Fire Retardancy of Polymeric Materials

The third edition of *Fire Retardancy of Polymeric Materials* provides a single source for all aspects of this highly challenging field of applied research. This authoritative book covers design and non-fire requirements that drive how these materials are fire protected.

Detailed study and consideration of chemistry, physics, materials science, economic issues, and fire safety science is necessary to address considerations of mechanical, thermal, environmental, and end-use requirements on top of fire protection means that the field requires. This thoroughly revised new edition continues to offer comprehensive coverage of the scientific approach for those developing fire safe materials. It covers new topics such as bio-based materials, regulatory issues, recycling, newer flame retardant chemical classes, and more details on how to flame retard materials for specific market applications.

Written by a team of experts, this book covers the fundamentals of polymer burning and combustion and how to apply fire protection or flame-retardant chemistries to specific material classes and applications. The book is written for material scientists and fire safety scientists who seek to develop new fire safe materials or understand why materials burn in our modern environment.

Features

- Connects fundamentals of material flammability to practical fire safety needs.
- Covers current fire safety requirements and regulations affecting flame retardant selection.
- Provides information on chemical structure-property relationships for flame retardancy.
- Provides practical guidance on how to design fire safe materials for specific fire risk scenarios.
- The new edition is expanded to 32 chapters, and all chapters are updated and revised with the newest information.

Fire Retardancy of Polymeric Materials

Third Edition

Alexander B. Morgan and Charles A. Wilkie

CRC Press
Taylor & Francis Group
Boca Raton London New York

CRC Press is an imprint of the
Taylor & Francis Group, an **informa** business

Designed cover image: Shutterstock Images

Third edition published 2024
by CRC Press
2385 NW Executive Center Drive, Suite 320, Boca Raton FL 33431

and by CRC Press
4 Park Square, Milton Park, Abingdon, Oxon, OX14 4RN

CRC Press is an imprint of Taylor & Francis Group, LLC

First edition published by CRC Press 2000
Second edition published by CRC Press 2010

Library of Congress Cataloging-in-Publication Data
Names: Wilkie, Charles A., 1941- editor. | Morgan, Alexander B., editor.
Title: Fire retardancy of polymeric materials / Charles A. Wilkie and Alexander B. Morgan.
Description: Third edition. | Boca Raton ; London : CRC Press, 2024. | Includes bibliographical references and index.
Identifiers: LCCN 2023056620 (print) | LCCN 2023056621 (ebook) | ISBN 9781032457543 (hardback) |
ISBN 9781032462332 (paperback) | ISBN 9781003380689 (ebook)
Subjects: LCSH: Polymers—Fire testing. | Polymers—Fires and fire prevention. | Fire resistant plastics.
Classification: LCC TH9446.5.P65 F56 2024 (print) | LCC TH9446.5.P65 (ebook) |
DDC 620.1/92—dc23/eng/20240128
LC record available at https://lccn.loc.gov/2023056620
LC ebook record available at https://lccn.loc.gov/2023056621

ISBN: 9781032457543 (hbk)
ISBN: 9781032462332 (pbk)
ISBN: 9781003380689 (ebk)

DOI: 10.1201/9781003380689

Typeset in Times
by Apex CoVantage, LLC

Contents

Preface

Greetings, fellow fire safety scientist/engineer:

In your hands you hold the third edition of "Fire Retardancy of Polymeric Materials." The first edition was published in 2000, the second edition was published in 2010, and now we have the third edition. Each time an edition has been published, the book has expanded, making it more comprehensive in scope. The book now covers every class of flame retardant chemistry, regulations, codes and standards, fire physics, emissions toxicity, aspects of fire safety engineering, testing, and how to apply fire protection to materials. Flame retardant material design was always multidisciplinary, and you'll see this reflected in the book.

This book includes a wide range of authors and emphasizes the multidisciplinary aspect and teamwork required to generate successful fire safe materials. Successful fire protection of materials requires chemistry, physics, engineering, and now more often, economics and environmental assessments to get it right. The field of fire protection, and indeed flame retardancy of polymeric materials, has changed greatly since the second edition was published in 2010, and this book captures those changes, the additional needs, and proposes the way forward so that we can continue to practice this field. It's important to realize that our field of fire protection is driven by laws and regulations and is a very old field of science. Fire protection of materials has been in place since the dawn of recorded history, and as long as we have carbon-based materials and oxygen, fire can and will be something we must protect against.

As with previous editions, this book is a mix of "how-to" guidance, distillates of technical knowledge, and reference material to enable the fire safety scientist/engineer to protect polymeric materials. We've included bio-based materials in this edition as well, so fire protection of wood (which is a polymeric material) is covered as well. We believe that this book is an excellent place to start to "dive in" if you're new to the field of fire safe materials, and for those of you with experience or who are experts, we believe you'll find something as well since THE experts in our field have contributed to this book. It should be noted though (and this will be obvious to some) that you will not be able to learn everything just from reading the book as some experimentation and experience is required in this field. For those reading this book hoping to find a quick answer to solving a specific material flame retardancy problem—you will find answers, but as you read the book, you will realize that there is no quick answer to material fire safety and material flammability. Flame retardancy of materials remains one of the most challenging materials science fields, as one must not only provide protection against fire for a polymeric material, but one must provide that protection while considering all the other balance of property issues (cost, mechanical, and thermal performance) and ever increasing regulations (banned chemistries, changing fire risks, recycling, end-of-life disposal). It sounds like we're underselling the book, but in reality, we're telling you that this book is a great place to start, and it's got all the starting information in it, but expect to put in some work if you really want to get to solutions.

Once again, we would like to thank all those who have helped make this book possible and helped build the foundation that has led to the 3rd edition. This includes all of our colleagues who contributed new chapters and those who revised chapters from the second edition. Additionally, we want to thank Taylor & Francis for agreeing to go with a third edition and providing insight on changes to the book market and e-books. Finally, we wish to thank our wives, Julie Ann Morgan and Nancy Wilkie, for their tireless support, editing skills, and putting up with us talking "shop" well after work hours have ended.

—Alexander B. Morgan, PhD
April, 2024. Dayton, Ohio USA

—Charles A. Wilkie, Professor Emeritus
April, 2024. Milwaukee, WI USA

Alexander B. Morgan received his BS in chemistry from Virginia Military Institute in 1994, his PhD in chemistry from the University of South Carolina in 1998. Dr. Morgan has over 27 years of experience in the areas of materials flammability, polymeric material flame retardancy, fire science, fire testing, thermal decomposition chemistry, and fire safety engineering with an emphasis on chemical structure property relationships and fire safe material design. He has helped academic, government, and industrial customers solve their flame retardant and fire safety needs in a wide range of applications. Dr. Morgan is on the editorial review board for Fire and Materials and is the editor-in-chief for the *Journal of Fire Science*. He is also a member of ASTM, Sigma Xi, International Association of Fire Safety Scientists, and the American Chemical Society. His current research areas include the following:

- New Flame Retardant Technology with Reduced Environmental Impact for Electronics, Building Materials, Bedding/Furniture, Wire and Cable, Transportation, Textiles
- Fire Testing Method Development
- Fire Safety/Survivability for Aircraft and Ground Vehicles
- Materials Development and Technology Assessment for Additive Manufacturing for Aerospace Applications

Charles A. Wilkie received his BS degree in chemistry from the University of Detroit in 1963 and his PhD in Inorganic Chemistry from Wayne State University in 1967. He subsequently joined Marquette University as an assistant professor and rose through the ranks until retiring in 2009; he now is a professor emeritus. In 1978, he read a paper on phosphorus-based flame retardants and decided to move into that area. He has worked in the field of flame retardancy for more than 30 years. His initial work was to attempt to understand mechanistically how a flame retardant interacts with a polymer. His research group also did quite a bit of work on nanocomposite based flame retardancy. Wilkie has served, and continues to serve, on the editorial boards of several journals and has been involved with organizing several meetings in this area.

Contributors

Walid H. Awad
SACO AEI Polymers, USA

Adrian Beard
Clariant Plastics and Coatings Deutschland GmbH (Corporation)

Aurelio Bifulco
Department of Chemical Materials and Industrial Production Engineering (DICMaPI)
University of Naples Federico II
Italy

Sergio Bocchini
DISMIC Politecnico di Torino sede di Alessandria
INSTM Research Unit
Italy

Serge Bourbigot
Univ. Lille, CNRS, INRAE, Centrale Lille, UMR 8207—UMET—Unité Matériaux et Transformations, F-59000 Lille
France
Institut Universitaire de France (IUF)

Giovanni Camino
DISMIC Politecnico di Torino sede di Alessandria, INSTM Research Unit, V
Italy

Federico Carosio
Dipartimento di Scienza Applicata e Tecnologia, Politecnico di Torino, Italy

Dushyant M. Chaudhari
Fire Safety Research Institute (FSRI)
UL Research Institute
USA

Jiuke Chen
Laboratory for Advanced Fibers, Empa Swiss Federal Laboratories for Materials Science and Technology
Switzerland
Department of Materials
ETH Zurich, Switzerland

Eyal Eden
ICL-IP
USA

Sabyasachi Gaan
Laboratory for Advanced Fibers, Empa Swiss Federal Laboratories for Materials Science and Technology
Switzerland

Ali Gooneie
Circular Plastics, Department of Circular Chemical Engineering, Faculty of Science and Engineering
Maastricht University
The Netherlands

Jaime C. Grunlan
Department of Mechanical Engineering
Texas A&M University
USA
Department of Materials Science Engineering
Texas A&M University
USA
Department of Chemistry
Texas A&M University
USA

Laura E. Hasburgh
Institute for Materials Research and Innovation
University of Bolton
UK

Qingliang He
Lanxess AG
Germany

Marcelo M. Hirschler
GBH International
USA

Peter Hornsby
Queen's University Belfast

A. Richard Horrocks
Fire Chemistry Laboratory
CMRI

Simo Hostikka
Aalto University
Espoo, Finland

T Richard Hull
Centre for Fire and Hazards Science
University of Central Lancashire
Preston, UK

Marc Janssens
Southwest Research Institute
San Antonio, USA

David D. Jiang
Organic Technology/Corporate Research
New York, USA

Paul Joseph
Institute of Sustainable Industries and Liveable Cities
Victoria University
Australia

Baljinder K. Kandola
Institute for Materials Research and Innovation
University of Bolton
UK

Grant T. Kirker
Institute for Materials Research and Innovation
University of Bolton
UK

Wenyu Wu Klingler
Laboratory for Advanced Fibers, Empa Swiss Federal Laboratories for Materials Science and Technology
Switzerland

Thomas J. Kolibaba
Applied Chemicals and Materials Division
National Institute of Standards and Technology
USA

Michael Kuepfert
Lanxess AG
Germany

Sergei Levchik
ICL-IP
USA

Zheng-Ming Li
The Collaborative Innovation Center for Eco-Friendly and Fire-Safety Polymeric Materials (MoE)
National Engineering Laboratory of Eco-Friendly Polymeric Materials (Sichuan)
College of Chemistry
Sichuan University
China

Bo-Wen Liu
The Collaborative Innovation Center for Eco-Friendly and Fire-Safety Polymeric Materials (MoE)
National Engineering Laboratory of Eco-Friendly Polymeric Materials
College of Chemistry
Sichuan University
China

Jing-Hong Liu
The Collaborative Innovation Center for Eco-Friendly and Fire-Safety Polymeric Materials (MoE)
National Engineering Laboratory of Eco-Friendly Polymeric Materials (Sichuan)
College of Chemistry
Sichuan University
China

Alessandra Lorenzetti
Department of Industrial Engineering
University of Padova
Italy

Michele Modesti
Department of Industrial Engineering
University of Padova
Italy

Alexander B. Morgan
University of Dayton Research Institute
Dayton, Ohio

Katie M. Ohno
Institute for Materials Research and Innovation
University of Bolton
UK

Rudolf Pfaendner
Fraunhofer Institute for Structural Durability and System Reliability LBF
Germany

D. Price
Centre for Research and Innovation
University of Bolton
UK

Guillermo Rein
BRE Centre for Fire Safety Engineering
The University of Edinburgh
UK

Bernhard Schartel
Bundesanstalt für Materialforschung und-prüfung (BAM)
Germany

Daniel F. Schmidt
Luxembourg Institute of Science and Technology (LIST)
Luxembourg

Arvindh Sekar
Laboratory for Advanced Fibers
Empa Swiss Federal Laboratories for Materials Science and Technology
Switzerland

Kelvin K. Shen
Rio Tinto/U.S. Borax, USA

Christopher Simpson
Lanxess AG
Germany

Anna A Stec
Centre for Fire and Hazards Science
University of Central Lancashire
UK

Stanislav I. Stoliarov
Department of Fire Protection Engineering
University of Maryland
College Park

Lein Tange
ICL IP
USA

Jose L. Torero
BRE Centre for Fire Safety Engineering
The University of Edinburgh
UK

Svetlana Tretsiakova-McNally
Belfast School of Architecture and the Built Environment
Ulster University

Xiu-Li Wang
The Collaborative Innovation Center for Eco-Friendly and Fire-Safety Polymeric Materials (MoE)
National Engineering Laboratory of Eco-Friendly Polymeric Materials (Sichuan)
College of Chemistry
Sichuan University
China

Yin-Long Wang
The Collaborative Innovation Center for Eco-Friendly and Fire-Safety Polymeric Materials (MoE)
National Engineering Laboratory of Eco-Friendly Polymeric Materials (Sichuan)
College of Chemistry
Sichuan University
China

Yu-Zhong Wang
The Collaborative Innovation Center for Eco-Friendly and Fire-Safety Polymeric Materials (MoE)
National Engineering Laboratory of Eco-Friendly Polymeric Materials (Sichuan)
College of Chemistry
Sichuan University
China

Carl-Eric Wilen
Faculty of Science and Engineering
Åbo Akademi University
Finland

Charles A. Wilkie
Marquette University
USA

Mauro Zammarano
Fire Research Division, Engineering Laboratory
National Institute of Standards and Technology
USA

1 An Introduction to Materials Flame Retardancy and Fire Protection

Alexander B. Morgan[a] *and Charles A. Wilkie*[b]
[a] University of Dayton Research Institute, Dayton, OH, 45469
[b] Marquette University, Milwaukee, WI, 53233

1.1 FIRE AND SOCIETY

As per the US National Fire Prevention Association (NFPA) annual report on fire loss in the United States, for the year 2021 there was 1.35 million reported fires resulting in 3,800 deaths, 14,700 injuries, and property losses of \$15.9 billion.[1] Fire losses vary by country outside the US, and some countries have larger fire losses than others, but regardless, fire losses continue in just about every country on the planet.[2] In the 14 years since the second edition of this book was published, there has been an increase in wildland-urban interface (WUI) fires, and some notable fire losses that have made the news due to both natural and manmade disasters. There is a lot to digest in all this fire loss data, and detailed study does give insight into where more protection against fire is needed. Statistics, though, are just numbers and do not really tell the whole story even though they attempt to. Most people are not swayed by numbers that try to tell them something, but humanity does react to stimuli, and one of those stimuli is an innate fear of fire, and with good reason. Fire is a unique destructive force of nature; what it touches cannot easily be repaired, rebuilt, or restored to its normal form. Scientifically, fire is thermo-oxidative decomposition of a material, which means it converts carbon-based materials and other combustible matter into CO_2 and water. In layman's terms, fire consumes what it touches and leaves behind nothing but ashes. Even non-carbon-based materials will suffer fire damage, from melting to thermal shock damage, to oxidation of metal surfaces caused by fire. Even concrete suffers damage under fire as water trapped in the concrete becomes superheated steam, which pushes out of concrete with explosive force, leading to cracking and, in some severe cases, structural failure of the concrete. Fire is fatal to almost all life on the planet (except for a few plants which germinate after fire has destroyed forest and underbrush) and is something that all of us know to avoid since we know it will burn and maim us, and if we cannot escape, it will kill us. But in modern society, fire is something we use to make life easier. It is a universal energy source for converting matter into heat energy, which is either used directly or is used to drive other devices which do work for us. It is not that modern society has lost its fear of fire, but instead we have grown accustomed to controlled, and controlling, fire, where we ignite the flame and control its use. Uncontrolled fire activates our instinctual fear mechanism, but until we see it, we really do not think about it. Only those who have encountered accidental fires in their lifetime become acutely aware of fire dangers and hazards—the rest of us tend to not think about it until it appears. So at first glance we might say that humanity has conquered fire except for the occasional outbreak, but when we return to the statistics, we see that this is not the case. Fire can be prevented in most cases, yet we still have catastrophic fire losses which result in loss of wealth, possessions, livelihood, and life. It is for these reasons that there are scientists and engineers who not only study fire but seek to provide passive fire protection for daily life so that we, as a society, will not have to worry about fire, and should an accidental fire begin, it can be rapidly extinguished before it grows. If you are reading this book, a lot of this may be very obvious to you, but it is not obvious to everyone for many of the reasons described previously. Yes, we all know fire is there, but what many people do not realize is the great potential for fire all around us. It is due to this potential that the discipline of flame retardancy exists.

1.2 DEFINITIONS OF FLAME RETARDANCY

Before one starts reading a book about flame retardancy, there are some terms which need to be defined for both the inexperienced and experienced reader, since one will see many terms throughout the book that at first glance look similar but have very different meaning when studying flame retardancy of materials. First of all, the terminology of "fire retardancy" and "flame retardancy" are used more or less interchangeably. There is some perception that flame retardancy involves a lower risk than fire retardancy; in this volume these terms are both used and mean the same thing. However, it is important to understand fire protection terms as well, as the discipline of fire protection engineering drives the requirements that may require a polymeric material to have flame retardant properties.

DOI: 10.1201/9781003380689-1

By understanding fire protection principles it becomes possible to know when to use flame retardants and when other engineering methods of fire protection are more appropriate.

1.2.1 Fire Protection

This is the overarching concept driving the purpose behind this book. Because fire happens, and it leads to losses, there is a need to protect against it. Fire protection engineering uses a wide range of concepts and multiple fields of science to engineer solutions that protect materials against fire. Fire protection can take many forms, including active approaches such as someone on fire watch looking for fires that they then put out, to reactive approaches that activate once some sort of fire stimulus require them. A good example of reactive fire protection would be water sprinklers for buildings or temperature activated extinguishers for restaurant exhaust hoods above cooking fryers. Then there are passive fire protection methods which assume fire will happen and the material needs to be there to slow fire growth/flame spread and allow time to escape. Flame retardants typically fall into the passive category, but depending upon how they get used, they can be part of reactive approaches as well.

1.2.2 Flame Retardancy

Flame retardancy means that something has been done to a material so that when exposed to a flame, either the material will retard the growth and propagation of that flame, or it will retard (slow) the growth and propagation of any flames that may come from the material once it has been ignited. Fire retardancy does not mean that the material will not burn, but rather that it will be more difficult burn. In some cases the flame retardant material may self-extinguish after being ignited if the external flame is removed, but in other cases, the flame retardant approach assumes the material will stay lit once ignited and will instead just burn slowly. What is required of a flame retardant material (self-extinguishment or slow burn rate) is dictated by the specific regulatory test under which the material is rated/sold, which in turn is governed by fire risk scenarios.

1.2.3 Fire Risk Scenario

A fire risk scenario is considering how a material/structure/assembly may be exposed to fire, and is it expected to lead to loss and injury. More specifically, it considers the probabilities of fires happening and the probability of that leading to loss, and how bad that loss could be. This is an important concept to understand as, once identified, it helps determine how a material will be fire retarded or made fire safe to resist that particular fire risk scenario, or at least not contribute negatively to fire growth. Fire risk scenarios are highly varied, and their relationship to regulatory tests is not always straightforward. For more details on this, the reader is encouraged to peruse Chapters 3, 17, 18, 21, and 22.

1.2.4 Flame Retardant

This term is used for any chemical that allows a material to retard a flame, or for any polymer that shows the ability to slow fire growth when ignited. It does not mean noncombustible or ignition resistant—these are very different terms and should not be used to describe a flame retardant material.[3] A material that is truly noncombustible or ignition resistant either cannot be combusted (no thermo-oxidative decomposition can occur) or cannot be ignited with a particular size flame/heat source. Here, things sometimes get confusing: a material could be flame retarded so that under a test which measures aspects of ignition or combustibility, the material is measured/assessed to be noncombustible/ignition resistant, but under another set of conditions it burns with ease (see Chapters 3, 7, 18, 21, and 22). It is important here to realize that the flame retardant effect of a chemical is a function of its particular chemical structure in the presence of a particular polymer. Something that may be a flame retardant for one polymeric material in one fire risk scenario may not provide any flame retardant effect for another fire risk scenario. As this book will make it clear, there is no generic class of "flame retardants." The ability of a chemical to have a flame retardant action is chemical structure dependent, and the chemical structures can be highly varied. Of final note, the term "flame retardant" is often abbreviated as "FR," such that flame retardant additives are described as "FR additives" or sometimes just as "FRs." A flame retardant plastic, though, may be listed as "FR plastic" or "FR resin," or described simply as an "FR material." When reading, look at how the author uses the term since one may use an FR additive to make a FR polymer.

1.2.5 Combustibility

This is defined by a series of conditions, as is the ability to resist ignition. It can be said that all carbon-based materials can be combusted (converted to CO_2 and H_2O) with enough heat and oxygen, but it may be hard to do so under some conditions, so a particular test may define certain carbon-based materials as noncombustible. Only inorganics (glasses, ceramics) and metals in their highest oxidation state would be noncombustible using the definition of thermo-oxidative decomposition. Again, Chapters 3, 17, 18, 21, and 22 address aspects of these definitions, but the reader is encouraged to understand that fire retardant does not always mean noncombustible.

1.2.6 Thermal Degradation/Decomposition

Another set of definitions with which one should be aware is the difference between thermal degradation and thermal decomposition, since they are not exactly the same thing.[3] *Thermal degradation* means that under heat exposure, some property of the material has degraded or become less than what it was before exposure to heat.

Thermal decomposition, on the other hand, means that the material has decomposed from its component parts upon exposure to heat, and for polymeric materials, this means that bonds have been broken, causing the polymer to depolymerize and/or break up into potential fuel molecules. It is important to remember that polymer decomposition is a free-radical process under fire conditions, and the chemical reactions of bond cleavage and subsequent pyrolysis of smaller molecular weight fragments from the polymer can be quite complex. A synopsis of these chemical pathways is discussed in more detail throughout the book, with primary emphasis in Chapters 2, 4, 5, 6, 7, 8, 9, 10, 11, 12, 13, 14, 15, 16, 19, and 20. Indeed, understanding how a polymer decomposes is essential to understanding its flammability behavior and how to flame retard it.

1.2.7 Other Relevant Fire Protection Definitions

Fire protection engineering uses other terms to determine if a flame retardant or other fire protection approach should be used, and while these terms are used mostly exclusively in the fire safety engineering profession, it's important to understand them here as they do get used in test methods and fire safety standards.

Fire Decision/Fire Safety Decision Tree: This is a decisional "tree," which provides guidance and concept about what fire protection measures to use and when to use them. The main guidance document for this work is NFPA 550: Guide to the Fire Safety Concepts Tree.[4] This document helps fire protection engineers decide how to provide protection against a fire risk scenario and would provide guidance on when a flame retardant would be best suited to use as either a source of primary or secondary protection in a fire.

Fire Performance: This the how a material, product, or structure/assembly responds to a particular fire other than a controlled test fire. To better explain, an analogy is useful. How a plastic behaves in a small flame ignition test to simulate an electric short circuit would be a "fire-test-response." How that same plastic behaves in a room fire where something else has caught fire and is radiating heat onto it, and it ignites and starts to burn, would be its fire performance. Sometimes, fire protection engineers do have to consider the fire performance and the fire-test-response of a material in their protection decisions. For the material scientist trying to design a fire safe product, it can be difficult to protect against all fire scenarios, and therefore, one should consider if the product may encounter fire outside of standard fire risk scenarios. If so, additional fire protection measures may be needed, or guidance to customers not to use that product outside the fire risk scenario it was designed for.

Fire Resistance: This is the ability of a material to withstand fire, or protect against it, for some period of time. A flame retarded material may be able to generate fire resistance, but the time-based aspect of this is important to understand, as something that may be flame retarded against a 60-second flame exposure could be very inappropriate for a one-hour flame resistance tests.

Gasification: This is the process by which solid and liquid flammable materials are converted into a gaseous state, which would enable ignition and flame propagation. Simplistically, it's the process where something is heated up, and it goes from solid to liquid to gas, but the processes can be much more complex depending upon the chemical structures of the fuels involved and the heating rates applied to the material.

Pyrolysis: This is a particular form of thermal decomposition that is very important to understand. Specifically, it's a gasification process that occurs under anaerobic conditions (non-oxidizing). Effectively, post ignition, everything under the flame is pyrolysis meaning the polymeric material is decomposing and gasifying in the absence of oxygen. This is why it's so important to understand thermal decomposition of materials measured under inert gases rather than oxidizing gases to really determine how it thermally decomposes.[5,6] Thermal decomposition studies in oxidizing conditions (such as air) are best saved for studies on how materials smolder, or how they decompose pre-ignition.

1.3 PROTECTION FROM FIRE—MODERN VS. HISTORICAL VIEWPOINTS

The need to protect materials against fire has been a scientific undertaking for a very long time. The use of flame retardant additives for polymeric materials is, of course, a more recent thing, with modern synthetic polymeric materials having been invented in the early 20th century. The earliest fire retardants for synthetic polymers were halogen-based, based on the discovery of halogenated hydrocarbons and waxes which effectively flame retarded Army tents.[7] New chemistries followed shortly thereafter, but each of these chemistries was created in response to a particular need to flame retard an object. It began with tents, but as more and more polymers moved into daily life, fire safety engineers and insurance companies found a need to push for additional fire safety. It is worth mentioning again that this field of research is regulation-driven, and many of the fire safety regulations in effect today began in the early 20th century when insurance companies, tired of paying on losses due to fire, began to push for greater fire safety. Eventually fire safety advocates took over pushing fire safety from the insurance companies, but this early start in the insurance industry is why so many fire safety standards were created by organizations like Underwriters Laboratories (UL) and Factory Mutual (FM); commercial products today are certified by these organizations before they can be sold in the US. Later the US federal government added its own regulations for particular fire safety needs, as did US state governments and other organizations worldwide. There is no one universal fire standard worldwide, which is why fire risk scenarios set the fire test that a material must pass and, in turn, why a particular flame retardant system is designed to pass that test. This is important to understand because no one system will pass all fire tests, and in general, today one designs a system to meet a particular test, not to be flame retardant against all fire situations. Since the intensity of a flame will vary from situation to situation, it is almost impossible to create a universal flame retardant system; instead, industry, and others, design flame retardant polymeric materials to pass the test which is related to a specific fire risk scenario. This approach, though, runs into trouble with the "reactive" approach of providing fire safety for a material.

The way that new fire safety regulations have come into being is from a reaction or response to a particular problem. There are many examples to choose from here, but we will focus on three: plastics in cars, polyurethane foam, and polymers for electronics.

For automobiles in the US in the 1970s, there was strong concern over the high number of fires in automobile crashes, but also over how the plastics in the interior of the car would respond to small ignition sources (cigarette ashes). A fire standard, Federal Motor Vehicle Safety Standard 302 (FMVSS 302), was created to assess the potential for flame spread in an automobile. At the time, the standard did its job, but as there is an ever increasing demand for fuel efficiency in cars and a desire to use new materials in automobile construction, the amount of plastics in a car crept up from 10kg in a 1970s car to 90–100 kg in a modern 1990s-2000s car.[8] The FMVSS 302 test, which is a very simple test to pass, really just assesses how well a flame will spread once a car interior material is ignited, and almost any plastic today will pass the test. This presents a problem in that a historical standard, which worked for its time, has not kept up with modern changes, so that now the primary fire risk in a collision is not just from the fuel tank rupturing and leading to fire, but the fact that there is about 90–100 kg of solid fuel *inside* the automotive compartment where the passengers are located. Indeed, studies have shown that modern cars do much worse in fires than their older counterparts,[9–15] but the standard has remained unchanged since only recently has this concern begun to be voiced. Indeed, the high level of polymeric materials present on both the inside and the outside of modern vehicles is leading not just to higher hazard for occupants but also increased fire hazard for buildings that these vehicles are located in. Specifically, modern vehicles release high levels of heat release that will damage structures, including parking garages.[16] It will likely again take catastrophic fire losses for people to react and demand a change in fire safety, but there is hope that regulators will listen to history and be proactive, rather than reactive, to the problem. If a change in fire safety performance for automotive plastics is enacted, then there will be a huge demand for flame retardant products in cars. It should be noted that with increased electrification of cars and use of high energy intensity battery packs (Lithium Ion) now being used in all-electric vehicles, the fire risk scenario changes again from crash + fuel release yielding fuel pool fires, to battery failures (in crash or during charging/short circuit failure) yielding very fast fire growth events and great difficulty in extinguishing lithium ion battery packs once they ignite.[17–20] Despite the clear and obvious issue behind Li Ion battery fires, it remains unclear when or if the fire safety standard for automotive plastics will change.

The flammability issues of polyurethane foam have been well understood for some time now, but the approaches to fire safety with this highly flammable material have been varied, including by country.[21] In the US, the fire risk scenario was identified as people smoking in bed, and the solution at the time was to educate the public about smoke detectors, as well as to make sure the fabrics used could resist a smoldering ember source, like that from a cigarette. In the United Kingdom, though, the problem was tackled at the polyurethane level and required high levels of flame retardancy to pass a very strict fire test, the Crib V test.[22] This test, named for the class V wooden "crib" fire source, is still the strictest in the world for fire safety of polyurethane foam used in furniture or bedding, and as such, the UK has practically no fire losses due to furniture or bedding ignition. The US, due to perceptions about the hazards of flame retardant chemicals vs. any potential benefit for fire protection,[23] has instead now moved to smolder-only testing even though smoking in the US has decreased. This began with a change to the California Technical Bulletin 117 (TB 117) in 2013, moving from open flame testing to cigarette testing, and then this approach was codified into US law via 16 CFR 1640 in 2021. This move to smolder-based testing has proven to be problematic, and fire loss statistics from 2013 to present are showing that when furniture is involved in fire events, it does lead to major losses and high probability of injuries and death.[24,25] Since polyurethane foam in furniture and bedding remains the number one source of flammable fuel in the US home,[26–28] it is clear that we need to fire protect this product either through flame retardants, use of sprinklers in home, or the use of barrier fabrics. The current political and regulatory environment in the US seems to indicate that the fire losses involving flexible polyurethane foam are only going to increase in the near future, and it will likely take ever-increasing losses of life and property for this to change. Smolder-based testing does not provide any protection against open flames or high-intensity ignition sources (such as failing lithium ion batteries from cellphones or laptops), and so we have a situation here where fire safety is heading in the wrong direction, even though there are solutions available.

Since flame retardants are used in response to changing fire scenarios, which develop as technology changes, it's noteworthy to speak about polymers for electronics and how much the fire risk scenarios have changed due to technological changes, even if the test methods to assess fire safety of electronics has changed very little. Electronics technology has changed a lot in the past 60+ years, and as it has changed, so has plastic usage for this technology. Originally, plastic components were a minor part of electronics, only used in some circuit boards and internal wiring and parts of keyboards. Today, they are a part of almost every component on a computer (laptop or desktop) as well as most modern electrical equipment and electronic appliances. Indeed, with many modern electronic appliances starting to have high levels of electronic control and, in some cases, wireless communication capability, there are far more electronic plastics in use today, and the amounts used continue to increase. Since electrical short circuits (failures) can be possible ignition sources, it is very important to flame retard the polymer if the plastic is anywhere near a power supply or electrical component that can give off heat when it fails (capacitors, heat sinks, computer chips, wiring, etc.). The insurance industry helped create the early regulations and fire testing standards to protect electrical and electronic equipment, and the best example of this, still in use today, is the UL-94 test.[29] This test exposes a small plastic specimen, in either a vertical or horizontal configuration, to a small calibrated Bunsen burner flame (25 to 125 mm in height) for short ten-second ignitions, and the time to extinguishment is recorded. The time to self-extinguishment determines the rating of the material, with higher ratings (V-0 or 5-V) being typically used for electronic equipment. This test originated in the early days of electronic equipment in the US home, specifically the 1950s, where many US homes had a television set in the living room. The UL-94 V test (V for vertical test) was designed to mimic the television (TV) set on wheels, which was often moved over the 1950s style carpet for typical American family gatherings. If the television set caught fire, flaming drips could fall from the TV set and ignite the carpet underneath. Since then, parts of the carpet have become flame retardant, and more importantly, the design of the television set has changed. Flat screen TVs can be mounted on walls, ceilings, or many other surfaces, and therefore if the power supply ignites and the polymer drips—where would it go? The flame spread may be much less if the plastic sticks to a painted wall (gypsum wallboard, plaster), but could be much worse if the wall is wooden. With other electronic equipment becoming so ubiquitous in the home, now the question arises about not only internal ignition (electrical failure inside the equipment) but also external ignition. There are increasing concerns about small flame sources igniting electrical equipment from the outside,[30] especially in Europe.[31–33] So when a candle flame touches a non-flame retardant plastic on the outside of an electrical equipment item, what will happen? The answer is very complex, and it depends upon the level of flame retardant in the plastic exposed to the candle-like flame. If the plastic is UL-94 rated V-1 or better, nothing will happen, as the flame will go out, but if the plastic has a lower flame retardant rating, the flame can spread rapidly, and in some cases can lead to very large fires. Extensive studies done by many laboratories have shown that a small candle-like flame can ignite a non-flame retardant plastic casing for a TV set or printer which can easily lead to room flashover.[34] The US National Institute of Standards and Technology (NIST) had a program called "Flammability Measures of Electronic Equipment," which studied several commercial plastics at both bench scale and large scale, and they found that parts made from UL-94 V-1 or better plastics would not ignite with a small candle-like flame.[35–37] However, if a non-flame retardant keyboard was put under a computer monitor case, and the keyboard was ignited, the heat release from the burning keyboard would overwhelm even the higher rated flame retardant plastics. Therefore the fire risk scenario of electronics is continuing to change as new technology enters the marketplace and the regulations are not keeping pace. Keeping materials with a UL-94 V rating in place appears to be a reasonable place to start, but many changes in how these items are mounted on walls or used in homes may result again in regulators reacting to change rather than being proactive in fire safety performance.

None of these scenarios described in the previous section are easy to address in a proactive manner. If regulators kept products from entering into use until they could determine all possible fire risk scenarios, many products might never come to the marketplace. Practically speaking, no regulator can envision all possible fire risk scenarios for a material. Some things cannot be envisioned until they are used by the consumers, and therefore some regulation changes must be reactive rather than proactive. Still, regulators and consumers need to be aware that fire risk never goes away, and so one should not be surprised when changes in technology lead to changes in consumer use which either defeats the fire protection in place for that material or makes the fire protection irrelevant for the ways the material is used. Perhaps now more than at any other time in human history, the rapid changes in technology require all of us to be flexible not just about how we use a material, but also how we think about that material from a fire safety perspective and a total life cycle view.

1.4 CHANGING PERCEPTIONS, REGULATIONS, AND CUSTOMER DEMANDS

Just as changes in technology result in new fire risk scenarios, societal changes can affect flame retardancy and fire safety as well. Some of the biggest changes in flame retardant materials have come from changes in waste disposal and recycling of materials that have reached the end of their usefulness. The best example of this is how the European Union (EU) disposes of its waste and the laws it enacted to deal with the waste. In the EU, landfills are not typically permitted since land in Europe is at a premium, with almost all existing land either used for agriculture or for living space and industry. So waste is incinerated rather than sent to a landfill. Early on in this process, it was found, in some cases, that the brominated flame retardants in flame retardant plastics would produce brominated dioxin compounds. When this was discovered, some EU states pushed to have these compounds removed from use in plastics, which led to a decade long legal struggle between EU governments, environmental non-government organizations (NGOs), fire safety regulators, and the makers of brominated flame retardants. Much of this story and the current state of these regulations are described later in Chapter 23, but to summarize, the change in a non-fire related activity has led to changes in how flame retardants are used and which ones are perceived to be safe. More importantly, it has led to a discussion about fire safety in general and whether fire safety/flame retardancy is still an acceptable environmental policy.

The environmental impact of waste disposal and of chemical use in Europe has led to three legislative actions that, in today's global economy, greatly affect flame retardant use and research. These actions go by the acronyms of RoHS (Reduction of Hazardous Substances),[38,39] WEEE (Waste Electrical and Electronic Equipment),[40,41] and REACH (Registration, Evaluation, Authorisation, and Restriction of Chemical substances).[42] These actions are discussed in detail in Chapter 23 but need to be mentioned here as they are clear examples of how changing regulations affect flame retardant use, selection, and new fire safety developments. The first one, RoHS, refers to how new items are manufactured and specifically bans chemicals and elements of environmental and toxicological concern in Europe. One fall-out item of RoHS is the move from lead-based solder on circuit boards to non-lead based solder. The new solders (usually tin based) are higher melting and, as such, require epoxy circuit boards of higher thermal stability. This has led to the some de-selection of brominated flame retardant epoxies simply because the C-Br bond on tetrabromobisphenol A is not thermally stable enough to survive exposure to the higher melting point solder. The organobromine bonds break down upon exposure to this solder, which degrades circuit board performance. So the selection of a material which at first glance has nothing to do with FR performance has led to a new paradigm in flame retardant research: higher thermal stability of the polymer itself to withstand high-temperature solder while maintaining flame retardant performance.

WEEE has had a direct effect on flame retardant use, since flame retardants are used in almost all electrical and electronic equipment to prevent fires from short circuits. This directive lays down rules for disposal and recycling of all electrical and electronic equipment which goes back to the previous incinerator discussion. In regards to flame retardants, this directive affects how the plastic parts, cable jackets, and enclosures are flame retarded. If the plastic cannot be reground and recycled, it must go to incineration, in which case it cannot form toxic by-products during incineration. This has led to the rapid de-selection of brominated FR additives in European plastics which are used in electronics, or the complete removal of FR additives from plastics used in electronics in Europe. This de-selection and removal led to increases in electrical fires in Europe, and now customers and fire safety experts demand low environmental impact and fire safety. However, the existing non-halogen flame retardant solutions brought in to replace bromine have their own balance-of-property issues, and so research continues to develop materials that can meet WEEE objectives. See Chapter 24 for more information on recycling of flame retardant polymers.

Of all the new regulations affecting flame retardant research, REACH is likely to have the greatest impact on any new flame retardant solution created to address RoHS and WEEE regulations. Since this regulation affects all new and existing chemicals in the EU, all flame retardant additives (each of which is a chemical compound) in use, or proposed for future use, must go through extensive testing and certification before they can be used. If they do not pass this battery of tests, then they are banned from future use, which means there would be fewer solutions available for fire safety. Therefore, REACH affects not only current flame retardant additive use, but also research to develop new flame retardant additives. Why spend any time and effort on the development of a new flame retardant solution if it has no chance of passing REACH? This makes it more difficult for the flame retardant researcher to figure out which chemical compounds to synthesize and test if they do not pay attention to potential environmental impact and toxicology first. Therefore, REACH may add yet another scientific discipline to flame retardant research—toxicology and environmental science. So one can ask—why did these regulations come into being in the EU and not everywhere at once? To understand this, one must again look at the "life-cycle" (creation, use, and disposal) of plastics.

Plastics that contain flame retardant additives have one of three fates at the end of their life cycle. One is to put the material into landfill; another is to incinerate it to recover energy from the material; and the final is to regrind, recycle, and reprocess the plastic for later use. All three approaches have their advantages and disadvantages, but overall, incineration and recycling are preferred. Within those two approaches though, the perception exists that flame retardants do not belong due to pollution they cause during incineration or pollution caused by recycling the plastic which may cause the additive to leach out of the plastic. There is data both for and against these perceptions of flame retardants in plastics, but the environmental NGO groups have been winning public opinion that it is better environmental policy to leave the flame retardants out completely. This argument, though, only holds true if that non-flame retardant plastic does not catch fire. Several studies have shown that the amount of pollution which enters the environment when a plastic catches fire is far greater than the pollution used in creation and disposal of a flame retardant plastic.[43–51] This is because when a non-flame retardant plastic catches fire, it tends to give off so much heat that it ignites other nearby objects, and those objects in turn quickly cause the heat release in the room to escalate, leading to major fire losses, large amounts of toxic emissions, and in some cases, loss of life (see Chapter 19 for more information on fire emission toxicity). So one has to make a decision—is it more acceptable to lose life and property to fire for the potential long-term benefit of the environment, or is it more acceptable to save lives and deal with potential pollution issues when the plastics are not dealt with correctly at the end of their lifetime? The answer to this question is complex, and both sides of the argument have made impassioned pleas for their side. It is possible to have both fire safety and good environmental policy, but it requires an understanding of the following:

1) Passive fire protection (flame retardants in plastics) is an important societal benefit.
2) Passive fire protection can be developed in a way that is environmentally friendly and is easily recycled.
3) The cost-benefit analysis of fire protection in relevant fire risk scenarios.
4) Both sides will need to set aside their preconceived notions about the other and be willing to work together.

Humanity is, by nature, emotional, and emotions on both sides of this issue make progress difficult for new fire safe materials, whether in response to new fire risk scenarios or in response to non-fire related issues. This last point should make it clear that flame retardancy of plastics is not a simple scientific issue, but one that must take into consideration societal, emotional, and non-fire related issues to make an acceptable new flame retardant material in the 21st century.

Along with the changes in perceptions of flame retardants, there are changes in fire safety requirements as well. Many of the changes were discussed in the previous section, and they are covered in detail in Chapter 23, but one of the changes is a move from pass/fail testing to performance-based testing. Performance-based testing allows for more freedom in how an item is constructed or assembled to meet fire safety needs. This tends to be final article testing rather than testing of a specific flame retardant material, and therefore modeling becomes quite important in performance-based methodology. These modeling approaches are discussed in Chapters 20 and 21, and modeling allows regulators and fire safety engineers to see how a construction performs in a particular fire scenario. While this approach yields high flexibility for the fire safety engineer, it can make things difficult for the scientist trying to flame retard a material for an application. With no guidance on what the flame retardant test should be for a particular component of the final article, the flame retardant materials scientist will need to understand fire safety engineering and these new models to better design their materials. It is likely that the change in performance-based codes, along with changes in environmental standards, will yield all sorts of new flame retardant materials used in new and very creative ways. However, this change means that the burden of flame retarding the polymeric material may fall upon the final item/assembly producer and no longer on the resin companies which compound flame retardants into plastics before they sell it. All of this is uncharted territory, and the correct path for polymeric fire-retardant materials remains to be seen. It is likely, though, that there are multiple paths forward where new engineering approaches and flame retardant chemistry will coexist to produce new fire-safe *and* environmentally friendly polymeric materials.

1.5 FLAME RETARDANT CHEMISTRIES

With this background, one begins to see why this field is so complex and that there is no simple answer to flame retardancy of polymers. This complexity is also why there are so many flame retardant chemistries to choose from since many flame retardant chemistries are tailored to a specific polymer for a specific test (see Chapters 25–31). However, there are some broad categories of flame retardants in use today, including the following:

- Halogenated compounds (Chapter 4)
- Phosphorus-based compounds (Chapter 5)
- Mineral fillers (Chapter 6)
- Intumescent protective systems (Chapter 7)
- Silicon and inorganic oxides (Chapter 8)
- Boron chemistry (Chapter 9)
- Nitrogen-based compounds (Chapter 10)
- Radical generators (Chapter 11)
- Polymer nanocomposite systems (Chapters 12)
- Bio-based flame retardants (Chapter 13)
- Conformal coatings (Chapter 14)
- Inorganic/Unusual chemistries (Chapter 15)
- Multi-component systems (Chapter 16)

For many of the reasons described in the last section, halogenated flame retardant chemistry is not as greatly researched in academic circles as it once was, although it continues to be a robust field of flame retardant chemistry, especially amongst the industrial manufacturers who produce these compounds. Non-halogenated flame retardant systems have grown by leaps and bounds in the open literature as well as patent literature, again for the reasons previously described.

Within each of these chemical approaches to flame retardancy, one will find a variety of chemical structures and compounds that yield specific effects of flame retardancy under specific conditions. Some compounds are fairly universal and reduce heat release under all conditions (halogen, phosphorus, mineral fillers), while others have unique activation conditions and protective schemes that can be very robust but are limited to a small number of polymers with which they can be used (intumescent, inorganic compounds, nanocomposites). Some materials by themselves enhance the flame retardancy of other compounds (boron, inorganic oxides, nanocomposites), while others not only enhance a wide range of flame retardants, but also bring multifunctional performance to a material. Polymer nanocomposite technology fits this last category in that the nanocomposite brings enhanced mechanical, thermal, and sometimes electrical properties while also reducing the heat/fuel release of that specific polymer.[52–58] Normally, flame retardant chemistry has involved the addition of a particular material only for enhancement of flammability performance to the material, so the promise of bringing something in for flame retardancy *and* thermal *and* mechanical *and* other properties has led to an explosion in new material science research for nanocomposites in general, and it certainly has been a boon to flame retardant research as well. However, it should be noted that there has been no commercialization of flame retardant polymer nanocomposites besides one example in wire and cable applications.* Therefore, it is not clear how well this technology will advance beyond its current use in coatings, as discussed in Chapter 14 of this book.

It is important to remember, again, that this field of research is driven by regulations, both fire and non-fire related, and with that the research sometimes is reactive to the regulations rather than proactive. Flame retardant chemistry is much more varied now than it has ever been, and it will continue to grow as chemists look for new alternatives to existing flame retardant systems.

1.6 FLAME RETARDANT RESEARCH

Research into flame retardancy of polymeric materials, and indeed fire safety of all materials in use today (wood, steel, glass, plastics, etc.), is truly a worldwide field of materials science study. Research is done in academia, government, and industry on every continent in the world today, and each group has its own areas of emphasis and expertise. Industrial work focuses largely on the development of new materials for sale that meet the regulations and, as such, is not broadly shown in the scientific literature and is almost completely limited to the patent literature. Therefore, it is hard to determine which companies are leading the way in new flame retardant material research other than by looking at their offerings of flame retardant products. Obviously, the main flame retardant additive manufacturers (ICL, Albemarle, Clariant, Lanxess) conduct the greatest amount of new flame retardant chemical research, but it is not unheard of for the end-users of these additives to conduct their own flame retardant chemical research. So in published research of access to everyone, most of the conducted research focuses on fire safety engineering more than new flame retardant chemistry, but still, the work is related as the new fire safety engineering research leads to new regulations which the flame retardant chemistry has to meet.

Listed in the next section are some of the pre-eminent fire safety and flame retardant research organizations worldwide. The list is by no means comprehensive and focuses on the organizations that are known to publish often in the fields of flame retardant research or fire safety engineering. The list here includes links to the organizations to learn more about the range of services and research they offer.

- National Institute of Standards and Technology Fire Research Division (NIST): www.nist.gov/el/fire-research-division-73300
- Underwriters Laboratories Fire Safety Research Institute (UL-FSRI): www.ul.com/ and https://fsri.org/
- FM Global (Factory Mutual): www.fmglobal.com/
- Southwest Research Institute (SwRI) Fire Technology Division: www.swri.org/industries/fire
- Federal Institute for Materials Research and Testing (Germany—BAM): www.bam.de/Navigation/EN/Topics/Infrastructure/FireScience/fire-science.html
- SP Technical Research Institute of Sweden: https://risefr.com/
- Bolton Institute for Materials Research and Innovation www.bolton.ac.uk/explore-our-schools/institute-for-materials-research-and-innovation/
- State Key Laboratory of Fire Science, Hefei, China http://fire.ustc.edu.cn/
- Commonwealth Scientific and Industrial Research Organization (CSIRO) Fire Safety & Control Division: www.csiro.au/en/work-with-us/services/testing-and-certification/fire-safety

1.7 CONCLUSIONS AND KEY ITEMS TO REMEMBER

As you proceed through this book, remember the importance of knowing the definitions of flame retardancy and fire protection, and then see how the book is organized. Flame retardant chemistries come first, followed by the fundamentals of how to assess material flammability (test methods, emissions, modeling), and then the regulations. The final chapters of the book explain how flame retardants are used in specific applications, and then the book ends with a potential view of the future and expected research/regulatory trends.

The field of polymeric flame retardancy is undergoing a great deal of change, reacting to new requirements, fire risk scenarios, and other requirements for polymeric materials. Those who pay attention to these coming changes, as well as being aware of the current technology available to flame retardancy, should be able to develop new materials in response to these changes, as well as come up with new innovations in flame retardancy. Fire will not just go away as our society advances, and it is very likely that we will see more, not fewer, polymers in use throughout society worldwide. With this increase it seems that there will be more need, not less, for those of us who practice this field of science. At the moment in some countries the level of flame retardant research funding is quite low (US), while in other countries the funding is higher (EU, Asia), but perhaps still not high enough to keep up with the changes. Since polymeric materials will only be used more in modern society, it falls upon fire scientists and fire safety engineers to educate the public that these materials are inherently flammable unless they are made flame retardant. Ideally, we would like to have the fire protection in place before catastrophic fires occur, but we have to keep in mind that there are many demands on the few resources that society has at the moment, and if we want those resources, we will have to voice the need for it. Until then, it is up to those of us who practice in this field to keep guard over the potential for fire in plastics and be willing to lend a hand when society asks for our help.

NOTE

* Products reported by Dr. Günter Beyer of Kabelwerk Eupen at many conferences up to 2016, but unclear if this product is still being sold. Cannot find product ID or number on Kabelwerk Eupen website. Information based upon personal communications between authors and Dr. Günter Beyer.

REFERENCES

1. "Fire Loss in the United States during 2021." National Fire Prevention Association. www.nfpa.org//-/media/Files/News-and-Research/Fire-statistics-and-reports/US-Fire-Problem/osFireLoss.pdf (accessed 10/12/22).
2. https://ctif.org/world-fire-statistics (accessed 10/12/22).
3. "ASTM E176–21ae: Standard Terminology of Fire Standards." American Society of Testing and Materials, West Conshohocken, PA, 2021.
4. www.nfpa.org/codes-and-standards/all-codes-and-standards/list-of-codes-and-standards/detail?code=550 (accessed 10/23/22).
5. "Recommendations on the Scientific Approach to Polymer Flame Retardancy: Part 1—Scientific Terms and Methods" Schartel, B.; Wilkie, C. A.; Camino, G. *J. Fire Sci.* **2016**, *34*, 447–467.
6. "Recommendations on the Scientific Approach to Polymer Flame Retardancy: Part 2—Concepts" Schartel, B.; Wilkie, C. A.; Camino, G. *J. Fire Sci.* **2017**, *35*, 3–20.
7. Sandholzer, M. W. "Flameproofing of Textiles", Circular of the National Bureau of Standards, C455 1946. U.S. Dept. of Commerce, National Bureau of Standards.
8. "Human Survivability in Motor Vehicle Fires" Digges, K. H.; Gann, R. G.; Grayson, S. J.; Hirschler, M. M.; Lyon, R. E.; Purser, D. A.; Quintiere, J. G.; Stephenson, R. R.; Tewarson, A. *Fire Mater.* **2008**, *32*, 249–258.
9. Ahrens, M. U.S. Vehicle fire trends and patterns. National Fire Protection Association, Quincy, MA, 2005.
10. U.S.F.A. Highway vehicle fires. Topical Fire Research Series, vol. 2, No. 4. U.S. Fire Administration, Emmitsburg, MD, 2002.
11. Friedman, K.; Holloway, E.; Kenney, T. Impact induced fires and fuel leakage: analysis of FARS and state data files (1978–2001). *Society of Automotive Engineers World Congress*, Detroit, MI, 11–14 April 2005. Paper 200501–1421.
12. "Guide on Methods for Evaluating Fire Hazard to Occupants of Passenger Road Vehicles—NFPA 556" National Fire Prevention Association (NFPA). www.nfpa.org/codes-and-standards/all-codes-and-standards/list-of-codes-and-standards/detail?code=556 (accessed 10/22/22).
13. "Investigation of Smoke Gases and Temperatures During Car Fire—Large-Scale and Small-Scale Tests and Numerical Investigations" Krüger, S.; Hofmann, A.; Berger, A.; Gude, N. *Fire Mater.* **2016**, *40*, 785–799.
14. "Post-Collision Fires in Road Vehicles between 2002 and 2015" Otxoterena, P.; Bjornstig, U.; Lindkvist, M. *Fire Mater.* **2020**, *44*, 767–775.
15. "Human Survivability in Motor Vehicle Fires" Digges, K. H.; Gann, R. G.; Grayson, S. J.; Hirschler, M. M.; Lyon, R. E.; Purser, D. A.; Quintiere, J. G.; Stephenson, R. R.; Tewarson, A. *Fire Mater.* **2008**, *32*, 249–258.
16. "Fire Hazard Analysis of Modern Vehicles in Parking Facilities" Boehmer, H. R.; Klassen, M. S.; Olenick, S. M. *Fire Technol.* **2021**, *57*, 2097–2127.
17. "Heat Release During Thermally-Induced Failure of a Lithium Ion Battery: Impact of Cathode Composition" Liu, X.; Wu, Z.; Stoliarov, S. I.; Denlinger, Ml; Masias, A.; Snyder, K. *Fire Saf J.* **2016**, *85*, 10–22.
18. "Analysis of Effectiveness of Suppression of Lithium Ion Battery Fires With a Clean Agent" Said, A. O.; Stoliarov, S. I. *Fire Saf J.* **2021**, *121*, 103296.
19. "Experimental Investigation of Suppression of 18650 Lithium Ion Cell Array Fires with Water Mist" Said, A. O.; Garber, A.; Peng, Y.; Stoliarov, S. I. *Fire Technol.* **2022**, *58*, 523–551.
20. "Risk Minimization of Electric Vehicle Fires in Underground Traffic Infrastructures" Mellert, L. D.; Welte, U.; Tuchschmid, M.; Held, M.; Hermann, M.; Kompatscher, M.; Tesson, M.; Nachef, L. https://plus.empa.ch/images/2020-08-17_Brandversuch-Elektroauto/AGT_2018_006_EMob_RiskMin_Undergr_Infrastr_Final_Report_V1.0.pdf (accessed 10/22/22).
21. "Revisiting Flexible Polyurethane Foam Flammability in Furniture and Bedding in the United States" Morgan, A. B. *Fire Mater.* **2021**, *45*, 68–80.
22. BS5852, Methods of test for assessment of the ignitability of upholstered seating by smoldering and flaming ignition sources, British Standards Institution, London, 2006.
23. "The Well-Meaning but Misguided Rollback of Fire Safety in the United States" Morgan, A. B. *J Fire Sci* **2022**, *40*, 249–253.
24. "Soft Furnishing Fires: They're Still a Problem" Ahrens, M. *Fire Mater.* **2021**, *45*, 8–16.
25. "Fire Safety in the United States since 1980: Through the Lens of the NFPA Fire and Life Safety Ecosystem" Ahrens, M.; Messerschmidt, B. National Fire Prevention Association, 2021. www.nfpa.org/~/media/Files/News%20and%20Research/Fire%20statistics%20and%20reports/US%20Fire%20Problem/osNFPAEcosystemFireSafetyReport2021.ashx?utm_source=emil&utm_medium=email_medium&utm_campaign=emil0358&utm_content=mbrs&order_src=e827 (accessed 10/22/22).
26. "Polyurethane Foam and Fire Safety" Hirschler, M. M. *Polym. Adv. Technol.* **2008**, *19*, 521–529.
27. "Aspects of the Fire Behavior of Thermoplastic Materials" Ohlemiller, T.; Shields, J. NIST Technical Note 1493, January, 2008.

28. "Flammability Reduction of Flexible Polyurethane Foams via Carbon Nanofiber Network Formation" Zammarano, M.; Kramer, R.H.; Harris, Jr., R.; Ohlemiller, T.J.; Shields, J.R.; Rahatekar, S.S.; Lacerda, S.; Gilman, J.W. *Polym. Adv. Technol.* **2008**, *19*, 588–595.
29. UL-94: Test for Flammability of Plastic Materials for Parts in Devices and Applications [Includes ASTM D635 (UL-94 HB), ASTM D3801 (UL-94 V), ASTM D4804 (UL-94 VTM), ASTM D5048 (UL-94 5V), ASTM D4986 (UL-94 HBF)].
30. "Fires Involving Appliance Housings—Is There a Clear and Present Danger?" Hall, J. R. *Fire Technol.* **2002**, *38*, 179–198.
31. "Full Scale Burn Tests of Television Sets and Electronic Appliances" Hoffmann, J. M.; Hoffmann, D. J.; Kroll, E. C.; Kroll, M. J. *Fire Technol.* **2003**, *39*, 207–224.
32. "Fire Performance of Selected IT-Equipment" Simonson, M.; Andersson, P.; Bliss, D. *Fire Tech.* **2004**, *40*, 27–37.
33. "Flame Retardancy Performance of Various UL94 Classified Materials Exposed to External Ignition Sources" Hong, S.; Yang, J.; Anh, S.; Mun, Y.; Lee, G. *Fire Mater.* **2004**, *28*, 25–31.
34. "Fire Behaviour of Plastic Parts in Electrical Appliances—Standards Versus Required Fire Safety Objectives" Babrauskas, V.; Simonson, M. *Fire Mater.* **2007**, *31*, 83–96.
35. "Cone Calorimeter Analysis of UL-94 V-Rated Plastics" Morgan, A. B.; Bundy, M. *Fire Mater.* **2007**, *31*, 257–283.
36. Bundy, M.; Ohlemiller, T. Full Scale Flammability Measures for Electronic Equipment. *NIST Technical Note 1461* **August 2004**, U.S. Dept. of Commerce.
37. Bundy, M.; Ohlemiller, T. Bench-Scale Flammability Measures for Electronic Equipment. *NISTIR (National Institute of Standards and Technology Internal Report) 7031* **July 2003**, U.S. Dept. of Commerce.
38. http://en.wikipedia.org/wiki/Restriction_of_Hazardous_Substances_Directive (accessed 10/23/22).
39. http://eur-lex.europa.eu/legal-content/EN/TXT/?qid=1399998664957&uri=CELEX:02011L0065-20140129 (accessed 10/23/22).
40. http://en.wikipedia.org/wiki/Waste_Electrical_and_Electronic_Equipment_Directive (accessed 10/23/22).
41. http://eur-lex.europa.eu/legal-content/EN/TXT/?uri=celex%3A32012L0019 (accessed 10/23/22).
42. https://en.wikipedia.org/wiki/Registration,_Evaluation,_Authorisation_and_Restriction_of_Chemicals (accessed 10/22/22).
43. Simonson, M.; Blomqvist, P.; Boldizar, A.; Möller, K.; Rosell, L.; Tullin, C.; Stripple, H.; Sundqvist, J. O. Fire-LCA model: TV case study. SP Report 2000:13, 2000.
44. "Emissions from Fires Part I: Fire Retarded and Non-Fire Retarded TV-Sets" Blomqvist, P.; Rosell, L.; Simonson, M. *Fire Tech.* **2004**, *40*, 39–58.
45. "Emissions from Fires Part II: Fire Retarded and Non-Fire Retarded TV-Sets" Blomqvist, P.; Rosell, L.; Simonson, M. *Fire Tech.* **2004**, *40*, 59–73.
46. Simonson, M.; Andersson, P.; van den Berg, M. Cost Benefit Analysis Model for Fire Safety and Deca BDE Case Study. *Proceedings of Fire and Materials 2007 Conference*, January 29–31, 2007, San Francisco, CA. Interscience Communications.
47. "Fire Emissions of Organics into the Atmosphere" Blomqvist, P.; Persson, B.; Simonson, M. *Fire Technol.* **2007**, *43*, 213–231.
48. "Combustion Characteristics of Flat Panel Televisions With and Without Flame Retardants in the Casing" Blais, M.; Carpenter, K. *Fire Technol.* **2015**, *51*, 19–40.
49. "Application of a Cost-benefit Analysis Model to the Use of Flame Retardants" Simonson-McNamee, M.; Andersson, P. *Fire Technol.* **2015**, *51*, 67–83.
50. "Flexible Polyurethane Foams: A Comparative Measurement of Toxic Vapors and Other Toxic Emissions in Controlled Combustion Environments of Foams With and Without Fire Retardants" Blais, M.; Carpenter, K. *Fire Technol.* **2015**, *51*, 3–18.
51. "Comparative Room Burn Study of Furnished Rooms from the United Kingdom, France and the United States" Blais, M. S.; Carpenter, K.; Fernandez, K. *Fire Technol.* **2020**, *56*, 489–514.
52. Morgan, A. B.; Wilkie, C. A. (Eds.). Flame retardant polymer nanocomposites. John Wiley & Sons, Hoboken, NJ, 2007.
53. "Flame Retarded Polymer Layered Silicate Nanocomposites: A Review of Commercial and Open Literature Systems" Morgan, A. B. *Polym. Adv. Technol.* **2006**, *17*, 206–217.
54. "Polymer/Layered Silicate Nanocomposites: A Review From Preparation to Processing" Ray, S. S.; Okamoto, M. *Prog. Polym. Sci.* **2003**, *28*, 1539–1641.
55. "Twenty Years of Polymer-Clay Nanocomposites" Okada, A.; Usuki, A. *Macromol. Mater. Eng.* **2007**, *291*, 1449–1476.
56. "Synthetic, Layered Nanoparticles for Polymeric Nanocomposites (PNCs)" Utracki, L. A.; Sepehr, M.; Boccaleri, E. *Polym. Adv. Technol.* **2007**, *18*, 1–37.
57. "A Review of Flame Retardant Nanocoatings Prepared Using Layer-by-Layer Assembly of Polyelectrolytes" Holder, K. M.; Smith, R. J.; Grunlan, J. C. *J. Mater. Sci.* **2017**, *52*, 12923–12959.
58. "Fire Retardancy in Nanocomposites by Using Nanomaterial Additives" Teles, F.; Martins, G.; Antunes, F. *J. Anal. Appl. Pyrolysis* **2022**, *163*, 105466.

2 Thermal Decomposition of Flame Retardant Polymers

Aurelio Bifulco[1], Sabyasachi Gaan[2], D. Price[3], and A. Richard Horrocks[4*]*

[1] *Department of Chemical Materials and Industrial Production Engineering (DICMaPI), University of Naples Federico II, Piazzale V. Tecchio 80, 80125, Naples, Italy*

[2] *Laboratory for Advanced Fibers, Empa Swiss Federal Laboratories for Materials Science and Technology, Lerchenfeldstrasse 5, 9014, St. Gallen, Switzerland*

[*] 2nd Edition Chapter Authors

[3] *Centre for Research and Innovation, University of Bolton, Deane Road, Bolton, BL3 5AB, UK*

[4] *Fire Chemistry Laboratory, CMRI*

2.1 STATE OF ART AND THERMAL DECOMPOSITION OF POLYMERS

In fires, polymeric materials are consumed by flaming combustion, which is a gas phase process. Thus, the polymer must decompose to yield volatile combustible species to fuel the conflagration. The conflagration involves a large release of combustion products, which move upward to the gas phase, causing the influx of more dense cold air to the condensed phase in the combustion zone (Figure 2.1). In the case of a very intense influx, the fire grows into a firestorm.[1,2] To begin, this chapter first considers the various processes by which pure polymer systems decompose. Then any influence by which the presence of oxygen can affect these processes is discussed. The different structures of the various polymer types influence the end consequence of any decomposition, and this will affect the resistance, if any, to combustion. At this point, the polymer combustion cycle will be described.

Pure polymeric materials decompose via one or more of the following simple processes:

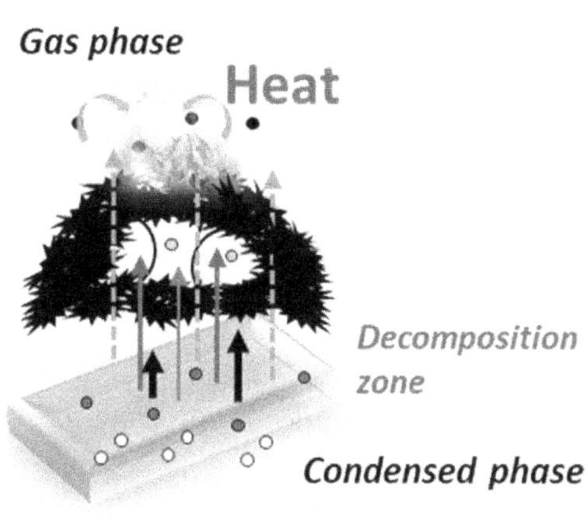

FIGURE 2.1 Schematic representation of polymer combustion cycle.

- End chain scission, individual monomer units successively cleaved from chain end;
- Random chain scission, scissions occur at random locations along the polymer chain;
- Chain stripping, atoms or groups not part of the polymer backbone are cleaved off;
- Cross linking, bonds created between polymer chains.

Table 2.1 collates various examples of each of these processes and the decomposition polymers obtained.

TABLE 2.1
Typical Decomposition Products for Each Generalized Mechanism of Polymer Decomposition

Mechanism	Examples of Polymer	Typical Products
Random Chain Scission	Polyethylene	Alkanes, alkenes, very little monomer
	Polypropylene	Alkanes, alkenes, very little monomer
	Polystyrene	Styrene monomer, dimer, and trimer.
	More generally	Monomers and oligomers
End Chain Scission	Polymethylmethacrylate	90–100% monomer
	Polytetrafluoroethylene	90–100% monomer
	More generally	Monomer
Chain Stripping	Polyvinyl chloride	Hydrogen chloride, aromatic hydrocarbons and char
	Polyvinyl alcohol	Water and char
	General	Small molecules and char
Cross-linking	Polyacrylonitrile	Char (and HCN)
	Poly(oxy-m-xylene)	Char
	General	Much char, few volatile products

Source: Adapted from reference 1.

DOI: 10.1201/9781003380689-2

Decomposition is also influenced by the chemical structure of the polymer, that is, straight chain, branched chain, or cross linked. In addition, synthetic polymers fall into three physical types, each of which will decompose differently when heated. These are thermoplastics, which will soften and melt before decomposing; thermoset (cross-linked), which do not melt and decompose yielding char and evolving volatiles; and elastomers, which are rubber-like materials. Recently, an additional class of plastics named "vitrimers" has been invented and is spreading in the scientific community and industry. Vitrimers are derived from thermosetting polymers and consist of molecular, covalent networks able to change their topology through thermally activated bond-exchange reactions.[3] They can flow like viscoelastic liquids at high temperatures, whereas the bond-exchange reactions are so slow at low temperatures that they behave like thermosets. Their characteristic behavior opens new horizons in the development of self-healing and highly reprocessable materials in a wide temperature range.[4] Lately, thermoplastics have been also modified to produce vitrimers and materials responding to different stimuli.[5] The decomposition of vitrimers involves the same steps as their precursors, though the operative conditions for a specific phenomenon may be changed.

In the main, the pure polymers decompose by some kinetic process. Madorsky[6] defined their relative thermal stability in terms of the temperature, T_h, at which their half-life equaled 30 minutes, see examples given in Table 2.2.[1,7]

From the T_h values, the effects of the various chemical structures on the thermal stability of these pure polymers can be deduced, see Table 2.3.

The Madorsky approach has the advantage of providing a simple method for comparing the thermal stabilities of a range of polymers by determining the temperature at which a polymer has lost half its initial weight in 30 minutes. However, it should be appreciated that it is based on the simplified assumption of a single first order decomposition reaction, which is not always valid. Thus, fundamental information concerning the decomposition should not be deduced from the T_h values determined.

However, the intrinsic thermal decomposition characteristics of any polymer may be influenced by impurity species present since polymers are rarely pure in the true chemical sense. Such impurities may include one or more of the following:

TABLE 2.2
Relative Thermal Stability of Selected Polymers Based on the Temperature at which Their Half-Life T_h Is 30 Minutes, Data Adapted from Madorsky[6]

Polymer	T_h (°C)
Polymethylmethacrylate A (molecular wt. 1.5×10^5)	283
Polymethylmethacrylate B (molecular wt. 1.5×10^6)	327
Poly alpha-styrene	287
Polyisoprene	323
Polymethylacrylate	328
Polyethylene oxide	345
Polyisobutylene	348
Polystyrene	364
Polypropylene	387
Polydivinyl benzene	399
Polyethylene	406
Polymethylene	415
Polytetrafluoroethylene	509

TABLE 2.3
Factors which Affect the Thermal Stability of Polymers, after Madorsky[6]

Polymer	Effect on Thermal Stability	Examples	T_h (°C)
Chain branching	Weakens	Polymethylene	415
		Polyethylene	406
		Polypropylene	387
		Polyisobutylene	348
Double bonds in polymer backbone	Weakens	Polypropylene	387
		Polyisoprene	323
Aromatic ring in polymer backbone	Strengthens	Polybenzyl	430
		Polystyrene	364
High molecular weight	Strengthens	PMMA B	327
		PMMA A	283
Cross-linking	Strengthens	Polydivinyl benzene	399
		Polystyrene	364
Oxygen in the polymer backbone	Weakens	Polymethylene	415
		Polyethylene oxide	345
		Polyoxymethylene	< 200

Notes: *included for comparison, no double bond in backbone **included for comparison, aromatic groups pendant

- Impurities already present in monomeric feeds to polymerization plants, although it may be generally stated that monomer purity is recognized as a critical variable by commercial polymer producers;
- Polymerization initiation or catalyst residues present in both addition and condensation polymers;
- Products of decomposition generated during polymerization and processing, often of a thermally-derived origin. These may include products of thermal oxidation (see next section) or compounds formed in the upcycling of melting polymers;
- Contaminants introduced during processing including atmospheric oxygen and metallic ions released from processing plant equipment. As for example in the case of thermoplastics, which may be contaminated by metallic species released during the extrusion process and manufacturing transformations.

These factors tend to be specific to each polymer type and its related polymerization history and will be referred to later when discussing individual polymer decomposition behavior, if relevant. It is sufficient to state at this point that the consequences of these impurities are usually one of promoting the overall decomposition, because of a catalytic function, and gives rise to the slow thermal decomposition and related deterioration in polymer properties often experienced when in use and exposed to service temperatures well below their normal rapid decomposition temperatures as defined by T_h in the previous section. In many cases, when exposing a polymer to its maximum service temperature, its effective lifetime is determined by the length of the induction period for these low temperature decomposition reactions to promote sufficient loss in properties so as to render it useless for its intended purpose. Such induction times may be quantified in terms of times to embrittlement, to lose 50% tensile strength, or to change its character (e.g., color) by a specified magnitude. Often, aging at temperatures above the service life temperature but below the polymer melting point, for example, enables an "apparent" activation energy to be determined based on assumed Arrhenius law behavior, from which service lives may be predicted.[7–9] Since the decomposition process is complex consisting of a series of simultaneous and possibly consecutive reactions, it should be stressed that such "apparent" activation energies determined in this manner are oversimplified values since they are composite values. Thus these "apparent" activation energies do not provide any fundamental information concerning the individual reactions which contribute to the overall complex decomposition process, though it remains an important tool for analyzing and understanding reaction rates, mechanistic details of complex reaction mechanisms, elementary-step energetics, catalytic activity, and reaction selectivity. Thermogravimetric analysis has been widely used to investigate polymer decompositions.[8,10] Such studies often indicate that the decomposition is a multistep process, the chemistry of the decomposition changing as the temperature is raised. To gain an insight into the chemistry occurring, evolved gas analysis techniques such as FTIR[11] and MS[12] are coupled to the TGA instrument. The microscale calorimetry technique developed by Lyon[13] combines the ability to study the pyrolysis of polymers in an inert gas flow followed by complete combustion (high-temperature oxidation) of the evolved volatile gases.[14,15] Oxygen depletion is used to measure the heat of combustion, which can be related to the flammability of the material under investigation. The technique provides a method of evaluating both the thermal behavior and the flammability of milligram quantities of polymer materials.[16,17] The microscale calorimetry is a non-flaming test, where the condensed phase and gas phase processes of flaming combustion occur separately, whereas the cone calorimeter simulates a real fire event, as the combustion of the tested sample takes place continuously and in the presence of air. Microscale combustion calorimetry can be very useful to study the mechanisms concerning the gas phase of polymer decomposition. Solid state NMR[18] and selective isotopic labelling[19] provide even more detailed insight into the decomposition chemistry. Most research has been carried out on single polymers. Sometimes the behavior of polymer mixtures is of interest. The distribution kinetics of a binary polymer mixture decomposition has been investigated by applying distribution kinetic theory based on molecular weight distributions.[20] Attempts have been made to apply quantum chemistry and molecular simulation techniques to determine the mechanisms and rates of polymer decompositions.[21,22] Also, some methodologies based on molar group contributions have been proposed to evaluate the polymer flammability.[23] It is hoped that this work will lead to methods for predicting material flammability and the development of strategies to improve fire resistance.

In addition to the challenges posed by the presence of an impurity, polymeric materials are rarely used in the "pure" or even stabilized state but are normally compounded with various compounds designed to enhance their properties – for example, flexibility, mechanical strength, color, stability, fire resistance – so that they can be deemed "fit for purpose." For example, stabilizing additives are usually included during the processing stages of any polymer, and because of the complexity and often interrelated nature of secondary reactions, such stabilizers may be required to minimize thermal decomposition and oxidation in a concerted manner (see next section). Among the additives, a plasticizer is a substance that is added to a polymeric material to make it softer and more flexible, to increase its plasticity, to decrease its viscosity, and/or to decrease friction during its handling in manufacture. As for example, plasticizers are commonly added to polyvinyl chloride to make it soft and pliable, making it suitable to be used as shower curtains, vinyl flooring, electric wire insulation/coating, etc.

This chapter and the subsequent chapters in the book are focused on the fire resistance aspect of modern-day polymeric materials, plastics, and textiles, as they are more commonly referred to.

2.1.1 Oxidative Decomposition

Polymer decomposition is almost always faster in the presence of oxygen or air due to the accelerating reactions between oxygen and carbon centered radicals (RO·) released from the initial decomposition products. These interactions with oxygen result in an increase in the concentration of polymer alkyl radicals (R·) leading to higher levels of scission and cross-linked products. Also, fragmentation reactions of oxygen-centered radicals yield new oxidation products with structures not found under an inert atmosphere. These radicals can proceed to undergo abstraction, fragmentation, and combination reactions both with the original polymer and other products from the decomposition. Such reactions can affect the polymer during processing, particularly if the temperature required is high, and also its performance during its end-use. For example, photo-oxidation reactions cause deterioration in the mechanical and physical properties of LLDPE during the early stages of exposure. Photo-oxidation studies of LLDPE-based films produced with the spherilene technology have demonstrated that the decomposition of LLDPE results in the formation of carbonyl and vinyl species.[24,25] The photo-oxidative exposition triggers chain scission, which makes the amorphous chains more mobile and freer for further crystallization, especially when the decomposition occurs simultaneously in many chains.[24,25] Antioxidants can be added to the plastic formulation to inhibit such effects. Antioxidants function by interfering with the radical reactions leading to polymer oxidation and decomposition.

To understand these reactions, the so-called Bolland and Gee reaction scheme[26,27] and its subsequent developments have been applied to explain the chain reaction characteristics of both thermal and photo-oxidation of polyolefins in the main. The scheme (Scheme 2.1) has been found to be

Initiation: $$RH \xrightarrow{\Delta} R^{\bullet} + {}^{\bullet}H$$

Propagation: $$R^{\bullet} + O_2 \rightarrow RO_2^{\bullet}$$

$$RO_2^{\bullet} + RH \rightarrow ROOH + R^{\bullet}$$

Termination: $$R^{\bullet} + R^{\bullet} \rightarrow R\text{-}R$$

$$RO_2^{\bullet} + R^{\bullet} \rightarrow ROOR$$

$$RO_2^{\bullet} + RO_2^{\bullet} \rightarrow \text{Products}$$

Chain branching: $$ROOH \rightarrow RO^{\bullet} + {}^{\bullet}OH$$

SCHEME 2.1 The general Bolland and Gee mechanism for the oxidation of polymers, RH.

a useful model for many other polymers comprising significant aliphatic character such as the aliphatic polyamides and polyesters and certain polyvinyls, including PVC.

2.2 Decomposition of Individual Polymer Types

When polymers are subjected to heat, it is generally the case that the weakest bonds will break first, and these determine the overall character of the subsequent decomposition pathways defined in section 2.1 and exemplified in Table 2.1. Since flammability is associated with the availability and ease of oxidation of volatile decomposition products, it is the decomposition pathways that form volatiles that are of importance in the first instance. However, since cross-linking reactions give rise to eventual char formation and thus may minimize volatile formation, these reactions are essential in determining the potential of a polymer to be rendered flame retardant by condensed phase flame retardants that may favor cross-linking. Char-forming behavior is highly desired to form a thermal shield and oxygen barrier, which may not only decrease heat coefficients at the boundary phase but also slow down the thermo-oxidative reactions during the burning.[28]

In the discussion in the next section, these reactions will be emphasized only within the overall context of the complex decomposition processes that most polymers exhibit when thermally decomposed.

2.2.1 Thermoplastics

2.2.1.1 Polyolefins

For both polyethylene and its many copolymeric variants and polypropylene, the main thermal degradative routes follow initial random chain scission. These reactions are only slightly affected by the differences in physical structure such as crystallinity but are influenced by the presence of impurities. However, it is largely true that while these may influence the processibility and long-term stability of respective polyolefins, they have little or no effect on the flammability.

In the case of polypropylene, pyrolysis is dominated by initial chain scissions, usually at the carbon-carbon bond adjacent to the labile tertiary hydrogen atom in the repeat group. Research has shown that heating the polymer, including waste polypropylene, generates a mixture of quite clean hydrocarbon fuels[29,30] and other valuable products such as lubricants.[31,32] This fuel-forming tendency explains the high flammability of polypropylene and the difficulty of generating high levels of flame-retardant properties while maintaining optimum polymer properties.

When heated under non-isothermal conditions, the maximum volatile product evolution temperature was 425°C for isotactic PP to yield volatile products comprising dienes, alkanes, and alkenes. Furthermore, the hydrogen content of pyrolysis products obtained by flash pyrolysis at 520°C indicates the magnitude of the flammability problem in terms of its fuel-forming potential.[33] The flammability of volatiles is further enhanced by the abundance of unsaturated less-volatile fuel fragments which behave as secondary fuel sources and which decompose further.[34]

The complete absence of cross-linking reactions prevents potential char-forming reactions from being favored in the presence of conventional condensed phase flame retardants, and so the most effective flame retardants for polyolefins are usually bromine-based so that flame inhibition in the vapor phase is affected or intumescent-based where char-promotion arises from the flame retardant itself. Recently, meltable alkyl sulfone bridged compounds have been synthesized and used as flame retardants for polypropylene.[35] Such compounds were able to be melt-processed with polypropylene, thanks to their high thermal stability (> 250°C).[36] Also, the incorporation of meltable alkyl sulfone bridged compounds into PP improved the thermo-oxidative stability and fire behavior of polypropylene, reducing the heat release rate by up to 48% compared to the blank polymer.[36]

2.2.1.2 Aliphatic Polyamides

The examples of PA 6 and 6.6 will illustrate the challenges that these polymers create. The classical research into the thermal decomposition occurred during the 1950–1970 period, and extensive reviews of this work include those by Peters and Still.[37,38] Essentially for all linear, aliphatic polyamides, thermal decomposition is influenced by one major factor: The strength of the weakest chain bonds around the amide group (Figure 2.2).

With bond cleavages occurring at the arrowed positions and preferential cleavage suggested to occur at the –NH–CH_2 bond.[6,38,39] These occur randomly and give rise to the gaseous products, NH_3, CO, and CO_2, low molecular weight fragments and subsequent decomposition products from these latter. Of the simple gases, only CO is flammable, but the volatiles generated from the smaller polymer chain fragments provide the major fuel components. The earliest PA-6.6 pyrolysis work published showed the products to comprise cyclopentanone, its derivatives, and various hydrocarbons, and this was supported by later work[38] although the former was unique to PA-6.6 and not PA-6. Thermal decomposition of nylon-6 involves the depolymerization of its monomer, caprolactam, which is not only faster at higher temperatures but is volatile.

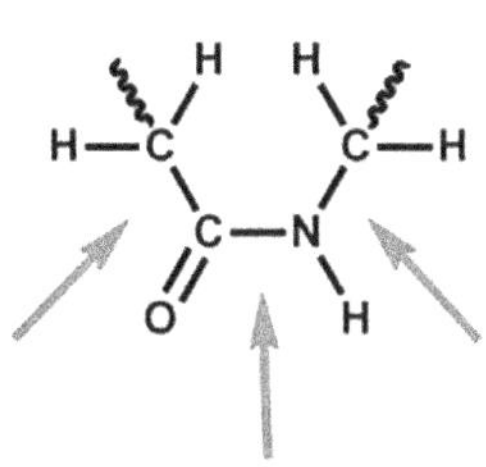

FIGURE 2.2 Representation of the chain bonds around the amide group.

Thermal lability of aliphatic polyamides in general is influenced by the potential for ring-forming during chain decomposition, and this is particularly the case with PA-6.6 in which the adipate repeat unit enables the formation of a six-membered intermediate along the polymer chains with eventual formation of cyclopentanone and its derivatives.[40]

The tendency of certain aliphatic polyamides to form 3-dimensional structures leads to gel formation. PA-6.6 is particularly prone to this and explains why melt extrusion processes often require more interruptions because of potential gel blockages than is the case with nylon 6, for example, PA-6.6 gels typically after six hours at 305°C while PA-6 may be heated for up to ten days at 281°C before it gels.[38] While gel formation mechanisms are not well understood, in nylon-6.6, the formation of cyclopentanone derivatives and their subsequent reaction products are believed to be involved.

It is thus apparent that the overall flammability of the simple polyamides is determined by their relative propensities to shrink and melt away from an ignition source in the first instance followed by the nature of the volatiles formed, which if ammonia and carbon dioxide are significantly present, will have a reduced fuel value. Any flame-retardant strategy may thus address this volatile formation or perhaps more interestingly accelerate gel formation, which could lead to a significant char-forming character. Unfortunately, to date, few flame retardants have been successfully commercialized for PA-6 and PA-6.6 partly because of the reactivity of polyamide melts to bromine-containing retardants and also the adverse effects of phosphorus-containing species on the molecular weight of melts during processing. Levchik and Weil[41] have reviewed this whole area and show that certain melamine salts in particular show promise. Exolit® OP 1230 is a flame retardant based on organic phosphinates and is one of the most used non-halogenated compounds for reinforced polyamide-6, polyamide-66, and high temperature polyamides.[42] Exolit exerts its flame-retardant mechanism through a combined gas phase and condensed phase mode of action. To give an example, the addition of a certain amount of Exolit into GF–PA6 provides self-extinguishing materials with very good thermal behavior.[42]

2.2.1.3 Polyesters

The principal linear polyester is PET, and so this will be the chosen exemplar. Studies of its thermal degradative behavior mirror those of the aliphatic polyamides in the previous section, in that during the commercial development of PET during the 1950s and 1960s, the basic research work was undertaken.[37] While some cross-linking tendency has been identified,[43] in the main, random chain scission dominates thermal decomposition with the major product being acetaldehyde formed at temperatures up to 290°C along with smaller amounts of CO, CO_2, and ethane and very small amounts of other fuels such as methane and benzene.[6,39] A simplified version of the primary stage appears to be as shown in Scheme 2.2 in which it is seen that acetaldehyde

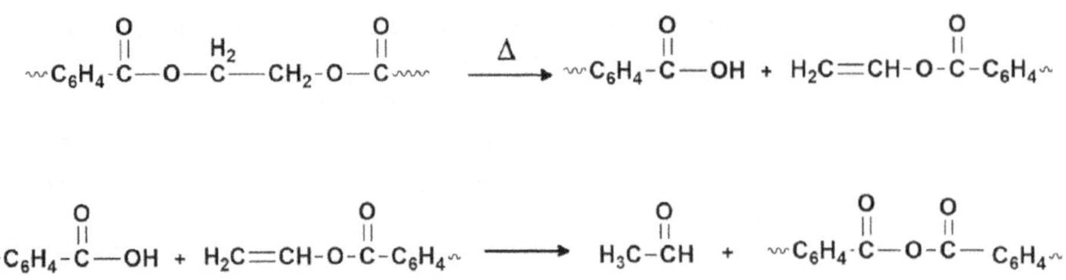

SCHEME 2.2 Initial stage of thermal decomposition of poly(ethylene terephthalate).

is formed as the major initial flammable volatile. Action of further heat causes polymerization of the vinyl ends coupled with the loss of CO and CO_2 as the anhydride links undergo further scission.

It is evident that any flame retardant must counteract the effect of or reduce the amount of acetaldehyde formed. While the actions of bromine- and phosphorus-containing species have achieved varying degrees of success, no successful flame retardant to date has managed to confer a significant char-forming character to the decomposition mechanism, and this is perhaps an indication in the challenges involved with effectively flame retarding linear polyesters in general. As PET is one of the "big five" recyclable plastics, the scientific community is looking for the development of always more efficient process to recycle poly(ethylene terephthalate) containing flame retardants. The issue is related to the downgrade during the thermo-mechanical recycling of flame retardant PET. Recently, Gooneie et al. found out that the addition of two phosphorous-based compounds, DOPO-PEPA and Aflammit PCO 900, to PET by extrusion enhanced lubrication and control melt rheology over long periods by stabilization.[44,45] Besides, Aflammit PCO 900 catalyzed the chain extensions and branching in PET, which could counter chain scissions to some extent. The analysis of evolved gases revealed that the decomposition of both flame retardants released active radicals and moieties that either inhibited other radicals such as oxygen radicals or initiated a reaction with the PET chains leading to chain scissions and/or branching. The addition of DOPO-PEPA into PET allowed to preserve ductile tensile behavior after recycling, whereas the incorporation of Aflammit PCO 900 conferred brittle characteristics to the polymeric material.[44,45] In the future, the synthesis of new phosphorous-based flame retardants may result in the development of PET products with improved recyclability and fire behavior.[44,45]

2.2.1.4 Polyacrylonitrile

Most commercial polymers comprising AN are copolymeric, and those containing the highest levels of AN monomer, usually 85 wt% or more, are for use in fiber end uses, including carbon fibers, where they are major precursors. It is as a consequence of their importance as carbon fiber precursors that most research on the thermal degradative and oxidative processes associated with acrylonitrile copolymers has focused in this area and took place over the 1960–1980 period in the main.[37,43,46] It is generally accepted that the pyrolysis of AN-containing copolymers of this type is dominated by the behavior of the AN monomeric unit itself and that this undergoes a cyclization reaction accompanied by an intense exotherm either in an inert atmosphere or in the presence of oxygen.[47] This gives rise to a so-called "ladder" structure as opposed to the random chain scission of chains into potential volatile product formation. The cyclization reaction occurs before any mass is lost and results in the formation of aromatic rings. The presence of head-to-head linkages within the polymer controls the extent of cyclization.[48] The analysis of evolved gases has demonstrated that ammonia and hydrogen cyanide are the first gases lost, whereas oligomers are lost from the uncyclized portion of the polymer lying between the cyclized portions and the amount of nonvolatile fraction. This latter is largely determined by the extent of oligomer loss.[48] In carbon fiber production, this cyclization is closely controlled by heating in an oxygenated atmosphere to produce so-called oxidized acrylic fibers which have acceptable fiber properties in their own right.[49] Furthermore, because they are highly carbonized, these fibers have a high inherent fire resistance with LOI values of 50–55 vol%. Subsequent heating in an inert atmosphere converts these fibers into carbon fibers having an essentially graphitic structure.

In parallel, however, has been the exploitation of fiber-forming acrylic copolymers in the textile area where they produce fabrics having similar levels of flammability as cotton with similar Limiting Oxygen Index values of about 18 vol%. This high level of flammability at first sight appears to be at odds with the cross-linking, carbonizing reactions observed in carbon fiber production. However, work in our own laboratories[50,51] showed that the pyrolysis mechanism is both temperature and heating rate dependent. Under the slow heating conditions and temperatures up to 400°C associated with carbon fiber production, the cyclization and cross-linking reactions prevail, whereas under the high heating rates and temperatures above 400°C associated with burning, volatilization and fuel-forming reactions predominate.

It, therefore, becomes evident that in order to flame retard polymers containing high levels of acrylonitrile, this tendency to volatilize at high heating rates must be overcome. In commercial terms, this has proved to be impossible to date and the only successful AN-containing, fiber-forming polymer is the group of modacrylics, containing between 35 and 85 wt% AN with the other comonomers being a halogen-containing species such as vinyl chloride or vinylidene chloride. These release chlorine atoms into the flame on heating and so act as vapor phase flame retardants. However, at the experimental level, Zhang et al. have also shown that the volatilization reactions may be suppressed in favor of char formation if ammonium polyphosphate and similar flame retardants are introduced, but as yet, these have not been commercialized.[50,51]

2.2.1.5 Polystyrene

Polystyrene is well-known for a multitude of general-purpose applications. Derivatives with superior properties for particular applications, in particular, ABS and rubber-modified, HIPS are used to replace PS or used in conjunction with it. General purpose polystyrene has been shown to decompose at 280°C.[52] The decomposition is initiated by main chain scission at a head-to-tail unit present in the polymer due to polymerization termination by radical coupling. The resulting macroradicals readily unzip to expel styrene monomer. A more thermally stable form of polystyrene can be produced via nitroxyl-mediated polymerization of styrene monomer followed by removal of the nitrosyl-end groups by a reductive technique.[53] Untreated PS decomposes above 300°C producing mainly styrene monomer plus lesser amounts of the dimer, trimer, and tetramer, all of which are highly flammable; as a consequence, the LOI value for PS is 19.0 vol%. The mechanism is dominated by chain scission, depolymerization, intramolecular hydrogen transfer, and bimolecular termination.[1,2,54] The main products are styrene monomer and its oligomers plus benzene and toluene. As would be expected from the aromatic nature of the PS structure, the principal flame retardant mechanism occurs in the condensed phase facilitating char formation. Post-polymerization modification to flame retard PS is easily achieved via electrophilic aromatic substitution of suitable flame retardant groups onto the phenyl rings. Successful methods include boronation,[55,56] sillation,[55,56] and phosphorylation.[57] PS, ABS, and HIPS are more often flame retarded using additives that are cost-effective and easy to process. Various halogenated-flame retardant/ATO combinations which evolve flame inhibitors under fire conditions are often used. Recently, AP has been largely exploited for the preparation of flame retardant PS.[58,59] Yan et al. obtained polystyrene based composites showing self-extinguishing capability through the incorporation of 25 wt% AP.[58,59] The flame retardant could exert a combined gas phase and condensed phase activity in the combustion and decomposition of the composites.

2.2.1.6 Poly(Vinyl Chloride)

Loss of the flame inhibitor HCl, via autocatalytic chain stripping, occurs from about 100°C. Thus PVC has its own "built-in" fire retardant, which is why PVC initially found extensive use in hazardous situations such as coal mines. This reaction yields other products which can be involved in other reactions such as cross-linking.[60,61] The conjugated double bonds resulting from the loss of HCl give rise to aromatic structures, for example, benzene, that burn producing significant quantities of hazardous smoke. Extensive studies have been undertaken to identify efficient smoke suppression systems for PVC, in particular those by Starnes[60] and Carty.[27] Alternatively, the polyene structures can continue to undergo cross-linking to produce the much less flammable char.

2.2.1.7 Ethylene/Vinyl Acetate Copolymers

EVA is a widely used material, particularly as a low cost, zero-halogen sheathing material in the electric cable industry. EVA is known to form a protective layer that can inhibit combustion.[62] In the literature, it is reported that the heating of EVA composites results in a two-step decomposition over the ranges 360–450°C and 450–550°C.[11,62] The first step is due to the evolution of acetic acid, while the second one to the release of a mixture composed by 1-butene, carbon dioxide, ethylene, methane, and carbon monoxide. A recent study[18] of the mechanism and kinetics of PVA and EVA decompositions has shown that the deacetylation process leaves a highly unsaturated polyene-type residue. The deacetylation of PVA is autocatalytic, but upon the incorporation of ethylene entities into the polymer backbone, this autocatalysis disappears. Between 400–500°C, the polyene will decompose further by chain scission reactions in inert conditions or aromatize in an oxidative environment into a char, and eventually CO_2 above 500°C. Under inert conditions, deacetylation is endothermic, but in the presence of oxygen, large exothermic effects are found for each decomposition step. This indicates the occurrence of additional oxidation reactions during deacetylation, an important reorganization of the polyene structure prior to char formation and oxidation of the latter to CO_2. Recently, Costache et al. studied the decomposition pathway of ethylene–vinyl acetate copolymers and their nanocomposites, confirming that the presence of a clay affects the decomposition pathway of EVA.[63] More specifically, the presence of a clay with hydroxyl groups on the edges accelerates the first step of the decomposition involving the loss of acetic acid by chain stripping, but also the filler may influence the quantity and the type of products formed in the second step of the decomposition.[63]

2.2.2 Foams

Thermal decomposition of foams is no different from that of the solid polymer except in that the foam structure imparts superior thermal insulation properties and low heat transfer so that the decomposition of the foam will be slower than that of the solid polymer. Almost every plastic can be produced with a foam structure, but only a few are commercially significant. Of these, flexible and rigid polyurethane foams, which have urethane links in the polymer chain, are the most important. The thermal decomposition products of polyurethane will depend on its composition, which can be chemically complex due to the wide range of starting materials and combinations which can be used to produce them and their required properties.[64] Basically, these involve the

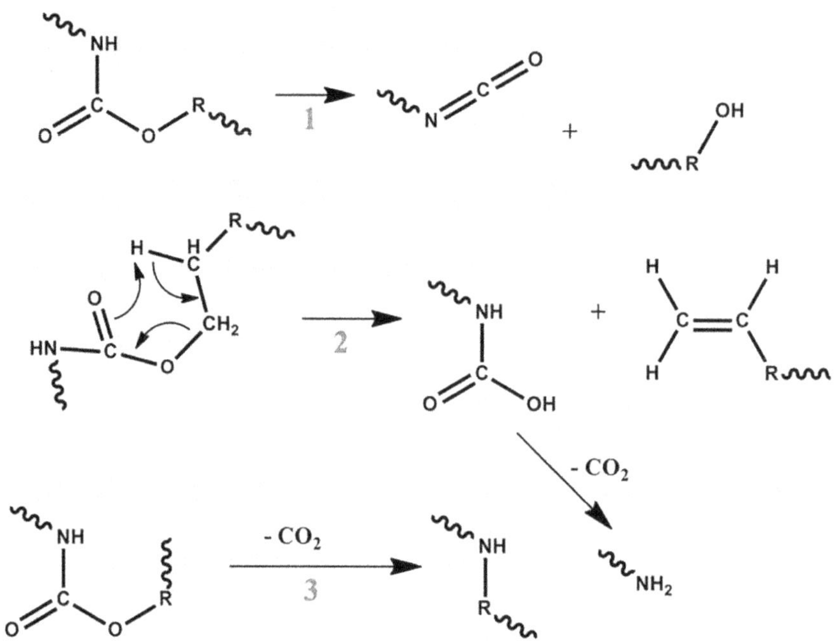

FIGURE 2.3 Decomposition mechanisms of urethane bonds: (1) reversal to the original isocyanate and alcohol; (2) formation of the primary amine, olefin, and CO_2 through the intermediate state of a six-membered ring; (3) formation of the secondary amine and CO_2 through the intermediate state of a four-membered ring.

reaction between isocyanates, such as TDI or MDI, and polyols (i.e., polymers showing an average of two or more hydroxyl groups per molecule). If the requirement is for greater heat stability and reduced brittleness, then MDI is favored over TDI. Polyurethanes are combustible, and their decomposition can produce significant amounts of hydrogen cyanide and carbon monoxide, together with nitrogen oxides, isocyanates, and other toxic products. Due to the high flammability of polyurethanes, these materials are normally treated with flame retardants.[65,66] Urethane linkages tend to dissociate above about 200°C. Fabris[67] indicated that urethanes from many isocyanates and primary and secondary alcohols begin to decompose at 150–200°C proceeding at a measurable rate above this range. Urethane bonds decompose by the mechanisms reported in Figure 2.3.[68]

Woolley[69] used GC/MS to investigate the thermal decomposition of commercial TDI-based flexible foams under nitrogen. The decompositions began with urethane bond scissions at 200–300°C to yield relatively nonvolatile polyol components and nitrogen-rich volatiles. The latter were termed "yellow smoke" and appeared to be polymerized or condensed forms of TDI with some free TDI. At higher temperatures, further decomposition of the polyol residue occurs to yield small organic species. Scheme 2.3 describes the mechanism outlined previously.

The major application of PU foam is for upholstered furniture. Because of their large surface area and high air permeability, polyurethane foams are highly flammable. As a consequence, it is essential that flame retarded PU foam be used in upholstered furniture.[70] Chlorinated phosphate esters are widely used to flame retard PU foams. These have

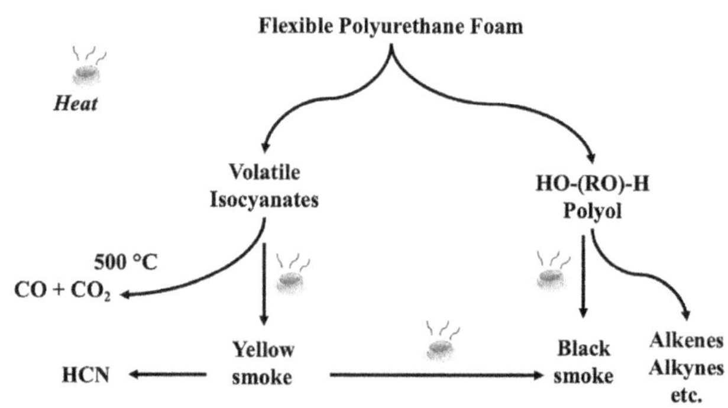

SCHEME 2.3 Thermal decomposition of flexible polyurethane foam.[69]

the disadvantage that they can increase smoke formation. An additive that can effectively trap the volatile isocyanate evolved during the thermal decomposition of the foam can lead to a reduction in the smoke and toxic gas yields. A common example is melamine. Price et al.[71] studied the reduction of smoke due to the presence of melamine in polyurethane foams. Overall, the interaction between melamine and the released isocyanate fraction arising from the decomposition of polyurethane foam is considered as the main reason for the smoke-suppression of melamine. Although no reaction is believed to occur between melamine and TDI during the manufacture of polyurethane foam at processing temperatures around 100°C, at higher temperatures interaction may occur. The melamine-NH_2 group is very reactive towards an isocyanate (–NCO) group. Thus, the reaction shown in Scheme 2.4 would be expected to occur when the temperature is over 250°C.

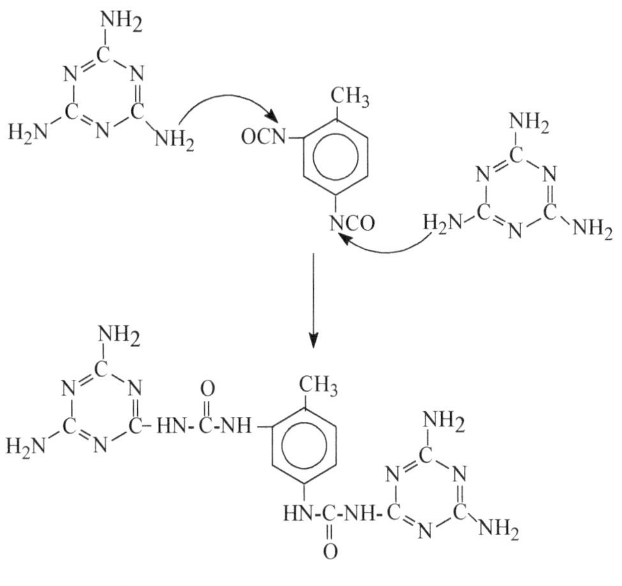

SCHEME 2.4 Melamine-isocyanate interaction above 250°C.[71]

The polymeric structure so formed would reduce the amount of aromatic smoke precursors volatilized, thus reducing the smoke released. This type of structure would decompose to a char, which will protect the remaining foam. Gaan et al. found that the addition of EDAB-DOPO into a polyurethane foam resulted in higher residual char and very low flammability due to a strong gas phase activity and inhibition.[72] The addition of PTFE had a synergistic effect, and the FR action was attributed to the condensed phase action of PTFE.[73]

2.2.3 Thermosets

Thermoset resins cover an extremely wide range, including phenol formaldehyde polymers, aminopolymers, polyurethanes, epoxies, and thermoset polyesters, which include alkyd and unsaturated vinyl ester resins. Of special interest at the present time are those that comprise the resin component of fiber-reinforced composites, which are finding increasing use in commercial and defense sectors where fire resistance is of paramount importance. Typical resins used here are those listed in Table 2.4 along with typical, respective LOI values in descending order of increased inherent fire resistance. Their general thermal stability and flame retardancy have been recently reviewed.[74–76]

2.2.3.1 Polyester Resins

Polyesters are probably the most commonly used polymeric resin materials and consist of a relatively low molecular weight unsaturated polyester chain dissolved in styrene, which on curing forms cross-links across unsaturated sites in the polyester. Figure 2.4 reports the typical formula for a resin.

Most polyesters start to decompose above 250°C, whereas the main step of weight loss occurs between 300

TABLE 2.4
Thermoset resins used in composites.

Resin Type	LOI, vol%
Polyester	20–22
Vinyl ester	20–23
Epoxy	23
Phenolic	25
Polyaromatic melamine	30
Bismaleimide	35

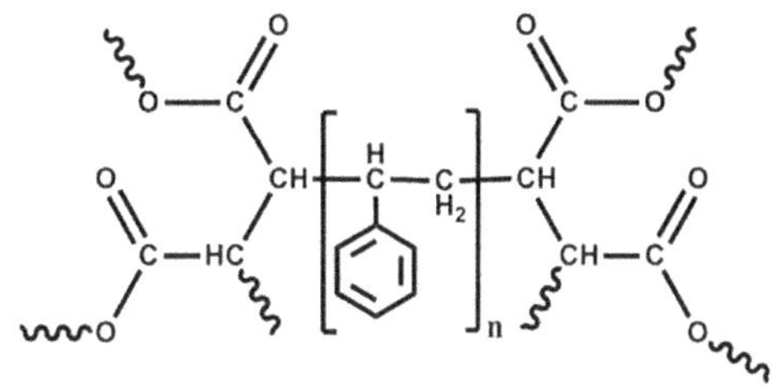

FIGURE 2.4 Typical chemical structure for a polyester resin.

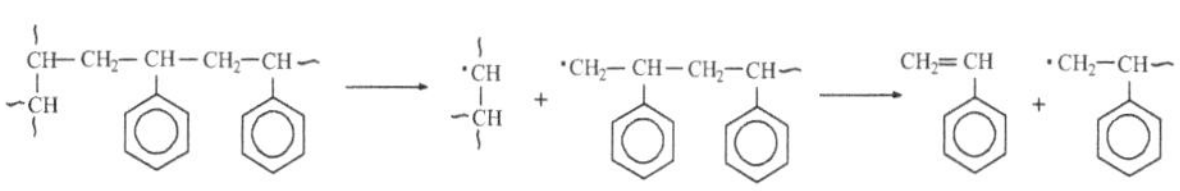

SCHEME 2.5 Initial chain scissions in polyester.[41,77]

and 400°C.[77] During thermal decomposition, polystyrene cross-links start to decompose first, and styrene is volatilized according to Scheme 2.5.

The linear polyester portion undergoes scission similar to thermoplastic polyesters, undergoing decarbonylation, decarboxylation, or splitting off of methylacetylene. Because of the ease of formation of these flammable pyrolysis products, polyesters have LOI values of 20–22 vol% (see Table 2.4) and, hence, burn readily and, because of the styrene content, give heavy soot formation. Because these resins are cured at room temperature, bromine-containing flame retardants, which would decompose in melt-processed, thermoplastic polymers, may be used with effectiveness here.

Bansal et al. observed that the cured bisphenol-A-based polyester resins are thermally more stable than the cured isophthalic acid-based ones, whereas the polyester resins cured with methyl ethyl ketone peroxide resulted in a thermally more stable matrix compared to the benzoyl peroxide-cured products.[78] The research group also found that all investigated polyester resins decompose in the first step involving the scission of cross-links/weak links with the liberation of free linear chains and subsequently a second step characterized by a random scission of the free linear chains into smaller fragments.[78]

2.2.3.2 Vinyl Ester Resins

These are mainly derived from the reaction of an epoxy resin, for example, bisphenol A diglycidyl ether, with acrylic or methacrylic acid. Figure 2.5 shows the general formula for vinyl ester resins.

Like unsaturated polyesters, they are copolymerized with diluents such as styrene using similar free-radical initiators. They differ from polyesters in that the unsaturation is at the end of the molecule and not along the polymer chains. Their burning behavior falls between that of polyester and epoxy resins (LOI = 20–23 vol%, Table 2.4).

2.2.3.3 Epoxy Resins

These resins, extensively used in the aerospace industry, consist of an epoxy resin component, often based on epichlorohydrin and a curing agent (e.g., aliphatic, cycloaliphatic, or aromatic compound) and comprising the epoxy or glycidyl group (Figure 2.6).[79]

This group will react typically with phenolic–OH groups and Bisphenol-A type resins to yield a general structure as shown in Figure 2.7, where X can be H and Y depends upon the structure of the curing agent. This yields a relatively thermally stable structure with the weakest bonds at the ether linkage, –O–.

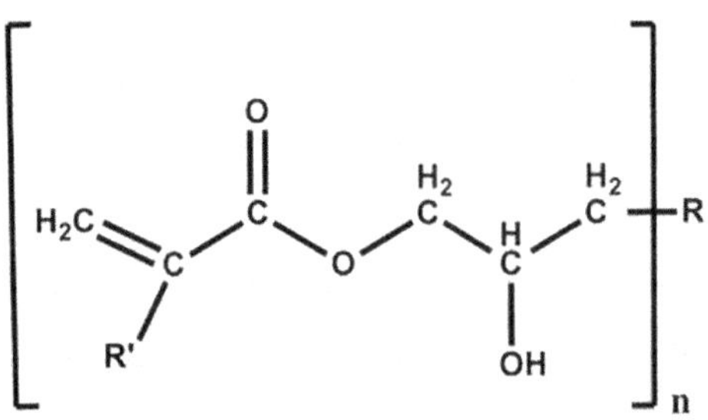

FIGURE 2.5 Typical chemical structure for a polyester resin. R is any aliphatic or aromatic residue and R' is typically either H or CH_3.

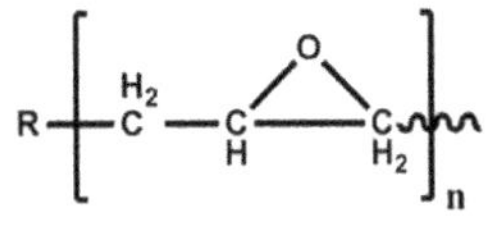

FIGURE 2.6 Chemical structure of an oxirane ring. R is an aliphatic or aromatic residue.

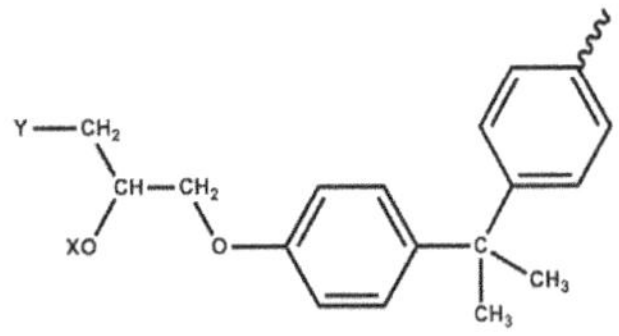

FIGURE 2.7 Typical chemical structure of a bisphenol-A type resin after reaction of oxirane ring.

During early stages of thermal decomposition, the reactions are mainly non-chain-scission type, whereas at higher temperatures, chain-scissions occur.[80] The most important non-scission reactions occurring in these resins are the competing dehydration and dehydrogenation reactions associated with secondary alcohol groups in the cured resin structures. The main products are methane, carbon dioxide, formaldehyde, and hydrogen.

During chain scission reactions, the aliphatic segments break down into methane and ethylene (and possibly propylene), acetone, acetaldehyde, and methane (and probably carbon monoxide and formaldehyde); all of which are flammable. From the aromatic segments of the polymer, phenol is liberated. For phthalic anhydride–cured resins, phthalic anhydride is regenerated together with CO and CO_2, benzene, toluene, o- and p-cresols and higher phenols. However, the flammable volatiles outlined above are produced only in relatively small quantities, and this, coupled with their cross-linked and related char-forming character, ensures that epoxy resins are less combustible than polyester resins with higher LOI values in the range of 22–23 vol%.

It is well-known that the flame retardation of aliphatic cured epoxy systems is not easy because of the high flammability of the polymer matrix, which contains a lower content of thermally stable aromatic moieties, with respect to resins cured with aromatic hardeners. Recently, Bifulco et al. proposed some hybrid strategies based on the sol-gel modification of epoxy and the addition of nonreactive DOPO-derivatives and other synergists.[81,82] The research group found out that the incorporation of DOPO-AAM or DOPO-PEPA, together with melamine, into silica-epoxy nanocomposites allows to achieve self-extinguishing capability, even with the use of an aliphatic hardener (isophorone diamine) and very low loadings of phosphorous.[81,82] The presence of melamine and in situ sol-gel silica in the epoxy matrix boosted the formation of an abundant and intumescent char, acting as an efficient thermal insulator. Also, the decomposition of DOPO-AAM and DOPO-PEPA resulted in the release of phosphorus flame inhibitors in the gas phase, suppressing oxygen radicals in the flame and in the production of phosphoric acid compounds accelerating the carbonization process (Figure 2.8).[81,82]

2.2.3.4 Phenolic Resins

Reaction of phenol with less than equimolar proportions of formaldehyde under acidic conditions gives so-called novolac resins containing aromatic phenol units linked predominantly by methylene bridges. These are thermally stable and can be cured by cross-linking with formaldehyde donors such as hexamethylenetetramine. However, the most widely used phenolic resins for composites are resoles manufactured by reacting phenol with a greater than equimolar amount of formaldehyde under alkaline conditions. Resoles are essentially hydroxymethyl functional phenols or polynuclear phenols with the general formula showed in Figure 2.9.

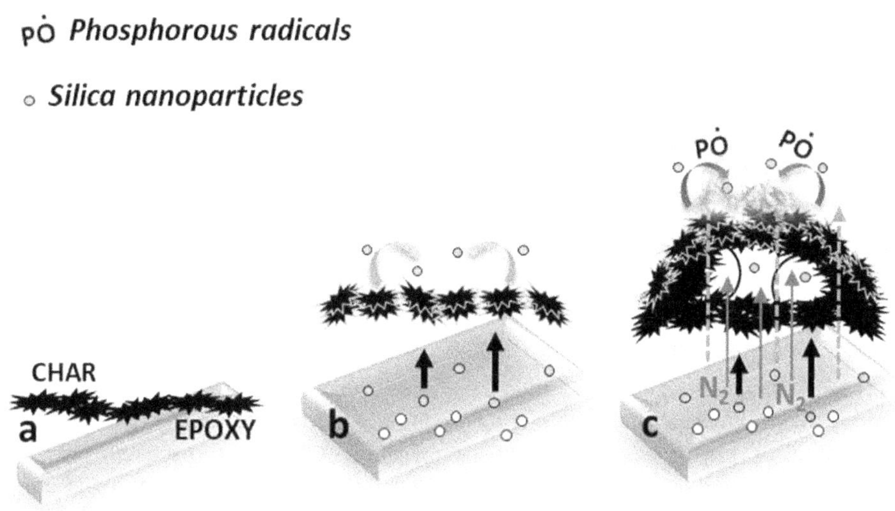

FIGURE 2.8 Fire behavior of flame retardant epoxy resins containing DOPO-derivatives. In particular: (a) shows that pristine resin completely burns, (b) indicates the formation of a silicon-containing char acting as thermal shield in the condensed phase, and (c) displays the combined flame retardant actions of an intumescent additive and phosphorus inhibitors in the gas phase.[81,82]

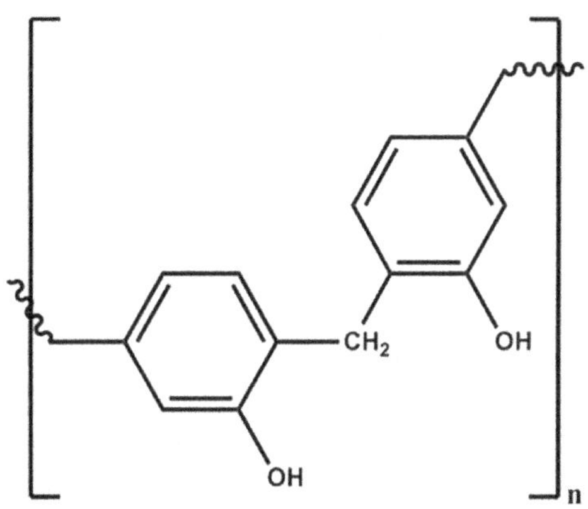

FIGURE 2.9 Typical chemical formula of hydroxymethyl functional phenols or polynuclear phenols.

Phenolics have LOI values of 25 vol% or so, and this high level of inherent flame resistance is associated with the general thermal stability and often means that no further flame retarding is necessary to create composites having required performance levels. During heating, water is generated chemically during the first step of thermal decomposition primarily because of phenol-phenol condensation by reactions of the type shown in Scheme 2.6.

This is followed by the oxidation of methylene groups by the released water to carbonyl linkages,[77] which then decompose further, releasing CO, CO_2, and other volatile products to yield ultimately char. Zhi Qin et al. found that the decomposition process of phenolic resins can be separated into three main stages[83]: firstly, the decomposition of ether bonds and unreacted terminal hydroxymethyl groups of the cured resin occurs; secondarily, methylene bridges

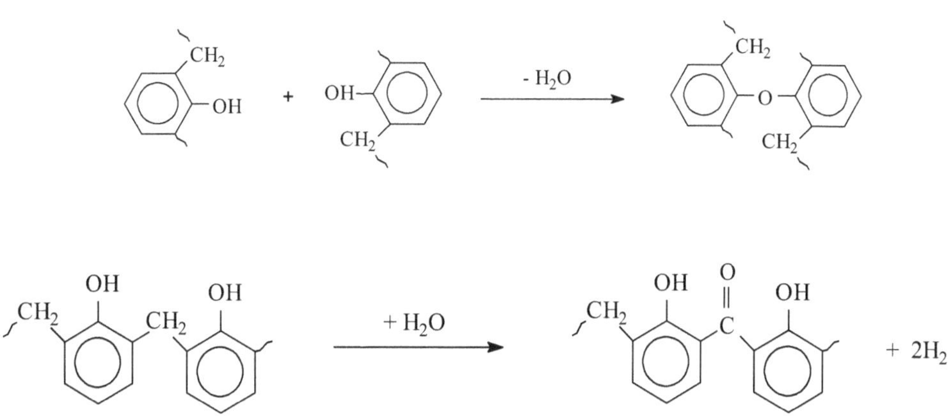

SCHEME 2.6 Condensation formation of methylene groups and their subsequent oxidation.[41,77]

decompose into methyl groups through the release of both phenol and cresol homolog; thirdly, phenolic hydroxyl undergoes cyclodehydration.[83]

In the case of highly cross-linked material, water is not released until above 400°C, and decomposition starts above 500°C, as confirmed using DTA.[80,84] The amount of char depends upon the structure of phenol, initial cross-links, and tendency to cross-link during decomposition. The main decomposition products may include methane, acetone, carbon monoxide, propanol, and propane.

2.2.3.5 Maleimide and Polyimide Resins

Their chemistry is often complex with a general formula for polyimide resins represented in Figure 2.10.

The aromatic structure of polyimides in particular ensures that they are thermally resistant and hence characterized by high char formation on pyrolysis, low flammability (LOI > 30 vol%), and low smoke production.

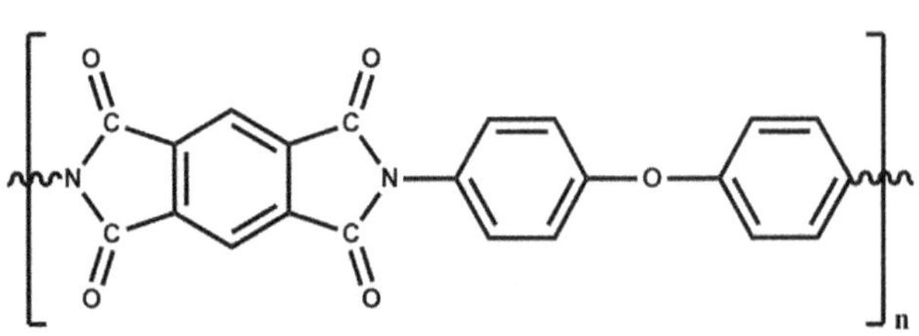

FIGURE 2.10 Typical chemical formula of polyimide resins.

2.2.4 Natural Polymers

2.2.4.1 Cellulose

Cellulose, either as the major component of wood or as the major textile fiber cotton, is frequently involved in fires. Thermal decomposition of cellulose results in the evolution of highly combustible volatiles, which will be consumed in a flame if ignited. Flame retardant treatments need to affect this decomposition process either by reducing the extent of volatile escape in favor of less flammable residue formation or evolve flame inhibitors such as Br· or Cl· species. Wang et al. applied a phosphorous-nitrogen synergistic additive, CTSGP, to flame retard cotton fabric.[85] CTSGP is able to react with the hydroxyl groups of cotton fabrics to form covalent bonds that bunch firmly to the surface. After the treatment with CTSGP, the cotton fabrics increased their LOI value from 18.0 to 29.0%; furthermore, the length of char residue was only 62 mm without after-flame time and after-glow time.[85] The chemical treatment did not change the color and handle of cotton fabrics, though the char-forming ability and fire behavior significantly improved compared to the untreated textile. Nazir et al. have recently developed an efficient methodology to flame retard cotton fabrics, simultaneously conferring antimicrobial properties.[86] The cellulose fibers were durably modified by forming in situ phosphine oxide macromolecular physical networks, where these latter were formed by treating the cellulose fibers with a mixture of Michael adducts (TVPO and cyclic amines, i.e., piperazine and 2,4,6-tri(piperazine-1-yl)-1,3,5-triazine).[86] The chemical treatment was durable against laundering and achieved a phosphorus retention higher than 95%, even after 50 laundry cycles. The treated cotton passed the Swiss vertical flammability test and showed a Limiting Oxygen Index value > 27% compared to the pristine fabric.[86]

Cellulose consists of long, linear chains of β-1,4-*D*(+)-glucopyranose units linked by 1,4-glucosidic bonds. The cellulose molecule is not planar but has a screw axis, each cellulose unit being at right angles to the previous one. Free rotation about the C-O-C link does not occur due to steric effects in the solid state. The decomposition of wood will not be discussed here because in addition to cellulose, wood and plant cells contain hemicellulose and lignin, which further complicate the decomposition process. Because of its wide usage in the textile and other industries, as a source of alternative fuels, the pyrolytic decomposition of cellulose has been extensively studied.[87] Whilst other more detailed mechanisms have appeared in the literature,[88,89] the basic processes proposed are all in line with that first suggested by Bradbury and Shafizadeh[90] who suggested that a precursor step in which an "activated" cellulose species Cellulose* which then undergoes further reaction depending on the temperature regime as presented in Scheme 2.7.

Whilst there is controversy as to whether or not this Cellulose* species exists, experimental evidence for the Cellulose* species was obtained by Price et al.,[91] who suggested it could be free radical in nature. At lower temperatures, oxygen plays a dominant role in cellulose decomposition; pyrolysis is faster in an oxidative atmosphere than in an inert one.[92] Oxygen catalyzes the formation of both volatiles and char promoting reactions.[93] At higher temperatures, the decomposition products are little affected.[92]

2.2.4.2 Protein Polymers

Proteins or poly(α-amino acids) feature the amide link common to the polyamides and may, in fact, be considered to be α-carbon substituted polyamide 2 variants. Thus, their potential thermal decomposition behavior might be expected to be similar to that of the aliphatic polyamides defined above. However, the α-substituents or -R groups are often quite reactive because of their functionalities, and so these will significantly influence, if not determine, the thermal decomposition behavior and potential flammability.

Commercial protein polymers were first developed during the early 20th century, a prime example being casein from milk. In reaction with formaldehyde, this gives a polymer which found much use as a synthetic horn or tortoiseshell material, and it still has some commercial presence in milk-producing countries like New Zealand. However, the most important protein polymers that require flame retardation are those associated with textiles with silk and wool as

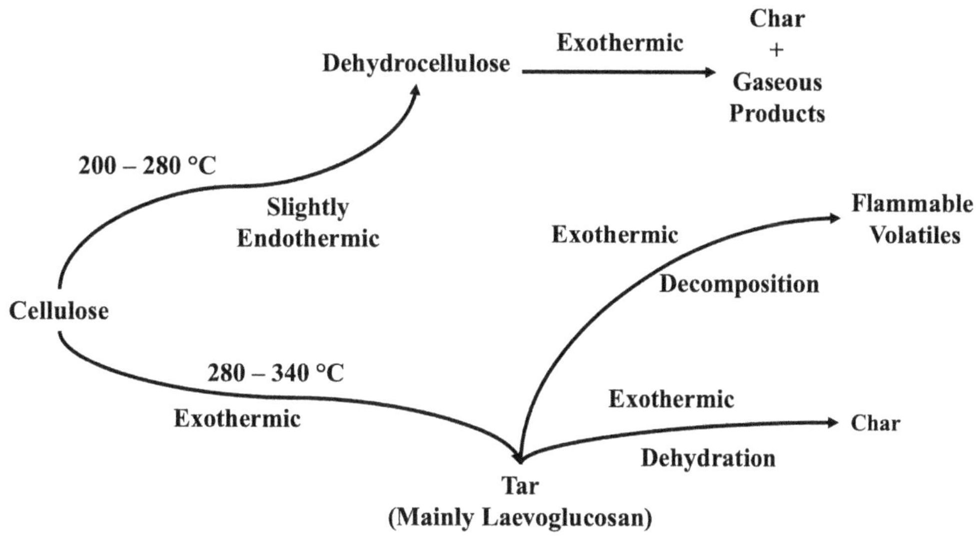

SCHEME 2.7 Basic scheme for cellulose decomposition process; after Bradbury et al.[90]

the principal examples. Whereas the aesthetics of silk define its commercial importance and the effect that most proprietary flame retardant treatments have on this fiber precludes it from being used in fire resistant textiles, there has been recent commercial interest in its use in executive jet aircraft interior dе́cor in which, as with normal commercial airliners, stringent fire standards are demanded. Kandola et al. have published work in this area to demonstrate the flame retardant challenges to be overcome.[94] Silk comprises 16 α-amino acids of which glycine (R = H), alanine (R = CH_3), and serine (R = CH_2OH) are the major comonomers present. When heated, silk starts to decompose above 250°C and forms a char. This charring characteristic is probably largely influenced by the dehydrating and cross-linking tendency of the hydroxyl group within the serine -CH_2OH α-substituent. Charring can be increased by the application of phosphorus-containing species, for example phosphate compounds.[95,96] The natural fiber LOI value is 22–23 vol% reflecting this higher char-forming tendency than the simple aliphatic polyamides which have LOI values of about 21 vol%.

Wool fibers and fabrics, however, have significantly greater commercial applications in products such as protective clothing and contract upholstery, where high levels of fire-resistant performance are demanded. Wool, while also comprising a large number (18) of α-amino acids, some of which are in common with silk, is uniquely identified by the presence of sulfur-containing α-substituents of which cystine (R = -CH_2-S-S-CH_2-) comprises nearly 10 wt% of the whole fiber and provides cross-links between adjacent polypeptide chains. This high sulfur content (3–4 wt%) coupled with the high nitrogen content (15–16 wt%) present in both chain and side groups contributes to the inherently low flammability of wool. The fiber also contains about 15 wt% of adsorbed moisture under normal atmospheric conditions, and LOI values are in the 25–26 vol% range. When wool is heated, it starts to give off its adsorbed moisture at 100°C and above and then starts to thermally decompose rapidly above 200°C, giving off gases which include H_2S following the cleavage of disulfide bonds above 230°C[97] alongside char formation.[98] The relatively nonflammable volatiles coupled with char formation are encouraged by cross-linking and dehydrating tendencies of the α-substituents present. The overall action of these is to give a relatively high ignition temperature of 570–600°C and a low flame temperature of about 680°C. The cystine disulfide link is particularly interesting here since it has highly reducing properties and so encourages subsequent oxidation by oxygen during the pyrolysis/combustion process. Pre-oxidation of the cystine to cysteic acid (R=CH_2. SO_3H) residues actually improves flame retardancy.

2.3 HIGH TEMPERATURE RESISTANT POLYMERS

These tend to be highly aromatic in character with rigid polymer chain backbones to yield polymers having very high second order transition values, absence of achievable melting transitions, and decomposition temperatures rarely below 400°C. It is generally the case that the lower the aliphatic content, the lower is the hydrogen to carbon ratio, and hence the lower is the flammability of any polymer. Aromatic chain polymers generally have H/C ratios < 1, and so their ability to generate volatile and flammable decomposition species at temperatures below 500°C or so is very limited. Consequently, they have LOI values generally above 30 vol% and are generally deemed to be sufficiently flame resistant for the applications for which they are selected.

Table 2.5 illustrates a selection of the more common high temperature, aromatic-structured polymers used for producing heat and flame resistant high performance fibers and their related thermal transitions and LOI values.[99]

TABLE 2.5
Thermal Transitions and LOI Values for Selected Aromatic, High Temperature Resistant Fiber-Forming Polymers

Fiber Genus	Second Order Temperature, °C	Melting Temperature, °C	Onset of Decomposition, °C	LOI, vol %
Phenol formaldehyde: Novoloid	NA	NA	> 150	30–34
m-Aramid	275	375–430 (decomposition)	425	28–31
p-Aramid	340	560 (decomposition)	> 590	29–31
Copolymeric p-aramid	-	-	500	25
Arimid (P84)	315	-	450	36–38
Aramid-arimid	< 315	-	380	32
Semicarbon	NA	NA	NA	55
Polybenzimidazole, PBI	> 400	NA	450/air; 1000/inert	> 41
Polybenzoxazole, PBO	-	-	650; > 700/inert	68

Source: Adapted from Horrocks.[99]
Notes: NA = not applicable; (decomposition) = with decomposition

These polymers may be compared with the more detailed discussion of thermal decomposition pathways for phenol-formaldehyde resins in a previous section above which more fully explains the reasons for generally low flammabilities in such highly aromatic structures. This same polymer in its novoloid form is commercially available as a fiber with properties defined in Table 2.5.

Addition of flame retardant species to these polymers is rarely undertaken since not only are they intractable during processing, but also the added value in terms of improved fire resistance is usually difficult to observe. The high cost of these polymers also negates the use of additional additives unless a real benefit is to be achieved.

HTRPs are heat-resistant plastics suitable in applications where the materials must maintain sufficient strength and stiffness at operating temperature. HTRPs can exceed 148°C in continuous operating temperature without any negative effect on their mechanical properties. Chemical resistance, friction, and wear performances and tensile strength are some of the characteristics considered to select the right HRTP for a specific application. Polytetrafluoroethylene is one of the best performing HRTPs, which is widely known for its non-stick behavior and very low coefficients of friction. PTFE has a large operating temperature range, as it can be used between –200°C and 260°C without decomposition.[100,101] Also, PTFE is chemically inert, hydrophobic, high electrical resistant, and has high flexural strength. Such material is practically insoluble in most chemicals and solvents. Another example of well-known HRTP is PEEK. PEEK is a colorless organic thermoplastic polymer belonging to the PAEK family and largely used in engineering applications, owing to its melting point over 371°C, superior hardness and continuous operating temperature of 260°C.[100,101] PEEK can withstand wear effects even under harsh chemical and physical conditions. This plastic material exhibits metal-like durability and the highest tensile and flexural strength of any high-performance polymer.[100,101]

2.4 POLYMER FIRE

Figure 2.11 is a schematic cross-section of a polymer fire indicating the important reaction zones.

The flame is fueled by combustible pyrolysis products escaping from the polymer surface due to heat being conducted from the flame in contact with the polymer surface and also radiated from the flame. The latter is the significant cause of flame spread and this process is modelled by the cone calorimeter.[102] The oxygen required to sustain the flame combustion diffuses in from the air environment. Flame combustion is a complex phenomenon taking place in three main phases: gas, boundary (at the interphase liquid/solid), and condensed.[103,104] The boundary phase represents the interface between the gas and condensed phase during the burning, and it is mainly influenced by the pyrolytic processes.[103,104] For example, the burning of a horizontal polymer slab with the release of a diffusion flame involves some physical processes: (1) energy transport by convention and radiation between the gas phase (i.e., flame) and the boundary phase and (2) mass transfer, vaporization of the pyrolysis gases, leading to energy loss from the boundary phase, and conduction into the solid.[103,104] At typical burning rates in polymer flaming combustion, fuel generation is the rate-limiting step, and it is strongly dependent on the rate at which heat and mass are transported to and from the polymer, respectively.

During the flaming combustion, various solid particles escape from the flame as smoke which is accompanied by gaseous species, some of which can be toxic.[105] As previously mentioned, the significant polymer decomposition reactions occur within a millimeter or so of the boundary phase between the flame and the solid polymer. Here, the temperature is high enough for condensed phase decomposition reactions to occur. These involve the polymer and any additive systems included in the polymer formulations.

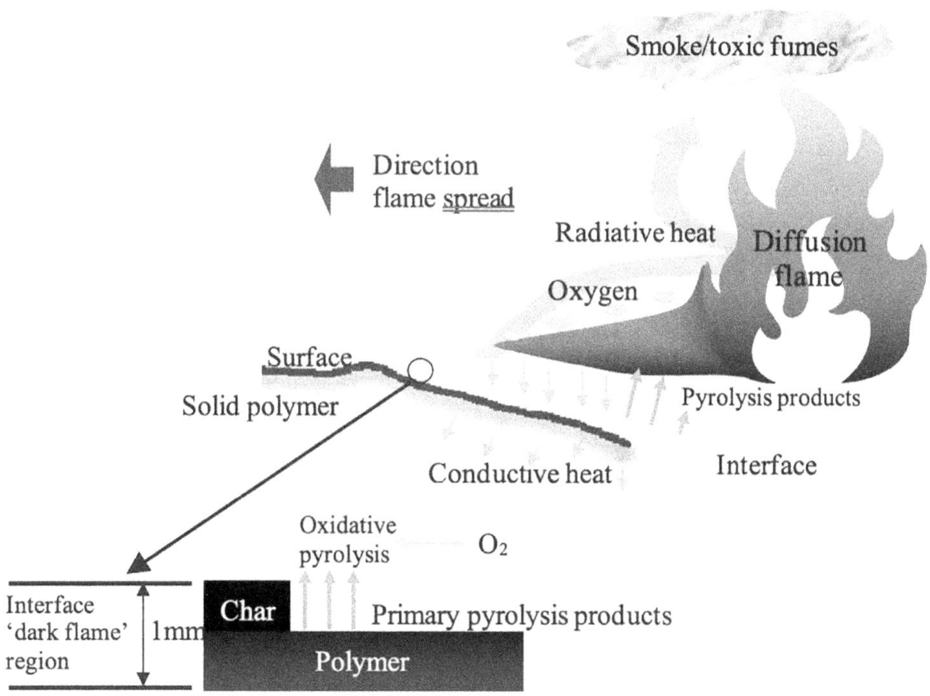

FIGURE 2.11 Schematic representation of a burning polymer.

Volatile species formed escape into the flame, whilst heavier species remain to undergo further reaction and may eventually decompose, leaving a char. This is where the significant condensed phase chemistry occurs. Experimental studies of this region have been undertaken by Price[106] and Marosi et al.[107]

2.4.1 Polymer Combustion Cycle

An account of the polymer combustion cycle is simplified by reference to the schematic representation given in Figure 2.11. In order for a polymer material to undergo flaming combustion, it must first decompose to evolve combustible volatiles which escape and mix with an oxidative atmosphere up to the lower flammability limit. Provided the temperature is above the ignition temperature or a suitable ignition source, such as a spark, is present, this mixture will ignite. The flames will yield gaseous products some of which may be toxic, smoke and fumes, as well as heat. Some of the heat will be conducted or radiated back to the original polymer to cause further decomposition and the release of additional combustible gases. Provided this heat is sufficiently intense, a combustion cycle will be established, as indicated schematically in Figure 2.12.

This scheme may now be used to understand how different flame retardant strategies may be designed and adopted in order to break the combustion cycle. For a given polymer, the strategy to be adopted will be largely dictated by its particular thermal degradative chemistry, which has been briefly reviewed in section 2.2 above for all the significant polymer types.

2.4.1.1 Flame Retardance

Most plastics and textiles are organic and thus vulnerable in a fire situation. A major concern of their manufacturers is, therefore, to render their products resistant to ignition or, if they are ignited, to burn less efficiently so that their rate of heat release is significantly reduced. The approach to achieving this is termed "flame retardance." Unless the polymer is inherently flame retarded, the various approaches indicated in Figure 2.12 can be used to reduce the fire threat of such materials. One method is to prevent access of oxygen to the flame, another is the introduction of flame inhibitors such as halogen atoms, Cl· and particularly Br·, or phosphorus into the flame. This can be accomplished by including additives, in the material's formulation, which release these flame inhibitors if the material is exposed to temperatures approaching the ignition temperature. An alternative approach is to introduce suitable chemical groups into the polymer structure, so-called "reactive flame retardants," which provide the same effect. The combustion can also be halted by reducing the heat flow back to the polymer, thus preventing further decomposition. This can be achieved by the introduction of a heat sink such as aluminum oxide trihydrate ($Al(OH)_3$) or magnesium hydroxide ($Mg(OH)_2$), which decompose with a large endothermicity. Formation of a heat barrier, for example, either a char or intumescent barrier as a result of exposure of the material to a fire is another successful method. Finally there is the option to modify the polymer decomposition chemistry so that the amount of flammables released is below the level required to fuel the flames whilst at the same time increasing the less combustible char formation. The char has the

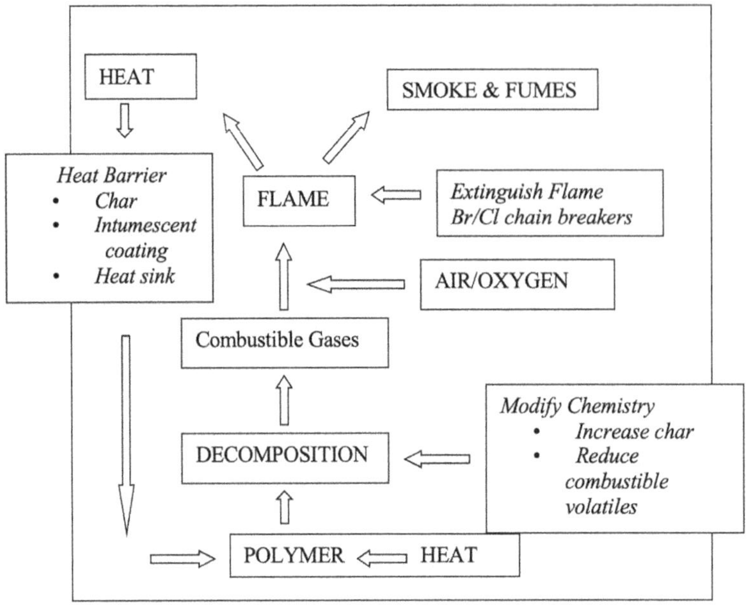

FIGURE 2.12 Schematic representation of the polymer combustion cycle; main approaches to flame retardancy are shown in italics.

beneficial effect of forming a barrier between the polymer surface and the flame. It is during these latter condensed phase processes that polymer decomposition plays a significant role in flame retardant action. It is on such processes that this chapter is focused.

As previously mentioned, some polymers can be said to be inherently flame retarded. Bourbigot and Duquesne[108] recently classified such polymers as having a continuous operating temperature range from 180°C to 300°C or above together with a decomposition temperature above 350°C. Such polymers can have high thermal stability due to their high aromatic and low hydrogen contents, for example, polyarylates and polycarbonates, phenolic resins and aromatic polyesters, polyethers and polyamides. Others, like PVC, decompose to evolve flame inhibitors such as HCl from PVC, or release sufficient quantities of nitrogen or similar heteroatoms, for example, polyamides and aminoresins, to blanket out any flames at high temperature. Poly(aryl ketone)s and ether ketones have above average thermal and thermo-oxidative stability at a high temperature, which will result in resistance to fire.

Few conventional synthetic polymers, however, are inherently resistant to heat and fire. The traditional method of rendering them fire retarded is to include a flame retarded additive in the formulation during polymer processing. The choice of fire retardant depends on whether the fire retardant is required to predominately function in the gas phase, for example, ATH, magnesium hydroxide ($Mg(OH)_2$), ATO systems, or the condensed phase via char formation enhancement, for example, APP. In addition, the chosen additive must be stable at the polymer processing temperature whilst being compatible with the polymer itself. In addition, cost is another significant factor, for example, ATH-containing plastic sheathing used as insulation for low-cost electrical wiring. Gas phase retardants function by releasing species which either blanket out the flames with noncombustible gases such as water from ATH or halogen flame inhibitors from the halogen/ATO type systems. The incorporation of additives however does have several disadvantages. The additive is often required in high loadings to be effective (typically 10–60 wt%), which may result in adverse changes to the physical and mechanical properties of the polymer rendering the polymer unsuitable for a particular end use. The alterative reactive fire retardant approach is to incorporate the fire retardant species, via copolymerization or some other chemical modification, to produce what is essentially an inherently fire retarded polymer.[109] The relatively low load required to achieve sufficient fire retardance, and careful selection of the comonomer, can keep detrimental changes to the physical and mechanical properties at an acceptable level. Also, because it is chemically incorporated into the polymer, the fire retardant will not be easily lost from the polymer. Thus, one of the major problems associated with additive systems is eliminated.

Because of the advantages and despite their higher costs, in recent years there has been a growing interest in the reactive approach to produce high value, high performance fire retarded polymers. Because of the environmental pressures to reduce/eliminate the use of halogen-containing systems, much interest has focused on phosphorus as the fire retardant moiety incorporated in the polymer chain. During the combustion of plastics, phosphorus compounds can exert a hybrid flame retardant mechanism in both gas and condensed phases.[75,76,110] For this reason and due to

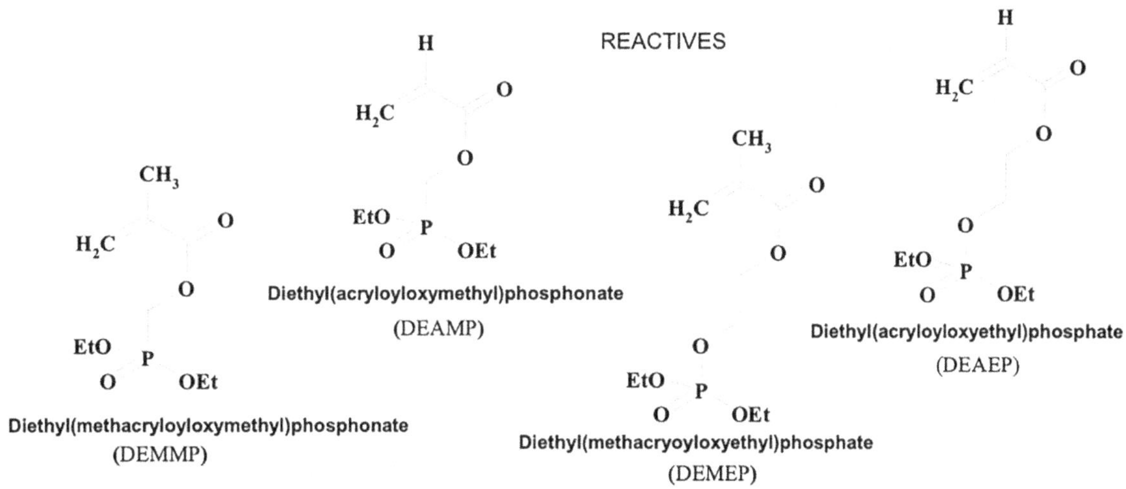

FIGURE 2.13 Structures of copolymers used for reactive fire retardant studies of Price, Ebdon et al.[115–117]

their lower toxicity compared to halogen-based species, phosphorus-based flame retardants are commonly used as functional additive or reactive compounds for plastics in several industrial applications.[111] It is well-known that the pyrolysis process feeds the flame with combustible gases during the decomposition. The formation of char is promoted by phosphorus-based flame retardants acting mostly in the solid phase of burning polymeric materials.[75,76,110] On the other side, intumescent flame retardants form a stable foam layer that acts as an insulating barrier between the flame and the solid phase. Also, phosphorus-based flame retardants can decompose, releasing phosphorus inhibitors (e.g., PO·, HPO·) able to suppress the flame by termination of oxygen radicals.[75,76,110,112–114] These flame retardant actions make phosphates, phosphonates, and phosphinates suitable alternative to brominated flame retardants in engineering plastics including polyesters, polyurethane foams, polyamides, and elastomers, even for electronic and automotive applications, where stringent fire regulations have been fulfilled. Besides, phosphorus compounds are frequently exploited as flame retardants in combination with nitrogen-containing compounds, metal hydroxides, and nanofillers.[75,76,110] As an example, to phase out halogenated additives, DOPO has been largely used in the flame retardation of printed circuit boards as well as cables and connectors. This phosphorus-based flame retardant is well recognized for its effectiveness in poisoning the flame by inhibition mechanism, though numerous derivatives of DOPO have been developed to boost its char-forming character.[75,76,110]

One example of phosphorus used as fire retardant moiety is the work of Price, Ebdon et al., who have synthesised[75] and studied the flammability and decomposition behavior[115–117] of poly(methyl methacrylate) and polystyrene polymers copolymerized with a range of phosphorus containing copolymers. Typical copolymers were DEAMP, DEMMP, DEAEP, DEMEP. Their structures are given in Figure 2.13.

The fire retardant mechanisms identified for the various phosphorus moieties in these PMMA and polystyrene copolymers investigated can be summarized:

- Vapor phase fire retardant action assumed for gas phase phosphorus species released from all polymers;
- The rate of volatile production was reduced for phosphorus-containing copolymers as compared to that of the corresponding additive system;[117]
- The normal unzipping process of PMMA decomposition was obstructed in the case of the acrylate copolymers, thus reducing the evolution of the flammable MMA monomer;
- Condensed phase cross-linking occurred as the copolymer containing phosphorus decomposed facilitating char formation and reducing flammable volatile evolution;
- Interference with the H-transfer reactions occurred during polystyrene decomposition for acrylate copolymers.

Other work in this area has been reported in recent years. Of particular note is that of Lyon has also been involved in synthesizing inherently flame retardant materials to improve the fire resistance of aircraft interiors.[16,17]

2.4.1.2 Relevance of Polymer Stabilization to Flame Retardant Processes

It might be assumed that, since condensed phase flame retardants function by modifying the normal thermal decomposition processes of polymers, they would also function as thermal stabilizers and that thermal antioxidant stabilizers would show flame retardant properties. However, these statements are rarely the case, and to understand why, it is necessary to compare the mechanistic aspects of flame retardance as discussed previously with those of thermal decomposition and thermal oxidation also briefly alluded to previously and, in the case of the latter, the Bolland and Gee mechanism[27] in Scheme 2.1.

Cursory comparison of the character and behavior of flame retardants and thermal stabilizers including antioxidants yields the following:

- Flame retardants are generally present at concentrations of greater than 10 wt% for them to be effective and relate to effective elemental concentrations in the case of phosphorus of the order of 2–4 wt% and in the case of bromine 5–10 wt% with respect to the polymer;
- Thermal and photo-antioxidants are often present and effective at concentrations of the order of 0.5–1.0 wt%;
- Flame retardants at low concentrations (≤1 wt%) are seldom, if ever, known to function as thermal stabilizers;
- Antioxidants when introduced at high concentrations (which would be very expensive given their relatively high costs), are not reported to be flame retardant.

There is only perhaps one significant case where low concentrations of an antioxidant shows flame retardant behavior, and that is in the case of certain hindered amine stabilizers (see Chapter 11 of this book for more details) that at the normally used concentrations (≤1 wt%) offer low levels of flame retardancy in polypropylene and show synergy with bromine-containing flame retardants.[118,119]

Before returning to this example, it is pertinent to review the mechanisms by which thermal antioxidant stabilizers work. Since thermal stability is determined inherently by the lability of bonds present within a polymer, the only means of developing thermal stabilization is to offer means of scavenging or rendering inert impurities present which might otherwise sensitize decomposition. So for example, in PVC where release of hydrogen chloride sensitizes further decomposition, the presence of a basic additive, such as metal carboxylates and even calcium carbonate, have thermal stabilizing properties. Similarly, the presence of radical scavengers such as hindered phenols may interact and terminate impurity-generated radicals which might otherwise promote eventual chain scission mechanisms, as shown in Scheme 2.1.

Most thermal stabilizers fall into one of two groups: they function either as antioxidants or in some other manner, such as buffers, to remove excess acidity as exemplified by the PVC example above. In the case of antioxidants, these are often characterized according to their means of operation[7,120–122]:

1. "Primary" antioxidants, also termed chain-breaking antioxidants, interfere with the chain reaction in Scheme 2.1 by trapping radicals or labile hydrogen atom donors. These are exemplified by hindered phenols and alkylarylamines. Scheme 2.8 schematically demonstrates the scavenging activity of a typical hindered phenol. R'-OH is a tertiary phenol, for example octadecyl 3,5-di(tert)-butyl-4-hydroxyhydrocinnamate or Irganox 1076, Ciba (see Table 2.6); R'-O· is a stable radical;

$$\dot{R} + R'{-}OH \longrightarrow RH + R'{-}\dot{O}$$

SCHEME 2.8 Stabilizing activity of chain-breaking, primary antioxidants.

$$R{-}OOH + (R'O)_3{-}P \longrightarrow ROH + (R'O)_3{-}P{=}O$$

SCHEME 2.9 Stabilizing activity of hydroperoxide-decomposing secondary antioxidants.

2. "Secondary" antioxidants or hydroperoxide decomposers (see Scheme 2.1) are typified by organosulfur species having reducing properties such as sulfides and thioethers. Tertiary phosphites also fall into this category (see Scheme 2.9). $(R'O)_3$-P is an organophosphite, for example tris(2,4-di-tert-butylphenyl) phosphite or Irgafos 168, Ciba (see Table 2.6);
3. Photoantioxidants are typified by the class of HAS which although they were developed for photostabilization of polyolefins, they also possess thermal antioxidant properties. They are generally assumed to function as "primary" antioxidants in that they scavenge radicals and, in particular, peroxy radicals;
4. Metal ions and particularly heavy metal ions tend to sensitize peroxy radical formation, and so the presence of metal scavenging or chelating species can offset this effect. This form of stabilization is particularly important for polymers in which metal-containing polymerization catalyst residues are present, such as polyolefins. While simple additives like calcium stearate may be used, more sophisticated ones based on bifunctional chelating species also are available commercially;
5. Some redox systems have been developed for certain polymers. The copper/iodine system is well-established for polyamide thermal stabilization and, in spite of introducing a heavy metal ion into the polymer, works well in an oxygen-free environment.[120]

Very often, antioxidants are used in combinations to ensure maximum activity, and typically a commercial additive system may comprise both a primary and secondary antioxidant species, although the total concentration remains ≤1 wt%. Scheme 2.10 shows schematically how a combination of primary and secondary antioxidants functions in a polyolefin matrix.[121]

Some metal chelate scavengers may also be based on a tertiary phenolic structure, thereby introducing two antioxidant properties into the same molecule.

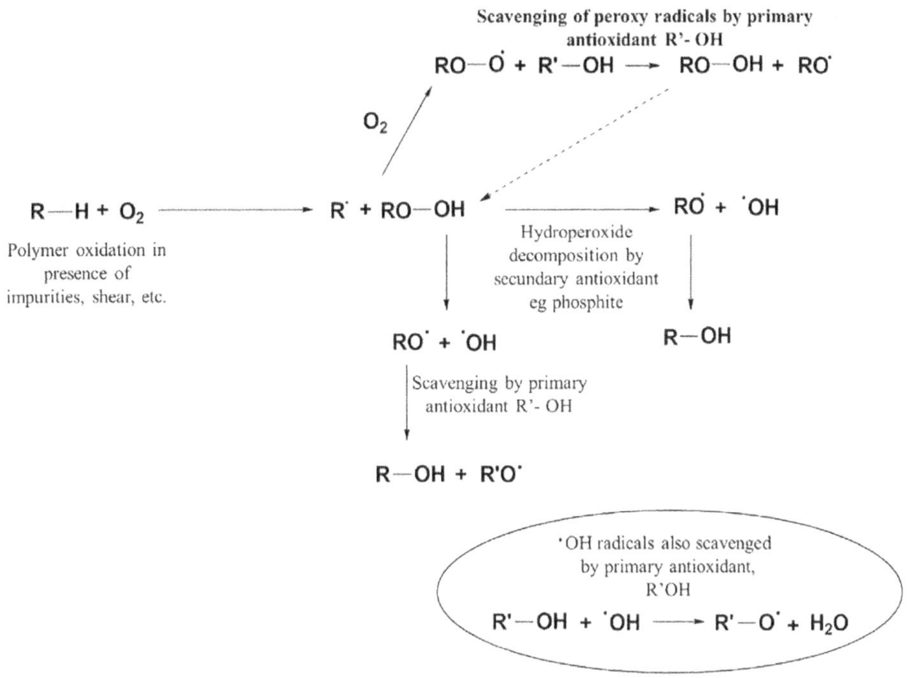

SCHEME 2.10 Combined stabilizing activity of primary and secondary antioxidants.

Table 2.6 provides typical examples of each of these antioxidants from which it may be seen that they bear little resemblance to the flame retardant molecular structures described in the remaining parts of this book.

However, notwithstanding this observation, it was briefly mentioned above that a recently developed HAS-based system, commercialized as NOR116 by Ciba, is marketed as both a photoantioxidant and a flame retardant for polypropylene.[119] While very little, if any, literature is available to explain its flame retardant activity, it is noteworthy that research into the burning behavior by Stuetz et al.[123] 30 years ago suggested that the burning process of polypropylene involves an oxidative pyrolytic step as a prerequisite for fuel formation, and so it is possible that the introduction of a HAS photoantioxidant will interfere with this stage, thus promoting a flame retardant effect. That there is also synergy observed between NOR116 and bromine-containing flame retardants like decabromodiphenyl ether,[118] for example, in which bromine radical formation and reaction determine flame retardant behavior. It is possible that the radical interacting character of the HAS may have a beneficial effect on the effectiveness of the Br· radicals in terminating flame chemical chain reactions.

2.4.1.3 Alternative Polymer Decomposition Processes, For Example, Photochemical, Plasma, Irradiation, and Their Potential Influence on Flame Retardant Behavior

In the previous section, comparison was made between the nature and means of operation of antioxidants (including some photoantioxidants), which are introduced into polymers to improve either processing or long-term stability, and flame retardants, which may interact with and modify the thermal decomposition process as well as the ensuing flame chemistry. It is evident that while there are considerable differences between the chemistries of flame retardancy and thermal (and photo-) stabilization, there are similarities in that the polymer decomposition pathways, while being specific for each polymer, are driven by thermal energy and involve the same thermal decomposition pathways. A major difference is, of course, the rates of heating, which during normal polymer processing and long-term exposures during service are generally low while in a fire are large. The effect of the rate of heating was noted for the acrylic copolymers above where low rates favor carbonization (and char-forming), whereas high rates favor volatilization.[51]

Since most polymer decomposition processes tend to lead to chain scission, cross-linking, or both, it might be assumed that flammability would be influenced by the decomposition history of any given polymer. Cursory analysis of the literature shows that little if anything has been published on the effects of aging or decomposition on resulting polymer flammability. This is not to be confused with the loss of flame retardants during service life by leaching, cleaning, or other process. Generally, however, the influence of polymer history of non-retarded polymers appears not to be an important issue since during the service life of a polymer, only a minority of polymer chains require to be broken before the polymer becomes unserviceable in terms of reduced tensile, impact, or other significant property. In fact, most decomposition processes comprising external agencies involve attack of polymer chains in the amorphous regions in the first instance. In highly crystalline polymers like polyethylene,

TABLE 2.6
Examples of primary, secondary, and hindered amine antioxidants marketed by Ciba for use with polypropylene.

Type	Commercial Name (Ciba)	Chemical Formula
Primary (radical scavenger)	Irganox 1076	Octadecyl 3,5-di(tert)-butyl-4-hydroxyhydrocinnamate) 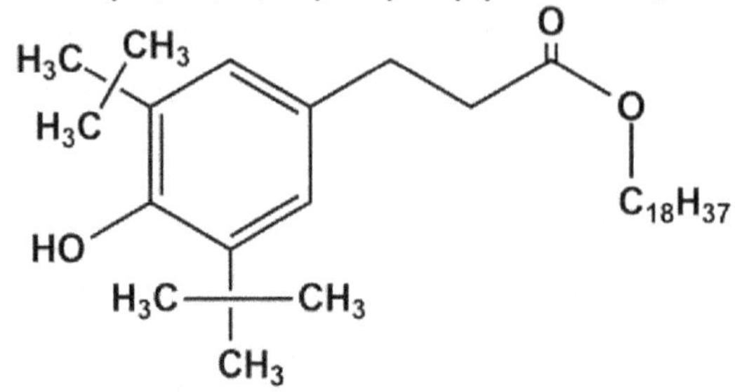
Secondary (hydroperoxide decomposer)	Irgafos 168	Tris(2,4-di-tert-butylphenyl)phosphite 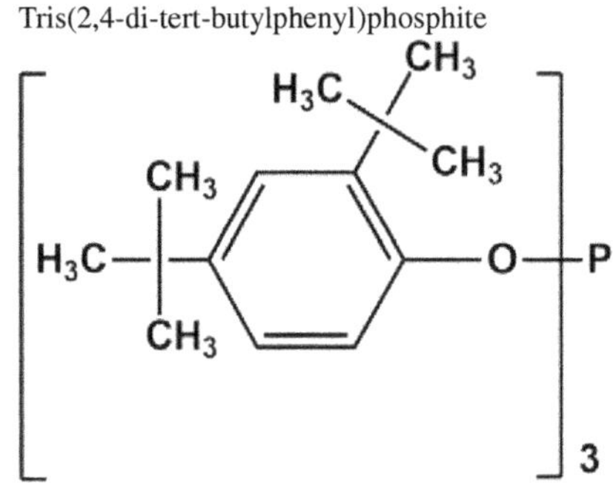
Hindered amine stabilizer (combined light and heat)	Chimassorb 944	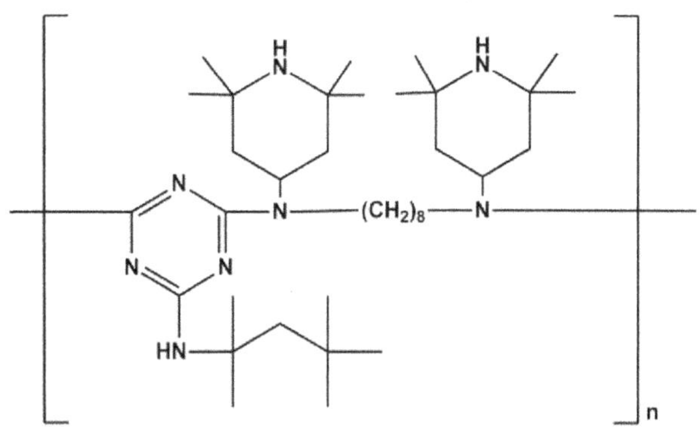
Hindered amine stabilizer with flame retardant properties	Flamstab NOR116	The reaction product of 2,4-bis[(1-cyclohexyloxy-2,2,6,6-piperidin-4-yl) butylamino]- 6-chloro-s-triazine with N,N′-bis(3-aminopropyl) ethylenediamine) [**CAS** Reg. No. **191680–81–6**]

polypropylene, and aliphatic polyamides, for example, these comprise less than 50 wt% of the total polymer and often much less, and so the majority of the polymer molecules present are non-decomposed even when the overall polymer serviceability has reduced. Furthermore, it is generally noted that during the weathering of linear crystalline polymers where the degrading agencies are a complex combination of heat, light, water, and possibly air polluting species, crystallinity increases following internal relaxation of polymer chains after scission reactions have occurred. For example, weathering of linear low density polyethylene increases its degree of crystallinity from just less than 40% to over 55% during a 12 month period.[124] In other even more highly crystalline polymers like polypropylene (> 70%), this means that nominally decomposed polymer actually comprises a higher proportion of pure polymer in the crystalline phase with products of decomposition concentrated in the minor amorphous phase. The effects of this concentrated but minor zone of decomposition appear to have little effect on the overall flammability although research in this area would be welcome. This would be especially relevant to the generally amorphous polymers such as the thermosets, polystyrene, and copolymers like EVA and ABS.

Conversely, degrading treatments that may modify a polymer surface in a manner that enables a subsequent or simultaneous modification to be undertaken may reduce polymer flammability. Such modifications could be seen to be potential flame retardant processes. Surface graft copolymerization of an activated underlying polymer surface comes to mind here where activation during surface chemical grafting,[125] radiation,[126] pre-irradiation,[127,128] and plasma[129] treatments are examples. This area may be considered to be an important and emerging means of conferring flame retardancy in a more sophisticated and both environmentally- and cost-effective manner than the traditional use of bulk flame retardants for a number of polymers. The current state of developments here along with the recent interest in depositing nanoparticles on to polymer surfaces with a view to improving overall flame retardancy needs also to be mentioned and have been recently reviewed by ourselves with an emphasis on textile substrates.[130]

2.5 CONCLUSIONS AND FUTURE OUTLOOKS

The flammability of polymers is one of the major aspects that needs to be considered in the design of a new polymeric material (PM) for fire safe application. The knowledge of fire behavior of polymers is crucial to develop high-performing polymer-based products. As new fire and environment regulations are currently under preparation, the scientific community needs to invest more efforts to study the key parameters affecting the fire performances of PMs, especially considering the chemical nature of polymers and their typical decomposition pathway. Thermoplastics require phosphorus-based flame retardants (FRs) showing high thermal stability during their extrusion, while thermosets demand for more sustainable FR strategies. The effectiveness of future technologies is linked to our knowledge of flame retardancy and polymer chemistry, as the constant evolution of fundamentals can foster the innovation in all sectors of our lives.

2.5.1 Author Statement

The manuscript was written through the contributions of all authors. All authors have approved the final version of the manuscript.

2.5.2 Declaration of Competing Interest

The authors declare no conflict of interest. We confirm that this submission is original, it is not published elsewhere, and not under consideration for the publication elsewhere.

2.5.3 List of Abbreviations

acrylonitrile-butadiene-styrene	**ABS**
acrylonitrile	**AN**
aluminum hypophosphite	**AP**
ammonium polyphosphate	**APP**
acrylonitrile-butadiene-styrene	**ABS**
alumina trihydrate	**ATH**
halogen/antimony III oxide	**ATO**
6-monochloro-1,3,5-triazine-4-p-aminobenzenesulfonate-2-methylphosphinic acid dimethyl ester	**CTSGP**
diethyl(acryloyloxyethyl)phosphate	**DEAEP**
diethyl(acryloyloxymethyl)phosphonate	**DEAMP**
diethyl(methacryloyloxyethyl)phosphate	**DEMEP**
diethyl(methacryloyloxymethyl)phosphonate	**DEMMP**
9,10-dihydro-9-oxa-10-phosphaphenanthrene-10-oxide	**DOPO**
DOPO-acrylamide	**DOPO-AAM**
6H-dibenz[c,e][1,2]oxaphosphorin,6-[(1-oxido-2,6,7-trioxa-1-phosphabicyclo[2.2.2]oct-4-yl)methoxy]-,6-oxide	**DOPO-PEPA**
differential thermal analysis	**DTA**
6,6′-(ethane-1,2-diylbis-(azanediyl)-bis-(6H-dibenzo[c,e][1,2]-oxaphosphine-6-oxide	**EDAB-DOPO**
ethylene/vinyl acetate	**EVA**
Fourier-transform infrared spectroscopy	**FTIR**
glass-fiber-reinforced aliphatic polyamides	**GF-PA6**
hindered amine stabilizers	**HAS**
high impact polystyrene	**HIPS**
high temperature resistant polymers	**HTRPs**
linear low-density polyethylene	**LLDPE**
Limiting Oxygen Index	**LOI**
diphenylmethane 4,3 diisocyanate	**MDI**
methyl methacrylate	**MMA**
mass spectrometry	**MS**
nuclear magnetic resonance	**NMR**
polyamide	**PA**
polyaryletherketone	**PAEK**
polyether ether ketone	**PEEK**
polyethylene terephthalate	**PET**
poly(methyl methacrylate)	**PMMA**
polypropylene	**PP**
polystyrene	**PS**
polytetrafluoroethylene	**PTFE**
polyurethane	**PU**
polyvinyl acetate	**PVA**
polyvinyl chloride	**PVC**
polyvinylchloride	**PVC**
toluene 2,4- and 2,6- diisocyanate	**TDI**
thermal gravimetric analysis	**TGA**
trivinyl phosphine oxide	**TVPO**

REFERENCES

1. Cullis, C. F.; Hirschler, M. M., The Combustion of Organic Polymers. Oxford University Press, Oxford: 1981; p 117.
2. Miao, J.; Fang, Y.; Yang, X.; Zhu, Y.; Hu, A.; Wang, G., Fabrication, flame retardancy and physical properties of

phosphorus containing porous organic polymers/epoxy resin composites. *Polym. Degrad. Stab.* **2020,** *176*, 109159.
3. Guerre, M.; Taplan, C.; Winne, J. M.; Du Prez, F. E., Vitrimers: Directing chemical reactivity to control material properties. *Chem. Sci.* **2020,** *11* (19), 4855–4870.
4. Röttger, M.; Domenech, T.; van der Weegen, R.; Breuillac, A.; Nicolaÿ, R.; Leibler, L., High-performance vitrimers from commodity thermoplastics through dioxaborolane metathesis. *Science* **2017,** *356* (6333), 62–65.
5. Klingler, W. W.; Bifulco, A.; Polisi, C.; Huang, Z.; Gaan, S., Recyclable inherently flame-retardant thermosets: Chemistry, properties and applications. *Compos. B: Eng.* **2023**, 110667.
6. Madorsky, S. L.; Straus, S., Thermal degradation of polymers at high temperatures. *J. Res. Natl. Bur. Stand. (U. S.). Section A, Physics and Chemistry* **1959,** *63* (3), 261.
7. Verdu, J.; Colin, X.; Fayolle, B.; Audouin, L., Methodology of lifetime prediction in polymer aging. *J. Test. Eval.* **2006,** *35* (3), 289–296.
8. British Standards Institution, Plastics: Thermogravimetry (TG) of polymers: General principles. BS EN ISO 11358:1997, British Standards Institution, UK, London: 1997.
9. Hamid, S. H., Lifetime prediction plastics. In Handbook of Polymer Degradation, CRC Press, New York: 2000; pp 725–741.
10. Wang, X.; Hu, Y.; Song, L.; Yang, H.; Xing, W.; Lu, H., Synthesis and characterization of a DOPO-substitued organophosphorus oligomer and its application in flame retardant epoxy resins. *Prog. Org. Coat.* **2011,** *71* (1), 72–82.
11. Çopuroğlu, M.; Şen, M., A comparative study of thermal ageing characteristics of poly (ethylene-co-vinyl acetate) and poly (ethylene-co-vinyl acetate)/carbon black mixture. *Polym. Adv. Technol.* **2004,** *15* (7), 393–399.
12. Draye, A. C.; Persenaire, O.; Brožek, J.; Roda, J.; Košek, T.; Dubois, P., Thermogravimetric analysis of poly (ε-caprolactam) and poly [(ε-caprolactam)-co-(ε-caprolactone)] polymers. *Polymer* **2001,** *42* (20), 8325–8332.
13. Huggett, C., Estimation of rate of heat release by means of oxygen consumption measurements. *Fire Mater.* **1980,** *4* (2), 61–65.
14. Salmeia, K.; Fage, J.; Liang, S.; Gaan, S., An overview of mode of action and analytical methods for evaluation of gas phase activities of flame retardants. *Polymers* **2015,** *7* (3), 504–526.
15. Gaan, S., Evaluation of Gas Phase: Mechanisms and Analyses. Elsevier, Amsterdam: 2022; pp 117–159.
16. Lyon, R. E.; Walters, R. N.; Beach, M.; Schall, F. P. Flammability screening of plastics containing flame retardant additives. ADDITIVES 2007, 16th International Conference, San Antonio, TX, USA, 2007.
17. Walters, R. N.; Lyon, R. E., Flammability of polymer composites. US Department of Transportation, Federal Aviation Administration. DOT/FAA/AR-08/18, May 2008. https://www.tc.faa.gov/its/worldpac/techrpt/ar0818.pdf (accessed 25/01/24)
18. Rimez, B.; Rahier, H.; Van Assche, G.; Artoos, T.; Biesemans, M.; Van Mele, B., The thermal degradation of poly (vinyl acetate) and poly (ethylene-co-vinyl acetate), Part I: Experimental study of the degradation mechanism. *Polym. Degrad. Stabil.* **2008,** *93* (4), 800–810.
19. Bernstein, R.; Thornberg, S. M.; Assink, R. A.; Irwin, A. N.; Hochrein, J. M.; Brown, J. R.; Derzon, D. K.; Klamo, S. B.; Clough, R. L., The origins of volatile oxidation products in the thermal degradation of polypropylene, identified by selective isotopic labeling. *Polym. Degrad. Stabil.* **2007,** *92* (11), 2076–2094.
20. Madras, G.; McCoy, B. J., Distribution kinetics for polymer mixture degradation. *Ind. Eng. Chem. Res.* **1999,** *38* (2), 352–357.
21. Sun, Q.; Ding, Y.; Stoliarov, S. I.; Sun, J.; Fontaine, G.; Bourbigot, S., Development of a pyrolysis model for an intumescent flame retardant system: Poly (lactic acid) blended with melamine and ammonium polyphosphate. *Compos. B: Eng.* **2020,** *194*, 108055.
22. Lyon, R. E.; Takemori, M. T.; Safronava, N.; Stoliarov, S. I.; Walters, R. N., A molecular basis for polymer flammability. *Polymer* **2009,** *50* (12), 2608–2617.
23. Walters, R. N.; Lyon, R. E., Molar group contributions to polymer flammability. *J. Appl. Polym. Sci.* **2003,** *87* (3), 548–563.
24. Guadagno, L.; Naddeo, C.; Vittoria, V.; Camino, G.; Cagnani, C., Chemical and morphologial modifications of irradiated linear low density polyethylene (LLDPE). *Polym. Degrad. Stabil.* **2001,** *72* (1), 175–186.
25. Therias, S.; Rapp, G.; Masson, C.; Gardette, J.-L., Limits of UV-light acceleration on the photooxidation of low-density polyethylene. *Polym. Degrad. Stabil.* **2021,** *183*, 109443–109443.
26. Chirinos-Padr6n, A. J.; Allen, N. S., Aspects of polymer stabilization. In Handbook of Polymer Degradation. Edited by AH Hamid, MB Amin, and AG Maadhah. Marcel Dekker Inc., New York: 1992; pp 261–303.
27. Carty, P.; White, S., A review of the role of basic iron (III) oxide acting as a char forming/smoke suppressing/flame retarding additive in halogenated polymers and halogenated polymer blends. *Polym. Polym. Compos.* **1998,** *6* (1), 33–38.
28. Kashiwagi, T., Polymer combustion and flammability—Role of the condensed phase. *Symp. (Int.) Combus.* **1994,** *25*, 1423–1437. https://www.sciencedirect.com/science/article/abs/pii/S0082078406807861
29. Ballice, L.; Reimert, R., Classification of volatile products from the temperature-programmed pyrolysis of polypropylene (PP), atactic-polypropylene (APP) and thermogravimetrically derived kinetics of pyrolysis. *Chem. Eng. Process* **2002,** *41* (4), 289–296.
30. Ballice, L., Classification of volatile products evolved during temperature-programmed co-pyrolysis of low-density polyethylene (LDPE) with polypropylene (PP). *Fuel* **2002,** *81* (9), 1233–1240.
31. Ochoa, R.; Van Woert, H.; Lee, W. H.; Subramanian, R.; Kugler, E.; Eklund, P. C., Catalytic degradation of medium density polyethylene over silica-alumina supports. *Fuel Process. Technol.* **1996,** *49* (1–3), 119–136.
32. Ding, W.; Liang, J.; Anderson, L. L., Thermal and catalytic degradation of high density polyethylene and commingled post-consumer plastic waste. *Fuel Process. Technol.* **1997,** *51* (1–2), 47–62.
33. Jha, N. K.; Misra, A. C.; Bajaj, P., Flame-retardant additives for polypropylene. *J. Macromol. Sci.* **1984,** *24* (1), 69–116.
34. Pearce, E. M.; Khanna, Y. P.; Raucher, D., Thermal characterization of polymeric materials. In Thermal Analysis in Polymer Flammability, Academic Press, London: 1981; pp 793–843.
35. Mountassir, A.; Tirri, T.; Sund, P.; Wilén, C.-E., Sulfenamides as standalone flame retardants for polystyrene. *Polym. Degrad. Stabil.* **2021,** *188*, 109588.
36. Nazir, R.; Gooneie, A.; Lehner, S.; Jovic, M.; Rupper, P.; Ott, N.; Hufenus, R.; Gaan, S., Alkyl sulfone bridged

phosphorus flame-retardants for polypropylene. *Mater. Design* **2021,** *200*, 109459.
37. Peters, R. H.; Still, R. H., Applied Fibre Science. Academic Press, London: 1979.
38. Palmer, R. J.; Staff, U. b., Polyamides, plastics. *Kirk-Othmer Encycl Chem Technol* **2000,** *19*, 772–797.
39. Wall, L. A.; Straus, S.; Florin, R. E.; Mohler, F. L.; Bradt, P., Phosphinoborine compounds: Mass spectra and pyrolysis. *J. Res. Natl. Bur. Stand. (U. S.). Section A, Physics and Chemistry* **1959,** *63* (1), 63.
40. Goodman, I., The thermal degradation of 66 nylon. *J. Polym. Sci.* **1954,** *13* (68), 175–178.
41. Levchik, S. V.; Weil, E. D., Combustion and fire retardancy of aliphatic nylons. *Polym Int.* **2000,** *49* (10), 1033–1076.
42. Zhao, B.; Hu, Z.; Chen, L.; Liu, Y.; Liu, Y.; Wang, Y. Z., A phosphorus-containing inorganic compound as an effective flame retardant for glass-fiber-reinforced polyamide 6. *J. Appl. Polym. Sci.* **2011,** *119* (4), 2379–2385.
43. Yoda, K.; Tsuboi, A.; Wada, M.; Yamadera, R., Network formation in poly (ethylene terephthalate) by thermooxidative degradation. *J. Appl. Polym. Sci.* **1970,** *14* (9), 2357–2376.
44. Gooneie, A.; Simonetti, P.; Salmeia, K. A.; Gaan, S.; Hufenus, R.; Heuberger, M. P., Enhanced PET processing with organophosphorus additive: Flame retardant products with added-value for recycling. *Polym. Degrad. Stabil.* **2019,** *160*, 218–228.
45. Bascucci, C.; Duretek, I.; Lehner, S.; Holzer, C.; Gaan, S.; Hufenus, R.; Gooneie, A., Investigating thermomechanical recycling of poly (ethylene terephthalate) containing phosphorus flame retardants. *Polym. Degrad. Stabil.* **2022,** *195*, 109783–109783.
46. Olive, G. H.; Olive, S., The chemistry of carbon fiber formation from polyacrylonitrile. *Adv. Polym. Sci.* **1983,** *51*, 1–60.
47. Houtz, R. C., "Orlon" acrylic fiber: Chemistry and properties. *Text. Res. J.* **1950,** *20* (11), 786–801.
48. Xue, T. J.; McKinney, M. A.; Wilkie, C. A., The thermal degradation of polyacrylonitrile. *Polym. Degrad. Stabil.* **1997,** *58* (1–2), 193–202.
49. Vosburgh, W. G., The heat treatment of orlon acrylic fiber to render it fireproof. *Text. Res. J.* **1960,** *30* (11), 882–896.
50. Hall, M. E.; Zhang, J.; Richard Horrocks, A., The flammability of polyacrylonitrile and its copolymers III. Effect of flame retardants. *Fire Mater.* **1994,** *18* (4), 231–241.
51. Horrocks, A. R.; Zhang, J.; Hall, M. E., Flammability of polyacrylonitrile and its copolymers II. Thermal behaviour and mechanism of degradation. *Polym. Int.* **1994,** *33* (3), 303–314.
52. Howell, B. A., The utilization of TG/GC/MS in the establishment of the mechanism of poly (styrene) degradation. *J. Therm. Anal. Calorim.* **2007,** *89*, 393–398.
53. Howell, B. A.; Cui, Y.; Chaiwong, K.; Zaho, H., Polymerization as a means of stabilization for polystyrene. *J. Vinyl Addit. Technol.* **2006,** *12* (4), 198–203.
54. Hirschler, M. M., Chemical aspects of thermal decomposition of polymeric materials. *Fire Retard Poly Mater.* **2000,** *2*.
55. Ebdon, J. R.; Hunt, B. J.; Jones, M. S.; Thorpe, F. G., Chemical modification of polymers to improve flame retardance—II. The influence of silicon-containing groups. *Polym. Degrad. Stabil.* **1996,** *54* (2–3), 395–400.
56. Armitage, P.; Ebdon, J. R.; Hunt, B. J.; Jones, M. S.; Thorpe, F. G., Chemical modification of polymers to improve flame retardance—I. The influence of boron-containing groups. *Polym. Degrad. Stabil.* **1996,** *54* (2–3), 387–393.
57. Liu, Y. L.; Hsiue, G. H.; Chiu, Y. S.; Jeng, R. J.; Ma, C., Synthesis and flame-retardant properties of phosphorus-containing polymers based on poly (4-hydroxystyrene). *J. Appl. Polym. Sci.* **1996,** *59* (10), 1619–1625.
58. Yan, Y.-W.; Huang, J.-Q.; Guan, Y.-H.; Shang, K.; Jian, R.-K.; Wang, Y.-Z., Flame retardance and thermal degradation mechanism of polystyrene modified with aluminum hypophosphite. *Polym. Degrad. Stabil.* **2014,** *99*, 35–42.
59. Faravelli, T.; Pinciroli, M.; Pisano, F.; Bozzano, G.; Dente, M.; Ranzi, E., Thermal degradation of polystyrene. *J. Anal. Appl. Pyrolysis* **2001,** *60* (1), 103–121.
60. Pike, R. D.; Starnes, W. H.; Jeng, J. P.; Bryant, W. S.; Kourtesis, P.; Adams, C. W.; Bunge, S. D.; Kang, Y. M.; Kim, A. S.; Kim, J. H., Low-valent metals as reductive cross-linking agents: A new strategy for smoke suppression of poly (vinyl chloride). *Macromolecules* **1997,** *30* (22), 6957–6965.
61. Tran, V. H., Polaron mechanism in the thermal degradation and stabilization of poly (vinyl chloride). *J. Macromol. Sci. – Rev. Macromol. Chem. Phys. C* **1998,** *38* (1), 1–52.
62. Hull, T. R.; Price, D.; Liu, Y.; Wills, C. L.; Brady, J., An investigation into the decomposition and burning behaviour of Ethylene-vinyl acetate copolymer nanocomposite materials. *Polym. Degrad. Stabil.* **2003,** *82* (2), 365–371.
63. Costache, M. C.; Jiang, D. D.; Wilkie, C. A., Thermal degradation of ethylene–vinyl acetate coplymer nanocomposites. *Polymer* **2005,** *46* (18), 6947–6958.
64. Chan, Y. Y.; Ma, C.; Zhou, F.; Hu, Y.; Schartel, B., Flame retardant flexible polyurethane foams based on phosphorous soybean-oil polyol and expandable graphite. *Polym. Degrad. Stabil.* **2021,** *191*, 109656–109656.
65. Papa, A. J., Reactive flame retardants for polyurethane foams. *Ind. Eng. Chem. Res.* **1970,** *9* (4), 478–496.
66. Zhang, G.; Lin, X.; Zhang, Q.; Jiang, K.; Chen, W.; Han, D., Anti-flammability, mechanical and thermal properties of bio-based rigid polyurethane foams with the addition of flame retardants. *RSC Adv.* **2020,** *10* (53), 32156–32161.
67. Takeichi, T.; Suefuji, K.; Inoue, K., Novel poly (urethane-amide) s from polyurethane prepolymer and reactive polyamide. Preparation and properties. *Polym. J.* **2002,** *34* (6), 455–460.
68. Hiltz, J. A., Analytical pyrolysis gas chromatography/mass spectrometry (py-GC/MS) of poly (ether urethane) s, poly (ether urea) s and poly (ether urethane-urea) s. *J. Anal. Appl. Pyrolysis* **2015,** *113*, 248–258.
69. Woolley, W. D., Nitrogen-containing products from the thermal decomposition of flexible polyurethane foams. *Br. Polym. J.* **1972,** *4* (1), 27–43.
70. Consumer Protection Act, The furniture and furnishing (fire)(safety) regulations, 1988 (amended 1989 and 1993), SI 1324. Her Majesty's Stationery Office, London: 1987.
71. Price, D.; Liu, Y.; Milnes, G. J.; Hull, R.; Kandola, B. K.; Horrocks, A. R., An investigation into the mechanism of flame retardancy and smoke suppression by melamine in flexible polyurethane foam. *Fire Mater.* **2002,** *26* (4–5), 201–206.
72. Gaan, S.; Liang, S.; Mispreuve, H.; Perler, H.; Naescher, R.; Neisius, M., Flame retardant flexible polyurethane foams from novel DOPO-phosphonamidate additives. *Polym. Degrad. Stabil.* **2015,** *113*, 180–188.
73. Przystas, A.; Jovic, M.; Salmeia, K. A.; Rentsch, D.; Ferry, L.; Mispreuve, H.; Perler, H.; Gaan, S., Some key factors influencing the flame retardancy of EDA-DOPO containing flexible polyurethane foams. *Polymers* **2018,** *10* (10), 1115.

74. Horrocks, A. R.; Kandola, B. K., Flammability and fire resistance of composites. In Design and Manufacture of Textile Composites, Elsevier, Cambridge: 2005; pp 330–363.
75. Varganici, C.-D.; Rosu, L.; Bifulco, A.; Rosu, D.; Mustata, F.; Gaan, S., Recent advances in flame retardant epoxy systems from reactive DOPO–based phosphorus additives. *Polym. Degrad. Stabil.* **2022**, 110020–110020.
76. Bifulco, A.; Varganici, C. D.; Rosu, L.; Mustata, F.; Rosu, D.; Gaan, S., Recent advances in flame retardant epoxy systems containing non-reactive DOPO based phosphorus additives. *Polym. Degrad. Stabil.* **2022**, 109962.
77. Troitzsch, J.; Antonatus, E., Plastics Flammability Handbook: Principles, Regulations, Testing, and Approval. Carl Hanser Verlag GmbH Co KG, München: 2021.
78. Bansal, R. K.; Mittal, J.; Singh, P., Thermal stability and degradation studies of polyester resins. *J. Appl. Polym. Sci.* **1989,** *37* (7), 1901–1908.
79. Guadagno, L.; Raimondo, M.; Vittoria, V.; Vertuccio, L.; Naddeo, C.; Russo, S.; De Vivo, B.; Lamberti, P.; Spinelli, G.; Tucci, V., Development of epoxy mixtures for application in aeronautics and aerospace. *RSC Adv.* **2014,** *4* (30), 15474–15488.
80. Bishop, D. P.; Smith, D. A., Combined pyrolysis and radiochemical gas chromatography for studying the thermal degradation of epoxy resins and polyimides. I. The degradation of epoxy resins in nitrogen between 400 C and 700 C. *J. Appl. Polym. Sci.* **1970,** *14* (1), 205–223.
81. Bifulco, A.; Parida, D.; Salmeia, K.; Lehner, S.; Stämpfli, R.; Markus, H.; Malucelli, G.; Branda, F.; Gaan, S., Improving flame retardancy of in-situ silica-epoxy nanocomposites cured with aliphatic hardener: Combined effect of DOPO-based flame-retardant and melamine. *Compos. C: Open Access* **2020,** *2*.
82. Bifulco, A.; Parida, D.; Salmeia, K. A.; Nazir, R.; Lehner, S.; Stämpfli, R.; Markus, H.; Malucelli, G.; Branda, F.; Gaan, S., Fire and mechanical properties of DGEBA-based epoxy resin cured with a cycloaliphatic hardener: Combined action of silica, melamine and DOPO-derivative. *Mater. Design* **2020,** *193*, 108862–108862.
83. Chen, Z. Q.; Chen, Y. F.; Liu, H. B., Study on thermal degradation of phenolic resin. *Appl. Mech. Mater.* **2013,** *422*, 24–28.
84. Kandola, B. K.; Horrocks, A. R.; Myler, P.; Blair, D., Thermal characterization of thermoset matrix resins. In Fire and Polymers; ACS Symposium Series; American Chemical Society, Washington, D.C.: 2001; Vol. 797, pp 344–360. https://pubs.acs.org/doi/10.1021/bk-2001-0797.ch027
85. Wang, S.; Liu, J.; Sun, L.; Wang, H.; Zhu, P.; Dong, C., Preparation of flame-retardant/dyed cotton fabrics: Flame retardancy, dyeing performance and flame retardant/dyed mechanism. *Cellulose* **2020,** *27*, 10425–10440.
86. Nazir, R.; Parida, D.; Borgstädt, J.; Lehner, S.; Jovic, M.; Rentsch, D.; Bülbül, E.; Huch, A.; Altenried, S.; Ren, Q., In-situ phosphine oxide physical networks: A facile strategy to achieve durable flame retardant and antimicrobial treatments of cellulose. *Chem. Eng. J.* **2021,** *417*, 128028–128028.
87. Kandola, B. K.; Horrocks, A. R.; Price, D.; Coleman, G. V., Flame-retardant treatments of cellulose and their influence on the mechanism of cellulose pyrolysis. *J. Macromol. Sci. C: Polym. Rev.* **1996,** *36* (4), 721–794.
88. Faroq, A. A.; Price, D.; Milnes, G. J.; Horrocks, A. R., Use of gas chromatographic analysis of volatile products to investigate the mechanisms underlying the influence of flame retardants on the pyrolysis of cellulose in air. *Polym. Degrad. Stabil.* **1991,** *33* (2), 155–170.
89. Price, D.; Horrocks, A. R.; Akalin, M.; Faroq, A. A., Influence of flame retardants on the mechanism of pyrolysis of cotton (cellulose) fabrics in air. *J. Anal. Appl. Pyrolysis* **1997,** *40*, 511–524.
90. Bradbury, A. G. W.; Sakai, Y.; Shafizadeh, F., A kinetic model for pyrolysis of cellulose. *J. Appl. Polym. Sci.* **1979,** *23* (11), 3271–3280.
91. Price, D.; Coleman, G.; Horrocks, A., Use of cyclic differential scanning calorimetry: Experiments to investigate the formation of an 'activated cellulose'species during cotton (cellulose) fabric pyrolysis in air and the influence of flame retarded treatments thereon. *J. Therm. Anal. Calorim.* **1993,** *40* (2), 649–656.
92. Shafizadeh, F.; Bradbury, A. G. W., Thermal degradation of cellulose in air and nitrogen at low temperatures. *J. Appl. Polym. Sci.* **1979,** *23* (5), 1431–1442.
93. Fairbridge, C.; Ross, R. A.; Sood, S. P., A kinetic and surface study of the thermal decomposition of cellulose powder in inert and oxidizing atmospheres. *J. Appl. Polym. Sci.* **1978,** *22* (2), 497–510.
94. Kandola, B. K.; Horrocks, A. R.; Padmore, K.; Dalton, J.; Owen, T., Comparison of cone and OSU calorimetric techniques to assess the flammability behaviour of fabrics used for aircraft interiors. *Fire Mater.* **2006,** *30* (4), 241–255.
95. Guan, J. P.; Chen, G. Q., Flame retardancy finish with an organophosphorus retardant on silk fabrics. *Fire Mater.* **2006,** *30* (6), 415–424.
96. Schartel, B., Phosphorus-based flame retardancy mechanisms—Old hat or a starting point for future development? *Materials* **2010,** *3* (10), 4710–4745.
97. Menefee, E.; Yee, G., Thermally-induced structural changes in wool. *Text. Res. J.* **1965,** *35* (9), 801–812.
98. Benisek, L., Flame retardance of protein fibers. In Flame Retardant Polymeric Materials, Springer, New York: 1975; pp 137–191.
99. Horrocks, A. R.; Eichhorn, H.; Schwaenke, H.; Saville, N.; Thomas, C., Thermally resistant fibres. In High Performance Fibres, CRC Press, Cambridge: 2001; pp 281–324.
100. Gupta, R.; Shinde, S.; Yella, A.; Subramaniam, C.; Saha, S. K., Thermomechanical characterisations of PTFE, PEEK, PEKK as encapsulation materials for medium temperature solar applications. *Energy* **2020,** *194*, 116921–116921.
101. Thiruchitrambalam, M.; Kumar, D. B.; Shanmugam, D.; Jawaid, M., A review on PEEK composites–Manufacturing methods, properties and applications. *Mater. Today: Proc.* **2020,** *33*, 1085–1092.
102. Dis, I. S. O., 5660, Fire tests–Reaction to fire–rate of heat release from building products. International Organization for Standardization, Geneva: 1990.
103. Shum-Kivan, F.; Santiago, J. M.; Verdier, A.; Riber, E.; Renou, B.; Cabot, G.; Cuenot, B., Experimental and numerical analysis of a turbulent spray flame structure. *Proc. Combust. Inst.* **2017,** *36* (2), 2567–2575.
104. Cui, G.; Dong, Z.; Wang, S.; Xing, X.; Shan, T.; Li, Z., Effect of the water on the flame characteristics of methane hydrate combustion. *Appl. Energy* **2020,** *259*, 114205–114205.
105. Hull, T. R., Challenges in fire testing: Reaction to fire tests and assessment of fire toxicity. In Advances in Fire Retardant Materials, Elsevier, Cambridge: 2008; pp 255–290.
106. Price, D.; Gao, F.; Milnes, G. J.; Eling, B.; Lindsay, C. I.; McGrail, P. T., Laser pyrolysis/time-of-flight mass spectrometry studies pertinent to the behaviour of flame-retarded

polymers in real fire situations. *Polym. Degrad. Stabil.* **1999,** *64* (3), 403–410.
107. Szabó, A.; MarosfŐi, B. B.; Anna, P.; Marosi, G., Complex micro-analysis assisted design of fire-retardant nanocomposites–contribution to the nanomechanism. In Flame Retardancy of Polymers: New Strategies and Mechanisms, The Royal Society of Chemistry, Cambridge: 2008; pp 74–91.
108. Bourbigot, S.; Duquesne, S., Fire retardant polymers: recent developments and opportunities. *J. Mat. Chem.* **2007,** *17* (22), 2283–2300.
109. Price, D.; Pyrah, K.; Hull, T. R.; Milnes, G. J.; Ebdon, J. R.; Hunt, B. J.; Joseph, P.; Konkel, C. S., Flame retarding poly (methyl methacrylate) with phosphorus-containing compounds: Comparison of an additive with a reactive approach. *Polym. Degrad. Stabil.* **2001,** *74* (3), 441–447.
110. Salmeia, K. A.; Gaan, S., An overview of some recent advances in DOPO-derivatives: Chemistry and flame retardant applications. *Polym. Degrad. Stabil.* **2015,** *113*, 119–134.
111. Hirsch, C.; Striegl, B.; Mathes, S.; Adlhart, C.; Edelmann, M.; Bono, E.; Gaan, S.; Salmeia, K. A.; Hoelting, L.; Krebs, A., Multiparameter toxicity assessment of novel DOPO-derived organophosphorus flame retardants. *Arch. Toxicol.* **2017,** *91* (1), 407–425.
112. Liang, S.; Hemberger, P.; Neisius, N. M.; Bodi, A.; Grützmacher, H.; Levalois-Grützmacher, J.; Gaan, S., Elucidating the thermal decomposition of dimethyl methylphosphonate by vacuum ultraviolet (VUV) photoionization: Pathways to the PO radical, a key species in flame-retardant mechanisms. *Chem. Eur. J.* **2015,** *21* (3), 1073–1080.
113. Liang, S.; Hemberger, P.; Steglich, M.; Simonetti, P.; Levalois-Grützmacher, J.; Grützmacher, H.; Gaan, S., The underlying chemistry to the formation of PO_2 radicals from organophosphorus compounds: A missing puzzle piece in flame chemistry. *Chem. Eur. J.* **2020,** *26* (47), 10795–10800.
114. Liang, S.; Hemberger, P.; Levalois-Grützmacher, J.; Grützmacher, H.; Gaan, S., Probing phosphorus nitride (p≡ n) and other elusive species formed upon pyrolysis of dimethyl phosphoramidate. *Chem. Eur. J.* **2017,** *23* (23), 5595–5601.
115. Price, D.; Cunliffe, L. K.; Bullet, K. J.; Hull, T. R.; Milnes, G. J.; Ebdon, J. R.; Hunt, B. J.; Joseph, P., Thermal behavior of covalently bonded phosphonate flame-retarded poly (methyl methacrylate) systems. *Polym. Adv. Technol.* **2008,** *19* (6), 710–723.
116. Bullett, K. J., 'Flame retarding acrylic and styrenic polymers by modification with phosphorus-containing compounds', PhD thesis, University of Salford, 2002.
117. Cunliffe, L. K., 'Decomposition studies of fire retardant action of phosphorus compounds in poly (methyl methacrylate), polystyrene and their copolymers', PhD thesis, University of Salford, 2005.
118. Kaprinidis, N.; Shields, P.; Leslie, G., Antimony free flame retardant systems containing Flamestab NOR 116 for polypropylene molding. In Flame Retardants 2002, Interscience Communications, London: 2002; pp 107–112.
119. Zhang, S.; Horrocks, A. R., A review of flame retardant polypropylene fibres. *Prog. Polym. Sci.* **2003,** *28* (11), 1517–1538.
120. Broadbelt, L. J.; Klein, M. T.; Dean, B. D.; Andrews, S. M., Thermal stability of aliphatic-aromatic polyamide model compounds: Structure–reactivity relationships in the catalyzed thermal reaction of benzamides in the presence of copper halides. *J. Polym. Sci.* **1997,** *35* (15), 3305–3322.
121. Malík, J.; Kröhnke, C., Polymer stabilization: present status and possible future trends. *C. R. Chim.* **2006,** *9* (11–12), 1330–1337.
122. Gugumus, F. L., Polyolefin stabilization: From single stabilizers to complex systems. *Environ. Sci. Pollut. Control Ser.* **2000**, 1–38.
123. Stuetz, D. E.; DiEdwardo, A. H.; Zitomer, F.; Barnes, B. P., *J. Polym. Sci. Polym. Chem. Ed.* **1975,** *13*, 585–621.
124. Qureshi, F. S.; Amin, M. B.; Maadhah, A. G.; Hamid, S. H., Weather-induced degradation of linear low-density polyethylene: Mechanical properties. *Polym.-Plast. Technol. Mater.* **1989,** *28* (7–8), 649–662.
125. Chiang, W.-Y.; Hu, C.-H., The improvements in flame retardance and mechanical properties of polypropylene/FR blends by acrylic acid graft copolymerization. *Eur. Polym. J.* **1996,** *32* (3), 385–390.
126. Friese, K.; Tannert, F., Radiation-induced grafting of chlorinated PVC for preparation of plastic foams. *Radiat. Phys. Chem.* **1999,** *55* (1), 47–54.
127. Zhang, S.; Wang, J.; Ding, Y.; Wu, S.; Xie, L.; Wen, C., The electron-beam (EB) irradiation and grafting of acrylic monomers onto EPDM copolymer. *Chinese Sci. Bull.* **2000,** *45*, 322–325.
128. Wang, J.; Zhang, S.; Xie, L.; Wen, C., Approaches to the flame retardancy of polymers. I.: Electron-beam pre-irradiation and grafting of acrylic monomers onto EVA copolymers. *J. Fire Sci.* **1997,** *15* (1), 68–87.
129. Laishun, S.; Jianqi, W., An approach to the flame retardancy of EVA copolymer by plasma treatmentc2. *J. Beijing Inst. Technol.* **1997,** *6* (4), 313–320.
130. Horrocks, A. R., Flame retardant/resistant textile coatings and laminates. In Advanced Fire Retardant Materials, Elsevier, Cambridge: 2008; pp 159–187.

3 Physical Parameters Affecting Fire Growth

Alexander B. Morgan[1], *Jose L. Torero*[2], *and Guillermo Rein*[2]
[1] 3rd Edition Revised Chapter Author:
University of Dayton Research Institute, Dayton, OH 45469
Alexander B. Morgan:
[2] 2nd Edition Chapter Authors:
BRE Centre for Fire Safety Engineering
The University of Edinburgh
United Kingdom

Nomenclature List

Area	A
Pre-exponential factor	A_i
Radiative absorptivity	a
Specific heat	C
Damköhler number	Da
Activation energy	E_i
Gravity acceleration	$\vec{g}$
Heat transfer coefficient	h
Heat of reaction	ΔH
Thermal conductivity	k
Mass flow	$\dot{m}''$
Pressure	P
Heat flux	$\dot{q}''$
Heat release rate	$\dot{Q}$
Universal gas constant	R
Spread	S
Temperature	T
Time	t
Velocity vector	$\vec{u}$
Gas velocity	U
Volume	V
Spatial location	x
Species concentration	Y

Greek Symbols

Thermal diffusivity	α
Emissivity	ε
Stephan-Boltzmann constant	σ
Density	ρ
Porosity	φ
Reaction rate	$\dot{\omega}'''$
Area	A
Viscosity	μ
Non-dimensional temperature	θ
Stoichiometric coefficient	Θ
Flame radiative fraction	χ

3.1 FIRE SAFETY

3.1.1 Objectives of Fire Safety

Because fire represents a threat to life, property, and the environment, there is a need to control its impact in such a way that life is fully protected, and damage to property and the environment are minimized. Fire safety is the means by which infrastructure is designed in a manner such that these goals are achieved. Fire safety engineering as a discipline uses a variety of concepts to minimize fire losses, whether they be human lives or property.[1] Fire safety engineering considers how fire grows in a building, compartment, vehicle, or other scenario, and so it is important to understand this fundamental process to enable fire safety engineering approaches.

The schematic presented in Figure 3.1 represents the possible sequence of events during a fire in a building. The safety objectives for a building can be quantified in terms of the different characteristic times of the events. It follows that the time needed to evacuate a particular compartment, $t_{e,i}$, is required to be much smaller than the time to reach untenable conditions in that compartment, $t_{f,i}$. The characteristic values of $t_{e,i}$ and $t_{f,i}$ can be established for different levels of containment, that is, room of origin (i = 1), floor (i = 2), and building (i = n). Furthermore, it is necessary for the time to evacuate the building to be much smaller than the time when structural integrity starts to be compromised (t_s).

In summary:

$$\forall i,\ \mathrm{i}=1 \text{ to } \mathrm{n};\ t_{e,i} << t_{f,i} \tag{3.1}$$

DOI: 10.1201/9781003380689-3

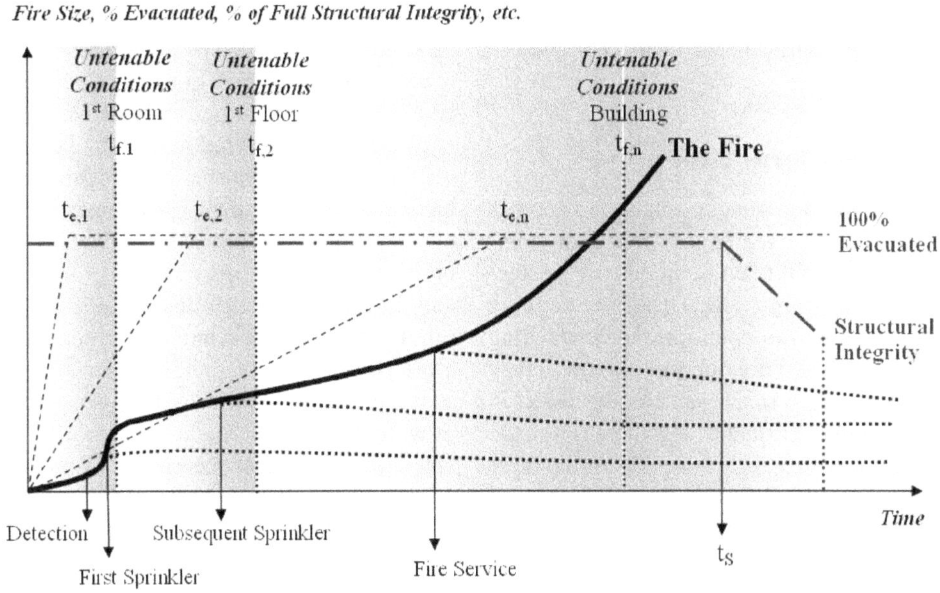

FIGURE 3.1 Schematic of the sequence of events following the onset of a fire in a multistorey building. The solid line corresponds to the "fire size," the dotted lines to the possible outcome of the different forms of intervention (sprinkler activation, fire service). The units of "fire size" could be defined as heat-release rate, area of fire, or any other means to quantify the magnitude of the event. The dashed lines are the percentage of people evacuated for the room, floor, and building respectively, with the ultimate goal of 100% represented by a horizontal dashed line. The dashed and dotted line corresponds to the percentage of the full structural integrity of the building.

$$\forall i,\ i = 1 \text{ to } n;\ t_{e,i} << t_S$$

It could be added to these objectives that full structural collapse is an undesirable event however long the fire lasts, therefore:

$$t_S \to \infty \tag{3.2}$$

Although these generalized criteria for safety times are a simplified statement, they describe well the main objectives of a fire safety strategy.

When designing for fire safety, a number of strategies are put in place aiming at achieving these objectives. These include those factors which are intended to increase t_s and $t_{f,i}$, such as active (e.g., sprinklers, or the intervention of the fire service) and passive systems (e.g., fire proofing or compartmentation). As shown by Figure 3.1 (the dotted lines branching off below the Fire curve), success of these strategies can result in control or suppression of the fire. Passive protection such as thermal insulation of structural elements becomes part of the design, with the purpose of increasing t_S. Finally, but most importantly, evacuation protocols and routes are designed to reduce $t_{e,i}$ at all stages of the building evacuation. It is important to note that the safe operation of the fire service within these times needs also to be included in the design.

Figure 3.1 makes clear that fire safety is the superposition of three different types of events occurring simultaneously. Two of these events, egress and structural behavior, are reactive events, while the rate of fire growth is the driving process. The structure will be designed, and it will respond to the fire. Some passive fire protection systems (detection, alarm) are designed and implemented to warn of the fire, and others are designed to affect the rate of growth (suppression). People within a building are located according to the general use of the premises but will change their behavior in response to the fire. Occupants will have mostly a passive role, while firefighters will have an active role attempting to control the growth of the fire.

Building design and firefighter intervention procedures are defined on the basis of one or more fire growth scenarios. In the case of prescriptive design (codes and standards), the fire growth scenarios are implicit, while in the case of performance-based design (engineering based methods), they are explicitly defined and are referred to as "design fires." Prescriptive design rules use knowledge on fire dynamics and empirical data to bound the fire growth for the specific conditions of the implied scenarios. Fire safety systems are designed to operate within these bounds and are deemed adequate for a range of building. But given that there are unavoidable and significant differences between buildings, there is a risk of extrapolating codes and standards outside its range of applicability. Therefore, to know the extent of the extrapolation of prescriptive solutions requires understanding the parameters that govern and bound the fire growth scenario. In the case of performance based design, knowledge on fire dynamics is used to predict fire growth under the particular conditions of the building. Thus the link between fire safety objective and understanding of the physical parameters controlling fire growth is important and explicit.

3.1.2 Basic Definition of Fire Growth

By understanding combustion fundamentals, it becomes possible for the material scientist or fire protection engineer to implement flame retardant solutions/fire protection systems that inhibit fire growth. Therefore, defining fire growth is essential to ensure a uniform understanding among fire safety scientists and fire safety engineers so they can work together toward the same goal. As discussed in the previous section, Figure 3.1 implies that there is a single variable to quantify fire growth, but the reality is that there are many different variables. The variable or variables of interest depend on the objective of the system under design.

At the core of a fire there is a flame or a reaction front that is effectively a combustion process and thus is governed by the mechanisms and variables controlling combustion.[2] The interaction between the fire and the environment determines the behavior of the flame and nature of the combustion processes. This is commonly referred to as fire dynamics. An extensive introduction to the topic is provided by Drysdale.[3]

As indicated by Drysdale,[3] fire dynamics involves a compendium of different sub-processes that start with the initiation of a fire and end with its extinction. The onset of the combustion process, that is, ignition, in a fire is a complex process that implies not only the initiation of an exothermic reaction but also a decomposition process that provides the fuel feeding the fire. In a fire it is common to have different materials involved, and given the nature of the fire growth many could be involved simultaneously but others sequentially. The sequence of ignitions of items in an enclosure will affect the nature of the combustion processes. Thus, ignition mechanisms set the dynamics of the fire and also are affected by the fire itself, creating a feedback loop.

Once a material is ignited, the flame propagates over the condensed fuels by transferring sufficient heat to the fuel until a subsequent ignition occurs. This process is commonly referred to as flame spread and is described in detail by Fernandez-Pello[4] and covered in Section 3.5 of this chapter. Flame spread defines the surface area of flammable material that is delivering gaseous fuel to the combustion process. The quantity of fuel produced per unit area is the mass burning rate. The mass burning rate multiplied by the surface area determines the total amount of fuel produced. If the total amount of fuel produced is multiplied by the effective heat of combustion (energy produced by combustion per unit mass of fuel burnt), it yields the heat release rate. The heat release rate is generally considered the single most important variable to describe fire growth.[5] Given the nature of the environment, the oxygen supply might not be enough to consume all the fuel, thus in many cases combustion is incomplete, and therefore the heat of combustion is not a material property but a function of the interactions between the environment and the fire. In these cases it is usually deemed appropriate to calculate the heat release rate as the energy produced per unit mass of oxygen consumed multiplied by the available oxygen supply. This is covered in more detail in Section 3.6.

A fire can end when it is extinguished or when oxygen or fuel supplies are depleted (oxygen starvation and burnout respectably). In all cases, extinction of the combustion process is brought by interactions of fuel and oxygen supply and the energy balance that permits the combustion reaction to remain self-sustained.[6–8] Suppression agents affect a fire by reducing fuel and oxygen supply or by removing heat (i.e., disturbing the fire triangle). At each stages of fire growth, it is more or less feasible to affect these three variables. Thus the effectiveness of a suppression system is dictated by its capability to affect the targeted variables at the moment of deployment.

The following sections describe all the above processes in more detail extracting at each stage the main material properties and physical parameters that affect fire growth and how they relate to the fire safety.

3.2 FUNDAMENTALS OF COMBUSTION

3.2.1 Reactive Gaseous Mixtures

A flame is the thin sheet where a gas phase combustion reaction is taking place, resulting in gases at high temperature. These gases could be hot enough to radiate their own light and thus be visible to the naked eye. An exothermic reaction is the essential chemical phenomena driving the combustion and involves the exchange of atoms between two reactants brought together, a fuel and oxidizer (typically the oxygen in air). It results in the release of heat and species products. In general, the process can involve millions of elementary chemical reactions, including very large numbers (6.022×10^{23}) of molecules and atoms, but the overall process can be represented by a reduced set of lumped pathways. For example, for methane, the overall combustion reaction for ideal and stoichiometric conditions can be simplified to the following:

$$CH_4 + 2\,O_2 \rightarrow CO_2 + 2\,H_2O$$

A reaction is said to be stoichiometric when the fuel and oxygen consume each other completely, forming only in ideal conditions carbon monoxide (CO_2) and water (H_2O). The equivalence ratio is the parameter relating a mixture proportions to stoichiometry. It is defined as the ratio of fuel to oxygen amounts times the stoichiometric ratio of oxygen to fuel amounts. If there is an excess of fuel, the mixture is called fuel-rich or rich, and the equivalence ration is > 1, and if there is an excess of oxygen, it is called fuel-lean or lean, and the equivalence ration is < 1.

Not all mixtures of reactants lead to combustion. There is a limiting amount of fuel per amount of oxygen above which the reaction does not take place. This is called the upper flammability limit (for methane/air mixtures this is 15% at atmospheric conditions). Likewise, there is a limiting amount of fuel per amount of oxygen below which the reaction does not take place. This is called the lower flammability limit (for methane/air mixtures this is 5% at atmospheric conditions). Tabulated values of these limits for different gaseous fuels can be found in.[1,2]

The net heat that is liberated in the overall combustion reaction is called heat of combustion and is equal to the heat released by the exothermic reactions minus the heat absorbed by the endothermic reactions. The heat of combustion is partially stored in the combustion product as sensible energy, thus increasing their temperature, and partially lost to the environment. The maximum temperature that the combustion products can ever reach is given for the ideal conditions of no heat losses or adiabatic combustion. This is called adiabatic temperature, and it sets for stoichiometric conditions the theoretical maximum limit for the temperature of a flame (for methane/air mixtures this is 1950°C at atmospheric conditions). Tabulated values of adiabatic temperature for different gaseous fuels can be found in.[1,2]

3.2.2 Premixed and Diffusion Flames

There are two basic flame structures depending on the mixing process of the gaseous reactants; premixed and diffusion (or non-premixed) flames. In premixed flames, fuel and oxygen are premixed before combustion. The process in a gasoline engine is a good example. The travelling of a freely premixed flame can be characterized by its flame speed, which is a function of the fuel, equivalence ratio, pressure and thermal properties of the reactants (for stoichiometric methane/air mixtures this is 36 cm/s). Tabulated values of flame speed for different gaseous fuels can be found in.[1] The flame speed is together with the heat of combustion and the adiabatic flame temperature is the most fundamental property of premixed flammable mixtures. If the speed at which the reactants arrive to the reaction front is lower than the flame speed, the reaction zone location is unstable, and the flame is said to blow off. Premixed flames are easier to study and control in laboratory conditions and thus are preferred for research purposes. This has led to the largest body of combustion knowledge being on premixed flames, especially in laminar flows.

In diffusion flames, fuel and oxygen are initially separated and mix at the flame location during the combustion process. Most combustion systems of technological interest, like diesel engines, gas burners, and accidental fires, are diffusion flames. The distinctive characteristic of a diffusion flame is that the burning rate is determined by the rate at which the fuel and the oxygen are brought together. Thus, the chemistry can be assumed infinitely fast, and the reaction be approximated by a one-step overall reaction (as the CH_4 example presented above). The mixing is controlled at the small-scale through molecular and turbulent diffusion (quantify by a variable called scalar dissipation rate[1]) and at large-scale through the flow conditions around the flame (convection transfer). There is a two-way coupling between the flame and the flow since the heat released induces buoyancy motion of combustion products and thus affects the flow environment. For this reason, flames that are larger than approximately 30 cm are turbulent, and in most cases of technological interest, the flame is turbulent.

3.2.3 Burning of Condensed Fuels

The fuel can be in the form of a gaseous jet or a condensed medium (either solid or liquid). Diffusion flames do not have a fundamental characteristic property that can be measured, such as flame speed. The equivalence ratio of the flame is not uniquely defined but cover a wide range of values in the spatial locations around the flame zone. The overall ratio of fuel and oxygen flows over the stoichiometric ratio of fuel and oxygen flow to the reaction zone are used to classification diffusion flames as overventilated and underventilated. Overventilated flames have enough oxygen supply to burn all the available gaseous fuel. Underventilated flames do not have enough oxygen supply to burn all the gaseous fuel.

The process via which a flame is started is called ignition. There are two types of ignition for gaseous mixtures; *induced ignition* via a hot spot, spark, or a small pilot flame, and *spontaneous ignition* via chemical process inherent to the mixture at particular conditions and where no external source is involved. The necessary condition for a thermal ignition is that the rate of energy release is greater than the rate of energy loss (dissipation). Ignition conditions are treated in more detail in Section 3.4 of this chapter. The process via which a flame is terminated is called extinction. A flame can be extinguished by quenching (excessive removal of heat by the presence of a solid) or be blown out (excessive removal of heat via convective flow). Both extinction mechanisms can be explained in terms of the dimensionless parameter Damköhler number (Da). This is defined as the ratio of a characteristics reaction time over a characteristic mixing time (or residence time). Large values of Da lead to extinction of the flame. Extinction conditions are treated in more detail in Section 3.7 of this chapter.

When a condensed phase fuel, either liquid or solid, is burning, a diffusion flame is established at some distance on top of the free surface of the fuel. The gaseous fuel supply feeding the flame is produced by evaporation, gasification, and pyrolysis of the condensed fuel. Pyrolysis is the chemical decomposition of organic materials by heating and not involving oxygen reactions. In solids, it leads to charring of the fuel, which is the chemical process of decomposition and incomplete combustion resulting in char as a solid residue. The gaseous fuel produced this way is then transported away from the surface into the air towards the flame as the oxygen is transported from the surroundings to the burning front. At the diffusion flame, the chemical process can be assumed to be infinitely fast and thus the burning rate of condense fuels is determined by heat and mass transfer rates alone. The rate at which the condense fuel is converted into gaseous form is generally determined by the rate of heat transfer from the flame to the fuel surface. This heat is in turn conducted into the condensed fuel and drive the evaporation or pyrolysis.

Before a flame is established, the ignition of the condensed fuel takes place. For the case of induced ignition with a pilot, the process has two conditions that are of interests and are generally considered material properties of the condensed fuels. Upon heating from initial ambient temperature, the material produces gaseous fuels that are transported away and mix with air, thus creating a spatial distribution of different fuel concentrations. The first point of interest is reached; the temperature of the material is high enough to produce a flammable gaseous mixture that reaches the location of the pilot. Then, the mixture ignites, and a flame propagates rapidly over the surface of the fuel consuming the gaseous fuel like a flash. This is termed the flash point and is quantified by the temperature of the fuel when the flash is observed in a standard test (flash point temperature) such as ASTM D56[9] or ASTM D93.[10] As the temperature of the condensed material keeps increasing, the supply of gaseous fuel is large enough that a stabled flame is anchored over the surface. This is termed the fire point and is quantified by the temperature of the fuel when the stable flame is observed in a standard test (fire point temperature). In some fuels, the flash point and the fire point are identical as measured in the test.

3.3 FUNDAMENTALS OF FIRE DYNAMICS

3.3.1 Open Fires

Accidental fires, the main objective of fire dynamics, mostly involve condensed fuels and thus are driven by diffusion flames. The single most important variable to describe and quantify fire is the heat released rate (HRR), which is the rate of thermal energy liberated at the combustion process. After ignition of the condensed fuel, if flame spread occurs, the surface area of burning fuel will increase in time and also ignition of other fuel packages might take place. The evolution in time of the HRR is related to the global fire growth from flame spread to burn out and extinction. In the early stages, the HRR typically follows a t^2 law in time with different strengths/speeds for different fire scenario. During the burn out phase, it follows an exponential decay in time.

The rates of productions of bulk combustion products (CO, CO_2, soot, and smoke in general) are proportional to the heat release rate since both of them are originated by the reaction rate at the total burning surface of fuel. The hot smoke emitted from a fire will rise up due to buoyancy forming the plume. As the plume rises, it entrains air from the surroundings in turn feeding the fire with oxygen. Also the diameter and mass flow rate of the plume increase with elevation as a result of the air entrainment. Air entrainment has the effect of increasing the mass flow of gas in the plume but also dilutes smoke concentration and lowers the smoke temperature.

The heating mechanisms from an open fire are the radiation from the flames that transport energy in rays going in all directions originated at the fire but is of relatively short range and the heat convection of the hot smoke that can travel significant distances. Only very close to the flame rim on the surface of the fuel heat transfer by conduction is important.

3.3.2 Compartment Fires

When the fire is located inside an enclosure bound by a ceiling and walls, the fire dynamics differ from those of an open fire. The products of combustion together with the air entrained by the fire flow up until they reach the ceiling. Upon reaching it, a ceiling jet is formed and smoke spreads concentrically from the fire axis. The smoke accumulates between the ceiling and the upper parts of the walls, eventually building a hot layer. The enclosure is then divided into two distinct horizontal layers filling the space. The hot products of combustion occupy the upper layer, while the cooler gas, mostly ambient air, occupies the lower layer. The hot gases can also flow through openings and smoke, and fire can spread to other compartments of the buildings.

In a similar manner to the classification of diffusion flames, the overall flows of fuel and oxygen are used to classify compartment fires in fuel controlled and ventilation controlled. Fuel controlled fires have enough oxygen supply from the opening to the compartment to burn all the available gaseous fuel. The size and evolution of the fire is then governed by the amount and configuration of the fuel load in the compartment. Ventilation controlled fires do not have enough oxygen supply to burn all the gaseous fuel and the combustion reaction is incomplete. The size and evolution of the fire is then governed by the amount and configuration of area of openings in the compartment. This leaves a significant fraction of unburned gaseous fuel in the smoke. This fuel-rich smoke could burn once it reaches a fresh supply of oxygen at the compartment openings resulting in external flaming from doors in corridors or windows on the façade of the building. Also, as the fire heating of window covers can lead to their breakage, the total area of openings in the compartment would increase and so will the fire size.

As a fuel-controlled fire grows in size and involves more fuel surface area and more fuel packages, it can reach a condition where the entire fuel load in the compartment is burning at once. When this condition takes place as a transition suddenly, it is termed flashover. The transition from fuel controlled to ventilation control conditions usually takes place at a flashover event.

The other fate of interest for compartment fires is vitiation. When some smoke instead of pure air is entrained into the flame, the combustion reaction is incomplete, releases less heat, and the flame could be locally extinguished. If the smoke layer keeps growing and the vitiation continues, the fire could be extinguished by this process. Alternatively, the fire could be extinguished by consumption of most of the oxygen in the compartment volume. Oxygen starvation and vitiation are possible in compartments with very small openings relative to the fire load.

3.3.3 Fire Modeling

Computer fire modeling was first developed as a research tool in the 1970s after the surge of computer resources. It reached its first applications to real fire engineering problems in the late 1980s and now is widely used in many aspects of fire science and engineering. Its current applications range from forensic investigations, risk assessments, life safety, smoke movement and detection, sprinkler performance, structural behavior, and design of fire safety. Modeling is among the fastest developing areas in fire safety science, and it continues to gain the ability to reproduce fire phenomena but still lags experimental understanding. The two are best used to complement one another, where modelling helps explain and visualize experimental results, or to help design experiments by modelling fires before the experiments are carried out.

There are three main approaches to model compartment fires.[2,3] The simplest is to use basic expressions and experimental correlations of the thermochemical and fluid processes occurring to produce an analytical model of the fire development. Analytical fire models are fast to set up and easy to use because of the few mechanisms involved,[2] however, the results are only correct in the order of magnitude because coupling of the different fire phenomena is difficult in these models. Nevertheless, they can serve as a baseline for more sophisticated computer modeling.

Of a more complete approach are the zone models,[3] which consider two (or more) distinct horizontal layers filling the compartment, each of which is assumed to be spatially uniform in temperature, pressure, and species concentrations, as determined by simplified transient conservation equations for mass, species, and energy. The hot gases tend to form an upper layer, and the ambient air stays in the lower layers. A fire in the enclosure is treated as a pump of mass and energy from the lower layer to the upper layer. As energy and mass are pumped into the upper layer, its volume increases causing the interface between the layers to move toward the floor. Mass transfer between compartments can also occur by means of vents such as doorways and windows. Heat transfer in the model occurs due to conduction to the various surfaces in the room. In addition, heat transfer can be included by radiative exchange between the upper and lower layers and between the layers and the surfaces of the room.

Zone models are in general computer-based, such as FIRST and CFAST, but can also be set up and solves by hand (i.e., using pen and paper). In a modern desktop computer, the computational time for the solution is in the order of seconds. Their main advantage is that this small computer-time allows for extensive parametric studies. Since they take into account more coupled physical mechanisms, they have potential to give better results than analytical models. The primary inputs to zone models are the geometry of the compartments, including all vents and windows, and the primary fire, in terms of the heat release rate, fire area, and the yields of products. The most important limitations of zone models are that radiation is elementary, that there is no significant horizontal layer growth, that the enclosure has to be simple and nearly of uniform cross-section, and that the plume-entrainment mechanism applies only simple scenarios.

The most complete approach is computational fluid dynamics (CFD) (also called field modelling),[3] which numerically solves the transient conservation equations of mass, momentum, energy, and species for the motion of a gas. It divides the three-dimensional space into small rectangular volumes. Within each volume the gas variables are assumed to be uniform but changing with time. For momentum conservation, fire dynamic simulator (FDS)[11] solves the Navier-Stokes equations using Reynolds-averaged (RANS) or large eddy simulations (LES) to account for turbulence. For the combustion reactions, CFD can use mixture fraction or flamelet models assuming infinitely fast chemistry, or more complex chemical models. Heat transfer to the solid surfaces and convection within the fluid are taken into account. In addition, the radiative transport equation for an absorbing/emitting and scattering medium can also be solved. Examples of field models are JASMINE, SOFIE, and FDS. FDS is a very powerful tool to simulate smoke and heat transport in fire events and is kept and maintained by the US National Institute of Standards and Technology (NIST). The software is free to download and install and is regularly updated by NIST. Users of this software are strongly encouraged to use the newest version, and if they make contributions, work with NIST to capture their additions/revisions to the models in this software rather than try to develop a parallel codes. Those interested in more aspects of fire safety modeling, such as FDS, should refer to Chapter 21 of this book.

The major advantage of CFD is that it gives spatial variation and temporal evolution of the fire variables and visual information about the complex fire development. They require a more significant setup effort and longer computational time, in the order of hours or days on a modern desktop PC. These models consider in more detail the most important fire mechanisms with the immediate advantage that its range of applicability widens, although it also implies an increase in the number of fundamental parameters required, and therefore further calibration and validation is needed. With the ever-increasing computer power available, it can be expected that field models will become more important in fire modelling.

Modelling approaches continue to greatly improve to meet simulation of HRR evolution and fire growth, for both bench-scale and full-scale efforts. There are recent good reviews on this topic and the advanced made, as well as the remaining work that needs to be done to capture the complexity of fire growth in real world enclosures/scenarios.[12–16] Readers can learn more about the current state of modeling in Chapters 20 and 21 of this book, which cover fire growth of polymeric solids (Chapter 20) and full-scale fire modeling (Chapter 21).

Every model, as an approximate representation of an actual phenomenon, has limitations that constrain its use and narrow its range of applicability. Generally speaking, a more complete model contains fewer simplifications that imply more freedom and fewer limitations. However, fewer simplifications imply the need to know a larger number of fundamental parameters, to be extracted in general from experiments, which in fire science often are associated to significant uncertainties. There has to be a consideration also towards whether the model has been validated for the particular circumstances of interest.

3.4 MATERIAL PARAMETERS CONTROLLING THE IGNITION OF SOLID FUELS

3.4.1 Spontaneous Ignition

Spontaneous ignition of a fuel results from the self-heating of the material induced by a chemical reaction that produces more heat than the heat that can be locally lost. Detailed analysis of this process can be found in Bowes[17] or Kanury.[18] The general requirement for spontaneous ignition to occur is that both fuel and oxygen are present in sufficient quantities at the same location. Thus a porous media or a material that can self-oxidize become a requirement. The solution requires the resolution of mass heat and momentum conservation equations within a porous media that include all chemical heat sources and sinks. A simple way of looking at this phenomenon is to assume that the porous media is a solid through which heat is gained and lost only through convection and conduction. The transient energy equation for a three-dimensional Cartesian control volume takes the following form:

$$\rho C\left(\frac{\partial T}{\partial t}+\vec{u}\cdot\nabla T\right)=\Delta H\frac{\dot{\omega}_O'''}{\varphi}-\nabla\cdot(-k\nabla T) \tag{3.3}$$

Where the reaction rate is given by $\dot{\omega}_O''' = AY_O^{n-j}Y_F^j\, e^{-E/RT}$ and in this particular case A, E, and R. For simplicity the reaction has been represented by a single step, but in most cases the reaction term will be described by the summation of multiple chemical reactions. The flow could be described by Darcy's law:

$$\vec{u}=-\frac{K}{\mu}(\nabla P-\rho\vec{g}) \tag{3.4}$$

If equation (3) is rearranged:

$$\underbrace{\rho C\left(\frac{\partial T}{\partial t}\right)}_{Change}=\underbrace{\Delta H\frac{\dot{\omega}_O'''}{\varphi}}_{Generation}-\underbrace{\left[\rho C(\vec{u}\cdot\nabla T)+\nabla\cdot(-k\nabla T)\right]}_{Losses} \tag{3.5}$$

The appropriate boundary conditions need to be included before a solution can be achieved. The solution to this problem remains complex, therefore simplifications are necessary.

The first approach was developed by Semenov, who treated the system as a lumped control volume V, which results in the following equation:

$$\underbrace{V\rho C\left(\frac{\partial T}{\partial t}\right)}_{Change}=\underbrace{V\Delta H\frac{\dot{\omega}_O'''}{\varphi}}_{Generation}-\underbrace{Ah_T(T-T_0)}_{Losses} \tag{3.6}$$

Where h_T is the total heat transfer coefficient and A the area for heat transfer. The analysis leads to two eigenvalues, the ignition temperature (T_C) and the critical ambient temperature that will lead to spontaneous ignition (T_0). These are found by making generation and loss terms and their derivatives equal. This is illustrated in Figure 3.2, where the curve

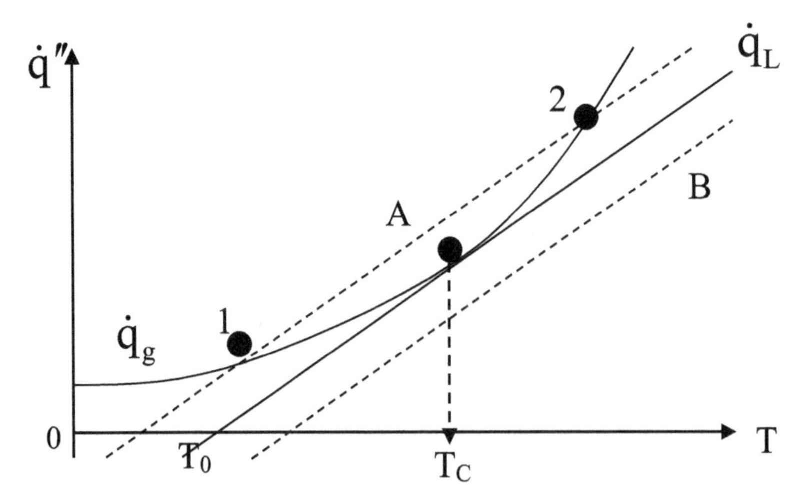

FIGURE 3.2 Schematic of the conditions leading to spontaneous ignition. The horizontal axis is the system temperature, and the vertical axis is the rate of heat generation/loss.

is the generation term ($\dot{q}_g$) and the line is the heat loss term ($\dot{q}_L$). T_0 is the minimum ambient temperature that will lead to heat generation being always greater than heat losses. The dotted line to the right of T_0 shows that if the ambient temperature increases, the condition of generation being greater than the losses is maintained. But the dotted line to the left of T_0 shows that if the ambient temperature decreases, the losses line intercept the heat generation at two points. Once point (1) is reached, stable thermal equilibrium will be attained. Point (2) is an unstable equilibrium.

A more elaborated approach is that of Frank-Kamenetski, who relaxed the assumption of homogeneous solid temperature allowing conduction within the solid. A similar Eigen value analysis will lead again to a critical ignition temperature (T_C) and the ambient temperature required for ignition to occur (T_0). Nevertheless, the ambient temperature, given that conduction of heat from the core to ambient is allowed, becomes a function of the volume of the solid, so for each T_0 a critical volume is obtained, V_C.

At a fundamental level, the process of spontaneous ignition depends strongly on the thermal properties (ρ,λ,C) and the reaction constants, and weakly on the viscosity (μ) and the permeability (K). The final parameter is the Eigen value of the problem corresponding to the ignition temperature, T_C. The critical ambient temperature, T_0, and the critical volume, V_C, are truly not physical parameters controlling spontaneous ignition but the result of the mathematical analysis.

3.4.2 Induced Ignition

The mechanisms leading to gas phase ignition from a solid fuel sample can be described as follows. The solid fuel sample is considered initially at ambient temperature, T_∞. After suddenly imposing an incident heat flux ($\dot{q}_i''$), the temperature of the solid fuel sample rises until the surface reaches the pyrolysis temperature (T_p). The time required for the fuel surface to attain T_p will be referred the pyrolysis time, Δt_p. After attaining T_p, the vapor (pyrolysate) leaves the surface, is diffused and convected outwards, mixes with the ambient oxygen, and creates a flammable mixture near the solid surface. This period will be referred here as the mixing time (Δt_m). The flow and geometrical characteristics determine the mixing time. If the mixture temperature is increased, the combustion reaction may become strong enough to overcome the heat losses to the solid and the ambient, thus becoming self-sustained, at which point flaming ignition will occur. This period corresponds to the induction time (Δt_i).

Extending the analysis proposed by Fernandez-Pello,[3] the ignition delay time (Δt_{ig}) is given by the following:

$$\Delta t_{ig} = \Delta t_p + \Delta t_m + \Delta t_i \tag{3.7}$$

Under ideal conditions, introducing a strong pilot reduces the induction time (Δt_i), making it negligible when compared to Δt_p and Δt_m. Here we will focus on piloted ignition as opposed to auto-ignition, where the absence of a pilot requires the heating of the gas phase mixture. For assessment of the physical variables controlling ignition, piloted ignition is preferred to spontaneous ignition because it reduces the uncertainty associated to the mass and heat transfer processes required to bring the mixture to ignition temperature.

A strong pilot ensures a minimal induction time (Δt_i), but due to the effect that scale has on the flow, the relative importance of the mixing time (Δt_m) sometimes needs to be evaluated,[19] thus, the ignition delay time could be presented as follows:

$$\Delta t_{ig} = \Delta t_p + \Delta t_m \tag{3.8}$$

Mixing is commonly considered to be a fast process compared to heating of the solid fuel sample, therefore, the fuel and oxygen mixture becomes flammable almost immediately after pyrolysis starts. Pyrolysis temperatures and times are thus commonly referred to as ignition temperature (T_{ig}) and ignition delay time (Δt_{ig})[20] and equation (7) finally simplifies to

$$\Delta t_{ig} = \Delta t_p \tag{3.9}$$

and T_{ig} can be defined as T_p. Although such a definition is not physically correct, it can be very useful in some practical applications since it provides a reference parameter that could serve to characterize ignition. In the case of liquids, the ignition temperature, T_{ig}, is referred to as the flash point temperature.

3.4.2.1 Pyrolysis Time (t_p)

The energy balance at the surface of the fuel sample under radiative heating is shown in Figure 3.3 and given by equation (10).

$$\dot{q}_s''(0,t) = a\dot{q}_i'' - \varepsilon\sigma(T^4(0,t) - T_\infty^4) - h_c(T(0,t) - T_\infty) \tag{3.10}$$

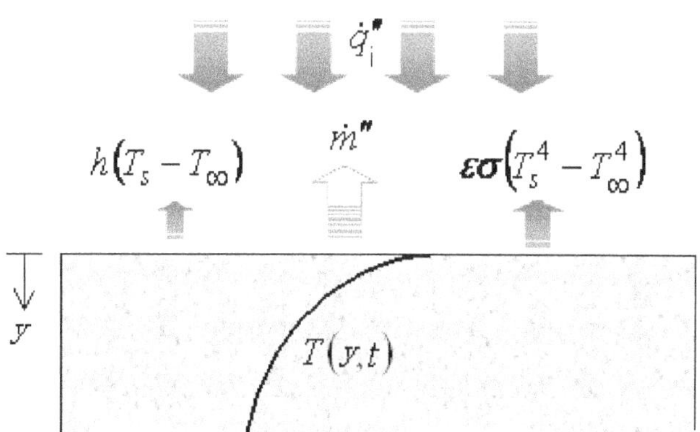

FIGURE 3.3 Energy flows at the surface of the solid fuel sample.

Where ($\dot{q}_s''$) is the net heat flux at the surface of the solid fuel sample, (a) is the absorptivity of the solid fuel sample, (ε) is the emissivity of the solid fuel sample, (σ) is the Stefan-Boltzmann constant, (T(0,t)) is the surface temperature at time (t), (h_c) is the convective heat transfer coefficient, and (T_∞) is the ambient temperature.

The classical analysis corresponding to the ignition process assumes a linear approximation for the surface re-radiation. The radiative term is then defined as follows:

$$\varepsilon\sigma\left(T^4(0,t)-T_\infty^{\ 4}\right)=h_r\left(T(0,t)-T_\infty\right) \tag{3.11}$$

This simplification allows an analytical solution of the one dimensional heat conduction energy equation. Substituting (11) into (10) and assuming that the total heat transfer coefficient (h_T) is equal to the sum of the convective heat transfer coefficient (h_c) and the radiative heat transfer coefficient (h_r), the following expression (Equation 12) defines the net heat flux ($\dot{q}_s''$) at the surface of the solid fuel sample.

$$\dot{q}_s''(0,t)=a\dot{q}''-h_T\left(T(0,t)-T_\infty\right) \tag{3.12}$$

A series of assumptions are regularly made to provide a solution to the heating problem arising from Equation 12. If the heat-flux is applied on only one face, then the heating problem can be treated in one dimension. Most materials relevant to fire can be assumed to be thermally thick and can be simplified to a semi-infinite solid, and the material remains inert until the pyrolysis time (t_P). This allows the elimination of reactive and convective terms and to solve the energy equation in one dimension. The differential formulation governing energy equation is then given by the following:

$$\frac{\partial^2 T}{\partial x^2}=\frac{1}{\alpha}\frac{\partial T}{\partial t} \qquad \text{Boundary Conditions} \quad x=0,\ -k\frac{\partial T}{\partial x}=\dot{q}_s''(0,t) \quad t=0 \quad x\to\infty \quad T=T_\infty \tag{3.13}$$

Where α is the thermal diffusivity ($k/\rho C$). The change of variable $\theta=(T-T_\infty)$ produces the following differential equation:

$$\frac{\partial^2\theta}{\partial x^2}=\frac{1}{\alpha}\frac{\partial\theta}{\partial t} \qquad \text{B.C.} \quad x=0,\ -k\frac{\partial\theta}{\partial x}=\dot{q}_s''(0,t) \quad t=0 \quad x\to\infty \quad \theta=0 \tag{3.14}$$

That can be solved (e.g., by means of a Laplace transformation) to provide a general solution for the temperature distribution in the sample as a function of the location (x) and time (t):

$$T-T_\infty=\frac{a\dot{q}_i''}{h_T}\left[erfc\left(\frac{x}{\sqrt{4\alpha t}}\right)-e^{\left(\frac{h_T}{k}\right)x+\left(\frac{h_T^{\ 2}}{k\rho C}\right)t}\, erfc\left(\frac{h_T}{\sqrt{k\rho C}}\sqrt{t}+\frac{x}{\sqrt{4\alpha t}}\right)\right] \tag{3.15}$$

To obtain the surface temperature (T_s), x is set equal to 0 and $T_s=T(0,t)$. Therefore equation (15) simplifies to the following:

$$T_s=T_\infty+\frac{a\dot{q}_i''}{h_T}\left[1-e^{\left(\frac{h_T^{\ 2}}{k\rho C}\right)t}\, erfc\left(\frac{h_T}{\sqrt{k\rho C}}\sqrt{t}\right)\right] \tag{3.16}$$

Details of this resolution can be found in any heat transfer book.[21] Figure 3.4 presents a typical set of data that shows the evolution of the surface temperature as a function of time. The peaks indicate the moment of ignition.

From equation (16),

$$\overline{T}=\frac{a\dot{q}_i''}{h_T} \tag{3.17}$$

can be defined as a characteristic temperature and

$$t_c=\frac{k\rho C}{h_T^{\ 2}} \tag{3.18}$$

is defined as a characteristic time. Equations (17) and (18) establish the physical parameters governing piloted ignition,

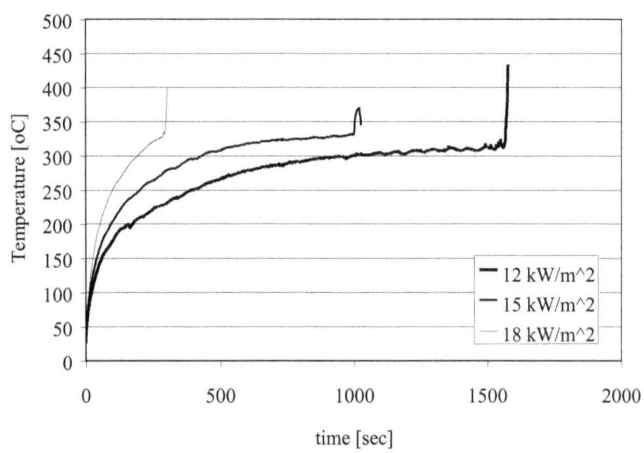

FIGURE 3.4 Characteristic surface temperature histories (thermocouple at center of ignition test specimen) for several heat fluxes. The material used is PMMA.

and these are the absorptivity (a), the thermal inertia ($k\rho C$), while the heat flux and total heat transfer coefficient represent the environmental variables.

Equation (16) is the general solution to the surface temperature at all levels of incident heat flux. To obtain the pyrolysis time (t_p), the surface temperature (T_s) is substituted by T_p, and equation (16) can be rewritten as follows:

$$T_P = T_\infty + \overline{T}\left[1 - e^{\frac{t_p}{t_c}} erfc\left(\left(\frac{t_p}{t_c}\right)^{\frac{1}{2}}\right)\right] \quad (3.19)$$

Where T_P is the final physical parameter to define ignition.

To avoid the complex form of the error function, simplified solutions have been proposed in the literature.[11] In order to solve for the ignition delay time ($t_p \approx t_{ig}$), a first order Taylor series expansion of equation (19) is conducted. The range of validity of this expansion is limited, thus cannot be used over a large range of incident heat fluxes. Therefore, the domain has to be divided at least in two.

The first domain corresponds to high incident heat fluxes where the pyrolysis temperature (T_p) is attained very fast, thus $t_p \ll t_c$. Application of the first order Taylor Series Expansion to equation (13) around $t_p / t_c \to 0$ yields the following formulation for the pyrolysis time (t_p):

$$\frac{1}{\sqrt{t_p}} = \frac{2}{\sqrt{\pi}} \frac{a}{\sqrt{\lambda \rho C}} \frac{\dot{q}_i''}{\left(T_p - T_\infty\right)} \quad (3.20)$$

As can be seen from equation (20), the short time solution for the pyrolysis time (t_p) is independent of the total heat transfer coefficient term ($h_T = (h_r + h_c)$). Thus the pyrolysis time (t_p) is only a function of the energy absorbed ($a\dot{q}_i''$) due to radiation from the radiant panel and the properties (k, ρ, C_p) of the solid fuel sample.

For low incident heat fluxes $t_p \gg t_c$, the Taylor series expansion is made around $t_p / t_c \to \infty$, where the first order approximation yields the following:

$$\frac{1}{\sqrt{t_p}} = \frac{\sqrt{\pi}\sqrt{\lambda \rho C}}{h_T}\left[1 - \frac{h_T\left(T_p - T_\infty\right)}{a\dot{q}_i''}\right] \quad (3.21)$$

Solving equation (20) and (21) for the pyrolysis time (t_p) will yield a theoretical value for the time at which the solid fuel sample begins to pyrolyze and produce fuel vapors. The use of the appropriate simplified solution will allow the evaluation of the pyrolysis time (t_p) over the entire domain of imposed incident heat fluxes.

Figure 3.5 shows $t_{ig}^{-0.5}$. This representation of the data corresponds to the dependencies established in equations (20) and (21). The dotted line corresponds to equation (21) while the full line to equation (20).

Equations (20) and (21) are used to extract the ignition temperature and the product $k\rho C$. A typical set of data is presented in Table 3.1.

3.5 MATERIAL PARAMETERS CONTROLLING FLAME SPREAD

3.5.1 DEFINITION

One of the most clearly defined fire safety issues is that of flame propagation. When a heat source is in contact with a combustible material (liquid or solid), ignition might occur. Once the fuel is ignited the flame propagates or spreads across its surface, establishing a diffusion flame over the fuel. The flame transfers heat to the surface, and the combustible material vaporizes, providing the necessary gaseous fuel to sustain the flame. Once the flame has propagated throughout the fuel, the diffusion flame releases heat that is partially used to vaporize the fuel that, in its turn, feeds the flame.

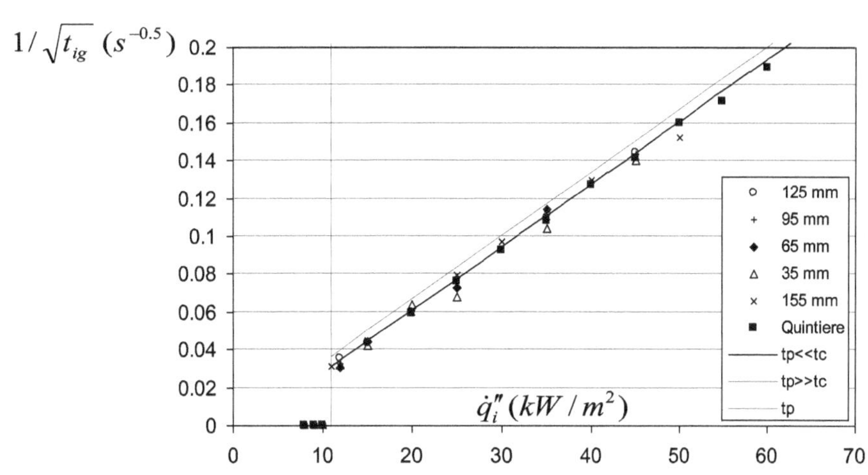

FIGURE 3.5 Ignition delay time ($t_{ig}^{-0.5}$) for different external heat fluxes. The material for the tests is PMMA.

TABLE 3.1
Ignition Data from ASTM E-1321 per Quintiere [13]

Material	T_{ig} [°C]	$k\rho C$ [(kW/m²K)².s]
Wood fiber board	355	0.46
Wood hardboard	365	0.88
Plywood	390	0.54
PMMA	380	1.00
Flexible foam plastic	390	0.32
Rigid foam plastic	435	0.03
Acrylic carpet	300	0.42
Wallpaper on plasterboard	412	0.57
Asphalt shingle	378	0.70
Glass reinforced plastic	390	0.32

The characteristics of the flame propagation process are determined by the fuel, oxygen flow structure, and orientation. If the fuel is vertical or a forced flow is imposed parallel to its surface, a boundary layer will form. If the fuel is horizontal and the oxygen is quiescent, a pool fire will develop. If the fuel is inclined or a flow non-parallel to the surface is imposed, a complex mixed flow problem will arise.

3.5.2 Classification

Different types of flame spread can be grouped into two main categories based on the relative direction of the propagating flame to the gas oxygen flow:

- Flame spread in gas flows moving in the opposite direction of the flame propagation (opposed flame spread)
- Flame spread in gas flows moving in the same direction as that of the flame propagation (concurrent flame spread)

In the first case, as shown in Figure 3.6, the flame front is usually distinctive and well behaved, and the flame propagates rather slowly because the heat transfer from the flame to the unburned fuel is hindered by the opposing gas flow. In the second case, Figure 3.7, the gas flow drives the diffusion flame ahead of the pyrolysis region and largely enhances the heat transfer from the flame to the fuel. Therefore, concurrent flame spread propagates much more rapidly than other cases, and it is known as the most dangerous type of flame spread.

Extensive studies have been conducted both analytically and experimentally in order to study opposed and concurrent flame spread.[22,23] One frequently used approach is to isolate the effects of individual parameters and concentrate on the influence of one parameter at a time for the sake of simplicity. Among most frequently studied is the effect of gas flow velocity [16],[24] external radiation,[11] oxygen concentration [17],[25] buoyancy [18],[26] and fuel sample dimension [19].[27] A relation to determine the flame spread rate can be obtained from a simple energy balance at the fuel surface:

$$\rho S \Delta h \delta = \dot{q}''_f \, l \qquad (3.22)$$

Where ρ is the fuel density, S the flame spread rate, δ the thickness of the pyrolysis layer, l the length of the pyrolysis region exposed to heat flux, $\dot{q}''_f$ the energy per unit area, per unit time transferred across the surface, and Δh is the critical enthalpy increase per unit mass of fuel needed for ignition to occur.

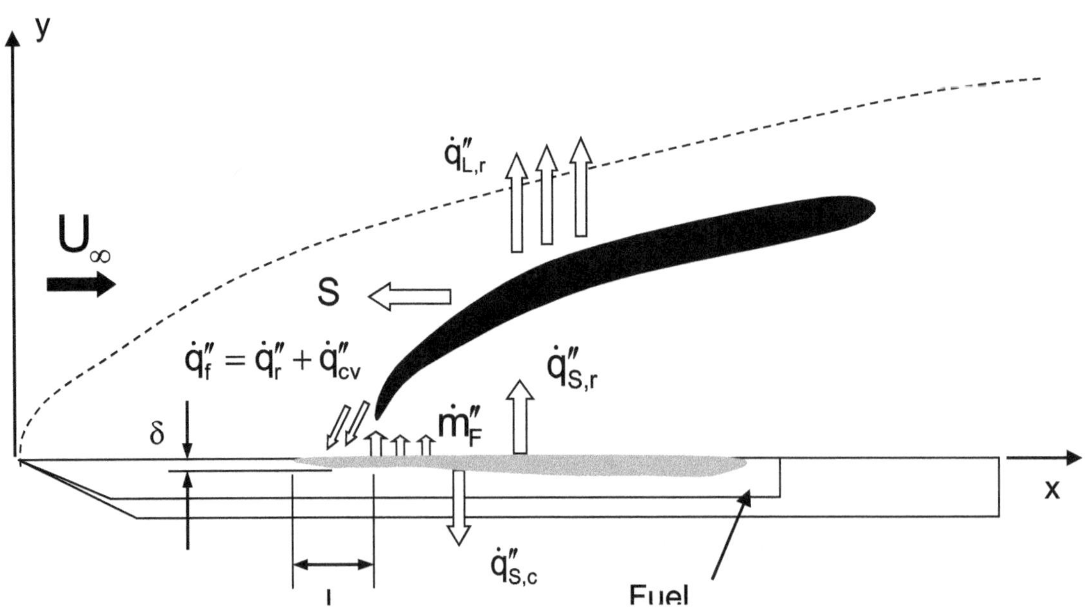

FIGURE 3.6 Schematic of opposed flow flame spread.

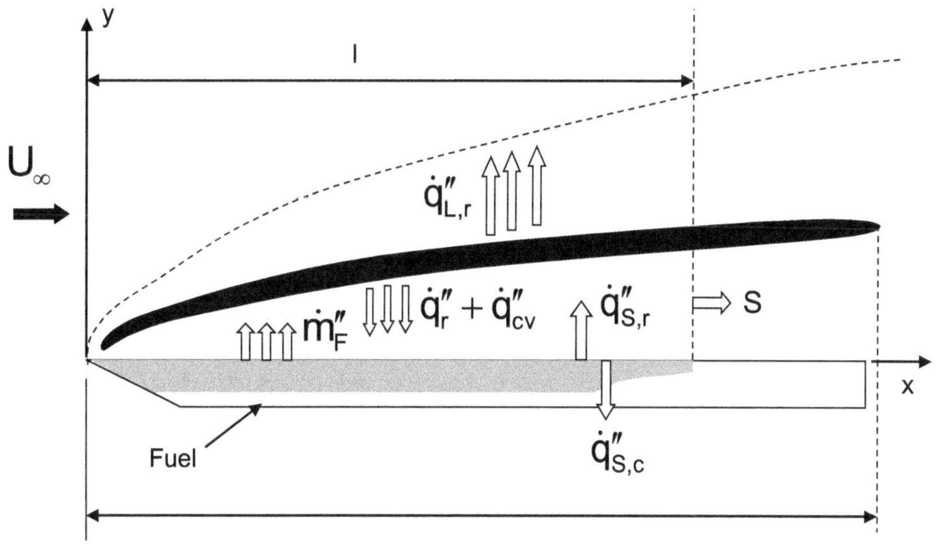

FIGURE 3.7 Schematic of concurrent flow flame spread.

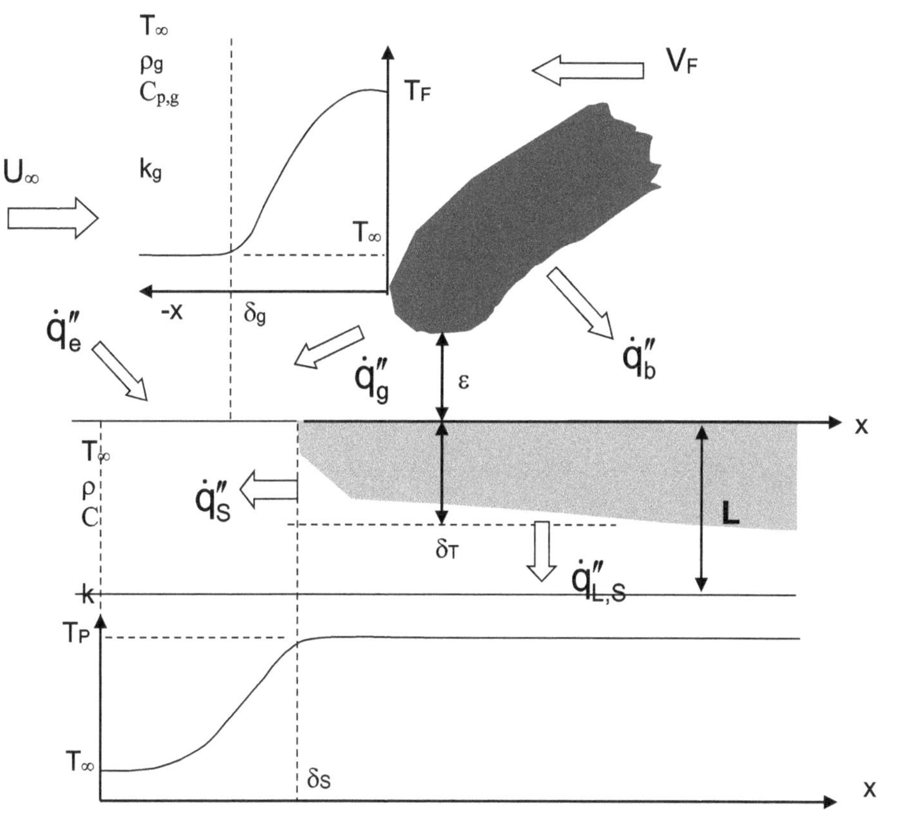

FIGURE 3.8 Schematic of the leading edge of the flame.

3.5.3 Opposed Flame Spread

There are many mechanisms involved in the propagation of a flame through a condensed fuel.[28] The problem is extremely complicated, therefore the simplest case will be used here, that of opposed-flow flame spread. The tip of the flame controls opposed-flow flame spread, and all the transport mechanisms involved are presented in Figure 3.8.

There are three independent sources of heat that will lead to the spread of a flame and that will define the boundary condition:

- Heat transfer from the flame through the gas: $\dot{q}''_g$
- Heat transfer through the solid: $\dot{q}''_S$
- Any external heat contribution: $\dot{q}''_e$

The external heat transfer can be independently controlled so will not be addressed here. Heat transfer through the solid is given by the following expression:

$$\dot{q}_s'' \approx k\frac{(T_P - T_\infty)}{\delta_S} \tag{3.23}$$

Heat transfer through the gas phase has been observed to be mainly due to conduction, therefore radiation is generally neglected. The following expression gives an estimate of the heat flux from the flame to the surface and how that heat is transferred inwards into the material:

$$\dot{q}_g'' \approx k_g\frac{(T_F - T_P)}{\varepsilon} \approx k\frac{(T_P - T_\infty)}{\delta_T} \tag{3.24}$$

Estimation of the relative importance of both these parameters was presented in reference.[14] It can be seen that for thin fuels, the heat transferred through the gas is dominant; for thick materials, it is the opposite.

The above results correspond to PMMA under downward burning conditions and are not general to all materials or configurations. In most cases heat transfer through the gas phase will be dominant for most conditions, mainly because as the fire grows, the contribution from flame radiation increases, enhancing the relative importance of gas phase heat transfer. A number of papers summarizing different results are referred to Fernandez-Pello and Hirano.[14]

As can be seen in Figure 3.9, only ~50% of the total available energy ($\dot{q}_T''$) is used to heat up the solid, the rest is lost from the surface either through convection or reradiation to the environment. As for ignition, both heat loss terms have generally been simplified to a total heat transfer coefficient:

$$\dot{q}_L'' = h_T(T_P - T_\infty) \tag{3.25}$$

As seen before, the material thickness plays a significant role on the rate of flame spread, either through the mode of heat transfer or through the losses that can occur through the solid. Materials are generally defined as thermally thin, where $\delta_T \approx L$, or thermally thick, where $\delta_T << L$. The thermal thickness (δ_T) needs to be carefully defined and generally is given by the following:

$$\delta_T = \sqrt{\alpha t_C} \tag{3.27}$$

Where the characteristic time is defined as follows:

$$t_C = \frac{\delta_S}{S}$$

Assuming that all the energy used for propagation comes through conduction in the gas phase, equation (22) can be rewritten as follows:

$$\rho CS\,(T_P - T_\infty)\,\delta_T = \dot{q}_g''\,\delta_S \tag{3.28}$$

And therefore

$$S = \frac{\dot{q}_g''\,\delta_S}{\rho C(T_P - T_\infty)\,\delta_T} \tag{3.29}$$

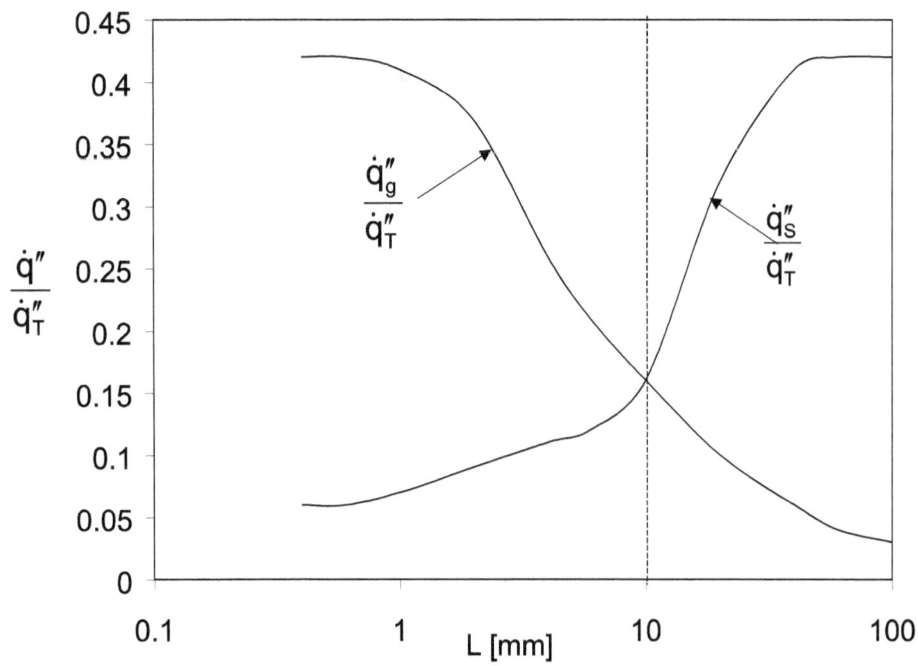

FIGURE 3.9 Relative magnitude of different modes of heat transfer (from [15]).

For thermally thick materials δ_T can be extracted from equation (24):

$$\delta_T = \frac{k(T_P - T_\infty)}{\dot{q}''_g}$$

And for thermally thin materials:

$$\delta_T = L$$

So the following expressions for the propagation velocity can be obtained for a thermally thick solid:

$$S = \frac{\left(\dot{q}''_g\right)^2 \delta_S}{k\rho C(T_P - T_\infty)^2} \tag{3.30}$$

At this point two different pathways can be followed. The first approach takes all the unknowns and brings them together into "global properties" that are evaluated experimentally. This is the approach followed by Quintiere.[11]

Flame Spread Parameter: $\varphi = \left(\dot{q}''_g\right)^2 \delta_S$

Thermal Inertia: $k\rho C$

Which gives the following expression for the flame spread velocity:

$$S = \frac{\varphi}{k\rho\, C(T_P - T_\infty)^2} \tag{3.31}$$

Where the pyrolysis temperature (T_p) and the thermal inertia ($k\rho C$) were described in the ignition section.

An alternate approach is to evaluate the unknown terms of equation (30) ($\delta_S, \dot{q}''_g$). Several expressions can be found in the literature, where the original solutions were presented by deRis[15] and Tarifa et al.[29] The expression presented here was obtained from Fernandez-Pello[30]:

$$S = c\frac{k_g \rho_g C_{P,g}(T_F - T_p)^2 (U_\infty / x)^{1/2}}{k\rho C(T_p - T_\infty)^2} \tag{3.32}$$

Equation (32) provides an expression for the flame spread velocity based only on the parameters of the problem and on the experimental conditions but still relies on the presence of an unknown constant that needs to be determined experimentally.

In a similar manner the flame spread velocity can be obtained for thermally thin materials:

$$S = \frac{\dot{q}''_g\, \delta_S}{\rho_g\, C_{P,g}(T_P - T_\infty)\, L} \tag{3.33}$$

Where similar global parameters can be extracted or the solution can be represented in terms of the properties and experimental parameters with the supplement of an empirical constant. The expression here was provided by Fernandez-Pello.[21]

$$S = c\frac{\left(k_g \rho_g Cp_g\right)^{1/2} (T_F - T_p)^2}{\rho CL(T_p - T_\infty)} \tag{3.34}$$

It is important to note that the physical parameters controlling opposed flame spread are the same as those controlling ignition, with the added parameter associated to the flame heat flux. Depending on the approach to be followed, the flame input can be either defined by the flame temperature, T_F or by q''_g.

3.5.4 Co-Current Flame Spread

The analysis for co-current flame spread is very similar to that of opposed flame spread. It is further complicated because the flame covers the fuel thus the flame length is a further parameter that needs to be analyzed. The flame length can be represented empirically as being proportional to the pyrolysis length or can be calculated using boundary layer theory and the assumption of infinitely fast gas phase chemistry.[31] Despite the added complexity, co-current flame spread is controlled by the same physical parameters as ignition or opposed flame spread.

3.5.5 Propagation of a Smouldering Front

Smouldering is characterized by an exothermic heterogeneous combustion reaction that occurs on the surface of a solid fuel and the interior of porous combustible materials. The heat released during the heterogeneous oxidation of the solid is transferred toward the un-reacted material by conduction, convection, and radiation, supporting the propagation of the smoulder reaction. The oxygen, in turn, is transported to the reaction zone by diffusion and convection. These transport mechanisms not only influence the rate at which the smoulder reaction propagates but also the limiting factors of the smoulder process, that is, ignition and extinction (lower bounds), and transition to flaming (upper bound). The propagation of the smoulder reaction is, therefore, a complexly coupled phenomenon involving processes related to the transport of heat and mass in a porous media, together with surface pyrolysis and combustion reactions.[1,32]

The main difference between smoulder and any of the other combustion processes is that oxidation does not occur in the gas but on the solid phase. Fuels that sustain smoulder consist, in general terms, of an aggregate and permeable medium formed by particulates, grains, fibers, or a porous matrix (Figure 3.10). These aggregate fuel elements facilitate the surface reaction with oxygen by providing a large surface area per unit volume. They also act as thermal insulation that reduce heat loss but, at the same time,

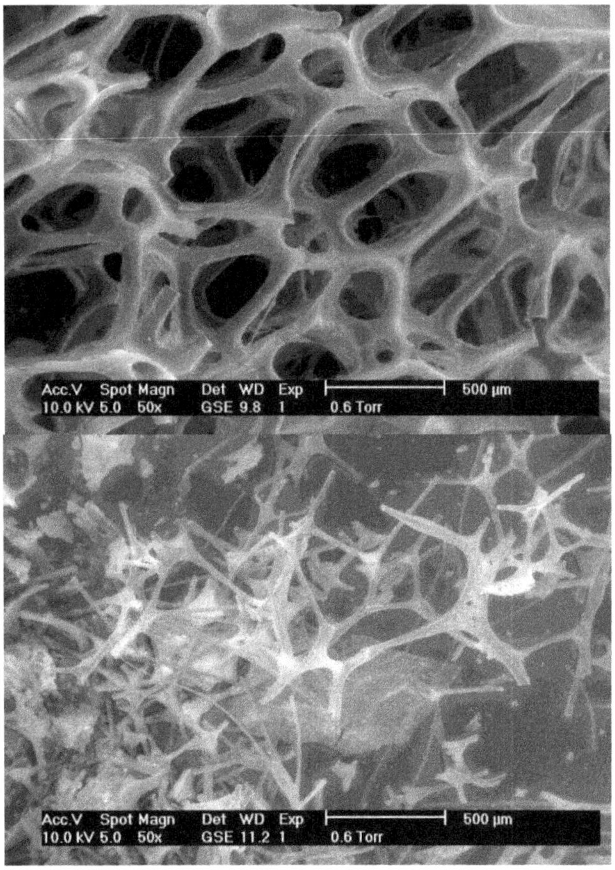

FIGURE 3.10 Scan Electron Microscopy imaging of the porous matrix of polyurethane foam; (top) virgin foam and (bottom) smouldered char.

permit oxygen transport to the reaction sites by convection and diffusion.

As the reaction propagates, the oxygen inside the porous matrix is completely consumed leaving residual char. Figure 3.11 shows a picture of smouldered polyurethane foam. The foam was ignited at the top of the sample, and the reaction was allowed to propagate downwards leaving a black char behind.

A smouldering reaction is not always established when heat is applied to a porous fuel susceptible to smoulder. Actually the conditions for the onset of smouldering are in some cases (i.e., polyurethane foam) very restrictive. The following possible pathways have been established as viable when heating a fuel susceptible to smoulder:

The first step of the decomposition process, as shown in Figure 3.12, corresponds to the different magnitudes of the net heat flux imposed to the material. A typical ignition plot is presented in Figure 3.13 for polyurethane foam. If the net heat flux is weak (i.e., for polyurethane foam < 6 kW/m^2 (Figure 3.13)), Figure 3.12 shows that the fuel is decomposed via pyrolysis. Nonetheless, a decomposed material is formed. This material is generally of liquid form and is commonly referred as tar. If the net heat flux is very strong (i.e., for polyurethane foam > 7 kW/m^2 (Figure 3.13)), a similar process is observed where a liquid tar remains as the product of the decomposition. Both decomposition branches are endothermic, and once the external heat source is withdrawn, extinction follows. The main difference between the two branches seems to be that the amount of heat determines the fraction of tar that will be evaporated, thus the production of airborne aerosols.

Smouldering of the material occurs only when the heat flux imposed on the material is in between the two limits described above. In the presence of an oxidative atmosphere, an exothermic surface reaction (smoulder) will lead to the release of heat and gaseous products and the formation of a residual char. Char is a solid matrix that generally conserves the structure of the original fuel. The char has a high carbon content and is combustible. The char can further react in the presence of oxygen if its temperature is high enough. Reaction temperatures of char have been

FIGURE 3.11 Photograph of a polyurethane foam sample through which a smolder reaction has propagated. Photograph by NASA

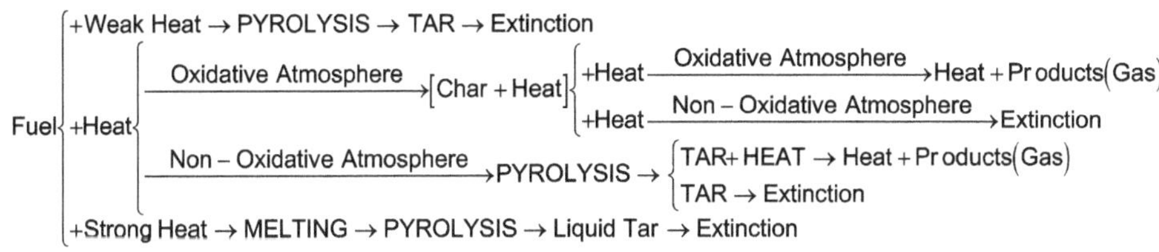

FIGURE 3.12 Possible reaction pathways for a smoulder reaction.

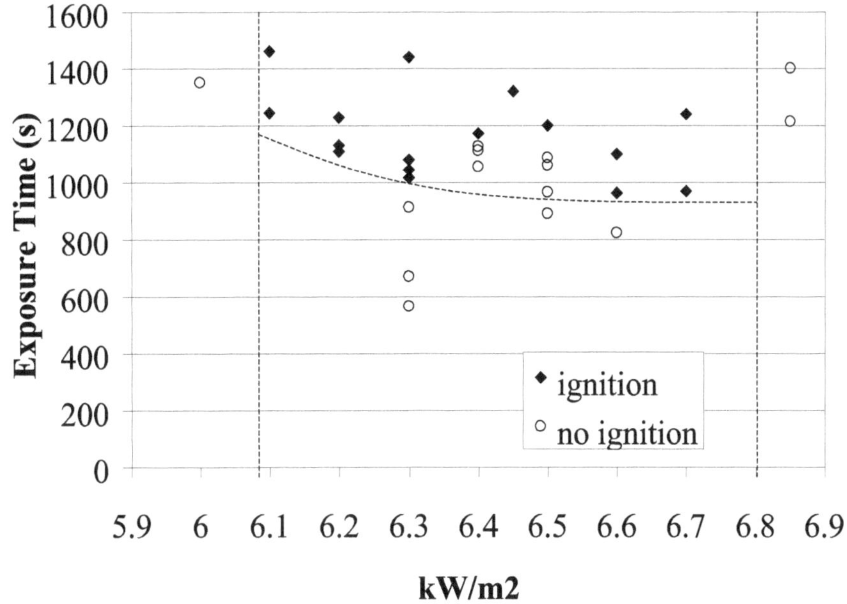

FIGURE 3.13 Ignition characteristics of polyurethane foam as a function of an external heat flux [24].

observed to be higher than the temperatures observed during direct smoulder of the fuel. The final products are in most cases in gaseous form and particulate (smoke), but some fuels lead to a residual noncombustible ash. If all the oxygen is consumed by the smoulder reaction, the char will not react and cool down will follow.

Even under the appropriate heating conditions, if not enough oxygen is available, the decomposition chemistry will privilege the endothermic pyrolysis of the fuel. This will again lead to the formation of tar and the consequent extinction. Attempts to identify the exothermic and endothermic decomposition processes have been made by means of thermogravimetric analysis (TGA) differential scanning calorimetry (DSC).[33] These methods can provide important and insightful information about the smoulder mechanism and can be used as input parameters in complex smouldering propagation models.[34]

Once the smoulder reaction has been initiated, the reaction front propagates across the porous matrix. When studying smoulder propagation, it is very common to use simplified scenarios. A frequent approach is that of treating the smoulder front as a flame spread problem and classify the smoulder reaction in two main groups: opposed and forward propagation. These are defined according to the direction in which the fuel and oxygen enter the reaction zone (Figure 3.14).

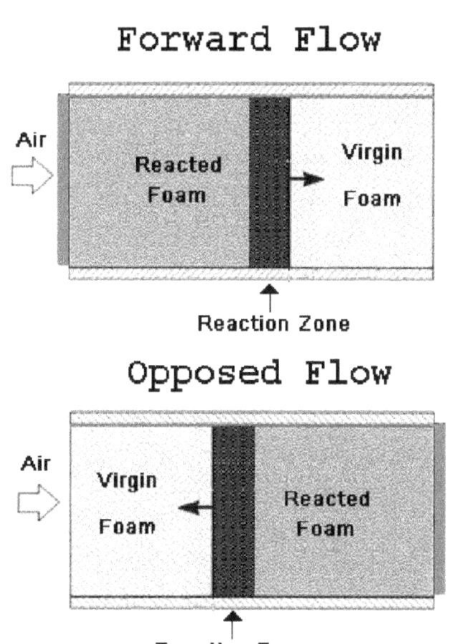

FIGURE 3.14 Schematic of Forward and Opposed flow smoulder.

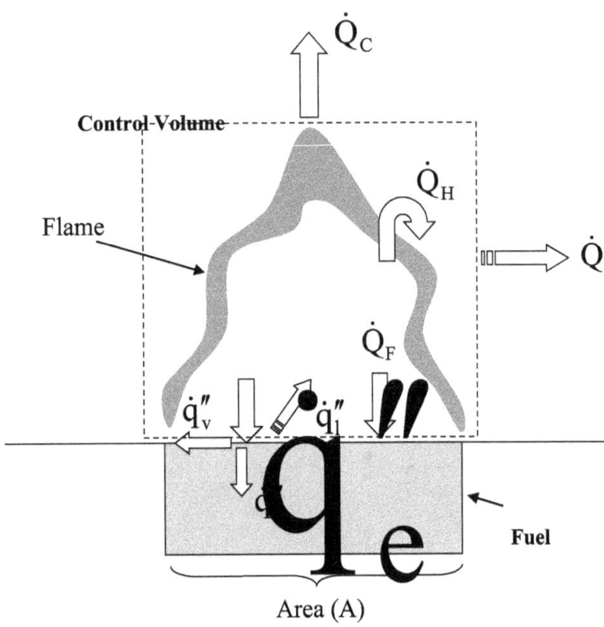

FIGURE 3.15 Repartition of energy within a diffusion flame.

3.5.5.1 Opposed Smouldering

In opposed (or reverse) smoulder, the reaction front propagates in a direction opposite to the oxygen flow. This configuration is also referred to as co-current, or premixed-flame-like smoulder, because with the coordinate system anchored at the reaction zone, fuel and oxygen enter the reaction zone from the same direction, albeit with different velocities. In forward smoulder, the smoulder reaction front moves in the same direction as the oxygen flow. This configuration is also referred to as counter-current, or diffusion-flame-like smoulder, because the fuel and the oxygen enter the reaction zone from opposite directions. Processes involved in the transport of heat and mass include radiation (heat), convection (natural and forced), and diffusion. In the presence of gravity, buoyancy interferes with both forced convection and diffusion and consequently with the transport processes controlling smouldering. In opposed smoulder propagation, the heat released by the heterogeneous oxidation (smoulder) reaction is transferred ahead of the reaction by conduction and radiation, heating the unreacted fuel and the incoming oxygen. The resulting increase of the virgin fuel temperature leads to the onset of the smoulder reaction and consequently gives way to its propagation through the fuel. The combustion process is generally oxygen deficient, and the propagating reaction leaves behind a char that contains a significant amount of unburned fuel. The rate of smoulder propagation is basically dictated by a balance between the rate of heat released by the reaction and the energy required to heat the solid fuel and gaseous oxygen to the smoulder reaction temperature. Increasing the oxygen flow rate increases the rates of fuel oxidation and heat release, and consequently the rate of smoulder propagation, until it reaches a point at which the rate of heat losses to the incoming oxygen overwhelm the heat released at the reaction and extinction occurs.

3.5.5.2 Forward Smouldering

Also known as counter-current smoulder, in this case the smoulder wave travels in the direction of the flow. In forward smoulder propagation, the oxygen reaches the reaction zone after passing through the hot char left behind by the propagating reaction, and the hot post-combustion gases flow through the virgin fuel ahead of the reaction, preheating it. Thus, increasing the oxygen flow rate increases the oxygen supply to the reaction zone, which enhances the reaction, and increases fuel preheating, which also tends to enhance the reaction. However, the post-combustion gases will also tend to dilute the oxygen contained inside the fuel pores and thus lower the oxygen concentration ahead of the reaction. Furthermore, oxidation of the char may deplete the oxygen in the gas flowing ahead of the reaction. As explained before, under these conditions, the fuel may undergo an endothermic, thermal decomposition reaction ahead of the oxidation reaction. These two reactions may propagate at different velocities depending on the oxygen flow and fuel characteristics. Another important process that may occur in forward smoulder is the transition from the heterogeneous surface reaction to a gas phase homogeneous reaction (flaming). This transition process may occur as the oxygen flow rate is increased and may be caused by the acceleration of the smoulder reaction (due to fuel preheating) or by vigorous oxidation of the char (due to the increased oxygen supply).

The transition from smolder to flaming combustion in polyurethane foam, as well as other materials, remains of great interest due to the fire safety aspects of furniture flammability and cigarette ignition, as well as wildland-urban-interface (WUI) fire events where smoldering embers from wildfires can cause flame to spread throughout the WUI built environment. In regards to furniture and polyurethane foam, Chapter 31 of this book covers this topic in detail, and there has been some recent experimental work on flexible polyurethane smolder-to-flaming transition as well.[35–37] Smolder-to-flaming in wildfires has also been recently reviewed and continues to be an area of active research.[38–41]

3.5.2.3 Physical Parameters Controlling Smoldering

Smolder propagation is generally treated as a flame spread problem, thus a similar thermal analysis to the one presented in the previous section is conducted for both opposed and concurrent smoldering. Many expressions for a smoldering propagation velocity can be found in the literature. Here we will use for illustration only the one presented by Torero et al.[42]

$$S = \frac{\rho_g \left[\Delta H_{O_2} Y_{O_2} - C_{pg}(T_S - T_i)\right]}{(1-\varphi)\rho C(T_S - T_i)} U \tag{3.35}$$

Where the physical parameters affecting the smolder propagation velocity are the heat of reaction per unit oxygen mass consumed (ΔH_{O_2}), the void fraction (φ), the smoldering temperature (T_s), the density (ρ), and the specific heat (C). All other parameters of equation (35) are environmental variables such as the ambient density (ρ_g), ambient oxygen concentration (Y_{O_2}), specific heat (C_{pg}), and temperature (T_i). The term U corresponds to the air flow velocity through the porous material and U is quantified using expressions of the form of equation (4).

3.6 MASS BURNING RATE—ENERGY RELEASE RATE

3.6.1 Definition

Once a flame is ignited and spreads over a surface, the area under the flames undergoes decomposition supported by the energy released by the flame. The production of fuel is known as the burning rate. The fuel emerging from the surface will react at the flame releasing energy; this is called the heat release rate. A brief summary of the different components of the mass burning and heat release rates and their association to physical material parameters will be presented here. Details on how to evaluate the burning rate are given by Drysdale[2] and Quintiere[15] and on the energy release rate and its estimation by Babrauskas and Grayson.[4]

3.6.2 Heat Release Rate ($\dot{Q}$)

The energy release rate is the amount of energy produced by the flame per unit time. In its simplest form is given by equation (36):

$$\dot{Q} = A\,\Delta H_{C,F}\,\dot{m}_F'' \tag{3.36}$$

Where A is the area, $\Delta H_{C,F}$ the heat of combustion per unit mass of fuel burnt, and $\dot{m}_F''$ the mass of fuel produced at the surface (mass burning rate). An alternate way to express the energy release rate is by means of the heat of combustion per unit mass of oxygen consumed ($\dot{m}_O$)

$$\dot{Q} = \Delta H_{C,O}\,\dot{m}_O \tag{3.37}$$

Where $\Delta H_{C,O}$ is the heat of combustion per unit mass of oxygen consumed (a constant value of 13.1 MJ/kg is generally used for most materials relevant to fire).[4] Note that the mass of oxygen consumed ($\dot{m}_O$) is not expressed per unit area, this is because the area through which the oxygen reaches the flame is difficult to determine. Equations (1) and (2) are equivalent and which one to use depends mostly on practical issues related to the quantification of the different values.

The total energy released by the flame can be divided into the different parts that are transferred by different modes of heat transfer or towards different directions:

$$\dot{Q} = \dot{Q}_C + \dot{Q}_R + \dot{Q}_H + \dot{Q}_F \tag{3.38}$$

$\dot{Q}_C$ Is the energy that is convected towards the plume above of the flame. Natural convection will drive the hot products (and consequently the energy) above the flame.
$\dot{Q}_R$ Is the energy that is radiated away from the flame towards external targets. The fraction of the energy radiated is generally given by a radiative fraction (χ) which is generally taken as approximately 30%, $\chi \approx 0.3$ [3] and thus $\dot{Q}_R \approx \chi\dot{Q}$.
$\dot{Q}_H$ Is the energy that remains within the control volume of the flame. Once flames have attained steady state conditions $\dot{Q}_H$ tends towards zero. The gas phase transient tends to be very short and thus neglected for most fire analyses.
$\dot{Q}_F$ Is the energy transmitted by the flame towards the surface of the fuel. This energy includes convection, conduction and radiation. Convection and conduction will be positive inputs towards the fuel but convection will vary depending on the temperature of the gases over the material at a specific location within the surface.

3.6.3 Energy Balance at the Material Surface

Conducting an energy balance per unit area allows a closer analysis of the fuel surface. This implies the assumption that the energy is distributed homogeneously across the surface. This assumption is justifiable for most of the surface area of the fuel. The energy input to the surface is given by:

$$\dot{Q} = A\dot{q}'' \tag{3.39}$$

The energy necessary to vaporize a unit mass of fuel varies depending on the nature of the gasification process. Most liquid fuels (light hydrocarbons) undergo gasification without any chemical decomposition. The energy necessary for the phase change of a unit mass of fuel is called the latent heat of vaporization (ΔH_V). For most solids and liquids formed by heavier hydrocarbons, the decomposition process implies a chemical breakdown of the molecules. This process is generally endothermic and the energy necessary to gasify a unit mass of fuel is called the heat of pyrolysis (ΔH_P). For the purposes of this analysis the distinctions in the decomposition pathway are not relevant, what is important is the endothermicity of the process, therefore a generic heat of gasification will be used (ΔH_G). The energy necessary to vaporize a unit mass of fuel per unit are is thus given by:

$$\dot{q}_V'' = \Delta H_G \dot{m}_{F,G}'' \tag{3.40}$$

Where $\dot{m}''_{F,G}$ is the mass of fuel generated per unit surface area.

As indicated in Figure 3.3, the heat feedback ($\dot{q}''_e$) is used for vaporization ($\dot{q}''_v$) and to compensate for in-depth conduction ($\dot{q}''_c$) and heat losses ($\dot{q}''_l$). The heat lost from the fuel surface includes convection and re-radiation and can be described by any of the formulations below:

$$\begin{aligned}\dot{q}''_L &= \dot{q}''_{CV} + \dot{q}''_{S,R} = h_C(T_P - T_\infty) + \varepsilon\sigma(T_P^4 - T_\infty^4) \\ &= h_C(T_P - T_\infty) + h_R(T_P - T_\infty)\end{aligned} \quad (3.41)$$

As it was done for ignition and flame spread, it is a common practice to simplify this term to a simple linearized total heat transfer coefficient

$$\dot{q}''_L = h_T(T_P - T_\infty) \quad (3.42)$$

Finally, in-depth conduction is given by the temperature gradient at the surface

$$\dot{q}''_C = -k\left.\frac{\partial T}{\partial x}\right|_{x=0} \quad (3.43)$$

This term varies significantly with the nature of the fuel. Liquid fuels are characterized by the presence of re-circulation currents induced by buoyancy. These currents homogenize the temperature distributions reducing in-depth conduction. Nevertheless, convective motion transfers heat to the interior of the pool leading to an additional term of in-depth convection. Convective heat transfer in polymer melts has a similar behaviour and could potentially have an important effect on burning rates, nevertheless, the impact of polymer melt rheology on the flammability of materials is still a matter of great controversy.

Assuming that in-depth heat transfer is only by conduction, the energy balance at the surface is then given by:

$$a\dot{q}''_e = \dot{q}''_V + \dot{q}''_C + \dot{q}''_L \quad (3.44)$$

For convenience, the right hand side of the equation is generally presented in a different form:

$$\begin{aligned}a\dot{q}''_e &= \Delta H_G \dot{m}''_{F,G} + \left(\frac{\dot{q}''_C + \dot{q}''_L}{\dot{m}''_{F,G}}\right)\dot{m}''_{F,G} \\ &= (\Delta H_G + Q_L)\dot{m}''_{F,G}\end{aligned} \quad (3.45)$$

Where Q_L represent the total heat not used to vaporize the fuel normalized per unit mass of fuel generated. In other words, of the total heat available a fraction goes to gasification (ΔH_G) and the rest goes away to the solid or the gas (Q_L). Assuming the absortivity to be unity (a=1) the mass of fuel generated is give by

$$\dot{m}''_{F,G} = \frac{\dot{q}''_e}{(\Delta H_G + Q_L)} \quad (3.46)$$

3.6.4 The Mass Transfer Number (B)

From equation (39) the heat release rate per unit area is given by the following expression:

$$\dot{q}''_g = \frac{\dot{Q}}{A} = \Delta H_{C,F}\ \dot{m}''_{F,B} \quad (3.47)$$

Where $\dot{m}''_{F,B}$ is the mass of fuel burnt. As shown before, the energy generated can be divided into the energy fed back to the fuel ($\dot{q}''_e$), the energy necessary to heat the gases that then will be lost due to convection,

$$\dot{q}''_{L,C} = \frac{\dot{Q}_C}{A} = (\dot{m}''_{F,B} + \dot{m}''_{O,B})C_{pg}(T_F - T_\infty) \quad (3.48)$$

and the energy lost due to radiation, which can be given by the radiative fraction (χ):

$$\dot{q}''_{L,r} = \frac{\dot{Q}_R}{A} = \chi\Delta H_{C,F}\ \dot{m}''_{F,B} \quad (3.49)$$

Therefore, an energy balance in the gas phase is given by:

$$\begin{aligned}\dot{q}''_g &= \Delta H_{C,F}\ \dot{m}''_{F,B} \\ &= \chi\Delta H_{C,F}\ \dot{m}''_{F,B} \\ &\quad + (\dot{m}''_{F,B} + \dot{m}''_{O,B})C_{pg}(T_F - T_\infty) + \dot{q}''_e\end{aligned} \quad (3.50)$$

Since diffusion flames are expected to burn at stoichiometric conditions, then:

$$\frac{\dot{m}''_{O,B}}{\dot{m}''_{F,B}} = \Theta$$

Where Θ is the stoichiometric coefficient. Substituting the above expression in equation (50) and rearranging an expression for the energy feedback per unit area can be obtained:

$$\dot{q}''_e = (1-\chi)\Delta H_{C,F}\ \dot{m}''_{F,B} + \dot{m}''_{F,B}(1+\varphi)C_{pg}(T_F - T_\infty) \quad (3.51)$$

And substituting into equation (50) leads to a phenomenological definition of the mass transfer number:

$$B = \frac{\dot{m}''_{F,G}}{\dot{m}''_{F,B}} = \frac{(1-\chi)\Delta H_{C,F} + (1+\varphi)C_{pg}(T_F - T_\infty)}{\Delta H_G + Q_L} \quad (3.52)$$

As can be noted, the mass transfer number represents the mass of fuel generated per unit mass of fuel burnt and is the additional physical parameter that controls the burning rate.

Although there are many definitions of the mass transfer number, mostly generated for specific conditions such as droplet burning or boundary layer burning. All retain the same physical concept, which is the capability of the flame to self-sustain by generating more fuel. If B > 1 the flame will produce more fuel than the necessary to sustain burning.

TABLE 3.2
Examples of the "B Number" for Different Liquids as Extracted from Drysdale [2]

Fuel	B Number
n-Pentane	8.1
n-Hexane	6.7
n-Heptane	5.8
n-Octane	5.2
n-Decane	4.3
Benzene	6.1
Toluene	6.1
Xylene	5.8
Methanol	2.7
Ethanol	3.3
Acetone	5.1
Kerosene	3.9
Diesel Oil	3.9

As observed from equation (52) the "B" number has several unknown or difficult to determine parameters, the first of them are the losses (Q_L). Since the losses depend on the particular conditions and could evolve in time [7], there is a tendency to neglect them when tabulating values for the "B" number. This is appropriate as a reference value but loses quantitative meaning when it is used to calculate the mass burning rate. A set of idealized values commonly reported in the literature is presented in Table 3.2. In most cases the "B" number will be presented for liquid fuels. It is not so common to have these values derived for solid materials. This is because the "B" number is considered to be representative of the steady mass burning rate where, in the case of liquids Q_L will tend to zero. In the case of solids, there is recognition that the role of Q_L will be more important and remain relevant throughout the entire sample burning.

3.7 MATERIAL PARAMETERS CONTROLLING EXTINCTION

Suppression is probably one of the most complex processes associated to fire. Extinction of fires implies the understanding of heat and mass transfer in the solid and gas phase as well as the concept chemical inhibition of the combustion reactions. Only a brief summary of the different material parameters controlling extinction will be presented here.

Extinction can be ultimately defined as the reduction of the combustion reaction rates below a critical threshold that do not permit a self-sustained reaction. This is ultimately achieved in the gas phase and is the result of a negative energy balance at the flame that will lead to the local reduction of the flame temperature. A detailed analysis of the energy equation leads to the definition of two dimensional groups:

$$\text{First Damköhler number: } Da_I = \frac{\tau_{Residence}}{\tau_{Conduction}} \tag{3.53}$$

$$\text{Second Damköhler number: } Da_{II} = \frac{\tau_{Residence}}{\tau_{Chemical}} \tag{3.54}$$

The first Damköhler number quantifies the ratio between the heat transferred from the flame (conduction time) and the energy required to heat the reactants to the ignition temperature (residence time). Extinction will occur when heat cannot be transferred fast enough. Equation (53) is written in terms of the ratio between conduction and convection (residence time) but, in a more general form, could include all forms of heat transfer like gas phase radiation. The second Damköhler number indicates if the reaction has sufficient time to proceed. In general, extinction is attained when either of the Damköhler groups is reduced below unity.

Other criteria can be used to establish the extinction condition and that are partially equivalent to the critical Damköhler number. Such criteria are a critical mass transfer numbers (B_{cr}),[21,43] critical mass flux of fuel,[2,6,44] or critical temperatures (T_{cr}).[2,5,45–47] The critical mass transfer number has a direct influence over the flame temperature, thus represents the link between the condensed phase (i.e., production of fuel) and the chemical time. The critical mass flux operates under the same principle but assumes a consistent heat input. Combustion reactions generally have high activation energy, therefore the reaction can be assumed to abruptly cease when the temperature reaches a critical value (T_{cr}).

A different way to look at extinction is by reducing the oxygen concentration and thus increasing the characteristic chemical time. This will result in a decrease of the second Damköhler number. This mechanism of extinction is analyzed in a standardized manner by the Limiting Oxygen Index (LOI),[27] although it should be noted that LOI measurements do not provide any meaningful prediction of real-world fire behavior.[48]

3.8 CONCLUSIONS

By understanding fire growth and the underlying physics of fire, it allows us to develop approaches to provide fire protection to polymeric materials involved in fire. These fundamentals of fire growth and fire physics are well understood, but there continues to be ongoing research to improve models that predict and understand fire growth, as well as ongoing research on fire growth involved in smoldering combustion. Therefore, readers are encouraged to continue to make themselves aware of advances in these areas, while being able to use the foundation of what is known today in regards to fire growth for combustible materials.

REFERENCES

1. Hurley, M.J., Editor-in-Chief. "SFPE Handbook of Fire Protection Engineering," 5th Edition, Springer, 2016. doi:10.1007/978-1-4939-2565-0
2. Williams, F.A. "Combustion Theory," Second Edition, The Benjamin/Cummings Publishing Company, Inc., 1985. doi:10.1201/9780429494055
3. Drysdale, D. "An Introduction to Fire Dynamics," John Wiley and Sons, 3rd ed., 2011. doi:10.1002/9781119975465
4. Cox, G., Editor. "Combustion Fundamentals of Fire," Academic Press, 1995.
5. Babrauskas, V., Grayson, S.J., Editors. "Heat Release in Fires," Elsevier Applied Science, 1992.
6. Quintiere, J. G.; Rangwala, A. S. A Theory for Flame Extinction Based on Flame Temperature. *Fire Mater.* 2004, 28 (5), 387–402.
7. Beyler, C. A Unified Model of Fire Suppression. *J. Fire Prot. Eng.* **1992**, *4* (1), 5–16.
8. Torero, J. L.; Vietoris, T.; Legros, G.; Joulain, P. Estimation of a Total Mass Transfer Number from Stand-off Distance of a Spreading Flame. *Combust. Sci. Technol.* **2002**, *174* (11–12), 187–203.
9. ASTM D56-22. "Standard Test Method for Flash Point by Tag Closed Cup Tester," ASTM International, West Conshohocken, PA, 2022.
10. ASTM D93-20. "Standard Test Methods for Flash Point by Pensky-Martens Closed Cup Tester," ASTM International, West Conshohocken, PA, 2020.
11. https://pages.nist.gov/fds-smv/ (accessed 09/11/23).
12. Nyazika, T.; Jimenez, M.; Samyn, F.; Bourbigot, S. Pyrolysis Modeling, Sensitivity Analysis, and Optimization Techniques for Combustible Materials: A Review. *J. Fire Sci.* **2019**, *37*, 377–433.
13. Stoliarov, S. I.; Ding, Y. Pyrolysis Model Parameterization and Fire Growth Prediction: The State of the Art. *Fire Saf. J.* **2023**, *140*, 103905.
14. Amokrane, A.; Alloschery, G.; Gautier, B. Experimental Study and Modelling of Real-Scale Vertical Cable Tray Fires. *J. Fire Sci.* **2023**, *41*, 169–185.
15. Marti, J.; Schartel, B.; Oñate, E. Simulation of the Burning and Dripping Cables in Fire Using the Particle Finite Element Method. *J. Fire Sci.* **2022**, *40*, 3.
16. Langot, J.; Pelzmann, T.; Chavez-Gomez, P.; Boanta, C. S.; Karsten, J.; Fiedler, B.; Levesque, M.; Robert, E. Multi- Physics Modeling of the Ignition of Polymer Matrix Composites Exposed to Fire. *Fire Saf. J.* **2021**, *122*, 103312.
17. Bowes, P.C. "Self-Heating: Evaluating and Controlling the Hazards," Elsevier, 1984.
18. Kanury, A.M. "Introduction to Combustion Phenomena," Combustion Science and Technology, Volume 2, Gordon and Breach Publishers, 1984.
19. Long, R. T.; Torero, J. L.; Quintiere, J. G.; Fernandez-Pello, A. C. Scale and Transport Considerations on Piloted Ignition of PMMA. *Fire Saf. Sci.* **1999**, *6*, 567–578.
20. Quintere, J. G. A Simplified Theory for Generalizing Results from a Radiant Panel Flame Spread Apparatus. J. *Fire Mater.* **1974**, *5*, 85.
21. Carslaw, H.S., Jaeger, J.C. "Conduction of Heat in Solids," Oxford University Press, 2nd ed., pp. 70–76, 1963.
22. Williams, F. A. Mechanisms of Fire Spread. *Proc. Combust. Inst.* **1976**, *16*, 1281.
23. Fernandez-Pello, A. C.; Hirano, T. Controlling Mechanisms of Flame Spread. *Combust. Sci. Technol.* **1983**, *32* (1).
24. De Ris, J. N. Spread of a Laminar Diffusion Flame. *Proc. Combust. Inst.* **1969**, *12*, 241.
25. Fernandez-Pello, A. C.; Ray, S. R.; Glassman, I. Flame Spread in Opposed Forced Flow: The Effect of Ambient Oxygen Concentration. *Proc. Combust. Inst.* **1980**, *8*, 579.
26. Altenkirch, R. A.; Eichorn, R.; Rizvi, A. R. Buoyancy Effects on Flame Spreading Down Thermally Thin Fuels. *Combust. Flame* **1980**, *37*, 71.
27. Perrins, L. E.; Pettet, K. Measurements of Flame Spread Velocities. J. *Fire Mater.* **1974**, *5*, 85.
28. Quintiere, J.G. "Principles of Fire Behavior," 2nd ed., CRC Press, 2016. doi:10.1201/9781315369655
29. Tarifa, C. S.; Perez del Notario, P.; Munoz Torralbo, A. Flame Spread. *Proc. Combust. Inst.* **1969**, *12*, 229–240.
30. Fernandez-Pello, A.C. "Combustion Fundamentals on Fire: The Solid Phase," Academic Press, pp. 31–100, 1995.
31. Pagni, P. J.; Shih, T. M. Excess Pyrolyzate. *Proc. Combust. Inst.* **1978**, *16*, 1329.
32. Ohlemiller, T. J. Modeling of Smoldering Combustion Propagation. *Prog. Energy Combust. Sci.* **1986**, *11*, 277–310.
33. Bilbao, R.; Mastral, J. F.; Ceamanos, J.; Aldea, M. E. Kinetics of the Thermal Decomposition of Polyurethane Foams in Nitrogen and Air Atmospheres. *J. Anal. Appl. Pyrolysis* **1996**, *39*, 69–82.
34. Rein, G.; Lautenberger, C.; Fernandez-Pello, A. C.; Torero, J. L.; Urban, D. L. Smoldering Combustion of Polyurethane Foam: Using Genetic Algorithms to Derive Its Kinetics. *Combust. Flame* **2006**, *146* (1–2), 95–108.
35. Morgan, A. B. Revisiting Flexible Polyurethane Foam Flammability in Furniture and Bedding in the United States. *Fire Mater.* **2021**, *45*, 68–80.
36. Morgan, A. B.; Knapp, G.; Stoliarov, S. I.; Levchik, S. V. Studying Smoldering to Flaming Transition in Polyurethane Furniture Subassemblies: Effects of Fabrics, Flame Retardants, and Material Type. *Fire Mater.* **2021**, *45*, 56–67.
37. Stoliarov, S. I.; Zeller, O.; Morgan, A. B.; Levchik, S. An Experimental Setup for Observation of Smoldering-to-Flaming Transition on Flexible Foam/Fabric Assemblies. *Fire Mater.* **2018**, *42*, 128–133. doi:10.1002/ fam.2464.
38. Santoso, M. A.; Christensen, E. G.; Yang, J.; Rein, G. Review of the Transition from Smouldering to Flaming Combustion in Wildfires. *Front. Mech. Eng.* **2019**. doi:10.3389/fmech.2019.00049.
39. Rinta-Paavola, A.; Sukhomlinov, D.; Hostikka, S. Modelling Charring and Burning of Spruce and Pine Woods During Pyrolysis, Smoldering and Flaming. *Fire Technol.* **2023**, *59*, 2751–2786.
40. Richter, F.; Jervis, F. X.; Huang, X.; Rein, G. Effect of Oxygen on the Burning Rate of Wood. *Combust. Flame* **2021**, *234*, 111591.
41. Kwon, B.; Liao, Y.-T. T. Effects of Spacing on Flaming and Smoldering Firebrands in Wildland–Urban Interface Fires. *J. Fire Sci.* **2022**, *40*, 155–174.
42. Torero, J. L.; Kitano, M.; Fernandez-Pello, A. C. Opposed Flow Smoldering of Polyurethane Foam. *Combust. Sci. Technol.* **1993**, *91* (1–3), 95–117.
43. ASTM-D2863-19. "Standard Test Method for Measuring the Minimum Oxygen Concentration to Support Candle-Like Combustion of Plastics (Oxygen Index)," ASTM International, West Conshohocken, PA, 2019.

44. Rasbash, D. J.; Drysdale, D. D.; Deepak, D. Critical Heat and Mass Transfer at Pilot Ignition and Extinction of a Material. *Fire Saf. J.* **1986**, *10*, 1–10.
45. Williams, F. A. A Review of Flame Extinction. *Fire Saf. J.* **1981**, *3*, 163–175.
46. Quintiere, J.G. "Fundamentals of Fire Phenomena," John Wiley and Sons, 2006.
47. Thomson, H. E.; Drysdale, D. D.; Beyler, C. L. An Experimental Evaluation of Critical Surface Temperature as a Criterion for Piloted Ignition of Solid Fuels. *Fire Saf. J.* **1988**, *13*, 185–196.
48. Weil, E. D.; Hirschler, M. M.; Patel, N. G.; Said, M. M.; Shakir, S. Oxygen Index: Correlations to Other Fire Tests. *Fire Mater.* **1992**, *16*, 159–167.

4 Halogen-Containing Flame Retardants

Sergei Levchik[1], *Sergio Bocchini*[2], *Giovanni Camino*[2]
[1] 3rd Edition Author:
ICL-IP America, 769 Old Saw Mill River Rd., Tarrytown, NY 10591, USA
[2] 2nd Edition Authors:
DISMIC Politecnico di Torino sede di Alessandria, INSTM Research Unit, V. Teresa Michel 5 15100 Alessandria, Italy

4.1 INTRODUCTION

Organic polymer materials may originate or propagate fire because they decompose under the action of heat evolving combustible products. The complex overall burning process can be schematically represented, as in Figure 4.1. Combustion begins when the heat from the ignition source leads to volatile combustion products whose concentration is within the flammability limits and at a temperature above the ignition temperature. The combustion, which is a radical chain thermal oxidation process producing heat and light, proceeds then as long as the heat supplied to the polymer is sufficient to sustain its thermal degradation at a rate exceeding that required to feed the flame. Otherwise, the flame extinguishes. When the heat supply from the ignition source is discontinued or is negligible, a self-sustaining process occurs, if the aforementioned heat requirements are satisfied by thermal oxidation taking place either in the gas phase (flame), in the condensed phase, or in both phases. Although the scheme in Figure 4.1 is of general applicability, some steps, such as charring or thermal oxidation in the condensed phase, might not take place depending on the type of polymer and the burning conditions. Please see Chapter 2 of this book for more fundamentals about when charring or thermal oxidation may take place in a burning polymer.

The purpose of fire-retardant systems is to reduce the heat supplied to the polymer below the critical level for flame stability. This can be achieved by modifying (generally decreasing) the rate of chemical or physical processes taking place in one or more of the steps of the burning process.

This chapter focuses on commercial flame-retardant systems presently available which are based upon halogen-containing components. Halogenated flame retardants tend to be the most effective by weight when compared to other flame retardant additives. Further, halogenated flame retardants are generally applicable across many classes of flammable materials because of their vapor phase activity. With aromatic compounds, their mechanism of action is related to the scission of the carbon-halogen bond, however with the aliphatic organohalogens it can be different because they may decompose through mechanisms other than free radical scission. The order of stability for this bond in halogen compounds, derived from Table 4.1, is F > CI > Br > I. The thermal stability of iodine compounds is below that required to withstand polymer processing, whereas fluorine

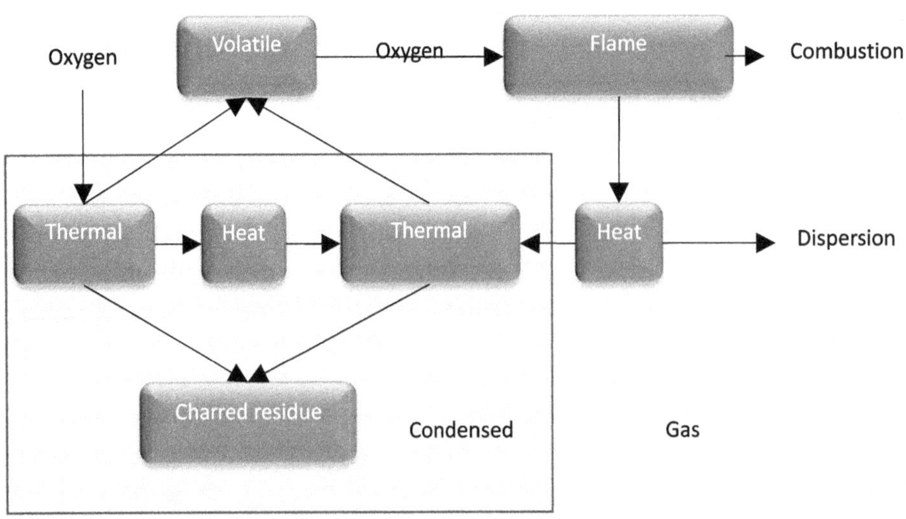

FIGURE 4.1 Self-sustained polymer combustion cycle.

DOI: 10.1201/9781003380689-4

TABLE 4.1
Bond Energies for Carbon-Halogen Bonds[4,5]

Bond	Bond Energy, kJ/mol	Temperature at which Degradation Begins, °C
C_{aliph}-F	443–450	> 500
C_{arom}-Cl	419	> 500
C_{aliph}-Cl	339–352	370–380
$C_{benzylic}$-Br	214	150
C_{aliph}-Br	285–293	290
C_{arom}-Br	335	360
C_{aliph}-I	222–235	180
C_{aliph}-C_{aliph}	330–370	400
C_{aliph}-H	390–436	> 500
C_{arom}-H	469	> 500

compounds are too thermally stable (the C-F bond will not readily break) to be generally useful. However, interestingly recent patent application shows peculiar use of iodine-based flame retardants in the coatings of 3D printed polyamide 12.[1] Some fluorine compounds have however been used as a component in a synergistic mixture of flame retardants.[2,3]

Typically, a brominated or chlorinated organic compound is added to the polymer or, in suitable cases, halogenated structures are introduced into the polymer chain by copolymerization to prepare fire-retardant polymer materials. Metal compounds, such as antimony trioxide, which do not, by themselves, impart significant fire-retardant properties to polymers, can strongly enhance the fire-retardant effectiveness of halogenated compounds (synergistic effect). On heating, these fire-retardant mixtures of antimony oxide and halogenated compounds evolve volatile metal halides, which are well-known flame inhibitors of much greater effectiveness as compared to hydrogen halides evolved in the absence of the metal compound. The detailed mechanism of the reactions which produce the volatile flame inhibitors are, however, not fully understood although these systems have been used for several decades.

These systems are of great importance despite of the current cautious attitude towards the use of halogen-based systems fueled by numerous academic publications about spread of halogenated flame retardants in the environment.[6–8] On the other hand, there is a large variety of naturally occurring organobromine molecules which can be found in the environment,[9] and some of them were even suggested as potential bio-sustainable flame retardants.[10] Please see Chapter 23 of this book for insight into the current regulatory situation around flame retardants of all chemistries.

Recent mechanistic results show that the active species formed by halogen compounds in the condensed phase may play a fundamental role in the fire-retardant mechanism which could be reproduced using halogen-free moieties. Some basic concepts of the mechanism of action of halogen-containing fire-retardant systems are reported below, including reactions of halogen products in the vapor and in the condensed phase, thermal decomposition of halogen-containing compounds, and the interaction of polymers with halogen-containing fire retardants and with metal compounds.

4.2 MECHANISM OF GAS PHASE ACTION OF HALOGEN COMPOUNDS

The purpose of this section is to briefly review the molecular level mechanistic aspects of the gas phase fire retardancy action of halogen-based compounds. The efficiency of fire extinguishing agents is often evaluated based on the critical concentration of particular species to achieve flame suppression. The action is attributed to the ability of the added substance to trap radicals responsible for the branching and propagating of thermal oxidation in the flame. Based on the literature data comparing halogenated compounds as the fire-extinguishing agents for n-heptane-air flame, expressed as minimum volume percentage (vol. %) to snuff out the flame,[11] the atomic ranking of halogens in terms of fire extinguishing effectiveness is as follows:

$$I > Br > Cl > F \tag{4.1}$$

In the hydrocarbon flames it has been shown that oxygen is consumed in the branching reaction,[12]

$$H\cdot + O_2 = \cdot OH + O\cdot \tag{4.2}$$

and the highly exothermic oxidation of CO to CO_2 is carried out by hydroxyl radicals:

$$HO\cdot + CO = CO_2 + H\cdot \tag{4.3}$$

The effective flame retardant acting in the gas phase should inhibit one or both reactions (4.2) and (4.3) because they have a paramount effect on the increase of the overall rate of thermal oxidation process occurring in the flame. Indeed, reaction (4.2) increases the concentration of radicals, while reaction (4.3) increases the temperature. From a mass spectrometry study of species sampled in the low-pressure flame,[13] it became evident that the introduction of halogen species into a premixed CH_4/O_2 flame leads to the production of hydrogen halide, HX, early in the flame. It was also observed that the production of H_2 is enhanced. This provides the evidence of the removal of H· radicals from the flame, and the predominant reaction is considered to be as follows:

$$H\cdot + HX = H_2 + X\cdot \tag{4.4}$$

Under the flame conditions, this is a fast reaction, and it effectively competes with the chain branching reaction

(2) in the preflame region.[14] Moreover, the observed degree of inhibition is greater than that which would be predicted considering the above reaction to reach equilibrium.

The mechanism by which halogen radical is removed from the system has been established (Reaction 4.5)[15]:

$$X\cdot + RH = HX + R\cdot \tag{4.5}$$

In which RH represents the combustible gas or volatile moieties from the polymer thermal degradation. Reaction (4.5) accounts for the high efficiency of flame inhibition by halogen species because it regenerates the flame inhibitor HX, which is not consumed and acts thus in a catalytic mode.

Note that this mechanism explains non-inhibiting properties of fluorides, because HF needs an excessively high activation energy to react with H· radicals to form F· radicals. The lower effectiveness of chloride compared to bromide inhibitors is probably because the HCl reaction is close to the thermodynamically neutral, hence it is likely that the reaction can also proceed in the opposite direction to generate H· radicals.

The scavenging effect of halogen radicals were also used to explain iodine action (4.6):

$$H\cdot + HI = H_2 + I\cdot \tag{4.6}$$

But in this case the weakness of the H-I bond means that HI is not readily formed by attack of RH groups; instead R· radicals can recombine with iodine radicals through the reaction (4.7)[16]:

$$R\cdot + I\cdot = RI \tag{4.7}$$

which explains the low effectiveness of iodine compounds as flame inhibitors.

HX can also trap the OH· radicals (4.8)[17–20]:

$$HX + \cdot OH = H_2O + X\cdot \tag{4.8}$$

In the study of low-pressure explosive combustion of styrene-oxygen mixtures, Petrella[21] observed that the production of OH· was delayed in the presence of HBr. The general understanding of the role of halogens in the trapping OH· radicals in the flames remains somewhat unsettled. The apparent difficulties may be related to the use of different flames for inhibition studies. For example, Wilson et al.[13] have obtained indirect evidence of the increase of maximum OH· level in the low pressure lean CH_4-O_2 flames containing HBr. However, using 1 atm lean propane fueled flames, indirect evidence was provided[22] for the reduction of maximum OH· concentration in the presence of HBr. It is known that in the higher temperature zone of the flame, reactions involving H·, OH·, and O· are balanced, and the pseudo-equilibrium exists between these species. Hence under these conditions, the argument as to whether OH·, O·, or H· are the inhibited species is not particularly critical, because the reduction in any of these radicals would serve to inhibit the chain branching.

An interesting alternative theory has been proposed,[23–25] which attempts to explain the inhibiting action of halogens in the purely physical terms. The proponents of the physical theory compare the halogen activity to that of inert gases, CO_2, and water.[26] Re-examination of flammability limits for halocarbons shows that over 70 wt% halogen is needed in a compound to prevent flame propagation, and only the total amount of halogen and not its nature is important. The peak flammability limit "k" of halocarbon-fuel mixtures, defined as minimum quantity of halogen to snuff out the flame, is a constant given by equation (9):

$$k = \frac{wt.halogen}{wt.halocarbon + wt.fuel} \times 100 = 69.8 \pm 3.5\ wt\% \tag{4.9}$$

Where wt.halogen is the weight of the halogen present in the flame, wt.halocarbon is the weight of the halogenated compound, and wt.fuel is the weight of the organic compound that feeds the flame. Furthermore, the relative effectiveness of the various halogens is directly proportional to the ratio of their atomic weights, namely (4.10):

$$F : Cl : Br : I = 1 : 1.9 : 4.2 : 6.7 \tag{4.10}$$

Therefore, the role of the halogens is simply increasing the total mass of material which must be vaporized per unit of time without an equivalent increase in the heat flux from the flame fuel.[27] However, these physical processes can be important only when the polymers are overwhelmingly converted to the gaseous fuels.[28]

Thus there appears to be no contradiction between the radical trap theory and the physical theory because the halogens can strongly inhibit the ignition of polymers but will probably not be very efficient in the preventing combustion when the external heat flux is large enough to volatilize most of the polymer.[27] Both approaches complement each other; it is difficult to determine in a general way the relative contribution of each of the two modes of activity. This will usually depend on the structure and properties of the polymer and of the flame retardant as well as on the conditions and parameters of the flame and, finally, on the size of the samples.

4.3 GAS PHASE MECHANISM: SYNERGISTIC EFFECT BASED ON HALOGEN-ANTIMONY COMPOUNDS

The flame-retardant properties of halogen compounds are often considerably enhanced when they are used in conjunction with antimony oxides, which used alone have little effect on the fire retardancy. This is a typical case of synergism, that is, a larger effect of combination of the additives than that expected from the individual contribution.[29] Among practicing fire retardancy researchers, achieving this synergism is one of the most desirable goals because it allows utilization of flame retardant additives at a lower concentrations than predicted on the basis of separate additives performances. Antimony oxide-halogen combination represents one of the most important synergistic fire-retardant combinations. The ability of antimony to enhance the effectiveness of halogen-based flame retardants was first demonstrated for cellulosic fabrics treated with chlorinated paraffins and antimony trioxide[30,31]; the synergism has also been shown in polyester resins,[32–34] polystyrene resins,[35] and polyolefins.[36] The use of antimony compounds in the conjunction with halogenated flame retardants is also documented for polyurethanes, polyacrylonitrile, and polyamides.[30,37]

The first studies on the effect of antimony oxides and various chlorinated compounds[30,31] on the flammability of cotton showed that the flame retardance depends on the ease of elimination of hydrogen halide from the chlorinated additive. Therefore, the active species inhibiting combustion were thought to be the antimony oxychloride liberated in the heating. Thus, most experiments were carried out at the atomic ratio of antimony to chloride 1:1. Attempts to determine the optimum ratio led to the discovery that higher chlorine levels improved the efficiency up to the atomic ratios of about 1:2.[30] However, this was attributed to incomplete volatilization of hydrogen chloride. At the same time Coppick et al.[31] found an optimal atomic ratio of about 1:3 for chlorinated alkanes.

The literature contains numerous speculations on the mechanism of antimony-halogen synergism. Some suggest a condensed phase mechanism, as in the case of cellulose it is suggested the formation in situ of antimony chloride which may react with the cellulose to alter the course of thermal decomposition of cellulose and/or a formation of heavy vapors which helps to extinguish the flame.[38] Others involve a physical gas phase mechanism such as formation of antimony containing solid or liquid particles in the flame whose surface provides a site for dissipation of the energy and change the chemistry of the flame (e.g., formation of $\cdot HO_2$ rather than $\cdot HO$ radicals). Gas phase chemical mechanisms have been also proposed based either on the oxygen-inhibiting effect of the heavy vapors of antimony chloride or oxychloride in addition to the physical inhibition of the oxidation chain reaction (wall effect) and chemical inhibition by chlorine[39] or based on the formation of an antimony oxygen halogen (antimony oxyhalide) intermediate compound which increases the presence of halogen radicals with resulting interference in the free radical mechanism of the flame propagation.[35]

It is known that most of the antimony is vaporized during burning[40] or char formation.[41] The ignition behavior of polyester resins inhibited by antimony halogen systems has been considered to indicate the likelihood of gas phase inhibition.[33,34]

An indication of whether a fire-retardant additive acts in the gas phase by inhibiting the radical chain reactions in the flame (flame poisoning) or in the condensed phase can be obtained by carrying out oxygen index measurements using nitrous oxide instead of oxygen as the oxidizer. If a fire-retardant additive acts by flame poisoning, its concentration dependent effectiveness in the nitrous oxide atmosphere would be negligible compared to that in the oxygen atmosphere. This criterion of distinguishment between flame poisoning and condensed phase mechanism of fire retardancy was supported by the finding that in hydrocarbons or hydrogen flames oxygen is mainly consumed in the branching reaction (4.2), while nitrous oxide is mostly consumed in the non-branching reaction (4.11)[42]:

$$H\cdot + N_2O = \cdot OH + N_2 \tag{4.11}$$

The main inhibition effect in the gaseous phase is due to partial indirect suppression of the branching step (4.2). Hence, N_2O flames would be less susceptible to poisoning because reaction (4.11) is a non-branching reaction.

The observation[40] that Sb_2O_3 inhibits the burning of chlorinated polyethylene in air, but not in nitrous oxide, suggests that inhibition of the gas phase reactions is specific to the fuel-oxidizer system. The same results were obtained with a ternary system polypropylene/chloroparaffin/antimony trioxide[43] where comparison of oxygen and nitrous oxide indexes (Figure 4.2) proves that the Sb-Cl synergism is due to the flame poisoning. In fact, there is no change of nitrous oxide index as a function of antimony trioxide concentration, whereas in the same range of Cl/Sb ratio, the oxygen index increases to a maximum followed by a decrease in the fire retardancy at Cl/Sb ratios less than 3. On the other hand, enhancement of the char formation[37,44] in the antimony containing systems can be credited to some involvement of antimony in the solid phase action.

It is clear that the synergistic action involves reaction of Sb_2O_3 with either the halogenated compound or a

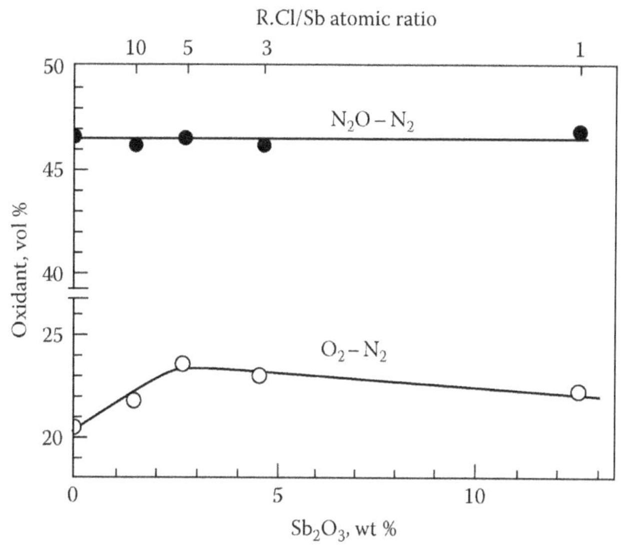

FIGURE 4.2 Oxygen and nitrous oxide index of PP-CP mixtures, 95–5 wt%, with added antimony trioxide.

decomposition product of the halogenated material, presumably HX, in fact optimum conditions for the retardancy depend on the ratio of antimony to chlorine and the ease of decomposition of the halogenated species.[37]

Studies on antimony oxides/fire retardant compounds containing chlorine/bromine confirm formation of gaseous species containing antimony and halides. It is believed that first some hydrogen halide is released from the halogen compound due to interaction with antimony trioxide or with polymer. The HX reacts with the Sb_2O_3 producing SbX_3.[18,45–48] Although the final product of halogenated additives/antimony reaction is antimony trihalide, which is volatile at the temperature of the polymer combustion, different mechanisms have been proposed for its formation. Literature data[47,49,50] favor the formation of SbX_3 through intermediate oxyhalides as compared to the direct complete halogenation,[51] as in the case of Sb_2O_3 reacting with chlorinated or brominated compounds (4.12).

The chemical halogenation of Sb_2O_3 leads to progressively halogen-richer oxyhalides up to the trihalide while the oxyhalides undergo thermal disproportionation with evolution of the trihalide and formation of the halogen-poorer oxyhalide. Therefore, the mechanism of the process depends on the temperature and on the interplay between thermal stability and chemical reactivity of the oxyhalides. Evidence for catalysis of the dehalogenation of halogenated compounds by metal oxyhalides or halides acting as Lewis acids has been reported in the literature.[46]

In the case of chlorinated additives, independently of whether they release HCl on heating or not, $Sb_4O_5Cl_2$ was the dominant oxychloride found in conditions close to those of polymer burning, although $Sb_8O_{11}Cl_2$ was expected to be the most stable oxychloride under these conditions.[52,53] This was explained assuming that $Sb_8O_{11}Cl_2$ is formed first but does not accumulate because it is a highly reactive dechlorinating agent and gives $Sb_4O_5Cl_2$. This last would give $SbCl_3$ at relatively lower rate either through thermal disproportionation or direct complete chlorination. $Sb_4O_5Cl_2$ and $Sb_8O_{11}Cl_2$ might also undergo chlorination to SbOCl which then rapidly disproportionates. It is not possible to demonstrate whether SbOCl is an intermediate in the process because of its high thermal instability under these conditions.

In the case of brominated additives, the process is less studied, for decabromodiphenyl oxide, a widely used brominated fire retardant additive, $Sb_8O_{11}Br_2$ is the dominant oxybromide, whereas $Sb_4O_5Br_2$ was not detected in measurable amounts. It was assumed that, if this last is formed by the bromination of $Sb_8O_{11}Br_2$, it eliminates $SbBr_3$ relatively rapidly either by thermal disproportionation or by chemical reaction with decabromodiphenyl oxide.[54]

The role of antimony halides as flame-retardant species in the gaseous phase is well-established: when SbX_3 (X = chlorine or bromine) is introduced into premixed methane-oxygen flames, atomic antimony and antimony monoxide are found in the flame.[55] Sb_2O_3 was shown by mass spectrometry to be present only in the pre-flame zone, and no antimony-halogen species could be detected in the flame itself.

The proposed sequence of reactions taking place in the flame includes the following steps, where X· is a halogen atom.

$$SbX_3 + H\cdot = SbX_2 + HX \tag{4.13}$$

$$SbX_2\cdot + H\cdot = SbX + HX \tag{4.14}$$

$$SbX\cdot + H\cdot = Sb + HX \tag{4.15}$$

$$Sb + \cdot OH = SbOH \tag{4.16}$$

$$Sb + O\cdot = SbO \tag{4.17}$$

$$SbO + H\cdot = SbOH \tag{4.18}$$

$$SbOH + H\cdot = SbO + H_2 \tag{4.19}$$

$$SbOH + HO\cdot = SbO + H_2O \tag{4.20}$$

Antimony halides are believed to have two functions in the flame. The first is to provide a ready source of hydrogen halide early on, and the second is to produce a "mist" of fine particles of solid SbO in the middle of the flame

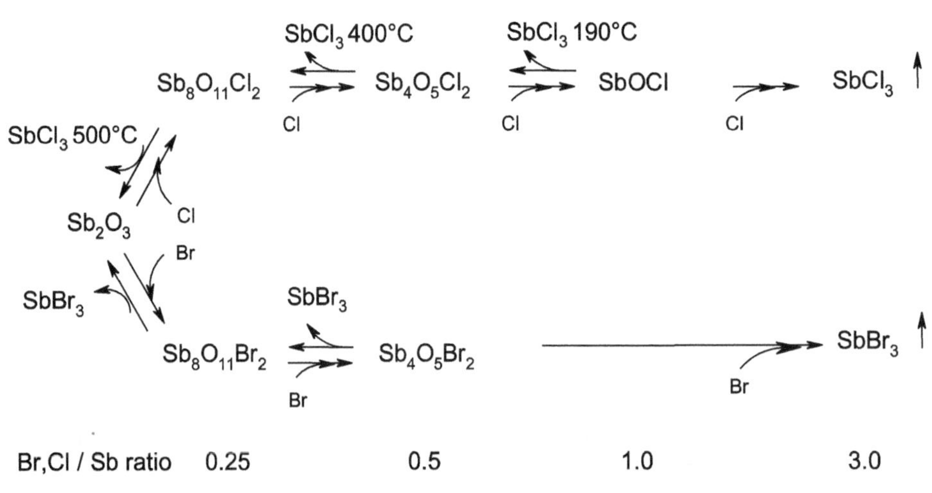

(4.12)

region. The function of SbO as an inhibitor independent of the presence of halogen species is verified by the effect on fire retardancy of triphenylantimony in the absence of halogen compounds.[56] Triphenylantimony is an efficient flame retardant for epoxy resins, even in the absence of halogen.[56] This can be explained in terms of the low volatility, but triphenylantimony can be easily oxidized so that it forms particles of antimony oxides in the gaseous phase. It is supposed by Hastie[55] that antimony monoxide is sufficiently stable in the flame to catalyze the recombination of H·, O··, and ·OH via the formation of transient species (4.16– 4.20) such as SbOH analogously to what was suggested to explain the catalytic effect of other oxides such as SnO.[57]

Over many years there was a significant effort to minimize use or find a replacement to antimony trioxide in industry because of the fluctuating prices and tight supply from China, which is the major producer of Sb_2O_3, and in academia because of the curiosity of the uniqueness of antimony trioxide as a synergist. Use of nanosized Sb_2O_3 in poly(butylene terephthalate) flame retarded with brominated epoxy oligomer[58,59] didn't show obvious advantage over typical micron-sized Sb_2O_3. Interestingly, it was found that concentration of antimony trioxide needed to achieve UL-94 V-0 rating more dependent on the nature of the brominated flame retardant than the particle size of Sb_2O_3. It was shown[60] that brominated flame retardants which melt during the extrusion and molding require two to three times less antimony trioxide. This was explained by more intimate contact between Sb_2O_3 particles and the brominated FR, which accelerates debromination because of Lewis acid effect of Sb_2O_3 and antimony oxybromides.

Zinc borates are well-known secondary synergists which help to decrease loading of antimony trioxide and keeping the similar flame retardant performance.[61,62] Other zinc compounds like zinc stannate,[63] hydrostannate,[63,64] and zinc oxalate[65] allow partial replacement of antimony trioxide or even full replacement but at higher loading. The mechanism of flame retardant action was attributed to the debromination effect of Lewis acid $ZnBr_2$ and possible volatilization of $SnBr_4$, but the last was never detected in the gas phase. In the effort to find a replacement to antimony trioxide, Shibata et al.[66] based on thermodynamic data estimated boiling points of numerous metal bromides and found out that some of them boil at the temperature lower or close to $SbBr_3$ so they can serve as a carrier of bromine to the flame (Table 4.2). Unfortunately, besides $BiBr_3$ all other bromides are tetra-, penta-, and hexabromides, and they are not the most thermodynamically stable compounds on the way from oxides to bromides. Like antimony trioxide (reaction scheme 4.12) these oxides on the way to bromides go through a number of oxybromides, which thermodynamically is more favorable, and complete transition to the bromides is impossible in the polymer burn scenario. In fact, Holdsworth et al.[67] studied Al, Sn(I), and Zn tungstates and found poor flame retardant efficiency, which they attributed to oxybromides.

Interestingly, a UL-94 V-0 rating can be achieved in glass-filled PBT without the use of antimony trioxide.[68,69] It requires an insignificant increase in the bromine content and aluminum hypophosphite or calcium hypophosphite, which replaces antimony trioxide. Further study using milligram scale flame calorimetry showed[70] no synergism between brominated FR and aluminum hypophosphite, but instead both had strong gas phase action, one due to bromine and another due to volatile phosphorus.

The evidence for the action of halogen and halogen antimony compounds in the gaseous phase is well established; however, halogen-containing flame retardant systems are often twofold systems providing radical action inhibition in the gaseous phase and, at the same time, inhibition in the condensed phase, as will be seen in the next section.

4.4 MECHANISM OF ACTION OF HALOGENATED COMPOUNDS IN THE CONDENSED PHASE

Measurement of the oxygen and nitrous oxide index of chlorinated polyethylene[36,40] suggests that, for this polymer, the flame-retardant influence of chlorine primarily takes place in the condensed phases. This suggestion is confirmed by the fact that the addition of quite large amounts of gaseous chlorine or hydrogen chloride to the nitrogen/oxygen atmosphere above burning polyethylene does not affect the oxygen index.

The studies of fire retardant action of halogenated compounds in the condensed phase were performed mostly on mixtures of chloroparaffin with vinyl polymers such as polyethylene, polypropylene and polystyrene.[46] The comparison of the oxygen index measured on chloroparaffin-polymer mixtures before and after HCl elimination by preliminary heat treatment (Table 4.3) demonstrate that chemical flame inhibition by HCl evolved from chloroparaffin appears to contribute to the fire retardation of PP but not of PE and PS. The oxygen index increase was suggested to be due to the condensed phase mechanism and explained by taking into account the mechanism of thermal degradation of chloroparaffin and the interaction with the polymers.

The thermal degradation of chloroparaffin occurs in two main steps. The first takes place below 400°C and consists of an endothermic HCl elimination, a chain elimination of

TABLE 4.2
Estimated Boiling Points of Low Boiling Metal Bromides

Metal Bromide	$SnBr_4$	$TiBr_4$	$SbBr_3$	$TaBr_5$	WBr_6	$ZrBr_4$	WBr_5	$NbBr_5$	$MoBr_4$	$BiBr_3$
Boiling point, °C	224	235	269	344	349	355	360	361	385	461

TABLE 4.3
Oxygen Indices of Polyethylene, Polypropylene, and Polystyrene and Their Mixtures with 30 wt% Chloroparaffin before and after Elimination of 50 wt% of Total Chlorine[46]

Sample	Oxygen Index	
	Original	Dehydrochlorinated
Polyethylene	18	-
Polyethylene/Chloroparaffin	22	22
Polypropylene	18	-
Polypropylene/Chloroparaffin	28	22
Polystyrene	18.5	-
Polystyrene/Chloroparaffin	24	24

TABLE 4.4
Molecular Weight after Treatment under Nitrogen Flow at 260°C[46]

Polymer	Mw		
		After Treatment	
	Original	Control	Mixture with 30%wt of Chloroparaffin
PE	138000	112000	Cross-linked
PP	285000	99000	12000
PS	137000	109000	1700

HCl (4.21). For example, in the case of a 70% Cl containing chloroparaffin[71]:

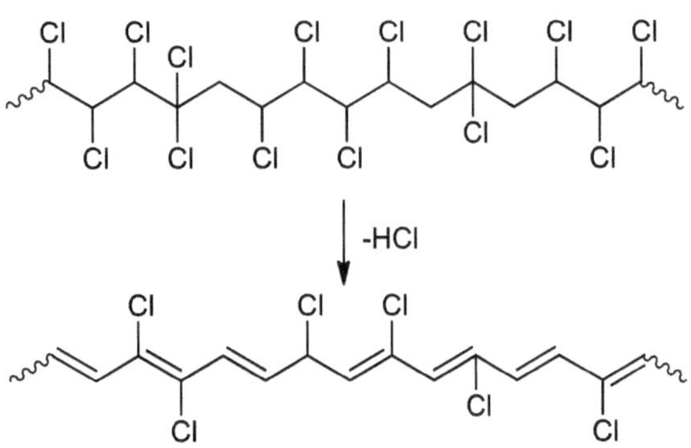

(4.21)

The second step, between 400°C and 800°C is exothermic. In this step, elimination of HCl takes place from the polyene at a lower rate since chlorine atoms linked to unsaturated carbon atoms are involved, which gives a more thermally stable bond than chlorine-saturated carbon of the original chloroparaffin. Intermolecular addition, responsible for the exothermic effect, takes place between double bonds leading to cross-linked products[71] and possibly to aromatization.

The curves of isothermal weight loss at 260°C for chloroparaffin/polymer mixtures show that the polymers are thermally destabilized by the presence of chloroparaffin. The destabilization increases in the sequence polyethylene, polypropylene, and polystyrene; chain scission is induced in polypropylene and polystyrene, whereas cross-linking is promoted in polyethylene, as shown in Table 4.4. At the same time the rate of elimination of HCl from chloroparaffin is decreased by the presence of the three polymers.

Finally, the chloroparaffin modifies the composition of volatile products. In the case of polyethylene, the main effect is to increase the ratio of saturated to unsaturated hydrocarbons.[72] A similar effect is observed for polypropylene in which an important reduction of the overall amount of the lighter hydrocarbons ($C_n \ll C_9$), in favor of longer oligomeric or chain fragments ($C_n > C_9$) was also detected.[42] In polystyrene the presence of chloroparaffin induces a decrease in the yield of styrene and an increase in that of oligomeric products.[73,74] Chloroparaffin also induces a noticeable increase in the yields of toluene, ethylbenzene, and several other fragments evolved in small amounts by pure polystyrene.

The thermal degradation behavior of the mixtures can be explained by the assumption that in chloroparaffin-polymer blends, chlorine radicals, that propagate the dehydrochlorination in the chloroparaffin domains, can migrate in the polymer phase where they abstract hydrogen atoms from the polymer backbone (4.22):

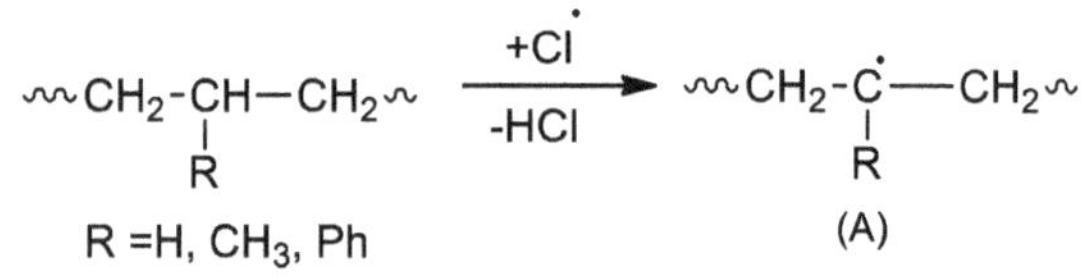

(4.22)

Further reactions of the resulting macroradicals (A) depend on the substituent R (4.23):

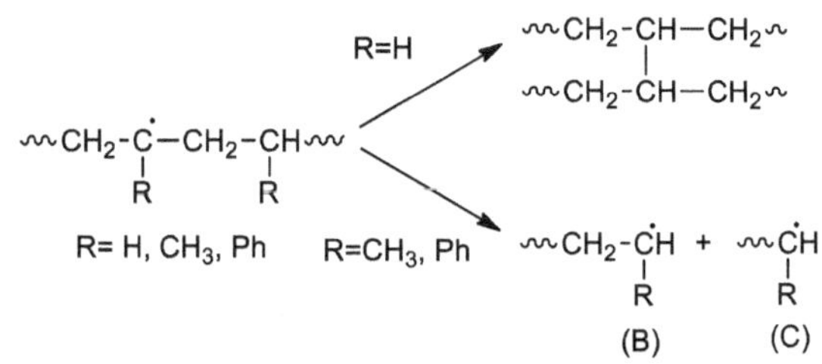

(4.23)

When the substituent R stabilizes radicals, as in (A) and (C), chain scission is more likely than termination by coupling. Radicals (C) then propagate the depolymerization process with volatilization of polypropylene and polystyrene at a temperature at which these polymers would not give significant amounts of volatile products when heated alone. Moreover, unsaturated chain ends such as (B) would also initiate the volatilization process because of the thermal instability of carbon/carbon bonds in β position to a double bond (4.24).

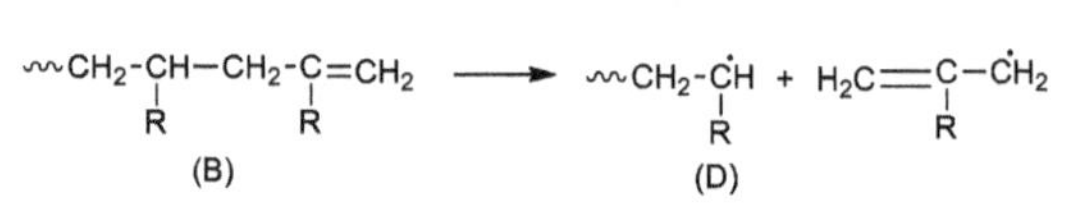

(4.24)

The dehydrochlorination rate of chloroparaffin is lowered by elimination of chlorine radical chain carrier of chloroparaffin dehydrochlorination, through the hydrogen abstraction process. The larger ratio of saturated to unsaturated hydrocarbons is explained based on interaction between the degrading polymer and the cross-linked polyene formed by chloroparaffin. For example, reactive allylic hydrogen atoms can react with the volatile hydrocarbons diffusing through the degrading mixture or with the radical propagating the volatilization process (4.25) increasing the hydrogen/carbon ratio in the polymers and their degradation products.

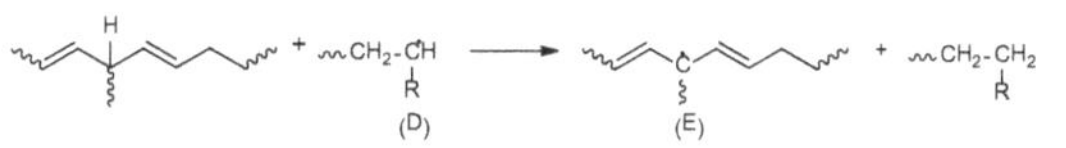
(4.25)

The resulting radical (E) is stabilized by the polyene sequence and may terminate by coupling. After evolution of part of the hydrogen chlorine from the chloroparaffin, the terminating effect of reaction (4.25) can compensate for the initial destabilization. This indeed happens in the case of polystyrene with the reduction of the relative rate of volatilization in the mixture as the degradation proceeds.[74] Moreover, because of reaction (4.25), the lifetime of radical (D) is reduced, and so reactions such as depolymerization by β-scission will be reduced, thus the further volatilization of the polymer will lead to a larger fraction of longer and saturated chain fragments instead of monomer produced by depolymerization.

The fire-retardant action of chloroparaffin through the condensed phase mechanism agrees with the expected somewhat lower flammability of the products evolved by thermal degradation of polyethylene, polypropylene, and polystyrene in combination with chloroparaffin as compared to those evolved from the polymer heated alone. The chain fragments produced in a larger amount are less flammable than the hydrocarbons of low molecular weight whose yield is correspondingly reduced. For example, for the polystyrene degradation products, styrene and ethyl benzene have oxygen indices of 14.4 and 14.7 respectively, whereas that of diphenyl methane is 16.4.[73] Furthermore, polystyrene also benefits from the reduction of pyrolysis rate attributed to interaction with the polyene. Thus, the polyene accumulating on the surface of the burning polystyrene could reduce the rate of products of degradation of polystyrene to the flame.

Concluding, the destabilization of the above polymers by chloroparaffin should be beneficial in terms of fire retardance because it induces the formation of fuel at the temperature at which HCl is evolved. Thus, the occurrence of a polymer-additive interaction in the condensed phase may simply optimize the gas phase retardant action of HCl. However, the composition of the volatile products of degradation of the polymers is modified, giving a mixture likely to be less flammable than that obtained from the pure polymers. Moreover, preventive elimination of the HCl released did not affect the oxygen index of polyethylene and polystyrene while it decreases that of polypropylene.[46] It is possible to conclude that chemical inhibition of the flame by HCl evolved from chloroparaffin should contribute to the fire retardation in polypropylene, whereas in polyethylene and polystyrene the reduced flammability is the direct result of the chloroparaffin-polymer interaction in the condensed phase, which overwhelms the contribution of flame poisoning by HCl.

Recent study[75] of the effect of poly(pentabromobenzyl acrylate) on combustion of PBT using microscale combustion calorimetry also confirmed significant contribution of brominated flame retardant in the condensed phase mechanism.

4.5 MECHANISM OF CONDENSED PHASE ACTION OF SYNERGISTIC SYSTEMS BASED ON HALOGEN-METAL COMPOUNDS

The mechanism of fire retardant action of synergistic halogen-metal fire retardant systems is not yet completely understood in spite of their wide use. In particular, besides the well-known flame inhibition activity of volatile metal halide resulting from halogen-metal compounds reaction in the degrading polymer, condensed phase reactions of halogen derivatives with the polymer have also been suggested to contribute to fire retardance.[46,76] Interaction of chloroparaffin with antimony[43,77,78] and bismuth[43,77,79] were extensively studied, and the studies were extended to other metal salts[80] in order to understand the mechanism of chloroparaffin/metal systems condensed phase action in polypropylene.

The comparison between oxygen and nitrous oxide index (Figure 4.2) of a ternary system polypropylene/chloroparaffin/antimony trioxide indicates that the Sb-Cl "synergism" depends on a flame poisoning effect.[43] For the ternary system polypropylene/chloroparaffin/bismuth carbonate (Figure 4.3), a maximum in both the oxygen and nitrous oxide indices is present at the same Cl/Sb ratio in both oxidant atmospheres, which indicates that the Bi-chlorine synergism must depend essentially on a condensed phase effect. The mixture at which maximum of oxygen index is observed is also that at which polypropylene pyrolyzes at the lowest rate.[43] Furthermore, the oxygen index increase produced by direct addition of the potential flame poison $BiCl_3$ to polypropylene is one order of magnitude smaller than that obtained for the ternary system.[77] Addition of $BiCl_3$ or BiOCl to polypropylene/chloroparaffin mixtures gives similar results as for the addition of bismuth carbonate.[77]

Thus, the flame poisoning activity of $BiCl_3$ evolved is overwhelmed by condensed phase effects. This agrees with the fact that the flame poisoning activity of $BiCl_3$ is only slightly better than that of HCl[81] and would not account for the large "synergistic" effect observed upon addition of bismuth carbonate. Further evidence of the condensed phase mechanism is shown by the higher effectiveness of bismuth carbonate

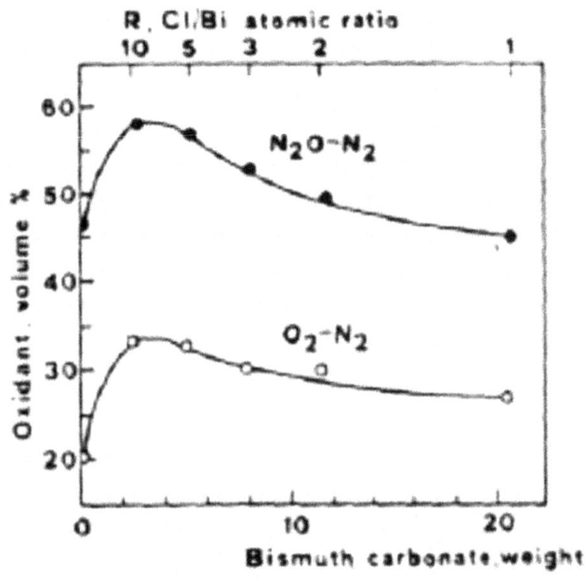

FIGURE 4.3 Oxygen and nitrous oxide index of PP-CP mixtures, 95–5 wt%, with added bismuth carbonate.[43]

compared to antimony oxide in the polypropylene/chloroparaffin ternary mixture,[46] which is in contrast with the higher efficiency of $SbCl_3$ respect to $BiCl_3$ in flame inhibition.[81]

The condensed phase mechanism was explained taking into account the decrease of the pyrolysis rate of polypropylene; $BiCl_3$ could catalyze the condensation between chloroparaffin and polypropylene by addition to chain end double bonds (4.26) formed either in reaction (4.24) or in chain scission occurring during volatilization of polypropylene[42]:

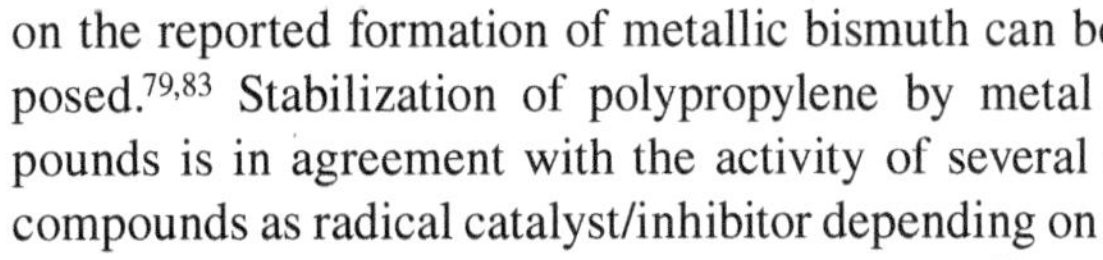
(4.26)

Reaction (4.26) would reduce the rate of volatilization of polypropylene by reducing the concentration of unsaturated chain ends which act as initiating structures.[82] As an alternative or in addition, a stabilization mechanism based on the reported formation of metallic bismuth can be proposed.[79,83] Stabilization of polypropylene by metal compounds is in agreement with the activity of several metal compounds as radical catalyst/inhibitor depending on metal concentration and/or temperature of the system.[84]

Reduction of bismuth compounds could take place by reaction with polymer radicals propagating the depolymerization of polypropylene, either by electron transfer or ligand transfer, which are typical redox reactions between alkyl radicals and metal compounds[85]:

$$R\cdot + M^{n+} = R^{+} + M^{(n-1)+} \tag{4.27}$$

$$R\cdot + MCl_n = RCl + MCl_{(n-1)} \tag{4.28}$$

Successive repetition of reactions (4.26) or (4.27) lead to formation of Bi^0. These redox reaction act as termination steps for polymer degradation. Radical terminations by metal compounds through redox reaction are well-known.[46,84] Further catalytic terminating activity may be displayed by the finely divided metallic bismuth within polypropylene by the so-called "wall-effect" in which radicals recombine at the active surface of the metal. The results obtained with bismuth carbonate was extended to other metals, which increase the oxygen index of PP, with a study of their interaction with polypropylene/chloroparaffin mixtures, which is reported in the work of Costa et al.[80]

4.6 MAIN APPLICATION OF HALOGENATED FLAME RETARDANTS BY FAMILIES OF POLYMERS

The names, structures, and physical properties of the most important halogen-containing fire retardant chemicals are reported in Appendix A of this chapter. This comprehensive list includes the chemical name, the chemical structure, the halogen content (percentage by weight), the melting or softening temperature range, and the polymer families with which they are used.

4.6.1 Styrenic Homopolymer and Copolymers

Halogen fire retardants are used in two members of this family: polystyrene homopolymer foams and styrenic copolymers, including high-impact polystyrene (HIPS) and acrylonitrile-butadiene-styrene (ABS). Foams used in the building industry for thermal insulation and for decorative profiles (extrusion foamed polystyrene XPS) and polystyrene foam (expanded polystyrene EPS) used in the packaging of electronic goods generally need to be fire retarded. Hexabromocyclododecane (HBCD) was a commonly used halogen-containing flame retardant for these applications, but it was banned in Europe in 2016 with other countries followed suit shortly afterwards. The mechanism of fire-retardant action, the aliphatic brominated flame retardants in the foam is due to a combination of chain scission that causes molecular weight reduction of the styrenic polymer, and this helps cool the flame by enhanced dripping and flame poisoning.[86,87] Two alternatives were introduced to the market well ahead of the ban of HBCD. Tetrabromobisphenol A bis(2,3-dibromo-2-methylpropyl ether)[88] is mostly used in Asia, and brominated butadiene-styrene block copolymer is mostly used in Europe and the USA.[89,90] The main properties of these compounds are summarized in Appendix A. Antimony trioxide is not used as a synergist in the fire retarded polystyrene foam. The loading of the fire retardant is typically between 0.8 and 4 wt%. The higher loadings are necessary to counteract the replacement of low flammability chlorofluorocarbon foaming agents with the much more flammable hydrocarbon foam blowing agents (e.g., cyclopentane/n-pentane).

Tetrabromobisphenol A bis(2,3-dibromo-2-methylpropyl ether) has a tertial aliphatic bromine in its structure.

Because of this it can partially debrominate during EPS polymerisation or extrusion of XPS. On the other hand brominated butadiene-styrene block copolymer apparently doesn't affect foaming of polystyrene[91] and is stable under the foam use conditions.[92] Despite being accessed as a sustainable alternative to HBCD,[93] there is still effort from the environmentalists to take out any flame retardants from the polystyrene foam.[94]

Decabromodiphenyl ether, which was widely used in HIPS was banned in Europe in 2008 because of fear that it can debrominate to more toxic octabromodiphenyl ether, which was banned even earlier in 2005. In 2010 major brominated flame retardant producers stopped production and sales of decabromodiphenyl ether in the USA. It has been replaced with decabromodiphenyl ethane, which is nontoxic[95] and more stable to photolytic debromination,[96,97] with debromination half-life in HIPS believed to be more than 200 years.[98] Another brominated flame retardant for HIPS which has even better UV stability is ethylene bis-tetrabromophthalimide.[99] Tetrabromobisphenol A (TBBA) was widely used in ABS because it is melt-blendable and easy to process. In 2022 TBBA was added to the RoHS list of hazardous substances and effectively banned for use in ABS. TBBA is still allowed for use as reactive flame retardant in epoxy resin and as a reactive intermediate for other flame retardants. After removal from the market, brominated diphenyl ethers and most recently TBBA tris(tribromophenyl) cyanurate gained significant use[100] because it melts at 230°C, which is below processing temperature of both HIPS and ABS, which translates into high-impact strength.[101] Recent publications[102,103] show a synergism between tris(tribromophenyl) cyanurate, antimony trioxide, and nitroxide type of heat stabilizers in HIPS. Despite relatively low bromine content end-capped and uncapped brominated epoxy oligomers and polymers with molecular weight of 1400–2000 are also widely used in HIPS and ABS.[104]

4.6.2 Thermoplastic Polyesters

Polymeric fire retardants, such as poly(pentabromobenzyl acrylate) or the phenoxy- or tribromophenol terminated carbonate oligomer of tetrabromobisphenol A, are used in poly(butylene terephthalate) (PBT) at bromine contents as low as 5–6% to produce short total after-flame times. Commercial glass-filled PBT generally contain between 7% and 10% bromine in order to provide a safety factor and to maintain the UL 94 V-0 rating after several recycling steps. Major use of glass-filled flame retardant PBT is in electrical and electronic connectors. Increasing the loading of poly(pentabromobenzyl acrylate) in a glass reinforced PBT has been shown to increase the oxygen index, but also to improve certain mechanical and impact properties for the resin, compared to the lower loading of FR. This is attributed to the efficient coupling effect of the FR additive between glass fibers and the polymeric matrix.[105]

High-molecular-weight brominated epoxy polymers are used in both non-reinforced and glass-reinforced PBT. They are used in conjunction with UV-stabilizer and UV-screener in order to provide light stability and sometimes are added with acid scavengers to improve corrosion resistance to copper.[106] Poly(ethylene terephthalate) (PET) is usually made fire retardant by the addition of brominated polystyrene, which has the necessary thermal stability to withstand processing temperatures that may exceed 300°C. Some compounders are using sodium antimonate as a synergist because it is believed that antimony trioxide contributes to the thermal decomposition of PET during processing.

4.6.3 Polycarbonate

The most widely used halogen flame retardant for polycarbonate is tetrabromobisphenol A that is copolymerized into the polycarbonate resin in the 3–5 wt% bromine concentration range. Phenoxy-terminated carbonated oligomers of tetrabromobisphenol A and brominated epoxy oligomers were also developed, which could be added to PC/ABS or PC/PBT in a compounding step. An occasionally overlooked problem with the use of halogen fire retardants for polycarbonate is that the use of antimony oxide, a common synergist with halogen FR systems, is not an option as it will degrade the PC molecular weight.[107] End-capped brominated epoxy is used in PC/ABS without antimony trioxide.[108] Interestingly, that tris(tribromophenyl) cyanurate in combination with small addition of potassium salt of diphenyl sulfone (KSS) results in a high transparency polycarbonate with UL-94 V-0 rating as at low as 0.4 mm thickness.[109]

4.6.4 Polyamides

Because of very good thermal stability, brominated polystyrene is the most used brominated flame retardant in polyamide 6, polyamide 6.6, and semi-aromatic polyamides. It also provides the highest comparative tracking index among all brominated flame retardants. Brominated polyphenylene oxide is made by polymerization of tribromophenol. It has exceptional thermal stability and chemical resistance. It is recommended for applications where minimal thermal discoloration is required. Poly(pentabromobenzyl acrylate), another polymeric fire retardant, is particularly suitable for use with fiberglass reinforced polyamide 6. It doesn't have high enough thermal stability to be used in polyamide 6.6, but its advantages over other fire retardant additives result from a combination of its polymeric nature and high bromine content.

High-molecular-weight brominated epoxy polymers (Mw = 15,000–68,000) are efficient fire retardants for polyamides, which offer the following advantages: high thermal stability and thermal aging, excellent processability, nonblooming, high UV stability, and low corrosivity.[106] It is important to mention that only brominated epoxies with negligible epoxy content are suitable for polyamide applications, in order to avoid adverse reactions between epoxies moieties and amine groups in the polyamide, which can happen during processing or thermal aging.[110] Although tris(tribromophenyl) cyanurate typically is not used in polyamides, it has been tested in long glass fibers filled with polyamide 6 because of good processing properties.[111]

The Diels-Alder adduct of hexachlorocyclopentadiene and 1,5-cyclooctadiene, also known as Dechlorane Plus, was the only commercial chlorinated fire retardant used in polyamides. It has even higher comparative tracking index compared to brominated polystyrene. However, it will be banned in Europe starting January 1st, 2024.

4.6.5 Polyolefins

Polypropylene has few fire retardant applications because of the difficulty to reach the high performances required for regulatory fire safety applications. The high fire retardant loading required tends to increase brittleness and decrease mechanical properties for polypropylene.[104] As already seen in the mechanism section, the primary fire retardant action of halogen for polypropylene is in the gaseous phase, thus the fire retardant additives for polypropylene are often based on aliphatic bromine compounds in order to release bromine at its low ignition temperature. Tetrabromobisphenol A bis(2,3-dibromopropyl ether) at 5–7 wt% loading with 1–2 wt% antimony trioxide is used for common UL-94 V-2 rating with the largest application in sockets for Christmas lights. At higher concentration this FR tends to bloom out and therefore is not used for V-0 rating. Very similar tetrabromobisphenol S bis(2,3-dibromopropyl ether) has better blooming performance, but not quite enough for V-0 rating. Another brominated FR used for V-2 polypropylene is 2,4,6-tris(2,3-dibromopropyl)-1,3,5-triazine.

To improve the fire retardancy of polypropylene, beyond the UL 94 V-2 level, it is necessary to use blends of aromatic bromine fire retardants with antimony trioxide as a synergist. Most commonly used is decabromodiphenyl ethane at the loading between 35 and 40 wt%, including antimony trioxide. It is used in aramid fiber/polypropylene composites[112] and in long glass fiber composites[113] where it demonstrates good thermal oxidative stability. Moreover, the presence of aromatic bromine increases the photooxidation instability of polypropylene,[114–116] inactivating hindered amines. To reduce the cost without losing in efficacy, the combination of brominated flame retardant/antimony trioxide system with magnesium hydroxide has been proposed[117] to provide a balance of properties at optimal cost. A typical loading might be 15% brominated flame retardant, 5 wt% antimony trioxide, and 20 wt% magnesium hydroxide. This combination also has better smoke characteristics compared to the system of only brominated FR and antimony trioxide. When magnesium hydroxide is used alone, a concentration of at least 60 wt% is necessary, affecting the impact and elongation properties of the polymer system. Poly(pentabromobenzyl acrylate) is the only suitable polymeric flame retardant for V-0 propylene, but its relatively high cost limits its use.

Tris(tribromoneopentyl) phosphate combines bromine and phosphorus in the same molecule. It has been successfully incorporated into polypropylene. Studies have dealt with the question of synergism between bromine and phosphorus present in the same molecule.[118,119] Because of its exceptional UV stability, it is used in plastic outdoor furniture and more specifically in the stadium seats. Fire-retardant efficiency without the need for antimony trioxide opens the door for this product in the field of PP fibers and textiles.

Decabromodiphenyl ethane is also used for polyethylene and EVA wire and cable applications included at a level of 20–24 wt%. Chlorinated paraffins with a loading of about 25 wt% are also used in cables. Because of good thermal stability, decabromodiphenyl ethane or ethylene bis(tetrabromophthalimide) are used in polyethylene/wood flour composites.[120] Ethylene bis(tetrabromophthalimide) with antimony trioxide has showed good thermal aging properties compared to chloroparaffin in ethylene-propylene-diene termonomer (EPDM).[121]

4.6.6 Poly(Vinyl Chloride)

Liquid chlorinated paraffins are the main halogen-containing fire retardant additives used for flexible poly(vinyl chloride) often in combination with a phosphate esters. In this case, the chlorinated paraffins have the secondary function of plasticizers.[122] The thermal degradation mechanism of chlorinated paraffins is similar to that of poly(vinyl chloride), so in this case poly(vinyl chloride) stabilizers have also the secondary function to stabilize chlorinated paraffins. For more demanding applications like plenum cables in the USA, di-(2-ethylhexyl) tetrabromophthalate is a thermally stable liquid fire-retardant additive with a bromine content of approximately 45 wt%. Decabromodiphenyl ethane is used for foamed soft PVC for thermal insulation.

4.6.7 Thermosetting Resins

In the case of thermosetting resins, the fire retardant is usually copolymerized with the resin. Tetrabromobisphenol A (TBBA) is used in epoxy resins especially for glass fiber reinforced printed circuit board. There are two ways how TBBA is applied to epoxy: (1) it is reacted with diglycidyl ether bisphenol A to produce chain extended resin or (2) reacted with epichlorohydrin to produce diglycidyl ether of TBBA. Nonreactive compounds such as decabromodiphenyl ethane or bis(tetrabromophthalimide) are used in non-epoxy typically polyphenylene oxide/styrene butadiene rubber printed wiring boards for high frequency applications. Antimony trioxide is not used in the printed wiring boards because it negatively affects electrical properties. Although not very common, glass fiber brominated epoxy composites can be produced by pultrusion process.[123]

Phenolic resins have a low flammability by themselves due to the high aromatic content which leads to a high char formation on thermal degradation. However, end-capped brominated epoxy oligomers are used when it is necessary.

Decabromodiphenyl ethane in combination with antimony trioxide can be also used.

The main reactive flame retardants compounds used in unsaturated polyesters[124] are tetrabromophthalic anhydride, dibromoneopentyl glycol, bis(2-hydroxyethyl ether of tetrabromobisphenol A, and chlorendic anhydride. Additive flame retardant compounds include brominated epoxy oligomers, chlorinated paraffins, and decabromodiphenyl ethane. Where transparency is not important, antimony trioxide can be used as a synergist to reduce the amount of the halogen required.

4.6.8 Polyurethanes

Rigid and flexible polyurethane foams often utilize fire-retardant chemicals. After ban of pentabromodiphenyl ether and voluntary removal of 2-ethylhexyl tetrabromobenzoate, flexible PU foams don't use brominated flame retardants. Most commonly used chlorinated phosphate esters are outside of the scope of this chapter. Brominated flame retardant compounds for rigid foams are primarily of the reactive type.[125] These include a mixed tetrabromophthalate ester of diethylene and propylene glycol, end-capped brominated epoxy oligomer (Mw = 1400) and tribromoneopentyl alcohol. All these flame retardant products are reactive with the isocyanate group and, hence, are incorporated into the polymer chain. Brominated flame retardants for rigid PU foams must be soluble in tris(chloropropyl) phosphate or other phosphate esters. More on the flame retardant solutions for polyurethanes can be found in Chapters 30 and 31 of this book.

4.6.9 Textiles

There are three primary methods for controlling the flammability of textile products. These include the following: incorporation of the fire retardant into the fiber during production, wet treatment of the fabric after production at the dying step, or back-coating the fabric.

Fiber modification is applicable to synthetic fibers only. Commonly treatment is performed during spinning of the fiber with melt-blendable fire-retardant compounds. The fire retardant must be sufficiently stable to withstand the extrusion conditions. A halogen-phosphorus-containing flame retardant,[126] tris(tribromoneopentyl) phosphate,[50] can be incorporated into fiber grade polypropylenes. This molecule has a high bromine content (70 wt%), which coupled with a 3 wt% phosphorus level ensures that it can be efficient as a flame retardant at 3–5 wt% concentrations sufficient for carpets, contract upholstery, office partitions, curtains, and drapes and wall coverings. Moreover, it has an excellent UV stability, and it is compatible with HALS stabilizers. Spinning must be performed at lowest possible temperature because tris(tribromoneopentyl) phosphate can release unpleasant smell even at minimal thermal decomposition. Brominated TBBA polymers can be incorporated in PET fibers during extrusion. In order to keep sufficient melt strength for the spinning process brominated epoxy polymers with Mw > 50,000 should be used.

Back-coating is the major area where bromine-based flame retardants compounds are used for textiles.[127] This technique is applied primarily in applications such as upholstery, curtains, and wall coverings in public building and seats in public transportation. The system would comprise a flame retardant additive with a latex binder, the latter often based on acrylic or ethylene vinylacetate. Use of vinylidene chloride-modified acrylics reduce the amount of flame retardant additive required. Depending on the composition of textile which often is polyester or polyamide or various polyester/cotton or polyamide/cotton blends, large variety of brominated flame retardants can be used typically in combination with antimony trioxide. For example, tetrabromobisphenol A bis(2,3-dibromopropyl ether) has high efficiency because of the presence of aliphatic bromine. Tris(tribromoneopentyl) phosphate has good UV stability and therefore most suitable for the curtains and blinds. Any bominated flame retardant for use in back-coatings needs to be finely milled. Two aliphatic brominated FRs mentioned above are difficult to mill because of low melting temperature. Aromatic decabromodiphenyl ethane and ethylene bis (tetrabromophtalimide) found broader application in the textiles. End-capped brominated epoxy polymers are used in upholstery because they are considered safer alternatives for frequent skin contact. All brominated flame retardants are permanent in back-coating and withstand at least 25 laundering. Precursor copolymers of acrylates and pentabromobenzyl acrylate after curing on the textile produces even more permanent coating withstanding 50 laundering. More details on textile flame retardancy can be found in Chapter 25 of this book.

4.7 CONCLUSIONS AND FUTURE TRENDS

Halogen-based flame retardants have served a great need for effective flame retardancy for many years. Due to recent environmental concerns (see Chapter 23 of this book), there is a continuing trend towards the development of non-halogenated materials to replace these systems. While this has been underway for quite some time, it does not appear that non-halogenated materials will be easily available to replace all applications where halogen is used today. Halogen-free flame retardants are efficient only in handful of polymers containing oxygen or nitrogen atoms in their structure. On the other hand, halogen and more specifically bromine-based flame retardants are more universal in their effectiveness across all polymers, and often selection of the specific flame retardant is determined by the compatibility with the polymer matrix but not efficiency. Hence it appears that there is still a need for these halogenated materials to prevent fires. Polymeric and reactive flame retardants based upon halogen are increasingly being offered for commercial use due to better environmental profiles and greatly reduced ability to migrate/bloom out of the polymer. For these reasons, it is highly likely that future halogenated flame retardants will be produced with chemical structures that allow for incorporation during polymerization, or as polymeric additives to enable polymer/polymer blend flame retardant materials.

Appendix A Common brominated flame retardants and their application

Chemical Name and CAS No.	Chemical Structure	Bromine or Chlorine Content, wt%	Melting/Softening Temperature °C	Main Applications
Medium (C_{14-17}) and long ($>C_{17}$) chain chlorinated paraffins, CAS No. 63449–39–8	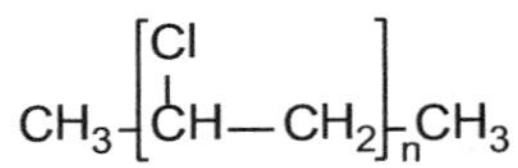	60–70	–45–105	Plasticizers and FRs for PVC and coatings.
Chlorendic anhydride CAS No. 115–27–5	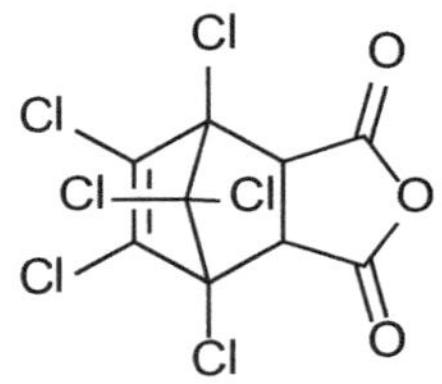	55	235–239	Reactive FR for UPE
Tribromoneopentyl alcohol CAS No. 36483–57–5	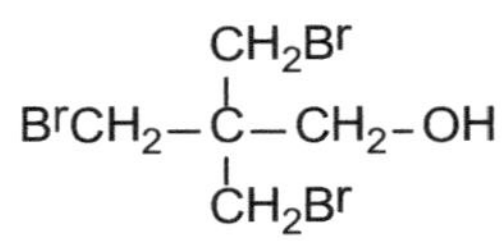	73	65	Reactive FR for rigid PU foams. Intermediate
Dibromoneopentyl alcohol CAS No. 3296–90–0	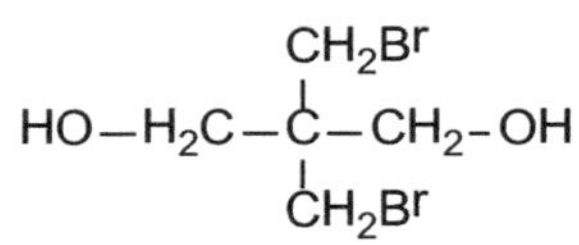	60	110	Reactive FR for UPE
Tris(tribromoneopentyl) phosphate CAS No. 19186–97–1	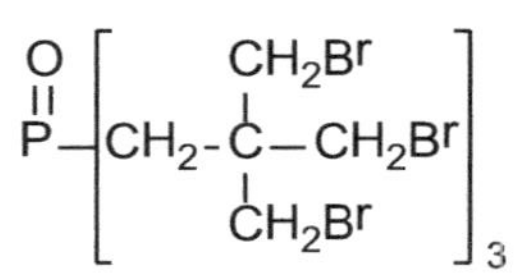	70 3.0 P	181	PP, PP fibers, HIPS, ABS, Adhesives
Brominated butadiene-styrene block copolymer CAS No. 1195978–93–8	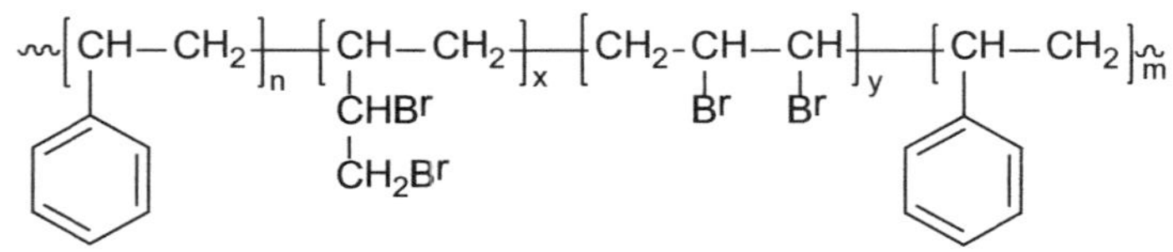	65	120–140	EPS, XPS

(Continued)

Appendix A (Continued)

Chemical Name and CAS No.	Chemical Structure	Bromine or Chlorine Content, wt%	Melting/Softening Temperature °C	Main Applications
Tetrabromobisphenol A CAS No. 79–94–7	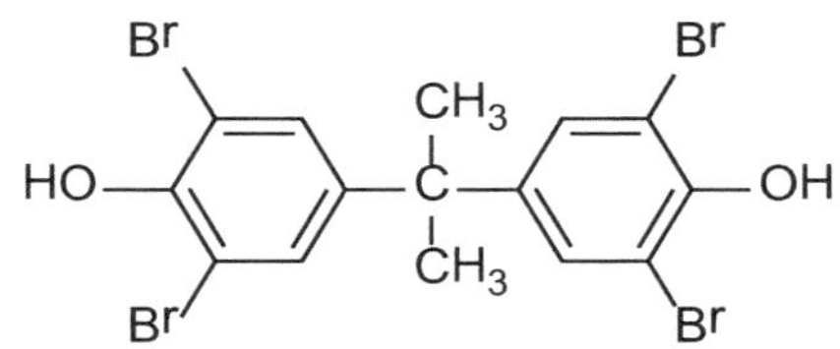	59	181	Reactive FR for epoxy, PC Intermediate
Bisphenol A bis(hydroxyethyl) ether CAS No. 4162–45–2	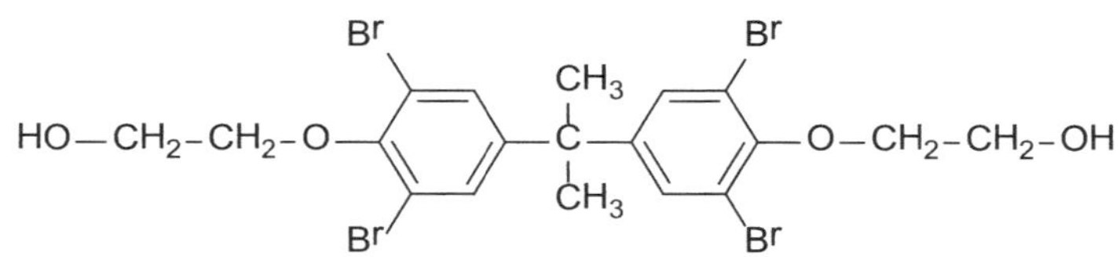	51	119	Reactive FR for UPE
Tetrabromobisphenol A bis(2,3-dibromopropyl ether) CAS No. 21850–44–2	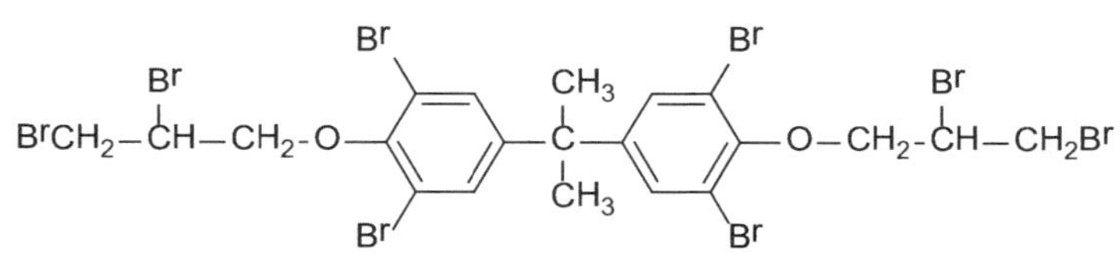	68	113–117	PP, HIPS, ABS
Tetrabromobisphenol A bis(2,3-dibromo-2-methylpropyl ether) CAS No. 97416–84–7	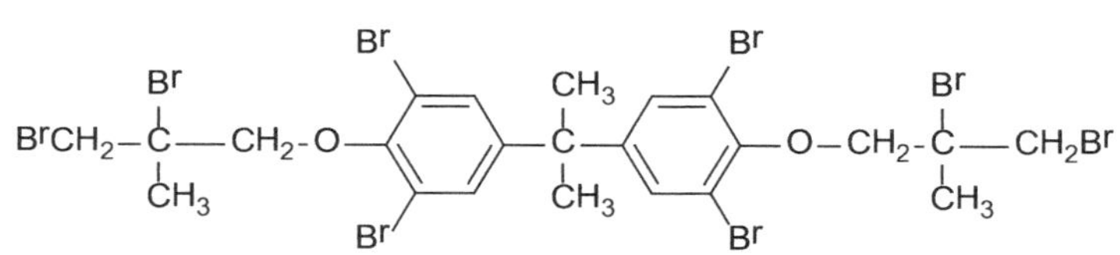	66	110–125	EPS, XPS
Tetrabromobisphenol S bis(2,3-dibromopropyl ether) CAS No. 42757–55–1	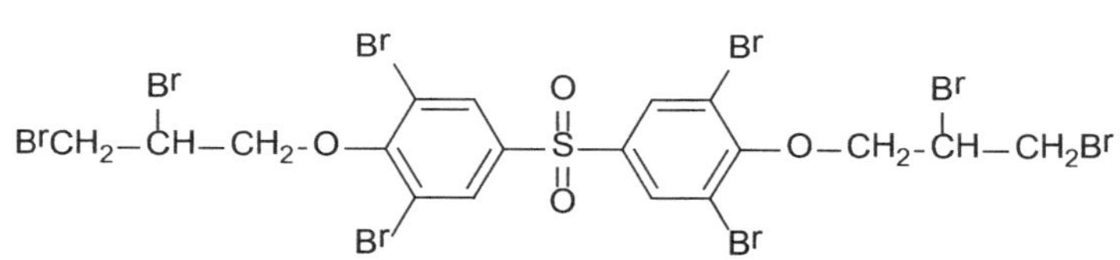	64	80–130	PP

(Continued)

Appendix A (Continued)

Chemical Name and CAS No.	Chemical Structure	Bromine or Chlorine Content, wt%	Melting/Softening Temperature °C	Main Applications
Decabromodiphenyl ethane CAS No. 84852–53–9	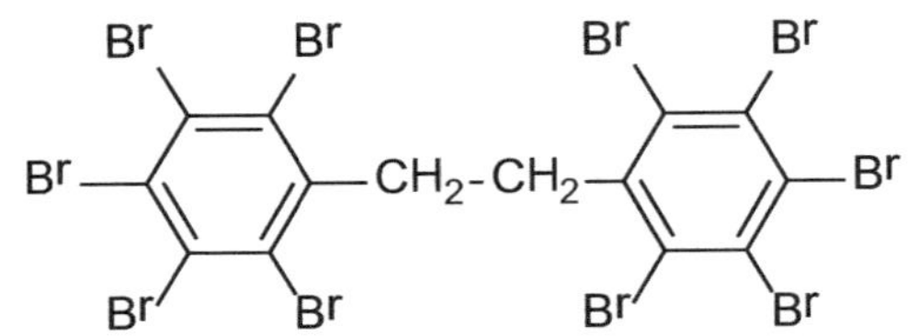	82	> 350	PP, PE, HIPS, PBT, Polyamides, UPE, Textile back-coating
N,N-Ethylenebis (3,4,5,6-tetrabromo-phtalimide) CAS No. 32588–76–4	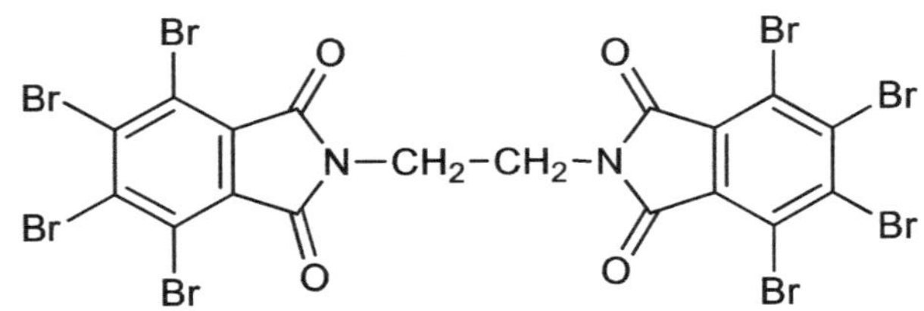	67	470	PP, PE, HIPS, PBT, Polyamides, Textile back-coating
2,4,6-Tris(2,4,6-tribromophenoxy)-1,3,5-triazine CAS No. 25713–60–4	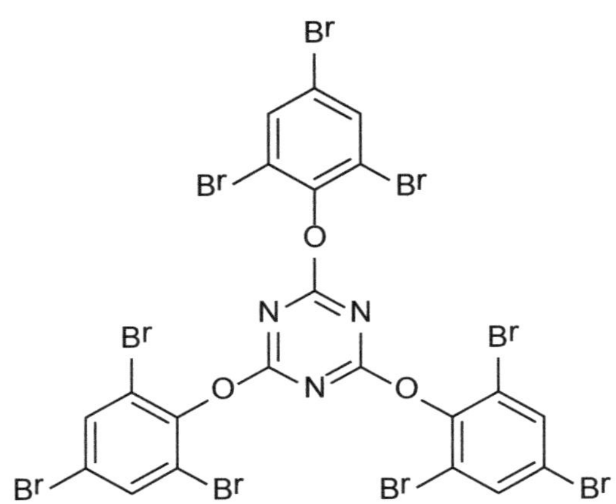	67	230	HIPS, ABS

(Continued)

Appendix A (Continued)

Chemical Name and CAS No.	Chemical Structure	Bromine or Chlorine Content, wt%	Melting/Softening Temperature °C	Main Applications
2,4,6-Tris(2,3-dibromopropyl)-1,3,5-triazine CAS 52434–90–9	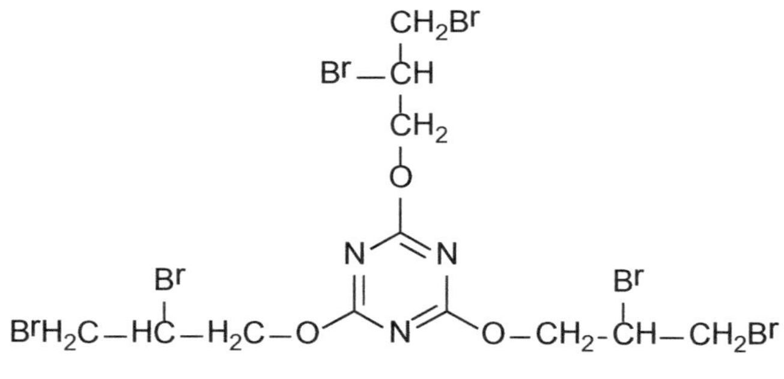	66	110	PP, Textile back-coating
Tetrabromophthalic anhydride CAS No. 632–79–1	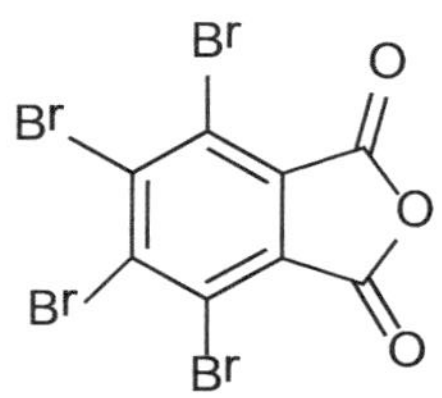	67	269–271	Reactive FR for UPE
Tetrabromophthalic acid mixed esters with diethylene glycol and propylene glycol CAS No.77098–7–8	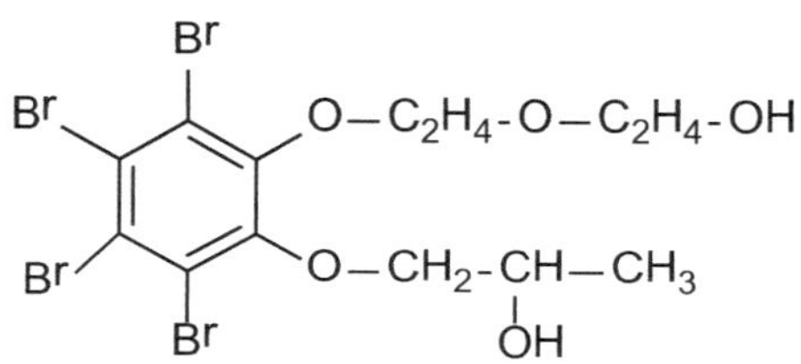	56		Reactive FR for rigid PU foams
Di-(2-ethylhexyl) tetrabromophthalate CAS No. 26040–51–7	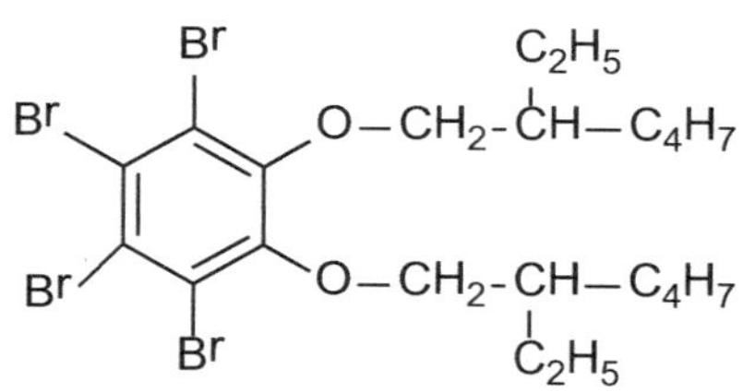	45	-27	Plasticizer for PVC

(Continued)

Appendix A (Continued)

Chemical Name and CAS No.	Chemical Structure	Bromine or Chlorine Content, wt%	Melting/Softening Temperature °C	Main Applications
Brominated epoxy oligomers and polymers CAS No. 68928–70–1	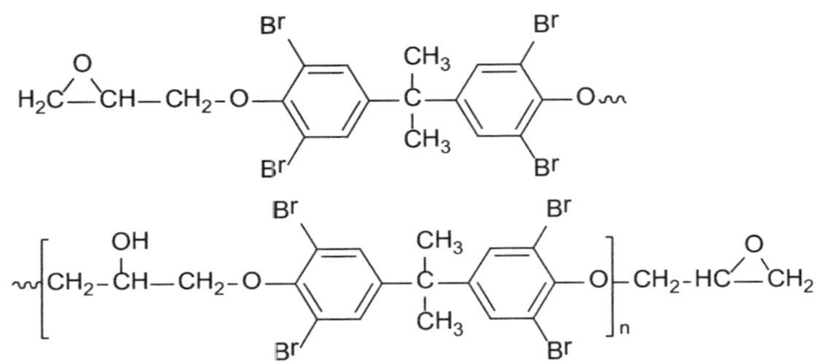	50–53	60–155	Depending on Mw HIPS, XPS, ABS, PBT, PET, Polyamides, TPU, PC/PBT, PC/ABS
End-capped brominated epoxy oligomers and polymers CAS No. 53458–61–7	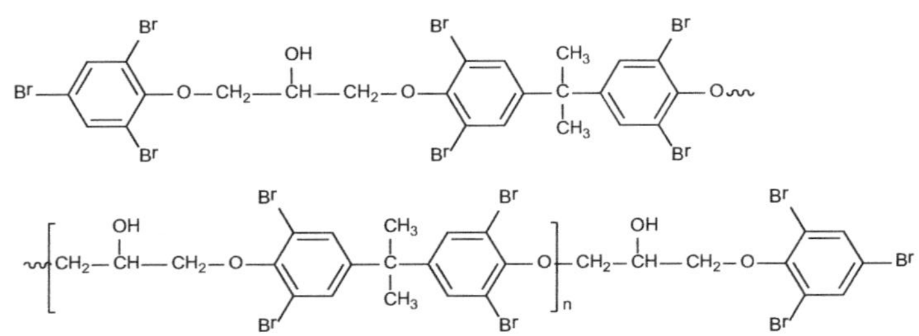	53–60	85–220	Depending on Mw HIPS, ABS, PBT, Polyamides
Phenol terminated brominated polycarbonate CAS No. 94334–64–2	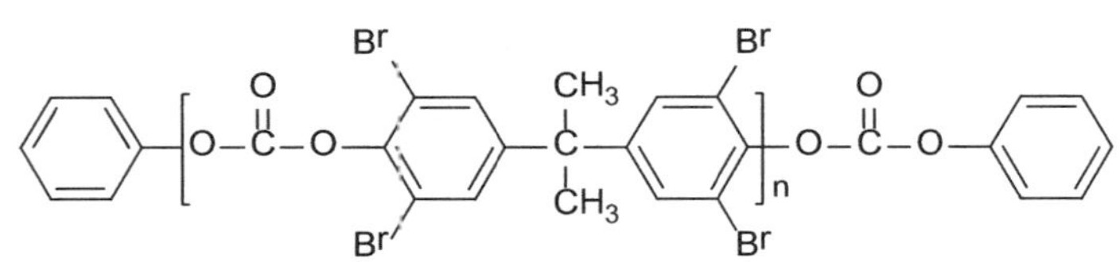	52	180–210	ABS, PC, PBT, PET
Tribromophenyl terminated brominated polycarbonate CAS No.	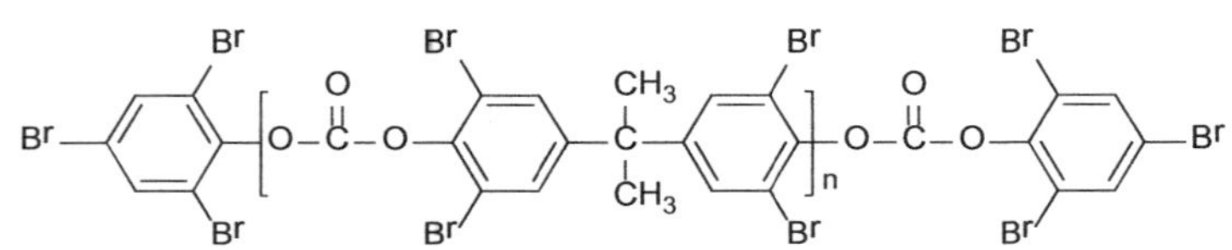	58	200–230	ABS, PC, PC/ABS, PBT, PC/PBT

(Continued)

Appendix A (Continued)

Chemical Name and CAS No.	Chemical Structure	Bromine or Chlorine Content, wt%	Melting/Softening Temperature °C	Main Applications
Poly(pentabromobenzyl acrylate) CAS No. 59447–57–3	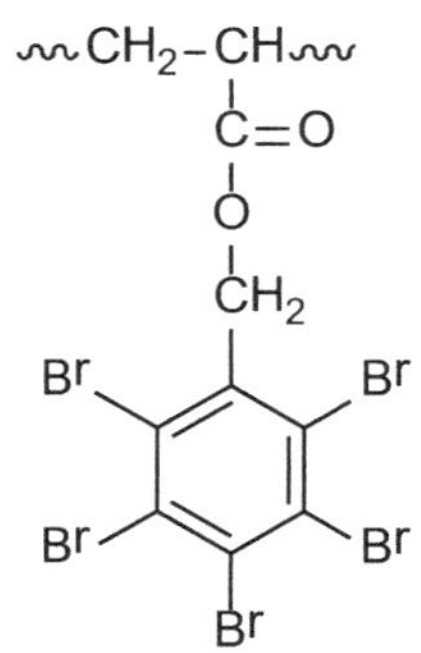	71	190–220	PBT, Polyamides
Brominated polystyrene CAS No. 88497–56–7	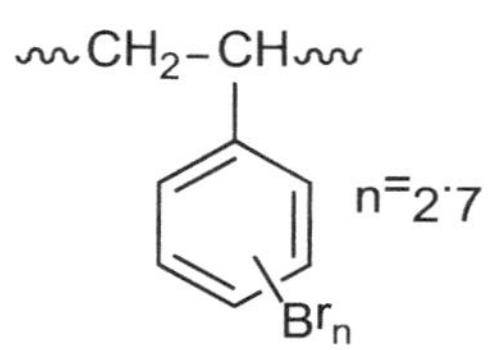	66	265–325	Polyamides, PBT, PET
Poly(2,6-dibromophenylene oxide) CAS No. 69882–11–7	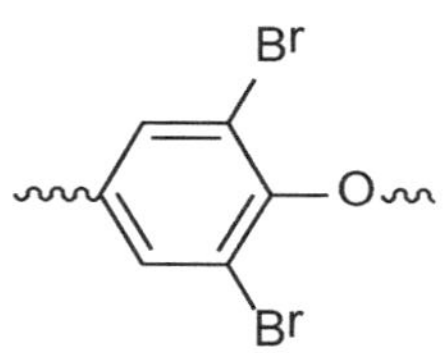	64	210–240	Polyamides, PBT, PET

REFERENCES

1. Woodruff, S.R., and Bhatt, J.C., (Hewlett Packard) Flame-resistant three-dimentional printed articles, PCT Patent Application WO 2023/163710: 2023.
2. Roma, P.; Camino, G.; Luda, M. P., Mechanistic Studies on Fire Retardant Action of Fluorinated Additives in ABS. *Fire Mater.* 1997, 21, 199–204. https://doi.org/10.1002/(SICI)1099-1018(199709/10)21:5 < 199::AID-FAM605 > 3.0. CO;2-O
3. Nishihara, H.; Tanji, S. High-Order Structure and Flame Retardation Effect of Poly(Tetrafluoroethylene) in Thermo-Plastic Resin Compositions. *Polym. J.* **1998**, *30*, 322–326. https://doi.org/10.1295/polymj.30.322
4. Van Krevelen, D.W., *Properties of Polymers: Their Correlation with Chemical Structure; Their Numerical Estimation and Prediction from Additive Group Contributions*, 3rd ed., Amsterdam: Elsevier, 1997.
5. Mita, I., Effect of structure on degradation and stability of polymers. In *Aspects of Degradation and Stabilisation of Polymers*, ed Jellinek, H.H.G., 1978, 247–294. New York: Elsevier.
6. De Wit, C. A. An Overview of Brominated Flame Retardants in the Environment. *Chemosphere.* **2002**, *46*, 583–624. https://doi.org/10.1016/S0045-6535(01)00225-9
7. Covaci, A.; Harrad, S.; Abdallah, M. A.-E.; Alia, N.; Law, R. J.; Herzke, D.; de Wit, C. A. Novel Brominated Flame Retardants: A Review of their Analysis, Environmental Fate and Behavior. *Environ. Int.* **2011**, *37*, 532–556. https://doi.org/10.1016/j.envint.2010.11.007
8. Jans, U. Emerging Brominated Flame Retardants in Sedi-Ments and Soils: A Review. *Curr. Pollut. Rep.* **2016**, *2*, 213–223. https://doi.org/10.1007/s40726-016-0041-5
9. Wan, Y.; Jones, P. D.; Wiseman, S.; Chang, H.; Chorney, D.; Kannan, K.; Zhang, K.; Hu, J.-Y.; Khim, J. S.; Tanabe, S.; Lam, M. H. W.; Giesy, J. P. Contribution of Synthetic and Naturally Occurring Organobromine Compounds to Bro-Mine Mass in Marine Organisms. *Environ. Sci. Technol.* **2010**, *44*, 6068–6073. https://doi.org/10.1021/es100914r
10. Horrocks, A. R. The Potential for Bio-Sustainable Organo-Bromine-Containing Flame Retardant Formulations for Textile Applications—A Review. *Polymers.* **2020**, *12*, 2160. https:// doi.org/10.3390/polym12092160
11. Lyons, J.W., *The Chemistry and Uses of Fire-Retardants*, 1970, 16. New York: Wiley-Interscience.
12. Minkoff, G.I., and Tipper, C.F.H., *Chemistry of Combustion Reactions*, 1962. London: Butterworth.
13. Wilson, W.E., O'Donovan, J.T., and Fristrom, R.M., Flame inhibition by halogen compounds. In *Twelfth Symposium (International) on Combustion*, 1969, 929–942. Pittsburgh, PA: The Combustion Institute.
14. Fristrom, R.M., and Sawyer, R.F., Flame inhibition chemistry. In *Advisory Group for Aerospace Research and Development Conference Proceedings*, No. 84 on Aircraft Fuels, Lubricants and Fire Safety, 1971 AGARD-CP-84-71, Section 12, Brussels: North Atlantic Treaty Organization.
15. Day, M.J., Stamp, D.V., Thompson, K., and Dixon-Lewis, G., Inhibition of hydrogen–air and hydrogen–nitrous oxide flames by halogen compounds. In *Thirteenth Symposium (International) on Combustion*, 1971, 705–712. Pittsburgh, PA: The Combustion Institute.
16. Noto, T., Babushok, V., Burgess, D.R., Hamins, A., Tsang, W., and Miziolek, A., Effect of halogenated flame inhibi- tors on C_1-C_2 organic flames. In *Twenty-Sixth Symposium (International) on Combustion*, 1996, 1377–1383. Pittsburgh, PA: The Combustion Institute.
17. Rosser, W.A., Wise, H., and Miller, J., Mechanism of combustion inhibition by compounds containing halogen. In *Seventh Symposium (International) on Combustion*, 1959, 175–182. London: Butterworth.
18. Hastie, J. W. Molecular Basis of Flame Inhibition. *Journal of Research of NBS-A Physics and Chemistry.* **1973**, *77A*, 733–754.
19. Levy,A., Droege, J. W., Tighe, J. J., and Foster, J. F., The inhibition of lean methane flame. In *Eighth Symposium (International) on Combustion*, 1962, 524–533. Philadelphia, PA: Williams and Wilkins.
20. Petrella, R.V., Factors affecting combustion of polystyrene and styrene. In *Flame Retardant Polymeric Materials*, eds. Lewin, M., Atlas, S.M., Pearce, E.M., Vol. 2, 1978, 159–201. New York: Plenum Press.
21. Petrella, R.V., Studies of the combustion of hydrocarbons by kinetic spectroscopy. II. The explosive combustion of styrene inhibited by halogen compounds, private communication, 1972.
22. Pownall, C., and Simmons, R.F., Some observation on the near flame limit. In *Thirteenth Symposium (International) on Combustion*, 1971, 585–592. Pittsburgh, PA: The Combustion Institute.
23. Larsen, E. R. Mechanism of Flame Inhibition. I: The role of halogens. *J. Fire Flammability/Fire Retardant Chem.* **1974**, *1*, 4–12.
24. Larsen, E. R. Mechanism of Flame Inhibition II: A New Principle of Flame Suppression. *J. Fire Flammability/Fire Retardant Chem.* **1975**, *2*, 5–20.
25. Larsen, E. R., Some Effects of Various Unsaturated Poly-Ester Resin Components Upon FR-Agent Efficiency. *J. Fire Flammability/Fire Retardant Chem.* **1975**, *2*, 209–223.
26. Larsen, E.R., Fire retardants (halogenated). In *Kirk-Othmer Encyclopedia of Chemical Technology*, 3rd ed., Vol. 10, 1980, 373–395. New York: Wiley.
27. Larsen, E. R.; Ludwing, R. B. On the Mechanism of Hal-Ogen's Flame Suppressing Properties. *J. Fire Flammability/Fire Retardant Chem.* **1979**, *10*, 69–77.
28. Hindersinn, R.R., and Witschard, G., The importance of intumescence and char in polymer fire retardance. In *Flame Retardancy of Polymeric Materials*, Vol. 4, eds. Kuryla, W.C., and Papa, A.J., 1978, 1–107. New York: Marcell Dekker.
29. Weil, E.D., Synergists, adjuvants, and antagonists in flame-retardant systems. In *Fire Retardancy of Polymeric Materials*, eds. Grand, A.F., and Wilkie, C.A., 2000, 115–145. New York: Marcel Dekker.
30. Little, R. W., Metallic oxide-chlorinated: Laboratory impregnations. In *Flameproofing Textile Fabrics*, ed. Little, R.W., 1947, 248–256. New York: Reinhold Publishing Company.
31. Coppick, S., Metallic oxide-chlorinated: Foundamentals of process. In *Flameproofing Textile Fabrics*, ed. Little R.W., 1947, 239–248. New York: Reinhold Publishing Company.
32. Nametz, R. C. Self Extinguishing Polyester Resins. *Ind. Eng. Chem. Res.* **1967**, *59*, 5, 99–116.
33. Learmonth, G. S.; Nesbitt, A.; Thwaite, D. C. Flammability of Plastics I. Relation Between Pyrolysis and Burning. *Br. Polym. J.* **1969**, *1*, 149–153.
34. Learmonth, G. S.; Thwaite, D. C. Flammability of Plastics II. Effect of Additives on the Flame. *Br. Polym. J.* **1969**, *1*, 154–160.
35. Lindemann, R. F. Flame-Retardants for Polystyrenes. *J. Ind. Eng. Chem.* **1969**, *61*, 5, 70–75.

36. Fenimore, C. P.; Martin, F. J. Flammability of Polymers. *Combust. Flame* **1966**, *10*, 135–139.
37. Lyons, J.W., *The Chemistry and Uses of Fire-Retardants*, 1970, 209–218. New York: Wiley-Interscience.
38. Read, N. J.; Heighway-Bury, E. G. Flameproofing of Textile Fabrics with Particular Reference to the Function of Antimony Compounds. *J. Soc. Dyers Colour.* **1958**, *74* (12), 823–829.
39. Thiery, P., *Fireproofing, Chemistry, Technology and Applications, English translation by Goundry, J.H.*, 1970. Barking: Elsevier.
40. Fenimore, C. P.; Jones, G. W. Modes of Inhibiting Polymer Flammability. *Combust. Flame* **1966**, *10*, 295–301.
41. Read, N. J.; Heighway-Bury, E. G. Flameproofing of Textile Fabrics with Particular Reference to the Function of Antimony Compounds. *J. Soc. Dyers Colour.* **1958**, *74*, 823–829.
42. Camino, G.; Costa, L.; Trossarelli, L. Thermal Degradation of Polymer-Fire Retardant Mixtures: Part III— Degradation Products of Polypropylene-Chlorinated Paraffin Mixtures. *Polym. Degrad. Stab.* **1982**, *4*, 133–144. https://doi.org/10.1016/0141-3910(82)90035-0
43. Costa, L.; Camino, G.; Luda, M. P. Effect of the Metal on the Mechanism of Fire Retardance in Chloropar-Affin-metal Compound-Polypropylene Mixtures. *Polym. Degrad. Stab.* **1986**, *14* (2), 113–123. https://doi.org/10.1016/0141-3910(86)90010-8
44. Hindersinn, R. R., and Wagner, G. M., Fire retardancy. In *Encyclopedia of Polymer Science and Technology*, Vol. 7. 1967, 1–57. New York: Interscience Communication.
45. Hastie, J.W. and McBee, C.L., Mechanistic studies of tri-phenylphosphine oxide poly(ethylentereftalate)and related flame retardant systems, *NBS Final Report NBSIR 75–742*, 1975. Gaithersburg, MD: National Bureau of Standards.
46. Camino, G., Mechanism of fire retardancy in chloropar-affin-polymer mixtures. In: *Developments in Polymer Degradation*, Vol 7, ed. Grassie, N., 1987, 221–269. London: Elsevier Applied Sciences.
47. Camino, G., Fire retardant polymeric materials. In *Atmospheric Oxidation and Antioxidants*, Vol. II, ed. Scott, G., 1993, 461–493. Amsterdam: Elsevier.
48. Lum, R. M. Antimony Oxide-PVC Synergism: Laser Pyrolysis Studies of the Interaction Mechanism. *J. Polym. Sci. Polym. Chem. Ed.* **1977**, *15*, 489–497.
49. Pitts, J. J.; Scott, P. H.; Powell, D. G. Thermal Decomposition of Antimony Oxychloride and Mode in Flame Retardancy. *J. Cell. Plast.* **1970**, *6* (1), 35–37.
50. Camino, G.; Costa, L.; Luda, M. P. Overview of Fire Retardant Mechanisms. *Polym. Degrad. Stab.* **1991**, *33*, 131–154. https://doi.org/10.1016/0141-3910(91)90014-I
51. Lum, R. M. Antimony Oxide-PVC Synergism: Laser Pyrolysis Studies of the Interaction Mechanism. *J. Polym. Sci. Polym. Chem. Ed.* **1977**, *15* (2), 489–497. https://doi.org/10.1002/pol.1977.170150223
52. Costa, L.; Goberti, P.; Paganetto, G.; Camino, G.; Sgarzi, P. Thermal Behaviour of Chlorine-Antimony Fire- Retardant Systems. *Polym. Degrad. Stab.* **1990**, *30*, 13–28. https://doi.org/10.1016/0141-3910(90)90114-M
53. Costa, L.; Paganetto, G.; Bertelli, G.; Camino, G. Thermal Decomposition of Antimony Oxyhalides I Oxychlorides. *J. Therm. Anal.* **1990**, *36*, 3, 1141–1153. https://doi.org/10.1007/BF01904648
54. Bertelli, G.; Costa, L.; Fenza, S.; Marchetti, E.; Camino, G.; Locatelli, R. Thermal Behavior of Bromine-Metal Fire Retardant Systems. *Polym. Degrad. Stab.* **1988**, *20*, 295–314. https://doi.org/10.1016/0141-3910(88)90074-2
55. Hastie, J. W. Mass Spectrometric Studies of Flame Inhibition: Analysis of Antimony Trihalides in Flames. *Combust. Flame* **1973**, *21*, 49–54.
56. Martin, F. J.; Price, K. R. Flammability of Epoxy Resins, *J. Appl. Polym. Sci.* **1968**, *12*, 143–158.
57. Bulewicz, E. M.; Padley, P. J. Photometric Observations on the Behaviour of Tin in Premixed H2+O2+N2 Flames. *J. Chem. Soc. Faraday Trans.* **1971**, *67*, 2337–2347.
58. Niu, L.; Xu, J.; Yang, W.; Ma, B.; Kang, C. Crystallization, Flame Retardancy and Mechanical Properties of Poly(Butylene Terephthalate)/Brominated Epoxy/ Nano-Sb_2O_3 Composites Dispersed by High Energy Ball Milling. *J. Macromol. Sci. Part B.* **2018**, *57*, 572–584. https:// doi.org/10.1080/00222348.2018.1493173
59. Niu, L.; Xu, J.; Yang, W.; Ma, J.; Zhao, J.; Kang, C.; Su, J. Study on the Synergetic Fire-Retardant Effect of Nano-Sb_2O_3 in PBT Matrix. *Materials.* **2018**, *11*, 1060. https://doi.org/10.3390/ma11071060
60. Levchik, S.; Georlette, P. Flame Retardancy without Antimony. *Specialty Chemical Magazine.* **2012**, September, 31–33.
61. Shen, K. K.; Kochesfahani, S.; Jouffret, F. Zinc Borates as Multifunctional Polymer Additives. *Polym. Adv. Technol.* **2008**, *19*, 469–474. https://doi.org/10.1002/pat.1119
62. Kaynak, C.; Ozkaraca, A. C. Use of Zinc Borate as the Third Component of a Traditional Brominated Flame Retardant System in Acrylonitrile Butadiene Styrene. *Fire Mater.* **2013**, *37*, 491–502. https://doi.org/10.1002/fam.2142
63. Horrocks, A. R.; Smart, G.; Kandola, B.; Holdsworth, A.; Price, D. Zinc Stannate Interactions with Flame Retardants in Polyamides; Part 1: Synergies with Organobromine-Containing Flame Retardants in Polyamides 6 (PA6) and 6.6 (PA6.6). *Polym. Degrad. Stab.* 2012, 97, 2503–2510. https://doi.org/10.1016/j.polymdegradstab.2012.07.002
64. Horrocks, A. R.; Smart, G.; Nazaré, S.; Kandola, B.; Price, D. Quantification of Zinc Hydroxystannate and Stannate Synergies in Halogen-Containing Flame-Retardant Polymeric Formulations, *J. Fire Sci.* **2010**, *28*, 217–248. https:// doi.org/10.1177/0734904109344302
65. Holdsworth, A. F.; Horrocks, A. R.; Kandola, B. K.; Price D. The Potential of Metal Oxalates as Novel Flame Retardants and Synergists for Engineering Polymers. *Polym. Degrad. Stab.* **2014**, *110*, 290–297. https://doi.org/10.1016/j.polymdegradstab.2014.09.007
66. Shibata, E.; Grabda, M.; Nakamura, T. Thermodynamic Consideration of the Bromination Reaction of Inorganic Compounds. A Study on Flame Resistant Mechanism and Pyrolysis Recycling of Bromine-Based Flame-Resisting Plastic. *Waste Manage. Res.* **2006**, *17*, 361–371.
67. Holdsworth, A. F.; Horrocks, A. R.; Kandola, B. K. Potential Synergism Between Novel Metal Complexes and Polymeric Brominated Flame Retardants in Polyamide 6.6. *Polymers.* **2020**, *12*, 1543. https://doi.org/10.3390/polym12071543
68. Hirchsohn, Y., Eden, E., and Epshtein Assor, Y., (ICL-IP America), Flame retardant polyester composition, PCT Patent Application WO2018/026621: 2018.
69. Hirchsohn, Y., Eden, E., and Epshtein Assor, Y., (Bromine compounds), Flame retardant polyester compositions, European Patent 3 529 306: 2022.
70. Raffan-Montoya, F.; Stoliarov, S. I.; Levchik, S.; Eden, E. Screening Flame Retardants Using Milligram- Scale Flame

Calorimetry. *Polym. Degrad. Stab.* **2018**, *151*, 12–24. https://doi.org/10.1016/j.polymdegradstab.2018.02.018
71. Camino, G.; Costa, L. Thermal Degradation of a Highly Chlorinated Paraffin Used as a Fire Retardant Additive for Polymers. *Polym. Degrad. Stab.* **1980**, *2*, 23–33. https://doi.org/10.1016/0141-3910(80)90013-0
72. Costa, L.; Camino, G. Thermal Degradation of Polymer-Fire Retardant Mixtures: Part V—Polyethylene- Chloroparaffin Mixtures. *Polym. Degrad. Stab.* **1985**, *12*, 105–116. https://doi.org/10.1016/0141-3910(85)90069-2
73. Costa, L.; Camino, G. Thermal Degradation of Polymer- Fire Retardant Mixtures: Part VII—Products of Degradation and Mechanism of Fire Retardance in Polystyrene-Chloropar- Affin Mixtures. *Polym. Degrad. Stab.* **1985**, *12*, 287–296. https://doi.org/10.1016/0141-3910(85)90120-X
74. Costa, L.; Camino, G.; Trossarelli, L. A Study of the Thermal Degradation of Polystyrene-Chloroalkane Mixture by Thermogravimetry-High Resolution Gas Chromatography. *J. Anal. Appl. Pyrolysis.* **1985**, *8*, 15–24. https://doi. org/10.1016/0165-2370(85)80011-5
75. Leventon, I.; Kraemer, R.; Stoliarov, S. The Impact of Bromine- and Phosphorus-Based Flame Retardants on Flame Stability and Heat Feedback from Laminar Wall Flames. *J. Phys.: Conf. Ser.* **2018**, *1107*, 032015. https://doi.org/10.1088/1742–6596/1107/3/032015
76. Pitts, J. J. Antimony-Halogen Synergistic Reactions in Fire Retardants. *J. Fire Flammability.* **1972**, *33*, 51–84.
77. Costa, L.; Camino, G.; Luda, M. P. Mechanism of Condensed Phase Action in Fire Retardant Bismuth Compound-Chloroparaffin-Polypropylene Mixtures: Part I—The Role of Bismuth Trichloride and Oxychloride. *Polym. Degrad. Stab.* **1986**, *14*, 159–164. https://doi.org/10.1016/0141-3910(86)90014-5
78. Bogdanova, V. V.; Fedeev, S. S.; Lesnikovich, A. I.; Klimovtsova, I. A.; Sviridov, V. V. The Formation of Antimony Oxychloride in Flame Retardant Mixtures and its Influence on Flame Retardant Efficiency. *Polym. Degrad. Stab.* **1985**, *11*, 205–210. https://doi.org/10.1016/0141-3910(85)90044-8
79. Costa, L.; Camino, G.; Luda, M. P. Mechanism of Con densed Phase Action in Fire Retardant Bismuth Compound-Chloroparaffin-Polypropylene Mixtures: Part II—The Thermal Degradation Behavior. *Polym. Degrad. Stab.* **1986**, *14*, 165–177. https://doi.org/10.1016/0141-3910(86)90015-7
80. Costa, L.; Luda, M. P.; Trossarelli, L. Mechanism of Condensed Phase Action in Flame Retardants. Synergistic Systems Based on Halogen-Metal Compounds. *Polym. Degrad. Stab.* **2000**, *68*, 67–74. https://doi.org/10.1016/S0141-3910(99)00166-4
81. Learmonth, G. S.; Thwaite, D. G. Flammability of Plastics: IV. An Apparatus for Investigating the Quenching Action of Metal Halides and Other Materials on Premixed Flames. *Brit. Polym. J.* **1970**, *2*, 249–253.
82. Camino, G.; Costa, L.; Trossarelli, L. Thermal Degradation of Polymer-Fire Retardant Mixtures—Part II: Mechanism of Interaction in Polypropylene-Chlorinated Paraffin Mixtures. *Polym. Degrad. Stab.* **1982**, *4*, 39–49. https://doi.org/10.1016/0141-3910(82)90004-0
83. Ballistreri, A.; Foti, S.; Montaudo, G.; Pappalardo, S.; Scamporrino, E. Thermal Decomposition of Flame Retardants: Mixtures of Chlorinated Polymers with Sb_2O_3 and $(BiO)_2CO_3$. *J. Polym. Sci.: Polym. Chem. Ed.* **1979**, *17*, 2469–2475. https://doi.org/10.1002/pol.1979.170170819
84. Laver, H.S., The influence of metal chelates on the oxidation of polymers. In *Developments in Polymer Stabilization*, vol 1, ed. Scott, G., 1979, 167–197. London: Applied Science Publisher.
85. Nonhebel, D.C., Tedder, J.M., and Walton, J.C., *Radicals*, 1979, 316. Cambridge: Cambridge University Press.
86. Kaspersma, J.; Doumen, C.; Munro, S.; Prins, A. M. Fire Retardant Mechanism of Aliphatic Bromine Compounds in Polystyrene and Polypropylene. *Polym. Degrad. Stab.* **2002**, *77*, 325–331. https://doi.org/10.1016/S0141-3910(02)00067-8
87. Beach, M. W.; Rondan, N. G.; Froese, R. D.; Gerhart, B. B.; Green, J. G.; Stobby, B. G.; Shmakov, A. G.; Shvartsberg, V. M.; Korobeinichev, O. P. Studies of Degradation Enhancement of Polystyrene by Flame Retardant Additives. *Polym. Degrad. Stab.* **2008**, *93*, 1664–1673. https://doi.org/10.1016/j.polymdegradstab.2008.06.010
88. Onishi, H., and Morishita, K., (Dai-Ichi Kogyo Seiyaki), Flame-retardaed foamed plastic composition and shaped articles, US Patent 8 668 850: 2014.
89. Beach, M. W.; Beaudoin, D. A.; Beulich, I.; Bloom, J. C.; Davis, J. W.; Hollnagel, H. M.; Hull, J. W.; King, B.; Kram, S.; Lukas, C.; Matteucci, M.; Morgan, T.; Stobby, B. New Class of Brominated Polymeric Flame Retardants for Use in Polystyrene Foams. *Cellular Polym.* **2013**, *32*, 229–236.
90. Beach, M. W.; Hull, J. W.; King, B. A.; Beulich, I. I.; Stobby, B. G.; Kram, S. L.; Gorman, D. B. Development of a New Class of Brominated Polymeric Flame Retardants Based on Copolymers of Styrene and Polybutadiene. *Polym. Degrad. Stab.* **2017**, *135*, 99–110. https://doi.org/10.1016/j.polymdegradstab.2016.11.008
91. Wang, Y.; Jiang, H.; Ni, J.; Chen, J.; Zhou, H.; Wang, X.; Xin, F. Study on the Effect of PolyFR and its FR System on the Flame Retardancy and Foaming Behavior of Polystyrene. *RSC Adv.* **2019**, *9*, 192–205.
92. Beach, M. W.; Kearns, K. L.; Davis, J. W.; Stutzman, J. R.; Lee, D.; Lai, Y.; Monaenkova, D.; Kram, S.; Hu, J.; Lukas, C. Stability Assessment of a Polymeric Brominated Flame Retardant in Polystyrene Foams Under Application-Relevant Conditions. *Environ. Sci. Technol.* **2021**, *55*, 3050–3058. https://doi.org/10.1021/acs.est.0c04325
93. Beulich, I.; Lukas, C. Flame Retardant with Molecular Anchor. Sustainable Alternative for Forbidden HBCD in Polystyrene Insulating Materials. *Kunstoffe Int.* **2017**, *12*, 53–56.
94. Babrauskas, V.; Lucas, D.; Eisenberg, D.; Singla, V.; Dedeo, M.; Blum, A. Flame Retardants in Building Insulation: A Case for Re-Evaluating Building Codes. *Building Research Info.* **2012**, *40*, 738–755. https://doi.org/10.1080/09613218.2012.744533
95. DeSchyver, D., DeSoto, T., Dawson, R., Landry, S.D., and Herbiet, R., Sustainable fire safety in electrical and electronic equipment. In *Frame Retardants 2002*, 2002, pp. 75–82. London: Intercsience Communications.
96. Kajiwara, N.; Noma, Y.; Takigami, H. Photolysis Studies of Technical Decabromodiphenyl Ether (DecaBDE) and Ethane (DeBDethane) in Plastics Under Natural Sunlight. *Environ. Sci. Technol.* **2008**, *42*, 4404–4409. https://doi.org/10.1021/es800060j
97. Wang, J.; Chen, S.; Nie, X.; Tian, M.; Luo, X.; An, T.; Mai, B. Photolytic Degradation of Decabromodiphenyl Ethane (DBDPE). *Chemosphere.* **2012**, *89*, 844–849. https:// doi.org/10.1016/j.chemosphere.2012.05.006

98. Mathur, R. S.; Banstola, B.; Lopez, K. M.; Beach, M. W.; DeSchryver, D. Assesing the Environmental Impact Due to Photolytic Degradation of Ethane Bis(Pentabromophenyl) in Plastics. *Chemosphere*. **2023**, *320*, 138063. https://doi.org/10.1016/j.chemosphere.2023.138063
99. DeSoto, T., Kumar, K., and Steukers, V., Sustainable flame retardeants—beyond fire safety ROHs, and WEEE compliance. In *Proceeding of the Conference Polymers in Electronics 2007*, Paper 6, Munich, January 30–31, 2007.
100. Borms, R., Blundell, C., Desikan, A., Georlette, P., and Bar Yaakov, Y., Implementation of higher fire safety requirements for european televisions. In *Flame Retardants 2008*, 2008, pp. 93–100. London: Interscience Communications.
101. Jia, L.; Li, B.; Shi, B.; Zhang, H.; Lan, Y.; Zhang, X.; Wang, Y.; Wang, L. Quantitative Analysis of Interfacial Tension Effect on the Impact Strength of Organic Flame Retardants and Acrylonitrile-Butadiene-Styrene Blends. *J. Appl. Polym. Sci.* **2012**, *124*, 1815–1823. https://doi.org/10.1002/app.35211
102. Wu, H.; Yang, J.; Liu, J.; Sun, M.; Wang, Y.; Tang, L. Synergistic Flame Retardancy of Tris(1-Methoxy-2,2,6,6-Tetramethyl-Piperidin-4-yl) Phosphite and Tris (2,4,6-Tribromophenoxy)-1,3,5-Triazine/Sb_2O_3 in High-Impact Polystyrene. *Fire Mater.* **2020**, *44*, 573–584. https://doi.org/10.1002/fam.2819
103. Wu, H.; Liu, J.; Zhang, Q.; Sun, M.; Yang, J.; Tang, L. Synergistic Fire Retardancy of Bis(1-Methoxy-2,2,6,6-Tetramethylpiperidin- 4-yl) Sebacate and Tris(2,4,6 - Tribromophenoxy)-1,3,5-Triazine/Sb_2O_3 in HIPS. *Adv. Mater. Sci. Eng.* **2020**, 6384530.
104. Smith, R.; Georlette, P.; Finberg, I.; Reznick, G. Development of Environmentally Friendly Multifunctional Flame Retardants for Commodity and Engineering Plastics. *Polym. Degrad. Stab.* **1996**, *54*, 167–173. https://doi. org/10.1016/S0141-3910(96)00040-7
105. Reznick, G., Bar Yaakov, Y., Utevskii, L., Georlette, P., and Lopez Cuesta, J.M., Optimization of flame retarded thermoplastics for engineering applications. In *Flame Retardants '98*, 1998, 125–137. London: Interscience Communications.
106. Bar Yaakov, Y., Finberg, I., Peled, M., Georlette, P., Stainmetz, A., and Plevinsky, D., Development of brominated flame retardant systems with improved corrosion resistance for application in plastics. In *Proceedings of the Fall FRCA Conference*, October 20–22, 2002, pp. 159–172. Cleveland, OH/Lancaster, PA.
107. Innes, J.; Innes, A. Flame Retardants for Polycarbonate—New and Classical Solutions. *Plast. Add. Compound.* **2006**, *8* (1), 26–29.
108. Geran, T., Finberg, I., Reznik, G., Hini, S., Plewinsky, D., and Bar Yaakov, Y., Development of fire retarded plastics with reduced or no presence of antimony trioxide. In *Proceedings of Conference BCC on Recent Advances in Flame Retardancy Polymeric Materials*, June 2–4, 2003, Stamford, CT/Norwalk, CT: BCC.
109. Eden, E., Hirschsohn, Y., and Epshtein Assor, E., (Bromine compounds), Flame-retarded trasparent polycarbonate compositions, US Patent 10 800 903: 2020.
110. Wang, M.; Song, H.; He, W.; Wang, J.; Zhou, D.; Guo, J. Thermal Oxidation and Thermal Degradation Kinetics of Brominated Epoxy Resin/Sb_2O_3 Flame Retardant PA10T/GF Composites. *Polym. Eng. Sci.* **2018**, *58*, 1583–1595. https:// doi.org/10.1002/pen.24747
111. Zuo, X.; Shao, H.; Zhang, D.; Hao, Z.; Guo, J. Effects of Thermal-Oxidative Aging on the Flammability and Thermal-Oxidative Degradation Kinetics of Tris(Tribromophenyl) Cyanurate Flame Retardant PA6/LGF Composites. *Polym. Degrad. Stab.* **2013**, *98*, 2774–2783. https://doi.org/10.1016/ j.polymdegradstab.2013.10.014
112. Chen, X.-S.; Xu, G.-Z.; Zhang, S.; Zhang, R.; Sun, J.; Dong, M.-Z.; Zhu, X.-J.; Liu, W. Improving the Flame Retardancy of the Polypropylene/Aramid Fiber Composites by the Introduction of Decabromodiphenyl Ethane and Antimony Trioxide. *J. Appl. Polym. Sci.* **2013**, *127*, 1446–1453. https://doi.org/10.1002/app.36789
113. Guo, J.; Wang, M.; Li, L.; Wang, J.; He, W.; Chen, X. Effects of Thermal-Oxidative Aging on the Flammability, Thermal Degradation Kinetics and Mechanical Properties of DBDPE Flame Retardant Long Glass Fiber Reinforced Polypropylene Composites. *Polym. Compos.* **2018**, *39*, E1733–E1741. https://doi.org/10.1002/pc.24720
114. Sinturel, C.; Philippart, J. L.; Lemaire, J.; Gardette, J. Photooxidation of Fire Retarded Polypropylene. I. Photoageing in Accelerated Conditions. *Eur. Polym. J.* **1999**, *35*, 1773–1781. https://doi.org/10.1016/S0014-3057(98)00280-8
115. Sinturel, C.; Lemaire, J.; Gardette, J. L. Photooxidation of Fire Retarded Polypropylene: II. Photooxidation Mechanism. *Eur. Polym. J.* **1999**, *35*, 1783–1790. https://doi. org/10.1016/S0014-3057(98)00281-X
116. Sinturel, C.; Lemaire, J.; Gardette, J. L. Photooxidation of Fire Retarded Polypropylene: III. Mechanism of HAS Inactivation. *Eur. Polym. J.* **2000**, *36*, 1431–1443. https://doi.org/10.1016/S0014-3057(99)00218-9
117. Montenzin, F.; Lopez-Cuesta, J. M.; Crespy, A.; Georlette, P. Flame Retardant and Mechanical Properties of a Copolymer PP/PE Containing Brominated Compounds/Antimony Trioxide Blends and Magnesium Hydroxide or Talc. *Fire Mater.* **1997**, *21*, 245–252. https://doi. org/10.1002/(SICI)1099-1018(199711/12)21:6 < 245::AID- FAM616 > 3.0.CO;2-F
118. Green, J., Phosphorous-bromine flame retardant synergy in engineering thermoplastics. In *Advances in Organobromine Chemistry II Proceedings, ORGABROM '93*, 1995, 341–354. Amsterdam: Elsevier.
119. Squires, G. E., Flame retardant polypropylene—A new approach that enhances form, function and processing. In *Flame Retardants'96*, 1996, pp. 107–114, London: Interscience Communications.
120. Zhang, J.; Li, G.; Wu, Q.; Li, M.-C.; Sun, X.; Ring, D. Synergistic Influence of Halogenated Flame Retardants and Nanoclay on Flame Performance of High Density Polyethylene and Wood Flour Composites. *RSC Adv.* **2017**, *7*, 24895. https://doi.org/10.1039/C7RA033270
121. Nah, C.; Oh, J.; Mensah, B.; Jeong, K.-U.; Ahn, D. U.; Kim, S.-J.; Lee, Y.; Nam, S.-H. Effects of Thermal Aging on Degradation Mechanism of Flame Retardant-Filled Ethylene-Propylene-Diene Termonomer Compounds. *J. Appl. Polym. Sci.* **2015**, *132*, 41324. https://doi.org/10.1002/app.41324
122. Weil, E. D.; Levchik, S. V.; Moy, P. Flame and Smoke Retardants in Vinyl Chloride Polymers—Commercial Usage and Current Developments. *J. Fire Sci.* **2006**, *24*, 211–236. https://doi.org/10.1177/0734904106057951
123. Chen, C.-H. Pultruded Glass Fiber Reinforced Brominated-Epoxy Composites: Dynamic Mechanical And Flame Retardant Properties. *Mod. Phys. Lett. B.* **2020**, *34*, 2040015. https://doi.org/10.1142/S0217984920400151

124. Weil, E. D.; Levchik, S. V. Commercial Flame Retardancy of Unsaturated Polyester and Vinyl Resins: Review. *J. Fire Sci.* **2004**, *22*, 293–303. https://doi.org/10.1177/0734904104041210
125. Silva, N.G.S., Zanini, N.C., de Souza, A.G., Barbosa, R.F.S., Rosa, D.S., and Mulinari, D.R., Halogen-based flame retardants in polyurethanes. In *Materials and Chemistry of Flame-Retardant Polyurethanes. Vol. 1. Fundamental Approach*, ed. Gupta, R.M., 2021, pp. 141–172, ACS Symposium Series 1399. Washington, DC: ACS.
126. Buszard, D.L., Inherently flame retadant polypropylene fibres. In *Proceedings of Conference Textile Flammability: Current and Future Issues*, 30 –31st March 1999. Manchester: Textile Institute.
127. Weil, E. D.; Levchik, S. V. Flame Retardants in Commercial Use or Development for Textiles. *J. Fire Sci.* **2008**, *26*, 243–281. https://doi.org/10.1177/0734904108089485

5 Phosphorus-Containing Flame Retardants

Svetlana Tretsiakova-McNally[1] *and Paul Joseph*[2]

[1] Belfast School of Architecture and the Built Environment, Ulster University, 2-24 York Street, Belfast, Northern Ireland, UK, BT15 1AP

[2] Institute of Sustainable Industries and Liveable Cities, Victoria University, P.O. Box 14428, Melbourne, VIC 8001, Australia

5.1 BACKGROUND

Elemental phosphorus and its various compounds have been used to flame retard a wide variety of polymer-based materials for several decades. Environmental considerations, especially concerning the use of halogen-based systems, have paved the way, especially in recent years, for the increased use of phosphorus-based flame retardants (FRs) as alternatives to the halogen-containing compounds. Furthermore, this had generated active research in identifying novel flame retardants based on phosphorus, as well as synergistic combinations with compounds of other flame retardant elements (such as nitrogen and halogens) and with several inorganic nano-fillers (for example, phyllosilicates and carbon nano-tubes).[1,2] This research on halogen alternatives has also resulted, especially in recent years, in a wealth of literature (including patents) and in the market acceptance of several phosphorus-based flame retardants.[3]

Although phosphorus compounds can be highly effective flame retardants, they are not effective with some major classes of polymers such as styrenic resins and polyolefins.[4] Furthermore, the basic mode of intervention of phosphorus compounds, regardless of the phase in which they are active, is to suppress the efficiency of combustion reactions, which are mostly radical in nature, occurring in the gas phase. This invariably involves less oxidation of a carbonaceous substrate to carbon dioxide and thus leads to the production of more soot/smoke and to a comparatively higher value of the CO/CO_2 ratio.[5] Needless to say, the increased production of smoke and toxic vapors (mainly CO) is a major concern, especially in real fire scenarios, where materials passively protected with phosphorus flame retardants are involved. However, from an ignitability point of view, albeit only a low-level indicator of the overall flammability hazard associated with a material, the use of phosphorus-containing flame retardants has proved to be of great value in that they generally increase the ignition resistance of materials significantly.[6]

The structure-toxicity relationships of organophosphorus compounds have been extensively studied and are relatively well understood.[7] Generally, phosphorus-containing flame retardants, as a class, exhibit only low to moderate toxicity, as gauged by their Lethal Dose-LD-values. However, there are exceptions to the above norm. These include the neurotoxicity of tricresyl phosphate (mainly owing to the presence of the *o*-isomer) and mutagenic/carcinogenic effects of tris(2,3-dibromopropyl) phosphate, a flame retardant used for polyester fabrics in the 1970s. The main effect of phosphorus fire retardants on the overall combustion toxicity of materials seems to stem from the increased yields of toxic gases such as CO and, in some instances, HCN. However, the results from various studies were inconclusive owing to the variables involved in these experiments, such as the fuel-air ratios, burning configurations, equipment design, heat input, and the particular toxicity criterion in question.[8] See Chapter 19 of this book for more information on smoke and fire toxicity and emissions.

5.2 ELEMENTAL PHOSPHORUS AND ITS COMPOUNDS

Phosphorus, which exists in various singly bonded allotropic forms, has vital biological functions and was first isolated from human urine. The element can be also volatilized from phosphate rock by heating with sand and coke. With an electronic configuration $[Ne]3s^23p^3$, phosphorus, like nitrogen, can complete its octet by forming three σ bonds, to give compounds such as PX_3, which can use the lone pair of electrons to act as Lewis bases. Unlike nitrogen, however, but like silicon, phosphorus can also expand its octet. Predictably, phosphorus is slightly smaller than silicon and is slightly more electronegative. The strengths of the P-H and Si-H bonds are about the same, with phosphorus having the closer approach but forming almost non-polar bonds, but the P-P bonds are weaker than Si-Si, and phosphorus forms somewhat weaker σ bonds than silicon to carbon, oxygen, fluorine, and chlorine.[9]

There are several solid forms of phosphorus, all containing single bonds. The low melting, white allotrope, which is soluble in carbon disulfide, is very toxic and reactive. The other major allotrope, red phosphorus, formed by heating the white form in the absence of air, is denser, harder, and is less toxic and less reactive. Heating the white

DOI: 10.1201/9781003380689-5

form at high pressures forms the more stable black form. Phosphorus can form a great variety of binary compounds with almost all elements, often by direct action. In addition, phosphorus may form a number of different binary compounds with a given element (for example, with Ni). It also forms a series of hydrides, P_nH_{n+2}, of which the first member, PH_3, is stable. Phosphorus forms about 12 possible halides and also some mixed derivatives. All can be hydrolyzed to the corresponding oxoacids. When three hydroxyl groups are attached to phosphorus, the particular oxoacid formed is phosphoric acid, H_3PO_4. A lower hydroxylated species, H_3PO_2, is a moderately strong monobasic acid. Two or more OPOH groups in oxoacids may be joined together, often through -P-O-P- links, to form polyphosphorus acids; species with chains containing up to 17 links are known, together with highly polymerized phosphoric acid.

Phosphorus also forms a wide variety of compounds in which it is joined to nitrogen. P-N bonds are short and strong, and many of these compounds are quite stable. There is a great tendency to form chains and rings, some of which are planar. For example, in $(NPCl_2)_x$, which can exist as a planar six membered cyclic trimer, all P-N distances are very similar, suggesting the presence of electron delocalization. In the case of the phosphazene unit (-N=P-), which is isoelectronic with the silicone unit (-O-Si-), a range of phosphazene polymers resemble the silicones, as they are water repellent, flexible, and stable to heat and solvents. There are also a vast number of compounds in which phosphorus is bound to carbon, since it can replace both nitrogen atoms and CH groups. The resulting compounds are collectively termed organophosphorus compounds. These include several classes such as aliphatic and aromatic phosphines, phosphine oxides, phosphites, phosphates, phosphinites, phosphinates, phosphonate esters, phosphonium salts, etc.[10] Several of the above classes of compounds have been very successfully used both as additives and as reactive flame retardants for a wide variety of polymer-based systems.

5.3 PHOSPHORUS-BASED INORGANIC ADDITIVE FLAME RETARDANTS

Phosphorus-containing flame retardants include red phosphorus, inorganic phosphates, organophosphorus compounds, and chlorophosphorus compounds.[11] From a practical point of view, adequate flame retardancy is achieved either by the mechanical blending of a flame retardant compound with the polymeric substrate (i.e., by introducing an *additive*), or by the chemical incorporation of the flame retardant into the polymer molecule by co- or ter-polymerization, or by chemical modification of the preformed polymer, that is, using a *reactive* component. Currently, synthetic polymers are usually made more flame retardant by incorporating additives. Such additives often have to be used at high loadings to achieve a significant effect, for example, 30% by weight or more, which occasionally can have a more detrimental effect on the physical and mechanical properties of polymers than that produced by reactive flame retardants. Nevertheless, additives are more generally used as they are usually cheaper and more widely applicable.[12,13]

It is common practice, especially from a commercial point of view, to use a combination of flame retardants for polymeric materials. In many cases, these flame-retardant mixtures can give an enhanced performance at a low cost and possibly at a lower overall loading. The synergistic effects of phosphorus-nitrogen and phosphorus-halogen are well-documented.[14,15] Certain inorganic phosphorus compounds (especially ammonium salts of phosphoric acid oligomers) have been extensively used as components in intumescent systems.[16] In recent years, there has been an intense research interest in the effects of nanoscopic inorganics (such as clays) in polymeric systems containing phosphorus compounds, both as additives and as reactives.[17]

5.3.1 Red Phosphorus

The red allotropic form of phosphorus is relatively nontoxic and, unlike white phosphorus, is not spontaneously flammable. However, red phosphorus can be easily ignited. It is a polymeric form of phosphorus, thermally stable up to *ca.* 450°C. In finally divided form, it has proved to be a powerful flame retardant additive.[18] Elemental red phosphorus is a highly efficient flame retardant, especially for oxygen-containing polymers such as polycarbonates and poly(ethylene terephthalate). Red phosphorus is particularly useful in glass-filled polyamide 6,6, where high processing temperature (about 280°C) excludes the use of most phosphorus compounds.[19] In addition, coated red phosphorus is used to flame-retard nylon electrical parts, mainly in Europe and Asia.[20]

Red phosphorus will react with atmospheric moisture to form toxic phosphine gas and will ignite readily in the air. As a result, the commercial product is often encapsulated in an appropriate polymer matrix. Suitable stabilization and encapsulation have led to commercial concentrates containing 50% of red phosphorus. Handling hazards and color of the final flame-retardant products may be a deterrent for wider use.

5.3.2 Ammonium Phosphates

Ammonium phosphates were first recommended for flame retarding theatre curtains by Gay-Lussac in 1821. Monoammonium and diammonium phosphates, or mixtures of the two, are widely used to impart flame resistance to a wide variety of cellulosic materials such as paper, cotton, and wood.[21] These salts have proven to be highly efficient at relatively low costs of application. They are also very effective in preventing afterglow. However, flame-retardant formulations based on these salts are generally nondurable since they are water soluble and, therefore, are easily susceptible to leaching out from the material matrix.

Ammonium polyphosphates, on the other hand, are relatively water insoluble, non-melting solids with very high phosphorus contents (up to about 30%). There are several crystalline forms, and the commercial products differ in molecular weights, particle sizes, solubilities, etc. They are also widely used as components of intumescent paints and mastics where they function as the acid catalysts (i.e., by producing phosphoric acid upon decomposition). These compounds are often used in paints with pentaerythritol (or with a derivative of pentaerythritol) as the carbonific component and melamine as the spumific compound.[22] In addition, the intumescent formulations typically contain resinous binders, pigments, and other fillers. These systems are highly efficient in flame retarding hydroxylated polymers.

A series of compounded flame retardants, based on finally divided insoluble ammonium phosphate together with char-forming nitrogenous resins, has been developed for thermoplastics.[23] These compounds are particularly useful as intumescent flame retardant additives for polyolefins, ethylene-vinyl acetate, and urethane elastomers. The char-forming resin can be, for example, an ethyleneurea-formaldehyde condensation polymer, a hydroxyethyl isocyanurate, or a piperazine-triazine resin. Commercial leach-resistant flame-retardant treatments for wood have also been developed based on a reaction product of phosphoric acid with urea-formaldehyde and dicyandiamide resins.

5.4 ORGANOPHOSPHORUS ADDITIVE FLAME RETARDANTS

5.4.1 Phosphates and Phosphonates

There are several classes of amine phosphates commercially available to flame retard a wide variety of polymeric substrates, both natural and synthetic.[24] A classic example is the three variations of melamine phosphate: melamine orthophosphate, dimelamine orthophosphate, and melamine pyrophosphate. Of these, the pyrophosphate is the least soluble and the most thermally stable. Melamine orthophosphate is converted to the pyrophosphate upon heating, with the loss of water. All the aforementioned variations are available as finally divided solids, are used commercially in coatings, and have utility in a wide variety of thermoplastics and thermosets (mostly presented in the patent literature).

Phosphate esters (alkyl or aryl, or mixed) of phosphoric acid (Structure **5.1**) constitute an important family of organophosphorus flame retardants.[25] Triethylphosphate, a colorless liquid boiling between 209 and 218°C, and containing 17 wt% phosphorus, has been used commercially as an additive for polyester resins/laminates and in cellulosics. In polyester resins, it functions as a viscosity depressant as well as a flame retardant. Trioctylphosphate is employed as a specialty flame-retardant plasticizer for vinyl composites where low temperature flexibility is critical. It can be also included in blends, along with general purpose plasticizers, such as phthalate esters, to improve low temperature flexibility.

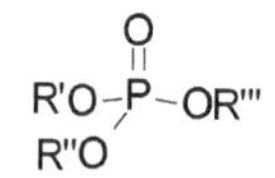

STRUCTURE 5.1 A phosphate ester (alkyl, aryl, or both) of phosphoric acid and a corresponding alcohol (R', R", and R"' are alkyl or aromatic substituents).

Aryl phosphates were introduced into commercial use early in the 20th century for flammable plastics such as cellulose nitrate and later for cellulose acetate.[26] In vinyls (plasticized), arylphosphates are frequently used with phthalate plasticizers. The principal applications are in wire and cable insulation, connectors, automotive interiors, vinyl moisture barriers, plastic greenhouses, furniture upholstery, and vinyl forms. Triarylphosphates are also used, on a large scale, as flame-retardant hydraulic fluids, lubricants, and as lubricant additives. Smaller amounts are used as nonflammable dispersing media for peroxide catalysts. Blends of triarylphosphates and pentabromodiphenyl oxide are extensively used as flame retardant additives for flexible urethane foams (however, pentabromodiphenyl oxide has been banned from use in the US, UK, and EU – see Chapter 23 of this book). It has been also used as a flame retardant additive for engineering thermoplastics such as polyphenylene oxide, high impact polystyrene, and ABS-polycarbonate blends.

Mixed esters, such as isopropylphenyl diphenyl phosphate and *tert*-butylphenyl diphenyl phosphate, are also widely used as both plasticizers/flame-retardants for engineering thermoplastics and hydraulic fluids.[11] These esters generally show slightly lower flame-retardant efficacy, when compared to triaryl counterparts; however, they have the added advantage of lower smoke production when burned. Some novel oligomeric phosphate flame retardants (based on tetraphenyl resorcinol diphosphate) are also employed to flame retard polyphenylene oxide blends, thermoplastic polyesters, polyamides, vinyls, and polycarbonates.

Dimethyl methylphosphonate, a water soluble liquid (boiling point: 185°C), made by the Arbuzov rearrangement of trimethyl phosphite, has a phosphorus content of 25 wt%, the highest possible for an alkyl phosphonate ester.[27] Applications of dimethyl methylphosphonate include the use as a viscosity depressant and flame retardant in alumina trihydrate filled polyester resins and as a flame suppressant for halogenated polyesters. It is also used in flame-retardant rigid polyurethane foams, and as a component in blends with triarylphosphates to make flame-retardant plasticizer for synthetic rubber and cellulosics. Diethyl ethylphosphonate, a higher member of this family of alkyl phosphonates, has a higher boiling point and is believed to be less susceptible to undesirable interactions with halogenated aliphatic components, such as blowing agents, or with amine catalysts.

5.4.2 Halogenated Phosphates and Phosphonates

In this important class of additives, the halogen contributes somewhat to the flame retardancy although this contribution is offset by the lower phosphorus content. The halogens generally reduce vapor pressure and water solubility, thus aiding retention of these additives. Thus, more efficient and effective blending/manufacturing processes involved usually lead to favorable economics of polymer flame retardation.[26]

Tris(2-choloroethyl)phosphate, tri(1-chloroethyl)phosphate, 1,3-dichloro-2-propanol phosphate, bis(2-chloroethyl) 2-chloroethylphosphonate are the cardinal members of this class of flame retardants and are quite compatible with polymers containing polar groups. These compounds have different viscosities, solubilities, hydrolytic stabilities, and boiling points. Their flame-retardant efficacies generally depend on the structural features as well as on the phosphorus-halogen ratios. Tris(2-choloroethyl)phosphate is widely used in rigid polyurethane and polyisocyanurate foams, most classes of thermosets, cast acrylics, and in wood-resin composites. Tri(1-chloroethyl)phosphate, owing to the presence of a branched alkyl group, has lower reactivity to water and bases than the 2-choloroethyl homologue. It is a preferred additive for rigid urethane foams where good storage stability in the isocyanate or the polyol-catalyst mixture is required.[25]

1,3-Dichloro-2-propanol phosphate has, compared to the mono-halogenated phosphates, high boiling point, much lower water solubility, and enhanced stability toward the amine catalyst used in the foam manufacture. Therefore, it is the leading additive for flexible polyurethane foams and can be added to the isocyanate or the polyol-catalyst mixture. This halogenated phosphate is also useful as a flame retardant in styrene-butadiene and acrylic lattices for textile back-coating and binding of nonwovens. Bis(2-chloroethyl) 2-chloroethylphosphonate, commercially available as a mixture of isomers, is not as stable as the corresponding phosphate; however, it is widely used as a flame retardant additive for rigid urethane foams, adhesives, and coatings.

5.4.3 Dimeric, Oligomeric, and Cyclic Phosphates and Phosphonates

Dimeric phosphates include several compounds amongst which three different 2-chloroethyl diphosphates have proved to be of commercial significance.[25] Generally, they have very low volatilities and reasonably good thermal stabilities, and thus are useful in those open cell (flexible) foams where there are requirements for improved resistance to dry and humid aging. The commercially available variants include tetrakis(2-chloroethyl) ethylenediphosphate and tetrakis(2-chloroethyl) ethyleneoxyethylenediphosphate (Structure 5.2). A third member of the family (Structure 5.3), having additional pendant chloromethylene groups in the bridging segment of the diphosphate, was introduced into the market later on, has enhanced hydrolytic stability, and is useful in many types of flexible foams, as well as in adhesives and in epoxy- or phenolic-based laminates.

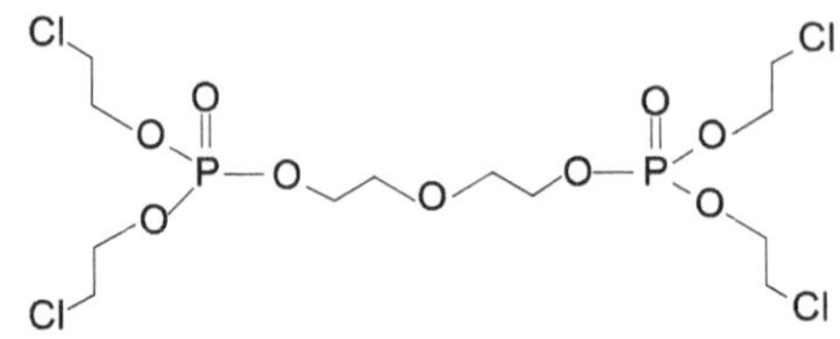

STRUCTURE 5.2 Tetrakis(2-chloroethyl) ethyleneoxyethylenediphosphate.

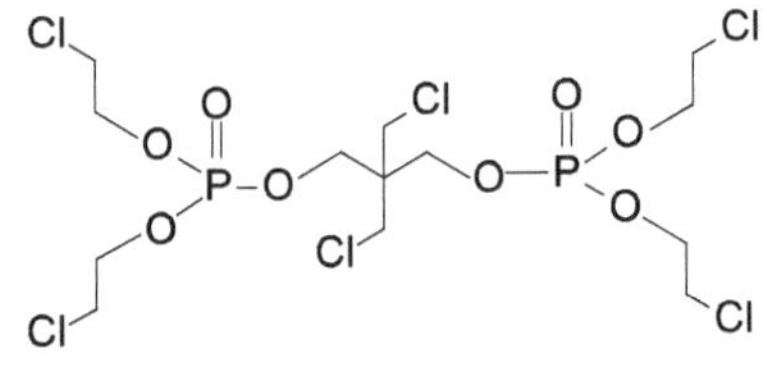

STRUCTURE 5.3 2,2-Bis(chloromethyl)trimethylene bis(bis(2-chloroethyl)phosphate).

Pentaerythritol phosphate has an excellent char-forming ability owing to the presence of the pentaerythritol structure. The bis-melamine salt of the bis-acid phosphate of pentaerythritol is also available commercially. This is a high melting solid that acts as an intumescent flame retardant additive for polyolefins. Synergistic combinations with ammonium polyphosphates have also been developed primarily for urethane elastomers. Self-condensation of tris(2-chloroethyl) phosphate produces oligomeric 2-chloroethylphosphate. It has a low volatility and is useful in resin-impregnated air filters, in flexible urethane foams, and in other structural foams.[11]

Cyclic oligomeric phosphonates with varying degrees of structural complexity (Structure 5.4) are also available in the market.[25] They are widely used as flame-retardant finishes for polyester fabrics. After the phosphonate is applied from an aqueous solution, the fabric is heated to swell and soften the fibers, thus allowing the phosphonate to be absorbed and strongly held. It is also a useful retardant in polyester resins, polyurethanes, polycarbonates, polyamide-6, and in textile back-coatings. A bicyclic pentaerythritol phosphate has been introduced into the market for the use in thermosets as well as for polyolefins (preferably, in combination with melamine or ammonium polyphosphate) (Structure 5.5). Cyclic neopentyl thiophosphoric anhydride (Structure 5.6), a solid additive, is used to flame retard viscose rayon, especially in Europe. In spite of the anhydride structure, it is remarkably stable, surviving addition to the highly alkaline viscose, the acidic

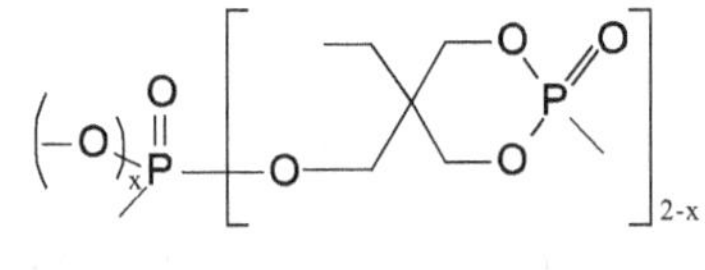

STRUCTURE 5.4 A cyclic oligomeric phosphonate.

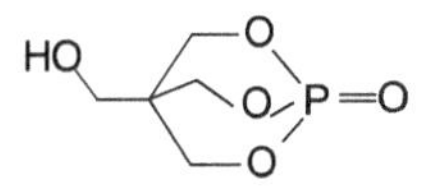

STRUCTURE 5.5 A bicyclic pentaerythritol phosphate.

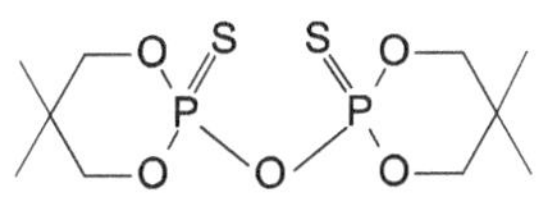

STRUCTURE 5.6 A neopentyl thiophosphoric anhydride.

coagulating bath, and also resisting the multiple launderings of the rayon fabric.

In the context of the use of polyphosphonates and their copolymer variants, especially for flame-retarding condensation polymers, it is worth mentioning several commercially available and highly effective polymeric additives from FRX Innovations, traded under the brand name of Nofia. Generally, these homopolymeric components are melt-processable and possess relatively high melt-flow indices. Given that these additives consist of molecular chains with sufficiently high degrees of polymerization, they are less likely to leach out during the service life of the finished products but at the same time impart an acceptable degree of flame retardance. So far, Nofia additive systems are extensively used for flame-retarding polycarbonates, polyurethanes, and several classes of polyesters—the base polymers for making textiles, carpets, transparent films, wires, cables, etc.

5.4.4 Phosphine Oxides

Phosphine oxides have hydrolytically stable P-C bonds, whereas the P-O-C bonds of phosphate esters hydrolyze more readily.[26] They also generally have higher phosphorus contents than the corresponding aromatic phosphate esters and, therefore, are more effective flame retardants. Triphenylphosphine oxide is disclosed in many patents as a flame retardant and may find some limited usage as such a vapor phase flame inhibitor. A phosphine oxide incorporating hydroxyl groups (Structure 5.7) has been proposed as a flame retardant for the use in polypropylene. In spite of its high hydroxyl content, it is a relatively stable, high melting solid additive.

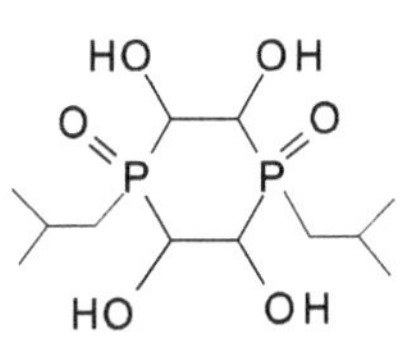

STRUCTURE 5.7 A phosphine oxide incorporating hydroxyl groups.

5.5 REACTIVE FLAME RETARDANTS

Reactive flame retardants, that is, those that are covalently attached to polymer chains, offer several advantages over those that are merely additives. For example, reactive flame retardants are inherently immobile within the polymer matrix and therefore will not be susceptible to loss during service through migration to the polymer surface (so-called blooming) or solvent leaching. Reactive flame retardants incorporated at the time of synthesis of the polymer also can be homogeneously dispersed throughout the polymer at the molecular level and because of this may be required at lower concentrations than comparable additives to achieve a desired level of flame retardance. The lower levels of incorporation of flame-retardant groups may bring with it an added advantage that the overall properties of the polymer (chemical, physical, and mechanical) are less likely to be adversely affected when compared with those of the non-flame-retarded counterpart. Moreover, reactive flame retardants are prevented from forming a separate phase within the polymer matrix; this is particularly important in uses of polymers as fibers in which a heterogeneous phase structure is likely to bring with it problems during fiber spinning and reductions in the modulus of the resultant fibers.

However, reactive incorporation of flame-retarding groups can also bring problems, not least the relative difficulty and expense of reactively modifying polymers for which commercially well-established methods for manufacturing the unmodified variants already exist. Furthermore, the extensive reactive modification of a partly crystalline polymer is likely to lead to a significant loss of crystallinity, whereas if an additive is introduced to a partly crystalline polymer, it will most probably end up in the amorphous phase and will have little impact upon crystallinity (although it may plasticize the amorphous regions). Moreover, in general, the reactive modification of chain-reaction polymers is less readily accomplished than for step-reaction polymers, unless the reactive modification is applied after the manufacture of the primary chains, say, through a post-polymerization grafting reaction. Grafting reactions may also be useful for covalently attaching flame-retardant groups to the surfaces of polymer-based plastics moldings, films, and fibers, where they will be particularly effective at the point of first impingement of any flame.

5.6 STEP-REACTION POLYMERS

5.6.1 Polyesters and Polycarbonates

The most commercially important linear polyester is poly(ethylene terephthalate) (PET), used extensively as a fiber and as a film. A P-containing comonomer can be introduced into PET either as a diacidic component (or its diester derivative) or as a diol. In commercial practice, the former strategy has been preferred. Thus, flame retardancy has been achieved in commercial PET fibers by incorporation of 2-carboxyethyl(methyl)phosphinic acid (Structure 5.8),

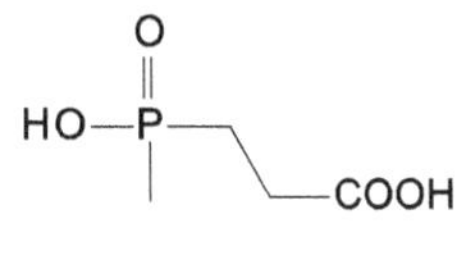

STRUCTURE 5.8 2-Carboxyethyl(methyl)phosphinic acid.

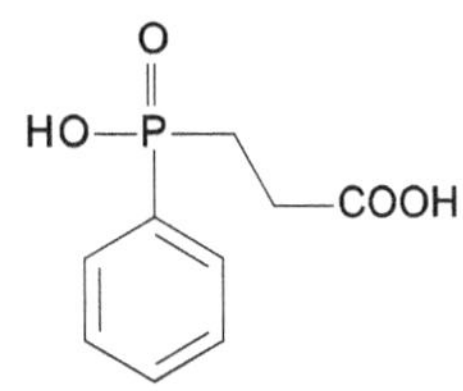

STRUCTURE 5.9 2-Carboxyethyl(phenyl)phosphinic acid.

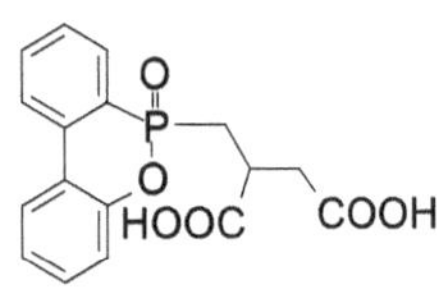

STRUCTURE 5.10 9,10-Dihydro-9-oxa-10-phosphaphenanthrenyl-10-oxide (DOPO) adduct of itaconic acid.

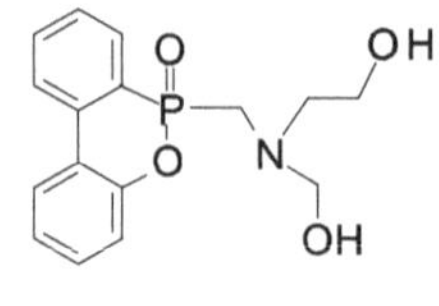

STRUCTURE 5.11 An alcoholamine derivative of DOPO.

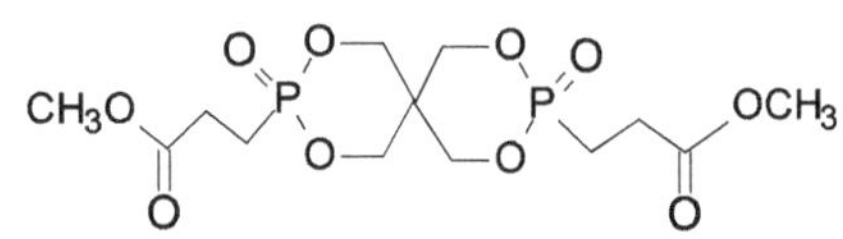

STRUCTURE 5.12 An alicyclic (spiro) bisphosphonate.

2-carboxyethyl(phenyl)phosphinic acid (Structure 5.9), or their cyclic anhydrides. PETs based on the former are marketed under the trade name Trevira CS. It has been suggested that fibers containing either Structure 5.8 or Structure 5.9 might be further improved by the additional incorporation of aromatic dicarboxylic acid monomers to act as charring agents.[28] Another commercially utilized P-containing comonomer is the 9,10-dihydro-9-oxa-10-phosphaphenanthrenyl-10-oxide (DOPO) adduct of itaconic acid[29] (Structure 5.10). Filament fibers and fabrics based on PET copolymers containing this flame-retardant comonomer are commercially available from Toyobo under the trade name HEIM, and have Limiting Oxygen Index (LOI) of up to 28.

It has been reported that the effectiveness of copolymerized DOPO-type monomers can be further improved if alcoholamine derivatives of DOPO, for example, Structure 5.11, are used rather than similar structures not containing nitrogen.[30] Of the FR fibers based on P-containing comonomers, it has been found that those based on Structure 5.10 are more hydrolytically stable, presumably because the P-containing group is in a cyclic structure, and also should the hydrolysis of the P-O bond occur, it will not lead automatically to a marked reduction in molecular weight.[31] All the P-modified PETs appear to be subject to both vapor phase and condensed phase mechanisms of flame retardance, with the former predominating.[32,33]

Other reactive P-containing comonomers reported to improve flame retardance in PET include aliphatic and alicyclic (spiro) bisphosphonates[34] such as Structure 5.12.

Poly(butylene terephthalate) (PBT) and poly(ethylene-2,6-naphthenate) (PEN) can be reactively flame retarded with the same types of reagents[35] as are used for PET, for example, Structure 5.10.

There have been some reports also of polycarbonates reactively modified with P-containing groups. For example, a novel P-containing copolycarbonate was prepared via melt polycondensation of diphenyl carbonate, bisphenol A, and 2-(6-oxo-6H-dibenz[c,e][1,2]oxaphosphorin-6-yl)-1,4-benzenediol. This polycarbonate exhibited a significantly better flame retardancy, a higher char yield, a higher degradation temperature, and a better thermal stability than conventional bisphenol A polycarbonate.[36]

5.6.2 Polyamides

Reactive strategies, that is, chemical modifications, appear not to have been explored to a large extent as a route to improve flame retardance in aliphatic polyamides, probably, because the chemical modification of aliphatic polyamides disrupts intermolecular hydrogen bonding and hence the crystallinity, thus reducing melting points. Thus, although the reactive incorporation of bis(4-carboxyphenyl) phenylphosphine oxide (Structure 5.13) at up to 20 mol% into polyamide-6,6 gives flame retardant polymers that are still partly crystalline, at 30 mol% incorporation, all crystallinity is lost.[37] Similar results have been observed with incorporated diphenylmethylphosphine oxide groups.[38]

Aromatic polyamides modified with phosphine oxide groups have been synthesized also by condensation of bis(4-carboxyphenyl)phenylphosphine oxide but with various aromatic diamines.[39,40]

As yet, it is not clear that there has been any successful commercial production of flame-retarded polyamides

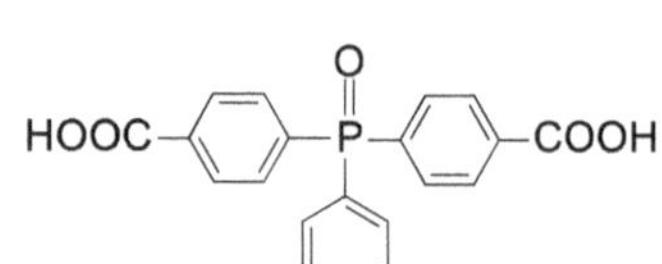

STRUCTURE 5.13 Bis(4-carboxyphenyl)phenylphosphine oxide.

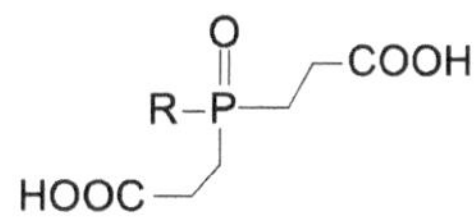

STRUCTURE 5.14 3,3'-(Alkylphosphoryl)dipropanoic acid (R is alkyl group).

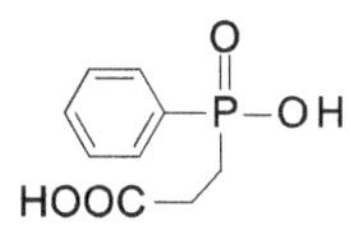

STRUCTURE 5.15 3-[Hydroxy(phenyl)phosphoryl]propanoic acid.

incorporating P-containing reactives, although both Monsanto and Solutia have patented P-containing diacids (Structures 5.14 and 5.15, respectively) designed to replace some of the adipic acid in the manufacture of flame-retardant polyamide-6,6 for fibers.[41,42]

5.7 CHAIN-REACTION POLYMERS

The most obvious way of reactively flame retarding a chain-reaction polymer is by incorporating a P-containing comonomer during polymerization. However, there have been no reports, to date, of significant commercial progress on this front, although a variety of chain-reaction polymers have been reactively modified with P-containing groups on the laboratory scale and have been shown to be significantly flame retarded when compared with their unmodified counterparts. Some of these are discussed below.

5.7.1 Acrylics

A variety of P-containing acrylates, methacrylates (Structure 5.16) and styrene derivatives (Structure 5.17) have been synthesized and copolymerized using free-radical initiators with methyl methacrylate (MMA)[43–45] and acrylonitrile (AN).[46,47] Copolymers of MMA with diethyl(methacryloyloxy)methyl phosphonate (DEMMP, Structure 5.16 with R = C_2H_5 and R' = CH_3), for example, have higher LOIs, the lower rates of heat release, the lower overall heats of combustion, and give a V-0 rating in a UL94 protocol.[44] Other P-containing comonomers demonstrate similar improvements. Moreover, these copolymers have transparencies comparable with those of PMMA, are hydrolytically reasonably stable, and have physical and mechanical properties not significantly inferior to those of PMMA. Studies of the thermal degradation and combustion of some of these reactively modified PMMAs indicate that flame retardance involves both the vapor and condensed phase activity, with the latter leading to significant char production. The char production, probably, involves transesterification reactions and interchain cyclizations catalyzed by phosphonic acid groups liberated during combustion. The reactive route to flame retardance has been shown

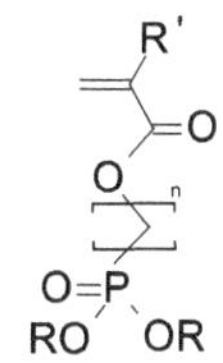

STRUCTURE 5.16 Dialkyl(acryloyloxy) (when R' = H) or dialkyl(methacryloyloxy) (when R' = CH_3) -alkylphosphonate (R is alkyl group).

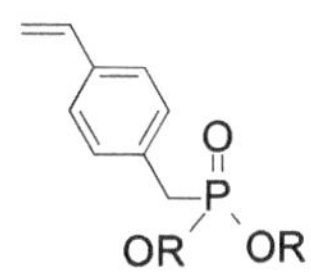

STRUCTURE 5.17 Dialkyl-*p*-vinylbenzyl phosphonate (R is alkyl group).

also to offer significant advantages over routes using analogous P-containing additives, in that additives tend to plasticize PMMA significantly and to give rise only to vapor phase modes of flame retardance.[5,27,43,45]

The incorporation of P-containing comonomers such as diethyl-*p*-vinylbenzene phosphonate (Structure 5.17 with R = C_2H_5) and acrylate and methacrylate phosphates and phosphonates (Structure 5.16) into polyacrylonitrile (PAN) also greatly increases char production during combustion. In this case, the P-containing comonomer appears to act by initiating the intramolecular cyclization of -C≡N groups along the PAN chain, leading to the well-known precursor structure to a graphitic char.[46,47]

5.7.2 Polystyrene

Monomers such as Structure 5.16 and Structure 5.17 copolymerize, free radically, as readily with styrene as they do with MMA and give rise to products that are also significantly flame retarded. Flame retardance here too involves both vapor and condensed phase action, with the condensed phase activity leading to a significant charring. In this case, char production is probably initiated by the Friedel-Crafts type condensation of phenyl groups, catalyzed by phosphorus acids liberated during combustion.[43,48,49]

5.7.3 Polyolefins

The reactive modification of polyolefins via copolymerization with P-containing monomers is not straightforward given the stringent conditions that pertain during the polymerization of olefin monomers, especially using organometallic coordination catalysts. However, both polyethylene (PE) and polypropylene (PP) can be chemically modified after polymerization so as to introduce P-containing side-groups. For example, PE has been oxidatively phosphonylated to give a polymer containing about 5 wt% P in -P(O)$(OCH_3)_2$ side groups. This polymer has an LOI significantly greater than that of unmodified PE, but unfortunately, the crystallinity is greatly reduced leading to the inferior physical and mechanical properties.[50]

PP is commercially modified by radical grafting with maleic anhydride to give maleated PPs used as compatibilizers for additives such as nanoclays. Similar strategies can be used to graft P-containing monomers to PPs. These copolymers, when blended with virgin PP, also allow compatibilization with nanoclays but in addition work synergistically with the clays to improve flame retardance.[51]

5.8 THERMOSETS

In thermosetting systems, the reactive flame retardant can be incorporated either in one or more of the principal chain-forming components, or in the crosslinking agent. Both strategies have been employed with P-containing flame retardants in a variety of thermosets.

5.8.1 Epoxy Resins

The major developments in reactive phosphorus-based flame retardants for epoxy resins to 2005 have been well reviewed.[52] It will suffice here, therefore, to outline just the major developments and to highlight the most recent work.

Since the P-H bond (like N-H, O-H and S-H) can add across an epoxy group with consequent ring-opening, hydrogen phosphonates, or phosphinates, can readily be incorporated covalently into epoxy resins. This reaction is currently widely used to incorporate DOPO (see Section 5.6.1) into epoxies destined for electronic applications.[53–56] Reacting DOPO first with reagents such as *p*-benzoquinone or naphthaquinone gives (after mild reduction) diphenolic derivatives (e.g., DOPO-HQ) that can be incorporated in an epoxy resin through a conventional chain-extension process with no net loss of functionality.[57] The itaconic derivative of DOPO (Structure 5.10) used in the manufacture of flame-retardant polyesters can be cured also into epoxies,[58,59] as can dicyanate ester,[60] tris(4-aminophenyl) methyl,[61] succinic anhydride,[62] epoxy silane,[63] and oligomeric tricyanurate derivatives.[64]

The major commercial alternative to DOPO and its derivatives for reactively flame retarding epoxy resins, especially for electronic applications, is an oligo(1,3-phenylene methylphosphonate) (Structure 5.18) marketed by Supresta under the tradename Fyrol PMP. It has a functionality higher than that of DOPO and can be used either as a hardener (a curing agent) or be precondensed with an epoxy.

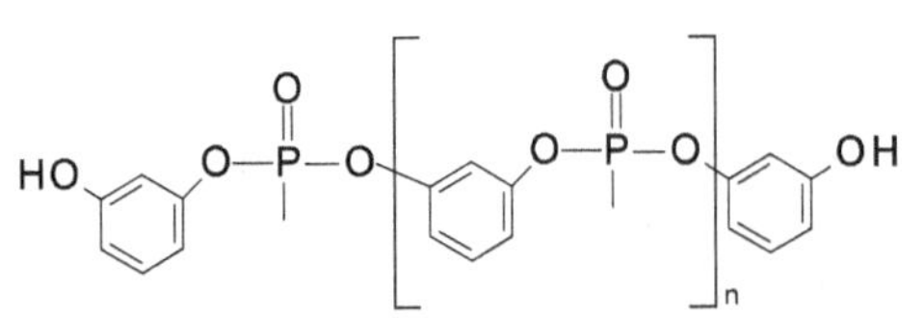

STRUCTURE 5.18 An oligo(1,3-phenylene methylphosphonate).

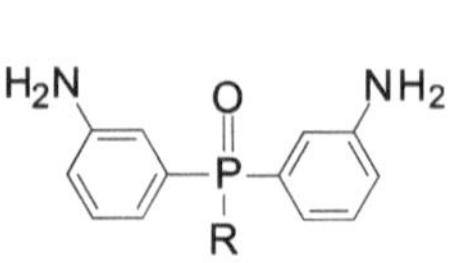

STRUCTURE 5.19 Bis(aminophenyl)alkyl- or bis(aminophenyl) aryl phosphine oxides (R is alkyl or aryl group).

Amino and functional aromatic phosphates, phosphonates, and phosphine oxides have also been used as reactive components to impart flame retardancy to cured epoxy resins. Phosphine oxides are particularly hydrolytically stable, and several studies have been reported, for example, of the curing of epoxy resins with bis(aminophenyl) alkyl and aryl phosphine oxides[65–67] (Structure 5.19). The relative performances of epoxies cured with aromatic diamine hardeners containing phosphine oxide, phosphinate, phosphonate, or phosphate units have recently been assessed.[68] Aromatic phosphine oxides have been functionalized also with maleimide and hydroxyl groups for the use as epoxy resin hardeners.[69,70]

5.8.2 Polyurethanes

Organophosphorus compounds, monomeric and oligomeric, additive and reactive, have a history of exploration, development, and use as flame retardants and anti-scorch agents in polyurethanes (PUs) dating back to the 1960s. This history, together with pertinent details of mechanisms of thermal degradation and combustion of polyurethanes, has been comprehensively reviewed.[71,72] It will thus suffice here merely to highlight the most important strategies that have been suggested and employed for reactively flame-retarding PUs with organophosphorus compounds.

A common approach with segmented PUs has been to incorporate a P-containing group (phosphate, phosphonate, phosphine oxide, or phosphazene) into a diol, which is then chain-extended with a diisocyanate (Structures 5.20 and 5.21).[73,74] In one such study,[75] phosphonate was found to be a more effective flame retardant than phosphate at

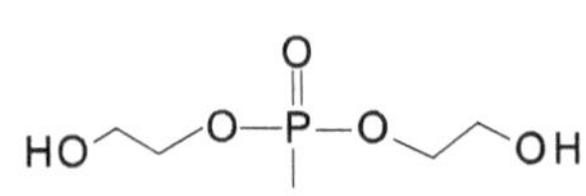

STRUCTURE 5.20 Bis(2-hydroxyethyl) methyl phosphonate.

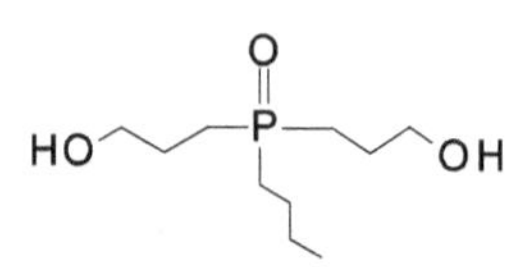

STRUCTURE 5.21 Bis(2-hydroxypropyl) *n*-butyl phosphonate.

the same phosphorus concentration and, in another study,[76] in-chain phosphorus to be more effective than phosphorus within a side chain. The UV curable PUs containing reactively incorporated phosphorus have been synthesized from acrylate-ended species.[77] The P-containing species can also be incorporated in PUs in curing. For example, P-containing aziridines have been suggested as curing agents for carboxylated PU aqueous dispersions used as surface coatings.[78–80]

Some of the most critical applications of PUs are as foams for which flammability can be particularly high owing to their high surface areas and, in the case of flexible foams, their open cell structures. Commercially, such foams are still flame retarded mainly with organophosphorus additives, albeit sometimes of high molecular weight. The use of high molecular weight additives is particularly important for the foams used in automotive applications where the use of more volatile, low-molecular weight additives can lead to excessive fogging of glass surfaces. Reactive organophosphorus flame retardants such as P-containing diols (e.g., Structure 5.20) have clear potential here.

5.8.3 Unsaturated Polyesters

Clearly, P-containing unsaturated polyesters can be readily synthesized using many of the same P-containing diols and diacids used for constructing P-containing saturated polyesters and polyurethanes. However, such reagents have not found commercial favor for the use in unsaturated polyester resins, either as components of the main polyester chain or as components of the cross-linking vinyl monomer system. This has been ascribed to the general poor hydrolytic stability of aliphatic polyesters based on P-containing diols and diacids and on the poor free radical copolymerizability of readily available P-containing cross-linking monomers, such as vinyl phosphonates.[81] Nevertheless, there are a number of P-containing acrylates, methacrylates, and styrenes now available, at least on a laboratory scale (for example, Structure 5.16 and 5.17), that copolymerize more readily than vinyl phosphonates, so this is an area that might repay further work.

5.9 SURFACE GRAFTING

Given that pyrolysis, ignition, and combustion affect first the surface of a polymeric material, it is sensible to ask whether or not flame-retardant treatments might optimally be applied to the surfaces of materials, especially when surface to volume ratios are high, for example, films, fibers, and textiles. A number of interesting studies have been reported in this regard in which, for example, radical graft-copolymerizations have been used to attach P-containing monomer units to the surfaces of cotton fabrics to impart flame retardancy.[82,83] Cotton surfaces have also been grafted with P-containing monomers by plasma polymerization as have the surfaces of PAN fabrics.[84,85]

5.10 MECHANISMS OF FLAME RETARDANCE: GENERAL CONSIDERATIONS

When subjected for a sufficient length of time to an external heat source, organic polymeric materials undergo thermal degradation, generating various products in varying concentrations over different temperature ranges. The chemical steps leading to the formation of volatiles may be homolytic or heterolytic, that is, be radical or ionic.[12] The three overall processes implicated in the thermal degradation of most thermoplastic polymers are as follows:

1. A random chain cleavage followed by chain unzipping is characterized by high monomer yields and by a slow decrease in the molecular weight of the polymer, for example, exhibited by PMMA, polystyrene, poly(α-methyl styrene), and polytetrafluoroethylene.
2. A random chain cleavage followed by further chain scission is characterized by very low monomer yields amongst the degradation products and a rapid drop in molecular weight, for example, exhibited by PE, PP, PMMA, and polychlorotrifluoroethylene.
3. An intra-chain chemical reaction followed by a cross-linking reaction and formation of a carbonaceous residue, or a random chain cleavage. This generates a relatively high yield of volatiles from the intra-chain reaction, but emits little monomer, and produces no, or only a very slight, reduction in molecular weight during the initial stages of degradation, for example, exhibited by poly(vinyl chloride), poly(vinyl alcohol), and PAN.

In some cases, several of these processes occur simultaneously, depending on the sample size, the heating rate, the pyrolysis temperature, the environment, and the presence of any additives. Although polymer degradation schemes can be greatly altered by the presence of comonomers, side-chain substituents, and other chemical constituent factors, the ultimate thermal stability is determined by the relative strengths of the main-chain bonds. Many additives and comonomers employed as flame retardants are thermally labile, and as a result the thermal stability of the polymer system is reduced. In order to reduce the observed effects of the flame retardant additives on the thermal stability of the polymeric materials, more thermally stable and hence inherently flame-resistant polymers are of increasing interest.

Thermosets, on the other hand, owing to their highly cross-linked three-dimensional nature, have a much lesser propensity to thermally degrade (especially at lower temperatures) to yield flammable volatiles. The rupture of bonds does not initially generate combustible gases, and carbonization is usually promoted. Ladder polymers, in which the main chains are bonded together at each repeat unit by a cross-link, show similar behaviors. Polymers with relatively strong main chain bonds and/or with aromatic and heterocylic structural units are also inherently thermally stable.[86] There are several classes of polymers, such as polyphenylenes, poly(*p*-phenylene oxide)s, polybenzimidazoles, polybenzamides, that have relatively high thermal decomposition temperatures coupled to low levels of fuel production on degradation.

The ability to form char is related to the flammability of a polymer. The higher the amount of residual char after combustion, the lower the amount of combustible material available to perpetuate the flame and the greater is the degree of flame retardance of the material.[87] Therefore, one of the ways to achieve high degrees of flame retardancy or non-combustibility of a polymeric material is to increase the amount of char produced on combustion. This is illustrated by the fact that aromatic polymers, for example, polycarbonate and poly(phenylene oxide), are less flammable than purely aliphatic polymers. The greater thermal stability of cross-linked and aromatic structures within thermosets gives rise to a greater degree of condensation into an aromatic char, and therefore only relatively low levels of flammable gases are available to feed the flame.

5.10.1 Mechanisms of Flame Retardance by Phosphorus Compounds

Phosphorus-containing fire retardants can be active in the condensed phase, or in the vapor phase, or in both phases.[88] The relative predominance of the different mechanisms actually operating in a particular system depends on the structural features of the base polymer as well as on the chemical environment of the phosphorus, that is, whether phosphorus is in the elemental state or, if present as a compound, the valency and the nature of the chemical moieties surrounding the P atoms. Physical and chemical actions have also been implicated in both phases. These could involve flame inhibition, heat loss owing to melt flow, surface protection by phosphorus-containing acids, acid catalyzed char accumulation, char enhancement, and protection of char from oxidation. Flame-retardant mechanisms can also be influenced by the presence of other additives, for example, the synergistic combinations of co-additives such as compounds of nitrogen or halogens.

A predominantly vapor phase mechanism of flame retardation for the flame retardants based on triphenylphosphine oxide, or triphenyl phosphate, has been proposed (Scheme 5.1). The salient features of these reactions are the promotion of hydrogen recombination and the scavenging of oxygen radicals by molecular phosphorus. This, in turn, will reduce the number of effective flame propagating radical species below the level at which the flame can be sustained.[26]

A classical and thoroughly investigated example, which amply illustrates the condensed- phase activity of phosphorus flame retardants, is the flame retardation of cellulosic substrates with phosphoric acid and/or its organic/inorganic

$$R_3PO \longrightarrow PO + P\cdot + P_2$$
$$H\cdot + PO\cdot + M \longrightarrow HPO + M$$
$$HO\cdot + PO\cdot \longrightarrow HPO + \cdot O\cdot$$
$$HPO + H\cdot \longrightarrow H_2 + PO\cdot$$
$$P_2 + \cdot O\cdot \longrightarrow P\cdot + PO\cdot$$
$$P\cdot + HO\cdot \longrightarrow PO\cdot + H\cdot$$

SCHEME 5.1 Elementary steps involved in vapor phase flame retardation by triphenylphosphine oxide (M is a third body species; R is aryl group).

esters.[89] In the presence of phosphoric acid and/or its organic and inorganic esters, or compounds having P-O-C bonds, phosphorylation of cellulose occurs predominantly at the C-6 hydroxyls. This, in turn, alters the degradation pathway of cellulose at the C-6 hydroxyls, where instead of the depolymerization reactions, the dehydration of the cellulose substrate occurs, thus, leading to an enhanced char production. Similar explanations can be given for the P-N synergism in cellulosics. The polar nature of P-N bonds, as compared with P-C and P-O bonds, would result in the enhanced rates of phosphorylation at the cellulose hydroxyls. Thus, the formation of P-N bonds in situ increases the reactivity of the phosphorus compound toward the cellulose hydroxyls, leading to a reduction in volatiles and a corresponding increase in the char formation.

Another mode of action in which phosphorus is important as a char promoter is in intumescent fire-retardant paints and mastics.[90] These typically have a phosphorus compound, such as ammonium phosphate; a char forming polyol, such as pentaerythritol; a blowing agent, such as melamine; and a binder. On decomposition, the phosphorus compound provides the phosphorylating agent that in turn reacts with the pentaerythritol to form polyol phosphates, which then break down to char through a series of elimination steps. The spumific agent will endothermally decompose to form gaseous products, which will expand the char already formed.

Phosphorylated poly(vinyl alcohol) and ethylene-vinyl alcohol copolymers were found to be significantly flame retarded as compared to their unmodified parent polymers.[50] The flame retardance in these systems is shown to arise primarily from a condensed phase mechanism involving dehydration, cross-linking, and char formation (Scheme 5.2). The authors have also reported that phosphonylation of PE, achieved through the reaction of molecular oxygen with a solution of the polymer in PCl_3, resulted in significant improvements in the flame retardance. However, higher degrees of phosphonylation of PE (above 10 mol% or above) resulted in the reduced crystallinity, and as a consequence, its physical and mechanical properties were substantially impaired. The flame retardance of the modified PE is believed to arise, at least in part, through a vapor phase mechanism.

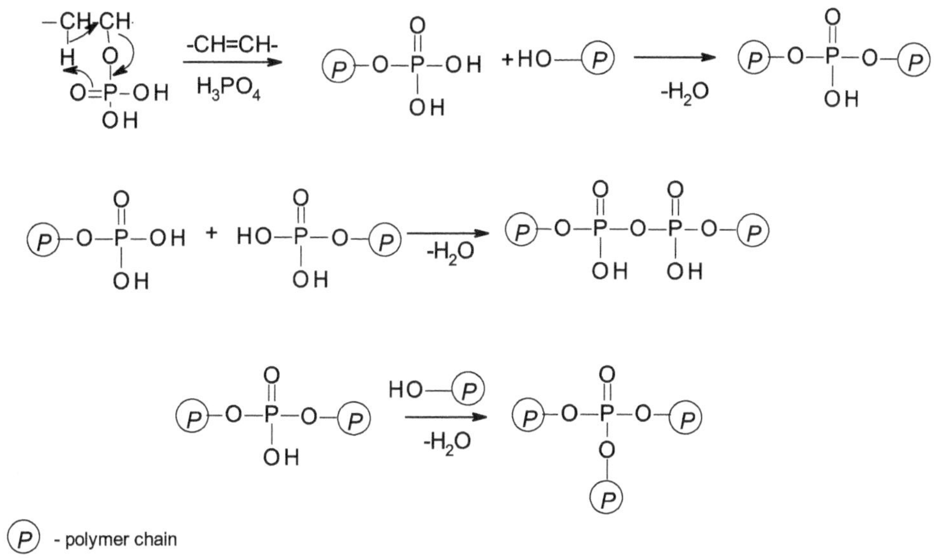

SCHEME 5.2 A formation of phosphorus anhydride cross-links in phosphorylated poly(vinyl alcohol) and ethylene-vinyl alcohol copolymers.

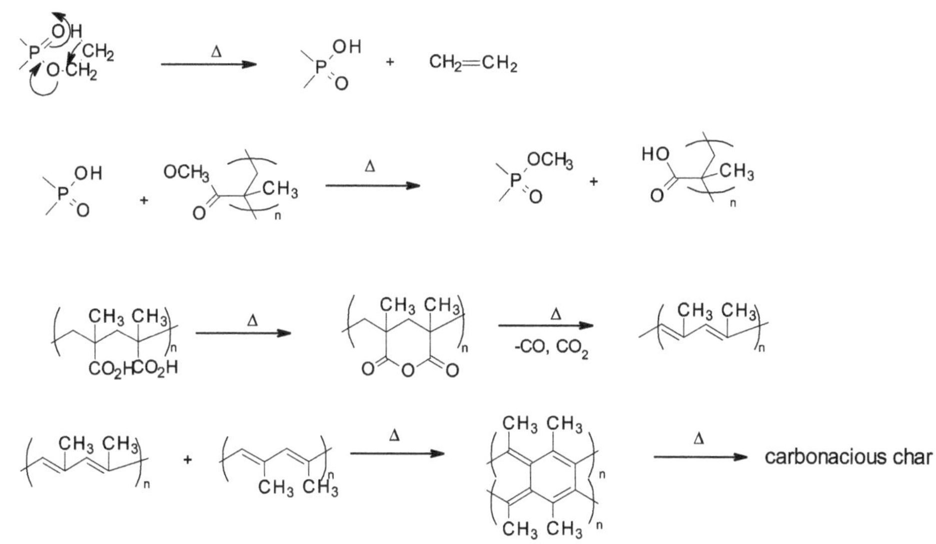

SCHEME 5.3 A suggested mechanism of char formation in methacrylate-phosphonate ester copolymer.

A systematic exploration of the effects of incorporating various phosphorus-containing comonomer units on the flame retardance, particularly of polystyrene and PMMA, has been reported.[6] It is found that the introduction of phosphorus, irrespective of its chemical environment within the comonomer unit, increases the flame retardance to a significant degree and also leads to an enhanced production of char. Phosphorus retention in the char residues of the modified systems was also found to be significant, suggesting a predominant element of condensed phase activity of the phosphorus-containing groups. In addition, copolymers of acrylamide with phosphorus-containing comonomers appear to display considerable phosphorus-nitrogen synergism.[91]

The mechanistic studies of thermal degradation and flame retardance in some copolymers of methyl methacrylate containing pendant phosphonate ester groups were performed using analyses of gaseous products of degradation and through spectroscopic examinations of solid residues.[43] Flame retardance in these systems is believed to occur by a combination of vapor phase processes and condensed phase activity, that is, char formation. The condensed phase mechanism is believed to involve the generation of phosphonic acid species and their subsequent reaction with the methyl methacrylate units to give methacrylic anhydride units, which then decarboxylate en route to forming char (Scheme 5.3).

Condensed phase activity of phosphorus-containing covalently bound groups is also reported for chemically modified PAN polymers.[47,92] It is shown that, in these systems, the intramolecular cyclization of the pendant nitrile groups is greatly favored, which in turn decreases the production of combustible volatiles including the monomer in the initial stages of the thermal degradation of the copolymer. The mechanistic pathway involves the thermal cracking of the pendant phosphorus groups to produce phosphorus acid species. These species later act as nucleophilic centers, aiding the cyclization processes (Scheme 5.4). Thus, the condensed phase activity of the modifying groups in PAN, in enhancing the extent of intramolecular cyclization, leads to the increased production of char.

The predominant mode of action of phosphorus-containing flame retardants (both additives and reactives), when present in thermoplastics or thermosets, is considered to be in the condensed phase. Generally, as with cellulose, flammable gas generation is reduced, and char formation is promoted. In some cases, the char cohesiveness is also enhanced. The retention of phosphorus in the chars in these systems is also generally high. Furthermore, vapor phase inhibitory mechanisms alone cannot account for the very high flame-retardant efficiencies observed, even in the polymers with nominal char formation tendency.[48] For example, in polystyrenes, phosphorus compounds can affect the burning behavior by altering the gas generation rate.[49]

The mechanistic details of flame-retardant synergism in phosphorus-halogen have not been as well studied as in phosphorus-nitrogen systems.[15] In the case of halogen-containing organophosphorus fire retardants, formation of phosphorus halides and oxyhalides is implicated. Other proposed mechanisms involve nominal condensed phase activity of halogens coupled to an enhanced vapor phase activity for the phosphorus in the presence of the halogen compounds. In the absence of any clearly established mechanistic pathways, it is sensible to assume that halogens and phosphorus will generally act more or less independently of each other when present together.

Physical effects such as those involving an increased heat capacity, the heat of vaporization, and an endothermic dissociation in the vapor phase may make a nominal, but recognizable, contribution to the overall efficacies of various phosphorus-based flame-retardant systems.[3] These types of mode of action, especially in the vapor phase, become more important in the case of readily volatile additives. An example is the case of tris(dichloroisopropyl) phosphate, which is used to flame retard flexible urethane foam. Phosphorus acid species, particularly phosphoric acid and its oligomers, can form a protective layer over a decomposing polymer matrix; this essentially acts as an oxygen-excluding physical barrier, thus protecting the underlying unburnt polymer.

5.11 CONCLUDING REMARKS AND FUTURE TRENDS

It has been seen that a wide variety of phosphorus compounds have been explored, both as additives and reactive components, for the flame retardation of a wide range of polymers, both thermoplastics and thermosets. However, the commercial exploitation of phosphorus-based flame-retardant systems is still somewhat in its infancy, with other classes of flame retardants still dominating the market. Largely this is a consequence of the generally higher cost of phosphorus-based materials, especially organophosphorus compounds, and the lower flame-retardant efficiency of many. As the environmental pressures to reduce the use of halogenated organic flame retardants increase, as they almost certainly will, it is likely that the commercial exploitation of phosphorus-containing alternatives will grow, bringing with it economies of scale and reduced costs. In this context, it is hardly surprising that the attention in recent years has turned to bio-based and more environmentally friendly versions.[93,94] The added impetus in such a quest has also stemmed from the fact that some toxic FRs have leached out into our potable water table and have also already found their way into the food chain, this posing serious health issues. However, in the interim period, the use of phosphorus compounds in combination with established halogenated flame retardants is likely to increase, leading at least to some reductions in overall use of the latter.[95,96]

The use of phosphorus-based flame retardants in combination with other, better established, flame retardants is most effective in situations in which the combination proves synergistic. However, our understanding of such synergistic effects is far from complete, and more fundamental work is required in this area; the work in which the gaseous and solid products of combustion, with and without

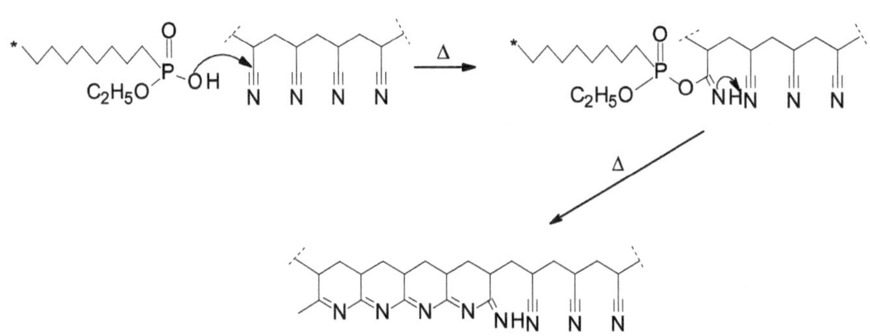

SCHEME 5.4 An intramolecular cyclization of nitrile groups of PAN initiated by nucleophilic phosphorus-containing species.

flame retardants present, are to be carefully analyzed. Such analyses can now be undertaken more readily than in the past, owing to the relatively recent development of hyphenated techniques such as gas phase FT-infrared spectroscopy and laser-pyrolysis time-of-flight mass spectrometry for the identification of volatiles, and solid-state NMR spectroscopy and X-ray photoelectron spectroscopy for the analysis of chars.

Perhaps the most exciting developments are in the increasing use of various nanoscopic additives, such nanoclays and carbon nanotubes (see Chapter 12 of this book), in combination with both halogenated and phosphorus-containing flame retardants. In addition, we are likely to see the increased emphasis on topical treatments of polymeric materials with flame retardant chemicals (see Chapter 14 of this book). Such treatments, involving conventional chemical or physical processes, are already well-established in the textile arena but will be augmented by more technologically advanced processes, such as those involving plasma polymerizations. Surface treatments in which a flame-retardant barrier layer is laid down by plasma polymerization could be particularly effective for polymeric materials having high surface-to-volume ratios, such as fibers and films. As mentioned earlier, the exploration of bio-based coating materials for protecting lignocellulosic substrates has also gained an added momentum in recent years given the frequency and intensity of wildland, or bush, fires.[97–100]

5.12 ACKNOWLEDGMENTS

We express our sincere gratitude to our colleagues, past and present, who have contributed to the research outputs from our group. They include, Professors John R. Ebdon, Richard Horrocks, T Richard Hull, and Baljinder K. Kandola; Drs Paul Wyman, Mavis Banks, Christopher Konkel, Kelly Bullette, Vincent Crooke, Ananya Thomas, Malavika Arun; and Ms Aloshy Baby.

REFERENCES

1. Gilman, J. W. Flammability and thermal stability studies of polymer layered-silicate (clay) nanocomposites. *Appl. Clay Sci.* **1999**, *15*, 31–49. DOI: 10.1016/S0169-1317(99)00019-8
2. Moniruzzaman, M. and Winey, K. I. Polymer nanocomposites containing carbon nanotubes. *Macromolecules* **2006**, *39*, 5194–5204. DOI: 10.1021/ma060733p
3. Levchik, S. V. and Weil, E. D. A review of recent progress in phosphorus-based flame retardants. *J. Fire Sci.* **2006**, *24*, 345–364. DOI: 10.1177/0734904106068426
4. Lewin, M. and Weil, E. D. Mechanisms and modes of action in flame retardancy of polymers. In *Fire Retardant Materials*, Horrocks, A. R. and Price, D. (Eds.). **2000**, Woodhead Publishing Limited, Cambridge, 31–68.
5. Price, D., Bullett, K. J., Cunliffe, L. K., Hull, T. R., Milnes, G. J., Ebdon, J. R., Hunt, B. J. and Joseph, P. Cone calorimetry studies of polymer systems flame retarded by chemically bonded phosphorus. *Polym. Degrad. Stab.* **2005**, *88*, 74–79. DOI: 10.1016/j.polymdegradstab.2003.09.025
6. Joseph, P. and Tretsiakova-McNally, S. Reactive modifications of some chain- and step-growth polymers with phosphorus compounds: effects on flame retardance – a review. *Polym. Adv. Technol.* **2011**, *22*, 395–406. DOI: 10.1002/pat.1900
7. van der Veen, I. and de Boer, J. Phosphorus flame retardants: Properties, production, environmental occurrence, toxicity and analysis. *Chemosphere* **2012**, *88*, 1119–1153. DOI: 10.1016/j.chemosphere.2012.03.067
8. Shirley, M., Stec, A. A. and Hull, T. R. The effect of gas phase flame retardants on fire effluent toxicity. *Polym. Degrad. Stab.* **2014**, 106, 36–46. DOI: 10.1016/j.polymdegradstab.2013.09.013
9. Rossotti, H. *Diverse Atoms: Profiles of the Chemical Elements*, **1998**, Oxford University Press, Oxford.
10. Greenwood, N. N. and Earnshaw, A. *Chemistry of Elements*, **1984**, Pergamon Press, Oxford, 546–636.
11. Chen, L. and Wang, Y. A review on flame retardant technology in China. Part I: development of flame retardants. *Polym. Adv. Technol.* **2009**, *21*, 1–16. DOI: 10.1002/pat.1550
12. Joseph, P. and Ebdon, J. R. Recent developments in flame-retarding thermoplastics and thermosets. In *Fire Retardant Materials*, Horrocks, A. R. and Price, D. (Eds.). **2000**, Woodhead Publishing Limited, Cambridge, 220–263.
13. Ebdon, J. R., Hunt, B. J., Joseph, P. and Konkel, C. S. Flame-retarding thermoplastics: additive versus reactive approach. In *Speciality Polymer Additives: Principles and Applications*, Al-Malaika, S., Golvoy, A. and Wilkie, C. A. (Eds.). **2001**, Blackwell Science, Oxford, 231–257.
14. Kannan, P. and Kishore, K. Novel flame retardant polyphosphoramide esters, *Polymer* **1992**, *33*, 412–422.
15. Green, J. A phosphorus-bromine flame-retardant for engineering thermoplastics – a review. *J. Fire Sci.* **1994**, *12*, 38–408.
16. Puri, R. G. and Khanna, A. S. Intumescent coatings: a review on recent progress. *J. Coat. Technol. Res.* **2017**, *14*, 1–20. DOI: 10.1007/s11998-016-9815-3
17. Ribeiro, S. P., Cescon, L. S., Ribeiro, R. R., Landesmann, A., Estevão, L. R. and Nascimento, R. V. Effect of clay minerals structure on the polymer flame retardancy intumescent process. *Appl. Clay Sci.* **2018**, *161*, 301–309. DOI: 10.1016/J.CLAY.2018.04.037
18. Gibertini, E., Carosio, F., Aykanat, K., Accogli, A., Panzeri, G. and Magagnin, L. Silica-encapsulated red phosphorus for flame retardant treatment on textile. *Surf. Interfaces* **2021**, *25*, 101252. DOI:10.1016/j.surfin.2021.101252.
19. Levchik, S. V., Levchik, G. F., Balanovich, A. I., Camino, G. and Costa, L. Mechanistic study of combustion performance and thermal decomposition behaviour of Nylon 6 with added halogen-free fire retardants. *Polym. Degrad. Stab.* **1996**, *54*, 217–222.
20. Levchik, G. F., Vorobyova, S. A., Gorbarenko, V. V., Levchik, S. V. and Weil, E. D. Some mechanistic aspects of the fire retardant action of red phosphorus in aliphatic nylons. *J. Fire Sci.* **2000**, *18*, 172–182. DOI: 10.1177/073490410001800302
21. Li, S., Zhong, L., Huang, S., Wang, D., Zhang, F. and Zhang, G. A novel flame retardant with reactive ammonium phosphate groups and polymerizing ability for preparing durable flame retardant and stiff cotton fabric. *Polym. Degrad. Stab.* **2019**, *164*, 145–156. DOI: 10.1016/j.polymdegradstab.2019.04.009
22. Lim, K., Bee, S., Sin, L. T., Tee, T., Ratnam, C. T., Hui, D. and Rahmat, A. R. A review of application of ammonium polyphosphate as intumescent flame retardant in

thermoplastic composites. *Compos. B Eng.* **2016**, *84*, 155–174. DOI: 10.1016/j.compositesb.2015.08.066
23. Cavdar, A. D., Torun, S. B., Ertas, M. and Mengeloglu, F. Ammonium zeolite and ammonium phosphate applied as fire retardants for microcrystalline cellulose filled thermoplastic composites. *Fire Saf. J.* **2019**, *107*, 202–209. DOI: 10.1016/j.firesaf.2018.11.008
24. Yang, Q., Wang, J., Chen, X., Yang, S., Huo, S., Chen, Q., Guo, P., Wang, X., Liu, F., Chen, W., Song, P. and Wang, H. A phosphorus-containing tertiary amine hardener enabled flame retardant, heat resistant and mechanically strong yet tough epoxy resins. *Chem. Eng. J.* **2023**, *468*, 143811. DOI: 10.1016/j.cej.2023.143811
25. Weil, E. D. Flame retardants: phosphorus compounds. In *Kirk-Othmer Encyclopedia of Chemical Technology*, 4th edn., Vol. 10, **1993**, Wiley-Interscience, New York, 976–979.
26. Green, J. Phosphorus-containing flame retardants. In *Fire Retardancy of Polymeric Materials*, Grand, A. F. and Wilkie, C. A. (Eds.). **2000**, Marcel Dekker, New York, 147–170.
27. Price, D., Pyrah, K., Hull, T. R., Milnes, G. J., Ebdon, J. R., Hunt, B. J. and Joseph. P. Flame retardance of poly(methyl methacrylate) modified with phosphorus-containing compounds. *Polym. Degrad. Stab.* **2002**, *77*, 227–233.
28. Asrar, J., Berger, P. A. and Hurlbut, J. Synthesis and characterization of a fire-retardant polyester: copolymers of ethylene terephthalate and 2-carboxyethyl(phenylphosphinic) acid. *J. Polym. Sci. A Polym. Chem.* **1999**, *37*, 3119–3128.
29. Endo, S., Kashihara, T., Osako, A., Shizuki, T. and Ikegami, T. Phosphorus-containing compounds. US 4 127 590, 1978.
30. Kim, Y. C. Phosphorus-containing nitrogen compounds as flame retardants and synthetic resins containing them. US 4 742 088, 1988.
31. Sato, M., Endo, S., Araki, Y., Matsuoka, G., Gyobu, S. and Takeuchi, H. The flame-retardant polyester fiber: improvement of hydrolysis resistance. *J. Appl. Polym. Sci.* **2000**, *78*, 1134–1138. DOI: 10.1002/1097-4628(20001031)78:5%3C1134::AID-APP230%3E3.0.CO;2-5
32. Day, M., Suprunchuk, T. and Wiles, D. M. Combustion and pyrolysis of poly(ethylene terephthalate) II. Study of the gas-phase inhibition reactions of flame retardant systems. *J. Appl. Polym. Sci.* **1981**, *26*, 3085–3098.
33. Chang, S. J. and Chang, F. C. Synthesis and characterization of copolyesters containing the phosphorus linking pendent groups. *J. Appl. Polym. Sci.* **1999**, *72*, 109–122.
34. Murayama, K. and Kashihara, T. Flame resistant polyesters. US 4 086 208, 1978.
35. Wang, C. S. and Lin, C. H. Synthesis and properties of phosphorus-containing PEN and PBN copolyesters. *Polymer* **1999**, *40*, 747–757.
36. Wang, C. S. and Shieh, J. Y. Synthesis and flame retardancy of phosphorus containing polycarbonate. *J. Polym. Res. (Taiwan)* **1999**, *6*, 149–154.
37. Wan, I. Y., McGrath, J. E. and Kashiwagi, T. Triarylphosphine oxide containing nylon-6,6 copolymers. *ACS Symp. Ser.* **1995**, *599*, 29–40.
38. Wan, I. Y. and McGrath, J. E. Synthesis and characterization of diphenylmethylphosphine oxide containing nylon-6,6 copolymers. *Abstr. Pap. Am. Chem. Soc.* **1995**, *209*(2), 119.
39. Delaviz, Y., Gungor, A., McGrath, J. E. and Gibson, H. W. Soluble phosphine oxide containing aromatic polyamides. *Polymer* **1993**, *34*, 210–213.
40. Zhang, Y. H., Tebby, J. C. and Wheeler, J. W. Polyamides incorporating phosphine oxide groups. 1. Wholly aromatic polymers. *J. Polym. Sci. A Polym. Chem.* **1996**, *34*, 1561–1566.
41. Picket, O. A., Jr. and Stoddard, J. W. Phosphorus-containing copolyamides and fibers thereof. US 4 032 517, 1977.
42. Asrar, J. Polymer-bound non-halogen fire resistant compositions. US 5 750 603, 1988.
43. Ebdon, J. R., Hunt, B. J., Joseph, P., Konkel, C. S., Price, D., Pyrah, K., Hull, T. R., Milnes, G. J., Hill, S. B., Lindsay, C. I., McCluskey, J. and Robinson, I. Thermal degradation and flame retardance in copolymers of methyl methacrylate with diethyl(methacryloyloxymethyl)phosphonate. *Polym. Degrad. Stab.* **2000**, *70*, 425–436. DOI: 10.1016/S0141-3910(00)00138-5
44. Price, D., Pyrah, K., Hull, T. R., Milnes, G. J., Ebdon, J. R., Hunt, B. J., Joseph, P. and Konkel, C. S. Flame retarding poly(methyl methacrylate) with phosphorus-containing compounds: comparison of an additive with a reactive approach. *Polym. Degrad. Stab.* **2001**, *74*, 441–447. DOI: 10.1016/S0141-3910(01)00184-7
45. Price, D., Pyrah, K., Hull, T. R., Milnes, G. J., Wooley, W. D., Ebdon, J. R., Hunt, B. J. and Konkel, C. S. Ignition temperatures and pyrolysis of a flame-retardant methyl methacrylate copolymer containing diethyl(methacryloyloxymethyl)-phosphonate units. *Polym. Int.* **2000**, *49*, 1164–1168. DOI: 10.1002/1097-0126(200010)49:10%3C1164::AID-PI513%3E3.0.CO;2-M
46. Joseph, P. and Tretsiakova-McNally, S. Combustion behaviours of chemically modified polyacrylonitrile polymers containing phosphorylamino groups. *Polym. Degrad. Stab.* **2012**, 97, 2531–2535. DOI: 10.1016/j.polymdegradstab.2012.07.014
47. Tretsiakova-McNally, S. and Joseph, P. Pyrolysis combustion flow calorimetry studies on some reactively modified polymers. *Polymers* **2015**, *7*, 453–467. DOI: 10.3390/polym7030453
48. Arun, M., Bigger, S., Guerrieri, M., Joseph, P. and Tretsiakova-Mcnally, S. Thermal and calorimetric investigations of some phosphorus-modified chain growth polymers 1: polymethyl methacrylate. *Polymers* **2022**, *14*, 1447. DOI: 10.3390/polym14071447
49. Arun, M., Bigger, S., Guerrieri, M., Joseph, P. and Tretsiakova-McNally, S. Thermal and calorimetric investigations of some phosphorus-modified chain growth polymers 2: polystyrene. *Polymers* **2022**, *14*, 1520. DOI: 10.3390/polym14081520
50. Banks, M., Ebdon, J. R. and Johnson, M. Influence of covalently bound phosphorus-containing groups on the flammability of poly(vinyl alcohol), poly(ethylene-co-vinyl alcohol) and low-density polyethylene. *Polymer* **1993**, *34*, 4547–4556.
51. Tretsiakova-McNally, S., Arun, M., Guerrieri, M. and Joseph, P. An overview of some reactive routes to flame-retardant fibre-forming polymers: polypropylene and polyacrylonitrile. *Organics* **2023**, *4*, 386–402. DOI: 10.3390/org4030028
52. Levchik, S., Piotrowski, A., Weil, E. and Yao, Q. New developments in flame retardancy of epoxy resins. *Polym. Degrad. Stab.* **2005**, *88*, 57–62. DOI: 10.1016/j.polymdegradstab.2004.02.019
53. Wang, C. S. and Lin, C. H. Synthesis and properties of phosphorus containing epoxy resins by novel method. *J. Polym. Sci. A Polym. Chem.* **1999**, *37*, 3903–3909.

54. Wang, C. S. and Lin, C. H. Epoxy resin rendered flame retardant by reaction with 9,10-dihydro-9-oxa-10-phosphaphenanthrene-10-oxide. US 6 291 627, 2001.
55. Lin, C. H. and Wang, C. S. Novel phosphorus-containing epoxy resins part 1. Synthesis and properties. *Polymer* **2001**, *42*, 1869–1878. DOI: 10.1016/S0032-3861(00)00447-X
56. Lin, C. H. and Wang, C. S. Halogen-free FR-4 printed circuit board. *ACS Polym. Prepr.* **2002**, *43*(2), 908–909.
57. Wang, C. S. and Lee, M. C. Synthesis and properties of epoxy resins containing 2-(6-oxid-6H-dibenz(c,e)(1,2) oxaphosphorin-6-yl) 1,4-benzenediol (II). *Polymer* **2000**, *41*, 3631–3638. DOI: 10.1016/S0032-3861(99)00541-8
58. Lin, C. H., Wu, C. Y. and Wang, C. S. Synthesis and properties of phosphorus-containing advanced epoxy resins II. *J. Appl. Polym. Sci.* **2000**, *78*, 228–235.
59. Hörold, S. and Kleiner, H. J. Phosphorus-containing dicarboxylic reaction product of epoxy resins and phosphorus acid (anhydride) with hardener. US 5 959 043, **1999**.
60. Lin, C. H. Synthesis of novel phosphorus-containing cyanate esters and their curing reaction with epoxy resin. *Polymer* **2004**, *45*, 7911–7926. DOI: 10.1016/j.polymer.2004.09.023
61. Lin, C. H., Cai, S. X. and Lin, C. H. Flame-retardant epoxy resins with high glass-transition temperatures. II. Using a novel hexafunctional curing agent: 9,10-dihydro-9-oxa-10-phosphaphenanthrene 10-yl-tris(4-aminophenyl) methane. *J. Polym. Sci. A Polym. Chem.* **2005**, *43*, 5971–5986. DOI: 10.1002/pola.21072
62. Cho, C. S., Fu, S. C., Chen, L. W. and Wu, T. R. Aryl phosphinate anhydride curing for flame retardant epoxy networks. *Polym. Int.* **1998**, *47*, 203–209.
63. Hou, M., Liu, W., Su, Q. and Liu, Y. Studies on the thermal properties and flame retardancy of epoxy resins modified with polysiloxane containing organophosphorus and epoxide groups. *Polym. J. (Japan)* **2007**, *39*, 696–702. DOI: 10.1295/polymj.PJ2006111
64. Perez, R. M., Sandler, J. K. W., Altstaedt, V., Hoffmann, T., Pospiech, D., Artner, J., Ciesielski, M., Doering, M., Balabanovich, A. I. and Schartel, B. Effective halogen-free flame retardancy for a monocomponent polyfunctional epoxy using an oligomeric organophosphorus compound. *J. Mat. Sci.* **2006**, *41*, 8347–8351. DOI: 10.1007/s10853-006-1079-3
65. Levchik, S. V., Camino, G., Luda, M. P., Costa, L., Muller, G. and Costes, B. Epoxy resins cured with aminophenylmethylphosphine oxide 1: combustion performance. *Polym. Adv. Technol.* **1996**, *7*, 823–830.
66. Shau, M. D. and Wang, T. S. Synthesis, structure, reactivity, and thermal properties of new cyclic phosphine oxide epoxy resins cured by diamines. *J. Polym. Sci. A Polym. Chem.* **1996**, *34*, 387–398.
67. Levchik, S. V., Camino, G., Luda, M. P., Costa, L., Muller, G. and Costes, B. Epoxy resins cured with aminophenymethylphosphine oxide: mechanism of thermal decomposition. *Polym. Degrad. Stab.* **1998**, *60*, 169–183.
68. Perez, R. M., Sandler, J. K. W., Altstaedt, V., Hoffmann, T., Pospiech, D., Artner, J., Ciesielski, M., Doering, M., Balabanovich, A. I., Knoll, U., Braun, U. and Schartel, B. Novel phosphorus-containing hardeners with tailored chemical structures for epoxy resins: synthesis and cured resin properties. *J. Appl. Polym. Sci.* **2007**, *105*, 2744–2759. DOI: 10.1002/app.26537
69. Jeng, R. J., Lo, G. S., Chen, C. P., Liu, Y. L., Hsiue, G. H. and Su, W. C. Enhanced thermal properties and flame retardancy from a thermosetting blend of a phosphorus-containing bismaleimide and epoxy resins. *Polym. Adv. Technol.* **2003**, *14*, 147–156. DOI: 10.1002/pat.343
70. Ren, H., Sun, J., Wu, B. and Zhou, Q. Synthesis and properties of a phosphorus-containing flame retardant epoxy resin based on bisphenoxy(3-hydroxy) phenyl phosphine oxide. *Polym. Degrad. Stab.* **2007**, *92*, 956–961. DOI: 10.1016/j.polymdegradstab.2007.03.006
71. Levchik, S. V. and Weil, E. D. Thermal decomposition, combustion and fire-retardancy of polyurethanes—a review of the recent literature. *Polym. Int.* **2004**, *53*, 1585–1610. DOI: 10.1002/pi.1473
72. Weil, E. D. and Levchik, S. V. Commercial flame retardancy of polyurethanes. *J. Fire Sci.* **2004**, *22*, 183–210. DOI: 10.1177/0734904104040259
73. Witte, A. and Krieger, W. Halogen-free, pentane-blown, flame-retardant rigid polyurethane foam and a process for its production. US 6 593 385, 2001.
74. Lee, F. T., Green, J. and Gibilisco, R. D. Recent developments using phosphorus-containing diol as a reactive combustion modifier for rigid polyurethane foams 3. *J. Fire Sci.* **1984**, *2*, 439–453.
75. Yanchuk, N. I. Fireproof phosphorus-containing polyurethanes. *Khim. Technol. (Kiev)*, **1981**, (1), 56–58.
76. Wang, T. L., Cho, Y. L. and Kuo, P. L. Flame-retarding materials II. Synthesis and flame-retarding properties of phosphorus-on-pendent and phosphorus-on-skeleton polyols and the corresponding polyurethanes. *J. Appl. Polym. Sci.* **2001**, *82*, 343–357. DOI: 10.1002/app.1858
77. Huang, W. K., Chen, K. J., Yeh, J. T. and Chen, K. N. Curing and combustion properties of a PU-coating system with UV-reactive phosphazene. *J. Appl. Polym. Sci.* **2002**, *85*, 1980–1991. DOI: 10.1002/app.10776
78. Shao, C. H., Huang, J. J., Chen, G. N., Yeh, J. T. and Chen, K. N. Thermal and combustion behaviors of aqueous-based polyurethane system with phosphorus and nitrogen containing curing agent. *Polym. Degrad. Stab.* **1999**, *65*, 359–371.
79. Wang, T. Z. and Chen, K. N. Introduction of covalently bonded phosphorus into aqueous-based polyurethane system via postcuring reaction. *J. Appl. Polym. Sci.* **1999**, *74*, 2499–2509.
80. Huang, W. K., Yeh, J. T., Chen, K. J. and Chen, K. N. Flame retardation improvement of aqueous-based polyurethane with aziridinyl phosphazene curing system. *J. Appl. Polym. Sci.* **2001**, *79*, 662–673. DOI: 10.1002/1097-4628(20010124)79:4%3C662::AID-APP100%3E3.0.CO;2-T
81. Weil, E. D. and Levchik, S. V. Commercial flame retardancy of unsaturated polyester and vinyl resins: review. *J. Fire Sci.*, **2004**, *22*, 293–303. DOI: 10.1177/0734904104041210
82. Abdel-Mohdy, F. A. Graft copolymerization of nitrogen- and phosphorus-containing monomers onto cellulosics for flame-retardant finishing of cotton textiles. *J. Appl. Polym. Sci.* **2003**, *89*, 2573–2578. DOI: 10.1002/app.12247
83. Kaur, I., Sharma, V. and Sharma, R. Development of flame retardant cotton fabric through grafting and post-grafting reactions. *Indian J. Fibre Textile Res.* **2007**, *32*, 312–318.
84. Tsafack, M. J. and Levalois-Grutzmacher, J. L. Plasma-induced graft-polymerization of flame retardant monomers onto PAN fabrics. *Surf. Coat. Technol.* **2006**, *200*, 3503–3510. DOI: 10.1016/j.surfcoat.2004.11.030
85. Ren, Y., Jianga, L., Tiana, T., Liu, X. and Hana, Z. Durable flame retardant polyacrylonitrile fabric via UV-induced grafting polymerization and surface chemical modification. *RSC Advances* **2018**, *8*, 41389–41396. DOI: 10.1039/C8RA09005J

86. Arnold, C., Jr. Stability of high temperature polymers. *J. Polym. Sci.: Macromol. Revs.* **1979**, *14*, 265–378.
87. Martel, B. Charring process in thermoplastic polymers: effect of condensed phase oxidation on the formation of chars in pure polymers. *J. Appl. Polym. Sci.* **1988**, *35*, 1213–1226.
88. Ebdon, J. R. and Jones, M. S. Flame retardants: an overview. In *Polymeric Materials Encyclopaedia*, Salamone, J. C. (Ed.). **1995**, CRC Press, Boca Raton, 2397–2411.
89. Kandola, B., Horrocks, A. R., Price, D. and Coleman, G. Flame retardant treatment of cellulose and their influence on the mechanism of cellulose pyrolysis. *Revs. Macromol. Chem. Phys.* **1996**, *C36*, 721–794.
90. Le Bras, M., Camino, G., Bourbigot, S. and Delobel, R. (Eds.). *Fire Retardancy of Polymers: The Use of Intumescence*, **1998**, Royal Society of Chemistry, London.
91. Banks, M., Ebdon, J. R. and Johnson, M. The flame-retardant effect of diethylvinyl phosphonate in copolymers with styrene, methyl methacrylate, acrylonitrile and acrylamide. *Polymer* **1994**, *35*, 3470–3473.
92. Ebdon, J. R., Hunt, B. J., Joseph, P. and Wilkie, T. K. Flame retardance of polyacrylonitiles covalently modified with phosphorus and nitrogen-containing groups. In *Fire Retardancy of Polymers: New Strategies and Mechanisms*, Hull, T. R. and Kandola, B. K. (Eds.). **2009**, Royal Society of Chemistry, Cambridge, 331–340.
93. Naiker, V. E., Mestry, S., Nirgude, T., Gadgeel, A. and Mhaske, S. T. Recent developments in phosphorous-containing bio-based flame-retardant (FR) materials for coatings: an attentive review. *J. Coat. Technol. Res.* **2023**, *20*, 113–139. DOI: 10.1007/s11998-022-00685-z
94. Mensah, R. A., Shanmugam, V., Narayanan, S., Renner, J. S., Babu, K., Neisiany, R. S., Försth, M., Sas, G. and Das, O. A review of sustainable and environment-friendly flame retardants used in plastics. *Polym. Test.* **2022**, *108*, 1–14. DOI: 10.1016/j.polymertesting.2022.107511
95. Schartel, B. Phosphorus-based flame retardancy mechanisms—old hat or a starting point for future development? *Materials* **2010**, *3*, 4710–4745. DOI: 10.3390/ma3104710
96. Özer, M. S. and Gaan, S. Recent developments in phosphorus based flame retardant coatings for textiles: synthesis, applications and performance. **2022**, *171*. DOI: 10.1016/j.porgcoat.2022.107027
97. Thomas, A., Joseph, P., Moinuddin, K., Zhu, H. and Tretsiakova-McNally, S. Thermal and calorimetric evaluations of some chemically modified carbohydrate-based substrates with phosphorus-containing groups. *Polymers* **2020**, *588*. DOI: 10.3390/polym12030588
98. Thomas, A., Moinuddin, K., Zhu, H. and Joseph, P. Passive fire protection of wood using some bio-derived fire retardants. *Fire Saf. J.* **2020**, *120*. DOI: 10.1016/j.firesaf.2020.103074
99. Bhat, T., Jones, M., Joseph, P., Thomas, A., Kandare, E., Yuen, R., John, S., Ma, J. and Wang, C. Thermal degradation and fire properties of fungal mycelium and mycelium-biomass composite materials. *Sci. Rep.* **2018**, *8*, 17583. DOI: 10.1038/s41598-018-36032-9
100. Wadhwani, R., Sutherland, D., Moinuddin, K. A. M. and Joseph, P. Kinetics of pyrolysis of litter materials from pine and eucalyptus forests. *J. Therm. Anal. Calorim.* **2017**, *130*, 2035–2046. DOI: 10.1007/s10973-017-6512-0

6 Flame Retardant Fillers

Charles A. Wilkie[1], A. Richard Horrocks[2], and Peter Hornsby[3]
[1] Marquette University
[2] A. Richard Horrocks, University of Bolton
[3] Original Chapter Author, 2nd Edition:
[3] Queen's University Belfast

6.1 INTRODUCTION

Inorganic hydroxides, mainly aluminum trihydroxide, $Al(OH)_3$, also known as alumina trihydrate, $Al_2O_3.3H_2O$, ATH, and magnesium dihydroxide, $Mg(OH)_2$, MDH, are the most used flame retardants (FRs) globally. This is due to their low-cost relative to antimony-halogen systems and phosphorus-containing flame retardants as well as their low toxicity and corrosivity. They can also contribute to the reduction of smoke emission during combustion. ATH is far and away the best-selling inorganic hydroxide and is used in a wide range of elastomers, thermoplastics, and thermosets processed below 200°C. The higher thermal stability of MDH in that decomposition occurs above 300°C means that MDH may be used in higher processing temperature polymers, including polypropylene, engineering thermoplastics, and some elastomers. Both materials are available in a number of different forms which can be critical to determine their suitability as flame retardants

There are also a number of other hydroxides and hydroxycarbonates that are used as alternatives to other types of systems because they too are believed to have less adverse environmental effects.

A unique feature of these fillers is the very high loadings that are required to ensure adequate fire protection. It is not unusual to see loadings of close to 50 wt% filler within a polymer. This can, of course, affect processability and melt rheology and, in load bearing situations, can result in deterioration in strength and toughness of the composite. Surface treatments can be used to lessen these limitations, however.

In this chapter, we shall first of all examine the mechanism by which these inorganic hydroxides function as flame retardants, secondly, provide an overview of the principal types of fillers, then examine the mechanisms in more detail. Finally, we will consider the potential of using co-additives to possibly increasing their effect.

6.2 AN INITIAL LOOK AT THE FLAME-RETARDANT MECHANISM

There are three major classifications of flame retardants, halogen (usually synergized with antimony or some other material), phosphorus-based FRs, and the aluminum hydroxide/magnesium hydroxide filler and examples considered in this chapter. The first two function through chemical reactions, radical trapping in the case of halogen FRs, and reaction between the degrading polymer and its FR in the case of phosphorus-based systems. The hydroxides considered here function primarily through thermodynamics. Both aluminum hydroxide and magnesium hydroxide decompose endothermically with the evolution of water.

$$Al_2O_3{\cdot}3H_2O(s) \rightarrow Al_2O_3(s) + 3\ H_2O(g)\ \Delta H = +\ 1300 kJ/mol$$
$$Mg(OH)_2(s) \rightarrow MgO(s) + H_2O(g)\ \Delta H = +145\ kJ/mol$$

Thus, any heat added to the polymer-mineral system is used to decompose the mineral and will have no effect on the polymer, as long as some mineral remains. For this reason, the typical filler levels are between 40 and 60% by weight in order to achieve acceptably high levels of flame retardancy. Lower levels may be acceptable when the mineral is combined with other FRs or if the mineral is used primarily to provide smoke suppression.

Additive levels are dependent on the type of polymer and its decomposition mechanism. In polypropylene, a loading typically of 60% by weight is needed to obtain a Limiting Oxygen Index (LOI) of 26%. In polyamide, at the same loading level, an oxygen index of 70% can be obtained [1]. The testing protocol will also influence the filler loading. In order to satisfy UL94 test requirements for dripping, polyamides require a significant filler level notwithstanding the high oxygen index, because this increases the viscosity of the decomposing polymer, thus inhibiting its tendency to drip [2].

The key aspect of polymer fire retardancy using hydrated fillers involves energy changes which occur on decomposition of the filler [3]. Related heat capacity effects also may influence the degradation profile of the polymer and so affect thermal barrier formation resulting from the residue from the degraded filler.

Differential scanning calorimetry (DSC) and thermogravimetric analysis (TGA) have been extensively used to study the thermal decomposition and endothermic breakdown of hydrated mineral fillers. As shown in Table 6.1 and in Figure 6.1, there are large differences in decomposition temperature with different filler types [4]. Different grades of the same filler type may also behave differently

DOI: 10.1201/9781003380689-6

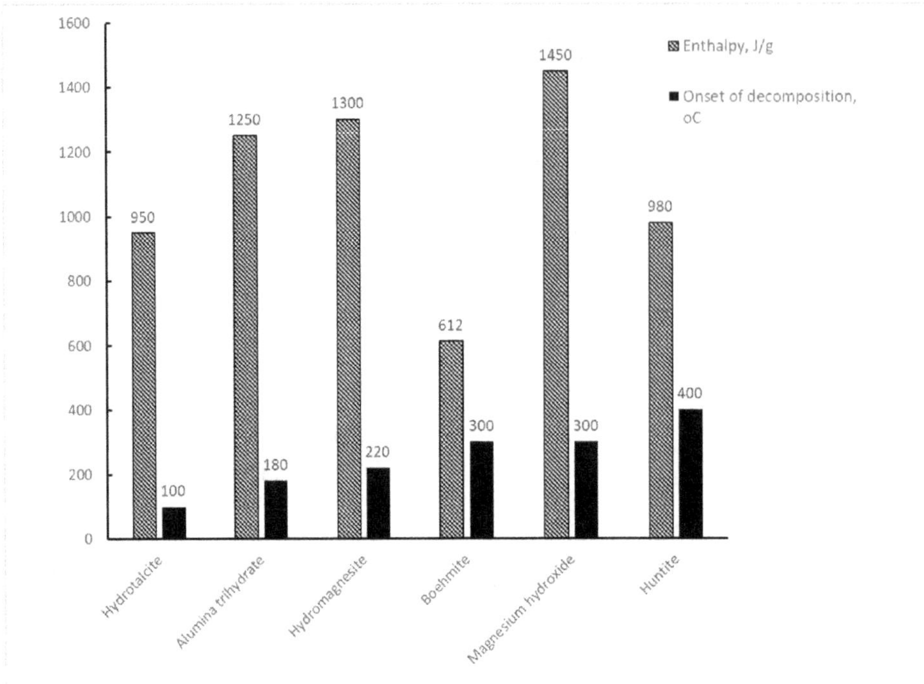

FIGURE 6.1 Onset temperatures (°C) and enthalpies (J/g) of decomposition of selected hydrated fillers.

in both decomposition temperature and the magnitude of the decomposition endotherm [5]. Furthermore, the rates of degradation of various grades of magnesium hydroxide will vary; both are dependent on the filler morphology and surface area [6]. It must be noted that the process of thermal analysis can affect the decomposition kinetics of the filler [3]. Differences have been observed depending on size of the sample that is used, the heating rate, inert gas flow, and sample pan closure.

The endothermic decomposition of aluminum hydroxide has been extensively studied giving an enthalpy for complete decomposition of about 1,300 kJ/kg (see Table 6.1). Thermal breakdown depends on the heating rate and on escape of decomposition products. The decomposition commences at about 200°C. According to reports, ATH is less thermally stable as the particle size increases [7]; precipitated grades are more stable than those produced by grinding [8].

Although the decomposition of aluminum hydroxide to the oxide is usually written as a one step process, it can proceed through the intermediate monohydroxide (boehmite, AlO(OH), see below), which has a much higher stability than does ATH, decomposing at about 450°C, and this may reduce the effectiveness. Several so-called transitional aluminas can be formed upon heating of hydroxides, a process which is temperature dependent [9]. When the escape of water is slowed, boehmite formation can occur. The alumina produced is very reactive early in the decomposition process and may easily combine with the evolved water to reform ATH. In large particles, water escaping from near the center of the particle has a larger diffusion path and thus more time to combine with alumina nearer to the surface. With increased particle size, the transition from gibbsite to pseudo-boehmite increases [10]. Hydrothermal treatment of a gibbsite suspension will produce both boehmite and pseudo-boehmite.

The heat capacity of these fillers, in addition to their strong endotherm, has an influence on the heat required for polymer decomposition and release of combustible volatiles [1]. This effect has been modelled, based on the oxygen index, using a heat balance approach which can be applied to the whole combustion process [11]. There is a good fit between model and experimental data, except for very fine fillers. At sufficiently high filler levels, hydrated fillers can reduce the mass burning rate by inhibiting the rates of heat transfer from the flame to the underlying matrix, causing the flame to extinguish due to fuel starvation [12]. Hence reductions in applied heat flux or increased surface heat losses will lead to a decrease in the mass burning rate of the polymer. This effect has been reported for polypropylene/aluminum hydroxide composites [13].

The particle size of the filler also plays a significant role in flammability resistance. For example, below a certain particle size (about 1–2μm), in many tests, including oxygen index, aluminum hydroxide shows enhanced fire retarding performance [14], which may be associated with the rate of filler decomposition and/or with the formation of a more stable ash. However, it has been found that the particle size effect is absent, or less evident, when tested in the cone calorimeter [15]. Similarly, particle size reduction has been shown to enhance fire retardancy in magnesium hydroxide-filled polypropylene; in this case samples were characterized by UL94 testing [16]. This raises the question as to whether further reductions in particle size to the

TABLE 6.1
Common Flame-Retardant Fillers and Thermal and Physical Properties

Filler	Formula	Specific Gravity	Onset of Decomposition, T_{on}, °C	Maximum Rate Temperature, T_{max}, °C	Enthalpy of Decomposition (or Formation), J/g	Theoretical Volatile Emissions, wt%
Alumina trihydrate	$Al_2O_3.3H_2O$ or $Al(OH)_3$	2.42	~180	312	1190–1300	34.5 (H_2O)
Magnesium hydroxide	$MgO.H_2O$ or $Mg(OH)_2$	2.34	~300	414	1450	31 (H_2O)
Hydromagnesite	$4MgCO_3.Mg(OH)_2.4H_2O$	2.16–2.2	~220(H_2O), 330 (H_2O), 350 (CO_2)		1300	19 (H_2O); 38 (CO_2)
Huntite	$Mg_3Ca(CO_3)_4$	2.70	> 400	NA	980	49.8 (CO_2)
Boehmite	AlO(OH)		> 300	504	612	15 (H_2O)
Calcium sulphate dihydrate	$CaSO_4.2H_2O$	2.64	> 120	NA	NA	21 (H_2O)
Hydrotalcite (layered double hydroxide (LDH))	$Mg_6Al_2CO_3(OH)_{16}{\cdot}4H_2O$	2.03–2.09	< 100	207, 291, 416	950	~36 (H_2O);~7 (CO_2)
Halloysite nanotubes	$Al_2Si_2O_5(OH)_4.nH_2O$	2.55–2.57	> 400	-	-	~14 (H_2O)
Vermiculite	$\{(Mg,Fe^{2+},Fe^{3+})_3[(Al,Si)_4O_{10}](OH)_2{\cdot}4H_2O\}_n$ (hydrous phyllosilicate mineral)	2.4–2.7	NA	NA	NA	-
Mica	$X_2Y_{4-6}Z_8O_{20}(OH, F)_4$, in which X is K, Na, or Ca or less commonly Ba, Rb, or Cs; Y is Al, Mg, or Fe or less commonly Mn, Cr, Ti, Li; *Z* is chiefly Si or Al, but also may include Fe^{3+} or Ti^{3+}.	2.8–3.0	NA	NA	NA	-

Sources: W. A. Deer, R. A. Howie and J. Zussman (1966) *An Introduction to the Rock Forming Minerals*, Longman, ISBN 0-582-44210-9; refs 4, 11 and 24.

nanoscale will lead to an additional increase in flammability performance, and perhaps enable filler overall levels to be significantly reduced. This aspect is considered in a later section.

Forced combustion studies provide a practical method for quantifying this thermal insulating effect and enabling measurement of rates of heat transfer through a fire-retardant polymer composition exposed to an externally applied ignition source. During thermal breakdown of polypropylene, magnesium and aluminum hydroxides decompose to their respective oxides, which together with any carbonaceous char produced, provide an effective thermal barrier, reducing heat transmission to the underlying substrate Figure 6.2 [17]. Similar behavior has been demonstrated with other polymer types, including modified-polyphenylene oxide (PPO), poly(butylene terephthalate) (PBT), and acrylonitrile-butadiene-styrene terpolymer (ABS) [18]. Further analysis of data from refs. 1 and 18, presented in Figure 6.2, in which the increase in Limiting Oxygen Index (ΔLOI) is plotted as a function of filler content, shows that for PP, polyamide 6 (PA6), PBT, and ABS, the increasing trends are generally similar. From these it may be

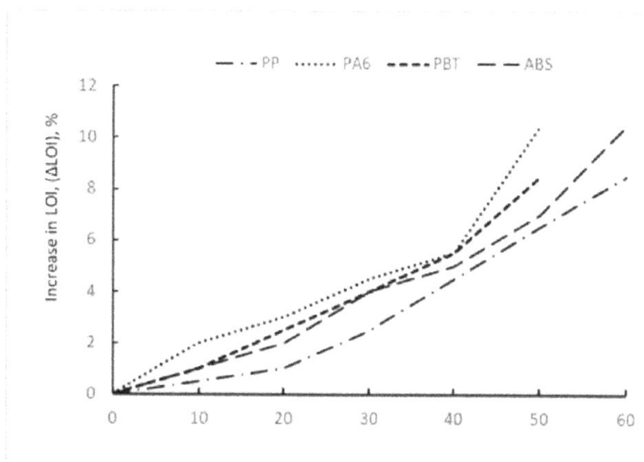

FIGURE 6.2 Increase in LOI (ΔLOI) versus magnesium hydroxide filler content in selected thermoplastic polymers.

concluded that not only is the magnesium hydroxide behaving similarly in each polymer but also that high loadings are required to produce significant increases in flame retardancy as stated previously.

In magnesium hydroxide-filled polypropylene compounds, the oxide/char residue formed during combustion has a morphology similar in form to the parent hydroxide [19]. In this regard, hexagonal platelets show preferred alignment and, in some cases overlap, in contrast to aggregated structures derived from hydroxide particles comprising an association of small crystallites. There is some evidence from oxygen index tests of increased crystal growth and that the coherency of the association between oxide particles contributes to the stability of the decomposition residue observed from combustion products, which may partially account for reported differences in performance observed between different grades of the same filler [11]. It is possible that partial sintering between oxide particles may also occur. In this regard, the strength of agglomerates containing magnesium hydroxide pseudomorphs has been estimated as 50MPa, arising from physio-chemical association between magnesium oxide and water [20].

The high filler loadings required using hydrated fillers create a significant dilution effect resulting from the presence of the filler, thereby reducing the amount of combustible polymer available. With magnesium hydroxide, the presence of 60% by weight of filler is equivalent to around 35% by volume reduction of polymer content. However, not all so-called inert fillers appear to act in the same way as matrix diluents.

With polypropylene compositions containing different grades of magnesium hydroxide, magnesium oxide and glass beads, values of heat release rate (HRR) were determined by cone calorimetry [19]. It was found in all cases that rates of heat release were significantly reduced, after allowing for the volume dilution of each of these fillers, but not to the same extent. Unexpectedly, magnesium oxide was far more effective than the glass beads in suppressing heat release rate, even though both are considered to be inert fillers. This suggests that even with thermally stable fillers showing no endothermic transition, other factors have a role in influencing fire retardancy, such as particle geometry, surface chemistry, chemical interactions between the filler, and polymer and possibly thermal conductivity.

6.3 FLAME RETARDANT FILLER TYPES

In order to have a useful commercial flame retardant filler, the following are required [11]:

1. The filler must undergo an endothermic decomposition in the range between about 100–300°C, with the release of at least 25% by weight of water and/or carbon dioxide, which, of course also depends on the polymer and its decomposition pathway and its inherent combustion resistance.
2. The filler must be readily available at low cost.
3. It must have low toxicity.
4. It must be available in small particle size, with a defined morphology and, ideally, low surface area.
5. It must have low water solubility and reasonable purity, that is, no detrimental impurities that may cause polymer degradation.
6. It must have no color.

The hydroxides, hydroxy carbonates, and hydrates of aluminum, calcium, and magnesium which may meet these requirements are shown in Table 6.1, along with their thermal properties and the gaseous products evolved on decomposition. Of these materials in commercial use, about 90% of the market by mass is aluminum hydroxide; magnesium hydroxide and basic magnesium carbonate are used in niche applications.

These materials may be produced in various ways which results in filler forms differing in particle size, surface area, purity, and morphology; these all have an influence on their performance as flame retardants. An overview of the major methods of their production is considered below.

6.3.1 Aluminum Hydroxide

Aluminum hydroxide ($Al(OH)_3$), sometimes also written as alumina trihydrate, $Al_2O_3.3H_2O$, is herein after referred to as ATH. It is commonly produced by the Bayer process from the mineral bauxite, which is a crude form of aluminum hydroxide containing 40–70% aluminum minerals, mainly gibbsite, a polymorph of $Al(OH)_3$, and boehmite, AlO(OH). The process involves dissolving bauxite with sodium hydroxide to form a solution of sodium aluminate, followed by controlled precipitation. The process is of low cost because it is linked to the production of alumina.

The ground Bayer hydrate is the largest volume of ATH sold as a flame retardant, with particle sizes ranging from 1.5 to 35 μm, and they may be further classified to have a tighter particle size distribution. These are low in cost and have platy, irregular particles, which may not be ideally suited for many applications. The second largest segment of the market is formed by precipitation from a purified sodium aluminate under conditions that control the shape and size to yield median particle size ranging from 0.25 to 3 μm.

There are a variety of surface treatments that are available for ATH. For applications at very high loadings, specially tailored particle size distributions may be used. Mechanical properties may suffer when fillers are used at high loadings, but this can be ameliorated by surface treatment and mixing techniques to optimize dispersion.

Basic aluminum oxalate can be produced by the reaction of aluminum hydroxide with oxalic acid. This is thermally stable up to 330°C; it loses 51% of its mass upon decomposition above 450°C. Because of its higher thermal stability, it can be used in polyamides and thermoplastic polyesters, and it does not cause hydrolytic degradation [21], which may occur when magnesium hydroxide is used.

6.3.2 Magnesium Hydroxide ($Mg(OH)_2$)

The second most widely used flame retardant filler is magnesium hydroxide, commonly called MDH. It has a higher decomposition temperature than ATH (near 300°C), so it is used in thermoplastics where higher processing temperatures are needed. It is also more expensive than ATH, and there are several routes to this product [22].

Where low cost is a requirement and color and thermal stability are not critical, a natural product, brucite, may be used; this is impure, of lower thermal stability than the refined MDH, and generally is colored.

Another possibility is MDH precipitated with lime from seawater with high magnesium content. The seawater is pretreated with sulfuric acid to remove bicarbonates, and the precipitate is washed, filtered, and dried. This material usually consists of small aggregates of crystallites with high surface areas.

The so-called Aman process uses concentrated magnesium chloride brine as a source of magnesium, which is hydrolyzed to produce a mixture containing magnesium hydroxide which is then filtered, washed, and dried.

Magnesium hydroxy silicate, $Mg_3(Si_2O_5)(OH)_4$, which is the ore serpentinite, can be converted into magnesium chloride by treatment with hydrochloric acid. This can then be hydropyrolyzed to magnesium oxide and then hydrolyzed to magnesium hydroxide. The result is particles with low surface area and defined platy morphologies.

There are many variations of these processes which exist to control surface area, shape, and purity. These properties control the flame retarding properties of MDH, which may be especially important in demanding applications where processability and mechanical properties are important.

Nano-sized MDH has also been produced, as has the development of nickel-doped forms of magnesium hydroxide, which are claimed to have superior fire-retardant properties [23].

6.3.3 Basic Magnesium Carbonates [24]

These comprise the mineral hydromagnesite, $4MgCO_3.Mg(OH)_2.4H_2O$, which is a mixture of magnesium carbonate and magnesium hydroxide. This decomposes over a range of temperatures, starting at about 220°C and with a total weight loss of 57%. At 220°C water initially generated is that of hydration and then at 330°C the magnesium hydroxide component dehydrates until at 350°C when CO_2 is evolved. The equation

$$4MgCO_3.Mg(OH)_2.4H_2O \rightarrow 5MgO + 4CO_2 + 5H_2O$$

represents this sequential decomposition.

Hydromagnesite is most usually found naturally mixed with the mineral huntite, itself a mixed magnesium calcium carbonate, $Mg_3Ca(CO_3)_4$, which decomposes to lose carbon dioxide initially at about above 410°C and then principally at about 570°C and a final mass loss at about 720°C:

$$Mg_3Ca(CO_3)_4 \rightarrow 3MgO + CaO + 4CO_2$$

Thus, a mixture of the two components will generate both water and carbon dioxide at temperatures over the range 220–720°C. The thermal decomposition temperatures reflect those of the component magnesium and calcium carbonates which alone decompose at about 440°C [25] and 800°C [26] respectively.

Not dissimilar to huntite is the mineral dolomite, which is often present with huntite and has the ideal formula $CaMg(CO_3)_2$. Recent research has studied the use of dolomite (DOL) at 5, 10, and 15 wt% levels together with mixtures with an intumescent (IFR) in various weight ratios DOL/IFR (5/5, 10/5, and 5/10 wt%) added to rigid polyurethane foams [27]. Results showed that up to 15 wt% DOL could be incorporated into the rigid polyurethane foams as filler without any problems in the foam production, and the thermal and combustion behaviors of the foams were slightly enhanced. DOL/IFR combinations, however, comprising 5 wt% DOL and 10 wt% IFR addition, satisfied the best performance in terms of the thermal stability and the fire resistance with reducing the smoke, CO, CO_2, and NO emissions, suggesting some degree of synergy between the components.

In principle, this group should be excellent fire retardants for many polymers, including some thermoplastics, but due to their complex morphology derived from the mixture of finer, plate-like particles of huntite with the larger, more block-like ones of hydromagnesite, processability problems may arise. Not only does the emission of both water and carbon dioxide dilute and so suppress polymer flame formation, but also the mixture forms a char layer with a platy structure on top of the underlying polymer, thereby isolating the fuel source. This char layer also reduces the release of smoke.

In practice natural deposits have to be converted into purer forms (~95%) by flotation processes, for example, if they are to be considered suitable for flame retardant applications [28]. Commercial forms are often coated with small amounts (e.g., 1–2 wt%) of dispersing agents such as stearic acid. While the huntite content can be up to 50 wt%, its platy morphology can affect processing [29]. Products approximating to the hydromagnesite composition can also be precipitated from solutions of magnesium salts, and this process has been used and is mainly sold for smoke suppressing applications, where it is effective at lower levels than those needed for fire retardancy.

Often mixtures of magnesite and huntite are also combined with aluminum hydroxide, which allows thermal decomposition and release of water to start at 180°C and which also will reduce overall flame-retardant cost.

6.3.4 Boehmite

Boehmite is partly decomposed aluminum hydroxide, where two-thirds of the water have been removed. It undergoes endothermic decomposition, commencing around 300°C, which is considerably higher than the temperature of breakdown of aluminum hydroxide. Results from this study yielded a heat absorption of 612 J/g relative to 1190 J/g for ATH [4]. In this study, the boehmite had a median particle size of 0.6 µm and surface area of 17 m^2/g. With a high theoretical residue of 85% it could potentially act in the condensed state by formation of an insulating layer but is not considered to have high fire retarding qualities, although it has been promoted as a fire retardant in its own right. However, it does seem to have some potential in mixtures with other fire-retardant fillers [30]. However, recent research has investigated the use of boehmite as a suitable flame retardant in lithium batteries [31]. Boehmite particles microencapsulated by a chemically stable poly(urea-formaldehyde) and embedded in the cathode showed better self-extinction efficiency when particle sizes were reduced from 0.9 mm and 0.35 mm.

Research has also focused on the incorporation of nanoparticulate boehmite to a number of polymers including polycarbonate(PC)/acrylonitrile-butadiene-styrene (PC/ABS) blends in the presence of selected aryl phosphates [32]. For instance, adding 5 wt% AlO(OH) decreases the peak heat release rate by releasing water, which enhances hydrolysis of PC with positive influences on char formation and LOI.

6.3.5 Calcium Sulfate Dihydrate (Gypsum)

This is a low-cost material with limited fire-retardant properties since as a dihydrate, it emits a limited amount of water when heated. It has a low onset of decomposition (under 100°C) and has been reported to find some use as a fire retardant in unsaturated polyester resins [33] and carpet backing coatings [34].

6.3.6 Hydrotalcites

These are a series of magnesium aluminum hydroxycarbonates with varying magnesium to aluminum ratios between 1.5 and 3.0 g-atoms of magnesium to 1 g-atom of aluminum. They have layers of magnesium hydroxide interspersed with aluminum cations and carbonate anions, which can potentially be chemically modified to allow intercalation of polymer chains and subsequent exfoliation into nanoplatelets in a similar manner to silicate layer nanofillers, for example, based on montmorillonite. DSC of hydrotalcite shows three peaks at 207, 291, and 416°C with corresponding decomposition enthalpies of 356 J/g associated with the first peak and 594 J/g from the second and third peak (see Table 6.1) [4]. The weight loss begins at a very low temperature (often as low as 50°C but < 100°C), which is attributed to water loss, but overall its thermal behavior in terms of heat absorption capacity and mass loss is comparable with that of magnesium and aluminum hydroxide and therefore might be expected to confer similar fire retarding properties to polymers.

They show similar flame-retardant activity and thermal stability to ATH, but their higher cost currently limits their potential use. In this regard, additions of only 30% by weight of hydrotalcite to unplasticized PVC gave an LOI value of 29% and UL94 V-0 rating [35]. However, it should be noted that in this study the initial hydrotalcite particles, made by chemical co-precipitation, were found to be very coarse with an average size of around 200 µm. This dimension was reduced substantially after treatment with a titanate coupling agent and by shearing with the polymer on a two-roll mill.

More recently, hydrotalcite has been used as a nanofiller at much lower levels than would be required as a filler [36]. Furthermore, it has received interest as a significant component in flame retardant polymer-layered double hydroxide (LDH) nanocomposites [37].

6.3.7 Halloysite Nanotubes

One of the more recent inorganic hydrate fillers is halloysite, which is an aluminosilicate, $Al_2Si_2O_5(OH)_4.nH_2O$, similar to kaolin, but instead of existing as sheets as in the latter, has a nanotubular structure. It loses its water of hydration at temperatures in excess of 400°C and because of its tubular structure has an extremely high specific surface area [38]. While it is classed as a nanofiller, it has the advantage, similar to those of other nanoparticulate materials, of improving fire performance of polymer composites at low concentrations, often less than 5–10 wt% [39]. However, this mineral has been shown to improve fire performance at levels as high as 30 wt% in polyamide 610 [40] and polypropylene [41], for example.

6.3.8 Inert Fillers

Inert fillers do not act as flame retardants in the mechanistic sense, but because they are used to at quite high levels (> 10 wt%) to reduce the cost of the resulting polymer composite as well as modify its physical properties, typically increased modulus/stiffness, they reduce flammability as a consequence of the reduced polymer content. Typical examples are talc, vermiculite, mica. Because of their inert activity, they require the additional presence of conventional flame retardants if high levels of flame retardancy are required. In the presence of other FRs, sometimes additional or even synergistic activity [42] is observed, as in the case of talc when added to an ammonium phosphate formulation in nylon 6, which improves flame retardancy by forming silicophosphates, which prevent the volatile loss of phosphorus [43]. Conversely, talc may react antagonistically and so reduce flame retardancy, as in the example where it reduces the effectiveness of RDP in polycarbonates [44].

Micas are a group comprising about 37 phyllosilicate minerals, which have unique physical properties in that the crystalline structure exists layers that can be split or delaminated into thin sheets. The most common and commercially available classes of mica are muscovite (hydrated silicate of aluminum and potassium) and phlogopite (magnesium and potassium hydrated silicate), although synthetic mica is also available. They can be ground into very fine laminar particles which when introduced into a polymer influences its flammability. These may then form a ceramifying barrier either alone or in combination with other additives. For instance, introducing up to 20 wt% mica in polystyrene, while not reducing the tendency to ignite, progressively reduces the total heat release during cone calorimetric studies, as expected as the polymer content reduces. However, and in addition, the peak heat release rate was more than half in value with respect to the pure polystyrene [45]. Mica presence has also been shown to reduce the levels of a conventional flame retardant such as an intumescent normally required when combined in polypropylene [46] and silicone rubber [47, 48].

Vermiculite is a hydrous phyllosilicate mineral which undergoes significant expansion when heated, and exfoliation occurs when the mineral is heated sufficiently, a process that significantly reduces particle size and density. Commercially the more common method of exfoliating vermiculite is by short-lived rapid, so-called "shock" or "flash," heating to temperatures of around 900°C for a few minutes. Table 6.1 shows the structure to contain water of hydration, and so the mechanism of thermal exfoliation of vermiculite is a consequence of the rapid production of steam during flash-heating, which forces open the layers of vermiculite [49]. Exfoliated vermiculate generally has much smaller particle sizes, sometimes < 2 μm, than the unexfoliated form and so is more amenable to inclusion as a polymer FR additive. Consequently, exfoliated vermiculite, like mica, when used a fire retarding element does not release any further water. Like mica, both unexfoliated and exfoliated vermiculite may be used in conjunction with other flame retardants, although evidence suggests the latter is superior [50]. Again, like mica, the presence of vermiculite alone in a polymer may reduce the peak heat release rate significantly [51].

6.4 SCOPE OF APPLICATION

There is widespread application of ATH in elastomers, thermosetting resins, and thermoplastics, its use being generally limited to polymers processed or cured below 200°C. This ranges from styrene-butadiene rubber latex for carpet backing, flame-retardant rubber-insulated cable, conveyor belting roofing and hoses, to flame-retardant unsaturated polyester resins in a wide range of molded products using sheet and dough molding compounds [52]. ATH is also used in epoxy and phenolic resins, especially in electrical and construction applications, and in cross-linked acrylic resins, for example in countertops, sinks, and bathroom panels. In thermoplastics, applications for ATH have been found in rigid and plasticized poly(vinyl chloride), high, low and linear low density polyethylene, ethylene-propylene rubber, ethylene-propylene-diene crosslinked rubbers, ethylene-ethyl acrylate copolymers, and ethylene-vinyl acetate copolymers [10, 53–55].

A major use is in low smoke, halogen-free wire, cable, and conduit applications, where there has been significant commercial activity [56–60].

Although it is claimed that ATH can also be used in polypropylene, the limited thermal stability of this filler generally necessitates special compounding and processing measures, which has inhibited its large-scale application in this polymer [61].

The greater thermal stability of MDH to temperatures in excess of 300°C permits its use in polymers such as polypropylene, polyamides, ABS, modified poly(phenylene oxide), and aliphatic polyketones, in addition to certain elastomers where increased thermal stability is a requirement. Major end uses are in wire and cable, appliance housings, and electrical components. A particularly important consideration in engineering components is the effect of the filler on mechanical properties, and as is the case with ATH, a variety of surface-treated grades are available to suit the physical demands of intended applications. However, use in thermoplastic polyesters is restricted by its tendency to hydrolyze and degrade the polymer during melt processing. Unlike ATH, in unsaturated polyester resins, MDH acts as a chain extender thus adversely affecting resin rheology. Although this effect can be ameliorated by using maleic acid-coated grades of MDH filler, there are long-term issues concerning the stability of these systems [62, 63].

Hydromagnesite (usually found in combination with huntite as stated previously) has intermediate thermal stability, decomposing between 220–240°C [64, 65]. Mixtures of these minerals are used in wire and cable applications, due to their higher thermal resistance than ATH and lower cost compared to MDH. They have also been considered for use in ethylene-propylene copolymers [66] and PVC formulations, where reduced smoke and acid gas emission are requirements [67].

Magnesite/huntite addition has also been reported to give superior heat stabilization and flame retardancy to plasticized PVC [68], elastomeric PU [69], EVA copolymers [70].

Boehmite has recently attracted interest in improving the fire resistance of silicon battery cathodes [71] as well as in epoxy resins [72], where results indicated that AlO(OH) can improve the flame retardancy and T_g of epoxy resin.

The inert fillers mica and vermiculite, as stated above, while having the ability to reduce the peak heat release rate when present alone in polymers, have been shown to have both positive and negative influences on the effectiveness of other flame retardants also present. However, one particular advantage of these two fillers is their ability to improve the ceramification of other inorganic additives present in the composite and even interact with silicone-containing polymer matrices such as silicone rubbers [47, 48].

6.5 A MORE DETAILED LOOK AT THE MECHANISM

As indicated earlier, a major drawback to the use of fire-retardant fillers, in relation to alternative flame retardants, including phosphorus-based intumescent and halogen-containing formulations, is their inefficiency, and hence their requiring significant addition levels in order to achieve acceptable combustion resistance. While typical filler levels required are between 40 and 60% by weight, lower levels may be acceptable in combination with more effective alternative fire retardants or where the purpose of their inclusion in a formulation is principally to aid smoke suppression.

It is difficult to generalize on the mechanism of fire retardancy in polymers modified with hydrated fillers, since additive levels required are very dependent on polymer type, whether or not they are thermoplastic and melt, and in particular, on their decomposition mechanism. For example, with polypropylene, 60 wt% would be required to achieve an oxygen index in excess of 26%. In polyamide 6 at the same addition level, however, an oxygen index of nearly 70% can be achieved as stated previously [73]. However, the testing mode can also influence the required filler loading, especially if the polymer melts and forms drips. For example, in order to satisfy UL94 test requirements in terms of dripping during combustion, polyamides, like polypropylene, require significant filler levels despite their high oxygen index, since this raises the viscosity of the decomposing polymer, inhibiting its tendency to drip [74].

When introducing such filers at the high levels required, there is the challenge posed by the individual particle-surface affinity, hence surface wetting, which is influenced by the particle characteristic such as size, shape, purity, and the presence of any surface treatments. Furthermore, the decomposition behavior and degradation mechanism of the polymer, together with any filler/polymer interactions which might occur, will affect the polymer thermal stability and possible char formation. This latter will most likely involve carbonaceous and inorganic contents possibly existing as an interacted, interpenetrating network.

The major flame retarding effects of hydrated fillers are determined by their thermophysical effects on the burning polymer mechanism, although tempered by polymer-filler interactions as well as effects of released gases, principally water and/or carbon dioxide, on the burning polymer flame chemistry and physics.

6.5.1 Thermal Effects from the Filler

Hull et al. identify four physical contributions to the overall fire retardant effects of mineral fillers, especially those that are hydrated [75]. These are the dilution of the polymer in the condensed phase, the endothermic decomposition of the filler, the presence of inert residue, and the diluent effect of released gases. These affect the energy changes associated with reduced polymer content, the decomposition of the filler, and related heat capacity effects and thermal barrier effects resulting from the residue remaining from degraded filler. Based on the known thermal decomposition behavior of a number of hydrated fillers, the relative contributions of estimations of the heat absorption in terms of the filler (% filler), the residue (% residue), the evolved water vapor and carbon dioxide (% gas), and the decomposition endotherm (wt% decomposition endotherm) are presented in Figure 6.3, using the data from ref. 75.

Here it is seen that broad similarities exist between each filler and that the greatest contributions come from the endothermic decomposition. In terms of energy absorbed for the respective endothermic transition per unit mass in order of decreasing value: magnesium hydroxide

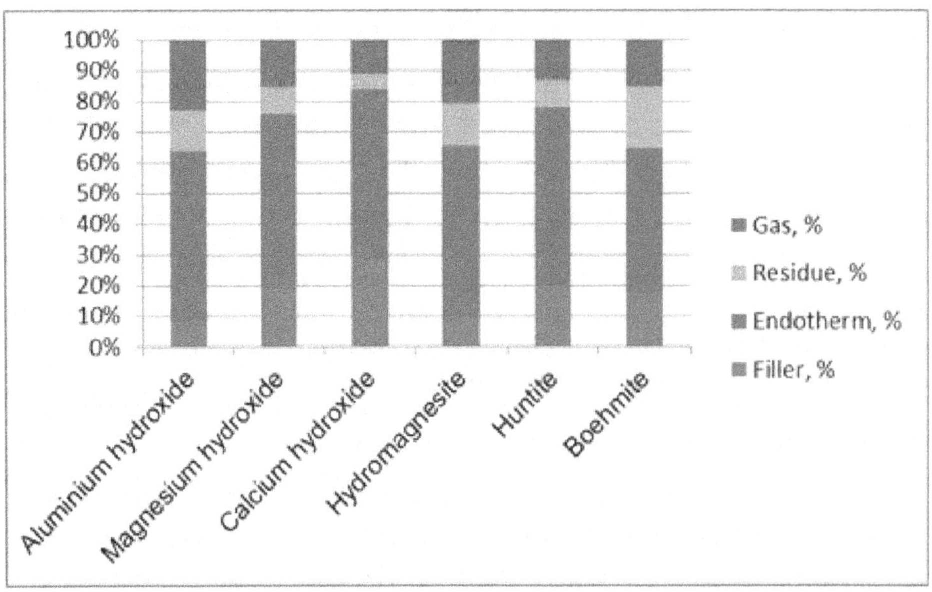

FIGURE 6.3 Relative contribution of heat-absorbing effects for mineral filler fire retardants [data from ref. 75].

(1.41 kJ/g) > aluminum hydroxide = hydromagnesite (1.27 kJ/g) > calcium hydroxide (1.12 kJ/g0 > huntite ((0.95 kJ/g) > boehmite (0.75 kJ/g). Thus the advantages of aluminum and magnesium hydroxides as endothermic fillers on a weight for weight basis are clearly evident.

Differential scanning calorimetry (DSC) and thermogravimetric analysis (TGA) have enabled full analysis of the thermal decomposition and endothermic breakdown of hydrated mineral fillers to be investigated with major transitions shown in Figure 6.2 and Table 6.1 [11]. Different grades of the same filler type may also behave very differently in terms of decomposition temperature and magnitude of the decomposition endotherm [5]. Furthermore, the rate of degradation of different grades of magnesium hydroxide may vary, depending on filler morphology and surface area [6]. A further complication results from the observation that filler decomposition kinetics can be influenced by the procedure adopted during thermal analysis [11]. For example, differences have been observed depending on sample size, rate of heating, rate of inert gas flow rate, and extent of sample pan closure. Isothermal studies have shown that while decomposition of ATH starts at about 200°C, it becomes less thermally stable as the particle size increases [21], with precipitated grades being significantly more stable than those produced by grinding [8].

As stated previously, there is evidence to show that particle size of the filler also plays a significant role in flammability resistance. While particle size reduction has been shown to enhance fire retardancy in magnesium hydroxide filled polypropylene, reductions in particle size to the nanoscale (see Section 6.7.2) offer novel possibilities.

6.5.2 Filler-Polymer Interactions

The changes in thermal degradation profile often observed in thermoplastic polymers in the presence of many fillers can also be a significant factor affecting the action of fire-retardant fillers. TGA and DSC have enabled such filler-polymer interactions to be studied especially when used in combination with evolved gas analysis (EGA) and online FTIR techniques. In studies with polyamides, it was shown that on decomposition, magnesium hydroxide promotes degradation of polyamide 6 (PA6) and polyamide 66 (PA66), which can be attributed to hydrolysis of the polymer resulting from water release from the filler [74]. This research showed that evolved gases detected by mass spectrometry from degradation of both filled and unfilled PA compositions contained water, carbon monoxide, carbon dioxide, ammonia, and various hydrocarbon fragments. Not surprisingly there was much greater similarity between evolved gas compositions of PA6-magnesium hydroxide combinations with respect to their components, where the polymer decomposed before the filler. This resulted in significantly improved fire retardancy in PA6 relative to PA66.

Filler-polymer interactions have also been observed in EVA copolymers yielding differences in fire retarding effectiveness between ATH and MDH [76]. In EVA with 30% vinyl acetate content, magnesium hydroxide had an oxygen index of 46%, whereas aluminum hydroxide gave a value of only 37%. From non-isothermal thermogravimetric analysis, it was proposed that in this polymer, water release is delayed from aluminum hydroxide, whereas it is accelerated from magnesium hydroxide as a consequence of reaction of acetic acid, evolved from the polymer, and this filler. This example again highlights the importance of relative decomposition temperatures of filler and polymer. Cone calorimetry data, however, may lead to different conclusions regarding the effectiveness between ATH and MDH in these kinds of compounds [77].

In EVA with 19% vinyl acetate content and a filler loading of 60 wt% of ATH or MDH grades with a similar particle size distributions and specific BET surface areas, the peak heat release rate (PHRR) measured at 35 kW/m^2 was significantly lower (121 kW/m^2) for ATH than for MDH (200 kW/m^2). There is evidence that filler-polymer interactions may occur between so-called inert fillers like mica and vermiculite, especially when present in silicon-containing polymers. For example, the previously mentioned effect of the improvement in ceramification of these in the presence of other inorganic additives such as ammonium polyphosphate, APP, and ATH in silicon rubber (SiR) composites. This work has indicated that compounds containing P-O-Al are initially formed due to the reaction between APP and $Al(OH)_3$ during the early combustion stages. Subsequently, a coherent, dense, and sealed structure is formed due to the reaction in mica, phosphates, and the thermal decomposition products of SiR during combustion in the later stage [47, 48].

Besides chemical filler-polymer interactions as exemplified above, the presence of large amounts of filler may influence and usually inhibit polymer melt dripping behavior as mentioned previously. Thus the physical influence of the filler on melt rheology of the decomposing polymer is of significance and may be influenced by filler physical properties such as particle size. For example, parallel plate rheometry has shown that different magnesium hydroxide filler variants influence the rheological behavior of thermally decomposing polyamides in different ways and hence their resistance to dripping [78]. Of note also, plate-like filler particles were observed to give the greatest increase in viscosity [79].

6.5.3 Vapor Phase Action

In addition to the endothermic effect exerted by hydrated fillers in the condensed phase, the release of water and/or inert gas into the vapor phase also contributes significantly to the overall fire retardation mechanism. While little detailed analysis has been undertaken on this aspect, it is generally considered that water release into the vapor phase exerts a beneficial action through dilution and cooling of flammable volatiles produced during polymer pyrolysis [73]. Of these factors, the heat capacity of the evolved gases, including carbon dioxide if present, will add to that from the oxide residue resulting in a significant cooling effect as indicated in Figure 6.3. Whether or not either evolved water or carbon dioxide interacts chemically with the polymer flame chemistry is not known and merits further research.

6.6 SMOKE SUPPRESSION

Whilst being more efficient than hydrated fillers in preventing flame ignition, in the event of a fire, halogenated and phosphorus-containing fire retardants have been claimed to increase the quantity and toxicity of smoke evolution [80] although evidence to the contrary also exists [81]. In contrast, hydrated fillers exert a strong smoke suppressing effect and are frequently chosen for this purpose in situations, such as cables used in underground railways or in ships, where legislation or specific demands of the application require it. Although there are many reports of the use of hydrated fillers as smoke suppressants [82], little detailed work has been undertaken on the mechanism.

An early example of the effect of calcium carbonate, ATH and MDH fillers on smoke production from styrene butadiene (SBR) foams has been reported [83]. It was evident that all the fillers reduced soot formation relative to unfilled foam with the hydrated fillers being more effective than the calcium carbonate, which was considered to act merely as matrix diluent. ATH and MDH were found to give enhanced char formation with promotion of solid-state cross-linking as opposed to pyrolytic degradation. An afterglow effect, occurring after extinction of the flame, was noted with MDH and attributed to slow combustion of carbon residues. A more recent example of smoke reduction is that of an inorganic flame-retardant filler containing magnesium hydroxide, aluminum hydroxide, inorganic phosphate, and melamine salt when present in an SBR binder in asphalt mastic while increasing both LOI and char formation [84].

Many other reports have demonstrated the smoke suppressing tendencies of hydrated fillers in various polymers, including ethylene-propylene-diene elastomers [83], polypropylene [18], polystyrene [85], modified polyphenylene oxide, polybutylene terephthalate, and ABS [18]. In addition to suppressing smoke generation, a delay in onset of smoke evolution is also achievable [1].

Analysis of smoke and soot formation from polymers during combustion has been extensively studied [86], however less is understood on how hydrated fillers influence this mechanism. It has been proposed that smoke reduction results from deposition of carbon onto the high surface area oxide surface, produced on decomposition of the filler [18]. Volatilization of carbonaceous residue as carbon oxides then occurs, reducing obscuration effects from the smoke.

It is well-known that freshly formed oxides have high surface areas and, in addition, can be catalytically active [87], thereby promoting both carbon deposition and subsequent oxidation processes [88]. The reduced combustion rate arising from the effects of the fire retardant filler also contributes to lowering the rate of smoke evolution and, by improving oxygen to fuel ratios, further limits levels of smoke density [11].

In this context, the role of evolved water from hydroxide decomposition is of interest, since water is also known to oxidize carbon. However, comparing smoke yields from polypropylene compounds containing magnesium hydroxide and magnesium oxide, little difference was seen in levels of smoke emitted [18]. There were also similar CO emissions from burning ABS containing oxide and hydrated forms of magnesium, suggesting that water has limited effect on the smoke suppression mechanism. This observation is supported by the fact that the so-called water-gas oxidation reaction occurs at temperatures and pressures well in excess of those normally found at the burning surface of a polymer [89].

Other aspects of fire-retardant performance include the effect of the additives on the interrelationship between smoke and afterglow or incandescence. Oxidation of trapped carbonaceous species on the oxide surface is the likely cause of after-glow commonly observed with fire retardant fillers [90]. In studies using EVA copolymer oxidation, self-ignition and incandescence were monitored as a function of temperature and filler loading. Both ATH and MDH were found to increase the self-ignition temperature, but to decrease the incandescence temperature of the EVA copolymer. Using TGA, in these systems it was concluded that the solid state after-glow effects observed were due to oxidation of carbonaceous residues.

6.7 ENHANCING THE EFFICIENCY OF FIRE RETARDANT FILLERS

While hydrated fillers have many attributes, including their low toxicity and smoke suppressing character, their use is restricted by their relative inefficiency as fire retardants, the high levels required, and the consequent reduction in polymer composite physical and mechanical properties. Thus means of improving their efficiencies with associated reductions in filler loadings and minimization of their impact on processibility and physical properties of the host polymer is a challenge worth addressing. However, use of processing additives and/or filler surface treatments are well established routes for minimizing these adverse effects [91], which are beyond the scope of this review.

Alternative means for improving fire retarding and mechanical performance of the composites are considered below, which include the use of synergistic co-additives, that may form multi-component structures during processing, thereby modifying the filler-polymer interface. Secondly and not unrelated are the use of fillers with significantly reduced particle size (at the nanoscale), which again will increase the effective areas of filler polymer interfaces.

TABLE 6.2
Examples of Synergists for Mineral Fillers

Co-Additive/Synergist	Filler	Polymer	Effects	Ref.
Borate compounds (Zn and Ca borate)	ATH/huntite-magnesite	EVA PE	Low co-additive and smoke levels. Enhanced flame resistance; increased char.	[92–94]
MDH/ATH combinations	MDH/ATH	PVC	Reduced flammability, increased endotherm width and water release. Oxide thermal barrier.	[95, 96]
Molybdenum compounds	ATH/MDH	PVC	Reduced flammability and smoke emission; increased char.	[97]
Red phosphorus	ATH MDH	PE/PP	Reduced filler levels; metal hydroxide suppresses phosphine formation. Coloured formulations. Low co-additive levels.	[98, 99]
Silicon-containing compounds	ATH MDH	PE/PP	Reduced flammability and smoke release. Improved processibility and physical properties.	[100–102]
Antimony III oxide	ATH MDH ABS	EVA	Sb_2O_3 inhibiting char incandescence; Synergistic effect of nano Sb_2O_3	[103, 104]
Transition (& rare earth) metal oxides	ATH MDH	PE/PP	Reduced filler levels. Increased oxidatively driven char formation (PP). Possible color and toxicity limitations.	[105, 106]
Metal nitrates (Cu and Fe)	ATH	PVC EVA	Increased flame resistance at low co-additive levels.	[107]
Introduction of intumescents	MDH Dolomite	PP Rigid PU	Increased flame resistance at low co-additive levels.	[27, 46, 107–109]
Melamine and/or novolac	ATH MDH	PP	Reduced flammability and afterglow.	[110, 111]
Tin compounds	ATH MDH	PVC Cl rubbers EVA	Reduced flammability and smoke emission especially with ZH/ZHS-coated variants.	[112–116]
Carbon/graphite/graphene species	ATH/MDH	PE/PP PBS	Reduced flammability and filler levels at low co-additive concentrations. Possible coloration.	[93, 109, 117–119]

6.7.1 Synergism

There are many reported examples of synergists used in combination with hydrated fillers. Examples are given in Table 6.2, although this list is not exhaustive. In many of these cases, although beneficial effects are observed, the mechanisms of interaction are not clear, or there are other reasons, such as cost, which limit their use industrially.

Combinations of MDH and ATH can give improved performance when used together, due to the increased range of endothermic reaction (180–400°C) and release of water in the vapor phase [95, 96]. The different metal oxides produced on dehydration may also contribute to this effect no doubt because the oxide layer production will occur over a longer temperature gradient than when either hydrate is used alone.

ATH and red phosphorus (3–5%) have also been used in synergistic mixtures to increase fire retardancy and enable lower filler loadings [98]. Recent research has shown that there is claimed synergy in composites comprising aluminum hydroxide, red phosphorus, and expandable graphite in polyethylene to yield LOI = 25.4% upon the incorporation of only 15 wt% of the mixture of three fillers with ATH:RP:EG mass ratio of 1:1:1 and a UL94/V-0 rating [99].

The addition of melamine and novolac (~ 1%) to PP/MDH mixtures has been found to reduce the burning time and give a UL94 V-0 rating at lower filler levels (30–50 wt%) as opposed to a more usual value of around 60 wt%, allowing the formulation to be mechanically more flexible. The novolac causes a structurally stabilizing effect above the melting point of PP. Thermal evidence suggests that a novolac magnesia gel may be formed [99].

Metal hydroxides in combination with various silicon-containing compounds have been used to reduce the amount of additive required to achieve a required level of flame retardancy in a variety of polymeric materials, including polyolefins [100, 101]. Systems which have been used contain a combination of reactive silicone polymers, a linear silicone fluid or gum, and a silicone resin, which is soluble in the fluid, plus a metal soap, in particular magnesium stearate. How these formulations work, however, is not clear given their complexity.

It has also been shown that the required loading levels of metal hydroxides to flame retard polyolefins could be reduced by addition of transition metal oxides as synergistic agents. For example, combination of 47.6 wt% MDH modified with nickel oxide in PP gave a UL94 V-0 flammability rating, which would require ~55 wt% of unmodified MDH [105]. These systems, however, can only be used where the color of the product is not important. Of similar interest is the effect of adding lanthanum oxide to an MDH-filled, PP composite. The significant enhancement in the flame retardancy with LOI > 31% at 2.5 wt% La_2O_3 and a UL94 V-0 rating, was attributed to the greatly improved quality of the condensed phase charred layers deposited on the surface of the formed magnesium oxide particles from the catalytic action of La_2O_3 on adjacent PP macromolecular chains [106].

The addition of metal nitrates to improve the flame retardancy of metal hydroxides and EVA has been reported [107]. Synergistic behavior was observed by the addition of 2 wt% of copper nitrate to EVA containing only 33 wt% ATH, and the LOI was raised from 19.9 to 30.0%.

The combination of fillers with intumescents can not only improve overall flame retardancy but enable reduced concentrations to be used. This was demonstrated previously with the introduction of dolomite/intumescent combinations into rigid polyurethane foam [27] and mica in the presence of an intumescent in polypropylene [46]. In the latter case, when the content of mica was 6 wt% and intumescent 24 wt%, the LOI value of PP compound reached 34.9% and yielded a V-0 rating in UL 94 flammability test. Much earlier work also showed that flammability properties of an intumescent fire-retardant PP formulation with added MDH had improved flammability properties but gave higher CO and smoke evolutions [108], the addition of MDH was found to reduce smoke density and CO emissions, in addition to giving superior fire resistance. More recently, addition of thermally reduced graphite oxide (TRGO), containing only four single carbon layers on average, together with ammonium polyphosphate (APP) and magnesium hydroxide (MDH), in polypropylene (PP) exhibited at ≤ 1 wt% TRGO a V-0 classification in the UL 94 test and LOI ≥ 31% [109].

Addition of silane cross-linkable PE copolymer to PE/metallic hydroxide systems can significantly improve the flame retardant properties of these materials, allowing lower filler levels to be used [102].

The combination of melamine with hydrated mineral fillers was shown to improve the fire retardancy behavior of PP, eliminating at the same time the after-glow phenomenon associated with these fillers used in isolation [111]. Similarly in EVA copolymer, antimony trioxide used in combination with metal hydroxides has been reported to reduce incandescence [103]. Chlorinated and brominated flame retardants are sometimes used in combination with metal hydroxides to provide a balance of enhanced fire retardant efficiency, lower smoke evolution, and lower overall filler levels. For example, in polyolefin wire and cable formulations, magnesium hydroxide in combination with chlorinated additives was reported to show synergism and reduced smoke emission [120].

In a similar manner, non-hydrated filler concentrations may be reduced by combination with halogenated flame retardants. For example, the natural mineral filler comprising mainly huntite and hydromagnesite, together with a blend of antimony trioxide and decabromodiphenyl oxide, reduced the flammability of an ethylene-propylene copolymer [121]. However, and rather surprisingly, the addition of nano-particulate antimony trioxide alone to a huntite/hydromagnesite-filled ABS composite has been claimed to be synergistic and to achieve a UL94/V-0 classification [104].

The addition of very small amounts of fine carbon fibers [117] or polyacrylonitrile fibers [122] have also been shown to reduce the level of inorganic hydroxide required to achieve UL94 V-0 flammability ratings in polyolefin compounds. These secondary additives are thought to function as char promoters.

The addition of low levels (~3 wt%) of zinc borate to metal hydroxides can give synergistic effects in EVA composites [92]. For example in an EVA/MDH formulation, LOI was found to increase from 39 to 43%, together with a significant reduction in heat release rate. Solid state carbon NMR of the residues showed that polymer fragments were in the char layer. It was suggested that zinc borate slows the degradation of the polymer, creating a vitreous protective physical barrier to combustion. However, no such effect of zinc borate was observed when added to a huntite-hydromagnesite filler present in a polyurethane elastomeric composite [93]. Boric acid in combination with antimony III oxide and huntite-hydromagnesite, on the other hand, have been reported to enable lower concentrations of the mineral filler to be used although decreased particle size is a significant factor in increasing flame retardancy [94].

While the additions of ATH and MDH are well-known to reduce smoke generation [18, 82–85], addition of molybdenum compounds have been reported over 40 years ago as additional smoke suppressants, possibly synergistically, in the presence of calcium carbonate and ATH in PVC [97].

Very recent research has shown that when expanded graphite (EG) was introduced as a synergist into poly(butylene succinate) (PBS)/$Mg(OH)_2$ system to investigate the results showed that addition of only 5 wt% EG into PBS/20 wt% $Mg(OH)_2$ gave excellent flame retardancy expressed as LOI = 29.4% and a UL-94 rating of V-0. Furthermore, the peak heat release rate, PHRR, decreased by 73% [118]. In addition, this formulation yielded much better thermal stability and mechanical properties than a PBS/40 wt% $Mg(OH)_2$ without EG, indicating a balanced improvement on the overall performances of PBS. Not dissimilar work showed that the level of ATH may be reduced in polypropylene following the introduction of benzoic acid functionalized graphene oxide where the latter behaves as a synergist within an $Al(OH)_3$/polypropylene composite [119].

6.7.2 Nano-Size Fire Retardant Fillers

The use of nano-particulate fillers as modifiers for polymers at low levels usually ≤ 5 wt% has increased in recent years because of their promotion of significant increases in mechanical properties, reduced gas and vapor transmission, and notably decreased flammability [123]. However, at these levels in terms of their flame retarding activity, they cannot be compared with the fillers of interest here, but there is evidence that their presence as a co-additive with the latter may improve their effectiveness with the aim of reducing overall filler loadings needed. Furthermore, when hydrated fillers themselves are used in nano-particulate form, they offer improvements in flame retardant efficiency. See Chapter 12 of this book for more information on nanoparticles and the effects polymer nanocomposites have on material flammability and flame retardancy.

6.7.2.1 Nano-Particulate Hydrated Fillers

In this regard, a number of reports consider the synthesis and application of magnesium hydroxide nanoparticles as flame retardants in polymers [124, 125]. Magnesium hydroxide nanoparticles with different morphological structures of needle-, lamellar-, and rod-like nanocrystals have been synthesized by solution precipitation reactions of magnesium chloride in the presence of water soluble polymer dispersants [126]. The size and morphologies of the magnesium hydroxide nanocrystals were controlled by the reaction conditions, in particular temperature, alkaline solution injection rate, and reactant concentration. Needle-like morphologies having dimensions of 10 × 100 nm^2, laminar particles around 50 nm in diameter and an estimated 10 nm in thickness, and rod-like particles 4 μm in length and 95 nm in diameter were studied. TGA-DTA analysis of the lamellar crystals gave a pronounced weight loss between 250 and 396°C with a corresponding endothermic peak near 354°C ascribed to decomposition of magnesium hydroxide. The overall weight loss for the MDH was 30.1%. A comparison of LOI and tensile strength (TS) data for EVA/MDH composites using three different types of MDH, nano-size filler and micron-sized filler, showed that tensile strength for the larger particles decreased with increasing filler level, whereas values for the nano-MDH/EVA composite increased. LOI results for the nano-MDH were also superior, especially at high filler levels. The enhancement in fire retardancy seen using these nano-scale fillers was attributed to a more compact char structure creating more effective gas barrier properties [127]. However, it should be noted that in this work high nano-MDH filler levels were still required to achieve reasonable resistance to ignition, in common with more conventional MDH fillers.

Nanotubes of magnesium hydroxide, synthesized by a solvothermal reaction from basic magnesium chloride and ethylenediamine solvent, were reported to have diameters of 80–150 nm, a wall thickness of 30–50 nm, and lengths of 5–10 μm [128]. Their use as polymer fire retardants was not considered, however, at the time of publication.

Nano-magnesium hydroxide (n-MDH) and three forms of micro-magnesium hydroxide filler (all commercially available in China) mixed with EPDM rubber have been analyzed for mechanical properties and fire resistance [129]. The particle size of the micro-MDH used was around 2.5 μm, and the nano-MDH had a hexagonal sheet-like structure with dimensions of 100 nm width by < 50 nm thick. Thermal analysis on the fillers showed that the nano-filler had a lower decomposition temperature and a larger endotherm. The LOI of these composites were very similar, although it was noted that the nano-filled material showed less tendency to drip and gave a more coherent char residue. The heat release rate (HRR) from the nanocomposite was substantially lower than the micro-filled systems (peak HHR 259 relative to 329–346 kW/m^2 respectively), and the time to ignition increased from 81–89s to 95s. It was noted, however, that filler levels for all the composites were again high with test samples containing between 60 and 100 phr of additive.

A further study considered the use of n-MDH on the mechanical and flame retarding properties of polypropylene [130]. Substantial improvements in mechanical properties were obtained by the addition of only small amounts of nano-magnesium hydroxide. For example the addition of 12 wt% of filler gave a 433% increase in Young's modulus and a 35% improvement in flame retardancy (expressed as a burning rate). However, no comparison was given with conventional micron-sized magnesium hydroxide filler.

Nano-magnesium hydroxide particles with a mean particle size of about 10 nm and a core-shell structure comprising a magnesium hydroxide inner region surrounded by an alkyl chain when present [131] in EVA at a 5 wt% loading were compared with conventional 1 μm micron-sized filler at addition levels of 5, 30, and 60 wt% for respective fire behaviors. It was shown that the peak HRR of the EVA/n-MDH composite was reduced by 60% compared to pure EVA even though only 5 wt% of n-MDH was present. The n-MDH also caused a significant reduction in smoke (specific extinction area) although LOI was not significantly affected. It was evident that the n-MDH significantly outperformed micro- or μ-MDH at equivalent filler loadings.

A comparison has also been given of the fire retarding performance in EVA of a commercially available hydrotalcite, with median particle size of 2.26 μm and surface area 17 m^2/g, and aluminum and magnesium hydroxides with similar particle sizes [132]. It was demonstrated that the most significant flame retardancy effect in terms of slowest heat release rate, lowest evolved gas temperature, and longest ignition time was in EVA filled with 50 wt%

hydrotalcite. This was attributed to the wide temperature range (200 to 500°C) for water loss and associated heat absorption, a delay in deacetylation of the EVA due to interlayer water loss at about 207°C and possible intercalation of acetate anions in the layer structure.

Very recent work on nano-MDH developments has been discussed in terms of a novel hydrous mechano-synthesis of magnesium hydroxide nano-fillers for flame-retardant polyester composites [133] and cites other recent research regarding their fire retarding and smoke-suppressing character [134, 135], selected use in textile coatings [136] and as fillers in fire-resistant polymeric composites [137].

6.7.2.2 Filler/Micro and Nano-Particulate Combinations

The addition of small amounts of a nanofiller together with a main filler has proved to be of benefit in improving the fire properties of the latter. For instance, in halogen-containing polymers, zinc hydroxystannate (ZHS) or zinc stannate-coated hydrated fillers can give significantly improved flame resistance and lower smoke emission compared with uncoated fillers, which suggests a synergistic activity and so may be included in Table 6.2 [112–114]. Since normal diameters of commercial ZS and ZHS are of the order of 1–2 μm, their use as nano-coatings has proved not to be a problem. The efficiency of ZHS-coated fillers was found to be superior to simple admixtures of these components, reflecting improved dispersion and possible synergism in these systems. Detailed characterization of the nature of the ZHS coating using XPS revealed the level of coverage and interaction between the coating and supporting substrate surfaces [115]. At levels of ZHS-contents ranging from 1 to 15 wt%, ATH was more effective than MDH in both uncoated and ZHS-coated forms in polyvinyl-chloride (PVC) based cable formulations [116].

One of the earliest commercial examples that used nanocomposite technology in combination with the hydrated filler, ATH, was the use by Kabelwerk Eupen AG specifically to improve fire performance of electrical cable EVA sheathings [138, 139]. For normal flame retardancy requirements, up to 65 wt%, ATH is required, which means that the overall cable sheathing physical properties may be significantly reduced. By incorporation of up to 5 wt% of a functionalized nanoclay (with functionalizing dimethyldistearylammonium salts), the ATH content may be reduced to about 45 wt% with a consequent improvement in sheath properties and saving on FR costs. Furthermore, the char formed by the nano-filled, ATH-containing formulation was far more rigid and less prone to crack development than the conventional analogue. The presence of nanoclay reduced the flame height from > 4 m to < 2 m and maximal flame temperature at 3.66 m from 1054 to 345°C in the cable sheathing test UL-1666.

A method for synthesizing nano-aluminum hydroxide with an ATH core and alkyl hydrocarbon chain shell structure has also been described [140]. In composites prepared using EVA, mechanical properties of the nano-filled material were almost the same as a conventional 1 μm sized ATH variant, however the heat release rate of the former composition was markedly lower. In this work 10 phr of filler was used.

Magnesium hydroxide sulphate hydrate (MHSH) whiskers have been used with and without microencapsulated red phosphorus synergist as fire retardants in low density polyethylene [141]. The MHSH whiskers were shown to degrade endothermically with release of water in a two-step process with DTG peaks at 301 and 411°C. The additive was shown to be an effective fire retardant and smoke suppressant but only at high filler levels. For example, to achieve a UL-94 V-0 rating required 60 wt% loading, similar to that for magnesium hydroxide. However, these fibers appear to have length dimensions in the micron range, which combined with their relatively inefficient fire retarding efficiency, makes their useful application unlikely.

It has also been reported that the interlayer spacing in hydrotalcite filler particles can be increased by intercalation with organic molecules, such as citric acid [142]. Results showed that for linear burning rates from a UL-94 horizontal burn test comparing nano- and micro-hydrotalcite modified epoxy resin, a 5 wt% addition level gave a reduction in burn rate of unmodified epoxy resin from around 22 mm/min to 18 mm/min, while the apparent nanocomposite variant, also at 5 wt%, gave 5 mm/min.

While there has been considerable interest in the fire properties of polymer composites containing carbon nanofillers such as multi-walled carbon nanotubes (MWCNTs), graphene and graphene oxide, the research regarding their action in the presence of micro-inorganic fillers has been surprisingly little [143]. An early paper has reported that combinations of MWCNT and micron-sized ATH in EVA function as very efficient fire retardants through enhanced char formation and coherency [144]. A later study describes the incorporation of a reduction graphene oxide (RGO) within a more complex $4ZnO{\cdot}B_2O_3{\cdot}H_2O$/RGO hybrid material present with magnesium hydroxide in a PVC composite. The incorporation of 2 pph ZB/RGO and 13 pph MDH yielded LOI = 29.5% and a 35% reduction in cone calorimetrically determined peak heat release rate relative to the PVC control (LOI = 24.5%) [145].

6.8 CONCLUSIONS

There is no doubt that the major effect from these mineral fillers is their endothermic decomposition. The effect of these fillers on flame retardancy depends on the nature of the filler, including its size, shape, purity, and behavior on decomposition but also on the degradation pathway of the polymer. Importantly, polymer-filler interactions may occur and can affect the thermal stability of the polymer and its ability to char. There will always be a need for an inexpensive, effective flame retardant, which at the present time, inorganic fillers satisfy.

REFERENCES

1. Hornsby, P. R. The Application of Magnesium Hydroxide as a Fire Retarding and Smoke Suppressing Additive for Polymers. *Fire Mater.* **1994**, *18*, 269.
2. Hornsby, P. R.; Wang, J.; Cosstick, K.; Rothon, R. N.; Jackson, G.; Wilkinson, G. Mechanism of Fire Retardancy in Polyamides Filled with Magnesium Hydroxide. *Prog. Rubber Plast. Recycl. Technol.* **1994**, *10* (3), 204–220.
3. Hornsby, P.R. and Rothon, R.N. Chapter 2 'Fire retardant fillers for polymers', in *Fire Retardancy of Polymers: New Applications of Mineral Fillers*, Le Bras, M., Wilkie, C.A., Bourbigot, S., Duquesne, S. and Jama, C., Eds., The Royal Society of Chemistry, Cambridge, 2005.
4. Camino, G.; Maffezzoli, A.; Braglia, M.; De Lazzaro, M.; Zammarano, M. Effect of Hydroxides and Hydroxycarbonate Structure on Fire Retardant Effectiveness and Mechanical Properties in Ethylene-Vinyl Acetate Copolymer. *Polym. Degrad. Stab.* **2001**, *74* (3), 457–464.
5. Hornsby, P.R. and Watson, C.L. 'Aspects of the mechanism of fire retardancy and smoke suppression in metal hydroxide filled thermoplastics', p 17 *IOP Short Meetings Series No 4*, Institute of Physics, London, April, 1997.
6. Hornsby, P.R. and Mthupha, A. 'Mechanism of fire retardancy in magnesium hydroxide filled polypropylene', *Proceedings from Society of Plastics Engineering Annual Technical Conference (ANTEC '93)*, New Orleans, LA, 1954–1956, May 9–13, 1993.
7. Hancock, M. and Rothon, R.N., Chapter 2 'Principal types of particulate fillers', in *Particulate Filled Polymer Composites Second Edition*, Rothon, R.N., Ed., Rapra Technology, UK, Shawbury, 88–90, 2003.
8. Dando, N.R., Clever, T.R., Pearson, A., Stinson, J.M., Kolok, P.L. and Martin, E.S. 'Aluminum trihydroxide (ATH) as a filler for polymer composites: Improvements in thermal stability by controlled precipitation', *Proceedings from 50th Annual Technical Conference, Composite Institute*, pp 1–4, Detroit, MI, Society of Plastics Industry Inc, Session 1-D, May 3–7, 1992.
9. Souza Santos, P.; Souza Santos, H.; Toledo, S. P. Standard Transition Aluminas. Electron Microscopy Studies. *Mater. Res.* **2000**, *3* (4), 104–114.
10. Sobolev, I. and Woychesin, E.A., Chapter 16 'Alumina Trihydrate', in *Handbook of Fillers for Plastics*, H.S. Katz and J.V. Milewski, Eds. Van Nostrand Reinhold, New York, 1987.
11. Rothon, R., Chapter 6 'Effects of particulate fillers on flame retardant properties of composites', in *Particulate-Filled Polymer Composites*, Rothon, R.N., Ed., 2nd Edition, p. 271. Rapra Technology Ltd, UK, Shawbury, 2003.
12. Tewarson, A., *Flame Retardant Polymeric Materials*, Lewin, M., Atlas, S.M. and Pearse, E.M., Eds., Plenum Press, New York, 1982.
13. Spilda, I.; Kosik, M.; Blazej, A. The Mass Burning Rate of Flame-Retarded Polypropylene in an Oxygen-Enriched Atmosphere. *J. Appl. Polym. Sci.* **1986**, *31*, 589.
14. Hughes, P.; Jackson, G. V.; Rothon, R. N. Particle Morphology Effects on the Performance of PMMA Filled with Aluminum Hydroxide in a Variety of Fire Tests. *Makromol Chem Makromol Symp.* **1993**, *74*, 179.
15. Herbert, M.J. 'Aluminum hydroxide for non-halogen compounds', *Proceedings of Flame Retardants '94*, p. 59, Interscience, London, 1994.
16. Miyata, S.; Imahashi, T.; Anabuki, H. 'Fire-retarding polypropylene with magnesium hydroxide', *J. Appl. Polym. Sci.* **1980**, *25*, 415.
17. Hornsby, P. R.; Watson, C. L. 'A study of the mechanism of flame retardancy and smoke suppression in polymers filled with magnesium hydroxide', *Polym. Degrad. Stab.* **1990**, *30*, 73–87.
18. Hornsby, P. R.; Watson, C. L. 'Mechanistic aspects of smoke suppression and fire retardancy in polymers containing magnesium hydroxide filler', *Plast. Rubber Process Appl.* **1989**, *11*, 45–51.
19. Hornsby, P. R.; Mthupha, A. Analysis of Fire Retardancy in Magnesium Hydroxide-Filled Polypropylene Composites. *Plast. Rubber Compos. Process. Appl.* **1996**, *25*, 347–355.
20. Itatani, K.; Kiyotaka, K.; Koizumi, K.; Howell, F. S.; Kishioka, A.; Kinoshita, M. Agglomeration of Magnesium Oxide Particles Formed by the Decomposition of Magnesium Hydroxide. *J. Mater. Sci.* **1988**, *23*, 3405.
21. Stinson, J.M. and Horn, W.E. 'Flame retardant performance of a modified aluminum hydroxide with increased thermal stability', *Proceedings from Society of Plastics Engineers 52nd Annual Technical Conference (ANTEC '94)*, Part 3, pp 2829–2833, 1–5 May, 1994.
22. Hancock, M. and Rothon, R.N. Chapter 2 'Principal types of particulate fillers', in *Particulate Filled Polymer Composites Second Edition*, Rothon, R.N., Ed., Rapra Technology, UK, Shawbury, 88–90, 2003.
23. Imahasi, T., Okada, A. and Abe, T. Flame retardant aid, flame retardant and flame-retardant composition. U.S. Patent 5,583,172, December 10, 1996.
24. Hollingbery, L. A.; Hull, T. R. The Thermal Decomposition of Huntite and Hydromagnesite—A Review. *Thermochim. Acta.* **2010**, *509* (1–2), 1–11.
25. Mahon, D.; Claudio, G.; Eames, P. An Experimental Study of the Decomposition and Carbonation of Magnesium Carbonate for Medium Temperature Thermochemical Energy Storage. *Energies* **2021**, *14* (50), 1316.
26. Xiang-Guo, L.; Yang, L.; Bao-Guo, M.; Wei-Qiang, W.; Shou-Wei, J. Decomposition Kinetic Characteristics of Calcium Carbonate Containing Organic Acids by TGA. *Arab. J. Chem.* **2017**, *10* (Supplement 2), 2534–2538.
27. Aydoğan, B.; Usta, N. Effects of Dolomite and Intumescent Flame Retardant Additions on Thermal and Combustion Behaviors of Rigid Polyurethane Foams. *J. Appl. Polym. Sci.* **2023**, *140* (16), e53739.
28. Kangal, O.; Kökkılıç, O.; Burat, F. Production of Huntite and Hydromagnesite with Flame Retardant Featured by Flotation. *Min., Metall. Explor.* **2009**, *26*, 109–113.
29. Doyle, M., Clemens, M., Lees, G., Briggs, C. and Day, R. *Proceedings of Flame Retardants '94*, Interscience, London, 193, 1994.
30. Sauerwein, R. *Proceedings of Fire and Materials 2001*, Interscience, San Francisco, CA, 395, 2001.
31. Pei-Hsuan, H.; Shinn-Jen, C.; Chia-Chen, L.; Chi-An, C. Boehmite-Based Microcapsules as Flame-Retardants for Lithium-ion Batteries. *Electrochim. Acta* **2017**, *228*, 597–603.
32. Pawlowski, K. H.; Schartel, B. Flame Retardancy Mechanisms of Aryl Phosphates in Combination with Boehmite in Bisphenol A Polycarbonate/Acrylonitrile-Butadiene-Styrene Blends. *Polym. Degrad. Stab.* **2008**, *93*, 657–667.
33. Rust, D.A. *Proceedings of Functional Fillers and Reinforcements '99*, Intertech, Atlanta, GA, Paper 22, 1999.
34. Anon, 'Use of calcium sulfate gypsum as a fire retardant for polyester spray-up and latex carpet backings; *C. E. Westbrook, SME Annual Meeting*, Reno, Nevada, February 15–18, 1993. www.911metallurgist.

com/use-of-calcium-sulfate-gypsum-as-a-fire-retardant/ (accessed 14 April 2023).
35. Wang, X.; Zhang, Q. Effect of Hydrotalcite on the Thermal Stability, Mechanical Properties, Rheology and Flame Retardance of Poly(Vinyl Chloride). *Polym. Int.* **2004**, *53*, 698–707.
36. Wang, H.; Huang, W.; Yang, L. Preparation and Flame Retardancy of Hydrotalcite/Montmorillonite Nanocomposite. *J. Phys. Conf. Ser.* **2020**, *1676* (1), 012060.
37. Gao, Y.; Wu, J.; Wang, Q.; Wilkie, C. A.; O'Hare, D. Flame Retardant Polymer/Layered Double Hydroxide Nanocomposites. *J. Mater. Chem. A* **2014**, *2*, 10996–11016.
38. Murray, H.H. *Applied Clay Mineralogy: Occurrences, Processing and Application of Kaolins, Bentonites, Palygorskite-Sepiolite, and Common Clays.* Elsevier, Amsterdam, 2007.
39. Jasinski, E.; Bounor-Legaré, V.; Taguet, A.; Beyou, E. 'Influence of Halloysite Nanotubes onto the Fire Properties of Polymer Based Composites: A Review. *Polym. Degrad. Stab.* **2021**, *183*, 109407.
40. Marset, D.; Dolza, C.; Fages, E.; Gonga, E.; Gutiérrez, O.; Gomez-Caturla J.; Ivorra-Martinez, J.; Sanchez-Nacher, L.; Quiles-Carrillo, L. The Effect of Halloysite Nanotubes on the Fire Retardancy Properties of Partially Biobased Polyamide 610. *Polymers* **2020**, *12* (12), 3050.
41. Mingliang, D.; Baochun, G.; Demin, J. Thermal Stability and Flame Retardant Effects of Halloysite Nanotubes on Poly(Propylene). *Eur. Polym. J.* **2006**, *42* (6), 1362–1369.
42. Lewin, M. Synergistic and Catalytic Effects in Flame Retardancy of Polymeric Materials—An Overview. *J. Fire Sci.* **1999**, *17* (1), 3–19.
43. Weil, E. D.; Levchik, S. V.; Ravey, M.; Zhu, W. A Survey of Recent Progress in Phosphorus-Based Flame Retardants and Some Mode of Action Studies. *Phosphorus, Sulfur Relat. Elem.* **1999**, *144* (1), 17–20.
44. Levchik, S. V.; Weil, E. D. Overview of Recent Developments in the Flame Retardancy of Polycarbonates. *Polym. Internat.* **2005**, *54* (7), 981–998.
45. Morgan, A.B., Chu, L.-L., and Harris, J.D. 'Flammability performance comparison between synthetic and natural clays in polystyrene nanocomposites', in *Proc. Flame Retardants 2004*, Interscience Communications, London, pp. 85–96, 2004.
46. Kahraman, M.; Kızılcan, N.; Oral, M. A. Influence of Mica Mineral on Flame Retardancy and Mechanical Properties of Intumescent Flame Retardant Polypropylene Composites. *Open Chem.* **2021**, *19* (1), 904–915.
47. Hu, S.; Chen, F.; Li, J.-G.; Shen, Q.; Huang, Z.-X.; Zhang, L.-M. The Ceramifying Process and Mechanical Properties of Silicone Rubber/Ammonium Polyphosphate/Aluminium Hydroxide/Mica Composites. *Polym. Degrad. Stab.* **2016**, *126*, 196–203.
48. Hu, S.; Tan, Z.-W.; Chen, F.; Li, J.-G.; Shen, Q.; Huang, Z.-X.; Zhang, L.-M. Flame-Retardant Properties and Synergistic Effect of Ammonium Polyphosphate/Aluminum Hydroxide/Mica/Silicone Rubber Composites. *Fire Mater.* **2020**, *44* (5), 673–682.
49. Hillier, S.; Marwa, E. M. M.; Rice, C. M. On the Mechanism of Exfoliation of Vermiculite. *Clay Miner.* **2013**, *48* (4), 563–582.
50. Ren, Q.; Zhang, Y.; Li, J.; Li, J. C. Synergistic Effect of Vermiculite on the Intumescent Flame Retardance of Polypropylene. *J. Appl. Polym. Sci.* **2011**, *120* (2), 1225–1233.
51. Macheca, A. D.; Focke, W. W.; Kaci, M.; Panampilly, B.; Androsch, R. Flame Retarding Polyamide 11 with Exfoliated Vermiculite Nanoflakes. *Polym. Eng. Sci.* **2018**, *58*, 1746–1755.
52. Horn, W.E. Chapter 9 'Inorganic hydroxides and hydroxycarbonates: Their function and use as flame-retardant additives', *Fire Retardancy of Polymeric Materials*, Grand, A.F. and C.A. Wilkie, C.A., Ed., Marcel Dekker, Basel, 2000.
53. Kirschbaum, G.S. 'Aluminum hydroxide for non-halogen compounds-well known- and still ever young', *Proceedings from Flame Retardants '94*, p 169, British Plastics Federation London, Interscience Communication Ltd., 1994.
54. Brown, S.C. and Herbert, M.J. 'New developments in ATH technology and applications', *Proceedings from Flame Retardants '92*, p 100, Plastics and Rubber Institute, Elsevier Applied Science, 1992.
55. Szablowska, B. and Pelica, J. 'Fire proofed floor coverings on the basis of polyolefins', *Proceedings from Flame Retardants '96*, p 241 S. J. Grayson (Ed.), British Plastics Federation, London, Interscience Communications Ltd., 1996.
56. Jow, J. and Gomolka, D. (Union Carbide Chemicals) 'Flame-retardant compositions useful for cable insulating and jacketing', US Patent 5482990A, 1996.
57. Sezaki, E., Akami, M. and Endo, H. 'Flame-retarded thermoplastic elastomer composition', European Patent 0331358, 1989.
58. Yamamoto, Y. and Tanmachi, M. 'Thin fire-resistant electrically insulated wire', Japanese Patent 06215644, 1994.
59. Hitachi Cable Ltd. 'Fire-resistant electrically insulating polyolefin compositions', Japanese Patent 04253748, 1992.
60. Hitachi Cable Ltd. 'Manufacture of halo-free flame-retardant polyolefin-covered electric wires', Japanese Patent 07122140, 1995.
61. Brown, S.C., Evans, K.A. and Godfrey, E.A. 'New Developments in Alumina Trihydrate', Paper 11 *Proceedings from BPF/PRI Conference Flame Retardants '87*, London, British Plastics Federation/Plastics and Rubber Institute, 1987.
62. Wan Hanafi, W. Z. A.; Hornsby, P. R. Smoke suppression and Fire Retardancy in Unsaturated Polyester Compositions Containing Hydrated Fillers. *Plast. Rubber Comp. Proc. Appl.* **1993**, *19*, 175–184.
63. Wan Hanafi, W.Z.A. 'A study of magnesium hydroxide as a flame retardant and smoke suppressant for unsaturated polyesters', M.Phil Thesis, Brunel University, July, 1988.
64. Bolger, R. Flame Retardant Materials. *Ind. Miner.* **1996**, 29.
65. Govanucci, K.-L. New Huntite Site Expands Possibilities for FR. *Functional Filler News* **1995**, *40* (1–2).
66. Toure, B.; Cuesta, J.-M. L.; Benhassaine, A.; Crespy, A. The Combined Action of Huntite and Hydromagnesite for Reducing Flammability of an Ethylene-Propylene Copolymer. *Intl. J. Polym. Anal. Charact.* **1996**, *2*, 193.
67. Briggs, C.C., Bhardwaj, B. and Gilbert, M. 'Flame retardant PVC cable compounds using huntite-hydromagnesite', Paper 9.51, *Proceedings of 5th BPF International Fillers Conference*, May 19–20, Manchester, 1992.
68. Molefe, D. M.; Labuschagne, J.; Focke, W. W.; van der Westhuizen, I.; Ofosu, O. The Effect of Magnesium Hydroxide, Hydromagnesite and Layered Double Hydroxide on the Heat Stability and Fire Performance of Plasticized Poly(Vinyl Chloride). *J. Fire Sci.* **2015**, *33* (6), 493–510.
69. Dike, A. S.; Tayfun, U.; Dogan, M. Influence of Zinc Borate on Flame Retardant and Thermal Properties of

Polyurethane Elastomer Composites Containing Huntite-Hydromagnesite Mineral. *Fire Mater.* **2017**, *47*, 890–897.
70. Witkowski, A., Hollingbery, L. and Hull, T.R. 'Fire retardancy of mineral fillers in EVA copolymers', *Fire Polymers VI: New Advances in Flame Retardant Chemistry and Science. ACS Symp. Ser.* 1118, 97–111, 2012.
71. Huang, P.-H.; Chang, S.-J.; Li, C.-C.; Chen, C.-A. Boehmite-Based Microcapsules as Flame-Retardants for Lithium-Ion Batteries. *Electrochim. Acta.* **2017**, *228*, 597–603.
72. Sun, T.; Zhuo, Q.; Chen, Y.; Wu, Z. Synthesis of Boehmite and its Effect on Flame Retardancy of Epoxy Resin. *High Perform. Polym.* **2015**, *27* (1), 100–104.
73. Hornsby, P. R. The Application of Magnesium Hydroxide as a Fire Retarding and Smoke Suppressing Additive for Polymers. *Fire Mater.* **1994**, *18*, 269.
74. Hornsby, P. R.; Wang, J.; Cosstick, K.; Rothon, R. N.; Jackson, G.; Wilkinson, G. Mechanism of Fire Retardancy in Polyamides Filled with Magnesium Hydroxide. *Prog. Rubber Plast. Technol.* **1994**, *10* (3), 204–220.
75. Hull, T. R.; Witkowski, A.; Hollingbery, L. Fire Retardant Action of Mineral Fillers. *Polym. Degrad. Stab.* **2011**, *96* (8), 1462–1469.
76. Rychlý, J.; Vesely, K.; Gal, E.; Kummer, M.; Jancar, J.; Rychla, L. Use of Thermal Methods in the Characterization of the High-Temperature Decomposition and Ignition of Polyolefins and EVA Copolymers Filled with $Mg(OH)_2$, $Al(OH)_3$ and $CaCO_3$. *Polym. Deg. Stab.* **1990**, *30*, 57.
77. Herbiet, R. Paper 17 'Nanocomposites based on boehemite alumina', *Proceedings of the Functional Fillers Conference*, Hamburg, September 15–17, 2004.
78. Hornsby, P.R., Wang, J., Jackson, G., Rothon, R.N., Wilkinson, G. and Cosstick, K. 'Analysis of fire retardancy in polyamides modified with magnesium hydroxide filler', *Proceedings from Society of Plastics Engineers Annual Technical Conference (ANTEC '94)*, San Francisco, CA, 2834–2839, 1–5 May, 1994.
79. Wang, J. 'Mechanism of flame retardancy in polyamides containing magnesium hydroxide', PhD thesis, Brunel University, 1994.
80. Stec, A. A. Fire Toxicity—The Elephant in the Room? *Fire Safe. J.* **2017**, *19*, 79–90.
81. Hirschler, M.M. 'Flame Retardants and the Associated Toxicity,' *PolyFlame*, Newsletter n°25, September 7–12, 2022.
82. Mochane, M. J.; Mokhothu, T. H.; Mokhena, T. C. Synthesis, Mechanical, and Flammability Properties of Metal Hydroxide Reinforced Polymer Composites: A Review. *Polym. Eng. Sci.* **2022**, *62* (1), 44.
83. Moseman, M.; Ingham, J. D. Smoke Properties of Highly Filled Ethylene-Propylene-Diene Terpolymer Rubbers. *Rubber Chem. Technol.* **1978**, *51*, 970.
84. Chen, X.; Ma, Z.; Zhou, J.; Wang, J.; Zhang, X.; Zhao, R.; Tong, J. Thermal Degradation Characteristics of Styrene-Butadiene-Styrene Copolymer Asphalt Binder Filled with an Inorganic Flame-Retarding Agent. *Polymers* **2022**, *14*, 3761.
85. Hirschler, M. M.; Thevaranjan, T. R. Effects of Magne Sium Oxide/Hydroxide on Flammability and Smoke Production Tendency of Polystyrene. *Eur. Polym. J.* **1985**, *21* (4), 371.
86. Lahaye, J. Mechanisms of Soot Formation. *Polym. Degrad. Stab.* **1990**, *30*, 111–121.
87. Krylov, O.V. *Catalysis by Non-Metals.* Academic Press, London, 1970.
88. McKee, D.W. 'The Catalyzed Gasification Reactions of Carbon', in *Chemistry and Physics of Carbon*, Vol 16, Walker, P.L. and Thrower, P.A., Eds., Marcel Dekker, New York, 1981.
89. Walker, P.L., Shelef, M. and Anderson, R.A. 'Catalysis of Carbon Gasification' in *Chemistry and Physics of Carbon*, Vol 4, Walker, P.L., Ed., Edward Arnold, London, 1968.
90. Delfosse, L.; Baillet, C.; Brault, A.; Brault, D. Combustion of Ethylene-Vinyl Acetate Copolymer Filled with Aluminum and Magnesium Hydroxides. *Polym. Degrad. Stab.* **1989**, *23*, 337.
91. Hornsby, P. R.; Watson, C. L. Interfacial Modification of Polypropylene Composites Filled with Magnesium Hydroxide. *J. Mat. Sci.* **1995**, *30*, 5347–5355.
92. Carpentier, F.; Bourbigot, S.; Le Bras, M.; Delobel, R.; Foulon, M. Charring of Fire Retarded Ethylene Vinyl Acetate Copolymer—Magnesium Hydroxide/Zinc Borate Formulations. *Polym. Degrad. Stab.* **2000**, *69*, 83–92.
93. Dike, A. S.; Tayfun, U.; Doğan, M. Influence of Zinc Borate on Flame Retardant and Thermal Properties of Polyurethane Elastomer Composites Containing Huntite-Hydromagnesite Mineral. *Fire Mater.* **2017**, *41*, 890–897.
94. Atay, H. Y.; Çelik, E. Flame Retardant Properties of Boric Acid and Antimony Oxide Accompanying with Huntite and Hydromagnesite in the Polymer Composites. *Polym. Polym. Comp.* **2016**, *24* (6), 419–428.
95. Kirschbaum, G.S. 'Recent developments in ATH and magnesium hydroxides-A challenge to traditional materials', *Proceedings of the 1995 Fall Conference of the Fire Retardant Chemical Association*, p 145, Rancho Mirage, CA, Technomic Publishing Co., October 29–November 1, 1995.
96. Kirschbaum, G. S. Halogenfreier Flammschutz. *Kunststoffe* **1989**, *79* (11), 62–64.
97. Moore, F. W.; Weber, T. R.; Tsigdinos, G. A. Advances in the Use of Molybdenum Compounds as Smoke Suppressants for PVC. *J. Vinyl Addit. Technol.* **1981**, *3*, 139–142.
98. Staendeke, H. *Proceedings of the Spring Conference of the Fire Retardant Chemical Association*, p 32, Grenelefe, FL, Technomic Publishing Co., March 20–23, 1988.
99. Thi, N. H.; Nguyen, T. N.; Oanh, H. T.; Trang, N. T. T.; Tham, D. Q.; Nguyen, H. T.; Nguyen, T. V.; Hoang, M. H. Synergistic Effects of Aluminum Hydroxide, Red Phosphorus, and Expandable Graphite on the Flame Retardancy and Thermal Stability of Polyethylene. *J. Appl. Polym. Sci.* **2021**, *138*, 50317.
100. Chavez, M.J. and Romanesco, D.J. 'Fire retarded plastics: Computer manipulation and video cone calorimeter data', *Proceedings of the 1995 Fall Conference of the Fire Retardant Chemical Association*, p 169, Rancho Mirage, CA, Technomic Publishing Co., October 29–November 1, 1995.
101. Huber, M.S., 'Non-halogenated flame retarded systems for olefins', *Proceedings of the Spring Conference of the Fire Retardant Chemical Association*, p 237, New Orleans, LA, Technomic Publishing Co., March 25–28, 1990.
102. Yeh, J. T.; Yang, H. M.; Huang, S. S. Combustion of Polyethylene Filled with Metallic Hydroxides and crosslinkable. Polyethylene. *Polym. Deg. Stab.* **1995**, *50*, 229–234.
103. Baillet, C.; Delfosse, L. The Combustion of Polyolefins Filled with Metallic Hydroxides and Antimony Trioxide. *Polym. Degrad. Stab.* **1990**, *30*, 89–99.
104. Yurddaskal, M.; Nil, M.; Ozturk, Y.; Celik, E. 'Synergetic Effect of Antimony Trioxide on the Flame Retardant and Mechanical Properties of Polymer Composites for Consumer Electronics Applications. *J Mater Sci: Mater Electron.* **2018**, *29*, 4557–4563.

105. Imahasi, T., Okada, A. and Abe, T. Flame retardant aid, flame retardant and flame-retardant composition, U.S. Patent 5,583,172, December 10, 1996.
106. Shen, L.; Chen, Y.; Li, P. Synergistic Catalysis Effects of Lanthanum Oxide in Polypropylene/Magnesium Hydroxide Flame Retarded System. *Compos. A Appl. Sci. Manuf.* **2012**, *43* (8), 1177–1186.
107. Zhu, W.; Weil, E. D. A Paradoxical Flame-Retardant Effect of Nitrates in ATH-Filled Ethylene-Vinyl Acetate Copolymer. *J. Appl. Polym. Sci.* **1995**, *56*, 925–933.
108. Chiu, S. H.; Wang, W. K. Dynamic Flame Retardancy of Polypropylene Filled with Ammonium Polyphosphate, Pentaerythritol and Melamine Additives. *Polymer* **1998**, *39*, 1951–1955.
109. Dittrich, B.; Wartig, K.-A.; Mülhaupt, R.; Schartel, B. Flame-Retardancy Properties of Intumescent Ammonium Poly(Phosphate) and Mineral Filler Magnesium Hydroxide in Combination with Graphene. *Polymers* **2014**, *6*, 2875–2895.
110. Weil, E. D.; Lewin, M.; Lin, H. S. Enhanced Flame Retardancy of Polypropylene with Magnesium Hydroxide, Melamine and Novolac. *J. Fire Sci.* **1998**, *16*, 383–404.
111. Bertelli, G., Goberti, P., Marchini, R., Camino, G. and Luda, M.P. 'Combined melamine/mineral fillers as fire retardants for polypropylene', *Proceedings from 6th European Meeting on Fire Retardancy of Polymeric Materials (FRPM '97)*, p 34, Lille, 1997.
112. Baggaley, R. G.; Hornsby, P. R.; Yahya, R.; Cusack, P. A.; Monk, A. W. The Influence of Novel Zinc Hydroxystannate-Coated Fillers on the Fire Properties of Flexible PVC. *Fire Mater.* **1997**, *21*, 179.
113. Cusack, P. A.; Hornsby, P. R. Zinc Hydroxystannate/ Zinc Stannate-Coated Fillers: Novel Flame Retardants and Smoke Suppressants for Polymeric Materials. *J. Vinyl Addit. Technol.* **1999**, *5*, 21–30.
114. Horrocks, A. R.; Smart, G.; Price, D.; Kandola, B. Zinc Stannates as Alternative Synergists in Selected Flame Retardant Systems. *J. Fire Sci.* **2009**, *27* (5), 495–522.
115. Mohai, M.; Tóth, A.; Hornsby, P. R.; Cusack, P. A.; Cross, M.; Marosi, G. XPS Analysis of Zinc Hydroxystannate Coated Fillers. *Surf. Interface Anal.* **2002**, *34*, 735–739.
116. Hornsby, P. R. G.; Cusack, P. A.; Cross, M.; Tóth, A.; Zelei, B.; Marosi, G. Zinc Hydroxystannate-Coated Metal Hydroxide Fire Retardants: Fire Performance and Substrate-Coating Interactions. *Mater. Sci.* **2003**, *38*, 2893–2899.
117. Namike, Y., Kato, Y., Kitano, Y., Kuriso, H. and Yokata, Y. 'Flame retardant polyolefin compound having low smoking and toxicity', US Patent No. 5,654,356, August 5th, 1997.
118. Chen, H.; Wang, T.; Wen, Y.; Wen, X.; Gao, D.; Yu, R.; Chen, X.; Mijowska, E.; Tang, T. Expanded Graphite Assistant Construction of Gradient-Structured Char Layer in PBS/$Mg(OH)_2$ Composites for Improving Flame Retardancy, Thermal Stability and Mechanical Properties. *Compos. B. Eng.* **2019**, *177*, 107402.
119. Wang, N.; Wang, S.; Chen, J.; Wang, S. Synergistic Effect of Benzoic Acid-Functionalized Graphene-$Al(OH)_3$ on Flame Retardant of Polypropylene. *Fuhe Cailiao Xuebao/ Acta Mater. Compos. Sinica* **2018**, *35* (9), 2434–24411.
120. Markezich, R. L.; Aschbacker, D. G. Chlorinated Flame-Retardant in Combination with Other Flame Retardants. *ACS Symp. Ser.* **1995**, *599*, 65–75.
121. Toure, B.; Cuesta, J. M. L.; Gaudon, P.; Benhassaine, A.; Crespy, A. Fire Resistance and Mechanical Properties of a Huntite/Hydromagnesite/Antimony Trioxide/Decabromo-Diphenyl Oxide Filled PP-PE Copolymer. *Polym. Degrad. Stab.* **1996**, *53*, 371–379.
122. Miyata, S. and Imahasi, T. 'Fire-retardant resin composition', US Patent No. 5,094,781, March 10th, 1992.
123. Morgan, A.B. and Wilkie, C.A. (Eds.). *Flame Retardant Polymer Nanocomposites*. John Wiley & Sons, Hoboken, NJ, 2007.
124. Gui, H.; Zhang, X.; Dong, W.; Gao, J.; Song, Z.; Lai, J.; Liu, Y.; Huang, F.; Qiao, J.; Wang, Q. Effect of Rubbers on the Flame Retardancy of EVA/Ultrafine Fully Vulcanized Powdered Rubber/Nanomagnesium Hydroxide Ternary Composites. *Polym. Compos.* **2007**, *28*, 479–483.
125. Liu, L.-H.; Chen, J.-M.; Song, Y.-H.; Guo, F.; Chen, J.-F. Application of Nano-Sized Magnesium Hydroxide Flame Retardant in Flexible PVC. *J. Chem. Eng. Chin. Univ.* **2004**, *18*, 339–343.
126. Lv, J.; Qie, L.; Qu, B. Controlled Synthesis of Magnesium Hydroxide Nanoparticles with Different Morphological Structures and Related Properties in Flame Retardant Ethyl-Ene–Vinyl Acetate Blends. *Nanotechnol.* **2004**, *15*, 1576–1581.
127. Qiu, L.; Xie, R.; Ding, P.; Qu, B. Preparation and Characterization of $Mg(OH)_2$ Nanoparticles and Flame Retardant Property of its Nanocomposites with EVA. *Compos. Struct.* **2003**, *62*, 391–395.
128. Fan, W.; Sun, S.; You, L.; Cao, G.; Song, X.; Zhang, W.; Yu, H. Solvothermal Synthesis of $Mg(OH)_2$ Nanotubes Using $Mg_{10}(OH)_{18}Cl_2.5H_2O$ Nanowires as Precursors. *J. Mater. Chem.* **2003**, *13*, 3062–3065.
129. Zhang, Q.; Tian, M.; Wu, Y.; Lin, G.; Zhang, L. Effect of Particle Size on the Properties of $Mg(OH)_2$-Filled Rubber Composites. *J. Appl. Polym. Sci.* **2004**, *94*, 2341–2346.
130. Mishra, S.; Sonawane, S. H.; Singh, R.; Bendale, A.; Patil, K. Effect of Nano-$Mg(OH)_2$ on the Mechanical and Flame-Retarding Properties of Polypropylene Composites. *J. Appl. Polym. Sci.* **2004**, *94*, 116–122.
131. Okobo, N., Okumura, H., Okoshi, M. and Hamada, H. 'Flame retardancy of EVA/nano magnesium hydroxide hybrid', *Proceedings from Additives 2004*, Clearwater Beach, FL, Executive Conference Management, March 22–24, 2004.
132. Camino, G.; Maffezzoli, A.; Braglia, M.; De Lazzaro, M.; Zammarano, M. Effect of Hydroxides and Hydroxycarbonate Structure on Fire Retardant Effectiveness and Mechanical Properties in Ethylene-Vinyl Acetate Copolymer. *Polym. Degrad. Stab.* **2001**, *74* (3), 457–464.
133. Santhosh, B.; Kumar, M.; Mathews, J. M.; Mohamed, A. A. P.; Solaiappan, A. A Facile Hydrous Mechano-Synthesis of Magnesium Hydroxide [Hy-$Mg(OH)_2$] Nano Fillers for Flame- Retardant Polyester Composites. *Chem. Eng. J. Adv.* **2023**, *14*, 100466.
134. Brostow, W.; Lohse, S.; Lu, X.; Osmanson, A. T. Nano-Al $(OH)_3$ and Mg $(OH)_2$ as Flame Retardants for Polypropylene Used on Wires and Cables. *Emergent Mater.* **2019**, *2*, 23–34.
135. Pilarska, A.; Bula, K.; Myszka, K.; Rozmanowski, T.; Szwarc-Rzepka, K.; Pilarski, K.; Chrzanowski, Ł.; Czaczyk, K.; Jesionowski, T. Functional Polypropylene Composites Filled with Ultra-Fine Magnesium Hydroxide. *Open Chem.* **2015**, *13* (1), 10151520150024.
136. Zhang, C.; Cheng, X.-W.; Guan, J.-P.; Chen, G. Preparation of Nano-$Mg(OH)_2$ for Surface Coating of Silk Fabric with Improved Flame Retardancy and Smoke Suppression. *Colloids Surf. Physicochem. Eng. Asp.* **2021**, *625*, 126868.
137. Naguib, H. M. Environmental-Friendly Recycled Polyester/ $Mg(OH)_2$ Nanocomposite: Fire-Retardancy and Thermal Stability. *Polym. Test.* **2018**, *72*, 308–314.

138. Beyer, G. Flame Retardant Properties of EVA-Nanocomposites and Improvements by Combination of Nanofillers will Aluminium Trihydrate. *Fire Mater.* **2001**, *25*, 193–197.
139. Beyer, G. Flame Retardancy of Nanocomposites—From Research to Technical Products. *J. Fire Sci.* **2005**, *23*, 75–87.
140. Okoshi, M.; Nishizawa, H. Flame Retardancy of Nano-Composites. *Fire Mater.* **2004**, *28*, 423–429.
141. Lu, H.; Hu, Y.; Yang, L.; Wang, Z.; Chen, Z.; Fan, W. Study of the Fire Performance of Magnesium Hydroxide Sulfate Hydrate Whisker Flame Retardant Polyethylene. *Macromol. Mater. Eng.* **2004**, *289*, 984–989.
142. Torno, O., Paper 19, 'Nanocomposites based on boehemite alumina', *Proceedings from Conference on Functional Fillers for Plastics*, 15–17 September, Hamburg, Interscience, 2004.
143. Mochane, M. J.; Mokhothu, T. H.; Mokhena, T. Synthesis, Mechanical, and Flammability Properties of Metal Hydroxide Reinforced Polymer Composites: A Review. *Polym. Eng. Sci.* **2022**, *62* (1), 44.
144. Beyer, G., Paper 16, 'Nanocomposites – A new class of flame retardants for polymer and cable applications', *Proceedings from Conference on Functional Fillers for Plastics*, 15–17 September, Hamburg, Interscience, 2004.
145. Zhang, Z.; Wu, W.; Zhang, M.; Qu, J.; Shi, L.; Qu, H.; Xu, J. Hydrothermal Synthesis of $4ZnO{\cdot}B_2O_3{\cdot}H_2O$/RGO Hybrid Material and its Flame Retardant Behavior in Flexible PVC and Magnesium Hydroxide Composites', *Appl. Surf. Sci.* **2017**, *425*, 896–904.

7 Intumescence-Based Flame Retardant

Serge Bourbigot[1,2]
[1] Univ. Lille, CNRS, INRAE, Centrale Lille, UMR 8207—UMET—Unité Matériaux et Transformations, F-59000 Lille, France
[2] Institut Universitaire de France (IUF)

7.1 INTRODUCTION

The word "intumescence" comes from Latin "intumescere" which means "to swell up." From the dictionary, intumescence is defined as the act or process of swelling or enlarging. Some definition also mentions the action of heat to expand the body. French Writers like Jules Verne or François-René de Chateaubriand used the word "intumescence" with different meanings: a hilly area on volcano (J. Verne in The Mysterious Island, 1874) or the waves' swell (F.R. de Chateaubriand in Mémoires d'Outre-Tombe, 1848). Taken "as is," those definitions are not directly linked to the flame retardancy of polymeric materials or to the fire protection. Nevertheless, the above definitions describe well the behavior of an intumescent material used for fire retardancy. When heating beyond a critical temperature, it begins to swell and then to expand. The result of this process is a foamed cellular charred layer on the surface which protects the underlying material from the action of the heat flux or the flame. Visually, the swelling and the expansion looks like "black waves" swollen at the surface of the material, and the final char exhibits hemispheric shape with a roughed or smooth surface.

It can be considered that intumescence was first reported in the scientific literature by J.L. Gay-Lussac in 1821 in the case of the flame retardancy of textiles.[1] The word "intumescence" was not mentioned, but we suspect an intumescent phenomenon (similar experiments were done by this author) when woven fabrics in hemp and flax were coated by a mixture of borate and ammonium phosphate: here there are the three ingredients of intumescence (it will be commented further in the paper): the acid source, namely, borate and phosphate; the char former, namely, flax or hemp; and the blowing agent, namely, the evolving ammonia from the phosphate. In 1912, intumescence was mentioned in a US patent.[2] It describes the formation of an intumescent material made from the expansion of silicates. The resulting material acts as heat insulator in the technical arts, but no application for fire protection was reported. In 1934, a German patent (extended in the US in 1938) claimed the fire protection of wood using a mixture of diammonium phosphate and formaldehyde.[3] It was reported the formation of a swollen char layer upon heating, protecting wood, but the word "intumescence" was not used in the text. Again in 1940, the word "intumescence" was used in a US patent where it was claimed the making of silicate-based intumescent coating to protect combustibles.[4] It is the same technology as that used in the aforementioned US patent of 1912[2] using alkali silicates. The first comprehensive paper was published in the early seventies by H. Vandersall and gives the fundamentals of intumescence.[5] In this latter paper, only coatings were considered, and it was only in the eighties that G. Camino applied with success this concept to bulk polymers (mainly thermoplastics).[6] This basic work was useful for the development of products which were launched by different companies in the nineties.

It may be thought that intumescence is an old concept since the first comprehensive paper was published in the early seventies, but a brief review of the literature shows that intumescence is still largely employed to make flame retardant (FR) polymers and FR paints and that some recent developments are very promising for fire protection[7–9] and for flame retardancy.[10–12] This concept of intumescence appears as an attractive topic as demonstrated by the number of papers or patents dealing with this topic, which has been growing since the fifties (Figure 7.1). The number of papers (all types) and patents was the highest between 2014 and 2022 evidencing the interest for this concept from the scientific community and the industry. It is then the purpose of this chapter to review the developments done in the reaction and resistance to fire with the intumescence concept.

This chapter is organized in four parts. The two first sections briefly survey intumescence to provide the reader the basic understanding on the mechanisms of action by intumescence and to give the main intumescent products available on the market associated with their typical application's fields. Reaction to fire of intumescent polymers and textiles, that is, the contribution of the material to fire growth, is examined in the third section. The latest developments of flame retardants as well as synergists are considered, and mechanism of action is commented based on published work and on our own experience. The fourth section is devoted to the resistance to fire of materials using intumescent paint or coating. Resistance to fire is defined as the ability of materials to resist the passage of fire and/or gaseous products of combustion and is capable of meeting specified performance criteria to those ends (the reader should make the distinction between reaction and resistance to fire). A short description of the normalized tests is first given. It also explains why bench-scale testing must be developed and correlation between small- and large-scale

DOI: 10.1201/9781003380689-7

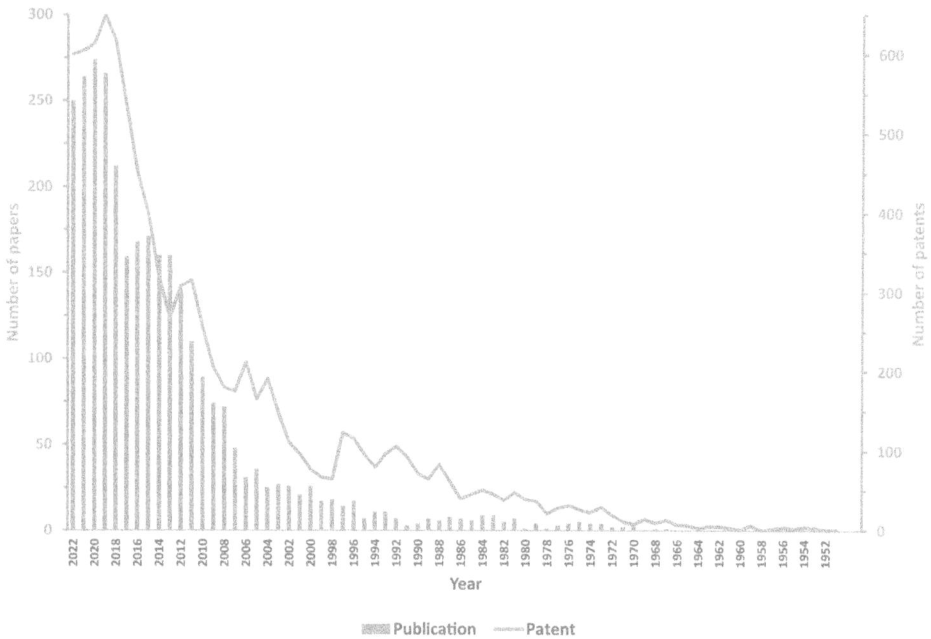

FIGURE 7.1 Number of publications (all types) and patents (extracted from the database of US patent office, Japan patent office, European patent office, world intellectual property organization, and UK intellectual property office) devoted to intumescence from 1951 to 2022 (Scopus database and Web of Science with the keywords combination: ["intumescence" OR "intumescent"] AND ["flame retardant" OR "flame retardancy" OR "fire retardant" OR "fire retardancy" OR "fire protection"]).

tests is shown. The mechanisms of action are discussed considering the chemical, physical, and thermal aspects.

7.2 FUNDAMENTALS OF INTUMESCENCE

Flame retarding polymers or textiles by intumescence are essentially a special case of a condensed phase mechanism.[13–17] Intumescent systems interrupt the self-sustained combustion of the polymer at its earliest stage, that is, the thermal decomposition with the evolution of gaseous fuels (Figure 7.2). The intumescence process results from a combination of charring and foaming at the surface of the substrate. So the charred layer acts as a physical barrier which slows down heat and mass transfer between gas and condensed phase. The formation of an effective char occurs via a semi-liquid phase, which coincides with gas formation and expansion of the surface. Gases released from the decomposition of the intumescent material, and in particular of the blowing agent, have to be trapped and to diffuse slowly in the highly viscous melt degraded material in order to create a layer with appropriate morphological properties. The viscosity of the degraded matrix in the blowing phase is therefore a critical factor. Another significant aspect of intumescent formulations is the mechanical strength of the intumescent char. In the conditions of a fire, char destruction can proceed not only by means of ablation and heterogeneous surface burning but also by means of an external influence such as wind, mechanical action of the fire, or convective air flows. The mechanical stability of the intumescent char depends both on the structure and porosity of the foamed intumescent material. If the structure of the char (morphology, distribution of voids inside the char) is appropriate, the thermal conductivity of intumescent chars can be very low, and it limits efficiently heat transfer from the heat source to the substrate. Finally, the construction of the whole intumescent structure is controlled by its chemical composition and its dynamics (kinetics).

FIGURE 7.2 Intumescent polylactide (PLA) during a cone calorimeter experiment. Note the small flames on the side of the intumescent "cake" showing how the intumescent coating smother the fire.

As mentioned above, the intumescence mechanism is that the intumescent material reacts upon heating to produce gases that are partially trapped in a viscoelastic matrix (rheological aspect). The matrix expands as gases are produced (from the blowing agent and/or from the decomposition products of the polymeric matrix), and at the same time, cross-linking reactions and charring cause the matrix to harden, thereby producing a coherent highly porous char (chemical aspects and formulation). The porosity of the char is generally extremely high that the resulting structure has extremely low thermal conductivity. The most important parameters of the expanded char affecting its thermal insulation performance are its heat conductivity and its ability to swell rapidly (thermal aspects). The thermal properties are fully dependent of the rheology and of the chemistry of the intumescent, and so, the appropriate values cannot be reached without considering them.

7.2.1 Chemistry of Intumescence

The usual chemistry of intumescence involves composition containing an inorganic acid or a material yielding acidic species, of a char former and of a component that decomposes to enable the expansion of the system (blowing agent). Typical examples of components used in intumescent systems are reported in Table 7.1.

The following sequences of events have been proposed to describe the development of the intumescent phenomenon:

- The inorganic acid is released at temperatures depending on its source and other components.
- The acid esterifies the carbon-rich components at temperatures slightly above the acid release temperature.
- The mixture of materials melts prior or during esterification.
- The ester decomposes via dehydration resulting in the formation of a carbon-inorganic residue.
- Released gases from the above reactions and decomposition products cause the carbonizing material to foam.
- As the reaction nears completion, gelation and finally solidification occurs, the resulting solid is in the form of multicellular foam.

A typical example is the case of polypropylene (PP)-ammonium polyphosphate (APP)/pentaerythritol (PER) system.[18,19] The reaction of the acidic species (APP and its decomposition products into orthophosphates and phosphoric acid) with the char former agent (PER) takes place in a first stage ($T < 280°C$) with formation of esters mixtures. The carbonization process then takes place at about 280°C (via the formation of double bonds followed by Diels Alder reaction and a free radical process increasing the size of the polyaromatic structure[20]). In a second step, the blowing agent decomposes to yield gaseous products (i.e., evolved ammonia from the decomposition of APP), which cause the char to swell ($280 < T < 350°C$). The intumescent material decomposes then at higher temperatures and loses its foamed character at about 430°C. Concurrently, the heat conductivity of the char decreases between 280°C and 430°C, and the insulation of the substrate is enhanced.[21]

The method to develop the intumescence phenomenon is mainly based on a series of chemical reactions occurring at the right time. Another way to make intumescence in a polymeric matrix is the physical expansion. In other words, the rapid sublimation of a molecule (and/or the decomposing products) creates the expansion of the top degraded layer of the polymer to make an intumescent coating. This mechanism occurs when using expandable graphite (EG). EG is a synthesized intercalation compound of graphite that expands or exfoliates when heated. A wide variety of chemical species can be used to intercalate graphite materials (e.g., sulfate, nitrate, various organic acids).[22] A typical example is the incorporation of 10 wt% EG in polypropylene (PP).[23] PP/EG was evaluated by cone calorimetry at an external heat flux of 35 kW/m² and compared to neat PP (Figure 7.3—(a)). When incorporated in PP, the peak of heat release rate (pHRR) is decreased by about 70%, and total heat release rate (THR) is also significantly reduced (by 35%). The formation of an expanded charred layer is

TABLE 7.1
Examples of Components of Intumescent Systems

(a) **Acid source**	(b) **Char former**
Phosphoric	Starch
Sulfuric	Dextrins
Boric	Sorbitol, mannitol, xylitol
Ammonium salts	Pentaerythritol, monomer, dimer, trimer
Phosphates, polyphosphates	Phenol-formaldehyde resins
Borates, polyborates	Methylol melamine
Sulfates	Isocyanurates
Phosphates of amine or amide	Char former polymers (PA-6, PA-6.6, PU, PC, PVC)
Products of reaction of urea or Guanidyl urea with phosphoric acids	(c) **Blowing agents**
Melamine phosphate	Urea
Product of reaction of ammonia with P_2O_5	Urea-formaldehyde resins
Organophosphorus compounds	Dicyandiamide
Tricresyl phosphate	Melamine
Alkyl phosphates	
Haloalkyl phosphates	

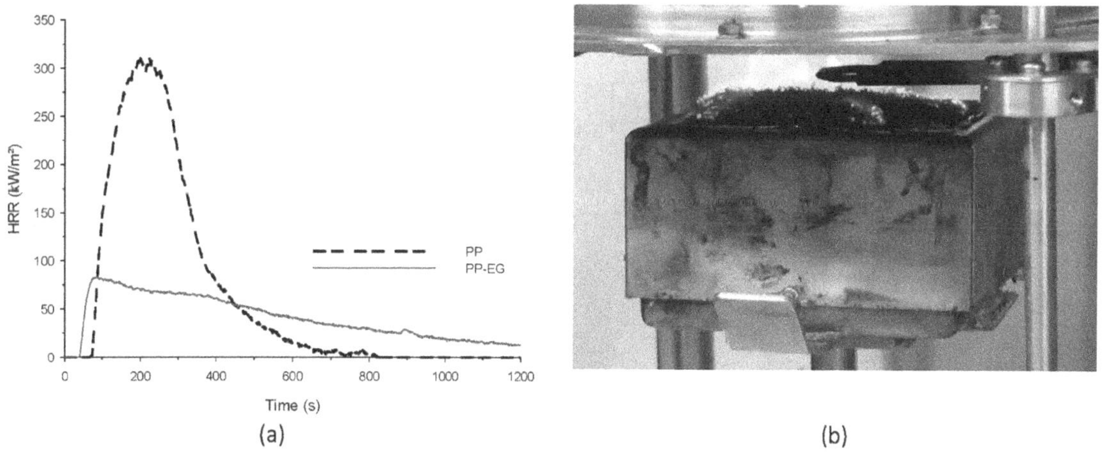

FIGURE 7.3 (a) HRR curves as a function of time of intumescent PP (external heat flux = 35 kW/m²) and (b) intumescent coating formed from PP/EG during the cone calorimetry experiment (note the formation of an entangled network made of graphite worms).

$$-\mathrm{Si}-\mathrm{OH} + \mathrm{HO}-\mathrm{Si}- \rightarrow -\mathrm{Si}-\mathrm{O}-\mathrm{Si}- + H_2O$$

SCHEME 7.1 Chemical reaction occurring during the expansion of silicates.

FIGURE 7.4 Intumescent silicate-based coating prepared in a furnace at high temperature.

observed at the surface of PP/EG evidencing an intumescent phenomenon. Not as conventional intumescent char which exhibits flat expanded carbonaceous layer, PP/EG shows an expanded "hairy" char during the experiment (Figure 7.3—(b)). The intercalation compounds contained in EG decompose rapidly into gaseous products, which blast off the graphite flakes. Those flakes then make worms to form an entangled network at the surface of the material. This network acts as a protective layer.

Alkali silicates can be also used to make intumescent coating.[4,24,25] The mechanism of intumescence is the rapid evolution of water at temperature higher than 100°C, creating a silicate-based foamed material (chemical reaction in Scheme 7.1) (Figure 7.4). This foamy material exhibits low heat conductivity and is appropriate for fire protection.[24] In addition to this, the evolution of water is strongly endothermic (4280 J/g), and hence it created an additional cooling effect and dilution.[26]

7.2.2 Rheology of Intumescence

Upon heating the intumescent material yields a viscoelastic char with the appropriate rheological properties (viscosity) to be expanded by the increasing internal pressure coming from the accumulation of the decomposition products and/or of the blowing agent. It must occur at the right time and at the right temperature to be efficient by controlling the kinetics of the decomposition reactions (dynamic of the intumescence). This phenomenon is relatively well understood, but its measurement has been reported only in the late nineties using a homemade device in a calibrated tube[27] and a rheometer.[28]

A typical example is the rheological measurements using a rheometer in a parallel plate configuration which were done to measure viscosity and expansion of an epoxy-based intumescent coating as a function of temperature.[29] It is clearly shown that expansion occurs when the viscosity is the lowest and when the temperature of charring is reached (about 250°C) (Figure 7.5-a). It is noteworthy that the viscosity must be high enough to accommodate internal stresses due to the internal pressure coming from the decomposition gases (for comparison the viscosity at 250°C of the virgin epoxy is twice lower than that of the intumescent coating). Expansion is also measured in the rheometer (convective heating in a furnace), and it gives useful indications on which temperatures intumescence takes place. Another method can be used to follow the expansion in situ during cone calorimetry experiment (the heat source is a radiative heat flux). It is done using an infrared camera combined with image analysis, and it provides the expansion as a function of time (Figure 7.5-b). Those protocols have been applied to numerous intumescent coatings, and the same trends were observed for each system evidencing the general behavior of intumescent material.[30–32]

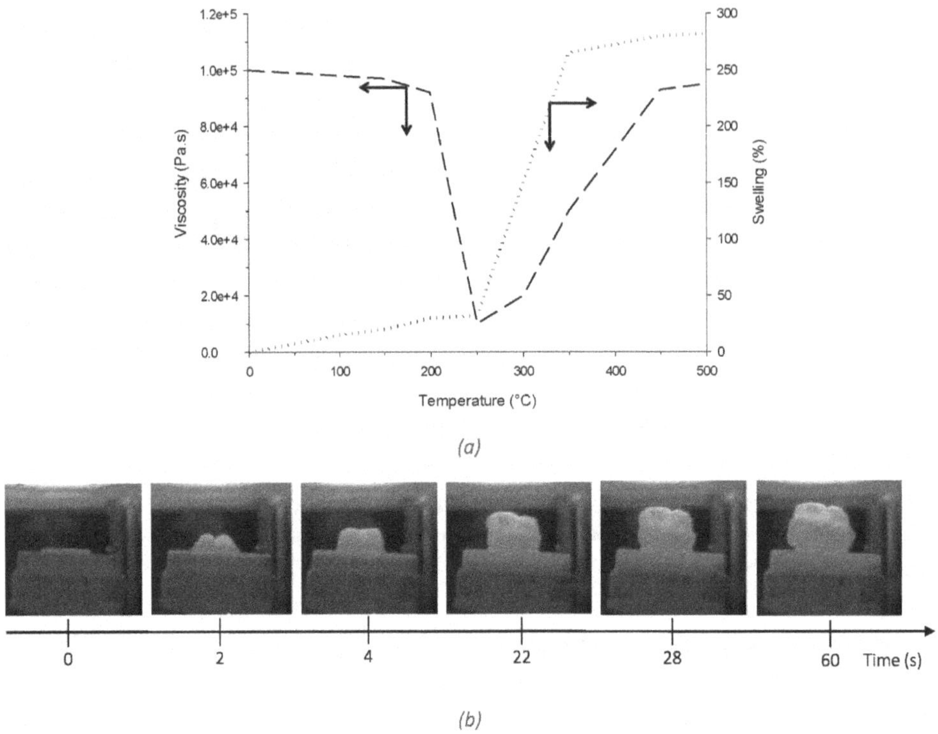

FIGURE 7.5 (a) viscosity and swelling as a function of temperature of an epoxy-based intumescent coating; (b) formation of an intumescent char as a function of time during cone calorimetry (external heat flux at 50 kW/m²) until the full expansion.

7.2.3 Thermal Behavior of Intumescence

The main purpose of an intumescence coating is to create an insulative material at the surface of the substrate to limit heat transfer and so to slow down the decomposition of the substrate and/or the diffusion of heat. If the structure of the char (morphology, distribution of voids inside the char) is appropriate[33] (Figure 7.6), the thermal conductivity of intumescent chars is very low, and hence, it limits efficiently heat transfer from the heat source to the substrate.

Heat conductivity (k) is one of the most influencing parameters determining the efficiency of an intumescent barrier. In a general way,[34] intumescent coating reacts around 250°C forming char and starting to swell. At higher temperatures (up to 350°C), the porosity of the intumescent char increases as well as its volume. Finally, between 400 and 600°C, intumescent char degrades slowly, and its expansion remains constant. Experimental results of k were determined with the transient plane source (TPS) method[35] up to 800°C for an epoxy-based intumescent paint.[31] At low temperature (T < 200°C), k increases, but when the material starts reacting and forming the char, k decreases dramatically (k drops down from 0.45 to 0.1 W/m.K). It is attributed to the formation of an expanded foamed char exhibiting low heat conductivity. At higher temperatures (T > 375°C), k increases considerably up to 500°C due to the shrinkage and the partial destruction of the char. At T > 500°C, k increases smoothly. From this experiment, it is then evidenced that intumescent coating exhibits low heat conductivity limiting heat transfer. This example can be extended to other intumescent systems,[36,37] and this trend can be generalized. It makes sense because the general mechanism of intumescence is based on the formation of multicellular structure from a bulk material at a given temperature, and hence, the structure changes lead to a specific k-changes as a function of temperature.

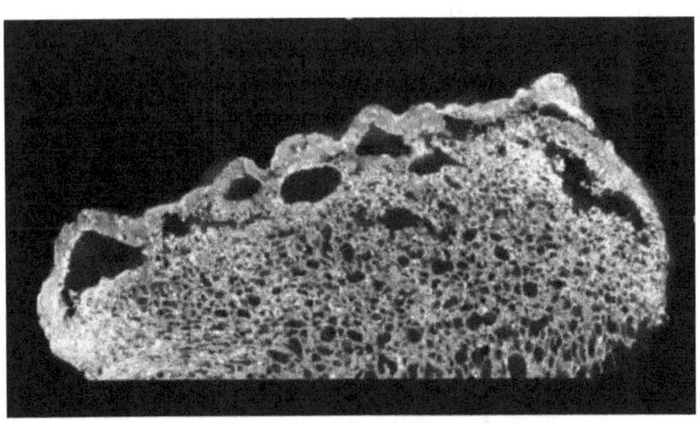

FIGURE 7.6 Internal structure of an intumescent char observed by X-ray tomography.

7.3 INTUMESCENCE ON THE MARKET

According to researchandmarkets.com, the market of halogen-free flame-retardants is projected to witness \$3,871.09 million in 2025 owing to the increasing government regulations for environmental safety and the flourishing electrical and electronics and automotive industries. This market is expected to increase at a highest growth rate between 2020 and 2025. Electrical and electronics (E&E) industry is the most significant end-use industry in the global halogen-free flame-retardant market. Phosphorus segment (phosphorus is one of the main ingredients to make intumescent systems) held the major revenue share of 57% in 2020, followed by minerals segment. Its global market is likely to log the fastest 2019–2026 CAGR (compounded annual growth rate) of 5.6% and reach a projected US$3.1 billion by 2026. Some of the key players operating in the global phosphorus-based flame retardants market are Clariant AG (Switzerland), Lanxess (Germany), Adeka (Japan), Huber Engineered Materials (USA), Israel Chemicals Limited (Israel), BASF SE (Germany), Italmatch Chemicals (Italy), Chang Chun Group (Taiwan), Rin Kagaku Kogyo Co (Japan), Jiangsu Liside (China), Shandong Ruixing (China), Shouguang Weidong Chemical Co (China), Shandong Moris Tech Co (China), Zhejiang Wansheng Co (China).

Typically, a phosphorus-based FR is designed to develop its activity just before the starting decomposition of the specific polymer it is used for. It can offer a partial gas phase contribution to the flame extinguishing effect, which is comparable to bromine- or chlorine containing FRs. However, the main feature is mostly char-forming activity like intumescence. The phosphorus-based FRs can be used in numerous polymers and resins such as polyamides, polyesters, polyolefins, epoxy, and styrenics. The main area of application for the compounded materials is injection-molded E&E parts. Besides the electrical market, a very important market is flame retardant fabrics for public buildings and public transportation seating. Other compounds as phosphorus can be used in intumescent systems (Table 7.2), but they are not widely employed except

TABLE 7.2
Examples of Intumescent Compounds Available on the Market

Type	Brand Name	Supplier	Examples of Application
Ammonium polyphosphate[(1)]	AP420 series	Clariant (Switzerland)	Intumescent gelcoats (Epoxy, UP or PUR based)
Ammonium polyphosphate combined with char former and nitrogen synergist[(2)]	AP740/750/760 series		FR polyolefins for E&E and building/transportation Applications—FR for lightweight composites
Aluminum phosphinate salt + melamine polyphosphate[(3)]	OP1310 series		FR polyamides for E&E applications
Ammonium polyphosphate[(1)] (crystal forms I and II)	FR CROS 580/750 series	Budenheim (Germany)	FR for PU, EP and paints
Melamine phosphate	Budit 310 series (310 to 312)		Unsaturated polyester resins, particularly for the SMC-process
Melamine borate	Budit 313		In combination with melamine phosphate for phenolic resin bound nonwoven based on cotton fibers
Coated ammonium polyphosphate[(2)]	Budit 660 series		FR polyolefins for E&E and building/transportation applications
Melamine polyphosphate	Melapur 200	BASF (Germany)	PA66/Glass fibers, epoxies, synergistic blends with other flame retardants
Hydrated sodium silicate	Palusol series		Intumescent seals which prevent the passage of smoke, heat and flames
Cresyl diphenyl phosphate[(4)]	Disflamoll DPK	Lanxess (Germany)	For thermosets like phenolic, polyester and epoxy resins, PC/ABS, HIPS/PPE blends, TPU compounds, PUR-foams (rigid and flexible) and rubbers
APP-based mixture containing char former and blowing agent	FlameOff	FlameOff (USA)	Intumescent powder for polyolefins
Piperazine pyrophosphate containing char former and blowing agent	ADK Stab FP 2000 series	Adeka (Japan)	Intumescent powder for polyolefins
Cresyl diphenyl phosphate[(4)]	CDP	Shandong Ruixing (China)	For thermosets, PVC and natural rubbers
Expandable graphite	GrafGuard	GrafTech (USA)	FR additive in plastics (processing up to 200°C), foams, putties, and coatings

Notes:
[(1)] *Typical acid source (dehydrating agent) used in numerous intumescent formulations*
[(2)] *This product contains all intumescent ingredients, that is, acid source, char former, blowing agent and a synergist*
[(3)] *In some specific grade, zinc borate is added as synergist and processing aid*
[(4)] *Intumescence is observed when used in char forming polymer (e.g., TPU, epoxy)*

boron-based compounds for nondurable textile applications and in intumescent coating for fire protection, and also expandable graphite and silicates for applications, including intumescent seals, foams, and strips. Table 7.2 gathers (it is not our intention to make an exhaustive list) intumescent compounds as typical examples of product available on the market. It shows that four main groups of additives are available and used industrially: phosphate, borate, and expandable graphite.

7.4 REACTION TO FIRE OF INTUMESCENT MATERIALS

7.4.1 Evaluation of the Reaction to Fire

This section intends to remind the reader the basics of fire testing of polymeric materials and to explain why certain physical parameters were considered to evaluate fire. Various organizations (e.g., ASTM, UL, CEN, ISO, IMO, etc.) throughout the world write fire standards. Their purpose was to create various types of fire standards, guidance documents, and test methods addressing most of the major fire properties. It is interesting that fire standards throughout the world are all tending in the same direction: away from simple ranking tests and geared towards test results that can be used as input into mathematical fire models or fire hazard assessment and research, and real-scale tests to be used for validation of the small-scale tests and models.[38,39]

The growth of fire follows distinct phases starting with ignition and early growth (it includes fire spread), to the fully developed stage, leading through to the eventual decay of the fire. These different stages are associated to fire scenarios mimicked by the standards depending on the application field (E&E, building, transportation, etc.). They are briefly commented hereafter, giving some fire tests as examples:

1. *Ignition*: All fires start with an ignition event, where a source of heat comes into contact with a fuel (here materials) in the presence of oxygen. Different standards can determine the ignition temperature or the ignition time on samples of small size (barrel or plaque of few centimeters and of few grams). The glow wire test (EN 60335–1) is one of them, and it is used to evaluate the flammability of material used within an appliance mimicking over-current or short circuit failures. It is an indirect method (no direct contact with the flame but with a heating glow wire element), and the ignition temperature can be determined. The well-known UL-94 test is a direct method where the flame is applied directly to a vertically or horizontally mounted specimen under controlled conditions. Test results from applying these methods provide a way to compare the materials' tendency to resist ignition, self-extinguish flames (ignition should occur), and to not propagate fire via dripping. The cone calorimeter (ISO 5660) is another test method which permits to vary the incident heat flux and to measure the time to ignition (among many other parameters, see below). It simulates a developing fire scenario.
2. *Flame spread*: An important parameter is the measurement of the rate at which a flame front tends to propagate along a flammable material. Usually, the tests measure the propagation of a flame away from the source of ignition across the surface of a material or assembly and evaluate the potential for spreading flames in the event of a fire. UL-94 test in its two configurations (horizontal and vertical burning) can estimate the flame spread. It provides preliminary indications on the flammability of polymeric materials. FMVSS 302 (or ISO 3795) is a horizontal burning test which measures the flame spread of samples used in passengers vehicles. The LIFT (Lateral Ignition and Flame Spread Test) apparatus (ISO 5658–2) provides ignitability and flame spread information on, typically, vertically oriented samples. It can generate fundamental data for fire modeling.
3. *Developing fire (or heat release rates measurement)*: The rate of heat release is arguably the most important fire parameter since it controls the rate of growth in the fire, including heat and ultimately the amount of smoke and toxic gas generated. It is measured by cone calorimetry setting heat flux resulting from an electrical heater rod, tightly wound into the shape of truncated cone. It is the so-called cone calorimeter, and it yields heat release rate information on small samples. The results of the test include the heat release rate (HRR) versus time at a specified external heat flux. Other tests can also provide HRR measurement such as Ohio State University apparatus (or OSU in short) used for cabin aircraft materials and EN 30399 to evaluate the burning behavior of bunched cables.
4. *Extinction*: Limiting Oxygen Index (LOI, ISO 4589 part 2) serves as a measure of the ease of extinction of the material. It determines the minimum oxygen concentration required to support candle-like downward flaming combustion.

All the measurements described previously can be applied on intumescent materials. Their behavior during the test does not create issues in general. However, it can be problematic with samples exhibiting high expansion during a cone experiment. The char layer formed upon heating can hit the electrical heater and, in some cases, can plug the extraction chimney (Figure 7.7). A way to avoid this is to adjust the distance between the cone heater and the sample holder but if the distance is too long then the homogeneity of the heat flux at the surface of the sample is lost. Another way is to use a grid as recommended in the standard, but in this case, the benefit of the expansion to create a heat barrier is lost.

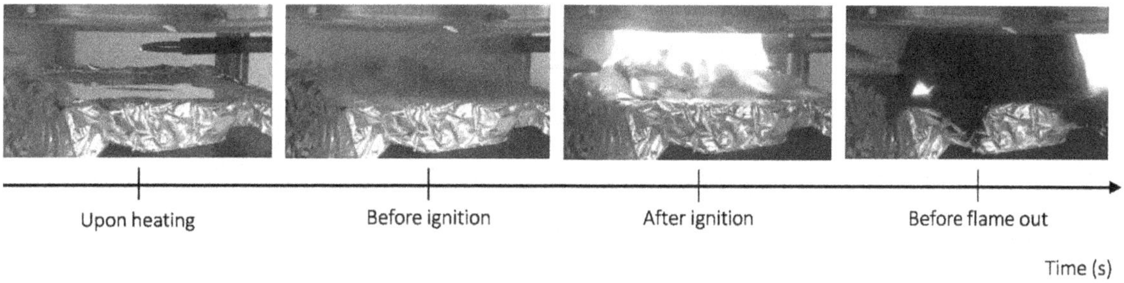

FIGURE 7.7 Snapshots as a function of time of a burning intumescent epoxy-based polymer during a cone experiment.

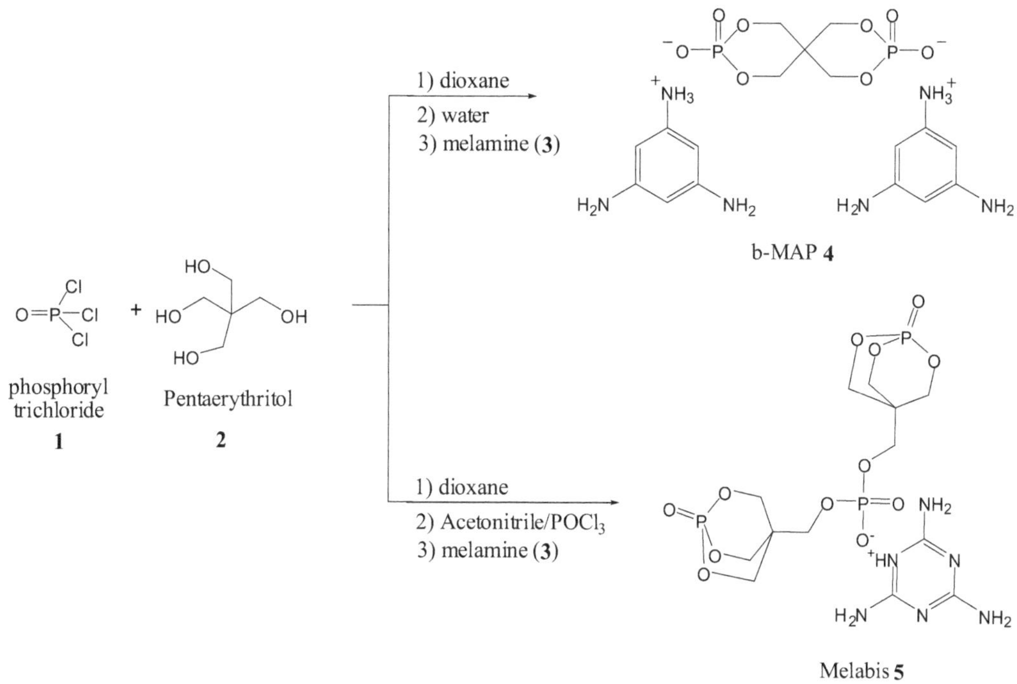

FIGURE 7.8 Melabis and b-MAP synthesis.

7.4.2 Intumescent Organic Polymeric Materials

7.4.2.1 Organic-Based Systems

In the eighties, intumescent systems (APP-based formulations) were used to flame retard polymer and in particular thermoplastics.[40–42] However, such systems present some drawbacks (water solubility, migration of the additives throughout the polymer) that can be prevented by synthesizing novel phosphorus- and/or nitrogen-containing systems. The last 40 years have seen the improvement of the intumescent concept applied to polymers in terms of performance, processability, and durability. A survey of the recent literature shows that numerous new intumescent compounds have been synthesized.[17] Basic concepts are used to combine in a single molecule the three main ingredients of intumescence, that is, the acid source, the char former, and the blowing agent (see the pioneering paper of Halpern[42]) and to develop new systems containing complex organic molecules (e.g., hyperbranched molecules[43]).

The use of polyols such as pentaerythritol, mannitol, or sorbitol as "conventional" char formers in intumescent formulations for thermoplastics is associated with migration and water solubility issues. Moreover, these additives are often not compatible with the polymeric matrix, and the mechanical properties of the formulations are then very poor. Those problems can be solved (at least partially) by the synthesis of additives that concentrate the three intumescent flame retardant elements in it as suggested by the pioneering work of Halpern.[42] The b-MAP (4) (melamine salt of 3,9-dihydroxy-2,4,8,10-tetraoxa-3,9-diphosphaspiro[5,5]-undecane-3,9-dioxide) and Melabis (5) (melamine salt of bis(1-oxo-2,6,7-trioxa-1-phosphabicyclo[2.2.2]octan-4-ylmethanol)phosphate) were synthesized from pentaerythritol (2), melamine (3), and phosphoryl trichloride (1) (Figure 7.8). They were found to be more effective to fire retard polypropylene than standard halogen-antimony fire retardant.

Based on this work and searching for environmental friendly process, Fontaine et al.[44] suggested the synthesis of PEPA (1-oxo-4-hydroxymethyl-2,6,7-trioxa-l-phosphabicyclo[2.2.2]octane) and bis(PEPA)phosphate-melamine salt derivatives prepared by a novel and safe protocol (one step, one pot process). These new salts were incorporated in PP and exhibit high fire performance via an intumescence mechanism (LOI of 32 vol% and V-0 rating at UL-94 on bars of 3.2 mm thick). Moreover, Liu et al.[45] proposed to use catalytic action of phosphotungstic acid in the synthesis of melamine salts of pentaerythritol phosphate (called MPP in their paper, but it is the same molecule as b-MAP) in order to solve the problems of conventional preparation methods (the use of $POCl_3$; the high temperature of the reaction and thus the high energy consumption) (Figure 7.9). It was shown that the acid catalysis can enhance the conversion degree of the reaction and decreases the reaction temperature while keeping a satisfactory conversion, thus greatly controlled the energy-consuming in the preparation process of MPP (or b-MAP). The results also indicated that as compared to non-catalyzed MPP, catalyzed MPP flame retardant system remarkably improved the flame-retardant properties reinforcing and stabilizing the char layer, and maintaining acceptable mechanical properties.

Other authors also suggested the use of organic compound '3 in 1' containing the three main ingredients of intumescence. A recent paper reports the synthesis of phytic acid functionalized MOF (Metal Organic Framework) tailored by carbon source β-cyclodextrin as FR for epoxy resin.[46] It provides a porous and foamy char acting as heat insulator and permitting to reduce by 65% HRR. Also, a clever application of a well-known natural product, namely, deoxyribose nucleic acid or in short DNA,[47] was reported by Alongi et al. in thermoplastics.[48,49] They showed that DNA could compete with conventional FRs when evaluated by cone calorimetry.

Wang et al. reported[50] the synthesis of a novel intumescent flame retardant, poly(2,2-dimethylpropylene spirocyclic pentaerythritol bisphosphonate) to yield polyethylene terephthalate (PET) with both excellent flame retardancy and anti-dripping properties. A novel phosphorus-nitrogen containing intumescent flame retardant was prepared via the reaction of a caged bicyclic phosphorus (PEPA) compound and 4,4-diamino diphenyl methane (DDM) in two steps (Figure 7.10).[51] Incorporated in PBT and in combination with thermoplastic polyurethane (TPU) as additional char former, the intumescent PBT exhibits V-0 rating at the UL-94 test (3.2 mm) and an enhanced thermal stability. It is suggested that P-N bonds detected in the charred structure might play a role in its efficiency.

To avoid dissolution or migration of the FR from the polymer, FR oligomer can be used instead of monomer. Using this approach, Ma et al.[52] reported the synthesis of phosphate-polyester copolymers from spirocyclic pentaerythritol di(phosphate acid monochloride)s. It was shown that LOI of the copolymer increases with increasing phosphate content to reach a maximum of 30 vol%. As the polyol-based char formers need to be substituted, Li et al.[53] reported the synthesis of a novel char former for intumescent system based on triazines and their derivatives. It is

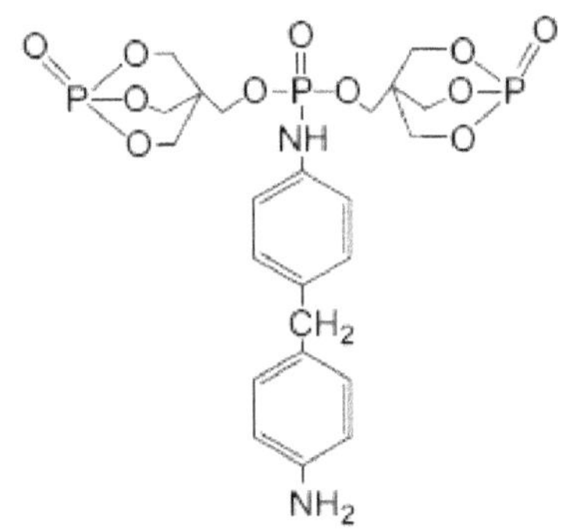

FIGURE 7.10 Phosphorus-nitrogen intumescent flame retardant (from[50]).

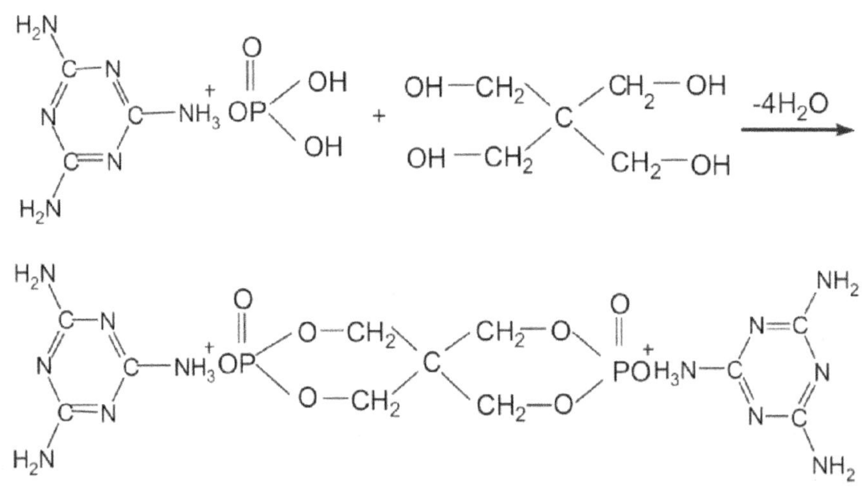

FIGURE 7.9 Synthesis of melamine salts of pentaerythritol phosphate (MPP or b-MAP) (from[45]).

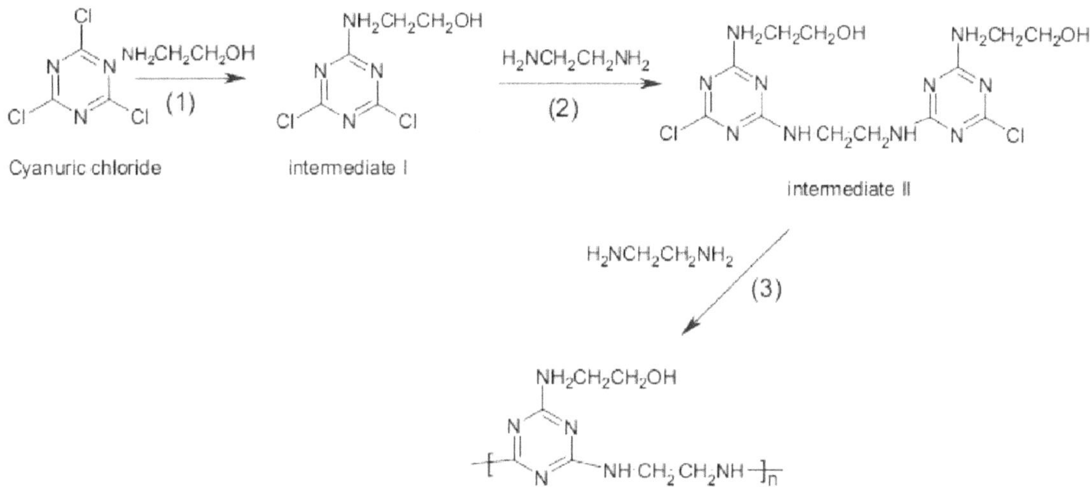

FIGURE 7.11 Synthesis of macromolecular triazines derivatives as char former for intumescent systems (from[53]).

a macromolecular triazine derivative containing hydroxyethylamino, triazine rings, and ethylenediamino groups (Figure 7.11). They showed that the new char former in an intumescent formulation containing APP and a zeolite as synergist can achieve low flammability at only 18 wt% loading in PP (LOI of 30 vol% and V-0 rating (3.2 mm) at the UL-94 test). No tentative of mechanism is commented, but we suspect that it should be close to what we described in a previous paper.[54]

Schartel's group followed a promising approach developing novel hyperbranched molecules as potential FRs.[55,56] They synthesized halogen-free aromatic and aliphatic hyperbranched polyphosphoesters (hbPPEs), which were prepared by olefin metathesis polymerization and investigated them in epoxy resins. The aliphatic version of hbPPEs exhibits a gas phase activity, and they are not intumescent. On the contrary, the aromatic version of hbPPEs increases the char yield and enhances the thermal stability of the epoxy-based material. Intumescence is not mentioned in the paper,[55] but the authors observed the formation of "voluminous char layer." So it can be suspected the formation of an intumescent material (it is also confirmed by the pictures of the cone residues). hbPPEs were then modified by sulfur moieties in their structure.[56] The authors reported that the presence of sulfur increased thermal stability of the flame retardants and introduced additional condensed phase action. Once again, the formation of char acting as protective layer was formed during cone testing. According to the pictures showed in the paper, it can be said an intumescent char was created upon heating. Other authors synthesized hyperbranched molecules as char former in intumescent systems. Examples are molecules containing s-triazine, diphenylmethane and urea groups,[57] and hyperbranched and phosphorus-containing triazine derivative.[58] They are very efficient in combination with the conventional APP in polypropylene (PP) as the pHRR is reduced by about 80% in the two cases.

7.4.2.2 Inorganic-Based Systems

The organic intumescent systems represent a large part of the studies dealing with intumescence while the processing of intumescent mineral systems is even older.[2] However, it is rarely commented in the literature. Mineral intumescent systems are based on alkali silicates. The swelling of the material upon heating or on contact with a flame is due to an endothermic process and is associated with the emission of water vapor that is ionically hydrated in the silicate system.[26] The solid foam formed is rigid and consists of hydrated silica. The structure remains solid until it reaches its glass-softening point. Since only water vapor is released, toxic fumes that may be released from organic-based systems are eliminated. But intumescent alkali silicates have serious limitation; in particular they are sensitive to carbon dioxide and water, which is present in the atmosphere, causing the silicate coating to gradually lose its intumescence, to become brittle, and to lose its adhesion.

Hermansson et al.[54] showed that incorporation of chalk filler and silicone can greatly improve the FR properties of ethylene-butyl acrylate (EBA) formulations. A filling of 30 wt% of chalk filler and 5 wt% silicone was employed. Compared to the pure polymer, an increase of LOI from 18 to 30 vol% and a decrease in the pHRR from 1300 to 330 kW/m² were measured. The improvements were assigned to an intumescent process. The suggested mechanism is as follows[59]: (i) ester pyrolysis occurs when decomposition starts at 300°C producing carboxylate ions on the copolymer backbone; (ii) volatile decomposition products of EBA copolymer lead to foaming of the melt, which is stabilized by the formation of carboxylate ions; and (iii) calcium ions from the chalk (Figure 7.12). Other copolymers (EBA blended with PP and poly(ethylene-co-methacrylic acid) (EMAA)) have been investigated by Krämer et al.,[60] and they found that EMAA formulation has the most effective intumescent process with a low heat release rate and good char stability.

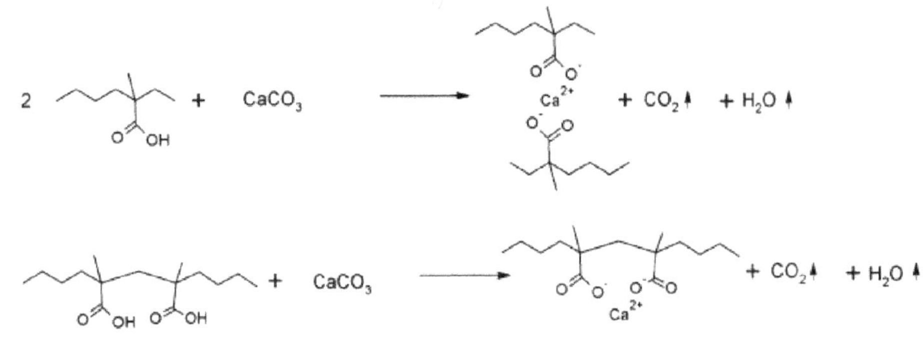

FIGURE 7.12 Reaction scheme for the calcium salt formation in EBA copolymer containing chalk (from[60]).

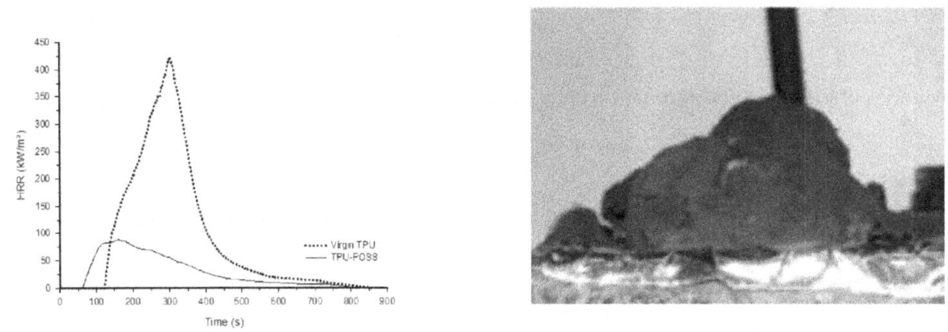

(a) (b)

FIGURE 7.13 HRR as a function of time of pure TPU and TPU/FQ-POSS composite (external heat flux = 35 kW/m²) (a) and intumescent char residue at the end of the cone experiment (b) (from[63]).

Polyhedral silsesquioxanes (POSS) are not known as intumescent FR but as nanoparticles for making polymer nanocomposites. POSS can provide low flammability to polymer via a mechanism in condensed phase.[61] In a previous work,[62] it was shown that the incorporation of POSS in thermoplastic polyurethane (TPU) used as coating on woven PET fabrics permitted 50% reduction in pHRR. The suggested mechanism is char formation at the surface of the material, which can act as an insulative barrier. It was also reported[63] that the incorporation of 10 wt% FQ-POSS (Poly(vinylsilsesquioxane) with the brand name Fire Quench from Hybrid Plastics—USA) in TPU (in bulk polymer) permits to decrease by 80% the pHRR (Figure 7.13-a) without any significant enhancement of LOI (22 vs. 23 vol%) and UL-94 (V-2 at 3.2 mm in the two cases). Visually, the mechanism of protection occurs via an intumescent phenomenon (Figure 7.13-b), and it was evidenced that the mechanism consists in the following steps: (i) charring of TPU and POSS taking place at similar temperatures with the formation of a "viscous" paste, (ii) expansion of a ceramified char (char reinforced by a silicon network) from the evolution of the degraded products and partial sublimation of POSS, and (iii) strong reduction of heat transfer to the substrate (measured experimentally) by the formation of the intumescent layer.

The incorporation of different types of POSS at low loading (about 5 wt%) in epoxy resins decreases their flammability via an intumescent mechanism. Gerard et al.[31] reports the formation of highly expanded intumescent coating during cone calorimetry, but the multilayered structure does permit a high reduction of pHRR (only −10%). With another type of POSS (reactive POSS vs not reactive) able to be chemically incorporated in the epoxy network, Raimondo et al.[64] showed that pHRR is decreased by 40%. It is assigned to the formation of an intumescent coating exhibiting a multilayered structure. According to the pictures published by the authors, the char created with the reactive POSS looks more cohesive than that formed with the nonreactive POSS. With another reactive POSSs in an aeronautic epoxy resin (specific chemistry), Laik et al.[65] showed up to 65% reduction of pHRR. An intumescent mechanism was reported, and it was evidenced the formation of a foamy and cohesive char. This foamy structure containing well-distributed voids explains the superior performance measured by cone calorimetry. In those papers, the mechanism was not clearly detailed, but it is reasonable to assume the formation of ceramified char by the interactions between the degrading epoxy resin and POSS. It is followed by the expansion of the char (viscous charred material) because of the evolving decomposition products of the epoxy resin and of the organic parts of POSS.

Silicates can be used as intumescent FRs in some polymers[66] like unsaturated polyester resins and silicones. More specific silicate systems with bound water were incorporated in polyvinyl chloride (PVC) and polyethylene (PE), and the resulting materials were evaluated by cone calorimetry.[67] The loading was quite high (50 wt%), but the purpose was to make a comparison with the conventional aluminum trihydroxide (ATH). It was shown that the silicate system is more efficient than ATH, and the pHRR was decreased by 70% for the two polymers. The reported mechanism is intumescence combined with the endothermal dehydration of the silicates and the dilution effect of the water release.

In an unconventional way, it was reported that calcium carbonate ($CaCO_3$) was an intumescent filler in PE.[68,69] To get this effect, stearic acid must be added to well disperse the particles of $CaCO_3$. Upon heating, a layer of $CaCO_3$ appears at the surface of the sample, and thanks to its good dispersion, the decomposition gases of the polymer are trapped in the cohesive structure, and the increasing internal pressure creates the expansion of the structure. The gases' release through the sample was slowed down thanks to the high tortuosity of the intumescent structure, which causes the reduction of the HRR during cone calorimetry (reduction by 80% compared to virgin PE).

Expandable graphite represents another class of inorganic intumescent systems as mentioned in Section 2 of this chapter. Expandable graphite is a graphite intercalation compound (GIC). It appears in the literature in 1840.[70] It is a layered crystal consisting of sheets of carbon atoms tightly bound to each other. Chemicals (such as sulfuric acid) may be inserted between the carbon layers. Upon heating EG expands and generates an entangled network of graphite worms providing fire performance of interest to the polymeric matrix. It creates an insulative protective coating exhibiting high anisotropy of its heat conductivity.[71] This anisotropy permits the dissipation of heat in-plane (heat dissipator) and poorly out of plane (heat insulator). EG was used advantageously in polyurethane coating to develop fire protective coating for polymeric substrates.[72] The intercalation compound can also be more sophisticated than the convention sulfuric acid. In combination with $KMnO_4$ and H_2SO_4, hydrated sodium silicate was employed.[73,74] It was shown that V-0 rating was achieved in ethylene vinyl acetate copolymer (EVA). The mechanism is an intumescent phenomenon with the formation of stable char.

7.4.3 Intumescent Textile

In this part, we should distinguish between natural and synthetic fibers because different methods are usually involved to provide them flame retardancy by intumescence. Very often the authors describe the mechanism of action of their materials as "mechanism via charring enhancement" or something similar (see[75] as general review on flame retardant for textiles and references therein). However, based on the chemical nature of the flame retardant used and by the described mechanism, we can sometimes suspect an intumescent behavior. A typical example is that mentioned in the introduction about the paper by J.L. Gay-Lussac in 1821.[1] It deals with the flame retardancy of curtains used in theater where fire hazard was high because of candles enlightening the scene and the room. Because of the ingredients used, it can then be suspected an intumescent behavior as protection mechanism against fire. In 2017, other authors repeated the same formulations on cotton and polyester, and they got acceptable performance.[76] No comments were made on the mechanism of action, but the formulation was similar.

Horrocks et al. ([77] and references therein) undertook considerable work on intumescents applied to textile structures, in particular substantive fiber treatments for cellulose. Based on the work of Halpern[42] on cyclic organophosphorus molecules, they developed a phosphorylation process for cotton fibers achieving intumescent cotton fabric with considerable durability. Char enhancement is as high as 60 wt% compared to the virgin cotton, which is associated with very low flammability. Coating (or back-coating) on fabric is another way to provide flame retardancy to cotton. Horrocks' group[78] used intumescent back-coatings based on APP as the main FR combined with metal ions as synergist. Metal ions promote thermal decomposition of APP at lower temperatures than in their absence, and this enables flame retardant activity to commence at lower temperatures in the polymer matrix thereby enhancing flame retardant efficiency. Giraud et al.[79,80] developed the concept of microcapsules of ammonium phosphate embedded in polyurethane and polyurea shells to make an intrinsic intumescent system compatible in normal polyurethane (PU) coating for textiles. The advantage of this concept is to reduce the water solubility of the phosphate and to produce textile backcoatings with good durability. The flame retarding behavior of these coated cotton fabrics was evaluated by cone calorimetry evidencing a significant FR effect. Development of intumescent char at the surface of the fabric was observed confirming the expected mechanism.

Layer-by-layer (LbL) assembly has been used as a surface treatment to impart flame resistance to cotton fabric. It is a simple versatile method to incorporate various polymers, colloids, or molecules into a thin film that is typically no thicker than 1 μm.[81,82] So it gives the opportunity to create intumescent systems selecting the right combination of ingredients. It was first evidenced by the Grunlan's group using LbL assembly of poly(sodium phosphate) (PSP), which acts as the acid source and is negatively charged in water, and poly(allylamine) (PAAm), which is used as the blowing agent.[83] From 5 to 20 bilayers (BL) were applied on cotton fabric and the coated fabrics were subjected to vertical flame testing (VFT). The uncoated control fabric was completely consumed by direct flame, while coated fabric with a 5 BL coating was preserved as a complete piece. For the 20 BL-coated fabric, the flame extinguished right after ignition while the 10 BL-coated fabric exhibits an intermediate behavior (Figure 7.14). It is evidenced an intumescent behavior protecting the fibers against the combustion. Other ingredients including chitosan (CH, acid source), APP (acid source and blowing agent), and branched polyethylenimine (BPEI) were used to coat cotton.[84] It is revealed an intumescent behavior during the VFT and the

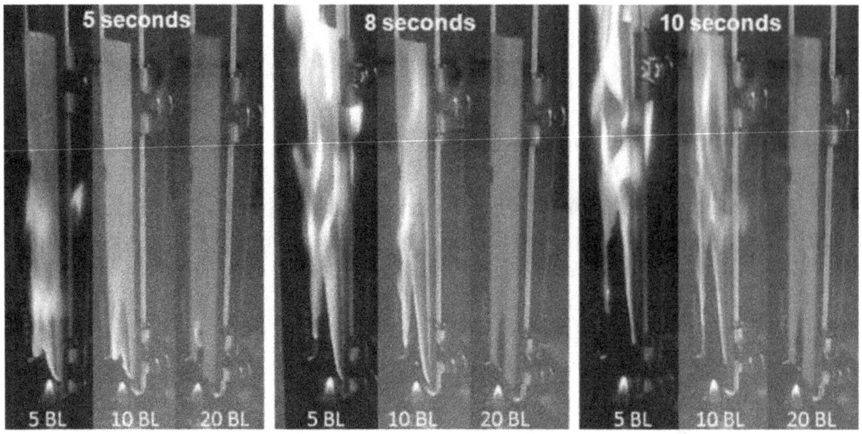

FIGURE 7.14 Cotton fabric coated with intumescent bilayers subjected to vertical flame test; images during the flame testing recorded at 5, 8, and 10 s (from[83]).

mechanism was completely elucidated. A combination of both condensed (formation of aromatic char) and gas phase (release of water and highly flammable gases) mechanisms impart the FR behavior promoting a kind of "microintumescence" phenomenon.

The application of DNA as all-in-one intumescent system (see the previous section) was first reported to impart flame retardancy to cotton fabric.[16,85,86] By combining DNA with chitosan, the authors made LbL assemblies on cotton fabric. The system is able to extinguish flame in horizontal flammability tests and strongly reduce the heat released during combustion. It was confirmed that DNA is an intrinsically intumescent material evidenced by charring and small expansion in the residues. It was reported that DNA layers promote the char formation of chitosan counterparts, by releasing phosphoric and polyphosphoric acid. The obtained char is thermally stable, and it imparts flame retardant character to cotton. Based on the idea of the all-in-one DNA, Jeong et al. used adenosine triphosphate (ATP) on cotton.[87] An intumescent char layer was formed at the surface of the textile VFT preventing fire spread, and a LOI of 32% was measured.

Within the area of natural fibers, wool has the highest inherent non-flammability. It exhibits a relatively high LOI of about 25 vol% and low flame temperature of about 680°C.[88] The inherent flame-retardant activity of the fiber can be associated with char-forming reactions, which may be enhanced by a number of flame retardants. Based on their fundamental work to enhance char formation, Horrocks et al.[89] offered intumescent formulations based on melamine phosphate to flame retard wool. From TGA and SEM characterization, they proposed a comprehensive model on the mechanism of protection via an intumescent process. It involves the formation of cross-linked char by P-N and P-O bonds resistant to oxidation. In another work, they used spirocyclic pentaerythritol phosphoryl chloride (SPDPC) phosphorylated wool to achieve intumescent wool which exhibits large char expansion and good flame retardancy.[90] More recently, a water-soluble polyelectrolyte complex (PEC) consisting of bio-resourced phytic acid and chitosan was prepared to impart flame retardancy to wool fabric.[91] The technology is close to LbL process, but it is faster. The coated wool exhibits an intumescent behavior upon testing using VFT and LOI. LOI jumps from 24% to 33%, and the char length (length of fire damage on the fabric) is short when the coating is present.

Polyester fibers are the main synthetic fibers used in the industrial manufacturing sector and can be found in several areas of application. Several flame retardants have also been designed for polyester extrusion (bisphenol-S-oligomer derivatives from Toyobo or phosphinate salts such as OP950, from Clariant). According to previous work on PET,[92] phosphinate salts provide flame retardancy to PET via an intumescent process, but it is never clearly commented in the literature when it is applied to textile. Chen et al.[93] proposed the use of a novel anti-dripping flame retardant, poly(2-hydroxy propylene spirocyclic pentaerythritol bisphosphonate) (PPPBP) to impart flame retardancy and dripping resistance to PET fabrics. Flammability of PET fabrics treated with PPPBP was investigated by VFT, which showed a significant enhancement of the flame retardancy (producing a non-ignitable fabric) and either a significant reduction of melt dripping at low levels or absence of dripping at higher levels of PPPBP. The same authors investigated in details the mechanism of flame retardancy of their FR PET fabrics.[94] They showed that it is a condensed phase mechanism via char promotion in which PPPBP produces phosphoric or polyphosphoric acid upon thermal decomposition leading to the formation of phosphorus-containing complexes at higher temperatures. They suggest that the high yields of char are protected from thermo-oxidation by the presence of phosphoric acid contained in the charred residue and because of the high thermal stability of C = C groups in the char. In their paper, the authors do not report any intumescent behavior, but it is suspected because of the type of the compound used and according to the described mechanism that an intumescent process should be involved.

Like polyester, polyamides are synthetic fibers made from semicrystalline polymers which find use in a variety of applications in textiles almost similar as those of polyester. Horrocks et al.[77] has investigated the effect of adding selected intumescent flame retardants based on APP, melamine phosphate, pentaerythritol phosphate, cyclic phosphonate, and similar formulations into nylon 6 and 6.6 in the presence and absence of nanoclay. They found that in nylon 6.6, all of the systems comprising the nanoclay demonstrated significant synergistic behavior except for melamine phosphate because of the agglomeration of the clay. Recently, it was reported that enzymatically modified nylon 6.6 fabric, coated via LbL deposition of an intumescent APP/CH coating, with and without urea or thiourea provide self-extinguishing behavior[95] (horizontal test). This result is not exceptional in terms of fire retardancy, but it is encouraging for further development.

Polypropylene (PP) is one of the fastest-growing fibers for technical end uses where high tensile strength coupled with low-cost are essential features. While polypropylene fibers may be treated with flame retardant finishes and back-coatings (intumescent or not) in textile form with varying and limited success,[96] the ideal flame retardant solution for achieving fibers with good overall performance demands that the property is inherent within the fiber. Presently, the use of phosphorus-based, halogen-free flame retardants in PP fibers is disfavored due to the need to have at least 15–20 wt% additive in the fiber. Since the latter are char-promoting while all halogen-based systems are essentially non-char-forming in polypropylene, the way forward for a halogen-free, char-forming flame retardant conferring acceptable levels of retardancy at low additive levels, 10 wt% will require either completely new FR chemistry or the development of a suitably synergistic combination. Based on this, Zhang et al.[96] reported that the flammability of polypropylene is reduced by the addition of small amounts of clay in conjunction with a conventional phosphorus-containing FR and a hindered amine. The authors suspect a P-N synergism to exist, and the LOI value for the best formulation is 22 vol% compared to 19 vol% for neat PP, with only 6 wt% total loading. Microcapsules containing phosphate embedded in different polymeric shells (polyurethane, melamine formaldehyde, and poly(hexamethylene adipate glycol)—melamine formaldehyde) were padded on PP fabrics.[97] The resulting fabrics were evaluated by cone calorimetry, and it was shown that pHRR was decreased by up to 50%. The mechanism involved is the formation of an expanded charred layer protecting the PP fabric.

7.4.4 Intumescent Inorganic Polymeric Materials

Few papers report the fire behavior of intumescent inorganic polymers. A possible explanation is that inorganic polymers are inherently flame retardant and are used "as is."[98] The particular case of inorganic-organic polymers is either they are already flame retardant or they are used in applications which do not require high level of flame retardancy. The only published work devoted to intumescent inorganic polymer concerns polyphosphazene.

While polyphosphazene exhibits some level of flame retardancy, its performance is not enough to fulfill the requirements of the FAA for aircraft interiors. In order to make ultra-fire-resistant elastomers, the use of expandable graphite was investigated.[99] In a cone calorimeter experiment, the addition of the expandable graphite to the polyurethane reduces its heat release rate to a level approximating that of the pure polyphosphazene rubber. The addition of the expandable graphite to the polyphosphazene rubber reduces its heat release rate to a level approximating that of the fire-resistant engineering plastics or of thermoset resins currently used in aircraft interiors (< 100 kW/m²). The post fire test photos (not shown) show that the polyurethane rubber is completely consumed in the fire test. In contrast, the polyphosphazene leaves a char residue equal to 35% of the original weight. The polyurethane and polyphosphazene formulated with the expandable graphite leave a light friable char on the order of 20–50 times the original sample volume. The expanded graphite char insulates the underlying polymer from burning. The high thermal efficiency of the expanded char results in a peak heat release rate that is five and seven times lower than that for the virgin polyphosphazene and polyurethane polymers, respectively.

7.4.5 Synergy in Intumescents

Interesting developments have occurred which involve unexpected "catalytic" effects in various intumescent systems. Performance in terms of LOI, UL-94, or cone calorimetry was enhanced dramatically adding small amount of an additional compound leading to a synergistic effect. In the following, we use the definition of synergy as "a synergistic effect occurs when the combined effects of two chemicals are much greater than the sum of the effects of each agent give alone."

In the literature, numerous synergists (micro- and nanofillers) were used in conventional "three-based ingredients" intumescent formulations. It covers the boron compounds (zinc borates,[44] borophosphate,[100] borosiloxane[101]), phosphorus compounds (phosphazene,[102] $ZrPO_4$[103]), silicon compounds (silica,[104] silicone,[105] silicalite[106]), aluminosilicate (mordenite,[107] zeolite,[106] montmorillonite[108]), rare earth oxides[109,110] (La_2O_3, Nd_2O_3), metal oxides (Al_2O_3,[104] MnO_2,[111] ZnO,[112] Ni_2O_3,[113] Bi_2O_3,[114] TiO_2,[115] ZrO_2,[116] Fe_2O_3[117]), and others (carbon nanotubes,[118] silsesquioxanes,[119] layered double hydroxides,[120] Cu,[121] Fe,[122] talc,[123] sepiolite,[124] zinc and nickel salts,[125] MOF[126]). The presence of this additional filler can modify the chemical (reactivity of the filler versus the ingredients of the intumescent system)[108] and physical (expansion, char strength, and thermophysical properties)[37] behavior of the intumescent char when undergoing flame or heat flux leading to enhanced performance. Those aspects are rarely described for the systems, and papers only report the description of the enhanced performance with few physical and chemical characterizations. The understanding is that the chemical reaction between the fillers and the acid source (mainly phosphate derivatives) yields to phospho-X compounds (e.g., phosphosilicate, zinc phosphate, borophosphate, etc.) reinforcing the structure and/or the action of the fillers (or its reaction products) as nucleating agent permits the formation of a homogeneous foamed structure with appropriate thermophysical properties (lower heat conductivity, lower emissivity at the surface, etc.). Note in the particular case of using phosphinates or phosphonates in intumescent formulations, no reaction occurs between the filler and the other ingredients, and in this case, the mechanism of action is mainly due to a physical effect.[127] Some examples are reported in the following to show how large can be the improvement and the mechanism involved in the synergy.

The first paper reporting synergy in intumescent systems was written by Scharf et al.[115] They used metal oxides (TiO_2 and SnO_2), and they found a dramatic increase of LOI when incorporating a small amount of oxides in APP/PER system. The explanation was a chemical and physical interaction stabilizing the char and modifying its morphology. It was not clearly evidenced, but it is correct because APP reacts with TiO_2 to yield titanium phosphate upon heating. This last molecule is stable at high temperatures, and it can prevent the char from oxidation (so it explains why high LOI can be achieved). Concurrently, Bourbigot et al. did all types of characterization to explain this phenomenon and to conceptualize the aspects of synergy in intumescent systems.[37,128] Based on the same type of mechanism, they showed that adding a small amount of minerals, such as zeolites,[106] natural clays,[107] and zinc borates[44,129] in intumescent systems, the FR performance (LOI, UL-94, and cone calorimetry) can be dramatically enhanced. It was confirmed by Levchik et al.[111,130] in the same years. They reported that the use of small amount of talc and manganese dioxide combined with APP in PA-6 could promote charring and enhance insulative properties of the intumescent coating leading to a significant improvement of the FR performance.

Large synergistic effects are observed when incorporating nanofillers in intumescent formulation, and most of the recent works on synergy are devoted to this. The basic mechanism remains similar to that described above, but the incorporation of nanofiller can bring additional properties (e.g., mechanical properties). It is noteworthy that the nanofiller is a crucial ingredient (reactant) in the development of intumescence forming additional species stabilizing the structure (it is also true with microfiller) and modifying the rheological behavior (more specific to nanofiller when the dispersion is really at the nanoscale). The nanofiller is incorporated at amount as low as 1 wt% (sometimes less, as in the case of the incorporation of nanoparticles of copper at amount as low as 0.1 wt% in epoxy resin containing APP[131]), and it permits the formation of active species selecting chemical reactions in the condensed phase and yielding to char with the dynamic properties of interest.

A typical example of using nanofillers as synergists in an intumescent system was given by Muller et al.[37] The polymeric matrix was a polyurethane (PU) synthesized in the lab by conventional methods. It was known that APP in PU provides an intumescent because PU is the char former, and APP is the acid source and the blowing agent. The nanofillers considered were MgO, SiO_2, and octamethyl polyhedral silsesquioxane (OMPOSS). The incorporation of 30 wt% APP in PU decreases by 62% the pHRR compared to virgin PU when measured by cone calorimetry (Figure 7.15). The time to ignition is similar for all the materials, and it occurs at about 45 seconds. The substitution of 2 wt% of APP by nanoparticles shows significant effects on the reaction to fire of the intumescent PU. PU/APP and PU/APP-SiO_2 exhibit similar HRR curves as a function of time. The incorporation of OMPOSS in PU/APP brings an additional decrease to pHRR: decrease by 71% vs. 62% compared to virgin PU. A larger synergistic effect is observed with the incorporation of MgO since pHRR is decreased by 83% (PU/APP-MgO) instead of 62% (PU/APP) compared to virgin PU. The HRR curves are stretched over time (pHRR occurs at 200s compared to 100s for PU-APP), and even if the time to ignition of PU/APP-MgO is not significantly modified, HRR remains quite low (pHRR < 100 kW/m²), and its slope increases significantly only from 100s. In all cases, the reduction of HRR is assigned to an intumescence phenomenon protecting the material (visual observation). The char of PU/APP exhibits an expansion of 1,100% associated to a residual mass of 44 wt%, while those containing the nanoparticles exhibit an expansion of 1,200% (SiO_2), 1,300% (OMPOSS), and 2,200% (MgO) associated to a residual mass of 44 wt% (SiO_2), 42 wt% (OMPOSS), and 52 wt% (MgO).

The nanofiller of MgO gave the largest synergistic effect, and the mechanism of action was investigated. It was shown that a reaction between APP and MgO occurred yielding to the formation of magnesium phosphates (evidenced by solid state NMR of ^{31}P and by XRD). Those latter stabilizes

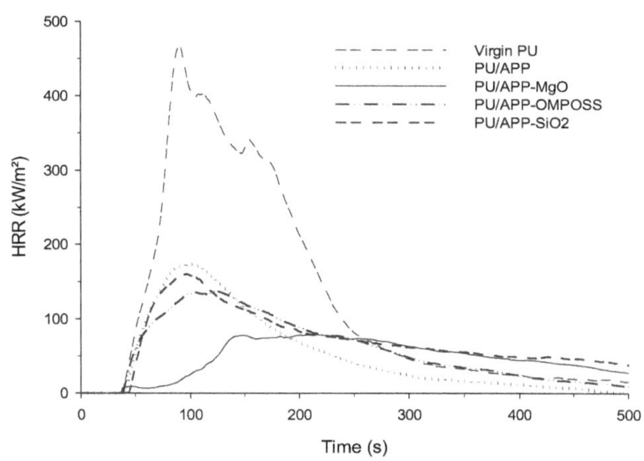

FIGURE 7.15 HRR curves as a function of time for PU, PU/APP, PU/APP-MgO, PU/APP-SiO_2, and PU/APP-OMPOSS (external heat flux = 50 kW/m²).

phosphates at high temperature which can prevent the oxidation of the intumescent char,[132] and hence, the protection is more efficient. The char morphology was also examined since it governs all the thermo-physical properties (heat conductivity for example) leading to an efficient thermal barrier. PU-APP char morphology exhibits an irregular surface, but it is more homogeneous inside the char. The inner structure is composed of big bubbles separated with thick porous walls (Figure 7.16). On the contrary, PU/APP-MgO char is highly expanded, and it exhibits a regular surface. Inside the char, two distinct structures are observed: at the bottom, there are many bubbles of various sizes separated by thin walls; on the top, only one or two huge bubbles are observed (Figure 7.17). Compared to the PU/APP char structure, the walls of PU/APP-MgO char are generally thinner than those of PU/APP char. In particular the crust is very thin at the top of the char where huge bubbles are observed, while the walls between bubbles are thicker at the bottom of the char where many cells are observed. The incorporation of MgO modifies the morphology of the char obtained with PU/APP. The expansion, the size, number and repartition of cells, the walls thickness, the aspect and size of the crust are indeed affected by the presence of MgO. It is observed there are two different parts in the char structure with many unorganized cells at the bottom and huge voids at the top. The morphology exhibited by the intumescent PU formulations is a result of a swelling and blowing mechanism leading to the formation of cells of different size. The fillers incorporated in PU participate in this bubbles development to a certain extent depending on the fillers or combination of filler used. Upon thermal stress, the fillers degrade and/or react leading to the release of gases which are "used to build the intumescent structure." The increasing internal pressure from decomposition gases allows the char to expand. It was evidenced that the incorporation of MgO affects the final char structure which is much more complex than that without MgO. The synergy mechanism between APP and MgO in PU is consequently partially due to the char morphology and the development of intumescence.

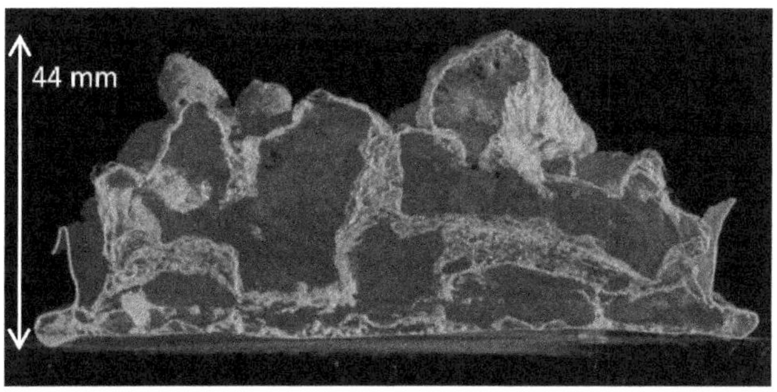

FIGURE 7.16 X-ray tomography picture of the inner structure of the char formed by PU/APP after cone calorimetry.

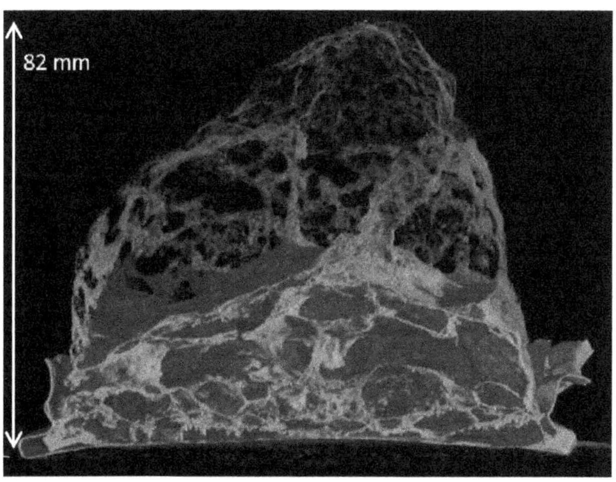

FIGURE 7.17 X-ray tomography picture of the inner structure of the char formed by PU/APP-MgO after cone calorimetry.

7.5 RESISTANCE TO FIRE OF INTUMESCENT MATERIALS

Intumescent coatings have had a large increase in use, as a method of passive fire protection, over the past few years. Their main purpose is to protect construction materials such as steel and wood in case of fire. Intumescent coatings swell when subjected to high temperatures, and they must resist the fire and avoid its passage and/or that of gaseous products of combustion. The goal is to keep the integrity and the functionality of the substrate as long as possible. That is why the main parameters considered to measure the resistance to fire of a material is the "time-temperature curve" and the determination of a "time to failure" (time to reach a given temperature, temperature depending on the specification). In this part, evaluation of the resistance to fire of materials are presented and commented in the first section. The second section is devoted to the development of bench scale tests and to recent protocols developed to measure key parameters of intumescence. The performance and the behavior of intumescence coatings are then surveyed in the third section.

7.5.1 Evaluation of the Resistance to Fire

In the event of a fire in buildings, vehicles, or means of transportation, the fire resistance time of the materials and products is the most critical variable and has a direct influence on the occupants' safety. Fire-resistant products should allow the structures to retain their minimum functions during the time required to evacuate the people, even despite the extreme conditions of heat and pressure to which they are subjected. The products must be evaluated according to the applicable regulation.

After about a century, the standard fire resistance test is the predominant means to characterize the response of structural elements in fires. In 1908 the American Society for Testing Materials (ASTM) published a standard test on the need to develop a common approach to evaluating the fire safety of construction materials. In 1918 a time-temperature curve was established as part of the standard based on the maximum temperatures experienced in real fires at the time (known as ASTM E119[133]). The curve was not based on the response of building components to a real fire but rather what the authors have described as a worst-case time-temperature relationship to be expected during a fire. This approach is still in use at this time even if some authors pointed out many shortcomings and showed it is sometimes unrealistic (see e.g., the review of Bisby et al.[134]).

Required levels of protection are specified in terms of time and temperature on the basis of one or more criteria, which may include statutory requirements, design considerations, and insurance cost implications. It can vary from a few minutes to several hours, but it usually takes the form of 15 minutes increments. The duration is established by a time rating which is determined by testing in accordance with an approved standard. The standards depend on the kind of fire the material should resist. There are three main categories of fires: the cellulosic or wood fire, the hydrocarbon fire, and the jet fire. However, a number of different fire test curves have been proposed, as shown in Figure 7.18.[135–137]

The cellulosic fire curve (ASTM E119) simulates the rate of temperature increase observed in a residential or commercial building fire where the main sources of combustion fuel are cellulosic in nature, such as wood, paper, furniture, and common building materials. In the development of new fireproofing products, this test method is currently used primarily to evaluate the relative performance of fireproofing on full scale standard sizes and shapes of structural steel under controlled conditions. Although this fire curve is still used, it is noteworthy that the burning rates for certain materials, for example, petrol gas, chemicals, etc., are well in excess of the rate at which, for instance, timber burns. As such, there was a need for an alternative fire test for the evaluation of structures and materials used within the petrochemical industry, and therefore the hydrocarbon test curve was developed.

The hydrocarbon test curve should duplicate or be indicative of the rapid temperature rise seen when a hydrocarbon fuel such as oil or natural gas burns: the temperature rises rapidly to 900°C within four minutes, and significantly higher overall temperatures are reached (between 1,100°C and 1,200°C). This hydrocarbon fire test curve, developed by the Mobil Oil Company in the early 1970s and adopted by a number of organizations and in particular Underwriters Laboratories (UL 1709 "Rapid Temperature Rise"), UK Dept. Of Energy, BSI, ISO, and the Norwegian Petroleum Directorate is now a common test method for high-risk environments such as petrochemical complexes and offshore platforms. Derived from the abovementioned hydrocarbon curve, the French regulators asked for a modified version, the so-called hydrocarbon modified curve (HCM). The maximum temperature of the HCM curve is 1,300°C instead of the 1,100°C standard HC curve. However, the temperature gradient in the first few minutes of the HCM fire is as severe as all hydrocarbon-based fires.

The RABT (Richtlinien für die Ausstattung und den Betrieb von strassen Tunnels) curves were developed in Germany.[138] In the RABT curve, the temperature rise is very rapid up to 1,200°C within five minutes, faster than the Hydrocarbon curve, which rises only to 1,150°C in ten minutes. The duration of the 1,200°C exposure is shorter than other curves with the temperature drop off starting to occur at 30 minutes. This test curve can be adapted to meet specific requirements. In testing to this exposure, the heat rise is very rapid but is only held for a period of 30 minutes: it is similar to the sort of temperature rise that would be expected from a simple truck fire, but with a cooling down period of 110 minutes.

The RWS (Rijks Water Staat) curve[138] was developed by the Ministry of Transport in the Netherlands. This curve assumes that in a worst-case scenario, a fuel, oil, or petrol

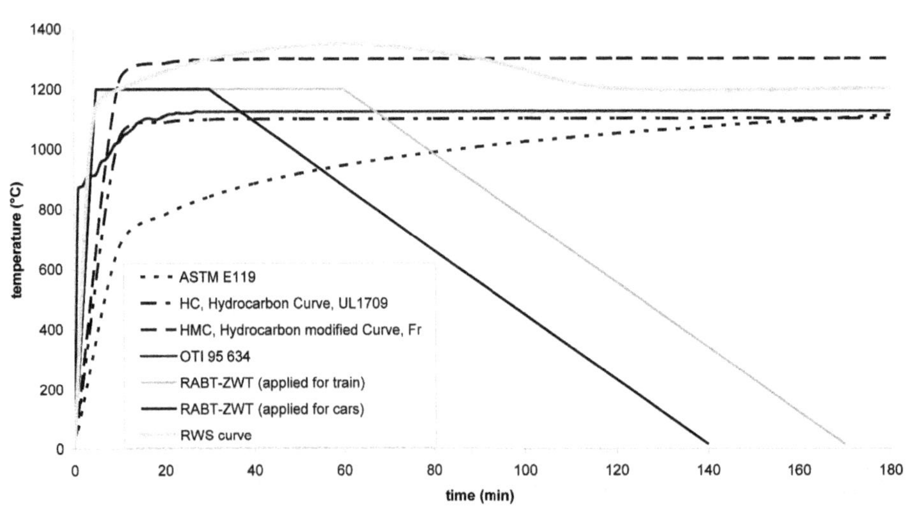

FIGURE 7.18 Standard fire test curves.

tanker fire with a fire load of 300MW could occur, lasting up to 120 minutes. The RWS curve was based on the results of testing carried out by TNO Centre for Fire Research in the Netherlands in 1979. The difference between the RWS and the hydrocarbon curve is that the latter is based on the temperatures that are expected from a fire occurring within a relatively open space, where some dissipation of the heat occurs, whereas the RWS curve is based on temperature found in a fire occurring in an enclosed area, such as a tunnel, where there is little or no chance of heat dissipating into the surrounding atmosphere. The RWS curve simulates the initial rapid growth of a fire using a petroleum tanker as the source, and the gradual drop in temperatures to be expected as the fuel load is burnt off.

All the tests of fire resistance are not based on time-temperature curve. Some of them involve jet fires to mimic a specific fire scenario. The properties of jet fires depend on the fuel composition, release conditions, release rate, release geometry, direction, and ambient wind conditions. Jet fires represent a significant element of the risk associated with major accidents on offshore installations or in plants.[139] The jet fire scenario assumes a flammable or explosive atmosphere when tank, vessel, or pipeline containing pressurized and/or liquefied gas fail, and hence, gas leakage entrains air with the gas effluent stream. The gas dilution starts at the leaking point because of the high pressure in the container. Therefore, a turbulent-free jet is formed, wherein the gas concentration and velocity along its axis depend on the orifice diameter. The heat fluxes released from these fires are very high, ranging from 200 to 400 kW/m^2 depending on the type of fuel. This range of heat fluxes is considered by the standard ISO 22899–1:2007 according to the Health & Safety Executive, Offshore Technology Report OTI 95 634: 1995. The applicability of this test was critically reviewed by Bradley et al.,[140] and they found that a single heat flux was inadequate means of specifying the severity of the hazard, but instead, a detailed description of the nature of the release should be provided.

In transportation such as aviation, jet fires can occur because of the large quantities of highly flammable fuel which can ignite in case of crash or during flight. So the threat is particularly high compared to other situations. Materials and composites used in the construction of transport category aircraft must comply with national and international regulations with most countries having adopted the US Federal Aviation Regulations (FAR). The fire resistance of components, equipment, and structure located in "fire zones" (e.g., compartments containing main engines and auxiliary power units) is conventionally evaluated according to the international standard ISO 2685.[141] The test is established on a pass/fail basis, thereby an element is considered to pass its fire test if it is capable to sustain its design function after the standard flame exposure for a specific lapse of time. It uses a specific burner fed by propane or kerosene with a flame temperature of 1,100°C and delivering a heat flux of 116 kW/m^2. In addition to this standard and to increase survivability, the Federal Aviation Administration (FAA) has developed a medium-scale laboratory test to analyze the burn through resistance of aircraft skin components using an impinging high heat flux ($\approx$ 180 kW/m^2) jet fuel flame from an oil burner to simulate conditions observed in post-crash aviation fires.[142]

The standards listed above are the most commonly used fire tests, but they are not exhaustive. They give an overview of fire testing in the field of resistance to fire, and they show the conditions can be very severe and much more than those involved in the case of reaction to fire. It is therefore the challenge of intumescence to provide the protection of interest in those severe conditions. Intumescent materials are classified as either thick or thin film intumescent coatings. Intumescent thick films are usually based on epoxy, vinyl, or other elastomeric resins. These films are particularly efficient in the case of hydrocarbon and jet fires. They are hard and durable and can provide excellent protection against corrosion. This is due to their very high adhesion to the substrate and resistance to impact, abrasion, and vibration damage. High tensile and compressive strengths can be obtained, and weather resistance is excellent. On the other hand, thin intumescent films are used for protection from cellulosic-type fires. They typically use thermoplastic acrylic or polyvinylacetate resin, and they respond rapidly and intumesce quickly when exposed to a cellulosic-type fire environment. One-hour protection can be achieved with between 1 and 3 mm of product. Thin film intumescents are often referred to as "fire retardant paints" rather than "fireproofing" materials due to their inferior fire resistance compared to thick films intumescents. They are mostly used inside buildings because of their poor durability.

7.5.2 Testing and Characterization of Intumescence at Reduced Scale

Resistance to fire of intumescent coating is typically measured by a curve "temperature as a function of time" or with the impingement of burner flame into the material (see previous section). In all cases, it requires large scale equipment, and those tests are very expensive and time-consuming. Even if due to the complexity of fire phenomenon, full-scale tests remain the main and the most credible tool for investigating fire-related issues, the cost of those tests significantly increases with scale, and it prevents the fast development of new fire-resistant materials and of new structures. This section examines the scale reduction of different tests to mimic fire scenarios at the reduced scale. Using selected representative examples, it is shown how to evaluate fire-resistant intumescent material.

The protection of metallic materials against fire has become a very important issue in the construction and petrochemical industries, as well as in the marine and military fields. Structural steel loses a significant part of its load-carrying ability when its temperature exceeds 500°C. Prevention of the structural collapse of a building is crucial to ensure the safe evacuation of people, and it

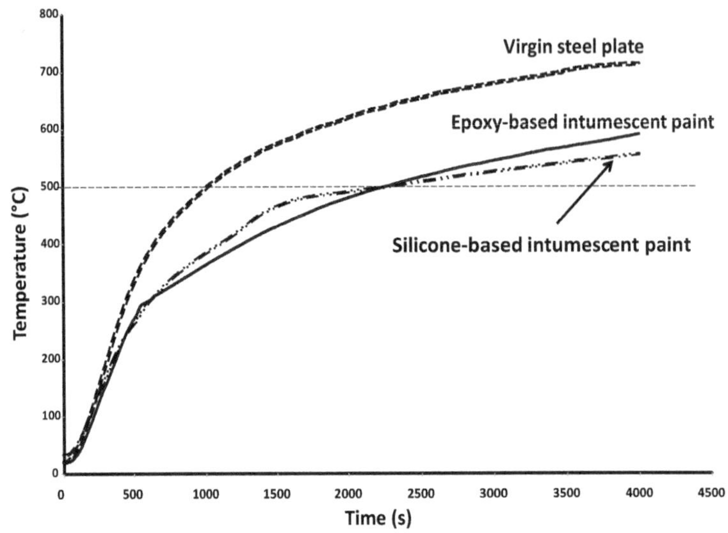

FIGURE 7.19 Temperature as function of time on the backside of steel plate protected or not by an intumescent paint when undergoing ISO 834 heating in the small furnace.

is a prime requirement of building regulations in many countries. A small-scale furnace test (internal volume of 40 dm^3) was built to evaluate the fire performance of steel protected by intumescent coatings in cellulosic and hydrocarbon fires. This test was designed to mimic the ISO834 (or ASTME119) and the UL1709 normalized temperature/time curves, respectively related to cellulosic fire and to hydrocarbon fire using propane burners.[143] Samples of 10 × 10 cm^2 are tested with this equipment, and the temperature as a function time is recorded on the backside of the steel plate. Intumescent paints based on silicone as polymeric matrix were applied on steel plates and were evaluated within the small-scale furnace according to ISO 834 temperature curve (Figure 7.19). The silicone-based intumescent paint was compared to a commercial epoxy-based intumescent paint. The failure temperature was taken at 500°C. Uncoated steel plate reaches 500°C at 1,000s, while the coated steel with intumescent coatings reaches the failure temperature at times higher than 2,400s. It shows that intumescence is an efficient way to protect steel since the time to failure temperature is twice longer than that with virgin steel. The silicone paint was compared to the commercial paint, and it appears that its behavior is similar as epoxy-based paint.

Similar approach was followed by Schartel et al. modifying a lab muffle furnace to comply with the time/temperature curve described in EN 1363–1.[144] The furnace is completely instrumented with in particular, a high temperature endoscope permitting to observe the foaming/expansion process of the intumescent coating (Figure 7.20). This specific instrumentation gives new insights on the formation of the intumescent coating, which allows the ranking of the intumescent materials and the development of new ones exhibiting higher performance. They also evaluate different intumescent coatings under four different but similar shaped heating curves with different maximum temperatures (standard time-temperature curve, hydrocarbon curve, and two self-designed curves with reduced temperature).[145] They found that the time/temperature curves affected the inner structures and the residue surface leading to difference thermal performance. They suggested that the coating should be appropriately selected according to those curves.

A novel fire testing method, named the Heat-Transfer Rate Inducing System (H-TRIS), was presented by Maluk et al.[146] (Figure 7.21-a). The method directly controls the thermal boundary conditions imposed on a test specimen by controlling a specified time-history of incident radiant heat flux at its exposed surface. It was used to examine the effectiveness of thin intumescent coatings (Figure 7.21-b).[147] It was found the onset of swelling is directly influenced by

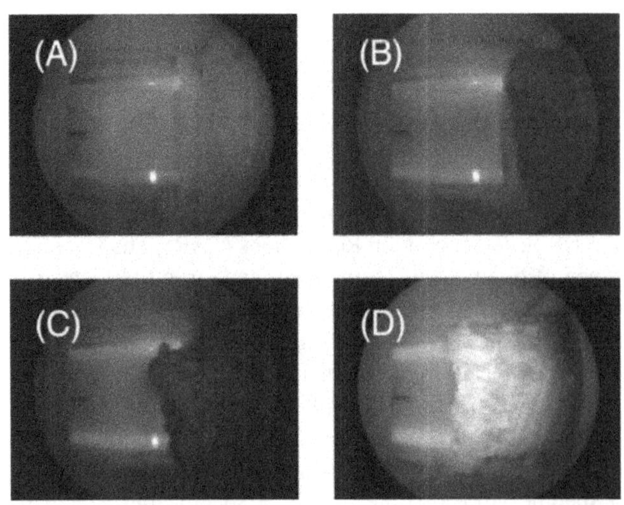

FIGURE 7.20 A–D, pictures as a function of time of the foaming specimens recorded by the high-temperature endoscope inside the furnace (from[144]).

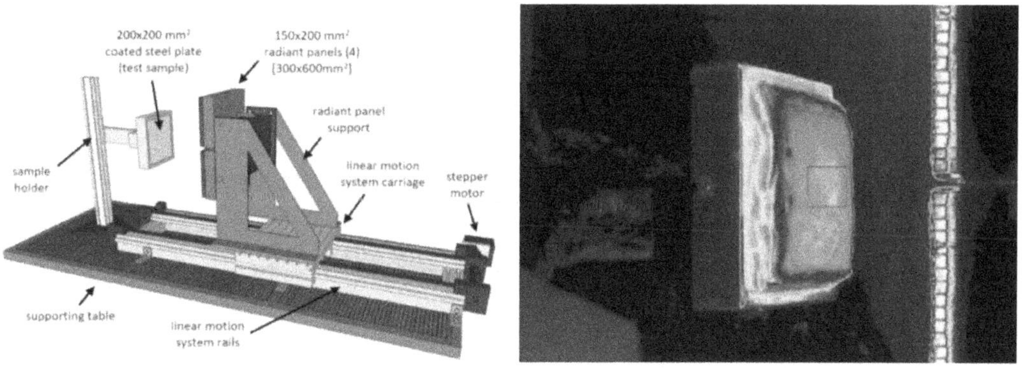

(a) (b)

FIGURE 7.21 (a) scheme of the H-TRIS bench-scale test and (b) analysis of an intumescent coating using infrared thermography undergoing radiative heat flux on H-TRIS test (from[147]).

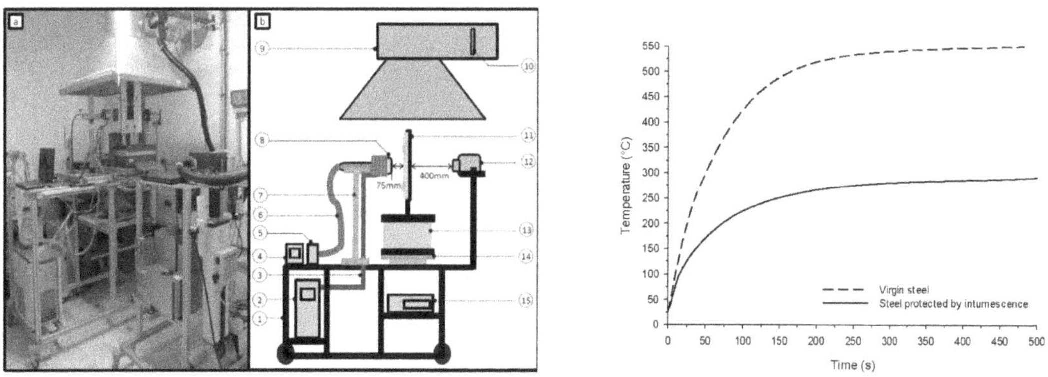

(a) (b)

FIGURE 7.22 (a) Picture and scheme of the horizontal burner test bench (from[148]) and (b) temperature as a function time at the burn through test of steel protected by intumescent coating.

the heating conditions at the exposed surface and the original applied dry film thickness.

A horizontal burner test was developed to mimic the burn through taking place in ISO 2685 (Figure 7.22 (a)). ISO 2685 is used to evaluate the fire resistance of components located in "fire zones" of aircraft.[148] The test consists of impacting a 100 × 100 mm² sample by a propane flame at 1,100°C and with a calibrated heat flux of 116 kW/m². Note that the propane burner can be substituted by a kerosene burner, as done by Coppalle et al. in another version of the test.[149,150] A commercial epoxy-based intumescent coating was evaluated for protecting steel when the flame directly impinges the intumescent paint. The comparison of the temperatures as a function of time on the backside of steel plate with and without paint shows the efficiency of the intumescent protection (Figure 7.22 (b)): in the steady state, the temperature only reaches 290°C with the intumescent protection, while the temperature is 450°C without protection.

Infrared images (Figure 7.23 (a) and (b)) show the development of the intumescent coating and the flame contour during the test. Note the infrared images were obtained using specific filters to observe the flame shape (Figure 7.23 (a)) and the surface of the sample through the flame (Figure 7.23 (b)). The burner flame engulfs the intumescent coating (Figure 7.23 (a)), and the expansion of the intumescent coating is limited. It does not form a regular expanded char, but the expansion is a distribution of small bubbles at the surface of the material (Figure 7.23 (b)). It is probably due to the velocity of the gas jet limiting the expansion. A crude assumption without any further measurement on the emissivity of the surface (here the emissivity is assumed to be one) permits the scientist to estimate the surface temperature as a function of time of the intumescent coating during the test (Figure 7.23 (c)). It is an average temperature measured in an elliptic zone, as shown in Figure 7.23 (b). The temperature increases sharply when applying the burner flame and reaches a pseudo steady state temperature

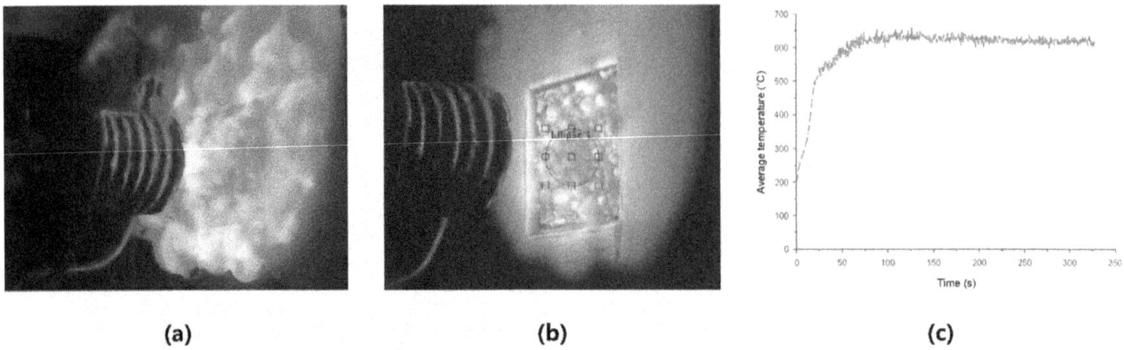

FIGURE 7.23 Intumescent coating on steel plate during the burn through test: (a) infrared image of flame contour and (b) infrared image of the intumescent surface through the flame; (c) average temperature at the surface of the intumescent coating on the zone defined by the ellipse.

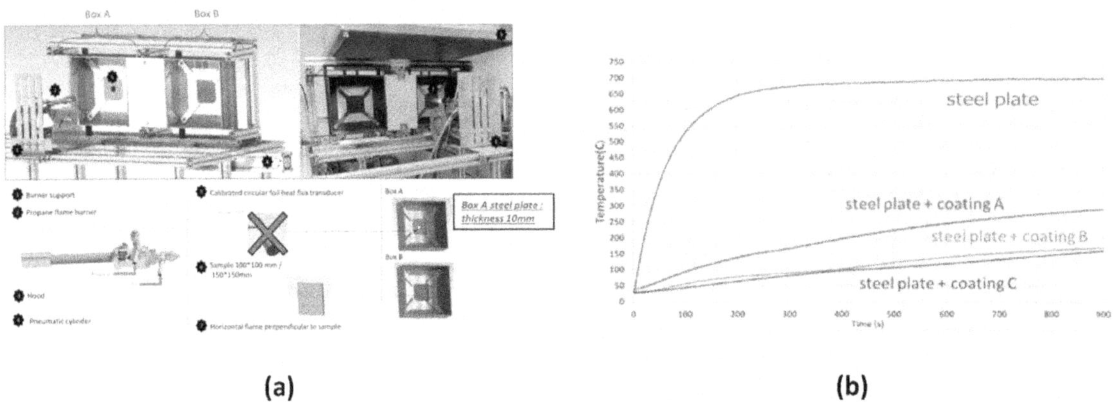

FIGURE 7.24 (a) Schematic description of the jetfire bench-scale test (note heat gauge is substituted by steel plate for day-to-day calibration) and (b) time/temperature curves of intumescent coating steel plates compared to virgin one.

of 615°C. The temperature gradient between the front side and the backside is therefore of 325°C on a small thickness. It evidences the efficiency of the intumescence for fire protection and the approach permits to get quantitative data on temperature gradient involved in the intumescent char.

Jetfire scenario involves extremely high heat flux (up to 450 kW/m²), and it is generally not investigated at the reduced scale. In our lab, we developed bench-scale test mimicking jetfire for testing intumescent paint on steel plate (Figure 7.24-a).[151] A powerful premixed propane/air burner provides a flame impinging the sample to be tested. Heat fluxes up to 450 kW/m² can be reached by adjusting the propane to air ratio of the burner as well as their mass flow. The sample holder is shaped like a box dimensioned in order to avoid backfiring of the flame. Opposite to the burner, two boxes are used respectively to perform the calibration of the burner (the heat flux and temperature of the flame are measured using respectively a water-cooled heat gauge and thermocouples) and to evaluate the behavior of materials. Three intumescent coatings based on different chemistries were evaluated: (i) coating A (silicone-based coating containing expandable graphite), (ii) coating B (water-based coating), and (iii) coating C (epoxy-based coating). They were applied on steel plate and submitted to a heat flux of 300 kW/m². Time-temperature curves were recorded during the experiments, and the results are shown in Figure 7.24-b. The profile obtained with the uncoated plate is compared to those obtained for coated plates whose basis weight was 1g/cm², leading to thicknesses of 6 mm, 7 mm, and 5 mm for coatings A, B, and C, respectively. All the coated plates show an improved behavior compared to the uncoated one thanks to the formation of an intumescent char. After a 15 minute test, the plates coated with A, B, and C coatings reach respectively 287°C, 169°C, and 154°C compared to 697°C for the uncoated one. It evidences that intumescent coating can resist high velocity jet and at high heat flux.

Schartel et al.[152] also build up a bench scale at very high heat flux using a powerful oxygen-propane burner. The heat flux was not estimated, but the temperature of the flame was extremely high in the range of 1,200–1,800°C. It was shown it was useful for evaluating intumescent coating in extreme conditions.

7.5.3 Fire Protective Intumescent Materials

Few recent scientific papers are devoted to the resistance to fire involving new developed intumescent coatings while existing a growing market for this application and while academics are active in understanding the mechanisms of protection (see the previous section). However, it should be cited the excellent review of E. Weil published in 2011[153] reporting recent developments for intumescent coatings and their application for fire protection. An updated version was then done by Mariappan in 2016[154] and by Sabee et al. in 2022.[155]

A rapid survey of the literature shows that the used formulations are always based on the well-known triplet APP/PER-Melamine as it was reviewed by Weil et al. in 1994.[156] The combination APP/PER-Melamine offers the possibility to substitute the char former (PER) by its dimer and trimer. Comparisons of temperature development, foaming ratios, and rheological behavior was performed by Andersson et al.[157] between formulations containing PER, di-PER, and tri-PER. A simulated fire test developed by the authors (the coated face is upside down and exposed to Bunsen burner and an infrared pyrometer records the temperature on the backside), in which the temperature increase during intumescence was studied. They showed that the formulations containing PER were considerably more efficient in keeping a low temperature throughout the process. A more rapid temperature development was displayed when using di-PER and tri-PER as char former. Rheometer tests indicate that PER formulations enter the intumescent process at a lower temperature and stays in it for a longer time than di-PER and tri-PER formulations. No explanation is given why PER is more efficient than the two others. It might be proposed it is because of the higher reactivity of PER to APP (reaction of esterification between APP and PER occurs at 190°C[158]) occurring at lower temperature permitting a faster development of the intumescent coating.

Other char formers in substitution to PER and in combination with APP can be found in epoxy-based commercial formulations such as Chartek (Akzo Nobel) and Pittchar (PPG) (see[153] and references therein). They are complex formulations involving numerous ingredients, but the triplet acid source/char former/blowing agent is still there in majority. The char former was tris(hydroxyethyl) isocyanurate (THEIC) in Chartek and bisphenol-A derivative in Pittchar. Those two formulations are used nowadays as thick coatings for steel protection in off-shore oil drilling platforms and petrochemical chemical installations. It is noteworthy the presence of boron molecules, namely, boric acid (Chartek) or zinc borate (Pittchar). They play a particular role reinforcing the char-forming borophosphate-type molecules upon heating.[159] In epoxy-based formulations, the role of char former can be the matrix itself.[31] Another interesting substitution of the char former is the use of ketone thermoplastic resin in epoxy-based intumescent coating.[153]

Zeng et al.[160] report the mapping performance of intumescent coatings as a function of concentration of ingredients. The char former was the polymeric matrix (namely, epoxy), and they examined the effects of APP, melamine, TiO_2, $CaCO_3$, and vitreous silicate fiber on the fire-resistance performance of zinc borate (ZB) containing and ZB-free hydrocarbon intumescent coatings with exposure to the hydrocarbon fire testing curve UL 1709. As previously reported, they found the coating with 25 wt% APP or 5 wt% MEL obtained the best performance. TiO_2 and $CaCO_3$ and the fiber do not significantly modify the performance of the ZB-containing intumescent coatings, but the time to failure (here 550°C) and char appearance were strongly affected for the ZB-free coatings. Increasing the content of any of the three inorganics led to reduced expansion of the ZB-free coatings but improved the chars with more mechanically stable structures. The longest time to failure were found with the ZB-free formulations containing 1.5 wt% TiO_2, 2.5 wt% $CaCO_3$, and 5 wt% vitreous silicate fiber.

The use of additional ingredients acting potentially as synergists (micro- and nanoparticles) was also investigated in intumescent coatings. Multi-walled carbon nanotube (MWCNT) was found to be effective as flame retardant[161] and Beheshti et al.[162] evaluated it in intumescent coating as a potential synergist. Using a burner test, they found it is ineffective for improving the protection in acrylic-based polymer containing APP/PER/MEL. Li et al.[163] suggested combining EG and/or molybdenum disilicide ($MoSi_2$) in an intumescent system based on APP/PER-Melamine. The results show that incorporating EG and $MoSi_2$, the time to failure (time required to reach 500°C in this case) is prolonged (heat insulation test similar as that described in the previous paragraph), and char formation rate is enhanced. The largest improvement is achieved combining the two ingredients $MoSi_2$ and EG in an appropriate ratio. The suggested mechanism of action is revealed using scanning electron microscopy (SEM) images and thermogravimetry showing that the synergistic effect is obtained through a ceramic-like layer produced by $MoSi_2$ covered on the surface of an "open-cellular" structural char and a better resistance to the thermooxidation. EG was also used as additional blowing agent in benzoxazine-based coating containing APP and boric acid to make an efficient intumescent coating.[164] In another paper, Wang et al.[165] report the use of organomodified layered double hydroxide (OLDH) as nanofiller combined with an intumescent system in acrylate resin. The intumescent paints were evaluated on steel plate measuring the temperature on the backside of the plate as a function of time when undergoing temperature (standard ISO-834). It was found that incorporating 1.5% OLDH the fire-resistant time (time to reach 300°C) jumps at 100 minutes compared to 60 minutes without OLDH (virgin coating). It is also noteworthy that the thickness of char layer of the formulation with 1.5% OLDH is similar to that of the formulation without. It confirms the results suggesting that the highest expansion is not necessary to get the best performance. The improvement of the performance is partially explained by the char strength. The morphology of char layer exhibits close holes, but the diameters of holes of the char containing OLDH are much smaller (10–30 μm) than those of the char without OLDH. Wang suggests that small holes reinforce the char strength and avoid the formation of cracks at the surface of the char. Indeed, the formation of close cells in the char structure like in foam evenly dispersed, reduces heat transfer, and then increases the efficiency of the char. When the cells are too big, char strength is reduced, and cracks can appear.

In addition to the oxides as synergists, Ahmad et al.[166] investigated the effect of nano-sized boron nitride reinforced with alumina in epoxy-based intumescent coatings in terms of improvement of char expansion and of the morphology. Epoxy plays the role of char former and the formulation contains boric acid and EG in addition with APP and MEL. They showed that the formulations reinforced with nano-alumina effectively protected steel and kept its temperature well below the time to failure (measured at the burner test). Nano-alumina decreases the expansion but forms an additional protection of alumina and of aluminophosphate at the surface of the char layer. It is also believed that the reduction of pore size in the char structure increases the insulation performance of the layer.

Environmental concerns promote the development of bio-based materials. Bio-fillers as additional ingredients in intumescent formulations were suggested in the recent literature. Nasir et al.[167] used rice husk ash (RHA) and chicken eggshell (CES) from waste product as bio-filler in intumescent coating. The formulation was based on APP/PER/MEL in acrylic polymer and were evaluated with a home-made burner test. When applied on steel plate, it was found that RHA acts as a synergist providing the lowest temperature in the steady state. Its mechanism of action is not clear, and it is only assigned to better insulation due to the high expanded cross-linked and foamed char layer. Lignin is a natural product and is abundant. Baldissera et al.[168] used it as char former in intumescent formulation in combination with polyaniline. They found that the new formulations gave similar protection at the burner test as those containing oil-based char formers. In addition to bio-based molecules, recycled materials might be interesting to solve our environmental issues. Liu et al.[169] used recycled powders containing flame-retardants from the precipitate of industrial effluents (the composition is not clearly defined in the paper) and apply them in an intumescent coating. The intumescent coating was based on APP/PER/MEL with styrene-acrylic polymer as binder in which the recycled powder was incorporated. The formulations were evaluated with a torch test, and it was shown the benefit of recycled powder as it enhances the char strength and the heat insulation of the intumescent layer.

Instead of the triplet APP/PER/MEL, other ingredients can be employed to make intumescent coating. It includes diglycidyl methylphosphonate, triglycidyl phosphate, bis[2-(methacryloyloxy) ethyl] phosphate, or dihydroxa-phosphaphenanthrene oxide (DOPO), or preferably an adduct of DOPO with an epoxy-novolac, in intumescent coatings especially for steel.[170] The purpose was to achieve a substantial reduction in the film thickness of intumescent coating required to protect a given substrate from the effects of fire (cellulosic and hydrocarbon fires). The mechanism was not described in the patent, but we can suspect an additional effect in the gas phase thanks to the action of DOPO enhancing the performance at low thickness.

Silicones have the properties to exhibit low thermal conductivity, to be water and heat resistant, to evolve few toxic gases during their decomposition, and to have high durability: they are therefore good candidate to make fire protective coating. Gardelle et al. showed that intumescent silicone-based-coating is very efficient to protect steel in the case of hydrocarbon fire[143] and cellulosic fire[171] scenarios. Expandable graphite (EG) was used as main ingredient in a silicone-based coating and provides good performance protecting steel (Figure 7.25). The performance increases as a function of EG loading in the coating, and at 25 wt% EG, the silicone coating exhibits higher performance (time to reach 500°C) than conventional commercial intumescent

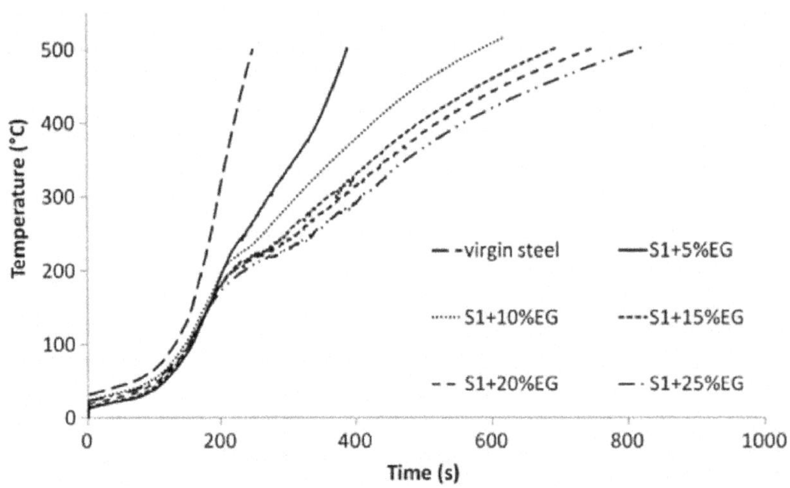

FIGURE 7.25 Temperature as a function of time measured on the backside of steel covered by an intumescent silicone-based coating at different loading of EG (hydrocarbon fire scenario according to UL1709 done in a furnace at a reduced scale, plaque of 10 × 10 cm²) (from[143]).

paint. This is explained by a high swelling velocity (18%/s), a high expansion (3,400%), an impressive cohesion of the char, and a low heat conductivity at high temperature (0.35 W/(K.m) at 500°C). This type of intumescent silicone-based coatings were also evaluated on carbon fiber reinforced polymer (CFRP) using a bench developed in the laboratory mimicking a jet fuel fire occurring at high heat flux (200 kW/m²).[172] A large intumescence was observed, and the composite does not pierce after 15 minutes. EG was also incorporated in thermoset resins for protecting wood composite and was successful at the Steiner tunnel (ASTM E84).[173]

Alkali silicates can be used to make intumescent inorganic coating for fire protection.[24] The expansion of the material upon heating over 100°C is due to the rapid release of water vapor.[26] It makes a foamy material which is rigid and is constituted of hydrated silica. It acts as thermal insulator and provides the fire protection of interest. Veinot et al.[26] evaluated the performance of different formulations of silicate-based coatings applied on aluminum according to the time/temperature curve of ASTM E-119 in a tubular furnace at the reduced scale. They showed that they could keep the temperature of the substrate below 300°C for 15 minutes with the appropriate thickness and the right amount of an acidic curing agent. This agent was not described in the paper, but it is reported it has a beneficial effect on the intumescent structure, which is solid and thick almost like cement in texture. Silicate-based coating was also applied on wood and gives satisfactory performance.[25] Additional magnesium-based ingredients were incorporated in the formulations providing higher expansion and more efficient structure limiting the diffusion combustion gases. An interesting version of silicate-based coating was to make silicate-polymer hydrogels.[174] The resulting coating is transparent, and hence, it can be used as varnish or for window protection. The hydrogel was incorporated between two glass panes and was evaluated in a furnace of fire resistance. It was shown that the hydrogel-based materials meet the requirements for fire tests in the construction. It was revealed the formation of an intumescent protective coating and that the temperature on the pane was kept below 180°C for 30 minutes for the optimized formulation of hydrogel.

A novel way to make intumescent coating is to use aluminosilicate composites or, in other words, geopolymers (GP). This approach was evaluated by Schartel et al.[175] to protect steel plates according to cellulosic fire scenario (ISO 834). Two aluminosilicate formulations were prepared based either on microsilica and alumina or on metakaolin and microsilica containing or not flame retardants including boron trioxide, borax, ATH, and MDH. It was observed an intumescent behavior during fire testing (up to 300% expansion) which is an unusual behavior for geopolymers (Figure 7.26). The best fire insulation was obtained with 10% borax in the formulation based on microsilica and alumina (original layer thickness of 6 mm). The fire protection can reach up to 30 minutes (temperature of failure of 500°C).

Other geopolymers were prepared consisting of metakaolin-based alkaline aluminosilicate with a swelling agent and calcium carbonate as intumescent coating for the fire protection of structural elements.[176] An intumescent behavior was observed, and the resulting material exhibits low heat conductivity and good mechanical stability. Conventional FRs can be incorporated in these types of geopolymer to further enhance the fire protection.[177] FRs were APP and EG, and it gave high expansion. In another work,[178] rice husk ash was incorporated in a metakaolin-based geopolymer to make fire resistant coating for structural insulated panel. The optimized geopolymer-based coating permits to resist 2h at the torch test before reaching 100°C. A relatively high degree of expansion was observed, but some cracks and holes can be distinguished on the residues after fire testing. A foam version of geopolymers also exists, which is made with metakaolin and by acid- or alkali-activation.[179] Geopolymer foams based on mixtures of alkali-activated metakaolin and silica fume were prepared in the lab.[180] Hydrogen peroxide (H_2O_2) was used as a porogen agent and a cationic surfactant (cetyltrimethyl ammonium bromide) as foam stabilizer (Figure 7.27—(a)). The foam was coated on steel plate and was evaluated at the burn through test mimicking ISO 2685 at the reduced scale (heat flux of 116 kW/m²).[148] The temperature reached on the backside of the steel plate is much lower for the geopolymer foam compared to the virgin steel (Figure 7.27—(b)). In the steady state,

FIGURE 7.26 Intumescent coating of geopolymer containing borax after fire testing (scale bar in cm) (from[175]).

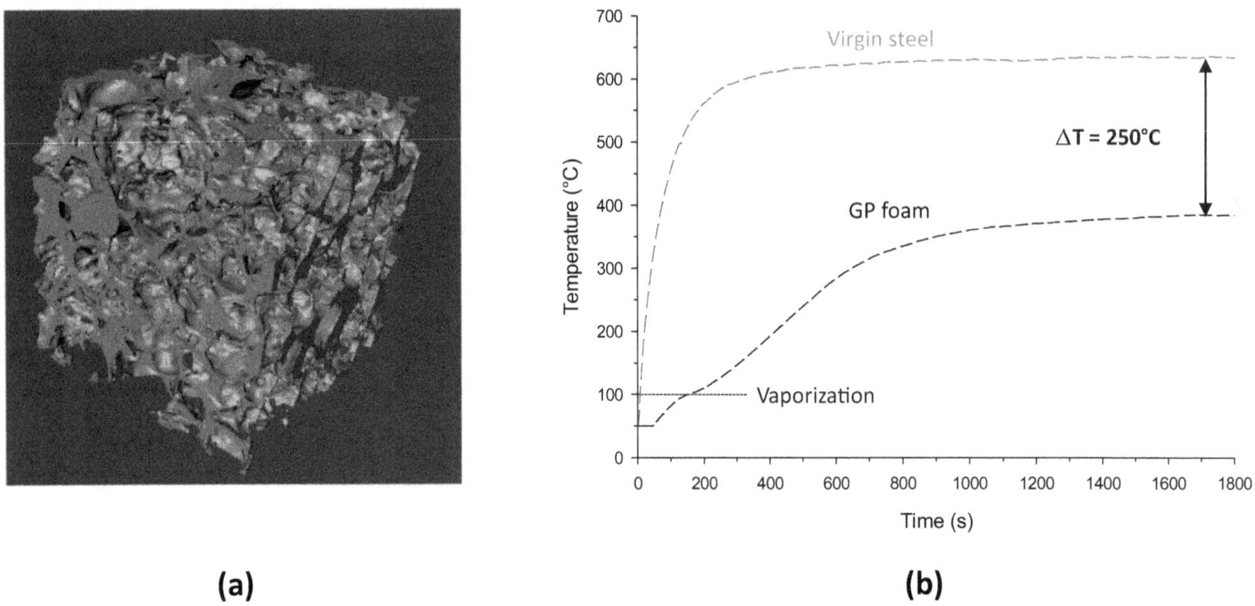

FIGURE 7.27 (a) X-ray tomography of GP foam showing the internal foamy structure and (b) temperature as a function of time of GP foam compared to virgin steel plate undergoing burn through test.

the temperature difference is about 250°C. It is also noteworthy that a plateau can be distinguished at about 100°C. It is assigned to H_2O molecules released from the foam, because of structure changes (amorphization) and vaporizing almost instantaneously. A robust numerical model was also made simulating completely the fire behavior of the geopolymer.[181] It provided some key information for further development and improvement.

An alternative approach to make novel intumescent coating is to develop novel design. Polymer metal laminates consisting of alternating thin aluminum foils and thin epoxy resin layers provide protection against fire due to the delamination between layers during burning (radiant panel at 50 kW/m²).[182] However, it is not enough when considering other and more severe scenarios, and a new concept called "Intumescent Polymer Metal Laminates," or IPML, was developed[183]: the thin epoxy resin layers were substituted by selected intumescent coatings. When evaluating at the burn through test, it was shown that the slopes of the time/temperature curves was dramatically decreased at the beginning of the test compared to the unmodified intumescent coating, but the temperature in the steady state was not lower. It was attributed to the delamination of the aluminum foils (Figure 7.28). The concept was then further enhanced combining the use of two different intumescent paints and a metal laminate structure.[184] The design with two aluminum foils and the overlay of both intumescent coatings reveals efficient fire protection with stabilization at low temperatures after 30 minutes of fire exposure at the burn through test. It was shown this type of design exhibits much higher performance than that of the conventional intumescent coating.

FIGURE 7.28 IPML before and after fire testing at the burn through test, delamination of the aluminum foils can be clearly seen.

Source: adapted from[183]

7.6 CONCLUSION AND FUTURE TRENDS

This chapter has discussed the developments of intumescent fire retarded materials in terms of reaction and resistance to fire. Research work in intumescence is still very active even if the first papers on this topic appeared at the beginning of the 20th century, and the concept of intumescence remains attractive. The quick overview of the mechanisms of action reveals that the formation of an expanded charred insulative layer acting as thermal shield is involved. The chemistry is very often based on the well-known triplet APP/PER/MEL. Numerous other char formers were suggested as well as other blowing agents, but the basic mechanism remains similar. New molecules (commercial molecules and new concepts like all-in-one molecules) appeared based on this. Micro- and/or nanofillers can be also combined with the conventional ingredients of intumescence to create synergistic effects. Reaction can take place between the additional filler and some ingredients of the intumescent formulation (e.g., the phosphate) in order to thermally stabilize the charred structure (chemical stabilization and prevention of oxidation). Physical interactions are also observed (e.g., additional "boost" of expansion). Those interactions permit the reinforcement of the char strength and to avoid the formation of cracks. The development rate and the quality of the layer are of the primary importance, and research work should be focused on this.

However, other chemistries cannot be ignored, namely, those based on a physical expansion. Silicate-based material can create an inorganic intumescent layer by the vaporization of water molecules. It is an old concept which is used in many applications. Few recent papers are devoted to this, but it should be explored further because of the advantages of having inorganic materials (e.g., thermal stability and resistance to oxidation of the intumescent layer). Expandable graphite provides an intumescence based on physical expansion (i.e., sublimation of the insertion molecules): it makes an insulative protective layer formed by entangled worms of graphite while the graphite is highly thermally conductive. The mechanism is not completely elucidated, and it opens the door to a novel way of research. Inorganic chemistry can be still considered with geopolymers giving the opportunity of novel chemistries to design new intumescent materials exhibiting high performance in extreme fire scenarios. Finally, the design of intumescent multi-materials should also be considered as it gives an exceptional opportunity to make materials of high performance, and it paves the way for novel concepts.

Environment is a main concern in our society. It can be considered making new intumescent materials with positive life cycle analysis (LCA). Bio-based and recycled ingredients and/or polymers offer this opportunity based on our knowledge on the diverse chemistries of intumescence. The resulting materials could exhibit higher performance, or at least a similar one, and hence, it is an alternative to be explored to make fire-safe and environmentally friendly intumescent materials.

REFERENCES

1. Gay Lussac, L. J., Note sur la propriété qu'ont les matières salines de rendre les tissus incombustibles. *Annales de Chimie et de Physique* **1821,** *18*, 211–218.
2. Arthur, W. Intumescent material and process of making the same. 1912-10-15, 1912.
3. Tramm, H.; Clar, C.; Kühnel, P.; Schuff, W. Fireproofing of wood. United States Patent office, Ruhrchemie Aktiengesellschaft, Alexandria, VA, USA: 1938; pp 1–2.
4. Sverre, Q. Silicate-containing product and process for making it. Sverre, Quisling, Alexandria, VA, USA: 1940.
5. Vandersall, H. J., Intumescent coating systems, their development and chemistry. *J. Fire Flammability* **1971,** *2*, 97–140.
6. Camino, G.; Costa, L.; Trossarelli, L., Study of the mechanism of intumescence in fire retardant polymers: Part I-Thermal degradation of ammonium polyphosphate-pentaerythritol mixtures. *Polym. Degrad. Stab.* **1984,** *6* (4), 243–252.
7. Gardelle, B.; Duquesne, S.; Vandereecken, P.; Bellayer, S.; Bourbigot, S., Resistance to fire of intumescent silicone based coating: The role of organoclay. *Prog. Org. Coat.* **2013,** *76* (11), 1633–1641.
8. Zeng, Y.; Weinell, C. E.; Dam-Johansen, K.; Ring, L.; Kiil, S., Exposure of hydrocarbon intumescent coatings to the UL1709 heating curve and furnace rheology: Effects of zinc borate on char properties. *Prog. Org. Coat.* **2019,** *135*, 321–330.
9. Nicoară, A. I.; Bădănoiu, A. I.; Voicu, G.; Dinu, C.; Ionescu, A., Intumescent coatings based on alkali-activated borosilicate inorganic polymers. *J. Coat. Technol. Res.* **2020,** *17* (3), 681–692.
10. Bourbigot, S.; Sarazin, J.; Samyn, F.; Jimenez, M., Intumescent ethylene-vinyl acetate copolymer: Reaction to fire and mechanistic aspects. *Polym. Degrad. Stab.* **2019,** *161*, 235–244.
11. Tawiah, B.; Zhou, Y.; Yuen, R. K. K.; Sun, J.; Fei, B., Microporous boron based intumescent macrocycle flame retardant for poly(lactic acid) with excellent UV protection. *Chem. Eng. J.* **2020,** *402*.
12. Liu, X.; Meng, X.; Sun, J.; Tang, W.; Chen, S.; Peng, X.; Gu, X.; Fei, B.; Bourbigot, S.; Zhang, S., Improving the flame retardant properties of polyester-cotton blend fabrics by introducing an intumescent coating via layer by layer assembly. *J. Appl. Polym. Sci.* **2020,** *137* (41).
13. Camino, G.; Costa, L.; Trossarelli, L., Study of the mechanism of intumescence in fire retardant polymers: Part V-Mechanism of formation of gaseous products in the thermal degradation of ammonium polyphosphate. *Polym. Degrad. Stab.* **1985,** *12* (3), 203–211.
14. Bourbigot, S.; Le Bras, M.; Delobel, R., Fire degradation of an intumescent flame retardant polypropylene. *J. Fire Sci.* **1995,** *13* (1), 3–22.
15. Zhang, S.; Horrocks, A. R., Substantive intumescence from phosphorylated 1,3-propanediol derivatives substituted on to cellulose. *J. Appl. Polym. Sci.* **2003,** *90* (12), 3165–3172.
16. Alongi, J.; Carletto, R. A.; Di Blasio, A.; Cuttica, F.; Carosio, F.; Bosco, F.; Malucelli, G., Intrinsic intumescent-like flame retardant properties of DNA-treated cotton fabrics. *Carbohydr. Polym.* **2013,** *96* (1), 296–304.
17. Alongi, J.; Han, Z.; Bourbigot, S., Intumescence: Tradition versus novelty. A comprehensive review. *Prog. Polym. Sci.* **2015,** *51*, 28–73.

18. Camino, G.; Costa, L.; Trossarelli, L., Study of the mechanism of intumescence in fire retardant polymers: Part II-Mechanism of action in polypropylene-ammonium polyphosphate-pentaerythritol mixtures. *Polym. Degrad. Stab.* **1984,** *7* (1), 25–31.
19. Camino, G.; Costa, L.; Luda di Cortemiglia, M. P., Overview of fire retardant mechanisms. *Polym. Degrad. Stab.* **1991,** *33* (2), 131–154.
20. Bourbigot, S.; Le Bras, M.; Delobel, R., Carbonization mechanisms resulting from intumescence association with the ammonium polyphosphate-pentaerythritol fire retardant system. *Carbon* **1993,** *31* (8), 1219–1230.
21. Bourbigot, S.; Duquesne, S.; Leroy, J. M., Modeling of heat transfer of a polypropylene-based intumescent system during combustion. *J. Fire Sci.* **1999,** *17* (1), 42–56.
22. Focke, W. W.; Badenhorst, H.; Mhike, W.; Kruger, H. J.; Lombaard, D., Characterization of commercial expandable graphite fire retardants. *Thermochim. Acta* **2014,** *584*, 8–16.
23. Bourbigot, S.; Sarazin, J.; Bensabath, T.; Samyn, F.; Jimenez, M., Intumescent polypropylene: Reaction to fire and mechanistic aspects. *Fire Safety Journal* **2019,** *105*, 261–269.
24. Bulewicz, E. M.; Pelc, A.; Kozlowski, R.; Miciukiewicz, A., Intumescent silicate-based materials: Mechanism of swelling in contact with fire. *Fire Mater.* **1985,** *9* (4), 171–175.
25. Kazmina, O.; Lebedeva, E.; Mitina, N.; Kuzmenko, A., Fire-proof silicate coatings with magnesium-containing fire retardant. *J. Coat. Technol. Res.* **2018,** *15* (3), 543–554.
26. Veinot, D. E.; Langille, K. B.; Nguyen, D.; Bernt, J. O., Soluble silicate-based coatings for fire protection. *Fire Technol.* **1989,** *25* (3), 230–240.
27. Zubkova, N. S.; Butylkina, N. G.; Chekanova, S. E.; Tyuganova, M. A.; Khalturinskii, N. A.; Reshetnikov, I. S.; Naganovskii, Y. K., Rheological and fireproofing characteristics of polyethylene modified with a microencapsulated fire retardant. *Fibre Chem.* **1998,** *30* (1), 11–13.
28. Bugajny, M.; Le Bras, M.; Bourbigot, S., New approach to the dynamic properties of an intumescent material. *Fire Mater.* **1999,** *23* (1), 49–51.
29. Jimenez, M.; Duquesne, S.; Bourbigot, S., Characterization of the performance of an intumescent fire protective coating. *Surf. Coat. Technol.* **2006,** *201* (3–4), 979–987.
30. Duquesne, S.; Magnet, S.; Jama, C.; Delobel, R., Thermoplastic resins for thin film intumescent coatings—Towards a better understanding of their effect on intumescence efficiency. *Polym. Degrad. Stab.* **2005,** *88* (1), 63–69.
31. Gérard, C.; Fontaine, G.; Bellayer, S.; Bourbigot, S., Reaction to fire of an intumescent epoxy resin: Protection mechanisms and synergy. *Polym. Degrad. Stab.* **2012,** *97* (8), 1366–1386.
32. Bodzay, B.; Bocz, K.; Bárkai, Z.; Marosi, G., Influence of rheological additives on char formation and fire resistance of intumescent coatings. *Polym. Degrad. Stab.* **2011,** *96* (3), 355–362.
33. Okyay, G.; Naik, A. D.; Samyn, F.; Jimenez, M.; Bourbigot, S., Fractal conceptualization of intumescent fire barriers, toward simulations of virtual morphologies. *Sci. Rep.* **2019,** *9* (1).
34. Bourbigot, S.; Le Bras, M.; Duquesne, S.; Rochery, M., Recent advances for intumescent polymers. *Macromol. Mater. Eng.* **2004,** *289* (6), 499–511.
35. Gustavsson, M. K.; Gustafsson, S. E. Different ways of evaluating thermal transport properties from measurements with the transient plane source (hot disk) method. In *29th International Thermal Conductivity Conference, ITCC29 and the 17th International Thermal Expansion Symposium, ITES17*, Birmingham, AL, 2008; pp 397–402.
36. Gardelle, B.; Duquesne, S.; Rerat, V.; Bourbigot, S., Thermal degradation and fire performance of intumescent silicone-based coatings. *Polym. Adv. Technol.* **2013,** *24* (1), 62–69.
37. Muller, M.; Bourbigot, S.; Duquesne, S.; Klein, R.; Giannini, G.; Lindsay, C.; Vlassenbroeck, J., Investigation of the synergy in intumescent polyurethane by 3D computed tomography. *Polym. Degrad. Stab.* **2013,** *98* (9), 1638–1647.
38. Hirschler, M. M., Survey of American test methods associated with fire performance of materials or products. *Polym. Degrad. Stab.* **1996,** *54* (2–3 Special issue), 333–343.
39. Troitzsch, J. H., The globalisation of fire testing and its impact on polymers and flame retardants. *Polym. Degrad. Stab.* **2005,** *88* (1), 146–149.
40. Brady, D. G.; Moberly, C. W.; Norell, J. R.; Waltrs, H., Intumescence: A novel effective approach to flame retarding polypropylene. *J Fire Retardant Chem* **1977,** *4*, 150–164.
41. Montaudo, G.; Scamporrino, E.; Vitalini, D., Intumescent flame retardants for polymers. II. The polypropylene-ammonium polyphosphate-polyurea system. *J. Polym. Sci. Part A-1, Polym. Chem.* **1983,** *21* (12), 3361–3371.
42. Halpern, Y.; Mott, D. M.; Niswander, R. H., Fire retardancy of thermoplastic materials by intumescence. *Ind. Eng. Chem. Prod. Res. Dev.* **1984,** *23* (2), 233–238.
43. Yan, J.; Xu, M., Design, synthesis and application of a highly efficient mono-component intumescent flame retardant for non-charring polyethylene composites. *Polym. Bull.* **2020,** *78* (2), 643–662.
44. Fontaine, G.; Bourbigot, S.; Duquesne, S., Neutralized flame retardant phosphorus agent: Facile synthesis, reaction to fire in PP and synergy with zinc borate. *Polym. Degrad. Stab.* **2008,** *93* (1), 68–76.
45. Liu, Y.; Wang, Q., Catalytic action of phospho-tungstic acid in the synthesis of melamine salts of pentaerythritol phosphate and their synergistic effects in flame retarded polypropylene. *Polym. Degrad. Stab.* **2006,** *91* (10), 2513–2519.
46. Zhang, J.; Li, Z.; Yin, G. Z.; Wang, D. Y., Construction of a novel three-in-one biomass based intumescent fire retardant through phosphorus functionalized metal-organic framework and beta-cyclodextrin hybrids in achieving fire safe epoxy. *Comp. Comm.* **2021,** *23*.
47. Alongi, J.; Di Blasio, A.; Milnes, J.; Malucelli, G.; Bourbigot, S.; Kandola, B.; Camino, G., Thermal degradation of DNA, an all-in-one natural intumescent flame retardant. *Polym. Degrad. Stab.* **2015,** *113*, 110–118.
48. Alongi, J.; Cuttica, F.; Bourbigot, S.; Malucelli, G., Thermal and flame retardant properties of ethylene vinyl acetate copolymers containing deoxyribose nucleic acid or ammonium polyphosphate. *J. Therm. Anal. Calorim.* **2015,** *122* (2), 705–715.
49. Alongi, J.; Cuttica, F.; Carosio, F., DNA coatings from byproducts: A panacea for the flame retardancy of EVA, PP, ABS, PET, and PA6? *ACS Sustainable Chem. Eng.* **2016,** *4* (6), 3544–3551.
50. Wang, D. Y.; Ge, X. G.; Wang, Y. Z.; Wang, C.; Qu, M. H.; Zhou, Q., A novel phosphorus-containing poly(ethylene terephthalate) nanocomposite with both flame retardancy and anti-dripping effects. *Macromol. Mater. Eng.* **2006,** *291* (6), 638–645.
51. Gao, F.; Tong, L.; Fang, Z., Effect of a novel phosphorous-nitrogen containing intumescent flame retardant on the fire

retardancy and the thermal behaviour of poly(butylene terephthalate). *Polym. Degrad. Stab.* **2006,** *91* (6), 1295–1299.
52. Ma, Z.; Zhao, W.; Liu, Y.; Shi, J., Synthesis and properties of intumescent, phosphorus-containing, flame-retardant polyesters. *J. Appl. Polym. Sci.* **1997,** *63* (12), 1511–1515.
53. Li, B.; Xu, M., Effect of a novel charring-foaming agent on flame retardancy and thermal degradation of intumescent flame retardant polypropylene. *Polym. Degrad. Stab.* **2006,** *91* (6), 1380–1386.
54. Bourbigot, S.; Duquesne, S., Fire retardant polymers: Recent developments and opportunities. *J. Mater. Chem.* **2007,** *17* (22), 2283–2300.
55. Markwart, J. C.; Battig, A.; Velencoso, M. M.; Pollok, D.; Schartel, B.; Wurm, F. R., Aromatic vs. aliphatic hyperbranched polyphosphoesters as flame retardants in epoxy resins. *Molecules* **2019,** *24* (21).
56. Battig, A.; Markwart, J. C.; Wurm, F. R.; Schartel, B., Sulfur's role in the flame retardancy of thio-ether–linked hyperbranched polyphosphoesters in epoxy resins. *Eur. Polym. J.* **2020,** *122.*
57. Yan, H.; Zhao, Z.; Ge, W.; Zhang, N.; Jin, Q., Hyperbranched polyurea as charring agent for simultaneously improving flame retardancy and mechanical properties of ammonium polyphosphate/polypropylene composites. *Ind. Eng. Chem. Res.* **2017,** *56* (30), 8408–8415.
58. Zhu, C.; He, M.; Cui, J.; Tai, Q.; Song, L.; Hu, Y., Synthesis of a novel hyperbranched and phosphorus-containing charring-foaming agent and its application in polypropylene. *Polym. Adv. Technol.* **2018,** *29* (9), 2449–2456.
59. Lundgren, A.; Hjertberg, T.; Sultan, B. A., Influence of the structure of acrylate groups on the flame retardant behavior of ethylene acrylate copolymers modified with chalk and silicone elastomer. *J. Fire Sci.* **2007,** *25* (4), 287–319.
60. Krämer, R. H.; Blomqvist, P.; Hees, P. V.; Gedde, U. W., On the intumescence of ethylene-acrylate copolymers blended with chalk and silicone. *Polym. Degrad. Stab.* **2007,** *92* (10), 1899–1910.
61. Gupta, S. K.; Schwab, J. J.; Lee, A.; Fu, B. X.; Hsiao, B. S. POSS™ reinforced fire retarding EVE resins. In *47th International SAMPE Symposium and Exhibition*, Long Beach, CA, 2002; pp 1517–1526.
62. Devaux, E.; Rochery, M.; Bourbigot, S., Polyurethane/clay and polyurethane/POSS nanocomposites as flame retarded coating for polyester and cotton fabrics. *Fire Mater.* **2002,** *26* (4–5), 149–154.
63. Bourbigot, S.; Turf, T.; Bellayer, S.; Duquesne, S., Polyhedral oligomeric silsesquioxane as flame retardant for thermoplastic polyurethane. *Polym. Degrad. Stab.* **2009,** *94* (8), 1230–1237.
64. Raimondo, M.; Russo, S.; Guadagno, L.; Longo, P.; Chirico, S.; Mariconda, A.; Bonnaud, L.; Murariu, O.; Dubois, P., Effect of incorporation of POSS compounds and phosphorous hardeners on thermal and fire resistance of nanofilled aeronautic resins. *RSC Adv.* **2015,** *5* (15), 10974–10986.
65. Laik, S.; Galy, J.; Gérard, J. F.; Monti, M.; Camino, G., Fire behaviour and morphology of epoxy matrices designed for composite materials processed by infusion. *Polym. Degrad. Stab.* **2016,** *127*, 44–55.
66. Horacek, H.; Pieh, S., Importance of intumescent systems for fire protection of plastic materials. *Polym. Int.* **2000,** *49* (10), 1106–1114.
67. Liu, T. M.; Baker, W. E.; Langille, K. B.; Nguyen, D. T.; Bernt, J. O., New silicate-based powders for fire protection of thermoplastics. *J. Vinyl Addit. Technol.* **1998,** *4* (4), 246–258.
68. Bellayer, S.; Tavard, E.; Duquesne, S.; Piechaczyk, A.; Bourbigot, S., Mechanism of intumescence of a polyethylene/calcium carbonate/stearic acid system. *Polym. Degrad. Stab.* **2009,** *94* (5), 797–803.
69. Bellayer, S. P.; Tavard, E.; Duquesne, S.; Piechaczyk, A.; Bourbigot, S., Natural mineral fire retardant fillers for polyethylene. *Fire Mater.* **2011,** *35* (3), 183–192.
70. Schafhaeutl, C., Ueber die Verbindungen des Kohlenstoffes mit Silicium, Eisen und anderen Metallen, welche die verschiedenen Gallungen von Roheisen, Stahl und Schmiedeeisen bilden. *J. Prakt. Chem.* **1840,** *21* (1), 129–157.
71. Bensabath, T.; Sarazin, J.; Bourbigot, S., Anisotropy of expandable graphite to explain its behavior as a flame-retardant. *J. Fire Sci.* **2021,** *39* (2), 109–118.
72. Duquesne, S.; Le Bras, M.; Bourbigot, S.; Delobel, R.; Vezin, H.; Camino, G.; Eling, B.; Lindsay, C.; Roels, T., Expandable graphite: A fire retardant additive for polyurethane coatings. *Fire Mater.* **2003,** *27* (3), 103–117.
73. Pang, X. Y.; Tian, Y.; Weng, M. Q., Preparation of expandable graphite with silicate assistant intercalation and its effect on flame retardancy of ethylene vinyl acetate composite. *Polym. Compos.* **2015,** *36* (8), 1407–1416.
74. Pang, X. Y.; Tian, Y.; Zhai, Z. X.; Duan, M. W., Preparation of expandable graphite intercalated by sulfuric acid and sodium silicate and its flame retardancy application for ethylene vinyl acetate copolymer. *Asian J. Chem.* **2014,** *26* (14), 4297–4302.
75. Weil, E. D.; Levchik, S. V., Flame retardants in commercial use or development for textiles. *J. Fire Sci.* **2008,** *26* (3), 243–281.
76. Buyukakinci, B. Y.; Yilmaz, A., Investigation of boric acid and sodium borate effect on flame retardancy of cotton and polyester fabrics. *Asian J. Chem.* **2017,** *29* (4), 893–895.
77. Horrocks, A. R.; Kandola, B. K.; Davies, P. J.; Zhang, S.; Padbury, S. A., Developments in flame retardant textiles—A review. *Polym. Degrad. Stab.* **2005,** *88* (1), 3–12.
78. Davies, P. J.; Horrocks, A. R.; Alderson, A., The sensitisation of thermal decomposition of ammonium polyphosphate by selected metal ions and their potential for improved cotton fabric flame retardancy. *Polym. Degrad. Stab.* **2005,** *88* (1), 114–122.
79. Giraud, S.; Bourbigot, S.; Rochery, M.; Vroman, I.; Tighzert, L.; Delobel, R., Microencapsulation of phosphate: Application to flame retarded coated cotton. *Polym. Degrad. Stab.* **2002,** *77* (2), 285–297.
80. Giraud, S.; Bourbigot, S.; Rochery, M.; Vroman, I.; Tighzert, L.; Delobel, R.; Poutch, F., Flame retarded polyurea with microencapsulated ammonium phosphate for textile coating. *Polym. Degrad. Stab.* **2005,** *88* (1), 106–113.
81. Decher, G.; Hong, J. D., Buildup of ultrathin multilayer films by a self-assembly process, 1 consecutive adsorption of anionic and cationic bipolar amphiphiles on charged surfaces. *Macromol. Sympos.* **1991,** *46* (1), 321–327.
82. Decher, G.; Lvov, Y.; Schmitt, J., Proof of multilayer structural organization in self-assembled polycation-polyanion molecular films. *Thin Solid Films* **1994,** *244* (1–2), 772–777.
83. Li, Y. C.; Mannen, S.; Morgan, A. B.; Chang, S.; Yang, Y. H.; Condon, B.; Grunlan, J. C., Intumescent all-polymer multilayer nanocoating capable of extinguishing flame on fabric. *Advanced Materials* **2011,** *23* (34), 3926–3931.
84. Jimenez, M.; Guin, T.; Bellayer, S.; Dupretz, R.; Bourbigot, S.; Grunlan, J. C., Microintumescent mechanism of flame-retardant water-based chitosan-ammonium polyphosphate multilayer nanocoating on cotton fabric. *J. Appl. Polym. Sci.* **2016,** *133* (32).

85. Alongi, J.; Carletto, R. A.; Di Blasio, A.; Carosio, F.; Bosco, F.; Malucelli, G., DNA: A novel, green, natural flame retardant and suppressant for cotton. *J. Mater. Chem. A* **2013,** *1* (15), 4779–4785.
86. Carosio, F.; Di Blasio, A.; Alongi, J.; Malucelli, G., Green DNA-based flame retardant coatings assembled through layer by layer. *Polymer (United Kingdom)* **2013,** *54* (19), 5148–5153.
87. Jeong, S. H.; Heo, J. H.; Lee, J. W.; Kim, M. J.; Park, C. H.; Lee, J. H., Bioinspired adenosine triphosphate as an "all-in-one" green flame retardant via extremely intumescent char formation. *ACS Appl. Mater. Interfaces* **2021,** *13* (19), 22935–22945.
88. Horrocks, A. R., High performance textiles for heat and fire protection. In *High Performance Textiles and Their Applications*, Elsevier Ltd: Sawston: 2014; pp 144–175.
89. Horrocks, A. R.; Davies, P. J., Char formation in flame-retarded wool fibres. Part 1. Effect of intumescent on thermogravimetric behaviour. *Fire Mater.* **2000,** *24* (3), 151–157.
90. Horrocks, A. R.; Zhang, S., Char formation in polyamides (nylons 6 and 6.6) and wool keratin phosphorylated by polyol phosphoryl chlorides. *Text. Res. J.* **2004,** *74* (5), 433–441.
91. Cheng, X. W.; Guan, J. P.; Yang, X. H.; Tang, R. C.; Yao, F., A bio-resourced phytic acid/chitosan polyelectrolyte complex for the flame retardant treatment of wool fabric. *J. Clean. Prod.* **2019,** *223*, 342–349.
92. Vannier, A.; Duquesne, S.; Bourbigot, S.; Alongi, J.; Camino, G.; Delobel, R., Investigation of the thermal degradation of PET, zinc phosphinate, OMPOSS and their blends-identification of the formed species. *Thermochim. Acta* **2009,** *495* (1–2), 155–166.
93. Chen, D. Q.; Wang, Y. Z.; Hu, X. P.; Wang, D. Y.; Qu, M. H.; Yang, B., Flame-retardant and anti-dripping effects of a novel char-forming flame retardant for the treatment of poly(ethylene terephthalate) fabrics. *Polym. Degrad. Stab.* **2005,** *88* (2), 349–356.
94. Liu, W.; Chen, D. Q.; Wang, Y. Z.; Wang, D. Y.; Qu, M. H., Char-forming mechanism of a novel polymeric flame retardant with char agent. *Polym. Degrad. Stab.* **2007,** *92* (6), 1046–1052.
95. Jordanov, I.; Kolibaba, T. J.; Lazar, S.; Magovac, E.; Bischof, S.; Grunlan, J. C., Flame suppression of polyamide through combined enzymatic modification and addition of urea to multilayer nanocoating. *J. Mater. Sci.* **2020,** *55* (30), 15056–15067.
96. Zhang, S.; Horrocks, A. R., A review of flame retardant polypropylene fibres. *Prog. Polym. Sci.* **2003,** *28* (11), 1517–1538.
97. Vroman, I.; Giraud, S.; Salaün, F.; Bourbigot, S., Polypropylene fabrics padded with microencapsulated ammonium phosphate: Effect of the shell structure on the thermal stability and fire performance. *Polym. Degrad. Stab.* **2010,** *95* (9), 1716–1720.
98. Ramgobin, A.; Fontaine, G.; Bourbigot, S., Thermal degradation and fire behavior of high performance polymers. *Polym. Rev.* **2019,** *59* (1), 55–123.
99. Lyon, R. E.; Speitel, L.; Walters, R. N.; Crowley, S., Fire-resistant elastomers. *Fire Mater.* **2003,** *27* (4), 195–208.
100. Dočan, M.; Ylmaz, A.; Bayraml, E., Synergistic effect of boron containing substances on flame retardancy and thermal stability of intumescent polypropylene composites. *Polym. Degrad. Stab.* **2010,** *95* (12), 2584–2588.
101. Anna, P.; Marosi, G.; Csontos, I.; Bourbigot, S.; Le Bras, M.; Delobel, R., Influence of modified rheology on the efficiency of intumescent flame retardant systems. *Polym. Degrad. Stab.* **2001,** *74* (3), 423–426.
102. Yang, S.; Wang, J.; Huo, S.; Wang, M.; Wang, J.; Zhang, B., Synergistic flame-retardant effect of expandable graphite and phosphorus-containing compounds for epoxy resin: Strong bonding of different carbon residues. *Polym. Degrad. Stab.* **2016,** *128*, 89–98.
103. Khanal, S.; Lu, Y.; Dang, L.; Ali, M.; Xu, S., Effects of α-zirconium phosphate and zirconium organophosphonate on the thermal, mechanical and flame retardant properties of intumescent flame retardant high density polyethylene composites. *RSC Advances* **2020,** *10* (51), 30990–31002.
104. Wei, P.; Hao, J.; Du, J.; Han, Z.; Wang, J., An investigation on synergism of an intumescent flame retardant based on silica and alumina. *J. Fire Sci.* **2003,** *21* (1), 17–28.
105. Wang, C.; Wei, P.; Qian, Y.; Liu, J., The synthesis of a novel flame retardant and its synergistic efficiency in polypropylene/ammonium polyphosphate system. *Polym. Adv. Technol.* **2011,** *22* (7), 1108–1114.
106. Bourbigot, S.; Le Bras, M.; Bréant, P.; Trémillon, J. M.; Delobel, R., Zeolites: New synergistic agents for intumescent fire retardant thermoplastic formulations—Criteria for the choice of the zeolite. *Fire Mater.* **1996,** *20* (3), 145–154.
107. Le Bras, M.; Bourbigot, S., Mineral fillers in intumescent fire retardant formulations—Criteria for the choice of a natural clay filler for the ammonium polyphosphate/pentaerythritol/polypropylene system. *Fire Mater.* **1996,** *20* (1), 39–49.
108. Bourbigot, S.; Le Bras, M.; Dabrowski, F.; Gilman, J. W.; Kashiwagi, T., PA-6 clay nanocomposite hybrid as char forming agent in intumescent formulations. *Fire Mater.* **2000,** *24* (4), 201–208.
109. Ren, Q.; Wan, C.; Zhang, Y.; Li, J., An investigation into synergistic effects of rare earth oxides on intumescent flame retardancy of polypropylene/poly (octylene-co-ethylene) blends. *Polym. Adv. Technol.* **2011,** *22* (10), 1414–1421.
110. Zhang, H.; Lu, X.; Zhang, Y., Synergistic effects of rare earth oxides on intumescent flame retardancy of Nylon 1010/ethylene-vinyl-acetate rubber thermoplastic elastomers. *J. Polym. Res.* **2015,** *22* (2), 1–10.
111. Levchik, S. V.; Levchik, G. F.; Camino, G.; Costa, L.; Lesnikovich, A. I., Mechanism of action of phosphorus-based flame retardants in nylon 6. III. Ammonium polyphosphate/manganese dioxide. *Fire Mater.* **1996,** *20* (4), 183–190.
112. Hu, S.; Zheng, H.; Shen, S.; Han, H.; Jing, H.; Li, S., *Synergistic effects of zinc oxide on intumescent flame-retarded polypropylene composites.* In 6th East Asian Symposium on Functional Ion Application Technology and 2010 International Forum on Ecological Environment Functional Materials and Industry, Shanghai, 2011; Vol. 178, pp 279–284.
113. Liu, G.; Zhou, Y.; Hao, J., Thermal degradation mechanism of Ni_2O_3 synergistic intumescent flame retardant polypropylenes. *Gaofenzi Cailiao Kexue Yu Gongcheng/Polym. Mater. Sci. Eng.* **2013,** *29* (9), 98–102 + 106.
114. Xu, Z.; Jia, H.; Yan, L.; Chu, Z.; Zhou, H., Synergistic effect of bismuth oxide and mono-component intumescent flame retardant on the flammability and smoke suppression properties of epoxy resins. *Polym. Adv. Technol.* **2020,** *31* (1), 25–35.
115. Scharf, D.; Nalepa, R.; Heflin, R.; Wusu, T., Studies on flame retardant intumescent char: Part I. *Fire Safety Journal* **1992,** *19* (1), 103–117.

116. Wang, X.; Wu, L.; Li, J., A study on the performance of intumescent flame-retarded polypropylene with nano-ZrO2. *J. Fire Sci.* **2011,** *29* (3), 227–242.
117. Lu, H.; Wilkie, C. A., Study on intumescent flame retarded polystyrene composites with improved flame retardancy. *Polym. Degrad. Stab.* **2010,** *95* (12), 2388–2395.
118. Bourbigot, S.; Duquesne, S.; Fontaine, G.; Bellayer, S.; Turf, T.; Samyn, F., Characterization and reaction to fire of polymer nanocomposites with and without conventional flame retardants. *Mol. Cryst. Liq. Cryst.* **2008,** *486.*
119. Vannier, A.; Duquesne, S.; Bourbigot, S.; Castrovinci, A.; Camino, G.; Delobel, R., The use of POSS as synergist in intumescent recycled poly(ethylene terephthalate). *Polym. Degrad. Stab.* **2008,** *93* (4), 818–826.
120. Ding, P.; Tang, S.; Yang, H.; Shi, L., PP/LDH nanocomposites via melt-intercalation: Synergistic flame retardant effects, properties, and applications in automobile industries. In *2009 Advanced Polymer Processing International Forum, APPF2009,* Qingdao, 2010; Vol. 87–88, pp 427–432.
121. Lewin, M.; Endo, M., Catalysis of intumescent flame retardancy of polypropylene by metallic compounds. *Polym. Adv. Technol.* **2003,** *14* (1), 3–11.
122. Chen, X. L.; Jiao, C. M.; Wang, Y., Synergistic effects of iron powder on intumescent flame retardant polypropylene system. *Express Polym. Lett.* **2009,** *3* (6), 359–365.
123. Duquesne, S.; Samyn, F.; Bourbigot, S.; Amigouet, P.; Jouffret, F.; Shen, K. K., Influence of talc on the fire retardant properties of highly filled intumescent polypropylene composites. *Polym. Adv. Technol.* **2008,** *19* (6), 620–627.
124. Huang, N. H.; Chen, Z. J.; Wang, J. Q.; Wei, P., Synergistic effects of sepiolite on intumescent flame retardant polypropylene. *Express Polymer Letters* **2010,** *4* (12), 743–752.
125. Wu, N.; Ding, C.; Yang, R., Effects of zinc and nickel salts in intumescent flame-retardant polypropylene. *Polym. Degrad. Stab.* **2010,** *95* (12), 2589–2595.
126. Zhang, J.; Fernandez-Blazquez, J. P.; Li, X. L.; Wang, R.; Zhang, X. Q.; Wang, D. Y., A facile technique to investigate the char strength and fire retardant performance towards intumescent epoxy nanocomposites containing different synergists. *Polym. Degrad. Stab.* **2022,** *202.*
127. Samyn, F.; Bourbigot, S., Protection mechanism of a flame-retarded polyamide 6 nanocomposite. *J. Fire Sci.* **2013,** *32* (3), 241–256.
128. Bourbigot, S.; Le Bras, M.; Delobel, R.; Trémillon, J. M., Synergistic effect of zeolite in an intumescence process: Study of the interactions between the polymer and the additives. *J. Chem. Soc. Faraday Trans.* **1996,** *92* (18), 3435–3444.
129. Samyn, F.; Bourbigot, S.; Duquesne, S.; Delobel, R., Effect of zinc borate on the thermal degradation of ammonium polyphosphate. *Thermochim. Acta* **2007,** *456* (2), 134–144.
130. Levchik, S. V.; Camino, G.; Costa, L.; Levchik, G. F., Mechanism of action of phosphorus-based flame retardants in nylon 6. I. Ammonium polyphosphate. *Fire Mater.* **1995,** *19* (1), 1–10.
131. Antonov, A.; Yablokova, M.; Costa, L.; Balabanovich, A.; Levchik, G.; Levchik, S., The effect of nanometals on the flammability and thermooxidative degradation of polymer materials. *Mol Cryst Liq Cryst Sci Technol Sect A Mol Crys Liq Cryst* **2000,** *353,* 203–210.
132. McKee, D. W.; Spiro, C. L.; Lamby, E. J., The inhibition of graphite oxidation by phosphorus additives. *Carbon* **1984,** *22* (3), 285–290.
133. E119–20, A., Standard test methods for fire tests of building construction and materials. In *ASTM E119–20,* ASTM International: West Conshohocken, PA, 2020.
134. Bisby, L.; Gales, J.; Maluk, C., A contemporary review of large-scale non-standard structural fire testing. *Fire Sci. Rev.* **2013,** *2* (1), 1.
135. Schmid, J.; Brandon, D.; Werther, N.; Klippel, M., Technical note—Thermal exposure of wood in standard fire resistance tests. *Fire Saf. J.* **2019,** *107,* 179–185.
136. Blagojević, M. D.; Pešić, D. J., A new curve for temperature-time relationship in compartment fire. *Therm. Sci.* **2011,** *15* (2), 339–352.
137. Barnett, C. R., Replacing international temperature-time curves with BFD curve. *Fire Saf. J.* **2007,** *42* (4), 321–327.
138. Khorasani, N. E.; Billittier, J.; Stavridis, A. Structural performance of a railway tunnel under different fire scenarios. In *2018 Joint Rail Conference, JRC 2018,* American Society of Mechanical Engineers (ASME): Pittsburgh, PA, April 18–20, 2018.
139. White, G. C.; Shirvill, L. C. Fire testing a review of past, current and future methods. In *Proceedings of the 14th International Conference on Offshore Mechanics and Arctic Engineering. Part 5 (of 5),* New York, Mercati, D.; Murray, A., Eds. ASME: New York, NY: 1995; pp 379–388.
140. Bradley, I.; Willoughby, D.; Royle, M., A review of the applicability of the jet fire resistance test of passive fire protection materials to a range of release scenarios. *Process Saf. Environ. Prot.* **2019,** *122,* 185–191.
141. Sikoutris, D. E.; Vlachos, D. E.; Kostopoulos, V.; Jagger, S.; Ledin, S., Fire burnthrough response of CFRP aerostructures. Numerical investigation and experimental verification. *Appl. Compos. Mater.* **2012,** *19* (2), 141–159.
142. Dierdorf, D.; Menchini, C.; Sellers, R.; Carr, Jr., V. J., Design and analysis of alternative high heat flux sources for materials fire testing. *Preprint.* **2007,** *2,* 5.
143. Gardelle, B.; Duquesne, S.; Vandereecken, P.; Bourbigot, S., Characterization of the carbonization process of expandable graphite/silicone formulations in a simulated fire. *Polym. Degrad. Stab.* **2013,** *98* (5), 1052–1063.
144. Morys, M.; Illerhaus, B.; Sturm, H.; Schartel, B., Revealing the inner secrets of intumescence: Advanced standard time temperature oven (STT Mufu+)—µ-computed tomography approach. *Fire Mater.* **2017,** *41* (8), 927–939.
145. Morys, M.; Häßler, D.; Krüger, S.; Schartel, B.; Hothan, S., Beyond the standard time-temperature curve: Assessment of intumescent coatings under standard and deviant temperature curves. *Fire Saf. J.* **2020,** *112.*
146. Maluk, C.; Bisby, L.; Krajcovic, M.; Torero, J. L., A heat-transfer rate inducing system (H-TRIS) test method. *Fire Saf. J.* **2019,** *105,* 307–319.
147. Lucherini, A.; Maluk, C., Assessing the onset of swelling for thin intumescent coatings under a range of heating conditions. *Fire Saf. J.* **2019,** *106,* 1–12.
148. Tranchard, P.; Samyn, F.; Duquesne, S.; Thomas, M.; Estèbe, B.; Montès, J. L.; Bourbigot, S., Fire behaviour of carbon fibre epoxy composite for aircraft: Novel test bench and experimental study. *J. Fire Sci.* **2015,** *33* (3), 247–266.
149. Vieille, B.; Coppalle, A.; Le Pluart, L.; Schuhler, E.; Chaudhary, A.; Rijal, B.; Alia, A.; Delpouve, N.; Bourdet, A., Influence of a flame-retardant on the fire-behaviour and the residual mechanical properties of C/PEKK composite laminates exposed to a kerosene flame. *Compos. A: Appl. Sci. Manuf.* **2022,** *152.*

150. Schuhler, E.; Chaudhary, A.; Vieille, B.; Coppalle, A., Fire behaviour of composite materials using kerosene burner tests at small-scales. *Fire Saf. J.* **2021,** *121.*
151. Adanménou, R. *Mesures à échelle réduite de paramètres pertinents issus de scénarios feu.* University of Lille: Lille: 2020.
152. Krüger, S.; Gluth, G. J. G.; Watolla, M. B.; Morys, M.; Häßler, D.; Schartel, B., New ways: Reactive fire protection coatings for extreme conditions. *Bautechnik* **2016,** *93* (8), 531–542.
153. Weil, E. D., Fire-protective and flame-retardant coatings—A state-of-the-art review. *J. Fire Sci.* **2011,** *29* (3), 259–296.
154. Mariappan, T., Recent developments of intumescent fire protection coatings for structural steel: A review. *J. Fire Sci.* **2016,** *34* (2), 120–163.
155. Sabee, M.; Itam, Z.; Beddu, S.; Zahari, N. M.; Kamal, N. L. M.; Mohamad, D.; Zulkepli, N. A.; Shafiq, M. D.; Hamid, Z. A. A., Flame retardant coatings: Additives, binders, and fillers. *Polym.* **2022,** *14* (14).
156. Weil, E.; McSwigan, B., Melamine phosphates and pyrophosphates in flame-retardant coatings: Old products with new potential. *J Coat Technol* **1994,** *66* (839), 75–82.
157. Andersson, A.; Landmark, S.; Maurer, F. H. J., Evaluation and characterization of ammoniumpolyphosphate-pentaerythritol-based systems for intumescent coatings. *J. Appl. Polym. Sci.* **2007,** *104* (2), 748–753.
158. Delobel, R.; Le Bras, M.; Ouassou, N.; Alistiqsa, F., Thermal behaviours of ammonium polyphosphate-pentaerythritol and ammonium pyrophosphate-pentaerythritol intumescent additives in polypropylene formulations. *J. Fire Sci.* **1990,** *8* (2), 85–108.
159. Jimenez, M.; Duquesne, S.; Bourbigot, S., Intumescent fire protective coating: Toward a better understanding of their mechanism of action. *Thermochim. Acta* **2006,** *449* (1–2), 16–26.
160. Zeng, Y.; Weinell, C. E.; Dam-Johansen, K.; Ring, L.; Kiil, S., Effects of coating ingredients on the thermal properties and morphological structures of hydrocarbon intumescent coating chars. *Prog. Org. Coat.* **2020,** *143.*
161. Kashiwagi, T.; Grulke, E.; Hilding, J.; Groth, K.; Harris, R.; Butler, K.; Shields, J.; Kharchenko, S.; Douglas, J., Thermal and flammability properties of polypropylene/carbon nanotube nanocomposites. *Polymer* **2004,** *45* (12), 4227–4239.
162. Beheshti, A.; Heris, S. Z., Is MWCNT a good synergistic candidate in APP-PER-MEL intumescent coating for steel structure? *Prog. Org. Coat.* **2016,** *90,* 252–257.
163. Li, G.; Liang, G.; He, T.; Yang, Q.; Song, X., Effects of EG and $MoSi_2$ on thermal degradation of intumescent coating. *Polym. Degrad. Stab.* **2007,** *92* (4), 569–579.
164. Beraldo, C. H. M.; da S Silveira, M. R.; Baldissera, A. F.; Ferreira, C. A., A new benzoxazine-based intumescent coating for passive protection against fire. *Prog. Org. Coat.* **2019,** *137.*
165. Wang, Z. Y.; Han, E. H.; Ke, W., Fire-resistant effect of nanoclay on intumescent nanocomposite coatings. *J. Appl. Polym. Sci.* **2007,** *103* (3), 1681–1689.
166. Ahmad, F.; Zulkurnain, E. S. B.; Ullah, S.; Al-Sehemi, A. G.; Raza, M. R., Improved fire resistance of boron nitride/epoxy intumescent coating upon minor addition of nano-alumina. *Mater. Chem. Phys.* **2020,** *256.*
167. Nasir, K. M.; Sulong, N. H. R.; Johan, M. R.; Afifi, A. M., Synergistic effect of industrial- and bio-fillers waterborne intumescent hybrid coatings on flame retardancy, physical and mechanical properties. *Prog. Org. Coat.* **2020,** *149.*
168. Baldissera, A. F.; Silveira, M. R.; Dornelles, A. C.; Ferreira, C. A., Assessment of lignin as a carbon source in intumescent coatings containing polyaniline. *J. Coat. Technol. Res.* **2020,** *17* (5), 1297–1307.
169. Liu, S.; Wang, C.; Hu, Q.; Huo, S.; Zhang, Q.; Liu, Z., Intumescent fire retardant coating with recycled powder from industrial effluent optimized using response surface methodology. *Prog. Org. Coat.* **2020,** *140.*
170. Bradford, S.; Hallam, S.; Taylor, A. P. Intumescent coating composition. No. 2,451,233, 2007.
171. Gardelle, B.; Duquesne, S.; Vandereecken, P.; Bourbigot, S., Resistance to fire of silicone-based coatings: Fire protection of steel against cellulosic fire. *J. Fire Sci.* **2014,** *32* (4), 374–387.
172. Bourbigot, S.; Gardelle, B.; Duquesne, S. Intumescent silicone-based coatings for the fire protection of carbon fiber reinforced composites. In *11th IAFSS Symposium*, Christchuch, New Zealand, International Association for Fire Safety Science (IAFSS), 2014.
173. Qureshi, S. P.; Krassowski, D. W. Intumescent resin system for improving fire resistance of composites. In *Proceedings of the 1997 29th International SAMPE Technical Conference*, Orlando, FL, 1997; pp 625–634.
174. Mastalska-Popławska, J.; Izak, P.; Wójcik, Ł.; Stempkowska, A.; Góral, Z.; Krzyżak, A. T.; Habina, I., Synthesis and characterization of cross-linked poly(sodium acrylate)/sodium silicate hydrogels. *Polym. Eng. Sci.* **2019,** *59* (6), 1279–1287.
175. Watolla, M. B.; Gluth, G. J. G.; Sturm, P.; Rickard, W. D. A.; Krüger, S.; Schartel, B., Intumescent geopolymer-bound coatings for fire protection of steel. *J. Ceram. Sci. Technol.* **2017,** *8* (3), 351–364.
176. Sotiriadis, K.; Guzii, S. G.; Mácová, P.; Viani, A.; Dvořák, K.; Drdácký, M., Thermal behavior of an intumescent alkaline aluminosilicate composite material for fire protection of structural elements. *J. Mater. Civ. Eng.* **2019,** *31* (6).
177. Chen, S. N.; Li, P. K.; Hsieh, T. H.; Ho, K. S.; Hong, Y. M., Enhancements on flame resistance by inorganic silicate-based intumescent coating materials. *Materials* **2021,** *14* (21).
178. Abdul Rashid, M. K.; Ramli Sulong, N. H.; Alengaram, U. J., Fire resistance performance of composite coating with geopolymer-based bio-fillers for lightweight panel application. *J. Appl. Polym. Sci.* **2020,** *137* (47).
179. Shuai, Q.; Xu, Z.; Yao, Z.; Chen, X.; Jiang, Z.; Peng, X.; An, R.; Li, Y.; Jiang, X.; Li, H., Fire resistance of phosphoric acid-based geopolymer foams fabricated from metakaolin and hydrogen peroxide. *Mater. Lett.* **2020,** *263.*
180. Sarazin, J.; Davy, C. A.; Bourbigot, S.; Tricot, G.; Hosdez, J.; Lambertin, D.; Fontaine, G., Flame resistance of geopolymer foam coatings for the fire protection of steel. *Compos. B. Eng.* **2021,** *222,* 109045.
181. Bourbigot, S.; Sarazin, J.; Davy, C. A.; Fontaine, G., Foamed geopolymers for fire protection: Burn-through testing and modeling. *Fire Mater.* **2021**.
182. Christke, S.; Gibson, A. G.; Grigoriou, K.; Mouritz, A. P., Multi-layer polymer metal laminates for the fire protection of lightweight structures. *Mater. Des.* **2016,** *97,* 349–356.
183. Geoffroy, L.; Samyn, F.; Jimenez, M.; Bourbigot, S., Intumescent polymer metal laminates for fire protection. *Polym.* **2018,** *10* (9).
184. Geoffroy, L.; Samyn, F.; Jimenez, M.; Bourbigot, S., Bilayer intumescent paint metal laminates: A novel design for a high-performance fire barrier. *Ind. Eng. Chem. Res.* **2020,** *59* (7), 2988–2997.

8 Silicon and Inorganic Flame Retardants

Walid H. Awad

8.1 INTRODUCTION

As pressure, due to environmental concern, on the halogenated flame retardants increases, the need and the search for flame retardant (FR) alternatives increases as well. Amongst the number of elements in the periodic table that proved to provide flame retardant characteristics, silicon presents strong possibilities. Even though most of the time, silicon compounds do not work as a standalone FR material, but the beautiful and unique thing about silicon is that this element can take a multitude of chemical structures, and all of them can provide FR performance in one way or another. Many monographs and reviews have been written in the areas of organic, inorganic, organometallic, polymer, and mechanistic silicon chemistry.[1–11] We will go through the different chemical forms of silicon and see how it can deliver the intended FR performance.

8.2 SILICON-BASED FLAME RETARDANTS

8.2.1 Silicones

Silicones are a class of synthetic polymers having the general formula $(R_mSi(O)_{4-m/2})_n$, where m = 1–3 and n ≥ 2. The most common example is polydimethylsiloxane (PDMS). This polymer has a repeating $(CH_3)_2SiO$ unit. Silicones are the subject of many reviews.[12–22] It is not possible to cover adequately the many uses to which silicones have been put. However, a summary of the most important properties and representative applications that utilize these properties are provided in the literature.[23,24]

Silicone materials exhibit relatively low rates of heat release, a uniquely low dependence of rate of heat release on external heat flux. One of the causes of the lower burning rate is attributed to the accumulation of the silica ash layer at the silicone fuel surface. This accumulation of amorphous silica ash at the surface results from the deposition of silica particles, one of the major combustion products of silicone oligomers (cyclic and/or linear structures) in the gas phase.

According to a gasification study conducted on silicones, two modes of gasification have been proposed: (1) the volatilization of molecular species native to the polymer and (2) the volatilization of molecules resulting from thermally induced degradation of the polymer via siloxane rearrangement. The former process is dominant for short-chain oligomers, whereas the latter one is dominant in all higher molecular weight polymers.[25,26]

Kashiwagi et al. investigated the flammability properties of various silicone-containing polycarbonates using the cone calorimeter.[27] The silicone was incorporated in the polycarbonate matrix as an additive or as a copolymer. Figure 8.1 shows a comparison between the heat release rate (HRR) of silicone/polycarbonate and pure polycarbonate. The data indicated that silicone has a significant effect in reducing the rate of the heat release. However, the delay in the ignition time for the silicone containing sample is shorter than that for the pure polycarbonate sample.

One effective approach in increasing the fire resistance of polyurethane (PU) is to incorporate PDMS into its structure.[28] Microphase-segregated PU block copolymers containing hydroxyl-terminated silicone as the soft segment have been synthesized. The enhanced siliconated surface of the block copolymer has been confirmed by ESCA and x-ray analyses.

Thermal analysis (TGA) data of these siliconated block copolymers revealed that they are thermally more stable than the reference materials, poly(tetrahydrofuran block urethane) and poly(ethyleneglycol block urethane) copolymers. The thermal stability was found to depend on the silicone content, with stability increasing as the silicone content in the structure increases.

The flammability behavior of these materials, as revealed by oxygen index, showed that siliconated PUs have higher oxygen index values compared to reference materials. The siliconated block copolymers with higher PDMS content have higher oxygen index values. The oxygen index values depend upon the diisocyanate used. For example, block copolymers made of hydroxy-terminated PDMS, H(12) MDI, and 1,6-hexanediol showed higher oxygen index values compared to block copolymers made of hydroxy-terminated PDMS, TDI, and 1,6-hexanediol. This difference is related to the extent of soft block segregation.

The observed enhancement in oxygen index could be attributed to phase segregation in these block copolymers, which leads to domination of siloxane on the polymer surface. Siloxanes have solid-phase activity rather than vapor phase activity and reduce flammability through increased formation of pyrolytic char.

Thermogravimetric analysis of PU/PDMS hybrid materials[29] indicated that no enhancement in the thermal stability of PU was obtained under the thermo-oxidative conditions, but under pyrolytic conditions, some improvement in the thermal stabilization of PU was observed. The thermal stability of PU is generally controlled by the thermally weakest link: the urethane bond. In the case of PU/PDMS

DOI: 10.1201/9781003380689-8

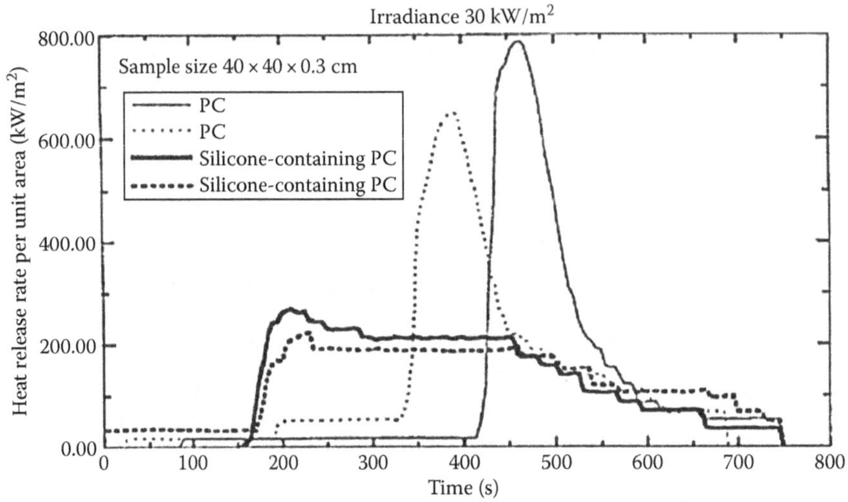

FIGURE 8.1 HRR for polycarbonate and polycarbonate-containing silicone (two repeated tests for each sample).

Source: From Kashiwagi, T. et al., A nonhalogenated, flame retarded polycarbonate, in *Conference of Advanced Fire Resistant Aircraft Interior Materials*, Atlantic City, NJ, February 9–11, 1993, 175.

hybrid system, any decrease in the thermal stability might be explained by thermal instabilities in PDMS structure.

In terms of flammability properties, both the Limiting Oxygen Index (LOI) (21% for PU/PDMS vs. 22% for PU) and UL 94 (no rating for PU and PU/PDMS) indicated that no benefit was achieved by incorporating PDMS into PU. However, the most interesting results were revealed by the cone calorimeter, where the data showed that there is significant reduction in the HRR of PU/PDMS compared to PU. This can be explained by the accumulation of silica on the char layer that provides protection to the underlying polymer from heat transfer with the flame. To examine the efficiency of this char layer in protecting the underlying polymer, a heat transfer measurement was carried out on both PU and PU/PDMS during heat exposure. The temperature profiles indicated that for pure PU the evolution of temperature was relatively rapid compared to PU/PDMS.

Hermansson et al. carried out extensive investigations on the fire-retardant behavior of ethylene–acrylate copolymer modified with chalk and silicone elastomer.[30–32] They have shown that incorporation of a silicone elastomer (at 5 wt%) and chalk filler (at 30 wt%) can greatly improve the flame-retardant properties of ethylene butyl acrylate formulations. The results showed that, compared to the pure polymer, an increase in the LOI from 18 to 30 and a decrease in the peak heat release rate (PHRR) from 1300 to 330 kW/m^2 were observed.

These property improvements were assigned to an intumescent process, which was proposed to consist of the following steps: (1) ester pyrolysis occurs when degradation starts at 300°C producing carboxylate ions on the copolymer backbone; (2) the foam structure started to form when the volatile degradation products of ethylene butyl acrylate (EBA) copolymer are produced; (3) the ionomers produced during the degradation process increase the melt viscosity by crosslinking reactions and provide stabilization to the foam structure.

The melt-dripping behavior of various ethylene–acrylate formulations with chalk and silicone have also been addressed by exposing these formulations to a Bunsen burner. It has been demonstrated that viscosity plays a crucial role in this process by affecting the transport of volatile gases, eventual dripping, and the formation of the intumescent structure.[32]

One of the most recent manifestations of the potential use of silicone as FR materials is the use of these materials to develop ceramifiable composition. In wire and cable applications, there is a need, as the cable burns during a fire, to develop a robust char structure, such as a ceramic, that protects the cable to do its function in delivering the electrical signals. Examples of this application can be found in recent literature.[33–35]

Polymer-based ceramifiable composites generally consist of polymer matrix with refractory silicate minerals that form the ceramic framework in combination with a flux system. A key characteristic of ceramifying polymers is their ability to form a self-supporting structure throughout the temperature range from ambient service temperature to over 1,000°C.

Typically, the reactions in the inorganic ceramic forming systems can commence from temperatures as low as 350°C and continue to 800°C or higher. This is achieved with fluxes which produce a controlled, low level of liquid phase at these temperatures.

The ceramification in these materials is not simply the bonding or fusing of the silicate particles by a viscous liquid phase, such as with relatively high levels of low melting point glasses. Such materials tend to collapse at high temperatures and are not self-supporting. Ceramification involves reaction sintering assisted by the controlled level of liquid phase.

In other words, the ceramification process relies on the formation of a eutectic liquid phase at the edges of the fillers, which penetrate into the matrix region with increasing temperature and exposure time. The extent of the eutectic formation is further influenced by factors such as particle size and chemical composition of the fillers.

Adjusting the ratio of the different components in the composition, in addition to the particle size (coarse vs. fine grade) will be crucial in forming ceramified layer with enhanced mechanical properties and help the ceramification process.

Functional additives may be added such as glass frits that contain a third category of oxides to strengthen, increase the viscosity, and widen the melting range of the material. The total ceramifying additive level may be required to be quite high, although the materials can still be processed like conventional polymers.

It should be noted that the first indication of the bridging effect of functional fillers in enhancing the flame retardancy was pointed out by Weil et al.,[36] where it was indicated that small amounts of high aspect ratio materials such as mica or wollastonite can improve fire-resistance properties of residues by a "bridging action."

8.2.2 Silanes

The strategy of incorporating silicon as a reactive component in the polymeric system to attain flame retardancy has been explored. For example, Ebdon et al. carried out silylation to the polystyrene using n-butyl lithium in the presence of tetramethylethylenediamine, followed by reaction with trimethylchlorosilane, dichlorodimethylsilane, or trichloromethylsilane, as shown in Scheme 8.1. Poly(vinyl alcohol) films have also been modified with chlorosilanes (Scheme 8.1).

The limiting oxygen indices of these systems showed a significant reduction in the flammability of the silylated polymers. In both systems, it appears that silicon has a much greater effect in the presence of halogen, probably through the evolution of silicon chlorides during the combustion process.[37]

The reactive approach has been employed to prepare various polymeric systems.[38,39] Silicon-containing polystyrenes and poly(methyl methacrylate)s (PMMAs) copolymers have been prepared by free radical polymerization. The LOI data indicated that a marginal improvement in flame retardancy has been observed compared to the parent homopolymers. The authors speculated that the nature of the silicon-containing group has an effect on the flame-retardant mechanism.[38]

Flame-retardant epoxy resins with different silicon contents were prepared using silicon-containing epoxides or silicon-containing prepolymers. The thermal stability and flame-retardant properties of the produced epoxide systems were evaluated and related to the silicon content. The char yields under nitrogen and air atmospheres increased with increase in silicon content. The authors pointed out that the silicon-containing resin has improved flame retardancy over the silicon-free resin, as evidenced by the LOI. LOI values increased from 24 for a standard commercial resin to 36 for silicon-containing resins.[39]

In wire and cable applications, silane has a major role to play in enhancing the thermal and flammability properties of the cable insulation. One of the major polymers used in wire and cable applications is the silane cross-linked polyethylene. The cross-linking is initiated by radical cross-linking of the silane grafted polyethylene. The silicone network developed after cross-linking provides a major boost in enhancing the thermal and flammability properties of the cable. This improvement in flammability performance comes from the combination of silicone content and the network structure developed. However, even though the role of silane here, as a graft or copolymer, does not act as a standalone flame retardant, its role in delivering the final flame rating cannot be underestimated.

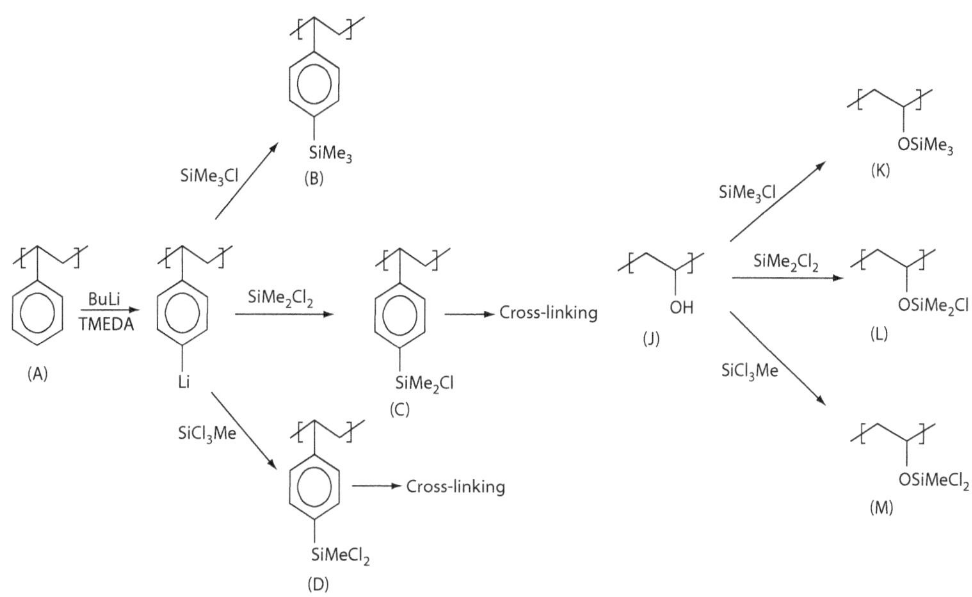

SCHEME 8.1 Silylation of polystyrene (left) and poly(vinyl alcohol) (right).

Source: From Ebdon, J.R. et al., *Polym. Degrad. Stab.*, 54, 395, 1996. With permission.

Lithium-ion batteries have many commercial applications. Recently, the potential of Vinyl-tris-(methoxydiethoxy) silane to act as flame retardant in lithium ion batteries has been explored.[40] This silane possesses similar structure and properties to phenyl-tris-(methoxydiethoxy)silane, which was reported to be a good additive to form efficient solid-electrolyte interface (SEI) films to prevent the decomposition of propylene carbonate on graphite surface.

8.2.3 Silsesquioxane

The term silsesquioxanes refers to silicon structures that have the empirical formula $RSiO_{3/2}$ where R is a hydrogen atom or a carbon moiety. Polyhedral oligomeric silsesquioxane (POSS) is a unique member of this group with its name derived from the noninteger (one and one-half or sesqui) ratio between oxygen and silicon organic substituent. Actually, POSS represents an intermediate structure between those of silicone and silica. Figure 8.2 shows the general structure of POSS macromers. The POSS molecules, which were first isolated along with other volatile compounds in 1946 through thermolysis of the polymeric products of methyltrichlorosilane co-hydrolysis, contain a polyhedral silicon-oxygen nanostructured skeleton.[41] The uniqueness of POSS molecules arises from the thermally stable silicon-oxygen framework and the flexible chemical composition; a variety of organic substituents can be attached to each corner silicon atom to give different functionality.

POSS molecules can be incorporated into polymer systems through blending, grafting, or copolymerization aiming at nanostructured polymeric materials, which have properties that bridge the property space between organic plastics and ceramics. Recent studies on POSS-containing hybrid polymers have been reported indicating reinforced mechanical, thermal, and flammability properties.[42–53]

In a detailed investigation about the thermolysis of POSS macromers and POSS-siloxane copolymers, Mantz et al.[43] investigated the pyrolysis of four POSS macromers: $Cy_8\text{-}Si_8O_{11}(OH)_2$, $Cy_8\text{-}Si_8O_{11}(OSiMe_3)_2$, $Cy_6\text{-}Si_6O_9$, and $Cy_8\text{-}Si_8O_{12}$ (where $Cy = c\text{-}C_6H_{11}$) (Figure 8.3) and two POSS siloxane copolymers ($Cy_8\text{-}Si_8O_{11}(OSiMe_2)_nO$–) (Figure 8.4).

They indicated that both fully and incompletely condensed POSS macromers have a tendency to sublime upon heating, assuming they contain functionalities which do not undergo cross-linking reactions. The original POSS structure was observed in the sublimates for both fully and incompletely condensed structures. However, once this POSS structure was incorporated into a polymeric form, these macromers were no longer observed in the sublimates. Rather, they decompose primarily through partial loss of their organic substituents followed by subsequent cross-linking reactions which incorporates the remaining composition into a SiO_xC_y network (char).

Extensive thermal degradation study was carried out on octaisobutyl POSS under both inert and air atmospheres.[48] The degradation under inert conditions led to almost complete weight loss regardless of the heating rate. More complex degradation was observed under the oxidative atmosphere. The authors proposed a possible competition between the evaporation and oxidation processes, which eventually led to the production of silica-like thermally stable char. Analysis of the char residue after thermo-oxidative conditions indicated that the organic content decreased by increasing the treatment temperature. This char was converted to amorphous silica by heating the POSS up to 800°C.

The same group reported on the thermal and thermo-oxidative degradation of a number of POSS cages with different substituent groups; namely, R = H, CH_3, C_4H_9, C_8H_{17}, C_6H_5, and vinyl (see Figure 8.5). Different thermal behavior was observed depending on the type of substituent. The basic difference between alkyl-substituted POSS and phenyl-substituted POSS is the intrinsic thermal stability of the substituent. For the alkyl substituent, cleavage of the Si-C and C-C bonds could occur, producing unsaturated moieties. On the other hand, phenyl POSS degradation would proceed by hydrogen abstraction of the phenyl group, and the resulting aromatic radicals would combine to form very stable polyaromatic structures. The highest ceramic yield

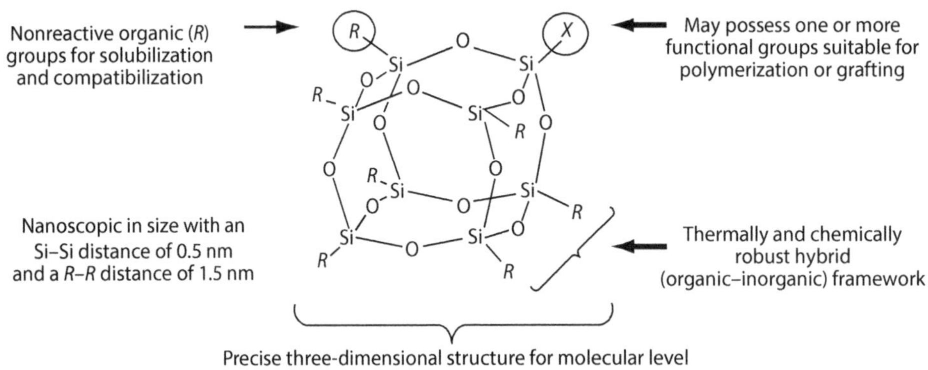

FIGURE 8.2 General structure of polyhedral oligomeric silsesquioxane.

Source: From http://www.hybridplastics.com

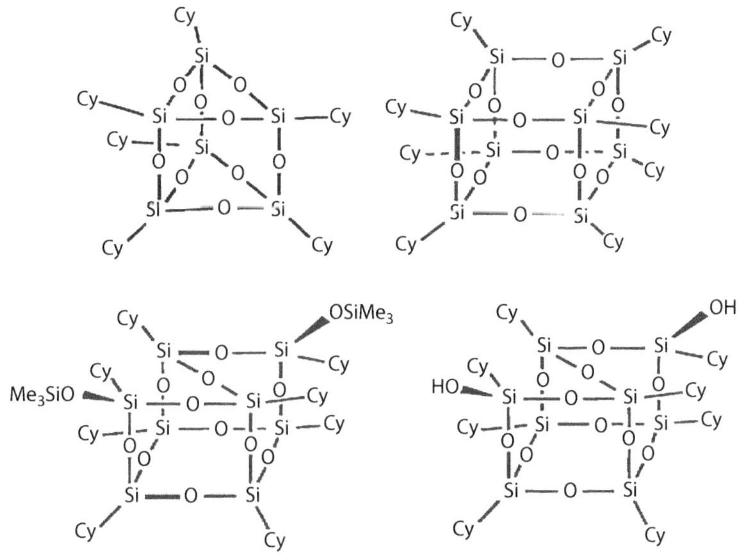

FIGURE 8.3 POSS macromers: Cy_6-Si_6O_9, Cy_8-Si_8O_{12}, and Cy_8-Si_8O_{11} $(OSiMe_3)_2$, Cy_8-Si_8O_{11} $(OH)_2$, (where Cy = c-C_6H_{11}).

Source: From Mantz, R.A. et al., *Chem. Mater.*, 8, 1250, 1996. With permission.

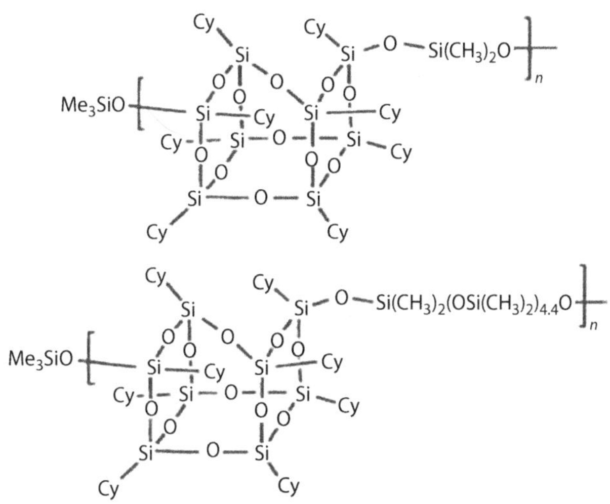

FIGURE 8.4 POSS-siloxane copolymer: (Cy_8-$Si_8O_{11}$$(OSiMe_2)_1$O-) and ($Cy_8$-$Si_8O_{11}$$(OSiMe_2)_{5.4}$O-).

Source: From Mantz, R.A. et al., *Chem Mater.*, 8, 1250, 1996. With permission.

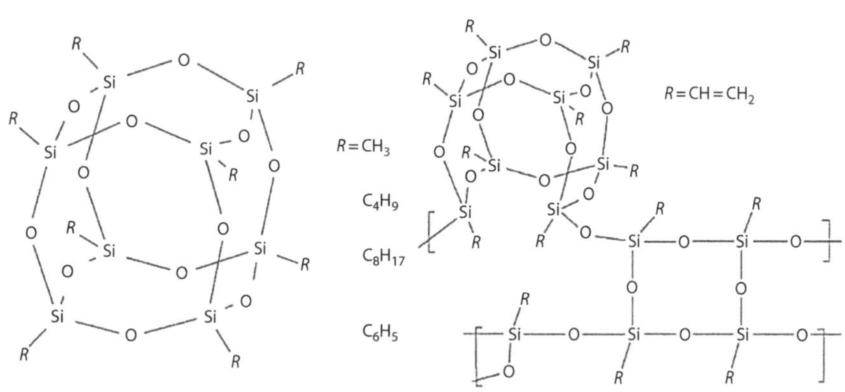

FIGURE 8.5 T_8 POSS structure and polyvinyl POSS.

Source: From Fina, et al., *Thermochim. Acta*, 440, 36, 2006. With permission.

was obtained for the polyvinyl POSS through material reorganization into O-Si-C_n-Si structure.[49]

The above thermal analysis studies demonstrated the enhanced thermal stability of POSS materials and suggested that there is potential to improve the flammability properties of polymers when compounded with these macromers. In a typical example of their application as flame retardants, a US patent[44] described the use of preceramic materials, namely, polycarbosilanes (PCS), polysilanes (PS), polysilsesquioxane (PSS) resins, and POSS (structures are shown in Figure 8.6) to improve the flammability properties of thermoplastic polymers such as polypropylene and thermoplastic elastomers such as Kraton (polystyrene-polybutadiene-polystyrene, SBS) and Pebax (polyether block–polyamide copolymer).

The resulting blends were shown to possess enhanced flammability and mechanical properties. The cone calorimeter data shown in Table 8.1 indicates that the highest reduction in the PHRR was obtained for the Pebax system where a reduction of 65%–70% in PHRR was reported. The reduction in heat release was obtained by enhanced char formation. The char reduces the escape of small volatile molecules to the gas phase, which in turn reduces the amount of heat release feedback to the polymer surface.

Another example of enhanced flammability properties was obtained by adding POSS to the coating of textile fabrics.[45] In this study, POSS was used as an additive for the PU resin, and the product was applied as a coating to the polyester fabrics. For comparison purposes, another kind of additive (Cloisite 30B) has been used. The structure of POSS materials employed in this investigation is shown in Figure 8.7a and b. The data in Figure 8.8 show the HRR curves for the knitted fabrics coated with PU alone and PU compounded with each of the following additives: POSS MS2, POSS FQ2, and 30B. It is apparent that POSS MS2 has no significant effect on the reduction of PHRR.

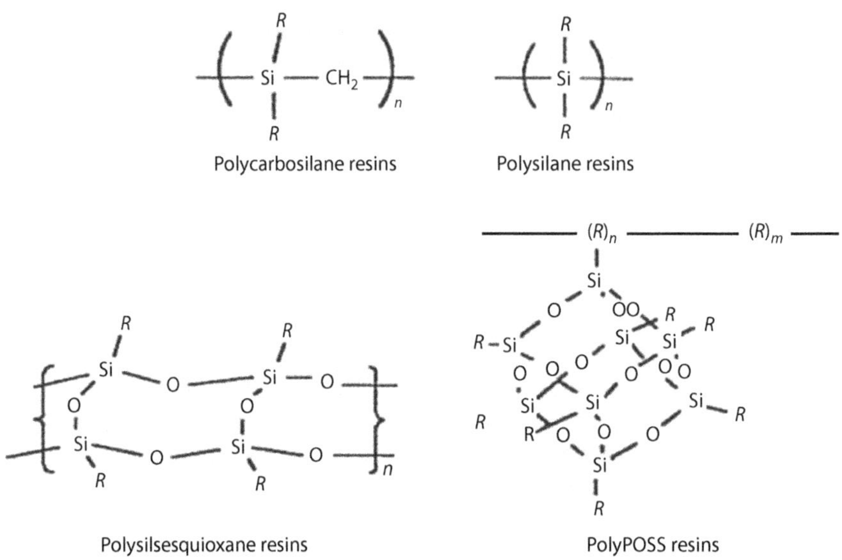

FIGURE 8.6 PCS, PS, PSS, and POSS structures.

Source: From Lichtenhan, J.D. and Gilman, J.W., Preceramic additives as fire retardants for plastics, US patent 6,362,279, 2002.

TABLE 8.1
Cone Calorimeter Data for Polymers and Pre-Ceramic Polymer-Polymer Blends

Sample	Char Yield (%)	Peak HRR (Δ%) (kW/m²)	Mean HRR (Δ%) (kW/m²)	Mean Heal of Combustion (MJ/kg)	Total Heat Released (MJ/m²)	Mean Ext Area (m²/kg)	Mean CO Yield (kg/kg)
PP	0	1.466	741	34.7	141	650	0.03
PP w/20%PSS	17	892 (40%)	432(42%)	29.8	106	821	0.03
PEBAX	0	2.020	780	29.0	332	187	0.02
PEBAX w/20%PCS	15	699 (65%)	419(46%)	28.5	272	260	0.02
PEBAX w/10%PSS	6	578 (72%)	437 (44%)	25.2	301	367	0.02
Kraton	1	1.405	976	29.3	351	1.750	0.08
Kraton w/20%PCS	20	825 (42%)	362 (63%)	26.4	266	1.548	0.07
Kraton w/10%PSS	6	1.027(27%)	755(23%)	26.9	324	1.491	0.07

Source: Lichtenhan, J.D. and Gilman, J.W., Preceramic additives as fire retardants for plastics, US Patent 6,362,279, 2002.

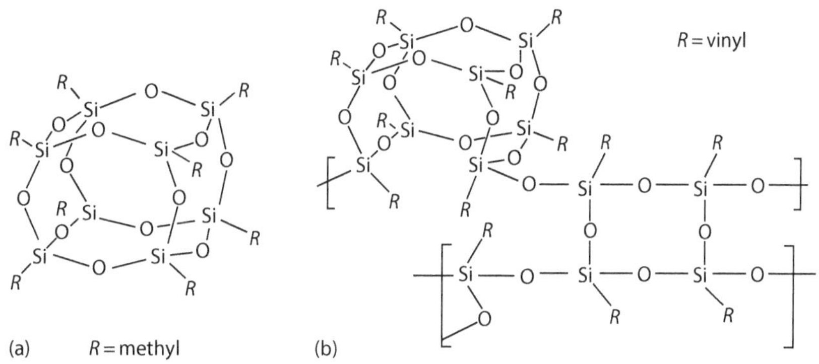

FIGURE 8.7 (a) Octamethyl POSS (POSS-MS) and (b) poly(vinylsilsesquioxane) (POSS-FQ).

Source: From Devaux, E. et al., *Fire Mater.*, 26, 149, 2002. With permission.

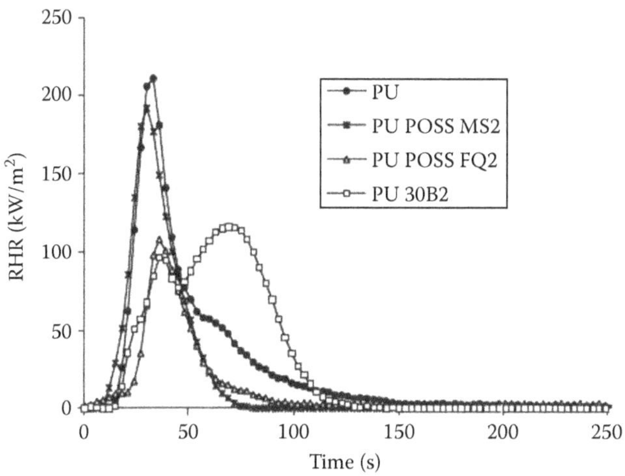

FIGURE 8.8 Rate of heat release curves of PU-nanocomposite coatings on PET knitted fabrics at 35 KW/m^2.

Source: From Devaux, E. al., *Fire Mater.*, 26, 149, 2002. With permission.

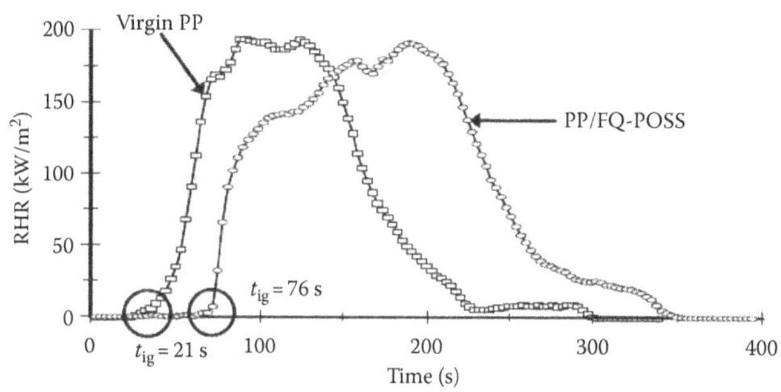

FIGURE 8.9 Rate of heat release curves of PP and PP/FQ-POSS knitted fabrics at 35 KW/m^2.

Source: From Bourbigot, S. et al., *Fire Retardancy of Polymers: New Applications of Mineral Fillers*, Bras, M. et al., Eds., The Royal Society of Chemistry, Cambridge, MA, 2005, 189. With permission.

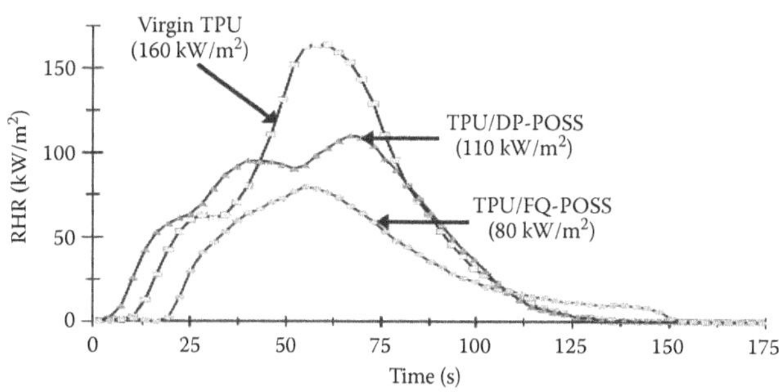

FIGURE 8.10 Rate of heat release curves of TPU, TPU/DP-POSS, and TPU/FQ-POSS as a coating of woven PET fabrics at heat flux 35 KW/m^2.

Source: From Bourbigot, S. et al., *Fire Retardancy of Polymers: New Applications of Mineral Fillers*, Bras, M. et al., Eds., The Royal Society of Chemistry, Cambridge, MA, 2005, 189. With permission.

However, 55% reduction in PHRR was obtained for POSS FQ2. This reduction was achieved coupled with increase in time to ignition (20 vs. 10 s for pure PU).

Bourbigot et al. reported on the fire behavior of the polypropylene-POSS multifilament yarns.[50] The material was prepared by melt spinning of both polypropylene and poly(vinylsilsesquioxane) (FQ-POSS). The HRRR data shown in Figure 8.9 indicates that no reduction in PHRR was obtained using FQ-POSS. However, the time to ignition of PP-POSS system was increased relative to the pure PP. Thermoplastic polyurethane (TPU) coating systems were also prepared using two types of POSS macromers, namely, dodecaphenyl-POSS (DP-POSS) and FQ-POSS. The HRR data (Figure 8.10) showed that the PHRR was reduced by 31% and 50% for TPU/DP-POSS and TPU/FQ-POSS systems, respectively.

The effects of metal-containing POSS on the combustion behavior of polypropylene have been investigated.[51] Metal-POSS was prepared from incompletely condensed structures by reaction with organometallic compounds. The dimeric and oligomeric of Al and Zn–isobutyl silsesquioxane (POSS) have been evaluated as potential flame retardants for polypropylene, and the results were compared

TABLE 8.2
Cone Calorimeter Data for Different PP/POSS Formulations

Sample	TTI (s)	PHRR (kW/m²)	MAHRE (kW/m²)	THR (MJ/m²)	EHC (MJ/kg)	Residual Weight (%)	TSR	Avg CO_2 Yield (kg/kg)	Avg CO Yield (g/kg)
PP	5615	1103 ± 50	509120	111 ± 2	4711	–0	1427 ± 30	3.39 ± 0.02	43 ± 5
PP/ T_8-POSS	5013	1325 ± 50	591 ± 20	112 ± 3	4811	0.2 ± 0.1	1564 ± 70	3.45 ± 0 04	42 ± 1
PP/AI POSS	37 ± 1	624 ± 20	399 ± 10	98 ± 1	41 ± 1	2.5 ± 0.1	1850 ± 20	3.05 ± 0.03	3512
PP/ Zn-POSS	54 ± 4	1069 ± 50	509120	108 ± 4	46 ± 1	3.3 ± 0.1	1525 ± 25	3.35 ± 0.08	4212

Source: Fina, A. et al., *Polym. Degrad. Stab.*, 91, 2275, 2006. With permission.
Note: TTI, time to ignition; PHRR, peak of heat release rate; MAHRE, maximum average rate of heat emission; THR, total heat release; EHC, effective heat of combustion; TSR, total smoke released.

with PP/octaisobutyl POSS. The cone calorimeter data (Table 8.2) revealed that the addition of octaisobutyl POSS had a deteriorating effect on the flammability properties of polypropylene, which was ascribed to the partial evaporation of T_8-POSS, and the subsequent oxidation of POSS organic substituents. Al-containing POSS can be an effective flame retardant, as evidenced by the significant reduction in PHRR. This enhanced performance was attributed to the catalytic effect of Al which promotes char formation. Zn-POSS had no effect on polypropylene combustion behavior.

It should be noted that the inconsistency in the flammability performance of various types of polymer/POSS hybrid systems could be ascribed to several factors, such as, the type of polymeric material, the structure of POSS macromer, and the degree of dispersion of POSS in the polymer matrix.

The potential of synergistic effects on the POSS materials has been illustrated recently.[52] Octamethyl POSS (OMPOSS) and Exolit OP950, a phosphinate-based compound, have been incorporated with the recycled poly(ethylene terephthalate) (PET) by melt blending. Exolit OP950 showed intumescent behavior. The LOI increased from the initial value 29 vol% for PET/10% OP950 to 36 vol%, when 1 wt% of OP950 is substituted by OMPOSS (Figure 8.11). This suggests that there is a synergistic effect between OP950 and OMPOSS.

The cone calorimeter data (Table 8.3) provided further evidence for this synergistic behavior. A reduction in PHRR of 30% and 37% was obtained for PET compounded with 20 wt% and 10 wt% of OP950, respectively. However, by partial substitution of OP950 with 2 wt% and 1 wt% of OMPOSS, the reduction observed in PHRR values was 65% and 61%, respectively.

The degradation and combustion behavior of polycarbonate/POSS hybrid system has been reported recently.[53] Different loading contents of trisilanolphenyl-POSS (TPOSS) were melt blended with polycarbonate matrix (PC). The data shown in Table 8.4 indicate that no improvement

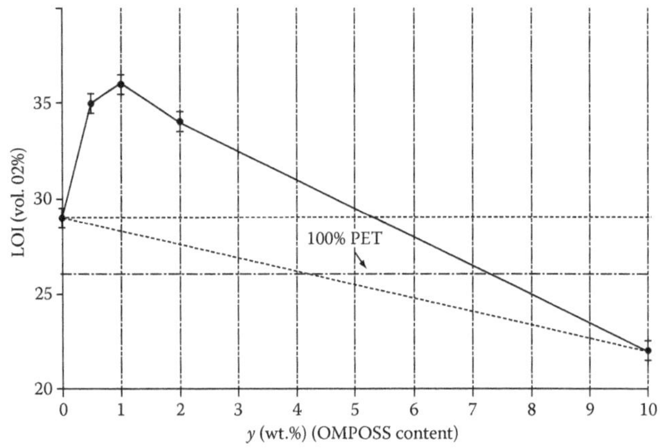

FIGURE 8.11 LOI values of formulations PET/OP950/OMPOSS (90/10—y/y).

Source: From Vannier, A. et al., Polym. Degrad. Stab., 93, 818, 2008. With permission.

in thermal stability parameters (i.e., onset decomposition temperature and peak decomposition temperature) was observed compared to the neat polycarbonate. The thermo-oxidative degradation process of the hybrid system proved to be a complicated process, which includes hydrolysis/alcoholysis of the carbonate linkage, free radical oxidative chain degradation, reformation, and branching and cross-linking reactions.

The flame-retardant properties as evaluated by the cone calorimeter indicate that adding TPOSS to polycarbonate leads to a significant reduction in the PHRR value, but no improvement in the time to ignition was observed (Figure 8.12). The decrease in the time to ignition was attributed to the evolution of small molecules produced from dissociation of TPOSS. The authors proposed that during the combustion process a char layer is formed on the surface of the sample and TPOSS undergoes a series of oxidation reactions to form SiO_2. The presence of SiO_2 can enhance the viscosity and the thermal oxidative stability of

TABLE 8.3
Cone Calorimeter Data for Various PET/OP950/OMPOSS Formulations (35 KW/m²)

Sample	T_{ign} (s)	PHRR (kW/m²) (% Reduction)	THR (MJ/m²)	AMLR (g/s)	ASEA (m²/kg)
PET	75	708	49	0.11	269
PET + OP950 20 wt%	145	496(30)	35	0.09	616
PET + OP950 10 wt%	109	448 (37)	37	0.10	454
PET + OP950 18 wt% + OMPOSS 2 wt%	62	247 (65)	44	0.06	481
PET + OP950 9 wt% + OMPOSS 1 wt%	73	279 (61)	46	0.06	440

Source: Vannier, A. et al., Polym. Degrad. Stab., 93, 818, 2008. With permission.
Note: T_{ign} = time to ignition, PHRR = peak heat release rate, THR = total heat release, AMLR = average mass loss rate, ASEA = average specific extinction area.

TABLE 8.4
Thermal Analysis (TGA, DTG, and DTA) Data for Different Formulations of PC/TPOSS, under Ar Atmosphere

Sample	$T_{-10\%}$(°C)	$T_{-50\%}$(°C)	$T_{-70\%}$(°C)	T_{max}(°C)	T_{max}(°C)	Solid Residue (%) at 800°C
TPOSS	436	—	—	43, 210, 568	684	56.1
PC	478	525	553	524	552	19.7
PC/1 wt% TPOSS	467	508	557	517	558	21.7
PC/2 wt% TPOSS	477	514	596	518, 671	540	14.3
PC/3 wt% TPOSS	469	520	600	518, 670	538	2.9

Source: Song, L. et al., *Polym. Degrad. Stab.*, 93, 627, 2008. With permission.

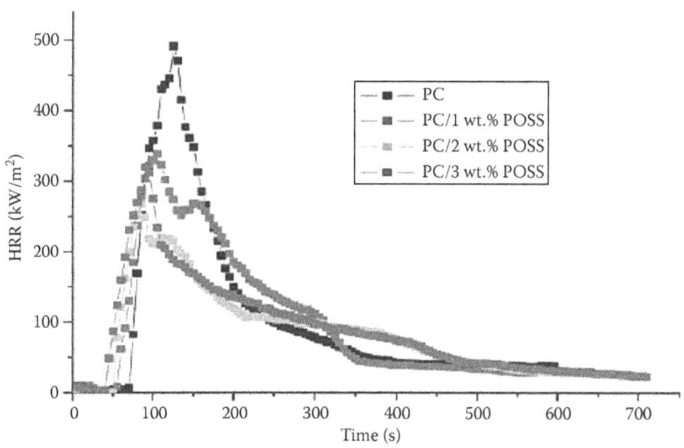

FIGURE 8.12 Rate of heat release curves of PC, PC/1 wt% TPOSS, PC/2 wt% TPOSS, and PC/3 wt% TPOSS.

Source: From Song, L. et al., *Polym. Degrad. Stab.*, 93, 627, 2008. With permission.

the char. This enhanced char structure protects and insulates the underlying polymer from any further degradation.

8.2.4 Silica

In an attempt to look for alternatives to the use of halogenated fire retardants, which function in the gas phase, an approach has been pursued which controls the polymer flammability by modifying the condensed phase chemistry. Silica gel combined with potassium carbonate have been reported to be an effective fire retardant for a wide variety of common polymers, such as polypropylene, nylon, poly(methylmethacrylate), poly(vinyl alcohol), cellulose, and to a lesser extent, polystyrene and styrene–acrylonitrile.[54] The cone calorimeter data shown in Table 8.5 indicate that the PHHR is reduced by up to 68% without significantly increasing the smoke or carbon monoxide levels during the combustion.

To further explore the influence of silica material properties (morphology, surface area, silanol concentration, and surface treatment) on the silica flame-retardant properties, various types of silicas (silica gel, fumed silicas, and fused silica) were investigated.[55,56] Material properties of the various silicas are summarized in Table 8.6. These different types of silicas were added to polypropylene and polyethylene oxide to determine their flame-retardant effectiveness and mechanisms. Polypropylene was chosen as a non-char-forming thermoplastic, and polyethylene oxide was chosen as a polar slightly char-forming thermoplastic. Flammability properties were measured in the cone calorimeter, and the mass loss rate was measured in the radiative gasification device in nitrogen to exclude any gas phase oxidation reactions.

TABLE 8.5
Cone Calorimeter Data for Different Polymers with Silica Gel and Potassium Carbonate

Sample Disk (75 mm × 8 mm)	Char Yield (%)	LOI (%)	Peak RHR (Δ) (kW/m²)	Mean RHR (kW/m²)	Mean Heat of Combustion (MJ/m²)	Tolal Heat Released (MJ/m²)	Mean Ext. Area (m²/kg)	Mean CO Yield (kg/kg)
PP	0	—	1761	803	37.9	357	689	0.04
PPw/6% SG & 4%PC	10	—	736 (58%)	512	33.1	297	710	0.04
PS	0	18	1737	1010	24.6	277	1422	0.07
PSw/6% SG & 4% PC	6	24	1190 (31%)	725	24.7	246	1503	0.07
PMMA	0	IK	722	569	23.1	319	210	0.01
PMMAw/3% SG & 1% PC	15	25	420 (42%)	246	20.9	231	199	0.05
PVA	4	—	609	381	17 0	221	594	0.03
PVAw/6% SG & 4%PC	43	—	194 (68%)	114	12.4	101	201	0.03
Cellulose	4	—	310	161	11.3	101	27	0.02
Cellulose w/6% SG & 4% PC	32	—	149 (52%)	71	5.3	34	20	0.04
SAN	2	—	1499	837	25.2	197	1331	0.07
SAN w/6% SG & 4% PC	3	—	1127 (25%)	772	23.0	169	1301	0.06
Nylon 6.6	1	30	1131	640	23.2	108	234	0.02
Nylon6,6w/3% SG & 2% PC	5	33	526 (53%)	390	22.0	105	171	0.02
Nylon6,6w/6% SG & 4% PC	6	30	546 (52%)	370	23.5	102	185	0.02

Source: Gilman, J.W. et al., Environmentally safe tire retardants for polymeric materials—I. Silica gel-potassium carbonate additive, in *Society for the Advancement of Material and Process Engineering (SAMPE)*. Anaheim. CA. Match 24-28. 1996. 708.
Note: SG. silica gel; PC, potassium carbonate.

TABLE 8.6
Material Properties of Various Silicas

Silica	Porosity (cm³/g)	Thermal Treatment (°C, h)	Silanol Surface Concentration (SiOH/nm²)	Surface Area (m²/g)	Primary Particle Size (μm)
Fused silica amorphous	–0	100°C (2h)	Loss	Low	7
Fumed silica hydrophilic	NA	None	3–4	255 ± 25	Aggregate length 0.2–03
Fumed silica hydrophobic	NA	100°C (15h)	1–2	140 ± 30	NA
Silica gel	2.0	900°C (15h)	0.4	400 ± 40	17

Source: Kashiwagi, T. et al., *Fire Mater.*, 24, 277, 2000. With permission.

Figures 8.13 and 8.14 show the HRR and mass loss rate of the polypropylene samples with each silica additive. The addition of low density, large surface area silica, such as fumed silicas and silica gel, to PP and PEO significantly reduced the HRR and mass loss rate. However, the addition of fused silica did not reduce the flammability properties as much as other silicas.

The authors proposed a mechanism that accounts for the reduction in flammability properties, which depends on physical processes in the condensed phase rather than chemical reactions. Three factors are critical in determining the silica behavior during the combustion process: the density and surface area of the additive, the melt viscosity of the polymer. The interplay between these factors can determine whether the silica will accumulate near the surface or sink through the polymer melt. Fumed silica and silica gel provide examples for the first case where the silica particles accumulated on the surface and formed an insulating layer that provide protection to the underlying polymer. This is in contrast to the other case where the fused silica particles sank through the polymer melt.

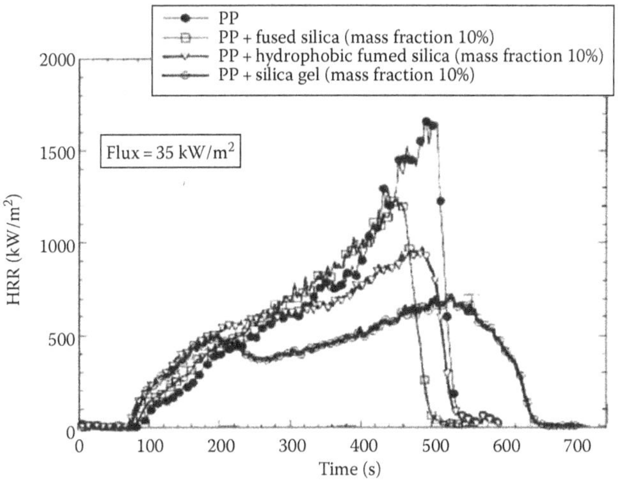

FIGURE 8.13 Effects of the addition of various silica types on the HRR of PP.

Source: From Kashiwagi, T. et al., *Fire Mater.*, 24, 277, 2000. With permission.

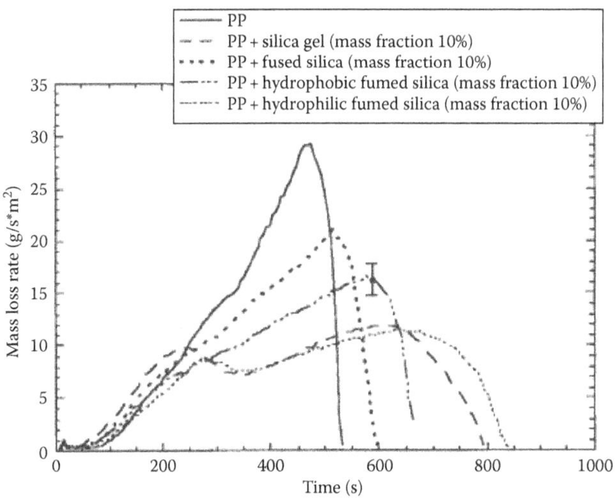

FIGURE 8.14 The effects of the addition of various silica types on the mass loss rate of PP in nitrogen.

Source: From Kashwagi, T. et al., *Fire Mater.*, 24, 277, 2000. With permission.

The proposed hypothesis mentioned above for the flame retardancy of silica was further examined by studying the effects of polymer melt viscosity on the flammability properties of PMMA with the addition of the various types of silica.[57] PMMA was chosen because of its large average zip length, which means that the molecular weight of PMMA does not significantly change during thermal degradation. The effects of melt viscosity were investigated by comparing the flammability properties of low molecular weight PMMA (LMWPMMA) with two types of silica (silica gel and fused silica) with that of high molecular weight PMMA (HMWPMMA) with the same types of silica.

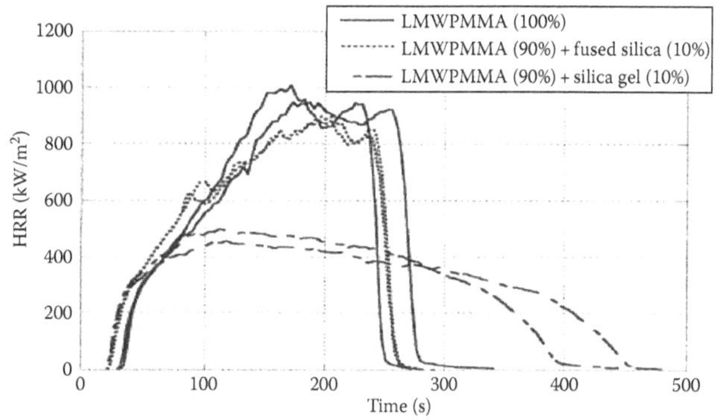

FIGURE 8.15 Effects of silica types on the HRR of LMWPMMA at 40 KW/m^2.

Source: From Kashwagi, T. et al., *J. Appl. Polym. Sci.*, 87, 1541, 2003. With permission.

The addition of fused silica to LMWPMMA did not bring a significant increase in the melt viscosity. However, adding silica gel (10% mass fraction) increased the melt viscosity by one order of magnitude compared to pure LMWPMMA. On the other hand, no significant increase in melt viscosity was observed for HMWPMMA by adding fused silica, but the addition of silica gel (10% mass fraction) increased the melt viscosity of HMWPMMA by a factor of five. The pyrolysis of both LMWPMMA and HMWPMMA with different silica types was investigated under nitrogen in the gasification apparatus. The results for the LMWPMMA system indicate mass loss rate curves similar to that obtained for both pure LMWPMMA and LMWPMMA/fused silica. However, half the mass loss rate can be observed for the LMWPMMA/silica gel system. As opposed to the negligible effect observed for fused silica with LMWPMMA, a significant effect was observed with HMWPMMA. Silica gel, however, was more effective in reducing the mass loss rate for HMWPMMA.

The burning behavior for LMWPMMA and HMWPMMA samples was investigated using the cone calorimeter at 40 kW/m^2. Figure 8.15 shows the heat release curves for LMWPMMA samples. The trends are similar to those of the mass loss rate in nitrogen. The data indicate that a roughly 50% reduction in heat release rate was observed by adding silica gel to LMWPMMA, but no significant reduction in heat release rate was obtained with fused silica.

The results for the HMWPMMA system are shown in Figure 8.16. Comparing these results with the mass loss rates for the same samples indicate that the trends are similar. Adding fused silica reduced the heat release rate for HMWPMMA by about 25% compared to the pure sample. However, the addition of silica gel reduced the heat release rate by about one-third of the peak rate of the pure HMWPMMA.

Based upon these results, the authors concluded that the two critical factors for reducing the mass loss rate and HRR for PMMA are the amount of accumulation of silica

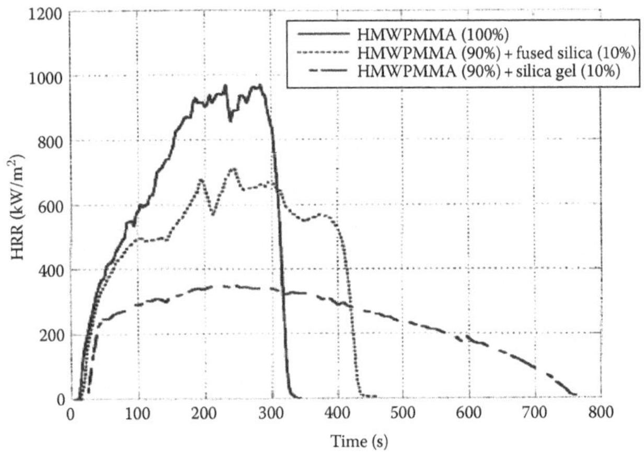

FIGURE 8.16 Effects of silica types on the HRR of HMWPMMA at 40 KW/m².

Source: From Kashwagi, T. et al., *J. Appl. Polym. Sci.*, 87, 1541, 2003. With permission.

TABLE 8.7
LOI and UL-94 Data for Different Formulations of EVA/MH/SiO$_2$

Sample Formulation	Composition (wt%)			LOI	UL-94
	EVA	MH	SiO$_2$		
A	100	0	0	18	Fail
B	40	60	0	35	V-0
C	40	58	2	37	V-0
D	40	55	5	39	V-0
E	40	52	8	39	V-0
F	40	48	12	37	V-1
G	45	55	0	32	Fail
H	45	53	2	34	V-2
I	45	50	5	35	V-0
J	45	47	8	35	V-0

Source: Fu. M. and Baojun. Q., *Polym. Degrad. Stub.*, 85. 633. 2004. With permission.

particles on the surface and their extent of surface coverage. These factors are determined not only by the silica characteristics, but also by the melt viscosity of PMMA.

Some reports have explored the potential of synergistic effect between silica and other flame retardants.[58–60] For example, silica showed synergistic effect with alumina in polypropylene (PP)/ammonium polyphosphate (APP)—pentaerythritol (PER) intumescent-based system. The data indicate that the HRR values improved by incorporating silica into the intumescent-based formulation, and the improvement was much more pronounced by combining both silica and alumina in the formulation.

A synergistic effect was also observed between silica and ATH in LLDPE/EVA composite.[59] Based upon the data collected from TGA and cone calorimetry, the report concluded that the synergistic mechanism of silica with ATH is mainly due to physical effect in the condensed phase. A more compact barrier is produced by combining silica with ATH, which leads to the reduction of HRR, MLR as revealed by cone calorimetry. The system combining fumed silica and magnesium hydroxide (MH) in ethylene vinyl acetate (EVA) showed synergistic behavior.[60] As can be seen in Table 8.7, partial substitution of magnesium hydroxide with fumed silica gave V-0 rating in UL-94 and increased the LOI value.

Recently, the mesoporous form of silica started to grab attention as another form of flame retardant material.[61] This interest stems from the fact that it has a pore structure with a high surface area. A number of studies indicated that combining the mesoporous silica (SBA-15) with intumescent system has the potential to improve the overall flame performance.[62,63] In one particular study, it was possible to achieve UL94 V-0 rating in polypropylene matrix.

One of the most promising examples of the use of silica was demonstrated recently by developing organic-inorganic nanostructure composite aerogel material. In the course of developing energy efficient building materials, phenolic-silica aerogel was prepared to act as insulation material that has flame retardant properties. The enhanced flammability performance was shown by the use of cone calorimeter.[64]

8.2.5 Silicate

Zeolite is a crystalline, porous aluminosilicate mineral with a unique interconnecting lattice structure. This lattice structure is arranged to form a honeycomb framework of consistent diameter interconnecting channels and pores. Negatively charged alumina and neutrally charged silica tetrahedral building blocks are stacked to produce the open three-dimensional honeycomb framework.

An extensive study was conducted on the effect of chemical and structural aspects of zeolites on the fire performance of the intumescent system, ammonium polyphosphate–pentaerythritol (APP–PER), where a marked improvement of the fire-retardant properties within different polymeric matrices has been observed.[65–67] The synergistic mechanism of zeolite 4A with the intumescent materials was investigated using solid-state NMR. Chemical analysis combined with cross-polarization dipolar-decoupled magic-angle spinning NMR revealed that the materials resulting from the thermal treatment of the APP–PER and APP–PER/4A systems were formed by carbonaceous and phosphocarbonaceous species, and that the zeolite enhances the stability of the phosphocarbonaceous species.

When evaluated by LOI, the system with zeolites has higher LOI than that without it. For example, compared with PP-APP–PER, which has an LOI of 30%, the LOI of PP-APP–PER system with zeolite 13X increases to 45%, an increase of 50%. Also the LOI of LRAM3.5 (ethylene–butyl acrylate–maleic anhydride terpolymer)–APP–PER system with 4A increases to 39%, relative to 29% for LRAM3.5–APP–PER system. All systems with zeolites obtained the UL-94 V-0 grade.

Wei et al. conducted investigation on the effects of zeolites 4A, 13X, mordenite, and ZSM-5 on the thermal degradation and charring of intumescent APP–pentaerythritol systems.[68] The residues formed with all types of zeolites at 600°C are higher (and more significantly with 4A and 13X) than that obtained with pure APP/PER. The authors argued that during the decomposition process of zeolite, the aluminum separates and migrates to the surface. This process leads to increasing the Si–Al ratio and enhancing the acid catalytic property, which improves the intumescent structure by retaining the volatile degradation products and promoting cross-linking reactions in the condensed phase.

Mizuno et al.[69] investigated the possibility of changing the pyrolysis behavior of polynorbornene (PNB) using the catalytic effect of zeolite. The flammability behavior of PNB compounded with zeolite was studied using a small-sized vertical combustion test. The results suggested that adding zeolite to PNB changed its primary ignition phenomena and increased the lower flammability limits (using Le Chatelier's equation) from 0.9 to 1.3.

In an attempt to provide compatibility between the FR and polymer matrix, Wang et al. reported on a novel microencapsulated intumescent system containing 4A zeolite as a potential flame retardant for natural rubber (NR).[70] The flame-retardant properties of NR composites loaded with different amounts of intumescent flame retardant (IFR), IFR-4A zeolite, and microencapsulated intumescent flame retardant (MIFR)-4A zeolite agents were studied and compared. The LOI data demonstrate that the NR composite filled with 50 phr of MIFR-4A zeolite agent and 50 phr of IFR-4A zeolite shows better FR properties as compared to NR and 50 phr of IFR-filled systems.

Talc is a natural magnesium silicate, which when compounded with plastics improves, among other properties, the flame retardancy. For example, talc combined with magnesium hydroxide or ATH improved the fire-retardance behavior of PP and EVA.[71,72]

Talc particles of different lamellarity and specific surface area have been incorporated into polyethylene vinyl acetate (EVA) copolymer/magnesium hydroxide (MH) composite system.[72] The fire retardancy of this system has been studied and compared with formulations containing only EVA and MH and formulations containing EVA, MH, and organomodified montmorillonites (oMMT). It was observed that talc with higher lamellar index showed fire behavior similar to that of EVA/MH/oMMT system with some intumescence.

One of the significant uses of silicate as FR materials was demonstrated in the application of silicate nanocomposite materials. Compared to the traditional flame retardants, this form of silicate nanocomposites initiated a completely new form of FR materials that grabbed a huge interest both from academia and industry.[73]

Montmorillonite is a well-known form of nano silicate materials. The ability to blend the montmorillonite clay with the polymeric system through melt blending, which is industrially feasible, creates materials that function as flame retardant.

Nanoclay composite materials started to attract commercial interest when Toyota patented nanoclay-polyamide system in 1988, which was probably the first time that the molecular layering of silicates was recognized to be a key in transferring nano properties to the macrocomposite.

Due to the polarity of montmorillonite material, it will be compatible only with polar resins, such as nylons. However, to develop nanocopmposite in hydrophobic systems, such as polyolefins would require coating the montmorillonite with organic coating with long hydrocarbon chain. This type of coating can be considered a disadvantage from the flammability standpoint due to its lower thermal stability.

Extensive literature can be found on the use of polymer-layered silicate nanocomposites as a flame retardant materials.[74,75] It was later realized that, combining the nanocomposite with the traditional flame retardant is better in order to get the flame rating for the specific application.

8.3 INORGANIC FLAME RETARDANTS

In addition to the silicon-based materials discussed previously, there are other classes of inorganic flame retardants that are based upon different elements; each one of them can deliver a special type of flame retardancy.[76,77] A brief review of these materials will be covered later.

8.3.1 Tin Compounds

Tin compounds are a class of inorganic flame retardants that can enhance fire safety and reduce the smoke emission of various materials, especially those containing halogens. Tin compounds can act in both the condensed and the gas phase, depending on the type of material and the halogen present.

This class of materials represented in different chemical structures, such as tin oxide, zinc stannate (ZS), and zinc hydroxyl stannate (ZHS), showed its potential to act as effective flame retardants and smoke suppressants. Both ZS and ZHS have the potential to partially replace antimony trioxide in halogenated systems or halogen-containing polymers, like PVC. From the mechanistic standpoint, these compounds behave both in condensed and vapor phase fire retardants in the presence of halogenated compounds.

ZS and ZHS also found applications as a coating to hydrated fillers, such as magnesium hydroxide. One of the effective tools to enhance the performance of hydrated fillers is to coat these materials with different types of coating materials, such as stearate, silane, etc. This enhanced performance means better dispersion and, hence, improved flammability.

Tin compounds can improve the thermal stability and char formation of materials, which can prevent or delay the ignition and combustion processes. They can also react with the halogens released from the materials to form tin halides, which can act as radical scavengers and quench the flame propagation in the gas phase.

Tin compounds have several advantages over other flame retardants, such as low toxicity, high thermal stability, good compatibility with polymers, and low loading levels. However, they also have some limitations, such as high cost, limited availability, and possible corrosion of metal components.

8.3.2 Molybdenum Compounds

Molybdenum compounds are a class of inorganic flame retardants that can synergize with halogens or other agents to improve the flame resistance and smoke suppression of polymeric materials. Molybdenum compounds can act in both the condensed and the gas phase, depending on the type of material and the flame retardant system.

Some of the common molybdenum compounds used as flame retardants are molybdic oxide (MoO3), ammonium octamolybdate ($(NH_4)_4Mo_8O_{26}$), and zinc molybdate ($ZnMoO_4$). These compounds can be used in combination with other flame retardants, such as bromine, chlorine, phosphorus, or intumescent agents. Molybdenum compounds can also be used as coatings or fillers for plastics, textiles, paints, and other materials.

In terms of smoke suppression additives, Molybdenum compounds have a unique position compared to other flame retardants. Char buildup is the mechanism proposed for their effectiveness as flame retardant and smoke suppressant. The major commercial application of molybdenum compounds is in PVC, where it can partially replace the antimony trioxide.

Ammonium octamolybdate is a major additive used in PVC for smoke reduction. It is proposed that the mode of action is condensed phase, as judged by the molybdenum content in the char residue. It is likely acts by catalyzing the dehydrochlorination to all-trans polyenes, thus preventing the cyclic aromatic structures that dominates the smoke. Also, different metal-molybdenum complexes demonstrated smoke suppressants capability in PVC wire and cable.

Molybdenum compounds have several advantages over other flame retardants, such as low toxicity, high thermal stability, good compatibility with plastics, and low loading levels. However, they also have some limitations, such as high cost, limited availability, and possible corrosion of metal components.

8.3.3 Iron Compounds

Iron compounds are a class of inorganic flame retardants that can synergize with halogens or other char-forming agents to improve fire safety and reduce the smoke emission of various polymeric materials. Iron compounds can act in both the condensed and the gas phase, depending on the type of material and the flame retardant system.

Some of the common iron compounds used as flame retardants are iron oxides (Fe_2O_3 and Fe_3O_4) and iron hydroxide ($Fe(OH)_3$). Iron compounds shown to be effective in both halogenated and non halogenated systems

Iron compounds can enhance the thermal stability and char formation of materials, which can prevent or delay the ignition and combustion processes. They can also react with the halogens released from the materials to form iron halides, which can act as radical scavengers and quench the flame propagation in the gas phase. Additionally, iron compounds can act as smoke suppressants by reducing the amount and toxicity of smoke generated by burning materials.

Iron compounds have several advantages over other flame retardants, such as low cost, high availability, good compatibility with polymers, and low loading levels. However, they also have some limitations, such as possible corrosion of metal components, coloration of materials, and incomplete combustion of char residues.

8.3.4 Metal Oxides

Metal oxides, including antimony trioxide, zinc oxide, magnesium oxide, tin oxide, etc., are not flame retardants by themselves. However, when combined with flame retardant packages; either halogenated or non-halogenated, their effect is typically demonstrated in the overall flame performance.

Other than the extensively documented data on the role of Sb_2O_3 in the halogenated system, there is not much data published on the mechanism of other metal oxides. Their overall loading in the formulation is typically low, but their effect can be demonstrated through enhancement in the char formation.

Metal oxides can enhance the thermal stability and char formation of materials, which can prevent or delay the ignition and combustion processes. They can also react with the halogens released from the materials to form metal halides, which can act as radical scavengers and quench the flame propagation in the gas phase. Additionally, metal oxides can act as smoke suppressants by reducing the amount and toxicity of smoke generated by burning materials.

Metal oxides have several advantages over other flame retardants, such as low toxicity, high thermal stability, good compatibility with polymers, and low loading levels.[23] However, they also have some limitations, such as possible corrosion of metal components, coloration of materials, and incomplete combustion of char residues.

8.3.5 Magnesium Carbonate-Based Materials

Magnesium carbonate in its pure form is rarely used as flame retardant. However, naturally occurring minerals such as Huntite and Hydromagnesite, which have magnesium carbonate-based chemical composition, have gained attention as low cost and effective flame retardants, especially for wire and cable applications.

Magnesium carbonate and its derivatives (Huntite and Hydromagnesite) have attracted attention as endothermic flame retardants that can be incorporated into thermoplastics without decomposition. Endothermic flame retardants are basically substances that absorb heat from the combustion zone and lower the temperature of the polymer, thus reducing its flammability. Magnesium carbonate decomposes into magnesium oxide and carbon dioxide upon heating, which can also dilute the oxygen and fuel gases in the flame.

The flame retardant mechanism of these minerals, which is available commercially as Ultracarb material, is based upon endothermic decomposition reaction and the release of water and carbon dioxide. That means the FR performance of these materials can be seen as a reduction in the flame energy due to the endothermic effect, dilution in the concentration of the combustible gases due to the release of water and carbon dioxide, in addition to reduction in smoke production.

8.4 CONCLUSION

The element silicon provides unique opportunities as a source of flame retardancy. This chapter has provided a concise account of the important types of flame retardants that are based upon silicon. These types include, but are not limited to, silicones, silanes, silsesquioxane, silica, and silicate. These classes of flame retardants have been covered briefly in this chapter along with some recent trends in using these materials.

Also, different classes of inorganic flame retardants have been covered, such as tin, molybdenum, iron, and magnesium compounds. Each class of these materials operates in the fire with different mechanism.

These classes of flame retardants provide a solution and opportunity to develop systems for fire retardancy that are environmentally friendly. There is a growing interest in this type of flame retardant, and this trend most likely will continue in the future, given the increasing concern over the release of the halogenated species into the environment.

REFERENCES

1. Eaborn, C., *Organosilicon Compounds*, Butterworths, London, 1960.
2. *The Chemistry of Organic Silicon Compounds*, Wiley, Chichester, Rappoport, Z.; Apeloig, Y., Eds., 1998, Vol. 2; Patai, S. and Rappoport, Z., Eds., 1989, Vol. 1.
3. Sommer, L.H., *Stereochemistry, Mechanism & Silicon*, McGraw Hill, New York, 1956.
4. Brook, M.A., *Silicon in Organic, Organometallic, and Polymer Chemistry*, John Wiley & Sons, Inc., New York, 2000.
5. Michl, J., Silicon chemistry, *Chem. Rev.* 1995, 95, 1135.
6. Jones, R.G., Ed., Silicon-containing polymers, *Proceedings of an RSC Macro Group Symposium* University of Kent, 1994, Royal Society of Chemistry, Cambridge, MA, 1995.
7. Marciniec, B., and Chojnowski, J., Eds., Progress in organosilicon chemistry, *Proceedings of the 10th International Symposium on Organosilicon Chemistry*, Poznan, Poland, 1993, Gordon & Breach, Basel, 1995.
8. Bassindale, A.R., and Gaspar, P.P., Eds., Frontiers of organosilicon chemistry, *Proceedings of the 9th International Symposium on Organosilicon Chemistry*, Edinburgh, 1990, Royal Society of Chemistry, Cambridge, MA, 1991.
9. Colvin, E.W., Silicon, in *Comprehensive Organometallic Chemistry II: A Review of the Literature 1982–1994*, Wilkinson, G., Stone, F.G.A., Abel, E.W., Eds.; McKillop, A., Vol. Ed., Pergamon, Oxford, 1982, Vol. 2, Chap. 1, p. 1.
10. Larson, G.L., Ed., *Advances in Silicon Chemistry*, JAI, Greenwich, CT, 1996, Vol. 3; 1993, Vol. 2; 1991, Vol. 1.
11. Kashiwagi, T., and Gilman, J.W., Silicon-based flame retardants, in *Fire Retardancy of Polymeric Materials*, Grand, A.F. and Wilkie, C.A., Eds., Marcel Dekker, New York, 2000, Chap. 10, p. 353.
12. Butts, M., Cella, J., Wood, C.D., Gillette, G., Kerboua, R., Leman, J., Lewis, L., Rajaraman, S., Rubinsztajn, S., Schattenmann, F., Stein, J., Wengrovius, J., and Wicht, D., Silicones, in *Encyclopedia of Polymer Science and Technology*, 3rd edn., Mark, H.F., Ed., John Wiley & Sons, Inc, Hoboken, NJ, 2003, Vol. 11.
13. Noll, W., *Chemistry and Technology of Silicones*, Academic Press, Inc., New York, 1978.
14. Clarson, S.J., and Semlyen, J.A., Eds., *Siloxane Polymers*, Prentice Hall, Englewood Cliffs, NJ, 1993.
15. Rochow, E.G., *Silicon and Silicones*, Springer-Verlag, Berlin, 1987.
16. Hunter, D., Carbide's silicones: The "sale of the century," *Chemical Week* 24, February 19, 1992.
17. Gorbunfov, A. I.; Belyi, A. P.; Filippov, G. G. Reactions of Silicon and Germanium with Halogens, Their Hydrides, and Organic Halogen Derivatives. *Russ. Chem. Rev.* **1974**, *43*, 291.
18. Petrov, A.D., Mironov, B.F., Ponomarenko, V.A., and Chernyshev, E.A., *Synthesis of Organosilicon Monomers*, Consultants Bureau, New York, 1964, p. 36.
19. Zeigler, J.M., and Fearon, F.W.G., Eds., *Silicon-Based Polymer Science*, ACS Advances in Chemistry, Series 224, American Chemical Society, Washington, D.C., 1990.
20. Stark, F.O., Falender, J.R., and Wright, A.P., Silicones, in *Comprehensive Organometallic Chemistry*, Wilkinson, G., Stone, F.G.A., and Abel, E.W., Eds., Pergamon Press, Oxford, 1982, Chapter 9.3, p. 305.
21. Eaborn, C., and Bott, R.W., *The Bond to Carbon*, MacDiarmid, A.G., Ed., Marcel Dekker, Inc., New York, 1968, Vol. I, Part 1, p. 105.
22. Drake, R., MacKinnon, I., and Taylor, R., Recent advances in the chemistry of siloxane polymers and copolymers, in *The Chemistry of Organic Silicon Compounds*, Rappoport, Z. and Apeloig, Y., Eds., Wiley, Chichester, 1998, Vol. 2, Chapter 38, p. 2217.
23. Arkles, B. Look What You Can Make Out of Silicones (Biomedical Applications of Silicones). *Chem. Tech.* **1983**, *13*, 542.
24. Tomanek, A., *Silicones and Industry*, Hanser (Wacker Chemie), Munich, Germany, 1991.
25. Austin, P. J.; Buch, R. R.; Kashiwagi, T. Gasification of Silicone Fluids Under External Thermal Radiation Part I. Gasification Rate and Global Heat of Gasification. *Fire Mater.* **1998**, *22*, 221.
26. Austin, P. J.; Buch, R. R.; Kashiwagi, T. Gasification of Silicone Fluids Under External Thermal Radiation Part 2. Gasification Products—Characterization and Quantitation. *Fire Mater.* **1998**, *22*, 239.

27. Kashiwagi, T., Cleary, T.G., Davis, G.C., and Lupinski, J.H., A non-halogenated, flame retarded polycarbonate, *Conference of Advanced Fire Resistant Aircraft Interior Materials*, Atlantic City, NJ, February 9–11, 1993, 175.
28. Benrashid, R., and Nelson, G.L., Flammability improvement of polyurethanes by incorporation of a silicone moiety into the structure of block copolymers, in *Fire and Polymers*, ACS Symposium, Series 599, American Chemical Society, Washington, D.C., 1995, p. 217.
29. Belva, F.; Bourbigot, S.; Duquesne, S.; Jama, C.; Le Bras, M.; Pelegris, C.; Rivenet, M. Heat and Fire Resistance of Polyurethane-Polydimethylsiloxane Hybrid Material. *Polym. Adv. Technol.* **2006**, *17*, 304.
30. Hermansson, A.; Hjertberg, T.; Sultan, B. A. The Flame Retardant Mechanism of Polyolefins Modified with Chalk and Silicone Elastomer. *Fire Mater.* **2003**, *27*, 51.
31. Hermansson, A.; Hjertberg, T.; Sultan, B. A. Linking the Flame-Retardant Mechanisms of Ethylene Acrylate Copolymer, Chalk and Silicone Elastomer System with Its Intumescent Behavior. *Fire Mater.* **2005**, *29*, 407.
32. Hermansson, A.; Hjertberg, T.; Sultan, B. A. Influence of the Structure of Acrylate Groups on the Flame Retardant Behavior of Ethylene Acrylate Copolymers Modified with Chalk and Silicone Elastomer. *J. Fire Sci.* **2007**, *25*, 287.
33. Li, Y.-M.; Hu, S.-L.; Wang, D.-Y. Polymer-Based Ceramifiable Composites for Flame Retardant Applications: A Review. *Comp. Comm.* **2020**, *21*, 100405.
34. Li, Z.; Liang, W.; Shan, Y.; Wang, X.; Yang, K.; Cui, Y. Study of Flame-Retarded Silicone Rubber with Ceramifiable Property. *Fire Mater.* **2020**, *44*, 487.
35. Mansouri, J.; Burford, R. P.; Cheng, Y. B.; Hanu, L. Formation of Strong Ceramified Ash from Silicone-Based Compositions. *J. Mater. Sci.* **2005**, *40*, 5741.
36. Weil, E. D. Interaction of Additives in FR Systems. *Plast. Comp.* **1987**, 31–40.
37. Ebdon, J. R.; Hunt, B. J.; Jones, M. S.; Thorpe, F. G. Chemical Modification of Polymers to Improve Flame Retardance-II. The Influence of Silicon-Containing Groups. *Polym. Degrad. Stab.* **1996**, *54*, 395.
38. Ebdon, J. R.; Hunt, B. J.; Joseph, P. Thermal Degradation and Flammability Characteristics of Some Polystyrenes and Poly(Methylmethacrylate)s Chemically Modified with Silicon-Containing Groups. *Polym. Degrad. Stab.* **2004**, *83*, 181.
39. Mercado, L. A.; Reina, J. A.; Galia, M. Flame Retardant Epoxy Resins Based on Diglycidyloxymethylphenylsilane. *J. Polym. Sci. Polym. Chem.* **2006**, *44*, 5580.
40. Zhang, H. P.; Xia, Q.; Wang, B.; Yang, L. C.; Wu, Y. P.; Sun, L. D. L.; Gan, C. L.; Luo, H. J.; Bebeda, A. W.; van Ree, T. Vinyl-Tris-(Methoxydiethoxy)Silane as an Effective and Ecofriendly Flame Retardant for Electrolytes in Lithium Ion Batteries. *Electrochem Commun.* **2009**, *11*, 526.
41. Scott, D. W. Thermal Rearrangement of Branched-Chain Methylpolysiloxanes. *J. Am. Chem. Soc.* **1946**, *68*, 356.
42. www.hybridplastics.com
43. Mantz, R. A.; Jones, P. F.; Chaffee, K. P.; Lichtenhan, J. D.; Gilman, J. W. Thermolysis of Polyhedral Oligomeric Silsesquioxane (POSS) Micromers and POSS-Siloxane Copolymers. *Chem Mater.* **1996**, *8*, 1250.
44. Lichtenhan, J.D., and Gilman, J.W., Preceramic additives as fire retardants for plastics, U.S. Patent 6,362,279, 2002.
45. Devaux, E.; Rochery, M.; Bourbigot, S. Polyurethane/Clay and Polyurethane/POSS Nanocomposites as Flame Retarded Coating for Polyester and Cotton Fabrics. *Fire Mater.* **2002**, *26*, 149.
46. *Proceedings of POSS Nanotechnology Conference*, Huntington Beach, CA, September 2002.
47. Lu, S.; Hamerton, I. Recent Developments in the Chemistry of Halogen-Free Flame Retardant Polymers, *Prog. Polym. Sci.* **2002**, *27*, 1661.
48. Fina, A., Tabuani, D., Frache, A., Boccaleri, E., and Camino, G., *Fire Retardancy of Polymers: New Applications of Mineral Fillers*, Le Bras, M., Wilkie, C., and Bourbigot, S., Eds., Royal Society of Chemistry, Cambridge, MA, 2005, p. 202.
49. Fina, A.; Tabuani, D.; Carniato, F.; Frache, A.; Boccaleri, E.; Camino, G. Polyhedral Oligomeric Silsesquioxanes (POSS) Thermal Degradation. *Thermochim Acta.* **2006**, *440*, 36.
50. Bourbigot, S., Le Bras, M., Flambard, X., Rochery, M., Devaux, E., and Lichtenhan, J.D., *Fire Retardancy of Polymers: New Applications of Mineral Fillers*, Bras, M., Bourbigot, S., Duquesne, S., Jama, C., and Wilkie, C.A., Eds., The Royal Society of Chemistry, Cambridge, MA, 2005, p. 189.
51. Fina, A.; Abbenhuis, H. C. L.; Tabuani, D.; Camino, G. Metal Functionalized POSS as Fire Retardants in Polypropylene. *Polym. Degrad. Stab.* **2006**, *91*, 2275.
52. Vannier, A.; Duquesne, S.; Bourbigot, S.; Castrovinci, A.; Camino, G.; Delobel, R. The Use of POSS as Synergist in Intumescent Recycled Poly(Ethylene Terephthalate). *Polym. Degrad. Stab.* **2008**, *93*, 818.
53. Song, L.; He, Q.; Hu, Y.; Chen, H.; Liu, L. Study on Thermal Degradation and Combustion Behaviors of PC/POSS Hybrids. *Polym. Degrad. Stab.* **2008**, *93*, 627.
54. Gilman, J.W., Kashiwagi, T., and Lomakin, S.M., Environmentally safe fire retardants for polymeric materials—I. Silica gel-potassium carbonate additive, *Society for the Advancement of Material and Process Engineering (SAMPE)*, Anaheim, CA, March 24–28, 1996, p. 708.
55. Gilman, J.W., Kashiwagi, T., Nyden, M., and Harris, R.H., New flame retardants consortium: Final report, Flame retardant mechanism of silica, NISTIR 6357, June 1999.
56. Kashiwagi, T.; Gilman, J. W.; Butler, K. M.; Harris, R. H.; Shields, J. R.; and Asano, A. Flame Retardant Mechanism of Silica Gel/Silica. *Fire Mater.* **2000**, *24*, 277.
57. Kashiwagi, T., Shields, J. R., Harris, R. H., Davis, R. D. Flame-Retardant Mechanism of Silica: Effects of Resin Molecular Weight. *J. Appl. Polym. Sci.* 2003, *87*, 1541.
58. Wei, P.; Hao, J.; Du, J.; Han, Z.; Wang, J. An investigation on Synergism of an Intumescent Flame Retardant Based on Silica and Alumina. *J. Fire Sci.* **2003**, *21*, 17.
59. Wei, P.; Han, Z.; Xu, X.; Li, Z. Synergistic Flame Retardant Effect of SiO_2 in LLDPE/EVA/ATH Blends. *J. Fire Sci.* **2006**, *24*, 487.
60. Fu, M.; Baojun, Q. Synergistic Flame Retardant Mechanism of Fumed Silica in Ethylene-vinyl Acetate/Magnesium Hydroxide Blends. *Polym. Degrad. Stab.* **2004**, *85*, 633.
61. Pinnavaia, T.J., Dulebohn, J.I., and Nanasy, B., Mesoporous silicate fire retardant composition, US20140179831A1.
62. Li, J.; Wei, P.; Li, L.; Qian, Y.; Wang, C.; Huang, N. H. Synergistic Effect of Mesoporous Silica SBA-15 on Intumescent Flame-Retardant Polypropylene. *Fire Mater.* **2011**, *35*, 83.
63. Matar, M.; Azambre, B.; Cochez, M.; Vahabi, H.; Fradet, F. Influence of Modified Mesoporous Silica SBA-15 on the Flammability of Intumescent High-Density Polyethylene. *Polym. Adv. Technol.* **2016**, *27*, 1363.
64. Thong, Y. X.; Li, X.; Yin, X. J. Determining the Best Flame Retardant for Rigid Polyurethane

Foam—Tris(2-Chloroisopropyl) Phosphate, Expandable Graphite, or Silica Aerogel, *J Appl Polym Sci.* **2022**, *139*, e51888.
65. Bourbigot, S.; Le Bras, M.; Delobel, R.; Breantb, P.; Tremillon, J. M. 4A Zeolite Synergistic Agent in New Flame Retardant Intumescent Formulations of Polyethylenic Polymers-Study of the Effect of the Constituent Monomers. *Polym. Degrad. Stab.* **1996**, *54*, 275.
66. Bourbigot, S.; Le Bras, M.; Delobel, R.; Decressainb, R.; Amoureuxb, J. P. Synergistic Effect of Zeolite in an Intumescence Process: Study of the Carbonaceous Structures Using Solid-State NMR. *J. Chem. SOC., Faraday Trans.* **1996**, *92*(1), 149.
67. Bourbigot, S.; Le Bras, M.; Delobel, R.; Tremillon, J. M. Synergistic Effect of Zeolite in an Intumescence process: Study of the Interactions between the Polymer and the Additives. *J. Chem. SOC., Faraday Trans.* **1996**, *92*(18), 3435.
68. Wei, P.; Jiang, P.; Han, Z.; Wang, J. An Investigation of the Effects of Zeolites on the Thermal Degradation and Charring of APP-PER by TGA-XPS. *J. Fire Sci.* **2005**, *23*, 173.
69. Mizuno, K.; Ueno, T.; Hirata, A.; Ishikawa, T.; Takeda, K. Thermal Degradation and Flame Retardancy of Polynor-Bornene by a Zeolite, *Polym. Degrad. Stab.* **2007**, *92*, 2257.
70. Wang, J.; Chen, Y. Effect of Microencapsulation and 4A Zeolite on the Properties of Intumescent Flame-Retardant Natural Rubber Composites. *J. Fire Sci.* **2008**, *26*, 153.
71. Jouffret, F., and Meli, G., Talc as a functional additive in flame retardant systems, in *Flame Retardants* 2004, London, January 27–28, 2004, Interscience Communications, Greenwich, U.K., pp. 129–132.
72. Clerc, L.; Ferry, L.; Leroy, E.; Lopez-Cuesta, J. M. Influence of Talc Physical Properties on the Fire Retarding Behavior of (Ethylene-Vinyl Acetate Copolymer/Magnesium Hydroxide/Talc) Composites, *Polym. Degrad. Stab.* **2005**, *88*, 504.
73. Morgan, A.B., and Wilkie, C.A., Eds., *Flame Retardant Polymer Nanocomposites*, John Wiley & Sons Inc., Hoboken, NJ, 2007.
74. Ray, S. S.; Okamoto, M. Polymer/Layered Silicate Nano-Composites: A Review from Preparation to Processing. *Prog. Polym. Sci.* **2003**, *28*(11), 1539.
75. Beall, G.W., and Pinnavaia, T.J., *Polymer- Clay Nanocomposites*, John Wiley & Sons Ltd, Chichester, 2000.
76. Dufton, P., *Flame Retardants for Plastics*, Rapra Technology Limited, Shawbury, Shrewsbury, Shropshire, 2003.
77. Wypych, G., *Handbook of Flame Retardants*, ChemTec Publishing, Toronto, ON, 2021.

9 Boron-Based Flame Retardants

Kelvin K. Shen
Rio Tinto/U.S. Borax (Consultant)
San Diego, California 92131
KelvinKShen@gmail.com

9.1 INTRODUCTION—MULTIFUNCTIONAL BORATES IN FLAME RETARDANCY

Boron is ubiquitous in nature. Sea water contains about 5 ppm of boron. The earth's crust contains only 3 ppm, but there are areas where borate salts (salts of the oxidized form of elemental boron) in substantial volume is concentrated for mining. Boron is the only electron-deficient nonmetallic element. Thus boron has a high affinity for electronegative atoms such as oxygen and forms strong B-O bonds in compounds known as borates. The boron atoms can bond to oxygen to form either planar trigonal BO_3 or negatively charged tetrahedral BO_{4-} units.

One of the many industrial applications of borates is fire retardancy. Boron compounds such as boric acid and borax are well-known fire retardants for wood/cellulosic products and coatings. Borates can function as flame retardants, smoke suppressants, afterglow suppressants, and anti-tracking agents in both halogen-containing and non-halogenated polymers.[1–7] The use of boron compounds, such as zinc borate, calcium borate, melamine borate, boron phosphate, ammonium pentaborate as flame retardants in polymers, became prominent only since the late 1990s. In recent years, the use of functionalized boron nitride, graphene derivatives containing boron, and boronic acid/borane derivatives as flame retardants have attracted considerable attention. In this chapter, the chemical/physical properties, end use applications, fire test performance, as well as modes of action of these boron-based flame retardants in many polymer systems will be reviewed. Synergistic/beneficial interaction between "boron and metal hydroxide," "boron and nitrogen," "boron and phosphorus," "boron-phosphorus-nitrogen," "boron and carbon," and "boron and silicon" in fire test performances will be illustrated.

In halogen-containing systems, boron-based flame retardants such as zinc borates can provide the following benefits:

- Synergists of halogen sources
- Synergists of antimony oxide
- Can either partially or completely replace antimony oxide
- Smoke suppressant
- Afterglow suppressant
- HCl/HBr suppressant
- Improve aged elongation properties of polyolefin
- Improve thermal stability of aromatic bromine/Sb system
- Provide anti-tracking (i.e., Comparative Tracking Index—CTI) and anti-arcing properties of base polymer

In non-halogen based systems:

- Reduce rate of heat release
- Promote char/residue formation
- Stabilize the char and prevent dripping; inhibit oxidation of the char (afterglow suppression)
- Most borates release significant amount of water to provide flame retardancy
- Promote sintering between filler particles—effective flux and glass former for ceramification
- Provide anti-tracking and anti-arcing properties of base polymer
- Display synergy with nitrogen, phosphorus, silicon compounds

In addition to acting as flame retardants, boron-based compound can provide an additional function depending on the polymer used:

- Provide corrosion inhibition function
- Function as a nucleating agent for polymer foam formation
- Provide UV protection
- IR screening
- Photocure catalyst
- Nuclear shielding for thermal neutrons

9.2 PROPERTIES OF BORON-BASED PRODUCTS AND APPLICATIONS

9.2.1 Boric Acid (H_3BO_3), Boric Oxide (B_2O_3), and Boric Acid Esters [$B(OR)_3$]

Boric acid (BA, commercially available as *Optibor*®) is soluble in water (5.46 wt% 25°C), alcohols, and glycerin. It is a weak acid and has a pH of 4 (saturated solution at room temperature). Upon heating from 75°C to about 125°C, it loses part of its hydration water to form metaboric acid

DOI: 10.1201/9781003380689-9

(HBO_2). The metaboric acid can be further dehydrated to boric oxide at approximately 260–270°C.

$$2H_3BO_3 \rightarrow 2\ HBO_2 + 2H_2O \rightarrow B_2O_3 + 3H_2O$$

Boric oxide, also known as anhydrous boric acid, is a hard glassy material that softens at about 325°C and melts at about 450–465°C. It is produced commercially by fusion of boric acid. Through this procedure, it generally contains up to 0.5% water. It can, however, absorb water and revert back to boric acid; however, this usually does not affect its fire retardancy performance. Boric oxide and boric acid are not recommended for use in non-polar hydrocarbon polymers because boric acid (or boric acid formed via boric oxide hydration by moisture) may migrate to the polymer surface. Some examples of boric acid/boric oxide usage as flame retardants are illustrated below.

Boric acid, working mostly in the condensed phase, is used extensively as a flame retardant in cellulosic products. For example, it is used in cotton batting for inhibiting smoldering combustion, in cellulosic insulation for inhibiting flame spread/afterglow combustion, and in various wood/paper product products for flammability control. In many applications, boric acid is used in combination with sodium borate.

9.2.1.1 Epoxy

BA is known to function as a flame retardant in epoxy for a variety of commercial applications. In particular, BA can function as a curing catalyst in a certain epoxy system and is known to enhance the thermal stability of epoxy. Incorporation of BA into epoxy can not only provide an increase in char residue but also reduce volatile products generation.[7]

9.2.1.1.1 Boron-Carbon Synergy

The addition of 0.5 wt% MWCNT (multiwall carbon nanotube) in epoxy (DGEBA, cured with polyethylene polyamine) did not improve the thermal stability of the epoxy composite. However, the use of the MWCNT and boric acid combination can improve significantly the thermal stability.[8]

9.2.1.1.2 Boron-Nitrogen Synergy

Schiff base is a condensation product of an amine and aldehyde with the functional imine group—$R_1R_2C = NR_3$. The use of Schiff bases as flame retardants has attracted some attention.[9] The boron-containing Schiff base was prepared with ethanolamine, 2,4-dihydroxybenzaldehyde, and BA.[10] This material, at 3 wt% loading, was reported to enhance the thermal stability of epoxy composite. In addition, it reduced peak heat release rate (pHRR), total heat release (THR), total smoke production (TSP), and CO production in the cone calorimeter test.

9.2.1.1.3 Boron-Phosphorus Synergy

Commercially, BA has been used extensively in conjunction with ammonium polyphosphate (APP) in epoxy intumescent coatings (See **9.2.2.3.2.** Zinc Borate).

DOPO (9, 10-dihydro-9-oxa-10-phosphaphenanthrene-10-oxide) and its derivatives has been a very unique class of phosphorus-containing flame retardant. It was reported that reacting DOPO with benzaldehyde to yield a special phosphorus compound (PFR). A combination of this PFR (2 wt% P) and BA (2 wt% B) decreased considerably the flammability of the epoxy (DGEBA).[7]

9.2.1.1.4 Boron-Phohsphorus-Nitrogen Synergy

Many boric acid ester derived from BA and alcohol containing P/N have been reported. One obvious advantages of this type material is that it takes advantage of synergy between boron and P/N, and most of all, it will provide a more uniform distribution of boron in epoxy. The hyperbranched material was prepared from DOPO, diethanolamine, paraformaldehyde, and BA[11] (Figure 9.1). At low loadings (< 3.0%), this material enhanced glass-transition temperature, mechanical, dielectric, fire, and smoke suppression performances.

9.2.1.1.5 Boron-Silicon Synergy

Polyhedral oligomeric silsesquioxanes (POSS, 1–3 nm) with cage-like structure is known due to its fire resistance and mechanical properties in polymers. A combination of tris(2-hydorxypropyl)borate (THPB, a boric acid ester) and octa-aminophenyl POSS (OapPOSS) has been used as curing agents in epoxy (DGEBA).[12] The LOI reached 30.5%, when the content of THPB and OapPOSS of boron and silicon in the cured epoxy were 1.5% and 0.5% respectively. MCC test showed that, relative to pure epoxy, the pHRR and THR of the treated epoxy were reduced by 69% and 46% respectively (see also **9.2.1.6. Wood/Cellulosic** for B/Si interaction).

9.2.1.2 Phenolics

Phenolics have been noted as good char sources in many FR formulations. BA is commonly used in phenolic to impart thermal stability (stabilizing both the phenolic and char formed) and fire retardancy. The chemical

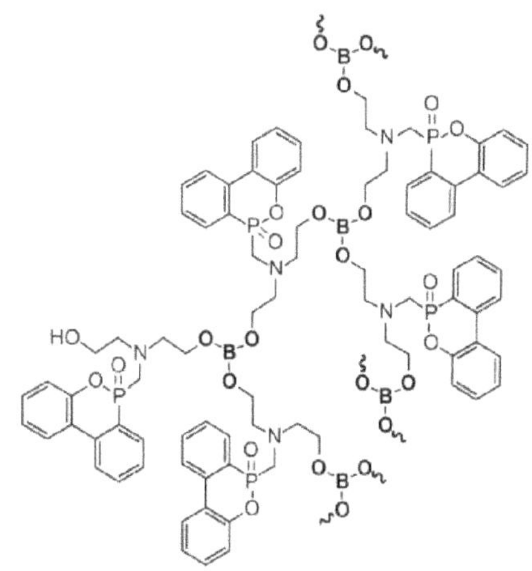

FIGURE 9.1 Hyperbranched product of DOPO, diethanolamine, formaldehyde, and boric acid.

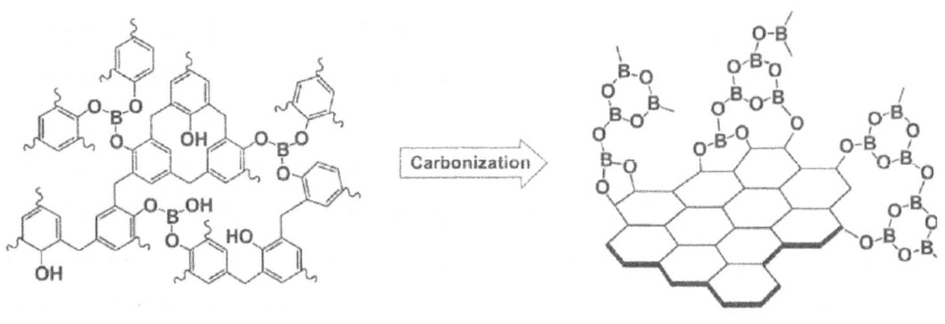

FIGURE 9.2 Left structure illustrates the boron-containing cured phenolics and the right one illustrates the pyrolysis/graphitization product.

Source: Wang, S., Jin X., et al. *Applied surface science*, 331, March 15, 2015, p. 519–529.

structure of the cured boron-containing phenolic and its structural evolution at high temperature were investigated (Figure 9.2).[13] High char yield of the system was mainly due to boron improving the graphitization and promoting the formation of a more ordered glassy carbon during pyrolysis The use of BA and aluminum trihydroxide (ATH) in phenolic for sandwich panel construction was reported.[14] A flame-retardant fiberglass plastics cover for conveyors was also reported.[15] The cover was made of phenolic resin (100 parts), BA (50–100 parts). It does not deform for 20 minutes in the cone calorimeter test (ISO 5660–1/50 kW/m^2) and has a pHRR of 13 KW/m^2 and THR of 6 MJ/m^2.

9.2.1.3 Polyimide and Polyetherimide (PI and PEI)

PIs possess outstanding thermal stability, mechanical properties, and chemical resistance. They are widely used in the fields of coatings, aerospace, filtrations, adhesives, as well as electric and photonic applications. In situ thermal conversion of BA to B_2O_3 nanoparticles (< 10 nm) in B_2O_3/PI composite films was reported.[16] Ultrafine B_2O_3 were homogeneously distributed in the PI and greatly enhanced the thermal and mechanical properties of the composite. The oxidation behavior of PI-derived carbon films doped with boron in low levels (up to 1.23×10^{-2} B/C) was also reported.[17] Interestingly, it was found that boron-doping of less than 7.09×10^{-3} B/C in the carbon film (formed at 1,500°C) had a catalytic effect on the oxidation of the carbon. The catalytic effect would be dominant up to some level of carbon burn off when less boron is doped; the contribution of the inhibiting effects become dominant when more doped boron is present.

Unlike PI, PEI offers many attributes such as high transparency, high heat resistance, impact strength, mechanical strength, good electrical properties, broad chemical resistance, etc.

It was reported that a small amount of BA (around 0.25 wt%) in glass-filled polyetherimide (PEI) can reduce peak heat release rate (pHRR) by almost 50% in the FAR Part 25 OSU Heat Release test for aircraft industry (Figure 9.3).[18]

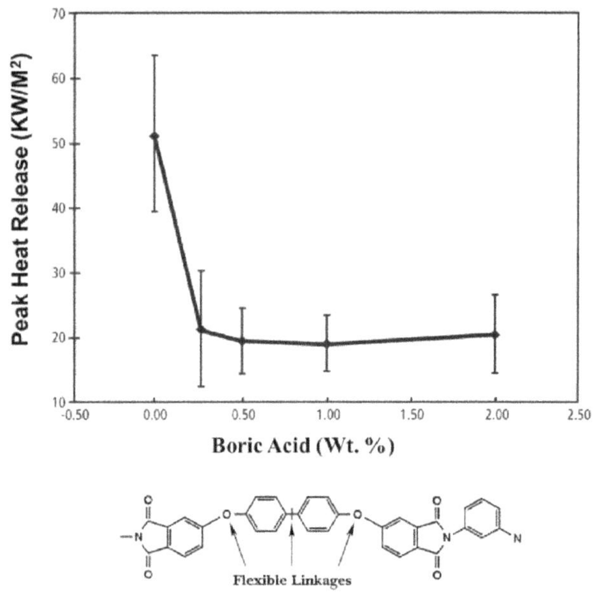

FIGURE 9.3 FAR Part 25 OSU vertical radiant panel heat release test of polyetherimide for aerospace application.

9.2.1.4 Polyphenylene Ether (PPE)

Boric oxide in conjunction with polyvinylidene difluoride (PVDF) is an effective flame retardant and smoke suppressant.[19,20] A combination of PPE (80%), pentaerythritol tetrabenzoate plasticizer (PETB) (20%), boric oxide (< 1%), and PVDF (2%) can resulted in V-0 (1.6 mm) and smoke density (ASTM E662) Ds 87 (4 min.). The benefits of having boric oxide and PVDF were also illustrated in systems such as PPE/HIPS, polycarbonate, polyether sulfone, polyphenylene sulfide. It was also reported that BA (1%) alone without PVDF can provide a drastic smoke reduction in PPE/PS and PPE/PETB.

9.2.1.5 Polystyrene Foam (PS Foam)

Flame retardant PS foam has always been a technical challenge. Fujimori reported that pre-treating polystyrene beads with BA/sodium silicate/phenolics resulted in low flammability and smoke generation.[21] XFLAM in Australia reported the use of a combination of phenolic resole, expandable thermoplastic microspheres, and BA (chloroalkyltrimethoxysilane treated) to treat pre-expanded PS beads.[22] The final PS foam produced has an average specific mass loss rate of around 1.27 g/m^2.s., whereas the untreated PS foam has the mass loss rate of 9.81 g/m^2.s. (ISO 17554 at 50 kW/m^2).

9.2.1.6 Wood/Cellulosics

Commercially, BA has been used in products such as cellulosic insulation, cotton batting for mattresses, various paper products, dimensional lumber/plywood, medium density fiber board (MDF), etc.[2] Silica sol/silica gel are attractive flame retardant for wood/cellulosics. For example, poplar pine treated with BA modified silica sol was recently reported.[23] Cone calorimeter and TGA analysis showed that wood treated with modified silica sol showed significant improvement of flame retardancy than untreated wood.

9.2.2 Metallo-Borate

9.2.2.1 Alkali Metal Borates (Sodium Borates)

The most common sodium borate is Borax Decahydrate (known as Borax or B-10, $Na_2O{\cdot}2B_2O_3{\cdot}10H_2O$), Borax Pentahydrate (*Neobor®*, or B-5, $Na_2O{\cdot}2B_2O_3{\cdot}5H_2O$), Anhydrous Borax, and Disodium Octaborate Tetrahydrate (DOT, also known as *Polybor®* or *Timbor®* -$Na_2O{\cdot}4B_2O_3{\cdot}4H_2O$). B-5 and B-10 are slightly soluble in cold water and very soluble in hot water. Their pHs is around 9.2, and they have excellent buffering properties. DOT is a borate produced by a spray-drying process. It is an amorphous material (pH around 8) and thus can be dissolved into water rapidly (solubility 9.7 wt% at room temperature and 21.9% at 30°C). It is particularly effective in reducing the flammability of wood/cellulose/paper products.

In many wood and paper products, it is common to use a combination of BA and sodium borate.

Cross-laminated timber (CLT) features multiple layers of two-dimensional lumber that is glued perpendicular to each other (mostly with urethane adhesive), compressed tight, tested to be fire rated. It is certified as structurally sound and approved by the 2021 International Building Code (IBC) for multiple stories tall. It was recently reported that the Douglas-fir wood samples treated with DOT and polyurethane adhesive resulted in not only better fire test performance as evidenced by one calorimeter test, but also higher termite and wood decay resistance than the untreated wood.[24] NordTreat of Finland claimed that boards treated with NT DECO™ (a water-based, environmentally friendly transparent coating) can meet Euroclass B-s1,d0 (EN13501-1) (i.e., the boards contribute to fire in a very limited extent, smoke production is very limited and has no flaming droplets).[25] Recently it was reported that reduced graphene oxide (rGO) and sodium metaborate synergistically display flame retardancy in sawdust pellets[26] (see section 9.2.6.4 Graphene).

9.2.2.2 Alkaline Earth Metal Borate

9.2.2.2.1 Calcium Borates ($xCaO{\cdot}yB_2O_3{\cdot}zH_2O$)

A variety of calcium borates can be prepared by reacting calcium hydroxide and BA.

In halogen-containing systems, calcium borates are generally not as effective as zinc borates, because zinc ion is a stronger Lewis acid than calcium ion.

9.2.2.2.1.1 Roofing Membrane Most calcium borates have low dehydration temperatures around 115°C–200°C except the natural mineral colemanite ($2CaO{\cdot}3B_2O_3{\cdot}5H_2O$) that has a dehydration temperature of 290–300°C. It is mostly used in low-sloped rubber modified roofing membrane (around 12–14% loading). Synthetic calcium borates with lower dehydration temperatures have been used in fire retardant grade sealants and caulks.

9.2.2.2.1.2 Epoxy Hitachi reported the use of synthetic calcium borate ($2CaO{\cdot}B_2O_3{\cdot}H_2O$) in epoxy filled with silica for semiconductor encapsulation. This calcium borate's dehydration temperature is in the range of 200–400°C.[27] Calcium metaborate is claimed to be advantageously incorporated into frits as fire retardant in fire doors particularly for those based on sodium silicate.

It is believed that calcium borosilicate will be formed upon heating in this composition.[28]

9.2.2.2.1.3 Polyester (PBT) The PBT composite comprising PBT (100 parts), brominated flame retardant (20), glass fiber (50), anhydrous calcium borate or metaborate (2–5) can effectively improve glow wire ignition temperature (GWIT) of the composite due to the synergistic effect between calcium borate and glass fiber.[29] In addition, the black spots in the product during processing can be alleviated. Interestingly, the combination of the proprietary calcium borate (0.5 μm)/calcium metaborate (0.5 μm) (such as 2.5:2.5 parts) resulted in the best GWIT (930°C) and UL-94 V-0 rating; whereas the formulation with calcium borate (0.5 parts) provides GWIT (825°C) and V-1 rating.

9.2.2.2.1.4 Wood Products Calcium borate whisker ($2CaO{\cdot}B_2O_3{\cdot}H_2O$) was prepared by a hydrothermal method. It releases water at a remarkably high temperature (370°C).[30] Recently it was demonstrated that the $2CaO{\cdot}B_2O_3{\cdot}H_2O$ nano-flake displays good flame retardancy with wood powder as shown by the LOI test.[31]

9.2.2.2.2 Magnesium Borates ($xMgO{\cdot}yB_2O_3{\cdot}zH_2O$)

Due to its high charge to size ratio, the Mg^{2+}cation has a strong tendency to include water in its coordination sphere. Thus, most synthetic magnesium borates contain non-hydroxyl water which can cause them to have low dehydration temperature. For use in plastics, these magnesium borates have to be fully or partially pre-dehydrated.

9.2.2.2.2.1 Epoxy Kyocera reported the use of an unspecified magnesium borate in silica-filled epoxy/

phenolic for electronic packaging. The addition of 25% magnesium borate resulted in a V-0 (1.6 mm) formulation with good moldability and high temperature reliability.[32]

9.2.2.2.2.2 Polyamide Anhydrous magnesium borate ($2MgO{\cdot}B_2O_3$) whisker has a melting point of 1,340°C and can replace glass fiber in thermoplatic ploymers.[33] A PA 66 (100 parts) formulation containing melamine cyanurate (8.8) and glass fiber (27.2) can only yield CTI 400 volts and UL 94 V-2 (0.8 mm) rating, whereas using magnesium borate whisker (diameter ~ 0.7 μm/length ~20 μm) to replace glass fiber results in CTI 500 volts and UL-94 rating V-0. Relative to glass fiber, the magnesium borate formulation displays only slight inferior flexural strength and flexural modulus.

9.2.2.2.2.3 Wood Products Preparation of magnesium borate ($2MgO{\cdot}B_2O_3{\cdot}1.5H_2O$) nano-rod and nanowire were reported. LOI of compressed pure wood dust, wood dust with regular magnesium borate (20 wt%), nano-rod (20%), and nano-wire (20%) are 23.8, 25.3, 29.0, 29.3, respectively.[34]

9.2.2.3 Transition Metal Borates and Miscellaneous Metal Borates

9.2.2.3.1 Aluminum Borate ($9Al_2O_3{\cdot}2B_2O_3$)

A variety of transition and other metal borates are reported in the literature. These include aluminum borate, silver borate, iron borate, copper borate, nickel borate, strontium borate, lead borate, zirconium borate, etc. In a PBT formulation containing brominated polystyrene and antimony trioxide, and glass fiber, anhydrous aluminum borate whiskers ($9Al_2O_3?2B_2O_3$, mp 1440°C) was used to replace glass fiber resulting in much improved CTI (from 275 to 375 volts) and some flexural modulus improvement and still maintain the V-0 (0.8 mm) performance.[35] Aluminum borate can also replace glass fiber in a non-halogen melamine cyanurate based PA 66 to achieve excellent mechanical strength, CTI, and better UL-94 test performance.[33]

9.2.2.3.2 Zinc Borates ($xZnO{\cdot}yB_2O_3{\cdot}zH_2O$)

Among all of the boron-containing fire retardants used in polymers, zinc borates have played the dominant role in commercial applications. The annual global volume requirement for flame retardancy application is estimated to be more than 15,000 MT/year.

9.2.2.3.2.1 Products Rio Tinto Minerals/U.S. Borax first patented and commercialized the unique form of zinc borate with a mineral formula of $2ZnO{\cdot}3B_2O_3{\cdot}3.5H_2O$ (ZB 2335), known in the trade as *Firebrake*®ZB.[36] In contrast to previously known zinc borates, this zinc borate is stable to 290–300°C (Figure 9.4). Thus, it can be used in engineering polymers such as glass reinforced polyamide 6,6. It should be noted that this borate does not have interstitial water molecules as shown in the mineral formula (Figure 9.5). Due to their low dehydration temperature, the usage of ZB-223 ($2ZnO{\cdot}2B_2O_3{\cdot}3H_2O$) and ZB-237 ($2ZnO{\cdot}3B_2O_3{\cdot}7H_2O$) are mostly limited to halogen-containing polymers such as flexible PVC and some elastomers. Due to the demand of high production throughput and thin-walled electrical parts, engineering plastics are processed at increasingly higher temperatures. To meet this market demand, US Borax also developed an anhydrous zinc borate, *Firebrake*®500 ($2ZnO{\cdot}3B_2O_3$), that is stable to at least 500°C and *Firebrake*®415 ($4ZnO{\cdot}B_2O_3{\cdot}H_2O$), that is stable to > 415°C.

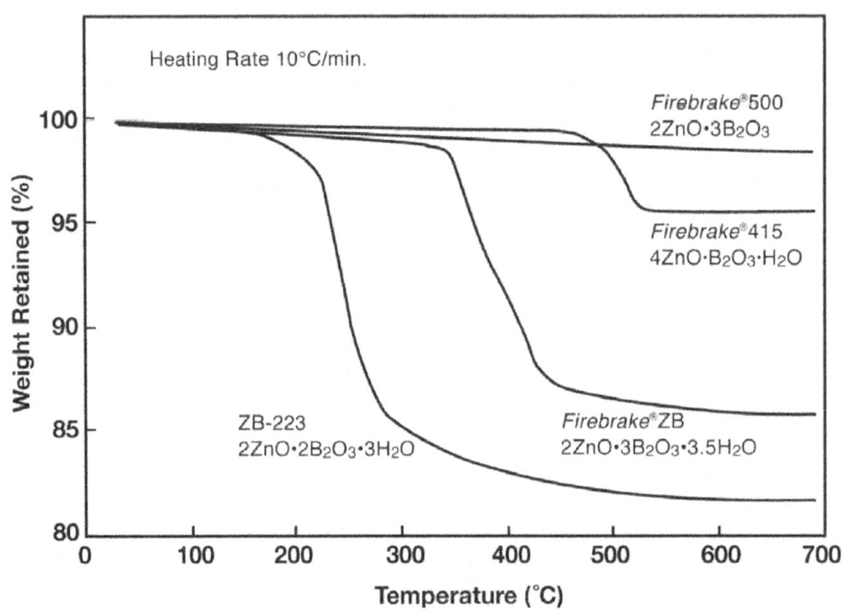

FIGURE 9.4 Thermogravimetric analyses of commercial zinc borates in air.

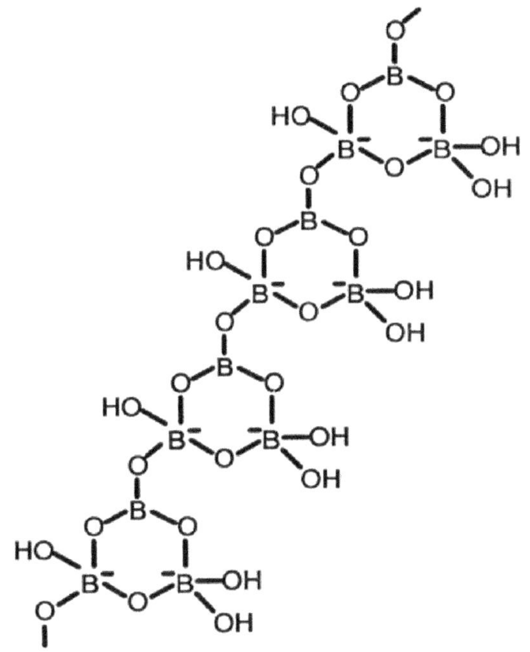

FIGURE 9.5 Molecular structure of ***Firebrake*** Zinc Borate (zinc ion complexing to dihydroxyborate anion is not displayed).

9.2.2.3.2.2 Epoxy In halogen-containing epoxy, *Firebrake* ZB (ZB) is an effective flame retardant and smoke suppressant, particularly with an aliphatic halogen source.

In non-halogenated epoxy, ZB has been used as the sole flame retardant in semiconductor encapsulation applications (i.e., high loading of spherical silica with ZB).[37] Zinc borate alone can outperform magnesium hydroxide and zinc carbonate in halogen-free epoxy/novolac for electrical insulator applications.[38] Zinc borate can not only provide a V-0 performance at 0.8 mm but also can improve the "stability temperature" and copper/epoxy adhesion, as evidenced by the shear tensile strength. This type of material is used as insulation material in electrical machines and converters, especially in transformers. They also have to meet a dielectric breakdown resistance of greater than 30 kV/mm.

In epoxy intumescent coatings for hydrocarbon fires, ZB is commonly used in conjunction with APP, silica, ATH, ammonium pentaborate, chloroalkyl phosphate, wollastonite, etc., for strong char formation (also see Section 9.2.3.1 Ammonium Pentaborate).[39,40] The effect of ZB on the char property was systematically evaluated. It was found that (1) ZB at 15 wt% showed the best efficacy, (2) higher level of ZB significantly increase the dynamic viscosity of incipient char that can result in a more uniform but less expanded char layer, and (3) ZB can help to reduce epoxy binder degradation and oxidation of the carbon-inorganic residue.[41]

The use of a combination of ATH and ZB in epoxy for aerospace applications was reported. In cone calorimeter tests, it was demonstrated that zinc borate promotes the formation of a strong char *via* sintering with alumina and results in substantial decrease of pHRR and aveHRR of epoxy resin. Synergy was observed when the total loading reaches 20% or higher at 1:1 ratio of ZB:ATH.[42]

Dow reported the use of a combination of ATH, zinc borate, and melamine in epoxy structural adhesives for bonding metals. The adhesive strongly resists ignition and does not interfere significantly with mid-frequency DC welding performance.[43] Boron compounds are known to inhibit thermal oxidation of carbon species. Interestingly enough, in a carbon fiber reinforced epoxy containing ATH or MDH, it was demonstrated that addition of zinc borate (almost independent of the loading level) can minimize the respirable carbon fiber formation during polymer combustion.[44]

9.2.2.3.2.3 Polyamides (PA) Zinc borates (*Firebrake* ZB and *Firebrake* 500) have been used extensively in both halogen-containing and non-halogen glass reinforced PA. Depending on the processing temperature, *Firebrake* 500 has been used in PA, particularly in high temperature PA (PPA) and PA 66.

9.2.2.3.2.3.1 Firebrake ZB can function as multifunctional flame retardants in both halogen-containing and halogen-free PAs. They can not only function as flame retardant synergists in fire retardancy, but also improve thermal stability, CTI, GWFI/GWIT, color stability, recyclability, corrosion resistance of processing equipment, mechanical properties, and laser marking.[6]

9.2.2.3.2.3.2 Halogen-Containing PA A combination of brominated polystyrene, zinc borate, and antimony trioxide can provide good CTI and V-0 performance. For an antimony-free formulation, one can still maintain V-0 rating with the use of zinc borate but at slightly higher bromine content. In high temperature PA, *Firebrake* 500 can replace antimony trioxide completely.

9.2.2.3.2.3.3 Non-halogen PA Zinc borate in PA 66 containing Red Phosphorus not only displays synergy in fire retardancy but also imparts corrosion resistance toward metals such as copper. Zinc borate can trap trace amounts of phosphine (PH_3) derived from red phosphorus. Aluminum diethylphosphinate (DEPAL) offered by Clariant is the dominant fire retardant on the market for non-halogen PA 66. DEPAL does not melt before it starts to decompose/vaporize at temperatures > 350°C. In 25% glass reinforced PA 66, OP 1312 (a blend of DEPAL, MPP, and zinc borate) can achieve V-0 at 1.6 mm with 16% total loading and 5VA with 18% loadings. A combination of DEPAL and aluminum ethylphosphinate in the presence of MPP/*Firebrake* ZB (or *Firebrake* 500) provides not only good flame retardancy, but also no discoloration, no extrudation, and low melt index (i.e., no degradation during processing). The zinc borate in OP1312 is also a thermal stabilizer. Zinc borate can also improve corrosion resistance of DEPAL in high temperature PPA. Clariant recently reported that PA 66 containing DEPAL (11.4%), MPP (5.7), talc (10), *Firebrake* 500, and glass fiber (20%) can achieve UL 94 V-0 at 0.6 mm and GWIT of 775°C at 1 mm and 800°C at 2 mm. The addition of MPP and zinc borate to DEPAL in PA causes the change of mode of action from gas phase to a mixed gas/condensed phase. The formation of boron phosphate and aluminum phosphate in the condensed phase was demonstrated.[45]

9.2.2.3.2.3.4 Polyamide 12 and Polyamide 12 (3D Printing) PA12 is a semi-crystalline material with longer aliphatic chain than polyamide 66 (PA 66). They have low moisture absorption/good dimensional stability, good flexibility, good abrasion, UV resistance, and good retention of performance over time. But they do have poorer impact resistance, lower dielectric strength, and higher cost relative to PA 66. Clariant reported the use of aluminum diethylphosphinate (18 wt%, Exolit 1230 from Clariant) and *Firebrake* ZB (2%,) in unreinforced PA12 to achieve UL 94 V-0 (1.5 mm). The use of Exolit 1312 (22%, a blend of DEPAL/MPP/zinc borate) in glass reinforced PA12 can achieve V-0 (0.8 mm).[6]

Among the different additive manufacturing techniques (i.e., 3D printing) available for commercial production, Selective Laser Sintering (SLS) allows the production of complex 3D shapes for aerospace and defense industries. The SLS is based on selective fusion and recrystallization of polymer powders layer by layer (generally with a CO_2 laser). Lopez-Cuesta and coworkers reported a systematic evaluation of flame retardant additives in semi-crystalline PA12 for powder flowability, thermal degradation when subjected to laser beam, and final porosity of the finished articles. APP, zinc diehtylphosphinate (Exolit OP 950), ZB and pentaerythritol ester of methane phosphonic acid (Aflammit PCO900) were evaluated individually, as well as compositions consisting of 2–3 FR components. It was found that, for compositions in which only one flame retardant was used, ZB gave the best fire test performance. Moreover, for binary and ternary compositions, the zinc borate was able to form zinc phosphate and borophosphate to retain nearly all of the phosphorus in the composition, and resulted in good fire test performances.[6]

9.2.2.3.2.4 Polycarbonates (PC) In PC, a combination of core-shell polymer (5.75%), BDP [bis-phenol-A-bis(diphenyl phosphate)] (12.5). PTFE-SAN (0.8), and zinc borate (1.48) provides UL-94 V-0 (1.5 mm), 5V 5A (2.3 mm), and good physical properties. Synergism of the zinc borate and polysiloxane in PC were reported to achieve V-0 at 1.6 mm. Zinc borate in PC containing BDP and with a core/shell modifier can lower the pHRR and provide a drastic smoke reduction per ASTM E662.[46]

9.2.2.3.2.5 Polyimide/Polyetherimide/Silicone Polyetherimide- A flame-retardant formulation of silicone polyetherimide containing zinc borate (*Firebrake* 415, 3%) and PTFE (1%) can achieve Flame Spread Index (ASTM E162) of 6 at 3.2 mm and with no dripping. It is suitable for transportation industry applications.[47]

9.2.2.3.2.6 Polyolefins

9.2.2.3.2.6.1 Halogen-Containing System When used in conjunction with halogen-based flame retardants, *Firebrake* ZB can partially replace antimony oxide (30–40%) and still maintain the same fire test performance. In addition, it can improve aged elongation properties, increase char formation, and decrease smoke generation. The B_2O_3 moiety in zinc borate can also provide afterglow suppression.[1]

9.2.2.3.2.6.2 Non-Halogen System Depending on the application, high loadings of ATH or MDH are generally required (30–70 wt%) in polyolefins. Recent developmental efforts have focused on co-additives of ATH or MDH with the aim of reducing total loading and developing stronger char/residue formation.[2]

- The main benefit of zinc borate is promoting the formation of a strong char/ceramic residue that prevents burning drips, delays oxidative pyrolysis, and protects underlying polymers.
- The use of co-additives such as silicone/silica, melamine polyphosphate, phosphate ester, red P, APP, and melamine cyanurate can augment the fire test performance of zinc borate/metal hydroxide system.
- Partial substitution of MDH with the fine grade of zinc borate (5–10 wt%) in EVA (28% VA) resulted in not only HRR reduction but also better electrical properties/physical properties (except tensile strength) and improvement of processability.
- Partial replacement of ATH with zinc borate in EVA reduces carbon monoxide yield under fuel-rich conditions in the Purser furnace.[48]

Polypropylene (PP)

In halogen-containing PP, zinc borate can partially replace antimony trioxide and provide good afterglow control. For example, the PP containing ZB has been used for producing electrical switch boxes and air-conditioning pan. In non-halogen system, the use of a combination of silicone, *Firebrake* ZB, ATH, and a zinc or sodium ionomer in PP achieves low HRR and smoke generation.[49]

Recently, Sun and Zhang et al. reported the preparation of a superior charring agent N-tris(hydroxymethyl)-amino-triazine-piperazine co-polymer (TIPC) (Figure 9.6). In PP, the test results showed that the presence of 21% of APP-TTPC-ZB with mass ratio of 16:4:1 significantly reduced pHRR from 1547 to 230 kW/m^2, increased LOI from 18.0 to 29.6%, and improved the UL-94 rating from no rating (NR) to V-0 at 3.2 mm/no dripping.[50] The results demonstrated that TTPC is an efficient charring agent and zinc borate is an indispensable synergist of the intumescent system.

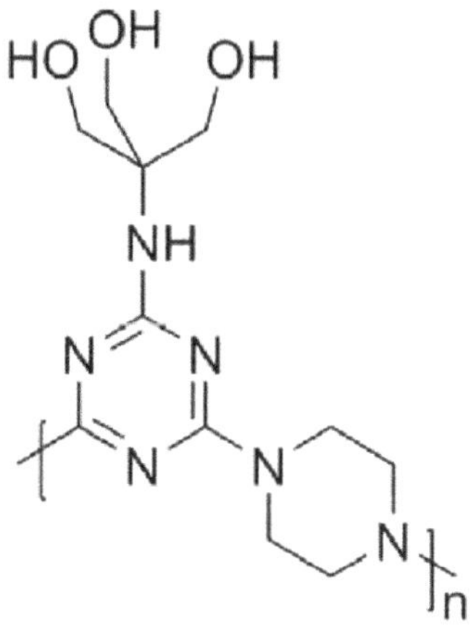

FIGURE 9.6 A polymeric reaction product of aminotriol/tiazine/piperazine—a charring agent.

Source: "Synthesis of a novel polydroxy triazine-based charring agent and its effects on improving the flame retardancy of polypropylene with ammonium polyphosphate and zinc borate" Wu, Q; Guo, J.; Fei, B.; Li, X.; Sun, J.; Gu, X.; Li, H.; Zhang, S. *Polym. Degrad. Stab.* 2020, *175*, 109123; with permission.

Fine grade of zinc borate is also known to be an effective nucleating agent for PP form production.

9.2.2.3.2.7 Polyphenylene Ether In a cross-linked PPE/PS printed wire board for satellite application, a combination of phosphate ester (PX200) and zinc borate (*Firebrake* ZB-Fine) can provide a V-0 at 1.6 mm, good dielectric constant, insulation resistance, and soldering heat resistance with a Tg of 165°C.[51]

9.2.2.3.2.8 PVC Zinc borate is widely used commercially in many PVC applications such as wall covering, wire and cable, coated fabrics/military tenting, carpet backing, conveyor belting/transparent mining curtain, protection screen, coatings for fiberglass sun screening mesh, etc.

9.2.2.3.2.8.1 Rigid PVC Firebrake ZB can outperform antimony trioxide as evidenced by LOI tests.

9.2.2.3.2.8.2 Flexible PVC Depending on the plasticizer, filler, and halogen content, ZB can either partially or completely replace antimony oxide. In contrast to antimony oxide, ZB also functions as a smoke suppressant and afterglow suppressant.[1]

9.2.2.3.2.9 Silicone/Elastomers (SBR, SEBS, SEPS, etc.) Zinc borates have commonly been used in both halogen and non-halogen elastomers (such as EPDM, EP, SBR, neoprene, and silicone). The applications include conveyor belting, flooring, wire and cable, electrical insulation, door seals, etc. The use of a combination of ATH, silica, zinc borate, and zeolite in styrene-ethylene-butylene-styrene (SEBS) and styrene-ethylene-propylene-styrene (SEPS) can achieve good dielectric breakdown strength, track and erosion resistances. In addition, good physical properties such as 650% elongation can be achieved.[52]

Commercially, zinc borate has been used in silicone wire and cables and insulating material applications. It is known that borate and silica (derived from silicone) or other mineral fillers (such as ATH) can result in ceramification during polymer combustion.[4,53] It was also demonstrated that zinc borate performed better than APB and borax in silicone/halloysite ceramification and achieved the best electrical properties.[54]

In many commercial polyolefin wire and cable applications, ZB is used in conjunction with metal hydroxide in the presence of silicone such as SFR100 (a viscous liquid) from Momentive or a silicone deposited on a powder carrier such as Dow's DC4–7081. Boric oxide from zinc borate can react with silica (derived from silicone combustion) to form borosilicate.

9.2.2.3.2.10 Unsaturated Polyesters/Vinyl Esters

9.2.2.3.2.10.1 Firebrake ZB can be used in halogen-containing polyester. For example, in Hetron 92 AT (an unsaturated polyester with chloredic anhydride and 3–5% Sb_2O_3), the zinc borate can replace antimony trioxide completely and result in good transparency and still meet the Class I in the E-84 tunnel test. Unlike antimony trioxide, *Firebrake* ZB has a refractive index similar to that of organic polymer; as a result, they do not induce opacity in polymer products.

In non-halogen UP/VE, bi- or multi-modal and surface treated ATH can be used with ZB in a high ratio (about 50–100 to 1) to achieve Class A in the E-84.[1,2] High loadings of ATH provides lower smoke generation during combustion. To lower smoke value even more, one could use methyl methacrylate to partially or completely replace styrene in the formulation.

9.2.3 Boron-Nitrogen-Containing Compounds

9.2.3.1 Ammonium Pentaborate [$(NH_4)_2O{\cdot}5B_2O_3{\cdot}8H_2O)$]

Upon thermal decomposition, ammonium pentaborate (APB) first gives off a large amount of water starting at about 120°C. At about 200°C, it starts to give off ammonia, and at about 450°C, it is all converted to boric oxide, which is a glass former. APB functions as both inorganic blowing agent and a glass-forming fire retardant.

$$(NH_4)_2O \cdot 5B_2O_3 \cdot 8H_2O) \rightarrow 2NH_3\uparrow + 8H_2O\uparrow + 5B_2O_3$$

APB solution can be sprayed on paper, or the paper can be dipped into the solution to yield fire retardant products. In addition to cellulose, it has also been used as a flame retardant in polyurethane foam (anechoic chamber) and epoxy intumescent coatings.

APB is an effective intumescent flame retardant in the thermoplastic polyurethane. Partially dehydrated APB at 5–10 phr loading in TPU can provide seven to tenfold improvement in a burn through test.

A recent patent by PPG disclosed the use of APB but with no zinc borate in epoxy intumescent coatings.[39,40] It provides excellent low temperature cracking resistance and good UL1709 fire test performance for coating structural steel application.[55,56]

It was reported that APB in urea formaldehyde adhesive for plywood can not only increase gel time but also decrease formaldehyde emission.[1,2]

9.2.3.2 Boron Nitride—2D (BN)

Hexagonal boron nitride (h-BN) can be prepared from reaction of boric acid and urea or melamine. For example, pyrolysis of melamine diborate can yield h-boron nitride. It is commonly referred to as "white graphite" because of its platy hexagonal structure similar to graphite. The h-BN is stable in inert or reducing atmosphere to about 2,700°C and in oxidizing atmosphere to 850°C. It has excellent anisotropic thermal conductivity and good electrical resistance. Furthermore, it can be chemically functionalized with numerous functional groups such as -OH, -NH_2, -NHR, and -COR. The h-BN has been frequently quoted as functional filler for fire retardant electrical/electronic applications (Figure 9.7).

The h-BN platelets can be exfoliated to generate boron nitride nanosheets (BNNSs). It can be exfoliated via mechanical exfoliation, solvent exfoliation, and other techniques.[57]

9.2.3.2.1 Acrylics

3M reported the use of h-BN platelets in acrylics (~73% loading) with 3D printing by fused filaments fabrication (FDM). Carefully controlled thickness of the fused filament allows the fabrication of electrical parts with either in-plane or through-plane orientation with maximum thermal conductivity.[58]

9.2.3.2.2 Epoxy

The h-BN was simultaneously exfoliated and flame-retardant functionalized to BNNSs via a one-step ball milling process. It is based on the synergistic effect of mechanical shear and chemical peeling of ammonium phosphate and sodium hydroxide. As one of the potential applications, this functionalized BNNSs could be used as a multifunctional additive for high performance thermally conductive epoxy composites. This material displayed superior flame retardancy and enhanced the thermal conductivity.[59]

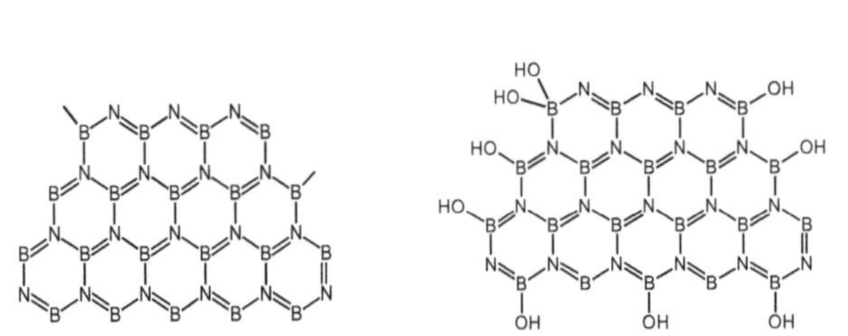

FIGURE 9.7 h-Boron nitride structure (left, nanosheet) and hydroxylated boron nitride (right).

9.2.3.2.3 Polyurethane Foam

It was reported that h-BN platelet can be partially hydrolyzed to hydroxylated nanosheet via a sonication-assisted process. This functionalized h-BN is water compatible. Grunlan et al. reported that the hydroxylated h-BN nanoparticles were conformally deposited on the cell wall of the flexible foam with polyethylenimine (PEI) in a single 82 nm bilayer via a layer-by-layer assembly (the foam was pretreated with polyacrylic acid). The treated foam resulted in reduction of pHRR (54%), THR (20%), and gas emission. The PEI/h-BN coating also protects PU foam from weathering and only modestly alters mechanical property of the foam.[60]

9.2.3.2.4 Wood

The use of exfoliated h-BN nano-sheet as a high performance binder-free fire-resistant coating for wood was claimed. It was demonstrated that the oxidation resistance property of h-BN wood coating is stable to 900°C in air.[61]

9.2.3.3 Boron Nitride Nanotubes (BNNT)

BNNT was reported around 2001. They are far more resistant to oxidation than carbon nanotubes

BNNT (0.01 wt%) was reported to improve thermal resistance of polybenzimidazole.[62] It was also claimed that a boron nitride coated calcium borate in combination with spherical alumina can be used in epoxy printed circuit boards having a high glass transition temperature, peel strength, moisture/heat resistance, and UL-94 V-0 performance at 0.4 mm.[63]

9.2.3.4 Guanidinium Borate [$x[C(NH_2)_3]O \cdot yB_2O_3 \cdot zH_2O$]

Several guanidinium borates have been reported in the literature. For example, guanidinium tetraborate [$C(NH_2)_3]_2O.2B_2O_3.4H_2O$] can be prepared by reacting guanidinium carbonate with BA in hot water or by reacting guanidinium chloride with borax in cold water.[64] A variety of nitrogen compound plus BA and guanidium borate are commonly used for flame retardant wood product, such as plywood. They provide better high temperature aging performance than nitrogen compounds plus phosphoric acid.

9.2.3.5 Melamine Diborate [$(C_3H_8N_6)O \cdot B_2O_3 \cdot 2H_2O)]/(C_3H_6N_6 \cdot 2H_3BO_3)$]

Melamine diborate (MB), known in the fire retardant trade as melamine borate, is a white powder which can be prepared readily from melamine and boric acid. It is partly soluble in water and acts as an afterglow suppressant and char promoter in cellulosic materials. Budenheim Iberica claims that, in a 1:1 combination with ammonium polyphosphate, MB (10–15%) can be used for phenolic-bound non-woven cotton fibers. In general, MB can be used as a char promoter in intumescent systems for various polymers, including polyolefins or elastomers. However, its low dehydration temperature (about 130°C) limits its application in thermoplastics that are processed at above 130°C. In EVA (60% VA), it was found that EVA-ATH-MB is protected by both the gas and condensed phase mechanism, whereas MB contributes to the formation of a protective layer due to the formation of boric oxide, boron nitride, and BNO at the surface. The boron species can decrease thermal conductivity in the ignition temperature range leading to delay of time to ignition relative to melamine or melamine phosphate.[65]

9.2.4 Boron-Phosphorus-Containing Compounds

9.2.4.1 Boron Phosphate (BPO_4)

Boron phosphate, an inorganic polymer of empirical formula BPO_4, can be prepared by heating phosphoric and boric acids to calcining temperature. It is a white infusible solid that vaporizes at 1,450–1,462°C without decomposition. As opposed to the tri-valency in boric oxide, most of the boron atoms in boron phosphate are tetra-coordinated (i.e., as a BO_{4-} group). Boron phosphate is a low cost, low toxicity flame retardant. An amorphous form of boron phosphate is commercially available from Budenheim. Commercially it is mostly used as an additive in polyphenylene ether alloy. In a PA-PPO alloy, a combination of boron phosphate (5%) and silicone oil (2%) resulted in a V-0 (1.6 mm) rating. The use of boron phosphate in conjunction with zinc borate prevents the UL-94 5VA burn through in PPE/HIPS/SEBS containing triphenyl phosphate. A patent by Nexans claimed the use of zinc borate or boron phosphate in electrical insulation, which is resistant to high temperature such as 1,100°C. The use of boron phosphate as a flame retardant in proton-conducting polyimide for fuel cell membrane application was reported. Recently it was reported that crystalline boron phosphate can function as a solid acid. In epoxy, it can catalyze the pyrolysis of epoxy at a lower temperature, reduce the release of flammable gas, and promote char formation.[4,6]

9.2.4.2 Metal Borophosphates

BASF reported the preparation of different metal borophosphates.[66] These compounds were found to exhibit high thermal stability and are useful as flame retardants for plastics. Partial replacement of DEPAL with the supposedly less expensive zinc or magnesium borophosphate can still maintain the UL-94 V-0 performance. The advantage of this additive is that it contains both boron and phosphorus. For easy processing, it could replace both melamine polyphosphate and zinc borate.[6]

9.2.4.3 Ammonium Borophosphate

It was claimed that boiling a mixture of boric acid and diammonium phosphate can result in a solid that has flame retardancy effect in unsaturated polyester (Sperber 2017). The solid product was not analyzed and could be some sort of "ammonium borophosphate" that has never been reported.[4,6]

9.2.5 Boron-Silicon-Containing Compounds

The use of borosilicate glass/ceramic frits and borosiloxane were previously reviewed.[3]

Borosilicate glass is a range of glasses based on boric oxide, silica, and metal oxide. It has excellent thermal shock and chemical resistance. Synergistic flame retardant effects of borosilicate hollow glass microsphere and magnesium hydroxide in EVA was reported.[67]

Thovotech in Germany reported a combination of PA 6 (90 wt%), MC (5), and porous amorphous glass particle (5) achieved a non-dripping V-0 (0.8 mm) rating. Whereas a typical polyamide 6 would require about 14–16% of MC. These porous borosilicate glass particles can sinter at heat treatment temperature around 360–400°C. This sintering action is believed to be responsible for the flame retardancy.[68]

9.2.6 Boron-Carbon-Containing Compounds

9.2.6.1 Arylboronic Acid Derivatives [$ArB(OH)_2$]

Boronic acids are known to release water on thermolysis, thereby leading to the formation of boroxines or boronic anhydrides. Unlike boric acid ester, B-C bond in boronic acid is not readily hydrolysable. Morgan and Benin's group reported the use of boronic acid derivatives as flame retardants for polyurethane materials. Fei et al. reported the preparation of a macromolecule with p-formylphenylboronic acid, pentaerythritol, and tris(2-aminoethyl)amine.[6] With a very low loading (~2%) in PLA, this microporous macromolecule can enhance remarkably the flammability performance such as LOI 31.5%, UL-94 V-0, and it showed reductions in pHRR, THR, total smoke release (TSR), and CO emissions based on the cone calorimeter test. In addition, it can improve UV protection. Although boronic acid derivatives are not yet used as flame retardants in commercial application, there have been many recent research reports and patents in the field of fire retardancy.[4]

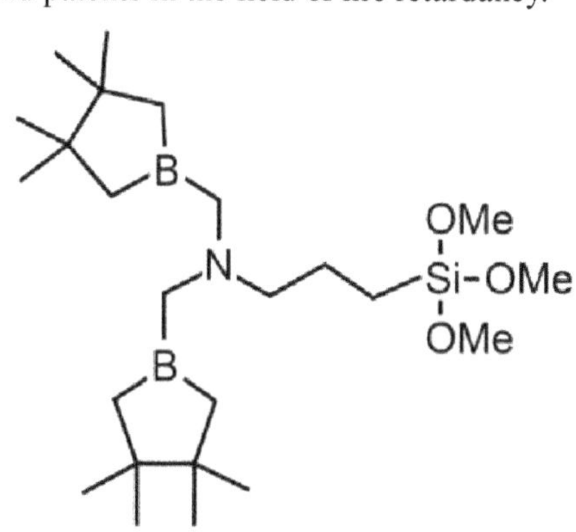

FIGURE 9.8 B/N/Si compound as a flame retardant/smoke suppressant in rayon/polypropylene.

9.2.6.2 Borane Derivatives [$B(CR)_3$]

T. Mayer-Gall, E. Ali, et al. of DTNW (German Textile Research Center North-West) reported the use of a combination of Borane/Nitrogen/Silane compound (Figure 9.9.) and DOPO in natural fiber reinforced composites such as rayon in PP or flax in epoxy. The borane compound functions as a smoke suppressant/char stabilizer and improves water leaching resistance due to the silane functionality.[69]

Polyphosphinoboranes $[RR'PBH_2]_n$ is a new class of inorganic polymers that are high in boron and phosphorus content. Interestingly, when cotton is treated with this material, the product becomes self-extinguishing and has good water non-leachability.[70]

9.2.6.3 Boron Carbide (B_4C)

Boron carbide is produced industrially by the carbo-thermal reduction of B_2O_3 in an electric arc furnace. It is a black powder and has a melting point of 2445°C. At around 400–500°C, B_4C can be oxidized to B_2O_3. At around 400–500°C, BC can be oxidized to B_2O_3 in dry air. Ranilli et al. reported the evaluation of the effect of nano-sized BC on phenolic/resol bulk molding compounds reinforced with carbon fiber.[71] It was found that, at 5% BC loading, the composite exhibited a remarkable dimensional stability and char residue structural integrity after combustion. This was attributed to BC being oxidized to B_2O_3 that are known to stabilize char residue and carbon fibers. Thus, BC is in essence acting like a B_2O_3 carrier that can release B_2O_3 slightly above polymer decomposition temperature. In water-air system, BC can be oxidized to B_2O_3, carbon, and hydrogen.

$$B_4C(s) + 4O_2(g) \rightarrow B_2O_3(l) + CO_2(g)$$

$$B_4C\ (s) + 6H_2O\ (g) \rightarrow 2B_2O_3 + C(s) + 6H_2O\ (g)$$

It was reported that the addition of boron carbide (10–15 wt%) in a variety of intumescent coatings containing ammonium polyphosphate and blowing agent resulted in the following benefits[6]:

- Forms a uniform and hard foamed layer.
- Prevents excessive expansion on heating.
- Retards oxidative weight loss of the foamed layer at 1,000°C or even higher.
- Improves weight retention, compression strength, and peel strength during fire.

9.2.6.4 Graphene (Boron–Doped)

Graphene is known to have excellent thermal/electrical conductivities and mechanical properties in polymer composites. Incorporation of graphene at < 5% in polymers can result in reduction in heat release and smoke reduction.[72] However, graphene can be oxidized at about 300°C in air, and exfoliated graphenes are prone to restacking. To improve its compatibility with polymers and boost its fire retardancy, graphene is functionalized. Normally it is converted to graphene oxide (GO)/reduced graphene (rGO) and their derivatives (Figure 9.9).

Boron is known to inhibit the oxidation of carbon species such as carbon fibers, carbon nanotubes (CNT). Substitutional boron doping of carbon materials is known. Wang and Hu et al. reported the preparation of boron-doped graphene oxide by annealing BA and GO at 900°C in argon. The boron content in BG (boron-doped graphene) is approximately 3% and is thermally stable to approximately 500°C.[4,6]

9.2.6.4.1 Epoxy

To overcome the brittleness and flammability of epoxy (nonreinforced), a cardanol-benzoxazine monomer (CBz) was prepared. Cardanol is a phenolic lipid. The incorporation of CBz can improve both the flame retardancy and impact strength of epoxy. The superior flame retardancy was ascribed to Bz enhancing the char formation/char stability.[4]

9.2.6.4.2 Rigid Polyurethane Foam

(RPUF): Recently, Gao et.al. reported the use of a combination of graphene oxide, boric acid, and 3-aminopropyltriethoxysilane to prepare a functionalized graphene oxide (fGO). The silanol reacted with OH group of graphene and acting as a bridge between boric acid and GO. This fGO was incorporated flame retardant RPUF as a synergist for expandable graphite (GF) and dimethyl methyltriphosphonate (DMMP). The LOI reached 28.1%

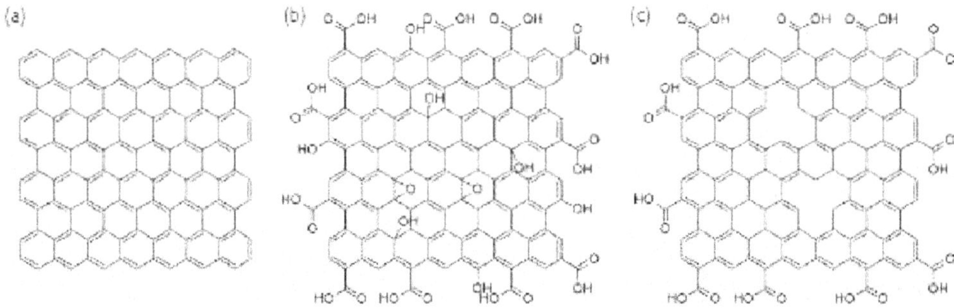

FIGURE 9.9 Schematic illustration of (a) graphene, (b) graphene oxide (GO), (c) reduced graphene oxide (rGO).

Source: Kim, J.; Kim, F.; Huang, J. *Materials Today*, 13, pp *28–38* (2010).

with 0.25 phr of fGO and 7.5 phr EG/2.5 phr DMMP, and the UL-94 test reached V-0. Remarkably, in the cone calorimeter test, the pHHR of the fGO foam relative to RPUF foam reduced by 33%, and the THR reduced by 25.1%. In addition, fGO enhanced the mechanical properties of RPUF.[73]

9.2.6.4.3 Cellulose/Particle Boards

Recently, Losic et al. reported using graphene-borate composite comprising hydrated sodium metaborate (Na2O·B2O3·xH2O) intercalating rGO layers for flame retarding cellulosic products.[26] The authors claimed it has potential for use in commercial particle boards and medium density fiberboard.

9.3 CONCLUSIONS

- Boron-based flame retardants have very broad applications and are multifunctional fire retardants. They can function as flame retardants, smoke suppressants, afterglow suppressants, anti-tracking/anti-arcing agents, etc.
- Borates function as flame retardants predominately in condensed phase.
- Borates have been commercially successful in both plastics and elastomers. The fire test performance of borates can be enhanced with the use of co-additives such as metal hydroxides, antimony trioxide, various N/P/Si-containing compounds, and different carbon species.
- Borates alone can also function as flame retardant in char-forming polymers such as polyethersulfone, polyimide, rigid PVC, silicone, etc.
- There is a tremendous growing demand for thermally conductive and electrically insulating polymer materials. Hexagonal boron nitride will play an increasingly important role in thermal management and flame retardancy in E/E applications.
- Graphene (boron-doped) will continue to be an interesting research subject.
- Non-halogen boronic acid/borane derivatives will attract considerable attention in future flame-retardant research.

REFERENCES

(1) "Zinc borates: 30 years of successful development as multifunctional fire retardants" Shen, K.K. in Nelson, G. L.; Wilkie, C.A., eds. *Fire and Polymers* (ACS symposium series 797); pp. 228, **2001**; Shen, K. K.; Kochesfahani, S; Jouffret, S. F. Zinc Borates as Multifunctional Polymer Additives. *Polym. Adv. Technology.* **2008,** *19*, 469–474.

(2) "Boron-based flame retardants and fire retardancy" Shen, K.K.; Kochesfahani, S.H.; Jouffret, F. in Wilkie, C.A.; Morgan, A., eds., *Fire Retardancy of Polymeric Materials-2nd Edition*, Taylor and Francis; pp. 207–237, **2010**.

(3) "Chapter 12. Review of recent advances on the use of boron-based flame retardants" Shen, K.K. in Papaspyrides, C.D.; Kiliaris, P., eds., *Green Flame Retardants for Polymers-A Comprehensive Guide to Additives and Their Applications*, Elsevier, **2014.**

(4) "Chapter 6. Recent advances in boron-based flame retardants", Shen, K.K. in Hu, Y.; Wang, X., eds., *Flame Retardant Polymeric Materials: A Handbook*, Taylor & Francis, pp. 97–1117, **2020**.

(5) Dogan, M.; Dogan, S. D.; Savas, L. A.; Ozcelik, G.; Tayfun, U. Flame Retardant Effect of Boron Compounds in Polymeric Materials. *Composites Part B.* **2021**, *222*, 109088.

(6) "Boron-based flame retardants in non-halogen based polymers" Shen, K.K. in Morgan, A. ed., *Non-halogenated Flame Retardant Handbook*, 2nd Edition, Wiley, **2022**, 309–336.

(7) "Chapter 6. Boron-containing flame retardants for epoxy thermosets and composites" Shen, K.K.; in Hu, Y.; Wang, X., eds., *Non-Halogenated Flame-Retardant Technology for Epoxy Resin Thermosets and Composites*, Elsevier, **2024** (in press).

(8) Nazarenko, O. B.; Amelkovich, Y. A.; Bannov, A. G.; Berdyugina, I. S.; Maniya V. P. Thermal Stability and Flammability off Epoxy Composites Filled with Multi-Walled Carbon Nanotube, Boric Acid, and Sodium Bicarbonate. *Polymers.* **2021,** *13,* 638.

(9) Naik, A. D.; Fontaine, G.; Bellayer, S.; Bourbigot, S. Salen Based Schiff Bases to Flame Retardant Thermoplastic Polyurethane Mimicking Operational Strategies of Thermosetting Resin. *RSC Adv.* **2015**, *5,* 48224–48235.

(10) He, T.; Guo, J.; Qi, C.; Feng, R.; Yang, B.; Zhou, Y.; Tian, L.; Cui, J. Synthesis of a Novel Type of High-Efficiency Boron-Containing Schiff Base Flame Retardant for Epoxy Resin. *Fire Mater.* **2022**, *47*(3), 341–351.(10)

(11) Huo, S.; Sai, T.; Ran, S.; Guo, Z.; Fang, Z.; Song, P. A Hyperbranched P/N/B-Containing Oligomers as Multifunctional Flame Retardant for Epoxy Resins. *Compos. B.* **2022**, *234*, 109701.

(12) Yang, H.; Wang, X.; Yu, B.; Lei, S.; Hu, Y.; Yuen, R. K. K. Effect of Borates on Thermal Degradation and Flame Retardancy of Epoxy Resins Using Polyhedral Oligomeric Silsesquioxane as a Curing Agent**.** *Thermochimica Acta.* **2012**, *535*, 771–778.

(13) Wang, S.; Wang, Y.; Bian, C.; Zhong, Y.; Jing X. The Thermal Stability and Pyrolysis Mechanism of Boron-Containing Phenolic Resins: The Effect of Phenyl Borates on the Char Formation**.** *Appl. Surf. Sci.* **2015**, *331*, 519–529.

(14) "Resol-type phenolic resin compound for manufacturing phenolic" Nobuyoki, T.; Yasuaki, N.; Takashi, H.; Hitoshi, T. 2002, Japan. Patent, 2002241530.

(15) "Cover for conveyor", Saratani, G. 2011, Japan. Patent, 2011026454.

(16) Jian, S.; Liu, S.; Chen, L.; Zhou, S.; Fan, P.; Zeng, Y.; Hou, H. Nano-Boria Reinforced Polyimide Composites With Greatly Enhanced Thermal and Mechanical Propertied via in-Situ Thermal Conversion of Boric Acid. *Comp. Comm.* **2017**, *3*, 14–17.

(17) Zhong, D. H.; Sano, H.; Uchiyama, Y.; Kobayashi, K. Effect of Low-Level Boron Doping on Oxidation Behavior of Polyimide-Derived Carbon Films. *Carbon.* **2000**, *38*(8): 1199–1206.

(18) "Non-halogen, flame retardant engineering thermoplastics" Bassett, W.H. *Third Annual BCC Conference on Flame Retardancy*, Stamford, CT, May 1992.

(19) "Thermoplastic composition with boron compound" Claesen, C.A.; Lohmejier, J.H.G.; Boogers, M.P.J.; Bussink,

J.; Savenije, H.R.; Hoeks, T.L.; Gosens, J.C. 1993, U.S. Patent, 5,182,325.
(20) "Polymer containing fluorinated polymer and boron compound" Hoeks, T.L.; Claesen, C.A.; Lohmeijer, J.H.G.; Pickett, J.E. 1997, U.S. Patent, 5,648,415.
(21) "Foamed polystyrene products and method for their production" Fujimori, T. 2002, U.S. Patent, 6,344,267.
(22) "Foam composites" Bracegirdle, B.; Monti, M.; Smyth, G.; Clarke, S.; McMillen, S.; Glyann, J.; Orpin, M. 2019, U.S. Patent Appl. 2019/0194411 A1.
(23) Liu, Q.; Chai, Y.; Ni, L.; Lyu, W. Flame Retardant Properties and Thermal Decomposition Kinetics of Wood Treated with Boric Acid Modified Silica Sol. *Materials.* **2020,** *13*(20), 4478. https://doi.org/10.3390/ma13204478
(24) Baghenri, S.; Alineja, M.; Ohno, K.; Hasburgh, L.; Arango, R.; Nejad, M. Improving Durability of Cross Laminated Timber (CLT) with Borate Treatment**.** *J. Wood Sci.* **2022**, *68*, 34.
(25) www.nordtreat.com/en/insights/clt-fire-protection-according-to-fire-class-b-s1-d0-requirements (NordTreat, Finland).
(26) Nine, J. M.; Tran, D. N. H.; Kabiri, S.; Losic., D. Graphene-Borate as an Efficient Fire Retardant for Cellulosic Materials with Multiple and Synergistic Modes of Action**.** *ACS Appl. Mater. Interface.* **2017**, *9*(11), 10160–10168.
(27) Ishii, T.; Kokaku, M.; Nagsi, A.; Nishita, T.; Kakimoto, M. Calcium Borate Flame Retardation System for Epoxy Molding Compound. *Polymer Eng. Sci.* **2006,** *46*(5): 799–806.
(28) "Materials for and manufacture of fire and heat resistant components" Crompton, G., U.S. Patent, 1992, 5,082,494.
(29) "Polyester composite material, preparation method therefor and application thereof" Chen, P.; Ye, N.; Wang, F.; Ding, C.; Feng, J.; Fu, X.; Wu, C.; Liu, X. 2023, WO 2023/116334 A1.
(30) Song, S.; Nai, X.; Li, W.; Zhu, C.; Meng, Q. Hydrothermal Synthesis of Calcium Borate Whiskers. *Adv. Mater. Res.* **2012,** *399–401*, 693–697.
(31) "Preparation and thermodynamic characterization of 2CaO?B_2O_3?H_2O nanomaterials with enhanced flame retardant properties" Geng, Y.-J.; Liu, Z.-H. **2017**, *Colloids and Surfaces A: Physical and Chemical Aspects*, 522: 563–568.
(32) "Epoxy resin composition and electrical/electronic part equipment" Haruomi, H. 2003, Japan. Patent 2003096272.
(33) "Flame retardant polyamide resin composition" Takeda, T.; Kirikoshi, K.H. 2002, U.S. Patent Appl. 2002/0002228.
(34) Zhang, L.; Liu, Z.-H. Preparation of $2MgO \cdot B_2O_3 \cdot 1.5H_2O$ Nanomaterial and Evaluation of Their Flame Retardant Properties**.** *J. Therm. Anal. Calorim.* **2017**, *129*, 715–719.
(35) "Reinforced and flame retarded thermoplastic resin composition and process for producing the same" Hirokana, H. 2001, U.S. Patent 6,221,947.
(36) "Zinc borate of low hydration and method for preparing same" Nies, N.; Hulbert, P. 1972, U.S. Patent 3,649,172.
(37) "Production of flame resistant epoxy resin composition" Kamikooryma, Y.; Hirose, M.; Koshibie S. 1994, Japan. Patent 06,107,914.
(38) "Halogen-free mixture, a self-extinguishing prepreg containing this mixture and the use of such a prepreg" Janowitz, O.; Walter, P. 1997, U.S. Patent 5,648,436.
(39) "Flexible intumescent coating composition" Nugent, Jr. R.M.; Ward, T.A.; Greiger, P.D.; Seiner, J.A. 1991, U.S. Patent 5,070, 119.
(40) "Low density light weight intumescent coating" Hanafin, J., Bertrand, D.C. 2000, U.S. Patent 6,096,812.
(41) Zeng, Y.; Erik Weinell, C.; Dam-Johansen, K.; Ring, L.; Kiil, S. Exposure of Hydrocarbon Intumescent Coatings to the UL1709 Heating Curve and Furnace Rheology: Effects of Zinc Borate on Char Properties. *Prog. Org. Coat.* **2019**, *135*, 321–330. https://doi.org/10.1016/j. porgcoat.2019.06.020
(42) Formicola, A. D.; De Fenzo, A.; Zarrelli, M.; Frache, A.; Giordano, M.; Camino, G. Synergistic Effects of Zinc Borate and Aluminum Trihydroxide on Flammability Behavior of Aerospace Epoxy System. *eXPRESS Polym. Lett.* **2009**, *3*(6), 376–384.
(43) "Flame retardant structural epoxy resin adhesives and process for bonding metal members" Eagle, G.G.; Lutz, A.; Jialanella, G.L. 2016, U.S. Patent 9,346,983.
(44) "Potential for the formation of respirable fibers in carbon fiber reinforced plastic materials after combustion" Eibl, S. **2017**, *Fire Mater.*, https://doi.org/10.1002/fam.2423.
(45) Braun, U.; Schartel, B.; Ficherra, M. A.; Jaeger, C. Flame Retardancy Mechanisms of Aluminum Phosphinate in Combination with Melamine Polyphosphate and Zinc Borate in Glass-Fiber Reinforced Polyamide 66. *Polym. Degrad. Stab.* **2007**, *92*, 1528–1545.
(46) "Flame retardant polycarbonate" Li, X. 2016, U.S. Patent Appl. 2016/0032097.
(47) "Silicone polyimide compositions with improved flame retardancy" Bhandari, Y.; Gallucci. R. 2013, U.S. Patent 8,349,933.
(48) Hull T. R.; Quinn R. E.; Areri, I. G.; Purser, D. A. Combustion Toxicity of Fire Retarded EVA. *Polym. Degrad. Stab.* **2002**, *77*, 235–242.
(49) "Flame retardant additive for polymers, free of halogens, antimony oxide, and phosphorus containing substance" Paul, S. 2015, U.S. Patent 8,975320.
(50) Wu, Q.; Guo, J.; Fei, B.; Li, X.; Sun, J.; Gu, X.; Li, H.; Zhang, S. Synthesis of a Novel Polyhydroxy Triazine-Based Charring Agent and Its Effects on Improving the Flame Retardancy of Polypropylene with Ammonium Polyphosphate and Zinc Borate. *Polym. Degrad. Stab.* **2020**, *175*, 109123.
(51) "Polyphenylene ether resin composition, prepreg containing the same, laminated plate, copper-clad laminated plate and printed circuit board" Satoshi, M.; Kazuhiko, O. 2009, Japan. Patent 4255759.
(52) "Electrical insulating material and method for preparing insulating material element" Wu, F.; Don, X.; Hillbourg, H.; Chen, J.; Rocks, J.; Fu, M.; Wang, S.; Weng, Y. 2017, U.S. Patent Appl. 2017/0250001.
(53) Song, J.; Huang, Z.; Qin Y.; Li, X. Thermal Decomposition and Ceramification of Ceramifiable Silicone Rubber Composite with Hydrated Zinc Borate. *Materials*, **2019**, *12*, 1591.
(54) Guo, H.; Chen, X.; Zhang, Y. Improving the Mechanical and Electrical Properties of Ceramifiable Silicone Rubber/Halloysite Composites and Their Ceramic Residue by Incorporating Different Borates. *Polymers.* **2018**, *10*, 388.
(55) "Intumescent coating composition", Wu, H.; Zhu, L.; Yang, L. 2020, U.S. Patent Appl. US 2020/0181429 A1.
(56) "Standard for rapid rise fire tests of protection materials for structural steel", UL 1709. 2017, Underwriters Laboratories, Northbrook, IL.
(57) Gautam, C.; Chelliah, S. Methods of Hexagonal Boron Nitride Exfoliation and Its Functionalization: Covalent and Non-Covalent Approaches. *RSC Adv.* **2021**, *11*, 31284.
(58) "3D printed component comprising a thermoplastic composite material and boron nitride, method for mak- ing a 3D printed component and use of 3D printed com- ponent"

Schadel, R.M.; Unibel, K.B.; Wildhack, S. WO 2020/004236 A1, assigned to 3M.

(59) Han, G.; Zhao, X.; Feng, Y.; Ma, J.; Zhou, K.; Shi, Y.; Liu, C.; Xie, X. Highly Flame-Retardant Epoxy-Based Thermal Conductive Composites with Functionalized Boron Nitride Nanosheets Exfoliated by One-Step Balling Milling. *Chem. Eng. J.* **2021**, *407*, 127099.

(60) Davesne, A.; Lazar S.; Bellayer S.; Qin S.; Grunlan J. C.; Bourbigot, S.; Jimenez, M.; Bourbigot, S.; Jimenez, M. Hexagonal Boron Nitride Platelet-Based Nano-Coating for Fire Protection. *ACS Appl. Nano Mater.* **2019**, *2*, 5450–5459.

(61) Liu, J.; Kutty, R.G.; Zheng, Q.; Eswarich, V.; Screejith, S.; Liu, Z. Hexagonal Boron Nanosheet as High-Performance Binder-Free Fire Resistant Wood Coatings. *Small.* **2017,** *13*, 1602456.

(62) "Heat resistant composite composition and method for making the same" Hiroaki, K.; Yoshio, B.; Chunyi, Z.; Tang, C.; Goldberg, D. **2008**, *Jpn. Kokai Tokkyo Koho*, JP 2008007699(A).

(63) "Resin composition, and prepreg and laminate sheet containing the same" Ohtsuka, H.; Ueyama, D.; Sogame, M. 2013, U.S. Patent Appl. U.S. 2013/0337269 A1.

(64) Weakley, T. J. R. Guanidinium Tetraborate. *Acta Cryst.* **1985**, *C41*, 377–379.

(65) Hoffendahl, C., Fontaine, G., Duquesne, S., Taschner, F., Mezger, M., Bourbigot, S. The Combination of Aluminum Trihydrate and Melamine Borate as Fire Retardant Additives for Elastomeric Ethylene Vinyl-Acetate. *Poly. Degrad. Stab.* **2015**, *115*, 77–88.

(66) "Borophosphate, borate phosphate, and metal borophosphate as flame proofing additives for plastics" Ewald, B., Hibst, H. 2014, U.S. Patent 8,841,373.

(67) Liu, L.; Hu, J.; Zhuo, J.; Jiao, C.; Chen, X.; Li, S. Synergistic Flame-Retardant Effects Between Hollow Glass Microsphere and Magnesium Hydroxide in Ethylene-Vinyl-Acetate. *Poly. Degrad. Stab.* **2014,** *104*, 87–94. "Flame retardant compositions for thermoplastic poly- mers consisting of porous, amorphous glass powder and melamine cyanurate" Hans-Juergen, V.; Ferner, U. 2015, (68) U.S. Patent Appl. US 2015/0018453 A1.

(69) "Nitrogen and phosphorus containing silanes as versatile flame retardant: not only for textiles" Mayer-Gall; Ali W.; Shabani, V.; Danielski, D.; Textor, T.; Guttmann, J.S. *FRPM23 Conference Proceedings in Switzerland,* June 28, 2023.

(70) Knights, K. W.; Nascimento, M. A.; Manners, I. An Investigation of Polyphosphinoboranes as Flame Retardant Materials. *Polymer.* **2022**, *247,* 124795.

(71) Ranilli, M.; Torre, L.; Kenny, J. M.; Natali, M. Effect of Boron Carbide Nanoparticles on the Thermal Stability of Carbon/Phenolic Composites**.** *Polym. Compos.* **2015,** *38*(9), 1819–1827.

(72) "Chapter 9. Graphene/polymer composites: a new class of fire retardant" Wang, X.; Guo, W., in Hu, Y.; Wang, X., eds., *Flame Retardant Polymeric Materials: A Handbook*, CRC Press, Taylor & Francis, pp. 151–180, **2020**.

(73) Gao, M.; Li, J.; Zhou, X. A Flame-Retardant Rigid Polyurethane Foam System Including Functionalized Graphene Oxide. *Polym. Compos.* **2018**, *40*(S2), 1–9.

10 Nitrogen-Based Flame Retardants

Christopher Simpson, Qingliang He, and Michael Kuepfert

10.1 INTRODUCTION

With the continued increase in usage of plastic/polymeric materials in our everyday lives conjoined with the need for such materials to be resistant to ignition, as well as fitting into the world of "sustainability," there is significant pressure and efforts to find new and useful flame retardant materials. Historically the main go-to flame retardant materials were based on mineral fillers (e.g., aluminum hydroxide ATH) and phosphorus or brominated small molecule compounds. However, as regulatory and NGO pressure grows for some of these classic materials, new developments are increasing towards the goal of "sustainability." The trends in the last decade or so are to migrate away from halogenated and/or small-molecule materials if possible, and on the market today are an array of polymeric and reactive flame retardant materials that help to satisfy the push towards more sustainable flame retardants and an area where nitrogen-containing compounds have some foothold and affords a deeper discussion. Nitrogen-containing compounds as flame retardant have shown significant importance in the plastic industry benefiting from versatile flame retardant mechanisms for a broad range of applications. Nitrogen-based flame retardants (NFRs) have shown to be advantageous in terms of low smoke generation, less toxic smoke, and less corrosive gas volatiles compared to halogenated flame retardants during combustion. As such, the various classes and compounds are discussed in detail below.

10.2 MAIN TYPES OF NITROGEN-BASED FLAME RETARDANTS

When identified by their mode of actions, NFRs are often categorized into two main groups. The first group includes chemical compounds that undergo endothermic decomposition reactions (absorb heat generated from the combustion) and releasing nonflammable gases such as nitrogen and ammonia (diluting oxygen and flammable gas volatiles generated from polymer degradation during a fire). The second group includes chemistries that during combusting proceed through characteristic condensation mechanisms and interactions with certain polymer matrices that result in polymer degradation or substantial charring (char formation to protect the underlying polymer from further degradation).[1] Either way, NFRs are efficient and effective flame retardants that interrupt the triangle of fire (fuel, oxygen, and heat) with advantages like delayed ignition, reduced heat release rate, lower smoke density, reduced smoke toxicity.

The most well-known NFRs are based on ammonia or melamine derivatives, and others from urea, guanidine, and phosphazene derivatives are also proven to be effective.[2] Even though melamine can be used as a flame retardant by itself (e.g., in flexible polyurethane foams) typically, ammonia and melamine based NFRs are in their salt forms, which results in higher thermal stability, lower solubility, and volatility. Typical examples include ammonium polyphosphate, melamine pyrophosphate, and melamine polyphosphate.

10.3 MELAMINE-BASED FLAME RETARDANTS

Melamine, melamine condensation products, and their respective salts are effective flame retardants in a variety of applications. Melamine and its derivatives combined with phosphorus-containing flame retardants (PFRs) demonstrate synergism that provides higher FR efficiency and less deterioration to the physicochemical properties of the polymer matrix like mechanical and electrical properties. The combination of NFR and PFR can also provide additional benefits like acid-base neutralization effects that lead to better processability during extrusion and injection molding conditions. Combinations of melamine and its derivatives used in conjunction with metal hydroxides[3] and halogenated flame retardants[4] can also be effective and show synergist effects.

Typically melamine undergoes endothermic decomposition thereby cooling the underlaying polymer and releasing nonflammable gases like water vapor, carbon dioxide, and ammonia.[5] These nonflammable gas volatiles dilute the combustible gas volatiles, effectively interrupting the sustainability of the flame. It is also known that this mode of action is independent from the polymer matrix type and can be broadly applied to many polymer systems. The rule of thumb is that the selection of such melamine derivatives depends on the processing temperature of different polymers and the sensitivity towards hydrolysis.[5] When melamine salts are used, the acid plays an important role in its overall flame retardant effectiveness. Typically, when melamine is reacted with acids like phosphoric acid, it provides strong dehydration reactions that promote char formation and polymer degradation *via* a catalytic effect of the acid.[6] The thermal stability of melamine salts strongly depends on the acid source,[5,6] which is an important criteria for FR selection. Synergistic FR systems can be obtained by combining melamine salts with other FRs of high thermal stability to withstand more stringent flame retardant requirement in high-end applications.

DOI: 10.1201/9781003380689-10

10.3.1 Melamine as Flame Retardant

Melamine (1,3,5-Triazine-2,4,6-triamine, CAS#: 108–78–1) is a white crystalline powder with a melting point of approximately 354°C and a density of 1.571 g/ml.[7] The manufacturing of melamine on large scale is based on heating dicyandiamide or urea in the presence of ammonia.[7] Melamine has low solubility in cold water and common solvents at room temperature but much higher solubility in hot water.[5,7]

Melamine undergoes sublimation at approximately 200°C, which limits its feasibility of being melt blended into many thermoplastics *via* extrusion processes at high temperatures. It does fit with other processes however, when melt blending is not necessary. A classic application of melamine as flame retardant is in flexible polyurethane foams as melamine can be dispersed in polyols. The incorporation of melamine into polyurethane occurs without excessive heating, so there is no concern for melamine volatilizing. When the melamine flame retarded polyurethane is ignited, the melamine sublimes and dilutes the fuel volatiles and oxygen.[5] Additionally, the sublimation energy of melamine and its endothermic decomposition acts as a heat sink during a fire scenario.

Typically high dosing of melamine (about 20–30 parts by weight polyol) is necessary to have meaningful flame retardance in flexible polyurethane foams used for upholstered furniture or mattresses, for instance, to pass the Crib 5 flammability test in the UK. PFRs such as dimethyl methyl phosphonate (DMMP)[8,9] or tris(2-chloroisopropylphosphate (TCPP)[10] are commonly used together with melamine in such application. Please see Chapter 31 for more information on flame retardancy of foam materials.

10.3.2 Melamine Condensates as Flame Retardants

As mentioned above, there is only limited usage of neat melamine as a flame retardant due to its sublimation at ~ 200°C; Which is a concern from environmental and occupational health perspectives. However, as an effective flame retardant with 67% nitrogen by mass, melamine serves as an important building block in the category of NFRs. When thermally treated at high temperature, melamine undergoes condensation and forms much more thermally stable products like melam, melem, melon, and ultimately graphitic carbon nitride.[11] Detailed chemical reactions of the melamine condensation processes are shown in Figure 10.1. Further details of the melamine condensation products and their salts was reviewed by Schwarzer.[12]

Melam, sometimes considered as the dimer of melamine, was first prepared in 1834 by Liebig.[12] The reaction used potassium thiocyanate and ammonium chloride. However, the crystal structure and spectroscopic characterization of melam and other melamine condensation products were not completed until recently by Schnick[13,14] with the commercial production of melam patented by DSM.[15–17] Briefly, in the commercial production process, melam is generated by thermally treating melamine in the presence of an acid, preferably toluene sulfonic acid at temperatures around 300°C.[17]

Melem, sometimes referred as the trimer of melamine, is another melamine condensation product upon heat treatment. Literature reports have demonstrated different routes for how melem is formed (e.g., thermolysis of melamine).[18,19] Alternatively, melam can be heated to generate melem with ammonia gas as a byproduct, shown in Figure 10.1.

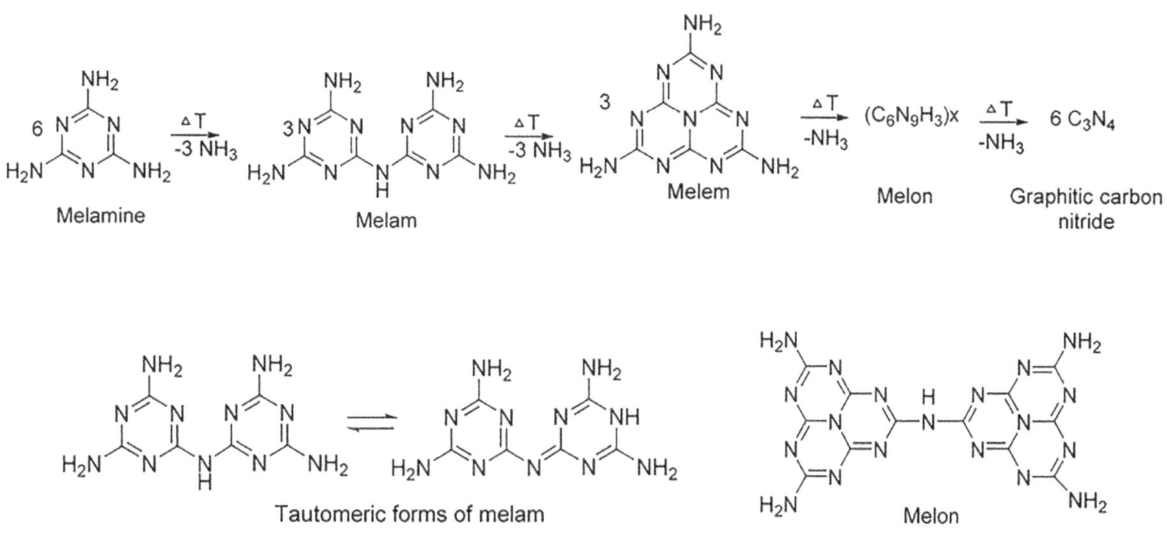

FIGURE 10.1 Thermal decomposition of melamine and formation of melamine condensates (melam, melem, and melon), and the formation of graphitic carbon nitride, tautomeric forms of melam, and chemical structure of melon.

Source: Adapted with permission from: Liu, X.; Hao, J.; Gaan, S. Recent Studies on the Decomposition and Strategies of Smoke and Toxicity Suppression for Polyurethane Based Materials. *RSC Adv.* 2016, 6 (78), 74742–74756. https://doi.org/10.1039/C6RA14345H.

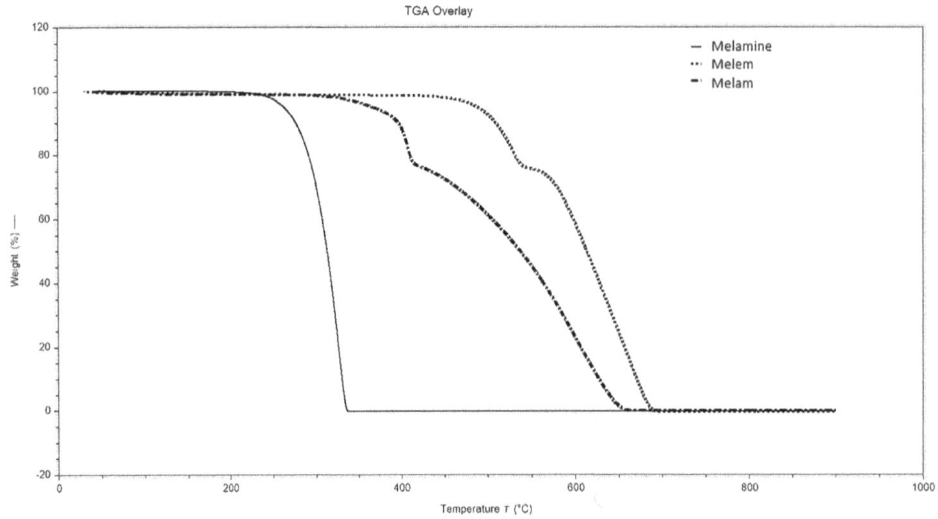

FIGURE 10.2 TGA results of melamine, melam, and melem (courtesy from LANXESS Corporation). Testing was done under nitrogen condition with a heating rate of 10°C/min.

TGA thermographs of melamine, melam, and melem are shown below in Figure 10.2. It is clear that the thermal stability of melem is higher than that of melam, which is much higher than melamine itself. Additionally, melam has a very low solubility in water, and melem is insoluble in water. The high thermal stability and low to no solubility of melam and melem make them suitable for high temperature flame retardant applications. Melam, melem, and their salts are used as flame retardants for engineering thermoplastics, especially polyamides. DSM reported the use of melam as a flame retardant in PA 6, PA 66, and PA 46.[20] Clariant reported the use of melam as synergist to aluminum diethylphosphinate[21] and aluminum phosphite[22] in glass filled PA 6, PA 66, and PBT. LANXESS also reported the use of melam and melem as synergists to a newly developed phosphonate based flame retardant product in glass filled engineering thermoplastics including PA6, PA66.[23,24]

Other patents also reported the use of melam and melem salts of different acids (phosphoric acid, polyphosphoric acid, or methane sulfonic acid) as flame retardants.[5] These salts were produced by thermally treating melamine and then neutralizing with the respective acid at temperatures above 300°C. A mixture of melamine and polyphosphoric acid, for example, was calcinated at 300°C–400°C for several hours to yield melem polyphosphate. Depending on the exact calcination conditions (time and temperature), predominately melam polyphosphate is formed.[5]

Melon, which can be formed by thermally treating melem, is another melamine condensation product, sometimes referred as dimer of melem. Due to its high thermal stability it has been rarely seen as an efficient flame retardant as it does not release when needed in fire events. Specifically, most of the polymer will have pyrolyzed before the melon has volatilized, thus creating a mismatch between polymer decomposition temperature and flame retardant activation temperature.

10.3.3 Melamine Salts as Flame Retardants

Aside from thermally treating melamine to form melamine condensation products with higher thermal stability, another method is to form melamine salts with different acid species to generate materials that can be processed at temperatures above 200°C (PP, TPU, PA, polyesters). The most well-known products are melamine polyphosphate (MPP) and melamine cyanurate (MCA).

When reacted with polyphosphoric acid or cyanuric acid, the formed salts MPP or MCA are much more thermally stable than melamine. MPP starts to decompose at about 350°C, and MCA shows initial decomposition around 300°C. Although they are quite similar in structure, the flame retardant mechanism between MPP and MCA are drastically different from each other.

10.3.3.1 Melamine Polyphosphate

Melamine phosphate flame retardants show endothermic cooling effects, fuel diluting capabilities, and dehydration effects during combustion with the potential for char formation. There are several melamine phosphorous salts like melamine phosphate, melamine pyrophosphate, and MPP. Detailed TGA results were summarized by Klatt.[5] Their thermal stability increases as follows: melamine phosphate < melamine pyrophosphate < melamine polyphosphate.

Melamine phosphate is the salt of melamine and phosphoric acid (chemical reaction is shown in Figure 10.3).[25] At around 200°C, melamine phosphate is converted to melamine pyrophosphate, and above 260°C, melamine polyphosphate is formed.[26] These transformations involve the release of water and lead to a heat sink effect. The chemical structure of MPP is shown in Figure 10.4.

When heated above 350°C, MPP undergoes endothermic decomposition with the release of polyphosphoric acid. The polyphosphoric acid serves as a barrier and prevents the underlying polymer matrix from fast decomposition.

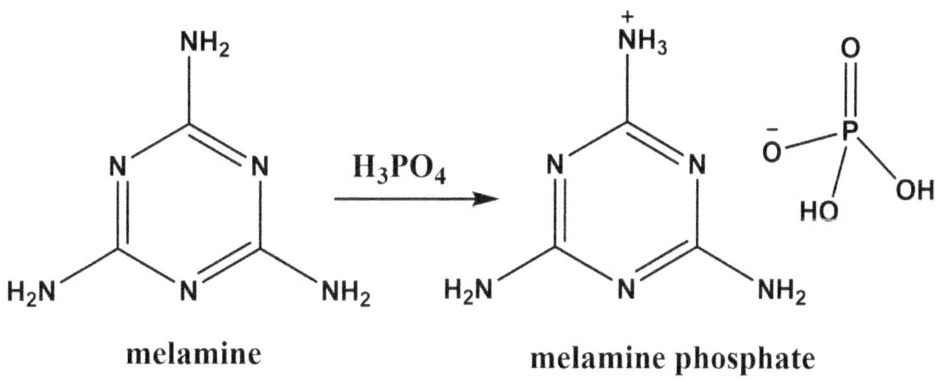

FIGURE 10.3 Melamine reacts with phosphoric acid to form melamine phosphate.

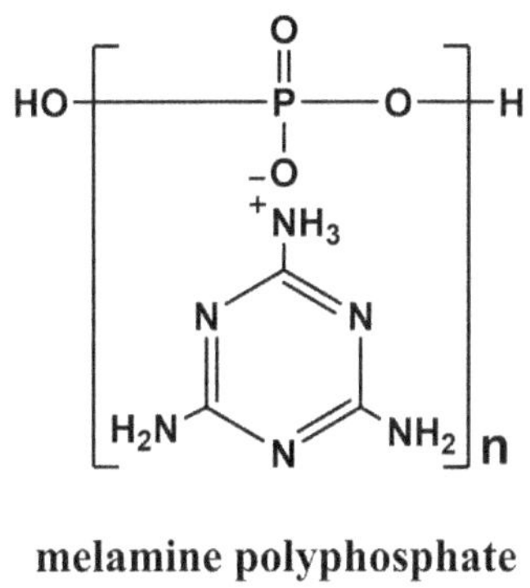

FIGURE 10.4 Chemical structure of melamine polyphosphate.

Additionally, char formation on the polymer surface reduces the amount of oxygen present at the combustion source and further protects the underlying polymer. Furthermore, melamine volatilizes and acts as a blowing agent, creating a protective intumescent char to further insulate the bulk polymer material. Benefiting from its high thermal stability, MPP can be used in both thermoplastic and thermoset applications. Commercial products are available from BASF (Melapur 200 with several different grades) and Budenheim (Budit® 341 and 342). Additionally, there are other MPP manufacturers in the Asian Pacific region. Hangzhou JLS Flame Retardants Chemical Co. is a manufacturer of both a neat powder-form of MPP and a MPP masterbatch. MPP masterbatches offer the advantages of being dust-free and having better dispersion capabilities, which improves the FR efficiency and decreases the chance of blooming compared with the neat powder. Overall better processability, improved performance (reduced energy consumption and higher throughput), and lower cost can be achieved with the usage of an MPP masterbatch.[27]

There have been many reports with MPP being utilized as a standalone flame retardant in polyamides, however, its efficiency is rather low. Others have reported synergistic effects between MPP and other P-containing flame retardant systems, including but not limited to aluminum diethylphosphinate,[21] aluminum phosphite,[22] and phosphonate[23] based FR systems.

Recently, Budenheim reported on a new type of MPP where the polyphosphate salt contains cations including melamine and at least one condensation product of melamine such as melam, melem, or melon.[28] They claimed that this new MPP improves thermal stability, processability, and has synergistic effects with other PFRs, for instance, aluminum diethylphosphinate (DEPAL).

Finally, there is another new class of melamine polyphosphate with the brand name Safire® that is based on melamine metal polyphosphate. This unique system contains both N- and P-elements and was originally developed by Floridienne Chimie, which is now part of Huber Engineered Materials. It was claimed that melamine poly(zinc phosphate) under the brand name Safire® 400 has synergistic effect with phosphinates and other halogen-free flame retardant systems with added benefits (reduced smoke generation and toxicity).[29]

10.3.3.2 Melamine Cyanurate

Melamine cyanurate [1,3,5-Triazinane-2,4,6-trione—1,3,5-triazine-2,4,6-triamine (1/1), with the chemical formula of $C_6H_9N_9O_3$ (or $C_3H_6N_6{\cdot}C_3H_3N_3O_3$); CAS#: 37640–57–6], also known as melamine-cyanuric acid adduct or melamine-cyanuric acid complex, is a crystalline compound formed from a 1:1 mixture of melamine and cyanuric acid.

The substance is not a salt despite its name. The complex is held together by an extensive two-dimensional network of hydrogen bonds between the two compounds, reminiscent to that seen in DNA base pairing, as illustrated in Figure 10.5. The network of hydrogen bonds in melamine cyanurate is responsible for its higher thermal stability compared to pure melamine and its insolubility in common solvents.[5] Additionally, the likelihood of melamine cyanurate migrating out of polymer matrices where this hydrogen bonding can occur between the polymer and melamine cyanurate is limited.

Melamine cyanurate is a suitable flame retardant for polymers with processing temperatures up to 300°C, including polyamides (PA6), TPU, and polyesters. Melamine cyanurate starts to decompose at approximately 300°C, which limits its use for polymer systems like PA66 with

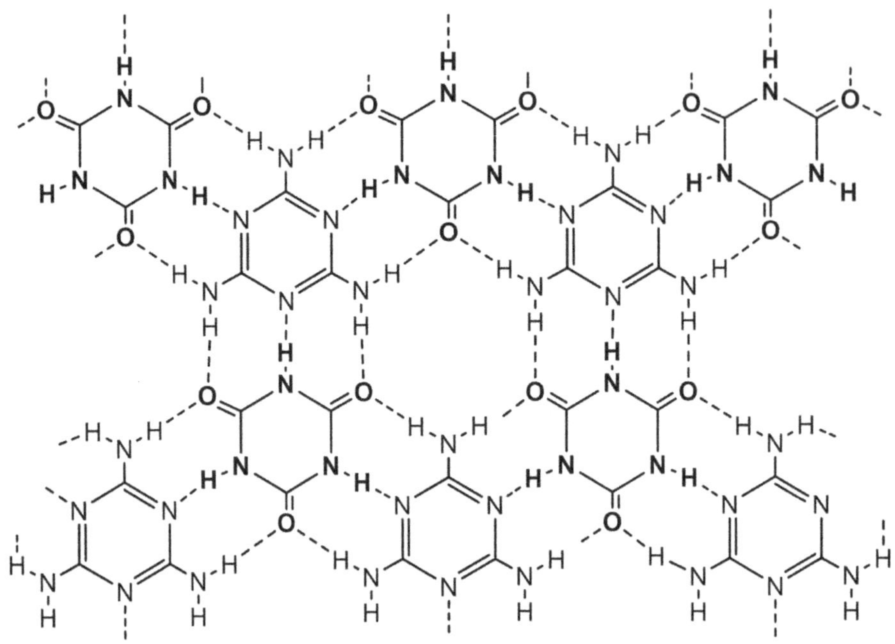

FIGURE 10.5 Chemical structure of melamine cyanurate and its hydrogen-bonding network.

processing temperatures that are known to reach 330°C.[5] Conversely, it can be used in flame retardant PA6 copolymers (PA6/PA66) or in PA6 polymer blends (PA6 with PA66). Commercial products are available from BASF (Melapur MC25 and MC50 as neat powder-form products and MCXL as a low dust granule version); other manufacturers like Hangzhou JLS Flame Retardants Chemical Co. offer JLS MC15, MC25, and MC50 in a neat powder-form, as well as a granule grade MCG.

Above 320°C melamine cyanurate undergoes endothermic decomposition to form melamine and cyanuric acid, which further decompose to ammonia, water, and carbon dioxide at higher temperatures. Melamine cyanurate acts as a heat sink during decomposition. The vaporized melamine and other inert gas products are capable of diluting oxygen and the fuel gas volatiles decomposing from the polymer during pyrolysis. Interestingly, melamine cyanurate has some additional effects if used in polyamides. When used as a stand-alone flame retardant in polyamides, it is one of the most efficient flame retardant systems in unreinforced PA6 and PA6/PA66 copolymers. At 6–10% dosing level, melamine cyanurate can achieve a UL-94 V-0 rating at thicknesses of 0.4 mm. It is theorized that this outstanding performance is related to a so-called "run-away" mechanism.[5]

During a typical UL-94 vertical test, melamine cyanurate starts to decompose immediately, with the developments of small gas bubbles indicating an intumescent effect, while no char formation is noticed under these conditions. Subsequently, small molten droplets start to form during the second flame application period. Due to the large amount of noncombustible gas volatiles (water, ammonia, and carbon dioxide) formed by the decomposing melamine cyanurate, there is no flaming on these small molten droplets. These droplets are capable of carrying away the energy so quickly that the flame is ceased by the heat sink/diluent effect of melamine cyanurate before reaching the cotton pad below, thus achieving a UL-94 V-0 rating. See Figure 10.6 for more details.[30]

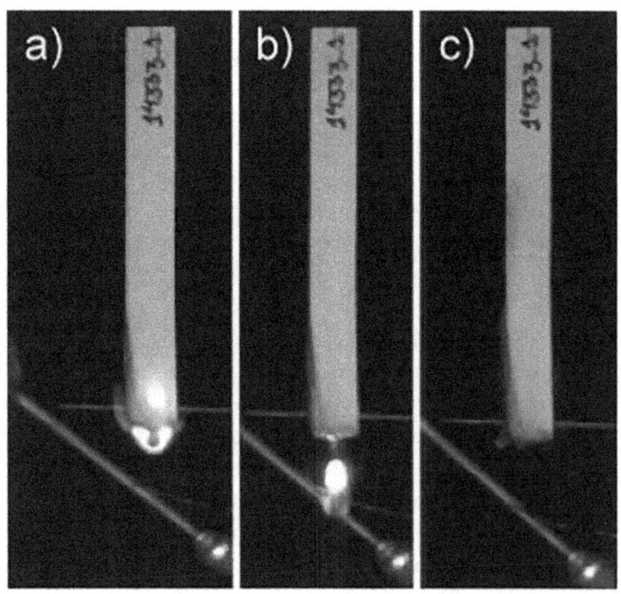

FIGURE 10.6 PA6 with melamine cyanurate dripping behavior: (a) pre-drop, (b) drop, (c) specimen extinguished.

Source: Adapted with permission from: Turski Silva Diniz, A.; Huth, C.; Schartel, B. Dripping and Decomposition under Fire: Melamine Cyanurate vs. Glass Fibers in Polyamide 6. *Polymer Degradation and Stability* 2020, *171*, 109048. https://doi.org/10.1016/j.polymdegradstab.2019.109048

This combination of the heat sink/diluent effect of melamine cyanurate and the runaway mechanism is very specific to polyamide, owning to the cyanuric acid degradation/chain scission.[5]

This differs considerably when discussing glass fiber reinforced polyamides. The fibers form entanglements (structural support) that limit the melt flowability of the polymer. As a result, much larger droplets are formed during the UL-94 testing. As these larger droplets contain high energy, the heat sink/diluent effect of melamine cyanurate is not sufficient enough to quench the flame before they reach the cotton pad during the UL-94 vertical burning test. One solution is to use short glass fiber, milled glass fibers, or fibrous fillers like wollastonite as reinforcing agent to polyamides. This way, the heat sink/diluent effect of melamine cyanurate is still sufficient enough to counteract the limited entanglement effect from short-length glass fibers or other short fibers to allow fast enough melt drippings. Therefore, at minimum a UL-94 V2 rating is expected.

Aside from the unique mechanism in polyamides, melamine cyanurate shows limited flame retardancy in other polymer systems. Synergistic effects can be achieved if melamine cyanurate is combined with PFRs. A typical example is the combination of melamine cyanurate with aluminum diethylphosphinate in polyesters[31] and TPU.[32]

Recent academic research has focused on utilizing melamine cyanurate in conjunction with novel flame retardants like polyphosphamide-containing triazine in PA6,[33] piperazine pyrophosphate in PP[34] and LLDPE,[35] as well as combination with magnesium hydroxide and phytic acid complex.[36] In the polyphosphamide-triazine combined with melamine cyanurate in PA6 system,[33] the optimal flame retardant effect was achieved by gas phase flame inhibition and physical shielding effects of char layers in the condensed phase. Particularly, in the gas phase, P-containing radicals generated by polyphosphamide pyrolysis are believed to quench the free-radical chain reaction during burning and lead to effective flame inhibition, as illustrated in Figure 10.7.

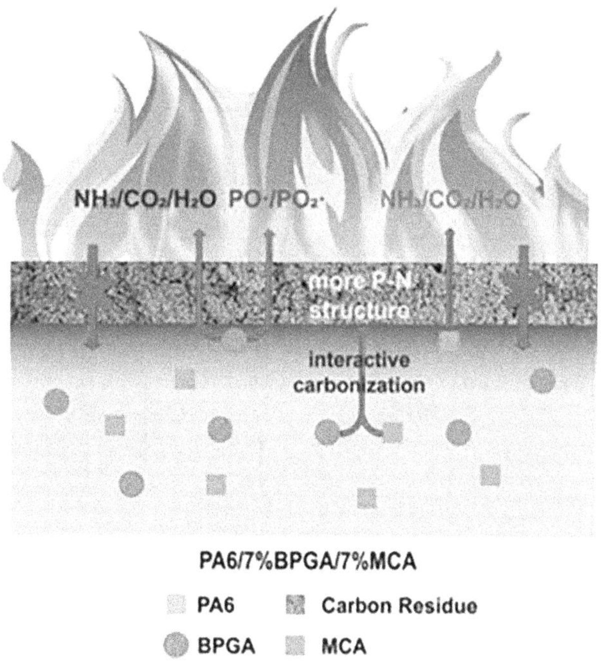

FIGURE 10.7 The flame retardant mechanism of polyphosphamide (BPGA)/triazine/melamine cyanurate system in PA6: gas and condensed phase flame retardant effect.

Source: Adapted with permission from: Shan, H.; Yan, L.; Xu, B.; Wang, D.; Wu, M. Polyphosphamide Containing Triazine and Melamine Cyanurate for Flame-Retardant PA6. *ACS Appl. Polym. Mater.* 2023, *5* (7), 5322–5333. https://doi.org/10.1021/acsapm.3c00732

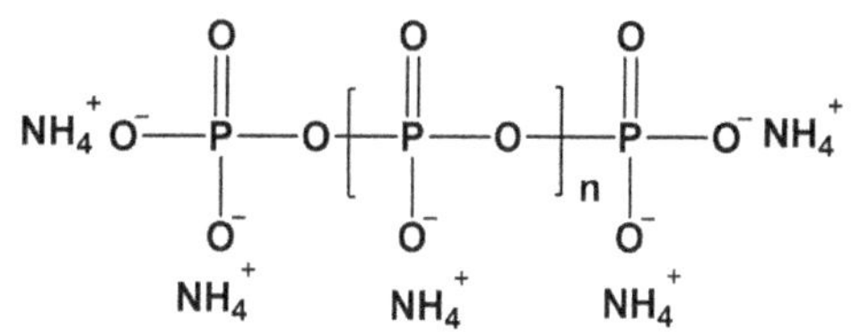

FIGURE 10.8 Chemical structure of linear APP.

10.4 AMMONIA-BASED FLAME RETARDANTS

Ammonia-based flame retardants are another efficient NFR class and are used in large volumes owing to market conditions and economics. Typically, ammonia-based flame retardants undergo endothermic decomposition while releasing ammonia and leaving behind an acid that causes charring.[5] There are only a few commercial ammonia-based flame retardants used today, the most prominent being ammonium polyphosphate (APP), followed by ammonium pentaborate and ammonium sulfamate.

10.4.1 Ammonium Polyphosphate

APP (chemical formula: $H[NH_4PO_3]_nOH$; CAS#: 68333-79-9) is the inorganic salt of ammonia and polyphosphoric acid with both linear and branched structures. APP consists of an orthophosphate anion neutralized by an ammonium cation leaving two hydroxy groups free to polymerize. In case of branching, some monomers have no ammonium anion and instead are linked to three other monomers.[5]

The properties of APP depend on the number of monomers in each molecule and the degree of branching. Shorter chains ($n < 100$) are more water sensitive and less thermally stable than longer chains ($n > 1000$).[37]

There are two main families of APP: Crystal phase I APP (APP I) and crystal phase II APP (APP II) (Figure 10.8 and Figure 10.9). Crystal phase I APP (APP I) is characterized by a variable linear chain length, possessing a lower decomposition temperature (around 150°C) and higher water solubility than APP II. In APP I, "n" is generally lower than 100.

As shown in Figure 10.9, the APP II structure can be cross-linked or branched. The molecular weight is much

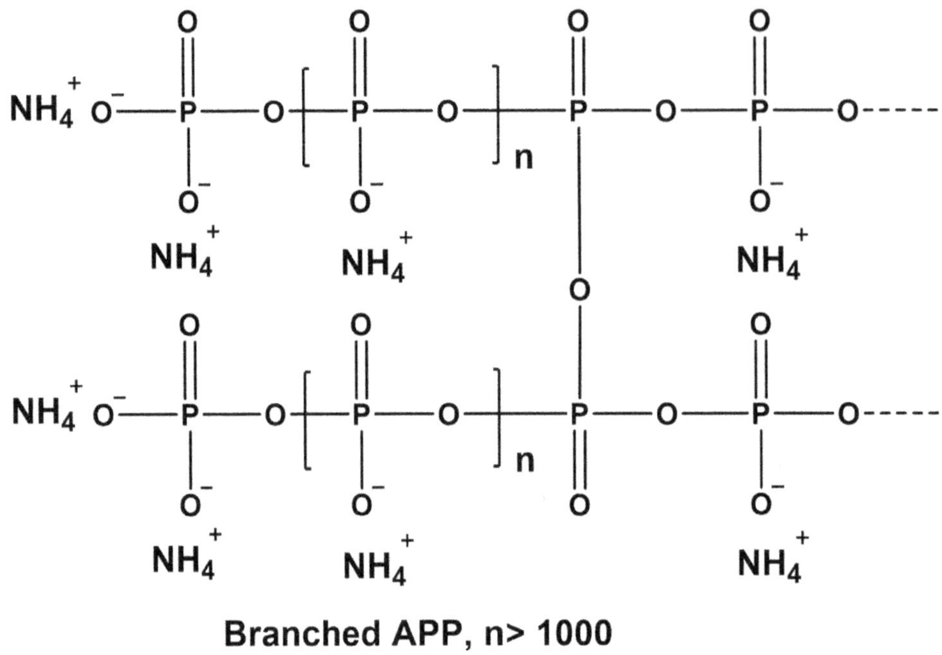

FIGURE 10.9 Chemical structure of branched APP.

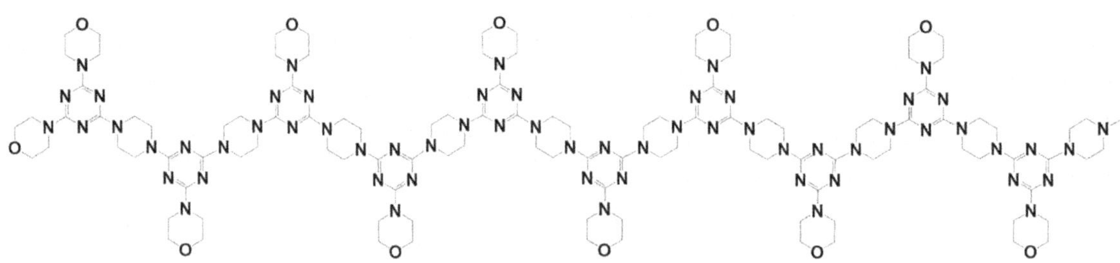

FIGURE 10.10 Chemical structure of MCA® PPM Triazine HF.

higher than APP I with "n" greater than 1,000. APP II has a higher thermal stability (decomposition starts at approximately 300°C) and lower water solubility than APP I.

Typically, APP acts as an intumescent flame retardant with broad application in thermoplastics, thermosets, foams, and coatings. One of the most important areas is the use of APP in polyolefins for wire and cable applications. APP starts to decompose at high temperatures to form polymeric phosphoric acid and ammonia. This endothermic decomposition process has a strong cooling effect, and the noncombustible gas volatiles (ammonia and water) subsequently dilute the combustible gases evolved from the polymer decomposition. Polyphosphoric acid then reacts with hydroxyl groups in the polymer matrices or added synergist to form an unstable phosphate ester. Subsequently, the dehydration of the phosphate ester continues, which leads to an intumescent char layer formed on the surface. The foamed char layer then acts as insulation, preventing further decomposition of the underlying polymer. Addition of synergistic products like pentaerythritol and its derivatives, carbohydrates, and foaming agents (e.g., melamine) significantly improve the flame retardant performance of APP.[5]

Although, APP II has much lower water solubility than APP I, further surface treatment is often necessary for end applications that require better water resistance. For example, melamine-formaldehyde resin,[38] silicone,[39] poly-siloxane,[40] silica gel,[41] and ethanolamine[42] are reported as coating materials to improve the water resistance of APP. Additionally, other functions have been introduced in APP flame retardant systems with multilayered coatings like nano-scale SiO_2/allyl polyoxypropylene ether/boron layers,[43] and zirconium-based metal-organic framework/tetraethyl orthosilicate/silane layers.[44] Overall, Si-, N-, and P-synergism can be achieved with these complex coating systems.

10.5 CYANURIC ACID–BASED FLAME RETARDANTS

Another unique class of flame retardants is based on cyanuric acid. Cyanuric chloride reacts stepwise with alcohols and amines to form the corresponding esters or amides.[5] The reaction with one mole of morpholine leads to a bifunctional cyanuric chloride derivative which can be polymerized with bifunctional amines or esters. MCA Technologies GmbH (Switzerland) produces a polymer with morpholine and piperazine with the following formula marketed under the trade name MCA® PPM Triazine HF (Figure 10.10).

PPM Triazine HF is reported to have a molecular weight of ~2755 g/mol and a nitrogen content of about 31%. Some work has been done on interactions and synergies with other flame retardants like aluminum diethyl phosphinate.[45]

10.6 NITROGEN-CONTAINING RADICAL GENERATORS

When examining nitrogen-containing flame retardant materials, one class has some interesting properties that have been leveraged for use in polymeric materials typically in polyolefins.[46,47] These are the nitrogen-containing radical generators, some examples shown Figure 10.11, including N-alkoxy hindered amine light stabilizers (e.g., Flamestab® NOR 116, TEMPO, Tinuvin 152) and azoalkane (e.g., Azobisisobutyronitrile AIBN) materials.

While these moieties and materials have a longstanding history of usage in polymer science as free radical initiators and controlling agents for radical polymerization, with the latter control achieved utilizing the persistent radical effect (PRE)[48,49] and use in the UV-stabilization of polymeric coatings, thin films, and fibers[50,51] (Figure 10.12). Their usage within the fire sciences is, relatively speaking, a young area of research, with currently very few commercially produced and sold materials available, Flamestab® NOR 116 from BASF (originally developed at Ciba Specialty Chemicals) being the most typical NOR type FR used today.

The NOR-type materials were seemingly designed or discovered to have FR efficacy for polypropylene (PP).[50] The prevailing mode of action discussed at the time was effectively to generate reactive radicals upon heating that would assist in the quick depolymerization (chain scission) of the PP, thus allowing the material to quickly shrink or

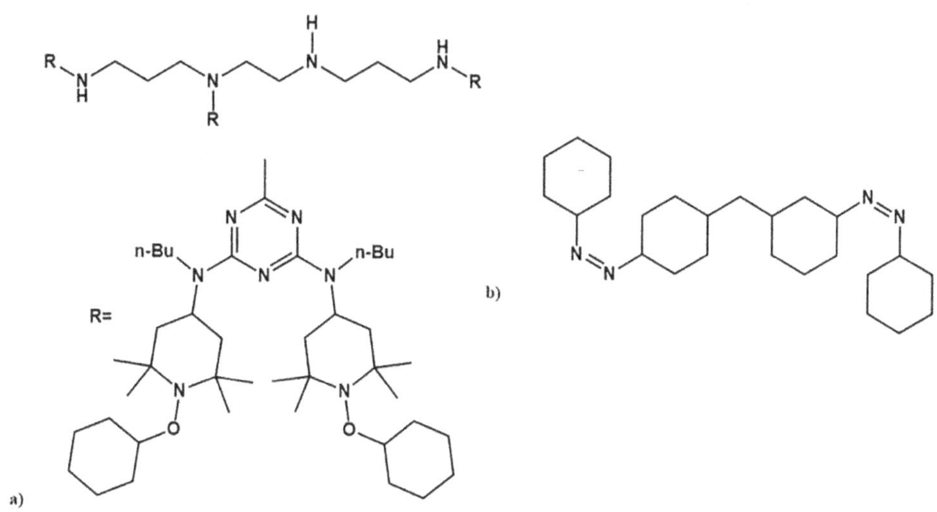

FIGURE 10.11 Examples of nitrogen-containing radical generators: (a) Flamestab® NOR® 116, (b) 4,4'-bis(cyclohexylazocyclohe xyl)methane.

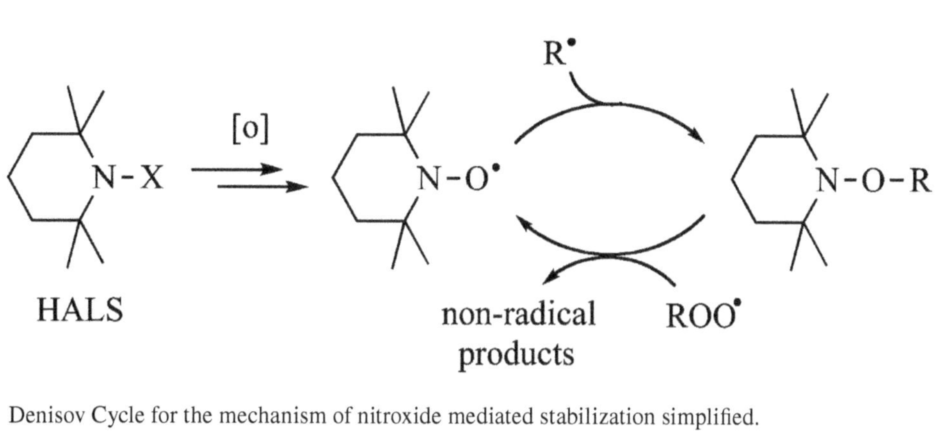

FIGURE 10.12 Denisov Cycle for the mechanism of nitroxide mediated stabilization simplified.

Source: Adapted with permission from: Hodgson, J. L.; Coote, M. L. Clarifying the Mechanism of the Denisov Cycle: How Do Hindered Amine Light Stabilizers Protect Polymer Coatings from Photo-Oxidative Degradation? Macromolecules 2010, 43 (10), 4573–4583. https://doi.org/10.1021/ma100453d. Copyright 2010 American Chemical Society.

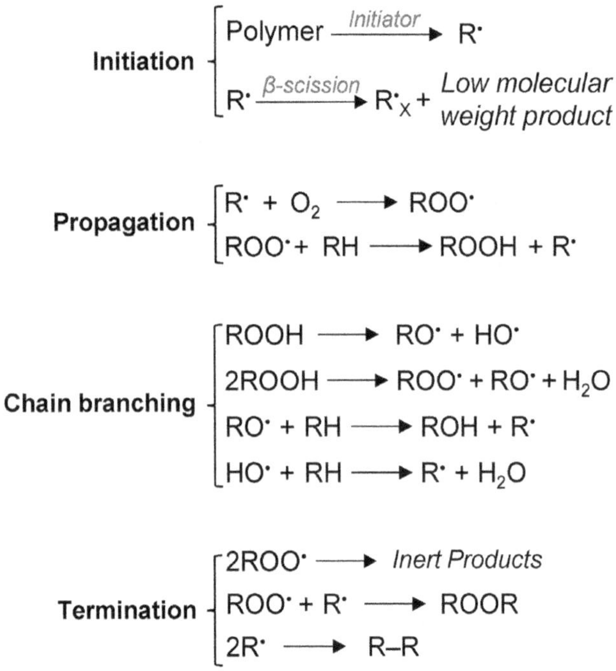

FIGURE 10.13 Degradation mechanism of polypropylene.

Source: Adapted with permission from: Nguyen, T. A.; Lim, H. M.; Kinashi, K.; Sakai, W.; Tsutsumi, N.; Okubayashi, S.; Hosoda, S.; Sato, T. Spin Trapping Analysis of Radical Intermediates on the Thermo-Oxidative Degradation of Polypropylene. Polymers 2023, 15 (1), 200. https://doi.org/10.3390/polym15010200.

melt drip away from the heat source without combustion occurring.[46,52] This theory being supported by the knowledge that alkoxy radicals can help to promote chain scission in polypropylene or cross-linking in polyethylene.[53–56] (Figure 10.13, where initiator can be alkoxy radicals.)

Previous studies discussed the relationship of NOR 116 concentration or sample thickness on FR efficacy. The former demonstrating in the NFPA 701 (1989) test on compression molded polypropylene films; 1% NOR load levels proved effective, while 5% and 10% failed the trails.[52] The supposition at the time, due to the organic nature of the material, higher loadings increased the amount of combustible materials present and therefore had a deleterious effect at higher concentrations. However, it is probable that at higher concentrations there is an increased likelihood of radical coupling in the insipient stages of the homolytic bond cleavage.

As there are two potential pathways for dissociation in Figure 10.14, it is possible that if the formation of alkyl radicals is slightly favored, generating a high concentration of alkyl radicals will lead to terminal irreversible coupling rather than the formation of alkoxy/nitroxyl radicals known to support the chain scission mechanism required for the polypropylene films to "run-away" from the heat source. The latter demonstrated that for thicker samples, the efficiency of chain scission and "run-away" are decreased; as such, hindered alkoxy amines cannot be utilized alone in these applications and are oftentimes combined with other flame retardant materials (e.g., PFR).[52] Conversely Aubert et al. have shown that certain cyclic azo compounds can beneficially be utilized in thick PP plaques as an effective flame retardant.[57] They examined a number of azo compounds looking for relationships between their structure and efficacy, the most efficient FR identified is seen in Figure 10.11b. These materials thermally decompose to release nitrogen and generate active radical species that participate in the decomposition of the polypropylene via the "run-away" effect.

Going beyond the chain scission and "run-away" effect, other researchers have attributed multiple roles played by

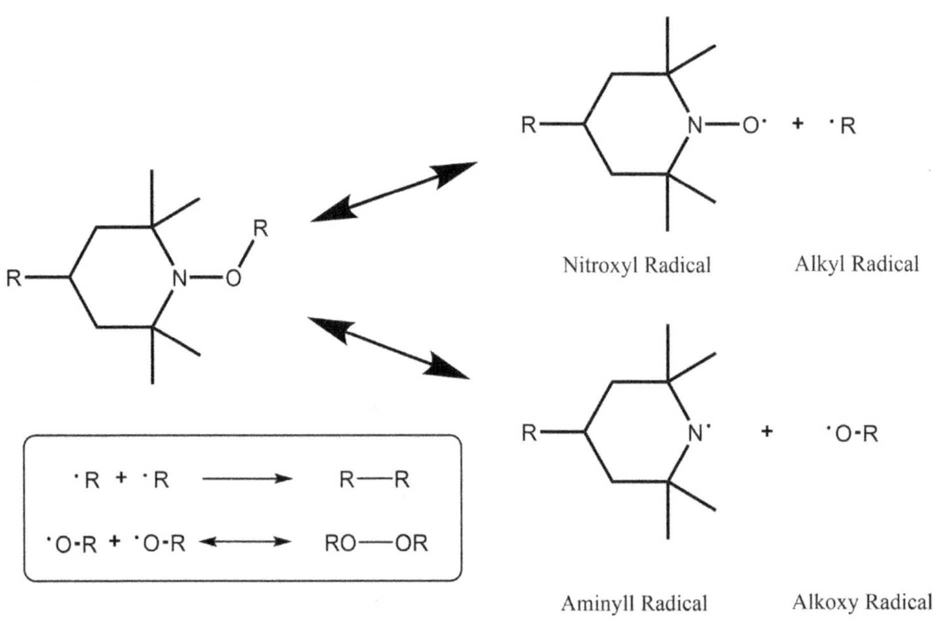

FIGURE 10.14 Degradation pathways.

NOR type materials in the FR space. These include gas phase activity as well as synergistic effects with halogenated or non-halogenated FRs.[58] Pfaendner had discussed and shown that the alkoxy/nitroxyl radicals play a dual role. First with the typical polypropylene degradation but also in the gas phase radical trapping mechanism, highlighting the synergistic effects when combined with halogenated flame retardants, allowing for decreased overall loading of flame retardants.[58] Xie *et al.* have also demonstrated the dual nature of the NOR type materials in combination with ammonium polyphosphate, where nitroxyl radicals contribute to gas phase radical trapping (PRE) in addition to promoting the formation of stable intumescent char in the condense phase.[59,60] There has been some research and indication that NOR type materials can also be utilized in conjunction with cyclic phosphonate compounds[61] (e.g., LANXESS' Amgard CU or Thor's Aflammit PCO 900).

Unique Si containing NOR materials[62] have been synthesized and utilized in recent studies. The authors were able to highlight the benefits of the nitroxyl radical in the gas phase mechanism in addition to the other benefits from the alkoxy radicals in the condensed phase.[63] These silane containing NOR materials were examined in as part of an intumescent flame retardant system in PP and demonstrated an increase in LOI values and overall improvement in the intumescent behavior as well as providing a UL-94 V-0 rating, and noted that at concentrations above 1% NOR compound, the system shifts to a UL 94 V-2, again holding that low levels of NOR type materials are needed. Their proposed mechanism is seen in Figure 10.15.

Beyond the traditional polyolefin usage, there has been an uptick in research and development via the intellectual property landscape to demonstrate combinations of nitrogen radical generators in other system such as in styrenics,[64–67] polyurethanes,[68] epoxy resins,[69–71] adhesives,[72] polyester,[73] polycarbonate,[74] as well as continued research into the polyolefin area.[75] The researchers have found successful combinations of NOR type materials with various phosphorus and bromine-containing flame retardants (e.g., phosphinate, phosphonate, APP, hypophosphite, et al.)

While there has been significant academic advancement in the area of nitrogen-containing radical generators in the last decade, inclusive of the extension to materials beyond polyolefins and as a synergist in combination with halogenated or non-halogenated FRs, there remains very few commercially available options for widespread adoption in the FR space at this time. As research and advances continue, hopefully so will the presence of other materials in the market.

Additional details on this class of flame retardants is covered in Chapter 11 of this book.

10.7 DIETHYL N,N-BIS(2-HYDROXYETHYL) AMINOMETHYLPHOSPHONATE

Diethyl N,N-bis(2-hydroxyethyl)aminomethylphosphonate was invented in the 1960s at Victor Chemical Works and is today commercially sold as Levagard® 4090N (LANXESS) and Fyrol® 6 (ICL) (Figure 10.16). The presence of the two

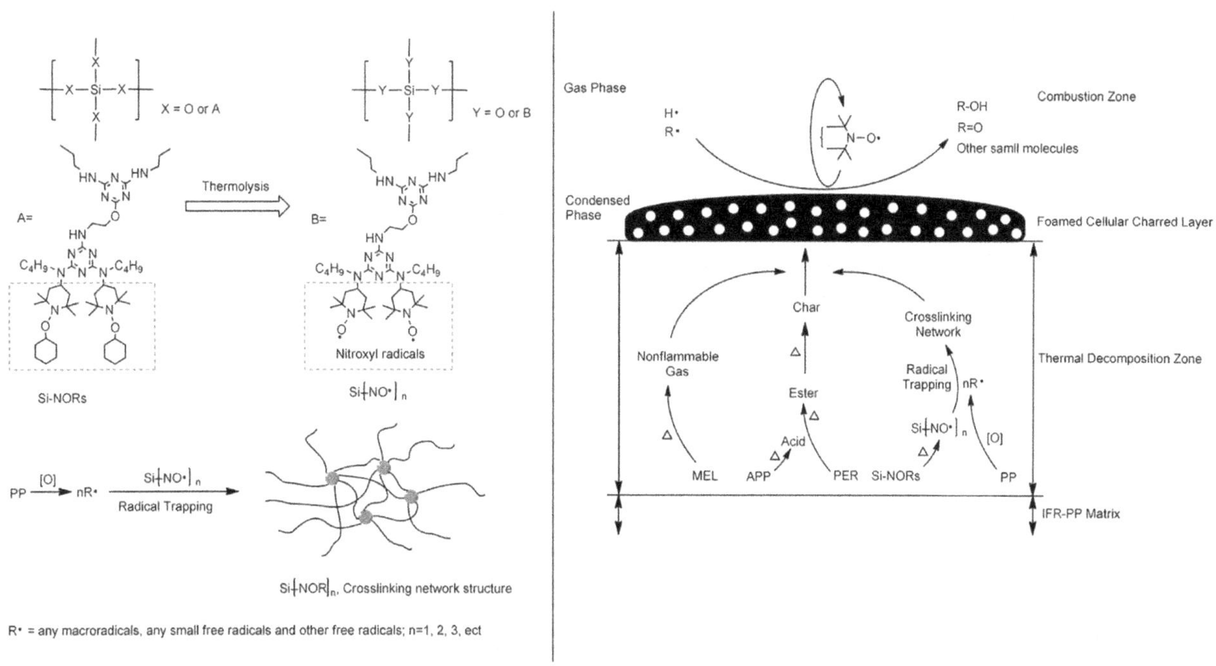

FIGURE 10.15 Proposed silane-NOR mechanism.

Source: Adapted with permission from: Cao, K.; Wu, S.; Qiu, S.; Li, Y.; Yao, Z. Synthesis of N-Alkoxy Hindered Amine Containing Silane as a Multifunctional Flame Retardant Synergist and Its Application in Intumescent Flame Retardant Polypropylene. Ind. Eng. Chem. Res. 2013, 52 (1), 309–317. https://doi.org/10.1021/ie3017048.

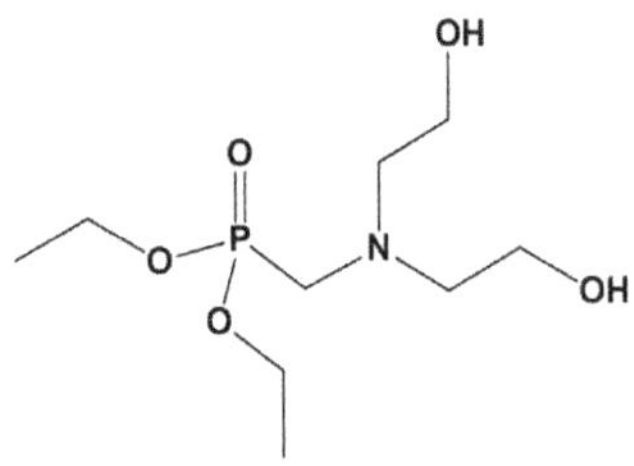

FIGURE 10.16 Chemical structure of diethyl N,N-bis(2-hydroxyethyl)aminomethyl phosphonate.

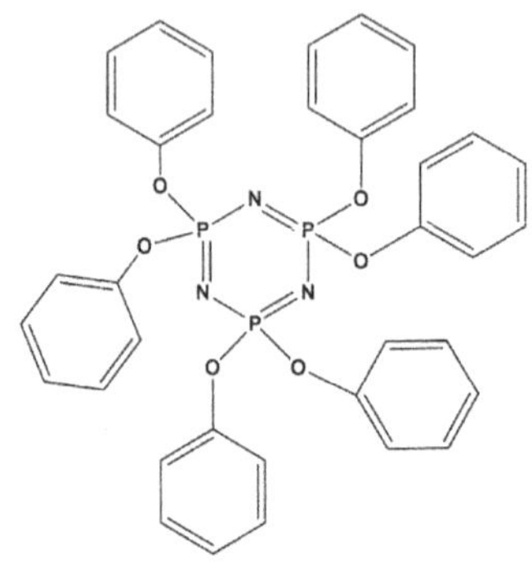

FIGURE 10.17 Chemical structure of hexaphenoxytricyclophosphazene.

primary hydroxyl groups renders this diol phosphonate reactive to be covalently incorporated into PU and thermosets. Because of the molecule's reactive nature, migration and leaching out of the polymer matrix is avoided, and therefore long-term retention of flame retardancy can be assured. The covalent incorporation of the FR is especially important in applications that are subjected to long-term heat exposure, as for example in PU foam in stadium domes or attics. Other applications include insulation foams for large refrigerators. The presence of the amino group contributes to the catalytic activity during foaming, which is a practical advantage in spray foams, and also may have some buffering (stabilizing) effect on the finished foam. Partially replacing TCPP with Levagard® 4090N in some spray foam formulations can lead to faster curing than in conventional spray foams.[76] Rigid polyurethane and isocyanurate foams are sometimes formulated with a blend of triethyl phosphate (LANXESS Levagard® TEP) and triaryl phosphates (LANXESS Disflamoll®) and can be combined with Levagard® 4090N. Due to the high OH number of the diol phosphonate, it is not used in flexible PU foams, except in high density foams at low concentrations.

Phosphorous-containing chemicals', such as 4090N, main mode of action is a condensed phase mechanism producing relatively incombustible carbonaceous char.[77] The flame retardant mechanism of the diol phosphonate has been studied in waterborne PU coatings for potential textile and wood applications.[78–80] For example, Sun et al. prepared a diol phosphonate-containing waterborne PU coating for polyester fabric and analyzed the char residue through energy dispersive X-ray spectroscopy (EDS).[79] While the analysis showed condensed phase activity as the char surface was enriched with carbon and phosphorous, the mass concentration of nitrogen decreased, indicating the formation of some nonflammable gases and synergistic activity. The diol-phosphonate also finds application in epoxy-based adhesives and coatings.[81,82]

10.8 PHOSPHAZENES

Cyclic phenoxyphosphazenes are compounds containing alternating phosphorus and nitrogen in their six-membered ring structures, which makes them attractive candidates as flame retardants with inherent synergistic properties in addition to exhibiting high thermal and hydrolytic stability. Hexaphenoxytricyclophosphazene is commercially available from Fushimi Pharmaceutical Co. or Otsuka Chemical Co. or can be synthesized from hexachlorocyclotriphosphazene (HCCP) (Figure 10.17). Cyclic phenoxyphosphazenes have low polarity and favorable electrical properties and therefore have found use in high end epoxy,[83] cyanate ester,[84] benzoxazine,[85] and PPE[86] based printed wiring boards. Loading levels of cyclic phenoxyphosphazenes are often low because of its low melting point (110°C) to avoid excessive resin flow in the lamination process, and therefore, other flame retardant types are often added to formulations. They have also been applied in PC/ABS blends at 15 wt% loading to achieve a UL 94 V-0 classification.[87] Due to the relatively high cost compared to traditional biphosphates, their use in PC/ABS is limited. Further formulation work with cyclic phenoxyphosphazenes has been done in PC,[88–91] PC/PLA/ABS blends,[92] PLA/PC and PLA/polyphosphonate-co-carbonate blends,[93] rigid polyurethanes,[94] PPE,[95] and polyamides.[96,97]

Extensive research has focused on generating cyclotriphosphazene-based epoxy resins, recently reviewed by Dagdad et al.[98] In subsequent studies, cyclotriphosphazenes were functionalized with vanillin-triazoles[99] and benzimidazolyl[100] to achieve UL 94 V-0 classification and a reduction in total heat release (THR) and peak heat release rate (PHRR) in epoxy resins. Selvaraj et al. prepared hydroxyl-terminated cyclotriphosphazene (HTCP) from HCCP and 4,4'-sulfonyl diol and incorporated it at 1, 3, and 5 wt% into hybrid composites based on renewable cardanol-benzoxazine blended epoxy resins.[101] Thermal characterization showed an increase in the glass transition temperature, thermal stability, and char yield with increasing HTCP content. A simple Bunsen burner flame test carried out on thin films showed a significant reduction in burn time as a function of higher weight percentages of HTCP. Benzoxazine-epoxy samples not containing HTCP also showed improved flame retardancy, indicating that benzoxazine itself also functions as flame retardant. Finally, corrosion studies with this epoxy resin coated on MS plates showed a decrease in corrosion rate, indicating possible application as a coating material with anti-corrosion properties. Cyclotriphosphazenes modified with siloxane[102] and phenolic hydroxyl groups[103] have been applied in epoxy resins. Further studies focused on functionalizing cyclic phenoxyphosphazene with cyanate groups for specific applications with cyanate esters,[104] or with allylic groups for use in acrylic terminated PPE laminates[105] and PC.[106]

10.9 POLYPHOSPHAZENES

Polyphosphazenes are high molecular weight inorganic-organic polymers with alternating phosphorous and nitrogen atoms in their main chain and organic groups in their side chains.[107] Polyphosphazenes can be synthesized by various methods, most commonly through thermal ring-opening polymerization of HCCP to produce polydichlorophosphazene $(NPCl_2)_n$.[108,109] Subsequent post-polymerization modification by nucleophilic substitution of the chlorines with various nucleophiles such as sodium alkoxides, aryloxides, or amines results in a -P = N- repeating unit that can form various high molecular weight linear, branched, dendritic, and cross-linked architectures (Figure 10.18).[110]

Linear and cross-linked architectures are widely studied and show excellent flame retardancy, high thermal stability, and enhanced mechanical properties.[110,111] Rothemund et al. reviewed chemical synthesis strategies to generate polyphosphazenes with controlled molecular weights and polydispersities, and well-defined architectures.[112] Polyphosphazene-based thermally conductive composites showed excellent thermal stability and flame retardancy (UL 94 V-0; LOI 30.2–35.5%) in combination with a variety of fillers, including aluminum oxide, boron nitride, silicon carbide, and expandable graphite.[111] Polyphosphazenes functionalized with phenolic hydroxyl groups achieved UL-94 V-0 in epoxy-based copper-clad laminates.[103]

10.10 OXYIMIDES

Oxyimides have found interests as flame retardants in PP and PA6 formulations due to their ability to generate radicals.[113–116] Oxyimide esters have been studied in PA6, but due to possible transesterification or transamidation concerns, oxyimide ethers were investigated more recently.[117] UL 94 V-0 and V-2 ratings were obtained at loadings of 1–3% depending on the specific oxyimide ether structure. SEM analysis of burned PA6 specimen revealed a smooth glassy surface layer rather than intumescence or char buildup. EDX analysis hinted towards gas phase activity of the oxyimide ether as decreased concentrations of nitrogen and oxygen in the burned regions of the specimen were detected compared to the non-burned areas.

10.11 TRIAZOLES

The application of 1,2,3-triazoles as flame retardants has recently been reviewed.[118] The 1,2,3-triazole moieties can be covalently incorporated through azide-alkyne click chemistry into various polymers such as polyacrylates, polyurethanes, polyimides, and PVC. A series of analyses (FTIR spectroscopy, TGA, EDX-elemental analysis, and XPS spectral analysis) indicated that the combustion of 1,2,3-triazoles led to the release of nitrogen to dilute the

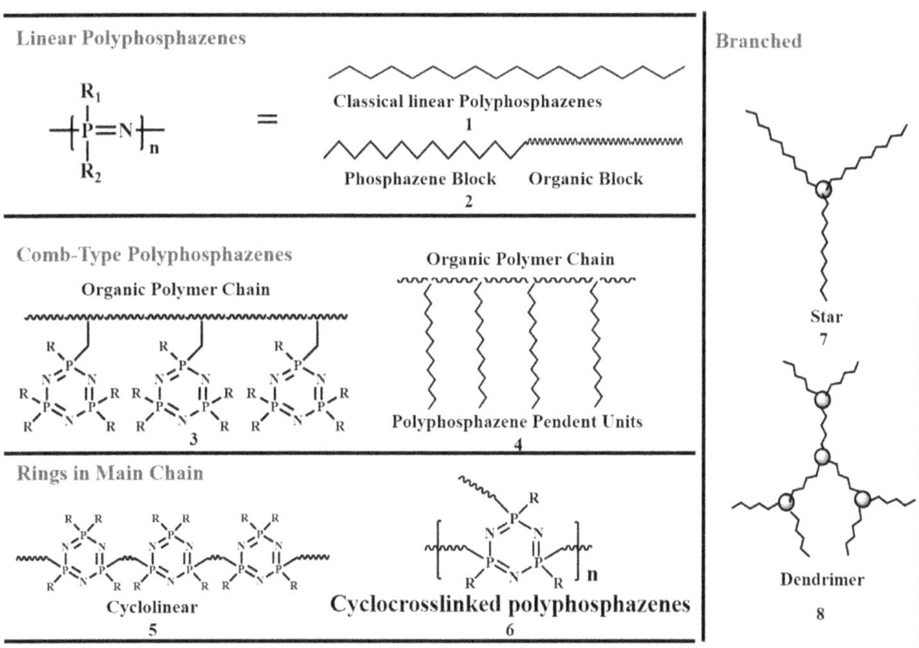

FIGURE 10.18 The structures of different polyphosphazenes.

Source: Adapted with permission from: Zhou, X.; Qiu, S.; Mu, X.; Zhou, M.; Cai, W.; Song, L.; Xing, W.; Hu, Y. Polyphosphazenes-Based Flame Retardants: A Review. *Composites Part B: Engineering* **2020**, *202*, 108397. https://doi.org/10.1016/j.compositesb.2020.108397

gas phase and also to produce intumescent char in combination with other flame retardants. Subsequent studies focused on derivatizing triazoles with vanillin derived cyclotriphosphazene,[99] diphenylphosphinyl vanillin,[119] and DOPO[77,120–122] for epoxy systems. A hydroxy functionalized triazole-diphenyl phosphate with inherent intumescent and char promotion properties was applied in corrosion-resistant PU coatings.[123] Triazole and triazine containing siloxane- and phosphorous derivatives were applied in flame retardant textiles through a sol-gel process.[124] A DOPO-triazole diallyl ester was synthesized and photocured into a thermoset with an LOI of 27.8% and a UL 94 V-0 rating.[125]

10.12 IMIDAZOLE AND IMIDAZOLIUM STRUCTURES

Recent studies have investigated the use of imidazoles as flame retardants in epoxy resins and composites.[126,127] For example, imidazolium diphenylphosphinate (IDPP) was synthesized and used as hardener for bisphenol A diglycidyl ether (DGEBA) and tetraglycidyl diaminodiphenylmethane (TGDDM) resins.[128] The resins had controllable latent curing behaviors as they showed good long-term stability in their liquid states at ambient temperature and rapid curing under heating. The cured resins at 15 wt% loading achieved UL 94 V-0 ratings and high LOIs of 37.0–38.0%. Other studies have prepared 2-methyl imidazolium conjugated HCCP as latent hardeners in epoxy resins that exhibited moderately improved flame retardancy due to P/N synergism compared to resins only containing 2-methyl imidazolium (LOI of 25.3% vs. 20.9%).[129]

Imidazoles have also been studied as ionic liquids in electrolytes for lithium ion batteries. Mixtures of lithium salt (LiPF6) and organic carbonate solvents are widely applied as liquid electrolytes in commercial lithium-ion batteries. The notion of flame retarding inherently flammable electrolytes, however, is rather in its infancy.[130–133]

10.13 FUTURE OUTLOOK AND CONCLUDING REMARKS

Nitrogen-based flame retardants constitute a unique class of substances with diverse applications. The majority of these compounds undergo an endothermic decomposition, producing inert volatiles that effectively hinder the flame spread through intumescence and heat dissipation. Combinations with phosphorous acid generators promotes char formation that results in enhancement of their fire-retardant capabilities. As discussed in Chapter 23, all flame retardants are under regulatory scrutiny, including nitrogen-based flame retardants. Ideally, reactive and polymeric chemistries should be developed and utilized to limit chances of leaching and migration from the polymeric matrices. While melamine can be covalently attached into polymers like polyurethane foam, keeping the content of free melamine below a certain threshold is important due to health and environmental concerns. A shift towards the thermally more stable condensation products of melamine and their salts can be expected due to a general trend towards engineering thermoplastics requiring higher thermal stability in electrical and electronic applications. Nitrogen-containing compounds like guanamine and amidine derivatives may find utility as efficient flame retardants in the future if requirements can be met. Progress has been made in academia with flame retardants like polyphophazenes, oxyimides, and imidazoles, their potential commercial success as cost-effective solutions will be interesting to see unfold in the future. As with all new advancements, commercialization and wide-spread adoption is a slow and arduous process. However, there is certainly enough beneficial aspects in the NFR space to warrant and expect increased adoption as components of flame retardant packages for the wide array of materials in use today.

REFERENCES

(1) Arastehnejad, N.; Sulaiman, M. R.; Gupta, R. K. Nitrogen-Based Ecofriendly Flame Retardants for Polyurethane Foams. In *Polyurethane Chemistry: Renewable Polyols and Isocyanates*; ACS Symposium Series; American Chemical Society, 2021; Vol. 1380, pp 167–185. https://doi.org/10.1021/bk-2021-1380.ch006.

(2) Coquelle, M.; Duquesne, S.; Casetta, M.; Sun, J.; Gu, X.; Zhang, S.; Bourbigot, S. Flame Retardancy of PA6 Using a Guanidine Sulfamate/Melamine Polyphosphate Mixture. *Polymers* **2015**, *7* (2), 316–332. https://doi.org/10.3390/polym7020316.

(3) Effect of Different Metallic Hydroxides on Flame-Retardant Properties of Low Density Polyethylene/Melamine Polyphosphate/Starch Composites. *Journal of Applied Polymer Science*. **2011**. https://onlinelibrary.wiley.com/doi/10.1002/app.34398 (accessed 2023–07–25).

(4) Arechederra, R. Flame Retardant Polymer Compositions. US10227482B2, March 12, 2019. https://patents.google.com/patent/US10227482B2/en?q=(melamine+halogenated+flame+retardant)&oq=melamine+halogenated+flame+retardant (accessed 2023–07–25).

(5) Klatt, M. Nitrogen-Based Flame Retardants. In *Non-Halogenated Flame Retardant Handbook*; John Wiley & Sons, Ltd, 2014; pp 143–168. https://doi.org/10.1002/9781118939239.ch4.

(6) Cichy, B.; Łuczkowska, D.; Nowak, M.; Władyka-Przybylak, M. Polyphosphate Flame Retardants with Increased Heat Resistance. *Industrial & Engineering Chemistry Research* **2003**, *42* (13), 2897–2905. https://doi.org/10.1021/ie0208570.

(7) Bann, B.; Miller, S. A. Melamine and Derivatives of Melamine. *Chemical Reviews* **1958**, *58* (1), 131–172. https://doi.org/10.1021/cr50019a004.

(8) Chan, Y. Y.; Ma, C.; Zhou, F.; Hu, Y.; Schartel, B. A Liquid Phosphorous Flame Retardant Combined with Expandable Graphite or Melamine in Flexible Polyurethane Foam. *Polymers for Advanced Technologies* **2022**, *33* (1), 326–339. https://doi.org/10.1002/pat.5519.

(9) Zhou, F.; Zhang, T.; Zou, B.; Hu, W.; Wang, B.; Zhan, J.; Ma, C.; Hu, Y. Synthesis of a Novel Liquid Phosphorus-Containing Flame Retardant for Flexible Polyurethane Foam: Combustion Behaviors and Thermal Properties. *Polymer Degradation and Stability* **2020**, *171*, 109029. https://doi.org/10.1016/j.polymdegradstab.2019.109029.

(10) Denecker, C.; Liggat, J. J.; Snape, C. E. Relationship between the Thermal Degradation Chemistry and Flammability of Commercial Flexible Polyurethane Foams. *Journal of Applied Polymer Science* **2006**, *100* (4), 3024–3033. https://doi.org/10.1002/app.23701.

(11) Liu, X.; Hao, J.; Gaan, S. Recent Studies on the Decomposition and Strategies of Smoke and Toxicity Suppression for Polyurethane Based Materials. *RSC Advances* **2016**, *6* (78), 74742–74756. https://doi.org/10.1039/C6RA14345H.

(12) Schwarzer, A.; Saplinova, T.; Kroke, E. Tri-s-Triazines (s-Heptazines)—From a "Mystery Molecule" to Industrially Relevant Carbon Nitride Materials. *Coordination Chemistry Reviews* **2013**, *257* (13), 2032–2062. https://doi.org/10.1016/j.ccr.2012.12.006.

(13) Lotsch, B. V.; Schnick, W. New Light on an Old Story: Formation of Melam during Thermal Condensation of Melamine. *Chemistry—A European Journal* **2007**, *13* (17), 4956–4968. https://doi.org/10.1002/chem.200601291.

(14) Kessler, F. K.; Burow, A. M.; Savasci, G.; Rosenthal, T.; Schultz, P.; Wirnhier, E.; Oeckler, O.; Ochsenfeld, C.; Schnick, W. Structure Elucidation of a Melam–Melem Adduct by a Combined Approach of Synchrotron X-Ray Diffraction and DFT Calculations. *Chemistry—A European Journal* **2019**, *25* (35), 8415–8424. https://doi.org/10.1002/chem.201901391.

(15) Stoelwinder, C. J. C.; Kwant, G. J. Process for the Preparation of Condensation Products of Melamine. US9096552B2, August 4, 2015. https://patents.google.com/patent/US9096552B2/en?oq=EP+2691378+B1 (accessed 2023–07–25).

(16) Stoelwinder, C. J. C.; Kwant, G. J. Process for the Preparation of Condensation Products of Melamine. EP2691378B1, April 22, 2015.

(17) Keijzer, A. E. H. D.; Dijkstra, B. J. H. Process for the Preparation of Condensation Products of Melamine. WO1996016948A1, June 6, 1996. https://patents.google.com/patent/WO1996016948A1/en?oq=wo9616948 (accessed 2023–07–25).

(18) Jürgens, B.; Irran, E.; Senker, J.; Kroll, P.; Müller, H.; Schnick, W. Melem (2,5,8-Triamino-Tri-s-Triazine), an Important Intermediate during Condensation of Melamine Rings to Graphitic Carbon Nitride: Synthesis, Structure Determination by X-Ray Powder Diffractometry, Solid-State NMR, and Theoretical Studies. *Journal of the American Chemical Society* **2003**, *125* (34), 10288–10300. https://doi.org/10.1021/ja0357689.

(19) Lotsch, B. V.; Schnick, W. The Crystal Structures of Two No-Metal Tricyanomelaminates: Diammonium Tricyanomelaminate Dihydrate $[NH_4]_2[C_6N_9H]$ 2 H_2O and Dimelaminium Tricyanomelaminate Melamine Dihydrate $[C_3N_6H_7]_2[C_6N_9H]$ $C_3N_6H_6$ 2 H_2O. *Zeitschrift für anorganische und allgemeine Chemie* **2007**, *633* (9), 1435–1441. https://doi.org/10.1002/zaac.200700147.

(20) Keijzer, A. E. H. D.; Sham, C. K.; Flippo, P.; Hulskotte, R. J. M. Flame Retardant Polyamide Composition. WO1996017013A1, June 6, 1996. https://patents.google.com/patent/WO1996017013A1/en?oq=wo9617013 (accessed 2023–07–25).

(21) Bauer, H.; Hoerold, S.; Sicken, M. Halogen-Free Solid Flame Retardant Mixture and Use Thereof. US10508238B2, December 17, 2019. https://patents.google.com/patent/US10508238B2/en?oq=US20160340588A1 (accessed 2023–07–25).

(22) Bauer, H.; Hoerold, S.; Sicken, M. Mixtures of Flame Protection Means Containing Flame Protection Means and Aluminium Phosphites, Method for Production and Use Thereof. US9481831B2, November 1, 2016. https://patents.google.com/patent/US9481831B2/en?q=(melam+clariant)&oq=melam+clariant (accessed 2023–07–25).

(23) Bonyhady, S. J.; Lee, J. Y.; He, Q.; Sharma, R. Method of Preparing Phosphorus-Containing Flame Retardants and Their Use in Polymer Compositions. US20220073547A1, March 10, 2022. https://patents.google.com/patent/US20220073547A1/en?q=(lanxess+melam)&oq=lanxess+melam (accessed 2023–07–25).

(24) Timberlake, L. D.; Hanson, M. V. Flame Retardant Resins Comprising Phosphorous Containing Flame Retardants. US9994696B2, June 12, 2018. https://patents.google.com/patent/US9994696B2/en?q=(flame+retardant)&q=(metal)&q=(alkyl)&q=(zinc)&q=(aryl)&before=priority:20181220&scholar (accessed 2023–07–25).

(25) Khalil, H.; Gläsel, H.-J.; Buchmeiser, M. R. High-Mechanical-Strength Flame-Retardant Nanocomposites Based on Novel Al(III)- and Zr(IV)-Melamine Phosphates and Sulfates. *Macromolecular Materials and Engineering* **2013**, *298* (6), 690–698. https://doi.org/10.1002/mame.201200168.

(26) Rakotomalala, M.; Wagner, S.; Döring, M. Recent Developments in Halogen Free Flame Retardants for Epoxy Resins for Electrical and Electronic Applications. *Materials* **2010**, *3* (8), 4300–4327. https://doi.org/10.3390/ma3084300.

(27) Rex, G. Advances in HFFRs via Efficient & Innovative Masterbatch Concentrates, Part III. In *AMI Fire Retardants in Plastics*, Philadelphia, PA, 2023.

(28) Futterer, T.; Martinez, D. G.; Moschel, S.; Rochholz, H.; Moss, T.; Abele, B. C. Neuartiges flammschutzmittel. WO2023006622A1, February 2, 2023. https://patents.google.com/patent/WO2023006622A1/en?oq=WO2023006622A1 (accessed 2023–07–27).

(29) *Safire® Nitrogen-Phosphorus Technology*. www.huber-advancedmaterials.com/solutions/products-brands/nitrogen-phosphorous-technology.

(30) Turski Silva Diniz, A.; Huth, C.; Schartel, B. Dripping and Decomposition under Fire: Melamine Cyanurate vs. Glass Fibres in Polyamide 6. *Polymer Degradation and Stability* **2020**, *171*, 109048. https://doi.org/10.1016/j.polymdegradstab.2019.109048.

(31) Borade, P. K.; Gallucci, R. R.; Konduri, R.; Lohmeijer, J. H. G. M.; Pal, S. K.; Srinivasan, V.; Weele, C. van der; de Wit, G. Polyester Compositions, Method of Manufacture, and Uses Thereof. US7812077B2, October 12, 2010. https://patents.google.com/patent/US7812077B2/en?q=(Melamine+cyanurate+polyesters+clariant)&oq=Melamine+cyanurate+polyesters+clariant (accessed 2023–07–27).

(32) Hao, F.; Chen, Y.; Sun, Z.; Qian, L. Component Ratio Effects of Melamine Cyanurate and Aluminum Diethylphosphinate in Flame Retardant TPU. *Journal of Polymer Research* **2022**, *30* (1), 25. https://doi.org/10.1007/s10965-022-03401-y.

(33) Shan, H.; Yan, L.; Xu, B.; Wang, D.; Wu, M. Polyphosphamide Containing Triazine and Melamine Cyanurate for Flame-Retardant PA6. *ACS Applied Polymer Materials* **2023**, *5* (7), 5322–5333. https://doi.org/10.1021/acsapm.3c00732.

(34) Liu, W.; Wang, Z.; Su, S.; Wu, H.; Sun, M.; Tang, L. Synergistic Flame Retardancy of ZnO and Piperazine Pyrophosphate/Melamine Cyanurate in Polypropylene. *Journal of Vinyl and Additive Technology* **2023**, *29* (1), 202–219. https://doi.org/10.1002/vnl.21956.

(35) Chen, K.; Bao, Q.; Chen, G.; Wu, S.; Liu, Y.; Wang, Q. Synergistic Flame Retardant Linear Low-Density Polyethylene Based on Piperazine Pyrophosphate and Melamine Cyanurate. *Fire and Materials* **2023**, *47*, 979–989. https://doi.org/10.1002/fam.3142.
(36) Feng, X.; Wang, Z.; Ma, H.; Dang, B.; Li, J. A New Bio-Based Flame Retardant Simply Synthesized: Improving the Flame Retardancy and Thermal Stability of Low-Density Polyethylene Together with Melamine Cyanurate. *Journal of Applied Polymer Science* **2023**, *140* (24), e53953. https://doi.org/10.1002/app.53953.
(37) Young, D. C. Ammonium Polyphosphate Production. US4041133A, August 9, 1977. https://patents.google.com/patent/US4041133A/en?oq=+US+4041133 (accessed 2023–07–27).
(38) Lin, G.; Fu, T.; Qiu, Y.; Li, F.; Sun, W. Flame Retardancy and Anti-Impact Performance of Polyurea Composite Coating Reinforced by Modified Ammonium Polyphosphate and Two-Dimensional Nano-Fillers. *Progress in Organic Coatings* **2023**, *180*, 107554. https://doi.org/10.1016/j.porgcoat.2023.107554.
(39) Li, S.; Gao, R.; Xue, H.; Zhang, Y.; Ye, Y.; Lin, Q.; Zeng, Q.; Liu, L.; Cao, C.; Xia, J. Sustainable Flame Retardant Coatings Derived from Chinese Lacquer Blend with Synergistic Flame Retardant Silicon-Modified Ammonium Polyphosphate. *Advances in Materials Science and Engineering* **2023**, *2023*, e9973474. https://doi.org/10.1155/2023/9973474.
(40) Ke, Q.; Bai, J.; Zhang, G.; Zhang, J.; Yang, M. Simultaneously Enhancing the Flame Retardancy, Water Resistance, and Mechanical Properties of Flame-Retardant Polypropylene via a Linear Vinyl Polysiloxane-Coated Ammonium Polyphosphate. *Polymers* **2023**, *15* (9), 2074. https://doi.org/10.3390/polym15092074.
(41) *Fire Behavior and Mechanical Properties of Phosphorus-Containing Silica Gel Coated Ammonium Polyphosphate as Fire Retardants in Thermoplastic Polyurethane (TPU).* https://doi.org/10.21203/rs.3.rs-3029148/v1.
(42) Sun, R.-Y.; Wang, F.; Li, C.-B.; Deng, Z.-P.; Song, F.; Wang, Y.-Z. Formulation of Environmentally Robust Flame-Retardant and Superhydrophobic Coatings for Wood Materials. *Construction and Building Materials* **2023**, *392*, 131873. https://doi.org/10.1016/j.conbuildmat.2023.131873.
(43) Jun, Y.; Jin, Y.; Lele, S.; Guilong, X.; Zhaohui, L. Preparation, Structure and Application of Styrene-Acrylic Emulsion/Modified Ammonium Polyphosphate in Flame Retardant Air Filter Paper. *Journal of Industrial Textiles* **2022**, *51* (4 Suppl.), 7223S-7238S. https://doi.org/10.1177/15280837211066629.
(44) Wan, M.; Shi, C.; Qian, X.; Qin, Y.; Jing, J.; Che, H.; Ren, F.; Li, J.; Yu, B.; Zhou, K. Design of Novel Double-Layer Coated Ammonium Polyphosphate and Its Application in Flame Retardant Thermoplastic Polyurethanes. *Chemical Engineering Journal* **2023**, *459*, 141448. https://doi.org/10.1016/j.cej.2023.141448.
(45) Endtner, J.; Bienmüller, M. Polyamidzusammensetzungen. EP3034554B1, August 12, 2020. https://patents.google.com/patent/EP3034554B1/en?q=(MCA%C2%AE+PPM+Triazine+HF)&oq=MCA%C2%AE+PPM+Triazine+HF (accessed 2023–07–31).
(46) Morgan, A. B., Ed. *Non-Halogenated Flame Retardant Handbook*, 2nd ed.; Wiley, 2021. https://doi.org/10.1002/9781119752240.fmatter.
(47) Wilkie, C. A., Morgan, A. B., Eds. *Fire Retardancy of Polymeric Materials*, 2nd ed.; CRC Press/Taylor & Francis: Boca Raton, FL, 2010.
(48) Greene, A. C.; Grubbs, R. B. Current Methods for N-Alkoxyamine Synthesis. In *Controlled/Living Radical Polymerization: Progress in RAFT, DT, NMP & OMRP*; ACS Symposium Series; American Chemical Society, 2009; Vol. 1024, pp 81–93. https://doi.org/10.1021/bk-2009-1024.ch006.
(49) Grishin, D. F.; Kolyakina, E. V.; Pavlovskaya, M. V.; Lazarev, M. A.; Shchepalov, A. A. New Nitroxide Mediators for Controlled Synthesis of Polystyrene, Poly(Meth) Acrylates and Their Copolymers. In *Controlled/Living Radical Polymerization: Progress in RAFT, DT, NMP & OMRP*; ACS Symposium Series; American Chemical Society, 2009; Vol. 1024, pp 95–114. https://doi.org/10.1021/bk-2009-1024.ch007.
(50) Kaprinidis, N.; Earhart, N.; Zingg, J. Overview of Recent Advances in Flame Retardant Compositions and UV Stable Flame Retardant Systems; Fully Formulated Antimony Free Flame Retardant Products for Polyolefins. In *Proceedings of the 51st International Wire and Cable Symposium*, 2002, pp 594–596. https://apps.dtic.mil/sti/tr/pdf/ADA417493.pdf.
(51) Hodgson, J. L.; Coote, M. L. Clarifying the Mechanism of the Denisov Cycle: How Do Hindered Amine Light Stabilizers Protect Polymer Coatings from Photo-Oxidative Degradation? *Macromolecules* **2010**, *43* (10), 4573–4583. https://doi.org/10.1021/ma100453d.
(52) Klatt, M. Nitrogen-Based Flame Retardants. In *Non-Halogenated Flame Retardant Handbook*; John Wiley & Sons, Ltd, 2014; pp 143–168. https://doi.org/10.1002/9781118939239.ch4.
(53) Triacca, V. J.; Gloor, P. E.; Zhu, S.; Hrymak, A. N.; Hamielec, A. E. Free Radical Degradation of Polypropylene: Random Chain Scission. *Polymer Engineering & Science* **1993**, *33* (8), 445–454. https://doi.org/10.1002/pen.760330802.
(54) Gugumus, F. Re-Examination of the Thermal Oxidation Reactions of Polymers1. New Views of an Old Reaction. *Polymer Degradation and Stability* **2001**, *74* (2), 327–339. https://doi.org/10.1016/S0141-3910(01)00165-3.
(55) Gugumus, F. Re-Examination of the Thermal Oxidation Reactions of Polymers3. Various Reactions in Polyethylene and Polypropylene. *Polymer Degradation and Stability* **2002**, *77* (1), 147–155. https://doi.org/10.1016/S0141-3910(02)00093-9.
(56) Nguyen, T. A.; Lim, H. M.; Kinashi, K.; Sakai, W.; Tsutsumi, N.; Okubayashi, S.; Hosoda, S.; Sato, T. Spin Trapping Analysis of Radical Intermediates on the Thermo-Oxidative Degradation of Polypropylene. *Polymers* **2023**, *15* (1), 200. https://doi.org/10.3390/polym15010200.
(57) Aubert, M.; Nicolas, R. C.; Pawelec, W.; Wilén, C.-E.; Roth, M.; Pfaendner, R. Azoalkanes—Novel Flame Retardants and Their Structure–Property Relationship. *Polymers for Advanced Technologies* **2011**, *22* (11), 1529–1538. https://doi.org/10.1002/pat.1642.
(58) Pfaendner, R. Nitroxyl Radicals and Nitroxylethers beyond Stabilization: Radical Generators for Efficient Polymer Modification. *Comptes Rendus Chimie* **2006**, *9* (11), 1338–1344. https://doi.org/10.1016/j.crci.2006.08.001.
(59) Xie, H.; Lai, X.; Zhou, R.; Li, H.; Zhang, Y.; Zeng, X.; Guo, J. Effect and Mechanism of N-Alkoxy Hindered Amine on the Flame Retardancy, UV Aging Resistance and Thermal Degradation of Intumescent Flame Retardant Polypropylene. *Polymer Degradation and Stability* **2015**, *118*, 167–177. https://doi.org/10.1016/j.polymdegradstab.2015.04.022.
(60) Xie, H.; Lai, X.; Li, H.; Zeng, X. Synthesis of a Novel Macromolecular Charring Agent with Free-Radical Quenching Capability and Its Synergism in Flame

Retardant Polypropylene. *Polymer Degradation and Stability* **2016**, *130*, 68–77. https://doi.org/10.1016/j.polymdegradstab.2016.05.029.
(61) Butz, V. Flame-Retardant Composition Comprising a Phosphonic Acid Derivative. US2011257310A1, October 20, 2011.
(62) Liao, F.; Zhou, L.; Ju, Y.; Yang, Y.; Wang, X. Synthesis of A Novel Phosphorus–Nitrogen-Silicon Polymeric Flame Retardant and Its Application in Poly(Lactic Acid). *Industrial & Engineering Chemistry Research* **2014**, *53* (24), 10015–10023. https://doi.org/10.1021/ie5008745.
(63) Cao, K.; Wu, S.; Qiu, S.; Li, Y.; Yao, Z. Synthesis of N-Alkoxy Hindered Amine Containing Silane as a Multifunctional Flame Retardant Synergist and Its Application in Intumescent Flame Retardant Polypropylene. *Industrial & Engineering Chemistry Research* **2013**, *52* (1), 309–317. https://doi.org/10.1021/ie3017048.
(64) Nodera, A. Flame-Retardant Styrene-Based Resin Composition, and Molding. JP2022069320A, May 11, 2022.
(65) Nodera, A. Styrenic Resin Composition and Molded Article. JP2022041091A, March 11, 2022.
(66) Nodera, A. Flame Retardant Master Batch, and Flame-Retardant Styrenic Resin Composition Including the Same. JP2021075613A, May 20, 2021.
(67) Nodera, A. Flame-Retardant Foam. JP2020117582A, August 6, 2020.
(68) Tamura, A.; Noda, O.; Kobayashi, T. Method for Producing Thermoplastic Polyurethane Elastomer Composition. JP2021080313A, May 27, 2021.
(69) Ogawa, S.; Masaki, T.; Kawamori, T. Resin Composition and Conductor Substrate. JP2022018376A, January 27, 2022.
(70) Wasano, T. Halogen-Free Flame-Retardant Curable Resin Composition, Prepreg, Metal-Clad Laminated Board, and Printed Wiring Board. WO2023053782A1, April 6, 2023.
(71) Ueno, T.; Emmei, C. Flame-Retardant Resin Composition. JP2022071496A, May 16, 2022.
(72) Felix, B.; Strebl-Pfarr, M.; Homann, V. Flame-Retardant Pressure-Sensitive Adhesive. US2021147725A1, May 20, 2021.
(73) Okunaga, K.; Sakai, H.; Tokimine, K. Flame-Retardant Polyester Elastomer Resin Composition and Flame-Retardant Polyester Elastomer Filament. JP2021070805A, May 6, 2021.
(74) Inasawa, Y.; Sato, H. Flame-Retardant Polycarbonate Resin Composition. JP2016138195A, August 4, 2016.
(75) Iwata, M.; Nakamura, S. Flame-Retardant Resin Composition, Flame-Retardant Resin Composition for Cable, Cable, Molded Article, and Flame Retardant Master Batch Using Same, and Flame Retardant. WO2020129871A1, June 25, 2020.
(76) Lacarte, C.; Gardin, G.; Harris, T. U.S Patent 6534556, 2003.
(77) Wang, X.-L.; Yang, K.-K.; Wang, Y.-Z. Physical and Chemical Effects of Diethyl N,N′-Diethanolaminomethylphosphate on Flame Retardancy of Rigid Polyurethane Foam. *Journal of Applied Polymer Science* **2001**, *82* (2), 276–282. https://doi.org/10.1002/app.1849.
(78) Huang, Y.; Ma, T.; Li, L.; Wang, Q.; Guo, C. Facile Synthesis and Construction of Renewable, Waterborne and Flame-Retardant UV-Curable Coatings in Wood Surface. *Progress in Organic Coatings* **2022**, *172*, 107104. https://doi.org/10.1016/j.porgcoat.2022.107104.
(79) Sun, Y.; Liu, C.; Hong, Y.; Liu, R.; Zhou, X. Synthesis and Application of Self-Crosslinking and Flame Retardant Waterborne Polyurethane as Fabric Coating Agent. *Progress in Organic Coatings* **2019**, *137*, 105323. https://doi.org/10.1016/j.porgcoat.2019.105323.
(80) Yin, X.; Luo, Y.; Zhang, J. Synthesis and Characterization of Halogen-Free Flame Retardant Two-Component Waterborne Polyurethane by Different Modification. *Industrial & Engineering Chemistry Research* **2017**, *56* (7), 1791–1802. https://doi.org/10.1021/acs.iecr.6b04452.
(81) Mahdi, S. Z.; Mattice, H. O.; Gibbons, E. E.; Quast, M. J. Crash Durable, Stress Durable and Weldable Epoxy Adhesives. WO 2022/187580 Al, 2022.
(82) Abecassis-Wolfovich, M.; Swissa, S. Antimony Free Flame-Retarded Epoxy Compositions. WO 2018/109766 Al, 2018.
(83) Taniguchi, M.; Tada, Y.; Nishioka, Y. Flame-Retardant Epoxy Resin Composition, Molded Article Thereof, and Electronic Part. US 6,797,750 B2, September 28, 2004. https://patentimages.storage.googleapis.com/18/09/c5/de9d9f3ccdd0b3/US6797750.pdf.
(84) Yu, L.-C.; Lee, T.-A. Halogen-Free Resin Composition and Copper Clad Laminate and Printed Circuit Board Using Same. US 9,279,051, March 8, 2016.
(85) He, Y.; Su, S. Halogen-Free Flame Retardant Resin Composition and the Use Thereof. US Patent 9745464, August 29, 2017.
(86) Sagara, T.; Kakiuchi, H.; Kashihara, K.; Kitai, Y.; Saito, H.; Yokoyama, D.; Fujiwara, H. Resin Composition, Resin Varnish, Prepreg, Metal-Clad Laminate, and Printed Wiring Board. US Patent 9528026, December 27, 2016.
(87) Maruyama, K.; Motoshige, R. Flame Retardant Thermoplastic Resin Composition. European Patent 0728811, September 17, 2003.
(88) Heuer, H. W.; Wehrmann, R.; Boumans, A.; Bauer, J. Multilayer Composite Material Containing Special Polycarbonate Compositions as a Matrix Material. WO 2019/121232 A1, June 27, 2019.
(89) Chen, L.; Shi, H.; Zheng, Y.; Zhang Chunjie. Flame Retardant Composition, Methods of Manufacture Thereof and Articles Comprising the Same. WO 2016/174592 A1, November 3, 2016.
(90) Huang, Z.; Han, H. Polycarbonate Compositions. WO 2019/062472 A1, April 4, 2019.
(91) Wen, L.; Shan, W.; Li, M.; Xu Jenny. Flame Retardant Polycarbonate Compositions, Methods of Manufacture Thereof and Articles Comprising the Same. WO 2013/175453 A1, November 28, 2013.
(92) Zhao, W.; Daga, V.; Kulkami, A. Flame Retardant Compositions, Articles Comprising the Same and Methods of Manufacture Thereof. WO 2013/177495 A1, November 28, 2013.
(93) Zhu, S.; Avakian, R. W. Flame Retardant Polylactic Acid Compounds. US 2015/0274966 A1, October 1, 2015.
(94) Niesten, M.; Welsch, N.; Hagen, T.; Albach, R. Flame-Retardant PUR-PIR Rigid Foams. WO 2019/011761 A1, January 17, 2019.
(95) Sharma, K. Flame Retardant Thermoplastic Compositions, Methods of Manufacture Thereof and Articles Comprising the Same. WO 2013/175454 A1, May 24, 2013.
(96) König, A. Flame-Retardant Thermoplastic Molding Composition. US 2012/0289634 A1, November 15, 2012.
(97) Lu, W.; Parappuveetil Sarangadharan, S.; Chuang, F. C. Polyamide Composition and the Article Thereof. WO 2021/069456 A1, April 15, 2021.
(98) Dagdag, O.; Bachiri, A. E.; Hamed, O.; Haldhar, R.; Verma, C.; Ebenso, E.; Gouri, M. E. Dendrimeric Epoxy Resins Based on Hexachlorocyclotriphosphazene as a Reactive Flame Retardant Polymeric Materials: A Review. *Journal of Inorganic and Organometallic Polymers and*

Materials **2021**, *31* (8), 3240–3261. https://doi.org/10.1007/s10904-021-01900-x.
(99) Nabipour, H.; Wang, X.; Rahman, M. Z.; Song, L.; Hu, Y. Improvement of the Flame Retardant and Thermomechanical Properties of Epoxy Resins by a Vanillin-Derived Cyclotriphosphazene-Cored Triazole Compound. *Polymer Degradation and Stability* **2022**, *204*, 110088. https://doi.org/10.1016/j.polymdegradstab.2022.110088.
(100) Hu, B.; Wang, J.; Wang, J.; Yang, S.; Li, C.; Wang, F.; Huo, S.; Song, P.; Fang, Z.; Wang, H. Flame-Retardant Single-Component Epoxy Resin Cured by Benzimidazolyl-Substituted Cyclotriphosphazene: Storage Stability, Curing Behaviors and Flame Retardancy. *Polymer Degradation and Stability* **2022**, *204*, 110092. https://doi.org/10.1016/j.polymdegradstab.2022.110092.
(101) Selvaraj, V.; Raghavarshini, T.; Alagar, M. Evaluation of Thermo-Mechanical, Dielectric and Corrosion Resistant Properties of Cardanol Benzoxazine-Epoxy Based Hybrid Composites: A Very Low Temperature Curing Pre-Polymer for High Performance Paint Related Applications. *High Performance Polymers* **2020**, *32* (5), 524–539. https://doi.org/10.1177/0954008319885006.
(102) Wang, Y.; He, Y.; Yang, Z.; Xu, Y. Siloxane-Modified Cyclotriphosphazene Halogen-Free Flame Retardant, Preparation Process and Use Thereof. EP3312186 A1, April 25, 2018.
(103) Pan, Q. Flame Retardant Compounds, Hardeners and Polyphenol-Based Epoxy Resins. US20160333137A1, November 17, 2016. https://patents.google.com/patent/US20160333137A1/en?oq=US20160333137+.
(104) Tada, Y.; Inoue, T. Cyanato Group-Containing Cyclic Phosphazene Compound and Method for Producing the Same. US7767739B2, August 3, 2010. https://patents.google.com/patent/US7767739B2/en?oq=US+Patent+7767739.
(105) Liu, S.-F.; Chen, M.-H. Resin Composition and Uses of the Same. US10196502B2, February 5, 2019. https://patents.google.com/patent/US10196502B2/en?oq=US+Patent+10196502.
(106) Zhou, C.; Avakian, R. W. Phosphazene Flame Retardant Polycarbonate Compounds. WO2015069642A1, May 14, 2015. https://patents.google.com/patent/WO2015069642A1/en?oq=WO+2015%2f069642+.
(107) Gleria, M.; De Jaeger, R. Polyphosphazenes: A Review. In *New Aspects in Phosphorus Chemistry V*; Majoral, J.-P., Ed.; Topics in Current Chemistry; Springer: Berlin, Heidelberg, 2005; pp 165–251. https://doi.org/10.1007/b100985.
(108) Amin, A. M.; Wang, L.; Wang, J.; Yu, H.; Huo, J.; Gao, J.; Xiao, A. Recent Research Progress in the Synthesis of Polyphosphazene and Their Applications. *Designed Monomers and Polymers* **2012**. https://doi.org/10.1163/138577209X12486896623373.
(109) Allcock, H. R.; Chen, C. Polyphosphazenes: Phosphorus in Inorganic–Organic Polymers. *The Journal of Organic Chemistry* **2020**, *85* (22), 14286–14297. https://doi.org/10.1021/acs.joc.0c01710.
(110) Zhou, X.; Qiu, S.; Mu, X.; Zhou, M.; Cai, W.; Song, L.; Xing, W.; Hu, Y. Polyphosphazenes-Based Flame Retardants: A Review. *Composites Part B: Engineering* **2020**, *202*, 108397. https://doi.org/10.1016/j.compositesb.2020.108397.
(111) Zou, W.; Basharat, M.; Dar, S. U.; Zhang, S.; Abbas, Y.; Liu, W.; Wu, Z.; Zhang, T. Preparation and Performances of Novel Polyphosphazene-Based Thermally Conductive Composites. *Composites Part A: Applied Science and Manufacturing* **2019**, *119*, 145–153. https://doi.org/10.1016/j.compositesa.2019.01.024.
(112) Rothemund, S.; Teasdale, I. Preparation of Polyphosphazenes: A Tutorial Review. *Chemical Society Reviews* **2016**, *45* (19), 5200–5215. https://doi.org/10.1039/C6CS00340K.
(113) Pfaendner, R.; Mazurowski, M. Use of Oxyimide-Comprising Copolymers or Polymers as Flame Retardants, Stabilisers, Rheology Modifiers for Plastic Materials, Initiators for Polymerisation- and Grafting Processes, Crosslinking- or Coupling Agents and Also Plastic Moulding Compounds Comprising Such Copolymers or Polymers. US10370537B2, 2015. https://patents.google.com/patent/US10370537B2/en.
(114) Pfaendner, R.; Mazurowski, M. Use of Organic Oxyimide Salts as Flame Retardants, Flame-Retardant Plastic Compositions, Method for the Production Thereof and Moulded Part, Paint or Coating. US10450442B2, October 22, 2019. https://patents.google.com/patent/US10450442B2/en?oq=US10450442B2 (accessed 2023–06–05).
(115) Pfaendner, R.; Mazurowski, M. Use of Organic Oxyimides as Radical Generators in Plastics, Method for Generating Radicals in Plastics and Use of Said Method. US10364340B2, July 30, 2019. https://patents.google.com/patent/US10364340B2/en?oq=US10364340B2 (accessed 2023–06–05).
(116) Pfaendner, R.; Metzsch-Zillingen, E.; Stec, M. Use of Organic Oxyimides as Flame Retardant for Plastic Materials and Also Flame-Retardant Plastic Material Composition and Moulded Parts Produced Therefrom. US20160052927A1, February 25, 2016. https://patents.google.com/patent/US20160052927A1/en?oq=US20160052927A1 (accessed 2023–06–05).
(117) Spieß, B.; Metzsch-Zilligen, E.; Pfaendner, R. A New Class of Oxyimides: Oxyimide Ethers and Their Use as Flame Retardants. *Macromolecular Materials and Engineering* **2021**, *306* (3), 2000650. https://doi.org/10.1002/mame.202000650.
(118) Sykam, K.; Donempudi, S.; Basak, P. 1,2,3-Triazole Rich Polymers for Flame Retardant Application: A Review. *Journal of Applied Polymer Science* **2022**, *139* (32), e52771. https://doi.org/10.1002/app.52771.
(119) Hua, Y.; Chen, J.; Liu, J.; Sun, J.; Gu, X.; Jiang, S.; Zhang, S. Fabrication of a Transparent, Flame Retardant, and Antimicrobial Epoxy Resin by a Novel Phosphorus-Containing Schiff Base Molecule. *Polymer Degradation and Stability* **2023**, *209*, 110274. https://doi.org/10.1016/j.polymdegradstab.2023.110274.
(120) Zhang, W.; Zhou, M.; Kan, Y.; Chen, J.; Hu, Y.; Xing, W. Synthesis and Flame Retardant Efficiency Study of Two Phosphorus-Nitrogen Type Flame Retardants Containing Triazole Units. *Polymer Degradation and Stability* **2023**, *208*, 110236. https://doi.org/10.1016/j.polymdegradstab.2022.110236.
(121) Zhao, W.-H.; Wang, J.-W.; Zhuang, G.-C.; Liu, D.-Y.; Sun, C.-C. Thermal Stability and Flame Retardancy of Epoxy Resin with DOPO-Based Compound Containing Triazole and Hydroxyl Groups. *Macromolecular Materials and Engineering* **2022**, *307* (11), 2200408. https://doi.org/10.1002/mame.202200408.
(122) Wang, H.; Yuan, J.; Wang, Y.; Ma, Y.; Lyu, S.; Zhu, Z. A Nitrogen Heterocyclic/Phosphaphenanthrene Derivative as a Reactive Additive for Simultaneous Improvement of Flame Retardancy, Mechanical and Dielectric Properties of Epoxy Resins. *Polymer Degradation and Stability* **2022**, *199*, 109909. https://doi.org/10.1016/j.polymdegradstab.2022.109909.

(123) Sykam, K.; Sivanandan, S.; Basak, P. 1,2,3-Triazole Mediated, Non-Halogenated Phosphorus Containing Protective Coatings from Castor Oil: Flame Retardant and Anti-Corrosion Applications. *Progress in Organic Coatings* **2023**, *178*, 107475. https://doi.org/10.1016/j.porgcoat.2023.107475.

(124) Shabani, V.; Ali, W.; Gutmann, J. S.; Mayer-Gall, T. Flammschutzmittel auf Basis von phosphorhaltigen Triazol- und Triazinverbindungen. DE102020132465A1, June 9, 2022. https://patents.google.com/patent/DE102020132465A1/en?oq=DE102020132465+ (accessed 2023–06–05).

(125) Sagdic, G.; Cakmakci, E.; Daglar, O.; Gunay, U. S.; Hizal, G.; Tunca, U.; Durmaz, H. Thermal and Mechanical Properties of Thiol-Ene Photocured Thermosets Containing DOPO-Based Liquid Reactive Flame Retardant Synthesized by Metal-Free Azide-Alkyne Click Reaction. *Progress in Organic Coatings* **2022**, *167*, 106825. https://doi.org/10.1016/j.porgcoat.2022.106825.

(126) Wang, J.; Chen, X.; Wang, J.; Yang, S.; Chen, K.; Zhu, L.; Huo, S.; Song, P.; Wang, H. High-Performance, Intrinsically Fire-Safe, Single-Component Epoxy Resins and Carbon Fiber Reinforced Epoxy Composites Based on Two Phosphorus-Derived Imidazoliums. *Polymer Degradation and Stability* **2023**, *208*, 110261. https://doi.org/10.1016/j.polymdegradstab.2023.110261.

(127) Xu, Y.-J.; Chen, L.; Rao, W.-H.; Qi, M.; Guo, D.-M.; Liao, W.; Wang, Y.-Z. Latent Curing Epoxy System with Excellent Thermal Stability, Flame Retardance and Dielectric Property. *Chemical Engineering Journal* **2018**, *347*, 223–232. https://doi.org/10.1016/j.cej.2018.04.097.

(128) Xu, Y.-J.; Shi, X.-H.; Lu, J.-H.; Qi, M.; Guo, D.-M.; Chen, L.; Wang, Y.-Z. Novel Phosphorus-Containing Imidazolium as Hardener for Epoxy Resin Aiming at Controllable Latent Curing Behavior and Flame Retardancy. *Composites Part B: Engineering* **2020**, *184*, 107673. https://doi.org/10.1016/j.compositesb.2019.107673.

(129) Zhu, S.; Zheng, T.; Li, F.; Wang, L.; Jiang, Q.; Wei, Y.; Liu, W. Synthesis of Cyclotriphosphazene-Containing Imidazole as a Thermally Latent Hardener for Epoxy Resins and Its Application in Carbon Fiber Reinforced Composites. *Journal of Applied Polymer Science* **2022**, *139* (41), e52980. https://doi.org/10.1002/app.52980.

(130) Zhu, X.; Fang, Z.; Deng, Q.; Zhou, Y.; Fu, X.; Wu, L.; Yan, W.; Yang, Y. Poly(Ionic Liquid)@PEGMA Block Polymer Initiated Microphase Separation Architecture in Poly(Ethylene Oxide)-Based Solid-State Polymer Electrolyte for Flexible and Self-Healing Lithium Batteries. *ACS Sustainable Chemistry & Engineering* **2022**, *10* (13), 4173–4185. https://doi.org/10.1021/acssuschemeng.1c08306.

(131) Chen, H.; He, P.; Li, M.; Wen, Y.; Wang, Y.; Qiu, J.; Cao, G.; Zhao, P.; Zhang, S.; Ming, H. Enabling Stable Cycling of 4.6 V High-Voltage $LiCoO_2$ with an in Situ-Modified PEGDA-Based Quasi-Solid Electrolyte. *ACS Applied Energy Materials* **2022**, *5* (4), 5170–5181. https://doi.org/10.1021/acsaem.2c00511.

(132) Zhu, Y.; Shen, H.; Ji, H.; Hua, M.; Pan, X.; Bussemaker, M.; Liu, L. Application of 1-Butyl-3-Menthylimidazolium-Hexafluorophosphate as Flame Retardant in Electrolyte of Lithium Ion Battery. *Journal of Loss Prevention in the Process Industries* **2022**, *79*, 104837. https://doi.org/10.1016/j.jlp.2022.104837.

(133) Song, J.; Liao, K.; Si, J.; Zhao, C.; Wang, J.; Zhou, M.; Liang, H.; Gong, J.; Cheng, Y.-J.; Gao, J.; Xia, Y. Phosphonate-Functionalized Ionic Liquid Gel Polymer Electrolyte with High Safety for Dendrite-Free Lithium Metal Batteries. *ACS Applied Materials & Interfaces* **2023**, *15* (2), 2901–2910. https://doi.org/10.1021/acsami.2c18298.

11 Non-Halogen Radical Generators for Enhanced Flame Retardancy of Plastics

Carl-Eric Wilen[1] *and Rudolf Pfaendner*[2]

[1] Faculty of Science and Engineering
Åbo Akademi University
Henriksgatan 2
20500 Åbo, Finland

[2] Fraunhofer Institute for Structural Durability and System Reliability LBF
Schlossgartenstr. 6, 64289 Darmstadt, Germany

11.1 INTRODUCTION

Thermal decomposition of plastics in a fire scenario is accompanied through the formation of molecular fragments with radical species. Many flame retardants interfere with these radicals to interrupt the self-accelerating process, for example, through formation of thermally stable intermediates or less burning reaction products. Therefore, the concept of using radical generators as potential flame retardants or synergists was proposed already in the sixties of the last century [1, 2], where it was shown that peroxides, azo-, sulfur-, bibenzyl-compounds, and hydrazones act synergistically with known brominated flame retardants. Consequently, combinations of hexabromocyclododecane (HBCD) and peroxides have been used as standard FR system in polystyrene foam applications until 2017 [3], and as of today HBCD has been replaced with novel polymeric brominated systems [4]. Alternative radical generators were found with cumyl and polycumyl acting in a similar way as peroxides with brominated compounds with the advantage of a somewhat higher thermal stability than peroxides [5, 6]. Furthermore, it was experimentally shown that radical generators of peroxides, hydroperoxides, and polysulfides do not alone act as flame retardants [7], and the established systems were not developed further for many years. At the end of the nineties, a novel area of radical generators acting as flame retardants showed up when alkoxyamines (NORs) were introduced into the market. These innovative findings were the nucleus for novel research activities for using radical generators as flame retardants and synergists on the basis of nitrogen compounds covering primarily alkoxyamines, azo compounds, sulfenamides, oxyimides, silylamines, and siloxyamines but as well sulfur compounds like disulfides.

11.2 RADICAL GENERATORS AND THEIR TENTATIVE MODES OF ACTION IN FLAME RETARDANCY

The entire combustion process of solid polymeric materials is a very complex coupling of energy feedback from a flame to the polymer surface with gasification of the polymer to generate combustible decomposition products. Combustion will occur when the above volatile substances accumulate to a certain concentration in air and heat.

The polymer combustion process begins when the polymer is subjected to sufficiently high temperature capable of breaking chemical bonds that form the polymer backbone or its side chains. The thermal decomposition of the polymer matrix gives rise to volatile fragments via pyrolysis of the solid polymer via random scission, depolymerization, and/or side group elimination reactions (see Chapter 2 of this book for more information). In general, the polymer decomposition reactions have been divided into three types of reactions: the primary reaction involves the formation of wax-like intermediates, condensable and non-condensable volatiles and/or char, whereas the second and third type of reactions involve cracking or polymerization of the primary volatile products. The combustible components evolved through pyrolysis are mixed with air and are potentially ignited to initiate a fire and thereby releasing free radical chain reactions and heat. A series of reactions among volatile free radicals, such as hydrogen radicals and hydroxyl radicals, and oxygen in the gaseous phase are mainly responsible for sustaining polymer combustion. As a result, the polymer combustion processes can be affected either by free radical reactions that alter the pyrolysis process, by reactions in the gaseous phase or in the condensed phase by forming protective carbonaceous structures.

DOI: 10.1201/9781003380689-11

Studies on the combustion process of thermoplastic polymers have normally been divided into four stages: (1) melting, (2) thermal decomposition, (3) ignition, and (4) flame spread. The combustion process of thermoplastic polymers will cause melt dripping or melt flow, which directly affects the flame spread behavior. Continuous heating on the molten mass will result in pyrolysis and combustion. The presence of melt flow and melt dripping behavior has the potential to accelerate or impede flame spread. The flame spread and combustion behaviors of solid combustibles will be affected by many environmental factors in real fire scenarios.

The mode of actions of flame retardants based on different radical generators of polymeric materials is yet not fully understood, and several different mechanisms have been postulated. The prevailing modes of actions reported so far are strongly dependent on the types of radical generators used and on the chemical and physical composition/chemical structure of the polymer matrix to be flame retarded. In Table 11.1 non-halogen radical generators used for flame retardancy of polymeric materials have been listed and the types of atom-centered radicals they give rise to upon thermal dissociation.

It has been proposed that radical generators may exert their flame retarded effect in the condensed phase, gaseous phase, or in the dual phases. For instance, it has been found that radical generators based on modified carbon-based and metal-derived metal nanomaterials can increase the pyrolysis temperature by free-radical scavenging in the solid phase, whereby they prevent or delay the generation of combustible gases and the thermal decomposition of the polymeric material [3]. The higher fire resistance has been proven experimentally, that is, the time to ignition (TTI) has been significantly prolonged and lower peak heat release rate (PHRR) values have been recorded in the presence of 1 wt% radical generator in the polymer matrix compared to without the addition of radical generator. The nanomaterial-based free radical capturing derivatives show a polymer-matrix-dependent characteristics [8, 9]. Whereas in contrast, radical generators based on, for example, alkoxyamines have been shown to promote depolymerization by decreasing the pyrolysis temperature, whereby the formed polymer fragments of low molecular weight are rapidly withdrawn from the flame zone in the form of droplets that either drip off or are immediately consumed in the combustion process. This type of melt away mechanism is also common for many other free radical generators such as oxyimides, peroxides, triazene [10], and azo [11] compounds.

On the other hand, radical generators containing sulfur such as elemental sulfur, sulfide, sulfenamides, and disulfide generate sulfur radicals, which besides aiding the polymer fragmentation also promote cross-linking at the polymer surface. Moreover, S-containing flame retardants during thermal decomposition generate sulfur dioxide (SO_2) in the gas phase, which acts not only as a fuel-diluent, but can also act as a radical-scavenger that suppress combustion efficiency in the gas phase [12, 13].

In general, it has been demonstrated that several radical generators exhibit a very strong synergistic effect and/or they act as adjuvants with conventional flame retardants based on halogen-, phosphor-, aluminum-containing flame retardants and in various intumescent flame retardant systems. Thus, the radical generator class of flame retardants also represents a promising strategy to decrease the amount of conventional flame retardants in polymeric materials to achieve desired high fire-retardant ratings such as UL94-V0. Moreover, the mentioned organic radical generators can replace problematic inorganic synergists such as antimony oxide in halogenated flame retardant formulations.

11.3 ALKOXYAMINES

Alkoxyamines (NORs or nitroxylethers) have been known since 1990 as improved versions of hindered amine light stabilizers (HALS) first for coatings [14] and later for agricultural greenhouse or mulch films even under severe acidic conditions, for example, in the presence of acidic pesticides.

TABLE 11.1
Different Halogen-Free Radical Generators and Thermolysis Thereof

Halogen-Free Radical Generators	Radicals Generated via Thermolysis
Peroxides	Oxygen and carbon-centered radicals
Dicumyl	Carbon-centered radicals
Azoalkanes	Carbon-centered radicals
Alkoxyamines	Carbon, oxygen, and nitrogen-centered radicals
Disulfides	Carbon and sulfur-centered radicals
Oxyimides	Carbon, oxygen, and nitrogen-centered radicals
Sulfenamides	Carbon, sulfur, nitrogen-centered radicals
Silylamines	Carbon, silyl, nitrogen-centered radicals
Siloxyamines	Carbon, silyl, oxygen, and nitrogen-centered radicals
Carbon-based nanomaterials	Carbon and oxygen-centered radicals
Phosphorus (where P is at low oxidation state of -3, 0, +1, or +3) flame retardants	Phosphorus-centered radicals (see Chapter 5)

The NOR structures show a lower basicity compared to classical -NH-HALS structures. In addition, advanced NOR structures outperform HALS as photo- and long-term thermal stabilizers. By working on novel alkoxyamines, researchers at Ciba Specialty Chemicals realized that selected structures act as flame retardants without any other flame retardant primarily in polypropylene fibers and films [15].

The first commercially promoted NOR flame retardant was Flamestab NOR 116 (Figure 11.1 (a), now BASF SE), which is still the only larger scale nitrogen-based radical generator flame retardant on the market. However, the flame-retardant performance of other commercial alkoxyamines is proven as well. For example, Tinuvin NOR 371 (Figure 11.1 (b), supplier BASF SE [16]) or ADK Stab LA-81 (Figure 11.1 (e), supplier Adeka [17]), Hostanox NOW (Figure 11.1 (c), supplier Clariant SE [18, 19] and various other NOR structures have been evaluated, e.g., hydroxy-methylpropyl HALS derivatives (Figure 11.1 (d), [16]). In addition, several other chemical structures related to alkoxyamines are claimed to provide flame retardancy, namely, hydroxylamine esters [20], hydroxylamines and nitroxyl radicals [21]. Another beneficial feature of NOR flame retardants is the contribution to UV stability and long-term thermal stability of the polymers [22]. Applications of NOR flame retarded polymers are often limited to polyolefins, that is, mainly polypropylene and polyethylene films, nonwovens, or fibers.

Efficient loadings to fulfil film (e.g., DIN 4102 B2) and fiber standards (e.g., NFPA 701 fiber test) are in the range of 0.5 to 2.0%. Weaknesses of the NOR approach are that more severe standards (e.g., UL 94 V-0) can be hardly achieved, and the overall thermostability is limited within the range up to 230°C, where early decomposition and color formation may start depending on the process conditions.

The activity of the alkoxyamine as flame retardant is based on the thermolysis, which leads to the formation of either alkoxy and aminyl radicals or alkyl and nitroxyl radicals (Figure 11.2). Aminyl and alkoxy radicals are very reactive and cause, on the one hand, decomposition of polypropylene (and cross-linking of polyethylene). On the other hand, they are involved in the free radical chemical reactions during the combustion process [23].

Performance improvements of alkoxyamine flame retardants have been addressed through identifying synergism

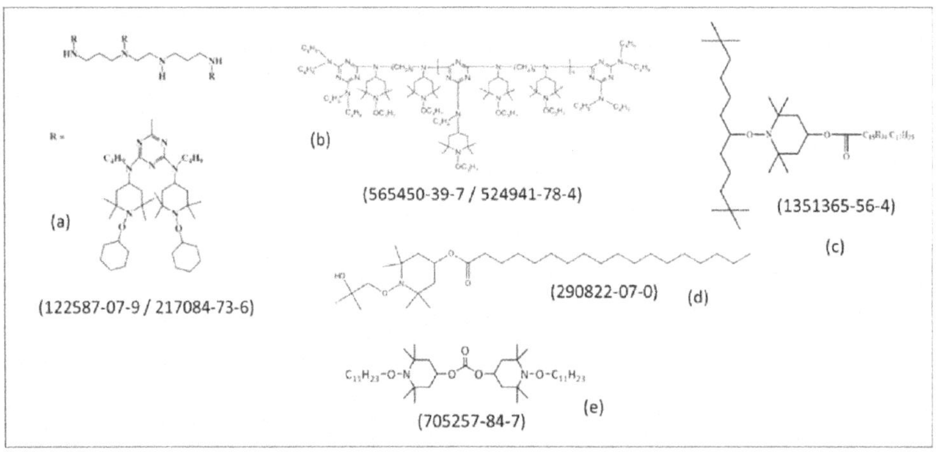

FIGURE 11.1 Examples of NOR flame retardants and their corresponding CAS numbers.

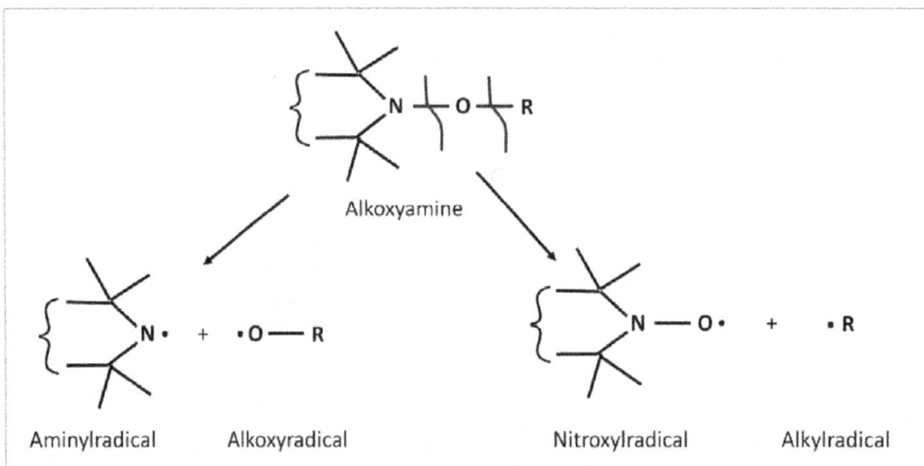

FIGURE 11.2 Thermolysis of alkoxyamines.

with other conventional flame retardants and through structural variations. The influence of the R-substitution pattern is not clearly expressed; the performance as flame retardant seems to be more related to the thermal stability. Therefore, NO-Methyl shows somewhat improved performance in PP versus the NO-Cyclohexyl of Flamestab 116 [24]. Substantially improved structural versions can be achieved through increasing the possibilities of forming radical species. This was achieved through combining azo structures with NOR in one molecule [25]. The resulting Azo-NOR structures clearly outperform concentration-wise the classical NOR structures in flame retardancy and pass more stringent fire tests, for example, UL 94 VTM-2 in concentrations between 0.25 and 1% in polypropylene.

Despite the synthetic approaches to enhance the fire performance of NOR molecules, commercial focus has been more on combining conventional flame retardants with NOR to identify synergistic behavior. For example, alkoxy amines can interact with brominated flame retardants and facilitate the release of bromine, consequently increasing the overall flame-retardant performance. In detailed analysis it could be experimentally shown that the alkoxy amine acts in the condensed and gas phase, modifies the decomposition pathway of phosphoric acid tris-(3-bromo-2,2-bis-bromomethylpropyl)ester and polypropylene, while lowering the temperature to release HBr as flame quenching species and the temperature of dripping of the polymer [26, 27]. Consequently, NOR can be used as antimony trioxide replacement in brominated systems.

Preferred synergistic partners of NORs today are phosphonates introduced through the company Thor [21], especially dimethylspirophosphonate (Aflammit PCO 900, Figure 11.3). In comparison to the individual substances, the combinations of dimethylspirophosphonate and Flamestab NOR 116 results in an increased LOI of 28.5% versus 25% at a concentration of 2% [28]. Moreover, an excess of the phosphonate seems to improve the performance further. On the same basis, three component mixtures with synergistic benefits have been published, for example, Flamestab NOR 116, phosphonate and melamine cyanurate [29] or silica [30].

With regard to circular economy and mechanical recycling, it could be shown that multiple extrusion of NOR flame retarded PP film material had no influence on mechanical properties and flame-retardant performance, meaning that such compositions have good potential for recyclability [31].

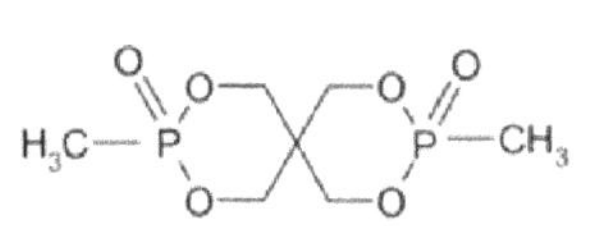

FIGURE 11.3 Dimethylphosphonate synergist for NOR flame retardant compositions and its CAS number.

11.4 SULFUR-CONTAINING RADICAL GENERATORS

Up to date, many studies have shown that flame retardant efficacy can be improved by introducing sulfur as building blocks in the flame-retardant structure in a variety of polymeric materials. Sulfur can show different FR mechanisms depending on the nature of the polymer where it has been utilized. For instance, investigations into the flame-retardant action of elemental sulfur, sulfides, and disulfides have revealed that these compounds decompose to form sulfur radicals, that induce a fast decomposition of the polymer chain, and flame retardancy is achieved by removing the substrate from the flame, or the sulfur radicals can take part in the flame inhibition process. They may also promote cross-linking reactions since sulfur-containing FR produces sulfonic acid species during combustion, which trigger decomposition and char formation. Whereas, sulfone-containing FRs have been shown to release sulfur dioxide into the gas phase, which besides diluting the flammable volatiles may also act as a radical-scavenger. It has been reported that disulfides themselves function as effective flame retardants for polypropylene films at already very low concentrations of 0.5 wt% [32].

Moreover, sulfur-containing systems have been exploited in combination with phosphorus-based flame retardants. It has an enhanced FR effect due to synergism between two atoms for epoxy resins, expanded polystyrene, polycarbonate, thermoplastic urethanes, and in various textiles. Other sulfur-based flame retardants include sulfamic acid-based salts, that is, ammonium sulfamate, or diammonium imidobisulfonate for flame retardancy of wood products.

11.5 OXYIMIDES

Oxyimides with the general structure as shown in Figure 11.4 have been developed with the idea to overcome some weaknesses of alkoxyamine structures like thermal instabilities. By introducing the imide ring, the range of thermal stabilities and adjustment of thermal radical formation through structural variations were considerably extended. Depending on the structure, the radical formation temperature can be tailor-made from below 200°C to more than 350°C (Figure 11.5). The chemical structures of oxyimides comprise aliphatic and aromatic esters [33], ethers [33], phosphorus- [34], sulfone- [35] derivatives, polymers [36], and salts [37]. Furthermore, structural variations allow the synthesis of meltable products or as well filler-like flame retardants.

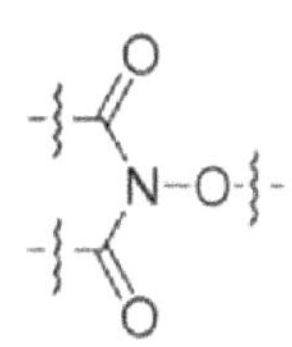

FIGURE 11.4 General structure of oxyimides.

Oxyimide	Thermal decomposition [°C] Decomposition onset [TGA, N_2]
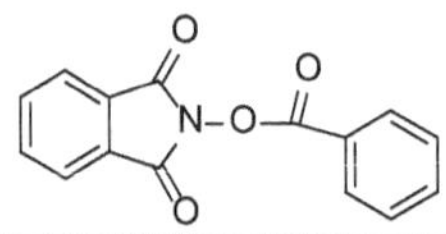	237
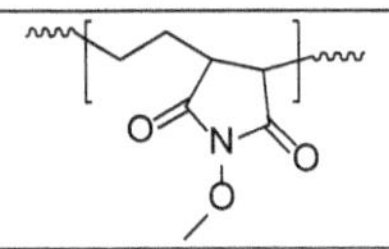	259
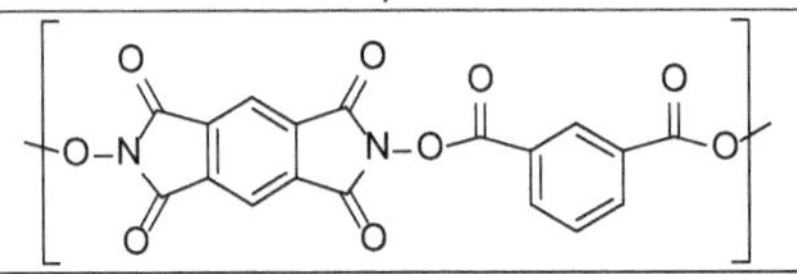	263
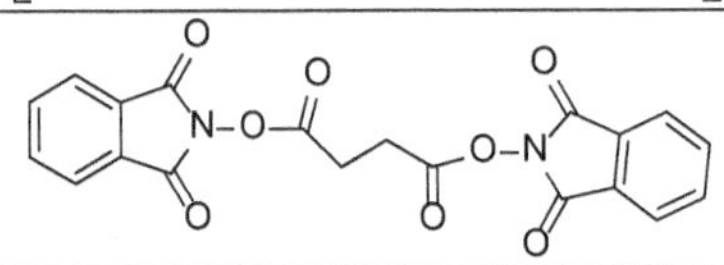	271
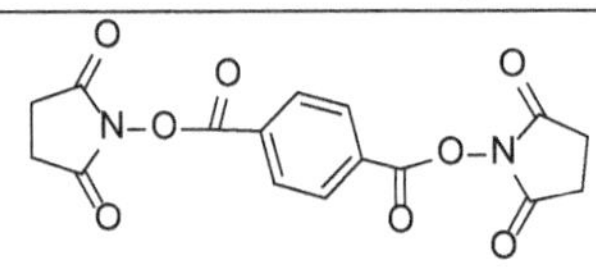	275
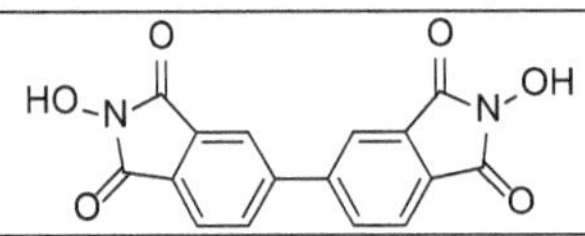	283
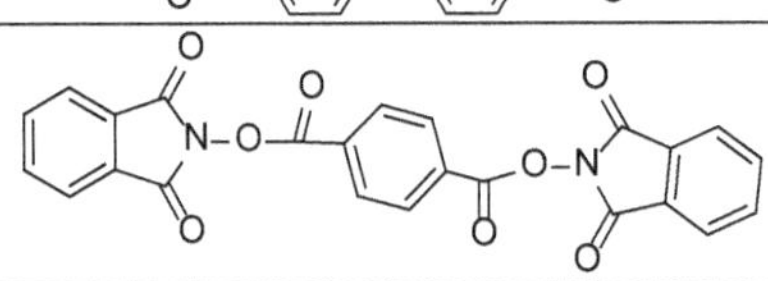	314
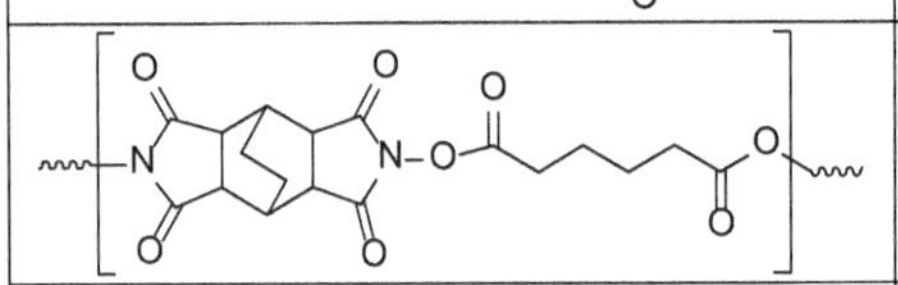	335
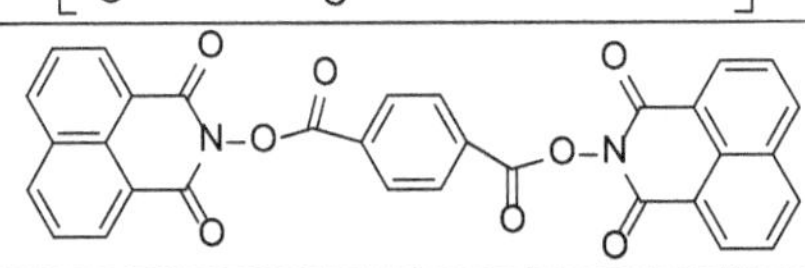	377
Flamestab NOR 116	250

FIGURE 11.5 Selected oxyimides and decomposition temperature.

Performance of oxyimides as single flame retardants in polyolefins is comparable to alkoxyamines with somewhat higher efficiency in some cases, for example, by using sulfone derivatives where DIN 4102 B2 standard is passed at a loading of 0.5% (230 μm film). In most case studies synergistic combinations with phosphorus flame retardants were used, for example, with dimethylspirophosphonate (Figure 11.3), whereas UL 94 V-0 (1.6 mm) for polyolefins can be achieved in considerably low concentrations around 6%. However, low melt flow PP needs higher concentrations as well as polyethylene (Figure 11.6, [38]). Further synergistic effects of oxyimides have been verified with phosphates, metal-phosphinates, phosphazenes, metalhypophosphites, and even aluminum hydroxide. Main polymer application area has been polyolefins of different structures, thermoplastic polyurethanes, ABS, engineering plastics, polylactic acid, and epoxides. The mechanism of oxyimides is again a dripping mechanism with accelerated decomposition of the polymer ("run-away effect"), whereas it is essential that the drips are non-burning. Therefore, all polymers with the trend of thermal decomposition into lower molecular weight fragments are preferred substrates to be flame protected through oxyimides. In polymers tending to cross-linking, carbonization or barrier formation oxyimides can synergistically support the formation process through radical reactions.

Because of the possibility to synthesize oxyimides with high thermal stability, engineering plastics such as polyamides [39] or polyesters [40] are accessible as potential substrates to be fire protected. As an example, polyamide-6 may be flame protected through oxyimide ethers compared

Polymer	Oxyimide[a)] [loading. %]	Phosphonate[b)] [loading %]	Total FR [loading %]	Burning time UL94 [s]	Classification
PP	1 [2]	8	10	1	V-0
PP	1 [4]	6	10	0	V-0
PP	1 [5]	5	10	1	V-0
PP	1 [2]	6	8	1.5	V-0
PP (MFR 2.3)	1 [2]	5	7	1.3	V 2
PP (MFR 2.3)	1 [2]	15	17	0	V-0
PP (MFR 13)	1 [1.5]	5	6.5	0	V-0
PP (MFR 35)	1 [1.5]	5	6.5	0	V-0
PP	2 [2]	4	5.5	0	V-0
HDPE	1 [4]	8	12	0	V-0
LDPE	1 [3]	8	11	0	V-0
LLDPE	1 [2]	15	17	0	V-0

a) b)

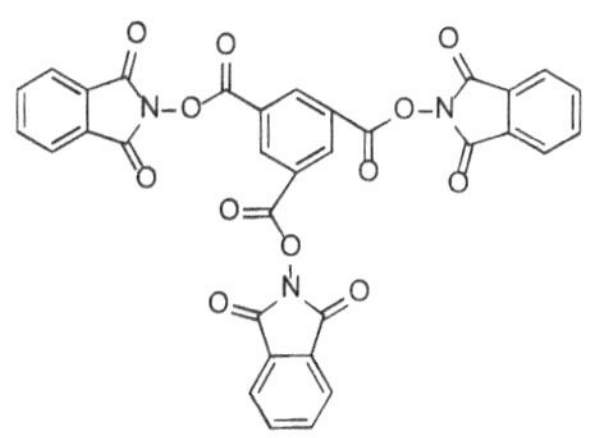

Oxyimide 1

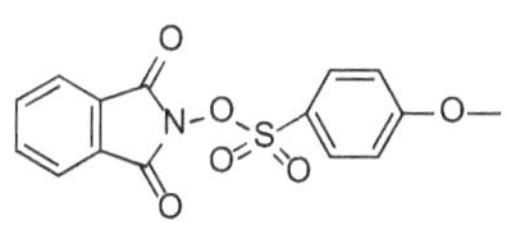

Oxyimide 2

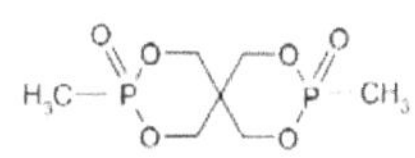

Phosphonate

FIGURE 11.6 Synergistic oxyimide flame retardant combinations for polyolefins.

to the classical melamine cyanurate protection (in both cases non-burning dripping mechanism is the main mode of flame-retardant action) at much lower concentrations resulting in V-0 (1.6 mm) classification at a loading of 3–5% [39].

11.6 SULFENAMIDES, SULFINAMIDES, AND SULFONAMIDES

In recent years, several radical generators have become an important branch of flame retardants and especially their utilization as synergists in combination with conventional flame retardants have grown in popularity. As of today, the efficacy of most radical generators is strongly dependent on the type of polymer matrix and its thickness. As a rule of thumb, radical generators need to be designed to match various polymers with inherent combustion characteristics.

Therefore, in many practical applications, there is still room for improvement and optimization of radical generator structures that would provide routes to more efficient flame retardancy of different polymeric materials and end use products.

Both academia and industry are still looking for more effective classes of radical generators. Especially, radical generators that contain organo-heteroatoms such as nitrogen, sulfur, oxygen, silicon, and phosphor in various combinations have received great attention [41–55]. In theory, such radical generators may have increased ability to inhibit the free radical reactions that initiate the thermal pyrolysis and sustain the fire of polymeric materials. It is foreseeable that the diverse chemistry provided by new generations of organo-heteroatom based radical generators could interfere with the entire combustion cycle of polymeric materials at various stages, that is, during thermal pyrolysis, ignition, flame propagation and spread, completed combustion, and flame attenuation.

Several sulfur-nitrogen compounds such as (1) sulfenamides, (2) sulfinamides, and (3) sulfonamides, as shown in Figure 11.7, have proven their potential as flame retardants for polymeric materials [56].

Despite the structural similarities, some differences in efficacy and mode of action of these sulfur-nitrogen compounds have been observed. A common nominator for all of them is a nitrogen-sulfur bond of low basicity, which is stable at normal polymer processing temperatures of polypropylene, but rapidly decomposes under fire conditions. Sulfenamides thermally dissociate to give simultaneously both aminyl- and sulfur-based radicals that display high flame retardant efficacy as standalone flame retardants in polystyrene (EPS) panels, polyacrylate coatings, and polypropylene

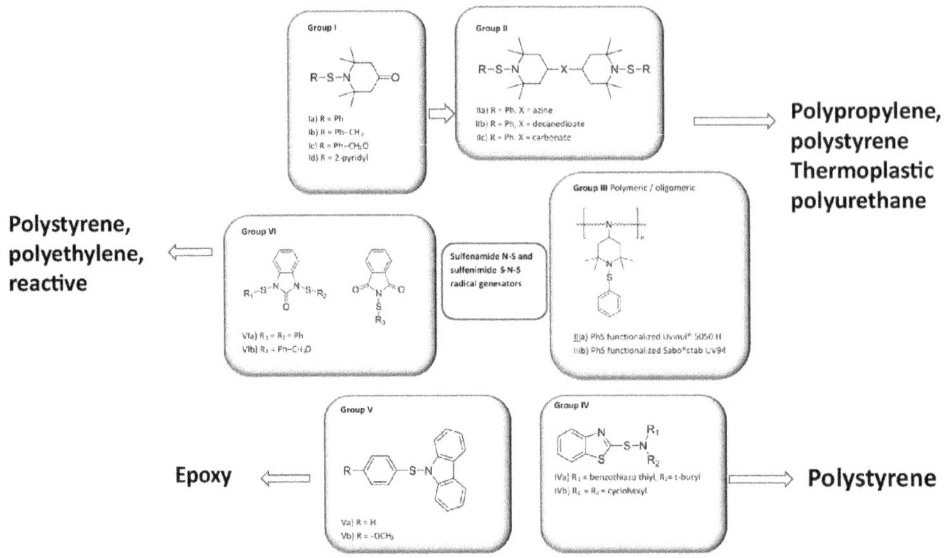

FIGURE 11.7 Road map for the selection of sulfenamide structures versus polymeric material to be fire retarded.

films [57]. The flame-retardant mechanism and efficacy have been investigated by performing standard fire tests such as DIN4102 and LOI tests in combination with thermal analysis (TGA/DSC). Moreover, chemical analysis of the decomposition products of the neat sulfur-nitrogen-based flame retardants in the presence of 9,10-dihydroanthracene (HA) have been conducted by NMR and GC MS analysis. Thermal decomposition has been performed for these compounds under an inert atmosphere in sealed pressure reactors at 310°C for ten minutes. The experiments revealed that both sulfenamides and sulfinamides initially form free radicals by facile cleavage of the sulfur-nitrogen bonds and the generated free radicals subsequently react with 9,10-dihydroanthracene (HA) to give anthracene. On the contrary, the experiments failed to provide evidence of the facile cleavage of sulfonamides into radical species at these experimental conditions, since 9,10-dihydroanthracene (HA) was not converted to anthracene under the used experimental conditions. In Scheme 11.1, tentative mechanisms for the decomposition of these compounds are presented.

It has been shown that sulfenamides exhibit a strong synergistic effect in various polymer formulations together with traditional halogen-free flame retardants such as alumina trihydrate (ATH), aluminum hypophosphite (AHP), phosphazene, and organic spirophosphonates (Figure 11.3) [58].

The synergistic efficacy is especially strong when the sulfenamides are combined with spirophosphonate, and the UL 94 V-0 classification can be reached at a total FR loading of only 10 wt%. The total FR loading can still be reduced to 9 wt% by shifting to a ternary FR system consisting of spirophosphonate (6 wt%), phosphazene (2 wt%), and SN-I (1 wt%). In the presence of 30 wt% of ATH, the UL 94 V-2 classification (with close to zero after flame burning times) has been reached when adding 1 wt% of SN-I.

Thermoplastic polyurethane (TPU) together with 40 wt% of ATH provides UL-94 V2 rating, whereas TPU in combination of 20% ATH together with 1 wt% of sulfenamide helps to ensure V0 in the UL-94 test. For epoxy resins based on bisphenol A diglycidyl ether (DGEBA)/isophorone diamine (IPDA), the combination between conventional phosphorous (P content 1.5 wt%) flame retardant together with 1 wt% of sulfenamide decreased the burning time to 2 seconds from 20 seconds when using the same phosphorous flame retardant by itself at the same loading.

Noteworthy is that it has been shown for the first time that multifunctional sulfenamides can even pass the UL-94 V2 criteria at a loading of only 4 wt% as a standalone flame retardant for polypropylene (first example of non-halogenated radical generator that can pass UL-94 V2 alone at 4 wt% loading) [59]. The structures of such multifunctional sulfenamides together with UL 94 V test results are shown in Table 11.1.

11.7 RADICAL GENERATORS BASED ON SILYLAMINE (N-SI) AND SILOXYAMINE (N-O-SI)

In the past it has been reported that silicon radicals can play a vital role in photochemically induced radical polymerizations by acting as important intermediates in organosilicon chemistry, as ionic liquids in fuel cells, and more recently as synergists in flame retardancy of various polymers. Silicon radicals can be generated from, for example, siloxanes or cyclic aminylsilanes. Silicon as a part of flame retarding system has been used either as a coating for plastics than within the polymer matrix itself or as a part of the matrix as a charring agent at high temperatures. Several studies have shown the improvement in flame retardancy by combining phosphorus-nitrogen and silicon moieties in flame retardant formulations. Noteworthy is that in these formulations the nitrogen and silyl atoms have not been chemically bonded to each other. Typically, the added silicon additives are in

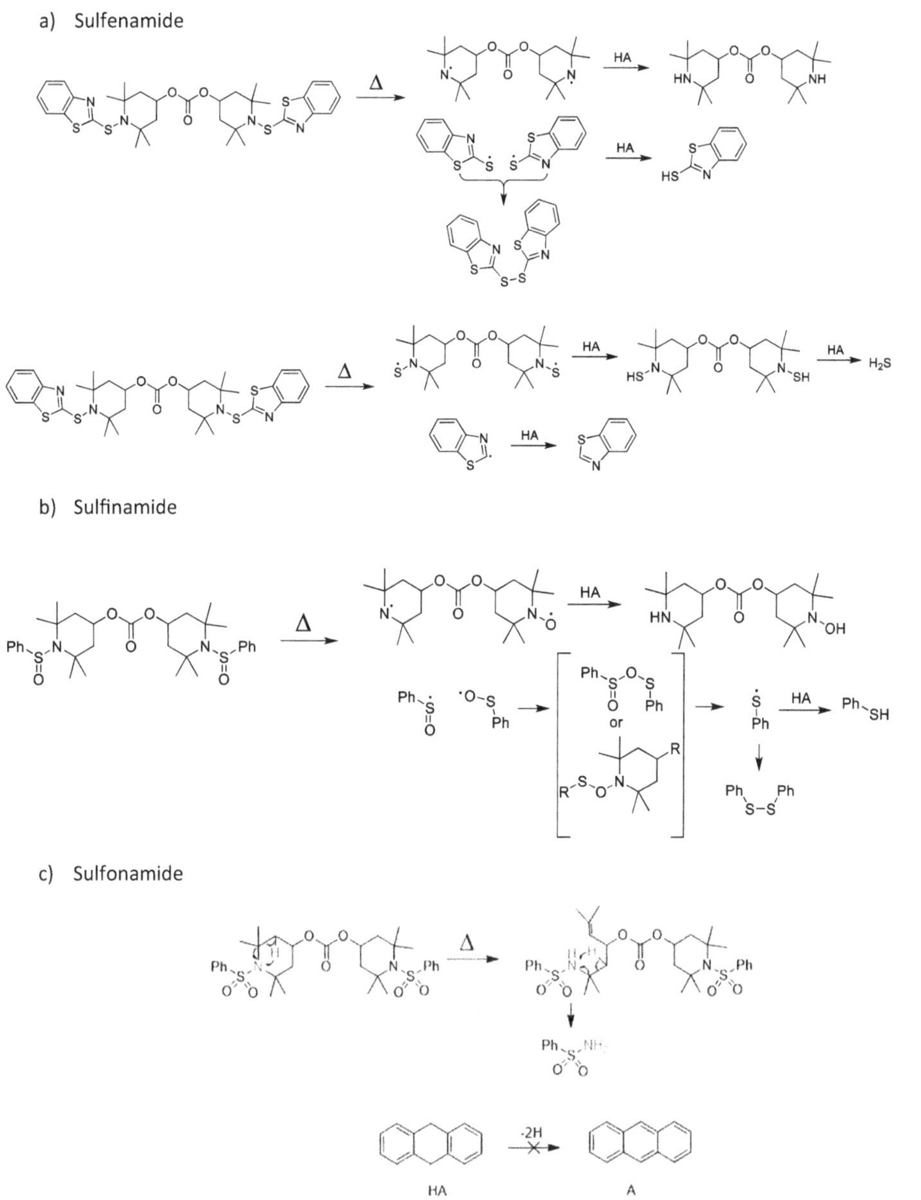

SCHEME 11.1 Thermal decomposition of (a) sulfenamide, (b) sulfinamide, and (c) sulfonamide in the presence of 9,10-dihydroanthracene (HA).

polymeric or oligomeric forms and require relatively high loadings of ca. 10–20 wt%. The silicon containing additives' contribution to flame retardancy has been mainly attributed to silicon compound's ability to form protective silica layer shields to protect the polymer from further thermal decomposition and its ability to retard oxidation of the char.

In general, for the phosphorus-nitrogen and silicon flame retardant formulations, it has been shown that by using the three diverse elements together, a high flame-retardant effect can be achieved, whereas when using the compounds as single elements, it has been very difficult to reach any good flame-retardant properties. Recently, the use of different aminylsilyl and aminylsiloxy compounds as synergistic radical generators at low loadings of 1–2 wt% together with traditional phosphorous flame retardants has been reported [60–62]. The results indicate that their thermal and hydrolytic stabilities could be adjusted by refining the silylamine (N-Si) and siloxyamine (N-O-Si) skeletons. Upon thermal dissociation, the N-Si and N-O-Si compounds were shown to yield aminyl, silyl, and oxygen-centered radicals that provided fire proofing effect in polypropylene (PP) films alone and a strong synergistic effect in combination with conventional phosphorous flame retardants in PP, linear low-density polyethylene (LLDPE), and epoxy resins (EP), as shown in Table 2.

The thermogravimetric analysis carried out under inert atmosphere revealed enhanced and earlier onset of

TABLE 11.1
Polypropylene Plaques of 1.6 mm Containing 4 wt% of Multifunctional Sulfenamide

Code	Polymer	Radical Generator [wt%]	UL 94V Rating (1.6 mm)
1	PP	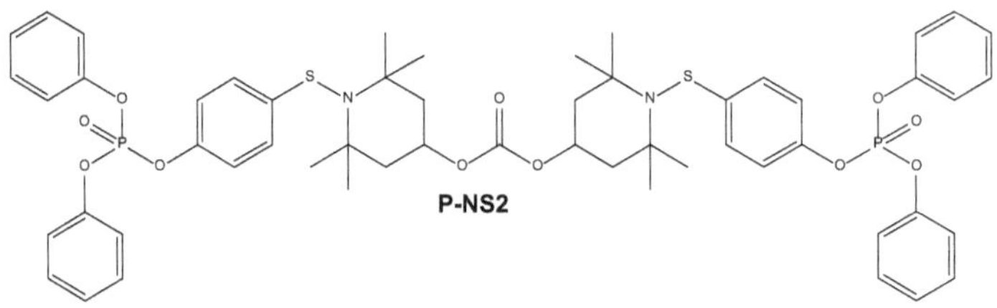 P-NS2 [4wt%]	V-2 rating
2	PP	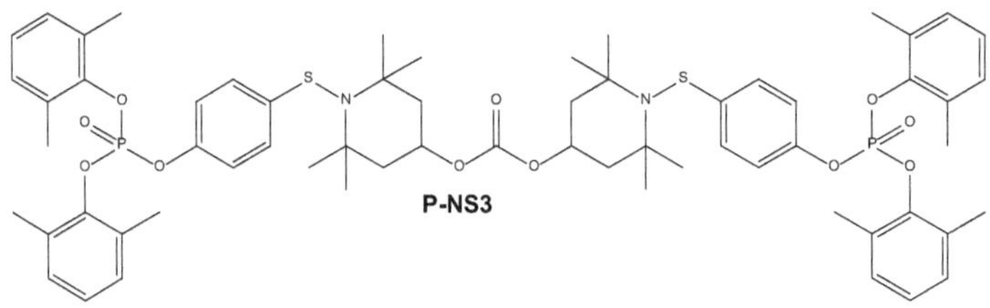 P-NS3 [4wt%]	V-2 rating

TABLE 11.2
UL-94 Results of Silylamine or Siloxyamine Compounds in Combination with Phosphonate (Figure 11.3) (P1) in PP, LLDPE, and Epoxy Resin

Sample	Radical Generator (Siloxyamine or Silylamine)	Loading of Silylamine/wt%	Phosphorous FR	Loading of Phosphorous FR /wt%	UL 94 Classification	Plaque Thickness/ mm
			Polypropylene (PP)			
1	-	-	-	-	Fail	1.6
2	-	-	P1	10	Fail	1.6
3	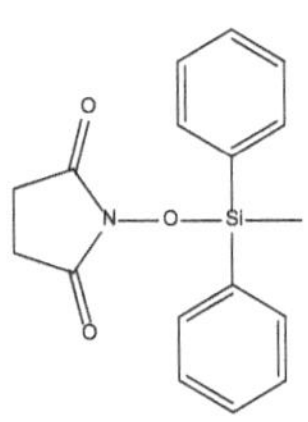	2	P1	10	V0	1.6
			Linear low-density polyethylene (LLDPE)			
4	-	-	-	-	Fail	3.2
5	-	-	P1	15	V2	3.2
6	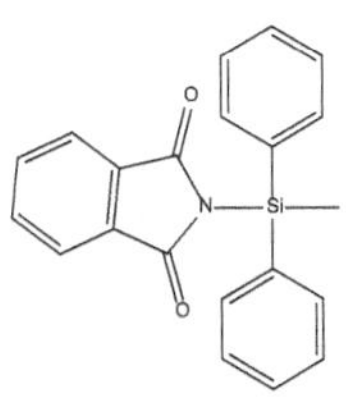	1	P1	15	V0	3.2

Sample	Radical Generator (Siloxyamine or Silylamine)	Loading of Silylamine/wt%	Phosphorous FR	Loading of Phosphorous FR /wt%	UL 94 Classification	Plaque Thickness/ mm
			Epoxy resin			
7	-	-	P1	2.5	Fail	1.6
8	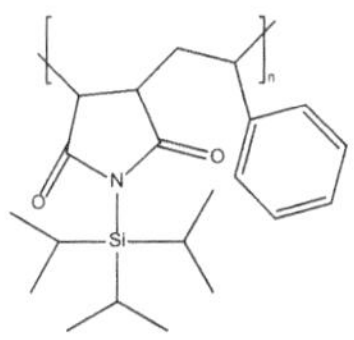	2	P1	2.5	V0	1.6

decomposition and thermal activation of phosphorous-based flame retardant in the presence of N-Si or N-O-Si in comparison to spirocyclic phosphonate, N-Si, or N-O-Si used alone, respectively.

11.8 CARBON-BASED NANOCOMPOSITES

In recent years several research groups have explored the potential of utilizing various carbon-based nanocomposites such as fullerene, carbon nanotubes (CNT), carbon black as fire retardants. Their efficacy has been ascribed to nanocomposites' ability to trap free radicals mainly in the condensed phase, whereby the polymer pyrolysis rate is retarded. Therefore, they do not work as radical generators and are often antagonistic with these types of flame retardants. While nanocomposites do reduce heat release when put into polymers, they fail to provide sufficient fire resistance in regulatory tests when used alone and must therefore be used in combination with traditional flame retardants, as discussed in Chapter 12 of this book.

11.9 SUMMARY

Radical generators are still a small but growing market in the landscape of flame retardants. There are well-established applications in films, fibers, and foams, and the future of radical generators as flame retardants looks promising. Chemistries investigated today cover a wide range of structures from alkoxyamines to oxyimides, sulfenamides, and silyl structures. The formed C, N, S, or Si-centered radicals act by different mechanisms covering accelerated decomposition and runaway effect, cross-linking in the condensed phase and radical reactions in the gas phase. As a common advantage of radical generators loadings to pass the relevant tests can be considerably lower compared to other solutions specifically in polyolefins. Combinations with synergistic conventional flame retardants, for example, phosphorus-based structures open additional opportunities for novel formulations in polyolefins, polystyrenes, polyurethanes, epoxy resins, and engineering plastics. Further growth of radical generators can be expected when chemistries are more refined, and their interactions with various polymers and other families of flame retardants are better understood.

REFERENCES

[1] Eichhorn, J. Synergism of Free Radical Initiators with Self-Extinguishing Additives in Vinyl Aromatic Polymers. *J. Appl. Polym. Sci.* **1964**, *8*, 2497–2524.

[2] US 3058928 (1962), inv. J. Eichhorn, S.I. Bates, assigned to Dow Chemical Corp.,

[3] M. Modesti, A. Lorenzetti in Fire Retardancy of Polymeric Materials, 2nd edition, C. A. Wilkie, A. B. Morgan (eds.), Boca Raton 2010, p. 763–781.

[4] Beach, M. W.; Hull, J. W.; King, B. A.; Beulich, I. I.; Stobby, B. G.; Kram, S. L.; Gorman, D. B. Development of a New Class of Brominated Polymeric Flame Retardants Based on Copolymers of Styrene and Polybutadiene. *Polym. Degrad. Stab.* **2017**, *135*, 99–110.

[5] US 3457204, inv.: H. Burger, G. Daumiller, J. Grohmann, E. G. Kastning, H. Weber, H. Willersinn, assigned to BASF AG.

[6] US 3420786, inv.: H. Weber, H. Burger, G. Daumiller, H. Willersinn, J. Grohmann, assigned to BASF AG.

[7] Chiotis, A.; Clouet, G.; Brossas, J. Effect of Free Radicals Generating Compounds Towards the Flammability of W-Phosphonated Polystyrenes. *Polym. Bull.* **1984**, *11*, 275–280.

[8] Joseph, P.; Tretsiakova-McNally, S. Melt-Flow Behaviours of Thermoplastic Materials Under Fire Conditions: Recent Experimental Studies and Some Theoretical Approaches. *Materials.* **2015**, *8*, 8793–8803.

[9] Sai, T.; Ran, S.; Guo, Z.; Song, P.; Fang, Z. Recent Advances in Fire-Retardant Carbon-Based Polymeric Nanocomposites Through Fighting Free Radicals. *SusMA.* **2022**, *2*, 411–434.

[10] Pawelec, W.; Aubert, M.; Pfaendner, R.; Hoppe, H.; Wilén, C.-E. Triazene Compounds as a Novel and Effective Class of Flame Retardants for Polypropylene. *Polym. Degrad. Stab.* **2012**, *97*(6), 948–954.

[11] Nicolas, R.; Wilén, C.-E.; Roth, M.; Pfaendner, R.; King, R. E. Azoalkanes: A Novel Class of Flame Retardants. *Macromol. Rapid Commun.* **2006**, *27*(12), 976–981.

[12] Battig, A.; Markwart, J. C.; Wurm, F. R.; Schartel, B. Sulfur's Role in the Flame Retardancy of Thio-Ether–Linked Hyperbranched Polyphosphoesters in Epoxy Resins. *Eur. Polym. J.* **2020**, *122*(5), 109390.

[13] Howell, B. A.; Daniel, Y. G. The Impact of Sulfur Oxidation Level on Flame Retardancy. *J. Fire Sci.* **2018**, *36*(6), 518–534.

[14] Bramer, D.; Holt, M. S.; Mar, A. Hindered Amino Ethers: A New Class of Light Stabilizers for Coatings. *Polym. Mater. Sci. Eng.* **1990**, *63*, 647–653.

[15] WO 99/00450, inv.: D.W. Horsey, S. M. Andrews, L. H. Davies, D. D. Dyas, R. L. Gray, A. Gupta, B. V. Hein, J.

S. Puglisi, R. Ravichandran, P. Shields, R. Srinivasan, assigned to Ciba Specialty Chemicals Holding Inc.
[16] WO 2017013028, inv.: G. Lips, D. Mueller, T. F. Thompson, S. M. Andrews, assigned to BASF SE.
[17] WO 2013136285, inv.: G. Lips, G. Huber, H. Hoppe, A. Le Gal, assigned to BASF Schweiz AG.
[18] WO 2019092163, inv.: S. Hoerold, T. Wang, J. P. Lens, Y. Jeong, assigned to Clariant Plastics and Coatings Ltd., FRX Polymers Inc.
[19] WO 2015010775, inv.: S. Hörold, M. Zäh, assigned to Clariant International Ltd.
[20] WO 2003087211 inv.: M. Roth, D. Simon, G. Leslie, P. Nesvadba, R. E. King, N. Kaprinidis assigned to Ciba Specialties Holding Inc.
[21] WO02/074847 inv.: M.V. Troutman, R. Ravichandran, R. Srinivasan, R. E. King, D. W. Horsey, assigned to Ciba Specialty Chemicals Holding Inc.
[22] Pilar, J.; Michalkova, D.; Sedenkova, I.; Pfleger, J.; Pospisil, J. NOR and Nitroxide-Based HAS in Accelerated Photooxidation of Carbon-Chain Polymers; Comparison with Secondary HAS: An ESRI and ATR FTIR Study. *Polym. Degrad. Stab.* **2011**, *96*, 847–862.
[23] N. Kaprinidis, J. Zingg, Presentation, Additives 2003, Executive Conference Management, San Francisco, USA, 2003.
[24] C. E. Wilen, R. Pfaendner, Polymer Green Flame Retardants ed. by C. D. Ppaspyrides, P. Kiliaris, Amsterdam 2014, 267–288.
[25] Aubert, M.; Wilen, C. E.; Pfaendner, R.; Kniesel, S.; Hoppe, H.; Roth, M. Bis(1-Propyloxy-2,2,6,6-Tetramethylpiperidin-4-yl)-Diazene, an Innovative Multifunctional Radical Generator Providing Flame Retardancy to Polypropylene Even After Extended Artificial Weathering. *Polym. Degrad. Stab.* **2011**, *96,* 328–333.
[26] Marney, D. C. O.; Russell, L. J.; Stark, T. M. The Influence of an N-alkoxy HALS on the Decomposition of a Brominated Fire Retardant in Polypropylene. *Polym. Degrad. Stab.* **2008**, *93,* 714–722.
[27] Marney, D. C. O.; Russell, L. J.; Soegang, T. M.; Dowling, V. P. Mechanistic Analysis of the Fire Performance of a Fire Retardant System. *J. Fire Sci.* **2007**, *25*, 471–497.
[28] WO 2010/026230, inv.: V. Butz, assigned to Thor GmbH.
[29] WO 2017/013028, inv.: G. Lips, D. Mueller, T. F. Thompson, S. M. Andrews, assigned to BASF SE.
[30] Üreyen, M. E.; Kaynak, E.; Yüksel, G. Flame-Retardant Effects of Cyclic Phosphonate with HALS and Fumed Silica in Polypropylene. *J. Appl. Polym. Sci.* **2020**, *137*, 48308.
[31] Schultheis, C.; Grosshauser, M.; Metzsch-Zilligen, E.; Pfaendner, R. Welche Materialien Stabil Bleiben. *Nachrichten aus der Chemie.* **2018**, *66*, 1067–1069.
[32] Pawelec, W.; Holappa, A.; Tirri, A. T.; Aubert, M.; Hoppe, H.; Pfaendner, R.; Wilén, C.-E. Disulfides–Effective Radical Generators for Flame Retardancy of Polypropylene. *Polym. Degrad. Stab.* **2014**, *110,* 447–456.
[33] WO 2014154636 inv.: R. Pfaendner, E. Metzsch-Zilligen, M. Stec, assigned to Fraunhofer Gesellschaft.
[34] WO 2016042040 inv.: R. Pfaendner, M. Mazurowski, assigned to Fraunhofer Gesellschaft.
[35] WO 2019020699 inv.: R. Pfaendner, T. Driever, A. Leipold, E. Metzsch-Zilligen, assigned to Fraunhofer Gesellschaft.
[36] WO 2015180888, inv.: R. Pfaendner, M. Mazurowski, assigned to Fraunhofer Gesellschaft.
[37] WO 2016042043 inv.: R. Pfaendner, M. Mazurowski, assigned to Fraunhofer Gesellschaft.
[38] A. Leipold, PhD thesis, Darmstadt 2021.
[39] Spiess, B.; Metzsch-Zilligen, E.; Pfaendner, R. A New Class of Oxyimides: Oxyimide Ethers and Their Use as Flame Retardants. *Macromol. Mater. Eng.* **2021**, *306*, 200650.
[40] Goedderz, D.; Weber, L.; Markert, D.; Schießer, A.; Fasel, C.; Riedel, R.; Altstädt, V.; Bethke, C.; Fuhr, O.; Puchtler, F.; Breu, J.; Döring, M. Flame Retardant Polyester by Combination of Organophosphorus Compounds and an NOR Radical Forming Agent. *J. Appl. Polym. Sci.* **2020**, *137*, 47876.
[41] Shao, Z. B.; Deng, C.; Tan, Y.; Chen, M. J.; Chen, L.; Wang, Y.-Z. An Efficient Mono-Component Polymeric Intumescent Flame Retardant for Polypropylene: Preparation and Application. *ACS Appl. Mater. Interfaces.* **2014**, *6*, 7363–7370.
[42] Lv, Q.; Huang, J. Q.; Chen, M. J.; Zhao, J.; Tan, Y.; Chen, Y.; Wang, Y.-Z. An Effective Flame Retardant and Smoke Suppression Oligomer for Epoxy Resin. *Ind. Eng. Chem. Res.* **2013**, *52*, 9397–9404.
[43] Deng, C.; Yin, H.; Li, R.-M.; Huang, S.-C.; Schartel, B.; Wang, Y.-Z. Modes of Action of a Mono-Component Intumescent Flame Retardant MAPP in Polyethylene-Octene Elastomer. *Polym. Degrad. Stab.* **2017**, *138,* 142–150.
[44] Yan, J. L.; Xu, M. J. Design, Synthesis and Application of a Highly Efficient Mono-Component Intumescent Flame Retardant for Non-charring Polyethylene Composites. *Polym. Bull.* **2021**, *78*, 643–662.
[45] Liu, L. B.; Xu, Y.; He, Y. T.; Xu, M. J.; Wang, W.; Li, B. A Facile Strategy for Enhancing the Fire Safety of Unsaturated Polyester Resins Through Introducing an Efficient Mono-Component Intumescent Flame Retardant. *Polym. Adv. Technol.* **2020**, *31*, 1218–1230.
[46] Xia, S. Y.; Zhang, Z. Y.; Leng, Y.; Li, B.; Xu, M. J. Synthesis of a Novel Mono-component Intumescent Flame Retardant and Its High Efficiency for Flame Retardant Polyethylene. *J. Anal. Appl. Pyrolysis.* **2018**, *134*, 632–640.
[47] Haiqiang, L.; Wenhui, R.; Zhao, P.; Wang, L.; Liu, Y.; Yu, C. An Efficient Organic/Inorganic Phosphorus–Nitrogen–Silicon Flame Retardant Towards Low-Flammability Epoxy Resin. *Polym. Degrad. Stab.* **2020**, *178,* 109195.
[48] L. Yi, Z. Huang, Y. Cao and Y. Peng, Synthesis of a Novel Spiro Phosphorus–Nitrogen Concerted Reactive Flame-Retardant Curing Agent and Its Application in Epoxy Resin, Frontiers in Materials, September 2020, 7, article 293.
[49] Zhu, Z.-M.; Xu, Y.-J.; Liao, W.; Xu, S.; Wang, Y.-Z. Highly Flame Retardant Expanded Polystyrene Foams from Phosphorus–Nitrogen–Silicon Synergistic Adhesives. *Ind. Eng. Chem. Res.* **2017**, *56*, 4649–4658.
[50] Chen, Y.; Peng, H.; Li, J.; Xia Z.; Tan, H. A Novel Flame Retardant Containing Phosphorus, Nitrogen, and Sulfur. *J. Therm. Anal. Calorim.* **2014**, *115*, 1639–1649.
[51] Song, K.; Wang, Y.; Ruan, F.; Liu, J.; Li, N.; Li, X. Effects of a Macromolecule Spirocyclic Inflatable Flame Retardant on the Thermal and Flame Retardant Properties of Epoxy Resin. *Polymers.* **2022**, *12*, 132.
[52] Shao, Z.-B.; Deng, C; Tan, Y.; Chen, M.-J.; Chen, L.; Wang, Y.-Z. An Efficient Mono-Component Polymeric Intumescent Flame Retardant for Polypropylene: Preparation and Application. *ACS Appl. Mater. Interfaces.* **2014**, *6*, 7363–7370.

[53] Yi, L.; Huang, Z.; Cao, Y.; Peng, Y. Novel Phosphorus–Nitrogen–Silicon Flame Retardants and Their Application in Cycloaliphatic Epoxy Systems. *Polym. Chem.* **2015**, *6*, 2977–2985.
[54] Li, Q.; Jiang, P.; Wei, P. S. Synthesis, Characteristic, and Application of New Flame Retardant Containing Phosphorus, Nitrogen, and Silicon. *Polym. Eng. Sci.* **2006**. DOI 10.10.1002/pen.20472
[55] Zhang, S.; Chu, F.; Xu, Z.; Zhou, Y.; Qiu, Y.; Qian, L.; Hu, Y.; Wang, B.; Hu, W. The Improvement of Fire Safety Performance of Flexible Polyurethane Foam by Highly-Efficient P-N-S Elemental Hybrid Synergistic Flame Retardant. *J. Colloid Interface Sci.* **2022**, *606*, 768–783.
[56] WO2015/067736 A1 inv.: C-E. Wilen, M. Aubert, T. Tirri, W. Pawelec, assigned to Songwon International AG.
[57] Mountassir, A.; Tirri, T.; Sund, P.; Wilén, C.-E. Sulfenamides as Standalone Flame Retardants for Polystyrene. *Polym. Degrad. Stab.* **2021**, *188*, 109588.
[58] Tirri, T.; Aubert, M.; Pawelec, W.; Holappa, A.; Wilén, C.-E. Structure–Property Studies on a New Family of Halogen Free Flame Retardants Based on Sulfenamide and Related Structures. *Polymers.* **2016**, *8*, 360.
[59] Ääritalo, T.; Brusentsev, Y.; Tirri, T.; Breton, L.; Wilen, C.-E. Highly Efficient Multifunctional Mono- component Flame Retardants Comprising of Sulfenamide and Phosphate Moieties for Polypropylene. *Polym. Degrad. Stab.* **2023**, *218*, 110568
[60] Ääritalo, T.; Tirri, T.; Aubert, M.; Wilen, C.-E. Highly Efficient Multifunctional Mono-Component Flame Retardants Comprising of Sulfenamide and Phosphate Moieties for Polypropylene. *Polym. Degrad. Stab.* **2023**, *211*, 110336.
[61] WO2020212377A1 inv.: T. Kröller, A. Meyer, M. Al-Masri, F. Piestert, C-E. Wilen, M. Aubert. T. Tirri, T. Ääritalo, assigned to BYK Chemie GmbH.
[62] WO2020212374A1 inv.: T. Kröller, A. Meyer, M. Al-Masri, F. Piestert, C-E. Wilen, M. Aubert. T. Tirri, T. Ääritalo, assigned to BYK Chemie GmbH.

12 Polymer Nanocomposites

Daniel F. Schmidt
Luxembourg Institute of Science and Technology (LIST)
5 Avenue des Hauts-Fourneaux
L-4362 Esch-sur-Alzette, Luxembourg

2nd Edition Chapter Author:
David D. Jiang
Organic Technology/Corporate Research
Corning Incorporated/Sullivan Park
Painted Post, NY 14870

12.1 INTRODUCTION

One of the most significant impacts of nanotechnology on polymeric materials is the preparation of polymer nanocomposites (PNs) through the addition of inorganic nanomaterials having at least one dimension measuring less than 100 nanometers (nm) (referred to as nanoadditives in this chapter) to polymer matrices through a variety of methods/processes. Figure 12.1 shows a generic illustration of how nanoadditives may be combined with polymer matrices to generate PNs, using clay as the example.

Nanoadditives used for fabricating PNs include a broad range of inorganic nanomaterials that may be generally classified into four categories according to dimensionality, as shown in Figure 12.2: nanoparticles (0D), nanofibers (1D), nanolayers (2D), and nano-networks (3D).

Examples of 0D nanoadditives include nano-metal oxides (such as silica,[1] titania,[2,3] alumina,[4,5] etc.), nano-metal carbides,[6] and polyhedral oligomeric silsesquioxanes (POSS),[7–9] to name just a few possibilities; 1D nanoadditives include carbon nanotubes (CNTs),[10–12] which may be single wall (SWCNTs), dual wall (DWCNTs) or multi-wall (MWCNTs), isostructural boron nitride nanotubes (BNNTs),[13] carbon nanofibers,[14,15] etc.; 2D nanoadditives include layered silicates,[16–19] layered double hydroxides (LDH),[20,21] layered zirconium phosphates and phosphonates,[22] layered titanates,[23] graphene and its derivatives,[24] hexagonal boron nitride,[13] metal dichalcogenides,[25] etc.; finally, examples of 3D nano-networks include IPN-type systems including an inorganic component, along the lines of ORMOSILs/ORMOCERs and related organic/inorganic hybrids.[26]

The impact of nanocomposite technology on polymers has been substantial and, when properly applied, is reflected in the enhanced performance of the resulting PNs, including improved mechanical, barrier, and ablative properties among others.[27] The effect of nanocomposite technology on the thermal and fire performance of polymers is primarily observed via changes in two important polymer characteristics: (1) the onset temperature (T_{onset}) in the thermogravimetric analysis (TGA) curve—one measure of the thermal stability of the polymer, and (2) in the peak heat release rate (PHRR) measured during cone calorimetry—a reflection of the combustion behavior (the flammability) of the polymer. In the ideal case, the T_{onset} is increased, and the PHRR is reduced when nanoscale dispersion of the nanoadditive is achieved in the polymer matrix.

In this chapter, an overview of the fundamentals of PNs is provided according to the authors' understanding and experience coupled with support from numerous references and review articles. The contents of this chapter cover all kinds of inorganic nanoadditives, but because the most widely studied and industrially applied nanoadditives for the purposes of PN formation are clays and CNTs, the focus of the chapter is primarily on the effect of these materials on the thermal and fire resistance of polymers. This includes the structure, properties, and surface treatment of the nanoadditives, design of the modifiers, synthesis, characterization of structure/morphology, thermal and fire performance of PNs, and the subsequent mechanisms of flammability reduction. A brief discussion of current trends in PN studies is also included. Since the focus here is to familiarize the reader with the basics of nanoadditives and their applications in polymer matrices, it is recommended that readers refer to key reviews, books, and papers in this area for more detailed information and/or a more complete picture concerning nanoadditives and their influence on polymer properties and behavior. A few such references are listed at the end of the chapter.

12.2 STRUCTURE AND PROPERTIES OF NANOADDITIVES

12.2.1 Structure and Properties of Clay

Clays, natural or synthetic, are the most widely investigated and understood nanoadditives used to enhance the flame retardancy of polymers through nanocomposite technology

DOI: 10.1201/9781003380689-12

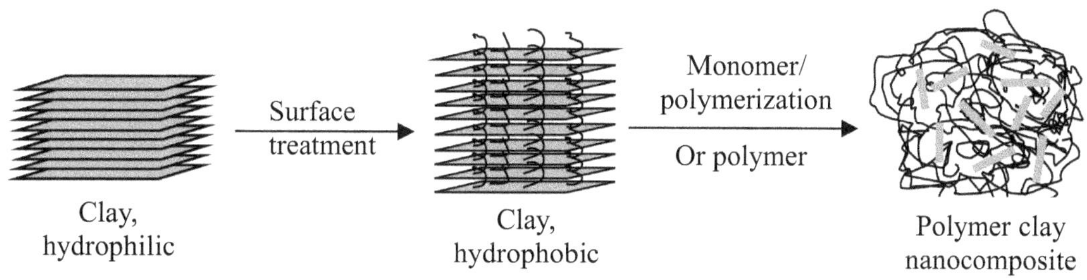

FIGURE 12.1 A schematic illustration of the general process for preparing PNs with clay as a nanoadditive.

Source: For illustration only.

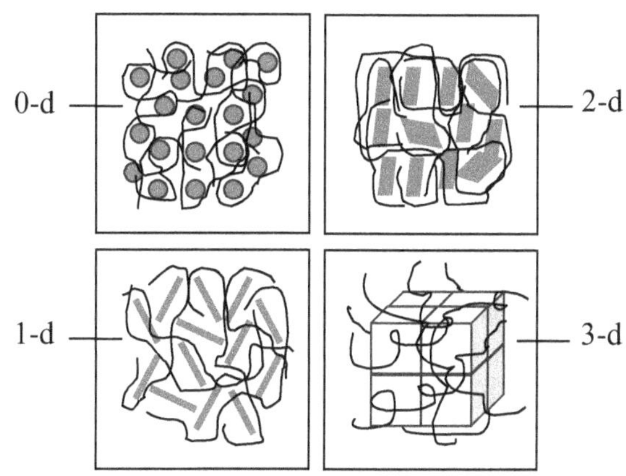

FIGURE 12.2 Nanoadditive classes used for fabrication of polymer nanocomposites.

Source: For illustration only.

because of their low cost, ease of surface treatment, and straightforward application in polymer matrices. Clays can be cationic or anionic in nature depending on the composition of the clay layers. In this chapter, the focus is on two kinds of clays: montmorillonite (MMT), a naturally occurring cationic clay that belongs to the smectite group of silicates; and layered doubly hydroxides (LDHs), anionic clays of which hydrotalcite is a naturally occurring example but for which synthetic forms are more common. Other clays will also be mentioned as appropriate.

Silicate Clay

Silicate clay[28] belongs to an important group of layered inorganic nanomaterials known as phyllosilicates that can be divided into four major sub-groups: kaolinite, montmorillonite (MMT)/smectite, illite (clay-mica), and chlorite. Structurally, phyllosilicates are made up of two types of sub-layers: tetrahedrally substituted (TS) and octahedrally substituted (OS). In the case of the 2:1 phyllosilicates, a single OS sub-layer is sandwiched between two TS sub-layers to form each clay platelet (Figure 12.3[29]). Here, each TS sub-layer consists of tetrahedrally coordinated silicon atoms fused to an edge-shared OS sub-layer consisting primarily of aluminum or magnesium hydroxide. The overall thickness of the platelet is around 1 nm, while the lateral dimensions of the layers vary dramatically, from tens of nanometers to several microns, depending on the type of clay. Due to isomorphic substitutions within the layers (such as the replacement of Al^{3+} by Mg^{2+} or Fe^{2+}, or Mg^{2+} by Li^{+}), the silicate layers generally bear negative charges that are counterbalanced by alkali and/or alkaline earth cations residing within the galleries. In contrast with other phyllosilicates where the presence of exposed OS sub-layers enables hydrogen-bonding between adjacent platelets, the lack of hydrogen bonding coupled with the presence of negative charges on the TS sub-layers that make up the platelet surfaces provide greater opportunities for polymer/silicate interactions. Furthermore, the metal cations residing in the galleries between the layers can be replaced by many other cationic species, enabling these clays to be readily modified with organo-cations. Such a replacement is characterized by the ion exchange capacity (IEC), an important parameter for clays, generally expressed as milliequivalents (meq)/100 g. Typical IECs for 2:1 layered silicates range from 60 to 200 meq/100 g, with the IEC of MMT tending towards ~100 meq/100 g. It is important to note, however, that the IEC is always an average value representative of a distribution of charges whose density varies both within and between platelets. Along the same lines, the type and concentration of counter-ions that reside in the interlayer galleries and counter-balance the charges on the silicate platelets can vary within and between platelets as well, allowing for the generation of a wide range of heterostructured/"partially modified" systems.[30,31]

Among all the silicate clays, MMT is the most widely used for preparing the PNs, referred to as polymer/clay nanocomposites (PCNs). Two characteristics of the silicate clays are generally considered in the fabrication of the PCNs: (1) the ability to fine-tune their surface chemistry through ion exchange reactions between organic and inorganic cations, and (2) the ability of the silicate particles to be dispersed at the level of individual layers in the matrix polymer. These two characteristics are co-related because the degree of dispersion for a given clay in a given polymer matrix is strongly dependent on the number and type of cation(s) present in the interlayer galleries.

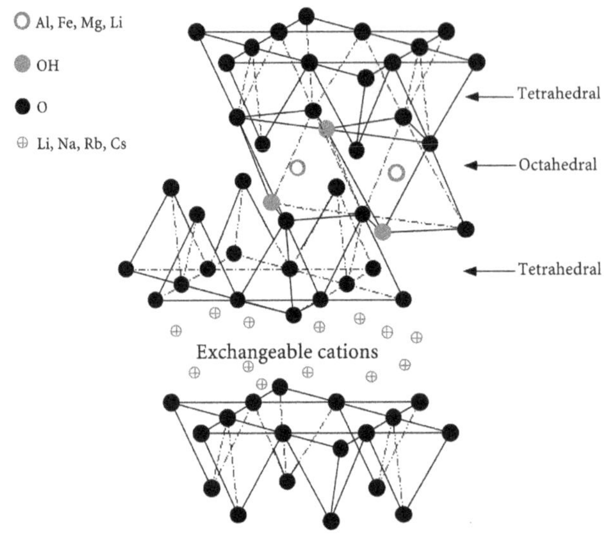

FIGURE 12.3 A schematic illustration of the structure of clay.

Source: Part of the figure is reprinted from ref. 29 with permission from Springer.

A clay with a small IEC may not have enough anchoring points to realize the extent of modification needed for compatibility with a given polymer, preventing the clay from dispersing effectively and hindering nanocomposite formation. Likewise, a clay with a large IEC may be modified so heavily as to entirely suppress polymer/silicate interactions, compromising nanocomposite properties as a result. The lateral dimensions of the clay platelets represent another important factor to consider when choosing a clay. The lateral dimensions of the clay platelets in MMT, the most commonly utilized clay, are generally larger than those of hectorite and saponite silicate clays but smaller than those of synthetic silicate clays such as synthetic fluoromica.[32]

LDH Clay

Structurally, the LDH clays,[33] natural or synthetic, have some similarity to the silicate clays. Both are layered inorganic nano-materials bearing surface charges with ions residing in the interlayers. However, according to composition, geometry, and thickness of the layers, the LDHs are quite different from the silicate clays. While as noted above the layers of the silicate clays have a sandwich structure—one octahedral sheet containing oxides of Fe, Al, Mg, etc. sandwiched between two silica tetrahedral sheets (that is the origin of the term "2:1 layered silicate")—LDH platelets consist of only a single octahedral metal hydroxide sheet. Another feature is that the silicate clays bear negative charges on the layers with exchangeable metal cations residing in the interlayer gallery, while the LDHs consist of brucite-like sheets and bear positive charges on the layers that are compensated by exchangeable (hydrated) anions residing in the interlayer gallery. The aforementioned differences in layer structure also mean that LDHs consist of platelets with about half the thickness of those in MMT and with a significantly lower rigidity as well.

Among the LDHs, synthetic LDHs are most commonly used. They are usually prepared via the co-precipitation method with M(II) and M(III) sources (in the form of chloride, nitrate, sulfate, carbonate, etc.) in basic solution, using the desired cations and anionic species to obtain the desired LDH structure as well as a tunable layer charge density through variations in composition. This is quite different from the silicate clays, which are mostly naturally occurring. The charge density of the LDHs, which determines the IEC, can be much higher, for instance doubled, compared to MMT and other silicate clays. As a result, LDHs can have charge densities similar to those seen in mica (0.32–0.34 C/m^2).[34]

The most common naturally occurring LDH clay is Mg-Al-LDH, a hydrotalcite having a typical chemical formula of $Mg_6Al_2(OH)_{16}CO_3.4H_2O$. Because of partial substitution of M^{II} by M^{III}, there are positive charges on the layers. The ideal formula of a generalized LDH is $[M^{II}{}_xM_{1-x}{}^{III}(OH)_2]_{intra}[A^{m-}{}_{x/m}\cdot nH_2O]_{inter}$, where M^{II} and M^{III} are metal cations, such as Mg^{2+}, Ca^{2+}, Zn^{2+}, etc. for M^{II} and Al^{3+}, Cr^{3+}, Fe^{3+}, Co^{3+}, etc. for M^{III}; A is the anion, which can be Cl^-, $CO_3{}^{2-}$, $NO_3{}^-$, etc.; while the "intra" and "inter" denote the intralayer and interlayer domains, respectively. When many LDH layers are stacked together one upon another through van der Waals interactions, the interlayer distance or basal spacing is ~0.77 nm for Mg-Al-LDH (Figure 12.4). Being hydrated (like $Mg(OH)_2$ and $Al(OH)_3$), LDHs show the potential to release water under fire conditions, a

FIGURE 12.4 A schematic illustration of the structure of LDHs

Source: For illustration only.

potentially useful behavior in the context of flame retardant nanoadditives.

12.2.2 Structure and Properties of Carbon Nanotubes and Other Nanoadditives

CNTs,[35] allotropes of carbon from the fullerene structural family that were first definitively identified in the literature in 1991,[36] are nano-cylindrical carbon molecules (with similarity to graphite in terms of the sp^2 hybridization of the carbon-carbon bonds they contain) that possess aspect ratios (length-to-diameter ratios) as high as 1,000,000:1 or more and that may be categorized as SWCNTs, which have a diameter of close to 1 nanometer and may possess various "zigzag," "armchair," or "chiral" structures that affect their electrical properties, DWCNTs, which consist of exactly two concentric nanotubes, one inside the other, and MWCNTs, which consist of multiple concentric nanotubes. Concentric tubes can be "fused" together when high pressure is exerted, leading to the conversion of sp^2 bonds to sp^3 bonds and producing stronger CNTs.[37] Because of van der Waals forces, CNTs tend to aggregate. Unlike layered materials that tightly stack together layer by layer, however, CNTs often organize themselves into rope-like structures.

Figure 12.5 shows a schematic illustration of the structure of a SWCNT and high-resolution transmission electron microscopy (HRTEM) images of MWCNTS.[38] Compared to SWCNTs, MWCNTs have significantly improved chemical resistance, a critical property when functionalizing CNTs. This is because CNT functionalization generally involves the formation of defects (in the form of broken carbon-carbon bond). In the case of SWCNTs, this compromises the structural integrity of the tube. In the case of MWCNTs, however, only the outer wall is involved. This has also inspired interest in DWCNTs, where the outer tube provides a basis for functionalization, while the inner tube remains intact and provides desired performance benefits while minimizing weight and avoiding inter-tube slip that may otherwise occur in MWCNTs.

Because of the unique combination of mechanical, electrical, and thermal properties, CNTs have been identified as candidates to substitute or complement other nanofillers in the fabrication of multifunctional PNs. The first CNT-based PNs were reported in 1994[39]; since then, CNTs have been second only to clays as far as nanoadditives investigated for polymer flammability reduction are concerned. One difficulty in the application of CNTs remains their effective dispersion in the matrix polymer, however, with their high cost representing a second problem. Questions have also been raised concerning toxicity, either due to CNTs themselves or to transition metal impurities they often contain as a result of the production process, as well as the potential for release in the context of machining operations and in fire scenarios.[40]

Other nanoadditives that have been studied include nanofibers (incl. carbon nanofibers,[41,42]), graphene and its derivatives,[24] boron nitride-based analogs to graphene and CNTs (i.e., planar BN and BN nanotubes),[13] metal dichalcogenides,[25] and various colloidal nanoparticles, such as POSS,[43,44] silica,[45] alumina,[46] etc., to name several examples, though as of this writing most tend to be less well-studied and understood. Among these nanoparticles, POSS is unique. Unlike other nanoadditives, POSS is an inorganic-organic hybrid material, containing an inorganic siloxane-like core, Si_8O_{12} (Figure 12.6 in which [2-(3,4-epoxycyclohexyl)ethyl]-heptaisobutyl substituted POSS[47] is shown), and organic substituents that can be tailored for compatibility with various matrix polymers, targeting applications ranging from mechanical reinforcement to flammability reduction in polymers.

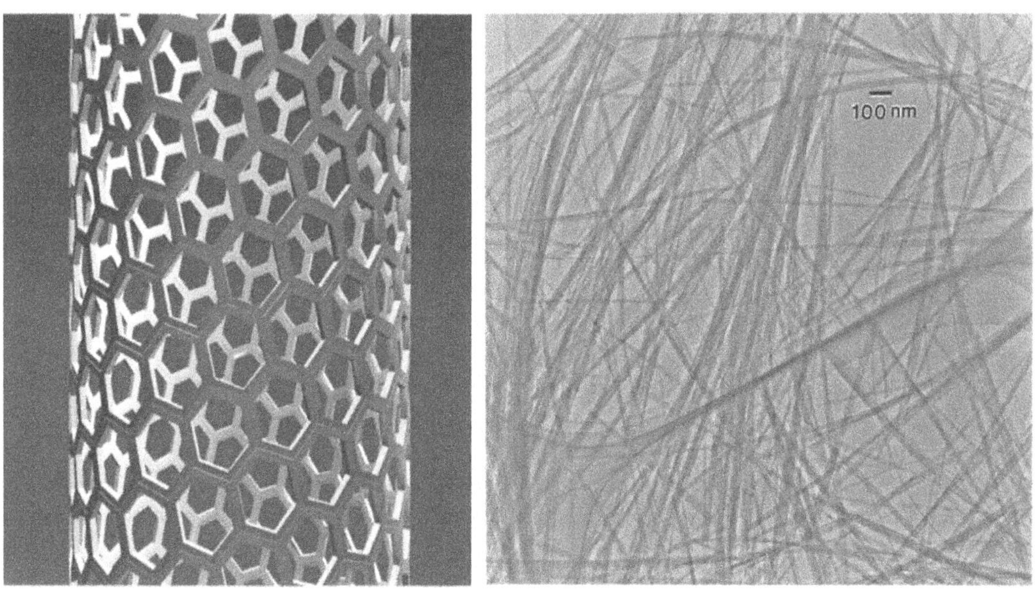

FIGURE 12.5 Left—A schematic illustration of a SWCNT; right—TEM image of MWCNTs.

Source: Reproduced from ref. 38 with permission from IOP Publishing Ltd.

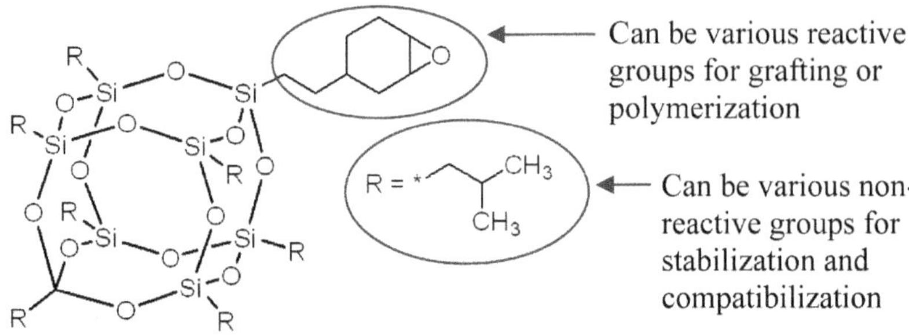

FIGURE 12.6 A representative structure of POSS.

Source: Reproduced from Sigma-Aldrich website, www.sigmaaldrich.com/catalog/search/ProductDetail/ALDRICH/560316, accessed Jul 31, 2023.

It has been shown in the literature that POSS may be utilized as a precursor of flame retardants.[43,44] The literature also reports observations of significantly reduced PHRR for a few polymers[48] and increased time to ignition[49] when using POSS as a nanoadditive. However, two significant practical disadvantages to the use of POSS are the high loading levels required to realize good flame retardant performance and the high cost of these materials. Finally, while other nanoadditives have also been used to enhance the thermal and fire performance of polymers, because they are less well-investigated and understood and have seen little or no commercial development, no detailed discussion is included in this chapter.

12.3 SURFACE TREATMENT OF NANOADDITIVES

The surfaces of common nanoadditives prior to chemical modification may be generally classified as metal oxide based or carbonaceous; unfortunately, neither lend themselves to compatibility with a broad range of polymers. Metal oxide-based surfaces tend to be hydrophilic, with nanoadditives possessing such surfaces showing compatibility only with highly polar polymers. For example, unmodified clays only form relatively well-dispersed PNs with poly(ethylene oxide),[50] poly(vinyl alcohol),[51] and a few other water-soluble polymers. Most other polymers possess poor compatibility with unmodified metal oxide surfaces. Likewise, the surfaces of carbon-based nanofillers such as CNTs and graphene, being graphitic in nature, tend to be compatible primarily with other polycyclic aromatic moieties found in very few polymers. While the means of surface modification of such disparate nanoadditives is quite different in practice, in the unmodified state they share the inability to form PNs with a high level of thermodynamically stable nanoscale dispersion in most cases. Consequently, for most nanoadditives used to prepare PNs, an important and necessary precondition is a surface treatment rendering them thermodynamically compatible with the polymer matrix of interest and thus capable of being uniformly dispersed and/or separated at the level of the primary particle.

12.3.1 Surface Treatment of Clays

12.3.1.1 Surface Treatment of Clays

In the case of cationic silicate clays such as MMT, modification is generally achieved through the introduction of a cationic surfactant, typically an organo-onium, to a suspension of inorganic clay. The resulting ion-exchange process anchors the cationic surfactant molecules onto the clay surface, rendering it organophilic and enabling polymer intercalation within the interlayer galleries. The oniums can be primary, secondary, tertiary, or quaternary based alkylammoniums[52] or alkylphosphoniums[16] as well as other oniums such as imidazolium,[53] pyridinium,[54] quinolinium,[54] tropylium,[55] stibonium,[56] etc. Modification of LDH clays is achieved in a parallel fashion, via ion-exchange with anionic surfactants such as sodium dodecylbenzenesulfonate (SDBS), lauric acid (LA), sodium dodecylsulfate (SDS), bis(2-ethylhexyl)hydrogen phosphate (BEHHP), etc.[57] Generic structures for selected cationic and anionic surfactants are shown in Figure 12.7, and a general procedure for lab-scale organo-modification (suitable for both silicate and LDH clays) is provided in Figure 12.8. In all cases a readily dispersed but unmodified form of the clay (i.e., containing small, monovalent counterions, for example Na^+ in the case of MMT, OH- in the case of LDHs, etc.) is first dispersed in (heated) water or an aqueous medium to

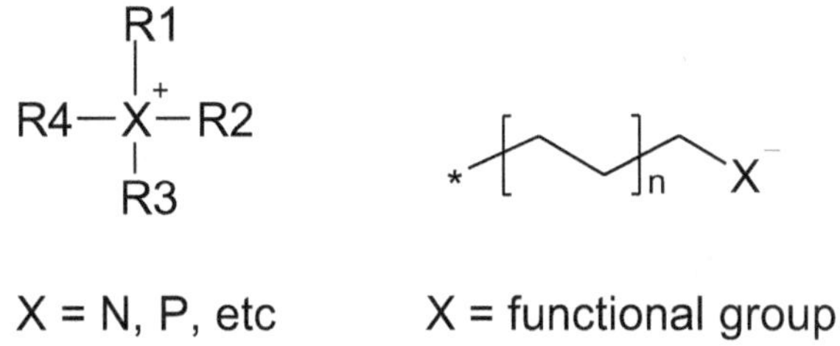

FIGURE 12.7 Generic structures of clay modifiers: left—an -onium type cationic surfactant representative of typical alkylammonium and alkylphosphonium modifiers for instance; and right—an anionic surfactant representative of alkylcarboxylates ($X = COO^-$) and alkylsulfates ($X = SO_4^-$) for instance.

Source: For illustration only.

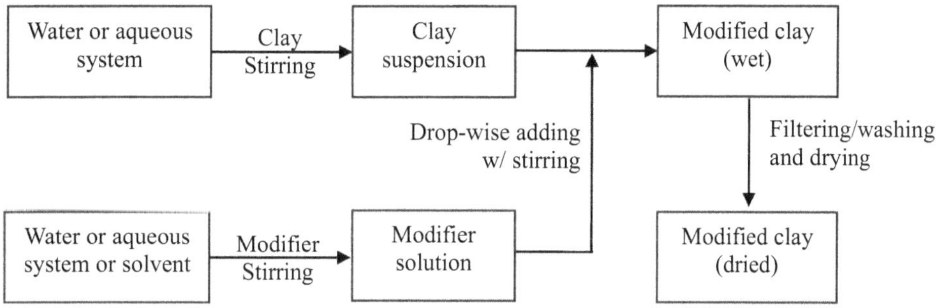

FIGURE 12.8 A schematic illustration of the ion-exchange-based organo-modification of clays.

Source: For illustration only.

form a dilute suspension. The clay concentration should be kept below the percolation threshold of the individual clay layers to enable complete dispersion at the level of the primary particle (typically a few wt%), and probe ultrasonication coupled with stirring and heating (60–80°C is typical for MMT modification for instance) may be used to ensure complete dispersion in practice. A (heated) solution of the organic modifier in water or an aqueous medium is then gradually added to the clay suspension, using a stoichiometric excess vs. the IEC if complete exchange is desired (as is usually the case). As ion-exchange tends to be quite rapid especially at elevated temperatures, the organo-modifier is anchored onto the surface of the clay platelets immediately following its addition, which may result in partial or total precipitation of the clay as the ion-exchange process proceeds. Nevertheless, such suspensions are typically maintained at temperature and stirred for several hours or even overnight to ensure complete and homogeneous ion exchange. Finally, care must be taken to remove the excess modifier and inorganic salts formed as a byproduct of ion-exchange. This is most effectively accomplished by centrifugation and decanting of the supernatant, followed by several additional washing cycles involving redispersion in a medium capable of dissolving both the excess modifier and the inorganic salts produced during ion-exchange, centrifugation, and decanting of the liquid. Finally, the drying of the modified clay should be carried out in such a way as to avoid the formation of hard aggregates, as tends to occur when the wet "cake" is directly air-dried or oven-dried; freeze-drying and spray drying are preferable to the aforementioned approaches in this context. Some representative modifiers (both cationic and anionic) that have been reported in the literature[16,53,57–65] as surface treatments for clays are shown in Figure 12.9.

12.3.1.2 Structure of Modified Clays

After drying, the modified clay layers will reform into multilayer stacks thanks to their high surface energy, effectively remembering their original structure prior to modification. In general, a single clay aggregate contains many primary clay particles with a multilayered structure. After organo-modification, the gap (commonly referred to d-spacing or basal spacing) between the clay layers increases due to the presence of the modifier in the interlayer gallery. Two major factors influence the interlayer spacing: (1) the size and chain length of the modifiers, with larger sizes/longer chain lengths leading to larger d-spacings; and (2) the extent of ion exchange, which is limited by the IEC and which determines the packing density of the modifiers in the interlayers of the modified clay. The former varies based on the choice of modifier, while the latter is influenced by the selection of the clay, given the variation in IEC as a function of clay type (see Section 12.2.1). The combined effect of these choices can greatly affect the efficiency of ion-exchange between the inorganic ions and the organo-ions, thus influencing the compatibility and dispersibility of the clay in the polymer matrix. In determining the d-spacing of the clay before and after modification, X-ray diffraction (XRD) has been the most used characterization method, thanks to the observation of one or more low-angle Bragg diffraction peaks associated with the presence of parallel, evenly spaced clay layers in the aforementioned multilayer stacks.

The structure of the modified clay generally refers to the orientation and alignment of the modifier in the clay interlayers. After anchoring onto the clay surface, the modifier is able to take different orientations or arrangements in the clay interlayer. The long chains of the modifier are posited to take one of two distinct orientations/arrangements with respect to the silicate layers, as shown in Figure 12.10[66]: (1) oriented parallel to the clay layers, typically in the form of mono or bilayers, and (2) radiating away from the clay layers at an angle, with mono and bimolecular arrangements possible. The orientation or the arrangement of the modifier is observed to be temperature-sensitive as well, with long chain modifiers confined between parallel clay layers sometimes displaying a thermal transition similar to a solid or liquid-crystalline to liquid-like transition that may be detected both via thermal analysis and X-ray diffraction.

The aforementioned discussions of organoclay structure refer to the modified clay in the dry state and assume an idealized model implying all-trans conformations for the long chains. This has been demonstrated to be unrealistic according to the results of FTIR analysis[67] and through molecular dynamics simulations.[68] It has been shown that the chains of the modifier possess varying degrees of order based on the frequency shifts of the asymmetric CH_2

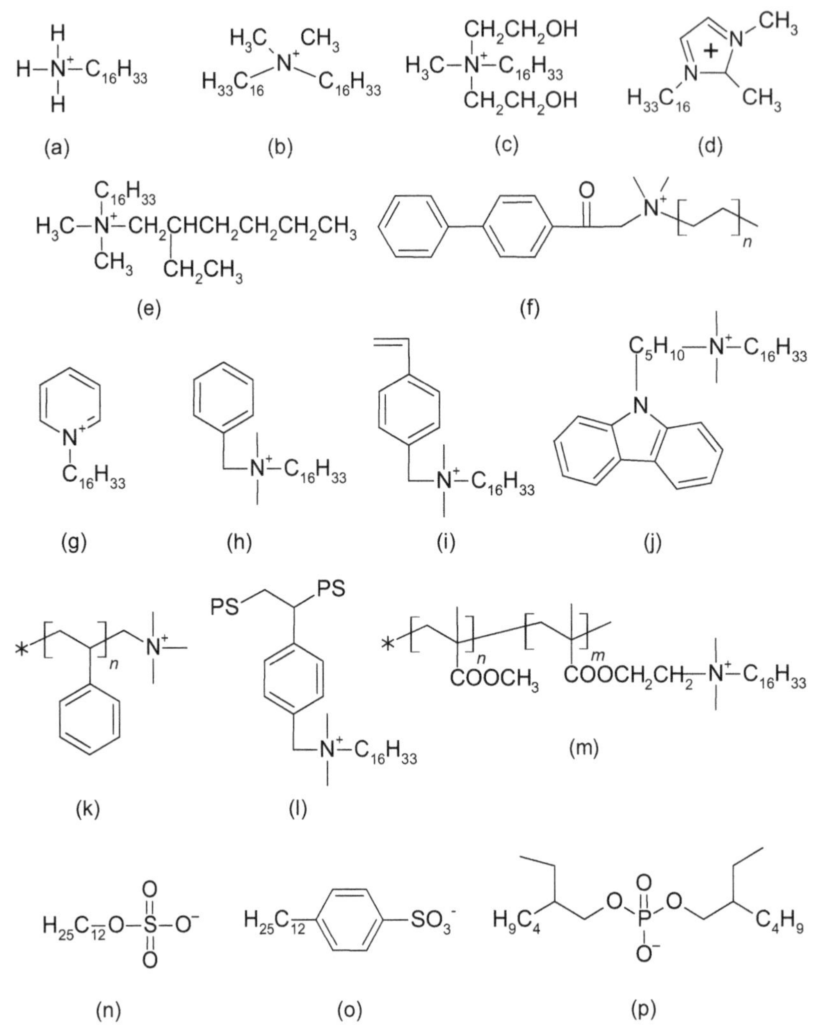

FIGURE 12.9 Representative organo-modifiers for clay surface treatment: (a) hexadecylammonium (ref. 58); (b) dimethyldihexadecylammonium (ref. 59); (c) methyl-bis-2-hydroxyethylhexadecylammonium (ref. 59); (d) 1,2-dimethyl-3-hexadecylimidazolium (ref. 53); (e) dimethyl-2-ethylhexyl-hydrogenated tallow (ref. 59); (f) phenylacetophenonedimethylhexadecyl ammonium (ref. 60); (g) hexadecylpyridium (ref. 61); (h) dimethylbenzylhydrogenatedtallow ammonium (ref. 59); (i) dimethylvinylbenzylhexdecylammonium (VB16, ref. 16); (j) carbazole-based ammonium (ref. 62); (k) trimethylammonium-terminated PS surfactant (ref. 63); (l) quaternary ammonium functional oligomeric copolymer of styrene and vinylbenzyl chloride (COPS, ref. 64); (m) quaternary ammonium functionalized methyl methacrylate oligomer (MAPS, ref. 65); (n) dodecylsulfate (ref. 57); (o) dodecylbenzenesulfonate (ref. 57); (p) bis(2-ethylhexyl) hydrogen phosphate (ref. 57).

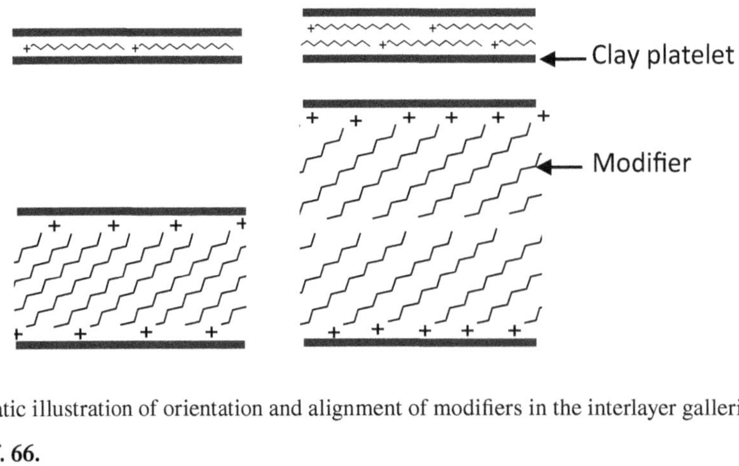

FIGURE 12.10 A schematic illustration of orientation and alignment of modifiers in the interlayer galleries of layered clays.

Source: Modified from ref. 66.

stretching and bending vibrations. In general, the chains of the modifier adopt a more disordered liquid-like structure with a decrease in packing density or chain length or an increase of temperature. When the available surface area/molecule is within a certain range, the chains are not completely disordered, but retain some orientational order similar to that of a liquid crystalline state. However, the modifiers display a different arrangement/degree of alignment in the "wet" state (for instance, when the modified clay is dispersed in a solvent), a phenomenon that may be studied via neutron scattering.[69,70]

12.3.1.3 Compatibility/Miscibility and Design of the Modifier

First, it is important to clarify the terms being used here. Compatibility[71] may generally be considered to be a measure of the stability of the substance when it is mixed with another material. Miscibility[72] refers to the property of two or more liquids to be mixed and form a clear, homogeneous solution in all proportions, for instance, a mixture of water and ethanol. Miscibility/immiscibility has been extended to mixtures of solids and of gases. However, one should be aware that a transparent "solution" does not mean that two materials are miscible. They may form an immiscible but visibly clear mixture if a refractive index match is achieved. Furthermore, the requirement for a precise refractive index match becomes less and less stringent as the domain size of either phase decreases to a value well below the wavelength of light. Aerogels offer a striking example of this phenomenon, where, thanks to the extremely small pore size (as low as a few nanometers), thin sections of silica aerogel can be almost entirely transparent despite the substantial refractive index mismatch between the empty pores and the silica skeleton ($n \sim 1$ vs. ~ 1.46 in the visible range). The same lesson applies to polymer nanocomposites. If an optically transparent polymer becomes opaque upon nanoparticle addition, barring significant crystallization of the polymer phase, this would imply poor nanoparticle dispersion. However, it does not follow that the retention of optical transparency on nanoparticle addition implies that the system is compatible/miscible.

As mentioned previously, modifiers provide for nanoadditive compatibility with the polymer matrix. For example, after anchoring the modifier onto the clay surface, the resulting organoclay can be dispersed at the nanoscale (intercalated and/or exfoliated) in the polymer matrix. It has been demonstrated in the literature that clay modifiers may be small-molecule-based or oligomer-based,[73–76] nonreactive or reactive,[77,78] including those containing reactive functional group(s), such as a vinyl groups in the modifier,[16] or generating reactive species during PN formation, for example through the production of radicals upon heating.[79,80] The discussion that follows is focused on cationic clays, but it is also relevant to LDHs. In this context it should be noted that, although quaternary alkylammoniums are the most frequently chosen modifiers for cationic clays, that such modifications are distinct from the use of processing aids or compatibilizers that help disperse the clay particles into the polymer matrix and further define the PN structure, but which are not necessarily attached to the clay layers in the same fashion as the clay modifiers. For instance, the maleic anhydride grafting of non-polar polymers such as polypropylene (to give PP-g-MA) has proven to be an effective means of improving compatibility in polyolefin/modified clay nanocomposites,[81,82] but such graft copolymers do not constitute an integral part of the modified clay.

In designing a modifier, the compatibility and the miscibility of the modifier with the polymer matrix are of significant importance. In particular, the structure of the modifier should be similar to that of the polymer in order to favor compatibility and miscibility between the modified clay and the polymer matrix. For instance, clay modified with a quaternary ammonium salt containing a benzyl group[16] shows excellent compatibility and miscibility with aromatic polymers such as polystyrene. However, other features of the molecular structure, such as alkyl chain length and number of alkyl chains, etc., also play an important role. It has been found, based on numerous experiments, that the length of the alkyl chains is critical. Longer chains seem to be able to expand the gallery space far enough to allow for intercalation of the polymer chains, leading to the general proposition that the modifier should have at least one alkyl-chain with 12 carbons (in practice, typically 12 to 18) in order to be effectively intercalated or exfoliated by polymer chains.[83,84] In some cases, two long chains will be more helpful.[85] Indeed, this observation relates to the behavior of a specific subset of modified clays—those where the modifier length and packing density enables their alignment parallel to the clay surface, as previously shown in Figure 12.10.[66] In this specific but important subset of modified clays (to which many commercially available organoclays belong), intercalation occurs when polymer-silicate interactions are more favorable than modifier-silicate interactions. In this scheme, the long alkyl tails of a typical alkylammonium modifier "stand up" and attempt to get as far away from the clay layer surface as possible in order to allow the polymer chains access to that surface in their place.[86] While polymer-modifier interactions remain important in these systems given the presence of an intimate mixture of modifier tails and polymer in the interlayer galleries, such a description implies that, in the absence of this additional driving force for intercalation, none should be expected. This observation points to the need to consider not only polymer-modifier interactions but, in specific cases, nanoparticle-modifier interactions as well. The loading of the modified clay in the polymer matrix is another factor which affects the miscibility and should be considered, in particular in light of the percolation threshold alluded to earlier. As the percolation threshold is approached, interparticle contacts become more and more likely, ensuring that nanoparticles can no longer be fully dispersed and randomly oriented. Typically, this gives rise to some level of preferred orientation (in the case of anisotropic particles) coupled with a reduction in the level of nanoscale dispersion. Given the impact of such changes on PN properties, it is noteworthy that the value of the percolation threshold is readily estimated given the aspect ratio of the nanoparticle of interest.[87]

Some attempts have been made to predict polymer/modified clay compatibility using solubility parameters.[88,89] Dispersing modified clay in a monomer or a solvent in which polymerization is carried out is another approach that has proven useful. Screening methods involving analysis of dispersion levels of modified clay in a monomer or solvent have been investigated in previous work as well.[90] In general, a modified clay that disperses well in good solvents for the matrix polymer is likely to disperse well in the polymer itself. While it will not compensate for a lack of thermodynamic compatibility, ultra-sonication can provide substantial benefits as well.[91] Other factors worth considering include the ability of the polymer chains to entangle with the modifier, hydrogen bonding, and alignment of the modifier on the clay surface.[92–95] Finally, as implied in previous discussions concerning the intercalation mechanism, the entropy and enthalpy of mixing between the polymer and the organic treatment on the clay play a role as well.

In addition to the role of modifiers as a means of realizing compatibility, the modifier may also be designed with additional roles in mind. Beyond their impact on compatibility in the absence of chemical reactions, for instance, the presence of potentially reactive functional groups in the modifier can also provide a chemical basis for enhanced dispersion in the polymer matrix, particularly in the case of nanocomposite formation via in situ polymerization but potentially during melt blending as well. The reactive functional groups can for instance provide a reactive site (such as a radical), initiating polymerization and/or grafting reactions at the clay surface; both can lead to better results with regard to clay exfoliation.[77–80] However, functionalized clays may themselves cause problems, for instance with respect to polymerization control in the case of thermosetting resins like polyurethanes and epoxies.[96,97] They may also induce unwanted cross-linking of the final polymer. An example of this is clay modified with *N,N*-dimethyl-*N*-vinylbenzylhexdecylammonium cations.[16] As the modifier in this case contains a vinyl group that can participate in polymerization reactions, a pseudo-grafted and potentially cross-linked polystyrene/clay nanocomposite (PSCN) is obtained. The use of a modifier without functional groups will not lead to grafting or cross-linking, but the interface between the clay and the polymer may not be as strong as in the case where the vinyl group is present.

The thermal stability of the modifier (and hence of the modified clay) is another important factor to consider. Pure clays such as MMT do not undergo any significant mass loss until temperatures well above the typical decomposition temperatures of organic polymers. Following modification, however, the situation changes, with the onium cation often decomposing at temperatures much lower than the onset temperature of the matrix polymer. Alkylammonium cations for instance typically decompose in the temperature range of 160–200°C,[98–102] either via a Hofmann elimination reaction to give an alpha olefin and amine or through an S_N2 nucleophilic substitution reaction to produce an amine and a substituted alkyl group; this occurs because the C-N bond is the weak link in the ammonium structure and breaks easily upon heating. The thermal degradation of the modifier, which can be monitored through TGA, can dramatically change the structure of the modified clay. This may be observed via XRD as demonstrated in Figure 12.11, in which the XRD patterns of Cloisite 30B, a commercial organoclay formerly produced by Southern Clay Products (now BYK) in the pristine state and after heat treatment

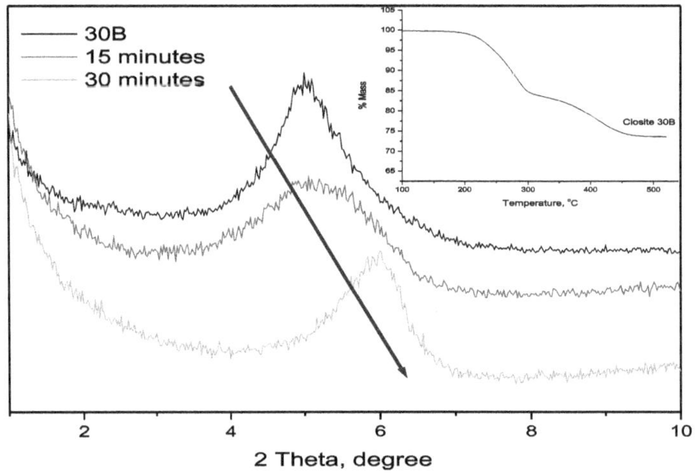

FIGURE 12.11 Thermal stability of Cloisite 30B (*N*-methyl-tallow-bis(hydroxyethyl) ammonium MMT, from Southern Clay Products, now BYK) seen in XRD patterns before and after heating at 190°C in air for 15 and 30 minutes; inset: TGA data in N_2 at 20°C/min.

Source: For illustration only.

at 190°C for 15 and 30 minutes, are shown. As is true for most organic species more generally, the thermal stability of modifiers is typically higher in inert atmospheres such as nitrogen, whereas most processing operations take place in the presence of air. With this in mind, it is critical to evaluate thermal stability in air as a "worst case" scenario and to keep in mind that TGA does not capture additional factors like mechanical shear and the subsequent localized heating that can occur when melt processing of polymer nanocomposites is carried out. In general, practical levels of thermal stability are almost always lower than what is indicated by typical TGA. While there is no substitute for an actual process trial, isothermal TGA in air at the highest process temperature of interest is the most realistic means of assessing thermal stability in this context.

Proper assessment of the sensitivity of the modifier to degradation is critical because it greatly restricts the use of the modified clays both in polymerizations, which require temperatures exceeding the degradation temperature of the modifier[103] and in melt processing operations as well, given that many PNs are prepared via melt-compounding and few polymers can be processed below 200°C. Indeed, even when low-temperature solution-based processes are used for combining polymer and modified clay, the subsequent forming of these PCNs into useful parts generally requires high temperature processing. As such, modifier degradation is a major concern for many thermoplastics because they are often injection/compression molded or extruded at elevated temperatures, and because actual melt temperatures can easily exceed the set points by 20–30°C or more due to the shearing and localized heating, which can further accelerate the decomposition of the modifier. Modifier decomposition, in turn, can lead to a range of effects on the PCN. For instance, thermal degradation of the organoclay through the Hofmann mechanism produces an olefin and an amine and leaves an acidic proton on the surface of the MMT.[98] This acidic site may have a catalytic effect on the initial stages of decomposition of organic material within the clay layers, decreasing the molecular weight of the matrix polymer by encouraging chain scission, thus exerting an impact on mechanical properties and flammability. Furthermore, if a significant fraction of the modifier degrades, this causes a loss of thermodynamic compatibility with the polymer matrix and re-aggregation of the clay platelets, resulting in the formation of a conventional composite, or microcomposite, instead of a nanocomposite.

Measures have been taken to improve the thermal stability of modified clays. Extraction of excess modifier can improve the thermal stability of alkyl ammonium based systems, sometimes up to 220°C.[104,105] Additionally, the use of sub-stoichiometric amounts of modifier can give some improvement in the thermal stability of the modified clay.[99] Furthermore, in melt compounding, careful experimental design and processing equipment choice can be of great importance in overcoming the thermal degradation of the modifier.[106–112] However, the structure of the modifier is the most important factor affecting the thermal stability. The use of modifiers based on aromatic amines can enhance thermal resistance.[113] Likewise, some studies indicate that alkylphosphoniums degrade at temperatures 20–30°C[16] higher than that of alkylammoniums, though in practice the thermal stability of such cations may be significantly reduced when they are attached to MMT.[114] Exotic options such as stibonium[56] modifiers have been investigated as well, with a stibonium modifier containing hexadecyl substituents displaying enhanced thermal stability and incomplete degradation even at elevated temperatures.[115] However, the most studied approach to the thermally stable modification of clays generally involves the use of modifiers containing cationic head groups with an aromatic character, most notably based on imidazolium cations,[53] with pyridinium,[54] quinolinium,[54] and tropylium[55] cations showing enhanced stability in this context as well. Application of oligomer-based modifiers with one or more cationic groups has been used as another means of improving the thermal stability of the organoclay.[116–119] The need for thermally stable modification applies to LDH clays also, though in such cases, the relatively low dihydroxylation temperature of the LDH layers themselves coupled with the relatively good thermal stability of the fatty acid carboxylates commonly used as LDH modifiers means modifier degradation is less likely to be the primary factor limiting their use. Finally, charged oligomeric species (in this case anionic) may be useful in the context of LDH modification.

12.3.2 Surface Treatment of Carbon Nanotubes and Other Nanoadditives

Similar to clays, surface treatment of CNTs, graphene, and other nanoadditives is necessary to establish an interface between the polymers and these nanoadditives. However, the method for modification or functionalization of these nanoadditives is dramatically different than that for clay modification in which a charged species is anchored to the clay surface via ionic interactions. The surface treatment of carbon-based nanofillers typically involves the breaking of carbon-carbon bonds to form more polar functional groups via oxidation with nitric acid[120] or other strong oxidizers, or through the realization of bond scission and the generation of reactive species through ion beam[121] or plasma treatment. In contrast, the modification of various mineral and metal oxide nanoparticles is typically achieved through treatment with silanes or siloxanes. **Figure 12.12** shows a schematic of such approaches. These surface treatments, as for clays, may be based on small molecules[122] or polymeric species.[120]

12.4 PREPARATION OF POLYMER NANOCOMPOSITES

A characteristic feature of PNs is the level of interaction between the polymer matrix and the nanoadditives and the associated nanoscale dispersion of the nanoadditives in the

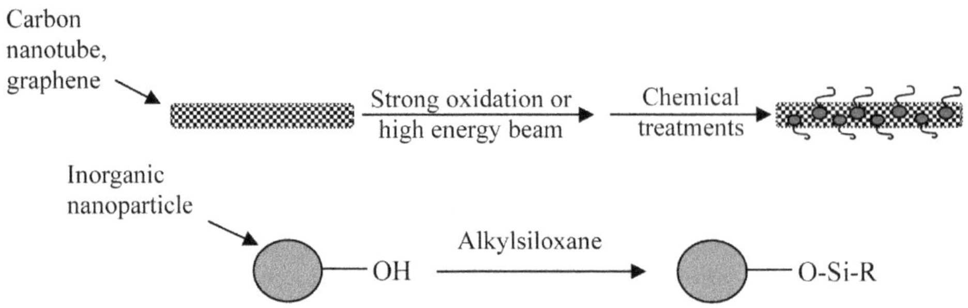

FIGURE 12.12 A schematic illustration showing the surface treatment of CNTs, graphene, and inorganic nanoparticles with hydroxyl-terminated surfaces.

Source: For illustration only.

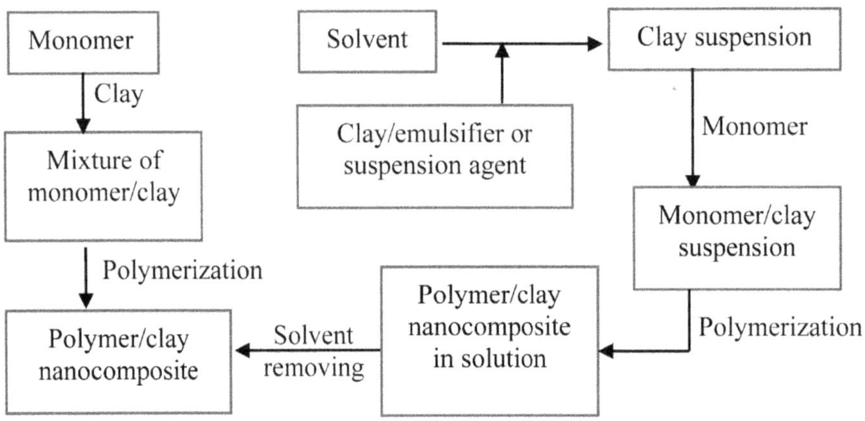

FIGURE 12.13 A schematic illustration of the in situ polymerization process for nanocomposite formation with clay used as an example nanoadditive.

Source: For illustration only.

polymer matrix. As a result, PNs exhibit unique properties that are not shared by their microscale counterparts—conventional polymeric composites.[29,123–126] However, PNs are not easy to obtain in practice. In most cases, the simple physical mixing of a polymer with a nanoadditive does not result in a PN but rather a more conventional composite with poor mechanical and thermal properties, with a lack of compatibility driving phase separation and resulting in poor physical interactions between the matrix polymer and the nanoadditives.

In order to successfully produce a PN, the first problem to be addressed is how to incorporate the nanoadditive into the polymer matrix. Not only is it necessary to break up agglomerates to form smaller agglomerates or single nanoparticles; additionally, these particles must be well dispersed at the nanoscale in the polymer matrix. This requires that the methods and processes used for fabricating PNs are compatible with the chemistry of both the polymer and the (modified) nanoparticle. A variety of methods/processes that are suitable for various nanoadditives have been and continue to be applied for the preparation of PNs. They are generally classified into three categories: in situ polymerization, solvent blending, and melt compounding. A simple discussion is given in the next section.

12.4.1 In Situ Polymerization

This method has been a widely applied to obtain PNs and involves polymerization either in a solvent-free system (i.e., a bulk phase) or in a solvent-based system (including aqueous systems) that may be either a solution, an emulsion,[127,128] or a suspension.[129] While the structure of the modifier is critical and strongly related to the need for compatibility and the miscibility of the nanoadditives with the matrix polymer, mechanical agitation is an important factor as well, providing shear forces that help to accelerate nanoparticle dispersion in the case of a compatible system. Ultrasonication can be an even more efficient means of achieving this, though it should be noted that ultrasonication of polymers in solution can also result in a loss of molecular weight.[130]

Figure 12.13 shows a schematic illustration of the in situ polymerization process with clay as example. An advantage of this process is that it favors the production of PNs with a well-dispersed morphology at both the micro and nanoscale, particularly in presence of well-chosen functional group(s) in the modifier. This process is particularly relevant for preparing PNs based on thermosetting polymers, given that thermosets are almost always made via in situ polymerization of liquid precursors into which nanoadditives may be readily added and mixed.

One feature of some nanoadditives, such as clays, is that they can significantly change, often increase, the viscosity of a fluid.[77,78] In some cases this may be beneficial (for instance, it may be used to adjust formulation viscosity to produce an ordered and well-aligned PN structure), but it can also be a problem (for instance, the processing of a thermoset formulation may be rendered much more difficult as a consequence). A disadvantage of in situ polymerization is that, depending upon the final interface between the matrix polymer and the nanoparticle, a well-dispersed but thermodynamically incompatible PN may be produced which may revert to a poorly dispersed morphology in the course of subsequent processing.[131] Furthermore, the process is not easily applied in industry, particularly at the multiton scale. In the case of clays, the PA-6/clay nanocomposite from Ube/Toyota is the most significant commercial example of the in situ process.

12.4.2 Solvent Blending

Solvent blending, also called solution intercalation in the case of clay and other nanolayers, involves dispersing the nanoadditive and dissolving the matrix polymer in a common solvent or solvent mixture. Three parameters have been considered to be important, particularly for clays, in choosing the surface treatment of the nanoadditives with this process in mind: the structure of the modifier, its miscibility with the polymer, and its thermal stability. The miscibility of the modifier here has two meanings: miscibility with both the final polymer and with the solvent chosen to dissolve the polymer. The modifier structure and its miscibility are perhaps more important than the thermal stability, given that thermal stability is primarily needed for post-processing (molding operations for instance) requiring elevated temperatures. Figure 12.14 shows a schematic illustration of the solvent intercalation process using clay as the example.

An advantage of using this process to obtain PNs is that the polymer can be dissolved and the nanoadditive can be dispersed in the same solvent/solvent mixture. Similar to the in situ process, a critical issue in obtaining PNs through this process is breaking apart nanoadditive agglomerates, with uniformity of mixing being of key importance as well. Again, ultra-sonication has proven to be a very useful tool to break apart such agglomerates, as shown in the case of clay.[90] It has been observed that a clay loading of ~1–3 wt% enables delamination of the solution and thus in the final PCN. This process is best used for preparing PNs based on thermosetting polymers or polymers showing good solubility in one or more solvents, thus allowing the polymer chains and the nanoadditives to mix freely.

Although solution blending has only been used at the lab scale at this time, compared to the situ polymerization, it may be more industrially relevant, particularly for primary polymer producers who have the capacity to recover and recycle solvents. However, as noted above, specific concentration ranges are preferred, and the process is quite dependent on the individual polymer. Some polymers have many solvents from which to choose while others do not. A typical example is polystyrene, which can dissolve in a variety of solvents, making it relatively easy to find a solvent that is compatible with both the clay and the polymer. Polyolefins, on the other hand, can only be dissolved in very specific solvents at elevated temperatures, with potential consequences as far as thermal degradation of the modifier is concerned.

12.4.3 Melt Compounding

Among all the methods/processes used for preparation of PNs in general and PCNs in particular, melt compounding (Figure 12.15) is perhaps the most widely studied approach[29,132–134] because of its simplicity and the ready availability of the relevant processing equipment. Preparation of PNs through melt compounding requires the feeding of the polymer and the nanoadditive into compounding equipment, such as an extruder, followed by

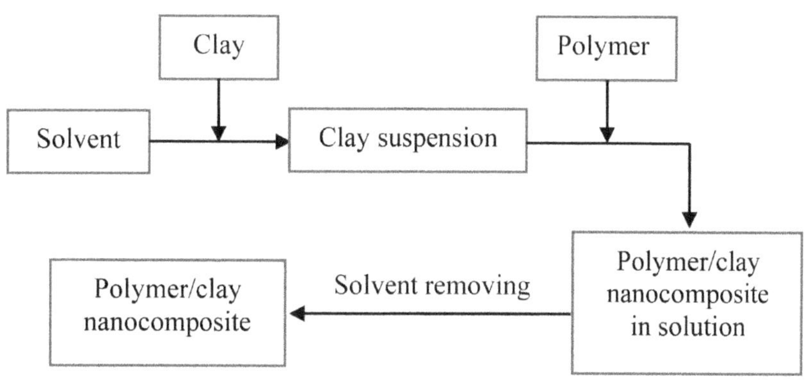

FIGURE 12.14 A schematic illustration of the solution processing of a nanocomposite with clay used as an example nanoadditive.

Source: For illustration only.

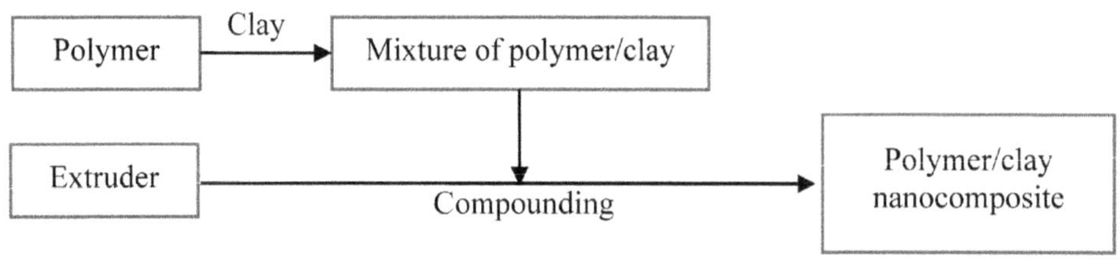

FIGURE 12.15 A schematic illustration of the melt compounding process for nanocomposite formation with clay used as example nanoadditive.

Source: For illustration only.

compounding at elevated temperatures required for the melt mixing of the matrix polymer. The modified clay should not only be compatible with the matrix polymer but also stable during the melt compounding operation as well. The need for functional modifiers that enable covalent bond formation between polymer and nanoparticle becomes less important with this process, but it can nonetheless enable in situ grafting of the polymer chains onto the surface of the nanoadditives and improve dispersion levels as a consequence.

In addition to its simplicity and the ready availability of the processing equipment, one advantage of melt compounding is that nanoadditives which may not disperse well in an in situ polymerization or solvent blending process may nevertheless be well-dispersed into the polymer matrix thanks to the ability of the significant applied shear forces to more efficiently accelerate the breakup of agglomerated nanoadditives into small aggregates and/or primary particles in compatible systems. However, this technique is generally restricted to thermoplastics.

12.5 STRUCTURE OF POLYMER NANOCOMPOSITES

In general, PN structure may be broadly classified by the level of dispersion of the nanoadditives in the polymer matrix: nanoscale dispersed PNs and micro/macroscale dispersed conventional polymer composites. Many factors play a role in achieving a particular morphology. For a given organo-modified nanoadditive, dispersion levels are a function of the matrix polymer, the structure of the organomodifier, and the exact method and process parameters used for fabricating the PN. In order to successfully obtain a PN, all of these factors must be considered.

12.5.1 Structure of Polymer Nanocomposites

In the case of PCNs (and for other PNs containing inorganic nanolayers), three kinds of commonly accepted terms have been widely used to describe the structures (ordered and disordered): immiscible, intercalated, and exfoliated.[135] In most cases, all three structures coexist in the same PCN, but one is generally dominant. A schematic illustration of the structures of PCNs is given in Figure 12.16. The immiscible structure (Figure 12.16a) indicates that the matrix polymer and the clay are incompatible following processing—either due to an initial incompatibility or to a change in the system during processing itself. The implication of this outcome is that phase separation has occurred in the PCN because either the chains of the matrix polymer were unable to intercalate into the galleries between the clay layers or the clay modifier has been degraded in the course of PCN fabrication. Such an outcome may mean that the chosen production process is not well-suited for the particular clay and matrix polymer chosen. The resulting materials tend to have similar properties to those of conventional composites.

The two kinds of common intercalated structures observed in PCNs are shown in Figure 12.16b and 12.16c. Both are considered to be nanocomposite materials, but PCNs with the intercalated structure shown in Figure 12.16c possess higher levels of dispersion vs. those shown in Figure 12.16b. The definition of "intercalated" derives from X-ray diffraction measurements, which indicate that the clay layers maintain parallel registry while moving further apart. The formation of intercalated structures in the PCN occurs because of the penetration of the polymer chains into the interlayer spacing of the clay platelets, which leads to an expansion of the d-spacing of these platelets (Figure 12.16b). Upon further intercalation of the matrix polymer into the interlayers of the clay platelets or under an external force such shear, some delamination occurs, leading to an intercalated structure based on stacks consisting of only a few clay layers with the potential for more uniform dispersion (**Figure 16c**).

With more and more polymer chains entering the interlayer galleries of the clay platelets, the d-spacing becomes increasingly large, and exfoliation becomes increasingly favorable (Figure 12.16d); this is often considered to be the most desirable of nanocomposite structures. In practice however, PCNs with the intercalated structure as shown in Figure 12.16c may be similar in performance to the PCNs with exfoliated structure. Furthermore, it should be noted that there is no standard criterion to describe what is meant by "exfoliated." Some researchers refer to "exfoliation" when the d-spacing of the clay platelets is larger than 6–8 nm, while others consider that the term "exfoliated" must be used only when the parallel registry of the clay platelets has been lost. An alternative to the term "exfoliated" is "delaminated" and both are used interchangeably.

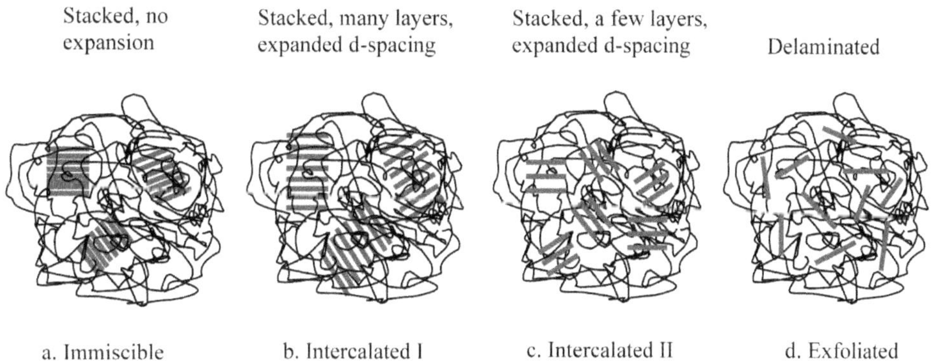

FIGURE 12.16 A schematic classifying typical PCN morphologies.

Source: For illustration only.

While the aforementioned terms are widely used to classify the degree of dispersion of clay in the polymer matrix, it should be pointed out that they are strictly qualitative, potentially controversial, and potentially dependent on the judgement of the individual. Because there are no established standards to follow, there is room for differences in interpretation of the morphology of PCNs. For instance, both delamination and exfoliation have been used to describe the morphology of the PCNs in which the clay platelets have a d-spacing larger than 6–8 nm or have been separated at the level of the primary particle; at the same time, other researchers argue that delamination is a dispersion state beyond exfoliation. Attempts to provide standard descriptions for levels of clay dispersion can be seen in a few papers.[136,137] Image analysis[137–139] has also been used to try to quantify the degrees of nanoadditive dispersion, but more work is needed in these areas.

In the case of PNs containing nanoadditives other than nanolayers, such as CNTs and POSS, the terms above are not used. Nonetheless, similar states of dispersion are described in these systems as well, including substantially aggregated (implying an immiscible or partially immiscible structure) and well-dispersed (implying small aggregates or single nanoparticles). Furthermore, such states of dispersion may coexist in the same PN.

12.5.2 Characterization

Various nanoscale and microscale techniques have been used to directly or indirectly characterize the structure and morphology of PNs. In the case of clays, these methods include x-ray diffraction (XRD),[90,140] transmission electron microscopy (TEM),[103,136] nuclear magnetic resonance (NMR) spectroscopy,[141,142] Fourier transmission infrared (FTIR) spectroscopy,[143] atomic force microscopy (AFM),[144] scanning electronic microscopy (SEM),[145,146] and the use of dyes as indicators of dispersion state,[147] to name just a few. Materials property testing can also be an option for indirectly confirming PCN morphology, for instance by measuring mechanical, thermal, and/or transport characteristics (e.g., gas barrier, thermal or electrical conductivity, etc.). If nano-dispersion is achieved, this can be inferred from changes in such properties vs. conventional composites or the base polymer.

Among the methods for the determination of morphology, XRD and TEM are among the most commonly used techniques. In this chapter, a simple introduction to these two techniques is provided. For more detailed information about these and other methods, the reader is referred to related papers/references.[103,136,137,148] At this point however it is important to note that it is generally not possible to assess the structure and morphology of a PN using any single method in isolation, and that a combination of multiple methods is typically necessary in order to understand the structure-properties relations active in a given materials system.

The use of XRD as one of the two primary tools to characterize the structure and morphology of PCNs relies on the fact that, in the case of immiscible and intercalated multilayer stacks of clay platelets, parallel registry and regular spacing are often maintained, enabling diffraction and allowing the determination of the interlayer distance (or d-spacing) between the clay platelets in these stacks using Bragg's law. This represents an application of wide angle x-ray diffraction (WAXD),[140] which is distinct from small angle X-ray scattering (SAXS),[63,90] though both can be complementary given their ability to probe different dispersion length scales. Nevertheless, as the instrumentation is more readily accessible and the data more easily interpreted, the use of the WAXD is much more common. WAXD is useful for measuring regular interlayer d-spacings of less than 6–8 nm, while SAXS gives information on the state of dispersion at larger interparticle distances undetectable by WAXD and does not require that the particles are regularly spaced. However, WAXD is enough in most cases to determine if intercalation has occurred.

The clay can be ordered or disordered in the polymer matrix, but only an ordered clay structure generates x-ray diffraction peaks. In this case, the diffraction peaks that appear in the WAXD pattern will provide information concerning the morphology of the PCN. A sharp and strong peak is consistent with the presence of larger/more highly ordered multilayer stacks of clay platelets in the polymer matrix (immiscible or intercalated), while a broad and weak peak is consistent with smaller and/or more disordered multilayer stacks and is usually taken as evidence of a mixture of intercalated (ordered) and disordered structures. When a completely exfoliated structure is achieved, no diffraction peaks associated with the interlayer spacing of the clay layers should be observed in the WAXD pattern because the d-spacing between the layers is too large (i.e., exceeding 6–8 nm)/too inconsistent and parallel registry between the layers has been lost. However, it should be pointed out that "exfoliation" only means that the clay is exfoliated relative to other clay platelets; this does not imply that a uniform distribution of clay has been achieved in the polymer matrix. Furthermore, while an exfoliated structure should not produce a diffraction peak, the absence of a diffraction peak does not prove that an exfoliated structure has been obtained in the PCN. Based on the authors' experiences, the absence of a peak only indicates an absence of ordering—in many cases, additional analyses such as TEM confirm that this is due to poorly dispersed but disordered clay layers, not true exfoliation. The PCN literature also highlights the fact that the absence of signal cannot be used as proof of the exfoliation of the clay.[136] In such cases, high resolution microscopy (TEM or AFM, most notably) must be used in order to obtain additional information about the morphology of the PCN. Another option is to use SAXS, as diffraction peaks occurring at angles too low to be detected via WAXD may be visible in SAXS, and because SAXS can additionally provide information about the size and spacing of scattering domains even when no diffraction peaks are observed. As even SAXS has limitations related to the length scales it is able to probe, however, this may not be sufficient to fully confirm the microstructure of a given PCN.

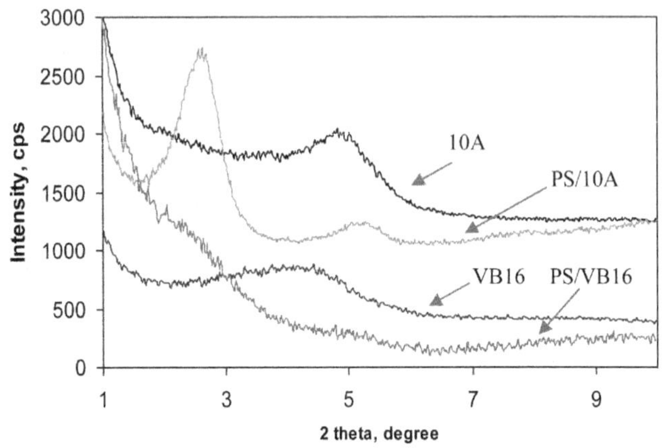

FIGURE 12.17 XRD patterns of two organoclays (Cloisite 10A, dimethyl-benzyl-hydrogenated tallowammonium MMT from Southern Clay Products, now BYK, and VB16 from ref. 16) and two PSCNs with intercalated (PS/10A) or exfoliated (PS/VB16) morphology.

Source: For illustration only.

With the XRD method, some practical experimental details[149] that need to be considered include scattering/diffraction mode (transmission or reflection), sample form (powder or solid), data collection parameters (step size, scan rate), typical locations and intensities of clay diffraction peaks/signals and equipment parameters (e.g., beam intensity and degree of collimation, beam-stop size and geometry, detection modes, etc.). Information on why these parameters matter and the kind of information X-ray data can provide about a nanocomposite material can be found in the references[140] as well as in books and manuscripts about X-ray diffraction.[148,150,151] Figure 12.17 provides examples of WAXD patterns for two organoclays and two PSCNs containing these two organoclays, either intercalated (showing a strong and sharp diffraction peak associated with the interlayer spacing of the clay platelets) or disordered (showing no such peak).

WAXD is easily used and generates data representative of the nanostructure of macroscopic volumes of material but can only give a partial view of dispersion state, and indirectly at that. In contrast, TEM is an expensive and difficult method to use and provides insights at such a local level that it can be challenging to ensure that the data is representative—but it nevertheless gives a direct (if qualitative) view on the nanostructure of the material. AFM is easier to carry out than TEM and requires less expensive equipment, but this is counterbalanced by the need for a skilled operator coupled with the right tip and the right imaging parameters in order to generate images equivalent to those taken via TEM.[152] For these reasons, the WAXD method is usually combined with TEM or (to a lesser extent) AFM to characterize the morphology of PCNs. Attempts have been made to perform quantitative image analysis in this context,[138,139] but in contrast with isotropic nanoparticles, where effective methodologies exist to quantify dispersion levels based on microscopy data,[153] in the case of clays and other anisotropic nanoparticles, while attempts have been made,[154,155] this remains a challenge.

TEM tends to be more qualitative than quantitative because of the lack of well-defined standards for exfoliation, intercalation, and distribution of the clay layers in the polymer matrix. Figure 12.18 shows TEM micrographs which, based on authors' understanding and experience with TEM, illustrate the following morphologies: immiscible, intercalated with large multilayer stacks, intercalated with small, uniformly distributed multilayer stacks, and exfoliated with uniformly distributed clay platelets. Other examples of such structures and morphologies may be found in the literature[135,136,156] and may serve as a useful basis with which researchers can compare their own imaging data.

Nonetheless, as noted above, it is better to apply multiple techniques in order to get an overall picture of the nanoscale structure of a nanocomposite. Furthermore, as each TEM image is focused on a very small portion of the overall specimen (typically microns wide by 100–200 nm thick), multiple images in both low magnification and high magnification and the analysis of several sections from different parts of a given nanocomposite specimen are necessary in

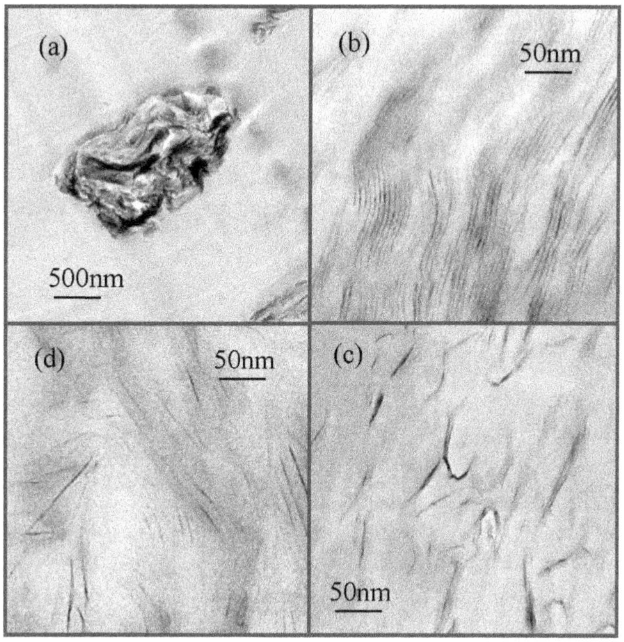

FIGURE 12.18 Representative TEM images for various PCN morphologies: (a) immiscible; (b) intercalated with large multilayer stacks; (c) intercalated and uniformly dispersed with small multilayer stacks; (d) exfoliated (delaminated) with a uniform dispersion.

Source: For illustration only.

order to make a well-informed judgement about the clay dispersion and morphology. Low magnification images are particularly important for determining the overall clay distribution and dispersion in the polymer matrix. Although this greatly increases the analysis time and cost, it provides significantly greater confidence when assessing nanocomposite morphology, particularly when used in combination with other techniques, such as WAXD, property measurements, etc. At the same time, when uniform dispersion is achieved (such as clay in polyamide-6 in most cases), a uniform distribution of the clay in the polymer matrix is seen in the low magnification TEM image and much less work is needed to collect representative images at higher resolutions given that the structures observed at any randomly chosen point within the specimen will appear similar.

The structures of the PNs based on CNTs and other nanoadditives have both similarities and differences with those seen for the PCNs because of the dramatic difference of nanoadditive geometry. Like clays, CNTs and other nanoadditives can be immiscible or aggregated with only a few primary particles dispersed in the polymer matrix, but intercalated and exfoliated are not suitable to describe the morphology of the PNs derived from CNTs and other nanoadditives as those terms are only suitable for describing the structure of layered nanoadditives as in the PCNs. The methods that are used for characterizing the nanostructure of the PCNs can be used to characterize the structure of the PNs that contain CNTs and other nanoadditives as

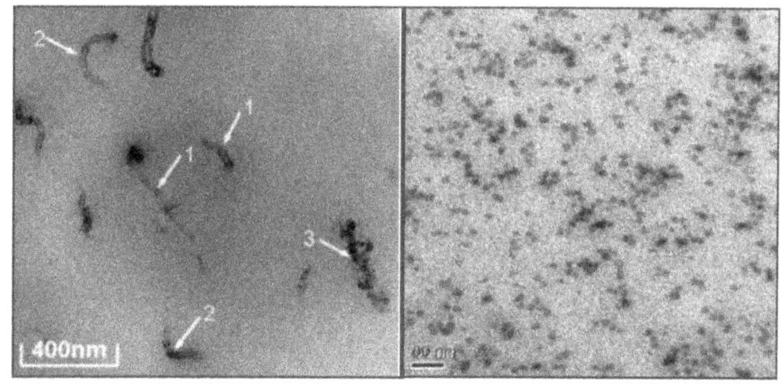

FIGURE 12.19 TEM images of carbon nanotubes in polypropylene (left, arrow 1, straight tubes, arrow 2, curved tubes); and arrow 3, bundled or entangled tubes (left, from ref. 157, with permission from Wiley); and nanosilica in acrylate polymer (right, from ref. 158, with permission from Elsevier).

well, albeit with some limitations related to WAXD in particular, given that only layered nanoadditives tend to give diffraction peaks indicative of dispersion state. With this in mind, high resolution microscopy (TEM or AFM) will be among the most useful tools for examining the morphology/dispersion of CNTs and other nanoadditives in the polymer matrices (Figure 12.19,[157,158]).

12.6 THERMAL AND FIRE PERFORMANCE OF POLYMER NANOCOMPOSITES AND ASSOCIATED MECHANISMS

12.6.1 Thermal and Fire Performance of Polymer Nanocomposites

The impact of nanocomposite technology on the thermal and fire resistance of polymers, as mentioned in the introductory section, is primarily reflected in two important parameters: (1) the T_{onset} in the TGA curve, representative of the thermal stability of the polymer, and (2) the peak heat release rate, or HRR, as measured by cone calorimetry analysis (CCA)—reflective of the combustion behavior (the flammability) of the polymer. The T_{onset} may be increased and the peak HRR may be reduced for a variety of polymers when nanodispersion of the nanoadditive is achieved in the polymer matrix, as seen in Figure 12.20 in which both the TGA curves and the CCA curves for pure polystyrene and a few PSCNs[16] obtained by in situ bulk polymerization are presented. In addition to these two methods, Limiting Oxygen Index (LOI), UL-94 testing, and other application-specific tests have also been used to assess fire performance. In most cases, PNs that contain no additional FR additives do not show an increased LOI, nor can they be classified in the UL-94 test. Detailed information about LOI and UL-94 is not provided in this chapter.

TGA is the most widely used tool for analyzing the thermal degradation behavior of PNs. A nanodispersed nanoadditive, as seen in Figure 12.20, has the potential to bring about a significant improvement of the thermal stability to the matrix polymer. More examples are given in Figure 12.21 in which TGA curves of polydimethylsiloxane

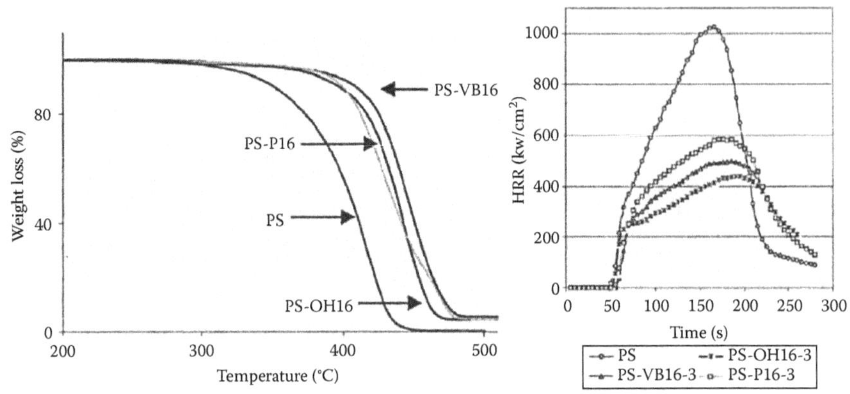

FIGURE 12.20 TGA (10°C/min in N_2) and CCA (35 kW/m^2) curves for pure polystyrene and three PS/clay nanocomposites obtained via in situ polymerization in bulk; VB16 and OH16 are ammonium modified MMTs, while P16 is a phosphonium modified MMT.

Source: Reprinted with permission from reference 16. Copyright 2001, American Chemical Society.

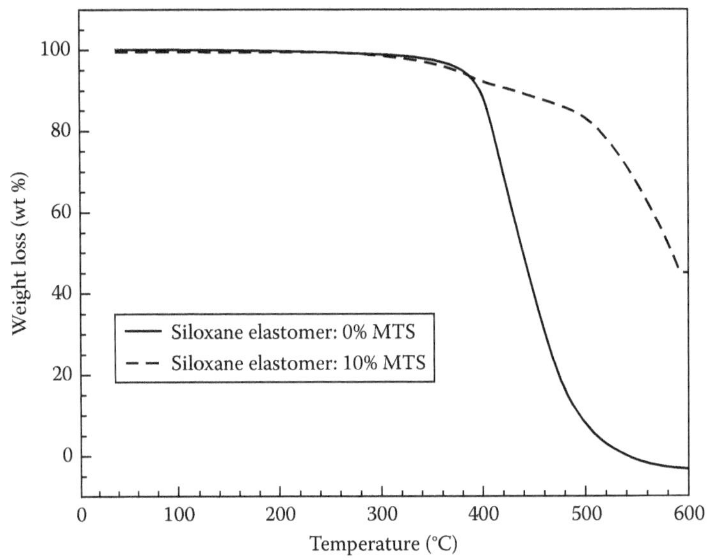

FIGURE 12.21 TGA traces for PDMS and its clay nanocomposite.

Source: Reprinted with permission from ref. 159. Copyright 1995, American Chemical Society.

(PDMS) and its clay nanocomposite (at 10% organo-MMT)[159] are shown—a drastic shift of the T_{onset} to a higher temperature (with a stabilization as high as 140°C at 50% weight loss) is observed and in **Figure 12.22** in which TGA curves of poly(1-caprolactone), PCL, and its clay nanocomposites[160] are provided. In all of these cases, the PCNs show significantly enhanced thermal stability. That said, it is important to note that the enhancement of the thermal stability due to PN formation is strongly dependent on the individual matrix polymers and nanoadditives. For a given nanoadditive, the resulting PNs based on one polymer matrix may show a significant enhancement of the thermal stability, while for another polymer matrix, such an enhancement may be less significant to nonexistent; decreases in thermal stability on addition of nanoadditives are possible as well. Similarly, for a given matrix polymer, one nanoadditive, when it is nanodispersed, may significantly enhance the thermal stability of this matrix polymer, while another nanoadditive may be ineffective. This strong composition-dependence is attributed to specific interactions between the polymer matrix (and/or its pyrolysis products) and the nanoadditives.

Examples illustrating the dependence of PN thermal stability on the polymer matrix (with MMT as the nanoadditive) are given here: a PSCN at 3% loading of organoclay (Figure 12.20) shows an increase of about 50°C in T_{onset} (at 5% mass loss); for PCNs based on polyolefins, however, a less significant enhancement in the thermal stability is observed (Figure 12.23,[161] with results for a polyethylene PCN shown); finally, for PCNs based on PA-6, no change in the TGA parameters is observed.

Examples of the dependence of the thermal stability of PNs on the nanoadditive (with PMMA as the matrix polymer) are given here: $ZnAl_2$, $CoAl_2$, and $NiAl_2$ LDHs

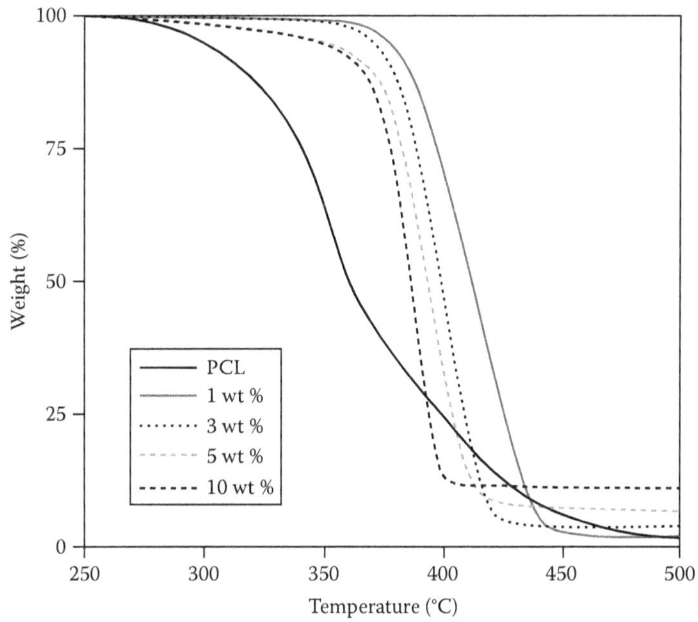

FIGURE 12.22 Relationship between weight loss and temperature in air for neat PCL and PCL nanocomposites containing 1, 3, 5, and 10 wt% clay.

Source: Reprinted from ref. 160 with permission from Elsevier.

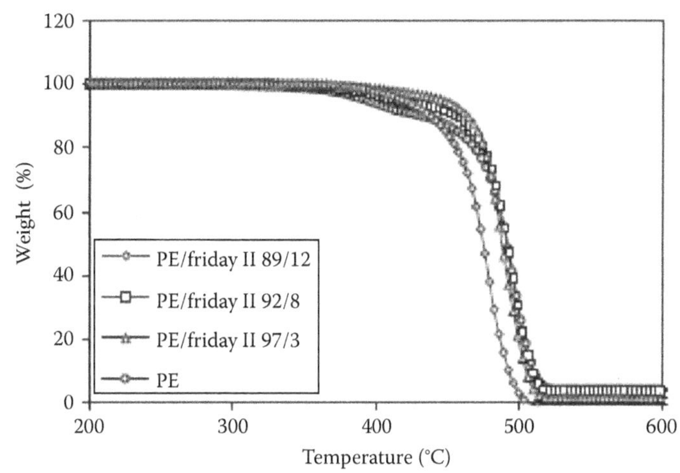

FIGURE 12.23 TGA curves of neat PE and its clay nanocomposites.

Source: Reprinted from ref. 161 with permission from Elsevier.

enhance the thermal stability of PMMA (Figure 12.24),[162] but $CuAl_2$ LDH significantly reduces thermal stability. It is also seen that use of CNTs can increase T_{onset}. This is seen for example in PAN/MWCNT composite fibers,[163] where a 24°C increase in T_{onset} at 5% MWCNT loading is observed vs. pure PAN, as opposed to much more limited enhancements when using clay as a nanoadditive.[164]

The thermal stability of PNs also shows a dependence on the surface treatment of the nanoadditive, as shown in Figure 12.25[165] in the form of TGA curves of PMMA and its clay nanocomposites (comprised of the same MMT but different surface treatments). It is evident in this case that the choice of surface treatment results in a different level of thermal stability in the resulting PCN. The presence of oxygen plays an important role in the thermal performance of the PNs as well, as seen in Figure 12.26[166] for PSCNs in both inert atmosphere and oxidative conditions. In the absence of oxygen, the thermal stability of the PSCN (as measured by the onset temperature of degradation) was significantly enhanced, increasing by 50°C as compared to pure PS. Under thermo-oxidative conditions, however, the degradation of pristine polymer and the PSCN occurred at

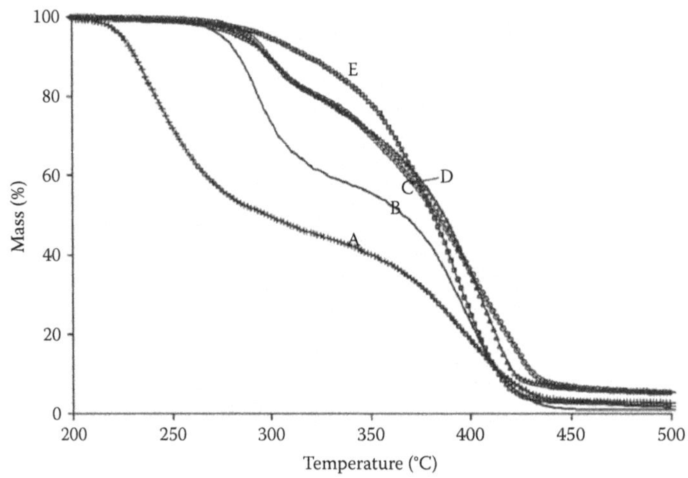

FIGURE 12.24 TGA curves of neat PMMA and various PMMA/LDH nanocomposites (5% loading) in N_2 at 20°C/min. A: PMMA/$CuAl_2$; B: PMMA; C: PMMA/$ZnAl_2$; D: PMMA-$CoAl_2$; E: PMMA/$NiAl_2$.

Source: From ref. 162, reproduced by permission of The Royal Society of Chemistry.

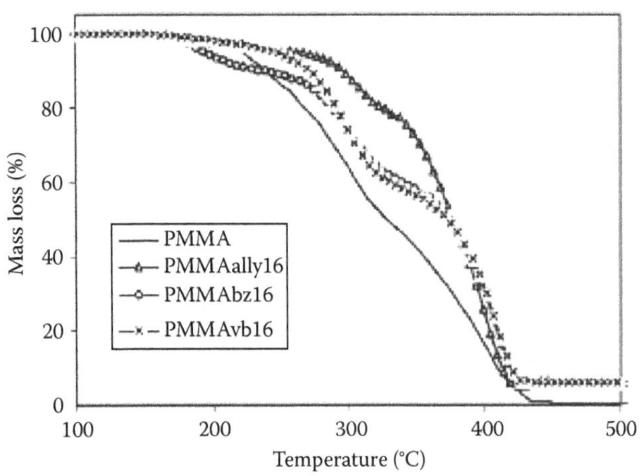

FIGURE 12.25 TGA curves for neat PMMA and various PMMA/MMT nanocomposites at 5% loading.

Source: Reprinted from ref. 165 with permission from Elsevier.

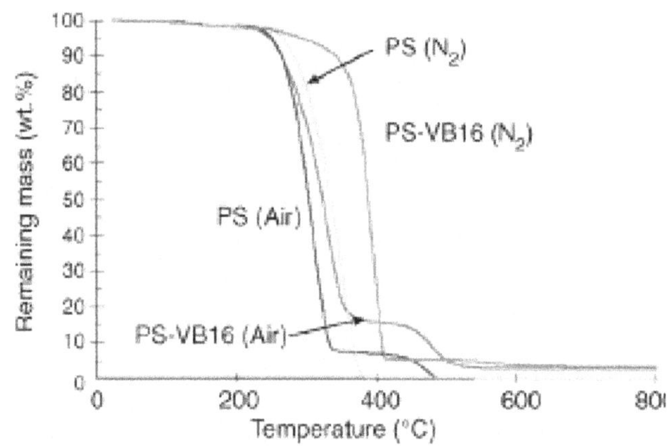

FIGURE 12.26 TGA curves of PS and its clay nanocomposites under pyrolytic and thermo-oxidative conditions at 1°C/min.

Source: Reprinted from ref. 166 with permission from Elsevier.

comparable temperatures. Nevertheless, regardless of atmosphere, greater char formation was consistently observed in the case of the PSCN, which displayed 15% residue at 400°C as compared to only 5% for the pure polymer.

The loading of the nanoadditive is another factor affecting the thermal stability of PNs. In particular, the most significant improvements in thermal stability are often observed at low loading levels, with the addition of larger amounts of nanofiller resulting in diminishing returns. This is shown in Figure 12.27[167] in which onset temperature as a function of clay loading is given for various PSCNs. Furthermore, it is noted that only the addition of surface treated MMTs results in enhanced thermal stability, in contrast with unmodified MMT, where no such enhancements are observed. These observations are fully in-line with trends generally observed in PNs based on high aspect ratio nanoparticles more broadly, where high levels of dispersion are only achieved when the nanofillers are appropriately surface modified and present at concentrations below the percolation threshold. As the percolation threshold is exceeded, filler aggregation becomes more and more likely, and performance gains are more and more limited as a consequence. The value of the percolation threshold differs as a function of filler aspect ratio but is readily estimated if the aspect ratio is known.[87]

In the case where high levels of dispersion are achieved, on the other hand, enhanced fire performance may be observed; in this context, CCA is arguably the most powerful tool for analyzing the flammability of the resultant PNs. In most cases (as seen in Figure 12.20), a significant reduction in peak HRR is observed in the CCA curve. Additional examples include a PA-6/clay nanocomposite which shows a 63% reduction in the peak HRR at 5% loading (Figure 12.28)[168] and in a poly(ethylene-co-vinyl

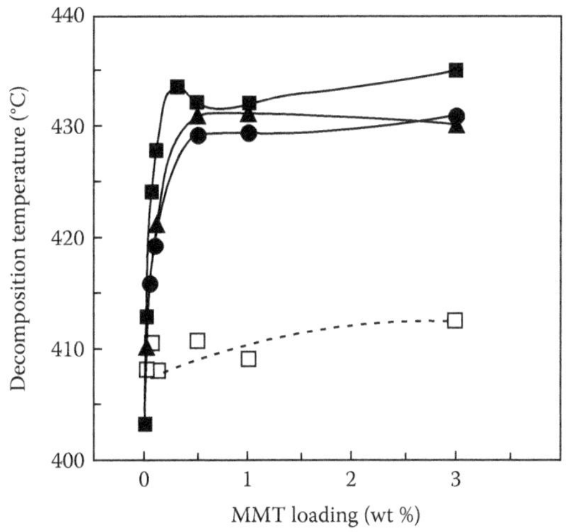

FIGURE 12.27 Decomposition temperature as a function of MMT loading in PS nanocomposites. The MMT is modified with the following: (filled squares) dimethyl(hydrogenated tallow alkyl)benzyl ammonium ions, (filled circles) dimethyl-di(hydrogenated tallow alkyl) ammonium ions, (filled triangles) dimethyl(hydrogenated tallow alkyl) 2-ethylhexyl ammonium ions, or (open squares) sodium ions (from ref. 167 with permission from Springer Science).

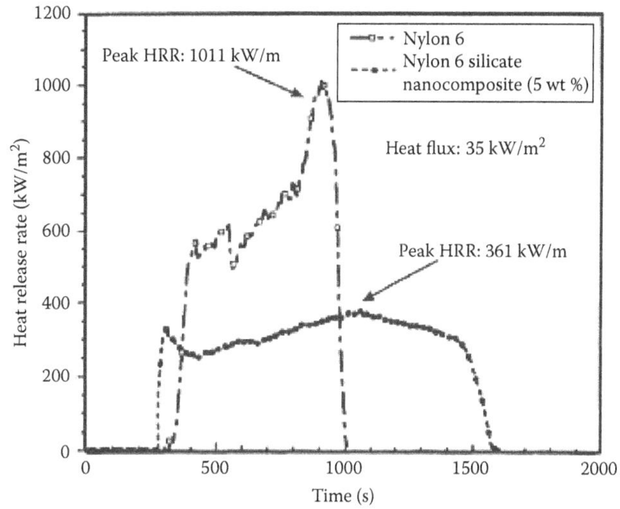

FIGURE 12.28 CCA curves of PA-6 and its clay nanocomposites (5 wt%) tested at a heat flux of 35 kW/m².

Source: Reprinted from ref. 168 with permission from Elsevier.

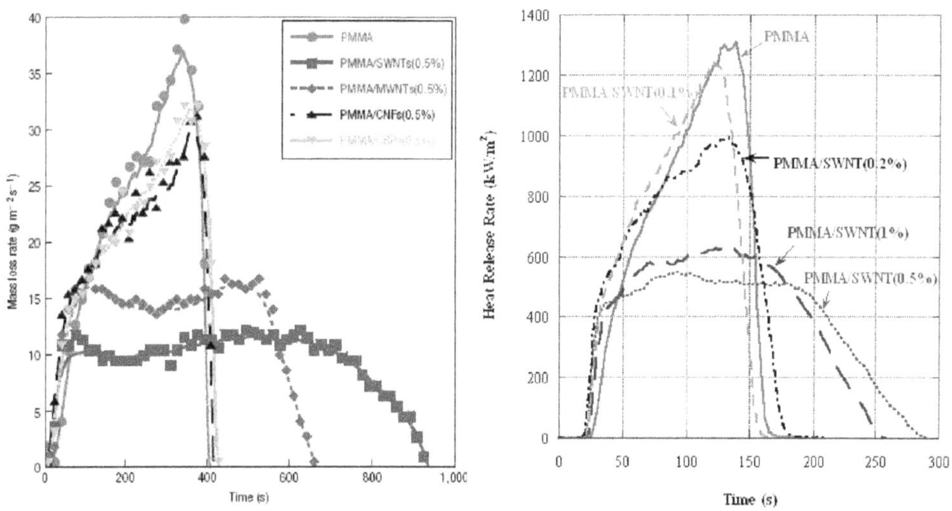

FIGURE 12.29 Left—mass loss rate as a function of gasfication time for several PMMA nanocomposites, at an external radiant flux of 50 kW/m² in nitrogen (reprinted from ref. 172 with permission from Macmillan Publishers Ltd); right—effects of SWNT concentration on heat release rate of PMMA at 50 kW/m².

Source: Reprinted from ref. 173 with permission from Elsevier.

acetate) (EVA)/clay nanocomposite[169] which shows a 50% reduction in the peak HRR at 5% organoclay loading. As with thermal stability, however, the magnitude of the reduction in the flammability of the PNs is strongly dependent on composition. Examples of matrix polymer dependence are readily illustrated with various MMT-based PCNs: the PCNs based on PS,[16] EVA,[169] and PA-6[168] show a significant reduction in the peak HRR in the CCA curve (40–70% at a clay loading about 5%), while PCNs based on polymers like polypropylene (PP) show a moderate reduction in the peak HRR (30–50% at a 5% clay loading) and PCNs based on poly(methyl methacrylate) (PMMA) and polyacrylonitrile (PAN) etc. show an even smaller reduction (10–20% at a 5% clay loading).[164] As far as the effects of nanofiller are concerned, CNTs have been observed to surpass both clays and other nanoadditives, as reflected in the lower CNT loadings needed to enhance the thermal and fire resistance.[170,171] As an example, the results for PMMA in the presence of different nanoadditives are seen in Figure 12.29.[172] A PMMA nanocomposite containing only 0.5% CNTs gives significant reductions in the peak mass loss rate, or MLR (55–70%). The effect of CNT loading on the HRR of the PMMA[173] is also

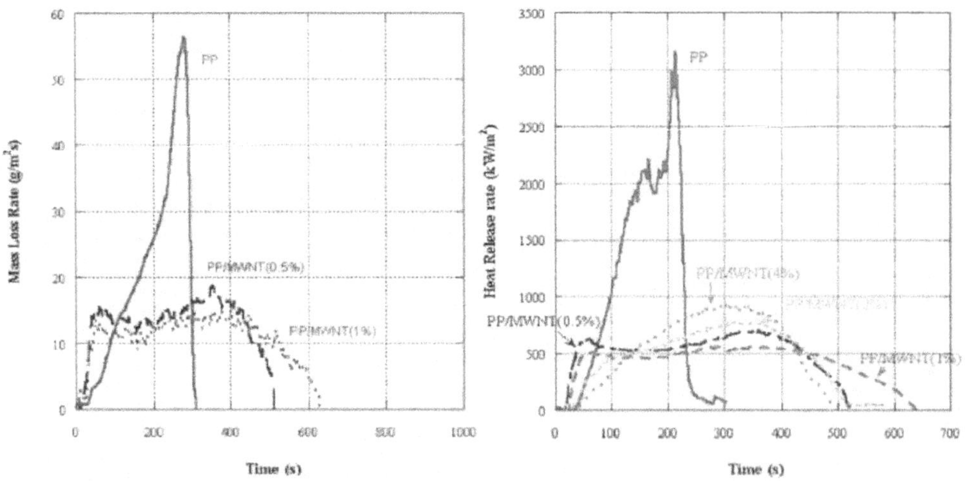

FIGURE 12.30 Left—Effects of MWNT concentration on mass loss rate of PP at 50 kW/m² in nitrogen; right—effects of MWNT concentration on heat release rate of PP at 50 kW/m².

Source: Reprinted from ref. 174 with permission from Elsevier.

shown in Figure 12.29, where a significant reduction in the peak HRR (~55%) is also seen. In contrast, PMMA nanocomposites containing other nanoadditives, clays included, tend to show more limited reductions in the peak HRR.[164] In the case of other polymers as well—for example, in PP, as shown in Figure 12.30[174]—CNTs can exert a significant effect on peak MLR and peak HRR values, while MMT[161] gives a more moderate reduction in the peak HRR.

In addition to the identity of the matrix polymer and nanoadditive, other factors (e.g., concentration, size, and degree of dispersion of the nanoadditive) also play a role. Consistent with the aforementioned arguments concerning percolation, the greatest flammability reductions in PCNs tend to occur at a clay loading of ~5 wt%, regardless of polymer type,[168,175,176] although exceptions are known. A more uniform dispersion of the nanoadditive in the PN is responsible for a larger reduction in peak HRR and more reproducible fire performance. This can be seen in Figure 12.31,[173] where CCA data for PMMA/SWCNT nanocomposites displaying either poor or good dispersion is shown; as is evident from the data, there is dramatic difference in the peak HRR between these two cases. As the reduction of the peak HRR in CCA is related to the morphology, therefore, CCA may be used as an indirect means of assessing whether or not a well-dispersed nanocomposite has been obtained.[149] For instance, when the peak HRR of a PN is less than the best that has been observed for a particular polymer, it may indicate that the PN is not as well-dispersed or that "defects" are present; usually this implies the presence of a significant amount of immiscible material in the PN.

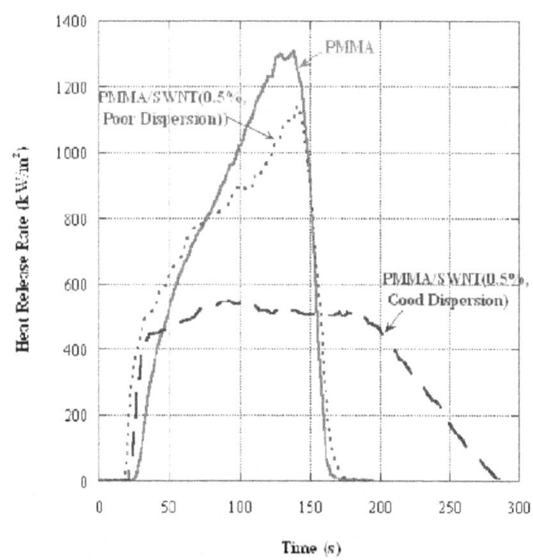

FIGURE 12.31 Effect of SWNT dispersion (at 0.5% loading) on heat release rate of PMMA at 50 kW/m².

Source: Reprinted from ref. 173 with permission from Elsevier.

Results[177–179] from the National Institute of Standards and Technology (NIST) show that the formation of a homogeneous carbonaceous char during the combustion of a PN is critical for suppressing the rate of combustion. This is generally associated with the realization of both effective dispersion and distribution of the nanoadditive in the polymer matrix, though even in such cases, variability in the dispersion and/or distribution of the nanofiller may nonetheless hinder the formation of a homogeneous char. Further insights may be drawn from comparisons of MLR data during gasification tests for three different polypropylene/clay nanocomposites (PPCNs) containing maleic anhydride grafted PP (PP-g-MA) and using either MMT, synthetic hectorite or synthetic fluoromica as the nanoadditive. In this system, the PP/PP-g-MA matrix produces little or no carbonaceous material after gasification. The PPCN with the synthetic fluoromica gives the largest reduction in

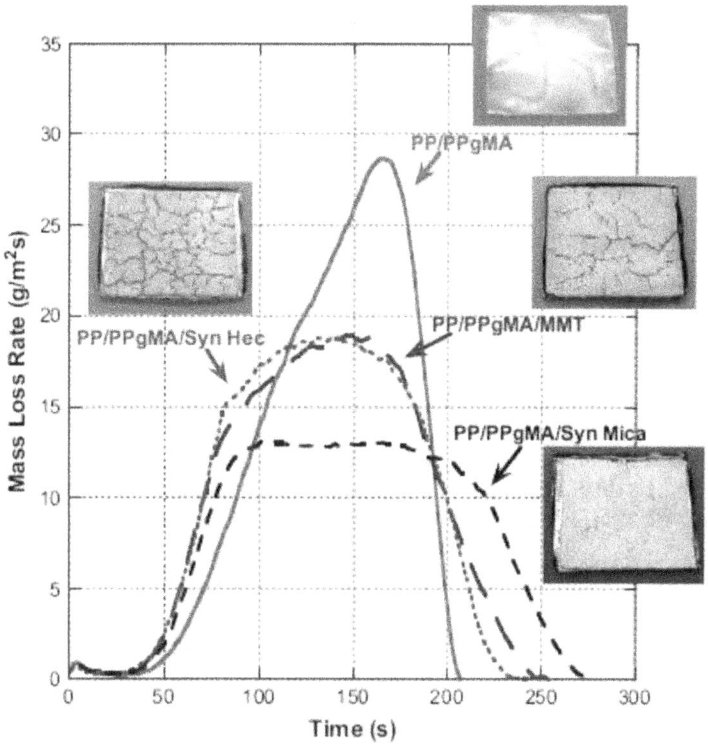

FIGURE 12.32 Effect of clay type on MLR of PP (84.6%)/PP-g-MA (7.7%)/clay (7.7%) at 50 kW/m^2 in N_2.

Source: Reprinted from ref. 179 with permission from Elsevier.

the peak MLR in the gasification test (> 50%), as shown in Figure 12.32.[179] This is attributed to the ability of the synthetic fluoromica to form a more continuous and crack-free residue (also shown in Figure 12.32) than the other two samples. This also highlights the importance of the aspect ratio of the clay in forming a continuous, defect-free char layer: the synthetic fluoromica possesses a much larger aspect ratio (about 1200) than MMT (about 200) and hectorite (about 50).

Although nanoadditives can enhance both the thermal stability and the fire performance of the matrix polymer, it should be noted that enhancements in fire performance do not typically parallel enhancements in thermal stability; in other words, observation of a reduction in the peak HRR of a given PN does not necessarily mean that an enhancement of the thermal stability of that PN will be observed via TGA, and vice versa. A typical example is PA-6, where the associated PCN shows a significantly reduced peak HRR[168] but does not show enhanced thermal stability.

In characterizing the fire performance of polymeric materials, microscale combustion calorimetry (MCC, also called pyrolysis combustion flow calorimetry (PCFC)) can also be a useful tool. MCC is a more recently introduced technique developed by scientists at the Federal Aviation Administration (FAA) laboratories[180] and written into a Standard Test Method.[181] The technique utilizes traditional oxygen depletion calorimetry. The specimen is first heated at a constant rate (typically 1°C/sec) in a pyrolyzer. The pyrolyzed products are swept by an inert gas, such as nitrogen, and then mixed with oxygen to form a gas stream that enters the combustor in which the decomposition products are completely oxidized at 900°C. Oxygen concentrations and flow rates of the combustion gases are used to determine the oxygen depletion involved in the combustion process, and heat release rates are determined from these measurements. With the MCC, various parameters can be determined in a short time with a very small specimen (1–50mg), including heat release rate (HRR, W/g), heat of combustion (HC, J/g), ignition temperature (T_i, °C), etc. While the MCC technique has some similarities to CCA in terms of the data generated, it is a distinctly different test method that cannot capture all aspects of the CCA testing of larger specimens (for instance, the formation of a robust macroscopic char layer which acts as a physical barrier to further combustion). Nonetheless, in addition to the advantages of requiring only a small amount of sample and obtaining results in a very short time, an important advantage is that the results from the MCC can be correlated to those from the other fire test testing methods, such as CCA, flammability tests (Limiting Oxygen Index (LOI), UL-94), combustion tests (bomb calorimeter), etc. For this reason, MCC represents a useful low-cost tool for screening and predicting the flammability of polymers and other materials. Figure 12.33 shows a typical MCC curve obtained with a commercial PE based FR composite.

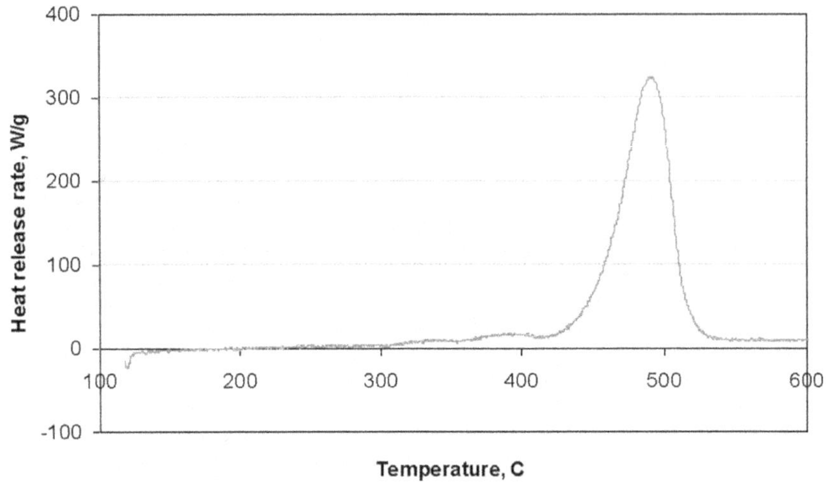

FIGURE 12.33 A representative MCC curve (relating heat release rate to temperature) for a polyethylene based FR composite.

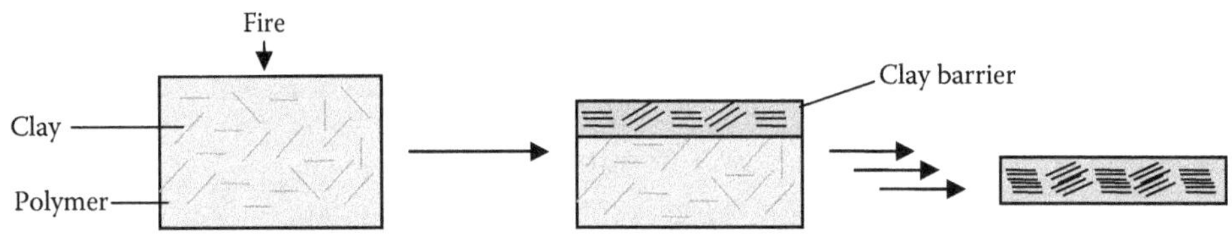

FIGURE 12.34 Schematic of combustion-induced formation of clay-based char (barrier) layer in a PCN.

Source: For illustration only.

12.6.2 Mechanisms Underlying the Effects of Nanoadditives on the Flammability of Polymers

The mechanisms underlying the effects of nanoadditives on the thermal and fire performance of the matrix polymers have been widely investigated, and various proposals have been made,[182] but a complete understanding is not yet available. Because the degree of enhancement of the thermal and fire performance observed in a given PN is dependent on the identity of the polymer, the effect of nanoadditives on the flammability of polymers is variable. For instance, in the case of clay, one hypothesis is that migration of clay particles to the surface of the matrix polymer occurs during burning.[183] In this scenario it is argued that the clay is transported by numerous rising bubbles of degradation products and the associated convective flow in the melt from the interior of the sample towards the sample surface, with pyrolysis of the matrix polymer meaning only de-wetted clay particles are left behind. These clay particles, in turn, show an increased tendency to aggregate, as degradation of the organic modifier renders them more hydrophilic and less compatible with the matrix polymer. The end result is the production of a clay-rich barrier at the sample surface that slows the rate of mass loss and, in turn, lowers the heat release rate. That said, the same outcome may also be explained by invoking the well-accepted barrier mechanism[175,184–186] in the condensed phase during burning.[187–191] The barrier mechanism (Figure 12.34) suggests that, under pyrolysis conditions, the clay forms a char-like material that acts as both a barrier to mass transport of the degradation products to the surface of the degrading polymer and as a thermal barrier, shielding the underlying polymer matrix from heat and oxygen. The clay-rich char can additionally provide thermal insulation for the condensed phase and thus increase the thermal stability of the matrix polymer. Finally, in addition to the above arguments, another mechanism that can also impact nanocomposite fire performance is paramagnetic radical trapping[192]; when the clay loading is low, paramagnetic radical trapping represents a stabilization mechanism that may be attributed to the presence of paramagnetic impurities within the clay. This has been confirmed in PSCNs prepared by bulk-polymerization using either an iron-containing or an iron-depleted clay, a specially designed experiment demonstrating the ability of paramagnetic iron impurities present in the clay layers to enhance thermal stability.

As implied by the aforementioned discussions, research shows that clay can qualitatively affect polymer degradation.[164] The presence of the clay can significantly change the degradation pathway of the matrix polymer, and clay is capable of promoting the production of oligomers rather than monomers.[193] In the case of PCNs that degrade by a radical pathway, the effect of clay on the enhancement of the thermal and fire resistance of matrix polymers is strongly related to the stability of

TABLE 12.1
Thermal Degradation Pathways of Some Polymers and Their Clay Nanocomposites and Resulting Reductions in Peak HRR

Polymer	Degradation Pathway of Virgin Polymer	Degradation Pathway in the Presence of Clay	Reduction in Peak HRR (%)
PA6	Intra-aminolysis/acidolysis, random scission	Inter-aminolysis/acidolysis, random scission	50–70
PS, HIPS	β-scission (chain end and middle)	Recombination, random scission	40–70
EVA	Chain stripping, disproportionation	Hydrogen abstraction, random scission	50–70
SAN, ABS	β-scission (chain end and middle)	Recombination, random scission	20–50
PE	Disproportionation	Hydrogen abstraction	20–40
PP	β-scission, disproportionation	Random scission	20–50
PAN	Cyclization, random scission	No change	< 10
PMMA	β-scission	No change	10–20

Source: Reprinted from ref. 164 with permission from Elsevier.

FIGURE 12.35 Cross-section of the char layer of a PP/MWNT (1%) nanocomposite.

Source: Reprinted from ref. 173 with permission from Elsevier.

the radicals that are produced during the thermal degradation of the matrix polymer. The more stable the polymeric radical that is produced, the larger the observed enhancement in the fire resistance (as reflected in the reduction of the peak HRR) of the matrix polymer. When a polymer such as PS degrades by more than one pathway, the presence of the clay generally favors one degradation pathway (most typically oligomer formation) over another, leading to the matrix polymer degrading more slowly than the pure polymer. However, this is not seen for polymers such as PMMA that degrade by only a single pathway and where the clay cannot promote the formation of different degradation products. Table 12.1 shows the thermal degradation pathways of some pure polymers and their clay nanocomposites and resulting reductions in peak HRR. The presence of clay also significantly affects the apparent activation energy of degradation processes.[194]

The mechanism by which clays enhance the thermal and fire resistance of the matrix polymers may also be used to inform discussions of the behavior of CNTs and other nanoadditives in a similar context. The formation of a continuous and robust char layer on the material surface, as seen in Figure 12.32 for PPCN[179] and in Figure 12.35 for PP/CNT nanocomposite,[173] is the primary pathway by which nanoadditives enhance the thermal and fire resistance of the matrix polymer, because such layers, when intact, physically stable and free of flaws, can provide the underlying polymer with significant protection from heat and oxygen, thus reducing the mass loss and heat release rates and slowing the burning process (Figure 12.36[195]).

12.7 CONCLUDING REMARKS AND FUTURE DIRECTIONS

Based on studies of PNs to date, it is quite clear that the nanoadditives can exert a definite effect on both thermal stability and fire properties, appearing in the form of an increase in the T_{onset} and a reduction in the peak HRR; this conclusion is supported by studies on a variety of thermoplastics, including PS, PA-6, EVA, and many others. An additional phenomenon that should be noted given its potential utility when designing flame retardant polymer systems is that, due to changes in the low shear rheology of PNs, under fire conditions PN specimens tend to be less likely to drip or flow and better able to retain their original shape. Nonetheless, production of a continuous, robust, and physical stable char layer at the surface of the material is critical to enhancing the fire performance of the matrix polymer, and this in turn is most readily achieved by ensuring that the nanoadditives are well-dispersed and distributed in the polymer matrix and that this situation is maintained for as long as possible in the polymer melt during pyrolysis. This is true whether the char layer is comprised of a mixture of organic degradation products (carbonaceous materials typically) and inorganic

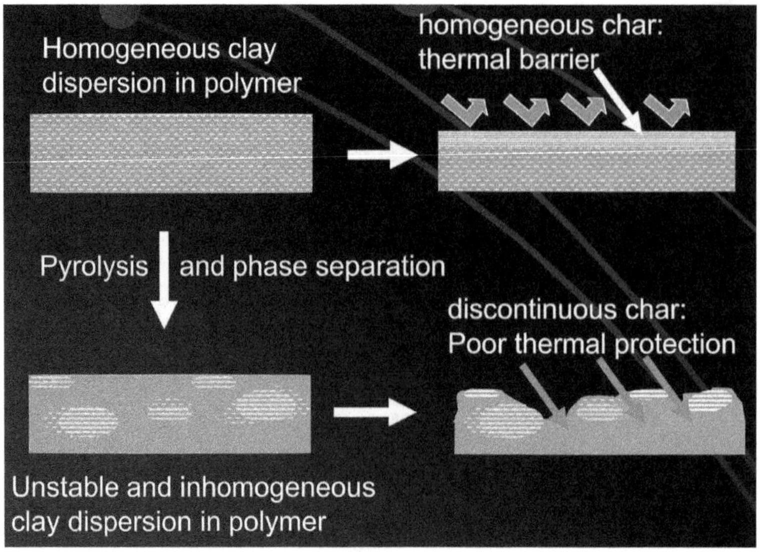

FIGURE 12.36 Schematic showing the impact of a stable, robust, homogeneous char later on the fire properties of the polymer matrix.[195]

nanoadditive residues or if nanoadditives are the primary contributors to char formation.

Having said this, a common lesson from the study of PN fire performance to date is that, even in the best cases, the addition of a single nanoadditive (e.g., clay) will only slow the combustion process (implying reductions in the peak HRR/MLR and the average HRR/MLR); it will not stop it altogether, nor will it alter the total heat released or mass lost compared to that of the same amount of base polymer. The only exception to this statement involves the use of nanoadditives capable of forming nonflammable compounds under fire conditions whose vaporization requires thermal energy; in such cases (i.e., in the case of the low temperature dihydroxylation of LDHs to form water), decreases in total heat release may be observed as well. Furthermore, an earlier time to ignition (and perhaps also the time to peak HRR) has also been observed in many PNs vs. base polymer via CCA experiments.[196–198] This can occur when the modifiers on the nanoadditives degrade to flammable vapors more rapidly than the polymer matrix itself (as is often the case with the alkylammonium treatments found on most commercially available modified MMT, given the tendency of such modifiers to begin to decompose at temperatures below 200°C[98,102]), or when the surface chemistry of the nanoadditive is able to catalyze the degradation of the matrix polymer.

The utilization of nanoadditives in polymer with relevant properties, in ways that are at the same time effective, economical, and scalable, is of high interest and remains a significant challenge. Many predictions concerning the utilization of nanoadditives in polymers to promote thermal and fire resistance as well and to enhance other properties have been mentioned in the literature.[149] A simple discussion, summarized from the literature, is provided below.

To enhance the thermal and fire resistance (as well as other properties) of the polymer matrix, one approach is the simultaneous addition of two or more types of nanoadditives in the same polymer matrix. In this way the advantages and disadvantages of the different nanoadditives may compensate for one another, and synergism may be observed as well, resulting in a PN with better performance than if either nanoadditive were used alone. Clay is of great interest as one component in such systems because it is easy to obtain at low cost and can often provide an array of performance benefits. As an example of this approach, it has been reported that use of a combination of clay and CNTs brings enhanced performance.[199] The use of synthetic clays,[32] such as synthetic fluoromica, magadiite, and LDHs, is of interest as well. LDHs, as a hydrated materials (very similar to $Mg(OH)_2$ or $Al(OH)_3$), are of particular interest because they combine a layered structure like that of the silicate clays with the potential to release water under fire conditions, thus providing additional benefits as noted above. However, the higher costs and more limited availability of these synthetic clays have slowed their adoption.

Another approach is the combination of nanoadditives with more conventional additives in order to enhance the overall fire resistance of the matrix polymer—for instance, the addition of an additive that can form a glassy barrier to seal the cracks in the char layer, resulting in a self-extinguishing PN. One common and practical way is to use nanoadditives in presence of traditional halogen-[200] or phosphorous-based[201] flame retardants, or as a nanoscale component in a traditional flame retardant system to achieve a fire resistant polymer composite; an example of the latter would be the use of nano-$Mg(OH)_2$ and nano-$Al(OH)_3$,[202–204] obtained from traditional inorganic flame retardants that have an original particle size in microns. Still, clay remains of significant interest because of its excellent performance, including drip resistance and robust char formation as well as potential synergism with many traditional fillers/flame retardants. For instance, the use of clay as a nanoadditive in flame retardant composites based on EVA and PP can enable significant reductions in the loading of $Mg(OH)_2$, the primary filler.[205] The authors have observed similar effects in unpublished work related to PA6 as well, where it was possible to design a halogen- and phosphorus-free glass reinforced PA6 formulation with a lower overall loading of inorganic material by replacing some amount of $Mg(OH)_2$ by a smaller amount of a well-chosen nanoclay. This industrially sponsored effort resulted in the development of a formulation with commercial relevance given its ability to achieve UL-94 V0 certification while at the same time meeting key requirements related to mechanical performance—in contrast with conventional systems, where either the material was too brittle, insufficiently flame retardant, or required the use of unwanted FR chemistries. Nanoadditives like CNTs and POSS, while expensive when used alone, can also be combined with traditional flame retardant systems, not only for the purposes of reducing costs but also in order to improve the fire performance of the resulting PNs. Closely related is a trend of using PNs to improve the fire resistance of more traditional composites.

In the context of future trends, two more topics are noted here: the precise control of nanocomposite structure and the realization of true multifunctional materials.[149] Control of nanocomposite structure in this context goes beyond achieving high levels of dispersion of the nanoparticles in the polymer matrix and refers to the ability to engineer a specific micro/nanostructure in these materials, such as an ordered alignment of the nanoadditives (e.g., the clay), to produce the desired levels of performance in the final nanocomposite. True multifunctional materials have been and will continue to be one of the targets for materials development as well, but this is somewhat elusive and premature in the PN area. Because such "performance" PNs may or may not show greater improvements in thermal and fire behavior than conventional PNs, they are not discussed in this chapter. Nevertheless, it is clear that, even when they are not designed with such characteristics in mind, that new multifunctional nanocomposite materials may already display enhanced thermal and fire properties simply because of their composition and structure.

FURTHER READING

a. Morgan, A. B.; Wilkie, C. A., Eds.; *Flame Retardant Polymer Nanocomposites*; Wiley-Interscience: Hoboken, NJ, 2007.

b. Leszczyńska, A.; Njuguna, J.; Pielichowski, K.; Banerjee, J. R. Polymer/Montmorillonite Nanocomposites with Improved Thermal Properties. *Thermochimica Acta*, **2007**, *453* (2), 75–96. https://doi.org/10.1016/j.tca.2006.11.002

c. Leszczyńska, A.; Njuguna, J.; Pielichowski, K.; Banerjee, J. R. Polymer/Montmorillonite Nanocomposites with Improved Thermal Properties. *Thermochimica Acta*, **2007**, *454* (1), 1–22. https://doi.org/10.1016/j.tca.2006.11.003

d. Bourbigot, S.; Duquesne, S. Fire Retardant Polymers: Recent Developments and Opportunities. *Journal of Materials Chemistry*, **2007**, *17* (22), 2283. https://doi.org/10.1039/b702511d

e. Beyer, G. Flame retardant properties of organoclays and carbon nanotubes and their combinations with alumina trihydrate. In *Flame Retardant Polymer Nanocomposites*; Morgan, A. B., Wilkie, C. A., Eds.; John Wiley & Sons, Inc.: Hoboken, NJ, 2007; pp 163–190. https://doi.org/10.1002/9780470109038.ch7

f. Pinnavaia, T. J.; Beall, G. W., Eds.; *Polymer-Clay Nanocomposites*; Wiley Series in Polymer Science; Wiley: Chichester and New York, 2000.

g. Alexandre, M.; Dubois, P. Polymer-Layered Silicate Nanocomposites: Preparation, Properties and Uses of a New Class of Materials. *Materials Science and Engineering: R: Reports*, **2000**, *28* (1–2), 1–63. https://doi.org/10.1016/S0927-796X(00)00012-7

h. Untrack, L. A. *Clay-Containing Polymeric Nanocomposites*; Rapra Technology Ltd: Shrewsbury, 2004.

i. Sinha Ray, S.; Okamoto, M. Polymer/Layered Silicate Nanocomposites: A Review from Preparation to Processing. *Progress in Polymer Science*, **2003**, *28* (11), 1539–1641. https://doi.org/10.1016/j.progpolymsci.2003.08.002

j. Alexandre, M.; Dubois, P. Nanocomposites. In *Macromolecular Engineering*; Matyjaszewski, K., Gnanou, Y., Leibler, L., Eds.; Wiley, 2007; pp 2033–2070. https://doi.org/10.1002/9783527631421.ch49

k. Leroux, F.; Besse, J.-P. Polymer Interleaved Layered Double Hydroxide: A New Emerging Class of Nanocomposites. *Chemistry of Materials*, **2001**, *13* (10), 3507–3515. https://doi.org/10.1021/cm0110268

l. Costa, F. R.; Saphiannikova, M.; Wagenknecht, U.; Heinrich, G. Layered double hydroxide based polymer nanocomposites. In *Wax Crystal Control Nanocomposites Stimuli-Responsive Polymers*; Springer: Berlin and Heidelberg, 2007; Vol. 210; pp 101–168. https://doi.org/10.1007/12_2007_123

m. Ahir, S. V.; Terentjev, E. M. Polymers containing carbon nanotubes: active composite materials. In *Polymeric Nanostructures and Their Applications*; Nalwa, H. S., Ed.; Nanotechnology Book Series; American Scientific Publication: Stevenson Ranch, CA, 2007; pp 153–200.

n. Coleman, J. N.; Khan, U.; Gun'ko, Y. K. Mechanical Reinforcement of Polymers Using Carbon Nanotubes. *Advanced Materials*, **2006**, *18* (6), 689–706. https://doi.org/10.1002/adma.200501851

o. Grossiord, N.; Loos, J.; Regev, O.; Koning, C. E. Toolbox for Dispersing Carbon Nanotubes into Polymers to Get Conductive Nanocomposites. *Chemistry of Materials*, **2006**, *18* (5), 1089–1099. https://doi.org/10.1021/cm051881h

p. Pielichowski, K.; Njuguna, J.; Janowski, B.; Pielichowski, J. Polyhedral Oligomeric Silsesquioxanes (POSS)-Containing Nanohybrid Polymers. In *Supramolecular Polymers Polymeric Betains Oligomers*; Springer: Berlin, Heidelberg, 2006; Vol. 201, pp 225–296. https://doi.org/10.1007/12_077

q. Moniruzzaman, M.; Winey, K. I. Polymer Nanocomposites Containing Carbon Nanotubes. *Macromolecules*, **2006**, *39* (16), 5194–5205. https://doi.org/10.1021/ma060733p

r. Joshi, M.; Butola, B. S. Polymeric Nanocomposites—Polyhedral Oligomeric Silsesquioxanes (POSS) as Hybrid Nanofiller. *Journal of Macromolecular Science, Part C: Polymer Reviews*, **2004**, *44* (4), 389–410. https://doi.org/10.1081/MC-200033687

s. Sang, B.; Li, Z.; Li, X.; Yu, L.; Zhang, Z. Graphene-Based Flame Retardants: A Review. *Journal of Materials Science*, **2016**, *51* (18), 8271–8295. https://doi.org/10.1007/s10853-016-0124-0
t. Yue, X.; Li, C.; Ni, Y.; Xu, Y.; Wang, J. Flame Retardant Nanocomposites Based on 2D Layered Nanomaterials: A Review. *Journal of Materials Science*, **2019**, *54* (20), 13070–13105. https://doi.org/10.1007/s10853-019-03841-w
u. Cai, W.; Wang, B.-B.; Wang, X.; Zhu, Y.-L.; Li, Z.-X.; Xu, Z.-M.; Song, L.; Hu, W.-Z.; Hu, Y. Recent Progress in Two-Dimensional Nanomaterials Following Graphene for Improving Fire Safety of Polymer (Nano) Composites. *Chinese Journal of Polymer Science*, **2021**, *39* (8), 935–956. https://doi.org/10.1007/s10118-021-2575-2
v. Hou, Y.; Xu, Z.; Chu, F.; Gui, Z.; Song, L.; Hu, Y.; Hu, W. A Review on Metal-Organic Hybrids as Flame Retardants for Enhancing Fire Safety of Polymer Composites. *Composites Part B: Engineering*, **2021**, *221*, 109014. https://doi.org/10.1016/j.compositesb.2021.109014
w. He, W.; Song, P.; Yu, B.; Fang, Z.; Wang, H. Flame Retardant Polymeric Nanocomposites through the Combination of Nanomaterials and Conventional Flame Retardants. *Progress in Materials Science*, **2020**, *114*, 100687. https://doi.org/10.1016/j.pmatsci.2020.100687
x. Beyer, G., Ed.; *Industry Guide to Polymer Nanocomposites*; A PID technical book; Plastics Information Direct: Bristol, 2009.

REFERENCES

1. Garnweitner, G.; Smarsly, B.; Assink, R.; Ruland, W.; Bond, E.; Brinker, C. J. Self-Assembly of an Environmentally Responsive Polymer/Silica Nanocomposite. *J. Am. Chem. Soc.*, **2003**, *125* (19), 5626–5627. https://doi.org/10.1021/ja0342648
2. Khaled, S. M.; Sui, R.; Charpentier, P. A.; Rizkalla, A. S. Synthesis of TiO_2–PMMA Nanocomposite: Using Methacrylic Acid as a Coupling Agent. *Langmuir*, **2007**, *23* (7), 3988–3995. https://doi.org/10.1021/la062879n
3. Xiong, H.-M.; Shen, W.-Z.; Wang, Z.-D.; Zhang, X.; Xia, Y.-Y. Liquid Polymer Nanocomposites PEGME–SnO_2 and PEGME–TiO_2 Prepared through Solvothermal Methods. *Chem. Mater.*, **2006**, *18* (16), 3850–3854. https://doi.org/10.1021/cm060432u
4. Guo, Z.; Pereira, T.; Choi, O.; Wang, Y.; Hahn, H. T. Surface Functionalized Alumina Nanoparticle Filled Polymeric Nanocomposites with Enhanced Mechanical Properties. *J. Mater. Chem.*, **2006**, *16* (27), 2800. https://doi.org/10.1039/b603020c
5. Ash, B. J.; Siegel, R. W.; Schadler, L. S. Mechanical Behavior of Alumina/Poly(Methyl Methacrylate) Nanocomposites. *Macromolecules*, **2004**, *37* (4), 1358–1369. https://doi.org/10.1021/ma0354400
6. Corriu, R. J. P.; Gerbier, P.; Guérin, C.; Henner, B. From Preceramic Polymers with Interpenetrating Networks to SiC/MC Nanocomposites. *Chem. Mater.*, **2000**, *12* (3), 805–811. https://doi.org/10.1021/cm991140e
7. Ni, Y.; Zheng, S. A Novel Photocrosslinkable Polyhedral Oligomeric Silsesquioxane and Its Nanocomposites with Poly(Vinyl Cinnamate). *Chem. Mater.*, **2004**, *16* (24), 5141–5148. https://doi.org/10.1021/cm049463k
8. Zheng, L.; Hong, S.; Cardoen, G.; Burgaz, E.; Gido, S. P.; Coughlin, E. B. Polymer Nanocomposites through Controlled Self-Assembly of Cubic Silsesquioxane Scaffolds. *Macromolecules*, **2004**, *37* (23), 8606–8611. https://doi.org/10.1021/ma048557c
9. Tabuani, D.; Frache, A.; Boccaleri, E.; Camino, G. Octaisobutyl POSS Thermal Degradation. In *Fire Retardancy of Polymers New Applications of Mineral Fillers*; Le Bras, M.; Wilkie, C. A.; Bourbigot, S.; Duquesne, S.; Jama, C., Eds.; Royal Society of Chemistry: Cambridge, 2005; pp 202–220.
10. Moniruzzaman, M.; Winey, K. I. Polymer Nanocomposites Containing Carbon Nanotubes. *Macromolecules*, **2006**, *39* (16), 5194–5205. https://doi.org/10.1021/ma060733p
11. Haggenmueller, R.; Fischer, J. E.; Winey, K. I. Single Wall Carbon Nanotube/Polyethylene Nanocomposites: Nucleating and Templating Polyethylene Crystallites. *Macromolecules*, **2006**, *39* (8), 2964–2971. https://doi.org/10.1021/ma0527698
12. Datsyuk, V.; Guerret-Piécourt, C.; Dagréou, S.; Billon, L.; Dupin, J.-C.; Flahaut, E.; Peigney, A.; Laurent, C. Double Walled Carbon Nanotube/Polymer Composites via in-Situ Nitroxide Mediated Polymerisation of Amphiphilic Block Copolymers. *Carbon*, **2005**, *43* (4), 873–876. https://doi.org/10.1016/j.carbon.2004.10.052
13. Joy, J.; George, E.; Haritha, P.; Thomas, S.; Anas, S. An Overview of Boron Nitride Based Polymer Nanocomposites. *Journal of Polymer Science*, **2020**, *58* (22), 3115–3141. https://doi.org/10.1002/pol.20200507
14. Dalmas, F.; Chazeau, L.; Gauthier, C.; Cavaillé, J.-Y.; Dendievel, R. Large Deformation Mechanical Behavior of Flexible Nanofiber Filled Polymer Nanocomposites. *Polymer*, **2006**, *47* (8), 2802–2812. https://doi.org/10.1016/j.polymer.2006.02.014
15. Kelarakis, A.; Yoon, K.; Somani, R. H.; Chen, X.; Hsiao, B. S.; Chu, B. Rheological Study of Carbon Nanofiber Induced Physical Gelation in Polyolefin Nanocomposite Melt. *Polymer*, **2005**, *46* (25), 11591–11599. https://doi.org/10.1016/j.polymer.2005.10.009
16. Zhu, J.; Morgan, A. B.; Lamelas, F. J.; Wilkie, C. A. Fire Properties of Polystyrene–Clay Nanocomposites. *Chem. Mater.*, **2001**, *13* (10), 3774–3780. https://doi.org/10.1021/cm000984r
17. Su, S. Polybutadiene-Modified Clay and Its Nanocomposites. *Polym. Degrad. Stab.*, **2004**, *84* (2), 279–288. https://doi.org/10.1016/j.polymdegradstab.2003.09.023
18. Wang, D.; Jang, B. N.; Su, S.; Zhang, J.; Zheng, X.; Chigwada, G.; Jiang, D. D.; Wilkie, C. A. Fire Retardancy of Polystyrene-Hectorite Nanocomposites. Editor(s): Le Bras, M. *Fire Retard. Polym.: New Appli. Miner. Fillers*, [9th European Meeting on Fire Retardancy and Protection of Materials], 2005, 126–138.
19. Maiti, P.; Yamada, K.; Okamoto, M.; Ueda, K.; Okamoto, K. New Polylactide/Layered Silicate Nanocomposites: Role of Organoclays. *Chem. Mater.*, **2002**, *14* (11), 4654–4661. https://doi.org/10.1021/cm020391b
20. Costache, M. C.; Heidecker, M. J.; Manias, E.; Camino, G.; Frache, A.; Beyer, G.; Gupta, R. K.; Wilkie, C. A. The Influence of Carbon Nanotubes, Organically Modified Montmorillonites and Layered Double Hydroxides on the Thermal Degradation and Fire Retardancy of Polyethylene, Ethylene–Vinyl Acetate Copolymer and Polystyrene. *Polymer*, **2007**, *48* (22), 6532–6545. https://doi.org/10.1016/j.polymer.2007.08.059
21. Leroux, F.; Besse, J.-P. Polymer Interleaved Layered Double Hydroxide: A New Emerging Class of Nanocomposites. *Chem. Mater.*, **2001**, *13* (10), 3507–3515. https://doi.org/10.1021/cm0110268

22. Sue, H.-J.; Gam, K. T.; Bestaoui, N.; Spurr, N.; Clearfield, A. Epoxy Nanocomposites Based on the Synthetic α-Zirconium Phosphate Layer Structure. *Chem. Mater.*, **2004**, *16* (2), 242–249. https://doi.org/10.1021/cm030441s
23. Sukpirom, N.; Lerner, M. M. Preparation of Organic–Inorganic Nanocomposites with a Layered Titanate. *Chem. Mater.*, **2001**, *13* (6), 2179–2185. https://doi.org/10.1021/cm0101226
24. Mittal, V. Functional Polymer Nanocomposites with Graphene: A Review. *Macromol. Mater. Eng.*, **2014**, *299* (8), 906–931. https://doi.org/10.1002/mame.201300394
25. Ahmadi, M.; Zabihi, O.; Jeon, S.; Yoonessi, M.; Dasari, A.; Ramakrishna, S.; Naebe, M. 2D Transition Metal Dichalcogenide Nanomaterials: Advances, Opportunities, and Challenges in Multi-Functional Polymer Nanocomposites. *J. Mater. Chem. A*, **2020**, *8* (3), 845–883. https://doi.org/10.1039/C9TA10130F
26. Sanchez, C.; Belleville, P.; Popall, M.; Nicole, L. Applications of Advanced Hybrid Organic–Inorganic Nanomaterials: From Laboratory to Market. *Chem. Soc. Rev.*, **2011**, *40* (2), 696. https://doi.org/10.1039/c0cs00136h
27. Giannelis, E. P. Polymer Layered Silicate Nanocomposites. *Adv. Mater.*, **1996**, *8* (1), 29–35. https://doi.org/10.1002/adma.19960080104
28. Clay. *Wikipedia*; 2023. https://en.wikipedia.org/wiki/Clay
29. Giannelis, E. P.; Krishnamoorti, R.; Manias, E. Polymer-Silicate Nanocomposites: Model Systems for Confined Polymers and Polymer Brushes. In *Polymers in Confined Environments*; Granick, S., Binder, K., de Gennes, P.-G., Giannelis, E. P., Grest, G. S., Hervet, H., Krishnamoorti, R., Léger, L., Manias, E., Raphaël, E., et al., Eds.; Advances in Polymer Science; Springer: Berlin and Heidelberg, 1999; pp 107–147. https://doi.org/10.1007/3-540-69711-X_3
30. Ijdo, W. L.; Pinnavaia, T. J. Staging of Organic and Inorganic Gallery Cations in Layered Silicate Heterostructures. *J. Solid State Chem.*, **1998**, *139* (2), 281–289. https://doi.org/10.1006/jssc.1998.7842
31. Ijdo, W. L.; Pinnavaia, T. J. Solid Solution Formation in Amphiphilic Organic–Inorganic Clay Heterostructures. *Chem. Mater.*, **1999**, *11* (11), 3227–3231. https://doi.org/10.1021/cm990294r
32. Utracki, L. A.; Sepehr, M.; Boccaleri, E. Synthetic, Layered Nanoparticles for Polymeric Nanocomposites (PNCs). *Polym. Adv. Technol.*, **2007**, *18* (1), 1–37. https://doi.org/10.1002/pat.852
33. Layered Double Hydroxides. *Wikipedia*; 2023. https://en.wikipedia.org/wiki/Layered_double_hydroxides
34. Meyn, M.; Beneke, K.; Lagaly, G. Anion-Exchange Reactions of Layered Double Hydroxides. *Inorg. Chem.*, **1990**, *29* (26), 5201–5207. https://doi.org/10.1021/ic00351a013
35. Carbon Nanotube. *Wikipedia*; 2023. https://en.wikipedia.org/wiki/Carbon_nanotube
36. Iijima, S. Helical Microtubules of Graphitic Carbon. *Nature*, **1991**, *354* (6348), 56–58. https://doi.org/10.1038/354056a0
37. Kumar, R. S.; Pravica, M. G.; Cornelius, A. L.; Nicol, M. F.; Hu, M. Y.; Chow, P. C. X-Ray Raman Scattering Studies on C60 Fullerenes and Multi-Walled Carbon Nanotubes under Pressure. *Diam. Relat. Mater.*, **2007**, *16* (4–7), 1250–1253. https://doi.org/10.1016/j.diamond.2006.11.103
38. Ajayan, P. M.; Ebbesen, T. W. Nanometre-Size Tubes of Carbon. *Rep. Prog. Phys.*, **1997**, *60* (10), 1025–1062. https://doi.org/10.1088/0034-4885/60/10/001
39. Ajayan, P. M.; Stephan, O.; Colliex, C.; Trauth, D. Aligned Carbon Nanotube Arrays Formed by Cutting a Polymer Resin—Nanotube Composite. *Science*, **1994**, *265* (5176), 1212–1214. https://doi.org/10.1126/science.265.5176.1212
40. Christou, A.; Stec, A. A.; Ahmed, W.; Aschberger, K.; Amenta, V. A Review of Exposure and Toxicological Aspects of Carbon Nanotubes, and as Additives to Fire Retardants in Polymers. *Critical Reviews in Toxicology*, **2016**, *46* (1), 74–95. https://doi.org/10.3109/10408444.2015.1082972
41. Lozano, K.; Yang, S.; Zeng, Q. Rheological Analysis of Vapor-Grown Carbon Nanofiber-Reinforced Polyethylene Composites. *J. Appl. Polym. Sci.*, **2004**, *93* (1), 155–162. https://doi.org/10.1002/app.20443
42. Zeng, J.; Saltysiak, B.; Johnson, W. S.; Schiraldi, D. A.; Kumar, S. Processing and Properties of Poly(Methyl Methacrylate)/Carbon Nanofiber Composites. *Compos. B. Eng.*, **2004**, *35* (3), 245–249. https://doi.org/10.1016/j.compositesb.2003.08.009
43. Fina, A.; Tabuani, D.; Carniato, F.; Frache, A.; Boccaleri, E.; Camino, G. Polyhedral Oligomeric Silsesquioxanes (POSS) Thermal Degradation. *Thermochim. Acta*, **2006**, *440* (1), 36–42. https://doi.org/10.1016/j.tca.2005.10.006
44. Fina, A.; Abbenhuis, H. C. L.; Tabuani, D.; Frache, A.; Camino, G. Polypropylene Metal Functionalised POSS Nanocomposites: A Study by Thermogravimetric Analysis. *Polym. Degrad. Stab.*, **2006**, *91* (5), 1064–1070. https://doi.org/10.1016/j.polymdegradstab.2005.07.013
45. Yang, F.; Nelson, G. L. Polymer/Silica Nanocomposites Prepared via Extrusion. *Polym. Adv. Technol.*, **2006**, *17* (4), 320–326. https://doi.org/10.1002/pat.695
46. Yang, F.; Bogdanova, I.; Nelson, G. L. Mechanism Study on the Flammability and Thermal Stability of Polymer/Alumina Nanocomposites via Extrusion. *PMSE Prepr.*, **2008**, *98*, 319–320.
47. www.sigmaaldrich.com/US/en/product/aldrich/560316 (accessed July 31, 2023).
48. Lichtenhan, J. D.; Gilman, J. W. Preceramic Additives as Fire Retardants for Plastics. US6362279B2, March 26, 2002. https://patents.google.com/patent/US6362279B2/en
49. Bourbigot, S.; Le Bras, M.; Flambard, X.; Rochery, M.; Devaux, E.; Lichtenham, J. D. Polyhedral Oligomeric Silsesquixones: Application to Flame Retardant Textiles. In *Fire Retardancy of Polymers New Applications of Mineral Fillers*; Le Bras, M.; Wilkie, C. A.; Bourbigot, S.; Duquesne, S.; Jama, C., Eds.; Royal Society of Chemistry: Cambridge, 2005; pp 189–201.
50. Aranda, P.; Ruiz-Hitzky, E. Poly(Ethylene Oxide)-Silicate Intercalation Materials. *Chem. Mater.*, **1992**, *4* (6), 1395–1403. https://doi.org/10.1021/cm00024a048
51. Greenland, D. J. Adsorption of Polyvinyl Alcohols by Montmorillonite. *J. Colloid Sci.*, **1963**, *18* (7), 647–664. https://doi.org/10.1016/0095-8522(63)90058-8
52. Wang, D.; Jiang, D. D.; Pabst, J.; Han, Z.; Wang, J.; Wilkie, C. A. Polystyrene Magadiite Nanocomposites. *Polym. Eng. Sci.*, **2004**, *44* (6), 1122–1131. https://doi.org/10.1002/pen.20105
53. Gilman, J. W.; Awad, W. H.; Davis, R. D.; Shields, J.; Harris, R. H.; Davis, C.; Morgan, A. B.; Sutto, T. E.; Callahan, J.; Trulove, P. C.; et al. Polymer/Layered Silicate Nanocomposites from Thermally Stable Trialkylimidazolium-Treated Montmorillonite. *Chem. Mater.*, **2002**, *14* (9), 3776–3785. https://doi.org/10.1021/cm011532x

54. Chigwada, G.; Wang, D.; Wilkie, C. A. Polystyrene Nanocomposites Based on Quinolinium and Pyridinium Surfactants. *Polym. Degrad. Stab.*, **2006**, *91* (4), 848–855. https://doi.org/10.1016/j.polymdegradstab.2005.06.016
55. Zhang, J.; Wilkie, C. A. A Carbocation Substituted Clay and Its Styrene Nanocomposite. *Polym. Degrad. Stab.*, **2004**, *83* (2), 301–307. https://doi.org/10.1016/S0141-3910(03)00275-1
56. Wang, D.; Wilkie, C. A. A Stibonium-Modified Clay and Its Polystyrene Nanocomposite. *Polym. Degrad. Stab.*, **2003**, *82* (2), 309–315. https://doi.org/10.1016/S0141-3910(03)00184-8
57. Costa, F. R.; Saphiannikova, M.; Wagenknecht, U.; Heinrich, G. Layered Double Hydroxide Based Polymer Nanocomposites. In *Wax Crystal Control Nanocomposites Stimuli-Responsive Polymers*; Springer: Berlin and Heidelberg, 2007; Vol. 210, pp 101–168. https://doi.org/10.1007/12_2007_123
58. Li, Y.; Ishida, H. Concentration-Dependent Conformation of Alkyl Tail in the Nanoconfined Space: Hexadecylamine in the Silicate Galleries. *Langmuir*, **2003**, *19* (6), 2479–2484. https://doi.org/10.1021/la026481c
59. Morgan, A. Multifunctional Polymer-Clay Nanocomposites. *Mater. Matt.*, **2007**, *2* (1), 20–23. www.sigmaaldrich.com/deepweb/assets/sigmaaldrich/product/documents/262/416/al_material_matters_v2n1.pdf#page=20 (accessed Jul 31, 2023).
60. Chigwada, G.; Wang, D.; Jiang, D. D.; Wilkie, C. A. Styrenic Nanocomposites Prepared Using a Novel Biphenyl-Containing Modified Clay. *Polym. Degrad. Stab.*, **2006**, *91* (4), 755–762. https://doi.org/10.1016/j.polymdegradstab.2005.05.026
61. Xiao, J.; Hu, Y.; Wang, Z.; Tang, Y.; Chen, Z.; Fan, W. Preparation and Characterization of Poly(Butylene Terephthalate) Nanocomposites from Thermally Stable Organic-Modified Montmorillonite. *Eur. Polym. J.*, **2005**, *41* (5), 1030–1035. https://doi.org/10.1016/j.eurpolymj.2004.11.025
62. Chigwada, G.; Jiang, D. D.; Wilkie, C. A. Polystyrene Nanocomposites Based on Carbazole-Containing Surfactants. *Thermochim. Acta*, **2005**, *436* (1–2), 113–121. https://doi.org/10.1016/j.tca.2005.02.033
63. Beyer, F. L.; Beck Tan, N. C.; Dasgupta, A.; Galvin, M. E. Polymer–Layered Silicate Nanocomposites from Model Surfactants. *Chem. Mater.*, **2002**, *14* (7), 2983–2988. https://doi.org/10.1021/cm011639k
64. Su, S.; Jiang, D. D.; Wilkie, C. A. Novel Polymerically-Modified Clays Permit the Preparation of Intercalated and Exfoliated Nanocomposites of Styrene and Its Copolymers by Melt Blending. *Polym. Degrad. Stab.*, **2004**, *83* (2), 333–346. https://doi.org/10.1016/S0141-3910(03)00278-7
65. Su, S.; Jiang, D. D.; Wilkie, C. A. Poly(Methyl Methacrylate), Polypropylene and Polyethylene Nanocomposite Formation by Melt Blending Using Novel Polymerically-Modified Clays. *Polym. Degrad. Stab.*, **2004**, *83* (2), 321–331. https://doi.org/10.1016/S0141-3910(03)00277-5
66. Lagaly, G. Interaction of Alkylamines with Different Types of Layered Compounds. *Solid State Ionics*, **1986**, *22* (1), 43–51. https://doi.org/10.1016/0167-2738(86)90057-3
67. Vaia, R. A.; Teukolsky, R. K.; Giannelis, E. P. Interlayer Structure and Molecular Environment of Alkylammonium Layered Silicates. *Chem. Mater.*, **1994**, *6* (7), 1017–1022. https://doi.org/10.1021/cm00043a025
68. Hackett, E.; Manias, E.; Giannelis, E. P. Molecular Dynamics Simulations of Organically Modified Layered Silicates. *J. Chem. Phys.*, **1998**, *108* (17), 7410–7415. https://doi.org/10.1063/1.476161
69. Ho, D. L.; Briber, R. M.; Glinka, C. J. Characterization of Organically Modified Clays Using Scattering and Microscopy Techniques. *Chem. Mater.*, **2001**, *13* (5), 1923–1931. https://doi.org/10.1021/cm0008617
70. Hanley, H. J. M.; Muzny, C. D.; Ho, D. L.; Glinka, C. J. A Small-Angle Neutron Scattering Study of a Commercial Organoclay Dispersion. *Langmuir*, **2003**, *19* (14), 5575–5580. https://doi.org/10.1021/la026342v
71. Compatibility (Chemical). *Wikipedia*; 2023. https://en.wikipedia.org/wiki/Compatibility_(chemical)
72. Miscibility. *Wikipedia*; 2023. https://en.wikipedia.org/wiki/Miscibility
73. Zhang, J.; Jiang, D. D.; Wang, D.; Wilkie, C. A. Styrenic Polymer Nanocomposites Based on an Oligomerically-Modified Clay with High Inorganic Content. *Polym. Degrad. Stab.*, **2006**, *91* (11), 2665–2674. https://doi.org/10.1016/j.polymdegradstab.2006.04.017
74. Zhang, J.; Manias, E.; Wilkie, C. A. Polymerically Modified Layered Silicates: An Effective Route to Nanocomposites. *J. Nanosci. Nanotechnol.*, **2008**, *8* (4), 1597–1615. https://doi.org/10.1166/jnn.2008.18226
75. Zheng, X.; Jiang, D. D.; Wilkie, C. A. Methyl Methacrylate Oligomerically-Modified Clay and Its Poly(Methyl Methacrylate) Nanocomposites. *Thermochim. Acta*, **2005**, *435* (2), 202–208. https://doi.org/10.1016/j.tca.2005.06.006
76. Su, S.; Jiang, D. D.; Wilkie, C. A. Methacrylate Modified Clays and Their Polystyrene and Poly(Methyl Methacrylate) Nanocomposites. *Polym. Adv. Technol.*, **2004**, *15* (5), 225–231. https://doi.org/10.1002/pat.456
77. Imai, Y.; Nishimura, S.; Abe, E.; Tateyama, H.; Abiko, A.; Yamaguchi, A.; Aoyama, T.; Taguchi, H. High-Modulus Poly(Ethylene Terephthalate)/Expandable Fluorine Mica Nanocomposites with a Novel Reactive Compatibilizer. *Chem. Mater.*, **2002**, *14* (2), 477–479. https://doi.org/10.1021/cm010408a
78. Paul, M.-A.; Alexandre, M.; Degée, P.; Calberg, C.; Jérôme, R.; Dubois, P. Exfoliated Polylactide/Clay Nanocomposites by In-Situ Coordination–Insertion Polymerization. *Macromol. Rapid Commun.*, **2003**, *24* (9), 561–566. https://doi.org/10.1002/marc.200390082
79. Weimer, M. W.; Chen, H.; Giannelis, E. P.; Sogah, D. Y. Direct Synthesis of Dispersed Nanocomposites by in Situ Living Free Radical Polymerization Using a Silicate-Anchored Initiator. *J. Am. Chem. Soc.*, **1999**, *121* (7), 1615–1616. https://doi.org/10.1021/ja983751y
80. Fan, X.; Zhou, Q.; Xia, C.; Cristofoli, W.; Mays, J.; Advincula, R. Living Anionic Surface-Initiated Polymerization (LASIP) of Styrene from Clay Nanoparticles Using Surface Bound 1,1-Diphenylethylene (DPE) Initiators. *Langmuir*, **2002**, *18* (11), 4511–4518. https://doi.org/10.1021/la025556+
81. Cui, L.; Paul, D. R. Evaluation of Amine Functionalized Polypropylenes as Compatibilizers for Polypropylene Nanocomposites. *Polymer*, **2007**, *48* (6), 1632–1640. https://doi.org/10.1016/j.polymer.2007.01.036
82. Kim, D. H.; Fasulo, P. D.; Rodgers, W. R.; Paul, D. R. Structure and Properties of Polypropylene-Based Nanocomposites: Effect of PP-g-MA to Organoclay Ratio. *Polymer*, **2007**, *48* (18), 5308–5323. https://doi.org/10.1016/j.polymer.2007.07.011
83. Reichert, P.; Nitz, H.; Klinke, S.; Brandsch, R.; Thomann, R.; Mülhaupt, R. Poly(Propylene)/Organoclay Nanocomposite Formation: Influence of Compatibilizer Functionality and Organoclay Modification. *Macromol. Mater. Eng.*,

2000, *275* (1), 8–17. https://doi.org/10.1002/(SICI)1439-2054(20000201)275:1<8::AID-MAME8>3.0.CO;2-6

84. Usuki, A.; Kawasumi, M.; Kojima, Y.; Okada, A.; Kurauchi, T.; Kamigaito, O. Swelling Behavior of Montmorillonite Cation Exchanged for ω-Amino Acids by ∈-Caprolactam. *J. Mater. Res.*, **1993**, *8* (5), 1174–1178. https://doi.org/10.1557/JMR.1993.1174
85. Fornes, T. D.; Hunter, D. L.; Paul, D. R. Nylon-6 Nanocomposites from Alkylammonium-Modified Clay: The Role of Alkyl Tails on Exfoliation. *Macromolecules*, **2004**, *37* (5), 1793–1798. https://doi.org/10.1021/ma0305481
86. Vaia, R. A.; Giannelis, E. P. Lattice Model of Polymer Melt Intercalation in Organically-Modified Layered Silicates. *Macromolecules*, **1997**, *30* (25), 7990–7999. https://doi.org/10.1021/ma9514333
87. Garboczi, E. J.; Snyder, K. A.; Douglas, J. F.; Thorpe, M. F. Geometrical Percolation Threshold of Overlapping Ellipsoids. *Phys. Rev. E.*, **1995**, *52* (1), 819–828. https://doi.org/10.1103/PhysRevE.52.819
88. Ishida, H.; Campbell, S.; Blackwell, J. General Approach to Nanocomposite Preparation. *Chem. Mater.*, **2000**, *12* (5), 1260–1267. https://doi.org/10.1021/cm990479y
89. Jang, B. N.; Wang, D.; Wilkie, C. A. Relationship between the Solubility Parameter of Polymers and the Clay Dispersion in Polymer/Clay Nanocomposites and the Role of the Surfactant. *Macromolecules*, **2005**, *38* (15), 6533–6543. https://doi.org/10.1021/ma0508909
90. Vaia, R. A.; Liu, W.; Koerner, H. Analysis of Small-Angle Scattering of Suspensions of Organically Modified Montmorillonite: Implications to Phase Behavior of Polymer Nanocomposites. *J. Polym. Sci. B Polym. Phys.*, **2003**, *41* (24), 3214–3236. https://doi.org/10.1002/polb.10698
91. Morgan, A. B.; Harris, J. D. Exfoliated Polystyrene-Clay Nanocomposites Synthesized by Solvent Blending with Sonication. *Polymer*, **2004**, *45* (26), 8695–8703. https://doi.org/10.1016/j.polymer.2004.10.067
92. Fornes, T. D.; Yoon, P. J.; Hunter, D. L.; Keskkula, H.; Paul, D. R. Effect of Organoclay Structure on Nylon 6 Nanocomposite Morphology and Properties. *Polymer*, **2002**, *43* (22), 5915–5933. https://doi.org/10.1016/S0032-3861(02)00400-7
93. McAlpine, M.; Hudson, N. E.; Liggat, J. J.; Pethrick, R. A.; Pugh, D.; Rhoney, I. Study of the Factors Influencing the Exfoliation of an Organically Modified Montmorillonite in Methyl Methacrylate/Poly(Methyl Methacrylate) Mixtures. *J. Appl. Polym. Sci.*, **2006**, *99* (5), 2614–2626. https://doi.org/10.1002/app.22582
94. Zeng, Q. H.; Yu, A. B.; Lu, G. Q.; Standish, R. K. Molecular Dynamics Simulation of Organic–Inorganic Nanocomposites: Layering Behavior and Interlayer Structure of Organoclays. *Chem. Mater.*, **2003**, *15* (25), 4732–4738. https://doi.org/10.1021/cm0342952
95. Heinz, H.; Koerner, H.; Anderson, K. L.; Vaia, R. A.; Farmer, B. L. Force Field for Mica-Type Silicates and Dynamics of Octadecylammonium Chains Grafted to Montmorillonite. *Chem. Mater.*, **2005**, *17* (23), 5658–5669. https://doi.org/10.1021/cm0509328
96. Dean, D.; Walker, R.; Theodore, M.; Hampton, E.; Nyairo, E. Chemorheology and Properties of Epoxy/Layered Silicate Nanocomposites. *Polymer*, **2005**, *46* (9), 3014–3021. https://doi.org/10.1016/j.polymer.2005.02.015
97. Ton-That, M.-T.; Ngo, T.-D.; Ding, P.; Fang, G.; Cole, K. C.; Hoa, S. V. Epoxy Nanocomposites: Analysis and Kinetics of Cure. *Polym. Eng. Sci.*, **2004**, *44* (6), 1132–1141. https://doi.org/10.1002/pen.20106
98. Xie, W.; Gao, Z.; Pan, W.-P.; Hunter, D.; Singh, A.; Vaia, R. Thermal Degradation Chemistry of Alkyl Quaternary Ammonium Montmorillonite. *Chem. Mater.*, **2001**, *13* (9), 2979–2990. https://doi.org/10.1021/cm010305s
99. He, H.; Ding, Z.; Zhu, J.; Yuan, P.; Xi, Y.; Yang, D.; Frost, R. L. Thermal Characterization of Surfactant-Modified Montmorillonites. *Clays Clay Miner.*, **2005**, *53* (3), 287–293. https://doi.org/10.1346/CCMN.2005.0530308
100. Dharaiya, D.; Jana, S. C. Thermal Decomposition of Alkyl Ammonium Ions and Its Effects on Surface Polarity of Organically Treated Nanoclay. *Polymer*, **2005**, *46* (23), 10139–10147. https://doi.org/10.1016/j.polymer.2005.08.027
101. Gelfer, M.; Burger, C.; Fadeev, A.; Sics, I.; Chu, B.; Hsiao, B. S.; Heintz, A.; Kojo, K.; Hsu, S.-L.; Si, M.; et al. Thermally Induced Phase Transitions and Morphological Changes in Organoclays. *Langmuir*, **2004**, *20* (9), 3746–3758. https://doi.org/10.1021/la035361h
102. Cervantes-Uc, J. M.; Cauich-Rodríguez, J. V.; Vázquez-Torres, H.; Garfias-Mesías, L. F.; Paul, D. R. Thermal Degradation of Commercially Available Organoclays Studied by TGA–FTIR. *Thermochim. Acta*, **2007**, *457* (1–2), 92–102. https://doi.org/10.1016/j.tca.2007.03.008
103. Morgan, A. B.; Gilman, J. W.; Jackson, C. L. Characterization of the Dispersion of Clay in a Polyetherimide Nanocomposite. *Macromolecules*, **2001**, *34* (8), 2735–2738. https://doi.org/10.1021/ma0008847
104. Davis, R. D.; Gilman, J. W.; Sutto, T. E.; Callahan, J. H.; Trulove, P. C.; De Long, H. C. Improved Thermal Stability of Organically Modified Layered Silicates. *Clays Clay Miner.*, **2004**, *52* (2), 171–179. https://doi.org/10.1346/CCMN.2004.0520203
105. Morgan, A. B.; Harris, J. D. Effects of Organoclay Soxhlet Extraction on Mechanical Properties, Flammability Properties and Organoclay Dispersion of Polypropylene Nanocomposites. *Polymer*, **2003**, *44* (8), 2313–2320. https://doi.org/10.1016/S0032-3861(03)00095-8
106. Hotta, S.; Paul, D. R. Nanocomposites Formed from Linear Low Density Polyethylene and Organoclays. *Polymer*, **2004**, *45* (22), 7639–7654. https://doi.org/10.1016/j.polymer.2004.08.059
107. Lee, H.; Fasulo, P. D.; Rodgers, W. R.; Paul, D. R. TPO Based Nanocomposites. Part 1. Morphology and Mechanical Properties. *Polymer*, **2005**, *46* (25), 11673–11689. https://doi.org/10.1016/j.polymer.2005.09.068
108. Ton-That, M.-T.; Perrin-Sarazin, F.; Cole, K. C.; Bureau, M. N.; Denault, J. Polyolefin Nanocomposites: Formulation and Development. *Polym Eng Sci*, **2004**, *44* (7), 1212–1219. https://doi.org/10.1002/pen.20116
109. Dennis, H. R.; Hunter, D. L.; Chang, D.; Kim, S.; White, J. L.; Cho, J. W.; Paul, D. R. Effect of Melt Processing Conditions on the Extent of Exfoliation in Organoclay-Based Nanocomposites. *Polymer*, **2001**, *42* (23), 9513–9522. https://doi.org/10.1016/S0032-3861(01)00473-6
110. Tanoue, S.; Utracki, L. A.; Garcia-Rejon, A.; Tatibouët, J.; Kamal, M. R. Melt Compounding of Different Grades Polystyrene with Organoclay. Part 3: Mechanical Properties. *Polym. Eng. Sci.*, **2005**, *45* (6), 827–837. https://doi.org/10.1002/pen.20336
111. Hasegawa, N.; Okamoto, H.; Kato, M.; Usuki, A.; Sato, N. Nylon 6/Na–Montmorillonite Nanocomposites Prepared by Compounding Nylon 6 with Na–Montmorillonite Slurry. *Polymer*, **2003**, *44* (10), 2933–2937. https://doi.org/10.1016/S0032-3861(03)00215-5
112. Kato, M.; Matsushita, M.; Fukumori, K. Development of a New Production Method for a Polypropylene-Clay

Nanocomposite. *Polym. Eng. Sci.*, **2004**, *44* (7), 1205–1211. https://doi.org/10.1002/pen.20115

113. Liang, Z.-M.; Yin, J.; Xu, H.-J. Polyimide/Montmorillonite Nanocomposites Based on Thermally Stable, Rigid-Rod Aromatic Amine Modifiers. *Polymer*, **2003**, *44* (5), 1391–1399. https://doi.org/10.1016/S0032-3861(02)00911-4
114. Xie, W.; Xie, R.; Pan, W.-P.; Hunter, D.; Koene, B.; Tan, L.-S.; Vaia, R. Thermal Stability of Quaternary Phosphonium Modified Montmorillonites. *Chem. Mater.*, **2002**, *14* (11), 4837–4845. https://doi.org/10.1021/cm020705v
115. Xie, W.; Gao, Z.; Pan, W. P.; Vaia, R.; Hunter, D.; Singh, A. Characterization of Organically Modified Montmorillonite by Thermal Techniques. *Polym. Mater Sci. Eng.*, **2000**, 83, 284.
116. Zhang, J.; Jiang, D. D.; Wang, D.; Wilkie, C. A. Mechanical and Fire Properties of Styrenic Polymer Nanocomposites Based on an Oligomerically-Modified Clay. *Polym. Adv. Technol.*, **2005**, *16* (11–12), 800–806. https://doi.org/10.1002/pat.655
117. Zhang, J.; Jiang, D. D.; Wilkie, C. A. Fire Properties of Styrenic Polymer–Clay Nanocomposites Based on an Oligomerically-Modified Clay. *Polym. Degrad. Stab.*, **2006**, *91* (2), 358–366. https://doi.org/10.1016/j.polymdegradstab.2005.04.040
118. Zheng, X.; Jiang, D. D.; Wang, D.; Wilkie, C. A. Flammability of Styrenic Polymer Clay Nanocomposites Based on a Methyl Methacrylate Oligomerically-Modified Clay. *Polym. Degrad. Stab.*, **2006**, *91* (2), 289–297. https://doi.org/10.1016/j.polymdegradstab.2005.05.007
119. Su, S. Study on the Thermal Stability of Polystyryl Surfactants and Their Modified Clay Nanocomposites. *Polym. Degrad. Stab.*, **2004**, *84* (2), 269–277. https://doi.org/10.1016/j.polymdegradstab.2003.10.021
120. Tao, L.; Chen, G.; Mantovani, G.; York, S.; Haddleton, D. M. Modification of Multi-Wall Carbon Nanotube Surfaces with Poly(Amidoamine) Dendrons: Synthesis and Metal Templating. *Chem. Commun.*, **2006**, No. 47, 4949. https://doi.org/10.1039/b609065f
121. Raghuveer, M. S.; Kumar, A.; Frederick, M. J.; Louie, G. P.; Ganesan, P. G.; Ramanath, G. Site-Selective Functionalization of Carbon Nanotubes. *Adv. Mater.*, **2006**, *18* (5), 547–552. https://doi.org/10.1002/adma.200500181
122. Yokoi, T.; Iwamatsu, S.; Komai, S.; Hattori, T.; Murata, S. Chemical Modification of Carbon Nanotubes with Organic Hydrazines. *Carbon*, **2005**, *43* (14), 2869–2874. https://doi.org/10.1016/j.carbon.2005.06.022
123. Okada, A.; Kawasumi, M.; Usuki, A.; Kojima, Y.; Kurauchi, T.; Kamigaito, O. Synthesis and Properties of Nylon-6/Clay Hybrids. In *Polymer Based Molecular Composites*; Schaefer, D. W., Mark, J. E., Eds.; MRS Symposium Proceedings, Pittsburgh, 1990; pp 171, 45–50.
124. LeBaron, P. Polymer-Layered Silicate Nanocomposites: An Overview. *Appl. Clay Sci.*, **1999**, *15* (1–2), 11–29. https://doi.org/10.1016/S0169-1317(99)00017-4
125. Vaia, R. Polymer/Layered Silicate Nanocomposites as High Performance Ablative Materials. *Appl. Clay Sci.*, **1999**, *15* (1–2), 67–92. https://doi.org/10.1016/S0169-1317(99)00013-7
126. Biswas, M.; Ray, S. S. Recent Progress in Synthesis and Evaluation of Polymer-Montmorillonite Nanocomposites. In *New Polymerization Techniques and Synthetic Methodologies*; Abe, A., Albertsson, A.-C., Cantow, H.-J., Dušek, K., Edwards, S., Höcker, H., Joanny, J.-F., Kausch, H.-H., Kobayashi, T., Lee, K.-S., et al., Series Eds.; Springer: Berlin, Heidelberg, 2001; Vol. 155, pp 167–221. https://doi.org/10.1007/3-540-44473-4_3
127. Voorn, D. J.; Ming, W.; Van Herk, A. M. Polymer–Clay Nanocomposite Latex Particles by Inverse Pickering Emulsion Polymerization Stabilized with Hydrophobic Montmorillonite Platelets. *Macromolecules*, **2006**, *39* (6), 2137–2143. https://doi.org/10.1021/ma052539t
128. Xu, M.; Choi, Y. S.; Kim, Y. K.; Wang, K. H.; Chung, I. J. Synthesis and Characterization of Exfoliated Poly(Styrene-Co-Methyl Methacrylate)/Clay Nanocomposites via Emulsion Polymerization with AMPS. *Polymer*, **2003**, *44* (20), 6387–6395. https://doi.org/10.1016/S0032-3861(03)00653-0
129. Wang, D.; Zhu, J.; Yao, Q.; Wilkie, C. A. A Comparison of Various Methods for the Preparation of Polystyrene and Poly(Methyl Methacrylate) Clay Nanocomposites. *Chem. Mater.*, **2002**, *14* (9), 3837–3843. https://doi.org/10.1021/cm011656
130. Price, G. J. Applications of High Intensity Ultrasound in Polymer Chemistry. In *Chemistry Under Extreme and Non-Classical Conditions*; van Eldik, R., Hubbard, C. D., Eds.; John Wiley & Sons: New York, 1997; Ch. 9, pp. 381–427.
131. Utracki, L. A. *Clay-Containing Polymeric Nanocomposites*; Rapra Technology Ltd: Shrewsbury, 2004; p 496.
132. Vaia, R. A.; Ishii, H.; Giannelis, E. P. Synthesis and Properties of Two-Dimensional Nanostructures by Direct Intercalation of Polymer Melts in Layered Silicates. *Chem. Mater.*, **1993**, *5* (12), 1694–1696. https://doi.org/10.1021/cm00036a004
133. Chigwada, G.; Jiang, D. D.; Wilkie, C. A. Fire Retardancy of Polystyrene Nanocomposites Using Naphthenate-Containing Clays. In *Fire and Polymers IV: Materials and Concepts for Hazard Prevention*; Wilkie, C. A., Nelson, G. L., Eds.; ACS Symposium Series; American Chemical Society: Washington, DC, 2005; Vol. 922, pp 103–116. https://doi.org/10.1021/bk-2006-0922.ch009
134. Zhang, J.; Jiang, D. D.; Wilkie, C. A. Polyethylene and Polypropylene Nanocomposites Based upon an Oligomerically Modified Clay. *Thermochim. Acta*, **2005**, *430* (1–2), 107–113. https://doi.org/10.1016/j.tca.2005.01.028
135. Alexandre, M.; Dubois, P. Polymer-Layered Silicate Nanocomposites: Preparation, Properties and Uses of a New Class of Materials. *Mater. Sci. Eng.: R: Rep.*, **2000**, *28* (1–2), 1–63. https://doi.org/10.1016/S0927-796X(00)00012-7
136. Morgan, A. B.; Gilman, J. W. Characterization of Polymer-Layered Silicate (Clay) Nanocomposites by Transmission Electron Microscopy and X-Ray Diffraction: A Comparative Study. *J. Appl. Polym. Sci.*, **2003**, *87* (8), 1329–1338. https://doi.org/10.1002/app.11884
137. Vermogen, A.; Masenelli-Varlot, K.; Séguéla, R.; Duchet-Rumeau, J.; Boucard, S.; Prele, P. Evaluation of the Structure and Dispersion in Polymer-Layered Silicate Nanocomposites. *Macromolecules*, **2005**, *38* (23), 9661–9669. https://doi.org/10.1021/ma051249
138. Perrin-Sarazin, F.; Ton-That, M.-T.; Bureau, M. N.; Denault, J. Micro- and Nano-Structure in Polypropylene/Clay Nanocomposites. *Polymer*, **2005**, *46* (25), 11624–11634. https://doi.org/10.1016/j.polymer.2005.09.076
139. Causin, V.; Marega, C.; Marigo, A.; Ferrara, G. Assessing Organo-Clay Dispersion in Polymer Layered Silicate Nanocomposites: A SAXS Approach. *Polymer*, **2005**, *46* (23), 9533–9537. https://doi.org/10.1016/j.polymer.2005.08.034
140. Vaia, R. A.; Liu, W. X-Ray Powder Diffraction of Polymer/Layered Silicate Nanocomposites: Model and Practice. *J. Polym. Sci. B Polym. Phys.*, **2002**, *40* (15), 1590–1600. https://doi.org/10.1002/polb.10214

141. Bourbigot, S.; VanderHart, D. L.; Gilman, J. W.; Awad, W. H.; Davis, R. D.; Morgan, A. B.; Wilkie, C. A. Investigation of Nanodispersion in Polystyrene-Montmorillonite Nanocomposites by Solid-State NMR. *J. Polym. Sci. B Polym. Phys.*, **2003**, *41* (24), 3188–3213. https://doi.org/10.1002/polb.10707
142. Gilman, J. W.; Bourbigot, S.; Shields, J. R.; Nyden, M.; Kashiwagi, T.; Davis, R. D.; Vanderhart, D. L.; Demory, W.; Wilkie, C. A.; Morgan, A. B.; et al. High Throughput Methods for Polymer Nanocomposites Research: Extrusion, NMR Characterization and Flammability Property Screening. *J. Mater. Sci.*, **2003**, *38* (22), 4451–4460. https://doi.org/10.1023/A:1027369115480
143. Vaia, R. A.; Vasudevan, S.; Krawiec, W.; Scanlon, L. G.; Giannelis, E. P. New Polymer Electrolyte Nanocomposites: Melt Intercalation of Poly(Ethylene Oxide) in Mica-Type Silicates. *Adv. Mater.*, **1995**, *7* (2), 154–156. https://doi.org/10.1002/adma.19950070210
144. Maiti, M.; Bhowmick, A. K. New Insights into Rubber–Clay Nanocomposites by AFM Imaging. *Polymer*, **2006**, *47* (17), 6156–6166. https://doi.org/10.1016/j.polymer.2006.06.032
145. Wu, D.; Wu, L.; Zhang, M.; Zhou, W.; Zhang, Y. Morphology Evolution of Nanocomposites Based on Poly(Phenylene Sulfide)/Poly(Butylene Terephthalate) Blend. *J. Polym. Sci. B Polym. Phys.*, **2008**, *46* (12), 1265–1279. https://doi.org/10.1002/polb.21461
146. Lutkenhaus, J. L.; Olivetti, E. A.; Verploegen, E. A.; Cord, B. M.; Sadoway, D. R.; Hammond, P. T. Anisotropic Structure and Transport in Self-Assembled Layered Polymer–Clay Nanocomposites. *Langmuir*, **2007**, *23* (16), 8515–8521. https://doi.org/10.1021/la700432p
147. Maupin, P. H.; Gilman, J. W.; Harris, R. H.; Bellayer, S.; Bur, A. J.; Roth, S. C.; Murariu, M.; Morgan, A. B.; Harris, J. D. Optical Probes for Monitoring Intercalation and Exfoliation in Melt-Processed Polymer Nanocomposites. *Macromol. Rapid Commun.*, **2004**, *25* (7), 788–792. https://doi.org/10.1002/marc.200300262
148. Drits, V. A.; Tchoubar, C. *X-Ray Diffraction by Disordered Lamellar Structures*; Springer: Berlin and Heidelberg, 1990. https://doi.org/10.1007/978-3-642-74802-8
149. Morgan, A. B.; Wilkie, C. A. Practical Issues and Future Trends in Polymer Nanocomposite Flammability Research. In *Flame Retardant Polymer Nanocomposites*; Morgan, A. B., Wilkie, C. A., Eds.; John Wiley & Sons, Inc.: Hoboken, NJ, 2007; pp 355–399. https://doi.org/10.1002/9780470109038.ch12
150. Bish, D. L.; Post, J. E.; Mineralogical Society of America, Eds.; *Modern Powder Diffraction*; Reviews in Mineralogy; Mineralogical Society of America: Washington, DC, 1989.
151. Roe, R. J. *Methods of X-Ray and Neutron Scattering in Polymer Science*; Topics in Polymer Science; Oxford University Press: New York, 2000.
152. Clément, F.; Lapra, A.; Bokobza, L.; Monnerie, L.; Ménez, P. Atomic Force Microscopy Investigation of Filled Elastomers and Comparison with Transmission Electron Microscopy—Application to Silica-Filled Silicone Elastomers. *Polymer*, **2001**, *42* (14), 6259–6270. https://doi.org/10.1016/S0032-3861(01)00058-1
153. Kim, D.; Lee, J. S.; Barry, C. M. F.; Mead, J. L. Microscopic Measurement of the Degree of Mixing for Nanoparticles in Polymer Nanocomposites by TEM Images. *Microsc. Res. Tech.*, **2007**, *70* (6), 539–546. https://doi.org/10.1002/jemt.20478
154. Basu, S. K.; Tewari, A.; Fasulo, P. D.; Rodgers, W. R. Transmission Electron Microscopy Based Direct Mathematical Quantifiers for Dispersion in Nanocomposites. *Applied Physics Letters*, **2007**, *91* (5), 053105. https://doi.org/10.1063/1.2760182
155. Basu, S. K.; Fasulo, P. D.; Rodgers, W. R. Stereology-Based Quantitative Characterization of Dispersion from TEM Micrographs of Polymer-Clay Nanocomposites. *J. Appl. Polym. Sci.*, **2011**, *119* (1), 396–411. https://doi.org/10.1002/app.32756
156. Sinha Ray, S.; Okamoto, M. Polymer/Layered Silicate Nanocomposites: A Review from Preparation to Processing. *Prog. Polym. Sci.*, **2003**, *28* (11), 1539–1641. https://doi.org/10.1016/j.progpolymsci.2003.08.002
157. Yang, J.; Zhang, Z.; Friedrich, K.; Schlarb, A. K. Creep Resistant Polymer Nanocomposites Reinforced with Multiwalled Carbon Nanotubes. *Macromol. Rapid Commun.*, **2007**, *28* (8), 955–961. https://doi.org/10.1002/marc.200600866
158. Qi, D.; Bao, Y.; Weng, Z.; Huang, Z. Preparation of Acrylate Polymer/Silica Nanocomposite Particles with High Silica Encapsulation Efficiency via Miniemulsion Polymerization. *Polymer*, **2006**, *47* (13), 4622–4629. https://doi.org/10.1016/j.polymer.2006.04.024
159. Burnside, S. D.; Giannelis, E. P. Synthesis and Properties of New Poly(Dimethylsiloxane) Nanocomposites. *Chem. Mater.*, **1995**, *7* (9), 1597–1600. https://doi.org/10.1021/cm00057a001
160. Lepoittevin, B.; Devalckenaere, M.; Pantoustier, N.; Alexandre, M.; Kubies, D.; Calberg, C.; Jérôme, R.; Dubois, P. Poly(ε-Caprolactone)/Clay Nanocomposites Prepared by Melt Intercalation: Mechanical, Thermal and Rheological Properties. *Polymer*, **2002**, *43* (14), 4017–4023. https://doi.org/10.1016/S0032-3861(02)00229-X
161. Zhang, J.; Jiang, D. D.; Wilkie, C. A. Thermal and Flame Properties of Polyethylene and Polypropylene Nanocomposites Based on an Oligomerically-Modified Clay. *Polym. Degrad. Stab.*, **2006**, *91* (2), 298–304. https://doi.org/10.1016/j.polymdegradstab.2005.05.006
162. Manzi-Nshuti, C.; Wang, D.; Hossenlopp, J. M.; Wilkie, C. A. Aluminum-Containing Layered Double Hydroxides: The Thermal, Mechanical, and Fire Properties of (Nano) Composites of Poly(Methyl Methacrylate). *J. Mater. Chem.*, **2008**, *18* (26), 3091. https://doi.org/10.1039/b802553c
163. Ge, J. J.; Hou, H.; Li, Q.; Graham, M. J.; Greiner, A.; Reneker, D. H.; Harris, F. W.; Cheng, S. Z. D. Assembly of Well-Aligned Multiwalled Carbon Nanotubes in Confined Polyacrylonitrile Environments: Electrospun Composite Nanofiber Sheets. *J. Am. Chem. Soc.*, **2004**, *126* (48), 15754–15761. https://doi.org/10.1021/ja048648p
164. Jang, B. N.; Costache, M.; Wilkie, C. A. The Relationship between Thermal Degradation Behavior of Polymer and the Fire Retardancy of Polymer/Clay Nanocomposites. *Polymer*, **2005**, *46* (24), 10678–10687. https://doi.org/10.1016/j.polymer.2005.08.085
165. Zhu, J.; Start, P.; Mauritz, K. A.; Wilkie, C. A. Thermal Stability and Flame Retardancy of Poly(Methyl Methacrylate)-Clay Nanocomposites. *Polym. Degrad. Stab.*, **2002**, *77* (2), 253–258. https://doi.org/10.1016/S0141-3910(02)00056-3
166. Bourbigot, S.; Gilman, J. W.; Wilkie, C. A. Kinetic Analysis of the Thermal Degradation of Polystyrene–Montmorillonite Nanocomposite. *Polym. Degrad. Stab.*, **2004**, *84* (3), 483–492. https://doi.org/10.1016/j.polymdegradstab.2004.01.006
167. Doh, J. G.; Cho, I. Synthesis and Properties of Polystyrene-Organoammonium Montmorillonite Hybrid. *Polym.*

Bull., **1998**, *41* (5), 511–518. https://doi.org/10.1007/s002890050395
168. Gilman, J. Flammability and Thermal Stability Studies of Polymer Layered-Silicate (Clay) Nanocomposites. *Appl. Clay Sci.*, **1999**, *15* (1–2), 31–49. https://doi.org/10.1016/S0169-1317(99)00019-8
169. Costache, M. C.; Jiang, D. D.; Wilkie, C. A. Thermal Degradation of Ethylene–Vinyl Acetate Coplymer Nanocomposites. *Polymer*, **2005**, *46* (18), 6947–6958. https://doi.org/10.1016/j.polymer.2005.05.084
170. Kashiwagi, T.; Grulke, E.; Hilding, J.; Harris, R.; Awad, W.; Douglas, J. Thermal Degradation and Flammability Properties of Poly(Propylene)/Carbon Nanotube Composites. *Macromol. Rapid Commun.*, **2002**, *23* (13), 761–765. https://doi.org/10.1002/1521-3927(20020901)23:13<761::AID-MARC761>3.0.CO;2-K
171. Kashiwagi, T. Progress in Flammability Studies of Nanocomposites with New Types of Nanoparticles. In *Flame Retardant Polymer Nanocomposites*; Morgan, A. B., Wilkie, C. A., Eds.; John Wiley & Sons, Inc.: Hoboken, NJ, 2007; pp 285–324. https://doi.org/10.1002/9780470109038.ch10
172. Kashiwagi, T.; Du, F.; Douglas, J. F.; Winey, K. I.; Harris, R. H.; Shields, J. R. Nanoparticle Networks Reduce the Flammability of Polymer Nanocomposites. *Nat. Mater*, **2005**, *4* (12), 928–933. https://doi.org/10.1038/nmat1502
173. Kashiwagi, T.; Du, F.; Winey, K. I.; Groth, K. M.; Shields, J. R.; Bellayer, S. P.; Kim, H.; Douglas, J. F. Flammability Properties of Polymer Nanocomposites with Single-Walled Carbon Nanotubes: Effects of Nanotube Dispersion and Concentration. *Polymer*, **2005**, *46* (2), 471–481. https://doi.org/10.1016/j.polymer.2004.10.087
174. Kashiwagi, T.; Grulke, E.; Hilding, J.; Groth, K.; Harris, R.; Butler, K.; Shields, J.; Kharchenko, S.; Douglas, J. Thermal and Flammability Properties of Polypropylene/Carbon Nanotube Nanocomposites. *Polymer*, **2004**, *45* (12), 4227–4239. https://doi.org/10.1016/j.polymer.2004.03.088
175. Gilman, J. W.; Jackson, C. L.; Morgan, A. B.; Harris, R.; Manias, E.; Giannelis, E. P.; Wuthenow, M.; Hilton, D.; Phillips, S. H. Flammability Properties of Polymer–Layered-Silicate Nanocomposites. Polypropylene and Polystyrene Nanocomposites. *Chem. Mater.*, **2000**, *12* (7), 1866–1873. https://doi.org/10.1021/cm0001760
176. Gilman, J. W.; Kashiwagi, T.; Morgan, A. B.; Harris, R. H. J.; Brassell, L.; VanLandingham, M.; Jackson, C. L. *Flammability of Polymer Clay Nanocomposites Consortium: Year One Annual Report*; NIST IR 6531; National Institute of Standards and Technology: Gaithersburg, MD, 2000. https://doi.org/10.6028/NIST.IR.6531
177. Kashiwagi, T. Flammability of Nanocomposites: Effects of the Shape of Nanoparticles. In *Fire Retardancy of Polymers: New Applications of Mineral Fillers*; Le Bras, M., Royal Society of Chemistry (Great Britain), Eds.; Royal Society of Chemistry: Cambridge, UK, 2005; pp 81–99.
178. Gilman, J. W. Flame Retardant Mechanism of Polymer–Clay Nanocomposites. In *Flame Retardant Polymer Nanocomposites*; Morgan, A. B., Wilkie, C. A., Eds.; John Wiley & Sons, Inc.: Hoboken, NJ, 2007; pp 67–87. https://doi.org/10.1002/9780470109038.ch3
179. Cipiriano, B. H.; Kashiwagi, T.; Raghavan, S. R.; Yang, Y.; Grulke, E. A.; Yamamoto, K.; Shields, J. R.; Douglas, J. F. Effects of Aspect Ratio of MWNT on the Flammability Properties of Polymer Nanocomposites. *Polymer*, **2007**, *48* (20), 6086–6096. https://doi.org/10.1016/j.polymer.2007.07.070
180. Lyon, R. E.; Walters, R. N.; Stoliarov, S. I. Screening Flame Retardants for Plastics Using Microscale Combustion Calorimetry. *Polym. Eng. Sci.*, **2007**, *47* (10), 1501–1510. https://doi.org/10.1002/pen.20871
181. Standard Test Method for Determining Flammability Characteristics of Plastics and Other Solid Materials Using Microscale Combustion Calorimetry www.astm.org/d7309-21b.html (accessed July 31, 2023).
182. Gilman, J. W.; Kashiwagi, T.; Brown, J. E. T.; Lomakin, S.; Giannelis, E. P.; Manias, E. Flammability Studies of Polymer Layered Silicate Nanocomposites: Proceedings of the 1998 43rd International SAMPE Symposium and Exhibition. Part 1 (of 2). *Int. SAMPE Symp. Exhib. (Proceedings)*, **1998**, *43* (1), 1053–1066.
183. Kashiwagi, T.; Harris, R. H.; Zhang, X.; Briber, R. M.; Cipriano, B. H.; Raghavan, S. R.; Awad, W. H.; Shields, J. R. Flame Retardant Mechanism of Polyamide 6–Clay Nanocomposites. *Polymer*, **2004**, *45* (3), 881–891. https://doi.org/10.1016/j.polymer.2003.11.036
184. Akelah, A. Nanocomposites of Grafted Polymers onto Layered Silicate. In *Polymers and Other Advanced Materials*; Prasad, P. N., Mark, J. E., Fai, T. J., Eds.; Springer: Boston, MA, 1995; pp 625–644. https://doi.org/10.1007/978-1-4899-0502-4_65
185. Xiao, P.; Xiao, M.; Gong, K. Preparation of Exfoliated Graphite/Polystyrene Composite by Polymerization-Filling Technique. *Polymer*, **2001**, *42* (11), 4813–4816. https://doi.org/10.1016/S0032-3861(00)00819-3
186. Tseng, C.-R.; Wu, J.-Y.; Lee, H.-Y.; Chang, F.-C. Preparation and Crystallization Behavior of Syndiotactic Polystyrene–Clay Nanocomposites. *Polymer*, **2001**, *42* (25), 10063–10070. https://doi.org/10.1016/S0032-3861(01)00568-7
187. Komori, Y.; Sugahara, Y.; Kuroda, K. Direct Intercalation of Poly(Vinylpyrrolidone) into Kaolinite by a Refined Guest Displacement Method. *Chem. Mater.*, **1999**, *11* (1), 3–6. https://doi.org/10.1021/cm9804721
188. Nisha, A.; Rajeswari, M. K.; Dhamodharan, R. Intercalative Redox Polymerization and Characterization of Poly(n-Vinyl-2-Pyrrolidinone) in the Gallery of Vermiculite: A Novel Inorganic-Organic Hybrid Material. *J. Appl. Polym. Sci.*, **2000**, *76* (12), 1825–1830. https://doi.org/10.1002/(SICI)1097-4628(20000620)76:12<1825::AID-APP13>3.0.CO;2-S
189. Fournaris, K. G.; Karakassides, M. A.; Petridis, D.; Yiannakopoulou, K. Clay–Polyvinylpyridine Nanocomposites. *Chem. Mater.*, **1999**, *11* (9), 2372–2381. https://doi.org/10.1021/cm981140z
190. Parfitt, R. L.; Greenland, D. J. The Adsorption of Poly(Ethylene Glycols) on Clay Minerals. *Clay Miner.*, **1970**, *8* (3), 305–315. https://doi.org/10.1180/claymin.1970.008.3.08
191. Zhao, X.; Urano, K.; Ogasawara, S. Adsorption of Polyethylene Glycol from Aqueous Solution on Montmorillonite Clays. *Colloid Polym. Sci.*, **1989**, *267* (10), 899–906. https://doi.org/10.1007/BF01410338
192. Zhu, J.; Uhl, F. M.; Morgan, A. B.; Wilkie, C. A. Studies on the Mechanism by Which the Formation of Nanocomposites Enhances Thermal Stability. *Chem. Mater.*, **2001**, *13* (12), 4649–4654. https://doi.org/10.1021/cm010451y
193. Jang, B. N.; Wilkie, C. A. The Thermal Degradation of Polystyrene Nanocomposite. *Polymer*, **2005**, *46* (9), 2933–2942. https://doi.org/10.1016/j.polymer.2005.01.098
194. Guo, B.; Jia, D.; Cai, C. Effects of Organo-Montmorillonite Dispersion on Thermal Stability of Epoxy Resin Nanocomposites. *Eur. Polym. J.*, **2004**, *40* (8), 1743–1748. https://doi.org/10.1016/j.eurpolymj.2004.03.027

195. https://web.archive.org/web/20150905194058/http://www.epa.gov/oppt/nano/p2docs/casestudy2_gilman.pdf (Gilman, J.W. "Flame retardant polymer nanocomposite")
196. Marosi, G.; Márton, A.; Szép, A.; Csontos, I.; Keszei, S.; Zimonyi, E.; Toth, A.; Almeras, X.; Le Bras, M. Fire Retardancy Effect of Migration in Polypropylene Nanocomposites Induced by Modified Interlayer. *Polym. Degrad. Stab.*, **2003**, *82* (2), 379–385. https://doi.org/10.1016/S0141-3910(03)00223-4
197. Szép, A.; Szabó, A.; Tóth, N.; Anna, P.; Marosi, G. Role of Montmorillonite in Flame Retardancy of Ethylene–Vinyl Acetate Copolymer. *Polym. Degrad. Stab.*, **2006**, *91* (3), 593–599. https://doi.org/10.1016/j.polymdegradstab.2005.02.026
198. Marosi, G.; Anna, P.; Márton, A.; Bertalan, G.; Bóta, A.; Tóth, A.; Mohai, M.; Rácz, I. Flame-Retarded Polyolefin Systems of Controlled Interphase: Flame-Retarded Polyolefin Systems. *Polym. Adv. Technol.*, **2002**, *13* (10–12), 1103–1111. https://doi.org/10.1002/pat.284
199. Kotaki, M.; Wang, K.; Toh, M. L.; Chen, L.; Wong, S. Y.; He, C. Electrically Conductive Epoxy/Clay/Vapor Grown Carbon Fiber Hybrids. *Macromolecules*, **2006**, *39* (3), 908–911. https://doi.org/10.1021/ma0522561
200. Chigwada, G.; Jash, P.; Jiang, D. D.; Wilkie, C. A. Synergy between Nanocomposite Formation and Low Levels of Bromine on Fire Retardancy in Polystyrenes. *Polym. Degrad. Stab.*, **2005**, *88* (3), 382–393. https://doi.org/10.1016/j.polymdegradstab.2004.12.002
201. Chigwada, G.; Jash, P.; Jiang, D. D.; Wilkie, C. A. Fire Retardancy of Vinyl Ester Nanocomposites: Synergy with Phosphorus-Based Fire Retardants. *Polym. Degrad. Stab.*, **2005**, *89* (1), 85–100. https://doi.org/10.1016/j.polymdegradstab.2005.01.005
202. Mishra, S.; Sonawane, S. H.; Singh, R. P.; Bendale, A.; Patil, K. Effect of Nano-$Mg(OH)_2$ on the Mechanical and Flame-Retarding Properties of Polypropylene Composites. *J. Appl. Polym. Sci.*, **2004**, *94* (1), 116–122. https://doi.org/10.1002/app.20750
203. Zhang, Q.; Tian, M.; Wu, Y.; Lin, G.; Zhang, L. Effect of Particle Size on the Properties of $Mg(OH)_2$-Filled Rubber Composites. *J. Appl. Polym. Sci.*, **2004**, *94* (6), 2341–2346. https://doi.org/10.1002/app.21037
204. Okoshi, M.; Nishizawa, H. Flame Retardancy of Nanocomposites. *Fire Mater.*, **2004**, *28* (6), 423–429. https://doi.org/10.1002/fam.842
205. Lan, T. Nanomer Nanoclay as Flame Retardation Additives. In *Environmentally Friendly Flame Retardants, Proceedings of the Intertech Pira Conference*, Baltimore, MD, March 28, 2024. https://www.researchgate.net/profile/Tie-Lan/publication/268363265_NanomerR_Nanoclay_as_Flame_Retardation_Additives/links/5aaf1f8ca6fdcc1bc0bc78d3/NanomerR-Nanoclay-as-Flame-Retardation-Additives.pdf (accessed July 31, 2023).

13 Bio-Based Flame Retardants

Xiu-Li Wang, Jing-Hong Liu, Yin-Long Wang, and Zheng-Ming Li
The Collaborative Innovation Center for Eco-Friendly and Fire-Safety Polymeric Materials (MoE), National Engineering Laboratory of Eco-Friendly Polymeric Materials (Sichuan), State, College of Chemistry, Sichuan University, Chengdu 610064, China

13.1 INTRODUCTION: WHY WE NEED BIO-BASED FLAME RETARDANTS

Flame retardants (FRs) can effectively suppress the fire hazards of polymeric materials via reducing ignitability, heat release, and overall combustion efficiency. By using FRs, the fire risk caused by the application of combustible polymeric materials has been greatly reduced. The most used FRs are mineral, halogenated, phosphorus-containing, nitrogen-containing, silicon-containing, and nanometric compounds. However, not all FRs are environmentally and human safe. Some commonly used efficient small molecule halogen-containing flame retardants have been banned by many countries and international organizations through laws and regulations (such as RoHS and WEEE instructions) because of their biological accumulation problems, which would cause harm to the ecological environment and human body. Therefore, researchers began to develop halogen-free flame retardants and macromolecular flame retardants that are not easy to bioaccumulate, and gradually replace the market occupied by the banned halogen-containing flame retardants.

Renewable resources are any type of resource that can be regenerated at a rate that is at least equal to the speed with which humanity can consume that resource. Today, some new development strategy of FRs using renewable resources (usually called as bio-based FRs) has received continuous attention due to their advantages of multiple sources, renewability, and the growing concerns of environmental issues linked to the increased use of fossil fuel feedstock. At the same time, the increased application of bio-based polymers in many technical fields also needs FRs fabricated by renewable resources to maintain their sustainability and good environmental impact. Bio-based resources for fabrication of FRs can be classified into polysaccharides-based products, protein, nucleic acid, and aromatic compounds, etc. In this chapter the biomass substances with high molecular weight such as polysaccharides-based products, protein, nucleic acid, lignin, DNA, alginates, et al. are ascribed as biomacromolecules. And those biomass substances with low molecular weight including phytic acid, cyclodextrins, tannin, vanillin, ferulic acid, cardanol, and others are called bio-based small molecule compounds.

13.2 BIOMACROMOLECULES AS FRS

Biomass refers to various organisms produced through photosynthesis, such as the atmosphere, water, and land, which are all living and growing organic substances collectively known as biomass. Biomass as the largest reservoir for renewable carbon mainly is composed of biomacromolecules. Most biomacromolecules such as chitosan, starch, cellulose, lignin, protein, etc. have high carbon contents and hydroxyl groups, and even nitrogen, can be used as carbonization agent or gas source in in intumescent flame retardant. Using these biomacromolecules alone as flame retardants, their flame-retardant efficiency is not sufficient. Usually, they are combined with other substances to achieve flame-retardant coating, flame-retardant additives, or intrinsic flame-retardant materials. In addition, introducing phosphorus, nitrogen, or other flame-retardant elements via chemical modification is also a widely used method.

13.2.1 CHITOSAN

Chitosan (CS), the only natural alkaline polysaccharide, can be obtained by deacetylation of chitin, which mainly exists in the shells of crustaceans, insects, and some fungi and is the second largest natural polymer after cellulose. Chitosan is a random copolymer of D-glucosamine and N-acetyl-D-glucosamine bonded by β-1–4 linkage. Its detailed chemical structure is shown in Figure 13.1. Because of rich sources and renewability, chitosan has been well investigated as a flame-retardant component. Since ring-opening reaction will occur at high temperature, self-agglutination will form aromatic ring cross-linking structure in the matrix, which is conducive to the inhibition of heat exchange in the matrix. At the same time, the amino group in chitosan is released into the gas phase in the form of NH_3 during the thermal decomposition process, which can dilute the concentration of combustible gas and promote the formation of expanded carbon layer.[1]

13.2.1.1 Chitosan-Based Coating

Because of the active groups in chitosan, chitosan can be modified to optimize its flame retardancy. When the pH of chitosan solution is less than 6, the amino group in its

DOI: 10.1201/9781003380689-13

structural unit is protonated to make the main chain of chitosan molecule positively charged so it can attract molecules with opposite charges through electrostatic interaction. With the development of layer-by-layer assembly (LbL) technology, more and more researchers are using chitosan to construct flame retardant coatings (see Chapter 14 of this book for more details). LbL is a simple, efficient, and universal surface modification technology and enables one to construct a multilayered coating on the surface of a substrate through the alternate deposition of oppositely charged polyelectrolytes or particles. The schematic diagram of the LbL preparation is shown in Figure 13.2. Usually, one bilayer (BL) is obtained by alternating the deposition of positively charged polyelectrolyte and negative polyelectrolyte. The electrostatic attraction between them is the most common and widely used interaction. In addition, other interactions, such as hydrogen bond, coordination bond, hydrophobic interaction, etc., are also feasible. Using the weak interaction between the molecules of each layer, spontaneous association between the layers has occurred and formed a molecular aggregate with stable performance, complete structure, controllable thickness and specific functions. These coatings with specific structures play an important role in delaying ignition and combustion of polymers by isolating the heat and volatile products derived from the thermal degradation of polymers. With the development of LbL technology, more and more biomacromolecules are used as flame-retardant components of coating; their poor thermal processing behavior can be neglected because only their solution is needed for constructing the layer.

Via this LbL technology, various flame-retardant systems based on chitosan have been prepared, include many inorganic particles such as clay,[2] titanate nanotube,[3] α-zirconium phosphate[4] (α-ZrP), graphene oxide,[5] metal hydroxides,[6] and hydroxyapatite[7] (HAP), etc. Chitosan can interact with ammonium polyphosphate (APP) to form polyelectrolyte complex, a classic intumescent flame-retardant (IFR) system, in which chitosan is an effective carbon source, APP molecule is as an acid source and gas source. This system can be used as an IFR coating for cotton fabric,[8] polyester,[9] polyamide,[10] polyurethane foam,[11] etc. The char yields of chitosan are higher than that of cellulose under the same pyrolytic conditions. The synergistic charring between chitosan and cellulose promotes substantial improvements in the char yields of cellulose fabrics. So chitosan usually is employed as a synergistic agent in the flame-retardant cotton fabrics. One bilayer (BL) is obtained by alternately depositing positively charged chitosan and negative APP. The phosphorus content increase with the bilayer number. For cotton fabric, 20 BL of chitosan-APP coating significantly reduces the burning time and effectively control the fire propagation with only 27 mm damaged length in the horizontal direction. The residual char of the coated fabric still perfectly retains the textile structure with considerable mechanical strength, due to the protection of the assembly coating.[8] The microscale combustibility test results show that the peak heat release rate and total heat release have been reduced compared to the uncoated material.

APP is not a biomass derived reagent but it has high phosphorus and nitrogen content. The other phosphorus-containing molecules, such as phytic acid (PA), sodium alginate,[12,13] et al. can also be used with chitosan to constitute the IFR system. Phytic acid, whose structure consists of six phosphate group, is the major storage form of phosphorus in cereal grains, beans, and oil seeds, and is environmentally friendly, biocompatible, as well as nontoxic.[14] It can interact with positively charged molecules (pKa 1.9–9.5),[15] illustrating it is a good candidate for LbL assembly. By changing pH values of the chitosan and PA solution, the coating thickness and the phosphorus-to-nitrogen ratio can be well controlled. Figure 13.3 shows images of the treated

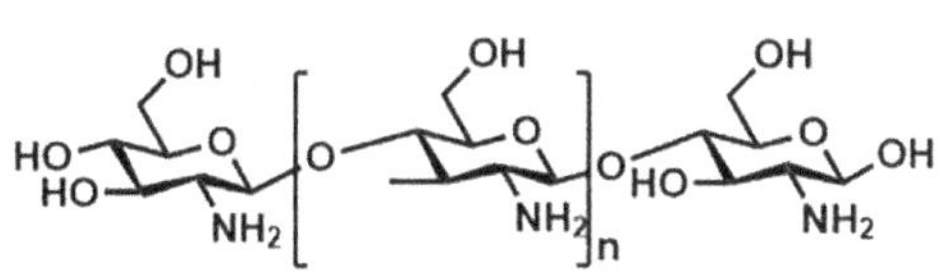

FIGURE 13.1 Chemical structure of chitosan.

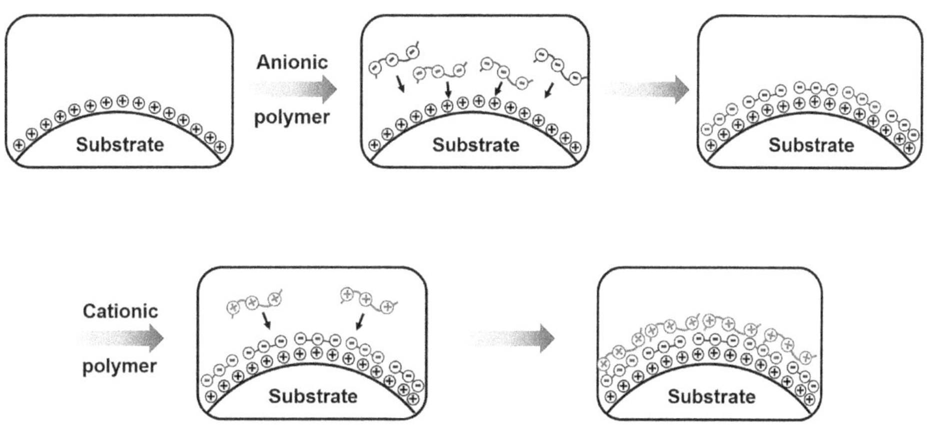

FIGURE 13.2 Schematic diagram of the LbL preparation.

fabric samples after a 12 s exposure to direct flame. It can be seen that uncoated control fabric is completely consumed during the test, while fabric coated with 30 BL of pH 5 coating completely stops the flame from propagating almost immediately after ignition and more than 90 wt% of the fabric is preserved. Fabric coated with pH 4 solutions also shows significant flame-retardant capacity, leaving about 80 wt% of the fabric, while adding only 16 wt% to the fabric weight before burning. Fabric coated with pH 6 solutions is burned much more extensively but still has some unburned white spots and a significant amount of residue left (42 wt%). When comparing the same coating weight (18 wt%), it is clear that higher PA content yields a more

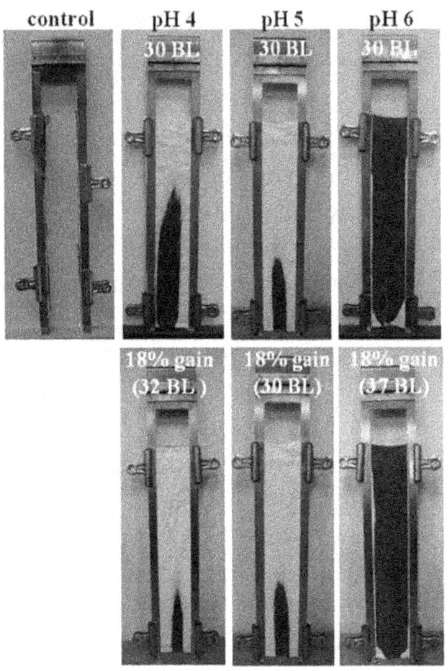

FIGURE 13.3 Images of uncoated control and fabrics coated with 30 BL of CH-PA (top) and 18 wt% CH-PA (bottom) deposited at varying pH level after 12 s exposure to direct flame.[16]

effective flame retardant. Micro-combustion calorimetry (MCC) shows that all coated fabric exhibits reduced total heat release by at least 70%.[16]

Durability must be addressed when LbL technology faces industrial application. Combining LbL with UV-curing process is a feasible method to improve durability of coating because highly cross-linked covalent networks can be formed through UV-curing. Cotton fabrics are pre-activated by a branched poly(ethylene imine) and then alternatively dipped in the UV-curable aliphatic acrylic polyurethane latex, APP emulsion (negative), and chitosan solution (positive).[17] The schematic preparation is shown in Figure 13.4. The process is repeated in order to deposit 3BL, and a durable flame-retardant coating has been generated. Even after washing in 65°C water for 1 h, the modified cotton fabric can still self-extinguish in the horizontal flame spread test. The flame-retardant fabric can withstand the attack of a 1 M solution of acetic acid or ammonia for 1 h without losing its original structure.

As one of the most available commercial lignins, lignosulfonate (LS) is a byproduct of sulfite pulping, and it has a hydrophobic backbone and hydrophilic branches (sulfonic, carboxyl, and phenolic hydroxyl groups). Because of the ionization of functional groups, LS has negative charges and exhibits polyelectrolyte behavior in aqueous solution. Therefore, it is possible to construct LbL multilayers with some cationic polyelectrolytes. The 8 BL CS/LS coating can endow polyurethane foam with flame retardancy, and the peak heat release rate (PHRR) is reduced 42% compared with untreated material, although the LOI value shows almost no change.[18] To further improve the flame retardancy of chiton/lignin coating, 10 and 20% of low-molecular-weight monoammonium phosphate (MAP) is added to the magnesium lignosulfonate solution. The coating comprising magnesium lignosulfonate, chitosan, and MAP is deposited on the cotton fabric using LbL assembly.

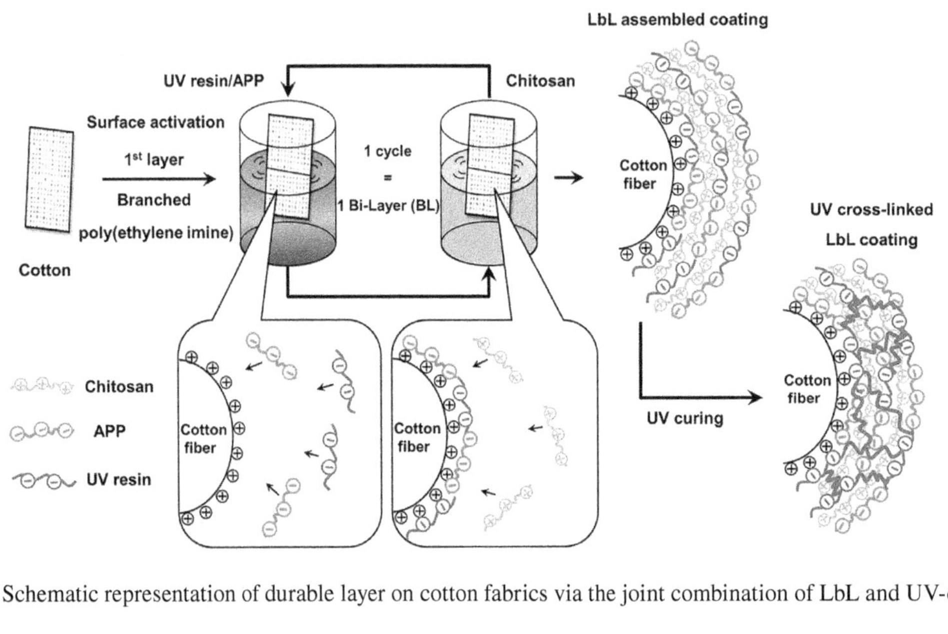

FIGURE 13.4 Schematic representation of durable layer on cotton fabrics via the joint combination of LbL and UV-curing.[17]

MAP is both a benign fertilizer and flame-retardant chemical, combining with lignin and chitosan, 5 bilayer coating (15.5 wt% add-on) imparts fire extinguishing behavior, and excellent UV and antimicrobial protection to the fabric.[19]

13.2.1.2 Chitosan-Based Flame-Retardant Additives

To further enhance the flame-retardant efficiency and widen its application, some chemical/physical modification especially phosphorylation has been carried out on the functional groups of chitosan. Considering the sustainability and good environmental impact, the demand for bio-based flame retardants is most urgent for bio-based polymers represented by poly(lactic acid) (PLA), which has been applied in textile industry, medical and agriculture field, food packaging, and disposable items. PLA is one of the most used bio-based polymers at present, and its physical properties are similar to that of petroleum-based plastics. Expanding the applications of PLA can reduce the use of non-degradable plastics and undoubtedly reduce the environmental pollution caused by plastics. But the inherent flammability and brittleness limit its further application in electronic and electrical appliances, vehicles, and rail transit materials.

Introducing chitosan together with APP into PLA by melt blending can give PLA good flame retardancy. A PLA sample containing 2% chitosan and 5% APP achieves the maximal LOI value of 33%, obtains the UL-94 V-0 rating, and decreases the peak heat release rate to 426 kW/m^2. It is found that chitosan acts as a carbon agent owning to its high content of carbon atoms and multi-hydroxyl groups, and the interaction between chitosan and APP provides synergistic effects in improving the flame retardancy of PLA.[20]

To further enhance the charring ability of chitosan, benzaldehyde, salicylaldehyde, and hydroxybenzaldehyde can be used to obtain derivatives.[21] The synthetic route for these derivatives is shown in Figure 13.5. These chitosan derivatives associated with APP are introduced into thermoplastic polyurethane (TPU) to prepare flame retardant TPU composites. The initial decomposition temperatures of theses CS derivatives are significantly increased to the range of 259–286°C, which are much higher than that of neat CS (222°C). The enhancement is due to the n-conjugated effects between C = N bond and benzene. Salicylaldehyde modified CS (SCS) presents the highest residue char amount among CS and the other derivatives. The possible reason for this phenomenon is that SCS contains o-hydroxyl, which can develop into o-quinone methide (o-QM) and resorcinol during the degradation processing. The sample containing 18.8% APP and 6.3% SCS achieves the maximal LOI of 30%, obtains the V-0 rating, and significantly decreases the PHRR from 841 kW/m^2 of neat TPU to 206 kW/m^2. SCS/APP promotes char formation to form more intact intumescent char structure in these CS derivatives-APP systems. It is proposed that the synergism between SCS and APP is responsible for the improvement of flame retardancy.

Via LbL technology, many chitosan derivatives can be prepared and used as flame retardant additives for PLA. Chitosan with opposite charge is coated on the surface of APP, and then a layer of sodium phytate (PA-Na) is also coated to fabricate flame retardant APP@CS@PA-Na. Introducing 10 wt% APP@CS @PA-Na into PLA, the material obtains UL 94 V0 grade, and LOI is enhanced from 20% of PLA to 31%. Its elongation at break is increased 28% compared with pure PLA.[22] Replacing PA-Na with sodium alginate, APP@CS@AA-3BL can be obtained.[23] The inclusion of 10 wt% APP@CS@AA-3BL gives rise to ~23% increase in the impact strength of PLA possibly because of interfacial hydrogen-bonding interactions. The PLA containing 10 wt% APP@CS@AA-3BL shows the highest LOI value (31%) and achieves a UL94 V-0 rating in the vertical burning test. Meanwhile, the cone calorimetry results demonstrate that the PHRR and total heat release (THR) are respectively decreased by 23% and 11% relative to the PLA bulk.

If carboxylated silicone oil (Si-COOH/Si) is used with APP and chitosan, a novel core-shell flame-retardant APP@CS@Si can be prepared in water, in which Si-COOH/Si is coated on the outside of APP and CS.[24] The results show that Si-COOH plays three roles, a water-resistant agent of the flame retardant, toughening agent of PLA, and the modifier of the char layer. Only 2 wt% APP@CS@Si can endow PLA with satisfying flame retardancy such as the UL-94 V-0 grade. Higher values of the fire propagation index (0.15) and flame retardancy index (1.33) can be also achieved by PLA with 4 wt% APP@CS@Si compared with those of

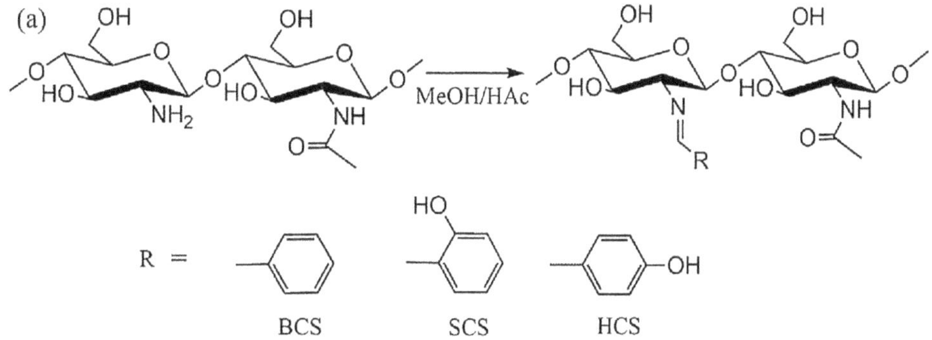

FIGURE 13.5 Synthetic route for different chitosan derivatives.[21]

PLA with the same loading of APP@CS. Interestingly, the total smoke release amount (TSR) of PLA/4APP@CS@Si is 60% lower than that of PLA/4APP@CS, exhibiting its excellent smoke suppression ability. This can be due to the promotion of Si-COOH for the carbonization process of APP@CS and PLA, a cross-linked and more compact char layer has formed with more P and Si contents. Si-COOH as a soft shell improves the toughness of PLA with increased impact strength and elongation at break of the PLA composite by 20 and 16%, respectively.

Both hydroxyl and amino groups of chitosan can undergo phosphoric acid modification to the preparation of phosphorus-containing bio-based flame retardants. Two novel CS-based flame retardants, phenylphosphorylated CS (PhPCS) and phenylphosphoramidated CS (PhPNCS), can be synthesized via the chemical modification of CS with phenylphosphoryl dichloride or tetraethylenepentamine, respectively.[25] Their synthetic routes are shown in Figure 13.6. Both of them exhibit excellent flame retardancy efficiency for PLA. The LOI values of the PLA/3 wt% PhPCS and PLA/3 wt% PhPNCS bio-composites increase to 29% and 27%, respectively, and they both achieve V-0 rating during the UL-94 vertical combustion test. Because of more CH_2 groups in PhPNCS, it shows lower flame-retardant efficiency than PhPCS. Owing to the incorporation of only several percent of PhPCS or PhPNCS, the formed char residue during the combustion is less and not dense enough to completely inhibit the melt-dripping of the PLA. The pHRR value of PLA/3 wt% PhPCS decreases from 472 kW/m^2 to 424 kW/m^2 indicating that the flame retardancy of the PLA is enhanced. The PHRR and THR values of PLA/3 wt% PhPNCS and PLA/5 wt% PhPNCS are slightly higher than those of the corresponding PLA/PhPCS, which agrees with the LOI results. PhPNCS decomposes to release more inert NH_3 in the gaseous phase than PhPCS, which dilutes the concentration of gases and inhibits the further decomposition of PLA. Therefore, the TSP values of PLA/PhPNCS is lower than those of PLA/PhPCS. The highly efficient flame retardancy of PhPCS and PhPNCS is attributed to the inhibition of the PO radical in the gas phase, and the synergistic effect in the condensed phase is an auxiliary flame-retardant mechanism. The mechanical properties of the PLA/PhPNCS are better than those of the PLA/PhPCS. The bending and tensile strengths of the PLA/3 wt% PhPNCS increase to 99 and 57 MPa, which are 28% and 11% higher than those of the PLA/3 wt% PhPCS, respectively. This finding is ascribed to the enhanced interfacial compatibility and even dispersion of PhPNCS nanoparticles in the PLA matrix.

For enhancing char-forming ability of phosphorylated chitosan (CS-P), some metal salts that can form ionic bonds are introduced. Cobalt, a typical transition metal element, can not only form ionic bonding with CS-P but also catalyze the char formation of polymer matrices, which is beneficial to the flame retardancy of polymer. CS-P-Co achieves a high char layer of 63 wt% with the incorporation of Co^{2+}, which may be attributed to the strong coordinative interaction between hydroxyl/amino groups in the CS-P and Co^{2+} cations, and the catalysis effect of cobalt ion on the degradation of CS. The PLA/CS-P-Co composites with small amounts of CS-P-Co ($\leq$ 4.0 wt%) show remarkably improved flame retardancy, such as decreased peak heat release rate and total heat release by 23% and 20%, respectively.[26]

13.2.2 Starch

Starch is one of the most common stored polysaccharides in plants, which is composed of amylase (chemical structure is shown in Figure 13.7), a linear polymer made of α-(1–4) glucose units, and amylopectin (chemical structure

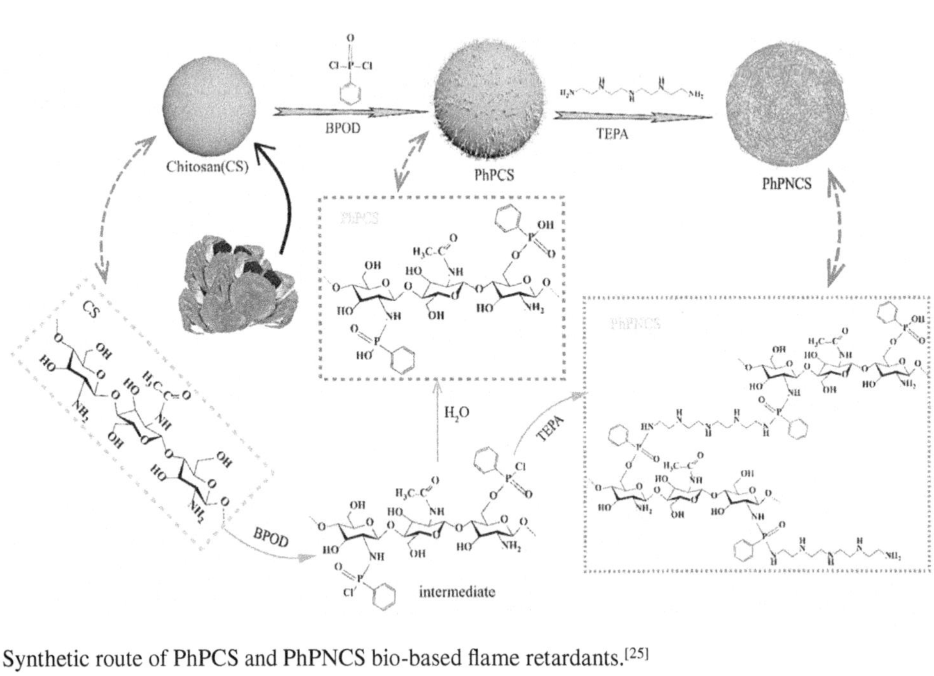

FIGURE 13.6 Synthetic route of PhPCS and PhPNCS bio-based flame retardants.[25]

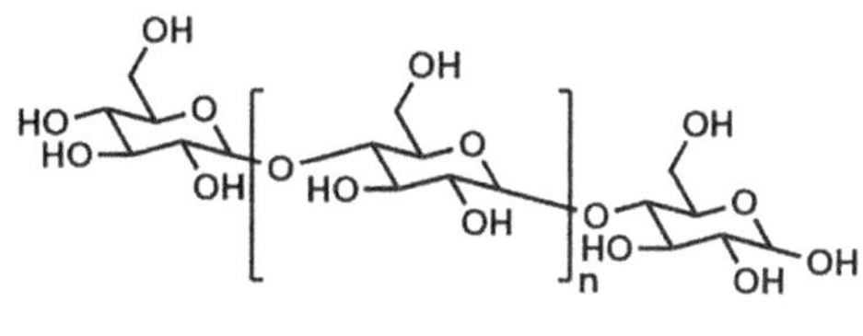

FIGURE 13.7 Chemical structure of amylose.

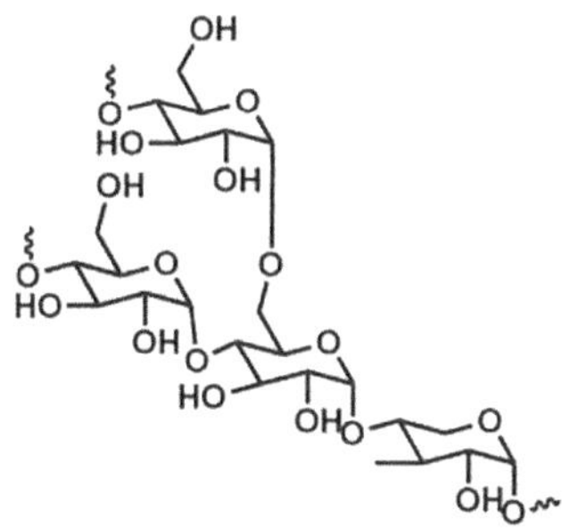

FIGURE 13.8 Chemical structure of amylopectin.

is shown in Figure 13.8), a highly branched polymer containing short glucose segments linked to the main glucose chain via α-(1–6) connections. The thermal degradation of starch occurs in three stages, including dehydration related to the release of water, formation of C = C bonds and carbonization reactions to form large conjugated aromatic structures.[27] Therefore, starch can be used as an intumescent flame retardant due to its good char forming ability.

It has been found that when 1–5 wt% starch is combined with phosphorus-containing commercial flame retardant in PLA foam, the LOI of PLA/FR/starch foam is raised to 31%, and melt-dripping is eliminated; starch acts as a natural charring agent. Compared to PLA composites, these PLA composites foams show significantly reduced LOI values, due to the volume expansion and the increased interface defects in the polymer matrix.[28] With starch as a carbon source and APP as an acid source for the IFR, the PHRR of PLA has a 66% reduction and leaving a residual mass of 43% (no residual mass left for pure PLA), UL 94 rating is V-0, at the same time the LOI is raised to 37%. From these results, we can make a conclusion that when starch shows great potential as a renewable carbonization agent.[29]

It is well-known that the thermoplastic processing of starch is poor because of the presence of intermolecular and intramolecular hydrogen bonds. Using plasticizers such as glycol, water, or urea that weaken the hydrogen bonds, the thermoplastic processing of starch can be enhanced, however, the addition of these plasticizers will lower the flame-retardant efficiency. Replacing starch with oxidized starch (OS) that is obtained by starch oxidation in the presence of oxidizing agents such as sodium hypochlorite, hydrogen peroxide, and ammonium persulfate can resolve the above problems. When OS is melt-blended with PLA, and then compounded with APP, the PLA composites can be melt-spun to multifilament fibers. Besides, the needle-punched non-woven fabrics from as prepared melt-spun PLA/OS/APP fibers (containing 7 wt% OS) show good fire resistance: PHRR and THR are 52% and 30% lower than those of pure PLA samples, respectively. For those with higher 5 wt% or 7 wt% OS, their effective heat combustion and total smoke production are significantly reduced.[30]

The coating built by LbL usually needs 20–40 deposition steps to achieve significant flame-retardant performances. If cationic starch is coupled with poly(phosphoric acid) (PPA), the flame-retardant efficiency of LbL coating is significantly raised. Figure 13.9 show the schematic process for cotton fabrics pre-activated by poly(acrylic acid) and then alternatively dipped in the starch (positive) and PAA (negative) solutions. Only 2 BL is needed to confer seld-extinguishability on the cotton fabric. Cone calorimetry results illustrate all treated cottons show significant reductions (up to 40%) in the total heat released during combustion, thus demonstrating the high efficiency achieved. It has found that for starch, its ability of forming thermally stable char is further improved by the dehydration ability of PPA.[31]

To achieve better flame retardancy, phosphorus can be introduced into the structure of starch via reacting with its hydroxyl group. Using starch, phosphoric acid and urea as raw materials, ammonium phosphate starch carbamates (APSC) are synthesized and coated on the surface of expandable polystyrene (EPS) particles.[32] The addition of 47 wt% of APSC on EPS enhances the LOI from 18% to 35% with V-0 rating in UL-94 and decreases the PHRR from 666 kW/m^2 of EPS to 316 kW/m^2. Moreover, the total smoke production (TSP) is also sharply decreased from 70 m^2 of EPS to 17 m^2. The decomposition of APSC can volatize nonflammable gases, like H_2O and NH_3, to dilute the oxygen as well as to expand the volume of the char barrier. During combustion, the intumescent char layer formed by APSC acts as a barrier to prevent the transfer of heat, oxygen, and flammable products. In this way, this bio-based flame-retardant coating enhances the flame retardancy of ESP.

If starch is modified by phytic acid via one-step esterification (FRS), this derivative also can endow EPS with excellent flame retardancy when it is used as a coating and adhesive for EPS. FRS coating has open pores, in combination with the closed cells formed by EPS beads, a hierarchically porous structure in FRS-EPS has created. The modified EPS has a superior thermal insulation with a lower thermal conductivity of 27 mW·(m·K)$^{-1}$. After being burnt with an alcohol lamp for 30 minutes, the top side temperature of the FRS-EPS remained at only 140°C with ignition, thereby exhibiting excellent fire resistance. Besides this, FRS-EPS also exhibits self-extinguishing behavior in vertical burning tests and has a high LOI value of 36%. A compact phosphorus-rich hybrid barrier brought by FRS coating, and the phosphorus-containing compounds that formed in the gas phase, contribute to the excellent flame retardancy and smoke suppression of FRS–EPS.[33]

As one of the lightest solid materials, aerogel has lower density and thermal conductivity than traditional foam materials. It is a "super" thermal insulation material emerging in recent years. Since inorganic aerogels, such

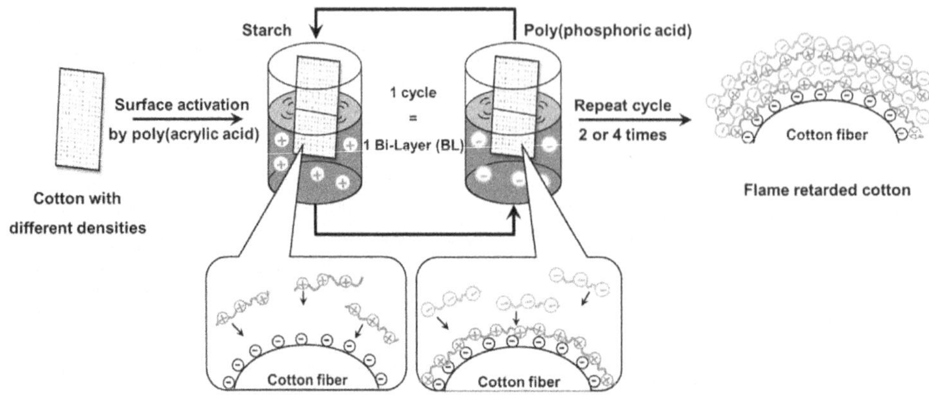

FIGURE 13.9 Schematic process for cotton fabrics pre-activated by poly(acrylic acid) and then alternatively dipped in the starch (positive) and PAA (negative) solutions. The process is repeated two or four times in order to deposit two or four bilayers (BL), respectively.[31]

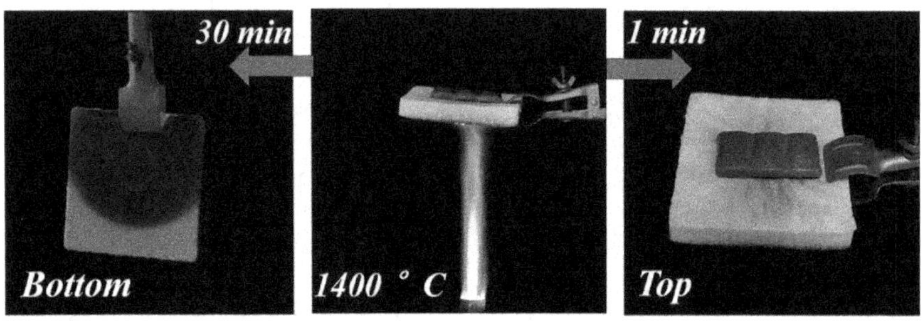

FIGURE 13.10 Optical images of $MMT_{3.37}\%/CAP_{0.63}$ aerogel under the fire of a butane blowtorch.[34]

as silica-based aerogels, are brittle, organic polymers are introduced in order to improve the toughness of the materials, which bring fire safety hazards. The preparation of aerogel usually consists of sol gel process and supercritical or freeze-drying process. Because some biomacromolecules are water soluble, they can be used to fabricate aerogel with flame retardance. Amylopectin can be chemically modified with cations (CAP) and then used as a binder to cross-link with montmorillonite (MMT), with negative charges, in water. The resultant aerogels with special microstructure of "brick-mortar-bridges" have low density (25 mg/cm^3), practical mechanical strengths (10–40 kNm/kg), low thermal conductivity, high-temperature insulation, and heat resistance up to 1,400°C. Figure 13.10 shows the chocolate on the top of the aerogel does not melt under the fire of a butane blowtorch (1,400°C) in one minute. All MMT/CAP aerogels are inherently nonflammable materials with an extremely high LOI value (> 60%) and with the highest rate (V-0) in UL-94 tests, which can be due to the low content of organic component, carbon source nature of starch, and protection and catalysis capability of clay.[34]

13.2.2 Cellulose

Cellulose is the most widely distributed and abundant polysaccharide in nature. Plants on Earth produce up to 7 billion tons of cellulose annually, accounting for over 50% of the carbon content in the plant kingdom. Cellulose is composed of β-D-glucopyranosyl with the degree of polymerization from 100 to 40,000, depending on both its botanical origin and the different treatments used for its extraction. Its chemical structure is shown in Figure 13.11. Cellulose chains are arranged in a tridimensional structure that is stabilized by hydrogen bonds, van der Waals bonds and hydrophobic interactions. The organization of cellulose chains results in the formation of microfibrils that are grouped together to form cellulose fibers (shown in Figure 13.12) containing both crystalline and amorphous regions.[27]

During thermal decomposition, cellulose can produce charred structures under some conditions. The amount of char formed is affected by the thermal degradation conditions as well as by the surrounding components during cellulose pyrolysis. In general, cellulose-based materials, such as cotton or wood, have always been considered to be materials with a high risk of ignition and fire propagation. Fireproof cellulosic materials can be prepared by enhancing their char forming ability by combining them with flame-retardant agents or chemically modifying them. Cellulose has the advantage of being easily functionalized because of its many hydroxyl groups. For example, phosphorylated cellulose can be obtained by the reaction of cellulose with phosphorus acid in molten urea. The phosphorylated cellulose and chitosan also can constitute fully bio-based IFR coating via LbL technology, in which chitosan acts as both carbon source and foaming agent, and phosphorylated cellulose as an acidic source. In the vertical flame test, the cotton fabric with 20 bilayers at the higher phosphorylated cellulose concentration (2 wt%) can extinguish the flame, and the char length is 11 cm.[35]

Using cellulose fibers with surface absorbed phosphorus-containing compound not only improve the flame retardancy of PLA but also enhance its mechanical properties. When a liquid FR resorcinol bis(diphenyl phosphate) (RDP) is introduced onto the cellulose fibers (CF), and then melt blended with PLA, the composite can self-extinguish in less than 2 s, easily obtaining the UL-94 V-0 criteria with LOI of 28%. The coexistence of RDP and CF has better flame-retardant effect than that of CF alone, indicating a synergistic effect. The dehydration process of CF-RDP, which is associated with the release of water, adsorption of heat, and enhancing PLA to form char, leads to an overall decrease in the HRR and an increase in LOI of the composite. When the CF is coated with RDP, they interact strongly with the PLA matrix, and this interaction reinforces the interface against crack propagation, leading to the enhancement both of impact strength (from 30 J/m in pure PLA to 35 J/m) and tensile strength of PLA (69 MPa *vs* 85 MPa).[36]

If phosphorus-nitrogen compounds are chemically grafted on the surface of cellulose nanofiber (CNF), the

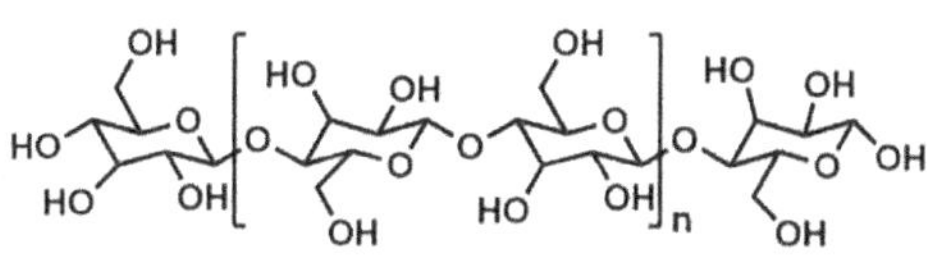

FIGURE 13.11 Chemical structure of cellulose.

flame-retardant efficiency and mechanical properties increase can be higher. Using CNF as a reinforcing template, via a three step chemical reaction, the flame-retardant shell is generated on the surface of CNF (PN-FR@CNF). And the detailed synthetic route is shown in Figure 13.13. Adding 10 wt% PN-FR@CNF enables PLA to achieve a V-0 flame resistance rating during vertical burning tests and to exhibit a dramatically reduced PHRR in cone calorimetry, indicating a significantly reduced flammability. Addition of 5 wt% PN-FR@CNF increases the impact strength from 8 kJ/m^2 for the PLA to 10 kJ/m^2, about a 28% increase, whereas the same loading level of CNF only leads to a negligible improvement (8 kJ/m^2). This illustrates 5 wt% PN-FR@CNF has unform dispersion and strong interface adhesion with PLA matrix, which facilitates the dissipation of impact energy. In addition, the tensile strength of PLA also increases by around 24% (about 72 MPa).[37]

To simplify the phosphating process, APP and CNF can be added together in water to prepare a mixture, then after ball-milling and filtration, the APP@CNF multifunctional additive is obtained. Due to the improved dispersion of APP@CNF and its smaller size within the PLA matrix along with their synergistic effect between APP and CNF, the flame retardancy and mechanical strength performance of PLA composite has been enhanced. The impact strength of PLA containing 5 wt% APP@CNF is significantly improved from 8 to 12 kJ/m^2 (increase of 54%), except the tensile strength has a slight reduction (reduced to 50 MPa). Only 5 wt% addition can give PLA a V-0 rating, illustrating a high flame-retardant efficiency for PLA.[38] Comparably, 10 wt% PN-FR@CNF is still required to enable PLA to achieve a V-0 rating.

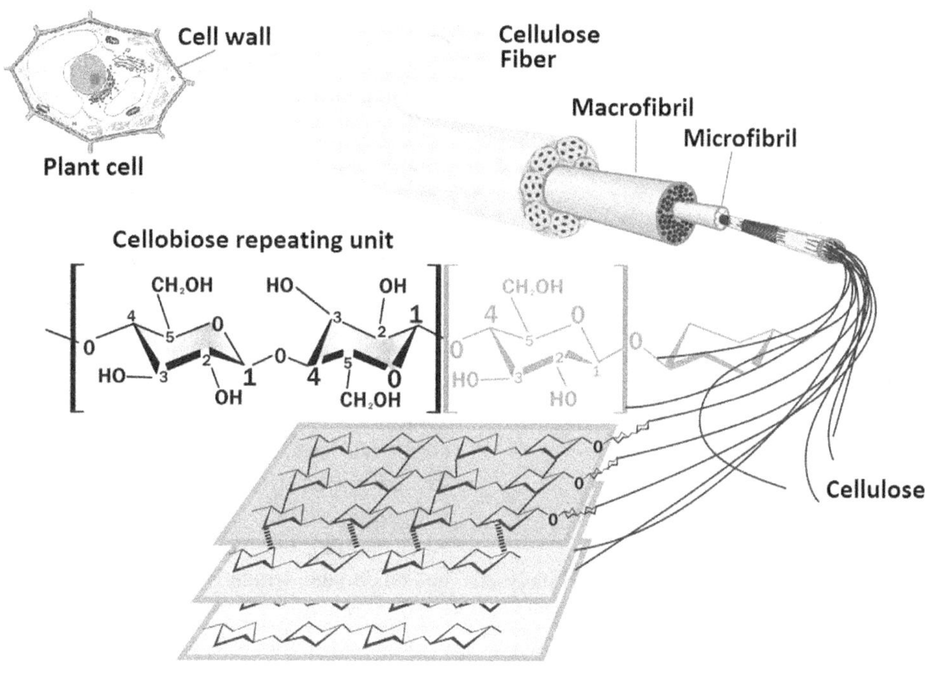

FIGURE 13.12 Representation of a cellulose fiber showing its organization.[27]

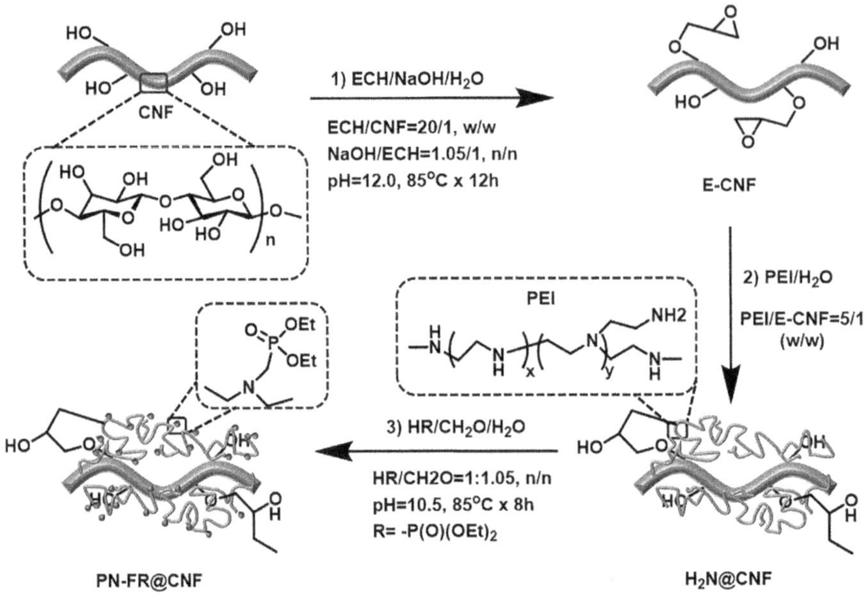

FIGURE 13.13 Synthetic route to core-shell nanofibrous flame retardants, PN-FR@CNF.[36]

Treating cellulose with a strong acid can result in the formation of cellulose nanocrystals. The highly cellulose crystalline regions can be extracted since acids can easily penetrate into the amorphous regions having a low level of order and hydrolyze them. Control of the acid strength, the temperature, and the time of hydrolysis are important for the production of micro or nano crystalline cellulose. Both microcrystalline cellulose (MCC) and nanocrystalline cellulose (NCC)can be obtained by the hydrolysis of cellulose by inorganic acid. When they are used together with aluminum phytate (Al-Phyt), they show different char-forming ability. The combination of MCC and Al-Phyt does not influence the pHRR value of the PLA, however, the combination of NCC and Al-Phyt reduce the pHRR value significantly (from 390 kW/m^2 to 240 kW/m^2, about 40%). The superior flame-retardant effect when NCC particles are used instead of MCC is attributed to their nanometric size, which provides them with high specific surface area and increases the interfacial interaction between cellulose and Al-Phyt. Because the crystalline structure and the nanometric size of NCC cannot be preserved during this phosphorylation process, only phosphorylated MCC has been prepared. When 20 wt% phosphorylated MCC is incorporated into PLA, PLA can reach V-0 grade in vertical burning, but the PHRR reduction is not obvious.[39]

13.2.3 Protein

Since proteins contain nitrogen, phosphorus, and sulfur, they can be used as an FR. Among the proteins derived from animal or microbial sources, caseins, hydrophobins, and whey protein et al. exhibit a great potential as flame retardant systems for textiles or thermoplastic bio-polymers.

Casein is the major milk protein (80%) and is obtained as a coproduct during the production of skim milk; it is made up of many components, among which αsl-casein, αs2-casein, β-casein, and k-casein are prevalent. αs-Caseins are the major casein proteins containing 8–10 seryl-phosphate groups, and β-casein contains about 5 phosphoserine residues.[40] Hydrophobins are produced by filamentous fungi and consist of a large family of small cysteine-rich proteins with a molecular mass of ca.10 kDa.[41] The eight cysteine residues form four non-sequential disulphide bonds stabilizing the tertiary structure of hydrophotins. When casein and hydrophobins are used as coatings for cotton, both phosphate groups of caseins and the disulphide bonds of hydrophobins lead the cellulose pyrolysis towards the formation of char, instead of the production of combustible volatile species. Thus, they prevent the combustion of cotton or at least delay the spread of fire after ignition. When approximately 20% caseins or hydrophobins are added, a significant reduction of the PHRR of cotton (~ 27 and ~ 45% for caseins and hydrophobins, respectively) is obtained.[42] When 20 wt% casein is coated on the surface of polyester (PET)-cotton fabrics, dripping cannot be totally inhibited, but a longer burning time is obtained. And the LOI value only increases slowly from 19% of pure polyester-cotton fabrics to 21%. In the cone (35 kW/m^2), the PHRR of treated polyester-cotton fabrics is reduced 15% compared with untreated one. Unfortunately, for pure polyester fabric, when casein is used as a flame-retardant coating, the PHRR has no change, which is due to the poor char-forming effect for pure polyester.[43]

Whey proteins also have the same function, whether its secondary and tertiary structure is folded or unfolded/denatured. The coating of whey proteins extends the burning time of cotton in the presence of a small flame or afterglow, together with a reduced burning rate. The high char residues may partially protect the cotton because they are able to hinder the oxygen diffusion and absorb the heat during the combustion.[44]

To enhance the durability of coating, UV-curable acrylic resin also can be combined with caseins. The coating containing the 15 wt% casein shows a decrease of PHRR and an increase of the time to peak as compared to the unfilled UV-cured counterparts. And the UV-curable coatings containing casein show good washing fastness. In fact, when the coated ethylene vinyl acetate copolymers (EVA) plates are subjected to washing cycles (i.e., operating in water at 60°C in the presence of a detergent for 1 h), their fire performances under the cone calorimeter do not change at all.[45]

Casein can be incorporated into PLA matrix by melt compounding in order to improve the fire resistance and sustain the biodegradable character of PLA simultaneously. The addition of 20% casein increases the LOI value of PLA composite (PLA-20) from 20% to 32%, upgrades the UL-94 from no rating to V-0, and decreases the PHRR from 779 to 639 kW/m^2. It is found that casein has an effect in both gas phase, by releasing nonflammable gases (such as NH_3 and H_2O), and condensed phase, by the formation of protective char layers. But the introduction of casein will induce an unavoidable deterioration in the mechanical performance of the composite. Both the tensile strength and elongation at break are reduced with the increase of casein. For PLA-20 casein composites, the tensile strength is reduced to 41.3 MPa. However, it is still higher than that of many widely used polymers such as polypropylene (35.3 MPa) and poly(butylene succinate).[46]

13.2.4 Lignin

Lignin, the second most available polymer next to cellulose, is a polyaromatic based amorphous polymer, which is normally available in large quantity as a byproduct of wood pulp. Lignin exists in plant cell walls and constitutes the basic skeleton of plants together with cellulose and hemicellulose. In light of lignin's merits of renewability, abundance and low cost, researchers have recently shifted their interest to the development of lignin-derived flame retardants due to its high thermal stability and char-forming capability.

Industrial lignin is normally isolated by two main approaches, namely, sulfur and sulfur-free processes. Both lignosulfonate lignin (LL) and Kraft lignin (KL) are usually obtained by sulfur processes of extraction with HSO^{3-} at 140°C and pH = 1–2, and NaOH with Na_2S at 170°C and pH = 13, respectively. In comparison, organosolv lignin (OL) and soda lignin (SL) are produced by a sulfur-free process of extraction with solvent pulping in the range of 150–200°C, and soda pulping in the range of 150–170°C at a pH value of 11–13, respectively.[47]

Compared with traditional carbon source pentaerythritol (PER) in IFR system, lignin shows higher efficiency because of the richness of phenylpropane repeat units, aromatic and aliphatic hydroxyl groups. When 3 wt% Kraft lignin and 20 wt% APP are used to constitute IFR and used in PLA, the LOI value of PLA composite is raised from 20% to 37 with UL 94 V-0 grade. Besides, the PHRR of PLA/APP20/KL3 is reduced to 250 kW/m^2, which is 50% less than pure PLA. For IFR consisting of APP and PER, using the same amount, the LOI value of PLA/APP20/PER3 is 34% with UL 94 V-1, and PHRR is only lowered to 310 kW/m^2. The combined effect of APP and KL make the melt viscosity of PLA composite increase which improves the swelling of the substrate, which generates more compact char structure, thereby hindering the free escape of pyrolysis gases, hence the ignition time is prolonged (increased from 63 s of pure PLA to 79 s~ 81 s).[48]

The flame-retardant efficiency of lignin has relationship with their extraction modes. For example, when Kraft lignin (obtained during the paper making process thanks to a mixture composed of sodium hydroxide and sodium sulfide), organosolv lignin (separated from other plant constituents by selective solubilization in organic solvents) are used in PLA matrix, they have different thermal stability and fire resistance. Due to higher carboxylic acid content in organosolv lignin, the 10% weight loss for PLA/organosolv lignin is lower than that of PLA/Kraft lignin. As far as PHRR and THR of PLA composites are concerned, a decrease of about 21% and 23% for PLA/20 kraft lignin, while these reduction values are 33% and 30% for PLA/20 organosolv lignin although both of them can make PLA form an insulating char layer at the surface during combustion. Unfortunately, TTI and the thermal stability of PLA are also decreased. To resolve this problem, these two kinds of lignin (Kraft lignin and organosolv lignin) are chemically modified by grafting phosphorus and nitrogen, and the synthetic route for lignin-$PONH_4$ is shown in Figure 13.14. After modification, they are proved to be highly effective to reduce the flammability of PLA composites. PLA/modified lignin composites achieve V0 in UL-94. The TTI and thermal stability of PLA are also maintained. Once modified, the effect of both lignins becomes similar, and neither the plant origin nor the

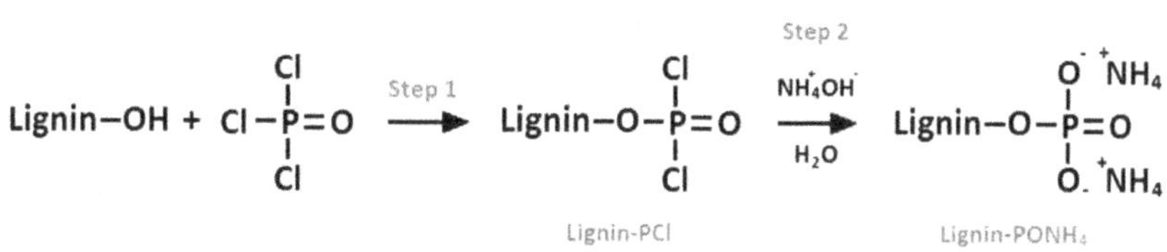

FIGURE 13.14 Two-step chemical modification of lignin.[49]

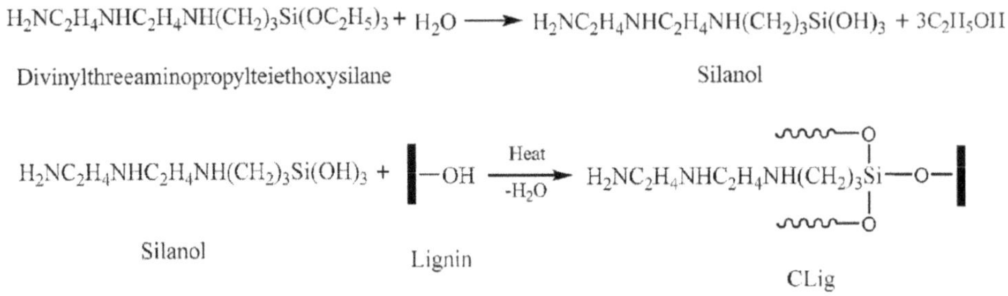

FIGURE 13.15 Synthetic route for CLig.[50]

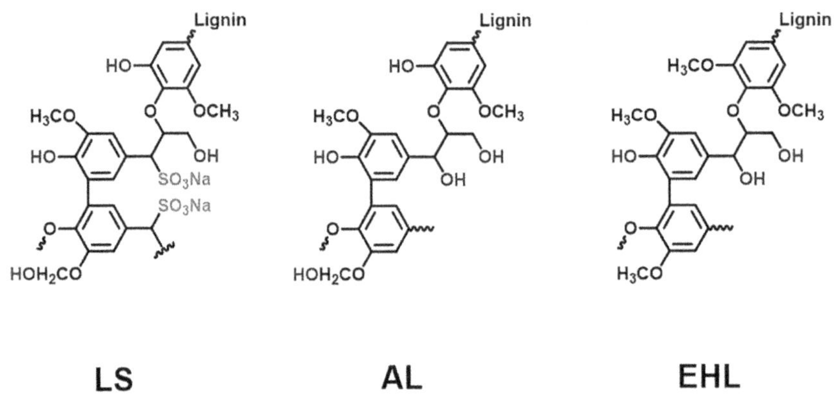

FIGURE 13.16 The chemical structure of three types of lignin.[53]

extraction process of the lignins affect the properties of the PLA composites.[49]

Lignin has a wide range of functional groups, including hydroxyl, phenyl, methoxy, and carbonyl groups, so it can be chemically modified and some flame-retardant elements such as phosphorus, nitrogen, and silicon and so on can be introduced, just as with Lignin-PONH$_4$. If lignin is reacted with divinyl-3-aminopropyltriethoxysilane (B201) CLig that simultaneously contains silicon and nitrogen has been obtained, the synthetic route is shown in Figure 13.15. When CLig combined with APP (mass ratio of 1:4) are used in PLA substrate, PLA obtains UL-94 V-0 rating, and its LOI value reaches 31%. Compared with PLA/APP/Lignin, the PHRR and THR of PLA/APP/CLig are largely reduced, and their char residues are increased. The synergistic effect between APP and CLig leads to thermal stability at high temperature and forms a continuous, compact, and intumescent char layer on the surface of materials during combustion. The tensile strength of PLA/APP/CLi is higher than that of PLA/APP/Lignin, which may be ascribed to the weaker intra or intermolecular hydrogen bond effects among the lignin caused by modification by B201 and the improvement of lignin's dispersion in PLA matrix.[50]

Compared to pure lignin, which is difficult to process and dissolve, lignin sulfonates have good water solubility. When sodium lignin sulfonate (SLS) is used as a flame-retardant finishing agent on cotton fabrics via a hot padding, 30% add-on can endow cotton fabric with good flame retardancy (LOI value of 29% with a minimum char length of 4 cm and a minimum afterglow for 9 s, while the control burns out with flame and afterglow within 1 min).[51]

Meanwhile, lignin-based flame-retardant coatings also can be used on various polymeric foams and paper.[52] Usually, when lignin is used as a flame retardant, it can change the thermal degradation and burning behavior of polymeric materials in the condensed phase (char forming) and gas phase by radical quenching (phenolic groups). Besides, lignin also can release inert gases to dilute the oxygen content to obstruct burning. It is found that the types of lignin have great effect on the flame retardancy. When alkaline lignin (AL), sodium lignosulfonate (LS), and enzymatic hydrolysis lignin (EHL), whose chemical structure is shown in Figure 13.16, are used alone for PU foam, AL (242% add-on) has the best performance in the horizontal burning test, and LOI of modified PU foam is increased to 28%. For the foam coated with EHL, even 277% uptake just reach 24% in LOI. LS also has a positive effect on the flame retardance of PU foam, but the discontinuous char is not as effective as the char formed by AL, leaving some positions for attack by fire and heat. Although EHL has the highest carbon percentage among the three types of lignin, the components of cellulose and hemicellulose block the flame retardancy of EHL.[53]

13.2.5 DNA

Deoxyribonucleic acid (DNA) is a molecule found in all living organisms that carries genetic information. Each molecule of DNA is formed by dehydration and condensation of a nitrogen-containing base, a phosphate, and a deoxyribose. The four bases that make up DNA are adenine (A), cytosine (C), guanine (G), and thymine (T). The double helix of DNA is stabilized by hydrogen bonds established between nitrogen-containing bases present on two strands. The chemical structure of DNA is exhibited in Figure 13.17. DNA possesses a special structure: phosphate group, deoxyribose unit, and nitrogen-containing bases making itself can form an intrinsically IFR system. As the temperature, increasing the phosphate group produces phosphoric acid, constituting the acid source. The deoxyribose unit is decomposed by a self-crosslinking process as a carbonizing agent to form an aromatic structure, and the nitrogenous base is released as a foaming agent ammonia.[54] Therefore, DNA can be used as IFR coating for polymers.[55]

To further enhance the efficiency, DNA and chitosan are used together to fabricate flame-retardant and green coating for cotton. A 20 BL treated cotton self-extinguishes in a shorter time (burning time from 104 s of untreated one to 56 s), and LOI value increases to 24%. Besides, the PHRR and THR values have 40% and 30% reduction in cone calorimetry. This decrease is higher than that of the chitosan/APP system. It is found that phosphoric acid released from DNA can catalyze the dehydration of cellulose, similarly to APP. P-N synergism in DNA is of benefit by forming a thermally stable aromatic structure in a more efficient way.[56]

When DNA, silver nitrate, and octadecyltriethoxysilane (ODTS) are used together, the treated cotton not only has self-extinguish ability but also shows superhydrophobicity.[57] For enhancing the washing fastness, durability, and thermal stability of DNA coating, DNA solution can be mixed with the opposite charged TiO_2 nanoparticles to form TiO_2@DNA systems. And this hybrid coating shows an improved washing fastness due to the cross-linking action of TiO_2 phase that strongly interacts both with DNA and cotton substrate. The nano-TiO_2@DNA coatings show very good flame-retardant properties, with a decrease of PHRR, THR, and the burning rate (even achieving self-extinction), and an increased final residue.[58]

DNA on the surface is more efficient than DNA in the bulk to reduce the total heat release during the combustion of ethylene vinyl acetate copolymers (EVA). As far as smoke is concerned, TSR does not change in a remarkable way, regardless of the presence of DNA. The rate of smoke release (RSR) as well as CO and CO_2 yields are lowered when DNA is present on EVA surface. The final residue turned out to be around 5%, regardless of the presence of the DNA in bulk or on the copolymer surface. It is worth noting that the DNA on surface can exhibit exceptional thermal shielding properties as it is capable of protecting the underlying material from a butane/propane torch during three consecutive applications for 5s.[59]

13.2.6 Alginates

Alginates refer to the derivatives of alginic acid and alginic acid itself. Alginic acid, a natural and renewable raw material, is extracted from marine plants such as brown algae, kelp, etc. It is a random copolymer of β-D-mannuronic acid (M) and α-L-guluronic acid (G), repeating units bonded by a β-1–4 linkage with varying proportions (shown in Figure 13.18). Just like most polysaccharides, alginic acid fibers without metal ions are flammable with an LOI value of about 24%. However, alginates can present inherent flame retardancy when some alkaline-earth and transition mental ions are incorporated.[60,61]

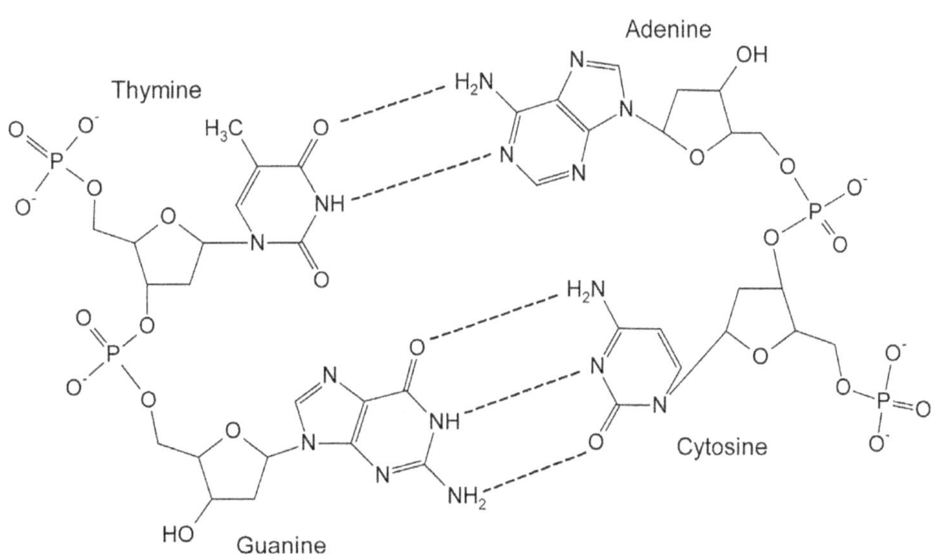

FIGURE 13.17 DNA molecular structures.

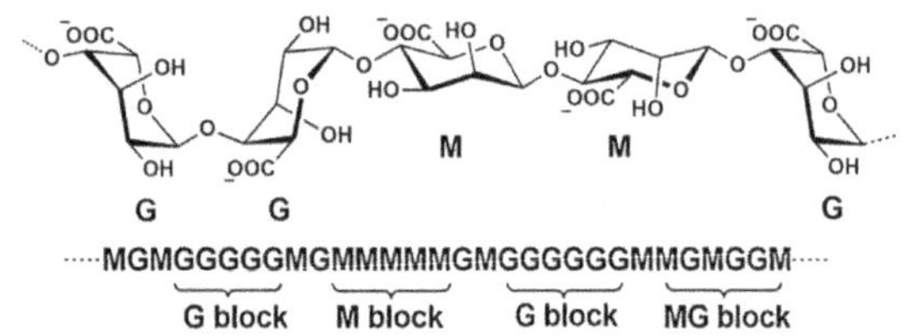

FIGURE 13.18 Chain conformation and block distribution of alginates.[12]

TABLE 13.1
Flammability Test Results of Alginate Film and Fabrics[12]

Ions	Ion content (wt%)	Thickness (mm)	LOI (%)	UL-94	Char (wt%)
H^+	~	~	~	~	-
Al^{3+}	~	0.4	42	V0	30
Na^+	6.8	0.1	25	NR	38
Mn^{2+}		0.1	31	NR	33
Co^{2+}	~	0.1	45	V0	22
Ca^{2+}	~	~	50	~	37
Ni^{2+}	13	0.8	50	V0	25
Cu^{2+}	15.3	0.8	23	NR	23
Zn^{2+}	8.7	0.2	-	V0	38
Ba^{2+}	11	2.0	52	V0	44
Fe^{3+}	~	0.1	35	V0	24

Notes: The sample in the work of "Ca^{2+}" was fabrics, and the others listed in the table were films.

Different metal ions have great influence on the alginate's pyrolysis behaviors and flame retardancy. Table 13.1 exhibits flammability test results of alginate film and fabrics with various metal ions. Sodium alginate is flammable with LOI value as low as 25%, although it has a relatively high char residue. This can be ascribed to the fact that there are no cross-linking networks among the sodium alginate chains, resulting in easy chain scission and degradation, and the formed char residues are loose and not good as an oxygen-resistant barrier to shield the matrix.

The alkaline earth metal ions, Ca^{2+} and Ba^{2+}, significantly enhance the flame retardancy of alginates because of their good affinity to alginate chains and high thermal stability of char residues. Unexpectedly, Co^{2+} and Zn^{2+} endow alginates with excellent flame retardancy, while Mn^{2+} and Cu^{2+} are ineffective under the same conditions. Some researchers suggest that Mn^{2+} has a low affinity to form cross-linked networks with alginates and that there are insufficient junctions for forming the char skeleton during burning, which are the reasons the manganese alginates showed unsatisfactory flame retardancy.[62] Copper alginates have very poor thermal stability and start to sharply lose weight at a very low temperature (below 150°C); meanwhile, Cu_2O and Cu nanoparticles are generated in situ and then act as catalysts to further accelerate the decomposition of alginates.[63]

Many efforts have been actively dedicated to use alginates for fire-retardant materials, such as fibers/fabrics,[64] aerogel composites, coating,[65] and flame retardant additives, et al., which have been introduced by Liu.[16] Calcium alginate fibers that can be prepared by wet spinning has been investigated as a flame-retardant fiber. Calcium alginate fibers have low resistance to salts and detergents caused by the ion exchange tendency of Ca^{2+} with Na^+ or other monovalent ions. Usually, incorporating polyvalent metal ions with a high affinity to alginates (e.g., Al^{3+}, Cu^{2+}, and Ba^{2+}) or using cross-linking agents (such as glutaraldehyde, borates, and tetraborates) can enhance the resistance of alginate fibers to salts and detergents. The blended fibers have been developed to utilize the char-forming feature of calcium alginate fibers. With an addition of 50 wt% alginate fibers, the alginate/cotton blended fabrics (cotton5/alginate5) quickly self-extinguish, with a damaged length of about 40 mm in the VFT. The blended fabrics exhibit a much longer TTI than cotton fabrics (100 s *versus* 21 s) and show sharp decreases in heat and smoke release compared with the reference sample, indicating excellent fire safety of the blended materials.[66] In addition to cotton, other synthetic fibers such as polyester, polyamide 5 can also be blended with alginate fibers to obtain flame-retardant nonwoven fabrics.[67,68]

Using intrinsically flame-retardant alginates to combine with clays can endow the aerogel with excellent flame retardancy, which is a valuable feature in applications requiring high fire safety. Many researchers have found that alginate/clay aerogels exhibit a very low density and excellent flame retardancy (e.g., high LOI, long TTI, low heat, and smoke releases).

A nonflammable aerogel based on ammonium alginate and MMT show a PHRR as low as 32 kW/m^2, while the aerogel after cross-linking with Ca^{2+} with a density of 0.104 g/cm^3 cannot be ignited in a cone test under a heat flux of 50 kW/m^2.[69] When magnesium hydroxide, aluminum hydroxide or layered double hydroxides (LDHs) are introduced in the ammonium alginate/MMT aerogel, the LOI value is raised to a very high level (< 60%) and satisfactory mechanical strength (0.7–3.5 MPa), as well as low thermal conductivity (0.024 ~ 0.046 W/m·K) is obtained.[70]

In this chapter, we will focus on the introduction of flame retardant additives or coating based on alginates. Alginates can be applied as an LbL assembly negatively charged component to construct a flame-retardant coating with different positively charged polyelectrolytes such as chitosan, branched polyethyleneimine (BPEI), clay, and carbon materials, etc. This coating can be used on the surface of polyurethane foam, cotton fabrics, polyester fabrics, and nylon 66 fabrics, etc. When some inorganic additives are introduced into LbL systems, the flame retardancy of coatings is improved greatly. Layered double hydroxide (LDH) has been added to chitosan and sodium alginate solution to constitute LbL coating for flexible PU foam, and the modified PU foam with 4.0% weight on coating present excellent fire-resistance in the cone: PHRR reduction of 66%, THR and TSP reduction of 24% and 74%, respectively.[71]

To improve the durability of LbL coating based on alginates, polyvalent metal ions are incorporated to cross-link the coating. The LbL coating prepared by alternately depositing BPET and sodium alginate can be cross-linked when Ba^{2+}, Ni^{2+}, and Co^{2+} are introduced. It is found that the Ba^{2+}-modified cotton fabric exhibits the best flame retardancy both before and after washing for 6 h in a detergent solution. In addition, these coatings possess excellent durability. The cotton fabrics with a 12 ± 3 wt% coating, as an example, show a burning rate of 2.1 ± 0.1 mm/s in a horizontal flame test; they feature a weight gain of 11 ± 2 wt% after washing and show a burning rate of 2.2 ± 0.2 mm/s in the flammability test.[72] The incorporation of hypophosphorous acid also can make covalent cross-links in LbL systems, and durable flame retardancy is achieved. The polyester-cotton blended fabrics with 4.6 wt% LbL coating (BPEI/oxidized sodium alginate/hypophosphorous acid) self-extinguish in a horizontal flame test and show a decrease in PHRR (77%) and THR (75%) compared with the uncoated sample. Moreover, the fabrics also exhibit good flame retardancy after 12 laundering cycles, indicating satisfactory wash durability of the cross-linking coating.[73]

The deposition of polysaccharide (oxidized sodium alginate and chitosan) along with the phosphorus-rich compound phytic acid and nitrogen-rich compound melamine on the phosphorylated chitosan (PCS) modified polyamide 6.6 (PA6.6) fabric surfaces can endow it with durable flame retardancy. The schematic representation of LbL assembled coatings on PCS-modified PA66 fabrics is shown in Figure 13.19. The treated textiles retain their flame-retardant performance even after 20 laundering cycles. This durability is ascribed to the dual interaction in the coating, that is, the covalent interaction between the amino groups of chitosan/melamine and the aldehyde groups of the oxidized sodium

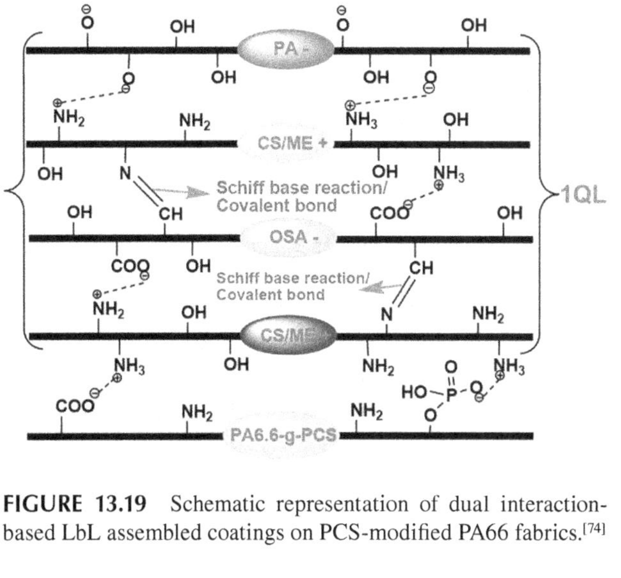

FIGURE 13.19 Schematic representation of dual interaction-based LbL assembled coatings on PCS-modified PA66 fabrics.[74]

alginate, the ionic interaction between positively charged chitosan/melamine and negatively charged phytic acid.[74]

Although alginates are nonflammable on account of their self-charring and self-cross-linking actions, alginates cannot be used as FR additives because they barely promote the formation of char. In addition, they have high water absorption and the low decomposition temperature (below 20°C), which do not meet the demand of most polymers processing. Chemically combining calcium alginate with a reactive compound containing phosphorus and silicon (CADPP, preparation route is shown in Figure 13.20) can overcome the disadvantages of the original alginate, including low thermal stability, high moisture absorption, and poor compatibility with the resin. At 30% loading, flame-retardant epoxy achieves a UL94 V-0 rating and 29% LOI, showing much better flame retardancy than the alginate or the FR containing phosphorus and silicon with the same loading. The highest charring amount and the most compact char layer of the former are the main factors resulting in the satisfactory flame retardance.[75]

13.2.7 Other Biomacromolecules-Based FR

Leather, an abundant biomass material, can be extensively obtained from animal skins. But only 15% of the raw materials can be converted into leather,[76] resulting in a large amount of byproduct, that is, leather solid wastes (LSW). Because of high nitrogen content in collagen, LSW can possess certain flame retardance. Introducing 70% LSW into PVA significantly enhances the Limiting Oxygen Index of PVA-based aerogels from 27 to 32%. During combustion, the amide group and other nitrogen-containing groups in LSW decompose endothermically, producing a large number of nonflammable gases, such as HCN, NH_3, CO_2, and HCNO, to prevent combustion. Besides, LSW can rapidly carbonize during combustion to form a char layer, which prevents the entry of oxygen and combustible gases and further inhibits the process of combustion.

Besides LSW, fish gelatin or porcine gelatin[77] (one of the structures of gelatin is shown in Figure 13.21) can also be added into the PVA/clay aerogel system to prepare a low-flammability materials. It is found that the aerogels containing 16.7 wt% gelatin exhibit LOI values greater than 28% and obtained a UL 94 V-0. With increasing gelatin concentration in the aerogel system, the PHRR of the PVA/MMT/gelatin aerogels decrease continuously. The low flammability of gelatin and the physical cross-linking structure between the gelatin and the PVA/clay system make aerogel composites difficult to ignite, especially in small-fire burning test. The char structure enabled by the interaction between the gelatin and the PVA/MMT system prevents the transfer of heat and combustible volatiles, limiting the further decomposition of the polymer matrix and keeping the heat-release rates of aerogels in the cone-calorimeter tests relatively low.

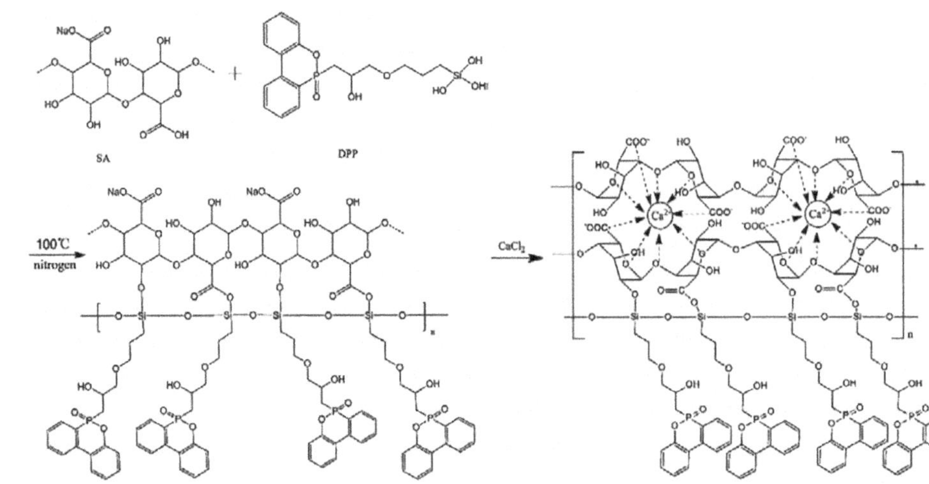

FIGURE 13.20 Preparation route of sodium alginate-based CADPP.[75]

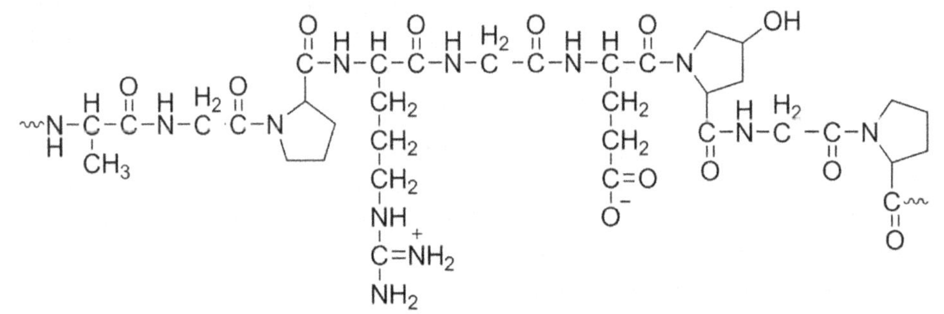

FIGURE 13.21 One of the structures of gelatin.

13.3 BIO-BASED SMALL MOLECULE COMPOUNDS AS FRS

Besides biomacromolecules, many bio-based small molecule compounds, such as phytic acid, tannin, vanillin, or ferulic acid et al., have been used in the flame-retardant industry because of renewability, environmental friendliness, and reactivity. Compared with biomacromolecules, these small molecule compounds have high reactivity because of their structures as well as good solubility in most common solvents.

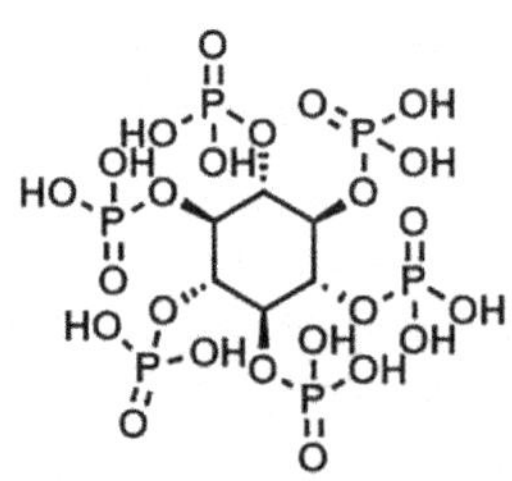

FIGURE 13.22 Chemical structure of phytic acid.

13.3.1 Phytic Acid

Phytic acid (PA) is a saturated cyclic acid and the main phosphorous storage molecule found in plant seeds, whose chemical structure is shown in Figure 13.22. It has ~28 wt% phosphorus and thus has great potential as suitable alternative phosphorus source.[78] For example, PA has been used to improve the flame retardancy of PLA nonwoven fabric by a pad-dry-cure technique. When the fabric is treated with 250 g/L PA, it obtains a high weight gain (around 45%) and exhibits a remarkably increased LOI value (36%).[79] PA also can be used as a phosphorous-based FR fixed on the surface of cotton fabric by a carbamate bond. It releases phosphoric acid/polyphosphoric acid upon thermal decomposition, which catalyzes the dehydration of cellulose, improving the char-forming ability.[80] Via LbL assembly, PA as part of a coating can be deposited on different substrates. The fabrics protected by the coating containing a high amount of PA are self-extinguishable and present better fire performance than those coated with materials with low PA content.

For nylon and cotton blend fabrics (nyco) that has been extensively used in the army combat uniforms and battle dress uniforms, the unique burning behavior called the "scaffold effect," making it difficult to impart flame retardancy. To endow nyco with flame retardancy, a combination of two bio-derived materials, tannic acid (TA), a naturally occurring polyphenol, and PA, is used. TA and PA are sequentially attached to nylon and cotton fibers through hydrogen bonding interactions and phosphorylation, respectively, and the reaction scheme illustrating attachment of tannic

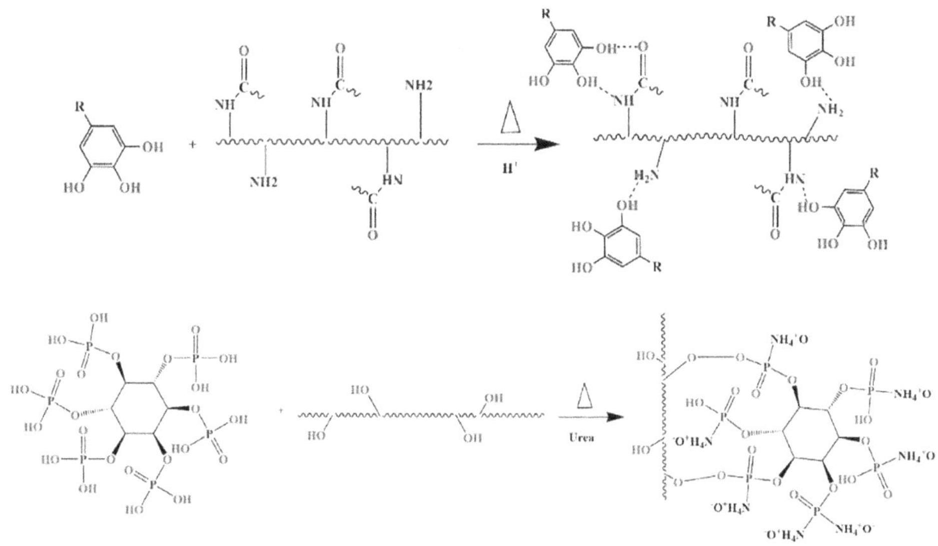

FIGURE 13.23 Reaction scheme illustrating attachment of tannic acid onto nylon (a) and phytic acid onto the cotton (b) in nyco fabric.[81]

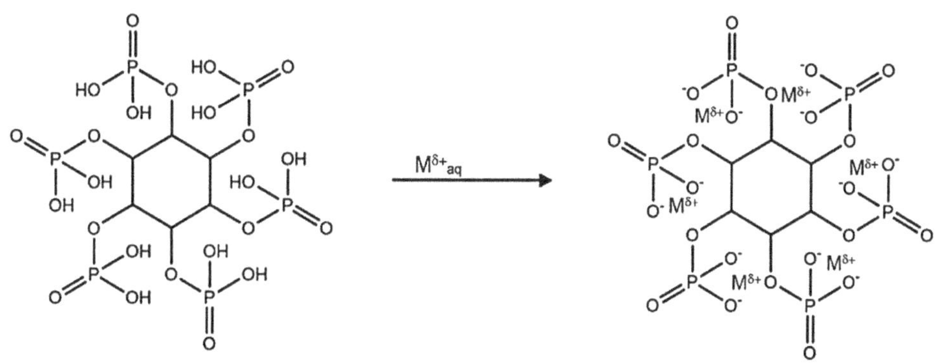

FIGURE 13.24 Phytic acid chemical structure and its chelation with metal ions.

acid onto nylon and phytic acid onto the cotton is shown in Figure 13.23. When nyco fabric is treated with either TA or PA alone and subjected to the vertical flame test, both samples burn completely. Amazingly, the TA-PA treated nyco exhibits a self-extinguishing behavior with an after-flame time of less than 2 s and no afterglow. Introducing both TA and PA into nyco fabric simultaneously enhances its char-forming ability, reducing the release of flammable volatiles, which in turn lowers the heat release and slows the flame spread.[81] It is worth mentioning that this FR system is not durable due to the water solubility of PA. The treated PLA fabric is found to lose the FR performance after several times of laundering.

Because of the inherent hydrophilicity of PA, a large quantity is often required to realize ideal flame resistance performance. Nonetheless, there is growing interest in using PA due to its high content of phosphorus in combination of other FRs and/or as part of IFR. This process does not only overcome PA limitations, but also results in materials with excellent flame resistance property. It has been recognized that a combination of metallic/phosphorus-containing compounds results in effective flame resistance performance. Salts are associated with gas phase, namely, it has flame poisoning effect, resulting from salts evolution from the condensed phase to the gas phase. Therefore, the most volatile salts lead to superior FR performance. Meanwhile, the presence of phosphorus linked to a metal cation results in thermally stable compound that can form a compact glassy isolation layer during combustion. As a rare multi toothed metal chelating agent, a phytic acid molecule contains six phosphate groups that can chelate with metal ions, each containing two hydroxyl groups and four oxygen atoms (chemical structure is shown Figure 13.24). Many metal ions such as sodium, aluminum, iron, magnesium, copper, nickel, tungsten, and lanthanum have been introduced into PA.[78]

When calcium and magnesium phytate (CaMg-Ph) together with carbon nanotubes (CNT) are applied in PLA, the PHRR and THR are greatly reduced. PLA/CaMg-Ph19/CNT1 shows lower PHRR (35%) and higher char yield (18 wt%) compared to PLA/CaMgPh20, suggesting preferable flame-retardant properties. More importantly, the CNT has a reinforcement effect on both of the tensile strength and char layers of PLA/CaMg-Ph.[82] Considering the limited flame retardancy of phytic acid metal salts, the LOI and UL-94 testing results of PLA/CaMg-P or PLA/CaMg-Ph19/CNT1 are not desirable. The main concern about the presence of these metals is that they cause polymer degradation during processing, especially for biopolymers. Therefore, there has been a paradigm shift towards combining PA with other compounds to overcome its limitations while improving the flame resistance performance.

When 1,6-hexanediamine is reacted with PA, a phosphorus-nitrogen containing (PA-HDA) is obtained, which can be used in the toughened PLA (TPLA) achieved by in situ formed cross-linked. With 5 wt% PA-HDA loading, TPLA can obtain UL-94 V-0 while its LOI value is 26%. The flame retardant TPLA still keeps good tensile and impact toughness, that is, its elongation at break and notched impact strength are enhanced to 207% and 5 kJ/m^2, which are 27 and 3 times of those of neat PLA, respectively; this is presented in Figure 13.25. By adjusting the PA-HDA loading, the balance between the flame retardancy and good mechanical properties of PLA can be achieved.[83]

When piperazine is used to replace 1,6-hexanediamine, this novel phytic acid salt PHYPI salt can achieve good flame retardancy in polypropylene (PP). The LOI value for PP containing 18 wt% PHYPI is 25%, showing a 39% increase compared to 18% of original PP. Moreover, it has the UL-94 V-0 rating in the vertical burning test. The flame-retarding efficiency of PHYPI is higher than that of typical traditional IFR containing APP or pentaerythritol. Together with the diluting action of the released H_2O and CO_2 during the thermally decomposing process of PHYPI, both condensed phase and gaseous phase FR played their role in the flame retardance of PP/PHYPI.[84]

A "three sources in one" IFR monomer (PAMAD) can be prepared in which PA as an acid source, melamine (MA) as a gas source, and dihydroxymethylpropionic acid (DMPA) as a carbon source, and a detailed synthetic route is shown in Figure 13.26. PAMAD contains two hydroxyl groups, which can react with -NCO, therefore, it can be used as a chain extender to participate in the synthesis of waterborne polyurethane (PWPU). The sample PWPU-5, with the content of PAMAD is 11 wt%, achieves a V-0 classification in UL-94 test and the residual carbon increases to 12%. Compared with the sample without flame retardant modification, the PHRR and the THR of PWPU-5 are reduced by 51% and 34%, respectively. Meanwhile, the release rates of CO and CO_2 decrease by 56% and 33%, respectively. The phosphorus-containing derivatives such as phosphoric acid derived from PAMAD can catalyze the backbone of polyurethane to char and cooperate with the nitrogen in the system to allow material to form a dense and swelling charcoal

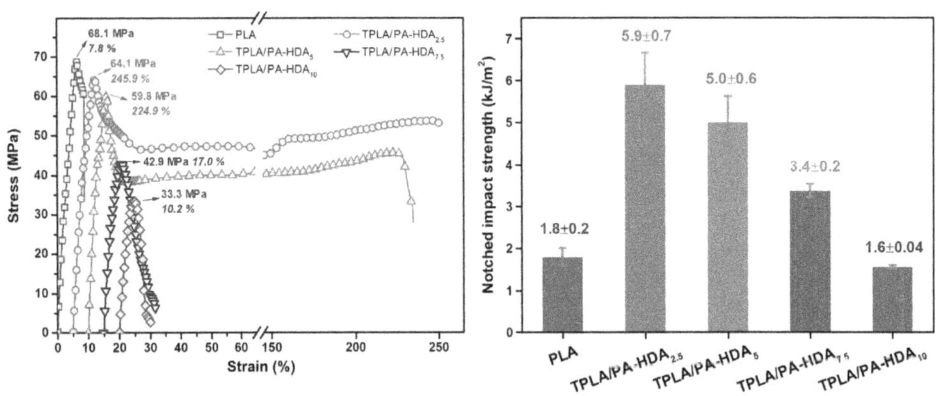

FIGURE 13.25 Stress-strain curves and bar graphs of notched Izod impact strength of PLA and TPLA/PA-HDAx.[83]

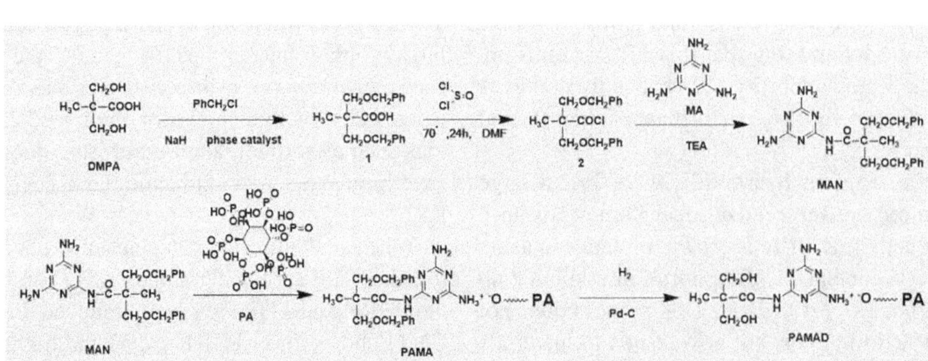

FIGURE 13.26 Preparation route of PAMAD using phytic acid, melamine, and dihydroxymethylpropionic acid as raw materials.[85]

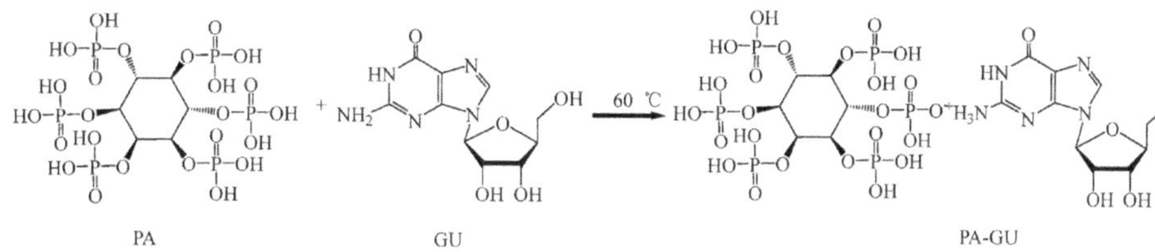

FIGURE 13.27 Synthetic route of PA-GU.[86]

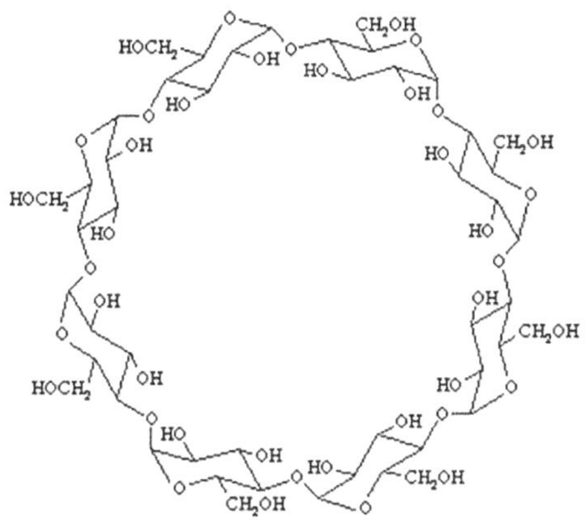

FIGURE 13.28 Chemical strucutre of cyclodextrins.

layer, which hindered the exchange of internal and external gases and heat, so as to play a flame-retardant effect.[85]

Guanosine (GU) extracted from natural plants can be used as charring and blowing agent, while phytic acid is used as the acid source to form a fully bio-based IFR (PA-GU) via convenient ionic reaction. The synthetic route for PA-GU is shown in Figure 13.27. When 10 wt% PA-GU is added to poly(butylene succinate), PBS, the PBS composite reaches a V-2 rating, and the LOI increased to 20% and its pHRR and THR decrease 59% (from 1413 kW/m^2 to 577 kW/m^2) and 38.6% (from 132 MJ/m^2 to 81 MJ/m^2), respectively.[86]

13.3.2 Cyclodextrins

Cyclodextrins (CDs), a cyclic oligosaccharide, consists of 6 ~ 12 α-1,4-linked α-D-glucopyranose units, whose chemical structure is presented in Figure 13.28. The molecules containing 6, 7, and 8 glucose units are called α-, β-, and γ-cyclodextrins, respectively, which have been well studied. CDs have a truncated cone (or toroid) structure because of the chair conformation of their glucopyranose units. In its hollow structure, the outer upper end (the larger open end) is composed of the secondary hydroxyl groups of C2 and C3, and the lower end (the smaller open end) is composed of the primary hydroxyl group of C6, which is hydrophilic. The hydrophobic zone is formed in the cavity due to the shielding effect of the C-H bond. Since the outer rim of CDs is hydrophilic while the inner cavity is hydrophobic, it can provide a hydrophobic binding site like an enzyme and act as a host to envelope various appropriate guests, such as organic molecules, inorganic ions, and gas molecules through van der Waals force, hydrophobic interaction force, etc.

If volatile or low melting flame retardants form inclusion complex (IC) with CDs, their drawbacks can be overcome. Triphenyl phosphate (TPP) is a widely used FR or plasticizer in polymeric materials; however, its melting temperature is only around 50°C, and it may volatilize under environmental conditions, leading to potential environmental hazards. When the IC based on β-CD and TPP is formed, the thermal stability of the guest TPP in the IC is enhanced (no weight loss is observed around the boiling point of TPP), while the flame retarding properties remains. Even though much less TPP is contained in IC, the treated polyethylene terephthalate (PET) films can self-extinguish at 3 s after removing the ignition source.[87] Besides β-CD, γ-CD can also generate stable IC with TPP with the molar ratio of TPP to γ-CD is 1:1 or 1:2. As far as α-CD is concerned, it cannot form stable IC with TPP, even though quantum chemistry calculations reveal that IC formation with both α-CD and γ-CD is feasible.[88]

As a cyclic oligosaccharide, CD also has charring ability, especially when it is modified by chain extension or crosslinking with other compounds, so in some research it has been used as IFR component in polymeric materials. For example, when β-CD is cross-linked in the presence of hexamethylene diisocyanate to fabricate a new char-forming agent, HDI-CD, whose chemical structure is shown in Figure 13.29; the combination of this with APP can constitute an effective IFR for PP. When the feed ratio β-CD:HDI is 1:4, the product HDI-CD(4) with suitable crosslinking degree obtains the maximum total pore volume and specific surface area, leading to PP/APP/HDI-CD(4) with 28 wt% IFR addition showing the best flame-retardant performances (LOI: 33%, UL94: V-0 grade). Since the -(CH_2)n- is thermally stable in early degradation, it can maintain the external cavity structure and trap more volatile degradation products, reduce the fuel supply of burning, and contribute to char formation. Besides, the suitable density of flexible linkage can help to accommodate the stress during combustion and form a more compact char layer.[89]

If CDs are modified by some phosphorus-containing compounds, their char-forming ability will be further enhanced. When phenyl phosphonic acid dichloride (BPOD) is used to modify CD to obtain phospholipidated b-cyclodextrin (PCD) through interfacial condensation, its char forming ability is greatly enhanced. PLA with 30 wt% PCD and APP loading presents the highest LOI value (43%)

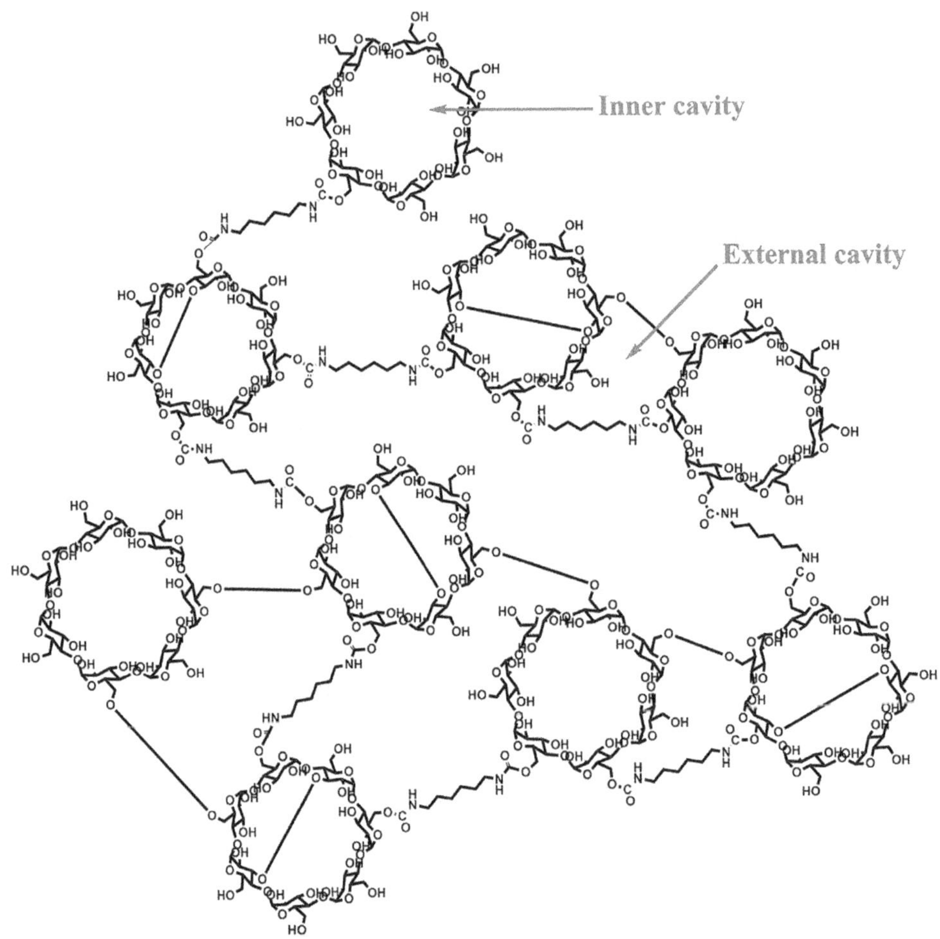

FIGURE 13.29 Chemical structure of HDI-CDS prepared by β-CD and hexamethylene diisocyanate[89]

and UL94 V-0. When the mass ratio of APP and PCD is 5:1, the highest char residue reaches 72 wt%, and the PHRR and THR show maximum decreases of 56 and 84%, respectively, compared with the pure PLA[90].

13.3.3 Aromatic Compounds from Biomass

Besides the above bio-based small molecule compounds, many aromatic compounds from biomass, like natural phenols (such as tannin), lignin derivatives (such as vanillin or ferulic acid, cardanol and others), have been used to obtain flame-retardant materials. Because of high reactivity, these bio-based aromatic compounds often are incorporated into polymer chain by covalent bonds. Among these polymeric materials, epoxy resin (EP) has been investigated extensively and in-depth due to the urgent demand for flame retardancy and the substitution of bisphenol A. Epoxy resin is the most widely used type of thermosetting resin, possessing many favorable qualities, including excellent mechanical strength, good chemical and electrical resistance, superior moisture resistivity, and high dimensional stability. Currently, 90% of the global demand for EP originates from petroleum-based diglycidyl ether bisphenol A (DGEBA)-type epoxy monomers, which are prepared from the reaction between epichlorohydrin (ECH) and bisphenol A (BPA) under alkaline conditions. The question of BPA on human health and the environment has led people to spend significant effort to find its substitutes. Most of the work has focused on replacing BPA with compounds from renewable resources. Just as with petroleum-based epoxy, bio-based EP also has high flammability, which hinders its applications in the fields of automobiles, trains and airplanes, construction, and electronic appliances. If the introduction of bio-based compounds not only avoids the use of bisphenol A, but also gives epoxy resins better flame retardancy, this will be a strategy of killing two birds with one stone. Usually, the bio-based compounds can be incorporated as epoxy compounds, curing agents, or flame retardants. The distinction between curing agents and additives in epoxy resins is a little hazy. Some flame retardants can react with reactive epoxy groups and participate in the resin network when used as curing agents with co-curing agents.

13.3.3.1 Bio-Based Flame-Retardant Epoxides

A number of bio-based flame-retardant epoxides, including bisphenol acid, phloroglucinol, cashew phenol, guaiacol, vanillin, and (iso) eugenol, have been synthesized. The flame-retardant properties of the EP based on these

TABLE 13.2
Bio-Based Flame-Retardant Epoxides and Flame Retardancy of the Corresponded EPs Reported in the Literature

Entry	Bio-Based Platform	Epoxy Monomer	Curing Agent	LOI	UL 94[a]	PHRR kW/m²	THR MJ/m²	References
1	Diphenolic acid	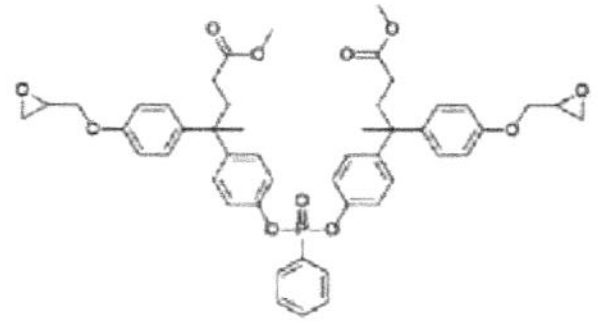 PCDGEDP	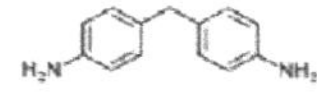 DDM	30	V –0	266 (– 63%)	41 (– 39%)	[92]
2	phloroglucinol	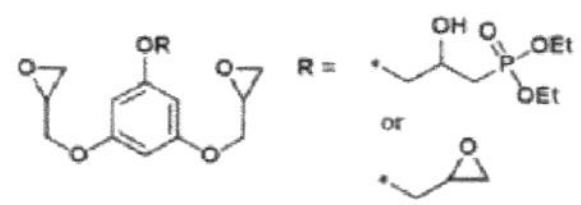 P2EP1P	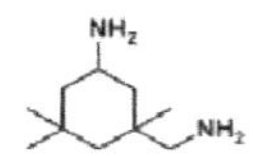 IPDA	-	-	166 (– 75%) [PCFC]	17 (– 50%) [PCFC]	[93]
				Phosphorus content = 3wt%				
3	Cardanol	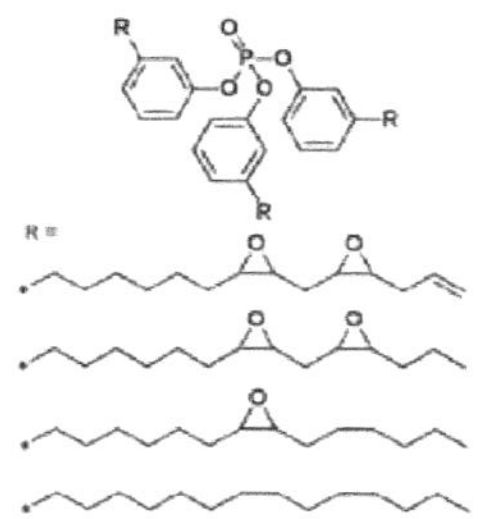 TECP	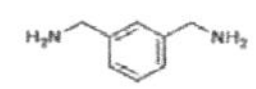 IPDA	-	-	708 (– 52%)	20 (– 4%)	[94]
4	Guaiacol &Vanillin	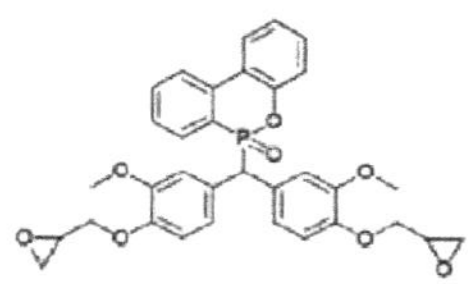 DEGEDB	DDM	33	V –0	-	18 (kJ/g, – 27%) [MCC]	[95]
				Mixture of monomer and DGBEA (7:3)				
5	Vanillin	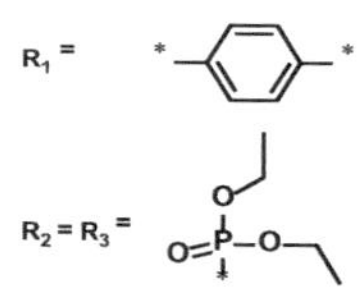 EP2	DDM	31	V –0	-	-	[96]

(Continued)

TABLE 13-2 (Continued)

Entry	Bio-Based Platform	Epoxy Monomer	Curing Agent	LOI	UL 94[a]	PHRR kW/m²	THR MJ/m²	References
6		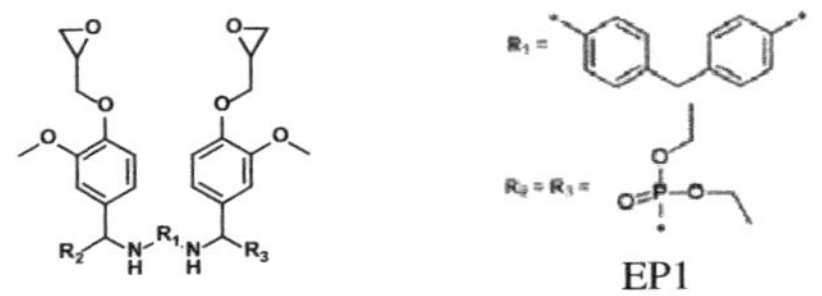 EP1	DDM	33	V –0	-	-	[96]
7		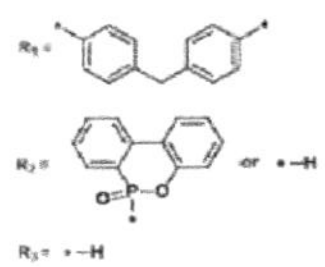 VAD-EP	DDM	29	V –0	467 (–48%)	67 (–32%)	[97]
					Molar ratio of R_2 (DOPO:H = 7:3)			
8		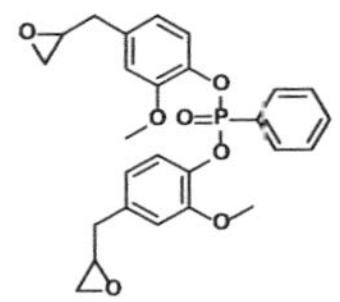 DEEP	MXDA	-	-	665 (–55%)	8 (–58%)	[94]
9	Eugenol	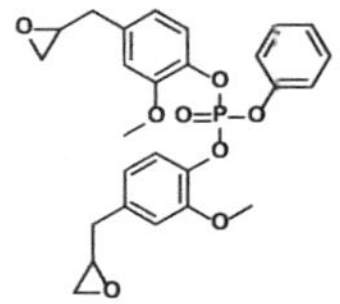 DEEP-Ph	MXDA	-	-	593 (–60%)	9 (–54%)	[94]
10		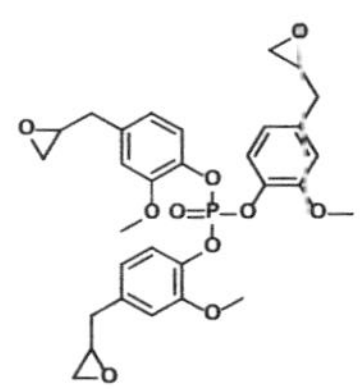 TEEP/TEUP-EP	MXDA	-	-	659 (– 55%)	10 (53%)	[94]
			DDM	31	V-0	119.3 (W/g, –63%) [MCC]	10 (KJ/g, –57%) [MCC]	[98]
11		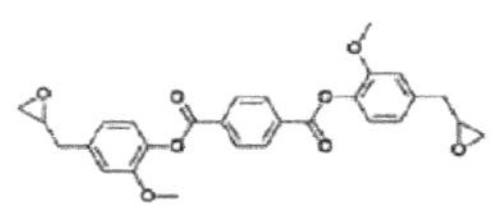 TPEU-EP	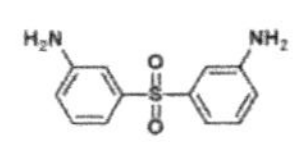 33DDS	27	V –1	860 (–20%)	50 (–29%)	[99]

(Continued)

TABLE 13-2 (Continued)

Entry	Bio-Based Platform	Epoxy Monomer	Curing Agent	LOI	UL 94[a]	PHRR kW/m²	THR MJ/m²	References
12		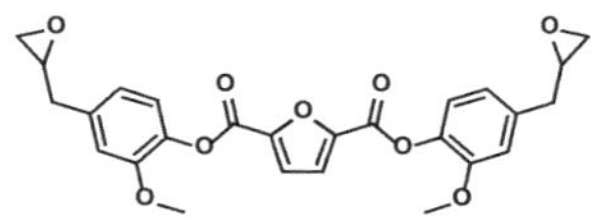 EUFU-EP	Methyl hexahydrophthalic anhydride (MHHPA)	-	-	291 (– 19%)	20 (– 19%)	[100]
13		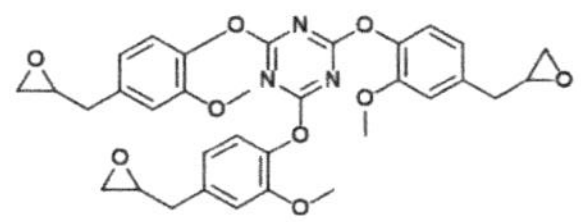 TEU-EP	33DDS	-	-	191 (W/g, –50%) [MCC]	18 (KJ/g, –29%) [MCC]	[101]
14		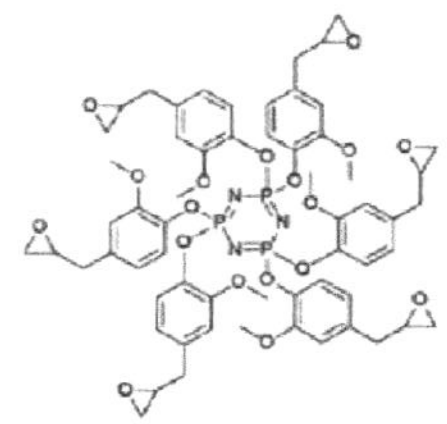 EHEP	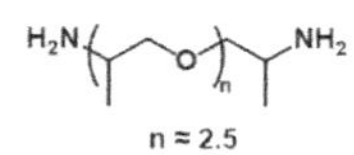 D230	31	V –0	426 (–66%)	34 (–65%)	[102]
15		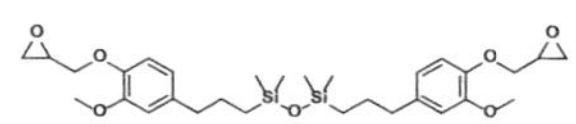 SIEEP2	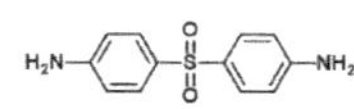 DDS	26	-	952 (–34%)	80 (–26%)	[103]
16		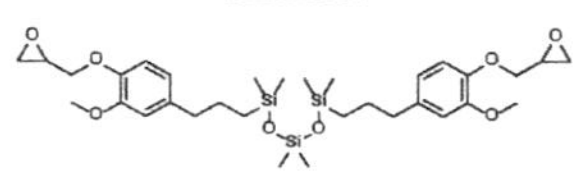 SIEEP4	DDS	28	-	662 (–62%)	74 (–32%)	[103]
17		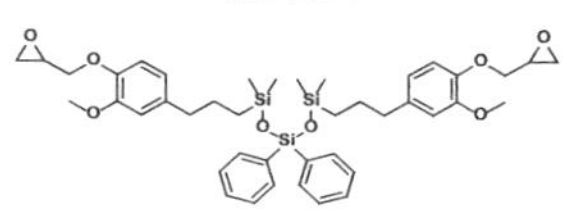 SIEPEP	DDS	31	-	781 (–42%)	79 (–27%)	[103]
18		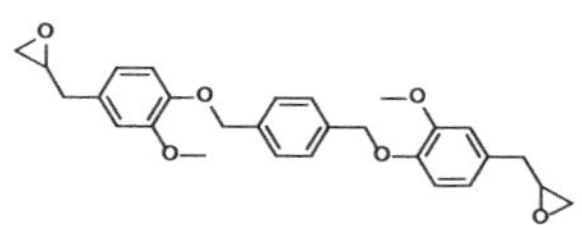 DEU-EP	DDM	-	-	201 (W/g, –55%) [MCC]	16 (KJ/g, –39%) [MCC]	[104]

(Continued)

TABLE 13-2 (Continued)

Entry	Bio-Based Platform	Epoxy Monomer	Curing Agent	LOI	UL 94[a]	PHRR kW/m²	THR MJ/m²	References
19	Isoeugenol	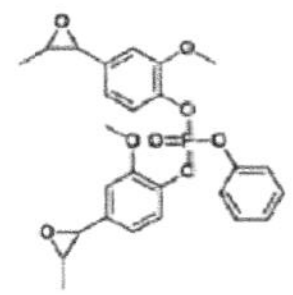 DEpiEPP	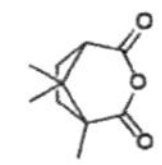 CA (Bio-based)	-	-	890 (−52%) Mixture resin of monomer and DGBEA (3 wt% P)	17 (−32%)	[105]
20	Ellagic Acid	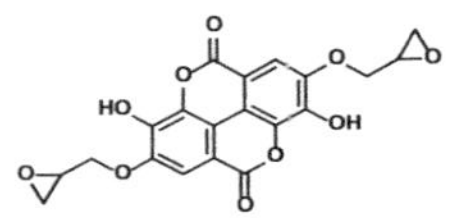 DGEEA	DDM	40	V −0	78 (W/g, −85%) [MCC]	10 (KJ/g, −70%) [MCC]	[106]
21	Guaiacol	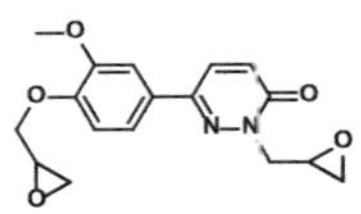 GSPZ-EP	DDM	-	V −1	185 (W/g, −61%) [MCC]	12 (KJ/g, −54%) [MCC]	[107]
22	Daidzein	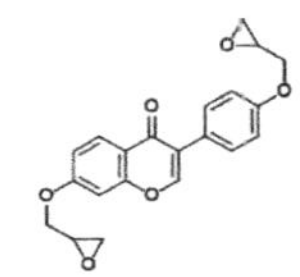	DDM	31	V −0	-	-	[108]
23	Ferulic acid	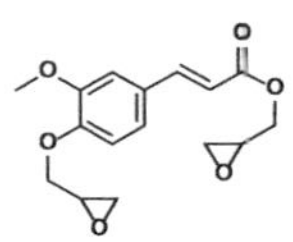 DGEFA	DDM	33	V −1	609 (− 39%)	49 (−54%)	[109]
24	Luteolin	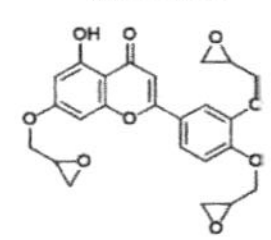 DGEL	DDS	32	V −0	108 (W/g, −75%) [MCC]	24 (KJ/g, −16%) [MCC]	[110]
25	Magnolol	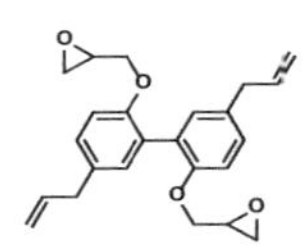 DGEM	DDS	-	V −0	138 (W/g, −70%) [MCC]	16 (KJ/g, −26%) [MCC]	[111]

bio-based flame-retardant epoxides are shown in Table 13.2. Diphenolic acid (DPA) is synthesized by condensation of phenol with bio-based levulinic acid (LA), which is obtained by the hydrolysis of cellulose and has been selected as one of the top 12 potential bio-based platform chemicals by the US Department of Energy.[91] The corresponding phospho-containing epoxy compound (PCDGEDP, Entry 1, Table 13.2) can subsequently be made from glycidyl ether. It is worth noting that PCDGEDP/DDM thermoset exhibits improved flame retardant with UL-94 V-0 and an LOI of 30%. Moreover, incorporating reactive phosphorus-containing epoxy monomer PCDGEDP into epoxy network can decrease the heat release rate, total heat release, smoke produce rate, total smoke release and, therefore, enhance the fire safety of EP.[92]

Phloroglucinols are widely found in plants, fungi and marine organisms, among which plants can be regarded as the main source. Phloroglucinol is chosen as a bio-based aromatic raw material because it exhibits a symmetrical structure with three hydroxyl groups having the same reactivity. It can be functionalized by glycidylation and/or phosphorylation to obtain building blocks that can be further used either as epoxy monomer or as phosphorus reactive flame retardant. The synthetic route for diglycidyl monophosphonated phloroglucinol is shown in Figure 13.30. When P3EP is used, the obtained FR-free epoxy resin (cured with isophorone diamine, IPDA) has a high glass transition temperature (T_g) of 177°C, compared to 157°C for the DGEBA–IPDA thermoset, and high char residue, because of the high aromatic rings content. The presence of P2EP1P is shown to decrease the thermal stability of the epoxy network, but it results in an increase in the char yield at elevated temperature. Pyrolysis/combustion flow calorimetry (PCFC) results show that when P3EP and P2EP1P (phosphorus content of 3%) are used together the PHRR value for the obtained EP has a reduction of around 75%[93] (Entry 2, Table 13.2).

Cardanol is non-edible phenolic oil obtained from cashew nuts, a side-product of the cashew industry, which is considered to be one of the most promising renewable aromatic by-products, relatively cost-effective and available. Cardanol, with its unique chemical structure, has the following characteristics: a benzene ring with polar hydroxyl group, with high temperature resistance and reactivity. The 15-carbon aliphatic chain with unsaturated double bonds in the meta-position can provide good toughness, excellent hydrophobicity, low permeability, and reactivity. When cardanol is used as a bio-based epoxy monomer, the phenol hydroxyl group can be substituted with phosphorus oxytrichloride, and the double bond is epoxidized with peroxic acid to obtain the corresponding epoxy compound and resin (synthetic pathway is shown in Figure 13.31). The resin reduces the PHHR by over 52% compared to DGEBA thermosets. But the THR value only drops 4%. This long aliphatic chain leads to flexible thermosets with low T_g (−26°C).[94]

Vanillin is 3-methoxy group-4-hydroxy benzaldehyde, a biological platform chemical, which is extracted from vanilla planifolia bean, a Rutaceae plant. Vanillin is the only monoaromatic compound produced on an industrial

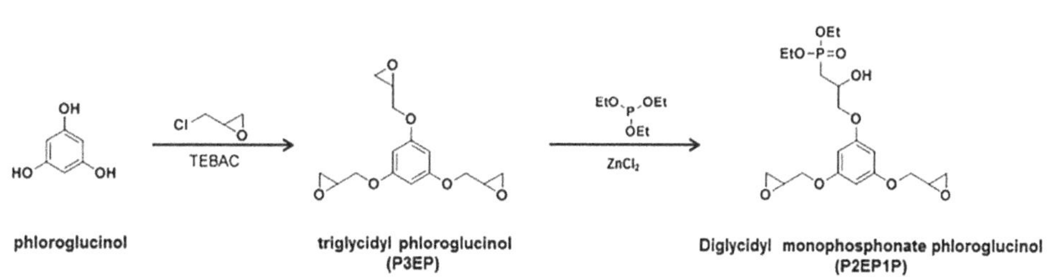

FIGURE 13.30 Synthetic route for diglycidyl monophosphonated phloroglucinol (P2EP1P).[93]

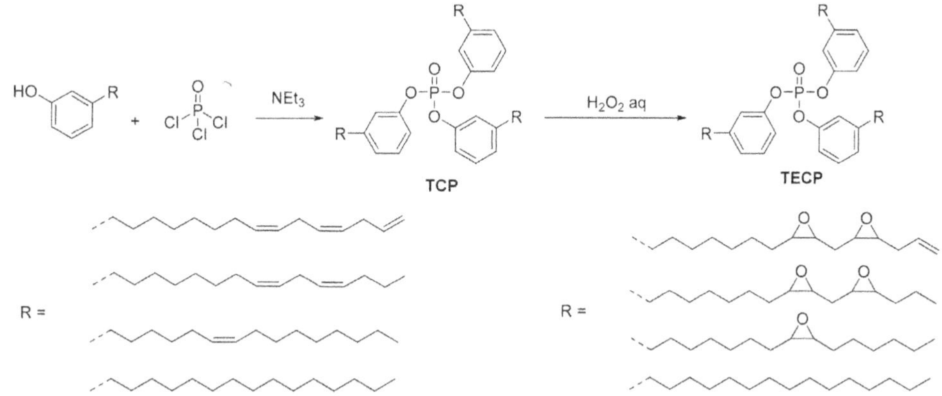

FIGURE 13.31 Synthetic pathways for cardanol-based epoxy monomer containing phosphorus. (Cardanol is a mixture of different molecules containing from 0 to 3 unsaturations.)[94]

scale from lignin and has exhibited tremendous potential to be used as a renewable building block for polymers. Due to the lack of flame retardants, the flame retardancy of these bio-based EP materials are poor. The presence of phenolic hydroxyl and aldehyde groups mean that vanillin can be easily converted into phosphorus-containing epoxy compounds; the aldehyde group of vanillin can react with amines to generate Schiff bases (C = N), which undergo an addition reaction with a P-H bond at high temperature.[95–97]

Guaiacol, 2-methoxyphenol, is the acidic component of wood distillation oil, which can also be used to synthesize the epoxy monomer. The P-H bond of DOPO and the vanillin aldehyde group combine to form a hydroxymethyl phenol, and then guaiacol is condensed with this hydroxymethyl phenol under acidic conditions to produce a very stiff bisphenol molecule for epoxidation. The synthetic route for phosphorus-containing epoxy monomer from guaiacol and vanillin (DGEBDB) is shown in Figure 13.32. DOPO, vanillin, and guaiacol may be readily converted into a bisphenol compound in a one-pot reaction since acids do not have a significant impact on the reactivity of P-H and aldehyde groups. This epoxy compound is employed with DGEBA with a ratio of 3:7 to obtain UL 94 V-0 flame retardant EP with LOI of 33% (Entry 4, Table 13.2). With the addition of DGEBDB, a slight decrement in the T_g value for cured resin is noticed (in the range of 173°C–155°C). Microscale combustion calorimetry (MCC) results show the flammability of EP containing DGEBDB is considerably lower than that of DGEBA/DDM.[95]

Using diamines as coupling agent, two vanillins may be connected, then by reaction with epichlorohydrin phosphorus-containing diepoxies are obtained. The synthetic route is shown in Figure 13.33. After curing with 4,4-diaminodiphenylmethane (DDM), these bio-based EP (EP1 and EP2) present excellent flame retardancy with UL-94 V0 rating and high LOI of 31% and 33% (Entry 5–6, Table 13.2), which are due to the outstanding intumescent and char formation ability. More interestingly, a high T_g of 183°C and

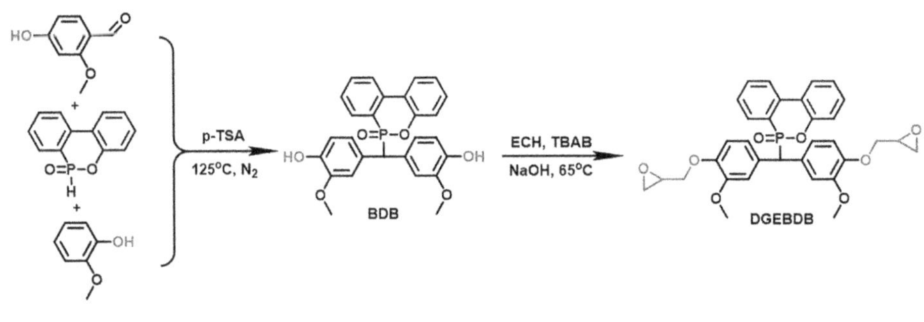

FIGURE 13.32 Synthesis of phosphorus-containing epoxy monomer from guaiacol and vanillin.[95]

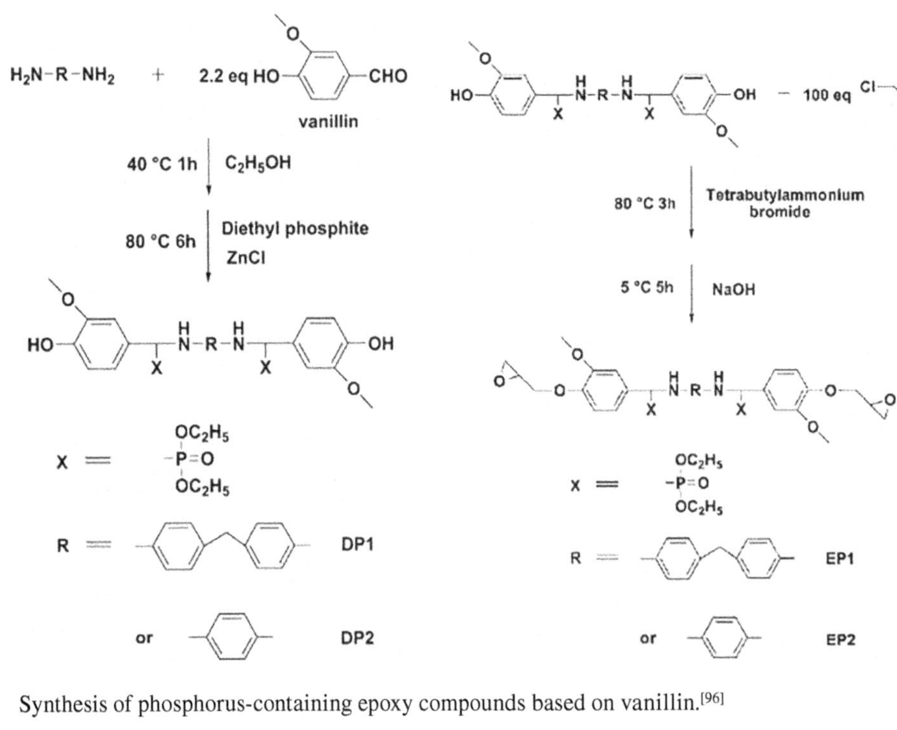

FIGURE 13.33 Synthesis of phosphorus-containing epoxy compounds based on vanillin.[96]

outstanding mechanical properties with tensile strength of ~ 80.3 MPa, much higher than those of the DGEBA system, are achieved for the bio-based EP1 due to their rigid structures.[96] Changing diethyl phosphate with 9, 10-dihydro-9-oxa-10-phosphaphenanthrene-10-oxide (DOPO), another phosphorus-containing epoxy monomer is obtained. Combination with epoxide, without DOPO, a flame retardant bio-based epoxy vitrimer can be obtained (VAD-EP, Entry 7, Table 13.2).[97] This epoxy vitrimer exhibits excellent self-healing capability, reprocessability, and degradation behavior.

Eugenol, 4-allyl-2-methoxyphenol, a bio-based phenol, is generally obtained either from cloves (of which the main product is clove oil) or from lignin by enzymatic reactions. It is an essential oil with relatively low cost, consisting of the rigid aromatic ring substituted by an allyl group, a hydroxyl group, and a methoxyl group. This makes its reactions easier and richer than vanillin (Entry 8–18, Table 13.2),[94–104] since both phenolic hydroxyl groups and allyl groups may be used for the introduction of epoxy groups, although they have different reactivity. In the presence of $POCl_3$, $POCl_2Ph$, or $POCl_2(OPh)$, a nucleophilic substitution of chloride by the phenol of the eugenol occurs, and the compounds with three or two ally groups are obtained. Then, meta-chloroperoxybenzoic acid (m-CPBA) is used for the epoxidation reaction of allylic function of eugenol and detailed synthetic routes are shown in Figure 13.34. These phosphorylated eugenols containing EP thermosets (cured with meta-xylylenediamine, MXDA) have intrinsically both lower pHRR and THR than the DGEBA system (Entry 8–10, Table 13.2). If the curing agent is DDM, the LOI value of TEEP-based EP is raised from 27% of DGEBA/DDM to 31% with UL 94 V-0.[98]

Using terephthaloyl chloride, a difunctional eugenol-based epoxy monomer (TPEU-EP, Entry 11, Table 13.2) is obtained. With 3,3'-diaminodiphenyl sulfone (33DDS) as the curing agent, the eugenol-base EP possesses high mechanical stiffness and low flammability because of its full aromatic structure. Even without phosphorus, TPEU-EP/33DDS can self-extinguish in a vertical burning test and has a much improved flame retardancy with notably lowered smoke production.[99] Replacing terephthaloyl chloride with furan-2,5-dicarbonyl dichloride, eugenol is esterified to produce an intermediate (EUFU) with high yield. Then, the target product with 2 functional epoxy groups is obtained by oxidization of the allyl group in EUFU by mCPBA (Entry 12, Table 13.2). The biomass content of EUFU-EP cured with methyl hexahydrophthalic anhydride (MHHPA) is as large as 93.3%. MCC results show that both peak heat release rate and total heat release are reduced by 19%.[100]

A triazine-bridged eugenol-based epoxy monomer has been developed using cyanuric chloride to react with the hydroxyl group of eugenol in a highly efficient phase-transfer catalytic reaction system to yield an intermediate bearing three allyl groups (TEU). Upon epoxidation of the allylic double bonds, a trifunctional epoxy monomer (TEU-EP) is obtained; the synthetic pathway is shown in Figure 13.35. TEU-EP/33DDS shows markedly decreased peak HHR (191 W/g) and total heat release rate (18 kJ/g) in

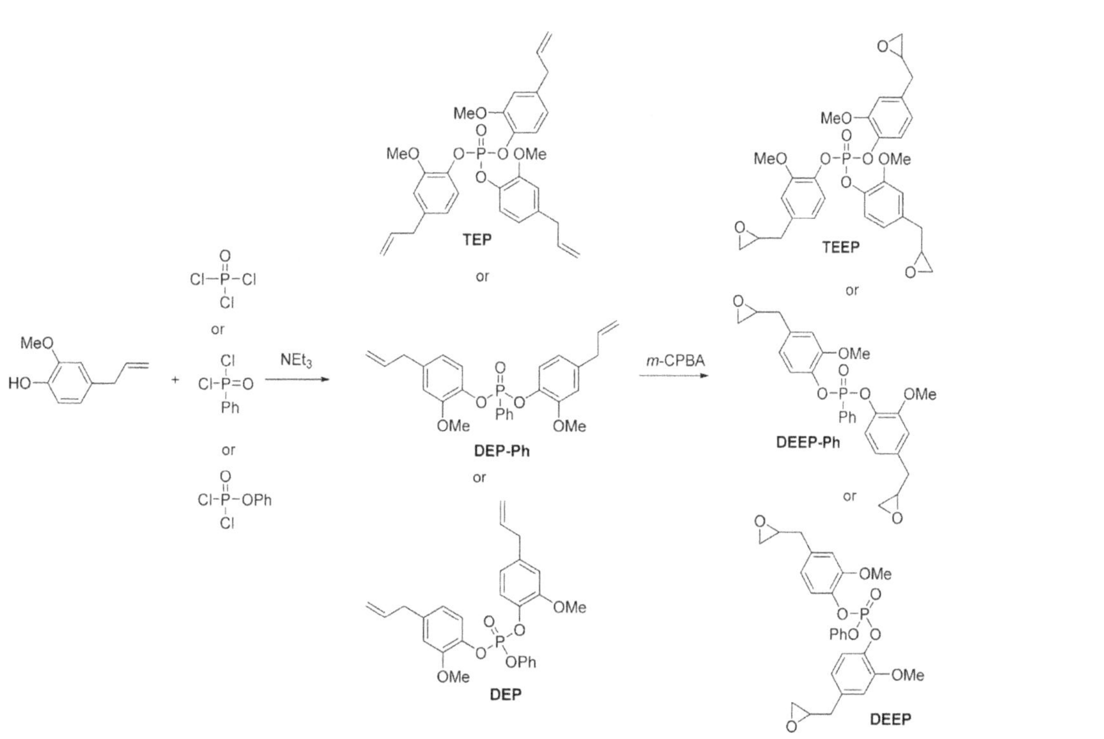

FIGURE 13.34 Synthesis of eugenol-based epoxy monomers.[94]

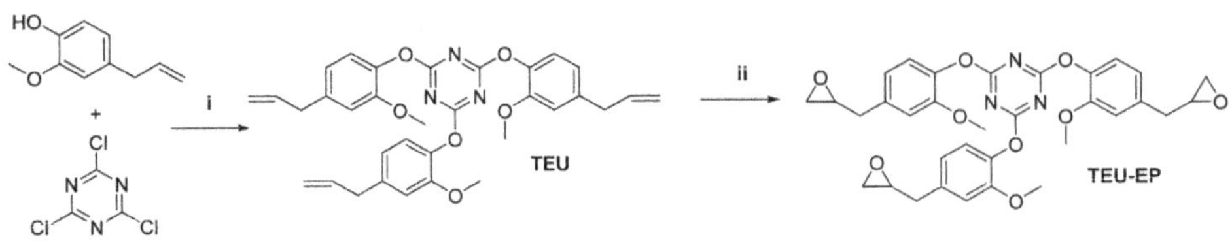

FIGURE 13.35 Synthesis of the triazine-bridged eugenol-based epoxy compound.[101]

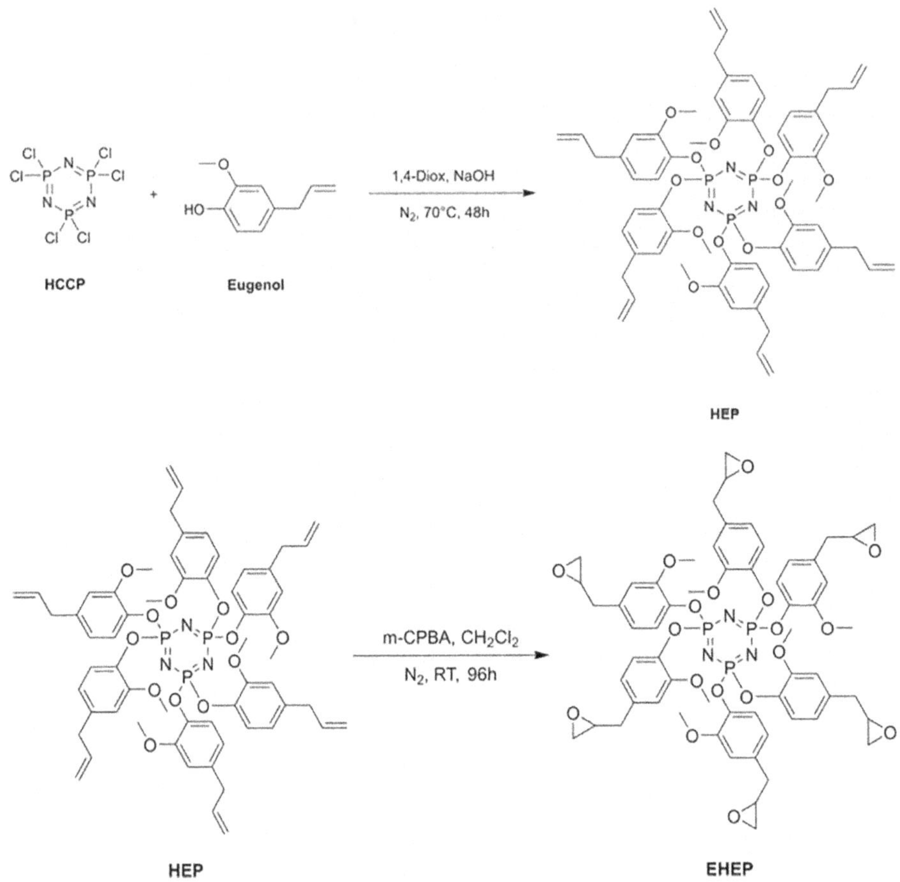

FIGURE 13.36 Synthetic route for eugenol-based epoxy monomer with six epoxy groups.[102]

the MCC, compared with DGEBA/33 DDS, the reductions are 52% and 29% (Entry 13, Table 13.2), respectively.[101]

If hexachlorocyclotriphosphazene (HCCP) is used as a connective unit, a eugenol-based epoxy monomer with six epoxy groups per molecule (EHEP) can be obtained; its synthetic route is presented in Figure 13.36. EHEP, with a star-like structure, not only introduces a cyclotriphosphazene ring into the epoxy resin network, but also endows the matrix with a high cross-linking density, therefore, the cured thermoset exhibits satisfactory comprehensive performance even when an aliphatic ether amine (D230) is used as the curing agent, which typically harm both the strength and flame retardancy. Compared with DGEBA/D230, EHEP/D230 possesses a higher tensile strength (57 to 64 MPa) and better thermal resistance. Moreover, EHEP/D230 exhibits an LOI value of 31%, and a UL94 V-0 without any after-flame time (Entry 14, Table 13.2). The PHRR, THR, and total smoke production for EHEP-D230 are reduced significantly by 66%, 65%, and 78%, respectively.[102]

Epoxidized eugenol (EPEU) also can be made through glycidyletherification between a phenolic hydroxyl group and epichlorohydrin (Figure 13.37), which is similar to the well-established industrial procedure to manufacture DGEBA-type epoxy resins. The terminal aliphatic C = C of EPEU can undergo highly selective and highly efficient hydrosilylation reaction to introduce different silicone linkages into the molecular backbone of the epoxy monomers under mild conditions; the synthetic pathways are shown in Figure 13.38. These silicone-bridged difunctional epoxy monomers exhibit much lower viscosity (< 2.5 Pa s) than commercial DGEBA epoxy (10.7 Pa s) suitable for composites and prepregs. After curing with DDS, they exhibit a dielectric permittivity as low as 2.8 and good intrinsic flame retardancy with LOI value in the range of

26 ~ 31%, far outperforming DGEBA. The markedly lowered PHRR and THR (Entry 15–17, Table 13.2) values indicate the decreased flammability of SIEEP2/DDS, SIEEP4/DDS and SIEPEP/DDS with respect to DGEBA/DDS. Incorporation of Si-O moieties into the epoxy network will reduce the release of the gaseous fuels during the pyrolysis of the epoxy network matrix. In addition, during the burning process, the Si-O-related moieties will migrate onto the burning surface, which functions as a barrier to slow the diffusion of oxidant and fuel and transport of heat to the polymer matrix.[103]

Eugenol has an isomer with very similar structures, called isoeugenol (2-methoxy-4-propenylphenol), but the tail double bond positions are different. Because the double bond of isoeugenol is in the center of the aliphatic chain, this potential bio-based platform molecule is more readily oxidized to epoxy. The majority of the chemical modifications of eugenol can be used for isoeugenol. For example, epichlorohydrin can react with isoeugenol to synthesize glycidyl ether epoxy of isoeugenol (GEEpiE, synthetic route is shown in Figure 13.39a). Phenyldichlorophosphate can react with isoeugenol to obtain di(isoeugenol) phenylphosphate (DiEPP), and then using oxone (potassium peroxomonsulfate) as an oxidant diepoxy-isoeugenol monomers (DEpiEPP) is synthesized (synthetic route is shown in Figure 13.39b). DEpiEP can copolymerize with DGEBA or GEEpiE to fabricate a thermoset in the presence of bio-based curing agent camphoric anhydride (CA). The resulting EP with a weight content of 2.0 wt% P, both GEEpiE-CA-DEpiEPP (95% bio-based content) and partially bio-based DGEBA-CA-DEpiEPP (57% bio-based content) epoxy thermosets possess high T_g (respectively 129 and 105°C) and bending elastic modulus and strength of respectively 4 GPa and 90 MPa and demonstrate good flame-retardant properties (Entry 19, Table 13.2) due to char promotion showing high promise for application.[105]

Incorporation of phosphorus-containing groups usually leads to the deterioration of T_g of these bio-based EP compared to the DGEBA-based thermoset. Furthermore, the phosphorus-containing groups are apt to absorb moisture and hydrolyze, which is not conducive for applications in the electronic and electrical fields. Therefore, phosphorus-free bio-based intrinsically flame retardant EP has received increasing attention. Researchers discovered that using some bio-based phenols, one may obtain a flame retardant UL-94 V-0 with an LOI > 28% novel bio-based EP without

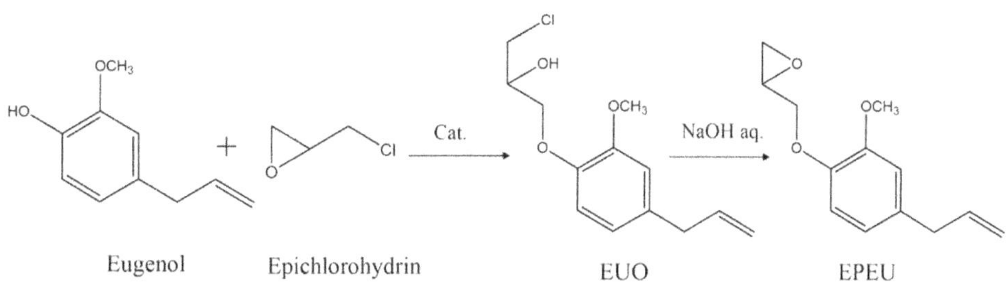

FIGURE 13.37 Synthesis of epoxidized eugenol.[103]

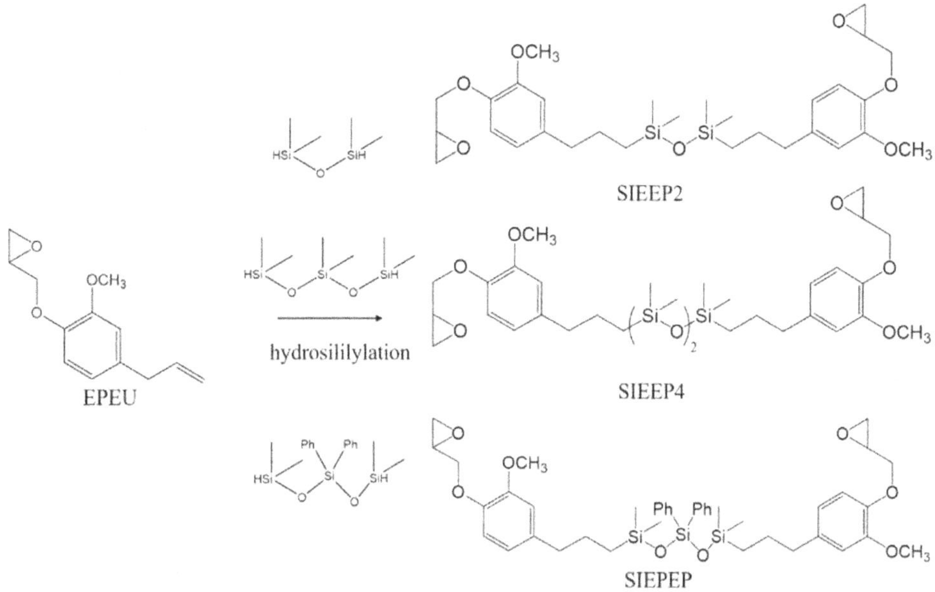

FIGURE 13.38 Synthetic routes to silicone-containing eugenol-based epoxy monomers.[103]

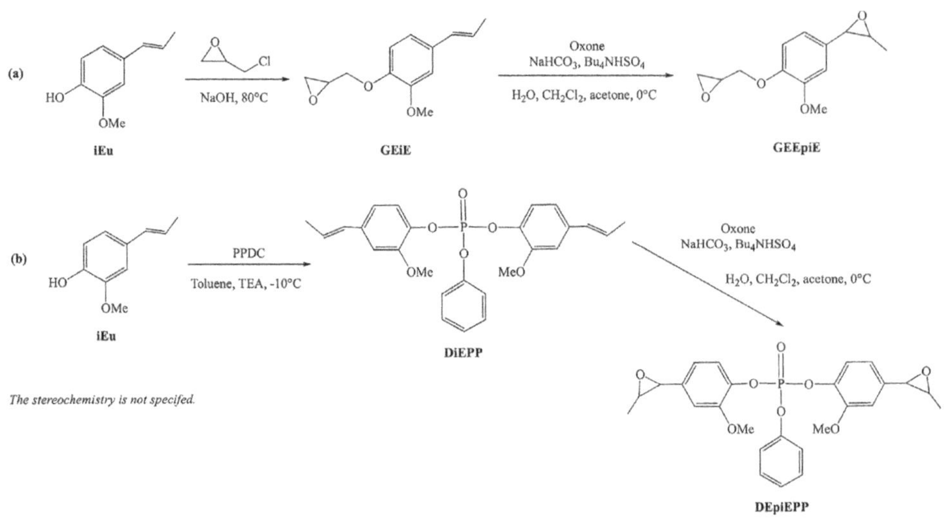

FIGURE 13.39 Synthetic route for isoeugenol-based diepoxy monomers.[105]

the addition of phosphorus-containing structures (Entries 20–25, Table 13.2).

Ellagic acid, a polyphenol dilactone, is a natural polyphenol component widely found in various soft fruits, nuts, and other plant tissues. Compared to bisphenol A, ellagic acid has more rigid conjugate structure and functional phenolic hydroxyl groups. The rigid conjugate structure facilitates the formation of a thermal-resistant char during combustion, while the functional phenolic hydroxyl groups are conducive to react with epichlorohydrin to prepare an epoxy monomer (DGEEA, synthetic route shown in Figure 13.40). The cured DGEEA/DDM shows a high T_g of 220°C, a high tensile strength of 76 MPa, and a high Young's modulus of 3 GPa, which outperforms the conventional petroleum-based DGEBA/DDM system. The resultant thermoset also exhibits an excellent intrinsic FR in terms of the significantly decreased PHRR (MCC results) and the notably improved LOI (40.0% *versus* 23.0%). Furthermore, it can also self-extinguish in the UL-94 vertical burning test (corresponding to the V-0 grade) (Entry 20, Table 13.2)[106]

IntroductionofaromaticN-heterocyclicintoEPcanenhance its properties or impart new functionality. A pyridazine-based compound 6-(4-hydroxy-3-methoxyphenyl)pyridazin-3(2H)-one (GSPZ) can be efficiently synthesized by the Friedel-Crafts reaction of guaiacol and succinic anhydride, both of which can be derived from biomass. Following treatment with epichlorohydrin (GSPZ-EP, Entry 21, Table 13.2) and then using DDM as a curing agent, the resulting bio-based epoxy resin possesses higher T_g (187°C *vs.* 173°C) and shows 61% and 54% reduction in PHRR and THR (MCC) compared with its petroleum-based counterpart (DGEBA/DDM). Unfortunately, only UL 94 V-1 grade is achieved.[107]

Daidzein (7-hydroxy-3-(4-hydroxyphenyl)-4H-chromen-4-one) is a natural polyphenol monomer that can be obtained from soybeans, whose chemical structure is similar to 4,4'-bishydroxydeoxybenzoin and BPA. The routine glycidylation reaction between daidzein and excess epichlorohydrin is conducted in the presence of sodium hydroxide (DGED, its structure shown in Table 13.2, Entry 22). The cured DGED/DDM system demonstrates excellent flame-retardant properties, showing a residual char of 43% at 800°C, LOI of 32%, and flammability rating of V-0 in UL94 test. And its T_g is up to 205°C (172°C for DGEBA/DDM), and tensile strength, tensile modulus, and flexural modulus are 83, 2972, and 2980 MPa, respectively, all much higher than those of cured DGEBA/DDM.[108]

Ferulic acid (FA), a plant-based phenolic acid, usually is directly obtained from ferula asafoetida, rattletops, spina date seeds, etc. FA is also an important derivative of lignin, regarded as one of the most abundant bio-based polymers. It is found that diglycidyl ether of ferulic acid (DGEFA) -DDM thermoset shows a char residue of 41% (in N_2), LOI of 33%, UL-94 rating of V-1, and PHRR of 609 kW/m^2 (Entry 23, Table 13.2), indicating a high carbonization and flame retardancy. The smoke density of DGEFA-DDM is 57% lower than DGEBA-DDM.[109]

Luteolin is a natural isoflavone compound derived from various plants, including onion leaves, celery, green pepper, and chamomile flower, which can be easily modified by epichlorohydrin (DGEL, synthetic route exhibited in Figure 13.41) in the presence of tetrabutylammonium bromide (TBAB). Unlike the other hydroxyls, the phenolic hydroxyl group close to carbonyl group in DGEL is not involved in the reaction, which is attributed to the formation of strong intramolecular hydrogen bond between -OH and oxygen atom of the carbonyl group that reduces the reactivity of phenolic hydroxyl group (Figure 13.41). DGEL/DDS can pass the V-0 rating in UL-94 test with a high LOI (33%) as well as extremely low peak heat release rate (108 W/g) (Entry 24, Table 13.2). Smoke density testing results show that the peak density of smoke for DGEL/DDS is 305, much lower than that for DGEBA/DDS (661) indicating the remarkable reduction of toxicity hazards.[110]

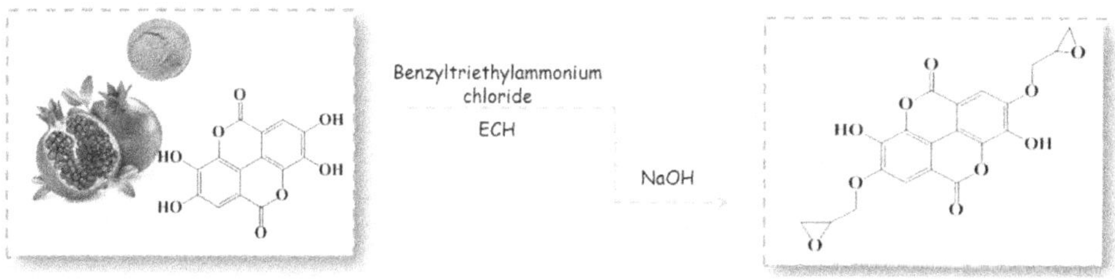

FIGURE 13.40 Synthetic routes to diglycidyl ether of ellagic acid.[106]

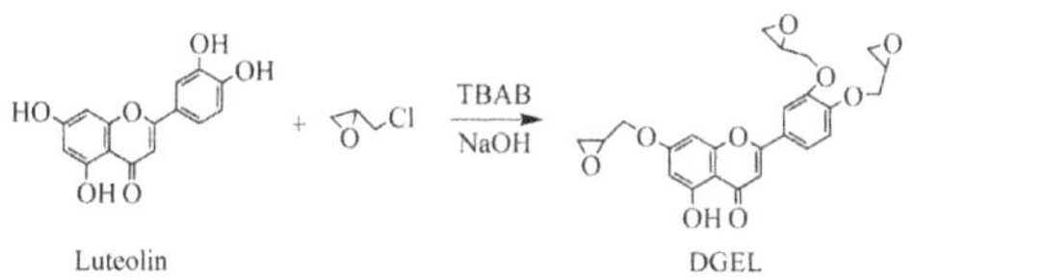

FIGURE 13.41 Synthetic route to luteolin-based epoxy compound.[110]

Magnolol is a bioactive compound that can be extracted from the bark of magnolia officinalis. Magnolol contains a symmetrical bisphenol and diallyl, which is very suitable for synthesis of various bio-based polymers. The biphenyl structure in magnolol belongs to a group of a highly aromatic compound, when incorporated in EP not only enhances the thermal stability but also forms a highly pyrolysis-resistant carbonaceous mass during combustion, thus, improving the flame retardancy of EP. Glycidylation reaction is carried out easily between magnolol and epichlorohydrin in alkaline conditions (DGEM, whose chemical structure is shown in Table 13.2 as Entry 25). After curing, DGEM/DDS reveals an extremely high T_g of 279°C, which is at the highest level among the bio-based epoxy resins reported so far. DGEM/DDS exhibit good flame retardancy with V-0 rating in UL-94. MCC signifies that DGEM/DDS has a 70% and 26% reduction of peak heat release rate and total heat release, respectively, compared to DGEBA/DDS, exhibiting both gas phase and condensed phase synergy flame retardant mechanism.[111]

The unsaturated ketone structure in the above bio-based aromatic compounds may experience mutual cross-linking reactions at high temperatures, enhancing carbonization. Additionally, allylic may cross-link, but to achieve cross-linking, it has to cure for a longer period of time and at a higher temperature (> 200°C).[111] This is because the resin breaks down over time in a high-temperature environment, causing a double-bond addition process that is initiated by free radicals.[112] The important factors for phosphorus-free bio-based flame retardant epoxy resin to achieve outstanding flame retardant performance may be high cross-linking density and stiff structure; however there are currently few examples, and its universality has to be further investigated.

Besides adopting biomass resources, the development of recyclable thermosets is also an effective way to solve the current energy and environmental issues. The cross-linked structures of the traditional thermosets determine that they cannot be reprocessed, recycled, reshaped, remolded. Recently, the introduction of dynamic covalent bonds into cross-linked networks to prepare malleable thermosets, named vitrimers or covalent adaptable networks (CANs), has attracted attention. A few researchers have found that intrinsically flame-retardant, bio-based thermosets with malleability can be fabricated via rational design. For example, using vanillin and phosphorus oxychloride as raw materials, trialdehyde monomer (TFMP) is synthesized and then cured with three different diamines, so the Schiff-based CANs are obtained. The obtained Schiff base CANs are readily recycled by two methods of reprocessing with short time and degradation to original monomer TFMP under mild condition and exhibit excellent flame retardancy as high as UL-94 V0 during vertical burning.[113] Epoxy monomer synthesized from itaconic acid can be cross-linked with phytic acid to achieve the phosphate-based epoxy CANs. The cross-linking reaction of phytic acid with itaconic acid-based diepoxide is shown in Figure 13.42. The epoxy CANs exhibit flame retardancy with UL-94 V0 rating and high LOI of 31–34%. When it is cut into small pieces and hot pressed at 160°C for 1.5 h, the sample with the original shape can be obtained.[114]

13.3.3.2 Bio-Based Flame-Retardant Curing Agent

For EP, flame retardants can be introduced in the main backbone of epoxy or can be introduced in the curing agent. Since curing agents for epoxy include amines (primary, secondary), anhydrides, carboxylic acids, phenols, alcohols, and mercaptans, there is a diversified choice. The bio-based curing ingredient is not as harmful as BPA, so bio-based hardeners have received some attention. Some bio-based curing agents reported in the literature are listed in the Table 13.3.

A phosphorus-sulfur-containing multifunctional carboxyl curing agent based on cardanol is synthesized by the reaction of cardanol with phosphorus oxychloride followed by thiol-ene reaction of thioglycolic acid across the unsaturation of cardanol, and the detailed synthetic route is shown in Figure 13.43. The commercial epoxy coating cured with P-II exhibits higher LOI value (24%) and can pass UL 94 vertical thin material (VTM) test VTM-0 rating, which is ascribed to higher sulfur content in P-II (Entry 1–2, Table 13.3).

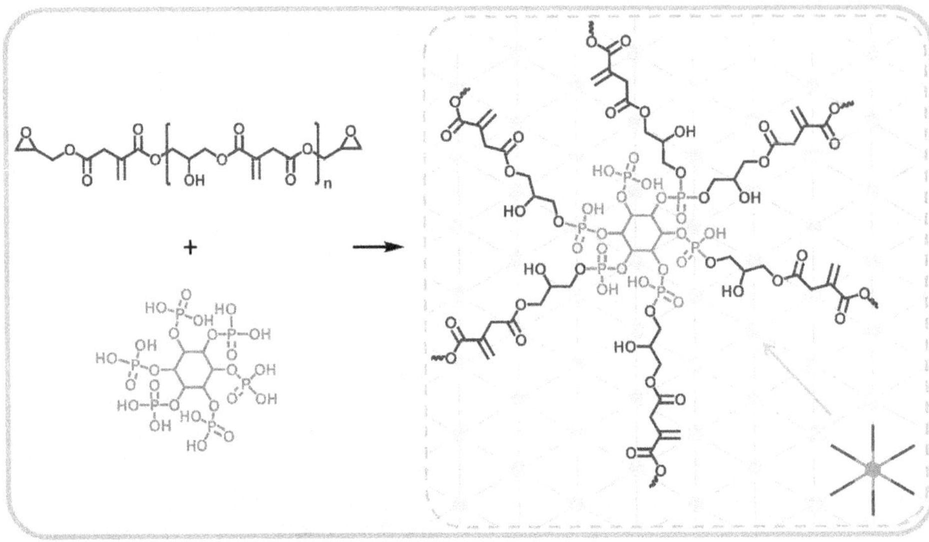

FIGURE 13.42 Cross-linking reaction of phytic acid with itaconic acid-based diepoxide.[114]

TABLE 13.3
Bio-Based Flame-Retardant Curing Agents and Flame Retardancy of the Corresponded EPs Reported in the Literature

Entry	Bio-Based Platform	Curing Agent	LOI	UL 94	PHRR kW/m²	THR MJ/m²	References
1	Cardanol	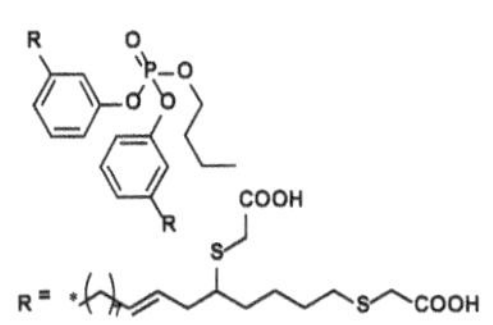 P-II	24	VTM-0 P (wt%) = 1.71%, S (wt%) = 7.07%	-	-	[115]
2		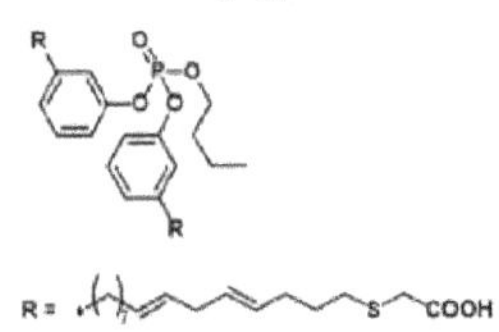 P-I	22	VTM-0 P (wt%) = 2.40%, S (wt%) = 4.95%	-	-	[115]
3	Castor oil	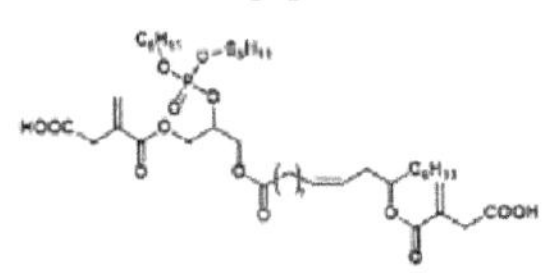 IDDRA	25	-	-	-	[116]
4	Acetovanillone	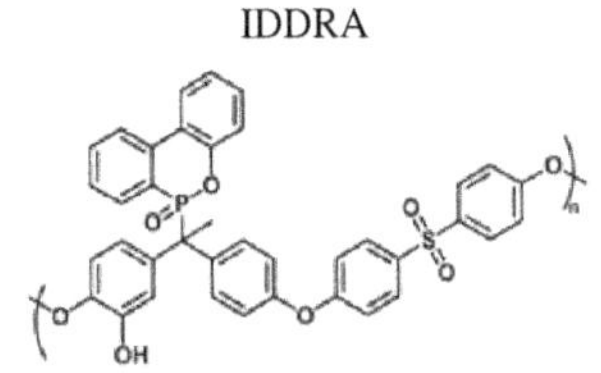 Hydroxyl-PES-1	-	VTM-0 P (wt%) = 2.40%, S (wt%) = 4.95%	-	-	[117]
5	Pytic acid	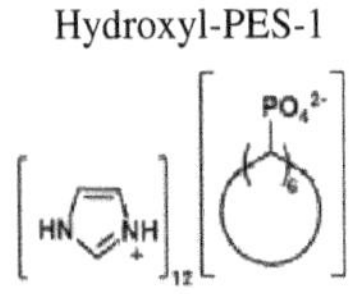 IMPA	35	V -0	344 (–43%)	47 (–32%)	[118]

Entry	Bio-Based Platform	Curing Agent	LOI	UL 94	PHRR kW/m²	THR MJ/m²	References
6		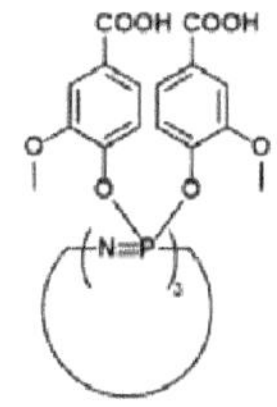 HCPVC	31	V -0	-	-	[119]
8		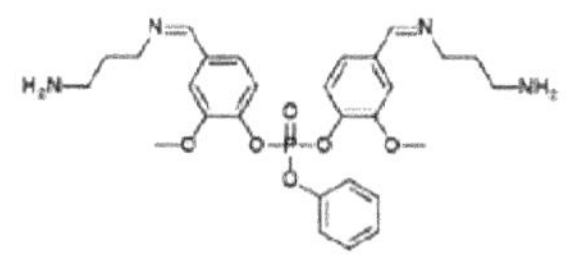 VDTS2	31	V -0	-	-	[120]
7	vanillin	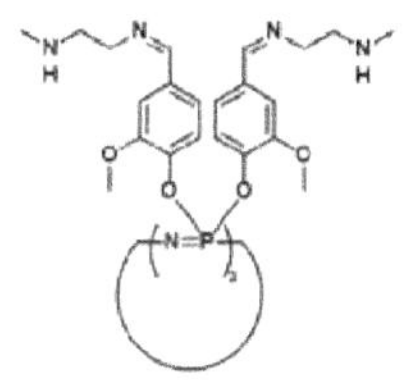 HVPA	29	V -0	352 (−42%)	31 (−32%)	[121]
9		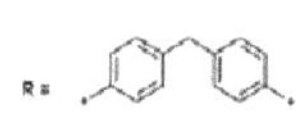	36	V -0 as a co-curing agent, P (wt%) = 1.09% with DDM	-	-	[122]
10		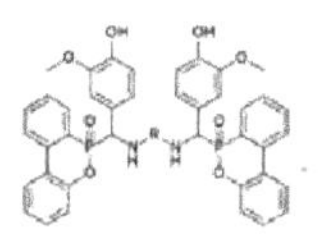 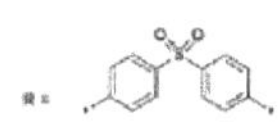	35	V-0 as a co-curing agent, P (wt%) = 0.73% with DDM	-	-	[123]
11		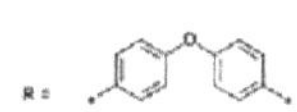	33	V-0 as a co-curing agent, P (wt%) = 0.75% with DDM	-	-	[123]
12	protocatechualdehyde	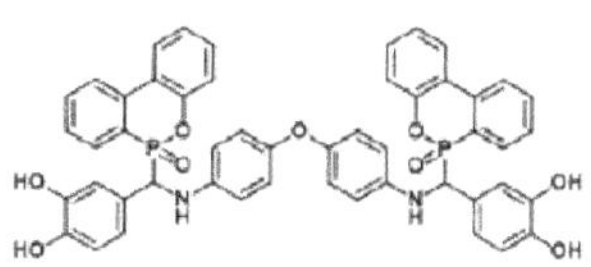 PDD	33	V -0 as a co-curing agent, P (wt%) = 1.09% with DDM	357 (−20%)	65 (−40%)	[124]
13	Furfural	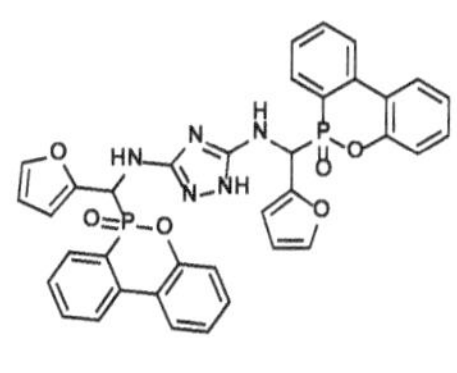 FGD	32	V -0 as a co-curing agent, P (wt%) = 0.45% with DDM	820 (−31%)	114 (−8%)	[125]

Notes: *Unless otherwise specified, it generally refers to bisphenol A epoxy oligomer: 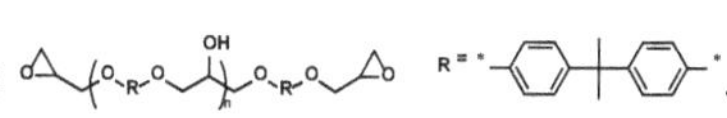.

Ricinoleic acid modified with itaconic anhydride (IDDRA) is synthesized from castor oil via glycidylation, epoxy ring-opening reaction and esterification (IDDRA), and the chemical structure of bio-based curing agent is shown in Table 13.3 (Entry 3). Unfortunately, the LOI of resultant EP (IDDRA/E51) is only 25%, meaning that long fatty chains are still detrimental to flame retardancy.[116] Acetovanillone, DOPO, and phenol can be catalyzed by protons to prepare highly rigid bisphenol compounds and then polymerized with 4,4'-difluorodiphenyl sulfone to fabricate phenolic hydroxyl poly(ether sulfone). Finally, a bio-based polyphenol oligomer curing agent containing phosphorus (hydroxyl-PES-1) is obtained by removing the methyl group from the methoxy group (Entry 4, Table 13.3). The hydroxyl-PES-1-cured commercially available epoxy films have higher T_g (> 200°C) and high transparency and reaches VTM-0 grade in the UL 94 VTM.[117] Phytic acid can undergo a neutralizing reaction with imidazole to form

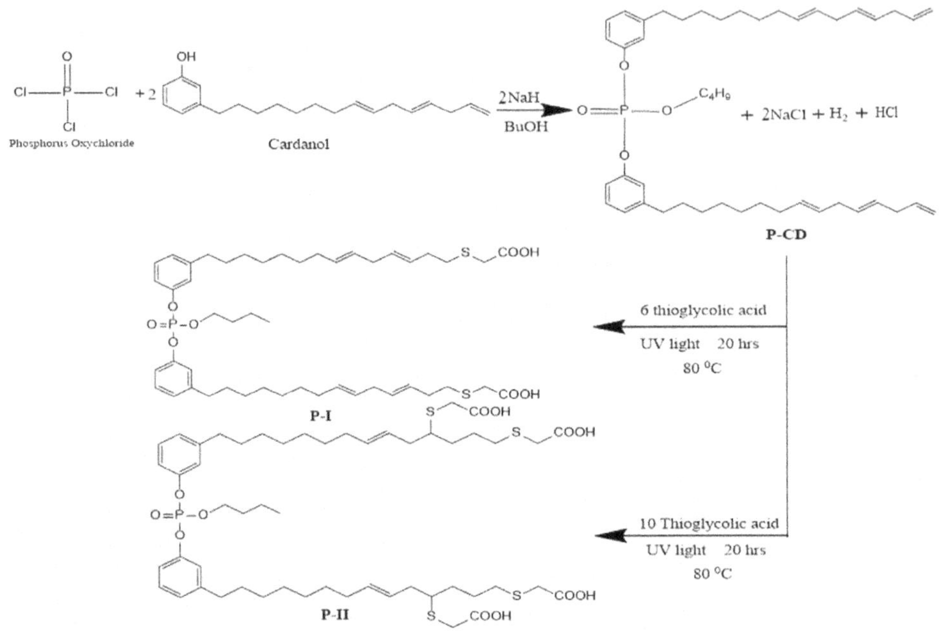

FIGURE 13.43 Schematic representation of preparation of P-S containing multifunctional carboxyl curing agent based on cardanol.[115]

a latent curing agent (IMPA, Entry 5, Table 13.3). For the corresponding bisphenol-A-type epoxy resin thermoset (E53/IMPA), its LOI is as high as 35%, and its PHHR, THR, and TSP have decreased 43%, 32%, and 32%, respectively compared with E53/imidazole. It is found that IMPA exerts a flame-retardant effect in both gaseous and condensed phases.[118]

Unlike cardanol and ricinoleic acid, which have long flexible aliphatic chains lignin derivatives are useful because of the aromatic skeleton. Vanillin-based flame retardant curing agents have been well investigated. Using vanillin and hexachlorocyclotriphosphazene as raw materials, polybasic carboxylic acid curing agent hexa-(4-carboxyl-2-methoxyphenoxy)-cyclotriphosphazene (HCPVC) is synthesized via the nucleophilic substitution and Pinnick oxidation reaction, and the synthetic pathway is shown in Figure 13.44. The obtained EP coating has a LOI of 31% and obtains UL 94 V-0, which is much superior to that cured with DDM or maleic anhydride.[119]

Besides polybasic carboxylic acid, multifunctional primary amine,[120] secondary amines[121–125] (Table 13.3, Entry 7–10) based on vanillin also can be designed. Via the Schiff base reaction between six-vanillin functionalized cyclophosphazene and N-methylethylenediamine, a multifunctional amine curing agent (HVPA, Entry 7, Table 13.3) is obtained. The resultant HVPA/glycerol diglycidyl ether (GDE) thermoset exhibits excellent flame retardancy: LOI value of 29%; UL 94 V-0 grade, PHRR, and THR reduction of 42% and 32% (compared to DDM/GDE due to the high content of N/P elements. The existence of dynamic imine endows HVPA/GED with fast self-healing, excellent shape memory, easy reprocessability, mild degradation, and superior bacterial resistance.[121]

Protocatechualdehyde (PH), which exists in the roots of the Chinese herbal medicines Salvia miltiorrhiza, is a natural phenolic acid compound containing two phenolic hydroxyl groups and an aldehyde group, which has been widely used medicinally due to its biological activity. Using PH, 4,4' -diaminodiphenylether and DOPO as raw materials, co-curing agent (PDD) is fabricated, and its synthetic route is shown in Figure 13.45. When it is used with DDM, the LOI value of epoxy resin thermoset reaches 33% with UL 94 V-0 grade, and the THR and TSP decrease by 40% and 61%, respectively (Entry 12, Table 13.3). PDD has flame-retardant action both in the gas and condensed phase, not only promoting formation of continuous and dense char layer but also releasing phosphorus radicals and nonflammable gases.[124]

Due to its many sources, low production cost, and wide applications, furfural is considered a very promising biomass material, which is obtained from fungal fermentation or biochemical treatment of plant cellulose. Furfural contains aldehyde and dienyl ether functional groups, and many chemical reactions can be carried out. For example, it can react with 3,5-diamino-1,2,4-triazole to form a Schiff base, and then an addition reaction occurs between the Schiff base and P-H bonds of DOPO (FGD, chemical structure shown in Table 13.3, Entry 13). Via this route, a furfural-based co-curing agent is obtained. Upon combination with DDM, low-phosphorus containing flame-retardant epoxy thermosets are obtained. For EP-5.0 whose phosphorus content is only 0.45 wt%, it achieves the UL-94 V-0 with an LOI value of 32%; the PHRR of EP-5.0 decreases by 31% compared with the control sample.[125]

In short, there are currently limited studies on bio-based flame-retardant curing agents, mostly concentrating

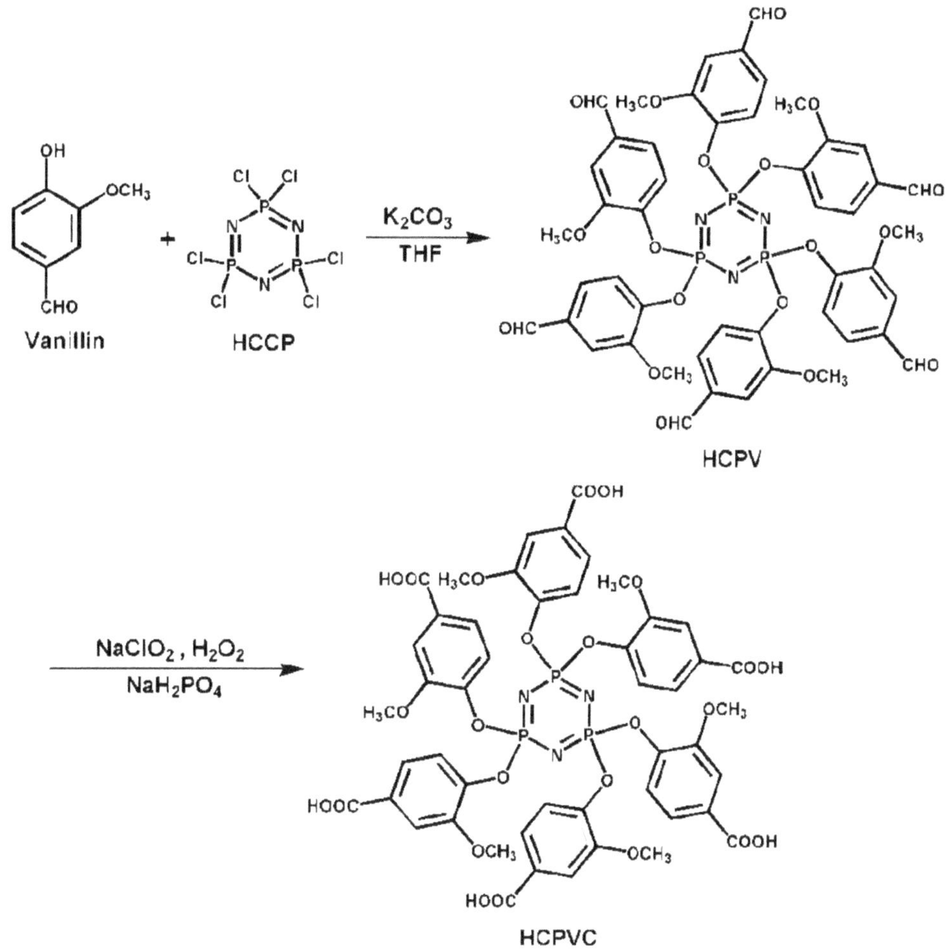

FIGURE 13.44 Synthetic route for polybasic carboxylic acid curing agent based on vanillin.[119]

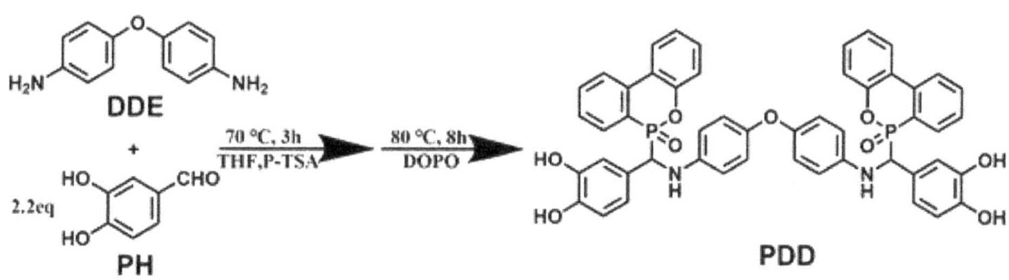

FIGURE 13.45 Synthetic route for co-curing agent based on protocatechualdehyde[124]

on secondary amines, phenols, and carboxylic acid curing agents, with very little research being done on primary amines and anhydrides.

13.3.3.3 Bio-Based Flame-Retardant Additives for EP

In addition to these reactive flame retardants, either as epoxide compounds or curing agent, some bio-based flame-retardant additives have been designed and used.[126–130] Bio-based compounds, like daidzein, vanillin, lignin, et al., have good char-forming ability during combustion, so combining them with phosphorus-containing bio-based FRs with high flame-retardant efficiency can be obtained. For example, using daidzein and diphenylphosphinyl chloride, which may work in the gas phase (effect of diphenyl phosphine oxide) and the condensed phase (effect of daidzein), a flame retardant additive (DPOD, whose chemical structure shown in Table 13.4 as Entry 1) suitable for EP is obtained. The EP blends with 6 wt% DPOD achieve the UL-94 V-0 rating, and an LOI value of 35%. The pHRR of EP/DPOD6 has also decreased.[126]

Commercial EP has poor toughness and its crack initiation and propagation happen easily, limiting its usage in

TABLE 13.4
Bio-Based Flame-Retardant Additives and Flame Retardancy of the Corresponded EPs Reported in the Literature

Entry	Bio-Based Platform	Bi-Based Flame Retardants	LOI	UL 94	PHRR kW/m²	THR MJ/m²	References
1	Daidzein	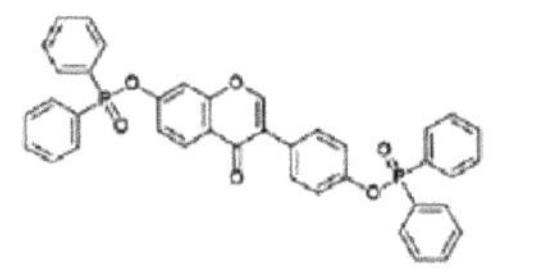 DPOD	35	V-0	544 (−34%)	75 (−23%)	[126]
2	Itaconic anhydride	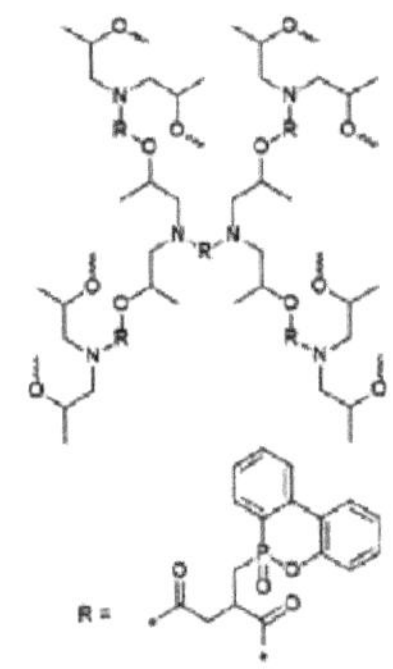 ITA-HBP	37	-	619 (−10%)	22(−14%)	[127]
3	Cardanol	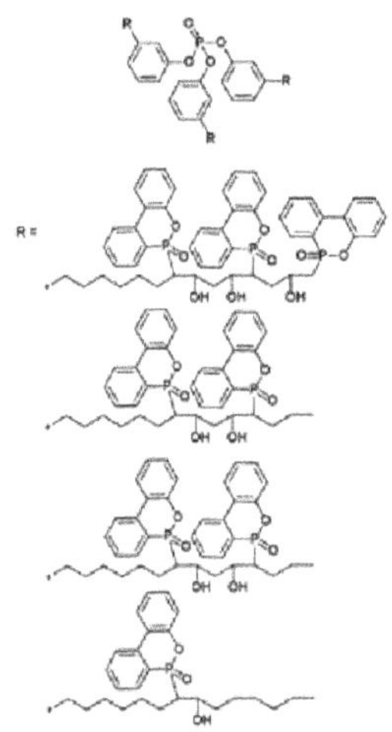 PTCP	31	-	748 (−50%)	63 (−27%)	[128]
4		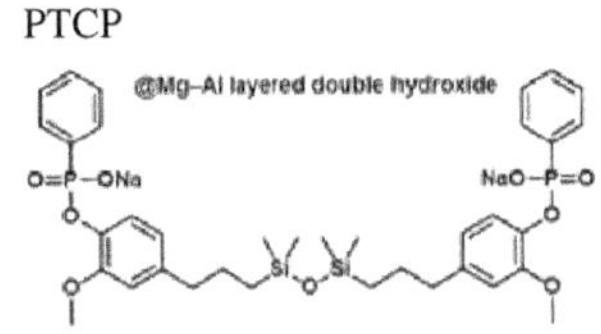 SIEPDP-LDH	28	V-0	658 (−28%)	87 (−4%)	[129]
5		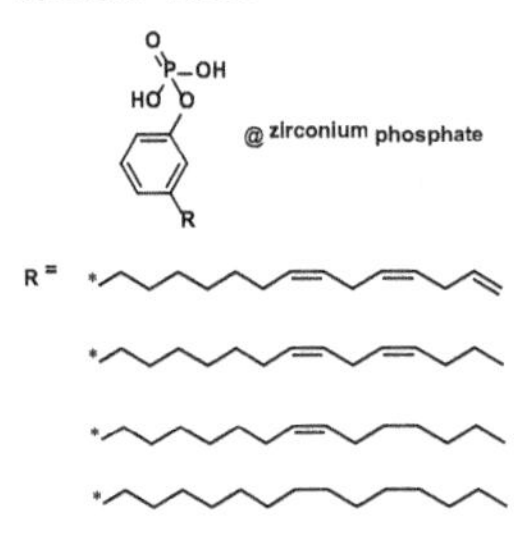 MCPA@ZrP	-	-	85 (−42%)	68 (−21%)	[130]

high-performance engineering materials. Some bio-based FRs which simultaneously endow EP with toughness have been designed. Itaconic acid (IA), also called methylene-succinic acid, is an unsaturated diacid produced by fermentation of carbohydrates using aspergillus terreus. Due to its great potential to replace the petrochemicals in the chemical industry, it has been selected as one of the top 12 potential bio-based platform chemicals by the US Itaconic anhydride, a derivative of itaconic acid, its double bonds can react with P-H of DOPO and N,N-dimethylformamide to obtain a hyperbranched oligomeric FR (ITA-HBP, Figure 13.46). With the addition of 5 phr ITA-HBP (0.26 wt% phosphorus content), the obtained epoxy thermoset (GEBA/DDM/ITA-HBP system) obtains a V-0 with an LOI of 37% (Entry 2, Table 13.4); PHRR, THR, and TSR are reduced as well. When ITA-HBP loading reaches 10 phr, the impact strength, fracture toughness, and fracture energy of the epoxy resins are significantly enhanced by 133%, 79%, and 125%, respectively, due to the in situ reinforcing and toughening effects induced by ITA-HBP.[127]

Cardanol-based flame-retardant toughening agent (phosphaphenanthrene groups-containing triscardanyl phosphate, PTCP) is synthesized via debydrochlorination, epoxidation, and ring opening reaction with phosphorus oxychloride, 3-chloroperbenzoic acid, and DOPO; the synthetic route is shown in Figure 13.47. The LOI value of EP (DGEBA/DDM)/PTCP-30% increases to 31% from 23% for neat EP (Entry 3, Table 13.4), while the PHRR, THR, and average effective heat of combustion values of EP/PTCP-30% are decreased by 50%, 27%, and 32%, respectively in comparison to those of neat EP. The impact strength of EP/PTCP-30% is enhanced from 15 kJ/m^2 in neat EP to 19 kJ/m^2, indicating the toughening effect of PTCP on EP.[128]

Nanocomposite technology provides a new idea for flame-retardant polymers without deterioration of their mechanical and thermal properties. In recent years, inorganic layered/polymer nanocomposites have received much attention to enhance the flame retardant properties of polymer composites. Combining inorganic layered compounds with bio-based flame reactants, these inorganic-organic hybrids FRs will provide EP with high flame retardancy while enhancing the mechanical properties (Entry 4–5, Table 13.4).[129,130] The cardanol-based zirconium phosphate (CZrP) is obtained by zirconium oxychloride octahydrate and monocardanyl phosphoric acid (MCPA). MCPA is synthesized from the reaction of cardanol and P_2O_5 in dry hexane (synthetic route is shown in Figure 13.48). Adding CZrP to EP not only the thermal stability but also the char yields have been raised. Compared with pure EP, the PHRR, THR, TSP, and CO production of EP/CZrP-6% decrease by 42%, 21%, 26%, and 36%, respectively.[130]

13.4 SUMMARY AND OUTLOOK

With the increasing concern on sustainable development, environmental protection, as well as human safety, many bio-based FRs have been fabricated to substitute petroleum-based counterparts over the past decades. This chapter reviews the fabrication and application research status of bio-based FRs based biomacromolecules (including

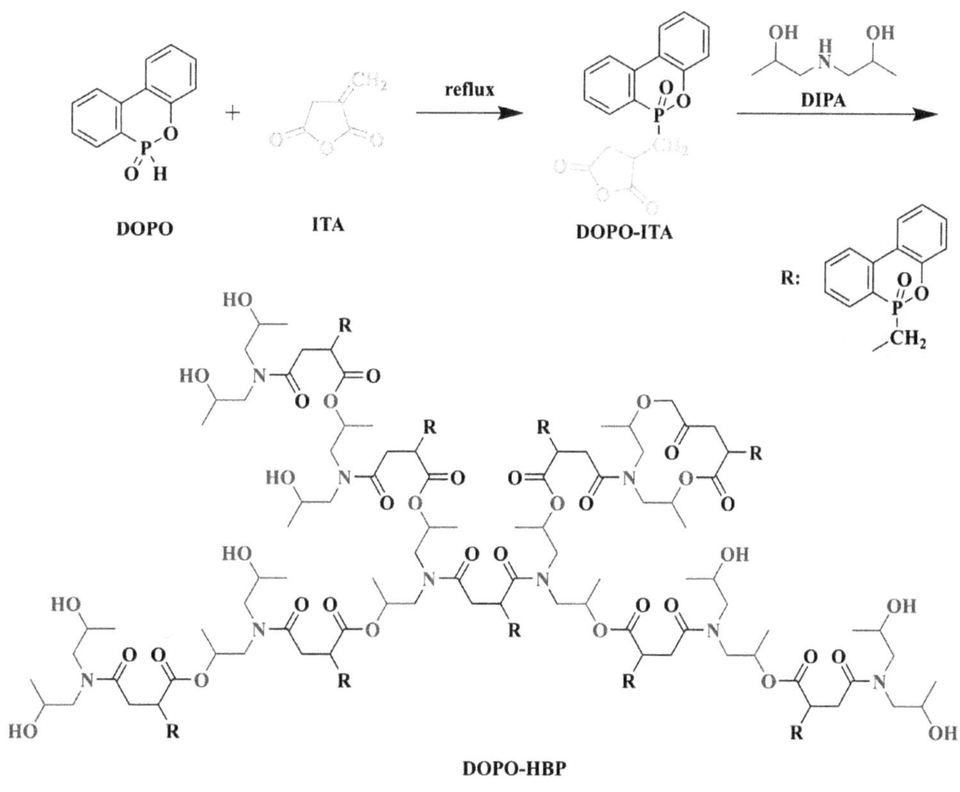

FIGURE 13.46 Synthetic route for itaconic anhydride-based hyperbranched FR.[127]

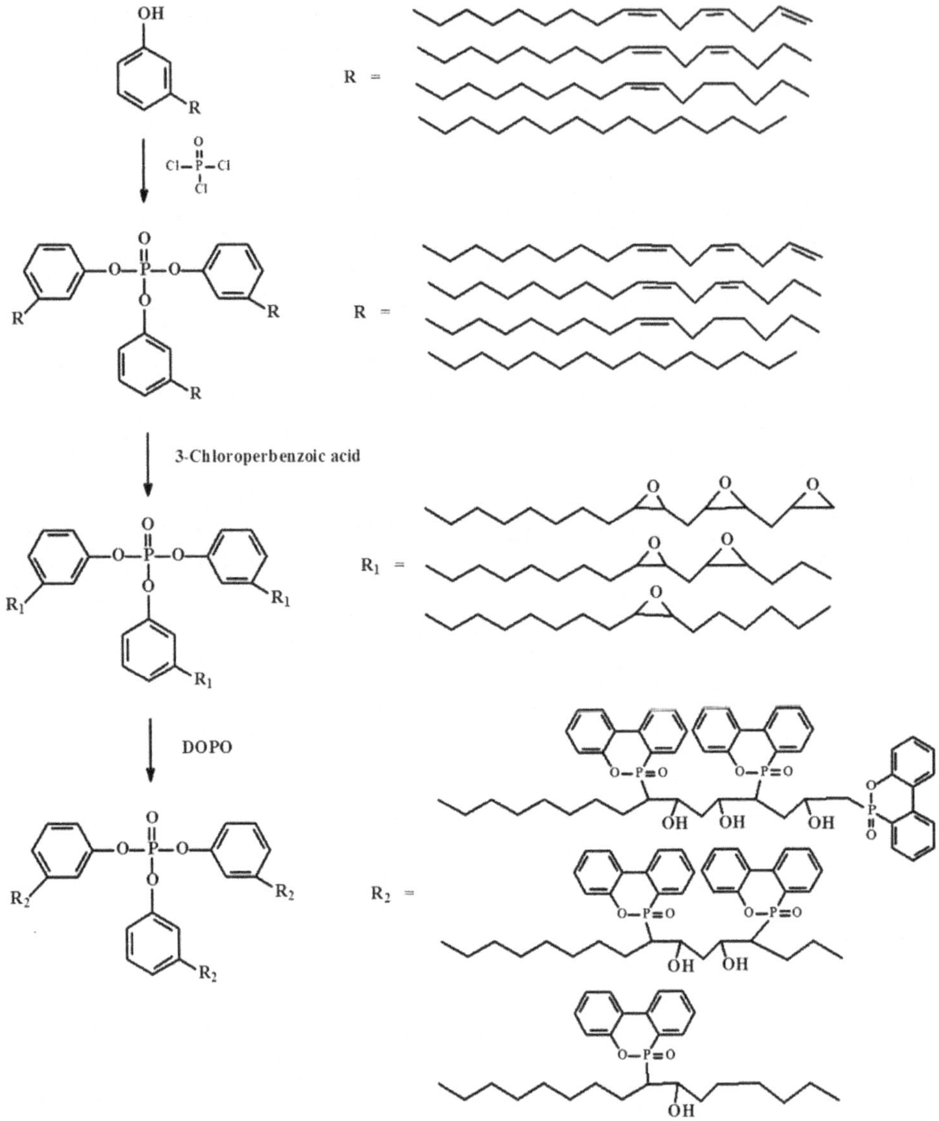

FIGURE 13.47 The synthetic route for cardanol-based flame-retardant toughening agent.[128]

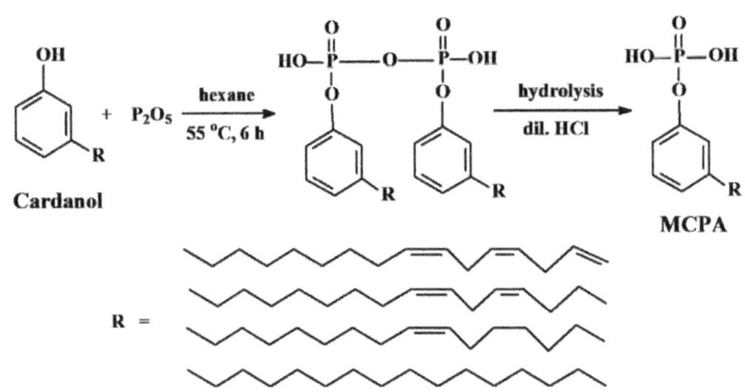

FIGURE 13.48 Synthetic route for MCPA.[130]

polysaccharides, protein, and lignin et al.) and bio-based small molecules (such as phytic acid, tannin, vanillin, or ferulic acid, et al.). Bio-based FRs show potential in flame-retardant applications, however, industrial scale up will not be feasible in the short term. Most biomacromolecules have complex composition and structure leading to difficulty in accurately controlling chemical modification. The poor melt processability make these biomacromolecules difficult to use as charring agents and introduced into polymer via melt blending. In terms of industrial application prospects, bio-based small molecules show potential. Via chemical decoration, they can be fabricated into different FR or intrinsic flame-retardant polymers. It is worth noting that some bio-based small molecules suffer from relatively high hydrophilicity and high cost, which will hamper their further application. Another thing that cannot be ignored is the production of these raw materials made from renewable substances which requires water and land use that are thus no longer available for food production, generating a competition for land use. The increasing market and interest in bio-based FRs are a positive trend, but during the replacement process, the production process of biomass, including the extraction of raw materials, are essential aspects that have to be considered in the evaluation of the life cycle assessment (LCA).

REFERENCES

[1] Qiu, X. Q.; Li, Z. W.; Li, X. H.; Zhang. Z. J. Flame Retardant Coatings Prepared Using Layer by Layer Assembly: A Review. *Chem. Eng. J.* **2018**, *334,* 108–122.

[2] Laufer, G.; Kirkland, C.; Cain, A. A. Grunlan, J. C. Clay–Chitosan Nanobrick Walls: Completely Renewable Gas Barrier and Flame-Retardant Nanocoatings. *ACS Appl. Mater. Interfaces.* **2012**, *4*, 1643–1649.

[3] Pan, H.; Wang, W.; Pan, Y.; Zeng, W.; Zhan, J.; Song, L.; Hu, Y.; Liew, K. M. Construction of Layer-by-Layer Assembled Chitosan/Titanate Nanotubes Based Nanocoating on Cotton Fabrics: Flame Retardant Performance and Combustion Behavior. *Cellulose.* **2015**, *22*, 911–923.

[4] Zhang, Z.; Ma, Z.; Wang, Y. Green and Facile Fabrication of α-Zirconium Phosphate-Based Core–Multishell Nanoparticles for Enhancing Fire Retardant Performance of Poly(Vinyl Alcohol). *ACS Appl. Nano Mater.* **2020**, *3*, 8708–8718.

[5] Maddalena, L.; Carosio, F.; Gomez, J.; Saracco, G.; Fina. A. Layer-by-Layer Assembly of Efficient Flame Retardant Coatings Based on High Aspect Ratio Graphene Oxide and Chitosan Capable of Preventing Ignition of PU Foam, *Polym. Degrad. Stab.* **2018**, *152*, 1–9.

[6] Wojtal, P.; Luo, D.; Chen, S.; Zhitomirsky, I. Composite Polymer–Metal Hydroxide Coatings with Flame-Retardant Properties. *Mater. Manuf. Process.* **2016**, *31*, 1201–1205.

[7] Nabipour, H.; Wang, X.; Song, L.; Hu. Y. A Fully Bio-Based Coating Made from Alginate, Chitosan and Hydroxyapatite for Protecting Flexible Polyurethane Foam from Fire. *Carbohydr. Polym.* **2020**, *246*, 116641.

[8] Fang, F.; Zhang, X.; Meng, Y.; Gu, Z.; Bao, C.; Ding, X.; Li, S.; Chen, X.; Tian, X. Intumescent Flame Retardant Coatings on Cotton Fabric of Chitosan and Ammonium Polyphosphate via Layer-by-Layer Assembly. *Surf. Coat. Technol.* **2015**, *25*, 9–14.

[9] Alongi, J.; Carosio, F.; Malucelli, G. Layer by Layer Complex Architectures Based on Ammonium Polyphos-Phate, Chitosan and Silica on Ppolyester-Cotton Blends: Flammability and Combustion Behaviour. *Cellulose.* **2012**, *19*, 1041–1050.

[10] Kundu, C. K.; Wang, X.; Hou, Y.; Hu, Y. Construction of Flame Retardant Coating on Polyamide 6.6 via UV Grafting of Phosphorylated Chitosan and Sol–Gel Process of Organosilane. *Carbohydr. Polym.* **2018**, *181*, 833–840.

[11] Tang, W.; Liang, G.; Wang, L.; Yuan, Y.; Dessie, W.; Liu, F.; Qin, Z.; Wang, Y.; Xiao, A.; Jin, X. Multi-Functional Flame Retardant Coatings Comprising Chitosan/Gelatin and Sodium Phytate for Rigid Polyurethane Foams. *J. Clean. Prod.* **2023**, *354*, 136371.

[12] Xu, Y.; Qu, L.; Liu, Y.; Zhu, P. An Overview of Alginates as Flame-Retardant Materials: Pyrolysis Behaviors, Flame Retardancy, and Applications. *Carbohyd. Polym.* **2021**, *260*, 117827.

[13] Wang, X.; Pan, Y.; Wan, J. Wang, D. An Eco-Friendly Way to Fire Retardant Flexible Polyurethane Foam: Layer-by-Layer Assembly of Fully Bio-Based Substances. *RSC. Adv.* **2014**, *4*, 46164–46169.

[14] Duskova, D.; Marounek, M.; Brezina, P. Determination of Phytic Acid in Feeds and Faeces of Pigs and Poultry by Capillary Isotachophoresis. *J. Sci. Food. Agric.* **2001**, *81*, 36–41.

[15] Evans, W. J.; McCourtney, E. J.; Shrager, R. I. Titration Studies of Phytic Acid. *J. Am. Oil. Chem. Soc.* **1982**, *59*, 189–191.

[16] Laufer, G.; Kirkland, C.; Morgan, A. B.; Grunlan, J. C. Intumescent Multilayer Nanocoating, Made with Renewable Polyelectrolytes, for Flame-Retardant Cotton. *Biomacromolecules.* **2012**, *13*, 2843–2848.

[17] Carosio, F.;Alongi, J. Few Durable Layers Suppress Cotton Combustion Due to the Joint Combination of Layer by Layer Assembly and UV-Curing. *RSC. Adv.* **2015**, *5*, 71482–71490.

[18] Pan, Y.; Zhan, J.; Pan, H.; Wang, W.; Tang, G.; Song, L.; Hu, Y. Effect of Fully Biobased Coatings Constructed via Layer-by-Layer Assembly of Chitosan and Lignosulfonate on the Thermal, Flame Retardant, and Mechanical Properties of Flexible Polyurethane Foam. ACS *Sustain. Chem. Eng.* **2016**, *4*, 1431–1438.

[19] Petkovska, J.; Mladenovic, N.; Marković, D.; Radoičić, M. ; Vest, N. A.; Palen, B.; Radetić, M.; Grunlan, J. C.; Jordanov, I. Flame-Retardant, Antimicrobial, and UV-Protective Lignin-Based Multilayer Nanocoating. *ACS Appl. Polym. Mater.* **2022**, *4*, 4528–4537.

[20] Chen, C.; Gu, X.; Jin, X.; Sun, J.; Zhang, S. The Effect of Chitosan on the Flammability and Thermal Stability of Polylactic Acid/Ammonium Polyphosphate Biocomposites. *Carbohyd. Polym.* **2017**, *157*, 1586–1593.

[21] Liu, X.; Gu, X.; Sun, J.; Zhang, S. Preparation and Characterization of Chitosan Derivatives and Their Application as Flame Retardants in Thermoplastic Polyurethane. *Cabohyd. Polym.* **2017**, *167*, 356–363.

[22] Xiong, Z.; Zhang, Y.; Du, X.; Song, P.; Fang, Z. Green and Scalable Fabrication of Core–Shell Biobased Flame Retardants for Reducing Flammability of Polylactic Acid. *ACS Sustain. Chem. Eng.* **2019**, 7, 8954–8963.

[23] Zhang, Y.; Xiong, Z.; Ge, H.; Ni, L.; Zhang, T.; Huo, S.; Song, P.; Fang, Z. Core–Shell Bioderived Flame Retardants Based on Chitosan/Alginate Coated Ammonia Polyphosphate for Enhancing Flame Retardancy of Polylactic acid. *ACS Sustain. Chem. Eng.* **2020**, *8*, 6402–6412.

[24] Liu, L.; Yao, M.; Zhang, H.; Zhang, Y.; Feng, J.; Fang, Z.; Song, P. Aqueous Self-Assembly of Bio-Based Flame Retardants for Fire-Retardant, Smoke-Suppressive, and Toughened Polylactic Acid. *ACS Sustain. Chem. Eng.* **2022**, *10*, 16313–16323.

[25] Ma, X.; Wu, N.; Liu, P.; Cui, H. Fabrication of Highly Efficient Phenylphosphorylated Chitosan Bio-Based Flame Retardants for Flammable PLA Biomaterial. *Carbohyd. Polym.* **2022**, *287*, 119317.[26] Shi, X.; Jiang, S.; Hu,

Y.; Peng, X.; Yang, H.; Qian, X. Phosphorylated Chitosan-Cobalt Complex: A Novel Green Flame Retardant for Polylactic Acid. *Polym. Adv. Technol.* **2018**, *29*, 860–866.

[27] Costes, L.; Laoutid, F.; Brohez, S.; Dubois, P. Bio-Based Flame Retardants: When Nature Meets Fire Protection. *Mater. Sci. Eng. R Rep.* **2017**, *117*, 1–25.

[28] Wang, J.; Ren, Q.; Zheng, W.; Zhai, W. Improved Flame-Retardant Properties of Poly(Lactic Acid) Foams Using Starch as a Natural Charring Agent. *Ind. Eng. Chem. Res.* **2014**, *53*, 1422–1430.

[29] Maqsood, M.; Seide, G. Investigation of the Flammability and Thermal Stability of Halogen-Free Intumescent System in Biopolymer Composites Containing Biobased Carbonization Agent and Mechanism of Their Char formation. *Polymers*, **2019**, *11*, 48.

[30] Maqsood, M.; Seide, G. Improved Thermal Processing of Polylactic Acid/Oxidized Starch Composites and Flame-Retardant Behavior of Intumescent Non-Wovens. *Coatings*. **2020**, *10*, 291.

[31] Carosio, F.; Fontaine, G.; Alongi, J.; Bourbigot, S. Starch-Based Layer by Layer Assembly: Efficient and Sustainable Approach to Cotton Fire Protection. *ACS Interfaces, Appl. Mater.* **2015**, *22*, 12158–12167.

[32] Ji, W.; Wang, D.; Guo, J.; Fei, B.; Gu, X.; Li, H.; Sun, J.; Zhang, S. The Preparation of Starch Derivatives Reacted with Urea-Phosphoric Acid and Effects on Fire Performance of Expandable Polystyrene Foams. *Carbohyd. Polym.* **2020**, *233*, 115841.

[33] Li, M.; Zhao, H.; Cheng, J.; Wang, T.; Fu, T.; Zhang, A.; Wang, Y. An Effective Green Porous Structural Adhesive for Thermal Insulating, Flame-Retardant, and Smoke-Suppressant Expandable Polystyrene Foam. *Engineering*. **2022**, *17*, 151–160.

[34] Ye, D.; Wang, T.; Liao, W.; Wang, H.; Zhao, H.; Wang, Y.; Xu, S.; Wang, Y. Ultrahigh-Temperature Insulating and Fire-Resistant Aerogels from Cationic Amylopectin and Clay via a Facile Route. *ACS Sustain. Chem. Eng.* **2019**, *7*, 11582–11592.

[35] Pan, H.; Song, L.; Ma, L.; Pan, Y.; Liew, K. M.; Y. Hu. Layer-by-Layer Assembled Thin Films Based on Fully Bio-Based Polysaccharides: Chitosan and Phosphorylated Cellulose for Flame-Retardant Cotton Fabric. *Cellulose*, **2014**, *21*, 2995–3006.

[36] Guo, Y.; He, S.; Zuo, X.; Xue, Y.; Chen, Z.; Chang, C.; Weil, E.; Rafailovich, M. Incorporation of Cellulose with Adsorbed Phosphates Into Poly (Lactic Acid) for Enhanced Mechanical and Flame Retardant Properties. *Polym. Degrad. Stabil.* **2017**, *144*, 24–32.

[37] Feng, J.; Sun, Y.; Song, P.; Lei, W.; Wu, Q.; Liu, L.; Yu, Y.; Wang, H. Fire-Resistant, Strong, and Green Polymer Nanocom-Posites Based on Poly(Lactic Acid) and Core–Shell Nanofibrous Flame Retardants. *ACS Sustain. Chem. Eng.* **2017**, *5*, 7894–7904.

[38] Yin, W.; Chen, L.; Lu, F.; Song, P.; Dai, J.; Meng, L. Mechanically Robust, Flame-Retardant Poly(Lactic Acid) Bio-Composites via Combining Cellulose Nanofibers and Ammonium Polyphosphate. *ACS Omega*, **2018**, *3*, 5615–5626.

[39] Costes, L.; Laoutid, F.; Khelifa, F.; Rose, G.; Brohez, S.; Delvosalle, C.; Dubois, P. Cellulose/Phosphorus Combinations for Sustainable Fire Retarded Polylactide. *Eur. Polym. J.* **2016**, *74*, 218–228.

[40] Phadungath, C. Casein Micelle Structure: A Concise Review. *Songklanakarin J. Sci. Technol.* **2005**, *27*, 201–212.

[41] Linder, M. B. Hydrophobins: Proteins That Self Assemble at Interfaces. *Curr. Opin. Colloid Interface Sci.* **2009**, *14*, 356–363.

[42] Alongi, J.; Carletto, R. A.; Bosco, F.; Carosio, F.; Blasio, A. D.; Cuttica, F.; Antonucci, V.; Giordano, M.; Malucelli, G. Caseins and Hydrophobins as Novel Green Flame Retardants for Cotton Fabrics. *Polym. Degrad. Stabli.* **2014**, *99*, 111–117.

[43] Carosio, F.; Blasio, A. D.; Cuttica, F.; Alongi, J.; Malucelli, G. Flame Retardancy of Polyester and Polyester–Cotton Blends Treated with Caseins. *Ind. Eng. Chem. Res.* **2014**, *53*, 3917–3923.

[44] Bosco, F.; Carletto, R. A.; Alongi, J.; Marmo, L.; Blasio, A. D.; G. Malucelli. Thermal Stability and Flame Resistance of Cotton Fabrics Treated with Whey Proteins. *Carbohyd. Polym.* **2013**, *94*, 372–377.

[45] Malucelli, G.; Barbalini, M. UV-Curable Acrylic Coatings Containing Biomacromolecules: A New Fire Retardant Strategy for Ethylene-Vinyl Acetate Copolymers. *Prog. Org. Coat.* **2019**, *127*, 330–337.

[46] Zhang, S.; Jin, X.; Gu, X.; Chen, C.; Li, H.; Zhang, Z.; Sun, J. The Preparation of Fully Bio-Based Flame Retardant Poly(Lactic Acid) Composites Containing Casein. *J. Appl. Polym.* **2018**, *135*, 46599.

[47] Yang, H.; Yu, B.; Xu, X.; Bourbigot, S.; Wang, H.; Song, P. Lignin-Derived Bio-Based Flame Retardants Toward High-Performance Sustainable Polymeric Materials. *Green Chem.* **2020**, *22*, 2129–2161.

[48] Maqsood, M.; Langensiepen, F.; Seide, G. The Efficiency of Biobased Carbonization Agent and Intumescent Flame Retardant on Flame Retardancy of Biopolymer Composites and Investigation of Their Melt-Spinnability. *Molecules*. **2019**, *24*, 1513–1530.

[49] Costes, L.; Laoutid, F.; Aguedo, M.; Richel, A.; Brohez, S.; Delvosalle, C.; Dubois, P. Phosphorus and Nitrogen Derivatization as Efficient Route for Improvement of Lignin Flame Retardant Action in PLA. *Eur. Polym. J.* **2016**, *84*, 652–667.

[50] Song, Y.; Zong, X.; Wang, N.; Yan, N.; Shan, X.; Li, J. Preparation of γ-Divinyl-3-Aminopropyltriethoxysilane Modified Lignin and Its Application in Flame Retardant Poly(Lactic Acid). *Materials*. **2018**, *11*, 1505.

[51] Shukla, A.; Sharma, V.; Basak, S.; Ali, S. W. Sodium Lignin Sulfonate: A Bio-Macromolecule for Making Fire Retardant Cotton Fabric. *Cellulose*. **2019**, *26*, 8191–8208.

[52] Palen, B.; Rabaey, M. G.; Danixa, R. M.; Iverson, E. T.; Kolibaba, T. J.; Grunlan, J. C. Polymeric Coacervate Coating for Flame Retardant Paper. *Cellulose*. **2022**, *29*, 4589–4597.

[53] Zhang, D.; Zeng, J.; Liu, W.; Qiu, X.; Qian, Y.; Zhang, H.; Yang, Y.; Liu, M.; Yang, D. Pristine Lignin as a Flame Retardant in Flexible PU foam. *Green Chem.* **2021**, *23*, 5972–5980.

[54] Alongi, J.; Cuttica, F.; Blasio, A. D.; Carosio, F.; Malucelli, G. Intumescent Features of Nucleic Acids and Proteins. *Thermochim. Acta*. **2014**, *591*, 31–39.

[55] Albite-Ortega, J.; Sanchez-Valdes, S.; Ramirez-Vargas, E.; Nunez-Figueredo, Y.; DeValle, L. F. R; Martínez-Colunga, J. G.; Graciano-Verdugo, A. Z.; Sanchez-Martínez, Z. V.; Espinoza-Martinez, A. B.; Rodriguez-Gonzalez, J. A.; Castaneda-Flores, M. E. Influence of Keratin and DNA Coating on Fire Retardant Magnesium Hydroxide Dispersion and Flammability Characteristics of PE/EVA Blends. *Polym. Degrad. Stabil.* **2019**, *165*, 1–11.

[56] Sousa, A. F.; Matos, M.; Freire, C. S. R.; Silvestre, A.; J. D.; J. F. J. Coelho. New Copolyesters Derived from Terephthalic and 2,5-Furandicarboxylic Acids: A Step Forward in the Development of Biobased Polyesters. *Polymer*, **2013**, *54*, 513–519.

[57] Suryaprabha, T.; Sethuraman, M. G. Fabrication of a Superhydrophobic and Flame-Retardant Cotton Fabric Using a DNA-Based Coating. *J. Mater. Sci.* **2020**, *55*, 11959–11969.

[58] Ortelli, S.; Malucelli, G.; Blosi, M.; Zanoni, I.; Costa, A. L. NanoTiO2@DNA Complex: A Novel Eco, Durable, Fire Retardant Design Strategy for Cotton Textiles. *J. Colloid Interface Sci.* **2019**, *546*, 174–183.

[59] Alongi, J.; Blasio, A. D.; Cuttica, F.; Carosio, F.; Malucelli, G. Bulk or Surface Treatments of Ethylene Vinyl Acetate Copolymers with DNA: Investigation on the Flame Retardant Properties. *Eur. Polym. J.* **2014**, *51*, 112–119.

[60] Kong, Q.; Wang, B.; Ji, Q.; Xia, Y.; Guo, Z.; Yu, J. Thermal Degradation and Flame Retardancy of Calcium Alginate Fibers. Chinese. *J. Polym. Sci.* **2009**, *27*, 807–812.

[61] Zhang, Y. Z.; Zhu, P.; Zhang, J. B.; Guo, X. Q. The Properties and Application of Alginate Fiber. *Text. Aux.* **2006**, *23*, 9–12.

[62] Agulhon, P.; Robizer, M.; Habbas, J.; Quignard, F. Influence of Both Cation and Alginate Nature on the Rheo-Logical Behavior of Transition Metal Alginate Gels. *Carbohyd. Polym.* **2014**, *112*, 525–531.

[63] Lu, Y.; Chen, J.; Wang, R.; Xu, P.; Zhang, X.; Gao, B.; Guo, C.; Yang, G. Bio-Inspired Cu-Alginate to Smartly Enhance Safety Performance and the Thermal Decomposition of Ammonium Perchlorate. *Appl. Surf. Sci.* **2019**, *15*, 269–275.

[64] Wang, Q.; Zhang, L.; Liu, Y.; Zhang, G.; Zhu, P. Characterization and Functional Assessment of Alginate Fibers Prepared by Metal-Calcium Ion Complex Coagulation Bath. *Carbohyd. Polym.* **2020**, *232*, 115693.

[65] Kundu, C. K.; Wang, W.; Zhou, S.; Wang, X.; Sheng, H.; Pan, Y.; Song, L.; Hu, Y. A green Approach to Constructing multilayered Nanocoating for Flame Retardant Treatment of Polyamide 66 Fabric from Chitosan and Sodium Alginate. *Carbohyd. Polym.* **2017**, *166*, 131–138.

[66] Wang, B. B.; Li, P.; Xu, Y.; Jiang, Z.; Dong, C.; Liu, Y.; Zhu, P. Bio-Based, Nontoxic and Flame-Retardant Cotton/Alginate Blended Fibres as Filling Materials: Thermal Degradation Properties, Flammability and Flame-Retardant Mechanism. *Compos. Part B-Eng.* **2020**, *194*, 108038.

[67] Li, P.; Wang, Q.; Wang, B.; Liu, Y.; Xu, Y.; Liu, Y.; Zhu, P. Blending Alginate Fibers with Polyester Fibers for Flame-Retardant Filling Materials: Thermal Decomposition Behaviors and Fire Performance. *Polym. Degrad. Stab.* **2021**, *183*, 109470.

[68] Zhang, F.; Wang, B.; Xu, Y.; Liu, Y.; Zhu, P. Convenient Blending of Alginate Fibers with Polyamide Fibers for Flame-Retardant Non-Woven Fabrics. *Cellulose.* **2020**, *27*, 8341–8349.

[69] Chen, H.; Wang, Y.; M. Sanchez-Soto and Schiraldi, D. A. Low Flammability, Foam-Like Materials Based on Ammonium Alginate and Sodium Montmorillonite Clay. *Polymer.* 2012, 53, 5825–5831.

[70] Shang, K.; Liao, W.; Wang, J.; Wang, Y.; Wang, Y.;Schiraldi, D. A. Nonflammable Alginate Nanocomposite Aerogels Prepared by a Simple Freeze-Drying and Post-Cross-Linking Method. *ACS Appl. Mater. Interfaces.* **2016**, *8*, 643–650.

[71] Liu, L.; Wang, W.; Hu, Y. Layered Double Hydroxide-Decorated Flexible Polyurethane Foam: Significantly Improved Toxic Effluent Elimination. *RSC Adv.* **2015**, *5*, 97458–97466.

[72] Pan, Y.; Wang, W.; Liu, L.; Ge, H.; Song, L.; Hu, Y. Influences of Metal Ions Crosslinked Alginate Based Coatings on Thermal Stability and Fire Resistance of Cotton Fabrics. *Carbohyd. Polym.* **2017**, *170*, 133–139.

[73] Pan, Y.; Liu, L.; Wang, X.; Song, L.; Hu, Y. Hypophosphorous Acid Cross-Linked Layer-by-Layer Assembly of Green Polyelectrolytes on Polyester-Cotton Blend Fabrics for Durable Flame-Retardant Treatment. *Carbohyd. Polym.* **2018**, 201, 1–8.

[74] Rahman, M. Z.; Kundu, C. K.; Wang, X.; Song, L.; Hu, Y. Bio-Inspired and Dual Interaction-Based Layer-by-Layer Assembled Coatings for Superior Flame Retardancy and Hydrophilicity of Polyamide 6.6 Textiles. *Eur. Polym. J.* **2021**, *145*, 110320.

[75] Chen, W.; Liu, Y.; Liu, Y.; Wang, Q. Preparation of Alginate Flame Retardant Containing P and Si and Its Flame Retardancy in Epoxy Resin. *J. Appl. Polym. Sci.* **2017**, *134*, 45552.

[76] Zhang, T.; Zeng, S.; Jiang, H.; Li, Z.; Bai, D.; Li, Y.; Li, J. Leather Solid Waste/Poly(Vinyl Alcohol)/Polyaniline Aero-Gel with mechanical Robustness, Flame Retardancy, and Enhanced Electromagnetic Interference Shielding. *ACS Appl. Mater. Interfaces.* **2021**, *13*, 11332–11343.

[77] Wang, Y.; Zhao, H.; Degracia, K.; Han, L.; Sun, H.; Sun, M.; Wang, Y.; Schiraldi, D. A. Green Approach to Improving the Strength and Flame Retardancy of Poly(Vinyl Alcohol)/Clay Aerogels: Incorporating Biobased Gelatin. *ACS Appl. Mater. Interfaces.* **2017**, *9*, 42258–42265.

[78] Mokhena, T. C.; Sadiku, E. R.; Ray, S. S.; Mochane, M. J.; Matabola, K. P.; M. Motloung. Flame Retardancy Efficacy of Phytic Acid: An Overview. *J. Appl. Polym. Sci.* **2022**, *139*, 52495.

[79] Cheng, X.; Guan, J.; Tang, R.; Liu, K. Phytic Acid as a Bio-Based Phosphorus Flame Retardant for Poly(Lactic Acid) Nonwoven Fabric. *J. Clean. Prod.* **2016**, *124*, 114–119.

[80] Thota, S.; Somisetti, V.; Kulkarni, S.; Kumar, J.; Negarajan, R.; Mosurkal, R. Covalent Functionalization of Cellulose in Cotton and a Nylon-Cotton Blend with Phytic Acid for Flame Retardant Properties. *Cellulose.* **2020**, *27*, 11–24.

[81] Kulkarni, S.; Xia, Z.; Yu, S.; Kiratitanavit, W.; Morgan, A. B.; Kumar, J.; Mosurkal, R.; Nagarajan, R. Bio-Based Flame-Retardant Coatings Based on the Synergistic Combination of Tannic Acid and Phytic Acid for Nylon–Cotton Blends. *ACS Appl. Interfaces, Mater.* **2021**, *13*, 61620–61628.

[82] Yang, W.; Tawiah, B.; Yu, C.; Qian, Y.; Wang, L.; Yuen, A. C.; Zhu, S.; Hu, E.; Chen, T. B.; Yu, B.; Lu, H.; Yeoh, G. H.; Wang, X.; Song, L.; Hu, Y. Manufacturing, Mechanical and Flame Retardant Properties of Poly(Lactic Acid) Biocomposites Based on Calcium Magnesium Phytate and Carbon Nanotubes. *Compos. Part A-Appl. S.* **2018**, *110*, 227–236.

[83] Li, D.; Zhao, X.; Jia, Y.; Wang, X.; Wang, Y. Tough and Flame-Retardant Poly(Lactic Acid) Composites Prepared via Reactive Blending with Biobased Ammonium Phytate and in Situ Formed Crosslinked Polyurethane. *Compos. Commun.* **2018**, *8*, 52–57.

[84] Gao, Y.; Deng, C.; Du, Y.; Huang, S.; Wang, Y. A Novel Bio-Based Flame Retardant for Polypropylene from Phytic Acid. *Polym. Degrad. Stabil.* **2019**, *161*, 298–308.

[85] Gu, L.; Shi, Y.; Zhang, A. Synthesis and Characterization of Bio-Based "Three Sources in One" Intumescent Flame Retardant Monomer and the Intrinsic Flame Retardant Water-Borne Polyurethane. *J. Polym. Res.* **2022**, *29*, 189.

[86] Chen, S.; Wu, F.; Hu, Y.; Lin, S.; Yu, C.; Zhu, F.; Wang, X. A Fully Bio-Based Intumescent Flame Retardant for Poly(Butylene Succinate). *Mater. Chem. Phys.* **2020**, *252*, 123222.

[87] Zhang, N.; Shen, J.; Pasquinelli, M. A.; Hinks, D.; Tonelli, A. E. Formation and Characterization of an Inclusion Complex of Triphenyl Phosphate and β-Cyclodextrin and Its Use as a Flame Retardant for Polyethylene Terephthalate. *Polym. Degrad. Stabil.* **2015**, *120*, 244–250.

[88] Zhang, N.; Yildirim, E.; Zane, C. P.; Shen, J.; Vinueza, N.; Hinks, D.; Tonelli, A. E.; Pasquinelli, M. A. Improved Eco-Friendliness of a Common Flame Retardant Through Inclusion Complexation with Cyclodextrins. *ACS Appl. Polym. Mater.* **2019**, *1*, 2768–2777.

[89] Ding, S.; Liu, P.; Zhang, S.; Gao, C.; Wang, F.; Ding, Y.; Yang, M. Crosslinking of β-Cyclodextrin and Combining with Ammonium Polyphosphate for Flame-Retardant Polypropylene. *J. Appl. Polym. Sci.* **2020**, *137*, 48320.

[90] Zhang, Y.; Han, P.; Fang, Z. Synthesis of Phospholipidated β-Cyclodextrin and Its Application for Flame-Retardant Poly(Lactic Acid) with Ammonium Polyphosphate. *J. Appl. Polym. Sci.* **2018**, *135*, 46054.

[91] T. Werpy and G. Petersen. Top Value Added Chemicals from Biomass: Volume I – Results of Screening for Potential Candidates from Sugars and Synthesis Gas. 2004. Available electronically at http://www.osti.gov/bridge.

[92] Gao, L.; Zheng, G.; Xie, X.; Wang, Y. Thermal Performance, Mechanical Property and Fire Behavior of Epoxy Thermoset Based on Reactive Phosphorus-Containing Epoxy Monomer. *J. Therm. Anal. Calorim.* **2017**, *127*, 1419–1430.

[93] Menard, R.; Negrell, C.; Fache, M.; Ferry, L.; Sonnier, R.; David, G. From a Bio-Based Phosphorus-Containing Epoxy Monomer to Fully Bio-Based Flame-Retardant Thermosets. *RSC Adv.* **2015**, *5*, 70856–70867.

[94] Ecochard, Y.; Decostanzi, M.; Negrell, C.; Sonnier, R.; Caillol, S. Cardanol and Eugenol Based Flame Retardant Epoxy Monomers for Thermostable Networks. *Molecules.* **2019**, *24*, 1818.

[95] Liu, J.; Dai, J.; Wang, S.; Peng, Y.; Cao, L.; Liu, X. Facile Synthesis of Bio-Based Reactive Flame Retardant from Van-Illin and Guaiacol for Epoxy Resin. *Compos. B. Eng.* **2020**, *190*, 107926.

[96] Wangm, S.; Ma, S.; Xu, C.; Liu, Y.; Dai, J.; Wang, Z.; Liu, X.; Chen, J.; Shen, X.; Wei, J.; Zhu, J. Vanillin-Derived High-Performance Flame Retardant Epoxy Resins: Facile Synthesis and Properties. *Macromolecules.* **2017**, *50*, 1892–1901.

[97] Yang, W.; Ding, H.; Zhou, W.; Liu, T.; Xu, P.; Puglia, D.; Kenny, J. M.; Ma, P. Design of Inherent Fire Retarding and Degradable Bio-Based Epoxy Vitrimer with Excellent Self-Healing and Mechanical Reprocessability. *Compos. Sci. Technol.* **2022**, *230*, 109776.

[98] Miao, J.; Yuan, L.; Guan, Q.; Liang, G.; Gu, A. Biobased Epoxy Resin Derived from Eugenol with Excellent Integrated Performance and High Renewable Carbon Content. *Polym. Int.* **2018**, *67*, 1194–1202.

[99] Wan, J.; Gan, B.; Li, C.; Molina-Aldareguia, J.; Li, Z.; Wang, X.; Wang, D. Y. A Novel Biobased Epoxy Resin with High Mechanical Stiffness and Low Flammability: Synthesis, Char-Acterization and Properties. *J. Mater. Chem. A.* **2015**, *3*, 21907–21921.

[100] Miao, J.; Yuan, L.; Guan, Q.; Liang, G.; Gu, A. Biobased Heat Resistant Epoxy Resin with Extremely High Biomass Content from 2,5-Furandicarboxylic Acid and Eugenol. *ACS Sustain. Chem. Eng.* **2017**, *5*, 7003–7011.[101] Wan, J.; Zhao, J.; Gan, B.; Li, C.; Molina-Aldareguia, J.; Zhao, Y.; Pan, Y.; Wang, D. Ultrastiff Biobased Epoxy Resin with High Tg and Low Permittivity: From Synthesis to Properties. *ACS Sustain. Chem. Eng.* **2016**, *5*, 2869–2880.

[102] Liu, J.; He, Z.; Wu, G.; Zhang, X.; Zhao, C.; Lei, C. Synthesis of a Novel Nonflammable eugenol-Based Phosp-Hazene Epoxy Resin with Unique Burned Intumescent Char. *Chem. Eng. J.* **2020**, *390*, 124620.

[103] Li, C.; Fan, H.; Aziz, T.; Bittencourt, C.; Wu, L.; Wang, D.; Dubois, P. Biobased Epoxy Resin with Low Electrical Permissivity and Flame Retardancy: From Environmental Friendly High-Throughput Synthesis to Properties. *ACS Sustain. Chem. Eng.* **2018**, *6*, 8856–8867.

[104] Wan, J.; Gan, B.; Li, C.; Molina-Aldareguia, J.; Kalali, E. N.; Wang, X.; Wang, D. Y. A Sustainable, Eugenol-Derived Epoxy Resin with High Biobased Content, Modulus, Hardness and Low Flammability: Synthesis, Curing Kinetics and Structure–Property Relationship. *Chem. Eng. J.* **2016**, *284*, 1080–1093.

[105] Pourchet, S.; Sonnier, R.; M. Ben-Abdelkader, Gaillard, Y.; Ruiz, Q.; Placet, V.; Plasseraud, L.; Boni, G. New Reactive Isoeugenol Based Phosphate Flame Retardant: Toward Green Epoxy Resins. *ACS Sustain. Chem. Eng.* **2019**, *7*, 14074–14088.

[106] Nabipour, H.; Qiu, S.; Wang, X.; Song, L.; Hu, Y. Phosphorus-Free Ellagic Acid-Derived Epoxy Thermosets with Intrinsic Antiflammability and High Glass Transition Temperature. *ACS Sustain. Chem. Eng.* **2021**, *9*, 10799–10808.

[107] Qi, Y.; Wang, J.; Kan, Y.; Pang, H.; Zhang, S.; Li, N.; Liu, C.; Weng, Z.; Jian, X. Synthesis of an Aromatic N-Heterocycle Derived from Biomass and Its Use as a Polymer Feedstock. *Nat Commun.* **2019**, *10,* 2107.

[108] Dai, J.; Peng, Y.; Teng, N.; Liu, Y.; Liu, C.; Shen, X.; Mahmud, S.; Zhu, J.; Liu, X. High-Performing and Fire-Resistant Bio-Based Epoxy Resin from Renewable Sources. *ACS Sustain. Chem. Eng.* **2018**, *6*, 7589–7599.

[109] Song, X.; Deng, Z.; You, C.; Sun, R.; Song, F.; Wang, X.; Chen, L.; Wang, Y. High-Performance and Fire-Resistant Epoxy Thermosets Derived from Plant-Derived Ferulic Acid. *Ind. Crops. Prod.* **2022**, *187*, 115445.

[110] Gao, T. Y.; Wang, ;Xu, Y.; Wei, C. X.; Zhu, S. E.; Yang, W.; Lu, H. D. Luteolin-Based Epoxy Resin with Exceptional Heat Resistance, Mechanical and Flame Retardant Properties. *Chem Eng. J.* **2022**, *428*, 131173.

[111] Qi, Y.; Weng, Z.; Zhang, K.; Wang, J.; Zhang, S.; Liu, C.; Jian, X. Magnolol-Based bio-Epoxy Resin with Acceptable Glass Transition Temperature, Processability and Flame Retardancy. *Chem. Eng. J.* **2020**, *387*, 124115.

[112] Liu, J.; Zhang, X.; Liu, S.; Lei, C. Allyl-Modified Epoxy Networks That Break and Reorganize at High Temperatures to Avoid T Reduction. *Chem. Eng. J.* **2023**, *455,* 140674.

[113] Wang, S.; Ma, S.; Li, Q.; Yuan, W.; Wang, B.; Zhu, J. Robust, Fire-Safe, Monomer-Recovery, Highly Malleable Thermosets from Renewable Bioresources. *Macromolecules.* **2018**, *51*, 8001–8012.

[114] Liu, Y.; Wang, B.; Ma, S.; Xu, X.; Qiu, J.; Li, Q.; Wang, S.; Lu, N.; Ye, J.; Zhu, J. Phosphate-Based Covalent Adaptable Networks with Recyclability and Flame Retardancy from Bio-Resources. *Eur. Polym. J.* **2021**,*144*, 110236.

[115] Wazarkar, K.; Sabnis, A. Synergistic Effect of P Single Bond S and Crosslink Density on Performance Properties of Epoxy Coatings Cured with Cardanol Based Multifunctional Carboxyl Curing Agents. *React. Funct. Polym.* **2018**, *128*, 74–83.[116] Mao, W.; Li, S.; Yang, X.; Cao, S.; Li, M.; Huang, K.; Xia, J. Preparation of a Flame-Retardant Epoxy Curing Agent Based on Castor Oil and Study on the Curing Reaction Kinetics. *J. Therm. Anal. Calorim.* **2017**, *130*, 2113–2121.

[117] Lin, C. H.; Chou, Y. C.; Shiao, W. F.; Wang, M. W. High Temperature, Flame-Retardant, and Transparent Epoxy Thermosets Prepared from an Acetovanillone-Based Hydroxyl Poly(Ether Sulfone) and Commercial Epoxy Resins. *Polymer.* **2016**, *97*, 300–308.

[118] Wang, J.; Huo, S.; Wang, J.; Yang, S.; Chen, K.; Li, C.; Fang, D.; Fang, Z.; Song, P.; Wang, H. Green and Facile Synthesis of Bio-Based, Flame-Retardant, Latent Imidazole Curing Agent for Single-Component Epoxy Resin. *ACS Appl. Polym. Mater.* **2022**, *4*, 3564–3574.

[119] Huang, Y.; Ma, T.; Wang, Q.; Guo, C. Synthesis of Bio-Based Flame-Retardant Carboxylic Acid Curing Agent and Application in Wood Surface Coating. *ACS Sustain. Chem. Eng.* **2019**, *7*, 14727–14738.

[120] Naiker, V. E.; Phalak, G. A.; Patil, D. A.; More, A. P.; Mhaske, S. T. Synthesis of Phosphorous-Containing Bio-Based Curing Agent for Flame Retardant Epoxy Resin System. *J. Coat. Technol. Res.* **2023**, *20*, 1325–1341.

[121] Peng, J.; Xie, S.; Liu, T.; Wang, D.; Ou, R.; Guo, C.; Wang, Q.; Liu, Z. High-Performance Epoxy Vitrimer with Superior Self-Healing, Shape-Memory, Flame Retardancy, and Anti-Bacterial Properties Based on Multifunctional Curing Agent. *Compos. B. Eng.* **2022**, *242*, 110109.

[122] Gu, L.; Qiu, J.; Sakai, E. A Novel DOPO-Containing Flame Retardant for Epoxy Resin: Synthesis, Nonflammability, and an Optimized Curing Procedure for High Performance. *High Perform. Polym.* **2017**, *29*, 899–912.

[123] Gu, L.; Chen, G.; Yao, Y. Two Novel Phosphorus–Nitrogen-Containing Halogen-Free Flame Retardants of High Performance for Epoxy Resin. *Polym. Degrad. Stabil.* **2014**, *108*, 68–75.

[124] Li, M.; Hao, X.; Hu, M.; Huang, Y.; Qiu, Y.; Li, L. Synthesis of Bio-Based flame-Retardant Epoxy Co-Curing Agent and Application in Wood Surface Coating. *Prog. Org. Coat.* **2022**, *167*, 106848.

[125] Niu, H.; Ding, H.; Huang, J.; Wang, X.; Song, L.; Hu, Y. A Furan-Based Phosphaphenanthrene-Containing Derivative as a Highly Efficient Flame-Retardant Agent for Epoxy Thermosets Without Deteriorating Thermomechanical Performances. *Chinese J. Polym. Sci.* **2020**, *40*, 233–240.

[126] Ma, C.; Li, J. Synthesis of an Organophosphorus Flame Retardant Derived from Daidzein and Its Application in Epoxy Resin. *Compos. B. Eng.* **2019**, *178*, 107471.

[127] Zhang, J.; Mi, X.; Chen, S.; Xu, Z.; Zhang, D.; Miao, M.; Wang, J. A Bio-Based Hyperbranched Flame Retardant for Epoxy Resins. *Chem. Eng. J.* **2020**, *381*, 122719.

[128] Wang, X.; Zhou, S.; Guo, W.; Wang, P.; Xing, W.; Song, L.; Hu, Y. Renewable Cardanol-Based Phosphate as a Flame Retardant Toughening Agent for Epoxy Resins. *ACS Sustain. Chem. Eng.* **2017**, *5*, 3409–3416.

[129] Li, C.; Wan, J.; Kalali, E. N.; Fan, H.; Wang, D. Y. Synthesis and Characterization of Functional Eugenol Derivative Based Layered Double Hydroxide and Its Use as a Nanoflame-Retardant in Epoxy Resin. *J. Mater. Chem. A.* **2015**, *3*, 3471–3479.

[130] Fu, X. L.; Wang, X.; Xing, W.; Zhang, P.; Song, L.; Hu, Y. Two-Dimensional Cardanol-Derived Zirconium Phosphate Hybrid as Flame Retardant and Smoke Suppressant for Epoxy Resin. *Polym. Degrad. Stab.* **2018**, *151*, 172–180.

14 Conformal Flame Retardant Coating Systems

Thomas J. Kolibaba[1], *Federico Carosio*[2], *and Jaime C. Grunlan*[3,4,5]

[1] Applied Chemicals and Materials Division, National Institute of Standards and Technology, Boulder, CO, USA.

[2] Dipartimento di Scienza Applicata e Tecnologia, Politecnico di Torino, Alessandria Campus, Viale Teresa Michel 5, 15121 Alessandria, Italy.

[3] Department of Mechanical Engineering, Texas A&M University, College Station, TX, USA.

[4] Department of Materials Science Engineering, Texas A&M University, College Station, TX, USA.

[5] Department of Chemistry, Texas A&M University, College Station, TX, USA.

14.1 MEETING THE FLAMMABILITY CHALLENGE AT THE INTERFACE

Flame retardant (FR) surface treatments have been a target of scientific and industrial focus for many decades.[1] The development of coating technologies was borne out of a combination of necessity and convenience. Necessity due to the fact that natural fibers such as cotton cannot be coextruded with a flame retardant FR molecule in a manner analogous to synthetic polyester or polyamide fabrics. Convenience in the sense that the processing equipment to finish textiles has been refined for centuries to apply dyes or other finishes to fabrics. Flame-retardant finishes are a natural extension of these conditions, and flame-retardant treatments for textiles have been an active research topic ever since. Please see Chapter 27 of this book for more information on textile flammability and fire protection.

Inadvertently, the development of flame retardant textile finishes stumbled upon a critical advantage in the fight against fire and flammability: surface area. In the decades since the first flame retardant textile finishes appeared, researchers have become keenly aware of the importance of surface area and localization of flame retardant chemistries to impart fire protection to materials. Combustion naturally begins at the exterior of a sample, meaning that surface treatments will yield materials with a higher ratio of FR:substrate in the early stages of combustion. This is critical to mitigate the hazards of material flammability. Additionally, the use of a surface treatment mitigates the degradation of mechanical properties brought on by incorporation of bulk additives.[2] When a surface treatment can be easily removed at the end of a product's life cycle, the underlying material can be more easily recycled. Increasing the potential for material recyclability and circularity makes surface treatments a core research focus. For these reasons, surface treatments for fire protection have become a major area of academic and industrial investment in recent years.[3]

In this chapter, we will outline some of the key substrates that benefit most from surface treatments for flammability reduction and introduce the major techniques to apply such treatments. Particular attention will be paid to the material and process sustainability of the outlined techniques. We note that many techniques that could broadly be classified as "covalent surface modification" insomuch as they yield a mostly unaltered (at both macro- and microscopic length scales) surface could be considered conformal treatments. Many of those chemistries are covered more thoroughly in other chapters and, as such, will only be minimally addressed in this chapter.

14.1.1 Common Substrates for Flame-Retardant Surface Treatments

14.1.1.1 Textiles

As previously noted, textiles were some of the first substrates where coating technologies for flammability reduction became commonplace.[1] Textiles remain a critical substrate to protect due to the widespread prevalence of severe injuries and deaths related to clothing burns, especially from synthetic fibers such as polyester and polyamide that are prone to melting.[4] Naturally derived cotton and synthetic fibers such as polyesters (most often polyethylene terephthalate, PET) or polyamides (most often polyamide-6,6 or simply abbreviated as PA), and their blends with cotton, are the subject of most textile flammability research. The degradation pathways of cotton, which is cellulosic, and synthetic fibers are markedly different and require different chemistries in order to reduce their flammability.

Cotton's thermal decomposition occurs in a two-stage process, beginning with a dehydroxylation and partial charring.[5] The initial degradation step preserves the structure of the cotton, and this is followed by a diffusion-controlled oxidation of the charred material. Cotton's relatively high char yield and low tendency to produce flammable volatiles

DOI: 10.1201/9781003380689-14

during its decomposition are the cause of its relatively low heat release. The tendency of cotton to form carbonaceous material is leveraged by intumescent treatments, which are designed to promote cross-linking and density of the charred area. The cross-linked char yields a material that is more resistant to the secondary oxidation step.[6] As a result, many FR treatments for cotton and other natural fibers are phosphorus-based, leveraging cotton's tendency to char to their advantage.

In contrast, synthetic fibers do not tend to form char as they decompose. Rather, they typically depolymerize into lower molecular weight oligomers during pyrolysis.[7,8] Instead of preserving the fibrous structure like cotton does, synthetic fibers are instead prone to melt dripping.[9] This dripping promotes flame spread but moreover poses a challenge in arresting flame spread since there is no easy scaffold for maintaining structure and promoting char formation. In addition to the depolymerization and melt behavior, synthetic fibers based upon aliphatic carbon chemistry (polyolefins, aliphatic nylons) have a higher total heat release (THR) and peak heat release rate (pkHRR) than cotton.[10] Blends of cotton and synthetic fibers such as polyester-cotton (PECO) or nylon-cotton (NYCO) are commonplace to take advantage of the robustness of synthetic fibers and the breathability and comfort of cotton. Blends pose a particular flammability challenge because the charring of the cotton in the blend provides a scaffold for faster flame spread as the synthetic component degrades, yielding extremely flammable fabrics.[11]

14.1.1.2 Wood

Home structure fires are responsible for approximately three quarters of fire deaths in the United States.[12] Wood is a common building component in home construction and as such is a critical substrate for fire protection. There has also been an added emphasis on the flammability of outdoor wood structures (e.g., fences) owing to their tendency to propagate wildfires.[13] Wood has a high fraction of cellulose and as such exhibits many of the same degradation properties as cotton. The early stages of wood degradation lead to the removal of aliphatic materials, which are lost as CO_2 or become aromatic char.[14] Despite this mass loss, the porosity of wood does not increase appreciably during the early stages of thermal degradation (up to ~250°C) but does increase notably at higher temperatures when aromatic carbon begins to degrade. Please see Chapter 26 of this book for more information on wood flammability and fire protection approaches.

Wood has slow diffusion kinetics owing to its very complex surface. These slow transport properties mean that, unlike in the cases of textiles or foam, coatings are unlikely to permeate the entirety of a treated piece of wood.[15] Much like cotton, char promotion via phosphorus-based FRs have been demonstrated to be effective. Since wood has fewer flexibility and comfort demands than cotton, there is also a greater tendency to incorporate inorganic treatments to reduce wood flammability.[16] Flame-retardant treatments for wood must consider carefully the end use. For example, flooring has high durability demands, while fencing or building siding will have higher demands for weathering and UV exposure. As with textiles, there is a strong existing market for surface treatments of wood, such as varnishes, paints, and even pressure treatments to impregnate wood with materials to prevent rot. Due to this existing infrastructure, it is ideal if surface treatments to reduce wood flammability are compatible with these technologies.

14.1.1.3 Foam

The United States alone produces over 1.9 million metric tons of polyurethane foam annually, split nearly evenly between rigid and flexible types.[17] Furniture and bedding account for over half of all flexible foam production, with the bulk of the remainder going to transportation. These materials are now ubiquitous in homes as furnishings and bedding, and in the transportation sector where they are used most heavily for seating. Polyurethane foams exhibit a high heat release rate (HRR) after ignition and produce a large amount of especially toxic smoke.[18] In addition, during combustion polyurethane foams tend to aggressively melt-drip, which can contribute to flame spread and flashover events. Coatings serves as a good means of imparting fire resistance to foams owing to their high surface area and porosity. Because of foam's tendency to melt-drip, foam FR coatings tend to focus first on providing structural support. This is often in the form of a ceramic treatment or shell and can sometimes be paired with other chemistries to slow and/or reduce heat release as the foam degrades. Due to the high levels of smoke release by most polyurethane foams, reduction of smoke release is especially important. Please see Chapter 31 of this book for more information on flammability and fire protection of foam materials.

Unlike textiles and wood, there are not currently commonplace coating technologies in use for foam. Due to their compressibility, conformal coatings can be applied to flexible foams, followed by manual expulsion of the solution carrying the coating materials. Rigid foams present a greater challenge and are not as amenable to coating treatments. The implementation of flow-through technologies could offer an automated way to coat foam at an industrial scale. Even so, the sheer volume of solution paired with the time and energy costs of drying coated foams have acted as a barrier to commercial implementation of many solution-based coating technologies for foam.

14.2 LAYER-BY-LAYER TREATMENTS

Layer-by-layer (LbL) assembly represents an interesting technique for the design and deposition of functional coatings and membranes. The fundamental concepts of this technique were disclosed in 1966 by Iler who reported the deposition of oppositely charged micro particles on glass.[19] Despite this early work, the practical application of the LbL was only realized decades later by Decher.[20] Since then, the LbL technique has been successfully applied in many research fields including the deposition of FR coatings.[21,22] Through the years, this multilayered deposition has proved to be an excellent FR strategy, as demonstrated by high number of published papers as well as by the many reviews available on this topic.[3,23–26] Within this context, this section is aimed at providing the fundamental operating principles of the technique and the most interesting FR results achieved on different substrates.

14.2.1 Practical Considerations of Layer-by-Layer Assembled Flame-Retardant Treatments

LbL assembly results in multilayered coatings by exploiting one or more interactions occurring between selected components in a specific solvent.[22] Although many possibilities exist (e.g., hydrophobic interactions, donor/acceptor interactions, or specific recognition), up to now most of the LbL assemblies have been fabricated through the electrostatic attraction that occurs between polyelectrolytes (PE) or nanoparticles (NP) in water.[27] A typical deposition procedure encompassing for example a positive PE and a negative NP can be performed by alternately dipping the substrate in the PE and NP containing aqueous solutions/suspensions (Figure 14.1). A rinsing step is normally performed in order to remove the excess PE before transferring the substrate into the NP suspension. A final rinsing step completes the deposition cycle. During each deposition step, polyanion-polycation complexes are formed at the substrate/water interface with an overall surface charge reversal that depends on the charge of the deposition bath.[28] By repeating these steps multiple times, a multilayered coating with controlled thickness and composition is generated.

The number of interacting layers within the assembly repetitive unit (deposited in a single cycle) is conventionally referred to as bilayer (BL), trilayer (TL), or quadlayer (QL) when two, three, or four components are employed, respectively. The chemistry of the selected components, the concentration, the pH, the ionic strength, temperature, and time are the main parameters controlling the outcome of the deposition.[22] This produces broad processing windows where it is always possible to find the conditions that allow for an optimal film build-up once the constituents have been identified. Thanks to this design freedom, the development of LbL nanocoatings in the field of flame retardancy started off by assembling components capable of mimicking the main FR action of conventional systems such as polymer nanocomposites and intumescent coatings.[29–32] Indeed, both systems can achieve substantial FR results by producing, during combustion, a protective barrier capable of limiting the mass/heat transfer from/to the flame. By combining synthetic or natural polyelectrolytes (e.g., polyethylenimine (PEI), chitosan (CS), etc.) with nanoparticles of different shapes (e.g., silica nanospheres (SNP), halloysite nanotubes (HN), montmorillonite nanoplatelets (MMT), etc.), it is possible to assemble a nanoparticle rich layer similar to the one produced during combustion by polymer-clay nanocomposites.[29] Similarly, if the nanoparticles are replaced by phosphate containing polyelectrolytes (e.g., ammonium polyphosphate (APP) and poly(sodium phosphate) (PSP, sometimes referred to as sodium hexametaphosphate)), it is possible to deposit assemblies with micro-intumescent features.[32]

To state that the FR LbL coatings fall in either one of the two described FR archetypes would be a great limitation. There are indeed many other classes of materials that can be incorporated within an LbL coating such as carbon-related materials (e.g., nanotubes, nanographites, graphenes, etc.), biomacromolecules (e.g., nucleic acids, proteins, etc.), and MXenes (a novel broad family of two-dimensional carbides and nitrides of transition metals).[22] With such a broad selection of components, it is possible to investigate different hybrid FR mechanisms as well as include additional functional characteristics such as super-hydrophobicity, electrical conductivity, and anti-bacterial properties.[33–36]

Up to now, dipping has been the most widely used approach to deposit FR LbL coatings. This is linked to the ease and accessibility of this method that represents the most studied choice for the assembly of LbL coatings in all research fields. Laboratory scale dipping methods can be easily scaled up to industrial pad-dry processing equipment and maintain coating performance.[37,38] In an effort to speed up the processing time, spray-assisted LbL deposition has also been demonstrated as viable option to deposit functional coatings on various substrates.[39–41] Contrary to the conventional dipping procedure, this latter approach is relatively new, and more work is needed in regard to the fundamental aspects of the process.

14.2.2 Use-Cases of Layer-by-Layer Assembled Flame-Retardant Coatings

Here we discuss the most efficient FR LbL treatments for textiles, foams, and other substrates. Examples were chosen based on the achieved FR performances, the adopted FR mechanism, and the particular features of the assembled LbL coating (e.g., first use of a certain chemical/FR mechanism/deposition approach, implementation of multifunctional characteristics, etc.).

14.2.2.1 Textiles

Textiles (natural, synthetic, and blends) were the first substrates to be coated using layer-by-layer assembly for flame retardant purposes. This can be linked to the already existing surface approaches and industrial-scale equipment conventionally employed to improve their FR properties. Indeed, the direct impregnation of the FR chemical represent the only options to confer FR properties to natural fabrics such as cotton, ramie, silk, and wool.[42] Synthetic fabrics such as PET, PA, and polyacrylonitrile (PAN) might also rely on additional approaches, such as the bulk inclusion of the FR additive during melt- or wet-processing and the copolymerization of FR monomer within the polymer chain.[43] These

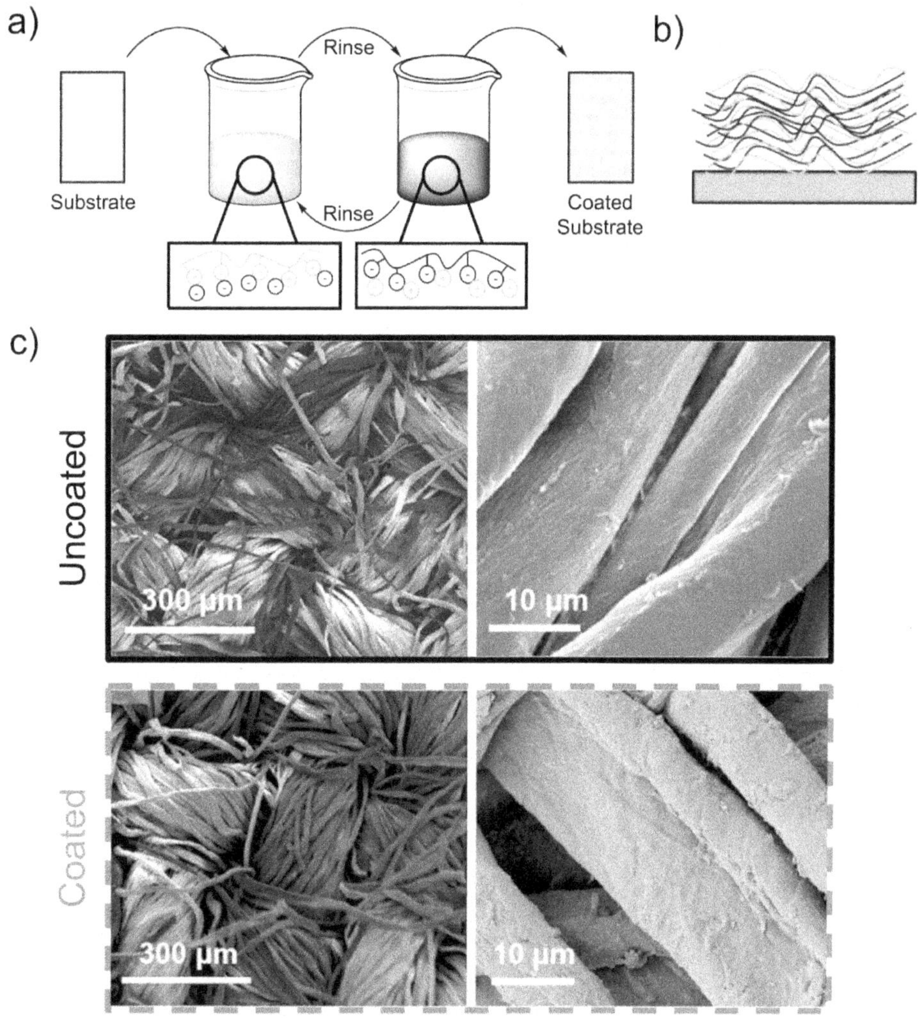

FIGURE 14.1 (a) Schematic of layer-by-layer assembly, where a substrate is first dipped into a solution containing a charged polymer (here shown as a polycation), followed by a rinse step and then immersion in a solution containing the oppositely charged polymer. (b) Cross-section cartoon showing the deposited coating atop a substrate. Interdiffusion between deposited polyelectrolytes is common, as shown here. (c) Scanning electron microscope images highlight the conformal nature of a CS/PSP coating on cotton fabric. The fibrous structure of the textile is preserved after coating even at the microscopic scale. Adapted from Ref. [44] with permission.

latter approaches suffer from constraints such as limited processability and decreased mechanical properties of the produced fibers. From this point of view LbL conformal coatings represent a viable choice as the treatment does not affect the fiber production outcome. Electron microscope images, such as those displayed in Figure 14.1c, highlight the conformal nature of LbL coatings, with treated fabric appearing nearly indistinguishable from untreated. If designed correctly, an LbL treatment should not modify the main fabric features related to mechanical and optical properties (i.e., comfort and design).[44] This is thanks to the already mentioned versatility (materials choice and processing parameters) of the LbL approach that allow for the proper design of the coating depending on the substrate of choice and the desired functional properties. Indeed, although there are some studies reporting a singular LbL treatment capable of conferring FR properties to cotton, PET, and their blend (i.e., PECO), the most efficient approach is to select the coating constituents and FR action in order to match the chemistry and burning behavior of the fabric.[45] To this end, coating formulations based on the concept of intumescence have proved most effective on natural fabrics such as cotton and ramie.

Intumescent-like formulations can be easily assembled in a BL fashion by combining polyphosphates (e.g., polyphosphoric acid PPA, APP, and PSP) with either synthetic or natural polyelectrolytes (e.g., polyethylenimine (PEI), chitosan (CS), or starch (ST)).[46–49] The polyanions act as the acid source, while the polycations act as the carbon source. The foaming agent, the third component of an intumescent formulation, is produced upon decomposition of the coating components. For example, ammonia can be released from either unexchanged ammonium ions in APP or the amino groups of CS.[50] Decomposition products coming from the polycation can also contribute to the formation of an expanded protective structure. A landmark article was published in 2011 by Li et al. where poly(allylamine) PAH was combined with PSP on cotton fabric.[46] A 20 BL coating imparted self-extinguishing behavior in a vertical flame spread test (VFST, ASTM D6413) and dramatically increased the amount of char produced upon cotton decomposition. The formation of micron-scale intumescent like structures was confirmed by post combustion residue SEM observations (Figure 14.2). Interestingly, the same PAH/PSP coating did not produce expanded structures when deposited at 10 BL, indicating that an intumescent system requires a minimum coating thickness in order to properly display its FR action. Similar FR results were achieved on ramie fabrics by using PEI and APP.[48] This system was deposited by either conventional dipping or spray, highlighting how the latter can reduce the number of BL required to achieve the desired FR properties to 10–15, while also reducing the overall processing time.[51] The durability of the coating and its FR properties to laundering represents an additional challenge for waterborne LbL coating.

In order to address the washing problem, different strategies can be exploited. For example, the inclusion of a UV-curable latex in an intumescent formulation based on CH and APP has been shown to improve the durability of the coating while maintaining good FR properties.[49] After photo-curing, a 3 BL coating exhibits self-extinguishing behavior in a horizontal flame spread test (HFST) after washing for 1h at 65°C. More recently, LbL coatings durable to 50 washing cycles (AATCC 143–2014 method) have been produced by assembling phenylphosphonic acid (PHA) and 3-aminopropyltriethoxysilane (APTES) on polydopamine (PDA)-coated cotton.[52] Fully renewable intumescent-like LbL coatings can also be assembled by replacing the synthetic acid source with bio-macromoleclules such as deoxyribonucleic acid (DNA), caseins (CA), vitamins, and phytic acid (PhA).[53–56] PhA has been recently used to create a high performing intumescent treatment deposited on cotton.[57,58] For example, PEI/melamine and PhA can self-extinguish in a vertical flame test with only 4 BL deposited. When post treated with a diluted solution of poly(dimethylsiloxane) (PDMS), the coated fabric also achieves good water repellent behavior.[58]

In parallel to synthetic or natural polyelectrolytes, nanoparticles have also been explored in FR assemblies. As far as FR properties are concerned, although nanoparticle-based LbL coatings were first to be investigated on cotton, the achieved results were limited compared with intumescent assemblies.[59,60] Hybrid coatings encompassing both approaches have recently demonstrated efficacy when the selected nanoparticle is employed as mechanical

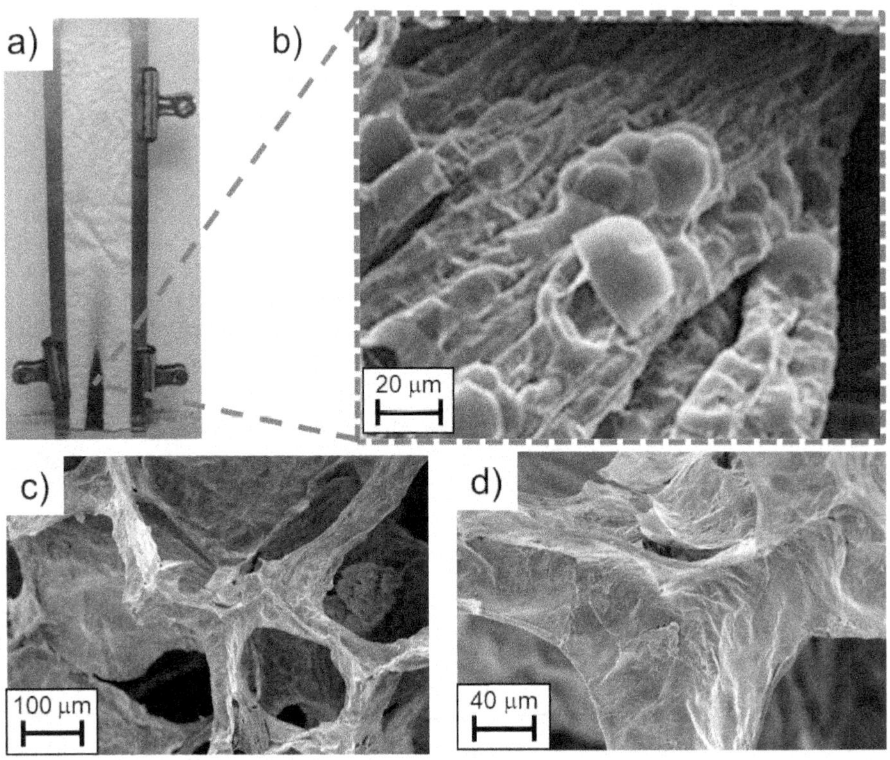

FIGURE 14.2 (a) Image of cotton fabric coated with an intumescent LbL film after vertical flame testing (ASTM D6413). (b) Inset shows the micro-intumescent effect, which yields self-extinguishing behavior and protects the fabric from further damage by fire. (c) Example of a nanoparticle-based LbL coating preserving the structure of open cell polyurethane foam after fire testing. (d) Electron micrograph showing struts of the foam are hollowed out, leaving an empty shell of graphitic material behind. Parts (a,b) adapted from Ref. [46].

reinforcement of the produced intumescent structure.[61] Besides improving the FR properties, the use of nanoparticles can provide additional properties to the deposited coating. For example, the use of Ag nanowires in combination with a PEI/PhA assembly has shown both self-extinguishing behavior and EMI shielding characteristics on cotton fabric.[62]

As far as synthetic fabrics are concerned, PET represents the most coated substrate followed by PA and PAN. LbL coatings designed for these fabrics must consider their inherently different burning behavior with respect to natural ones. Indeed, the melting processes that occur during combustion can have detrimental effects on the ability of the coating to produce a protective barrier and thus its overall FR behavior.[63] For these reasons, the compositions of most of the LbL coatings developed for synthetic fabrics differ significantly from those already discussed for natural ones. This peculiarity is certainly highlighted by the results achieved by one of the first FR LbL assemblies developed for PET.[64] Similar to early attempts on cotton, nanoparticles were employed to achieve a result typical of polymer nanocomposites.[60] Positively charged alumina coated silica and negatively charged silica were assembled on PET fabric. A 5 BL coating achieved self-extinguishing behavior in VFST, while also prolonging the time required to ignite the fabric during cone calorimetry tests (35 kW/m^2).[64] Interestingly, the increase in BL number did not produce additional benefit as the increased thickness of the deposited coating led to its instability during combustion due to the synthetic fiber melting. Spray assisted deposition of this silica/silica LbL was also attempted on PET mats (density 490 g/m^2).[65] Horizontal and vertical spray configurations were also investigated, highlighting the former as the most efficient, as supported by a dedicated study on cotton.[41] When compared, the results achieved by horizontal spray deposition on both cotton and PET confirms the good FR potential of nanoparticles for synthetic fabrics.

Despite the good results achieved, a major problem of nanoparticle-based coatings is poor washing durability. To address this issue, polyelectrolytes can be used with different cross-linking strategies that can be applied to reduce the coating's swelling in water. Cross-linked polysaccharide-based coatings have been developed with either an improved char forming ability or an intumescent behavior.[66–68] The assembly of CS with sodium alginate (SA) and Cu^{2+} ions on PET produces dense charred residues capable of preventing melt-dripping and self-extinguishing the flame during VFST with just 2 QL deposited.[66] Similarly, Pan et al. exploited partially oxidized SA in combination with PEI and a post cross-linking treatment with hypophosphorous acid on PET fabric subjected to a UV-grafting pretreatment.[67] Fabric treated with 15 BL are capable of self-extinguishing the flame during HFST even after 12 laundering cycles (GB/T 17595–1998).

Intumescent-like assemblies have been achieved by including a phosphate containing component such as APP or PhA.[68–70] When deposited on PET, 10 BL of a CS/APP assembly can only suppress the melt-dripping phenomenon without being able to stop flame spread during flammability tests.[69] This deficiency can be remedied to achieve self-extinguishing behavior by including guanidine sulfamate within the CS solution, without increasing the number of deposited BL.[69] Similarly, a QL architecture encompassing PhA, CS, and SA deposited on PA66 fabric can achieve a UL-94 V1 classification with only 5 QL (i.e., 20 deposition steps), whereas only 5BL of a CH/PhA assembly (i.e., 10 deposition steps) are required to reach the same classification.[71,72] This difference can be ascribed to the increase in PhA content within the coating when moving from a QL to a BL architecture. In addition, the CS/PhA assembly also exhibits durable FR properties (5 washing cycles according to AATCC test method 61(2A)-1996) when post cross-linked with sodium tetraborate decahydrate.[72]

PET and PA fibers are also typically employed in blends with cotton in order to target specific characteristics such as durability, reduced cost, and comfort.[73] From an FR point of view, this further complicates the task as blends tends to combine the weakness of their components (e.g., ignitability of natural fibers and melting of synthetic fibers).[74] The best way to address this problem seems to be the use of intumescent-like formulations or hybrid coatings based on Si-P interactions.[75–77] For example, a CS/PhA coating assembled at 15BL was reported to impart self-extinguishing behavior to PECO during VFST (ASTM D6413).[75] Similar results were achieved when 15 QL encompassing CS, TA, and PhA were deposited on NYCO.[78] It should be noted that the number of deposition steps required to achieve satisfactory results on blends is, on average, higher than those related to a single fabric type (e.g., 15 BL vs 5 BL for PECO vs PET, respectively). This observation further highlights the challenges associated with textile blends. Besides PET and PA, studies have been performed on PAN fabrics as well.[79] Coatings encompassing silica precursors and phosphate-containing components have been developed achieving a hybrid FR effect.[80]

14.2.2.2 Foams

A few years after the publication of the first studies dealing with FR LbL coated fabric, a study reported of a LbL coating capable of reducing heat release rates of open cell flexible polyurethane foam (PUF).[81] The study of PUF has been growing ever since and has formed a significant enough body of research to have a review solely on this topic.[82] Polyurethane foam has a favorable surface/bulk ratio and the open cells allow for an efficient and straightforward deposition of a LbL coating. Indeed, each solution/dispersion can easily penetrate inside the 3D structure of the foam and eventually yield a continuous coating that extends through the entire volume of the foam.[83] The well-known structural collapse of polyurethane foam is often considered one of the most dangerous phenomena at the early stages of a developing fire and, as a consequence, has been a key target of FR LbL assemblies.[84]

Melt-dripping suppression and conspicuous HRR reduction is normally observed if the performed surface modification is capable of either delaying or preventing foam collapse. The literature suggests that the best way to achieve this is by depositing a nanoparticle-based assembly, where the presence of nanoparticles is likely to provide the mechanical strength required to withstand the foam collapsing (Figure 14.2).[23] The plethora of LbL coatings developed up to now can be roughly classified according to the shape/aspect ratio of the nanoparticle employed: spherical, needle-like, and platelet-like. While the last two categories have been widely investigated, very few studies use spherical nanoparticles,[85–87] which might be linked to a reduced FR efficiency with respect to aspect ratio. The deposition of needle-like particles such as sepiolite rods, halloysite, and titanate nanotubes in combination, paired with different polyelectrolytes, has shown effective reductions in combustion rates and smoke production during cone calorimetry test.[88–90]

Indeed, the dense fibrous network resulting from LbL deposition has proven capable of preventing foam collapsing. Additionally, combustible volatiles can be either trapped by the tubular structure of the nanoparticles or diluted over time as they have to pass through a dense inorganic fibrous network.[90] The use of multi-walled carbon nanotubes (MWCNT) clearly demonstrates these two FR effects.[91,92] MWCNT can be dispersed in water by employing stabilizing polyelectrolytes such as pyrene functionalized-PEI and PAA.[91] The resulting dispersions are then capable of yielding steady LbL growth, where the MWCNT are found embedded within a PEI/PAA matrix. From a flame retardant point of view, foam coated with 9 BL self-extinguishes the flame in VFST while also achieving 67% and 80% reductions in pkHRR and TSR, respectively by cone calorimetry (35 kW/m^2). It is important to note that this performance represents FR properties seldom achieved by other LbL systems. A recent work exploited silver microwires to simultaneously achieve FR and EMI shielding properties, further highlighting the versatility of these multilayered coatings in delivering multifunctional properties.[93]

Nanoplatelets are the most common morphology of nanoparticles employed in FR LbL coatings applied to polyurethane foam.[23] This is related to their shape that, during LbL deposition with a polyelectrolyte, forces them to assemble parallel to the surface of the substrate, recreating a nacre-like or brick and mortar ("nanobrick wall") structure that is well-known for its FR behavior.[94] An early study demonstrated the potential of this structure by employing chitosan and montmorrilonite (MMT) clay.[95] The resulting stratified coating produced a 52% reduction in pkHRR in cone calorimetry testing at a flux of 35 kW/m^2. The use of MMT-based coatings was also evaluated in a large scale test with promising results.[96] The influence of different platelet sizes has been investigated, highlighting how high aspect ratio particles can achieve the best FR performance due to the increased probability of particle overlapping even when few bilayers are deposited.[97,98] This is clearly shown in a study where a direct comparison between MMT and the larger diameter vermiculite (VMT) has been performed, with the latter performing best, even with only 1 or 2 deposited BL.[97] Within this context, the use of graphene-related materials such as graphene oxide nanoplatelets (GO) produced interesting results.[99,100] In particular, the use of GO in combination with either CS or poly(diallyldimethylammonium chloride) (PDAC) has been shown to influence the release of volatiles to the point of reaching a non-igniting behavior during cone calorimetry tests (35 kW/m^2).[101,102] The 6 BL PDAC/GO coatings assembled in the presence of a phosphate salt, as an ionic strength modifier, exhibit heat-shielding characteristics close to inorganic silica aerogels during flame penetration tests.[102] Recently, MXenes have also been employed with interesting results.[103–105]

Coatings containing only organic polyelectrolytes have also been developed for foam by targeting either a char forming or an intumescent behavior.[106,107] Char forming LbL coatings have been shown to produce substantial FR properties at the expense of a relatively high number of deposited layers (10 BL or 5 QL) when compared with nanoparticle-based coatings. Interestingly, the deposition of intumescent-like formulations was found insufficient to confer FR characteristics to PUF.[108,109] This phenomenon has been related to the early collapsing of the foam structure and its impact on preventing the proper intumescent expansion to occur.[110] A workaround for this problem is the deposition of hybrid coatings encompassing a stacked configuration, where an intumescent system is deposited on top of a nanoparticle containing assembly. Self-extinguishing behavior in HFST and substantial reductions in heat release rates by cone calorimetry can be achieved with systems encompassing either CS/APP or CS/PhA supported by a MMT-rich layer.[111]

By evaluating the literature on PUF it is possible to understand how the research efforts have been primarily directed towards the development of LbL formulations capable of meeting certain FR properties. Particular attention has been paid to reducing the number of deposition steps required for the coating to be effective. There is, however, a limited amount of information on the durability of the deposited coating, where multiple compression cycles and/or wear/abrasion processes occur. A few studies highlight how an LbL coating cracks but does not detach from the surface of PUF upon multiple deformations when a proper adhesion between the substrate and the coating is achieved.[112,113] Despite being cracked, the deposited coating remains flame retardant. Within this context, the plasticizing effect of water must be considered. It has been indeed demonstrated that damaged LbL coatings might display self-healing characteristics when exposed to a high humidity environment with subsequent recovery of functional properties.[114] Self-healing is certainly dependent on the LbL formulation employed and thus should be evaluated systematically for each coating composition.

14.2.2.3 Other Substrates

The excellent FR results achieved on textiles and polyurethane foam led to the use of FR LbL coatings on many other substrates. While the number of published papers is not enough to justify a dedicated sub-chapter, it is certainly worth highlighting some of the most interesting papers. In regard to polymer cellular solids, LbL assemblies have been deposited on open cell melamine and silicone foams as well as closed cell PET foam.[115–117] This latter work demonstrated that LbL coatings can be effective on cellular solids even if they are not able to penetrate the internal pores. A 4 QL of an intumescent system encompassing PDAC, PAA, and APP self-extinguishes the flame in HFST and achieves a 25% reduction in pkHRR during cone calorimetry at a flux of 35 kW/m^2.[117] Cellulose-based materials such as paper, cellulose fiber networks, and nanocellulose aerogels have also been coated by LbL formulations derived from those already employed on natural textiles.[118–120] Of particular interest is the approach employed for the production of FR paper, where the cellulose fibers are LbL-coated before being assembled in the final material.[118] Paper sheets achieving self-extinguishing behavior in HFST were produced with fibers coated with 3.5 BL of a PEI/PSP assembly.[121]

Wet-stable and cross-linked cellulose porous materials made of either nanofribils or fibers were also LbL modified by flowing the solution/suspension through the open pores of the substrate in a filter-like configuration. This approach turned out to be extremely efficient, producing porous cellulose-based materials characterized by self-extinguishing during HFST and very low HRR (< 100 kW/m^2) in cone calorimeter testing at a flux of 35 kW/m^2.[119,120] Thin polymer films (polycarbonate, polystyrene, polylactide, and polyamide), with thicknesses ranging from several hundred microns to a few millimeters, have also been LbL-coated in order to impart FR behavior.[122–125] These film substrates pose the great challenge of having a highly unfavorable surface to bulk ratio. As a major consequence, the thickness of the film has been demonstrated to play a key-role in determining whether the deposited coating can effectively exert an FR effect.[122] Thick samples (≥ 1 mm) would require the deposition of a relatively high number of BL (~ 60) in order to achieve substantial FR properties.[123,124] This problem has been addressed by the implementation of thickening agents such as salts and short chain amines in order to increase the amount of coating deposited per layer.[125] Using this thickening approach, a CS/clay assembly is capable of shielding a 3.2 mm PS film from a flame torch with only 3 BL deposited.

14.2.3 Outlook and Future Research Directions

In just 15 years, LbL assembly applied to the FR field has arisen as one of the most advanced and versatile techniques to impart FR properties to numerous substrates. The favorable processing conditions (room temperature and pressure in aqueous media), and the potential of the coating ingredients to be either environmentally benign or even biologically sourced, have been the keys to this technique's success. Despite the advantages of having impressive FR properties with an environmentally friendly approach, the high number of deposition steps required to achieve the desired behavior, as well as the production of large volumes of exhausted deposition baths, currently represent the main constraints that are holding back LbL commercialization. Technological and scientific advances are needed to minimize the economic impact of the LbL process, which are currently being addressed by many research groups. The durability of the coating to different stresses (e.g., washing, aging, deformation, and abrasion) must be evaluated and improved. Aside from these practical considerations, it is worth noting how the development of LbL flame retardant treatments introduced the field of fire protection to a number of new chemistries. Before the "LbL era," the properties of polyelectrolyte/nanoparticle complexes in terms of thermal stability and flame retardancy were not known. After years of development, this multilayered technique has also inspired a secondary subfield of preformed polyelectrolyte complexes, which is described in Section 14.3.

14.3 POLYELECTROLYTE COMPLEX COATINGS

A frequent criticism levied at layer-by-layer coatings is their very time intensive processing due to the many layers. Depositing a single bilayer in LbL requires at least four liquid exposure steps (one for each solution and its associated rinse), with the possibility of a drying step either in between bilayers or after each rinse. Considering that most published coatings utilize 8–20 BL, the time and energy-intensive nature of the rinse and dry steps have led to research into alternative methods to deposit the charged networks yielded by LbL coatings. Further examination into the fundamental interactions of the charged species in LbL coatings is thus necessary to design improved coatings/processes.

14.3.1 Controlling Polyelectrolyte Complexation to Deposit Coatings

Polyelectrolyte complexes (PECs) are assemblies of oppositely charged polymers (i.e., polycations and polyanions), whose formation is driven by primarily entropic factors.[126] The individual charged polymeric species exist in an equilibrium with the associated complex that can be tuned by a variety of factors that include salinity, charge density (driven by pH changes), polymer molecular weight, and stoichiometry.[127–131] Many of these parameters can be changed either ex situ or in situ, yielding control over how, when, and where a complex is formed. The first noted case of a PEC for flame retardancy was in 2014 when delayed complexation was first noticed following inadvertent mixing of two solutions intended for a layer-by-layer film.[132] This initial use-case for the deposition of a single-step PEC was unprecedented at the time, and that work was quickly built upon and has yielded a rich field of active research that is described here.

The first class of PEC coatings to be published are what we will describe herein as buffer-cured PECs, demonstrated schematically in Figure 14.3a.[133] In order to replicate the material deposition of layer-by-layer assembled coatings, PEC coatings must begin with significantly higher solution concentrations. A typical LbL coating solution will contain ≤ 1 wt% charged species, while PEC coatings often have upwards of 10 wt% solids content. The PEC solution content includes a mixture of cationic and anionic species, at least one of which is a weak (i.e., charge density tied to solution pH) polyelectrolyte. The use of literature pKas enables the preparation of solutions at a pH such that one of the species is uncharged (or at least low enough charge to prevent complexation), removing the driving force for complexation and yielding a "soluble PEC." After deposition on a substrate and drying, the PEC is then exposed to a buffer that "cures" the PEC, charging the uncharged species while minimally altering the charge state of the other species. This process yields a complexed film on the substrate that is resistant to rinsing (owing to the water-excluding PEC equilibrium).

Newer classes of PEC treatments have been developed more recently by delving further into methods of tuning the PEC equilibrium. One such example is the use of a flame retardant polyelectrolyte complex coacervate (Figure 14.3b). Polyelectrolyte complexes exist along a continuum of states, being an insoluble complex at one end of the continuum (high charge density and low salinity) or a polymeric solution on the other end (low charge density and/or high salinity). The coacervate phase is a form of liquid-liquid phase separation at moderate charge/salinity that yields the formation of an elastic liquid of high (> 40 wt%) polymer content. This liquid is amenable to bar coating and can then be deposited as a flame retardant treatment on film/sheet-like substrates.[134] The full potential of coacervate treatments has yet to be realized, but they have all of the characteristics to be effective FR treatments at

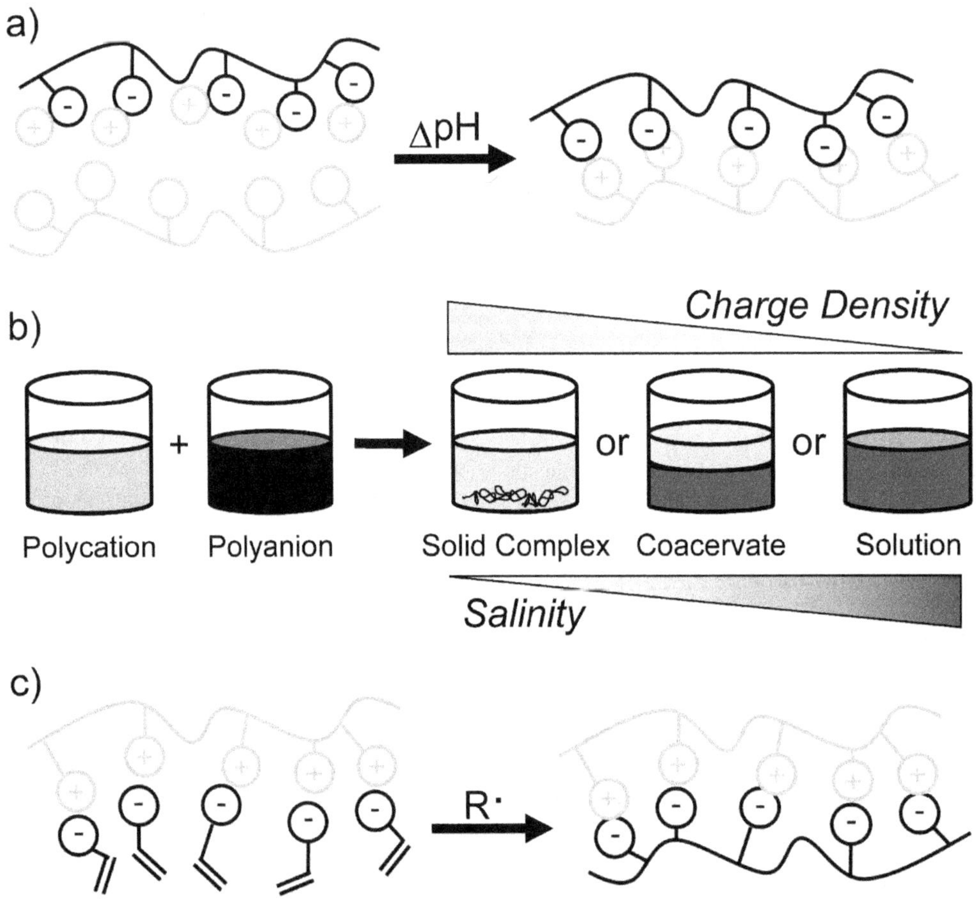

FIGURE 14.3 (a) Schematic demonstrating the use of buffer curing to yield a pH-induced change in charge density between two uncomplexed polymers. The result is an insoluble complex that is resistant to water. (b) Schematic of the polyelectrolyte complex-coacervate-solution continuum. At high charge density and low salinity, a solid complex precipitates from solution after mixing. As charge density decreases and/or salinity increases, complexation changes to a liquid-liquid phase separation (coacervation), ultimately yielding a soluble polyelectrolyte solution at very low charge density/very high salinity. (c) Schematic demonstrating how polymerization can be utilized to turn an uncomplexed polycation/monomeric anion pair into a polyelectrolyte complex by polymerizing the monomeric species. Heat or light can be used to initiate the polymerization reaction.

the industrial scale with little modification of existing coating infrastructure. Another emerging class of PEC coating, shown in Figure 14.3c, makes use of polymerization to yield a complex in situ on the surface of a treated material. In this case, a polyelectrolyte is paired with an oppositely charged monomer in solution with a thermal initiator (e.g., potassium persulfate) or a photoinitiator. After deposition of the liquid coating, an external stimulus corresponding to the chosen initiator is applied to polymerize the monomeric component, yielding a polyelectrolyte complex film.[135] Photopolymerization is often a more energy-efficient coating method since heating is not necessary but poses the risk of substrate/coating degradation and incomplete polymerization due to light absorption issues.

14.3.2 Use-Cases of Polyelectrolyte Complex Flame-Retardant Coatings

Polyelectrolyte complex flame retardant coatings have found applications across a wide variety of substrates. Even accounting for the fact that PEC treatments have been an active research area for approximately half as long as layer-by-layer assembled FR treatments, the breadth of PEC FR literature is significantly smaller than LbL. This is likely due to a greater number of constraints (e.g., high concentrations, adjustable charge densities, material cost, etc.) placed on this type of treatment. The following subsections summarize some of the seminal work on flame retardant PEC treatments for a variety of commonly studied substrates.

14.3.2.1 Textiles

The earliest polyelectrolyte complex coatings were deposited on textiles. As noted previously, the first serendipitous discovery of a "single step" coating represented a significant paradigm shift in the deposition of polyelectrolyte treatments for fire protection.[132] This first treatment mixed polyethyleneimine [PEI] and poly(sodium phosphate) [PSP] and allowed them to complex over time on an immersed fabric substrate. While this time-dependent deposition saved on a number of processing steps, as compared to LbL deposition, the processing time was still considerable and generated a significant amount of waste because the solution as not reusable. Haile et al. later developed a different deposition system where PEI and PSP were mixed at a high pH such that PEI had low charge density.[133] This system could then be complexed through a subsequent buffer curing step after deposition. The buffer curing step is critical to the deposition of an intumescent coating. Without it, there is no acid source for the intumescence process and coated/uncured fabric performs very poorly. The PEI/PSP PEC coating is self-extinguishing through an intumescent-type mechanism in VFST (ASTM D6413).[136] Later work has adopted this technology for nylon-cotton fabric,[137] including wash-durable chemistries,[138,139] and for other natural textiles.[140]

More recent iterations of buffer cured PECs have made use of additives, to either the polymer solutions or to the buffers.[141] On nylon-cotton fabric, it was found that the incorporation of a melamine deposition step prior to the ultimate buffer curing of the PEC could yield self-extinguishing behavior after forming melamine polyphosphate in situ (Figure 14.4a).[137] The melamine-containing PEC coating is self-extinguishing in VFST (ASTM D6413) and exhibits a negative pyrolytic energy balance in differential scanning calorimetry. Importantly, the incorporation of metal ions affects PEC interactions and thus can enable tunable viscosity, which is important for controlling weight gain and processability.[141,142] Other studies have included inorganic clays in PEC-style motifs to protect high heat release textiles that lack the char-forming potential of cotton, such as acrylic fabric.[143] In this study, Carosio et al. developed a chitosan/montmorillonite treatment that formed a complexed gel that could be applied via blade coating. Despite not using LbL processing, this coating exhibited the nanobrick wall structure common to LbL treatments (Figure 14.4b), which allows the coating to act as an effective thermal barrier and reduce flammability. More recent work has sought to increase the breadth of available chemistries (e.g., inclusion of biomolecules) and incorporate multifunctionality, such as water repellency, to the PEC coating on fabric.[144–146]

14.3.2.2 Foams

As noted throughout the discussion of layer-by-layer treatments, flame retardants for polyurethane foam most often requires a high percentage of inorganic content to counteract its high heat release and tendency to form fire-spreading melt pools. The need for inorganic content poses a challenge for most PEC-based FR treatments since the single-step deposition requires a generally much higher concentration of dissolved solids. Clays and other inorganic species are challenging to load to 10+ wt% in solution due to their large size and poor stability. To date, very little published work exists that successfully reduces foam flammability using PEC treatments. Conveniently, polyphosphates have long been known to act as deflocculants/dispersants for a variety of inorganic particles.[147] Since polyphosphates like APP and PSP have good utility as FR agents, it seems natural to pair them for improved clay loading in PEC FR treatments. The first work to pursue this line of reasoning made use of the strong ability of polyphosphates to stabilize halloysite to prepare clay/polyelectrolyte solutions with a very high clay loading.[148] This clay allowed both improved fire resistance by the PEC/clay system, but also counteracted the pure intumescent PEC's tendency to increase the degrading foam's smoke production through the sequestration mechanisms noted in the LbL subsection. The halloysite-filled PEC system reduced the pkHRR of foam by 53% (35 kW/m^2) and reduced the total smoke release (TSR) by 17% compared to the unfilled PEC, although it still had a higher TSR than neat foam. A similar work utilized graphene oxide (GO) deposited as a second step after coating the foam in an intumescent PEC precursor solution.[149] Thermal reduction of the GO layer led to improved char yield and fire protection, with coated foam passing a vertical burning test

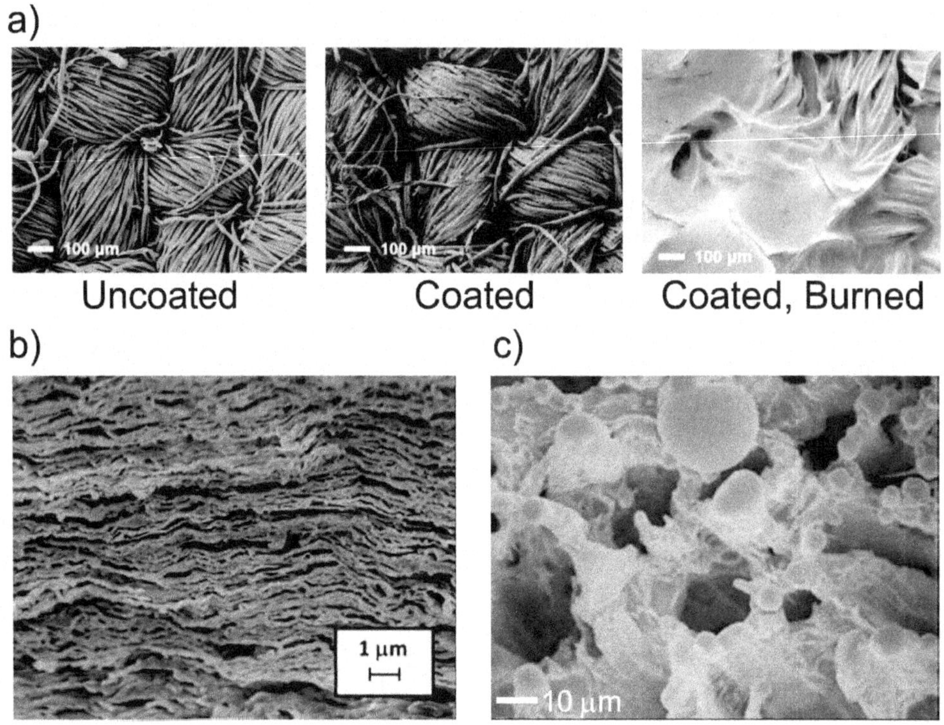

FIGURE 14.4 (a) Scanning electron microscope (SEM) images of uncoated nylon-cotton, NYCO coated with a melamine-containing PEC, and the coated NYCO after vertical flame testing, respectively. Despite utilizing high concentrations, PEC coatings remain highly conformal to their substrate. (b) Cross-sectional SEM image of a chitosan/montmorillonite PEC coating on acrylic fabric. The coating self-stratifies into a layered structure after blade coating. (c) Image of a PEC-coated wood article following fire exposure. Much like cotton fabric, wood's charring nature leads to a microintumescent effect with polyamine/polyphosphate PEC treatments. Part (a) reproduced from Ref. [137] with permission, part (b) reproduced from Ref. [143] under a creative commons license, and part (c) reproduced from Ref. [151] with permission.

(TB 117–2000). The dynamic nature of PEC ionic bonding yielded a coating that was also found to be self-healing.

14.3.2.3 Wood

The processing advantages of PEC coatings has been most pronounced and widespread in wood. As mentioned previously, wood's complex surface yields extremely slow rates of diffusion and adsorption. As a result, traditional polyelectrolyte treatments making use of LbL deposition take prohibitively long to deposit.[150] The influence of immersion time on the deposition of buffer cured PECs was investigated on wood and yielded self-extinguishing wood articles through an intumescent polyelectrolyte complex system (Figure 14.4c).[151] It was found that the diffusion front of the curing buffer could carry the coating further inside the wood, ultimately leading to a 1 mm depth of penetration. Treatment time remains a challenge because of diffusion of polymeric species and buffer into the bulk of the wood. Pressure treatment with buffer cured PECs dramatically reduces processing time, representing an important step toward the commercialization of polyelectrolyte FR technologies for wood.[152]

A critical flaw in the deposition of buffer-cured PECs is their reliance on diffusion to deposit a coating. While this is unavoidable for the flame retardant species, diffusion of a buffer to cure the complex presents a time barrier that diminishes the potential of PECs for wood fire protection. Recent efforts have turned towards photopolymerization and coacervates to further reduce immersion steps.[135,153] In the case of photopolymerization, a monomeric phosphate was paired with a cationic polymer and post-deposition photopolymerization formed the complex on the surface of the wood. Despite the short penetration depth of the UV light, early studies indicate that the coated material is "sealed" by the polymerization of the exterior material.[135,154] The photopolymerized PEC treatments were more effective on plywood, where they exhibited a pkHRR decrease of nearly 50% (at a flux of 35 kW/m^2). Further chemical alterations will need to be investigated to determine optimal treatment conditions for other forms of wood. Although significantly less common in the literature, the use of coacervates as a flame retardant treatment allow for truly single-step deposition of a flame retardant coating. A similar coacervation approach was also recently used to flame retard paper in a single step.[134]

14.3.3 Outlook and Future Research Directions

Since their discovery a decade ago, polyelectrolyte complex treatments for fire protection have rapidly matched the performance of layer-by-layer systems, while requiring significantly fewer processing steps. While many promising steps have been taken toward commercializing PECs, future work must seek to understand and address the lingering questions/flaws associated with these treatments. First, the "potlife" of the deposition solutions must be evaluated. As noted previously, PEC deposition solutions are significantly (up to 10×) more concentrated than those used for LbL deposition. Treatments must remain effective after numerous deposition cycles to avoid waste and ensure economic viability of these coatings. Second, the field should continue investing in ultraviolet curing. The buffers utilized for curing PEC treatments are prone to fouling and generate a large amount of solution waste. Switching to UV-based curing approaches will avoid an immersion/drying step and generate less waste. Finally, as with LbL systems, the durability of these treatments to washing/weathering for textiles and wood, respectively, must be assessed. While the field is still young, PEC-based approaches hold significant promise for flame retardant treatments that are environmentally benign and scalable for widespread implementation.

14.4 SOL-GEL COATINGS

The sol-gel process provides a wet-chemical route for the synthesis of either inorganic or hybrid organic–inorganic chemical structures.[155] Very generally, hydrolysis and polycondensation reactions are used to prepare well controlled hybrid materials with a broad selection of processing routes and parameters.[156] The sol-gel procedure had been initially developed for the synthesis of homogeneous and pure metal oxide particles.[157] Throughout the years, the technique has been extended to the surface modification of different substrates.[158–160] The intrinsic fire retardant characteristics of sol-gel coatings have been known since the nineties, as evidenced by patents focusing on the production of composite coatings for patching/sealing of substrates.[161] Sol-gel based FR coatings for porous materials (e.g., synthetic and natural woven and nonwoven fabrics, fibers etc.) were demonstrated in the early 2000s.[162] During the past two decades, the sol-gel technique has been widely developed in the field of FR coatings capable of meeting increasingly demanding fire safety criteria.[3,163,164]

14.4.1 Sol-Gel Chemistry

In a sol-gel process a colloidal solution ("sol") is obtained starting from liquid precursors. This is commonly achieved through hydrolysis and condensation reactions of metal alkoxide precursors, although a sol can be generally linked to a colloidal system to encompass different systems under the general "sol-gel" term.[155] The so produced "sol" can be further developed into a network structure ("gel") defined as non-fluid 3D network that extends through a fluid phase (Figure 14.5). Gels can be grouped into five classes that are relevant in the sol-gel synthesis of materials: colloidal, metal-oxane polymer, metal complex, polymer complex I, and polymer complex II.[165] The polymer complexes are classified based on the type of bonding occurring between the polymers. The sol-gel chemistry was first reported in the 19th century when a gel was produced from a silicon alkoxide through hydrolysis and condensation reactions favored by atmospheric moisture.[166] This general concept has been widely employed and investigated through years by exploiting, for example, acid or base catalysis in order to form gels with different structures. Although this hydrolysis and condensation chemistry, often referred as "traditional sol-gel chemistry," presents some limitations (e.g., high reactivity of the precursors and phase separation when combining precursor with different rates of hydrolysis), it still represents the most widely used sol-gel approach in the field of materials chemistry. Here we will focus on traditional sol-gel that is also the most common employed in the flame retardant field.

Metal alkoxides are the most commonly employed reactants for sol-gel processes. The most common examples involve the use of transition group metals such as Ti and Zr or p-block elements such as Al and Si. There are, however, many other examples of metal alkoxides produced by different synthesis strategies.[167,168] The ionic character of the M-O bond and the stability of the alkoxy groups can influence the gel structure by affecting the relative rates of hydrolysis and condensation and, eventually, the degree of polymerization.[169] On the other hand, the volatility and viscosity of the alkoxides might affect its suitability for a sol-gel process.[170] Hydrolysis and condensation reactions represent the core of the sol-gel chemistry and are strongly affected by many processes parameters such as the nature of the R group, the water to alkoxide ratio, and the presence of a catalyst.[155] This latter, in the form of either an acid or base catalyst, is needed for the sol-gel chemistry of silica because the reaction carried over at neutral pH is notoriously slow. During hydrolysis, the alkoxy group is replaced by a hydroxyl. The number of alkoxy groups hydrolyzed depends on the reaction conditions and the Si/H_2O ratio. The subsequent condensation step produces siloxane bonds (or metaloxane bonds for other elements). The degree of hydrolysis, which is affected by pH, has a direct influence on the progression of the condensation. Under basic conditions, hydrolysis is generally completed before the condensation step, whereas in acidic conditions condensation starts before the completion of hydrolysis. The former likely ends up in a colloidal gel, while the latter produces network-like gels. Besides pH, there are other factors controlling the structure of the gel. The presence of solvents in addition to water might enable or improve mixing (many metal alkoxides are not miscible with water) or interact with the metal center. The alkoxide/water ratio is also important as it can be tailored to control hydrolysis. This latter step can be directly influenced by the nature of the R group. In addition to the chemistry linked to gel formation, there are other ways to affect the outcome of a sol-gel process. The drying step is, for instance, capable of impacting the gel structure.[171] Uncontrolled evaporation

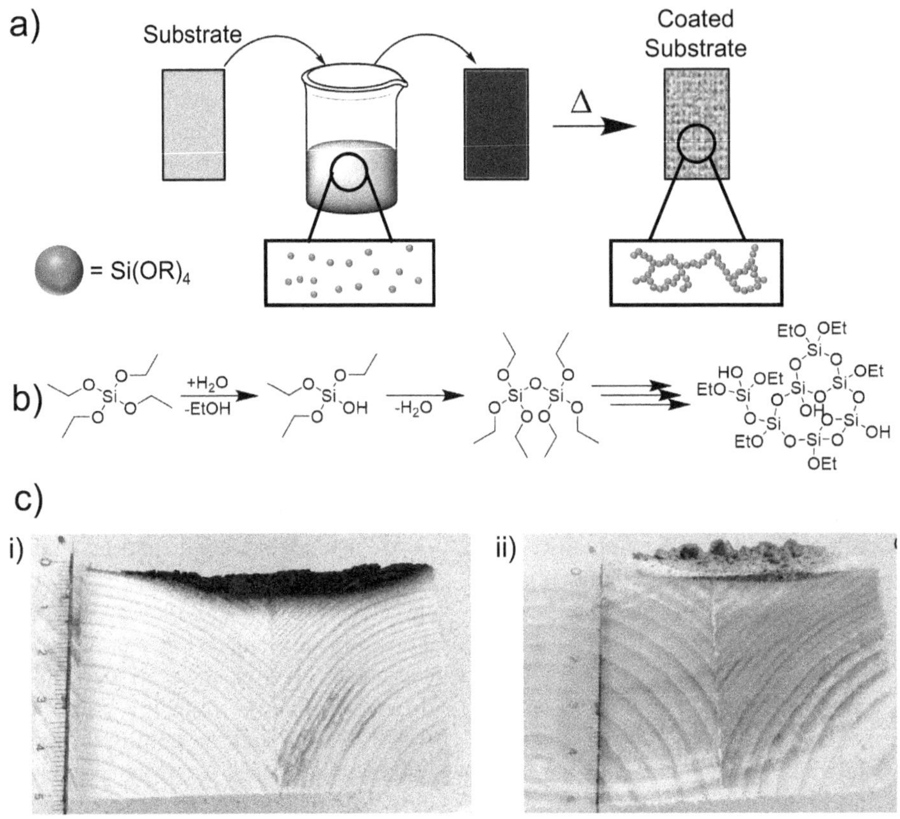

FIGURE 14.5 (a) Schematic of the deposition process for sol-gel coatings. (b) Chemical scheme for the hydrolysis/condensation process that yields a siloxane network. (c) Photos of sol-gel coated wood after single burning item testing, reproduced from Ref. [201] with permission.

produces a collapse of the network into a dense xerogel, which often shows high surface area due to the presence of disordered porosity. Conversely, drying under controlled conditions (e.g., by supercritical drying or freeze-drying) might preserve a large pore volume yielding aerogels or cryogels characterized by low densities.[172–174]

14.4.2 Use-Cases of Sol-Gel Flame-Retardant Coatings

With all the abovementioned possibilities, the use of the sol-gel approach in the FR field has been widely studied for almost 20 years. Although conformal coatings represent the majority of the published papers, the use of sol-gel has been also extended to the synthesis of new chemicals or surface modification of particles used for bulk inclusion in polymers.[163]As far as FR coatings are concerned, the first attempts were performed using a traditional sol-gel chemistry (conventional approach). Additives have been included within the formed gel (use of dopants) in order to improve the achieved FR properties. Recently, the synthesis of sol-gel precursors encompassing FR functionality has been proposed.[3] Some of the best performing and unique sol-gel based FR solutions are discussed below.

14.4.2.1 Textiles

As already noted throughout this chapter, textiles represent an optimal starting substrate for the evaluation of coating treatments aimed at FR properties. The first sol-gel treatment with FR properties was attempted on regenerated cellulose fibers using a tetraethylorthosilicate (TEOS) precursor.[175] The treatment produced, under basic conditions, an improved thermal-oxidative stability of the fibers. Subsequently, the influence of different TEOS/water ratios on the FR properties of coatings deposited on cotton, PET, and their blends was evaluated.[176] While the treatment had limited influence on PET fabric, PET-rich blends displayed a remarkable increase in TTI (up to +93%) and a reduction in pkHRR (–34%) in cone calorimetry (at a flux of 35 kW/m^2). Further studies performed on cotton evaluated the influence of different alkoxysilane or metal-alkoxide precursors, highlighting how TEOS-based sol-gel coatings can achieve the best FR performance in terms of TTI increase and pkHRR and THR reductions.[177] Interestingly, despite the limited reduction in pkHRR when using tetraethylortho-titanate or tetraethylortho-zirconate, and the increase in pkHRR achieved with aluminum isopropylate, a reduction in specific extinction area (i.e., the optical density of smoke per unit mass) and TSR values was observed for all different metal-alkoxide precursors. This result has been ascribed to the increased water content of the treated fabric as well as the barrier effect exerted by the metal-oxide coating. These preliminary works clearly demonstrated the potential of the sol-gel approach and inspired further developments.

A straightforward way to improve FR properties is represented by the inclusion of FR components within the sol-gel network. The use of phosphorus-based molecules such as ethyl dichlorophosphate or phosphorus acid in a TEOS-based sol-gel process on cotton fabric has been shown to reduce flammability. In particular, the presence of phosphoric acid was found to grant a self-extinguishing behavior during flame tests (TS 5569 EN ISO 6941) with a partial durability to ten washing cycles.[178] Similar results were also achieved on polyacrylonitrile fabric.[179] Commercially available P-based FR formulations and nanoparticles have also been investigated by Alongi and coworkers.[180] AP Exolit® OP1230 (aluminum phosphinate), APMP Exolit® OP1312 (aluminum phosphinate 63.5% by mass, melamine polyphosphate 32% by mass, and zinc and boron oxide 4.5% by mass), and α-Zirconium dihydrogen phosphate were included within a sol-gel process exploiting tetramethyl orthosilicate (TMOS) as silica precursor. Cotton fabric treated with doped sol-gel coatings required multiple flame applications to achieve ignition. In addition, the inclusion of AP Exolit® OP1230 within a silica network considerably limited the removal of the FR during washing, thus resulting in a semi-durable treatment. Similar properties where achieved with the use of γ-aminopropyltriethoxysilane (APTES) in combination with diethylphosphite.[181] Bio-sourced FR such as phytic acid have also been included in sol gel processes.[35,182,183] When deposited on silk in combination with TEOS, the presence of phytic acid imparts self-extinguishing behavior in vertical flame testing (ASTM D6413–2008).[182] The use of different coupling agents to increase hydrophobicity of the coating imparted wash durability.

The combination of phosphorus-based and nitrogen-based compounds has also been attempted.[184,185] For example, the use of (3-aminopropyl)triethoxysilane doped with phytic acid sol-gel coatings has been shown to confer remarkable FR properties on cotton by achieving a self-extinguishing behavior during flammability tests (ASTM D6413–2008) and LOI values as high as 28.5%.[185] A post-treatment with n-dodecyltrimethoxysilane further endowed the fabric with hydrophobic properties without affecting the FR efficiency. Similar results have been achieved by combining TEOS, hydroxyl-terminated polydimethylsiloxane, and APP on cotton.[186] Besides Si, other metal alkoxides have been evaluated.[187] For example, titanium (IV) butoxide and boric acid (employed with different molar ratios) have been deposited on cotton by a pad/dry/cure process, achieving self-extinguishing behavior during vertical flame tests. In addition, this titania-based coating also produced an increase in hydrophobicity that was not sufficient to impart washing durability. Although the use of doping agents has been demonstrated to improve the FR properties of the treated fabric, this approach might suffer from a poor compatibility of the doping agent with the adopted sol-gel chemistry. In order to address this problem, recent research has been focused on the modification of the siloxane precursor.[188–190] An early study compared the FR performance of a hybrid precursor with those achievable with a traditional organophosphate. Only the latter solution was capable of self-extinguishing the flame during vertical flammability tests (DIN 53906).[188] Notwithstanding this, the use of a hybrid precursor allowed for a formaldehyde-free process and improved hydrophobicity thus showing the potential for the synthesis of new sol-gel precursors with specific FR functionalization.[191]

More recent work has employed a di-(triethoxysilylpropyl) phenylphosphamide hybrid precursor that produces a self-extinguishing behavior on cotton after prolonged impregnation times. Similar results have been achieved with a 2,2-dimethyl-1,3-propanediol-(3′-triethoxysilanepropyl) phosphoramidate precursor that also produces an increase in LOI (up to 30.1%). Surprisingly, the same treatment produces detrimental results in cone calorimetry (i.e., increased pkHRR and TSR values).[189] Conventional FR chemicals have been also modified in order to be exploited in sol-gel chemistry. A 9,10-dihydro-9-oxa-10-phosphaphenanthrene-10-oxide-modified vinyl trialkoxysilane (DOPO-VTS) has been synthetized and combined with TEOS on PA 6 fibers.[192] There is a synergistic FR effect in the condensed phased resulting in the complete suppression of the melt dripping.

With the FR ability of hybrid sol-gel precursors demonstrated, the next step has been related to the combination of multiple precursors to target multifunctional coatings. For example, 1H,1H,2H,2H-perfluorooctyltriethoxysilane and P,P-diphenyl-N-(3-(trimethoxysilyl)propyl)phosphinic amide have been combined on cotton fabric in order to achieve an FR coating with superhydrophobicity and high oleophobicity, as well as passive antibacterial activity.[193] The use of ionic liquids in combination with sol-gel has also been proposed.[187] The 1-methylimidazolium chloride propyltriethoxysilane and pyridinium chloride propyltriethoxysilane were applied to cotton, followed by a subsequent anion exchange step by dipping the coated fabric in solutions containing hexafluorophosphoric acid, bis(trifluoromethane) sulfonimide lithium, sodium tetrafluoroborate, or sodium acetate. Of all the different counterions evaluated, the hexafluorophosphoric acid was found to provide the optimal FR properties. P,P-diphenyl-N-(3-(trithoxysilyl)propyl)phospinic amide and the hydrophobic finishing agent polydimethylsiloxane were combined on cotton.[194] The so treated fabrics achieved a self-extinguishing behavior during vertical flame testing (GB/T 5455–1997), while also displaying superhydrophobic, self-cleaning, and abrasion resistance properties.

14.4.2.2 Other Substrates

Following the good results achieved on textiles, the sol-gel approach has been extended to other substrates such as polymeric foam and wood. An early study by Bellayer et al.

designed a sol-gel formulation encompassing tetraethoxysilane, methyltriethoxysilane, 3-aminopropyl triethoxysilane, and diethyl phosphite for flexible PU foam.[195] The treated foam self-extinguishes the flame during UL-94 tests in a vertical configuration. In addition, a conspicuous reduction in pkHRR (-60%) was observed in cone calorimetry (at a flux of 50 kW/m^2). The same authors evaluated the FR mechanism behind the achieved results, revealing a condensed phase effect involving an intumescing layer, comprising SiO_2 and Si-O-P networks, as well as a gas phase effect related to the release of diethyl phosphite as a free radical scavenger.[196] Shi et al. exploited TEOS in combination with a polyacrylate resin in order to deposit a protective coating filled with silica nanoparticles on the surface of a melamine-formaldehyde foam. The presence of the coating further improved upon the already good FR properties of the uncoated foam, as demonstrated by an increase of the LOI (up to 43.7%).[197]

Recently, a sol-gel approach combined with freeze-drying has been employed to produce a silica aerogel to be exploited as either filler or surface coating for rigid PU foams.[198] When employed as a coating, the aerogel achieves the best FR properties, with LOI up to 29.4%, V0 classification by UL-94, and considerably reduced pkHRR values. Similar to open cell PU foam, the sol-gel approach has also been exploited to coat wood substrates.[199] An early study employed titanium(IV) isopropoxide on pine sapwood wood.[200] The treatment penetrated inside the wood structure, achieving different results based on the precursor concentration. Within this context, it is important to note how the possibility to perform a sol-gel process by mixing different solvent might reduce the surface tension of the sol and thus improve wettability. The impregnated wood displays improved char forming ability and an increase in LOI up to 38% (as well as a decrease in the pkHRR in cone calorimetry). The flammability, evaluated by UL-94 shows limited improvement with coated samples, achieving an HB classification while failing the test in a vertical configuration. Recently, Bellayer et al. reported a sol-gel formulation encompassing tetraethoxysilicate, methyltriethoxysilicate, 3-aminopropyl triethoxysilane, and diethyl phosphite that was either impregnated into the wood structure or deposited as thick coating.[201] The latter process was achieved by partially drying the solution in order to achieve a viscosity suitable for a doctor blade application. The adopted sol-gel formulation improved the FR properties of wood only when deposited as thick coating. The intumescing behavior of the coating also thermally insulated the underlying wood during a small-scale furnace test (Figure 14.5c).

14.4.3 Outlook and Future Research Directions

The sol-gel approach has been clearly demonstrated as a viable tool for conferring flame retardant properties to different substrates. The ease of design and the versatility of the sol-gel chemistry is a key factor in the successful development of these kind of coatings in the FR field. The ability to tune the polarity of sol-gel precursors and the use of different solvents further allow adjusting the overall polarity of the sol based on the substrate choice. On the other hand, the need for additional solvents besides water might lead to waste management problems if large volumes are employed. Another drawback, mostly related to textiles, is represented by the durability of the coating to washing or abrasion. Although the use of different cross-linking strategies or the implementation of hydrophobic components might improve the wash-fastness of the coating, the literature shows that the FR efficiency of the coating is partially lost after few laundering cycles. A last problem to be addressed is related to scale-up. As discussed in the previous chapters, the best performing sol-gel solutions rely on the use of novel synthesized precursors that will require cost-effective production for commercial use.

14.5 PLASMA AND VAPOR BASED TREATMENTS

While layer-by-layer assembly, polyelectrolyte complexation, and sol-gel coatings represent the most active areas for flame retardant coating research, other approaches are worth noting in the context of this subfield. Of particular note are plasma- and vapor-based treatments, which make use of vapor deposition and/or plasma assisted polymerization to deposit or graft a flame retardant treatment to a substrate.[2,202,203] These approaches not only yield highly conformal coatings but also modify the surface of the substrate to covalently attach the polymerized monomers onto the surface. A chief advantage of this technique is the installation of the FR agent through a covalent bond. Covalent attachment generally yields treatments which are more durable to washing, which is especially important for textile finishes.

Plasma, the so-called fourth state of matter, is a quasi-neutral gas with a high degree of free ionic and radical species. Exposure of gaseous monomers to a plasma environment can lead to polymerization, yielding networked polymers of irregular topology due to the incidence of molecular rearrangement in the plasma environment.[204] The plasma can cause radical propagation in the gas state as well as on the surface of a material through adsorption and subsequent reaction of "activated" species. After decades of study, this once-challenging state of matter has been made significantly more available to researchers and industry through the advent of atmospheric plasmas.[205,206] The network-forming and surface-grafting tendencies of plasma polymerization are conceptually appealing because of the durability of networks to dissolution, which can impart wash resistance.

Plasma-driven polymerization for flame protection has been studied for many decades, with the earliest work polymerizing polysiloxanes onto the surface of polyacrylonitrile and polyamides.[207–209] The plasma process is especially amenable to utilizing mixed or sequential treatments of gases to deposit a coating. The modularity of the plasma treatment process has been used to deposit multifunctional

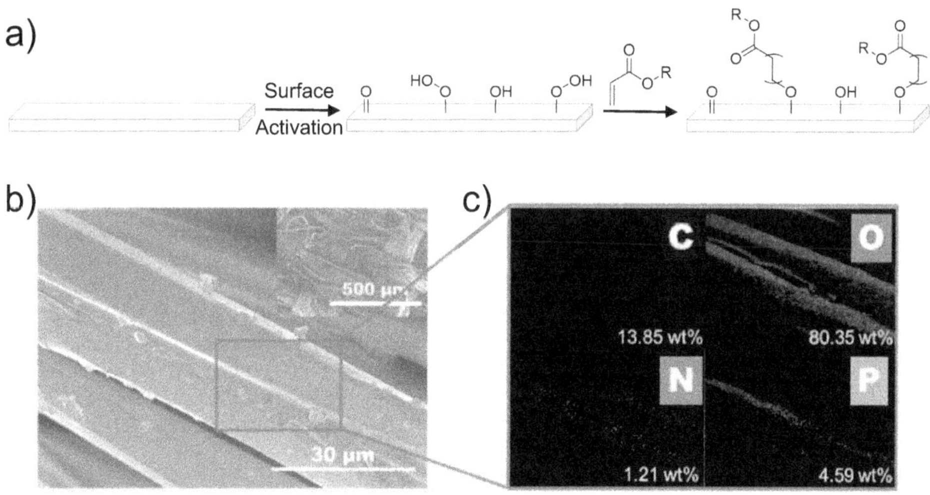

FIGURE 14.6 (a) Schematic of the activation of surface functional groups from a plasma treatment, followed by the surface-initiated polymerization after exposure to a generic acrylate monomer. Most radically-polymerizable groups can be substituted for or incorporated with the acrylate shown. (b) Electron microscopy along with (c) energy dispersive spectroscopy (EDS) of phosphate-acrylate coated polyester fibers. The coating is clearly highly conformal to the fibers along with a uniform distribution of constituent elements along the fibers. Parts (b) and (c) reproduced from Ref. [216] with permission.

(typically flame retardant and hydrophobic) treatments through either the use of two monomers (e.g., a phosphonated acrylate and a fluorinated acrylate, see Figure 14.6a) or through sequential treatment by a flame retardant monomer followed by SF_6 to add a water-repellant layer.[210,211] Plasma treatment has also been used historically to improve the adhesion of nanoparticles, such as clays or vapor-deposited carbon to textile fibers, that improve coating durability.[212–214] Recently, efforts have been made to afford the ability for plasma treatments for textile processing, offering hope for widespread adoption of these technologies.[205]

Plasma-assisted treatments have continued to be an active research area in recent years. An especially useful development has been the unique crosslinking network topologies. Recent work in plasma coatings has focused on multiplexing the capabilities of UV curing for rapid coating deposition along with plasma activation for covalent surface grafting by the photopolymerized species. This motif has led to wash-durable treatments for cotton and pure polyester fabric (Figure 14.6b,c).[215,216] Success in these research efforts could lead to more overlap between plasma treatments and other popular forms of flame retardant coating research (e.g., LbL, PEC, sol-gel, etc.) to improve durability. Plasma treatments have to this point focused primarily on textile treatment, owing to the volume of that market and the relative ease of implementation on commercially relevant scales. There have been promising results with plasma treatments on polyurethane foams,[217] but more work needs to be done to make this technology viable for commercial production.

An important advantage of plasma treatments, as compared to the other techniques discussed in this chapter, is that the gaseous nature of the polymerized species removes the need for any immersions of the substrate. By eliminating solution immersion steps, coating processes can be faster and avoid time and energy intensive rinsing and drying steps. As plasma treatment technologies become cheaper and more widespread, they will continue to represent an important segment of FR-application technologies. Future research to develop multifunctional (e.g., antimicrobial/FR or hydrophobic/FR) treatments from the same monomer feedstock will further improve the utility of plasma treatments. Considerable work remains to be done for non-textile substrates. Foam and wood in particular could achieve significant improvements in fire safety, albeit while facing scaling issues owing to how plasmas are formed.

14.6 CONCLUSIONS

In this chapter we have described the most active current research areas in conformal flame retardant treatments: layer-by-layer assembly, polyelectrolyte complexation, sol-gel coatings, and plasma treatments. Layer-by-layer and polyelectrolyte complex treatments are the most active research areas in this subfield of fire protection currently, owing to their modularity, advantageous environmental profile, and high efficacy on a variety of substrates. Each of the described methodologies have significant hurdles to overcome prior to widespread commercialization and use. A chief concern for most of the described techniques is durability to laundering and weathering. Furthermore, the environmental ramifications of both the materials used and the process of deposition must be carefully considered if the global effort towards a circular economy is to be realized. Continued research and investment in these technologies promises to yield highly effective and environmentally benign fire protection for nearly any flammable material encountered in everyday life.

REFERENCES

(1) Little, R. W.; Church, J. M.; Coppick, S. Commercial Application of Flame-Resistant Finishes. *Industrial & Engineering Chemistry* **1950**, *42* (3), 432–440. https://doi.org/10.1021/ie50483a015.

(2) Liu, B.; Zhao, H.; Wang, Y. Advanced Flame-Retardant Methods for Polymeric Materials. *Advanced Materials* **2022**, 2107905. https://doi.org/10.1002/adma.202107905.

(3) Lazar, S. T.; Kolibaba, T. J.; Grunlan, J. C. Flame-Retardant Surface Treatments. *Nature Reviews Materials* **2020**, *5* (4), 259–275. https://doi.org/10.1038/s41578-019-0164-6.

(4) Hoebel, J. F.; Damant, G. H.; Spivak, S. M.; Berlin, G. N. Clothing-Related Burn Casualties: An Overlooked Problem? *Fire Technology* **2010**, *46* (3), 629–649. https://doi.org/10.1007/s10694-009-0113-9.

(5) Bourbigot, S.; Chlebicki, S.; Mamleev, V. Thermal Degradation of Cotton under Linear Heating. *Polymer Degradation and Stability* **2002**, *78* (1), 57–62. https://doi.org/10.1016/S0141-3910(02)00119-2.

(6) Kandola, B. K.; Horrocks, S.; Horrocks, A. R. Evidence of Interaction in Flame-Retardant Fibre-Intumescent Combinations by Thermal Analytical Techniques. *Thermochimica Acta* **1997**, *294* (1), 113–125. https://doi.org/10.1016/S0040-6031(96)03151-6.

(7) Buxbaum, L. H. The Degradation of Poly(Ethylene Terephthalate). *Angewandte Chemie International Edition in English* **1968**, *7* (3), 182–190. https://doi.org/10.1002/anie.196801821.

(8) Holland, B. J.; Hay, J. N. Thermal Degradation of Nylon Polymers. *Polymer International* **2000**, *49* (9), 943–948. https://doi.org/10.1002/1097-0126(200009)49:9<943::AID-PI400>3.0.CO;2-5.

(9) Hilado, C. J. An Overview of the Fire Behavior of Polymers. *Fire Technology* **1973**, *9* (3), 198–208. https://doi.org/10.1007/BF02624777.

(10) Yang, C. Q.; He, Q. Textile Heat Release Properties Measured by Microscale Combustion Calorimetry: Experimental Repeatability. *Fire and Materials* **2012**, *36* (2), 127–137. https://doi.org/10.1002/fam.1093.

(11) Lawton, E. L.; Setzer, C. J. Flame-Retardant Polyethylene Terephthalate Fibers. In *Flame-Retardant Polymeric Materials*; Lewin, M., Atlas, S. M., Pearce, E. M., Eds.; Springer: Boston, MA, 1975; pp 193–221. https://doi.org/10.1007/978-1-4684-2148-4_4.

(12) Hall, S.; Evarts, B. *Fire Loss in the United States During 2021*; National Fire Protection Agency: Quincy, MA, 2022.

(13) Dubrulle, L.; Zammarano, M.; Davis, R. D. *Effect of Fire-Retardant Coatings and Accelerated-Weathering on the Flammability of Wood-Based Materials in Wildland-Urban Interface (WUI) Communities*; National Institute of Standards and Technology, 2020. https://doi.org/10.6028/NIST.TN.2094.

(14) Rutherford, D. W.; Wershaw, R. L.; Cox, L. G. *Changes in Composition and Porosity Occurring During the Thermal Degradation of Wood and Wood Components*; Scientific Investigations Report 2004–5292; United States Geological Survey: Reston, VA, 2005.

(15) Hill, C. A. S. *Wood Modification: Chemical, Thermal and Other Processes*; John Wiley & Sons, Ltd: Chichester, UK, 2006. https://doi.org/10.1002/0470021748.

(16) Lowden, L. A.; Hull, T. R. Flammability Behaviour of Wood and a Review of the Methods for Its Reduction. *Fire Science Reviews* **2013**, *2* (1), 4.

(17) Liang, C.; Gracida-Alvarez, U. R.; Gallant, E. T.; Gillis, P. A.; Marques, Y. A.; Abramo, G. P.; Hawkins, T. R.; Dunn, J. B. Material Flows of Polyurethane in the United States. *Environmental Science & Technology* **2021**, *55* (20), 14215–14224. https://doi.org/10.1021/acs.est.1c03654.

(18) Yadav, A.; de Souza, F. M.; Dawsey, T.; Gupta, R. K. Recent Advancements in Flame-Retardant Polyurethane Foams: A Review. *Industrial & Engineering Chemistry Research* **2022**, *61* (41), 15046–15065. https://doi.org/10.1021/acs.iecr.2c02670.

(19) Iler, R. K. Multilayers of Colloidal Particles. *Journal of Colloid and Interface Science* **1966**, *21* (6), 569–594. https://doi.org/10.1016/0095-8522(66)90018-3.

(20) Decher, G.; Hong, J.-D. Buildup of Ultrathin Multilayer Films by a Self-Assembly Process, 1 Consecutive Adsorption of Anionic and Cationic Bipolar Amphiphiles on Charged Surfaces. *Makromolekulare Chemie. Macromolecular Symposia* **1991**, *46* (1), 321–327. https://doi.org/10.1002/masy.19910460145.

(21) Zhang, X.; Xu, Y.; Zhang, X.; Wu, H.; Shen, J.; Chen, R.; Xiong, Y.; Li, J.; Guo, S. Progress on the Layer-by-Layer Assembly of Multilayered Polymer Composites: Strategy, Structural Control and Applications. *Progress in Polymer Science* **2019**, *89*, 76–107. https://doi.org/10.1016/j.progpolymsci.2018.10.002.

(22) Decher, G.; Schlenoff, J. B. *Multilayer Thin Films: Sequential Assembly of Nanocomposite Materials*, 2nd ed.; John Wiley & Sons, Ltd.: Weinheim, Germany, **2012**.

(23) Yang, H.; Yu, B.; Song, P.; Maluk, C.; Wang, H. Surface-Coating Engineering for Flame Retardant Flexible Polyurethane Foams: A Critical Review. *Composites Part B: Engineering* **2019**, *176*, 107185. https://doi.org/10.1016/j.compositesb.2019.107185.

(24) Qiu, X.; Li, Z.; Li, X.; Zhang, Z. Flame Retardant Coatings Prepared Using Layer by Layer Assembly: A Review. *Chemical Engineering Journal* **2018**, *334*, 108–122. https://doi.org/10.1016/j.cej.2017.09.194.

(25) Liang, S.; Neisius, N. M.; Gaan, S. Recent Developments in Flame Retardant Polymeric Coatings. *Progress in Organic Coatings* **2013**, *76* (11), 1642–1665. https://doi.org/10.1016/j.porgcoat.2013.07.014.

(26) Kundu, C. K.; Li, Z.; Song, L.; Hu, Y. An Overview of Fire Retardant Treatments for Synthetic Textiles: From Traditional Approaches to Recent Applications. *European Polymer Journal* **2020**, *137*, 109911. https://doi.org/10.1016/j.eurpolymj.2020.109911.

(27) Ariga, K.; Ahn, E.; Park, M.; Kim, B. Layer-by-Layer Assembly: Recent Progress from Layered Assemblies to Layered Nanoarchitectonics. *Chemistry: An Asian Journal* **2019**, *14* (15), 2553–2566. https://doi.org/10.1002/asia.201900627.

(28) van der Gucht, J.; Spruijt, E.; Lemmers, M.; Cohen Stuart, M. A. Polyelectrolyte Complexes: Bulk Phases and Colloidal Systems. *Journal of Colloid and Interface Science* **2011**, *361* (2), 407–422. https://doi.org/10.1016/j.jcis.2011.05.080.

(29) Gilman, J. W.; Harris, R. H.; Shields, J. R.; Kashiwagi, T.; Morgan, A. B. A Study of the Flammability Reduction Mechanism of Polystyrene-Layered Silicate Nanocomposite: Layered Silicate Reinforced Carbonaceous Char. *Polymers for Advanced Technologies* **2006**, *17* (4), 263–271. https://doi.org/10.1002/pat.682.

(30) Gilman, J. W.; Jackson, C. L.; Morgan, A. B.; Harris, R.; Manias, E.; Giannelis, E. P.; Wuthenow, M.; Hilton, D.; Phillips, S. H. Flammability Properties of Polymer–Layered-Silicate Nanocomposites. Polypropylene and Polystyrene Nanocomposites. *Chemistry of Materials* **2000**, *12* (7), 1866–1873. https://doi.org/10.1021/cm0001760.

(31) Schartel, B.; Bartholmai, M.; Knoll, U. Some Comments on the Main Fire Retardancy Mechanisms in Polymer Nanocomposites. *Polymers for Advanced Technologies* **2006**, *17* (9–10), 772–777. https://doi.org/10.1002/pat.792.
(32) Alongi, J.; Han, Z.; Bourbigot, S. Intumescence: Tradition versus Novelty. A Comprehensive Review. *Progress in Polymer Science* **2015**, *51*, 28–73. https://doi.org/10.1016/j.progpolymsci.2015.04.010.
(33) Zhang, L.; Li, Y.; Sun, J.; Shen, J. Mechanically Stable Antireflection and Antifogging Coatings Fabricated by the Layer-by-Layer Deposition Process and Postcalcination. *Langmuir* **2008**, *24* (19), 10851–10857. https://doi.org/10.1021/la801806r.
(34) Li, Y.; Liu, F.; Sun, J. A Facile Layer-by-Layer Deposition Process for the Fabrication of Highly Transparent Superhydrophobic Coatings. *Chemical Communications* **2009**, 19, 2730. https://doi.org/10.1039/b900804g.
(35) Li, P.; Wang, B.; Xu, Y.-J.; Jiang, Z.; Dong, C.; Liu, Y.; Zhu, P. Ecofriendly Flame-Retardant Cotton Fabrics: Preparation, Flame Retardancy, Thermal Degradation Properties, and Mechanism. *ACS Sustainable Chemistry & Engineering* **2019**, *7* (23), 19246–19256. https://doi.org/10.1021/acssuschemeng.9b05523.
(36) Chen, X.; Fang, F.; Zhang, X.; Ding, X.; Wang, Y.; Chen, L.; Tian, X. Flame-Retardant, Electrically Conductive and Antimicrobial Multifunctional Coating on Cotton Fabric via Layer-by-Layer Assembly Technique. *RSC Advances* **2016**, *6* (33), 27669–27676. https://doi.org/10.1039/C5RA26914H.
(37) Ahmed, H. B.; Emam, H. E. Layer by Layer Assembly of Nanosilver for High Performance Cotton Fabrics. *Fibers and Polymers* **2016**, *17* (3), 418–426. https://doi.org/10.1007/s12221-016-5814-3.
(38) Uğur, Ş. S. Sustainable Dyeing and Finishing of Cotton Fabrics with Layer-by-Layer Technique. *Coatings* **2023**, *13* (6), 1129. https://doi.org/10.3390/coatings13061129.
(39) Izquierdo, A.; Ono, S. S.; Voegel, J.-C.; Schaaf, P.; Decher, G. Dipping versus Spraying: Exploring the Deposition Conditions for Speeding Up Layer-by-Layer Assembly. *Langmuir* **2005**, *21* (16), 7558–7567. https://doi.org/10.1021/la047407s.
(40) Dierendonck, M.; De Koker, S.; De Rycke, R.; De Geest, B. G. Just Spray It—LbL Assembly Enters a New Age. *Soft Matter* **2014**, *10* (6), 804–807. https://doi.org/10.1039/C3SM52202D.
(41) Alongi, J.; Carosio, F.; Frache, A.; Malucelli, G. Layer by Layer Coatings Assembled through Dipping, Vertical or Horizontal Spray for Cotton Flame Retardancy. *Carbohydrate Polymers* **2013**, *92* (1), 114–119. https://doi.org/10.1016/j.carbpol.2012.08.086.
(42) Alongi, J.; Malucelli, G. Cotton Flame Retardancy: State of the Art and Future Perspectives. *RSC Advances* **2015**, *5* (31), 24239–24263. https://doi.org/10.1039/C5RA01176K.
(43) Zhao, H.-B.; Wang, Y.-Z. Design and Synthesis of PET-Based Copolyesters with Flame-Retardant and Antidripping Performance. *Macromolecular Rapid Communications* **2017**, *38* (23), 1700451. https://doi.org/10.1002/marc.201700451.
(44) Guin, T.; Krecker, M.; Milhorn, A.; Grunlan, J. C. Maintaining Hand and Improving Fire Resistance of Cotton Fabric through Ultrasonication Rinsing of Multilayer Nanocoating. *Cellulose* **2014**, *21* (4), 3023–3030. https://doi.org/10.1007/s10570-014-0286-3.
(45) Alongi, J.; Carosio, F.; Malucelli, G. Influence of Ammonium Polyphosphate-/Poly(Acrylic Acid)-Based Layer by Layer Architectures on the Char Formation in Cotton, Polyester and Their Blends. *Polymer Degradation and Stability* **2012**, *97* (9), 1644–1653. https://doi.org/10.1016/j.polymdegradstab.2012.06.025.
(46) Li, Y.-C.; Mannen, S.; Morgan, A. B.; Chang, S.; Yang, Y.-H.; Condon, B.; Grunlan, J. C. Intumescent All-Polymer Multilayer Nanocoating Capable of Extinguishing Flame on Fabric. *Advanced Materials* **2011**, *23* (34), 3926–3931. https://doi.org/10.1002/adma.201101871.
(47) Carosio, F.; Fontaine, G.; Alongi, J.; Bourbigot, S. Starch-Based Layer by Layer Assembly: Efficient and Sustainable Approach to Cotton Fire Protection. *ACS Applied Materials & Interfaces* **2015**, *7* (22), 12158–12167. https://doi.org/10.1021/acsami.5b02507.
(48) Zhang, T.; Yan, H.; Wang, L.; Fang, Z. Controlled Formation of Self-Extinguishing Intumescent Coating on Ramie Fabric via Layer-by-Layer Assembly. *Industrial & Engineering Chemistry Research* **2013**, *52* (18), 6138–6146. https://doi.org/10.1021/ie3031554.
(49) Carosio, F.; Alongi, J. Few Durable Layers Suppress Cotton Combustion Due to the Joint Combination of Layer by Layer Assembly and UV-Curing. *RSC Advances* **2015**, *5* (87), 71482–71490. https://doi.org/10.1039/C5RA11856E.
(50) Apaydin, K.; Laachachi, A.; Ball, V.; Jimenez, M.; Bourbigot, S.; Toniazzo, V.; Ruch, D. Intumescent Coating of (Polyallylamine-Polyphosphates) Deposited on Polyamide Fabrics via Layer-by-Layer Technique. *Polymer Degradation and Stability* **2014**, *106*, 158–164. https://doi.org/10.1016/j.polymdegradstab.2014.01.006.
(51) Zhao, L.; Yan, H.; Fang, Z.; Wang, J.; Wang, H. On the Flameproof Treatment of Ramie Fabrics Using a Spray-Assisted Layer-by-Layer Technique. *Polymer Degradation and Stability* **2015**, *121*, 11–17. https://doi.org/10.1016/j.polymdegradstab.2015.08.007.
(52) Fu, C.; Xu, X.; Yin, G.-Z.; Xu, B.; Li, P.; Ai, B.; Zhai, Z.; Gao, F.; Zhai, J.; Wang, D.-Y. Surface Engineering for Cellulose as a Boosted Layer-by-Layer Assembly: Excellent Flame Retardancy and Improved Durability with Introduction of Bio-Based "Molecular Glue." *Applied Surface Science* **2022**, *585*, 152550. https://doi.org/10.1016/j.apsusc.2022.152550.
(53) Carosio, F.; Di Blasio, A.; Alongi, J.; Malucelli, G. Green DNA-Based Flame Retardant Coatings Assembled through Layer by Layer. *Polymer* **2013**, *54* (19), 5148–5153. https://doi.org/10.1016/j.polymer.2013.07.029.
(54) Zhang, W.; Wang, M.; Guan, J.-P.; Tang, R.-C.; Qiao, Y.-F. Casein Phosphopeptide-Metal Salts Combination: A Novel Route for Imparting the Durable Flame Retardancy to Silk. *Journal of the Taiwan Institute of Chemical Engineers* **2019**, *101*, 1–7. https://doi.org/10.1016/j.jtice.2019.04.038.
(55) Lv, Z.; Hu, Y.-T.; Guan, J.-P.; Tang, R.-C.; Chen, G.-Q. Preparation of a Flame Retardant, Antibacterial, and Colored Silk Fabric with Chitosan and Vitamin B2 Sodium Phosphate by Electrostatic Layer by Layer Assembly. *Materials Letters* **2019**, *241*, 136–139. https://doi.org/10.1016/j.matlet.2019.01.005.
(56) Mokhena, T. C.; Sadiku, E. R.; Ray, S. S.; Mochane, M. J.; Matabola, K. P.; Motloung, M. Flame Retardancy Efficacy of Phytic Acid: An Overview. *Journal of Applied Polymer Science* **2022**, *139* (27). https://doi.org/10.1002/app.52495.
(57) Magovac, E.; Jordanov, I.; Grunlan, J. C.; Bischof, S. Environmentally-Benign Phytic Acid-Based Multilayer Coating for Flame Retardant Cotton. *Materials* **2020**, *13* (23), 5492. https://doi.org/10.3390/ma13235492.
(58) Liu, L.; Huang, Z.; Pan, Y.; Wang, X.; Song, L.; Hu, Y. Finishing of Cotton Fabrics by Multi-Layered Coatings to

Improve Their Flame Retardancy and Water Repellency. *Cellulose* **2018**, *25* (8), 4791–4803. https://doi.org/10.1007/s10570-018-1866-4.
(59) Li, Y.-C.; Schulz, J.; Mannen, S.; Delhom, C.; Condon, B.; Chang, S.; Zammarano, M.; Grunlan, J. C. Flame Retardant Behavior of Polyelectrolyte–Clay Thin Film Assemblies on Cotton Fabric. *ACS Nano* **2010**, *4* (6), 3325–3337. https://doi.org/10.1021/nn100467e.
(60) Laufer, G.; Carosio, F.; Martinez, R.; Camino, G.; Grunlan, J. C. Growth and Fire Resistance of Colloidal Silica-Polyelectrolyte Thin Film Assemblies. *Journal of Colloid and Interface Science* **2011**, *356* (1), 69–77. https://doi.org/10.1016/j.jcis.2010.12.072.
(61) Wang, D.; Ma, J.; Liu, J.; Tian, A.; Fu, S. Intumescent Flame-retardant and Ultraviolet-blocking Coating Screen-Printed on Cotton Fabric. *Cellulose* **2021**, *28* (4), 2495–2504. https://doi.org/10.1007/s10570-020-03669-7.
(62) Zhang, Y.; Tian, W.; Liu, L.; Cheng, W.; Wang, W.; Liew, K. M.; Wang, B.; Hu, Y. Eco-Friendly Flame Retardant and Electromagnetic Interference Shielding Cotton Fabrics with Multi-Layered Coatings. *Chemical Engineering Journal* **2019**, *372*, 1077–1090. https://doi.org/10.1016/j.cej.2019.05.012.
(63) Carosio, F.; Di Pierro, A.; Alongi, J.; Fina, A.; Saracco, G. Controlling the Melt Dripping of Polyester Fabrics by Tuning the Ionic Strength of Polyhedral Oligomeric Silsesquioxane and Sodium Montmorillonite Coatings Assembled through Layer by Layer. *Journal of Colloid and Interface Science* **2018**, *510*, 142–151. https://doi.org/10.1016/j.jcis.2017.09.059.
(64) Carosio, F.; Laufer, G.; Alongi, J.; Camino, G.; Grunlan, J. C. Layer-by-Layer Assembly of Silica-Based Flame Retardant Thin Film on PET Fabric. *Polymer Degradation and Stability* **2011**, *96* (5), 745–750. https://doi.org/10.1016/j.polymdegradstab.2011.02.019.
(65) Carosio, F.; Di Blasio, A.; Cuttica, F.; Alongi, J.; Frache, A.; Malucelli, G. Flame Retardancy of Polyester Fabrics Treated by Spray-Assisted Layer-by-Layer Silica Architectures. *Industrial & Engineering Chemistry Research* **2013**, *52* (28), 9544–9550. https://doi.org/10.1021/ie4011244.
(66) Liu, J.; Xiao, C. Fire-Retardant Multilayer Assembled on Polyester Fabric from Water-Soluble Chitosan, Sodium Alginate and Divalent Metal Ion. *International Journal of Biological Macromolecules* **2018**, *119*, 1083–1089. https://doi.org/10.1016/j.ijbiomac.2018.08.043.
(67) Pan, Y.; Liu, L.; Song, L.; Hu, Y.; Wang, W.; Zhao, H. Durable Flame Retardant Treatment of Polyethylene Terephthalate (PET) Fabric with Cross-Linked Layer-by-Layer Assembled Coating. *Polymer Degradation and Stability* **2019**, *165*, 145–152. https://doi.org/10.1016/j.polymdegradstab.2019.05.009.
(68) Fang, Y.; Liu, X.; Tao, X. Intumescent Flame Retardant and Anti-Dripping of PET Fabrics through Layer-by-Layer Assembly of Chitosan and Ammonium Polyphosphate. *Progress in Organic Coatings* **2019**, *134*, 162–168. https://doi.org/10.1016/j.porgcoat.2019.05.010.
(69) Jordanov, I.; Magovac, E.; Fahami, A.; Lazar, S.; Kolibaba, T.; Smith, R. J.; Bischof, S.; Grunlan, J. C. Flame Retardant Polyester Fabric from Nitrogen-Rich Low Molecular Weight Additives within Intumescent Nanocoating. *Polymer Degradation and Stability* **2019**, *170*, 108998. https://doi.org/10.1016/j.polymdegradstab.2019.108998.
(70) Guo, J.; Cang, D.; Zhao, Z.; Yin, Y.; Yang, Z.; Saiyin, B. Fabrication of Superhydrophobic and Flame-Retardant Polyethylene Terephthalate Fabric through a Fluorine-Free Layer-by-Layer Technique. *International Journal of Chemical Reactor Engineering* **2022**, *20* (12), 1283–1295. https://doi.org/10.1515/ijcre-2022-0010.
(71) Kumar Kundu, C.; Wang, W.; Zhou, S.; Wang, X.; Sheng, H.; Pan, Y.; Song, L.; Hu, Y. A Green Approach to Constructing Multilayered Nanocoating for Flame Retardant Treatment of Polyamide 66 Fabric from Chitosan and Sodium Alginate. *Carbohydrate Polymers* **2017**, *166*, 131–138. https://doi.org/10.1016/j.carbpol.2017.02.084.
(72) Kundu, C. K.; Wang, X.; Song, L.; Hu, Y. Borate Cross-Linked Layer-by-Layer Assembly of Green Polyelectrolytes on Polyamide 66 Fabrics for Flame-Retardant Treatment. *Progress in Organic Coatings* **2018**, *121*, 173–181. https://doi.org/10.1016/j.porgcoat.2018.04.031.
(73) Yoon, H. N.; Buckley, A. Improved Comfort Polyester: Part I: Transport Properties and Thermal Comfort of Polyester/Cotton Blend Fabrics. *Textile Research Journal* **1984**, *54* (5), 289–298. https://doi.org/10.1177/004051758405400502.
(74) Alongi, J.; Carosio, F.; Kiekens, P. Recent Advances in the Design of Water Based-Flame Retardant Coatings for Polyester and Polyester-Cotton Blends. *Polymers* **2016**, *8* (10), 357. https://doi.org/10.3390/polym8100357.
(75) Fang, Y.; Sun, W.; Li, J.; Liu, H.; Liu, X. Eco-Friendly Flame Retardant and Dripping-Resistant of Polyester/Cotton Blend Fabrics through Layer-by-Layer Assembly Fully Bio-Based Chitosan/Phytic Acid Coating. *International Journal of Biological Macromolecules* **2021**, *175*, 140–146. https://doi.org/10.1016/j.ijbiomac.2021.02.023.
(76) Wang, B.; Xu, Y.-J.; Li, P.; Zhang, F.-Q.; Liu, Y.; Zhu, P. Flame-Retardant Polyester/Cotton Blend with Phosphorus/Nitrogen/Silicon-Containing Nano-Coating by Layer-by-Layer Assembly. *Applied Surface Science* **2020**, *509*, 145323. https://doi.org/10.1016/j.apsusc.2020.145323.
(77) Narkhede, M.; Thota, S.; Mosurkal, R.; Muller, W. S.; Kumar, J. Layer-by-Layer Assembly of Halogen-Free Polymeric Materials on Nylon/Cotton Blend for Flame Retardant Applications: Layer-by-Layer Assembly of Halogen-Free Polymeric Materials. *Fire and Materials* **2016**, *40* (2), 206–218. https://doi.org/10.1002/fam.2280.
(78) Smith, D. L.; Vest, N. A.; Rodriguez-Melendez, D.; Palen, B.; Grunlan, J. C. Bio-Sourced Intumescent Nanocoating. *Advanced Engineering Materials* **2023**, *25* (4), 2200911. https://doi.org/10.1002/adem.202200911.
(79) Carosio, F.; Alongi, J. Influence of Layer by Layer Coatings Containing Octapropylammonium Polyhedral Oligomeric Silsesquioxane and Ammonium Polyphosphate on the Thermal Stability and Flammability of Acrylic Fabrics. *Journal of Analytical and Applied Pyrolysis* **2016**, *119*, 114–123. https://doi.org/10.1016/j.jaap.2016.03.010.
(80) Ren, Y.; Huo, T.; Qin, Y.; Liu, X. Preparation of Flame Retardant Polyacrylonitrile Fabric Based on Sol-Gel and Layer-by-Layer Assembly. *Materials* **2018**, *11* (4), 483. https://doi.org/10.3390/ma11040483.
(81) Kim, Y. S.; Davis, R.; Cain, A. A.; Grunlan, J. C. Development of Layer-by-Layer Assembled Carbon Nanofiber-Filled Coatings to Reduce Polyurethane Foam Flammability. *Polymer* **2011**, *52* (13), 2847–2855. https://doi.org/10.1016/j.polymer.2011.04.023.
(82) Liu, Q.; Gao, S.; Zhao, Y.; Tao, W.; Yu, X.; Zhi, M. Review of Layer-by-Layer Self-Assembly Technology for Fire Protection of Flexible Polyurethane Foam. *Journal of Materials Science* **2021**, *56* (16), 9605–9643. https://doi.org/10.1007/s10853-021-05904-3.
(83) Carosio, F.; Ghanadpour, M.; Alongi, J.; Wågberg, L. Layer-by-Layer-Assembled Chitosan/Phosphorylated Cellulose

Nanofibrils as a Bio-Based and Flame Protecting Nano-Exoskeleton on PU Foams. *Carbohydrate Polymers* **2018**, *202*, 479–487. https://doi.org/10.1016/j.carbpol.2018.09.005.

(84) Krämer, R. H.; Zammarano, M.; Linteris, G. T.; Gedde, U. W.; Gilman, J. W. Heat Release and Structural Collapse of Flexible Polyurethane Foam. *Polymer Degradation and Stability* **2010**, *95* (6), 1115–1122. https://doi.org/10.1016/j.polymdegradstab.2010.02.019.

(85) Haile, M.; Fomete, S.; Lopez, I. D.; Grunlan, J. C. Aluminum Hydroxide Multilayer Assembly Capable of Extinguishing Flame on Polyurethane Foam. *Journal of Materials Science* **2016**, *51* (1), 375–381. https://doi.org/10.1007/s10853-015-9258-8.

(86) Pan, Y.; Zhan, J.; Pan, H.; Wang, W.; Ge, H.; Song, L.; Hu, Y. A Novel and Effective Method to Fabricate Flame Retardant and Smoke Suppressed Flexible Polyurethane Foam. *RSC Advances* **2015**, *5* (83), 67878–67885. https://doi.org/10.1039/C5RA09553K.

(87) Lazar, S.; Shen, R.; Quan, Y.; Palen, B.; Wang, Q.; Ellison, C. J.; Grunlan, J. C. Mixed Solvent Synthesis of Polydopamine Nanospheres for Sustainable Multilayer Flame Retardant Nanocoating. *Polymer Chemistry* **2021**, *12* (16), 2389–2396. https://doi.org/10.1039/D1PY00111F.

(88) Pan, Y.; Liu, L.; Cai, W.; Hu, Y.; Jiang, S.; Zhao, H. Effect of Layer-by-Layer Self-Assembled Sepiolite-Based Nanocoating on Flame Retardant and Smoke Suppressant Properties of Flexible Polyurethane Foam. *Applied Clay Science* **2019**, *168*, 230–236. https://doi.org/10.1016/j.clay.2018.11.014.

(89) Pan, H.; Wang, W.; Pan, Y.; Song, L.; Hu, Y.; Liew, K. M. Formation of Layer-by-Layer Assembled Titanate Nanotubes Filled Coating on Flexible Polyurethane Foam with Improved Flame Retardant and Smoke Suppression Properties. *ACS Applied Materials & Interfaces* **2015**, *7* (1), 101–111. https://doi.org/10.1021/am507045g.

(90) Smith, R. J.; Holder, K. M.; Ruiz, S.; Hahn, W.; Song, Y.; Lvov, Y. M.; Grunlan, J. C. Environmentally Benign Halloysite Nanotube Multilayer Assembly Significantly Reduces Polyurethane Flammability. *Advanced Functional Materials* **2018**, *28* (27), 1703289. https://doi.org/10.1002/adfm.201703289.

(91) Holder, K. M.; Cain, A. A.; Plummer, M. G.; Stevens, B. E.; Odenborg, P. K.; Morgan, A. B.; Grunlan, J. C. Carbon Nanotube Multilayer Nanocoatings Prevent Flame Spread on Flexible Polyurethane Foam. *Macromolecular Materials and Engineering* **2016**, *301* (6), 665–673. https://doi.org/10.1002/mame.201500327.

(92) Kim, Y. S.; Davis, R. Multi-Walled Carbon Nanotube Layer-by-Layer Coatings with a Trilayer Structure to Reduce Foam Flammability. *Thin Solid Films* **2014**, *550*, 184–189. https://doi.org/10.1016/j.tsf.2013.10.167.

(93) Zhang, L.; Liu, B.-W.; Wang, Y.-Z.; Fu, T.; Zhao, H.-B. P-Doped PANI/AgMWs Nano/Micro Coating towards High-Efficiency Flame Retardancy and Electromagnetic Interference Shielding. *Composites Part B: Engineering* **2022**, *238*, 109944. https://doi.org/10.1016/j.compositesb.2022.109944.

(94) Carosio, F.; Kochumalayil, J.; Cuttica, F.; Camino, G.; Berglund, L. Oriented Clay Nanopaper from Biobased Components—Mechanisms for Superior Fire Protection Properties. *ACS Applied Materials & Interfaces* **2015**, *7* (10), 5847–5856. https://doi.org/10.1021/am509058h.

(95) Laufer, G.; Kirkland, C.; Cain, A. A.; Grunlan, J. C. Clay–Chitosan Nanobrick Walls: Completely Renewable Gas Barrier and Flame-Retardant Nanocoatings. *ACS Applied Materials & Interfaces* **2012**, *4* (3), 1643–1649. https://doi.org/10.1021/am2017915.

(96) Kim, Y. S.; Li, Y.-C.; Pitts, W. M.; Werrel, M.; Davis, R. D. Rapid Growing Clay Coatings to Reduce the Fire Threat of Furniture. *ACS Applied Materials & Interfaces* **2014**, *6* (3), 2146–2152. https://doi.org/10.1021/am405259n.

(97) Cain, A. A.; Plummer, M. G. B.; Murray, S. E.; Bolling, L.; Regev, O.; Grunlan, J. C. Iron-Containing, High Aspect Ratio Clay as Nanoarmor That Imparts Substantial Thermal/Flame Protection to Polyurethane with a Single Electrostatically-Deposited Bilayer. *Journal of Materials Chemistry A* **2014**, *2* (41), 17609–17617. https://doi.org/10.1039/C4TA03541K.

(98) Lazar, S.; Carosio, F.; Davesne, A.-L.; Jimenez, M.; Bourbigot, S.; Grunlan, J. Extreme Heat Shielding of Clay/Chitosan Nanobrick Wall on Flexible Foam. *ACS Applied Materials & Interfaces* **2018**, *10* (37), 31686–31696. https://doi.org/10.1021/acsami.8b10227.

(99) Maddalena, L.; Gomez, J.; Fina, A.; Carosio, F. Effects of Graphite Oxide Nanoparticle Size on the Functional Properties of Layer-by-Layer Coated Flexible Foams. *Nanomaterials* **2021**, *11* (2), 266. https://doi.org/10.3390/nano11020266.

(100) Pan, H.; Yu, B.; Wang, W.; Pan, Y.; Song, L.; Hu, Y. Comparative Study of Layer by Layer Assembled Multilayer Films Based on Graphene Oxide and Reduced Graphene Oxide on Flexible Polyurethane Foam: Flame Retardant and Smoke Suppression Properties. *RSC Advances* **2016**, *6* (115), 114304–114312. https://doi.org/10.1039/C6RA15522G.

(101) Maddalena, L.; Carosio, F.; Gomez, J.; Saracco, G.; Fina, A. Layer-by-Layer Assembly of Efficient Flame Retardant Coatings Based on High Aspect Ratio Graphene Oxide and Chitosan Capable of Preventing Ignition of PU Foam. *Polymer Degradation and Stability* **2018**, *152*, 1–9. https://doi.org/10.1016/j.polymdegradstab.2018.03.013.

(102) Carosio, F.; Maddalena, L.; Gomez, J.; Saracco, G.; Fina, A. Graphene Oxide Exoskeleton to Produce Self-Extinguishing, Nonignitable, and Flame Resistant Flexible Foams: A Mechanically Tough Alternative to Inorganic Aerogels. *Advanced Materials Interfaces* **2018**, *5* (23), 1801288. https://doi.org/10.1002/admi.201801288.

(103) Lin, B.; Yuen, A. C. Y.; Li, A.; Zhang, Y.; Chen, T. B. Y.; Yu, B.; Lee, E. W. M.; Peng, S.; Yang, W.; Lu, H.-D.; Chan, Q. N.; Yeoh, G. H.; Wang, C. H. MXene/Chitosan Nanocoating for Flexible Polyurethane Foam towards Remarkable Fire Hazards Reductions. *Journal of Hazardous Materials* **2020**, *381*, 120952. https://doi.org/10.1016/j.jhazmat.2019.120952.

(104) Lin, B.; Yuen, A. C. Y.; Oliver, S.; Liu, J.; Yu, B.; Yang, W.; Wu, S.; Yeoh, G. H.; Wang, C. H. Dual Functionalisation of Polyurethane Foam for Unprecedented Flame Retardancy and Antibacterial Properties Using Layer-by-Layer Assembly of MXene Chitosan with Antibacterial Metal Particles. *Composites Part B: Engineering* **2022**, *244*, 110147. https://doi.org/10.1016/j.compositesb.2022.110147.

(105) Lin, B.; Yuen, A. C. Y.; Chen, T. B. Y.; Yu, B.; Yang, W.; Zhang, J.; Yao, Y.; Wu, S.; Wang, C. H.; Yeoh, G. H. Experimental and Numerical Perspective on the Fire Performance of MXene/Chitosan/Phytic Acid Coated Flexible Polyurethane Foam. *Scientific Reports* **2021**, *11* (1), 4684. https://doi.org/10.1038/s41598-021-84083-2.

(106) Wang, X.; Pan, Y.-T.; Wan, J.-T.; Wang, D.-Y. An Eco-Friendly Way to Fire Retardant Flexible Polyurethane Foam: Layer-by-Layer Assembly of Fully Bio-Based

Substances. *RSC Advances* **2014**, *4* (86), 46164–46169. https://doi.org/10.1039/C4RA07972H.

(107) Pan, Y.; Zhan, J.; Pan, H.; Wang, W.; Tang, G.; Song, L.; Hu, Y. Effect of Fully Biobased Coatings Constructed via Layer-by-Layer Assembly of Chitosan and Lignosulfonate on the Thermal, Flame Retardant, and Mechanical Properties of Flexible Polyurethane Foam. *ACS Sustainable Chemistry & Engineering* **2016**, *4* (3), 1431–1438. https://doi.org/10.1021/acssuschemeng.5b01423.

(108) Laufer, G.; Kirkland, C.; Morgan, A. B.; Grunlan, J. C. Exceptionally Flame Retardant Sulfur-Based Multilayer Nanocoating for Polyurethane Prepared from Aqueous Polyelectrolyte Solutions. *ACS Macro Letters* **2013**, *2* (5), 361–365. https://doi.org/10.1021/mz400105e.

(109) Carosio, F.; Di Blasio, A.; Cuttica, F.; Alongi, J.; Malucelli, G. Self-Assembled Hybrid Nanoarchitectures Deposited on Poly(Urethane) Foams Capable of Chemically Adapting to Extreme Heat. *RSC Advances* **2014**, *4* (32), 16674–16680. https://doi.org/10.1039/C4RA01343C.

(110) Cain, A. A.; Nolen, C. R.; Li, Y.-C.; Davis, R.; Grunlan, J. C. Phosphorous-Filled Nanobrick Wall Multilayer Thin Film Eliminates Polyurethane Melt Dripping and Reduces Heat Release Associated with Fire. *Polymer Degradation and Stability* **2013**, *98* (12), 2645–2652. https://doi.org/10.1016/j.polymdegradstab.2013.09.028.

(111) Weldemhret, T. G.; Menge, H. G.; Lee, D.-W.; Park, H.; Lee, J.; Song, J. I.; Park, Y. T. Facile Deposition of Environmentally Benign Organic-Inorganic Flame Retardant Coatings to Protect Flammable Foam. *Progress in Organic Coatings* **2021**, *161*, 106480. https://doi.org/10.1016/j.porgcoat.2021.106480.

(112) Li, Y.-C.; Kim, Y. S.; Shields, J.; Davis, R. Controlling Polyurethane Foam Flammability and Mechanical Behaviour by Tailoring the Composition of Clay-Based Multilayer Nanocoatings. *Journal of Materials Chemistry A* **2013**, *1* (41), 12987. https://doi.org/10.1039/c3ta11936j.

(113) Carosio, F.; Fina, A. Three Organic/Inorganic Nanolayers on Flexible Foam Allow Retaining Superior Flame Retardancy Performance Upon Mechanical Compression Cycles. *Frontiers in Materials* **2019**, *6*. https://doi.org/10.3389/fmats.2019.00020.

(114) Song, Y.; Meyers, K. P.; Gerringer, J.; Ramakrishnan, R. K.; Humood, M.; Qin, S.; Polycarpou, A. A.; Nazarenko, S.; Grunlan, J. C. Fast Self-Healing of Polyelectrolyte Multilayer Nanocoating and Restoration of Super Oxygen Barrier. *Macromolecular Rapid Communications* **2017**, *38* (10), 1700064. https://doi.org/10.1002/marc.201700064.

(115) Yang, J.-C.; Cao, Z.-J.; Wang, Y.-Z.; Schiraldi, D. A. Ammonium Polyphosphate-Based Nanocoating for Melamine Foam towards High Flame Retardancy and Anti-Shrinkage in Fire. *Polymer* **2015**, *66*, 86–93. https://doi.org/10.1016/j.polymer.2015.04.022.

(116) Deng, S.-B.; Liao, W.; Yang, J.-C.; Cao, Z.-J.; Wang, Y.-Z. Flame-Retardant and Smoke-Suppressed Silicone Foams with Chitosan-Based Nanocoatings. *Industrial & Engineering Chemistry Research* **2016**, *55* (27), 7239–7248. https://doi.org/10.1021/acs.iecr.6b00532.

(117) Carosio, F.; Cuttica, F.; Di Blasio, A.; Alongi, J.; Malucelli, G. Layer by Layer Assembly of Flame Retardant Thin Films on Closed Cell PET Foams: Efficiency of Ammonium Polyphosphate versus DNA. *Polymer Degradation and Stability* **2015**, *113*, 189–196. https://doi.org/10.1016/j.polymdegradstab.2014.09.018.

(118) Köklükaya, O.; Carosio, F.; Grunlan, J. C.; Wågberg, L. Flame-Retardant Paper from Wood Fibers Functionalized via Layer-by-Layer Assembly. *ACS Applied Materials & Interfaces* **2015**, *7* (42), 23750–23759. https://doi.org/10.1021/acsami.5b08105.

(119) Köklükaya, O.; Carosio, F.; Durán, V. L.; Wågberg, L. Layer-by-Layer Modified Low Density Cellulose Fiber Networks: A Sustainable and Fireproof Alternative to Petroleum Based Foams. *Carbohydrate Polymers* **2020**, *230*, 115616. https://doi.org/10.1016/j.carbpol.2019.115616.

(120) Köklükaya, O.; Carosio, F.; Wågberg, L. Superior Flame-Resistant Cellulose Nanofibril Aerogels Modified with Hybrid Layer-by-Layer Coatings. *ACS Applied Materials & Interfaces* **2017**, *9* (34), 29082–29092. https://doi.org/10.1021/acsami.7b08018.

(121) Köklükaya, O.; Carosio, F.; Wågberg, L. Tailoring Flame-Retardancy and Strength of Papers via Layer-by-Layer Treatment of Cellulose Fibers. *Cellulose* **2018**, *25* (4), 2691–2709. https://doi.org/10.1007/s10570-018-1749-8.

(122) Carosio, F.; Di Blasio, A.; Alongi, J.; Malucelli, G. Layer by Layer Nanoarchitectures for the Surface Protection of Polycarbonate. *European Polymer Journal* **2013**, *49* (2), 397–404. https://doi.org/10.1016/j.eurpolymj.2012.10.032.

(123) Laachachi, A.; Ball, V.; Apaydin, K.; Toniazzo, V.; Ruch, D. Diffusion of Polyphosphates into (Poly(Allylamine)-Montmorillonite) Multilayer Films: Flame Retardant-Intumescent Films with Improved Oxygen Barrier. *Langmuir* **2011**, *27* (22), 13879–13887. https://doi.org/10.1021/la203252q.

(124) Apaydin, K.; Laachachi, A.; Ball, V.; Jimenez, M.; Bourbigot, S.; Toniazzo, V.; Ruch, D. Polyallylamine–Montmorillonite as Super Flame Retardant Coating Assemblies by Layer-by Layer Deposition on Polyamide. *Polymer Degradation and Stability* **2013**, *98* (2), 627–634. https://doi.org/10.1016/j.polymdegradstab.2012.11.006.

(125) Guin, T.; Krecker, M.; Milhorn, A.; Hagen, D. A.; Stevens, B.; Grunlan, J. C. Exceptional Flame Resistance and Gas Barrier with Thick Multilayer Nanobrick Wall Thin Films. *Advanced Materials Interfaces* **2015**, *2* (11), 1500214. https://doi.org/10.1002/admi.201500214.

(126) Fu, J.; Schlenoff, J. B. Driving Forces for Oppositely Charged Polyion Association in Aqueous Solutions: Enthalpic, Entropic, but Not Electrostatic. *Journal of the American Chemical Society* **2016**, *138* (3), 980–990. https://doi.org/10.1021/jacs.5b11878.

(127) Mattison, K. W.; Dubin, P. L.; Brittain, I. J. Complex Formation between Bovine Serum Albumin and Strong Polyelectrolytes: Effect of Polymer Charge Density. *The Journal of Physical Chemistry B* **1998**, *102* (19), 3830–3836. https://doi.org/10.1021/jp980486u.

(128) Chen, J.; Heitmann, J. A.; Hubbe, M. A. Dependency of Polyelectrolyte Complex Stoichiometry on the Order of Addition. 1. Effect of Salt Concentration during Streaming Current Titrations with Strong Poly-Acid and Poly-Base. *Colloids and Surfaces A: Physicochemical and Engineering Aspects* **2003**, *223* (1–3), 215–230. https://doi.org/10.1016/S0927-7757(03)00222-X.

(129) de Vasconcelos, C. L.; Bezerril, P. M.; dos Santos, D. E. S.; Dantas, T. N. C.; Pereira, M. R.; Fonseca, J. L. C. Effect of Molecular Weight and Ionic Strength on the Formation of Polyelectrolyte Complexes Based on Poly(Methacrylic Acid) and Chitosan. *Biomacromolecules* **2006**, *7* (4), 1245–1252. https://doi.org/10.1021/bm050963w.

(130) Chollakup, R.; Beck, J. B.; Dirnberger, K.; Tirrell, M.; Eisenbach, C. D. Polyelectrolyte Molecular Weight and Salt Effects on the Phase Behavior and Coacervation of Aqueous Solutions of Poly(Acrylic Acid) Sodium Salt and Poly(Allylamine) Hydrochloride. *Macromolecules* **2013**, *46* (6), 2376–2390. https://doi.org/10.1021/ma202172q.

(131) Lounis, F. M.; Chamieh, J.; Leclercq, L.; Gonzalez, P.; Geneste, A.; Prelot, B.; Cottet, H. Interactions between Oppositely Charged Polyelectrolytes by Isothermal Titration Calorimetry: Effect of Ionic Strength and Charge Density. *The Journal of Physical Chemistry B* **2017**, *121* (12), 2684–2694. https://doi.org/10.1021/acs.jpcb.6b11907.
(132) Cain, A. A.; Murray, S.; Holder, K. M.; Nolen, C. R.; Grunlan, J. C. Intumescent Nanocoating Extinguishes Flame on Fabric Using Aqueous Polyelectrolyte Complex Deposited in Single Step: Intumescent Nanocoating Extinguishes Flame on Fabric. *Macromolecular Materials and Engineering* **2014**, *299* (10), 1180–1187. https://doi.org/10.1002/mame.201400022.
(133) Haile, M.; Fincher, C.; Fomete, S.; Grunlan, J. C. Water-Soluble Polyelectrolyte Complexes That Extinguish Fire on Cotton Fabric When Deposited as PH-Cured Nanocoating. *Polymer Degradation and Stability* **2015**, *114*, 60–64. https://doi.org/10.1016/j.polymdegradstab.2015.01.022.
(134) Palen, B.; Rabaey, M. G.; Rodriguez-Melendez, D.; Iverson, E. T.; Kolibaba, T. J.; Grunlan, J. C. Polymeric Coacervate Coating for Flame Retardant Paper. *Cellulose* **2022**, *29* (8), 4589–4597. https://doi.org/10.1007/s10570-022-04594-7.
(135) Kolibaba, T. J.; Vest, N. A.; Grunlan, J. C. Polyelectrolyte Photopolymer Complexes for Flame Retardant Wood. *Materials Chemistry Frontiers* **2022**, *6*, 1630–1636. https://doi.org/10.1039/D2QM00257D.
(136) Jimenez, M.; Guin, T.; Bellayer, S.; Dupretz, R.; Bourbigot, S.; Grunlan, J. C. Microintumescent Mechanism of Flame-Retardant Water-Based Chitosan-Ammonium Polyphosphate Multilayer Nanocoating on Cotton Fabric. *Journal of Applied Polymer Science* **2016**, *133* (32), 43783. https://doi.org/10.1002/app.43783.
(137) Leistner, M.; Haile, M.; Rohmer, S.; Abu-Odeh, A.; Grunlan, J. C. Water-Soluble Polyelectrolyte Complex Nanocoating for Flame Retardant Nylon-Cotton Fabric. *Polymer Degradation and Stability* **2015**, *122*, 1–7. https://doi.org/10.1016/j.polymdegradstab.2015.10.008.
(138) Haile, M.; Leistner, M.; Sarwar, O.; Toler, C. M.; Henderson, R.; Grunlan, J. C. A Wash-Durable Polyelectrolyte Complex That Extinguishes Flames on Polyester–Cotton Fabric. *RSC Advances* **2016**, *6* (40), 33998–34004. https://doi.org/10.1039/C6RA03637F.
(139) Li, L.; Qi, P.; Peng, A.; Sun, J.; Cui, Z.; Liu, W.; Li, H.; Gu, X.; Zhang, S. Preparation of Durable Flame Retardant Nylon-Cotton Blend Fabrics by 3-Glycidyloxypropyl Trimethoxy Silane Associated with Polyethyleneimine and Phytic Acid. *Cellulose* **2022**, *29* (13), 7413–7430. https://doi.org/10.1007/s10570-022-04693-5.
(140) Cheng, X.-W.; Guan, J.-P.; Yang, X.-H.; Tang, R.-C.; Yao, F. A Bio-Resourced Phytic Acid/Chitosan Polyelectrolyte Complex for the Flame Retardant Treatment of Wool Fabric. *Journal of Cleaner Production* **2019**, *223*, 342–349. https://doi.org/10.1016/j.jclepro.2019.03.157.
(141) Wang, Z.-H.; Zhang, A.-N.; Liu, B.-W.; Wang, X.-L.; Zhao, H.-B.; Wang, Y.-Z. Durable Flame-Retardant Cotton Fabrics with Tannic Acid Complexed by Various Metal Ions. *Polymer Degradation and Stability* **2022**, *201*, 109997. https://doi.org/10.1016/j.polymdegradstab.2022.109997.
(142) He, J.-L.; Luo, W.; Wang, T.; He, L.; Deng, J.-N.; Fu, Z.-C.; Grunlan, J. C.; Chen, M.-J. Polyelectrolyte Complex with Controllable Viscosity by Doping Cu $^{2+}$ Protects Nylon–Cotton Fabric against Fire. *ACS Applied Materials & Interfaces* **2022**, *14* (48), 54225–54232. https://doi.org/10.1021/acsami.2c16343.
(143) Carosio, F.; Alongi, J. Flame Retardant Multilayered Coatings on Acrylic Fabrics Prepared by One-Step Deposition of Chitosan/Montmorillonite Complexes. *Fibers* **2018**, *6* (2), 36. https://doi.org/10.3390/fib6020036.
(144) Kundu, C. K.; Wang, X.; Song, L.; Hu, Y. Chitosan-Based Flame Retardant Coatings for Polyamide 66 Textiles: One-Pot Deposition versus Layer-by-Layer Assembly. *International Journal of Biological Macromolecules* **2020**, *143*, 1–10. https://doi.org/10.1016/j.ijbiomac.2019.11.220.
(145) Rao, W.; Shi, J.; Yang, M.; Yang, F.; Wu, T.; Yu, C.; Chen, M.-J.; Zhao, H.-B. One-Pot Fabrication of Bio-Based Hydrophobic and Flame-Retardant Coating for Cotton Fabric. *Cellulose* **2023**, *30* (3), 1943–1955. https://doi.org/10.1007/s10570-022-05005-7.
(146) Wang, T.; Xu, J.; Zhan, Y.-J.; He, L.; Deng, J.; Fu, Z.-C.; Zhao, H.-B.; Chen, M.-J. Eco-Friendly and Facile Integrated Intumescent Polyelectrolyte Complex Coating with Universal Flame Retardancy and Smoke Suppression for Cotton and Its Blending Fabrics. *ACS Sustainable Chemistry & Engineering* **2023**, *11* (12), 4838–4849. https://doi.org/10.1021/acssuschemeng.3c00028.
(147) Papo, A.; Piani, L.; Ricceri, R. Sodium Tripolyphosphate and Polyphosphate as Dispersing Agents for Kaolin Suspensions: Rheological Characterization. *Colloids and Surfaces A: Physicochemical and Engineering Aspects* **2002**, *201* (1–3), 219–230. https://doi.org/10.1016/S0927-7757(01)01024-X.
(148) Palen, B.; Kolibaba, T. J.; Brehm, J. T.; Shen, R.; Quan, Y.; Wang, Q.; Grunlan, J. C. Clay-Filled Polyelectrolyte Complex Nanocoating for Flame-Retardant Polyurethane Foam. *ACS Omega* **2021**, *6* (12), 8016–8020. https://doi.org/10.1021/acsomega.0c05354.
(149) Meng, D.; Liu, X.; Wang, S.; Sun, J.; Li, H.; Wang, Z.; Gu, X.; Zhang, S. Self-Healing Polyelectrolyte Complex Coating for Flame Retardant Flexible Polyurethane Foam with Enhanced Mechanical Property. *Composites Part B: Engineering* **2021**, *219*, 108886. https://doi.org/10.1016/j.compositesb.2021.108886.
(150) Bellanger, H.; Casdorff, K.; Muff, L. F.; Ammann, R.; Burgert, I.; Michen, B. Layer-by-Layer Deposition on a Heterogeneous Surface: Effect of Sorption Kinetics on the Growth of Polyelectrolyte Multilayers. *Journal of Colloid and Interface Science* **2017**, *500*, 133–141. https://doi.org/10.1016/j.jcis.2017.02.048.
(151) Kolibaba, T. J.; Grunlan, J. C. Environmentally Benign Polyelectrolyte Complex That Renders Wood Flame Retardant and Mechanically Strengthened. *Macromolecular Materials and Engineering* **2019**, *304* (8), 1900179. https://doi.org/10.1002/mame.201900179.
(152) Soula, M.; Samyn, F.; Duquesne, S.; Landry, V. Innovative Polyelectrolyte Treatment to Flame-Retard Wood. *Polymers* **2021**, *13* (17), 2884. https://doi.org/10.3390/polym13172884.
(153) Gao, S.; Qi, J.; Qi, P.; Xu, R.; Wu, T.; Zhang, B.; Huang, J.; Yan, Y. Unprecedented Nonflammable Organic Adhesives Leading to Fireproof Wood Products. *ACS Applied Materials & Interfaces* **2023**, *15* (6), 8609–8616. https://doi.org/10.1021/acsami.2c19072.
(154) Fisher, S. G.; Rodriguez-Melendez, D.; Iverson, E. T.; Kolibaba, T. J.; Grunlan, J. C. Fire Protection of Wood with an Environmentally Benign UV-Cured Polyelectrolyte Complex. *Polymer Degradation and Stability* **2023**, 110467. https://doi.org/10.1016/j.polymdegradstab.2023.110467.
(155) Danks, A. E.; Hall, S. R.; Schnepp, Z. The Evolution of 'Sol–Gel' Chemistry as a Technique for Materials Synthesis. *Materials Horizons* **2016**, *3* (2), 91–112.
(156) Brinker, C. J.; Scherer, G. W. *Sol-Gel Science: The Physics and Chemistry of Sol-Gel Processing*; Academic Press: Cambridge, MA, **2013**.

(157) Hench, L. L.; West, J. K. The Sol-Gel Process. *Chemical Reviews* **1990**, *90* (1), 33–72.
(158) Avnir, D.; Coradin, T.; Lev, O.; Livage, J. Recent Bio-Applications of Sol–Gel Materials. *Journal of Materials Chemistry* **2006**, *16* (11), 1013–1030.
(159) Figueira, R. B. Hybrid Sol–Gel Coatings for Corrosion Mitigation: A Critical Review. *Polymers* **2020**, *12* (3), 689.
(160) Mackenzie, J. D.; Bescher, E. P. Physical Properties of Sol-Gel Coatings. *Journal of Sol-Gel Science and Technology* **2000**, *19*, 23–29.
(161) Wicks, G. G. *Ceramic Composite Coatings*; Du Pont de Nemours (EI) and Co.: Aiken, SC, **1991**.
(162) Holman, M. E. Gel-Coated Materials with Increased Flame Retardancy. US 6,696,145 B2, February 24, **2004**. https://patentimages.storage.googleapis.com/6e/ed/2c/da2ab6b183f58f/US6696145.pdf.
(163) Bifulco, A.; Imparato, C.; Aronne, A.; Malucelli, G. Flame Retarded Polymer Systems Based on the Sol-Gel Approach: Recent Advances and Future Perspectives. *Journal of Sol-Gel Science and Technology* **2022**, 1–25.
(164) Periyasamy, A. P.; Venkataraman, M.; Kremenakova, D.; Militky, J.; Zhou, Y. Progress in Sol-Gel Technology for the Coatings of Fabrics. *Materials* **2020**, *13* (8), 1838.
(165) Kakihana, M. Invited Review "Sol-Gel" Preparation of High Temperature Superconducting Oxides. *Journal of Sol-Gel Science and Technology* **1996**, *6*, 7–55.
(166) Ebelmen. Untersuchungen Über Die Verbindungen Der Borsäure Und Kieselsäure Mit Aether. *Justus Liebigs Annalen der Chemie* **1846**, *57* (3), 319–355.
(167) Mazdiyasni, K. S.; Brown, L. M. Synthesis and Some Properties of Yttrium and Lanthanide Isopropoxides. *Inorganic Chemistry* **1970**, *9* (12), 2783–2786.
(168) Turova, N. Y.; Korolev, A. V.; Tchebukov, D. E.; Belokon, A. I.; Yanovsky, A. I.; Struchkov, Y. T. Tantalum (V) Alkoxides: Electrochemical Synthesis, Mass-Spectral Investigation and Oxoalkoxocomplexes. *Polyhedron* **1996**, *15* (21), 3869–3880.
(169) Livage, J.; Henry, M.; Sanchez, C. Sol-Gel Chemistry of Transition Metal Oxides. *Progress in Solid State Chemistry* **1988**, *18* (4), 259–341.
(170) Guglielmi, M.; Carturan, G. Precursors for Sol-Gel Preparations. *Journal of Non-Crystalline Solids* **1988**, *100* (1–3), 16–30.
(171) Stöber, W.; Fink, A.; Bohn, E. Controlled Growth of Monodisperse Silica Spheres in the Micron Size Range. *Journal of Colloid and Interface Science* **1968**, *26* (1), 62–69.
(172) Sui, R.; Charpentier, P. Synthesis of Metal Oxide Nanostructures by Direct Sol–Gel Chemistry in Supercritical Fluids. *Chemical Reviews* **2012**, *112* (6), 3057–3082.
(173) Pirzada, T.; Ashrafi, Z.; Xie, W.; Khan, S. A. Cellulose Silica Hybrid Nanofiber Aerogels: From Sol–Gel Electrospun Nanofibers to Multifunctional Aerogels. *Advanced Functional Materials* **2020**, *30* (5), 1907359.
(174) Loy, D. A.; Russick, E. M.; Yamanaka, S. A.; Baugher, B. M.; Shea, K. J. Direct Formation of Aerogels by Sol– Gel Polymerizations of Alkoxysilanes in Supercritical Carbon Dioxide. *Chemistry of Materials* **1997**, *9* (11), 2264–2268.
(175) Alongi, J.; Ciobanu, M.; Tata, J.; Carosio, F.; Malucelli, G. Thermal Stability and Flame Retardancy of Polyester, Cotton, and Relative Blend Textile Fabrics Subjected to Sol–Gel Treatments. *Journal of Applied Polymer Science* **2011**, *119* (4), 1961–1969.
(176) Alongi, J.; Ciobanu, M.; Malucelli, G. Sol–Gel Treatments on Cotton Fabrics for Improving Thermal and Flame Stability: Effect of the Structure of the Alkoxysilane Precursor. *Carbohydrate Polymers* **2012**, *87* (1), 627–635.
(177) Alongi, J.; Ciobanu, M.; Malucelli, G. Thermal Stability, Flame Retardancy and Mechanical Properties of Cotton Fabrics Treated with Inorganic Coatings Synthesized through Sol–Gel Processes. *Carbohydrate Polymers* **2012**, *87* (3), 2093–2099.
(178) Cireli, A.; Onar, N.; Ebeoglugil, M. F.; Kayatekin, I.; Kutlu, B.; Culha, O.; Celik, E. Development of Flame Retardancy Properties of New Halogen-free Phosphorous Doped SiO_2 Thin Films on Fabrics. *Journal of Applied Polymer Science* **2007**, *105* (6), 3748–3756.
(179) Yaman, N. Preparation and Flammability Properties of Hybrid Materials Containing Phosphorous Compounds via Sol-Gel Process. *Fibers and Polymers* **2009**, *10* (4), 413–418.
(180) Alongi, J.; Ciobanu, M.; Malucelli, G. Novel Flame Retardant Finishing Systems for Cotton Fabrics Based on Phosphorus-Containing Compounds and Silica Derived from Sol–Gel Processes. *Carbohydrate Polymers* **2011**, *85* (3), 599–608.
(181) Brancatelli, G.; Colleoni, C.; Massafra, M. R.; Rosace, G. Effect of Hybrid Phosphorus-Doped Silica Thin Films Produced by Sol-Gel Method on the Thermal Behavior of Cotton Fabrics. *Polymer Degradation and Stability* **2011**, *96* (4), 483–490.
(182) Cheng, X.-W.; Liang, C.-X.; Guan, J.-P.; Yang, X.-H.; Tang, R.-C. Flame Retardant and Hydrophobic Properties of Novel Sol-Gel Derived Phytic Acid/Silica Hybrid Organic-Inorganic Coatings for Silk Fabric. *Applied Surface Science* **2018**, *427*, 69–80.
(183) Cheng, X.-W.; Tang, R.-C.; Guan, J.-P.; Zhou, S.-Q. An Eco-Friendly and Effective Flame Retardant Coating for Cotton Fabric Based on Phytic Acid Doped Silica Sol Approach. *Progress in Organic Coatings* **2020**, *141*, 105539.
(184) Grancaric, A. M.; Colleoni, C.; Guido, E.; Botteri, L.; Rosace, G. Thermal Behaviour and Flame Retardancy of Monoethanolamine-Doped Sol-Gel Coatings of Cotton Fabric. *Progress in Organic Coatings* **2017**, *103*, 174–181.
(185) Nie, S.; Jin, D.; Yang, J.; Dai, G.; Luo, Y. Fabrication of Environmentally-Benign Flame Retardant Cotton Fabrics with Hydrophobicity by a Facile Chemical Modification. *Cellulose* **2019**, *26*, 5147–5158.
(186) Lin, D.; Zeng, X.; Li, H.; Lai, X.; Wu, T. One-Pot Fabrication of Superhydrophobic and Flame-Retardant Coatings on Cotton Fabrics via Sol-Gel Reaction. *Journal of Colloid and Interface Science* **2019**, *533*, 198–206.
(187) Bentis, A.; Boukhriss, A.; Gmouh, S. Flame-Retardant and Water-Repellent Coating on Cotton Fabric by Titania–Boron Sol–Gel Method. *Journal of Sol-Gel Science and Technology* **2020**, *94*, 719–730.
(188) Vasiljević, J.; Hadžić, S.; Jerman, I.; Černe, L.; Tomšič, B.; Medved, J.; Godec, M.; Orel, B.; Simončič, B. Study of Flame-Retardant Finishing of Cellulose Fibres: Organic–Inorganic Hybrid versus Conventional Organophosphonate. *Polymer Degradation and Stability* **2013**, *98* (12), 2602–2608.
(189) Liu, Y.; Pan, Y.-T.; Wang, X.; Acuña, P.; Zhu, P.; Wagenknecht, U.; Heinrich, G.; Zhang, X.-Q.; Wang, R.; Wang, D.-Y. Effect of Phosphorus-Containing Inorganic–Organic Hybrid Coating on the Flammability of Cotton Fabrics: Synthesis, Characterization and Flammability. *Chemical Engineering Journal* **2016**, *294*, 167–175.
(190) Castellano, A.; Colleoni, C.; Iacono, G.; Mezzi, A.; Plutino, M. R.; Malucelli, G.; Rosace, G. Synthesis and Characterization of a Phosphorous/Nitrogen Based Sol-Gel Coating as a Novel Halogen-and Formaldehyde-Free

Flame Retardant Finishing for Cotton Fabric. *Polymer Degradation and Stability* **2019**, *162*, 148–159.

(191) Yang, Z.; Fei, B.; Wang, X.; Xin, J. H. A Novel Halogen-free and Formaldehyde-free Flame Retardant for Cotton Fabrics. *Fire and Materials* **2012**, *36* (1), 31–39.

(192) Šehić, A.; Tomšič, B.; Jerman, I.; Vasiljević, J.; Medved, J.; Simončič, B. Synergistic Inhibitory Action of P-and Si-Containing Precursors in Sol–Gel Coatings on the Thermal Degradation of Polyamide 6. *Polymer Degradation and Stability* **2016**, *128*, 245–252.

(193) Vasiljević, J.; Tomšič, B.; Jerman, I.; Orel, B.; Jakša, G.; Kovač, J.; Simončič, B. Multifunctional Superhydrophobic/Oleophobic and Flame-Retardant Cellulose Fibres with Improved Ice-Releasing Properties and Passive Antibacterial Activity Prepared via the Sol–Gel Method. *Journal of Sol-Gel Science and Technology* **2014**, *70*, 385–399.

(194) Qi, L.; Qiu, S.; Xi, J.; Yu, B.; Hu, Y.; Xing, W. Construction of Super-Hydrophobic, Highly Effective Flame Retardant Coating for Cotton Fabric with Superior Washability and Abrasion Resistance. *Journal of Colloid and Interface Science* **2022**, *607*, 2019–2028.

(195) Bellayer, S.; Jimenez, M.; Barrau, S.; Bourbigot, S. Fire Retardant Sol–Gel Coatings for Flexible Polyurethane Foams. *RSC Advances* **2016**, *6* (34), 28543–28554.

(196) Bellayer, S.; Jimenez, M.; Prieur, B.; Dewailly, B.; Ramgobin, A.; Sarazin, J.; Revel, B.; Tricot, G.; Bourbigot, S. Fire Retardant Sol-Gel Coated Polyurethane Foam: Mechanism of Action. *Polymer Degradation and Stability* **2018**, *147*, 159–167.

(197) Shi, Y.; Wang, N.; Liu, L.; Liu, Y. Surface Sedimentation and Adherence of Nano-SiO_2 to Improve Thermal Stability and Flame Resistance of Melamine-formaldehyde Foam via Sol-gel Method. *Fire and Materials* **2018**, *42* (2), 183–189.

(198) Bo, G.; Xu, X.; Tian, X.; Wu, J.; He, X.; Xu, L.; Yan, Y. Enhancing the Flame Retardancy for Castor Oil-Based Rigid Polyurethane Foams via Silica Aerogel. *Journal of Non-Crystalline Solids* **2021**, *562*, 120783.

(199) Qu, L.; Rahimi, S.; Qian, J.; He, L.; He, Z.; Yi, S. Preparation and Characterization of Hydrophobic Coatings on Wood Surfaces by a Sol-Gel Method and Post-Aging Heat Treatment. *Polymer Degradation and Stability* **2021**, *183*, 109429.

(200) Shabir Mahr, M.; Hübert, T.; Sabel, M.; Schartel, B.; Bahr, H.; Militz, H. Fire Retardancy of Sol–Gel Derived Titania Wood-Inorganic Composites. *Journal of Materials Science* **2012**, *47*, 6849–6861.

(201) Bellayer, S.; Gossiaux, A.; Duquesne, S.; Dewailly, B.; Bachelet, P.; Jimenez, M. Transparent Fire Protective Sol-Gel Coating for Wood Panels. *Polymer Testing* **2022**, *110*, 107579.

(202) Zille, A.; Oliveira, F. R.; Souto, A. P. Plasma Treatment in Textile Industry: Plasma Treatment in Textile Industry. *Plasma Processes and Polymers* **2015**, *12* (2), 98–131. https://doi.org/10.1002/ppap.201400052.

(203) Tudoran, C.; Roşu, M.; Coroş, M. A Concise Overview on Plasma Treatment for Application on Textile and Leather Materials. *Plasma Processes and Polymers* **2020**, *17* (8), 2000046. https://doi.org/10.1002/ppap.202000046.

(204) Bell, A. T. The Mechanism and Kinetics of Plasma Polymerization. In *Plasma Chemistry III*; Vepřek, S., Venugopalan, M., Eds.; Topics in Current Chemistry; Springer-Verlag: Berlin and Heidelberg, **1980**; Vol. 94, pp 43–68. https://doi.org/10.1007/BFb0048586.

(205) Horrocks, A.; Eivazi, S.; Ayesh, M.; Kandola, B. Environmentally Sustainable Flame Retardant Surface Treatments for Textiles: The Potential of a Novel Atmospheric Plasma/UV Laser Technology. *Fibers* **2018**, *6* (2), 31. https://doi.org/10.3390/fib6020031.

(206) Jang, H. J.; Jung, E. Y.; Parsons, T.; Tae, H.-S.; Park, C.-S. A Review of Plasma Synthesis Methods for Polymer Films and Nanoparticles under Atmospheric Pressure Conditions. *Polymers* **2021**, *13* (14), 2267. https://doi.org/10.3390/polym13142267.

(207) Akovali, G.; Gundogan, G. Studies on Flame Retardancy of Polyacrylonitrile Fiber Treated by Flame-Retardant Monomers in Cold Plasma. *Journal of Applied Polymer Science* **1990**, *41* (910), 2011–2019. https://doi.org/10.1002/app.1990.070410907.

(208) Bourbigot, S.; Charaffedine, J.; Le Bras, M.; Delobel, R.; Dessaux, O.; Goudmand, P. New Approach to Flame Retardancy Using Plasma Assisted Surface Polymerisation Techniques. *Polymer Degradation and Stability* **1999**, *66*, 153–155.

(209) Schartel, B.; Kühn, G.; Mix, R.; Friedrich, J. Surface Controlled Fire Retardancy of Polymers Using Plasma Polymerisation. *Macromolecular Materials and Engineering* **2002**, *287*, 579–582.

(210) Tsafack, M. J.; Levalois-Grützmacher, J. Towards Multifunctional Surfaces Using the Plasma-Induced Graft-Polymerization (PIGP) Process: Flame and Waterproof Cotton Textiles. *Surface and Coatings Technology* **2007**, *201* (12), 5789–5795. https://doi.org/10.1016/j.surfcoat.2006.10.027.

(211) Kamlangkla, K.; Hodak, S. K.; Levalois-Grützmacher, J. Multifunctional Silk Fabrics by Means of the Plasma Induced Graft Polymerization (PIGP) Process. *Surface and Coatings Technology* **2011**, *205* (13–14), 3755–3762. https://doi.org/10.1016/j.surfcoat.2011.01.006.

(212) Horrocks, A. R.; Nazaré, S.; Masood, R.; Kandola, B.; Price, D. Surface Modification of Fabrics for Improved Flash-Fire Resistance Using Atmospheric Pressure Plasma in the Presence of a Functionalized Clay and Polysiloxane. *Polymers for Advanced Technologies* **2011**, *22* (1), 22–29. https://doi.org/10.1002/pat.1707.

(213) Carosio, F.; Alongi, J.; Frache, A. Influence of Surface Activation by Plasma and Nanoparticle Adsorption on the Morphology, Thermal Stability and Combustion Behavior of PET Fabrics. *European Polymer Journal* **2011**, *47* (5), 893–902. https://doi.org/10.1016/j.eurpolymj.2011.01.009.

(214) Caschera, D.; Toro, R. G.; Federici, F.; Riccucci, C.; Ingo, G. M.; Gigli, G.; Cortese, B. Flame Retardant Properties of Plasma Pre-Treated/Diamond-Like Carbon (DLC) Coated Cotton Fabrics. *Cellulose* **2015**, *22* (4), 2797–2809. https://doi.org/10.1007/s10570-015-0661-8.

(215) Ayesh, M.; Horrocks, A. R.; Kandola, B. K. The Effect of Combined Atmospheric Plasma/UV Treatments on Improving the Durability of Flame Retardants Applied to Cotton. *Molecules* **2022**, *27* (24), 8737. https://doi.org/10.3390/molecules27248737.

(216) Qi, L.; Wang, B.; Zhang, W.; Yu, B.; Zhou, M.; Hu, Y.; Xing, W. Durable Flame Retardant and Dip-Resistant Coating of Polyester Fabrics by Plasma Surface Treatment and UV-Curing. *Progress in Organic Coatings* **2022**, *172*, 107066. https://doi.org/10.1016/j.porgcoat.2022.107066.

(217) Jimenez, M.; Lesaffre, N.; Bellayer, S.; Dupretz, R.; Vandenbossche, M.; Duquesne, S.; Bourbigot, S. Novel Flame Retardant Flexible Polyurethane Foam: Plasma Induced Graft-Polymerization of Phosphonates. *RSC Advances* **2015**, *5* (78), 63853–63865. https://doi.org/10.1039/C5RA08289G.

15 Flame Retardants without Conventional Flame-Retardant Elements

Bo-Wen Liu and Yu-Zhong Wang
The Collaborative Innovation Center for Eco-Friendly and Fire-Safety Polymeric Materials, National Engineering Laboratory of Eco-Friendly Polymeric Materials (Sichuan), State Key Laboratory of Polymer Materials Engineering, College of Chemistry, Sichuan University, Chengdu 610064, China; yzwang@scu.edu.cn

15.1 INTRODUCTION

In the conventional understanding, flame retardancy is achieved through the use of flame-retardant elements such as halogens, phosphorus, silicon, nitrogen, and so on,[1] as is covered in multiple chapters of this book. Flame retardancy is usually realized in either or both of the gas phase and the condensed phase. In the gas phase, flame retardation is mainly achieved by capturing combustion free radicals, diluting combustible or combustion-supporting gases. In the condensed phase, flame retardation is mainly achieved through heat absorption/heat dissipation, dripping promotion, and barrier protection effects. However, some flame retardants, especially halogen-containing ones represented by polybrominated diphenyl ethers (PBDEs), tetrabromobisphenol A (TBBPA) and hexabromocyclododecane (HBCD), can cause significant harm to human health and the ecological environment due to the migration or release of toxic small molecules during their use or combustion.[2] With the increasing public awareness of safety and environmental protection, many countries have enacted regulations (represented by RoHS) prohibiting the use of some halogen-containing flame retardants.[3] While phosphorus- and nitrogen-containing flame retardants are relatively safer and more environmentally friendly than halogen-based ones, they still pose certain hazards to the environment. There is ongoing legislation in the USA and EU to ban the use of organophosphorus, organonitrogen, and other flame retardants in select applications (please see Chapter 23 of this book for more details). Faced with such a challenging situation, achieving flame retardancy in polymers without the use of conventional flame-retardant elements undoubtedly represents the optimal solution to address these issues. One approach is to take advantage of thermal reactions, specifically, to use special structures to promote char formation through catalysis, self-cross-linking/rearrangement reactions at high temperatures, as this offers a promising avenue to achieve the aforementioned objectives.

This chapter provides a detailed exploration of flame retardants without conventional flame-retardant elements from various aspects, including its chemical fundamentals, development and perspectives. In Section 15.2, the chemical fundamentals, underlying principles, and flame-retardant mechanism are extensively discussed. Section 15.3 examines establishment and development of the flame-retardant method without conventional flame-retardant elements through metal ion promoted char formation, high-temperature chemical self-cross-linking, and thermal rearrangement. Section 15.4 introduces the combination of flame-retardant method with and without the use of conventional flame-retardant elements. Sections 15.5 provides a summary of the flame-retardant-free method and offer insights into its future development trends.

15.2 CHEMICAL FUNDAMENTALS

Upon heating, polymers will undergo thermal decomposition, "a process of extensive chemical species change caused by heat,"[4] resulting in the formation of combustible gaseous fuel vapor, which will burn on the top of polymers in the presence of heat and oxygen. The combustion of gaseous fuel vapor generates a large amount of heat, part of which feeds back to the polymers, leading to further decomposition and continuous generation of combustible gaseous fuel vapor. Please see Chapters 2 and 3 of this book for more information on the fundamentals of polymer decomposition. Therefore, if the decomposition process of polymers can be altered without using conventional flame-retardant elements, thereby suppressing the formation of combustible small molecules, flame retardancy may be achieved.[1]

To achieve this goal, researchers initially utilized the catalytic charring effect of metal ions in combination with conventional flame retardants.[5,6] Then, in subsequent research, it was found that alginate (a natural polysaccharide salt) has very high flame-retardant efficiency without using conventional flame-retardant elements like halogen, phosphorus, etc. Our research found a high-temperature self-cross-linking/rearrangement charring flame-retardant method, in which some specific structures can achieve flame retardancy using only C, H, and O, representing a new flame-retardant method which does not use conventional flame-retardant elements. Through the utilization of high-temperature chemical reactions such as self-cross-linking

DOI: 10.1201/9781003380689-15

and rearrangement, the char-forming ability of polymers is significantly enhanced, leading to high flame retardancy without the use of conventional flame-retardant elements.

15.3 ESTABLISHMENT AND DEVELOPMENT

15.3.1 Metal Ion Promoted Char Formation

The catalytic char-forming effect of metal ion compounds has long been used as synergistic compound for conventional flame-retardant systems. Cullis et al.[5] first reported the char-forming effect of metal chelates on polyolefins in the presence of chlorinated additives. Then, Levchik et al.[7] investigated the catalytic flame-retardant effect of manganese dioxide (MnO_2) on ammonium polyphosphate (APP) in polyamide 6 (PA 6). Following this, researchers have conducted extensive investigations on metal-catalyzed flame retardancy and achieved a series of remarkable results.[5,8–18] Unfortunately, the primary flame-retardant action of this method still relies on conventional flame retardants represented by intumescent flame retardants. In a similar concept, Wilkie and his coworkers[19] used metal ion catalysts to enable Freidel-Crafts cross-linking of styrenic polymers for flame retardant applications.

15.3.1.1 Alginate

Alginates are naturally occurring polysaccharide materials that are mostly extracted from brown algae and are characterized by high biocompatibility, sustainability, low toxicity, and relatively low cost. Most polysaccharides are flammable despite the fact that they have great char-forming ability, thus alginic acid fibers without metal ions are highly flammable. With the incorporation of alkaline-earth and transition metal ions, "egg-box" junctions are formed between the guluronic acid blocks (G-block) and the metal ions (**Figure 15.1**). It was reported by Zhu, Xia, and their coworkers[20,21] that the inherent flame retardancy of alginates was mainly attributed to the incorporation of polyvalent metal ions like Ca^{2+}, which can change the decomposition process of alginates, enhance the char-forming ability, and reduce the formation of flammable gases, thus improve the flame retardancy. In recent years, numerous endeavors have been fervently directed towards utilizing alginates for flame-retardant applications, including: (1) producing inherently flame-retardant fibers and fabrics, (2) crafting fire-resistant alginate/clay aerogel composites, (3) formulating functional coatings and films, and (4) modifying inorganic flame retardants through core-shell structures, intercalation hybrids, and additive strategies.

Alginates have been successfully utilized to produce flame-retardant fibers through wet spinning in a coagulation bath without the need for toxic solvents. Zhu and coworkers[23] reported a series of hybrid alginate fibers by combining Al^{3+}, Cu^{2+}, Zn^{2+}, and Ba^{2+} with Ca^{2+} in a coagulation bath, among which the Ca^{2+}/Al^{3+}-containing ones exhibited high mechanical properties, flame retardancy, and resistance to salts. Specially that, the breaking strength and elongation of the resultant metal-alginate fibers were greatly improved by 40–75%. Alginate fibers can also be applied to be a component of flame-retardant fiber blends. Recent researches[24,25] have interwoven alginate fibers with cotton or viscose fibers to prepare fully bio-based flame-retardant blended fibers and fabrics. With an introduction of 50 wt% alginate fibers, the alginate/cotton blended fabrics (cotton5/alginate5) exhibited a self-extinguishing behavior, with a damaged section of around 40 mm in the ASTM D6413 vertical flame test (VFT). This approach also exhibited great applicability in synthetic fabrics, for which natural/synthetic blended nonwoven fabrics can be prepared by physically mixing synthetic fibers with alginate fibers in different ratios. Polyester/alginate hybrid fibers,[26] with a content of 20 wt% alginate fibers, realized self-extinguishing without dripping in both the VFT test and the Pennsylvania stuffed toy regulations assessment (Figure 15.2). Meanwhile, the addition of 50 wt% alginate fibers (polyester5/alginate5) led to a pronounced reduction in heat and smoke release in the cone calorimeter (ISO5660) test. Moreover, when applied in polyamide,[27] the polyamide5/alginate5 (with a 1:1 mass ratio of the two fibers) exhibited self-extinguishing performance without dripping in the VFT test, as well as substantial reductions in heat, smoke, and toxic gas release.

Utilizing its capacity to form films, alginates were applied in the creation of flame-retardant coatings, incorporating both organic and inorganic additives. Wang and coworkers[28] reported the construction of a fully renewable flame-retardant chitosan-alginate coating on flexible polyurethane foams (FPUFs) by layer-by-layer (LbL) assembly. The obtained coating dramatically improved the char-forming ability of FPUF, thus resulting in significant reduction of heat and smoke release. Specially, in the cone calorimeter test, PUF-10 BL (the FPUF foam with 10 bilayers of the chitosan-alginate coating) exhibited a 66%, 11%, and 6% reduction in peak HRR (PHRR), total heat release (THR), and total smoke produce (TSP) values, respectively. Researchers further improved the flame-retardant efficiency of the chitosan-alginate coating by incorporating inorganic additives such as titanate nanotubes (TNT),[29] α-zirconium phosphate (α-ZrP),[30] graphene oxide (GO),[31] metal hydroxides,[32] and hydroxyapatite (HAP).[33] As expected, when combined with LDHs,[34] the hybrid coating help the FPUF foam achieve significant reductio in PHRR (66%), THR (24%), and TSP (74%) in the cone calorimeter test at a coating content of 4.0 wt%.

The distinct properties of alginates have also garnered attention in the field of flame-retardant aerogels. Alginate/clay aerogels have been reported to exhibit both low density and excellent flame-retardant properties such as high Limiting Oxygen Index (LOI) value, delayed time-to-ignition (TTI), and low heat and smoke release. Wang, Schiraldi, and coworkers[35] reported nonflammable aerogel based on ammonium alginates (AA) and MMT, and further a cross-linked aerogel using Ca^{2+}. The resultant alginate/clay aerogel (A5C5) showed a low density of 0.085 g/cm^3 and low

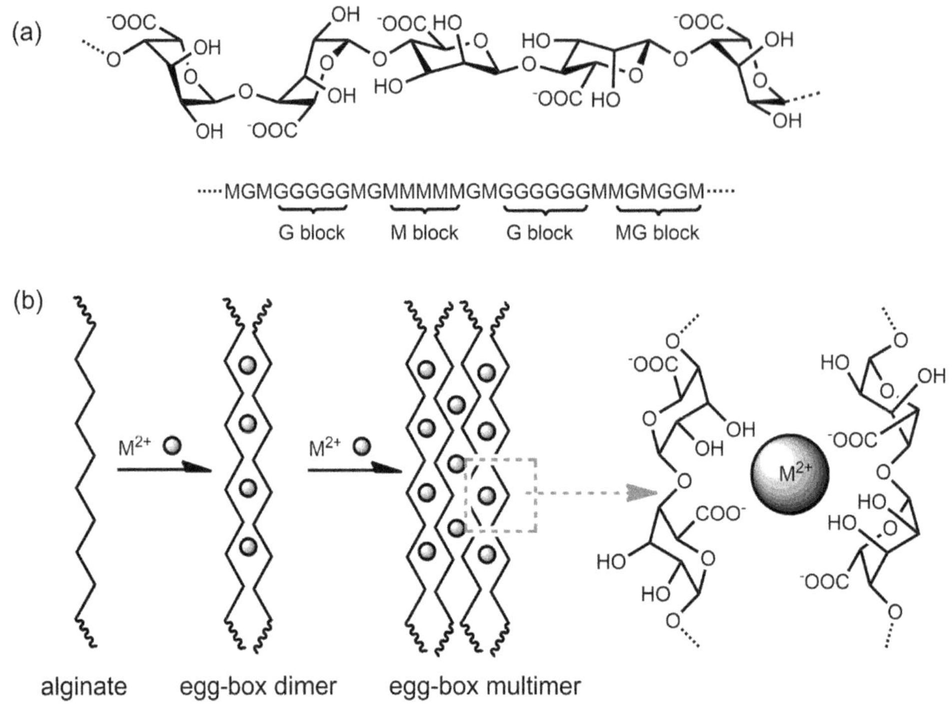

FIGURE 15.1 Schematic illustration of (a) chain conformation and block distribution of alginates and (b) egg-box junctions of divalent metal ions and the G blocks of alginates.

Source: Reproduced with permission.[22] Copyright 2021, Elsevier.

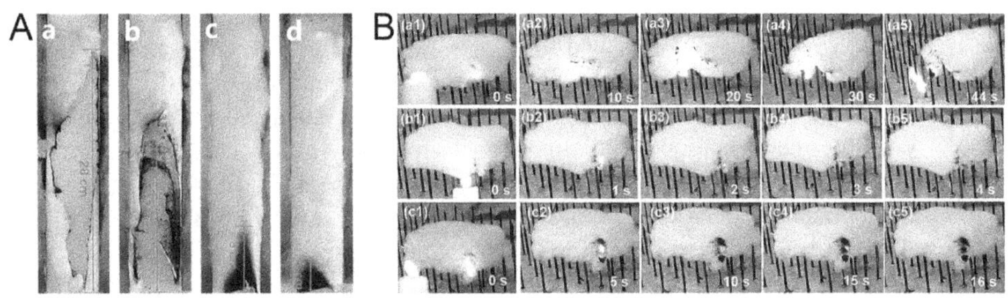

FIGURE 15.2 Photographs of the polyester/alginate blended fibers (A) after the VFT test: (a) polyester, (b) polyester$_{80}$/alginate$_{20}$, (c) polyester$_{50}$/alginate$_{50}$, and (d) alginate fibers; and (B) during the test according to Pennsylvania stuffed toy regulations: (a1–a5) polyester, (b1–b5) polyester$_{80}$/alginate$_{20}$, and (c) polyester$_{50}$/alginate$_{50}$ fiber.

Source: Reproduced with permission.[26] Copyright 2021, Elsevier.

PHRR value of 32 kW/m^2 in the cone calorimeter test under a heat flux of 50 kW/m^2. After cross-linked with Ca^{2+}, the density of the aerogel increased to 0.104 g/cm^3, which could not be ignited throughout the cone calorimeter test. Wang and coworkers[36] also developed a series of AA-based aerogels by mixing, freeze-drying, and post-cross-linking with Ca^{2+}. With the incorporation of nanoparticles like magnesium hydroxide (MH) and aluminum hydroxide (ATH), MMT, and layered double hydroxides (LDHs), the obtained aerogels all exhibited low density (0.064–0.116 g/cm^3), low thermal conductivity (0.024–0.046 W/m K), and satisfactory mechanical strength (0.7–3.5 MPa). More importantly, the aerogels exhibited high LOI values (> 60%) and could not be ignited in the cone calorimeter test under a heat flux of 50 kW/m^2.

15.3.1.2 Novel Metal-Based Flame Retardants

Recently, Wang and his coworkers[38,39] developed a kind of novel metal-based flame retardants containing no conventional flame-retardant elements such as halogens, P, N, and Si. Those flame retardants showed excellent char-forming ability and expansibility and can expand to several times their original volume at relatively low temperatures of 250°C. This expandable charring behavior is very similar to expandable graphite (EG). However, compared with EG, this kind of flame retardants have much better compatibility

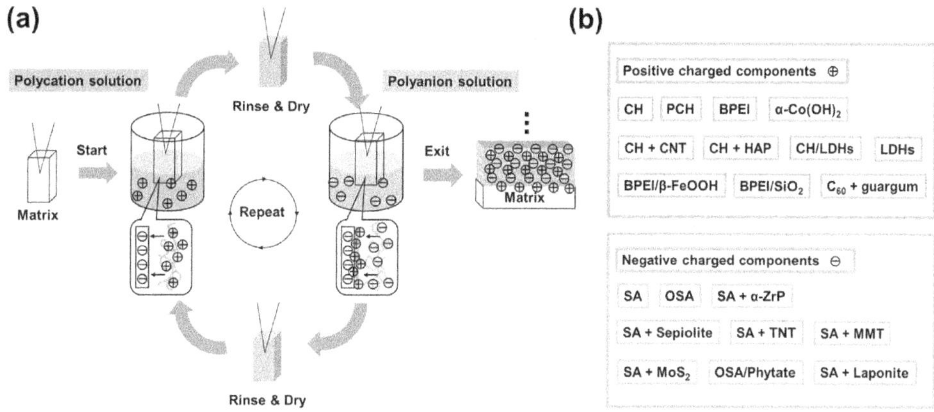

FIGURE 15.3 Schematic illustration of (a) LbL assembly depositions by an alternating immersion in the aqueous solutions and (b) the components of the reported LbL systems involving alginates.

Source: Reproduced with permission.[37] Copyright 2017, Elsevier.

with polymers and, especially, showed excellent compatibility and dispersion in flexible polyurethane foams (FPUF). However, different metal-based flame retardants show different flame-retardant efficiency. A highly efficient flame retardant can increase the LOI value of FPUF to 27% at a content of 10 wt%, and the resulting flame-retardant FPUF exhibits 34% and 48% reduction of peak HRR and SPR, respectively, and most importantly, only 1.2% loss of compression deformation at 75%. Moreover, the flame-retardant FPUF foams also possessed great resistance against the butane torch flame.

15.3.2 Chemical Self-Cross-Linking

Chemical cross-linking through the formation of irreversible stable chemical bonds can effectively increase the melt viscosity and melt strength of polymers during combustion. More importantly, cross-linking of some structures will lead to aromatization and carbonization, thus enhancing the char-forming ability and flame retardancy. However, introducing irreversible chemical cross-linking networks into thermoplastic polymers can compromise their processability. Therefore, such cross-linked structures should not exist during the preparation and processing of polymers but can be rapidly formed during burning to achieve flame retardancy. More importantly, the cross-linking reaction should be able to enhance the char-forming ability of the polymers.

The use of functional groups which activate at desired temperatures to improve the thermal stability of materials has been known since the 1990s, but it was not exploited for flame retardant effect beyond some simple studies until recently. Based on this design concept, Wang and his coworkers[40,41] proposed the concept of "smart high-temperature self-cross-linking" flame-retardancy. The flame-retardant polymers can be prepared by introducing high-temperature self-cross-linkable groups (Figure 15.4) into the polymer molecular chain. These obtained polymers do not undergo cross-linking reactions during polymerization and processing; however, rapid self-cross-linking reactions occur at higher temperatures or during burning, leading to the formation of three-dimensional networks. These cross-linked networks can significantly increase the melt viscosity of the copolymers and further promote the aromatization and carbonization, resulting in the formation of stable and compact char layer covering the material surface. These char layers inhibit further degradation of the substrate, act as a barrier to heat transfer and oxygen exchange, and ultimately achieve flame retardancy.

15.3.2.1 Phenylacetylene

The key to realizing such smart self-cross-linkable polymers is to design appropriate cross-linkable monomers, which should (1) be chemically reactive for copolymerization, (2) whose derived chemical structures in the copolymers be inactive at the temperature of copolymerization and processing, and (3) be able to cross-link rapidly at higher temperatures, and to promote char formation.

Phenylacetylene represents a self-cross-linkable functional group with the ability for high-temperature cross-linking, typically occurring above 300°C, thereby satisfying the aforementioned criteria for polyethylene terephthalate (PET). Based on this, Wang and coworkers[42] first designed and synthesized a novel flame-retardant monomer with the abbreviation of "PEPE" containing pendant phenylacetylene structures (Figure 15.5), then prepared the corresponding PET copolyesters P(ET-*co*-P)$_n$. For pure PET, a typical pseudoplastic fluid, the chain movability becomes easier with increasing temperature, resulting in a gradual decrease in the complex viscosity as observed in dynamic rheological test. However, the complex viscosities of P(ET-*co*-P)$_n$ exhibit a "U"-shaped curve, which first decreased and then increased with increasing temperature. This is attributed to the occurrence of cross-linking reactions between phenylacetylene units (Figure 15.5), which connect neighboring polymer chains and form a cross-linked network structure, significantly enhancing the melt viscosity and strength of the copolyesters. The cross-linking reaction

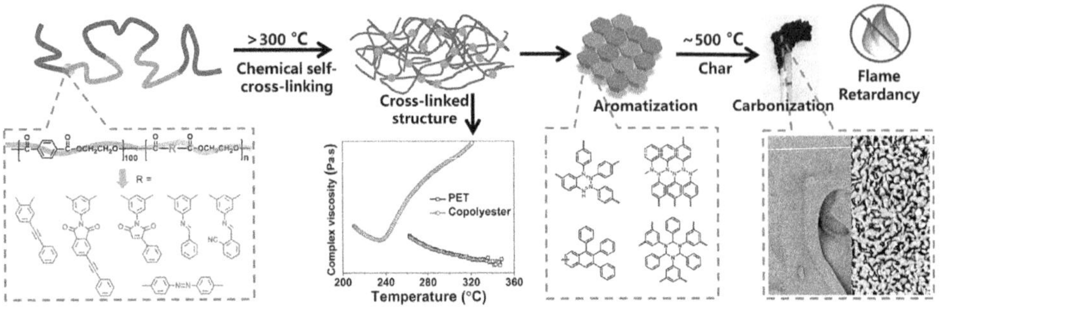

FIGURE 15.4 High-temperature self-cross-linkable flame-retardant mode of action.

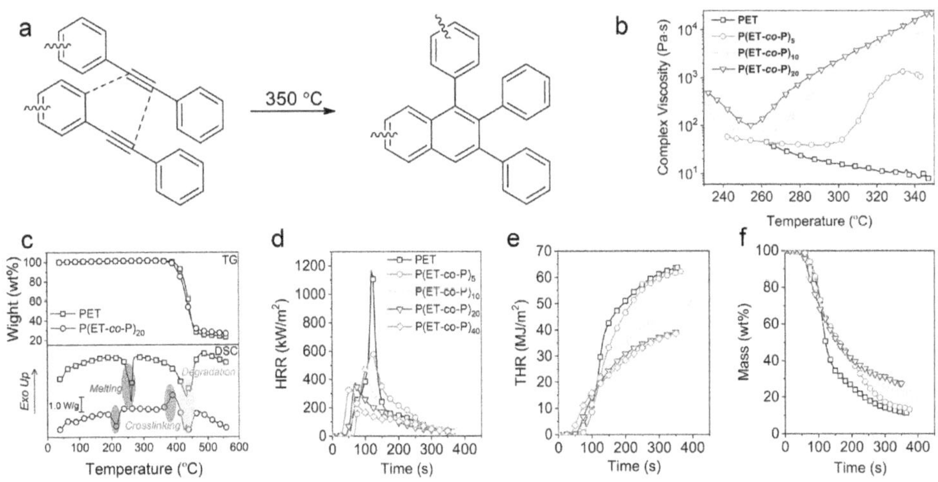

FIGURE 15.5 (a) Self-cross-linking reaction of PEPE, (b) change of complex viscosity, (c) TG and DSC curves obtained from the simultaneous TG-DSC test, and (d) HRR curves, (e) THR curves, and (f) MLR curves of PET and P(ET-*co*-P)$_n$.

of diphenylacetylene also enhances the char-forming ability of the copolyesters. In the thermogravimetric analysis (TGA) test under nitrogen atmosphere, P(ET-*co*-P)$_{20}$ and P(ET-*co*-P)$_{80}$ exhibit increased residual char contents of 15.2 wt% and 28.5 wt%, respectively, compared with the 9.0 wt% of pure PET. Consequently, the copolyesters exhibit enhanced flame retardancy and anti-dripping performance. At a PEPE content of 20 mol%, P(ET-*co*-P)$_{20}$ achieves an LOI value of 24.5% and a 3.2 mm UL-94 V-2 rating. The higher diphenylacetylene content in P(ET-*co*-P)$_{80}$ elevates the LOI value to 30.0% and enables it to pass the 3.2 mm UL-94 V-0 rating without dripping. In the cone calorimeter test, the PHRR and THR values all gradually decreased with the increase of PEPE content. This work represents the first successful attempt to achieve flame retardancy through high-temperature self-cross-linking, demonstrating the feasibility of this method.

However, due to the large rigid structure and high content of PEPE units, the crystallization ability of P(ET-*co*-P)$_n$ copolyesters was severely compromised. To address this issue, the authors prepared the block copolyesters containing diphenylacetylene structures in the side chains via solid-phase polymerization.[43] The resulting copolyesters exhibit significantly improved crystallinity while maintaining excellent flame retardancy. In addition, the high content of diphenylacetylene structures leads to a significant deterioration in the mechanical properties of P(ET-*co*-P)$_n$, limiting its practical applications. To further enhance the flame-retardant efficiency and mitigate the influence of high monomer content on mechanical properties, Wang and coworkers[44] integrated the phenylacetylene and phenylimide structures to synthesize a novel monomer with synergistic cross-linking effects (Figure 15.6). Phenylimide can undergo an "imide-isoimide" rearrangement at high temperature to form an isoimide structure containing unsaturated C=N group, which further decomposed to form a benzonitrile structure. Besides, they found that the phenylacetylene structure can capture the benzonitrile structure produced by the decomposition of phenylimide, leading to the synergistic cross-linking effect (Figure 15.6). The resulting copolyester, P(ET-*co*-PN)$_n$, exhibits higher cross-linking efficiency and char-forming abilities. At a low monomer content of 13.0 mol%, the copolyester P(ET-*co*-PN)$_{15}$ achieves a 3.2 mm UL-94 V-0 rating

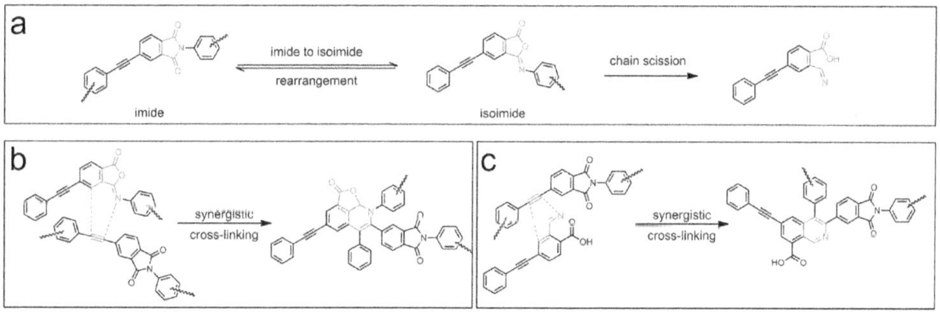

FIGURE 15.6 (a) The imide-isoimide rearrangement process; (b–c) the synergistic cross-linking processes between.

Source: Reproduced with permission.[45] Copyright 2021, Elsevier.

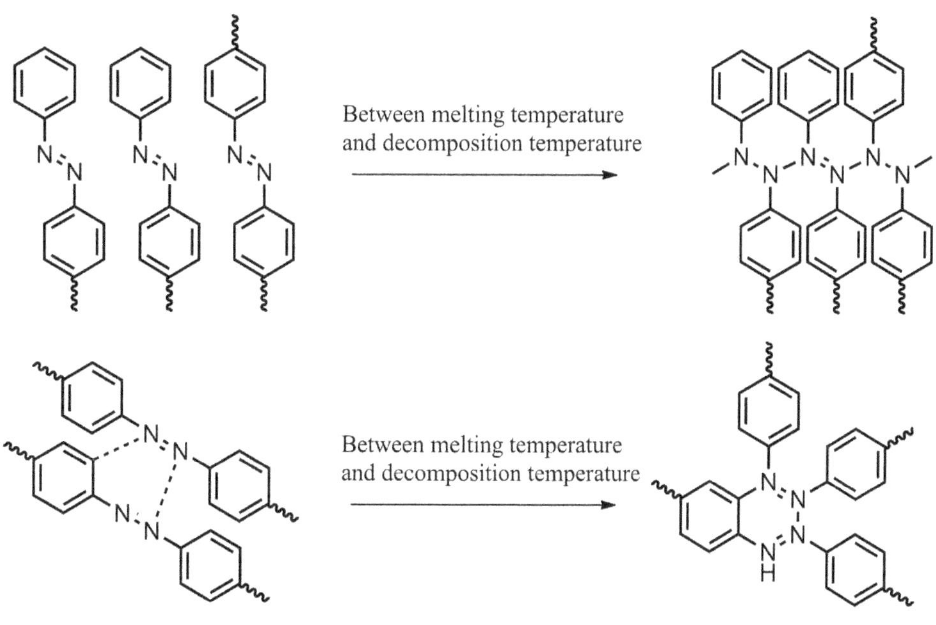

FIGURE 15.7 Possible cross-linking structure of azobenzene groups.

Source: Reproduced with permission.[48] Copyright 2013, Royal Society of Chemistry.

without dripping, with a correspondingly high LOI value of 33.5%. Furthermore, the phenylacetylene units in the P(ET-*co*-PN)$_n$ structure form physical π-π stacking interactions, which not only enhance intermolecular forces and improve the mechanical strength of the copolyester (up to a 45.0% increase in tensile strength compared to pure PET) but also serve as fixed phases in shape memory and dynamic cross-linking points in self-healing processes. Additionally, the copolyester exhibits excellent shape-memory performance, self-healing properties, and is suitable for 3D printing applications. This synergistic cross-linking strategy also exhibited great applicability in polytrimethylene terephthalate (PTT).[45]

Following this concept, Yang and coworkers[46] synthesized an intrinsically flame-retardant thermoplastic polyurethane elastomer (TPU) by incorporating PEPE as a chain extender. The incorporation of PEPE led to a substantial reduction in the PHRR and THR values, decreasing by 46.2% and 24.5%, respectively. Scanning electron microscopy (SEM) analysis of the residual char revealed the formation of a cross-linked structure during burning. Moreover, the addition of PEPE significantly improved the tensile strength of TPU, reaching 39.2 MPa at a PEPE content of 1.35 wt%, nearly double that of neat TPU.

15.3.2.2 Azobenzene

Azobenzene-containing polymers have attracted significant attention due to the microcosmic cis-trans isomerization of the azobenzene group, which leads to macroscopic changes in geometry or polarity.[47] Additionally, the "N=N" group in azobenzene can undergo π-π* electronic transitions and form a large conjugated system with the benzene ring, resulting in high thermal stability. Moreover, the azobenzene structure can undergo high-temperature self-cross-linking reactions, forming stable nitrogen-containing cross-linked structures (Figure 15.7)[48] that contribute to solid-phase charring. Conversely, the non-cross-linked azobenzene groups decompose to release nitrogen gas that can dilute the concentration of flammable volatiles and combustion-supporting gases in the combustion zone. Consequently, azobenzene group exhibits both condensed phase and gas phase flame-retardant effects.

Based on these characteristics of azobenzene groups, Wang and coworkers[48] designed and synthesized a novel copolyester, called PEAT in this paper, by incorporating azobenzene structures into the main chain. With a 15 mol% azobenzene content, PEAT15 achieved an elevated LOI value of 29.0%. However, due to the relatively low cross-linking efficiency of azobenzene structures, PEATs only achieved a 3.2 mm V-2 rating in the UL-94 test, although the dripping phenomenon was improved. In the cone calorimeter test, PEATs exhibited a reduction of 57.1% in PHRR and 34.7% in TSP compared with pure PET. Furthermore, the conjugation effect of the azobenzene structure enhanced the van der Waals forces between the copolyester chains, which promoted the crystalline growth rate and improved the crystallization temperature and enthalpy, resulting in good crystallinity.

The azobenzene structure was further utilized for the synthesis of a self-cross-linkable epoxy monomer named Azo-EP, which was cured with either phthalic anhydride (PA) or anhydrides containing phenylethynyl groups (3-phenylethynyl phthalic anhydride—PEPA) to prepare high-temperature self-cross-linkable epoxy resins.[49] The presence of a co-cross-linking network of azobenzene and phenylethynyl significantly enhanced the char-forming ability of the resultant epoxy resins (Figure 15.8). The obtained epoxy resin, Azo-EP-PEPA, exhibited a residual mass of 48.2 wt% at 700°C in TGA test and a char yield of 52 wt% after cone calorimeter test. The improved char-forming ability effectively suppressed the heat release, as evidenced by a 58% reduction in PHRR and a 49% reduction in THR for Azo-EP-PEPA in the cone calorimeter test.

15.3.2.3 Aromatic Schiff Base

Schiff base compounds, characterized by the C = N functional group in their molecular structure, exhibit significant electron effects, making them of great research value in various fields such as biomedical, smart materials, catalytic materials, corrosion inhibitors, and photochromic materials.[50,51] Additionally, phenyl Schiff bases can undergo self-cross-linking reactions at high temperatures, where the "CH=N" moieties engage in addition reactions to form stable nitrogen-containing six-membered rings (Figure 15.9a). Based on this, Wang and coworkers[52] designed and synthesized a flame-retardant monomer containing Schiff base structures in the side chain and prepared the corresponding copolyester with the acronym of BAnPET (shown in Figure 15.9). Dynamic rheological test demonstrated that due to the cross-linking reactions between Schiff base groups, BAnPET exhibited a sharp increase in complex viscosity with increasing temperature. With further temperature elevation, the nitrogen-containing six-membered ring structures formed by cross-linking transformed into compact char layers, acting as barriers to heat sources and oxygen. These nitrogen-containing char layers also inhibited heat release and the volatilization of organic small molecules. When the content of phenyl Schiff base units was 9.1 mol%, the copolyester BA10PET achieved a 3.2 mm UL-94 V-0 rating without dripping, and the corresponding LOI value reached 31.0%. In the cone calorimeter test, both the PHRR and TSR values of BAnPET were reduced by over 30%.

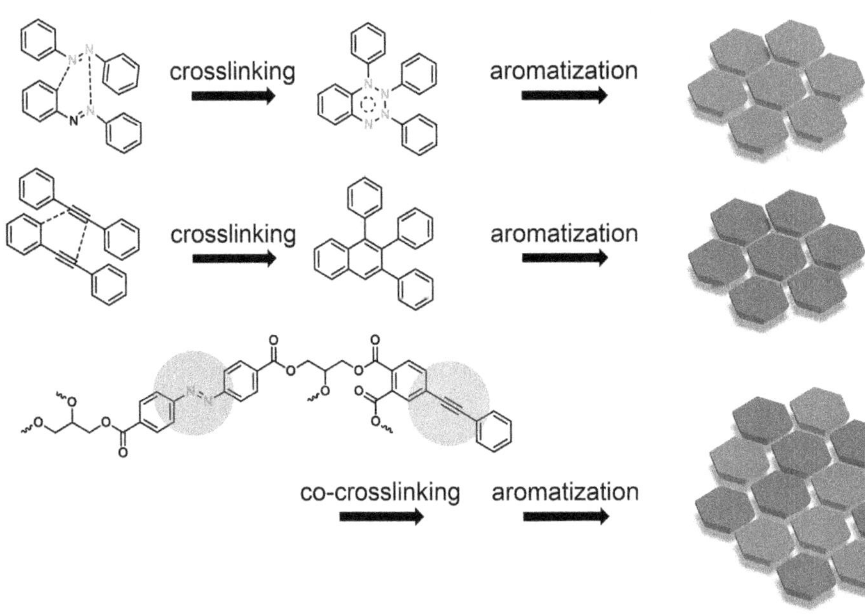

FIGURE 15.8 The co-cross-linking reaction between the azobenzene and phenylethynyl groups.

Source: Reproduced with permission.[49] Copyright 2015, Elsevier.

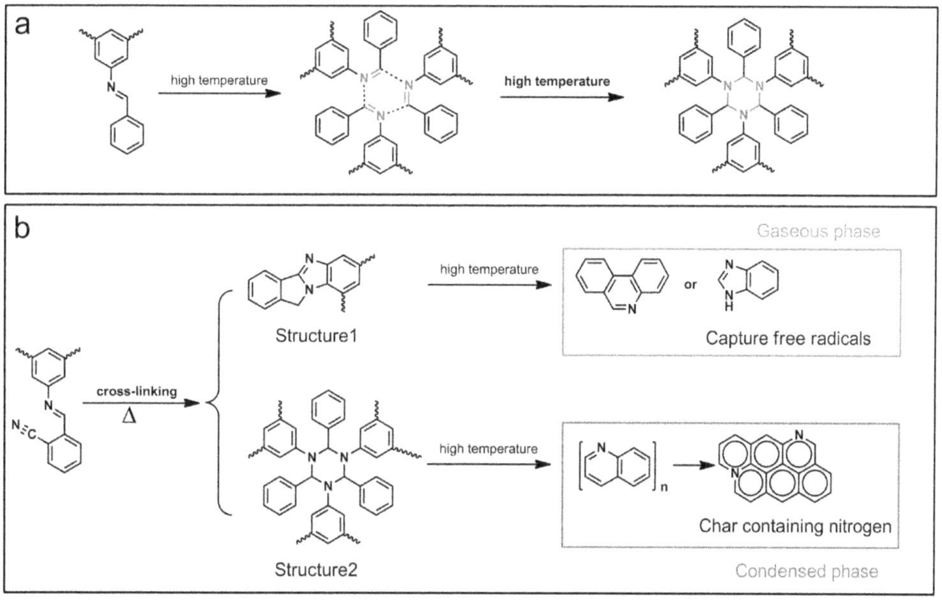

FIGURE 15.9 Self-cross-linking process of (a) Schiff base groups and (b) Schiff base groups with cyano group at the ortho position. **Source: Reproduced with permission.[53] Copyright 2018, Elsevier.**

Because the self-cross-linking temperature of the phenyl Schiff base is in close proximity to the processing temperature of the polyester, the BAnPET copolyesters exhibit a limited processing temperature range. To improve the processability, the authors further proposed a strategy to reduce the cross-linking reactivity between "CH=N" moieties by introducing electron-withdrawing cyano groups at the ortho position of the benzene ring connected to the Schiff base structure. A novel monomer named CA containing phenyl Schiff base and cyano structures were designed and synthesized, and the corresponding copolyester CBAAnPET were prepared.[53] The introduction of ortho cyano groups not only effectively raises the cross-linking temperature of the copolyester from 250°C to 365°C but also enables the cross-linking reactions between "C=N" groups and "CH=N" groups at elevated temperatures, resulting in a dual cross-linking effect that significantly enhances the melt viscosity and char-forming ability of the copolyester (structure 2 in Figure 15.9b). The formation of less stable nitrogen-containing five-membered rings between "CH=N" and "C≡N" moieties (structure 1 in Figure 15.9b) leads to a thermally unstable structure that further decomposes, releasing nitrogen-containing radicals. These radicals then capture active radicals from the combustion process in the gas phase, thereby exerting a flame-retardant effect in the gas phase. When the content of cyano-Schiff base units was 9.1 mol%, the copolyester CBAA10PET achieved an LOI value of 31.0% and passed the 3.2 mm UL-94 V-0 rating without dripping. Furthermore, CBAAnPET exhibited better inhibition effect on heat and smoke release than BAnPET and exhibited 52.9% reduction of PHRR and 32.5% reduction of TSR compared to pure PET when this was measured by cone calorimeter.

Following a similar approach, Zhang et al.[54] synthesized a Schiff base monomer named NBA containing a $-NO_2$ group and prepared the corresponding PET copolyester NBA_nPET. The findings demonstrated that the presence of the $-NO_2$ group in NBA reduced the cross-linking reactivity of the Schiff base, thereby improving the cross-linking temperature of NBAnPET from the range of 250–348°C to the range of 347–372°C, attributed to the substituent effect of $-NO_2$. Moreover, with the NBA content of 7.9 mol%, $NBA_{10}PET$ achieved an LOI of 32% and a 3.2 mm UL-94 V-0 rating without dripping.

15.3.2.4 Phenylmaleimide

The phenylmaleimide structure can undergo a $2\pi + \pi$ cycloaddition cross-linking reaction during burning (Figure 15.10), which promotes solid-phase aromatization and char formation, significantly enhancing the flame-retardant and anti-dripping properties of the copolymer.[55] Even at a low molar content of 10 mol%, the phenylmaleimide-containing copolyester named PET-*co*-DPDPI10 passed the 3.2 mm UL-94 V-0 rating without dripping, and achieved a high LOI value of 30.4% (**Table 1**). In the cone calorimeter test, both the PHRR and TSR value of PET-*co*-DPDPI gradually decreased with the increase of the phenylmaleimide content. PET-*co*-DPDPI10 exhibits a 47.8% reduction in PHRR and 64.2% reduction in TSR compared to pure PET. However, the incorporation of phenylmaleimide structures simultaneously deteriorated the regularity and symmetry of the copolyester chains, significantly reducing the crystallinity of the copolyester. It is noteworthy that PET-co-DPDPI10 maintains good mechanical properties, with a tensile strength of 49.5 MPa.

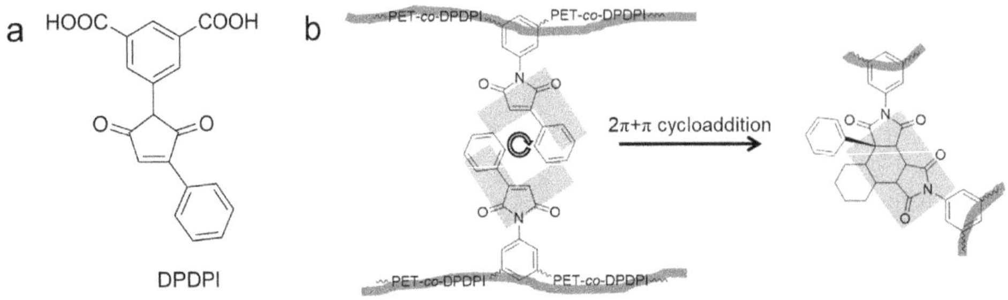

FIGURE 15.10 The 2π + π cycloaddition cross-linking reaction of the phenylmaleimide structure.

Source: Reproduced with permission.[55] Copyright 2016, Royal Society of Chemistry.

15.3.3 Thermal Rearrangement and Others

Similar to chemical cross-linking, thermal rearrangement of certain aromatic structures, such as "Ar-CH_2-Ar," "Ar-O-Ar," "Ar-O-Ar-O-Ar," and "Ar-O-Ar-CO-Ar-O-Ar," can facilitate the formation of a char layer with conjugated aromatic structures. Incorporating these structures into the molecular chain of polymers can also realize the improvement of flame retardancy through rearrangement-based char formation. Based on this, various polymers based on high-temperature rearrangement reactions have been developed in recent years.

15.3.3.1 Phenylate-Based Thermal Rearrangement

Based on the thermal rearrangement of the above aromatic structures, a novel copolymer monomer, PBPBD, containing a phenyl ether structure was designed by Wang and coworkers.[56] Then the monomer was introduced into the main chain of PET through melt polycondensation to prepare the corresponding copolyester BD_xPET. TGA results showed that the introduction of the phenyl ether structure improved the thermal stability and char-forming ability of the copolyester. Under nitrogen atmosphere, when 20 mol% PBPBD was introduced, the char residue at 700°C increased to 22.5 wt% for BD_{20}PET from 9.9 wt% of pure PET. More importantly, pyrolysis gas chromatography/mass spectrometry (Py-GC/MS) results indicated that PBBPD underwent a unique rearrangement reaction and formed conjugated heteroaromatic structures, resulting in a dense char layer with a "stick-like" microstructure (Figure 15.11). Due to the rearrangement and subsequent carbonization process of PBBPD, the BD_xPET copolyesters achieved high flame retardancy, smoke suppression, and anti-dripping performance simultaneously. BD_{20}PET exhibited significantly reduced PHRR of 541.9 kW/m^2 and TSP of 13.0 m^2 compared to pure PET (1004.5 kW/m^2 and 18.8 m^2, respectively) when this was measured by cone calorimeter testing. Additionally, BD_{20}PET had a high LOI value of 32%, and an intumescent char layer was formed at the top of the sample, preventing the generation of melt drips during the LOI test.

Wang and coworkers[57,58] further designed four flame-retardant monomers containing "Ar-CO-Ar," "Ar-O-Ar," "Ar-O-Ar-O-Ar," and "Ar-O-Ar-CO-Ar-O-Ar" moieties and prepared four corresponding semi-aromatic polyesters (named PKT, POT, PEET, and PEKET) through melt polycondensation with dimethyl terephthalate as model compounds to investigate their rearrangement and flame-retardant processes. These polyesters exhibited significantly improved flame retardancy and anti-dripping performance compared to pure PET. The LOI values of the polyesters containing "Ar-O-Ar-O-Ar" and "Ar-O-Ar-CO-Ar-O-Ar" moieties were 29.9% and 30.9%, respectively, and the introduction of the "Ar-O-Ar-O-Ar" moiety allowed the corresponding polyester to pass the 3.2 mm UL-94 V-0 rating without dripping. The significant difference between the residual char yield of the polyester at high temperatures and the calculated values suggests that the functional groups in the polyester undergo direct interactions during the burning process, rather than simple molecular chain scission. Py-GC/MS results revealed that the benzene ether/ketone structures could combine with small molecules and form char layers with different conjugated aromatic structures through rearrangement reactions (Figure 15.12), thereby enhancing the flame retardancy and anti-dripping performance of the polyester.

To further investigate the influence of rearrangement reactions on the flame retardancy and anti-dripping performance of copolyesters, Wang and coworkers further designed two derivative monomers based on bisphenol F (BHBPF) and bisphenol A (BHBPA) and introduced them into the PET molecular chain through melt polycondensation to prepare the corresponding F_xPET and A_xPET copolyesters.[59] The chemical difference between BHBPF and BHBPA lies only in the presence of substituents on the carbon atom linking the phenyl rings. Interestingly, this subtle structural variation led to different thermal decomposition processes and flame-retardant properties in F_xPET and A_xPET copolyesters. Py-GC/MS test results revealed that the "Ar-CH_2-Ar" group in BHBPF underwent rearrangement to form fluorene structures at high temperatures, while the "Ar-$C(CH_3)_2$-Ar" group in BHBPA did not undergo this rearrangement reaction (Figure 15.13). The rearrangement reaction of BHBPF promoted the formation of a char layer with a unique microstructure during the combustion process of the copolyester, thereby endowing F_xPET with good flame retardancy and anti-dripping

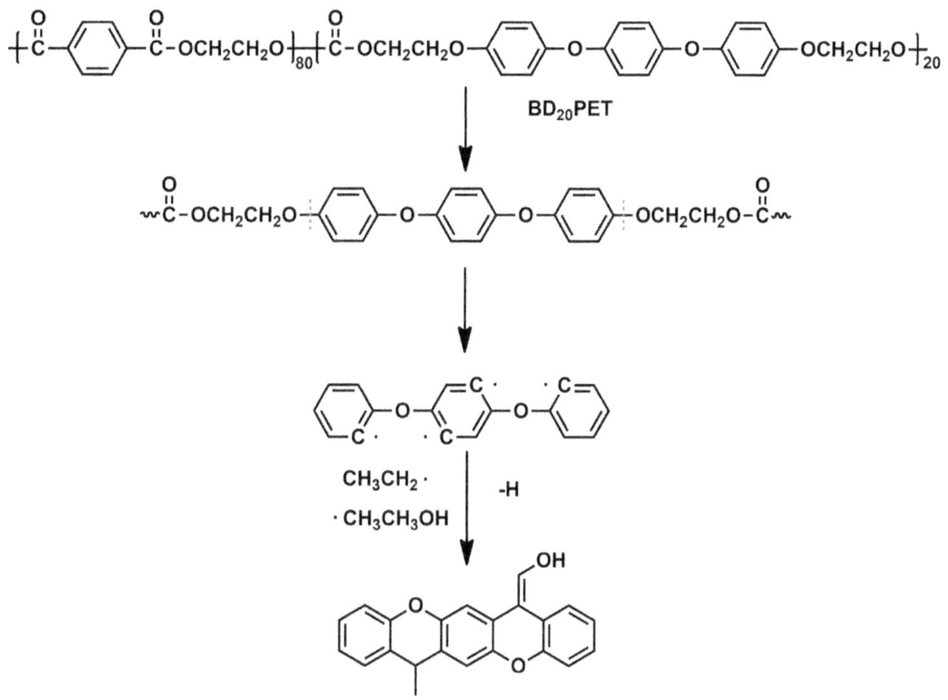

7'

FIGURE 15.11 Proposed pyrolysis process for PBPBD structures.

Source: Reproduced with permission.[56] Copyright 2015, Elsevier.

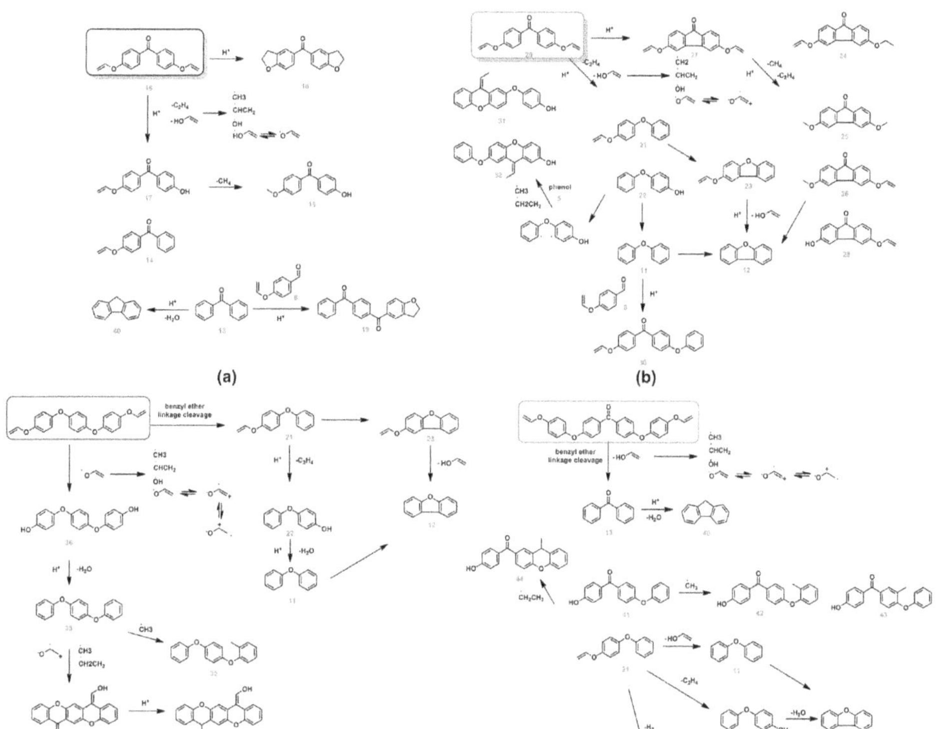

FIGURE 15.12 Pyrolysis processes of "Ar-CO-Ar," "Ar-O-Ar," "Ar-O-Ar-O-Ar," and "Ar-O-Ar-CO-Ar-O-Ar" structures.

Source: Reproduced with permission.[57] Copyright 2016, Royal Society of Chemistry.

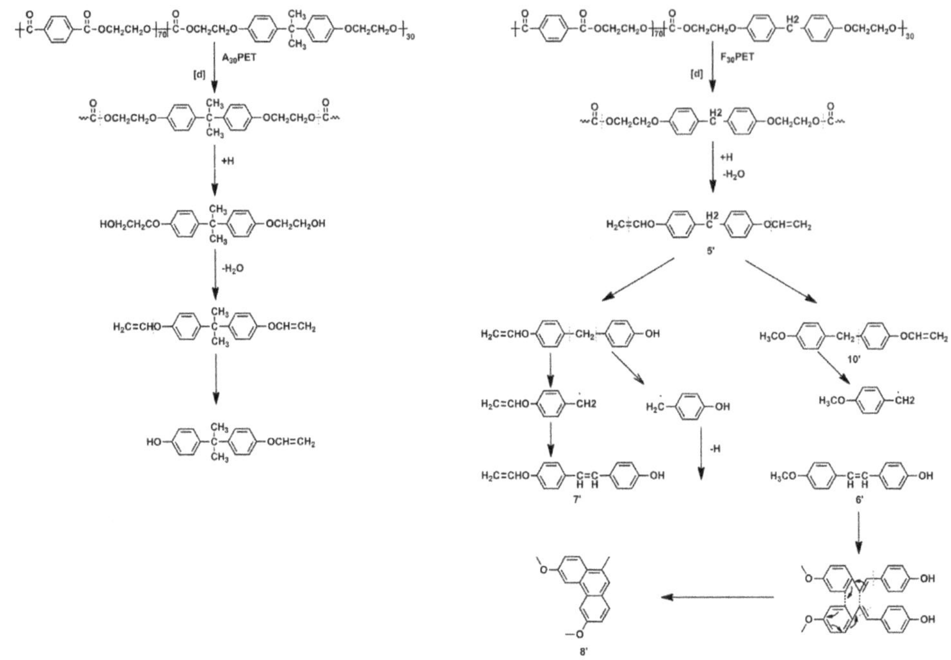

FIGURE 15.13 Pyrolysis process of the AxPET and FxPET copolyesters.

Source: Reproduced with permission.[59] Copyright 2015, Elsevier.

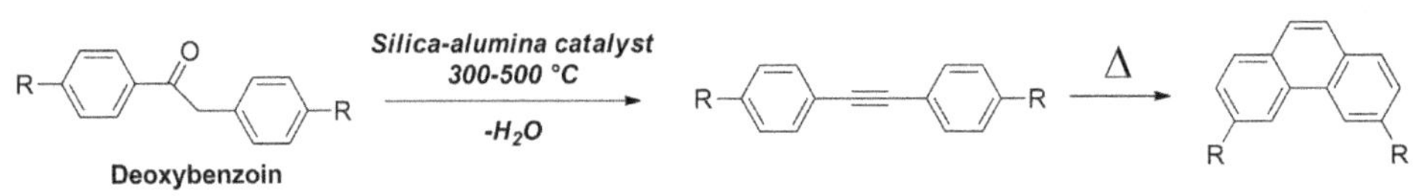

FIGURE 15.14 The deoxybenzoin-to-diphenylacetylene thermal rearrangement process.

Source: Reproduced with permission.[61] Copyright 2006, American Chemical Society.

properties. Compared to the A_xPET copolyester containing bisphenol A, the F_xPET copolyester containing bisphenol F exhibited higher char formation capability, better flame retardancy, and fewer melt drips due to the thermal rearrangement reactions.

15.3.3.2 Deoxybenzoin-Based Thermal Rearrangement

The catalytic conversion of deoxybenzoin to diphenylacetylene (Figure 15.14) at temperatures ranging from 300 to 500°C was reported by van der Waals et al.[60] The observed temperature range for the transformation of deoxybenzoin into diphenylacetylene highlights its potential applicability in promoting char formation of polymeric materials. Taking advantage of this, Emrick and coworkers[61] synthesized a deoxybenzoin-based monomer BHDB and explored its impact on the thermal stability and flame retardancy of the polyarylates. They discovered that the thermal rearrangement of BHDB facilitated aromatization and enhanced the char-forming capability. Consequently, the char yield in the TGA test increased from 25.8 wt% for the BPA-polyarylate to 42.2 wt% for the BHDB-polyarylate. Notably, the microscale combustion calorimeter (MCC, ASTM D7309) results revealed a decline in the heat release capacity (HRC) from 456.3 J/g·K to 61.5 J/g·K and a reduction in the total heat release (THR) from 17.7 kJ/g to 7.5 kJ/g. However, the BHDB-polyarylate showed rather low solubility, which severely limited their molecular weight and processability. To solve this problem, they synthesized novel BHDB-containing polyphosphonates through the interfacial polycondensation between BHDB and phenylphosphonic dichloride (PPDC).[62] BHDB-PPDC polyphosphonates exhibit high molecular weight (Mw) of ~20 000 g/mol and polydispersity indices (Mw/Mn) of ~2.6; at the same time, they possess good-to-excellent solubility at room temperature in many solvents, including tetrahydrofuran (THF) (70 mg/mL), chloroform (100 mg/mL), and dimethyl formamide (DMF) (400 mg/mL). MCC analysis of the BHDB-polyphosphonate revealed exceptionally low HRC values of < 100 J/g·K. In addition, they developed and prepared a novel monomer containing allyl-functionalized deoxybenzoin and employed it in interfacial polycondensation reactions with isophthaloyl or terephthaloyl chloride.[63] The incorporation of the allyl group conferred excellent solubility, high molecular weight, and remarkable flame retardancy to the aromatic polyesters.

Furthermore, they successfully developed a deoxybenzoin-containing epoxy compound called BEDB[64] and utilized it to create a range of epoxy resins through the curing reaction with various diamines. The resulting epoxy resins exhibited low HRC values comparable to halogen-containing counterparts, while significantly outperforming conventional non-halogenated alternatives. A unique diisocyanate monomer, named DBDI, which was derived from deoxybenzoin, was successfully synthesized to facilitate the production of flame-retardant polyurethanes.[65] Incorporating DBDI into polyurethane structures resulted in the development of highly flame-retardant materials, as evidenced by their exceptionally low heat release rates and high char yields. Additionally, by incorporating innovative oligomers based on BHDB into the polyol formulation, polyurethane foams were synthesized, showcasing a remarkable propensity for char formation. Following a similar approach, novel poly(arylene ether) sulfones, sulfoxides, and sulfides incorporating deoxybenzoin subunits were successfully synthesized through step-growth polymerization, utilizing BHDB and the corresponding sulfur-containing monomers.[66] Remarkably, all the resulting polymers exhibited outstandingly low flammability properties, characterized by exceptionally low THR values of 6 J/g·K, and remarkably high char yield values, up to 54 wt%.

15.3.3.3 Triazole-Based Thermal Rearrangement

The thermal rearrangement of bisphenyl-1,2,3-triazole structures into phenylindole (Figure 15.15) at elevated temperatures was first reported by Gilchrist *et al.*[67–69] Inspired by this finding, Emrick and coworkers prepared novel aromatic polyesters with the incorporation of the bisphenol-1,2,3-triazole (BPT) structure.[70] The resulting BPT-containing polymers and copolymers exhibited TGA curves with a two-step weight loss process, corresponding to the elimination of nitrogen (N_2) during the conversion of triazole units to indoles (Figure 15.15). Remarkably, the char yield of the 3-BPT polymer at 850°C reached 56 wt%, significantly higher than that of the 4-BPT materials (47 wt%). This difference was attributed to the lower crystallinity and closer proximity of the phenyl rings in 3-BPT, rendering it more favorable for thermally induced aromatization compared to the para-substituted structures. This reaction greatly enhanced the char formation of the copolyesters, consequently improving their flame retardancy and anti-dripping performance. Additionally, the MCC results of these 3-BPT aromatic polyesters exhibited low HRC and THR values of 23 J/g·K and 4.5 kJ/g, respectively. These exceptionally low heat release values for hydrocarbon-based polymers even surpass those of inorganic-based polyphosphazenes, and the performance of BPT structures in terms of heat release is comparable to that of the aromatic polyimide Kapton™. In addition, they further investigated the application of BPT in epoxy resins, and the research revealed that epoxy resins containing BPT structures also exhibited exceptional flame-retardant properties.[71]

Following this, Wei et al.[72] synthesized 5-chloro-1-(4-chlorophenyl)benzotriazole-containing conjugated microporous materials (with the acronym of ZCMPs) through Sonogashira-Hagihara cross-coupling polymerization, using 1,3,5-triethynylbenzene and 1,4-diethynylbenzene as alkynes. The resulting ZCMPs, denoted as ZCMP-1 and ZCMP-2, exhibited porous structures and demonstrated

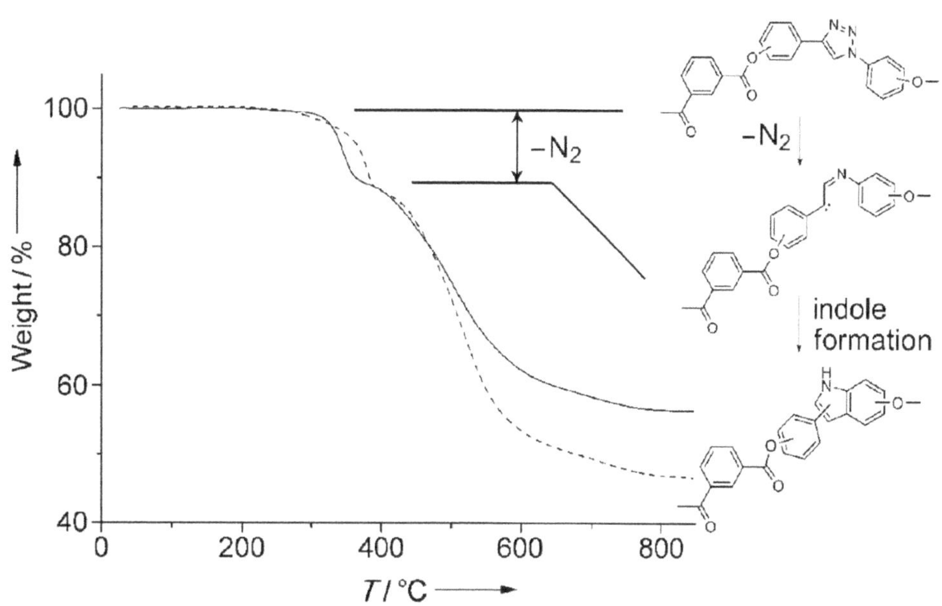

FIGURE 15.15 The triazole to phenylindole thermal rearrangement.

Source: Reproduced with permission.[70] Copyright 2010, Royal Society of Chemistry.

excellent flame retardancy. The MCC results exhibited low PHRR values of 11.9 and 9.9 W/g, as well as low THR values of 6.7 and 5.9 kJ/g, for ZCMP-1 and ZCMP-2, respectively. The flame-retardant behavior of ZCMPs was evaluated by incorporating them into epoxy composites with a loading of only 0.25 wt%. The ZCMP-2/epoxy composites showed a significant 17.42% reduction in PHRR values and 34.9% in THR values compared to the control sample as measured by cone calorimeter test. The incorporation of 1,2,3-triazole moieties into polyimide systems to prepare 1,2,3-triazole-containing polytriazoleimides (PTAIs) was reported by Balasubramanian et al.[73] They initially synthesized aromatic diamines with 1,2,3-triazole moieties through a CuAAC reaction, which were then polymerized with various commercially available aromatic dianhydrides via a polycondensation reaction. The resulting PTAIs exhibited displayed high initial decomposition temperatures exceeding 350°C, high char yields at 800°C (ranging from 53.5 wt% to 57.8 wt%), and high LOI values ranging from 38.9% to 40.5%.

15.3.3.4 Hydrogen Bond and π-π Stacking

Except for chemical process like cross-linking and rearrangement, physical interactions like hydrogen bonds and π-π stacking may also be capable of enhancing the melt viscosity and melt strength, thus promoting the aromatization and char formation of polymers. Following this concept, Wang and coworkers[74,75] designed and synthesized a novel benzimidazole-based monomer PBM (Figure 15.16) and prepared the corresponding copolyester PET-co-PBMs. The incorporation of the benzimidazole structure endowed the copolyesters with a dual physical cross-linking network: the "N–H" groups of the benzimidazole units can form hydrogen bonds with the carbonyl groups in the polyester backbone, as well as hydrogen bonds with the "C=N" groups of the imidazole units; moreover, there exist strong π-π stacking interactions between the benzimidazole moieties. The presence of this dual physical cross-linking network restricted the mobility of the copolyester chains, significantly increasing the high-temperature melt viscosity and melt strength of the copolyesters. Furthermore, the excellent char-forming ability of the benzimidazole groups themselves enhances the char-forming ability of the copolyester during combustion, thereby endowing excellent flame retardancy to the copolyesters. The PET-co-PBM20 copolyester achieved a 3.2 mm UL-94 V-0 rating without dripping and possessed a high LOI value of 36%. In the cone calorimeter test, PET-co-PBM20 exhibited a 74.8% reduction in PHRR and a 38.8% reduction in THR compared with pure PET.

Recently, combining the hydrogen bonds and π-π stacking interactions with the ionic aggregation, Wang and coworkers[76,77] designed and synthesized two types of ion monomers, BIS-Na and MS-K, containing both sulfonic acid salts and benzimidazole structures.[76,77] Due to the synergistic effect of ionic aggregation and char-forming ability of benzimidazole, the corresponding PET copolyesters, PETB-NaX and PETMn-K, exhibited extremely high melt viscosity and excellent char-forming ability at high temperatures. When the content of BIS-Na monomer was 8 mol%, PETB-Na8 achieved a 3.2 mm UL-94 V-0 rating without dripping. Additionally, the LOI value reached 33.0%. Compared to PET, the copolyesters showed via cone calorimeter testing a 45.2% reduction in total smoke released (TSR), 60.5% in PHRR, and 75.0% in peak carbon monoxide production (PCOP) values. Furthermore, the copolyesters possessed great spinnability, and the exhaustion rate of cationic dyes increased from 6.5% in PET to 99.6%.

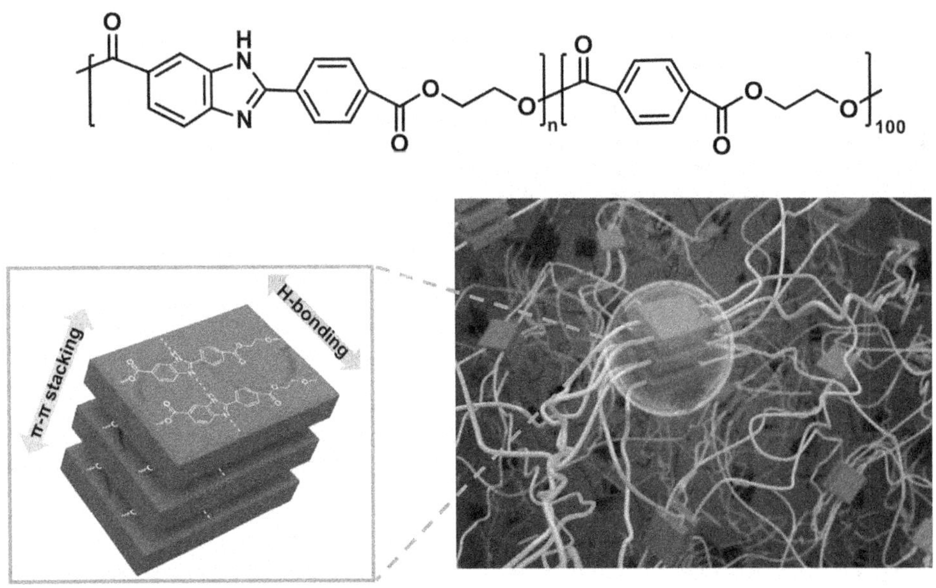

FIGURE 15.16 The chemical structure of PET-*co*-PBM and its dual physical cross-linking network.

Source: Reproduced with permission.[74] Copyright 2019, Elsevier.

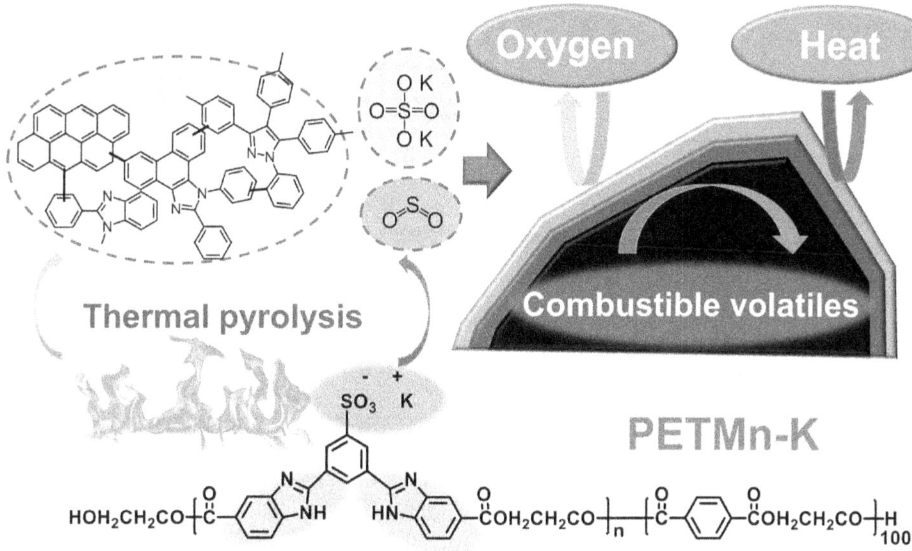

FIGURE 15.17 Synergistic effect of ionic aggregation and char-forming ability of benzimidazole.

Source: Reproduced with permission.[77] Copyright 2022, Elsevier.

15.3.3.5 End-Group-Capturing

The thermal rearrangement of hydroxy-containing phthalimide (HPI) into benzoxazole (Figure 15.18) with the release of carbon dioxide was first reported by Mathias et al.[78,79] The products generated from the thermal rearrangement, such as the nonflammable carbon dioxide and water vapor, as well as the thermally stable (mono or bis)benzoxazole structures, all have positive contributions to the flame retardancy of polymers. Wang and coworkers[80] employed the unique thermal rearrangement of HPI structure (Figure 15.18) to design and synthesize a novel monomer containing the HPI structure and the corresponding copolyester PET-*co*-HPI. Additionally, by changing the substituent from hydroxy group in the HPI structure to methyl (MPI) and hydrogen (PI), two comparative copolyesters, PET-*co*-MPI and PET-*co*-PI were prepared. The study revealed that PET-*co*-HPI exhibited outstanding flame retardancy and anti-dripping performance. At a HPI content of 20 mol%, PET-*co*-HPI$_{20}$ achieved a 3.2 mm UL-94 V-0 rating without dripping and high LOI value of 33% and showed 62%, 59%, and 75% reduction of PHRR, peak smoke production rate (PSPR), and PCOP values when measured via cone calorimeter, respectively, compared with pure PET. In contrast, the two comparative samples, PET-*co*-MPI and PET-*co*-PI, only achieved a 3.2 mm UL-94 V-2 rating and generated flaming melt drips that ignited the underlying absorbent cotton. Further mechanism studies revealed that the carboxyl-, hydroxyl-, and carbonyl-terminated structures, as well as hydrocarbons generated during the decomposition process of the PET-*co*-HPI were significantly reduced compared to pure PET, indicating the "end-group-capturing" effect. Furthermore, the benzoxazole structures formed through rearrangement were detected from the chromatography spectrometry obtained from the Py-GC/MS test, indicating the occurrence of the HPI-to-benzoxazole rearrangement during decomposition, while such structures could not be detected for the MPI and PI copolyesters. Consequently, the thermal rearrangement of HPI improved the melt viscosity and enhanced the char-forming ability of the copolyester, thus led to the improvement of flame retardancy.

The influence of substitution position on the efficiency of this flame-retardant method was further investigated.[81] It was observed that the para-substitution copolymer, named PET-*co*-*p*HPI, exhibited significantly higher flame-retardant efficiency compared to the meta-substitution copolymer, PET-*co*-HPI, for which a 25% lower monomer content was required to achieve 3.2 mm UL-94 V-0 rating, high LOI value, and low smoke and toxic gas release as measured by cone calorimeter. This phenomenon can be attributed to the disparity in thermal rearrangement efficiency between the two copolymers. Specifically, PET-*co*-HPI rearranged at a temperature range that did not align well with its decomposition temperature, resulting in a slower "end-group-capturing" rate. Conversely, PET-*co*-*p*HPI exhibited rearrangement within a temperature range that better matched its decomposition temperature, leading to a faster "end-group-capturing" speed and better flame retardancy. The feasibility of employing the flame-retardant method in PTT (polytrimethylene terephthalate) and PBT (polybutylene terephthalate) was further investigated.[82] The results revealed a gradual decline in flame-retardant efficiency as the number of methylene groups in the polymer chain increased. Specifically, the flame retardancy of PET-*co*-HPI, PTT-*co*-HPI, and PBT-*co*-HPI exhibited a gradual reduction. Mechanism studies unveiled that the "end-group-capturing" efficiency gradually decreased due

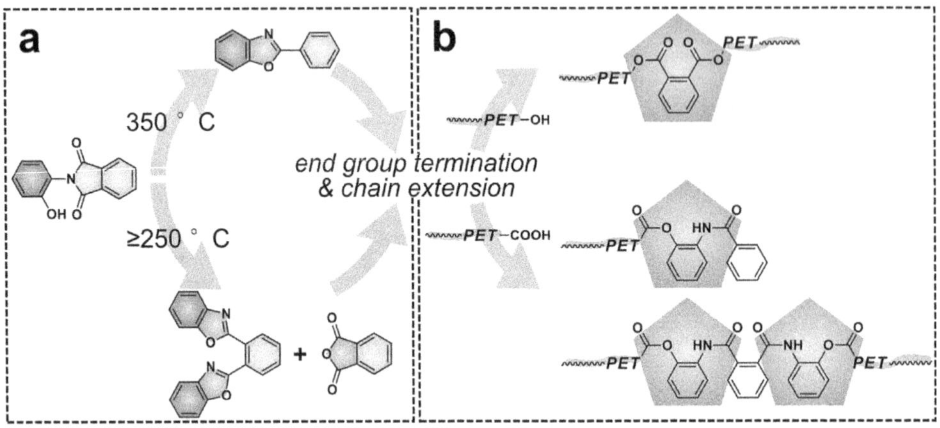

FIGURE 15.18 The thermal rearrangement of HPI (a) and the corresponding "end-group-capturing" process (b).

Source: Reproduced with permission.[80] Copyright 2019, John Wiley and Sons.

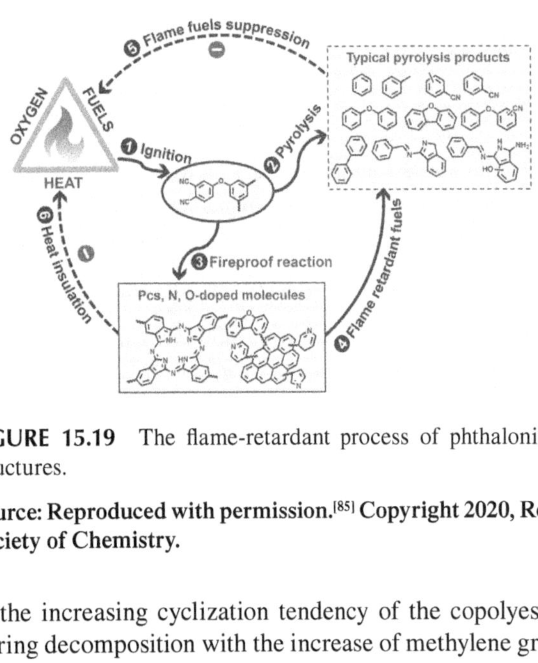

FIGURE 15.19 The flame-retardant process of phthalonitrile structures.

Source: Reproduced with permission.[85] Copyright 2020, Royal Society of Chemistry.

to the increasing cyclization tendency of the copolyesters during decomposition with the increase of methylene group numbers.

15.3.3.6 Other Related Chemistries

Phthalonitriles have been reported to undergo self-cross-linking reactions at temperatures above 200°C,[83] leading to the formation of phthalocyanine structures. These structures possess potential as aromatization precursors and exhibit the ability to enhance the char formation. Inspired by this, Wang and coworkers[84,85] chemically incorporate aryl ether nitrile structures into the PET chain to improve the flame retardancy. Through flame-responsive cyclotrimerization and aliphatic fragment capture (Figure 15.19), the flammability risks and hazards of heat, smoke, toxicity, and dripping were significantly reduced. The modified PET exhibited an increased LOI value of 31% and significant reductions of 49.0%, 31.1%, and 52.6% in PHRR, TSR, and COP, respectively, when measured by cone calorimeter. Moreover, the modified PET successfully achieved a 3.2 mm UL-94 V-0 rating without dripping. Wang et al.[86] synthesized an additive flame retardant named DBA-Ph containing nitrile groups and a double bond for improving the flame retardancy of EPs. The introduction of DBA-Ph into the EP matrix facilitates multiple cross-linking reactions, thereby significantly improving the flame-retardant properties. Remarkably, with 10 wt% addition of DBA-Ph, the resultant EP achieved a 3.2 mm UL-94 V-0 rating. Furthermore, a substantial 40% reduction in PHRR was observed when the DBA-Ph content was increased to 20 wt%.

Recently, taking advantage of the multifunctionality of natural polyphenols, Wang and coworkers[87] reported a novel polyphenol-inspired polymer, PIP, by incorporating polyphenol structures into the polymer chains through a simple melt polymerization. At low polyphenol content, PIP maintained high transparency in the visible light range while exhibiting excellent UV shielding efficiency. With a polyphenol content of only 2 mol%, PIP_2 achieved complete blocking of UV-A radiation; further increasing the polyphenol content results in PIP_3 effectively shielding high-energy short-wavelength blue light (SWB) at a wavelength of 425 nm. Due to the enhanced hydrogen bond from the polyphenol structure, PIP_5 exhibits an 88% increase in tensile strength compared to commercial PET materials. Additionally, the free radical scavenging and char-forming effects of polyphenols significantly enhance the flame retardancy of PIP. When the polyphenol content is increased to 30 mol% or higher, PIP_{30} successfully achieved a high LOI value of 30.5%, a 3.2 mm UL-94 V-0 rating without dripping, and significant reduction of toxic gas and smoke release when measured by cone calorimeter.

15.4 THE COMBINATION OF CONVENTIONAL AND UNCONVENTIONAL FLAME-RETARDANT METHOD

The aforementioned flame-retardant method, based on thermal self-cross-linking/rearrangement reactions without conventional flame-retardant elements, exhibits excellent

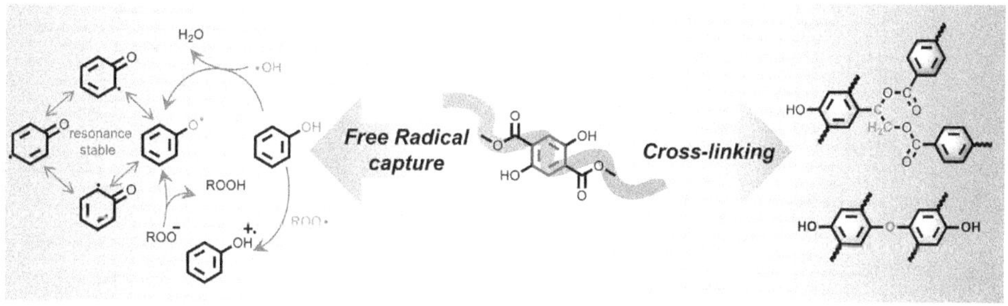

FIGURE 15.20 The free radical capturing and thermal cross-linking process during the burning of PIP.

Source: Reproduced with permission. Copyright 2023, Elsevier.

flame retardancy both when used independently and in combination with conventional flame-retardant elements such as phosphorus.

In this regard, Morgan and Tour[88] synthesized several thermally stable alkyne-containing compounds from commercially available brominated flame retardants. High thermal stability and great char-forming ability were confirmed by the DSC and TGA results. They found that the compounds containing *o*-dialkyne substitutions on the aromatic ring had lower cross-linking temperature and higher char yields. The addition of just 0.5 wt% of such alkyne- and bromide-containing flame retardant enabled polycarbonate to achieve a UL-94 V-0 rating with a thickness of 3.2 mm. Following this, a series of alkyne- and phosphorus-containing compounds, including phosphine oxides, phosphates, diphosphates, and oligophosphates, are synthesized and the thermal and char-forming properties were investigated in detail.[89] The DSC results confirmed the cross-linking reaction, while the TGA results demonstrated that the inclusion of alkyne groups significantly enhanced the char-forming capability of these compounds. Thanks to the dual cross-linking mechanism between alkyne and phosphorus, these compounds exhibited great flame-retardant efficiency, for which a loading amount of 10 wt% helped polycarbonate pass the 3.2 mm UL-94 V-0 rating.

Wang and coworkers[90] successfully designed and synthesized an alkynyl-containing phosphorus-based oligomer with the acronym of PPBP, along with a control sample PPBOP without alkynyl group (Figure 15.21). TGA results demonstrated that the presence of the alkynyl group substantially enhanced the char-forming ability of PPBP. The residual char content increased from 1.6 wt% for PPBOP to 45.0 wt% for PPBP, representing a remarkable increase of 2,713%. These results established that the decomposition of PPBP was accompanied by a cross-linking reaction, wherein the cross-linking of the alkynyl group, the dehydration and carbonization effects of phosphorus, and the unique thermal decomposition behavior of PPBP collectively played a pivotal role in its notably superior char-forming capability. Upon incorporation of PPBP into thermoplastic polyurethane (TPU), PPBP not only conferred excellent flame retardancy and anti-dripping performance to TPU, enabling it to achieve a LOI value of 28.3% and a 3.2 mm UL-94 V-0 rating at a PPBP content of 3 wt%, but also resulted in remarkably low heat release rate and smoke production rate with a reduction of 70.7% in PHRR and 53.9% in PSPR, respectively, as measured by cone calorimeter. Furthermore, a novel piperazine-containing additive named PPBPP was designed for the flame-retardancy of TPU.[91] Experimental results showed that 3.0 wt% PPBPP endowed TPU with a 3.2 mm UL-94 V-0 rating, and anti-dripping performance was realized at 12.0 wt% PPBPP. Compared with that of pure TPU, the LOI value of TPU/PPBPP12 was increased to 35.0%, and the PHRR and THR values were drastically reduced by 76.7 and 40.7%. The flame-retardant TPU/PPBPP also exhibited great mechanical properties, with the elongation at break reaching 812.1% at 12 wt% PPBPP according to the ISO 37 standard.

The combination of aromatic Schiff base and silicon-containing flame retardants was favored by researchers. Fan et al.[92] synthesized a novel aromatic Schiff base-containing branched polysiloxane (PCNSi) and prepared corresponding PA 6/PCNSi composites by melt blending. The addition of only 6 wt% PCNSi resulted in a remarkable 48.9% reduction in the PHRR value compared with pure PA 6. Additionally, the anti-dripping behavior of PA 6/PCNSi was completely suppressed with the incorporation of 4 wt% PCNSi. The enhanced flame retardancy of the composites was attributed to the formation of a protective char layer, facilitated by the cross-linking reaction of the Schiff base structure and the thermo-oxidative degradation of polysiloxane units in PCNSi. In another work, Li et al.[93] synthesized a novel flame-retardant nanoparticle with the acronym of SiDP, by incorporating nano SiO_2, tetra-brachial Schiff base, and quaternary ammonium salt (N, N'-dimethyl-N-(3-(trimethoxy silyl) propyl) octadecan-1-aminium chloride—DMOAC), aiming for imparting flame retardancy and antibacterial properties to cellulose-based materials. Thanks to the high thermal stability of nano SiO_2 and the cross-linking structure of aromatic Schiff base, the modified cotton fabrics treated with SiDP exhibited remarkable improvement in flame retardancy, whose LOI value was increased to 27.6% and PHRR valued reduced by 34.2%. Furthermore, SiDP demonstrated high antibacterial efficacy, with up to 99.9% inhibition against Staphylococcus aureus and Escherichia coli through cell

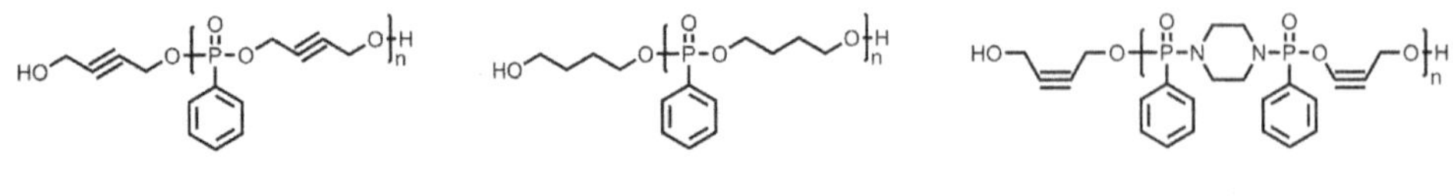

FIGURE 15.21 Chemical structures of PPBP, PPBOP, and PPBPP.

membrane disruption. Related to this concept, some hydrazides have also shown the ability to release nitrogen during fire events and showed some mixed condensed phase and gas phase flame retardant effects, but only when phosphorus was present as well.[94] Please see Chapter 10 of this book for more chemistry on nitrogen-releasing flame retardants.

Aromatic Schiff base was also obtained from fully bio resources, for which a fully bio-based epoxy monomer, having the acronym SH-BDA-EP, was synthesized from naturally occurring syringaldehyde and 1,4-butanediamine.[95] When cured with 4,4'-diaminodiphenylmethane (DDM), the resulting bio-epoxy thermoset (SH-BDA-EP/DDM) exhibited outstanding mechanical and flame-retardant properties. The self-cross-linking of the Schiff base structure facilitated the formation of a stable protective char layer, thus enhancing the flame retardancy. Specially, SH-BDA-EP/DDM successfully passed the 3.0 mm UL-94 V-0 rating, achieved a high LOI value of 39.6%, and showed significantly 82.2% and 52.4% reduction in PHRR and THR values, respectively, when measured by cone calorimeter. Additionally, the aromatic Schiff base structure endowed SH-BDA-EP/DDM with rigidity and reversibility, resulting in highly integrated performance, including excellent mechanical properties and acid-catalyzed degradation.

Ammonium polyphosphate (APP), a common and well-studied commercial intumescent flame retardant, was combined with aromatic Schiff base to prepare an organic-inorganic flame retardant, having the acronym of SP-APP, by coating APP with a Schiff base-containing branched polysiloxane.[96] This enhanced the water resistance and char-forming capacity of APP, leading to excellent flame retardancy and anti-dripping properties in polyamide 6 (PA6). The PA6/SP-APP composite with 12 wt% SP-APP achieved a high LOI value of 32.5% and passed the 3.2 mm UL-94 V-0 rating without dripping. Furthermore, PA6/SP-APP exhibited a significant reduction of 50.9% and 29.4% in PHRR and THR values compared to PA6, respectively, as measured by cone calorimeter. The mechanism behind the improved flame retardancy were extensively studied, revealing the synergistic effects of catalytic degradation of APP, self-cross-linking behavior of the Schiff base structure, and thermo-oxidative degradation of the branched polysiloxane.

Sykam et al.[97] recently developed phosphorus and triazole-containing flame-retardant polyurethane triazole foams (PUTFs) using the solvent-free and catalyst-free polymerization via azide-alkyne "click" reaction between urethane diazide monomers and triprop-2-ynyl phosphate (Figure 15.22). Notably, PUTFs achieved a 2 mm V-1 rating in the UL-94 Vertical burning test, while the control PUF failed to pass the test owing to the serious dripping phenomenon. PUTF-100 demonstrated a 6% and 76% reduction in PHRR and THR values, respectively, compared to the control PUF, when measured by cone calorimeter. They confirmed the role of triazole in the generation of nitrogen gas, as both PUTFs self-extinguished upon ignition and formed a dense intumescent char layer.

It has been reported by Schartel and coworker[98] that certain phosphorus-containing flame retardants, such as bisphenol A bis(diphenyl phosphate) (BDP) and resorcinol bis(diphenyl phosphate) (RDP), can catalyze the Fries rearrangement of PC, thereby enhancing the char-forming ability. The combination of gas and condensed phase action of BDP results in a slightly superior performance in terms of fire load and decreasing the flame spread. Phosphorus was combined with deoxybenzoin to synthesize an additive flame retardant named BDPP using a Friedel-Crafts acylation reaction.[99] BDPP exhibited synergistic char-formation mechanism in the condensed phase and the flame-inhibition mechanism in the gas phase. When applied in PTT, the introduction of 5% BDPP increased the LOI from 22.5% to 25.0% and reduced the THR and PHRR values (measured by cone calorimeter) by 25 and 24% in comparison to neat PTT, respectively. Furthermore, the addition of BDPP enabled the PTT–BDPP(95/5) blend to be processed at lower temperatures and exhibit superior mechanical properties and thermal stability. A cross-linkable N/P flame retardant (acronym of APN-DOPO) was synthesized and used in epoxy resin (EP) composites.[100] The nitrile groups in APN-DOPO underwent thermal cross-linking reaction, which significantly improved the flame-retardant properties. With the incorporation of 20 wt% APN-DOPO, the LOI value of the resulting EP increased to 46.2%, and the PHRR value decreased by 77.4% compared to neat EP when measured by cone calorimeter. Furthermore, 3.0 mm UL-94 V-0 rating was achieved with only 5 wt% APN-DOPO incorporated.

15.5 CONCLUSION AND PERSPECTIVES

In this chapter, we provide a detailed overview of flame retardants without conventional flame-retardant elements, including their chemical fundamentals, development, and the combination with conventional flame-retardant approaches. Unlike conventional methods that rely on specific flame-retardant effect of elements such as halogens, phosphorus, silicon, and boron, this chapter has outlined several alternative and promising flame-retardant methods which use different chemical functionalities. Specifically,

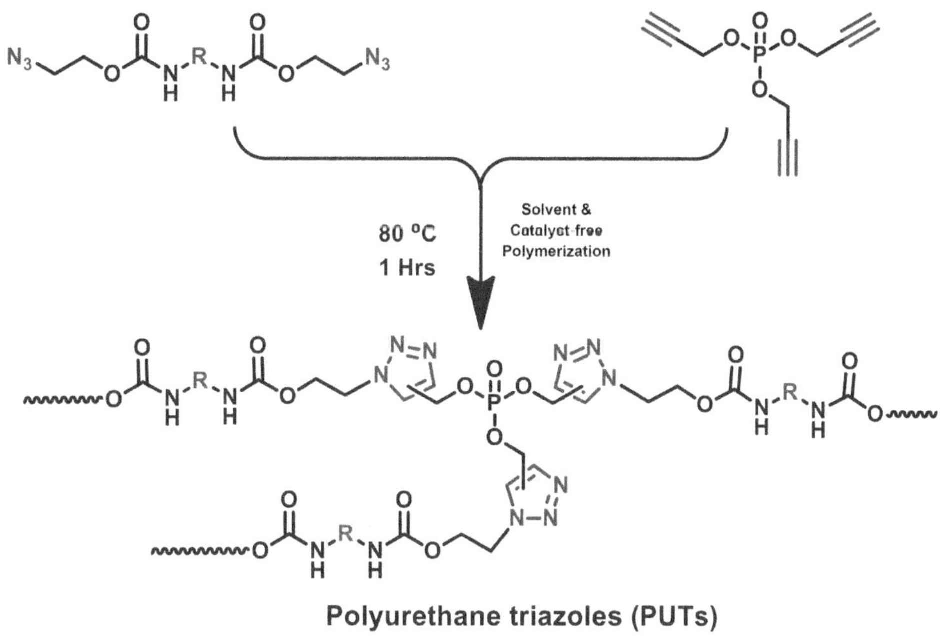

FIGURE 15.22 The preparation process of PUTs.

Source: Reproduced with permission.[97] Copyright 2021, John Wiley and Sons.

using metal ion-promoted char formation, and special chemical reactions of certain structures at high temperatures (such as self-cross-linking and rearrangement) to promote the char formation of polymers, achieves flame retardancy, anti-dripping performance (for thermoplastic polymers), and suppression of heat and toxic gas release. This method has shown significant potential for enhancing the comprehensive properties of polymers, including flame retardancy, mechanical performance, and multifunctionality, while avoiding the use of conventional flame retardants, some of which are currently under regulatory scrutiny.

Looking ahead, the future development of the chemistries and approaches discussed in this chapter shows promising prospects in the following areas:

1) High-performance flame-retardant polymers. The incorporation of special structures into the molecular chains of polymers has the potential to improve the intermolecular interactions through π-π stacking, hydrogen bond, etc., thereby enhancing mechanical properties. Thus, these methods hold tremendous potential in the field of high-performance flame-retardant polymers.
2) Heat-protection materials. The formation of the compact char layers at high temperatures provides potential thermal-insulating performance, making these methods promising for extreme heat protection applications.
3) Advanced multifunctional flame-retardant polymers. The incorporation of these special structures can impart additional functionalities to polymers, such as UV shielding, antimicrobial properties, self-healing capabilities, and 3D printing, expanding their potential in advanced multifunctional flame-retardant polymers.
4) Flame-retardant materials meeting current regulatory requirements. By avoiding the use of conventional flame-retardant elements such as phosphorus, halogen and boron, these methods present the optimal approach for pursuing the development of maintaining flame-retardant performance while meeting regulatory needs that ban the use of these conventional flame retardant elements.

REFERENCES

[1] B.W. Liu; H.B. Zhao; Y.Z. Wang, Advanced flame-retardant methods for polymeric materials. *Advanced Materials* **2022,** *34* (46), 2107905.

[2] A. Blum, The Fire Retardant Dilemma. *Science* **2007,** *318* (5848), 194–195.

[3] T.E. Parliament; T.C.O.T.E. Union, The restriction of the use of certain hazardous substances in electrical and electronic equipment. In *Directive 2002/95/EC of the European Parliament and of the Council*, CRC Press, Taylor & Francis Group: Boca Raton, FL, 2023.

[4] D. Price; A.R. Horrocks, Polymer Degradation and the Matching of FR Chemistry to Degradation. In C.A. Wilkie; A.B. Morgan (eds.) *Fire Retardancy of Polymeric Materials*. 2nd ed.; CRC Press: Boca Raton, FL, **2010,** 823.

[5] C.F. Cullis; M.M. Hirschler, Char formation from polyolefins. Correlations with low-temperature oxygen uptake and with flammability in the presence of metal halogen systems. *European Polymer Journal* **1984,** *20* (1), 53–60.

[6] A.B. Morgan, A Review of Transition Metal-Based Flame Retardants: Transition Metal Oxide/Salts, and Complexes. In *Fire and Polymers V*, 1st ed.; American Chemical Society: Washington, DC, **2009**, Vol. 1013, 312–328.

[7] S.V. Levchik; G.F. Levchik; G. Camino; L. Costa; A.I. Lesnikovich, Mechanism of Action of Phosphorus-based Flame Retardants in Nylon 6. III. Ammonium Polyphosphate/Manganese Dioxide. *Fire and Materials* **1996,** *20* (4), 183–190.

[8] G.F. Levchik; S.V. Levchik; P.D. Sachok; A.F. Selevich; A.S. Lyakhov; A.I. Lesnikovich, Thermal behaviour of ammonium polyphosphate-inorganic compound mixtures. Part 2. Manganese dioxide. *Thermochimica Acta* **1995,** *257*, 117–125.

[9] M. Lewin; M. Endo, Catalysis of intumescent flame retardancy of polypropylene by metallic compounds. *Polymers for Advanced Technologies* **2003,** *14* (1), 3–11.

[10] M. Lewin, Synergism and catalysis in flame retardancy of polymers. *Polymers for Advanced Technologies* **2001,** *12* (3–4), 215–222.

[11] P.J. Davies; A.R. Horrocks; A. Alderson, The sensitisation of thermal decomposition of ammonium polyphosphate by selected metal ions and their potential for improved cotton fabric flame retardancy. *Polymer Degradation and Stability* **2005,** *88* (1), 114–122.

[12] F. Xie; Y.-Z. Wang; B. Yang; Y. Liu, A novel intumescent flame-retardant polyethylene system. *Macromolecular Materials and Engineering* **2006,** *291* (3), 247–253.

[13] D.-L. Wang; Y. Liu; D.-Y. Wang; C.-X. Zhao; Y.-R. Mou; Y.-Z. Wang, A novel intumescent flame-retardant system containing metal chelates for polyvinyl alcohol. *Polymer Degradation and Stability* **2007,** *92* (8), 1555–1564.

[14] R.P.R. Ranaweera; G. Scott, Mechanisms of antioxidant action: Antioxidant behaviour of nickel complex U.V. stabilisers. *European Polymer Journal* **1976,** *12* (12), 825–830.

[15] C.X. Zhao; Y. Liu; D.Y. Wang; D.L. Wang; Y.Z. Wang, Synergistic effect of ammonium polyphosphate and layered double hydroxide on flame retardant properties of poly(vinyl alcohol). *Polymer Degradation and Stability* **2008,** *93* (7), 1323–1331.

[16] C. Manzi-Nshuti; D. Wang; J.M. Hossenlopp; C.A. Wilkie, Aluminum-containing layered double hydroxides: The thermal, mechanical, and fire properties of (nano)composites of poly(methyl methacrylate). *Journal of Materials Chemistry* **2008,** *18* (26), 3091–3102.

[17] C. Manzi-Nshuti; D. Wang; J.M. Hossenlopp; C.A. Wilkie, The role of the trivalent metal in an LDH: Synthesis, characterization and fire properties of thermally stable PMMA/LDH systems. *Polymer Degradation and Stability* **2009,** *94* (4), 705–711.

[18] Z. Li; J.H. Zhang; F. Dufosse; D.Y. Wang, Ultrafine nickel nanocatalyst-engineering of an organic layered double hydroxide towards a super-efficient fire-safe epoxy resin via interfacial catalysis. *Journal of Materials Chemistry A* **2018,** *6* (18), 8488–8498.

[19] J.-Q. Wang; J.-X. Du; H.-Y. Yao; C.A. Wilkie, XPS characterization of Friedel-Crafts cross-linked polystyrene. *Polymer Degradation and Stability* **2001,** *74* (2), 321–326.

[20] Y.-Z. Zhan; P. Zhu; J.-B. Zhang; X.-Q. Guo, The properties and application of alginate fiber. *Textile Auxiliaries* **2006,** *23* (6), 9–12.

[21] Q.-S. Kong; B.-B. Wang; Q. Ji; Y.-Z. Xia; Z.-X. Guo; J. Yu, Thermal degradation and flame retardancy of calcium alginate fibers. *Chinese Journal of Polymer Science* **2009,** *27* (6), 807.

[22] Y.-J. Xu; L.-Y. Qu; Y. Liu; P. Zhu, An overview of alginates as flame-retardant materials: Pyrolysis behaviors, flame retardancy, and applications. *Carbohydrate Polymers* **2021,** *260*, 117827.

[23] Q.-Q. Wang; L. Zhang; Y.-Y. Liu; G.-Q. Zhang; P. Zhu, Characterization and functional assessment of alginate fibers prepared by metal-calcium ion complex coagulation bath. *Carbohydrate Polymers* **2020,** *232*, 115693.

[24] Y. Tao; B. Wang; C. Liu; P. Li; Y.-J. Xu; Z.-M. Jiang; Y. Liu; P. Zhu, Cotton/alginate blended knitted fabrics: Flame retardancy, flame-retardant mechanism, water absorption and mechanical properties. *Cellulose* **2021,** *28* (7), 4495–4510.

[25] Y. Liu; Y. Tao; B. Wang; P. Li; Y.-J. Xu; Z.-M. Jiang; C.-H. Dong; P. Zhu, Fully bio-based fire-safety viscose/alginate blended nonwoven fabrics: Thermal degradation behavior, flammability, and smoke suppression. *Cellulose* **2020,** *27* (10), 6037–6053.

[26] P. Li; Q.-Z. Wang; B. Wang; Y.-Y. Liu; Y.-J. Xu; Y. Liu; P. Zhu, Blending alginate fibers with polyester fibers for flame-retardant filling materials: Thermal decomposition behaviors and fire performance. *Polymer Degradation and Stability* **2021,** *183*, 109470.

[27] F.-Q. Zhang; B. Wang; Y.-J. Xu; P. Li; Y. Liu; P. Zhu, Convenient blending of alginate fibers with polyamide fibers for flame-retardant non-woven fabrics. *Cellulose* **2020,** *27* (14), 8341–8349.

[28] X. Wang; Y.-T. Pan; J.-T. Wan; D.-Y. Wang, An eco-friendly way to fire retardant flexible polyurethane foam: Layer-by-layer assembly of fully bio-based substances. *RSC Advances* **2014,** *4* (86), 46164–46169.

[29] H. Pan; W. Wang; Y. Pan; L. Song; Y. Hu; K.M. Liew, Formation of layer-by-layer assembled titanate nanotubes filled coating on flexible polyurethane foam with improved flame retardant and smoke suppression properties. *ACS Applied Materials & Interfaces* **2015,** *7* (1), 101–111.

[30] Y. Pan; H. Pan; B. Yuan; N. Hong; J. Zhan; B. Wang; L. Song; Y. Hu, Construction of organic–inorganic hybrid nano-coatings containing α-zirconium phosphate with high efficiency for reducing fire hazards of flexible polyurethane foam. *Materials Chemistry and Physics* **2015,** *163*, 107–115.

[31] X. Zhang; Q. Shen; X. Zhang; H. Pan; Y. Lu, Graphene oxide-filled multilayer coating to improve flame-retardant and smoke suppression properties of flexible polyurethane foam. *Journal of Materials Science* **2016,** *51* (23), 10361–10374.

[32] X. Mu; B. Yuan; Y. Pan; X. Feng; L. Duan; R. Zong; Y. Hu, A single α-cobalt hydroxide/sodium alginate bilayer layer-by-layer assembly for conferring flame retardancy to flexible polyurethane foams. *Materials Chemistry and Physics* **2017,** *191*, 52–61.

[33] H. Nabipour; X. Wang; L. Song; Y. Hu, A fully bio-based coating made from alginate, chitosan and hydroxyapatite for protecting flexible polyurethane foam from fire. *Carbohydrate Polymers* **2020,** *246*, 116641.

[34] L. Liu; W. Wang; Y. Hu, Layered double hydroxide-decorated flexible polyurethane foam: Significantly improved toxic effluent elimination. *RSC Advances* **2015,** *5* (118), 97458–97466.

[35] H.-B. Chen; Y.-Z. Wang; M. Sánchez-Soto; D.A. Schiraldi, Low flammability, foam-like materials based on ammonium alginate and sodium montmorillonite clay. *Polymer* **2012,** *53* (25), 5825–5831.

[36] K. Shang; W. Liao; J. Wang; Y.-T. Wang; Y.-Z. Wang; D.A. Schiraldi, Nonflammable alginate nanocomposite aerogels prepared by a simple freeze-drying and post-cross-linking method. *ACS Applied Materials & Interfaces* **2016,** *8* (1), 643–650.

[37] X. Wang; E.N. Kalali; J.-T. Wan; D.-Y. Wang, Carbon-family materials for flame retardant polymeric materials. *Progress in Polymer Science* **2017,** *69*, 22–46.

[38] Y.-Z. Wang, *Fire Retardation of Polymeric Materials without Conventional Flame-Retardant Elements.* 5th Asia-Oceania Symposium on Fire Safety Materials Science and Engineering, Sydney, Australia, **2023**.

[39] D.-X. Fang; H.-B. Zhao; M.-J. Chen; Y.-Z. Wang, *Flame-Retardant and Recyclable Flexible Polyurethane Foams.* 5th Asia-Oceania Symposium on Fire Safety Materials Science and Engineering, Sydney, Australia, **2023**.

[40] H.B. Zhao; Y.Z. Wang, Design and synthesis of PET-based copolyesters with flame-retardant and antidripping performance. *Macromolecular Rapid Communications* **2017,** *38* (23), 1700451.

[41] L. Chen; B. Liu; T. Fu; Y. Ni; X. Wang; Y. Wang, New methods for flame-retarding PET without melt dripping. *Chinese Science Bulletin* **2020,** *65* (28–29), 3160–3172.

[42] H.-B. Zhao; L. Chen; J.-C. Yang; X.-G. Ge; Y.-Z. Wang, A novel flame-retardant-free copolyester: Cross-linking towards self extinguishing and non-dripping. *Journal of Materials Chemistry* **2012,** *22* (37), 19849.

[43] H.B. Zhao; X.L. Wang; Y. Guan; X.L. Wang; L. Chen; Y.Z. Wang, Block self-cross-linkable poly(ethylene terephthalate) copolyester via solid-state polymerization: Crystallization, cross-linking, and flame retardance. *Polymer* **2015,** *70*, 68–76.

[44] L. Chen; H.B. Zhao; Y.P. Ni; T. Fu; W.S. Wu; X.L. Wang; Y.Z. Wang, 3D printable robust shape memory PET copolyesters with fire safety via π-stacking and synergistic crosslinking. *Journal of Materials Chemistry A* **2019,** *7* (28), 17037–17045.

[45] B.-W. Liu; H.-B. Zhao; L. Chen; L. Chen; X.-L. Wang; Y.-Z. Wang, Eco-friendly synergistic cross-linking flame-retardant strategy with smoke and melt-dripping suppression for condensation polymers. *Composites Part B: Engineering* **2021,** *211*, 108664.

[46] M. Xie; D. Jia; J. Hu; J. He; X. Li; R. Yang, Fabrication of enhanced mechanical properties and intrinsic flame-retardant polyurethane elastomer containing 4-(phenylethynyl) di(ethylene glycol) phthalate. *Polymers* **2021,** *13* (15), 2388.

[47] T. Ikeda, Photomodulation of liquid crystal orientations for photonic applications. *Journal of Materials Chemistry* **2003,** *13* (9), 2037–2057.

[48] X.-K. Jing; X.-S. Wang; D.-M. Guo; Y. Zhang; F.-Y. Zhai; X.-L. Wang; L. Chen; Y.-Z. Wang, The high-temperature self-crosslinking contribution of azobenzene groups to the flame retardance and anti-dripping of copolyesters. *Journal of Materials Chemistry A* **2013,** *1* (32), 9264.

[49] B.-W. Liu; H.-B. Zhao; Y. Tan; L. Chen; Y.-Z. Wang, Novel crosslinkable epoxy resins containing phenylacetylene and azobenzene groups: From thermal crosslinking to flame retardance. *Polymer Degradation and Stability* **2015,** *122*, 66–76.

[50] Z.Q. Lei; P. Xie; M.Z. Rong; M.Q. Zhang, Catalyst-free dynamic exchange of aromatic Schiff base bonds and its application to self-healing and remolding of crosslinked polymers. *Journal of Materials Chemistry A* **2015,** *3* (39), 19662–19668.

[51] N. Ji; J. Yuan; M. Liu; T. Lan; W. He, Novel chiral Schiff base/Ti(IV) catalysts for the catalytic asymmetric epoxidation of N-alkenyl sulfonamides. *Chemical Communications* **2016,** *52* (49), 7731–7734.

[52] J.N. Wu; L. Chen; T. Fu; H.B. Zhao; D.M. Guo; X.L. Wang; Y.Z. Wang, New application for aromatic Schiff base: High efficient flame-retardant and anti-dripping action for polyesters. *Chemical Engineering Journal* **2018,** *336*, 622–632.

[53] J.-N. Wu; Z.-H. Qin; L. Chen; B.-W. Liu; X.-L. Wang; Y.-Z. Wang, Tailoring Schiff base cross-linking by cyano group toward excellent flame retardancy, anti-dripping and smoke suppression of PET. *Polymer* **2018,** *153*, 78–85.

[54] X.-X. Zhang; Q.-Y. Wang; S.-Y. Liu; L.-L. Zhang; G.-Y. Wang, Synthesis and characterization of fire-safety PET by Schiff base with nitro group. *European Polymer Journal* **2021,** *145*, 110230.

[55] X. Dong; L. Chen; R.T. Duan; Y.Z. Wang, Phenylmaleimide-containing PET-based copolyester: Cross-linking from 2π + π cycloaddition toward flame retardance and anti-dripping. *Polymer Chemistry* **2016,** *7* (15), 2698–2708.

[56] D.M. Guo; T. Fu; C. Ruan; X.L. Wang; L. Chen; Y.Z. Wang, A new approach to improving flame retardancy, smoke suppression and anti-dripping of PET: Via arylene-ether units rearrangement reactions at high temperature. *Polymer* **2015,** *77*, 21–31.

[57] T. Fu; D.M. Guo; J.N. Wu; X.L. Wang; X.L. Wang; L. Chen; Y.Z. Wang, Inherent flame retardation of semi-aromatic polyesters via binding small-molecule free radicals and charring. *Polymer Chemistry* **2016,** *7* (8), 1584–1592.

[58] T. Fu; X.L. Wang; Y.Z. Wang, Flame-responsive aryl ether nitrile structure towards multiple fire hazards suppression of thermoplastic polyester. *Journal of Hazardous Materials* **2021,** *403*, 123714.

[59] D.M. Guo; X.Q. Chen; L. Tang; J.N. Wu; T. Fu; X.L. Wang; X.L. Wang; L. Chen; Y.Z. Wang, PET-based copolyesters with bisphenol A or bisphenol F structural units: Their distinct differences in pyrolysis behaviours and flame-retardant performances. *Polymer Degradation and Stability* **2015,** *120*, 158–168.

[60] A.C.L.M. Van Der Waals; A.J.H. Klunder; F.R. Van Buren; B. Zwanenburg, Novel dehydration of stilbene oxides to diphenylacetylene using flash vacuum thermolysis with solid acid catalysts. *Journal of Molecular Catalysis A: Chemical* **1998,** *134* (1), 179–189.

[61] K.A. Ellzey; T. Ranganathan; J. Zilberman; E.B. Coughlin; R.J. Farris; T. Emrick, Deoxybenzoin-based polyarylates as halogen-free fire-resistant polymers. *Macromolecules* **2006,** *39* (10), 3553–3558.

[62] T. Ranganathan; J. Zilberman; R.J. Farris; E.B. Coughlin; T. Emrick, Synthesis and characterization of halogen-free antiflammable polyphosphonates containing 4,4'-bishydroxydeoxybenzoin. *Macromolecules* **2006,** *39* (18), 5974–5975.

[63] U. Choudhary; A.A. Mir; T. Emrick, Soluble, allyl-functionalized deoxybenzoin polymers. *Macromolecules* **2017,** *50* (10), 3772–3778.

[64] B.Y. Ryu; S. Moon; I. Kosif; T. Ranganathan; R.J. Farris; T. Emrick, Deoxybenzoin-based epoxy resins. *Polymer* **2009,** *50* (3), 767–774.

[65] T. Ranganathan; P. Cossette; T. Emrick, Halogen-free, low flammability polyurethanes derived from deoxybenzoin-based monomers. *Journal of Materials Chemistry* **2010,** *20* (18), 3681.

[66] A.A. Mir; S. Wagner; R.H. Krämer; P. Deglmann; T. Emrick, Deoxybenzoin-containing polysulfones and

polysulfoxides: Synthesis and thermal properties. *Polymer* **2016,** *84*, 59–64.

[67] T.L. Gilchrist; G.E. Gymer; C.W. Rees, Reactive intermediates. Part XXIV. 1H-Azirine intermediates in the pyrolysis of 1H-1,2,3-triazoles. *Journal of the Chemical Society, Perkin Transactions* **1975,** *1* (1), 1–8.

[68] T.L. Gilchrist; C.W. Rees; C. Thomas, Reactive intermediates. Part XXV. Investigation of the pyrolysis of 1,4- and 1,5-diphenyl-1,2,3-triazoles by use of 13C-labelled compounds. *Journal of the Chemical Society, Perkin Transactions* **1975,** *1* (1), 8–11.

[69] T.L. Gilchrist; C.W. Rees; C. Thomas, Reactive intermediates. Part XXVI. Flash vacuum pyrolysis of phenyl-substituted 1,2,4-triazoles; A new synthesis of isoindoles. *Journal of the Chemical Society, Perkin Transactions* **1975,** *1* (1), 12–18.

[70] B.Y. Ryu; T. Emrick, Thermally induced structural transformation of bisphenol-1,2,3-triazole polymers: Smart, self-extinguishing materials. *Angewandte Chemie* **2010,** *122* (50), 9838–9841.

[71] B.Y. Ryu; T. Emrick, Bisphenol-1,2,3-triazole (BPT) epoxies and cyanate esters: Synthesis and self-catalyzed curing. *Macromolecules* **2011,** *44* (14), 5693–5700.

[72] H. Wei; F. Wang; H. Sun; Z. Zhu; C. Xiao; W. Liang; B. Yang; L. Chen; A. Li, Benzotriazole-based conjugated microporous polymers as efficient flame retardants with better thermal insulation properties. *Journal of Materials Chemistry A* **2018,** *6* (18), 8633–8642.

[73] B. Rukmanikrishnan; S. Muthusamy, Preparation and properties of polyimides containing 1,2,3-triazole moieties. *Advances in Polymer Technology* **2018,** *37* (1), 50–59.

[74] Y.P. Ni; Q.T. Li; L. Chen; W.S. Wu; Z.H. Qin; Y. Zhang; L. Chen; X.L. Wang; Y.Z. Wang, Semi-aromatic copolyesters with high strength and fire safety via hydrogen bonds and π-π stacking. *Chemical Engineering Journal* **2019,** *374*, 694–705.

[75] Y.P. Ni; W.S. Wu; L. Chen; X. Zhao; Z.H. Qin; X.L. Wang; Y.Z. Wang, How hydrogen bond interactions affect the flame retardancy and anti-dripping performances of PET. *Macromolecular Materials and Engineering* **2020,** *305* (1), 1900661.

[76] W.-S. Wu; P.-H. Duan; Y.-L. Wang; L. Chen; X.-L. Wang; Y.-Z. Wang, High-fire-safety thermoplastic polyester constructed by novel sulfonate with benzimidazole structure. *Science China Materials* **2021,** *64* (8), 2067–2080.

[77] W.-S. Wu; Y.-P. Ni; L. Chen; T. Fu; X.-L. Wang; Y.-Z. Wang, Trinity effect of potassium sulfonate-benzimidozale towards self-intumescent flame-retarded polyester with low fire hazards. *Chemical Engineering Journal* **2022,** *429*, 132121.

[78] G.L. Tullos; L.J. Mathias, Unexpected thermal conversion of hydroxy-containing polyimides to polybenzoxazoles. *Polymer* **1999,** *40* (12), 3463–3468.

[79] G.L. Tullos; J.M. Powers; S.J.J. And; L.J. Mathias, Thermal conversion of hydroxy-containing imides to benzoxazoles: Polymer and model compound study. *Macromolecules* **1999,** *32* (11), 3598–3612.

[80] B.W. Liu; L. Chen; D.M. Guo; X.F. Liu; Y.F. Lei; X.M. Ding; Y.Z. Wang, Fire-safe polyesters enabled by end-group capturing chemistry. *Angewandte Chemie International Edition* **2019,** *58* (27), 9188–9193.

[81] B.W. Liu; Y.F. Lei; X.F. Liu; L. Chen; D.M. Guo; Y.Z. Wang, Small change, big impact: Simply tailoring the substitution position towards significant improvement of flame retardancy. *Composites Part B: Engineering* **2021,** *223*, 109109.

[82] B.W. Liu; Y.F. Lei; X.F. Liu; D.M. Guo; L. Chen; Y.Z. Wang, Thermally induced end-group-capturing as an eco-friendly and general method for enhancing the fire safety of semi-aromatic polyesters. *Polymer* **2021,** *218*, 123430.

[83] N.B. Mckeown, *Phthalocyanine Materials: Synthesis, Structure and Function*. Cambridge University Press: Cambridge, **1998**, 222.

[84] T. Fu; X.-L. Wang; Y.-Z. Wang, Flame-responsive aryl ether nitrile structure towards multiple fire hazards suppression of thermoplastic polyester. *Journal of Hazardous Materials* **2021,** *403*, 123714.

[85] T. Fu; D.M. Guo; X.-L. Wang; Y.-Z. Wang, In situ phthalocyanine synthesis chemistry in flames towards molecular fireproof engineering. *Chemical Communications* **2020,** *56* (66), 9525–9528.

[86] R. Wang; P.-L. Zheng; J.-W. Li; J.-C. Sun; H.-Y. Liu; X. Li; Q.-Y. Liu, An efficient cross-linked phosphorus-free flame retardant for epoxy resins. *ACS Omega* **2022,** *7* (42), 37170–37179.

[87] G.-Q. Zheng; B.-W. Liu; L. Yuan; L. Chen; X.-L. Wang; H.-B. Zhao; Y.-Z. Wang, Natural polyphenol-inspired polymer towards multifunction and high performance. *Chemical Engineering Journal* **2023,** *462*, 142212.

[88] A.B. Morgan; J.M. Tour, Synthesis and testing of nonhalogenated alkyne-containing flame-retarding polymer additives. *Macromolecules* **1998,** *31* (9), 2857–2865.

[89] A.B. Morgan; J.M. Tour, Synthesis and testing of nonhalogenated alkyne/phosphorus-containing polymer additives: Potent condensed-phase flame retardants. *Journal of Applied Polymer Science* **1999,** *73* (5), 707–718.

[90] H. Chen; C. Deng; Z.-Y. Zhao; S.-C. Huang; Y.-X. Wei; Y.-Z. Wang, Novel alkynyl-containing phosphonate ester oligomer with high charring capability as flame retardant additive for thermoplastic polyurethane. *Composites Part B: Engineering* **2020,** *199*, 108315.

[91] H. Chen; C. Deng; Z.-Y. Zhao; L. Wan; A.-H. Yang; Y.-Z. Wang, Novel piperazine-containing oligomer as flame retardant and crystallization induction additive for thermoplastics polyurethane. *Chemical Engineering Journal* **2020,** *400*, 125941.

[92] S. Fan; B. Peng; R. Yuan; D. Wu; X. Wang; J. Yu; F. Li, A novel Schiff base-containing branched polysiloxane as a self-crosslinking flame retardant for PA6 with low heat release and excellent anti-dripping performance. *Composites Part B: Engineering* **2020,** *183*, 107684.

[93] N. Li; G. Kang; H. Liu; M. Li; W. Qiu; Q. Wang; L. Liu; J. Yu; B. Li; F. Li; D. Wu, Hybrid nanoparticles based on novel Schiff base for durable flame retardant and antibacterial properties. *Composites Part B: Engineering* **2022,** *238*, 109905.

[94] A.B. Morgan; V. Benin; D.A. Klosterman; A.B. Sulayman; M. Mukhtar; M.L. Galaska, Organophosphorus-hydrazides as potential reactive flame retardants for epoxy. *Journal of Fire Sciences* **2020,** *38* (1), 28–52.

[95] W. Xie; S. Huang; S. Liu; J. Zhao, Imine-functionalized biomass-derived dynamic covalent thermosets enabled by heat-induced self-crosslinking and reversible structures. *Chemical Engineering Journal* **2021,** *404*, 126598.

[96] S. Fan; Y. Sun; X. Wang; J. Yu; D. Wu; F. Li, A novel organic-inorganic flame retardant of ammonium polyphosphate chemically coated by Schiff base-containing branched polysiloxane for polyamide 6. *Polymers for Advanced Technologies* **2020,** *31* (11), 2763–2774.

[97] K. Sykam; P. Harika; S. Donempudi, Flame-retardant, phosphorous-based polyurethane triazoles via solvent-free and catalyst-free azide–alkyne cycloaddition and their cure kinetics. *Polymers for Advanced Technologies* **2021,** *32* (4), 1636–1653.
[98] K.H. Pawlowski; B. Schartel, Flame retardancy mechanisms of triphenyl phosphate, resorcinol bis(diphenyl phosphate) and bisphenol A bis(diphenyl phosphate) in polycarbonate/acrylonitrile–butadiene–styrene blends. *Polymer International* **2007,** *56* (11), 1404–1414.
[99] X. Hu; Y. Wang; J. Yu; J. Zhu; Z. Hu, Synthesis of a deoxybenzoin derivative and its use as a flame retardant in poly(trimethylene terephthalate). *Journal of Applied Polymer Science* **2018,** *135* (8), 45904.
[100] X. Peng; Z. Li; D. Wang; Z. Li; C. Liu; R. Wang; L. Jiang; Q. Liu; P. Zheng, A facile crosslinking strategy endows the traditional additive flame retardant with enormous flame retardancy improvement. *Chemical Engineering Journal* **2021,** *424*, 130404.

16 Multi-Component Flame-Retardant Systems

Bernhard Schartel
Bundesanstalt für Materialforschung und -prüfung (BAM), Unter den Eichen 87, 12205 Berlin, Germany

16.1 THE IMPERATIVE FOR MULTI-COMPONENT SOLUTIONS

Most modern flame-retardant polymeric materials are based on multi-component systems. Two flame retardants are combined, or flame retardants are used in combination with adjuvants or synergists to enhance efficiency, reduce the amount of flame retardant required, or in the interest of reducing costs. Since adding flame retardants usually tends to worsen the material properties, such as mechanical strength, elongation to break, glass transition temperature, or processability, multi-component systems are intensively used as an approach to optimize property profiles. Fibers, several fillers, and some additives also make a crucial contribution to the fire behavior of a polymeric material. This chapter will illuminate the concept of multi-component systems in flame retardancy by explaining the need for using these systems and the main phenomena they achieve. Multi-component flame retardants are discussed as an overall powerful strategy for achieving and optimizing actual and future flame-retardant polymeric materials. It should be noted that this chapter utilizes the same concepts published in a similar chapter underlining the importance of multi-component systems in halogen-free flame retardancy[1] and therefore has some similarity in text because the fundamentals are the same for halogen-free systems as well as halogenated systems.

Flame-retardant polymeric materials are the clear preference for technical or high-performance materials used in flame-retarded lightweight solutions in the transportation sector, for instance for shipping, railway vehicles, and aviation. Flame-retardant polymeric materials are dominant in electrical engineering and electronics for housings, cables, switches, and circuit boards. Large amounts of flame-retardant materials are used as cost-efficient flame-retardant building products. The market increasingly demands efficient, cost-efficient, and environmentally friendly solutions and in the future will also require increased sustainability. What is more, the markets in question are highly competitive, and several are extremely innovative. There is constant pressure to improve the flame-retardant materials on offer, which should ideally be systems that simultaneously reduce flammability, smoke production, and the release of toxic fire effluents, while reducing costs, improving mechanical properties, prolonging durability (mainly stability against hydrolysis and photo-oxidation), enabling recyclability or biodegradability, adjusting color, and improving processability. The importance of applying multi-component systems to optimize the property profile cannot be overemphasized when it comes to commercial flame-retardant polymeric materials.

Clearly, the possible combinations of established commercially available flame retardants, synergists, adjuvants, and fillers as well as their concentration, particle size distribution, and so forth present seemingly infinite possibilities to proceed and advance in thousands of different directions. Yet frankly, it is generally not these infinite possibilities that drive innovation in the application of multi-component systems, but the needs of the market. However, several of the most important flame retardants determine the need for multi-component systems, as the efficiency of their flame-retardant effects shows a non-linear dependence on increasing concentrations. In other words, it is not always profitable or in some cases it is even not possible to achieve the desired flame retardancy by merely increasing the amount of a certain flame retardant. Figure 16.1 illustrates some characteristic tasks for flame retardants by showing the linear and non-linear dependencies of their efficiency on the concentration. Hereby, the characteristic non-linear behavior is sketched for phosphorous flame retardants functioning mainly in the gas phase, for intumescent systems, and for metal hydrates. For flame retardants with a linearly increasing effect, the use of synergistic multi-component systems to decrease the necessary amount of flame retardants is obviously rather straightforward (Figure 16.1a). Multi-component systems are applied to increase the efficiency per concentration to achieve the desired flame retardancy with a lower amount of flame retardants. However, most of the flame retardants probably do not show a linear increase in flame retardancy over a wider range of concentrations. Phosphorous flame retardants usually show a flame retardancy that levels off with increasing concentrations, when flame inhibition is the main flame-retardant mode of action. Flame retardancy may even pass through a maximum (Figure 16.1b) before somewhat fire-accelerating effects of phosphorus set in in the flame. As red phosphorus

DOI: 10.1201/9781003380689-16

has been reported to show such Limiting Oxygen Index (LOI) behavior in different polymers,[2,3] it makes little sense to add more than 7.5–10 wt% red phosphorus. Similar behavior was reported for aluminum diethylphosphinate in polyester, indicating that the improvement to LOI levels off at around 12–14 wt%, an effect that is also seen in epoxy resin and in polyolefins.[3–5] As the gap to the flame retardancy required is much larger for polyolefins than for technical or high-performance polymers, in some polymers, phosphorous flame retardants must be combined with other flame retardants to reach the demanded flame retardancy effect.[5,6] An important, well-known task is determined by the non-linear dependency of the flame retardancy achieved with metal hydrates.[7,8] Metal hydrates, such as aluminum trihydrate, are inexpensive compared to other flame retardants, so that their use in polyolefins are their main application. Nevertheless, quite a large amount (see Chapter 6 in this book), up to 65 wt%, must be applied to achieve UL 94 V-0 classifications, which harms the mechanical properties of the material (Figure 16.1c). The increase in flame retardancy until these very large amounts are added is moderate, so that the entire curve resembles a nearly exponential-like behavior. Several dozen different approaches are proposed to shift the strong increase to lower concentrations,[7–9] many of which have been commercialized. The combination with synergist in multi-component systems is often crucial to reduce the filler content and thus to achieve an acceptable property profile. Charring, intumescent, or protective layer systems typically show an S-shape or a step function over concentration (Figure 16.1d). At low amounts the systems function as simple charring systems. Total fuel release and fuel release rate are reduced since carbon remains in the carbonaceous char. At a certain threshold concentration, the protective layer function of the fire residue becomes effective enough to prevent the specimen from complete pyrolysis.[3,10,11] In some cases, thermal insulation by the fire residue becomes strong enough to prevent the pyrolysis front from passing through the entire specimen.[12,13] Depending on the flame retardancy demanded, it may be necessary to optimize the multi-component intumescent system to reduce the concentration needed to reach the threshold or increase the level of flame retardancy after the

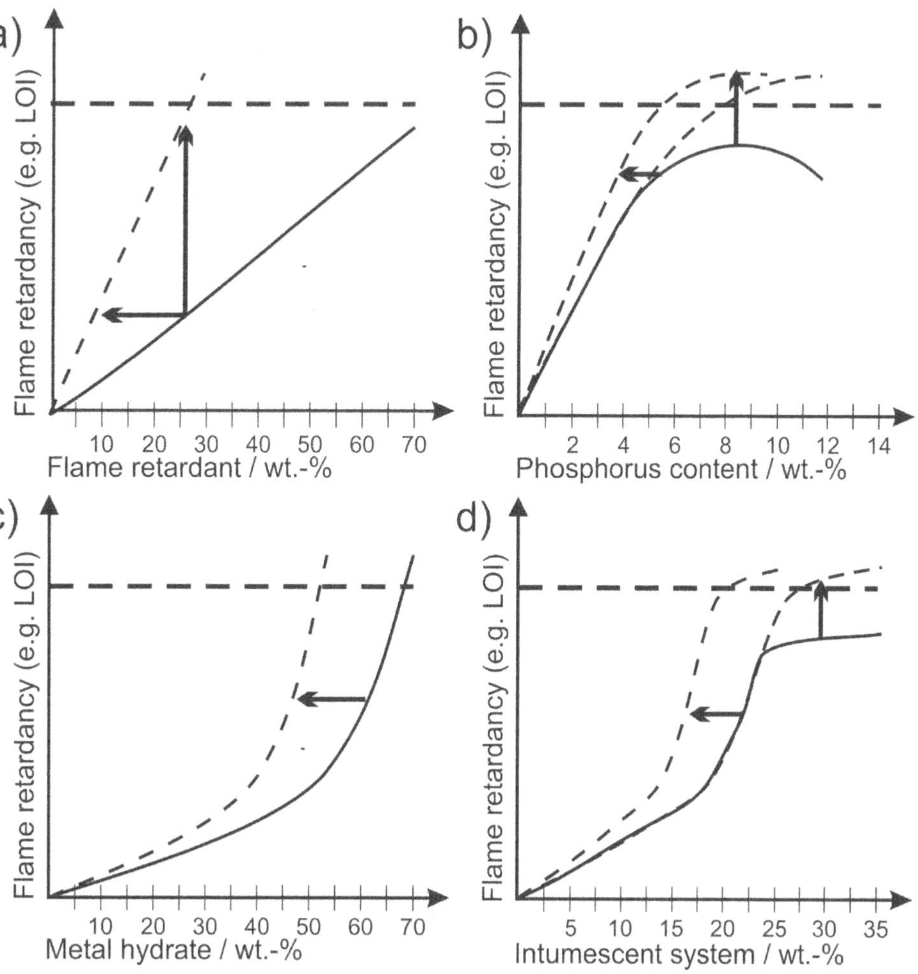

FIGURE 16.1 (a) Increasing efficiency of flame retardants with linear dependence on the concentration, and (b-d) flame retardants with characteristic non-linear dependence on the concentration; (b) phosphorous flame retardant working in the gas phase, (c) metal hydrate, and (d) intumescent systems. Dashed lines represent the desired flame retardancy effect and required flame retardancy levels, respectively; arrows, the goals for applying multi-component systems.

threshold (Figure 16.1d). Indeed, in flame retardant polyolefin materials such as polypropylene (PP)/charring agent/ammonium polyphosphate (APP), one of the most important research goals is to decrease the amount of additives needed by applying synergists. Thus, plenty of synergists are proposed for intumescent systems in polymeric materials, including several of the flame retardants discussed throughout this book, such as boron compounds, phosphorous compounds, silicone, nanoparticles, metal oxides, and fibers and fillers.[14–24] More details on intumescent systems can be found in Chapter 7 of this book.

The second main motivation for multi-component systems is to balance the property profiles of the flame-retardant technical or high-performance polymeric materials, since a dozen properties must fulfill multiple demanding technical specifications at any given time. Approaches using only one flame retardant are often detrimental to other key properties such as mechanical strength, ductility, glass transition temperature, color, transparency, processability, or durability. Tailored and synergistic multi-component systems provide routes to reduce the flame retardant or overall filler content needed. This reduction of the content contributes to a better property profile. Further, multi-component systems are used as alternatives to replacing certain flame retardants. Several halogen-free multi-component systems were ultimately applied mainly to replace former halogenated solutions. For instance, the remarkable story of success of metal phosphinate salts originates exclusively from adding flame-retardant mixtures to polyamides and polyesters for electrical engineering.[25–35] However, most of the halogen flame retardants are also applied within multi-component systems.[36–39] The combination of a halogen-containing flame retardant with antimony oxide is one of the most successful synergistic combinations and used widely.[40–42] Other halogen-containing flame retardants are combined with charring or dripping agents.[43]

The third main force driving the development and application of multi-component flame retardant polymers is cost reduction. Material costs or cost-efficiency are the most important factors in the area of commodities and technical polymers, as demonstrated well, when plotting fire hazard over the bulk price of polymers.[44] An important component of the flame retardancy of polymers is how well they compete with more expensive polymers that have higher intrinsic flame retardancy. Flame-retarded commodities were developed to compete with technical polymers, and flame-retarded technical polymers to compete with high-performance polymers. Consequently, flame retardants must be very economical for polyolefins and low-priced for technical polymers; only for high-performance polymers does the role of the bulk price of the flame retardant lose some of its dominance. Multi-component systems exploiting the combination with less expensive flame retardants or inexpensive fillers are a successful way to reduce costs. Synergistic systems reducing the necessary amount of expensive flame retardant are a prevalent way to reduce costs as well, for example, synergistic combinations make rather expensive halogen-containing flame retardants so successfully in commodities and technical polymers.

Clearly, several of these reasons overlap and coincide with each other, such that cost is usually the most important factor within any given property profile. For instance, for decades, nearly unbelievably low amounts of hexabromocyclododecane (HBCD) were used together with drip-promoting agents in flame-retardant polystyrene foams.[41] See Chapter 11 of this book for more information on drip-promoting agents/radical generator flame retardants. Some flame inhibition and a pronounced retreat effect was exploited to pass the fire tests for building products. The combination allows very low concentrations, resulting in reduced costs while retaining foamability.

When it comes to developing a flame-retardant polymeric material the huge number of possible variations and combinations of flame retardants, synergists, adjuvants, and fillers, along with their concentration, particle size distribution, and so forth, determines typical research projects in the field. Even following one distinct idea still requires that a wide range of different variations be compared and combinations and concentrations optimized. Research and development approaching application usually entail much screening of several alternative approaches, combinations, and additives. For instance, during research projects performed in close cooperation with industrial partners we investigated in my working group, various inorganic additives of different kinds, including layered materials, metal hydroxides, metal oxides/carbonate, and metal borates in bisphenol A polycarbonate/silicon rubber/bisphenol A bis(diphenyl phosphate) (PC/SiR/BDP),[45] multi-component flame retardant systems containing aluminum diethylphosphinate and another flame retardant and synergist in thermoplastic styrene-ethylene-butylene-styrene elastomers,[5] different combinations of flame retardant and adjuvants in wood plastic composites,[46] and different double and triple combinations of flame retardants in thermoplastic polyurethane,[47] and so forth. Even if an ingenious empirical concept reduces the number of systems to be investigated, a comprehensive screening of a variety of systems generally remains. Thus, in a typical example for industrial research and development, the patent DE 42 31 774 A1 [48] protects a thermoplastic polymer flame retarded with a combination of 0.1–30 parts per weight phosphorous flame retardant and 0.1–15 parts per weight aluminosilicate. The special idea proposed by this patent covers an overwhelming multitudinous matrix of components: all kinds of thermoplastics, including homopolymers and copolymers, polymer blends, and different composites, all kinds of phosphorous flame retardants combined with all kinds of zeolites and layered silicates.[48] When it comes to commercial systems, complex, synergistic, multifunctional multi-component systems are the standard, not the exception.

The almost infinite feasible variations and combinations demand preselection, guidelines, rapid screening, and accelerated assessment, so that "high-throughput" approaches are proposed.[49–58] Because fire behavior is not a material property but the response of a defined test specimen in a certain fire scenario, some intrinsic problems become obvious, such as limited possibilities to reduce the specimen size or to perform parallel testing. Nevertheless, developing multi-component systems generates strong demands to accelerate tools in one way or the other.

16.2 CONCEPTS: SYNERGY, TAILORED SYSTEMS, AND RESIDUE DESIGN

In practice, several combinations showing superposition or even slightly antagonistic combinations are used with commercial success. They are prolifically applied to balance or optimize property profiles, including cost efficiency. However, the dominant concept for multi-component flame retardant is synergy. The flame-retardant effect achieved by the combination of additives or measures is greater than the simple sum of the effects (superposition) caused by the individual contributions (Figure 16.2a). Thus, using synergistic combinations is the most effective route to reduce the content of flame retardant needed. The term "synergy" is not synonym for any kind of "improvement," but the concept of synergy in flame retardancy has been well defined and reviewed several times at length over recent decades.[36,37,40,59,60] Synergism is defined and thus observed in the measure of a distinct fire performance and can be due to quite different physical and/or chemical phenomena. There are specific molecular effects that cause synergy, for instance, chemical reactions which are possible only in the multi-component system, such as the formation of metal phosphates remaining in the residue rather than the release of phosphorus into the gas phase. In other systems, material properties that control the response in distinct fire tests are optimized. Efficiency is increased due to chemical interactions or physical phenomena that adjust key parameters such as the decomposition temperature range, dispersion, or viscosity of the melt. Enhancing flame-retardant modes of action play a role by mechanically reinforcing fire residues or by closing the surface of the fire residue through glass formation. Indeed, many synergies are exploited in today's flame-retardant polymeric products: halogen-containing flame retardant + antimony oxide, metal hydroxides + synergists, all kinds of phosphorus–nitrogen combinations, flame retardant + anti-dripping agent, flame inhibition working in the gas phase + free radical generators causing enhanced decomposition/melt flow (retreat effect), halogen-phosphorus combinations, and so on. Consequently, several groups of additives are used regularly as adjuvants or synergists to achieve synergy in performance, such as zinc additives (zinc borates, zinc sulfide, zinc oxide, zinc stannate, zinc hydroxy stannate), boron substances, metal/metal oxides, silica/silicone/siloxane/silicate fillers, sulfur compounds, and poly(tetrafluoroethylene) fibers.[41,61–63]

Further, combining condensed phase flame-retardant modes of action and gas phase flame-retardant modes of action is concluded to be a general route to synergy.[60,64] One model to describe the burning of plastics are two strongly coupled chemical reactions: exothermic reaction with oxygen (combustion) in the flame heating the condensed phase and the endothermic decomposition (pyrolysis) in the condensed phase feeding the gas phase with fuel.[65] In this model, the synergistic interaction of gas phase and condensed phase is an intrinsic feature of the burning. What is more, the use of systems exploiting "pseudo-synergy" is prevalent.[40] Although no additional synergistic mechanism occurs, the efficiency of flame retardancy is crucially increased for a combination of flame retardants just by optimizing the concentration and mixing ratio when the single flame retardants show non-linear behavior as their concentrations are increased. This effect becomes most obvious when systems are compared in which the total amount of flame retardant is fixed at a certain level (Figure 16.2b). An adjusted mixture can outperform one with the same amount of flame retardants. Synergism is appraised to go along with surprising performance, and thus a bargain to get a patent exploiting the great potential for commercial systems.

A second general concept is the tailored multi-component system. The main flame-retardant modes of action such as charring, protection layer formation/intumescence, and

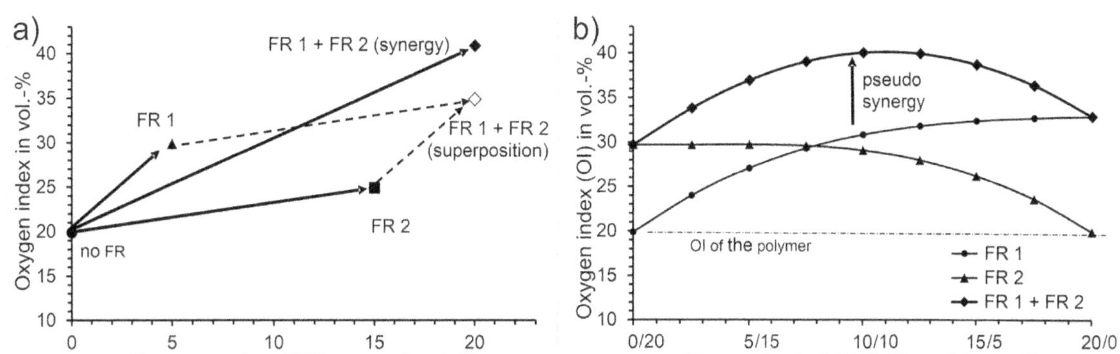

FIGURE 16.2 (a) Synergy based on an additional mechanism so that the performance of the mixture outperforms the linear superposition of the individual contributions, and (b) pseudo synergy = the superposition of the individual non-linear contribution outperforms the linear mixing rule, for a constant overall flame-retardant content (FR1 + FR2 = const).

flame inhibition originate from specific chemical reactions during pyrolysis. The efficiency of flame retardancy can be enhanced by increasing the reactivity and yield through adjustments to the systems. Corresponding decomposition temperature ranges of the flame retardant and the polymer are a key criterion for selecting promising flame retardants, not only for efficient flame inhibition, but also for charring.[66,67] Further, multi-component mixtures can be used to shift the decomposition temperature range or to broaden it.[68] Synergists like metal oxides are reported to change the reactivity of chemical reactions in the condensed phase due to catalysis, salt formation, or stabilization, or to change the conditions for chemical reactions, such as the pH value.[41,69]

To optimizing the performance profile, often multi-component flame retardant systems are used to realize multifunctional performance in the polymeric material of interest. Flame retardants are combined with smoke suppressant to reduce flammability and smoke production.[41,63,70] Carbon fillers are proposed to achieved flame retardancy and electric/thermal conductivity. Several fibers, fillers such as carbon black or talc, and nanoparticles are applied or proposed to enhance both flame retardancy and mechanical properties.

In many flame-retardant polymers, it turns out that synergistic systems are based on fire residue design. Charring itself, or in other words, reducing the amount of carbon-based fuel released by forming a carbonaceous residue, is often only half the story. Adjuvant or synergistic combinations are used to improve the protective properties of the fire residue. These protective properties are controlled by the properties of the fire residue, not by its amount. Most important fire residue properties are strongly linked to the morphology of the residue, such as closed surface, multicellular or other thermal insulating structures (Figure 16.3), and the thermal and mechanical stability of the residue. Charring aims to reduce the total mass loss and thus the fire load = total heat evolved (THE) (Equation 2) in a fire test as well as the mass loss rate and heat release rate (Equation 2) via the factor (1-μ), where μ is the char yield between 0 and 1.[71–73] The

$$THE \sim \chi h_c^0 (1-\mu) m_0 \tag{16.1}$$

and the

$$HRR(t) \sim \theta(t) \chi h_c^0 (1-\mu) \tag{16.2}$$

with the effective heat of combustion of the volatiles h_c^0, and the combustion efficiency of the flame χ. At first glance, protective layer properties reduce only the rates, such as mass loss rate, heat release rate (HRR), or the peak of heat release rate (PHRR) by a factor θ(t) without changing the fire load. The total heat evolved (total heat release, THR, at flameout or end of test = total heat evolved, THE) remain the same. The sample burns with a reduced intensity for a longer time. Nanocomposites such as polyolefins/layered silicate are good examples illustrating this description.[74] 5–10 wt% nanoparticles, well dispersed in non-charring polyolefins, switch the systems from burning without any relevant residue to materials that form a protective residual layer. The PHRR can be reduced by 50–70%, whereas the reduction in THE is restricted to 5–10%.[75] In many systems, residue design to improve the residual protective layer, combined with other flame-retardant modes of action such as fuel dilution, cooling, charring, or flame inhibition, is a most powerful approach to optimize efficient flame-retardant systems. General residue design approaches for creating an enhanced protective layer entail inorganic-organic, phosphorous-carbonaceous, or nitrogenous carbonaceous fire residues with better thermal stability. Mechanically reinforced chars, crusts, and glassy coatings provide for closed surfaces. Moreover, the better the protective layer, the greater the chance that the mass loss rate or HRR of a burning specimen falls below the critical value for self-extinguishing. The burning stops before all of the material is pyrolyzed. All kinds of intumescent systems

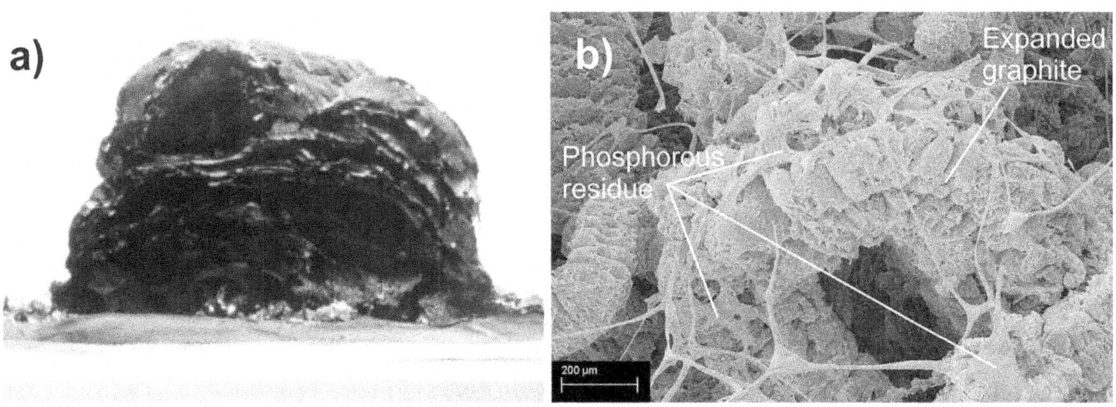

FIGURE 16.3 Fire residue design: (a) intumescence forming a multicellular, thermally insulating char (cross-section by cutting the fire residue in two halves),[88] residue 11 wt%, volume increase > 10 times, with a closed surface and (b) scanning electron microscopy image proposed to illustrate the stabilization of the residue of expandable graphite by the residue of a phosphorous flame retardant as synergist.[11]

are based on multi-component systems that incorporate the residue design of thermally insulating morphologies. (See Chapter 7 of this book). Usually nonflammable gases are released, some of which are used to form a multicellular carbonaceous or silicate layer structure.[76] Intumescent systems are complex multi-component systems often consisting of several ingredients such as an acid source, a charring agent, a blowing agent, and several adjuvants and used in a polymer matrix such as polyolefins or with a binder in intumescent coatings.[14,77,78] Several chemical and physical phenomena act together, and most of them are activated thermally. There is plenty of room to optimize the reactivity and yield of the chemical reactions as well as to adjust the viscosity and optimize the residue properties and morphology to produce clearly improved thermal insulation.[14,79–83] It should be noted that efficient thermal insulation is delivered not only by foam-like multicellular carbonaceous systems, but also by inorganic sponges, namely, layered structures such as those yielded by expandable graphite on the microscopic level, but also other layered structures on the scale of delaminating long-fiber-reinforced composites or multi-component laminates.[84–87]

It should not escape our notice that synergistic phenomena are usually of a specific nature, and their specific nature may yield considerable consequences. Synergists that adjust physical properties, for instances nanoparticles changing the viscosity of the pyrolyzing melt during intumescence, may work only at a distinct concentration in a distinct matrix; the same additive can be a synergist in one system and strongly antagonistic in another.[89] Synergism originating from chemical reactions may be dependent on factors such as temperature, reactivity, or residence times. The fruitful design of residues may be very sensitive to a complex interplay of morphology and properties;[90] thus synergy and antagonism may be close neighbors for the same combination of materials. Further, whatever the origin of synergism in flame retardancy is, it is defined by the performance in a specific fire test. Consequently, the synergy can differ strongly in different fire tests that check different fire properties in different fire scenarios, such as ignitability, flame spread, or fire load.

16.3 FIBERS, FILLERS, NANOPARTICLES

The fire behavior of fiber-reinforced composites clearly differs from one of their corresponding polymer matrix.[4,91–95] Even though glass fibers and carbon fibers are rather inert with respect to the mainly anaerobic pyrolysis taking place during burning, they strongly influence flammability. They change heat absorption, thermal conduction in the condensed phase, and convective heat transfer in the melt, so that the changed heating behavior of the materials influences ignition and flame spread.[96,97] Composites show a dramatically changed melt flow and dripping of pyrolyzing melts during burning; in some setups wicking (transportation of fuel due to capillary forces) may even occur.[98,99] Performance in several fire tests is changed significantly, with one notable example being how composites can affect the classification in UL 94, a reaction to small flame test. The amount of fire residue of a composite increases dramatically; in other words, the fire load decreases due to fuel replacement by inert fiber. The morphology and properties of the fire residue are changed. In sum, flame retardancy concepts tailored to composites are needed. Furthermore, tasks that are specific for composites, such as structural integrity in fire, come into focus.[100–102] Achieving both the demanded flame retardancy and superior mechanical properties is a further intrinsic challenge that demands outstanding solutions, sometimes even thinking-out-of-the-box approaches such as phase-selective filled multi-component laminates.[103]

Halogen-free flame-retarded short-fiber-reinforced technical polymers such as polyesters and polyamides play a major role in electrical engineering. Short glass fibers are typically used in the range of 20–40 wt%. According to their higher density of around 2.6 g/cm^3, the vol% of the filling is less than half of the wt%; the fire load of a test specimen is reduced to just 8–15%. The increased viscosity of the melt impedes any retreat effect, thus more material is trapped in the pyrolysis zone; wicking may also occur. Usually, poorer performance was observed in fire tests monitoring the reaction to small flame, such as UL 94 and Limiting Oxygen Index (LOI) or glow wire testing. Reduced melt flow and reduced dripping worsen the performance in other tests as well, including for PP-GF in comparison to PP in FMVSS 302. In other fire tests, such as in the cone calorimeter, short fibers contribute crucially to the amount of fire residue; in non-charring systems nearly the entire polymer matrix is pyrolyzed, with the residue consisting of 80–97 wt% glass fiber. The remaining fiber mat changes burning behavior, as a PHHR occurs at the beginning of burning before the residue drastically reduces the HRR. The PHRR is drastically reduced. In charring systems the fibers increase not only the amount of residue but also may improve its properties. Successful flame-retardant systems combine flame inhibition with a relevant protective layer, based, for instance, on a mechanically stabilized glass fiber mat functioning as barrier for heat;[104] alternatively, they target residue design in the condensed phase, for instance, using the glass fiber to stabilize thermally insulating morphologies.[12]

Long-fiber-reinforced materials are usually characterized by higher filling grades of around 60–70 vol%, anisotropic properties, and laminate behavior. The reduction in fire load becomes a major effect as well as the increase in residue. Controlled delamination of single layers can be used as a thermally insulating structure, with woven fiber resulting in a superior residue.[105] In terms of the morphology and properties of the fire residue, the mats of long fibers become dominant. The tight fiber layers are reported to be so dense that the release of gas is disrupted, temporally and spatially. Such a disruptive gas release together with flame inhibition can yield to phenomenon in UL 94 called the "blowing out" effect, where the pyrolysis gases are released as a fuel-rich/oxygen starved jet stream to "blow out" the flame.[106–108] Successful flame retardants usually combine flame inhibition with minor charring accompanied by the formation of a protective layer, ideally one that has been optimized.[109–111]

Large amounts of fibers or fillers are used to replace fuel and thus reduce the fire load. This strategy is sometimes used for polyolefins with char yields close to nil and high effective heat of combustions (39–44 MJ/kg). The combination of flame retardants with fibers/fillers that reduce fire load is crucial to enabling a material to achieve the demanded flame retardancy, such as a V0 rating in UL 94. Further, there are materials such as sheet molding composites (SMC) and bulk molding compounds (BMC), but also other materials in which only a rather small amount of polymeric binder is used to glue incombustible fillers together; such materials may even approach a classification as incombustible when glass fibers and metal hydrates are combined to filling grades most probably > 90 wt%. SMC and BMC systems with a Limiting Oxygen Index of 80 vol% and even approaching 100 vol% have been commercialized.[112–115]

There is increasing demand for flame-retardant natural-fiber-reinforced polymers [116–118] and biocomposites in the transportation sector,[119,120] but also in building construction.[121] Natural fibers are not inert under the pyrolytic conditions of a burning specimen. Nevertheless, since they usually show charring, and their effective heat of combustion is typically between 12–19 MJ/kg, they cause relevant fuel dilution in the condensed phases as well as in the gas phase, since their heat of combustion is crucially lower than most of the commodity and technical plastics. Tailored flame retardancy attempts to use them as charring agents and the fiber residue to reinforce the char. Flame-retardant coatings and soaking fibers have been reported.[122–124] In biocomposites, synergistic combinations that remain compostable are demanded,[125] and in the future, increasingly sustainable solutions that utilize industrial waste.[126–128]

Inorganic fillers are commonly used in a wide range of thermoplastics for mechanical reinforcement. Their impact on fire behavior is often significant and sometimes relevant.[129] They are often inert fillers with respect to decomposition but change heat capacity, replace fuel, and contribute to the amount and properties of the residue. Some show sintering (fusing into glassy/ceramic residues) or relevant mechanical reinforcement of the char. Some inorganic fillers release water or CO_2 when heated to higher temperatures, and nearly all of them change the viscosity of melts, melt flow, and dripping behavior in flammability tests. Talc is one of the most typical substances in this class. Originally used for mechanical reinforcement, talc has a crucial influence on fire behavior in most systems; combinations of talc and phosphorous flame retardant can show synergistic performance.[41,130] Melt flow and dripping behavior are strongly hindered, fire residue mechanically reinforced, and in some systems chemical interactions are reported.[131,132] The lamellar behavior of talc is believed to form barriers.[133,134]

Most polymers in most fire tests burn based on anaerobic pyrolysis feeding the flame with volatile fuel. Combustion is the exothermic reaction with oxygen taking place rather solely in the flame. Thus, all kinds of carbon fibers and fillers function as inert fillers during burning,[135–137] due to their thermal stability in anaerobe conditions. Carbon black is used in high amounts in various elastomers; however, the impact on fire properties is reported to be limited.[138] Some years ago carbon nanotubes were introduced in flame retardancy applications.[139–142] Like other anisotropic nanoparticles, well dispersed carbon nanotubes change the melt viscosity, usually preventing dripping in UL 94. The dispersion of nanoparticles in non-charring polymers works as prerequisite for forming a protective fire residue layer.[143] Burning behavior is changed, dramatically reducing the PHRR in the cone calorimeter. In charring polymers, nanoparticles can improve the properties of fire residues. More recently graphene,[136,144] graphene oxides, functionalized graphene, and functionalized graphene oxide have been proposed as nanofillers with great potential in flame retardancy applications.[145–147] Hardly any of these graphene materials consist of single graphene sheets; instead, they are used as few-layer or multi-layer graphene, with stacks of 5 to 20 graphene layers, comprising extraordinary nanometer-thick plates with an impressive aspect ratio. Well dispersed, these graphene nanofillers are believed to outperform nanotubes in performance and should become available at a lower price in the future. Carbon nanoparticles usually show a relevant impact on fire behavior in terms of maximum heat release rate and are reported to exhibit flame-retardant effects; their use as synergists is proposed in combination with flame retardants. Their filler character is usually obvious through a limited reduction in fire load.[148,149] Although carbon nanoparticles alone are not flame retardants, when cleverly functionalized or used in combination with flame retardants they can be impressive multifunctional adjuvants and synergists with respect to mechanical properties, thermal and electrical conductivity, and flame retardancy.[150–154]

Well dispersed all kinds of anisotropic inorganic nanoparticles are proposed as a synergist for flame retardancy, such as layered silicates,[155] oligomeric silsesquioxane,[156–158] layered double hydroxides,[159] sepiolite,[160] halloysite,[161,162] boron nitride,[163] zirconium phosphate,[164] or boehmite nanocrystals,[165] as they change the melt viscosity considerably; for instance, dripping in UL 94 is usually prevented. The dispersion of the nanoparticles in non-charring polymers works as a prerequisite for forming residual protective layers.[166–168] Burning behavior is changed; in the cone calorimeter the PHRR of non-charring polymers is reduced dramatically.[74,169,170] In some charring polymers and intumescent systems, nanoparticles may improve the properties of the fire residues.[171,172] Nanocomposites exhibit a clear nanoscience character and outperform the analogous microcomposites by far. Although they are proposed as a revolutionary new approach for flame retardancy,[173] here they are discussed also as fillers, in keeping with their inert filler character with respect to fire load, time to ignition, and reaction to small flame. Nevertheless, nanoparticles often show a distinct impact on fire behavior; their flame-retardant effects are apparent, and they are successfully implemented as synergists in combination with flame retardants.[174–179] Although nanoparticles alone do not provide sufficient flame retardancy, and combinations are not beneficial in every application, in some multi-component systems nanoparticles are applied as impressive multifunctional synergists with respect to mechanical properties and flame retardancy.[180] Although proposed right at the start of nanocomposite development in the second half of the 1990s, the combination with phosphorous flame retardants has generated all kinds of outcomes: commercial systems, innovative ideas, and deleterious combinations.[89,181–184] At the beginning of our century, the combination of layered silicate and ATH was commercialized.[185,186] This combination enabled the reduction of filler content from 65 wt% to < 55 wt% in flame retardant cable sheet materials.

16.4 ADDITIVES, ADJUVANTS, SYNERGISTS

Most of the commercially successful multi-component systems applying well-established adjuvants are also strongly synergistic. Indeed, several of the groups of additives discussed contain examples for adjuvants and synergists with respect to flame retardancy. The performance of some of the fillers and nanoparticles mentioned, as well as most of the examples of adjuvants and nearly all synergists discussed is quite material-specific; in other words, the same additive or combination of groups may work synergistically in one system, while their effectivity might be limited in another. Several classes of additives contain crossover materials and include examples of typical fillers, adjuvants, and synergists. Thus, discussing groups of additives under the terms "adjuvants," synergists," or in the previous paragraph under "fillers" is somewhat arbitrary. Nevertheless, I have tried to discuss each group only once in detail to avoid redundancy, summarizing additives that mainly add or enable flame-retardant modes of action under "adjuvants," and additives that mainly optimize a flame-retardancy function under the term "synergists."

Melt flow and dripping are crucial properties that influence the response of test specimens in several of the most important fire tests for polymeric materials. When a small flame or a glowing wire is applied, removing fuel and heat by melt flow and dripping can result in efficient extinguishing.[187] There are not only V-2 and V-0 thin foils and fibers, but also several successful bulk materials on the market with V-0 non-flaming dripping behavior in the UL 94 test. Also, in other mandatory fire tests using a small burner, such as for building products or in the automotive industry, what is sometimes called the "retreat" effect is used to pass. Drip promoting agents are often used as adjuvants, with flame retardants causing flame inhibition or strong fuel dilution.[188–191] The combination of adjuvants enhancing dripping and flame inhibition is a very old idea that was applied to halogenated systems,[43] the popularity of which is increasing in phosphorus-based multi-component systems. The most successful dripping agents are radical generators in the condensed phase. They enhance early decomposition reactions yielding chain scission, thus changing the intermediate product balance and increasing the fraction of low-viscosity liquid. The mass loss rate in thermogravimetry does not necessarily show any relevant change, but the viscosity of the pyrolyzing melt does. The impact of the retreat effect differs crucial from fire test to fire test, tests such as the cone calorimeter may show hardly any relevant change, whereas the UL 94 classification or LOI changes dramatically. Additives based on N-alkoxy amine, with a chemical structure closely related to antioxidants,[192,193] particularly Flamestab ®NOR 116 a monomeric N-alkoxy hindered amine stabilizer, have been widely introduced as dripping agents in not only halogen-containing flame retardants, but also in several halogen-free flame-retardant combinations. Cyclic dimethyl spirophosphonate AFLAMMIT® PCO 900, for instance, shows a synergistic effect with N-alkoxy hindered amines like Flamestab® NOR 116.[194–197] Several new systems are proposed, such as symmetrical and asymmetrical azoalkanes, along with related azoxy, hydrazone, and azine derivatives,[198,199] disulfide,[200] sulfenamides,[201,202] and oxyimide[203,204] systems. Recently, ever more combinations have been proposed, some of which show synergistic effects that demand more complex modes of action, such as additional activity in the gas phase. See Chapter 11 of this book for more about these drip promoting agents.

In most of the flame-retardant technical thermoplastics, such as polyamides, polyesters, or polycarbonate blends, dripping must be avoided to achieve a UL 94 V-0 non-dripping classification. Sometimes enhanced charring and cross-linking prevent dripping in fire tests,[205–209] sometimes inorganic fillers[210] such as talc[130] prevent dripping, and more recently nanocomposites have been proposed.[45,74,211] In a multitude of flame-retarded thermoplastics, 0.1–0.8 wt% microfibrillated polytetrafluoroethylene (PTFE) is added as an anti-dripping agent.[40,41] In some systems where the flame retardant also works as plasticizer, combination with an anti-dripping agent becomes essential to enable the flame retardant to work in the condensed phase.[212,213] Anti-dripping agents such as microfibrillated polytetrafluoroethylene (PTFE) provide quite clever tailored solutions. The plasticizing effect, or generally low viscosity due to shear thinning, is often desired during the processing of thermoplastics, such as during extrusion and injection molding. Thus, anti-dripping agents introduce flow limits in the pyrolyzing melts at low shear rates to prevent dripping without increasing the low viscosity of the melt needed for the high shear rates at the elevated temperatures typical for processing.[213]

One of the most important classes of adjuvant is charring agents. Some flame retardants are typically used together with charring agents or adjuvants that combine charring and blowing properties, in particular ammonium polyphosphate

(APP) in polyolefins.[41] In some systems that do not provide any interactions with phosphoric acid, without a charring agent, fire residue would consist only of phosphate, not phosphate-stabilized carbonaceous char. Of course, the combinations of phosphorous flame retardant and charring agent in these systems are strongly synergistic. One of the standard charring agents used with APP is pentaerythritol (PER) and its derivatives.[214–216] Another charring agent is polyphenylether (PPE), formerly called polyphenyleneoxide (PPO), synthesized via self-condensation of 2,6-xylenol, which is processed as a blend with polystyrene to yield high-impact polystyrene HIPS/PPE.[41] Several sources propose adding novolac to thermoplastics,[217,218] some adding polyamides or polyimides.[219] Charring hyperbranched structures and dendrimers have been proposed as well as condensation products of melamine, morpholine, and piperazine.[63,220–223] Triazines and piperazine are proposed as charring groups in different molecules.[224–227] Following the trend toward replacing petrochemicals with natural renewable resources, bio-based charring agents are increasingly favored, such as lignin, vanillin, tannic acid, starch, cellulose, or cyclodextrin and their derivatives.[228–240]

Blowing agents such as melamine and additives containing melamine and its derivatives are a very successful class of additives, because these substances often provide manifold benefits in multi-component systems (see Chapters 7 and 10 of this book).[241,242] The conversion to nonvolatile char-enhancing products and the release of NH_3 spread over a certain temperature range generate much more than simple fuel dilution in the condensed and gas phases but make melamine the most important blowing agent in commercial intumescent systems. Melamine cyanurate and melamine salts are reported to combine fuel dilution with enhanced melt flow.[191,243,244]

There is a strong need to add smoke suppressants for applications in transportation, particularly because several flame retardants efficiently reduce ignitability, flame spread, and heat release via flame inhibition but intrinsically exhibit increases in yields of CO and smoke. In some systems the total smoke release is even increased under forced flaming conditions. Although poly(vinyl chloride) (PVC) is widely used for building products, the use of smoke suppressants in PVC is the most comprehensive, intensively studied, and best understood benchmark for applying smoke suppressants;[245,246] moreover, several smoke suppressants started their carrier in PVC and are now widely used in all kinds of materials. Metal hydrates are applied to reduce smoke production,[247,248] as well as zinc borate, zinc stannate, all kinds of catalytic materials such as metal oxides,[249] and nanoparticles are also believed to harbor potential as a smoke suppressant.[70,163,250–252] The understanding of smoke suppression is often insufficient because the production of smoke is complex and controlled by both the pyrolysis of the polymeric material and the combustion of the volatiles in the flame. Both pyrolysis and combustion change under different fire scenarios such as ventilation, heat flux, and temperature, thus the smoke assessment depend strongly on the used test methods. The interactions reducing smoke production may be quite specific and differ from one system to the next. Catalysis and perhaps specific chemical reactions are also believed to play an important role in reducing aromatic volatiles by charring or reducing the size of volatiles through further decomposition.[253–255] As changing the residence time and temperature in the pyrolysis zone changes pyrolysis products, protective layers also influence smoke production.[256,257]

Several synergistic combinations with phosphorus are reported. The most popular rule of thumb is the phosphorus-nitrogen (P-N) synergy. However, as it is an oversimplification to call P and N flame-retardant elements, there is hardly a general P-N synergism. Specific phosphorous and nitrogenous groups are applied to realize specific chemical reactions (mechanisms) that cause flame-retardant modes of action, such as fuel dilution, charring, and flame inhibition. The impression of a general P-N synergism is most probably artificial due to the fact that so many different combinations of phosphorous and nitrogenous structures are successfully used as synergistic combinations; indeed, different phenomena contribute to the concept of P-N synergism. Phosphorous flame retardants enhance the charring of amides quite efficiently.[41] Amides are even proposed as charring agents for phosphorous flame retardants.[171] Several molecules containing N cause the endothermic formation and release of NH_3.[258,259] In flame-retardant combinations, nitrogenous groups often support phosphorous flame retardants crucially through fuel dilution in the gas phase,[260] the combination of melamine cyanurate with phosphorous flame retardants in thermoplastic polyurethane may be a good example.[41] Nitrogenous molecules such as melamine are widely used as the blowing agents in intumescent systems. For instance, combining ammonium polyphosphate with melamine is the synergistic combination of ingredients in most of the intumescent coatings used. P-N bonds or the formation of P-N bonds essentially provide for thermal stabilization of the char in charring systems.[261] Phosphoramidates and phosphazene are interesting classes of flame retardant similar to phosphine oxides, phosphinates, phosphonates, and phosphates.[63,240–266]

Although phosphorus-sulfur (P-S) synergy and phosphorus-silicone (P-Si) synergy are well-known, attempts to exploit these phenomena seem to be recently increased.[267–269] P-Si systems cover a wide range of performance. In some systems, whenever molecules containing Si were applied, they behaved like inert fillers, with a negligible influence on fire behavior. Further, in oligomeric or polymeric Si-containing substances such as silicone oils and polysiloxanes, degrees of branching and side groups control the pyrolysis into volatiles or into inorganic residue. Nevertheless, in some systems, the potential of inorganic-organic residues with superior protective layer properties are spawning synergistic solutions.[270–276] In several combinations the formation of silicophosphates is observed, and thus such combinations are proposed to be crucial for improving the stability of fire residue.[277] Special silicone synergists are under development to improve the barrier properties of the fire residues, for example, were proposed for application in glass-fiber-reinforced polyamides flame-retarded with aluminum diethylphosphinate, but they also work in natural-fiber-reinforced biocomposites, in synergism with aluminum diethylphosphinate.[278,279] Incidentally, the key flame-retardant mode of action of the successful nanocomposites is not the inorganic residue itself, but the formation of an inorganic-organic residual protective layer.[280,281] P-S synergism is obvious and impressive when the flame inhibition of a phosphorous flame retardant is combined with disulfides as a radical generator in the condensed phase, analogous to the combination of halogen and disulfides in polystyrene.[282] Analogously, sulfenamides are proposed as extremely efficient synergists in combination with phosphorous flame retardants in polyolefins.[201] However, sulfurous groups harbor the potential for several flame-retardant modes of action and synergistic impacts in combinations with phosphorous flame retardants.[283] Organo-/polysulfones are known to release SO_2, functioning to dilute fuel in the gas phase, while the remaining structure simultaneously contributes to charring.[284–287]

All kinds of boron compounds (see Chapter 9 of this book), including boric acid, boric oxide, and borates,[63,288–290] particularly zinc borate, are applied as synergists in multi-component systems.[291–293] Zinc borate is used together with metal hydrates such as ATH in EVA and other polymers to improve the protective layer properties of the fire residue,[294] but also together with phosphorous flame retardants. Borates release water, sinter, and form glasses. In combination with phosphorous flame retardants, the formation of borate networks and borophosphates are observed; the inorganic-organic fire residue is proposed to improve flame retardancy.[295,296] Improved protective residual layers are observed when zinc borate is combined with phosphorous flame retardants.[297] Melamine borates have also been proposed.[298]

Metal compounds, mainly metal oxides, are used as synergists in flame-retardant polymeric materials.[250,299–302] Although antimony oxide can cause cancer and should not be released into the environment, the combination of antimony oxide and halogen-containing flame retardant was and is one of the most successful synergistic flame-retardant combinations known and widely used in large amounts.[41] Compared to other metal oxides, the function of antimony oxides in combinations with halogen-containing flame retardants is not only outstanding but also unique. The formation of antimony halides particularly supports the flame inhibition, increasing efficiency dramatically (see Chapter 4 of this book). Sometimes metal oxides are multifunctional, improving thermal stability and flame retardancy. Zinc oxide of less than 1 wt% is often used in thermoplastics, such as in combination with red phosphorus in polyamides and polyesters.[41,69] However, all kinds of metal oxides harbor the potential to catalyze charring, for instance, in polyesters, or to enhance the condensed phase action of phosphorous flame retardant. These distinct effects can be enhanced further using nanoparticles with larger specific surfaces.[303–308] A large variety of metal compounds, such as oxides, acetates, acetyl acetonates, borates, and sulfates, are proposed as synergists in intumescent systems applied in polypropylene.[309] TiO_2 has become a standard ingredient in intumescent coatings and intumescent systems, due to its beneficial reaction with APP to form metal phosphates, stabilizing the intumescent residue as well as improving its properties.[310–313] Sophisticated, more complex structures such as metal oxide hybrids, metal complexes. and metal organic frameworks have been proposed more recently, however.[314,315]

All kinds of synergistic combinations of conventional flame retardants and nanoparticles are proposed.[316–319] Carbon and inorganic nanofillers were already discussed in this chapter. The few successful commercialized flame retardant nanocomposites synergy is exploited. The best example is the abovementioned combination of organically modified layered silicate with ATH, which enables a beneficial reduction in the amount of filler needed to achieve the demanded flame retardancy.[185,186] Layered plate-like nanoparticles, tube-shaped nanoparticles, zeolites, but also talc and fibers are reported to act as synergists in intumescent systems when the changed viscosity favors intumescence, and the particles stabilize the residue due to physical or chemical interactions.[23,83,320–330] Although usually reported as success, the beneficial use of nanoparticles in intumescent systems is somewhat tricky, because the viscosity of the pyrolysis melt must be adjusted, reinforcement can also effectively prevent successful intumescent char formation.

Glass-forming substances are proposed as synergists in flame-retardant multi-component systems. They are applied to optimize the properties of the fire residue by closing its surface. Borates and phosphates form glasses, too, so that glass formation is a phenomenon occurring as minor mode of action for several flame retardants. A few products are commercialized as glass-forming flame retardants, consisting of distinct mixtures of glass frits and fumed glass used as synergists.[41,331–333] The combination of melamine cyanurate and porous glass powder is proposed for unfilled nylon 6. It has been argued that adjusting the softening temperatures is the key to achieving efficient systems.[334–336]

Apart from the P-Si synergy, silicone, siloxane, fumed silica, and amorphous SiO_2 (see Chapter 8 of this book) have been proposed as synergists.[337–341] Yet additives containing Si exhibit quite a wide range of performance, from decomposing into volatiles, to inert filler character, to very efficient synergism. Side groups and branching make major differences with respect to decomposing in volatiles and residue formation. In some systems the potential of inorganic-organic residues with superior protective layer properties are spawning synergistic solutions. Sub-micrometer spherical particles of amorphous SiO_2 are used intensively as multifunctional synergist to ATH in the cable industry and in rubbers for conveyor belts.[342,343]

Metal hydrates (see Chapter 6 of this book) are economical, environmentally friendly, and reduce not only fire risks but also smoke production. Obviously, these are very good reasons to apply metal hydrates in polyolefins and more recently in biopolymers. Hereby, replacing fuel in polyolefins by high content of filler is a crucial part of the flame retardancy. Considering both the high flame-retardant efficiency demanded in polyolefins and the limited effectiveness of metal hydrates, filling content of over 60 wt% is necessary to achieve the demanded flame retardancy. With a density of 2.42 g/cm^3, over 25% of the fire load is thus reduced by fuel dilution in the condensed phase. What is more, as the high content of filler deteriorates the mechanical property profile of the materials, metal hydrates like ATH and magnesium hydrate are frequently used together with synergists to reduce the necessary amount of filler. Mixtures with other fillers or fibers are proposed, including layered silicate, nanotubes, amorphous silicon dioxide, silicones, silanes, metal borates, zinc stannate/zinc hydroxystannates, carbonates, diatomite, melamine, triazine, and charring agents.[7,8,344–349] Most of these multi-component systems show improved protective properties in the fire residue; closed layers capping entire surfaces and ceramification are identified to be vital. Modification and various coatings are proposed,[350] such as silane, melamine-formaldehyde resin, melamine/novolac resin, poly-methyl-methacrylate, and other sophisticated ideas.[351]

In practice what is proposed is often not the combination of one flame retardant and one adjuvant, or one flame retardant and one synergist, but multi-component systems consisting of flame retardants, fillers, adjuvants, and synergists.[352]

16.5 COMBINING DISTINCT FLAME RETARDANTS

There is proof of phosphorus-phosphorus (P-P) synergism.[40,353] The synergism is achieved mainly due to the combination of different flame-retardant modes of action in the gas and condensed phases, flame inhibition, and charring/protective layer formation, for instance, aryl phosphates with ammonium polyphosphate, or P-containing copolymers, polymeric P-containing flame retardants, phosphates, and phosphonates with phosphazene.[354–356] Aluminum diethylphosphinate is combined with melamine polyphosphate or ammonium polyphosphate[41] and aluminum hypophosphite with resorcinol diphosphate (RDP). Enhancing the charring in the condensed phase is usually controlled by specific chemical reactions between the partly decomposed polymer and the partly decomposed flame retardant.[66] Further phosphorous flame retardants are known to be sensitive to chemical interactions that shift their pyrolysis products from inorganic phosphates to phosphates, enhancing the amount and thermal stability of char or to the release of phosphorous volatiles. Thus, in several systems, not only simple superpositions of the effects of the different flame retardants occur but also "new" chemical interactions. In some systems these interactions contribute to synergistic performance or are the main reason for synergy. For instance, the combination of aluminum diethylphosphinate and melamine polyphosphate is reported to show heavy reactions between the flame retardants, improving the flame-retardant mode of action in the condensed phase.[357–359] A reaction between two phosphorous additives improving the effectiveness of the condensed phase has been observed in other systems as well.[360] In the case of arylphosphate-arylphosphate synergism in polycarbonate (PC) blends, the intermediate products exhibiting beneficial higher thermal stability are proposed to cause the synergy.[361]

As metal hydrates are economical, environmentally friendly, provide water for desired hydrolysis, reduce smoke, and provide inorganic residue, metal ions, and metal oxides, they harbor plenty of features that make them interesting for multi-component flame retardant mixtures. In combinations of metal hydrates and phosphorous flame retardants functioning mainly in the gas phase, the synergistic interaction of different flame-retardant modes of action is encouraging: release of phosphorus resulting in flame inhibition in the gas phase, release of water resulting in cooling and fuel dilution, and formation of an inorganic residue. However, metal hydrates also react with phosphorous flame retardants to form metal phosphates,[362] improving in some systems the protective residue properties, for instance in red phosphorus/$Mg(OH)_2$ or APP/$Mg(OH)_2$ combinations.[6,297,363,364] Due to its higher decomposition temperature, magnesium hydrate may exhibit this phenomenon much more strongly than ATH. The combination of ATH with a small amount of phosphorous flame retardant is also reported, for instance, ATH with RDP in thermoplastic polyurethane (TPU) for power and communication cables.[41,365] The combination of metal hydrate and a phosphorous flame retardant to decrease the filler content can also be proposed for other polymers such as EVA, as well as combination with the mineral huntite/hydromagnesite.[366,367] Combinations of higher content 30 wt% of boehmite or ATH with reasonably small amounts of reactive organophosphorus, 9,10-dihydro-9-oxo-10-phosphaphenanthrene-10-oxide (DOPO) derivatives, APP, aluminum diethylphosphinate, zinc diethylphosphinate, or hypophosphite are proposed for epoxy resins and polyester resins.[368–370] Although the combination of metal hydrates and phosphorous flame retardants may be prevalent, other combinations such as with expandable graphite have also been proposed.[371,372] The combination of a large amount of metal hydrate and other flame retardants and synergists is proposed for ceramifying systems.[373,374]

Expandable graphite itself is already an efficient flame retardant, often showing extraordinary performance in fire tests such as the cone calorimeter.[375] Yet in other fire tests, and sometimes even in the cone calorimeter, the fire residue is too fluffy and detaches from the sample. Expandable graphite is widely used in combination with phosphorous flame retardants; the combination is often proposed for biopolymers and wood plastic composites.[376–380] Massive synergy occurs, mainly due to the fixation of the expanded graphite protective layer, sometimes also due to the combination of the thermal insulating residue and flame inhibition.[381,382] The combination of expandable graphite and phosphorous flame retardants is a prevalent approach in hard and flexible polyurethane foams.[383–390] Literature reports on the combination of expandable graphite (EG) with additive and reactive phosphorous flame retardants.

There are also examples for combining three and more flame retardants, for instance expandable graphite, phosphorous flame retardant and metal hydrate, or expandable graphite and two phosphorous flame retardants.[372,391,392] The multi-component synergistic flame retardant system of melamine polyphosphate (MPP), melamine cyanurate (MC), and aluminum diethylphosphinate (AlPi) is proposed for thermoplastic polyurethane (TPU) as well as other similar combinations.[393,394]

Synergistic combinations are proposed which exploit different particle shapes or sizes and also broaden the decomposition temperature range. The latter is reported to be a key factor for the success of natural mixtures of huntite-hydromagnesite.[395] Mixtures of ATH, Mg(OH), and boehmite are proposed. With respect to protective layer formation, synergistic combinations of different filler sizes and shapes are reported. ATH with a bimodular size distribution is commercialized; the combination of layered silicate with carbon nanotubes is reported[396] as well as synergy in layered silicate and spherical amorphous SiO_2.[397]

Flame-retarded materials can be refined with an intumescent coating or gel coatings, but also with other non-intumescent flame retardant coatings.[398–408] A successful commercial flame retardant composite is reported to consist of fiber mats soaked with flame retardants, thermoset resins modified with flame retardant, and intumescent coatings applied on the surface.[409,410]

16.6 FLAME RETARDANTS WITH DIFFERENT FLAME-RETARDANT GROUPS

Although proposed as a new concept recently,[411] and spawning innovative developments of commercial flame retardants,[412–416] the combination of different flame retardant groups in one flame retardant is well established, such as in ammonium polyphosphate and melamine (poly)phosphate. Flame retardants containing halogen and phosphate such as tris(2-chloroethyl) phosphate and tris(1-chloro-2-propyl) phosphate are intensively used in hard and flexible polyurethane foams.[41] Their outstandingly high efficiency is believed to originate from a phosphorus-halogen synergy. In melamine polyphosphate or ammonium polyphosphate combinations, phosphorus working in the condensed phase is combined with nitrogenous groups causing fuel dilution in the gas phase or working as a blowing agent.[41] The P-N combination is established also for charring groups, for instance with bis(melamine) salt of pentaerythritol bis(acid phosphate).[416,417] The idea is to achieve intumescence with a single flame retardant. Replacing multi-component intumescent systems with one flame retardant functioning as the acid source, charring agent and blowing agent is indeed very interesting.[418–421] Several mono-component intumescent flame retardants have been proposed.[421–425] The idea of slightly upgrading phosphorous flame retardants by adding functional groups such as ethylenediamine to achieve more efficient intumescence is a very successful concept that has been proposed and used over the last four decades and discussed in recent papers.[355,356,426–428]

The functionalization of bio-based aromatic polymers with phosphorus, such as phosphorylated lignin, can be described as combining P with a charring agent.[329–434] Phosphorylated biopolymers are proposed not only for biopolymers such as PLA, but also for non-charring polyolefins. With natural phosphorous substances such as DNA or phytic acid, the door is open to completely bio-based intumescent flame retardants.[435–441] Phosphorylated vegetable oils are believed to combine flame inhibition and dripping agents in polyester.[442] 9,10-dihydro-9-oxa-10-phosphaphenanthrene-10-oxide—cellulose—acrylate has been proposed for transparent flame retardant coatings.[443]

Analogous to combining two different phosphorous flame retardants acting mainly in the gas and condensed phase, two different phosphorous groups can be combined in a single flame retardant. The combination of phosphine oxide or phosphinate with phosphonate or phosphate, respectively, harbors the potential for combining efficient flame inhibition and charring.[444–446] Several phosphine oxides and phosphinate groups, such as side groups consisting of DOPO derivatives, efficiently release phosphorus volatiles that generate radical trapping in the flame. Several phosphonate and phosphate groups are precursors for acids that catalyze and stabilize carbonaceous char.

The combination of phosphorus and sulfur is proposed.[447–450] The combination of P and N is widely used, not only in the form of melamine derivatives and salts, but also molecules containing triazine–phosphorus,[451–455] or structures based on P-N bonds such as phosphazene, phosphoramines, or phosphoramidates.[413,456–460] P-N combinations in flame retardants have always attracted much attention in academic and industrial research. Compounds containing P-N harbor benefits such as high thermal stability of the flame retardant or the fire residue, and potential for flame inhibition, fuel dilution, and charring.[268] Incorporating silicone or borate groups together with phosphorus were proposed.[461–464] Due to the higher costs of high performance matrices, for example, epoxy resins, all kinds of complex structures have been proposed.[322] Recently, molecules containing complex combinations of P with N, S, Si, and boron have been reported.[465–472]

More recently, sophisticated nanoparticles, such as hierarchical and functionalized carbon nanoparticles to promote them to full-fledged flame retardants have been proposed.[150,473–475] In particular, functionalized graphene and graphene oxides are proposed to work as flame retardants;[10,63,145,476–480] however, upgrading synergistic nanofillers to stand-alone flame retardants is also being proposed for other nanoparticles.[481]

16.7 SUMMARY

Although some mechanisms between distinct components and some interactions between phenomena are known, in practice, our detailed scientific understanding usually still falls short when it comes to complex and synergistic multi-component systems, crucial minor optimizations, or quantification in terms of specific fire properties. The description of the flame retardancy yielded by multi-component systems usually remains sketchy or fragmentary. Nevertheless, this chapter has attempted to summarize some of the relevant literature, provide thought-provoking impulses on how multi-component systems can be utilized, and direct the development of future flame-retardant polymer products.

REFERENCES

[1] B. Schartel. Multicomponent Flame Retardant. In: A. B. Morgan, Ed. Non-Halogenated Flame Retardant Handbook, 2nd Edition. Scriver Publishing, Beverly, 2022, chap. 9, 413–473.

[2] G. F. Levchik, S. V. Levchik, G. Camino, E. D. Weil. Fire Retardant Action of Red Phosphorus in Nylon 6. In: M. Le Bras, G. Camino, S. Bourbigot, R. Delobel, Eds. Fire Retardancy of Polymers the Use of Intumescence. The Royal Society of Chemistry, London, 1998, 304–315.

[3] Rabe, S.; Chuenban, Y.; Schartel, B. Exploring the Modes of Action of Phosphorus-Based Flame Retardants in Polymeric Systems. *Mater.* **2017**, *10*, 455. https://doi.org/10.3390/ma10050455

[4] Brehme, S.; Köppl, T.; Schartel, B.; Altstädt, V. Competition in Aluminium Phosphinate-Based Halogen-Free Flame Retardancy of Poly(Butylene Terephthalate) and its Glass-Fibre Composites. *e-Polym.* **2014**, *14*, 193–208. https://doi.org/10.1515/ epoly-2014-0029

[5] Langfeld, K.; Wilke, A.; Sut, A.; Greiser, S.; Ulmer, B.; Andrievici, V.; Limbach, P.; Bastian, M.; Schartel, B. Halogen-free Fire Retardant Styrene-Ethylene-Butylene-Styrene-Based Thermoplastic Elastomers Using Synergistic Aluminum Diethylphosphinate-Based Combinations. *J. Fire Sci.* **2015**, *33*, 157–177. https://doi.org/10.1177/0734904114565581

[6] Braun, U.; Schartel, B. Flame Retardant Mechanisms of Red Phosphorus and Magnesium Hydroxide in High Impact. *Polystyrene. Macromol. Chem. Phys.* **2004**, *205*, 2185–2196. https://doi.org/10.1002/macp.200400255

[7] W. E. Horn Jr. Inorganic Hydroxides and Hydroxycarbonates: Their Function and Use as Flame-Retardant Additives. In: A. F. Grand, C. A. Wilkie, Eds. Fire Retardancy of Polymeric Materials. Marcel Dekker, New York, 2000, chap. 9, 285–352.

[8] P. Hornsby. Fire-Retardant Fillers. In: C. A. Wilkie, A. B. Morgan, Eds. Fire Retardancy of Polymeric Materials, 2nd Edition. CRC Press, Boca Raton, FL, 2010, chap. 7, 163–185.

[9] Hull, T. R.; Witkowski, A.; Hollingbery, L. Fire Retardant Action of Mineral Fillers. *Polym. Degrad. Stab.* **2011**, *96*, 1462–1469. https://doi.org/10.1016/j.polymdegradstab.2011.05.006

[10] Wang, X.; Xing, W.; Feng, X.; Yu, B.; Song, L.; Hu, Y. Functionalization of Graphene with Grafted Polyphosphamide for Flame Retardant Epoxy Composites: Synthesis, Flammability and Mechanism. *Polym. Chem.* **2014**, *5*, 1145–1154. https://doi.org/10.1039/C3PY00963G

[11] Chan, Y. Y.; Schartel, B. It Takes Two to Tango: Synergistic Expandable Graphite–Phosphorus Flame Retardant Combinations in Polyurethane Foams. Polymers **2022**, *14*, 2562. https://doi.org/10.3390/polym14132562

[12] Braun, U.; Schartel, B.; Fichera, M. A.; Jäger, C. Flame Retardancy Mechanisms of Aluminium Phosphinate in Combination with Melamine Polyphosphate and Zinc Borate in Glass-Fibre Reinforced Polyamide 6,6. *Polym. Degrad. Stab.* **2007**, *92*, 1528–1545. https://doi.org/10.1016/j. polymdegradstab.2007.05.007

[13] Chan,Y. Y.; Schartel, B. It Takes Two to Tango: Industrial Benchmark PU-Foams with Expandable Graphite/P-Flame Retardant Combinations. *KGK*, **2022**, *75* (6), 39–46.

[14] Alongi, J.; Han, Z. D.; Bourbigot, S. Intumescence: Tradition Versus Novelty. A Comprehensive Review. *Prog. Polym. Sci.* **2015**, *51*, 28–73. https://doi.org/10.1016/j.progpolymsci.2015.04.010

[15] Bourbigot, S.; Le Bras, M.; Bréant, P.; Trémillon, J. M.; Delobel, R. Zeolites: New Synergistic Agents for Intumescent Fire Retardant Thermoplastic Formulations—Criteria for the Choice of the Zeolite. *Fire Mater.* **1996**, *20*, 145–154. https://doi.org/10.1002/(SICI)1099-1018(199605)20:3<145::AID-FAM569>3.0.CO;2-L

[16] Vannier, A.; Duquesne, S.; Bourbigot, S.; Castrovinci, A.; Camino, G.; Delobel, R. The Use of POSS as Synergist in Intumescent Recycledpoly(Ethylene Terephthalate). *Polym. Degrad. Stab.* **2008**, *93*, 818–826. https://doi.org/10.1016/j.polymdegradstab.2008.01.016

[17] Anna, P.; Marosi, G.; Csontos, I.; Bourbigot, S.; Le Bras, M.; Delobel, R. Influence of Modified Rheology on the Efficiency of Intumescent Flameretardant Systems. *Polym. Degrad. Stab.* **2001**, *74*, 423–426. https://doi.org/10.1016/S0141-3910(01)00180-X

[18] Qian, Y.; Wei, P.; Jiang, P.; Zhao, X.; Yu, H. Synthesis of a Novel Hybrid Synergistic Flame Retardant and Its Application in PP/IFR. *Polym. Degrad. Stab.* **2011**, *96*, 1134–1140. https://doi.org/10.1016/j.polymdegradstab.2011.02.017

[19] Liu, Y.; Wang, J. S.; Deng, C. L.; Wang, D. Y.; Song, Y. P.; Wang, Y.-Z. The Synergistic Flame-Retardant Effect of O-MMT on the Intumescent Flame-Retardant PP/CA/APP Systems. *Polym. Adv. Technol.* **2010**, *21*, 789–796. https://doi.org/10.1002/pat.1502

[20] Ren, Q.; Wan, C.; Zhang, Y.; Li, J. An Investigation Into Synergistic Effects of Rare Earth Oxides on Intumescent Flame Retardancy Of Polypropylene/Poly (Octylene-Co-Ethylene) Blends. *Polym. Adv. Technol.* **2011**, *22*, 1414–1421. https://doi.org/10.1002/pat.1624

[21] Lin, M.; Li, B.; Li, Q.; Li, S.; Zhang, S. Synergistic Effect of Metal Oxides on the Flame Retardancy and Thermal Degradation of Novel Intumescent Flame-Retardant Thermoplastic Polyurethanes. *J. Appl. Polym. Sci.* **2011**, *121*, 1951–6190. https://doi.org/10.1002/app.33759

[22] Wang, X.; Wu,L.; Li, J. A Study on the Performance of Intumescent Flame-Retarded Polypropylene with Nano-ZrO2. *J. Fire Sci.* **2011**, *29*, 227–242. https://doi.org/10.1177/0734904110388923

[23] Liu, Y.; Zhao, J.; Deng, C. L.; Chen, L.; Wang, D. Y.; Wang, Y.-Z. Flame-Retardant Effect of Sepiolite on an Intumescent Flame-Retardant Polypropylene System. *Ind. Eng. Chem. Res.* **2011**, *50*, 2047–2054. https://doi.org/10.1021/ie101737n

[24] Wang, J. S.; Liu, Y.; Zhao, H. B.; Liu, J.; Wang, D. Y.; Song, Y. P.; Wang, Y.-Z. Metal Compound-Enhanced Flame Retardancy of Intumescent Epoxy Resins Containing Ammonium Polyphosphate. *Polym. Degrad.*

Stab. **2009**, *94*, 625–631. https://doi.org/10.1016/j.polymdegradstab.2009.01.006

[25] W. Herwig, H. J. Kleiner, H. D. Sabel. Flammschutzmittel und ihre Verwendung zur Herstellung schwerentflammbarer Thermoplasten. European Patent EP 0006568 A1, 1980.

[26] H. J. Kleiner, W. Budzinsky, G. Kirsch. Low-flammability polyamide molding materials. US Patent US 5773556A, 1998.

[27] H. J. Kleiner, W. Budzinsky, G. Kirsch. Flameproofed polyester molding composition. US Patent US 5780534A, 1998.

[28] H. J. Kleiner, W. Budzinsky, G. Kirsch. Salts of phosphoric acids and use therof as flame retardants in plastics. US Patent US 5891226A, 1999.

[29] H. J. Kleiner. Process for preparing aluminum salts of dialkylphosphinic acids and diphosphinic acids. US Patent US 6194605B1, 2001.

[30] H. J. Kleiner. Phosphinic and aluminum salts. US Patent US 6211402 B1, 2001.

[31] E. Schlosser, B. Nass, W. Wanzke. Flame-retardant combination. US Patent US 6,255,371B1, 2001.

[32] S. Hörold. Flame-retarding thermosetting compositions. US Patent US 6420459B1, 2002.

[33] E. Jenewein, H. J. Kleiner, W. Wanzke, W. Budzinsky. Synergistic flame protection agent combination for thermoplastic polymers. US Patent US 6365071B1, 2002.

[34] M. Klatt, B. Leutner, M. Nam, H. Fisch. Flame-retardant polyester molding compositions containing flame retardant nitrogen compounds and diphosphinates. US Patent US 6503969B1, 2003.

[35] S. Knop, M. Sicken, S. Hoerold. Flammwidrige duroplastische Massen. European Patent EP 1403309 A1, 2004.

[36] Lewin, M. Synergistic and Catalytic Effects in Flame Retardancy of Polymeric Materials—An Overview. *J. Fire Sci.* **1999**, *17*, 3–19. https://doi.org/10.1177/073490419901700101

[37] Lewin, M. Synergism and Catalysis in Flame Retardancy of Polymers. *Polym. Adv. Technol.* **2001**, *12*, 215–222. https://doi.org/10.1002/pat.132

[38] Kaynak, C.; Isitman, N. A. Synergistic Fire Retardancy of Colemanite, a Natural Hydrated Calcium Borate, in High-Impact Polystyrene Containing Brominated Epoxy and Antimony Oxide. *Polym. Degrad. Stab.* **2011**, *96*, 798–807. https://doi.org/10.1016/j.polymdegradstab.2011.02.011

[39] Horrocks, A. R.; Smart, G.; Nazare, S.; Kandola, B.; Price, D. Quantification of Zinc Hydroxystannate** and Stannate** Synergies in Halogen-Containing Flame-Retardant Polymeric Formulations. *J. Fire Sci.* **2010**, *28*, 217–248. https://doi.org/10.1177/0734904109344302

[40] E. D. Weil. Synergists, Adjuvants, and Antagonists in Flame-Retardant Systems. In: A. F. Grand, C. A. Wilkie, Eds. Fire Retardancy of Polymeric Materials. Marcel Dekker, New York, 2000, chap. 4, 115–145.

[41] E. D. Weil, S. V. Levchik. Flame Retardants for Plastics and Textiles, 2nd Edition. Hanser, Munich, 2016.

[42] Weil, E. D.; Levchik, S. V. Flame Retardants in Commercial Use or Development for Polyolefins. *J. Fire Sci.* **2007**, *26*, 5–43, https://doi.org/10.1177/0734904107083309

[43] Eichhorn, J. Synergism of Free Radical Initiators with Self-Extinguishing Additives in Vinyl Aromatic Polymers. *J. Appl. Polym. Sci.* **1964**, *8*, 2497–2524. https://doi.org/10.1002/ app.1964.070080604

[44] R. E. Lyon, M. L. Janssens. Flammability. In: H. F. Mark, Ed. Encyclopedia of Polymer Science and Technology. Wiley, Hoboken, NJ, 2005.

[45] Wawrzyn, E.; Schartel, B.; Karrasch, A.; Jäger, C. Flame-retarded Bisphenol A Polycarbonate/Silicon Rubber/Bisphenol A Bis(Diphenyl Phosphate): Adding Inorganic Additives. *Polym. Degrad. Stab.* **2014**, *106*, 74–87. https://doi.org/10.1016/j. polymdegradstab.2013.08.006

[46] Yin, H.; Sypaseuth, F. D.; Schubert, M.; Schoch, R.; Bastian, M.; Schartel, B. Routes to Halogen-Free Flame-Retardant Polypropylene Wood Plastic Composites. *Polym. Adv. Technol.* *30*, **2019**, 187–202. https://doi.org/10.1002/pat.4458

[47] Sut, A.; Metzsch-Zilligen, E.; Großhauser, M.; Pfaendner, R.; Schartel, B. Rapid Mass Calorimeter as a High-Throughput Screening Method for the Development of Flame-Retarded TPU. *Polym. Degrad. Stab.* **2018**, *156*, 43–58. https://doi.org/10.1016/j.polymdegradstab.2018.08.004

[48] T. Eckel, P. Ooms, D. Wittmann, H.-J. Buysch. Thermoplastische Polymerformmassen mit flammwidrigen Eigenschaften. German Patent DE 42 31 774 A1, 1994.

[49] Gilman, J. W.; Bourbigot, S.; Shields, J. R.; Nyden, M.; Kashiwagi, T.; Davis, R. D.; Vanderhart, D. L.; Demory, W.; Wilkie, C. A.; Morgan, A. B.; Harris, J.; Lyon, R. E. High Throughput Methods for Polymer Nanocomposites Research: Extrusion, NMR Characterization and Flammability Property Screening. *J. Mater. Sci.* **2003**, *38*, 4451–4460. https://doi.org/10.1023/A:1027369115480

[50] Chigwada, G.; Wilkie, C. A. Synergy Between Conventional Phosphorus Fire Retardants and Organically-Modified Clays Can Lead to Fire Retardancy of Styrenics. *Polym. Degrad. Stab.* **2003**, *80*, 551–557. https://doi.org/10.1016/S0141-3910(03)00156-3

[51] Wilkie, C. A.; Chigwada, G.; Gilman, J. W.; Lyon, R. E. High-Throughput Techniques for the Evaluation of Fire Retardancy. *J. Mater. Chem.* **2006**, *16*, 2023–2029. https://doi.org/10.1039/B514598H

[52] J. W. Gilman, L. D. Brassell, R. D. Davis, J. R. Shields, R. H. Harris, D. Wentz, R. Reitz, A. Becker, C. Chi. Evaluation of Iron Carbonate as a Flame Retardant for Polyolefins. In: B. Schartel, Ed. Advances in the Flame Retardancy of Polymeric Materials: Current Perspectives Presented at FRPM'05. Books on Demand, Verlag, 2007, 131–140.

[53] R. D. Davis, R. E. Lyon, M. T. Takemori, N. Eidelman. High Throughput Techniques for Fire Resistant Materials Development. In: C. A. Wilkie, A. B. Morgan, Eds. Fire Retardancy of Polymeric Materials, 2nd edition. CRC Press, Boca Raton, FL, 2010, 421–451.

[54] Lyon, R. E.; Walters, R. N. Pyrolysis Combustion Flow Calorimetry. *J. Anal. Appl. Pyrolys.* **2004**, *71*, 27–46. https://doi.org/10.1016/S0165-2370(03)00096-2

[55] Lyon, R. E.; Walters, R. N.; Stoliarov, S. I. Screening Flame Retardants for Plastics Using Microscale Combustion Calorimetry. *Polym. Eng. Sci.* **2007**, *47*, 1501–1510. https://doi.org/10.1002/pen.20871

[56] Schartel, B.; Pawlowski, K. H.; Lyon, R. E. Pyrolysis Combustion Flow Calorimeter: A Tool to Assess Flame Retarded PC/ABS Materials? *Thermochim. Act.* **2007**, *462*, 1–14. https://doi.org/10.1016/j.tca.2007.05.021

[57] Rabe, S.; Schartel, B. The Rapid Mass Calorimeter: Understanding Reduced-Scale Fire Test Results. *Polym. Test.* **2017**, *57*, 165–174. https://doi.org/10.1016/j.polymertesting.2016.11.027

[58] Rabe, S.; Schartel, B. The Rapid Mass Calorimeter: A Route to High Throughput Fire Testing. *Fire Mater.* **2017**, *41*, 834–847. https://doi.org/10.1002/fam.2420

[59] E. D. Weil. Additivity, Synergism and Antagonism in Flame Retardancy. In: W. C. Kuryla, A. J. Papa, Eds. Flame

Retardancy of Polymeric Materials. Marcel Dekker, New York, 1975, 185–243.

[60] Schartel, B.; Wilkie, C. A.; Camino, G. Recommendations on the Scientific Approach to Polymer Flame Retardancy: Part 2—Concepts. *J. Fire Sci.* **2017**, *35*, 3–20. https://doi.org/10.1177/0734904116675370

[61] Kind, D. J.; Hull, T. R. A Review of Candidate Fire Retardants for Polyisoprene. *Polym. Degrad. Stab.* **2012**, *97*, 201–213. https://doi.org/10.1016/j.polymdegradstab.2011.12.008

[62] Kashiwagi, T.; Shields, J. R.; Harris, R. H.; Davis, R. D. Flame-Retardant Mechanism of Silica: Effects of Resin Molecular Weight. *J. Appl. Polym. Sci.* **2003**, *87*, 1541–1553. https://doi.org/10.1002/app.11967

[63] J. K. Fink. Flame Retardants. Scrivener Publishing and Wiley & Sons, Hoboken, NJ, 2020.

[64] Y. P. Khanna, E. M. Pearce. Synergism and Flame Retardancy. In: M. Lewin, S. M. Atlas, E. M. Pearce, Eds. Flame—Retardant Polymeric Materials, Vol. 2. Plenum Press, New York, 1978, 43–61.

[65] Camino, B.; Camino, G. The Chemical Kinetics of the Polymer Combustion Allows for Inherent Fire Retardant Synergism. *Polym. Degrad. Stab.* **2019**, *160*, 142–147. https://doi.org/10.1016/j.polymdegradstab.2018.12.018

[66] Schartel, B.; Perret, B.; Dittrich, B.; Ciesielski, M.; Krämer, J.; Müller, P.; Altstädt, V.; Zang, L.; Döring, M. Flame Retar dancy of Polymers: The Role of Specific Reactions in the Condensed Phase. *Macromol. Mater. Eng.* **2016**, *301*, 9–35. https://doi.org/10.1002/mame.201500250

[67] Perret, B.; Pawlowski, K. H.; Schartel, B. Fire Retardancy Mechanisms of Arylphosphates in Polycarbonate (PC) and PC/Acrylonitrile-Butadiene-Styrene. *J. Therm. Anal. Calorim.* **2009**, *97*, 949–958. https://doi.org/10.1007/s10973-009-0379-7

[68] Hollingbery, L. A.; Hull, T. R. The Fire Retardant Behaviour of Huntite and Hydromagnesite—A Review. *Polym. Degrad. Stab.* **2010**, *95*, 2213–2225. https://doi.org/10.1016/j.polymdegradstab.2010.08.019

[69] U. Braun, U. Knoll, D. Neubert, B. Schartel. Fire Retarding Mechanisms of Phosphorus in Polybutylenterephthalat (PBT) in Combination with Melamine Cyanurat or Zinc Oxide. In: B. Schartel, Ed. Advances in the Flame Retardancy of Polymeric Materials: Current Perspectives Presented at FRPM'05. Books on Demand, Norderstedt, 2007, 35–49.

[70] Liu, X.; Hao, J. W.; Gaan, S. Recent Studies on the Decomposition and Strategies of Smoke and Toxicity Suppression for Polyurethane Based Materials. *RSC Adv.* **2016**, *6*, 74742–74756. https://doi.org/10.1039/C6RA14345H

[71] R. E. Lyon. Plastics and Rubbers. In: C. A. Harper, Ed. Handbook of Building Materials for Fire Protection. McGraw-Hill, New York, 2004, chap. 3, 3.1–3.51.

[72] van Krevelen, D. W. Some Basic Aspects of Flame Resistance of Polymeric Materials. *Polymer* **1975**, *16*, 615–620. https://doi.org/10.1016/0032-3861(75)90157-3

[73] B. Schartel, K. Kebelmann. Fire Testing for the Development of Flame Retardant Polymeric Materials. In: Y. Hu, X. Wang, Eds. Flame Retardant Polymeric Materials A Handbook. CRC Press, Boca Raton, FL, 2020, chap. 3, 35–55.

[74] Schartel, B.; Bartholmai, M.; Knoll, U. Some Comments on the Main Fire Retardancy Mechanisms in Polymer Nanocomposites. *Polym. Adv. Technol.* **2006**, *17*, 772–777. https://doi.org/10.1002/pat.792

[75] Bartholmai, M.; Schartel, B. Layered Silicate Polymer Nanocomposites: New Approach or Illusion for Fire Retardancy? Investigations on the Potential and on the Tasks Using a Model System. *Polym. Adv. Technol.* **2004**, *15*, 355–364. https://doi.org/10.1002/pat.483

[76] Morys, M.; Illerhaus, B.; Sturm, H.; Schartel, B. Variation of Intumescent Coatings Revealing Different Modes of Action for Good Protection Performance. *Fire Technol.* **2017**, *53*, 1569–1587. https://doi.org/10.1007/s10694-017-0649-z

[77] Vandersall, H. L. Intumescent Coating System, Their Development and Chemistry. *J. Fire Flamm.* **1971**, *2*, 97–140.

[78] Weil, E. D. Fire-Protective and Flame-Retardant Coatings— A State-of-the-Art Review. *J. Fire Sci.* **2011**, *29*, 259–296. https://doi.org/10.1177/0734904110395469

[79] Morys, M.; Illerhaus, B.; Sturm, H.; Schartel, B. Revealing the Inner Secrets of Intumescence: Advanced Standard Time Temperature Oven (STT Mufu+)—μ-Computed Tomography Approach. *Fire Mater.* **2017**, *41*, 927–939. https://doi.org/10.1002/fam.2426

[80] Müller, P.; Morys, M.; Sut, A.; Jäger, C.; Illerhaus, B.; Schartel, B. Melamine Poly(Zincphosphate) as Flame Retardant in Epoxy Resin: Decomposition Pathways, Molecular Mechanisms and Morphology of Fire Residues. *Polym. Degrad. Stabil.* **2016**, *130*, 307–319. https://doi.org/10.1016/j.polymdegradstab.2016.06.023

[81] Muller, M.; Bourbigot, S.; Duquesne, S.; Klein, R.; Giannini, G.; Lindsay, C.; Vlassenbroeck, J. Investigation of the Synergy in Intumescent Polyurethane by 3D Computed Tomography. *Polym. Degrad. Stabil.* **2013**, *98*, 1638–1647. https://doi.org/10.1016/j.polymdegradstab.2013.06.018

[82] Horacek, H.; Pieh, S. The Importance of Intumescent Systems for Fire Protection of Plastic Materials. *Polym. Int.* **2000**, *49*, 1106–1114. https://doi.org/10.1002/1097-0126(200010)49:10<1106::AID-PI539>3.0.CO;2-I

[83] Yuan, B. H.; Fan, A.; Yang, M.; Chen, X. F.; Hu, Y.; Bao, C. L.; Jiang, S. H.; Niu, Y.; Zhang, Y.; He, S.; Dai, H. M. The Effects of Graphene on the Flammability and Fire Behavior of Intumescent Flame Retardant Polypropylene Composites at Different Flame Scenarios. *Polym. Degrad. Stab.* **2017**, *143*, 42–56. https://doi.org/10.1016/j.polymdegradstab.2017.06.015

[84] Duquesne, S.; Le Bras, M.; Bourbigot, S.; Delobel, R.; Vezin, H.; Camino, G.; Eling, B.; Lindsay, C.; Roels, T. Expandable Graphite: A Fire Retardant Additive for Polyurethane Coatings. *Fire Mater.* **2003**, *27*, 103–117. https://doi.org/10.1002/fam.812

[85] Parlevliet, P.; Geistbeck, M. Investigations into Lightweight Solutions for Epoxy Composite Fire Property Improvement. *Plast. Rubber Compos.* **2015**, *44*, 104–110. https://doi.org/10.1179/1743289815Y.0000000001

[86] Vogelesang, L. B.; Vlot, A. Development of Fibre Metal Laminates for Advanced Aerospace Structures. *J. Mater. Process. Technol.* **2000**, *103*, 1–5. https://doi.org/10.1016/S0924-0136(00)00411-8

[87] Sturm, H.; Schartel, B.; Weiß, A.; Braun, U. SEM/EDX: Advanced Investigation of Structured Fire Residues and Residue Formation. *Polym. Test.* **2012**, *31*, 606–619. https://doi.org/10.1016/j.polymertesting.2012.03.005

[88] Brehme, S.; Schartel, B.; Goebbels, J.; Fischer, O.; Pospiech, D.; Bykov, Y.; Döring, M. Phosphorus Polyester Versus Aluminium Phosphinate in Poly(Butylene Terephthalate) (PBT): Flame Retardancy Performance and Mechanisms. *Polym. Degrad. Stab.* **2011**, *96*, 875–884. https://doi.org/10.1016/j.polymdegradstab.2011.01.035

[89] Schartel, B.; Weiß, A.; Mohr, F.; Kleemeier, M.; Hartwig, A.; Braun, U. Flame Retarded Epoxy Resins by Adding Layered Silicate in Combination with the Conventional Protection-layer-Building Flame Retardants Melamine

Borate and Ammonium Polyphosphate. *J. Appl. Polym, Sci.* **2010**, *118*, 1134–1143. https://doi.org/10.1002/app.32512

[90] Bourbigot, S.; Sarazin, J.; Bensabath, T.; Samyn, F.; Jimenez, M. Intumescent Polypropylene: Reaction to Fire and Mechanistic Aspects. *Fire Safety J.* **2019**, *105*, 261–269. https://doi.org/10.1016/j.firesaf.2019.03.007

[91] A. R. Horrocks, B. K. Kandola. 9 — Flammability and Fire Resistance of Composites. In: A. C. Long, Ed. Design and Manufacture of Textile Composites. Woodhead Publishing Series in Textiles, Woodhead Publishing, Cambridge, 2005.

[92] Casu, A.; Camino, G.; De Giorgi, M.; Flath, D.; Laudi, A.; Morone, V. Effect of Glass Fibres and Fire Retardant on the Combustion Behaviour of Composites, Glass Fibres Poly(Butylene Terephthalate). *Fire Mater.* **1998**, *22*, 7–14. https://doi.org/10.1002/(SICI)1099-1018(199801/02)22:1<7::AID-FAM623>3.0.CO;2-3

[93] Casu, A.; Camino, G.; Luda, M. P.; Degiorgi, M. Mechanisms of Fire Retardantion in Glass-Fiber Polymer Composites. *Makromol. Chem. Macromol. Symp.* **1993**, *74*, 307–310. https://doi.org/10.1002/masy.19930740140

[94] Köppl, T.; Brehme, S.; Wolff-Fabris, F.; Altstädt, V.; Schartel, B.; Döring, M. Structure-Property Relationships of Halogen-Free Flame Retarded Poly(Butylene Terephthalate) (PBT) and Glass Fibre Reinforced PBT. *J. Appl. Polym. Sci.* **2012**, *124*, 9–18. https://doi.org/10.1002/app.34910

[95] Perret, B.; Schartel, B.; Stöß, K.; Ciesielski, M.; Diederichs, J.; Döring, M.; Krämer, J.; Altstädt, V. Novel DOPO-Based Flame Retardants in High-Performance Carbon Fibre Epoxy Composites for Aviation. *Eur. Polym. J.* **2011**, *47*, 1081–1089. https://doi.org/10.1016/j.eurpolymj.2011.02.008

[96] Eibl, S. Influence of Carbon Fibre Orientation on Reaction-to-fire Properties of Polymer Matrix Composites. *Fire Mater.* **2012**, *36*, 309–324. https://doi.org/10.1002/fam.1112

[97] Eibl, S.; Swanson, D. Influence of Out-of-Plane Fiber Orientation on Reaction-to-Fire Properties of Carbon Fiber Reinforced Polymer Matrix Composites. *Fire Mater.* **2018**, *42*, 234–243. https://doi.org/10.1002/fam.2484

[98] Ding, Y.; Swann, J. D.; Sun, Q.; Stoliarov, S. I.; Kraemer, R. H. Development of a Pyrolysis Model for Glass Fiber Reinforced Polyamide 66 Blended with Red Phosphorus: Relationship Between Flammability Behavior and Material Composition. *Compos. Part B-Eng.* **2019**, *176*, 107263. https://doi.org/10.1016/j.compositesb.2019.107263

[99] Ding, Y.; Stoliarov, S. I.; Kraemer, R. H. Pyrolysis Model Development for a Polymeric Material Containing Multiple Flame Retardants: Relationship Between Heat Release Rate and Material Composition. *Combust. Flame* **2019**, *202*, 43–57. https://doi.org/10.1016/j.combustflame.2019.01.003

[100] Feih, S.; Mathys, Z.; Gibson, A. G.; Mouritz, A. P. Modelling the Tension and Compression Strengths of Polymer Laminates in Fire. *Compos. Sci. Technol.* **2007**, *67*, 551–564. https://doi.org/10.1016/j.compscitech.2006.07.038

[101] A. P. Mouritz, A. G. Gibson. Fire Properties of Polymer Composite Materials. Book Series: Solid Mechanics and Its Applications, Vol. 143. Springer, Dordrecht, 2006, 1–401. Https://doi.org/10.1007/978-1-4020-5356-6

[102] Gibson, A. G.; Wright, P. N. H.; Wu, Y. S.; Mouritz, A. P.; Mathys, Z.; Gardiner, C. P. The Integrity of Polymer Composites During and After Fire. *J. Compos. Mater.* **2004**, *38*, 1283–1307. https://doi.org/10.1177/0021998304042733

[103] Gallo, E.; Schartel, B.; Acierno, D.; Cimino, F.; Russo, P. Tailoring the Flame Retardant and Mechanical Performances of Natural Fiber-Reinforced Biopolymer by Multi-Component Laminate. *Compos. B Eng.* **2013**, *44*, 112–119. https://doi.org/10.1016/j.compositesb.2012.07.005

[104] Braun, U.; Bahr, H.; Sturm, H.; Schartel, B.. Flame Retardancy Mechanisms of Metal Phosphinates and Metal Phosphinates in Combination with Melamine Cyanurate in Glass-Fiber Reinforced Poly(1,4-Butylene Terephthalate): The Influence of Metal Cation. *Polym. Adv. Technol.* **2008**, *19*, 680–692. https://doi.org/10.1002/pat.1147

[105] Eibl, S. Influence of Unwoven Roving and Woven Fabric Carbon Fiber Reinforcements on Reaction-to-Fire Properties of Polymer Matrix Composites. *Fire Mater.* **2020**, *44*, 557–572. https://doi.org/10.1002/fam.2815

[106] Zhang, W. C.; Li, X. M.; Yang, R. J. Blowing-Out Effect in Epoxy Composites Flame Retarded by DOPO-POSS and Its Correlation with Amide Curing Agents. *Polym. Degrad. Stab.* **2012**, *97*, 1314–1324. https://doi.org/10.1016/j.polymdegradstab.2012.05.020

[107] Zhang, W. C.; Li, X. M.; Li, L. M.; Yang, R. J. Study of the Synergistic Effect of Silicon and Phosphorus on the Blowing-Out Effect of Epoxy Resin Composites. *Polym. Degrad. Stab.* **2012**, *97*, 1041–1048. https://doi.org/10.1016/j.polymdegradstab.2012.03.008

[108] Zhang, W. C.; Li, X. M.; Yang, R. J. Blowing-Out Effect and Temperature Profile in Condensed Phase in Flame Retarding Epoxy Resins by Phosphorus-Containing Oligomeric Silsesquioxane. *Polym. Adv. Technol.* **2013**, *24*, 951–961. https://doi.org/10.1002/pat.3170

[109] Braun, U.; Balabanovich, A. I.; Schartel, B.; Knoll, U.; Artner, J.; Ciesielski, M.; Döring, M.; Perez, R.; Sandler, J. K. W.; Altstädt, V.; Hoffmann, T.; Pospiech, D. Influence of the Oxidation State of Phosphorus on the Decomposition and Fire Behaviour of Flame-Retarded Epoxy Resin Composites. Polymer **2006**, *47*, 8495–8508. https://doi.org/10.1016/j.polymer.2006.10.022

[110] Schartel, B.; Balabanovich, A. I.; Braun, U.; Knoll, U.; Artner, J.; Ciesielski, M.; Döring, M.; Perez, R.; Sandler, J. K. W.; Altstädt, V.; Hoffmann, T.; Pospiech, D. Pyrolysis of Epoxy Resins and Fire Behaviour of Epoxy Resin Composites Flame-Retarded with 9,10-Dihydro-9-Oxa-10-Phosphaphenanthrene-10-Oxide Additives. *J. Appl. Polym. Sci.* **2007**, *104*, 2260–2269. https://doi.org/10.1002/app.25660

[111] Perret, B.; Schartel, B.; Stöß, K.; Ciesielski, M.; Diederichs, J.; Döring, M.; Krämer, J.; Altstädt, V. A New Halogen-Free Flame Retardant Based on 9,10-Dihydro-9-Oxa-10-Phosphaphenanthrene-10-Oxide for Epoxy Resins and Their Carbon Fiber Composites for the Automotive and Aviation Industries. *Macromol. Mater. Eng.* **2011**, *296*, 14–30. https://doi.org/10.1002/mame.201000242

[112] P. Ooms, T. Lorenz. World Premiere: Flame Retardant Plastic BMC 0204 Sets New Standards. 25th Annual Conference on Recent Advances in Flame Retardancy of Polymeric Materials, May 19–21, 2014.

[113] T. Lorenz, Lomix BMC 0204—neues flammenfestes BMC für den weltweiten Einsatz. 2. AVK-Fachtagung—„Flammschutz bei Composites-Anwendungen"10th December 2013, Frankfurt.

[114] Duroplast SMC 0208 für die Pressverarbeitung, Produktmeldung 12.05.2016, article-253228, Kunststoffe.de, Hanser, München 2016

[115] N. Stoess. Neue Generation flammgeschützter SMC-Materialien. 2. AVK-Fachtagung—„Flammschutz bei Composites-Anwendungen"10th December 2013, Frankfurt.

[116] R. Sonnier, A. Taguet, L. Ferry, J. M. Lopez-Cuesta. Flame Retardancy of Natural Fibers Reinforced Composites In: R. Sonnier, A. Taguet, L. Ferry, J.-M. LopezCuesta, Eds. Towards Bio-Based Flame Retardant Polymers. Book Series:

Springer Briefs in Molecular Science, Springer, Cham, 2018, 73–98. Https://doi.org/10.1007/978-3-319-67083-6_3

[117] Chapple, S.; Anandjiwala, R. Flammability of Natural Fiber-reinforced Composites and Strategies for Fire Retardancy: A Review. *J. Thermoplast. Compos. Mater.* **2010**, *23*, 871–893. https://doi.org/10.1177/0892705709356338

[118] Schartel, B.; Braun, U.; Schwarz, U.; Reinemann, S. Fire Retardancy of Polypropylene/Flax Blends. Polymer **2003**, *44*, 6241–6250. https://doi.org/10.1016/S0032-3861(03)00692-X

[119] Kozłowski, R.; Władyka-Przybylak, M. Flammability and Fire Resistance of Composites Reinforced by Natural Fibers. *Polym. Adv. Technol.* **2008**, *19*, 446–453. https://doi.org/10.1002/pat.1135

[120] Shah, A. U. R.; Prabhakar, M. N.; Song, J.-I. Current Advances in the Fire Retardancy of Natural Fiber and Bio-Based Composites–A Review. *Int. J. Prec. Eng. Manuf. Green Technol.* **2017**, *4*, 247–262. https://doi.org/10.1007/s40684-017-0030-1

[121] Hapuarachchi, T. D.; Ren, G.; Fan, M.; Hogg, P. J.; Peijs, T. Fire Retardancy of Natural Fibre Reinforced Sheet Moulding Compound. *Appl. Compos. Mater.* **2007**, *14*, 251–264. https://doi.org/10.1007/s10443-007-9044-0

[122] Zhang, L.; Li, Z.; Pan, Y.-T.; Pérez Yáñez, A.; Hu, S.; Zhang, X.-Q.; Wang, R.; Wang, D.-Y. Polydopamine Induced Natural Fiber Surface Functionalization: A Way Towards Flame Retardancy of Flax/Poly(Lactic Acid) Biocomposites. *Compos. B Eng.* **2018**, *154*, 56–63. https://doi.org/10.1016/j.compositesb.2018.07.037

[123] Schirp, A.; Su, S. Effectiveness of Pre-Treated Wood Particles and Halogen-Free Flame Retardants Used in Wood-Plastic Composites. *Polym. Degrad. Stab.* **2016**, *126*, 81–92. https://doi.org/10.1016/j.polymdegradstab.2016.01.016

[124] Hamalainen, K.; Karki, T. Effects of Wood Flour Modification on the Fire Retardancy of Wood-Plastic Composites. *Eur. J. Wood Wood Prod.* **2014**, *72*, 703–711. https://doi.org/10.1007/s00107-014-0833-5

[125] Gallo, E.; Sánchez-Olivares, G.; Schartel, B. Flame Retarded Starch-Based Biocomposites—Aluminum Hydroxide-Coconut Fiber Synergy. *Polimery* **2013**, *58*, 395–402.

[126] Sain, M.; Park, S. H.; Suhara, F.; Law, S. Flame Retardant and Mechanical Properties of Natural Fibre–PP Composites Containing Magnesium Hydroxide. *Polym. Degrad. Stab.* **2004**, *83*, 363367. https://doi.org/10.1016/S0141-3910(03)00280-5

[127] Sanchez-Olivares, G.; Sanchez-Solis, A.; Calderas, F.; Alongi, J. Keratin Fibres Derived from Tannery Industry Wastes for Flame Retarded PLA Composites. *Polym. Degrad. Stab.* **2017**, *140*, 4254. https://doi.org/10.1016/j.polymdegradstab.2017.04.011

[128] Rabe, S.; Sánchez-Olivares, G.; Pérez-Chávez, R.; Schartel, B. Natural Keratin and Coconut Fibres from Industrial Wastes in Flame Retarded Thermoplastic Starch Biocomposites. *Materials* **2019**, *12*, 344. https://doi.org/10.3390/ma12030344

[129] Batistella, M. A.; Sonnier, R.; Otazaghine, B.; Petter, C. O.; Lopez-Cuesta, J.-M. Interactions Between Kaolinite and Phosphinate-Based Flame Retardant in Polyamide 6. *Appl. Clay Sci.* **2018**, *157*, 248256. https://doi.org/10.1016/j.clay.2018.02.021

[130] B. Schartel, K. H. Richter, M. Böhning. Synergistic Use of Talc in Halogen-free Flame Retarded Polycarbonate/Acrylonitrile-Butadiene-Styrene Blends. In: A. B. Morgan, C. A. Wilkie, G. L. Nelson, Eds. Fire and Polymers VI: New Advances in Flame Retardant Chemistry and Science. ACS Symposium Series 1118. ACS, Washington, DC, 2012, chap. 2, 15–36. Https://doi.org/10.1021/bk-2012-1118.ch002

[131] Levchik, S. V.; Levchik, G. F.; Camino, G.; Costa, L. Mechanism of Action of Phosphorus-Based Flame Retardants in Nylon-6. 2. Ammonium Polyphosphate Talc. *J. Fire Sci.* **1995**, *13*, 43–58. https://doi.org/10.1177/073490419501300103

[132] Laoutid, F.; Ferry, L.; Lopez-Cuesta, J.-M.; Crespy, A. Flame-retardant Action of Red Phosphorus/Magnesium Oxide and Red Phosphorus/Iron Oxide Compositions in Recycled PET. *Fire Mater.* **2006**, *30*, 343–358. https://doi.org/10.1002/fam.914

[133] Longerey, M.; Lopez-Cuesta, J.-M.; Gaudon, P.; Crespy, A. Talcs and Brominated Trimethylphenyl Indane/Sb2O3 Blend in a PP-PE Copolymer. *Polym. Degrad. Stab.* **1999**, *64*, 489–496. https://doi.org/10.1016/S0141-3910(98)00147-5

[134] Durin-France, A.; Ferry, L.; Lopez-Cuesta, J.-M.; Crespy, A. Magnesium Hydroxide/Zinc Borate/Talc Compositions as Flame-Retardants in EVA Copolymer. *Polym. Int.* **2000**, *49*, 1101–1105. https://doi.org/10.1002/1097-0126(200010)49:10<1101::AID-PI523>3.0.CO;2-5

[135] Wang, X.; Kalali, E. N.; Wan, J. T.; Wang, D. Y. Carbon-Family Materials for Flame Retardant Polymeric Materials. *Prog. Polym. Sci.* **2017**, *69*, 22–46. https://doi.org/10.1016/j.progpolymsci.2017.02.001

[136] Dittrich, B.; Wartig, K.-A.; Hofmann, D.; Mülhaupt, R.; Schartel, B. Flame Retardancy Through Carbon Nanomaterials: Carbon Black, Multiwall Nanotubes, Expanded Graphite, Multi-Layer Graphene and Graphene in Polypropylene. *Polym. Degrad. Stab.* **2013**, *98*, 1495–1505. https://doi.org/10.1016/j.polymdegradstab.2013.04.009

[137] Liu, S.; Chevali, V. S.; Xu, Z. G.; Hui, D.; Wang, H. A Review of Extending Performance of Epoxy Resins Using Carbon Nanomaterials. *Compos. B Eng.* **2018**, *136*, 197–214. https://doi.org/10.1016/j.compositesb.2017.08.020

[138] Wen, X.; Wang, Y. J.; Gong, J.; Liu,J.; Tian, N. N.; Wang, Y. H.; Jiang, Z. W.; Qiu, J.; Tang, T. Thermal and Flammability Properties of Polypropylene/Carbon Black Nanocomposites. *Polym. Degrad. Stab.* **2012**, *97*, 793–801. https://doi.org/10.1016/j.polymdegradstab.2012.01.031

[139] Beyer, G. Short Communication: Carbon Nanotubes as Flame Retardants for Polymers. *Fire Mater.* **2002**, *26*, 291–293. https://doi.org/10.1002/fam.805

[140] Kashiwagi, T.; Grulke, E.; Hilding, J.; Harris, R.; Awad, W.; Douglas, J. Thermal Degradation and Flammability Properties of Poly(Propylene)/Carbon Nanotube Composites. *Macromol. Rapid Commun.* **2002**, *23*, 761–765. https://doi.org/10.1002/1521-3927(20020901)23:13<761::AID-MARC761>3.0.CO;2-K

[141] Kashiwagi, T.; Grulke, E.; Hilding, J.; Groth, K.; Harris, R.; Butler, K.; Shields, J.; Kharchenko, S.; Douglas, J. Thermal and Flammability Properties of Polypropylene/Carbon Nano-tube Nanocomposites. *Polymer* **2004**, *45*, 4227–4239. https://doi.org/10.1016/j.polymer.2004.03.088

[142] Schartel, B.; Pötschke, P.; Knoll, U.; Abdel-Goad, M. Fire Behaviour of Polyamide 6/Multiwall Carbon Nanotube Nanocomposites. *Eur. Polym. J.* **2005**, *41*, 1061–1070. https://doi.org/10.1016/j.eurpolymj.2004.11.023

[143] Kashiwagi, T.; Du, F. M.; Douglas, J. F.; Winey, K. I.; Harris, R. H.; Shields, J. R. Nanoparticle Networks Reduce the Flammability of Polymer Nanocomposites. *Nature Mater.* **2005**, *4*, 928–933. https://doi.org/10.1038/nmat1502

[144] Wang, X.; Song, L.; Yang, H. Y.; Lu, H. D.; Hu, Y. Synergistic Effect of Graphene on Antidripping and Fire

Resistance of Intumescent Flame Retardant Poly(Butylene Succinate). *Composites. Ind. Eng. Chem. Res.* **2011**, *50*, 5376–5383. https://doi.org/10.1021/ie102566y

[145] Yu, B.; Shi, Y. Q.; Yuan, B. H.; Qiu, S. L.; Xing, W. Y.; Hu, W. Z.; Song, L.; Lo, S. M.; Hu, Y. Enhanced Thermal and Flame Retardant Properties of Flame-Retardant-Wrapped Graphene/Epoxy Resin Nanocomposites. *J. Mater. Chem. A* **2015**, *3*, 8034–8044. https://doi.org/10.1039/c4ta06613h

[146] Xu, W. Z.; Liu, L.; Zhang, B. L.; Hu, Y.; Xu, B. L. Effect of Molybdenum Trioxide-Loaded Graphene and Cuprous Oxide-Loaded Graphene on Flame Retardancy and Smoke Suppression of Polyurethane Elastomer. *Ind. Eng. Chem. Res.* **2016**, *55*, 4930–4941. https://doi.org/10.1021/acs.iecr.6b00383

[147] Yu, B.; Wang, X.; Qian, X. D.; Xing, W. Y.; Yang, H. Y.; Ma, L. Y.; Lin, Y.; Jiang, S. H.; Song, L.; Hu, Y.; Lo, S. M. Functionalized Graphene Oxide/Phosphoramide Oligomer Hybrids Flame Retardant Prepared Via in Situ Polymerization for Improving the Fire Safety of Polypropylene. *RSC Adv.* **2014**, *4*, 31782–31794. https://doi.org/10.1039/c3ra45945d

[148] Zirnstein, B.; Tabaka, W.; Frasca, D.; Schulze, D.; Schartel, B. Graphene/Hydrogenated Acrylonitrile-Butadiene Rubber Nanocomposites: Dispersion, Curing, Mechanical Reinforcement, Multifunctional Filler. *Polym. Test.* **2018**, *66*, 268–279. https://doi.org/10.1016/j.polymertesting.2018.01.035

[149] Frasca, D.; Schulze, D.; Böhning, M.; Krafft, B.; Schartel, B. Multilayer Graphene Chlorine Isobutyl Isoprene Rubber Nanocomposites: Influence of the Multilayer Graphene Concentration on Physical and Flame-Retardant Properties. *Rubber Chem. Technol.* **2016**, *89*, 316–334. https://doi.org/10.5254/rct.15.84838

[150] Ma, H. Y.; Tong, L. F.; Xu, Z. B.; Fang, Z. P. Functionalizing Carbon Nanotubes by Grafting on Intumescent Flame Retardant: Nanocomposite Synthesis, Morphology, Rheology, and Flammability. *Adv. Funct. Mater.* **2008**, *18*, 414–421. https://doi.org/10.1002/adfm.200700677

[151] Dittrich, B.; Wartig, K.-A.; Mülhaupt, R.; Schartel, B. Flame-Retardancy Properties of Intumescent Ammonium Poly(Phosphate) and Mineral Filler Magnesium Hydroxide in Combination with Graphene. *Polymers* **2014**, *6*, 2875–2895. https://doi.org/10.3390/polym6112875

[152] Dittrich, B.; Wartig, K.-A.; Hofmann, D.; Mülhaupt, R.; Schartel, B. Carbon Black, Multiwall Carbon Nanotubes, Expanded Graphite and Functionalized Graphene Flame Retarded Polypropylene Nanocomposites. *Polym. Adv. Technol.* **2013**, *24*, 916–926. https://doi.org/10.1002/pat.3165

[153] Zhang, T.; Yan, H. Q.; Peng, M.; Wang, L. L.; Ding, H. L.; Fang, Z. P. Construction of Flame Retardant Nanocoating on Ramie Fabric Via Layer-by-Layer Assembly of Carbon Nano-Tube and Ammonium Polyphosphate. *Nanoscale* **2013**, *5*, 3013–3021. https://doi.org/10.1039/c3nr34020a

[154] Strommer, B.; Battig, A.; Frasca, D.; Schulze, D.; Huth, C.; Böhning, M.; Schartel, B. Multifunctional Property Improvements by Combining Graphene and Conventional Fillers in Chlorosulfonated Polyethylene Rubber Composites. *ACS Appl. Polym. Mater.* **2022**, *4*, 1021–1034. https://doi.org/10.1021/acsapm.1c01469

[155] Gilman, J. W. Flammability and Thermal Stability Studies of Polymer Layered-Silicate (Clay) Nanocomposites. *Appl. Clay Sci.* **1999**, *15*, 31–49. https://doi.org/10.1016/S0169-1317(99)00019-8

[156] Zhang, W. J.; Camino, G.; Yang, R. J. Polymer/Polyhedral Oligomeric Silsesquioxane (POSS) Nanocomposites: An Overview of Fire Retardance. *Prog. Polym. Sci.* **2017**, *67*, 77–125. https://doi.org/10.1016/j.progpolymsci.2016.09.011

[157] Bourbigot, S.; Turf, T.; Bellayer, S.; Duquesne, S. Polyhedral Oligomeric Silsesquioxane as Flame Retardant for Thermo-Plastic Polyurethane. *Polym. Degrad. Stab.* **2009**, *94*, 1230–1237. https://doi.org/10.1016/j.polymdegradstab.2009.04.016

[158] Fina, A.; Tabuani, D.; Frache, A.; Camino, G. Polypropylene-polyhedral Oligomeric Silsesquioxanes (POSS) Nanocomposites. *Polymer* **2005**, *46*, 7855–7866. https://doi.org/10.1016/j.polymer.2005.06.121

[159] Zammarano, M.; Franceschi, M.; Bellayer, S.; Gilman, J. W.; Meriani, S. Preparation and Flame Resistance Properties of Revolutionary Self-Extinguishing Epoxy Nanocomposites Based on Layered Double Hydroxides. *Polymer* **2005**, *46*, 9314–9328. https://doi.org/10.1016/j.polymer.2005.07.050

[160] Alongi, J. Investigation on Flame Retardancy of Poly(Ethylene Terephthalate) for Plastics and Textiles by Combination of an Organo-Modified Sepiolite and Zn Phosphinate. *Fibers Polym.* **2011**, *12*, 166–173. https://doi.org/10.1007/s12221-011-0166-5

[161] Du, M. L.; Guo, B. C.; Jia, D. M. Thermal Stability and Flame Retardant Effects of Halloysite Nanotubes on Poly(Propylene). *Eur. Polym. J.* **2006**, *42*, 1362–1369. https://doi.org/10.1016/j.eurpolymj.2005.12.006

[162] Marney, D. C. O.; Russell, L. J.; Wu, D. Y.; Nguyen, T.; Cramm, D.; Rigopoulos, N.; Wright, N.; Greaves, M. The Suitability of Halloysite Nanotubes as a Fire Retardant for Nylon 6. *Polym. Degrad. Stab.* **2008**, *93*, 1971–1978. https://doi.org/10.1016/j.polymdegradstab.2008.06.018

[163] Yu, B.; Xing, W. Y.; Guo, W. W.; Qiu, S. L.; Wang, X.; Lo, S. M.; Hu, Y. Thermal Exfoliation of Hexagonal Boron Nitride for Effective Enhancements on Thermal Stability, Flame Retardancy and Smoke Suppression of Epoxy Resin Nanocomposites Via Sol-Gel Process. *J. Mater. Chem. A* **2016**, *4*, 7330–7340. https://doi.org/10.1039/c6ta01565d

[164] Yang, D. D.; Hu, Y.; Song, L.; Nie, S. B.; He, S. Q.; Cai, Y. B. Catalyzing Carbonization Function of Alpha-ZrP Based Intumescent Fire Retardant Polypropylene Nanocomposites. *Polym. Degrad. Stab.* **2008**, *93*, 2014–2018. https://doi.org/10.1016/j.polymdegradstab.2008.02.012

[165] Pawlowski, K. H.; Schartel, B. Flame Retardancy Mechanisms of Aryl Phosphates in Combination with Boehmite in Bisphenol a Polycarbonate/Acrylonitrile-Butadiene-Styrene Blends. *Polym. Degrad. Stab.* **2008**, *93*, 657–667. https://doi.org/10.1016/j.polymdegradstab.2008.01.002

[166] J. W. Gilman. Flame Retardant Mechanism of Polymer-Clay Nanocomposites. In: A. B. Morgan, C. A. Wilkie, Eds. Flame Retardant Polymer Nanocomposites. John Wiley and Sons, Hoboken, NJ, 2007, 67–87.

[167] Kashiwagi, T.; Harris, R. H.; Zhang, X.; Briber, R. M.; Cipriano, B. H.; Raghavan, S. R.; Awad, W. H.; Shields, J. R. Flame Retardant Mechanism of Polyamide 6-Clay Nanocomposites. *Polymer* **2004**, *45*, 881–891. https://doi.org/10.1016/j.polymer.2003.11.036

[168] Schartel, B.; Weiß, A.; Sturm, H.; Kleemeier, M.; Hartwig, A.; Vogt, C.; Fischer, R. X. Layered Silicate Epoxy Nanocomposites: Formation of the Inorganic-Carbonaceous Fire Protection Layer. *Polym. Adv. Technol.* **2011**, *22*, 1581–1592. https://doi.org/10.1002/pat.1644

[169] Gilman, J. W.; Jackson, C. L.; Morgan, A. B.; Harris, R.; Manias, E.; Giannelis, E. P.; Wuthenow, M.; Hilton, D.;

Phillips, S. H. Flammability Properties of Polymer—Layered-Silicate Nanocomposites. Polypropylene and Polystyrene Nanocomposites. *Chem. Mater.* **2000**, *12*, 1866–1873. https://doi.org/10.1021/cm0001760

[170] B. Schartel. Considerations Regarding Specific Impacts of the Principal Fire Retardancy Mechanisms in Nanocomposites. In: A. B. Morgan, C. A. Wilkie, Eds. Flame Retardant Polymer Nanocomposites. John Wiley & Sons, Hoboken, 2007, chap. 5, 107–129.

[171] Bourbigot, S.; Le Bras, M.; Dabrowski, F.; Gilman, J. W.; Kashiwagi, T. PA-6 Clay Nanocomposite Hybrid as Char Forming Agent in Intumescent Formulations. *Fire Mater.* **2000**, *24*, 201–208. https://doi.org/10.1002/1099-1018(200007/08)24:4<201::AID-FAM739>3.0.CO;2-D

[172] Pappalardo, S.; Russo, P.; Acierno, D.; Rabe, S.; Schartel, B. The Synergistic Effect of Organically Modified Sepiolite in Intumescent Flame Retardant Polypropylene. *Eur. Polym. J.* **2016**, *76*, 196–207. https://doi.org/10.1016/j.eurpolymj.2016.01.041

[173] Gilman, J. W.; Kashiwagi, T.; Lichtenhan, J. D. Nanocomposites: A Revolutionary New Flame Retardant Approach. *SAMPE J.* **1997**, *33*, 40–46.

[174] He, W. T.; Song, P. A.; Yu, B.; Fang, Z. P.; Wang, H. Flame Retardant Polymeric Nanocomposites Through the Combination of Nanomaterials and Conventional Flame Retardants. *Prog. Mater. Sci.* **2020**, *114*, 100687. https://doi.org/10.1016/j.pmatsci.2020.100687

[175] Liu, Y.; Gao, Y. S.; Wang, Q.; Lin, W. R. The Synergistic Effect of Layered Double Hydroxides with Other Flame Retardant Additives for Polymer Nanocomposites: A Critical Review. *Dalton Transact.* **2018**, *47*, 14827–14840. https://doi.org/10.1039/c8dt02949k

[176] Lu, H. D.; Song, L.; Hu, Y. A. A Review on Flame Retardant Technology in China. Part II: Flame Retardant Polymeric Nanocomposites and Coatings. *Polym. Adv. Technol.* **2011**, *22*, 379–394. https://doi.org/10.1002/pat.1891

[177] J.-M. Lopez-Cuesta, F. Laoutid. Multicomponent FR Systems: Polymer Nanocomposites Combined with Additional Materials. In: C. A. Wilkie, A. B. Morgan, Eds. Fire Retardancy of Polymeric Materials. 2nd Edition. CRC Press, Boca Raton, FL, 2010, chap. 12, 301–328.

[178] Bourbigot, S.; Duquesne, S.; Fontaine, G.; Bellayer, S.; Turf, T.; Samyn, F. Characterization and Reaction to Fire of Polymer Nanocomposites with and without Conventional Flame Retardants. *Mol. Cryst. Liquid Cryst.* **2008**, *486*, 1367–1381. https://doi.org/10.1080/15421400801921983

[179] J. W. Gilman, T. Kashiwagi. Polymer-Layered Silicaste Nanocomposites With Conventional Flame Retardants. In: T. J. Pinnavaia, G. W. Beall, Eds. Polymer-Clay Nanocomposites. John Wiley & Sons, Chichestrer, 2000, chap. 10, 193–206.

[180] B. Schartel. Fire Retardancy Based on Polymer Layered Silicate Nanocomposites. In: M. Okamoto, Ed. Advances in Polymeric Nanocomposite. CMC Publishing, Osaka, 2004, chap. 2–3, 242–257.

[181] M. Bödiger T. Eckel, D. Wittmann, H. Alberts. German Patent DE 19530200 A1, 1997.

[182] Schartel, B.; Knoll, U.; Hartwig, A.; Pütz, D. Phosphonium-modified Layered Silicate Epoxy Resins Nanocomposites and Their Combinations with ATH and Organo-Phosphorus Fire Retardants. *Polym. Adv. Technol.* **2006**, *17*, 281–293. https://doi.org/10.1002/pat.686

[183] Chigwada, G.; Jash, P.; Jiang, D. D.; Wilkie, C. A. Fire Retardancy of Vinyl Ester Nanocomposites: Synergy with Phosphorus-Based Fire Retardants. *Polym. Degrad. Stab.* **2005**, *89*, 85–100. https://doi.org/10.1016/j.polymdegradstab.2005.01.005

[184] M. Klatt, S. Grutke, T. Heitz, V. Rauschenberger, T. Plesnivy, P. Wolf, J. R. Wünsch, M. Fischer. German Patent DE 19705998 A1, 1998.

[185] Beyer, G. Flame Retardant Properties of EVA-Nanocomposites and Improvements by Combination of Nanofillers with Aluminium Trihydrate. *Fire Mater.* **2001**, *25*, 193–197. https://doi.org/10.1002/fam.776

[186] N. Schall, T. Engelhardt, H. Simmler-Hübenthal, G. Beyer. German Patent DE 19921 472 A1, 2000.

[187] M. Matzen, J. Marti, E. Oñate, S. Idelsohn, B. Schartel. Advanced experiments and particle finite element mod- elling on dripping V-0 polypropylene. Conference Proceedings, Fire Mater. 2017, 15th International Conference, Interscience Communications, 2017; 57–62.

[188] Buczko, A.; Stelzig, T.; Bommer, L.; Rentsch, D.; Heneczkowski, M.; Gaan, S. Bridged DOPO Derivatives as Flame Retardants for PA6. *Polym. Degrad. Stab.* **2014**, *107*, 158–165. https://doi.org/10.1016/j.polymdegradstab.2014.05.017

[189] Beach, M. W.; Rondan, N. G.; Froese, R. D.; Gerhart, B. B.; Green, J. G.; Stobby, B. G.; Shmakov, A. G.; Shvartsberg, V. M.; Korobeinichev, O. P. Studies of Degradation Enhancement of Polystyrene by Flame Retardant Additives. *Polym. Degrad. Stab.* **2008**, *93*, 1664–1673. https://doi.org/10.1016/j.polymdegradstab.2008.06.010

[190] Liu, Y.; Wang, Q. The Investigation on the Flame Retardancy Mechanism of Nitrogen Flame Retardant Melamine Cyanurate in Polyamide 6. *J. Polym. Res.* **2009**, *16*, 583–589. https://doi.org/10.1007/s10965-008-9263-6

[191] Casu, A.; Camino, G.; De Giorgi, M.; Flath, D.; Morone, V.; Zenoni, R. Fire-Retardant Mechanistic Aspects of Melamine Cyanurate in Polyamide Copolymer. *Polym. Degrad. Stab.* **1997**, *58*, 297–302. https://doi.org/10.1016/S0141-3910(97)00061-X

[192] R. Srinivasan, A. Gupta, D. Horsey. A revolutionary UV stable flame retardant system for polyolefins. International Conference on Additives for Polyolefins. Book Group Soc Plast Engineers 69–83, 1998, Document Type: Proceedings Paper Conference: International Conference on Additives for Polyolefins Location: Houston, TX Date: FEB 23–25, 1998.

[193] Ureyen, M. E.; Kaynak, E.; Yuksel, G. Flame-Retardant Effects of Cyclic Phosphonate with HALS and Fumed Silica in Polypropylene. *J. Appl. Polym. Sci.* **2020**, *137*, 48308. https://doi.org/10.1002/app.48308

[194] M. Klatt. Nitrogen-Based Flame Retardants. In: A. B. Morgan, C. A. Wilkie, Eds. Non Halogenated Flame Retardant Handbook, 1st Edition. Scrivener Publishing, Beverly, MA, 2014, chap. 4, 143–168.

[195] V. Butz. Flame retardant composition comprising a phosphonic acid derivative. Patent WO 002010026230 A1, 2010.

[196] S. M. Andrews, T. F. Thomson. Flame-retardant polyolefin systems. Patent WO 002017015338 A1, 2017.

[197] P. Van Ravenstyn. Flame retardant pre-expanded polymer material, in-mold or extruded foamed article and process for producing the material. European Patent EP 000 003002311 A1, 2016.

[198] Aubert, M.; Nicolas, R. C.; Pawelec, W.; Wilén, C. E.; Roth, M.; Pfaendner, R. Azoalkanes-Novel Flame Retardants and their Structure-Property Relationship. *Polym. Adv. Technol.* **2011**, *22*, 1529–1538. https://doi.org/10.1002/pat.1642

[199] Nicolas, R. C.; Wilén, C. E.; Roth, M.; Pfaendner, R.; King, R. E. Azoalkanes: A Novel Class of Flame Retardants. *Macromol. Rapid Commun.* **2006**, *27*, 976–981. https://doi.org/10.1002/marc.200600094

[200] P. Frank, U. Jonas, M. Luksin, S. Fuchs. Flammgeschützte Copolymere und Formmassen. German Patent DE 10 2019 135 325 A1, 2021.

[201] Tirri, T.; Aubert, M.; Aziz, H.; Brusentsev, Y.; Pawelec, W.; Wilén, C.-E. Sulfenamides in Synergistic Combination with Halogen Free Flame Retardants in Polypropylene. *Polym. Degrad. Stab.* **2019**, *164*, 75–89. https://doi.org/10.1016/j.polymdegradstab.2019.03.021

[202] C.-E. Wilen. Sulfenamides as Flame Retardants. European Patent EP 3066153 B1, 2014.

[203] Spiess, B.; Metzsch-Zilligen, E.; Pfaendner, R. A New Class of Oxyimides: Oxyimide Ethers and their Use as Flame Retardants. *Macromol. Mater. Eng.* **2021**, 2000650. https://doi.org/10.1002/mame.202000650

[204] R. Pfaendner, M. Mazurowski. Verwendung von phosphorhaltigen organischen Oxyimiden als Flammschutzmittel, Radikalgeneratoren und/oder als Stabilisatoren für Kunststoffe, flammgeschützte Kunststoffzusammensetzung, Verfahren. German Patent DE 102014218811 A1, 2016.

[205] Zhang, N.; Zhang, J.; Yan, H.; Guo, X.; Sun, Q.; Guo, R. A Novel Organic-Inorganic Hybrid K-HBPE@APP Perform-Ing Excellent Flame Retardancy and Smoke Suppression for Polypropylene. *J. Hazard. Mater.* **2019**, *373*, 856–886. https://doi.org/0.1016/j.jhazmat.2019.04.016

[206] Ke, C. H.; Li, J.; Fang, K. Y.; Zhu, Q. L.; Zhu, J.; Yan, Q.; Wang, Y.-Z. Synergistic Effect Between a Novel Hyperbranched Charring Agent and Ammonium Polyphosphate on the Flame Retardant and Anti-Dripping Properties of Polylactide. *Polym. Degrad. Stab.* **2010**, *95*, 763–770. https://doi.org/10.1016/j.polymdegradstab.2010.02.011

[207] Fu, T.; Wang, X. L.; Wang, Y.-Z. Flame-Responsive Aryl Ether Nitrile Structure Towards Multiple Fire Hazards Suppression of Thermoplastic Polyester. *J. Hazard. Mater.* **2021**, *403*, 123714. https://doi.org/10.1016/j.jhazmat.2020.123714

[208] Jing, X. K.; Wang, X. S.; Guo, D. M.; Zhang, Y.; Zhai, F. Y.; Wang, X. L.; Chen, L.; Wang, Y.-Z. The High-Temperature Self-Crosslinking Contribution of Azobenzene Groups to the Flame Retardance and Anti-Dripping of Copolyesters. *J. Mater. Chem. A* **2013**, *1*, 9264–9272. https://doi.org/10.1039/c3ta11267e

[209] Zhao, H. B.; Chen, L.; Yang, J. C.; Ge, X. G.; Wang, Y.-Z. A Novel Flame-Retardant-Free Copolyester: Cross-Linking Towards Self Extinguishing and Non-Dripping. *J. Mater. Chem.* **2012**, *22*, 19849–19857. https://doi.org/10.1039/c2jm34376b

[210] Zhan, J.; Wang, L.; Hong, N.; Hu, W.; Wang, J.; Song, L.; Hu, Y. Flame-Retardant and Anti-Dripping Properties of Intumescent Flame-Retardant Polylactide with Different Synergists. *Polym. Plast. Technol. Eng.* **2014**, *53*, 387–394. https://doi.org/10.1080/03602559.2013.844826

[211] Li, X. H.; Shao, L. B.; Song, N.; Shi, L. Y.; Ding, P. Enhanced Thermal-Conductive and Anti-Dripping Properties of Polyamide Composites by 3D Graphene Structures at Low Filler Content. *Compos. Part A Appl. Sci. Manuf.* **2016**, *88*, 305–314. https://doi.org/10.1016/j.compositesa.2016.06.007

[212] Pawlowski, K. H.; Schartel, B. Flame Retardancy Mechanisms of Triphenyl Phosphate, Resorcinol Bis(Diphenyl Phosphate) and Bisphenol a Bis(Diphenyl Phosphate) in Polycarbonate/Acrylonitrile-Butadiene-Styrene Blends. *Polym. Int.* **2007**, *56*, 1404–1414. https://doi.org/10.1002/pi.2290

[213] Kempel, F.; Schartel, B.; Marti, J. M.; Butler, K. M.; Rossi, R.; Idelsohn, S. R.; Oñate, E.; Hofmann, A. Modelling the Vertical UL 94 Test: Competition and Collaboration Between Melt Dripping, Gasification and Combustion. *Fire Mater.* **2015**, *39*, 570–584. https://doi.org/10.1002/fam.2257

[214] G. Camino, R. Delobel. Intumescence. In: A. F. Grand, C. A. Wilkie, Eds. Fire Retardancy of Polymeric Materials. Marcel Dekker, New York, 2000, chap. 7, 217–243.

[215] G. Yu, C. Ma, J. Li. Flame retardant effect of cyto- sine pyrophosphate and pentaerythritol on polypropyl- ene. Compos. B Eng., 180, 107520, 2020. Https://doi.org/10.1016/j.compositesb.2019.107520

[216] Guo, S.; Xu, J.; Ni, X. Synthesis, Structures, and Properties of a New Pentaerythritol-Derived Flame Retardant Used in Polyamide 66. *ACS Omega* **2012**, *6* (19), 12887–12897. https://doi.org/10.1021/acsomega.1c01385

[217] Harashina, H.; Tajima, Y.; Itoh, T. Synergistic Effect of Red Phosphorus, Novolac and Melamine Ternary Combination on Flame Retardancy of Poly(Oxymethylene). *Polym. Degrad. Stab.* **2006**, *91*, 1996–2002. https://doi.org/10.1016/j.polymdegradstab.2006.02.010

[218] Zhong, Y. H.; Wu, W.; Wu, R.; Luo, Q. L.; Wang, Z. The Flame Retarding Mechanism of the Novolac as Char Agent with the Fire Retardant Containing Phosphorous-Nitrogen in Thermoplastic Poly(Ether Ester) Elastomer System. *Polym. Degrad. Stab.* **2014**, *105*, 166–177. https://doi.org/10.1016/j.polymdegradstab.2014.04.013

[219] Le Bras, M.; Bourbigot, S.; Felix, E.; Pouille, F.; Siat, C.; Traisnel, M. Characterization of a Polyamide-6-Based Intumescent Additive for Thermoplastic Formulations. *Polymer* **2000**, *41*, 5283–5296. https://doi.org/10.1016/S0032-3861(99)00736-3

[220] Chen, T.; Xiao, X.; Wang, J. K.; Guo, N. Fire, Thermal and Mechanical Properties of TPE Composites with Systems Containing Piperazine Pyrophosphate (PAPP), Melamine Phosphate (MPP) and Titanium Dioxide (TiO2). *Plast. Rubber Compos.* **2019**, *48*, 149–159. https://doi.org/10.1080/14658011.2019.1582200

[221] Hu, Z.; Zhong, Z. Q.; Gong, X. D. Flame Retardancy, Thermal Properties, and Combustion Behaviors of Intumescent Flame-Retardant Polypropylene Containing (Poly) and Melamine Polyphosphate. *Polym. Adv. Technol.* **2020**, *31*, 2701–2710. https://doi.org/10.1002/pat.4996

[222] Feng, C. M.; Zhang, Y.; Liu, S. W.; Chi, Z. G.; Xu, J. R. Synthesis of Novel Triazine Charring Agent and Its Effect in Intumescent Flame-Retardant Polypropylene. *J. Appl. Polym. Sci.* **2012**, *123*, 3208–3216. https://doi.org/10.1002/app.34993

[223] Zhu, C. J.; He, M. S.; Liu, Y.; Cui, J. G.; Tai, Q. L.; Song, L.; Hu, Y. Synthesis and Application of a Mono-Component Intumescent Flame Retardant for Polypropylene. *Polym. Degrad. Stab.* **2018**, *151*, 144–151. https://doi.org/10.1016/j.polymdegradstab.2018.03.007

[224] Li, B.; Xu, M. J. Effect of a Novel Charring-Foaming Agent on Flame Retardancy and Thermal Degradation of Intumescent Flame Retardant Polypropylene. *Polym. Degrad. Stab.* **2006**, *91*, 1380–1386. https://doi.org/10.1016/j.polymdegradstab.2005.07.020

[225] Hu, X. P.; Li, Y. L.; Wang, Y.-Z. Synergistic Effect of the Charring Agent on the Thermal and Flame Retardant Properties of Polyethylene. *Macromol. Mater. Eng.* **2004**, *289*, 208–212. https://doi.org/10.1002/mame.200300189

[226] Yang, K.; Xu, M. J.; Li, B. Synthesis of N-Ethyl Triazine-Piperazine Copolymer and Flame Retardancy and Water Resistance of Intumescent Flame Retardant Polypropylene. *Polym. Degrad. Stab.* **2013**, *98*, 1397–1406. https://doi.org/10.1016/j.polymdegradstab.2013.03.023

[227] Liu, H. C.; Li, S.; Zhang, Z. Y.; Li, B.; Xu, M. J. An Efficient and Convenient Strategy Toward Fire Safety and Water Resistance of Polypropylene Composites Through Design and Synthesis of a Novel Mono-Component Intumescent Flame Retardant. *Polym. Adv. Technol.* **2019**, *30*, 1543–1554. https://doi.org/10.1002/pat.4585

[228] De Chirico, A.; Armanini, M.; Chini, P.; Cioccolo, G.; Provasoli, F.; Audisio, G. Flame Retardants for Polypropylene Based on Lignin. *Polym. Degrad. Stab.* **2003**, *79*, 139–145. https://doi.org/10.1016/S0141-3910(02)00266-5

[229] Cayla, A.; Rault, F.; Giraud, S.; Salaun, F.; Fierro, V.; Celzard, A. PLA with Intumescent System Containing Lignin and Ammonium Polyphosphate for Flame Retardant Textile. *Polymers* **2016**, *8*, 331. https://doi.org/10.3390/polym8090331

[230] Mandlekar, N.; Malucelli, G.; Cayla, A.; Rault, F.; Giraud, S.; Salaun, F.; Guan, J. P. Fire Retardant Action of Zinc Phosphinate and Polyamide 11 Blend Containing Lignin as a Carbon Source. *Polym. Degrad. Stab.* **2018**, *153*, 63–74. https://doi.org/10.1016/j.polymdegradstab.2018.04.019

[231] Luda, M. P.; Zanetti, M. Cyclodextrins and Cyclodextrin Derivatives as Green Char Promoters in Flame Retardants Formulations for Polymeric Materials. A Review. *Polymers* **2019**, *11*, 664. https://doi.org/10.3390/polym11040664

[232] Alongi, J.; Poskovic, M.; Frache, A.; Trotta, F. Novel Flame Retardants Containing Cyclodextrin Nanosponges and Phosphorus Compounds to Enhance EVA Combustion Properties. *Polym. Degrad. Stab.* **2010**, *95*, 2093–2100. https://doi.org/10.1016/j.polymdegradstab.2010.06.030

[233] Li, L.; Liu, X. L.; Shao, X. M.; Jiang, L. C.; Huang, K.; Zhao, S. Synergistic Effects of a Highly Effective Intumescent Flame Retardant Based on Tannic Acid Functionalized Graphene on the Flame Retardancy and Smoke Suppression Properties of Natural Rubber. *Compos. A Appl. Sci. Manufact.* **2020**, *129*, 105715. https://doi.org/10.1016/j.compositesa.2019.105715

[234] Kong, F.-B.; He, Q.-L.; Peng, W.; Nie, S.-B.; Dong, X.; Yang, J.-N. Eco-Friendly Flame Retardant Poly(Lactic Acid) Composites Based on Banana Peel Powders and Phytic Acid: Flame Retardancy and Thermal Property. *J. Polym. Res.* **2020**, *27*, 204. https://doi.org/10.1007/s10965-020-02176-4

[235] Gómez-Fernández, S.; Günther, M.; Schartel, B.; Corcuera, M. A.; Eceiza, A. Impact of the Combined Use of Layered Double Hydroxides, Lignin and Phosphorous Polyol on the Fire Behavior of Flexible Polyurethane Foams. *Ind. Crops Prod.* **2018**, *125*, 346–359. https://doi.org/10.1016/J.INDCROP.2018.09.018

[236] Chu, F.; Ma, C.; Zhang, T.; Xu, Z.; Mu, X.; Cai, W.; Zhou, X.; Ma, S.; Zhou, Y.; Hu, W.; Song, L. Renewable Vanillin-Based Flame Retardant Toughening Agent with Ultra-Low Phosphorus Loading for the Fabrication of High-Performance Epoxy Thermoset. *Compos. B Eng.* **2020**, *190*, 107925. https://doi.org/10.1016/j.compositesb.2020.107925

[237] Basak, S.; Raja, A. S. M.; Saxena, S.; Patil, P. G. Tannin Based Polyphenolic Bio-Macromolecules: Creating a New Era Towards Sustainable Flame Retardancy of Polymers. *Polym. Degrad. Stab.* **2021**, *189*, 109603. https://doi.org/10.1016/j.polymdegradstab.2021.109603

[238] Venezia, V.; Matta, S.; Lehner, S.; Vitiello, G.; Costantini, A.; Gaan, S.; Malucelli, G.; Branda, F.; Luciani, G.; Bifulco, A. Detailed Thermal, Fire, and Mechanical Study of Silicon-modified Epoxy Resin Containing Humic Acid and Other Additives. *ACS Appl. Polym. Mater.* **2021**, *3*, 5969–5981. https://doi.org/10.1021/acsapm.1c01240

[239] Wang, X.; Cai, W.; Ye, D.; Zhu, Y.; Cui, M.; Xi, J.; Liu, J.; Xing, W. Bio-Based Polyphenol Tannic Acid as Universal Linker Between Metal Oxide Nanoparticles and Thermoplastic Polyurethane to Enhance Flame Retardancy and Mechanical Properties. *Compos. B Eng.* **2021**, *224*, 109206. https://doi.org/10.1016/j.compositesb.2021.109206

[240] Battig, A.; Sanchez-Olivares, G.; Rockel, D.; Maldonado-Santoyo, M.; Schartel, B. Waste Not, Want Not: The Use of Leather Waste in Flame Retarded EVA. *Mater. Design* **2021**, *210*, 110100. https://doi.org/10.1016/j.matdes.2021.110100

[241] Weil, E. D.; Choudhary, V. Flame-Retarding Plastics and Elastomers with Melamine. *J. Fire Sci.* **1995**, *13*, 104–126. https://doi.org/10.1177/073490419501300202

[242] Liu, L. B.; Xu, Y.; He, Y. T.; Xu, M. J.; Shi, Z. X.; Hu, H. C.; Yang, Z. C.; Li, B. An Effective Mono-Component Intumescent Flame Retardant for the Enhancement of Water Resistance and Fire Safety of Thermoplastic Polyurethane Composites. *Polym. Degrad. Stab.* **2019**, *167*, 146–156. https://doi.org/10.1016/j.polymdegradstab.2019.07.006

[243] Levchik, S. V.; Levchik, G. F.; Balabanovich, A. I.; Camino, G.; Costa, L. Mechanistic Study of Combustion Performance and Thermal Decomposition Behaviour of Nylon 6 with Added Halogen-Free Fire Retardants. *Polym. Degrad. Stab.* **1996**, *54*, 217–222. https://doi.org/10.1016/S0141-3910(96)00046-8

[244] Turski Silva Diniz, A.; Huth, C.; Schartel, B. Dripping and Decomposition Under Fire: Melamine Cyanurate vs. Glass fibres in Polyamide 6. *Polym. Degrad. Stab.* **2020**, *171*, 109048. https://doi.org/10.1016/j.polymdegradstab.2019.109048

[245] Levchik, S. V.; Weil, E. D. Overview of the Recent Literature on Flame Retardancy and Smoke Suppression in PVC. *Polym. Adv. Technol.* **2005**, *16*, 707–716. https://doi.org/10.1002/pat.645

[246] Green, J. Mechanisms for Flame Retardancy and Smoke Suppression—A Review. *J. Fire Sci.* **1996**, *14*, 426–442. https://doi.org/10.1177/073490419601400602

[247] Kim, S. Flame Retardancy and Smoke Suppression of Magnesium Hydroxide Filled Polyethylene. *J. Polym. Sci. B—Polym. Phys.* **2003**, *41*, 936–944. https://doi.org/10.1002/polb.10453

[248] T. R. Hull, C. L. Wills, T. Artingstall, D. Price, G. J. Milnes. Mechanisms of Smoke and CO Suppression from EVA Composites. In: M. Le Bras, C. A. Wilkie, S. Bourbigot, S. Duquense, C. Jama, Eds. Fire Retardancy of Polymers: New Applications of Mineral Fillers. RSC, Cambridge, 2005, chap. 28, 372–385.

[249] Hirschler, M. M. Reduction of Smoke Formation from and Flammability of Thermoplastic Polymers by Metal-Oxides. *Polymer* **1984**, *25*, 405–411. https://doi.org/10.1016/0032-3861(84)90296-9

[250] Levchik, G. F.; Vorobyova, S. A.; Gorbarenko, V. V.; Levchik, S. V.; Weil, E. D. Some Mechanistic Aspects of the Fire Retardant Action of Red Phosphorus in Aliphatic Nylons. *J. Fire Sci.* **2000**, *18*, 172–182.

[251] Sheng, H. B.; Zhang, Y.; Ma, C.; Yang, L.; Qiu, S. L.; Wang, B. B.; Hu, Y. Influence of Zinc Stannate and Graphene Hybrids on Reducing the Toxic Gases and Fire Hazards During Epoxy Resin Combustion. *Polym. Adv. Technol.* **2019**, *30*, 666–674. https://doi.org/10.1002/pat.4502

[252] Zhou, K. Q.; Yang, W.; Tang, G.; Wang, B. B.; Jiang, S. H.; Hu, Y.; Gui, Z. Comparative Study on the Thermal Stability, Flame Retardancy and Smoke Suppression Properties of Polystyrene Composites Containing Molybdenum Disulfide

and Graphene. *RSC Adv.* **2013**, *3*, 25030–25040. https://doi.org/10.1039/c3ra43297a

[253] Price, D.; Liu, Y.; Milnes, G. J.; Hull, T. R.; Kandola, B. K.; Horrocks, A. R. An Investigation into the Mechanism of Flame Retardancy and Smoke Suppression by Melamine in Flexible Polyurethane Foam. *Fire Mater.* **2002**, *26*, 201–206. https://doi.org/10.1002/fam.810

[254] Chen, X. L.; Jiang, Y. F.; Jiao, C. M. Smoke Suppression Properties of Ferrite Yellow on Flame Retardant Thermoplastic Polyurethane Based on Ammonium Polyphosphate. *J. Hazard. Mater.* **2014**, *266*, 114–121. https://doi.org/10.1016/j.jhazmat.2013.12.025

[255] Xu, Q. W.; Zhai, H. M.; Wang, G. J. Mechanism of Smoke Suppression by Melamine in Rigid Polyurethane Foam. *Fire Mater.* **2015**, *39*, 271–282. https://doi.org/10.1002/fam.2249

[256] Shi, Y. Q.; Qian, X. D.; Zhou, K. Q.; Tang, Q. B.; Jiang, S. H.; Wang, B. B.; Wang, B.; Yu, B.; Hu, Y.; Yuen, R. K. K. CuO/graphene Nanohybrids: Preparation and Enhancement on Thermal Stability and Smoke Suppression of Polypropylene. *Ind. Eng. Chem. Res.* **2013**, *52*, 13654–13660. https://doi.org/10.1021/ie401535h

[257] Rao, W. H.; Liao, W.; Wang, H.; Zhao, H. B.; Wang, Y. Z. Flame-Retardant and Smoke-Suppressant Flexible Polyurethane Foams Based on Reactive Phosphorus-Containing Polyol and Expandable Graphite. *J Hazard Mater.* **2018**, *360*, 651–660. https://doi.org/10.1016/j.jhazmat.2018.08.053

[258] Costa, L.; Camino, G. Thermal-Behaviour of Melamine. *J. Therm. Anal.* **1988**, *34*, 423–429. https://doi.org/10.1007/BF01913181

[259] L. Costa, G. Camino, M. P. L. Dicortemiglia. Mechanism of Thermal-Degradation of Fire-Retardant Melamine Salts. In: G. L. Nelson, Ed. Fire and Polymers: Hazards Identification and Prevention. Book Series: ACS Symposium Series, Vol. 425. ACS, Washington, DC, 1990, 211–238.

[260] Braun, U.; Schartel, B. Flame Retardancy Mechanisms of Aluminium Phosphinate in Combination with Melamine Cyanurate in Glass-Fibre Reinforced Poly (1,4-Butylene Terephthalate). *Macromol. Mater. Eng.* **2008**, *293*, 206–217. https://doi.org/10.1002/mame.200700330

[261] Gaan, S.; Sun, G.; Hutches, K.; Engelhard, M. H. Effect of Nitrogen Additives on Flame Retardant Action of Tributyl Phosphate: Phosphorus-Nitrogen Synergism. *Polym. Degrad. Stabil.* **2008**, *93*, 99–108. https://doi.org/10.1016/j.polymdegradstab.2007.10.013

[262] Neisius, M.; Liang, S. Y.; Mispreuve, H.; Gaan, S. Phosphoramidate-Containing Flame-Retardant Flexible Polyurethane Foams. *Ind. Eng. Chem. Res.* **2013**, *52*, 9752–9762. https://doi.org/10.1021/ie400914u

[263] Liang, S. Y.; Neisius, M.; Mispreuve, H.; Naescher, R.; Gaan, S. Flame Retardancy and Thermal Decomposition of Flexible Polyurethane Foams: Structural Influence of Organophosphorus Compounds. *Polym. Degrad. Stab.* **2012**, *97*, 2428–2440. https://doi.org/10.1016/j.polymdegradstab.2012.07.019

[264] Neisius, N. M.; Lutz, M.; Rentsch, D.; Hemberger, P.; Gaan, S. Synthesis of DOPO-Based Phosphonamidates and Their Thermal Properties. *Ind. Eng. Chem. Res.* **2014**, *53*, 2889–2896. https://doi.org/10.1021/ie403677k

[265] Markwart, J. C.; Battig, A.; Zimmermann, L.; Wagner, M.; Fischer, J.; Schartel, B.; Wurm, F. R. Systematically Controlled Decomposition Mechanism in Phosphorus Flame Retardants by Precise Molecular Architecture: P–O vs P–N. *ACS Appl. Polym. Mater.* **2019**, *1*, 1118–1128. https://doi.org/10.1021/acsapm.9b00129

[266] Nazir, R.; Gaan, S. Recent Developments in P(O/S)-N Containing Flame Retardants. *J. Appl. Polym. Sci.* **2020**, *137*, 47910. https://doi.org/10.1002/app.47910

[267] Levchik, S. V.; Weil, E. D. Thermal Decomposition, Combustion and Flame-Retardancy of Epoxy Resins—A Review of the Recent Literature. *Polym. Int.* **2004**, *53*, 1901–1929. https://doi.org/10.1002/pi.1473

[268] Chen, X. L.; Jiao, C. M. Synergistic Effects of Hydroxy Silicone Oil on Intumescent Flame Retardant Polypropylene System. *Fire Safety J.* **2009**, *44*, 1010–1014. https://doi.org/10.1016/j.firesaf.2009.06.008

[269] M. Ciesielski, B. Burk, C. Heinzmann, M. Döring. Fire-Retardant High-Performance Epoxy-Based Materials. In: D.-Y. Wang, Ed. Novel Fire Retardant Polymers and Composite Materials. Book Series: Woodhead Publishing Series in Composites Science and Engineering, Vol. 73, 2017, 3–51. Https://doi.org/10.1016/B978-0-08-100136-3.00002-9

[270] Bonnet, J.; Bounor-Legare, V.; Boisson, F.; Melis, F.; Camino, G.; Cassagnau, P. Phosphorus Based Organic-Inorganic Hybrid Materials Prepared by Reactive Processing for EVA Fire Retardancy. *Polym. Degrad. Stab.* **2012**, *97*, 513–522. https://doi.org/10.1016/j.polymdegradstab.2012.01.018

[271] Zhu, Z.-M.; Xu, Y.-J.; Liao, W.; Xu, S.; Wang, Y.-Z. Highly Flame Retardant Expanded Polystyrene Foams from Phosphorus-Nitrogen-Silicon Synergistic Adhesives. *Ind. Eng. Chem. Res.* **2017**, *56*, 4649–4658. https://doi.org/10.1021/acs.iecr.6b05065

[272] Häublein, M.; Peter, K.; Bakis, G.; Mäkimieni, R.; Altstädt, V.; Möller, M. Investigation on the Flame Retardant Properties and Fracture Toughness of DOPO and Nano-SiO2 Modified Epoxy Novolac Resin and Evaluation of Its Combinational Effects. *Materials* **2019**, *12*, 1528. https://doi.org/10.3390/ma12091528

[273] Qiu, Y.; Wachtendorf, V.; Klack, P.; Qian, L. J.; Liu, Z.; Schartel, B. Improved Flame Retardancy by Synergy Between Cyclotetrasiloxane and Phosphaphenanthrene/Triazine Compounds in Epoxy Thermoset. *Polym, Int.* **2017**, *66*, 1883–1890. https://doi.org/10.1002/pi.5466

[274] Wang, X.; Hu, Y. A.; Song, L.; Xing, W. Y.; Lu, H. D. Thermal Degradation Behaviors of Epoxy Resin/POSS Hybrids and Phosphorus-Silicon Synergism of Flame Retardancy. *J. Polym. Sci. Part B Polym. Phys.* **2010**, *48*, 693–705. https://doi.org/10.1002/polb.21939

[275] Wu, C. S.; Liu, Y. L.; Chiu, Y. S. Epoxy Resins Possessing Flame Retardant Elements from Silicon Incorporated Epoxy Compounds Cured with Phosphorus or Nitrogen Containing Curing Agents. *Polymer* **2002**, *43*, 4277–4284. https://doi.org/10.1016/S0032-3861(02)00234-3

[276] Qian, X. D.; Song, L.; Yu, B.; Wang, B. B.; Yuan, B. H.; Shi, Y. Q.; Hu, Y.; Yuen, R. K. K. Novel Organic-Inorganic Flame Retardants Containing Exfoliated Graphene: Preparation and Their Performance on the Flame Retardancy of Epoxy Resins. *J. Mater. Chem. A* **2013**, *1*, 6822–6830. https://doi.org/10.1039/c3ta10416h

[277] Sut, A.; Greiser, S.; Jäger, C.; Schartel, B. Aluminium Diethylphosphinate Versus Ammonium Polyphosphate: A Comprehensive Comparison of the Chemical Interactions During Pyrolysis in Flame-Retarded Polyolefine/Poly(Phenylene Oxide). *Thermochim. Act.* **2016**, *640*, 74–84. https://doi.org/10.1016/j.tca.2016.08.004

[278] Product Information Dow Corning® 43–821 Additive, Form No. 26-2357-01, © 2017 Dow Corning Corporation.

[279] Sanchez-Olivares, G.; Rabe, S.; Pérez-Chávez, R.; Calderas, F.; Schartel, B. Industrial-Waste Agave Fibres in Flame-Retarded Thermoplastic Starch Biocomposites. *Compos. Part B* **2019**, *177*, 107370. https://doi.org/10.1016/j.compositesb.2019.107370

[280] Gilman, J. W.; Harris, R. H.; Shields, J. R.; Kashiwagi, T.; Morgan, A. B. A Study of the Flammability Reduction Mechanism of Polystyrene-Layered Silicate Nanocomposite: Layered Silicate Reinforced Carbonaceous Char. *Polym. Adv. Technol.* **2006**, *17*, 263–271. https://doi.org/10.1002/pat.682

[281] Morgan, A. B.; Harris, R. H.; Kashiwagi, T.; Chyall, L. J.; Gilman, J. W. Flammability of Polystyrene Layered Silicate (Clay) Nanocomposites: Carbonaceous Char Formation. *Fire Mater.* **2002**, *26*, 247–253. https://doi.org/10.1002/fam.803

[282] Wagner, J.; Deglmann, P.; Fuchs, S.; Ciesielski, M.; Fleckenstein, C. A.; Döring, M. Flame Retardant Synergism of Organic Disulfides and Phosphorous Compounds. *Polym. Degrad. Stab.* **2016**, *129*, 63–76. https://doi.org/10.1016/j.polymdegradstab.2016.03.023

[283] Howell, B. A.; Daniel, Y. G. The Impact of Sulfur Oxidation Level on Flame Retardancy. *J. Fire Sci.* **2018**, *36*, 518–534. https://doi.org/10.1177/0734904118806155

[284] Balabanovich, A. I.; Engelmann, J. Fire Retardant and Charring Effect of Poly(Sulfonyldiphenylene Phenylphosphonate) in Poly(Butylene Terephthalate). *Polym. Degrad. Stab.* **2003**, *79*, 85–92. https://doi.org/10.1016/S0141-3910(02)00258-6

[285] Macocinschi, A.; Grigoriu, A.; Filip, D. Aromatic Polysulfones for Flame Retardancy. *Eur. Polym. J.* **2002**, *38*, 1025–1031. https://doi.org/10.1016/S0014-3057(01)00253-1

[286] Braun, U.; Knoll, U.; Schartel, B.; Hoffmann, T.; Pospiech, D.; Artner, J.; Ciesielski, M.; Döring, M.; Perez-Gratero, R.; Sandler, J. K. W.; Altstädt, V. Novel Phosphorus-Containing Poly(Ether Sulfone)s and Their Blends with an Epoxy Resin: Thermal Decomposition and Fire Retardancy. *Macromol. Chem. Phys.* *207*, 1501–1514, **2006**. https://doi.org/10.1002/macp.200600182

[287] Battig, A.; Markwart, J.-C.; Wurm, F. R.; Schartel, B. Sulfur's Role in the Flame Retardancy of Thio-Ether–Linked Hyperbranched Polyphosphoesters in Epoxy Resins. *Eur. Polym. J.* **2020**, *122*, 109390. https://doi.org/10.1016/j.eurpolymj.2019.109390

[288] Tang, S.; Qian, L. J.; Qiu, Y.; Dong, Y. P. Synergistic Flame-Retardant Effect and Mechanisms of Boron/Phosphorus Compounds on Epoxy Resins. *Polym. Adv. Technol.* **2018**, *29*, 641–648. https://doi.org/10.1002/pat.4174

[289] Dogan, M.; Unlu, S. M. Flame Retardant Effect of Boron Compounds on Red Phosphorus Containing Epoxy Resins. *Polym. Degrad. Stab.* **2014**, *99*, 12–17. https://doi.org/10.1016/j.polymdegradstab.2013.12.017

[290] Unlu, S. M.; Dogan, S. D.; Dogan, M. Comparative Study of Boron Compounds and Aluminum Trihydroxide as Flame Retardant Additives in Epoxy Resin. *Polym. Adv. Technol.* **2014**, *25*, 769–776. https://doi.org/10.1002/pat.3274

[291] K. K. Shen, T. S. Griffin. Zinc Borate as a Flame Retardant, Smoke Suppressant, and Afterglow Surpressant in Polymers. In: G. L. Nelson, Ed. Fire and Polymers: Hazards Identification and Prevention. Book Series: ACS Symposium Series, Vol. 425. ACS, Washington, DC, 1990, 157–177.

[292] K. K. Shen, H. S. Kochesfahani, F. Jouffret. Boron- Based Flame Retardants and Flame Retardancy. In: C. A. Wilkie, A. B. Morgan, Eds. Fire Retardancy of Polymeric Materials, 2nd Edition. CRC Press, Boca Raton, FL, 2010, chap. 9, 207–237.

[293] Bourbigot, S.; Le Bras, M.; Leeuwendal, R.; Shen, K. K.; Schubert, D. Recent Advances in the Use of Zinc Borates in Flame Retardancy of EVA. *Polym. Degrad. Stab.* **1999**, *64*, 419– 425. https://doi.org/10.1016/S0141-3910(98)00130-X

[294] Formicola, C.; De Fenzo, A.; Zarrelli, M.; Frache, A.; Giordano, M.; Camino, G. Synergistic Effects of Zinc Borate and Aluminium Trihydroxide on Flammability Behaviour of Aerospace Epoxy System. *Express Polym. Lett.* **2009**, *3*, 376–384. https://doi.org/10.3144/expresspolymlett.2009.47

[295] Pawlowski, K. H.; Schartel, B.; Fichera, M. A.; Jäger, C. Flame Retardancy Mechanisms of Bisphenol A Bis(Diphenyl Phosphate) in Combination with Zinc Borate in Bisphenol a Polycarbonate/Acrylonitrile-Butadiene-Styrene Blends. *Thermochim. Act.* **2010**, *498*, 92–99. https://doi.org/10.1016/j. tca.2009.10.007

[296] Karrasch, A.; Wawrzyn, E.; Schartel, B.; Jäger, C. Solid-State NMR on Thermal and Fire Residues of Bisphenol a Polycarbonate/Silicone Acrylate Rubber/Bisphenol a Bis(Diphenyl-Phosphate)/(PC/SiR/BDP) and PC/SiR/BDP/Zinc Borate (PC/SiR/BDP/ZnB)—Part I: PC Charring and the Impact of BDP and ZnB. *Polym. Degrad. Stab.* **2010**, *95*, 2525–2533. https://doi.org/10.1016/j.polymdegradstab.2010.07.034

[297] Sut, A.; Greiser, S.; Jäger, C.; Schartel, B. Interactions in Multicomponent Flame-Retardant Polymers: Solid-state NMR Identifying the Chemistry Behind It. *Polym. Degrad. Stab.* **2015**, *121*, 116–125. https://doi.org/10.1016/j.polymdegradstab.2015.08.018

[298] Hoffendahl, C.; Fontaine, G.; Duquesne, S.; Taschner, F.; Mezger, M.; Bourbigot, S. The Combination of Aluminum Trihydroxide (ATH) and Melamine Borate (MB) as Fire Retardant Additives for Elastomeric Ethylene Vinyl Acetate (EVA). *Polym. Degrad. Stab.* **2015**, *115*, 77–88. https://doi.org/10.1016/j.polymdegradstab.2015.03.001

[299] Carty, P.; White, S. Smoke/Char Relationships in PVC Formulations. *Fire Mater.* **1994**, *18*, 151–166. https://doi.org/10.1177/073490419501300404

[300] Kuljanin, J.; Marinovic-Cincovic, M.; Zec, S.; Comor, M.; Nedeljkovic, C. Influence of Fe2O3-Filler on the Thermal Properties of Polystyrene. *J. Mater. Sci. Lett.* **2003**, *22*, 235–237. https://doi.org/10.1023/A:1022270514220

[301] Aufmuth, W.; Levchik, S. V.; Levchik, G. F.; Klatt, M. Poly(Butylene Terephthalate) Fire Retarded by 1,4-Diisobutylene-2,3,5,6-Tetraxydroxy-1, 4-Diphosphine Oxide. I. Combustion and Thermal Decomposition. *Fire Mater.* **1999**, *23*, 1–6. https://doi.org/10.1002/(SICI)1099-1018(199901/02)23:1<1::AID-FAM660>3.0.CO;2-K

[302] Peng, Y.; Niu, M.; Qin, R. H.; Xue, B. X.; Shao, M. Q. Study on Flame Retardancy and Smoke Suppression of PET by the Synergy Between Fe_2O_3 and New Phosphorus-Containing Silicone Flame Retardant. *High Perform. Polym.* **2022**, *32*, 871–882. https://doi.org/10.1177/0954008320914365

[303] A. Laachachi, M. Cochez, M. Ferriol, E. Leroy, J.-M. Lopez Cuesta. Thermal Degradation and Flammability of Poly(Methyl Methacrylate) Containing TiO_2 Nanoparticles and Modified Montmorillonite. In: C. A. Wilkie, G. L. Nelson, Eds. Fire and Polymers IV: Materials and Concepts for Hazard Prevention. Book Series: ACS Symposium Series, Vol. 922. ACS, 2006, 36–47. https://doi.org/10.1021/bk-2006- 0922.ch004

[304] Laachachi, A.; Cochez, M.; Leroy, E.; Ferriol, M.; Lopez-Cuesta, J.-M. Fire Retardant Systems in Poly (Methyl Meth-Acrylate): Interactions Between Metal Oxide Nanoparticles

and Phosphinates. *Polym. Degrad. Stab.* **2007**, *92*, 61–69, https://doi.org/10.1016/j.polymdegradstab.2006.09.011

[305] Gallo, E.; Braun, U.; Schartel, B.; Russo, P.; Acierno, D. Halogen-Free Flame Retarded Poly(Butylene Terephthalate) (PBT) Using Metal Oxides/PBT Nanocomposites in Combination with Aluminium phosphinate. *Polym. Degrad. Stab.* **2009**, *94*, 1245–1253. https://doi.org/10.1016/j.polymdegradstab.2009.04.014

[306] Gallo, E.; Schartel, B.; Braun, U.; Russo, P.; Acierno, D. Fire Retardant Synergisms Between Nanometric Fe_2O_3 and Aluminium Phosphinate in Poly(Butylene Terephthalate). *Polym. Adv. Technol.* **2011**, *22*, 2382–2391. https://doi.org/10.1002/ pat.1774

[307] Gallo, E.; Schartel, B.; Acierno, D.; Russo, P. Flame Retardant Biocomposites: Synergism Between Phosphinate and Nano-Metric Metal Oxides. *Eur. Polym. J.* **2011**, *47*, 1390–1401. https://doi.org/10.1016/j.eurpolymj.2011.04.001

[308] Xu, L. F.; Wu, X. D.; Li, L. S.; Chen, Y. J. Synthesis of a Novel Polyphosphazene/Triazine Bi-Group Flame Retardant in Situ Doping Nano Zinc Oxide and Its Application in Poly (Lactic Acid) Resin. *Polym. Adv. Technol.* **2019**, *30*, 1375–1385. https://doi.org/10.1002/pat.4570

[309] Lewin, L.; Endo, M. Catalysis of Intumescent Flame Retardancy of Polypropylene by Metallic Compounds. *Polym.* Adv. Technol. **2003**, 14, 3–11, **2003**. https://doi.org/10.1002/pat.265

[310] Duquesne, S.; P. Bachelet, S. Bellayer, S. Bourbigot, W. Mertens. Influence of inorganic fillers on the fire protection of intumescent coatings. *J. Fire Sci.* **2013**, *31*, 258–275. https://doi.org/10.1177/0734904112467291

[311] Gu, J. W.; Zhang, G. C.; Dong, S. L.; Zhang, Q. Y.; Kong, J. Study on Preparation and Fire-Retardant Mechanism Analysis of Intumescent Flame-Retardant Coatings. *Surf. Coat. Technol.* **2007**, *201*, 7835–7841. https://doi.org/10.1016/j. surfcoat.2007.03.020

[312] Zhou, Y.; Hao, J. W.; Liu, G. S.; Du, J. X. Influencing Mechanism of Transition Metal Oxide on Thermal Decomposition of Ammonium Polyphosphate. *Chin. J. Inorg. Chem.* **2013**, *29, 1115–1122.* https://doi.org/*10.3969/j. issn.1001-4861.2013.00.193*

[313] Mariappan, T.; Agarwal, A.; Ray, S. Influence of Titanium Dioxide on the Thermal Insulation of Waterborne Intumescent Fire Protective Paints to Structural Steel. *Prog. Org. Coat.* **2017**, 111, 67–74. https://doi.org/10.1016/j.porgcoat.2017.04.036

[314] Pan, Y. T.; Zhang, Z. D.; Yang. R. J. The Rise of MOFs and Their Derivatives for Flame Retardant Polymeric Materials: A Critical Review. *Compos. B Eng.* **2020**, *199*, 108265. https://doi.org/10.1016/j.compositesb.2020.1082650

[315] Xu, Z. S.; Jia, H. Y.; Yan, L.; Chu, Z. Y.; Zhou, H. Synergistic Effect of Bismuth Oxide and Mono-Component Intumescent Flame Retardant on the Flammability and Smoke Suppression Properties of Epoxy Resins. *Polym. Adv. Technol.* **2020,** *31*, 25–35. https://doi.org/10.1002/ pat.4744

[316] Morgan, A. B. Flame Retarded Polymer Layered Silicate Nanocomposites: A Review of Commercial and Open Literature Systems. Polym. Adv. Technol. 2006, 17, 206–217. https://doi.org/10.1002/pat.685

[317] Modesti, M.; Lorenzetti, A.; Besco, S.; Hrelja, D.; Semenzato, S.; Bertani, R.; Michelin, R. A. Synergism Between Flame Retardant and Modified Layered Silicate on Thermal Stability and Fire Behaviour of Polyurethane-Nanocomposite Foams. *Polym. Degrad. Stab.* **2008**, *93*, 2166–2171. https://doi.org/10.1016/j.polymdegradstab.2008.08.005

[318] Gomez-Fernandez, S.; Ugarte, L.; Pena-Rodriguez, C.; A. Corcuera, M. A.; Eceiza, A. The Effect of Phosphorus Containing Polyol and Layered Double Hydroxides on the Properties of a Castor Oil Based Flexible Polyurethane Foam. *Polym. Degrad. Stabil.* **2016**, *132*, 41–51. https://doi.org/10.1016/j. polymdegradstab.2016.03.036

[319] Yan, W.; Yu, J.; Zhang, M. Q.; Qin, S. H.; Wang, T.; Huang, W. J.; Long, L. J. Flame-Retardant Effect of a Phenethyl-Bridged DOPO Derivative and Layered Double Hydroxides for Epoxy Resin. *RSC Adv.* **2017**, *7, 46236–46245.* https://doi.org/*10.1039/c7ra08173a*

[320] Naderi Kalali, E.; Montes, A.; Wang, X.; Zhang, L.; Shabestari, M. E.; Li, Z.; Wang, D.-Y. Effect of Phytic Acid–Modified Layered Double Hydroxide on Flammability and Mechanical Properties of Intumescent Flame Retardant Poly-Propylene System. *Fire Mater.* **2018,** *42*, 213–220. https://doi.org/10.1002/fam.2482

[321] Bourbigot, S.; LeBras, M.; Delobel, R.; Tremillon, J. M. Synergistic Effect of Zeolite in an Intumescence Process—Study of the Interactions Between the Polymer and the Additives. *J. Chem. Soc. Faraday Transact.* **1996**, *92*, 3435–3444. https://doi.org/10.1039/ft9969203435

[322] Demir, H.; Arkis, E.; Balkose, D.; Ulku, S. Synergistic Effect of Natural Zeolites on Flame Retardant Additives. *Polym. Degrad. Stab.* **2005**, *89*, 478–483. https://doi.org/10.1016/j. polymdegradstab.2005.01.028

[323] Liu, Y. C.; Xu, W. Z.; Chen, R.; Cheng, C. M.; Hu, Y. Z.. Effect of Different Zeolitic Imidazolate Frameworks Nanoparticle-Modified Beta-FeOOH Rods on Flame Retardancy and Smoke Suppression of Epoxy Resin. *J. Appl. Polym. Sci.* **2021**, *138*, 49637. https://doi.org/10.1002/ app.49637

[324] Huang, G. B.; Chen, S. Q.; Song, P. G.; Lu, P. P.; Wu, C. L.; Liang, H. D. Combination Effects of Graphene and Layered Double Hydroxides on Intumescent Flame-Retardant Poly(Methyl Methacrylate) Nanocomposites. *Appl. Clay Sci.* **2014**, *88–89*, 78–85. https://doi.org/10.1016/j. clay.2013.11.002

[325] Duquesne, S.; Samyn, F.; Bourbigot, S.; Amigouet, P.; Jouffret, F.; Shen, K. Influence of Talc on the Fire Retardant Properties of Highly Filled Intumescent Polypropylene Composites. *Polym. Adv. Technol.* **2008**, *19*, 620–627. https://doi.org/10.1002/pat.1127

[326] Almeras, X.; Le Bras, M.; Hornsby, P.; Bourbigot, S.; Marosi, G.; Keszei, S.; Poutch, F. Effect of Fillers on the Fire Retardancy of Intumescent Polypropylene Compounds. *Polym. Degrad. Stab.* **2023**, *82*, 325–331. https://doi.org/10.1016/ S0141-3910(03)00187-3

[327] Clerc, L.; Ferry, L.; Leroy, E.; Lopez-Cuesta, J.-M. Influence of Talc Physical Properties on the Fire Retarding Behaviour of (Ethylene-Vinyl Acetate Copolymer/Magnesium Hydroxide/Talc) Composites. *Polym. Degrad. Stab.* **2005**, *88*, 504–511. https://doi.org/10.1016/j.polymdegradstab.2004.12.010

[328] Bertelli, G.; Marchetti, E.; Camino, G.; Costa, L.; Locatelli, R. Intumescent Fire Retardant Systems—Effect of Fillers on Char Structure. *Angew. Makromol. Chem.* **1989**, *172*, 153–163. https://doi.org/10.1002/apmc.1989.051720113

[329] Bourbigot, S.; Le Bras, M.; Duquesne, S.; Rochery, M. Recent Advances for Intumescent Polymers. *Macromol. Mater. Eng.* **2004**, *289*, 499–511. https://doi.org/10.1002/ mame.200400007

[330] Huang, G. B.; Wang, S. Q.; Song, P. A.; Wu, C. L.; Chen, S. Q.; Wang, X. Combination Effect of Carbon Nanotubes with Graphene on Intumescent Flame-Retardant Polypropylene

Nanocomposites. Compos. A Appl. Sci. Manufact. **2014**, 59, 18–25. https://doi.org/10.1016/j.compositesa.2013.12.010

[331] Wu, G. M.; Schartel, B.; Kleemeier, M.; Hartwig, A. Flammability of Layered Silicate Epoxy Nanocomposites Combined with Low-Melting Inorganic Ceepree Glass. *Polym.* Eng. Sci. **2012**, 52, 507–517. https://doi.org/10.1002/pen.22111

[332] Kroenke, W. J. Low-Melting Sulfate Glasses and Glass Ceramics, and Their Utility as Fire and Smoke Retarder Additives for Poly(Vinyl-Chloride). J. *Mater. Sci.* **1986**, *21*, 1123–1133. https://doi.org/10.1007/BF00553241

[333] Myers, R. E.; Licursi, E. Inorganic Glass Forming Systems as Intumescent Flame Retardants for Organic Polymers. *J. Fire Sci.* **1985**, *3*, 415–431. https://doi.org/10.1177/073490418500300603

[334] Yu, D.; Kleemeier, M.; Wu, G. M.; Schartel, B.; Liu, W. Q.; Hartwig, A. A Low Melting Organic-Inorganic Glass and Its Effect on Flame Retardancy of Clay/Epoxy Composites. *Polymer.* **2011**, *52*, 2120–2131. https://doi.org/10.1016/j.polymer.2011.03.033

[335] Yu, D.; Kleemeier, M.; Wu, G. M.; Schartel, B.; Liu, W. Q.; Hartwig, A. Phosphorous and Silicon Containing Low-Melting Organic-Inorganic Glasses Improve Flame Retardancy of Epoxy/Clay Composite. *Macromol. Mater. Eng.* **2011**, *296*, 952–964. https://doi.org/10.1002/mame.201100014

[336] Wu, G. M.; Schartel, B.; Yu, D.; Kleemeier, M.; Hartwig, A. Synergistic Flame Retardancy in Layered-Silicate Epoxy Nanocomposite Combined with Low-Melting Phenysiloxane Glass. *J. Fire Sci.* **2012**, *30*, 69–87. https://doi.org/10.1177/0734904111422417

[337] T. Kashiwagi, J. W. Gilman. Silicon-Based Flame Retardants. In: A. F. Grand, C. A. Wilkie, Eds. Fire Retardancy of Polymeric Materials. Marcel Dekker, New York, 2000, chap. 10, 353–390.

[338] Hamdani, S.; Longuet, C.; Perrin, D.; Lopez-Cuesta, J.-M.; Ganachaud, F. Flame Retardancy of Silicone-Based Materials. *Polym. Degrad. Stab.* **2009**, *94*, 465–495. https://doi.org/10.1016/j. polymdegradstab.2008.11.019

[339] Chen, Y. J.; Zhan, J.; Zhang, P.; Nie, S. B.; Lu, H. D.; Song, L.; Hu, Y. Preparation of Intumescent Flame Retardant Poly(Butylene Succinate) Using Fumed Silica as Synergistic Agent. Ind. Eng. Chem. Res. **2010**, 49, 8200–8208. https://doi.org/10.1021/ie100989j

[340] Wei, P.; Han, Z.; Xu, X.; Li, Z. Synergistic Flame Retardant Effect of SiO_2 in LLDPE/EVA/ATH Blends. *J. Fire Sci.* **2006**, *24*, 487–498. https://doi.org/10.1177/0734904106060954

[341] Fu, M. Z.; Qu, B. J. Synergistic Flame Retardant Mechanism of Fumed Silica in Ethylene-Vinyl Acetate/Magnesium Hydroxide Blends. *Polym. Degrad. Stab.* **2004**, *85*, 633–639. https://doi.org/10.1016/j.polymdegradstab.2004.03.002

[342] G. Schmaucks, B. Friede, H. Schreiner, J. Q. Roszinski. Amorphous Silicon Dioxide as Additive to Improve the Fire Retardancy of Polyamides. In: T. R. Hull, B. K. Kandola, Eds. Fire Retardancy of Polymers: New Strategies and Mechanisms. RSC, Cambridge, 2009, chap. 3, 35–48.

[343] Gallo, E.; Schartel, B.; Schmaucks, G.; von der Ehe, K; Böhning, M. The Effect of Well-Dispersed Amorphous Silicon Dioxide in Flame-Retarded Styrene Butadiene Rubber. Plas. Rubber Compos. **2013**, 42, 34–42. https://doi.org/10.1179/1743289812Y.0000000012

[344] Laoutid, F.; Duriez, V.; Brison, L.; Aouadi, S.; Vahabi, H.; Dubois, P. Synergistic Flame-Retardant Effect Between Lignin and Magnesium Hydroxide in Poly(Ethylene-Co-Vinyl Acetate). *Flame Retard. Therm. Stab. Mater.* **2019**, *2*, 9–18. https://doi.org/10.1515/flret-2019-0002

[345] Cavodeau, F.; Otazaghine, B.; Sonnier, R.; Lopez- Cuesta, J.-M.; Delaite, C. Fire Retardancy of Ethylene-Vinyl Acetate Composites—Evaluation of Synergistic Effects Between ATH and Diatomite Fillers. *Polym. Degrad. Stab.* **2016**, *129*, 246– 259. https://doi.org/10.1016/j.polymdegradstab.2016.04.018

[346] Hoffendahl, C.; Duquesne, S.; Fontaine, G.; Taschner, F.; Mezger, M.; Bourbigot, S. Decomposition Mechanism of Fire Retarded Ethylene Vinyl Acetate Elastomer (EVA) Containing Aluminum Trihydroxide and Melamine. *Polym. Degrad. Stab.* **2015**, *113*, 168–179. https://doi.org/10.1016/j. polymdegradstab.2014.09.016

[347] Ye, L.; Wu, Q. H.; Qu, B. J. Synergistic Effects and Mecha nism of Multiwalled Carbon Nanotubes with Magnesium Hydroxide in Halogen-Free Flame Retardant EVA/MH/ MWNT Nanocomposites. *Polym. Degrad. Stab.* **2009**, *94*, 751– 756. https://doi.org/10.1016/j.polymdegradstab.2009.02.010

[348] Wan, C.; Liu, J.; Wang, Y.; Han, Z. Enhanced Flame Retar dance in Polyethylene/Magnesium Hydroxide/Polycarbosilane Blends. *Mater. Chem. Phys.* **2020**, *253*, 123373. https://doi.org/10.1016/j.matchemphys.2020.123373

[349] Sanchez-Olivares, G.; Sanchez-Soli, A.; Manero, O.;Pérez-Chávez, R.; Jaramillo, M.; Alongi, J.; Carosio, F. Improving Mechanical Properties and Reaction to Fire of EVA/LLDPE Blends for Cable Applications with Melamine Triazine and Bentonite Clay. *Materials.* **2019**, *12*, 2393. https://doi.org/10.3390/ma12152393

[350] Cárdenas, M. A.; García-López, D.; Gobernado-Mitre, I.; Merino, J. C.; Pastor, J. M.; de D. Martinez, J.; Barbeta, J.; Calveras, D. Mechanical and Fire Retardant Properties of EVA/ clay/ATH Nanocomposites—Effect of Particle Size and Surface Treatment of ATH Filler. *Polym. Degrad. Stab.* **2008**, *93*, 2032–2037. https://doi.org/10.1016/j.polymdegradstab.2008.02.015

[351] Wang, H.; Zhou, X.; Abro, M.; Gao, M.; Deng, M.; Qin, Z.; Sun, Y.; Yue, L.; Zhang, X. Mussel-Inspired General Interface Modification Method and Its Application in Polymer Rein-Forcement and as a Flame Retardant. *ACS Omega*, **2018**, *3*, 4891–4898. https://doi.org/10.1021/acsomega.8b00182

[352] Zhang, R.; Xiao, X.; Tai, Q.; Huang, H.; Yang, J.; Hu, Y. The Effect of Different Organic Modified Montmorillonites (OMMTs) on the Thermal Properties and Flammability of PLA/MCAPP/Lignin Systems. *J. Appl. Polym, Sci.* **2013**, *127*, 4967–4973. https://doi.org/10.1002/app.38095

[353] Pomázi, Á.; Krecz, M.; Toldy, A. The Effect of the Combined Application of Solid-and Gas-Phase Flame Retardants in Epoxy Gelcoats on the Thermal Stability, Fire Performance and Adhesion of Coated Carbon Fibre–Reinforced Epoxy Composites, *J. Therm. Anal. Calorim.* **2023**, *148*, 257–270. https://doi.org/10.1007/s10973-022-11770-1

[354] Braun, U.; Schartel, B.. Effect of Red Phosphorus and Melamine Polyphosphate on the Fire Behavior of HIPS. *J.* Fire Sci. **2005**, 23, 5–30. https://doi.org/10.1177/0734904105043451

[355] Pomázi, Á.; Szolnoki, B.; Toldy, A. Flame Retardancy of Low-Viscosity Epoxy Resins and Their Carbon Fibre Reinforced Composites via a Ccombined Solid and Gas Phase Mechanism. *Polymers.* **2018**, *10*, 1081. https://doi.org/10.3390/polym10101081

[356] Szolnoki, B.; Marosi, K. B. G.; Toldy, A. Flame Retardancy of Sorbitol Based Bioepoxy via Combined Solid and Gas Phase Action. *Polymers.* **2016,** *8*, 322. https://doi.org/10.3390/ polym8090322

[357] Braun, U.; Bahr, H.; Schartel, B. Fire Retardancy Effect of Aluminium Phosphinate and Melamine Polyphosphate in Glass Fibre Reinforced Polyamide 6. *e-Polymers.* **2010**, *41.* https://doi.org/*10.1515/epoly.2010.10.1.443*

[358] Samyn, F.; Bourbigot, S. Thermal Decomposition of Flame Retarded Formulations PA6/Aluminum Phosphinate/ Melamine Polyphosphate/Organomodified Clay: Interactions Between the Constituents? *Polym. Degrad. Stabil.* **2012**, *97*, 2217–2230. https://doi.org/10.1016/j.polymdegradstab.2012.08.004

[359] Orhan, T.; Isitman, N. A.; Hacaloglu, J.; Kaynak, C. Thermal Degradation Mechanisms of Aluminium Phosphinate, Melamine Polyphosphate and Zinc Borate in Poly(methyl methacrylate). *Polym. Degrad. Stabil.* **2011**, *96*, 1780–1787. https://doi.org/10.1016/j.polymdegradstab.2011.07.019

[360] Xu, Z. Z.; Huang, J. Q.; Chen, M. J.; Tan, Y.; Wang, Y.-Z. Flame Retardant Mechanism of an Efficient Flame-Retardant Polymeric Synergist with Ammonium Polyphosphate for Polypropylene. *Polym. Degrad. Stab.* **2011**, 98, 2011–2020. https://doi.org/10.1016/j.polymdegradstab.2013.07.010

[361] Despinasse, M.-C.; Schartel, B. Aryl Phosphate—Aryl Phosphate Synergy in Flame-Retarded Bisphenol a Polycar-Bonate/Acrylonitrile-Butadiene-Styrene. *Thermochim. Act.* **2013**, *563*, 51–61. https://doi.org/10.1016/j.tca.2013.04.006

[362] Fichera, M. A.; Braun, U.; Schartel, B.; Sturm, H.; Knoll, U.; *Jäger, C.* Solid-State NMR Investigations of the Pyrolysis and Thermo-Oxidative Decomposition Products of a Poly-Styrene/Red Phosphorus/Magnesium Hydroxide System. J. Anal. Appl. Pyrolys. **2007**, 78, 378–386. https://doi.org/10.1016/j. jaap.2006.09.013

[363] Camino, A. Riva, D. Vizzini, A. Castrovinci, P. Amigouet, P. Bras Pereira. Effect of Hydroxides on Fire Retardance Mechanism of Intumescent EVA Compositions. In: M. Le Bras, C. A. Wilkie, S. Bourbigot, S. Duquense, C. Jama, Eds. Fire Retardancy of Polymers: New Applications of Mineral Fillers. RSC, Cambridge, 2005, chap 18, 248–263.

[364] J. C. Liu, S. Peng, Y. B. Zhang, H. B. Chang, Z. L. Yu, B. L. Pan, C. Lu, J. Y. Ma, Q. S. Niu. Influence of micro- encapsulated red phosphorus on the flame retardancy of high impact polystyrene/magnesium hydroxide composite and its mode of action. Polym. Degrad. Stab., 121 208–221, 2015. https://doi.org/10.1016/j.polymdegradstab.2015.09.011

[365] O. S. Henze, O. Muehren, A. Bertels, S. Conrad, T. Flug, H. Rudolph, C Beckmann, D. Meier. Flame retardant thermoplastic polyurethane comprising coated metal hydroxides, phosphorus-containing flame retardants and/or hydrotalcite or phyllosilicate. US Patent US 9688841 B2, 2017.

[366] Shen, L. G.; Shao, C. R.; Li, R. J.; Xu, Y. C.; Li, J. X.; Lin, H. J. Preparation and Characterization of Ethylene-Vinyl Acetate Copolymer (EVA)-Magnesium Hydroxide (MH)-Hexaphenoxycyclotriphosphazene (HPCTP) Composite Flame-Retardant Materials. *Polym. Bull.* **2019**, *76,* 2399–2410. https://doi.org/10.1007/s00289-018-2500-1

[367] Savas, L. A.; Deniz, T. K.; Tayfun, U.; Dogan, M. Effect of Microcapsulated Red Phosphorus on Flame Retardant, Thermal and Mechanical Properties of Thermoplastic Polyurethane Composites Filled with Huntite&Hydromagnesite Mineral. *Polym. Degrad. Stab.* **2017**, *135*, 121–129. https://doi.org/10.1016/j.polymdegradstab.2016.12.001

[368] M. Döring, M. Ciesielski, C. Heinzmann. Synergistic Flame Retardant Mixtures in Epoxy Resins. In: A. B. Morgan, C. A. Wilkie, G. L. Nelson, Eds. Fire and Polymers VI: New Advances in Flame Retardant Chemistry and Science. Book Series: ACS Symposium Series 1118. ACS, Washington, DC, 2012, 295–309. https://doi.org/10.1021/bk-2012-1118.ch020

[369] Reuter, J.; Greiner, L.; Schonberger, F.; Döring, M. Synergistic Flame Retardant Interplay of Phosphorus Containing Flame Retardants with Aluminum Trihydrate Depending on the Specific Surface Area in Unsaturated Polyester Resin. *J. Appl.* Polym, Sci. **2019**, 136, 47270. https://doi.org/10.1002/app.47270

[370] Yan, W.; Wang, K.; Huang, W. J.; Wang, M.; Wang, T.; Tu, C. Y.; Tian, Q. Synergistic Effects of Phenethyl-bridged DOPO Derivative with Al(OH)(3) on Flame Retardancy for Epoxy Resins. *Polym. Plast. Technol. Mater.* **2020**, *59*, 797–808. https://doi.org/10.1080/25740881.2019.1695269

[371] Li, Z. Z.; Qu, B. J. Flammability Characterization and Synergistic Effects of Expandable Graphite with Magnesium Hydroxide in Halogen-Free Flame-Retardant EVA Blends. *Polym. Degrad. Stab.* **2003**, *81*, 401–408. https://doi.org/10.1016/ S0141–3910(03)00123-X

[372] Wang, Z. Z.; Qu, B. J.; Fan, W. C.; Huang, P. Combustion Characteristics of Halogen-Free Flame-Retarded Polyethylene Containing Magnesium Hydroxide and Some Synergists. *J. Appl. Polym, Sci.* **2001**, *81,* 206–214. https://doi.org/*10.1002/*app.1430

[373] Hu, S.; Tan, Z.-W.; Chen, F.; Li, J.-G.; Shen, Q.; Huang, Z.-X.; Zhang, L.-M. Flame-Retardant Properties and Synergistic Effect of Ammonium Polyphosphate/Aluminum Hydroxide/Mica/Silicone Rubber Composites. Fire Mater. **2020**, 44, 673–682. https://doi.org/10.1002/fam.2831

[374] Li, Y. M.; Hu, S. L.; Wang, D. Y. Polymer-Based Ceramifiable Composites for Flame Retardant Applications: A Review. *Compos. Commun.* **2020**, *21*, 100405. https://doi.org/10.1016/j. coco.2020.100405

[375] Modesti, M.; Lorenzetti, A. Flame Retardancy of Polyiso-Cyanurate-Polyurethane Foams: Use of Different Charring Agents. *Polym. Degrad. Stabil.* **2002**, *78*, 341–347.

[376] Sypaseuth, F. D.; Gallo, E.; Çiftci, S.; Schartel, B.; Polylactic Acid Biocomposites: Approaches to a Completely Green Flame Retarded Polymer. *e-Polymers.* **2017**, *17*, 449–462. *https://doi.org/*10.1515/epoly-2017-0024

[377] Tang, G.; Zhang, R.; Wang, X.; Wang, B.; Song, L.; Hu, Y.; Gong, X. Enhancement of flame retardant performance of bio-based polylactic acid composites with the incorporation of aluminum hypophosphite and expanded graphite. J. Macromol. Sci. Part A Pure Appl. Chem. **2013**, 50, 255–269. https://doi.org/10.1080/10601325.2013.742835

[378] Yang, S.; Wang, J.; Huo, S. Q.; Wang, M.; Wang, J. P.; Zhang, B. Synergistic Flame-Retardant Effect of Expandable Graphite and Phosphorus-Containing Compounds for Epoxy resin: Strong Bonding of Different Carbon Residues. *Polym. Degrad. Stabil.* **2016**, *128*, 89–98. https://doi.org/10.1016/j. polymdegradstab.2016.03.017

[379] Guo, C. G.; Zhou, L.; Lv, J. X. Effects of Expandable Graphite and Modified Ammonium Polyphosphate on the Flame-Retardant and Mechanical Properties of Wood Flour-Polypropylene Composites. *Polym. Polym. Compos.* **2013**, *21*, 449–456. https://doi.org/10.1177/096739111302100706

[380] Zhu, H. F.; Zhu, Q. L.; Li, J. A.; Tao, K.; Xue, L. X.; Yan, Q. Synergistic Effect Between Expandable Graphite and Ammonium Polyphosphate on Flame Retarded Polylactide. *Polym. Degrad. Stab.* **2011**, *96*, 183–189. https://doi.org/10.1016/j. polymdegradstab.2010.11.017

[381] Wilke, A.; Langfeld, K.; Ulmer, B.; Andrievici, V.; A. Hörold, Limbach, P.; Bastian, M.; Schartel, B. Halogen-Free Multicomponent Flame Retardant Thermoplastic styrene-Ethylene-Butylene-Styrene Elastomers Based on Ammonium Polyphosphate—Expandable graphite Synergy. *Ind. Eng. Chem. Res.* **2017**, *56*, 8251–8263. https://doi.org/10.1021/acs. iecr.7b01177

[382] Seefeldt, H.; Braun, U.; Wagner, M. H. Residue Stabilization in the Fire Retardancy of Wood–Plastic Composites: Combination of Ammonium Polyphosphate, Expandable Graphite, and Red Phosphorus. *Macromol. Chem. Phys.* **2012**, *213*, 2370–2377. https://doi.org/10.1002/macp.201200119

[383] Li, J.; Mo, X. H.; Li, Y.; Zou, H. W.; Liang, M.; Chen, Y. Influence of Expandable Graphite pParticle Size on the Synergy Flame Retardant Property Between Expandable Graphite and Ammonium Polyphosphate in Semi-Rigid Polyurethane Foam. *Polym. Bull.* **2018**, *75*, 5287–5304. https://doi.org/10.1007/s00289-018-2309-y

[384] Meng, X. Y.; Ye, L.; Zhang, X. G.; Tang, P. M.; Tang, J. H.; Ji, X.; Li, Z. M. Effects of Expandable Graphite and Ammonium Polyphosphate on the Flame-Retardant and Mechanical Properties of Rigid Polyurethane Foams. *J. Appl. Polym. Sci.* **2009**, *114*, 853–863. https://doi.org/10.1002/app.30485

[385] Shi, L.; Li, Z. M.; Xie, B. H.; Wang, J. H.; Tian, C. R.; Yang, M. B. Flame Retardancy of Different-Sized Expandable Graphite Particles for High-Density Rigid Polyurethane Foams. *Polym. Int.* **2006**, *55*, 862–871. https://doi.org/10.1002/ pi.2021

[386] Modesti, M.; Lorenzetti, A. Halogen-Free Flame Retardants for Polymeric Foams. *Polym. Degrad. Stab.* **2002**, *78*, 167–173. https://doi.org/10.1016/S0141-3910(02)00130-1

[387] Zhang, L. Q.; Zhang, M.; Zhou, Y. H.; Hu, L. H. The Study of Mechanical Behavior and Flame Retardancy of Castor Oil Phosphate-Based Rigid Polyurethane Foam Composites Containing Expanded Graphite and Triethyl Phosphate. *Polym. Degrad. Stab.* **2013**, *98*, 2784–2794. https://doi.org/10.1016/j. polymdegradstab.2013.10.015

[388] Modesti, M.; Lorenzetti, A.; Simioni, F.; Camino, G. Expandable Graphite as an Intumescent Flame Retardant in Polyisocyanurate-Polyurethane Foams. *Polym. Degrad. Stab.* **2002**, *77*, 195–202. https://doi.org/10.1016/S0141-3910(02)00034-4

[389] Chan, Y. Y.; Ma, C.; Zhou, F.; Hu, Y.; Schartel, B. Flame Retardant Flexible Polyurethane Foams Based on Phosphorous Soybean-Oil Polyol and Expandable Graphite. *Polym. Degrad. Stab.* **2021**, *191*, 109656. https://doi.org/10.1016/j. polymdegradstab.2021.109656

[390] Chan, Y. Y.; Korwitz, A.; Pospiech, D.; Schartel, B. Flame Retardant Combinations with Expandable Graphite/Phosphorus/CuO/Castor Oil in Flexible Polyurethane Foams. *ACS Appl. Polym. Mater.* **2023**, *5*, 1891–1901. https://doi.org/10.1021/ acsapm.2c01969

[391] Thi, N. H.; Nguyen, T. N.; Oanh, H. T.; Trang, N. T. T.; Tham, D. Q.; Nguyen, H. T.; Nguyen, T. V.; Hoang, M. H. Synergistic Effects of Aluminum Hydroxide, Red Phosphorus, and Expandable Graphite on the Flame Retardancy and Thermal Stability of Polyethylene. *J. Appl. Polym, Sci.* **2020**, *138*, 50317. https://doi.org/10.1002/app.50317

[392] A. Mehl. Mixture for optical or electrical cable sheaths. European Patent EU 1167429A1, 2000.

[393] Sut, A.; Metzsch-Zilligen, E.; Großhauser, M.; Pfaendner, R.; Schartel, B. Synergy Between Melamine Cyanurate, Melamine Polyphosphate and Aluminum Diethylphosphinate in Flame Retarded Thermoplastic Polyurethane. *Polym. Test.* **2019**, *74*, 196–204. https://doi.org/10.1016/j.polymertesting.2019.01.001

[394] C. Hackl, F. Lehrich, G Bittner. Selbstverlöschende thermoplastische Polyurethane sowie Verfahren zur ihrer Herstellung DE 4309194 A1, 1994.

[395] Hollingbery, L. A.; Hull, T. R. The Thermal Decomposition of Huntite and Hydromagnesite—A Review. *Polym. Degrad. Stab.* **2010**, *95*, 2213–2225. https://doi.org/10.1016/j.polymdegradstab.2010.08.019

[396] Peeterbroeck, S.; Alexandre, M.; Nagy, J. B.; Pirlot, C.; Fonseca, A.; Moreau, N.; Philippin, G.; Delhalle, J.; Mekhalif, Z.; Sporken, R.; Beyer, G.; Dubois, P. Polymer-Layered Silicatecarbon Nanotube Nanocomposites: Unique Nanofiller Synergistic Effect. *Compos. Sci. Technol.* **2004**, *64*, 2317–2323. https://doi.org/10.1016/j.compscitech.2004.01.020

[397] Schartel, B.; Schmaucks, G. Flame Retardancy Synergism in Polymers Through Different Inert Fillers' Geometry. *Polym. Eng. Sci.* **2017**, *57*, 1099–1109. https://doi.org/10.1002/pen.24485

[398] Schartel, B.; Beck, U.; Bahr, H.; Hertwig, A.; Knoll, U.; Weise, M. Short Communication: Sub-Micrometer Coatings as an IR Mirror: A New Route to Flame Retardancy. *Fire Mater.* **2012**, *36*, 671–677. https://doi.org/10.1002/fam.1122

[399] Geoffroy, L.; Davesne, A. L.; Parent, F.; Sanchette, F.; Samyn, F.; Jimenez, M.; Bourbigot, S. Combining Low-Emissivity Thin Coating and 3D-Printed Original Designs for Superior Fire-Protective Performance. *ACS Omega.* **2020**, *5*, 27857–27863. https://doi.org/10.1021/acsomega.0c02902

[400] Jimenez, M.; Gallou, H.; Duquesne, S.; Jama, C.; Bourbigot, S.; Couillens, X.; Speroni, F. New Routes to Flame Retard Polyamide 6,6 For Electrical Applications. *J. Fire Sci.* **2012**, *30*, 535-551.

[401] Forsth, M.; Zhao, S. X.; Roos, A. Spectrally Selective and Adaptive Surfaces for Protection Against Radiative Heating: ITO and VO2. *Fire Mater.* **2014**, *38*, 111–124. https://doi.org/10.1002/ fam.2167

[402] Jimenez, M.; Duquesne, S.; Bourbigot, S. Enhanced Fire Retardant Properties of Glass-Fiber Reinforced Polyamide 6,6 by Combining Bulk and Surface Treatments: Toward a Better Understanding of the Fire-Retardant Mechanism. *Polym. Degrad. Stab.* **2013**, *98*, 1378–1388. https://doi.org/10.1016/j. polymdegradstab.2013.03.026

[403] Bourbigot, S.; Jama, C.; M. Le Bras, Delobel, R.; Dessaux, O.; Goudmand, P. New Approach to Flame Retardancy Using Plasma Assisted Surface Polymerisation Techniques. *Polym. Deg. Stab.* **1999**, *66*, 153–155. https://doi.org/10.1016/ S0141-3910(99)00059-2.

[404] Gallo, E.; Fan, Z.; Schartel, B.; Greiner, A. Electrospun Nanofibre Mat Coating—New Route to Flame Retardancy. *Polym. Adv. Technol.* **2011**, *22*, 1205–1210. https://doi.org/10.1002/pat.1994

[405] Schartel, B.; G. Kühn, Mix, R.; Friedrich, J. Surface Controlled Fire Retardancy of Polymers Using Plasma Polymerisation Macromol. *Mater. Eng.*

2002, *287*, 579–582. https://doi.org/10.1002/1439-2054(20020901)287:9<579::AID-MAME579>3.0.CO;2-6

[406] Davesne, A. L.; Bensabath, T.; Sarazin, J.; Bellayer, S.; Parent, F.; Samyn, F.; Jimenez, M.; Sanchette, F.; Bourbigot, S. Low-Emissivity Metal/Dielectric coatings as Radiative Barriers for the Fire Protection of Raw and Formulated Polymers. *ACS Appl. Polym. Mater.* **2020**, *2*, 2880–2889. https://doi.org/10.1021/acsapm.0c00399

[407] Pomázi, A.; Krecz, M.; Toldy, A. Thermal Behaviour and Fire and Mechanical Performance of Carbon Fibre-Reinforced Epoxy Composites Coated with Flame-Retardant Epoxy Gelcoats. *J. Therm. Anal. Calorim.* **2022**, *148*, 2685–2702. https://doi.org/10.1007/s10973-022-11710-z

[408] Pomázi, A.; Toldy, A. Development of Fire Retardant Epoxy based Gelcoats for Carbon Fibre Reinforced Epoxy Resin Composites. *Prog. Organ. Coat.* **2021**, *151*, 106015. https://doi.org/10.1016/j.porgcoat.2020.106015

[409] Bünker, J.; Elfgen, M. Fire-Retardant Composite Panel for Shipbuilding. *Lightweight Design Worldwide*, **2018**, *11*, 36–41. https://doi.org/10.1007/s41777-018-0030-z

[410] Hörold, A.; Schartel, B.; Trappe, V.; Korzen, M.; Bünker, J. Fire Stability of Glass-Fibre Sandwich Panels: The Influence of Core Materials and Flame Retardants. *Compos. Struct.* **2017**, *160*, 1310–1318. https://doi.org/10.1016/j.compstruct.2016.11.027

[411] Xi, W.; Qian, L. J.; Qiu, Y.; Chen, Y. J. Flame-Retardant Behavior of Bi-Group Molecule Derived from Phosphaphen-Anthrene and Triazine Groups on Polylactic Acid. *Polym. Adv. Technol.* **2016**, *27*, 781–788. https://doi.org/10.1002/pat.3714

[412] Chen, L.; Y.-Z. Wang. A Review on Flame Retardant Technology in China. Part 1: Development of Flame Retardants. *Polym. Adv. Technol.* **2010**, *21*, 1–26. https://doi.org/10.1002/pat.1550

[413] Chen, Y. J.; Wang, W.; Qiu, Y.; Li, L. S.; Qian, L. J.; Xin, F. Terminal group Effects of Phosphazene-Triazine Bi-Group Flame Retardant Additives in Flame Retardant Polylactic Acid Composites. *Polym. Degrad. Stab.* **2017**, *140*, 166–175. https://doi.org/10.1016/j.polymdegradstab.2017.04.024

[414] Sag, J.; Kukla, P.; Goedderz, D.; Roch, H.; Kabasci, S.; Döring, M.; Schonberger, F. Synthesis of Novel Polymeric acrylate-Based Flame Retardants Containing Two Phosphorus Groups in Different Chemical Environments and Their Influence on the Flammability of Poly (Lactic Acid). *Polymers.* **2020**, *12*, 778. https://doi.org/10.3390/polym12040778

[415] Müller, P.; Schartel, B. Melamine poly(Metal Phosphates) as Flame Retardant in Epoxy Resin: Performance, Modes of Action and Synergy. *J. Appl. Polym, Sci.* **2016**, *133*, 43549. https://doi.org/10.1002/app.43549

[416] Halpern, Y.; Mott, D. M.; Niswander, R. H. Fire Retardancy of Thermoplastic Materials by Intumescence. *Ind. Eng. Chem. Prod. Res. Dev.* **1984**, *23*, 233–238. https://doi.org/10.1021/ i300014a011

[417] Wang, X.; Hu, Y.; Song, L.; Xing, W. Y.; Lu, H. D. A.; Lv, P.; Jie, G. X. Flame Retardancy and Thermal Degradation Mechanism of Epoxy Resin Composites Based on a DOPO Substituted Organophosphorus Oligomer. *Polymer.* **2010**, *51*, 2435–2445. https://doi.org/10.1016/j.polymer.2010.03.053

[418] Wang, C.; Wu, Y. C.; Li, Y. C.; Shao, Q.; Yan, X. R.; Han, C.; Wang, Z.; Liu, Z.; Guo, Z. H. Flame-Retardant Rigid Polyurethane Foam with a Phosphorus-Nitrogen Single Intumescent Flame Retardant. *Polym. Adv. Technol.* **2018**, *29*, 668–676. https://doi.org/10.1002/pat.4105

[419] Ma, H. Y.; Tong, L. F.; Xu, Z. B.; Fang, Z. P.; Jin, Y. M.; Lu, F. Z. A Novel Intumescent Flame Retardant: Synthesis and Application in ABS Copolymer. *Polym. Degrad. Stab.* **2007**, *92*, 720–726. https://doi.org/10.1016/j.polymdegradstab.2006.12.009

[420] Lv, Q.; Huang, J. Q.; Chen, M. J.; Zhao, J.; Tan, Y.; Chen, L.; Y.-Z. Wang. An Effective Flame Retardant and Smoke Suppression Oligomer for Epoxy resin. *Ind. Eng. Chem. Res.* **2013**, *52*, 9397–9404. https://doi.org/10.1021/ie400911r

[421] Shao, Z. B.; Deng, C.; Tan, Y.; Chen, M. J.; Chen, L.; Y.-Z. Wang. An Efficient Monocomponent Polymeric Intumescent Flame Retardant for Polypropylene: Preparation and Application. *ACS Appl. Mat. Inter.* **2014**, *6*, 7363–7370. https://doi.org/10.1021/am500789q

[422] Deng, C.; Yin, H.; Li, R.-M.; Huang, S.-C.; Schartel, B.; Wang, Y.-Z. Modes of Action of a Mono-cComponent Intumescent Flame Retardant MAPP in Polyethylene-Octene Elastomer. *Polym. Degrad. Stab.* **2017**, *138*, 142–150. https://doi.org/10.1016/j. polymdegradstab.2017.03.006

[423] Yan, J. L.; Design, M.; Xu, J.; Synthesis and Application of a Highly Efficient Mono-Component Intumescent Flame Retardant for Non-Charring Polyethylene Composites. *Polym. Bull.* **2021**, *78*, 643–662. https://doi.org/10.1007/s00289-020-03130-6

[424] Liu, L. B.; Xu, Y.; He, Y. T.; Xu, M. J.; Wang, W.; Li, B. A Facile Strategy for Enhancing the Fire Safety of Unsaturated Polyester Resins Through Introducing an Efficient Monocomponent Intumescent Flame Retardant. *Polym. Adv. Technol.* **2020**, *31*, 1218–1230. https://doi.org/10.1002/pat.4852

[425] Xia, S. Y.; Zhang, Z. Y.; Leng, Y.; Li, B.; Xu, M. J. Synthesis of a Novel Mono-Component Intumescent Flame Retardant and Its High Efficiency for Flame Retardant Polyethylene. *J. Anal. Appl. Pyrolys.* **2018**, *134*, 632–640. https://doi.org/10.1016/j. jaap.2018.08.017

[426] Shao, Z. B.; Deng, C.; Tan, Y.; Chen, M. J.; Chen, L.; Wang, Y.-Z. Flame Retardation of Polypropylene via a Novel Intumescent Flame Retardant: Ethylenediamine Modified Ammonium Polyphosphate. *Polym. Degrad. Stabil.* **2014**, *106*, 88–96. https://doi.org/10.1016/j.polymdegradstab.2013.10.005

[427] Shao, Z. B.; Deng, C.; Tan, Y.; Yu, L.; Chen, M. J.; Chen, L.; Wang, Y.-Z. Ammonium Polyphosphate Chemically-Modified with Ethanolamine as an Efficient Intumescent Flame Retardant for Polypropylene. *J. Mat. Chem. A.* **2014**, *2*, 13955–13965. https://doi.org/10.1039/C4TA02778G

[428] P. Anna, M. G. Marosi, A, Toldy. Process for the preparation of amine functional phosphoric amides. Patent WO 2009/077796 A2, 2009.

[429] Zhu, H. B.; Peng, Z. M.; Chen, Y. M.; Li, G. Y.; Wang, L.; Tang, Y.; Pang, R.; Khan, Z. U.; Wan, P. Y. Preparation and Characterization of Flame Retardant Polyurethane Foams Containing Phosphorus-Nitrogen-Functionalized Lignin. *RSC Adv.* **2014**, *4*, 55271–55279. https://doi.org/10.1039/c4ra08429b

[430] Costes, L.; Laoutid, F.; Aguedo, M.; Richel, A.; Brohez, S.; Delvosalle, C.; Dubois, P. Phosphorus and Nitrogen Derivatization as Efficient Route for Improvement of Lignin Flame Retardant action in PLA. *Eur. Polym. J.* **2016**, *84*, 652–667. https://doi.org/10.1016/j.eurpolymj.2016.10.003

[431] Yu, Y.; Fu, S.; Song, P.; Luo, X.; Jin, Y.; Lu, F.; Wu, Q.; Ye, J. Functionalized Lignin by Grafting Phosphorus-Nitrogen Improves the Thermal Stability and Flame Retardancy of Polypropylene. *Polym. Degrad. Stab.* **2012**, *97*, 541–546. https://doi.org/10.1016/j.polymdegradstab.2012.01.020

[432] Prieur, B.; Meub, M.; Wittemann, M.; Klein, R.; Bellayer, S.; Fontaine, G.; Bourbigot, S. Phosphorylation of Lignin to Flame Retard Acrylonitrile Butadiene Styrene (ABS). *Polym. Degrad. Stab.* **2016**, *127*, 32–43. https://doi.org/10.1016/j. polymdegradstab.2016.01.015

[433] Zhang, R.; Xiao, X.; Tai, Q.; Huang, H.; Hu, Y. Modification of Lignin and Its Application as Char Agent in Intumescent Flame-Retardant Poly(Lactic Acid). *Polym. Eng. Sci.* **2012**, *52*, 2620–2626. https://doi.org/10.1002/pen.23214

[434] Xing, W. Y.; Yuan, H. X.; Zhang, P.; Yang, H. Y.; Song, L.; Hu, Y. Functionalized Lignin for Halogen-Free Flame Retardant Rigid Polyurethane Foam: Preparation, Thermal Stability, Fire Performance and Mechanical Properties. *J. Polym. Res.* **2013**, *20*, 234. https://doi.org/10.1007/ s10965-013-0234-1

[435] Zhang, T.; Yan, H. Q.; Shen, L.; Fang, Z. P.; Zhang, X. M.; Wang, J. J.; Zhang, B. Y. Chitosan/Phytic Acid Polyelectrolyte Complex: A Green and Renewable Intumescent Flame Retardant System for Ethylene Vinyl Acetate Copolymer. *Ind. Eng. Chem. Res.* **2014**, *53*, 19199–19207. https://doi.org/10.1021/ ie503421f

[436] Cheng, X. W.; Guan, J. P.; Tang, R. C.; Liu, K. O. Phytic Acid as a Bio-Based Phosphorus Flame Retardant for Poly(Lactic Acid) Nonwoven Fabric. *J. Clean. Produc.* **2016**, *124*, 114–119. https://doi.org/10.1016/j.jclepro.2016.02.113

[437] Fang, F.; Huo, S. Q.; Shen, H. F.; Ran, S. Y.; Wang, H.; Song, P. G.; Fang, Z. P. A Bio-Based Ionic Complex with Different Oxidation States of Phosphorus for rReducing Flammability and Smoke Release of Epoxy Resins. *Compos. Commun.* **2020**, *17*, 104–108. https://doi.org/10.1016/j.coco.2019.11.011

[438] Gao, Y. Y.; Deng, C.; Du, Y. Y.; Huang, S. C.; Wang, Y.-Z. A Novel Bio-Based Flame Retardant for Polypropylene from Phytic Acid. *Polym. Degrad. Stab.* **2019**, *161*, 298–308. https://doi.org/10.1016/j.polymdegradstab.2019.02.005

[439] Alongi, J.; R. Andrea Carletto, A. Di Blasio, Carosio, F.; Bosco, F.; Malucelli, G. DNA: A Novel, Green, Natural Flame Retardant and Suppressant for Cotton. *J. Mater. Chem. A.* **2013**, *1*, 4779–4785. https://doi.org/10.1039/ c3ta00107e

[440] Carosio, F.; A. Di Blasio, Alongi, J.; Malucelli, G. Green DNA-Based Flame Retardant Coatings Assembled Through Layer by Layer. Polymer. **2013**, *54*, 5148–5153. https://doi.org/10.1016/j. polymer.2013.07.029

[441] Jin, X.; Cui, S.; Sun, S.; Gu, X.; Li, H.; Liu, X.; Tang, W.; Sun, J.; Bourbigot, S.; Zhang, S. The Preparation of a Biopolyelectrolytes Based Core-Shell Structure and Its Application in Flame Retardant Polylactic Acid Composites. *Compos. A Appl. Sci. Manuf.* **2019**, *124*, 105485. https://doi.org/10.1016/j. compositesa.2019.105485

[442] Chang, B. P.; Thakur, S.; Mohanty, A. K.; Misra, M. Novel Sustainable Biobased Flame Retardant from Functionalized Vegetable Oil for Enhanced Flame Retardancy of Engineering Plastic. *Sci. Rep.* **2019**, *9*, 15971. https://doi.org/10.1038/ s41598-019-52039-2

[443] Chen, Z. Y.; Xiao, P.; Zhang, J. M.; Tian, W. G.; Jia, R. N.; Nawaz, H. F.; Jin, K.; Zhang, J. A Facile Strategy to Fabricate Cellulose-Based, Flame-Retardant, Transparent and Anti-Dripping Protective Coatings. *Chem. Eng. J.* **2020**, *379*, 122270. https://doi.org/10.1016/j.cej.2019.122270

[444] Zhang, Y.; Yu, B.; Wang, B.; Liew, K. M.; Song, L.; Wang, C.; Hu, Y. Highly Effective P–P Synergy of a Novel DOPO-Based Flame Retardant for Epoxy Resin. *Ind. Eng. Chem. Res.* **2017**, *56*, 1245–1255. https://doi.org/10.1021/acs. iecr.6b04292

[445] Peng, H. Q.; Zhou, Q.; Wang, D. Y.; Chen, L.; Wang, Y.-Z. A Novel Charring Agent Containing Caged Bicyclic Phosphate and Its Application in Intumescent Flame Retardant Polypropylene Systems. *J. Ind. Eng. Chem.* **2008**, *14*, 589–595. https://doi.org/10.1016/j.jiec.2008.05.011

[446] T. Zich, L. Babiniec, K. A. M. Salmeia, S Gaan. A process for preparation of DOPO derivatives. European Patent EP 3421479 A1, 2017.

[447] Liang, W. J.; Zhao, B.; Zhao, P. H.; Zhang, C. Y.; Liu, Y. Q. Bisphenol-S Bridged Pent2a(Anilino)Cyclotriphosphazene and Its Application in Epoxy Resins: Synthesis, Thermal Degradation, and Flame Retardancy. *Polym. Degrad. Stab.* **2017**, *135*, 140–151. https://doi.org/10.1016/j. polymdegradstab.2016.11.023

[448] Huang, J. Q.; Zhang, Y. Q.; Yang, Q.; Liao, X.; Li, G. X. Synthesis and Characterization of a Novel Charring Agent and Its Application in Intumescent Flame Retardant Polypropylene System. *J. Appl. Polym, Sci.* **2012**, *123*, 1636-1644. https://doi.org/10.1002/app.34578

[449] Jian, R. K.; Wang, P.; Duan, W. S.; Wang, J. S.; Zheng, X. L.; Weng, J. B. Synthesis of a Novel P/N/S-Containing Flame Retardant and Its Application in Epoxy Resin: Thermal Property, Flame Retardance, and Pyrolysis Behavior. *Ind. Eng. Chem. Res.* **2016**, *55*, 11520–11527. https://doi.org/10.1021/acs. iecr.6b03416

[450] Huo, S. Q.; Wang, J.; Yang, S.; Chen, X.; Zhang, B.; Wu, Q. L.; Zhang, B. Flame-Retardant Performance and Mechanism of Epoxy Thermosets Modified with a Novel Reactive Flame Retardant Containing Phosphorus, Nitrogen, and Sulfur. *Polym. Adv. Technol.* **2018**, *29*, 497–506. https://doi.org/10.1002/ pat.4145.

[451] Chen, M. J.; Shao, Z. B.; Wang, X. L.; Chen, L.; Y.-Z. Wang. Halogen-free flame-retardant flexible polyurethane foam with a novel nitrogen-phosphorus flame retardant. Ind. Eng. Chem. Res. **2012**, *51*, 9769–9776. https://doi.org/10.1021/ ie301004d

[452] Thirumal, M.; Khastgir, D.; Nando, G. B.; Naik, Y. P.; Singha, N. K. Halogen-Free Flame Retardant PUF: Effect of Melamine Compounds on Mechanical, Thermal and Flame Retardant Properties. *Polym. Degrad. Stab.* **2010**, *95*, 1138–1145. https://doi.org/10.1016/j.polymdegradstab.2010.01.035

[453] Tang, S.; Qian, L. J.; Liu, X. X.; Dong, Y. P. Gas-Phase Flame-Retardant Effects of a Bi-Group Compound Based on Phosphaphenanthrene and Triazine-Trione Groups in Epoxy Resin. *Polym. Degrad. Stab.* **2016**, *133*, 350–357. https://doi.org/10.1016/j. polymdegradstab.2016.09.014

[454] Qiu, Y.; Qian, L. J.; Xi, W. Flame-Retardant Effect of a Novel Phosphaphenanthrene/Triazine-Trione Bi-Group Compound on an Epoxy Thermoset and Its Pyrolysis Behaviour. *RSC Adv.* **2016**, *6*, 56018–56027. https://doi.org/10.1039/c6ra10752d

[455] Mishra, N.; Vasava, D. Recent Developments in S-Triazine Holding Phosphorus and Nitrogen Flame-Retardant Materials. *J. Fire Sci.* **2020**, *38*, 552–573. https://doi.org/10.1177/0734904120952347

[456] Allen, C. W. The Use of Phosphazenes as Fire Resistant Materials. *J. Fire Sci.* **1993**, *11*, 320–328. https://doi.org/10.1177/ 073490419301100404

[457] Zhou, X.; Qiu, S. L.; Mu, X. W.; Zhou, M. T.; Cai, W.; Song, L.; Xing, W. Y.; Hu, Y. Polyphosphazenes-Based Flame Retardants: A Review. *Compos. B Eng.* **2020**, *202*, 108397. https://doi.org/10.1016/j.compositesb.2020.108397

[458] Klinkowski, C.; Wagner, S.; Ciesielski, M.; Döring, M. Bridged Phosphorylated Diamines: Synthesis, Thermal Stability and Flame Retarding pProperties in Epoxy Resins. *Polym. Degrad. Stab.* **2014**, *106*, 122–128. https://doi.org/10.1016/j. polymdegradstab.2014.04.002

[459] Schafer, A.; Seibold, S.; Walter, O.; Döring, M. Novel High T-G Flame Retardancy Approach for Epoxy Resins. *Polym. Degrad. Stab.* **2008**, *93*, 557–560. https://doi.org/10.1016/j. polymdegradstab.2007.11.016

[460] Garth, K.; Klinkowski, C.; Fuhr, O.; Döring, M. Synthesis of a New Phosphorylated Ethylamine, Thereon Based Phosphonamidates and Their Application as Flame Retardants. *Heteroatom Chem.* **2017**, *28*, e21407. https://doi.org/10.1002/ hc.21407

[461] Tang, S.; Qian, L. J.; Qiu, Y.; Dong, Y. P. High-Performance Flame Retardant Epoxy Resin Based on a bi-Group Molecule Containing Phosphaphenanthrene and Borate Groups. *Polym. Degrad. Stab.* **2018**, *153*, 210–219. https://doi.org/10.1016/j. polymdegradstab.2018.04.037

[462] Zhang, P. K.; Fan, H. J.; Tian, S. Q.; Chen, Y.; Yan, J. Synergistic Effect of Phosphorus-Nitrogen and Silicon-Containing Chain Extenders on the Mechanical Properties, Flame Retardancy and Thermal Degradation Behavior of Waterborne Polyurethane. *RSC Adv.* **2016**, *6*, 72409–72422. https://doi.org/10.1039/c6ra15869b.

[463] Shao, L. S.; Xu, B.; Ma, W.; Wang, J. Y.; Liu, Y. T.; Qian, L. J. Flame Retardant Application of a Hypophosphite/ Cyclotetrasiloxane Bigroup Compound on Polycarbonate. *J. Appl. Polym, Sci.* **2020**, *137*, 48699. https://doi.org/10.1002/ app.48699

[464] Peng, Y.; Xue, B. X.; Song, Y. H.; Wang, J.; Niu, M. Preparation of a Novel Phosphorus-Containing Organosilicon and Its eEffect on the Flame Retardant and Smoke Suppression of Polyethylene Terephthalate. *Polym. Adv. Technol.* **2019**, *30*, 1279–1289. https://doi.org/10.1002/ pat.4561

[465] Luo, H. Q.; Rao, W. H.; Liu, Y. L.; Zhao, P.; Wang, L.; Yu, C. B. Novel Multi-Element DOPO Derivative Toward Low-Flammability Epoxy Resin. *J. Appl. Polym, Sci.* **2020**, *137*, e49427. https://doi.org/10.1002/app.49427

[466] Zhang, W. C.; Li, X. M.; Yang, R. J. Flame Retardant Mechanisms of Phosphorus-Containing Polyhedral Oligomeric Silsesquioxane (DOPO-POSS) in Polycarbonate Composites. *J. Appl. Polym, Sci.* **2012**, *124*, 1848–1857. https://doi.org/10.1002/app.35203

[467] Qian, X. D.; Song, L.; Yuan, B. H.; Yu, B.; Shi, Y. Q.; Hu, Y.; R. K. K. Yuen. Organic/Inorganic Flame Retardants Containing Phosphorus, Nitrogen and Silicon: Preparation and Their Performance on the Flame Retardancy of Epoxy Resins as a Novel Intumescent Flame Retardant System. *Mater. Chem. Phys.* **2014**, *143*, 1243–1252. https://doi.org/10.1016/j. matchemphys.2013.11.029

[468] Liu, C.; Chen, T.; Yuan, C. H.; Song, C. F.; Chang, Y.; Chen, G. R.; Xu, Y. T.; Dai, L. Z. Modification of Epoxy Resin Through the Self-Assembly of a Surfactant-Like Multi-Element Flame Retardant. *J. Mater. Chem. A.* **2016**, *4*, 3462–3470. https://doi.org/10.1039/c5ta07115a

[469] Ai, L.; Yang, L.; Hu, J.; Chen, S.; Zeng, J.; Liu, P. Synergistic Flame Retardant Effect of Organic Phosphorus-Nitrogen and Inorganic Boron Flame Retardant on Polyethylene. *Polym. Eng. Sci.* **2020**, *60*, 414–422. https://doi.org/10.1002/ pen.25296

[470] Yang, S.; Zhang, Q. X.; Hu, Y. F. Synthesis of a Novel Flame Retardant Containing Phosphorus, Nitrogen and Boron and Its Application in Flame-Retardant Epoxy Resin. *Polym. Degrad. Stab.* **2016**, *133*, 358–366. https://doi.org/10.1016/j. polymdegradstab.2016.09.023

[471] Zhang, T.; Liu, W. S.; Wang, M. X.; Liu, P.; Pan, Y. H.; Liu, D. F. Synthesis of a Boron/Nitrogen-Containing Compound Based on Triazine and Boronic Acid and Its Flame Retardant Effect on Epoxy Resin. *High Perform. Polym.* **2017**, *29*, 513–523. https://doi.org/10.1177/0954008316650929

[472] Jin, W. Q.; Yuan, L.; Liang, G. Z.; Gu, A. J. Multifunctional Cyclotriphosphazene/Hexagonal Boron Nitride Hybrids and Their Flame Retarding Bismaleimide Resins with High Thermal Conductivity and Thermal Stability. *ACS App. Mater. Interf.* **2014**, *6*, 14931–14944. https://doi.org/10.1021/ am502364k

[473] Song, P. A.; Xu, L. H.; Guo, Z. H.; Zhang, Y.; Fang, Z. P. Flame-Retardant-Wrapped Carbon Nanotubes for Simultaneously Improving the Flame Retardancy and Mechanical Properties of Polypropylene. *J. Mater. Chem.* **2008**, *18*, 5083–5091. https://doi.org/10.1039/b808309f

[474] Yan, L.; Xu, Z. S.; Deng, N.; Chu, Z. Y. Synergistic Effects of Mono-Component Intumescent Flame Retardant Grafted with Carbon Black on Flame Retardancy and Smoke Suppression Properties of Epoxy rResins. *J. Therm. Anal. Calorim.* **2019**, *138*, 915–927. https://doi.org/10.1007/ s10973-019-08298-2

[475] Yan, L.; Xu, Z. S.; Wang, X. H.; Deng, N.; Chu, Z. Y. Preparation of a Novel Mono-Component Intumescent Flame Retardant for Enhancing the Flame Retardancy and Smoke Suppression Properties of Epoxy Resin. *J. Therm. Anal. Calorim.* **2018**, *134*, 1505–1519. https://doi.org/10.1007/ s10973-018-7810-x

[476] Yue, X. P.; Li, C. F.; Ni, Y. H.; Xu, Y. J.; Wang, J. Flame Retardant Nanocomposites Based on 2D Layered Nanomaterials: A Review. *J. Mater. Sci.* **2019**, *54*, 13070–13105. https://doi.org/10.1007/s10853-019-03841-w

[477] Fang, F.; Ran, S. Y.; Fang, Z. P.; Song, P. G.; Wang, H. Improved Flame Resistance and Thermo-Mechanical Properties of Epoxy Resin Nanocomposites from Functionalized Graphene Oxide via Self-Assembly in Water. *Compos. Part B-Eng.* **2019**, *165*, 406–416. https://doi.org/10.1016/j.compositesb.2019.01.086

[478] Bao, C. L.; Guo, Y. Q.; Yuan, B. H.; Hu, Y.; Song, L. Functionalized Graphene Oxide for Fire Safety Applications of Polymers: A Combination of Condensed Phase fFlame Retardant Strategies. *J. Mater. Chem.* **2012**, *22*, 23057–23063. https://doi.org/10.1039/c2jm35001g

[479] Guo, Y. Q.; Bao, C. L.; L, Song, Yuan, B. H.; Hu, Y. In Situ Polymerization of Graphene, Graphite Oxide, and Function-Alized Graphite Oxide into Epoxy Resin and Comparison Study of on-the-Flame Behavior. *Ind. Eng. Chem. Res.* **2011**, *50*, 7772–7783. https://doi.org/10.1021/ ie200152x

[480] Cai, W.; Feng, X. M.; Wang, B. B.; Hu, W. Z.; Yuan, B. H.; Hong, N. N.; Hu, Y. A Novel Strategy to Simultaneously Electrochemically Prepare and Functionalize Graphene with a Multifunctional Flame retardant. Chem. Eng. J. **2017**, *316*, 514–524. https://doi.org/10.1016/j.cej.2017.01.017

[481] Cai, W.; Guo, W. W.; Pan, Y.; Wang, J. L.; Mu, X. W.; Feng, X. M.; Yuan, B. H.; Wang, B. B.; Hu, Y. Polydopamine-Bridged Synthesis of Ternary h-BN@PDA@SnO2 as Nanoenhancers for Flame Retardant and Smoke Suppression of Epoxy Composites. *Compos. A Appl. Sci. Manuf.* **2018**, *111*, 94–105. https://doi.org/10.1016/j. compositesa.2018.05.015

17 Uses of Fire Tests for Flame-Retardant Material Development

Bernhard Schartel
Bundesanstalt für Materialforschung und -prüfung (BAM),
Unter den Eichen 87, 12205 Berlin, Germany; bernhard.schartel@bam.de

17.1 INTRODUCTION

For economic reasons, flame-retarded commodity and engineering plastics are used widely, whereas only a few niche products are made from inherently flame-resistant high-performance polymers. There is increasing demand for sustainable materials, for materials that pass fire regulations with increasingly strict tests, and for materials that are suitable for new applications that require advanced property combinations. Thus, developing new flame-retarded polymeric materials is a current challenge for material science, research, and industrial development. Although flame-retarded polymeric materials are clearly mass-produced products, the field is characterized by quite a high level of innovation, meaning that a large number of commercialized materials has been developed within the last 20 years. Most of the development is based on the optimization of established approaches. These include, for instance, the modification of flame retardants using advanced technologies such as micro-encapsulation to increase stability against hydrolysis or avoid agglomeration, adjusting combinations of flame retardants with synergists or adjuvants, and proposing new combinations of additives in order to reduce the load or costs while increasing the efficiency of flame retardancy. Some work focuses on the synthesis of new derivates, adjusting properties of flame retardants such as molecular weight, thermal stability, or compatibility with the polymer. Ground-breaking new approaches such as nanocomposites,[1–4] the incorporation of 9,10-dihydro-9-oxa-10-phosphaphenanthrene-10-oxide (DOPO) structures in thermosets,[5–11] or metal phosphinate salts,[12–15] have been advanced in the last 25 years. While these advancements are rare, they are also a focus of discussion due to the promise they show for fulfilling future demands with respect to sustainability and efficiency. All of these developments are based primarily on rather empirical concepts and screening procedures. It should be noted that empirically-based development includes "trial and error" as well as decades of experience and ingenious sophisticated approaches. However, a detailed scientific understanding of the flame-retardant modes of action or structure-property relationships is usually lacking, even where successfully commercialized products are concerned. This fact is due mainly to two inherent characteristics of the task of developing flame-retarded materials.

First, it is rather typical for the area that the plethora of parameters that need to be optimized to obtain a product, such as the number and kind of components (polymer, flame retardant, synergist, adjuvant, additive), the concentration of each compound, the morphology of both the material (in particular the distribution of the components) and of each component (particle size, shape, and particle-size distribution), and the modification of each compound through processes such as coating and micro-encapsulation, but also molecular weight and branching, make for a very large multidimensional matrix. Analytical description of the matrix is impossible as long as the manifold interactions between the different compounds and their influences on fire behavior are unknown, apart from an empirical understanding of the main mechanisms. Indeed, even these are often described in detail and fully understood only for rather simple or model systems. Thus, it has been concluded that typical working packages in the area of developing flame-retarded polymeric materials demand high-throughput screening or combinatorial methods.[16–20] Even so, if an ingenious empirical concept reduces the number of systems to be investigated, the comprehensive screening of a variety of systems generally remains. So in a typical example, patent DE 42 31 774 A1 from 1994 protects a thermoplastic polymer flame-retarded with a combination of 0.1–30 parts per weight of a phosphorous flame retardant and 0.1–15 parts per weight aluminosilicate.[21] This special idea proposed by the patent covers an overwhelming number of possible components; all kinds of different polymeric matrices, distinct phosphorous compounds, and a plethora of aluminosilicates.

The second reason is that flame-retardancy mechanisms[22–26] on the molecular or microscopic scale are used to protect macroscopic specimens so that they pass a distinct fire test conceived with respect to a distinct application.[27] How distinct flame-retardancy mechanisms influence the flame-retardant modes of action and thus the performance in a distinct fire test is not a trivial question and cannot be answered generally, since a whole set of materials properties such as char yield (μ), effective heat of combustion of the volatiles (h_c), heat of gasification (h_g), thermal conductivity,

DOI: 10.1201/9781003380689-17

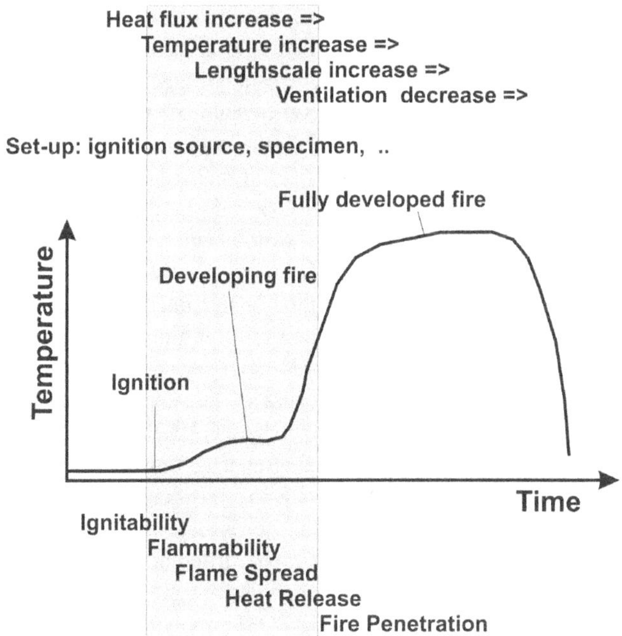

FIGURE 17.1 Schematic temperature development during a (room) fire; illustrating the fire scenarios: ignition, developing fire and fully developed fire with their characteristics heat flux, temperature, length scale involved, and ventilation; and the predominant fire properties ignition, flammability, flame spread, heat release, and fire penetration. The grey area marks the main area in which the flame retardancy of polymers is applied.

melt viscosity, and surface emissivity as well as the properties of the protective fire residue formed during burning usually determines the fire behavior of a specimen. Furthermore, the role played by each material property differs from one fire test to the next. Fire tests do not focus on certain material properties; instead, each fire test simulates a distinct fire scenario to check the response of a defined test specimen to assess whether a certain protection goal is achieved. The fire scenarios and protection goals considered are very different with respect to different applications. Generally, fires differ crucially in their heat and mass transport, defined by characteristics such as applied heat flux ($\dot{q}''_{ex}$), temperature, length scales, and ventilation (Figure 17.1). Their behavior can be divided into three main general stages of fire, or three general fire scenarios:

- Ignition: Piloted ignition is the onset of flaming combustion, characterized by an ignition source (flame, cigarette, glow wire, etc.), small length scale (cm), ambient temperatures in the range of ignition temperatures (600–700 K), and high ventilation.
- Developing fire: The continuation of flaming combustion occurring in this stage of fire growth is characterized by an external heat flux $\dot{q}''_{ex}$ = 20–60 kW m^{-2}, larger length scales (dm to m), ambient temperatures above the ignition temperature (700–900 K), and continuing high ventilation.
- Fully developed fire: The penultimate stage of fire growth is characterized by a high external heat flux $\dot{q}''_{ex} > 50$ kW m^{-2}, large length scales > m, ambient temperatures above auto-ignition temperatures (> 900 K), and low ventilation.

These different fire scenarios highlight different fire properties.

Ignition:

- Ignitability (time to ignition (t_{ig}), critical mass loss rate or heat flux for ignition, ignition temperature (T_{ig}), etc.)
- Flammability (extinction behavior under ambient conditions after removing an ignition source, reaction to a small flame)

Developing fire:

- Flame spread (opposed flow and wind-aided flame spread, fire growth rate, etc.)
- Heat release rate (HRR)
- Total heat release (THR)/total heat evolved (THE)/fire load (THR is the total heat released up to certain time point of testing and thus a function of time; THE = THR after burning and thus the fire load of the specimen monitored by the applied fire test)

Fully developed fire:

- Heat penetration
- Fire penetration/fire resistance
- THE/fire load
- Structural integrity/fire stability

As mentioned previously, these properties such as reaction to small flame (flammability), flame spread, and fire penetration are not material properties, but merely the performance of a certain specimen in a certain fire scenario. By the way, terms such as "fire-proof," "flame resistant," and "self-extinguishing," but also "flammable" and "(in) combustible" often have different meanings depending on the standard or regulation applied; this situation can cause confusion. It is strongly recommended to use the scientific terms according to the established state of the art summarized in ISO 13943 and ASTM E 176.[28] "Flammability" is used in this chapter in the specific sense of "showing sustained flaming after removal of a small ignition source" and not in the general sense of "fire behavior."

Most fire science and thus most fire testing focus on specific protection goals in the different applications, for good reason. Common protection goals include preventing sustained ignition (beginning of a fire), limiting the contribution to fire propagation (developing fire), or acting as a fire barrier (fully developed fire). Established fire tests try to simulate a specific fire scenario and monitor a specific fire risk or hazard from a specific specimen within that scenario. However, fire tests simulating similar fire scenarios do not necessarily deliver equal results for a material nor an equivalent ranking of materials. Even rather small differences in the fire scenario can play a crucial role and highlight different materials properties. For instance, the ignition and flammability hazards often change crucially depending on whether a pilot flame, irradiation, a glow wire, or a cigarette is used.[29,30] Further, some materials properties are important for the performance of the tested specimen in certain tests, but not in others. The melt viscosity of the material, for instance, and the resulting heat removal through dripping or melt flow have a crucial influence on some flammability tests,[31] whereas other fire tests are hardly influenced by these phenomena at all. For some applications, the market is controlled by national standards rather than international ones,[27,32] so that a material may pass the national test necessary in country A and fail the corresponding national test in country B, or vice versa. In fact, it should be noted that, in practice, developing flame-retarded polymeric materials basically means developing a material for the sole purpose of ensuring that a defined specimen can pass a distinct fire test. Sometimes this is achieved with little or no improvement in performance in another fire test. In most cases, this aspect also favors screening and empirical approaches rather than attempting to understand all of the structure-property relationships or mechanisms in detail. However, it is also the objective of this chapter to discuss the understanding of flame-retardant modes of action and structure-property relationships controlling the performance of polymeric materials in different fire tests. In particular, the utility of understanding flame retardancy and comprehensively assessing fire behavior are indicated as scientific fundamentals for the future, more tailored development of innovative materials.

Only the use of those fire tests most important for the development of flame-retarded polymers is discussed. As such, the chapter abstains from describing test apparatuses or the testing procedures themselves, since the relevant international standards can be used to ensure accurate quality assurance, and the comprehensive assemblies have already been described in established textbooks.[27] Instead, the main aim of this chapter is to arouse the reader's interest in advanced uses of fire tests for the development of new materials. Further, the chapter tries to give comprehensive insight into how fire testing can be used for the tailored development of flame-retarded materials, not a comprehensive review what has been achieved so far.

17.2 IGNITION: REACTION TO A SMALL IGNITION SOURCE

Considering the necessary amounts of material and time for investigation, which, taken together, constitute the costs, in practice, small- or bench-scale methods (specimen size up to 1 dm^2) are required to develop flame-retarded polymers. This length scale is typical only for the beginning of a fire and hence ignition, flammability, or response to an ignition source such as a small flame. Correspondingly, small-scale or bench-scale fire tests have become established, like limiting oxygen index (LOI, ISO 4589), glow wire tests (glow wire ignition temperature = GWIT, IEC 60695–2–13; glow wire flammability index = GWFI, IEC 60695–2–12), and reaction to a small flame—in particular, vertical and horizontal flame test methods according to IEC 60695–11–10/ASTM D3801 (= UL 94), which monitor ignition or sustained flammability after a defined contact with a certain ignition source. Tests such as the glow wire test represent a very specific scenario and are used only in the process of developing materials intended for a given application, whereas UL 94 and LOI are very common, widespread, and often used universally.

Whereas UL 94 delivers only a classification based on a pass-and-fail system, LOI can be used to rank and compare the flammability behavior of different materials. In Figure 17.2 the increasing LOI values are presented for different polymers as an example: POM = poly(oxymethylene), PEO = poly(ethyl oxide), PMMA = poly(methyl methacrylate), PE = poly(ethylene), PP, ABS, PS, PET = poly(ethylene terephthalate), PVA = poly(vinyl alcohol), PBT, PA = poly(amide), PC, PPO = poly(phenylene oxide), PSU, PEEK = poly(ether ether ketone), PAEK = poly(aryl ether ketone), PES, PBI = poly(benzimidazole), PEI = poly(ether imide), PVC = poly(vinyl chloride), PBO = poly(aryl ether benzoxazole), PTFE. The higher the LOI, the better the intrinsic flame retardancy. Apart from rigid PVC, nearly all commodity and technical polymers are flammable. Only a few high-performance polymers show self-extinguishing. Table 17.1 shows an example of how the LOI is used in the development of flame-retarded materials. The flame retardant red phosphorus (P_{red}) increases flame retardancy in HIPS and glass-fiber-reinforced PBT (PBT-GF), but only up to a certain amount of P_{red}. This observation corresponds to similar studies reported in the literature.[33–35] P_{red} is reported not only to cause flame inhibition, but also to harbor the propensity to enhance fire in the gas phase,[36] which may explain the behavior observed in HIPS and PBT-GF.

LOI and UL 94 are so common for plastics that in practise their results are often discussed as intrinsic fire properties of materials, which is not correct.[31,37–39] Even such small-scale fire tests are specific fire scenarios. This becomes obvious when, for instance, the dimensions of the specimen are changed,[40,41] as is shown in Figure 17.3 for carbon-fiber (60 vol%) reinforced epoxy composites (EP-CF). The LOI and the UL 94 are not intrinsic material properties but describe the ease of extinction of a specific specimen, which is of course crucially dependent on parameters such as the thickness of the specimen. Indeed, the LOI is reported to depend on every parameter of the setup such as temperature, mode of ignition, column size and diameter, gas flow rate, sample condition, and sample dimension.[40–44] It becomes obvious that tests which differ from the UL 94 or LOI standard, respectively, must not be performed and called UL 94 or LOI, respectively, because they lack comparability. Furthermore, comparing these two

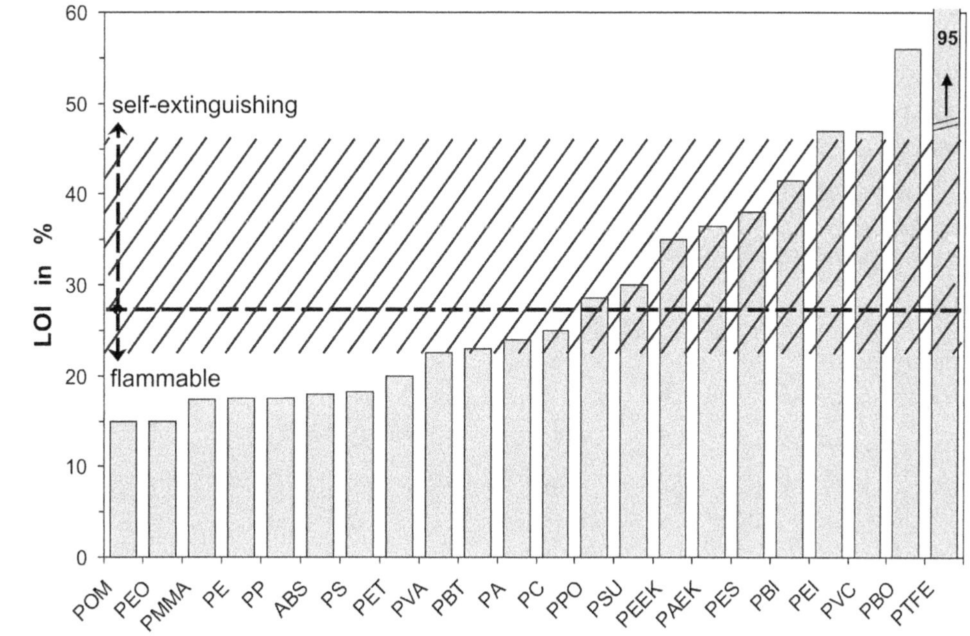

FIGURE 17.2 Limiting oxygen index (LOI) of several polymers. Often LOI values of up to roughly 27% (dashed line) signify flammability; higher LOI values, self-extinguishing behavior in the UL 94 test. There is no clear correlation between LOI and UL 94, and some extremes down to 22% and up to 47% (hatched zone) are reported for polymers that are self-extinguishing or flammable, respectively.

TABLE 17.1
LOI Results with Increasing P_{red} Content in HIPS and PBT-GF

	LOI/% (± 1%)		LOI/% (± 1%)
HIPS	17.2	PBT-GF	20.1
HIPS/7.5 wt% P_{red}	22.2	PBT-GF/7 wt% P_{red}	26.7
HIPS/10 wt% P_{red}	22.5	PBT-GF/10 wt% P_{red}	26.8

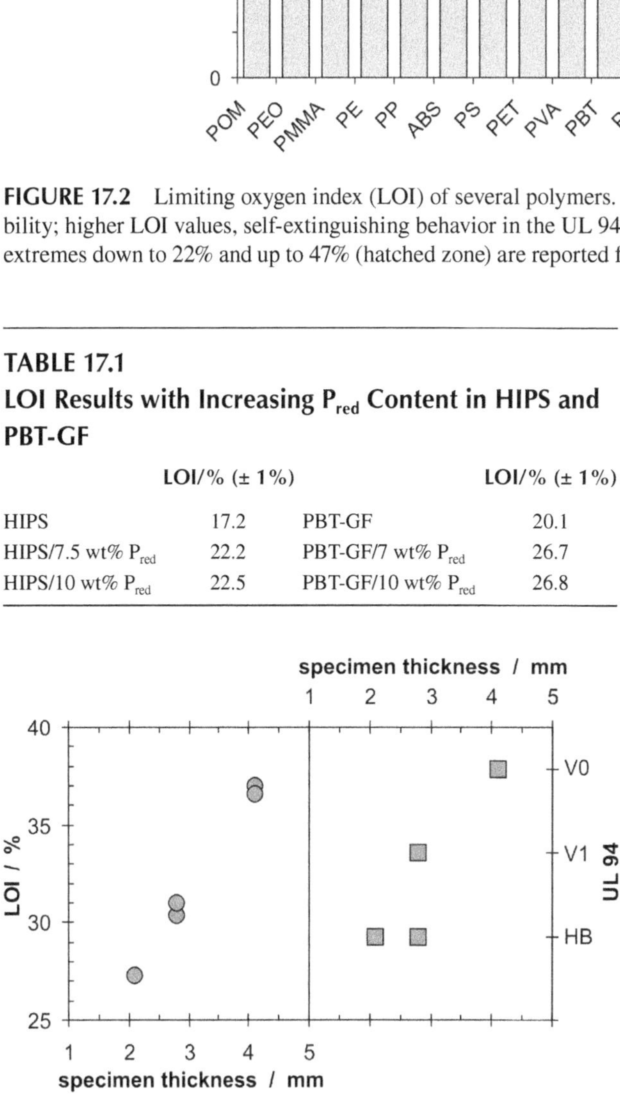

FIGURE 17.3 Influence of specimen thickness (2.1–4.2 mm) of EP-CF on the results of LOI and UL 94.

different specific fire scenarios, LOI and UL 94, shows that there is no clear general correlation between them, as indicated above in Figure 17.2. The main difference is not that a critical oxygen concentration is monitored in the LOI and an afterburning time in UL 94, as both results are based on observing the critical condition for extinguishing. The main difference is that LOI focuses on a downward flame spread and UL 94 on an upward flame spread.[45–48] Indeed it was concluded that the inability of flame propagation is the explicit cause of extinction in the LOI and should not be interpreted as a limit below which burning of the material cannot take place.[49] Modified LOI with upward burning was proposed[50] and showed better correlation with the UL 94.[51]

In general, LOI values of up to roughly 27% signify flammability[37]; higher LOI values self-extinguishing behavior in the UL 94 test (Table 17.2). However, there is no clear correlation between LOI and UL 94, and some extremes down to 22% and up to 47% (Figure 17.2, Table 17.2) are reported for polymers that are self-extinguishing or flammable in the UL 94, respectively.[38,39,52–56] The correlation seems to disappear when a set of materials is compared such as glass-fiber-reinforced PA 66 (PA 66-GF), PA 66-GF/melamine polyphosphate (MPP), PA 66-GF/aluminum phosphinate ($AlPO_2$), and PA 66-GF/AlPO2/MPP, in which the flame-retardant modes of action alternate between flame inhibition, charring, and barrier formation (Table 17.2).[56] The protection of the char is sensitive to ignition and burning from the top (LOI) or from the bottom (UL 94: V-classifications).[38]

Table 17.3 presents a rough empirical approach summarizing the LOI improvement achieved by adding 5–15 wt% flame retardant to some polymer systems.[52,56–61] LOI seems to be very sensitive to flame inhibition and efficient char formation. The important role of char yield (μ) was noted back in the 1970s by van Krevelen, who reported on the correlation between the char yield (μ) of halogen-free

TABLE 17.2
Flammability of Different Materials (APP = Ammonium Polyphosphate, DOP-Et = 10-Ethyl-9,10-Dihydro-9-Oxa-10-Phosphaphenanthrene 10-Oxide, DOP-Cy = 1,3,5-Tris[2-(9,10-Dihydro-9-Oxa-10-Phosphaphenanthrene 10-Oxide-10-)Ethyl]1,3,5-Triazine-2,4,6(1H,3H,5H)-Trione)

Material		EP	+ 10 wt% APP	+ 15 wt% APP
LOI/% (± 1%)		20.5	23.8	28.9
UL 94		HB	HB	V-0
	HIPS	HIPS/P_{red}	HIPS/$Mg(OH)_2$	HIPS/P_{red}/$Mg(OH)_2$
LOI/% (± 1%)	17.2	22.2	19.5	22.8
UL 94	HB	V-2	HB	V-1
		EP-CF	EP-CF/DOP-Et	EP-CF/DOP-Cy
LOI/% (± 1%)		31	44	49
UL 94		HB	V-1	V-0
	PA 66-GF	+ MPP	+ AlPO2	+ AlPO2/MPP
LOI/% (± 1%)	21.5	28.9	37.9	28.2
UL 94	HB	HB	HB	V-0

polymers—measured by thermogravimetry under nitrogen—and their fire properties like limiting oxygen index (LOI).[62,63] This approach was refined to take char yield (μ) and the effective heat of combustion of the pyrolysis products (h_c) divided by the heat of gasification (h_g) into account.[64–67] Indeed, the effective heat of combustion of volatile pyrolysis products is observed to lie between 3 and 45 kJ g^{-1}.[32,64,66,68] However, fire behavior is controlled not by the effective heat of combustion of the volatiles (h_c), but by its product with combustion efficiency ($\chi \cdot h_c$). Whereas combustion efficiency is near a value of $\chi = 1$ in the well-ventilated LOI or UL 94 setup, it typically changes to values below $\chi < 0.8$ for effective flame inhibition. Further, in some systems the best results in flammability seem to be achieved when flame inhibition is accompanied by some stable residue formation (Table 17.3).[52,56–61]

Furthermore, melt flow and dripping are sensitive to ignition and burning from the top (LOI) or from the bottom (UL 94: V classifications)[38] as well as to a horizontal (UL 94: HB classifications) or vertical (LOI, UL 94: V classifications) specimen position during testing. Both melt flow and dripping are significant phenomena to remove heat from the pyrolysis zone, thus reducing heat release, flame spread, and enhancing self-extinction. In the LOI two kinds of burning for a thermoplastic material were reported, clean burning and burning with melting and dripping, different in the resulting LOI values.[69] The left side of Figure 17.4 illustrates the influence of dripping on an upward-burning fire behavior for the investigated polypropylene graft maleic anhydride (PP-g-MA), which extinguishes due to dripping, but the burning drops of which probably ignite the surroundings, and a corresponding layered silicate (LS) PP-g-MA nanocomposite (PP-g-MA/LS), for which the dripping and hence the extinction of the specimen is inhibited.[57,70] The right side of Figure 17.4 shows the residues of PC/ABS, PC/ABS/PTFE (PTFE content < 0.5 wt%), PC/ABS/bisphenol A bis(diphenyl phosphate) (BDP), and PC/ABS/PTFE/BDP after UL 94 testing.[71] The anti-dripping effect of adding a small amount of PTFE becomes apparent,[26,72] whereas the flame retardant BDP also shows a clear plasticization effect. In Table 17.4 the corresponding LOI and UL 94 results are summarized. Not only PC/ABS shows self-extinction in the vertical UL 94 test due to burning dripping, but also PC/ABS/BDP. Indeed, it shows a clearly enhanced propensity for dripping (Figure 17.4). Adding PTFE inhibits dripping so that no self-extinction is observed in PC/ABS/PTFE (= HB classification). PTFE is a very important synergist for the effectiveness of aryl phosphates in PC/ABS. Avoiding dripping in combination with aryl phosphates, PC/ABS/PTFE/BDP achieves V-0, the prerequisite for challenging applications. Only when PTFE stabilizes the pyrolysis zone in the vertical UL 94 test is the flame-retardant action of BDP effective in the UL 94 V test for PC/ABS.[71]

TABLE 17.3
LOI Performance of the Materials (PP-g-MA = Poly(Propylene-Graft Maleic Anhydride), LS = Layered Silicate, ZnPi = Zinc Phosphinate)

	LOI/% (± 1%)		LOI/% (± 1%)
PP-g-MA	19.2	PA 66-GF	22.4
PP-g-MA/5% LS	19.3	PA 66-GF/7% P_{red}	26.0
Only barrier		Charring	
	LOI/% (± 1%)		LOI/% (± 1%)
HIPS	17.2	PBT-GF	19.7
HIPS/15% $Mg(OH)_2$	19.5	PBT-GF/13.3% ZnPi	27.0
Barrier + additional effects		Flame inhibition	
	LOI/% (± 1%)		LOI/% (± 1%)
PA 66-GF	21.5	PBT-GF	19.7
PA 66-GF/11.4% AlPi	37.9	PBT-GF/13.3% AlPi	42.0
Flame inhibition + charring		Flame inhibition + charring	

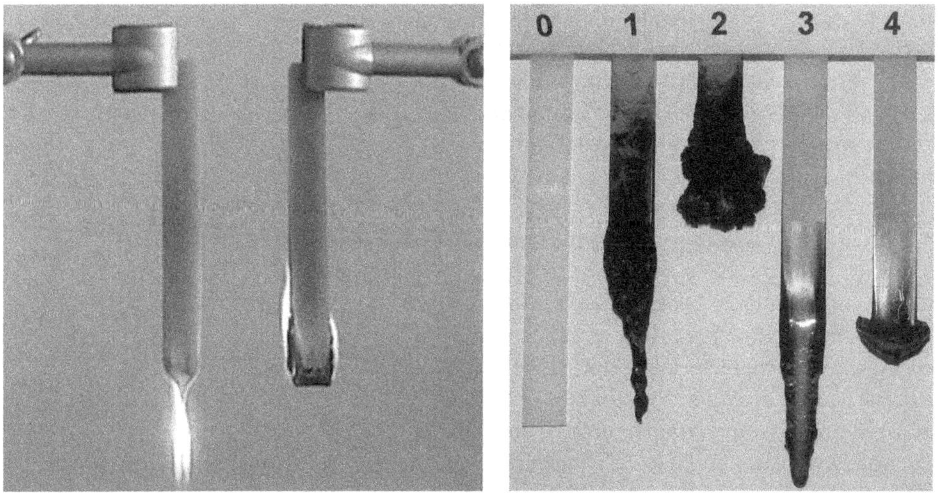

FIGURE 17.4 Influence of dripping on the fire behavior of PP-g-MA and PP-g-MA/LS nanocomposite (on left) and UL 94 residues (specimen 0 before UL 94 test) of PC/ABS (1), PC/ABS/PTFE (2), PC/ABS/BDP (3), PC/ABS/PTFE/BDP (4), PC/ABS and PC/ABS/BDP show extinction due to dripping (V-2); PC/ABS/PTFE HB behavior; PC/ABS/PTFE/BDP shows self-extinction (V-0).

TABLE 17.4
Flammability (LOI, UL 94) for PC/ABS, PC/ABS/PTFE, PC/ABS/BDP, PC/ABS/BDP/PTFE

	PC/ABS	PC/ABS/ PTFE	PC/ABS/ BDP	PC/ABS/ BDP/PTFE
LOI/% (± 1%)	23.2	23.6	26.0	28.1
UL 94	V-2	HB	V-2	V-0

Fibers such as glass fibers in glass-fiber-reinforced composites also inhibit dripping and melt flow [73] and thus often decrease the LOI or UL 94 performance, at least for the usual glass-fiber content. Consequently, flame-retarding such composites is generally more challenging than for their corresponding pure polymers and often requires different flame-retardancy approaches than for the polymer without glass fibers.

UL 94 and LOI are discussed in detail, since these two are used most commonly and universally in materials' flammability development. However, there are many more small heat source ignition tests. First of all, it should be noted that there are further tests based on a small flame as ignition source, similar to LOI and UL 94, such as small burner/flame tests (building products, textiles) or the needle flame test (electrical engineering). There are also two further tests based directly on the LOI setup and test procedure: LOI for elevated temperatures and determination of a nitrous oxide index. Both deliver quite interesting results with respect to flame-retardant modes of action[40] but are used extremely rarely. The FMVSS 302 test for motor vehicle interior components is another common Bunsen burner–type test, but its specific protection goal is so low that it is probably not relevant for development challenges. As indicated before, there are also tests which are crucially different as they use not a pilot flame but irradiation, a glowing wire, or a cigarette as the ignition source.[30] Often they simulate specific scenarios that are important for a given application, such as cigarette tests for upholstery and mattresses, and glow-wire tests and high-current/voltage-arc ignition tests in electrical engineering and electronics.

LOI (of minor importance in building, railway-vehicle codes) and UL 94 (important test for electronics and electrical engineering regulations) are commonly and successfully used far beyond their mandatory performance for the screening and estimation of fire behavior in flame-retarded materials development. However, due to their specific fire scenario, the explanatory power of LOI and UL 94 is restricted, particularly when the fire scenario or the protection goal is changed. Due to its advantages, the use of LOI in particular has been questioned, since its minor importance in genuine fires and its less demanding downward-burning configuration yields no or only rough correlations with other fire tests.[38,39,74] However, probably because the V-1 and V-0 classification is quite demanding in UL 94 and achieved only through a substantial flame-retardancy approach, the results correlate surprisingly well with fire tests covering the fire scenarios of a developing fire.[75]

17.3 DEVELOPING FIRE: FORCED FLAMING COMBUSTION

The cone calorimeter (ASTM E1354, ISO 5660) is a rather performance-based bench-scale fire testing apparatus.[76–79] It is a reaction-to-fire test and realizes forced-flaming conditions[66] and thus simulates a developing fire scenario with a very small specimen size. Therefore, it is not surprising that beyond its use in mandatory codes (transportation, building in Japan) the cone calorimeter has found increasing implementation as a characterization tool in the research and development of fire-retarded polymeric materials. When characteristics such as peak HRR (PHRR), THR(t), or

THE are discussed with respect to the development of new flame-retarded materials, they are used as a means of comparing systematically varied samples, whereas the absolute numbers are often of minor interest. Thus, some important aspects and approaches are addressed in the following, which received consideration in evaluating cone calorimeter data, interpreting the results and drawing conclusions.

Samples 100 mm × 100 mm square are used, most of a thickness between 3–10 mm; therefore the cone calorimeter is described as a small-scale or bench-scale fire test. Its length scale corresponds to the upper limit of the flammability tests typically used to characterize performance in ignition scenarios. The sample size is of the smallest order of magnitude discussed in fire engineering and of the largest used in polymer analysis. Hence, the cone calorimeter constitutes an important link between fire engineering and polymer science, which is crucial in the interdisciplinary area of the flame retardancy of polymers. Furthermore, it provides comprehensive insight into not only several fire risks such as HRR, THR(t), THE, and t_{ig} but also fire hazards such as smoke release and CO production. Further, the mass loss (ML) rate and ML are monitored. The cone calorimeter setup was developed thoroughly[78] to provide comprehensive and meaningful insight into the fire properties of materials,[79] rather than to correspond to a special real-scale fire scenario. However, the cone calorimeter setup also defines a specific fire scenario. As is typical for all fire tests, samples' performance in the cone calorimeter depends on the characteristics of the test, including ignition source, ventilation, irradiance (external heat flux, $\dot{q}''_{ex}$), temperature, and the geometry of the specimen. Strictly speaking, the cone calorimeter test characterizes the performance resulting from an interaction of material properties, specimen, and the defined fire scenario. The meaning of the results may have little relevance for other fire scenarios or fire tests that differ essentially in their setup.

After ignition, the cone calorimeter represents a well-defined flaming condition, forced by external radiation, typical for a developing fire scenario. The external radiant heat flux is varied; typically $\dot{q}''_{ex}$ = 25, 35 or 50 kW m^{-2} are chosen, corresponding to the heat fluxes typical for developing fires,[80,81] in the interest of replicating performance in developing fires. Thus, one of the key features of cone calorimeter testing is that reasonable insight into the developing fire behavior of a material can be obtained from a small specimen, reducing development time and cost. Although this is a significant advantage of the cone calorimeter, it is not without some important limitations. For instance, there is ignition by radiation accompanied by a spark, burning is well-ventilated, the specimens are usually in a horizontal position to rule out melt flow and dripping, no real flame spread over the surface of the test specimen occurs, and burning is one-dimensional (the flame front penetrates only the depth of the sample). This dimension, the thickness, can be varied in the cone calorimeter, and the thickness used has a significant influence on the results, as

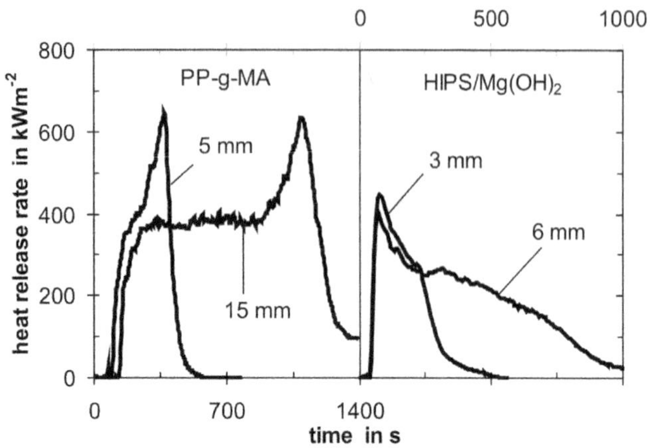

FIGURE 17.5 Comparing the influence of thickness for a non-charring (PP-g-MA) and a residue-forming (HIPS/$Mg(OH)_2$) polymeric material with respect to HRR.

is also typical of performance in authentic fires (fire load, flame spread, flammability, and ignition) and hence most fire tests. The thickness of the investigated samples has to be taken into account thoroughly in the discussion of cone calorimeter results. Thermally thin samples, especially, show a decreased ignition time, as it is no longer dependent on the conductivity controlling the heat transport into the sample, but on the thickness characterizing the limited volume that is heated up. Further, samples that are thermally thin or of intermediate thickness show a higher PHRR than do thermally thick samples.[82] There are no general limits for the different regimes, but they are dependent on the irradiation[83] and thermal properties of the specimen during burning, such as thermal conductivity and heat capacity. Further, it should be noted that, according to ISO 5660, all specimens less than 6 mm thick are considered to be thin specimens, which does not necessarily mean that they are thermally thin, but provides useful guidance about the type of behavior to be expected. Furthermore, any variation in specimen thickness results in a different typical set of HRR curves for different burning behavior. This influence is shown for non-charring and residue-forming material (Figure 17.5).[84,85] Most obviously, time to PHRR is shifted for the non-charring material by varying the thickness, whereas it remains constant for the residue-forming material. This result shows that specimen thickness is a key parameter which needs to be chosen carefully, in accordance with the aim of the cone calorimetric measurement. For applied research and development, as well as for regulatory testing, the end-use condition dictates the thickness to be tested. Thicker samples may be more advantageous to study material properties, but have little relevance with respect to end use conditions. In any case, a comparison of cone calorimeter results obtained for different materials requires testing with the same specimen thickness.

Higher irradiation levels give better reproducibility, more clearly defined ignition, and shorter measurement times, but correspond to more developed fires. Thus, particularly for flame-retarded polymers, a smaller irradiation level often corresponds better to the fire protection goals addressed. Cone calorimeter results for the HRR at small irradiances correspond to flammability tests such as LOI and UL 94, if a reasonable set of materials are compared and the behavior is not dominated by dripping effects. Thus, different considerations govern the choice of external heat flux [80,81]

The external heat flux ($\dot{q}''_{ex}$) from the cone heater does not exclusively determine the heat flux important for samples' pyrolysis in the cone calorimeter, since the re-radiation from the hot sample surface ($\dot{q}''_{rerad}$), the loss by thermal conductivity into the specimen and the surroundings ($\dot{q}''_{loss}$), and the heat flux from the flame ($\dot{q}''_{flame}$) are also of the same order of magnitude.[86–89] Thus, the heat flux effective with respect to pyrolysis during a cone calorimeter run ($\dot{q}''_{eff}$) is the result of the external heat flux and the material's response ($\dot{q}''_{eff} = \dot{q}''_{ex} + \dot{q}''_{flame} - \dot{q}''_{rerad} - \dot{q}''_{loss}$).

The cone heater is designed to provide rather homogeneous irradiation over the entire sample surface. It is reasonably optimized for a particular distance (2.5 cm) between the defined sample surface (a square 10 × 10 cm) and the cone heater. The irradiance is constant for an area larger than 5 cm × 5 cm at a distance of 2.5 cm, as specified by the ISO 5660 standard. There are considerable deviations in the heat flux when the distance is changed with respect to both the vertical and the horizontal characteristics.[84,85] Thus, strictly speaking, the cone calorimeter must not be used for samples that exhibit significant deformation, such as foams that collapse before ignition, some polymers showing extensive deformation, or intumescent materials. Changing the initial distance by the user or the change in distance due to deformation of the sample during the test will have a significant influence on the fire scenario and hence on the cone calorimeter results.

The maximum surface temperatures reached without ignition are around 570–790 K (300–520°C) for low external heat fluxes ($\dot{q}''_{ex}$ = 15–35 kW m^{-2}),[84,85] which corresponds to the critical radiant flux required to ignite many combustible organic materials; $\dot{q}''_{ex}$ = 50 kW m^{-2} causes temperatures up to around 880 K (610°C), and $\dot{q}''_{ex}$ = 70 kW m^{-2} up to 970 K (700°C). A direct measurement of the response to a small flame, like the ignition source targeted by a flammability test (UL 94, LOI), is not possible since the combination of small irradiation and a high temperature is not available in the cone calorimeter. Temperatures reached for high irradiation are lower than temperatures on the standard time temperature curve typically used in testing with respect to fully developed fires. Thus, no correlations can be drawn between the cone calorimeter and testing with the standard time temperature curve when high temperature processes are crucial.[90] After ignition, however, the materials, decomposition, pyrolysis and vaporization exert cooling effects on the surface. Indeed, for non-charring materials the pyrolysis temperature equals the surface temperature as the pyrolysis zone moves through the sample, consuming the material.

During the cone calorimeter experiment, the main result of heat release, especially HRR[85,91] and THR, is determined by oxygen consumption calorimetry.[92,93] Different types of typical burning behavior give rise to characteristic curves of HRR over time. Some are illustrated schematically in Figure 17.6[30,85]:

(a) Thermally thick non-charring (and non-residue forming) samples show a strong initial increase after ignition up to a quasi-static HRR value, corresponding to the HRR_{st}. This plateau remains until near the end of the test when the PHRR occurs. This peak is caused by diminution of $\dot{q}_{loss}$ as the glass wool supporting the sample prevents heat transfer to the sample holder as the pyrolysis zone approaches it.
(b) For non-charring samples of intermediate thickness, the plateau vanishes. The HRR_{st} is marked only by a shoulder. The PHRR increases more than for thermally thick non-charring samples since its origin is halfway between the behavior of thermally thick non-charring and thermally thin samples, as discussed below, or because the burning stage towards the end of the test becomes a kind of pool fire.
(c) Thermally thick charring (residue-forming) samples show an initial increase in HRR until an efficient char layer is formed. As the char layer thickens, this results in a decrease in HRR. The maximum reached at the beginning equals both the HRR_{st} and the PHRR.
(d) Some thermally thick charring materials, such as (but not exclusively) wood, tend to show a HRR peak at the beginning, prior to charring, and a second HRR peak at the end of the measurements. The second peak may be caused by cracking char, a second pyrolysis front with a higher pyrolysis temperature running through the test specimen, or just an increase in the effective pyrolysis zone analogous to that observed for the thick non-charring materials.
(e) Thermally thin samples are characterized by a sharp peak in HRR, since the whole sample is pyrolyzed at the same time. In this case, the PHRR is dependent on their total fire load.
(f) Some samples show HRR curves characterized by a kind of unsteady development of combustion. Reasons for this can be flashing (ignition followed by self-extinction) before a sustained flame or during the whole measurement, or deformation during burning, which changes the surface area and/or distance to the cone heater.

Obviously, this everyday interpretation of HRR curves is mainly empirical, but it clearly demonstrates that the full information on the fire behavior in terms of its HRR is available only by looking at the complete HRR curve

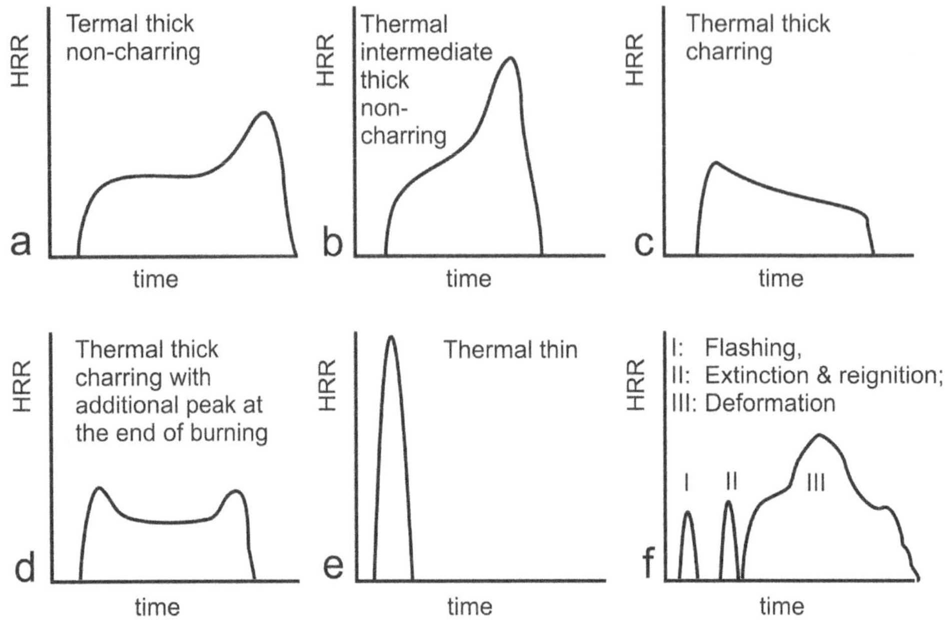

FIGURE 17.6 HRR in the cone calorimeter for typical burning behaviors (a) thermally thick non-charring; (b) intermediate thermally thick non-charring; (c) thermally thick charring; (d) thermally thick charring with additional heat release maximum at the end of burning; (e) thermally thin samples; and (f) samples that show flashing, extinction and reignition, and deformation.

over the full duration. Only the whole HRR curve can adequately represent the fire behavior controlled by the material-specific properties (char yield (μ), effective heat of combustion (h_c), heat of gasification (h_g), etc.), the influences from the specimen (thickness, deformation, etc.), and the physical and chemical mechanisms active during burning (increasing and cracking char layer, endothermic reactions, release of different pyrolysis products, afterglow, etc.). However, for practical reasons, often only characteristic values are used to describe burning behavior, such as averaged HRR or PHRR. These values do not contain the complete information and may sometimes be misleading.

The PHRR in the cone calorimeter is strongly dependent on the test setup and the specimen, as well as the intrinsic fire properties of the materials. To obtain comparable results, it is essential that the specimen and sample holder be used as defined in the standard. This is illustrated in Figure 17.7, where the cone calorimeter HRR curves are compared with results using a modified sample holder.[84,85]

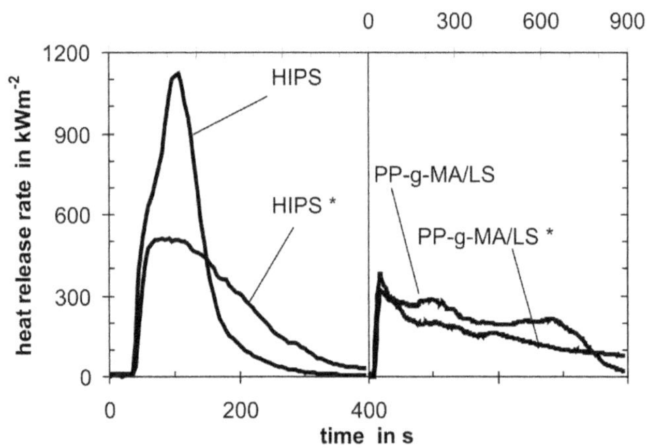

FIGURE 17.7 HRR for HIPS (non-charring) and PP-g-MA/LS (char/residue-forming) if the standard or a modified sample holder is used in the cone calorimeter. In the modified setup the sample is placed on a conductive 1-cm thick Cu plate.

The modified sample holder allows heat to be conducted away from the back of the specimen. Consequently, the large PHRR disappears at the end of burning of non-charring materials (around 100 s for a 3 mm-thick HIPS specimen in Figure 17.7), whereas the PHRR is barely affected for residue-forming materials at the beginning of the HRR curve (PP-g-MA/5 wt% LS material) (Figure 17.7). This difference is explained by the different origins of the PHRR, as described above. The PHRR of a specimen in the cone calorimeter is clearly not an "intrinsic" material property. This is also the reason why correlations with other tests such as LOI and UL 94 usually show some deficits when PHRR is used.[85] However, "non-intrinsic" material properties are not artefacts and may be quite important with regard to the assessment of genuine fire hazards.[91] Indeed, the thermal insulation at the back of the specimen in the cone calorimeter setup was originally chosen to represent free-standing material applications, and the PHRR is probably the most widely used result from cone calorimeter investigations.

To simplify the interpretation of cone calorimeter data, indices have been introduced to assess the hazard of developing fires,[94] such as $FPI = t_{ig} / PHRR$ (fire performance index),[95] PHRR/t(PHRR) or $PHRR / t_{ig}$ as a fire growth index or measure for the flashover propensity,[96] the FIGRA (fire growth rate = maximum of HRR(t)/t, sometimes but not

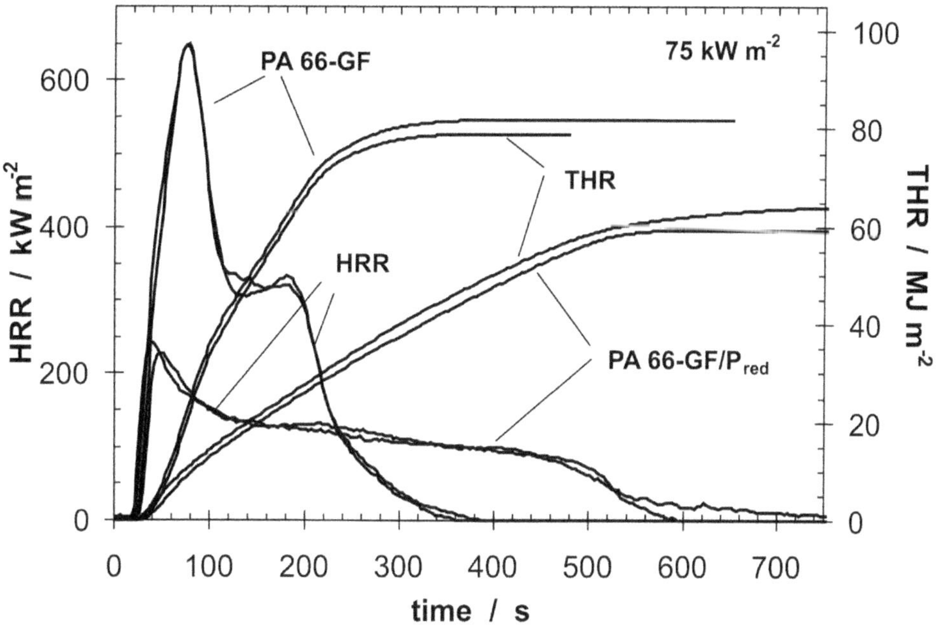

FIGURE 17.8 HRR and THR of PA 66-GF and PA 66-GF/P_{red} for irradiation = 75 kW m^{-2}. P_{red} reduces the THE and the PHRR.

always equalling PHRR/t(PHRR)),[97] parameter b (b=0.01 $HRR_{average}$—(1—t_{ig}/burning time ($t_{burning}$)),[98] $PHRR^2/t_{ig}$,[99] and the MARHE (maximum average rate of heat emission); more recently, the product of THR and PHRR/t_{ig} was proposed.[100] Some of these indices are used for regulatory purposes, such as FIGRA for the single burning item test (SBI; EN 13823) with respect to building products, and MARHE for cone calorimeter tests with respect to fire protection in railway vehicles according to EN 45545. However, these indices try to concentrate the relevant information into a single number, which has limited physical meaning with regard to fire behavior, and thus clearly oversimplify and can even be misleading. It should be noted that the flame spread in developing fires is controlled by several characteristics, such as the heat release of the burning part of the specimen, the specimen (thickness, orientation, etc.), the ignitability of the surface material, the fire scenario (ventilation, air flow, etc.), the direction in which the flame spread is considered (opposed flow, wind-aided, horizontal, vertical, etc.), and so on. Obviously, real flame spread cannot be measured in the cone calorimeter, and hence it is not incorporated directly into any of the fire growth indices.

The THR(t) during a cone calorimeter test is the integral of the HRR with respect to time—the total heat output up to that point. The THR at the end of the test is the THE and is, therefore, the fire load of the specimen in the cone calorimeter fire scenario. The THE and the HRR are mathematically related but monitor quite independent fire hazards.

Figure 17.8 and Figure 17.9 illustrate examples of how cone calorimeter data can be used in the development of flame-retarded materials. PA 66-GF without P_{red} showed typical fire behavior for non-charring polymers containing inorganic glass fiber as inert filler,[73] when high external heat flux is applied. The shape of the HRR curve is divided into two different parts. In the beginning the surface layer pyrolysis shows a sharp peak, followed by a reduced pyrolysis rate when the pyrolysis zone becomes covered by the glass-fiber network residue layer. When P_{red} was added, the PA 66-GF samples were transformed into carbonaceous char-forming materials, which led to a reduction in the THE and a reduction in the PHRR, accompanied by a prolongation of the burning time.[58] The reduction in the THE correlates with the increase in residue. The apparent effective heat of combustion (EHC) determined by the cone calorimeter software (dividing heat release by ML) is not changed, which means the ratio THE/ML ~ $\chi \cdot h_c$ is constant, indicating the absence of relevant gas phase flame-retardant modes of action such as flame inhibition or fuel dilution. The cone calorimeter investigations highlight P_{red} as an interesting halogen-free char-inducing flame retardant for PA 66 with respect to HRR and THE.

The influence of different flame retardants ($Mg(OH)_2$ and P_{red} in HIPS) and their combination is shown in Figure 17.9.[52] The most obvious effect of the 15 wt% $Mg(OH)_2$ used in the cone calorimeter is not the release of incombustible cooling agents such as water but barrier formation in the condensed phase. HIPS and HIPS/P_{red} show the typical HRR curve of a non-charring material, whereas HIPS/$Mg(OH)_2$ and HIPS/P_{red}/$Mg(OH)_2$ show the typical HRR curve of a charring—or better, residue-forming—one. The clear change in the HRR curve was accompanied by a clear decrease in PHRR and prolongation of burning time, but the amount added, 15 wt% $Mg(OH)_2$, is far too little to exert a significant influence on the THE (Figure 17.9). The barrier properties of the inorganic residue layer in HIPS/P_{red}/$Mg(OH)_2$ were clearly improved in comparison to HIPS/$Mg(OH)_2$.[52] With respect to barrier properties, the combination of P_{red} and $Mg(OH)_2$ shows clear synergy, since MgO is replaced by

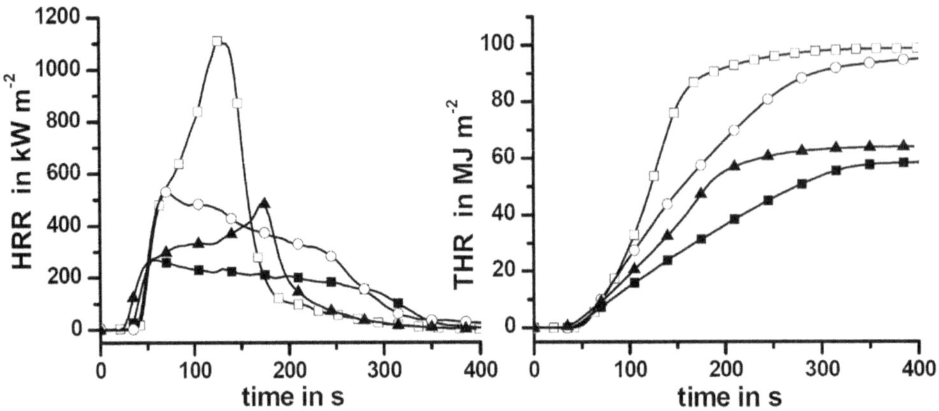

FIGURE 17.9 HRR and THR of HIPS (open squares), HIPS/P_{red} (filled triangles), HIPS/$Mg(OH)_2$ (open circles), and HIPS/P_{red}/$Mg(OH)_2$ (filled squares). HIPS and HIPS/P_{red} show the typical HRR curve of a non-charring material; HIPS/$Mg(OH)_2$ and HIPS/P_{red}/$Mg(OH)_2$, that of residue-forming material. Adding P_{red} reduces the HRR and THE by flame inhibition; $Mg(OH)_2$ reduces the HRR, mainly by introducing a barrier.

Mg-phosphates. Of course, in the cone calorimeter experiment, the formation of Mg-phosphates is indicated only by the increased residue. Unambiguous evidence was supplied by solid-state NMR investigations proving the formation of glassy Mg-phosphates.[101] Adding P_{red} to HIPS or HIPS/$Mg(OH)_2$, respectively, significantly reduced the THE/ML, and thus also the THE, through flame inhibition in the gas phase (Figure 17.9). P_{red} does not show a significant interaction with the HIPS decomposition. No additional carbonaceous char was produced. Detailed data evaluation indicates that the combination shows a slight antagonism with respect to flame inhibition.[52] However, overall the fire retardancy of HIPS containing P_{red} and $Mg(OH)_2$ was significantly better than when either additive was used alone. Since the flame-retardancy effect cannot be increased efficiently by using more P_{red} (Table 17.1), and because a very large amount of $Mg(OH)_2$ is needed to obtain performance similar to that of the combination, the combination is the far more interesting system with respect to fire retardancy.

As indicated in the introduction, not only the identification of promising polymer-flame retardant systems and promising additive combinations is at the focus of materials development, but also the amount and the optimization of the substances used. Such examples are shown in Figure 17.10 and Figure 17.11.[57,102] In Figure 17.10 the HRR and the THR of PP-g-MA/LS nanocomposites are shown with increasing LS content. In this system the additional carbonaceous char formation is negligible. The increase in residue corresponds to the inert filler effect of LS. As a consequence, the change in THE is below 10% and thus still in the same order of magnitude as the typical error. The only relevant flame-retardant mode of action that influences fire behavior in the cone calorimeter is the change in the HRR shape from a non-charring material to a residue-forming one.[70] The PHRR at the end of burning vanishes more and more and thus is replaced by the PHRR in the beginning,

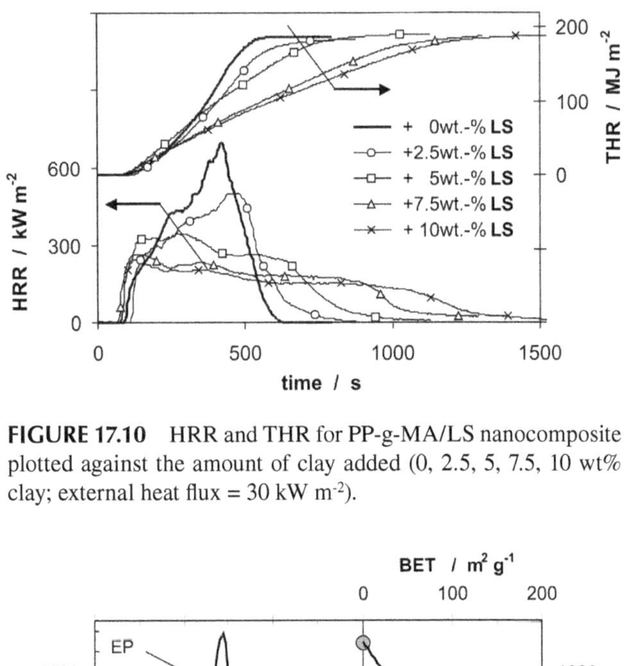

FIGURE 17.10 HRR and THR for PP-g-MA/LS nanocomposite plotted against the amount of clay added (0, 2.5, 5, 7.5, 10 wt% clay; external heat flux = 30 kW m^{-2}).

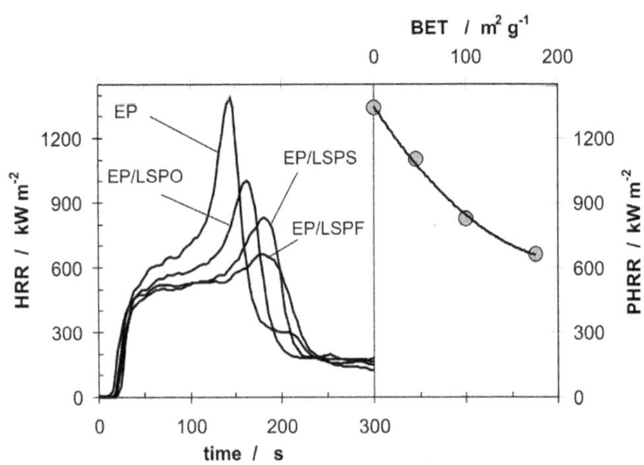

FIGURE 17.11 HRR (irradiation = 70 kW m^{-2}) of EP, EP/LSPO, EP/LSPS, EP/LSPF plotted against time and PHRR plotted against the BET of the LSP used. BET as well as the reduction in PHRR increase in the order: oven-dried (LSPO) < spray-dried (LSPS) < freeze-dried (LSPF).

FIGURE 17.12 Fire residue of EP + 15 wt% APP (left) and EP + LS + 10 wt% APP (right). Combining APP with LS inhibited intumescence, causing strong antagonism.

typical for residue-forming materials. The PHRR drops dramatically—by more than 50%—even when only small amounts of LS are added, but this effect is not linear in keeping with LS concentration, levelling out for amounts larger than 7.5 wt%. Only very high loadings are expected to show a larger impact, when the replacement of polymer results in a crucial additional reduction in fuel.

Figure 17.11 illustrates an example of the kind of development tasks that can exist and be addressed within the same materials system. The polymer and the additive as well as their concentration are the same for the layered silicate epoxy resin nanocomposites (EP/LS) investigated.[102] Only the morphology of the LS was varied, due to the different drying procedures during processing. The tetraphenyl phosphonium-modified LS was dried in a ventilated oven and subsequently ground in a ball mill (LSPO), spray-dried (LSPS), or freeze-dried (LSPF). The reduction in PHRR compared to the EP increases significantly with the increasing BET surface of the additives used. The reason for this is the increased homogeneous distribution of LS as monitored by TEM.[102]

During a cone calorimeter run, visual observation is often a crucial step in understanding the burning behavior of a specimen.[85] Bubbling, char formation, surface rise, deformation due to residual stresses, intumescence, char cracking, collapse of structures, and afterglow are examples of important observations which are made only during the test. These phenomena profoundly influence the HRR, but cannot necessarily be deduced unambiguously from the HRR curves. In Figure 17.12 such an example is shown by means of fire residues.

Combining 5 wt% LS with 10 wt% ammonium polyphosphate (APP) in EP (EP/LS/APP) inhibited the intumescence due to APP and causes a strong antagonism.[55] It is assumed that adding LS to any polymer/APP systems influences the viscosity in the pyrolysis zone, the amount, and the mechanical strength of the fire residue. These interactions can be deleterious, as in the investigated system, but can also be advantageous.[103] It should be noted that according to ISO 5660 (and ASTM E1354), visual observations are an essential part of the test results and must be recorded.

Apart from the cone calorimeter, fire testing within the fire scenario of a developing fire is usually based on larger specimens, generally ruling these tests out for comprehensive systematic investigations or screening approaches. For different applications (building, ships, railway vehicles), radiant panel tests are common, in which a real flame spread is measured based on samples 230 mm × 1,050 mm or 155 mm × 800 mm in size. Further, building codes require tests such as the European SBI test (EN 13823), which simulates fire propagation in a room corner, and the US Steiner tunnel test based on specimens of 2.25 m^2 and 3.73 m^2, respectively. Thus, within the topic development of fire-retardant materials, truly reliable procedures to predict performance in the SBI or Steiner tunnel test are needed and under development, in particular using the cone calorimeter data.[104–108]

17.4 FULLY DEVELOPED FIRE

Apart from some specific areas such as fuel tanks or intumescent coatings, the performance in a fire test simulating a fully developed fire has often not been of interest with respect to the development of fire-retarded polymers, since common flame-retarded polymeric materials are used in applications that demand the inhibition of fire hazards only with respect to the beginning (electrical engineering, electronics, etc.) and the development of a fire (building, transportation, etc.) (as illustrated in Figure 17.1). However, since recent developments like the increasing use of carbon and glass-fiber-reinforced polymer composites to replace metallic materials in load-bearing components of structural building and transportation applications, the focus of interest is shifting to passing tests simulating fully developed fire scenarios.

Testing codes within the scenario of a fully developed fire are based on intermediate, large, or full-scale testing. Specimens are typically in the dimension of several m^2 and often authentic components such as building columns are tested, or the whole product in the case of gas bottles. Tests like the small-scale test furnace based on specimens of 500 mm × 500 mm are the exception. Intensive flame application or the use of furnaces realizing standard time-temperature curves are used to simulate the characteristics of fully developed fires. Thus, in particular the heat impact of convection and the surface temperature are clearly greater than in the tests discussed above. The fire properties investigated are often resistance to fire, or the fire and/or temperature penetration.

The time effort and specimen size needed for these tests rule them out for comprehensive systematic investigations or screening approaches within the development of new materials. Thus, bench-scale setups are common to estimate performance in the larger tests. Using high external heat fluxes ($\dot{q}''_{ex} \geq 70$ kW m^{-2}) in the cone calorimeter combined with temperature measurements has been proposed, in particular for intumescent coatings,[109,110] even though this approach has some deficits with respect to the maximum surface temperature reached.[90] Furnaces that can realize the standard-time temperature curve for samples smaller than 250 mm × 250 mm are rare[111–114] but show clear advantages in simulating the fire scenario. In most cases a flame is applied directly to a smaller specimen by using Bunsen burner–type burners or burner areas, or similar but somewhat larger bench-scale burner approaches. Burners realizing heat fluxes around 180 kW/m^2 and temperatures above 1,000°C have been proposed together with compression loads in the range of several hundreds of kN to address the fire stability within the development of fire protection systems for composite materials.[115–118] All of these bench-scale approaches have some intrinsic limitations as compared to the larger test, in particular with respect to monitoring the structural integrity during fire, which often plays an important role for the fire stability component.[119] Nevertheless, bench-scale fire resistance testing is applied successfully to screen and rank materials, protective coatings, and to assist in component design.

17.5 OTHER FIRE HAZARDS

So far, the chapter has focused on preventing sustainable ignition or subsequent fire growth, which are, of course, the two main goals in fire protection. However, it is often not the heat or the flame that prevents people from escaping from a fire, but the smoke. In the majority of cases, toxic gases are the actual cause of the loss of human life. In general, combustion is never complete in a veritable fire, so that apart from products such as H_2O and CO_2, other products of incomplete combustion such as soot, smoke, HCN, and CO are quite characteristic and, in fact, important fire hazards. What is more, as a consequence, flame inhibition that reduces combustion efficiency (χ) is accompanied by an increase in CO yield and often smoke yield as well. Of course, this yield is lower than the absolute amount of material, and a burning item generally produces more smoke and toxic gases than a non-burning item; further, self-extinguishing during an early stage of a fire reduces smoke and toxic gases. However, particularly in some transportation applications like aviation, shipping, and railway vehicles, the combination of reducing fire growth and smoke yield (and/or the production and yield of toxic gases) is required and thus constitutes a crucial goal in the development of flame-retarded materials. Typical approaches are the use of a combination of a flame retardant and a smoke suppressant or implementing a change in the flame retardant or polymer material itself.

The ventilation condition has a significant effect on fire hazards like HRR, fire growth rate, smoke, and CO production; in the latter cases, this crucially influences both the absolute amount and the ranking of different materials. CO production and smoke production depend not only on the material, but also strongly on the fire scenario.[120,121] The most important influence is ventilation, but other parameters such as temperature, irradiation, residence time, and quenching effects are important. Comprehensive studies in which more than one of the key influences were systematically varied[122] are rare and show rather complex landscapes of dependencies. There are no easy correlations between different fire scenarios or fire tests, respectively. The absolute amount and the ranking of different materials in one test are of limited use in predicting the results in another. Thus, the development of flame-retarded materials in this area is based on the established smoke or toxicity tests that have to be passed for the different applications. The most important tests are probably the ones based on the NBS smoke density chamber.

Efficient fire protection is also based on the consideration of product or scenario-specific hazards, which may lead to very specific materials development goals. Examples are the combination of impacts, such as vandalism and ignition source for seats in railway vehicles, or a preceding shock wave before the fire impact in navy applications. Some such more product-specific phenomena are related directly to material properties, such as building up an increased risk for pool fires through burning thermoplastic plastics or dripping foams, and thus have become topics in the development of some flame-retarded materials.[123]

Additional goals become relevant, particularly with respect to the protection of larger properties not subjected to direct fire influences, such as corrosion induced by the fire effluents. Hereby, the long-term corrosion damage to the building and to equipment (industrial machinery as well as electrical and electronic installations) is of great interest.

17.6 UNDERSTANDING FIRE BEHAVIOR AND FIRE-RETARDANT MODES OF ACTION

Figure 17.13 illustrates the fundamental difference between the fire-retardant modes of action of P_{red} used in HIPS and PA 66-GF evaluated by cone calorimeter testing.[52,58,59,85] The THE for both materials with and without P_{red} as a fire retardant are plotted against the ML. The lack of significant change in the mass of residue shows that the phosphorus in HIPS vaporizes nearly completely, and no additional charring occurs. The THE is reduced through flame inhibition, reducing the product of the effective heat of combustion of the volatiles and combustion efficiency ($\chi \cdot h_c$), which is indicated by the reduced slope. For P_{red} in PA 66-GF $\chi \cdot h_c$ remains constant, whereas the reduction in THE is caused by a decreased ML. This shows that P_{red} induces char formation in the condensed phase.

It should not escape our notice that the cone calorimeter not only identifies flame-retardant modes of action but also monitors several of the most important flame-retardant modes of action directly and in a quantitative way. The cone calorimeter measures the apparent effective heat of combustion (EHC) during burning as a product of the effective heat of combustion of the volatiles h_c^0 and the combustion efficiency of the flame χ. Hereby the cone calorimeter simulates a well-ventilated developing fire; thus χ is almost 1 for most of the polymers. Fuel dilution in the gas phase reduces h_c^0; flame inhibition, the χ of the actual flame. Since the THE is the total mass loss multiplied by the EHC, it is proportional to χh_c^0. Charring and fuel dilution in the condensed phase are taken into account by the factor (1-μ), where μ is the char yield between 0 and 1. Thus the total mass loss is the mass of the specimen (m_0) multiplied by (1-μ) and the THE is[30]

$$THE \sim \chi h_c^0 (1-\mu) m_0 \qquad (17.1)$$

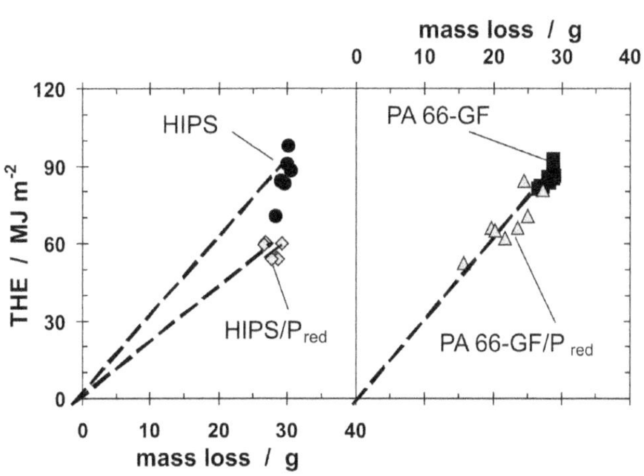

FIGURE 17.13 THE plotted against ML for HIPS, HIPS/P_{red}, PA 66-GF, and PA 66-GF/P_{red}, illustrating the pure flame inhibition effect of P_{red} in HIPS and the pure charring effect in PA 66-GF.

Equation 1 can be used to quantify the flame-retardant effects of charring, fuel dilution, and flame inhibition when comparing the results of the apparent EHC and the char yield of a flame-retarded polymeric material to the results of its corresponding polymer matrix.[124–126] Assuming the same factors, amount of fuel and quality of fuel, are also governing the HRR, any additional protective layer effect reducing the rate but not the total fire load can be estimated using an empirical factor θ(t) in a quantitative way (Equation 2):

$$PHRR \sim HRR(t) \sim \theta(t) \chi h_c^0 (1-\mu), \qquad (17.2)$$

It is concluded that the difference in flame-retardant modes of action of phosphorous flame retardants is due to the different decomposition pathways and interactions between flame retardant and polymer or their decomposition products, respectively. The release of phosphorous molecules into the gas phase competes with char-inducing or inorganic residue-forming interactions in the solid state. Thus, the combination of cone calorimeter results and thermal analysis becomes interesting. Figure 17.14 illustrates the potential of evolved gas analysis, here Fourier-transform infrared spectroscopy (FTIR), to identify volatile pyrolysis products as well as the release rate during decomposition. TG-MS or pyrolysis GC-MS are also proposed as methods targeting the volatile pyrolysis products. The cone calorimeter data reveal the flame-retardant modes of action, indeed, the reduction in the apparent effective heat of combustion EHC and the char yield can be used to quantify flame inhibition, fuel dilution, and the reduction in fuel release, respectively. The thermal analysis coupled with the evolved gas analysis can be used to illuminate the mechanisms, the chemical reactions occurring during pyrolysis. Solid-state products are often tackled by solid-state NMR, ATR-FTIR, or XPS. Table 17.5 shows an example of how the results of combined cone calorimeter and thermal analysis can be used to improve our understanding of a system like PC/ABS/PTFE, PC/ABS/PTFE/TPP (TPP = triphenylphosphate), and PC/ABS/PTFE/BDP.[71] Even though TPP and BDP are very similar in their structure, thermal analysis monitors an earlier and slightly larger phosphorus release for PC/ABS/PTFE/TPP than for PC/ABS/PTFE/BDP, and only for PC/ABS/PTFE/BDP is there an increase in residue compared to PC/ABS/PTFE.[71] The same trends are monitored by cone calorimeter for the apparent effective heat of combustion of volatiles EHC = χ h_C and the fire residue. Using both thermal analysis and fire testing reveals an unambiguous understanding of the mechanisms and modes of action controlling the change in PHRR and THE.

Combining data obtained by the cone calorimeter with pyrolysis combustion flow calorimeter (PCFC; sometimes called microscale combustion calorimeter: MCC) results also was proposed to increase the understanding of flame retardancy and flame-retardant modes of action.[127,128] Dividing the fraction of the apparent effective heat of combustion of

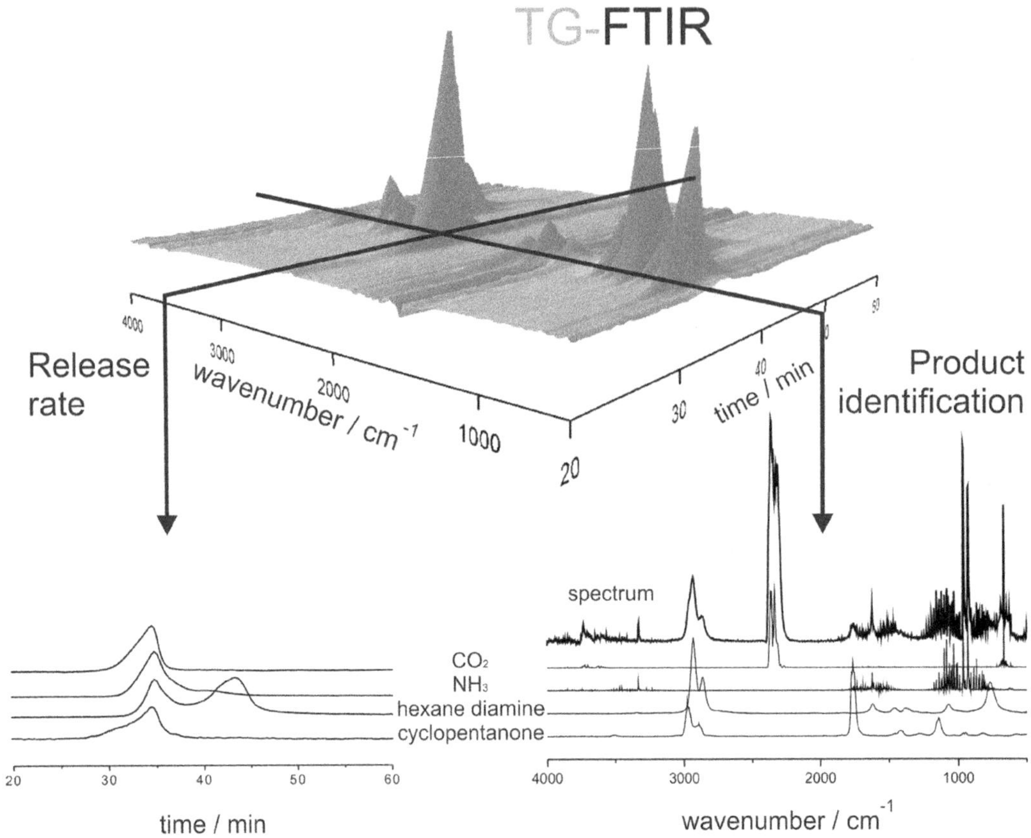

FIGURE 17.14 FTIR monitored online during a TG-FTIR experiment (this means several hundreds of FTIR spectra are taken) of PA 66-GF/MPP enables the identification of volatile pyrolysis products (H_2O, NH_3, CO_2, cyclopentanone, amines, and hydrocarbons) at distinct stages of the decomposition and the investigation of the product release rates.

the volatiles (THE/ML) obtained from the cone calorimeter by the heat of complete combustion of the volatiles obtained from PCFC yields the combustion efficiency χ. Thus, the combination of fire test and PCFC enables a quantitative evaluation of ventilation but also of the flame inhibition that reduces χ. Figure 17.15 shows an example for PC/ABS/PTFE, PC/ABS/PTFE/BDP, PC/ABS/PTFE/RDP (RDP = resorcinol bis(diphenyl phosphate), PC/ABS/PTFE/TPP, and PC/ABS/PTFE/BDP/ZB (ZB = zinc borate).

For PC/ABS/PTFE a combustion efficiency close to 1 ($\chi = 0.98$) was observed. Almost all volatile pyrolysis products were completely oxidized. This is in good accordance to the absence of a flame inhibition effect and the well-ventilated cone calorimeter fire scenario.[85] Corresponding results have been reported previously for different polymers in the cone calorimeter.[129] Adding BDP, RDP, and TPP results in a very similar flame inhibition activity in PC/ABS/PTFE (χ =~ 0.8), increasing slightly in the order BDP < RDP < TPP.[127] Using the combination of BDP and ZB results in the formation of zinc phosphates in the solid state. This competing mechanism reduces the release of phosphorus in the gas phase. The combination showed a significant antagonism in terms of flame inhibition.

The additional power of charring becomes clear when results using a bomb calorimeter and PCFC (Table 17.6) are compared, as was proposed before.[130] The bomb

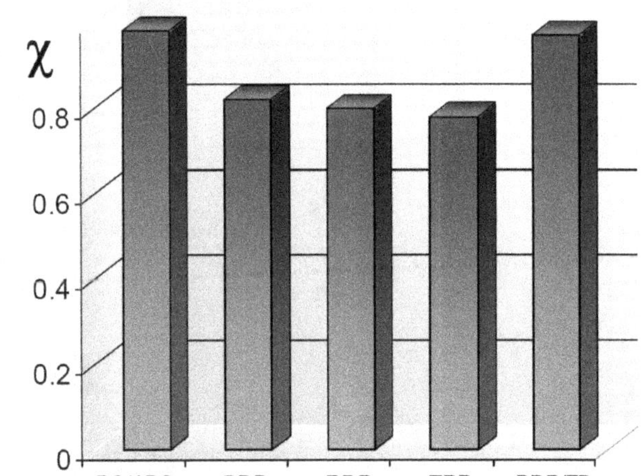

FIGURE 17.15 Combustion efficiency (χ) of various PC/ABS materials calculated using the THE/ML measured in the cone calorimeter and the heat release per ML for the complete combustion of the volatiles monitored in the PCFC. Systems that do not show flame inhibition show combustion efficiencies of around 1, according to the well-ventilated fire scenario of the cone calorimeter. Systems in which adding aryl phosphates results in flame inhibition show combustion efficiencies of around 0.8. When the release of phosphorus is reduced by competing reactions in the solid state, combustion efficiencies of between 0.8 and 1 are observed.

TABLE 17.6
Comparison between Results Obtained with the PCFC and the Bomb Calorimeter

	HR (PCFC)	h_c (PCFC)	HOC_{bomb}
Beech wood	11.1 +/- 0.4	12.7 +/- 0.6	17.84 +/- 0.05
PC	20.7 +/- 0.5	26.5 +/- 1.1	30.92 +/- 0.02
PESU	9.6 +/- 0.3	15.4 +/- 0.8	25.32 +/- 0.18

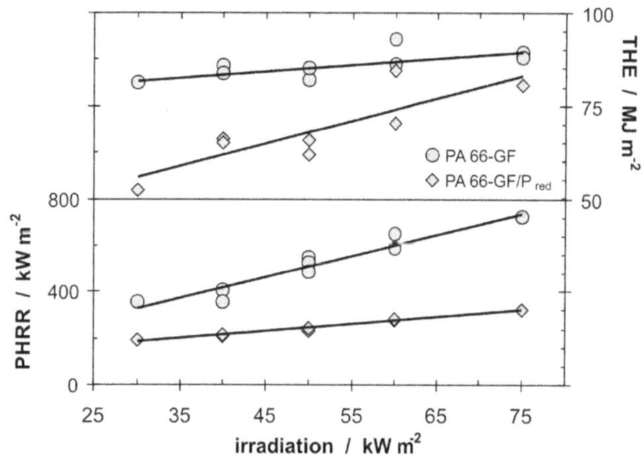

FIGURE 17.16 THE and PHRR of PA 66-GF and PA 66-GF/P_{red} plotted against irradiation. With increasing irradiation, the flame-retardancy effect vanishes with respect to THE and increases with respect to PHRR. The THE correlates to the ML during burning, which converges for PA 66-GF and PA 66-GF/P_{red} with increasing irradiation. The PHRR is determined more by the barrier properties of the char than by its quantity.

calorimeter delivers the heat of combustion per mass for the complete oxidation of a material (HOC_{bomb}). The PCFC determines the heat of combustion per mass for the complete oxidation of the volatiles (HR), whereas the char formed during anaerobe decomposition is not oxidized in the PCFC. Considering the char yield, one can calculate the effective heat of combustion of the volatiles h_c. Subtracting the HR from HOC_{bomb} results delivers the heat of combustion held in the condensed phase by carbonaceous charring. This value is zero for inert fillers, ash, or any inorganic residue. Inert fillers and inorganic residue are accompanied by a reduction in heat release only to the extent that polymeric mass per volume is replaced, but do not cause a reduction by storing the fire load in the residue. The ratio of the heat of combustion of the carbonaceous char and the total heat of combustion does not equal the char yield (μ). The heat of combustion per unit mass of char may be estimated to be around 33 kJ g^{-1}, as can be expected for a graphite-like material. This estimation also corresponds with investigations of the chemical composition of char reported in the literature,[131,132] demonstrating 90% carbon for PC char. Thus, the heat of combustion per unit mass of char is significantly higher than the effective heat of combustion of technical polymers (20–31 MJ/kg) or wood (14–19 MJ/kg) measured in the bomb calorimeter for complete oxidation (HOC_{bomb}) and clearly lower than for polyolefins where h_c is between 39 and 44 MJ/kg polyolefins. The enrichment of carbon in the char becomes apparent; as a consequence, the increased charring of technical polymers containing hetero-atoms is very efficient, because the reduction in heat release outperforms the reduction in mass loss. On the other hand, charring falls behind fuel replacement in polyolefins when the h_c of the char is less than the h_c of the polymer. Charring is believed to change the h_c of the volatiles and the oxidation of char release CO with a high yield, yet these two effects must not be interpreted as additional fuel dilution or flame inhibition. Further, differing in the formation of carbonaceous char and inorganic residue is key to properly describe not only the flame-retardancy mechanisms but also the flame-retardant modes of action.

Flame retardants or flame-retardant modes of action, respectively, influence different fire properties quite differently and, what is more, show different effectiveness in different fire scenarios and thus fire tests. In extreme cases, flame retardancy with respect to a specific fire property or specific test is achieved with little or no improvement in performance in another fire property or fire test. This fundamental understanding in fire science sometimes may be overlooked in materials development but is worth addressing. The influence on different fire risks and the dependency of effectiveness on the scenario addressed is discussed in the following based on the flame-retardant modes of action accompanying charring and barrier formation.

Char or inert residue influences fire behavior through two quite different modes of action. As discussed previously, char especially, but also inorganic fillers that replace polymeric material per volume, result in a reduction of the total fuel release and hence the THE. In a first approximation, the increase in carbonaceous char yield also yields a proportional decrease in HRR. Further, every fire residue also works as a kind of barrier to heat or mass transfer, suppressing the HRR in particular. Although the same char is the origin of both effects, these modes of action must be considered as to some degree independent. Again, this conclusion is illustrated exemplarily by the investigations on PA 66-GF/P_{red}. Figure 17.16 summarizes the THE and the PHRR for PA 66-GF/P_{red} in comparison to PA 66-GF.[58,59] With increasing irradiance, an increasingly complete decomposition of the polymer resulted so that fire retardancy vanished in terms of THE. In contrast to this, the reduction of PHRR increased for higher external heat fluxes, which was due to barrier properties reducing the rates but not the total amounts.

It is striking that this system shows such a clear differentiation between the behavior of the two probably most important fire risks PHRR and THE when the external heat flux is varied. The flame-retardancy effect with respect to the THE decreases with increasing irradiation, whereas the relative flame-retardancy effect with respect to the PHRR increases. The latter clearly indicates the predominant influence of the barrier effect on the HRR.

Taking into account the different heat fluxes in the cone calorimeter setup ($\dot{q}''_{\text{eff}} = \dot{q}''_{\text{ex}} + \dot{q}''_{\text{flame}} - \dot{q}''_{\text{rerad}} - \dot{q}''_{\text{loss}}$), Equation 3 was proposed for the idealized "steady-state" HRR (HRR_{st}) during "steady-state" burning[64,66,132–134]:

$$\begin{aligned} HRR_{st} &= \chi(1-\mu)\frac{h_c}{h_g}\dot{q}''_{\text{eff}} \\ &= \chi(1-\mu)\frac{h_c}{h_g}\left(\dot{q}''_{\text{ex}} + \dot{q}''_{\text{flame}} - \dot{q}''_{\text{rerad}} - \dot{q}''_{\text{loss}}\right) \end{aligned} \tag{17.3}$$

$$\begin{aligned} HRR_{st} &= \chi(1-\mu)\frac{h_c}{h_g}\dot{q}''_{\text{ex}} \\ &\quad + \chi(1-\mu)\frac{h_c}{h_g}\left(\dot{q}''_{\text{flame}} - \dot{q}''_{\text{rerad}} - \dot{q}''_{\text{loss}}\right) \\ &= HRP\dot{q}''_{\text{ex}} + HRR_0 \end{aligned}$$

Hence, a linear dependency of the HRR_{st} on the external heat flux is expected (Figure 17.17 based on[59]). The slope (referred to as the heat release parameter, HRP) is interpreted as a fire response parameter,[64,133] whereas the intercept (HRR_0) is considered a flammability parameter,[66,132] at least on this length scale.

Both values are based on materials properties and are, or approximate, "intrinsic" materials properties. Some ideas of how an idealized HRR such as the HRR_{st} can be deduced from experimental cone calorimeter HRR data have been described,[132] but in practise the proposed methods do not work satisfactorily for many HRR curves.[75] However, the improvement achieved by using HRR_{st} instead of PHRR was concluded through attempted correlation with other tests.[75,135] In fact, the flammability parameter was found to correlate quite reasonably with UL 94 performance.[66] A critical heat release rate of 125 ± 25 kW m^{-2} was proposed for self-extinguishing in the absence of the application of external heat.[64,132] It must be concluded that the data error is generally so large that the influence of thickness (1.5–3 mm) in the UL 94 classification and the differentiation between the classification as V-0, V-1, or V-2 cannot be considered. The results for different PA 66-GF and HIPS materials investigated are summarized in Figure 17.17 following this approach.[59] Some of the selected data are quite close to self-extinction (V-0, V-1) or flammable (HB) behavior, respectively. Thus, the data clearly challenge the practical use of the concept and hence validate it. However, the precise prediction of UL 94 performance through droves of cone calorimeter data is, of course, academic claptrap. The important conclusion is that the inter- and extrapolation of results obtained under forced-flaming conditions in the cone calorimeter can be used to get a reasonable and realistic idea of fire behavior in different fire scenarios such as flammability. Using several irradiances was proposed to obtain comprehensive information on material performance for different fire scenarios.[57–59,85]

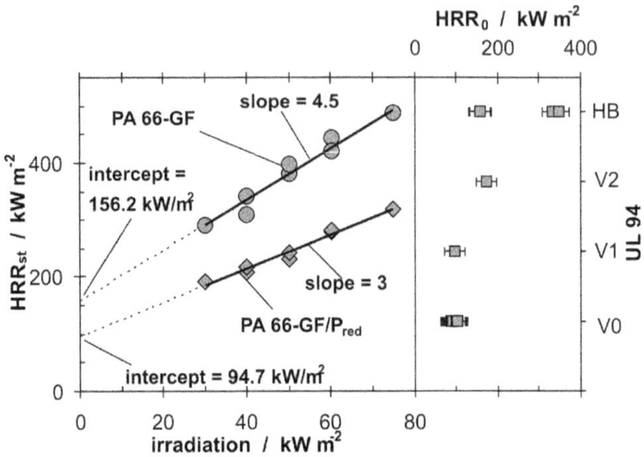

FIGURE 17.17 On left: HRR_{st} for PA 66-GF and PA 66-GF/P_{red} at 30, 40, 50, 60, and 75 kWm^{-2} and the extrapolation to irradiation = 0. The slope is the heat release parameter (HRP); the intercept a flammability parameter (HRR_0). On right: UL 94 performance plotted against HRR_0 for a variety of PA 66-GF and HIPS materials. The critical value of 125 ± 25 kW m^{-2} is a good determination of the point when self-extinguishing behavior is reached.

What is more, the use of several irradiances is proposed, to provide further valuable insight into the active flame-retardant modes of action themselves and their efficiency in different fire scenarios.[59,70] To a certain degree, different flame-retardancy effects show specific, characteristic behavior depending on the irradiation.[57,59] Typically the PHRR of non-charring plastics shows a strong linear increase with irradiation. When flame inhibition or charring is induced, the PHRR decreases, and the ratio between the PHRR with and without flame retardant is rather constant, corresponding to the model that generally only the factor combustion efficiency (χ) or the factor "(1–μ)" is changed in Equation 3. The efficiency is proposed to increase slightly with increasing irradiation due to a kind of self-reinforcing flame-retardancy effect, caused by additionally reduced thermal feedback from the flame. Inducing an efficient barrier results in a negligent increase in PHRR at higher irradiation. Thus, the relative reduction is large for high irradiation but can vanish for low irradiation. Such extreme behavior was found for some nanocomposites,[57,70,102] as illustrated in Figure 17.18. The vanishing flame retardancy for low irradiations corresponds to the flammability performance measured for these systems in the LOI.[57]

In principle, all char or residue formed during a fire also acts as a barrier to heat and mass transport. The effectiveness of the barrier depends not only on the amount of char, but also on the properties of the char, such as its morphology, which determines gas permeability and thermal conductivity. Further, the properties of char may be tailored by adding elements such as inorganic adjuvants and synergists.[56,136–140] Consequently, the heterogeneous, gradually changing, or structured morphology of fire residues plays an important role in terms of fire behavior and fire retardancy. Indeed, the design of a fire residue consisting of multicellular structures or closed glassy surface layers, respectively, rather than merely separated loose particles, is a promising approach for flame-retarded materials. Such approaches often play a crucial role, particularly in halogen-free systems containing phosphorus. Pursuing the objectives of the understanding and directed development of flame-retarded materials shifts the focus to the accurate investigation of heterogeneous or complex fire residues with respect to their morphology and chemical composition, as well as the investigation of their formation during fire. Scanning electron microscopy is proposed as a suitable and powerful tool for the advanced characterization of such complex fire residues, since it offers high resolution in combination with good depth of field and, simultaneously, when SEM-EDX is used, an analysis of the chemical composition at the same location. Because fire residues in the form of sticky smut, char, or slug obviously are not the preferred samples for investigation in an SEM apparatus, such SEM investigations are probably used more rarely than the multitude of fire tests and focus mainly on the characterization of special surface layers.[56,141–146]

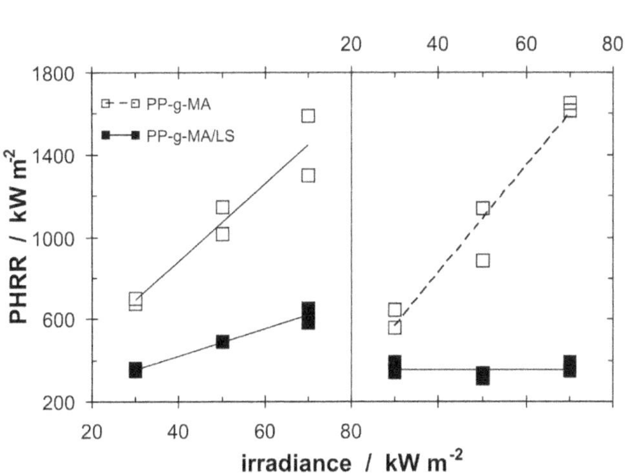

FIGURE 17.18 PHRR of PP-g-MA and PP-g-MA/5 wt% LS plotted against irradiance. The materials were compounded by extrusion and subsequently pressed into specimens (left) or injection-molded into specimens (right). The flame retardancy increases with increasing irradiance. Additional injection molding increased the nano-dispersion even further until the increase in PHRR as a function of irradiance vanished.

The surface temperature during burning and the gradient inside the sample are interesting for the development of materials, especially for materials that are flame-retarded by intumescent barrier layers.[147,148] In different groups sample holders were developed to address this objective for the cone calorimeter (Figure 17.19)[149] but also for other tests, especially to gain deeper insight into the actual pyrolysis conditions and flame-retardancy mechanism.

Shielding effects are observed, especially in the case of effective insulation accompanied by increased re-radiation due to high surface temperatures.[150] Indeed, optimizing the barrier properties so that the HRR drops below a critical value characteristic for passing a test, or even reducing the HRR to cause self-extinction, is a self-contained flame-retardancy approach. Most successful is intumescence, where applied heat results in expanding—often multicellular—char, insulating underlying material.[151,152] Recent reviews of this concept and its use in polymers, and precise descriptions of the chemistry involved have been published.[103,145,146,153–162] Temperature differences of up to several hundred degrees Kelvin are observed (Figure 17.20).[109] Intumescence in polymeric materials can be used to reduce HRR by slowing down the pyrolysis front velocity,[163] or even to achieve extinction before the pyrolysis front goes through the whole specimen.[164] In the latter case, the insulation of the char decreases the temperature in the underlying material to keep it from reaching the decomposition temperature of the polymer.

Since different flame-retardant approaches influence different fire properties quite differently and show different effectiveness in different fire tests, the focus of interest shifts to a comprehensive assessment of fire retardancy. The

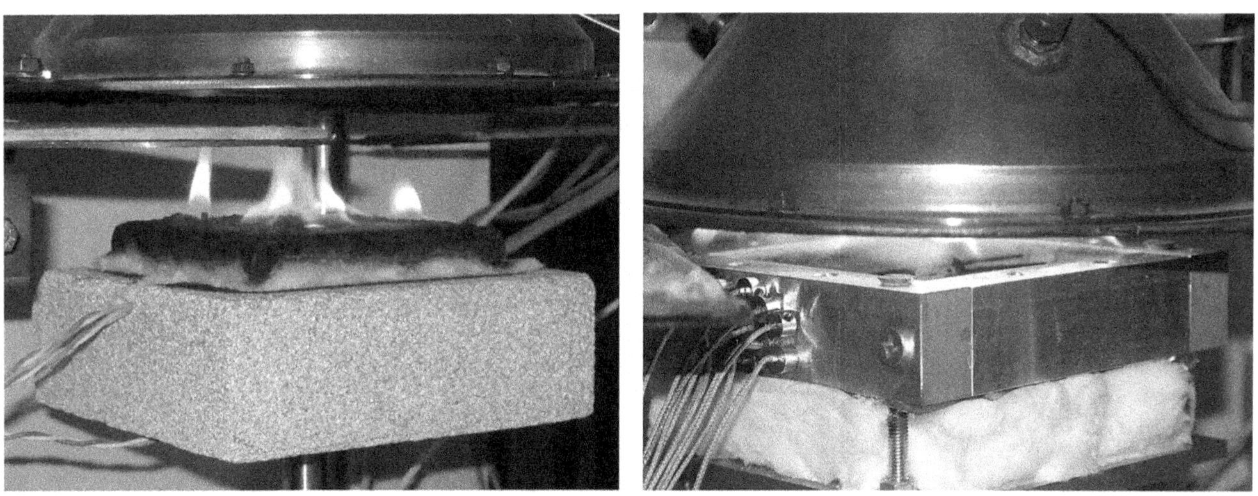

FIGURE 17.19 Sample holders used to measure the temperature on the back of coated steel plates (left) and at different positions within polymer samples (right) during flaming combustion in a cone calorimeter.

fire load and the flame spread or HRR are believed to have the greatest influence on the fire hazard. Petrella proposed plotting THE (as fire load) against $PHRR / t_{ig}$ (as a fire growth index) to give a comprehensive assessment of the two most important fire hazards of a material.[165] The use of $PHRR / t_{ig}$ as an index for the fire growth was reported in the 1980s.[96] Even though the index $PHRR / t_{ig}$ may be a better way to address flame spread than FIGRA, it still suffers from the disadvantages of oversimplified empirical indices discussed above. However, Petrella's idea of the assessment of both fire load and fire growth in a graphic at the same time was successfully developed further to achieve a comprehensive and reasonable scientific assessment of the flame retardancy of a material by comparing different approaches, or by comparing the effects for different irradiations. Examples for such a data evaluation are given in Figure 17.21 for the examples discussed above based on HIPS with P_{red} and $Mg(OH)_2$) and P_{red} in PA 66-GF.[52,58]

The fire hazard is reduced by the flame retardant, either by reducing the THE or by reducing the fire growth rate. The best fire retardancy would yield reduction in both, represented by approaching the origin. Overall (synergism in barrier properties and antagonism in flame inhibition), the fire retardancy of HIPS containing P_{red} and $Mg(OH)_2$ was approximated quite well by a superposition of the effects achieved by each of the flame retardants alone (Figure 17.21 left) so that the combination is the far more interesting system with respect to fire retardancy.

Figure 17.21 (right) uses the Petrella approach to assess the fire risks of both PA 66-GF and PA 66-GF/P_{red} at different external heat fluxes.[58,59,85] Thus the assessment is valuable for different applications, fire scenarios and fire tests, as these correspond to different external heat fluxes and define different demands on fire retardancy in terms of the duration and growth of a fire. The data for both materials show that with increasing heat flux the fire hazards increase in terms of fire growth, as expected, since a higher irradiance results in an increase in fuel production rate. The THE of PA 66-GF is fairly constant, since the polymer is combusted nearly completely for all irradiances used, whereas the THE decreases most for the flame-retarded material at low irradiance, because the char formation is greatest. This effect diminishes with increasing irradiance. In the case of P_{red} in PA 66-GF, combustion is complete at the highest external heat flux, with or without flame retardant, so the THE is almost the same, but the fire growth index is almost halved. Conversely, at low irradiance, not only is the fire growth index reduced, but the THE is almost halved as well. The change in the fire scenario changes the effectiveness of P_{red} added to PA 66-GF in two of the most important fire properties. P_{red} in PA 66-GF works best for low external heat flux. Flammability tests like LOI and, much more important, UL 94, are fire scenarios with low external heat flux.

It should not escape our notice that such a comprehensive assessment of a specific system sometimes amounts to the essential assessment of a whole fire retardancy approach and immediately blazes the trail for progress in the field. Such an example is shown in Figure 17.22, assessing the barrier effect in layered silicate-based nanocomposites with non-charring materials, which was also discussed previously.[57,70] Apart from the impressive reduction in fire growth, comparing these to an ideal flame retardancy revealed the deficits of nanocomposites, which must be remedied by combining the concept with traditional flame retardants or by designing special systems that show additional crucial carbonaceous charring. A discussion of the

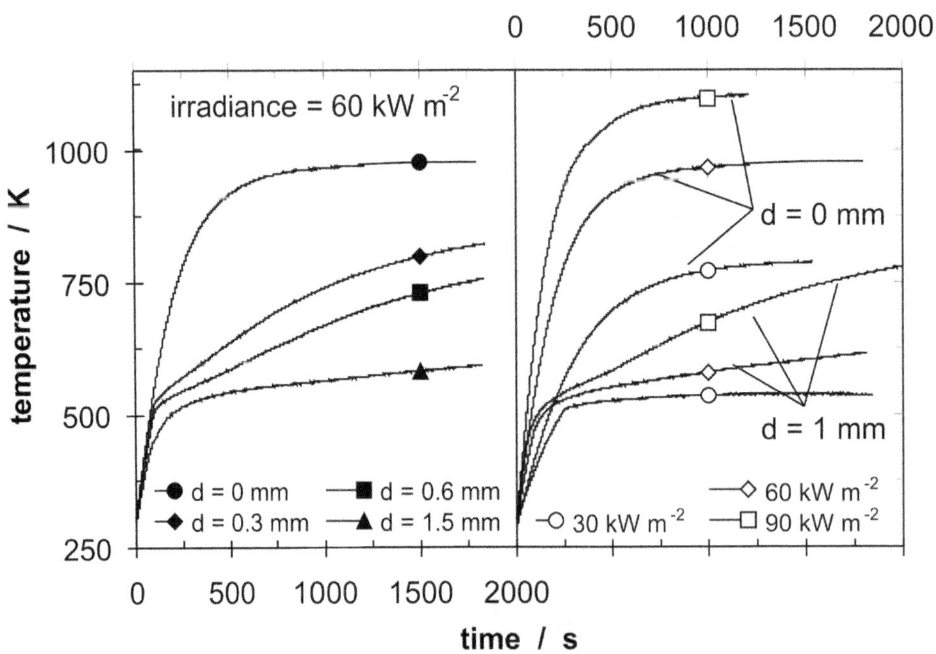

FIGURE 17.20 Temperature on the back of coated (system A) steel plates plotted against time, for different intumescent coating thicknesses (d) and external heat fluxes.

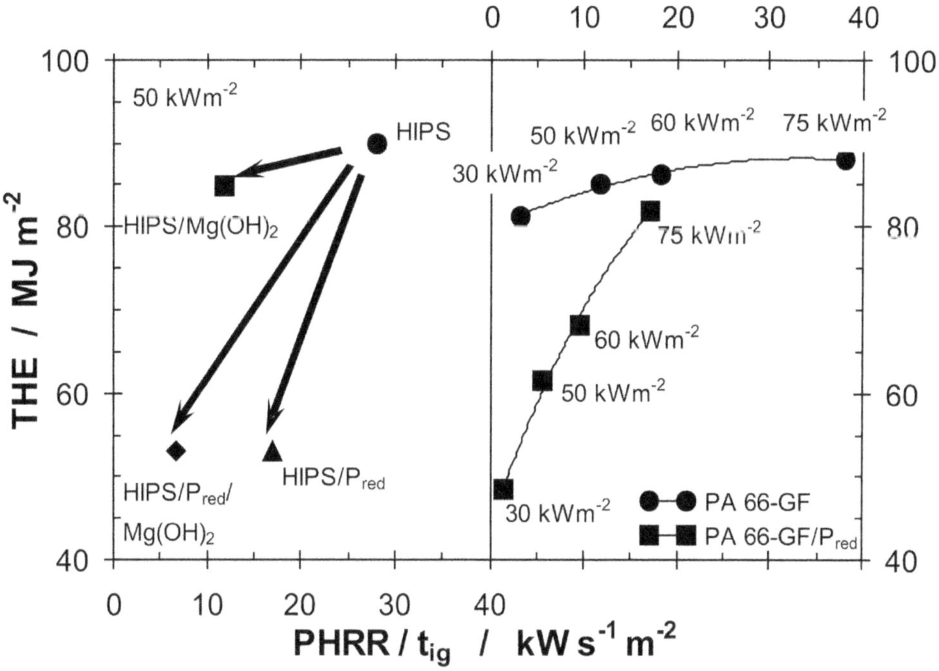

FIGURE 17.21 Using the Petrella plot for comprehensive and reasonable scientific assessment of flame retardancy by comparing different approaches, or by comparing the effects for different irradiations. THE stands for the fire load and $PHRR / t_{ig}$ for the fire growth rate; hence, the two most important fire risks are monitored at the same time. An ideal flame retardancy would decrease both hazards significantly, as is the case for the combination of both flame retardants on the left (comparison of HIPS), HIPS/P_{red}, HIPS/$Mg(OH)_2$, and HIPS/P_{red}/$Mg(OH)_2$) and for low external heat flux on the right (comparison of PA 66-GF and PA 66-GF/P_{red} for different irradiations).

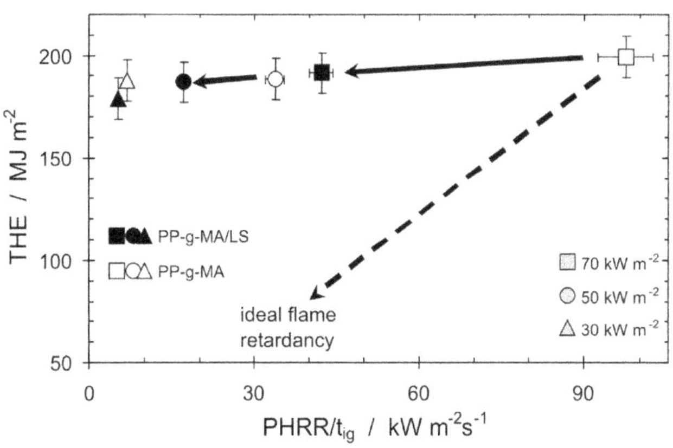

FIGURE 17.22 THE (fire load) plotted against the PHRR/t_{ig} (fire growth index) for PP-g-MA and the corresponding 5 wt% LS PP-g-MA nanocomposites (PP-g-MA/LS) at different irradiances: 30, 50, and 70 kW m^{-2}.

flame-retardancy performance for different irradiations used makes clear that the pronounced shielding effect in the cone calorimeter may have less impact in flammability tests like UL 94 or LOI.

17.7 CONCLUSIONS

This work outlined a great number of approaches for advanced investigation of fire retardancy and evaluated these for use in flame-retarded materials development. Understanding the key to fire retardancy and fire-retardancy mechanisms was deepened in keeping with the golden rule of using several methods of description. Fire behavior and fire retardancy were studied to ensure a comprehensive characterization. Several fire scenarios as well as several fire properties were discussed to work out in detail the fire retardancy, but also the potentials and limitations with respect to applications. Tailored experimental approaches and targeted data evaluation were not the focus of this chapter as ends in themselves, but as means to an end. Thus, neither a new fire testing apparatus nor a detailed theoretical description of fire behavior was evaluated, but merely small steps that could facilitate improved monitoring and understanding. While each of these steps may be not impressive itself, together they constitute a crucial force to tackle the challenge successfully and promise a significant scientific and commercial impact.

17.7.1 Acknowledgments

The examples used in this chapter are based on studies performed in the author's research group—thus, thanks to A. I. Balabanovich, H. Bahr, M. Bartholmai, U. Braun, J. Chacon Castro, P. Klack, U. Knoll, K. H. Richter, A. Weiß—and some of them were performed in cooperation with other research institutes or companies—thus, thanks to A. Hartwig (IFAM in Bremen), R. E. Lyon (FAA, US), M. Doering (KIT in Karlsruhe), V. Altstädt (University

Bayreuth), BASF AG, Clariant GmbH, and Bayer MaterialScience AG. Financial support was granted by the German Research Foundation (DFG, Scha 730/6–1, Scha 730/8–1, Scha 730/10–1, BR 3376/1–1), the Volkswagen Foundation (I/77 974), BASF AG, Bayer MaterialScience AG, and Clariant Produkte (Deutschland) GmbH.

REFERENCES

[1] Bödiger M, Eckel T, Wittmann D, Alberts H. Feinstteilige anorganische Pulver als Flammschutzmittel in thermoplastischen Formmassen. German patent. DE 195 30 200 A1, 1997.

[2] Gilman, J. W.; Kashiwagi, T.; Lichtenhan, J. D. Nanocomposites: A Revolutionary New Flame Retardant Approach. *Sampe J.* **1997**, *42*, 1078–1089.

[3] Gilman, J. W. Flammability and Thermal Stability Studies of Polymer Layered-Silicate (Clay) Nanocomposites. *Appl. Clay Sci.* **1999**, *15*, 31–49. https://doi.org/10.1016/S0169-1317(99)00019-8

[4] Zanetti, M.; Camino, G.; Reichert, P.; Mülhaupt, R. Thermal Behaviour of Poly(Propylene) Layered Silicate Nanocom- posites. *Macromol. Rapid Commun.* **2001**, *22*, 176–180. https://doi.org/10.1002/1521-3927(200102)22:3< 176::AID-MARC176 >3.0.CO;2-C

[5] Shieh, J. Y.; Wang, C. S. Effect of the Organophosphate Structure on the Physical and Flame-Retardant Properties of an Epoxy Resin. *J. Polym. Sci. Part A: Polym. Chem.* **2002**, *40*, 369–378. https://doi.org/10.1002/pola.10121

[6] Lin, C. H.; Wang, C. S. Novel Phosphorus-Containing Epoxy Resins Part I. Synthesis and Properties. Polymer. 2001, 42, 1869–1878. Https://doi.org/10.1016/S0032-3861(00)00447-X

[7] Hussain, M.; Varley, R. J.; Mathys, Z.; Cheng, Y. B.; Simon, G. P. Effect of Organo-Phosphorus and Nano-Clay Materials on the Thermal and Fire Performance of Epoxy Resins. *J. Appl. Polym. Sci.* **2004**, *91*, 1233–1253. https://doi.org/10.1002/app.13267

[8] Shieh, J. Y.; Wang, C. S. Synthesis of Novel Flame Retardant Epoxy Hardeners and Properties of Cured Products. *Polymer.* **2001**, *42*, 7617–7625. https://doi.org/10.1016/S0032-3861(01)00257-9

[9] Wang, C. S.; Lin, C. H. Synthesis and Properties of Phosphorus- Containing Epoxy Resins by Novel Method. *J. Polym. Sci. Part A: Polym. Chem.* **1999**, *37*, 3903–3909. https://doi.org/10.1002/(SICI)1099-0518(19991101)37:21 < 3903::AID-POLA4>3.0.CO;2-X

[10] Ciesielski M, Burk B, Heinzmann C, Doring M. Fire-Retardant High-Performance Epoxy-Based Materials. In: Novel Fire Retardant Polymers and Composite Materials. Wang D.-Y., Ed. Woodhead Publishing. Duxford 2017; chap. 2:3–51. Https://doi.org/10.1016/B978-0-08-100136-3.00002-9

[11] Liu, Y. L. Flame-Retardant Epoxy Resins from Novel Phosphorus-Containing Novolac. Polymer. 2001, 42, 3445–3454. https://doi.org/10.1016/S0032-3861(00)00717-5

[12] Kleiner, H-J, Verfahren zur Herstellung von Aluminiumsalzen von Phosphinsäuren. German Patent DE 196 16 025 C1, 1997.

[13] Kleiner H-J, Budzinsky W, Kirsch G. Schwerentflammbare Polyamidformmassen. German Patent DE 196 07 635 A1, 1997.

[14] Kleiner H-J, Budzinsky W, Kirsch G. Formteile aus schwerentflammbarer Polyesterformmasse. German Patent DE 196 08 008 A1, 1997.

[15] Jenewein E, Naß B, Wanzke W. Flammschutzmittel-Kombination. German Patent DE 199 33 901 A1, 2001.

[16] Gilman, J. W.; Bourbigot, S.; Shields, J. R.; Nyden, M.; Kashiwagi, T.; Davis, R. D.; Vanderhart, D. L.; Demory, W.; Wilkie, C. A.; Morgan, A. B.; Harris, J.; Lyon, R. E. High Throughput Methods for Polymer Nanocomposites Research: Extrusion, NMR Characterization and Flammability Property Screening. *J. Mater. Sci.* **2003**, *38*, 4451–4460. https://doi.org/10.1023/A:1027369115480

[17] Wilkie, C. A.; Chigwada, G.; Gilman, J. W.; Lyon, R. E. High-Throughput Techniques for the Evaluation of Fire Retardancy. *J. Mater. Chem.* **2006**, *16*, 2023–2029. https://doi.org/10.1039/b514598h

[18] Jimenez, M.; Duquesne, S.; Bourbigot, S. High-Throughput Fire Testing for Intumescent Coatings. *Ind. Eng. Chem. Res.* **2006**, *45*, 7475–7481. https://doi.org/10.1021/ie0608410

[19] Rabe, S.; Schartel, B. The Rapid Mass Calorimeter: A Route to High Throughput Fire Testing. *Fire Mater.* **2017**, *41*, 834–847. https://doi.org/10.1002/fam.2420

[20] Rabe, S.; Schartel, B. The Rapid Mass Calorimeter: Understanding Reduced-Scale Fire Test Results. *Polym. Test.* **2017**, *57*, 165–174. https://doi.org/10.1016/j.polymertesting.2016.11.027

[21] Eckel, T, Ooms P, Wittmann D, Buysch H-J. Thermoplastische Polymerformmassen mit f lammwidrigen Eigenschaften. German patent. DE 42 31 774 A 1, 1994.

[22] Bourbigot S, Le Bras M. Flame Retardants. In: Plastics Flammability Handbook. 3rd Edition. Troitzsch J, Ed. Hanser. Munich 2004; chap. 5:133–157.

[23] Camino, G.; Costa, L.; Luda di Cortemiglia, M. P. Overview of Fire Retardant Mechanisms. *Polym. Degrad. Stab.* **1991**, *33*, 131–154. https://doi.org/10.1016/0141-3910(91)90014-I

[24] Lewin M. Physical and Chemical Mechanisms of Flame Retarding of Polymers. In: Fire Retardancy of Polymers. The Use of Intumescence. Le Bras M, Camino G, Bourbigot S, Delobel R, Eds. Royal Society of Chemistry. London 1998:3–32.

[25] Lewin M, Weil ED. Mechanisms and Modes of Action in Flame Retardancy of Polymers. In: Fire Retardant Materials. Horrocks AR, Price D, Eds. Woodhead Publishing Limited. Cambridge 2001; chap. 2:31–68.

[26] Levchick SV. Introduction to Flame Retardancy and Polymer Flammability. In: Flame Retardant Polymer Nanocomposites. Morgan AB, Wilkie CA, Eds. John Wiley & Sons. Hoboken, NJ 2007; chap. 1:1–29.

[27] Troitzsch J, Antonatus E, Eds. Plastics Flammability Handbook Principles, Regulations, Testing, and Approval, 4th Edition. Hanser. Munich 2021.

[28] Schartel, B.; Wilkie, C. A.; Camino, G. Recommendations on the Scientific Approach to Polymer Flame Retardancy: Part 1—Scientific Terms and Methods. *J. Fire Sci.* **2016**, *34*, 447–467. https://doi.org/10.1177/0734904116675881

[29] Babrauskas V. Ignition Handbook. Fire Science Publishers and SFPE. Issaquah 2003.

[30] Schartel B, Kebelmann K. Fire Testing for the Development of Flame Retardant Polymeric Materials. In: Flame Retardant Polymeric Materials A Handbook. Hu Y, Wang X, Eds. CRC Press, Taylor & Francis Group. Boca Raton, FL 2020; chap. 3:35–55.

[31] Reimschuessel, H. K.; Shalaby, S. W.; Pearce, E. M. On the Oxygen Index of Nylon 6. *J. Fire Flammability.* **1973**, *4*, 299–308.

[32] DiNenno PJ, Drysdale D, Beyler CL, Walton WD, Custer RLP, Hall JR Jr, Watts JM Jr, Eds. The SFPE Handbook of Fire Protection Engineering, 3rd Edition. National Fire Protection Association, Inc. Quincy, MA 2002.

[33] Levchik GF, Levchik SV, Camino G, Weil ED. Fire Retardant Action of Red Phosphorus in Nylon 6. In: Fire Retardancy of Polymers. The Use of Intumescence. Le Bras M, Camino G, Bourbigot S, Delobel R, Eds. Royal Society of Chemistry. London 1998:304–315.

[34] Yeh, J. T.; Hsieh, S. H.; Cheng, Y. C.; Yang, M. J.; Chen, K. N. Combustion and Smoke Emission Properties of Poly(Ethylene Terephthalate) Filled with Phosphorous and Metallic Oxides. *Polym. Degrad. Stab.* **1998**, *61*, 399–407. https://doi.org/10.1016/S0141-3910(97)00225-5

[35] Balabanovich, A. I.; Zevaco, T. A.; Schnabel, W. Fire Retardance in Poly(Butylene Terephthalate). The Effects of Red Phosphorus and Radiation-Induced Cross-Links. *Macromol. Mater. Eng.* **2004**, *289*, 181–190. https://doi.org/10.1002/ mame.200300153

[36] Hastie, J. W. Molecular-Basis of Flame Inhibition. *J. Res. Nat. Bur. Stand. Sect. A Phys. Chem.* **1973**, *77A*, 733–754.

[37] Isaacs, J. L. The Oxygen Index Flammability Test. *J. Fire Flammability.* **1970**, *1*, 36–47.

[38] Weil, E. D.; Hirschler, M. M.; Patel, N. G.; Said, M. M.; Shakir, S. Oxygen Index: Correlations to Other Fire Tests. *Fire Mater.* **1992**, *16*, 159–167. https://doi.org/10.1002/fam.810160402

[39] Wharton, R. K. Correlation Between the Critical Oxygen Index Test and Other Fire Tests. *Fire Mater.* **1981**, *5*, 93–102. https://doi.org/10.1002/fam.810050304

[40] Camino, G.; Costa, L.; Casorati, E.; Bertelli, G.; Locatelli, R. The Oxygen Index Method in Fire Retardance Studies of Polymeric Materials. *J. Appl. Polym. Sci.* **1988**, *35*, 1863–1876. https://doi.org/10.1002/app.1988.070350712

[41] Wharton, R. K. The Effect of Sample Size on the Burning Behaviour of Thermoplastic Materials in the Critical Oxygen Index Test. *Fire Mater.* **1981**, *5*, 73–76. https://doi.org/10.1002/fam.810050207

[42] Wharton, R. K. Factors That Influence the Critical Oxygen Index of Various Solids. *Fire Mater.* **1979**, *3*, 39–48. https://doi.org/10.1002/fam.810030108

[43] Wharton, R. K. The Relative Performance of Critical Oxygen Index Testing Columns Under Ambient Conditions. *J. Fire Flammability.* **1981**, *12*, 266–271.

[44] Wharton, R. K. Comments on the Effect of Gas Velocity on Oxygen Index Measurements. *J. Fire Sci.* **1983**, *1*, 459–464. https://doi.org/10.1177/073490418300100605

[45] Sibulkin, M.; Lee, C. K. Flame Propagation Measurements and Energy Feedback Analysis for Burning Cylinders. *Combust. Sci. Technol.* **1974**, *9*, 137–147. https://doi.org/10.1080/00102207408960349

[46] Sibulkin, M.; Kim, J.; Creeden, J. V. Jr. The Dependence of Flame Propagation on Surface Heat Transfer I. Downward Burning. *Combust. Sci. Technol.* **1976**, *14*, 43–56. https://doi.org/10.1080/00102207608946745

[47] Sibulkin, M.; Kim, J. The Dependence of Flame Propagation on Surface Heat Transfer II. Upward Burning. *Combust. Sci. Technol.* **1977**, *17*, 39–49. https://doi.org/10.1080/00102209708946811

[48] Tewarson, A.; Ogden, S. D. Fire Behavior of Polymethylmethacrylate. *Combust. Flame.* **1992**, *89*, 237–259. https://doi.org/10.1016/0010-2180(92)90013-F

[49] Sibulkin, M.; Little, M. W. Propagation and Extinction of Downward Burning Fires. *Combust. Flame.* **1978**, *31*, 197–208. https://doi.org/10.1016/0010-2180(78)90129-3

[50] Arcand, C. G. Jr. The Bottom Ignition Oxygen Index Test. *Text. Res. J.* **1972**, *42*, 328–330. https://doi.org/10.1177/004051757204200604

[51] Ševěček, P.; Filipi, B.; Zapletalová, I. Modification of the Oxygen Index Test for Better Evaluation of the Relative Flammability. *Fire Mater.* **1981**, *5*, 39–40. https://doi.org/10.1002/ fam.810050107

[52] Braun, U.; Schartel, B. Flame Retardant Mechanisms of Red Phosphorus and Magnesium Hydroxide in High Impact Polystyrene. *Macromol. Chem. Phys.* **2004**, *205*, 2185–2196. https://doi.org/10.1002/macp.200400255

[53] Schartel, B.; Balabanovich, A. I.; Braun, U.; Knoll, U.; Artner, J.; Ciesielski, M.; Döring, M.; Perez, P.; Sandler, J. K. W.; Altstädt, V.; Hoffmann, T.; Pospiech, D. Pyrolysis of Epoxy Resins and Fire Behaviour of Epoxy Resin Composites Flame-Retarded with 9,10-Dihydro-9-oxa-10-Phosphaphenanthrene-10-Oxide Additives. *J. Appl. Polym. Sci.* **2007**, *104*, 2260–2269. https://doi.org/10.1002/app.25660

[54] Braun, U.; Balabanovich, A. I.; Schartel, B.; Knoll, U.; Artner, J.; Ciesielski, M.; Döring, M.; Perez, R.; Sandler, J. K. W.; Altstädt, V.; Hoffmann, T.; Pospiech, D. Influence of the Oxidation State of Phosphorus on the Decomposition and Fire Behaviour of Flame-Retarded Epoxy Resin Composites. *Polymer.* **2006**, *47*, 8495–8508. https://doi.org/10.1016/j.polymer.2006.10.022

[55] Schartel, B.; Weiß, A.; Kleemeier, M.; Hartwig, A.; Vogt, C.; Wendschuh, M.; Fischer, R. X. Flame Retarded Epoxy Resins by Adding Layered Silicate in Combination With the Conventional Protection Layer Building Flame Retardants Melamine Borate and Ammonium Polyphosphate. *J. Appl. Polym. Sci.* **2010**, *118*, 1134–1143. https://doi.org/10.1002/app.32512

[56] Braun, U.; Schartel, B.; Fichera, M. A.; Jäger, C. Flame Retardancy Mechanisms of Aluminium Phosphinate in Combination with Melamine Polyphosphate and Zinc Borate in Glass-Fibre Reinforced Polyamide 6,6. *Polym. Degrad. Stab.* **2007**, *92*, 1528–1545. https://doi.org/10.1016/j.polymdegradstab.2007.05.007

[57] Bartholmai, M.; Schartel, B. Layered Silicate Polymer Nanocomposites: New Approach or Illusion for Fire Retardancy? Investigations on the Potential and on the Tasks Using a Model System. *Polym. Adv. Technol.* **2004**, *15*, 355–364. https://doi.org/10.1002/pat.483

[58] Schartel, B.; Kunze, R.; Neubert, D. Red Phosphorus Controlled Decomposition for Fire Retardant PA 66. *J. Appl. Polym. Sci.* **2002**, *83*, 2060–2071. https://doi.org/10.1002/app.10144

[59] Schartel, B.; Braun, U. Comprehensive Fire Behaviour Assessment of Polymeric Materials Based on Cone Calorimeter Investigations. *e-Polymers* **2003**, *3*, 13. https://doi.org/10.1515/epoly.2003.3.1.177

[60] Braun, U.; Schartel, B. Flame Retardancy Mechanisms of Aluminium Phosphinate in Combination with Melamine Cyanurate in Glass-Fibre Reinforced Poly (1,4-Butylene Terephthalate). *Macromol. Mater. Eng.* **2008**, *293*, 206–217. https://doi.org/10.1002/mame.200700330

[61] Braun, U.; Bahr, H.; Sturm, H.; Schartel, B. Flame Retardancy Mechanisms of Metal Phosphinates and Metal Phosphinates in Combination with Melamine Cyanurate in Glass-fiber Reinforced Poly(1,4-Butylene Terephthalate): The Influence of Metal Cation. *Polym. Adv. Technol.* **2008**, *19*, 680–692. https://doi.org/10.1002/pat.1147

[62] van Krevelen, D. W. Some Basic Aspects of Flame Resistance of Polymeric Materials. *Polymer.* **1975**, *16*, 615–620. https://doi.org/10.1016/0032-3861(75)90157-3

[63] van Krevelen DW. Properties of Polymers: Their Correlation with Chemical Structure. Elsevier. Amsterdam 1990.

[64] Tewarson A. Generation of Heat and Chemical Compounds in Fires. In: The SFPE Handbook of Fire Protection Engineering, 3rd Edition. DiNenno PJ, Drysdale D, Beyler CL, Walton WD, Custer RLP, Hall JR Jr, Watts JM Jr, Eds. National Fire Protection Association, Inc. Quincy, MA 2002; chap. 3.4:3-82–3-161.

[65] Lyon RE, Janssens ML. Polymer Flammability. In: Encyclopedia of Polymer Science & Technology, 3rd Edition. Mark HF, Ed. Wiley. Hoboken 2004.

[66] Lyon RE. Plastics and Rubber. In: Handbook of Building Materials for Fire Protection. Harper CA, Ed. McGraw-Hill. New York 2004; chap. 3:3.1–3.51.

[67] Walters, R. N.; Lyon, R. E. Molar Group Contributions to Polymer Flammability. *J. Appl. Polym. Sci.* **2003**, *87*, 548–563. https://doi.org/10.1002/app.11466

[68] Quintiere JG. Fundamentals of Fire Phenomena. Wiley. Chichester 2006.

[69] Wharton, R. K. Identification of Two Pass Regimes for Nylon-66 Based Rods in the Critical Oxygen Index Test. *J. Appl. Polym. Sci.* **1982**, *27*, 3193–3197. https://doi.org/10.1002/ app.1982.070270901

[70] Schartel, B.; Bartholmai, M.; Knoll, U. Some Comments on the Main Fire Retardancy Mechanisms in Polymer Nanocomposites. *Polym. Adv. Technol.* **2006**, *17*, 772–777. https://doi.org/10.1002/pat.792

[71] Pawlowski, K. H.; Schartel, B. Flame Retardancy Mechanisms of Triphenyl Phosphate, Resorcinol Bis(Diphenyl Phosphate) and Bisphenol A Bis(Diphenyl Phosphate) in Polycarbonate/ Acrylonitrile-Butadiene-Styrene Blends. Polym. Int. 2007, 56, 1404–1414. https://doi.org/10.1002/pi.2290

[72] Weil ED. Synergists, Adjuvants, and Antagonists in Flame-Retardant Efficacy in Polymeric Materials. In: Fire Retardancy of Polymeric Materials. Grand AF, Wilkie CA, Eds. Marcel Dekker. New York 2000; chap. 4:115–145.

[73] Casu, A.; Camino, G.; De Giorgi, M.; Flath, D.; Laudi, A.; Morone, V. Effect of Glass Fibres and Fire Retardant on the Combustion Behaviour of Composites, Glass Fibres Poly(Butylene Terephthalate). *Fire Mater.* **1998**, *22*, 7–14. https://doi.org/10.1002/(SICI)1099-1018(199801/02)22:1<7::AID-FAM623>3.0.CO;2-3

[74] Day, A. G. Oxygen Index Test: Temperature Effect and Comparison with Other Flammability Tests. *Plast. Polym.* **1975**, *43*, 64–67.

[75] Morgan, A. B.; Bundy, M. Cone Calorimeter Analysis of UL-94 V-Rated Plastics. *Fire Mater.* **2007**, *31*, 257–283. https://doi.org/10.1002/fam.937

[76] Karlsson B. Performance-Based Test Methods for Material Flammability. In: Fire Retardant Materials. Horrocks AR, Price D, Eds. Woodhead Publishing Limited. Cambridge 2001; chap. 12:355–377.

[77] Babrauskas V. The Cone Calorimeter. In: Heat release in Fires. Babrauskas V, Grayson SJ, Eds. Elsevier Science Publishers Ltd. Barking 1992; chap. 4:61–91.

[78] Babrauskas, V. Development of the Cone Calorimeter—a Bench-Scale Heat Release Rate Apparatus Based on Oxygen-Consumption. *Fire Mater.* **1984**, *8*, 81–95. https://doi.org/10.1002/fam.810080206

[79] Babrauskas V. Fire Test Methods for Evaluation of Fire- Retardant Efficacy in Polymeric Materials. In: Fire Retardancy of Polymeric Materials. Grand AF, Wilkie CA, Eds. Marcel Dekker Inc. New York 2000; chap. 3:81–113.

[80] Scudamore, M. J.; Briggs, P. J.; Prager, F. H. Cone Calorimetry—a Review of Tests Carried Out on Plastics for the Association of Plastic Manufacturers in Europe. *Fire Mater.* **1991**, *15*, 65–84. https://doi.org/10.1002/fam.810180507

[81] Babrauskas, V. Specimen Heat Fluxes for Bench-Scale Heat Release Rate Testing. *Fire Mater.* **1995**, *19*, 243–252. https://doi.org/10.1002/fam.810190602

[82] Babrauskas V. Heat Release Rates. In: The SFPE Handbook of Fire Protection Engineering, 3rd Edition. DiNenno PJ, Drysdale D, Beyler CL, Walton WD, Custer RLP, Hall JR Jr, Watts JM Jr, Eds. National Fire Protection Association, Inc. Quincy, MA 2002; chap. 3.1:3-1–3-37.

[83] Scudamore, M. J. Fire Performance Studies on Glass-Reinforced Plastic Laminates. *Fire Mater.* **1994**, *18*, 313–325. https://doi.org/10.1002/fam.810180507

[84] Schartel, B.; Bartholmai, M.; Knoll, U. Some Comments on the Use of Cone Calorimeter Data. *Polym. Degrad. Stab.* **2005**, *88*, 540–547. https://doi.org/10.1016/j.polymdegradstab.2004.12.016

[85] Schartel, B.; Hull, T. R. Development of Fire Retarded Materials—Interpretation of Cone Calorimeter Data. *Fire Mater.* **2007**, *31*, 327–354. https://doi.org/10.1002/fam.949

[86] Drysdale DD. Fire Safety Design Requirements of Flame-Retarded Materials. In: Fire Retardant Materials. Horrocks AR, Price D, Eds. Woodhead Publishing. Cambridge 2001; chap. 13:378–397.

[87] Tewarson, A.; Pion, R. F. Flammability of Plastics. 1. Burning Intensity. *Combust. Flame.* **1976**, *26*, 85–103. https://doi.org/10.1016/0010-2180(76)90059-6

[88] Tewarson A. Flammability of Polymers. In: Plastics and the Environment. Andrady AL, Ed. John Wiley & Sons, Hoboken, NJ 2003; chap. 11.

[89] Hopkins, D. Jr.; Quintiere, J. G. Material Fire Properties and Predictions for Thermoplastics. Fire Safety J. 1996, 26, 241–268. https://doi.org/10.1016/S0379-7112(96)00033–1

[90] Bartholmai, M.; Schartel, B. Assessing the Performance of Intumescent Coatings Using Bench-Scaled Cone Calorimeter and Finite Difference Simulations. *Fire Mater.* **2007**, *31*, 187–205. https://doi.org/10.1002/fam.933

[91] Babrauskas, V.; Peacock, R. D. Heat Release Rate—the Single Most Important Variable in Fire Hazard. *Fire Safety J.* **1992**, *18*, 255–272. https://doi.org/10.1016/0379-7112(92)90019-9

[92] Janssens M, Parker WJ. Oxygen Consumption Calo- rimetry. In: Heat Release in Fires. Babrauskas V, Gray- son SJ, Eds. Elsevier Science Publishers. Barking 1992; chap. 3:31–59.

[93] Huggett, C. Estimation of Rate of Heat Release by Means of Oxygen-Consumption Measurements. *Fire Mater.* **1980**, *4*, 61–65. https://doi.org/10.1002/fam.810040202

[94] Schartel, B.; Wilkie, C. A.; Camino, G. Recommendations on the Scientific Approach to Polymer Flame Retardancy: Part 2—Concepts. *J. Fire Sci.* **2017**, *35*, 3–20. https://doi.org/10.1177/0734904116675370

[95] Hirschler, M. M. How to Measure Smoke Obscuration in a Manner Relevant to Fire Hazard Assessment: Use of Heat Release Calorimetry Test Equipment. *J. Fire Sci.* **1991**; 9,183–222. https://doi.org/10.1177/073490419100900302

[96] Babrauskas V. Bench-Scale Methods for Prediction of Full-Scale Fire Behaviour of Furnishings and Wall Linings. SFPE Technology Report 84–10. Society of Fire Protection Engineers. Boston, MA 1984.

[97] Sundström B, van Hees P, Thureson P. Results and analysis from fire tests of building products in ISO 9705, the room corner test. SP report 1998; 11.

[98] Cleary TG, Quintiere JG. Framework for Utilizing Fire Property Tests. In: Cox G, Langford B, Eds. Fire Safety Science: Proceedings of the 3rd International Symposium on Fire Safety Science (International Association for Fire Safety Science). Elsevier Applied Science. New York 1991:647–656.

[99] Delichatsios, M. Application of Upward Flame Spread for the Prediction of SBI and ISO Room Corner (and Parallel Wall) Experiments and Classification. *Therm Sci.* **2007**, *11*, 7–22. https://doi.org/10.2298/TSCI0702007D

[100] Vahabi, H.; Kandola, B. K.; Reza Saeb, M. Flame Retardancy Index for Thermoplastic Composites. *Polymers.* **2019**, *11*, 407. https://doi.org/10.3390/polym11030407

[101] Fichera, M. A.; Braun, U.; Schartel, B.; Sturm, H.; Knoll, U.; Jäger, C. Solid-State NMR Investigations of the Pyrolysis and Thermo-Oxidative Decomposition Products of a Polysty- rene/Red Phosphorus/Magnesium Hydroxide System. *J. Anal. Appl. Pyrolysis.* **2007**, *78*, 378–386. https://doi.org/10.1016/j. jaap.2006.09.013

[102] Schartel, B.; Knoll, U.; Hartwig, A.; Pütz, D. Phosphonium-Modified Layered Silicate Epoxy Resins Nanocomposites and their Combinations with ATH and Organo-Phosphorus Fire Retardants. *Polym. Adv. Technol.* **2006**, *17*, 281–293. https://doi.org/10.1002/pat.686

[103] Bourbigot, S.; Le Bras, M.; Duquesne, S.; Rochery, M. Recent Advances for Intumescent Polymers. *Macromol. Mater Eng.* **2004**, *289*, 499–511. https://doi.org/10.1002/mame.200400007

[104] Van Hees P, Axelsson J. Modelling of Euroclass Test Results by Means of the Cone Calorimeter. In: Multifunctional Barriers for Flexible Structure Textile, Leather, and Paper. Duquesne S, Magniez C, Camino G, Eds. Springer. Berlin 2007.

[105] Messerschmidt B, Van Hees P, Wickström U. Prediction of SBI (Single Burning Item) Test Results by Means of Cone Calorimeter Test Results. Interflam 1999, Proceedings of the Eighth International Conference. Interscience Communication Limited. London 1999:11–22.

[106] Hakkarainen, T.; Kokkala, M. A. Application of a One-dimensional Thermal Flame Spread Model on Predicting the Rate of Heat Release in the SBI Test. *Fire Mater.* **2001**, *25*, 61–70. https://doi.org/10.1002/fam.760

[107] Hansen, A. S. Prediction of Heat Release in the Single Burning Item Test. *Fire Mater.* **2002**, *26*, 87–97. https://doi.org/10.1002/fam.789

[108] Janssens M, Huczek J, Sauceda A. Development of a Model of the ASTM E84 Steiner Tunnel Test. Ninth International IAFSS Symposium. Interscience Communications Ltd. Greenwich. London 2008.

[109] Bartholmai, M.; Schriever, R.; Schartel, B. Influence of External Heat Flux and Coating Thickness on Thermal Insulation Properties of Two Different Intumescent Coatings Using Cone Calorimeter and Numerical Analysis. *Fire Mater.* **2003**, *27*, 151–162. https://doi.org/10.1002/fam.823

[110] Duquesne, S.; Magnet, S.; Jama C, Delobel R. Thermoplastic Resins for Thin Film Intumescent Coatings—Towards a Better Understanding of Their Effect on Intumescence Efficiency. *Polym. Degrad. Stab.* **2005**, *88*, 63–69. https://doi.org/10.1016/j.polymdegradstab.2004.01.026

[111] Morys, M.; Illerhaus, B.; Sturm, H.; Schartel, B. Revealing the Inner Secrets of Intumescence: Advanced Standard Time Temperature Oven (STT Mufu+)—μ-Computed Tomography Approach. *Fire Mater.* **2017**, *41*, 927–939. https://doi.org/10.1002/fam.2426

[112] Dreyer, J. A. H.; Weinell, C. E.; Dam-Johansen, K.; Kiil, S. Review of Heat Exposure Equipment and In-Situ Characterisation Techniques for Intumescent Coatings. *Fire Safety J.* **2021**, *121*, 103264. https://doi.org/10.1016/j.firesaf.2020.103264

[113] Gardelle B, Duquesne S, Vandereecken P, Bourbigot S, Characterization of the Carbonization Process of Expandable Graphite/Silicone Formulations in a Simulated. *Fire. Polym. Degrad. Stab.* **2013**, *98*, 1052–1063. https://doi.org/10.1016/j.polymdegradstab.2013.02.001

[114] Morys, M.; Illerhaus, B.; Sturm, H.; Schartel, B. Size Is Not All That Matters: Residue Thickness and Protection Performance of Intumescent Coatings Made From Different Binders. *J. Fire Sci.* **2017**, *35*, 284–302. https://doi.org/10.1177/0734904117709479

[115] Schartel, B.; Humphrey, J. K.; Gibson, A. G.; Hörold, A.; Trappe, V.; Gettwert, V. Assessing the Structural Integrity of Carbon-Fibre Sandwich Panels in Fire: Bench-Scale Approach. *Compos. B Eng.* **2019**, *164*, 82–89. https://doi.org/10.1016/j.compositesb.2018.11.077

[116] Tabaka, W.; Timme, S.; Lauterbach, T.; Medina, L.; Berglund, L. A.; Carosio, F.; Duquesne, S.; Schartel, B. Bench-Scale Fire Stability Testing—Assessment of Protective Systems on Carbon Fibre Reinforced Polymer Composites. *Polym Test.* **2021**, *102*, 107340. https://doi.org/10.1016/j.polymertesting.2021.107340

[117] Gibson, A. G.; Browne, T. N. A.; Feih, S.; Mouritz, A. P. Modelling Composite High Temperature Behavior and Fire Response Under Load. *J. Compos. Mater.* **2012**, *46*, 2005–2022. https://doi.org/10.1177/0021998311429383

[118] La Delfa, G.; Luinge, J. W.; Gibson, A. G. Integrity of Composite Aircraft Fuselage Materials Under Crash Fire Conditions. *Plast. Rubber Compos.* **2009**, *38*, 111–117. https://doi.org/10.1179/174328909X387900

[119] Hörold, A.; Schartel, B.; Trappe, V.; Korzen, M.; Naumann M. Structural Integrity of Sandwich Structures in Fire: An Intermediate-Scale Approach. *Compos. Inter.* **2013**, *20*, 741–759. https://doi.org/10.1080/15685543.2013.816620

[120] Hull, T. R.; Carman, J. M.; Purser, D. A. Prediction of CO Evolution from Small-Scale Polymer Fires. *Polym. Int.* **2000**, *49*, 1259–1265. https://doi.org/10.1002/1097-0126(200010)49:10<1259::AID-PI573>3.0.CO;2-D

[121] Babrauskas, V. The Generation of CO in Bench-Scale Fire Tests and the Prediction for Real-Scale Fires. *Fire Mater.* **1995**, *19*, 205–213. https://doi.org/10.1002/fam.810190503

[122] Wittbecker F-W, Bansemer B. A Model for the Scenario-Related Assessment of the Smoke Toxic Potency. Interflam 2004 Proceedings of the Tenth International Conference, Edinburgh. Interscience Communication Limited. London 2004:1479–1490.

[123] Zammarano, M.; Kramer, R.H.; Harris, R.; Ohlemiller, T. J.; Shields, J. R.; Rahatekar, S. S.; Lacerda, S.; Gilman, J. W. Flammability Reduction of Flexible Polyurethane Foams Via Carbon Nanofiber Network Formation. Polym. *Adv. Technol.* **2008**, *19*, 588–595. https://doi.org/10.1002/pat.1111

[124] Brehme, S.; Schartel, B.; Goebbels, J.; Fischer, O.; Pospiech, D.; Bykov, Y.; Döring, M. Phosphorus Polyester Versus Aluminium Phosphinate in Poly(Butylene Terephthalate) (PBT): Flame Retardancy Performance and Mechanisms. *Polym. Degrad. Stab.* **2011**, *96*, 875–884. https://doi.org/10.1016/j. polymdegradstab.2011.01.035

[125] Deng, C.; Yin, H.; Li, R.-M.; Huang, S.-C.; Schartel, B.; Wang, Y.-Z. Modes of Action of a Mono-Component Intumescent Flame Retardant MAPP in Polyethylene-Octene Elastomer. *Polym. Degrad. Stab.* **2017**, *138*, 142–150. https://doi.org/10.1016/j. polymdegradstab.2017.03.006

[126] Rabe, S.; Chuenban, Y.; Schartel, B. Exploring the Modes of Action of Phosphorus-Based Flame Retardants in Polymeric Systems. *Materials.* **2017**, *10*, 455. https://doi.org/10.3390/ ma10050455

[127] Schartel, B.; Pawlowski, K. H.; Lyon, R. E. Pyrolysis Combustion Flow Calorimeter: A Tool to Assess Flame Retarded PC/ ABS Materials? *Thermochim. Act.* **2007**, 462:1–14. https://doi.org/10.1016/j.tca.2007.05.021

[128] Sonnier, R.; Ferry, L.; Longuet, C.; Laoutid, F.; Friederich, B.; Laachachi, A.; Lopez Cuesta, J.-M. Combining Cone Calorimeter and PCFC to Determine the Mode of Action of Flame-Retardant Additives. *Polym. Adv. Technol.* **2011**, *22*, 1091–1099. https://doi.org/10.1002/pat.19891091

[129] Lyon, R. E.; Walters, R. N.; Gandhi, S. Combustibility of Cyanate Ester Resins. *Fire Mater.* **2006**, *30*, 89–106. https://doi.org/10.1002/fam.889

[130] Price D, Anthony G, Carty P. Introduction: Polymer Combustion, Condensed Phase Pyrolysis and Smoke Formation. In: Fire Retardant Materials. Horrocks AR, Price D, Eds. Woodhead Publishing. Cambridge 2001; chap. 1:1–30.

[131] Factor A. Char Formation in Aromatic Engineering Polymers. In: Fire and Polymers, ACS Symposium Series 425. Americal Chemical Society. Washington, DC 1990; chap. 19:274–287.

[132] Lyon RE. Ignition Resistance of Plastics. In: Recent Advances in Flame Retardancy of Polymers, Vol. 13. Lewin M, Ed. Business Communications Co. Inc. Norwalk 2002:14–25.

[133] Tewarson, A. Flammability Parameters of Materials—Ignition Combustion, and Fire Propagation. J. Fire Sci. 1994, 12, 329–356.

[134] Schartel B. The Burning of Plastics. In: Plastics Flammability Handbook, 4th Edition. Troitzsch J, Antonatus E, Eds. Hanser. Munich 2021; chapter 2:23–52.

[135] Bundy M, Ohlemiller T. Bench-Scale Flammability Measures for Electronic Equipment, NISTR 7031. National Institute of Standards and Technology, NIST. Gaithersburg 2003.

[136] Anna, P.; Marosi, G.; Csontos, I.; Bourbigot, S.; Le Bras, M.; Delobel, R. Influence of Modified Rheology on the Efficiency of Intumescent Flame Retardant Systems. *Polym. Degrad. Stab.* **2001**, *74*, 423–426.

[137] Laoutid, F.; Gaudon, P.; Taulemesse, J. M.; Lopez Cuesta, J. M.; Velasco, J. I.; Piechaczyk, A. Study of Hydromagnesite and Magnesium Hydroxide Based Fire Retardant Systems for Ethylene-Vinyl Acetate Containing Organo-Modified Montmorillonite. *Polym. Degrad. Stab.* **2006**, *91*, 3074–3082.

[138] Bourbigot, S.; LeBras, M.; Delobel, R.; Tremillon, J.-M. Synergistic Effect of Zeolite in an Intumescence Process—Study of the Interactions Between the Polymer and the Additives. *J. Chem. Soc. Faraday Trans.* **1996**, *92*, 3435–3444.

[139] Samyn, F.; Bourbigot, S.; Duquesne, S.; Delobel, R. Effect of Zinc Borate on the Thermal Degradation of Ammonium Polyphosphate. *Thermochim Act.* **2007**, *456*, 134–144.

[140] Pawlowski, K. H.; Schartel, B. Flame Retardancy Mechanisms of Aryl Phosphates in Combination with Boehmite in Bisphenol A Polycarbonate/ Acrylonitrile-Butadiene-Styrene Blends. *Polym. Degrad. Stab.* **2008**, *93*, 657–667.

[141] Song, L.; Hu, Y.; Lin, Z.; Xuan, S.; Wang, S.; Chen, Z.; Fan, W. Preparation and Properties of Halogen-Free Flame-Retarded Polyamide 6/Organoclay Nanocomposite. *Polym. Degrad. Stab.* **2004**, *86*, 535–540.

[142] Gentilhomme, A.; Cochez, M.; Ferriol, M.; Oget, N.; Mieloszynski, J. L. Thermal Degradation of Methyl Methacrylate Polymers Functionalized by Phosphorus-Containing Molecules. III: Cone Calorimeter Experiments and Investigation of Residues. *Polym. Degrad. Stab.* **2005**, *88*, 92–97.

[143] Tang, Y.; Hu, Y.; Song, L.; Zong, R. W.; Gui, Z.; Fan, W. C. Preparation and Combustion Properties Polypropylene-Polyamide-6 of Flame Retarded Alloys. *Polym. Degrad. Stab.* **2006**, *91*, 234–241. https://doi.org/10.1016/j. polymdegradstab.2005.05.016

[144] Laoutid, F.; Ferry, L.; Lopez-Cuesta, J. M.; Crespy, A. Flame- Retardant Action of Red Phosphorus/Magnesium Oxide and Red Phosphorus/Iron Oxide Compositions in Recycled PET. *Fire Mater.* **2006**, *30*, 343–358. https://doi.org/10.1002/fam.914

[145] Bertelli, G.; Marchetti, E.; Camino, G.; Costa, L.; Locatelli, R. Intumescent Fire Retardant Systems—Effect of Fillers on Char Structure. Angew. *Makromol. Chem.* **1989**, *172*, 153–163. https://doi.org/10.1002/apmc.1989.051690112

[146] Bertelli, G.; Camino, G.; Marchetti, E.; Costa, L.; Locatelli, R. Structural Studies on Chars from Fire Retardant Intumescent Systems. Angew. *Makromol. Chem.* **1989**, *169*, 137–142.

[147] Le Bras M, Bourbigot S, Siat C, Delobel R. Comprehensive Study of Protection of Polymers by Intumescence—Application to Ethylene Vinyl Acetat Copolymer Formulations. In: Fire Retardancy of Polymers: The Use of Intumescence. Le Bras M, Camino G, Bourbigot S, Delobel R, Eds. The Royal Chemical Society. Cambridge 1998:266–279.

[148] Bourbigot S, Leroy J-M. Modelling of Thermal Diffusivity During Combustion—Application to Intumescent Materials. In: Fire Retardancy of Polymers: The Use of Intumescence. Le Bras M, Camino G, Bourbigot S, Delobel R, Eds. The Royal Chemical Society. Cambridge 1998:129–139.

[149] Schartel, B.; Weiß, A. Temperature Inside Burning Polymer Specimens: Pyrolysis Zone and Shielding. *Fire Mater.* **2010**, *34*, 217–235. https://doi.org/10.1002/fam.1007

[150] Wu, G. M.; Schartel, B.; Bahr, H.; Kleemeier, M.; Yu, D.; Hartwig A. Experimental and Quantitative Assessment of Flame Retardancy by the Shielding Effect in Layered Silicate Epoxy Nanocomposites. *Combust Flame.* **2012**, *159*, 3616–3623. https://doi.org/10.1016/j.combustflame.2012.07.003

[151] Vandersall, H. L. Intumescent Coating Systems, Their Development and Chemistry. *J. Fire Flammability.* **1971**, *2*, 97–140.

[152] Kay, M.; Price, A. F.; Lavery, I. Review of Intumescent Materials, with Emphasis on Melamine Formulations. *J. Fire Retard. Chem.* **1979**, *6*, 69–91.

[153] Camino, G.; Costa, L.; Trossarelli, L. Study on the Mechanism of Intumescence in Fire Retardant Polymers. 1. Thermal- Degradation of Ammonium Polyphosphate Pentaerythritol Mixtures. *Polym. Degrad. Stab.* **1984**, *6*, 243–252. https://doi.org/10.1016/0141-3910(84)90004-1

[154] Camino, G.; Costa, L.; Trossarelli, L. Study on the Mechanism of Intumescence in Fire Retardant Polymers 2. Mechanism of Action in Polypropylene-Ammonium Polyphosphate

Pentaerythritol Mixtures. *Polym. Degrad. Stab.* **1984**, *7*, 25–31. https://doi.org/10.1016/0141-3910(84)90027-2

[155] Camino, G.; Costa, L.; Trossarelli, L. Study on the Mechanism of Intumescence in Fire Retardant Polymers. 3. Effect of Urea on the Ammonium Polyphosphate Pentaerythritol System. *Polym. Degrad. Stab.* **1984**, *7*, 221–229. https://doi.org/10.1016/0141-3910(84)90098-3

[156] Camino, G.; Costa, L.; Trossarelli, L.; Costanzi, F.; Landoni, G. Study on the Mechanism of Intumescence in Fire Retardant Polymers. 4. Evidence of Ester Formation in Ammonium Polyphosphate Pentaerythritol Mixtures. *Polym. Degrad. Stab.* **1984**, *8*, 13–22. https://doi.org/10.1016/0141-3910(84)90068-5

[157] Camino, G.; Costa, L.; Trossarelli, L. Study on the Mechanism of Intumescence in Fire Retardant Polymers. 5. Mechanism of Formation of Gaseous Products in the Thermal-Degradation of Ammonium Polyphosphate Pentaerythritol Mixtures. *Polym. Degrad. Stab.* **1985**, *12*, 203–211. https://doi.org/10.1016/0141-3910(84)90004-1

[158] Camino, G.; Costa, L.; Trossarelli, L.; Costanzi, F.; Pagliari, A. Study on the Mechanism of Intumescence in Fire Retardant Polymers. 6. Mechanism of Ester Formation in Ammonium Polyphosphate Pentaerythritol Mixtures. *Polym. Degrad. Stab.* **1985**, *12*, 213–228. https://doi.org/10.1016/0141-3910(85)90090-4

[159] Camino, G.; Costa, L.; Martinasso, G. Intumescent Fire-Retardant Systems. *Polym. Degrad. Stab.* **1989**, *23*, 359–376. https://doi.org/10.1016/0141-3910(89)90058-X

[160] Camino, G, Delobel R. Intumescence. In: Fire Retardancy of Polymeric Materials. Grand AF, Wilkie CA, Eds. Marcel Dekker. New York 2000; chap. 7:217–243.

[161] Camino G, Lomakin S. Intumescent Materials. In: Fire Retardant Materials. Horrocks AR, Price D, Eds. Woodhead Publishing. Cambridge 2001; chap. 10:318–336.

[162] Marchal, A.; Delobel, R.; Le Bras, M.; Leroy, J.-M.; Price, D. Effect of Intumescence on Polymer Degradation. *Polym. Degrad. Stab.* **1994**, *44*, 263–272. https://doi.org/10.1016/0141-3910(94)90088-4

[163] Wilke, A.; Langfeld, K.; Ulmer, B.; Andrievici, V.; Hörold, A.; Lim-bach, P.; Bastian, M.; Schartel, B. Halogen-Free Multicomponent Flame Retardant Thermoplastic Styrene-Ethylene-Butylene-Styrene Elastomers Based on Ammonium Polyphosphate—Expandable Graphite Synergy. *Ind. Eng. Chem. Res.* **2017**, *56*, 8251–8263. https://doi.org/10.1021/acs.iecr.7b01177

[164] Chan, Y. Y.; Schartel, B. It Takes Two to Tango: Synergistic Expandable Graphite–Phosphorus Flame Retardant Combinations in Polyurethane Foams. *Polymers.* **2022**, *14*, 2562. https://doi.org/10.3390/polym14132562

[165] Petrella, R. V. The Assessment of Full-Scale Fire Hazards from Cone Calorimeter Data. *J. Fire Sci.* **1994**, *12*, 14–43.

18 Material Flammability Measurements and Test Methods

Marc Janssens
Southwest Research Institute, San Antonio, TX, USA

18.1 FUNDAMENTALS OF FIRE TESTING

18.1.1 Typical Development of a Compartment Fire

The fire tests discussed in this chapter are used to evaluate one or several aspects of the behavior of a material in a compartment fire. This behavior is strongly affected by the environmental conditions in the compartment, that is, the surrounding gas temperature, the incident heat flux to the surface of the material and the amount of oxygen available for combustion. These conditions change as the fire develops. Compartment fires that progress to burnout typically consists of four stages: initiation, the pre-flashover growth stage, flashover, and the post-flashover stage. A typical sequence of fire development involving these four stages is briefly described in the next four sections.

18.1.1.1 Initiation

A compartment fire usually starts with the ignition of a small amount of combustible contents or interior finishes due to, for example, a lighter flame, an electrical fault, or a smoldering cigarette. This is illustrated in Figure 18.1a (upper left), which shows an upholstered chair shortly after the onset of flaming combustion due to exposure to a small open flame ignition source.

The fire remains limited in size for some time, during which only one item or a small area is involved. A single person can easily extinguish the flames with a small amount of water or a portable extinguisher, but the fire may not be detected at this time. The environment inside the compartment is not yet affected, and there is no major threat to occupants. The fire may be detected when flames are large enough to be visible or when smoke or heat is produced in sufficient quantities to activate a detector.

Some types of materials will first start smoldering when exposed to heat without an open flame ignition source. Smoldering combustion is a slow exothermic surface reaction between a solid fuel and oxygen in the air. Oxygen is needed to support smoldering combustion, but it is consumed at a much smaller rate than in flaming fires. Smoldering fires involve a low rate of mass loss per unit time, but a larger share of lost mass is released as products of incomplete combustion, in particular carbon monoxide (CO), than in flaming fire conditions.

A smoldering upholstered chair fire in a closed room can cause untenable conditions in approximately one to two hours, depending on the size of the room.[1] The heat produced by a smoldering fire is usually insufficient to activate a sprinkler head until it transitions to flaming combustion. It is difficult to predict whether or when this transition will occur, but it usually happens after conditions near the fire's point of origin have already become untenable due to the elevated concentration of carbon monoxide. However, the number of fire fatalities in the US due to smoldering fires that do not transition to flaming fires is relatively insignificant.[2]

18.1.1.2 Pre-Flashover Stage

As the fire grows, a hot smoke layer accumulates beneath the ceiling and temperatures gradually increase. The fire is still limited to the first item ignited, and it becomes increasingly difficult to see exit signs through the smoke layer, as illustrated in Figure 18.1b (upper right).

The fire eventually starts spreading beyond the first item ignited (note the burning curtain on the floor in Figure 18.1c, lower left), and visibility through the upper smoke layer drops to zero. Conditions become untenable when the heat flux to the lower part of the compartment exceeds a critical level, or when people become exposed to the hot toxic smoke. At this point it is no longer possible to control the fire with a portable extinguisher.

18.1.1.3 Flashover

When heat fluxes to the lower part of the compartment are high enough to ignite common combustible materials, a rapid transition occurs to a fully developed fire. This transition usually takes less than a minute and is referred to as flashover. When flashover occurs, it is no longer possible to survive in the fire compartment. All exposed combustible materials become involved in the fire (note burning rug and table top in Figure 18.1d, lower right). Commonly used criteria for the onset of flashover are a hot smoke layer temperature of 600°C and an incident heat flux at floor level of 20 kW/m^2.

DOI: 10.1201/9781003380689-18

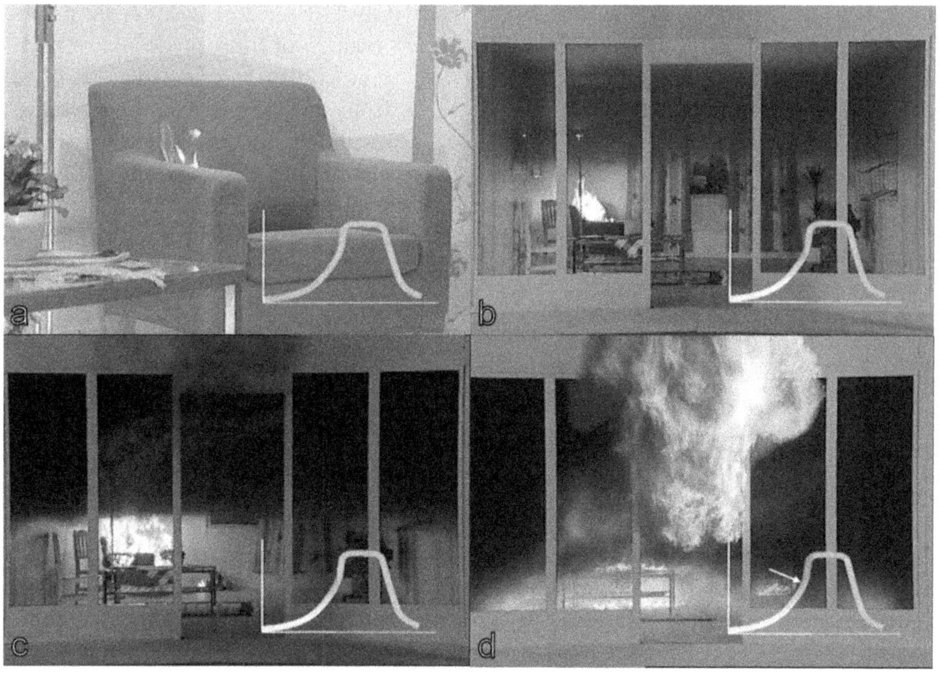

FIGURE 18.1 Compartment fire stages from initiation through flashover.

Source: Courtesy SP, Borås, Sweden.

18.1.1.4 Post-Flashover Stage

Flashover leads to the fully developed stage of a fire in which all exposed combustibles in the compartment are involved. The temperatures and heat fluxes in the compartment and the types of combustible materials that are present control the generation rate of fuel volatiles. Typical temperatures in a fully developed fire are 800–1,000°C, and corresponding incident heat fluxes range from 75 to 150 kW/m^2. The flow rate of air into the compartment is primarily determined by the size and shape of the ventilation openings. The supply of air in a fully developed fire is usually below what is needed to burn all fuel volatiles inside the compartment. The fire is therefore ventilation-limited, and some fuel volatiles burn outside the compartment; in other words, flames emerge from doors and windows (see Figure 18.1d, lower right).

Once a fire reaches the post-flashover stage, it becomes a threat to the entire building. Occupants remote from the fire compartment may be affected and evacuation of the entire building is necessary to avoid casualties. Flames can propagate to other compartments through interior or exterior pathways, and smoke may travel over long distances and pose a threat to occupants in remote parts of the building. According to US fire statistics post-flashover fires account for approximately 70% of fire fatalities, with the majority of deaths remote from the fire room of origin. Without intervention, the fire will eventually decay and burn out when all combustibles in the compartment are consumed.

18.1.2 The NFPA 550 Fire Concepts Tree

NFPA 550, *Fire Safety Concepts Tree*, was first developed in 1974. The document presents a systematic approach and identifies fundamental strategies for achieving a fire-safe environment. NFPA 550 suggests two strategies that pertain specifically to materials whose fire performance can be improved with flame retardants. The first strategy involves preventing, or at least minimizing the likelihood of ignition. Because in practice it is not possible to completely eliminate ignition, the second strategy comprises managing the impact of a subsequent fire. The latter is accomplished in part by delaying and reducing the likelihood of flashover, which in turn requires controlling the intensity with which a material burns and releases heat once ignited, its propensity to spread fire, and the rate at which it generates smoke and toxic or corrosive combustion products during gasification and burning.

18.1.3 Reaction to Fire of Materials

The two strategies discussed in the previous section are implemented in fire safety codes and regulations. The requirements are based on performance in "flammability" tests, which are intended to evaluate one or multiple aspects of the fire behavior of a material. A distinction can be made between two types of flammability tests. The first type pertains to the initiation stage and evaluates the ignition propensity of a material exposed to a small heat source such as a small flame, a cigarette, or a glowing wire. The second type assesses how a material reacts to the thermal exposure conditions in the pre-flashover stage of a fire and contributes to the fire's growth toward flashover. The different elements of the reaction to fire of a material are described below.

18.1.3.1 Ignition

When a combustible material is exposed to an external radiant heat source, its surface temperature starts to rise. The temperature inside the solid also increases with time, but at a slower rate. Provided the net heat flux into the material is sufficiently high, the surface temperature eventually reaches a level at which thermal decomposition begins. The fuel gases and vapors generated emerge through the exposed surface and mix with air in the gas phase. Under certain conditions, this mixture exceeds the lower flammability limit and ignites.

The initiation of flaming combustion as described in the previous paragraph is termed flaming ignition. Piloted ignition refers to the case when a pilot (small flame, spark igniter, hot wire, etc.) is used to ignite the fuel gases and vapors generated by a pyrolyzing specimen. Autoignition occurs when there is no pilot and the gas mixture is ignited at the hot surface of the material. A specimen of a specific material exposed to a specified irradiance will ignite faster when a pilot is present. Tests to evaluate the ignitability of a material exposed to radiant heat therefore typically include a pilot.

18.1.3.2 Heat Release

The rate at which heat is released in a compartment is the most important factor affecting fire growth towards flashover and the severity of subsequent post-flashover fire conditions. This can be illustrated as follows.

- When large surfaces are involved in the early stages of a fire, pre-flashover growth is determined primarily by the increase of the burning area. As discussed in section 18.1.3.3, this increase is primarily due to wind-aided flame spread, the rate of which depends in part on the flame length. The latter is a direct function of the local heat release rate of the pyrolyzing fuel downstream of the area being heated by the flame.[3]
- The rate of heat release of a fire is the driving force for a buoyant plume. When the plume hits the ceiling, it turns into a ceiling jet whose characteristics determine when heat detectors respond[4] or when sprinklers are activated.[5]
- Temperature rise in the upper layer in pre-flashover compartment fires has been correlated with the energy release rate.[6–8]
- The heat from the fire source and the upper layer determines when conditions become untenable inside the compartment and consequently how much time there is for safe egress.[9]
- Given the heat release rate in a compartment as a function of time, models have been developed to predict the transport of hot smoke and toxic gases from this compartment to neighboring compartments.[10,11] It has been demonstrated, using such models, that heat release rate is the single most important variable in characterizing fire hazard.[12]
- An upper layer temperature of 600°C is frequently used as a criterion for flashover. This criterion and the aforementioned upper layer temperature correlations resulted in expressions for the heat release rate required for flashover as a function of the geometry of the compartment, the thermal properties of the walls, and the ventilation opening.[8,13,14]
- Post-flashover fire temperatures also depend mainly on the energy release rate inside the compartment.[15] If the fire is ventilation limited, the heat release rate inside the compartment is controlled by the ventilation factor, which is equal to the area times the square root of the height of the ventilation opening.[16]

18.1.3.3 Surface Flame Spread

Flames can spread over a solid surface in two modes. In the wind-aided flame spread mode, flames spread in the same direction as the surrounding airflow or buoyant forces. The second mode is referred to as opposed-flow flame spread, which occurs when flames spread in the opposite direction of the surrounding airflow or buoyant forces. These two modes are illustrated for flame spread over a vertical wall surface in Figure 18.2.

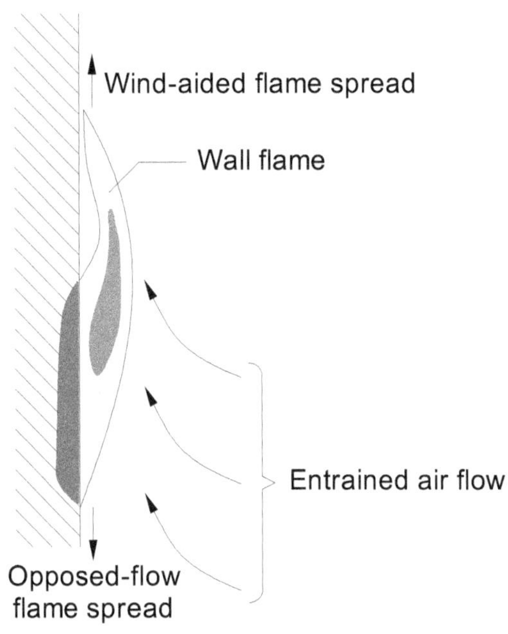

FIGURE 18.2 Modes of flame spread.

Flame spread in the upward direction over a vertical wall surface is concurrent with the surrounding airflow and is therefore wind-aided. Flame spread in the downward direction is against the entrained airflow and is of the opposed-flow type. The height of the region that is heated by the flame above the pyrolyzing region is much greater than the height of the heated region below the pyrolyzing region. The former is comparable to the height of the pyrolyzing region and is typically on the order of 1 m. The latter is only a few millimeters at most. The result is that upward or wind-aided flame spread is much faster than downward or opposed-flow flame spread.

18.1.3.4 Production of Smoke and Toxic and Corrosive Products of Combustion

Fires generate particulate matter, which reduces the intensity of light transmitted through smoke. The distance at which an exit sign can be seen through a smoke layer is a direct function of the concentration of particulates in the smoke.[17]

Fires also generate toxic products of combustion, primarily in gaseous form. There are two types of toxic gases: narcotic gases such as carbon monoxide (CO) and hydrogen cyanide (HCN), and irritant gases such as hydrochloric acid (HCl) and hydrogen bromide (HBr). Narcotic gases are absorbed into the bloodstream and reduce the intake of oxygen, which can lead to loss of consciousness and death. Irritant gases cause respiratory distress and indirectly contribute to incapacitation and lethality during exposure to fire gases. Inflammation of the respiratory tract may result in death within days or even weeks after the fire. Please see Chapter 19 of this book for more information on toxicity issues from fires.

Acid gases can also cause corrosion damage to structural steel and electrical equipment.

18.1.4 Types of Flammability Tests

This section provides a brief overview of the three different types of flammability tests that are in use. A more detailed discussion with specific examples follows in subsequent sections.

18.1.4.1 Small Heat Source Ignition Tests

A large variety of combustible materials are typically used throughout buildings and in transportation vehicles. A major concern is for materials that easily ignite when exposed to a small heat source and, thus, support flame propagation that quickly leads to a catastrophic fire. This concern can be addressed by requiring that materials used in significant quantities do not ignite and/or support flame propagation when exposed to a small heat source. A large number of small heat source ignition tests have been developed for this purpose.[18]

For some materials, or under certain conditions, combustion is not in the gas phase but in the solid phase. In such cases, no flame can be observed, and the surface is glowing. This very different phenomenon is termed smoldering ignition. Flaming combustion can be preceded by glowing ignition for char-forming materials exposed to low heat fluxes. Several small heat source ignition tests evaluate the smoldering ignition propensity of materials. The most common of these test methods are used to evaluate the ignition propensity of upholstered furniture components and composites exposed to the heat from a smoldering cigarette. Performance in these tests is assessed on the basis of specimen mass loss or the extent of charring and thermal degradation.

18.1.4.2 Bench-Scale Reaction-to-Fire Tests

Bench-scale reaction-to-fire tests are used to characterize the behavior of materials under more severe thermal exposure conditions that are representative of the growing pre-flashover stage of a compartment fire. These tests essentially determine how a material responds to the temperatures and heat fluxes in a growing fire. In these tests the fire conditions are simulated with a radiant panel or by inserting the specimen into a small furnace. A pilot may be used to ignite the flammable gases and vapors that are generated as a result of thermal decomposition of the specimen. Performance of a material is quantified on the basis of one or several of the different reaction-to-fire characteristics, which are typically obtained over a range of thermal exposure conditions.

18.1.4.3 Large-Scale Reaction-to-Fire Tests

In some cases, it is not possible to evaluate a material or product (combination of materials) in a bench-scale test in a manner that is representative of its end use. For example, it is difficult to use a bench-scale test method to evaluate the effect of joints on the fire performance of a thick sandwich panel that consists of a plastic foam core and metal skins. In this case a room test is used to assess the reaction to fire of the materials. Although models and correlations are available to predict the burning rate of upholstered furniture from small-scale reaction-to-fire test data, it is more practical (and generally more accurate) to measure the burning rate in a full-scale test.

18.2 SMALL HEAT SOURCE IGNITION TESTS

18.2.1 General Concepts

Tests of this type expose a relatively small specimen (linear dimensions on the order of centimeters) to a small heat source (Bunsen burner–type flame, glowing wire, smoldering cigarette, etc.) for a short duration (seconds). Pass/fail criteria are based on ignition of the specimen during exposure, formation of flaming droplets, extent or rate of flame propagation over the specimen surface and/or sustained flaming or smoldering after removal of the heat source. A few commonly used small heat source ignition tests are described in the next section.

18.2.2 Examples of Small Heat Source Ignition Tests

18.2.2.1 UL 94 20-mm Vertical Burning Test

The UL 94 standard specifies bench-scale test methods to determine the acceptability of plastic materials for use in appliances or other devices with respect to flammability under controlled laboratory conditions. The test method that is used depends on the intended end use of the material and its orientation in the device. The standard outlines two horizontal burning tests, three vertical burning tests, and a radiant panel flame spread test. The most commonly used test method described in the UL 94 standard is the "20-mm Vertical Burning Test; V-0, V-1, or V-2." The method is also described in ASTM D3801. A schematic of the test setup is shown in Figure 18.3.

The flame is maintained for ten seconds, and then removed to a distance of at least 150 mm. Upon flame removal, the specimen is observed for flaming and its duration time recorded. As soon as the flame ceases, the burner flame is reapplied for an additional ten seconds, then removed again. Duration of flaming and/or glowing after the second flame application are recorded. Ignition of the cotton by dripping particles from the test specimen is noted. Based on the results for five specimens tested, the material is classified as either V-0, V-1, or V-2.

The results of the UL 94 20 mm vertical burning test are very sensitive to the exact duration of flame application and to the technique that is used to apply and remove the flame.[19] The magnitude of the effect depends on the type of material that is being tested. Loosely following the test procedure might result in an unwarranted V-0 classification.

Specimen thickness affects performance in the test, as was demonstrated in a study involving 18 plastics.[20] The materials were tested in two thicknesses: 1.6 and 3.2 mm. Specimen thickness did not affect the rating for most materials, but for some the rating changed from V-0 to V-2 (two materials), from V-0 to not rated (one material), and from V-1 to not rated (one material). Small heat source ignition test standards therefore nearly always require that at least the minimum end use thickness be tested.

18.2.2.2 Limiting Oxygen Index Test

The limiting oxygen index (LOI) test is standardized in North America as ASTM D2863 and internationally as ISO 4589–2. The LOI apparatus consists of a glass tube of 75 to 100 mm in diameter and 450 to 500 mm in height. Figure 18.4 shows a schematic of the apparatus.

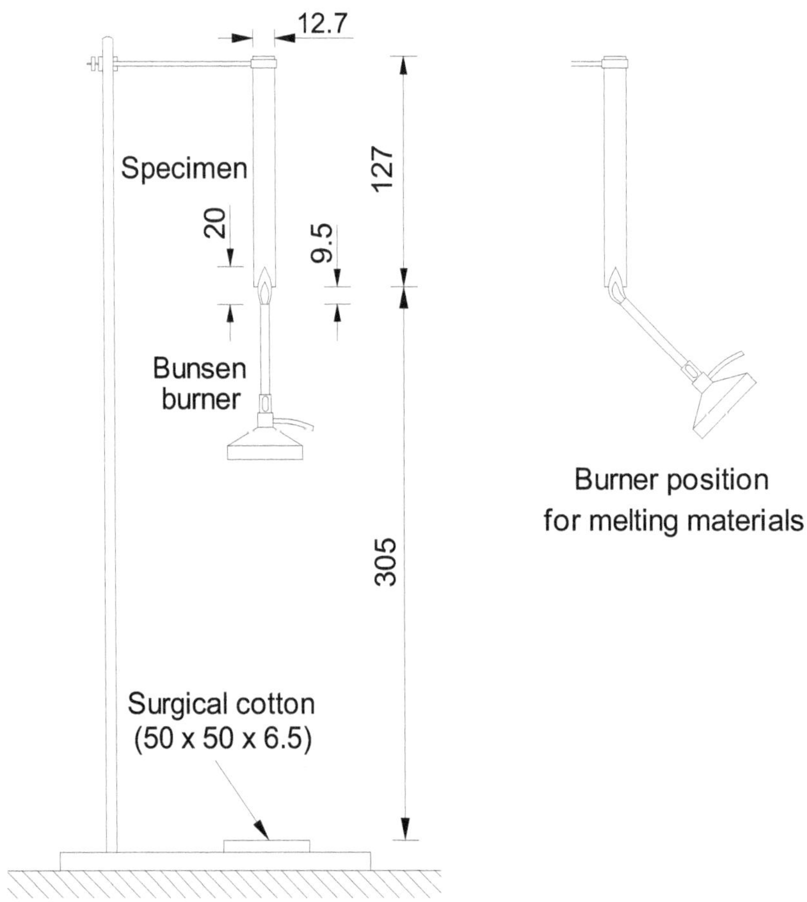

FIGURE 18.3 UL 94 20 mm vertical burning test (dimensions in mm).

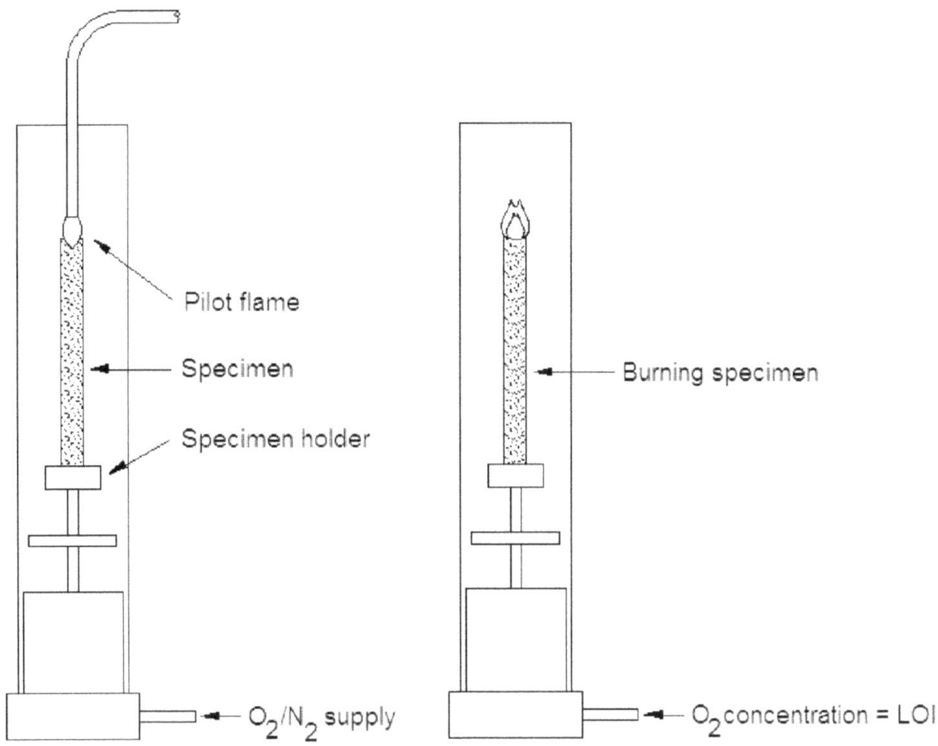

FIGURE 18.4 Limiting oxygen index test (LOI).

A specimen with a height between 70 and 200 mm and a width of 7 to 52 mm is supported inside the glass tube. A gas mixture of oxygen and nitrogen is supplied at the bottom of the tube, and a small candle-like flame is applied to the top of the specimen in an attempt to ignite it. The objective is to find the minimum oxygen concentration in nitrogen that will result in sustained combustion for at least three minutes or excessive flame propagation down the specimen. The dimensions of the specimen and the limit of excessive flame propagation depend on the type of material that is being tested. Although the LOI index may have some practical use, it does not correlate with performance in other reaction-to-fire tests that are more reliable predictors of a material's potential contribution to rapid fire growth because the flame in the LOI test spreads in the downward direction, that is, opposite to the direction of buoyant forces, instead of in the upward direction or aided by wind (see Section 18.1.3.3 for a discussion of flame spread modes).[21]

18.2.2.3 Other Small Heat Source Ignition Tests

A large number of flammability tests have been developed to evaluate the ignition propensity of a wide range of materials exposed, usually for a short duration, to a small heat source. For example, there are at least 30 ASTM fire standards that include small heat source ignition tests.[22] When adding methods that use an electric arc, hot wire, hot surface, etc. instead of a flame and those developed by other standards organizations (NFPA, UL, ISO, IEC, etc.), the total number of tests is in the hundreds.

Conceptually all small heat source ignition tests are very similar. The examples described in the preceding two sections illustrate these concepts. Other examples of commonly used small heat source ignition tests are CAL TB 117, FMVSS 302, and IEC 60695–2–13. California Technical Bulletin 117–2013 describes a test procedure for evaluating cigarette ignition resistance of resilient filling materials for use in upholstered furniture, and this test is now codified under federal law as 16 CFR 1640 (see Chapter 31 of this book for more details on this test). Federal Motor Vehicle Safety Standard 302 describes a small flame ignition test method for exposed materials in the passenger compartment of motor vehicles. The IEC test method assesses the susceptibility of electrical insulating materials to ignition as a result of exposure to a glowing wire.

18.2.3 Use and Limitations of Small Heat Source Ignition Tests

Fire safety codes and regulations generally include requirements that are based on small heat source ignition tests. The objective of these requirements is to greatly reduce the probability of a relatively benign ignition source causing a major catastrophic fire (first strategy in the NFPA 550 *Fire Concepts Tree*).

It is easy to see that the aforementioned requirements have a noticeable and positive impact on fire statistics. For example, the significantly lower number of fatalities per capita in fires involving television sets in the US compared to Europe can be attributed to the UL 94 V-0 requirement for the plastic housing in North America.[23]

However, good performance in a small heat source ignition test can give a false sense of confidence as far as a material's behavior in real fires is concerned. Performance in the test depends on how it is measured, that is, how the pass/fail or classification criteria are defined. Whether a material will ignite in the test depends on many factors, such as the following:

- Specimen size, in particular thickness.
- Specimen orientation and direction of flame propagation.
- Specimen mounting and substrate.
- Type and intensity of the heat source and duration of its application.

A material that does very well in the test might perform very differently in a real fire if any of these factors are different. For example, a V-0 rated material is not expected to ignite when subjected in a real fire to a heat source similar to that in the test. But what would happen when the real source is more severe or persists beyond the exposure time in the test? The results could be dramatically different. Ignition might occur and flames might subsequently propagate over the surface and quickly result in a catastrophic fire. There are numerous examples of materials that pass a flammability test with flying colors but perform miserably under slightly more severe real fire conditions.

In 1974 the US Federal Trade Commission issued a consent order that resulted in the withdrawal of ASTM D1692. The order was motivated by the fact that some ASTM flammability test standards used terms such as "non-burning" and "self-extinguishing" to label materials that perform well in the test. The Commission determined that these terms give the public a false sense of confidence in the actual fire performance of these materials. ASTM subsequently drafted a policy to ensure that the limitations of using test results for assessing the fire hazard and risk of materials, products, and assemblies be clearly stated in its fire test standards. The policy is no longer in effect, but the issue is addressed by the requirement that the following caveat be included in the scope section of every ASTM fire test standard: "This standard is used to measure and describe the response of materials, products, or assemblies to heat and flame under controlled conditions, but does not by itself incorporate all factors required for fire hazard or fire risk assessment of the materials, products, or assemblies under actual fire conditions." The caveat concisely sums up the main limitation of all flammability tests.

Small heat source ignition tests generally appear to be very sensitive to the composition of the material and some are therefore ideally suited to serve as a tool for formulation development and quality assurance of fire-retardant-treated products and materials. The equipment is inexpensive, only a small quantity of material is needed, the results are usually reasonably repeatable and reproducible, and a qualified laboratory technician can run many tests in a short time.

18.3 BENCH-SCALE REACTION-TO-FIRE TESTS

The purpose of bench-scale reaction-to-fire tests is to measure the flammability characteristics of materials, that is, ease of ignition, flame spread propensity, heat release, and production of smoke and toxic combustion products. Some tests are designed to measure only one of these characteristics. Other tests are more sophisticated and can be used to measure several characteristics at the same time.

18.3.1 General Concepts

Bench-scale reaction-to-fire tests characterize the behavior of materials under thermal exposure conditions that are representative of the growing pre-flashover stage of a compartment fire. The fire conditions are typically simulated in these bench-scale reaction-to-fire tests by exposing the specimen to the heat flux from a gas-fired radiant panel or an electrical heater. The range of heat fluxes that can be obtained in a particular test apparatus depends on the specimen-heater geometry and the type of radiant heat source that is used. Between different tests, heat fluxes vary over a broad range, from approximately 1 kW/m^2 to more than 100 kW/m^2. The heat flux to the specimen is specified in some test standards and is variable in others. Tests for measuring flame spread characteristics generally expose the specimen to a heat flux that varies in the direction of its largest dimension.

An important aspect of the heater is its ability to maintain the emitted radiant heat flux at a constant level during a test. If the heater is operated at a constant power level, incident radiant heat flux changes during testing. At the start of a test, a cold specimen is inserted. The specimen acts as a heat sink, resulting in a decrease of the heater temperature, and consequently a decrease of the incident radiant heat flux. After ignition, the heat released by the specimen results in an increase of the heater temperature and incident radiant heat flux. To maintain the incident radiant heat flux during a test, it is therefore necessary to keep the temperature of the heater constant. This is not trivial with a gas-fired radiant panel, but relatively straightforward for electrical heating elements.

Two types of electrical heaters are used, high and low temperature. The former are commonly tungsten filament lamps that operate at temperatures close to 2600 K. According to Wien's displacement law, peak radiant heat flux from such lamps is at a much shorter wavelength than for real fires, with temperatures in the range of 600 to 1400 K. Piloted ignition studies on plastics and wood have shown that these materials absorb much less radiation in the visible and near-infrared range than at higher wavelengths.[24,25] Specimens that are tested in an apparatus with tungsten filament lamps therefore have to be covered with a thin black coating. A procedure is described in the Fire Propagation Test standard ASTM E2058.

The heat flux can also be provided with a gas burner flame in contact with the specimen. Incident heat flux from impinging gas burner flames is primarily convective and can only be adjusted over a relatively narrow range. A gas burner flame is therefore not a suitable heat source if the response of a material over a range of heat fluxes is to be determined.

Finally, a small furnace is sometimes used to create the desired thermal exposure conditions. A furnace arrangement is ideal when the objective is to create a constant or specified time-varying temperature environment.

A pilot flame, spark plug, or hot wire is often used to ignite the gases and vapors that are generated by the pyrolysis of the heated specimen. When a pilot flame is used, it is either located in the gas phase or impinging on the specimen surface. The latter is less desirable because it locally increases the heat flux to the specimen by an unknown amount.

A pilot flame is sometimes extinguished by fire retardants or halogens in the fuel volatiles. An electric spark remains stable when fire retardants or halogens are present. However, it occupies a small volume, so that the positioning of the spark plug is more critical than with a pilot flame. A glowing wire is not an efficient method for igniting fuel volatiles, leading to poor repeatability.

18.3.2 Ignition

The performance of a material in a growing pre-flashover fire is affected by its ignition characteristics in primarily two ways. First, flame spread over the surface of the material can be viewed as a series of subsequent ignitions of incremental areas. Second, the flame spread process is typically initiated in an area that is heated by an impinging flame from a burning object. In both cases a flame is present in the vicinity of the material that is heated to ignition. The focus of this section is therefore on piloted ignition.

18.3.2.1 Pertinent Material Properties

The propensity for piloted ignition is characterized by two intrinsic material properties. The first property quantifies a critical condition for ignition. Piloted ignition occurs when the lower flammability limit is reached in the fuel-air mixture around the pilot. Consequently, a critical mass flux criterion appears to be logical. This has been proposed,[26] but unfortunately it is not very practical. The most common criterion is based on the assumption that ignition occurs when a critical temperature at the surface, T_{ig}, is reached. The surface temperature at ignition of a thermoplastic is reasonably constant and independent of heat flux.[27,28] A number of investigators measured T_{ig} for a range of wood products.[29–36] Reasonably constant values were found for each material at heat fluxes ≥ 25 kW/m^2. All studies reported a significant increase of T_{ig} at lower heat fluxes (50–150°C at 15 kW/m^2). This is due to the fact that pyrolysis and char formation at the surface are no longer negligible for ignition times exceeding three minutes. Under those conditions one of the basic assumptions of thermal ignition theory, that is, that the specimen behaves as an inert solid, is no longer valid.

The second property is the thermal inertia, kρc. This property is a measure of how fast the surface temperature of a material rises when exposed to heat. A material with lower kρc will ignite faster than a material with higher kρc and the same T_{ig} exposed to the same heat flux.

The minimum heat flux for ignition, $\dot{q}''_{min}$, is closely related to T_{ig}. This heat flux is just sufficient to heat the material surface to T_{ig} for very long exposure times (theoretically ∞). The minimum heat flux is not a true material property because it depends on the rate of convective cooling from the surface. This, in turn, depends primarily on the orientation, size, and flow field around the exposed surface. Since these are different in a bench-scale test vs. a real fire, the minimum heat flux determined based on test data is an approximate value. To make the distinction, it is referred to as the critical heat flux for ignition, $\dot{q}''_{cr}$. Since the rate of convective cooling in a bench-scale test is generally smaller than in a real fire, $\dot{q}''_{cr}$ is a conservative estimate of $\dot{q}''_{min}$.

18.3.2.2 Ignition Measurements

Ignition properties can be determined by direct measurements. T_{ig} can be measured with fine thermocouples attached to the exposed surface of ignition test specimens, or by using an optical pyrometer. ASTM D1929 is a bench-scale furnace test method to determine piloted ignition and auto-ignition temperatures of plastics (see section 18.3.2.3.1). The latter are reasonably accurate, but the former are not because the pilot is located too far above the specimen.[18]

$\dot{q}''_{cr}$ can be determined by bracketing, that is, by conducting experiments are incrementally decreasing heat flux levels until ignition does not occur within a specified period (usually 10 or 20 min). kρc can be determined by measuring thermal conductivity, density, and specific heat separately. However, since k and c are temperature-dependent, measurements at elevated temperature are needed.

Because it is very tedious to measure T_{ig} and kρc directly, it is much more common to determine ignition properties on the basis of an analysis of time-to-ignition data obtained over a range of heat fluxes. The analysis is usually based on a simple heat conduction model, which assumes that the solid is inert (negligible pyrolysis prior to ignition) and thermally thick (heat wave does not reach the back surface prior to ignition). An example of this type of analysis is discussed in Section 18.3.2.3.2.

Ignition is determined on the basis of visual observation. This can be tricky when the material exhibits extensive flashing before sustained flaming. An alternative method based on the second time derivative of the mass of the specimen has been suggested to alleviate this problem.[37,38]

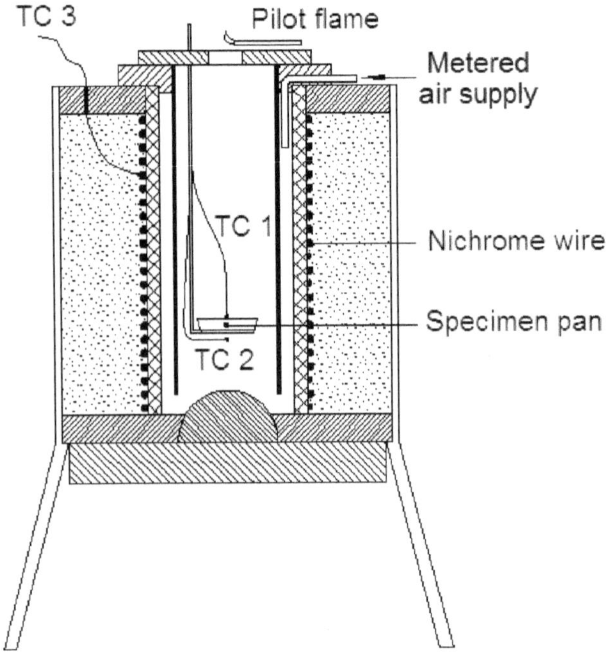

FIGURE 18.5 ASTM D 1929 test apparatus.

18.3.2.3 Examples of Ignition Tests

18.3.2.3.1 ASTM D1929

The test apparatus consists of a small tubular furnace, with an inside diameter of 76 mm and a height of 210 to 250 mm (see Figure 18.5). A controlled flow of ambient air is supplied at the bottom of the apparatus. A specimen with an area of 20 × 20 mm is inserted into the furnace from the top. Specimen height is 50 mm for materials with a density of 100 kg/m^3 or less. For materials with a higher density, the height of the specimen is adjusted so that its mass is equal to 3 g. The air temperature in the furnace is increased at increments 10°C and tests are repeated until ignition is observed. Tests are performed with and without pilot flame to determine the flash ignition temperature (FIT) and spontaneous ignition temperature (SIT), respectively. The former is equal to the piloted ignition temperature, T_{ig}.

18.3.2.3.2 Lateral Ignition and Flame Spread Test (ASTM E1321)

The lateral ignition and flame spread test (LIFT) apparatus was developed primarily for lateral flame spread measurements. The apparatus, test procedures, and methods for data analysis are described in ASTM E1321. A sample of 155 mm × 800 mm is exposed to the radiant heat of a gas-fired panel. The panel measures 280 mm × 483 mm. The heat flux is not uniform over the specimen but varies along the long axis as a function of distance from the hot end, as shown in Figure 18.6. The flux distribution is an invariant of distance when normalized to the heat flux at the 50 mm position. When burning methane or natural gas, the upper limit of the radiant heat flux is 60 to 65 kW/m^2. The lower limit is approximately 10 kW/m^2 because the porous ceramic tile surface of the panel is only partly covered with flame at lower heat fluxes.

In flame spread tests, the specimen is ignited at the hot end by a non-impinging premixed acetylene-air or impinging propane pilot flame. Flame spread rate over the surface is then monitored as a function of distance x. Thus, one experiment yields information on flame spread rate over a whole range of heat flux levels (or surface temperatures). This much information can be obtained in one run due to the particular shape of the flux invariant, which is the result of the specific geometry and specimen-panel arrangement shown in Figure 18.7.

The heat flux distribution is fairly uniform over the first 100–150 mm. Therefore, the same apparatus, with some slight modifications, may also be used for ignitability measurements. A 155 mm × 155 mm specimen is positioned at the hot end in the flame spread sample holder. An acetylene-air or propane pilot is located in the boundary layer above the specimen. A steel plate is attached to the sample holder to extend the boundary layer. The specimen is wrapped in aluminum foil and backed by a low-density ceramic fiber board. The LIFT apparatus as used for piloted ignition testing is illustrated in Figure 18.7.

Quintiere and Harkleroad developed a practical method for analyzing ignition data obtained with the LIFT

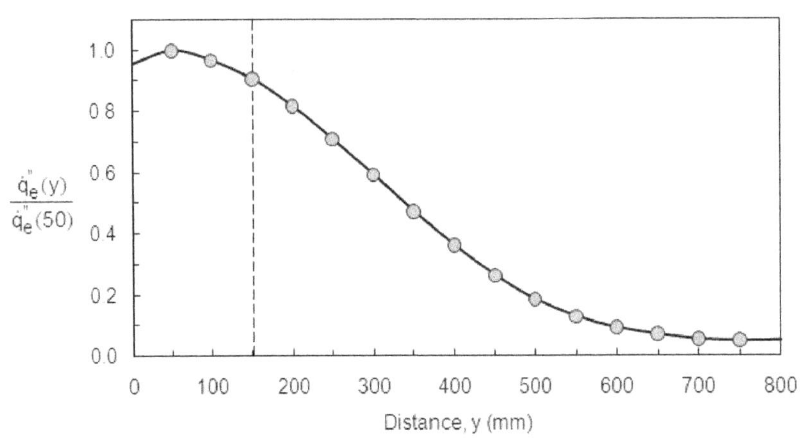

FIGURE 18.6 Radiant heat flux profile in the LIFT.

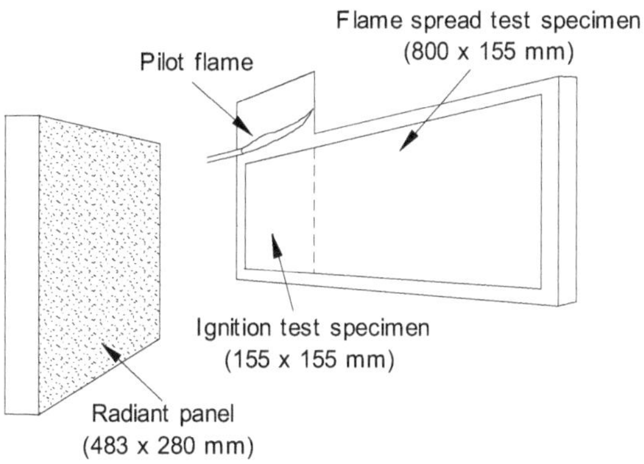

FIGURE 18.7 Schematic of the LIFT.

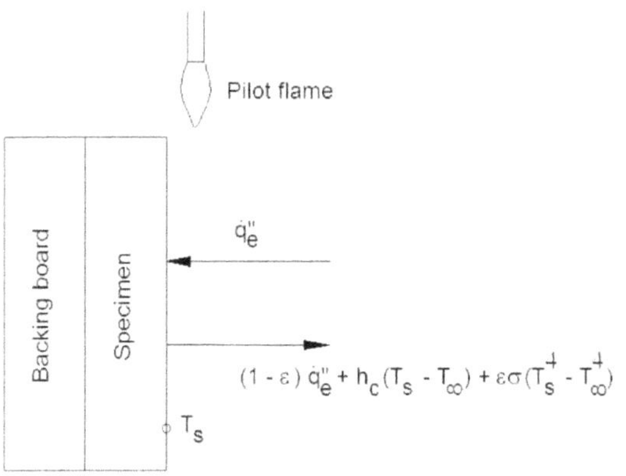

FIGURE 18.8 Schematic of the LIFT.

apparatus.[39] The method is described in ASTM E1321. The first step of the method consists of conducting ignition tests starting at a radiant heat flux level near the maximum for the apparatus (60–65 kW/m²). Time to ignition is obtained at heat flux levels in descending order at intervals of 5 to 10 kW/m², preferably with some replicates. When ignition time becomes sufficiently long (of the order of ten minutes), data is obtained at heat flux levels more closely together (1.5 to 2 kW/m² intervals). At a certain level, ignition will no longer occur within the (arbitrary) maximum test duration of 20 minutes. The critical heat flux is taken to be slightly above this level. Usually, a few more tests are conducted around this level to confirm its value. Once the critical heat flux is known, T_{ig} can be calculated from a heat balance at the surface (see Figure 18.8) after very long exposure, because heat conduction into the specimen then becomes negligible:

$$\begin{aligned}\varepsilon\dot{q}''_{cr} &\equiv h_c\left(T_{ig} - T_\infty\right) + \varepsilon\sigma\left(T_{ig}^4 - T_\infty^4\right)\\ &\equiv h_{ig}\left(T_{ig} - T_\infty\right)\end{aligned} \qquad [18\text{–}1]$$

where
ε = surface emissivity (—);
h_c = convection coefficient (kW/m²·K);
T_∞ = ambient temperature (K);
σ = Boltzmann constant ($5.67 \cdot 10^{-11}$ kW/m²·K⁴); and
h_{ig} = total heat transfer coefficient at ignition (kW/m²·K).

Surface temperature measurements under steady state conditions for a number of inert materials and some combustible materials resulted in the following fit[39]:

$$\begin{aligned}\dot{q}''_{cr} &= 0.015\left(T_{ig} - T_\infty\right) + \sigma\left(T_{ig}^4 - T_\infty^4\right)\\ &\equiv h_{ig}\left(T_{ig} - T_\infty\right)\end{aligned} \qquad [18\text{–}2]$$

Thus, if specimens are heated for a sufficiently long time in the LIFT apparatus, it may be assumed that $\varepsilon = 1$ and that h_c = 15 W/m²·K. Once T_{ig} is calculated from the empirical value for $\dot{q}''_{cr}$ via Equation 18–2, a total heat transfer coefficient from the surface at ignition can be obtained by rearranging this equation as follows:

$$\begin{aligned}h_{ig} &\equiv \frac{\dot{q}''_{cr}}{T_{ig} - T_\infty} = 0.015 + \sigma\frac{T_{ig}^4 - T_\infty^4}{T_{ig} - T_\infty} = 0.015\\ &\quad + \sigma\left(T_{ig} + T_\infty\right)\left(T_{ig}^2 + T_\infty^2\right)\end{aligned} \qquad [18\text{–}3]$$

The surface temperature at ignition for exposure to a constant radiant heat flux is approximated by the following:

$$T_{ig} \approx T_\infty + \frac{\dot{q}''_e}{h_{ig}}F\left(t_{ig}\right) \quad \text{with}$$
$$F(t) = \begin{cases}\dfrac{2h_{ig}\sqrt{t}}{\sqrt{\pi k\rho c}} & \text{small t}\\ 1 & \text{large t}\end{cases} \qquad [18\text{–}4]$$

where
t_{ig} = time to ignition at incident heat flux $\dot{q}''_e$ (s); and
F = function of time.

This leads to the following expression for correlation of piloted ignition data:

$$\frac{\dot{q}''_{cr}}{\dot{q}''_e} = F\left(t_{ig}\right) = \begin{cases}\dfrac{2h_{ig}\sqrt{t_{ig}}}{\sqrt{\pi k\rho c}} & t_{ig} \le t^*\\ 1 & t_{ig} > t^*\end{cases} \qquad [18\text{–}5]$$

where
t* = time to reach steady conditions (s).

Thus, all data are plotted in a graph of $\dot{q}''_{cr} / \dot{q}''_e$ versus $\sqrt{t_{ig}}$. An "apparent" value for $k\rho c$ can be calculated from the slope of the line through zero that best fits the data. This line crosses $\dot{q}''_{cr} / \dot{q}''_e = 1$ at t*, the time needed to reach "steady state" conditions. The functional form of Equation 18–5 for small times is identical to that of the solution of the one-dimensional heat conduction equation for a semi-infinite solid exposed to a constant heat flux without heat losses from the surface. Consequently, kρc values obtained with this procedure are higher than actual average values. The same procedure can be used to analyze piloted ignition data obtained with the Cone Calorimeter (see section 18.3.3.2.1), provided an adjustment is made to h_c to account for the differences in convective cooling conditions (use h_c. = 12 W/m²·K for tests in the Cone Calorimeter in the horizontal orientation, for example).

18.3.2.3.3 Other Ignition Tests

The time to ignition as a function of incident radiant heat flux can also be measured in the ISO ignitability test apparatus. This apparatus and its use are described in ISO 5657. Bench-scale heat release calorimeters such as the cone calorimeter (see 18.3.3.2.1) and the fire propagation apparatus (see 18.3.3.2.3) can also be used to obtain this kind of data.

18.3.3 Heat Release Rate

18.3.3.1 Pertinent Material Properties

There are two intrinsic material characteristics that are related to heat release rate. These two properties are the effective heat of combustion, ΔH_c (MJ/kg), and the heat of gasification, L (MJ/kg). The effective heat of combustion is the ratio of heat release rate to mass loss rate measured in a bench-scale calorimeter:

$$\Delta H_c = \frac{\dot{Q}''}{\dot{m}''} \quad [18\text{–}6]$$

where
$\dot{Q}''$ = heat release rate per unit exposed area (kW/m²); and
$\dot{m}''$ = mass loss rate per unit exposed area (g/m²·s).
ΔH_c is different from the lower calorific value measured in an oxygen bomb calorimeter. The latter is measured in a small container under high pressure and in pure oxygen, conditions that are not representative of real fires. The conditions in bench-scale calorimeters such as the cone calorimeter resemble those in real fires much more closely. For some fuels, in particular gases, both values are nearly identical. However, for charring solids such as wood, ΔH_c is significantly lower and equal to the heat of combustion of the volatiles during flaming combustion.

The second material property is heat of gasification, L, defined as the net heat flow into the material required to convert one unit mass of solid material to volatiles. The net heat flux into the material can be obtained from an energy balance at the surface of the specimen. Typically, a sample exposed in a bench-scale calorimeter is heated by external heaters and by its own flame. Heat is lost from the surface in the form of radiation. Due to the small sample size, the flame flux is primarily convective, and flame absorption of external heater and specimen surface radiation can be neglected. Hence, L can be defined as

$$L \equiv \frac{\dot{q}''_{net}}{\dot{m}''} = \frac{\dot{q}''_e + \dot{q}''_f - \dot{q}''_l}{\dot{m}''} \quad [18\text{–}7]$$

where
L = heat of gasification (MJ/kg);
$\dot{q}''_e$ = heat flux from external sources (kW/m²);
$\dot{q}''_f$ = heat flux from the flame (kW/m²); and
$\dot{q}''_l$ = heat losses from the exposed surface (kW/m²).

18.3.3.2 Heat Release Rate Measurements

A variety of calorimeters have been developed to determine the heat release rate and related properties under a range of conditions. The heat release rate is generally measured on the basis of the oxygen consumption technique,[40,41] although the generation of carbon dioxide is also used.[42,43] The oxygen consumption technique is based on the observation that for a wide range of materials undergoing complete combustion, a nearly constant amount of heat is released per unit mass of oxygen consumed.[44,45] The value is 13.1 kJ/g ± 0.7 kJ/g of O_2 consumed. This observation, generally referred to as Thornton's rule, implies that it is sufficient to measure the flow rate and oxygen concentration in the exhaust duct of a fire test apparatus to determine the heat release rate of the burning test specimen. In addition, bench-scale calorimeter specimens are usually placed on a load cell to measure the mass loss (rate) during a test.

Some materials exhibit nearly steady mass loss rates when exposed to a fixed radiant heat flux. The surface temperature for these materials reaches a steady value after a short initial transient period, and all terms in Equation 18–7 are approximately constant at a specified heat flux level. L can then be obtained by measuring steady mass loss rates at different radiant heat flux levels, and by plotting $\dot{m}''$ as a function of $\dot{q}''_e$. The reciprocal of the slope of a straight line fitted through the data points is equal to L. The intercept of the line with the abscissa is equal to $\dot{q}''_f - \dot{q}''_l$. Several investigators have used this technique to obtain average L values for a large number of materials.[46,47]

18.3.3.2.1 Cone Calorimeter

The cone calorimeter was developed at the National Bureau of Standards (NBS), currently the National Institute of Standards and Technology, or NIST, by Dr. Vytenis Babrauskas in the early 1980s.[48] It is presently the most commonly used bench-scale calorimeter. The apparatus and test procedure are standardized in the United States as ASTM E1354 and internationally as ISO 5660 Part 1. A schematic of the apparatus is shown in Figure 18.9.

A square specimen of 100 × 100 mm is exposed to the radiant flux of an electric heater. The heater has the shape of a truncated cone (hence the name of the instrument) and

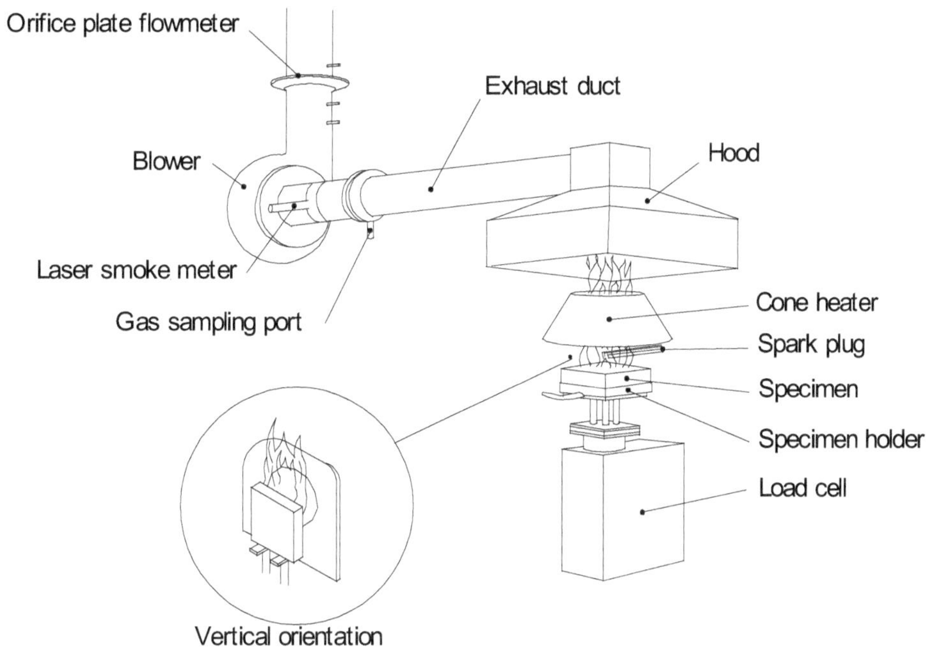

FIGURE 18.9 Cone calorimeter.

is capable of providing heat fluxes to the specimen in the range of 0–110 kW/m². Prior to testing, the heater temperature is set at the appropriate value resulting in the desired heat flux. At the start of a test, the specimen in the appropriate holder and the retainer frame (if used) is placed on the load cell, which is located below the heater. An electric spark is used to ignite the pyrolysis products released by the specimen. As soon as sustained flaming is observed, the electric spark igniter is removed. All combustion products and entrained air are collected by an exhaust hood. At a sufficient distance downstream from a mixing orifice, a gas sample is taken and analyzed for oxygen concentration. Measurements of the gas temperature and differential pressure across the orifice plate are used for calculating the mass flow rate of the exhaust gases. The heat released rate can be determined on the basis of the oxygen depletion and the mass flow rate in the exhaust duct.

Most cone calorimeters include instrumentation for measuring light extinction in the exhaust duct, using a laser light source, described in ASTM E1354 and ISO 5660–1 (see Section 18.3.5.3.2). Instrumentation to measure concentrations of soot, carbon dioxide, carbon monoxide and other gases are commonly added. Some laboratories have used a modified version of the standard apparatus to conduct studies in vitiated or oxygen-enriched atmospheres.[49–51]

18.3.3.2.2 FAA Microscale Combustion Calorimeter (ASTM D7309)

The US Federal Aviation Administration (FAA) developed the Microscale Combustion Calorimeter to assist with the development of fire-resistant polymers for use

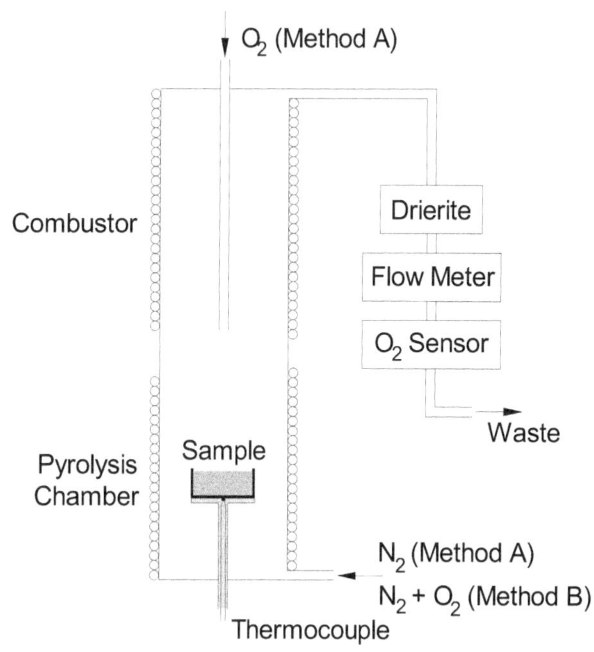

FIGURE 18.10 FAA microflow combustion calorimeter.

in commercial passenger aircraft. A schematic of this micro-scale calorimeter is shown in Figure 18.10. The apparatus is described in ASTM D7309.

A 1–10 mg specimen (typically between 2 and 5 mg) is heated at a constant rate between 0.2 and 2 K/s in the lower chamber. Decomposition can take place in nitrogen (method A) or in a mixture of nitrogen and oxygen (method B). When method A is used, char-forming specimens do not decompose completely and leave a solid residue. In this case the volatiles are mixed with a metered supply of oxygen in the combustor to obtain the heat release rate of the volatiles. When method B is used, the specimen is completely consumed with the exception of any non-combustible components. The total heat released in this case is comparable to that measured in an oxygen bomb calorimeter. The temperature of the combustor is set at approximately 900°C to ensure that all specimen gases are completely oxidized.

Oxygen consumption calorimetry with E = 13.1 kJ/g is used to measure heat release rate as a function of time. The specific heat release rate, Q(t), in W/g is equal to the heat release rate divided by the initial specimen mass, m_0. The following five parameters are calculated when method A is used:

1. The heat release capacity $\eta_c \equiv Q_{max} / \beta$ in J/g·K, where Q_{max} is the maximum value of $Q(t)$ and β is the heating rate in K/s.
2. The heat release temperature T_{max} in K as the pyrolysis chamber temperature at which $Q(t) = Q_{max}$.
3. The specific heat release h_c in J/g as the area under the $Q(t)$ curve.
4. The pyrolysis residue $Y_p \equiv m_p / m_o$ in g/g, where m_p is the residual mass of the specimen at the end of the test.
5. The specific heat of combustion of the specimen gases $h_{c,gas} \equiv h_c / (1 - Y_p)$ in J/g.

For method B only three parameters are calculated:

1. The combustion temperature T_{max} in K as the pyrolysis chamber temperature at which the specific heat release rate is a maximum, that is, $Q(t) = Q^0_{max}$.
2. The combustion residue $Y_c \equiv m_c / m_o$ in g/g, where m_c is the residual mass of the specimen at the end of the test.
3. The net calorific value h^0_c in J/g as the area under the $Q(t)$ curve.

The thermal combustion properties measured in the test are related to flammability characteristics of the material.[52–56] For example, the heat release temperature from method A approximates the surface temperature at ignition (see 18.3.2.1). The net calorific value from method B approximates the net heat of combustion measured in an oxygen bomb calorimeter.

18.3.3.2.3 Other Heat Release Rate Calorimeters

The apparatus designed by Prof. Ed Smith at Ohio State University is one of the most widely used and best known bench-scale calorimeters.[57] The test method based on this apparatus was first published as a proposed ASTM standard in 1980. In 1983 it was adopted as ASTM E906. The standard was later amended to include two configurations of the test apparatus. Configuration A is that which is used by the Federal Aviation Administration (FAA) for assessing aircraft cabin materials at a radiant heat flux of 35 kW/m². The test procedure in this configuration is also described in the FAA Aircraft Material Fire Test Handbook.[58] Configuration B is the original configuration. In both configurations heat release rate is determined on the basis of the temperature rise of the exhaust gases.

Factory Mutual Research Corporation (FMRC, currently FM Global Research) developed the Fire Propagation Apparatus (originally referred to as the Combustibility Apparatus) to measure heat release rate and generation rates of smoke and combustion products.[59] Originally, only convective heat release rate was measured on the basis of temperature rise of the exhaust gases. Test results reported since the late 1970s also include total heat release rates calculated from oxygen consumption or carbon dioxide generation. The standard test method based on the fire propagation apparatus is described in ASTM E2058.

18.3.4 Surface Flame Spread

18.3.4.1 Pertinent Material Properties

As described in Section 18.1.3.3, flames can spread over a surface in two modes, that is, concurrent with and in the opposite direction of the surrounding air flow. The rate of wind-aided flame spread is typically much higher than that of opposed-flow flame spread and is therefore of primary concern in fire hazard assessment. To predict the rate of wind-aided flame spread over the surface of a material, it is necessary to determine the heat flux distribution downstream of the pyrolysis front and the thermal response of the material as its surface temperature increases and eventually reaches the ignition threshold. The former depends primarily on the length of the flame, which, in turn, is a function of the heat release rate of the pyrolyzing fuel and the geometry, for example, the flame height from a fire with a specified heat release rate is shorter for a flat surface than in a corner.[60,61] The latter is determined by the ignition properties. In other words, there are no specific material properties pertaining to wind-aided flame spread other than those discussed in Sections 18.3.2.1 and 18.3.3.1.

The rate of opposed-flow flame spread can be determined from the following equation, derived theoretically by deRis.[62]

$$V_p = \frac{\phi}{k\rho c\left(T_{ig} - T_s\right)^2} \qquad [18\text{–}8]$$

Where
V_p = flame spread rate (m/s);
ϕ = flame heating parameter (kW²/m³); and
T_s = surface temperature ahead of the flame front (K).

It is of interest to know under what conditions opposed flow flame spread ceases. Quintiere and Harkleroad proposed the minimum surface temperature for spread, $T_{s,min}$, as a convenient criterion.[39] If the surface temperature just ahead of the pyrolysis front is lower than $T_{s,min}$, the gas phase heat conduction from the flame is insufficient to raise the fuel temperature to T_{ig}. Both ϕ and $T_{s,min}$ are material properties that can be measured in the LIFT apparatus as described in ASTM standard E1321 (see Section 18.3.4.3.2).

18.3.4.2 Surface Flame Spread Measurements

By far the most common and most practical approach to measure the rate of flame spread over a flat surface involves recording the location of the flame tip (wind-aided spread) or flame front (opposed-flow spread) as a function of time based on visual observations. In the case of wind-aided flame spread, however, it is very difficult to track propagation of the pyrolysis front (boundary between the pyrolyzing and non-pyrolyzing fuel) as it is hidden by the flame. This problem can be solved by attaching fine thermocouples to the surface at specified locations as ignition results in an abrupt rise of the surface temperature. This approach is very tedious and not suitable for routine use. An infrared video camera has been used to look through the flame and monitor the upward advancement of the pyrolysis front in a corner fire.[63]

18.3.4.3 Examples of Flame Spread Tests

Conceptually there are two types of flame spread tests. The length of the specimen is generally much greater than the width and flame propagation is measured in the direction of the longest dimension. In the first type of flame spread tests, the specimen is exposed to a gas burner flame at one end. In the second type, the specimen is exposed to a radiant panel producing a heat flux that varies from one end of the specimen to the other. A pilot flame is used to ignite the specimen at the hot end. An example of each of the two types is described below.

18.3.4.3.1 Steiner Tunnel Test

As mentioned earlier, the fire hazard of interior finish materials is primarily due to the potential for rapid wind-aided flame spread over the surface. It is therefore not a surprise that reaction-to-fire requirements for interior finish materials in US building codes are primarily based on performance in a wind-aided flame spread test. The apparatus of this test is often referred to as the Steiner tunnel. The Steiner tunnel test is described in ASTM E84. Although the test does not measure any material properties that can be used in a model-based hazard assessment, a discussion of the test is included here due to its practical importance for the passive fire protection of buildings in the US.

A schematic of the test apparatus is shown in Figure 18.11. It consists of a long tunnel-like enclosure measuring 8.7 × 0.45 × 0.31 m. The test specimen is 7.6 m long and 0.51 m wide and is mounted in the ceiling position. It is exposed at one end, designated as the burner end, to a 79 kW gas burner. There is a forced draft through the tunnel from the burner end with an average initial air velocity of 1.2 m/s.

FIGURE 18.11 ASTM E 84 Steiner tunnel test apparatus.
Left insert: burner flame viewed from tunnel inlet.
Right insert: initial flame tip location is 1.37 m (4.5 ft) from the bvurner.

Source: Photographs Courtesy Southwest Research Institute®

The measurements consist of flame spread over the surface and smoke obscuration in the exhaust duct of the tunnel. Test duration is ten minutes. A flame-spread index (FSI) is calculated on the basis of the area under the curve of flame tip location versus time as follows:

$$\text{FSI} = \begin{cases} 0.0282A_T & \text{for } A_T \leq 1783 \text{ m}\cdot\text{s} \\ \dfrac{89611}{3566 - A_T} & \text{for } A_T > 1783 \text{ m}\cdot\text{s} \end{cases} \qquad [18\text{–}9]$$

where
FSI = ASTM E84 Flame Spread Index; and
A_T = area under the curve of flame tip location versus time (m·s).
A_T is calculated based on the flame extension, that is, the distance of the flame tip from the burner minus the length of the burner flame. It is important to note that flame recession is ignored in the calculation of A_T.

The classification of linings in the US model building codes is based on the FSI and SDI (Smoke Developed Index). The latter is based on the area under the light transmission vs. time curve normalized to the area for a small heptane pool fire, which by definition has an SDI of 100. There are three classes: Class A for products with FSI ≤ 25, Class B for products with 25 < FSI ≤ 75, and Class C for products with 75 < FSI ≤ 200. In all cases the SDI must be 450 or less. Class A products are generally permitted in enclosed vertical exits. Class B products can be used in exit access corridors, and Class C products are allowed in other rooms and areas.

18.3.4.3.2 Lateral Ignition and Flame Spread Test (LIFT)

The LIFT apparatus can be used for obtaining opposed flow flame spread properties. The apparatus is briefly discussed in 18.3.2.3.2 and is described in detail in ASTM E1321. A schematic is shown in Figure 18.7.

The gas flow to the radiant panel is set to obtain a heat flux at the 50 mm location that is 5 to 10 kW/m^2 above the critical heat flux for ignition (see Section 18.2.3.2.3). The heat flux is verified with a heat flux meter inserted into a calcium silicate dummy specimen. The heat flux profile is shown in Figure 18.6 and can be used to determine the incident radiant heat flux at any location along the specimen centerline. The flux invariant may vary slightly and is usually determined for each apparatus during the initial calibration.

A horizontal line is drawn along the center of the 155 × 800 mm specimen, and vertical marks are made on the line at 25 mm intervals. A marked specimen backed by a ceramic fiber insulation board is mounted in the specimen holder. The holder is inserted when the panel reaches equilibrium and the heat flux at the 50 mm position is at the desired level. After preheating to steady conditions (see Equation 18–5 and subsequent discussion), the specimen is ignited at the hot end with a non-impinging premixed acetylene flame or an impinging propane flame. The time for the flame to travel to each mark is recorded. The flame spread rate at the location of every mark (except for the two extreme locations) is then calculated from a simple running 3-point least squares fit. The distance from the hot end of the specimen where the flame spread stops is also recorded.

The main objective of conducting LIFT flame spread experiments is to obtain material properties for predicting opposed-flow flame spread. Based on the analysis by deRis (see Equation 18–8), the flame heating parameter ϕ can be determined from the slope of a linear fit of $1/\sqrt{V_p(y)}$ plotted as a function of $\dot{q}_e''(y)$, where y is the distance from the hot end of the specimen. The minimum surface temperature for spread, $T_{s,min}$, is calculated according to an expression akin to Equation 18–4 from the incident heat flux at the maximum distance from the hot end reached by the flame.

18.3.4.3.3 Other Flame Spread Tests

The LIFT apparatus is used in conjunction with an instrumented exhaust stack to qualify finish materials for use on ships that sail on international voyages and need to comply with the SOLAS (Safety of Life at Sea) regulations developed by the International Maritime Organization (IMO). The method is described in ASTM E1317 and in Part 5 of Annex 1 to the IMO Fire Test Procedures (FTP) code.

The National Bureau of Standards (NBS, currently the National Institute of Standards and Technology, or NIST) conducted a series of full-scale fire tests in the 1970s to investigate the fire hazard of floor coverings.[64] The main concern was flame spread from a fire room to a connected corridor. This work resulted in the development of the radiant flooring panel test. This test is described in ASTM E648 and EN ISO 9239–1.

The test method described in ASTM E162 also evaluates opposed-flow flame spread characteristics of a product and is referred to in regulations that pertain to various modes of transportation.

18.3.5 Smoke and Toxicity

18.3.5.1 Pertinent Material Properties

The generation rate of smoke (soot) can be written as

$$\dot{m}_s = \dot{m}Y_s \qquad [18\text{–}10]$$

where
$\dot{m}_s$ = mass generation rate of soot (g/s);
$\dot{m}$ = mass loss rate of the fuel (g/s); and
Y_s = soot yield (g/g).

Similarly, the generation rate of toxic gas species can be expressed as

$$\dot{m}_i = \dot{m}Y_i \qquad [18\text{–}11]$$

where
$\dot{m}_i$ = mass generation rate of species i (g/s); and
Y_i = yield of species i (g/g).

The yields of soot and toxic gas species are material properties that may vary as a function of vitiation and/or ventilation. The latter is characterized by the equivalence ratio, that is, the actual mass ratio of fuel to air divided by the stoichiometric ratio. This implies, for example, that under-ventilated fires are characterized by an equivalence ratio greater than one.

18.3.5.2 Smoke and Toxicity Measurements

18.3.5.2.1 Smoke

There are essentially two different experimental configurations that are used for quantifying the attenuation of light in fire effluents. The test methods that are based on these two configurations are referred to as static and dynamic, respectively. Different units have been used to report the results of these two types of tests.

The intensity of light passing through smoke is attenuated due to scattering and absorption by the suspended particulates. Light attenuation by smoke is described by the following equation, which is known as the Bouguer or Lambert-Beer law

$$I = I_0 \exp(-k\ell) \qquad [18\text{–}12]$$

where
I = light intensity measured at the detector (mW);
I_0 = light intensity at the source (mW);
k = extinction coefficient (1/m); and
ℓ = path length (m).

The extinction coefficient is the most fundamental smoke measurement in dynamic test methods. The products of combustion generated in these test methods are collected in a hood and extracted through an exhaust duct. The smoke meter consists of a light source on one side of the duct and a detector on the opposite side of the duct. Assuming that the light source is stable, I_0 can be measured at the start of a test. Alternatively, the intensity of the source can be measured during the test if the smoke meter comprises a beam splitter at the light source to divert a fraction of the beam to a reference detector. This arrangement is typically used in laser light source systems. The path length is equal to the diameter of the duct. The extinction coefficient can be calculated from the measured light intensity during a test and is a direct measure of the concentration of particulates in the exhaust duct. This implies that k is a function of the generation rate of particulates in the fire and the volumetric flow rate in the duct. The smoke production rate is independent of the duct flow rate and is defined as follows:

$$\dot{s} \equiv k\dot{V}_s \qquad [18\text{–}13]$$

where
$\dot{s}$ = smoke production rate (m^2/s); and
$\dot{V}_s$ = actual volumetric flow rate at the smoke meter (m^3/s).

The fact that $\dot{s}$ is independent of duct flow rate can be illustrated as follows. Suppose the fire is generating particulates at a constant rate. At some point during a test the exhaust fan speed is adjusted so that the duct flow rate is doubled. As a result, additional ambient air is drawn into the hood, and the concentration of particulates in the exhaust is reduced by 50%. The extinction coefficient is therefore also cut in half, and the smoke production rate is unchanged.

Sometimes it is helpful to express the smoke production rate per unit of exposed specimen surface area:

$$\dot{s}'' = \frac{k\dot{V}_s}{A_s} = \frac{\dot{s}}{A_s} \qquad [18\text{–}14]$$

where
$\dot{s}''$ = smoke production rate per unit exposed specimen surface area (1/s); and
A_s = exposed area of the specimen (m^2).

The smoke production rate can also be expressed per unit of specimen mass loss by dividing the RHS in Equation 18–14 by the mass loss rate. The result is referred to as the specific extinction area because it has the units of area divided by mass.

$$\sigma_f = \frac{\dot{s}}{\dot{m}} = \frac{k\dot{V}_s}{\dot{m}} \qquad [18\text{–}15]$$

where
σ_f = specific extinction area (m^2/g); and
$\dot{m}$ = mass loss rate of the specimen (g/s).

The ratio of the light extinction coefficient of flame-generated smoke to the mass concentration of soot in the smoke is approximately constant and equal to $\sigma_s = 8.7 \pm 1.1$ m^2/g.[65] The implication of this nearly universal value is that the soot yield, that is, the mass of soot generated per mass unit of fuel burnt, can be estimated from light extinction measurements as follows:

$$Y_s = \frac{\dot{m}_s}{\dot{m}} = \frac{\left(\dfrac{k}{\sigma_s}\right)\dot{V}_s}{\dot{m}} = \frac{\dot{s}}{\sigma_s \dot{m}} = \frac{\sigma_f}{\sigma_s} \qquad [18\text{–}16]$$

where
σ_s = ratio of extinction coefficient to soot concentration ($\sigma_s = 8.7 \pm 1.1$ m^2/g).

The extinction coefficient in static methods is typically expressed per unit path length using base 10 logarithms:

$$D_u = \frac{1}{\ell} D = \frac{1}{\ell}\log\left(\frac{I_0}{I}\right) = \frac{1}{\ell}\log\left(\frac{100}{T}\right) \qquad [18\text{–}17]$$

where
D_u = optical density per unit distance or path length (1/m);
D = optical density; and
T = light transmittance (%).

The relationship between k and D is as follows:

$$k \approx 2.303\ D_u = \frac{2.303 D}{\ell} \qquad [18\text{–}18]$$

The specific optical density is a dimensionless quantity that is defined as follows:

$$D_s \equiv \frac{DV}{A_s \ell} \qquad [18\text{–}19]$$

where
D_s = specific optical density; and
V = volume of the chamber in which the smoke is collected (m^3).

Slight variations of the units described in this section have been used. Conversion factors can easily be derived on the basis of the definition of these alternative units.

18.3.5.2.2 Toxicity

Toxic potency of fire effluents can be quantified by evaluating the response of live animals to smoke generated from fires in a controlled atmosphere. Mice, rats, and primates have been used for this purpose. The results are expressed as an LC_{50} value. The LC_{50} is the concentration (by mass) estimated to produce lethality in 50% of the animals within the specified time (typically 30 minutes). Under pressure from animal rights groups and the general public, bioassay methods have fallen out of favor. Over 20 years of research using bioassay methods supplemented by gas analysis have resulted in an extensive database defining the concentration-response relationship for various materials and toxic gases. This has greatly eliminated the need for animal testing and made it possible to rely primarily on analytical techniques to obtain toxic potency data for fire hazard analysis.

Researchers at the National Institute of Standards and Technology (NIST) developed a concept to minimize the usage of animals for the assessment of the toxic potency of a material in fires. This concept is based on the well-established hypothesis that a small number (N) of gases in the smoke accounts for a large percentage of the observed toxic potency. Research at NIST of toxicologically important gases and their interactions resulted in the development of the N-Gas Model,[66] which is expressed by the following equation:

$$FED = \frac{a[CO]}{[CO_2]-b} + \frac{[HCN]}{LC_{50}(HCN)} + \frac{21-[O_2]}{21-LC_{50}(O_2)} + \frac{[HCl]}{LC_{50}(HCl)} + \frac{[HBr]}{LC_{50}(HBr)} \qquad [18\text{–}20]$$

where
FED = fractional effective exposure dose;
m = constant (m = −0.0018 if $[CO_2] \leq 5\%$ and m = 0.0023 if $[CO_2] > 5\%$);
[CO] = concentration of CO (ppm);
b = constant (b = 12.2% if $[CO_2] \leq 5\%$ and b = −3.86% if $[CO_2] > 5\%$);
$[CO_2]$ = concentration of CO_2 (%);
[HCN] = concentration of HCN (ppm);
$[O_2]$ = concentration of O_2 (%);
[HCl] = concentration of HCl (ppm);
[HBr] = concentration of HBr (ppm); and
$LC_{50}(X)$ = lethal concentration of species X resulting in 50% mortality (% or ppm).

The constants in Equation 18–20 are based on laboratory tests with deaths within a 30-minute exposure period and a 14-day post-exposure period. The LC_{50} values are 150 ppm for HCN, 5.4% for O_2, 3,800 ppm for HCl and 3,000 ppm for HBr. The first term on the RHS of Equation 18–20 indicates that CO and CO_2 interact in a non-linear way. Theoretically, 50% of the animals should die at FED = 1. Instead, due to slight non-linearities, the 50% lethality level corresponds to FED = 1.1.[67]

Equation 18–20 with FED = 1.1 can be used in a bench-scale toxicity test to determine the LC_{50} of a material. The test involves burning a specified amount of material and collecting the effluents in a closed chamber. Tests are conducted with varying amounts of material to find the mass loss Δm_{50} that is expected to result in 50% lethality of test animals exposed over a 30-minute test duration to the atmosphere in the chamber. The LC_{50} can then be calculated as follows:

$$LC_{50} = \frac{\Delta m}{V} \qquad [18\text{–}21]$$

where
LC_{50} = lethal concentration (g/m^3);
Δm_{50} = mass loss resulting in FED = 1.1 in the chamber after 30 minutes (g); and
V = volume of the camber (m^3).

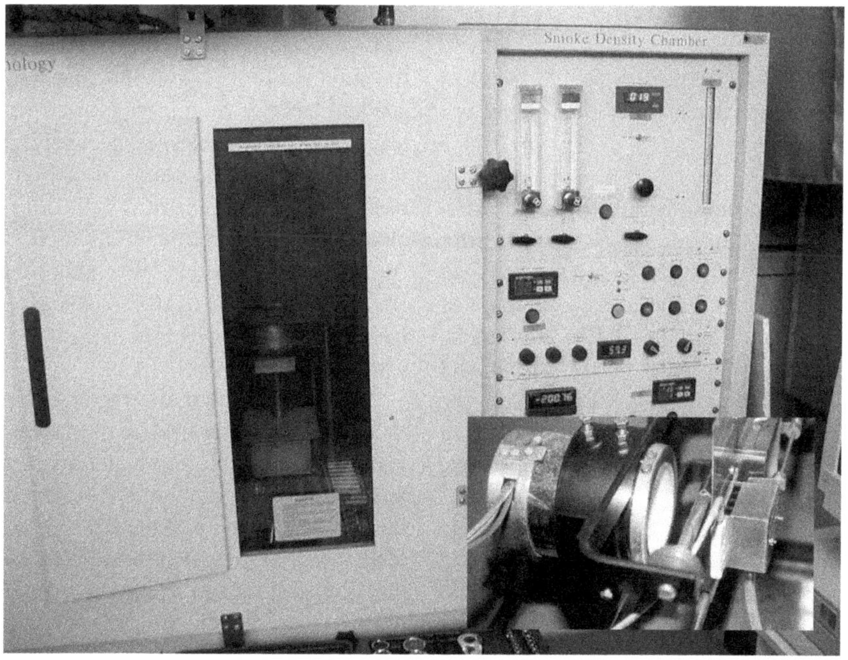

FIGURE 18.12 NBS smoke density chamber with heater described in IMO FTP code.
Insert: close-up view of ASTM E 622 heater.

Source: Photographs Courtesy Southwest Research Institute®

The total mass loss of a material in a pre-flashover fire that will result in a lethal mixture can theoretically be determined as the product of the LC_{50} and the volume in which the products of combustion are distributed. For post-flashover fires it is necessary to correct the LC_{50} measured in the bench-scale test to account for the increased CO production in these fires.[68] NIST conducted a series of full-scale validation tests and determined that the smoke from the bench-scale method reproduces that from room fires to within a factor of 3.[67]

$$(LC_{50})_{corrected} = \frac{1}{\frac{1}{LC_{50}} + 0.044 - \frac{0.00005[CO]}{\Delta m}} \quad [18\text{–}22]$$

The correction is important because the majority of US fire deaths occur remote from the fire room, overall and especially for fires that have proceeded past flashover.[69]

Various methods and instruments are available for measuring gas concentrations in fire effluents, for example, gas detector tubes, ion-selective electrodes, GC/MS (gas chromatography/mass spectrometry), and Fourier Transform Infrared (FTIR) spectroscopy. An extensive review of the methods is provided in ASTM E800 and ISO 19701.

18.3.5.3 Examples of Smoke and Toxicity Tests

18.3.5.3.1 NBS Smoke Chamber

The NBS smoke chamber is the most commonly used bench-scale test apparatus for measuring the optical density of smoke. The apparatus and test procedure are described in ASTM E662. The method was developed at the National Bureau of Standards (NBS, currently the National Institute of Standards and Technology or NIST) in the 1960s.[70]

The test apparatus consists of a 0.914 × 0.610 × 0.914-m enclosure (see Figure 18.12). A radiant heater with a diameter of 76 mm is used to provide a constant radiant heat flux of 25 kW/m^2 to the specimen surface. The specimen measures 76 × 76 mm and is oriented vertically.

Tests are conducted in two modes. A six-tube premixed pilot burner is used in the flaming mode. The burner flames impinge on the lower half of the specimen surface. The burner is removed to conduct tests in the non-flaming mode. Three replicate tests are typically conducted in each mode. An additional set of three specimens are tested if there is a high variation in the results from the first set.

Smoke density is measured based on the attenuation of a light beam by the smoke accumulating in the closed chamber. A white light source is located at the bottom of the enclosure, and a photomultiplier tube is mounted at the top. A modified version of Equation 18–19 is used to calculate the specific optical density from the measured transmittance:

$$D_s = \frac{V}{A_s L}\left[\log\left(\frac{100}{T}\right) + F\right] = 132\left[\log\left(\frac{100}{T}\right) + F\right] \quad [18\text{–}23]$$

where
V = volume of the smoke chamber (0.51 m^3);
L = interior height of the chamber (0.91 m); and
F = optical density of the range extension filter if moved out of the light path.

A test is terminated three minutes after the light transmittance value reaches a minimum value or 20 minutes after the start of the test, whichever occurs first. D_s calculated according to Equation 18–20 is reported as a function of time. In addition, the maximum value of the specific smoke density, D_m, is reported separately after correction for soot deposits on the windows of the optical system during the test. The correction is equal to the apparent optical density that is recorded after the test is terminated and the smoke has been cleared from the chamber.

The test method described in ASTM E662 appears to have some significant limitations, although it is hard to find objective evidence for that. To make the user aware of these limitations, the ASTM committee that developed the standard added the following caveat to the scope section of the document: "This test method is intended for use in research and development and not as a basis for ratings for regulatory purposes." In spite of this caveat, the method is referenced in several fire safety regulations and specifications for materials used in passenger aircraft, navy ships, and rail transportation vehicles. The acceptance criteria in these regulations and specifications are typically based on limits for D_s at specified times (typically 1 ½ or 4 minutes) and/or D_m.

The smoke chamber method described in ASTM E662 is often supplemented with toxic gas analysis. A PTFE-lined stainless steel tube is used to take a gas sample from the geometric center of the chamber at a specified time. This time can be fixed, for example 1 ½ or 4 minutes into the test, or variable, for example immediately following the maximum specific optical density. Regulations and specifications that call for these measurements require that the concentration of a predefined set of gases be determined. The product is acceptable if the concentration of every gas is within specified limits. These limits have been established from experience based on data for products that are deemed to be acceptable or not acceptable.

TABLE 18.1
Toxic Gas Concentration Limits for Three Static Methods

	Airbus	Bombardier	IMO
CO_2	NA	90000	NA
CO	1000	3500	1450
HF	100	100	600
HCl	150	500	600
HBr	NA	100	600
NO_x	100	100	350
HCN	150	100	140
SO_2	100	100	120

For example, Airbus Industrie ABD 0031 and Bombardier SMP 800-C are industry specifications that place limits on the amount of toxic gases that may be generated in the NBS smoke chamber by materials used in aircraft and ground transportation vehicles, respectively. Part 2 of Annex 1 to the IMO FTP Code places limits on the amount of toxic gases that may be generated in the ISO 5659–2 single smoke chamber method by materials used in ships. ISO 5659–2 specifies the same chamber and smoke meter as ASTM E662 but uses a scaled-down version of the cone calorimeter heater. The gas concentration limits for these three methods are given in Table 18.1.

Static smoke chamber methods have major limitations in terms of being indicative of the fire hazard due to smoke toxicity of products and materials in actual fires. As combustion products accumulate in the chamber during a test, the burning behavior of the test specimen may have a significant effect on the level of vitiation (oxygen concentration) and temperature rise in the chamber. Consequently, rather than simulating a specific fire scenario, conditions in smoke chamber tests vary over time and are not well-defined. A relatively new tube furnace method greatly alleviates this problem and, for that reason, is gaining increased international acceptance. The method is described in ISO TS 19700.

18.3.5.3.2 Cone Calorimeter

The standard cone calorimeter (see 18.3.3.3.1) section described in ASTM E1354 includes a smoke photometer to measure light extinction in the exhaust duct. The system is based on a laser light source. The same system is also standardized internationally and described in the main cone calorimeter standard (ISO 5660–1). Smoke measurements are reported in terms of the average specific extinction area (ASTM E1354 and ISO 5660–1) and the smoke production rate and total smoke production for the period prior to ignition and the flaming period (ISO 5660–2).

The cone calorimeter and other bench- and large-scale calorimeters have also been used to obtain dynamic smoke toxicity data. A continuous gas sample is taken through the oxygen consumption probe or through a separate gas sampling probe and analyzed for the gases of interest. Since the gas concentrations are a function of the flow rate in the exhaust duct, they need to be converted to generation rates or yields. For example, if the gas concentrations are measured on a wet basis (as in an FTIR spectrometer), the generation rate and yield of a particular gas species can be calculated from

$$\dot{m}_i = \frac{X_i \dot{V}_s M_i}{8.2 \cdot 10^{-5} T_s} \qquad [18\text{–}24]$$

and

$$Y_i = \frac{\dot{m}_i}{\dot{m}} \qquad [18\text{–}25]$$

where
X_i = mole fraction of species i in the exhaust duct on a wet basis;
$\dot{V}_e$ = actual volumetric flow rate at the sampling point (m^3/s);
M_i = molecular mass of species i (g/mol); and
T_e = absolute gas temperature at the sampling point (K).

If the gases are measured on a dry basis, the equations need to be adjusted to account for moisture. The derivation of the equations is then similar to those for calculating heat release rate based on oxygen consumption.

The cone calorimeter is also used to quantify the corrosivity of products of combustion as described in ASTM D5485. The "Cone Corrosimeter" uses the same load cell, specimen holder, retainer frame, spark igniter, conical heater, and exhaust system as the cone calorimeter. A heated stainless steel sampling tube is connected to a funnel placed on top of the conical heater. A gas sample is continuously drawn from the tube at a rate of 4.5 L/min. The sampling tube is connected with silicone rubber tubing to the pump via an 11.2 L exposure chamber, a filter and a flow meter. A target is placed in the exposure chamber at the start of the test and exposed to the corrosive atmosphere of the gas sample for 60 minutes or until the specimen has lost 70% of its total mass loss, whichever occurs first.

The target consists of a printed circuit board with two circuits. One circuit is exposed. The second circuit is protected and is used to compensate for the effect of elevated temperature on the electrical resistance of the circuit. The target is removed from the chamber one hour after the start of a test and the reduction of the electrical resistance of the exposed circuit is measured. The target is then exposed to an atmosphere of 23°C and relative humidity of 75% for 24 hours. The electrical resistance is measured again after the 24 hour post-test exposure. The electrical resistance measurements are converted to the equivalent reductions in metal thickness expressed in nm.

18.3.5.3.3 Other Smoke Toxicity Tests

Two smoke toxicity tests that involve animal exposure were developed but are no longer required for regulatory in the US. The University of Pittsburgh, or UPitt, method at one time was required to demonstrate compliance with the requirement in the New York City building code that no construction product shall be more toxic than wood. A small specimen of the product is heated in a muffle furnace, and four mice are exposed to the products of combustion diluted with air. The furnace temperature is ramped at a rate of 5°C/minute. The test is terminated after 30 minutes. The objective is to find the LC_{50}, defined in this case as the quantity of the product in grams that results in 50 percent mortality of the test animals. A product is acceptable if the LC_{50} is equal to or greater than 19.5 g, which is the value assigned to wood.

The test procedure described in ASTM E1678 minimizes the number of animal tests. In this test procedure, a specimen is exposed to a radiant heat flux of 50 kW/m^2, and the products of combustion are collected in a 0.2 m^3 chamber. Test duration is 30 minutes. Additional tests are performed with specimens of different size to find the exposed area that is expected to result in 50 percent lethality of test animals exposed over the 30-minute test duration to the atmosphere in the chamber. The lethality is determined on the basis of analytical measurements of the composition of the contents of the chamber and the N-Gas model (see Equation 18–20). To verify the results, two additional tests are conducted with 70% and 140% of that specimen area and six rats exposed to the gases in the chamber. For some time, after the UPitt method was abandoned, the New York City building code required smoke toxicity testing for specific construction products according to NFPA 269, which describes the same apparatus and test procedure as in ASTM E1678. This requirement is no longer in effect.

18.4 LARGE-SCALE REACTION-TO-FIRE TESTS

Often it is very difficult to determine the burning behavior of complex objects on the basis of the performance of its individual components in bench-scale reaction-to-fire tests. It is much more practical to measure the heat release rate and related properties for the complete object. This requires a large-scale test. In other cases it is not possible to capture certain aspects of real fire behavior such as melting, delamination, joint effects, etc. in a bench-scale test. A large-scale test is needed to assess these effects. Two commonly used large-scale reaction-to-fire tests methods are discussed next.

18.4.1 Furniture Calorimeter

It is very hard to determine the burning behavior of upholstered furniture on the basis of the fire characteristics of the foam, fabric, and framing materials and to account for the geometry and configuration of the furniture and how it is ignited. It is much easier to test the entire furniture item. The calorimeter described in the section was developed for this purpose.

A furniture calorimeter consists of a weighing platform that is located on the floor of the laboratory, beneath a hood connected to an instrumented exhaust duct (see Figure 18.13). The object is placed on the platform and ignited with the specified ignition source. The products of combustion are collected in the hood and extracted through the exhaust duct. Measurements of oxygen concentration, flow rate, and light transmission in the exhaust duct are used to determine the heat release rate and smoke production rate from the object as a function of time.

Furniture calorimeters were developed in the 1980s in several laboratories to obtain this kind of data.[71,72] The first furniture calorimeter test standard was published in 1987 in the Nordic countries as NT Fire 032. Furniture calorimeter test standards have been developed by ASTM for

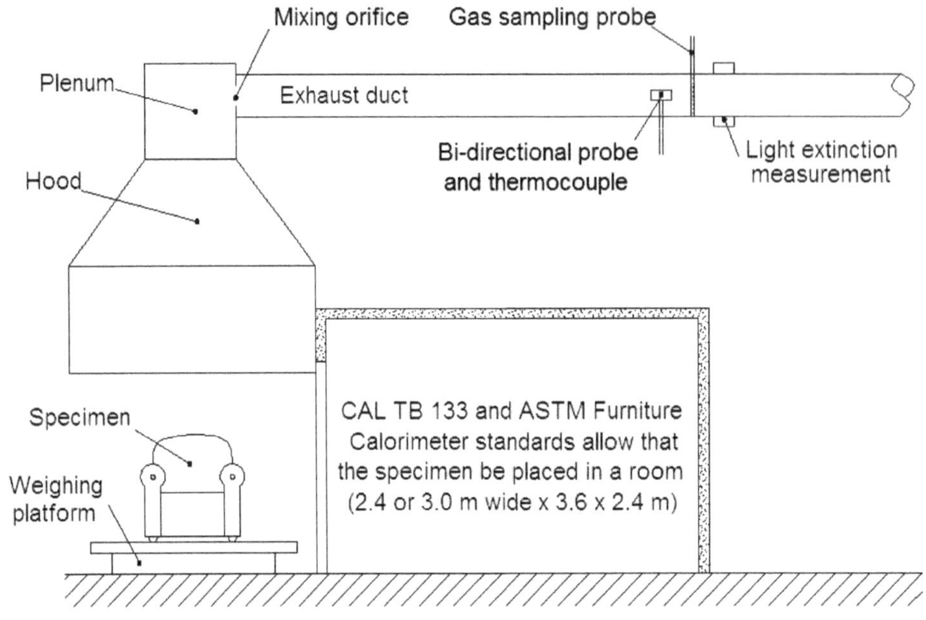

FIGURE 18.13 Furniture calorimeter.

chairs, mattresses, and stacked chairs. The corresponding designations are ASTM E1537, ASTM E1590, and ASTM E1822, respectively. The California Bureau of Home Furnishings and Thermal Insulation (CBHFTI) developed California Technical Bulletins (CAL TB) 133 and 603. These documents describe fire test procedures to qualify seating furniture and mattresses, respectively, for use in public occupancies in California. CAL TB 133 has been withdrawn, and CAL TB 603 has been superseded by the Federal CPSC standard 16 CFR 1633. The primary difference between the various chair and mattress tests is the ignition source.

18.4.2 Room/Corner Test

Several standard room/corner test protocols are now available and are specified in codes and regulations for qualifying interior finishes. For example, US model building codes require that textile wall coverings for use in unsprinklered compartments meet specific performance requirements when tested according to NFPA 265. The principal requirement of these tests is that flashover does not occur. The same codes also require that all other interior wall and ceiling finish materials comply with requirements based on NFPA 286, including a limit on the total smoke released.

SOLAS permits the use of combustible bulkhead and ceiling linings on high speed craft, provided they meet stringent fire performance requirements based on assessment according to ISO 9705. ASTM E2257 is the American version of ISO 9705.

The test apparatus and instrumentation described in the NFPA and ISO room/corner test standards are very similar (see Figure 18.14). However, there are some significant differences in terms of specimen configuration and ignition source. The apparatus consists of a room measuring 3.6 m deep by 2.4 m wide by 2.4 m high, with a single ventilation opening (open doorway) measuring approximately 0.8 m wide by 2 m high in the front wall. Walls and ceiling are lined for tests according to ISO 9705. For tests according to the NFPA standards, the interior surfaces of all walls (except the front wall) are covered with the test material. NFPA 286 is also suitable for evaluating ceiling finishes (see below).

The test material is exposed to a propane burner ignition source, located on the floor in one of the rear corners of the room opposite the doorway. The burner is placed directly against (ISO 9705 and NFPA 286) or at a distance of 50 mm (NFPA 265) from the walls. The ISO burner consists of a steel sandbox measuring 0.17 × 0.17 × 0.135 m. Propane is supplied to the burner at a specified rate such that a net heat release rate of 100 kW is achieved for the first ten minutes of the test, followed by 300 kW for the remaining ten minutes (20-minute test duration, unless terminated when flashover occurs). The NFPA burner consists of a steel sandbox measuring 0.305 × 0.305 m × 0.152 m, with the top surface positioned 0.305 m above the floor of the room. Propane is supplied at a specified rate so that a net heat release rate of 40 kW is achieved for the first five minutes of the test, followed by 150 kW (NFPA 265) or 160 kW (NFPA 286) for the remaining ten minutes (15 minute test duration unless terminated when flashover occurs).

A fundamental difference between NFPA 265 and NFPA 286 is the fact that the flame from the burner alone just touches the ceiling in NFPA 286. This makes it suitable for assessing the fire performance of interior ceiling finish, an application for which NFPA 265 is unsuitable.

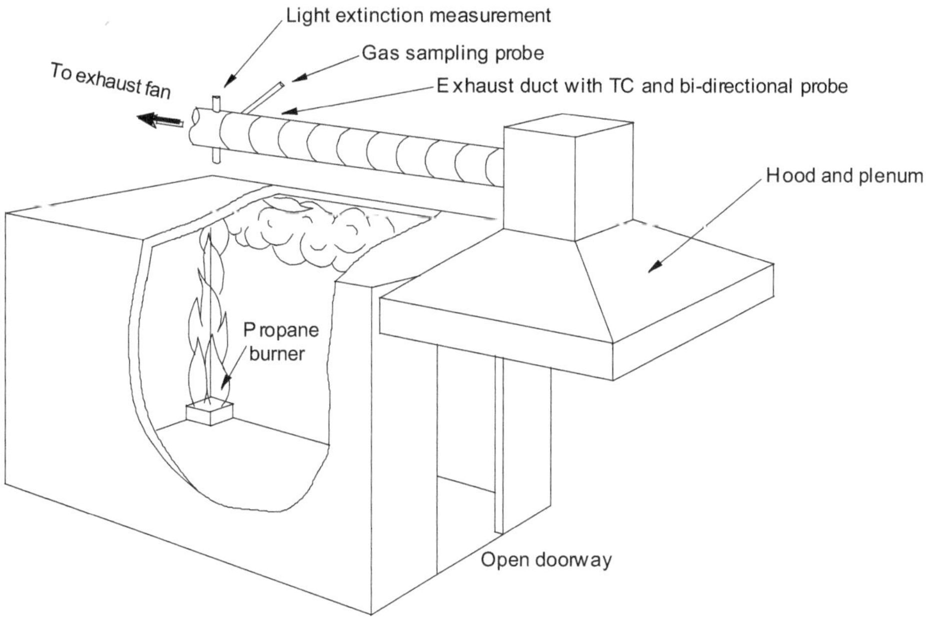

FIGURE 18.14 Room/Corner test.

This effect is partly due to the higher energy release rate of the NFPA 286 burner, but primarily because of the burner being in direct contact with the walls, thereby reducing the area over which the flames can entrain air and increasing the overall flame height.

All combustion products emerging from the room through the open doorway are collected in the standard hood. Instrumentation is provided in the exhaust duct for measuring heat release rate based on the oxygen consumed (ISO and NFPA standards) and smoke production rate (ISO 9705 and NFPA 286 only). The room contains a single heat flux meter located in the center of the floor. The NFPA standards also specify that seven thermocouples be installed in the upper part of the room and doorway to measure the temperature of hot gases that accumulate beneath the ceiling and exit through the doorway. In addition to quantitative heat release and smoke production rate measurements, time to flashover (if it occurs) is one of the main results of a room/corner test. Different criteria are commonly used to define flashover, e.g., upper layer temperature of 600°C, flames emerging through the doorway, heat flux to the floor of 20 kW/m², heat release rate of 1 MW, etc.

18.4.3 Other Large-Scale Reaction-to-Fire Tests

The intermediate-scale calorimeter (ICAL), described in ASTM E1623, is essentially a scaled-up version of the cone calorimeter. The apparatus consists of a vertical gas-fired radiant panel with a height of approximately 1.33 m and width of approximately 1.54 m. The standard test specimen measures 1 × 1 m and is positioned parallel to the radiant panel. The heat flux to the specimen is preset in the range of 10 to 60 kW/m² by adjusting the distance to the panel. The products of pyrolysis from the specimen are ignited with hot wires located near the top and bottom of the specimen. The specimen is placed on a load cell to measure mass loss during testing. Panel and specimen are positioned beneath the ISO 9705 hood and exhaust duct.

The European reaction-to-fire classification system described in EN 13501 is based primarily on performance in the single burning item test (SBI) for construction products except floor coverings. The apparatus and test method are described in EN 13823. The classification is based primarily on fire growth (FIGRA) and smoke development (SMOGRA) indices that are equal to the peak heat release and smoke production rate, respectively, divided by the time to reach the peak. FIGRA and SMOGRA limits were established based on performance in the ISO 9705 room/corner test as the reference scenario.[73]

18.5 CONCLUSIONS

There are literally hundreds of test methods to evaluate material flammability, and this chapter can only scratch the surface. The intent was to identify and describe the different aspects of material flammability and to illustrate how they can be measured. Several commonly used flammability tests not covered in this chapter are described elsewhere in the book. The reader is referred to the end of this chapter for a short list of more comprehensive surveys of flammability tests.

Each test method is unique and provides information concerning the specimen's response to the specific fire conditions simulated by the test. A material that does very well in a test might perform very poorly in a real fire if, for example, the thermal exposure conditions in the fire are more severe than those in the test. For the same reason, good performance in one test does not guarantee good performance in another. This helps to explain inconsistencies between fire codes and regulations that rely on different standard tests. Each test method has merits but also limitations. It is important to recognize the limitations and not to draw any unwarranted conclusions from the test results.

There is an increased use of flammability tests which measure fundamental properties as opposed to tests that simulate a specific fire scenario. The former can be used in conjunction with mathematical models to predict the performance of a material in a range of fire scenarios. This approach has become feasible due to the significant progress that has been made in the past few decades in our understanding of the physics and chemistry of fire, mathematical modeling of fire phenomena and measurement techniques. However, there will always be materials that exhibit a behavior that cannot be captured in bench-scale tests and computer models. The fire performance of those materials can only be determined in full-scale tests.

BIBLIOGRAPHY

Codes and Standards

'ABD 0031: Airbus Directives (ABD) and Procedures for Fire—Smoke—Toxicity (FST).' Airbus Industries.

'ASTM D1692: Method of Test for Rate of Burning or Extent and Time of Burning of Cellular Plastics Using a Specimen Horizontal (Withdrawn 1976).' ASTM International, West Conshohocken, PA.

'ASTM D1929: Test Method for Determining Ignition Temperature of Plastics.' ASTM International, West Conshohocken, PA.

'ASTM D2863: Test Method for Measuring the Minimum Oxygen Concentration to Support Candle-Like Combustion of Plastics (Oxygen Index).' ASTM International, West Conshohocken, PA.

'ASTM D3801: Test Method for Measuring the Comparative Burning Characteristics of Solid Plastics in a Vertical Position.' ASTM International, West Conshohocken, PA.

'ASTM D5485: Standard Test Method for Determining the Corrosive Effect of Combustion Products Using the Cone Corrosimeter.' *Annual Book of Standards, Vol. 04.07*, ASTM International, West Conshohocken, PA.

'ASTM D7295: Standard Practice for Sampling and Determination of Hydrogen Cyanide (HCN) in Combustion Effluents and Other Stationary Sources.' *Annual Book of Standards, Vol. 11.07*, ASTM International, West Conshohocken, PA.

'ASTM D7309: Standard Test Method for Determining Flammability Characteristics of Plastics and Other Solid Materials Using Microscale Combustion.' Annual Book of Standards, Vol. 08.03, ASTM International, West Conshohocken, PA.

'ASTM E84: Standard Test Method for Surface Burning Characteristics of Building Materials.' *Annual Book of Standards, Vol. 04.07*, ASTM International, West Conshohocken, PA.

'ASTM E162: Standard Test Method for Surface Flammability of Materials Using a Radiant Heat Energy Source.' Annual Book of Standards, Vol. 04.07, ASTM International, West Conshohocken, PA.

'ASTM E648: Standard Test Method for Critical Radiant Flux of Floor Covering Systems Using a Radiant Heat Energy Source.' Annual Book of Standards, Vol. 04.07, ASTM International, West Conshohocken, PA.

'ASTM E662: Standard Test Method for Specific Optical Density of Smoke Generated by Solid Materials.' ASTM International, West Conshohocken, PA.

'ASTM E800: Standard Guide for Measurement of Gases Present or Generated During Fires.' *Annual Book of Standards, Vol. 04.07*, ASTM International, West Conshohocken, PA.

'ASTM E906: Standard Test Method for Heat and Visible Smoke Release Rates for Materials and Products.' Annual Book of Standards, Vol. 04.07, ASTM International, West Conshohocken, PA.

'ASTM E1317: Standard Test Method for Flammability of Marine Surface Finishes.' Annual Book of Standards, Vol. 04.07, ASTM International, West Conshohocken, PA.

'ASTM E1321: Test Method for Determining Material Ignition and Flame Spread Properties.' ASTM International, West Conshohocken, PA.

'ASTM E1354: Standard Test Method for Heat and Visible Smoke Release Rates for Materials and Products Using an Oxygen Consumption Calorimeter.' *Annual Book of Standards, Vol. 04.07*, ASTM International, West Conshohocken, PA.

'ASTM E1537: Standard Test Method for Fire Testing of Upholstered Furniture.' *Annual Book of Standards, Vol. 04.07*, ASTM International, West Conshohocken, PA.

'ASTM E1623: Test Method for Determination of Fire and Thermal Parameters of Materials, Products, and Systems Using an Intermediate Scale Calorimeter (ICAL).' ASTM International, West Conshohocken, PA.

'ASTM E1590: Standard Test Method for Fire Testing of Mattresses.' *Annual Book of Standards, Vol. 04.07*, ASTM International, West Conshohocken, PA.

'ASTM E1678: Standard Test Method for Measuring Smoke Toxicity for Use in Fire Hazard Analysis.' ASTM International, West Conshohocken, PA.

'ASTM E1822: Standard Test Method for Fire Testing of Stacked Chairs.' *Annual Book of Standards, Vol. 04.07*, ASTM International, West Conshohocken, PA.

'ASTM E1995: Standard Test Method for Measurement of Smoke Obscuration Using a Conical Radiant Source in a Single Closed Chamber, With the Test Specimen Oriented Horizontally.' ASTM International, West Conshohocken, PA.

'ASTM E2058: Standard Test Methods for Measurement of Synthetic Polymer Material Flammability Using a Fire Propagation Apparatus.' *Annual Book of Standards, Vol. 04.07*, ASTM International, West Conshohocken, PA.

'ASTM E2257: Standard Test Method for Room Fire Test of Wall and Ceiling Materials and Assemblies.' *Annual Book of Standards, Vol. 04.07*, ASTM International, West Conshohocken, PA.

'CAL TB 117: Requirements, Test Procedures and Apparatus for Testing the Flame Retardance of Resilient Filling Materials Used in Upholstered Furniture,' Bureau of Home Furnishings and Thermal Insulation, Sacramento, CA.

'EN 13501: Fire classification of construction products and building elements—Part 1—Classification using test data from reaction-to-fire tests.' European Committee for Standardization, Brussels.

'EN 13823: Reaction to fire tests for building products—Building Products Excluding Flooring Exposed to the Thermal Attack of a single burning item." European Committee for Standardization, Brussels.

'FMVSS 302: Motor Vehicle Interior Materials Fire Test.' National Highway Traffic Safety Administration, Washington, DC.

'IEC 60695–2–13: Fire hazard testing—Part 2–13: Glowing/hot-wire based test methods—Glow-wire ignitability test method for materials.' International Electrotechnical Commission, Geneva.

'IMO FTP Code: International Code for Application of Fire Test Procedures, Annex 1. Part 2.' International Maritime Organization, London.

'IMO FTP Code: International Code for Application of Fire Test Procedures, Annex 1. Part 5.' International Maritime Organization, London.

'ISO 4589–2: Plastics—Determination of burning behavior by Oxygen Index—Part 2: Ambient-temperature test.' International Organization for Standardization, Geneva.

'ISO 5657: Fire tests—Reaction to fire—Ignitability of building products.' International Organization for Standardization, Geneva.

'ISO 5659–2: Plastics—Smoke generation –Part 2: Determination of optical density by a single-chamber test.' International Organization for Standardization, Geneva.

'ISO 5660–1: Reaction-to-fire tests—Heat release, smoke production and mass loss rate—Part 1: Heat Release Rate (Cone Calorimeter Method).' International Organization for Standardization, Geneva.

'ISO 9239–1: Reaction to fire tests for floorings—Part 1: Determination of the Burning Behaviour Using a Radiant Heat Source.' International Organization for Standardization, Geneva.

'ISO 9705: Fire tests—Reaction-to-fire—Room fire test.' International Organization for Standardization, Geneva.

'ISO TS 19700: Fire tests—Controlled equivalence ratio method for the determination of hazardous components of fire effluents.' International Organization for Standardization, Geneva.

'ISO 19701—Methods for analysis and sampling of fire effluents.' International Organization for Standardisation, Geneva.

'MIL-STD-2031: Fire and Toxicity Test Methods and Qualification Procedure for Composite Material Systems Used in Hull, Machinery and Structural Applications Inside Naval Submarines.'

'NFPA 265: Standard Method of Tests for Evaluating Room Fire Growth Contribution of Textile Wall Coverings.' National Fire Protection Association, Quincy, MA.

'NFPA 269: Standard Test Method for Developing Toxic Potency Data for Use in Fire Hazard Modeling.' National Fire Protection Association, Quincy, MA.

'NFPA 271: Standard Method of Test for Heat and Visible Smoke Release Rates for Materials and Products Using an Oxygen Consumption Calorimeter.' National Fire Protection Association, Quincy, MA.

'NFPA 286: Standard Method of Tests for Evaluating Contribution of Wall and Ceiling Interior Finish to Room Fire Growth.' National Fire Protection Association, Quincy, MA.

'NFPA 550: Guide to the Fire Safety Concepts Tree.' National Fire Protection Association, Quincy, MA.

'SMP 800-C: Bombardier Material and Process Specification, Toxic Gas Generation.' Bombardier, Montréal.

'SOLAS: Consolidated Edition, 2001, Chapter II-2–Construction: Fire protection, fire detection and fire extinction.' International Maritime Organization, London.

'UL 94: Test for Flammability of Plastic Materials for Parts in Devices and Appliances.' Underwriters Laboratories, Northbrook, IL.

REFERENCES

1. Quintiere, J., *Principles of Fire Behavior*, 2nd Edition. CRC Press, Taylor & Francis Group: Boca Raton, FL, 2016.
2. Gann, R.; Babrauskas, V.; Peacock, R.; Hall, J., Fire Conditions for Smoke Toxicity Measurement. *Fire and Materials* **1994,** *18*, 193–199.
3. Tu, K.; Quintiere, J., Wall Flame Heights With External Radiation. *Fire Technology* **1991,** *27*, 195–203.
4. Evans, D.; Stroup, D., Methods to Calculate the Response Time of Heat and Smoke Detectors Installed Below Large Unobstructed Ceilings. *Fire Technology* **1986,** *22*, 54–65.
5. Evans, D., Calculating Sprinkler Actuation Time in Compartments. *Fire Safety Journal* **1985,** *9*, 147–155.
6. Deal, S.; Beyler, C., Correlating Preflashover Room Fire Temperatures. *Journal of Fire Protection Engineering* **1990,** *2*, 33–48.
7. Foote, K.; Pagni, P.; Alvarez, N. *Temperature Correlations for Forced Ventilation Compartment Fires*, First International Symposium on Fire Safety Science, Gaithersburg, MD, Hemisphere Publishing Corporation, New York, NY, 1986; pp 139–148.
8. McCaffrey, B.; Quintiere, J.; Harkleroad, M., Estimating Room Fire Temperatures and the Likelihood of Flashover Using Fire Test Data Correlations. *Fire Technology* **1981,** *17*, 98–119.
9. Cooper, L., A Mathematical Model for Estimating Available Safe Egress Time in Fires. *Fire and Materials* **1982,** *6*, 135–144.
10. Jones, W.; Peacock, R.; Forney, G.; Reneke, P. *CFAST: Consolidated Model of Fire Growth and Smoke Transport (Version 5). Technical Reference Guide.* NIST Special Publication 1030; National Institute of Standards and Technology: Gaithersburg, MD, 2004; p 153.
11. McGrattan, K. B. H. S.; Floyd, J. E.; Baum, H. R.; Rehm, R. G. *Fire Dynamics Simulator (Version 5): Technical Reference Guide.* NIST Special Publication 1018–5; National Institute of Standards and Technology: Gaithersburg, MD, 2007; p 100.
12. Babrauskas, V.; Peacock, R., Heat Release Rate: The Single Most Important Variable in Fire Hazard. *Fire Safety Journal* **1992,** *18*, 255–272.
13. Thomas, P., Testing Products and Materials for Their Contribution to Flashover in Rooms. *Fire and Materials* **1980,** *5*, 103–111.
14. Babrauskas, V., Estimating Room Flashover Potential. *Fire Technology* **1980,** *16*, 94–104.
15. Babrauskas, V., A Closed Form Approximation for Post-Flashover Compartment Fire Temperatures. *Fire Safety Journal* **1981,** *4*, 63–73.
16. Thomas, P., SFPE Classic Paper Review: Fire Behavior In Rooms by Kunio Kawagoe. *Journal of Fire Protection Engineering* **2004,** *14*, 5–8.
17. Jin, T., Visibility Through Smoke. *Journal of Fire and Flammability* **1978,** *9*, 135–155.
18. Babrauskas, V., *Ignition Handbook*. Fire Science Publishers: Issaquah, WA, 2003.
19. Blaszkiewicz, M.; Bowman, P.; Masciantonio, M. *Understanding the Repeatability and Reproducibility of UL 94 Testing*, 18th BCC Conference on Flame Retardancy, Stamford, CT, May 21–23, 2007; Business Communications Co.: Stamford, CT, 2007.
20. Morgan, A.; Bundy, M., Cone Calorimeter Analysis of UL 94 V-Rated Plastics. *Fire and Materials* **2007,** *31*, 257–283.
21. Weil, E. D.; Hirschler, M. M.; Patel, N. G.; Said, M. M.; Shakir, S., Oxygen Index: Correlations to Other Fire Tests. *Fire and Materials* **1992,** *16*, 159–169.
22. *ASTM Fire Standards and Related Technical Material*, 7th Edition. ASTM International: West Conshohocken, PA, 2007.

23. Roed, J. *Low Voltage Directive*, FIRESEL 2003, Borås, Sweden, May 27–28, 2003; SP, Borås, Sweden, 2003.
24. Hallman, J. Ignition of Polymers by Radiant Energy. Ph.D. Thesis, University of Oklahoma, Norman, OK, 1971.
25. Koohyar, A. Ignition of Wood by Flame Radiation. Ph.D. Thesis, University of Oklahoma, Norman, OK, 1967.
26. Rasbash, D.; Drysdale, D.; Deepak, D., Critical Heat and Mass Transfer at Pilot Ignition and Extinction of a Material. *Fire Safety Journal* **1986,** *10*, 1–10.
27. Thomson, H.; Drysdale, D., Flammability of Plastics, I: Ignition Temperatures. *Fire and Materials* **1987,** *11*, 163–172.
28. Thomson, H.; Drysdale, D.; Beyler, C., An Experimental Evaluation of Critical Surface Temperature as a Criterion for Piloted Ignition of Solid Fuels. *Fire Safety Journal* **1988,** *13*, 185–196.
29. Atreya, A. Pyrolysis, Ignition and Fire Spread on Horizontal Surfaces of Wood. Ph.D. Thesis, Harvard University, Cambridge, MA, 1983.
30. Atreya, A.; Carpentier, C.; Harkleroad, M. *Effect of Sample Orientation on Piloted Ignition and Flame Spread*, First International Symposium on Fire Safety Science, Gaithersburg, MD, Hemisphere Publishing Co., New York, NY, 1985; pp 97–109.
31. Abu-Zaid, M. Effect of Water on Ignition of Cellulosic Materials. Ph.D. Thesis, Michigan State University, East Lansing, MI, 1988.
32. Fangrat, J.; Hasemi, Y.; Yoshida, M.; Hirata, T., Surface Temperature at Ignition of Wooden Based Slabs. *Fire Safety Journal* **1996,** *27*, 249–259.
33. Janssens, M. *Thermal Model for Piloted Ignition of Wood Including Variable Thermophysical Properties*, Third International Symposium on Fire Safety Science, Edinburgh, Scotland, Cox, G.; Langford, B., Eds. Elsevier Applied Science, New York, NY, 1991; pp 167–176.
34. Moghtaderi, B.; Novozhilov, V.; Fletcher, D.; Kent, J., A New Correlation for Bench-Scale Piloted Ignition Data of Wood. *Fire Safety Journal* **1997,** *29*, 41–59.
35. Urbas, J.; Parker, W., Surface Temperature Measurements on Burning Wood Specimens in the Cone Calorimeter and Effect of Grain Orientation. *Fire and Materials* **1993,** *17*, 205–208.
36. Yudong, L.; Drysdale, D., Measurement of the Ignition Temperature of Wood. *Fire Safety Science* **1992,** *1*, 25–30.
37. Khan, M.; deRis, J. *Operator Independent Ignition Measurements*, Eighth International Symposium on Fire Safety Science, Beijing, International Association of Safety Science: Beijing, 2005; pp 163–174.
38. Khan, M.; deRis, J. *Device and Method for Measuring Absorbed Heat Flux*, 10th Fire and Materials Conference, San Francisco, CA, Interscience Communications, London, 2007; pp 25–26 (abstract, full paper on CD).
39. Quintiere, J.; Harkleroad, M. *New Concepts for Measuring Flame Spread Properties*; NBSIR 84–2943; National Bureau of Standards: Gaithersburg, MD, 1984.
40. Janssens, M., Measuring Rate of Heat Release by Oxygen Consumption. *Fire Technology* **1991,** *27*, 234–249.
41. Parker, W., Calculations of the Heat Release Rate by Oxygen Consumption for Various Applications. *Journal of Fire Sciences* **1984,** *2*, 380–395.
42. Beaulieu, P.; Dembsey, N. *Enhanced Equations for Carbon Dioxide and Oxygen Calorimetry*, 9th International Fire and Materials Conference, San Francisco, CA, Interscience Communications Limited, London, 2005; pp 51–62.
43. Tewarson, A., Generation of Heat and Chemical Compounds in Fires. In *The SFPE Handbook of Fire Protection Engineering*, National Fire Protection Association: Quincy, MA, 2002; Section 3, pp 82–161.
44. Huggett, C., Estimation of the Rate of Heat Release by Means of Oxygen Consumption. *Fire and Materials* **1980,** *12*, 61–65.
45. Thornton, W., The Relation of Oxygen to the Heat of Combustion of Organic Compounds. *The London, Edinburgh and Dublin Philosophical Magazine and Journal of Science* **1917,** *33*, 196–203.
46. Petrella, V., The Mass Burning Rate of Polymers, Wood and Liquids. *Journal of Fire and Flammability* **1980,** *11*, 3–21.
47. Tewarson, A.; Pion, R., Flammability of Plastics. I. Burning Intensity. *Combustion and Flame* **1976,** *26*, 85–103.
48. Babrauskas, V., Development of the Cone Calorimeter—A Bench-Scale Heat Release Rate Apparatus Based on O_2 Consumption. *Fire and Materials* **1984,** *8*, 81–95.
49. Babrauskas, V.; Twilley, W.; Janssens, M.; Yusa, S., A Cone Calorimeter for Controlled Atmosphere Studies. *Fire and Materials* **1992,** *16*, 37–43.
50. Leonard, J.; Bowditch, P.; Dowling, V., Development of a Controlled-Atmosphere Cone Calorimeter. *Fire and Materials* **2000,** *24*, 143–150.
51. Hshieh, F.; Buch, R., Controlled-Atmosphere Cone Calorimeter Studies of Silicones. *Fire and Materials* **1997,** *21*, 265–272.
52. Lyon, R., Heat Release Kinetics. *Fire and Materials* **2000,** *24*, 179–186.
53. Lyon, R.; Walters, R.; Stoliarov, S., A Thermal Analysis Method for Measuring Polymer Flammability. *Journal of ASTM International* **2006,** *3*, 1–18.
54. Lyon, R.; Janssens, M., Polymer Flammability. In *Encyclopedia of Polymer Science & Engineering (On-line Edition)*, John Wiley & Sons: New York, NY, 2005.
55. Lyon, R., Plastics and Rubber. In *Handbook of Building Materials for Fire Protection*, Harper, C., Ed. McGraw-Hill: New York, NY, 2004; pp 3.1–3.51.
56. Lyon, R.; Walters, R.; Safronava, N.; Stoliarov, S. *A Statistical Model for the Results of Flammability Tests*, 11th International Fire and Materials Conference, San Francisco, CA, Interscience Communications Limited, London, 2009; pp 141–159.
57. Smith, E., Heat Release Rate of Building Materials. In *Ignition, Heat Release and Noncombustibility of Materials, ASTM STP 502*, American Society of Testing and Materials, Philadelphia, PA: October 6, 1971, Washington, DC, 1972; pp 119–134.
58. *Aircraft Material Fire Test Handbook*; DOT/FAA/CT-89/15; U.S. Department of Transportation, Federal Aviation Administration: Atlantic City, NJ, 1990.
59. Tewarson, A., Heat Release Rates from Burning Plastics. *Journal of Fire and Flammability* **1977,** *8*, 115–130.
60. Hasemi, Y., Experimental Wall Flame Heat Transfer Correlations for the Analysis of Upward Wall Flame Spread. *Fire Science and Technology* **1984,** *4*, 75–90.
61. Hasemi, Y.; Tokunaga, T. In *Modeling of Turbulent Diffusion Flames and Fire Plumes for the Analysis of Fire Growth*, 21st National Heat Transfer Conference, Fire Dynamics and Heat Transfer, Seattle, WA, 1983; pp 37–45.
62. deRis, J. *Spread of a Laminar Diffusion Flame*, Twelfth Symposium (International) on Combustion, Combustion Institute, Pittsburgh, PA: 1969; pp 241–252.
63. Qian, C.; Ishida, H.; Saito, K. *Upward Flame Spread Along the Vertical Corner Walls*; NIST-GCR-94-648; National

Institute of Standards and Technology: Gaithersburg, MD, 1994.
64. Hartzell, G. *Development of a Radiant Panel Test for Flooring Materials*; NBSIR 74–495; National Bureau of Standards: Washington, DC, 1974.
65. Mulholland, G.; Croarkin, C., Specific Extinction Coefficient of Flame Generated Smoke. *Fire and Materials* **2000,** *24*, 227–230.
66. Babrauskas, V.; Levin, B.; Gann, R.; Paabo, M.; Harris, R.; Peacock, R.; Yusa, S. *Toxic Potency Measurements for Fire Hazard Analysis*; Special Publication 827; National Institute of Standards and Technology: Gaithersburg, MD, 1991.
67. Babrauskas, V.; Gann, R.; Levin, B.; Paabo, M.; Harris, R.; Peacock, R.; Yusa, S., A Methodology for Obtaining and Using Toxic Potency Data for Fire Hazard Analysis. *Fire Safety Journal* **1998,** *31*, 345–358.
68. Babrauskas, V., The Generation of CO in Bench-Scale Fire Tests and the Prediction for Real-Scale Fires. *Fire and Materials* **1995,** *19*, 205–213.
69. Gann, R.; Babrauskas, V.; Peacock, R., Fire Conditions for Smoke Toxicity Measurement. *Fire and Materials* **1994,** *18*, 193–199.
70. Gross, D.; Loftus, J.; Robertson, A. *Method for Measuring Smoke From Burning Materials*, Fire Test Methods: Restraint and Smoke, ASTM STP 422, June 26–July 1, 1966, Atlantic City, NJ, American Society for Testing and Materials, Philadelphia, PA: 1967; pp 166–204.
71. Ames, S.; Rogers, S. *Large and Small Scale Fire Calorimetry Assessment of Upholstered Furniture*, 5th Interflam Conference, Canterbury, September 17–19, 2001; Interscience Communications, London, England: Canterbury, England, 1990; pp 221–232.
72. Babrauskas, V.; Lawson, J.; Walton, W.; Twilley, W. *Upholstered Furniture Heat Release Rates Measured with a Furniture Calorimeter*; NBSIR 82–2604; National Bureau of Standards: Gaithersburg, MD, 1982.
73. Smith, D.; Marshall, N.; Shaw, K.; Colwell, S. *Correlating Large-Scale Fire Performance with the Single Burning Item Test*, 9th Interflam Conference, Edinburgh, Interscience Communications, London: Edinburgh, 2001; pp 531–542.

Sources for Additional Information and Further Reading

Comprehensive compilations and an extensive discussion of various uses of flammability test methods can be found in the following two books:

Apte, V. *Flammability Testing of Materials Used in Construction, Transport and Mining*, 2nd Edition, Woodhead Publishing (an imprint of Elsevier), Cambridge, England, 2022.

Troitzsch, J., ed. *Plastics Flammability Handbook—Principles, Regulations, Testing, and Approval*, 4th Edition, Hanser Publisher, Munich, 2021.

All standards developed and maintained by ASTM Committee E05 on fire standards are in Volume 04.07 of the Annual Book of ASTM Standards. Members of ASTM have access free of charge to one volume of their choosing (see www.astm.org/astm-bos-04.07.html for details).

ASTM Fire Standards and Related Technical Material, 7th Edition, ASTM International, West Conshohocken, PA, 2007.

An extensive discussion of the principles of fire physics and chemistry and the behavior of materials in fire can be found in:

Fire Protection Handbook, 21st Edition, National Fire Protection Association, Quincy, MA, 2023 (https://catalog.nfpa.org/).

SFPE Handbook of Fire Protection Engineering. Springer, 5th Edition, Springer (https://link.springer.com/book/10.1007/978-1-4939-2565-0).

Finally, a tremendous amount of useful information can be found on the Internet. Instead of providing a list of web sites in this chapter, the reader is referred to the site of Dr. Vyto Babrauskas' company, Fire Science and Technology (URL: http://doctorfire.com). Not only is there a lot of useful information on this site, but you will also find what is, to the author of this chapter's knowledge, the longest list of fire-related links on the web (URL: http://doctorfire.com/links.html).

19 Fire Emissions and Smoke Toxicity

T Richard Hull and Anna A Stec
Centre for Fire and Hazards Science, University of Central Lancashire, Preston, UK

19.1 INTRODUCTION

The toxicity of fire effluents is known to be the biggest cause of death and injury from unwanted fires.[1] Compared with natural materials (wood, wool, cotton, leather, etc.), widely used synthetic polymers (derived from oil) burn more quickly; they also generate more smoke and toxic effluents. When compounded with flame retardants acting in the gas phase, the toxic product yield increases.[2, 3] Figure 19.1 shows the fire fatalities in Great Britain by cause of death. It is clear that "overcome by gas, smoke or toxic fumes" is a bigger category (31%) than burns (28%). However, badly burned fire victims who have inhaled toxic gas do not show a single, clear cause of death, making the combination of the third largest category (20%) difficult to interpret, although many fire victim's bodies continue to be burnt after breathing has stopped. The "unspecified" fire fatalities (16%) are awaiting a coroner's verdict and likely to fall proportionately into one of the other four categories. Proportionate redistribution of the "unspecified" category would increase the fire-related fatalities caused by smoke inhalation alone to 37% and the combination of burns and smoke to 23%. This shows that smoke toxicity is the major cause of all fire deaths.[1]

Over the period from 1955 to 2019 the overall number of fire deaths in Great Britain (GB) has decreased.[4] However, there has been a progressive shift in cause of death from "burns" to "overcome by toxic gas or smoke" ("Smoke"), as shown in Figure 19.2.

In Great Britain, and also in the United States,[5] most injuries in fires are also due to inhalation of toxic smoke, as shown in Figure 19.3, for all fires in Great Britain[1] from 1955 to 2019. In 2019/2020 "overcome by gas, smoke and toxic fumes" was the major contributor to nonfatal casualties resulting in 51% of injuries.

Recently there has been a resurgence of interest in fire toxicity. This has been driven by three factors:

1. The progressive replacement of prescriptive codes by performance-based design approaches to ensuring fire safety.
2. The development of tools to make meaningful assessments of fire toxicity.
3. The gradual recognition that the problem of fire toxicity had not been addressed by the emphasis on heat release in fire safety engineering.

The asphyxiants carbon monoxide (CO) and hydrogen cyanide (HCN) are known to cause the death of most fire victims in the short term. CO and HCN yields are much higher in under-ventilated flaming, which is more difficult to replicate on a bench scale. The recent findings that firefighters have twice the rate of cancer deaths of the civilian population, and that the international agency for research into cancer IRC have recognised firefighting as an occupation which causes cancer,[6] highlights the significance of carcinogens as longer-term fire toxicants. There is currently no requirement to quantify carcinogens from burning materials for products and no restriction on products emitting lethal quantities of carcinogens during a fire.

Further, it is becoming increasingly clear that airborne particulates result in a very large loss of human lives. In 2021, 238,000 European lives were lost from inhalation of airborne particulates.[7] These deaths had been attributed to diesel vehicles, but recent evidence shows that wood burning in homes now produces more small particle pollution than all road traffic in the UK.[8] A Swedish study[9] demonstrated that unwanted fires (*excluding* wildland fires) were responsible for around 10% of the emissions of polycyclic aromatic hydrocarbons (PAH) and atmospheric particulates; the emission of particulates and unburnt hydrocarbons from one ton of fuel from an unwanted fire is equivalent to that from burning almost 200 tons of solid fuel in a power station. The heterogeneity of unwanted fires and their high burning rates and low combustion efficiency are therefore highly likely to produce particles of widely different and potentially much greater toxicity than those from diesels or other combustion.

19.1.1 Fire Scenarios

The factors affecting fire gas toxicity[10–12] have been investigated in terms of the yields of toxic gases and other combustion products (smoke, particulates)[13, 14] as a function of fire condition for a wide range of polymeric materials. This requires realistic recreation of individual fire stages and robust analyses to correlate the concentrations of individual toxicants in fire gases.

The generalized development of a fire has been recognized and used to classify fire growth into a number of stages, from smoldering combustion and early well-ventilated flaming, through to fully developed under-ventilated flaming.[15]

DOI: 10.1201/9781003380689-19

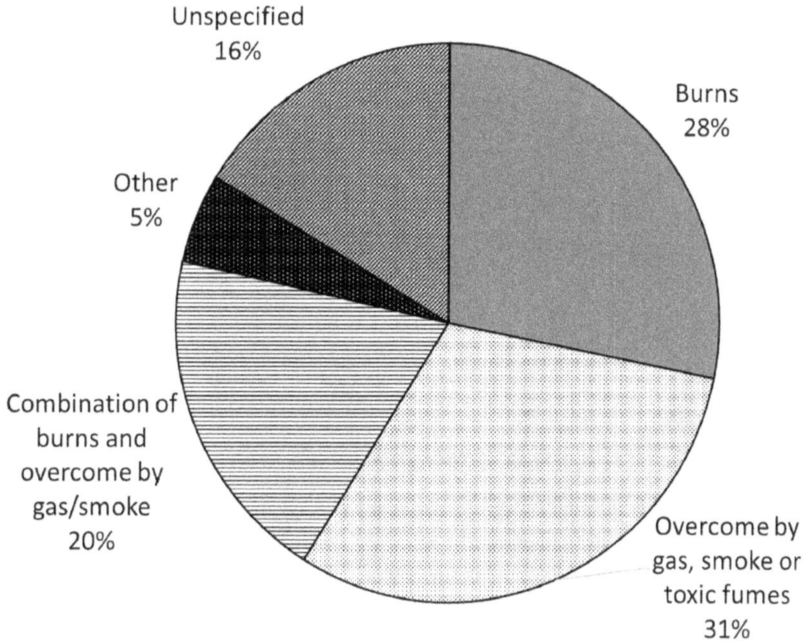

FIGURE 19.1 The cause of fatalities from fires in Great Britain 2019/20.

Source: Using data from ref 1

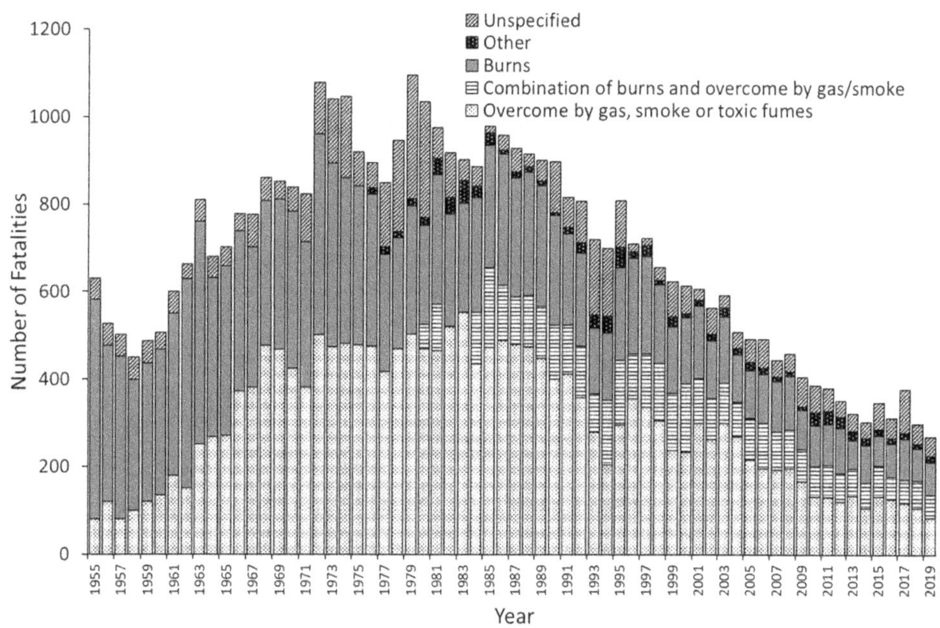

FIGURE 19.2 GB fire fatalities for all incident locations by cause of death, data from ref 1. A new category of combination of burns and overcome by gas/smoke was introduced in 1980. Grenfell fatalities are shown in 2017 as "unspecified," as they are awaiting the coroner's verdict.

Source: Using data from ref 1.

These are shown in Table 19.1. Although on some occasions smoldering (oxidative pyrolysis) can generate toxicologically significant quantities of effluent (for example smoldering cotton, or polyurethane foam), the rate of reaction, and hence the amount of toxic species generated, will be small, so unlikely to affect anyone outside the immediate vicinity. Similarly, well-ventilated fires are generally small, so extinguishment or escape is still feasible, while the fire effluent moves across the ceiling above head height. However, as they grow, fires become ventilation controlled, and fires in enclosures such as buildings rapidly change from well-ventilated to under-ventilated. These fires are larger and therefore produce greater volumes of effluent, affecting occupants over a much wider part of any building.

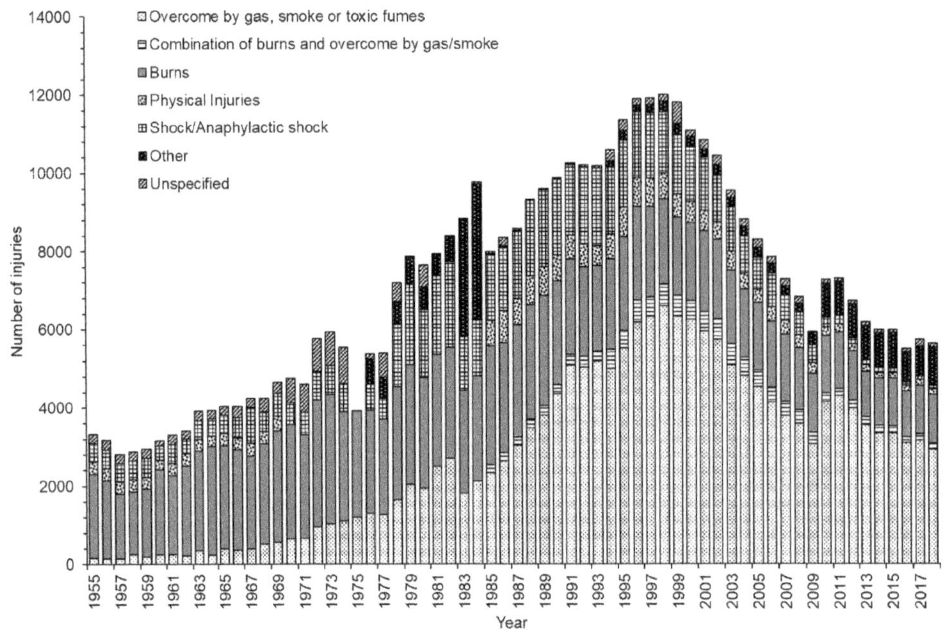

FIGURE 19.3 Nature of GB fire injuries for all incident locations from 1955 to 2019.

Source: Data taken from ref 1

TABLE 19.1
ISO Classification of Fire Stages, Based on ISO 19706[15]

Fire Stage	Heat/ kW m^{-2}	Max Temp/°C Fuel	Max Temp/°C Smoke	Oxygen % In	Oxygen % Out	Equiv-alence Ratio ϕ	$\frac{V_{CO}}{V_{CO_2}}$	Comb-ustion Efficiency %
Non-flaming								
1a. Self-sustained smoldering	n.a.	450–800	25–85	20	0–20	-	0.1–1	50–90
1b. Oxidative, external radiation	-	300–600		20	20	-		
1c. Anaerobic external radiation	-	100–500		0	0	-		
Well-ventilated flaming								
2. Well-ventilated flaming	0 to 60	350–650	50–500	~20	0–20	< 0.8	< 0.05	> 95
Under-ventilated flaming								
3a. Low vent. room fire	0 to 30	300–600	50–500	15–20	5–10	> 1.2	0.2–0.4	70–80
3b. Post flashover	50 to 150	350–650	> 600	< 15	< 5	> 1.2	0.1 0.4	70–90

While well-ventilated fire scenarios are routinely used for assessment of flammability, because the object is to stop the fire growing out of control, where fire toxicity is concerned, the important fire stages are under-ventilated. There are two reasons for this:

1. The volume of effluent is much greater.
2. The yields of the major toxic products (carbon monoxide and hydrogen cyanide, and other products of incomplete combustion) will be much greater.[2, 16]

The fire effluent composition can be obtained by interpretation of gas phase Fourier transform infrared (FTIR) spectra, alongside those from complementary analyses (high performance ion chromatography (HPIC), spectrophotometry, chemiluminescence, non-dispersive infrared (NDIR), gas chromatography-mass spectrometry (GC-MS), and wet chemical analysis).[17] The generation, sampling, and quantification of fire effluents has been described in detail elsewhere.[18] For many common materials CO is not the only significant toxicant in fire gas. HCN from burning nylon (polyamide or PA),[11] or hydrogen chloride (HCl) from burning PVC; are both produced in yields of greater toxicological significance than CO.[19]

The use of CO/CO_2 ratios can only be used to characterize fire stages for materials which do not contain chlorine or bromine since these elements significantly increase the CO yield in well-ventilated fires.

Studies such as these show that four factors multiply together to affect smoke toxicity, in roughly decreasing order of importance.

1. The flammability of the material. The more it burns, the more smoke and toxic gases will be produced.
2. The presence of a flame. Non-flaming fires (smolder) produce small quantities of highly toxic smoke, but flaming fires produce much larger volumes of usually less toxic smoke.
3. Ventilation condition. As the availability of oxygen decreases, so the smoke will contain more products of incomplete combustion. These include carbon monoxide, hydrogen cyanide, organics, and particulate matter. Most fire deaths result from inhalation of smoke from under-ventilated fires.
4. The material composition.
   a. If material only contains carbon, hydrogen, and oxygen, like many of the common polymers, the main toxicant will be carbon monoxide.
   b. If the material contains nitrogen (such as polyamides, polyacrylonitrile, polyurethanes, or polyisocyanurates), then hydrogen cyanide will also be produced. Hydrogen cyanide is approximately 20 times more toxic than carbon monoxide.
   c. Certain flame retardants, containing halogens (chlorine or bromine) or organophosphates, act as gas phase inhibitors which increase yields of all products of incomplete combustion, carbon monoxide, hydrogen cyanide (if nitrogen present), organics, and particulate matter, etc. Typically, their presence makes burning in well-ventilated flaming conditions as toxic as under-ventilated flaming, while burning in under-ventilated conditions is two to three times more toxic than without these flame retardants. In addition, the acid gases produced are potent respiratory tract irritants which will impede escape.

The most significant differences in fire conditions arise between flaming and non-flaming combustion. The high reactivity of the radicals in the flame drives the oxidation of fuel forward as far as it can go, with the available oxygen. Once flaming combustion is established, the ventilation condition, or more specifically the fuel/air ratio has the greatest effect on the yields. As a fire in a building develops, the temperature increases, and oxygen concentration decreases. Research predicting the carbon monoxide evolution from flames of simple hydrocarbons, reviewed by Pitts,[20] has shown the importance of the equivalence ratio, ϕ, for predicting the CO yield from the oxygen depletion in flaming conditions (Table 19.2).

In a fully developed fire, with low ventilation, ϕ can be as large as 5. For many hydrocarbon polymers, CO yield increases rapidly with increase in ϕ, almost independent of polymer.[21] In addition, a close correlation between CO formation and HCN formation has been established in studies of full-scale fires,[22, 23] as the formation of both species appear to be favorable under the same poorly ventilated fire conditions. The stochiometric condition provides just enough oxygen for complete conversion of fuel to CO_2 and water (and molecular nitrogen N_2 if N is present in the fuel).

19.1.2 Fire Hazards

Fires usually start small and grow slowly at first. Lethal fires often involve smoldering furnishing (bed or chair) where there is a gradual increase in the local concentrations of smoke and toxic gases. During this time carbon monoxide (CO) and a large number of organic products are generated in the smoke. This can incapacitate victims or put them into a deep sleep so they do not escape. Once *flaming* starts, these products of incomplete combustion are oxidized to carbon dioxide (CO_2) and water, and the effluent toxicity of the well-ventilated fire decreases briefly. However, if there is enough fuel, the fire quickly grows until it is limited by the availability of oxygen: it becomes under-ventilated. This results in an increase in smoke, reducing visibility as the concentrations of the asphyxiant (oxygen depriving) gases, CO and HCN, increase by factors of 10 to 50 with under-ventilation, creating a greater hazard than the lack of oxygen. At the same time, irritant gases (hydrogen chloride (HCl), hydrogen bromide (HBr), nitrogen dioxide (NO_2), ammonia (NH_3), acrolein (CH_3CHO) etc.) and deep lung irritants and particulates will also add to the noxious properties of the smoke.

TABLE 19.2
Relationship of Equivalence Ratio to Ventilation Condition, CO Yield and CO_2/CO Ratio for Flaming Fires

$$\varphi = \frac{actual\ fuel\ to\ air\ ratio}{stoichiometric\ fuel\ to\ air\ ratio}$$

Equivalence Ratio, ϕ	Type of Flaming	Typical CO Yield g/g	CO_2/CO Ratio
0.7	Well-ventilated	0.01	100
1.0	stoichiometric	0.05	20
1.5	Under-ventilated	0.20	10

TABLE 19.3
Hazards from Fire, Typical Origins, and Their Effect on Humans

Toxic Products	Origin	Effects	Tenability/Thresholds Limits (within 30 min)
Carbon monoxide (CO)	Incomplete combustion of any organic material	Functional asphyxia, narcosis tissue hypoxia, organ failure, death	5,700 ppm
Hydrogen cyanide (HCN)	Polyurethane, polyamides, acrylics (polyacrylonitrile), wool, etc.	Cellular asphyxia, tissue hypoxia, organ failure, death	165 ppm
Halogen acids (HCl, HBr)	Polyvinyl chloride, halogenated flame retardants	Respiratory tract irritation	100 ppm to 1,000 ppm
HF	Li-ion batteries, fluoropolymers	Respiratory tract irritation, lung damage, oedema, cell death	100 ppm
Nitrogen dioxide (NO_2)	High temperature combustion—severe fires, car exhausts	Respiratory tract irritation, mucus membrane irritant	170 ppm
Smoke	Organic material forming tar droplets or soot particles (especially if carbon rich or under-ventilated)	Airway obstruction, visual obscuration	0.33 OD·m^{-1} (OD is optical density)
Oxygen depletion (O_2)	Combustion/Pyrolysis	Asphyxia, laryngotracheitis	10–12%

19.2 ASSESSMENT OF FIRE TOXICITY

Exposure to toxic fire effluents can initially lead to a combination of physiological and behavioral effects of which physical incapacitation, loss of consciousness, motor coordination, or disorientation may be life-threatening because they can impair safe escape.

Quantification of fire effluent composition from bench-scale laboratory tests enables prediction of both incapacitation (inability to escape unaided) and lethality (based on rat exposure data) using ISO 13571[24] and ISO 13344[25] respectively.

The toxic effects of fire effluent can be expressed as a fractional effective dose (FED), based on the chemical composition of the effluent. When the FED is equal to one, 50% of a healthy adult exposed population would be predicted to succumb either to incapacitation or death. In both cases CO_2 increases respiration rate, and so a multiplication factor for CO_2-driven hyperventilation, V_{CO_2} in Equation 2, is included in the calculation. ISO 13571[24] considers the four major hazards from fire which may prevent escape (toxic gases, irritant gases, heat, and smoke obscuration). ISO 13571[25] stipulates the use of Equation 1 for estimation of incapacitation by the asphyxiant gases carbon monoxide (CO) and hydrogen cyanide (HCN):

$$FED = \left\{ \sum_{t_1}^{t_2} \frac{[CO]}{35000} \Delta t + \sum_{t_1}^{t_2} \frac{[HCN]^{2.36}}{1.2 \times 10^6} \Delta t \right\} \times V_{CO_2} \quad (19.1)$$

$$V_{CO_2} = 1 + \frac{\exp(0.14[CO_2]) - 1}{2} \quad (19.2)$$

Gas concentrations in [] are expressed in μL L^{-1} or ppm; time (*t*) is in minutes. The C·t product corresponds to the ratio of actual over incapacitating dose for CO of 35,000 μL L^{-1} min (approximately equal to ppm min), equal to around 1,170 ppm for each 30 minutes of exposure. The FED value is calculated using the exposed dose relationship (concentration-time product, C·t) for CO and an exponential relationship for HCN (because inhalation of HCN induces hyperventilation due to the oxygen deprivation). The concentration of HCN is raised to the power 2.36 to predict incapacitation.

Since the effect of irritant gases is more or less instantaneous (a victim for whom breathing was too painful would not attempt escape), only the concentration is required to predict a fractional effective concentration. The FEC for incapacitation by irritants is shown in Equation 3.

$$FEC = \frac{[HCl]}{IC_{50,HCl}} + \frac{[HBr]}{IC_{50,HBr}} + \frac{[HF]}{IC_{50,HF}} + \frac{[SO_2]}{IC_{50,SO_2}} + \frac{[NO_2]}{IC_{50,NO_2}} + \frac{[acrolein]}{IC_{50,acrolein}} + \frac{[fomaldehyde]}{IC_{50,fomaldehyde}} + \sum \frac{[irritant]}{IC_{50,irritant}} \quad (19.3)$$

Equation 4, from ISO 13344,[25] provides an estimation of lethality for a 30-minute exposure, using the ratio of each toxicant concentration to its lethal concentration (LC_{50}). An acidosis factor, A, is included to account for the toxicity of CO_2. AGI are acid gas irritants and OI are organic irritants.

$$FED = \left\{ \frac{[CO]}{LC_{50,CO}} + \frac{[HCN]}{LC_{50,HCN}} + \frac{[AGI]}{LC_{50,AGI}} + \frac{[OI]}{LC_{50,OI}} \ldots \right\} \times V_{CO_2} + A + \frac{21 - [O_2]}{21 - 5.4} \quad (19.4)$$

The acidosis factor A = $[CO_2] \times 0.05$. O_2 and CO_2 concentrations are expressed in vol %. The concentration of the other toxicants is in μL L^{-1} or ppm.

The fire toxicity of a material can also be expressed as a Material-LC_{50}, which in this case is the specimen mass M of a burning polymeric material which would yield an FED equal to one within a volume of 1 m^3. The relation to the FED is given in Equation 5.

$$\text{Material-LC}_{50} = \frac{M}{\text{FED} \times V} \quad (19.5)$$

V is the total volume of diluted fire effluent in m^3 at STP. An illustration of the use of Material-LC_{50} values is given later in the chapter, in Table 19.6.

It is important to recognize that hazard prediction describes an unacceptable outcome of incapacitation or lethality. In any hazard or risk assessment, a factor of 0.3 is applied for a healthy adult population to ensure that around 90% of the trapped fire victims do not suffer the outcomes above, and another factor of 0.3 is applied to that figure if typical human populations, such as office workers, are considered, in order to account for individuals of greater susceptibility. Thus, a value of FED or FEC greater than 0.1 for effluent flowing into an apartment or workplace would be considered an unacceptable hazard.

To illustrate how this translates into a toxic effect, the major LDPE, PMMA, PS, PA 6.6, and PVC product yields, obtained from the steady state tube furnace, have been translated into an overall estimation of the fire effluent toxicity, using the methods described in ISO 13344 to show the contribution of each toxicant towards the overall fire hazard (Figure 19.4). The toxicity is expressed as FED for a fuel mass charge concentration of 20 g/m^3.

Most polymers without heteroatoms follow the trend shown by LDPE and PMMA of fire toxicity increasing from a very low value in well-ventilated conditions, to a much higher value in under-ventilated flaming. As can be seen from Figure 19.4, there is large variation in FED values for materials containing chlorine or nitrogen. For well-ventilated tests, the largest FED value is for PVC. PVC is one of the few materials to show a predicted combustion toxicity which is almost independent of the equivalence ratio, showing unusually high toxicity in well-ventilated conditions. As an irritant gas, HCl will have the greatest impact on toxicity. However, its gas phase inhibition effect also increases the yield of CO. In well-ventilated flaming, neither CO nor HCN make a significant contribution to the toxicity. In contrast, for under-ventilated fires, HCN from PA 6.6 makes the most significant contribution to the toxicity, and a high dependency on fire conditions is observed. HCN generated during small and large under-ventilated flaming tests makes the most significant contribution to the toxicity. For PA 6.6 in under-ventilated conditions, and for PVC, HCN and HCl make the greatest contribution to the final FED, showing the importance of two toxicants other than CO. PMMA, LDPE, and PS are hydrocarbons without any halogens or nitrogen, and this is reflected in FED values well below 1 for the well-ventilated fire scenario. The fire effluent from polystyrene has relatively low toxicity and shows characteristically low dependence on fire conditions. The toxic contribution of CO under fuel-rich conditions is remarkably similar to those generated under fuel lean conditions, as shown in Figure 19.4. This has been attributed to the high stability of the aromatic ring.[26]

19.3 COMPOSITION OF FIRE EFFLUENTS

The heat generated during a flaming fire volatilizes solid and liquid fuels. In the vapor phase they can react with oxygen, releasing heat. In small fires, such as a match or gas stove, almost all the fuel is converted to CO_2 and water. Products

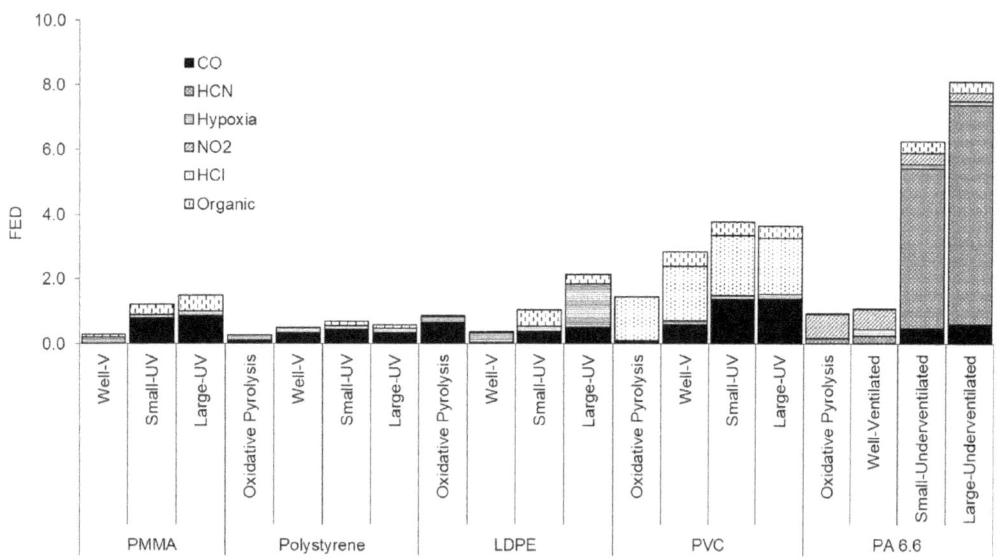

FIGURE 19.4 Contributions of individual toxicants to FED for lethality of common polymers using ISO 13344.[11]

such as water, CO, and CO_2 are almost always present in a fire effluent. In large fires, the volatiles are produced too quickly for complete reaction with oxygen, resulting in many products of incomplete combustion. These include CO, small molecules of carbon, hydrogen, and sometimes oxygen, described as volatile organic compounds, larger molecules such as polycyclic aromatic hydrocarbons, tar, and soot. The composition of VOCs and PAHs, as deposited, is somewhat dependent on the fuel, but mainly on the fire conditions.

Unwanted fires will produce many more products of incomplete combustion, including particulates, carcinogens, such as polycyclic aromatic hydrocarbons (PAH), respiratory sensitizers such as isocyanates, and persistent, bioaccumulative, and toxic (PBT) compounds such as polychloro- and polybromodibenzo dioxins and dibenzofurans (PCDD/F and PBDD/F) ("dioxins") from fuels containing halogens.

19.3.1 Volatile Organic Compounds

In fires, VOCs either originate from incomplete combustion, and therefore the VOC composition will depend on both the material burning (as partially decomposed fuel molecules), and the fire conditions, or in the absence of sufficient heat or oxygen, by radical recombination. Important examples of these compounds are benzene, styrene, and phenol, as these are all formed in the majority of fires.[27] To illustrate this process, the reactions occurring in a methane flame are shown schematically in Figure 19.5.[28]

Benzene is of particular toxicological significance as it is a precursor of carcinogenic polycyclic aromatic hydrocarbons (PAHs) as well as being a known carcinogen in its own right. The hydrophobic structures of VOCs increase their environmental persistence and bioaccumulativity in fatty tissues.

19.3.2 Incapacitating Irritants

The irritant chemicals released in fires are formed during the pyrolysis and partial oxidation of materials, and the combinations of products from different materials are often remarkably similar.[29, 30] However, for many organic materials and particularly for simple hydrocarbon polymers such as polypropylene or polyethylene, the main pyrolysis products, which consist of various hydrocarbon fragments, are innocuous. Thus when polypropylene is pyrolyzed in nitrogen, the product such as ethylene, formaldehyde, acetaldehyde, styrene, etc. are produced, and such an atmosphere was found to have no effect upon primates.[31] However, when

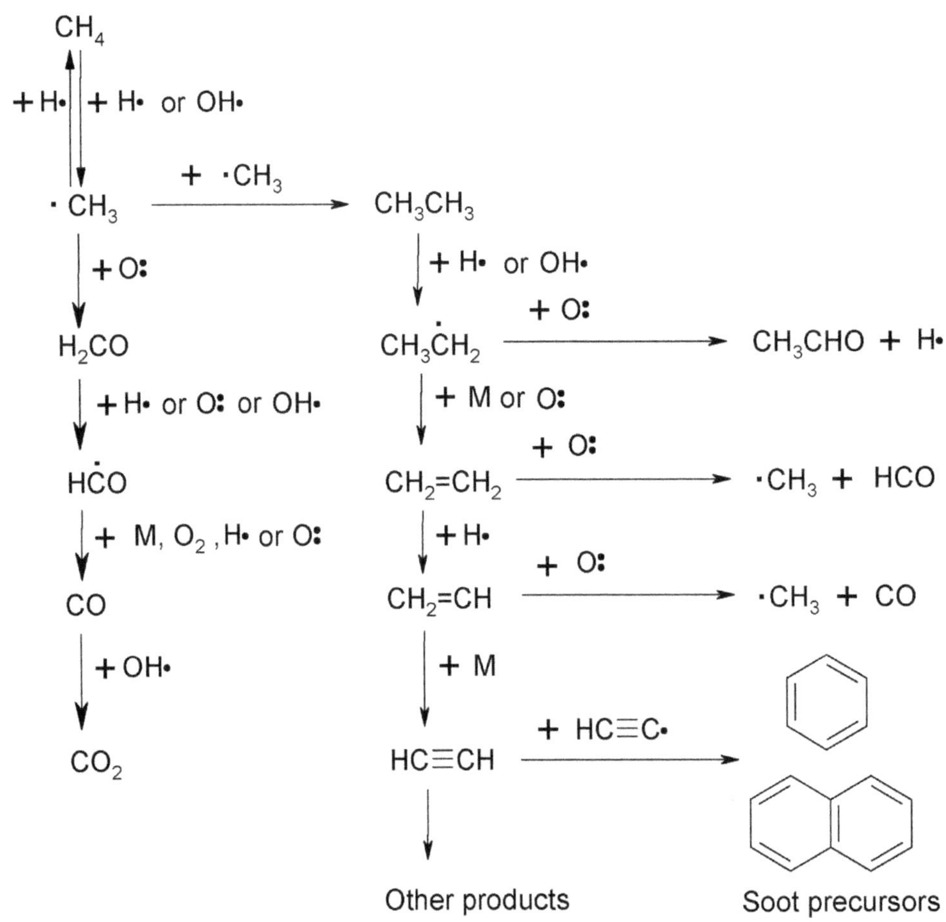

FIGURE 19.5 Reactions occurring in a methane flame (after Ref 28).

TABLE 19.4
Common Organo Irritants Found in Fire Gas with IDLH Values[33]

Substance	IDLH Value/ppm
Acetaldehyde	2,000
Acrolein	2
Benzene	500
Crotonaldehyde	50
Formaldehyde	20
Phenol	250
Toluene	500

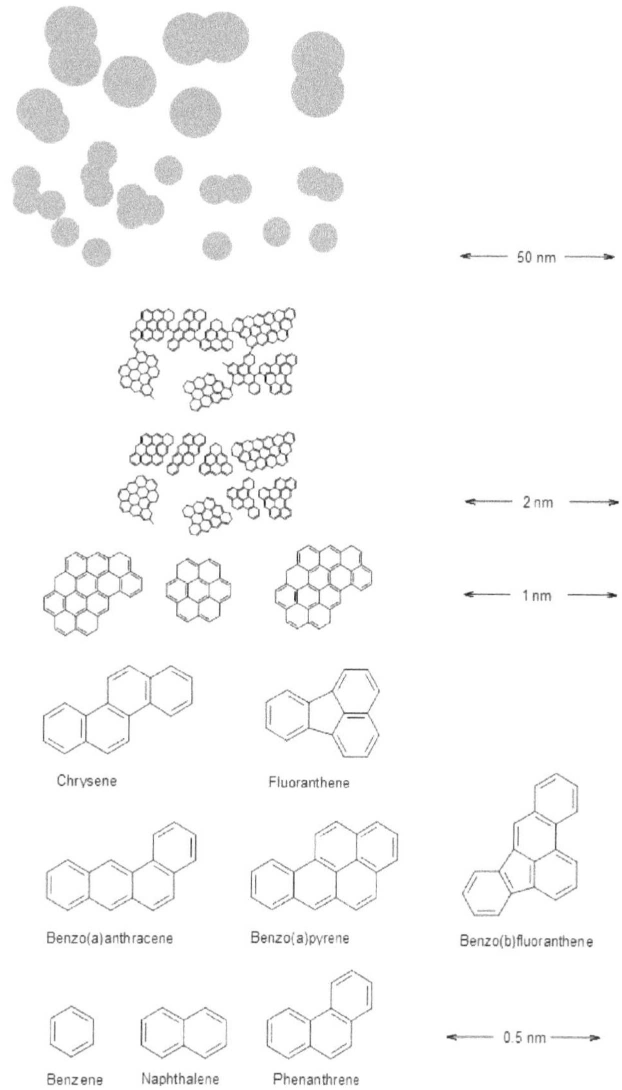

FIGURE 19.6 The process of agglomeration of PAHs to form soot particles.

these products are oxidized during non-flaming decomposition in air (oxidative pyrolysis), some are converted to highly irritant products, and such atmospheres were indeed found to be highly irritant to both mice and primates. In reports of mouse exposure experiments, some fire retarded materials, which could be induced to flame only intermittently, with considerable smoke production, were found to produce atmospheres up to 300 times more irritant than the same polymer in its non-fire retarded state, which burned cleanly.[32] Table 19.4 shows some of more toxic, commonly encountered organic species in fire gas, with the concentration considered by NIOSH to be Immediately Dangerous to Life or Health (IDLH).[33]

An approximation by Purser[34] predicts that an overall organic concentration of 10 mg L^{-1} is sufficient to cause incapacitation. For comparison, the 500 ppm IDLH value for benzene corresponds to 1.6 mg L^{-1}.

19.3.3 Polycyclic Aromatic Hydrocarbons

Most atmospheric PAHs are derived from the burning or combustion process of carbonaceous material (e.g., methane, wood, plastic, petrol, diesel, or other fuel). Tar and soot are mostly composed of agglomerations of PAHs. The process of agglomeration is illustrated in Figure 19.6. Individual PAHs, such as chrysene, phenanthrene, etc., attach themselves to other PAHs, either fusing to form structures with more rings, or loosely attaching, keeping their molecular identity. As more PAHs stick together, the accumulated gas phase molecules grow to become microscopic particulates and ultimately tar droplets and soot particles. In a small fire, as the fuel leaves the solid or liquid to the gas, it will be broken down into small molecules (around 1 nm in diameter). The formation of PAHs and the composition of PAHs in the fire effluent are dependent on the fire conditions. In particular, the yield of PAHs will increase with underventilation (where products of incomplete combustion become favored over carbon dioxide and water) and the composition of PAHs in the sample will vary with temperature, time, and turbulent mixing, particularly in the flame zone (larger PAH molecules will be favored by lower temperatures and longer residence times in the flame zone). Although detailed studies have been undertaken on PAH production in combustion systems (petrol or diesel engines etc.), the conditions in unwanted fires vary so much from fire to fire that meaningful predictive statements cannot be made.[35]

PAHs are human toxicants, carcinogens, and environmental pollutants: when agglomerated into combustion-derived particulates, they are estimated to kill three million people, globally per year.[36] They originate from various sources, including coal coking, diesel emissions, domestic heating, cooking, and from unwanted fires.[37] The molecular weight of PAHs strongly influences their distribution between the gas and condensed phases and is of direct relevance to their toxicity. Those in the low molecular weight category (fewer than four fused rings, e.g., naphthalene, phenanthrene) mainly exist in the gas phase. The tendency to adsorb onto organic carbon (e.g., soot particles) increases with increasing molecular weight, so the higher molecular weight compounds (more than four fused rings, such as benzo(a)pyrene or benzo(b)fluoranthrene) occur in the atmosphere mainly as particulates.[38]

The Environmental Protection Agency (EPA) have identified and classified 16 chronically toxic priority pollutant PAHs. Some PAHs are recognized carcinogens and mutagens, while those not yet classified are considered to be potentially harmful. Many PAH studies have found potential links to certain cancers but have not explicitly proven that these agents are causative as there is insufficient epidemiological or experimental data on individual PAHs and their mixtures. Most PAHs are not only pro-carcinogens but also are listed as genotoxic and mutagenic. The carcinogenicity of PAHs, as well as their toxic equivalency factor (TEF), the relative toxicity of different PAHs, is related to one of the most significant of the pyrogenic carcinogens, benzo[*a*]pyrene (B[*a*]P). Benzo(a) pyrene (BaP) has been identified as the most toxic PAH species and is used as a reference toxicant for other PAH, although benzo(a)anthracene is also of great concern.[39, 40] However, recent environmental studies identified PAHs of considerably higher toxicity than the US Environmental Protection Agency (EPA) 16 priority PAHs. From a regulatory perspective many of these have not been studied in sufficient detail with respect to frequency of occurrence in the environment and to their toxicity to different organisms to justify their inclusion in routine measurements.[41] For example, 3-methylcholanthrene (3-MCA) possesses a carcinogenic potential approximately 5.7-fold that of B[*a*] P, meaning that low concentrations of them may contribute significantly to the toxic effects of exposure. Alkyl derivatives of PAHs, for example 7,12-dimethylbenzo[*a*] anthracene, have a 20-fold higher TEF than its parent compound and twice that of B[*a*]P.[42] Compelling arguments have been made for the EPA list to be expanded to cover a wider group of toxic PAH.[42, 43]

In the US it has been reported that cancer accounts for 56% of line-of-duty deaths amongst firefighters,[44] around twice the national average. Only recently have these cancers (myelomas, lymphomas, melanomas, leukemia, prostate, testicular, brain, digestive tract, and skin cancers), and the carcinogenicity of the PAH and other compounds identified in fire effluents, been combined to estimate the carcinogenicity of fire effluents.[45–48]

Table 19.5 lists the 16 major PAHs together with carcinogenic classifications as defined by the International Agency for Research on Cancer (IARC).[49] IARC classifications are as follows:

- Group 1—carcinogenic to humans,
- Group 2A—probably carcinogenic to humans,
- Group 2B—possibly carcinogenic to humans,
- Group 3—unclassifiable as to carcinogenicity to humans,
- Group 4—probably not carcinogenic to humans.

A large number of PAH structures have been identified in fire effluents, some with greater carcinogenicity than the well-known benzo-a-pyrene (BaP). While relationships have been established to show the links between exposures to PAH and cancer, and atmospheric particulates ($PM_{2.5}$) and respiratory and cardiovascular diseases (acute lower respiratory illness, cerebrovascular disease, ischaemic heart disease, chronic obstructive pulmonary disease, and lung cancer)[50], the specific agents responsible have not been positively identified. The shape, size, and composition of particulates determine their toxicity.[51, 52] Organic species are chemisorbed onto their surface through transition metal oxides.[53–55] This results in surface-stabilized, environmentally persistent free radicals (EPFR), which continue to exist in ambient air from hours to months and can last at least 12 h in biological fluids. In laboratory animals and human lung cell lines they have been shown to cause the increased oxidative stress resulting in regeneration of reactive oxygen species (ROS) in biological systems, and associated with cancer, pulmonary, and other diseases.[56, 57] The formation of potentially toxic particulates, PAHs, and isocyanates and their distribution between gas phase and solid phase for different building material fires has been investigated.[38]

TABLE 19.5
The 16 Priority PAHs as Defined by the EPA

Name	Toxicity Data
Benzo(a)pyrene	IARC Group 1
Dibenzo(ah)anthracene	IARC Group 2A
Naphthalene **Benz(a)anthracene** **Chrysene** **Benzo(b)fluoranthene** **Benzo(k)fluoranthene** **Indeno(1.2.3-dc)pyrene**	IARC Group 2B
Acenaphthene **Fluorene** **Phenanthrene** **Anthracene** **Fluoranthene** **Pyrene** **Benzo(ghi)perylene**	IARC Group 3
Acenaphthylene	No IARC classification

Analysis of VOCs/SVOCs and PAH in a series of experimental house fires where cooking oil or a single sofa were used as initial fuel packages has been reported.[58] The initial fuel packages were either cooking oil or a single sofa; these were burned both alone and in furnished surroundings. Experiments were also carried out where the fire compartment was furnished. The ventilation conditions were varied between experiments by having the door and/or window to the room open or closed for the duration of the burn. Benzo(a)pyrene, the PAH of greatest toxicological concern, was detected in the majority of fires and most frequently in the developing and developed stages of fire. This is a particular concern due to the known mutagenic and carcinogenic effects of this molecule.

19.3.4 Isocyanates

The potent respiratory sensitizers, isocyanates, triggering asthma, and sometimes death after a single exposure, have been identified in fire effluents.[59, 60] Isocyanates are polymerized with polyols to make flexible polyurethane (PUF) foams for soft furniture and rigid polyurethane (PUR) foams for insulation in buildings, refrigerators, and freezers. Due to their versatility and wide range of applications, isocyanate-containing products can be found in significant quantities throughout the built environment. The two main products in the isocyanate market, with an approximate market share of 90%, are both di-isocyanates: isomers of toluene-di-isocyanate (TDI), mostly used in PUF, and diphenylmethane-di-isocyanate (MDI), mostly used in PUR. Inhalation of isocyanates is the most important human exposure route and may result in respiratory sensitization.[61] There is also limited evidence that dermal contact may result in human respiratory sensitization.[62] The chemical reactivity of isocyanates appears to be reflected in their toxicity, as exposure to isocyanates is one of the most common causes of occupational asthma in developed countries. Additionally, low concentrations of airborne isocyanates are acutely toxic following inhalation, and the production of isocyanates and products containing isocyanates is therefore heavily regulated from a health and safety at work perspective, particularly polyurethane (PU) based foams, fillers, and coatings. However, airborne isocyanates have also been sampled in fire effluent from building materials during the early stages of fire development.[59]

Airborne isocyanates from 23 common materials found in buildings, including PUR, particle board, and melamine-formaldehyde laminate, were sampled after being generated by a cone calorimeter.[59] The main findings from these initial small-scale tests were that all nitrogen-containing materials included in the study produced isocyanates upon thermal degradation. Airborne isocyanates were also sampled from an ISO 9705 room, burning a sofa containing flexible PU foam. The experiment went to flashover. The relatively high amounts of isocyanates measured, mainly isocyanic acid (ICA), but also methyl isocyanate (MIC), phenyl isocyanate (PHI), and 2,4-TDI, from this test were considered a potential health hazard.

Using the steady state tube furnace (SSTF) (ISO/TS 19700), the generation of isocyanates from materials such as PUF mattresses, PUR insulation, PVC carpets, and wood board were investigated.[38, 63] Similar yields for isocyanates from the polyurethane materials were found from the well-ventilated stage of the SSTF and from the cone calorimeter, with no significant difference between the well-ventilated and under-ventilated conditions. The smaller, more volatile mono-isocyanates were found almost exclusively in the gas phase, whereas the less-volatile MDI was only found in the particulate phase of the fire effluent.

19.3.5 Halogenated Dioxins

Polychlorinated dibenzo-p-dioxins (PCDD) and polychlorinated dibenzofurans (PCDFs) and their brominated homologues (PBDD/F) are toxic, chemically and thermally stable, and have a tendency to be strongly adsorbed on the surface of particulate matter. The formation of PCDD/Fs is a function of temperature, the availability of halogens, and the quantity of unburnt carbon, which depends upon the ventilation conditions.[64–66] Numerous studies on animals have also confirmed that some dioxin congeners are carcinogenic and produce mutagenic effects in certain species. However, it is difficult to extrapolate data from different fire test methods as there are significant differences in both the combustion conditions (such as large variations in temperature and oxygen environments) and also varying concentrations of precursors and chlorine in the product gases.[67] Therefore, at present, our knowledge of the distribution and quantities of these species in fire gases is fairly limited.

19.3.6 Particulates

Smoke not only impairs escape by visual obscuration but also contains particulate matter which is sufficiently small to pose a respiratory hazard. In spite of large amounts of particulates being generated by fire, relatively few investigations have been made into their size, distribution, and composition from unwanted fires.[68]

The particle size distribution is dependent on the material, temperature, fire conditions, and effluent flow. The particle size of the spherical droplets from smoldering combustion is generally of the order of 1μm, while the size of the irregular soot particulates from flaming combustion has a much wider range. This makes particle size distribution harder to determine and dependent on measuring technique and sampling position. The deposition zones for humans as a function of particle size are presented in Figure 19.7.

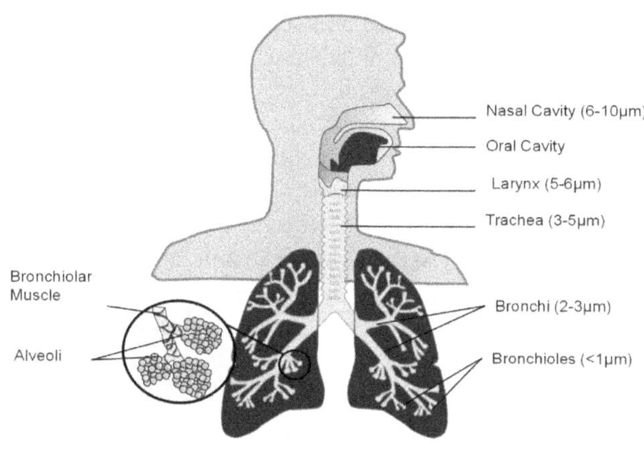

FIGURE 19.7 Deposition zones of particulates in the human respiratory tract.[69]

PAH agglomerate into small spherical particles, which then adhere to one another, like a string of beads, eventually tangling to form soot particles which are large enough to be trapped on a filter. Further, PAHs can then be adsorbed onto the growing soot surface. Both the volatile PAH and the soot particles will remain airborne almost indefinitely but have the potential to cause significant damage to the lungs. Although a large amount of research has focused on the toxicity of airborne particulates,[70] very little work has been reported on particulate emissions from unwanted fires.[68] Different particle size ranges are deposited in different parts of the respiratory tract. $PM_{2.5}$s (particles with diameter less than 2.5 μm) result in inflammation of the terminal bronchioles and may result in their complete blockage.

Oedema fluid disrupts the dispersion of the lung surfactant, causing collapse of the alveolae from higher surface tension of the fluid. The smallest particles (< 0.5 μm) penetrate the lung interstitium (between the surface of the alveolae and the blood capillaries), where they cause interstitial and luminal oedema. They can also cross the air-blood barrier and enter the bloodstream, triggering dangerous immune responses from the white blood cells, including polymer fume fever, leading to increased platelet stickiness and heart attacks. Attempts to characterize pyrogenic particulates, and the toxicants bound to them, are hampered by the establishment of dynamic equilibrium between adsorption and desorption on smoke particle surfaces, when trapped on a filter.

19.4 TOXICITY ASSESSMENT FROM SMALL AND LARGE-SCALE TEST METHODS

Assessment of toxic hazard is increasingly being recognized as an important factor in the assessment of fire hazard. Prediction of toxic fire hazard depends on two parameters.

1. Time-concentration profiles for major products. These depend on the fire growth curve and the yields of toxic products.
2. Toxic potency of the products, based on estimates of doses likely to impair escape efficiency, cause incapacitation or death.

Some criteria for the assessment of physical fire models have been published in ISO 16312-1 and reviewed elsewhere.[71] Essentially, these require toxic product yields from bench-scale tests to be replicated in large-scale fires. For a bench scale method to differentiate between the stages of a fire, it must allow the equivalence ratio and/or the combustion efficiency to be quantified so that the individual fire stages can be identified, and the product yields correlated to each stage.

Most bench-scale methods used for generation of toxic effluents can only reproduce a single well-ventilated fire stage or combustion condition. This is because they are based on flammability tests, where the worst-case scenario is indeed well-ventilated flaming. These methods can also be classified as those with constant combustion conditions, often achieving a prolonged steady-state period, and those with non-constant combustion conditions. Those with constant combustion conditions are best suited to producing reproducible data relating to specific fire stages suitable for fire toxicity assessment, since most effluent analysis is not instantaneous and benefits from constant fire effluent composition. Those bench-scale methods in closed chambers where the sample is exposed to a fixed heat flux have non-constant combustion conditions, including the toxicity tests derived from the smoke density chamber (SDC) (ISO 5659), where the oxygen concentration decreases during the test, and static tube furnace tests, such as the NF X 70–100, where the sample temperature, and hence the rate of pyrolysis, increases during the test. Intermediate between these two extremes are those that can produce quasi-steady combustion conditions such as the cone calorimeter (ISO 5660–1) replicating well-ventilated flaming. The controlled atmosphere cone calorimeter[72] (CACC) (ISO/TS 5660–5) and the Fire Propagation Apparatus (FPA)[73] (ASTM E 2058 and ISO 12136) have approximately constant combustion conditions and the facility to control the ventilation condition.

The steady state tube furnace (SSTF) also has constant combustion conditions. It has been developed from a British Standard (BS 7990)[74] into the first international standard (ISO/TS 19700)[75] for bench-scale measurement of fire toxicity[76], specifically as a tool for fire safety engineering. It is now commercially available from Concept Equipment, UK, and Phinix Instruments, China. The difficulty of replicating the conditions of fully developed under-ventilated flaming on a bench-scale is caused by two practical problems. In fully developed fires in under-ventilated conditions, the high heat flux is large enough for burning to continue at low oxygen concentrations (e.g., ~5%); in bench-scale experiments the heat flux is usually constant and often insufficient to sustain flaming at such low oxygen concentrations. In fully developed fires, mixing and recirculation of the fire effluent ensures that all the oxygen present passes into the fire plume; in many bench-scale apparatuses, such as the controlled atmosphere cone calorimeter (CACC) or smoke chamber, an unknown quantity of fresh air bypasses the fire plume so the ventilation condition remains undefined. The repeatability and reproducibility off the SSTF have been quantified,[77, 78] relating its yields to other national and international standards (the smoke density chamber (SDC), ISO 5659 modified for toxicity measurements as ISO draft DIS 21489, the French railway test NF X 70–100, the FPA and DIN 53436)[79] and finally to comparable data from the limited data available from large scale tests.[80, 81] This work showed that under controlled conditions, robust quantification of fire toxicity was achievable.

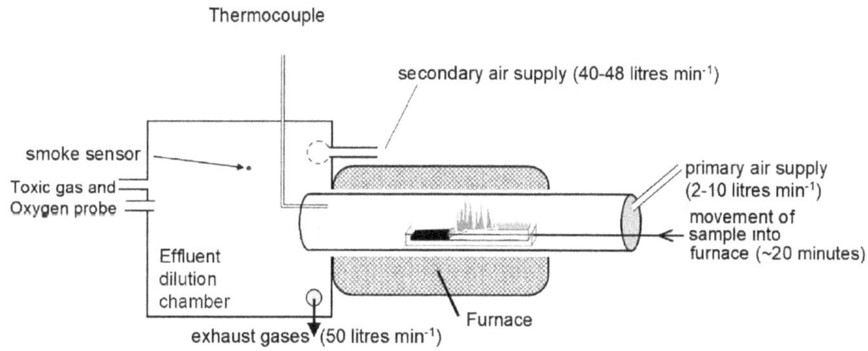

FIGURE 19.8 The steady state tube furnace apparatus, ISO/TS 19700.

19.4.1 The Steady State Tube Furnace (SSTF)

The steady state tube furnace (ISO/TS 19700), shown in Figure 19.8, feeds a linearly uniform 800 mm sample into its hot zone at a fixed rate, under a controlled air supply, inside a horizontal silica tube of diameter 48 mm, allowing adequate mixing of fuel and oxidant. It forces combustion by driving the sample into a furnace, where the heat flux is greatest in the centre, at a fixed rate, so that, by running several tests with the same material with different furnace temperatures and ventilation conditions, each fire stage can be replicated by steady state burning. The products generated in the flame zone then pass through the heated furnace tube, maintaining a high temperature, as in the upper layer of a compartment fire. The toxic product yields may be quantified from the gas concentrations and mass feed rate during the steady state burn period, after the initial peak. It has been designed to generate data for input to fire hazard assessments, using the methodology in ISO 13571[24] and ISO 13344[25], particularly in relation to the fire stages in ISO 19706.[15]

ISO/TS 19700 replicates the full range of fire conditions, characterizing the fire behavior of materials under controlled and well-defined laboratory conditions, in terms of the equivalence ratio (ϕ) or more roughly the CO_2/CO ratio in the fire effluent.[79] It was shown that for many common materials, the yield of toxic products such as CO, HCN, organo irritants, and smoke increases by a factor between 10 and 50 as the fire changes from well-ventilated ($0.5 < \phi < 0.7$) to under-ventilated ($1 < \phi < 5$). Figure 19.9 A shows how the CO yield (itself a useful marker of incomplete combustion) of most simple polymers (LDPE, PP, PMMA, EVA, and PA 6) rises rapidly as ϕ increases from 0.5 to 1.5 in the SSTF. In contrast, the CO yield from unplasticized PVC is consistently high, resulting from the flame quenching of HCl, and also does not follow the same trend for polystyrene (PS), as a result of the enhanced stability of the aromatic ring. Figure 19.9 B shows how the CO yield in the SSTF and FPA replicate the large-scale test data (ISO room and 1/3 scale ISO room), where the CACC does not, even when re-calculated using the effective value of ϕ (ϕ_{eff}).[82, 83]

From Figure 19.9 (B), the SSTF and the FPA show acceptable agreement with large-scale data over the range of fire conditions. The CACC cannot replicate the high yields of toxic gases (CO reported here, and HCN reported elsewhere[81]), which occur in under-ventilated fires. The SSTF and FPA are capable of forcing controlled combustion in under-ventilated conditions, allowing the relationship between equivalence ratio and the toxic product yields to be investigated in order to assess the toxic fire hazard of materials.[79, 81]

19.4.2 Use of the Smoke Density Chamber for Assessment of Fire Toxicity

To their credit, the mass transport industry (air, rail, and sea) recognized that their passengers could not escape and therefore regulated the components of their craft for smoke toxicity. Unfortunately, variations of the SDC were adopted as regulatory assessment tool for fire toxicity. In the SDC most of the toxic gas production should occur towards the end of the test, when the oxygen is most depleted, provided flaming combustion is maintained. However, it is this stage where it is most difficult to maintain flaming combustion. While the SDC is well-suited to the measurement of visual obscuration, during the early stages of flaming, where the smoke can be quantified non-destructively and is independent of the smoke layer height and distribution, it is unsuitable for toxic gas generation and measurement. The ventilation condition, or equivalence ratio, corresponding to the time when the products are generated is unknown, because an unknown fraction of the available oxygen will be accessible to the fire plume. Indeed, the circulation within the chamber will depend primarily on the heat release rate. When sampling a stratified fire effluent from a fixed point, such as a tube 20 cm below the chamber roof, this could be too low, so missing the main effluent plume altogether, or sample from its center, without dilution. The act of sampling, and cleanup prior to gas analysis, of the fire effluent will change its composition—if the fire gases are sampled in real time, it may be possible to isolate individual fire stages, but then the effluent may change during analysis or may not be returned to the chamber

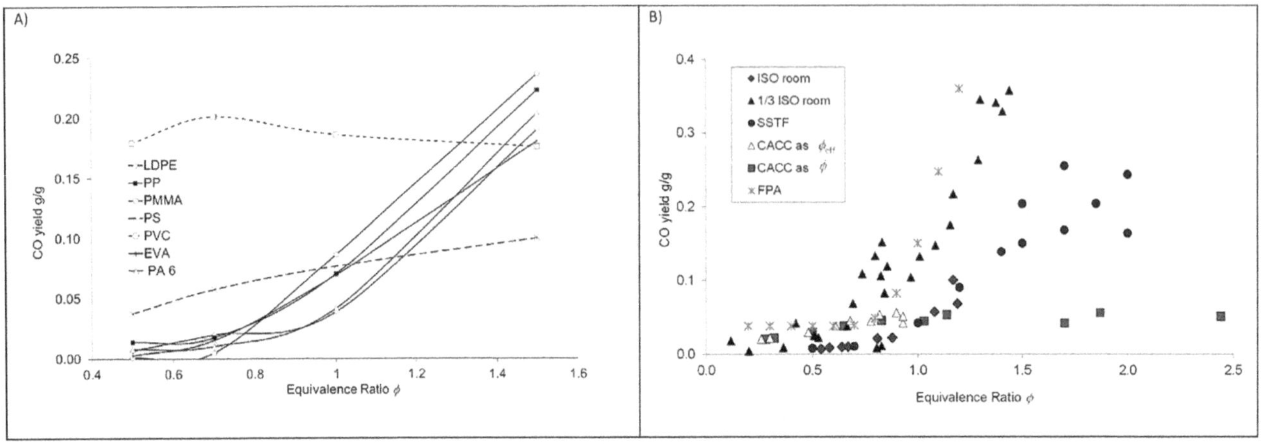

FIGURE 19.9 (A) Carbon monoxide yields for various polymers and products as a function of equivalence ratio and (B) comparison of CO yield for polyamide 6.6 from steady state tube furnace (SSTF), controlled atmosphere cone calorimeter (CACC), fire propagation apparatus (FPA), with ISO Room, and ⅓ ISO Room as a function of equivalence ratio ϕ.[81]

after analysis. If the analyte is returned to the chamber, it will usually have been filtered removing both the particulates and the gases absorbed onto them, to protect the analysers, while acid gases may have deposited onto the sampling line. Similarly, the high surface area of the chamber wall will promote deposition and disproportionately larger losses of certain analytes, such as hydrogen chloride (HCl) or hydrogen bromide (HBr). However, of greatest significance in quantification of the toxicity of fire effluents is that the most important under-ventilated conditions cannot be replicated.[81]

19.5 EFFECT OF FIRE RETARDANTS ON SMOKE TOXICITY

ISO defines a *fire retardant* as substance added, or a treatment applied, to a material in order to delay ignition or to reduce the rate of combustion.[84] They define a *flame retardant* as a substance added, or a treatment applied, to a material in order to suppress or delay the appearance of a flame and/or reduce the flame spread rate.[84] Thus *fire retardants* can act anywhere to reduce flammability, by releasing water, absorbing heat, forming a protective char or barrier layer, but also include the subset *flame retardants* which act by interfering with reactions in the flame. In the context of smoke toxicity, this distinction is important: by interfering with the flame chemistry, flame retardants produce more smoke and products of incomplete combustion, where many other types of fire retardant keep the fuel out of the flame zone.

19.5.1 Gas Phase Action

The widespread use of synthetic polymers in Europe and the US led to an increase in fires and fire fatalities, as shown in Figure 19.2. In the 1970s halogenated *flame* retardants started to be used to make products less flammable. While halogen-based fire retardants are very effective in reducing fire risk, that is, the probability of occurrence of a fire, they show a high fire hazard, that is, the probability of producing toxic, corrosive, obscuring smokes or when involved in a developed fire, and the fire is too big to be extinguished. Gas phase flame retardants, such as those based on organohalogen or organophosphorus compounds interfere with the gas phase free radical reactions, typically producing more CO, smoke, and other products of incomplete combustion.[26] This results in incomplete oxidation of vapour phase fuel molecules, leading to higher yields of all products of incomplete combustion. These are all more toxic than the cleaner products of complete combustion (CO_2 and water) and include CO, HCN, hydrocarbons, oxygenated organics (including organo irritants, such as acrolein and formaldehyde), and larger cyclic molecules such as PAHs and soot particulates. Fire toxicity increases as combustion becomes more incomplete, which can arise from chemical quenching (for example by gas phase *flame* retardants), insufficient heat (for example during smoldering), or when the fire becomes under-ventilated, and there is insufficient oxygen for complete combustion.[11, 79]

Thus in a fire, gas phase *flame* retardants will act to increase the gas phase products of incomplete combustion, and hence the yields of toxic gases and smoke, when compared to either the non-flame retarded polymer, or the same polymer incorporating a condensed phase *fire* retardant, which increases the char yield. As most fire deaths and most fire injuries result from toxic gas inhalation, the use of gas phase flame retardants is a compromise, between supressing ignition and increasing the fire toxicity (or decreasing the fire risk at the expense of increasing the fire hazard). In one case, it has been shown that the phosphorus flame retardants which act in the gas phase have a smaller influence on CO and HCN yields than the corresponding brominated flame retardants: that was in industry standard formulations of PA 6.6 meeting UL 94 V-0 at 0.8 mm.[3]

19.5.2 Condensed Phase Action

In contrast, modern *fire* retardants reduce fuel release to the gas phase, often by formation of a protective barrier layer, which acts as a radiation shield, and inhibits the flow of fuel and oxygen.[85] Barriers can be formed or reinforced with inorganic nanoparticles, such as montmorillonite.[86] The barrier is often carbonaceous, and increasing the char yield is synonymous with reducing the flammability. Moreover, on reducing the volatile evolution, the amount of toxic effluent is reduced, and for a given ventilation rate, the ventilation condition becomes more well-ventilated, also reducing the toxicity. Char formation has been enhanced in polyethylene, using functionalized graphene oxide[87]; in PMMA using graphene oxide and a nickel-aluminum layered double hydroxide.[88] Enhancement of the barriers can be achieved in intumescent systems, where gas is released within the molten polymer, causing significant swelling, and so increasing the effectiveness of the thermal barrier,[89] for example in polyethylene,[90] polypropylene,[91] and polystyrene.[92] Carbon nanotubes (CNT) have also been used as potential fire retardant additives, given their similarity to carbonaceous char, with mixed success.[93] That study, in the SSTF, also found that polypropylene samples with intumescent fire retardant and/or CNTs, or both, produced much more CO compared to the pure polymer during all fire stages, and hydrocarbon yield from the pure polymer increased more with the equivalence ratio, indicating that CNTs promote the formation of smoke particulates from hydrocarbons. Questions have also been raised about the long-term safety of CNTs, and whether airborne nanotubes (of similar dimensions to asbestos particles) could be equally hazardous.[94]

19.5.3 Assessment of Smoke Toxicity of Product Classes

19.5.3.1 Prediction of Lethal Fire Toxicity of Thermal Insulation Materials

Six commercial insulation products for use by the construction industry were tested for smoke toxicity in the SSTF at four fire conditions, non-flaming, well-ventilated, small slightly under-ventilated, and small very under-ventilated flaming.[95] Figure 19.10 shows the contributions to toxicity, expressed as a fractional effective dose (FED) of the individual toxicants for each fire condition (presented at a loading of 20 g m^{-3} or 1 kg of material in 50 m^3). The last two products, glass wool (GW) and stone wool (SW), are predominantly noncombustible, with around 5% of an organic binder. It was not possible to achieve flaming of those two products, so a non-flaming test was conducted at 825°C. The results for PIR and PUR show significantly higher toxicity, mostly from the HCN, but also from CO, particularly in under-ventilated conditions. The relatively high yields of CO and HCN in well-ventilated conditions result from the presence of a gas phase flame retardant, believed to be a tris(chloroalkyl)phosphate. The data show that different insulation materials have very different fire toxicities.

Table 19.6 shows the Material-LC_{50} values for the different insulation products shown in Figure 19.10. The Material-LC_{50} is the mass of material in grams that would create a lethal toxic effluent when dispersed into 1 m^3 after burning under the stated conditions. It provides a simple means to communicate to designers when the use of combustible products presents a toxic hazard. For example,

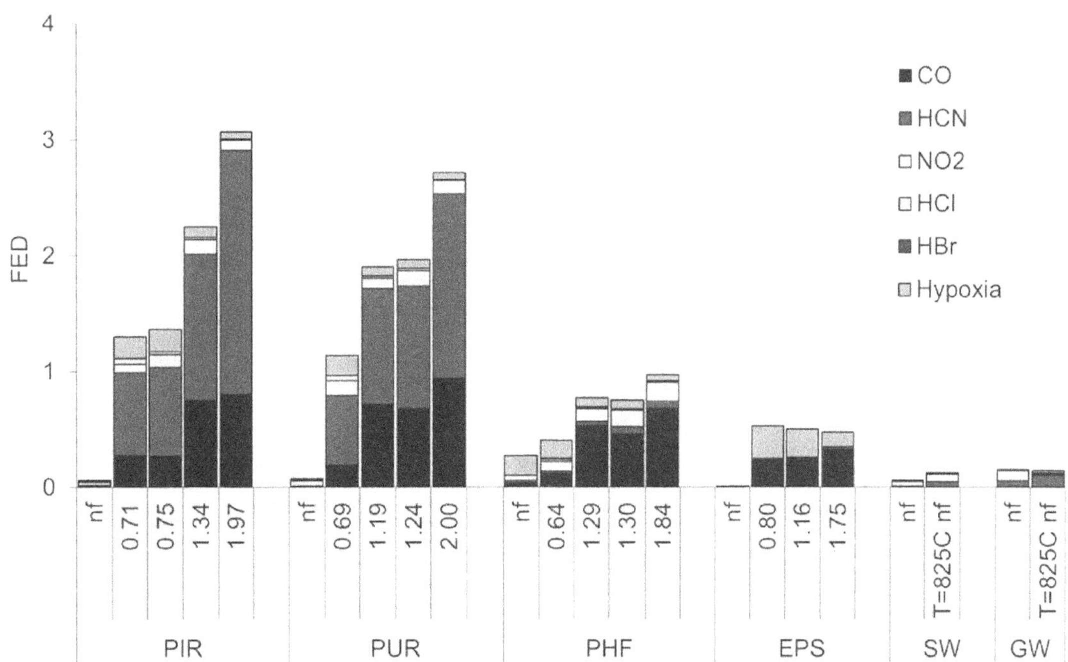

FIGURE 19.10 Smoke toxicity, expressed as a fractional effective dose (FED) (Equation 3) for construction insulation products, as a function of fire condition. Key: polyisocyanurate foam (PIR), polyurethane foam (PUR), expanded polystyrene foam (EPS), phenolic foam (PhF), stone wool (SW), glass wool (GW).

TABLE 19.6
Material-LC_{50} for Different Products

Material	Fire Conditions	Equivalence Ratio ϕ	Material-LC_{50} g/m³
GW	Smoldering	—	163.6
	T = 825°C	—	129.5
	No flaming		
SW	Smoldering	—	388.1
	T = 850°C	—	172.1
	No flaming		
	Smoldering	—	186.3
PHF	Well-ventilated	0.64	43.3
	Under-ventilated	1.29	22.3
	Under-ventilated	1.84	21.0
	Smoldering	—	5648.5
EPS	Well-Ventilated	0.80	28.4
	Under-ventilated	1.16	27.9
	Under-ventilated	1.75	27.6
	Smoldering	—	337.2
PUR	Well-Ventilated	0.69	15.7
	Under-ventilated	1.24	10.3
	Under-ventilated	2.00	11.4
	Smoldering	—	498.4
PIR	Well-Ventilated	0.75	16.5
	Under-ventilated	1.34	10.7
	Under-ventilated	1.97	8.3

this shows that 8 g of PIR or 11 g of PUR foam burning in under-ventilated conditions would make 1 m³ of air toxic, or 1 kg of such foam burning in under-ventilated conditions would provide lethal concentration of toxicants in a 100 m³ room. Engineers would normally divide this value by 10 to incorporate a reasonable safety margin (see Section 2).

19.5.3.2 Prediction of Lethal Fire Toxicity of Electric Cable Insulation Materials

Ten commercial cables were investigated for fire toxicity[96] using the conditions specified in the precursor International Electrotechnical Commission (IEC) standard, IEC 60695–7–50,[97] which uses the steady state tube furnace, but a simpler methodology for ensuring the setting up of the tube furnace ventilation condition to obtain a particular fire stage. Six cables, three for data and three for power distribution with halogen-free formulations were compared to 4 PVC cables. Typical halogen-free cable sheathing formulations use an elastomeric polymer such as ethylene vinyl acetate (EVA) and around 65% mineral filler such as aluminum or magnesium hydroxide, while PVC sheathing formulations are approximately equal masses of PVC, plasticizer and chalk.

For each cable, the FED is shown in Figure 19.11 for each of three fire stages (presented at a loading of 20 g m^{-3} or 1 kg of material in 50 m³). In most cases (8/10) oxidative pyrolysis is the least toxic, presumably due to the low mass loss of the polymers at 350°C.[96] However, this temperature is high enough to release HCl from PVC in two cases. In general, the FED, based on the main toxic gases analyzed, increases from oxidative pyrolysis, to well-ventilated, to developed flaming. In common with most materials, the most hazardous fire condition is confirmed to be the under-ventilated fully developed fire.

The two major toxicants are seen to be CO and HCl. Most clearly, it can be seen that hydrogen chloride increases the yield of carbon monoxide. Therefore, the CO yield from PVC cables is greater than from halogen-free cables. It is also noticeable that the CO yield from the PVC cables increases with under-ventilation, whereas pure PVC (see Figure 19.4), the yield is almost independent of ventilation.

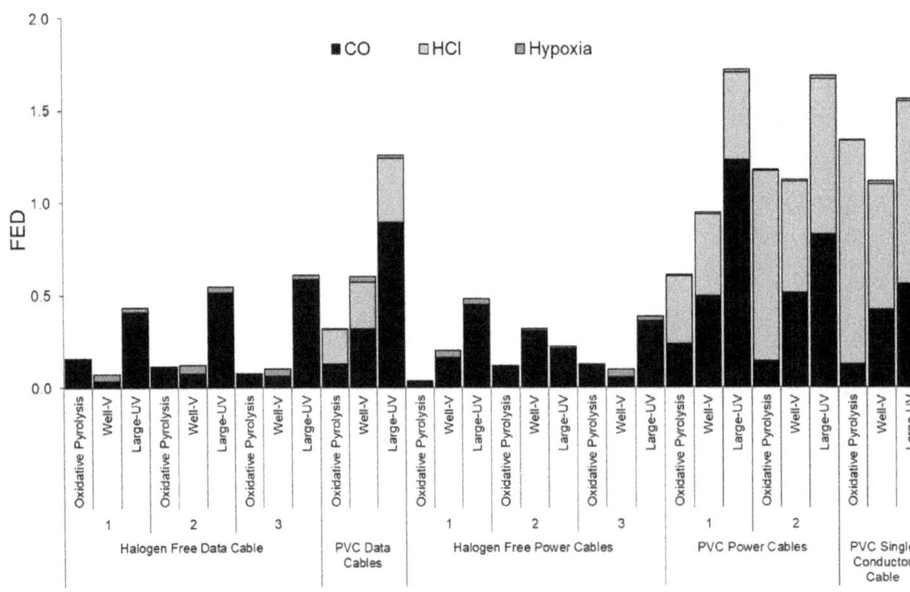

FIGURE 19.11 Contribution of individual gases to toxicity from burning cables.

19.5.3.3 Effect of Different Fire Retardants on Lethal Toxicity of Burning Polyamides

A comparison was made of industry-standard glass-fiber-reinforced formulations of PA 6 and PA 6.6, which had been optimized to meet UL94 V-0 at 0.8 mm. To meet the V-0 requirement, the formulation used either a BrSb combination (brominated polystyrene flame retardant in combination with antimony oxide) or a AlPiM formulation (diethyl aluminium phosphonate/melamine polyphosphate combination).

Figure 19.12 shows how the individual yields may be translated into estimates of toxicity[3] expressed as fractional effective dose (FED) (presented at a loading of 20 g m^{-3} or 1 kg of material in 50 m^3). At this loading, many of the material/conditions show FEDs well below the critical value of 1 (representing lethality to 50% of the population during a 30 minute exposure). The two exceptions are the materials formulated with AlPiM, when burning at the lower furnace temperature of 650°C, and all the materials formulated with BrSb at both furnace temperatures. In each case, where the FEDs are large, the major contribution to the toxicity comes from HCN, with a smaller contribution from CO. There is also a minor contribution from NO_2 from all materials, in all fire conditions. It is notable that the fire toxicity of the PA 6/BrSb materials is a factor of 10 larger than that of PA 6 and a factor of 5 larger than the AlPiM material at 650°C; at 825°C it is a factor of 30 larger than the PA 6, and 17 larger than the AlPiM material. The BrSb materials all show high fire toxicity under the normally least toxic, well-ventilated fire conditions. The lower fire toxicity of the AlPiM materials at higher temperatures suggest a reduction in fire hazard. It is well-known that under more severe conditions than a small flame ignition test, flame-retarded materials will burn and frequently produce more toxic effluents as a result of quenching of the gas phase flame reactions. It appears that the AlPiM system switches off this flame quenching process as the fire becomes more severe, reducing the toxicity of the effluent. These surprising findings led to the detailed investigation of the chemistry of the processes, which are discussed in more detail in reference 3.

Thus for the base polymers with glass fiber reinforcement, and for the four comparably flame retarded samples, an LC_{50} may be calculated (Table 19.7).

The LC_{50} is inversely proportional to the toxicity (higher values indicate lower toxicity). If the mass of material equal to the LC_{50} is burnt in well-ventilated conditions, this would be fatal to 50% of the occupying population. Therefore, if a 1 kg fuse box made of glass-reinforced polyamide 6, flame retarded to UL94 V-0 with brominated polystyrene and antimony oxide, was to burn completely in conditions corresponding to a small, well-ventilated fire

TABLE 19.7
LC_{50} for each material and furnace temperature (the mass required to generate a lethal effluent in grams per cubic meter).

	Furnace Temperature	
	650°C	825°C
Sample	Material-LC_{50}/g m^{-3}	Material-LC_{50}/g m^{-3}
PA 6	44	83
PA 6/AlPiM	15	40
PA 6/BrSb	3.3	2.4
PA 6.6	80	84
PA 6.6/AlPiM	16	66
PA 6.6/BrSb	4.3	4.1

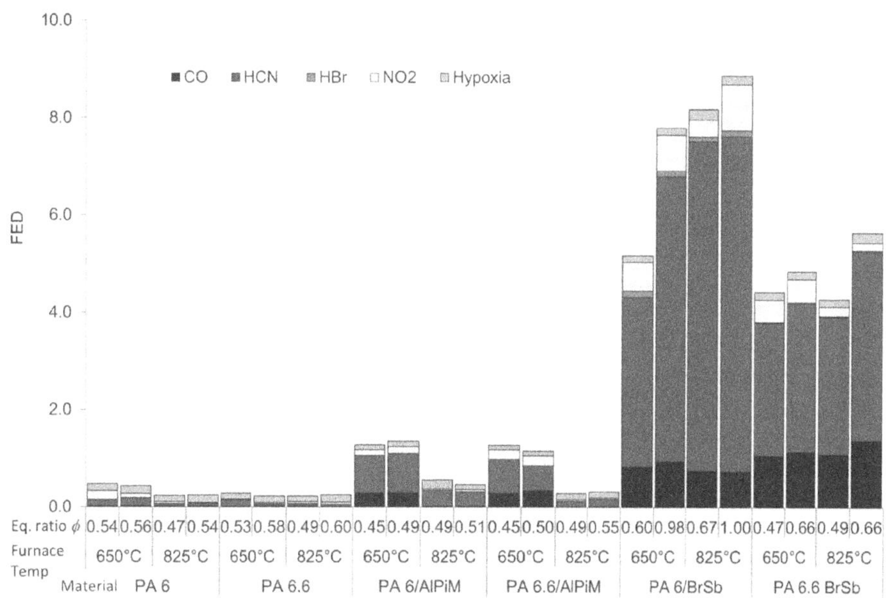

FIGURE 19.12 Contributions to effluent toxicity from fire retarded PA 6 and PA 6.6 based materials.

(650°C), it would generate a volume of effluent of 300 m^3 that would be lethal for 30 minute exposure, whereas if it were flame retarded with AlPiM, it would only generate a volume of lethal effluent of 67 m^3 under the same conditions.

19.5.3.4 Prediction of Incapacitation from Fire-Retarded Timber

The examples in sections 19.5.1 to 3 of this chapter have shown how lethality may be predicted from laboratory measurements of smoke toxicity. However, designers cannot rely on the presence of emergency responders at the scene, and therefore incapacitation, or inability to effect one's own escape, is probably a more meaningful end point to work from. An investigation of two commercial treatments for fire protection of timber, vacuum-pressure impregnation (FRP), and surface coating with an intumescent paint (FRC) have been compared.[98] All samples were made from the same batch of pine plywood (P). Assessment of smoke toxicity from the asphyxiants yields of CO and HCN were used to predict incapacitation as a fractional effective dose (FED) according to Equations 1 and 2. Prediction of incapacitation after 5 minute exposure to fire effluent, generated from 1 kg timber and distributed in a volume of 50 m^3 for all fire stages is shown in Figure 19.13. The FRP gave a much higher CO yield (784 mg g^{-1}) and a high HCN yield (7.5 mg g^{-1}) for fire stage 3a/3b.

Figure 19.13 shows that *FRP* burning with under-ventilated flaming (fire stages 3a and 3b) creates a far greater volume of incapacitating effluent than the other sample/fire condition combinations; fractional effective doses for samples during well-ventilated flaming (fire stage 2) were all below 0.01.

19.5.3.5 Prediction of Incapacitation from Li-Ion Batteries with a LiFePO4 (LFP) Cathode

Although not strictly related to polymers, the fire hazards of lithium ion batteries have been shown to be significant

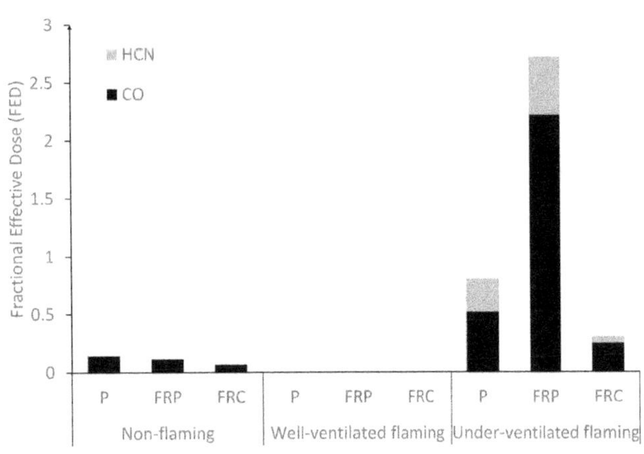

FIGURE 19.13 Fractional effective dose (FED) for incapacitation according to ISO 13571 by 5 min exposure to fire effluent generated by combustion of 1 kg timber (*Pine* (=P), *FRP* or *FRC*) distributed in a volume of 50 m^3.

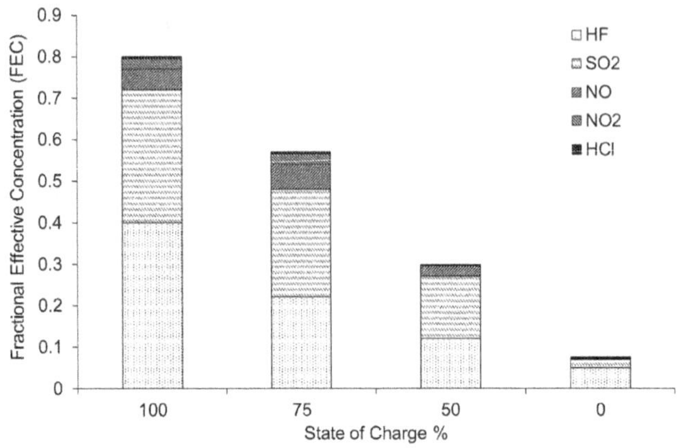

FIGURE 19.14 The peak FEC values contributed from each irritant gas of LIBs with different SOCs.

in their own right, as well as potential ignition sources for other fuels. Using an example reported in the literature, Figure 19.14 illustrates the peak FEC contributions from each irritant gas to show the relative importance of the individual toxicants[99] at different states of charge (SOC). Although the whole battery is combustible, with around 10 times greater heat release than the energy of the fully charged battery,[100] the high state of charge acts as a ready promoter of ignition. The sum of contributions from hydrogen fluoride (HF) and sulfur dioxide (SO_2) accounts for most of the irritant gases released at each SOC. It is notable that in other studies of lithium-ion battery fires which used the nickel manganese cobalt oxide (NMC) cathode, the emissions of HF were around a tenth of those from the $LiFePO_4$ (LFP) cathode cells.[101]

19.5.4 Fire Retardancy and Performance in Large-Scale Fires

The fire toxicity of large-scale fire tests has been investigated and demonstrated good correlations with the results from the SSTF[81, 102]. Other large scale fires have been investigated for their toxicity, showing life-threatening conditions in an entire house when a single sofa was burning[103] and identifying the presence of significant quantities of tris(chloropropyl) phosphate flame retardants in the residues of another fire involving a single sofa in a domestic property.[58]

19.6 CONCLUSIONS

Fire effluent toxicity is the largest killer and greatest cause of injury in fires. Assessment of fire gas toxicity is an essential component of fire hazard analysis. The hazard generally changes with fire scenario, and the toxic product yields are lower in well-ventilated flaming fires (at least with non-halogen-containing materials). CO is generally considered to be the most toxicologically significant component in fire gases, at least in part because of the ease of measurement of CO compared to other toxicants (particularly HCN) in the bodies of fire victims. However, it has been shown that CO is not the only toxicologically significant component of fire effluents. Laboratory investigations show the much greater toxicological significance of HCl from PVC, and HCN from PA 6.6 particularly in under-ventilated conditions, while epidemiological studies point to the importance of carcinogens (e.g., PAH and particulates). Change in ventilation conditions changes the relative toxicities by up to two orders of magnitude. For fire hazard assessments, fire toxicity must be determined as a function of both the material and fire condition, especially combustibility, presence of flame, oxygen availability in the fire zone, and the presence of heteroelements (P, N, Cl, and Br). The steady state tube furnace (SSTF) and the fire propagation apparatus (FPA) show acceptable agreement with large scale data over the range of fire conditions. The controlled atmosphere cone calorimeter (CACC) and the smoke density chamber (SDC) fail to replicate the high yields of asphyxiant gases, CO and HCN, which occur in under-ventilated fires. This shows that they are unsuitable for assessment of toxic hazard of fire effluents. The advantage of the SSTF over other techniques is its capability to replicate the whole range of fire conditions. The disadvantage is that it burns all materials at the same rate, irrespective of their flammability. To address this shortcoming a simplified methodology has been proposed[104] for combining the extent of burning (quantified using the results of the European SBI test, EN13823, required for all construction products, so all the data is publicly available) with toxic product yields (measured using the steady state tube furnace) allowing a quick assessment of the mass of material which would produce a lethal atmosphere in a given volume. Although the approach is simplistic, it provides fire safety engineers with a tool for estimating fire toxicity in a building.

The adoption of variations of the SDC to quantify fire toxicity in passenger aircraft, international shipping,[105] and across European rail network[106] is unfortunate. The rejection by ISO of the SDC as a toxicity test (DIS 21489) due to its irreproducibility ignores the more serious issue. The apparent inability of the smoke density chamber to replicate under-ventilated flaming renders it unsuitable for the assessment of fire effluent toxicity.

19.6.1 Future Prospects

19.6.1.1 Assessment of Acute Toxicity

Fire effluents contain such rich cocktails of asphyxiants, irritant gases and vapors, and particulates that full characterization could not be undertaken. Acute toxicity in well-ventilated conditions is assessed for materials used in mass transport applications; for performance-based engineering (PBE) assessments, ISO stipulate that the concentrations of eight acute toxic gases (CO, CO_2, NO, NO_2, SO_2, HCN, HBr, and HCl) should be quantified and related to rat lethality data or estimates of human incapacitation. In the ISO assessment of fire toxicity, no account need be taken of organic irritants ("unless their presence is known or suspected"), while particulate matter or chronic toxicants or carcinogens are not quantified at all. With the development of highly sensitive, lower cost instrumentation, a more complete assessment of fire effluent toxicity can be achieved, providing a complete set of data for a fire safety assessment. Building occupants who are trapped by a fire should not then be exposed to carcinogens during their escape, or while they await rescue.

The increased use of lightweight insulation materials will help to meet targets for carbon emissions but could create a time-bomb in terms of fire safety and particularly fire toxicity. BT Three recent examples stand out:

- The Grenfell Tower fire in 2017 where 72 people died as a result of fire rapidly spreading over polyethylene filled aluminum composite material and igniting the underlying PIR and phenolic foam insulation which lay behind it.[107, 108] The combination of the rapid burning polyethylene and the slower-burning PIR and phenolic foams led to such large numbers of deaths. In particular, the PIR releases of significant quantities of hydrogen cyanide when burning, sufficient to incapacitate victims, who then lost consciousness and were unable to attempt escape.
- The Colectiv Nightclub fire in Bucharest in October 2015, where most of the 64 people died as a result of inhaling HCN from burning polyurethane foam.[109]
- The fire at the Cuba Libre Club in Rouen where 13 people died from smoke inhalation from burning EPS in Aug 2016.[110]

19.6.1.2 Atmospheric and Particulate Toxicity

Almost all air pollution continues to be caused by combustion processes. In most cases combustion efficiency is key to limiting pollutant emissions: although vastly more fuel is burned in controlled conditions, unwanted fires contribute 10% of overall PAH and particulate emissions, because of their much lower combustion efficiency.[111] The increasing prevalence and severity of wildland and wildland urban interface WUI fires has caused much more attention to be focused on predicting and understanding smoke toxicity.[112, 113] In the wildland fires that surrounded Moscow in the 2010 heatwave, the average PM_{10} levels exceeded 300 μg/m^3 on several days, resulting in an 11,000 excess of deaths from cardiovascular, respiratory, genitourinary, and nervous system diseases, with the combination of high temperatures and air pollution contributing to a further 2,000 deaths.[114] In each case robust analysis of combustion products is the key to assessing their hazards. Combustion-derived particulates have been attributed to 3 million deaths per year across the globe, mostly from open cooking and heating within homes in India and China.[115] Since the particulates are composed of agglomerated PAH and hydroxy-PAH, with other toxicants adsorbed onto them, and possibly carrying EPFRs, the major challenge is to characterize the detailed chemistry and composition of these pyrogenic particles.

Across Europe, as demands for meaningful end-of-life processing increase, large waste storage sites are increasingly igniting and burning for long periods. In the UK, 300 of the 400 large fires each year occur at waste facilities, many burning for several months, blowing smoke of unknown toxicity into people's homes. Rapidly deployable techniques are only just being developed for monitoring such incidents, but they are clearly major sources of toxic emissions. Large waste fires have already included tires, furniture, plastic and recyclate, cables, white goods (fridges and washing machines etc.), and carpets.

Wildland fires are a growing European problem, and a significant number of deaths have been recorded as a result of the fire toxicity of wildland fires.

REFERENCES

1. Fire Statistics GOV.UK www.gov.uk/government/collections/ fire-statistics Scotland www.firescotland.gov.uk/about-us/fire-and-rescue-statistics.aspx and Wales https://gov.wales/fire-and-rescue-incident-statistics and preceding editions.
2. Stec, A.A., Hull, T.R. (2010) Fire Toxicity, Woodhead and Elsevier, Cambridge. https://doi.org/10.1533/9781845698072.1
3. Molyneux, S.; Hull, T. R.; Stec, A. A. The Effect of Gas Phase Flame Retardants on Fire Effluent Toxicity. *Polym. Degrad. Stab.* **2014**, *106*, 36–46. https://doi. org/10.1016/j.polymdegradstab.2013.09.013
4. Stull, J.O. (2008) *U.S. Fire Administration Firefighter Autopsy Protocol*, Federal Emergency Management Agency, Washington DC. https://www.usfa.fema.gov/downloads/pdf/publications/firefighter_autopsy_protocol.pdf
5. Reporter's Guide: The Consequences of Fire, NFPA, United States. www.nfpa.org/News-and-Research/Publications-and-media/Press-Room/Reporters-Guide-to-Fire-and-NFPA/Consequences-of-fire (accessed 11/01/2023)
6. Occupational Exposure as a Firefighter, *IARC Monographs on the Identification of Carcinogenic Hazards to Humans*, Volume 132, WHO, Geneva. 2023. https://publications.iarc.fr/615
7. Air Quality in Europe—2022 Report, European Environmental Agency, 2022 www.eea.europa.eu/publications/air-quality-in-europe-2022
8. Emissions of Air Pollutants in the UK—Particulate Matter (PM10 and PM2.5) National Statistics. GOV.UK. Updated 22 February 2023. www.gov.uk/government/statistics/emissions-of-air-pollutants/emissions-of-air-pollutants-in-the-uk-particulate-matter-pm10-and-pm25
9. Persson, P.; Simonson, M. Fire Emissions Into the Atmosphere. *Fire Technol.* **1998**, *34*, 266–279. https://doi.org/ 10.1023/A:1015350024118
10. Hull, T. R.; Quinn, R. E.; Areri, I. G.; Purser, D. A. Combustion Toxicity of Fire Retarded EVA. *Polym. Degrad. Stab.* **2002**, *77*, 235–242. https://doi.org/10.1016/s0141-3910(02)00039-3
11. Hull, T. R.; Stec, A. A.; Lebek, K.; Price, D. Factors Affecting the Combustion toxicity of Polymeric Materials. *Polym. Degrad. Stab.* **2007**, *92*, 2239–2246. https:// doi.org/10.1016/j.polymdegradstab.2007.03.032
12. Paul, K. T.; Hull, T. R.; Lebek, K.; Stec, A. A. Fire Smoke Toxicity: The Effect of Nitrogen Oxides. *Fire Saf. J.* **2008**, *43*, 243–251. https://doi.org/10.1016/j.firesaf.2007.10.003
13. Rhodes, J.; Smith, C.; Stec, A. A. Characterisation of Soot Particulates from Fire Retarded and Nanocomposite Materials, and Their Toxicological Impact. *Polym. Degrad. Stab.* **2010**, *96*(3), 277–284. https://doi.org/10.1016/j.polymdegradstab.2010.07.002
14. Stec, A. A.; Rhodes, J. Smoke and Hydrocarbon Yields from Fire Retarded Polymer Nanocomposites. *Polym. Degrad. Stab.* **2010**, *96*(3), 295–300. https:// doi.org/10.1016/j.polymdegradstab.2010.03.032
15. ISO 19706:2011, Guidelines for assessing the fire threat to people. ISO, Geneva. www.iso.org/standard/56864.html
16. Stec, A. A. Fire Toxicity—The Elephant in the Room? *Fire Saf. J.* **2017**, *91*, 79–90. https://doi.org/10.1016/j.firesaf.2017.05.003
17. Stec, A. A.; Fadell, P.; Blomqvist, P.; Bustamante-Valencia, L.; Sargoza, L.; Guillaume, E. Quantification of Fire Gases by FTIR: Experimental Characterisation of Calibration Systems. *Fire Saf. J.* **2011**, *46*, 225–233. https://doi.org/10.1016/j.firesaf.2011.02.004
18. Hull, T. R.; Stec, A. A. Chapter 5: Generation, Sampling and Quantification of Toxic Combustion Products. *Issues Toxicol.* **2016**, *23*, 108–138. Royal Society of Chemistry, https://doi.org/10.1039/9781849737487-00108
19. Lebek, K., Hull, R., Price, D. (2005) Products of Burning Rigid PVC Burning Under Different Fire Conditions. In *Fire and Polymers: Materials and Concepts for Hazard Prevention*, ACS Symposium Series No. 922, pp 334–347, Oxford University Press.
20. Pitts, W. M. Global Equivalence Ratio Concept and the Prediction of Carbon Monoxide Formation in Enclosure Fires. *Prog. Energ. Combust.* **1995**, *21*, 197–237. https://doi.org/10.1016/0360-1285(95)00004-2
21. Hull, T. R.; Carman, J. M.; Purser, D. A. Prediction of CO Evolution from Small-Scale Polymer Fires. *Polym Int.* **2000**, *49*, 1259. https://doi.org/10.1002/1097- 0126(200010)49:10 < 1259::aid-pi573 > 3.3.co;2-4
22. Purser, D. A. Toxic Product Yields and Hazard Assessment for Fully Enclosed Design Fires. *Polym Int.* **2000**, *49*, 1232–1255. https://doi.org/10.1002/1097-0126(200010)49:10%3C1232::aid-pi543%3E3.3.co;2-k
23. Molyneux, S. A.; Stec, A. A.; Hull, T. R. The Correlation Between Carbon Monoxide and Hydrogen Cyanide in Fire Effluents of Flame Retarded Polymers. *Fire Saf. Sci.* **2014**, *11*, 389–403. https://doi.org/10.3801/IAFSS. FSS.11-389
24. ISO 13571:2012, Life-threatening components of fire—Guidelines for the estimation of time to compromised tenability in fires, ISO, Geneva. www.iso.org/standard/56172.html
25. ISO 13344:2015, Estimation of the lethal toxic potency of fire effluents, ISO, Geneva. www.iso.org/standard/68029.html
26. Kaczorek, K.; Stec, A. A.; Hull, T. R. Carbon Monoxide Generation in Fires: Effect of Temperature on Halogenated and Aromatic Fuels. *Fire Saf. Sci.* **2011**, *10*, 253–263. https://doi.org/10.3801/IAFSS.FSS.10-253
27. Blomqvist P. Emissions from Fires. Ph. D thesis, Lund University; 2005.
28. Pilling, M.J., Seakins, P.W. (1996) *Reaction Kinetics*, 2nd Edition, Oxford Science Publications, Oxford University Press, Oxford.

29. Purser, D. A.; Wolley, W. D. Biological Studies of Combustion Atmospheres. *J. Fire Sci.* **1983**, *1*, 118–144. https://doi.org/10.1177/073490418300100204
30. Wolley, W. D.; Fardell, P. J. Basic aspects of combustion toxicology. *Fire Saf. J.* **1982**, *5*, 29–48. https://doi.org/10.1016/0379-7112(82)90005-4
31. Purser, D. A.; Buckley, P. Lung Irritation and Inflammation During and After Exposure to Thermal Decomposition Products from Polymeric Materials. *Med. Sci. L.* **1983**, *23*, 142–150.
32. D.A. Purser, unpublished data, 1984.
33. Documentation for Immediately Dangerous to Life or Health Concentrations (IDLH): NIOSH Chemical Listing and Documentation of Revised IDLH Values (as of 3/1/95), NTIS Publication No. PB-94–195047, U.S. Department of Health and Human Services, Public Health Service, Centers for Disease Control, National Institute for Occupational Safety and Health, Cincinnati, Ohio. https://doi.org/10.1016/b0-12-369400-0/01054-1
34. Purser, D.A. (2016) Assessment of Hazards to Occupants From Smoke, Toxic Gases, and Heat. In *SFPE Handbook of Fire Protection Engineering*, 5th Edition, pp. 2308–2428. https://doi.org/10.1007/978-1-4939-2565-0_63
35. Richter, H.; Howard, J. B. Formation of Polycyclic Aromatic Hydrocarbons and Their Growth to Soot-a Review of Chemical Reaction Pathways. *Prog. Energ Combust. Sci.* **2000**, *26*(4), 565–608.
36. World Health Organization (2016) Burden of disease from ambient and household air pollution. www.who.int/phe/health_topics/outdoorair/databases/en/
37. Blomqvist, P.; Simonson McNamee, M.; Andersson, P.; Lönnermark, A. Polycyclic Aromatic Hydrocarbons (PAHs) Quantified in Large-Scale Fire Experiments. *Fire Technol.* **2011**, *48*, 513–528. https://doi.org/10.1007/s10694-011-0242-9
38. Stec, A. A.; Readman, J. E.; Blomqvist, P.; Gylestam, D.; Karlsson, D.; Wojtalewicz, D.; Dlugogorski, B. Z. Analysis of Toxic Effluents Released from PVC Carpet Under Different Fire Conditions. *Chemosphere*, **2013**, *90*, 65–71. https://doi.org/10.1016/j.chemosphere.2012.07.037
39. Simonson-McNamee, M., Tullin, C., Stripple, H. Fire-LCA Study of TV Sets with V0 and HB Enclosure Material. *Chemosphere*, **2002**, *46*, 737–744. https://doi.org/10.1016/s0045-6535(01)00238-7
40. Kim, K. H.; Jahan, S. A.; Kabir, E.; Brown, R. J. A Review of Airborne Polycyclic Aromatic Hydrocarbons (PAHs) and Their Human Health Effects. *Environ Int.* **2013**, *60*, 71–80. https://doi.org/10.1016/j.envint.2013.07.019
41. Van Rooij, J. G.; Bodelier-Bade, M. M.; Jongeneelen, F. J. Estimation of Individual Dermal and Respiratory Uptake of Polycyclic Aromatic Hydrocarbons in 12 Coke Oven Workers. *Br. J. Sports Med.* **1993**, *50*(7), 623–632.
42. Achten, C.; Andersson, J. T. Time to Say Goodbye to the 16 EPA PAHs? Towards an Up-to-Date Use of PACs for Environmental Purposes. *Polycycl. Aromat. Compd.* **2015**, *35* (2–4), 330–354. https://doi.org/10.1080/10406638.2014.99 1042.
43. Samburova, V.; Zielinska, B.; Khlystov, A. Do 16 Polycyclic Aromatic Hydrocarbons Represent PAH Air Toxicity? *Toxics.* **2017**, *5*(3), art. no. 17. https://doi.org/10.3390/toxics5030017
44. Daniels, R. D.; Kubale, T. L.; Yiin, J. H.; Dahm, M. M.; Hales, T. R.; Baris, D.; Zahm, S. H.; Beaumont, J. J.; Waters, K. M.; Pinkerton, L. E. Mortality and Cancer Incidence in a Pooled Cohort of US Firefighters from San Francisco, Chicago and Philadelphia (1950–2009). *Occup. Environ. Med.* **2014**, *71*, 388–397 https://doi. org/10.1136/oemed-2013-101662
45. Golden, A. L.; Markowitz, S. B.; Landrigan, P. J. The Risk of Cancer in Firefighters. *Occup Med.* **1995**, *10*(4):803–820.
46. Ide, C. W. Cancer Incidence and Mortality in Serving Whole-Time Scottish Firefighters 1984–2005. *Occup. Med. Lond.* **2014**, *64*(6), 421–427 https://doi. org/10.1093/occmed/kqu080
47. Pukkala, E.; Martinsen, J. I.; Weiderpass, E.; Kjaerheim, K.; Lynge, E.; Tryggvadottir, L.; Sparén, P.; Demers, P. A. Cancer Incidence Among Firefighters: 45 Years of Follow-Up in Five Nordic Countries. *Occup. Environ. Med.* **2014**, *71*(6), 398–404. https://doi.org/10.1136/ oemed-2013-101803
48. Boffetta, P.; Jourenkova, N.; Gustavsson, P. Cancer Risk from Occupational and Environmental Exposure to Polycyclic Aromatic Hydrocarbons. *Cancer Causes Control.* **1997**, *3*, 444–472.
49. International Agency for Research on Cancer (2016) *Agents classified by the IARC monographs, volume 1–115.* http://monographs.iarc.fr/ENG/Classification/latest_classif.php
50. Burnett, R. T.; Pope, C. A.; 3rd.; Ezzati, M.; Olives, C.; Lim, S. S.; Mehta, S.; Shin, H. H.; Singh, G.; Hubbell, B.; Brauer, M.; Anderson, H. R.; Smith, K. R.; Balmes, J. R.; Bruce, N. G.; Kan, H.; Laden, F.; Prüss-Ustün, A.; Turner, M. C.; Gapstur, S. M.; Diver, W. R.; Cohen, A. An Integrated Risk Function for Estimating the Global Burden of Disease Attributable to Ambient Fine Particulate Matter Exposure. *Environ. Health Perspect.* **2014**, *122*, 397–403. https:// doi. org/10.1289/ehp.122-a235
51. Valavanidis, A.; Fiotakis, K.; Vlachogianni, T. Air-borne Particulate Matter and Human Health: Toxicological Assessment and Importance of Size and Composition of Particles for Oxidative Damage and Carcinogenic Mechanisms. *J. Environ. Health.* **2008**, *26*(4), 339–362. https://doi.org/10.1080/10590500802494538
52. Franck, U.; Herbarth, O.; Roder, S.; Schlink, U.; Borte, M.; Diez, U.; Kramer, U.; Lehmann, I. Respiratory Effects of Indoor Particles in Young Children Are Size Dependent. *Sci. Total Environ.* **2011**, *409*(9), 1621–1631. https://doi. org/10.1016/j.scitotenv.2011.01.001
53. Dellinger, B.; Pryor, W. A.; Cueto, R.; Squadrito, G. L.; Hegde, V.; Deutsch, W. A. Role of Free Radicals in the Toxicity of Airborne Fine Particulate Matter. *Chem. Res. Toxicol.* **2001**, *14*(10), 1371–1377. https://doi. org/10.1021/tx010050x
54. Lomnicki, S.; Truong, H.; Vejerano, E.; Dellinger, B. Copper Oxide-Based Model of Persistent Free Radical Formation on Combustion-Derived Particulate Matter. *Environ. Sci. Tech.* **2008**, *42*(13), 4982–4988. https://doi.org/10.1021/es071708h
55. Dugas, T. R.; Lomnicki, S.; Cormier, S. A.; Dellinger, B.; Reams, M. Addressing Emerging Risks: Scientific and Regulatory Challenges Associated with Environmentally Persistent Free Radicals. *Int. J. Environ. Res. Public Health.* **2016,** *13*(6), 573. https://doi.org/10.1021/es071708h
56. Thevenot, P. T.; Saravia, J.; Jin, N.; Giaimo, J. D.; Chustz, R. E.; Mahne, S.; Kelley, M. A.; Hebert, V. Y.; Dellinger, B.; Dugas, T. R.; Demayo, F. J.; Cormier, S. A. Radical-Containing PM0.2 Initiates Epithelial-to-Mesenchymal Transitions in Airway Epithelial Cells. *Am. J. Respir. Cell Mol. Biol.* **2013,** *48*(2), 188–197. https://doi.org/10.1165/rcmb.2012-0052oc
57. Balakrishna, S.; Saravia, J.; Thevenot, P.; Ahlert, T.; Lominiki, S.; Dellinger, B.; Cormier, S. A. Environmentally Persistent

Free Radicals Induce Airway Hyperresponsiveness in Neonatal Rat Lungs. *Part. Fibre Toxicol.* **2011**, *8*, 11. https://doi.org/10.1186/1743-8977-8-11

58. Hewitt, F.; Christou, A.; Dickens, K.; Walker, R.; Stec, A. A. Release of Volatile and Semi-Volatile Toxicants During House Fires. *Chemosphere.* **2017**, *173*, 580–593. https:// doi.org/10.1016/j.chemosphere.2016.12.079
59. Blomqvist, P.; Hertzberg, T.; Dalene, M.; Skarping, G. Isocyanates, Aminoisocyanates and Amines from Fires—A Screening of Common Materials Found in Buildings. *Fire Mater.* **2003**, *27*, 275–294. https://doi. org/10.1002/fam.836
60. Bengtström, L.; Salden, M.; Stec, A. A. The Role of Isocyanates in Fire Toxicity. *Fire Sci. Rev.* **2016**, *5*, 4. https://doi.org/10.1186/s40038-016-0013-2
61. Kimber, I.; Dearman, R. J. Chemical Respiratory Allergy: Role of IgE Antibody and Relevance of Route of Exposure. *Toxicology.* **2002**, *181–182*, 311–315. http://dx.doi. org/10.1016/S0300-483X(02)00299-8
62. Pronk, A. Dermal, Inhalation, and Internal Exposure to 1,6-HDI and Its Oligomers in Car Body Repair Shop Workers and Industrial Spray Painters. *Occup. Environ. Med.* **2006**, *63*(9), 624–631. http://dx.doi. org/10.1136/oem.2005.023226
63. Blomqvist, P., Persson, B., Simonson, M. Fire Emissions of Organics into the Atmosphere. *Fire Technol.* **2007**, *43*(3), 213–231. https://doi.org/10.1007/s10694-007-0011-y
64. Altarawneh, M.; Dlugogorski, B. Z.; Kennedy, E. M.; Mackie, J. C. Mechanisms for Formation, Chlorination, Dechlorination and Destruction of Polychlorinated Dibenzo-p-Dioxins and Dibenzofurans (PCDD/Fs). *Prog. Energy Combust. Sci.* **2009**, *35*(3), 245–274. https://doi. org/10.1016/j.pecs.2008.12.001
65. Tame, N. W.; Dlugogorski, B. Z.; Kennedy, E. M. Formation of Dioxins and Furans During Combustion of Treated Wood. *Prog. Energy Combust. Sci.* **2007**, *33*(4), 384–408. https://doi.org/10.1016/j.pecs.2007.01.001
66. Huang, H.; Buekens, A. On the Mechanisms of Dioxin Formation in Combustion Process. *Chemosphere.* **1995**, *31*, 4099–4117. https://doi.org/10.1016/0045-6535(95)80011-9
67. Blomqvist, P.; Rosell, L.; Simonson, M. Emissions from Fires Part II: Simulated Room Fires. *Fire Technol.* **2004**, *40*, 59–73. https://doi.org/10.1023/b:fire.0000003316.63475.16
68. Blomqvist, P.; Hertzberg, T.; Tuovinen, H.; Arrhenius, K.; Rosell, L. Detailed Determination of Smoke Gas Contents Using a Small-Scale Controlled Equivalence Ratio Tube Furnace Method. *Fire Mater.* **2007**, *31*(8), 495. https://doi. org/10.1002/fam.946
69. Hull, T.R., Stec, A.A. (2010) Introduction to Fire Toxicity. In *Fire Toxicity*, Chapter 1, pp. 3–25. Woodhead/Elsevier. https://10.1533/9781845698072.1.3
70. Heal, M. R.; Kumar, P.; Harrison, R. M. Particles, Air Quality, Policy and Health. *Chem. Soc. Rev.* **2012**, 41, 6606–6630. https://doi.org/10.1039/c2cs35076a
71. Hull, T. R.; Paul, K. T. Bench-Scale Assessment of Combustion Toxicity—A Critical Analysis of Current Protocols. *Fire Saf. J.* **2007**, *42*, 340–365. https://doi. org/10.1016/j.firesaf.2006.12.006
72. ISO/TS 5660-5:2020, Reaction-to-fire tests—Heat release, smoke production and mass loss rate—Part 5: Heat release rate (cone calorimeter method) and smoke production rate (dynamic measurement) under reduced oxygen atmo- spheres, ISO, Geneva www.iso.org/standard/71346.html
73. ISO 12136:2011, Reaction to fire tests—Measurement of material properties using a fire propagation apparatus, ISO, Geneva. www.iso.org/standard/51237.html
74. BS 7990:2003, Tube furnace method for the determination of toxic product yields in fire effluents, British Standards Institution, London.
75. ISO/TS 19700:2016, Controlled equivalence ratio method for the determination of hazardous components of fire effluents—Steady-state tube furnace, ISO, Geneva. www.iso.org/standard/70630.html
76. Stec, A. A.; Hull, T. R.; Lebek, K. Characterisation of the Steady State Tube Furnace. *Polym. Degrad. Stab.* **2008**, *93*, 2058–2065. https://doi.org/10.1016/j.polymdegradstab.2008.02.020
77. Stec, A. A.; Hull, T. R.; Purser, J. A.; Blomqvist, P.; Lebek, K. A Comparison of Toxic Product Yields Obtained from Five Laboratories Using the Steady State Tube Furnace (ISO TS 19700). *Fire Saf. Sci.* **2008**, *9*, 653–663. https://doi. org/10.3801/IAFSS.FSS.9-653
78. Purser, J. A.; Purser, D. A.; Stec, A. A.; Moffatt, C.; Hull, T. R.; Su, J. Z.; Bijloos, M.; Blomqvist, P. Repeatability and Reproducibility of the ISO/TS 19700 Steady State Tube Furnace. *Fire Saf. J.* **2013**, *55*, 22–34. https://doi. org/10.1016/j.firesaf.2012.10.002
79. Stec, A. A.; Hull, T. R.; Lebek, K.; Purser J. A.; Purser, D. A. The Effect of Ventilation Condition on the Toxic Product Yields from Burning Polymers. *Fire Mater.* **2008**, *32*, 49–60. https://doi.org/10.1002/fam.955
80. Stec, A. A.; Hull, T. R.; Purser, D. A. Comparison of Toxic Product Yields from Bench-Scale to ISO Room. *Fire Saf. J.* **2009**, 44:62–70. https://doi.org/10.1016/j. firesaf.2008.03.005
81. Stec, A. A.; Hull, T. R.; Purser, D. A.; Purser, J. A. Fire Toxicity Assessment: Comparison of Asphyxiant Yields From Laboratory and Large Scale Flaming Fires. *Fire Saf. Sci.* **2014**, *11*, 404–418. https://doi.org/10.3801/IAFSS. FSS.11-404
82. Andersson, B.; Markert, F.; Holmstedt, G. Combustion Products Generated by Hetero-Organic Fuels on Four Different Fire Test Scales. *Fire Saf. J.* **2005**, *40*(5), 439–465. https://doi.org/10.1016/j.firesaf.2005.03.002
83. Hietaniemi, J.; Kallonen, R., Mikkola, E. Burning Characteristics of Selected Substances: Production of Heat, Smoke and Chemical Species. *Fire Mater.* **1999**, *23*(4), 171–185. https://doi.org/10.1002/(SICI)1099-1018(199907/08)23:4<171::AID-FAM680>3.0.CO;2-C
84. ISO 13943:2017 fire safety—vocabulary, ISO, Geneva, www.iso.org/standard/63321.html
85. Stec, A.A., Hull, T.R., Torero, J.L., Carvel, C., Rein, G., Bourbigot, S., Samym, F., Camino, G., Fina, A., Nazare, S., Delichatsios, M. (2009) Effects of Fire Retardants and Nanofillers on the Fire Toxicity. In *Fire and Polymers V— Materials and Concepts for Fire Retardancy*, Chapter 21, pp. 342–366. ACS Symposium Series 1013. American Chemical Society, Washington, DC.
86. Hull, T. R.; Stec, A. A.; Nazare, S. Fire Retardant Effects of Polymer Nanocomposites. *J. Nanosci. Nanotechnol.* **2009**, *9*, 4478–4486. https://doi. org/10.1166/jnn.2009.M80
87. Hu, W.; Zhan, J.; Wang, X.; Hong, N.; Wang, B.; Song, L.; Stec, A. A.; Hull, T. R.; Wang, J.; Hu, Y. Effect of Functionalized Graphene Oxide with Hyper-Branched Flame Retardant on Flammability and Thermal Stability of Cross-Linked Polyethylene. *Ind. Eng. Chem. Res.* **2014**, *53*(8), 3073–3083. https://doi.org/10.1021/ ie4026743
88. Hong, N.; Song, L.; Wang, B.; Stec, A. A.; Hull, T. R.; Zhan, J.; Hu, Y. Co-Precipitation Synthesis of Reduced Graphene Oxide/NiAl-layered Double Hydroxide Hybrid and Its Application in Flame Retarding Poly(Methyl Methacrylate). *Mater. Res. Bull.* **2014**, *49*, 657–664. https://doi.org/10.1016/j.materresbull.2013.09.051

89. Stec, A.A., Hull, T.R. (2009) Assessment of fire toxicity from polymer nanocomposites. In *Fire Retardancy of Polymers: New Strategies and Mechanisms*. Edited by T.R. Hull, B.K. Kandola, Chapter 25, pp 405–418. Royal Society of Chemistry, Cambridge.
90. Zhang, Y.; Li, X.; Cao, Z.; Fang, Z.; Hull, T. R.; Stec, A. A. Synthesis of Zinc Phosphonated Poly(Ethylene Imine) and Its Fire-Retardant Effect in Low-Density Polyethylene. *Ind. Eng. Chem. Res.* **2015**, *54*(13), 3247–3256. https://doi.org/10.1021/ie504200y
91. Zhang, Y.; Li, X.; Fang, Z.; Hull, T. R.; Kelarakis, A.; Stec, A. A. Mechanism of Enhancement of Intumescent Fire Retardancy By Metal Acetates in Polypropylene. *Polym. Degrad. Stab.* **2017**, *136*, 139–145. https:// doi.org/10.1016/j.polymdegradstab.2016.12.018
92. Wu, Z.; Yang, L.; Zhan, J.; Chen, Y.; Zhou, X.; Hu, Y.; Stec, A. A.; Hull, T. R. Experimental Study on Polystyrene With Intumescent Flame Retardants From Different Scale Experiments. *Fire Mater.* **2016**, *40*(1), 18–26. https://doi.org/10.1002/fam.2251
93. Zhang, Q.; Zhan, J.; Zhou, K.; Lu, H.; Zeng, W.; Stec, A. A.; Hull, T. R.; Hu, Y.; Gui, Z. The Influence of Carbon Nanotubes on the Combustion Toxicity of PP/Intu- mescent Flame Retardant Composites. *Polym. Degrad. Stab.* **2015,** *115*, 38–44. https://doi.org/10.1016/j.polymdegradstab.2015.02.010
94. Christou, A.; Stec, A. A.; Ahmed, W.; Aschberger, K.; Amenta, V. Review of Exposure and Toxicological Aspects of Carbon Nanotubes, and as Additives to Fire Retardants in Polymers. *Crit. Rev. Toxicol.* **2016**, *46*(1), 74–95. https://doi.org/10.3109/10408444.2015.1082972
95. Stec, A. A.; Hull, T. R. Assessment of The Fire Toxicity of Building Insulation Materials. *Energy Build.* **2011**, *43*, 498–506. https://doi.org/10.1016/j.enbuild.2010.10.015
96. Hull, T. R.; Lebek, K.; Robinson, J. E. Acidity, Toxicity and European Cable Regulation. *Trans. Int. Wire Cable Symp. (Trans IWCS).* **2008**, *1*, 111–117.
97. IEC TS 60695-7-50 Fire hazard testing Part 7–50: Toxicity of fire effluent estimation of toxic potency Apparatus and test method, IEC, Geneva.
98. Hansen-Bruhn, I.; Hull, T. R. Smoke Toxicity of Fire Protecting Timber Treatments. *Fire Saf. J.* **2023**, *141*, 103977. https://doi.org/10.1016/j.firesaf.2023.103977
99. Peng, Y.; Yang, L.; Ju, X.; Liao, B.; Ye, K.; Li, L.; Cao, B.; Ni, Y. A Comprehensive Investigation on the Thermal and Toxic Hazards of Large Format Lithium-Ion Batteries With $LiFePO_4$ Cathode. *J. Hazard. Mater.* **2020**, *381*, 120916. https://doi.org/10.1016/j.jhazmat.2019.120916
100. Sun, P.; Bisschop, R.; Niu, H.; Huang, X. A Review of Battery Fires in Electric Vehicles. *Fire Technol.* **2020**, *56*(4), 1361–1410. https://doi.org/10.1007/s10694-019-00944-3
101. Willstrand, O.; Pushp, M.; Andersson, P.; Brandell, D. Impact of Different Li-ion Cell Test Conditions on Thermal Runaway Characteristics and Gas Release Measurements. *J. Energy Storage.* **2023**, *68*, 107785. https://doi.org/10.1016/j.est.2023.107785
102. Crewe, R. J.; Stec, A. A.; Walker, R. G.; Shaw, J. E. A.; Hull, T. R.; Rhodes, J.; Garcia-Sorribes, T. Experimental Results of a Residential House Fire Test on Tenability: Temperature, Smoke and Gas Analyses. *J. Forensic Sci.* **2014**, *59*, 139–154. https://doi.org/10.1111/1556–4029.12268
103. Crewe, R. J.; Stec, A. A.; Walker, R. G.; Shaw, J. E. A.; Hull, T. R.; Rhodes, J.; Garcia-Sorribes, T. Experimental Results of a Residential House Fire Test on Tenability: Temperature, Smoke and Gas Analyses. *J. Forensic Sci.* **2014**, *59*, 139–154. https://doi.org/10.1111/1556-4029.12268
104. Hull, T. R.; Brien, D.; Stec, A. A. Quantification of Toxic Hazard from Fires in Buildings. *J. Build. Eng.* **2016**, *8*, 313–318. https://doi.org/10.1016/j.jobe.2016.02.014
105. 2010 Fire Test Procedure Code, Maritime Safety Committee, (MSC 87/26/Add.3) Annex 34, Part 2 Smoke and Toxicity Test, International Maritime Organisation, London.
106. EN 45545-2:2013 Railway applications—Fire protection on railway vehicles—Part 2: Requirements for fire behaviour of materials and components.
107. McKenna, S. T.; Jones, N.; Peck, G.; Dickens, K.; Pawelec, W.; Oradei, S.; Harris, S.; Stec, A. A.; Hull, T. R. Fire Behaviour of Modern Façade Materials—Understanding the Grenfell Tower Fire. *J. Hazard. Mater.* **2019**, *368*, 115–123. https://doi.org/10.1016/j.jhazmat.2018.12.077
108. Peck, G.; Jones, N.; McKenna, S. T.; Glockling, J. L. D.; Harbottle, J.; Stec, A. A.; Hull, T. R. Smoke Toxicity of Rainscreen Façades. *J. Hazard. Mater.* **2021**, *403*, 123694. https://doi.org/10.1016/j.jhazmat.2020.123694
109. Wikipedia article on Colectiv nightclub fire. https://en.wikipedia.org/wiki/Colectiv_nightclub_fire (accessed 28/08/2023).
110. Rouen Fire: at least 13 dead in birthday disaster at bar in French city. www.theguardian.com/world/2016/aug/06/rouen-explosion-at-least-13-dead-after-blast-and-fire-at-bar-in-french-city-reports (accessed 28/08/2023).
111. Persson, B.; Simonson, M. Fire Emissions Into the Atmosphere. *Fire Technol.* **1998**, *34*(3), 266–279. https://doi. org/10.1023/a:1015350024118
112. Harries, M. E.; Allen, D. T.; Adetona, O.; Bell, M. L.; Black, M. S.; Burgess, J. L.; Dryer, F. L.; Holder, A. L.; Mascareñas, A.; Rosario-Ortiz, F. L.; Stec, A. A.; Turpin, B. J.; Zelikoff, J. T. A Research Agenda for the Chemistry of Fires at the Wildland-Urban Interface: A National Academies Consensus Report. *Environ. Sci. Tech.* **2022**, *56*(22), 15189–15191. https://doi.org/10.1021/acs.est.2c07015
113. *The Chemistry of Fires at the Wildland-Urban Interface* (2023) pp. 1–214. National Academies Press, Washington, DC. https://doi.org/10.17226/26460
114. Shaposhnikov, D.; Revich, B.; Bellander, T. Mortality Related to Air Pollution With the Moscow Heat Wave and Wildfire of 2010. *Epidemiology.* **2014**, *25*(3), 359–364. https://doi.org/10.1097/ede.0000000000000090
115. Lelieveld, J.; Evans, J. S.; Fnais, M.; Giannadaki, D.; Pozzer, A. The Contribution of Outdoor Air Pollution Sources to Premature Mortality on a Global Scale. *Nature.* **2015**, *525*, 367–371. https://doi.org/10.1038/nature15371

20 Toward a Quantitative Relation between the Molecular Structure and Fire Growth on Polymeric Solids

Dushyant M. Chaudhari[1] *and Stanislav I. Stoliarov*[2]
[1]Fire Safety Research Institute (FSRI), UL Research Institutes
[2]Department of Fire Protection Engineering, University of Maryland, College Park

20.1 INTRODUCTION

Between 1960 and 2019, polymer production has increased exponentially to more than 460 million tons.[1] The use of polymeric and composite materials is increasing in the built environment, propelled by several factors such as a relatively low cost of production, ability to design and manufacture materials with varying physical properties for wider applicability, and aesthetic demands of modern architecture. About 19% of these produced polymers are used in the building and construction industry.[1,2] Development of materials is also increasingly directed towards more environmentally benign alternatives or additives due to increasing concerns of toxic and greenhouse gas emissions during material degradation caused by ageing or burning.[3] Most of the polymers, however, being hydrocarbon in nature impose fire and toxic hazards in the built environment, sometimes higher than the fire and toxic hazards associated with traditional materials.[4] In 2020, incidents involving fires in structures accounted for about 35% of the total 1.4 million fire incidents reported in the United States of America (USA).[5] Over the same period of time, a home fire was estimated to occur once every 89 seconds, with one death approximately every 3 hours and 24 minutes accounting for about 74% of total deaths due to fire.

Prescriptive standards require any new polymeric materials to be tested in apparatuses carefully designed to emulate scenario-specific thermal conditions to which the material may be exposed in its final application. These tests are classified into reaction-to-fire tests that examine the response of a material to fire and resistance-to-fire tests that examine the ability of the material to contain a fully developed fire. Reaction-to-fire tests typically evaluate specific metrics such as ignition, thermal decomposition, flammability, flame spread, and gaseous emissions. Materials are classified based on their fire hazard parameters on a relative ranking scale. Fire hazard parameters such as heat release rate (HRR), ignition time, minimum temperature for ignition, effective heat of combustion, and fire growth index are important indicators of the propensity of a material to support early fire growth. However, standardized tests that are used to quantify such parameters are specific to test configuration and condition. Karlsson et al.[6] and Bill and Croce[7] argued that the data available from such tests are non-fundamental in nature and thus cannot be reliably used to predict results of another fire scenario, especially for complex materials. Bill and Croce[7] also suggested that the ultimate transition to computer-based predictive models can be achieved by promoting application of small-scale property measurements to inform first-principle models of larger-scale experiments. A step forward is to fundamentally understand the relationship between molecular structure and material properties and the dynamics of the flame spread problem and to apply available scientific methods to improve the prediction capabilities. Such prediction capability can potentially help guide the development and application of new fire-retardant materials and drastically reduce the monetary cost of standardized testing of new materials. Consequently, the fire incidents and their impact on life and public safety can be minimized.

20.1.1 Fire Growth Phenomenon

Fundamentals of flame spread phenomenon have been studied theoretically and experimentally for more than 50 years. Researchers have also attempted to assess material performance using semi-empirical models as well as comprehensive computational fluid dynamic models for about 30 years. Williams[8] in 1977 provided an in-depth theoretical review of flame spread phenomenon wherein was discussed the fundamental equation of fire spread, which relates the net energy per unit area incident on the unburnt fuel to the product of fuel density, velocity of flame spread, and enthalpy of unburnt fuel to raise its temperature to an ignition temperature.

Generally, when a combustible, solid material is heated, it undergoes a thermal degradation process, influenced by the surrounding aerobic or anaerobic environment, resulting in release of pyrolyzates (combustible or noncombustible gases). This thermal degradation along with associated mass and heat transfer through the degrading material is referred as *pyrolysis*. On a vertical wall, such as illustrated

DOI: 10.1201/9781003380689-20

for a concurrent flame spread scenario in Figure 20.1, the mass flux of pyrolyzates generated during pyrolysis from pyrolyzing region y_p is given as $\dot{m}_F''$. The *gas phase combustion* of pyrolyzates (if combustible) can result in a continuous flame, of length y_f, that may extend over the unburnt region of length y_e. The unburnt region is then subjected to heating via *flame heat feedback* expressed as flux, $\dot{q}''$, from the extending flame and simultaneously heated through conduction from the burning region, $\dot{q}_{k,p}''$. Concurrent, buoyancy-driven vertical flame spread is a result of nonlinear growth of the flame height and the extension region.

20.1.2 Fire Growth Modeling

Flame spread is thus governed by cyclic feedback[8,9] between three interdependent phenomena: (1) Gas phase combustion of pyrolyzates, (2) Heat transfer to the unburnt region, and (3) Condensed phase pyrolysis. Controlling mechanisms for flame spread over continuous surfaces for concurrent[8] are (1) flame to surface heat flux and (2) chemistry that controls the pyrolyzate production rate. Therefore, for a specific flame spread scenario, understanding these controlling mechanisms is important and each has been studied individually for decades. Modeling of the flame spread process thus involves modeling of these three key physical phenomena. Generally, a separate model is necessary for each of these phenomena which then can be coupled to create a flame spread model. Depending on the details that each sub-model attempts to resolve, a flame spread model can be generally classified into three categories:

1. Empirical model—Such models use parameters defined per unit area based on small-scale experiments and extrapolate those for larger scale scenarios. For example, ConeTools[10] is an empirical model that predicts transient heat release rate (HRR) of a material in a Single Burning Item test (EN-13823[11]) scenario based on extrapolation of heat release rate (HRR) per unit area obtained from cone calorimetry tests and an empirical burning area growth function. These models are heavily reliant on the accuracy of empirical functions and extrapolations to larger scales generally have limitations.
2. Semi-empirical model—Such models usually couple an empirical representation for some phenomena and comprehensive modeling of the rest. Several pyrolysis solvers such as ThermaKin[12,13] aim to represent pyrolysis comprehensively. Such models have great potential to improve accuracy compared to empirical models and have relatively smaller computational expense compared to field models.
3. Field (computational fluid dynamics (CFD)-based) models—These models, such as Fire Dynamics Simulator (FDS),[14] explicitly solve Navier-Stokes equation and comprehensively model all the phenomena and has the potential for greater accuracy and applicability to broader range of materials compared to other types of models. A drawback of such models is the computational cost required to comprehensively model all phenomena.

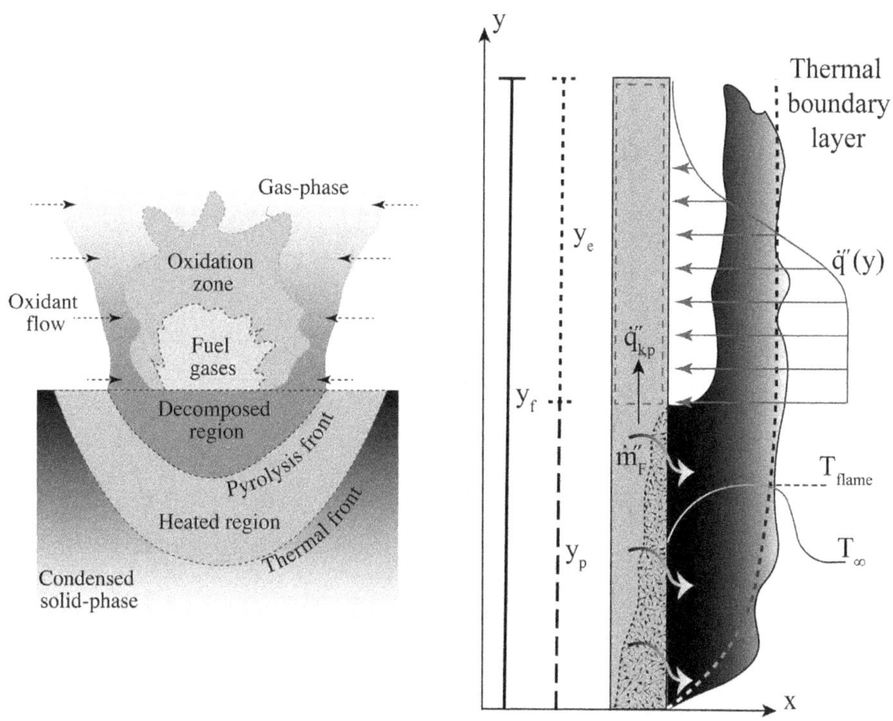

FIGURE 20.1 Illustration of gas phase and solid phase coupling (left). Schematic of a vertical flame spread process (right).

An overview of the semi-empirical and CFD-based model, that are later used for a fire growth case study in this chapter, is provided in Section 20.3.

20.2 OVERVIEW OF PYROLYSIS AND PYROLYSIS MODELING

Of the three phenomena that govern the concurrent, vertical flame spread over a polymer after its ignition, gas phase combustion and flame heat feedback (heat transfer) have been widely studied and modeled compared to the pyrolysis process. The pyrolysis process and heat transfer mechanism are arguably the controlling phenomena in the fire growth process and have been traditionally simplified in modeling of fire growth. More importance is gradually given to comprehensively model the pyrolysis phenomenon in a fire growth process. This section first provides an overview of the pyrolysis phenomenon and identifies key input parameters for comprehensive pyrolysis model development.

20.2.1 PYROLYSIS PHENOMENON

Pyrolysis is defined as the process by which the solid transforms into gas phase fuel and generally implies the breakdown of the molecules into different, generally smaller, molecules.[15] Solid gasification is different from liquid gasification primarily because a solid gasification, referred also as thermal degradation or decomposition, usually accompanies chemical change governed by multiple chemical reactions while liquids primarily undergo a phase-change without a chemical change.[15]

A detailed schematic of various processes occurring during a simplified, one-dimensional pyrolysis phenomenon is illustrated in Figure 20.2. The solid is external heated by a radiation source, q_e'', which could change with time, t. The temperature of the solid, initially at T_0, increases at the surface. As the thermal wave propagates when the solid is heated, temperature within the solid increases, and at around the thermal depth, $x = \delta_T$, the temperature is very close to the initial temperature. This depth is usually referred as the thermal front. The surface defines the heat and mass transfer in and out of the solid fuel, or mathematically, the boundary conditions. The surface could absorb the external radiative heat flux in-depth or only on its surface, depending on the absorption coefficient (α) of the solid. The re-radiative heat flux (q_r''), which is dependent on the emissivity (ε), and the convective heat flux (q_{conv}''), which is dependent on temperature gradient across the boundary layer, and the heat transfer coefficient are also important in defining the heat transfer boundary condition. The conductive heat flux, q_c'', depends on the temperature gradient within the solid and thermal conductivity (k), density (ρ), and specific heat capacity (C_p) of the polymeric solid.

A gaseous pyrolyzate, formed during a degradation reaction, could escape from the solid surface, at a rate of m_p'', primarily depending on the reaction kinetics and the gas transfer coefficient of the solid and the pyrolyzate gas. The volume fraction of this pyrolyzate gas, $\varphi_{F,g}$, would be the highest at the surface of the solid and decrease away from the solid as ambient airflow dilutes the concentrations. However, at a sufficient distance away from the solid surface, where the flammable gas mixture is within the flammability limits, and in presence of a piloted ignition source, these pyrolyzate gases could ignite and result in sustained flaming. This would then initiate the solid-gas coupling described earlier in Section 20.1.1.

The thermal degradation process could be influenced by the presence of oxygen immediately in the vicinity of the solid surface. Depending on the diffusion coefficient of oxygen,

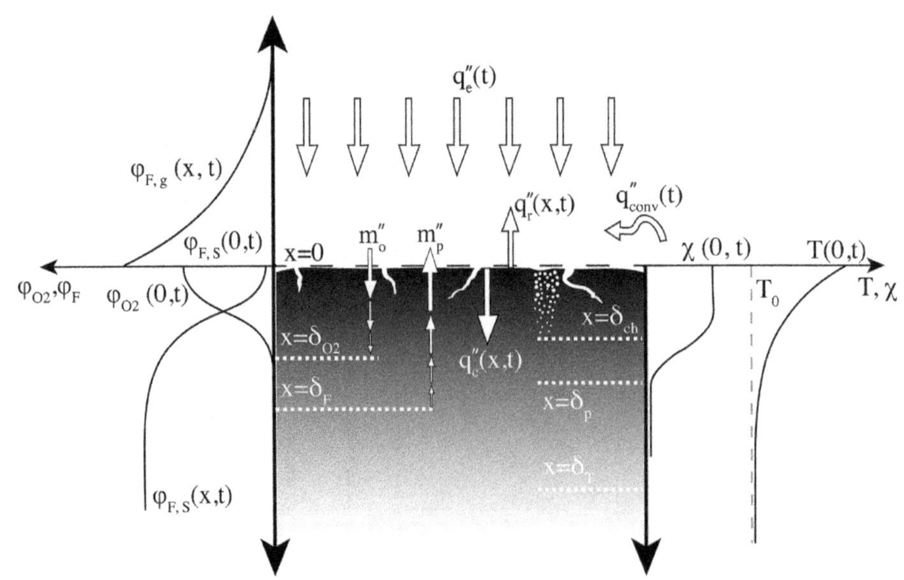

FIGURE 20.2 Schematic of various phenomena occurring during a simplified one-dimensional pyrolysis process.

oxygen could diffuse at a rate, $\dot{m}''_O$, and the volume fraction of oxygen, φ_{O2}, would be highest at the surface and decrease with depth inside the solid until depth, $x = \delta_{O2}$. Simultaneously, the volume fraction of undecomposed initial solid material, $\varphi_{F,S}$, would be smallest at the surface and increase deeper inside the solid, depending primarily on parameters that influence the conductive heat flux. Any solid below a depth, $x = \delta_F$, where $\varphi_{F,S}$ is equal or very close to 1, is undecomposed.

It is likely that a material degradation reaction would result in a chemical change which partly produces solid carbonaceous residue that require much higher energy than available during a fire scenario to completely transform into gas phase. This solid carbonaceous residue is often referred as char. This solid residue could change the permeability, χ, of the solid which is highest at the surface and lowest at and below depth, $x = \delta_{ch}$. The char formation may result in fractures and increased porous structure, and thus can affect permeability of gas in and out of the solid. Char formation, which can accompany intumescence (swelling) of the solid residue, can create a thermal and mass barrier that retards further degradation or flaming of the material. Below a depth of $x = \delta_p$, the pyrolysis process (or degradation reaction) can be considered insignificant, and this depth is considered as the pyrolysis front. As the solid is continuously heated, the transient pyrolysis process continues, and rates of propagation of the thermal front, pyrolysis front, and rates of pyrolyzate production, oxygen diffusion, char production change.

20.2.2 Comprehensive Pyrolysis Modeling

In general, a model that describes a pyrolysis process, as illustrated in Figure 20.2, can be classified into the following four categories,[16] listed with increasing order of complexity—(i) Empirical models, (ii) Thermal models, (iii) Polynomial pyrolysis models, and (iv) Comprehensive pyrolysis models. Empirical models are based on small-scale experimental data that quantify fire hazard parameters (commonly expressed per unit area) and extrapolated to large-scale assuming a critical ignition temperature. Thermal models do not consider chemical kinetics but solve for mass loss rate considering energy balance at a thin pyrolysis front, assuming that the onset of degradation occurs at a critical material-dependent pyrolysis temperature. Polynomial pyrolysis models have simplifications similar to that of thermal models and utilize polynomial equations that consider the impact of environment (such as incident heat flux, oxygen concentration) on the mass loss rate. Comprehensive pyrolysis models consider chemical kinetics by describing the degradation process as a reaction rate dependent on a local temperature and species' concentration.

Most pyrolysis reactions are given by Arrhenius-type reaction:

$$r''' = A \cdot \exp\left(-\frac{E}{RT}\right)[F]^{\eta} \tag{20.1}$$

Where r''' is the rate of reaction per unit volume of the material (kg m^{-3} s^{-1}), A is a pre-exponential factor (s^{-1} for a first order reaction), E is the activation energy (J mol^{-1}), R is the universal gas constant (8.314 J mol^{-1} K^{-1}), T is the solid temperature (K), $[F]$ is the mass of the fuel per unit volume (kg m^{-3}), and η is the reaction order. A, E, and η are generally referred as kinetic triplet and an apparent value for these parameters is usually estimated to characterize the solid degradation reaction. Each reaction could have an associated heat of degradation, ΔH_r (J kg^{-1}), which could be exothermic or endothermic.

Parameters that govern the pyrolysis process, and thereby are input parameters for comprehensive pyrolysis models, are given in Table 20.1. The list also includes some examples of possible estimation methods for determining the corresponding input parameters. This list of parameters does not include parameters that influence permeability, oxygen diffusion, and structural change of the solid such as melting and dripping. For a solid covered with a sustained continuous diffusion flame, the surface oxygen content is found to be negligible,[17] and its effect can be assumed to be negligible for the scenario in this chapter. Similarly, permeability (affected by material or char porosity) and structural changes are beyond the scope of this chapter. However, material porosity can be indirectly considered by the gas diffusion coefficient within the solid.

Measuring all relevant input parameter values is limited by the availability of experimental apparatus or simply the inability to measure some key properties (such as parameters describing reaction kinetics for solid phase thermal degradation). An alternative (or complimentary) approach involves inverse analysis of experimental data for deriving such properties. Inverse analysis, a technique used in the model-fitting approach, can be defined as estimating property value(s) that can describe an observed natural phenomenon, considering the measured quantity.[16]

A framework for inverse analysis is shown in Figure 20.3. This framework follows a hierarchical modeling approach of building a pyrolysis model using multiple experiments conducted across a range of scales. Individual properties of interest are isolated and calibrated through inverse analysis of the data from separate experiments. The boxes on the left side of this figure show experiments that are conducted in such hierarchical modeling approach to obtain data that can be used to optimize corresponding parameter types shown on the right. The data from simultaneous thermal analysis, a milligram-scale test, is generally used for determining reactions kinetics and thermodynamics. Microscale combustion calorimeter, a milligram-scale test, data is generally used for determining heats of combustion of pyrolyzate gases. A gram scale gasification test, such as the controlled atmosphere pyrolysis apparatus (CAPA), data can be used to quantify thermo-physical properties. Additionally, some properties can be obtained from other experiments. For example, broadband absorption coefficients can be derived by using a methodology developed by Linteris et al.,[19] where

TABLE 20.1
Parameters that Need to be Quantified for Input to Comprehensive Pyrolysis Models and Possible Sources for Estimating Respective Parameters

Type	Parameter	Variable	Units	Estimation Methods
Reaction kinetics and thermodynamics	Pre-exponential factor	A	s^{-1} for first order reaction	• Model-fitting approach using small-scale/milligram-scale experimental data
	Activation energy	E	$J\ mol^{-1}$	• Iso-conversional/model-free methods using thermogravimetric analysis
	Reaction order	η	-	• Population balance approach for linear molecules[18]
	Heat of degradation	ΔH_r	$J\ kg^{-1}$	• Calorimeter • Differential scanning calorimeter
Thermo-physical	Specific heat capacity	C_p	$J\ kg^{-1}K^{-1}$	• Differential scanning calorimeter • Instruments used for thermal conductivity
	Thermal conductivity	k	$W\ m^{-1}K^{-1}$	• Steady or transient measurements with Heat flow meter such as Modified Transient Plane Source (MTPS), Hot disk®, Laser Flash
	Density	ρ	$kg\ m^{-3}$	• Measurement of mass and volume of a sample
Optical	Absorption coefficient	α	m^{-1}	• Broadband absorption (such as a gasification device) • Integrating sphere (coupled with Fourier-transform infrared spectroscopy) for spectral absorption
	Emissivity	ε	-	• Calorimetric (total hemispherical) and radiometric (spectral) methods • Reflectometric methods such as integrating sphere (spectral emissivity)
Gas transport	Gas diffusion coefficient	D_G	$m^2\ s^{-1}$	• Available literature • Molecular dynamics simulations utilizing Leonnard-Jonnes potential

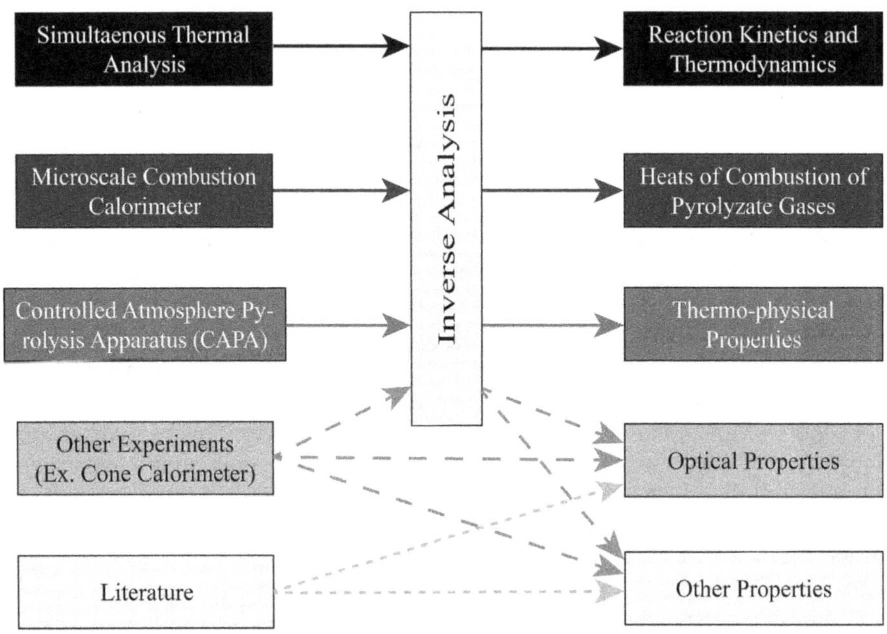

FIGURE 20.3 Framework for comprehensive pyrolysis model development.

a known heat flux is irradiated on a millimeter-scale thick sample and transmitted heat flux is measured. Existing literature is also a valuable source for pyrolysis properties, but the specifics of the material, property determination method, assumptions, and the applicability of existing values should be considered. The hierarchical approach shown here has provided pyrolysis property sets that are applicable to wide range of pyrolysis and fire scenarios.[20] Some studies have also attempted to develop comprehensive pyrolysis model based only on a single experiment, such as the cone calorimeter, an approach which limits the application of derived pyrolysis property sets to the scenario used for development of the model.[20–22] This framework is later applied in Sec. 20.4 for developing pyrolysis model of poly(isocyanurate) foam.

20.3 OVERVIEW OF MODELING TOOLS

In the past two decades, key computational solvers that are developed and used by the fire community include ThermaKin,[12,13] Fire Dynamics Simulator (FDS),[23] GPyro,[24] and FireFoam.[25] ThermaKin and Gpyro are comprehensive pyrolysis solvers that can model pyrolysis of only a solid phase for variety of external user-defined heating conditions. On the other hand, FireFoam and FDS can numerically solve for gas phase phenomena such as combustion and radiation along with the solid-phase comprehensive pyrolysis. Each program has its advantages and limitations. Amongst these, ThemaKin has been used to develop validated comprehensive pyrolysis models for charring,[26] non-charring,[27] and intumescing[28] polymers. FDS is widely used in research and by fire protection professionals and has been widely validated for a large variety of fire scenarios including pyrolysis and gas phase combustion.[29] Therefore, an overview of these two programs is provided below. These programs are applied later for developing a pyrolysis model in Section 20.4 and for fire growth case study in Section 20.6.

20.3.1 ThermaKin

ThermaKin, a comprehensive pyrolysis solver, solves unsteady mass and energy conservation equations for any condensed phase material that may undergo chemical or physical changes in response to user-defined external boundary conditions. Each material is defined as a mixture of components, where each component can degrade following a first or second order Arrhenius-type reaction kinetics. When ThermaKin was developed in 2008,[13] conduction, in-depth absorption, chemical reactions, and diffusion of gases from a thin layer were systematically verified. Later, in 2014, the ThermaKin2Ds was released to expand the computation to a two-dimensional space. This facilitated coupling of pyrolysis with an analytical representation of energy feedback from surface flames. Stoliarov et al. and Leventon et al. demonstrated application of such a model for laminar flame spread over a 0.15 m tall poly(methyl methacrylate) panel.[12,30] In 2019, a new module capable of simulating 2D axisymmetric material specimen (to accurately model intumescence) was introduced in ThermaKin2Ds.[31] ThermaKin2Ds with capability of modeling 2D axisymmetric scenario was used here to explain the development of comprehensive pyrolysis models. ThermaKin2Ds was also used to develop a semi-empirical flame spread model for the fire growth case study discussed later.

20.3.2 Fire Dynamics Simulator (FDS)

Fire Dynamics Simulator (FDS) is an open-source computational fluid dynamics (CFD) model, maintained by the National Institute of Standards and Technology (NIST), that numerically solves Navier-Stokes equations for low Mach number ($Ma < 0.3$) flows.[23,32] The emphasis of this software is thermally driven flow and mass and heat transfer from fires. Large eddy simulation (LES) is used for simulating turbulence, with default Deardorff model to describe turbulent viscosity, and the model utilizes an explicit predictor-corrector scheme for the hydrodynamic model. Combustion of a species is, by default, modeled as a single-step, mixing-controlled chemical reaction for three lumped species: air, fuel, and products. Flame extinction is modeled using the critical flame temperature and oxygen index concept, following which, for a cell temperature above approximately 600°C, no flame extinction is assumed unless the oxygen concentration drops to zero.[23] Radiation is modeled by spatial (adjusted via number of radiation angles) and temporal (adjusted via time step increment for radiation solver) discretization of radiation transport equation, and a radiative fraction (fraction of the fire's heat release rate released in the form of thermal radiation) is prescribed by default for the combusting region (defined based on volumetric heat release rate). A gray gas model is used by default for radiation absorption coefficient, where a RadCal narrow band model is utilized for obtaining temperature dependent absorption coefficient values for all gases present in the simulation plus soot for a default or user-defined path length. An optional weighted sum of gray gas (WSGG) model or the wide band (box) model can be used for modeling radiation, and their detail can be found elsewhere.[23,32] Convective heat transfer to the solid is, by default, modeled using empirical correlations and a limit of conduction for relatively small cell size. Convective heat transfer coefficient using empirical correlations for a corresponding geometry is calculated by considering the temperature differential of surface temperature and immediately adjacent gas cell temperature. FDS v.6.8, released on April 18th, 2023, was used here in the fire growth case study.

20.4 QUANTIFYING PARAMETERS TO DEVELOP COMPREHENSIVE PYROLYSIS MODELS

The importance of material properties on the fire growth was emphasized as early as 1994 by Kashiwagi,[33] where it was highlighted that the fundamental understanding of chemical and physical mechanisms in the condensed phase during combustion of polymers is severely lacking compared with that in the gas phase. In that study, Kashiwagi highlighted three challenges that fundamentally limit the understanding of the thermal degradation process: lack of chemical kinetic constants for polymer degradation reactions, role of bubbles and of oxygen in the gasification process; and chemical mechanisms of char formation and chemical and physical structures of chars. This chapter primarily addresses the first challenge and discusses the methodology to characterize the reaction kinetics and thermodynamics of polymer degradation, heats of combustion of evolving gaseous pyrolyzates, and the thermal transport properties, which together form a comprehensive pyrolysis model.

The pyrolysis model development framework was introduced earlier in Figure 20.3. This framework is applied in this section for an example material—poly(isocyanurate) (PIR) foam. ThermaKin2Ds was utilized to inversely analyze milligram and gram-scale experimental data.

20.4.1 Materials

The PIR foam was supplied by DuPont de Nemours Inc. and sold under the trademark Tuff-R.[34] Tuff-R insulation panels were 5.08×10^{-2} m thick and had a density (measured at room temperature) of 32 kg m^{-3}. The panels consisted of a closed-cell PIR foam core reinforced with non-woven glass-fiber and manufactured using iso-pentane blowing agent. The foam-core was sandwiched between two facers, each comprised of aluminum foil and Kraft paper. For all tests, the facer material was removed for sample preparation.

The PMMA selected for this study was cast black PMMA which was manufactured by Evonik industries under a tradename Acrylite cast black 9H01 GT. PMMA sheets were purchased and had an average thickness of 5.8×10^{-3} m and density (measured at room temperature) of 1210 ± 30 kg m^{-3}. The manufacturer, formulation, and batch of this PMMA is the same as that used by the Measurement and Computation of Fire Phenomenon (MaCFP) Condensed phase working group[35] round robin experiment and modeling exercise.

The PMMA pyrolysis is known to be relatively simple,[36] compared to PIR foam.[37] To discuss the pyrolysis model development framework in the context of a wall-lining insulation material, the framework is applied here for the PIR foam and not for the PMMA. The pyrolysis model for PMMA following the same framework has been developed and can be found elsewhere.[36]

20.4.2 Milligram-Scale Experiments and Modeling

20.4.2.1 Simultaneous Thermal Analysis (STA)

Simultaneous thermal analysis (STA), which involves thermogravimetric analysis (TGA) and differential scanning calorimetry (DSC), assume that a milligram-scale sample is uniformly heated. TGA and DSC are widely used to study the pyrolysis process of a material subjected to a controlled heating profile in a controlled atmosphere. Samples in the order of milligrams are placed in a crucible, placed within a tubular furnace, which is heated with a user-defined profile. The gases evolved during the pyrolysis process may be further analyzed when the instrument is coupled to additional gas phase sensors. The mass loss, mass loss rate, and heat flow data obtained from these tests can be inversely analyzed to characterize the kinetics and thermodynamics of the pyrolysis process. Nearly uniform heating conditions in STA allows simplifying the energy conservation equation such that reaction kinetics and thermodynamic parameters can be obtained for a global user-defined thermal degradation reaction mechanism. These parameters are usually optimized so that the predicted mass loss and heat flow for these tests match the experimental data.

TGA and DSC were performed simultaneously using a Netzsch F3 Jupiter Simultaneous Thermal Analyzer. Cryogenically grounded PIR foam core samples weighing 4–6 mg were used for testing. The samples were dried for at least 24 h in a desiccator in the presence of Drierite prior to testing. Ten tests were performed at 10 K min^{-1} in a nitrogen atmosphere; the results of these tests were averaged and used for the development of the thermal degradation model. The solid residue left behind at the end of these tests was collected and used to perform five additional tests under the same conditions to determine the heat capacity of the residue. Six additional tests were performed in nitrogen on the ground foam core samples at 5 and 20 K min^{-1} (three at each heating rate) to obtain the data that were used to examine whether the thermal degradation model was able to predict degradation at alternate heating conditions. All these tests were carried out in Pt-Rd crucibles with lids to maximize thermal contact and heat flow sensitivity. Gaseous degradation products vented through a small opening which was located at the center of each lid. Finally, two tests were performed in air at 10 K min^{-1} using ceramic crucibles without lids. The sole purpose of these tests was to determine the mass fraction of non-oxidizable constituents of the foam core, which were assumed to be the glass fibers. In all these tests conducted in inert atmosphere, the nitrogen flow was maintained at 80 ml min^{-1}.

20.4.2.2 Milligram-Scale Modeling

Considering the controlled atmosphere, small sample size, highly conductive crucible, and nearly uniform heating of the sample in these tests, it is justified to assume that the temperature is uniform across all the sample and that thermal degradation is a zero-dimensional process in these tests. This assumption facilitates inverse analysis of the data. Thus, during the modeling of these tests, all samples were modeled as zero-dimensional and represented by a single spatial element in ThermaKin2Ds. The element temperature was forced to follow the experimental temperature profile by defining a sufficiently high convection coefficient, 1×10^5 W m^{-2} K^{-1}, at the boundary. The experimental temperature profile was prescribed by expressing the heating rate via an exponentially decaying sinusoidal function. The mass flow boundary conditions were defined such that the gaseous pyrolyzate instantaneously escaped the element.

20.4.2.3 Modeling of TGA

The TGA data, shown in Figure 20.4, indicated that the mass loss rate (MLR) of the PIR foam thermal degradation had multiple local peaks at different temperatures and the onset of the first degradation process was approximately at 400 K.

The mass loss rate (MLR) peaked when the temperature was around 600 K but was insignificant after 900 K. The PIR foam left behind approximately 32 ± 1 wt% of solid residue at 900 K.

The PIR foam thermal degradation mechanism was assumed to consist of consecutive reactions; the number of reactions was determined from the inspection of the MLR data obtained in the 10 K min^{-1} TGA experiments performed in nitrogen. The foam material was represented by a mixture of 95.9 wt% reactive PIR component and 4.1 wt% of inert glass fiber component. The PIR component was assumed to decompose through seven consecutive first-order reactions. Each reaction product comprised of one condensed phase, PIT_Int_#, and one gaseous product, GAS_#, where # denote the number for the reaction. Table 20.2 shows the reaction mechanism. The kinetic parameters of these reactions were determined through fitting of the TGA data using a multiparameter optimization with a hill-climbing algorithm. Details of this algorithm implementation can be found elsewhere.[36] Figure 20.4 compares experimental TGA data with model predictions.

20.4.2.4 Modeling of DSC

The normalized heat flow data, shown in Figure 20.5, represented a multi-peak degradation and small net heat flow magnitude with maximum endothermic heat flow of 0.7 W g^{-1}. Higher uncertainty is observed in the heat flow data at higher temperature due to higher uncertainty and fluctuations in the baseline data at these temperatures in the apparatus. An onset of exothermic reaction was identified around 580 K, which is characteristic of PIR thermal degradation.[37]

The temperature-dependent heat capacity, C_p, of the PIR component was estimated by fitting the curve resulting after dividing the heat flow data by the instantaneous heating rate before the onset of the first degradation reaction (400 K). A similar procedure was applied to estimate temperature-dependent heat capacity for the final residue (PIR_Int_7) using the DSC tests conducted on the fully decomposed samples. The heat capacities of intermediate condensed phase components were assumed to be equal to the heat capacity of the PIR component at 420 K. Sensible enthalpy of the degradation process was then modeled by forcing the heats of reaction to be zero. Heats of reaction were then estimated by subtracting the DSC heat flow data from the baseline. The final heats of reaction, shown in Table 20.2, were selected to fully reproduce the experimental heat flow and its integral. Heat capacity of all gas components were assumed to be 2100 J kg^{-1} K^{-1}, which was the mean heat capacity of collection of C1 to C8 hydrocarbons at 600 K.[31] Figure 20.5 compares experimental DSC data with model predictions.

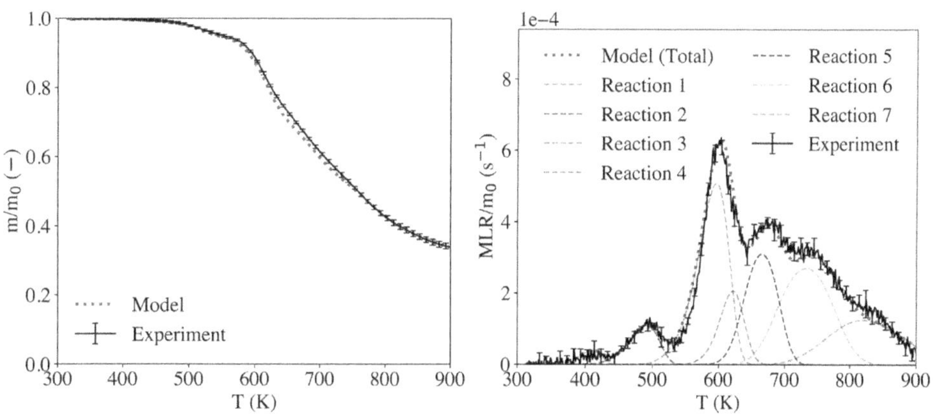

FIGURE 20.4 Measured mass loss and mass loss rate compared with model predictions for TGA tests conducted using PIR foam at 10 K min^{-1} nominal heating rate.

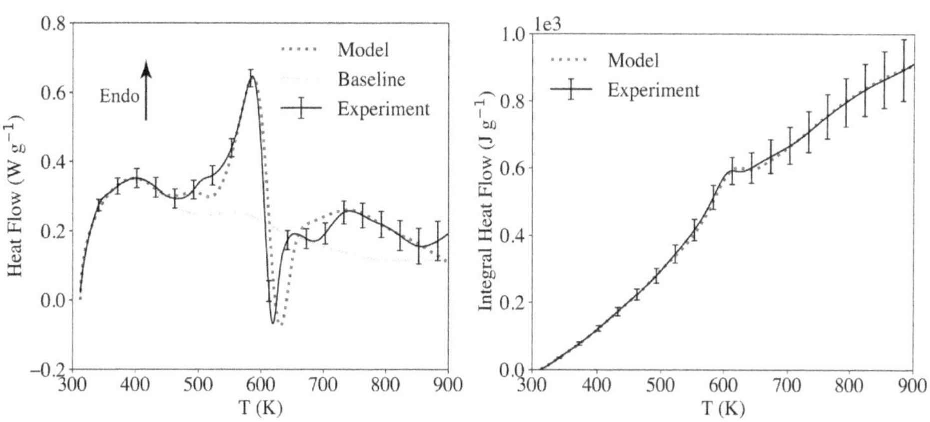

FIGURE 20.5 Measured DSC heat flow compared with model predictions for PIR foam STA tests conducted at 10 K min^{-1} nominal heating rate.

20.4.2.5 Microscale Combustion Calorimeter (MCC)

Similar to the STA, the milligram-scale samples in Microscale Combustion Calorimeter (MCC) are subjected to a user-defined heating rate and pyrolyze in a controlled atmosphere. Additionally, the pyrolyzates are forced to undergo complete oxidation in 20 vol% oxygen in a tubular flow reactor maintained at 1223 K. The flow rate and volumetric oxygen concentration are monitored, and the instantaneous heat release rate of the material is determined. Using this data, heats of combustion for the pyrolyzates corresponding to reactions identified using the TGA data can be estimated.

The MCC tests were carried out at a nominal heating rate of 10 K min^{-1} using the same foam core samples that were used in the TGA and DSC tests. The samples were pyrolyzed in nitrogen (Method A[38]). The HRR was determined using oxygen consumption calorimetry and recorded as a function of time and material sample temperature. These tests were repeated five times to estimate the uncertainty of the measurement.

20.4.2.6 Modeling of MCC

MCC test data, shown in Figure 20.6, showed similar multiple peaks of the HRR at different temperatures as that seen in the TGA MLR data. The mean heat of combustion was estimated to be 13.3 ± 0.4 kJ per gram of initial mass. Residue yield in MCC was 30.4 ± 0.4%, which was slightly lower to that of the TGA tests and is likely due to uneven heating of the sample attributed to radiation from the hot tubular reactor.

The heats of combustion of the gaseous pyrolyzates were determined via simulation of the MCC test by using the reaction model developed from the TGA test data. The heats of combustion were first assumed to be the same, equal to the mean heat of combustion of the gasified material obtained through the integration of the MCC HRR curve and were subsequently adjusted (if needed) to capture the shape of the MCC curve. The MCC instantaneous HRR and integral HRR normalized by initial mass of the PIR foam are shown in Figure 20.6 along with the model predictions.

20.4.2.7 Pyrolysis Model Parameters Determined from Milligram-Scale Tests

The reaction mechanism for the PIR foam is shown below, along with kinetic and thermodynamic parameters for each reaction, in Table 20.2. The specific heat capacities for the condensed phase component and heats of combustion for the gas phase component are shown in Table 20.3. This model was validated by predicting the experimental data obtained from STA tests performed at 5 and 20 K min^{-1} heating rates.[39] This provided confidence that the pyrolysis model developed until now can be used to predict mass loss rate at different heating rates.

20.4.3 Gram-Scale Experiments and Modeling

20.4.3.1 Controlled Atmosphere Pyrolysis Apparatus II (CAPA II)

Controlled Atmosphere Pyrolysis Apparatus II (CAPA II) is an experimental apparatus developed to study gasification of gram-scale, axisymmetric solid samples in a controlled atmosphere. In this apparatus, the sample is exposed to a radiant heat flux and its mass loss rate (MLR), back surface temperature, and cross-sectional profile are tracked simultaneously in time. The sample is placed on a copper foil inside a sample holder that rests on a high-resolution mass balance. The sample's front surface is heated with a nearly uniform radiant heat flux, while its sides are insulated to ensure nearly one-dimensional heating in the axial direction. The sample is supported by a copper foil whose back surface is coated with a black paint of known emissivity. A calibrated infrared camera, which does not interfere with the operation of the mass balance, is focused on this surface to provide an accurate temperature measurement. A video recording through an observation window allows for tracking cross-sectional profile of the material. Further details on the CAPA II design, characterization, and diagnostics can be found in an earlier publication.[28,31] The obtained back surface temperature and thickness change data can then be used to optimize effective thermal conductivity and density respectively for the solid components in the user-defined

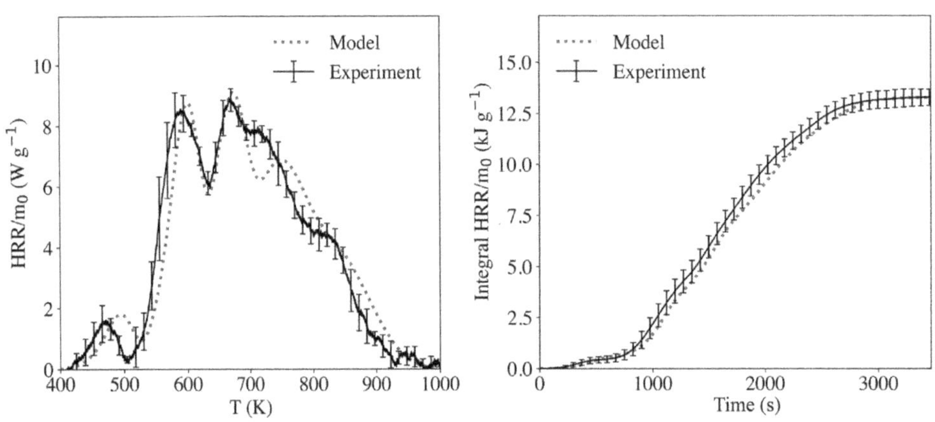

FIGURE 20.6 Measured HRR and integral HRR (both normalized by initial mass) compared with pyrolysis model predictions.

TABLE 20.2
Thermal Degradation Reaction Mechanism and Parameters for PIR Foam

#	Reaction Equation	A_r (s^{-1})	E_r (J mol^{-1})	ΔH_r (J kg^{-1})$^+$
1	PIR → 0.992 PIR_Int_1 + 0.008 Gas_1	2.42×10^8	8.09×10^4	0
2	PIR_Int_1 → 0.96 PIR_Int_2 + 0.04 Gas_2	2.27×10^7	8.96×10^4	2.16×10^4
3	PIR_Int_2 → 0.80 PIR_Int_3 + 0.20 Gas_3	5.63×10^8	1.24×10^5	2.15×10^5
4	PIR_Int_3 → 0.92 PIR_Int_4 + 0.08 Gas_4	1.00×10^8	1.19×10^5	-1.6×10^5
5	PIR_Int_4 → 0.83 PIR_Int_5 + 0.17 Gas_5	2.84×10^7	1.24×10^5	9.37×10^3
6	PIR_Int_5 → 0.71 PIR_Int_6 + 0.29 Gas_6	6.85×10^3	8.85×10^4	1.06×10^5
7	PIR_Int_6 → 0.73 PIR_Int_7 + 0.27 Gas_7	5.47×10^1	6.89×10^4	2.06×10^5

Notes: $^+$Positive values correspond to an endothermic reaction.

TABLE 20.3
Specific Heat Capacity of Condensed Phase Components and Heats of Combustion of Gas Phase Components

Condensed Phase Component	C_p (J kg^{-1} K^{-1})	Gas Phase Component*	ΔH_c (J kg^{-1})
Glass fiber	$1.24 \times T + 442$[40]	Gas_1	0
PIR	$4.86 \times T–357$	Gas_2	1.5×10^7
PIR_Int_1	1684	Gas_3	1.5×10^7
PIR_Int_2	1684	Gas_4	0.8×10^7
PIR_Int_3	1684	Gas_5	2.4×10^7
PIR_Int_4	1684	Gas_6	2.0×10^7
PIR_Int_5	1684	Gas_7	3.3×10^7
PIR_Int_6	1684		
PIR_Int_7	$0.41 \times T–883$		

Notes: *Specific heat capacity of all gas phase components was 2100 J kg^{-1} K^{-1}.

thermal degradation reaction mechanism. Eventually, the mass loss rate experimental data can be used for validating the pyrolysis model.

The PIR foam samples used for the CAPA II experiments were 1.27×10^{-2} m thick, 7.00×10^{-2} m diameter disks, which were cut from the core of the Tuff-R insulation panels. The samples were dried for at least 24 h in a desiccator in the presence of Drierite prior to testing. The initial mass and room temperature density of the samples were 1.6×10^{-3} kg and 32 kg m^{-3}, respectively. The samples were pyrolyzed in nitrogen atmosphere with oxygen content below 0.6 vol% O_2 in the vicinity of the front sample surface. Two tests were performed at a set radiant heat flux of 40 kW m^{-2} and three at 70 kW m^{-2}.

20.4.3.2 Modeling of CAPA II

Almost all the observed mass loss of PIR foam in CAPA II occurred in the first 200 s after being exposed to either of the heat fluxes. The back surface temperature reached a maximum of around 520 K for 40 kW m^{-2} and 620 K for 70 kW m^{-2}. CAPA II experimental data for 70 kW m^{-2} external heat flux is shown in Figure 20.7. The sample thickness was estimated for the 70 kW m^{-2} test and was observed to first increase by about 25% and subsequently contract to 15% of the original thickness. The back surface temperature and thickness change data from the 70 kW m^{-2} CAPA II experiments were used as target for optimization because the PIR samples underwent complete thermal degradation (throughout the entire depth) at this heat flux.

CAPA II tests were modeled as a one-dimensional process using ThermaKin2Ds because the experimental results indicated that both sample thickness and temperature remained radially uniform throughout the experiments. The convective and radiative boundary conditions of the CAPA II tests were carefully characterized in an earlier study[28] and were utilized in this model setup. The front sample surface was modeled to have no resistance to gas flow. The back surface, covered with copper foil, was set to be impenetrable to gas flow. The radiation absorption coefficients of all solid-phase components were set to 3,200 m^{-1}, which was estimated as reciprocal of the size of open pores at the surface of the sample. The densities of the condensed phase degradation products of the PIR foam were prescribed and then optimized to capture the changes in the sample thickness observed in the 70 kW m^{-2} tests. The thermal conductivities of the foam and products of its degradation were assumed to be independent of temperature, to minimize the number of adjustable parameters, and were optimized to capture the experimental back sample surface temperature profile obtained at 70 kW m^{-2}. As in the case of the TGA data, the optimization of densities followed by thermal conductivities was carried out using a hill climbing algorithm. Details of this algorithm implementation can be found in an earlier publication.[36] All simulations were performed with 5×10^{-5} m spatial resolution and 0.01 s time step. Increasing or reducing these integration parameters by a factor of two did not produce any significant changes in the simulation results, indicating convergence. Figure 20.7 compares CAPA II test data with simulation results. It is important to note that mass fraction and MLR profiles were not used in parameter optimization, and the results shown here are model predictions and thus provide model validation. Additionally, data from CAPA II experiments performed at 40 kW m^{-2} were used for model validation. Optimized model parameters are listed in Table 20.4.

Mass loss (mass fraction profile in Figure 20.7) was underpredicted in CAPA II simulations. This was attributed to the inability to account for release of the blowing agent located in the foam pores. This inability was a consequence of the fact that the blowing agent was lost during the preparation of the milligram-scale test samples, which were used to parameterize the thermal degradation mechanism.

20.4.3.3 Cone Calorimeter

The cone calorimeter is a fire test apparatus, standardized in ASTM E1354, that is used to determine instantaneous HRR and smoke release rate of a gram-scale sample having an exposed surface area of 0.1 m × 0.1 m. The sample is usually wrapped on the sides and bottom with aluminum foil and placed on an insulation. The front surface of the sample is exposed to a radiant heat flux, and its mass-loss and heat release rate are obtained as a function of time. In this study, the difference between cone calorimeter and CAPA II is that cone tests are performed in air atmosphere, and a continuous flame (dependent on the material) is present on top of the pyrolyzing material. Data obtained from cone calorimeter experiments can be useful in further validating and calibrating the pyrolysis model.

Cone calorimetry was performed on the PIR foam core to elucidate the impact of the blowing agent present in the foam on its pyrolysis model. It was hypothesized that the blowing agent contributed significantly to the burning process, and the effects of the blowing agent were not captured in the milligram-scale experiments and therefore were not characterized prior to modeling the cone calorimeter experiments. The cone calorimeter samples, which were chosen to be twice as thick as the CAPA II samples, were more likely to retain blowing agent in the foam pores during sample preparation and conditioning process. This test was carried out in accordance with the ASTM E1354 standard [35] at 50 kW m^{-2} of radiant heat flux and was repeated three times. The foam samples used in these tests were 0.1 m × 0.1 m × 2.54×10^{-2} m cuboids and sample MLR and HRR were recorded.

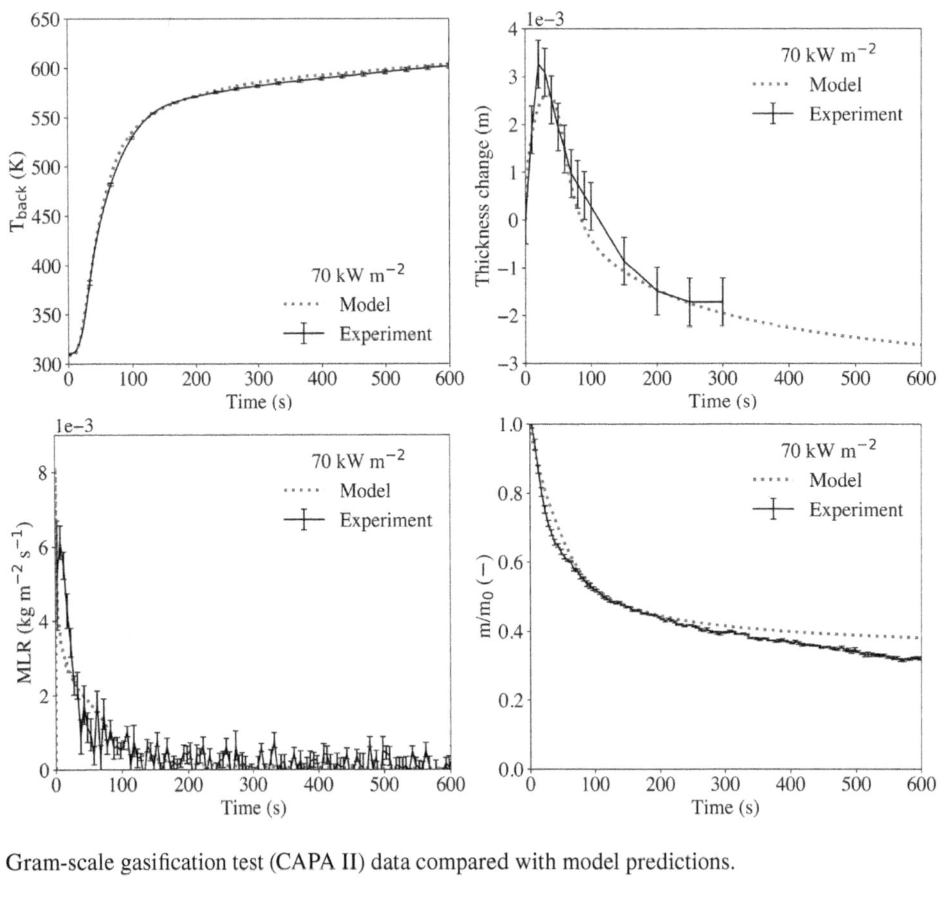

FIGURE 20.7 Gram-scale gasification test (CAPA II) data compared with model predictions.

20.4.3.4 Modeling of Cone Calorimeter

Cone calorimeter test results showed sustained flaming of the PIR foam immediately after its exposure to 50 kW m^{-2} heat flux. The sample expanded slowly, and continuous flaming transitioned to fragmented flames after about 70 s. Flaming ceased completely after 200 s and negligible mass loss was observed after that. The residue yield was estimated to be 18 ± 3 wt%, and effective heat of combustion was calculated as 13.9 ± 1 kJ per gram of initial mass. This heat of combustion value compares well with the value observed in MCC, which is unexpected due to known incomplete combustion in the cone calorimeter compared to complete combustion in the MCC tubular reactor.[41] However, this discrepancy is attributed to the iso-pentane blowing agent, which was likely lost during MCC sample (mg-scale) preparation but was contained in the cone calorimeter samples and contributed to the HRR.

The cone calorimetry experiments were also simulated using the one-dimensional version of ThermaKin2Ds. The pyrolysis model developed so far using previous tests was used in this simulation. The sample was simulated by prescribing 7 wt% iso-pentane blowing agent[34] in the interior 0.0234 m of the 0.0254 m sample based on observation of open pores on the surface of the samples. Heat capacity and heat of combustion of the blowing agent were set to the corresponding properties of iso-pentane[42] at 300 K, 1667 J kg^{-1} K^{-1}, and 49.5 kJ g^{-1}, respectively. The initial component PIR in the reaction mechanism was set to have a gas transfer coefficient of 1×10^{-30} m^2 s^{-1}, to ensure that it was impenetrable to gas flow and that the blowing agent was retained in the foam matrix until the first component decomposed to PIR_Int_1. Gas transfer coefficient of all other solid components (PIR_Int_1 through PIR_Int_7) was set as 2×10^{-5} m^2 s^{-1} to simulate uninhibited transport of gaseous pyrolyzate evolving from subsequent reactions. The front sample boundary condition was modeled as a constant radiative heat flux before ignition (50 kW m^{-2}) and a flame heat feedback model after ignition, which was developed and validated for the cone calorimeter by McCoy et al.[43] The ignition was assumed to take place once the gaseous fuel mass flux reached the magnitude corresponding to a critical value of HRR[44] of 21 kW m^{-2}. All simulations were performed with 5×10^{-5} m spatial resolution and 0.01 s time step. Prediction of MLR and HRR before and after the inclusion of blowing agent in the pyrolysis model is shown in Figure 20.8. Thus, including the blowing agent in the pyrolysis model was necessary to accurately capture its contribution in the thermal degradation of the PIR foam and subsequently improved integral HRR and integral MLR predictions of cone calorimeter tests.

20.4.3.5 Pyrolysis Model Parameters Determined from Gram-Scale Tests

Thermal transport properties, namely, emissivity (ε), absorption coefficient (α), density (ρ), and thermal conductivity (k) are tabulated below in Table 20.4. The emissivity value for condensed phase components was estimated by conducting CAPA II tests with and without a layer of paint, having an emissivity of 0.95, on the exposed surface. Both the tests showed negligible difference in the back surface and mass loss rate evolution with time, and thus the emissivity of all condensed phase components was fixed at 0.95. The absorption coefficients for all condensed phase components were set to 3200 m^{-1}, which was approximately equal to the reciprocal of the depth of open pores located on the sample surface. Gas transport coefficient for all condensed phase components, except the initial PIR component, were set to 2×10^{-5} m^2 s^{-1} to ensure uninhibited gas transport. The gas transport coefficient for the PIR component was modified to 1×10^{-30} m^2 s^{-1}, as described in Section 20.4.3.2, to retain blowing agent in the polymer matrix before the PIR component degraded.

20.4.4 State of the Art of Pyrolysis Modeling

Table 20.2, 20.3, and 20.4 together form a comprehensive pyrolysis model for the PIR foam. Such a model was also developed for the cast black PMMA[36] mentioned in

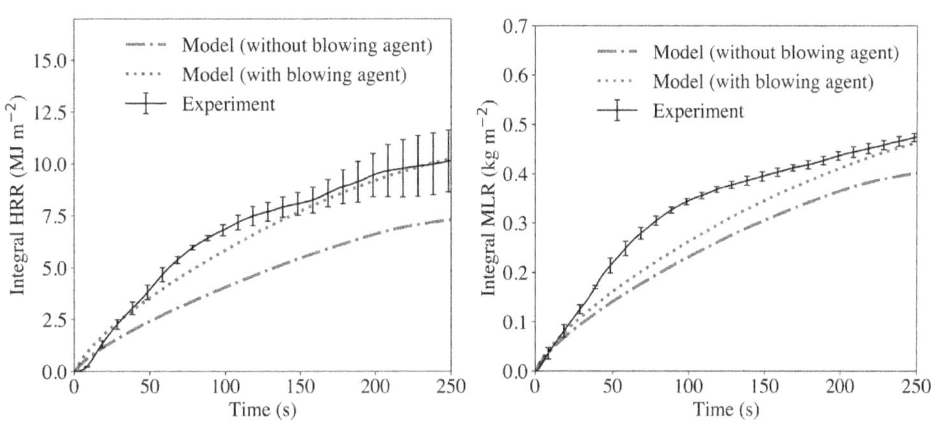

FIGURE 20.8 Integral HRR and MLR measured in cone calorimeter for PIR foam compared with predictions of model with and without the consideration of blowing agent.

TABLE 20.4
Thermal Transport Properties of Condensed Phase Components of the PIR Foam

Component	ε(-)	α(m^{-1})	ρ(kg m^{-3})	k (W m^{-1} K^{-1})
Glass fiber[40,45]	0.81	4160	2600	0.36
PIR	0.95	3200	31.4	0.05
PIR_Int_1	0.95	3200	15	0.09
PIR_Int_2	0.95	3200	30	0.04
PIR_Int_3	0.95	3200	18	0.08
PIR_Int_4	0.95	3200	17	0.06
PIR_Int_5	0.95	3200	16	0.14
PIR_Int_6	0.95	3200	14	0.15
PIR_Int_7	0.95	3200	14	0.25

Section 20.4.1. Li and Stoliarov measured kinetics and thermodynamics of thermal degradation of non-charring polymers[27] (such as poly(oxymethylene), high-impact polystyrene, polyamide 6,6, polypropylene, poly(lactic acid), and poly(acrylonitrile butadiene styrene)), and charring polymers[26] (such as poly(methyl methacrylate) epoly(vinyl chloride) alloy (Kydex), polymerized diglycidylether of bisphenol A, poly(ethylene terephthalate), poly(paraphenylene terephthalamide) (Kevlar), polymerized bisphenol A cyanate ester, poly(phenylene sulfide), polyetherimide, and poly(ether ether ketone)). Single-step and multi-step mechanisms describing thermal degradation of materials often require optimization of model parameters such that an optimization function (also called as evaluation function or goodness of fit criterion), that determines error between experimental and predicted data, is minimized. This was done in this study (as described earlier) using hill-climbing algorithm,[36] but other algorithms such as genetic algorithm (GA)[21,22] and shuffled complex evolution (SCE)[16] can also be used. However, it is essential to constrain optimization to physically meaningful values of various properties so that the pyrolysis model can be reliably used for heating scenarios not considered during model development.

The methodology shown in Figure 20.3 was also utilized to incorporate the effect of flame retardants and inert glass fiber additives[46–49] and intumescence.[50] In 2016, Ding et al. parametrized reaction kinetics, thermodynamics, and heats of combustion of gaseous pyrolyzates of poly(lactic acid) containing active flame retardants (intumescence-inducing ingredients), Melamine and ammonium polyphosphate (APP), with varying compositions. In 2019, Ding et al.[45] investigated the relationship between flammability behavior and material composition for glass fiber (GF) reinforced polyamide 66 (PA66 or Nylon 66) blended with red phosphorus (RP). In that study, a comprehensive pyrolysis model developed for varying composition of PA66, GF, and RP concentrations was used to correctly predict the pyrolyzate production rate for compositions where RP concentrations (in wt%) were between and significantly above those used for model development. Although gas phase inhibition effects in HRR predictions of these compositions exposed to 50 kW m^{-2} external heat flux was not considered, they found out that addition of 2 wt% RP reduces HRR significantly primarily due to the thermal barrier formed by the products of the chemical interactions between RP and PA66. The impact of chemical structure on thermal degradation have been observed experimentally.[37] The series of studies performed by Ding et al.,[46–49] Jayarama Krishna et al.,[51] Coimbra et al.[52] to model such interactions further demonstrates the ability of such methodology (Figure 20.3) to develop comprehensive pyrolysis models to assist with development of fire-safe polymeric solids.

20.5 RELATING MATERIAL PROPERTIES AND MOLECULAR STRUCTURE

Deriving the properties of materials from inverse analysis of data obtained from milligram and gram-scale experiments, as illustrated with an example of PIR foam in Section 20.4, is a reasonable approach when the polymeric material is available to produce such a range of samples. Polymer scientists, however, are constantly challenged with producing new formulations of polymers to address the needs of society such as reducing cost and environmental impact or formulating a polymer matrix with high resilience for a specific field of application. Polymeric properties, during their developmental phase cannot be known a priori, unless material scientists utilize some predictive models. Additive methods (utilizing group contributions), the backbone of *Quantitative Structure Activity Relationships* (QSAR), have been used for decades to successfully predict chemical reactivity and biological activity of molecules in medicinal and agricultural chemistry. Such methods are also used in *Quantitative Structure Property Relationships* (QSPRs) mathematically link physical or chemical properties with the structure of a molecule.

20.5.1 Group Contributions to Characterize Material Properties

D. Van Krevelen and K. Te Nijenhuis[53] were among the first to establish a structure-property relationship that correlated properties of known polymers with their chemical structure. The fundamental assumption in the proposed approach was that the physical property of a compound, for example, a polymer, can be estimated by the sum of contributions made by the structural and functional groups in the repeating unit of the polymer. This additive principle expresses a molar polymer property, ***P***, as a summation of either atomic, group, or bond contributions.

$$P = \sum_i n_i P_i \tag{20.2}$$

In Equation 20.2, P_i is the numerical contribution of component i, and n_i is the number of contributing components of the type i. These properties include thermo-physical and optical parameters and parameters describing thermal degradation, such as temperature at half degradation and char residue.

For example, Table 5.1 in the book *Properties of Polymers—Their Correlation With Chemical Structure; Their Numerical Estimation And Prediction From Additive Group Contributions*[53] provides molar specific heat capacity at 298 K for molar groups ($-CH_3$), ($-CH_2$), ($>C<$), and ($-COO-$), which are the main groups in the poly(methyl methacrylate) (PMMA) monomer shown below in Figure 20.9.

Considering 100% crystalline PMMA polymer, the specific heat capacity of PMMA (C_P) can be determined by contribution of the molar groups to molar specific heat capacity. The molar specific heat capacity (in $J\,mol^{-1}K^{-1}$) are 25.35, 30.9, 6.2, 46 for ($-CH_2$), ($-CH_3$), ($>C<$), and ($-COO-$), respectively. Following Equation 20.2 and considering that there are two molar group contributions from ($-CH_3$), $C_P = 139.35\,J\,mol^{-1}K^{-1} \approx 1393\,J\,kg^{-1}K^{-1}$. This value is in close agreement with experimental value[53] of 1340 $J\,kg^{-1}K^{-1}$. Temperature dependence of heat capacity was also correlated for various molar groups as a linear function of temperature. Similarly, thermal conductivity (k) was correlated as a function of specific heat capacity (C_P), density (ρ), velocity of sound (elastic waves) (U), and average free path (L) shown in Equation 20.3. Each of these parameters have corresponding molar group contribution values. Temperature dependent functions are also available to determine thermal conductivities at elevated temperatures.

$$k = C_p \rho U L \tag{20.3}$$

Bicerano[54] generalized the additive method using graph theory to develop atomic connectivity indices for describing the monomer unit topology of a polymer. This method can be used for prediction of the polymer properties by the summation of additive contributions over atoms and bonds, instead of groups. Connectivity indices can be obtained from Lewis diagrams which represent valence shell electrons in a molecule. However, obtaining the connectivity indices is also inherently empirical and dependent on experimental data or available group contribution data.

Modeling presents a viable alternative to experiments, and molecular modeling and data-centered approaches have gained attention in recent years. Molecular modeling

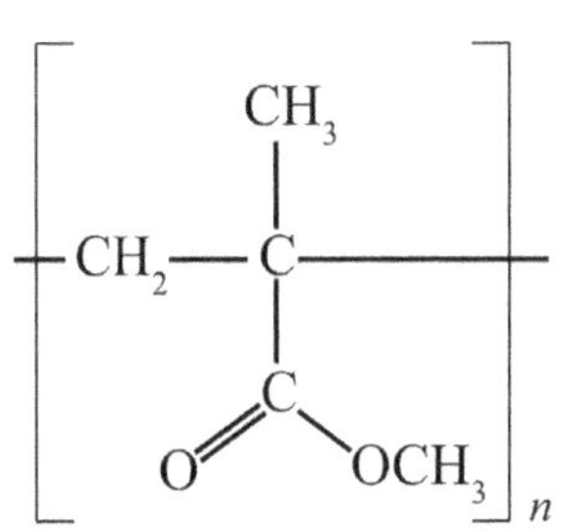

FIGURE 20.9 Monomer of poly(methyl methacrylate) (PMMA).

(of force field or quantum mechanical calculations) can be combined with group additivity theory to yield models that are not restricted by lack of requisite group additive parameters and take advantage of three-dimensional molecular information.[55] Reactive molecular dynamics (RMD) has also been applied to successfully predict the thermal degradation mechanism and product distribution of simple polymers and polymer composites[56–59] and studying the effect of fire-retardant filler on thermal degradation.[60] However, such simulations can be computationally expensive, especially when the reaction system involves long chain polymers.

20.5.2 State of the Art of Quantitative Structure-Property Relationships

Several QSPR studies have attempted to obtain polymers properties describing flammability using the group contribution approach. Walters et al.,[61] in 2000, compared heats of combustion, of 49 developmental and commercial polymers, measured in oxygen bomb calorimeter with those obtained from thermochemical calculations for complete combustion. The heats of combustion were calculated considering oxygen consumption principle and the gross heats of combustion from group additivity of the heats of formation of products and reactants. The calculated gross and net heats of combustion were within 5% of the experimentally measured values. However, the ability to perform these predictions was limited by availability of experimental data of group contribution to heats of formation. In the study conducted by Walters et al.,[61] molar group contribution of the phosphine oxide group ($-OPX_3$) to the heat of formation of the Arylether of hexafluorobisphenol-A and triphenylphosphine oxide (Tradename 6F-ETPP™), silicon monoxide (SiO) group to the heat of formation of polydimethylsiloxane (PDMS), and the naphthyl group to the heat of formation of polyethylene naphthalate (PEN) were not available, and the corresponding heats of combustion for these polymers could not be calculated.

In 2009, Lyon et al.[62] correlated material properties rooted in the polymer structure, for 73 synthesized and 11 commercial polymers, to standardized fire test results. Heats of combustion, heat release parameter, and char yield was determined using additive molar group contributions and compared with experimental data obtained using Microscale Combustion Calorimeter and vertical UL 94-V standard test.[63] The derived properties showed reasonable agreement with the measured quantities. In particular, the char yield and heat of combustion correlated well with qualitative and quantitative standardized test results, suggesting a molecular basis for flammability.

Later, in 2012, Sonnier et al.[64] studied the relationship of molecular structure and material flammability considering total heat released (THR) and heat release capacity (HRC) of polymers containing phosphorous, an element known for flame retardant properties. They quantified interaction indices of phosphonate functions with carboxylic acid or ester groups, whose interactions reduce the material flammability and promote char formation. The predictions of the material flammability properties improved after considering these interactions. An improved model was later proposed,[65] where a guideline was developed for fragmenting groups, by considering about 100 polymers. Based on this established methodology to quantify group contribution to THR and HRC, Sonnier et al.[66] further quantified the contributions to flammability from 14 new chemical groups from a set of 30 thermoset polymers. They also discussed the impact of aromaticity, number of covalent bonds between the chemical group and neighboring groups, and presence of heteroatoms (N, O) on flammability. The contribution of two phosphorus groups, particularly 9,10-dihydro-9-oxa-10-phosphaphenanthrene-10-oxide (DOPO) and PO3, were also explored and the calculated contribution to THR and HRC were consistent with the main modes of action of these two groups.[67]

Recent exploration using the amount of available data have also led to development of machine learning algorithms that can help speed up the process of material property estimation. Development of QSPRs consist of four steps[68]:

1. Calculation of molecular descriptors.
2. Descriptor selection for model building.
3. Finding an optimal relationship between selected descriptors and target activity.
4. Validation of that relationship's predictive capabilities and applicable domain.

Molecular descriptors could be as simple as atom count and structural fragment lists which can be determined from the text representing the molecular structure. Higher order descriptors characterize topology (2D) or topography (3D).[68] Polymer Genome project, developed by Rampi et al.,[69,70] numerically represents contributions from groups, atoms, bonds, etc. in such descriptors or a fingerprint. Currently, Polymer Genome supports two databases—linear and ladder polymers—for predicting a variety of material properties. Data centered approaches could potentially utilize available additive methods and facilitate faster prediction of material properties.

Although the group or atomic contribution approach, which could be coupled with data-centered approaches, can be a promising for predicting material properties, further research still needs to be conducted to facilitate a quantitative link between molecular structure and material properties. Some of the fundamental limitations are given here:

- The availability of high-quality experimental data for atomic or group contributions to a particular property.
- Group contribution formulations that link chemical structure of additives to their impact on the matrix material performance don't exist, and the data that are needed to develop such formulations are very sparse.
- Expanding such an approach is challenging for composites where the material property is not only a function of molecular structure but also of macrostructure. For example, thermal conductivity of a roof shingle cannot be calculated from the thermal conductivity of constituent polymers.
- Pyrolysis models for even simple monomer structure, such as PMMA, have not yet been developed using QSPR. Such an endeavor can help in determining whether the current QSPR correlations can be used for comprehensive pyrolysis modeling and predicting behavior of a material in a fire growth scenario.

20.6 FIRE GROWTH MODELING CASE STUDY

Apart from creating a link between molecular structure and material properties which consequently can be used for comprehensive pyrolysis model development, it is also important to understand whether these pyrolysis models can be used to predict large-scale flame spread scenarios. Therefore, a fire growth modeling case study is discussed in this section where a semi-empirical model and a CFD-based model were developed to predict fire growth over materials in a corner geometry. The experimental setup was based on a Single Burning Item (SBI) EN 13823 standardized test[11] which classifies wall-lining materials according to their fire growth and smoke production rate. A brief description of the setup and the instrumentation is provided here, but additional details of the experimental setup and other instrumentation can be found elsewhere.[71]

20.6.1 Experiment Overview

The experimental setup, shown in Figure 20.10, used to perform flame spread tests on the materials was built at the University of Maryland, College Park, based on the Single Burning Item (SBI) test[11] scenario. The SBI scenario is an important standardized test which assesses the flame spread propensity of wall mounting materials. The setup consisted of two identical panels of the desired test material rigidly mounted onto a 0.013 m thick Marinite I calcium silicate board in a symmetrical corner geometry. The panels were mounted such that the vertical distance from the top of the burner sidewalls to the top of the panels was 1.425 m. A triangular sandbox burner having cross-section of an isosceles right-angled triangle with equal sides being 0.25 m in length, fueled by propane, was placed 0.04 m away from both the panels. This burner was fueled with a constant propane flowrate of 6.47×10^{-4} kg s^{-1}, corresponding to the HRR of 30 kW. The entire corner assembly was placed under a 2 m × 2 m exhaust collection hood, whose bottom edge was 2.2 m above the floor. A flame-resistant curtain extended 1.6 m below the bottom of the hood. A fabric mesh covered the remaining 0.6 m to the concrete floor to homogenize the incoming air flow. The exhaust duct was a galvanized steel pipe having a circular cross-section and an inner diameter of 0.28 m. The combustion products were extracted continuously at a constant flowrate of 0.56 $m^3 s^{-1}$.

20.6.1.1 Measurements

A diagnostic section in the exhaust duct continuously monitored temperatures and flowrate using three K-type thermocouple probes and a multipoint averaging pitot tube, respectively. Exhaust gases were sampled, cooled by natural convection, filtered, dried, and desorbed of carbon dioxide before they were sent to a fast-response PM1111E Hummingbird paramagnetic oxygen sensor. The intrinsic error in O_2 measurements were < 0.01 vol%. The response time of the gas sampling system, defined as the time it takes to reach 63% of the full response to a step change in the O_2 concentration, was determined to be about 3 s. These measurements were used to perform oxygen consumption calorimetry, which was formulated to account for fuel molecular structure and correct for changes in ambient conditions and extracted moisture and CO_2,[71,72] to determine instantaneous HRR of the flame spread process. Corrections were made to account for transport delay of 10 s through the gas sampling system.

In addition to the measurements that facilitated HRR calculation, eight Schmidt-Boelter water-cooled total heat flux gauges (9.53×10^{-3} m diameter) were installed flush with the panel surface to measure flame heat feedback during the flame spread process. The cooling water temperature was maintained at 291 ± 7 K. The sensors were cleaned, repainted, and recalibrated against a standard gauge traceable to a NIST reference before every experiment. The gauge locations are defined as distance from the corner, *x*, and distance from top edge of the burner, *y*. The symmetry of the heat feedback was verified by placing some gauges at the same location on the opposite panel. Total flame heat flux was measured twice at each of the twenty-eight selected locations across 7 tests performed with PMMA. Total flame heat flux was measured at seven locations for the PIR foam. Experiments were conducted with PIR foam without the foil facers.

20.6.1.2 Experimental Heat Release Rate (HRR)

Instantaneous HRR for PMMA and PIR foam are plotted below in Figure 20.11. Here, $t = 0$ s corresponds to propane burner ignition and the HRR includes contribution of 30 kW from the propane burner. The HRR for PIR foam increased to the maximum of around 70 kW immediately after its ignition. PIR foam ignition was followed by a formation of a single fire vortex, attributed to quick release of the blowing agent, that rolled to the top of the corner, and the flame tip extended above the panels at the maximum HRR. The flames over the PIR foam gradually subsided and HRR nearly flattened around 35 kW after 200 s. Ignition of PMMA panels was observed to take place 45 s after propane burner ignition while PIR foam was seen to ignite within 5 to 15 s after the burner was turned on. The HRR for PMMA increased exponentially after its ignition corresponding to flame spread on the panels and the HRR rise rate became nearly linear as the flames reached the top of the panels.

20.6.1.3 Experimental Flame Heat Feedback

Two datasets of flame heat flux at each of the 28 locations (for PMMA) were first averaged, then post processed by binning them in 10 s intervals. Average error was calculated for each of the 10 s bins as two standard deviations of the mean plus 3% systematic error of the calibration. The average uncertainty was determined to be about 10%. The measured total flame heat flux was found to have a maximum of 65 kW m^{-2} near the corner (0.05 m away from the corner) at multiple distances above the burner surface. Generally, the heat fluxes were higher near the corner and decreased away from the corner to the outside of the panels. The data showed a persistent reduction in the value of the heat flux at about 0.50 m above the burner. This height approximately corresponds to the length of the burner flame, and this

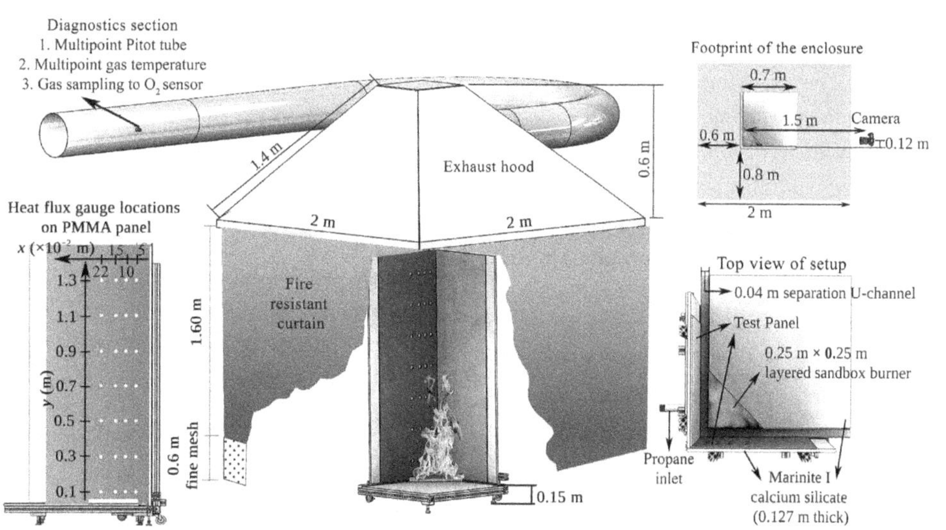

FIGURE 20.10 Schematic of large-scale experimental setup.

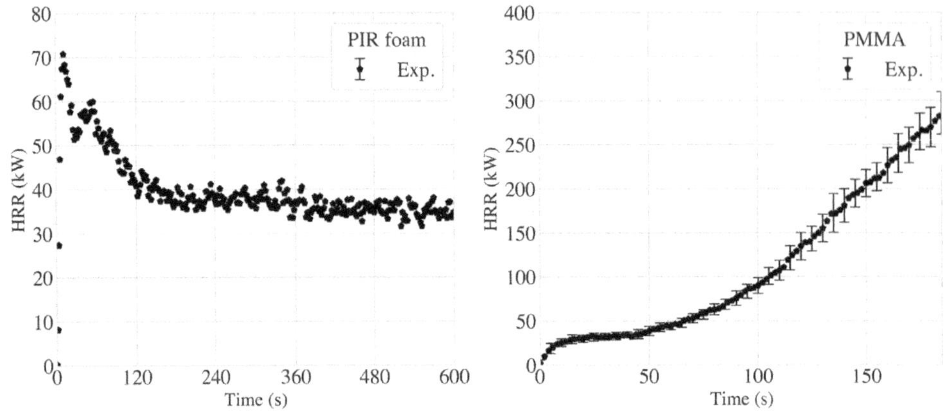

FIGURE 20.11 HRR evolution during flame spread over PIR foam (left) and PMMA (right).

reduction is speculated to be associated with air entrainment promoted by the burner flame that narrows the flame in this region and cools the solid surface. Further details of the data can be found elsewhere.[71,72]

Knowing that the wall flame heat flux profiles are observed to be a function of HRR for buoyancy-driven flames,[73,74] measured total flame heat flux at 28 locations in the PMMA tests were correlated with total measured HRR. An example of the relationship is shown in Figure 20.12 for four locations at 0.10 m above the burner surface. At most locations, the relationship reaches a plateau except at greater distance away from the corner (x = 0.22 m) where the flame does not fully spread. The flame heat flux measured at these locations in the PMMA tests were correlated with HRR using the following exponential function.

$$\dot{q}'' = \frac{e - a \cdot \exp(-b \cdot HRR) - c \cdot \exp(-d \cdot HRR)}{f - a \cdot \exp(-b \cdot HRR) - c \cdot \exp(-d \cdot HRR)} \quad (20.4)$$

The function parameters a, b, c, d, e, and f were optimized for each location such that the coefficient of determination was greater than 0.97. Parameters b and d are in kW^{-1} and the rest are in kW m^{-2}. The experimental data for PMMA measured 0.30 m above the burner surface at four locations away from the corner, compared with the Equation 20.4 fit for these locations are shown in Figure 20.12.

Flame heat flux obtained for the PIR foam tests at the seven locations follow a similar trend with changing HRR as that observed in the PMMA data at those locations. However, there were some differences in the measured total flame heat flux (compared to the PMMA data) during the initial fireball vortex event immediately following ignition. However, since comprehensive data for more locations was not available for PIR foam, this functional correlation of flame heat flux with HRR (Eq. 20.4) for 28 locations was used as a quantitative basis to formulate the flame heat feedback model for the semi-empirical flame spread model.

20.6.2 Fire Growth Modeling Framework

Modeling framework for a semi-empirical flame spread model and Fire Dynamics Simulator (FDS) model are discussed here. A semi-empirical flame spread model was formulated by coupling the flame heat feedback model and the pyrolysis model and used to predict fire growth over the two materials. Additionally, fire growth over these materials were also simulated in Fire Dynamics Simulator (FDS). The framework of these two models is presented below, and their predictions are compared and discussed in Section 20.6.3.

20.6.2.1 Semi-Empirical Model

The methodology for the developed semi-empirical model is illustrated in Figure 20.13. One panel area was divided into 28 areal elements, such that each element contained a heat flux gauge at or close to its geometric center. Considering the functional correlation between flame heat flux and HRR developed from the measurements over the PMMA panels, a time-dependent heat-flux profile was generated for each element corresponding to an arbitrary time-dependent input HRR profile. This resulting heat-flux profile, plus a small background contribution corresponding to average gauge temperature of 291 K, was applied to each element as a time-dependent external heating boundary condition in ThermaKin2Ds, by three segment piecewise-linear fit of the profile. The linear segments were fit such that the heating boundary condition could be modeled using either purely radiative heat flux model or hybrid heat flux model. In the radiation heat flux model, the entire heating profile was assumed to be completely radiative in nature. In the hybrid heat flux model, the linear segments were fit such that heat flux was modeled as a transition to purely radiative above 30 kW m^{-2} from purely convective heat flux below this threshold. The identified threshold was 2.5 times the convective flame heat flux estimate. One-dimensional transient simulation for 28 elements were then performed to model pyrolysis of the test material and heat transfer through the Marinite I calcium silicate board. The insulation board had nearly uniform properties across the temperature range of 290 to 550 K. The properties are

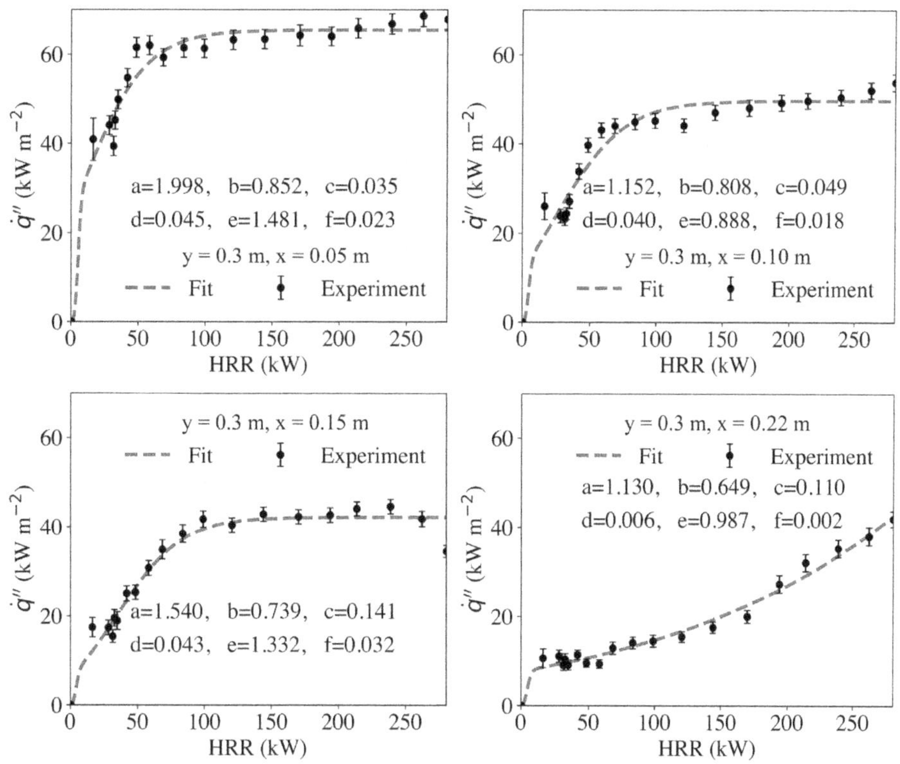

FIGURE 20.12 Example of measured total flame heat flux and corresponding fit during flame spread over PMMA. Experimental data and fits for other locations can be found elsewhere.[71,72]

density, $\rho = 737$ kg m^{-3}, specific heat capacity, $C_p = 1260$ J kg^{-1} K^{-1}, and thermal conductivity, $k = 0.13$ W m^{-1} K^{-1}, across the temperature range of 290 K to 550 K. Total simulated HRR was obtained by considering the propane contribution, HRR_b, and symmetric contribution from both panels in the corner geometry (consistent with the experimental observations). The contribution to total HRR from one panel was calculated by summation of heat released per unit area for each element, $HRRPUA_j$, multiplied by corresponding surface area of that element, A_j. The $HRRPUA_j$ was calculated, using Equation 20.4, by summation of the predicted pyrolyzate production rate, $\dot{m}''_{ij}$, of gaseous decomposition product i from an element j, assumed to be uniformly distributed across the area of an element, multiplied by the heats of combustion of respective pyrolyzate gas (ΔH_c^i) obtained from the MCC tests, for N_g gaseous pyrolyzates. The following Equations 20.5 and 20.6 summarizes the simulated HRR calculation.

$$HRRPUA_j(t) = \sum_{i=1}^{N_g} \dot{m}''_{ij}(t)\Delta H_c^i \quad (20.5)$$

$$Simulated\ HRR(t) = HRR_b(t) + 2 \sum_{j=1}^{28}\left[A_j \cdot HRRPUA_j(t)\right] \quad (20.6)$$

The simulated HRR was then used as an input HRR for the next iteration until the time-normalized L1-norm loss function until any time t_e ($t_e > 1$s), L_{SE}, defined by the following Eq. 20.7, was less than 5 kW.

$$L_{SE} = \frac{1}{t_e}\sum_{t'=0}^{t_e}\left|Input\ HRR(t') - Simulated\ HRR(t')\right| \quad (20.7)$$

The performance of these iterations was facilitated by development of a MATLAB script that was used to generate boundary condition parameters and post-process the results of individual ThermaKin2Ds simulations. Simulations for PMMA were conducted using hybrid heat flux model, and PIR foam were conducted with radiative heat flux model. The use of purely radiative heat flux for the PIR foam is justified by considering the pyrolyzate gases that are produced by PIR pyrolysis which are better precursor to soot. Flame supported over the PIR foam also showed significantly larger radiative intensity compared to the flame over PMMA. Convergence was achieved within six iterations for PMMA and three iterations for PIR foam, regardless of the assumed input HRR profile. These simulations were performed on a single core of an Intel Core i5 processor with max turbo frequency of 3.4 GHz.

20.6.2.2 Computational Fluid Dynamics Model

The flame spread model built in ThermaKin simulates the gas phase as a transient semi-empirical flame feedback process. However, CFD-based models such as FDS can numerically solve gas phase combustion and flame feedback to

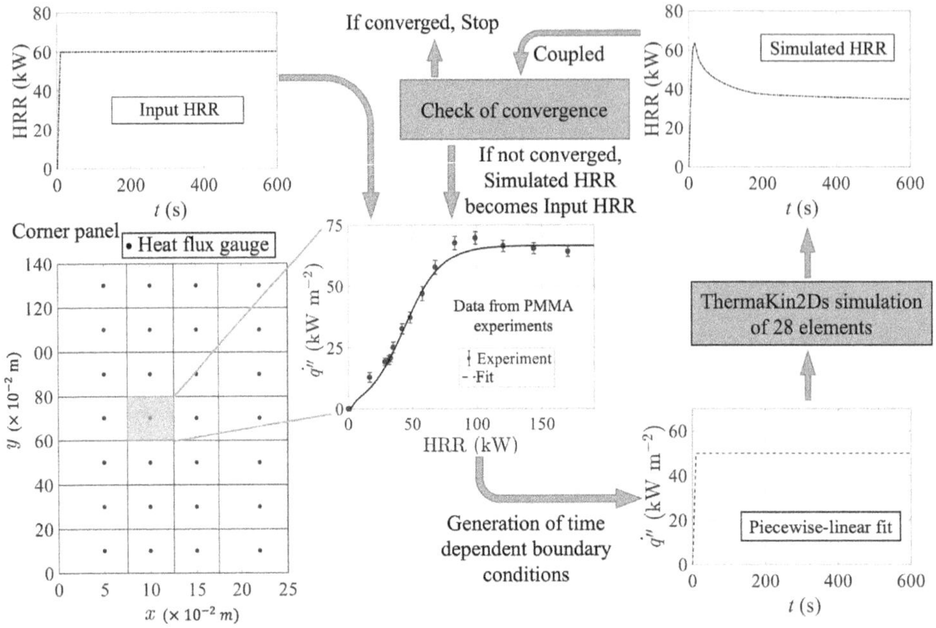

FIGURE 20.13 Framework of semi-empirical flame spread model.

the panel as desired in the case of a spreading flame. As discussed earlier, the gas phase combustion and heat transfer has been studied more compared to the condensed phase pyrolysis process. FDS simulations were conducted using FDS v.6.8, released on April 18, 2023, to investigate the state-of-the-art CFD modeling approach and understand the impact of coupling condensed phase modeling with gas phase calculations on fire growth predictions. FDS simulations were conducted for both PMMA and PIR foam. The reader is referred to Section 20.4.2 for an overview of default FDS model for turbulence, combustion, radiation, and convection which were used for setting up the FDS model.

First, the developed pyrolysis model for both the materials considered here was validated within the FDS framework by simulating the milligram-scale (TGA-DSC) and gram-scale gasification (CAPA II) tests. The experimental setup for the flame spread tests, shown in Figure 20.10, was then closely replicated, including the conical hood, in FDS. Here, the cartesian co-ordinates are defined such that x-direction is same as that used in the semi-empirical model definition, z-direction is in vertical direction, and y-direction is perpendicular to the xz-plane in the direction away from the corner. The exhaust was modeled through a single vent outlet, prescribed a flow rate of 0.56 $m^3\ s^{-1}$, having dimensions 0.2 m × 0.2 m placed at z = 2.88 m above the floor (approximate height at which the exhaust inlet was located). The base calcium silicate board was modeled as 0.04 m thick obstruction whose bottom surface was 0.08 m above the concrete floor. The finest mesh covered the region of height 1.68 m above the bottom surface of the base calcium silicate board and encompassed 0.8 m × 0.8 m area which consisted of 0.68 m in front of the panel surface and the remainder (0.12 m) extended behind the Marinite I backing board. Flame spread simulations were conducted with uniform cell sizes of 5×10^{-3} m, 0.01 m, 0.02 m, and 0.04 m for the finest mesh. The vertical panel and the backing board were modeled as 0.04 m thick obstruction. The coordinate (x, y, z) at the corner in the plane of the burner surface was (0, 0, 0.12). The remainder of the region was modeled with a uniform mesh having cell dimension of 0.04 m. The burner with triangular cross-section was modeled as a collection of 0.02 m thick rectangles and prescribed a mass loss rate per unit area equivalent to 30 kW burner flame.

Radiation was modeled using radiative fractions of 0.30 for propane, 0.35 for PMMA, and 0.40 pyrolyzate emanating from PIR foam pyrolysis, all which were obtained from literature.[75] The radiative fraction for the blowing agent was fixed at a value of 0.31 obtained from literature for butane,[75] gas having closest molecular structure to that of iso-pentane. The number of radiation angles (for spatial discretization of radiation transport equation) and path length (for RadCal narrow band model calculation of temperature-dependent absorption coefficient calculation for all gases and soot) were prescribed to be 400 and 0.5 m respectively for radiation calculation. These numbers were selected based on preliminary analysis and provided reasonable balance between accuracy and computational cost.

All combustion reactions were assumed to follow infinitely fast, single-step mixing-controlled combustion chemistry. Flame extinguishment was modeled using critical flame temperature and oxygen index concept, following which, for cell temperature above approximately 600°C, no flame extinction is assumed unless the oxygen concentration drops to zero.[23] Propane combustion was modeled using following parameter values[75]—heat of combustion of 43.8 MJ kg^{-1}, soot yield of 0.024 kg per kg of fuel, and CO yield of 0.005 kg per kg of fuel. Pyrolyzate emanating from PMMA pyrolysis were assumed to have a molecular formula of PMMA monomer ($C_3H_8O_2$) with a heat of combustion of 24.1 MJ kg^{-1}, soot and CO yield of 0.015 kg, and 0.005 kg per kg of fuel, respectively. For PIR foam, pyrolyzate gas emanating from all reactions was assumed to have the same molecular formula of methyl isocyanate ($H_3C-N{=}C{=}O$) based on the observed intensities of emitted gases measured by Marquis et al.[76] at around 600 K during PIR pyrolysis in nitrogen at 5 K min^{-1}. Pyrolyzate (methyl isocyanate) combustion was modeled such that the heat of combustion for each thermal degradation reaction corresponded to the value found from MCC analysis (Table 20.2). Soot yield and CO yield for methyl isocyanate combustion reaction was fixed at 0.13 kg and 0.031 kg per kg of fuel, respectively. These values were the values for GM29 rigid PU foam,[75] which had monomer molecular structure closest to estimated PIR foam monomer. Combustion of blowing agent for PIR foam (thermophysical properties of isopentane provided in Section 20.4.3.4) was modeled using properties for butane[75] (a hydrocarbon closest to isopentane for which the properties were available), with heat of combustion of 45.4 MJ kg^{-1} and soot and CO yield of 0.029 kg and 0.007 kg per kg of fuel, respectively.

Background air composition was fixed at the default of N_2, O_2, H_2O, CO_2 at 78.25, 20.76, 0.95, and 0.04 vol% respectively. Background temperature and background relative humidity was fixed at 20°C and 40% respectively. Simulations were conducted for 200 s for both PMMA and the PIR foam on FSRI computational cluster, which utilizes Intel Xeon Platinum 8358 processor with max turbo frequency of 3.40 GHz for computation.

20.6.3 Results and Discussion

Instantaneous heat release rate predicted by the semi-empirical (SE) model and CFD-based FDS model (0.01 m grid size) are compared in Figure 20.14. Here, the semi-empirical model used for the PIR foam simulated flame heat flux as purely radiative in nature and for PMMA it was simulated using the hybrid heat flux model. The details for these flame heat flux models are described earlier.

The model accuracy is quantified by calculating the HRR_{error}, which was formulated with the following expression.

$$HRR_{error} = \frac{1}{t_e - t_i} \int_{t'=t_i}^{t_e} \frac{(Simulated\ HRR - Experimental\ HRR)}{Peak\ Experimental\ HRR} dt' \quad (20.8)$$

Here, t_i is the initial time to calculate the error statistic which was $t_i = 0\,s$ for PIR foam and $t_i = 45\,s$ for PMMA, and t_e is the end time to calculate error statistics which was

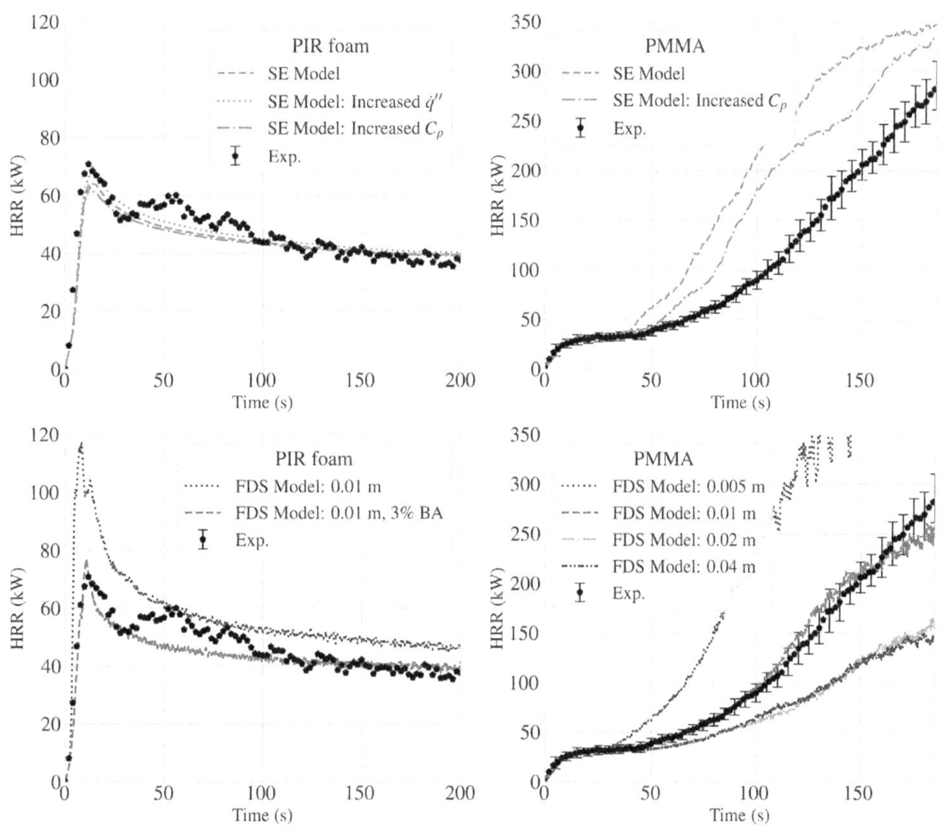

FIGURE 20.14 Predictions of heat release rate using semi-empirical (SE) model and fire dynamic simulator for PIR foam and PMMA.

$t_e = 200\,s$ for PIR foam and $t_e = 190\,s$ for PMMA. The initial time was chosen to calculate error in prediction of HRR evolution after the ignition of the respective panels. The HRR_{error} for both the models (for base case simulations) are tabulated in Table 20.5 below.

The semi-empirical (SE) model slightly underpredicted the HRR (still within the experimental uncertainty) for PIR foam by about 3% and overestimated the HRR for PMMA by about 33%. The peak heat release for PIR foam was predicted qualitatively well by the semi-empirical model. The semi-empirical model over-predicted the HRR evolution after the PMMA ignition. The onset of PMMA ignition was correctly predicted by both models to occur at around the experimental value of 45 s. A strong feedback between flame heat flux and the HRR in the simulations was more evident for larger fire growth (higher HRR). A slight overprediction (or underprediction) at an earlier iteration resulted in larger overprediction (or underprediction) in HRR in subsequent iterations due to the strong interaction. Thus, the excellent predictions for the PIR foam could be attributed to lack of significant flame spread that tends to amplify modeling errors. However, the SE model predictions for PMMA are outside the experimental uncertainties. It is possible that these discrepancies are possible due to the crude nature of the flame heat feedback model or uncertainties in the pyrolysis properties. To understand whether the uncertainties of pyrolysis properties can explain the discrepancies in predictions observed here, specific heat capacity was identified as a property whose variation had a large impact on pyrolyzate production rate. The uncertainty of heat capacity was determined to be 10% and increasing this value for the solid-phase components by 10% did not compromise on the CAPA II mass loss rate predictions. However, this increase produced a discernable impact on the HRR predictions, as shown in Figure 20.14. Similarly, increasing the heat flux profile values by its uncertainty (10%) for PIR foam simulation, in accordance with the observed flame being higher in radiation intensity than that of the flame supported over PMMA, further improved the HRR predictions. This indicates that the flame heat feedback model, which was developed using total flame heat flux data measured for experiments conducted with PMMA, needs to be calibrated for flames having different radiant fractions. Another parameter whose uncertainties were identified to affect predictions of HRR for PMMA was the split between convective and radiative portions of the flame heat feedback to the solid fuel.

TABLE 20.5
Error in Predictions of HRR during Flame Spread over PIR Foam and PMMA

Model Type	HRR_{error} (%)	
	PIR Foam	PMMA
Semi-empirical (SE) model	−3	+33
FDS model: 0.01 m	+14	+1

The FDS model (0.01 m grid size) overpredicts the HRR PIR foam by about 14% and predicts HRR for PMMA on an average within 1% of the experimental data. The FDS model overpredicted PIR foam HRR is primarily attributed to the uncertainty of the contribution of the blowing agent in the pyrolyzates. Although cone calorimeter data was used to optimize the model to consider blowing agent contribution in ThermaKin2Ds and validated in FDS, the FDS model was found to be sensitive to the uncertainty of the blowing agent composition. Another simulation conducted with 3 wt% blowing agent (BA) reduced the prediction error to 4%, as seen in Figure 20.14. The FDS model (0.01 m grid size) predicted HRR during flame spread over PMMA within the experimental uncertainty of the measured HRR data. The flame heat flux predicted by the FDS model (0.01 m grid size) were comparable to the measured values for PMMA at 28 locations (Figure 20.10), but these flame heat flux predictions were dependent on the resolution of the finest grid. As seen in Figure 20.14, the HRR predictions are also sensitive to the grid size of the finest mesh which results from the sensitivity of the flame heat feedback to the grid resolution and thereby affects the pyrolysis process. Part of the grid-dependency is hypothesized to result from convective heat flux model, which uses empirical correlations for estimating convective heat flux to the wall surface. The free stream temperature (for determining temperature gradient driving convection) of the boundary layer used for these calculations was assumed to be temperature of the gas immediately adjacent to the wall surface, which was naturally affected by the grid size. The good prediction for FDS model with 0.01 m grid size was a consequence of the selection of the grid size that probably represents the boundary layer for this scenario well and predicted HRR within experimental uncertainties. Apart from the pyrolysis properties and processes governing the flame spread process, the FDS model predictions were also sensitive to the composition of ambient gas (which affect ambient gas absorption coefficients) and the properties of surrounding materials that take part in the heat transfer (such as fire curtains and concrete floor).

It is also important to note that the time required for computation for these models was significantly different. The FDS model for PMMA and PIR foam with uniform 0.01 m cell size near the fire and panel region took approximately 68 hours each on 32 cores of the FSRI computational cluster while each semi-empirical model iteration took approximately 40 minutes on a single core of the Intel Core i5 processor. The semi-empirical model converged within five iterations for PMMA and two iterations for PIR foam.

The availability of comprehensive data for the presented experimental setup is valuable for making reliable predictions with CFD-based models as shown here. The semi-empirical flame spread model, which used the empirical flame heat feedback correlations developed using PMMA data, can be further improved by potentially calibrating it to account for the impact of flames with different radiative properties. The coupling of the pyrolysis model with the flame heat feedback model, as done in the SE model, amplified the sensitivity of the model predictions to the uncertainties in the input parameters. The amplification of uncertainties is a feature of any model attempting to capture the physics of the flame spread process. The uncertainties of key material parameters and convection and radiation split of the total flame heat flux, which rarely fall below 10%, need to be therefore substantially reduced to deliver measurements comparable to the measurement accuracy. FDS model predictions of HRR were also found to be sensitive to grid size, because of grid dependent flame heat feedback that affect the pyrolysis process. It is evident that sensitivity of the model to uncertainties in property values and model parameters affect HRR predictions need to be quantified. The next section attempts to quantify the SE model sensitivity to input parameters.

20.7 SENSITIVITY TO UNCERTAINTIES IN PROPERTY VALUES

The input parameters for a fire growth model, as described in Section 20.4, have uncertainties associated with them that are likely to contribute to uncertainty in the predictions themselves, as briefly discussed in Section 20.6.3. Several approaches have been identified that quantify the propagation of uncertainty of the input parameters in model predictions. However, a simplified nominal range sensitivity analysis[77] was performed here by varying individual model parameter one at a time and holding all other parameters at their nominal or base values. The effect on the model output due to a change in each parameter was then quantified. This analysis was performed using only the semi-empirical flame spread model because of relatively fewer model parameters compared to the FDS model.

The input parameters considered were all pyrolysis properties and the external heat flux boundary condition. The parameter value was increased one at a time by its uncertainty, which are listed in Table 20.6. The uncertainty values used here for α, ε, ΔH_c were measured in other studies, the values for ρ, $\dot{q}''$ were measured from the data obtained from experiments conducted with PMMA, and values for all other parameters were calculated from sensitivity analysis of inverse analysis techniques used to calculate the parameters. For degradation mechanisms with multiple reactions, the sensitivity analysis was performed by changing the property value, for example activation energy or density, for all reactions or solid-phase components respectively in the degradation mechanism. The impact on the model predictions was quantified by considering the change in HRR evolution from the base case.

The impact of change of a specific parameter on the HRR evolution was evaluated by calculating the sensitivity coefficient, S_{HRR}, using the following expressions,

TABLE 20.6
Uncertainties of Input Parameters Used for the Sensitivity Analysis

Type	Parameter	Variable	Uncertainty (±)
Reaction kinetics and thermodynamics	Pre-exponential factor	A	40% [27]
	Activation energy	E	2% [27]
	Heat of degradation	ΔH_r	5% [26,27]
Thermo-physical	Specific heat capacity	c_p	10% [26,27,36]
	Thermal conductivity	k	10% [26]
	Density	ρ	5% [26,36]
Optical	Absorption coefficient	α	30% [78]
	Emissivity	ε	2% [78]
Gas phase combustion	Heat of combustion	ΔH_c	5% [71,79]
Heat feedback	Heat flux	$\dot{q}''$	10% [71]

$$S_{HRR} = \frac{\Delta HRR_{error}}{\delta} \tag{20.9}$$

$$\Delta HRR_{error} = HRR_{error}^{P+\delta P} - HRR_{error}^{P} \tag{20.10}$$

Where δ is the fractional change in an input parameter corresponding to the parameter uncertainty given in Table 20.6, $HRR_{error}^{P+\delta P}$ is the HRR_{error} evaluated for simulation (Eq. 20.8) with one parameter base-value, P, increased by δP, and HRR_{error}^{P} represent base-case HRR_{error}. Thus, the parameter S_{HRR} provides a quantitative value for the change in HRR_{error} for a fractional change in the parameter value. For example, a 0.01 fractional increase in a parameter having a sensitivity coefficient of −0.5 indicates that the error in HRR for the base-case (HRR_{error}^{P}) would decrease by a magnitude of 0.005 (or 0.5%).

The results of sensitivity analysis, performed to quantify the impact of changing input parameters on HRR predictions, are shown in Figure 20.15 for converged simulations conducted using radiation heat flux model for PMMA and PIR foam. An additional third data-point is added here for Oriented Strand Board (OSB) to facilitate correlating sensitivity coefficient with fire size. Three experiments on OSB were conducted in the same corner setup. The HRR for OSB increased somewhat linearly above the propane burner contribution, after 10 s of propane ignition, to a peak of about 130 at 150 s. The HRR is mostly steady, around 120 ± 8 kW, for the rest of the experimental duration (300 s). Further details of the OSB material and the experiments can be found elsewhere.[72] The sensitivity coefficients for the three materials are listed in Table 20.7.

The parameters with smallest and highest magnitude (absolute value) of S_{HRR} are α and E respectively. The second largest magnitude of S_{HRR} was estimated for the flame heat flux parameter. The order of these three parameters, from lowest to highest magnitude of S_{HRR}, is the same for all the three materials studied. Pre-exponential factor, emissivity, and heat of combustion have positive S_{HRR} values for all cases. Heat of reaction, thermal conductivity, and density are parameters which have negative S_{HRR} value for PMMA and positive value for OSB and PIR foam.

The S_{HRR} also showed a linear correlation with peak HRR (80, 130, and 280 kW for PIR foam, OSB, and PMMA respectively) with a coefficient of determination greater than 0.95. The linear correlation parameter values, ψ_p (kW^{-1}) and β_S (–), are given in Table 20.7, which relate S_{HRR} with peak HRR, HRR_p, by the Equation 20.11. The parameter values are obtained by only interpolating between the known extents of peak HRR for the three materials studied here.

$$S_{HRR} = \psi_p \times HRR_p + \beta_S \tag{20.11}$$

S_{HRR} did not correlate linearly for the fire growth rate index (kW s^{-1}), calculated as the ratio of difference between the peak HRR and the propane burner contribution (30 kW) and the time to peak HRR from propane burner ignition (13, 145, and 185 s for PIR foam, OSB, and PMMA respectively). Thus, it can be inferred that the input parameters have higher sensitivity coefficient values for fires with higher peak HRR.

It is important to note that the impact of a parameter uncertainty on the model results (change in HRR_{error}) can be estimated by multiplying S_{HRR} with the fractional change in the parameter value, δ. From Figure 20.15, the impact of fractional change of model parameter value (which in this case was by its uncertainty) can help identify the parameter which uncertainties have highest impact on the HRR predictions. Flame heat flux was identified as the parameter which uncertainties impact the predictions the most. Reducing the flame heat flux by 10% reduced the HRR_{error} predictions for PMMA from 33% of the original simulation to 11% for the reduced flame heat flux simulation. However, as indicated by the sensitivity coefficients and the heat flux uncertainty value, the impact of increasing flame heat flux on the HRR prediction for flame spread over PIR foam was insignificant.

TABLE 20.7
Sensitivity Coefficients (S_{HRR}) for Parameters for PIR, OSB, and PMMA

Type	Variable	PIR	OSB	PMMA	Linear Fit (ψ_p, β_S)[a]
Reaction kinetics and thermodynamics	A	0.042	0.070	0.095	(0.00024, 0.02942)
	E	–0.950	–1.750	–2.845	(–0.008973, –0.38273)
	ΔH_r	0.070	0.020	–0.320	(–0.00202, 0.25377)
Thermo-physical	C_p	–0.035	–0.290	–0.790	(–0.00367, 0.22827)
	k	0.117	0.120	–0.240	(–0.00192, 0.31373)
	ρ	0.204	0.120	–0.240	(–0.00226, 0.39738)
Optical	α	0.008	0.003	0.027	(0.00011, –0.00530)
	ε	0.225	0.350	1.700	(0.00775, –0.50750)
Gas phase combustion	ΔH_c	0.212	0.020	0.900	(0.00400, –0.27600)
Heat feedback	$\dot{q}''$	0.143	0.560	1.850	(0.00855, –0.54550)

Notes: [a] The linear fit parameters are shown for (Eq. 20.11)

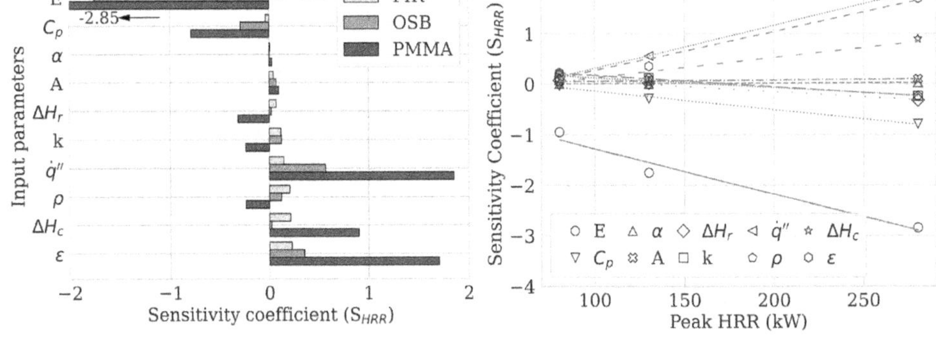

FIGURE 20.15 Sensitivity coefficients (S_{HRR}) for input parameters for simulations conducted using the radiation heat flux model.

It is important to note that parameter interactions that might affect the model sensitivity are not studied here. For example, it is known that parameters *A* and *E* have a compensation effect, and those interactions are not considered here in the sensitivity study. The sensitivity of a CFD-based model should also be investigated in the future. The sensitivity analysis conducted here, however, is the first of its kind and can potentially provide valuable information for the material development process of flame-retardant polymers.

20.8 CONCLUSIONS

Material scientists are constantly challenged with the need to develop new materials that have fire-retardant characteristics for many applications. The ability to predict material properties would potentially guide in development and application of new fire-retardant materials. For this to be realized, a quantitative link between molecular structure and fire growth on a polymeric solid needs to be established. This chapter discussed the importance of understanding interdependent phenomena of pyrolysis, flame heat feedback, and gas phase combustion that govern the fire growth process. Amongst these three phenomena, pyrolysis and flame heat feedback are limiting processes. Pyrolysis involves chemical change of a polymer when exposed to heating and releasing combustible or noncombustible gases called pyrolyzates and leaving behind solid carbonaceous residue called char. Comprehensive pyrolysis models, whose importance is gradually being recognized, describe thermal degradation reaction kinetics, associated thermodynamics, heats of combustion of pyrolyzates, thermo-physical properties such as thermal conductivity, density, and optical properties. A methodology to develop comprehensive pyrolysis models was discussed in detail with an example of poly(isocyanurate) (PIR) foam. Although the comprehensive pyrolysis models developed here were based primarily on anaerobic thermogravimetric and gasification tests, it is important to note that the presence of oxygen has been known to affect the pyrolysis process, and its effect can be incorporated during the pyrolysis model development. The developed comprehensive models, coupled with a flame heat feedback model (either semi-empirical or first-principle CFD-based), were shown to successfully predict heat release rate evolution during a fire growth scenario over PIR foam and poly(methyl methacrylate) in a Single Burning Item (SBI) test-like corner geometry. However, the fire growth predictions were also found to be sensitive to uncertainties in the flame heat flux treatment (convection-radiation split), to the uncertainties in model input parameters, and in the case of CFD model, to the grid size. A sensitivity study performed using the semi-empirical fire growth model to quantify the impact of various input parameters on fire growth predictions showed that the top three impactful parameters were activation energy of the degradation reactions, followed by flame heat feedback and emissivity.

A methodology illustrated using Figure 20.16 could potentially serve as a guidance to incorporate fire safety during the material development process. This methodology assumes that polymer property values can be estimated from its molecular structure using group contributions, molecular modeling, or data-centered approaches to parameterize pyrolysis of new fire-retardant materials. Multiple fire scenarios, guided by the end use of the material, then can be simulated to predict fire growth hazard of a potential new fire-retardant material. For example, for wall lining materials, the established semi-empirical flame spread model could then be used to estimate heat release predictions for a polymeric solid subjected to a corner flame spread test like the SBI test scenario. Once an initial estimate of fire growth is obtained, parameter values can then be optimized with the help of presented sensitivity analysis. Finally, when desired polymer structure is obtained, the synthesized material should be validated to produce predicted fire-retardant behavior.

Although this framework for material development could be potentially empowering, there are several factors that still need to be better understood to improve the quantitative connection between the molecular structure and fire growth. They are listed here:

1. Group contribution derived properties of simple polymers have yet to be tested within the framework of pyrolysis modeling. There is a need for

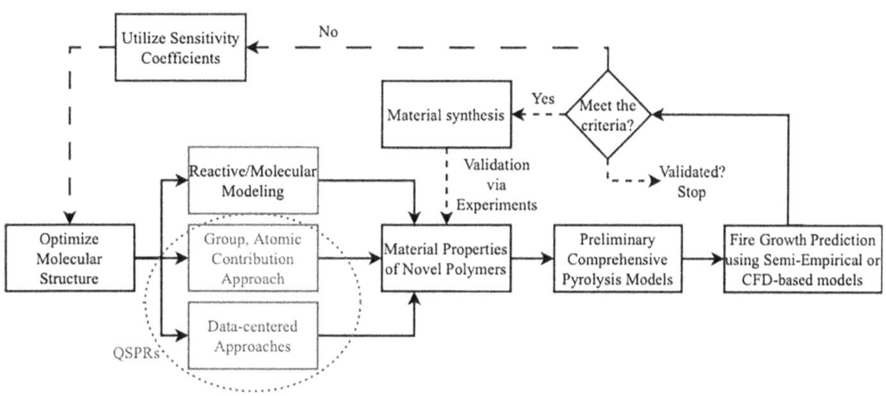

FIGURE 20.16 A proposed methodology for developing new materials by utilizing the quantitative structure property relationships (QSPRs) and sensitivity of scenario-specific fire growth rate to material property values.

verifying the applicability of existing QSPRs to derive pyrolysis properties even for polymers with relatively simple monomer molecular structure.

2. High-quality data for correlating group contributions and relevant properties required in fire growth models need to be obtained and made available. Group contributions of fire-retardant additive structures to polymer properties are not studied, and their contribution needs to be established.
3. Expanding group contribution approach is challenging for composites where the material property is not only a function of molecular structure but also of the macrostructure.
4. Calibrating flame heat feedback model that was used here for flames with a wide range of radiative fractions could help improve the fire growth predictions.
5. Reducing uncertainty in the measured and derived pyrolysis properties to improve fire growth predictions. A continuously evolving dataset that provides milligram-scale to gram-scale data of material degradation and flammability, such as the Materials and Products Database,[80] is a valuable resource for reliable, repeatable data that can be used to derive and validate pyrolysis properties.
6. Improving CFD-based sub-models to obtain more accurate, grid-independent flame heat feedback for better predictions of the fire growth hazard in a fire scenario of interest.

20.8.1 Acknowledgments

We thank all the partners, DuPont De Nemours Inc. and the Federal Aviation Administration, who provided funding for part of this project. We thank Dr. Alexander B. Morgan for the invitation to write this chapter. D.C is thankful to Dr. Steve Kerber, Dr. Daniel Madrzykowski, Dr. Craig Weinschenk, Dr. Mark McKinnon from Fire Safety Research Institute (FSRI) for their continued support and review of the final document. D.C is grateful to Dr. Jason Floyd for helpful discussions related to FDS.

BIBLIOGRAPHY

1. Richtie, H.; Samborska, V.; Roser, M. Plastic Pollution. In *Our World in Data*; **2023**. Published online at Our WorldInData.org. https://ourworldindata.org/plastic-pollution (accessed 2024-1-29).
2. Geyer, R.; Jambeck, J. R.; Law, K. L. Production, Use, and Fate of All Plastics Ever Made. *Sci Adv* **2017**, *3* (7), 19–24. https://doi.org/10.1126/sciadv.1700782.
3. Dasari, A.; Yu, Z. Z.; Cai, G. P.; Mai, Y. W. Recent Developments in the Fire Retardancy of Polymeric Materials. *Prog Polym Sci* **2013**, *38* (9), 1357–1387. https://doi.org/10.1016/j.progpolymsci.2013.06.006.
4. Kerber, S. Analysis of Changing Residential Fire Dynamics and Its Implications on Firefighter Operational Timeframes. *Fire Technol* **2012**, *48* (4), 865–891. https://doi.org/10.1007/s10694-011-0249-2.
5. Ahrens, M.; Evarts, B. Fire Loss in the United States During 2020. National Fire Protection Association, September 2021.
6. Karlsson, B.; North, G.; Gojkovic, D. Using Results from Performance-Based Test Methods for Material Flammability in Fire Safety Engineering Design. *J. Fire Prot. Eng.* **2002**, *12* (2), 93–108. https://doi.org/10.1177/1042391502012002313.
7. Bill, R. G.; Croce, P. A. The International FORUM of Fire Research Directors: A Position Paper on Small-Scale Measurements for Next Generation Standards. *Fire Saf J* **2006**, *41* (7), 536–538. https://doi.org/10.1016/j.firesaf.2006.05.005.
8. Williams, F. A. Mechanisms of Fire Spread. *Symp. Intern. Combust.* **1977**, *16* (1), 1281–1294. https://doi.org/10.1016/S0082-0784(77)80415-3.
9. Rangwala, A. S.; Buckley, S. G.; Torero, J. L. Upward Flame Spread on a Vertically Oriented Fuel Surface: The Effect of Finite Width. *Proc. Combust. Inst.* **2007**, *31 II*, 2607–2615. https://doi.org/10.1016/j.proci.2006.07.235.
10. van Hees, P.; Andersson, P.; Hjohlman, M.; Wenne, N.; Hassan, M. A. Use of Cone Calorimeter and Cone Tools Software for Development of Innovative Intumescing Graphite Systems. *Fire Mater* **2010**, *34*, 367–384. https://doi.org/10.1002/fam.1032.
11. *EN-13823 — Reaction to Fire Tests for Building Products—Building Products Excluding Floorings Exposed to the Thermal Attack by a Single Burning Item*; 2004.
12. Stoliarov, S. I.; Leventon, I. T.; Lyon, R. E. Two-Dimensional Model of Burning for Pyrolyzable Solids. *Fire Mater* **2014**, *38* (3), 391–408. https://doi.org/10.1002/fam.2187.
13. Stoliarov, S. I.; Lyon, R. E. *Thermo-Kinetic Model of Burning*; 2008.
14. McGrattan, K.; Hostikka, S.; McDermott, R.; Floyd, J.; Weinschenk, C.; Overhold, K. *Sixth Edition Fire Dynamics Simulator User 's Guide (FDS v 6.7.9).* NIST Special Publication 1019, 2016.
15. Torero, J. Flaming Ignition of Solid Fuels. In *SFPE Handbook of Fire Protection Engineering*, 5th Edition; National Fire Protection Association, Quincy, MA; **2016**, 633–661. https://doi.org/10.1007/978-1-4939-2565-0_21.
16. Rogaume, T. Thermal Decomposition and Pyrolysis of Solid Fuels: Objectives, Challenges and Modelling. *Fire Saf J* **2019**, *106*, 177–188. https://doi.org/10.1016/j.firesaf.2019.04.016.
17. Korobeinichev, O. P.; Karpov, A. I.; Bolkisev, A. A.; Shaklein, A. A.; Gonchikzhapov, M. B.; Paletsky, A. A.; Tereshchenko, A. G. An Experimental and Numerical Study of Thermal and Chemical Structure of Downward Flame Spread over PMMA Surface in Still Air. *Proc Combust Inst* **2019**, *37*, 4017–4024. https://doi.org/10.1016/j.proci.2018.06.005.
18. Staggs, J. E. J. Population Balance Models for the Thermal Degradation of PMMA. *Polymer (Guildf)* **2007**, *48* (13), 3868–3876. https://doi.org/10.1016/j.polymer.2007.04.075.
19. Linteris, G.; Zammarano, M.; Wilthan, B.; Hanssen, L. Absorption and Reflection of Infrared Radiation by Polymers in Fire-Like Environments. *Fire Mater* **2012**, *36* (7), 537–553. https://doi.org/10.1002/fam.1113.
20. Ding, Y.; Leventon, I. T.; Stoliarov, S. I. An Analysis of the Sensitivity of the Rate of Buoyancy-Driven Flame Spread on a Solid Material to Uncertainties in the Pyrolysis and Combustion Properties. Is Accurate Prediction Possible? *Polym Degrad Stab* **2023**, *214* (January), 110405. https://doi.org/10.1016/j.polymdegradstab.2023.110405.

21. Lautenberger, C.; Rein, G.; Fernandez-Pello, C. The Application of a Genetic Algorithm to Estimate Material Properties for Fire Modeling from Bench-Scale Fire Test Data. *Fire Saf J* **2006**, *41* (3), 204–214. https://doi.org/10.1016/j.firesaf.2005.12.004.
22. Chaos, M.; Khan, M. M.; Krishnamoorthy, N.; De Ris, J. L.; Dorofeev, S. B. Evaluation of Optimization Schemes and Determination of Solid Fuel Properties for CFD Fire Models Using Bench-Scale Pyrolysis Tests. *Proceedings of the Combustion Institute* **2011**, *33* (2), 2599–2606. https://doi.org/10.1016/j.proci.2010.07.018.
23. McGrattan, K.; Hostikka, S.; Floyd, J.; McDermott, R.; Vanella, M.; Mueller, E. *Sixth Edition Fire Dynamics Simulator User 's Guide (FDS)*; 2023. 6th ed. https://doi.org/10.6028/NIST.SP.1019.
24. Lautenberger, C. Gpyro3D: A Three Dimensional Generalized Pyrolysis Model. *Fire Saf. Sci.* **2014**, *11*, 193–207. https://doi.org/10.3801/IAFSS.FSS.11-193.
25. *OpenFOAM: User Guide v2112 — The open source CFD toolbox.* www.openfoam.com/documentation/guides/latest/doc/guide-applications-solvers-combustion-fireFoam.html (accessed 2023-04-24).
26. Li, J.; Stoliarov, S. I. Measurement of Kinetics and Thermodynamics of the Thermal Degradation for Charring Polymers. *Polym Degrad Stab* **2014**, *106*, 2–15. https://doi.org/10.1016/j.polymdegradstab.2013.09.022.
27. Li, J.; Stoliarov, S. I. Measurement of Kinetics and Thermodynamics of the Thermal Degradation for Non-Charring Polymers. *Combust Flame* **2013**, *160*, 1287–1297. https://doi.org/10.1016/j.combustflame.2013.02.012.
28. Swann, J. D.; Ding, Y.; McKinnon, M. B.; Stoliarov, S. I. Controlled Atmosphere Pyrolysis Apparatus II (CAPA II): A New Tool for Analysis of Pyrolysis of Charring and Intumescent Polymers. *Fire Saf J* **2017**, *91* (March), 130–139. https://doi.org/10.1016/j.firesaf.2017.03.038.
29. McGrattan, K.; Hostikka, S.; Floyd, J.; McDermott, R.; Vanella, M.; Mueller, E. *Fire Dynamics Simulator Technical Reference Guide Volume 3: Validation.* Gaithersburg, MD 2023.
30. Leventon, I. T.; Korver, K. T.; Stoliarov, S. I. A Generalized Model of Flame to Surface Heat Feedback for Laminar Wall Flames. *Combust Flame* **2017**, *179*, 338–353. https://doi.org/10.1016/j.combustflame.2017.02.007.
31. Swann, J. D.; Ding, Y.; Stoliarov, S. I. Characterization of Pyrolysis and Combustion of Rigid Poly(Vinyl Chloride) Using Two-Dimensional Modeling. *Int J Heat Mass Transf* **2019**, *132*, 347–361. https://doi.org/10.1016/j.ijheatmasstransfer.2018.12.011.
32. McGrattan, K. B.; Hostikka, S.; Floyd, J.; McDermott, R.; Vanella, M.; Mueller, E. *Fire Dynamics Simulator Technical Reference Guide Volume 1: Mathematical Model.* Gaithersburg, MD; 2023; Vol. 3. https://doi.org/10.6028/NIST.SP.1018.
33. Kashiwagi, T. Polymer Combustion and Flammability-Role of the Condensed Phase. *Symposium (International) on Combustion.* Elsevier; 1994, pp 1423–1437. https://doi.org/10.1016/S0082-0784(06)80786-1.
34. Safety Data Sheet. *TUFF-R^{TM} C 2.00 Inch Commercial Insulation Sheathing,* The Dow Chemical Company, Midlands, MI, March 2015.
35. Leventon, I.; Batiot, B.; Bruns, M.; Hostikka, S.; Nakamura, Y.; Reszka, P.; Rogaume, T.; Stoliarov, S. *Measurement and Computation of Fire Phenomena (MaCFP) Condensed Phase Material Database.* https://doi.org/10.18434/mds2-2586.
36. Fiola, G. J.; Chaudhari, D. M.; Stoliarov, S. I. Comparison of Pyrolysis Properties of Extruded and Cast Poly(Methyl Methacrylate). *Fire Saf J* **2021**, *120* (May 2020), 103083. https://doi.org/10.1016/j.firesaf.2020.103083.
37. Dick, C.; Dominguez-Rosado, E.; Eling, B.; Liggat, J. J.; Lindsay, C. I.; Martin, S. C.; Mohammed, M. H.; Seeley, G.; Snape, C. E. The Flammability of Urethane-Modified Polyisocyanurates and Its Relationship to Thermal Degradation Chemistry. *Polymer (Guildf)* **2001**, *42* (3), 913–923. https://doi.org/10.1016/S0032-3861(00)00470-5.
38. Lyon, R. E.; Walters, R. N.; Stoliarov, S. I.; Safronava, N. *Principles and Practice of Microscale Combustion Calorimetry.* DOT/FAA/TC-12/53, Federal Aviation Administration, NJ, April 2013.
39. Chaudhari, D. M.; Stoliarov, S. I.; Beach, M. W.; Suryadevara, K. A. Polyisocyanurate Foam Pyrolysis and Flame Spread Modeling. *Applied Sciences (Switzerland)* **2021**, *11* (8). https://doi.org/10.3390/APP11083463.
40. *MatWeb-Material Property Data, E-glass Fiber, Generic.* www.matweb.com/search/datasheet.aspx?MatGUID=d9c18047c49147a2a7c0b0bb1743e812 (accessed 2023-3-21).
41. Schartel, B.; Pawlowski, K. H.; Lyon, R. E. Pyrolysis Combustion Flow Calorimeter: A Tool to Assess Flame Retarded PC/ABS Materials? *Thermochim Acta* **2007**, *462* (1–2), 1–14. https://doi.org/10.1016/j.tca.2007.05.021.
42. *NIST Chemistry WebBook, SRD69, 2-Methyl-Butane.* https://webbook.nist.gov/cgi/cbook.cgi?Name=isopentane&Units=SI (accessed 2023-3-21).
43. Mccoy, C. G.; Tilles, J. L.; Stoliarov, S. I. Empirical Model of Flame Heat Feedback for Simulation of Cone Calorimetry. *Fire Saf J* **2019**, *103*, 38–48. https://doi.org/10.1016/j.firesaf.2018.11.006.
44. Lyon, R. E.; Quintiere, J. G. Criteria for Piloted Ignition of Combustible Solids. *Combust Flame* **2007**, *151* (4), 551–559. https://doi.org/10.1016/j.combustflame.2007.07.020.
45. Ding, Y.; Swann, J. D.; Sun, Q.; Stoliarov, S. I.; Kraemer, R. H. Development of a Pyrolysis Model for Glass Fiber Reinforced Polyamide 66 Blended with Red Phosphorus: Relationship between Flammability Behavior and Material Composition. *Compos B Eng* **2019**, *176*, 107263. https://doi.org/10.1016/j.compositesb.2019.107263.
46. Ding, Y.; McKinnon, M. B.; Stoliarov, S. I.; Fontaine, G.; Bourbigot, S. Determination of Kinetics and Thermodynamics of Thermal Decomposition for Polymers Containing Reactive Flame Retardants: Application to Poly(Lactic Acid) Blended with Melamine and Ammonium Polyphosphate. *Polym Degrad Stab* **2016**, *129*, 347–362. https://doi.org/10.1016/j.polymdegradstab.2016.05.014.
47. Ding, Y.; Stoliarov, S.; Kraemer, R. Development of a Semiglobal Reaction Mechanism for the Thermal Decomposition of a Polymer Containing Reactive Flame Retardants: Application to Glass-Fiber-Reinforced Polybutylene Terephthalate Blended with Aluminum Diethyl Phosphinate and Melamine Polypho. *Polymers (Basel)* **2018**, *10* (10), 1137. https://doi.org/10.3390/polym10101137.
48. Ding, Y.; Stoliarov, S. I.; Kraemer, R. H. Pyrolysis Model Development for a Polymeric Material Containing Multiple Flame Retardants: Relationship between Heat Release Rate and Material Composition. *Combust Flame* **2019**, *202*, 43–57. https://doi.org/10.1016/j.combustflame.2019.01.003.
49. Ding, Y.; Kwon, K.; Stoliarov, S. I.; Kraemer, R. H. Development of a Semi-Global Reaction Mechanism for Thermal Decomposition of a Polymer Containing Reactive Flame Retardant. *Proc. Combust. Inst.* **2019**, *37* (3), 4247–4255. https://doi.org/10.1016/j.proci.2018.05.073.
50. Swann, J. D.; Ding, Y.; Stoliarov, S. I. Comparative Analysis of Pyrolysis and Combustion of Bisphenol A Polycarbonate

and Poly(Ether Ether Ketone) Using Two-Dimensional Modeling: A Relation between Thermal Transport and the Physical Structure of the Intumescent Char. *Combust Flame* **2020**, *212*, 469–485. https://doi.org/10.1016/j.combustflame.2019.11.017.

51. Jayarama Krishna, J. V.; Srivatsa Kumar, S.; Korobeinichev, O. P.; Vinu, R. Detailed Kinetic Analysis of Slow and Fast Pyrolysis of Poly(Methyl Methacrylate)-Flame Retardant Mixtures. *Thermochim Acta* **2020**, *687*, 178545. https://doi.org/10.1016/j.tca.2020.178545.
52. Coimbra, A.; Sarazin, J.; Bourbigot, S.; Legros, G.; Consalvi, J. L. A Semi-Global Reaction Mechanism for the Thermal Decomposition of Low-Density Polyethylene Blended with Ammonium Polyphosphate and Pentaerythritol. *Fire Saf J* **2022**, *133* (June), 103649. https://doi.org/10.1016/j.firesaf.2022.103649.
53. Van Krevelen, D. W.; Te Nijenhuis, K. *Properties of Polymers: Their Correlation with Chemical Structure; Their Numerical Estimation and Prediction from Additive Group Contributions*, 4th ed.; 2009. https://doi.org/10.1016/b978-0-08-054819-7.00027-3.
54. Bicerano, J. *Prediction of Polymer Properties*, 3rd ed.; Bicerano, J., Ed.; 2002. https://doi.org/10.1201/9780203910115.
55. Hopfinger, A. J.; Koehler, M. G.; Pearlstein, R. A.; Tripathy, S. K. Molecular Modeling of Polymers. IV. Estimation of Glass Transition Temperatures. *J Polym Sci B Polym Phys* **1988**, *26* (10), 2007–2028. https://doi.org/https://doi.org/10.1002/polb.1988.090261001.
56. Stoliarov, S. I.; Westmoreland, P. R.; Nyden, M. R.; Forney, G. P. A Reactive Molecular Dynamics Model of Thermal Decomposition in Polymers: I. Poly(Methyl Methacrylate). *Polymer (Guildf)* **2002**, *44* (3), 883–894. https://doi.org/10.1016/S0032-3861(02)00761-9.
57. Nyden, M. R.; Coley, T. R.; Mumby, S. Application of Molecular Dynamics to the Study of Thermal Degradation in Aromatic Polymers. I: Polystyrene. *Polym Eng Sci* **1997**, *37* (9).
58. Stoliarov, S. I.; Lyon, R. E.; Nyden, M. R. A Reactive Molecular Dynamics Model of Thermal Decomposition in Polymers. II. Polyisobutylene. *Polymer (Guildf)* **2004**, *45* (25), 8613–8621.
59. Chen, T. B. Y.; Yuen, A. C. Y.; Lin, B.; Liu, L.; Lo, A. L. P.; Chan, Q. N.; Zhang, J.; Cheung, S. C. P.; Yeoh, G. H. Characterisation of Pyrolysis Kinetics and Detailed Gas Species Formations of Engineering Polymers via Reactive Molecular Dynamics (ReaxFF). *J Anal Appl Pyrolysis* **2020**, *153* (November). https://doi.org/10.1016/j.jaap.2020.104931.
60. Vaari, J.; Paajanen, A. Evaluation of the Reactive Molecular Dynamics Method for Research on Flame Retardants: ATH-Filled Polyethylene. *Comput Mater Sci* **2018**, *153*, 103–112. https://doi.org/10.1016/j.commatsci.2018.06.032.
61. Walters, R. N.; Hackett, S. M.; Lyon, R. E. Heats of Combustion of High Temperature Polymers. *Fire Mater* **2000**, *24* (5), 245–252. https://doi.org/10.1002/1099-1018(200009/10)24:5<245::AID-FAM744>3.0.CO;2-7.
62. Lyon, R. E.; Takemori, M. T.; Safronava, N.; Stoliarov, S. I.; Walters, R. N. A Molecular Basis for Polymer Flammability. *Polymer (Guildf)* **2009**, *50* (12), 2608–2617. https://doi.org/10.1016/j.polymer.2009.03.047.
63. ASTM. *Standard Test Method for Measuring the Comparative Burning Characteristics of Solid Plastics in a Vertical Position*; ASTM International; 2020. https://doi.org/10.1520/D3801-19.
64. Sonnier, R.; Negrell-Guirao, C.; Vahabi, H.; Otazaghine, B.; David, G.; Lopez-Cuesta, J. M. Relationships between the Molecular Structure and the Flammability of Polymers: Study of Phosphonate Functions Using Microscale Combustion Calorimeter. *Polymer (Guildf)* **2012**, *53* (6), 1258–1266. https://doi.org/10.1016/j.polymer.2012.01.055.
65. Sonnier, R.; Otazaghine, B.; Iftene, F.; Negrell, C.; David, G.; Howell, B. A. Predicting the Flammability of Polymers from Their Chemical Structure: An Improved Model Based on Group Contributions. *Polymer (Guildf)* **2016**, *86*, 42–55. https://doi.org/10.1016/j.polymer.2016.01.046.
66. Sonnier, R.; Otazaghine, B.; Dumazert, L.; Ménard, R.; Viretto, A.; Dumas, L.; Bonnaud, L.; Dubois, P.; Safronava, N.; Walters, R.; Lyon, R. Prediction of Thermosets Flammability Using a Model Based on Group Contributions. *Polymer (Guildf)* **2017**, *127*, 203–213. https://doi.org/10.1016/j.polymer.2017.09.012.
67. Sonnier, R.; Otazaghine, B.; Vagner, C.; Bier, F.; Six, J. L.; Durand, A.; Vahabi, H. Exploring the Contribution of Two Phosphorus-Based Groups to Polymer Flammability via Pyrolysis-Combustion Flow Calorimetry. *Materials* **2019**, *12* (18). https://doi.org/10.3390/ma12182961.
68. Colin Haser, J.; Herring, R. H.; Datta, S.; Eden, M. R. Development of QSPR Model Relating Solvent Structure to Crystal Morphology. In *Proceedings of the 8th International Conference on Foundations of Computer-Aided Process Design*; Eden, M. R., Siirola, J. D., Towler, G. P., Eds.; Computer Aided Chemical Engineering; Elsevier, 2014; Vol. 34, pp 321–326. https://doi.org/10.1016/B978-0-444-63433-7.50038-9.
69. Ramprasad, R.; Batra, R.; Pilania, G.; Mannodi-Kanakkithodi, A.; Kim, C. Machine Learning in Materials Informatics: Recent Applications and Prospects. *NPJ Comput Mater* **2017**, *3* (1), 1–13. https://doi.org/10.1038/s41524-017-0056-5.
70. Doan Tran, H.; Kim, C.; Chen, L.; Chandrasekaran, A.; Batra, R.; Venkatram, S.; Kamal, D.; Lightstone, J. P.; Gurnani, R.; Shetty, P.; Ramprasad, M.; Laws, J.; Shelton, M.; Ramprasad, R. Machine-Learning Predictions of Polymer Properties with Polymer Genome. *J Appl Phys* **2020**, *128* (17), 1–13. https://doi.org/10.1063/5.0023759.
71. Chaudhari, D. M.; Fiola, G. J.; Stoliarov, S. I. Experimental Analysis and Modeling of Buoyancy-Driven Flame Spread on Cast Poly(Methyl Methacrylate) in Corner Configuration. *Polym Degrad Stab* **2021**, *183*, 109433. https://doi.org/10.1016/j.polymdegradstab.2020.109433.
72. Chaudhari, D. M. Experiments and Semi Empirical Modeling of Buoyancy Driven, Turbulent Flame Spread over Combustible Solids in a Corner Configuration. PhD Dissertation, University of Maryland, College Park, MD, 2021.
73. Quintiere, J. G.; Cleary, T. G. Heat Flux from Flames to Vertical Surfaces. *Fire Technol* **1994**, *30* (2), 209–231. https://doi.org/10.1007/BF01040003.
74. Lattimer, B. Y.; Hunt, S. P.; Wright, M.; Beyler, C. Corner Fire Growth in a Room with a Combustible Lining. *Fire Safety Science* **2003**, 419–430. https://doi.org/10.3801/IAFSS.FSS.7-419.
75. Tewarson, A. Generation of Heat and Chemical Compounds in Fires. In *SFPE Handbook of Fire Protection Engineering*, 3rd Edition; National Fire Protection Association, Quincy, MA; **2002**; pp 3 (82–161).
76. Marquis, D. M.; Batiot, B.; Guillaume, E.; Rogaume, T. Influence of Reaction Mechanism Accuracy on the Chemical Reactivity Prediction of Complex Charring

Material in Fire Condition. *J Anal Appl Pyrolysis* **2016**, *118*, 231–248. https://doi.org/10.1016/j.jaap.2016.02.007.

77. Frey, H. C.; Patil, S. R. Identification and Review of Sensitivity Analysis Methods. *Risk Analysis* **2002**, *22*, 553–578. https://doi.org/10.1111/0272-4332.00039.
78. Linteris, G.; Zammarano, M.; Wilthan, B.; Hanssen, L. Absorption and Reflection of Infrared Radiation by Polymers in Fire-Like Environments. *Fire Mater* **2012**, *36* (7), 537–553. https://doi.org/10.1002/fam.1113.
79. Walters, R. N.; Hackett, S. M.; Lyon, R. E. Heats of Combustion of High Temperature Polymers. *Fire Mater* **2000**, *24* (5), 245–252. https://doi.org/10.1002/1099-1018(200009/10)24:5<245::AID-FAM744>3.0.CO;2-7.
80. McKinnon, M.; Weinschenk, C.; Dow, N.; DiDomizio, M.; Madrzykowski, D. *Materials and Products Database* (Version 1.0.0). Fire Safety Research Institute, UL Research Institutes, Columbia, MD; 2023.

21 Full-Scale Fire Modeling and Simulation

Simo Hostikka
Aalto University, Espoo, Finland

List of Symbols and Abbreviations

A	Pre-exponential factor
E	Activation energy
$\mathbf{f}$	Force vector
$\mathbf{g}$	Acceleration of gravity
h	Enthalpy
I	Radiation intensity (radiance)
N	Reaction order
q	Energy
$\mathbf{q}$	Heat flux vector
m	Mass
$\mathbf{s}$	Direction vector
T	Temperature
t	Time
$\mathbf{u}$	Velocity vector
$\mathbf{x}$	Position vector
Y	Mass fraction
Z	Mixture fraction

Greek	
χ_r	Local radiative fraction
ΔH	Heat of reaction (subscript r), heat of combustion (subscript c)
ε	Enthalpy dissipation rate, surface mean hemispherical emissivity
Φ	Scattering phase function
κ	Absorption coefficient
λ	Wavelength
Ω	Solid angle
θ	Polar angle
ρ	Density, mass concentration
σ	Scattering coefficient, Stefan-Boltzmann constant
τ	Mixing time coefficient, shear

Abbreviations

CFD	Computational Fluid Dynamics
CFL	Courant-Friedrichs-Lewy
DNS	Direct Numerical Simulation
DOM	Discrete Ordinates Methods
EBU	Eddy Break-Up
EDC	Eddy Dissipation Concept
DT	Discrete Transfer
FDS	Fire Dynamics Simulator
FVM	Finite Volume Method
LBL	Line-by-Line Solution of Radiation
LES	Large Eddy Simulation
PMC	Photon Monte Carlo
RANS	Reynolds-Averaged Navier-Stokes
RTE	Radiation Transport Equation

21.1 INTRODUCTION

Fire safety engineering is used to ensure that the risk of fire-induced losses for humans, property, environment, and the surrounding society associated with the target of the analysis is acceptable by the common standards of the society. Additional objectives may be imposed, for example by economic goals and needs to protect environment or cultural heritage. Fire safety engineering is typically used to design the fire safety measures of new buildings or transportation vehicles. Traditionally, the design is based on a set of *prescriptive requirements* for the physical characteristics, such as the dimensions of the fire compartments, classification of structures, and width and length of evacuation routes. These requirements are described in national (or state) building codes and are based mainly on experimental findings and accumulated lessons learned from past fires. An alternative design methodology is *performance-based design*, where the effectiveness of fire safety measures is assessed by considering the performance of an entire system, not as fulfillment of individual prescriptive requirements given by the building code. An essential part of the design process is analysis of the risks associated with fires and assessment of the efficiency of proposed fire safety measures. Full-scale fire modeling is used to assess the consequences of the identified fire scenarios.

During the last 20 years, computational fluid dynamics (CFD) has become the most widely used technique for the simulation of smoke transport and fire spread. Simpler techniques, such as hand-calculation formulas and two-zone models, still have an important role in engineering because they are faster and easier to use than CFD, but most fire simulations are currently performed using CFD. One application of the CFD-based fire modeling, that is becoming more commonplace, is fire development modeling (sometimes called flame spread modeling or fire growth modeling). In this type of simulation, the overall heat release rate and species production rates are predicted as the fire develops, rather than specified (prescribed) beforehand.

Over the years, several reviews of the CFD fire modelling have been published [1–5]. Early reviews [1, 2] described the state-of-the-art from rather general fire CFD viewpoint, whereas McGrattan and Miles [3] presented a comprehensive review of the modern CFD techniques and applications in the context of fire protection engineering. Since the publication of their chapter, only few things have changed regarding the state of the art of the fire engineering

DOI: 10.1201/9781003380689-21

use of CFD. Most importantly, the routine use of high-performance computing (HPC) has enabled the computations using larger models, partially making the simulations more realistic in terms of the problem size, and partially improving the numerical resolution. Yuen et al. [4] focused on multiphase modelling in the context of fire-CFD, discussing the coupling of CFD with pyrolysis models and the sub-grid scale models of suppression agents. According to their review, published in 2022, it was still not possible to perform predictive simulations of asphyxiant gases production, which is known to be responsible for the lethal hazard in most fires. See Chapter 19 of this book for more details on asphyxiant gas production. Fundamental understanding of the thermal-chemical breakdown process of the solid fuels was therefore identified as a key development target. According to the review of CFD-based wildfire studies by Morvan et al. [5], CFD is a promising approach for improving the knowledge about wildfire behavior and fire propagation mechanisms, and to propose solutions for fire safety engineering in wildland-urban interface, but the models are still computationally too expensive for fire front propagation forecasting or other regional scale computations.

This chapter reviews the aspects of physical and numerical modeling necessary to simulate fire development with present CFD fire models. The emphasis is put on physical phenomena such as fluid dynamics, combustion, radiation, solid phase heat transfer, and flame spread. The challenges of radiation modeling are discussed in detail. Models designed for special applications, such as explosions or direct numerical simulation of combustion processes, are not discussed. Examples of CFD-based predictions of fire development are presented to give an idea of current capabilities and limitations.

21.2 GAS PHASE PHYSICS REQUIRED FOR FIRE GROWTH PREDICTIONS

In their 2016 chapter of the Society of Fire Protection Engineers (SFPE) Handbook concerning CFD fire modeling, McGrattan and Miles [3] presented the governing equations, models for source terms, most important numerical methods, aspects of verification and validation, and applications of modern CFD in fire engineering. The reader is encouraged to study their chapter for an overview of fire CFD, or some of the many textbooks for details of the CFD methods in general [6, 7].

What makes a CFD model a fire model? In addition to the numerical solver for the Navier-Stokes equations, the computer code must have models for the essential physical processes and the phenomena of interest. Fire energy is produced by combustion, which obviously needs to be modelled in some way. The transport of the combustion products and heat is also necessary, and this is where the requirements of fire modelling start to deviate from the capabilities commonly available in general-purpose CFD solvers. For instance, the conjugate heat transfer between gas phase and its solid boundaries is always important in fires, but not necessarily included in more general programs. Also, while most CFD packages can model multiphase flows, where mass, momentum, and energy are exchanged between the fluid and condensed phase particles (or droplets), defining the source of water droplets in a way they are generated from a sprinkler head, as an example, may be very tedious unless using a CFD code that is dedicated for fire problems.

In connection with polymer materials, the main topic of this book, fire modeling can be used in (at least) two different ways. In the first, and less demanding way of coupling the gas phase CFD and polymeric solid, the fire simulation provides information on the conditions which the polymeric object would experience during a hypothesized fire, and this information (temperature, heat flux, chemical atmosphere) can be used for setting the design requirements for the material performance, such as fire retardancy. In the second, and the more demanding way of coupling, fluid flow and solid surfaces are tightly coupled—not only through the conjugate heat transfer, but also through the thermolysis process which turns the solid boundary into a source of flammable and nonflammable gases. These gases then contribute to combustion reactions driving the flow. In fact, the numerical predictions of the flame and fire spread represent the most demanding applications of fire modelling, both in terms of necessary physics, but also for the numerical resolution requirements.

21.2.1 Navier-Stokes Equations and Turbulence

Most physical phenomena in fires can be described as transfer of heat and mass and chemical reactions in either gas or condensed phases. The physics of the fluid flow is governed by the conservation equations of mass, momentum, and enthalpy:

$$\frac{\partial \rho}{\partial t} + \nabla \cdot \rho \boldsymbol{u} = \dot{m}''' \tag{21.1}$$

$$\frac{\partial}{\partial t}(\rho \boldsymbol{u}) + \nabla \cdot \rho \boldsymbol{u}\boldsymbol{u} + \nabla p - \rho \boldsymbol{g} = \boldsymbol{f} + \nabla \cdot \tau \tag{21.2}$$

$$\frac{\partial}{\partial t}(\rho h) + \nabla \cdot \rho h u = \frac{Dp}{Dt} + \dot{q}''' - \nabla \cdot \boldsymbol{q} + \varepsilon \tag{21.3}$$

where enthalpy is a function of the gas temperature

$$h(T) = \int_{T_0}^{T} c_p dT. \tag{21.4}$$

These equations are called the Navier-Stokes equations, and when supplemented by the state equation for fluid pressure, transport equations for individual gas species, and appropriate initial and boundary conditions, they form the basis for any computational model describing the flows in fires.

Flow solver can be either compressible or incompressible, which is sometimes called "low Mach number" formulation. The compressible formulation is more generic and computationally more expensive, while the low Mach number formulation reduces the number of unknowns in Equations 21.1–21.3 from six to five, as temperature can be calculated from known density, and simplifies the solution process as the pressure in Equation 21.3 and in the equation of state can be approximated by a time averaged value. Term "weakly compressible" is sometimes used for such solvers because the fluid density varies, not because of the pressure, but because of temperature variations. The low-Mach number approximation is adequate in typical fire application, such as a building fire, where the flow speed rarely exceeds 20 m/s. It is not valid when the flow speed exceeds roughly one third of the speed of sound, that is, 100 m/s, or when the gas compression by pressure is significant (detonation wave). However, sometimes fire engineers must deal with very high fluid velocities, for example during the activation of high-pressure suppression systems or opening a pressure relief valve of a hydrogen tank. Also, the release of the aviation fuel during aircraft impacts takes place at a speed which is roughly two times the aircraft speed, that is, quite close to the speed of sound. Despite the theoretical problem related to the use of low-Mach number CFD for simulating these events, the simulations can still provide useful information because the "too" high speeds are limited to a small portion of the entire computational domain, and there are other physical processes than the very high-speed phenomena that significantly contribute to the overall outcome of the event.

Fire-specific features of Equations 21.1–21.3 are included by modelling the right-hand side terms. In Equation 21.1, $\dot{m}'''$ is a mass source term originating from the evaporation of liquid droplets or thermolysis of solid particles, such as fire brands. In Equation 21.2, force term **f** describes the momentum transferred from condensed phase particles and droplets to gas, containing both drag and the momentum of the evaporated gases$\nabla \cdot \tau$. includes the viscous stresses and, in combination with the sub-grid scale turbulent dissipation, is the subject of turbulence modelling. And it is the turbulence modelling that estimates the production of sub-grid kinetic energy and, correspondingly, its dissipation to heat, appearing as ε on the right-hand side of the enthalpy equation, Equation 21.3. This term, as well as the pressure term, Dp/Dt, are usually negligible. The enthalpy source, $\dot{q}'''$, mainly comes from the combustion, but the heat exchange with droplets and particles would appear in this term too. Finally, the divergence of the heat flux vector, $\nabla \cdot \boldsymbol{q} = \nabla \cdot \boldsymbol{q}_c + \nabla \cdot \boldsymbol{q}_r$, contains contributions from gas phase heat conduction and emission/absorption of thermal radiation.

Regarding the turbulence modelling, practical CFD fire models have classically fallen into one of two major categories: Reynolds-averaged Navier-Stokes (RANS) or Large Eddy Simulation (LES). The difference of these two categories is the nature of the starting equations: In RANS, the Navier-Stokes equations are *time averaged* before the derivation of the discrete form suitable for programming as a solver algorithm. The solver then finds a *steady state* or *quasi steady state* solution for the equations. Time-dependent flows can be solved if the time scale of the mean flow is large compared to the time scale of the turbulent fluctuations. In LES, the time averaging is not performed, and the solutions can be considered "accurate" in time, meaning that the variations in the solution correspond to the motions resolvable by the numerical grid. The marching in time takes place using short time steps Δt, which are usually defined by the following stability criterion:

$$\Delta t < \min_{ijk}\left(\Delta x_{ijk} / u_{ijk}\right) \tag{21.5}$$

where Δx and u are the grid cell size and velocity, respectively, and the minimum is found over the entire computational domain. As the maximum velocity changes over time, so will the time step. Equation 21.5 is called Courant-Friedrichs-Lewy (CFL) condition [8, 9] after the German mathematicians who proposed it in 1928—well before modern computers. In explicit numerical solvers, the CFL condition must be met to ensure solution stability. For implicit solvers, the constraint for the time step is not caused by the stability criteria but the requirement of temporal accuracy, often leading to a time step of the same order as given by Equation 21.5.

In LES, the filtering is performed in space (not time), although the actual filtering is usually limited to the length scales below the grid cell size. This may be confusing since in the fluid dynamics literature, the derivation of LES algorithms usually involves the spatial filtering operation:

$$\bar{\varphi}(\boldsymbol{x},t) = \int_{\Omega} G(\boldsymbol{x}-\boldsymbol{x}')\varphi(\boldsymbol{x}',t)\,d\boldsymbol{x}' \tag{21.6}$$

where $G(x-x')$ is a filtering window. In the actual models, the filtering volume Ω is usually set equal to the numerical grid cell, and the cell value simply represents a volumetric average in the cell. Additional equations are needed for modelling the sub-grid scale turbulence.

The practical difference between RANS and LES is depicted in Figure 21.1, which shows the temperature fields of a pool fire flame. While the RANS result shows smooth variations and looks like a laminar flame, the LES result clearly illustrates the large-scale eddies. Both results are correct solutions of the corresponding equations. However, the time accuracy of LES is also essential for the quantitative accuracy of the buoyancy driven flows. As Baum and Rehm [10], and numerous following studies have shown, the dynamic motions or "eddies" are responsible for most of the air entrainment into the fire plumes. Since these motions cannot be captured by RANS, LES is practically always better suited for fire-driven flow and—without a doubt—the

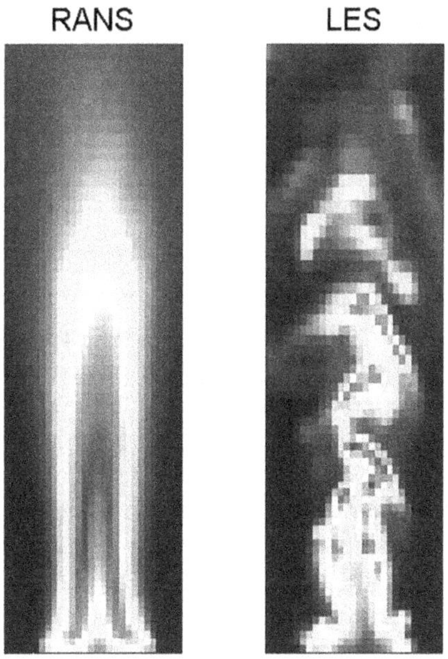

FIGURE 21.1 A comparison of temperature fields in a pool fire flame simulations using RANS and LES.

most common technique of turbulence modelling in fire simulations. LES is the only or preferred turbulence model in most CFD codes used for fire simulations, including Fire Dynamics Simulator (FDS) [11, 12], and FireFOAM [13].

There are certain applications where RANS may have an advantage over LES. RANS modelers can exploit a priori knowledge of the mean flow direction by accepting high aspect ratio grid cells, such as tunnels and pipes. The grid cells of a RANS model can be made long and thin, giving good accuracy in the direction of the largest gradients, that is, normal to the wall, but saving cells in the direction of the tunnel. In LES, all the velocity components are present with similar likelihood, and the cell aspect ratios must be close to unity. For this reason, tunnel simulations using LES are computationally expensive. RANS models have also been used in wind engineering and wildland fire applications, although LES is better suited for cases where a strong wake is formed behind a building [14]. In the simulation of long, quasi-steady or steady state fires, RANS with an implicit solver allows fast marching in time using long (~ 1 s) time steps, while LES time step is bound by the CFL condition, being typically order of 0.01 s or less.

The accuracy of LES technique in fire simulation has been studied extensively. Early validation of FDS predecessors was performed by comparing simulations against salt water experiments [15–17], fire plumes [18, 19], and room fires [20]. The first FDS versions were validated for fire plumes [21] and fires in enclosures in the context of the World Trade Center investigation [22, 23] and the fire model validation project sponsored by the U.S. Nuclear Regulatory Commission [24]. Since 2007, FDS program has had a dedicated Validation Guide, which currently contains comparisons against data from close to a hundred experimental campaigns [25]. Being continuously updated and maintained, it is the best source of information if an FDS user needs an estimate for the model uncertainty in the prediction of a specific fire-related output quantity, such as a fire plume temperature.

In RANS simulations, boundary layers have traditionally been handled using wall functions that assume a logarithmic velocity profile on the wall. In simple applications, these functions work well, having the most serious problems in situations involving separation and reattachment. Similar "sub-grid scale wall functions" can be derived for LES, or the effect of the wall can be taken into account in the sub-grid scale model of viscosity [26]. In their comparison of measured and predicted turbulence statistics, Zhang et al. [27] showed that good room-scale predictions can be achieved without any wall-functions, simply applying a no-slip boundary condition. However, in many cases the wall effects can be important, and wall boundary layer must be modelled in the sub-grid-scale for the wall shear and convective heat flux [28]. Wall function–type corrections have also been proposed for the temperature gradient and boundary layer mass transfer [29].

21.2.2 Combustion

21.2.2.1 Why Combustion Modeling?

The most important difference between the majority of CFD applications and the fire CFD is what drives the flow. In typical, non-fire CFD, the boundary conditions such as inflow velocity drive the flow. Fire problems, in turn, are always driven by the combustion source terms, that is $\dot{q}''$ in Equation 21.3. The accuracy of the combustion model is therefore essential for the quality of the whole simulation. Depending on the nature of the simulation, different aspects of combustion modeling become important. In the prediction of heat transfer from the flames to the nearby surfaces, the *spatial distribution* of the heat release rate is important because that determines the temperature field. Flame spread simulation is an important example where the fine details of the temperature field determine the heat fluxes and heating of the condensed phase. For simulation of quantities that are far from the flame zone, such as smoke filling of a large building, it is important that the combustion model can accurately predict the *global* heat release rate which acts as a source of buoyancy and energy. For the simulation of toxic hazards, on the other hand, the (in) completeness of the chemical reactions and the production of gaseous species, such as CO, are important.

In essence, combustion modelling aims at calculating the rate(s) of the chemical reaction(s) between reactants. For a single-step combustion reaction, this is the rate of fuel consumption $\dot{m}_F'''$. The local rate of heat release per unit volume is then calculated by multiplying with the heat of combustion:

$$\dot{q}''' = \Delta H_c \cdot \dot{m}_F''' \quad (21.7)$$

Fire science has always been a small field compared to combustion science, which is clearly the closest relative. Through the history of fire CFD, combustion models that have been developed for other combustion problems have been directly applied to fire problems. When doing that, the abovementioned aspects of fire simulation should be kept in mind. In particular, the requirement of an accurate global/integrated heat release rate value must be ensured (this is rarely found interesting in combustion science). The theoretical challenge for the modelling comes from the range of combustion phenomena relevant to large scale fire development: lean premixed combustion, partially premixed combustion, non-premixed combustion, and rich combustion during post-flashover burning.

Over the last 40 years, considerable progress has been made in understanding elementary chemical reactions relevant to combustion and pollutant formation. Even for simple hydrocarbons, hundreds of reactions and intermediate species are involved in the oxidation process. For example, the GRI-Mech 3.0 mechanism (intended primarily for combustion of methane, a relatively simple hydrocarbon) includes 325 elementary reactions and 53 species [30]. Additional reactions and intermediate species must be considered to describe the oxidation of higher hydrocarbons produced in fires. It is not yet possible to predict the chemical composition of these gases; even if this was possible, the combustion mechanism required to describe the oxidation of these gases would involve such a large number of species that computer power limitations would preclude the use of such a mechanism for practical except for highly simplified (i.e., 1D steady state) calculations. In fire growth modeling, complex combustion processes are usually relegated to either a single "global" irreversible reaction with infinite (mixing limited) or finite (temperature limited) rate, single or two-step reactions with an empirical "mixing time" as a rate-determining property, or a flamelet library representing the numerous reactions that cannot be explicitly calculated.

In addition to the inherent complexity of microscale combustion processes, the range of length (and time) scales relevant to combustion and turbulence also presents significant computational challenges for fire growth modeling. The thickness of the reaction zone in a diffusion flame and the Kolmogorov microscale are on the order of 1 mm or finer, but currently available computer power limits grid sizes used in 3D simulations of room-scale fire growth to approximately 10 mm. For these reasons, direct numerical simulation (DNS) of large-scale fire growth is not yet feasible. Consequently, CFD-based fire growth modeling requires the use of simplified descriptions of combustion and turbulence involving sub-grid-scale models [31]. The consequences of these turbulence and combustion modeling approaches on flame spread predictions have not been thoroughly explored.

21.2.2.2 Solving for Gas Species

In the formulation of numerical models for combustion, a number of simplifying assumptions and approximations are typically made. In Eulerian framework, combustion modeling requires solution of the gas phase transport equation for some scalar representing the fuel concentration. In real fires, a range of different fuels are produced from different materials, but in gas phase combustion modeling the different fuels are often represented by a single fuel. With most commonly used CFD codes, this is simply a cost-saving approximation, not a limitation by the code capabilities. The effect of this approximation has not been studied extensively, but if we are mainly interested in far-field quantities, it is reasonable to assume that the effect is small, as long as the correct amount of energy is released by taking into account the differences in heat of combustion. For example, Wang et al. simulated hydrogen/methane jet flames using FireFOAM [32] and included separate combustion reactions for the two fuels. The results, however, focused on the global radiative characteristics rather than the influence of combustion modelling. Errors may be introduced by the simplified reactions when simulating combustion of fuels with clearly disparate combustion characteristics, for example, a compartment fire with both methane (lightly sooting) and acetylene (highly sooting with luminous flames).

In full-scale fire modeling, a diffusion flame structure is usually assumed. However, in many fire situations, such as under-ventilated fires, premixed or partially premixed flame theory may be more appropriate. For the first tens of years since the beginning of the combustion CFD at 1970s, it was common to use the Burke-Schumann description of the diffusion flames to conveniently represent the transport of gaseous species by a single scalar quantity called *mixture fraction*. For a simple one-step reaction:

$$\nu_F \text{Fuel} + \nu_O \, \text{O}_2 \rightarrow \sum_i \nu_{P,i} \text{Products} \tag{21.8}$$

the mixture fraction is defined as a fraction of the mass that originated in the fuel stream:

$$Z = \frac{sY_\text{F} - \left(Y_\text{O} - Y_\text{O}^\infty\right)}{sY_\text{F}^I + Y_\text{O}^\infty} \quad ; \quad s = \frac{\nu_\text{O} M_\text{O}}{\nu_\text{F} M_\text{F}} \tag{21.9}$$

where Y_O and Y_F are the mass fractions of oxygen and fuel, respectively, $\text{Y}_\text{F}^\text{I}$ is the mass fraction of fuel in the fuel stream, and Y_O^∞ is the ambient mass fraction of oxygen; s denotes the stoichiometric oxygen to fuel mass ratio. The quantities M_F and M_O are the fuel and oxygen molecular weights, respectively. The transport equation for mixture fraction is

$$\rho \frac{DZ}{Dt} = \nabla \cdot \rho D \nabla Z \tag{21.10}$$

where D is the diffusion coefficient and D/Dt is the material derivative. This equation shows that the mixture fraction is a conserved scalar—the transport equation has no source or sink terms.

The benefits of using the mixture fraction were two-fold: First, instead of solving separate transport equations for fuel, oxygen, nitrogen, and products, it was sufficient to solve one single transport equation for mixture fraction and calculate individual species concentrations from the value of mixture fraction using so called state relations [33]. An example of state relations is shown in Figure 21.2 for ethanol. From the figure, we see that once the value of Z is known, it is very easy to obtain the local concentrations of all gas species. Second, since the mixture fraction is a perfectly conserved scalar, it was not necessary to calculate species formation/destruction rate source terms, which can be expensive to evaluate and cannot be accurately evaluated on coarse grids.

Despite the obvious benefits of the mixture fraction formulation, it is not used in FDS code because it limits the possibility for extending the model for more advanced but necessary details, such as multiple fuels, partial reactions, and non-trivial initial and boundary conditions. The current FDS formulation still makes use of the reduced variable set by specifying the reaction as

$$\text{Fuel} + \text{Air} \rightarrow \text{Products} \tag{21.11}$$

where Fuel, Air, and Products are lumped species, that is, groups of primitive species (fuel, oxygen, nitrogen, carbon dioxide, water vapor, etc.). Using three lumped species instead of mixture fraction increases the number of transport equations (to be solved) from one to three but provides more flexibility for the numerical solution.

21.2.2.3 Computation of Reaction Rate

The combustion reaction rate is controlled by the availability of fuel and oxygen, the rate at which they mix, and by temperature. In computations of very fine spatial and temporal resolution, local chemical equilibrium cannot be assumed, and the chemical reactions are described as *finite rate reactions*. The temperature dependence of the reaction (fuel consumption) rate is presented as an Arrhenius reaction:

$$\dot{m}_{\mathrm{F}}''' \propto e^{-E/RT} \tag{21.12}$$

The Arrhenius form of the reaction results from the Maxwell speed distribution and the rate at which molecular bonds in gas phase species are broken [34]. In full-scale fire modeling, the resolvable length and time scales are usually much larger than those associated with the scales of the chemical combustion reaction, and it is common to assume that the reactions are infinitely fast. The local reaction rate then depends on the rate at which oxygen and fuel are transported toward the reaction zone, shown in Figure 21.2 as a point of stoichiometric mixture fraction, where both oxygen and fuel mass fractions go to zero.

Vast majority of the CFD combustion simulations have used some variant of the same model family, carrying names like eddy breakup model (EBU), which was originally developed for pre-mixed combustion by Spalding [35], eddy dissipation method (EDM) [36], or eddy dissipation concept (EDC) [37]. The main assumptions of these models are that the chemical reactions take place at fine structures similar to Kolmogorov scales, and that the reaction rate can be calculated from the rate of mixing between the fine structures and the surrounding fluid. In each CFD cell with filtered gas density $\tilde{\rho}$, the fuel and oxygen mass concentrations are $\tilde{\rho}Y_{\mathrm{F}}$ and $\tilde{\rho}Y_{\mathrm{O}}$, respectively. As the reaction rate depends on both concentrations, the amount of reaction is limited by whichever is consumed first, and the fuel reaction rate can be expressed as

$$\dot{m}_F''' = -C\frac{\tilde{\rho}\min\left[Y_{\mathrm{F}}, \frac{Y_{\mathrm{O}}}{s}\right]}{\tau} \tag{21.13}$$

where s is the stoichiometric constant, τ is the mixing time, and C is a model constant to be calibrated.

Developments in different model versions have focused on determining the mixing rate, which is strongly related to the transfer of turbulent energy from integral (e.g., fire plume) scale to Kolmogorov scale. EBU model assumes very fast chemistry where the combustion process is limited by the rate at which unburned structure break up to smaller ones. Model was initially developed for the RANS simulations, where the turbulent time scale can be calculated as a ratio of the turbulent kinetic energy k and its dissipation rate ε, leading to the following reaction rate equation:

$$\dot{m}_{F,\mathrm{EBU}}''' = \tilde{\rho}\frac{\varepsilon}{k}\min\left[C_{\mathrm{R}}Y_{\mathrm{F}}, \frac{C_R'Y_O}{s}\right] \tag{21.14}$$

In the previous section, C_{R} and C_R' are empirical constants. EDM extended the EBU for non-premixed combustion.

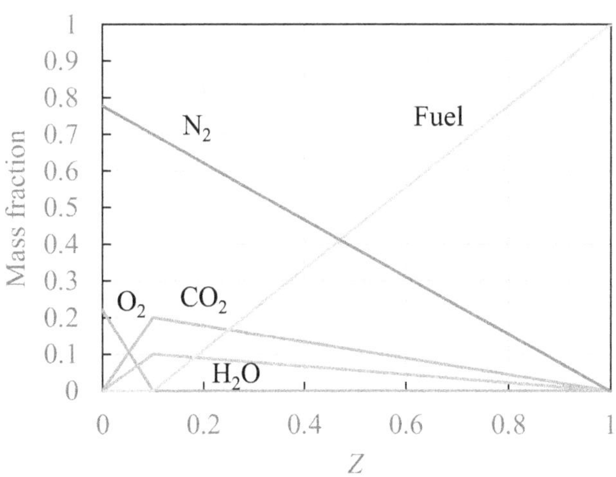

FIGURE 21.2 State relations for ethanol.

EDC then extended EDM for more detailed reaction kinetics by combining Arrhenius kinetics with mixing-limited rates and by considering the fraction of the reacting mixture and its mixing with surrounding flow, that is, an attempt to model so-called turbulence-chemistry interactions.

Extensions of the abovementioned combustion models to the LES framework have been problematic, partly because the eddy characteristic time scale cannot be easily determined in LES, where the kinetic energy k is not readily available. Modelers (actually, model developers) need to determine the position of the LES filter width Δ within the length scales of the energy cascade and formulate models that predict both the time scale and the fraction of the non-reacted reactants that will react during the time step [38]. Numerous ways to calculate the reaction time scale were developed in the field of turbulent combustion. For the applicability in fire simulations, it has been important to consider the practical range of different geometrical, flow and turbulence characteristics. Chen et al. [38] studied two different ways to calculate the reacting fraction of the SGS fine structures and observed that what works for pre-mixed combustion may not work for buoyancy-dominated diffusion flames. Ren et al. demonstrated how wall flames had both fully turbulent and partially turbulent or almost laminar regions [28]. These regions would have very different time scales at which fuel and oxygen mix with each other. In the turbulent regions, mixing would be a consequence the sub-grid scale turbulence, for which the SGS turbulence model of LES would provide the relevant time scale τ_{SGS}. In partially turbulent regions, it would be diffusion that drives the mixing, and the mixing time scale within each cell would be proportional to square of the cell size Δ^2 divided by thermal diffusivity. In each cell, minimum of the two was used:

$$\tau = \min\left(\tau_{SGS}, C_{diff}\left(\frac{\Delta^2}{D_{th}}\right)\right) \tag{21.15}$$

This idea has been extended in the combustion model of FDS to account for five different time scales that may appear in fire simulations [12, 39]. The relevance of each time scale in a specific simulation and fluid cell depends mainly on the LES filter width (cell size), as depicted in Figure 21.3. The smallest time scale is a chemical time scale $\tau_{chem} D_F / s_L^2$, where D_F is the fuel diffusivity and s_L is the flame speed. At slightly larger scale, the mixing is controlled by molecular diffusion and the mixing time, τ_d, should vary as the square of the cell size. Such simulations would be classified as DNS. As the cell size increases beyond the Kolmogorov scale, turbulent advection starts to control the mixing if the Reynolds number is sufficiently high, and the time scale varies as $\tau_u \Delta^{2/3}$. This is the regime where most LES submodels are valid. In the next scale, the grid cells are larger than the flames, and the time scale is controlled by buoyancy. Assuming a constant acceleration by gravity, we obtain a time scale τ_g which scales with the

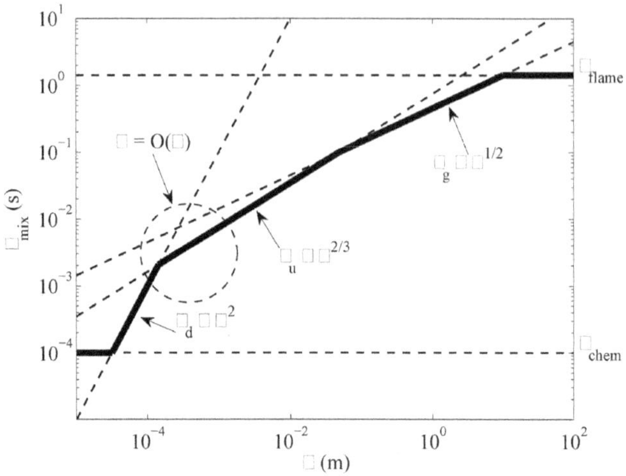

FIGURE 21.3 Combustion reaction time scale model in FDS [12, 39].

square root of the cell size. For very large cells, the upper limit of the time scale comes from the flame length because we know that all the combustion should occur during the time it takes for fuel molecule to travel from the flame base to the flame tip. Combining the different time scales into a single formula gives us the reaction time model of FDS:

$$\tau = \max\left(\tau_{chem}, \min\left(\tau_d, \tau_u, \tau_g, \tau_{flame}\right)\right) \tag{21.16}$$

Now that the mixing time scale, and thus the reaction rate, are known, the amount of fuel consumption can be calculated, taking into account the fraction of the cell contents that has already been mixed. In the beginning of each time step, a grid cell may contain both mixed and unmixed regions of fuel and air. Within the mixed region, reaction can be modelled as a batch reactor model with finite or infinite reaction rate. Rather than formulating a model for the fixed reactive fraction, like Chen et al. [38], in FDS the fraction of the mixed mass is solved numerically from an ordinary differential equation for mixed mass.

The LES combustion models developed by combustion scientists are often based on probabilistic and flamelet-based methods. They rely on good spatial accuracy where both temperature and concentration fields are well captured in the vicinity of the reaction zone. The sub-grid scale turbulence and combustion are related by assuming a statistical distribution of conditions inside a grid cell—instead of uniform conditions indicated by the averaged cell values. For simplicity, some presumed shape for the probability density function (pdf) representing the statistical distribution of conditions is assumed (i.e., beta function, clipped Gaussian, etc.). Then, the shape of the pdf is determined based on the mean, variance, and sometimes higher order moments of the pdf as determined from the filtered temperature/species fields. Unfortunately, in a typical fire simulation, neither the temperature nor the species concentrations are accurately captured, and the necessary pdf parameters may be difficult to determine. Combining the models with radiation effects can also be difficult. Robustness can thus only be achieved by simplicity.

Laminar flamelet-based combustion models within FDS were studied by Yang et al. [40] and Kang and Wen [41], but the performance or advantage over the simple flame-sheet model in large-scale fire simulation could not be demonstrated. In full-scale calculations, the concentration and temperature fields close to the flame sheet have overshoots, caused by the second-order transport scheme. Flamelets have also been used for the prediction of secondary species, such as CO, HCl, and soot [42]. The flamelets (and state relations) can be determined either experimentally [43] or computationally, using detailed models for combustion chemistry that incorporate strain rate effects [44]. More recently, combustion models based on steady [45] and unsteady [46] flamelets have been coupled with a LES-based CFD and radiation model using FireFOAM. The authors observed significant discrepancies between simulated and measured temperatures, when using steady flamelets and associated them with overestimated values of the width of the presumed probability density function representing sub-grid-scale variations of mixture fraction. Using unsteady flamelets was able to capture the physics of the turbulent line burner flame, where part of the flame has long residence time, that is, flame is physically and optically thick.

In full-scale fire modeling, the finite reaction rates may be needed if one attempts to model things such as CO and soot production and oxidation, or ignition and extinction. However, then the simple mixture fraction or lumped gas formulations must be supplemented by additional variables keeping track of the reaction progress. Even with the infinitely fast reactions, considering partial reactions requires more transport equations for gas species and more reaction rate calculations. McGrattan et al. [47] proposed a two-step combustion model to predict CO and soot in under-ventilated fires. This model provides the modeler an additional level of control over the in-flame soot concentration without Arrhenius-rate-dependent chemical models for the soot particle formation and oxidation. For propane, as an example, the two reaction steps would be the following:

$$\begin{aligned} &C_3H_8+\left(\frac{3}{2}\gamma+2\right)O_2 \rightarrow 3\gamma CO+3(1-\gamma) \\ &C+4H_2O \end{aligned} \quad (21.17)$$

$$\begin{aligned} &3\gamma CO+3(1-\gamma)C+4H_2O+\left(3-\frac{3}{2}\gamma\right) \\ &O_2 \rightarrow 3CO_2+4H_2O \end{aligned} \quad (21.18)$$

where γ denotes the fraction of the fuel's carbon that is converted to CO versus soot (here taken to be pure carbon, C) in the first reaction sub-step. The model postulates that CO and soot are produced in a fast initial step followed by oxidation to CO_2 in a slower second step. It assumes that while the second oxidation step is slow relative to the first step, it is nevertheless still faster than the time step used in the combustion sub-model. Thus, each reaction is treated as infinitely fast, but processed in serial, with the first reaction given priority for available oxygen. In practice, the post-flame yields of soot and CO are added to the reaction equation, but that would complicate the formula. For now, there is no theoretical or direct empirical justification for a particular value of γ, and to some extent, γ can be seen as a calibration coefficient. Nevertheless, this kind of formulation may serve as an important starting point for the fire engineers capability to predict the outcomes of under-ventilated combustion.

21.2.3 Radiative Heat Transfer

21.2.3.1 The Radiative Transfer Equation

In enclosure fires, radiation may be the dominant mode of heat transfer. For flames burning in an open atmosphere, the radiative fraction of overall heat transfer ranges from less than 0.1 to 0.4, depending both on the fuel type and the fire diameter [48]. Due to the important role that radiation plays in fires, all fire CFD models have a radiation model that solves the radiation transport equation (RTE) [49–51]:

$$\begin{aligned} \underbrace{s\cdot\nabla I_\lambda(x,s)}_{\text{rate of change}} &= \underbrace{-\left[\kappa_\lambda(x)+\sigma_\lambda(x)\right]I_\lambda(x,s)}_{\text{rate of attenuation}} \\ &\quad \underbrace{+\kappa_\lambda(x)I_{b,\lambda}(x)}_{\text{esmission source}}+\underbrace{\frac{\sigma_\lambda(x)}{4\pi}\int_{4\pi}\Phi_\lambda(s,s')\, I_\lambda(x,s')\,d\Omega'}_{\text{in-scattering}} \end{aligned} \quad (21.19)$$

where **s** is the unit direction vector; I_λ is the intensity at wavelength λ; $\kappa_\lambda(x)$ and $\sigma_\lambda(x)$ are the local absorption and scattering coefficients at λ, respectively; I_b is the emissive power of the medium; Φ(**s,s'**) is the scattering phase function giving the scattered intensity from direction **s'** to **s**. The terms of the RTE have the following interpretations: The left hand side is the rate of change of the intensity in direction **s**; the first right hand side (rhs) term describes the attenuation by absorption and scattering to other directions; the second rhs term is the emission source term; the last rhs term is the in-scattering integral, describing how much intensity is gained by scattering from all the other directions to the present direction. The intensity depends on location, direction, and wavelength.

Unlike the Navier-Stokes equations, the RTE normally does not have time derivative on its left hand side. At every moment, intensity is assumed to be in equilibrium state because the radiation propagates at the speed of light, that is, much faster than the other processes of a system. Intensity changes over time only when the other state variables (temperature, pressure, gas, and soot concentration) change.

The radiation field connects with the energy equation (Eq. 21.3) through the divergence of the radiant heat flux vector $\mathbf{q}_r$ (ignoring here the wavelength dependence for simplicity):

$$q_r(x) = \int_{4\pi} I(x,s)\,s\,d\Omega \tag{21.20}$$

$$-\nabla \cdot q_r(x) = \kappa(x)\left[U(x) - 4\pi I_b(x)\right]; \quad U(x) = \int_{4\pi} I(x,s)\,d\Omega \tag{21.21}$$

In words, the net radiant energy gained by a grid cell is the difference between that which is absorbed and that which is emitted.

To solve the RTE for intensity, one needs to do the following:

1) Discretize the domain and direction and formulate a numerical RTE solver to calculate the transport of intensity over the domain and in different directions.
2) Model the source term, that is, product $\kappa_\lambda(x) I_{b,\lambda}(x)$.
3) Model the absorption coefficient $\kappa_\lambda(x)$ in each location of the domain, relying on the available information on the local gas temperatures, gas concentrations, and soot concentrations.
4) Specify boundary conditions for the emitted, absorbed, and reflected intensity at all domain boundaries.
5) Model the absorption and scattering coefficients and the scattering phase functions of any scattering media like soot particles, liquid droplets, and vegetation.

Step 3 includes the fundamental choice of handling the wavelength dependence. Indeed, most practical calculations are performed using "gray approximation," where all spectral properties are averaged into corresponding mean properties. This modelling choice will obviously have strong implications on steps 2, 4, and 5, but not so much for the numerical implementation (Step 1). Next, we discuss the steps 1–3 of the above list. See, for example, [12] for the details of the boundary conditions in the context of fire simulations, and [52] for scattering.

21.2.3.2 Numerical Solvers for the RTE

The low-order spherical harmonics-based models like P-1 and six flux models [49] were popular in the early years of combustion CFD. In P-1, the diffusion approximation of RTE is adopted, and spherical harmonics are used to develop the intensity. It is best suited for optically thick cases where intensity fields have smooth variations. The six-flux model in turn is related to the use of Cartesian grid system; the intensity is solved in the six co-ordinate directions. It can be derived from the spherical harmonics solution with six ordinate directions.

The ray tracing models such as discrete transfer (DT) [53] and the photon Monte Carlo (PMC) method are theoretically good for fires but may become computationally expensive. The RTE is integrated along imaginary rays (lines of sight), and the number of these rays determines the angular accuracy of the solution. In DT, the ray directions are predetermined, and the tracing starts from the domain boundaries. PMC methods mimic the real electromagnetic processes by generating a statistical distribution of computational "photons" which represent different wavelengths and propagate in random directions. Tracing starts either from the boundary or from within the medium and ends when the photon is absorbed by gas or boundary. DT method has lost its popularity, mainly due to the very strong inhomogeneity of the predicted intensity field (ray effect). PMC methods, in turn, have been used extensively in the thermal radiation literature to generate accurate "benchmark" solutions, against which other solver can be compared. Thanks to the increasing computational resources, it has become possible to integrate PMC methods with transient CFD [54].

Discrete ordinates methods (DOM) [50] and finite volume method (FVM) [55] are currently the most popular ways to solve the RTE. In these models, the solid angle is first divided to small control angles or directions, and the flux of intensity for each direction is solved separately in space. DOM and FVM are very similar techniques. In DOM, the angular distributions are defined by generalized S_N and T_N quadratures. In FVM, the polar/azimuthal discretization is model-specific, but the angular integration is performed exactly. Since it is, indeed, the angular discretization, not spatial, it would be more accurate and less confusing to call FVM a "finite angle method" (FAM) instead of FVM [56].

The numerical accuracy of the FVM-DOM method depends on how the local propagation of intensity is solved [57, 58]. As usual, higher-order numerical methods provide less diffusive and thus more accurate solutions, but with increased computational cost. Figure 21.4 demonstrates the numerical diffusion of the total intensity $U(\mathbf{x})$ (see Eq. 21.21) when using a first-order upwind FVM. The model represents a L-shaped domain with cold ($T = 0$ K) boundaries, except one hot, emitting wall on the left side. In exact solution, radiation would not be observed below the diagonal from the domain center to the bottom-right corner. The diffusive method allows some radiation to "blend" in this region. On the other hand, like ray tracing, also the flux methods suffer from the "ray effect," as demonstrated in the second part of Figure 21.4;. At relatively large distances from the emission source, the ray effect manifests itself as fluctuations in radiative heat flux. To some extent, numerical diffusion may be exploited to "even out" the intensity gradients and thus reduce the ray effect. This fact has reduced the interest towards implementing the higher-order RTE solvers. Various modeling approaches for radiative heat transfer in pool fires were compared in [59]. The

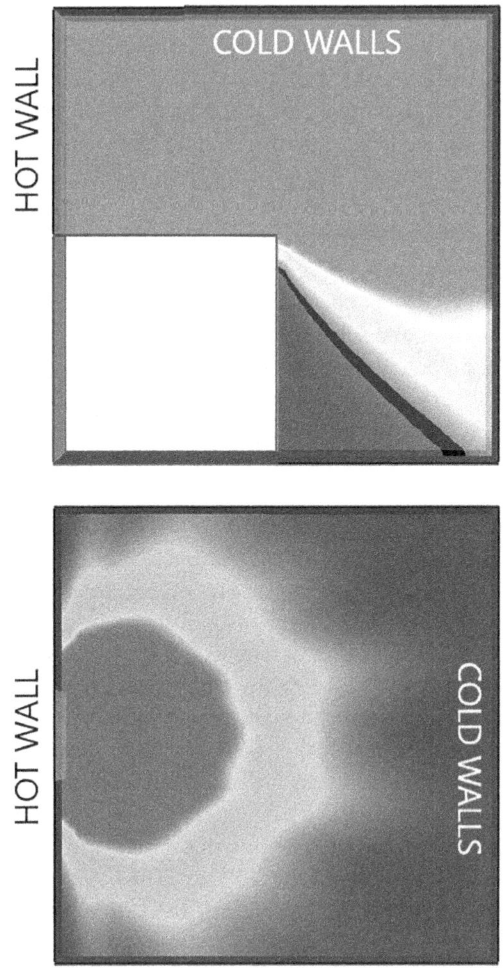

FIGURE 21.4 Examples of numerical errors in RTE solution of integrated intensity U using FVM. Top: diffusion. Bottom: ray effect.

results showed that, within the isotropic region (inside a flame), all methods gave comparable results, but in the anisotropic regions, angular treatments played an important role. The radiation schemes used in the two most popular fire-CFD modes FDS (FVM) and FireFOAM (DOM) have been compared in [60].

The inherent time dependence of fires sets strong requirements for the computational efficiency of the RTE solvers. The time accuracy of the radiation field can be relaxed by solving the FVM equations every N^{th} time step and only part of the directions at the time. This is the approach taken in FDS and FireFOAM, with typical values of $N = 3 \ldots 10$.

21.2.3.3 Modelling the Radiative Source Term

The emissive power at individual wavelength and absolute temperature T can be calculated using a Planck function. Integrating the emission source over the wavelength gives the usual engineering form:

$$\int_0^\infty \kappa_\lambda(\boldsymbol{x}) I_{b,\lambda}(\boldsymbol{x}) d\lambda = \tilde{\kappa}_p(\boldsymbol{x}) \cdot \frac{\sigma T^4(\boldsymbol{x})}{\pi} \quad (21.22)$$

where $\tilde{\kappa}_p = \int \kappa_\lambda I_{b,\lambda} d\lambda / \int I_{b,\lambda} d\lambda$ is the (Planck) mean absorption coefficient and σ is the Stefan-Boltzmann constant.

Despite the apparent simplicity of Equation 21.22, several problems appear in the practical CFD calculations. Due to the T^4-dependence, the emission source term is extremely sensitive to errors in temperature. For example, a 15% underestimation of temperature would lead to a source term that is 48% too small. In large-scale fire simulation, this kind of errors in the predicted gas temperature can rarely be avoided. Even if the CFD method conserves the convective enthalpy, the internal sub-grid scale temperature distribution within large grid cells would influence the predicted emission power. Taking a mean over the cell volume of a local temperature to fourth power is not straightforward as $\overline{T^4} \neq \overline{T}^4$, where the overbar denotes spatial average. In words, the mean of the fourth powers is not equal to the fourth power of the mean temperature. Yet another complexity follows from the sub-grid scale correlations between the absorption coefficient and temperature: $\overline{\kappa T^4} \neq \overline{\kappa} \cdot \overline{T^4}$.

A simple and robust way to calculate the emission source term is to use a prescribed radiative fraction χ of the combustion heat release. Coupling with the energy equation takes place through the source term (which in the case of net emission becomes a "sink" term):

$$-\nabla \cdot \boldsymbol{q}_r(\boldsymbol{x}) = -4\pi\kappa(\boldsymbol{x}) I_b(\boldsymbol{x}) \approx -\chi \cdot \dot{q}''' \quad (21.23)$$

In FDS, in its default mode, this approach is used whenever a non-zero radiative fraction is assigned for the combustion reaction and when the local emission power $\chi \cdot \dot{q}''' >$ 10 kW/m^3. It is important to note that the radiative fraction in above is *local* fraction of emitted energy, which may be different (usually larger) than the *global* counterpart. Global radiative fractions of diffusion flames have been reported in fire literature, and for most combustibles they are between 0.3 and 0.4. As the fire size increases, some of the emitted radiation is absorbed by the soot surrounding the high-temperature regions. This makes the global radiative fraction smaller than the local values. The local property should be treated independent of the problem scale, but there are little experimental observations of the local radiative fractions.

As the sub-grid scale fluctuations of κ and T are closely related to the turbulent nature of the flow, their inclusion in the calculation of $\kappa(\boldsymbol{x}) I_b(\boldsymbol{x})$ are commonly called *Turbulence-Radiation Interaction* (TRI). TRI can be directly modeled using Taylor expansion methods [61, 62], transported filtered density function [63], or presumed probability density function of the sub-grid scale distributions [46, 64]. Chatterjee et al. [65] used a laminar smoke point method to determine in-flame soot concentrations and then applied Taylor expansion of the temperature field to close TRI. Snegirev et al. [66] investigated TRI in methane and heptane flames by using a simple approximation for grid cell temperature when flamelet structure is not spatially resolved. Most recently,

Rashidzadeh et al. [67] introduced a two-zone model sub-grid flame model, attempting to reach grid-independent predictions of the radiative fraction in a line-burner flame with varying ambient oxygen concentration.

21.2.3.4 Modeling the Radiative Properties

The absorption coefficient $\kappa(\mathbf{x},\lambda)$ is an important parameter of RTE. It depends on the local gas and soot concentrations, as well as temperature. As with combustion modeling, the development of the fire-CFD radiation models has largely adapted techniques that were originally developed for combustion system simulations. In fires, however, the challenges of the radiation modeling may be even more pronounced due to the greater optical thickness, the relatively higher importance of absorption (combustion system studies almost exclusively focus on emission only) and a wider range of possible fuels.

The simplest way to account for radiation in a CFD code is to calculate the emission source term as a prescribed radiative fraction χ of the combustion heat release using Equation 21.23 and to ignore the gas phase emission and absorption in the rest of the domain. This optically thin approximation can provide a fast way to account for the most essential radiation physics, that is, the radiative heat loss inside a flame. The distribution of the emitted radiation on surfaces can be either ignored or solved from the RTE for a transparent medium (zero attenuation). The latter approach has been the default mode in FireFOAM and has been exploited in, for example [46, 68]. However, the authors of those studies concluded that obtaining satisfactory accuracy for the temperature distributions requires accounting for the reabsorption of the emitted radiation.

In fires, soot is usually the dominant emitter and absorber of radiation. The modeling of soot formation and oxidation processes is therefore important for the accurate prediction of radiation exchange. Over the years, detailed models that solve the fields of soot number density and mass fraction have been developed and implemented in fire CFD models [69–71]. "Engineering models" for soot formation and oxidation, relying on the laminar smoke point height, have been postulated [72–74] and extended for EDC combustion model and LES [75]. Fukumoto et al. [76] used the soot model in the simulation of vertical flame spread and concluded that reasonable or good accuracy was achieved despite the errors in the underlying models, such as soot formation and oxidation, because the two errors compensated each other.

Unfortunately, the soot formation and oxidation processes are sensitive to the temperature and the same problems appear as in detailed combustion modeling. Simple partial-combustion reaction models (Eqs. 21.17–21.18) may help to enhance the radiative emissions and reabsorption of the flame zone, but more work is needed to find general basis for the inputs of such models. Alternatively, the numerical accuracy of the scalar field solutions should achieve an accuracy required by the complex soot formation and oxidation models. Meanwhile, the standard engineering practice is to prescribe a post-combustion reaction yield of soot, and to use the transported concentration in the calculation of $\kappa(\mathbf{x},\lambda)$.

Although predicting the soot concentration has turned out to be difficult, its radiative characteristics are relatively simple to account for. Soot absorption coefficient is often assumed to be directly proportional to soot volume fraction and inversely proportional to wavelength, $\kappa_{\lambda,soot} \propto f_v / \lambda$. The experimentally observed wavelength dependency of the soot absorption coefficient [77] is mathematically easy to account for and is being used, for example, in FDS.

Unlike soot, the wavelength dependencies of the gaseous species' absorption coefficients are far from simple. Figure 21.5 shows the spectral absorption coefficients of pure water vapor and carbon dioxide at 300 K temperature. The data is shown as a function of wave number $\eta = 1/\lambda$. As can be seen, both gases show about five order of magnitude variation in the κ_λ values and fluctuations that seem to appear in two different scales. First, there are very narrow absorption lines with few cm^{-1} (sub-micron) width. Second, the water vapor lines are grouped in bands. The relative importance of the different wavenumbers depends on temperature according to the Planck distribution. Similar plots could be made for many of the other gases present in the practical combustion systems, including carbon monoxide and many of the fuels.

The main challenge of radiative property modelling is the integration of RTE (Eq. 21.19) over the wavelength spectrum in the presence of the extremely complex λ-dependency of $\kappa(\mathbf{x},\lambda)$. Direct, "brute force" integration of the spectral data that takes into account millions of wavelengths separately, is called a "line-by-line" method (LBL). LBL integration of Figure 21.5 is very expensive and used only to generate benchmark solutions, against which the practical engineering methods are fitted or tested.

A classical engineering approximation to the abovementioned integration problem is to assume gray media (gas and boundaries). The spectral dependency of κ is not considered during the RTE solution; instead $\kappa\ \lambda$ is replaced with an averaged $\tilde{\kappa}$ that can be obtained *in advance* either by integrating the spectraκ_λl over the entire wavelength spectrum or by determining some effective value. Due to the non-spectral nature, such models are often called *global*. The cost saving comes from the fact that the spectral integration is performed before the (thousands of) RTE solutions, storing the results in a table or a correlation-type model, from which the gray absorption coefficients are then fetched during the actual computation. *Weighted-Sum-of-Gray-Gases* (WSGG) models [78] are commonly used for this purpose. WSGG models are essentially polynomial functions that have been fitted to experimental or LBL predictions of emissivity from gas layers with different path lengths and H_2O-CO_2 ratios. The polynomial degrees of freedom are taken as fictitious gases. During the last years, WSGG models have been extended for wider range of H_2O-CO_2 ratios [79], to enable modelling of pure water vapor and carbon dioxide [80] and to account for fuel vapors [81].

Gray radiation is the default model operation in FDS too. To calculate the effective mean absorption coefficients, FDS exploits a "narrow band" model RadCal [82] which basically contains the essential parts of Figure 21.5 and can calculate the spectral integrals very accurately. The narrow band method is still too expensive to be used within the RTE solution, during the CFD simulation. Instead, it is run in the beginning of every simulation to generate a table of mean absorption coefficients for a range of concentrations and temperatures. During the CFD simulation, local values are found quickly from the table. The exploited RadCal output is an effective mean absorption coefficient $\tilde{\kappa}$, which is a solution to the following equation

$$I_{\mathrm{NB}}(L) = \frac{\sigma}{\pi}\left[e^{-\tilde{\kappa}L}T_{\mathrm{rad}}^4 + \left(1-e^{-\tilde{\kappa}L}\right)T^4\right] \quad (21.24)$$

where I_{NB} is the narrow-band solution of a line-of-sight intensity through a gas layer with thickness L, T_{rad} is a characteristic radiation temperature (wall temperature), and T is the gas temperature. Calculating $\tilde{\kappa}$ thus requires a specification of some characteristic temperature, for which a typical flame temperature is a good candidate, and a path length L. The main problem of this (and all gray methods in essence) is that, in a typical fire simulation scenario, it is not possible to define a single path length that would characterize all the optical paths appearing in the simulation. In FDS, the default $L = 10$ cm, a compromise between typical flame and domain length scales.

The performance of different gray models, implemented in FDS, was studied in [83] by comparing the predicted heat fluxes against the experimental measurements from liquid fuel flames. The RadCal-based and WSGG formulations performed similarly, providing in general more accurate results than the approach using the Planck-mean absorption coefficient.

Using a gray approximation can be accurate in homogeneous medium with isotropic radiation and only one characteristic length scale, over which the RTE needs to be integrated. Problems arise with strongly inhomogeneous scalar fields and several lengths scales. To overcome these limitations, non-gray or spectral global models have been developed. So called *k-distribution* methods were the first to introduce a physical basis for low-cost spectral integration. The main observation behind these models was that the individual spectral lines can be treated independently, and consequently, the order of their processing is unimportant. The spectral lines can thus be reordered in increasing order, from small to high value of absorption coefficient. After normalization, we have a cumulative density function of absorption coefficient, shown in Figure 21.6 for water vapor and carbon dioxide. These functions are obviously much easier to integrate numerically:

$$I_{tot} = \int_0^\infty I_\lambda(\kappa_\lambda)\,d\lambda = \int_0^\infty I_\eta(\kappa_\eta)\,d\eta = \sum_{k=1}^{k} w_k I_k(\kappa_k) \quad (21.25)$$

In above, the integral of a spectral intensity is calculated as (relatively short) sum of intensities I_k, determined at pre-selected values κ_k, multiplied by weights w_k, corresponding to some commonly used mathematical integration quadrature.

The Full Spectrum Correlated-k method (FSCK) [84] used this idea and enabled the spectral integration of the RTE with just few quadrature points (consequent RTE solutions)—a reasonable increase in cost over the single solution when using the gray method. The model assumes that the absorption coefficient spectra with different thermodynamic states are mutually correlated. The performance of the correlated-k method has been studied regarding the radiation emission in combustion systems [85] and the prediction of radiation attenuation within a fuel-rich core of laboratory-scale pool fires [86]. The FSCK was found to provide a good engineering performance, except when the strong attenuation by methane broke the

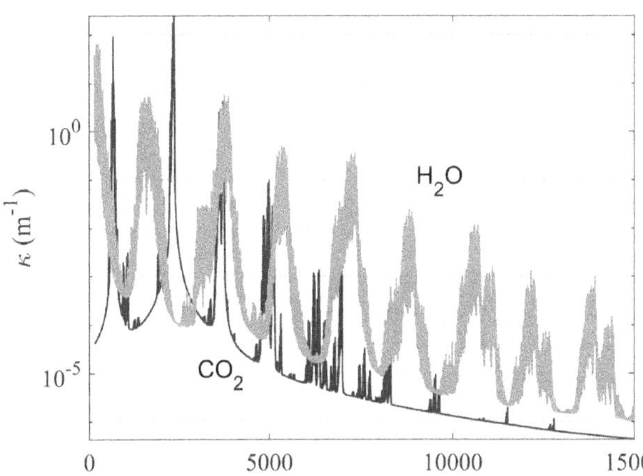

FIGURE 21.5 Spectral absorption coefficient of carbon dioxide and water vapor at 300 K and 1 bar.

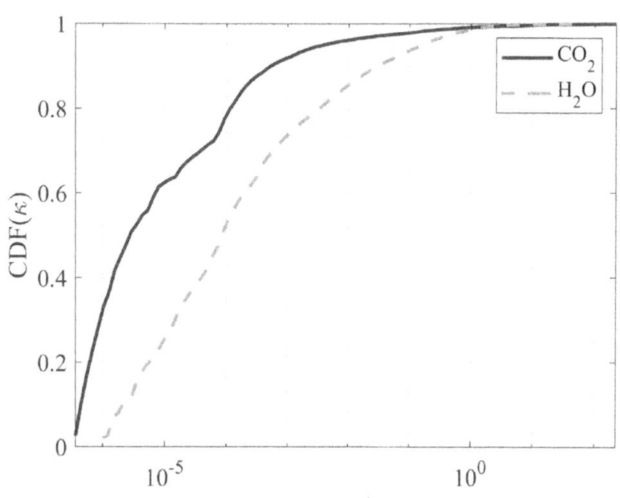

FIGURE 21.6 Cumulative density functions of the water vapor and carbon dioxide absorption coefficients at 300 K and 1 bar.

correlation-assumption, in which case a multi-scale version of the same method was needed.

Another non-gray global model is the Spectral-Line base Weighted-Sum-of-Gray-Gases (SLW) model by Denison and Webb [87]. The physical basis of the model was different from that of FSCK, but mathematically, the resulting procedures were almost identical, especially with the latest developments of both methods [56]. In fact, WSGG, SLW, and FSCK are technically very similar. They require a handful of RTE solutions which are then summed to get the total values of the radiative intensity and heat flux.

During their development, all the abovementioned non-gray models have been calibrated by experimental or LBL data of *emissivity* leaving a layer of hot combustion products. This is understandable, as predicting the flame emission is the key objective of a simulation for an industrial combustion system. In natural fires, however, the reabsorption by the relatively cold combustion products is almost equally important. Sometimes it can be even more important to be able to predict absorption, especially when the radiative emission is modeled using a prescribed radiative fraction but the RTE withing the absorbing parts is solved using modeled radiative properties. In near future, non-gray global models should be optimized for fire simulations to obtain better accuracy with reasonable computational cost.

21.2.4 Convective Heat Transfer

Convective heat transfer takes place between the fluid phase and solid walls and obstructions. Zooming in very close to a solid boundary, heat is transferred between the gas and the solid via conduction, and the rate of heat transfer can be calculated using Fourier's law:

$$\dot{q}'' = k\frac{dT}{dn} \tag{21.26}$$

where k is the thermal conductivity and n is the distance from the wall.

In very well-resolved simulations, the heat conduction across the viscous sub-layer can be calculated directly from the Fourier's law (Eq. 21.26). Such CFD simulations of turbulent wall fires have been reported using FireFOAM [28,76] and FDS [88]. When computing a flame spread on a one-meter tall vertical wall, Ren et al. [28] found out that the relevant length scale describing the boundary layer of a weakly turbulent wall fire was the flame-to-wall distance, which in their scenario was about 3 cm. Grid convergent convective heat fluxes were achieved within the pyrolysis (vertical burner) zone when there were ten cells across this distance, that is, with 3 mm cell size, but 1.5 mm cells were needed to obtain grid convergence in the preheating zone, for example, above the burner. As an indication of the practical cost of such computing, we can take a note of the 1,500 h CPU time that was required for 15 s simulation using 36 Xeon cores. Fukumoto et al. [76] simulated vertical flame spread on poly(methyl methacrylate) (PMMA) wall and, relying on the 3 mm resolution requirement by Ren et al. [28], they adopted 1.9 mm gas phase resolution in the direction normal to the wall. Li and Hostikka [88] simulated the vertical flame spread on thin birch wood rods using axi-symmetric, two-dimensional model. In these calculations, smallest Kolmogorov length scales were estimated to be order of 0.2 mm, but grid convergent flame spread rates were obtained with 1 mm numerical resolution in the gas phase.

From the above examples, we learn that the direct use of Equation 21.26 in a CFD simulation requires resolving the velocity and temperature profiles within the boundary layer. As the thickness of the viscous sub-layer in typical flame spread problems is only order of one centimeter, resolving the near-wall profiles would require about 1 mm grid cells. In the mentioned studies, focused scientific objective and significant computational resources enabled the use of sufficiently fine resolutions, but in full-scale fire engineering simulations, it would be practically impossible to compute the convective heat transfer directly. The domain sizes are typically more complex and larger and the computational cost too high. To overcome the problem of prohibitive resolution requirement, it is common to calculate the convective heat fluxes using conventional engineering approach, as product of the convective heat transfer coefficient h and temperature difference between the solid and gas phases:

$$\dot{q}_c'' = h\cdot\left(T_g - T_w\right) \tag{21.27}$$

where h is determined from experimental correlations that can be found in most heat transfer texts. These correlations are typically developed for Nusselt number:

$$\text{Nu} = \frac{hL}{k}, \tag{21.28}$$

from which h can be calculated when the convective length scale L and gas thermal conductivity k are known. The engineering correlations relate Nu with Reynolds, Grashof, and Prandtl numbers, depending on the nature of the flow. Maragkos and Beji [89] reviewed the convection models used in LES simulations and discussed the challenges of heat and mass transfer correlations in the context of liquid pool and turbulent wall fires. The lack of heat flux validation data was identified as one of the main reasons limiting the development in this field.

The first modelling step is usually to define whether the convective flow is driven by momentum or buoyancy. For a CFD model developer, it is difficult to know in advance what kind of flow situations the convection model will be applied. One solution, used in FDS, is to calculate Nu assuming both momentum and buoyancy-driven regimes, and then choose the value that is greater. The model user must confirm that appropriate correlations are being used.

Another challenge of the empirical convection correlations is the choice of the length scale L that appears in the formulas for most non-dimensional numbers of fluid dynamics, including the Nusselt, Reynolds, and Grashof numbers. Hong et al. [90] showed that, in a relatively well-defined scenario, such as a liquid pool fire, the convective heat fluxes can be calculated based on local conditions only, but often this is not the case. In the empirical setting behind the heat transfer correlations, L has usually some well-defined physical meaning, such as the distance from the leading edge of a flat plate or the diameter of a cylinder or sphere, around which the convection takes place. Now, using such a correlation for an arbitrary boundary cell of a large CFD model leads to obvious difficulty of assigning a value for L because such a characteristic length scale may not exist, and even if it does, it would be very difficult to develop an algorithm that would automatically identify and evaluate such a non-local quantity. The numerical methods used in generalized flow solvers rely on local information only, that is, the values of the solved and state variables in the cell itself and some finite neighborhood. Making use of geometrical or flow-related parameters will always limit the generality of the solver.

An example of an algorithm using a global length scale was developed for the vertical flame spread situation by Li and Hostikka [88]. They replaced the coarse-resolution heat fluxes with pre-calculated heat flux profiles, based on the non-dimensional vertical position relative to the ignition and extinction fronts. In their two-dimensional simulations, it was possible to keep track of the ignition and burn-out front locations and thus calculate a non-dimensional position, that is, the necessary global parameter. The method was shown to reduce the grid dependency of the vertical flame spread predictions, but it would be difficult to generalize to three dimensional cases due to the difficulty of defining the front locations.

According to the review of Maragkos and Beji [89], in well-resolved LES, it is possible to exploit the law of the wall for boundary layer flows and employ wall functions that define the non-dimensional speed u^+ and temperature T^+ within the viscous sublayer and log regions as

$$\begin{aligned} u^+ &= y^+ && \text{for } y^+ < 11.81 \\ u^+ &= \frac{1}{\kappa}\ln\left(y^+\right) + B && \text{for } y^+ \geq 11.81 \\ T^+ &= \Pr y^+ && \text{for } y^+ < 11.81 \\ T^+ &= \frac{\Pr_t}{\kappa}\ln\left(y^+\right) + B_T && \text{for } y^+ \geq 11.81 \end{aligned} \tag{21.29}$$

where $\kappa = 0.41$ is von Kármán constant, $B = 5.2$, Pr and $\Pr_t$ are molecular and turbulent Prandtl numbers. B_T, an integration constant from the relation between velocity and temperature gradients, has been determined experimentally. These velocity and temperature relations rely on the non-dimensional distance from the wall, $y+$, which is calculated by dividing the real distance with viscous length scale. This is a local quantity and thus appropriate for numerical implementation. Finally, the convective heat transfer coefficient would be calculated as

$$h = \frac{\rho c_p u_\tau}{T^+} \tag{21.30}$$

where u_τ is the friction velocity.

21.3 CONDENSED PHASE PYROLYSIS

In a spreading or growing fire, combustion occurs when gaseous fuel liberated from condensed phase (solid or liquid) materials mixes with the surrounding oxidizer and reacts with oxygen, thereby releasing heat and combustion products. This heat supports further fuel heating and gasification. Therefore, condensed phase processes are among the primary factors controlling the ignition and burning. Realistic simulation of flame spread and fire growth requires accurate simulation of condensed phase fuel generation processes.

Generically, a pyrolysis model is an algorithm that calculates the rate at which solid combustibles generate gaseous fuel (pyrolysate, CO) or other gases (CO_2, H_2O) when thermally stimulated. Pyrolysis models range from simple empirical formulations, that rely heavily on fire test data, to complex models that simulate microscale physical and chemical processes. The reader is referred to Chapter 20 for the details regarding the condensed phase thermal degradation modeling. In this chapter, we provide only a brief review of the different pyrolysis modeling approaches that have been applied in the simulations of large-scale fire growth and focus on the modeling strategies needed when coupling the condensed phase model with gas phase CFD. From the modeling literature, one can easily find reviews of the specific topics, such as the modeling of wood pyrolysis [91], the condensed phase heat and mass transfer processes [92], and the optimization techniques [93].

21.3.1 Conjugate Heat Transfer

A heat transfer process involving both gas and condensed phases is called conjugate heat transfer. The temperature of one phase (or domain) cannot be solved independently from the another. In fire safety, conjugate heat transfer problems are found in structural fire engineering, where the temperatures of the compartment gas space and solid boundaries are tightly coupled, and when simulating flame spread on solid surfaces, where the reactive boundary layer interacts with a solid wall. Sometimes the coupling is weak, and the conjugate heat transfer can be simplified to one-way coupling. For instance, the compartment temperature may be driven by the heat release, and the effect of an individual solid object on the large-scale temperature field is so small that it can be ignored. In structural engineering, an example of such a scenario could be a steel beam or column heated/

cooled by the surrounding gas space, so that the effect of *one* structural element on the fire itself is negligible. In the simulations of polymeric materials, predicting the ignition of some individual polymer component under the heating from already developing fire could be considered an uncoupled problem, but as soon as the material under investigation ignites, it starts to contribute to the fire, and the coupling becomes tighter.

Over the years, different computational techniques have been developed for solving the conjugate heat transfer problems:

1. The simplest technique is to use separate numerical solvers for the fluid and solid phases and to exchange information through the boundary conditions. Technically, the solvers can be part of the same computer code, though. The use of separate solvers allows a flexible gridding inside the solid phase, which is required due to the three orders of magnitude difference in thermal conductivities. It is also easy to include various physical phenomena such as charring and moisture transfer. Quite often, one-dimensional solution of the heat conduction equation on each wall cell is sufficiently accurate. This technique is implemented as an internal subroutine in FDS, FireFOAM, and ISIS.
2. Separate solvers of three-dimensional heat conduction can be linked to the CFD solver, either as an external program or internal subroutine. External programs are often based on the Finite Element Method (FEM). The link works in one direction: information from the gas phase (CFD) is transferred to the boundary conditions of the solid-phase solver. Specialized algorithms may be needed to model the connection between the gas and solid phases due to the disparity in length and time scales. In the context of fire simulations, Adiabatic Surface Temperature (AST) concept is commonly used to transfer the heat exposure information from CFD to FEM [94]. Examples of tools used for this purpose are Firestruc [95], FDS2FEM [96], FDS2FTI [97], FDS-FEM [98], and OpenFire [99].
3. A full coupling of the solid and fluid phases can be achieved by solving a single enthalpy equation, common for both phases. FDS currently contains a three-dimensional solid heat conduction solver with two-way coupling. Similar approach has been reported in codes SOFIE [100] and C-SAFE [101].

From the above categories, 1D solvers (option 1) are commonly used for full-scale fire simulations with conjugate heat transfer and pyrolysis. This technique is illustrated in Figure 21.7 for an arbitrary system with four gas phase CFD cells against a horizontal solid surface. The condensed phase geometry is thus broken into four "patches," each having a surface area corresponding to that of the gas phase cell face to which it abuts. For each patch, a 1D calculation is performed to determine the conduction heat flux $-k(dT/dz)$, internal temperature profile, and gaseous fuel generation rate. Each patch serves as a boundary condition to the gas phase CFD model. Coupling is achieved by passing the radiative and convective heat fluxes between gas and solid, and the mass flux of gaseous pyrolysate $\dot{m}''$ at solid temperature from the condensed phase solver to the gas phase CFD solver. Each condensed phase patch knows nothing of its neighboring patches, that is, no heat is transferred through the condensed phase in a direction parallel

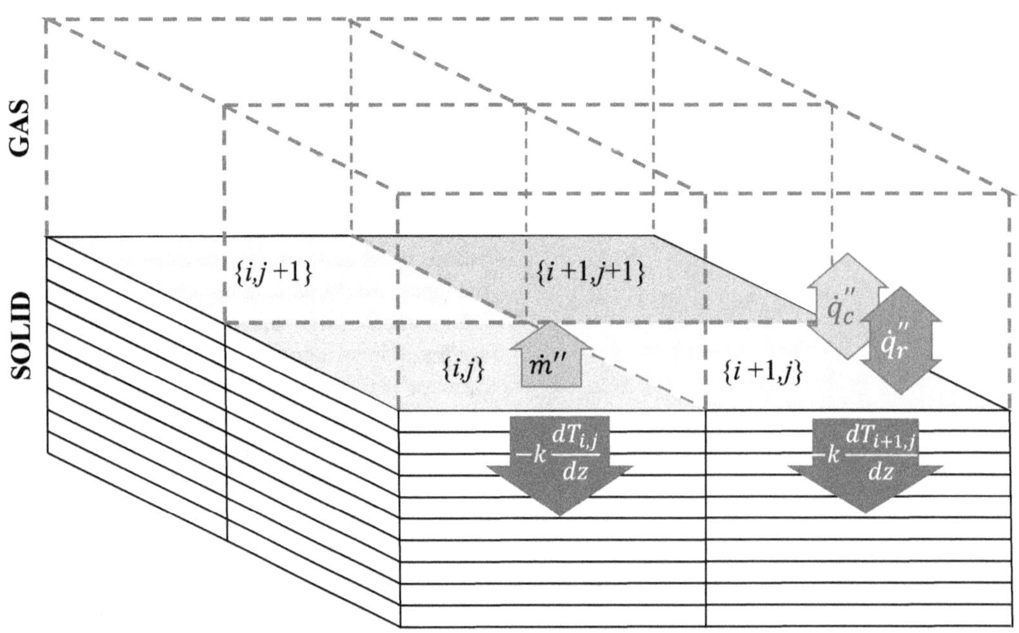

FIGURE 21.7 Coupling the 1D solid conduction solvers with 3D CFD for gas phase.

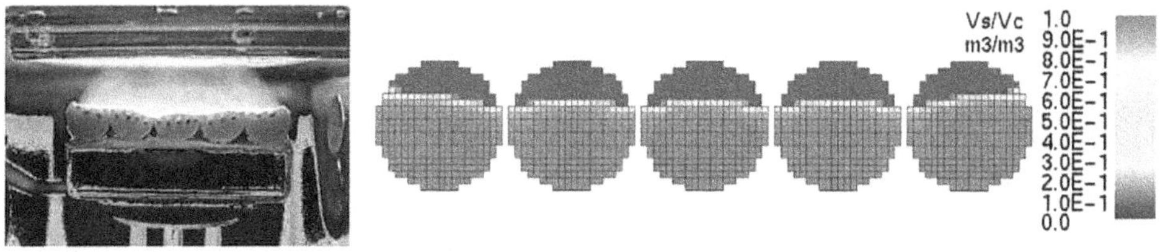

FIGURE 21.8 Gasification of five PMMA rods under external radiation. Thermal image from an experiment (left) and two-dimensional simulation using FDS, showing the solid mass fraction. [102]

to the solid surface. This approximation can fail if the thermal conductivity of the solid is high or if the heat conduction length scale is of the same order with the gas phase heat transfer, such as the opposed (downward) flame spread.

While the 1D approximation may work very well for planar surfaces (most construction products), it will obviously fail when modelling non-rectangular objects in a scale where the length scales of the condensed phase temperature field are of the same order with the modelled geometry. For example, the numerical investigation of flaming and gasification of PMMA cylinders in cone calorimeter in [102] made use of the 3D heat transfer and pyrolysis solver in FDS [103] (Figure 21.8). While the heat conduction was solved in 2D/3D geometry using 1 mm resolution, the mass transfer was not explicitly solved inside the solid. Instead, the pyrolysis products were released from the solid surface nearest to the point of origin. As the material was non-charring thermoplastic, the influence of this approximation was supposed to be small. A specific feature of the simulations in [102] was the reduction of the sample size as the polymer burned away.

21.3.2 Empirical Models for Solid Fuels

Among the different models for predicting the fuel generation rates in CFD-based fire growth calculations, the empirical models are the most simple ones. Instead of attempting to model the physical processes that lead to gaseous fuel production inside decomposing solids, empirical data that can be measured (transient heat release or mass loss rate) or inferred (heat of gasification) from common bench-scale fire tests such as the cone calorimeter are used to characterize the fuel generation processes.

With empirical pyrolysis models, material's burning rate is zero until its surface is heated to its ignition temperature (T_{ig}). Once the ignition temperature is reached, at time $t_{ig,mod}$, material's mass loss rate follows a transient measurement from a bench-scale fire tests:

$$\dot{m}''\left(t-t_{ig,mod}\right)=\dot{m}''_{exp}\left(t-t_{ig,mod}+t_{ig,exp}\right) \tag{21.31}$$

where a subscript "mod" denotes model, and a subscript "exp" denotes experiment. Since the net heat flux to the fuel surface or temperature do not appear in Equation 21.31, fuel generation is decoupled from the radiative and convective environment that develops in the simulation. The method is called a "simple pyrolysis model" in FDS and "thermal model" by some authors [104].

One step towards a predictive model can be taken by calculating the mass loss rate as the net heat flux to the fuel's surface ($\dot{q}''_{net}$) divided by the effective heat of gasification (ΔH_g):

$$\dot{m}''(t)=\frac{\dot{q}''_{net}(t)}{\Delta H_g} \tag{21.32}$$

The effective heat of gasification in Equation 21.32 is a material's fire property, a quantity of heat required to generate unit mass of volatiles at temperature T_{ig} from unit mass of solid initially at T_0.

The primary shortcoming of Equation 21.31 is the assumption that the net heat flux to the model is identical to the experimental net heat flux history. Laboratory experiments usually involve instantaneous application of high heat flux levels (50 kW/m^2 or higher). This type of heat flux history does not necessarily occur in spreading fires, where applied heat flux levels may change over time as the fire develops. Bench-scale laboratory experiments are conducted on small fuel samples burning under constant conditions, and real fires are unlikely to have the same ventilation conditions. In actual fires, fuel packages interact with one another and the fire compartment, so different parts of the same material may burn differently because they are exposed to different thermal conditions. This effect cannot be captured by a model that directly applies bench-scale fire test data without considering the applied heat flux levels, so model results are sensitive to the irradiance level at which experimental data are collected. As an attempt to account for some of the deficiencies pointed out here, an acceleration function was introduced in [105] by relating the modeled heat release rate to the total heat flux absorbed by the solid. Currently, FDS has a simple model to extrapolate burning rate data to the heat feedback occurring during an FDS simulation [106]. The measured heat release rate (HRR) is scaled both in magnitude and in time by the ratio of the realized incident heat flux and the reference heat flux in the experiment. Figure 21.9, copied from the FDS User's guide [107], demonstrates this method, where a material tested at 50 kW/m^2 heat flux is exposed to 25, 50, and 75 kW/m^2 heat fluxes. At 25 kW/m^2, the test data is stretched in time by a factor of 2 with reduction of burning rate with

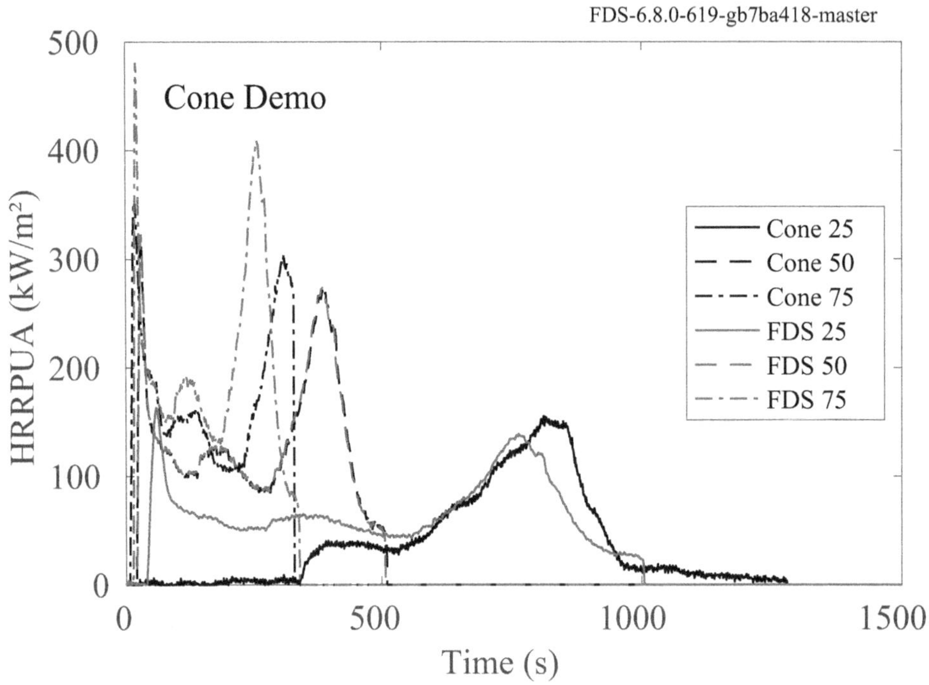

FIGURE 21.9 Demonstration of extrapolating cone test data according to the changing heat flux [107].

factor of 2. Similarly, at 75 kW/m^2 the data is collapsed by 50% and burning rate increased by 50%.

Empirical approaches are useful when macro-scale HRR measurements are available, but little or no information is available regarding the thermophysical properties, kinetic parameters, and heats of reaction that would be necessary to apply a more comprehensive pyrolysis model. Although these modeling approaches are crude in comparison to the comprehensive models presented in the next section, one advantage is that all required input parameters can be obtained from widely used bench-scale fire tests using well-established data reduction techniques. As greater levels of complexity are added, establishing the required input parameters (or "material properties") for different materials becomes an oncrous task.

21.3.3 Complex Pyrolysis Models for Solid Fuels

When the empirical models cannot adapt to the varying heating and other conditions of the transient simulation, more comprehensive and complex pyrolysis models are needed. These models are almost exclusively based on the calculation of kinetically controlled reaction rates. The solid is assumed to be a mixture of N_s condensed components, and each component is volatilized as function of its temperature at a local volumetric reaction rate $\dot{\omega}_i'''$ (kg/m^3-s), following an Arrhenius type reaction. By assuming that all volatiles escape instantaneously to the gas phase with no internal resistance, the fuel mass flux $\dot{m}''$ is obtained by integrating this volumetric reaction rate in depth. The thickness of the pyrolysis front is controlled by decomposition kinetics and heat transfer rates. A kinetically limited model for a solid with thickness δ is embodied by Equation 21.33:

$$\rho c \frac{\partial T}{\partial t} = \frac{\partial}{\partial z}\left(k \frac{\partial T}{\partial z}\right) - \sum_{i=1}^{N_s} \dot{\omega}_i''' \Delta H_{r,i} \text{ for } 0 < z < \delta \quad (21.33a)$$

$$\frac{\partial \rho_i}{\partial t} = -\dot{\omega}_i''' + \sum_{j=1}^{N_s} \nu_{j,i} \dot{\omega}_j''' \quad (21.33b)$$

$$\dot{\omega}_i''' = \rho^{n_i} A_i \exp\left(-\frac{E_i}{RT}\right) \quad (21.33c)$$

$$\dot{m}_i''' = \int_0^{\delta} \dot{\omega}_i''' dz \quad (21.33d)$$

Boundary conditions at $z = 0$ and $z = \delta$ are also required to complete the problem. Here, we have assumed that surface regression is negligible, material's density does not depend on temperature (except when as a consequence of charring reactions), the solid is opaque to thermal radiation, transfer of gases to the surface is instantaneous, and heat transfer from the condensed phase to the gas phase inside the decomposing solid is negligible.

While surface regression may be an important phenomenon for predicting mass burning rates long after ignition, it does not necessarily have a large effect on calculated flame spread rates (e.g., the surface regression velocity is at most 1% of the upward flame spread rate on a non-charring polymer [108]). From the numerical viewpoint, in one-dimensional models (Figure 21.7), surface regression and material shrinkage/swelling are easy to account for, as the solid phase meshing can be freely evolved over time. To avoid the problem of time-dependent gas-solid interface geometry, the changing thickness δ can be implemented by allowing the (virtual) back-side to move [76]. More realistic model using arbitrary Lagrangian-Eulerian (ALE) formulation was presented by Kacem et al. [109]. Even with charring polymers, some shrinking and swelling can occur, which can be taken into account by the numerical scheme with re-meshing [110].

Despite the higher burden associated with determining the required material properties, this type of pyrolysis model offers greater flexibility than the approaches that use the heat of gasification or measured burning rate approaches discussed in the previous section to estimate burning rates. Since the CPU time required for the solid-phase pyrolysis routines is small in comparison to the CPU time required for the gas phase routines in CFD-based simulations of large-scale flame spread, there is little computational advantage to using empirical pyrolysis models instead of a finite-rate pyrolysis model.

21.3.4 Liquid Fuels

In many situations involving liquid-fueled fires, the burning rate is specified, for example, using correlations from [111], rather than calculated. However, under some circumstances it may be desirable to predict the place and time dependent burning rate of the liquid. In FDS, this is can be accomplished in a manner analogous to the convective heat transfer: Fuel mass flux at the face of each liquid pool surface cell is

$$\dot{m}'' = h_{\text{m}} \rho_{\text{film}} \ln(1+B)\left(Y_{\text{sv}} + \frac{Y_{\text{sv}} - Y_{\infty}}{B} \right); \qquad B = \frac{Y_{\text{sv}} - Y_{\infty}}{1 - Y_{\text{sv}}} \tag{21.34}$$

where h_{m} is a mass transfer coefficient, ρ_{film} is the vapor film density, B is the Spalding mass transfer number (common in droplet evaporation modelling), Y_{sv} and Y_{∞} are the fuel mass fractions at the surface and in the center of the first CFD grid cell. Y_{sv} is calculated using Clausius-Clapeyron equation for the equilibrium vapor pressure, thus making the evaporation rate dependent on the surface temperature, and steeply increasing when approaching the boiling point. Temperature itself is calculated using one-dimensional heat conduction solver, as described in the previous chapter for solid fuels.

Several challenges are associated with the evaporation rate model Equation (21.34): (i) A natural choice for Y_{∞} is the value in the first gas phase grid cell. As Y_{∞} is dependent on $\dot{m}''$, Equation 21.34 must be solved through iteration. (ii) As Y_{∞} is also dependent on the cell size (volume where the evaporated fuel is diluted in), it is difficult to achieve mesh-independent solution. (iii) As expected from the analogy to the convective heat transfer, the mass transfer coefficient h_{m} depends on a global length scale, which is difficult to define for each liquid cell.

Internal convection within the liquid layer increases the heat dissipation significantly in comparison to pure heat conduction. In small hydrocarbon pools, this factor is 10–20 [112]. The internal convective motions cannot be solved by the current CFD sub-models. Instead, the liquid temperature is solved using a 1D heat conduction solver, as if the material was solid, and the internal convective motions are considered by artificially increasing thermal conductivity [113].

The calculated burning rates are obviously sensitive to the predicted heat fluxes, including the convective heating and the radiative flux from the flame, for which the model should capture the radiation attenuation by fuel vapor and soot within the fuel-rich core [86]. Many liquid fuels are semi-transparent to thermal radiation, so in-depth attenuation of thermal radiation must be accurately captured to accurately predict the internal temperature distribution and evaporation rate. Sikanen and Hostikka [113] determined gray absorption coefficients for hydrocarbon liquids and solved the internal RTE using a two-flux method, but some unphysical temperatures were obtained due to the inaccuracy of the gray method. Isojärvi et al. [114] and Alinejad et al. [115] developed non-gray *full-spectrum k*-methods for the radiation penetration into liquids, and implemented the method in FDS with a computationally efficient accounting for surface reflection and refraction [116].

While the majority of fire literature has studied single-component fuels, real pool fire scenarios often involve multi-component fuels, such as fuel blends. Yip et al. [117] developed a multi-component evaporation model for FDS and validated it using steady-state data of ethanol-water, ethanol-isopropanol and ethanol-hexane mixtures, and transient ethanol-heptane pool fire. The simulations reproduced the relationship between the fuel composition and maximum burning rate, as well as the temporal variations in transient burning rate due to the distillation effects.

21.3.5 In-Depth Radiation

In above, we have described the boundary condition of the condensed phase heat conduction equation as convective and radiative heat fluxes, both acting at the surface. If the radiative heat flux is divided into incident and emitted parts, the boundary condition states as

$$-k \left.\frac{\partial T}{\partial z}\right|_{z=0} = \varepsilon \dot{q}''_{\text{in}} - \varepsilon \sigma T_s^4 + \dot{q}''_{\text{c}} \tag{21.35}$$

where ε is the mean hemispherical emissivity of the surface. But real polymers and flammable liquids are not fully opaque to thermal radiation. Instead, radiation may penetrate through the surface and get absorbed within a finite (non-zero) distance of the material. The absorption distance may be short—about two-thirds of radiative energy is attenuated within the first 0.33 mm of black PMMA [118], which has led to the assumption that the surface absorption ($\kappa_\lambda \rightarrow \infty$) is a good approximation for the treatment of the thermal radiation boundary condition in pyrolysis models [119]. The validity of this assumption has, however, been questioned by studies that quantified the role of the in-depth penetration [120, 121]. In-depth absorption has been shown to be essential in the prediction of ignition time, especially at heat fluxes higher than 70 kW/m² [120, 122].

The in-depth radiation process is illustrated in Figure 21.10. Wavelength dependencies are not shown in the notation for simplicity, but in principle, all the properties should be treated as spectral properties. The gas phase radiation solver provides the incident intensities I_{in} and the corresponding incident radiative heat flux $\dot{q}''_{\text{in}}$ at the surface. Part of the incident energy is reflected as $I_{\text{in,r}} = \rho I_{\text{in}}$, where $\rho = (1-\varepsilon)$. The rest of the incident intensity, $\varepsilon I_{\text{in}}$, penetrates through the interface. The direction of the penerated ray of radiation depends on the ratio of the spectral refractive indices n_λ according to Snell's law.

Inside the material, the transmitted part of intensity can be calculated by Beer's law:

$$I_{\text{in,t}}(z') = (1-\rho) I_{\text{in}} \cdot \exp(-\kappa z') \tag{21.36}$$

where z' is the distance from the point of penetration along the new path of traveling. If the material is thick enough, practically all the transmitted radiation will be absorbed before the ray of radiation reaches the back side.

Besides absorbing, the condensed material emits radiation in all directions. The emitted intensity reaching the surface from inside would be

$$I_{\text{em}}(z=0) = \int_{\text{bottom}}^{\text{top}} \left(\kappa I_b(z') - \kappa_\lambda I_{\text{em}}(z')\right) dz' \tag{21.37}$$

where $I_b(z')$ is the emissive power at distance z'. In the case of a homogenous layer with constant temperature and absorption coefficient, this would lead to the well-known emissivity equation $I_{\text{em}}(0) = \kappa I_b\left(1-e^{-\kappa L}\right)$, L being the distance travelled. Part of the emitted intensity would reflect back into the solid, $I_{\text{em,r}} = \rho_2 I_{\text{em}}$, where ρ_2 is the internal reflectivity, and the rest would be transmitted to the gas phase. This transmitted part ($I_{\text{em,t}}$) would be considered as "emitted intensity" by the gas phase. Outgoing heat flux, $\dot{q}''_{\text{em}}(z)$ would then be an integral of all emitted intensities in different directions.

Taking into account all the penetrating directions, that is, intensities in all positive z-directions, we get the radiative heat flux to positive z-direction inside the material $\dot{q}''^{+}(z)$, which is a function of depth z. Correspondingly, integrating over all the intensities in negative z-direction, we get the radiative heat flux towards the surface $\dot{q}''^{-}(z)$. The difference ($\dot{q}''^{+} - \dot{q}''^{-}$) would be a net radiative heat flux, and its gradient would comprise an additional source term in the solid-phase energy equation (Eq. 21.33a).

As mentioned above, the first approximation to the problem is to assume opaque material, which means that the solid absorption coefficient κ is equal or close to infinity, $I_{\text{in},t}(z'>0) = 0$ (radiation is absorbed at surface) and $I_{\text{em,t}} = (1-\rho_2)\sigma T_s^4/\pi$, where T_s is the surface temperature. When the semi-transparent nature of a realistic material is deemed important, the in-depth radiation must be modeled in some way. The simplest way to approximate the in-depth heat flux is to use a diffusion approximation by increasing the thermal conductivity with a factor proportional to the third power of absolute temperature $k' = k + \omega T_{\square}^3$ [49]. This method has been used, for example, in [123, 124] to account for radiation inside the voids of a porous material.

The first level of actually modelling the in-depth radiation transfer is to approximate the inwards heat flux using Beer's law:

$$\dot{q}''^{+}(z) = \varepsilon\, \dot{q}''_{\text{in}} \cdot \exp(-\tilde{\kappa} z) \tag{21.38}$$

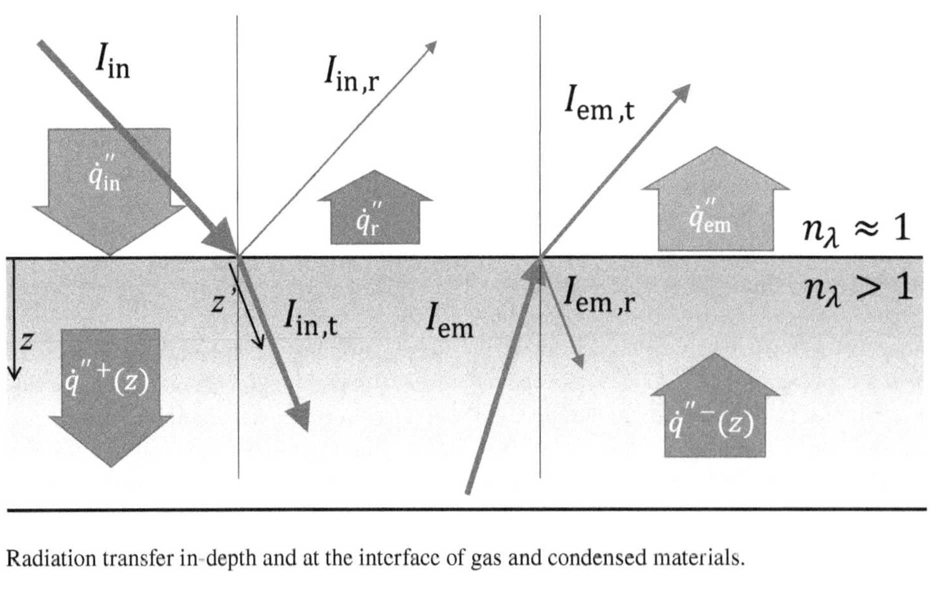

FIGURE 21.10 Radiation transfer in-depth and at the interface of gas and condensed materials.

where $\tilde{\kappa}$ is a mean absorption coefficient. The equation is sometimes combined with the diffusion approximation to have a separate treatment for the continuum and porous parts of the material. Eq. 21.38 ignores the facts that Beer's law is only valid for a monochromatic radiation (single λ) and that the rays propagating into different directions have travelled different distances z' at the same depth z. Due to the latter, the analytical solutions for the heat flux involve more complex functions than just the exponential function [49]. The mean absorption coefficient $\tilde{\kappa}$ in Eq. 21.38 must therefore be interpreted as an effective property. Nevertheless, this method is used, for example, in ThermaKin [124] and FireFOAM programs due to its simplicity. Slightly more complex form was used in [123], taking into account the variation of $\tilde{\kappa}$ over depth.

The next level of sophistication in the in-depth radiation calculation is so-called two-flux method, where the inwards heat flux is solved from the following ordinary differential equation

$$\frac{1}{2}\frac{d\dot{q}''^{+}(z)}{dz} = \tilde{\kappa}\pi I_b(z) - \tilde{\kappa}\dot{q}''^{+}(z) \qquad (21.39)$$

with boundary condition $\dot{q}''^{+}(0) = \pi\left[(1-\rho)I_{\text{in}} + \rho_2 I_{\text{em}}(0)\right]$. Similar equation can be derived for the outwards-flux $\dot{q}''^{-}$. The radiative source term in the heat conduction equation can then be calculated as

$$\dot{q}'''(z) = \frac{d\dot{q}''^{+}(z)}{dz} + \frac{d\dot{q}''^{-}(z)}{dz} \qquad (21.40)$$

Two-flux method is used in the FDS pyrolysis model.

More detailed solutions for the in-depth RTE are of course possible, but the benefit of any additional increase in complexity has not been demonstrated. On the contrary, Alinejad et al. [125] have shown that more accurate treatment of the directional dependence (three-dimensional FVM) provides only marginal improvement in the predicted radiative source terms. The two-flux method was found an optimal way for in-depth RTE solution in polymer pyrolysis simulations.

In above, the in-depth radiation equations have included an effective absorption coefficient $\tilde{\kappa}$, and solving just a single RTE implies assuming a gray medium. The gray coefficients can be obtained by measuring the transmitted heat flux [111], optimized for the mass loss rate predictions [123], or using a spectral technique. So far, spectral properties of polymeric materials have been reported for clear [126] and black [127, 128]. PMMA, which by far are the most commonly used polymers in fire science. These experimental results showed that the clear and black PMMA had almost identical absorption coefficients, with differences limited to wavelengths shorter than 2.23 μm, where the black carbon increases the absorption coefficient with four orders of magnitude [127]. The black PMMA sample in [128] was from the same batch used in the round robin study of the Measurement and Computation of Fire Phenomena (MaCFP) working group of International Association of Fire Safety Science (IAFSS) [129]. This makes the radiative properties directly applicable for the modelling of the MaCFP experiments.

The need for the spectral treatment in solving in-depth radiation was highlighted by Bal et al. [121]. Accordingly, non-gray models were developed by Pizzo et al. [130], Kacem et al. [109], and Alinejad et al. [115], allowing these modelers to reach predictions of the heat source distribution. But with additional complexity comes two problems: need for more input data (radiative properties) and the increased computational cost. Alinejad et al. [125] compared the relative accuracy of different in-depth radiation modeling options against their computational cost and concluded that the most efficient approach for coupling in-depth radiation with large-scale CFD is to use a gray absorption coefficient which is a function of depth and source temperature $\tilde{\kappa}(z, T_{\text{source}})$, so that the gray model can capture the non-gray effects of the fast attenuation and dependence on the type of heating. The method was implemented in FDS but is not yet in wide use among the fire simulation community.

21.3.6 Estimation of the Pyrolysis Model Parameters

The pyrolysis model parameters are difficult to measure directly for several reasons: First, the range of conditions (temperatures, gas concentrations, moisture levels) found in the possible simulations is wider than what usually can be covered in laboratory experiments. Second, many of the parameters are effectively model-specific parameters, not fundamental material properties. Kim and Dempsey [131] identified three groups of parameters: those related to thermal decomposition, material properties (either intrinsic or effective), and fitting-parameters. Yet one can argue that even many of the thermal decomposition parameters are model dependent, because they are only relevant in the context of a specific reaction pathway or scheme. The third challenge is the size of the problem: there are just so many parameters. Indeed, for a simple charring material with temperature-invariant thermal properties that degrades by a single-step first order reaction, there are ~11 parameters that must be specified (two kinetic parameters, one heat of reaction, two thermal conductivities, two specific heat capacities, two emissivities, and two in-depth radiation absorption coefficients). Letting the thermal properties change with temperature, adopting more complex reaction scheme, or pursuing non-gray radiation treatment would further increase the number of unknown parameters.

Model parameter estimation has been an active topic of pyrolysis modelling during the last 15 years. These methods were described in Chapter 20, which we refer for the details of the condensed phase pyrolysis models and the methods of obtaining their parameters. In this chapter, we try to cover some aspects of parameter estimation that are specific to coupled models, where the gas phase combustion plays a role.

The key principles of pyrolysis model parameter estimation have been summarized in a small number of review and guidance articles [131–133]. The ongoing MaCFP effort aims at improving the predictive capabilities of numerical simulations of thermal decomposition and pyrolysis, currently turning the focus on fully coupled simulations of vertical flame spread on non-charring wall [129]. Considering the examples of individual studies and the few summarizing presentations, the recommended practice of model parameter estimation seems to include the following steps:

1. Direct measurement or appropriate literature value should be used when possible. Typical properties for direct measurement include material density, thermal conductivity at or close to room temperature, and possibly specific heat.
2. Perform thermogravimetric analysis (TGA) at several heating rates and in different gas atmospheres.
3. Construct a reaction scheme and estimate the reaction kinetic parameters (A_i, E_i, n_i) as well as the yields of reaction products from the TGA data. Most parameters can be obtained using direct methods but inverse modelling may be needed especially when the model is more complex than a single step. Some authors seem to prefer first-order reactions ($n_i = 1$) and to capture the complex reaction rates through more complicated reaction schemes. This is a viable option especially when the computational cost is not an issue.
4. Perform differential scanning calorimetry (DSC) either as a stand-alone experiment or simultaneously with the TGA (STA). Estimate temperature dependent specific heat and the reaction enthalpies from the DSC data either analytically or by inverse modeling, using the reaction scheme from step 3 and the exact pyrolysis model in question.
5. Perform bench-scale experiments for the mass loss and heat release rates using fire propagation apparatus (FPA), cone calorimeter, or similar. Use several heat fluxes.
6. Construct a model of the bench scale experiment and start simulations from inert atmosphere experiments. Estimate the unknown thermal parameters using inverse modeling. Save some experimental conditions for validation.
7. Include the effects appearing in oxidizing atmosphere experiments, such as flaming and smoldering. Let the model predict HRR and estimate remaining unknown parameters.
8. Validate the model using bench-scale experiments which were not used for the model calibration.

Although several specialized laboratory tests are available to measure some properties (e.g., specific heat capacity can be determined by differential scanning calorimetry), many of these tests are still research tools and standard procedures to develop material properties for fire modeling have not yet been developed. Even if standard procedures were available, it would likely be so expensive to conduct 5+ different specialized laboratory tests for each material that practicing engineers would be unable to apply this approach to real-world projects in an economically viable way. Furthermore, there is no guarantee that properties measured independently from multiple laboratory tests will provide accurate predictions of pyrolysis behavior in a slab pyrolysis/combustion experiment such as the cone calorimeter or fire propagation apparatus.

To remedy the lack of direct measurements, inverse modeling is commonly used for finding model parameter values that reproduce the targeted experiments with highest accuracy. Inverse modeling can be based on direct methods [134], optimization using gradient-based methods or evolutionary algorithms, such as genetic algorithm (GA) [135–138], shuffled complex evolution (SCE) [139], fitness scaled chaotic artificial bee colony (FSCABC) [140], or a brute force stochastic search. Dedicated software have been developed for carrying out the inverse modelling with pyrolysis models, such as GPyro utilizing GA [123].

While many of the pyrolysis modelling studies and associated parameter estimations have been based on gasification experiments, the application of the model in the context of fire-CFD requires that the model also works in the presence of oxygen. This may be straightforward for some materials, such as homogenous thermoplastics, but can be difficult when dealing with charring, bio-based or composite materials because the composition of the pyrolysis products, associated reactions and, consequently, the heat of combustion may vary over time. Also, the experimental data may consist of test performed at larger scale than the typical bench scale experiments. In such tests, the flames from the sample itself may dominate the heat transfer, unlike in the cone or gasification apparatus, where the dominating source of heat is the radiant heater. To capture the variable heats of combustion and the coupled flame dynamics in the model being optimized, the gas phase phenomena must be included when performing the inverse modelling. It means that the inverse modelling tool must be able to call the actual CFD code for obtaining the model results, not just a stand-alone, one-dimensional pyrolysis solver. Matala and Hostikka [141] developed a Matlab-based PyroPlot tool that was able to run FDS simulations during the GA process and applied the tool for modelling electrical cables. Following their example, Arnold et al. [140] have developed a tool called PROPTI that has gained popularity among the researchers of CFD-based pyrolysis. PROPTI enables communication between arbitrary simulation models, optimization algorithms, and experimental datasets as optimization targets. It has been applied for modelling fire spread on electrical cables in [142].

21.3.7 Modeling Bench-Scale Experiments

In their review of pyrolysis models, Shi and Chew [132] observed that most pyrolysis models aim at predicting the production of gas volatiles. As suggested in the parameter estimation process description above, the role of the gas phase combustion reactions in the pyrolysis modelling utilizing CFD is mainly to produce heat. In cone calorimeter and other bench-scale experiments, where oxygen is available for combustion, the flame heat release will increase the heat flux to the material surface. Normally, the increased heat flux needs to be taken into account in the heat flux boundary condition, but ignoring it may be justified if the external heat flux level is very high, such as the 100 kW/m^2 heat flux in the study of Grange et al. [143] who modeled the decomposition of carbon-based composites with FireFOAM. Further details, such as the production of the incomplete combustion products, toxic gases, or soot are usually ignored.

Two common approaches for modelling the flame heat flux are illustrated in Figure 21.11 for a fictitious example of a horizontal cone calorimeter sample heated by 35 kW/m^2 radiative heat flux and burning at 150 kW/m^2 heat release rate. An instantaneous temperature field, shown on the right, was computed using FDS. The straightforward approach for flame heat flux modelling is to presume an experimental or fitted value for the flame heat flux and add it to the external flux boundary condition without explicitly modeling the gas phase combustion. For example, Lautenberger et al. [135] added 30 kW/m^2 to the sample surface based on experimental observations. Stoliarov et al. [144] suggested *a posteriori* fitted flame heat fluxes of 11 kW/m^2 for HIPS and HDPE and 12 kW/m^2 for PMMA. The nature of these heat fluxes was considered purely radiative, with convective heat fluxes set to zero after ignition. Before ignition, the role of the convective heat flux was to cool down the radiated surface of the polymer sample. To consider the spatial variability of the heat flux, McCoy et al. [145] divided the 10 × 10 cm^2 sample into center and edge regions with different heat fluxes. In this approach, the flame heat flux to the center was dominated by radiation, and close to the edges it was convection. The presumed flame heat flux is a common approach in pyrolysis modeling when using stand-alone models like ThermaKin or GPyro, which do include a model for the gas phase.

When modelling the pyrolysis using a pyrolysis model embedded in a CFD tool (FDS, FireFOAM, ISIS), it is possible to simulate the gas phase combustion and the associated flame heat fluxes explicitly. For example, Hostikka and Axelsson [146] used an early version of FDS to produce the spatial radiative heat flux fields at different values of heat release rate on a burner-like sample. The results demonstrated that the spatial distribution could be presented as a function of a single, normalized distance from the sample center. Yuen et al. [147] carried out 3D simulations of the cone calorimeter tests of wood samples, focusing on the prediction of the pilot ignition process. The transient and spatially distributed features of the flame heat flux were not reported, though.

Boyer simulated the flame heat fluxes in cone calorimeter experiments of three polymers using the gas phase combustion and radiation models of ISIS software [148]. His results indicated that the radiative fluxes strongly depend on the burning material, being relatively stable during the whole pyrolysis process: 4.5 kW/m^2 for PMMA, 5.5 kW/m^2 for HDPE and 11 kW/m^2 for HIPS. The radiative fluxes were only marginally affected by the imposed flux or sample thickness. Convective fluxes, in turn, decreased with increasing external heat flux, almost cancelling out at 75 kW/m^2 heat flux.

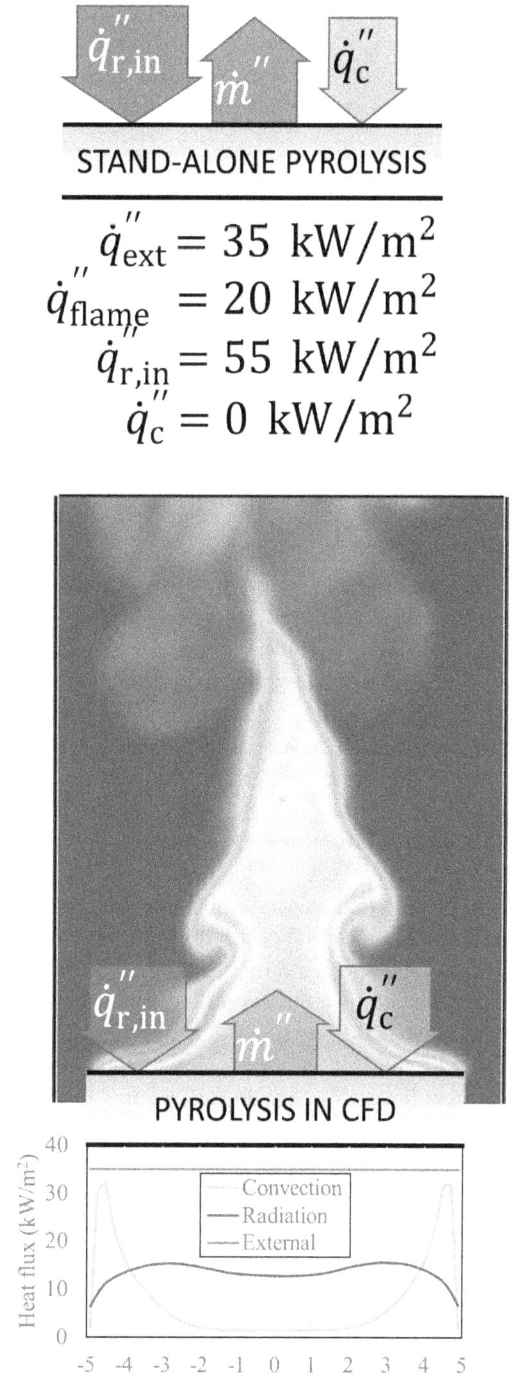

FIGURE 21.11 Two approaches for modelling flame heat fluxes for a fictitious cone calorimeter experiment: prescribed heat fluxes (top) and computed with coupled CFD (bottom).

CFD model with detailed gas phase solution has also been used for understanding the ambient pressure effects on piloted ignition [149]. A two-dimensional model was used with 1–2 mm grid cells to enable finite-rate combustion modelling.

As we see, detailed simulations of the coupled fluid and solid phase physics of bench-scale experiments can be carried out. Such simulations are, however, extremely time-consuming due to the large number of computational cells and small time-step, imposed by the small cell size, and as such, practically useless for parameter estimation. To be usable in any type of inverse modelling, each individual simulation must be executed in few seconds or minutes at most. This is why most parameter estimation studies using pyrolysis models, that are embedded in a CFD tool, use either the prescribed approach or an effective CFD solution, discussed below.

The idea of the effective flame modelling is to use very coarse numerical resolution, such as 5–10 cm mesh, that does not increase the computing time significantly in comparison to pure pyrolysis solution. Although such a resolution cannot provide a credible numerical solution to the underlying discretized equation, it may provide a useful approximation to the heat fluxes because the mixing-controlled combustion model enforces the desired heat release within the first one or two cells above the sample, and because the prescribed radiative fraction assigns a correct amount of energy to the radiation field [141]. For example, in the cone calorimeter models of birch wood, Hostikka and Matala [150] observed that a simulation with 5 cm resolution provided flame heat fluxes in range 4–12 kW/m^2, while refining the resolution to 1 cm magnified the flame heat fluxes to 10–15 kW/m^2 range, consistent with experimental observations. Despite this difference in flame heat fluxes, the simulations using 5 cm grid cells predicted heat release rates that were only few percent smaller than those with fine resolution, mainly because the external heat flux from the cone radiation was 50 kW/m^2, reducing the relative importance of few kW/m^2 differences in flame heat fluxes.

Alonso et al. [151] investigated the feasibility of the effective flame modelling by using two different models to simulate the same experiment. Electrical cables were used as a fuel. They first developed a simplified model with 1.0 cm resolution and slab-type idealization for the inverse modelling of cable samples and then tested the estimated pyrolysis model in simulations with (only slightly) refined mesh (6.7 mm resolution) but more realistic geometry. In general, the detailed model predicted the cone calorimeter results with higher accuracy than the simple model, but the simple model, which was used in the estimation, was more accurate in some key features like the time to ignition.

Rinta-Paavola et al. [110] used a hybrid between the fine-resolution and effective approaches. They first computed the flame heat fluxes for several different specified heat release rates and surface temperatures, averaged the computed heat fluxes over the sample area, and finally, added the tabulated heat fluxes to the cone heat flux during pyrolysis simulations. The averaged heat fluxes were in range 18.8–24.4 kW/m^2 and were found to be almost constant over time. The role of the cone radiation attenuation within the fuel vapors and flame was mentioned as a source of uncertainty.

Obviously, models with very coarse resolution cannot capture the spatial distributions of the radiative and convective fluxes. But based on the above examples, we can conclude that when such details are not detrimental, coarse and fast models can be used for the parameter estimation, avoiding the need to presume some ignition-triggered constant heat flux.

21.4 SIMULATION OF LARGE-SCALE BURNING AND FIRE GROWTH

21.4.1 Coupled Problem

Flame spread and fire growth are probably the most challenging modeling problems in fire CFD because they require accurate consideration of several subprocesses:

1. Flame temperature and thermal radiation. In large fires, flame radiation dominates heat transfer to unburned and burning fuel.
2. Flame structure and heat transfer in the region close to the wall. In small fires, the near field flames are responsible for most of the heat transfer to the wall. Larger pool fires, in turn, have a fuel rich flame core which effectively absorbs the flame radiation and acts as a buffer between the heat source and liquid surface. Nevertheless, in a typical fire simulation, this near-wall region is covered by just few grid cells at best, making it impossible to capture the flame structure, radiation profile, and temperature distribution in detail. Some kind of sub-grid scale model of this region is needed in fire CFD models to accurately model the flame spread. The sub-grid scale model might use the ideas of wall functions and boundary layer flame structures [152].
3. Heat and mass transfer and pyrolysis inside the condensed material. Wide range of possible natural and synthetic materials, including composites, makes it difficult to build and evaluate models that are efficient to use and accurate in prediction. Depending on the state of the fire, physical processes that cannot be characterized in bench-scale experiments may play a role.

It may be necessary to distinguish between *flame* and *fire* spread simulations. Flame spread simulation means the ability to predict the fire growth starting from a *small initial fire* or ignition point, where all three sub-processes are important, but the direct flame-to-surface heat transfer acts as a driving force. Heat flux distribution has high gradients, and they need to be captured and transferred to the condensed phase model. Fire spread, in turn, means consecutive ignition of solid surfaces in the presence of a relatively *large initial fire* dominating the heat transfer by radiation. Heat flux is relatively uniform across the surfaces. Flame spread depends mainly on the ignition source and only indirectly on the surrounding environment, while fire spread involves the surrounding compartment, accumulation of heat in surfaces and formation of hot, radiating smoke layers.

Typically, a fire growth model is evaluated by comparing its calculations (predictions) of large-scale behavior to experimental HRR measurements, thermocouple temperatures, or pyrolysis front position. The overall predictive capabilities of fire growth models depend on the pyrolysis model, treatment of gas phase fluid mechanics, turbulence, combustion chemistry, and convective/radiative heat transfer. Unless simulations are truly blind, some model "calibration" (adjusting various input parameters to improve agreement between model calculations and experimental data) is usually inherent in published results, so model calculations may not truly be predictions.

In order to rigorously assess a model's predictive capabilities for flame spread modeling or make suggestions as to how it could be enhanced by improving various subroutines, some level of deconstruction is necessary. This involves independently analyzing turbulence, combustion, flame heat transfer, and solid-phase phenomena before bringing them together to simulate large-scale flame spread. If a model reproduces well large-scale experimental flame spread data, it does not necessarily mean that all physics relevant to the flame spread process are accurately modeled. Instead, the good agreement between model calculations and experimental data may be due to compensating effects. A model that overestimates flame heat fluxes may be able to reproduce experimental flame spread data if the pyrolysis model underestimates the mass loss rate. This should be kept in mind when viewing the large-scale flame spread modeling results.

21.4.2 Burning Rate Prediction for Stationary (Non-Spreading) Fires

Before considering spreading fires, we first examine the simpler case of a stationary solid or liquid pool fire where the pyrolysis area does not change with time. Physically, the burning rate (more precisely, mass loss or heat release rate) of a stationary fire is controlled by condensed phase fuel generation processes and the rate of heat feedback from the flames (or external heat sources) to the burning fuel. Accurate prediction of burning rates with CFD-based fire modeling requires accurate prediction of flame heat transfer and condensed phase pyrolysis (or evaporation for liquid fuels). Since these processes drive fire growth rates in spreading fires, prediction of burning rates of stationary fires is of direct relevance to fire growth modeling.

The studies that have been conducted to date fall into two primary categories: (1) prediction of burning rates in small-scale flammability experiments such as the cone calorimeter, and (2) prediction of burning rates in liquid pool fires. Early predictions of the pool burning rates assuming opaque surface and nominal thermal conductivity were able to reach ~25% accuracy for heptane, toluene, and ethanol [153] and ~35%–100% overprediction for methanol [154]. More recent studies have included in-depth radiation using effective, gray absorption coefficients, and effective thermal conductivity. The effective thermal conductivity tries to account for the convective dissipation of thermal gradients [155]. More detailed physics was obtained, but the results were not always satisfactory. Sikanen and Hostikka [156] obtained 15% accuracy in the steady-state burning rates of hydrogenated tetra-propylene (TPH) pool fires under vitiated conditions despite reporting problems in obtaining realistic temperature distributions within the liquid layer. Attempts have been made to include non-gray radiation in the liquid [114, 115] and to provide non-gray models for the attenuation of radiation within the fuel-rich core [81], but significant improvement cannot be expected until a common, non-gray framework for the multiphase radiation heat transfer becomes available.

One of the main difficulties in predicting the pool burning rates is that many of the engineering heat and mass transfer correlations include a global length scale *L*. Hong et al. studied the natural convection correlation using FDS and showed that the dependence on the global length scale can be avoided with appropriate convective heat transfer and mass-transfer number correlations [90]. The natural convection correlations were found to produce more realistic surface temperatures, yet the burning rates were similar to those obtained using forced convection.

21.4.3 Flame Spread on Flat Surfaces

One of the simplest large-scale geometries relevant to real world fire growth modeling is vertical upward flame spread on a free-standing wall, meaning that the wall is not part of a compartment. Compartment effects, such as the accumulation of a hot ceiling layer, do not come into play.

Vertical flame spread on a 5 m tall PMMA wall, measured by Wu et al. [157] has been modeled by several authors. Liang [158] used FDS Version 2, which included a heat transfer limited pyrolysis model. Kwon et al. [159] used FDS4, which included a high-heat flux form of the finite-rate pyrolysis model. In the FDS4 study, the calculated heat release rate initially increased at a rate that is slower than that observed experimentally, but then jumped to a higher heat release rate. This behavior was found to be sensitive to a feature of FDS4 gas phase combustion model that automatically adjusted the stoichiometric value of mixture fraction on coarse grids. Markus et al. [160] and Li and Wang [161] used FDS version 6 with finite rate pyrolysis models. Zhang et al. [162] modeled the same configuration using SOFIE and obtained good agreement between

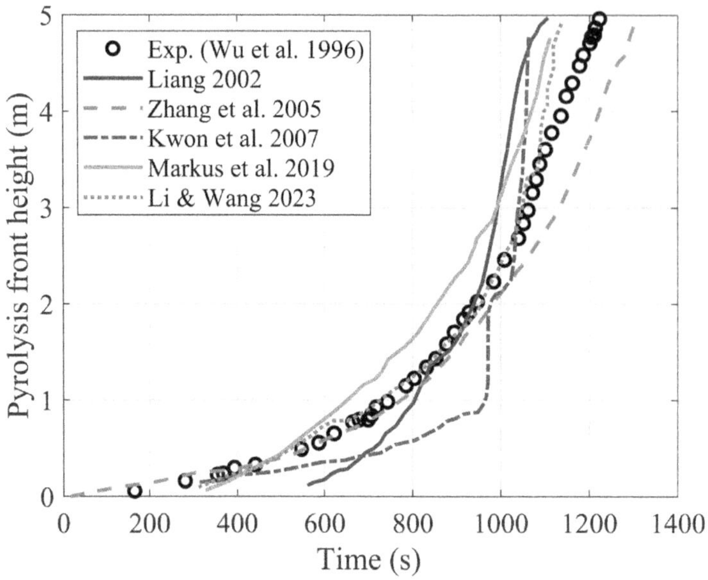

FIGURE 21.12 Comparison of measured and modeled pyrolysis front height for vertical flame spread on 5 m PMMA panel.

their model calculations and the experimental data. In Figure 21.12, the experimental flame spread data [157] are compared to the model calculations of Liang [158], Kwon et al. [159], Markus et al. [160], Li and Wang [161], and Zhang et al. [162] The differences in the calculations shown in Figure 21.12 are attributed to different gas phase solvers and resolutions, solid phase models, user assumptions, and model input parameters; no conclusions should be drawn regarding the accuracy of one model compared to another on the basis of Figure 21.12. For example, Markus et al. [160] demonstrated that the predicted flame spread rate was very sensitive to approximations in the ignition source model. Similar approximations and simplifications are embedded in all predictions.

Markus et al. [163] compared the simple (thermal) and complex (finite rate) pyrolysis models within FDS in their capability to predict vertical flame spread on the abovementioned 5 m tall PMMA slab. They concluded that the main difference in the results was observed in the HRR after an induction (preheating) period. While the fire growth (flame spread) using a thermal model was very abrupt after the induction period and spatially constant in the ignited region, the finite-rate model produced more gradual growth, finite width of pyrolysis development region, and almost linearly increasing mass burning rate within the pyrolysis region. The last observation was then used to develop a modified thermal model, improving its accuracy in the early part of the fire.

Ding et al. have studied the sensitivity of vertical flame spread predictions on combustible material and observed that the reaction kinetics, material heat capacity, and the gas phase heat of combustion had the highest influence on the predicted spread rate. [164]

Fukumoto et al. [76] used FireFOAM 2.2.x to simulate vertical flame spread on PMMA slabs of different heights using LES, EDC combustion, WSGG-based radiation model and a kinetically controlled pyrolysis model with Beer's law for in-depth radiation. The numerical resolution of the gas phase mesh was less than 2 mm in the direction normal to the wall. They obtained good accuracy for the flame spread rate, and in the largest scale also for the pyrolysis front position and flame length. They found out that the accurate calculation of the combustion reaction and, consequently, the convective heat flux at the flame leading edge (lowest position) required special modeling to avoid excessive resolution requirements, that is, the use of < 2 mm cells. Attempts to overcome the requirement of very good resolution inside the reactive boundary layer have utilized local [165] and global [88] relationships between the thermodynamic state and the surface heat flux. These relationships were "embedded" into the CFD solution to prescribe, rather than predict the details of the heat flux distribution, and will raise the problems of model generality, as discussed in Section 21.2.4.

Fire growth in a combustible corner (e.g., ISO 9705 or similar) has received considerable attention from modelers. One of the earliest three-dimensional models of large-scale fire growth is due to Yan and Holmstedt [166] who used CFD to model fire growth on particle board. Generation of gaseous fuel was modeled in two different ways. In the first, the heat release rate of a solid fuel element was zero until its surface was raised to its ignition temperature. Then, the heat release rate per unit area of that fuel element was specified as that measured in the Cone Calorimeter. With the second method, a heat transfer limited pyrolysis model was used to calculate the pyrolysis rate. Interestingly, the results obtained from both methods were similar. The high level of agreement between model calculations and experimental data was remarkable considering this was one of the earliest three-dimensional simulations of large-scale fire spread and that the grid spacing was coarse by today's standards. Later, Theuns et al. [167] conducted similar simulations of the same experiments, with a focus on the effect of material properties. The influence of the grid spacing in predicting the HRR of spruce wood panels in room corner test was demonstrated by [168]. The charring reaction kinetics were adjusted to reproduce the large-scale behaviors, as at that time, the model parameter estimation process exploiting small scale experiments like TGA had not been developed. Also, the grid resolutions were extremely coarse (6.7–15 cm) in comparison to the modern flame spread studies. Moghaddam et al. [169] used FDS (Version 3 or 4) to model flame spread on plywood in a room/corner fire test. These authors noted sensitivity to grid size and selection of the fuel reaction.

Hietaniemi et al. [170] used a pre-release version of FDS4 to model fire spread on several materials in several different configurations and compared the calculated results with experimental data. The materials simulated include spruce timber (single burning item (SBI), room/corner, and 6 m cavity), medium density fiber board (SBI and room/corner), polyvinyl chloride (PVC) wall carpet on gypsum board (SBI, room/corner), upholstered furniture (furniture calorimeter and ISO room), and polyethylene-sheathed cables in 6 m cavity. The model reproduced well most of the experiments conducted on spruce, but the calculated heat release rate was sensitive to the back-face boundary condition in the SBI test and the observed decay in the heat release rate in the cavity experiment was not captured by the model. Modeled heat release rate for medium density fiberboard closely matched room/corner test data; heat release rate in the SBI test was not reproduced as closely. The heat release rate of the PVC wall carpet was reproduced reasonably well in the SBI test but was over-predicted in the room corner test, probably due to discrepancy between the back-face boundary condition in reality and in the model. For the upholstered chair, FDS underestimated the time to ignition and the peak heat release rate compared with experimental data for both the furniture calorimeter and ISO room cases. Finally, for the polyethylene-sheathed cables in the 6 m cavity, the modeled heat release rate matched the experimental data fairly closely.

In addition to the different models and materials, the abovementioned CFD studies have used very different numerical resolutions. While Fukumoto et al. [76] resolved the details of the reactive boundary layer using < 2 mm meshing, Markus et al. [160] did not capture the details but resolved the large-scale eddies with ~ 2 cm resolution, and many of the room-corner predictions used grid cells large enough to retain most of the boundary layer within just one or two cells. The use of the finer resolutions in the most recent of the cited works reflects the increased recognition of the wall flow heat transfer challenges and more rigorous attempts to capture the physics. In addition, the availability of high-performance computing resources and simulation programs' better support for parallel processing have acted as enablers.

Finding the reasons that limit the predictive capability of CFD models in wall and corner flame spread situations is difficult due to the fully coupled nature of the problem. Zhang et al. [171] used FDS4 to simulate flame heat transfer in Single Burning Item (SBI) tests. They concluded that although FDS accurately predicted the gas burner flame heights, the surface heat fluxes were significantly underpredicted at higher heat release rates. These validation tests are included in today's FDS validation repository and shown in Figure 21.13 for wall, ceiling, and floor heat flux predictions [25]. In average, FDS is seen to underestimate heat fluxes by just 3%, and the model relative standard deviation

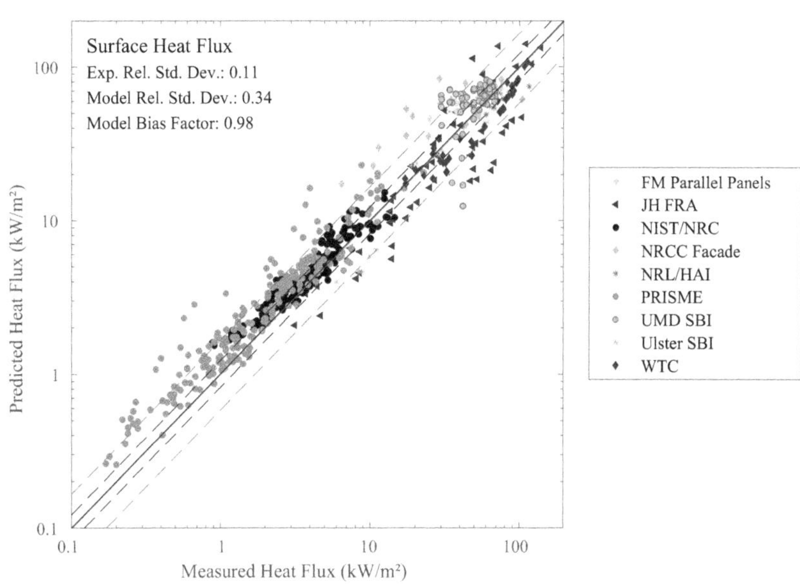

FIGURE 21.13 Summary of FDS validation for wall, ceiling, and floor heat fluxes [25, compiled April 13, 2023].

is only 33%. Still, individual heat flux predictions in the range relevant for flame spread (10–50 kW/m^2) may show much greater errors, which would naturally propagate to the predicted flame spread and heat release rate.

21.4.4 Predicting the Fire Growth on Electrical Cables

Electrical cables can act as ignition sources and form an important class of fire loads in many industrial systems. They are the most significant fire load within nuclear power plants and an important subject of nuclear safety research. Predictive simulations of cable fires are used by the technical support organizations carrying out probabilistic risk assessments and fire hazard analyses for power plants and other nuclear installations.

Early simulations of cables fires, whether using CFD or other modelling tools, relied on experimental data or simple analytical models on the fire size and growth. The simulations tried to predict the consequence from a hypothetical fire, making the results conditional to the quality of the initial assumptions and data. Uncertainties in the input data would be treated like any input uncertainty, that is, using probabilistic methods. Significance of the factors influencing fire growth, such as fuel layout or flammability characteristics, on the expected risk could not be estimated. In 2008 summary report of the International Collaborative Fire Modeling Project (ICFMP) [172], the status of the cable modelling was characterized "elementary" and the need for models predicting cable burning and dysfunction was identified.

Since the time of the ICFMP report [172], several studies have targeted in the predictive capability of cable fire simulations. Models have made use of the simple and complex pyrolysis models, but even the simple pyrolysis models may seem too demanding for large-scale industrial applications because they—despite being simple in formalism—require significant amount of bench-scale experimental data on the ignition temperature and heat release rate per unit area (HRRPUA). Considering the plethora of different cables and geometrical layouts in any practical application, obtaining these data may seem unfeasible. From the viewpoint of the practical usability, great results have been obtained using so called FLASH-CAT method [173]. It is an empirical fire spread model for horizontal cable trays, developed by the U.S. Nuclear Regulatory Commission in co-operation with the National Institute of Standards and Technology. FLASH-CAT assumes that fire starts from the bottom, requires the number and physical dimensions of the cable trays, their plastic mass content, and the heat of combustion as inputs, and provides the rates of fire spread in vertical and horizontal directions as outputs. The fire spread pattern looks like an upwards opening "V" shape which widens over time due to the horizontal spreading. Combining the burning area with assumed HRRPUA, one obtains the cable fire HRR as a function of time, and the model result can be specified in, for example, FDS simulation as fire boundary. The model is fully empirical, and therefore restricted to scenarios similar to the experimental situation. Jang et al. [174] compared the HRR predictions of FLASH-CAT and using the simple (thermal) pyrolysis model of FDS. They concluded that obtaining accurate results may be difficult if the inputs are based on literature values, thus indicating the need for some project specific experiments. Once such experimental data is available, the simple pyrolysis model of FDS was found to provide necessary tunability to reach HRR predictions that are close to the experimental data. Siemon et al. [175] used a similar technique in FDS by specifying the cable tray ignitions and fire spread rates rather than any of the pyrolysis parameters.

The original FLASH-CAT model has been modified by Tyas et al. [176] for closed compartments by tuning the spread rates and burning times to take into account the increased heat feedback from nearby walls and the reduced burning rate by under-ventilation. Verma et al. [177] combined the FLASH-CAT burning time predictions with the ignition temperature-based pyrolysis model of FDS, being able to capture fire spread in geometrical conditions that did not comply with the assumptions of the original model. Figure 21.14 visualizes the flame simulated in [177] and

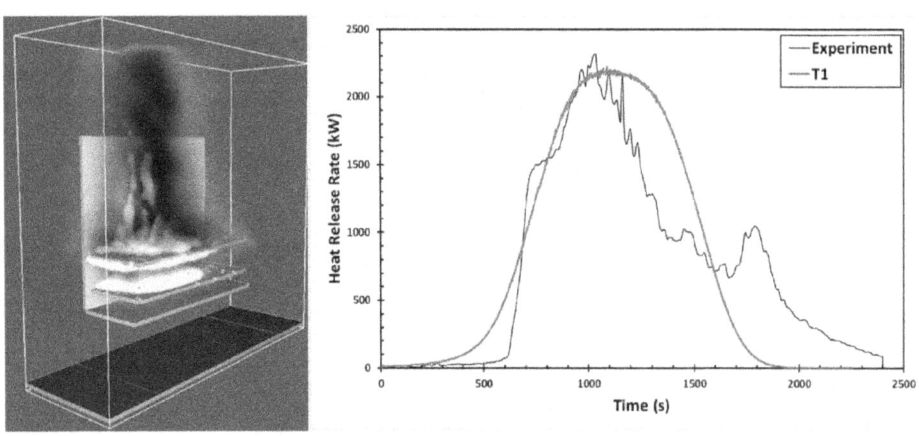

FIGURE 21.14 Simulation of PVC/PE cable tray fire using FDS and adapted FLASH-CAT. Left: visualization of instantaneous flame and smoke plume. Right: comparison of the experimental and simulated (T1) heat release rates. [177]

shows comparison of the measured and simulated HRR in a fire test with 15.9 mm diameter polyethylene (PE)/PVC cables.

Cable polymers range from simple polyolefins, like polyethylene, to those containing flame retardants. From the viewpoint of pyrolysis modelling, the treatment of these polymers does not differ from the other fire engineering applications, except for the shape. Classical flame retardant cable material is flexible PVC, but non-corrosive cables with alumina trihydrate (ATH) or magnesium hydroxide (MDH) fillers have become popular during the last 15 years. Pyrolysis modeling of PVC cable materials should include the plasticizers that evaporate from the polymer in relatively low temperatures and are usually highly combustible [141]. The flame retardant action by mineral fillers have been simulated in [176] using FLASH-CAT and in [178] using a complex pyrolysis model.

A special challenge in cable fire modeling comes from the handling of the detailed cable geometry. While most construction materials appear in planar form, and are thus well suited for one-dimensional pyrolysis modeling, cables have a cylindrical appearance but detailed internal structure. Cables consist of one or several metal conductors, each surrounded by a layer of insulation. Insulated conductors can be surrounded by a filler material and kept together by polymer jacket (sheath), as shown in Figure 21.15. Cables are typically laid out in bundles and bound on horizontal or vertical trays made of metal. In such an open configuration, cables can be heated effectively by an underneath flame, unless physically protected.

Detailed internal geometry is not important when using the simple pyrolysis model, for which the effective HRRPUA will be specified rather than predicted. What are really predicted are the ignition times of surface elements, and accurate calculation is required only for the heat conduction into the cable surface.

Geometrical representativeness is, however, desired when using complex pyrolysis models. Yet the detailed form and internal structure of cables cannot be resolved by the mesh used in the CFD simulation. Instead, the cables are commonly described as obstructions resolvable by the CFD mesh, so that one obstruction typically describes one tray or bundle of cables. Alternatively, cables can be modeled using sub-grid elements (particles). In either case, one-dimensional pyrolysis models can be exploited. Modeling the complex structure of multiple multi-component cables using one-dimensional pyrolysis models obviously requires simplifying assumptions [179]. Figure 21.15 shows two alternative approaches: in the upper, the inner structure is homogenized into a mixture of filler, insulator, and metal. In the second approach, the material components remain separated, in the form of layers. Both approaches include the cable jacket as a separate surface layer because it is the jacket's flammability that determines the ignition and early flame spread behavior. Metal, in turn, is often ignored due to its negligible role in the pyrolysis process [141, 142, 177]. In [179, 180], the layered approach was used with layer specific kinetic parameters, found by fitting the postulated reaction schemes to TGA data, and also layer specific thermal parameters, optimized from cone calorimeter experiments that were performed for each cable component material separately. This was possible as the material samples could be physically cut from a complete cable, large enough (diameter several cm) for forming thin 10 × 10 cm^2 samples. With smaller cables, it may not be possible to form planar samples from the pieces of detached cable components. The comparison of model predictions against medium-scale radiative panel tests in [180] indicates that the cable tray HRR predictions can be sufficiently accurate for engineering use when there is already sufficient (> 25 kW/m^2) external heating. At lower heat fluxes, there were factor of two or three errors in peak HRR and the time of ignition.

In addition to the layering and homogenization of the cable structure, the larger-scale structure of a cable bundle may also need some modelling. The assumption that the horizontal bundle of cables forms a uniform, non-permeable slab can be unrealistic if the actual cables are loose and not fully aligned. The modelling challenges arising from the very different fire growth behaviors seen with loose and tight cable layouts have been discussed by Jang et al. [174] and Beji and Merci [181]. They observed that the consequences of using the slab-model include (i) ignoring the penetration of underneath flames in between the cables, (ii) underprediction of heat flux to the tray above, (iii) too small total burning area, (iv) and underpredicted overall fire growth. To overcome these problems, Beji and Merci [181] introduced longitudinal gaps into the obstructions representing the trays to enhance flame penetration. Viitanen et al. [182], in turn, increased the burning area parametrically (area adjustment) and introduced randomly generated holes into the slab to enhance flame penetration. Combined with a finite-rate pyrolysis model and layered cable model, their predictions overpredicted the peak HRR by 8.3% in average, and the uncertainties of the simulated ignition times and flame spread velocities were 22.4% and 37.4%, respectively. Verma et al. [177] utilized the same approach in their FLASH-CAT modelling.

Practically all the works carrying out cable pyrolysis modelling have optimized their models using cone calorimeter experiments. In this scale, the consequences from approximating the (externally) cylindrical objects as a slab are particularly severe: while the nominal cone heat flux can well be expected on the top surface of a cable, the incident heat flux certainly decreases when going down the cable surface ($\theta > 0°$, Figure 21.15) and approaches zero in point between the substrate and the cable ($\theta = 180°$). Additionally, the convective heating from flame is going to vary over time and place, and the combustion propagates from the ignition point to cover the entire sample. This process was studied by Fard and Hostikka [102] using fine-resolution CFD of 20-mm PMMA cylinders tested in cone calorimeter. Figure 21.16 shows instantaneous radiative (incident) and convective heat fluxes on the perimeter of a single PMMA

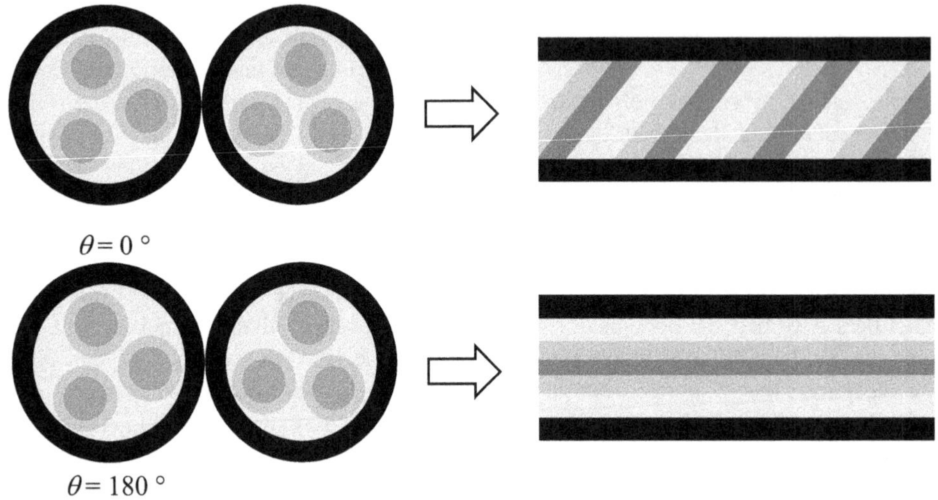

FIGURE 21.15 Slab-description of electrical cables for pyrolysis modelling using homogenization (top) and layering (bottom) of the internal parts.

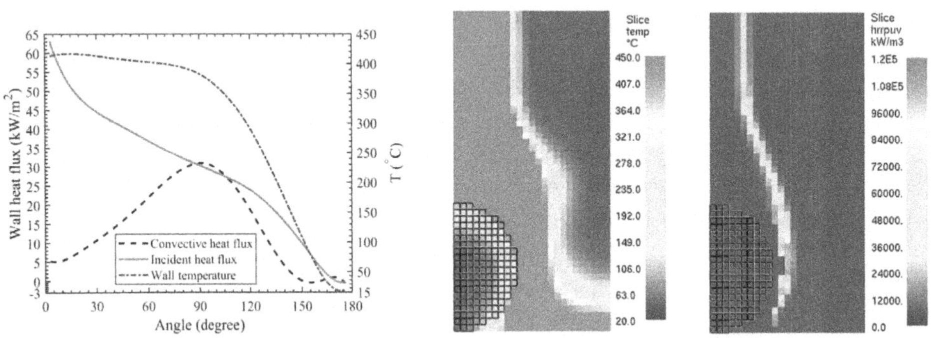

FIGURE 21.16 Simulated surface heat fluxes and temperature distributions (left), medium temperature (middle) and flame location (right) soon after the ignition of a 20 mm PMMA rod under cone calorimeter at 50 kW/m^2 nominal heat flux. [102]

rod placed in the middle of the cone calorimeter sample holder. The incident heat flux shows a decreasing trend with direction angle, as discussed above, and a detailed comparison with an earlier snapshot from the same simulation shows that the flame contribution to the incident radiative flux varies from about +15 kW/m^2 at $\theta = 0°$ to about +5 kW/m^2 at $\theta = 90°$ and is practically zero at $\theta = 180°$. The convective heat flux, in turn, shows a wide 30 kW/m^2 peak at $\theta = 90°$ (side of the cable), mainly due to the flame attached to the side of the cylinder. In the future, this kind of results may be used to refine the presumed temporal and spatial distributions of the heat exposure at cable surfaces during the cone calorimeter test simulations. However, building such inputs for a model used in the optimization can be laborious, and the benefits for the full-scale simulation cannot be known in advance. Nevertheless, as cables remain a topic of pyrolysis and fire spread modelling due to their outmost importance for nuclear and other industries, more detailed models and optimization procedures will certainly appear in the future, hopefully exploiting the emerging understanding of the conditions on the surfaces of tested cables.

21.4.5 CFD Simulations Involving Fire-Retardant Polymers

CFD can be used for demonstrating the benefits of fire retardant polymers in real applications if the modelling tool can reproduce the key fire retardancy mechanisms of the polymer material or product in question. The degree of complexity and detail in the sub-models then largely determine which mechanisms can be modelled, which can be prescribed as user inputs, so that the simulation reflects the effect of the polymer's flammability, and which cannot be investigated at all.

Gas phase fire retardancy mechanisms are difficult to reproduce for most, if not all models. The use of the infinitely fast chemistry may allow for modelling very basic dilution processes: replacing oxygen with an inert gas would reduce or prevent heat release and ultimately cut the feedback loop behind fire growth. Emission of the inert gases, such as water vapor, from a degrading polymer may have a positive effect on the fire retardancy in principle, but the effect may be too subtle or it may be influential only in combination with other mechanisms, so that it could be quantified through a CFD simulation. Furthermore, the gas phase mechanisms based on radical chemistry are clearly beyond the capabilities of most engineering combustion models. Detailed kinetic schemes for combustion could work in theory, but typically they require well-resolved scalar fields that cannot be achieved in the simulations of real-scale applications.

Among the various condensed phase mechanisms, the increased heat capacity by inert fillers is the most straightforward fire retardant feature to include in a pyrolysis model. Increased char yield and endothermicity of the polymer's degradation process are possible to reproduce, as long as the reaction and mass fractions can be observed in TGA, and the kinetics of the decomposition reaction estimated. These mechanisms are important for metal hydroxides, and electrical cables are a common application for polymers with such doping. For instance, Hostikka and Matala [183] used the FDS pyrolysis model for the simulation of TGA and cone calorimeter tests of an electrical cable with MDH-based flame retardant. The flame retardant action of MDH was captured through the inclusion of inert mass component (reduced flammable content, increased char yield) and through the endothermic decomposition reaction of MDH to its oxide and water vapor. Gas phase dilution of fuel vapor by the water vapor was technically included, but it did not cause any measurable impact on the coarse resolution CFD simulation of the cone calorimeter tests, especially because the combustion was modelled using the infinitely fast reaction.

Condensed phase mechanisms that influence through the reaction inhibition or improved transport properties are clearly more difficult to include in pyrolysis models. Considering the former would require complex heterogenous reactions, for which the kinetic coefficients would be difficult to determine. Transport properties regarding heat transfer can be included through the material composition-based thermal conductivity and porosity (enhanced/suppressed radiation), but the gas transfer resistance is still beyond the capabilities of the current models. A special case of the heat transfer reduction-based fire retardants are the intumescent coatings and additives. Although the detailed acid chemistry of gas source generation and foaming may be too complex for the pyrolysis models, effective models of the intumescence have been implemented using stand-alone pyrolysis codes [184, 185]. Nevertheless, coupling such models with CFD is still rare. De Cachinho Cordeiro et al. [186] studied the pyrolysis behavior of high density polyethylene (HDPE)—ammonium polyphosphate (APP) samples using ANSYS Fluent. The model described both the cone calorimeter sample and the conical heater with very fine resolution, and it was validated through the predictions of HDPE and HDPE/APP heat release rates. Unfortunately, the model description and results were not sufficiently detailed to explain why exactly the model did capture the HRR reduction.

21.4.5.1 Acknowledgments

The original version of this chapter was written for the second edition with Dr. Chris Lautenberger. Many parts of the current text were originally from his keyboard, and I am thankful for every word that I still could use. Thanks to Dr Randy McDermott, Dr Kevin McGrattan, Dr Jason Floyd, and Dr Jonathan Hodges who created the plots appearing in the FDS guides that were borrowed here.

REFERENCES

1. Cox, G. Compartment Fire Modelling. In: Combustion Fundamentals of Fire, Ed. G. Cox. Academic Press, London, 1995.
2. Novozhilov, V. Computational Fluid Dynamics Modeling of Compartment Fires. *Prog. Energy Combust. Sci.* **2001**, *27*, 611–666.
3. McGrattan, K., Miles, S. Modeling Fire Using Computational Fluid Dynamics (CFD). In: SFPE Handbook of Fire Protection Engineering, Ed. Hurley, M.J., 5th Edition, Society of Fire Protection Engineering. Springer, New York, 2016, pp. 1034–1065. https://doi.org/10.1007/978-1-4939-2565-0_32
4. Yuen, A. C. Y.; De Cachinho Cordeiro, I. M.; Chen, T. B. Y.; Chen, Q.; Liu, H.; Yeoh, G. H. Multiphase CFD Modelling for Enclosure Fires—a Review on Past Studies and Future Perspectives. *Exp. Comp. Mul. Flow.* **2022**, *4*(1), 1–25. https://doi.org/10.1007/ s42757-021-0116-4
5. Morvan, D.; Accary, G.; Meradji, S.; Frangieh N. Fifty Years of Progress in Wildland Fire Modelling: From Empirical to Fully Physical CFD Models. *C. R.-Mec.* **2022**, *350*. https://doi.org/10.5802/crmeca.133
6. Hirsch, C. Numerical Computations of Internal & External Flows, 2nd edition, Elsevier, 2007.
7. Zikanov, O. Essential Computational Fluid Dynamics, 2nd edition, Wiley, 2019.
8. Courant, R.; Friedrichs, K. O.; Lewy, H. Über Die Partiellen Differentialgleichungen Der Mathematischen Physik. *Math Ann.* **1928**, *100*, 32–74.
9. Courant, R.; Friedrichs, K.; Lewy, H. On the Partial Difference Equations of Mathematical Physics. *IBM J.* **1967**, 215–234. English translation of the 1928 German original paper.
10. Baum, H. R.; Rehm, R. G. Calculations of Three Dimensional Buoyant Plumes in Enclosures. *Combust. Sci. Technol.* **1984**, *40*, 55–77.
11. McGrattan, K. B.; McDermott, R.; Floyd, J.; Hostikka, S.; Forney, G.; Baum, H. Computational Fluid Dynamics Modelling of Fire. *Int. J. Comput. Fluid Dyn.* **2012**, *26*(6–8), 349–361. https://doi.org/10.1080/10618562.2012.659663
12. McGrattan, K., Hostikka, S., McDermott, R., Floyd, J., Weinschenk, C., Overholt, K. Fire Dynamics Simulator, Technical Reference Guide, Volume 1: Mathematical Model, No. 1018-1, 6th ed. National Institute of Standards and Technology, Gaithersburg, MD and VTT Technical Research Centre of Finland, Espoo, 2013.
13. Wang, Y.; Chatterjee, P.; de Ris, J. L. Large Eddy Simulation of Fire Plumes. *Proc. Combust. Inst.* **2011**, *33*(2), 2473–2480. https://doi.org/10.1016/j. proci.2010.07.031
14. Hadane, A.; Redford, J. A.; Gueguin, M.; Hafid, F.; Ghidaglia, J.-M. CFD Wind Tunnel Investigation for Wind Loading on Angle Members in Lattice Tower Structures. *J. Wind. Eng.*

Ind. Aerodyn. **2023**, *236*, 105397. https://doi.org/10.1016/j.jweia.2023.105397
15. McGrattan, K. B.; Rehm, R. G.; Baum, H. R. Fire-Driven Flows in Enclosures. *J. Comput. Phys.* **1994**, *110*(2), 285–291.
16. Rehm, R.G., McGrattan, K.B., Baum, H.R., Cassel, K.W. Transport by Gravity Currents in Building Fires. In: Fire Safety Science—Proceedings of the Fifth International Symposium. Hasemi, Y. (ed.). International Association for Fire Safety Science, March 3–7, Melbourne, 1997, pp. 391–402.
17. Clement, J.M., Fleischmann, C.M. Experimental Verification of the Fire Dynamics Simulator Hydrodynamic Model. In: Fire Safety Science—Proceedings of the Seventh International Symposium. Hasemi, Y. (ed.). International Association for Fire Safety Science, March 3–7, Melbourne, 1997, pp. 839–862.
18. Baum, H.R., McGrattan, K.B., Rehm, R.G. Three Dimensional Simulations of Fire Plume Dynamics. In: Proceedings of Fifth (5th) International Symposium on Fire Safety Science. Hasemi, Y. (ed.). International Association for Fire Safety Science, March 3–7, Melbourne, 1997, pp. 511–522.
19. Baum, H.R., McCaffrey, B.J. Fire Induced Flow Field—Theory and Experiment. In: Proceedings of the Second International Symposium on Fire Safety Science. Wakamatsu, T. et al. (eds.). International Association for Fire Safety Science, June 13–17, Tokyo, 1989, pp. 129–148.
20. McGrattan, K. B.; Baum, H. R.; Rehm, R. G. Large Eddy Simulations of Smoke Movement. *Fire Saf. J.* **1998**, *30*, 161–178.
21. Ma, T. G., Quintiere, J. G. Numerical Simulation of Axisym-Metric Fire Plumes: Accuracy and Limitations. *Fire Saf. J.* **2003,** *38*(5), 467–492.
22. Hamins, A., Maranghides, A., McGrattan, K.B., Johnsson, E., Ohlemiller, T., Donnelly, M., Yang, J., Mulholland, G., Prasad, K., Kukuck, S., Anleitner, R., McAllister, T. Report on Experiments to Validate Fire Dynamic and Thermal-Structural Models for Use in the World Trade Center Investigation. NIST Special Publication 1000-B. National Institute of Standards and Technology, Gaithersburg, MD, 2004.
23. Hamins, A., Maranghides, A., McGrattan, K.B., Ohlemiller, T., Anletiner, R. Experiments to Validate Models of Fire Growth and Spread for Use in the World Trade Center Investigation. NIST Special Publication 1000–10E. National Institute of Standards and Technology, Gaithersburg, MD, 2004.
24. McGrattan, K.B. Verification and Validation of Selected Fire Models for Nuclear Power Plant Applications. Volume 7. Fire Dynamics Simulator (FDS). NUREG-1824, EPRI 1011999. National Institute of Standards and Technology, Gaithersburg, MD, Nuclear Regulatory Commission, Washington, DC and Electric Power Research Institute, Palo Alto, CA, 2007, 205 p.
25. McGrattan, K., Hostikka, S., McDermott, R., Floyd, J., Weinschenk, C., Overholt, K. Fire Dynamics Simulator, Technical Reference Guide, Volume 3: Validation. National Institute of Standards and Technology, Gaithersburg, MD and VTT Technical Research Centre of Finland, Espoo, 6th Edition, 2013.
26. Pope, S. B. Ten Questions Concerning the Large-Eddy Simulation of Turbulent Flows. *New J. Phys.* **2004**, *6*, 1–24.
27. Zhang, W.; Hamer, A.; Klassen, M.; Carpenter, D.; Roby, R. Turbulence Statistics in a Fire Room Model by Large Eddy Simulation. *Fire Saf. J.* **2002**, *37*(8), 721–752.
28. Ren, N.; Wang, Y.; Vilfayeau, S.; Trouvé, A. Large Eddy Simulation of Turbulent Vertical Wall Fires Supplied with Gaseous Fuel Through Porous Burners. *Combust Flame.* **2016**, *169*, 194–208. https://doi.org/10.1016/j.combustflame.2015.12.008
29. Ren, N.; Wang, Y. A Convective Heat Transfer Model for LES Fire Modelling. *Proc. Combust. Inst.* **2021**, *38*(3), 4535–4542. https://doi.org/10.1016/j.proci.2020.07.077
30. Smith, G.P., Golden, D.M., Frenklach, M., Moriarty, N.W., Eiteneer, B., Goldenberg, M., Bowman, C.T., Hanson, R.K., Song, S., Gardiner, W.C., Lissianski, V.V., Qin, Z. GRI Mech 3.0, http://combustion.berkeley.edu/gri-mech/, accessed September 17, 2023.
31. Poinsot, T.; Veynante, D. Theoretical and Numerical Combustion. Edwards, Inc., Philadelphia, PA, 2005.
32. Wang, C. J.; Wen, J. X.; Chen, Z. B.; Dembele, S. Predicting Radiative Characteristics of Hydrogen and Hydrogen/methane Jet Fires Using Fire Foam. *Int. J. Hydrogen Energy.* **2014**, *39*(35), 20560–20569. https://doi.org/10.1016/j.ijhydene.2014.04.062
33. Bilger, R. W. Reaction Rates in Diffusion Flames. *Combust Flame.* **1997**, *30*, 277–284.
34. Williams, F.A. Combustion Theory: The Fundamental Theory of Chemically Reacting Flow Systems. Benjamin/Cummings, Menlo Park, CA, 2nd Edition, 680 p., 1985.
35. Spalding, D. B. Mixing and Chemical Reaction in Steady Confined Turbulent Flames. *Symp. (International) Combust.* **1971**, *13*(1), 649–657. https://doi.org/10.1016/S0082-0784(71)80067-X
36. Magnussen, B.F., Hjertager, G.H. On Mathematical Modelling of Turbulent Combustion With Special Emphasis on Soot Formation and Combustion. In: Proceedings of 16th International Symposium on Combustion. The Combustion Institute, Pittsburgh, PA, 1976, pp. 719–729.
37. Magnussen, B. F.; Hjertager, G. H.; Olsen, J. G.; Bhaduri, D. Effects of Turbulent Structure and Local Concentrations on Soot Formation and Combustion in C H diffusion flames. *Symp. (International) Combust.* **1979**, *17*(1), 1383–1393.
38. Chen, Z.; Wen, J.; Xu, B.; Dembele, S. Large Eddy Simulation of a Medium-Scale Methanol Pool Fire Using the Extended Eddy Dissipation Concept. *Int. J. Heat Mass Transf.* **2014,** *70*, 389–408. https://doi.org/10.1016/j.ijheatmasstransfer.2013.11.010
39. McDermott, R.; McGrattan, K.; Floyd, J. A Simple Reaction Time Scale for Under-Resolved Fire Dynamics. *Fire Saf. Sci.* **2011**, *10*, 809–820. https://doi.org/10.3801/IAFSS.FSS.10-809
40. Yang, R.; Weng, W. G.; Fan, W. C.; Wang, Y. S. Subgrid Scale Laminar Flamelet Model for Partially Premixed Combustion and Its Application to Backdraft Simulation. *Fire Saf. J.* **2005**, *40*(2), 81–98.
41. Kang, Y.; Wen, J. X. Large Eddy Simulation of a Small Pool Fire. *Combust. Sci. Technol.* **2004**, *176*(12), 2193–2223.
42. Peters, N. Laminar Diffusion Flamelet Model in Non-Premixed Turbulent Combustion. *Prog. Energy Combust. Sci.* **1984**, *10*, 319–339.
43. Köylü, U. O.; Feath, G. M. Carbon Monoxide and Soot Emissions from Liquid-Fueled Buoyant Turbulent Diffusion Flames. *Combust Flame.* **1991**, *87*, 61–76.
44. Tuovinen, H.; Blomqvist, P.; Saric, F. Modelling of Hydrogen Cyanide Formation in Room Fires. *Fire Saf. J.* **2004**, *39*(8), 737–755.
45. Le, V. M.; Marchand, A.; Verma, S.; Xu, R.; Wite, J.; Marshall, A.; Rogaume, T.; Richard, F.; Luche, J.; Trouvé,

A. Simulations of a Turbulent Line Fire with a Steady Flamelet Combustion Model Coupled with Models for Non-Local and Local Gas Radiation Effects. *Fire Saf. J.* **2019**, *106*, 105–113. https://doi.org/10.1016/j.firesaf.2019.04.011

46. Xu, R.; Le, V. M.; Marchand, A.; Verma, S.; Rogaume, T.; Richard, F.; Luche, J.; Trouvé, A. Simulations of the Coupling between Combustion and Radiation in a Turbulent Line Fire using an Unsteady Flamelet Model. *Fire Saf. J.* **2021**, *120*, 103101. https://doi.org/10.1016/j.firesaf.2020.103101
47. McGrattan, K., McDermott, R., Floyd, J. A Simple Two-Step Reaction Scheme for Predicting CO in Under-Ventilated Compartments. In: Proceedings of the Tenth International Symposium on Fire and Explosion Hazards, Oslo, 2022.
48. Beyler, C.L. Fire Hazard Calculations for Large, Open Hydrocarbon Fires. In: SFPE Handbook of Fire Protection Engineering. Society of Fire Protection Engineers, Quincy, MA, 2002, 3rd Edition, pp. 3-268–3-314.
49. Siegel, R., Howell, J.R. Thermal Radiation Heat Transfer. Hemisphere Publishing Corporation, New York, NY, 1981, 2nd Edition, 862 p.
50. Özisik, M.N. Radiative Transfer and Interactions With Conduction and Convection. John Wiley & Sons Inc., New York, NY, 1973, 575 p.
51. Modest, M.F., Mazumder, S. Radiative Heat Transfer, 4th ed., Academic Press, London, 2021, 1016 p. https://www.sci-encedirect.com/science/article/pii/B9780128181430000031
52. Hostikka, S.; McGrattan, K. Numerical Modeling of Radiative Heat Transfer in Water Sprays. *Fire Saf. J.* **2006**, *41*(1), 76–86. https://doi.org/10.1016/j.firesaf.2005.09.003
53. Lockwood, F.C., Shah, N.G. A New Radiation Solution Method for Incorporation in General Combustion Prediction Procedures. In: Proceedings of the 18th Symposium (International) on Combustion. The Combustion Institute, Pittsburgh, PA, 1981, pp. 1405–1416.
54. Wu, B.; Roy, S. P.; Zhao, X. Detailed Modeling of a Small-Scale Turbulent Pool Fire. *Combust Flame.* **2020**, *214*, 224–237. https://doi.org/10.1016/j.combustflame.2019.12.034
55. Raithby, G. D., Chui, E. H. A Finite-Volume Method for Predicting Radiant Heat Transfer in Enclosures with Participating Media. *J. Heat Transf.* **1990**, *112*(2), 415–423.
56. Modest, M. F. Radiative Heat Transfer in Turbulent Combustion 2021 Max Jakob memorial award paper. *The ASME J. Heat Mass Transf.* **2023**, *145*(7), 073101. https://doi.org/10.1115/1.4056402
57. Raithby, G. D. Discussion of the Finite-Volume Method for Radiation, and Its Application using 3D Unstructured Meshes. *Numer. Heat Transf.* **1999**, *35*, 389–405.
58. Chai, J. C.; Lee, H. S.; Patankar, S. V. Finite Volume Method for Radiation Heat Transfer. J. Thermophys. Heat Transf. 1994, *8*(3), 419–425.
59. Jensen, K. A.; Ripoll, J.-F.; Wray, A. A.; Joseph, D.; El Hafi, M. On Various Modeling Approaches to Radiative Heat Transfer in Pool Fires. *Combust Flame.* **2007**, *148*(4), 263–279. https://doi.org/10.1016/j.combustflame.2006.09.008
60. Runefors, M., Anderson, J., Wahlqvist, J., Huang, C., Husted, B. A Comparison of Radiative Transfer Models in FireFoam and FDS. In: 14th International Conference and Exhibition on Fire Science and Engineering (Interflam 2016), Windsor, Interscience Communications Ltd., Hampshire, 2016, pp. 4–6.
61. Snegirev, A. Y. Statistical Modeling of Thermal Radiation Transfer in Buoyant Turbulent Diffusion Flames. *Combust Flame.* **2004**, *136*(1–2), 51–71. https://doi.org/10.1016/j.combustflame.2003.09.005
62. Gupta, A.; Modest, M. F.; Haworth, D. C. Large-Eddy Simulations of Turbulence Radiation Interactions in a Turbulent Planar Channel Flow. *J. Heat Transf.* **2009**, *131*(1), 61704. https://doi.org/10.1115/1.3085875
63. Gupta, A.; Haworth, D. C.; Modest, M. F. Turbulence-Radiation Interactions in Large-Eddy Simulations of Luminous and Nonluminous Nonpremixed Flames. *Proc. Combust. Inst.* **2013**, *34*(1), 1281–1288. http:// dx.doi.org/10.1016/j.proci.2012.05.052
64. Nmira, F.; Ma, L.; Consalvi, J.-L. Assessment of Subfilter-Scale Turbulence-Radiation Interaction in Non-Luminous Pool Fires. *Proc. Combust. Inst.* **2021**, *38*(3), 4927–4934. https://doi.org/10.1016/j.proci.2020.06.271
65. Chatterjee, P.; Wang, Y.; Meredith, K.V.; Dorofeev, S.B. Application of a subgrid soot-radiation model in the numerical simulation of a heptane pool fire. *Proc. Combust. Inst.* **2015**, *35*(3), 2573–2580. http:// dx.doi.org/10.1016/j.proci.2014.05.045
66. Snegirev, A.; Markus, E.; Kuznetsov, E.; Harris, J.; Wu, T. On Soot and Radiation Modeling in Buoyant Turbulent Diffusion Flames. *Heat Mass Transf.* **2018**, *54*, 2275–2293. http://link.springer.com/article/10.1007/s00231-017-2198-x
67. Rashidzadeh, S.; Hostikka, S.; McDermott, R. A Two-Zone Subgrid Flame Model for Predicting Radiant Emission from Fires. *Fire Saf. J.* **2023**, *140*, 103865. https://doi.org/10.1016/j.firesaf.2023.103865
68. Ahmed, M.; Trouvé, A. Large Eddy Simulation of the Unstable Flame Structure and Gas-to-Liquid Thermal Feedback in a Medium-Scale Methanol Pool Fire. *Combust Flame.* **2021**, 225, 237–254. https://doi.org/10.1016/j.combustflame.2020.10.055
69. Moss, J. B.; Stewart, C. D. Flamelet-Based Smoke Properties for the Field Modelling of Fires. *Fire Saf. J.* **1998**, *30*, 229–250.
70. Yeoh, G. H.; Yuen, R. K. K.; Chueng, S. C. P.; Kwok, W. K. On Modelling Combustion, Radiation and Soot Processes in Compartment Fires. *Build. Environ.* **2003**, *38*(6), 771–785.
71. Wen, J. X.; Huang, L. Y.; Roberts, J. The Effect of Microscopic and Global Radiative Heat Exchange on the Field Predictions of Compartment Fires. *Fire Saf. J.* **2001**, *36*(3), 205–223.
72. Lautenberger, C. W.; de Ris, J. L.; Dembsey, N. A.; Barnett, J. R.; Baum, H. R. A Simplified Model for Soot Formation and Oxidation in CFD Simulation of Non Premixed Hydrocarbon Flames. *Fire Saf. J.* **2005**, *40*(2), 141–176.
73. Beji, T.; Zhang, J. P.; Delichatsios, M. Determination of Soot Formation Rate from Laminar Smoke Point Measurements. *Combust. Sci. Technol.* **2008**, *180*, 927–940.
74. Beji, T.; Zhang, J.; Delichatsios, M. Soot Formation and Oxidation in Fires from Laminar Smoke Point Measurements. Fire Saf. Sci. **2008**, *9*, 219–230.
75. Chen, Z.; Wen, J.; Xu, B.; Dembele, S. Extension of the Eddy Dissipation Concept and Smoke Point Soot Model to the LES Frame for Fire Simulations. *Fire Saf. J.* **2014**, *64*, 12–26. https://doi.org/10.1016/j.firesaf.2014.01.001
76. Fukumoto, K.; Wang, C.; Wen, J. Large Eddy Simulation of Upward Flame Spread on PMMA Walls with a Fully Coupled Fluid–Solid Approach. *Combust Flame.* **2018**, *190*, 365–387. https://doi.org/10.1016/j.combustflame.2017.11.012
77. Chang, H.-C.; Charalampopoulos, T. T. Determination of the Wavelength Dependence of Refractive Indices of Flame Soot. *Proc. R. Soc. Lond.* **1990**, *430*, 577–591. https://doi.org/10.1098/rspa.1990.0107
78. Modest, M. F. The Weighted-Sum-of-Gray-Gases Model for Arbitrary Solution Methods in Radiative

Transfer. *J. Heat Transf.* **1991**, *113*, 650–656. https://doi.org/10.1115/1.2910614
79. Bordbar, M. H.; Wecel, G.; Hyppänen, T. A Line by Line Based Weighted Sum of Gray Gases Model for Inhomogeneous CO_2-H_2O Mixture in Oxy-Fired Combustion. *Combust Flame.* **2014**, *161*, 2435–2445. https://doi.org/10.1016/j.combustflame.2014.03.013
80. Bordbar, H.; Fraga, G. C.; Hostikka, S. An Extended Weighted-Sum-of-Gray-Gases Model to Account for All CO_2-H_2O Molar Fraction Ratios in Thermal Radiation. *Int. Commun. Heat. Mass. Transf.* **2020**, *110*, 104400. https://doi.org/10.1016/j.icheatmasstransfer.2019.104400
81. Sadeghi, H.; Hostikka, S.; Fraga, G. C.; Bordbar, H. Weighted-Sum-of-Gray-Gases Models for Non-Gray Thermal Radiation of Hydrocarbon Fuel Vapors, CH4, CO and Soot. *Fire Saf. J.* **2021**, *125*, 103420. https://doi.org/10.1016/j.firesaf.2021.103420
82. Grosshandler, W. RadCal: A Narrow Band Model for Radiation Calculations in a Combustion Environment. NIST Technical Note TN 1402. National Institute of Standards and Technology, Gaithersburg, MD, 1993.
83. Fernandes, C. S.; Fraga, G. C.; França, F. H. R.; Centeno, F. R. Radiative Transfer Calculations in Fire Simulations: An Assessment of Different Gray Gas Models Using the Software FDS. *Fire Saf. J.* **2021**, *120*, 103103. https://doi.org/10.1016/j.firesaf.2020.103103
84. Modest, M. F.; Zhang, H. The Full-Spectrum Correlated-K Distribution for Thermal Radiation from Molecular Gas-Particulate Mixtures. *J. Heat Transf.* **2002**, *124*(1), 30–38. https://doi.org/10.1115/1.1418697
85. Dembele, S.; Wen, J. X. Evaluation of a Fast Correlated-K Approach for Radiation Calculations in Combustion Systems. *Numer. Heat Transf. Part B.* **2003**, *44*(4), 365–385.
86. Consalvi, J. L.; Liu, F. Radiative Heat Transfer Through the Fuel-Rich Core of Laboratory-Scale Pool Fires. *Combust. Sci. Technol.* **2014**, *186*(4–5), 475–489. http:// dx.doi.org/10.1080/00102202.2014.883221
87. Denison, M. K.; Webb, B. W. The Spectral-Line-Based Weighted-Sum-of-Gray-Gases Model in Nonisothermal Nonhomogeneous Media. *ASME J. Heat Mass Transf.* **1995**, *117*(2), 359–365. https://doi.org/10.1115/1.2822530
88. Li, K.; Hostikka, S. Embedded Flame Heat Flux Method for Simulation of Quasi-Steady State Vertical Flame Spread. *Fire Saf. J.* **2019**, *104*, 117–129. https://doi.org/10.1016/j.firesaf.2019.01.011
89. Maragkos, G.; Beji, T. Review of Convective Heat Transfer Modelling in CFD Simulations of Fire-Driven Flows. App. Sci. (Switzerland). **2021**, *11*(11), 5240. https://dx.doi.org/10.3390/app11115240
90. Hong, M. C.; Merci, B.; Beji, T. On the Use of Natural and Forced Convection Correlations in Predictive CFD Simulations of Liquid Pool Fires. *Fire Saf. J.* **2023**, *137*, 103765. https://dx.doi.org/10.1016/j.firesaf.2023.103765
91. Di Blasi, C. Modeling chemical and physical processes of wood and biomass pyrolysis. *Prog. Energy Combust. Sci.* **2008**, *34*, 47–90.
92. Lautenberger, C., Fernandez-Pello, C. 2008. Pyrolysis Modeling, Thermal Decomposition, and Transport Processes in Combustible Solids. In: Transport Phenomena in Fires, Ed. Faghri, M. and B. Sundén. WIT Press, Boston, MA, pp. 209–259.
93. Nyazika, T.; Jimenez, M.; Samyn, F.; Bourbigot, S. Pyrolysis Modeling, Sensitivity Analysis, and Optimization Techniques for Combustible Materials: A Review. *J. Fire Saf.* **2019**, *37*(4–6), 377–433. https://doi.org/10.1177/0734904119852740
94. Wickström, U., Duthinh, D., McGrattan, K. Adiabatic Surface Temperature for Calculating Heat Transfer to Fire Exposed Structures. In: Proceedings of the 11th International Interflam Conference (Interflam '07), London, 2007.
95. Welch, S.; Miles, S.; Kumar, S.; Lemaire, A.; Chan, A. Firestruc–integrating advanced three-dimensional modelling methodologies for predicting thermo-mechanical behaviour of steel and composite structures subjected to natural fires. *Fire Saf. Sci.* **2008**, *9*, 1315–1326. http://dx.doi.org/10.3801/IAFSS.FSS.9-1315
96. Paajanen, A.; Korhonen, T.; Sippola, M.; Hostikka, S.; Malendowski, M.; Gutkin, R. FDS2FEM—A Tool for Coupling Fire and Structural Analyses. In: Safety, Failures and Robustness of Large Structures, IABSE Report *Int. Assoc. for Bridge and Struct. Eng.* (IABSE). **2013**, *100* 218–224. https://doi.org/10.2749/222137813807054629
97. Zhang, C.; Silva, J. G.; Weinschenk, C.; Kamikawa, D.; Hasemi, Y. Simulation Methodology for Coupled Fire-Structure Analysis: Modelling Localized Fire Tests on a Steel column. Fire Techn. **2016**, *52*, 239–262.
98. Tondini, N.; Morbioli, A.; Vassart, O.; Lechêne, S.; Franssen, J.-M. An Integrated Modelling Strategy between a CFD and an FE Software: Methodology and Application to Compartment Fires. *J. Struct. Fire Eng.* **2016**, *7*(3), 217–233. https://doi.org/10.1108/JSFE-09-2016-015
99. Khan, A. A.; Khan, M. A.; Zhang, C.; Jiang, L.; Usmani, A. Open-Fire: An Open Computational Framework for Structural Response to Real Fires. Fire Technol. 2022, *58*, 1011–1038. https://doi.org/10.1007/s10694-021-01184-0
100. Welch, S., Rubini, P. Three-Dimensional Simulation of a Fire-Resistance Furnace. In: Proceedings of Fifth (5th) International Symposium on Fire Safety Science. March 3–7, 1997, Melbourne, Ed. Hasemi, Y. International Association for Fire Safety Science, Boston, MA, 1997, pp. 1009–1020.
101. de St. Germain, J.D., McCorquodale, J., Parker, S.G., Johnson, C.R. Uintah: A Massively Parallel Problem Solving Environment. In: Ninth IEEE International Symposium on High Performance and Distributed Computing, Pittsburg, PA. IEEE, 2000, pp. 33–41.
102. Gholami Haghighi Fard, M.; Hostikka, S. Combustion characteristics of non-charring polymer cylinders—experimental and numerical study. *Combust Flame.* **2023**, *249*, 112587. https://doi.org/10.1016/j.combustflame.2022.112587
103. Bruns, M., McDermott, R., Benkorichi, S., Hostikka, S. Development of 3d pyrolysis in FDS. In: Spring Technical Meeting of the Eastern States Section of the Combustion Institute (ESSCI 2018), State College, Pennsylvania, 2018.
104. Markus, E.; Snegirev, A.; Kuznetsov, E.; Tanklevskiy. L. Application of the Thermal Pyrolysis Model to Predict Flame Spread Over Continuous and Discrete Fire Load. *Fire Saf. J.* **2019**, *108*, 102825. https://doi.org/10.1016/j.firesaf.2019.102825
105. Mitler, H. E. Predicting the spread rates of fires on vertical sur- faces. *Symp. (International) Combust.* **1991**, *23*(1), 1715–1721. https://doi.org/10.1016/S0082-0784(06)80447-9
106. Hodges, J. L.; Lattimer, B. Y.; Kapahi, A.; Floyd, J. E. An Engineering Model for the Pyrolysis of Materials. *Fire Saf. J.* **2023**, *141*, 103980. https://doi.org/10.1016/j.firesaf.2023.103980
107. McGrattan, K., Hostikka, S., McDermott, R., Floyd, J., Weinschenk, C., Overholt, K. Fire Dynamics Simulator, User's Guide. National Institute of Standards and

Technology, Gaithersburg, MD and VTT Technical Research Centre of Finland, Espoo, 6th Edition, 2013.
108. Tewarson, A.; Ogden, S. D. Fire Behavior of Polymethyl-Methacrylate. *Combust Flame.* **1992**, *89*, 237–259.
109. Kacem, A.; Mense, M.; Pizzo, Y.; Boyer, G.; Suard, S.; Boulet, P.; Parent, G.; Porterie, B. A Fully Coupled Fluid/Solid Model for Open Air Combustion of Horizontally-Oriented PMMA Samples. *Combust Flame.* **2016,** *170*, 135–147. https://doi.org/10.1016/j.combustflame.2016.04.009
110. Rinta-Paavola, A.; Sukhomlinov, D.; Hostikka, S. Modelling Charring and Burning of Spruce and Pine Woods During Pyrolysis, Smoldering and Flaming. *Fire Technol.* **2023**. https://doi.org/10.1007/s10694-023-01458-9
111. Babrauskas, V. Heat Release Rates. In: The SFPE Handbook of Fire Protection Engineering, 3rd Edition. National Fire Protection Association, Quincy, MA, 2002, pp. 3–26.
112. Beji, T.; Helson, O.; Rogaume, T.; Luche, J. Experimental and Numerical Study on the Evaporation Rates of Liquid Fuels Using a Controlled Atmosphere Cone Calorimeter. *Fire Saf. J.* **2021**, *121*, 103317. https://doi.org/10.1016/j.firesaf.2021.103317
113. Sikanen, T.; Hostikka, S. Modeling and Simulation of Liquid Pool Fires with in-Depth Radiation Absorption and Heat Transfer. *Fire Saf. J.* **2016**, 80, 95–109. https://doi.org/10.1016/j.firesaf.2016.01.002
114. Isojärvi, T.; Bordbar, H.; Hostikka, S. Spectrally Resolved Calculation of Thermal Radiation Penetration into Liquid N-Heptane in Pool Fires. *Int. J. Heat. Mass. Transf.* **2018**, *127*(C), 1101–1109. https://doi.org/10.1016/j.ijheatmasstransfer.2018.08.108
115. Alinejad, F.; Bordbar, S.; Hostikka, S. Improving the Modelling of Spectral Radiation Penetration into the Condensed Materials with the Separated Full Spectrum Correlated-K Method. *Int. J. Heat. Mass. Transf.* **2021**, *176*, 121448. https://doi.org/10.1016/j.ijheatmasstransfer.2021.121448
116. Alinejad, F.; Bordbar, S.; Hostikka, S. The Ordinate Weighting Method for Solving Radiative Heat Transfer Through a Fresnel Interface. *J. Quant. Spectrosc. Radiat. Transf.* **2021**, *270*, 107685. https://doi.org/10.1016/j.jqsrt.2021.107685
117. Yip, A.; Haelssig, J. B.; Pegg, M. J. Simulating Fire Dynamics in Multicomponent Pool Fires. *Fire Saf. J.* **2021**, *125*, 103402. https://doi.org/10.1016/j.firesaf.2021.103402
118. Linteris, G.; Zammarano, M.; Wilthan, B.; Hanssen, L. Absorption and Reflection of Infrared Radiation by Polymers in Fire-Like Environments. *Fire and Mater.* **2012**, *36*(7), 537–553. https://doi.org/10.1002/fam.1113
119. Gong, J.; Chen, Y.; Jiang, J.; Yang, L.; Li, J. A Numerical Study of Thermal Degradation of Polymers: Surface and In-Depth Absorption. *Appl. Therm. Eng.* **2016**, *106*, 1366–1379. https://doi.org/10.1016/j.applthermaleng.2016.06.114
120. Bal, N.; Rein, G. Numerical Investigation of the Ignition Delay Time of a Translucent Solid at High Radiant Heat Fluxes. *Combust Flame.* **2011**, *158*(6), 1109–1116. http://doi.org/10.1016/j.combustflame.2010.10.014
121. Bal, N.; Raynard, J.; Rein, G.; Torero, J.; Försth, M.; Boulet, P.; Parent, G.; Acem, Z.; Linteris, G. Experimental Study of Radiative Heat Transfer in a Translucent Fuel Sample Exposed to Different Spectral sources. *Int. J. Heat. Mass. Transf.* **2013**, *61*, 742–748. https://doi.org/10.1016/j.ijheatmasstransfer.2013.02.017
122. Chaos, M. Application of Sensitivity Analyses to Condensed-phase Pyrolysis Modeling. *Fire Saf. J.* **2013**, *61*, 254–264. https://doi.org/10.1016/j.firesaf.2013.09.016
123. Lautenberger, C.; Fernandez-Pello, C. Generalized Pyrolysis Model for Combustible Solids. *Fire Saf. J.* **2009**, *44*(6), 819–839. https://doi.org/10.1016/j.firesaf.2009.03.011
124. Stoliarov, S. I.; Crowley, S.; Walters, R. N.; Lyon, R. E. Prediction of the Burning Rates of Charring Polymers. *Combust Flame.* **2010**, *157*(11), 2024–2034. http://doi.org/10.1016/j.combustflame.2010.03.011
125. Alinejad, F.; Bordbar, H.; Hostikka, S. On the Importance and Modeling of in-Depth Spectral Radiation Absorption in the Flammability Analyses of Black PMMA. *Fire Saf. J.* **2023**, *135*, 103706. https://doi.org/10.1016/j.firesaf.2022.
126. Boulet, P.; Gérardin, J.; Acem, Z.; Parent, G.; Collin, A.; Pizzo, Y.; Porterie, B. Optical and Radiative Properties of Clear PMMA Samples Exposed to a Radiant Heat Flux. *Int. J. Therm. Sci.* **2014**, *82*, 1–8. https://doi.org/10.1016/j.ijthermalsci.2014.03.013
127. Alinejad, F.; Bordbar, H.; Makowska, M.; Hostikka, S. Spectroscopic Determination of the Optical Constants and Radiative Properties of Black PMMA for Pyrolysis Modeling. *Int. J. Therm. Sci.* **2022**, *176*, 107501. https://doi.org/10.1016/j.ijthermalsci.2022.107501
128. Alinejad, F.; Bordbar, H.; Makowska, M.; Hostikka, S. A Dataset for Spectral Radiative Properties of Black Poly(Methyl Methacrylate). *Data Br.* **2022**, *42*, 108097. https://doi.org/10.1016/j.dib.2022.108097
129. Leventon, I., Batiot, B., Bruns, M., Hostikka, S., Nakamura, Y., Reszka, P., Rogaume, T., Stoliarov, S. The MaCFP Condensed Phase Working Group: A Structured, Global Effort Towards Pyrolysis Model Development, ASTM Selected Technical Papers (STP), Atlanta, GA, 2021. [online], https://tsapps.nist.gov/publication/get_pdf.cfm?pub_id=933681 (Accessed March 3, 2022).
130. Pizzo, Y.; Lallemand, C.; Kacem, A.; Kaiss, A.; Gerardin, J.; Acem, Z.; Boulet, P.; Porterie, B. Steady and Transient Pyrolysis of Thick Clear PMMA Slabs. *Combust Flame.* **2015**, *162*(1), 226–236. https://doi.org/10.1016/j.combustflame.2016.04.009
131. Kim, E.; Dembsey, N. Parameter Estimation for Comprehensive Pyrolysis Modeling: Guidance and Critical Observations. *Fire Technol.* **2015**, *51*, 443–477. https://doi.org/10.1007/s10694-014-0399-0
132. Shi, L., Chew, M.Y.L. A Review of Fire Processes Modeling of Combustible Materials Under External Heat Flux. *Fuel* **2013**, *106*, 30–50. https://doi.org/10.1016/j.fuel.2012.12.057
133. Nyazika, T.; Jimenez, M.; Samyn, F.; Bourbigot, S. Pyrolysis Modeling, Sensitivity Analysis, and Optimization Techniques for Combustible Materials: A Review. *J. Fire Sci.* **2019**, *37*(4–6), 377–433. https://doi.org/10.1177/0734904119852740
134. Lyon, R. E.; Safronava, N.; Oztekin, E. A Simple Method for Determining Kinetic Parameters for Materials in Fire Models. *Fire Saf. Sci.* **2011**, *10*, 765–778. https://doi.org/10.3801/IAFSS.FSS.10-765
135. Lautenberger, C.; Rein, G.; Fernandez-Pello, A. C. The Application of a Genetic Algorithm to Estimate Material Properties for Fire Modeling from Bench-Scale Fire Test Data. *Fire Saf. J.* **2006**, *41*, 204–214.
136. Rein, G.; Lautenberger, C.; Fernandez-Pello, A. C., Torero, J. L.; Urban, D. L. Application of Genetic Algorithms and Thermogravimetry to Determine the Kinetics of Polyurethane Foam in Smoldering Combustion. *Combust Flame.* **2006**, *146*, 95–108.
137. Saha, B.; Karthik Reddy, P.; Ghosal, A. K. Hybrid Genetic Algorithm to Find the Best Model and the Globally

Optimized Overall Kinetics Parameters for Thermal Decomposition of Plastics. *J. Chem. Eng.* **2008**, *138*, 20–29.
138. Matala, A., Hostikka, S., Mangs, J. 2008. Estimation of Pyrolysis Model Parameters for Solid Materials Using Thermogravimetric Data. In: Fire Safety Science—Proceedings of the Ninth International Symposium. Karlsson, B. (ed.). International Association for Fire Safety Science, September 21–26, 2008, pp. 1213–1224. DOI:10.3801/IAFSS.FSS.9-1213
139. Chaos, M.; Khan, M. M.; Krishnamoorthy, N.; de Ris, J. L.; Dorofeev, S. B. Evaluation of Optimization Schemes and Determination of Solid Fuel Properties for CFD Fire Models Using Bench-Scale Pyrolysis Tests. *Proc. Combust. Inst.* **2011**, *33*(2), 2599–2606. https://doi.org/10.1016/j.proci.2010.07.018
140. Arnold, L.; Hehnen, T.; Lauer, P.; Trettin, C.; Vinayak, A. Application Cases of Inverse Modelling with the PROPTI Framework. *Fire Saf. J.* **2019**, *108*, 102835. http://doi.org/10.1016/j.firesaf.2019.102835
141. Matala, A.; Hostikka, S. Pyrolysis Modelling of PVC Cable Materials. *Fire Saf. Sci.* **2011**, *10*, 917–30. http://dx.doi.org/10.3801/iafss.fss.10-917
142. Hehnen, T.; Arnold, L.; La Mendola, S. Numerical Fire Spread Simulation Based on Material Pyrolysis—An Application to the Christifire Phase 1 Horizontal Cable Tray Tests. *Fire.* **2020**, *3*(3), 33. https://doi.org/10.3390/fire3030033
143. Grange, N.; Chetehouna, K.; Gascoin, N.; Coppalle, A.; Reynaud, I.; Senave, S. One-Dimensional Pyrolysis of Carbon Based Composite Materials Using Fire Foam. *Fire Saf. J.* **2018**, *97*, 66–75. https://doi.org/10.1016/j.firesaf.2018.03.002
144. Stoliarov, S. I.; Crowley, S.; Lyon, R. E.; Linteris, G. T. Prediction of the Burning Rates of Non-Charring Polymers. *Combust Flame.* **2009**, *156*(5), 1068–1083. http://doi.org/10.1016/j.combustflame.2008.11.010
145. McCoy, C. G.; Tilles, J. L.; Stoliarov, S. I. Empirical Model of Flame Heat Feedback for Simulation of Cone Calorimetry. *Fire Saf. J.* **2019**, *103*, 38–48. https://doi.org/10.1016/j.firesaf.2018.11.006
146. Hostikka, S., Axelsson, J. Modelling of the Radiative Feedback from Flames in Cone Calorimeter. NORDTEST technical report 540. Nordtest, Espoo. 41 p. 2003. www.nordtest.info/wp/2003/10/02/modelling-of-the-radiative-feedback-from-the-flames-in-cone-calorimeter-nt-tr-540/
147. Yuen, R. K. K.; Yeoh, G. H.; de Vahl Davis, G.; Leonardi, E. Modelling the Pyrolysis of Wet Wood—II. Three-Dimensional Cone Calorimeter Simulation. *Int. J. Heat Mass Transf.* **2007**, *50*(21–22), 4387–4399. https://doi.org/10.1016/j.ijheatmasstransfer.2007.01.018
148. Boyer, G. Fully Coupled CFD Simulation of the Pyrolysis of Non-Charring Polymers: A Predictive Approach. *Fire Saf. J.* **2017**, *91*, 208–217. https://doi.org/10.1016/j.firesaf.2017.03.070
149. Fereres, S.; Lautenberger, C.; Fernandez-Pello, A. C.; Urban, D. L.; Ruff, G. A. Understanding Ambient Pressure Effects on Piloted Ignition Through Numerical Modeling. *Combust Flame.* **2012**, *159*(12), 3544–3553. http://doi.org/10.1016/j.combustflame.2012.08.006
150. Hostikka, S.; Matala, A. Pyrolysis Model for Predicting the Heat Release Rate of Birch Wood. *Combust. Sci. Technol.* **2017**, *189*(8), 1373–1393. https://doi.org/10.1080/00102202.2017.1295959
151. Alonso, A.; Lázaro, D.; Lázaro, M.; Alvear, D. Numerical Prediction of Cables Fire Behaviour Using Non-Metallic Components in Cone Calorimeter. *Combust. Sci. Technol.* **2023**, *195*(7), 1509–1525. https://doi.org/10.1080/00102202.2023.2182198
152. Atreya, A., Baum, H.R. Model of Opposed-Flow Flame Spread Over Charring Materials. In: Proceedings of the 29th Symposium (International) on Combustion, Eds. Chen, J.H., Colket, M.D. July 21–25, 2002, Sapporo. Combustion Institute, Pittsburgh, PA, 2002, Vol. 29, Part 1, pp. 227–236.
153. Novozhilov, V.; Koseki, H. CFD Prediction of Pool Fire Burning Rates and Flame Feedback. *Combust. Sci. Technol.* **2004**, *176*, 1283–1307.
154. Hostikka, S., McGrattan, K.B., Hamins, A. Numerical Modeling of Pool Fires Using LES and Finite Volume Method for Radiation. Fire Safety Science-Proceedings of the Seventh International Symposium, 2003, pp. 383–394.
155. Sikanen, T.; Hostikka, S. Modeling and Simulation of Liquid Pool Fires with in-Depth Radiation Absorption and Heat Transfer. *Fire Saf. J.* **2016**, *80*, 95–109. https://doi.org/10.1016/j.firesaf.2016.01.002
156. Sikanen, T.; Hostikka, S. Predicting the Heat Release Rates of Liquid Pool Fires in Mechanically Ventilated Compartments. *Fire Saf. J.* **2017**, *91*, 266–275. https://doi.org/10.1016/j.firesaf.2017.03.060
157. Wu, P.K., Orloff, L., Tewarson, A. Assessment of Material Flammability With the FSG Propagation Model and Laboratory Test Methods. 13th Joint Panel Meeting of the UJNR Panel on Fire Research and Safety, NIST Report NISTIR 6030. Gaithersburg, MD, 1996.
158. Liang, M. Evaluation Studies of the Flame Spread and Burning Rate Predictions by the Fire Dynamics Simulator. MS Thesis, University of Maryland, 2002.
159. Kwon, J.-W.; Dembsey, N. A.; Lautenberger, C. W. Evaluation of FDS V.4: Upward Flame Spread. *Fire Technol.* **2007**, *43*, 255–284.
160. Markus, E. S.; Kuznetsov, E. A.; Snegirev, A. Y. Natural Buoyant Turbulent Diffusion Flame Near a Vertical Surface. *Combust. Explos. Shock. Waves.* **2018**, *54*, 284–293. http://doi.org/10.1134/S0010508218030048
161. Li, C. C.; Wang, Y. C. Simplified Modelling of Transient Upward Flame Spread on Combustible Element. *Fire Saf. J.* **2023**, *136*, 103749. https://doi.org/10.1016/j.firesaf.2023.103749
162. Zhang, J.; Dembele, S.; Karwatzki, J.; Wen, J. X. Effect of Radiation Models on CFD Simulations of Upward Flame Spread. *Fire Saf. Sci.* **2005**, *8*, 421–432.
163. Markus, E., Snegirev, A., Kuznetsov, E., Tanklevskiy, L. Fire Growth in a High-Rack Storage. In: Proceedings of the Ninth International Seminar on Fire and Explosion Hazards, Eds. Snegirev A. et al., 2019, Vol. 2, pp. 796–807. https://doi.org/10.18720/spbpu/2/k19-70
164. Ding, Y.; Leventon, I. T.; Stoliarov, S. I. An Analysis of the Sensitivity of the Rate of Buoyancy-Driven Flame Spread on a Solid Material to Uncertainties in the Pyrolysis and Combustion Properties. Is Accurate Prediction Possible? *Polym. Degrad. Stab.* **2023**, *214*, 110405. https://doi.org/10.1016/j.polymdegradstab.2023.110405
165. Xie, W.; DesJardin, P. E. An Embedded Upward Flame Spread Model Using 2D Direct Numerical Simulations. Combust Flame. 2009, 156(2), 522–530. https://doi.org/10.1016/j. combustflame.2008.11.011
166. Yan, Z.; Holmstedt, G. CFD and Experimental Studies of Room Fire Growth on Wall Lining Materials. *Fire Saf. J.* **1996**, *27*, 201–238.
167. Theuns, E.; Merci, B.; Vierendeels, J.; Vandevelde, P. Influence of Solid Material Properties on Numerical

Large Scale Flame Spread Calculations. *Interflam*. **2004**, 1245–1256.
168. Hostikka, S.; McGrattan, K. B. Large Eddy Simulation of Wood Combustion. *Interflam*. **2001**, 755–762.
169. Moghaddam, A.Z., Moinuddin, K., Thomas, I.R., Bennets, I.D., Culton, M. Fire Behavior Studies of Combustible Wall Linings Applying Fire Dynamics Simulator. In: Proceedings of the 15th Australasian Fluid Mechanics Conference. Behnia, M. (ed.). The University of Sydney, December 13–17, Sydney, 2004. AFMC00168
170. Hietaniemi, J., Hostikka, S., Vaari, J. FDS Simulation of Fire Spread—Comparison of Model Results with Experimental Data. VTT Technica Research Centre of Finland, Espoo. VTT Working Papers 4, 2004. 45 p. + app. 6 p. www.vttresearch.com/sites/default/files/pdf/working papers/2004/W4.pdf.
171. Zhang, J.; Delichatsios, M.; Colobert, M. Assessment of Fire Dynamics Simulator for Heat Flux and Flame Heights Predictions from Fires in SBI Tests. *Fire Technol*. **2010**, *46*, 291–306. https://doi.org/10.1007/s10694-008-0072-6
172. Röwekamp, M., Klein-Heßling, W., Dreisbach, J., McGrattan, K., Miles, S., Plys, M., Riese, O. International Collaborative Fire Modeling Project (ICFMP). Summary of Benchmark. Gesellschaft für Anlagenund Reaktorsicherheit (GRS) mbH, 2008, GRS—227.
173. McGrattan, K.B., Lock, A.J., Marsh, N.D., Nyden, M.R. Cable Heat Release, Ignition, and Spread in Tray Installations During Fire (CHRISTIFIRE): Phase 1-Horizontal Trays. Nuclear Regulatory Commission, NUREG 7010 V1, 2012.
174. Jang, H.-Y.; Han, H.-S.; Hwang, C.-H.; Lee, S. K. Application of FDS to Predicting Fire Spread in a Horizontal Cable Tray. *Int. J. Fire Sci. Eng*. **2022**, *36*(4), 34–44. https://doi.org/10.7731/KIFSE.030659d0
175. Siemon, M.; Riese, O.; Forell, B.; Krönung, D.; Klein-Heßling, W. Experimental and Numerical Analysis of the Influence of Cable Tray Arrangements on the Resulting Mass Loss Rate and Fire Spreading. *Fire Mater*. **2019,** *43*, 497–513. https://doi.org/10.1002/fam.2689
176. Tyas, D.; Bagshaw, D.; Plummer, J.; Nyogeri, L. Modelling the Heat Release Rate of Prisme Experimental Cable Fires in a Confined, Ventilation Controlled, Environment Using Flash-CAT and FDS. *Fire Saf. J*. **2023**, *139*, 103828. https://doi.org/10.1016/j.firesaf.2023.103828
177. Verma, N.; Hostikka, S.; Vaari, J.; Korhonen, T. Adapted FlashCAT Methodology to Model Horizontal Cable Tray Fires Using Computational Fluid Dynamics. *Fire Saf. J*. **2023**, 103814. https://doi.org/10.1016/j. firesaf.2023.103814
178. Matala, A., Hostikka, S. Fire Modelling of Flame Retardant Electrical Cable. In: 12th International Fire and Materials Conference, 31.1.-2.2.2011 San Francisco, CA. Interscience Communications Ltd., Hampshire, 2011, pp. 583–595.
179. Matala, A.; Hostikka, S. Probabilistic Simulation of Cable Performance and Water Based Protection in Cable tunnel Fires. *Nucl. Eng. Des*. **2011**, 241(12), 5263–5274. https://doi.org/10.1016/j.nucengdes.2011.09.014
180. Matala, A., Hostikka, S. Predicting Fire Spread of a Cable Tray Using Methods of Pyrolysis Modelling. Paper presented at International Workshop on Fire Probabilistic Risk Assessment (PRA), Garching, 2014. www.oecd-nea.org/nsd/docs/2015/csni-r2015-12.pdf
181. Beji, T.; Merci, B. Numerical Simulations of a Full-Scale Cable Tray Fire Using Small-Scale Test Data. *Fire Mater*. **2019**, *43*(5), 486–496. https://doi.org/10.1002/fam.2687
182. Viitanen, A., Hostikka, S., Vaari, J. CFD simulations of fire propagation in horizontal cable trays using a pyrolysis model with stochastically determined geometry. *Fire Technol*. 2022. https://doi.org/10.1007/s10694-022-01291-6
183. Hostikka, S., Matala, A. Modelling the Fire Behaviour of Electrical Cables. 11th International Seminar on Fire Safety in Nuclear Power Plants and Installations. Post-Conference Seminar of the 20th International Conference on Structural Mechanics in Reactor Technology (SMiRT 20), August 17–19, 2009. Helsinki, Radiation and Nuclear Safety Authority STUK. https://publications.vtt.fi/julkaisut/muut/2009/SMiRT_20_Hostikka_Mat.pdf
184. Sun, Q.; Ding, Y.; Stoliarov, S. I.; Sun, J.; Fontaine, G.; Bourbigot, S. Development of a Pyrolysis Model for an Intumescent Flame Retardant System: Poly(Lactic Acid) Blended with Melamine and Ammonium Polyphosphate. *Compos. B Eng*. **2020**, 194, 108055. http://doi.org/10.1016/j.compositesb.2020.108055
185. Shi, J.; Boyer, G.; Mourzenko, V.; Thovert, J.-F. A Comprehensive Numerical Model for the Pyrolysis of Intumescent Polymers: Application to EVA-ATH Compounds. *Chem. Eng. Sci*. **2023**, *268*, 118385. https://doi.org/10.1016/j.ces.2022.118385
186. De Cachinho Cordeiro, I. M.; Chen, T. B. Y.; Yuen, A. C. Y.; Wang, C.; Chan, Q. N.; Zhang, J.; Heng Yeoh, G. Pyrolysis and Combustion Characterisation of HDPE/APP Composites via Molecular Dynamics and CFD Simulations. *J. Anal. Appl. Pyrolysis*. **2022**, *163*, 105499. https://doi.org/10.1016/j.jaap.2022.105499
186. De Cachinho Cordeiro, I.M., Chen, T.B.Y., Yuen, A.C.Y., Wang, C., Chan, Q.N., Zhang, J., Heng Yeoh, G. Pyrolysis and combustion characterisation of HDPE/APP composites via molecular dynamics and CFD simulations. Journal of Analytical and Applied Pyrolysis, 163, 2022, 105499. https://doi.org/10.1016/j.jaap.2022.105499

22 Regulations, Codes, and Standards Relevant to Fire Issues in the United States

Marcelo M. Hirschler (GBH International)

22.1 BACKGROUND: REGULATIONS, CODES, AND STANDARDS

North American fire safety requirements are based on an assortment of regulations, codes, and standards and are not, and probably will never be, uniform. It is important therefore to start by defining the terms to be used and then discussing the various organizations involved.

(a) Regulations are documents with the force of law which list general objectives and act as a framework for more detailed requirements. One example could be a national law requiring sprinkler protection of a certain type of occupancy.
(b) Codes are documents connected with regulations but which comprise more specific requirements and are valid for particular cases such as hospitals or schools. If they also offer practical solutions, they may be called codes of practice. An example associated with the national law above would be a building code detailing the layout of the sprinklers in each type of building.
(c) Standards are documents referred to either in regulations or in codes and which put forward special techniques to quantify results. Typical examples of standards are test methods and specifications. Examples associated with the national law above would be specifications describing sprinkler construction test methods assessing their effectiveness.
(d) Specifications are documents that describe requirements for a particular material or product for use by a specific purchaser. This is a document that is often part of an open solicitation or of an agreement and can be between two parties or a generic request. An example associated with the scheme above could involve requirements for the type of sprinkler pipe and/or for the discharge density.
(e) Guides are documents that explain the concepts associated with particular issues, such as test methods or properties. An associated example would be a guide explaining why sprinklers help in fires and what the effect of discharge density could be.

These definitions are generic enough that they can be applied worldwide. However, there are some distinctions in that codes and regulations can be (and in some instances are) issued together by some authorities.

In the European Union (EU), for example, the European Commission (EC) issued the Construction Products Directive (CPD) in 1989,[1] which all member states were required to comply with, and which contained "essential requirements" associated with fire safety. The CPD was replaced in 2011, by the Construction Products Regulation (CPR), most recently updated in 2022.[2] The Construction Products Regulation lays down harmonized rules for the marketing of construction products in the EU. The Regulation provides a common technical language to assess the performance of construction products. It ensures that reliable information is available to professionals, public authorities, and consumers, so they can compare the performance of products from different manufacturers in different countries. The governments of the individual member states (countries) issue regulations adopting certain requirements and incorporating them into their own local codes. These codes can then reference national or international standards. It is important to point out, however, that the national codes and standards are issued by organizations associated with the governments of the individual countries.

In the US, some states, most notably California, have a code development process that is combined with the adoption of that code into regulation.

In the United States, the Federal Government is entitled to regulate the fire safety of consumer products and of transportation vehicles but rarely does. In fact, there are very few consumer products the fire safety of which is being regulated (via the Consumer Product Safety Commission, CPSC). CPSC is an organization governed by a set of commissioners appointed by the US President, with congressional approval. Until recently, the only ones were the following: wearing apparel fabrics, children's sleepwear, carpets and rugs, mattress sets, and cellulose loose-fill insulation.

CPSC long ago (in 1994) published a notice of its intent to regulate upholstered furniture, but this did not come to fruition for many years due to political considerations (more details later), until a federal regulation was issued in late 2020. On December 27, 2020, the "COVID-19 Regulatory

DOI: 10.1201/9781003380689-22

Relief and Work From Home Safety Act," became law and included Public Law 116–260, Section 2101(c), that mandated that CPSC regulate the flammability of upholstered furniture based on a smoldering ignition test. In fact, CPSC had been required, under section 4 of the Flammable Fabrics Act (15 U.S.C. 1193), to regulate all consumer products that contain "flammable fabrics," but it had not acted.

CPSC also regulates, indirectly, the use of hazardous materials, which include flammable solids (this indirectly affects toys, as will be discussed later). On the other hand, CPSC is explicitly prohibited from regulating cigarettes.

Various separate agencies have, as a mandate, the regulation of the fire safety of transportation vehicles. All other materials and products are regulated via the adoption by a state or local government of one of various codes (issued by private organizations) or (rarely) by their issuance of their own special regulations.

Canada has a system that is intermediate between the European Union regulatory system (which is intended to be, at least ultimately, based on regulations from the European Commission) and the US system (where codes and standards are both developed primarily by private companies and can be adopted, or not, by the various state and local governments). The National Building Code of Canada can be different from provincial building codes and standards, which can be issued by private companies.

In the United States, there are two primary organizations that develop codes: ICC (International Code Council) and NFPA (National Fire Protection Association); both are private. ICC develops a full family of codes, including the following: International Building Code (IBC),[3] International Fire Code (IFC),[4] International Mechanical Code (IMC),[5] International Wildland Urban Interface Code (IWUIC),[6] International Residential Code (IRC)[7] and International Existing Building Code (IEBC),[8] which have been adopted widely throughout the country. NFPA also develops a full family of codes, with many NFPA codes being equivalent to many ICC codes. However, two NFPA codes (National Electrical Code, or NEC or NFPA 70,[9] and Life Safety Code, or NFPA 101[10]) are unique and have no ICC equivalent. NFPA is also one of the premier developers of fire safety standards, including a few key ones, which can almost be considered regulatory by their application: NFPA 90A (air conditioning standard),[11] NFPA 13 (sprinkler standard),[12] and NFPA 130 (rail standard).[13] Fire tests (standards) are developed primarily by ASTM International and by NFPA, although FM Global (formerly Factory Mutual) and Underwriters Laboratories have also developed some widely used tests. In some cases, states (like California) or cities (like Boston, Massachusetts) have issued their own standard fire tests.

The individual US states have always been entitled to issue their own regulation, as long as there is no contradiction with federal legislation, normally nonexistent. Thus, California has long had regulation for the fire performance of upholstered furniture and mattresses, based on requirements developed at the California Bureau of Home Furnishings and Thermal Insulation (CBHF) (which, in 2023, is called California Bureau of Household Goods and Services [CBHGS]) and this has, on occasion, been emulated by some other states. California also issues its own building and fire codes (albeit based on the corresponding ICC codes date the year before) and its very own wildland area code. The California wildland code is Chapter 7A of the California Building Code, which regulates construction in wildland urban interface areas and is independent of the IWUIC and not based on the ICC code at all. In 2023, there is a project to convert California Chapter 7A so that it has the structure of the IWUIC, with the intent of, eventually, getting both code more in concert with each other.

22.2 REGULATIONS

22.2.1 How Does Regulation for Fire Safety Work in the United States

As explained above, the US system works primarily on the basis of the actions of individuals who are not employees of codes and standards organizations but are codes and standards "volunteers," working within private codes and standards organizations to both develop new standards and also revise existing standards, which are then incorporated into new codes (or into revisions of existing codes). These people are called "volunteers" because their participation in the process is neither funded by the codes and standards organizations within which they are active nor by the US government (unless they are government employees). Each code development organization has a unique consensus process by which public proposals and comments are processed into new or revised documents. A successful process always consists of two stages before new documents or substantial revisions are approved. Having stated that, the development of the 2027 ICC codes will consist of three stages.

The process of adoption of codes and standards is tightly controlled by each organization so that no unfair advantage is provided to any individual interest and so that all concerns are addressed. For example, the International Code Council requires that the approval of all code changes be completed at a public meeting where the voting is restricted to public officials (although this is followed by an online vote open to all officials, even those not present at the meetings). In the case of organizations like ASTM International (formerly the American Society for Testing and Materials, ASTM) and the National Fire Protection Association (NFPA), the number of official voting members in each committee is strictly balanced, by combining representatives from producers or manufacturers and representatives from various other interests or stakeholders.

In some cases, standards developed by trade associations, instead of those developed by standards making bodies, are included within codes. However, such types of standards are frowned upon in recent years, and it is rare that new ones are being included. They are, however, often "grandfathered" [i.e., allowed to remain because they have been in the relevant code or in its predecessor for years]. Such standards must still comply with some consensus rules, and many of them tend to be eventually replaced by standards issued by the major standards development organizations.

As pointed out earlier, there are very few products the fire safety of which is actually mandated by the federal government in the US. The fire safety of the following products is covered by federal regulation.

(1) Wearing apparel is regulated by CPSC and is covered since the promulgation of the Flammable Fabrics Act [FFA], in 1953, by a 45° angle flammability test for apparel fabrics (CS 191–53), which became effective in 1954, and is still valid, as 16 CFR 1610.[14] The FFA later started covering other products. ASTM has developed a parallel standard (ASTM D1230[15]), which is substantially the same as 16 CFR 1610, but provides more guidance to labs.
(2) Ignitability of carpets and rugs is regulated by CPSC and is covered, also as a result of the FFA, since the development of the requirements in 1970 and 1971, by regulations based on the methenamine pill test, as 16 CFR 1630[16] and 16 CFR 1631.[17] The ASTM equivalent standard is ASTM D2859.[18]
(3) Flammability of children's sleepwear is regulated by CPSC and is covered, also as a result of the FFA, since a small-scale vertical open flame ignition test was developed in 1971, as 16 CFR 1615.[19] This was expanded to older children in 1974, as 16 CFR 1616.[20] The ASTM equivalent standard is ASTM D6545.[21]
(4) Smoldering ignition of mattresses is regulated by CPSC and is covered, once again via the FFA, since the smoldering test in 16 CFR 1632[22] was introduced in 1972. There is no ASTM equivalent standard.
(5) Flammable solids, products which are considered "dangerous goods" for transportation purposes and "hazardous materials" for storage and display purposes, have been regulated under the Federal Hazardous Substances Act (FHSA), by means of rules governing their amounts for storage and transport based on a candle test in 16 CFR 1500.44.[23] Hazardous materials (including flammable solids) are regulated by both CPSC and the Department of Transport. A flammable solid is often defined as follows:

    Flammable solid—A solid, other than a blasting agent, or explosive, that is liable to cause fire through friction, absorption or moisture, spontaneous chemical change, or retained heat from manufacturing or processing, or which can be ignited readily and when ignited burns so vigorously and persistently as to create a serious hazard.

(6) Open flame ignition of mattresses is regulated by CPSC and is covered, once again via the FFA, since the fairly severe open flame test for mattress sets in 16 CFR 1633[24] came into effect in 2007.
(7) Flammability of toys is regulated by CPSC and is covered since 2008 by the application of ASTM F963,[25] a standard specification for toys which includes the requirements that: (a) "materials other than textiles (excluding paper) used in toys shall not be flammable, as defined under 16 CFR 1500.3 (c) (6) (vi)" under the FHSA and (b) any textile fabrics used in toys must comply with 16 CFR 1610 (see previous section).
(8) Proposed fire safety regulations for upholstered furniture have existed since the CPSC issued various advanced notices of proposed rulemaking in the 1990s and early 2000s and a notice of proposed rulemaking in 2007.[26] None of them were implemented. As mentioned previously, a new regulation for flammability of upholstered furniture composites and components came into effect in 2021, following the 2020 federal law, based on a smoldering test (16 CFR 1640).[27]
(9) The fire safety of a variety of transportation vehicles (but not all) is covered by other federal government agencies, mostly agencies that are subsidiaries of the Department of Transportation, including National Highway Traffic Safety Administration (NHTSA, for highway vehicles), Federal Aviation Administration (FAA, for aircraft), Federal Railroad Administration (FRA, for interstate and intercity trains only), and US Coast Guard (USCG, for ships). The Federal Transit Administration (FTA) has jurisdiction over US buses and underground fixed guideway vehicles (meaning subway trains or underground trains), but it has never issued requirements and still simply recommends "guidelines" issued in the 1970s. It is notable, for example, that the FRA does not regulate urban or suburban trains (e.g., most commuter trains), whether above ground or below ground. As briefly mentioned previously, NFPA 130[13] is often used to regulate such trains, albeit on a voluntary basis.
(10) Fire safety of mine conveyor belts is covered by the Mine Safety and Health Administration (MSHA), subsidiary to the Department of Labor (responsible for the safety of miners). It has instituted some requirements, in accordance with an Act from 1969, which then became the 1977 Federal Mine Safety and Health Act. This regulation mandates the use of "flame-resistant conveyor belts." Follow-up legislation was introduced in 2007.

With regard to state and local authorities having jurisdiction, they are not permitted to issue regulation that conflicts with federal regulation, because of the issue of "preemption." In other words, it is typically not legal for a state or local authority to require that a particular product be less safe than is required by the federal regulations. On the other hand it is normally acceptable for a state or local jurisdiction to impose requirements that exceed those issued by the federal government. Therefore, since the information above shows that there are very few examples of federal fire safety regulation, state and local authorities have wide opportunities for action. One example is given by the adoption of various codes, which may include requirements that are not addressed at the federal level.

22.2.2 Federal Regulations

In order for the federal (national) government to introduce a new regulatory requirement, it must show the need for it. Once the need has been demonstrated to the satisfaction of the government agency about to issue the regulation, a notice of proposed rulemaking is published in the Federal Register. A notice of proposed rulemaking, or NPRM, must be issued by law when a regulatory agency of the United States Federal Government wishes to add, remove, or change a rule (or regulation) as part of the rulemaking process. The NPRM procedure is required and defined by the Administrative Procedure Act and not the Constitution. It was created by the US Congress to force agencies to listen to comments and concerns of people who the regulation will likely affect. Examples of agencies subject to these procedures are CPSC, FAA, FRA, NHTSA, USCG, MSHA, and EPA (Environmental Protection Agency). The NPRM is published in the Federal Register and typically gives 60 days for public comment from any interested party and an additional 30 days for reply comments. Original comments may still be filed in the reply comments "window." In fact, complex regulation often winds up being preceded by an ANPRM (Advance Notice of Proposed Rulemaking). Each notice includes the following:

- A statement of the time, place, and nature of the proposed rulemaking proceeding.
- A reference to the authority under which it is issued.
- A description of the subjects and issues involved or the substance and terms of the proposed regulation.
- A statement of the time within which written comments must be submitted.
- A statement of how, and to what extent, interested parties are allowed to participate in the proceeding.

At least in theory, the relevant authority having jurisdiction (AHJ) must start by making a cost-benefit analysis which investigates the history of losses and assesses the benefits to result from the regulation. In fact, to some extent at least, local and state authorities must go through a similar process.

This complex set of rules, which can be derailed at many stages of the process, explain why federal regulations are rare in the area of fire safety, where the impetus to introduce regulation is rarely front page news. Typically, changes in regulations follow flashy headlines, associated with a specific large fire loss. Also, federal regulations on fire safety are often based on old editions of fire test standards because of the awkwardness of updating them.

22.2.3 State Regulations

It is rare for individual states to issue specific regulations associated with fire safety, but some states have done so. In particular, the California Bureau of Home Furnishings and Thermal Insulation (CBHF, now California Bureau of Goods and Services, or BGHS, in 2023) has issued a number of Technical Bulletins addressing various furnishings fire issues (Table 22.1). No other state has been as pro-active as California. However, some activities have earlier taken place in Massachusetts, where the Board of Fire Prevention Regulations had also issued a number of fire safety requirements, most of which are no longer applicable.

In recent years US states have issued regulations governing the sale of so-called "fire-safe cigarettes." These are cigarettes with lower ignition propensity, as a result of containing "bands" of less porous material (speed bumps) that slow the combustion inside the cigarette and lead to extinction if left unattended. Cigarettes are assessed by means of the ASTM E2187 (Standard Test Method for Measuring the Ignition Strength of Cigarettes)[28] fire test and those that meet the test requirements have been labeled "fire-safe" (although some ignitions of fabrics are still possible, but to a much lower degree). The regulation to require "fire-safe cigarettes" was first adopted by New York State, in 2003, and was then followed by many other states. New York passed legislation in 2003 and implemented the law in 2004. Following that, regulations have become effective by August 26, 2011, in all US states, and the District of Columbia (Washington, DC). Thus, all US states have now passed legislation requiring that cigarettes sold there be "fire safe." The Coalition for Fire-Safe Cigarettes, which was headed by the president of NFPA, was the organization that spearheaded the effort to have this legislation adopted. The objective was, of course, to make people safer from fire by encouraging states to require that all cigarettes be designed so that they are less likely to start a fire. At the time the ASTM E2187 standard was being developed, between 700 and 900 people had been dying each year in the United States as a result of fires caused by cigarettes, according to NFPA. It should be noted that Native American nations may live in sovereign areas that have their own rules, and some of them continued (at least for a while) producing cigarettes that do not comply with the ASTM E2187 requirements.

However, the most typical (and important) way for individual states to apply fire safety regulations is by adoption of codes. Table 22.2 (based on ICC information) shows the adoption of the ICC codes (or International Codes, issued by the International Code Council).

TABLE 22.1
California Technical Bulletins and Some Other Fire Safety Regulations

Number	Last Issued	Withdrawn	Title
26	1987		Requirements for Record Keeping and Prototype Testing of Mattresses for Compliance with State and Federal Flammability Laws. Questions and Answers about the Amended Mattress Flammability Standard 16 CFR 1632 (FF 4–72)
106	1986	2007	Requirements, Test Procedures and Apparatus for Testing the Resistance of a Mattress or Mattress Pad to Combustion Which May Result from a Smoldering Cigarette
116	1980		Requirements, Test Procedure and Apparatus for Testing the Flame Retardance of Upholstered Furniture
117	2000	2013	Requirements, Test Procedures and Apparatus for Testing the Flame Retardance of Resilient Filling Materials Used in Upholstered Furniture
117–2013	2013		Requirements, Test Procedure and Apparatus for Testing the Smolder Resistance of Materials Used in Upholstered Furniture
121	1980		Flammability Test Procedure for Mattresses for Use in High Risk Occupancies
129	1992		Flammability Test Procedure for Mattresses for Use in Public Buildings
133	1991	2019	Flammability Test Procedure for Seating Furniture for Use in Public Occupancies
603	2003	2007	Requirements and Test Procedure for Resistance of a Mattress/Box Spring Set to a Large Open Flame
16 CFR 1632	2007		Standard for the Flammability (Cigarette Ignition Resistance) of Mattresses and Mattress Pads (Replaces CA TB 106)
16 CFR 1633	2007		Standard for the Flammability (Open Flame) of Mattress Sets (replaces CA TB 603)
16 CFR 1640	2021		Standard for the Flammability of Upholstered Furniture
Title 19 Ch 2 Art 4	2007		Tents, Awnings and other Fabric Enclosures—Flame Resistance and Labeling
Title 19 Ch 7 Art 1	2007		Standards of Flammability: Wearing Apparel—prohibition of Tris (2,3-dibromopropyl) phosphate
Title 19 Ch 7 Art 2/3	2007		Standards of Flammability: Hospital Fabrics, Sheets and Pillowcases, Sleepwear

TABLE 22.2
ICC Codes-Adoption by State (January 2023)

STATE	Jurisdiction	IBC	IRC	IFC	IMC	IPC	IFGC	IECC R	IECC C	IPMC	IEBC	ICCPC	IWUIC	IZC
AL	Alabama	2021	2015	2021	2021	2021	2021	2015	2015	L	2021	L	L	L
AK	Alaska	2021	L	2021	2021		2021	2018	2018		2021			
AZ	Arizona	L	L	L	L	L	L	L	L	L	L	L	L	L
AR	Arkansas	2012	2012	2012	2021	2018	2018	2009	2009	L	L			
CA	California	2021	2021	2021				State	State	L	2021		L	
CO	Colorado	L	L	L	2018	L	2018	L	L	L	L	L	L	L
CT	Connecticut	2021	2021	2021	2021	2021	2021	2021	2021	L	2021	L	L	L
DE	Delaware	L	L		2018	2018	2018	2018	2018	L	L			
DC	District of Columbia	2015	2015	2015	2015	2015	2015	2015	2015	2015	2015			
FL	Florida	2018	2018		2018	2018	2018	2018	2018	L	2018			
GA	Georgia	2018	2018	2018	2018	2018	2018	2015	2015	2018	2018		L	
HI	Hawaii	2018	2018			L		2018	2018		2018			
ID	Idaho	2018	2018	2018	2018		2018	2018	2018	L	2018		L	
IL	Illinois	L	L	L	L	L	L	2018	2018	L	L	L	L	L
IN	Indiana	2012	2018	2012	2012	2006	2012	State	State					
IA	Iowa	2015	2015	2015	2021	L	L	2012	2012	L	2015			L
KS	Kansas	L	L	L	L	L	L	L	L	L	L			
KY	Kentucky	2015	2015	2012	2015			2009	2012	L	2015			
LA	Louisiana	2015	2015	L	2015	2015		2009	2008	L	2015			
ME	Maine	2015	2015		2015			2015	2015	L	2015			
MD	Maryland	2018	2018		2018	2018	2018	2018	2018	L	2018			
MA	Massachusetts	2015	2015		2015			2018	2018		2015			
MI	Michigan	2015	2015	L	2015	2018	2015	2015	2015	L	2015		L	
MN	Minnesota	2018	2018	2018	2018		2018	2012	2018	L	2018		L	
MS	Mississippi	2018	2018	2018	2018	2018	2018	2018	2018	L	2018	L		

STATE	Jurisdiction	IBC	IRC	IFC	IMC	IPC	IFGC	IECC R	IECC C	IPMC	IEBC	ICCPC	IWUIC	IZC
MO	Missouri	L	L	L	L	L	L	L	L	L	L	L	L	L
MT	Montana	2021	2021	2012	2021		2021	2021	2021		2021		2021	
NE	Nebraska	2018	2018	L	L	L	L	2018	2018	L	2018		L	L
NV	Nevada	L	L	L	L	L	L	2021	2021	L	L	L	L	
NH	New Hampshire	2018	2018		2018	2018		2018	2018	L	2018			
NJ	New Jersey	2021	2021	2015	2021		2021	2021	State	L				
NM	New Mexico	2015	2015	2015	L	L	L	2018	2018	L	2015	L	L	
NY	New York	2018	2018	2018	2018	2018	2018	2018	2018	2018	2018			
NC	North Carolina	2015	2015	2015	2015	2015	2015	2015	2015		2015			
ND	North Dakota	2021	2021	2021	2021		2021	2021	L	L	L			
OH	Ohio	2015	2015	2015	2015	2015	2015	2018	2012	L				L
OK	Oklahoma	2018	2018	2018	2018	2018	2018	2009	2006	L	2018	L	L	L
OR	Oregon	2021	2018	2018	2021		2021	2021	State				L	
PA	Pennsylvania	2018	2018	2018	2018	2018	2018	2018	L	2018	2018	2018	2018	
RI	Rhode Island	2018	2018		2018	L	2018	2018	2018					
SC	South Carolina	2021	2021	2021	2021	2021	2009	2009	L	L	L			
SD	South Dakota	2021	L	2015	2015		L		L	2021	L	L	L	
TN	Tennessee	2012	2018	2012	2012	2012	2012	2009	2012	2012	2012	L		L
TX	Texas	2012	2012	L	L	L	L	2015	2015	L	L	L	L	L
UT	Utah	2018	2015	2018	2018	2018	2018	2018	2018	2018	2018			
VT	Vermont	2015	L			2021		2018	2018		2015			
VA	Virginia	2018	2018	2018	2018	2018	2018	2018	2018	2018	2018			
WA	Washington	2018	2018	2018	2018	L		2018	2018	L	2018	L	2018	
WV	West Virginia	2018	2018		2018	2018	2018	2015	State	2018	2018	2018		
WI	Wisconsin	2015		L	2015		2015	2009	2015		2015			
WY	Wyoming	2021	L	2021	2021	L	L	L	L	L	2021	L	L	L
TY	USTerritories													
PR	Puerto Rico	2018	2018	2018	2018	2018	2018				2018			
VI	US Virgin Islands	2018	2018	2018	2018						2018			

Source: L = One or more jurisdictions have adopted an edition of the specific code. However, the particular code is not used as a standard for all buildings.

Blank space: The specific code has not been adopted by any state or local jurisdiction in the state. States often adopt codes with local amendments.

As far as NFPA codes are concerned, as stated previously there are two key codes of relevance to fire safety that have no ICC equivalent. They are the National Electrical Code (NEC, NFPA 70), and the National Life Safety Code (NFPA 101), which have both been adopted by virtually all 50 US states (in whole or in part). However, the adoption may be partial or may involve older editions.

In the 1980s New York State incorporated into its code (New York State Uniform Fire Prevention and Building Code), into Article 15, Part 1120, a requirement for "Combustion Toxicity Testing"[29] of a variety of commercial products, based on a smoke toxicity test developed at the University of Pittsburgh.[30] The test method was never standardized, by either ASTM or NFPA because of severe concerns about its scientific validity.[31–34] The New York State regulation did not include pass/fail criteria and has since been rescinded. No further regulations have been developed (either by individual states or by the federal government) for smoke toxicity of individual commercial products, following an understanding that such regulation is not suitable for decreasing fire hazard. This will be discussed further later in this chapter.

22.2.4 Local Regulations

It is even less common in the United States for local jurisdictions to adopt specific regulations that differ from the federal (or state) requirements. A fairly unique set of stringent upholstered furniture fire safety requirements were managed by the Boston Fire Department for the city, based on recommendations from the fire department's Chemistry Department in the 1970s and 1980s. Another unusual requirement was put in place by the Department of Buildings in the city of New York, for smoke toxicity testing of some commercial products, also based on the University of Pittsburgh test. This requirement, as opposed to the one in New York State, did have pass/fail criteria.

It is not unusual for cities, especially the larger ones, such as New York (NY), Los Angeles (CA), Chicago (IL), and Phoenix (AZ), to adopt a citywide code or set of codes. Until recently some of these cities used their own unique local codes. More recently, most cities (including those just mentioned) have generally based their local building and fire codes on modifications of the applicable state codes (which, themselves, tended to be based on an ICC code, albeit not necessarily the most updated version). It is not unusual for specific events or personalities to influence local bans on certain products. One example is a requirement in Chicago that all cables installed in plenums must be enclosed in metal raceways, while most codes and standards (including the NEC) allows the use of cables that are fire tested for plenum use and also "listed."

22.2.5 Regulations of Specific Items

The key items regulated are components in vehicles and items covered under the flammable fabrics act. NFPA has developed statistics that show average annual US vehicular losses (1980–1998): 433,000 fires, with 679 civilian fire fatalities, 2,990 civilian fire injuries, and $959 millions in total fire losses.[35] In structures, the numbers, from 1980 to 1998, are 682,200 fires, with 4,440 civilian fire fatalities, 23,014 civilian fire injuries, and $6,438.3 millions in total fire losses.[36] This indicates that the number of vehicular fires is almost two thirds of the number of structure fires. More recent vehicle fire statistics (2018) showed: 212,500 fires, with 560 civilian fire fatalities, 1,500 civilian fire injuries, and $1.9 billion in total fire losses.[37] In 2018, only fires in one- and two-family homes claimed more lives than vehicle fires and vehicle fires caused 4.5 times the number of deaths as nonresidential structure fires and 1.6 times the number of apartment fire deaths. The heat release rate obtained from a burning automobile has been shown to be in the range of 1.5 to 8.0 MW, roughly the same order as heat released from a fully involved room in a home.[38]

In view of the concern that insufficient attention was being placed on the problem of transportation vehicle fire safety, the NFPA Fire Protection Research Foundation (FPRF) created, in late 2002, a Research Advisory Council on Fire and Transportation Vehicles, with Marcelo M. Hirschler as technical coordinator. The Transportation Council addressed vehicle fire safety in five modes of transportation (road, rail, water, air, and underground fixed guideway) and investigated the requirements (or guidelines) in the areas of fire safety. The Transportation Council decided that, although there are some unique characteristics depending on the type of vehicle, all modes of transportation require adequate fire safety and have some commonality. It also showed clearly that private cars (including minivans and sports utility vehicles), represent by far the mode of transportation with the highest fire losses. The issues studied were reaction-to-fire (furnishings and contents), fire resistance (structural fire protection), detection and suppression, egress, ignition sources (including wiring and arson), propulsion systems (fuels and fuel tanks), and ventilation. Although the Transportation Council focused primarily on the US fire problem, it became apparent that (a) most of the fire issues will be the same irrespective of where the vehicle operates and (b) that many vehicles cross international borders continually, so that the fire problem is "exported" from one country to another. In some areas, this is already being addressed by international organizations or by "de facto" arrangements whereby a certain organization is known to develop rules which are followed by many (or most countries). Transportation Council research found fire safety regulation only for trains traveling between cities and across state lines (Federal Railroad Administration, FRA), ships (US Coast Guard), aircraft (Federal Aviation Administration, FAA), and road vehicles (National Highway Traffic Safety Administration, NHTSA). The only mandatory fire safety regulation for road vehicles (including private cars) is a requirement that materials within 13 mm of the passenger compartment have a flame spread rate of less than 11 mm/min when tested according to FMVSS 302,[39] something that applies to cars, lorries or trucks, buses, and school buses (see also Section 2.5.4). Cars usually have to meet the same requirements throughout the world, although the standard called out has different designations (including ISO 3795 in Europe).[40] A comprehensive white paper was published in 2004 by the FPRF council.[41] NFPA also developed a guide for how to evaluate fire hazard in vehicles (NFPA 556),[42] which is now widely used to understand how to mitigate the hazard of fires in vehicles, but no new regulation has been issued.

In 1993 the Federal Transit Administration (responsible for fire safety of transit buses and vans) issued guidelines for the use of materials in such public transportation vehicles.[43] No requirements have ever been issued, but the guidelines are similar to those later used for trains.

22.2.5.1 Aircraft

The FAA has developed a series of requirements for many of the materials and products contained in airplanes. Its regulations, contained in FA 25.853, cover four categories of aircraft: transport category airplanes (Federal Aviation Regulation [FAR] Part 25), small airplanes (FAR Part 23), rotorcraft of all categories (FAR Parts 27 and 29), and engines and propellers (FAR Part 33 and 35). All the relevant fire tests are contained in the "Aircraft Materials Fire Test Handbook,"[44] for which periodic updates are issued. The FAA has also published a notice[45] that advises the public that the FAA considers the material flammability tests described in the latest version of that document to be the preferred acceptable test methods for showing compliance with the relevant regulations. The required test methods (non propulsion) are described in Chapters 1 through 10 and Chapter 15, and they are acceptable methods for showing compliance with, or providing an equivalent level of safety to, the required regulations as outlined in the chapter. Although these test methods cannot—and do not—supersede any method specified by and described in the regulations, they represent an acceptable means of compliance with the relevant regulation and, in some cases, a preferred option over the specified method. However, the FAA will consider other alternative methods that demonstrate an equivalent level of safety on a case-by-case basis along with the necessary supportive data. Chapters 11 through 14 contain required test methods (propulsion), and they remain as the name indicates. The test methods described in Chapters 18 through 22 are non-required test methods; they are included in the handbook for use as test methods applications where there currently are no requirements, and no process is required for their modification.

The principal fire tests in the Aircraft Materials Fire Test Handbook are the following:

1. Chapter 5: Heat release rate test. The test is based on the Ohio State University rate of heat release calorimeter,[46] which has also been standardized as ASTM E906.[47] The test exposes a vertical test specimen ca. 150 mm × 150 mm (6 in. × 6 in.) to an incident radiant heat flux of 35 kW/m^2, from four horizontal glow bars, for five minutes, in the presence of an open flame pilot burner. The test applies to the vast majority of nonmetallic materials contained inside the aircraft, typically vertical panels (which are often laminated and have multiple layers), ceiling and wall linings, luggage compartment racks, carts, and so on. It does not cover foams and fabrics for seats and draperies, windows floor coverings (carpets), electrical wiring, insulation, or firewalls. The acceptance criteria are an average maximum heat release rate during the five-minute test that does not exceed 65 kW/m^2 and an average total heat released during the first two minutes of test that does not exceed 65 kW min/m^2.
2. Chapter 6: Smoke test. The test is based on the NBS (National Bureau of Standards, now NIST, National Institute of Standards and Technology) smoke density chamber, which has also been standardized as ASTM E662.[48] The test exposes a vertical test specimen ca. 75 mm × 75 mm (3 in. × 3 in.) to an incident radiant heat flux of 25 kW/m^2, from a radiant heat burner, for four minutes, in the presence of an open flame pilot burner. The test applies to the same materials as the heat release rate test. The acceptance criterion is an average maximum specific optical density of smoke that does not exceed 200 (no units).
3. Chapter 7: Oil burner test for seat cushions. The test exposes a seat cushion to an oil burner (kerosene or fuel oil) at a rate of 126 mL/min (2 gallons/hour) gas flow for two minutes. The acceptance criteria are a burn length not exceeding 432 mm (17 inches), and a weight loss not exceeding 10%.
4. Chapters 1 through 3: Bunsen burner tests for miscellaneous materials. The tests are all conducted in cabinets, and they are a vertical test (Chapter 1: 12 or 60 seconds), a horizontal test (Chapter 2: 30 seconds), and a 45 degree angle test (Chapter 3: 15 seconds). These tests apply primarily to the materials and products in smaller non-commercial aircraft.
5. Chapter 4: 60° angle Bunsen burner test for electric wire insulation. The test is also conducted in a cabinet, and the flame is applied for 30 seconds. Manufacturers of commercial aircraft tend to require higher fire performance than this test provides, but many of the electrical services incorporated into the aircraft by the airlines use this test.
6. Chapter 18: Horizontal Bunsen burner test for blankets. This is similar to the test in Chapter 2 but is applied for 12 seconds.
7. Chapter 23: Ignitability test for thermal/acoustic insulation. This test is based on the flooring radiant panel test, for carpets and other floor coverings, which has also been standardized as ASTM E648.[49] The test exposes a bag of insulation material to a radiant panel at an angle of 30°, so that the incident radiant heat gradually decreases the further away it is measured from the exposure point. The test assesses whether the bag of insulation ignites and spreads flame for ca. 50 mm (2 in.).
8. Chapter 24: Burn through test for thermal/acoustic insulation blankets used in the fuselage. This test uses the same type of oil burner as the one for seat cushions (Chapter 7), but the fuel flow rate is 378 mL/min (6 gallons/hour) gas flow for four minutes. The acceptance criteria require that there be no flame penetration through the insulation blanket and that the heat released by the insulation not exceed 22.7 kW/m^2.

The requirements were obtained as a result of full-scale aircraft tests conducted by the FAA and by airframe manufacturers, which showed the key importance of heat release criteria for proper fire safety. The criteria were based on the concept of ensuring that, in a survivable post-crash scenario, aircraft evacuation can be completed in five minutes. Moreover, fires should not happen or self-extinguish in hidden areas. The FAA was asked by the aircraft industry to add the smoke obscuration test because that had been a test required by manufacturers in the past. The FAA does not have any regulatory smoke toxicity or material composition tests (because the research conducted showed that they were not needed), but some aircraft manufacturers and airlines do have such requirements.

22.2.5.2 Ships

The US Coast Guard regulates all ships that moor in US ports (both on the exterior coasts and on rivers or lakes). The US is a signatory to the International Convention for the Safety of Life at Sea (SOLAS),[50] and the USCG will enforce the corresponding regulations, which are issued by the International Maritime Organization (IMO).[51]

All ships that engage in international trade and fly the flag of a country that has signed the SOLAS convention must comply with the regulations of IMO, detailed in the SOLAS book, periodically amended by "Resolutions" of the IMO committees, and ratification by the signatory states. Details of the fire issues are given in the IMO Fire Test Procedures Code (FTP),[52] which is also reissued regularly. Some special vessels are regulated separately: high speed craft that is never too far from shore is regulated by the IMO High Speed Craft Code (HSC).[53] The High Speed Craft Code applies to small vessels, within certain parameters, that operate at a minimum specified high speed and carry a maximum number of passengers and make no overnight trips. However, in the US many ships never leave inland waters, or simply hug coastal areas and often choose not to be covered by SOLAS. The USCG regulates all ships that use US ports and enforces SOLAS or HSC requirements for those ships that need to meet those rules. All ships that sail in US waters must comply with the requirements specified by the US Coast Guard, laid out in US Federal Government—Coast Guard: Title 46, Shipping, Code of Federal Regulations, Parts 1–199,[54] and in a NVIC (US Coast Guard Guide to Structural Fire Protection).[55] The Coast Guard is also the authority having jurisdiction over ships engaging in international trade and sailing into US waters or US ports; such ships must comply with IMO regulations and need not also comply with separate Coast Guard requirements. With the instructions from the US federal government that standards should, whenever possible, be delegated to private organizations, the Coast Guard and NFPA agreed to develop a consensus code, NFPA 301, Code for Safety to Life from Fire on Merchant Vessels,[56] for ships that do not sail in international waters. NFPA 301 references both SOLAS and HSC for the corresponding types of ships. The USCG has a policy of permitting ships to comply with NFPA 301 instead of its requirements, but this code is rarely used, because it contains very onerous sprinkler requirements which exceed those of the USCG.

The IMO regulations, via the FTP code, cover reaction-to-fire requirements for cables (electric wires and cables), carpeting (floor coverings), deck coverings, fabrics, interior finish, mattresses and bedding, piping, and upholstered furniture. They also cover fire resistance requirements for bulkheads, dampers, doors, windows, and other divisions. The fire resistance requirements are based on testing to the time temperature curve in ISO 834[57] instead of the one used in the US, from ASTM E119.[58] Although the differences are of minor relevance technically, they do mean that testing must be accomplished with the correct standard. The fire safety philosophy used by SOLAS for materials and products in ships is the limitation on the amounts of combustible materials, while trying to build the ships as much as possible out of noncombustible materials (often stated as "steel or equivalent"). The HSC code has deviated considerably from this concept by developing heat release and smoke release criteria for "fire restrictive materials," which are combustible but have excellent fire performance, as assessed by a very severe room-corner heat release fire test, ISO 9705-1.[59] The rationale for this change in philosophy has been the fact that composites and other materials with excellent fire performance are much lighter than steel and can permit the ships to travel much faster.

The USCG regulations include reaction-to-fire requirements for the same products as the FTP code, but the tests required tend to be more realistic or severe. In particular upholstered furniture and mattresses need to meet open flame tests and not just smoldering tests, for example. Smaller boats are usually intended to comply with NFPA 302, Fire Protection Standard for Pleasure and Commercial Motor Craft,[60] but are not actually regulated.

Finally, the US Navy also has its own set of requirements for military ships and for submarines. It has, for some time now, been pursuing a similar type of approach to the HSC Code approach.[61] In this case there is an additional advantage in moving away from steel products: the alternative materials are much less detectable by enemy radar.

NFPA 301 has much more restrictive reaction-to-fire test requirements for furnishings and contents than SOLAS. However, its latest edition (dated 2023) permits the omission of many, if not most, of the reaction-to-fire requirements if the vessel is sprinklered or has a water mist system installed. This means that the requirements can be omitted for passenger ships, as all Group I passenger vessels (i.e., those with either over 3,000 day passengers or over 300 overnight passengers) and Group II passenger vessels (i.e., those with either over 1,000 and under 3,000 day passengers or over 150 and under 300 overnight passengers), and all passenger vessels with overnight accommodations for passengers must have a sprinkler or water mist system, to protect "all accommodation, service, and storage spaces." This makes reaction-to-fire requirements for passenger ships mostly optional.

22.2.5.3 Trains and Underground Rail Vehicles

As a result of the 1973 Urban Mass Transportation Administration (UMTA) program to improve fire safety of public transport, "Guidelines for Flammability and Smoke Emission Specifications" for materials used in transportation vehicles were generated (they are virtually the same as those reissued in 1993[43]). Although these guidelines have developed somewhat over the years, they are still based on the fire testing technology available in the 1970s, but the FRA has actually made them mandatory regulations for trains connecting cities (intercity trains) or for trains crossing state lines (interstate trains),[62,63] while the Federal Transit Administration (FTA) still uses them as guidelines for urban trains (including underground fixed guideway vehicles) and buses. Table 22.3 shows the FRA rulemaking to that effect, as issued in 2002 (which has not been updated). The table is the basis of the regulations for reaction-to-fire of materials and products. The two principal fire tests used are a radiant panel test for flame spread (ASTM E162)[64] and the static smoke obscuration test referenced earlier (ASTM E662[48]). The table contains detailed instructions and pass/fail criteria for all types of materials and products in trains (except for wire and cable products, which were eliminated between the 1999 and 2002 editions). These rules have ensured that the seating materials used no longer lead to dramatic fires, but some of the

TABLE 22.3
FRA Requirements for Passenger Rail Materials (2002)

		Flammability		Smoke Emission	
Category	Function of Material	Test Procedure	Performance Criteria	Test Procedure	Performance Criteria
Cushions, Mattresses	All (1, 2, 3, 4, 5, 6, 7, 8)	ASTM D3675	Is ≤ 25	ASTM E662	D_s (1.5) ≤ 100 D_s (4.0) ≤ 175
Fabrics	Seat upholstery, mattress ticking and covers, curtains, draperies, wall coverings, and window shades (1, 2, 3, 6, 7, 8)	14 CFR 25 Appendix F, Part I (vertical test)	Flame time ≤ 10 s; Burn length ≤ 6 in.	ASTM E662	D_s (4.0) ≤ 200
Other Vehicle Components (9, 10, 11, 12)	Seat and mattress frames, wall and ceiling panels, seat and toilet shrouds, tray and other tables, partitions, shelves, opaque windscreens, end caps, roof housings, and component boxes and covers (1, 2)	ASTM E162	Is ≤ 35	ASTM E662	D_s (1.5) ≤ 100 D_s (4.0) ≤ 200
	Flexible cellular foams used in armrests and seat padding (1, 2, 4, 6)	ASTM D3675	Is ≤ 25	ASTM E662	D_s (1.5) ≤ 100 D_s (4.0) ≤ 175
	Thermal and acoustic insulation (1, 2)	ASTM E162	Is ≤ 25	ASTM E662	D_s (4.0) ≤ 100
	HVAC ducting (1, 2)	ASTM E162	Is ≤ 25	ASTM E662	D_s (4.0) ≤ 100
	Floor covering (12, 13)	ASTM E648	CRF ≥ 5 kW/m^2	ASTM E662	D_s (1.5) ≤ 100 D_s (4.0) ≤ 200
	Light diffusers, windows, and transparent plastic windscreens (2, 14)	ASTM E162	Is ≤ 100	ASTM E662	D_s (1.5) ≤ 100 D_s (4.0) ≤ 200
Elastomers (1, 10, 11)	Window gaskets, door nosings, inter-car diaphragms, roof mats, and seat springs	ASTM C1166	Average flame propagation ≤ 4 in.	ASTM E662	D_s (1.5) ≤ 100 D_s (4.0) ≤ 200
Structural Components (15)	Flooring (16), other (17)	ASTM E119	Pass		

Notes to Table 22.3

1. Materials tested for surface flammability shall not exhibit any flaming running or dripping.

2. The ASTM E662 maximum test limits for smoke emission (specific optical density) shall be measured in either the flaming or non-flaming mode, utilizing the mode which generates the most smoke.

3. Testing of a complete seat assembly (including cushions, fabric layers, upholstery) according to ASTM E1537 using the pass/fail criteria of Cal TB 133, and testing of a complete mattress assembly (including foam and ticking) according to ASTM E1590 using the pass/fail criteria of Cal TB 129 shall be permitted in lieu of the test methods prescribed herein, provided the assembly component units remain unchanged or new (replacement) assembly components possess equivalent fire performance properties to the original components tested. A fire hazard analysis must also be conducted that considers the operating environment within which the seat or mattress assembly will be used in relation to the risk of vandalism, puncture, cutting, or other acts which may expose the individual components of the assemblies to an ignition source. Notes 5, 6, 7, and 8 apply.

4. Testing is performed without upholstery.

5. The surface flammability and smoke emission characteristics shall be demonstrated to be permanent after dynamic testing according to ASTM D3574, Test I 2 (Dynamic Fatigue Test by the Roller Shear at Constant Force) or Test I 3 (Dynamic Fatigue Test by Constant Force Pounding) both using Procedure B, except that the test samples shall be a minimum of 6 inches (154 mm) by 18 inches (457 mm) by the thickness of the material in its end use configuration, or multiples thereof. If Test I 3 is used, the size of the indentor described in paragraph 96.2 shall be modified to accommodate the specified test specimen.

6. The surface flammability and smoke emission characteristics shall be demonstrated to be permanent by washing, if appropriate, according to FED-STD–191A Textile Test Method 5830.

7. The surface flammability and smoke emission characteristics shall be demonstrated to be permanent by dry-cleaning, if appropriate, according to ASTM D2724.

8. Materials that cannot be washed or dry-cleaned shall be so labeled and shall meet the applicable performance criteria after being cleaned as recommended by the manufacturer.

9. Signage is not required to meet any flammability or smoke emission performance criteria specified in this Appendix.

10. Materials used to fabricate miscellaneous, discontinuous small parts (such as knobs, rollers, fasteners, clips, grommets, and small electrical parts) that will not contribute materially to fire growth in end use configuration are exempt from flammability and smoke emission performance requirements, provided that the surface area of any individual small part is less than 16 square inches (100 cm^2) in end use configuration and an appropriate fire hazard analysis is conducted which addresses the location and quantity of the materials used, and the vulnerability of the materials to ignition and contribution to flame spread.

11. If the surface area of any individual small part is less than 16 square inches (100 cm^2) in end use configuration, materials used to fabricate such a part may be tested in accordance with ASTM E1354 as an alternative to both (a) the ASTM E162 flammability test procedure, or the appropriate flammability test procedure otherwise specified in the table, and (b) the ASTM E662 smoke generation test procedure. Testing shall be at 50 kW/m^2 applied heat flux with a retainer frame. Materials tested in accordance with ASTM E1354 shall meet the following performance criteria: the average heat release rate over 180 seconds shall be less than or equal to 100 kW/m^2, and the average specific extinction area shall be less than or equal to 500 m^2/kg over the same 180-second period.

12. Carpeting used as a wall or ceiling covering shall be tested according to ASTM E162 and ASTM E662 and meet the respective criteria of Is less than or equal to 35 and D_s (1.5) less than or equal to 100 and D_s (4.0) less than or equal to 200. Notes 1 and 2 apply.

13. Floor covering shall be tested with padding in accordance with ASTM E648, if the padding is used in the actual installation.

14. For double window glazing, only the interior glazing is required to meet the requirements specified herein. (The exterior glazing is not required to meet these requirements.)

15. Penetrations (ducts, etc.) shall be designed against acting as passageways for fire and smoke and representative penetrations shall be included as part of test assemblies.

16. A structural flooring assembly separating the interior of a vehicle from its undercarriage shall meet the performance criteria during a nominal test period as determined by the railroad. The nominal test period must be twice the maximum expected time period under normal circumstances for a vehicle to stop completely and safely from its maximum operating speed, plus the time necessary to evacuate all the vehicle's occupants to a safe area. The nominal test period must not be less than 15 minutes. Only one specimen need be tested. A proportional reduction may be made in the dimensions of the specimen provided it serves to truly test the ability of the structural flooring assembly to perform as a barrier against under-vehicle fires. The fire resistance period required shall be consistent with the safe evacuation of a full load of passengers from the vehicle under worst-case conditions.

17. Portions of the vehicle body which separate major ignition sources, energy sources, or sources of fuel-load from vehicle interiors shall have sufficient fire endurance as determined by a fire hazard analysis acceptable to the railroad which addresses the location and quantity of the materials used, as well as vulnerability of the materials to ignition, flame spread, and smoke generation. These portions include equipment-carrying portions of a vehicle's roof and the interior structure separating the levels of a bi-level car, but do not include a flooring assembly subject to Note 16. A railroad is not required to use the ASTM E119 test method.

---

tests are less than fully adequate, so that their results may have random ability to predict real fire behavior.[65] FRA has realized that heat release rate is particularly critical as it is the single property that indicates the intensity of the fire, and controls whether a fire in a particular product will become large enough to ignite the next item. In fact, it has been shown that heat release rate is much more important than smoke toxicity or ignitability in determining the time remaining for occupants to escape. In view of this, FRA now permits approvals to be based on the use of alternative approaches to the prescriptive table, using heat release for hazard assessment. Both ASTM and NFPA have developed standard documents that address the way to do this: ASTM E2061[66] and NFPA 130; a review of these standards has been published.[67] Both standards consider critical the reaction-to-fire properties of wire and cable products, and the use of these documents is prevalent whenever a specifier (usually a local transportation authority) wants to exceed FRA regulations or is not bound by them (for example, outside of the US). ASTM E2061 is a guide on how to assess fire hazard in rail transportation vehicles, which gives detailed instructions and can be followed in two ways: by complying with a version of "the table" (which gets updated every time there is a new edition of NFPA 130) or by analyzing the fire hazard of the rail car with modern fire modeling technology. NFPA 130 is a standard with broader applicability because it includes the entire rail system but also has both a version of "the table" and a section on a performance-based approach. Table 22.4 contains the rail vehicle materials requirements from NFPA 130–2023 (with the associated notes in Table 22.5). The NFPA 130 standard also requires that all cables meet flame spread and smoke obscuration criteria, using a vertical cable tray test (UL 2556, vertical tray flame test method 2, FT4 and limited smoke, ST1[68]). That test is technically equivalent to the Canadian CSA FT4[69] test. In both cases, the requirements are for a char height less than 1.5 m (5 ft) when measured from the lower edge of the burner face, a total smoke release rate over 20 minutes that does not exceed 150 m^2, and a peak smoke release rate that does not exceed 0.40 m^2/s. As an alternative to that vertical tray flame test method, wires and cables are permitted to comply with the NEC requirements based on NFPA 262,[70] which are that the flame travel distance not exceed 1.5 m, the maximum peak optical density of smoke not exceed 0.50, and the maximum average optical density of smoke not exceed 0.15.

22.2.5.4 Motor Vehicles

The statistics of fire losses in vehicles show that over two thirds of all fires in vehicles occur in road vehicles, and over 95% of those occur in automobiles (a category that includes, of course, the popular private minivans and sports utility vehicles).[35,37] Moreover, the very mild fire safety requirements for cars, namely FMVSS 302,[39] were set by NHTSA in 1969 to address the problem that people discarded lit cigarettes in cars. This means that requirements have not changed in over 50 years, when there were many fewer than the almost 300 million cars in the US in 2020, and when cars had much more steel and much less plastic. The increased use of plastics has made cars much more fuel-efficient (lower weight) and allowed more flexibility in design and esthetics, but at the expense of increasing

TABLE 22.4
NFPA 130–2023 Requirements for Passenger Vehicle Rail Materials

Category	Function of Material	Flammability: Test Procedure	Flammability: Performance Criteria	Smoke Emission: Test Procedure	Smoke Emission: Performance Criteria
Cushioning	All individual flexible cushioning materials used in seat cushions, mattresses, mattress pads, armrests, crash pads, and grab rail padding (1, 2, 3, 4, 5)	ASTM D3675	Is ≤ 25	ASTM E662	D_s (1.5) ≤ 100 D_s (4.0) ≤ 175
Fabrics	Seat upholstery, mattress ticking and covers, curtains, draperies, window shades, and woven seat cushion suspensions (1, 2, 3, 6, 7, 8)	14 CFR 25 Appendix F, Part I (vertical test)	Flame time ≤ 10 s; Burn length ≤ 6 in.	ASTM E662	D_s (4.0) ≤ 200
Other Vehicle Components	Seat and mattress frames, wall and ceiling lining and panels, seat and toilet shrouds, toilet seats, trays and other tables, partitions, shelves, opaque windscreens, combustible signage, end caps, roof housings, articulation bellows, exterior shells, nonmetallic skirts, battery case material, and component boxes and covers (1, 2, 9, 10, 11)	ASTM E162	Is ≤ 35	ASTM E662	D_s (1.5) ≤ 100 D_s (4.0) ≤ 200
	Thermal and acoustical insulation (1, 2)	ASTM E162	Is ≤ 25	ASTM E662	D_s (4.0) ≤ 100
	HVAC ducting (1, 2)	ASTM E162	Is ≤ 25	ASTM E662	D_s (4.0) ≤ 100
	Floor covering (2, 11, 12)	ASTM E648	CRF ≥ 5 kW/m²	ASTM E662	D_s (1.5) ≤ 100 D_s (4.0) ≤ 200
	Light diffusers, windows, and transparent plastic windscreens (1, 2, 9)	ASTM E162	Is ≤ 100	ASTM E662	D_s (1.5) ≤ 100 D_s (4.0) ≤ 200
	Adhesives and sealants (1, 2, 14)	ASTM E162	Is ≤ 35	ASTM E662	D_s (1.5) ≤ 100 D_s (4.0) ≤ 200
Elastomers (1, 2, 9, 10)	Window gaskets, door nosings, intercar diaphragms, seat cushion suspension diaphragms, and roof mats	ASTM C1166	Flame propagation ≤ 100 mm (4 in.)	ASTM E662	D_s (1.5) ≤ 100 D_s (4.0) ≤ 200
Structural Components (13)	Flooring (17), other (15)	ASTM E119	Pass		
Wire and cable	All			See # 16	See # 16

TABLE 22-5
NFPA 130-2023 Requirements for Passenger Vehicle Rail Materials Notes to Table 22.4

1. Materials tested for surface flammability shall not exhibit any flaming running or flaming dripping.
2. The ASTM E662 maximum test limits for smoke emission (specific optical density) shall be based on both the flaming and the non-flaming modes.
3. Testing of a complete seat assembly (including cushions, fabric layers, and upholstery) according to ASTM E1537 using the pass/fail criteria of California Technical Bulletin 133 and testing of a complete mattress assembly (including foam and ticking) according to ASTM E1590 using the pass/fail criteria of California Technical Bulletin 129 shall be permitted in lieu of the test methods prescribed herein, provided the assembly component units remain unchanged or new (replacement) assembly components possess fire performance properties equivalent to those of the original components tested.
   3.1 A fire hazard analysis shall also be conducted that considers the operating environment within which the seat or mattress assembly will be used in relation to the risk of vandalism, puncture, cutting, introduction of additional combustibles, or other acts that potentially expose the individual components of the assemblies to an ignition source.
   3.2 The requirements of items 4 through 8 shall be met.
4. Testing shall be performed without upholstery.
5. The surface flammability and smoke emission characteristics shall be demonstrated to be permanent after dynamic testing according to ASTM D 3574, Test I_2 or Test I_3, both using Procedure B, except that the test samples shall be a minimum of 150 mm (6 in.) × 450 mm (18 in.) × the thickness used in end-use configuration, or multiples thereof. If Test I_3 is used, the size of the indentor described in Section 96.2 of ASTM D 3574 shall be modified to accommodate the specified test specimen.
6. The surface flammability and smoke emission characteristics shall be demonstrated to be permanent by washing, if appropriate, in accordance with the manufacturer's recommended procedure. If a washing procedure is not provided by the manufacturer, the fabric shall be washed in accordance with ASTM E2061, Annex A1.

(Continued)

TABLE 22-5 Continued

7. The surface flammability and smoke emission characteristics shall be demonstrated to be permanent by dry cleaning, if appropriate, according to ASTM D2724.
8. Materials that cannot be washed or drycleaned shall be so labeled and shall meet the applicable performance criteria after being cleaned as recommended by the manufacturer.
9. Combustible operational and safety signage shall not be required to meet flame spread or smoke emission requirements if the combustible mass of a single sign does not exceed 500 g (1.1 lb) and the aggregate area of combustible signage does not exceed 1 ft^2 per foot of car length.
10. Materials used to fabricate miscellaneous, discontinuous small parts (such as knobs, rollers, fasteners, clips, grommets, and small electrical parts) where the surface area of any individual small part is less than 100 cm^2 (16 $in.^2$) in end use configuration and that will not contribute materially to fire growth in end use configuration shall comply with either 10.1 or 10.2.
    - 10.1 The materials shall be exempt from flammability and smoke emission performance requirements provided that an appropriate fire hazard analysis is conducted that addresses the location and quantity of the materials used and the vulnerability of the materials to ignition and contribution to flame spread.
    - 10.2 The materials shall be tested in accordance with ASTM E1354 at an initial test heat flux of 50 kW/m^2 (4.4 Btu/sec·ft^2) in the horizontal orientation with a retainer frame and shall meet the performance criteria of a 180-second average heat release rate not exceeding 100 kW/m^2 (8.8 Btu/sec·ft^2) and a test average smoke extinction area not exceeding 500 m^2/kg (2441.2 ft^2/lb).
11. Carpeting used as a wall or ceiling covering shall be tested according to ASTM E162 and ASTM E662 and shall meet the respective criteria of Is ≤ 35, Ds (1.5) ≤ 100, and Ds (4.0) ≤ 200. (See Items 1 and 2)
12. The floor covering shall be tested as installed with padding, adhesives, or a combination of the two, as applicable, in accordance with NFPA 253 or ASTM E648.
13. Penetrations (ducts, etc.) shall be designed against acting as passageways for fire and smoke, and representative penetrations of each type shall be included as part of test assemblies.
14. Adhesives and sealants shall be tested in accordance with both ASTM E162 and ASTM E662 by applying the adhesive or sealant to the smooth face of 6.4 mm (1⁄4 in.) thick inorganic, reinforced cement board, nominal density 1762 kg/m^3 ± 160 kg/m^3 (110 lb/ft^3 ± 10 lb/ft^3), using recommended (or practical) application techniques and coverage rates.
15. Portions of the vehicle body that separate the major ignition source, energy sources, or sources of fuel load from vehicle interiors shall have fire resistance as determined by a fire hazard analysis acceptable to the authority having jurisdiction that addresses the location and quantity of the materials used, as well as vulnerability of the materials to ignition, flame spread, and smoke generation. These portions shall include equipment-carrying portions of a vehicle's roof and the interior structure separating the levels of a bi-level car but do not include a flooring assembly subject to Section 8.5. In those cases, the use of the ASTM E119 or UL 263 test procedure shall not be required.
16. Wires and cables. wires and cables must be "resistant to the spread of fire" and have reduced smoke emissions by complying with one of the following:
    - (1) They must exhibit a height less than 1.5 m (5 ft) when measured from the lower edge of the burner face, a total smoke release rate over 20 minutes that does not exceed 150 m^2 (1615 ft^2), and a peak smoke release rate that does not exceed 0.40 m^2/sec (4.3 ft^2/sec) when tested to the FT4/IEEE 1202 exposure of UL 1685.
    - (2) Wires and cables listed as having adequate fire-resistant and low-smoke-producing characteristics, by having a flame travel distance that does not exceed 1.5 m (5 ft) and generating a maximum peak optical density of smoke of 0.50 and a maximum average optical density of smoke of 0.15 when tested in accordance with NFPA 262 are permitted as an alternative to those in item 1.

    Wires and cables used for heat, smoke, or other detection systems must also comply with one of the following:
    - (1) Be capable of having 15-minute circuit integrity when tested in accordance with IEC 60331-11.
    - (2) Have fire alarm circuit integrity cable in accordance with NFPA 70.
17. Vehicle floor assembly. Vehicle floor assemblies must be tested in accordance with ASTM E119 using test assemblies that are representative of the vehicle construction and that are tested in a configuration to demonstrate that a fire will not extend into the passenger and crew areas during the fire exposure duration. The exposed portion of the floor assembly must be at least 3.7 m (12 ft) long by the normal width of the vehicle floor and be tested with a representative loading consistent with the vehicle design. The minimum fire exposure duration must be the greatest of the following: (1) twice the maximum expected time period under normal circumstances for a vehicle to stop completely and safely from its maximum operating speed, plus the time necessary to evacuate a full load of passengers from the vehicle under approved conditions or (2) 15 minutes for automated guideway transit vehicles and low floor vehicles and 30 minutes for all other passenger-carrying vehicles.

the potential for fire losses. Significant amount of recent work has highlighted some critical fire safety deficiencies of the materials used in this area.[71–74] In particular, such work showed problems in terms of the reaction-to-fire properties of automobile materials (from the passenger and engine compartments) in private cars (many of which are highly combustible and have poor fire performance) and in terms of the fire resistance properties of the so-called barrier materials used. For example, real-scale tests on actual cars showed that small fires in the engine compartment (not even involving liquid fuels) would penetrate into the passenger compartment through the "engine cover" in less than 2 min and that the "engine cover" would actually burn at low heat fluxes, and not offer any type of real barrier.[72] Figure 22.1 shows the temperature reached in the front and rear seats, headliner, duct, vent and carpet following ignition with a small ignition source in the engine compartment: temperatures exceeding 700°C were reached quickly

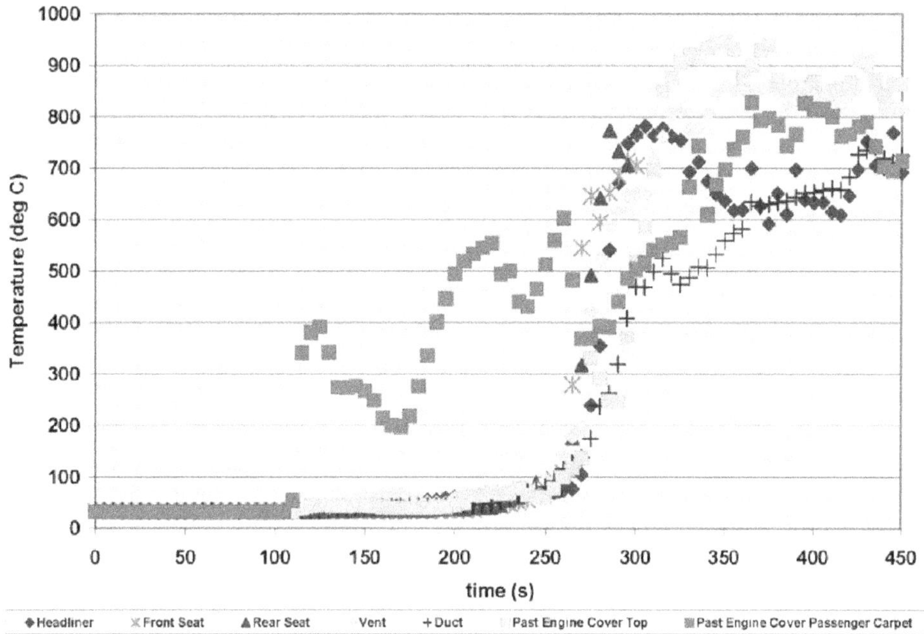

FIGURE 22.1 Temperatures in fire tests of real vans. [67]

in all cases.[72] It was also shown that materials used for the headliner, the foam padding and, especially, the duct were of particularly poor fire performance, with high heat release rates. Furthermore, the NHTSA web site contains extensive information on research conducted on fire safety of private cars as a result of the agreement between the US Federal Government and General Motors.[75] NHTSA has recognized that there is a serious fire safety problem and has undertaken various research programs to investigate alternative tests for material regulation, probably the cone calorimeter (ASTM E1354),[76] a source of heat release information. Two examples of such research have been published by Jason Huczek and others, at the Southwest Research Institute.[77,78] No regulatory action ensued, and none appears to be imminent.

One NFPA technical committee, titled "Hazard and Risk of Contents and Furnishings," is responsible for providing guidance to other NFPA technical committees on areas associated with fire hazard and fire risk of contents and furnishings. In 1996 the committee developed NFPA 555, "Guide on Methods for Evaluating Potential for Room Flashover." Subsequently, NFPA developed a document for vehicles (especially highway vehicles) intended to be parallel to NFPA 555, to focus on ways to lower fire hazard for occupants of such vehicles. The document addressed input from the automotive industry and other public sources, and it was published as NFPA 556[42] ("Guide on Methods for Evaluating Fire Hazard to Occupants of Passenger Road Vehicles"). It consists of various chapters, including administrative ones. The key concept is that the guide includes a discussion of the types of vehicles, a general description of passenger road vehicle fires and background information, an approach to evaluating vehicle fire hazard, how to develop objectives and design criteria for vehicle designs and an investigation of individual typical fire scenarios, evaluation methods and tools (including test methods and fire models), and further guidance. It also includes recommended types of mitigation strategies, such as decreasing ignition propensity of materials, decreasing heat release of materials, and designing improvements in vehicle engineering, while ensuring that the changes do not affect the vehicle safety or performance in terms of issues other than fire. In terms of fire properties, clearly it is principally important to reduce ignitability, heat release, and the potential for flame spread, which is why the test most widely considered for assessing material and product fire properties would be the cone calorimeter, ASTM E1354. The document recommends the use of tests based on heat release, at various levels: full-scale, fuel package scale, and individual material scale, so as to obtain data that are quantitative enough for fire hazard assessment. Guidance stresses the performance-based approach and contrasts this with the continued use of FMVSS 302 as the sole fire safety tool, since the exclusive use of FMVSS 302 is inconsistent with significant decreases in fire losses associated with road vehicles. Particularly with the growing use of combustible materials in road vehicles, a mild ignition test such as FMVSS 302 is clearly insufficiently severe to ensure that materials meeting it would allow enough time for escape to passengers and drivers in the case of a fire. It is important to recognize that, although full-scale tests are the most accurate way of understanding both deficiencies in road vehicle fire safety and to develop mitigation strategies, their high cost means that other strategies need to be used also. For example, testing of sections, such as individual road vehicle compartments or fuel packages in a furniture calorimeter (i.e., under a hood, above a load cell), is a way to develop an understanding of the interactions between the materials and products contained in the various sections of the road vehicle. Testing the fire properties of materials or products searching for an individual fire property must always be accompanied by an overall analysis that indicates that a resulting improvement, or apparent improvement, in the fire property assessed will be accompanied by an actual improvement in fire safety in the road vehicle. Battipaglia et al.[77] used the ASTM fire hazard assessment framework to assess the fire hazard of automotive materials in the engine compartment of a passenger road vehicle following a collision. The NFPA committee expected NHTSA to consider this work and NFPA 556 to develop considerations for highway vehicle fire safety that is adequately predictive of real-scale fire performance. This has not happened.

The ASTM committee on plastics (ASTM D20) has developed a standard, ASTM D5132,[79] which is a consensus standard version of the FMVSS 302 requirement. The development of such consensus standards resembling regulation is not uncommon. In fact, although the consensus standard cannot be used for approval of a material or product, it is often easier for a lab to follow and can be helpful for incorporating improvements, even if they cannot be recognized by the AHJ. In other words, such standards are basically intended for use by regulatory authorities outside the US, by AHJs, or codes, that go beyond federal regulations. Of course this is important principally when the test is of use for improving fire safety.

22.2.5.5 School Buses

School buses: The authority having jurisdiction is NHTSA and the fire safety requirements are included, together with all other NHTSA vehicle safety requirements in FMVSS 301[80] and FMVSS 302. FMVSS 301 deals with the structural integrity of the school bus and with the needed exits.

With regard to FMVSS 302, the same concerns expressed above about the validity of the fire tests on materials included in FMVSS 302 have also been expressed by the National Institute of Standards and Technology about school buses.[81] In fact, the report showed that school bus seat materials complying with FMVSS 302 would easily cause flashover on the bus with a small ignition source. In the case of school buses the report resulted from the investigation into a horrific school bus accident. On May 14, 1988, the worst drunk driving crash in US history occurred; it was the Carrollton, Kentucky, school bus fire. Drunk driver Larry Mahoney was northbound in the southbound lanes of Highway I-71 when he hit a school bus carrying 67 people. The bus caught fire trapping and killing 27 people, 24 of them children, and injuring 34 others.

Fortunately for schoolchildren, in spite of the inactivity by the federal government, local education authorities in many US states and Canadian provinces have decided to adopt a very simplistic but effective fire test, namely the National Safety Council Standard "School bus seat upholstery fire block test," which was approved by the National Conference on School Transportation as part of the National Standards for School Buses and National Standards for School Bus Operations.[82] The test involves putting a few sheets of newspaper in a paper bag, placing the bag on a school bus seat and igniting it with a match. If the seat is made of adequate materials, most of it will not burn, and if the materials are unsafe, the seat padding and cover will be completely destroyed. Unfortunately, this test has not been adopted by all local education authorities, and it is being applied in differing ways. ASTM developed a version of this test as ASTM E2574/E2574M.[83] In ASTM E2574, the paper bag is replaced by a gas burner and the acceptance criteria of the National Safety Council Standard are contained as information in an appendix (as is typical in most ASTM fire tests).

22.2.5.6 Cigarettes

A key development throughout North America is the requirements for cigarettes to comply with ASTM E2187[28] so that they are less likely to cause ignition of upholstery fabrics. These cigarettes are often designated as "fire-safe cigarettes," an incorrect designation, since they are still capable of igniting cellulosic fabrics, in some 25% of the cases. The test assesses whether a cigarette is capable of igniting filter paper, by burning past a certain mark, with multiple replicates. A cigarette complies with the regulation if it does not cause ignition > 25% of the time. This level of ignition power represents a dramatic decrease when compared to "normal" cigarettes, which ignite almost all cellulosic fabrics, unless the fabric has been specially treated. The cigarettes should more properly be designated as reduced ignition propensity cigarettes.

The key factor in determining smoldering combustion of cigarettes is their paper. The paper-modification approach most commonly used for manufacturing reduced ignition propensity cigarettes is the use of "banded" paper. This is roughly equivalent to increasing the weight of the paper or decreasing its porosity, but instead of changing these properties across the entire length of the cigarette, the paper is designed to have intermittent bands with altered properties. The intent is for the cigarette to self-extinguish when the ignited end hits a band and the cigarette is not being smoked, but to retain its normal smoking properties when it is not burning through a band.

On October 1, 2005, Canada became the first country to adopt the requirement for reduced ignition propensity cigarettes nationwide. In the United States, as stated earlier, reduced ignition propensity cigarettes are required in all states, as well as Washington, DC.

Outside of North America, "reduced ignition propensity cigarette" regulation has gone into effect in Australia and in the European Union. In the European Union the tobacco industry must manufacture cigarettes with special paper that will cause a cigarette to self-extinguish after 60 seconds if left unattended. Data from a number of European countries showed that cigarette-related fires account for some 11,000 fires each year, with 520 deaths and 1,600 injuries, with the elderly disproportionately affected.

Requirements for the sale of reduced ignition propensity cigarettes means that the probability of cigarette (smoldering) ignition of upholstery fabrics is decreased, even if the upholstery fabrics don't "pass" the smoldering ignition tests. As a result of the success of the campaign for "fire-safe" cigarettes, a discussion has started as to whether smoldering ignition (and even flaming ignition) tests should be eliminated. This would be a serious fire safety mistake. There are three reasons for still needing smoldering ignition tests in North America: (1) "fire-safe" cigarettes can still ignite fabrics, (2) traditional cigarettes are still commercially available (for example from the areas where there is no regulation, including the autonomous American Indian nations), and (3) the treatments used for improving the flaming performance of fabrics are usually ineffective in improving the smoldering ignition performance.

22.2.5.7 Flammable Fabrics

22.2.5.7.1 Wearing Apparel

As discussed previously, since the 1950s, wearing apparel (or clothing) is regulated via the flammability test for apparel fabrics in 16 CFR 1610.[14] In the test, samples tested are placed in a sample holder at a 45 degree angle, and an igniter flame is imposed on the upper surface of the sample. The test method requires that replicate, pre-conditioned samples of fabrics used in clothing apparel comply with one of the following criteria:

(a) no ignition when subjected to a small gas diffusion flame emitted from a burner based on a hypodermic needle during an exposure of 1.0 s, or
(b) if the fabric sample ignites, the flames shall not spread 127 mm (5 inches) in less than 3.5 seconds.

The regulation addresses the sensitivity of this test method to fabric weight (or areal density) by providing that fabrics with an areal density exceeding 88.3 g/m^2 (2.6 oz/yd^2) are permitted to be excluded from testing. These are considered too heavy to ignite under the test conditions.[†]

A typical criticism of this test method, for example, is that ordinary newsprint, and even tissue paper, has been shown to meet its requirements. That is a valid criticism. However, there seems to be general agreement that, in spite of its lack of sophistication, this test method has been successful in eliminating the fabrics with the poorest fire performance from the general population of fabrics in use for apparel in the US. Thus, fabric types such as the fibrous "torch" sweaters with raised surface fibers (very popular for boys in the 1950s) that ignite readily and spread flame quickly are no longer legally sold in the United States due to the test requirements. The test has also been able to screen out of use very sheer (i.e., lightweight) fabrics and cotton garments that are ultra-light. Thus, it has effectively removed extremely flammable fabrics from the marketplace. However, significant numbers of children and older adults are injured (or killed) from the effects of fire when their lightweight, loose fitting clothing made from 100% cellulosic or polyester cellulosic blend fabrics catch fire. In fact, fire statistics indicate that fires for which "worn apparel" is the item first ignited are more likely than those starting in other consumer products to result in fire fatality per unit fire.

Some forensic experts ignore the added frequency of fire injuries or fire fatalities among the very young and the very old when defending the 16 CFR 1610 test as sufficient, suggesting that additional requirements are unnecessary. Note that the ASTM committee on textiles (ASTM D13) has developed a standard, ASTM D1230,[15] which is a consensus standard version of the 16 CFR 1610 requirement. A couple of studies[84,85] have shown that changes in the regulation would be helpful. Testing was conducted to search for possible correlations between fabric test modes, fabric composition, and fire-related properties. The general trend found demonstrated that, as areal density (a representation of weight) of fabrics increase, their times to forced ignition and their times to spread flame across their surface to the top of a vertical sample both also increase, leading to improved fire performance. The most important consequence of this observation is that better fire performance in heavier fabrics is largely, *but not fully*, independent of fabric composition. In consequence, the hazard to an individual wearing a garment composed of a specific fabric type is far more complex an issue than can be simply assessed based on whether fabric composition is of a thermoplastic material, a charring material, or a blend. Moreover, in terms of the regulatory implications it appears that the regulation of very lightweight fabrics should be an important consideration and that the US regulatory cut off value, set at 88.3 g/m^2 (2.6 oz/yd^2), is relatively arbitrary.

21.2.5.7.2 Carpets and Rugs

The ignitability of carpets and rugs is regulated by requiring compliance with the methenamine pill test, included in 16 CFR 1630[6] and 16 CFR 1631.[17] As is often the case with US regulatory tests, the test is more properly described by an ASTM consensus standard, in this case ASTM D2859,[18] under the jurisdiction of the ASTM committee on fire standards (ASTM E05). It is important to point out that (a) some imported small carpets and rugs (usually those made of cotton and other cellulosics) are often not tested and may not comply with the requirement and (b) codes in the US typically require testing beyond just methenamine pill results, for occupancies covered by a building code (i.e., not for 1- and 2-family dwellings) and have additionally focused on critical radiant flux requirements, based on ASTM E648.[49]

22.2.5.7.3 Children's Sleepwear

The flammability of children's sleepwear is regulated via the small-scale vertical open flame ignition tests in 16 CFR

1615[19] and in 16 CFR 1616.[20] Also, once more, an ASTM D13 standard exists as an alternative, namely, ASTM D6545.[21] The critical part of the regulation is that children's sleepwear must either be tight-fitting (and the regulation contains strict measurements) or they must pass the test. The regulation has exceptions for sleepwear used by very young children (babies). A key concern associated with the regulation is the fact that garments that do not meet the requirements are often offered for sale as "active wear" and are sold at lower prices than sleepwear, even though they are clearly designed to be used as sleepwear. Outside of the US, regulation on the surface burning characteristics of children's pajamas and of their snug fit is also in place in Canada and New Zealand.

22.2.5.7.4 Mattresses, Mattress Sets, and Sleep Products (Bedding)

Smoldering ignition of mattresses is regulated, both by CPSC and by codes, via the smoldering test in 16 CFR 1632.[22] Since 2007, open flame ignition of mattresses and mattress sets is also regulated in the US via the fairly severe open flame test for mattress sets in 16 CFR 1633.[24] On March 28, 2000, attorney Whitney Davis, director of the Children's Coalition for Fire-Safe Mattresses submitted four petitions to CPSC concerning mattress flammability. The petitions proposed four options: (1) an open flame standard similar to the full-scale test set forth in California Technical Bulletin 129 (CA TB 129);[86] (2) an open flame standard similar to the component test set forth in British Standard BS 5852,[87] using a fairly severe crib 5 ignition source (see ignition sources described in BS 5852 in Table 22.6); (3) a warning label for mattresses warning of polyurethane foam fire hazards; and (4) a permanent, "fireproof mattress!" identification tag. Whitney Davis had previously been working with California Assemblyman John Dutra, as a result of which the California legislature passed a law, Assembly Bill AB 603, in 2001. AB 603 required that all mattresses manufactured or sold in California be "fire retardant," without details. Before AB 603 was issued, the state of California has been requiring, since 1975, that all foam used for mattresses to comply with California Technical Bulletin 117 (CA TB 117; last edition: 2000).[88] That edition of CA TB 117 includes a vertical fire test for foam that could not be "passed" unless the padding used is either polyurethane foam, containing some level of flame retardants, or a different type of padding with better fire performance. There is no consensus as to whether foam complying with CA TB 117 resulted in mattresses with highly increased fire safety. It is clear, however, that such foam would delay ignition of the mattress from small open flames and would even inhibit ignition from very small open flames. California also had in place a voluntary standard for high risk occupancies known as CA TB 121,[89] which used a metal container containing newspaper as the ignition source. The ignition source used in CA TB 129, a gas burner, is a T-shaped burner that applies propane gas for 180 seconds at a flow rate of 12 L/min. The test is severe enough that it can usually not be met unless the padding used for the mattress is either a well flame-retarded polyurethane foam or another padding with improved fire performance. CA TB 129 became a consensus standard through committee E05 as ASTM E1590,[90] but the ASTM standard does not have its own pass/fail criteria.

AB 603 was initially expected to require that mattresses comply with CA TB 129. After a research project was conducted at the National Institute of Standards and Technology (NIST) on behalf of the mattress industry, a new test was developed, based on using a new dual burner.[91,92] This resulted in the issuance of California Technical Bulletin 603 (CA TB 603),[93] which went into effect for all mattresses sold in the state of California from January 1, 2005. It is important to point out that CA TB 603 is a significantly less severe test than CA TB 129 and mattresses sold as complying with the test often use barrier materials under the mattress ticking (or as the mattress ticking) instead of adding flame retardants into the foam. In fact, if the barrier is pierced or damaged (by the effect of vandalism and/or of children playing with sharp objects) experience indicates that mattresses can comply with CA TB 603 but release large amounts of heat. However, in the absence of vandalism, clearly mattresses complying with CA TB 603 offer a very significant improvement in fire safety over non-compliant mattresses. Eventually, CPSC issued their rulemaking (law) and developed 16 CFR 1633,[24] which is technically equivalent to CA TB 603, but has some added requirements, in terms of labeling and record keeping. Once 16 CFR 1633 went into effect throughout the US, on July 1, 2007, CA TB 603 was withdrawn. The 16 CFR 1633 requires that all mattresses, mattress/foundation sets and futons sold in the US meet the following pass/fail criteria: (a) the peak heat release rate does not exceed 200 kW during the 30 minute test, and (b) the total heat release does not exceed 25 MJ in the first 10 minutes of the test. The propane gas flames from the dual burner are applied for 70 seconds (top burner, at a flow rate of 12.9 L/min) and 50 seconds (side burner, at a flow rate of 6.6 L/min), giving a total applied heat release of 27 kW. The total observation period from the start of the test is 30 minutes.

The codes in the US have long interpreted that the regulation based on 16 CFR 1633 applied to residential mattresses only and that higher fire safety requirements should apply to high risk environments, such as jails or prisons, hospitals, and university dormitories. NFPA 101[10] contains requirements that mattresses in health care occupancies and in detention and correctional occupancies that are not sprinklered must comply with a peak heat release rate of 100 kW and a total heat release no more than 25 MJ in the first 10 minutes of test when tested to ASTM E1590 (or CA TB 129). The IFC[4] has similar requirements, except that there is no sprinkler exception to the requirements for mattresses in detention occupancies; mattresses must comply with the requirements irrespective of the presence of sprinklers (because sprinklers are ineffective to handle

TABLE 22.6
Ignition Sources Described in BS 5852[87]

Ignition source 0:	Smoldering cigarette
Tubular sources:	A burner tube consisting of a length of stainless steel tube, with 8.0 ± 0.1 mm (5/16 ± 0.004 in.) outside diameter and 6.5 ± 0.1 mm (0.26 ± 0.004 in.) internal diameter and 200 ± 5 mm (7.9 ± 0.2 in.), connected by a flexible tube via a flow meter, fine control valve, an optional on-off valve, and a cylinder regulator providing a nominal outlet pressure of 2.8 kPa containing butane. The flow meter shall be calibrated to supply butane gas flow rates as required in 6.5.1 through 6.5.3. The flexible tube connecting the output of the flow meter to the burner tube shall be 2.5–3.0 m (8–10 ft) in length with an internal diameter of 7.0 ±1.0 mm (0.28 ± 0.004 in.).
	Gas: Butane gas, with a known net heat of combustion of 49.6 ± 0.5 MJ/kg, is the fuel for this ignition source. Meter the flow rate of butane and keep it constant throughout the test. Measure the gas flow rate at a pressure of 101 ± 5 kPa (standard atmospheric pressure, measured at the flow gage) and a temperature of 20 ± 5°C.
Ignition source 1:	45 mL/min at 25°C (77°F), corresponding to a flame height of 35 mm (1.4 in.), measured from the top of the burner tube, when held vertically upwards and when the flames are burning freely in air.
Ignition source 2:	160 mL/min at 25°C (77°F), corresponding to a flame height of 145 mm (5.7 in.), measured from the top of the burner tube, when held vertically upwards and when the flames are burning freely in air.
Ignition source 3:	350 mL/min at 25°C (77°F), corresponding to a flame height of 240 mm (9.4 in.), measured from the top of the burner tube, when held vertically upwards and when the flames are burning freely in air.
Wood crib sources:	The wood crib is constructed using wood (pine lumber conditioned to constant humidity) and cotton (long-fiber, pure, dry, untreated, surgical grade cotton not more than 6 mm (0.25 in.) and not less than 4 mm (0.16 in.) thick). The wood cribs are constructed so that the sticks in each layer are parallel to one another and at right angles to the sticks in the adjacent layer. The sticks in each layer are placed as far away from each other as possible (except for the two main cribs forming the base of crib 6), without overlap at the ends, forming a square-sectioned crib. The sticks are glued together with wood adhesive. Cotton is placed at the bottom of the cribs, with the "fluffy" side up. The cotton is glued to the sticks at the bottom of the crib with minimal amounts of wood adhesive.
Ignition source 4:	The crib is constructed with ten sticks, each of which is 40 ± 2 mm (1.6 ± 0.08 in.) long and has a square section of 6.5 ± 0.5 mm (1/4 ± 0.02 in.). There are 5 layers of 2 sticks each. The total mass of sticks is 8.5 ± 0.5 g. The bottom of the crib is covered by a flat horizontal layer of cotton approximately 40 × 40 mm (1.6 × 1.6 in.) in area.
Ignition source 5:	The crib is constructed with 20 sticks, each of which is 40 ± 2 mm (1.6 ± 0.08 in.) long and has a square section of 6.5 ± 0.5 mm (1/4 ± 0.02 in.). There are 10 layers of 2 sticks each. The total mass of sticks is 17.0 ± 1.0 g. The bottom of the crib is covered by a flat horizontal layer of cotton approximately 40 × 40 mm (1.6 × 1.6 in.) in area.
Ignition source 6:	The crib consists of an outside crib and an inside crib. The outside crib is constructed with eight sticks, each of which is 80 ± 2 mm (3.2 ± 0.08 in.) long and has a square section of 12.5 ± 0.5 mm (1/2 ± 0.02 in.). There are 4 layers of 2 sticks each. The inside crib is constructed with ten sticks, each of which is 40 ± 2 mm (1.6 ± 0.08 in.) long and has a square section of 6.5 ± 0.5 mm (1/4 ± 0.02 in.). There are 5 layers of 2 sticks each. The total mass of sticks is 60.0 ± 2.0 g. The bottom of the inside crib is covered by a flat horizontal layer of cotton approximately 40 × 40 mm (1.6 × 1.6 in.) in area.
Ignition source 7:	The crib consists of an outside crib and an inside crib. The outside crib is constructed with eighteen sticks, each of which is 80 ± 2 mm (3.2 ± 0.08 in.) long and has a square section of 12.5 ± 0.5 mm (1/2 ± 0.02 in.). There are 9 layers of 2 sticks each. The inside crib is constructed with eight sticks, each of which is 40 ± 2 mm (1.6 ± 0.08 in.) long and has a square section of 6.5 ± 0.5 mm (1/4 ± 0.02 in.). There are 5 layers of 2 sticks each. The total mass of sticks is 126.0 ± 4.0 g. The bottom of the inside crib is covered by a flat horizontal layer of cotton approximately 40 × 40 mm (1.6 × 1.6 in.) in area.

fires under the steel bunkers common in detention facilities). This is an important distinction since the vast majority of health care and detention facilities are fully sprinklered. The IFC also requires that mattresses in college and university dormitories meet the same criteria as those in hospitals.

As a result of the same petition that led to CA TB 603 and 16 CFR 1633, California AB 603 also mandated that filled bed products, such as comforters, pillows, and bedspreads, would also have to exhibit improved fire performance. CBHF studied the flammability of filled bed products and found that these products significantly contribute to mattress fires. In collaboration with the federal government, industry, and commercial laboratories, CBHF started working on California Technical Bulletin 604 (CA TB 604) in order to address this hazard. CBHF intended CA TB 604 to be a performance-based standard that would not prescribe the use of any specific additives or flame retardants but would encourage the use of innovative methods and products to increase fire performance without affecting the environment. In effect, the draft of CA TB 604 issued in 2007 consisted of three tests, all to be conducted in a small test chamber (or a small hood). They are as follows: a flat filled/bed clothing products test, a pillow/cushion and loose fill materials test, and a mattress pad test. The pass/fail criteria were intended to be based on weight loss (mass loss) for the first two tests and/or the generation of a hole for the flame to penetrate through (as seen by visual observation) in the third test. The ignition source was a propane gas burner tube (a 250–300 mm long stainless steel tube, 8 mm in outside diameter, 6.5 mm internal diameter). In the first test multiple layers of flat filling materials were to be inserted into a case made of standard sheeting fabric or of the product's actual cover fabric. The test specimen was to be placed on a horizontal cement board on a weighing device and ignited at one corner. In the second test the filling materials were to be encased in a case made of the standard sheeting fabric or of the ticking/fabric used in the actual bed product to encase the loose filling materials. The test specimen was to be placed on a horizontal cement board on a weighing device and ignited at one corner. In the last test, a specimen of the mattress pad or its filling material was to be sandwiched between layers of standard sheeting fabrics. The specimen was to be subjected to a small open flame applied to the center point of the top surface, and observations of the burning behavior and burn patterns are used to assess fire performance. The regulation was never issued, and is unlikely to proceed, in spite of the fact that the regulatory timeline had started both in California and at CPSC, who issued an Advanced Notice of Proposed Rulemaking.

22.2.5.7.5 Upholstered Furniture

In 1994, the National Association for State Fire Marshals (NASFM) submitted a petition to CPSC to enact mandatory regulation for the flammability of upholstered furniture. The recommendation by NASFM was that the requirements for upholstered furniture flammability used in California or those used in the United Kingdom be implemented. CPSC initially, in 2001, recommended an ignition test for the upholstery fabrics similar to the small gas burner sources in BS 5852.[87] This was widely derided as inappropriate, since it did not (and could not) address the issue of the heat release from the padding materials. Following other concerns expressed by interested parties, CPSC conducted studies of effectiveness of flame retardants, of toxicity of flame retardants, and many others. In the meantime, CBHF developed a draft revision of CA TB 117 (often known as CA TB 117 plus), which was made public in February 2002. This draft regulation addressed most of the concerns that had been expressed about CA TB 117, particularly regarding the fabric test and regarding the lack of substantive foam and composite tests. However, it was never implemented. In 2005 CPSC staff proposed a fire test scheme of 14 options, including smoldering and open flame tests as well as individual component and composite tests, but this was also never approved, as a result of both technical and political inputs. At a briefing of CPSC commissioners on December 6, 2007, a series of four options was presented to the commissioners: (a) the adoption of CA TB 117 plus, (b) the adoption of the 2001 recommendation by CPSC staff, (c) the adoption of the 2005 recommendation by CPSC staff, or (d) the adoption of a recommendation addressing only smoldering combustion for cover fabrics. In a Notice of Proposed Rulemaking (NPRM), dated March 4, 2008, the commissioners chose to go with a smoldering combustion test for cover fabrics and a barrier test for fabrics failing the smoldering test, which would become draft 16 CFR 1634. This was also never implemented.

However, in spite of the fact that CPSC had not implemented mandatory smoldering requirements for upholstered furniture fire safety, there had been widespread voluntary use of smoldering tests for components (ASTM E1353,[94] from ASTM committee E05) by the trade association for residential furniture manufacturers (Upholstered Furniture Action Council, UFAC) and for composites (ASTM E1352,[95] from ASTM committee E05) by the trade association for contract furniture (Business and Institutional Furniture Manufacturers Association, BIFMA). In essence, what CPSC staff had proposed as a smoldering test was to retain virtually the same requirements the vast majority of upholstered furniture items had been complying with, via their trade associations since the late 1970s. In fact, the NFPA committee on fire tests had already developed two standards which were technically equivalent to ASTM E1353 (namely NFPA 260[96]) and to ASTM E1352 (namely NFPA 261[97]), which were less widely used by manufacturers than the ASTM test methods. The test proposed by CPSC, and contained in draft 16 CFR 1634, is different from the one in ASTM E1353 but would have a very similar outcome for the applicable fabrics: most cellulosic cover fabrics would fail, and most thermoplastic fabrics would pass. It is of interest, however, that the draft 16 CFR 1634 applied only to cover fabrics and not to any other upholstered furniture components, while NFPA 260 (and ASTM E1353) applied to all individual components and NFPA 261 (and ASTM E1352) applied to mock-ups also containing all components. This means that the proposed draft 16 CFR 1634 regulation would have left unregulated not only foams (or any other fillings) but also welt cords and all other interior fabrics.

In the 2013 editions of NFPA 260 and NFPA 261, the ignition source for the tests was revised. In view of the fact that commercial cigarettes in the US were of the so-called "fire-safe cigarette" type, the likelihood of the ignition of fabrics using that type of cigarette was low (since that was the reason that commercial cigarettes were being required to be of that type). Therefore, NIST developed a batch of cigarettes, designated SRM 1196, that were very similar to the typical commercial cigarettes that had been used for ASTM E1352, ASTM E1353, NFPA 260, and NFPA 261, before the introduction of "fire-safe cigarettes." All 4 standards (NFPA 260, NFPA 261, ASTM E1352, and ASTM E1353) incorporated a requirement that the ignition source would have to be an SRM 1196 cigarette. However, when the original batch of SRM 1196 cigarettes was exhausted, NIST started developing a series of cigarettes (designated as: "the current supply of SRM 1196 series cigarettes") for use in the test methods. This change, and some other needed technical improvements, were implemented in NFPA 260 and in NFPA 261 but they were not incorporated into either ASTM E1352 or into ASTM E1353. As a result of this, in 2022, ASTM E1352 and ASTM E1353 were withdrawn by ASTM committee E05. The codes (IFC and NFPA 101) continue referencing both NFPA 260 and NFPA 261 and do not reference the ASTM standards (although the annex of NFPA 101 explains that they were formerly equivalent). Both IFC and

NFPA 101 require that upholstered furniture in some regulated occupancies comply with a smoldering ignition test.

In 2013, the state of California withdrew the open flame test contained in CA TB 117 and replaced it by a smoldering ignition test for all upholstery components, based on (but not identical to) the 2008a edition of ASTM E1353, while using either the SRM 1196 cigarette (or one that meets similar specifications) as the ignition source (CA TB 117–2013[98]). The standard does not include any open flame component. The CA TB 117–2013 test is the fire test included in 16 CFR 1640,[27] which applies to all upholstered furniture manufactured, imported, or reupholstered in the US as of June 25, 2021. It does not apply to mattresses (which are still regulated by means of 16 CFR 1632 and 16 CFR 1633). The regulation includes a clause that says that "no state or any political subdivision of a state may establish or continue in effect any provision of a flammability law, regulation, code, standard, or requirement that is designed to protect against the risk of occurrence of fire, or to slow or prevent the spread of fire, with respect to upholstered furniture." This preemption does not apply to the codes (IFC or NFPA 101), which still contain added requirements, as described in the next paragraph.

The state of California introduced requirements for upholstered furniture flammability for public occupancies when it developed California Technical Bulletin 133 (CA TB 133),[99] which was also made into a consensus standard by ASTM committee E05 as ASTM E1537,[100] albeit without acceptance criteria. The gas burner used as the ignition source in CA TB 133 is a square-shaped burner that applies propane gas for 80 seconds at a flow rate of 13 L/min. The test is severe enough that it can usually not be met unless: (a) the foam contained in the upholstered furniture item is either well flame-retarded polyurethane foam or an alternate padding with good fire performance, or (b) an adequate fire barrier is used. The pass/fail criteria of CA TB 133 are a peak heat release rate of 80 kW and a total heat released that does not exceed 25 MJ over the first ten minutes of test, and both IFC and NFPA 101 have parallel requirements to those discussed above for mattresses. ASTM E1537 (with the criteria shown above) applies to detention facilities, health care occupancies, and college and university dormitories, where upholstered furniture also needs to comply with smoldering ignition requirements based on NFPA 260 or NFPA 261. However, the open flame test is not required if the facility is fully sprinklered, and both detention facilities and health care facilities are required to be sprinklered, and usually are. The State of California declared that Technical Bulletin CA TB 133 was repealed effective January 22, 2019, but this does not affect ASTM E1537.

22.2.5.7.6 Mine Conveyor Belts

The fire safety of mine conveyor belts is covered by the Mine Safety and Health Administration (MSHA). The convoluted history of requirements is described by Verakis.[101] The regulation mandates the exclusive use of "flame-resistant conveyor belts," without details. The actual test used is a Bunsen burner–type test, based on ASTM D635 (UL 94 HB),[102] which has been shown to be inappropriate for the associated fire hazard. Originally, large scale tunnels were used to classify the flammability of conveyor belts, but those tunnels have since been destroyed. There have been fire testing research projects addressing the correlation of the various proposed tests with standard tests, but nothing has been implemented to date.

The Technical Study Panel on the Utilization of Belt Air and the Composition and Fire Retardant Properties of Belt Materials in Underground Coal Mining was created under Section 11 of the Mine Improvement and New Emergency Response Act of 2006 (MINER Act) (Public Law 109–236) and was chartered under the provisions of the Federal Advisory Committee Act (FACA). The panel's charge was to prepare and submit this report concerning the utilization of belt air and the composition and fire retardant properties of belt materials in underground coal mining to the Secretary of Labor, the Secretary of Health and Human Services, the Committee on Health, Education, Labor, and Pensions of the Senate, and the Committee on Education and the Workforce of the House of Representatives. A final report was published in December 2007.[103] The recommendations of the panel regarding a fire test for conveyor mine belts was the use of the BELT test for regulatory purposes, replacing the Bunsen burner test currently in 30 CFR §18.65, which was based on ASTM D635 (or UL 94 HB). The BELT test was developed by the US Bureau of Mines and is conducted in a 1.7 m (5.5 ft long by 0.2 m^2) (1.5 ft^2) ventilated tunnel. The belt material sample size is 1.5 m (5 ft) long by 230 mm (9 in.) wide. The sample is ignited by applying a gas burner to the front edge of the belt sample with the flames distributed equally on the top and bottom surfaces of the sample. After five minutes, the burner is removed, and the belt sample is allowed to burn until the flames are out. A belt passes the BELT if, in three separate trials, there remains a portion of the conveyor mine belt sample that is undamaged across its entire width. If, in any of the three trials, fire damage extends to the end of the sample, the conveyor belt formulation fails the test. Comparison testing showed excellent agreement for 19 of the belts between the pass/fail results of the large-scale fire gallery test and of the BELT. Agreement was reached that this would be implemented, after considering public input. On June 19, 2008, the Mine Safety and Health Administration (MSHA) published in the Federal Register a Request for Information (*www.msha.gov/REGS/FEDREG/RFI/E8-13633.pdf*) as to whether tests exist that can be used for assessing conveyor belt combustion toxicity and smoke density. The information collected resulted in the development of a final rule on December 31, 2008, with a requirement that conveyor mine belts must be tested to the BELT fire test,[104] consistent with the technical study panel report. No requirements were added for smoke density or smoke toxicity testing, but a recommendation was made to add smoke sensors along the belt entry, once adequate ones have been assessed.

22.2.5.7.7 Other Materials and Products

Flammable solids are regulated under the Federal Hazardous Substances Act (FHSA) in terms of storage and transport. That is picked up by fire codes, like the IFC, which lay out the maximum allowable quantities of hazardous materials permitted in various occupancies, depending on the type of construction. The quantity of flammable solids that can be stored is very limited, unless the storage location has been constructed as a "hazardous location," which means that it is built to extreme safety. An unusual added wrinkle is presented by the fact that CPSC defines what constitutes a flammable solid as "a material that exhibits a flame spread rate exceeding 2.5 mm per second when exposed to the flame in the test contained within 16 CFR 1500.44."[23] It has been found that non-flame-retarded polyurethane foam can be classified as a flammable solid.[105] The consequence of this finding should have been that upholstered furniture that is not flame-retarded cannot be stored (or kept) in showrooms or storage facilities that are not built as hazardous locations. This would have been a severe limitation since the majority of the upholstered furniture commercially available is not flame retarded (and the state of California, for example, prohibits the use of flame retardants in furniture). The IFC has therefore added a requirement that facilities used for the manufacture, display, sale, and storage of upholstered furniture and mattresses must be fully sprinklered.

ASTM F963[25] is a standard specification for toys which includes the requirements that (a) "materials other than textiles (excluding paper) used in toys shall not be flammable, as defined under 16 CFR 1500.3 (c) (6) (vi)" under the FHSA and (b) any textile fabrics used in toys must comply with 16 CFR 1610 (see above). Legislation in the US, intended to prevent the use of lead in toys, has mandated that all toys comply with ASTM F963. This leads to the unusual fact that children's toys (such as dolls and their furniture) must have higher fire safety than the upholstered furniture on which the child may sit.

22.2.6 Comparison with International Regulations

It is important to point out that the above regulations do not include any requirements in the US for construction products (building products) or for electrical wires and cables (or their optical fiber equivalents) other than in some transportation vehicles. As explained earlier, these requirements are contained in codes (most of the key ones having been issued by companies, such as ICC or NFPA, in the US), which may be (and typically are) adopted into regulation. Most other developed countries have government-mandated regulations for such products.

In the case of the European Union, the European Commission has developed "directives" including one requiring "safety in case of fire."[1] These directives were later converted into Construction Products Regulations in 2011,[106] and the regulations were updated with a road map in 2020[107] and further details were incorporated in 2022.[2]

The consequence of this was the development of a scheme for the classification of products,[108] starting with "construction products other than floorings," and "floorings." Basically, four tests are used for the "construction products other than floorings": a noncombustibility test (EN ISO 1182),[109] an oxygen bomb calorimeter (EN ISO 1716)[110] an intermediate scale heat release apparatus (the single burning item test, or SBI, EN 13823)[111] and a small ignitability apparatus (EN ISO 11925–2).[112] For floorings, the SBI test is replaced by the flooring radiant panel test (EN ISO 9239–1).[113] It is important to note that the tests used for regulation in the European Union are issued by CEN (European Committee for Standardization), but that the "Vienna Agreement" between ISO and CEN mean that the only difference between those EN standards where the word ISO follows the prefix EN and the corresponding ISO standards is the cover sheet. Every country adopting this classification scheme (including all European Union countries or member states) is entitled to choose the limits that apply to each application, but they must all adopt the overall scheme. In order to help with the development and implementation of this scheme, the European Community has developed a "reference scenario" fire test, namely, the ISO 9705–1 room-corner test,[59] and uses other tests as surrogates. Table 22.7 shows the concept of what each class is intended to represent in the European Community standard testing approach. The specific classification schemes for construction products (excluding floorings) and for floorings are shown in Tables 22.8 and 22.9, respectively. Afterwards, the European Commission addressed fire testing of linear pipe thermal insulation products[114] and, later, of electric cables.[115] The details are shown in Tables 22.10 and 22.11, respectively. The abbreviations used in the classification schemes above are shown in Table 22.12. The scheme for electric cables is based on the European Commission FIPEC project,[116] and the test is a vertical cable tray heat release test, described in the EN 50399 standard.[117] It is the sole classification that includes measurements of acid gas emissions (described as "acidity). These requirements are included as regulation by all European member states (and some countries that are not European Union members), with potential caveats and exceptions. The additional declarations in the European system (smoke obscuration, flaming droplets, and acidity) are not regulatory requirements but are often included in specifications, especially in the case of acid gas emissions from electric cables. The individual European countries have had different regulatory approaches to fire safety regulation, but most of them have since been withdrawn.

In the European Union there are, on the other hand, no regulatory requirements for upholstered furniture or

TABLE 22.7
European Community Concept for Classes Relating to Fire

Fire Class	Test Simulations		Time to Flashover	Contribution to Fire Growth
A1	Fully developed fire in a room	Fully developed fire in a room	Assumed to correspond to after flashover	None
A2	Fully developed fire in a room	Fully developed fire in a room	No flashover	None
B	Small fire attack on limited area	Single Burning Item in a Room	No flashover	Extra small
C	Small fire attack on limited area	Single Burning Item in a Room	After 10 minutes	Small
D	Small fire attack on limited area	Single Burning Item in a Room	2–10 minutes	Medium
E	Small fire attack on limited area	Single flame (i.e., match/lighter)	Before 2 minutes	Large
F	Does not comply with Classes A1 through E			

Source: Essential definitions associated with European Community regulations:
"Material": a single basic substance, or uniformly dispersed mixture of substances, such as metal, stone, timber, concrete, mineral wool with a uniformly dispersed binder, or polymers.
"Product": a material, element, or component about which fire safety information is required.
"Homogeneous product": a product consisting of a single material, which exhibits uniform density and composition throughout the product.
"Non-homogeneous product": a product that is not a homogeneous product. A non-homogeneous product is composed of one or more components, substantial and/or non-substantial, and does not have uniform density or composition.
"Substantial component": a material that constitutes a significant fraction of a non-homogeneous product. A substantial component must have a layer with either a mass per unit area ≥ 1.0 kg/m2 or a thickness ≥ 1.0 mm.
Two or more non-substantial layers that are adjacent to each other (i.e., with no substantial component(s) in-between the layers) are regarded as one non-substantial component and, therefore, must altogether comply with the requirements for a layer being a non-substantial component.
For non-substantial components, distinction is made between internal non-substantial components and external non-substantial components, as follows.
"Internal non-substantial component": a non-substantial component that is covered on both sides by at least one substantial component.
"External non-substantial component": a non-substantial component that is not covered on one side by a substantial component.

TABLE 22.8
European Community Classes of Reaction to Fire Performance for Construction Products Excluding Floorings, Linear Pipe Thermal Insulation Products, and Electrical Cables

Class	Test Method(s)	Classification Criteria	Additional Classification (Not a Requirement)
A1	EN ISO 1182[a] and	$\Delta T \leq 30°C$ and $\Delta m \leq 50\%$ and $t_f = 0$ (i.e., no sustained flaming)	-
	EN ISO 1716	PCS ≤ 2.0 MJ/kg [a] and PCS ≤ 2.0 MJ/kg [b, c] and PCS ≤ 1.4 MJ/m^2 [d] and PCS ≤ 2.0 MJ/kg [e]	-
A2	EN ISO 1182[a]	$\Delta T \leq 50°C$ and $\Delta m \leq 50\%$ and $t_f \leq 20s$	-
	EN ISO 1716 and	PCS ≤ 3.0 MJ/kg [a] and PCS ≤ 4.0 MJ/kg [b] and PCS ≤ 4.0 MJ/m^2 [d] and PCS ≤ 3.0 MJ/kg [e]	-
	EN 13823 (SBI)	FIGRA ≤ 120 W/s and LFS < edge of specimen and $THR_{600s} \leq 7.5$ MJ	Smoke production, [f] and Flaming droplets/particles [g]

(Continued)

TABLE 22-8 Continued

Class	Test Method(s)	Classification Criteria	Additional Classification (Not a Requirement)
B	EN 13823 (SBI) and	FIGRA ≤ 120 W/s and LFS < edge of specimen and THR_{600s} ≤ 7.5 MJ	Smoke production, [f] and Flaming droplets/particles [g]
	EN ISO 11925–2[i] (Exposure = 30 seconds)	F_s ≤ 150 mm within 60 seconds	
C	EN 13823 (SBI) and	FIGRA ≤ 250 W/s and LFS < edge of specimen and THR_{600s} ≤ 15 MJ	Smoke production, [f] and Flaming droplets/particles [g]
	EN ISO 11925–2[i] (Exposure = 30 seconds)	F_s ≤ 150 mm within 60 seconds	
D	EN 13823 (SBI) and	FIGRA ≤ 750 W/s and THR_{600s} ≤ 100 MJ	Smoke production, [f] and Flaming droplets/particles [g]
	EN ISO 11925–2[i] (Exposure = 30 seconds)	F_s ≤ 150 mm within 60 seconds	
E	EN ISO 11925–2[i] (Exposure = 15 seconds)	F_s ≤ 150 mm within 20 seconds	Flaming droplets/particles [h]
F	No performance determined		

Source: **(a)** For homogeneous products and substantial components of non-homogeneous products.
(b) For any external non-substantial component of non-homogeneous products.
(c) Alternatively, any external non-substantial component having a PCS ≤ 2.0 MJ/m², provided that the product satisfies the following criteria of EN 13823 (SBI): FIGRA ≤ 20 W/s; and LFS < edge of specimen, and THR_{600s} ≤ 4.0 MJ and s1 and d0.
(d) For any internal non-substantial component of non-homogeneous products.
(e) For the product as a whole.
(f) In the last phase of the development of the test procedure, modifications of the smoke measurement system have been introduced, the effect of which needs further investigation. This may result in a modification of the limit values and/or parameters for the evaluation of smoke production.
s1 = SMOGRA ≤ 30m²/s² and TSP_{600s} ≤ 50 m²; s2 = SMOGRA ≤ 180 m²/s² and TSP_{600s}≤ 200 m²; s3 = not s1 or s2.
(g) d0 = No flaming droplets/particles in EN 13283 (SBI) within 600 s; d1 = No flaming droplets/particles persisting longer than 10 s in EN 13823 (SBI) within 600 s; d2 = not d0 or d1. Ignition of the paper in EN ISO 11925–2 results in a d2 classification.
(h) Pass = no ignition of the paper (no classification).
Fail = ignition of the paper (d2 classification).
(i) Under conditions of surface flame attack and, if appropriate to the end-use application of the product, edge flame attack.

TABLE 22.9
European Community Classes of Reaction to Fire Performance for Floorings

Class	Test Method(s)	Classification Criteria	Additional Classification (Not a Requirement)
A1fl	EN ISO 1182[a] and	ΔT ≤30°C and Δm ≤50% and t_f = 0 (i.e., no sustained flaming)	-
	EN ISO 1716	PCS ≤ 2.0 MJ/kg [a] and PCS ≤ 2.0 MJ/kg [b] and PCS ≤1.4 MJ/m² [c] and PCS ≤ 2.0 MJ/kg [d]	-
A2fl	EN ISO 1182[a] and	ΔT ≤50°C and Δm ≤50% and t_f ≤ 20 s	-
	EN ISO 1716 and	PCS ≤ 3.0 MJ/kg [a] and PCS ≤ 4.0 MJ/m² [b] and PCS ≤ 4.0 MJ/m² [c] and PCS ≤ 3.0 MJ/kg [d]	-
	EN ISO 9239–1 [e]	Critical flux [f] ≥ 8.0 kW/m²	Smoke production [g]

Class	Test Method(s)	Classification Criteria	Additional Classification (Not a Requirement)
Bfl	EN ISO 9239–1 [e] and	Critical flux [f] ≥ 8.0 kW/m²	Smoke production [g]
	EN ISO 11925–2 [h] (Exposure = 15 sec)	F_s ≤ 150 mm within 20 seconds	-
Cfl	EN ISO 9239–1 [e] and	Critical flux [f] ≥ 4.5 kW/m²	Smoke production [g]
	EN ISO 11925–2 [h] (Exposure = 15 sec)	F_s ≤ 150 mm within 20 s	-
Dfl	EN ISO 9239–1 [e] and	Critical flux [f] ≥ 3.0 kW/m²	Smoke production [g]
	EN ISO 11925–2 [h] (Exposure = 15 sec)	F_s ≤ 150 mm within 20s	-
Efl	EN ISO 11925–2 [h] (Exposure = 15 sec)	F_s ≤ 150 mm within 20s	-
Ffl	No performance determined		

Source: **(a)** For homogeneous products and substantial components of non-homogeneous products.
(b) For any external non-substantial component of non-homogeneous products.
(c) For any internal non-substantial component of non-homogeneous products.
(d) For the product as a whole.
(e) Test duration = 30 minutes.
(f) Critical flux is defined as the radiant flux at which the flame extinguishes or the radiant flux after a test period of 30 minutes, whichever is the lower (i.e., the flux corresponding with the furthest extent of spread of flame).
(g) s1 = Smoke ≤ 750% min; s2 = not s1.
(h) Under conditions of surface flame attack and, if appropriate to the end use application of the product, edge flame attack.

TABLE 22.10
European Community Classes of Reaction-to-Fire Performance for Linear Pipe Thermal Insulation Products

Class	Test Method(s)	Classification Criteria	Additional Classification (Not a Requirement)
$A1_L$	EN ISO 1182[a] and	ΔT ≤ 30°C and Δm ≤ 50% and t_f = 0 (i.e., no sustained flaming)	-
	EN ISO 1716	PCS ≤ 2.0 MJ/kg [a] and PCS ≤ 2.0 MJ/kg [b] and PCS ≤1.4 MJ/m² [c] and PCS ≤ 2.0 MJ/kg [d]	-
$A2_L$	EN ISO 1182[a] and	ΔT ≤ 50°C and Δm ≤ 50% and t_f ≤ 20s	-
	EN ISO 1716 and	PCS ≤ 3.0 MJ/kg [a] and PCS ≤ 4.0 MJ/kg [b] and PCS ≤ 4.0 MJ/m² [c] and PCS ≤ 3.0 MJ/kg [d]	-
	EN 13823 (SBI)	FIGRA ≤ 270 W/s and LFS < edge of specimen and THR_{600s} ≤ 7.5 MJ	Smoke production, [e] and Flaming droplets/particles [f]
B_L	EN 13823 (SBI) and	FIGRA ≤ 270 W/s and LFS < edge of specimen and THR_{600s} ≤ 7.5 MJ	Smoke production, [e] and Flaming droplets/particles [f]
	EN ISO 11925–2[h] Exposure = 30 seconds	F_s ≤ 150 mm within 60 seconds	
C_L	EN 13823 (SBI) and	FIGRA ≤ 460 W/s and LFS < edge of specimen and THR_{600s} ≤ 15 MJ	Smoke production, [e] and Flaming droplets/particles [f]
	EN ISO 11925–2[h] Exposure = 30 seconds	F_s ≤ 150 mm within 60 seconds	

(Continued)

TABLE 22-10 Continued

Class	Test Method(s)	Classification Criteria	Additional Classification (Not a Requirement)
D_L	EN 13823 (SBI)	FIGRA ≤ 2,100 W/s and THR $_{600\,s}$ ≤ 100 MJ	Smoke production, [e] and Flaming droplets/particles [f]
	EN ISO 11925–2[h] Exposure = 30 seconds	F_s ≤ 150 mm within 60 seconds	
E_L	EN ISO 11925–2[h] Exposure = 15 seconds	F_s ≤ 150 mm within 20 seconds	Flaming droplets/particles [f]
F_L	No performance determined		

Source: (a) For homogeneous products and substantial components of non-homogeneous products.
(b) For any external non-substantial component of non-homogeneous products.
(c) For any internal non-substantial component of non-homogeneous products.
(d) For the product as a whole.
(e) s1 = SMOGRA ≤ 105 m^2/s^2 and TSP_{600s} ≤ 250 m^2; s2 = SMOGRA ≤ 580 m^2/s^2 and TSP_{600s} ≤1600 m^2; s3 = not s1 or s2.
(f) d0 = No flaming droplets/particles in EN13823 (SBI) within 600 s; d1 = No flaming droplets/particles persisting longer than 10 s in EN 13283 (SBI) within 600 s; d2 = not d0 or d1. Ignition of the paper in EN ISO 11925–2 results in a d2 classification.
(g) Pass = no ignition of the paper (no classification); fail = ignition of the paper (d2 classification).
Fail = ignition of the paper (d2 classification).
(h) Under conditions of surface flame attack and, if appropriate to end use application of product, edge flame attack.

TABLE 22.11
European Community Classes of Reaction to Fire Performance for Electric Cables

Class	Test Method(s)	Classification Criteria	Additional Classification (Not a Requirement)
A_{ca}	EN ISO 1716	PCS ≤ 2.0 MJ/kg [a]	
$B1_{ca}$	EN 50399 (30 kW flame source), and	FS ≤ 1.75 m and $THR_{1200sec}$ ≤ 10 MJ and Peak HRR ≤ 20 kW and FIGRA ≤ 120 W/sec	Smoke production[b,e] and flaming droplets/particles[c] and acidity[d]
	EN 60332–1–2	H ≤ 425 mm	
$B2_{ca}$	EN 50399 (20.5 kW flame source), and	FS ≤ 1.5 m and $THR_{1200sec}$ ≤ 15 MJ and Peak HRR ≤ 30 kW and FIGRA ≤ 150 W/sec	Smoke production[b,f] and flaming droplets/particles[c] and acidity[d]
	EN 60332–1–2	H ≤ 425 mm	
C_{ca}	EN 50399 (20.5 kW flame source), and	FS ≤ 2.0 m and $THR_{1200sec}$ ≤ 30 MJ and Peak HRR ≤ 60 kW and FIGRA ≤ 300 W/sec	Smoke production[b,f] and flaming droplets/particles[c] and acidity[d]
	EN 60332–1–2	H ≤ 425 mm	
D_{ca}	EN 50399 (20.5 kW flame source), and	$THR_{1200sec}$ ≤ 70 MJ and Peak HRR ≤ 400 kW and FIGRA ≤ 1300 W/sec	Smoke production[b,f] and flaming droplets/particles[c] and acidity[d]
	EN 60332–1–2	H ≤ 425 mm	
E_{ca}	EN 60332–1–2	H ≤ 425 mm	
F_{ca}	No performance determined		

Source: (a) For the product as a whole, excluding metallic materials, and for any external component (i.e., sheath) of the product.
(b) Smoke declarations: s1, s1a, s1b, s2, or s3
s1 = TSP_{1200} ≤ 50 m^2 *and* Peak SPR ≤ 0.25 m^2/s
s1a = **s1** and transmittance in accordance with EN IEC 61034–2 ≥ 80%
s1b = **s1** and transmittance in accordance with EN 61034–2 ≥ 60% < 80%
s2 = TSP_{1200} ≤ 400 m^2 *and* Peak SPR ≤ 1.5 m^2/s
s3 = not **s1** or **s2**
(c) Droplets declarations: d0, d1, or d2

d0 = No flaming droplets/particles within 1,200 sec
d1 = No flaming droplets/particles persisting longer than 10 sec within 1,200 sec
d2 = not **d0** or **d1**.
(d) Conductivity declarations: a1, a2, or a3 (in accordance with EN IEC 60754–2). Testing is conducted on each component of a cable excluding the metallic ones, and the results are evaluated for the whole cable.
a1 = conductivity < 2.5 μS/mm *and* pH > 4.3
a2 = conductivity < 10 μS/mm *and* pH > 4.3
a3 = not **a1** or **a2**.
(e) The smoke class declared for class $B1_{ca}$ cables must originate from the test according to EN 50399 (30 kW flame source).
(f) The smoke class declared for class $B2_{ca}$, C_{ca}, and D_{ca} cables must originate from the test according to EN 50399 (20.5 kW flame source).

TABLE 22.12
Abbreviations in European Community Classifications Relating to Fire

ΔT	temperature rise
Δm	mass loss
t_f	duration of flaming
PCS	gross calorific potential
FIGRA	fire growth rate
THR_{600s}	total heat release
LFS	lateral flame spread
SMOGRA	smoke growth rate
TSP_{600s}	total smoke production
F_s	flame spread
H	char height
EN IEC 60754–2	Test on gases evolved during combustion of materials from cables—Part 2: Determination of acidity (by pH measurement) and conductivity (International Electrotechnical Commission)
EN IEC 61034–2	Measurement of smoke density of cables burning under defined conditions—Part 2: Test procedure and requirements (International Electrotechnical Commission)

mattresses, but the United Kingdom and Ireland have adopted regulations as far back as 1988 (for furniture) and 1989 (for mattresses).[118,119] These regulations, based on the use of BS 5852,[87] have proved to be very effective in improving fire safety, as shown by a research project commissioned by the UK Department of Trade and Industry, after the regulations had been in place for over 10 years.[120] The European Commission also undertook a research project (known as CBUF) to investigate how such a scheme could be used in Europe,[121] but that successful research has not been implemented in Europe. The CBUF project recommended the use of full scale heat release tests similar to those used in the US for upholstered furniture and mattresses in codes (such as ASTM E1537[100] and ASTM E1590[90]) and recommended the use of a small-scale heat release test, the cone calorimeter (ASTM E1354,[76] standardized by ISO as ISO 5660–1[122]), for predicting full-scale test results. In fact, CBUF developed small scale tests with both individual component materials and with composites and models for using small-scale test data.

In Canada the system is somewhat similar to the US in the development of building codes and an electrical code, but the organization developing them is a part of the National Research Council and not a separate standards developing organization as in the US.

22.3 CODES

22.3.1 International Code Council Codes

As explained earlier, in the United States, there are two primary organizations that develop codes: ICC (International Code Council) and NFPA (National Fire Protection Association); they are both private companies. ICC develops a full family of codes, including the following: International Building Code (IBC),[3] International Fire Code (IFC),[4] International Mechanical Code (IMC),[5] International Wildland Urban Interface Code (IWUIC),[6] International Residential Code (IRC),[7] and International Existing Building Code (IEBC),[8] which have been adopted widely throughout the country, albeit not always the latest edition (see Table 22.2).

Until the mid-1990s, the US had three regional code development companies: (1) ICBO (International Conference of Building Officials),[123] which issued the Uniform Building Code or UBC, primarily used in the 25 states West of the Mississippi; (2) BOCA (Building Officials and Code Administrators),[124] which issued the National Building Code or NBC, primarily used in the Northeast and Midwest; and (3) SBCCI (Southern Building Code Conference International),[125] which issued the Standard Building Code or SBC, primarily used in the Southeast.

The three organizations chose, in 1994, to join up to form ICC, which has since developed (and maintained) the ICC set of codes in a uniform fashion and to discontinue the old codes (also called the legacy codes). At least some of the legacy codes also had parallel documents dealing with the other issues that need codification, but not all of them had complete sets.

For all ICC codes, the first edition (dated 2000 for most of them) was developed by a series of committees who worked to combine the best aspects of all legacy codes. Those drafts were then made available to the public for submission of code development proposals. ICC then set up a series of code development committees dealing with each one of the codes [in the case of the IBC and the IRC more than one committee was set up]. Proposals were submitted by the public (and ICC staff was not permitted to make proposals) and considered by each committee at a code development hearing meeting. The committees made decisions (and their options were to accept a proposal as submitted, make modifications to any proposal, or disapprove it), following opportunities for proponents and opponents to explain the pros and cons of each proposal. The public was then allowed to comment on the decisions and the hearings were held without code committees, and with the code officials present voting on the public comments (for proposals without comments the committee decisions were final). New codes are issued every three years (the latest editions as of the publication of this chapter, were dated 2024), but initially there were interim changes every 18 months. This system has continued, with some minor modifications, from the time the first set of ICC codes were issued in the year 2000 until the 2024 codes. Further details on the ICC code development process and on the development processes at NFPA (codes and standards) and at ASTM (standards only) have been published in 2017.[126] With regard to the 2027 editions, the codes will be developed with two sets of committee hearings before one public hearing where public officials, and no other interested parties, vote on the public comments. Jurisdictions are entitled to adopt the full codes or the codes with local (state or city) amendments.

All documents (mostly standards) referenced in an ICC code must comply with the ICC policy on referenced documents. This policy has been applied strictly to new referenced documents but not as strictly to "grandfathered" documents originating in the legacy codes. Two key provisions of this policy are the following:

(1) All documents referenced by ICC must be fully written in mandatory language (including the fact that they cannot contain the words "may" or "should" other than in non-enforceable sections) to be permitted to be referenced. This has been a key reason that many standards issued by ISO or IEC (as well as some ASTM standards) have not been able to be referenced, as they are often written including nonmandatory language. This has also resulted in extensive changes in the way consensus standards (by ASTM, NFPA, and others) have been written and revised. In fact, NFPA demands that all its documents (other than guides or recommended practices) be written in mandatory language and so does the ASTM committee on fire standards (ASTM E05), as well as some other committees with standards intended for code application. However, most other ASTM committees and many other organizations (including ISO and IEC) have not yet adopted this policy.

(2) All referenced documents must have been issued as a result of a full consensus process. This has been the reason that many documents issued by trade associations and similar organizations, or even regulations, have not been adopted by the codes.

22.3.1.1 International Building Code

The IBC applies to the "construction, alteration, movement, enlargement, replacement, repair, equipment, use and occupancy, location, maintenance, removal and demolition of every building or structure or any appurtenances connected or attached to such buildings or structures." However, the IBC does not apply to "detached one- and two-family dwellings and multiple single-family dwellings (townhouses) not more than three stories above grade plane in height with a separate means of egress and their accessory structures," to which the IRC applies.

The IBC consists of 34 chapters, plus a chapter on referenced standards and a few appendices, which are not a mandatory portion of the code and can be adopted (or not) separately by individual AHJs. The key chapters within the IBC associated with fire safety are the following:

- Chapter 2: Definitions of terms.
- Chapter 3: Use and occupancy classification of buildings. Requirements are usually based on what the type of use is for a building (or portion of a building).
- Chapter 4: Detailed requirements for some special occupancy buildings.
- Chapter 5: General building heights and areas.
- Chapter 6: Types of construction. The permission to use combustible materials or the type of fire resistance ratings are a function of whether the building is constructed completely, or primarily, of noncombustible materials.
- Chapter 7: Fire-resistant-rated construction. This chapter addresses the separation requirements between rooms/compartments in buildings and between buildings and also deals with insulation.
- Chapter 8: Interior finish and trim.
- Chapter 9: Fire protection systems: suppression and detection systems (including sprinklers and smoke detectors).

- Chapter 14: Exterior walls: dealing with exterior wall assemblies, including exterior wall coverings, and siding.
- Chapter 15: Roof assemblies and roof structures.
- Chapter 23: Wood (including fire-retardant-treated wood).
- Chapter 26: Plastics: this includes foam plastic insulation, plastic veneer, and light transmitting plastics.
- Chapter 27: Electrical: this contains primarily a reference to the National Electrical Code[9] but also some provisions on emergency and standby power systems, based on the IFC. It originally referenced a code named the ICC Electrical Code, which was soon discontinued.
- Chapter 31: Special construction: this deals with membrane structures, temporary structures, and various other structures, some of which are combustible and require fire testing.
- Chapter 33: Safeguards during construction: this deals with special needs during construction and demolition, since there have been many instances of fires during construction.
- Chapter 34: This chapter used to deal with existing buildings, but all its contents have been transferred to the IEBC (International Existing Building Code).

Interior wall and ceiling finish materials (or systems), addressed in Chapter 8 of the IBC, must be tested for heat release, flame spread, and smoke release in accordance with a room-corner test, NFPA 286.[127] In the NFPA 286 room-corner test, the materials line three walls and the ceiling of a room (with dimensions of 2.4 m × 3.7 m × 2.4 m high) and are exposed to a gas burner ignition source (5 minutes at 40 kW and 10 minutes at 160 kW). The burner is placed flush against one corner of the room. In order to comply with the code requirements, the material tested cannot reach flashover, and there are limits on heat release (800 kW), smoke release (1,000 m^2), and flame spread (flames are not allowed to spread to the outer extremity of the sample on any wall or ceiling). Alternatively, most (but not all) interior finish materials or systems can be tested for flame spread and smoke development in accordance with the traditional Steiner tunnel test (ASTM E84,[128] a 7.3 m (25 ft) tunnel with the test specimen on the underside of the ceiling). The Steiner tunnel provides results as a flame spread index (FSI) and a smoke developed index (SDI). There is clear evidence that the Steiner tunnel test is appropriate for many building materials but not for all of them, but that NFPA 286 is a suitable test for any wall or ceiling interior finish material. Some materials (textile and expanded vinyl wall coverings) are permitted to be tested by using a less severe room-corner test (NFPA 265),[129] but, if materials tested to NFPA 265 are used, the area where they are installed must be fully sprinklered. The differences between NFPA 265 and NFPA 286 are: (a) that the second exposure period uses a 150 kW source and (b) that the burner is not placed flush against the wall; the effect is that the room ceiling is not exposed to the burner flame.

Some materials (particularly foam plastics, high-density polyethylene, and polypropylene) are only permitted to be used if they have been fire tested for interior finish use via the NFPA 286 test (and have complied with the requirements above) and not via ASTM E84, because of evidence that the results from the Steiner tunnel test are misleading. Materials, such as most foam plastics, that are not allowed to be used exposed if tested only to ASTM E84, are allowed, however, to be used when covered by a thermal barrier. The IBC recognizes as a thermal barrier one of three options: (a) gypsum wallboard, ½ inch (12. 7 mm) thick, (b) heavy timber (as defined in the code), or (c) materials that comply with the NFPA 275[130] thermal barrier test.

The NFPA 275 test assesses whether the thermal barrier prevents the underlying material (which is often a foam plastic or a building product containing a foam plastic) from causing a large fire by means of two tests: one assessing reaction-to-fire test (which is typically NFPA 286) and one assessing fire resistance (for 15 minutes), using the same time-temperature curve as in the traditional ASTM E119[58] test, but with a much smaller test specimen.

Very thin (< 0.9 mm thick) materials coated directly onto the surface of walls or ceilings are exempt from testing because they are believed to be likely to produce insufficient heat or flames to cause serious problems (this exception does not apply if the thin material is applied to a combustible wall or ceiling surface). Very thick structural members are exempt from testing because they are believed to be of low enough ignitability that they do not provide a serious hazard. Interior trim (which is not allowed to occupy more than 10% of any wall or ceiling surface) is tested by the same means as other interior finish but is allowed greater leniency (and is often also tested using thin strips of material rather than full thickness test specimens).

Recent trends have incorporated information on specific mounting practices for the use of ASTM E84 with a variety of materials, such as wall coverings, to ensure consistent testing results among labs (so everyone uses the same methods). These standard practices (see further in Section 22.4.2.5) contain requirements for specimen preparation and mounting in the ASTM E84 test. Table 22.13 describes the key fire tests used to regulate most interior finish materials in the IBC (which includes references [as must use] to the relevant standard practices, when the code requires their use).

The IBC also has a table, in Chapter 8, which describes those applications, for each occupancy, which must have materials that meet the requirements of Class A, B, or C. Note that materials that comply with the requirements based on NFPA 286 are permitted anywhere that a material with a Class A, B, or C is required.

Interior floor finish materials must meet requirements based on both ignition, via the methenamine pill test (ASTM D2859[18]), and a critical radiant flux via the ASTM

TABLE 22.13
IBC Fire Test Requirements for Interior Wall and Ceiling Finish Materials

Material	Test Method	Acceptance Criterion
Interior wall and ceiling finish materials, except as shown elsewhere in this table	NFPA 286	No flashover; no flames to extremities of samples; Pk HRR ≤ 800 kW; TSR ≤ 1,000 m² Complying materials are also accepted as Class A materials
	ASTM E84	Class A: FSI ≤ 25—SDI ≤ 450
	ASTM E84	Class B: 75 ≤ FSI < 25—SDI ≤ 450 (If Class A or B are permitted)
	ASTM E84	Class C: 200 ≤ FSI < 75—SDI ≤ 450 (If Class A, B, or C are permitted)
Materials having a thickness less than 0.036 inch (0.9 mm) applied directly to the surface of walls or ceilings	No testing required	
Exposed portions of structural members	No testing required	
Foam plastics	NFPA 286	No flashover; no flames to extremities of samples; Pk HRR ≤ 800 kW; TSR ≤ 1,000 m² Must comply with NFPA 286 plus ASTM E84 Class B or better
	FM 4880	Pass and must also comply with ASTM E84 Class B or better
	UL 1040	Pass and must also comply with ASTM E84 Class B or better
	UL 1715	Pass and must also comply with ASTM E84 Class B or better
Textile wall coverings	NFPA 286	No flashover; no flames to extremities of samples; Pk HRR ≤ 800 kW; TSR ≤ 1,000 m²
	NFPA 265, Method B	No flashover; no flames to extremities of samples; TSR ≤ 1,000 m²
	ASTM E84	ASTM E84, Class A, but also requires sprinklers—Must use ASTM E2404
Expanded vinyl wall coverings	NFPA 286	No flashover; no flames to extremities of samples; Pk HRR ≤ 800 kW; TSR ≤ 1,000 m²
	NFPA 265, Method B	No flashover; no flames to extremities of samples; TSR ≤ 1,000 m²
	ASTM E84	ASTM E84, Class A, but also requires sprinklers—Must use ASTM E2404
Textile ceiling coverings	NFPA 286	No flashover; no flames to extremities of samples; Pk HRR ≤ 800 kW; TSR ≤ 1,000 m²
	ASTM E84	ASTM E84, Class A, but also requires sprinklers—Must use ASTM E 2404
Expanded vinyl ceiling coverings	NFPA 286	No flashover; no flames to extremities of samples; Pk HRR ≤ 800 kW; TSR ≤ 1,000 m²
	ASTM E84	ASTM E84, Class A, but also requires sprinklers—Must use ASTM E2404
Site-fabricated stretch systems	NFPA 286	No flashover; no flames to extremities of samples; Pk HRR ≤ 800 kW; TSR ≤ 1,000 m²
	ASTM E84	ASTM E84, Class A, but also requires sprinklers—Must use ASTM E2573
High density polyethylene	NFPA 286	No flashover; no flames to extremities of samples; Pk HRR ≤ 800 kW; TSR ≤ 1,000 m²
Polypropylene	NFPA 286	No flashover; no flames to extremities of samples; Pk HRR ≤ 800 kW; TSR ≤ 1,000 m²
Laminated products factory produced with a wood substrate	NFPA 286	No flashover; no flames to extremities of samples; Pk HRR ≤ 800 kW; TSR ≤ 1,000 m² Complying materials are also accepted as Class A materials
	ASTM E84	Class A: FSI ≤ 25—SDI ≤ 450 — Must use ASTM E2579
	ASTM E84	Class B: 75 ≤ FSI < 25—SDI ≤ 450 (If Class A or B are permitted)—Must use ASTM E2579
	ASTM E84	Class C: 200 ≤ FSI < 75—SDI ≤ 450 (If Class A, B, or C are permitted)—Must use ASTM E2579
Facings or wood veneers intended to be applied on site over a wood substrate	NFPA 286	No flashover; no flames to extremities of samples; Pk HRR ≤ 800 kW; TSR ≤ 1,000 m² Complying materials are also accepted as Class A materials
	ASTM E84	Class A: FSI ≤ 25—SDI ≤ 450 — Must use ASTM E2404
	ASTM E84	Class B: 75 ≤ FSI < 25—SDI ≤ 450 (If Class A or B are permitted)—Must use ASTM E2404
	ASTM E84	Class C: 200 ≤ FSI < 75—SDI ≤ 450 (If Class A, B, or C are permitted)—Must use ASTM E2404
Interior trim (other than foam plastic)	NFPA 286	No flashover; no flames to extremities of samples; Pk HRR ≤ 800 kW; TSR ≤ 1,000 m²
	ASTM E84	Class C: 200 ≤ FSI < 75—SDI ≤ 450
Interior trim (foam plastic)	NFPA 286	No flashover; no flames to extremities of samples; Pk HRR ≤ 800 kW; TSR ≤ 1,000 m²
	ASTM E84	Class B: 200 ≤ FSI < 75

E648[49] test. Compliance with the ASTM D2859 (or its equivalent tests 16 CFR 1630 and 16 CFR 1631, depending on the size of the carpet) is required for all carpets and rugs sold in the US, but compliance with ASTM E648 depends on the application and is covered only by the codes.

Chapter 14 contains all the requirements for exterior walls. A key requirement states that all exterior wall assemblies containing combustible materials must be fire tested to NFPA 285.[131] This requirement was clarified in the 2024 IBC, since earlier editions of the code contained some potential ambiguity.

NFPA 285 is a large-scale (two stories) fire test that exposes exterior wall assemblies, which means everything that is part of the exterior wall other than the structural element of the wall, to ensure that there is no significant flame spread upwards, no penetration into a second (or next) story window, and no flame spread sideways. The need for such a test was highlighted in 2017 by the tragic fire, on June 14, in the Grenfell tower in London (United Kingdom) that caused 72 fatalities; the exterior wall assembly had not been properly fire tested as an assembly, and it was subsequently proven to have failed the appropriate fire safety requirement. Note, however, that full-scale fire testing of exterior wall assemblies was not a regulatory requirement in the United Kingdom at the time. Chapter 14 also deals with all the various types of components and products contained in exterior wall assemblies, many of which need to undergo individual fire tests as well as fire tests of the entire assembly.

The IBC defines exterior wall assembly as "A system of exterior wall covering, framing and components, such as weather-resistive barriers, air barriers, and insulating materials. This system provides protection of the building structural members and conditioned interior space, from the detrimental effects of the exterior environment." It also defines exterior wall covering as

> a material or assembly of materials applied on the exterior side of exterior walls for the purpose of providing a weather-resisting barrier, insulation or for aesthetics, including but not limited to, veneers, siding, exterior insulation and finish systems, architectural trim and embellishments such as cornices, soffits, facias, gutters and leaders.

The major types of exterior wall coverings addressed in Chapter 14 are metal composite materials (or MCMs), exterior insulation and finish systems (or EIFS), high pressure decorative laminates (or HPLs), and insulated metal panels (IMPs, which were added for the first time in the 2024 edition). Exterior wall coverings, and veneers, also include a variety of materials, such as a variety of sidings. It is important to highlight that any exterior wall assembly that contains a foam plastic must be tested to NFPA 285, and the foam plastic itself must also comply with testing to ASTM E84 (and achieve a Class B) and must also be separated by a thermal barrier (as discussed elsewhere in this chapter). It is also essential to explain that exterior wall assemblies that contain as the sole combustible a water-resistive barrier (which involves very little mass) must still be tested to NFPA 285, unless the water-resistive barrier itself complies with requirements based on both ASTM E84 and ASTM E1354, because experiments have shown that the water-resistive barrier alone can spread fire upwards quite rapidly.

Chapter 15 deals with roofing materials and requires that all roof assemblies must be fire tested to ASTM E108.[132] Again, the emphasis is on testing the entire roof assembly and not just the top layer (or roof covering). This requirement is very important for fire safety since it has been shown that the roof assembly is one of the areas where exposure to burning embers is most likely to cause severe penetration of fire into the building. Moreover, ASTM E108 also contains a requirement for weathering of the roof assembly, to ensure that the fire performance of the roof assembly does not deteriorate after exposure to rain and sunshine (UV).

Chapter 26 deals with plastics and, in particular, foam plastics and light-transmitting plastics. Foam plastics (as discussed previously) are not permitted to be used exposed when tested to ASTM E84 but must be tested to NFPA 286. However, even when the foam plastic is protected (i.e., separated from the habitable space) by a thermal barrier, the foam plastic must be tested to ASTM E84 and achieve a flame spread index not exceeding 75 and a smoke developed index not exceeding 450 (i.e., a Class B, unless it meets the requirements based on NFPA 286). Furthermore, all foam plastics must be "listed" by a national-recognized laboratory in order to be permitted to be used in construction. A study by Evans and Hirschler[133] includes an overview of the requirements for foam plastics in building construction. Light-transmitting plastics must be fire tested for ignitability (via ASTM D1929[134]), flame spread (via ASTM D635[102]), and smoke release (via ASTM D2843[135]). All three of these fire tests for light-transmitting plastics are very mild, so the use of these plastics also has severe size limitations.

Chapter 26 also deals with some specific composite materials (plastic lumber and wood-plastic composites), in section 2612, when used as deck boards, stair treads, handrails, and guards, meaning deck boards and associated applications. When used in this way, the materials need to meet an ASTM E84 flame spread index of 200 (which is what untreated wood products typically meet) but have no smoke development requirements (which is typical for outdoor applications). In some applications in the 21st century it has been found that these same decking materials have been used as exterior wall coverings, which is a vertical application. In order to develop the proper requirements for those plastic lumber and wood-plastic composite materials used as exterior wall coverings, ASTM developed a new specification in 2023, namely, ASTM D8484.[136]

One unusual code requirement, in Chapter 4, deals with the fire safety of kiosks (typically contained within malls) and indoor play structures (for both adults and children). In both instances, there are heat release requirements for foam plastics (based on either UL 1975[137] or NFPA 289[138]) and for rigid plastic materials (based on ASTM E1354), as well as various other requirements.

Insulation is handled in the IBC in Chapter 7. Fire safety is addressed in two ways: (a) based on whether the insulation is exposed or unexposed and (b) based on the type of insulation. Exposed insulation must meet the requirements of a Class A when tested in accordance with ASTM E84.[128] Of course, as described above, foam plastic insulation is not permitted to be used exposed unless it complies with the requirements based on NFPA 286,[127] while it must otherwise be covered by a thermal barrier. However, foam plastic insulation used in concealed ceiling spaces and attic spaces is permitted to be used when protected by one of a series of acceptable fireblocking materials, which includes gypsum board, lumber, and other approved materials. Cellulosic loose-fill insulation is permitted to be used exposed if it meets smoldering requirements set out by CPSC (CPSC 16 CFR Parts 1209[139] and 1404[140]) as well as a smoke developed index of ≤ 450 (based on ASTM E84). It can also be used concealed if it meets a smoke developed index of ≤ 450 when tested to ULC S102.2[141] (which is very similar to ASTM E84, but with the test specimen mounted on the tunnel floor). Attic floor insulation (whatever the composition) must also meet a critical radiant flux of ≥ 0.12 W/cm^2 when tested in accordance with ASTM E970.[142]

Traditionally, there were five types of construction in the IBC, described broadly in the following sentences. Types I and II were noncombustible construction, with varying degrees of permission for some combustible building elements. Type III construction requires that the exterior walls be constructed of noncombustible materials, with the interior building elements permitted to be of combustible materials and fire-retardant-treated wood framing and sheathing being permitted. Type IV construction permitted the use of mass timber as construction elements, and Type V construction is combustible construction. In recent editions of the IBC, three new types of construction were added: Types IVA, IVB, and IVC. They all involve mass timber (typically cross-laminated timber, CLT) construction. CLT is a wood panel product made from gluing together at least three layers (and typically an odd number of layers) of solid-sawn lumber (lumber cut from a single log). Each layer of boards is usually oriented perpendicular to adjacent layers and glued on the wide faces of each board, usually in a symmetric way so that the outer layers have the same orientation. It has been shown to have much better fire performance (both in terms of reaction-to-fire and fire resistance) than traditional lumber. CLT construction, which must have additional protection, is allowed to rise to much higher numbers of stories than normal Type IV construction.

This section cannot describe all the provisions in the code, but the code contains specific requirements for a variety of materials, both based on the type of material and on its application. For example requirements exist for insulation materials (loose-fill, foamed, and others), siding, combustible materials in exterior walls, roofs, and decks, and for glazing materials, among others. Chapter 6 of the IBC contains a list of the combustible materials that are permitted within noncombustible construction and also describes the requirements associated with each one of them (by reference to the appropriate section). That same chapter also specifies that all combustible materials contained within plenums must meet the requirements of the IMC.[5]

Chapter 23, on wood, deals with a specific material, namely, fire-retardant-treated wood (FRTW) which is required to be a wood material impregnated with fire-retardant chemicals that is listed to exhibit a flame spread index not exceeding 25 (based on ASTM E84) and a flame front that does not progress more than 3.2 m beyond the centerline of the burner at any time during the test (when the test is extended from ten minutes). The code also explicitly states that coatings are not permitted to be used to develop FRTW products. The same requirements are also contained in the IFC and the IWUIC (see the corresponding sections).

It is important to highlight a section in Chapter 1 of the IBC (which is consistent with the equivalent chapter in all other ICC codes) that deals with "alternative materials and methods." This section governs how to replace materials (or methodologies) required by the code with others that are neither specifically allowed (nor prohibited) by the code, by providing evidence to the AHJ that the alternative is satisfactory and complies with the intent of the provisions of the code and is not less safe than the one explicitly required. One new section states that, if a material (or methodology) is explicitly prohibited by the code, the use of an alternative is not allowed.

22.3.1.2 International Fire Code

The IFC code establishes regulations affecting or relating to structures, processes, premises, and safeguards regarding the following:

1. The hazard of fire and explosion arising from the storage, handling or use of structures, materials, or devices.
2. Conditions hazardous to life, property, or public welfare in the occupancy of structures or premises.
3. Fire hazards in the structure or on the premises from occupancy or operation.
4. Matters related to the construction, extension, repair, alteration, or removal of fire suppression or alarm systems.

The IFC applies to the following:

1. Structures, facilities, and conditions arising after the adoption of the relevant IFC edition.
2. Existing structures, facilities, and conditions that were not legally in existence at the time of adoption of the relevant IFC edition.
3. Existing structures, facilities, and conditions when identified in specific sections of the IFC.
4. Existing structures, facilities, and conditions which, in the opinion of the fire code official, constitute a distinct hazard to life or property.

The IFC also applies to the following:

1. Conditions and operations arising after the adoption of the relevant IFC edition.
2. Existing conditions and operations.

The IFC does not allow any change to be made in the use or occupancy of any structure that would place the structure in a different division of the same group or occupancy or in a different group of occupancies, unless such structure is made to comply with the requirements of both the IFC and the IBC. Subject to the approval of the fire code official, the use or occupancy of any existing structure can be changed, and the structure can be occupied for purposes in other groups without conforming to all the requirements of both the IFC and the IBC for those groups, provided the new or proposed use is less hazardous, based on life and fire risk, than the existing use.

The design and construction of new structures must comply with the IBC, and any alterations, additions, changes in use or changes in structures required by the IFC, which are within the scope of the IBC, must be made in accordance with the IBC.

With regard to historic buildings, the provisions of the IFC relating to the construction, alteration, repair, enlargement, restoration, relocation, or moving of buildings or structures are not mandatory for existing buildings or structures identified and classified by the state or local jurisdiction as historic buildings when such buildings or structures do not constitute a distinct hazard to life or property. Fire protection in designated historic buildings and structures is provided in accordance with an approved fire protection plan.

The IFC consists of 44 chapters, plus a chapter on referenced standards and a large number of appendices, which are not a mandatory portion of the code and can be adopted (or not) separately by individual AHJs. The key chapters associated with fire safety are the following:

- Chapter 2: Definitions of terms
- Chapter 8: Interior finish, decorative materials, and furnishings
- Chapter 9: Fire protection systems: suppression and detection systems (including sprinklers and smoke detectors)
- Chapter 27: Hazardous materials
- Chapter 31: Tents, temporary structures, and membrane structures
- Chapter 34: Flammable and combustible liquids
- Chapter 35: Flammable gases
- Chapter 36: Flammable solids

The requirements for interior finish (including interior floor finish) and interior trim materials in Chapter 8 of the IFC are coordinated with those in the IBC, with some exceptions. Since the IFC applies to both new and existing buildings, the requirements in Chapter 8 of the IFC for the smoke released by textile or expanded vinyl wall coverings apply only to "newly introduced" ones. The IFC also contains, in Chapter 8, requirements for upholstered furniture and mattresses in some regulated occupancies, principally hospitals, nursing homes, board and care facilities, college and university dormitories, prisons and jails. The requirements are based on the tests in ASTM E1537[100] and ASTM E1590,[90] with the pass/fail criteria developed by CBHF and incorporated into CA TB 133[99] (before this standard was repealed by the state of California) and CA TB 129[86] respectively (as the test methods are equivalent to the ASTM standards). The materials must also meet smoldering test requirements (see also Sections 22.2.5.7.4 and 22.2.5.7.5). An alternative test to the heat release test for mattresses in detention environments is the very severe mass loss test known as the Michigan Roll test (ASTM F1085, Annex A3),[143] in which a rolled-up mattress is exposed to fire from sheets of newsprint burning inside the roll, and it cannot lose more than 15% of the initial mattress mass. Additional requirements exist for curtains and drapes, decorations (including combustible vegetation), all of which need to be tested for flame propagation in accordance with NFPA 701.[144]

It is important to highlight that NFPA 701 contains two fire test methods (named, simply, Test Method 1 and Test Method 2), which apply, respectively, to lightweight and heavyweight fabrics or materials. Of interest, however, is that until the 1989 edition, NFPA 701 contained a "small-scale test" and a "large-scale test" (the latter of which is basically the same as Test Method 2). The "small-scale test" was found to be misleading and was eliminated from NFPA 701, but it is common to find vendors offering fabrics tested to the "small-scale test of NFPA 701," which is why most codes refer to "Test Method 1 or Test Method 2, as appropriate," or specifically to the one test applicable to the product in question. For example, the IFC requires tent fabrics (and those used in bounce houses, or "inflatable amusement devices") to meet the flame propagation performance of NFPA 701, Test Method 2. In the case of tents (typically used outdoors) it is essential that they also undergo a weathering test contained in NFPA 701, to ensure any fabric treatment does not lose its efficacy after exposure to rain or sunlight.

The 2024 edition of the IFC consolidated requirements for heat release rate of materials used for constructing large waste containers and linen containers, and similar products, when tested to ASTM E1354, now included in Chapter 3. The 2024 edition includes new requirements (in a nonmandatory appendix) for a business that has been in existence for many years but was never regulated: valet trash collection. This involves apartments and town houses, where residents place garbage cans outside their doors (in corridors) at night, and the garbage is collected by the next morning. The garbage cans must be made of materials meeting a severe heat release rate requirement (namely, a peak heat release rate of 300 kW/m^2, when tested in accordance with the cone calorimeter in a horizontal orientation and at an incident heat flux of 50 kW/m^2) and they may not affect the width of the exit corridor. The 2021 edition also introduced a requirement for testing the efficiency of fire retardant treatments for natural trees, using ASTM E3082.[145]

The IFC explicitly permits the use of fire retardant coatings for upgrading the fire performance of existing interior finish, something not permitted in the IBC (i.e., it can be done in existing construction but not in new construction).

The IFC contains requirements for installation of residential sprinklers in all new one and two family dwellings and townhouses. This requirement is parallel to what is also required (see below) in NFPA 101 and NFPA 5000. However, in actual practice, most states have excluded that section from their local codes. The IFC also contains requirements for installation of carbon monoxide detectors.

22.3.1.3 International Residential Code

The provisions of the IRC[7] apply to the construction, alteration, movement, enlargement, replacement, repair, equipment, use and occupancy, location, removal, and demolition of detached one- and two-family dwellings and townhouses not more than three stories above-grade in height with a separate means of egress and their accessory structures. The IRC is intended to provide minimum requirements to safeguard the public safety, health, and general welfare through affordability, structural strength, means of egress facilities, stability, sanitation, light and ventilation, energy conservation, and safety to life and property from fire and other hazards attributed to the built environment.

The legal occupancy of any structure existing on the date of adoption of a particular edition of the IRC will be allowed to continue without change, except if it is deemed necessary by the building official for the general safety and welfare of the occupants and the public. Additions, alterations, or repairs to any structure must conform to the requirements for a new structure without requiring the existing structure to comply with all of the requirements of the code, unless otherwise stated. Additions, alterations, or repairs cannot cause an existing structure to become unsafe or adversely affect the performance of the building.

The 2024 edition of the IRC consists of 43 chapters, plus a chapter on referenced standards and a large number of appendices, which are not a mandatory portion of the code and can be adopted (or not) separately by individual AHJs. The key chapters associated with fire safety are the following:

- Chapter 2: Definitions of terms
- Chapter 3: Building planning, this chapter contains virtually all the requirements associated with fire safety, including those for interior finish, foam plastics, insulation, and siding
- Chapter 7: Wall covering
- Chapter 8: Roof-ceiling construction
- Chapter 9: Roof assemblies
- Chapter 16: Duct systems

General requirements for residential construction tend to use, when they exist, the same tests as those used in the IBC, but often with milder requirements. Chapter 3 has consolidated, in the 2024 edition, all the information associated with fire safety requirements (including details on fire-retardant-treated wood).

22.3.1.4 International Mechanical Code

The IMC regulates the design, installation, maintenance, alteration, and inspection of mechanical systems that are permanently installed and utilized to provide control of environmental conditions and related processes within buildings. It does not regulate the installation of fuel gas distribution piping and equipment, fuel-gas-fired appliances and fuel-gas-fired appliance venting systems. Also, mechanical systems in detached one- and two-family dwellings and multiple single-family dwellings (townhouses) not more than three stories high with separate means of egress and their accessory structures are regulated by the IRC. Other exclusions are historical buildings and existing buildings.

The key fire safety provisions covered by this code are those for materials in ducts and plenums and for piping systems. Plenums are enclosed portions of the building structure that are not occupiable and that have been designed to allow air movement and serve as part of an air distribution system. These concealed spaces can move air (and smoke) easily between building compartments without the occupants being aware. Therefore, typically fairly severe fire safety requirements govern all materials permitted in plenums. Traditionally, the only materials permitted in plenums were those that are noncombustible or that meet a flame spread index of no more than 25 and a smoke developed index of no more than 50 when tested to the ASTM E84 Steiner tunnel test. However, a number of other products are also permitted, based on specific tests. They are: wires and cables (that either meet criteria based on the NFPA 262[70] plenum cable fire test or that are enclosed in noncombustible raceways), combustible pneumatic tubing (that meet the UL 1820[146] test), plastic fire sprinkler piping (that meet the UL 1887[147] test), combustible optical fiber and communication raceways (that meet the UL 2024[148] test), and combustible electrical, mechanical, or plumbing products (that meet the UL 2043[149] test, a heat release test). There is an exception for plastic plumbing and tubing intended for water distribution, to be discussed later (see 22.4.2.5). Foam plastic wall and ceiling insulation must meet the same requirements as interior finish, based on the NFPA 286 room-corner test, or be covered by a thermal barrier or by a steel skin. The ASTM E84 requirements, which also apply to pipe and duct insulation, must meet a much more severe smoke criterion (a smoke developed index of 50) than that for non-plenum areas, where the smoke developed index is allowed to reach 450.

22.3.1.5 International Existing Building Code

The provisions of the IEBC[8] apply to the repair, alteration, change of occupancy, addition, and relocation of existing buildings. The intent is to provide flexibility to permit the use of alternative approaches to achieve compliance with minimum requirements to safeguard the public health, safety, and welfare insofar as they are affected by the repair, alteration, change of occupancy, addition, and relocation of existing buildings. A building or portion of a building that has not been previously occupied or used for its intended purpose in accordance with the laws in existence at the time of its completion must comply with the provisions of the IBC or of the IEBC, as applicable, for new construction or with any current permit for such occupancy.

This code contains relatively few fire safety provisions, unless a problem has been found and such provisions are deemed necessary by the code official for the general safety and welfare of the occupants and the public. However, when buildings are being renovated, the IEBC often requires that the provisions of either the IBC or the IFC apply. An example of fire safety criteria added to the IEBC is that exterior wall assemblies (and exterior wall coverings) must meet the IBC criteria if added or replaced (unless the changes involve less than 15% of the wall area of any side of the building and only a single building story). The IEBC also explicitly permits the use of fire retardant coatings for upgrading the fire performance of existing interior finish.

22.3.1.6 International Wildland Urban Interface Code

The objective of the IWUIC is to establish minimum regulations consistent with nationally recognized good practice for the safeguarding of life and property, so as to mitigate the risk to life and structures from intrusion of fire from wildland fire exposures and fire exposures from adjacent structures and to mitigate structure fires from spreading to wildland fuels. The extent of the regulation is intended to be tiered commensurate with the relative level of hazard present, since the unrestricted use of property in wildland-urban interface areas is a potential threat to life and property from fire and resulting erosion. The IWUIC code is intended to supplement the jurisdiction's building and fire codes, if such codes have been adopted, to provide for special regulations to mitigate the fire- and life-safety hazards of the wildland-urban interface areas.

Originally the IWUIC code contained requirements for fire resistance-rated construction, but, for several editions, it also contains alternate requirements based on "ignition-resistant building materials." The IWUIC defines "ignition-resistant building materials" as ones that resist ignition or sustained flaming combustion sufficiently so as to reduce losses from burning embers and small flames (note that the concept is to reduce losses and not to prevent them, something typically unlikely in the event of a severe wildfire). Such materials have to meet one of a number of reaction-to-fire tests, principally the 30 minute version of the ASTM E84 test (now also standardized as ASTM E2768[150]) that is used for defining fire-retardant-treated wood (or FRTW), for example in Chapter 23 of the IBC or in NFPA 703.[151] In order to meet ASTM E2768 a material must comply with a flame spread index of 25 or less and exhibit a flame front that does not progress more than 3.2 m beyond the centerline of the burner at any time during the entire 30-minute test. Chapter 23 of the IBC (and its equivalent in the IRC) is explicit that the addition of coatings or paints to a wood material does not generate an FRTW material, since FRTW materials require that the fire retardant treatment be "impregnated" into the wood. The code explicitly describes requirements for ignition-resistant building materials made out of fire-retardant-treated wood and ones made out of wood-plastic composites or plastic lumber. They all must meet the same fire test, as well as mechanical and weathering requirements.

With regard to coated wood materials, the IWUIC explicitly prohibits such use on the walking surfaces of decks but is silent on other uses. However, the section on ignition resistant building materials requires that they be tested in accordance with the extended ASTM E84 or with ASTM E2768, and pass, (a) on the front and back faces and (b) that panel products be tested with a ripped or cut longitudinal gap of 3.2 mm (1/8 inch). These requirements (from which metal panel products are excluded) implicitly mean that coated wood products are not allowed, since the coating is unlikely to penetrate that deeply into the wood and to be applied on both faces. The concern about coated wood materials is based on two factors: (a) no evidence exists that commercial wood coatings can survive the twin effects of rain and UV exposure for many years (and of course, buildings in wildland areas are exposed to weather) and (b) many coatings are intended to be applied on site, which means that there is no assurance that the applicator will apply the coating as intended by the manufacturer.

The IWUIC requires that decks and walls be covered by an ignition-resistant building material (or a noncombustible material), or an assembly rated for one-hour fire resistant construction. There is also permission for heavy timber or log wall construction.

It is of interest that the state of California has developed its own wildland urban interface code (Chapter 7A of the California Building Code). When comparing the two codes,[152] it is clear that they have many of the same requirements, with California Chapter 7A being more severe in a few areas, but with one major difference. In California, decks and walls are allowed to be covered by untreated wood, while that is not the case in the IWUIC. Also, Chapter 7A is explicit in prohibiting the use of coated wood products.

Controversies have arisen with regard to requirements for fire resistance-rated construction. There is no doubt that if walls are required to meet both a severe reaction-to-fire test (such as ASTM E2768) and also exhibit a fire resistance rating, they are more efficiently protected against both upward flame spread and fire penetration than if there is no fire resistance rating requirement. However, it is likely that the most immediate fire hazard is rapid flame spread upwards towards the roof, vents, or eaves, from which it can penetrate into the building (which is covered by the flame spread requirements for ignition-resistant building materials). That would lower the need for a fire resistance rating requirement and make it a potential burden towards the adoption of a wildland urban interface code, which needs to be made explicitly by regulation.

22.3.1.7 International Performance Code

This code (International Code Council Performance Code for Buildings and Facilities or ICCPC[153]) was developed to provide a performance-based alternative to the use of the prescriptive IBC and IFC. It requires adequate fire safety but is not explicit in the way that this is obtained. For example the chapter on fire prevention has language such as the following:

> The objective of the code is to limit or control the likelihood that a fire will start because of the design, operation or maintenance of a facility or its systems so as to minimize impacts on people, property, processes, and the environment. Facility services, systems and activities that represent a potential source of ignition or can contribute fuel to an incipient fire must be designed, operated, managed, and maintained to reduce the likelihood of a fire starting. Electrical, mechanical, and chemical systems or processes and facility services capable of supplying sufficient heat under normal operating conditions or anticipated failure modes to ignite combustible system components, facility elements or nearby materials must be designed, operated, managed, and maintained to prevent the occurrence of fire. The quantities, configurations, characteristics, or locations of combustible materials, including components or facility systems, facility elements, facility contents and accumulations of readily ignitable waste or debris must be managed or maintained to prevent ignition by facility service equipment and other ignition sources associated with processes normally present or expected to be present within the facility. Ignition and fuel source interactions must be designed, operated, and maintained so as to prevent the occurrence of ignition or to control the extent of atmospheres likely to pose an ignition hazard.

This code is still relatively rarely used and has not undergone too many revisions. The 2024 edition has several key changes, many of them intended to reinforce the concept, stated previously, that the ICCPC code is to be used as another pathway for evaluating alternative materials, designs, and methods of construction. The concepts presented for the changes were that, when projects are designed per the prescriptive requirements of any ICC code, there are situations where a single material, element, or system cannot conform to the prescriptive requirements and that new materials or systems are entering the construction market at a pace that the prescriptive codes cannot keep up. This has the potential to create a competition between improved (or retained) fire safety and improved environmental benefits.

22.3.2 National Fire Protection Association Codes and Standards

NFPA (National Fire Protection Association) is the other major developer of codes in the US. Table 22.14 shows the NFPA codes and standards (excluding the fire tests issued by the NFPA committee on fire tests that are of key regulatory importance even though they are not codes). This includes the NFPA 13 series of sprinkler standards (NFPA 13, 13D and 13R), which form the bases for all the requirements for sprinkler installations in the US (and probably worldwide), NFPA 130 (a standard for trains and underground systems), NFPA 99 (a code which governs all health care occupancies, in conjunction with the Life Safety Code), and NFPA 90A (a standard for air conditioning, which regulates materials in ducts and plenums). NFPA documents are revised every three to five years, with the period being strictly three years for all codes and the major standards.[126] With regard to standards, this puts NFPA standard test methods at a disadvantage over potentially competing ASTM standards, which are able to be revised multiple times each year, if necessary (and in the case of ASTM E84, that is often the case). In order to revise an NFPA document between official revision cycles, this must be done by means of a TIA (Tentative Interim Amendment) which requires that the technical committee confirm that such a change is both technically important and of an emergency nature (which is often difficult to prove to the satisfaction of the committee).

22.3.2.1 National Electrical Code

The National Electrical Code (NFPA 70, NEC[7]) regulates all electrical installations and also all electrical and optical fiber cables. Chapter 2 describes overall wiring requirements (including, in particular, grounding in Article 250). Chapter 3 describes wiring methods and materials. That includes Article 300, which is a type of general guide through the code with the requirements for many applications and materials (including plenums, ducts and other spaces used for environmental air, in section 300.22). Article 310 (Conductors for General Wiring) contains a table that describes the materials to be used and the standard that addresses the material composition and the fire performance of the material, usually based on a small scale UL fire test. Other articles in Chapter 3 address the different types of cables and raceways. Thus, in Article 334, NM cables (or nonmetallic sheathed cables) are described, and they are one example of cables that need to have flame retarded outer sheaths (or jackets). Other examples include Power and Control Tray Cables (Article 336) and Service Entrance Cables (Article 338, unless they are used underground). Similarly, conduits and raceways (enclosed channels designed expressly for holding wires, cables, or busbars) have specific articles, with additional functions as permitted in this Code, and many of those require the materials (or the final products) to meet some type of fire test (generally not included in the NEC but in the specific product standard). Thus, for example, high density polyethylene conduit (HDPE, article 353) has no fire performance requirements but is intended only for underground installations, while Reinforced Thermosetting resin conduit (RTRC, article 355) does have fire performance requirements and can be used exposed and concealed in a variety of locations (but not in plenums). PVC conduit is made into various types of products, including rigid PVC conduit (RNC, article 352), liquidtight nonmetallic conduit (LFNC, article 356), and electrical nonmetallic tubing (ENT, article 362). PVC has inherently improved fire performance and its conduits can generally be used in a variety of locations, but again usually not in plenums, unless specifically formulated for that use.

NM cables (Article 334) have seen abundant controversy. These cables are the ones most commonly used in

TABLE 22.14
Key NFPA Codes and Standards (Excluding Fire Test Methods)

Codes	
NFPA 1	Fire Code
NFPA 30	Flammable and Combustible Liquids Code
NFPA 30A	Code for Motor Fuel Dispensing Facilities and Repair Garages
NFPA 42	Code for the Storage of Pyroxylin Plastic
NFPA 52	Vehicular Fuel Systems Code
NFPA 54	National Fuel Gas Code
NFPA 58	Liquefied Petroleum Gas Code
NFPA 59	Utility LP Gas Plant Code
NFPA 70	National Electrical Code (NEC)
NFPA 72	National Fire Alarm Code
NFPA 85	Boiler and Combustion Systems Hazard Code
NFPA 99	Health Care Facilities Code
NFPA 101	Life Safety Code
NFPA 301	Code for Safety to Life from Fire on Merchant Vessels
NFPA 400	Hazardous Materials Code
NFPA 430	Code for the Storage of Liquid and Solid Oxidizers
NFPA 432	Code for the Storage of Organic Peroxide Formulations
NFPA 434	Code for the Storage of Pesticides
NFPA 490	Code for the Storage of Ammonium Nitrate
NFPA 495	Explosive Materials Code
NFPA 900	Building Energy Code
NFPA 909	Code for the Protection of Cultural Resources Properties—Museums, Libraries, and Places of Worship
NFPA 914	Code for Fire Protection of Historic Structures
NFPA 1122	Code for Model Rocketry
NFPA 1123	Code for Fireworks Display
NFPA 1124	Code for the Manufacture, Transportation, Storage, and Retail Sales of Fireworks and Pyrotechnic Articles
NFPA 1125	Code for the Manufacture of Model Rocket and High Power Rocket Motors
NFPA 1127	Code for High Power Rocketry
NFPA 5000	Building Construction and Safety Code
Standards	
NFPA 13	Standard for the Installation of Sprinkler Systems
NFPA 13D	Standard for the Installation of Sprinkler Systems in One- and Two-Family Dwellings and Manufactured Homes
NFPA 13R	Standard for the Installation of Sprinkler Systems in Residential Occupancies up to and Including Four Stories in Height
NFPA 37	Standard for the Installation and Use of Stationary Combustion Engines and Gas Turbines
NFPA 70E	Standard for Electrical Safety in the Workplace
NFPA 75	Standard for the Fire Protection of Information Technology Equipment
NFPA 76	Standard for the Fire Protection of Telecommunications Facilities
NFPA 80	Standard for Fire Doors and Other Opening Protectives
NFPA 90A	Standard for the Installation of Air-Conditioning and Ventilating Systems
NFPA 90B	Standard for the Installation of Warm Air Heating and Air-Conditioning Systems
NFPA 130	Standard for Fixed Guideway Transit and Passenger Rail Systems
NFPA 220	Standard on Types of Building Construction
NFPA 221	Standard for High Challenge Fire Walls, Fire Walls, and Fire Barrier Walls
NFPA 225	Model Manufactured Home Installation Standard
NFPA 318	Standard for the Protection of Semiconductor Fabrication Facilities
NFPA 501	Standard on Manufactured Housing (withdrawn)
NFPA 502	Standard for Road Tunnels, Bridges, and Other Limited Access Highways
NFPA 520	Standard on Subterranean Spaces
NFPA 855	Standard for the Installation of Stationary Energy Storage Systems
NFPA 1140	Standard for Wildland Fire Protection

residences, and their fire performance is relatively poor. Even though they are required to have a "flame retardant covering" (without an explicit explanation of the meaning of the term) they are required to be "listed" and that means that they must comply with some fire testing, typically of the components. They are not required to be installed within metal conduits and are, therefore, limited to the following: (a) one- and two-family dwellings and their attached or detached garages, and their storage buildings, (b) multi-family dwellings and their detached garages permitted to be of Types III, IV, and V construction and (c) other structures permitted to be of Types III, IV, and V construction. These cables must be concealed within walls, floors, or ceilings that provide a thermal barrier of material that has at least a 15-minute finish rating as identified in listings of fire-rated assemblies. When the IBC introduced Types IV A, IV B, and IV C construction, debate ensued as to whether the use of NM cables needed to be restricted further but this did not occur for the 2023 NEC.

Chapter 4 of the NEC addresses flexible cords and cables and many of them are also required to meet some fire tests, at least in terms of their materials of composition. Chapter 5 includes a variety of special occupancies and is similar to a building code in that it describes the materials and products permitted in those applications. Special occupancies include hazardous locations, healthcare facilities (including hospitals, article 517) and places of assembly (article 518), both of which have very restrictive requirements. Another occupancy, with much more lenient requirements, is associated with Carnivals, Circuses, Fairs, and Similar Events (article 525). Article 550 addresses wiring in manufactured homes and is coordinated with the 2017 edition of NFPA 501[154] and with Part 3280, Manufactured Home Construction and Safety Standards, of the Federal Department of Housing and Urban Development, or HUD (see section 22.3.2.7; note that the NFPA 501 standard has subsequently been withdrawn and the regulation of manufactured homes is under the jurisdiction of HUD). Chapter 6 addresses special equipment and the key article of importance to fire safety is article 645, on information technology equipment, which regulates wiring in computer rooms. Of particular interest is the fact that cables located in an area under a raised floor in an information technology equipment room that is fully sprinklered do not need to meet the severe plenum cable test, in spite of the fact that such locations are often deemed to be plenums, but only need to meet a less severe vertical cable tray fire test (see next paragraph).

Chapter 7 is the chapter dealing with "Special Conditions," and it addresses most of the cables with highly improved fire performance. Thus, articles 725 (Class 1, Class 2, and Class 3 Remote-Control, Signaling, and Power-Limited Circuits), 760 (Fire Alarm Systems), and 770 (Optical Fiber Cables and Raceways) all use the same two schemes for fire performance of cables, as shown in Figures 22.2 and 22.3. The figures show the top being NFPA 262,[70] a cable fire test for flame spread and smoke, conducted in a modified Steiner tunnel (88 kW or 300,000 BTU/hr), for which the requirements in the NEC are that the maximum peak optical density not exceed 0.5, the maximum average optical density not exceed 0.15 and the maximum allowable flame travel distance not exceed 1.52 m (5 ft). The next test, in order of decreasing severity is UL 1666,[155] known as the "riser cable test," which requires that, when exposed to a 155 kW (530,000 BTU/hr) flame, the cable must be capable of preventing the carrying of fire vertically from floor to floor. The next rung in the severity scale is the vertical cable tray test, and there are two parallel tests, listing to both of which allows the cable to be used in the same application, even though the tests are not of equal severity. The more severe one is the Canadian CSA FT4 test,[69] equivalent to the IEEE 1202 test,[156] and the other one is the vertical cable tray test with a perpendicular burner (which was formerly contained in UL 1581,[157] a standard that now refers to UL 2556[68] for the fire test details). Both of these tests are now contained with UL 1685[158] and in UL 2556.[68] The concept in these cable tray tests is that the cable does not spread flame vertically far enough. The cable test specimen is roughly 2.4 m high, and the requirement is that

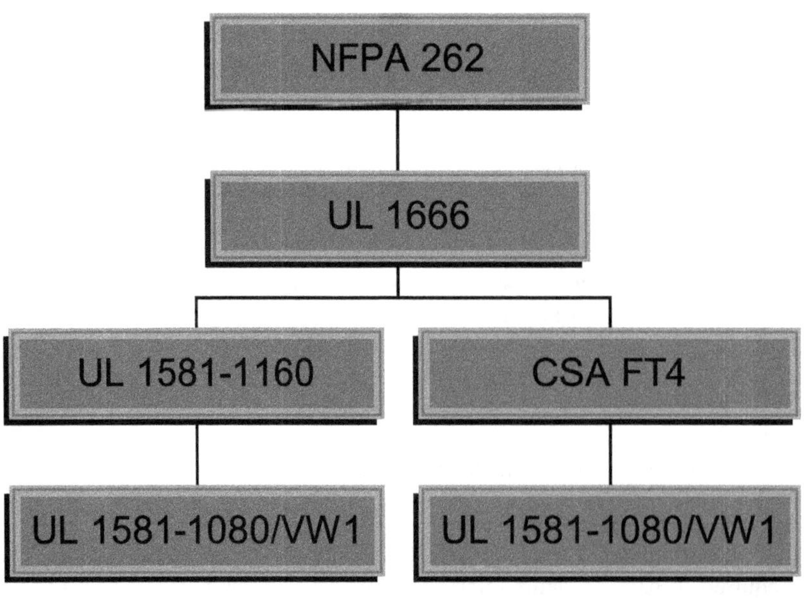

FIGURE 22.2 Flame spread requirements for cables in the NEC, in order of decreasing flame spread severity.

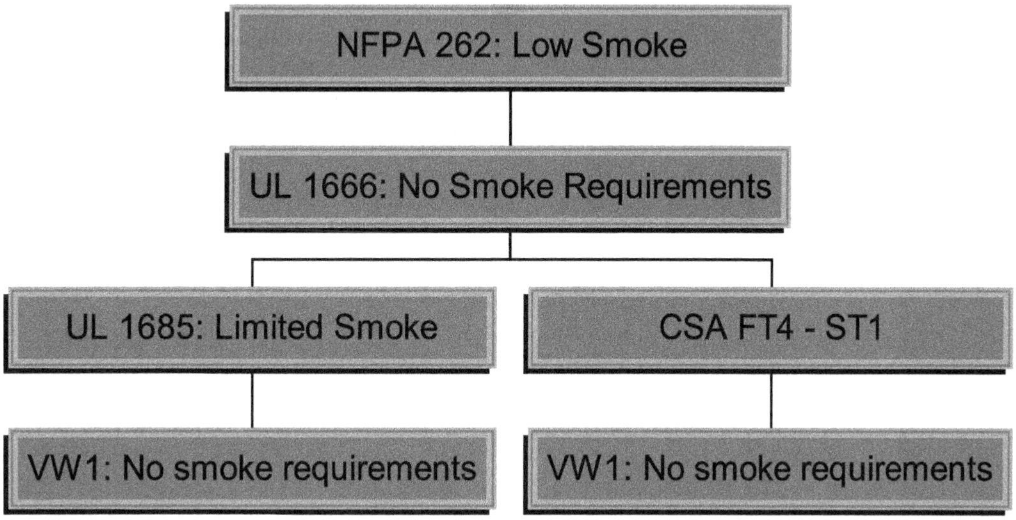

FIGURE 22.3 Smoke release requirements for cables in the NEC, with corresponding hierarchy (in order of decreasing flame spread severity).

the char length (cable destruction) does not go all the way to the top (in the vertical test) or does not reach 1.5 m (in the CSA FT 4 test). The flame source used is approximately 20 kW (70,000 BTU/hr.), applied for ten minutes, with the burner perpendicular to the cable tray (in the vertical test) and at a 20 degree angle (in the CSA FT 4 test). The loading of cables in the tray is also different between the tests, with a more severe loading in the CSA FT 4 test. The NEC does not require smoke measurements for these tests, but UL 1685 (as well as IEEE 1202 and CSA FT4) does have smoke pass/fail requirements. The fourth and minimal test in the NEC hierarchy is the small-scale Bunsen burner UL VW-1 test (or vertical wire test). Here a flame of only 500 W is applied, and the test simply requires the wire not to burn all the way; it is also included in UL 2556.

Chapter 8, on communications systems, has the same fire safety scheme as that for articles 725, 760, and 770, discussed previously. It has recently been reorganized with article 800 containing the general requirements for communications systems, which apply to articles 805 (communications circuits), 820 (Community Antenna Television and Radio Distribution Systems), 830 (Network-Powered Broadband Communications Systems), and 840 (Premises-Powered Broadband Communications Systems), albeit with some differentiations. Of interest is the fact that Chapter 8 is not subject to the requirements of Chapters 1 through 7 except where the requirements are specifically referenced in Chapter 8.

The independence of Chapter 8 is particularly important when dealing with plenum cables, for which the charging section is 300.22, which describes three types of ducts: (a) ducts for dust, loose stock, or vapor removal; (b) ducts specifically fabricated for environmental air; and (c) other spaces used for environmental air (plenums). Type (a) ducts are not allowed to contain any type of wiring.

Type (b) ducts are permitted to contain a very limited amount of wiring. The NEC states,

> Only wiring methods consisting of Type MI cable without an overall nonmetallic covering, Type MC cable employing a smooth or corrugated impervious metal sheath without an overall nonmetallic covering, electrical metallic tubing, flexible metallic tubing, intermediate metal conduit, or rigid metal conduit without an overall nonmetallic covering shall be installed in ducts specifically fabricated to transport environmental air. Flexible metal conduit shall be permitted, in lengths not to exceed 1.2 m (4 ft), to connect physically adjustable equipment and devices permitted to be in these fabricated ducts. The connectors used with flexible metal conduit shall effectively close any openings in the connection.

It then provides an exception, as follows:

> Wiring methods and cabling systems, listed for use in other spaces used for environmental air (plenums), shall be permitted to be installed in ducts specifically fabricated for environmental air-handling purposes under both of the following conditions: (1) The wiring methods or cabling systems shall be permitted only if necessary to connect to equipment or devices associated with the direct action upon or sensing of the contained air, and (2) The total length of such wiring methods or cabling systems shall not exceed 1.2 m (4 ft).

Type (c) ducts, which are known as “Other Spaces Used for Environmental Air (Plenums)” are the ones typically referred to as plenums. The NEC requirements are the same as those described earlier for plenums in the IMC. The NEC wording specifically reads as follows:

> The wiring methods for other spaces used for environmental air shall be limited to totally enclosed, nonventilated,

insulated busway having no provisions for plug-in connections, Type MI cable without an overall nonmetallic covering, Type MC cable without an overall nonmetallic covering, Type AC cable, or other factory-assembled multiconductor control or power cable that is specifically listed for use within an air-handling space, or listed prefabricated cable assemblies of metallic manufactured wiring systems without nonmetallic sheath. Other types of cables, conductors, and raceways shall be permitted to be installed in electrical metallic tubing, flexible metallic tubing, intermediate metal conduit, rigid metal conduit without an overall nonmetallic covering, flexible metal conduit, or, where accessible, surface metal raceway or metal wireway with metal covers. Nonmetallic cable ties and other nonmetallic cable accessories used to secure and support cables shall be listed as having low smoke and heat release properties.

When the NEC refers to "factory-assembled multiconductor control or power cable that is specifically listed for use within an air-handling space," it simply means "plenum cables." The NFPA Standards Council (which oversees the NFPA codes and standards development activities, administers the rules and regulations, and acts as an appeals body) determined, many years ago, that NFPA 90A[11] is responsible for the fire safety requirements for all materials and products in ducts and plenums. NFPA 90A is under the jurisdiction of a small committee, namely, the Technical Committee on Air Conditioning. Therefore, the NEC is restricted to regulating issues other than fire safety for wiring in plenums, but it regulates fire safety of wiring in all other applications in buildings. Similarly, neither NFPA 101[10] (Life Safety Code) nor NFPA 5000[159] (Building Construction and Safety Code) can assign their own fire safety requirements to materials or products in plenums, but must refer to NFPA 90A. The 2023 edition of NFPA 90A changed the terminology associated with wires and cables in plenums to correlate with NEC wording more directly.

Section 300.22 of the NEC also addresses information technology equipment. In particular it explains that electrical wiring in air-handling areas beneath raised floors for information technology equipment are regulated in accordance with Article 645 (with further details in NFPA 75[160]). The NEC article refers to section 300.2 (c) (meaning plenums) for the fire safety requirements of wiring but allows less severe fire safety requirements for wires and cables if the area is fully sprinklered.

22.3.2.2 National Life Safety Code

The 2024 edition of the Life Safety Code (NFPA 101) consists of 43 chapters, plus a few annexes, which are not mandatory portions of the code and can be adopted (or not) separately by individual AHJs. The key chapters within NFPA 101 are the following:

- Chapter 2: Mandatory references
- Chapter 3: Definitions of terms
- Chapter 4: General
- Chapter 5: Performance-based code alternative
- Chapter 6: Classification of occupancy and hazard of contents
- Chapter 7: Means of egress
- Chapter 8: Features of fire protection (such as compartmentation, barriers, and partitions)
- Chapter 9: Building service and fire protection equipment (such as sprinklers, alarms, and detectors)
- Chapter 10: Interior finish, contents, and furnishings (key chapter on reaction to fire requirements)
- Chapters 11–43: New and existing occupancy chapters

The requirements for interior finish, contents, and furnishings are very similar to those in the IBC and IFC, except that NFPA 101 is more detailed in some cases. For example, while the IFC requires polypropylene and high-density polyethylene (HDPE) materials when used as interior finish to be tested in accordance with NFPA 286 (and not ASTM E84), NFPA 101 expands this to "solid thermoplastics including, but not limited to, polypropylene, high-density polyethylene (HDPE), solid polycarbonate, solid polystyrene, and solid acrylic materials that melt and drip when exposed to flame."

Each of the occupancy chapters in NFPA 101 makes their independent decision as to whether the corresponding committee wants to incorporate requirements for interior finish, upholstered furniture, mattresses, or decorations. One consequence of this is that several occupancy chapters have decided that upholstered furniture and mattress requirements (for both smoldering and flaming ignition) should not be applied to their occupancies.

Annex A of NFPA 101 contains a table with a summary of the generic requirements for interior finish for each type of material.

22.3.2.3 Fire Code

The Fire Code issued by NFPA (NFPA 1)[124] is a blend of two documents, the NFPA Fire Prevention Code (also known as NFPA 1) and the Uniform Fire Code previously issued by the Western Fire Chiefs, which was a partner of ICBO when ICBO issued the Uniform Building Code. The 2024 edition consists of 75 chapters, although several of them are "reserved" for future use.

The purpose of the Fire Code is to prescribe minimum requirements necessary to establish a reasonable level of fire and life safety and property protection from the hazards created by fire, explosion, and dangerous conditions. It applies to both new and existing conditions. If there is a conflict between a general requirement and a specific requirement for a particular occupancy or situation, the specific requirement applies. If two or more classes of occupancy occur in the same building or structure and are so intermingled that separate safeguards are impracticable, means of egress facilities, construction, protection, and other safeguards must comply with the most restrictive fire safety requirements of the occupancies involved. The code applies also to vehicles and vessels when they are in a fixed location and are occupied as buildings. The code applies to buildings permitted for construction after the adoption of a particular edition of NFPA 1, but only with the provisions for new buildings. Buildings in existence or permitted for construction prior to the adoption of a particular edition of NFPA 1 must comply with the provisions for existing buildings. Repairs, renovations, alterations, reconstruction, change of occupancy, and additions to buildings must comply with both NFPA 101 and the building code. Newly introduced equipment, materials, and operations regulated by NFPA 1 need to comply with the requirements for new construction or processes.

The scope of the Fire Code includes the following:

(1) Inspection of permanent and temporary buildings, processes, equipment, systems, and other fire and related life safety situations
(2) Investigation of fires, explosions, hazardous materials incidents, and other related emergency incidents
(3) Review of design and construction plans, drawings, and specifications for life safety systems, fire protection systems, access, water supplies, processes, and hazardous materials and other fire and life safety issues
(4) Fire and life safety education of fire brigades, employees, responsible parties, and the general public
(5) Existing occupancies and conditions, the design and construction of new buildings, remodeling of existing buildings, and additions to existing buildings
(6) Design, alteration, modification, construction, maintenance, and testing of fire protection systems and equipment
(7) Access requirements for fire department operations
(8) Hazards from outside fires in vegetation, trash, building debris, and other materials
(9) Regulation and control of special events including, but not limited to, assemblage of people, exhibits, trade shows, amusement parks, haunted houses, outdoor events, and other similar special temporary and permanent occupancies
(10) Interior finish, decorations, furnishings, and other combustibles that contribute to fire spread, fire load, and smoke production
(11) Storage, use, processing, handling, and on-site transportation of flammable and combustible gases, liquids, and solids
(12) Storage, use, processing, handling, and on-site transportation of hazardous materials
(13) Control of emergency operations and scenes
(14) Conditions affecting firefighter safety

Chapter 10 addresses various fire safety concerns, including combustible vegetation (which is usually not permitted in most occupancies) and other combustible contents. In particular Chapter 10 has a section (10.20) that addresses indoor children's playground structures, which exceed 3.1 m (10 ft) in height and 14.9 m^2 (160 ft^2) in area. The materials of construction of these playgrounds must be among the following: fire-retardant-treated wood, light-transmitting plastics, foam plastics and balls with a maximum heat release rate ≤ 100 kW when tested in accordance with UL 1975, aluminum composite materials which are Class A in ASTM E84 or via NFPA 286, textiles, and films complying with the flame propagation performance criteria of NFPA 701 and rigid plastic materials with a peak rate of heat release ≤ 400 kW/m^2 when tested in accordance with ASTM E1354 at an incident heat flux of 50 kW/m^2 in the horizontal orientation at a thickness of 6 mm. The floor covering under the children's playground structure must meet a Class I interior floor finish classification in accordance with ASTM E648 (flooring radiant panel).

Chapter 20 addresses occupancy fire safety, and this is where most requirements for reaction-to-fire testing are included, usually by "extraction" of the requirements from the Life Safety Code (NFPA 101), for interior finish, furnishings, and decorations.

Chapter 19 addresses combustible waste and refuse and includes the specific requirement that nonmetallic rubbish containers exceeding a capacity of 0.15 m^3 (5 ft^3 or 40 gallons) to meet a peak rate of heat release not exceeding 300 kW/m^2 at a flux of 50 kW/m^2 when tested in the horizontal orientation, at a thickness as used in the container but not less than of 6 mm (0.25 in. (6 mm), in accordance with ASTM E1354 (cone calorimeter).

Chapters 21 through 34 address individual occupancies. Chapter 60 addresses Hazardous Materials and all subsequent chapters address some specific hazardous materials, including aerosols, compressed gases, corrosives, explosives, fireworks, flammable and combustible liquids, flammable solids, and toxics. Much of the old misleading information about some of these hazardous materials has been taken out of recent editions.

22.3.2.4 NFPA Building Code

The NFPA Building Construction and Safety Code or NFPA 5000[159] is the alternate building code to the IBC. In structure it is similar to the Life Safety Code (NFPA 101) in that it starts with general requirements and then includes chapters on occupancies. It is also similar to NFPA 1 in that, after chapter 40 it has specific chapters on individual materials, such as plastics, wood, glass, and gypsum board (which the IBC also has). It also contains a performance option alternative in Chapter 5.

NFPA 5000 addresses those construction, protection, and occupancy features necessary to minimize danger to life and property. The purpose of NFPA 5000 is to provide minimum design regulations to safeguard life, health, property, and public welfare and to minimize injuries by regulating and controlling the permitting, design, construction, quality of materials, use and occupancy, location, and maintenance of all buildings and structures within the jurisdiction and certain equipment specifically regulated herein. The provisions of the Code apply to the construction, alteration, repair, equipment, use and occupancy, maintenance, relocation, and demolition of every building or structure, or any appurtenances connected or attached to such buildings or structures within the jurisdiction. The provisions of the Code also apply to existing buildings and structures if any one of the following conditions applies:

(1) A change of use or occupancy classification occurs.
(2) A repair, renovation, modification, reconstruction, or addition is made.
(3) The building or structure is relocated.
(4) The building is considered damaged, unsafe, or a fire hazard.
(5) A property line that affects compliance with any provision of this Code is created or relocated.

The key chapters within NFPA 5000 are the following:

- Chapter 2: Mandatory references
- Chapter 3: Definitions of terms
- Chapter 4: General
- Chapter 5: Performance-based code alternative
- Chapter 6: Classification of occupancy and hazard of contents
- Chapter 7: Construction types and heights and areas
- Chapter 8: Fire resistive construction
- Chapter 10: Interior finish, contents, and furnishings (key chapter on reaction to fire requirements)
- Chapter 11: Means of egress
- Chapters 16–33: New occupancy chapters
- Chapters 41–48: Materials (concrete, aluminum, masonry, steel, glass and glazing, gypsum board and plastics)
- Chapter 55: Fire protection systems and equipment

Chapter 10 of NFPA 5000 is virtually identical to the corresponding chapter of NFPA 101, except that it does not address, of course, furniture, mattresses, or decorations (which are not building construction materials). One interesting addition is the permission to incorporate coated wood panels in new construction. NFPA 5000 states that field applied fire-retardant coatings are not permitted to obtain compliance with the interior finish requirements. But it also states that surfaces of walls, partitions, columns, and ceilings are permitted to be finished with factory-applied fire-retardant-coated products that have been listed and labeled to demonstrate compliance with the requirements of ASTM E2768 on the coated surface (this is not included in the IBC). NFPA 5000 also has a similar table to that in NFPA 101, in the annex, describing the requirements for each type of interior finish. The same technical committee is responsible for the Life Safety Code and Building Code chapters on interior finish, contents, and furnishings.

22.3.2.5 Buildings of Historic or Cultural Interest

There are two codes addressing buildings of particular interest: NFPA 909 (Code for the Protection of Cultural Resource Properties—Museums, Libraries, and Places of Worship)[161] and NFPA 914 (Code for Fire Protection of Historic Structures).[162] NFPA 909 and NFPA 914 include definitions for various terms, which are important to understand what the codes intend:

Cultural Resource Properties. Buildings, structures, sites, or portions thereof that are culturally significant or that house culturally significant collections for museums, libraries, and places of worship.

Historic Building. A building that is designated, or deemed eligible for such designation, by a local, regional, or national jurisdiction as having historical, architectural, or cultural significance.

Historic Fabric. Original or added building or construction materials, features, and finishes that existed during the period that is deemed to be most architecturally or historically significant, or both.

Historic Preservation. A generic term that encompasses all aspects of the professional and public concern related to the maintenance of a historic structure, site, or element in its current condition, as originally constructed, or with the additions and alterations determined to have acquired significance over time.

Historic Site. A place, often with associated structures, having historic significance.

Historic Structure. A building, bridge, lighthouse, monument, pier, vessel, or other construction that is designated or that is deemed eligible for such designation by a local, regional, or national jurisdiction as having historical, architectural, or cultural significance.

Therefore, the intent is for NFPA 909 to address buildings, both new and old, that have cultural significance, either because of their contents (museums, libraries, and the like) or because they are used for religious ceremonies. On the other hand, NFPA 914 deals with older buildings (or structures) that have some historic significance because of some particular historic association with the building. In many countries buildings or structures are entitled to be declared of historic significance if they are > 100 years old. However, not every structure that has survived for 100 years is of historic significance. Usually there is a need for the authority having jurisdiction to make a determination that the building or structure is associated with some historic event or represents some very unusual characteristic that makes it worth preserving (and often expending the necessary funds to update it for fire safety). Libraries, museums, and places of worship that are housed in historic structures need to comply with the requirements of both NFPA 909 and NFPA 914. In the NFPA system, the three key codes: the building code, the life safety code and the fire code all incorporate special considerations for historic buildings which comply with NFPA 914. On the other hand only NFPA 5000 (the building code) and NFPA 1 (Uniform Fire Code) incorporate requirements for NFPA 909; they say that buildings, structures, or spaces that are or contain cultural resource properties must be constructed in accordance with NFPA 909. In the ICC set of codes there are no references to either of these codes because the concept of buildings of particular significance is not considered.

The requirements contained in both of these codes tend to represent a lower level of passive fire protection than in other buildings (especially for NFPA 914) because of the interest in preserving the historic "look."

22.3.2.6 Ships

NFPA has issued NFPA 301[56] to allow regulation of merchant vessels (ships) to be based on a consensus document. The code covers three types of vessels: passenger, cargo, and tank vessels, and it is intended to apply to ocean-going towing vessels in future editions, in anticipation of an expected US law mandating fire protection upgrades. It was originally envisaged that future editions might also add other types of vessels; however this code is rarely, if ever, used for new ships, and it is almost superfluous. A key reason it is not used is that includes requirements for sprinkler protection that are more onerous than those of the regulatory authority, the US Coast Guard. Chapter 8 contains general fire safety requirements for any vessel, but they will only apply if referenced in the specific vessel or occupancy chapter. This chapter includes requirements for interior finish, upholstered furniture, mattresses, case furniture, draperies, stacking chairs, nonmetallic rubbish containers, and contains the overall requirements for electrical and optical fiber wires and cables. Interior wall and ceiling finish must comply with a flame spread index and a smoke developed index of 20 and 10 respectively (based on ASTM E84),[128] or they must meet the same requirements as in building codes when tested based on the room-corner test (NFPA 286).[127] They are also allowed to comply with the requirements of the IMO FTP code,[52] which involves testing for smoke obscuration (based on ASTM E1995[163] or NFPA 270[164] or ISO 5659–2[165]; all equivalent), for flame spread (based on the apparatus also used for the Lateral Ignition and Flame Spread Test, or LIFT test, per ASTM E1317)[166] and for heat content (based on ISO 1716).[167] Upholstered furniture must meet the requirements of the ASTM E1537 full scale heat release test,[100] with the requirements of CA TB 133.[99] Mattresses must meet the requirements of the ASTM E1590 full scale heat release test,[90] with the requirements of CA TB 129.[86] The requirements for furnishings in this chapter would result in better fire safety than those associated with the IMO SOLAS fire tests,[50–52] but unfortunately the code includes statements that exclude requirements if the space is fully protected by a sprinkler or water mist system. The code includes a requirement that stacking chairs must exhibit fire performance such that each stacking chair is incapable of causing flashover in the space on its own; there is an annex statement such chairs should comply with ASTM E1822,[168] but that is not an actual requirement. Nonmetallic rubbish containers must meet a cone calorimeter pass-fail criterion. The requirements for wire and cable are based on those of the National Electrical Code,[7] except that the minimum requirements are that cables meet the vertical cable tray fire test with a perpendicular burner from UL 2556,[68] and they need not be grounded (a logical requirement in ships).

Chapters 10 through 17 address the individual spaces within each type of vessel; the key chapters associated with fire safety are the ones for accommodation spaces (chapter 10) and for medical spaces (chapter 11). Chapter 21 is the overall chapter for passenger vessels, and it is here that the requirements from earlier chapters are called for or not. A key requirement in this chapter is the one that requires that on Group I and Group II passenger vessels and on all passenger vessels with overnight accommodations for passengers, an automatic sprinkler system or a water mist system must be installed to protect all accommodation, service, and storage spaces. Group I passenger vessels are those with either over 3,000 day passengers or over 300 overnight passengers, and Group II passenger vessels are those with either over 1,000 and under 3,000 day passengers or over 150 and under 300 overnight passengers.

Neither cargo vessels nor tank vessels have too many requirements associated with fire safety, and the key issues covered in their chapter (Chapter 20) are construction specifications and active fire protection. However the code also states that interior finish and furnishings in exits and exit accesses must meet the requirements of Chapter 8 and that accommodation spaces must meet the requirements of Chapter 10 (and these vessels are rarely sprinklered).

22.3.2.7 Manufactured Housing

Manufactured homes are built and installed to the US Department of Housing and Urban Development's (HUD) Manufactured Home Construction and Safety Standards (MHCSS). The standards address structural, fire safety, and energy-efficiency issues and require adequate ventilation. This regulation supersedes local and state building codes and is the current minimum standard that all HUD-code homes are required to meet. NFPA used to periodically update NFPA 501, Standard on Manufactured Housing,[149] because NFPA 501 was the standard approved by industry and other stakeholders, but it was never officially adopted by HUD, and has been withdrawn in 2022. The NFPA standard never had authority over the older MHCSS regulation but rather provided recommendations to HUD. Research conducted in 2004 by US Department of Energy (DOE) authors contributed to NFPA 501's improved stringency of thermal efficiency. NFPA 501 also incorporated improvements over the HUD-code based provisions, at the time, based on the experiences of energy-efficient manufactured home programs and on fire safety, an area in which improvements continue, as happens with all consensus standards. Subsequently, the HUD documents, which regulate the construction and transport of mobile homes can be found in the Code of Federal Regulations Title 24, Housing and Urban Development Subtitle B, most notably Part 3280[169] and Part 3285.[170] The former provides "Manufactured Home Construction and Safety Standards" while the latter provides "Manufactured Home Installation Standards." One problem associated with the use of the HUD requirements is that editions of referenced standards are not updated (for example, the reference to ASTM E84 is to the 2001 edition, long superseded).

In terms of fire safety, there are no fire resistance requirements (other than for parapets for attached manufactured homes) and all interior surfaces must comply with a flame spread index of 200 in the Steiner tunnel test, ASTM E84[128] or a radiant panel index of 200 in the radiant panel test, ASTM E162.[64] (In fact, the regulation describes the requirements in terms of the obsolete designations of "flame spread rating.") Thermal insulation materials, other than foam plastics and loose-fill insulation, must meet an ASTM E84 Class A requirement (i.e., FSI ≤ 25 and SDI ≤ 450). Cellulose loose-fill insulation must meet requirements based on CPSC smoldering tests only.[139,140] Foam plastic insulation must be treated the same way as it is handled in the building codes (see Table 22.13): it cannot be used exposed unless it complies with the requirements based o NFPA 286 and must also meet an ASTM E84 Class B requirement behind the thermal barrier. Note, however, that some of the less common foams that meet the requirements based on the NFPA 286 test are rarely used in manufactured housing.

22.3.2.8 Air Conditioning Standard

NFPA 90A[11] is a key document because it controls all the fire safety requirements for ducts and plenums in the NFPA system. The requirements are basically very similar to those in the IMC. The only materials permitted in plenums are: wires and cables that meet criteria based on the NFPA 262[70] plenum cable fire test or that are enclosed in noncombustible raceways, pneumatic tubing, based on UL 1820,[146] nonmetallic sprinkler pipes, based on UL 1887,[147] fiber optic raceway systems, based on UL 2024[148] and combustible electrical equipment, based on UL 2043,[149] so that each one is covered by a special test for that product. There is an exception for plastic plumbing and tubing intended for water distribution, to be discussed later (see 22.4.2.5). Foam plastic wall and ceiling insulation must meet the NFPA 286 room-corner test or be covered by a thermal barrier or by a steel skin. All other materials, including pipe and duct insulation, must meet the requirements of a flame spread index of 25 and a smoke developed index of 50 in the ASTM E84 Steiner tunnel test (a much more severe smoke criterion than that for non plenum areas, where the smoke developed index is allowed to reach 450). The standard also notes that electrical wires and cables listed to a more severe requirement, UL Subject 2424[171] can be used without further testing; this is very important because the UL Subject 2424 requirement (now obsolete) was based on "limited combustible" cables, that have a very low heat content (when tested to NFPA 259)[172] and a very low smoke release (when tested to ASTM E84); they would have to be almost invariably constructed of fluoropolymer materials, while other plenum cables tend to be constructed of fire retarded materials. It also notes that electrical wires and cables and optical fiber cables installed in metal raceways or metal sheathed cable are not considered to be exposed to the airflow and need not meet the requirements of NFPA 262. Plenum cables are permitted to be installed in three of the types of plenums covered by the standard: ceiling cavity plenums, raised floor plenums and air-handling room plenums, but not in apparatus casing plenums or in ducts. The materials of construction of the plenums themselves, and all other plenum or duct contents, must be made of "limited combustible" materials. Finally, ducts and plenums are not permitted to be used for storage. It is of interest to point out that there has been much debate regarding the use of plenum cables, due to various attempts to replace them by cables complying with UL Subject 2424. These efforts were unsuccessful because they were seen as an attempt by certain manufacturers to gain market share without a justification based on fire safety or fire losses, and UL 2424 is no longer referenced in NFPA 90A.

22.3.2.9 Other NFPA Codes and Standards

As discussed earlier, NFPA 130[12] is another important document because it is used for regulation of train and underground system, by authorities having jurisdiction over local rail systems in some US systems, in some Canadian ones and in some Asian ones. Table 22.4 shows the reaction-to-fire requirements for rail cars, from NFPA 130–2023, and Table 22.5 has the associated notes. The approach used by NFPA 130 for accepting vehicle materials is that materials intended for use in a limited area of the vehicle, and not meeting the requirements of the table, are permitted only after an appropriate fire hazard analysis establishes, within the limits of precision, that the material produces a contribution to fire hazard equal to or less than a material meeting the appropriate criteria of the table, where the alternative material is used in the same location to fulfill a function similar to the candidate material. NFPA 130 also includes requirements for stations and trainways as well as rail vehicles. The key issues to be considered in stations are the same ones as in other assembly occupancies, namely electrical, interior finish, upholstered furniture, decorations, insulation, and trash disposal. With regard to trainways the standard looks at the effect of areas that are potentially concealed spaces and considers that factor.

The NFPA 13 series of sprinkler standards (NFPA 13,[12] 13D,[173] and 13R[174]) form the bases for all the requirements for sprinkler installations in the US (and probably worldwide). These standards do not contain any requirements for passive fire protection but explain how to install sprinklers and in what spaces the sprinkler protection is needed. The recent adoption of requirements for sprinklers in residences and townhouses in the IRC is based on NFPA 13D systems, but most nonresidential occupancies are sprinklered based on the more stringent NFPA 13 systems.

NFPA 99[175] governs all health care occupancies, in conjunction with the Life Safety Code and the National Electrical Code. In fact virtually all hospitals, and the Joint Commission for the Accreditation of Healthcare Organizations (JCAHO), which regulates hospitals, use the combination above almost without any other consideration. It must be pointed out that JCAHO does not necessarily adopt the most recent editions of the codes; they are frequently dealing with code editions that are 2 or 3 editions old.

Other NFPA documents that are important for specific applications are NFPA 30 (Flammable and Combustible Liquids Code), NFPA 54 (National Fuel Gas Code), NFPA 75[160] (Standard for the Protection of Information Technology Equipment), and NFPA 318 (Standard for the Protection of Semiconductor Fabrication Facilities).

22.3.3 IAPMO Codes

Beyond the Western Fire Chiefs Association, issuing the Uniform Fire Code, now NFPA 1, another organization was a partner to ICBO, namely the International Association of Plumbing and Mechanical Officials (IAPMO) who issued the Uniform Mechanical Code (UMC)[176] and the Uniform Plumbing Code (UPC). They continue to do so, but now have an agreement with NFPA so that the UMC and the UPC are NFPA's equivalents to the ICC codes. The UPC has little, if any, importance with regard to fire issues.

22.3.3.1 Uniform Mechanical Code

The UMC applies to the addition to or erection, installation, alteration, repair, relocation, replacement, use, or maintenance of heating, ventilating, cooling, refrigeration systems, incineration, or other miscellaneous heat-producing appliances. It also covers design. In effect there is a considerable amount of overlap with NFPA 90A.[11]

The key provisions covered by this code are those for materials in ducts and plenums and for hydronic piping systems. Just like in the IMC and in NFPA 90A, the only materials permitted in plenums are wires and cables that meet criteria based on the NFPA 262[70] plenum cable fire test or that are enclosed in noncombustible raceways, pneumatic tubing, based on UL 1820,[146] sprinkler pipes, based on UL 1887,[147] fiber optic raceway systems, based on UL 2024[148] and loudspeakers and recessed lighting fixtures, based on UL 2043,[149] so that each one is covered by a special test for that product. Foam plastic wall and ceiling insulation must meet the NFPA 286 room-corner test or be covered by a thermal barrier or by a steel skin. All other materials, including pipe and duct insulation, must meet the requirements of a flame spread index of 25 and a smoke developed index of 50 in the ASTM E84 Steiner tunnel test (a much more severe smoke criterion than that for non plenum areas, where the smoke developed index is allowed to reach 450). However, insulation of hydronic pipe systems is allowed to be tested in accordance with NFPA 274[177] instead of using ASTM E84 (with the 25/50 pass fail criteria). NFPA 274 is a heat release test in which a set of pipes is insulated in a vertical chase, exposed to a gas burner and the heat and smoke release are measured. The pass/fail criteria used by the UMC are a maximum peak heat release rate of 300 kW, a maximum total heat release of 50 MJ, a maximum total smoke release of 500 m^2, and the flames cannot extend 305 mm or more above the top of the vertical portion of the apparatus at any time during the test. The UMC also explicitly permits (if listed) a series of specific piping materials permitted for use in these systems.

22.4 STANDARDS

22.4.1 Organizations and Committees Issuing Fire Standards or Standards with Fire Tests

As discussed earlier, fire test standards are developed primarily in the US by ASTM International and by NFPA, although FM Global (formerly Factory Mutual) and Underwriters Laboratories have also developed some widely used tests. Within ASTM the following committees have developed fire tests:

- ASTM E05: fire standards (primary fire standards development committee, with subcommittees on smoke and combustion products (which includes heat release), surface burning, combustibility, external fire exposures, furnishings and contents, transportation, large scale tests and fire resistance)
- ASTM C16: thermal insulation
- ASTM D07: wood (primarily subcommittee D07.07)
- ASTM D09: electrical and electronic insulating materials (primarily subcommittee D09.17)
- ASTM D13: textiles (primarily subcommittee D13.52)
- ASTM D20: plastics (primarily subcommittee D20.30)
- ASTM F08: sports equipment, playing surfaces, and facilities (primarily subcommittee F08.22)
- ASTM F15: consumer products
- ASTM F23: personal protective clothing and equipment (primarily subcommittee F23.80)
- ASTM F25: ships and marine technology (primarily subcommittee F25.03)
- ASTM F24: amusement rides and devices (subcommittee F24.10)
- ASTM F33: detention and correctional facilities (primarily subcommittee F33.05)

These same committees have also developed some specifications and practices which reference fire tests (and pass/fail criteria) and occasionally also contain some new fire tests. The NFPA Technical Committee on Fire Tests is the developer of the fire tests for NFPA (but some fire tests dealing with protective clothing, such as for firefighters, have been issued by other technical committees). Two other international organizations are systematic developers of fire tests, most of which are primarily used in Europe: ISO (International Organization for Standardization) and IEC (International Electrotechnical Commission). Within ISO and IEC, the key technical committees are as follows:

- ISO TC92: fire safety (with subcommittees on fire initiation and growth, fire containment (or fire resistance), fire threat to people and the environment (or smoke toxicity) and fire safety engineering)
- ISO TC61: plastics (primarily subcommittee SC4)
- IEC TC20: electric cables
- IEC TC89: fire hazard testing

There is an agreement between ASTM Committee E05, the NFPA Fire Tests Committee and UL that duplication of fire test standards between those organizations must be minimized. This has resulted in each one of the organizations withdrawing certain tests if they are suitably covered by a test from one of the other organizations.

22.4.2 Standard Test Methods

22.4.2.1 Noncombustibility

It is often thought that materials are basically of two types: combustible ones and noncombustible ones, and that the distinction is absolute. That is a flawed assumption. It is essential to clearly distinguish between the theoretical (or absolute) concept of noncombustibility, for a material or product, and the practical consideration, namely, that certain materials are suitable for actual use in construction when they do not contribute to a decrease in fire safety. In theory, a material will only be noncombustible if, when exposed to any applied heat source, the material will undergo no change, release no heat, or flame, and lose no mass. In practice, regulations and codes have established practical rules for what can be considered a noncombustible material for building construction. Therefore, a material will only meet the statutory requirements when assessed based on a fire test method. Of course, that does not mean that materials that are obviously noncombustible (such as steel or concrete) would need to be tested, but it applies to many other construction materials. The difference between theoretical noncombustibility and practical applications of noncombustibility of materials for use in building construction has been discussed.[178]

In the US, the test that has been used since the 1950s has been ASTM E136,[179] which exposes materials to a temperature of 750°C, in furnace, known as the Setchkin furnace, named after its inventor (Nicholas Setchkin).[180] It is critical to be aware that there are two thermocouples inside the furnace that are used to assess noncombustibility: one on the specimen surface and one inside the specimen (in its geometric center). A material "passes" the test (and can be considered a noncombustible material) when (and there are two options, both of which allow some burning):

(a) If the weight loss of the test specimen is 50% or less, the material passes the test when the two following criteria are met. (1) The recorded temperatures of either the surface or the interior thermocouple does not, at any time during the test, rise more than 30°C (54°F) above the stabilized furnace temperature measured prior to the test. (2) There is no flaming from the test specimen after the first 30 s.

(b) If the weight loss of the specimen exceeds 50%, the material passes the test when the two following criteria are met. (1) The recorded temperature of either the surface or the interior thermocouple does not, at any time during the test, rise above the stabilized furnace temperature prior to the test. (2) No flaming from the test specimen is observed at any time during the test.

Following extensive debate as to whether a material "passed" the test if only one of the thermocouples exceeded the limit, agreement was reached that, if the temperature at either thermocouple rose beyond the limit, the material fails the test. These accepted criteria are based on research conducted by Setchkin and published in 1952.[181-183]

The debate on these criteria continues and proposals are being made to develop new approaches which would allow more materials to be classified as noncombustible. An example of materials that could become classified as noncombustible with more lenient criteria are noncombustible materials that use combustible binders when commercialized. The percentage of binder included with such materials will affect whether the criteria are met.

In 2018, ASTM E136 was amended to add, as an alternative, a second furnace and procedure, as contained in ASTM E2652,[184] which was based on a furnace used in the ISO 1182 noncombustibility test.[109] (Note that when an ISO standard exists, the equivalent EN standard with the same designation is identical.) The requirement was that the acceptance criteria of ASTM E136 must be the same, irrespective of which furnace and procedure is used. It was determined that the results with either procedure were substantially equivalent.

A later complication must be described: the ISO 1182 standard was revised in 2020 to create a procedure with more precise temperature control, by having two thermocouples instead of only one controlling the furnace temperature (not the thermocouples by the specimen). However, neither ASTM E136 nor ASTM E2652[184] were amended in that fashion, meaning that the furnace used in both of the ASTM test methods is different from the one used in the ISO test method.

The acceptance criteria used in European regulation are different than those used in the US. The European regulation has created a Class A1, which requires materials to comply with the following 5 criteria: a temperature increase ≤ 30°C, a mass loss ≤ 50%, and no sustained flaming (all 3 in accordance with ISO 1182[109]). A gross calorific potential ≤ 2 MJ/kg and ≤ 1.4 MJ/m^2 (both in accordance with ISO 1716).[110] For comparison purposes, a material deemed complying with Class A1 may not always be deemed noncombustible in accordance with ASTM E136, since the latter criteria are more severe.

The IBC contains an exception that allows materials, such as gypsum board, to be acceptable as noncombustible materials. The exception applies to materials having a structural base of noncombustible material (as determined in accordance with ASTM E136), with a surfacing of not more than 0.125 inch (3.18 mm) in thickness having a flame spread index not greater than 50 when tested in accordance with ASTM E84.

Both the NFPA and the European classification systems include different approaches to the one in the IBC when considering the acceptance of materials that provide very little contribution to fire hazard but do not comply with strict noncombustibility requirements.

NFPA uses the designation of "limited combustible materials" for such materials, and they are frequently permitted for use together with strictly noncombustible materials. A material is considered to be a limited combustible material if it complies with either one of two sets of criteria. The first set of criteria states:

> The material is a limited combustible material if, in the form in which it is used, (a) exhibits a potential heat value ≤ 3500 Btu/lb. (8141 kJ/kg) where tested in accordance with NFPA 259[172]; (b) has a structural base of a noncombustible material with a surfacing not exceeding a thickness of 1/8 in. (3.2 mm) where the surfacing exhibits a flame spread index ≤ 50 when tested in accordance with ASTM E84, and (c) is composed of materials which, in the form and thickness used, neither exhibit a flame spread index > 25 nor evidence of continued progressive combustion when tested in accordance with ASTM E84 and are of such composition that all surfaces that would be exposed by cutting through the material on any plane would neither exhibit a flame spread index > 25 nor evidence of continued progressive combustion when tested in accordance with ASTM E84.

The second set of criteria states:

> The material is a limited combustible material if, where tested in accordance with ASTM E2965, at an incident heat flux of 75 kW/m^2 for a 20-minute exposure, both of the following conditions are met: the peak heat release rate does not exceed 150 kW/m^2 for longer than 10 seconds and the total heat released does not exceed 8 MJ/m^2.

In the European classification system, Class A2 materials are variously (depending on the member country) designated as either noncombustible materials or as materials of limited combustibility. They are required to comply with the following criteria: a temperature increase ≤ 50°C; a mass loss ≤ 50%; sustained flaming that does not continue for longer than 20 seconds (all 3 in accordance with ISO 1182[109]); a gross calorific potential ≤ 3 MJ/kg and ≤ 4 MJ/m2 (both in accordance with ISO 1716[110]); and additional criteria based on heat release, flame spread and ignitability (with a property known as FIGRA) (based on EN 13823[174]).

22.4.2.2 Ignitability

Unless a material ignites, there is no fire. Therefore, low ignitability is a key property. All combustible materials ignite, but from the point of view of ignitability, the higher the temperature a material has to reach before it ignites (or the longer time until it reaches that temperature), the safer it is. Therefore, measures of ignition are of three types:

(1) the minimum temperature for ignition (in the absence or presence of a pilot source of ignition),
(2) the time to ignition under a certain applied thermal insult (typically a heat flux or a flame of a particular size), and
(3) the minimum (or critical) heat flux required to reach ignition (because temperature and heat flux can be correlated with the Stefan-Boltzman constant and black-body radiation).

The traditional method of assessing ignition temperatures is done with a test, such as the Setchkin chamber (ASTM D1929[134]), which has been used for many years to determine self-ignition temperatures and flash ignition temperatures (with an applied pilot flame) of many materials. The furnace in the ASTM D1929 Setchkin chamber is the same one used in the test for assessing combustibility (ASTM E136), but the chamber temperature is not held constant, and there are other changes. The test exposes a small test specimen inside a vertical furnace tube, heated electrically. Its results are no longer considered very indicative of fire hazard, because the exposure is not truly representative of real fires. However, results from this test are frequently required in specifications and quoted in data sheets, and the test is referenced in building codes to determine the suitability of some plastics for use in construction (plastic signs and light transmitting plastics), because it represents an easily understandable single number.

The time to ignition at a certain incident heat flux is an excellent way of comparing material properties, but it will give a numerical result only, of course, if the heat flux is high enough to cause ignition. The advantage of the critical heat flux for ignition is that it can always be assessed; it results from extrapolation after measurements made at a variety of heat fluxes.

There are four other standard test methods used purely for assessing ignitability: a radiant cone heater with an intermittent gas flame igniter for building materials (ISO 5657,[185] (Figure 22.4) sometimes called the dipping duck test because of the shape of the igniter), a hot furnace test for plastics (ISO 871,[186] which contains two options: one based on the Setchkin chamber, virtually identical to ASTM D1929, and one based on the 2020 edition of the ISO 1182 chamber), and two tests for electrical materials coated onto wires: ASTM D3874[187] or IEC 60695–11–5[188] (needle flame test) and ASTM D6194[189] or IEC 60695–2–10[190] (glow wire test).

The glow wire test involves bringing an electrically heated wire in contact with the material and simulates the thermal stresses resulting from faults or overload conditions during cable use. The needle flame test simulates the flame that can result from a fault or overload condition. The needle flame test can also be used to obtain information on flame spread. In the case of the glow wire ignition test (ASTM D6194 or IEC 60695–2–10), the information obtained is the temperature (in °C) at which the coated wire ignites, and the thickness of the coating required for ignition. In the case of the needle flame ignition test (ASTM D3874 or IEC 60695–11–5), the information obtained is whether ignition occurred and whether it happened with edge flames (E) or with facing flames (F), as well as the thickness of the coating required for ignition. Edge flames are a more severe ignition source than the facing flames, so that if there is no ignition with edge flames, ignition will not occur with facing flames. Similarly, ignition becomes more likely the thinner the sample, so that if ignition does not occur at 1 mm, it will not occur at 3 mm thickness, but the reverse is not the case.

Other test methods can also be used to assess ignitability, together with other properties. Some important ones are the cone calorimeter (ASTM E1354,[76] (Figure 22.5) which has as its primary purpose the assessment of heat and smoke release); the Ohio State University calorimeter (ASTM E906,[47] (Figure 22.6) which also has as its primary purpose the assessment of heat and smoke release, and which is used by the FAA for regulatory purposes, as discussed above), and the lateral ignition and flame spread test (LIFT, ASTM E1321,[191] (Figure 22.7) which has as its primary purpose the assessment of flame spread). Both ASTM and ISO have issued standards with compilations of ignition sources: ASTM E3020[192] and ISO 10093.[193]

A study has shown the time-to-ignition (at 3 incident heat fluxes) in the cone calorimeter for a set of 35 materials (all plastics and Douglas fir wood) (Table 22.15).[194] Materials that do not ignite under the heat flux to which they have been exposed are shown as having an (arbitrary) time to ignition of 10,000 s. Table 22.15 also shows the minimum ignition flux to cause ignition within 100 s or 600 s, from the same test method. While a higher time to ignition represents better ignitability performance, the opposite is true for the critical ignition flux: the higher the better. It should be noted that the value presented for the extrapolation of the critical flux for ignition is not reliable under 15 kW/m^2, so all such data are shown as "≤ 15 kW/m^2."

FIGURE 22.4 ISO 5657 ignitability test apparatus.

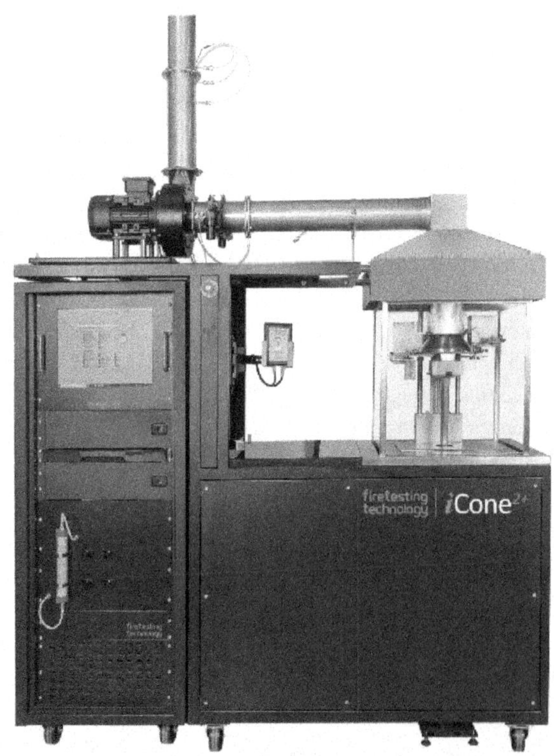

FIGURE 22.5 Cone calorimeter test apparatus.

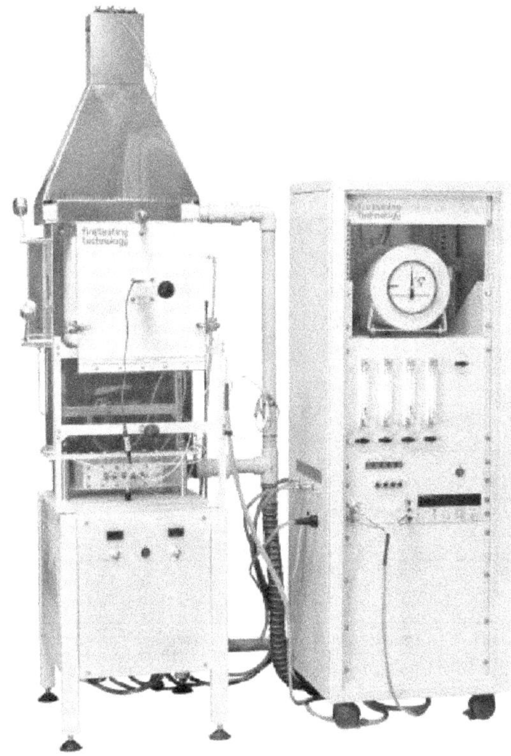

FIGURE 22.6 Ohio State University heat release rate calorimeter test apparatus.

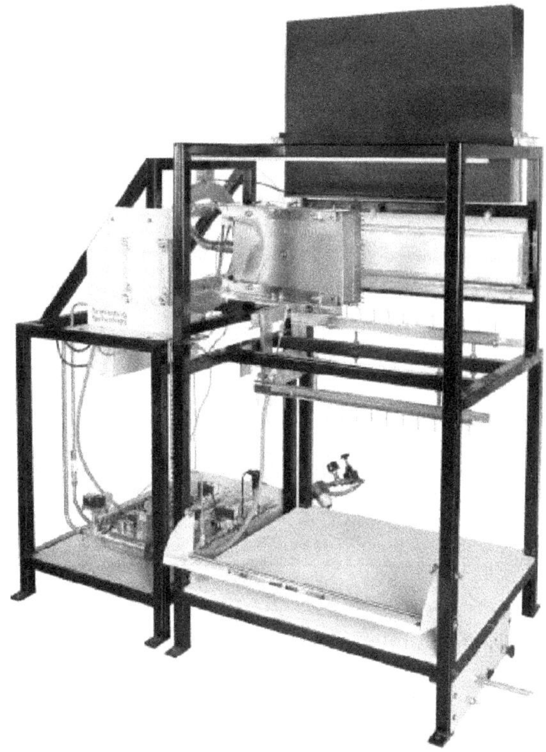

FIGURE 22.7 Lateral ignition and flame spread test (LIFT) apparatus.

TABLE 22.15
Ignitability of Materials in the Cone Calorimeter

	Time to Ignition (in s) at Heat Flux			Heat Flux (in kW/m²) for a Time to Ignition of	
	20 kW/m²	**40 kW/m²**	**70 kW/m²**	**600 s**	**100 s**
PTFE NV	10,000	10,000	252	63	83
VTE 3 V	10,000	1,212	17	45	64
VTE 2 V	10,000	1,253	424	60	110
VTE 4 V	10,000	10,000	1,583	86	115
PCARB NV	10,000	182	75	34	43
VTE 1 V	10,000	1,271	60	47	65
CPVC V	10,000	621	372	42	90
PVC CIM V	5,159	73	45	30	39
PVC WC FR V	236	47	12	≤ 15	31
PVC LS V	5171	187	43	33	44
XLPE NV	750	105	35	22	40
PVC WC SM V	176	36	14	≤ 15	27
PVC EXT V	3591	85	48	30	39
PVC WC V	117	27	11	≤ 15	22
ACR FR NV	200	38	12	≤ 15	28
PCARB B NV	6400	144	45	32	42
PPO GLAS NV	465	45	35	18	33
PPO/PS NV	479	87	39	17	38
ABS FV NV	5198	61	39	30	38
ABS FR NV	212	66	39	≤ 15	33
FL PVC V	102	21	15	≤ 15	20
DFIR NV	254	34	12	≤ 15	29

(Continued)

TABLE 22-15 Continued

	Time to Ignition (in s) at Heat Flux			Heat Flux (in kW/m²) for a Time to Ignition of	
	20 kW/m²	40 kW/m²	70 kW/m²	600 s	100 s
PS FR NV	244	90	51	≤ 15	38
ACET NV	259	74	24	≤ 15	35
PU NV	12	1	1	≤ 15	≤ 15
PMMA NV	176	36	11	≤ 15	27
THM PU NV	302	60	38	≤ 15	34
NYLON NV	1,923	65	31	27	37
ABS NV	236	69	48	≤ 15	34
PS NV	417	97	50	15	40
EPDM/SAN NV	486	68	36	18	36
PBT NV	609	113	59	20	41
PET NV	718	116	42	22	42
PE NV	403	159	47	≤ 15	50
PP NV	218	86	41	≤ 15	37

Description of materials in Table 21.14:

Abbreviation	Description and source (including trade name)
PTFE	Polytetrafluoroethylene sheet (samples were two sheets at 3 mm thickness each, Du Pont)
VTE 3	Flexible vinyl thermoplastic elastomer alloy wire and cable jacket experimental compound, example of the third of several families of VTE alloys
VTE 2	Flexible vinyl thermoplastic elastomer alloy wire and cable jacket experimental compound, example of the second of several families of VTE alloys
VTE 4	Semi flexible vinyl thermoplastic elastomer alloy wire and cable jacket experimental compound, example of a family of VTE alloys containing CPVC
PCARB	Polycarbonate sheeting (Lexan 141–111, General Electric)
VTE 1	Flexible vinyl thermoplastic elastomer alloy wire and cable jacket experimental compound, example of the first of several families of VTE alloys
CPVC	Chlorinated poly(vinyl chloride) sheet compound (BFGoodrich)
PVC CIM	Poly(vinyl chloride) general purpose rigid custom injection molding compound with impact modifier additives (BFGoodrich)
PVC WC FR	Flexible wire and cable poly(vinyl chloride) compound (containing fire retardants) (BFGoodrich)
PVC LS	Poly(vinyl chloride) rigid experimental sheet extrusion compound with smoke suppressant additives (BFGoodrich)
XLPE	Black non-halogen flame retardant, irradiation cross-linkable, polyethylene copolymer cable jacket compound (DEQD-1388, Union Carbide)
PVC WC SM	Flexible wire and cable poly(vinyl chloride) compound (containing minimal amounts of fire retardants) (BFGoodrich)
PVC EXT	Poly(vinyl chloride) rigid weatherable extrusion compound with minimal additives (BFGoodrich)
PVC WC	Flexible wire and cable poly(vinyl chloride) compound (non fire retarded) (BFGoodrich)
ACR FR	Kydex: fire retarded acrylic paneling, blue, (samples were 4 sheets at 1.5 mm thickness each, Kleerdex)
PCARB B	Commercial polycarbonate sheeting (Commercial Plastics)
PPO GLAS	Blend of polyphenylene oxide and polystyrene containing 30% fiberglass (Noryl GFN-3–70, General Electric)
PPO/PS	Blend of polyphenylene oxide and polystyrene (Noryl N190, General Electric)
ABS FV	Polymeric system containing acrylonitrile butadiene styrene and some poly(vinyl chloride) as additive
ABS FR	Cycolac KJT acrylonitrile butadiene styrene terpolymer fire retarded with bromine compounds (Borg Warner)
FL PVC	Standard flexible poly(vinyl chloride) compound (non-commercial; similar to a wire and cable compound) used for various sets of testing (including Cone Calorimeter RHR ASTM round robin; it contains PVC resin 100 phr; diisodecyl phthalate 65 phr; tribasic lead sulphate 5 phr; calcium carbonate 40 phr; stearic acid 0.25 phr
DFIR	Douglas fir wood board
PS FR	Fire retarded polystyrene, Huntsman 351 (Huntsman)
ACET	Polyacetal: polyformaldehyde (Delrin, Commercial Plastics)
PU	Polyurethane flexible foam, non fire retarded (Jo-Ann Fabrics)
PMMA	Poly(methyl methacrylate) (25 mm thick, lined with cardboard, standard RHR sample)
THM PU	Thermoplastic polyurethane containing fire retardants (estane, BFGoodrich)
NYLON	Nylon 6,6 compound (Zytel 103 HSL, Du Pont)
ABS	Cycolac CTB acrylonitrile butadiene styrene terpolymer (Borg Warner)
PS	Polystyrene, Huntsman 333 (Huntsman)
EPDM	Copolymer of ethylene propylene diene rubber (EPDM) and styrene acrylonitrile (SAN) (Rovel 701)
PBT	Polybutylene terephthalate sheet (Celanex 2000–2 polyester, Hoechst Celanese)
PET	Polyethylene terephthalate soft drink bottle compound
PE	Polyethylene (Marlex HXM 50100)
PP	Polypropylene (Dypro 8938)

22.4.2.3 Ease of Extinction

Once ignited, the easier it is to extinguish a material, the lower the resulting fire hazard. ASTM D2863[195] (or ISO 4589–2[196]) is used to determine the oxygen index (or LOI, Figure 22.8), which is the minimum oxygen concentration (in a flowing mixture of oxygen and nitrogen) required to support candle-like downward flaming combustion. This means that, in theory, a material with limiting oxygen index values above 25 to 27 should only burn under extreme conditions. In reality, the test serves principally as a measure of the ease of extinction of the material and is not properly predictive. The small specimen is placed vertically inside a glass column and ignited at the top with a small gas flame. This test method has excellent repeatability and reproducibility and can generate numerical data covering a very broad range of fire performance. However, once more, the test method is inappropriate as a predictor of real scale fire performance, mainly because of the low heat input and the artificiality of the high oxygen environments used. It is widely required in specifications and quoted in data sheets and is suitable as a quantitative quality control tool during manufacturing, and as a semi-qualitative indicator of the effectiveness of additives, during research and development, for low incident energy situations.

An application of this test method is ISO 4589–3,[197] a method which assesses the LOI at a variety of temperatures (Figure 22.9). The combination of these results allows a prediction of the flammability temperature, at which the LOI has a predicted value of 21, and the percentage of ash remaining after the test.

22.4.2.4 Flammability

Once ignited, the greater the flammability of a material, the greater will be the hazard associated with it. A small-scale flammability test which is very extensively used for plastic materials is the family of UL 94 tests[198] (also standardized in ASTM, ISO, and IEC, but most widely known by the UL standard designation). In this test, a small sample of material (127 mm × 13 mm, or 5 in × 0.5 in) is exposed vertically to a small Bunsen burner–type flame from underneath (in the UL 94 V test), and the results show a rating, ranging from V-0 (best), through V-1, V-2 to "B" (for Burn). The flame (using methane gas at a flow rate of 105 mL/min) is 20 mm high. Materials with a "B" rating (i.e., which fail) on the UL 94 Vertical test can also be tested in the less severe UL 94 HB (for Horizontal burning), where the assessment is whether a flame spread rate of 4 inches per minute is achieved. The UL 94 standard also contains a 500 W vertical test, with the test specimen being a plaque that is 150 mm × 150 mm, which assigns a classification of 5VA or 5VB. The UL 94 test is the most widely used fire test specified for plastic materials, especially fire retarded ones, and forms the basis of the famous "Yellow Card" used by Underwriters Laboratories to list plastic materials. Results from these tests are almost invariably found in a variety of specifications and in data sheets.

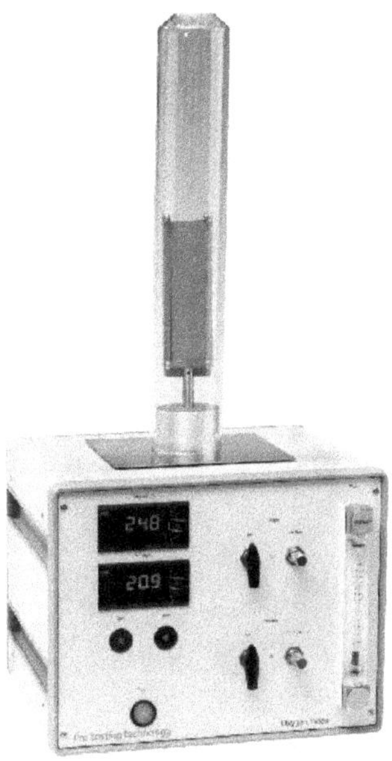

FIGURE 22.8 Oxygen index fire test apparatus.

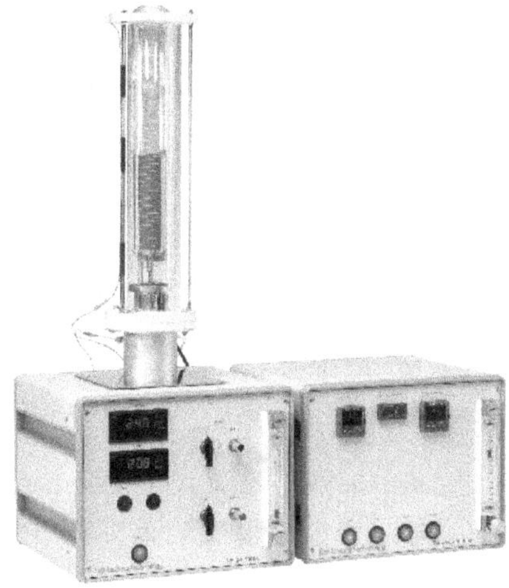

FIGURE 22.9 Temperature index test apparatus.

22.4.2.5 Flame Spread

The tendency of a material to spread a flame away from the fire source is critical to understand the potential fire hazard. Flame spread tests may refer to organic polymers themselves or to materials in diverse applications (such as textiles or electrical insulation sleeving), or to whole structures (such as furniture or building components). Most of the tests can be classified in terms of the angle between the exposed surface and the horizontal. Thus this angle may vary between 0°, for a horizontal material burning on its upper surface, through 180°, for a horizontal material burning on its lower surface, and 240°, for a material burning on its lower surface but rotated a further 60° from the horizontal, up to 270°, for a material burning vertically downwards.[199] This surface angle is closely related both to the extent to which the gaseous products of combustion heat the surface before the flame front reaches it and to the extent to which the hot gases penetrate into the polymer. Sample sizes can range very widely, from very small to quite large (7.3 m × 0.56 m, or 24 ft × 22 in, ASTM E84).[128]

Two other test apparatuses are also suitable to assess flame spread of materials: the ones in ASTM E162,[200] radiant panel (which is also used for cellular plastics, as ASTM D3675[201]), and the one in ASTM E1321,[191] also known as the LIFT test. A variation of the latter is used by the marine industry (ASTM E1317).[192]

The Steiner tunnel test method (ASTM E84)[128] was developed by Al Steiner for building materials,[202] such as wood or gypsum board. It has been adopted by every building and fire code in the USA. It is perhaps the most widely accepted test for surface flammability in North America, and there are a large number of applications tests developed from it. The most famous is the one used for electrical cables needed for plenum use (NFPA 262,[70] formerly also known as UL 910, now withdrawn), and there are also application tests for pneumatic tubing (UL 1820[146]), sprinkler piping (UL 1887[147]), optical-fiber and communications cable raceway (UL 2024[148]), and plastic plumbing and piping (UL 2846[203]). The specimen in ASTM E84 is a building material (normally up to 0.15 m thick), either in one unbroken length or in separate sections joined end to end, which is mounted face downwards so as to form the roof of a horizontal tunnel 305 mm high. The fire source, two gas burners, ignites the sample from below with an 89 kW fire source. When plastics started being used in construction, this test was applied to them, in spite of the fact that it is not always appropriate. For example, samples that cannot be retained in place above the tunnel floor, or which melt and continue burning on the tunnel floor (typical behavior for most thermoplastics) are still being tested with this equipment in spite of the results being misleading. It can also produce wrong results for thin materials.[204] The normal output is a flame spread index (FSI), calculated based on the area under the flame spread distance vs. time curve. The FSI index is a relative number based on an FSI = 0 for an almost noncombustible fiber cement board and FSI = 100 for red oak flooring. The SDI is also a relative number, but it is based on the same fire cement board (for SDI = 0) and heptane (for SDI = 100). The use of heptane improves the precision from the previously used red oak flooring. Results are used, in the codes, as classes, with Class A being an FSI of up to 25, Class B being an FSI of 26 up to 75 and Class C being an FSI of 76 up to 200. The test method also assesses smoke obscuration, and assigns a smoke developed index (SDI), based on a similar concept to the FSI. To be classified as Class A, B, or C, a material must have an SDI not exceeding 450. The requirements for tests in accordance with ASTM E84 for materials for use in plenums are an FSI of 25 and an SDI of 50.

Several of these test methods that are variations of ASTM E84 (NFPA 262, UL 1820, UL 1887, UL 2024, and UL 2846) use a tray inside the tunnel on which the test specimens are placed (instead of them being placed on the tunnel ledges). These tests have pass/fail criteria as follows: a flame spread distance not greater than 5 feet (1524 mm), a peak optical density not greater than 0.5, and an average optical density not greater than 0.15, and do not assess flame spread index or smoke developed index. UL 2846 is a test method for individual pairs of plastic water distribution plumbing piping that is permitted, as an exception, in both the UMC and NFPA 90A, for piping materials that are not tested to ASTM E84. It uses two pipes (without water) placed in the middle of the tray.

Since 2002, ASTM committee E05 has developed a series of standard practices to provide mandatory methods for specimen preparation and mounting materials and products in the ASTM E84 test method. The objective of developing these standard practices was to ensure that testing is being conducted in a standardized fashion at each laboratory, without local variability. The following standard practices have been issued between 2002 and 2022: ASTM E2231[205] (for pipe and duct insulation materials), ASTM E2404[206] (for textile, paper or polymeric (including vinyl) and wood wall or ceiling coverings, facings and veneers), ASTM E2573[207] (for site-fabricated stretch systems), ASTM E2579[208] (for wood products), ASTM E2599[209] (for reflective insulation, radiant barrier and vinyl stretch ceiling materials), ASTM E2688[210] (for tapes), ASTM E2690[211] (for caulks and sealants), ASTM E2988[212] (for flexible fibrous glass insulation for metal buildings), ASTM E3202[213] (for plastic composites for use as deck boards, stair treads, guards, or handrails), and ASTM E3287[214] (for fenestration profiles intended to support noncombustible infill materials). Also, many of the nonmandatory appendix sections have been revised to include clear instructions for improved standardization.

ASTM Test Method E162[64] (Figure 22.10) also assesses a flame spread index, but since the measurements are different from what is assessed by the Steiner tunnel test, the results are known as the radiant panel index. The radiant panel index is calculated as the product of a "flame spread factor" and a "heat evolution factor," using techniques

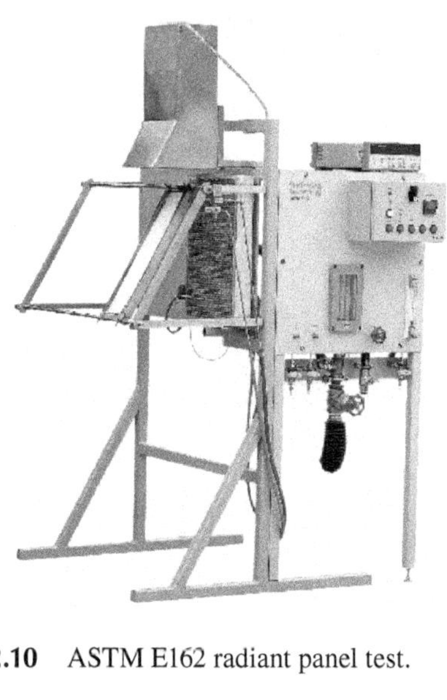

FIGURE 22.10 ASTM E162 radiant panel test.

described in the standard. The test apparatus consists of a gas-fed radiant panel in front of which an inclined (at a 30° angle) specimen (150 × 460 mm) is exposed to a radiant flux equivalent to a black body temperature of 670°C (943°K), namely approx. 45 kW/m², in the presence of a small gas pilot flame. The maximum thickness that can be tested in the normal specimen holder is 25 mm, but alternative specimen holders can accommodate thicker specimens. The ignition is forced near the upper edge of the specimen and the flame front progresses downward. The radiant panel index (I_s) (formerly flame spread index) is calculated as the product of a flame spread factor, which results from the measurements of flame front position and time, and a heat evolution factor, which is proportional to the maximum temperature measured in the exhaust stack. This test has also not been shown to be an adequate predictor of real scale fire performance. If the specimen melts or causes flaming drips, this is likely to affect the flame spread in a way that is uneven. The test method simply requires that such events be reported. Moreover, if flame spread is very rapid, the flame spread value is potentially lost unless recording is continuous. This apparatus is often referred to as the radiant panel, and results from this test are frequently required in regulations, particularly for transportation environments (both foamed and rigid materials) and detention environment specifications (foamed materials) and are quoted in data sheets.

The LIFT apparatus (lateral ignition and flame spread test, ASTM E1321)[191] was developed as an improvement on the apparatus in ASTM E162.[64] The specimen size for flame spread studies is 155 × 800 mm by a maximum thickness of 50 mm (and a smaller specimen is used for ignition studies). The test method determines the critical flux for flame spread, the surface temperature needed for flame spread, and the thermal inertia or thermal heating property (product of the thermal conductivity, the density, and the specific heat) of the material under test. These properties are mainly used for assessment of fire hazard and for input into fire models. A flame spread parameter is also determined, and this can be used as a direct way of comparing the responses of the specimens. This test method appears well-suited for materials (or composites), which do not melt, and which can be ignited without raising the incident flux to potentially dangerous limits. It has been used successfully for predictions of full-scale flame spread performance.[215]

22.4.2.6 Heat Release

The key question to ask in a fire is: "How big is the fire?" The critical fire property that presents the answer to that question is the maximum rate of heat release. A burning product will spread a fire to nearby products only if it gives off enough heat to ignite them. Moreover, the heat has to be released fast enough not to be dissipated or lost while traveling through the cold air surrounding any product that is not on fire. Therefore, fire hazard is dominated by the rate of heat release.[216–224] In fact, the rate of heat release has been shown to be much more important than either ease of ignition, smoke toxicity, or flame spread in controlling the time available for potential victims of a fire to escape.[225] In the late 1960s, Edwin Smith, at The Ohio State University (OSU), developed the first bench-scale test instrument, the OSU rate of heat release (RHR) calorimeter (ASTM E906),[47] to measure rates of heat release.[226]

In the early 1980s, Vytenis Babrauskas, at the National Institute of Standards and Technology (NIST, then National Bureau of Standards), developed a more advanced test method to measure rate of heat release: the cone calorimeter (ASTM E1354).[76,227] This fire test instrument can also be used to assess other fire properties, the most important of which are the ignitability (as discussed above), the mass loss, and the smoke released. Moreover, results from this instrument correlate with those from full-scale fires.[228–233] In order to obtain the best overall understanding of the fire performance of materials, it is important to test the materials under a variety of conditions. Therefore, tests are often conducted at a variety of incident heat fluxes. The peak rates of heat release (and total heat released) of the same materials included in Table 22.15 at each incident flux are shown in Table 22.16.[194]

A number of modern full-scale fire test methods have been developed for products, relying on rate of heat release measurements, such as those involving testing of upholstered furniture (ASTM E1537[100] and CA TB 133[99]), mattresses (ASTM E1590,[90] CA TB 129,[86] CA TB 603,[93] 16 CFR 1633,[24] and ASTM F1085 (Annexes A and C)[234]), stacking chairs (ASTM E1822[168]), electrical cables (ASTM D5424,[235] ASTM D5537,[236] and UL 1685[158]), plastic display stands (UL 1975),[237] other decorative items (NFPA 289,[238] a generic furniture calorimeter test), electrical equipment (UL 2043)[149] or wall lining products (NFPA 265,[129] NFPA 286,[127] ASTM E2257,[239] and ISO 9705–1[59]). In fact, room-corner tests are now being used in the codes, as the preferred alternatives to replace the ASTM E84 Steiner tunnel test, thus generating more useful results. Figure 22.11 shows a photograph of the equipment and a room-corner test layout. The cone calorimeter fire performance index (with tests conducted at 50 kW/m²)[240] was shown to be a good predictor of time to flashover in Federal Aviation Administration (FAA) full aircraft fires[233,241] and in the ISO 9705–1 room-corner test.[59] In addition, the same cone calorimeter tests, but using only heat release criteria, have been shown to have almost perfect predictability of ISO 9705–1 room-corner test rankings.[242]

22.4.2.7 Smoke Obscuration

Smoke obscuration (reduction in light transmitted) is a serious concern in a fire, because a decrease in visibility reduces the light available, thus hindering both escape from the fire and rescue by safety personnel. The main way in which a fire causes visibility decreases is through smoke emission. This decrease in visibility is the result of a combination of two factors: how much material is burnt in the real fire (which will be less if the material has better fire performance) and how much smoke is released per unit material burnt.

TABLE 22.16
Heat Release (of the Materials in Table 22.15) in the Cone Calorimeter (ASTM E1354) at Three Incident Heat Fluxes

	Flux 20 kW/m²		Flux 40 kW/m²		Flux 70 kW/m²	
Material	Pk RHR (kW/m²)	THR (MJ/m²)	Pk RHR (kW/m²)	THR (MJ/m²)	Pk RHR (kW/m²)	THR (MJ/m²)
PTFE	3	0.3	13	11.7	161	69.1
VTE 3	4	5.1	43	31.5	70	48.8
VTE 2	9	5.7	64	66.1	100	39.0
VTE 4	14	13.2	87	25.9	66	57.4
PCARB	16	0.1	429	119.2	342	121.7
VTE 1	19	12.2	77	48.1	120	63.4
CPVC	25	14.7	84	37.4	93	44.9
PVC CIM	40	3.0	175	24.3	191	93.0
PVC WC FR	72	36.5	92	51.7	134	65.5
PVC LS	75	6.6	111	73.6	126	75.5
XLPE	88	87.6	192	126.2	268	129.2
PVC WC SM	90	49.0	142	75.4	186	73.4
PVC EXT	102	2.9	183	90.8	190	96.5
PVC WC	116	47.3	167	95.7	232	94.4
ACR FR	117	20.5	176	86.7	242	77.2
PCARB B	144	35.4	420	134.7	535	143.5
PPO GLAS	154	111.0	276	125.8	386	125.7
PPO/PS	219	103.6	265	128.5	301	134.3
ABS FV	224	80.7	291	108.5	409	114.1
ABS FR	224	38.3	402	70.3	419	61.0
FL PVC	233	116.4	237	98.2	252	86.3
DFIR	237	46.5	221	64.1	196	50.0
PS FR	277	93.0	334	94.5	445	82.0
ACET	290	143.9	360	141.3	566	167.1
PU	290	9.4	710	13.2	1221	13.3
PMMA	409	691.5	665	827.9	988	757.1
THM PU	424	110.0	221	119.3	319	120.1
NYLON	517	188.0	1313	226.3	2019	233.8
ABS	614	160.0	944	162.5	1311	162.5
PS	723	202.6	1101	210.1	1555	197.8
EPDM	737	213.1	956	199.8	1215	215.7
PBT	850	96.7	1313	169.9	1984	197.4
PET	881	93.3	534	113.7	616	125.5
PE	913	161.9	1408	221.0	2735	227.5
PP	1170	231.3	1509	206.9	2421	231.1

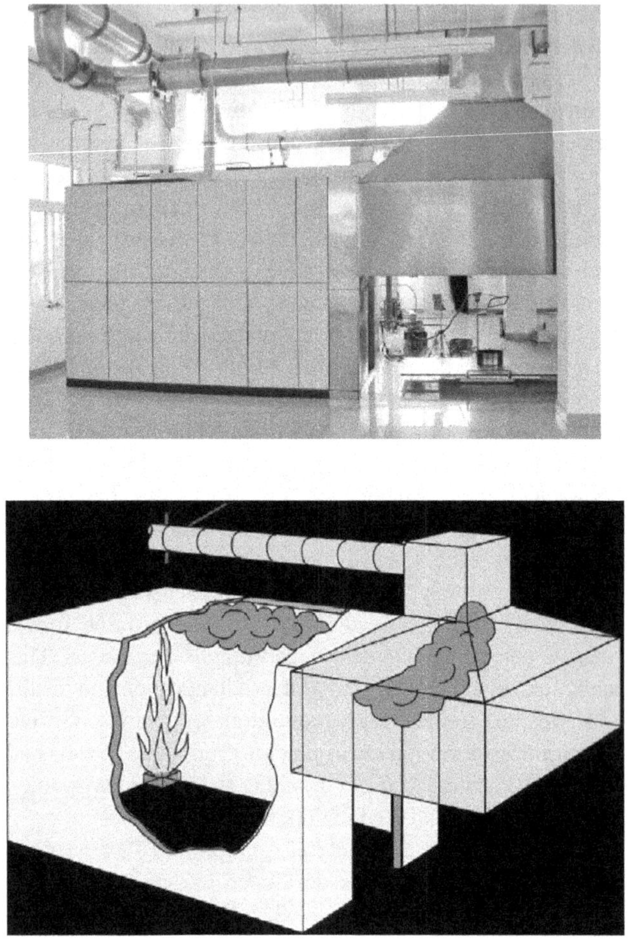

FIGURE 22.11 Photograph and schematic of room-corner fire test.

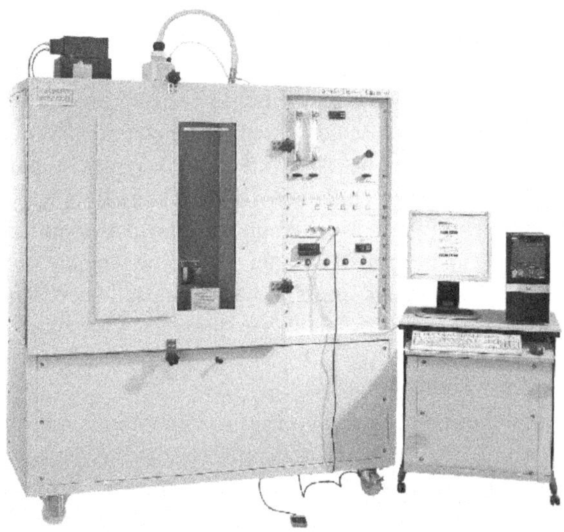

FIGURE 22.12 ASTM E662 smoke chamber.

Despite understanding that smoke obscuration ought to be measured in a large scale, or by a method which can predict large scale smoke release, the most common small scale test method for measuring smoke from burning products is the traditional smoke chamber in the vertical mode (ASTM E662)[48] (Figure 22.12). The test results are expressed in terms of a quantity called "specific optical density," which is defined in the test standard. This test has now been shown to have some serious deficiencies. The most important problem is misrepresentation of the smoke obscuration found in real fires and found for melting materials, such as thermoplastics.[243–247] When materials which melt or drip when exposed to flame, are exposed vertically in the smoke chamber test, the molten portions will have escaped the effect of the radiant heat source.[248] This means that some of the material does not burn during the test (and does not give off smoke), suggesting a low test result. In a real fire, all of the molten material will burn and generate smoke. If these dripping products are exposed horizontally, the entire sample will be consumed, and much more smoke will be released. The problem associated with the spread of fire from the flaming melt that occurs when thermoplastic materials have not been properly fire retarded has been highlighted by NIST work.[249] An alternate fire test exists also that assesses smoke obscuration in a closed chamber, ASTM E1995[163] or NFPA 270,[164] or ISO 5659–2,[165] it uses a conical heater instead of the traditional ASTM E662 heater (Figure 22.13).

The cone calorimeter,[76] which is a dynamic flow-through fire test, can also be used to assess smoke obscuration. The rankings tend to be quite different from those found with the static smoke chamber and are much more realistic. Several empirical parameters have been proposed to make this compensation for incomplete sample consumption, including one called the smoke factor (SmkFct), determined in small-scale rate of heat release calorimeters.[250] It combines the two aspects mentioned above: the light obscuration (as the total smoke released) and the peak rate of heat release.

The majority of materials with low flame spread (or low heat release) also exhibit low smoke release. However, it has been shown in a total of 5 projects involving series of room-corner tests (with the tested material lining either the walls or the walls and the ceiling), that approximately 10% of the materials tested (8 out of 84) exhibited adequate heat release (or fire growth) characteristics but have very high smoke release (Table 22.17 and Figure 22.14).[251,252] These materials would cause severe obscuration problems if used in buildings. A combination of this work, and the concept that a visibility of 4 m is reasonable for people familiar with their environment,[253] has led all US codes to include smoke pass/fail criteria when room-corner tests are used as alternatives to the ASTM E84 Steiner tunnel test.

22.4.2.8 Smoke Toxicity

Toxic potency values are most often assessed from the most suitable small-scale smoke toxicity test (NIST radiant test, using rats as the animal model, but only for confirmatory purposes, standardized in ASTM E1678[254] and NFPA 269[255]). The results from this test have been well validated with regard to toxicity in full-scale fires. However, such validation cannot be done to a better approximation than a factor of 3. This is illustrated by the fact that the range of the toxic potency of the smoke of almost all materials is so small that it pales in comparison with the ranges of toxic potencies of typical poisons.[256] Consequently, all smoke is extremely toxic, virtually irrespective of the material that is burning. The majority of fire fatalities are a result of inhalation of smoke and combustion products, rather than being the consequence of burns. Various organizations have been trying for many years to develop test

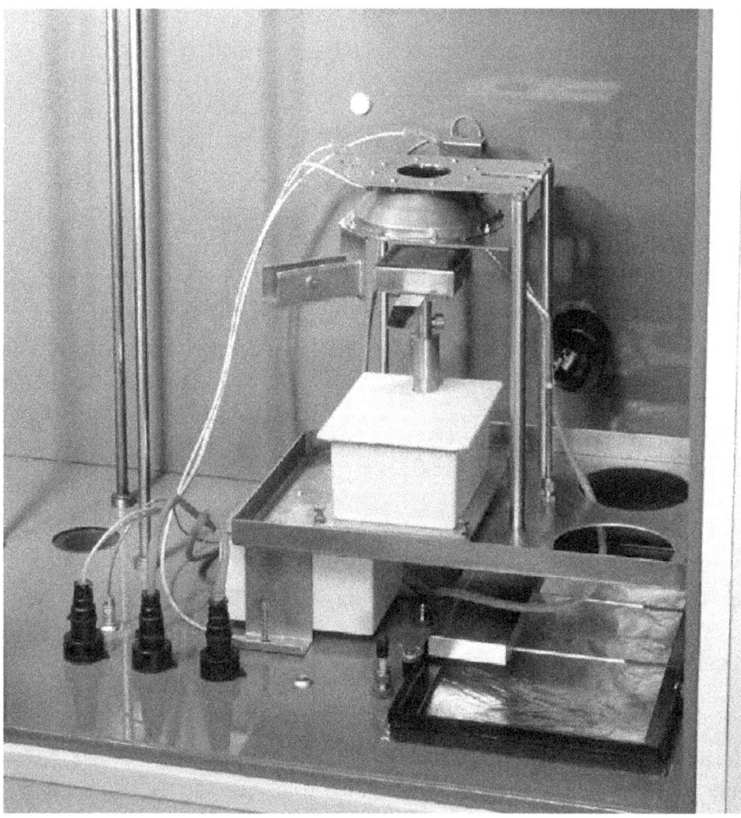

FIGURE 22.13 ASTM E1995 smoke chamber conical heater.

TABLE 22.17
Results of Six Series of Room-Corner Fire Tests

Series of Room-Corner Tests Conducted By	Materials Reaching Early Flashover	Materials with Adequate Heat and Low Smoke	Materials with Adequate Heat and High Smoke	Number of Materials Tested
Southwest Research Institute (SwRI)	1	8	1	10
Eurefic	14	12	2	28
SBI	12	15	3	30
US Coast Guard (USCG)	3	5	1	9
BFGoodrich (BFG)	1	5	1	7
Overall	**31**	**45**	**8**	**84**

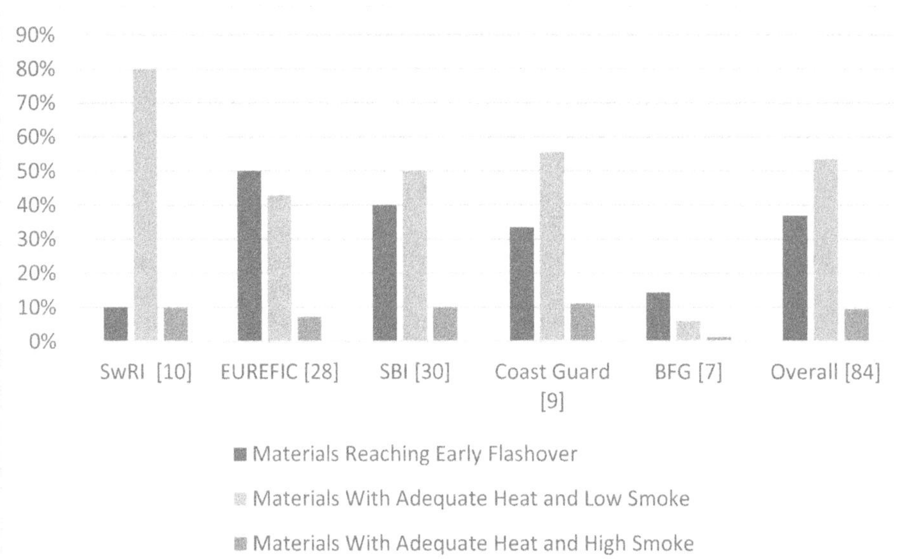

FIGURE 22.14 Heat and smoke release in room corner tests in six laboratories.

methods and guidance documents on smoke toxicity, but emotional responses arise from discussions on interpretation of results or requirements for the use of animals as test surrogates. The following is now accepted by many fire scientists[218,222,257–263] and is critical to understanding how to assess fire hazard:

- Most fire fatalities occur in fires that become very large. In fact, US statistics indicate that such fires account for over six times more fatalities than all other fires. This is illustrated by NFPA statistics of US fires in the 1986 to 1990 time period.[264,265] Similar information would be obtained from other time periods.
- Carbon monoxide concentrations in the atmospheres of flashover fires (the fires most likely to produce fatalities) are determined by geometric variables and oxygen availability but are virtually unaffected by chemical composition of fuels.
- CO yields in full-scale flashover fires are approximately 0.2 g/g, which can be calculated to correspond to a toxicity of 25 mg/l.[218,263] This consistent yield of CO is illustrated by a set of 24 studies where such results are shown is in Table 22.18.[266–271]
- The consequence of this is that any toxic potency (LC_{50}) higher than 8 mg/l will be subsumed within the toxicity of the atmosphere and is of no consequence. Thus, values 8 or greater should be converted to 8 mg/l for reporting purposes. Moreover, therefore, almost all common materials have virtually the same smoke toxicity; their associated fire hazard will not be a function of smoke toxic potency.
- A comprehensive study of fire (and non-fire) fatalities associated with CO showed that carbon monoxide inhalation statistically tracks fire fatalities.[262,263,266]

Consequently, smoke toxicity measurements are often of minimal consequence to fire hazard assessment.

TABLE 22.18
Yields of Carbon Monoxide in Fires, as Reported in the Literature

Material	Yield	Reference	Organization
Plywood walls, wood fiberboard ceiling	0.35	Budnick 1978	HUD
FR plywood walls, wood fiberboard ceiling	0.42	Budnick 1978	HUD
Plywood walls, no ceiling	0.10	Budnick 1978	HUD
Upholstered chair, bed FR plywood walls	0.36	Budnick et al. 1978	HUD
Plywood walls, bed	0.31	Budnick et al. 1978	HUD
Plywood walls, wood cribs, cellulosic ceiling tile	0.29	Levine and Nelson 1990	NIST
Non FR chair, TV cabinets, cables, etc.	0.22	Babrauskas et al. 1988	FRCA
FR chair, TV cabinets, cables, etc.	0.23	Babrauskas et al. 1988	FRCA
PMMA walls	0.28	NIST unpublished	NIST
Wood cribs	0.15	Gottuk et al. 1992	VPI
Flexible PU foam	0.25	Gottuk et al. 1992	VPI
PMMA	0.30	Gottuk et al. 1992]	VPI
Hexane	0.23	Gottuk et al. 1992	VPI
Propane	0.23	Beyler 1986	Harvard
Propene	0.20	Beyler 1986	Harvard
Hexanes	0.20	Beyler 1986	Harvard
Toluene	0.11	Beyler 1986	Harvard
Methanol	0.24	Beyler 1986	Harvard
Ethanol	0.22	Beyler 1986	Harvard
Isopropanol	0.17	Beyler 1986	Harvard
Acetone	0.30	Beyler 1986	Harvard
Polyethylene	0.18	Beyler 1986	Harvard
PMMA	0.19	Beyler 1986	Harvard
Pine	0.14	Beyler 1986	Harvard
Average CO yield	**0.236**	24 Cases	5 Studies

22.4.2.9 Microcalorimetry

A new fire test instrument was developed in 2004, the pyrolysis combustion flow calorimeter (PCFC), or microcalorimeter.[272,273] This instrument (Figure 22.15) was developed by Richard Lyon and coworkers at the Federal Aviation Administration laboratories. It enables the determination of parameters such as specific heat release rate (W/g), heat of combustion (J/g) and ignition temperature (°C) to be quickly determined from very small (1–50 mg) test specimens. The technique has been standardized by ASTM as ASTM D7309. Data from the PCFC has been shown to be capable of being correlated with data from standard heat release instrument fire tests (such as the cone calorimeter or the OSU calorimeter) and with other flammability results (such as the LOI or the UL 94). The data is also consistent with data obtained from an oxygen bomb calorimeter, so as to assess and predict flammability properties.

The PCFC technique utilizes traditional oxygen depletion calorimetry. The specimen is first heated at a constant rate of temperature rise (typically 1–5 °K/s) in a pyrolyzer. The thermal decomposition products are swept from the pyrolyzer by an inert gas. The gas stream is mixed with oxygen and enters a combustor at 900°C, where the decomposition products are completely oxidized. Oxygen concentrations and flow rates of the combustion gases are used to determine the oxygen depletion involved in the combustion process and the heat release, as well as the heat release capacity, is determined from these measurements. Then, the ignition temperature and the heat release capacity

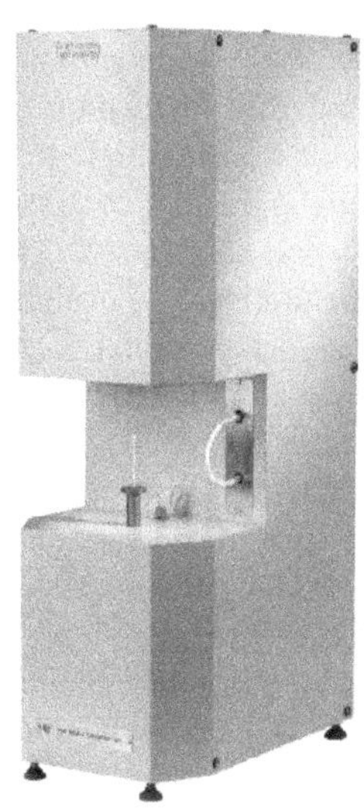

FIGURE 22.15 Microcalorimeter.

parameter (HRC) can be determined and used to compare PCFC data with data from other test methods. The HRC is defined as the ratio of the heat release rate and the heating rate. The peak heat release rates determined in cone calorimeter experiments correlate well with peak heat release capacity data from PCFC experiments. In terms of other tests, results from the limiting oxygen index (LOI, ASTM D2863) test method exhibit a reciprocal correlation with heat release capacity values, while heat release capacity can also be a rough indicator for UL 94 ratings. In approximate terms, it has been said that HRC results can classify materials into three ranges of material flammability, as follows:

- Materials with HRC > 400 J/g-K: poor fire performance
- Materials with 400 J/g-K < HRC < 200 J/g-K: average fire performance
- Materials with 200 J/g-K > HRC: superior fire performance

The advantage this instrument offers over traditional calorimeters is the very small sample size and the ease of performing tests, so that up to 50–60 tests can be conducted in a single day. The test is under consideration as a screening test for heat release tests such as those used for aircraft.

22.4.2.10 Cigarette Ignition

The vast majority of the tests discussed above are basically flaming ignition tests. However, test methods also exist for assessing the ignitability from cigarette ignition and the capability of cigarettes to ignite fabrics. Thus, NFPA 260[96] (and also formerly ASTM E1353[94]) assesses the capability of individual fabrics and foams from being ignited by cigarettes, with the intent of using them for controlling smoldering ignition of upholstered furniture and mattress components. As discussed earlier, this test is very similar to the UFAC (Upholstered Furniture Action Council, associated with residential furniture manufacturers) test that was long used voluntarily by many (or most) manufacturers of US residential upholstered furniture. NFPA 261[97] (and formerly ASTM E1352[95]) is similar to NFPA 260, but it addresses smoldering ignition of composites, and it is similar to the voluntary test that used to be used by BIFMA (Business and Institutional Furniture Manufacturers Association, trade association for US contract upholstered furniture). The tests in 16 CFR 1632[22] and in 16 CFR 1640[27] are federally mandated approaches to smoldering ignition of mattress and upholstered furniture components, respectively. In all these tests a standard cigarette is used and applied on fabrics, foams, or composites.

In view of the fact that regulatory assessment of fire performance of upholstered furniture flammability is based on smoldering tests, it is of interest that a test has been developed to predict the probability of transition from smoldering to flaming.[274] However it has not become a standard and is not used for regulation.

In contrast, ASTM E2187[28] is a test for assessing the ignition power of cigarettes. In the test cigarettes are assessed for their capability of igniting filter paper (as a surrogate for a cellulosic fabric). In this case, the cigarette is the item to be assessed and not the fabric or the foam.

22.4.3 Tests for Specific Products or Materials

As discussed earlier, a number of organizations and technical committees develop fire test methods or specifications that are specific to some particular material or product. It is not possible to cover all of them in this work and they are usually publicly available from the responsible organization via their website.

Within ASTM, technical committees associated with plastics, electrical materials, textiles, protective clothing, thermal insulation, consumer products, detention, and correctional facilities, sleeping bags, tents, and ships have developed tests that are often applications tests that are of specific interest to the products involved. One fire test has spawned more applications standards than any other, primarily because of its vast use in the US: ASTM E84 (Steiner tunnel). Thus, NFPA 262, UL 1820, UL 1887, UL 2024, UL 2846, ULC S102, ULC S102.2, and ASTM E2768, are all test methods based on the Steiner tunnel test. In some cases the base apparatus is being modified (thus, although it is possible to conduct the ASTM E84 test in the apparatus of some of the other tests, excluding the ULC ones, it is often not possible to conduct the other test in any apparatus strictly complying with the ASTM E84 apparatus). The ULC S102[275] and ULC S102.2[141]standards use a tunnel with slightly different construction characteristics than ASTM E84; the only difference between the two ULC standards is the procedure, since ULC S102.2 exposes the test specimen on the tunnel floor. The other test method that has resulted in many applications standards is the cone calorimeter; the applications standards issued by ASTM are ASTM D5485, ASTM D6113, ASTM E1474, ASTM E1740, ASTM E2102, ASTM E2965, and ASTM F1550 (see appendix).

Organizations such as FM Global and Underwriters Laboratories compete with the broader-based standards development organizations (ASTM and NFPA) in developing some fire tests. However, in many cases FM Global tests are being replaced by ones from ASTM and/or NFPA for use in codes and regulations. A number of trade associations also develop test methods, but these are also being replaced when intended for use in codes or regulations. The appendix (Section 22.8) contains the key fire test standards from ASTM, NFPA, and UL.

22.5 PRODUCT LIABILITY

In the US there is an added factor that is an alternate consideration to regulations: product liability. A number of specifications require materials suppliers to ensure that their materials meet certain criteria with regard to flammability (and sometimes smoke) performance, often exceeding regulatory requirements. Therefore, companies will often strive to meet the enhanced criteria for the materials or products they supply, perhaps in order to avoid liability issues.

In some cases regulation can have the result of stifling innovation and/or lowering fire safety. The earlier mentioned US regulation on upholstered furniture that resulted in 16 CFR 1640 has made furniture manufacturers somewhat shielded from product liability associated with open flame ignition of their products, so that they are less likely to introduce less fire safe products.

In cases where some manufacturers meet the letter (and not always the spirit) of the law, product liability is a critical component of safety (including fire safety). Product liability represents an added line of defense for the consumer over and above regulation, but only when regulation is flawed. Attorneys representing potential plaintiffs may be quite sophisticated and hire experts with extensive experience in fire investigation and flammability testing. It is not unusual for an expert (or the lawyer who uses said expert) to be able to use complex fire models and to understand the fundamental implications of test results from a broader perspective than the technical personnel in some manufacturing companies.

Once product liability "kicks in," in a particular incident, the situation facing a manufacturer involves the financial aspect of minimizing expenses, and technical issues start playing a minor role, although all arguments are presented in technical language. Furthermore, at this stage, compliance with all regulations and specifications is no longer considered an issue: failure to comply is an automatic admission of "guilt," but compliance does not bring automatic "acquittal." The issue that will be considered by legal counsel (on each side) will be whether "due diligence" was used by defendant, and whether state-of-the-art concepts were employed when the product that was involved in the fire was manufactured. The history of decision-making by the manufacturer will be brought into the open, and the methods relied upon to make recommendations and reach conclusions will be analyzed.

A manufacturer who can show that the materials (or products) made cannot be faulted because nothing better is available to meet the usage guidelines put forward by the final customer (and not necessarily the specifications), using the most modern experimental techniques and mathematical models, is most likely to be successful in preventing product liability concerns, and may even not face many lawsuits of this kind. The best defense for a manufacturer is to ensure that the precautions taken at every stage were such that a manufacturer has "gone the extra mile" and investigated the materials or products being sold beyond the specifications. In particular, when deficiencies of certain techniques (for example fire tests) are known, it is still essential for a manufacturer to conduct all needed testing to meet specifications. However, additional testing or modeling must also be conducted to ensure that the safety of the user of the product is not compromised. If this is not possible, for either practical or theoretical reasons, document trails are needed to show what course of action was considered, and the rationale for the decisions made. A good basic rule is always that manufacturers should be more aware of the shortcomings of their materials or products than their customers and should share the knowledge with the public. Product liability lawyers make it their business to care how manufacturers test for flammability to ensure that the public is adequately protected.

22.6 CONCLUSIONS

This work describes the regulatory (and semi-regulatory) situation in North America, in the area of fire safety. In recent years there have been considerable improvements in fire safety, but they are often relatively difficult to notice because of the fact that codes and standards are often not adopted for years by local or state governments. Codes have significantly improved requirements for fire safety. Examples where codes provide higher levels of protection include the wildland-urban interface area and the exterior claddings of tall buildings, as well as rail transportation vehicles (underground and above ground) and indoor playgrounds. In terms of consumer products, positive changes have occurred in the areas of mattresses and in the requirements for cigarettes with lower ignition propensity, so as to cause fewer ignitions of furnishings. However, the area of fire safety of upholstered furniture has moved backwards. This is particularly worrisome because upholstered furniture continues to be the item first ignited responsible for most fire deaths in the US and, moreover, the number of deaths per unit population from such fires has increased in the most recent data.[276]

In conclusion, the promotion of improved fire safety should continue to be a key societal goal, and it can be achieved, in the US, by work in the codes and standards arena.

22.7 ABBREVIATIONS

22.7.1 Regulations, Codes, Standards, and Organizations

AFSC	American Fire Safety Council (US)
AHJ	Authority Having Jurisdiction (US)
ANPRM	Advance Notice of Proposed Rulemaking (US)
ASTM	American Society for Testing and Materials (now ASTM International)
BIFMA	Business and Institutional Furniture Manufacturers Association (US)
BOCA	Building Officials and Code Administrators (US)
BRE	Building Research Establishment (UK)
BSEF	Bromine Science and Environmental Forum
BSI	British Standards Institution
CA TB	California Technical Bulletin
CBHF	California Bureau of Home Furnishings and Thermal Insulation
CBUF	Combustion Behaviour of Upholstered Furniture EC Research Project

CEN	European Committee for Standardization
CO	Carbon Monoxide
CPD	European Commission Construction Products Directive
CPSC	Consumer Product Safety Commission (US)
CSA	Canadian Standards Association
EC	European Commission
EN	Euro Norm (European Standard)
EPA	Environmental Protection Agency (US)
EU	European Union
FAA	Federal Aviation Administration (US)
FAR	Federal Aviation Regulation (US)
FFA	Flammable Fabrics Act (US)
FHSA	Federal Hazardous Substances Act (US)
FIPEC	Fire Performance of Electrical Cables EC Research Project
FM	FM Global (formerly Factory Mutual)
FPRF	NFPA Fire Protection Research Foundation (US)
FMVSS	Federal Motor Vehicle Safety Standard (US)
FRA	Federal Railroad Administration (US)
FTA	Federal Transit Administration (US)
FTP	IMO Fire Test Procedures Code
FSC	Coalition for Fire Safe Cigarettes FFA Flammable Fabrics Act (US)
FSI	Flame Spread Index (ASTM E84)
HRC	Heat Release Capacity (from Microcalorimeter)
HSC	IMO High Speed Craft Code
HUD	Department of Housing and Urban Development (US)
IAFC	International Association of Fire Chiefs
IAPMO	International Association of Plumbing and Mechanical Officials (US)
IBC	International Building Code (US)
ICBO	International Conference of Building Officials (US)
ICC	International Code Council (US)
IEC	International Electrotechnical Commission
IEBC	International Existing Building Code (US)
IEEE	Institute of Electrical and Electronics Engineers (US)
IFC	International Fire Code (US)
IMC	International Mechanical Code (US)
IMO	International Maritime Organization
IRC	International Residential Code (US)
ISO	International Organization for Standardization
ISPA	International Sleep Products Association (US)
IWUIC	International Wildland Urban Interface Code (US)
LIFT	Lateral Ignition and Flame Spread Test, ASTM E 1321
LOI	Limiting Oxygen Index
LPCB	Loss Prevention Control Board (UK)
LPS	Loss Prevention Standard (UK)
MSHA	Mine Safety and Health Administration (US)
NASFM	National Association of State Fire Marshals (US)
NBS	National Bureau of Standards (US)
NEC	National Electrical Code, NFPA 70 (US)
NEMA	National Electrical Manufacturers Association (US)
NFPA	National Fire Protection Association
NFPA 101	National Life Safety Code (US)
NFPA 5000	National Building Construction and Safety Code (US)
NHTSA	National Highway Traffic Safety Administration (US)
NIST	National Institute of Standards and Technology (US)
NPRM	Notice of Proposed Rulemaking (US)
NVIC	Navigation and Vessel Inspection Circular (US)
ODPM	Office of the Deputy Prime Minister (UK)
OSU	Ohio State University
PCFC	Pyrolysis Combustion Flow Calorimeter (Microcalorimeter)
RoHS	EC Directive on Restriction of Hazardous Substances
SBCCI	Southern Building Code Conference International (US)
SBI	Single Burning Item (European Norm EN 13823)
SDI	Smoke Developed Index (ASTM E 84)
SFPE	Society of Fire Protection Engineers (US)
SOLAS	International Convention for the Safety of Life at Sea (from IMO)
SP	Swedish National Testing and Research Institute
SRM	Standard Reference Material
UFAC	Upholstered Furniture Action Council (US)
UFC	Uniform Fire Code (also known as NFPA 1)
UL	Underwriters Laboratories
UKAS	United Kingdom Accreditation Service
UMC	Uniform Mechanical Code (US)
UMTA	Urban Mass Transportation Administration (US)
UPC	Uniform Plumbing Code (US)
USFA	United States Fire Administration
USCG	United States Coast Guard
VW-1	UL Small Scale Vertical Wire Fire Test in UL 1581
WEEE	EC Directive on Waste Electrical and Electronic Equipment

22.8 APPENDIX: FIRE TEST METHODS FROM ASTM, NFPA, AND UL

ASTM Fire Tests (ASTM International, 100 Bar Harbor Drive, West Conshohocken, PA, 19428–2959, US):

ASTM C1485: Standard Test Method for Critical Radiant Flux of Exposed Attic Floor Insulation Using an Electric Radiant Heat Energy Source

ASTM D568: Standard Test Method for Rate of Burning and/or Extent and Time of Burning of Flexible Plastics in a Vertical Position
ASTM D635: Standard Test Method for Rate of Burning and/or Extent and Time of Burning of Self-Supporting Plastics in a Horizontal Position
ASTM D777: Standard Test Method for Flammability of Treated Paper and Paperboard
ASTM D1230: Standard Test Method for Flammability of Apparel Textiles
ASTM D1360: Standard Test Method for Fire Retardancy of Paints (Cabinet Method)
ASTM D1929: Standard Test Method for Determining Ignition Temperature of Plastics
ASTM D2843: Standard Test Method for Density of Smoke from the Burning or Decomposition of Plastics
ASTM D2859: Standard Test Method for Flammability of Finished Textile Floor Covering Materials
ASTM D2863: Standard Test Method for Measuring the Minimum Oxygen Concentration to Support Candle-Like Combustion of Plastics (Oxygen Index)
ASTM D3014: Standard Test Method for Flame Height, Time of Burning, and Loss of Weight of Rigid Cellular Plastics in a Vertical Position
ASTM D3659: Standard Test Method for Flammability of Apparel Fabrics by Semi-Restraint Method
ASTM D3675: Standard Test Method for Surface Flammability of Flexible Cellular Materials Using a Radiant Heat Energy Source
ASTM D3801: Standard Test Method for Measuring the Comparative Extinguishing Characteristics of Solid Plastics in a Vertical Position
ASTM D3874, Standard Test Method for Ignition of Materials by Hot Wire Sources
ASTM D4151: Standard Test Method for Flammability of Blankets
ASTM D4804: Standard Test Methods for Determining the Flammability Characteristics of Nonrigid Solid Plastics
ASTM D4986: Standard Test Method for Horizontal Burning Characteristics of Cellular Polymeric Materials
ASTM D5048: Standard Test Method for Measuring the Comparative Burning Characteristics and Resistance to Burn-Through of Solid Plastics Using a 125-mm Flame
ASTM D5132: Standard Test Method for Horizontal Burning Rate of Flexible Cellular and Rubber Materials Used in Occupant Compartments of Motor Vehicles
ASTM D5424: Standard Test Method for Smoke Obscuration of Insulating Materials Contained in Electrical or Optical Fiber Cables When Burning in a Vertical Cable Tray Configuration
ASTM D5485: Standard Test Method for Determining the Corrosive Effect of Combustion Products Using the Cone Corrosimeter
ASTM D5537: Standard Test Method for Heat Release, Flame Spread, Smoke Obscuration, and Mass Loss Testing of Insulating Materials Contained in Electrical or Optical Fiber Cables When Burning in a Vertical Cable Tray Configuration
ASTM D6113: Standard Test Method for Using a Cone Calorimeter to Determine Fire-Test-Response Characteristics of Insulating Materials Contained in Electrical or Optical Fiber Cables
ASTM D6194: Standard Test Method for Glow Wire Ignition of Materials
ASTM D6413: Standard Test Method for Flame Resistance of Textiles (Vertical Test)
ASTM D6545: Standard Test Method for Flammability of Textiles Used in Children's Sleepwear
ASTM D7016: Standard Test Method to Evaluate Edge Binding Components Used in Mattresses After Exposure to An Open Flame
ASTM D7140: Standard Test Method to Measure Heat Transfer Through Textile Thermal Barrier Materials
ASTM D7309: Standard Test Method for Determining Flammability Characteristics of Plastics and Other Solid Materials Using Microscale Combustion Calorimetry
ASTM D7936 Standard Test Method for Flammability of Electrical Insulating Materials Intended for Wires or Cables When Burning in Horizontal Configuration
ASTM D8354 Standard Test Method for Flammability of Electrical Insulating Materials Intended for Wires or Cables When Burning in a Vertical Configuration
ASTM D8355 Standard Test Methods for Flammability of Electrical Insulating Materials Used for Sleeving or Tubing
ASTM E69: Standard Test Method for Combustible Properties of Treated Wood by the Fire-Tube Apparatus
ASTM E84: Standard Test Method for Surface Burning Characteristics of Building Materials
ASTM E108: Standard Test Methods for Fire Tests of Roof Coverings
ASTM E119: Standard Test Methods for Fire Tests of Building Construction and Materials
ASTM E136: Standard Test Method for Assessing Combustibility of Materials Using a Vertical Tube Furnace at 750°C
ASTM E162: Standard Test Method for Surface Flammability of Materials Using a Radiant Heat Energy Source
ASTM E648: Standard Test Method for Critical Radiant Flux of Floor-Covering Systems Using a Radiant Heat Energy Source
ASTM E662: Standard Test Method for Specific Optical Density of Smoke Generated by Solid Materials
ASTM E814: Standard Test Method for Fire Tests of Through-Penetration Fire Stops

ASTM E906: Standard Test Method for Heat and Visible Smoke Release Rates for Materials and Products

ASTM E970: Standard Test Method for Critical Radiant Flux of Exposed Attic Floor Insulation Using a Radiant Heat Energy Source

ASTM E1317: Standard Test Method for Flammability of Marine Surface Finishes

ASTM E1321: Standard Test Method for Determining Material Ignition and Flame Spread Properties

ASTM E1352: Standard Test Method for Cigarette Ignition Resistance of Mock-Upholstered Furniture Assemblies (withdrawn)

ASTM E1353: Standard Test Methods for Cigarette Ignition Resistance of Components of Upholstered Furniture (withdrawn)

ASTM E1354: Standard Test Method for Heat and Visible Smoke Release Rates for Materials and Products Using an Oxygen Consumption Calorimeter

ASTM E1474: Standard Test Method for Determining the Heat Release Rate of Upholstered Furniture and Mattress Components or Composites Using a Bench Scale Oxygen Consumption Calorimeter

ASTM E1529: Standard Test Methods for Determining Effects of Large Hydrocarbon Pool Fires on Structural Members and Assemblies

ASTM E1537: Standard Test Method for Fire Testing of Upholstered Furniture

ASTM E1590: Standard Test Method for Fire Testing of Mattresses

ASTM E1623: Standard Test Method for Determination of Fire and Thermal Parameters of Materials, Products, and Systems Using an Intermediate Scale Calorimeter (ICAL)

ASTM E1678: Standard Test Method for Measuring Smoke Toxicity for Use in Fire Hazard Analysis

ASTM E1822: Standard Test Method for Fire Testing of Stacked Chairs

ASTM E1966: Standard Test Method for Fire-Resistive Joint Systems

ASTM E1995: Standard Test Method for Measurement of Smoke Obscuration Using a Conical Radiant Source in a Single Closed Chamber, With the Test Specimen Oriented Horizontally

ASTM E2010: Standard Test Method for Positive Pressure Fire Tests of Window Assemblies (withdrawn)

ASTM E2058: Standard Test Methods for Measurement of Synthetic Polymer Material Flammability Using a Fire Propagation Apparatus (FPA)

ASTM E2074: Standard Test Method for Positive Pressure Fire Tests of Window Assemblies (withdrawn)

ASTM E2187: Standard Test Method for Measuring the Ignition Strength of Cigarettes

ASTM E2257: Standard Test Method for Room Fire Test of Wall and Ceiling Materials and Assemblies

ASTM E2405: Standard Test Method for Determination of Fire and Thermal Parameters of Materials Using an Intermediate Scale Test with Vertically Oriented Specimen (withdrawn)

ASTM E2574: Standard Test Method for Fire Testing of School Bus Seat Assemblies

ASTM E2632/E2632M Standard Test Method for Evaluating the Under-Deck Fire Test Response of Deck Materials

ASTM E2652 Standard Test Method for Assessing Combustibility of Materials Using a Tube Furnace with a Cone-shaped Airflow Stabilizer, at 750 °C

ASTM E2707 Standard Test Method for Determining Fire Penetration of Exterior Wall Assemblies Using a Direct Flame Impingement Exposure

ASTM E2726/E2726M Standard Test Method for Evaluating the Fire-Test-Response of Deck Structures to Burning Brands

ASTM E2768 Standard Test Method for Extended Duration Surface Burning Characteristics of Building Materials (30 min Tunnel Test)

ASTM E2816 Standard Test Methods for Fire Resistive Metallic HVAC Duct Systems

ASTM E2837 Standard Test Method for Determining the Fire Resistance of Continuity Head-of-Wall Joint Systems Installed Between Rated Wall Assemblies and Nonrated Horizontal Assemblies

ASTM E2874 Standard Test Method for Determining the Fire-Test Response Characteristics of a Building Spandrel-Panel Assembly Due to External Spread of Fire

ASTM E2886/E2886M Standard Test Method for Evaluating the Ability of Exterior Vents to Resist the Entry of Embers and Direct Flame Impingement

ASTM E2912 Standard Test Method for Fire Test of Non-Mechanical Fire Dampers Used in Vented Construction

ASTM E2957 Standard Test Method for Resistance to Wildfire Penetration of Eaves, Soffits and Other Projections

ASTM E2965 Standard Test Method for Determination of Low Levels of Heat Release Rate for Materials and Products Using an Oxygen Consumption Calorimeter

ASTM E3048 Standard Test Method for Determination of Time to Burn-Through Using the Intermediate Scale Calorimeter (ICAL) Radiant Panel

ASTM E3082 Standard Test Methods for Determining the Effectiveness of Fire Retardant Treatments for Natural Christmas Trees

ASTM E3367 Standard Test Method for Determining the Combustion Behavior of Layered Assemblies using a Cone Calorimeter

ASTM F955 Standard Test Method for Evaluating Heat Transfer through Materials for Protective Clothing Upon Contact with Molten Substances

ASTM F1060 Standard Test Method for Thermal Protective Performance of Materials for Protective Clothing for Hot Surface Contact

- ASTM F1085: Standard Specification for Mattress and Box Springs for Use in Berths in Marine Vessels (Annexes A1 and A3)
- ASTM F1358 Standard Test Method for Effects of Flame Impingement on Materials Used in Protective Clothing Not Designated primarily for Flame Resistance
- ASTM F1534: Standard Test Method for Determining Changes in Fire-Test-Response Characteristics of Cushioning Materials After Water Leaching
- ASTM F1550: Standard Test Method for Determination of Fire-Test-Response Characteristics of Components or Composites of Mattresses or Furniture for Use in Correctional Facilities after Exposure to Vandalism, by Employing a Bench Scale Oxygen Consumption Calorimeter
- ASTM F1930 Standard Test Method for Evaluation of Flame Resistant Clothing for Protection Against Flash Fire Simulations Using an Instrumented Manikin
- ASTM F1939 Standard Test Method for Radiant Heat Resistance of Flame Resistant Clothing Materials with Continuous Heating
- ASTM F2700 Standard Test Method for Unsteady-State Heat Transfer Evaluation of Flame Resistant Materials for Clothing with Continuous Heating
- ASTM F2701 Standard Test Method for Evaluating Heat Transfer through Materials for Protective Clothing Upon Contact with a Hot Liquid Splash
- ASTM F2702 Standard Test Method for Radiant Heat Performance of Flame Resistant Clothing Materials with Burn Injury Prediction
- ASTM F2703 Standard Test Method for Unsteady-State Heat Transfer Evaluation of Flame Resistant Materials for Clothing with Burn Injury Prediction

NFPA Fire Tests (National Fire Protection Association, 1 Battery March park, Quincy, MA, 02269, US).

- NFPA 251: Standard Methods of Tests of Fire Endurance of Building Construction and Materials (withdrawn)
- NFPA 252: Standard Methods of Fire Tests of Door Assemblies
- NFPA 253: Standard Method of Test for Critical Radiant Flux of Floor Covering Systems Using a Radiant Heat Energy Source
- NFPA 255: Standard Method of Test of Surface Burning Characteristics of Building Materials (withdrawn)
- NFPA 256: Standard Methods of Fire Tests of Roof Coverings (withdrawn)
- NFPA 257: Standard on Fire Test for Window and Glass Block Assemblies
- NFPA 258: Recommended Practice for Determining Smoke Generation of Solid Materials (withdrawn)
- NFPA 259: Standard Test Method for Potential Heat of Building Materials
- NFPA 260: Standard Methods of Tests and Classification System for Cigarette Ignition Resistance of Components of Upholstered Furniture
- NFPA 261: Standard Method of Test for Determining Resistance of Mock-Up Upholstered Furniture Material Assemblies to Ignition by Smoldering Cigarettes
- NFPA 262: Standard Method of Test for Flame Travel and Smoke of Wires and Cables for Use in Air-Handling Spaces
- NFPA 263: Standard Method of Test for Heat and Visible Smoke Release Rates for Materials and Products (withdrawn)
- NFPA 265: Standard Methods of Fire Tests for Evaluating Room Fire Growth Contribution of Textile Coverings on Full Height Panels and Walls
- NFPA 266: Standard Method of Test for Fire Characteristics of Upholstered Furniture Exposed to Flaming Ignition Source (withdrawn)
- NFPA 267: Standard Method of Test for Fire Characteristics of Mattresses and Bedding Assemblies Exposed to Flaming Ignition Source (withdrawn)
- NFPA 268: Standard Test Method for Determining Ignitability of Exterior Wall Assemblies Using a Radiant Heat Energy Source
- NFPA 269: Standard Test Method for Developing Toxic Potency Data for Use in Fire Hazard Modeling
- NFPA 270: Standard Test Method for Measurement of Smoke Obscuration Using a Conical Radiant Source in a Single Closed Chamber
- NFPA 271: Standard Method of Test for Heat and Visible Smoke Release Rates for Materials and Products Using an Oxygen Consumption Calorimeter
- NFPA 272: Standard Method of Test for Heat and Visible Smoke Release Rates for Upholstered Furniture Components or Composites and Mattresses Using an Oxygen Consumption Calorimeter (withdrawn)
- NFPA 274: Standard Test Method to Evaluate Fire Performance Characteristics of Pipe Insulation
- NFPA 275: Standard Method of Fire Tests for the Evaluation of Thermal Barriers Used Over Foam Plastic Insulation
- NFPA 285: Standard Fire Test Method for Evaluation of Fire Propagation Characteristics of Exterior Wall Assemblies Containing Combustible Components
- NFPA 286: Standard Methods of Fire Tests for Evaluating Contribution of Wall and Ceiling Interior Finish to Room Fire Growth
- NFPA 287: Standard Test Methods for Measurement of Flammability of Materials in Cleanrooms Using a Fire Propagation Apparatus (FPA)
- NFPA 288: Standard Methods of Fire Tests of Floor Fire Door Assemblies Installed Horizontally in Fire Resistance–Rated Floor Systems

NFPA 289: Standard Method of Fire Test for Individual Fuel Packages
NFPA 290: Standard for Fire Testing of Passive Protection Materials for Use on LP-Gas Containers
NFPA 701: Standard Methods of Fire Tests for Flame Propagation of Textiles and Films
NFPA 703: Standard for Fire Retardant Impregnated Wood and Fire Retardant Coatings for Building Materials
NFPA 705: Recommended Practice for a Field Flame Test for Textiles and Films

Key UL Fire Test Standards (Underwriters Laboratories Inc., 333 Pfingsten Road, Northbrook, IL 60062–2096, US):

UL 9, Standard for Fire Tests of Window Assemblies
UL 10, Fire Doors
UL 10B, Standard for Fire Tests of Door Assemblies
UL 214, Standard for Safety Tests for Flame-Propagation of Fabrics and Films (withdrawn)
UL 263 Standard for Fire Tests of Building Construction and Materials
UL 555, Standard for Fire Dampers
UL 555C, Standard for Ceiling Dampers
UL 555S, Standard for Leakage Rated Dampers for Use in Smoke Control Systems
UL 723, Standard for Test for Surface Burning Characteristics of Building Materials
UL 790, Standard for Standard Test Methods for Fire Tests of Roof Coverings (2004)
UL 910, Standard for Safety Test for Flame-Propagation and Smoke-Density Values for Electrical and Optical-Fiber Cables Used in Spaces Transporting Environmental Air (withdrawn)
UL 1040, Standard for Fire Test of Insulated Wall Construction
UL 1479, Standard for Fire Tests of Through-Penetration Firestops
UL 1056, Standard for Safety Fire Test of Upholstered Furniture (withdrawn)
UL 1666, Test for Flame Propagation Height of Electrical and Optical-Fiber Cables Installed Vertically in Shafts
UL 1685, Standard Vertical-Tray Fire-Propagation and Smoke-Release Test for Electrical and Optical-Fiber Cables
UL 1715, Standard for Fire Test of Interior Finish Material
UL 1820, Standard for Safety Fire Test of Pneumatic Tubing for Flame and Smoke Characteristics
UL 1887, Standard for Safety Fire Test of Plastic Sprinkler Pipe for Visible Flame and Smoke Characteristics
UL 1895, Standard for Safety Fire Test of Mattresses (withdrawn)
UL 1975, Standard for Fire Tests for Foamed Plastics Used for Decorative Purposes
UL 2024, Standard for Cable Routing Assemblies and Communications Raceways
UL 2043, Standard for Safety Fire Test for Heat and Visible Smoke Release for Discrete Products and Their Accessories Installed in Air-Handling Spaces
UL 2079, Standard for Tests for Fire Resistance of Building Joint Systems
UL 2335, Standard for Fire Tests of Storage Pallets
UL 2556, Wire and Cable Test Methods
UL 2596, Test Method for Thermal and Mechanical Performance of Battery Enclosure Materials
UL 2846, Fire Test of Plastic Water Distribution Plumbing Pipe for Visible Flame and Smoke Characteristics

NOTES

† 16 CFR 1610 states: "1610.62 (4) Note 2 - Some textiles never exhibit unusual burning characteristics and need not be tested. 16 CFR 1610.37(d). Such textiles include plain surface fabrics, regardless of fiber content, weighing 2.6 oz. or more per sq. yd., and plain and raised surface fabrics made of acrylic, modacrylic, nylon, olefin, polyester, wool, or any combination of these fibers, regardless of weight.

1. The term "fireproof" is deprecated by the fire science community, since it is not possible that a combustible material or product will not catch fire when exposed to any heat source.

REFERENCES

1. The Construction Products Directive (CPD) as Regards the Classification of the Reaction-to-Fire Performance of Construction Products, Council Directive 89/106/EEC, 21 December 1988, Brussels, Belgium, 1988, European Commission, Enterprise and Industry Directorate-General, Construction.
2. Construction Products Regulation (CPR), European Commission, 2022. https://single-market-economy.ec.europa.eu/sectors/construction/construction-products-regulation-cpr_en
3. International Building Code (IBC), International Code Council, Washington, DC.
4. International Fire Code (IFC), International Code Council, Washington, DC.
5. International Mechanical Code (IMC), International Code Council, Washington, DC.
6. International Wildland Urban Interface Code (IWUIC), International Code Council, Washington, DC.
7. International Residential Code for One- and Two-Family Dwellings (IRC), International Code Council, Washington, DC.
8. International Existing Building Code (IEBC), International Code Council, Washington, DC.
9. NFPA 70, National Electrical Code (NEC), National Fire Protection Association, Quincy, MA.
10. NFPA 101, National Life Safety Code, National Fire Protection Association, Quincy, MA.
11. NFPA 90A, Standard for the Installation of Air-Conditioning and Ventilating Systems, National Fire Protection Association, Quincy, MA.

12. NFPA 13, Standard for the Installation of Sprinkler Systems, National Fire Protection Association, Quincy, MA.
13. NFPA 130, Standard for Fixed Guideway Transit and Passenger Rail Systems, National Fire Protection Association, Quincy, MA.
14. 16 CFR 1610, Title 16, Commercial Practices, Chapter II, Consumer Product Safety Commission, Part 1610, "Standard for the Flammability of Clothing Textiles", Flammable Fabrics Act Regulations, US Federal Government.
15. ASTM D1230, "Standard Test Method for Flammability of Apparel Textiles", American Society for Testing and Materials, West Conshohocken, PA.
16. 16 CFR Part 1630, Title 16, Commercial Practices, Chapter II, Consumer Product Safety Commission, Part 1630, "Standard for the Surface Flammability of Carpets and Rugs (FF 1–70)", Flammable Fabrics Act Regulations, US Federal Government.
17. 16 CFR Part 1631, Title 16, Commercial Practices, Chapter II, Consumer Product Safety Commission, Part 1631, "Standard for the Surface Flammability of Small Carpets and Rugs (FF 2–70)", Flammable Fabrics Act Regulations, US Federal Government.
18. ASTM D2859, "Standard Test Method for Flammability of Finished Textile Floor Covering Materials", American Society for Testing and Materials, West Conshohocken, PA.
19. 16 CFR 1615, Title 16, Commercial Practices, Chapter II, Consumer Product Safety Commission, Part 1615, "Standard for the Flammability of Children's Sleepwear: Sizes 0 Through 6X (FF 3–71)", Flammable Fabrics Act Regulations, US Federal Government.
20. 16 CFR 1616, Title 16, Commercial Practices, Chapter II, Consumer Product Safety Commission, Part 1616, "Standard for the Flammability of Children's Sleepwear: Sizes 7 Through 14 (FF 5–74)", Flammable Fabrics Act Regulations, US Federal Government.
21. ASTM D6545, "Standard Test Method for Flammability of Textiles Used in Children's Sleepwear", American Society for Testing and Materials, West Conshohocken, PA.
22. 16 CFR 1632, Title 16, Commercial Practices, Chapter II, Consumer Product Safety Commission, Part 1632, "Standard for the Flammability of Mattresses (FF 4–72)", Flammable Fabrics Act Regulations, US Federal Government.
23. 16 CFR 1500.44, Title 16, Commercial Practices, Chapter II, Consumer Product Safety Commission, Part 1500, "Hazardous Substances and Articles; Administration and Enforcement Regulations: Method for Determining Extremely Flammable and Flammable Solids", US Federal Government.
24. 16 CFR 1633, Title 16, Commercial Practices, Chapter II, Consumer Product Safety Commission, Part 1633, "Standard for the Flammability (Open Flame) of Mattress Sets", Flammable Fabrics Act Regulations, US Federal Government.
25. ASTM F963, "Standard Consumer Safety Specification for Toy Safety", American Society for Testing and Materials, West Conshohocken, PA.
26. Proposed 16 CFR Part 1634, "Standard for the Flammability of Residential Upholstered Furniture", www.cpsc.gov/businfo/frnotices/fr08/cpscfr08.html, March 4, 2008.
27. 16 CFR 1640, "Standard for the Flammability of Upholstered Furniture", www.ecfr.gov/current/title-16/chapter-II/subchapter-D/part-1640.
28. ASTM E2187, "Standard Test Method for Measuring the Ignition Strength of Cigarettes", American Society for Testing and Materials, West Conshohocken, PA.
29. New York State Combustion Toxicity Regulations: "Combustion Toxicity Testing", Article 15, Part 1120 (9 NYCRR 1120)—Effective December 16, 1986.
30. Alarie, Y. C.; Anderson, R. C. Toxicologic and Acute Lethal Hazard Evaluation of Thermal Decomposition Products of Synthetic and Natural Polymers. *Toxicol. Appl. Pharmacol.* **1979**, *51*, 341.
31. Norris, J.C., 1990, "Investigation of the Dual LC_{50} Values in Woods Using the University of Pittsburgh Combustion Toxicity Apparatus", in "Characterization and Toxicity of Smoke", ASTM STP 1082, Amer. Soc. Testing and Materials, Philadelphia, PA, Ed. H.J. Hasegawa, pp. 57–71.
32. Hirschler, M.M., 1990, "General Principles of Fire Hazard and the Role of Smoke Toxicity", in "Fire and Polymers: Hazards Identification and Prevention" (Ed. G.L. Nelson), ACS Symposium Series 425, Amer. Chem. Soc., Washington, DC, Chapter 28, pp. 462–478.
33. Hirschler, M. M. Smoke Toxicity Measurements Made So That the Results Can Be Used for Improved Fire Safety. *J. Fire Sci.* **1991**, *9*, 330–347.
34. Hirschler, M. M.; Toxicity of the Smoke From PVC Materials: New Concepts. *Prog. Rubber Plast. Technol.* **1994**, *10–12*, 154–169.
35. Ahrens, M., U.S. Vehicle Fire Trends and Patterns, NFPA, Quincy, MA, August 2001.
36. Ahrens, M., The U.S. Fire Problem Overview Report: Leading Causes and Other Patterns and Trends, NFPA, Quincy, MA, June 2001.
37. Ahrens, M., Vehicle Fires, NFPA, Quincy, MA, March 2020.
38. Janssens, M.L., "Database on Full-Scale Calorimeter Tests on Motor Vehicles", HAR-AAA TC Meeting, Detroit, MI, October 23–24, 2006.
39. FMVSS 302, "Motor Vehicle Safety Standard No. 302, Flammability of Materials—Passenger Cars, Multipurpose Passenger Vehicles, Trucks and Buses", National Highway Traffic Safety Administration, Washington, DC. [Code of Federal Regulations § 571.302, Originally Federal Register 34, No. 229, pp. 20434–20436 (December 31, 1969)].
40. ISO 3795, "Road Vehicles, and Tractors and Machinery for Agriculture and Forestry—Determination of Burning Behaviour of Interior Materials", International Organization for Standardization (ISO), Geneva.
41. "Fire and Transportation Vehicles—State of the Art: Regulatory Requirements and Guidelines—A White Paper", M.M. Hirschler (Technical Coordinator) et al., Fire Protection Research Foundation Research Advisory Council on Transportation Vehicles, October 2004.
42. NFPA 556, Guide on Methods for Evaluating Fire Hazard to Occupants of Passenger Road Vehicles, National Fire Protection Association, Quincy, MA.
43. Federal Transit Administration, "Recommended Fire Safety Practices for Transit Bus and Van Materials Selection", Federal Register, pp. 54249–54254, October 20, 1993.
44. "Aircraft Materials Fire Test Handbook", www.fire.tc.faa.gov/, US Department of Transportation, Federal Aviation Administration, Washington, DC.
45. Federal Register, Vol. 66, No. 36/Thursday, February 22, 2001, FAA Policy on Use of the "Aircraft Materials Fire Test Handbook", pp. 11197–11199, 2001.
46. Smith, E.E., "Ignition, Heat Release and Noncombustibility of Materials, ASTM STP 502", (Ed. A.F. Robertson), Am. Soc. Test. Mats, Philadelphia, PA, 1972, p. 119.
47. ASTM E906, "Standard Test Method for Heat and Visible Smoke Release Rates for Materials and Products",

American Society for Testing and Materials, West Conshohocken, PA.
48. ASTM E662, "Standard Test Method for Specific Optical Density of Smoke Generated by Solid Materials", American Society for Testing and Materials, West Conshohocken, PA.
49. ASTM E648, "Standard Test Method for Critical Radiant Flux of Floor-Covering Systems Using a Radiant Heat Energy Source", American Society for Testing and Materials, West Conshohocken, PA.
50. International Convention for the Safety of Life at Sea (SOLAS), 1974, with Periodic Amendments, Consolidated Edition, International Maritime Organization, London.
51. International Maritime Organization (IMO), London.
52. International Code for Application of Fire Test Procedures (FTP Code), International Maritime Organization, London.
53. International Code of Safety for High-Speed Craft (HSC Code), International Maritime Organization, London.
54. Code of Federal Regulations, US Federal Government—Coast Guard: Title 46, Shipping, CFR, Parts 1–199, Subchapter J: Electrical Engineering (Parts 110–113).
55. US Coast Guard, NVIC (Navigation and Vessel Inspection Circular), US Coast Guard Guide to Structural Fire Protection.
56. NFPA 301, Code for Safety to Life from Fire on Merchant Vessels, National Fire Protection Association, Quincy, MA.
57. ISO 834–1, "Fire-Resistance Tests—Elements of Building Construction—Part 1: General Requirements", International Organization for Standardization (ISO), Geneva.
58. ASTM E119, "Standard Test Methods for Fire Tests of Building Construction and Materials", American Society for Testing and Materials, West Conshohocken, PA.
59. ISO 9705–1, "Reaction to Fire Tests—Room Corner Test for Wall and Ceiling Lining Products—Part 1: Test Method for a Small Room Configuration", International Organization for Standardization (ISO), Geneva.
60. NFPA 302, "Fire Protection Standard for Pleasure and Commercial Motor Craft", National Fire Protection Association, Quincy, MA.
61. Sorathia, U., "Materials in Military Applications", in Business Communications Company Fourteenth Ann. Conference on Recent Advances in Flame Retardancy of Polymeric Materials, June 2–4, Stamford, CT, Ed. M. Lewin, Norwalk, CT, 2003.
62. Federal Register, Vol. 64, No. 91/Wednesday May 12, 1999, "Department of Transportation, Federal Railroad Administration, 49 CFR Part 216 et al., Passenger Equipment Safety Standards; Final Rule", pp. 25539–25705, 1999.
63. Federal Register, Vol. 67, No. 122/Tuesday, June 25, 2002, "Department of Transportation, Federal Railroad Administration, 49 CFR Part 238, [FRA Docket No. PCSS-1, Notice No. 8] RIN 2130–AB48, Passenger Equipment Safety Standards", pp. 42892–42912, 2002.
64. ASTM E162, "Standard Test Method for Surface Flammability of Materials Using a Radiant Heat Energy Source", American Society for Testing and Materials, West Conshohocken, PA.
65. Peacock, R.D., Bukowski, R.W., Jones, W.J., Reneke, P.A., Babrauskas, V., and Brown, J.E., "Fire Safety of Passenger Trains: A Review of U.S. and Foreign Approaches", National Institute of Standards and Technology (Under Contract to US Dept. of Transportation, John A. Volpe National Transportation Systems Center, Cambridge, MA), Gaithersburg, MD (NIST Technical Note 1406), 1994.
66. ASTM E2061, "Guide for Fire Hazard Assessment of Rail Transportation Vehicles", American Society for Testing and Materials, West Conshohocken, PA.
67. Hirschler, M.M., "Fire Safety in Rail Transportation Vehicles: Special Focus on Recent Activities at the Federal Railroad Administration, NFPA 130 and ASTM E2061", Fire Protection Research Foundation Fire Risk and Hazard Assessment Research Application Symposium, July 9–11, 2003, Baltimore, MD.
68. UL 2556, "Wire and Cable Test Methods", Vertical Tray Flame Test Method 2 and ST1 Limited Smoke," Underwriters Laboratories, Northbrook, IL.
69. CSA FT4, "Vertical Cable Tray Test in CSA 22.2, Test Methods for Electrical Wires and Cables (C22.2 No. 0.3)", Canadian Standards Association, Rexdale, ON.
70. NFPA 262, "Standard Method of Test for Fire and Smoke Characteristics of Wires and Cables," National Fire Protection Assn., NFPA, Quincy, MA.
71. Grayson, S.J., and Hirschler, M.M., "Fire Performance of Plastics in Car Interiors", Flame Retardants 2002, February 5–6, London, pp. 197–207, Interscience Communications, London, 2002.
72. Hirschler, M.M., Hoffmann, D.J., Hoffmann, J.M., and Kroll, E.C., "Fire Hazard Associated with Passenger Cars and Vans", Fire and Materials Conf., San Francisco, CA, Jan. 27–28, 2003, Interscience Communications, London, pp. 307–319.
73. Janssens, M.L., and Huczek, J.P., "Comparison of Fire Properties of Automotive Materials", in Business Communications Company Fourteenth Ann. Conference on Recent Advances in Flame Retardancy of Polymeric Materials, June 2–4, Stamford, CT, Ed. M. Lewin, Norwalk, CT, 2003.
74. Hirschler, M.M., "Improving the Fire Safety of Road Vehicles", Chapter 17 in "Advances in Fire Retardant Materials", Ed. R. Horrocks and D. Price, Woodhead Publishing Ltd., London, 2008.
75. National Highway Traffic Safety Administration (NHTSA), "Car and Van Fire Research Work Conducted under the General Motors Corporation—Settlement Agreement—Section B. Fire Safety Research" at the web site: http:// dms.dot.gov/, Docket # 3588.
76. ASTM E1354, "Standard Test Method for Heat and Visible Smoke Release Rates for Materials and Products Using an Oxygen Consumption Calorimeter (Cone Calorimeter)", American Society for Testing and Materials, West Conshohocken, PA.
77. Battipaglia, K., Huczek, J., Janssens, M., and Miller, M., 'Development of a Method to Assess the Fire Hazard of Automotive Materials', in Proc. Interflam '04, 10th Int. Fire Safety Conference, July 5–7, 2004, Interscience Communications, Edinburgh, pp. 1587–1596.
78. Huczek, J.P., and Blais, M.S., "Motorcoach Fire Safety" (Final Report. Report No. DOT HS 812 213). National Highway Traffic Safety Administration, Washington, DC, November 2015.
79. ASTM D5132, "Standard Test Method for Horizontal Burning Rate of Polymeric Materials Used in Occupant Compartments of Motor Vehicles", American Society for Testing and Materials, West Conshohocken, PA.
80. Code of Federal Regulations, Title 49, Transportation, Chapter V, National Highway Traffic Safety Administration, Department of Transportation, Part 571, Federal Motor Vehicle Safety Standards, 571.301 Standard No. 301; Fuel System Integrity.

81. Braun, E., Davis, S., Klote, J., Levin, B., and Paabo, M., "Assessment of the Fire Performance of School Bus Interior Components", NISTIR 4347, Nat Inst. Stand. Tech., Gaithersburg, MD, 1990.
82. "School Bus Seat Upholstery Fire Block Test", Approved by the National Conference on School Transportation as Part of the National Standards for School Buses and National Standards for School Bus Operations (Available from National Safety Council, 444 North Michigan Avenue, Chicago, IL).
83. ASTM E2574/E2574M, "Standard Test Method for Fire Testing of School Bus Seat Assemblies", American Society for Testing and Materials, West Conshohocken, PA.
84. Hirschler, M. M.; Piansay, T. Survey of Small Scale Flame Spread Test Results of Modern Fabrics. *Fire Mater.* **2007**, *31*, 373–386.
85. Hirschler, M.M., Zicherman, J.B., and Umino, P.Y., "Forensic Evaluation of Clothing Flammability", Fire and Materials (2008).
86. CA TB 129, California Technical Bulletin 129, "Flammability Test Procedure for Mattresses for Use in Public Buildings", California Bureau of Home Furnishings and Thermal Insulation, North Highlands, CA, 1992.
87. BS 5852, "Methods of Test for Assessment of the Ignitability of Upholstered Seating by Smoldering and Flaming Ignition Sources", British Standards Institution, London.
88. CA TB 117, California Technical Bulletin 117, "Requirements, Test Procedures and Apparatus for Testing the Flame Retardance of Resilient Filling Materials Used in Upholstered Furniture", California Bureau of Home Furnishings and Thermal Insulation, North Highlands, CA, 2000.
89. CA TB 121, California Technical Bulletin 121, "Flammability Test Procedure for Mattresses for Use in High Risk Occupancies", California Bureau of Home Furnishings and Thermal Insulation, North Highlands, CA, 1980.
90. ASTM E1590, "Standard Method for Fire Testing of Mattresses", American Society for Testing and Materials, West Conshohocken, PA.
91. Ohlemiller, T., Shields, J., McLane, R., and Gann, R., "Flammability Assessment Methodology for Mattresses," National Institute of Standards and Technology NISTIR 6497, Gaithersburg, MD, June, 2000.
92. Ohlemiller, T., and Gann, R., "Estimating Reduced Fire Risk Resulting From an Improved Mattress Flammability Standard," National Institute of Standards and Technology Technical Note 1446, Gaithersburg, MD, August, 2002.
93. CA TB 603, California Technical Bulletin 603, "Requirements and Test Procedure for Resistance of a Mattress/Box Spring Set to a Large Open Flame", California Bureau of Home Furnishings and Thermal Insulation", North Highlands, CA, 2003.
94. ASTM E1353, "Standard Test Methods for Cigarette Ignition Resistance of Components of Upholstered Furniture", American Society for Testing and Materials, West Conshohocken, PA (withdrawn).
95. ASTM E1352, "Standard Method of Test for Cigarette Ignition Resistance of Mock-Up Upholstered Furniture Assemblies", American Society for Testing and Materials, West Conshohocken, PA (withdrawn).
96. NFPA 260, Standard Methods of Tests and Classification System for Cigarette Ignition Resistance of Components of Upholstered Furniture, National Fire Protection Association, Quincy, MA.
97. NFPA 261, Standard Method of Test for Determining Resistance of Mock-Up Upholstered Furniture Material Assemblies to Ignition by Smoldering Cigarettes, National Fire Protection Association, Quincy, MA.
98. CA TB 117–2013, California Technical Bulletin 117–2013, Requirements, Test Procedure and Apparatus for Testing the Smolder Resistance of Materials Used in Upholstered Furniture, California Bureau of Electronic & Appliance Repair, Home Furnishings and Thermal Insulation, Sacramento, CA, 2013.
99. CA TB 133, California Technical Bulletin 133, Flammability Test Procedure for Seating Furniture for Use in Public Occupancies, California Bureau of Home Furnishings and Thermal Insulation, North Highlands, CA, 1991.
100. ASTM E1537, "Standard Method for Fire Testing of Upholstered Furniture", American Society for Testing and Materials, West Conshohocken, PA.
101. Verakis, H., "Conveyor Belt Flammability", Mine Safety & Health Administration, www.msha.gov/beltair/msha/january/verakis 2007 presentation.ppt
102. ASTM D635, "Standard Test Method for Rate of Burning and/or Extent and Time of Burning of Self-Supporting Plastics in a Horizontal Position", American Society for Testing and Materials, West Conshohocken, PA.
103. "The Final Report of the Technical Study Panel on the Utilization of Belt Air and the Composition and Fire Retardant Properties of Belt Materials in Underground Coal Mining", by J.M. Mutmansky, J.F. Brune, F. Calizaya, T.P. Mucho, J.C. Tien, and J.L. Weeks (2007), www.cdc.gov/niosh/mining/mineract/pdfs/BeltAirFinalReport122007.pdf.
104. www.federalregister.gov/documents/2008/12/31/E8-30639/flame-resistant-conveyor-belt-fire-prevention-and-detection-and-use-of-air-from-the-belt-entry.
105. Hirschler, M.M., and Earl, T., "Is Upholstered Furniture a Flammable Solid?", Business Communications Company Seventeenth Ann. Conference on Recent Advances in Flame Retardancy of Polymeric Materials, June 9–11, Stamford, CT, Ed. M. Lewin, Norwalk, CT, 2008.
106. "Regulation (EU) No 305/2011 of the European Parliament and of the Council of 9 March 2011 Laying Down Harmonised Conditions for the Marketing of Construction Products and Repealing Council Directive 89/106/EEC", https://eur-lex.europa.eu/legal-content/EN/TXT/?uri=CELEX:32011R0305—European Union Website, 2011.
107. "Roadmap for the Implementation of the Construction ProductsRegulation(CPR)—DocumentDate:29/02/2020—Created by GROW.I.2—Last Update: 29/03/2021", https://ec.europa.eu/docsroom/documents/45224—European Union Website, 2020.
108. EN 13501–1, "Fire Classification of Construction Products and Building Elements—Part 1: Classification Using Test Data from Reaction to Fire Tests", CEN (European Committee for Standardization), Brussels.
109. EN ISO 1182, "Reaction to Fire Tests for Building Products—Non-Combustibility Test", CEN (European Committee for Standardization), Brussels.
110. EN ISO 1716, "Reaction to Fire Tests for Building Products—Determination of the Heat of Combustion", CEN (European Committee for Standardization), Brussels.
111. EN 13823, "Reaction to Fire Tests for Building Products. Building Products Excluding Floorings Exposed to the Thermal Attack by a Single Burning Item", CEN (European Committee for Standardization), Brussels.
112. EN ISO 11925–2, "Reaction to Fire Tests—Ignitability of Building Products Subjected to Direct Impingement

of Flame—Part 2: Single-Flame Source Test", CEN (European Committee for Standardization), Brussels.
113. EN 9239–1, "Reaction to Fire Tests for Floorings—Part 1: Determination of the Burning Behaviour Using a Radiant Heat Source", CEN (European Committee for Standardization), Brussels.
114. Pipe Insulation: Commission Decision of 26 August 2003 Amending Decision 2000/147/EC Implementing Council Directive 89/106/EEC as Regards the Classification of the Reaction-to-Fire Performance of Construction Products (2003/632/EC).
115. Cables: Commission Decision of 27 October 2006 Amending Decision 2000/147/EC Implementing Council Directive 89/106/EEC as Regards the Classification of the Reaction-to-Fire Performance of Construction Products (2006/751/EC).
116. Grayson, S., Van Hees, P., Vercellotti, U., Breulet, H., and Green, A., "Fire Performance of Electrical Cables—New Test Methods and Measurement Techniques. Final Report of EU SMT Project SMT4-CT96–2059," Interscience Communications, London, 2000.
117. EN 50399, "Common Test Methods for Cables Under Fire Conditions—Heat Release and Smoke Production Measurement on Cables During Flame Spread Test—Test Apparatus, Procedures, Results." European Committee for Electrotechnical Standardization (CENELEC), Brussels.
118. United Kingdom Furniture Regulations, "The Furniture and Furnishings (Fire) (Safety) Regulations 1988", Statutory Instrument 1988 No. 1324—Amended 1993 by Statutory Instrument 1993 No. 207.
119. United Kingdom Mattress Regulations, "The Furniture and Furnishings (Fire) (Safety) Regulations 1988", Statutory Instrument 1988 No. 1324—Amended 1989 by Statutory Instrument 1989 No. 2358.
120. UK Government Consumer Safety Research, "Effectiveness of the Furniture and Furnishings (Fire) (Safety) Regulations 1988", Consumer Affairs Directorate, Dept. Trade and Industry, London, June 2000 [Research Conducted by Professor Gary Stevens, Univ. of Surrey, Guildford].
121. CBUF Report, "Fire Safety of Upholstered Furniture—the Final Report on the CBUF Research Programme", Ed. B. Sundstrom, EUR 16477 EN, European Commission, Measurements and Testing Report, Contract No. 3478/1/0/196/11-BCR-DK(30), 1995, Interscience Communi- cations, London.
122. ISO 5660–1, "Reaction to-Fire Tests—Heat Release, Smoke Production and Mass Loss Rate—Part 1: Heat Release Rate (Cone Calorimeter Method) and Smoke Production Rate (Dynamic Measurement)", International Organization for Standardization (ISO), Geneva.
123. ICBO (International Conference of Building Officials), Whittier, CA.
124. BOCA (Building Officials and Code Administrators), Country Club Hills, IL.
125. SBCCI (Southern Building Code Conference International), Birmingham, AL.
126. Hirschler, M. M. Procedures for Development and Revision of Codes and Standards Associated with Fire Safety in the USA and Its Fire Properties. *Fire Mater. J.* **2017**, *41*(8), 1058–1071.
127. NFPA 286, Standard Methods of Fire Tests for Evaluating Contribution of Wall and Ceiling Interior Finish to Room Fire Growth, National Fire Protection Association, Quincy, MA.
128. ASTM E84, "Standard Test Method for Surface Burning Characteristics of Building Materials", American Society for Testing and Materials, West Conshohocken, PA.
129. NFPA 265, Standard Methods of Fire Tests for Evaluating Room Fire Growth Contribution of Textile Coverings on Full Height Panels and Walls, National Fire Protection Association, Quincy, MA.
130. NFPA 275, Standard Method of Fire Tests for the Evaluation of Thermal Barriers, National Fire Protection Association, Quincy, MA.
131. NFPA 285, Standard Fire Test Method for Evaluation of Fire Propagation Characteristics of Exterior Wall Assemblies Containing Combustible Components, National Fire Protection Association, Quincy, MA.
132. ASTM E108, "Standard Test Methods for Fire Tests of Roof Coverings", American Society for Testing and Materials, West Conshohocken, PA.
133. Evans, D.H., and Hirschler, M.M., "Foam Plastics in Building Construction", Session T44 at NFPA Conference and Expo, June 9–12, 2014, Las Vegas, NV.
134. ASTM D1929, "Standard Test Method for Determining Ignition Temperature of Plastics", American Society for Testing and Materials, West Conshohocken, PA.
135. ASTM D2843, "Standard Test Method for Density of Smoke from the Burning or Decomposition of Plastics", American Society for Testing and Materials, West Conshohocken, PA.
136. ASTM D8484, "Standard Specification for Plastic Lumber Materials and Wood-Plastic Composite Materials Used as Exterior Wall Coverings", American Society for Testing and Materials, West Conshohocken, PA.
137. UL 1975, Standard for Fire Tests for Foamed Plastics Used for Decorative Purposes", Underwriters Laboratories, Northbrook, IL.
138. NFPA 289, "Standard Method of Fire Test for Individual Fuel Packages", National Fire Protection Association, Quincy, MA.
139. CPSC 16 CFR Parts 1209, "Interim Safety Standard for Cellulose Insulation", www.ecfr.gov/current/title-16/chapter-II/subchapter-B/part-1209?toc=1.
140. CPSC 16 CFR Parts 1404, "Cellulose Insulation", www.ecfr.gov/current/title-16/chapter-II/subchapter-B/part-1404.
141. CAN/ULC-S102.2, "Standard Method of Test for Surface Burning Characteristics of Flooring, Floor Covering, and Miscellaneous Materials and Assemblies", Underwriters Laboratories of Canada, Richmond, BC.
142. ASTM E970, "Standard Test Method for Critical Radiant Flux of Exposed Attic Floor Insulation Using a Radiant Heat Energy Source", American Society for Testing and Materials, West Conshohocken, PA.
143. ASTM F1085, "Standard Specification for Mattress and Box Springs for Use in Berths in Marine Vessels (Annexes A1 and A3)", American Society for Testing and Materials, West Conshohocken, PA.
144. NFPA 701, "Standard Methods of Fire Tests for Flame-Propagation of Textiles and Films", National Fire Protection Association, Quincy, MA.
145. ASTM E3082, "Standard Methods for Determining the Effectiveness of Fire Retardant Treatments for Natural Christmas Trees", American Society for Testing and Materials, West Conshohocken, PA.
146. UL 1820, "Standard for Safety Fire Test of Pneumatic Tubing for Flame and Smoke Characteristics", Underwriters Laboratories, Northbrook, IL.

147. UL 1887, "Standard for Safety Fire Test of Plastic Sprinkler Pipe for Visible Flame and Smoke Characteristics", Underwriters Laboratories, Northbrook, IL.
148. UL 2024, "Standard for Cable Routing Assemblies and Communications Raceways", Underwriters Laboratories, Northbrook, IL.
149. UL 2043, "Standard for Safety Fire Test for Heat and Visible Smoke Release for Discrete Products and Their Accessories Installed in Air-Handling Spaces", Underwriters Laboratories, Northbrook, IL.
150. ASTM E2768, "Standard Test Method for Extended Duration Surface Burning Characteristics of Building Materials (30 min Tunnel Test)", American Society for Testing and Materials, West Conshohocken, PA.
151. NFPA 703, "Standard for Fire Retardant Impregnated Wood and Fire Retardant Coatings for Building Materials", National Fire Protection Association, Quincy, MA.
152. Hirschler, M.M., "Wildland Urban Interface Code Comparisons: IWUIC (ICC) and Chapter 7A (California)", Wildfire Resilient Structures (WiReS) Conference and Tradeshow, February 7–10, 2023, San Diego, CA.
153. International Code Council Performance Code for Buildings and Facilities (ICCPC), International Code Council, Washington, DC.
154. NFPA 501, "Standard on Manufactured Housing", National Fire Protection Association, Quincy, MA (withdrawn).
155. UL 1666, "Standard Test for Flame Propagation Height of Electrical and Optical-Fiber Cables Installed Vertically in Shafts", Underwriters Laboratories, Northbrook, IL.
156. IEEE 1202, "IEEE Standard for Flame-Propagation Testing of Wire & Cable", Institute of Electrical and Electronics Engineers, New York, NY.
157. UL 1581, "Reference Standard for Electrical Wires, Cables, and Flexible Cords", Underwriters Laboratories, Northbrook, IL.
158. UL 1685, "Standard Vertical-Tray Fire-Propagation and Smoke-Release Test for Electrical and Optical-Fiber Cables", Underwriters Laboratories, Northbrook, IL.
159. NFPA 5000, "Building Construction and Safety Code", National Fire Protection Association, Quincy, MA.
160. NFPA 75, "Standard for the Fire Protection of Information Technology Equipment", National Fire Protection Association, Quincy, MA.
161. NFPA 909, "Code for the Protection of Cultural Resource Properties—Museums, Libraries, and Places of Worship", National Fire Protection Association, Quincy, MA.
162. NFPA 914, "Code for Fire Protection of Historic Structures", National Fire Protection Association, Quincy, MA.
163. ASTM E1995, "Standard Test Method for Measurement of Smoke Obscuration Using a Conical Radiant Source in a Single Closed Chamber, With the Test Specimen Oriented Horizontally", American Society for Testing and Materials, West Conshohocken, PA.
164. NFPA 270, "Standard Test Method for Measurement of Smoke Obscuration Using a Conical Radiant Source in a Single Closed Chamber", National Fire Protection Association, Quincy, MA.
165. ISO 5659-2, "Plastics—Smoke Generation—Part 2: Determination of Optical Density by a Single-Chamber Test", International Organization for Standardization (ISO), Geneva.
166. ASTM E1317, "Standard Test Method for Flammability of Marine Surface Finishes", American Society for Testing and Materials, West Conshohocken, PA.
167. ISO 1716, "Reaction to Fire Tests for Building Products—Determination of the Heat of Combustion", International Organization for Standardization (ISO), Geneva.
168. ASTM E1822, "Standard Method for Fire Testing of Stacked Chairs", American Society for Testing and Materials, West Conshohocken, PA.
169. 24 CFR Part 3280, "Manufactured Home Construction and Safety Standards", www.ecfr.gov/current/title-24/subtitle-B/chapter-XX/part-3280.
170. 24 CFR Part 3285, "Manufactured Home Installation Standards", www.ecfr.gov/current/title-24/subtitle-B/chapter-XX/part-3285.
171. UL 2424, "Outline of Investigation for Cable Marked Limited Combustible", Underwriters Laboratories, Northbrook, IL (withdrawn).
172. NFPA 259, "Standard Test Method for Potential Heat of Building Materials", National Fire Protection Association, Quincy, MA.
173. ASTM E2965, "Standard Test Method for Determination of Low Levels of Heat Release Rate for Materials and Products Using an Oxygen Consumption Calorimeter", American Society for Testing and Materials, West Conshohocken, PA.
174. EN 13823, Reaction to Fire Tests for Building Products—Building Products Excluding Floorings Exposed to the Thermal Attack by a Single Burning Item, European Committee for Normalization (CEN), Geneva.
175. NFPA 13D, "Standard for the Installation of Sprinkler Systems in One- and Two-Family Dwellings and Manufactured Homes", National Fire Protection Association, Quincy, MA.
176. NFPA 13R, "Standard for the Installation of Sprinkler Systems in Residential Occupancies up to and Including Four Stories in Height", National Fire Protection Association, Quincy, MA.
177. NFPA 99, "Standard for Health Care Facilities", National Fire Protection Association, Quincy, MA.
178. UMC, Uniform Mechanical Code, International Association of Plumbing and Mechanical Officials (IAPMO), Ontario, CA.
179. NFPA 274, "Standard Test Method to Evaluate Fire Performance Characteristics of Pipe Insulation", National Fire Protection Association, Quincy, MA.
180. ASTM E136, "Standard Test Method for Assessing Combustibility of Materials Using a Vertical Tube Furnace at 750°C", American Society for Testing and Materials, West Conshohocken, PA.
181. Hirschler, M. M. Noncombustibility: Testing and Regulatory Requirements *Fire Mater.* **2023**, *47*(2), 182–198. https://doi.org/10.1002/fam.3087198
182. Setchkin, N. P. "A Method and Apparatus for Determining the Ignition Characteristics of Plastics," Research Paper RP 2052. *J. Res. Nat. Bureau Stand.* **1949,** *43*, 591–608.
183. Setchkin, N.P., "Combustibility Tests of 47 ASTM Material Samples", National Bureau of Standards (NBS) Project 1002-43-1029, Report No. 1454, February 6, 1952, Washington, DC.
184. ASTM E2652, "Standard Test Method for Assessing Combustibility of Materials Using a Tube Furnace With a Cone-Shaped Airflow Stabilizer at 750°C", American Society for Testing and Materials, West Conshohocken, PA.
185. ISO 5657, "Reaction to Fire Tests—Ignitability of Building Products Using a Radiant Heat Source", International Organization for Standardization, Geneva.
186. ISO 871, "Plastics—Determination of Ignition Temperature Using a Hot-Air Furnace", International Organization for Standardization, Geneva.

187. ASTM D3874, "Standard Test Method for Ignition of Materials by Hot Wire Sources", American Society for Testing and Materials, West Conshohocken, PA.
188. IEC 60695–11–5, "Fire Hazard Testing—Part 11–5: Test Methods—Needle-Flame Test Method—Apparatus, Confirmatory Test Arrangement and Guidance", International Electrotechnical Commission, Geneva.
189. ASTM D6194, "Standard Test Method for Glow Wire Ignition of Materials", American Society for Testing and Materials, West Conshohocken, PA.
190. IEC 60695–2–10, "Fire Hazard Testing—Part 2–10: Glowing/Hot-Wire Based Test Methods—Glow-Wire Apparatus and Common Test Procedure", International Electrotechnical Commission, Geneva.
191. ASTM E1321, "Standard Test Method for Determining Material Ignition and Flame Spread Properties", American Society for Testing and Materials, West Conshohocken, PA.
192. ASTM E3020, "Standard Practice for Ignition Sources", American Society for Testing and Materials, West Conshohocken, PA.
193. ISO 10093, "Plastics—Fire Tests—Standard Ignition Sources", International Organization for Standardization, Geneva.
194. Hirschler, M.M., "Heat Release from Plastic Materials," Chapter 12A, in "Heat Release in Fires", Eds. V. Babrauskas and S.J. Grayson, pp. 375–422, Elsevier, London, 1992.
195. ASTM D2863, "Standard Test Method for Measuring the Minimum Oxygen Concentration to Support Candle-Like Combustion of Plastics (Oxygen Index)", American Society for Testing and Materials, West Conshohocken, PA.
196. ISO 4589–2, "Plastics—Determination of Burning Behaviour by Oxygen Index—Part 2: Ambient-Temperature Test", International Organization for Standardization, Geneva.
197. ISO 4589–3, "Plastics—Determination of Burning Behaviour by Oxygen Index—Part 3: Elevated-Temperature Test", International Organization for Standardization, Geneva.
198. UL 94, "Standard for Test for Flammability of Plastic Materials for Parts in Devices and Appliances", Underwriters Laboratories, Northbrook, IL.
199. Cullis, C.F., and Hirschler, M.M., "The Combustion of Organic Polymers", Oxford University Press, Oxford, 1981.
200. ASTM E162, "Standard Test Method for Surface Flammability of Materials Using a Radiant Heat Energy Source)", American Society for Testing and Materials, West Conshohocken, PA.
201. ASTM D3675, "Standard Test Method for Surface Flammability of Flexible Cellular Materials Using a Radiant Heat Energy Source)", American Society for Testing and Materials, West Conshohocken, PA.
202. Steiner, A.J., Underwriters Laboratories, Research Bulletin No. 32 (1944).
203. UL 2846, "Standard Fire Test of Plastic Water Distribution Plumbing Pipe for Visible Flame and Smoke Characteristics", Underwriters Laboratories, Northbrook, IL.
204. Belles, D.W., Fisher, F.L., and Williamson, R.B., "How Well Does the ASTM E84 Predict Fire Performance of Textile Wallcoverings?", Fire J., 82(1), 24–30, 74 (1988).
205. ASTM E2231, "Standard Practice for Specimen Preparation and Mounting of Pipe and Duct Insulation Materials to Assess Surface Burning Characteristics", American Society for Testing and Materials, West Conshohocken, PA.
206. ASTM E2404, "Standard Practice for Specimen Preparation and Mounting of Textile, Paper or Polymeric (Including Vinyl) and Wood Wall or Ceiling Coverings, Facings and Veneers, to Assess Surface Burning Characteristics", American Society for Testing and Materials, West Conshohocken, PA.
207. ASTM E2573, "Standard Practice for Specimen Preparation and Mounting of Site-Fabricated Stretch Systems to Assess Surface Burning Characteristics", American Society for Testing and Materials, West Conshohocken, PA.
208. ASTM E2579, "Standard Practice for Specimen Preparation and Mounting of Wood Products to Assess Surface Burning Characteristics", American Society for Testing and Materials, West Conshohocken, PA.
209. ASTM E2599, "Standard Practice for Specimen Preparation and Mounting of Reflective Insulation, Radiant Barrier and Vinyl Stretch Ceiling Materials for Building Applications to Assess Surface Burning Characteristics", American Society for Testing and Materials, West Conshohocken, PA.
210. ASTM E2688, "Standard Practice for Specimen Preparation and Mounting of Tapes to Assess Surface Burning Characteristics", American Society for Testing and Materials, West Conshohocken, PA.
211. ASTM E2690, "Standard Practice for Specimen Preparation and Mounting of Caulks and Sealants to Assess Surface Burning Characteristics", American Society for Testing and Materials, West Conshohocken, PA.
212. ASTM E2988, "Standard Practice for Specimen Preparation and Mounting of Flexible Fibrous Glass Insulation for Metal Buildings to Assess Surface Burning Characteristics", American Society for Testing and Materials, West Conshohocken, PA.
213. ASTM E3202, "Standard Practice for Specimen Preparation and Mounting of Plastic Composites for Use as Deck Boards, Stair Treads, Guards or Handrails to Assess Surface Burning Characteristics", American Society for Testing and Materials, West Conshohocken, PA.
214. ASTM E3287, "Standard Practice for Specimen Preparation of Fenestration Profiles Intended to Support Non-Combustible in-Fill Materials to Assess Surface Burning Characteristics", American Society for Testing and Materials, West Conshohocken, PA.
215. Quintiere, J.G., and Harkleroad, M., "New Concepts for Measuring Flame Spread Properties", US National Bureau Standards, Gaithersburg, MD, NBSIR 84-2943, 1984.
216. Babrauskas, V. "Effective Measurement Techniques for Heat, Smoke and Toxic Fire Gases", Int. Conf. "FIRE: Control the Heat-Reduce the Hazard," Fire Research Station, October 24–25, 1988, London, # 4.
217. Babrauskas, V., and Grayson, S.J., "Heat Release in Fires", Elsevier, London, 1992.
218. Hirschler, M.M., "Fire Retardance, Smoke Toxicity and Fire Hazard", in Proc. Flame Retardants '94, British Plastics Federation Editor, Interscience Communications, London, January 26–27, pp. 225–237 (1994).
219. CBUFReport,"FireSafetyofUpholsteredFurniture—TheFinal Report on the CBUF Research Programme", Ed. B. Sundstrom, EUR 16477 EN, European Commission, Measurements and Testing Report, Contract No. 3478/1/0/196/11-BCR-DK(30), 1995, Interscience Communications, London.
220. Hirschler, M. M. Analysis of and Potential Correlations Between Fire Tests for Electrical Cables, and How to Use This Information for Fire Hazard Assessment. *Fire Technol.* **1997**, *33*, 291–315.
221. Hirschler, M.M., "How to Decide If a Material Is Suitable for an Application Where Fire Safety Is Required", Flame Retardants 2002, February 5–6, London, pp. 45–56, Interscience Communications, London, 2002.

222. Hirschler, M.M. "Fire Safety, Smoke Toxicity and Acidity", Flame Retardants 2006, February 14–15, London, pp. 47–58, Interscience Communications, London, 2006.
223. Hirschler, M. M. Flame Retardants and Heat Release: Review of Traditional Studies on Products and on Groups of Polymers. *Fire Mater.* **2015**, *39*, 207–231.
224. Hirschler, M. M. Flame Retardants and Heat Release: Review of Data on Individual Polymers. *Fire Mater.* **2015**, *39*, 232–258.
225. Babrauskas, V.; Peacock, R. D. Heat Release Rate: The Single Most Important Variable in Fire Hazard. *Fire Safety J.* **1992**, *18*, 255–272.
226. Smith, E.E., "Heat Release Rate of Building Materials," in "Ignition, Heat Release and Noncombustibility of Materials, ASTM STP 502," (A.F. Robertson, editor), p. 119, American Society for Testing and Materials, Philadelphia, PA (1972).
227. Babrauskas, V., "Development of the Cone Calorimeter. A Bench-Scale Heat Release Rate Apparatus Based on Oxygen Consumption," National Bureau of Standards, NBSIR 82–2611 (1982), Gaithersburg, MD.
228. Hirschler, M. M. The Measurement of Smoke in Rate of Heat Release Equipment in a Manner Related to Fire Hazard. *Fire Safety J.* **1991**, *17*, 239–258.
229. Babrauskas, V., "Bench-Scale Methods for Prediction of Full-Scale Fire Behavior of Furnishings and Wall Linings," Society of Fire Protection Engineers, Boston, MA, Technology Report 84–10 (1984).
230. Babrauskas, V. Upholstered Furniture Room Fires—Measurements, Comparison with Furniture Calorimeter Data, and Flashover Predictions. *J. Fire Sci.* **1984**, *2*, 5–19.
231. Babrauskas, V., and Krasny, J.F., "Prediction of Upholstered Chair Heat Release Rates from Bench-Scale Measurements", in "Fire Safety. Science and Engineering, ASTM STP 882" (T.Z. Harmathy, editor), p. 268, American Society for Testing and Materials, Philadelphia, PA (1985).
232. Hirschler, M.M., "Tools Available to Predict Full Scale Fire Performance of Furniture", in "*Fire and Polymers II. Materials and Tests for Hazard Prevention*" (G.L. Nelson, editor), ACS Symposium Series 599, Developed from ACS Symp. in 208th ACS Natl Mtg, Aug. 21–25, 1994, Amer. Chem. Soc., Washington, DC, Chapter 36, pp. 593–608, Amer. Chem. Soc. Washington, DC, 1995.
233. Lyon, R.E., "Fire-Safe Aircraft Cabin Materials", in "*Fire and Polymers II. Materials and Tests for Hazard Prevention*" (G.L. Nelson, editor), ACS Symposium Series 599, Developed from ACS Symp. in 208th ACS Natl Mtg, Aug. 21–25, 1994, Washington, DC, Chapter 36, pp. 618–638, Amer. Chem. Soc. Washington, DC, 1995.30.
234. ASTM F1085, "Standard Specification for Mattress and Box Springs for Use in Berths in Marine Vessels (Annexes A1 and A3)", American Society for Testing and Materials, West Conshohocken, PA.
235. ASTM D5424, "Standard Test Method for Smoke Obscuration of Insulating Materials Contained in Electrical or Optical Fiber Cables When Burning in a Vertical Cable Tray Configuration", American Society for Testing and Materials, West Conshohocken, PA.
236. ASTM D5537, "Standard Test Method for Heat Release, Flame Spread, Smoke Obscuration, and Mass Loss Testing of Insulating Materials Contained in Electrical or Optical Fiber Cables When Burning in a Vertical Cable Tray Configuration", American Society for Testing and Materials, West Conshohocken, PA.
237. UL 1975, Standard for Fire Tests for Foamed Plastics Used for Decorative Purposes", Underwriters Laboratories, Northbrook, IL.
238. NFPA 289, "Standard Method of Fire Test for Individual Fuel Packages", National Fire Protection Association, Quincy, MA.
239. ASTM E2257, "Standard Test Method for Room Fire Test of Wall and Ceiling Materials and Assemblies", American Society for Testing and Materials, West Conshohocken, PA.
240. Hirschler, M. M. Use of Heat Release Rate to Predict Whether Individual Furnishings Would Cause Self Propagating Fires. *Fire Safety J.* **1999**, *32*, 273–296.
241. Sarkos, C.P., and Hill, R.G., "Evaluation of Aircraft Interior Panels Under Full-Scale Cabin Fire Test Conditions", AIAA-85–0393, AIAA 23rd Aerospace Sciences Meeting, Reno, NV (1985).
242. Wickstrom, U., (Programme Manager) Proc. Int. EUREFIC Seminar 1991, Sept. 11–12, 1991, Copenhagen, Interscience Commun., London, 1991.
243. Babrauskas, V. Applications of Predictive Smoke Measurements. *J. Fire Flamm.* **1981**, *12*, 51.
244. Quintiere, J. G. Smoke Measurements: An Assessment of Correlations Between Laboratory and Full-Scale Experiments. *Fire Mat.* **1982**, *6*, 145.
245. Babrauskas, V., "Use of the Cone Calorimeter for Smoke Prediction Measurements," in SPE RETEC Conference on "PVC: THE ISSUES," Atlantic City, NJ, p. 41 (1987).
246. Hirschler, M.M., "Smoke in Fires: Obscuration and Toxicity", Plenary Lecture, Business Communications Company Conference on Recent Advances in Flame Retardancy of Polymeric Materials, May 15–17, Stamford, CT, Eds. G.S. Kirshenbaum and M. Lewin, pp. 70–82, Norwalk, CT, 1990.
247. Hirschler, M. M. How to Measure Smoke Obscuration in a Manner Relevant to Fire Hazard Assessment: Use of Heat Release Calorimetry Test Equipment, *J. Fire Sci.* **1991**, *9*, 183–222.
248. Breden, L. H.; Meisters, M. The Effect of Sample Orientation in the Smoke Density Chamber. *J. Fire Flamm.* **1976**, *7*, 234.
249. Ohlemiller, T.J., and Shields, J.R., "Aspects of the Fire Behavior of Thermoplastic Materials", NIST Technical Note 1493, National Institute of Standards and Technology, Gaithersburg, MD, January 2008.
250. Hirschler, M. M. Smoke and Heat Release and Ignitability as Measures of Fire Hazard From Burning of Carpet Tiles. *Fire Safety J.* **1992**, *18*, 305–324.
251. Hirschler, M.M., and Janssens, M.L., "Smoke Obscuration Measurements in the NFPA 265 Room-Corner Test", Proc. 6th Fire and Materials Conf., Feb. 22–23, 1999, San Antonio, TX; Interscience Commun., London, Ed. S.J. Grayson, pp. 179–198, 1999.
252. Hirschler, M.M., "Fire Performance of Organic Polymers, Thermal Decomposition, and Chemical Composition", American Chemical Society, Fire and Polymers—Materials and Solutions for Hazard Prevention, ACS Symposium Series 797, Eds. G.L. Nelson and C.A. Wilkie. Amer. Chem. Soc., Washington, DC, 2001, pp. 293–306.
253. Jin, T., "Studies of Emotional Instability in Smoke from Fires", J. Fire Flamm., 12, 130–142 (1981).
254. ASTM E1678, "Standard Test Method for Measuring Smoke Toxicity for Use in Fire Hazard Analysis", American Society for Testing and Materials, West Conshohocken, PA.
255. NFPA 269, "Standard Test Method for Developing Toxic Potency Data for Use in Fire Hazard Modeling", National Fire Protection Association, Quincy, MA.

256. Hirschler, M. M. Fire Hazard and Toxic Potency of the Smoke from Burning Materials. *J. Fire Sci.* **1987**, *5*, 289–307.
257. Babrauskas, V., Harris, R.H., Gann, R.G., Levin, B.C., Lee, B.T., Peacock, R.D., Paabo, M., Twilley, W., Yoklavich, M.F., and Clark, H.M., 1988, "Fire Hazard Comparison of Fire-Retarded and Non-Fire-Retarded Products," NBS Special Publ. 749, National Bureau of Standards, Gaithersburg, MD.
258. Mulholland, G.W., 1989, in W.M. Pitts, "Executive Summary for the Workshop on Developing a Predictive Capability for CO Formation in Fires", NISTIR 89-4093, National Institute of Standards and Technology, Gaithersburg, MD, p. 25.
259. Babrauskas, V., Harris, R.H., Braun, E., Levin, B.C., Paabo, M., and Gann, R.G., 1991, "The Role of Bench-Scale Data in Assessing Real-Scale Fire Toxicity", NIST Tech. Note # 1284, National Inst. Standards Technology, Gaithersburg, MD.
260. Hirschler, M.M., 1990, "General Principles of Fire Hazard and the Role of Smoke Toxicity", in "Fire and Polymers: Hazards Identification and Prevention" (Ed. G.L. Nelson), ACS Symposium Series 425, Amer. Chem. Soc., Washington, DC, Chapter 28, pp. 462–478.
261. Babrauskas, V., Levin, B.C., Gann, R.G., Paabo, M., Harris, R.H., Peacock, R.D., and Yusa, S., 1991, "Toxic Potency Measurement for Fire Hazard Analysis", NIST Special Publication # 827, National Inst. Standards Technology, Gaithersburg, MD.
262. Debanne, S.M., Hirschler, M.M., and Nelson, G.L., 1992, "The Importance of Carbon Monoxide in the Toxicity of Fire Atmospheres", in "Fire Hazard and Fire Risk Assessment", Ed. M. M. Hirschler, ASTM STP 1150, Amer. Soc. Testing and Materials, Philadelphia, PA, pp. 9–23.
263. Hirschler, M.M. (Editor-in-Chief), and Debanne, S.M., Larsen, J.B., and Nelson, G.L., "Carbon Monoxide and Human Lethality—Fire and Non-Fire Studies", Elsevier, London, 1993.
264. Gann, R. G.; Babrauskas, V.; Peacock, R. D.; Hall, J. R. Fire Conditions for Smoke Toxicity Measurement. *Fire Mater.* **1994**, *18*, 193–199.
265. Hall, J.R., "The US Fire Problem Overview Report—Leading Causes and Other Patterns and Trends", National Fire Protection Association, March 1998, Quincy, MA.
266. Hirschler, M.M., "The Role of Carbon Monoxide in the Toxicity of Fire Atmospheres", in Proc. 19th. Int. Conf. on Fire Safety, Product Safety Corp., San Francisco, CA, Ed. C.J. Hilado, January 10–14, pp. 163–84 (1994).
267. Budnick, E.K., "Mobile Home Living Room Fire Studies", NBSIR 78-1530, National Bureau of Standards, Gaithersburg, MD,1978.
268. Budnick, E.K., Klein, D.P., and O'Laughlin, R.J., "Mobile Home Bedroom Fire Studies: The Role of Interior Finish", NBSIR 78-1531, National Bureau of Standards, Gaithersburg, MD,1978.
269. Levine, R.S., and Nelson, H.E., "Full Scale Simulation of a Fatal Fire and Comparison of Results With Two Multiroom Models", NISTIR 90-4268, National Inst. Standards and Technology, Gaithersburg, MD,1990.
270. Gottuk, D. T.; Roby, R. J.; Peatross, M. J.; Beyler, C. L. CO Production in Compartment Fires. *J. Fire Protection Eng.* **1992**, *4*, 133–150.
271. Beyler, C. L. Major Species Production by Diffusion Flames in a Two-Layer Compartment Fire Environment. *Fire Safety J.* **1986**, *10*, 47–56.
272. Lyon, R. E.; Walters, R. N. Pyrolysis Combustion Flow Calorimetry. *J. Anal. Appl. Pyrolysis.* **2004**, *71*(1), 27–46.
273. ASTM D7309, "Standard Test Method for Determining Flammability Characteristics of Plastics and Other Solid Materials Using Microscale Combustion Calorimetry", American Society for Testing and Materials, West Conshohocken, PA.
274. Stoliarov, S. I.; Zeller, O.; Morgan, A. B.; Levchik, S. An Experimental Setup for Observation of Smoldering-to-Flaming Transition on Flexible Foam/Fabric Assemblies. *Fire Mater.* **2018**, *42*, 128–133. https://doi.org/10.1002/fam.2464.
275. CAN/ULC-S102, "Standard Method of Test for Surface Burning Characteristics of Building Materials and Assemblies", Underwriters Laboratories of Canada, Richmond, BC.
276. Hall, S., and Messerschmidt, B., "Civilian Fire Death Rate Trends: 1980–2021", at NFPA 2023 Conference and Exposition, June 2023, Las Vegas, NV. National Fire Protection Association, Quincy, MA.

23 Chemical Regulation of Flame Retardants

Adrian Beard
Clariant Plastics and Coatings Deutschland GmbH (Corporation)

23.1 INTRODUCTION

Disclaimer notice: laws and regulations change over time. The information presented in this section represents the status of summer 2023, and interpretations of legal texts are done with best efforts of the author, but without warranty. Please always consult the current legal documents for latest up to date status.

Flame retardants (FRs) have come under regulatory scrutiny since the 1990s when several brominated FRs were detected in the environment.[1,2] In the meantime, they have been found in all environmental compartments (water, air, soil) as well as in animals and even humans, including human milk.[3] The Human Biomonitoring for Europe (HBM4EU)[4] testing program was launched in 2016, which also included FRs based on bromine, chlorine, and phosphorus. Several small molecule FRs were detected in blood and urine, also in children.[5] However, there is a mismatch between modern chemical analytical tools which can detect extremely small traces of chemicals on the one hand and our ability to assessment the health impact of such trace amounts. It is very difficult to prove impacts across a highly variable population, which is exposed to many natural and man-made chemicals and other environmental impacts.

In order to assess the hazards of a particular chemical, both the physical and health hazards of the chemical must be considered. Some FRs do have toxic properties as neat chemicals. Whether a chemical which, for example, is neurotoxic as the pure chemical also exhibits this effect in real life depends on which quantity people are really exposed to, to which extent they come in touch with the chemical. The potential toxic effect is also very much dependent on the dose. Only the combination of hazard and exposure results in the risk to a person: risk = hazard * exposure. In everyday life we accept risks, in particular when they are far outweighed by benefits (which can also be very subjective like downhill skiing, driving a motorcycle). FRs provide important societal benefit of making many materials and everyday products (fire) safe and to protect us. However, some have been found to have detrimental environmental and health effects so that they have been regulated in many regions of the world. Small molecule, additively used (nonreactive) brominated (and chlorinated) FRs have been widely found in the environment, some also have bioaccumulative and toxic properties and therefore have been restricted under the UN Stockholm Convention or national laws, see Figure 23.1 for examples. Some of these products are precursors for the formation of chlorinated or brominated dibenzodioxins and dibenzofurans in accidental fires or uncontrolled combustion situations like in crude recycling of electronic waste. However, it should be noted that the combustion of any halogenated substance is likely to form some levels of halogenated dioxin,[6,7] so the issue of dioxin formation is not unique to just halogenated flame retardants.

More restrictions and regulation are to be expected, because of growing scrutiny and evidence around FRs. Nevertheless, industry has been developing improved FRs with a better environmental and health profile. New trends include bio- and renewable-based FRs[8] (see Chapter 13 of this book). Figure 23.2 shows the three essential elements of any successful regulation: without proper enforcement and sanctions—which are often lacking in the chemical sector—laws will have limited effect.

Increasingly, additional criteria beyond the chemical hazards are being considered when it comes to the question "what is a sustainable flame retardant (or chemical)?" Figure 23.3 presents an overview of criteria including life cycle aspects up to social impacts or the question whether critical raw materials are being used. This concept will be further elaborated in 23.7.3 on safe and sustainable-by-design and is also discussed in Chapter 24 of this book.

23.2 REGULATORY APPROACH FOR CHEMICALS

Chemical regulations are national or regional laws, with the exceptions of the treaties ("Conventions") mentioned below under the United Nations auspices. Regulations may even apply on a federal state level or down to regions or municipalities (e.g., in the USA). Countries have historically embarked on chemical regulation by setting up "inventories," which are a register of chemicals which are allowed to be manufactured and marketed. Whereas existing chemicals were often grandfathered into the system without further requirements, specific information including toxicity

DOI: 10.1201/9781003380689-23

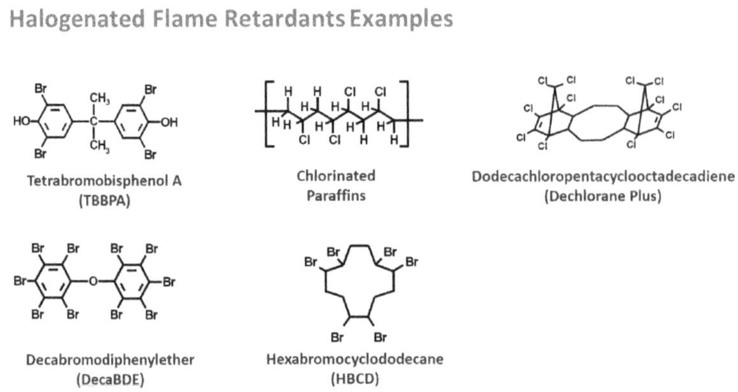

FIGURE 23.1 "First generation" halogenated flame retardants which have come under regulatory scrutiny and are restricted in many jurisdictions. TBBPA has become a SVHC candidate in Europe and is listed for regulatory action in Europe by the European Chemicals Agency.

FIGURE 23.2 Necessary elements of any successful regulation.

data were required for new chemicals. If a chemical was suspected to have detrimental effects on people or the environment, it was up to the authorities to prove so. Only then could a chemical be put on the list of restricted substances. This burden of proof principle was fundamentally reversed with the introduction of REACH (v.i.) in Europe. There are still some countries in the world without an effective chemical legislation in place, not even an inventory (e.g., in African countries, India CMSR is still in draft stage). Figure 23.4 gives a high-level overview of regulations and key drivers by regions, whereas Figure 23.5 highlights the impact of regulations on the different types of FR chemistries.

23.3 INTERNATIONAL

23.3.1 UN Stockholm Convention

Only very few flame retardants are regulated on a global level by the United Nations under the "persistent organic pollutants" (POP) rule.[9] The main reason is that it takes high efforts to reach a consensus with such a large group of nations with different national and economic interests. Only after several years of mounting evidence of detrimental effects of chemicals to humans and the environment will a restriction be passed. The UN Stockholm Convention restricts these flame retardants:

Sustainable FRs

HAZARDS

- PBT: Persistence, Bioaccumulation, Toxicity
- CMR: Carcinogenicity, Mutagenicity, Reproductive Toxicity
- Endocrine effects, Mobility (new in 2023 in Europe)

LIFE CYCLE

- Carbon footprint (global warming potential) and other Product Environmental Footprint criteria
- Recycling properties
- Production (value chain): impact on workers
- Use Phase: impact on consumers

OTHER

- Fires: Smoke formation / toxicity
- Critical raw materials
- Social impacts

FIGURE 23.3 Criteria to be considered for the sustainability of flame retardants.

- **Europe** clearly regulatory driver with Chemicals Strategy for Sustainability: REACH revision, Generic Risk Assessment, polymer registrations, Safe and Sustainable-by-Design, EcoDesign, Substances of Concern …
- **North America**:
    - USA: slow action on Federal level (TSCA), many activities on State level, with California, New York, Washington leading with initiatives
    - Canada: comprehensive approach with Domestic Substances List and evaluations
- **Asia**: lagging behind, slowly following
- **Rest of world**: very slow to no action

FIGURE 23.4 Perspective on the global regulatory horizon for flame retardants by geography.

- HBCD
- PBDE (tetra, penta, hexa, hepta, deca)
- SCCP

They have also recently included Perfluorooctanoic acid (PFOA), its salts and PFOA-related compounds, as well as Perfluorohexane sulfonic acid (PFHxS), its salts and PFHxS-related compounds.

23.3.2 UN Basel and Rotterdam Conventions

The "Basel Convention on the Control of Transboundary Movements of Hazardous Wastes and their Disposal"[10] now also including plastic waste. This brings up the issue of legacy flame retardants in waste plastic: too strict limit values for residual content of unwanted FRs is an impediment to more (mechanical) recycling of plastics.

Bromine and **chlorine** based:
- historically under most pressure, several legacy products restricted
- More restrictions very likely (see ECHA strategy in EU)

Phosphorus based:
- Very diverse chemistry
- Only few restrictions so far

Nitrogen based:
- melamine identified as Substance of Very High Concern (SVCH) in EU
- Limited impact for melamine derivatives

Inorganics:
- Some FR / synergists with issues, e.g. Zn-Borate

FIGURE 23.5 Perspective on the global regulatory horizon for flame retardants by chemistries.

FIGURE 23.6 As an example, the symbol for substances hazardous to the human health as implemented by GHS.[14]

The Rotterdam Convention[11] (formally, the Rotterdam Convention on the Prior Informed Consent Procedure for Certain Hazardous Chemicals and Pesticides in International Trade) puts restrictions on Polybrominated biphenyls (PBB), PBDEs, HBCD, SCCP, Tris(2,3-dibromopropyl)phosphate (TRIS) as well as perfluorooctanoic acid (PFOA), its salts and PFOA-related compounds.

23.3.3 Globally Harmonized System

Like many other chemicals, flame retardants can have dangerous properties as a neat chemical. Once incorporated into a target polymer or even reacted into the polymeric chains there is generally low to negligible chance of exposure to workers or consumers. However, to handle chemicals safely the Globally Harmonized System of Classification and Labelling of Chemicals (GHS) was developed by the United Nations as a standard (not a treaty or convention). This standard is implemented in national laws in most regions of the world starting around 2010 and mostly completed by 2017.[12] In the Europe Union the respective law is the Classification, Labelling and Packaging (CLP) Regulation[13] and in the United States there are the Occupational Safety and Health Administration standards. Figure 23.6 shows an example of a hazard symbol used on packing of dangerous goods.

23.4 EUROPE

23.4.1 REACH

In 2006, the European Union introduced the regulation concerning the Registration, Evaluation, Authorization, and Restriction of Chemicals (REACH[15]). By 2018, all chemicals manufactured of imported into EU (at > 1 metric ton per legal entity, except polymers and intermediates) had to be registered to the European Chemicals Agency (ECHA), with the obligation to provide data on physical-chemical, toxicological, and environmental properties ("no data—no market"). The ECHA database of registered substances is publicly available on their website.[16]

REACH in Articles 55 ff. also introduced the concept of Substances of Very High Concern for chemicals critical hazard properties like carcinogenic, mutagenic, or reprotoxic (CMR) or persistent, bioaccumulative, and toxic (PBT) properties. Endocrine disruption (ED) and mobility, that is, the potential to spread in the environment via aqueous media, are also coming into focus with the amended regulation on classification, labelling and packaging (CLP[17]). ECHA must be notified of the presence of SVHCs in articles if the total quantity is more than one ton per year, and the SVHC is present at more than 0.1% of the mass of the object. Some uses of

SVHCs may be subject to prior authorization from ECHA. Applicants for authorization have to include plans to substitute the use of the SVHC with a safer alternative. As of July 2023, there are 235 substances on the Candidate List for Authorisation.[18] Once such candidates have been confirmed, they are entered into Annex XIV[19] of REACH, the list for Authorization, or AnnexXVII,[20] the Restriction list. Table 23.1 shows which flame retardants are on these lists. The latest flame retardant entries to the Candidate List were tetrabromobisphenol-A (TBBPA), one of the most widely used brominated FRs, and melamine. The latter is being contest by industry, because health concern for melamine seem to be very specific and of low concern for its flame-retardant uses.

Whereas Table 23.1 lists the already regulated flame retardants, ECHA also maintains the Community Rolling Action Plan[21] (*CoRAP*). It contains the "to do" list for Evaluation of substances based on proposals from Member State authorities or ECHA. The outcome can be harmonized classification and labelling of a substance or other regulatory measure like a restriction or SVHC proposal. Examples of FRs on CoRAP are triphenylphosphate (TPP), tribromophenol, and antimony trioxide.

In 2020, the European Commission published its "Chemicals Strategy for Sustainability"[22] which included the announcement of a REACH revision. One new element is the registration and notification of polymers. For reactive FRs this might lead to a situation where users and processors of such FRs would have to notify or register each of their flame-retarded polymers, because reacting the FR in with monomers creates new polymers. Another new concept with possibly wide-ranging effects is the "Generic Risk Approach" (GRA, to be implemented in Art. 68 of REACH). It means that chemicals could be banned for certain uses purely based on their hazard properties (with CMR and PBT in the focus) without an assessment of potential exposure and therefore a real assessment of risk. Another new concept is that of "Essential Use" which means that hazardous chemicals may get a derogation from restriction, if it can be demonstrated that they have an essential use for society and no suitable alternative exists. How exactly this would be assessed and implemented is still being debated as of summer 2023.

One learning from assessing and restricting one substance at a time by authorities has been that it takes very long to cover a range of even similar substances, several years at least. In addition, banning one substance has sometimes led to a very similar one being introduced into the market which might also have similar risks when compared to the original one. This replacement flame retardants with molecules of similar structures (and therefore assumed risks) has been called "regrettable substitution." An example often cited by environmental groups is the replacement of decabromo diphenylether by decabromo diphenylethane (see Figure 23.7), also mentioned in the ECHA strategy.[22]

Authorities see the solution to slow substance-by-substance evaluation and regrettable substitution in a new approach where groups of similar substances are regulated together. ECHA has published a Restriction Roadmap[23] based on grouping and issued reports on individual groups as "Assessments of Regulatory Needs."[24] The European association for phosphorus, inorganic, and nitrogen FRs (pinfa.eu) has commissioned a study on how phosphorus-based FRs could be grouped (see Figure 23.8). One important finding was that even though many physico-chemical and toxicity properties are similar within a chemical group, there can be outliers and exceptions which need to be properly addressed.

TABLE 23.1
Flame Retardants Covered by Reach in Europe (as of 2023-07)

Flame retardants "Restricted" under Annex XVII of REACH (that is effectively banned except in specific "authorized" uses):
• For skin contact (e.g., textiles):
  • Tris(aziridinyl phosphinoxide),
  • Tris(2,3 dibromopropyl)phosphate (TRIS),
  • Polybromobiphenyls (PBB)
• In cellulose insulation materials:
  • ammonium salts
• Not authorized for any use:
  • Polybrominated diphenyls: Penta-, Octa-, and DecaBDE were on Annex XVII, but now covered under Stockholm POPs convention

Flame retardants which are candidates for "Substances of Very High Concern" (SVHC) or have been confirmed* and included in Annex XIV of REACH (the Authorisation list). These substances will be subject to Restriction, if not already restricted: • TBBPA = Tetrabromobisphenol-A (carcinogenic) • BEHTBP = Bisethylhexyltetrabromophthalate (vPvB) • BTBPE = Bistribomophenoxyethane (vPvB) • HBCD* = Hexabromocyclododecane (PBT) • TBNPA = Tribomoneopentylalcohol (carcinogenic) • SCCP = Short chain chlorinated paraffins (PBT, vPvB) • MCCP = Medium chain chlorinated paraffins (PBT, vPvB) • Dechlorane plus (vPvB) • TCEP* = Tris(2-chloroethyl)phosphate (toxic for reproduction) • Boric acid and *sodium salts (toxic for reproduction) • TXP* = Trixylyl phosphate (toxic for reproduction) • Melamine (equiv. level of concern)

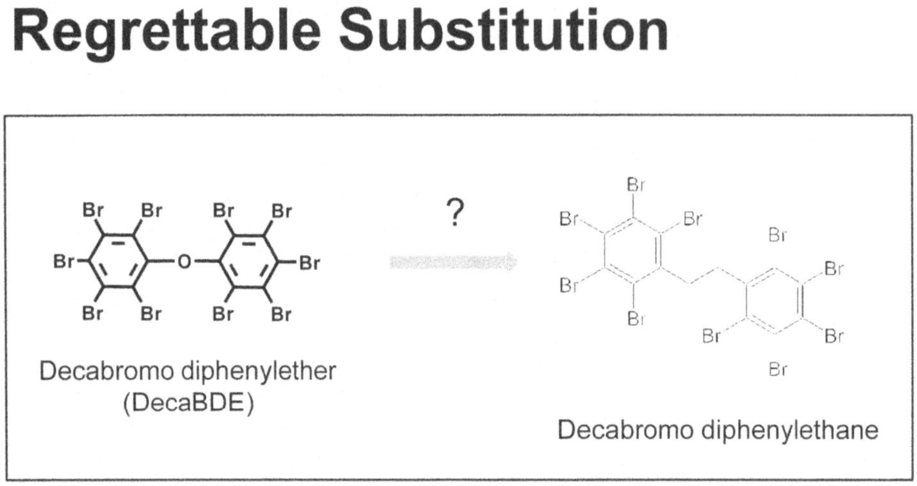

FIGURE 23.7 A debated example of regrettable substitution, replacing DecaBDE by Decabromo diphenylethane.

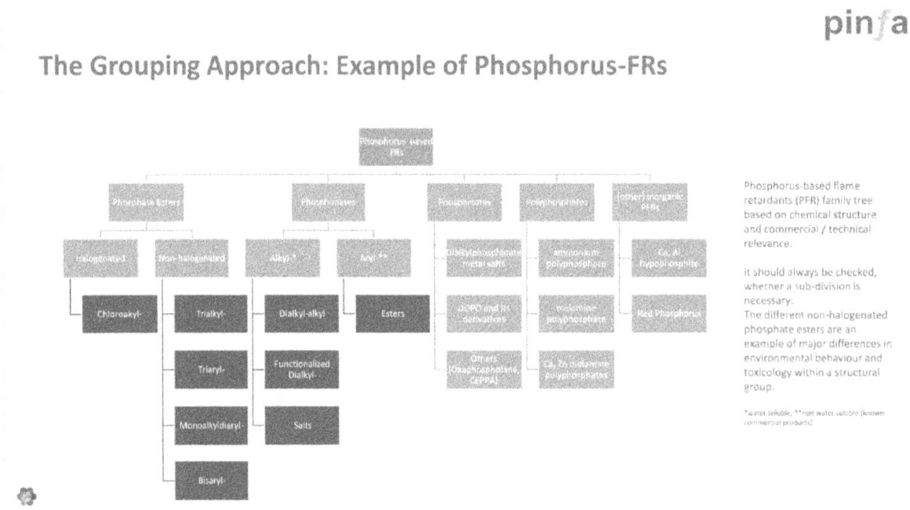

FIGURE 23.8 Example of grouping approach for phosphorus-based FRs by pinfa.[25]

ECHA expanded the concept of grouping to all flame retardants by publishing in early 2023 its Regulatory Strategy for Flame Retardants[26] in which it identified aromatic brominated FRs as candidates for EU-wide restriction because of their PBT properties. However, further information on end-of-life stages (waste, recycling operations) and availability of alternatives is deemed necessary. For many aliphatic brominated and some organophosphorus-based flame retardants, ECHA needs more data to determine if a restriction would be necessary. They expect such data to be available from 2024 onwards, and ECHA will subsequently assess those groups. For chlorinated paraffins and chlorinated alkyl-phosphate esters, regulatory measures are already in place or initiated. No regulatory action is recommended for several non-halogenated subgroups of flame retardants, see Figure 23.9.

23.4.2 Restriction of Hazardous Substances in Electronics (RoHS and WEEE)

Because of the massive increasing amount of waste from electric and electronic products, the European Union introduced two directives (= legal acts which need to be implemented into national laws) in 2003: the Directive on Waste of Electric and Electronic Equipment[27] which covers the collection and recycling of end-of-life products and the Directive on the Restriction of certain Hazardous Substances in Electric and Electronic Equipment (RoHS, 2002/95/EC).[28] WEEE has a provision that plastics containing brominated FRs need to be separated from the waste stream before further recycling. The intention is to not "contaminate" recyclates streams with legacy brominated FRs.[29] However, in practice this often leads to the incineration of

"Thus, based on the current information, groups of low or unlikely hazards can be identified such as:

- *Non-cyclic alkyl aryl esters of phosphoric acid,*
- *Other hydrogenphosphonates and alkyl phosphonates, their salts and esters, and*
- *Dibenzo oxaphosphorine oxide derivatives (DOPO derivatives)*
- *Hydrocarbyl phosphinates"*

"Inorganic phosphates, polyphosphates, phosphinates, phosphonates and phosphorus and its inorganic salts have been assessed with low hazard and no further EU regulatory risk management need."

FIGURE 23.9 Phosphorus Flame retardants identified as low concern by the ECHA Strategy, quote.[22]

the whole plastics fraction from WEEE, because sorting is too challenging or costly. The WEEE Directive was recast as 2012/19/EU with higher recycling quotas and additional product groups covered.

RoHS bans the use of the heavy metals cadmium, lead, chromium (VI), and mercury as well as the flame retardants polybrominated biphenyls and polybrominated diphenylethers since 2006. The Directive "recast" as 2011/65/EU widened the scope of covered products and added four phthalates as of 2019–07 to the list of restricted substances. RoHS has a procedure for granting (time-limited) exemptions for defined applications where no substitute exists, which is welcomed by the electronics industry and has been mentioned as the reason why they prefer RoHS over REACH for regulating chemicals in electronics (DigitalEurope[30]).

In 2019 the Oekoinstitute in Germany was tasked with assessing three more flame retardants for potential further restriction under RoHS: Tetrabromobisphenol-A, medium-chain chlorinated paraffins, and antimony trioxide (ATO).[31,32] When considering a ban on ATO the authors realized that this would have major implications for the use of brominated FRs, because ATO is generally used as a synergist. The approach was not further pursued, perhaps also in the light of the upcoming Chemicals Strategy for Sustainability and REACH revision.

Other countries and regions of the world have adopted similar legislation to RoHS and WEEE with different scopes and requirements, for example, China, Japan, California, and some other US Federal States, Taiwan, Korea, Norway, Turkey, and India.[33]

23.4.3 Swedish Chemical Tax

Sweden took a different approach to regulate chemicals: it introduced an additional tax ("exise duty") on electric and electronic products in 2017 (Act 2016:1067). The idea was to promote the phase-out of problematic flame retardants and their substitution by more benign substances by charging this surtax, which is based on the type of flame retardant used and the weight of the whole product (excluding packaging). In its *latest version* as of July 2023 the tax rates are 11 SEK/kg (1.04 USD/kg) for white goods and 160 SEK/kg (15 USD/kg) for other electronic products with a maximum rate of 489 SEK (46 USD) per item.[34] A 95% tax reduction will be applied if the product does not contain chlorine, bromine, or phosphorus, and a 50% reduction if it does not contain bromine or chlorine (but does contain phosphorus), whether these elements are in flame retardants or in other chemicals or polymers (see Figure 23.10).

Previously, the tax differentiated between reactive and additive FRs with its own "legal" definition of the terms. The respective annex will be deleted as of July 2023 because it is considered unworkable. Because the tax fails to promote environmentally preferable FRs it has been much criticized, not only by industry and business associations,[35,36] but also by environmental groups like ChemSec or the Swedish ecolabel organization TCO Certified. One main criticism is that there are many phosphorus flame retardants with a good environmental and health profile which are still penalized with a 50% tax rate, whereas all FRs which do not contain bromine, chlorine, or phosphorus get a blanket tax rate of only 5% without further proof of their properties. Perhaps, the harsh criticism of the Swedish tax has kept other countries from following this approach. In Europe, as part of the Chemicals Strategy for Sustainability also other policy instruments should be rolled out which should more effectively promote the transition to chemicals with improved environmental and health properties, including flame retardants.

23.5 NORTH AMERICA

23.5.1 United States

The Toxic Substances Control Act (TSCA)[37] from 1976 was the major federal law on chemicals for almost 40 years.

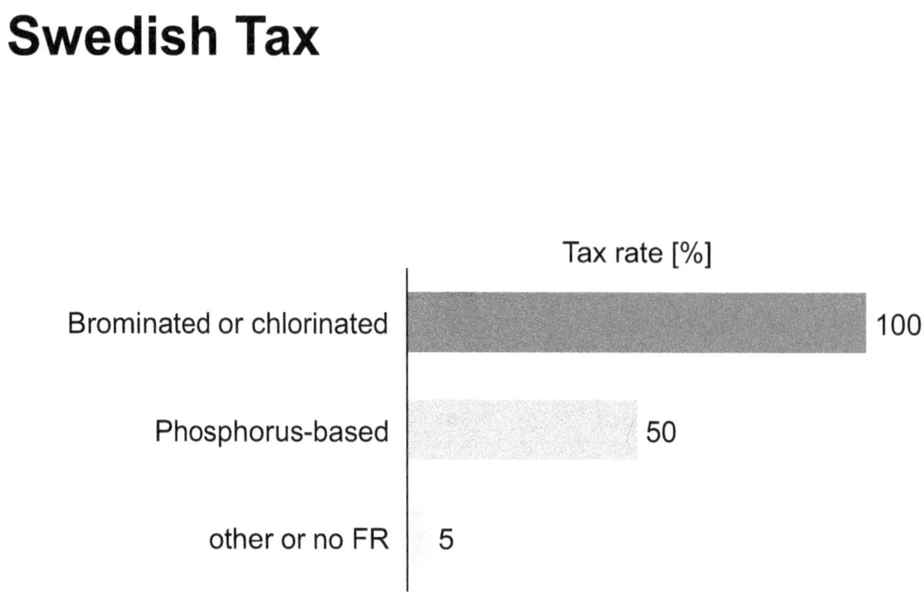

FIGURE 23.10 Tax rates of the Swedish chemical tax based on hich FRs are used in the piece of equipment.

Like in other countries, at the time of its implementation TSCA also grandfathered existing chemicals into its inventory and by default regarded them as safe. New commercial chemicals had to registered with a premanufacturing notice. Only few hazardous substances like asbestos, mercury, lead, polychlorinated biphenyls (PCB) were restricted under TSCA, not any flame retardants. The other instrument under TSCA to regulate substances is the Significant New Use Rule (SNUR) under which any new use of a chemical needs to be approved by EPA.

In 2016, the Frank R. Lautenberg Chemical Safety Act for the 21st Century[38] amended and updated TSCA. It introduced a duty on EPA to evaluate existing chemicals with clear and enforceable deadlines. Chemicals had to be assessed against a risk-based safety standard, unreasonable risks identified in the risk evaluation had to be eliminated. It also gave EPA expanded authority to more quickly require development of chemical information when needed.

The US EPA included these flame retardants in its workplan and risk evaluation/management under TSCA:

- 2019: "high priority Substances" for risk evaluation as "PBT Chemicals" under TSCA 6(h): TBBP-A, TCEP, HBCD, TPP
- 2021: "PBT Chemicals" under Risk management phase: HBCD, DecaBDE, IPTPP-PIP(3:1), ATO to be re-evaluated TSCA (re. new data)
- 2021: EPA extends compliance date to protect supply chains [IPTPP-PIP(3:1)]
- 2022: HBCD—restrictions imposed to eliminate unreasonable risk

The US Consumer Product Safety Commission as a federal institution has also been dealing with the regulation of flame retardants. In 2017 it granted a petition to ban organohalogen FRs (OHFRs) in specific consumer products.[39] Since this would have a very broad and diverse group of chemicals, the National Academy of Science (NAS) was tasked with assessing such a group-wise regulation. In 2019, they came to the conclusion that 14 subclasses should be defined for further hazard assessment.[40] To date, CPSC has not acted further on banning of these 14 subclasses.

In a separate workstream, in 2019 CPSC advised against adopting the Californian fire standard for upholstered furniture TB 117–2013 as a national standard. US Congress ignored this advice and instead adopted this test via legislation in 2021 so that TB 117–2013 became the national furniture flammability standard. This is peculiar, because this flammability standard was expressively promoted to reduce the need for flame retardants. It is an example where a fire safety requirement is lowered to reduce the consumption of critical FRs, rather than restricting the chemicals in question directly and letting industry find safer alternatives to maintain the safety of products. Please see Chapter 31 of this book for more information on this test method and the issues behind it.

In addition to federal laws, many states and even regions or municipalities have passed restrictions on flame retardants. Table 23.2 provides an overview. These local laws often target legacy brominated FRs, sometime also include organo-phosphate FRs and focus on upholstery or electronic applications. One latest example is the Washington State Regulations for FRs in external plastic enclosures for electronic and electric equipment and recreational polyurethane foam products.[41] It defines ten chemical-product categories where intentionally added organohalogen FRs (OFRs) are no longer permitted as of 2025 to 2028. However, due to strong industry lobbying, many exemptions were granted, and more can still be sought.

TABLE 23.2
Local US Laws on Flame Retardants

State/District/Citty	Law/Regulation Reference #	Table Narrative
California	AB 302, PROP 65, SB 1019, AB 2998, CBSC SFM 02/18, TB 117–2013, BEARHFTI OMN 2018–1210–03S, AB 267	The frequency of specific **FR chemical class** referenced in the 55 Local U.S. Laws/ Regulations in Table 23.2 are: • CFR: cited within 52% of the Laws • BFR: 78% • OPFR: 26% • ONFR: 21% • Other FR: 24% Frequency of **end applications**: • Upholstered furniture: 40% • Juvenile products: 45% • Electric & Electronic: 16% • Other applications: 21% Acronyms for Table 23-2: • CFR: chlorinated flame retardant • BFR: brominated flame retardant • OPFR: organophosphorus flame retardant (halogen-free) • ONFR: organonitrogen flame retardant • Other FR: miscellaneous (e.g., nanoscale, antimony trioxide)
Connecticut	HB 6369	
Delaware	HB 77/6 DE Code §2502G	
Hawaii	HCR 235/SR 107, HB 2013	
Illinois	HB 2572	
Iowa	HF 62	
Maine	LD 2048, LD 1129, LD 1568, EO 12, LD 1790, LD 1658, LD 182, LD 1662/HP 1233	
Maryland	SB 556, HB 83, HB 229, HB 99, SB 447	
Massachusetts	H 4900, HD 3912	
Michigan	HB 4406, SB 1458	
Minnesota	HF 2123, SF 2096, SF 1215, HF 359	
Nevada	AB 97	
New Hampshire	SB 193	
New York	EO 4, S 7621, A 6195, SB 3703, A 6296, SB 4630	
Oregon	SB 596, SB 962, SB 478/HB 3473	
Rhode Island	HB 7917, H 5082	
Vermont	H 444/S 109, S 81	
Washington	HB 1024, EO 04–01, HB 2545, SB 5135, Pub 23–03–015	
Washington, DC	ESHB 2545	
Anchorage, Alaska	AO No. 2019–15	
San Francisco, California	SF Ord. No. 211–17	

23.5.2 Canada

The Canadian Environmental Protection Act (CEPA, 1999)[42] is the fundamental federal law governing chemical. Section 93 of the act grants the federal government authority to make regulations to restrict and manage the Canadian List of Toxic Substances (LOTS).[43] Toxic substances have characteristics outlined in Section 64. These FRs have been evaluated and partially regulated under Canadian law:

- melamine, tricresyl phosphate (TCP), Dechlorane Plus (DP), TCPP, TDCPP, EBTBP, ATE (BFR), DBDPE, TBB, TBPH, ATO under the Chemical Management Plan (CMP, 2006)[44]
- pentaBDE and OctaBDE under the PBDE Regulations (2008)
- TCEP (foam products—children), TDBPP (textile wearing apparel) under Canada Consumer Product Safety Act (CCPSA, 2010)[45]
- HBCD, PBDEs (7), Dechlorane Plus and Decabromodiphenyl ethane were proposed amendments to the "Prohibition of Certain Toxic Substances" (2022)[46]; DBDPE may contribute to the formation of persistent, bioaccumulative, and inherently toxic transformation products, such as lower brominated BDPEs, in the environment. A ban on the manufacture, sale or import of the brominated FR DBDPE has been proposed.

23.6 ASIA, SOUTH AMERICA, AND AFRICA

In general, chemical legislation in these world regions lags behind developments in Europe and North America, only slowly following their regulatory approach. Rarely is a law or standard simply copied as they are adapted to local "circumstances." Weaker restrictions on chemicals might also be perceived as a business advantage (more freedom to use any chemical). Indeed, in China many flame retardants banned for use in the EU or US can be found being produced for sale. Enforcement of chemical regulation is another matter and resources for such activities are scarce in the Global South. China and Japan were chosen for further details, because of their market relevance in flame retardants consumption and production of final consumer or industrial articles like electronic equipment or electric vehicles.

23.6.1 China

Overall, China restricts the same "legacy" halogenated FRs as other regions with a mix of hard bans and simple labelling requirements (see Table 23.3). Only very few phosphorus based FRs are regulated.

Industrial chemicals are mainly regulated by Decree 591—Regulations on Safe Management of Hazardous Chemicals; MEE Order 12—The Measures for Environmental Administration Registration of New Substances; SAWS Order

TABLE 23.3
Flame Retardant Restrictions in China (as of July 2023)

Regulation Banning or Restricting FRs	Halogenated FRs Affected	PIN FRs Affected
List of Strictly Restricted Toxic Chemicals in China (2020) and the amendment (2022) on the basis of UN Stockholm Convention POPs47	HBCD, SCCPs (chlorinated alkanes), PCT (effective from 1st Jan. 2020) DecaBDE (effective from 6th June 2023)	none
Implementation of the conformity assessment system for restricted use of hazardous substances in electrical and electronic products. AKA China RoHS 2 (6th Jan. 2016)48	PBBs, PBDE (effective from 1st July 2016)	none
Technical requirements for environmental labeling products: small household appliances (HJ 1159–2021)49	Green label can only be applied if: • plastic parts > 25g: no, TCEP, TCPP, TDCP; • product shells, circuit board substrates and power cord: no HBCD, SCCPs (effective from 23rd April 2021)	none
Technical Requirements for Environmental labelling Products: Microcomputer, Monitor (HJ 2536–2014)50	Green label can only be applied if: • plastic parts > 25 g: no SCCPs, MCCP, HBCD; • Computer main board substrate: no HBCD (1st July 2014)	none
Technical requirement for environmental labeling products: Textile products (HJ 2546–2016)51	PBBs, TRIS, PeBDE, OcBDE, DecaBDE, HBCD, TCEP, SCCPs	TEPA
Chemicals Subject to Prioritized Control (1st batch, 27th Dec. 2017 & 2nd batch, 2nd Nov. 2022)52,53	Revision of national mandatory standards to restrict the use of SCCPs, HBCD, PBDE in certain products (27th Dec. 2017) restrict the use of TCEP	IPTPP (Phenol isopropylated phosphate)
Guidelines for use and control of key chemical substances in consumer product (19th Nov. 2020)54	Content/Migration control: PBBs, HBCD, PBDES, SCCPs, TCEP, TCP (1st June 2021)	TPP

53—The Measures for the Administration of Registration of Hazardous Chemicals; MEP Order 22—The Measures for Environmental Administration Registration of Hazardous Chemicals (revoked in July 2016); and other supporting regulations or national standards issued by various ministries.

23.6.2 Japan

The Japan Chemical Substances Control Law (CSCL) goes back to 1973 and is often referred to as the Japanese REACH. It is administered by the Ministry of Economy, Trade, and Industry (METI). The law is intended to manage the risks to the environment and human health of industrial chemicals manufactured in or imported to Japan. Under this law only a few of the "usual suspects" legacy brominated FRs are restricted, by designation of as Class 1 Specified Chemical Substances, and related to the UN Stockholm Convention: HBCD is prohibited since 2014[55] (except when businesses have obtained a prior permit from the competent authorities). Likewise for SCCPs and DecaBDE since 2018.[56] Dechlorane Plus will be banned as of 2024,[57] as it also listed in the Stockholm Convention.

Japan has also taken action of perfluorinated substances which include some that are used as FRs or synergists, for example, as anti-drip additives. PFOA and its salts are designated as Class I Specified Chemical Substances under METI law from 2021. In principle, import/manufacture of the products containing these substances is banned. (> 0%). As an exception, products containing trace amounts of these substances as impurities can be imported by submitting a "BAT report" to the authorities. PFOA-related substances (56 substance group) and PFHxS are planned to be designated as Class I Specified Chemical Substances under METI law.

23.7 NONREGULATORY DRIVERS

23.7.1 Alternatives Assessments

In the late 1990s, the United States Environmental Protection Agency (US-EPA) started "Design for Environment" projects (Figure 23.11) which assessed the impact of certain chemicals on the environment and established a methodology to compare different chemicals based on their hazard properties. Four large studies were done on flame retardants: "Flame Retardants Used in Flexible Polyurethane Foam"[58] (2005, updated 2015), "Flame Retardants in Printed Circuit Boards"[59] (2015), "Flame Retardant Alternatives to DecaBDE"[60] (2014), and "Flame Retardant Alternatives to

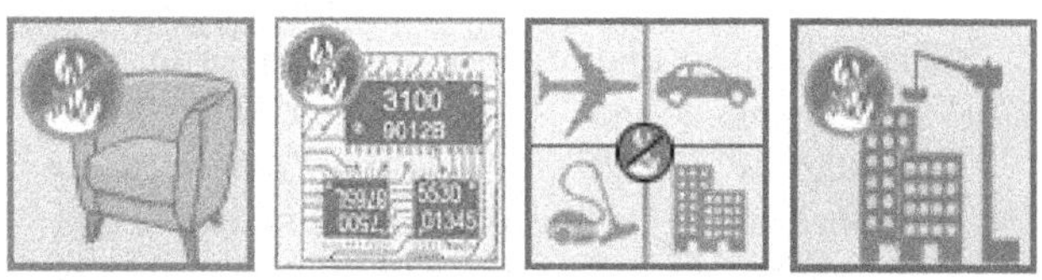

FIGURE 23.11 Logos of the US-EPA DfE reports on flame retardants (source[62]).

HBCD"[61] (2014). Overall, these reports concluded that viable alternatives for many applications of the legacy brominated FRs exist and that their hazard profile was often much better, even though some halogen-free alternative have hazardous properties.

Based on the DfE criteria, the nongovernmental organization CleanProductionAction developed the GreenScreen™[63] methodology to not only assess chemicals (FRs), but also rank them into four groups called "Benchmarks": Benchmark 1 "Avoid—chemical of high concern" is the worst and Benchmark 4 "Prefer—safer chemical." This simplified result of a detailed hazard assessment allows non-experts to choose safer chemicals, because even the Benchmark 1 group may contain chemicals which are still legally allowed. The system also differentiates the "better" FRs into Benchmarks 2 to 4 (with 4 hardly achievable due to very ambitious criteria). See Figure 23.12 for an example of an assessment of 48 phosphorus-based FRs assessed by the company ToxServices in 2018.[64]

The GreenScreen approach was in a way taken one step further by a group called ChemForward[65] by establishing a database of third-party verified assessments for chemicals. They extended the concept beyond the pure chemical defined, for example, by CAS registry number to trade-named commercial products and formulations and maintaining the confidentiality of sensitive data. Their database of "Safer" chemicals is accessible for a fee.

In Europe, the question of "do we know enough about the chemical alternatives to the legacy brominated FRs?" was answered in the ENFIRO[66] project funded by the European Commission. It concluded that viable alternatives exist with good environmental and health profiles, a summary video was also produced.[67] Table 23.4 from the final ENFIRO report demonstrates their finding that viable alternatives to several key brominated FRs exist.

23.7.2 Eco-Labels and Green Procurement

The German Blue Angel[68] ("Blauer Engel") was introduced in 1978 as one of the first ecolabels. It was one of the first ecolabels to restrict certain halogenated FRs in certified electronic products and later also for textiles. The EU Green Flower,[69] the Nordic Swan,[70] and other ecolabels followed with similar requirements. However, the details differ between different ecolabel systems and their product categories. There is a tendency to move away from "blacklisting" individual FRs to excluding them based generic hazard properties, like CMR or PBT properties. In Sweden,

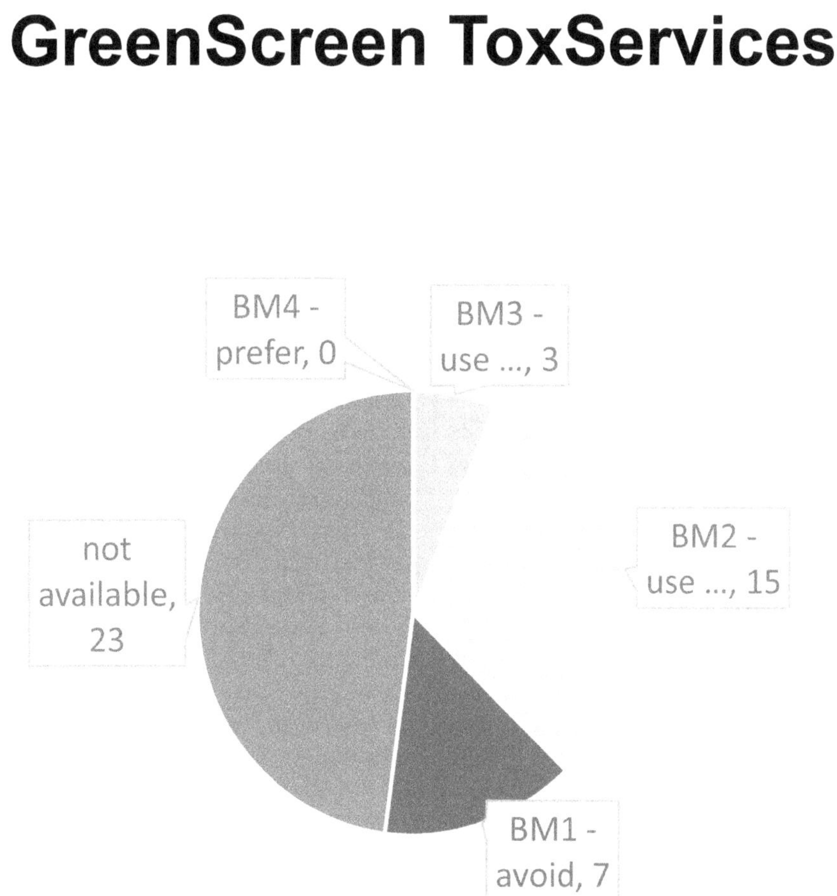

FIGURE 23-12 GreenScreen™ Benchmarks assigned by ToxServices[59] to 48 phosphorus-based FRs in 2018. For 23 FRs no assessments were available.

TABLE 23.4
List of Evaluated Non-Halogenated FRs in the ENFIRO Project, Showing Viable Alternatives to Halogenated FRs

Concern Level	Flame Retardant Name	Reason for Concern Level
No immediate concern	Ammonium polyphosphate (APP) Aluminium diethylphosphinate (Alpi) Aluminium hydroxide (ATH) Melamine polyphosphate (MPP) Dihydrooxaphosphaphenanthrene (DOPO) Zinc stannate (ZS) Zinc hydroxystannate (ZHS)	• Inorganic and organic substances with low acute (eco-)toxicity and no bioaccumulation potential. • Chemical stability required for application results in limited degradation (persistence). • Stannates: the (neuro-)tox effects found with in vitro cell based systems were not confirmed with animal studies (in-vivo, mice studies), probably due to low bioavailability, therefore no immediate concern.
Some concern for potential environment and humans	Resorcinol bisphosphate (RDP) Bisphenol-A bisphosphate (BDP)	• RDP toxicity to aquatic organisms is main concern, may be linked to by-products (TPP). Low and high toxicity are found for same test species, which is may be due to batch differences in the amount of TPP present as by-product. • BDP is persistent.
Of concern, risk assessment necessary	Triphenyl phosphate (TPP) Nanoclay (Cloisite)	• TPP very toxic to aquatic organisms is main concern, potential endocrine effects. • Nanoclay showed strong in vitro neurotoxicity. May be due to the nanoparticle coating, additional studies needed. Information on the leaching behavior of nanoclays from polymers is also needed.

TCO Certified has taken a different approach by restricting halogenated FRs and mandating any non-halogenated FRs to have a GreenScreen™ benchmark of at least 2 or higher. This is a criterion for inclusion on their "accepted substances list.[71]" This "white list" approach ensures FRs have undergone a thorough assessment and that critical FRs are not replaced with similar ones which happen to not have caught regulatory attention (yet) (see regrettable substitution in 23.4.1 REACH).

The US based ecolabel EPEAT (Electronic Product Environmental Assessment Tool) is run by the non-profit Global Electronics Council® (GEC). They reward the use of FRs which have undergone positive alternatives assessments with extra points [halogen-free FRs in housing materials[72]]. The total score of points determines the achieved ecolabel level (bronze, silver, gold). This system creates an incentive for using non-halogenated and more benign FRs without setting hard requirements which can in some cases be difficult to meet for technical reasons.

23.7.3 Safe and Sustainable-by-Design Concept in Europe

As an element of the Chemicals Strategy for Sustainability[19] launched by the European Commission in 2020, the concept of "Safe and Sustainable-by-Design" chemicals was launched. The Commission asked its Joint Research Centre (JRC) to compile an overview of existing methodologies for sustainability assessments as well as drafting a guideline for a harmonized European approach to assess the sustainability of chemicals,[73] taking into account all the aspects shown in Figure 23.13. This goes beyond the chemical alternatives assessments described previously (23.7.1) because in addition to hazard properties life cycle criteria, information on raw materials, the production process, impact on consumers, end-of-life considerations, product performance and social value are reflected. The result is a set of scores for the different sustainability dimensions, allowing a comparison of different chemicals. Background information including a case study with phosphinate FRs are available on a Commission website.[74] After a testing and refinement period until 2025, the concept is expected to become mandatory for research and innovation projects in Europe (at the minimum for publicly funded research).

23.8 CONCLUSIONS

Since several flame retardants (initially mostly brominated flame retardants) have been widely found in the environment, animals and even humans, they have become a focus of chemical regulatory activity. Whereas the "first generation" brominated FRs attracted most attention and their most prominent members are now restricted on a global level, other FR chemistries are being scrutinized—therefore, expect more restrictions to come, starting in the most active regions Europe and North America. Japan, China, South Korea, and the rest of world will follow at a slower pace. The FR industry has been developing safer alternatives and will have to keep doing so. However, in some cases these alternatives lack market acceptance because of higher price and often one-time switching cost. What is a sustainable FR? There is no definitive answer, because new criteria come into play: in addition to new hazard criteria like mobility or endocrine disruption also life cycle and end-of-life considerations. Therefore, this will remain a dynamic field. For now, inorganic FRs like aluminum hydroxide (ATH) and magnesium hydroxide (MDH) look "safe"; also phosphorus chemistry offers great versatility so that PFRs can be tailored to not only fit the target

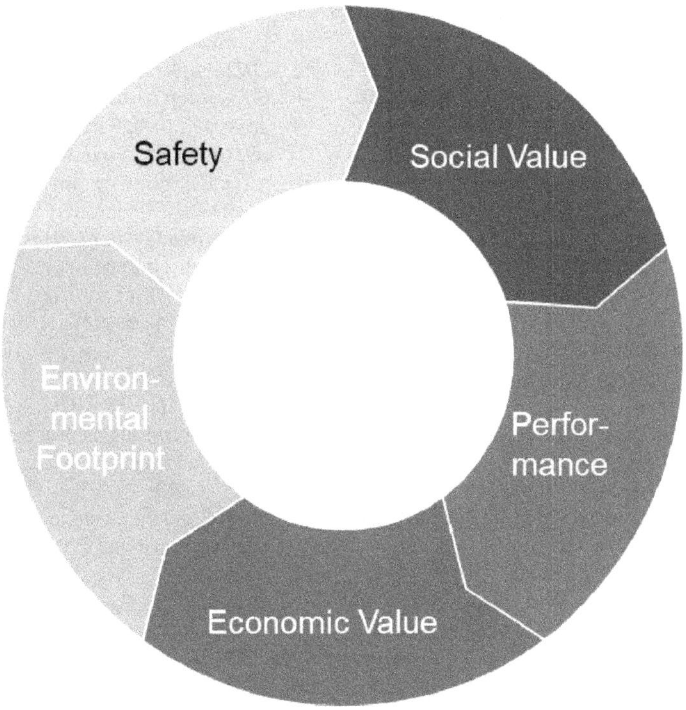

FIGURE 23.13 Elements of safe and sustainable-by-design evaluation.

polymer or textile but also sustainability criteria. Bio-based (see Chapter 13 of this book) and polymeric/polymer-bound flame retardants are options with certain (eco-)toxicity advantages, but often have technical limitations.

REFERENCES

1. DeBoer, J.; Stapleton, H.: Toward Fire Safety Without Chemical Risk. *Science.* **2019**. *364*(6437), 231–232. http://doi.org/10.1126/science.aax2054
2. Ethel Eljarrat, D. Barceló (2011): Brominated Flame Retardants (The Handbook of Environmental Chemistry 16), Springer, Berlin, Heidelberg
3. Schreder, E.; Zheng, G.; Sathyanarayana, S.; Gunaje, N.; Hu, M. Amina Salamova: Brominated Flame Retardants in Breast Milk From the United States: First Detection of Bromophenols in U.S. Breast Milk. *Environ Pollut.* **2023**, *334*, 122028. https://doi.org/10.1016/j.envpol.2023.122028
4. www.hbm4eu.eu/ accessed 2023–8–27.
5. van der Schyff, V.; Kalina, J.; Govarts, E.; Gilles, L.; Schoeters, G.; Castaño, A.; Esteban-López, M.; Kohoutek, J.; Kukučka, P.; Covaci, A.; Koppen, G.; Andrýsková, L.; Piler, P.; Klánová, J.; Jensen, T. K.; Rambaud, L.; Riou, M.; Lamoree, M.; Kolossa-Gehring, M.; Vogel, N.; Weber, T.; Göen, T.; Gabriel, C.; Sarigiannis, D. A.; Sakhi, A. K.; Haug, L. S.; Murinova, L. P.; Fabelova, L.; Tratnik, J. S.; Mazej, D.; Melymuk, L.. Exposure to Flame Retardants in European Children—Results from the HBM4EU Aligned Studies. *Int. J. Hyg. Environ. Health.* **2023**, *247*, 114070. https://doi.org/10.1016/j. ijheh.2022.114070.
6. J. Aurell, B. K. Gullet: "Emission Factors from Aerial and Ground Measurements of Field and Laboratory Forest Burns in the Southeastern U.S.: $PM_{2.5}$, Black and Brown Carbon, VOC, and PCDD/PCDF." Environmental Science & Technology, 2013. https://doi.org/10.1021/es402101k.
7. www.epa.gov/saferchoice/partnership-evaluate-flame-retar- dants-printed-circuit-boards-publications accessed 2023–9–19.
8. Wang, M.; Yin, G.-Z.; Yang, Y.; Fu, W.; Palencia, J. L. D.; Zhao, J.; Wang, N.; Jiang, Y.; Wang, D.-Y. Bio-Based Flame Retardants to Polymers: A Review. *Adv. Ind. Eng. Polym. Res.* **2023**, *6*(2), 132–155. https://doi.org/10.1016/j.aiepr.2022.07.003.
9. https://chm.pops.int/ accessed 2023–8–27.
10. www.basel.int/ accessed 2023–8–27.
11. www.pic.int/ accessed 2023–8–27.
12. "GHS Implementation—Transport—UNECE." www.unece.org accessed 2017–10–18.
13. European Union (1272/2008): Regulation on Classification, Labelling and Packaging of Substances and Mixtures, Amending and Repealing Directives 67/548/EEC and 1999/45/EC, and Amending Regulation (EC) No 1907/2006.
14. www.osha.gov/hazcom/pictograms accessed 2023–9–15.
15. Regulation (EC) No 1907/2006 of the European Parliament and of the Council of 18 December 2006 Concerning the Registration, Evaluation, Authorisation and Restriction of Chemicals (REACH). Official Journal of the European Union L 396/1.
16. echa.europa.eu accessed 2023–9–15.
17. Commission Delegated Regulation (EU) 2023/707 of 19 December 2022 Amending Regulation (EC) No 1272/2008 as Regards Hazard Classes and Criteria for the Classification, Labelling and Packaging of Substances and Mixtures.
18. https://echa.europa.eu/candidate-list-table accessed 2023–7–31.
19. https://echa.europa.eu/authorisation-list accessed 2023–9–15.
20. https://echa.europa.eu/substances-restricted-under-reach accessed 2023–9–15.

21. https://echa.europa.eu/information-on-chemicals/evaluation/community-rolling-action-plan/corap-table accessed 2023–7–31.
22. European Commission COM(2020) 667: Chemicals Strategy for Sustainability. https://eur-lex.europa.eu/legal-content/EN/TXT/?uri=COM%3A2020%3A667%3AFIN accessed 2023–9–1.
23. European Commission SWD(2022) 128: Commission Staff Working Document—Restrictions Roadmap under the Chemicals Strategy for Sustainability.
24. https://echa.europa.eu/assessment-regulatory-needs accessed 2023–9–15.
25. www.pinfa.eu/wp-content/uploads/2022/08/P-Fisk_organophosphorus_flame_retardants_grouping_for_pinfa_2022-07_final.pdf accessed 2023–9–15.
26. https://echa.europa.eu/documents/10162/2082415/flame_retardants_strategy_en.pdf accessed 2023–8–1.
27. European Commission Directive 2002/96/EC: Waste Electrical and Electronic Equipment Directive, replaced by 2012/19/EU.
28. European Commission 2002/95/EC (RoHS 1) Directive on the Restriction of the Use of Certain Hazardous Substances in Electrical and Electronic Equipment, Replaced by Directive 2011/65/EU, 2013.
29. Shockley, K. R.; Cora, M. C.; Malarkey, D. E.; Jackson-Humbles, D.; Vallant, M.; Collins, B. J.; Mutlu, E.; Robinson, V. G.; Waidyanatha, S.; Zmarowski, A.; Machesky, N.; Richey, J.; Harbo, S.; Cheng, E.; Patton, K.; Sparrow, B.; Dunnick, J. K. Comparative Toxicity and Liver Transcriptomics of Legacy and Emerging Brominated Flame Retardants Following 5-Day Exposure in the Rat. Toxicol Lett. **2020**, *332*, 222–234. https://doi.org/10.1016/j.toxlet.2020.07.016.
30. https://cdn.digitaleurope.org/uploads/2023/06/DE-position-on-proposal-for-EU-REACH-restriction-of-flame-retardants.pdf.
31. https://rohs.exemptions.oeko.info/fileadmin/user_upload/RoHS_Pack_15/Final_Results/Diantimony_trioxide_RoHS_Dossier_V3_final.pdf accessed 2023–9–15.
32. https://rohs.exemptions.oeko.info/substance-review-2018/interim-results accessed 2023–9–15.
33. www.rohsguide.com/rohs-future.htm accessed 2023–8–8.
34. www.skatteverket.se/servicelankar/otherlanguages/inenglishengelska/businessesandemployers/startingandrunningaswedishbusiness/payingtaxesbusinesses/taxonchemicalsincertainelectronics.4.5c281c7015abecc2e2019351.html accessed 2023–8–8.
35. "Näringslivets gemensamma position kring skatten på elektronik" (The Common Position of the Business Community Around the Tax on Electronics), APPLiA, ElektronikBranschen, IT & Telekomföretagen, Svensk Handel, Svenskt Näringsliv, Teknikföretagen, September 2021. https://applianytt.se/wp-content/uploads/2021/10/Naringslivets-gemensamma-position-kring-skatten-p%C3%A5-elektronik.pdf.
36. www.pinfa.eu/mediaroom/proposed-chemical-tax-revision- is-appreciated-necessary-steps-in-the-right-direction-but- complete-withdrawal-should-still-be-considered/#gsc.tab=0 accessed 2023–8–9.
37. United States of America, 94th Congress: Toxic Substances Control Act, Public Law 94–469—Oct. 11, 1976 (S. 3149).
38. United States of America, 114th Congress: Frank R. Lautenberg Chemical Safety for the 21st Century Act, Public Law 114–182—JUNE 22, 2016, 130 STAT. 448.
39. www.cpsc.gov/Business--Manufacturing/Business-Education/Business-Guidance/flame-retardants accessed 2023–9–15.
40. https://nap.nationalacademies.org/catalog/25412/a-class-approach-to-hazard-assessment-of-organohalogen-flame-retardants accessed 2023–9–15.
41. https://ecology.wa.gov/Regulations-Permits/Laws-rules-rulemaking/Rulemaking/WAC-173-337 accessed 2023–8–28.
42. www.canada.ca/en/environment-climate-change/services/canadian-environmental-protection-act-registry/related-documents.html accessed 2023–8–28.
43. www.canada.ca/en/environment-climate-change/services/canadian-environmental-protection-act-registry/substances-list.html accessed 2023–8–28.
44. www.canada.ca/en/health-canada/services/chemical-substances/chemicals-management-plan.html accessed 2023–8–28.
45. https://laws-lois.justice.gc.ca/eng/acts/c-1.68/ accessed 2023–8–28.
46. https://gazette.gc.ca/rp-pr/p1/2022/2022-05-14/html/reg2-eng.html accessed 2023–8–28.
47. www.mee.gov.cn/xxgk2018/xxgk/xxgk01/201912/t20191231_756318.html accessed 2023–7–1.
48. www.miit.gov.cn/jgsj/jns/qjsc/art/2020/art_0ffb0deb9185404aa94afe3b852dcc6f.html accessed 2023–7–1.
49. www.mee.gov.cn/xxgk2018/xxgk/xxgk06/201910/t20191016_737822.html accessed 2023–7–1.
50. www.mee.gov.cn/gkml/hbb/bgg/201404/t20140403_270117.htm accessed 2023–7–1.
51. www.mee.gov.cn/ywgz/fgbz/bz/bzwb/other/hjbz/201612/W020161202321802029261.pdf accessed 2023–7–1.
52. www.mee.gov.cn/gkml/hbb/bgg/201712/t20171229_428832.htm accessed 2023–7–1.
53. www.mee.gov.cn/xxgk2018/xxgk/xxgk01/202011/t20201102_805937.html accessed 2023–7–1.
54. GB/T 39498–2020, China National Standard (Voluntary).
55. METI, JP: Amendments to the Order for Enforcement of the Act on the Evaluation of Chemical Substances and Regulation of Their Manufacture, 19 March 2014.
56. METI, JP: Amendment to the Enforcement Order of the Law on the Evaluation of Chemical Substances and Regulation of their Manufacture (Cabinet Order No. 35 of 21 February 2018) + Amendment to the Public Notice of the Export Approval of Chemical Substances (Public Notice No. 3 of 16 March 2018) + Amendment to the Public Notice of the Implementation of the Export Trade Control Order.
57. METI, JP: Substances Newly Designated for Elimination at the 11th Conference of the Parties to the Stockholm Convention Will Be Designated as Class 1 Specified Chemical Substances under the Chemical Substances Control Law (Notice). www.meti.go.jp/policy/chemical_management/kasinhou/oshirase_230810.pdf.
58. www.epa.gov/saferchoice/flame-retardants-used-flexible-polyurethane-foam#finalreport accessed 2023–8–30.
59. www.epa.gov/saferchoice/partnership-evaluate-flame-retar-dants-printed-circuit-boards-publications accessed 2023–8–30.
60. www.epa.gov/saferchoice/partnership-evaluate-flame-retardant-alternatives-decabde accessed 2023–8–30.
61. www.epa.gov/saferchoice/partnership-evaluate-flame-retardant-alternatives-hbcd accessed 2023–8–30.
62. www.epa.gov/saferchoice/design-environment-alternatives-assessments accessed 2023–9–4.
63. www.greenscreenchemicals.org/accessed 2023–8–30.

64. ToxServices: Comparative Hazard Assessment of Reactive and Additive Flame Retardants (Promoting the Use of Safer Flame Retardants Under Sweden's Tax on Chemicals in E&E Products), 2018.
65. www.chemforward.org/safer accessed 2023–8–30.
66. www.flameretardants-online.com/images/itempics/8/8/4/item_18488_pdf_1.pdf and www.enfiro.eu/ accessed 2023– 8–30.
67. https://youtu.be/ouKg81NW2ow?feature=shared accessed 2023–8–30.
68. www.blauer-engel.de/en accessed 2023–9–4.
69. https://environment.ec.europa.eu/topics/circular-economy/eu-ecolabel-home_en accessed 2023–9–4.
70. www.nordic-swan-ecolabel.org/accessed 2023–9–4.
71. https://tcocertified.com/industry/accepted-substance-list/ accessed 2023–9–4.
72. EPEAT Imaging Equipment Criteria: Based on IEEE 1680.2–2012 — Standard for Environmental Assessment of Imaging Equipment and IEEE 1680.2a-2017 — Standard for Environmental Assessment of Imaging Equipment—Amendment 1.
73. Carla Caldeira, Lucian R. Farcal, Irantzu Garmendia Aguirre, Lucia Mancini, Davide Tosches, Antonio Amelio, Kirsten Rasmussen, Hubert Rauscher, Juan Riego Sintes, Serenella Sala, JRC: Safe and Sustainable by Design Chemicals and Materials, Framework for the Definition of Criteria and Evaluation Procedure for Chemicals and Materials, 2022. https://doi.org/10.2760/487955.
74. https://research-and-innovation.ec.europa.eu/research-area/ industrial-research-and-innovation/key-enabling-technologies/ chemicals-and-advanced-materials/ safe-and-sustainable- design_en.

24 Recycling of Flame-Retardant Polymeric Materials

State of the Art and Future Perspectives

Aurelio Bifulco[1], *Jiuke Chen*[2,3], *Arvindh Sekar*[2], *Wenyu Wu Klingler*[2], *Ali Gooneie*[4], *Sabyasachi Gaan*[2]

[1] *Department of Chemical Materials and Industrial Production Engineering (DICMaPI), University of Naples Federico II, Piazzale V. Tecchio 80, 80125, Naples, Italy*

[2] *Laboratory for Advanced Fibers, Empa Swiss Federal Laboratories for Materials Science and Technology, Lerchenfeldstrasse 5, 9014, St. Gallen, Switzerland*

[3] *Department of Materials, ETH Zurich, 8092 Zurich, Switzerland*

[4] *Circular Plastics, Department of Circular Chemical Engineering, Faculty of Science and Engineering, Maastricht University, Urmonderbaan 22, 6162 AL Geleen, The Netherlands*

24.1 INTRODUCTION: BACKGROUND AND DRIVING FORCES

According to the World Health Organization, the annual fire-related deaths are around 300,000 throughout the world.[1] To limit the growing number of fire-related victims, the most used polymers require the addition of suitable FRs, due to their poor fire resistance and high flammability.[2] During a fire scenario, FRs are designed to affect the chance of saving lives,[3] extending the combustion of PMs at the end of their lifetime and the time to flashover.[4] In 2015, the demand for FRs to be applied in EEE, construction, textile, furniture, and transportation was over 600,000 tons only in Europe.[5,6] A recent research work has estimated that the European FRs market is expected to bolster at a CAGR of ~5.08% from 2019 to 2024.[7] This CAGR increase is mainly linked to the large investments of the Asia-Pacific region in infrastructure activity, which demands a huge amount of FR products imported from Europe, due to the stringent fire-safety regulations in electronics manufacturing, automotive (ELV), and building construction. Based on market analysis,[8–10] the global HFFR market is foreseen to increase at a CAGR of ~8.50% by 2028, starting from a value of USD 4.129 billion in 2021.

Many types of FR are commercially available that can be classified by their chemical composition and modes of action.[11] Halogenated FRs (e.g., PBDEs and PBBs) produce halogen atoms able to terminate oxygen radicals in the flame during the combustion. On the other side, P-containing FRs (e.g., DOPO and APP) generate polyphosphoric acids, forming a thermal shield against heat transfer and P radicals poisoning radicals in the flame gas phase.[12] N-containing FRs are usually used in combination with P-based ones to form an insulating intumescent char and dilute the combustible gases in the surrounding environment.[13,14] Inorganic FRs exert the so-called "sink effect," lowering the heat exchange by endothermic decomposition. Also, the degradation of some inorganic FRs releases water, diluting the flammable gases, and metal oxides acting as barriers toward oxygen and heat.[15,16] The type of polymer, including the thermal decomposition profile of both polymer and FR, potential fire scenario, regulatory requirements, and the overall costs are various factors that need to be considered when selecting the suitable FR for a specific application.[17,18] Many of the chapters in this book provide high levels of detail on the above FR classes and their mode of action. The growing attention on topics related to sustainability and EoL recycling is fostering a better understanding of the environmental consequences arising from the use of FR products. Adopting a sustainable approach is getting a critical requirement in the preparation of both FRs and pristine polymeric materials.[19] Moving to an effective recycling of EoL PMs, composed of thermoplastics (or plastics) and thermosets is of paramount importance for contemporary society, especially for the sake of future generations. This transition would help to reduce the discarding of daily-use polymeric products, and thus the white pollution, into the environment and around the world. The largest amount of PMs produced from 1950 to 2015 has been landfilled or incinerated, while only ~6% of those have been recycled,[20] as displayed in Figure 24.1. Up to now, most of the products containing FRs went through such destiny rather than being recycled, causing greenhouse gas emissions and soil contamination.[21] The recent economic and political developments are making metals, CFs, and rare earth elements even more precious additives,[22] which are generally lost as

DOI: 10.1201/9781003380689-24

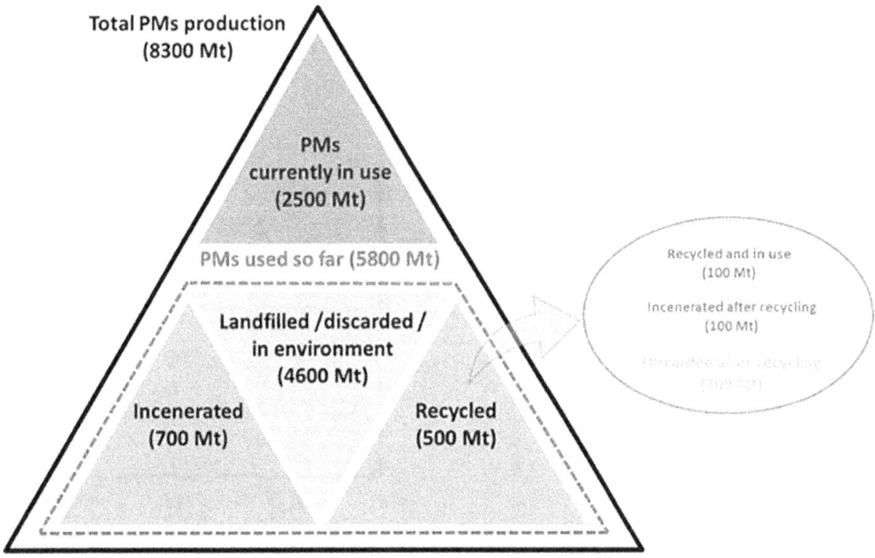

FIGURE 24.1 Fabrication to discarding of PMs from 1950 to 2015.[20]

waste together with potentially reusable FR polymers.[23] Thus, the expected transition to recycling FR polymeric products has motivated the present book chapter.

As alternatives to energy recovery from waste and incineration, different recycling approaches can be used to recycle EoL PMs. The direct reuse of wastes without any modification can be defined as primary recycling or secondhand use. Mechanical (or secondary) recycling allows for the conversion of the waste into a secondary raw material then upcycled for the fabrication of new products.[5] Chemical (or tertiary) recycling is a top-down approach transforming polymeric wastes back to their individual monomers or oligomers by thermal or solvent-assisted depolymerization (solvolysis).[24] Tertiary recycling may also be performed by advanced thermochemical recycling processes, for example pyrolysis.[25] Figure 24.2 reports that based on a recent market analysis, mechanical recycling, solvent washing treatment, and pyrolysis represent the three main routes of recycling for EoL PMs. Among solvent-based recycling methods, the solvent washing treatment (or solvent-based purification) involves extraction processes (e.g., accelerated solvent extraction) finalized to the removal of additives (e.g., HFRs) from plastics before being mechanically recycled.[26] Yet solvolysis is a solvent-assisted depolymerization, usually carried out at 70–340°C and 0.1–4 MPa, and consists of the chemolysis/depolymerization of polymer chains into their monomers.[27,28] Today, most of the thermosets are going to smelter or recycled by solvolysis, while plastic wastes are mainly recycled by thermochemical and mechanical recycling.[9,29] As previously mentioned, the depletion of minerals and energy resources, together with the enormous amount of EoL PMs, have impelled the industry and the scientific community to develop more chemical and thermochemical recycling technologies, as highlighted by a relevant increase in global revenue (see Figure 24.2) from the recycling of PMs over the next decade.[30]

FRs exert their action in case of a fire, though a small amount of them does not end up burning and is often still present in the EoL PMs. This residual content of FRs may be responsible for various environmental issues during the recycling of EoL PMs, including the release of toxic by-products, critical operative conditions, and leaching into groundwater sources. To give an example, in an urban watershed of Singapore, BTBPE (i.e., recognized as persistent and hazardous in the aquatic environment) has been found with concentrations ranging from 0.0039–0.49 ng/L.[31,32] Starting from the 1970s, PBDE and PBB were employed as primary FR molecules. Later, these BFRs were substituted by other halogen-based and phosphorous-containing FRs.[33] In 2003, some PBDEs, octaBDE, and pentaBDE (EC 251–087–9 or OBDPO) were restricted due to their health and environmental concerns, and also decaBDE soon after (EC 214–604–9). Even though, in 2010, tetra-, penta-, hexa-, and heptaBDE have been banned by the Stockholm Convention under the POPs Regulation,[33] such compounds could still be detected above all in plastic WEEE.[34] In 2011, the European Community introduced the Restriction of Hazardous Substances Directive to promote a safer recycling of WEEE, limiting the use of PBDE and PBB in EEE.[33] Based on a recent study, considering 220 kilotons of plastic WEEE, only around 2 kilotons undergoes effective recycling.[35] Besides, Vahabi et al. revealed that FR EoL polymers containing mineral fillers and halogen-based additives are less sensitive to aging than the ones filled with P-based compounds, which are susceptible to hydrolysis. The degradation of HCs may still occur during the recycling of the polymeric waste, negatively affecting the reprocessing procedures.[36] All these aspects need to be seriously considered, as for example, BFRs are largely used in the construction and automotive sectors, and chiefly in the EEE market. As shown in Figure 24.3, in Europe, the production of WEEE has particularly increased in the last two years.

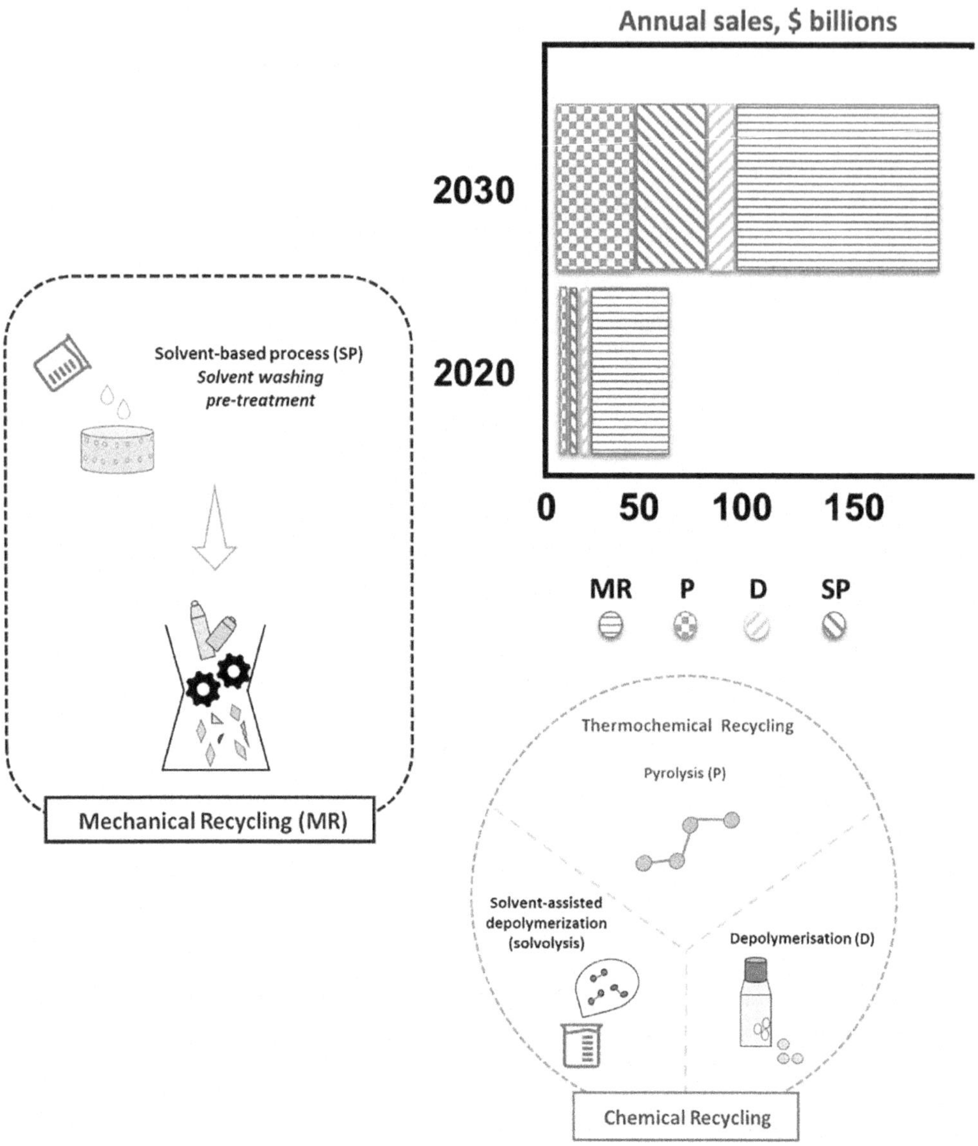

FIGURE 24.2 Global revenue from PMs recycling and main recycling technologies over the next decade.[30]

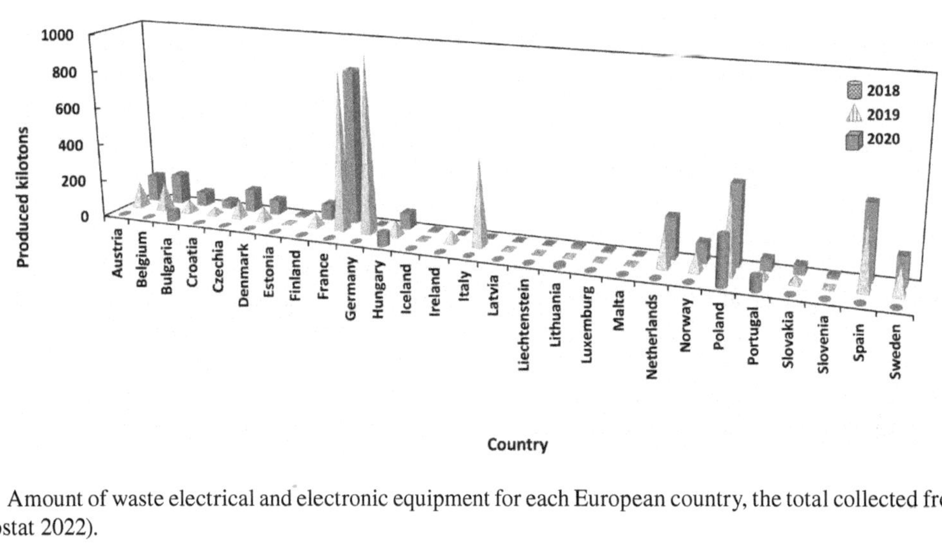

FIGURE 24.3 Amount of waste electrical and electronic equipment for each European country, the total collected from 2018 onwards (data from Eurostat 2022).

The stricter legislations (i.e., REACH and POPs safety regulations) and the huge volume of WEEE are promoting the development of new FRs, able to replace conventional ones, and the design of recycling methods taking into consideration the presence of BFRs or other additives in EoL PMs. Recently, some strategies have been proposed for a more sustainable recycling of FR thermoplastics.[37] It is well-known that FR PET is largely employed in packaging and textile industries, though the recycling of items made of such material is quite difficult, as the polymer matrix partially degrades lowering its overall properties during thermomechanical recycling. To overcome such limitation, Bascucci et al. incorporated DOPO-PEPA and AF, two PFRs, into PET by extrusion. AF promoted branching and chain extensions in PET, countering chain scissions, while DOPO-PEPA could boost control melt rheology and lubrification of PET for long-term.[38] In particular, the addition of DOPO-PEPA into the PET matrix, PET/DP samples, allowed to keep unvaried the ductile tensile behavior after recycling.[38] Also, Morita et al. sequestered FR bromine molecules using a hydrotalcite, while converting PP and PS mixed plastic, filled with TBBP-A, into oil via a thermolysis route.[39] Thermosetting polymers are composed of permanent cross-linked networks making these materials unable to flow by thermal stimulation, indeed they decompose at high temperatures. Pyrolysis-oxidation and solvolysis (e.g., acidolysis) are the most commonly used recycling methodologies for thermosetting polymers, resulting in the production of energy from the polymer matrix, small low-weight compounds, or in the recovery of potentially reusable reinforcing fillers (e.g., carbon fibers). Intriguingly, Vasile et al. applied a catalytic hydrogenation route to thermosetting parts of printed circuit boards and keyboards and obtained liquid products, also capturing some toxic BFRs along the process.[40] As a valuable alternative to permanently linked networks, many researchers are preparing inherently reprocessable and FR thermosets based on CANs, which exhibit excellent fire performances and can be easily reshaped and recycled without the need for extreme conditions. Following a sustainable approach, Wang et al. synthesized self-extinguishing thermosetting composites based on Schiff base CANs starting from vanillin as raw material.[41] The research group demonstrated that the monomer could be fully recovered through the immersion of the material in acidic water solutions at mild conditions.[41] Fostering the production of inherently reprocessable FR thermosets may notably reduce the amount of FR plastic waste and environmental damage.

Still-functional FRs in EoL PMs affect the design and the operative conditions of the recycling process in terms of effectiveness and sustainability. Thus, the scope of this book chapter is to give detailed insights into the development of more efficient recycling methods for FR PMs. The book chapter is mostly focused on theoretical aspects and the recycling methodologies are mainly discussed from an academic point of view. Given the most recent research efforts, this book chapter provides a comprehensive perspective of the recent advances inherent the recyclability and the reprocessing strategies of various FR polymer-based matrices (with a special focus on thermoplastics, thermosets, and covalent adaptable networks) and their composites.

24.2 RECYCLING OF THERMOPLASTICS CONTAINING HALOGENATED FLAME RETARDANTS

Packaging, textiles, electrical, construction, and automotive are some of the application fields of thermoplastics. Based on Plastics Europe,[42] fossil-based thermoplastics account for 83.1% of the global plastics production in 2021, which means around 390.7 Mt. To pass stringent requirements and deal with high fire risk, thermoplastics need the usage of specific FRs lowering their high flammability. The flame retardation of plastics by halogen-based species can involve the addition of halogenated molecules in the polymer matrix or the insertion of halogen-containing moieties in the backbone.[43] One of the most representative examples is PVC, which was produced on a scale of 5.5 million metric tons (as of 2018) and exhibits excellent mechanical and thermal properties, strong resistance toward chemicals, low carbon footprint and biological attacks, and poor air permeability.[44] Even now, due to their low cost and high availability, HCs are still largely used to confer enhanced FR properties to many plastics and composites. Chlorides and bromides are widely available in nature, owing to their presence in seawater, where their average concentrations are ~42,000 ppm[45] and ~65 ppm,[46] respectively.

Figure 24.4 shows some commonly used FRs. Unfortunately, such HCs have been found to act as endocrine disruptors,[49,50] unlike the more inert PVC. At the end of their life cycle, the possible degradation at elevated temperatures (> 220°C) of halogenated FRs may release chlorinated or brominated furans/dioxins, which cause toxicity in close human proximity.[51] Because of that, international organizations and governments are banning and phasing out several halogenated compounds from everyday use.[52] The high toxicity of possible by-products is also ruling out EoL methods like direct incineration or landfilling.[53] Recent legislative actions of the European Union are finalized to reduce the landfill to less than 10% by 2035.[54] Direct incineration of plastics containing HCs or halogenated PMs cannot be considered a suitable disposal method, as it would generate volatile organic species like chloro- or bromocarbons,[55] which either may be responsible for the formation of corrosive acids and the depletion of the ozone layer.[56] Some halogenated FRs, as for example PBDDs and PBDFs, do not undergo degradation upon incineration, but they persist in highly toxic flue gases,[57] though some R1 incinerator installations in Europe are fully equipped to handle these substances and are in compliance with the emissions.[58] In view of the above, tertiary recycling involving pyrolysis has to be performed by trapping any halogenated by-products. To date, only a few reports have been published wherein an

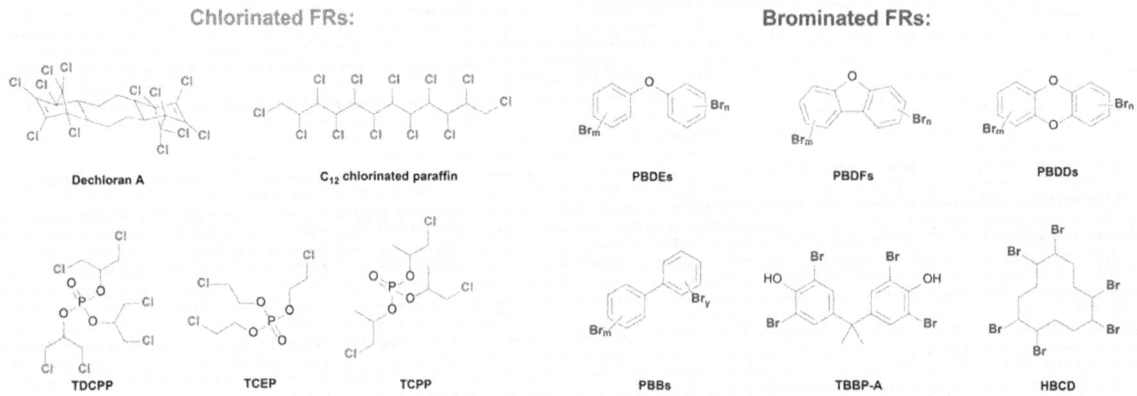

FIGURE 24.4 Commonly used HFR additives. (Left side) Chlorinated FRs: TDCPP, TCEP, TCPP are phosphate derivatives with 1,3-dicholoropropyl, 2-chloroethyl or 1-chloro-2-propyl groups. Dechloran A and chlorinated paraffins are highly regulated substances in many countries.[47] (Right side) Brominated FRs: PBDEs, PBDFs, PBDDs, PBBs, TBBP-A and HBCD.[48]

effective recycling of PMs and composites filled with halogenated FRs has been carried out. The following sections summarize the main achievements of such few studies.

24.2.1 Mechanical Recycling of Thermoplastics Containing Halogenated Flame Retardants

Mechanical recycling of the plastic waste is usually performed by extrusion, melting, or moulding processes. The waste does not undergo any dehalogenation treatment; therefore the HCs are still present in the PMs. In the case of PVC, as an inert halogenated PMs, dehalogenation treatments are not necessary and thus the recycling of such polymer has a minimal environmental impact and a maximum economic utility.[59] Indeed, it is reported in the literature that recycled PVC shows almost unvaried mechanical strength compared to the original polymer.[60] Owing to the restrictions imposed on HCs, it is mandatory to separate halogen-containing plastics from wastes like composites and WEEE before mechanical recycling[61] till it is proven that the POP concentration is below specific limits. Concerning brominated plastics, the ones showing a total bromine content higher than 2,000 ppm are subjected to recycling in proper plants for WEEE plastics and partial incineration. On the other side, plastics characterized by a total bromine content lower than 2,000 ppm can be reused to produce recycled items following the requirements of POPs and REACH regulations. For an efficient sorting of plastics, high-resolution near infrared devices plus XRF can be employed to screen wastes containing halogenated FRs before their reprocessing.[62] Cagnetta et al. used iron and quartz sand as co-milling reagents in a dry-ball milling apparatus to dehalogenate plastic wastes by mechanochemical process.[63] This was the first demonstration of plastic waste dehalogenation, showing a technology readiness level (TRL) of 7, as in some previous works, the ball milling was only used to debrominate[64] and dechlorinate[65] organic materials. Basically, the degradation of halo-organic molecules to amorphous carbon and halides can occur thanks to the mechanical energy derived from the milling.[66]

24.2.2 Chemical Recycling of Thermoplastics Containing Halogenated Flame Retardants

Chemical recycling allows for the conversion of plastic wastes into lower molecular weight compounds. To obtain such conversion, covalent chemical bonds need to break through energy-intensive and quite endothermic reactions. On the basis of the primary chemical process, chemical recycling can be classified as thermolysis, solvent washing, photolysis, gasification, etc. In 2003, in Japan, Yamawaki et al.[67] experimented with an early/small scale industrial setup to obtain usable raw materials from the gasification of WEEE, without producing a large amount of brominated toxic dioxins as by-products. However, the economical aspect was a significant downside, as the process required extremely high temperatures (~1200°C) and quite expensive reactors to depolymerize the wastes. To dechlorinate plastics, like PVC, Kakuta et al. in 2008 adopted a combination of hot water dissolution and melt processing.[68] During the recycling, the release of corrosive chlorinated gases was inhibited by using converter dust and coal tar, both containing inexpensive Fe_2O_3. This methodology allowed for a 98% removal of chlorine from plastics through its conversion to $FeCl_3$.[69] In comparison to thermal pyrolysis, the photolytic degradation of BFRs results in lower efficiency in terms of removal rate. Ling et al. demonstrated that UV radiation and natural sunlight can degrade BFRs to hazardous organobromides.[70] Also, Mathur et al. proved that the photolytic degradation of FR PP and FR PS occurred with quite slow debromination rates.[71] Further, Khaled et al. showed that the phototransformation of PS films containing BFRs was responsible for the production of brominated volatile photoproducts when exposed to sunlight, therefore wastes of such films need to be considered a possible source of hazardous contaminants.[72] Later on, Koch et al. found out that the treatment in the presence of water of a relatively inert brominated plastic (PolyFR) by both UV and thermal radiation led to the release of organobromides (i.e., Br_2 and HBr) along the time.[73]

To prevent problems arising from the pyrolysis of PMs containing brominated or chlorinated species, the treatment of these latter by solvents may be a valuable strategy. Unfortunately, solvent washing usually is not able to fully break down the plastics into lower molecular weight products, but it is still effective to remove halogenated additives from plastic wastes. As displayed in Figure 24.5, the additive migrates into the solvents because it dissolves better than the polymer matrix, which is insoluble in many commercial solvents.

For example, in 2019, Evangelopoulos et al. demonstrated (TRL: 7) that the solvent washing of WEEEs with toluene and isopropanol before pyrolysis results in the removal of TBBP-A and in the reduction of the bromine content in the WEEE by 36.5%, though this treatment was not able to reach the requirement like below 500 ppm of POPs.[74] Charitopoulou et al. applied the solvent washing methodology by alcohol/hexane mixtures, coupled with microwave irradiation, on polymer blends to reduce the bromine content of 73% and selectively extract TBPP-A.[75] Due to the commercial viability of the solvent washing technique, Fraunhofer IVV in Germany trademarked the CreaSolv® Process (registered trademark of CreaCycle GmbH).[76,77] This CreaSolv® Process (TRL: 4–6) is a combination of precipitation methods and dissolution able to selectively extract legacy halogenated additives, especially BFRs, as well as recover large fractions of PE and PS (i.e., polymer matrices), which are industrially significant.[78] Also, Fraunhofer IVV has trademarked the Centrevap® process to remove insoluble components at a sub-micron scale from other PMs and WEEE.[79] In 2012, Wang et al. obtained an oil with 60 wt% loading of brominated by-products by solvent washing of a BFR-containing waste computer housing plastic. Supercritical isopropanol was used as a washing agent and a debromination efficiency of ~97.6% was achieved with POP concentration of PBDEs < 500 ppm and for HBCD < 100 ppm.[80] One of the major downsides linked to the use of supercritical fluids like water, alcohol, or acetone for recycling is the the huge loss of the organic components, such as H_2O and CO_2.[81] However, Zhang et al. almost fully dibrominated and dehalogenated electronic housing display plastics, composed of BFRs and PVC, by employing supercritical CO_2, though the removal efficiency was far from REACH compliance.[82]

Finally, in order to recycle WEEE in the right way, the scientific community needs to spend more efforts in developing effective dehalogenation processes. Today, the chemical recycling of POP-containing waste is not allowed, as not approved in the technical guidelines. However, considering that mechanical recycling focuses more on re-melting or re-extruding the plastics, chemical recycling represents a promising methodology to obtain dehalogenation grades approved by the Basel Convention.[83] Solvent washing of native plastics is a valuable technique to isolate halogenated additives, before proceeding with their mechanical recycling. Even now, thermal energy, using gasification, pyrolysis (thermochemical recycling), or friction, is the only way to degrade the HCs.

24.3 RECYCLING OF THERMOPLASTICS CONTAINING HALOGEN-FREE FLAME RETARDANTS

The regulations restricting the use of HCs as flame retardants for plastics have increased up to 70% until 2019 the consumption around the world of HFFRs, including N-based, organophosphorus, and other chemicals (e.g., magnesium hydroxide).[84] Figure 24.6 shows some of the most common HFFRs. In the fields of furniture, EEE, textile, wire and cables, building construction, and transportation,

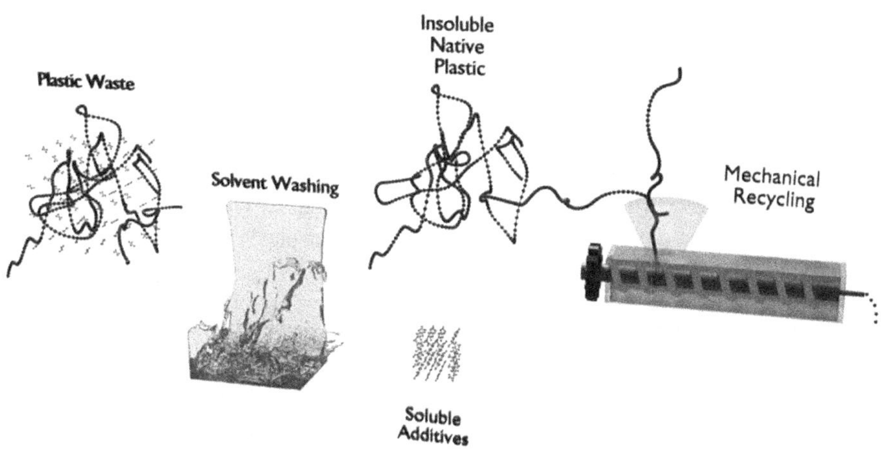

FIGURE 24.5 Solvent-washing treatments separate halogen-based additives from the native thermoplastics, before being reprocessed by mechanical recycling.

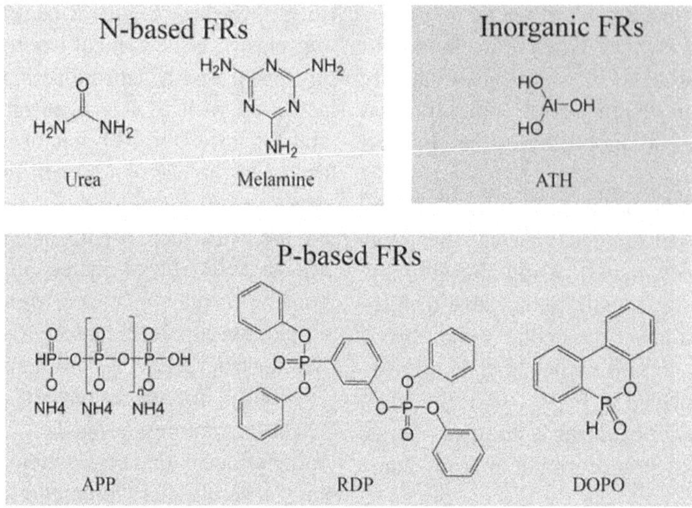

FIGURE 24.6 Chemical structures of HFFRs. P-based FRs: RDP, DOPO, and APP are resorcinol bis(diphenylphosphate), 9,10-dihydro-9-oxa-10-phosphaphenanthrene-10-oxide, ammonium polyphosphate, respectively. N-based FRs: urea and melamine are often used in combination with P-based FRs. Inorganic FRs: ATH is aluminum hydroxide.

N-containing and/or P-containing organic compounds are usually employed as HF alternatives.

The different physico-chemical properties (e.g., thermal behavior) of both polymer matrix and FR confers a heterogeneous composition to plastic wastes containing HFFRs, which makes quite difficult the recycling of these materials.[85] Such aspects strongly limit the effective recycling of HFFR thermoplastics, and thus they represent some of the major challenges that the scientific community needs to face.[86–88] Several methodologies have been developed to manage halogen-containing contaminants present in HFFR thermoplastics, for example, density sorting is used in recycling plants to separate brominated FR plastics from the waste stream.[89] Most of the recovered thermoplastics from WEEE waste are contaminated by brominated species, which are very resistant to aging compared to P compounds. Beigbeder et al. applied the near-infrared technique to sort brominated plastic wastes from different plastics and demonstrated that the impurities negatively impact the fire performances of recycled thermoplastics.[62] The removal of brominated residuals and a good compatibility between the additives and the EoL plastics in the waste stream allow for a good mechanical recycling of HFFR thermoplastics. Recent works have reported that ~20% of the total weight of WEEE streams is approximately composed of 15 different types of thermoplastics, especially including HIPS, PC, ABS, and PE.[90] Also, PC/ABS and HIPS/PPO blends can be found in the recycled thermoplastics from WEEE streams. As mentioned earlier, the stricter regulations of the European WEEE Directive have restricted the recycling and the use of HFRs, also defining a set of limits named "unintentional trace contaminant".[91] These restrictions have boosted the production of PC/ABS-containing PFRs, and thus further research needs to be considered to improve the recyclability of such plastics.[92]

The separation of halogen-containing components from WEEE by means of several techniques is crucial before implementing a recycling method, otherwise, the quality of the reprocessed material will be very poor.[93–95] Also, more research efforts should be devoted to effectively separating HFFR plastics containing P- or N-based FRs. Peeters et al. evaluated the economic and technical feasibility concerning the recycling of a waste stream based on EoL of LCD TVs through two procedures: (1) manual disassembly and (2) downsizing process.[96] PC/ABS are the main components of EoL LCD, which contains around 31 wt% PFRs and 18 wt% BFRs, respectively. Peeters et al. performed some laboratory-scale experiments and proved the feasibility of a closed-loop recycling for PC/ABS-containing PFRs, leading to a significant reduction of costs related to pollution, energy issues and degradation.[96]

24.3.1 Mechanical Recycling of Thermoplastics Containing Halogen-Free Flame Retardants

Mechanical recycling of HFFR thermoplastics is usually one of the preferred methodologies to manage wastes, due to its low costs and sustainable characteristics, together with the possibility to keep almost unvaried the properties (e.g., fire behavior, mechanical performance) of the original material. There are some aspects to be considered in order to perform an effective mechanical recycling of HFFR thermoplastics, which are linked to their high sensitivity toward hydrolysis and thermal degradation. For example, the hydrolysis of PFRs present in HFFR thermoplastics is responsible for the production of phosphoric acid, which combined with the presence of water causes the partial degradation of the polymer matrix and the decrease of FRs concentration. These phenomena may occur during both the recycling and service life of the FRs.[97]

To investigate the influence of FRs on the mechanical recycling of HFFR plastic wastes, repeated extrusion cycles on the material are usually carried out.[38,89,96,98–102] Imai et al. studied using multiple extrusion steps the recyclability of different commercial HFFR thermoplastics.[103] After only one recycling step, PC/ABS and PC/HIPS containing organic phosphate esters could not pass UL-94 5VB tests but still exhibited good mechanical properties. The research group assessed the effect of humidity on the recyclability of both PC/ABS and PC/HIPS by hydrolyzing (up to ~100%) HFFR thermoplastics before their extrusion. All the recycled HF thermoplastics could not pass the UL-94 5VB flammability tests, owing to the different degrees of hydrolysis.[103] Unlike HF thermoplastics, hydrolyzed ABS and virgin samples incorporating brominated FRs (i.e., TBBP-A and brominated epoxy oligomers) could pass UL-94 tests.

It looks evident that the mechanical recycling of HF thermoplastic waste strongly depends on the water content, chemical composition, and physico-chemical properties of the plastic waste stream,[100] as all these aspects affect the quality of the recycled materials. Almeras et al. investigated the flame retardancy and mechanical properties of recycled APP/PA-6/PP blends containing different compatibilizers.[98,99] The blends after four extrusion recycling cycles showed a slight decrease in elongation at break and an increase in Young's modulus. As proved by ^{31}P NMR performed on reprocessed materials, the second extrusion caused the degradation of APP and the consequent generation of orthophosphate and pyrophosphate, which supported the notable decline in fire behavior. On the other side, the aging of blends in the presence of compatibilizers led to a significant degradation of APP chains.[98] In case of anti-aging agents are consumed, recycling can be seen as a sort of aging to some extent. Indeed, accelerated aging can be employed to evaluate the effectiveness of a recycling process toward a specific plastic waste stream. Turku and Kärki studied the effects resulting from the accelerated aging of PP-based WPC containing titanium dioxide, melamine, aluminum trihydrate, graphite, and borate as commercial inorganic FRs.[104] The additives affect the surface morphology, mechanical properties, and color stability of WPC.[104] Statler et al. demonstrated that, similarly to PC/bromine samples, recycled HFFR PC containing KSS preserved its original mechanical properties (e.g., Young's modulus) and still exhibited a V-0 rating in the UL-94 test after eight cycles of extrusion.[101] The use of PFRs may influence the performances of the recycling and processing of thermoplastics. For example, DOPO-PEPA enables reprocessing at lower temperatures of PET, reducing its melt viscosity during extrusion.[105,106] As mentioned above, Bascucci et al. proved that PET/DOPO-PEPA keeps unvaried mechanical performance after an injection molding step and three extrusion cycles.[38] Unlike PET/DOPO-PEPA, the same recycling attempts performed on PET containing Aflammit PCO 900 (i.e., a commercially available organophosphate compound) resulted in a complete loss of its properties,[38] due to the differences between the chemical reactions that occurred by the PFRs in PET at elevated temperature. In the presence of different FRs (DOPO-PEPA, zinc diethyl phosphinate, phosphinate salt, Aflammit PCO 910, and an HFR), Bethke et al. upcycled bottle-grade PET through a reactive foam extrusion process using PMDA as chain extruder.[107] PET/HFR showed lower fire performance than other formulations containing PFRs but improved processability. The results may contribute to develop a potential alternative finalized to the production of high intrinsic viscosity PET foam with boosted FR performance.

24.3.2 Chemical Recycling of Thermoplastics Containing Halogen-Free Flame Retardants

The composition of FRs and the polymer matrix influence the optimal recycling strategy for a specific plastic waste stream. HFFR thermoplastics are not easy to be recycled by mechanical processes and tertiary recycling (i.e., gasification, depolymerization, dissolution, and pyrolysis) represents a valuable alternative to deal with such materials. Recycled plastics are expected to increase to 3.4 Mt, for this reason, the European Community has planned to invest 7.2 billion euros in chemical recycling plants by 2030.[108]

Pyrolysis is widely used for the thermochemical recycling of WEEE plastics and typically leads to the production of fuel-like products.[109–112] Perret and Schartel evaluated the effects of impact modifiers on the fire performances of PC blends containing BDP flame retardant.[113] In WEEE streams, the sorted heavy fraction up, are usually present significant amounts of brominated FRs, and thus many research efforts have been devoted to studying the effects of pyrolysis on halogen-based plastics, paying less attention to HFFR plastics. When mechanical recycling does not involve any effective separation of plastics and additives from waste streams, characterized by heterogeneous compositions, the properties of the recycled plastics are negatively affected. To overcome such limitations, chemical recycling strategies have been acquiring growing attention, including depolymerization processes, solvolysis, and other methods.[114,115] Wang et al. used DMCHA to selectively extract organophosphates (i.e., BDP) from commercial plastics. These latter were mainly composed of PC/ABS resin, which could be also recovered through precipitation.[116] This precipitation/dissolution approach offers relevant insights into exploring efficient and novel recycling methodologies. Currently, the waste management of HF thermoplastics is mostly performed by mechanical recycling. The increasing requirements in terms of scope and recycling quality are fostering the development of always more efficient thermochemical and chemical recycling techniques.

24.4 RECYCLING OF FLAME-RETARDED THERMOSETS: FROM CONVENTIONAL DISPOSAL METHODS TO RECENT APPROACHES

Thermosets are extensively used in several applications and thus the amount of waste resulting from such materials is huge. Nowadays, only a small percentage of thermoplastic materials undergo an effective recycling, while most of PMs are not properly disposed.[117] EoL conventional thermosets are characterized by permanent covalent crosslinks, making the recycling of these materials very difficult, even more when fillers (e.g., carbon fibers) or FRs are present in the polymer matrix.[118] However, due to their high stiffness and strength, thermosets are considered an ideal choice for the manufacturing of different lightweight components and fiber-reinforced composites, especially in the aerospace sector.[17] The high chemical and thermal stability of thermosets contribute to make challenging the recycling of such materials. Currently, with a CAGR of ~65 million tons, thermosets represent ~20% of global fossil-based PMs. In particular, owing to their excellent properties, epoxy resins account for around 70% of the global thermoset market with a CAGR of ~6.46% by 2024.[9,10] Unfortunately, the most common disposal methods for thermosets are incineration, sea dumping, and landfill.[119,120] Taking this further, based on the Environmental Protection Agency, white pollution will shortly become one of the most serious issues both outside and within Europe.[121] Depending on the particular need or request, thermoset wastes are usually mechanical or chemically recycled. During mechanical recycling, the thermoset waste is usually ground and re-applied as filler, without any chemical treatment or modification, for the manufacturing of composites with lower quality.[9,122] On the other side, chemical recycling involves pyrolysis-oxidation processes (thermochemical recycling) or solvent-assisted depolymerization techniques, allowing the recovery of energy by degradation of the polymer matrix or the regain of the precious filler from the thermosetting composite.[123,124] Both methodologies are considered partial recycling approaches, as the thermoset matrix is fully lost and only fibers or fillers are generally regained. This is a significant downside for these recycling methods, considering that the largest part of a composite material is the polymer matrix. Also, the presence of FRs in the matrix of some thermoset wastes is not envisaged in the current processes adopted in mechanical or chemical recycling. Thus, the FRs present in the polymer matrix of the waste are typically burned or reused as still active additives in the ground thermoset. As HCs represent the largest amount of FRs, the mechanical or chemical recycling of the thermoset waste may be responsible for the release of secondary pollutants or toxic flue gases.[5,122] The increasing development of thermosets composed by CANs may inspire the production of always more inherently FR recyclable and reprocessable materials, for which many of the abovementioned concerns would be overcome.[125,126] The growing interest of the industry toward such dynamic covalently crosslinked systems has been highlighted in recent published reviews, where some examples of recyclable thermosets approaching the market have been described in detail.[127–130]

24.4.1 Waste Treatment of Conventional Flame-Retarded Thermosets

Recent European restrictions are devoted to reduce the landfilling of thermosetting waste, promoting the development of novel lightweight and saving-energy thermosets and the design of more sustainable recycling technologies.[131] Glass and carbon fibers are typically used as fillers in thermosetting composites. Carbon fibers are generally more expansive than the polymer matrix, and thus great attention is given to the recovery of such fibers at high-quality standards and the recycling of fiber-reinforced composites.[122,132]

According to the European Union's Waste Framework Directive, the direct reuse of high-performance EoL thermosets and related components in lower quality applications, for example housing equipment, represents the most advantageous recycling strategy.[122,133] This sustainable approach demands lower reprocessing steps and costs, though some thermoset wastes are difficult to be adapted for a specific need due to geometry reasons.[132,134] Primary recycling, involving direct reuse processes, may be a valuable alternative for the recycling of flame retardant thermoset wastes, as FRs will still exert their function in the recycled product. However, the downsides linked to this waste-to-wealth approach do not allow it to be included among the most effective solutions for the recycling of thermoset waste. The most commercial and studied waste management technologies can be split into mechanical and chemical processing.[135] An overview of the applied recycling strategies for the treatment of wastes derived from conventional thermosetting composites is reported in Scheme 24.1. Most of these technologies are targeted to regain fillers or fibers from the composite, and only a few involve particular physical/chemical treatments for the recovery of the FR or the polymer matrix, resulting in negative repercussions on both the environment and the economy.

24.4.1.1 Mechanical and Chemical Recycling of Conventional Thermosets Containing Flame Retardants

By using a speed-cutting or crushing mill, mechanical recycling converts scrap thermosetting wastes with heterogeneous shapes into small sized flakes of ~50–100 nm. Indeed, such a process is usually carried out for both thermosetting and thermoplastic wastes, even filled with FR additives or other materials.[5,123] Unlike thermoplastic flakes, the thermoset ones require an additional grinding by high-speed mills to achieve products showing a micrometre-size.[135] As reported in the literature,[136,137] a proper tailoring of the ground products, before being upcycled as reinforcing fillers for the manufacturing of polymeric composites, would drive down costs, reduce the energy demand, and contribute to obtain materials with excellent interphases and overall properties.

Anaerobic and aerobic combustions represent the most used processes for the thermochemical recycling of thermosetting wastes. Pyrolysis is an anaerobic combustion involving the decomposition under an inert atmosphere of the organic polymer matrix. Most of these waste treatment methods do not separate additives and fillers by the matrix, therefore they may degrade producing by-products and harmful compounds.[138] Incineration is the widest employed aerobic thermal route, which does not allow the

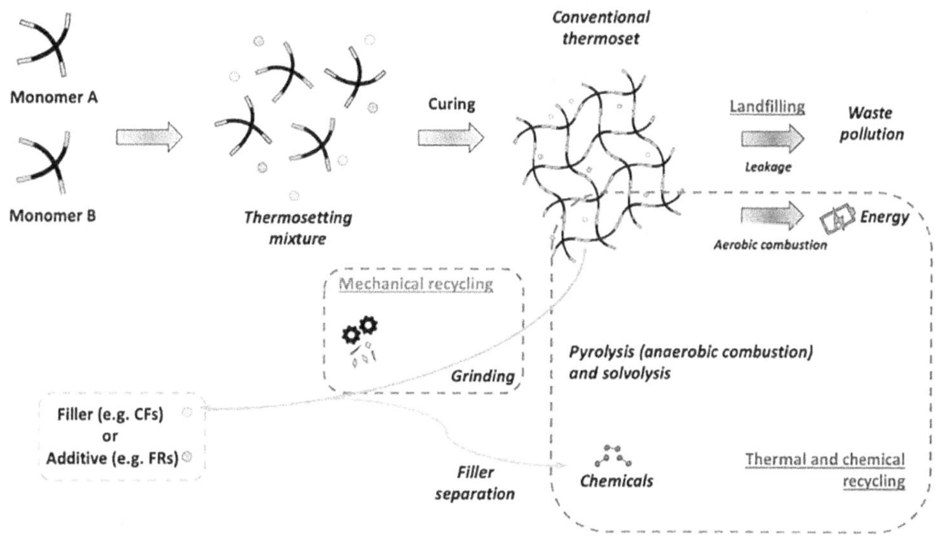

SCHEME 24.1 Recycling methodologies of conventional thermoset composites.[122,132]

recycle of the waste, but the oxidative combustion of this latter into recovered energy.[139,140] Anaerobic combustion or pyrolysis of EoL thermosetting products coverts the polymer matrix into value-added chemicals, consisting of different types of lower molecular weight organic species (see Scheme 24.1).[109] Sealants and putties are extensively used in integrated circuits and thus are some of the main components forming EoL WEEE items. For this reason, Luda et al. investigated the released gases from the pyrolysis recycling of such materials.[141] The research group found out that the thermochemical recycling of sealants and putties, composed of brominated anhydride-cured epoxies, led to the debromination of aromatic fractions, together with the release of hardener, bromine-based compounds, and glycidyl groups.[141] All of them are well-known for their adverse impacts on human health and the environment.[50] In 2013, Terakado et al. carried out the pyrolysis of printed circuit boards consisting of ET composites. The research group achieved over 80% suppression of both brominated secondary products and HB performing the anaerobic combustion in the presence of ZnO, La_2O_3, and other metal oxides.[142] Later, Altarawneh et al. demonstrated that TBBP-A could be co-pyrolyzed using Fe_2O_3 in order to convert the formed HBr into lower irritating iron oxybromides and iron bromides.[143] However, it is clear that in the design of future pyrolysis technologies for an effective recycling of EoL thermosetting wastes filled with FRs (e.g., HCs), the dehalogenation of streams, the limitation of pollutant emissions, and the reduction of energy request need to become the most important features.[144]

As previously mentioned, the polymer matrix represents one of the most abundant component of a thermoset composite waste, therefore its recovery and conversion into chemicals would significantly reduce costs and add value to the recycling methodology (see Scheme 24.1).[145] However, these chemicals need further processes before being implemented as raw materials for the synthesis of new polymers.[146] Most of the solvolysis technologies already available on the market and applied on thermosetting composite wastes allow for the regain of monomers, organic byproducts, and fibers. Unfortunately, it occurs frequently that the recycled fibers and matrix are not reusable, because partially damaged especially on the surface.[147–149] Compared to thermochemical treatments, recycling by solvolysis show many advantages, for example it does not lead to the formation of char lowering the quality of the recycled fiber surface. Besides, solvolysis is often conducted by low impact supercritical water and alcohols requiring for less expensive overall operative conditions, with respect to thermochemical recycling processes demanding for high-cost reactors.[150] Nevertheless, the extreme pH conditions and high temperatures (above 200°C) that are usually necessary for many solvolysis procedures cause often the decomposition of FR, resulting in the release of toxic compounds and hazardous molecules into the environment.[5,150] The scientific community needs to focus growing efforts on the development of lower energy-demanding strategies able to effectively regain not only the polymer matrix with a proper solvent, but also other organic species with added-value (e.g., FR moieties or FRs). Some recent works paved the way for future developments in this direction. Following a waste-to-wealth approach, Kim et al. prepared (TRL: 6) a recyclable carbon-fiber-reinforced EC using TA as a multifunctional filler. Indeed, TA exerted the role of reinforcing agent, smoke-suppressant and char-formed in the final products.[151] The research group performed the chemical recycling of TA-based ECs by oxidative degradation and supercritical water, also recovering reusable CFs.[151] Panda et al. found a recycling methodology for EoL waste deriving from printed circuit boards. They could separate BER, with 97% of efficiency, and regain ~93% of Cu as copper-containing species,[152] which is crucial to counter and limit the depletion of natural. Thanks to the use of dimethylformamide, the research group could dissolve and then remove

BER, preventing the production of harmful dibenzofurans and PBDDs during the regain of Cu.[152] Ma et al. synthesized a fully degradable DGEBA-based resin starting from DGEP as raw material and using DDM as a curing agent.[153] Unlike DGEBA/DDM system, DGEP/DDM exhibited self-extinguishing capability (V-0 rating in the UL-94 vertical flame test) and could degrade after 40 min of treatment in a methanol/sodium hydroxide solution kept at 50°C.[153]

UPR and its GFRCs have been largely used in the fabrication of windmills and in the aeronautical industry. Due to their extensive application, the waste UPR, end-of-life components, and unused prepeg are causing several concerns to the environment.[154] Thus, the development of effective recycling methodologies for UPR-based products is becoming urgent. However, due to the presence of a permanent three-dimensional network structure, the recycling of UPR and its composites is extremely tough. Currently, available recycling strategies include mechanical, thermal, and chemical methods.[155] Chemical methodologies show the highest effectiveness, though long reaction times, high temperatures, and the need for strong corrosive solvents (e.g., nitric acid) are still indispensable to cleave the chemical bonds of UPR. Recently, UPR and its GFRCs have been chemically recycled by selective cleavage of C-O bonds using the $AlCl_3/CH_3COOH$ system as a promoter for the acyl exchange reaction between CH_3COOH and UPR.[156] This process led to the recovery of monomers and oligomers from UPR, together with the regain of glass fibers. Wang et al. cleaved ester bonds by using hydrolysis of 80% hydrazine hydrate along with NaOH.[157] To achieve a complete degradation of UPRs, this mild hydrolysis-oxidation synergistic catalytic strategy was conducted at 100°C and accomplished via the Fenton reaction. The research group could recover carbon fibers showing tensile strength, chemical structure, and morphology very similar to the virgin carbon fibers.[157] Yao et al. prepared an inherently recyclable UPR-based material by modification of dimethacrylate monomer and its self-polymerization. A hindered C–O bond with a carbonyl group in the β-position of steric ester was introduced during the synthesis of the monomer.[158] The obtained resin exhibited very good properties and could be degraded by selective cleavage within 3.5 h at 90°C by immersion in an aqueous solution of hydrazine hydrate. The research group found out that the degradation products, characterized by benzene and carboxyl/carboxylate groups, could be easily separated according to their water solubility and then reused as functional additives and flocculants.[158] More research efforts need to be devoted toward the implementation of technologies involving the recycling of flame retardant UPRs and related composites, as most of the wastes derived from such materials contain FR additives due to the high flammability of virgin products.

PUs are widely used in several industrial applications, thanks to their durability and low-cost processing. They are applied in the manufacturing of bedding, polyurethane resin wind turbine blades, molded polyurethane foam, insulation materials, electronic components, and other goods.[159,160] In 2020, the consumption of PUs in Europe was estimated at around 3 Mt/year, including ~32% as rigid foam, ~36% as flexible foams, and ~32% as elastomers and reaction injection molding.[161,162] However, due to the chemical characteristics of PUs, conferring them high chemical stability and lightweight properties, these materials are difficult to recycle. Most of the waste (~40%) derived from PUs is converted into energy through thermochemical processes, while only ~30% is recycled, and the rest is landfilled.[163] The recycling of PUs involves mechanical, chemical, and thermochemical approaches.[164] Chemical recycling represents the most effective and promising route, as it breaks down the PU backbone to regain raw materials that can be upcycled into new PU-based products.[163] Glycolysis, aminolysis, and hydrolysis are methodologies commonly used to perform the chemical recycling of PUs, while pyrolysis and gasification are carried out to transform the PU waste into added-value monomers or syngas, respectively.[164] PUs are easily flammable and thus products based on such materials are usually flame retarded with several functional compounds, for example TCPP is one of the largely employed additive.[165] Also, flame retardants based on phosphorus, nitrogen, and sulfur are added to rigid urethane foams by mechanical mixing to improve their fire behavior. Alternatively, the fire resistance of urethane foams can be enhanced by chemical reactions between their hydroxyl groups and chlorine and/or phosphorus-containing polyols.[160] Unfortunately, huge amounts of chlorine compounds are usually released during the pyrolysis of PU waste from home appliances at 500–700°C, due to the presence of Cl-based FRs in the polymer matrix.[166] As an example, Eschenbacher et al. performed the thermochemical conversion of PUs (rigid foam, flexible foam, and semirigid) and PIRs, with and without TCPP flame retardant, via pyrolysis to obtain valuable compounds as volatiles for chemical recycling.[161] The research group tried to achieve high values of yields in terms of semi-volatile organic products. The main volatile pyrolysis vapors were ethylene, styrene, propylene, oxygenates (e.g., polyether species), and nitrogen compounds (e.g., 4,4′-methylenedianiline). It is worth to mention that the pyrolysis of TCPP-containing PIR was responsible for a significant release of Cl-containing molecules, as for example allyl chlorides.[161] Covestro has recently marketed an innovative process enabling the recovery of flexible PU foam and other core raw materials from used mattresses. This chemical recycling methodology aims to reduce the waste produced by mattresses at the end of their life, as these materials contain 15 to 20 kilograms of foam. Covestro has developed this process working in the PUReSmart project,[164,167] which is coordinated by Recticel company (i.e., a division of Wolfgang Binder GmbH), and has already started operating a pilot plant for flexible foam recycling at its Leverkusen site. Besides, Covestro has parallelly introduced an intelligent sorting solution based on algorithms for separating the different PU foams from post-consumer mattresses.

Electronic devices are always more used in our daily life and almost all electronic items are composed of a PWB. As these items are widely employed, stringent fire requirements need to be fulfilled for the fire resistance of PWBs, especially the UL94 vertical burning test is the most performed specification. The incorporation of additive FR systems into PWBs or the use of FR reactive species (e.g., dihydrooxaphosphaphenantrene, poly(1,3-phenylene methylphosphonate), phosphonate oligomers) for the matrix is commonly required to achieve UL94 V0 (i.e., self-extinguishing). PWB base materials can be manufactured by several polymers, however, epoxy resins usually represent the backbone of such products. In addition to the aspects related to FRs, some recent European directives are focusing the attention on the content of lead (Pb) in PWB base materials. Since July 2006, the European Directive on Restriction of certain Hazardous Substances in E&E (RoHS, 2002/95/EC and RoHS 2, 2011/65/EU) has banned PWB base products that contain Pb, unless their use is specifically exempted.[168] To produce lead-free soldering materials for the fabrication of PWB base products, very high process temperatures are needed, and thus the polymer matrix, as one of the main component, should be characterized by high thermal stability. For the manufacturing of halogen-free PWB base materials, epoxy-novolac resins are often selected due to a glass transition temperature higher than 165°C.[169,170] The higher thermal stability of epoxy-novolac resins than diglycidyl ether of bisphenol-A (DGEBA) ones allows PWB base materials to achieve technical requirements and simplify recycling procedures, even replacing halogen-containing FRs with more sustainable additives composed by P- or N-structures. This possibility is fostering the development of several halogen-free NEMA classes of fire retardant laminate materials for PWBs.[171,172] There are several categories of PWBs, including FR-4, FR-2, FR-1, CEM-1, and CEM-3 along with their application to electronics and electrics. The chemical composition of such categories is very different and influences the recycling of related products.[173–175] To give some examples, FR-2 is made from a phenolic resin reinforced with paper, which is commonly flame retarded using triphenyl phosphate as HFFR. For simple applications, FR-3 boards based on epoxy systems and paper are still employed. FPCs can be composed of FR PUs, brominated epoxy resins and phenoxy resins, and are used for mobile devices (i.e., cell phones) and computer peripherals. PWBs belonging to the FR-4 classification are usually made from epoxy resins with glass fiber reinforcement and in order to be applied in the electronics industry, these materials should not decompose or delaminate during lead-free soldering.[174,175] The growing environmental concerns, the urgent demand for more sustainable products and the need for easier recyclable goods are fostering the transition toward PWBs containing HFFRs, for these reasons the scientific community and printed wiring board manufacturers are testing the technical performances (e.g., thermomechanical and flammability properties) and suitability of halogen-free laminates compared to a benchmark brominated material.[176] Today, most of the FR PWBs are going to smelter, that is, ~1 ton of plastics provides 450 kg of cokes and BFRs are converted into bromide salts. An effective recycling of PWBs should involve not only the recycling of the polymer matrix but also the recovery of fibers (e.g., glass-wool) and precious metals such as Au, Ag, Cu, Ni, Sn, etc. The mechanical recycling of PWB base products allows the separation of resin, which can be converted into polyols by chemical wet processes conducted at specific operative conditions.[173,177] Recently, Krishnamoorthy et al. have used organic acids such as citric acid, itaconic acid, and oxalic acid, synthesized by Aspergillus niveus, to bioleach in 15 days metals (e.g., Zn, Ni, Cu) from waste printed wiring boards.[178] Pyrolysis and leaching are the main methodologies employed for the recycling of WPBs. Unfortunately, the pyrolysis of WPBs releases hydrogen bromide, together with a large amount of pollutants and brominated pyrolysis oils. Luyima studied a strategy for reducing the amount of brominated oils and HBr gas that evolved during the pyrolysis of WPBs.[179] This method involved the use of powder inorganic chemicals (CaO, $CaCO_3$, Fe_2O_3, Al_2O_3, YZeolite, and ZSM-5) to trap HBr and other brominated gases, resulting in more than 90% reduction in total bromine evolved. Also, Luyima found out that the leaching of WPBs by HCl, nitric acid and aqua regia could selectively recover precious metals contained in waste PWBs.[179]

In the following section, some recent advances related to synthesis procedures and characteristics of inherently recyclable thermosets will be summarized. Inherently FR and recyclable thermosets represent one of the most promising solutions to overcome downsides and limitations linked to the waste management of conventional flame retardant EoL thermosets.

24.4.2 Reprocessable Flame-Retardant Thermosets Based on CANs

Thermoset materials are characterized by a permanent cross-linked network and can be in cured in different shapes based on the need. The presence of permanent covalent crosslinks limits the options to dispose of EoL thermosets. For this reason, Leibler et al. have recently synthesized a class of materials called "vitrimers", which are composed of associative CANs.[180] The attention of the scientific community toward vitrimers and other types of CANs is growing,[181] owing to their chemical characteristics, making them able to be rigid at ambient temperature, and "flow" at a higher temperature. This flowing behavior is due to topological rearrangements, which can be triggered providing heat or other stimuli.[182] All these features enable either pristine vitrimers or the ones containing FRs and other fillers to be reshaped, recycled, and repaired by using a wide range of operating conditions. Similarly to thermoplastics, vitrimers can be easily reprocessed, because of their malleability and thermal ductility. However, vitrimers exhibit chemical resistance, mechanical performances and high thermostability like thermosetting materials.[183,184] Reversible ester bonds, disulfide exchanges, metathesis of dioxaborolanes, transalkylation, and -Si-O-Si- exchange may constitute CANs of reprocessable thermosets.[185] These dynamic linkages, for which Scheme 24.2 reports a simplified representation, offer several functional groups in the chemical structure of the polymer network, which can be used to insert the FR moieties as part of the polymer network or as a pendent group.

Inherently recyclable and FR thermosets can be reshaped and reprocessed by external stimuli (e.g., pH, heat and UV irradiation), without any detrimental impact in terms of waste production or environment pollution.[186] Bio-based derived thermosetting vitrimers highlight the potential of such materials, which are able to combine low flammability with high sustainability and recyclability.[187,188]

In the next section, the recycling of thermosets based on CANs is discussed, with a special focus on the main strategies for the recovery of both filler and polymer network.

24.4.2.1 Chemical and Mechanical Recycling of Flame-Retardant Thermosets Based on CANs

Owing to their main properties, EoL vitrimers can be directly reprocessed by traditional manufacturing methodologies, or ground into flakes to be used as fillers for the preparation of new products.[189] Physical recycling of vitrimers is performed through three thermomechanical techniques: injection molding, extrusion, and hot pressing (i.e., compression molding). Figure 24.7 reports a representation of some main features of FR vitrimers. As displayed in Figure 24.7, hot press recycling involves a down-sizing process of vitrimers to small pieces, which are subsequently hot pressed to obtain a recycled material.[190]

The final properties of the recovered material are strongly depending on the operative conditions selected for the hot pressing process. To avoid the thermal oxidative degradation of the FR polymer matrix, and the deterioration of the filler (e.g., short or long fibers) during the recycling of the EoL FR vitrimer, proper conditions of pressure, temperature and powder size are necessary.[191] These represent important limitations in view of a higher spreading of the hot pressing technology, especially because the FR part of many FR vitrimers exhibits lower thermal stability at the operating temperatures, with respect to the polymer matrix and other fillers.[192,193] Nowadays, limited compression molding is the most used methodology to recycle and reprocess vitrimers, though this latter is energy-intensive and often leads to recycled resin products showing lower mechanical performances.[126,194] Recently, some lower energy-demanding technologies have been developed, which are easy to be applied and allow for an effective reprocessing of the EoL vitrimer.[195,196] CompPair is an EPFL spin-off based in Switzerland, which has introduced on the market a technology called HealTech™. This latter enables the production of composite components able to repair on-site at low temperatures and within a few minutes.[195] Due to the downsides of thermomechanical recycling, the chemical recycling of FR thermosets composed of CANs is preferred.[130]

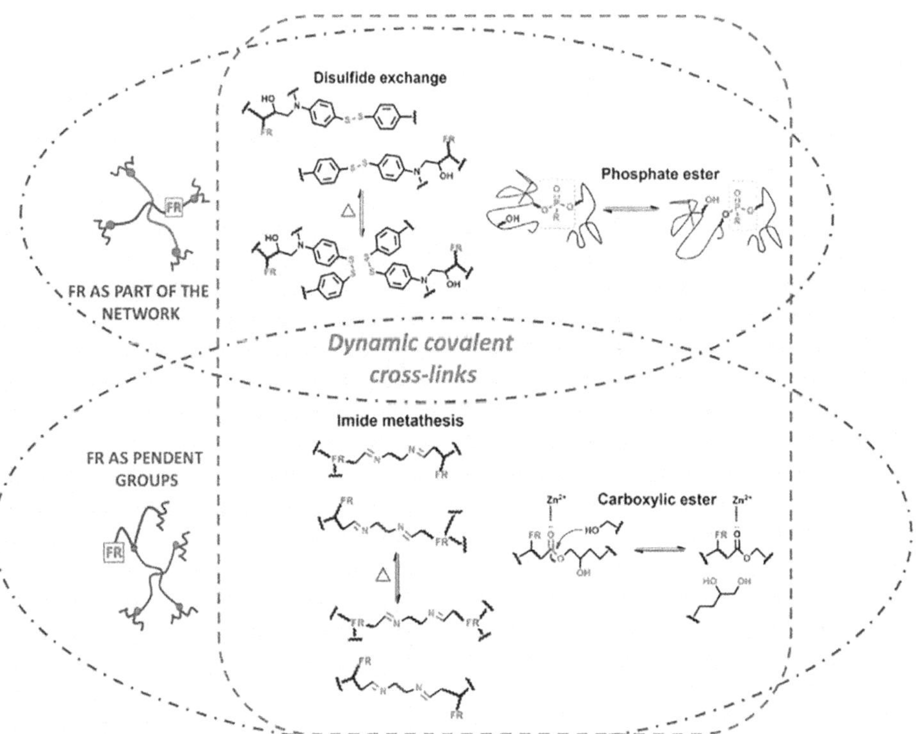

SCHEME 24.2 Current chemical moieties enabling reversable topological rearrangement in inherently recyclable and flame retardant thermosets. The covalent adaptable network in the polymer matrix of inherently recyclable and FR thermosets may be composed by FR as pendant group or as part of the network.

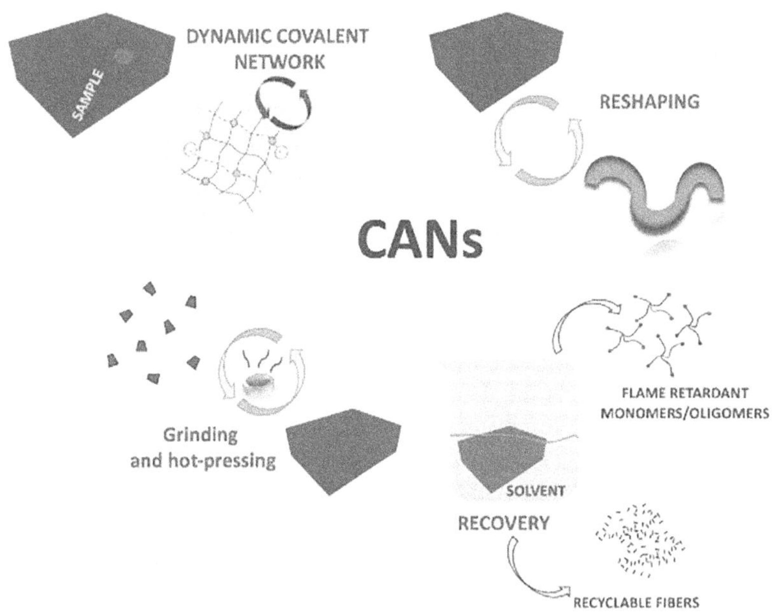

FIGURE 24.7 Some main features of FR vitrimers.

The chemical recycling of vitrimer as a single component leads to the recovery of monomers, with exchangeable functional groups, by means of a proper solvent. The same functional groups make possible the degradation of vitrimers into recyclable oligomers. On the other side, the chemical treatment of composite vitrimers results in multiple materials, FR monomers/oligomers and filler, which are separated in the last steps of recycling (see Figure 24.7).[197]

24.4.2.2 Recycling Methodologies of Flame-Retardant Thermosets Based on CANs

In electronics and aerospace, FR fiber-reinforced polyimides are widely employed to prepare composite materials with excellent flame retardancy and lightweight characteristics.[198] Due to their large use, EoL polymer waste based on polyimides is significant and needs to be reduced by the introduction on the market of fully chemical recyclable FR thermosetting polyimides based on CANs. Thermosets based on CANs, and composed of imine bonds can rearrange their polymer network through reversible metathesis reactions.[199] Owing to their features such materials can undergo chemical recycling and provide low-cost and easy methodologies for the regain of precious fillers, as for example CFs when these strategies are applied on EoL CFRPs (see Figure 24.7).[200] To synthesize a self-extinguishing polyimine binder material, Liu et al. used HVP, which is a monomer obtained from vanillin, D230 as a curing agent, and a catalyst-free Schiff base reaction. Then, the prepared system was exploited as a matrix for the fabrication of fully recyclable CFRPs.[201] More in detail, the research group could recover almost 100 wt% of the HVP by immersing HVP/CF-D230 into a mixed acidic solution containing THF. Also, unmodified HVP/D230 could be reduced into small pieces and reshaped by hot pressing at 120°C and 18 MPa for 5 min. At the same operative conditions and within 15 min, the resulting laminates could be easily repaired and reprocessed in complete films.[201] The high reprocessability of HVP/CF-D230 and its bio-based nature offer a clear example of recyclable fire-safe high-performance material. More extensive development of similar systems may significantly reduce the production of wastes related to CFRPs.

Among thermosets, EP is very well-known and widely used in several applications, due to its excellent properties. DGEBA-based ERs are very common and offer high performances, though the monomer is one of the most toxic endocrine-disrupting molecules. One drawback linked to ERs is their poor flame resistance, therefore the addition of halogenated compounds or P-based molecules is mandatory to pass specific requirements.[202] As for other polymer systems, the incorporation of FRs in the epoxy matrix strongly impacts the recycling strategy and its overall procedure. Among inherently FR and recyclable vitrimers, epoxy ones exhibit good recyclability and performances in terms of fire behavior and mechanical properties.[10] The introduction of disulfide bonds, as reversible covalent bonds in the polymer network, confers healing capability to ER even at low temperatures.[203,204] Zhou et al. obtained repairable self-extinguishing CTP-EP/DTDA by the modification of an epoxy matrix with exchangeable disulfide bonds. The research group found that CTP-EP/DTDA could be repaired by hot pressing at 200°C and 10 MPa, resulting in 97.5% of healing efficiency after 10 min of thermomechanical recycling.[205] Similarly to a thermoplastic, CTP-EP/DTDA could be reshaped many times without any detrimental impact on mechanical behavior, which is highly desirable for several applications in aerospace or automotive industries. In 2012, the company Connora Technologies marketed Recyclamine™, a product consisting of ammine reagents composed of sulfide points, which may undergo selective cleavage at mild conditions in an aqueous solution.[206] The use of Recyclamine™ guaranteed the recovery of cleaned fibers, and the possibility to obtain thermoplastic matrix from the reprocessing of the epoxy network in composites.[146,206] Later, the sustainability and cost-efficiency of exploiting Recyclamine™ as a curing agent for bio-based epoxy monomers were also assessed by life cycle analysis.[146,196] The introduction of ester bonds as reversible covalent linkages into a thermoset matrix allows the preparation of highly reprocessable vitrimers, as the dynamic network can topologically rearrange by transesterification,[207] boronic ester bonds or β-hydroxy ester exchange.[208] Such reversible exchanges are usually activated by thermal treatment in the presence of peculiar catalysts.[209] Feng et al. cured a DGEBA resin (EPON-826) using HEMAP (i.e., phosphate ester) as a hardener to synthesize PE-CANs.[210] PE-CANs did not capture the flame even after 10 min of flame application and could be thermomechanically recycled by hot pressing. Thanks to a dynamic β-hydroxy phosphate esters based covalent adaptable network, PE-CANs could be reduced into small pieces and then reshaped at 110°C for 1 h.[210] On the other side, the chemical recycling of PE-CANs involved a free-catalyst route concerning their dissolution in polar solvents (e.g., ethanol) at 80°C for 4 h.[210] The overall features of PE-CANs demonstrate that recyclable vitrimers exhibiting good mechanical and fire performances can be prepared by catalyst-free methodologies based on dynamic β-hydroxy phosphate ester chemistry.

However, limitations and downsides related to CANs are still notable. For example, suitable FRs are necessary to significantly improve the fire behavior of Schiff base CANs and the recovery of polymer matrix still requires low sustainable organic solvents (e.g., THF).[41] CANs based on disulfide exchangeable bonds exhibit good fire performances and mechanical behavior, though their oxidative degradation or chemical recycling may generate the release of toxic sulphureous anhydrides.[211] Similarly to vitrimers made of both boroxines and boronates, CANs composed by phosphate-ester-based networks may undergo hydrolysis, which negatively affects the overall effectiveness of recycling and reprocessing procedures, as water damages crosslinks in the matrix and FR monomers.[212,213]

In a timely manner, the scientific community needs to face such downsides to foster the industrial upscaling of inherently recyclable FR thermosetting CANs and break new ground with innovative recycling strategies.

24.5 CONCLUSIONS AND FUTURE OUTLOOKS

The scientific community is investing a lot of time and efforts in developing effective technologies for the recycling of flame retardant (FR) polymeric wastes. The currently used industrial-scale recycling methodologies for FR EoL plastics and thermosets are energy-intensive and do not always take into account the negative effects arising from the presence of FR moieties in the polymer network, which could lead to toxic pollution.

The direct incineration of plastic wastes containing halogen-based FRs (HFRs) can cause the release of toxic flue gases in the environment and thus does not represent a viable solution. Mechanical recycling of FR plastic waste involves its reuse as plasticizer or upcycling as filler in a new manufacturing process leading to poor quality composite products, exhibiting both low tensile strength and adhesion at the interphase. Furthermore, thermomechanical recycling of waste by melting, molding, or extrusion is often performed without any dehalogenation (e.g., debromination) process. To proceed with mechanical recycling, the removal of halogenated plastic wastes, for example, containing an operational and detectable threshold of 2,000 ppm of total bromine, is now mandatory to fulfill stringent restrictions on the use of HFRs. High-resolution near-infrared devices have been found very useful in screening plastics containing HFRs, becoming crucial tools during the sorting process. In addition to this separation, the native plastics can be dehalogenated by solvent-washing treatments exploiting specific and selective solvents for end-of-life products. Tertiary recycling allows for the conversion of free-POP halogenated plastic wastes into energy or lower molecular weight products. There are several tertiary recycling processes (e.g., photolysis, thermolysis, gasification, solvent-assisted depolymerization) that can be implemented for the reprocessing of plastic wastes, though one of the major challenges of effective trapping of any residual halogenated by-products during pyrolysis (thermochemical recycling) still remains. Halogen-free FR (HFFR) thermoplastic wastes are usually mechanically recycled as this strategy allows to keep low costs and high sustainability, and the fire behavior and mechanical performances of the reprocessed products appear almost unvaried. However, HFFR thermoplastics may undergo thermal degradation during recycling, resulting in a detrimental impact on the mechanical properties and flame retardancy of such materials. To design more efficient mechanical recycling processes of HFFR thermoplastics, repeated extrusion cycles were carried out to study the influence of FRs on reprocessing. The optimal operative conditions for the recycling of a specific plastic waste stream depend on the chemical composition of the polymer matrix and its FRs. Tertiary recycling involving chemical and thermochemical methods may be a valuable strategy to overcome the difficulties linked to the mechanical recycling of mixed waste streams composed of HFFR thermoplastics.

A better understanding of the influence of FRs on the recycling of thermoset waste is getting crucial. Currently, down-scaling treatments transform the thermosetting waste into small sized flakes, which are mechanically recycled by upcycling them as reinforcing fillers for the manufacturing of polymer-based composites via thermomechanical processes. However, thermochemical recycling is energy-demanding and often does not allow the recovery of the polymer matrix, which impedes the design of closed-loop strategies. Solvolysis of FR thermosetting wastes enables the separation of more resistant fibers (e.g., carbon fibers) but requires extreme operative conditions, converting the matrix and FRs into different chemical species that are not easy to be separated and reused as raw materials. On the contrary, inherently FR and recyclable EoL thermosets based on CANs make the full recovery of FR polymer matrix and its additives possible, but their rapid steady-state processing and upscaling still require further detailed investigations. In conclusion, these crucial challenges need to be addressed, however, development of sustainable FR polymers designed for recycling is the key to meet the needs of future sustainable society.

24.5.1 Author Statement

The manuscript was written through the contributions of all authors. All authors have approved the final version of the manuscript.

24.5.2 Declaration of Competing Interest

The authors declare no conflict of interest. We confirm that this submission is original, it is not published elsewhere, and not under consideration for the publication elsewhere.

List of Abbreviations

acrylonitrile-butadiene-styrene	**ABS**
aflammit PCO 90	**AF**
ammonium polyphosphate	**APP**
bisphenol A bis(diphenyl phosphate),	**BDP**
brominated epoxy resin	**BER**
brominated FRs	**BFRs**
bis(2,4,6-tribromophenoxy)ethane	**BTBPE**
compound annual growth rate	**CAGR**
covalent adaptable networks	**CANs**
carbon fibers	**CF**
cyclotriphosphazene-based epoxy vitrimer	**CTP-EP/DTDA**
copper	**Cu**
commercial polyetherimide	**D230**
4,4-diamino diphenylmethane,	**DDM**
bisphenol A diglycidyl ether	**DGEBA**
phenylphosphonate based epoxy monomer	**DGEP**
dimethylcyclohexylamin	**DMCHA**
9,10-dihydro-9-oxa-10-phosphaphenanthrene-10-oxide	**DOPO**
6-((1-oxido-2,6,7trioxa-1-phosphabicyclo[2.2.2] octan-4-yl)methoxy)dibenzo[c,e][1,2] oxaphosphinine 6-oxide	**DOPO-PEPA**
epoxy composite	**EC**
electric and electronic equipment	**EEE**
end-of-life	**EoL**
epoxy resin	**ER**
epoxy thermoset	**ET**
flame retardant	**FR**
acrylonitrile-butadiene-styrene	**ABS**
flexible printed circuits	**FPCs**
glass-fiber-einforced composites	**GFRCs**
Hexabromocyclododecane	**HBCD**
halogenated compound	**HCs**
phosphoric acid 2-hydroxyethyl methacrylate ester	**HEMAP**
halogen-free	**HF**
halogen-free FRs	**HFFRs**
halogenated FR	**HFR**
high impact polystyrene	**HIPS**
hexasubstituted cyclotriphosphazene	**HVP**
fiber-reinforced polyimine composite	**HVP/CF-D230**
potassium diphenylsulfone sulfonate	**KSS**
liquid crystal display	**LCD**
million tons	**Mt**
National Electrical Manufacturers Association	**NEMA**
nitrogen	**N**
phosphorus	**P**
Polyamide-6	**PA-6**
polybrominated biphenyls	**PBBs**
polybrominated dibenzo-p-dioxins	**PBDDs**
polybrominated diphenyl ethers	**PBDEs**
polybrominated dibenzofurans	**PBDFs**
polycarbonate	**PC**
polyethylene	**PE**
β-hydroxy phosphate-esters-based thermoset polymers	**PE-CANs**
phosphorus FR	**PFRs**
polyethylene terephthalate	**PET**
polyisocyanurates	**PIRs**
pyromellitic dianhydride	**PMDA**
polymeric materials	**PMs**
persistent organic pollutants	**POPs**
polypropylene	**PP**
poly(p-phenylene oxide)	**PPO**
polystyrene	**PS**
polyurethanes	**PUs**
polyvinyl chloride	**PVC**
printed wiring board	**PWB**
resorcinol bis(diphenylphosphate	**RDP**
tannic acid	**TA**
tetrabromobisphenol A	**TBBP-A**
tris(2-chloropropyl) phosphate	**TCPP**
tetrahydrofurane	**THF**
technology readiness level	**TRL**
unsatured polyester resin	**UPR**
waste from electrical and electronic equipment	**WEEE**
waste printed wiring boards	**WPBs**
wood-plastic composites	**WPC**

REFERENCES

1. Laoutid, F.; Ferry, L.; Lopez-Cuesta, J. M.; Crespy, A., Flame-retardant action of red phosphorus/magnesium oxide and red phosphorus/iron oxide compositions in recycled PET. *Fire Mater.* **2006,** *30* (5), 343–358.
2. Troitzsch, J.; Antonatus, E., *Plastics Flammability Handbook: Principles, Regulations, Testing, and Approval*, Carl Hanser Verlag GmbH Co KG, München: 2021.

3. Camino, G.; Costa, L., Performance and mechanisms of fire retardants in polymers—A review. *Polym. Degrad. Stabil.* **1988,** *20* (3–4), 271–294.
4. Bourbigot, S.; Duquesne, S., Fire retardant polymers: Recent developments and opportunities. *J. Mater. Chem.* **2007,** *17* (22), 2283–2300.
5. Delva, L.; Hubo, S.; Cardon, L.; Ragaert, K., On the role of flame retardants in mechanical recycling of solid plastic waste. *Waste Manage.* **2018,** *82*, 198–206.
6. Majewski, P.; Florin, N.; Jit, J.; Stewart, R. A., End-of-life policy considerations for wind turbine blades. *Renew. Sust. Energ. Rev.* **2022,** *164*, 112538.
7. Ngo, T., Development of sustainable flame-retardant materials. *Green Mater.* **2020,** *8* (3), 101–122.
8. Kim, Y. O.; Jung, Y. C., Flame Retardancy of Bioepoxy Polymers, Their Blends, and Composites. In *Bio-Based Epoxy Polymers, Blends and Composites: Synthesis, Properties, Characterization and Applications*, WILEY-VCH GmbH, Weinheim: 2021; pp 285–307.
9. Post, W.; Susa, A.; Blaauw, R.; Molenveld, K.; Knoop, R. J. I., A review on the potential and limitations of recyclable thermosets for structural applications. *Polym. Rev.* **2020,** *60* (2), 359–388.
10. Memon, H.; Wei, Y.; Zhu, C., Recyclable and reformable epoxy resins based on dynamic covalent bonds–present, past, and future. *Polym. Test.* **2022,** *105*, 107420–107420.
11. Dasari, A.; Yu, Z.-Z.; Cai, G.-P.; Mai, Y.-W., Recent developments in the fire retardancy of polymeric materials. *Prog. Polym. Sci.* **2013,** *38* (9), 1357–1387.
12. Schartel, B., Phosphorus-based flame retardancy mechanisms—Old hat or a starting point for future development? *Materials* **2010,** *3* (10).
13. Hörold, S., Phosphorus-Based and Intumescent Flame Retardants. In *Polymer Green Flame Retardants*, Elsevier, Amsterdam: 2014; pp 221–254.
14. Shan, X.; Jiang, K.; Li, J.; Song, Y.; Han, J.; Hu, Y., Preparation of β-cyclodextrin inclusion complex and its application as an intumescent flame retardant for epoxy. *Polymers* **2019,** *11* (1), 71–71.
15. Rothon, R. N.; Hornsby, P. R., Flame retardant effects of magnesium hydroxide. *Polym. Degrad. Stabil.* **1996,** *54* (2–3), 383–385.
16. Martins, M.; Pereira, C. *A Study on the Effect of Nano-Magnesium Hydroxide on the Flammability of Epoxy Resins*, Trans Tech Publication, Bäch: 2009; pp 72–78.
17. Guadagno, L.; Raimondo, M.; Vittoria, V.; Vertuccio, L.; Naddeo, C.; Russo, S.; De Vivo, B.; Lamberti, P.; Spinelli, G.; Tucci, V., Development of epoxy mixtures for application in aeronautics and aerospace. *RSC Adv.* **2014,** *4* (30), 15474–15488.
18. Toldy, A.; Szolnoki, B.; Marosi, G., Flame retardancy of fiber-reinforced epoxy resin composites for aerospace applications. *Polym. Degrad. Stab.* **2011,** *96* (3), 371–376.
19. Mensah, R. A.; Shanmugam, V.; Narayanan, S.; Renner, J. S.; Babu, K.; Neisiany, R. E.; Försth, M.; Sas, G.; Das, O., A review of sustainable and environment-friendly flame retardants used in plastics. *Polym. Test.* **2022**, 107511.
20. Shanmugam, V.; Das, O.; Neisiany, R. E.; Babu, K.; Singh, S.; Hedenqvist, M. S.; Berto, F.; Ramakrishna, S., Polymer recycling in additive manufacturing: An opportunity for the circular economy. *Mater. Circ. Econ.* **2020,** *2* (1), 1–11.
21. Broeren, M. L. M.; Molenveld, K.; van den Oever, M. J. A.; Patel, M. K.; Worrell, E.; Shen, L., Early-stage sustainability assessment to assist with material selection: A case study for biobased printer panels. *J. Clean. Prod.* **2016,** *135*, 30–41.
22. Salfate, G.; Sánchez, J., Rare earth elements uptake by synthetic polymeric and cellulose-based materials: A review. *Polymers* **2022,** *14* (21), 4786.
23. Al-Salem, S. M.; Antelava, A.; Constantinou, A.; Manos, G.; Dutta, A., A review on thermal and catalytic pyrolysis of plastic solid waste (PSW). *J. Environ. Manage.* **2017,** *197*, 177–198.
24. Charitopoulou, M. A.; Kalogiannis, K. G.; Lappas, A. A.; Achilias, D. S., Novel trends in the thermo-chemical recycling of plastics from WEEE containing brominated flame retardants. *Environ. Sci. Pollut. Res.* **2021,** *28* (42), 59190–59213.
25. Al-Salem, S. M.; Lettieri, P.; Baeyens, J., The valorization of plastic solid waste (PSW) by primary to quaternary routes: From re-use to energy and chemicals. *Prog. Energy Combust. Sci.* **2010,** *36* (1), 103–129.
26. Ügdüler, S.; Van Geem, K. M.; Roosen, M.; Delbeke, E. I. P.; De Meester, S., Challenges and opportunities of solvent-based additive extraction methods for plastic recycling. *Waste Manage.* **2020,** *104*, 148–182.
27. Moscatelli, D.; Pelucchi, M., Towards Circular Economy: Closing the Loop with Chemical Recycling of Solid Plastic Waste. In *Advances in Chemical Engineering*, Academic Press, Elsevier, Amsterdam: 2022; pp 2–313.
28. Kol, R.; Nachtergaele, P.; De Somer, T.; D'hooge, D. R.; Achilias, D. S.; De Meester, S., Toward more universal prediction of polymer solution viscosity for solvent-based recycling. *Ind. Eng. Chem. Res.* **2022,** *61* (30), 10999–11011.
29. Tabelin, C. B.; Park, I.; Phengsaart, T.; Jeon, S.; Villacorte-Tabelin, M.; Alonzo, D.; Yoo, K.; Ito, M.; Hiroyoshi, N., Copper and critical metals production from porphyry ores and E-wastes: A review of resource availability, processing/recycling challenges, socio-environmental aspects, and sustainability issues. *Resour. Conserv. Recycl.* **2021,** *170*, 105610.
30. Tullo, A. H., Companies are placing big bets on plastics recycling. Are the odds in their favor. *Chem. Eng. News* **2020,** *98*, 39.
31. Xiong, P.; Yan, X.; Zhu, Q.; Qu, G.; Shi, J.; Liao, C.; Jiang, G., A review of environmental occurrence, fate, and toxicity of novel brominated flame retardants. *Environ. Sci. Technol. Lett.* **2019,** *53* (23), 13551–13569.
32. Zuiderveen, E. A. R.; Slootweg, J. C.; de Boer, J., Novel brominated flame retardants-A review of their occurrence in indoor air, dust, consumer goods and food. *Chemosphere* **2020,** *255*, 126816.
33. Agency, E. C., *Regulatory Strategy for Flame Retardants*, European Chemical Agency (ECHA), Helsinki: 2023; pp 1–84. https://echa.europa.eu/documents/10162/2082415/flame_retardants_strategy_en.pdf/9dd56b7e-4b62-e31b-712f-16cc51d0e724?t=1679045593845 (accessed 26/01/2024).
34. Betts, K. S., *New Thinking on Flame Retardants*. National Institute of Environmental Health Sciences, EHP Publishing, Durham: 2008; Vol. 116, pp A210–A213. https://doi.org/10.1289/ehp.116-a210.
35. Lase, I. S.; Ragaert, K.; Dewulf, J.; De Meester, S., Multivariate input-output and material flow analysis of current and future plastic recycling rates from waste electrical and electronic equipment: The case of small household appliances. *Resour. Conserv. Recycl.* **2021,** *174*, 105772.
36. Vahabi, H.; Sonnier, R.; Ferry, L., Effects of ageing on the fire behaviour of flame-retarded polymers: A review. *Polym. Int.* **2015,** *64* (3), 313–328.

37. Nishida, H.; Fan, Y.; Mori, T.; Oyagi, N.; Shirai, Y.; Endo, T., Feedstock recycling of flame-resisting poly (lactic acid)/aluminum hydroxide composite to L, L-lactide. *Ind. Eng. Chem. Res.* **2005,** *44* (5), 1433–1437.
38. Bascucci, C.; Duretek, I.; Lehner, S.; Holzer, C.; Gaan, S.; Hufenus, R.; Gooneie, A., Investigating thermomechanical recycling of poly(ethylene terephthalate) containing phosphorus flame. *Polym. Degrad. Stabil.* **2022,** *195.*
39. Morita, N.; Kawabata, Y.; Wajima, T.; Saito, A. T.; Nakagome, H. Effect of the Bromine-Based Flame Retardant Plastic Pyrolysis of Hydrotalcite. In *MATEC Web of Conferences, 3rd International Conference on Chemical and Food Engineering*, EDP Sciences, Les Ulis: 2016; Vol. 62, p 5002. https://doi.org/10.1051/matecconf/20166205002.
40. Vasile, C.; Brebu, M. A.; Karayildirim, T.; Yanik, J.; Darie, H., Feedstock recycling from plastics and thermosets fractions of used computers. II. Pyrolysis oil upgrading. *Fuel* **2007,** *86* (4), 477–485.
41. Wang, S.; Ma, S.; Li, Q.; Yuan, W.; Wang, B.; Zhu, J., Robust, fire-safe, monomer-recovery, highly malleable thermosets from renewable bioresources. *Macromolecules* **2018,** *51* (20), 8001–8012.
42. Phonthamachai, N.; Chia, H.; Li, X.; Wang, F.; Tjiu, W. W.; He, C., Solvent-free one-pot synthesis of high performance silica/epoxy nanocomposites. *Polymer* **2010,** *51* (23), 5377–5384.
43. Geyer, R.; Jambeck, J. R.; Law, K. L., Production, use, and fate of all plastics ever made. *Sci Adv* **2017,** *3* (7), e1700782.
44. Miliute-Plepiene, J.; Fråne, A.; Almasi, A. M., Overview of polyvinyl chloride (PVC) waste management practices in the Nordic countries. *Clean. Eng. Technol.* **2021,** *4*, 100246.
45. Siddique, S.; Jang, J. G., Mechanical properties, microstructure, and chloride content of alkali-activated fly ash paste made with sea water. *Materials* **2020,** *13* (6).
46. Yalçin, H.; Koç, T.; Pamuk, V., Hydrogen and bromine production from concentrated sea-water. *Int. J. Hydrog. Energy* **1997,** *22* (10), 967–970.
47. Susan, S., Halogenated flame retardants: Do the fire safety benefits justify the risks? *Rev. Environ. Health* **2010,** *25* (4), 261–306.
48. Zhang, M.; Buekens, A.; Li, X., Brominated flame retardants and the formation of dioxins and furans in fires and combustion. *J. Hazard. Mater.* **2016,** *304*, 26–39.
49. Leonetti, C.; Butt, C. M.; Hoffman, K.; Hammel, S. C.; Miranda, M. L.; Stapleton, H. M., Brominated flame retardants in placental tissues: Associations with infant sex and thyroid hormone endpoints. *Environ. Health* **2016,** *15* (1), 113.
50. Imparato, C.; Bifulco, A.; Silvestri, B.; Vitiello, G., Recent advances in endocrine disrupting compounds degradation through metal oxide-based nanomaterials. *Catalysts* **2022,** *12* (3), 289–289.
51. Suzuki, G.; Someya, M.; Takahashi, S.; Tanabe, S.; Sakai, S.; Takigami, H., Dioxin-like activity in Japanese indoor dusts evaluated by means of in vitro bioassay and instrumental analysis: Brominated dibenzofurans are an important contributor. *Environ. Sci. Technol.* **2010,** *44* (21), 8330–8336.
52. Sharkey, M.; Harrad, S.; Abou-Elwafa Abdallah, M.; Drage, D. S.; Berresheim, H., Phasing-out of legacy brominated flame retardants: The UNEP Stockholm convention and other legislative action worldwide. *Environ. Int.* **2020,** *144*, 106041.
53. Gottschall, N.; Topp, E.; Edwards, M.; Payne, M.; Kleywegt, S.; Lapen, D. R., Brominated flame retardants and perfluoroalkyl acids in groundwater, tile drainage, soil, and crop grain following a high application of municipal biosolids to a field. *Sci. Total Environ.* **2017,** *574*, 1345–1359.
54. Directorate-General for, C., *Towards a Sustainable Europe by 2030: Reflection Paper.* Publications Office of the European Union, Wetstraat, Brussels: 2019; pp 1–132. https://op.europa.eu/en/publication-detail/-/publication/3b096b37-300a-11e9-8d04-01aa75ed71a1/language-en/format-PDF (accessed 25/01/24).
55. Glarborg, P., Chapter 11 — Detailed Kinetic Mechanisms of Pollutant Formation in Combustion Processes. In *Computer Aided Chemical Engineering*, Faravelli, T.; Manenti, F.; Ranzi, E., Eds. Computer Aided Chemical Engineering Series, Elsevier, Amsterdam: 2019; Vol. 45, pp 603–645. https://doi.org/10.1016/B978-0-444-64087-1.00011-5.
56. Oram, D. E.; Ashfold, M. J.; Laube, J. C.; Gooch, L. J.; Humphrey, S.; Sturges, W. T.; Leedham Elvidge, E. C.; Forster, G. L.; Harris, N. R. P.; Mead, M. I.; Samah, A. A.; Phang, S. M.; Ou-Yang, C.-F.; Lin, N.-H.; Wang, J.-L.; Baker, A. K.; Brenninkmeijer, C. A. M.; Sherry, D., A growing threat to the ozone layer from short-lived anthropogenic chlorocarbons. *Atmospheric Chem. Phys.* **2017,** *17* (19), 11929–11941.
57. Söderström, G.; Marklund, S., PBCDD and PBCDF from incineration of waste-containing brominated flame retardants. *EEnviron. Sci. Technol.* **2002,** *36* (9), 1959–1964.
58. Persson, U.; Münster, M., Current and future prospects for heat recovery from waste in European district heating systems: A literature and data review. *Energy* **2016,** *110*, 116–128.
59. Summers, J. W.; Mikofalvy, B. K.; Boo, H. K.; Krogstie, J. M.; Sell, W. A.; Rodriguez, J. C., Examples of recycled vinyl products. *J. Vinyl Technol.* **1992,** *14* (3), 166–170.
60. Janajreh, I.; Alshrah, M.; Zamzam, S., Mechanical recycling of PVC plastic waste streams from cable industry: A case study. *Sustain. Cities Soc.* **2015,** *18*, 13–20.
61. Delva, L.; Hubo, S.; Cardon, L.; Ragaert, K., On the role of flame retardants in mechanical recycling of solid plastic waste. *Waste Manage.* **2018,** *82*, 198–206.
62. Beigbeder, J.; Perrin, D.; Mascaro, J.-F.; Lopez-Cuesta, J.-M., Study of the physico-chemical properties of recycled polymers from waste electrical and electronic equipment (WEEE) sorted by high resolution near infrared devices. *Resour. Conserv. Recycl.* **2013,** *78*, 105–114.
63. Cagnetta, G.; Zhang, K.; Zhang, Q.; Huang, J.; Yu, G., Mechanochemical pre-treatment for viable recycling of plastic waste containing haloorganics. *Waste Manage.* **2018,** *75*, 181–186.
64. Cagnetta, G.; Liu, H.; Zhang, K.; Huang, J.; Wang, B.; Deng, S.; Wang, Y.; Yu, G., Mechanochemical conversion of brominated POPs into useful oxybromides: A greener approach. *Sci Rep.* **2016,** *6* (1), 28394.
65. Birke, V.; Schütt, C.; Ruck, W. K. L., Small Particle Size Magnesium in One-Pot Grignard-Zerewitinoff-Like Reactions Under Mechanochemical Conditions: On the Kinetics of Reductive Dechlorination of Persistent Organic Pollutants (POPs). In *Environmental Applications of Nanoscale and Microscale Reactive Metal Particles*, American Chemical Society, ACS Symposium Series, Washington, DC: 2009; Vol. 1027, pp 39–54.
66. Baláž, P.; Achimovičová, M.; Baláž, M.; Billik, P.; Cherkezova-Zheleva, Z.; Criado, J. M.; Delogu, F.;

Dutková, E.; Gaffet, E.; Gotor, F. J.; Kumar, R.; Mitov, I.; Rojac, T.; Senna, M.; Streletskii, A.; Wieczorek-Ciurowa, K., Hallmarks of mechanochemistry: From nanoparticles to technology. *Chem. Soc. Rev.* **2013,** *42* (18), 7571–7637.
67. Yamawaki, T., The gasification recycling technology of plastics WEEE containing brominated flame retardants. *Fire Mater.* **2003,** *27* (6), 315–319.
68. Kakuta, Y.; Hirano, K.; Sugano, M.; Mashimo, K., Study on chlorine removal from mixture of waste plastics. *Waste Manage.* **2008,** *28* (3), 615–621.
69. Bernal, J. D.; Dasgupta, D. R.; Mackay, A. L., The oxides and hydroxides of iron and their structural inter-relationships. *Clay Miner. Bull.* **1959,** *4* (21), 15–30.
70. Ling, S.; Huang, K.; Tariq, M.; Wang, Y.; Chen, X.; Zhang, W.; Lin, K.; Zhou, B., Photodegradation of novel brominated flame retardants (NBFRs) in a liquid system: Kinetics and photoproducts. *Chem. Eng. J.* **2019,** *362*, 938–946.
71. Mathur, R. S.; Banstola, B.; Lopez, K. M.; Beach, M. W.; De Schryver, D., Assessing the environmental impact due to photolytic degradation of ethane-bis(pentabromophenyl) in plastics. *Chemosphere* **2023,** *320*, 138063.
72. Khaled, A.; Rivaton, A.; Richard, C.; Jaber, F.; Sleiman, M., Phototransformation of plastic containing brominated flame retardants: Enhanced fragmentation and release of photoproducts to water and air. *Environ. Sci. Technol.* **2018,** *52* (19), 11123–11131.
73. Koch, C.; Nachev, M.; Klein, J.; Köster, D.; Schmitz, O. J.; Schmidt, T. C.; Sures, B., Degradation of the polymeric brominated flame retardant "polymeric FR" by heat and UV exposure. *Environ. Sci. Technol.* **2019,** *53* (3), 1453–1462.
74. Evangelopoulos, P.; Arato, S.; Persson, H.; Kantarelis, E.; Yang, W., Reduction of brominated flame retardants (BFRs) in plastics from waste electrical and electronic equipment (WEEE) by solvent extraction and the influence on their thermal decomposition. *Waste Manage.* **2019,** *94*, 165–171.
75. Charitopoulou, M. A.; Papadopoulou, L.; Achilias, D. S., Microwave-assisted extraction as an effective method for the debromination of brominated flame retarded polymeric blends with a composition that simulates the plastic part of waste electric and electronic equipment (WEEE). *Sustain. Chem. Pharm.* **2022,** *29*, 100790.
76. Yuan, Y.; Yu, B.; Shi, Y.; Mao, L.; Xie, J.; Pan, H.; Liu, Y.; Wang, W., Insight into hyper-branched aluminum phosphonate in combination with multiple phosphorus synergies for fire-safe epoxy resin composites. *Polymers (Basel, Switz.)* **2020,** *12* (1), 64.
77. Schlummer, M.; Maurer, A.; Wagner, S.; Berrang, A.; Fell, T.; Knappich, F., Recycling of flame retarded waste polystyrene foams (EPS and XPS) to PS granules free of hexabromocyclododecane (HBCDD). *Adv. Recycl. Waste Manage.* **2017,** *2* (5).
78. Cardamone, G. F.; Ardolino, F.; Arena, U., Can plastics from end-of-life vehicles be managed in a sustainable way? *Sustain. Prod. Consum.* **2022,** *29*, 115–127.
79. Nnorom, I. C.; Osibanjo, O., Sound management of brominated flame retarded (BFR) plastics from electronic wastes: State of the art and options in Nigeria. *Resour. Conserv. Recycl.* **2008,** *52* (12), 1362–1372.
80. Wang, Y.; Zhang, F.-S., Degradation of brominated flame retardant in computer housing plastic by supercritical fluids. *J. Hazard. Mater.* **2012,** *205–206*, 156–163.
81. Xiu, F.-R.; Qi, Y.; Zhang, F.-S., Recovery of metals from waste printed circuit boards by supercritical water pretreatment combined with acid leaching process. *Waste Manage.* **2013,** *33* (5), 1251–1257.
82. Zhang, C. C.; Zhang, F. S., Enhanced dehalogenation and coupled recovery of complex electronic display housing plastics by sub/supercritical CO(2). *J. Hazard. Mater.* **2020,** *382*, 121140.
83. Ma, C.; Yu, J.; Wang, B.; Song, Z.; Xiang, J.; Hu, S.; Su, S.; Sun, L., Chemical recycling of brominated flame retarded plastics from e-waste for clean fuels production: A review. *Renew. Sust. Energ. Rev.* **2016,** *61*, 433–450.
84. Mascia, L.; Prezzi, L.; Haworth, B., Substantiating the role of phase bicontinuity and interfacial bonding in epoxy-silica nanocomposites. *J. Mater. Sci.* **2006,** *41* (4), 1145–1155.
85. Morgan, A. B.; Gilman, J. W., An overview of flame retardancy of polymeric materials: Application, technology, and future directions. *Fire Mater.* **2013,** *37* (4), 259–279.
86. Zhang, G.; Ding, P.; Zhang, M.; Qu, B., Synergistic effects of layered double hydroxide with hyperfine magnesium hydroxide in halogen-free flame retardant EVA/HFMH/LDH nanocomposites. *Polym. Degrad. Stabil.* **2007,** *92* (9), 1715–1720.
87. Basel, C. L., *Clariant Making Its Additives Fit for Reuse, Halogen-Free Flame Retardant Allowing Re- and Up-Cycling of PA6 & PA66*, Clariant Ltd, Muttenz: 2019. https://www.clariant.com/en/Corporate/News/2019/08/Clariant-making-its-Additives-fit-for-reuse-halogenfree-flame-retardant-allowing-re-and-upcycling-of (accessed 25/01/24).
88. Quattrosoldi, S.; Lotti, N.; Soccio, M.; Schick, C.; Androsch, R., Stability of crystal nuclei of poly (butylene isophthalate) formed near the glass transition temperature. *Polymers* **2020,** *12*, 1099–1099.
89. Delva, L.; Hubo, S.; Cardon, L.; Ragaert, K., On the role of flame retardants in mechanical recycling of solid plastic waste. *Waste Manage.* **2018,** *82*, 198–206.
90. Vilaplana, F.; Karlsson, S., Quality concepts for the improved use of recycled polymeric materials: A review. *Macromol. Mater. Eng.* **2008,** *293* (4), 274–297.
91. Sharkey, M.; Harrad, S.; Abdallah, M. A.-E.; Drage, D. S.; Berresheim, H., Phasing-out of legacy brominated flame retardants: The UNEP Stockholm convention and other legislative action worldwide. *Environ. Int.* **2020,** *144*, 106041.
92. Wäger, P. A.; Schluep, M.; Müller, E.; Gloor, R., RoHS regulated substances in mixed plastics from waste electrical and electronic equipment. *Environ. Sci. Technol.* **2012,** *46* (2), 628–635.
93. Ragaert, K.; Delva, L.; Van Geem, K., Mechanical and chemical recycling of solid plastic waste. *Waste Manage.* **2017,** *69*, 24–58.
94. Tange, L.; Van Houwelingen, J. A.; Peeters, J. R.; Vanegas, P., Recycling of flame retardant plastics from WEEE, technical and environmental challenges. *Adv. Prod. Eng. Manag.* **2013,** *8* (2), 67–77.
95. Taurino, R.; Pozzi, P.; Zanasi, T., Facile characterization of polymer fractions from waste electrical and electronic equipment (WEEE) for mechanical recycling. *Waste Manage.* **2010,** *30* (12), 2601–2607.
96. Peeters, J. R.; Vanegas, P.; Tange, L.; Van Houwelingen, J.; Duflou, J. R., Closed loop recycling of plastics containing Flame Retardants. *Resour. Conserv. Recycl.* **2014,** *84*, 35–43.
97. Van der Veen, I.; de Boer, J., Phosphorus flame retardants: Properties, production, environmental occurrence, toxicity and analysis. *Chemosphere* **2012,** *88* (10), 1119–1153.
98. Almeras, X.; Bras, M. L.; Hornsby, P.; Bourbigot, S.; Marosi, G.; Anna, P.; Delobel, R., Artificial weathering and recycling effect on intumescent polypropylenebased blends. *J. Fire Sci.* **2004,** *22* (2), 143–161.

99. Almeras, X.; Le Bras, M.; Hornsby, P.; Bourbigot, S.; Marosi, G.; Anna, P.; Delobel, R., Effect of recycling on fire retardancy of intumescent polypropylene based blends. *Prog. Rubber Plast. Technol.* **2004,** *20* (1), 25–45.
100. Dawson, R. B.; Landry, S. D. Recyclability of Flame Retardant HIPS, PC/ABS, and PPO/HIPS Used in Electrical and Electronic Equipment. In *Proceedings of the 2005 IEEE International Symposium on Electronics and the Environment*, IEEE, New York: 2005; pp 77–82.
101. Statler, D.; Stajduhar, E.; Gupta, R. K., Flame retardancy of polycarbonate upon repeated recycling. *J. Fire Sci.* **2008,** *26* (4), 331–350.
102. Casetta, M.; Delaval, D.; Traisnel, M.; Bourbigot, S., Influence of the recycling process on the fire-retardant properties of PP/EPR blends. *Macromol. Mater. Eng.* **2011,** *296* (6), 494–505.
103. Imai, T.; Hamm, S.; Rothenbacher, K. P., Comparison of the recyclability of flame-retarded plastics. *Environ. Sci. Technol.* **2003,** *37* (3), 652–656.
104. Turku, I.; Kärki, T., Accelerated weathering of fire-retarded wood–polypropylene composites. *Compos. Part A Appl. Sci.* **2016,** *81*, 305–312.
105. Gooneie, A.; Simonetti, P.; Rupper, P.; Nazir, R.; Jovic, M.; Gaan, S.; Heuberger, M. P.; Hufenus, R., Stabilizing effects of novel phosphorus flame retardant on PET for high-temperature applications. *Mater. Lett.* **2020,** *276.*
106. Gooneie, A.; Simonetti, P.; Salmeia, K. A.; Gaan, S.; Hufenus, R.; Heuberger, M. P., Enhanced PET processing with organophosphorus additive: Flame retardant products with added-value for recycling. *Polym. Degrad. Stabil.* **2019,** *160*, 218–228.
107. Bethke, C.; Goedderz, D.; Weber, L.; Standau, T.; Döring, M.; Altstädt, V., Improving the flame-retardant property of bottle-grade PET foam made by reactive foam extrusion. *J. Appl. Polym. Sci.* **2020,** *137* (35), 49042.
108. Rothon, R. N. *Magnesium Hydroxide: New Products, Processes and Applications*, Intertech, Portland, ME/ Functional Effect Filler, Berlin: 2000.
109. Soni, V. K.; Singh, G.; Vijayan, B. K.; Chopra, A.; Kapur, G. S.; Ramakumar, S. S. V., Thermochemical recycling of waste plastics by pyrolysis: A review. *Energy Fuels* **2021,** *35* (16), 12763–12808.
110. Yang, X.; Sun, L.; Xiang, J.; Hu, S.; Su, S., Pyrolysis and dehalogenation of plastics from waste electrical and electronic equipment (WEEE): A review. *Waste Manage.* **2013,** *33* (2), 462–473.
111. Vollmer, I.; Jenks, M. J. F.; Roelands, M. C. P.; White, R. J.; van Harmelen, T.; de Wild, P.; van der Laan, G. P.; Meirer, F.; Keurentjes, J. T. F.; Weckhuysen, B. M., Beyond mechanical recycling: Giving new life to plastic waste. *Angew. Chem. Int. Ed.* **2020,** *59* (36), 15402–15423.
112. Perret, B.; Schartel, B., The effect of different impact modifiers in halogen-free flame retarded polycarbonate blends—I. Pyrolysis. *Polym. Degrad. Stabil.* **2009,** *94* (12), 2194–2203.
113. Perret, B.; Schartel, B., The effect of different impact modifiers in halogen-free flame retarded polycarbonate blends—II. Fire behaviour. *Polym. Degrad. Stabil.* **2009,** *94* (12), 2204–2212.
114. Xu, G.; Wang, Q., Chemically recyclable polymer materials: Polymerization and depolymerization cycles. *Green Chem.* **2022,** *24* (6), 2321–2346.
115. Miao, Y.; von Jouanne, A.; Yokochi, A., Current technologies in depolymerization process and the road ahead. *Polymers* **2021,** *13* (3), 449.
116. Wang, L.; Liu, Y.; Lu, H.; Huang, Z., Recycling of phosphorus-containing plastic based on the dual effects of switchable hydrophilicity solvents. *Chemosphere* **2020,** *259*, 127402.
117. Bucknall, D. G., Plastics as a materials system in a circular economy. *Philos. Trans. Royal Soc. A* **2020,** *378* (2176), 20190268–20190268.
118. Arrabiyeh, P. A.; May, D.; Eckrich, M.; Dlugaj, A. M., An overview on current manufacturing technologies: Processing continuous rovings impregnated with thermoset resin. *Polym. Compos.* **2021,** *42* (11), 5630–5655.
119. Rajmohan, K. V. S.; Ramya, C.; Viswanathan, M. R.; Varjani, S., Plastic pollutants: Effective waste management for pollution control and abatement. *Curr. Opin. Environ. Sci. Health.* **2019,** *12*, 72–84.
120. Qureshi, J., A review of recycling methods for fibre reinforced polymer composites. *Sustainability* **2022,** *14* (24), 16855–16855.
121. Lazarevic, D.; Aoustin, E.; Buclet, N.; Brandt, N., Plastic waste management in the context of a European recycling society: Comparing results and uncertainties in a life cycle perspective. *Resour. Conserv. Recycl.* **2010,** *55* (2), 246–259.
122. Morici, E.; Dintcheva, N. T., Recycling of thermoset materials and thermoset-based composites: Challenge and opportunity. *Polymers* **2022,** *14* (19), 4153–4153.
123. Pickering, K. L.; Beckermann, G. W.; Alam, S. N.; Foreman, N. J., Optimising industrial hemp fibre for composites. *Compos. Part A Appl. Sci.* **2007,** *38* (2), 461–468.
124. Guo, L.; Xu, L.; Ren, Y.; Shen, Z.; Fu, R.; Xiao, H.; Liu, J., Research on a two-step pyrolysis-oxidation process of carbon fiber-reinforced epoxy resin-based composites and analysis of product properties. *J. Environ. Chem. Eng.* **2022,** *10* (3), 107510.
125. Ma, S.; Webster, D. C., Degradable thermosets based on labile bonds or linkages: A review. *Prog. Polym. Sci.* **2018,** *76*, 65–110.
126. Krishnakumar, B.; Sanka, R. V. S. P.; Binder, W. H.; Parthasarthy, V.; Rana, S.; Karak, N., Vitrimers: Associative dynamic covalent adaptive networks in thermoset polymers. *Chem. Eng. J.* **2020,** *385*, 123820–123820.
127. Alabiso, W.; Schlögl, S., The impact of vitrimers on the industry of the future: Chemistry, properties and sustainable forward-looking applications. *Polymers* **2020,** *12* (8), 1660–1660.
128. Kissounko, D. A.; Taynton, P.; Kaffer, C., New material: Vitrimers promise to impact composites. *Reinf. Plast.* **2018,** *62* (3), 162–166.
129. Klingler, W. W.; Bifulco, A.; Polisi, C.; Huang, Z.; Gaan, S., Recyclable inherently flame-retardant thermosets: Chemistry, properties and applications. *Compos. B: Eng.* **2023**, 110667.
130. Zheng, N.; Xu, Y.; Zhao, Q.; Xie, T., Dynamic covalent polymer networks: A molecular platform for designing functions beyond chemical recycling and self-healing. *Chem. Rev.* **2021,** *121* (3), 1716–1745.
131. Ferdous, W.; Manalo, A.; Siddique, R.; Mendis, P.; Zhuge, Y.; Wong, H. S.; Lokuge, W.; Aravinthan, T.; Schubel, P., Recycling of landfill wastes (tyres, plastics and glass) in construction–A review on global waste generation, performance, application and future opportunities. *Resour. Conserv. Recycl.* **2021,** *173*, 105745–105745.
132. Khalid, M. Y.; Arif, Z. U.; Ahmed, W.; Arshad, H., Recent trends in recycling and reusing techniques of different plastic polymers and their composite materials. *Sustain. Mater. Technol.* **2021**, e00382–e00382.

133. de Jong, A. M.; Mellquist, A.-C., The potential of plastic reuse for manufacturing: A case study into circular business models for an on-line marketplace. *Sustainability* **2021,** *13* (4), 2007.
134. Bank, L. C.; Arias, F. R.; Yazdanbakhsh, A.; Gentry, T. R.; Al-Haddad, T.; Chen, J.-F.; Morrow, R., Concepts for reusing composite materials from decommissioned wind turbine blades in affordable housing. *Recycling* **2018,** *3* (1), 3–3.
135. Utekar, S.; Suriya, V. K.; More, N.; Rao, A., Comprehensive study of recycling of thermosetting polymer composites–Driving force, challenges and methods. *Compos. B: Eng.* **2021,** *207*, 108596–108596.
136. Zhang, J.; Chevali, V. S.; Wang, H.; Wang, C.-H., Current status of carbon fibre and carbon fibre composites recycling. *Compos. B: Eng.* **2020,** *193*, 108053–108053.
137. Ferrari, F.; Carallo, G. A.; Greco, A., Innovative closed-loop recyclable bio-based composites from epoxidized waste flour and recycled carbon fibers. *Polymers* **2022,** *14* (18), 3878.
138. Luda, M. P.; Balabanovich, A. I.; Zanetti, M.; Guaratto, D., Thermal decomposition of fire retardant brominated epoxy resins cured with different nitrogen containing hardeners. *Polym. Degrad. Stabil.* **2007,** *92* (6), 1088–1100.
139. Pickering, S. J., Recycling technologies for thermoset composite materials—Current status. *Compos. Part A Appl. Sci.* **2006,** *37* (8), 1206–1215.
140. Xue, X.; Liu, S.-Y.; Zhang, Z.-Y.; Wang, Q.-Z.; Xiao, C.-Z., A technology review of recycling methods for fiber-reinforced thermosets. *J. Reinf. Plast. Compos.* **2022,** *41* (11–12), 459–480.
141. Luda, M. P.; Balabanovich, A. I.; Zanetti, M., Pyrolysis of fire retardant anhydride-cured epoxy resins. *J. Anal. Appl. Pyrolysis* **2010,** *88* (1), 39–52.
142. Terakado, O.; Ohhashi, R.; Hirasawa, M., Bromine fixation by metal oxide in pyrolysis of printed circuit board containing brominated flame retardant. *J. Anal. Appl. Pyrolysis* **2013,** *103*, 216–221.
143. Altarawneh, M.; Ahmed, O. H.; Jiang, Z.-T.; Dlugogorski, B. Z., Thermal recycling of brominated flame retardants with Fe_2O_3. *J. Phys. Chem. A* **2016,** *120* (30), 6039–6047.
144. Naqvi, S. R.; Prabhakara, H. M.; Bramer, E. A.; Dierkes, W.; Akkerman, R.; Brem, G., A critical review on recycling of end-of-life carbon fibre/glass fibre reinforced composites waste using pyrolysis towards a circular economy. *Resour. Conserv. Recycl.* **2018,** *136*, 118–129.
145. Shen, M.; Cao, H.; Robertson, M. L., Hydrolysis and solvolysis as benign routes for the end-of-life management of thermoset polymer waste. *Annu. Rev. Chem. Biomol. Eng* **2020,** *11* (1), 183–201.
146. La Rosa, A. D.; Greco, S.; Tosto, C.; Cicala, G., LCA and LCC of a chemical recycling process of waste CF-thermoset composites for the production of novel CF-thermoplastic composites. Open loop and closed loop scenarios. *J. Clean. Prod.* **2021,** *304*, 127158–127158.
147. Fortunato, G.; Anghileri, L.; Griffini, G.; Turri, S., Simultaneous recovery of matrix and fiber in carbon reinforced composites through a Diels–Alder solvolysis process. *Polymers* **2019,** *11* (6), 1007–1007.
148. Oliveux, G.; Bailleul, J.-L.; La Salle, E. L. G., Chemical recycling of glass fibre reinforced composites using subcritical water. *Compos. Part A Appl. Sci.* **2012,** *43* (11), 1809–1818.
149. Shetty, S.; Pinkard, B. R.; Novosselov, I. V., Recycling of carbon fiber reinforced polymers in a subcritical acetic acid solution. *Heliyon* **2022,** *8* (12), e12242-e12242.
150. Khalil, Y. F., Sustainability assessment of solvolysis using supercritical fluids for carbon fiber reinforced polymers waste management. *Sustain. Prod. Consum.* **2019,** *17*, 74–84.
151. Kim, Y.-O.; Cho, J.; Kim, Y. N.; Kim, K. W.; Lee, B. W.; Kim, J. W.; Kim, M.; Jung, Y. C., Recyclable, flame-retardant and smoke-suppressing tannic acid-based carbon-fiber-reinforced plastic. *Compos. B: Eng.* **2020,** *197*, 108173–108173.
152. Panda, R.; Pant, K. K.; Bhaskar, T.; Naik, S. N., Dissolution of brominated epoxy resin for environment friendly recovery of copper as cupric oxide nanoparticles from waste printed circuit boards using ammonium chloride roasting. *J. Clean. Prod.* **2021,** *291*, 125928–125928.
153. Ma, C.; Guo, Z.; Fang, Z.; Li, J., Flame retardancy and chemical degradation of epoxy containing phenylphosphonate group under mild conditions. *Compos. B: Eng.* **2022**, 109967–109967.
154. Castro, A. C. M.; Carvalho, J. P.; Ribeiro, M. C. S.; Meixedo, J. P.; Silva, F. J. G.; Fiúza, A.; Dinis, M. L., An integrated recycling approach for GFRP pultrusion wastes: Recycling and reuse assessment into new composite materials using Fuzzy Boolean Nets. *J. Clean. Prod.* **2014,** *66*, 420–430.
155. Oliveux, G.; Dandy, L. O.; Leeke, G. A., Current status of recycling of fibre reinforced polymers: Review of technologies, reuse and resulting properties. *Prog. Mater. Sci.* **2015,** *72*, 61–99.
156. Wang, Y.; Cui, X.; Yang, Q.; Deng, T.; Wang, Y.; Yang, Y.; Jia, S.; Qin, Z.; Hou, X., Chemical recycling of unsaturated polyester resin and its composites via selective cleavage of the ester bond. *Green Chem.* **2015,** *17* (9), 4527–4532.
157. Wang, B.; Wang, X.; Xu, N.; Shen, Y.; Lu, F.; Liu, Y.; Huang, Y.; Hu, Z., Recycling of carbon fibers from unsaturated polyester composites via a hydrolysis-oxidation synergistic catalytic strategy. *Compos. Sci. Technol.* **2021,** *203*, 108589.
158. Yao, Y.; Wang, B.; Zhao, F.; Hu, Z.; Huang, Y., Degradable and recyclable unsaturated polyester resin based on the selective cleavage of a hindered C–O bond. *ACS Appl. Polym. Mater.* **2022,** *4* (2), 999–1009.
159. Włoch, M., Recycling of Polyurethanes Containing Flame-Retardants and Polymer Waste Transformed into Flame-Retarded Polyurethanes. In *Materials and Chemistry of Flame-Retardant Polyurethanes Volume 1: A Fundamental Approach*, ACS Publications, Washington, DC: 2021; pp 265–284.
160. Szycher, M., Waterborne Polyurethanes. In *Handbook of Polyurethanes*, 2nd ed.; Szycher, M., Ed. CRC Press, Boca Raton, FL: 2012; pp 417–448.
161. Eschenbacher, A.; Varghese, R. J.; Weng, J.; Van Geem, K. M., Fast pyrolysis of polyurethanes and polyisocyanurate with and without flame retardant: Compounds of interest for chemical recycling. *J. Anal. Appl. Pyrolysis* **2021,** *160*, 105374.
162. Deng, Y.; Dewil, R.; Appels, L.; Ansart, R.; Baeyens, J.; Kang, Q., Reviewing the thermo-chemical recycling of waste polyurethane foam. *J. Environ. Manage.* **2021,** *278*, 111527.
163. Gadhave, R. V.; Srivastava, S.; Mahanwar, P. A.; Gadekar, P. T., Recycling and disposal methods for polyurethane wastes: A review. *Open J. Polym. Chem.* **2019,** *9* (2), 39–51.
164. Eling, B.; Tomović, Ž.; Schädler, V., Current and future trends in polyurethanes: An industrial perspective. *Macromol. Chem. Phys.* **2020,** *221* (14), 2000114.

165. Gürkan, E. H.; Yaman, B., Comparative evaluation of flame retardant performance in rigid polyurethane foams: TCPP, TDCP MP, and ATH as promising additives. *J. Taibah Univ. Sci.* **2023,** *17* (1), 2233757.
166. Stančin, H.; Růžičková, J.; Mikulčić, H.; Raclavská, H.; Kucbel, M.; Wang, X.; Duić, N., Experimental analysis of waste polyurethane from household appliances and its utilization possibilities. *J. Environ. Manage.* **2019,** *243*, 105–115.
167. Schultz, F. C.; Everding, S.; Pies, I., Circular supply chain governance: A qualitative-empirical study of the European polyurethane industry to facilitate functional circular supply chain management. *J. Clean. Prod.* **2021,** *317*, 128445.
168. Deubzer, O., Reduction of Hazardous Materials in Electrical and Electronic Equipment. In *Waste Electrical and Electronic Equipment (WEEE) Handbook*, Woodhead Publishing, Elsevier, Amsterdam: 2019; pp 207–230.
169. Ciesielski, M.; Schäfer, A.; Döring, M., Novel efficient DOPO-based flame-retardants for PWB relevant epoxy resins with high glass transition temperatures. *Polym. Adv. Technol.* **2008,** *19* (6), 507–515.
170. Mariappan, T.; Wilkie, C., Flame retardant epoxy resin for electrical and electronic applications. *Fire Mater.* **2014,** *38*.
171. Ciesielski, M.; Diederichs, J.; Döring, M.; Schäfer, A., Advanced Flame-Retardant Epoxy Resins for Composite Materials. In *Fire and Polymers V*, American Chemical Society, ACS Symposium Series, Washington, DC: 2009; Vol. 1013, pp 174–190.
172. Sood, B.; Pecht, M., The effect of epoxy/glass interfaces on CAF failures in printed circuit boards. *Microelectron. Reliab.* **2018,** *82*, 235–243.
173. Shin, S.-R.; Mai, V. D.; Lee, D.-S., Chemical recycling of used printed circuit board scraps: Recovery and utilization of organic products. *Processes* **2019,** *7* (1), 22.
174. Awasthi, A. K.; Zeng, X., Recycling Printed Circuit Boards. In *Waste Electrical and Electronic Equipment (WEEE) Handbook*, Woodhead Publishing, Elsevier, Amsterdam: 2019; pp 311–325.
175. Li, J.; Zeng, X., Recycling Printed Circuit Boards. In *Waste Electrical and Electronic Equipment (WEEE) Handbook*, Woodhead Publishing, Elsevier, Amsterdam: 2012; pp 287–311.
176. Fu, H.; Tisdale, S.; Rausch, M.; Davignon, J.; Hall, S. H.; Pfahl, R. C. *iNEMI HFR-free Program Report*; IEEE, NC: 2009; pp 627–632.
177. Sontakke, S.; Parvat, B.; Bhagat, S., Waste plastic to liquid fuel with process optimization. *Pravara J. Sci. Technol.* **2017,** *1* (2).
178. Krishnamoorthy, S.; Ramakrishnan, G.; Dhandapani, B., Recovery of valuable metals from waste printed circuit boards using organic acids synthesised by Aspergillus niveus. *IET Nanobiotech.* **2021,** *15* (2), 212–220.
179. Luyima, A. 'Recycling of electronic waste: Printed wiring boards', PhD thesis, Missouri University of Science and Technology, USA, 2013. https://core.ac.uk/download/pdf/229061617.pdf.
180. Lucherelli, M. A.; Duval, A.; Avérous, L., Biobased vitrimers: Towards sustainable and adaptable performing polymer materials. *Prog. Polym. Sci.* **2022**, 101515.
181. Kloxin, C. J.; Bowman, C. N., Covalent adaptable networks: Smart, reconfigurable and responsive network systems. *Chem. Soc. Rev.* **2013,** *42* (17), 7161–7173.
182. Hayashi, M., Implantation of recyclability and healability into cross-linked commercial polymers by applying the vitrimer concept. *Polymers* **2020,** *12* (6), 1322.
183. Debsharma, T.; Amfilochiou, V.; Wróblewska, A. A.; De Baere, I.; Van Paepegem, W.; Du Prez, F. E., Fast dynamic siloxane exchange mechanism for reshapable vitrimer composites. *J. Am. Chem. Soc.* **2022,** *144* (27), 12280–12289.
184. Rashid, M. A.; Liu, W.; Wei, Y.; Jiang, Q., Review of reversible dynamic bonds containing intrinsically flame retardant biomass thermosets. *Eur. Polym. J.* **2022**, 111263.
185. Majumdar, S.; Zhang, H.; Soleimani, M.; Van Benthem, R. A. T. M.; Heuts, J. P. A.; Sijbesma, R. P., Phosphate triester dynamic covalent networks. *ACS Macro Lett.* **2020,** *9* (12), 1753–1758.
186. Shieh, P.; Zhang, W.; Husted, K. E. L.; Kristufek, S. L.; Xiong, B.; Lundberg, D. J.; Lem, J.; Veysset, D.; Sun, Y.; Nelson, K. A., Cleavable comonomers enable degradable, recyclable thermoset plastics. *Nature* **2020,** *583* (7817), 542–547.
187. Nabipour, H.; Wang, X.; Kandola, B.; Song, L.; Kan, Y.; Chen, J.; Hu, Y., A bio-based intrinsically flame-retardant epoxy vitrimer from furan derivatives and its application in recyclable carbon fiber composites. *Polym. Degrad. Stabil.* **2023,** *207*, 110206.
188. Zhang, Y.; Ma, F.; Shi, L.; Lyu, B.; Ma, J., Recyclable, repairable and malleable bio-based epoxy vitrimers: Overview and future prospects. *Curr. Opin. Green Sustain. Chem.* **2022**, 100726.
189. Yu, K.; Taynton, P.; Zhang, W.; Dunn, M. L.; Qi, H. J., Reprocessing and recycling of thermosetting polymers based on bond exchange reactions. *RSC Adv.* **2014,** *4* (20), 10108–10117.
190. Zhang, H.; Cui, J.; Hu, G.; Zhang, B., Recycling strategies for vitrimers. *Int. J. Smart Nano Mater.* **2022,** *13* (3), 367–390.
191. Zhang, B.; Li, H.; Yuan, C.; Dunn, M. L.; Qi, H. J.; Yu, K.; Shi, Q.; Ge, Q., Influences of processing conditions on mechanical properties of recycled epoxy-anhydride vitrimers. *J. Appl. Polym. Sci.* **2020,** *137* (41), 49246–49246.
192. Delahaye, M.; Winne, J. M.; Du Prez, F. E., Internal catalysis in covalent adaptable networks: Phthalate monoester transesterification as a versatile dynamic cross-linking chemistry. *J. Am. Chem. Soc.* **2019,** *141* (38), 15277–15287.
193. Montarnal, D.; Capelot, M.; Tournilhac, F.; Leibler, L., Silica-like malleable materials from permanent organic networks. *Science* **2011,** *334* (6058), 965–968.
194. Zhang, B.; Yuan, C.; Zhang, W.; Dunn, M. L.; Qi, H. J.; Liu, Z.; Yu, K.; Ge, Q., Recycling of vitrimer blends with tunable thermomechanical properties. *RSC Adv.* **2019,** *9* (10), 5431–5437.
195. Scazzoli, C.; Trigueira, R.; Cohades, A.; Michaud, V., A novel method to quantify self-healing capabilities of fiber-reinforced polymers. *Front. Mater.* **2022**, 461.
196. Cicala, G.; Mannino, S.; La Rosa, A. D.; Banatao, D. R.; Pastine, S. J.; Kosinski, S. T.; Scarpa, F., Hybrid biobased recyclable epoxy composites for mass production. *Polym. Compos.* **2018,** *39* (S4), E2217–E2225.
197. Yang, X.; Ke, Y.; Chen, Q.; Shen, L.; Xue, J.; Quirino, R. L.; Yan, Z.; Luo, Y.; Zhang, C., Efficient transformation of renewable vanillin into reprocessable, acid-degradable and flame retardant polyimide vitrimers. *J. Clean. Prod.* **2022,** *333* (June 2021), 130043.
198. Takekoshi, T.; Terry, J. M., High-temperature thermoset polyimides containing disubstituted acetylene end groups. *Polymer* **1994,** *35* (22), 4874–4880.
199. Lee, S.-H.; Shin, S.-R.; Lee, D.-S., Self-healing of cross-linked PU via dual-dynamic covalent bonds of a Schiff base from cystine and vanillin. *Mater. Design* **2019,** *172*, 107774.

200. Wang, S.; Ma, S.; Li, Q.; Xu, X.; Wang, B.; Yuan, W.; Zhou, S.; You, S.; Zhu, J., Facile in situ preparation of high-performance epoxy vitrimer from renewable resources and its application in nondestructive recyclable carbon fiber composite. *Green Chem.* **2019,** *21* (6), 1484–1497.
201. Liu, T.; Peng, J.; Liu, J.; Hao, X.; Guo, C.; Ou, R.; Liu, Z.; Wang, Q., Fully recyclable, flame-retardant and high-performance carbon fiber composites based on vanillin-terminated cyclophosphazene polyimine thermosets. *Compos. B: Eng.* **2021,** *224* (June), 109188.
202. Bifulco, A.; Varganici, C. D.; Rosu, L.; Mustata, F.; Rosu, D.; Gaan, S., Recent advances in flame retardant epoxy systems containing non-reactive DOPO based phosphorus additives. *Polym. Degrad. Stabil.* **2022**, 109962.
203. Matxain, J. M.; Asua, J. M.; Ruipérez, F., Design of new disulfide-based organic compounds for the improvement of self-healing materials. *Phys. Chem. Chem. Phys.* **2016,** *18* (3), 1758–1770.
204. Di Mauro, C.; Genua, A.; Mija, A., Kinetical study, thermo-mechanical characteristics and recyclability of epoxidized camelina oil cured with antagonist structure (aliphatic/aromatic) or functionality (acid/amine) hardeners. *Polymers* **2021,** *13* (15), 2503.
205. Zhou, L.; Zhang, G.; Feng, Y.; Zhang, H.; Li, J.; Shi, X., Design of a self-healing and flame-retardant cyclotriphosphazene-based epoxy vitrimer. *J. Mater. Sci.* **2018,** *53* (9), 7030–7047.
206. Cicala, G.; La Rosa, A. D.; Latteri, A.; Banatao, R.; Pastine, S., The Use of Recyclable Epoxy and Hybrid Lay up for Biocomposites: Technical and LCA Evaluation. In *Proceedings of the CAMX 2016-Composites and Advanced Materials Expo*, Anaheim, CA, 2016; pp 1–10. https://vvorpal.com/whitepapers/CAMX2016_CICALA.pdf.
207. Chen, J.-H.; An, X.-P.; Li, Y.-D.; Wang, M.; Zeng, J.-B., Reprocessible epoxy networks with tunable physical properties: Synthesis, stress relaxation and recyclability. *Chinese J. Polym. Sci.* **2018,** *36* (5), 641–648.
208. Röttger, M.; Domenech, T.; van der Weegen, R.; Breuillac, A.; Nicolaÿ, R.; Leibler, L., High-performance vitrimers from commodity thermoplastics through dioxaborolane metathesis. *Science* **2017,** *356* (6333), 62–65.
209. Liu, T.; Zhao, B.; Zhang, J., Recent development of repairable, malleable and recyclable thermosetting polymers through dynamic transesterification. *Polymer* **2020,** *194*, 122392–122392.
210. Feng, X.; Li, G., Catalyst-free β-hydroxy phosphate ester exchange for robust fire-proof vitrimers. *Chem. Eng. J.* **2021,** *417* (November 2020), 129132–129132.
211. Nazir, R.; Gooneie, A.; Lehner, S.; Jovic, M.; Rupper, P.; Ott, N.; Hufenus, R.; Gaan, S., Alkyl sulfone bridged phosphorus flame-retardants for polypropylene. *Mater. Design* **2021,** *200*, 109459.
212. Gosecki, M.; Gosecka, M., Boronic acid esters and anhydrates as dynamic cross-links in vitrimers. *Polymers* **2022,** *14* (4), 842–842.
213. Liu, Y.; Wang, B.; Ma, S.; Xu, X.; Qiu, J.; Li, Q.; Wang, S.; Lu, N.; Ye, J.; Zhu, J., Phosphate-based covalent adaptable networks with recyclability and flame retardancy from bioresources. *Eur. Polym. J.* **2021,** *144* (December 2020).

25 Flame-Retardant Design and Protection for Fiber-Reinforced Composites

Alexander B. Morgan[1] *and Baljinder K. Kandola*[2]

[1] University of Dayton Center for Flame Retardant Materials Science, Dayton, OH 45469 USA

[2] Institute for Materials Research and Innovation, University of Bolton, Deane Road, Bolton BL3 5AB United Kingdom

2nd Edition Chapter Authors: Michael G. Stevens and Alexander B. Morgan

25.1 INTRODUCTION TO FIBER-REINFORCED COMPOSITES AND THEIR CURRENT USES

Depending upon the inherent mechanical properties of a particular polymer in a specific application, the polymer may be able to be used "as is" with no additional reinforcement. However, some applications require higher levels of strength/stiffness, and this results in incorporation of fibers to reinforce the mechanical properties of polymer, thus yielding a fiber-reinforced composite. Polymers containing fibers are commonly referred to as polymer composites, fiber reinforced polymers (FRP), or simply "composites." Polymer composites can be reinforced with a variety of fibers, including carbon fiber, glass, aramid (Nomex, Kevlar), and natural plant fibers (jute, sisal). Even the glass fibers can be of different compositions, with glasses such as E-glass and S-2 glass being most common. These polymer composites are used in many applications where lightweight, good corrosion/environmental resistance and high strength-to-weight ratio are desired. Many applications use these advantages to replace other materials such as steel, aluminum, and wood. Indeed, these enhanced properties has led to wide spread use of polymer composites in aerospace, maritime, chemical industry, and mass transport applications. The heightened increase and sensitivity to fuel costs has led to a huge push to put composites into as many vehicle applications as possible to save weight and therefore improve fuel efficiency. Some recent examples of this trend are naval vessels built entirely out of polymer composite, train cars with significant portions of polymer composites in their interior, aircraft where both interior cabin components and parts of the wings and fuselage are increasingly being made out of FRP (typically carbon-fiber based) composite, and even automobiles, although this is currently limited to higher end sports cars.

Polymer composites are made up of the polymer and reinforcing fibers. In most polymer composites, the polymer is typically a thermoset polymer but thermoplastics can be used as well. Some typical thermoset polymers used are epoxy resins, unsaturated polyester resins, epoxy vinyl ester resins, phenolic resins, and high performance aerospace resins such as cyanate esters, polyimides, and bismaleimides. Thermoplastic polymers used in polymer composites can vary widely, but polyolefins, polyesters, polyamides, and selected engineering polymers (such as polyaryletherketones) are known and used commercially. Since the resin in the composite holds the fiber form together, it dominates the toughness, corrosion resistance, and fire performance of the composite. The fibers provide the strength and stiffness to the composite. So while these are different components, the resin and fibers have to be looked at together in the application. Varying the resin type and content or varying the type, orientation, and amount of fibers will vary the performance greatly. There are many guidelines for polymer composite design, but these are always set by the performance requirements of the application. This will determine the type of resin to use, the type of reinforcements to use, and the amount of reinforcing fibers. This chapter does not propose to cover all aspects of polymer composite design but rather to focus on the flammability of polymer composites and how to design fire protection for these materials. As has been alluded to in many places in the book, it is important to consider the end use application and its fire risk scenarios as this will determine the fire safety requirements, and the needed fire protection, for the material. This is true of polymer composites as well.

25.1.1 AEROSPACE

Polymer composite use has grown greatly in aerospace due to its benefits in increasing fuel efficiency (polymer composites are lighter than aerospace grade metals) and its resistance to corrosion (lower maintenance cost). Polymer composites used in aerospace are mostly thermoset resins (epoxies, phenolics, cyanate esters, vinyl esters, etc.) with carbon fiber, but some fiberglass composites can be found as well in interior cabin applications. For all aerospace applications, flammability requirements begin with the US Federal Aviation Regulations (FAR) 25.853, and the battery

DOI: 10.1201/9781003380689-25

of tests located within that requirement.[1] Depending upon where the polymer composite is located on the aircraft, such as the passenger cabin, wing, fuselage, engine compartments, or cargo hold, different flammability requirements will apply. For example, one area that has used polymer composites for many years is the cargo area. All cargo areas must contain a liner that is not part of the aircraft wall. The most common material of construction for the cargo liners is halogenated unsaturated polyester resins. These can be formulated to meet the requirements of this application. The requirement for this application is the 60 second Bunsen burner test. There is no smoke or smoke toxicity requirement for the cargo liners. The requirement for lining the inside of the passenger compartment is much more stringent if the plane is designed to carry more than 20 passengers. The passenger cabin requires that heat release rate, smoke, and smoke toxicity all be measured. For polymer composites materials to pass these stricter requirements in interior applications, more inherently fire retardant resin systems such as phenolic resins have to be used, although some low heat release thermoplastic resins such as polyetheretherketones (PEEK) can be used as well,[2,3] along with polyetherimides. For interior applications, the OSU rate of heat release test and the ASTM E662 smoke density test continue to be used for qualifying materials for these interior applications. The OSU test is similar to the cone calorimeter but has significant differences which must be considered when comparing the two techniques.[4] The Federal Aviation Administration has funded work on trying to develop more inherently fire retardant resins for use in aircraft interiors and has made great progress in this area, with several recent crashes allowing the passengers ample time to escape from the aircraft before the fuel and fire penetrated the fuselage. For exterior components (fuselage and wing) that would be exposed to the post-crash fuel pool fire scenario, there is a kerosene-fueled "oil burner" test which forces a large jet of flame onto the composite and time to burn through is measured. As will be discussed in the next section, this severe fire risk scenario presents some additional fire hazards for polymer composites that burn when exposed to these intense fire sources.

25.1.2 Mass Transit

Moving from aerospace to mass transit (trains, subways), the mass transit industry is growing rapidly on a global basis due to increasing population in urban centers, as well as increasing fuel costs. The test requirements for polymer composites in mass transit vary around the world. The United States, for instance, has fire regulations for passenger trains traveling between cities described in 49 CFR part 216[5]. The National Fire Protection Association (NFPA) also has a fire code for mass transit vehicles called NFPA 130.[6] Also, most South American countries follow the requirements of NFPA 130 for their mass transport (rail) systems. Finally, there are European standards for subway cars and trains which can be very strict on fire safety of materials used in mass transit construction. These standards though are a mix of local (national) standards and attempted EU "harmonized" standards, but further harmonization continues and those selling into this application should continue to check for what the current requirements may be. All of these standards have flame spread and smoke requirements to protect passengers from fire hazards and give them enough time to evacuate cars safely. In the US, the flame spread requirement is based on the ASTM E162 radiant panel test,[7] and the smoke requirement is based on ASTM E662 smoke density test.[8] Under this US standard, the requirement for all vehicle walls and ceilings is a maximum flame spread index of 35 and maximum specific smoke density at 90 seconds of 100 and a maximum specific smoke density at 4 minutes of 200. Some of the train car manufactures have more stringent requirements and include a smoke toxicity test requirement as well. Because of the smoke and smoke toxicity requirements in this application, halogen flame retardants are not generally used (see Chapter 4 for why halogenated FR induces more smoke). Most of these parts are made using unsaturated polyester resins, epoxy vinyl ester resins, or modified acrylic resins that are filled with aluminum trihydrate or intumescent FR packages (such as ammonium polyphosphate), but other flame retardants may also be present if they are compatible with the polymer composite manufacturing process. These resins are chosen due to their ease of manufacture and relatively low cost given the large size of the applications in question. These applications include seats, interior and exterior wall panels, lavatories, engineer's consoles, and other large parts. Some parts such as flooring are made out of phenolic resin to meet the ASTM E119 requirement of 15 minutes without the temperature exceeding 200°F on the back side of the panel for floors in passenger trains. Depending upon the size and dimensions of the part in question, particular processes will favor which will dictate resin choice which will in turn dictate base fire performance. Parts in mass transport are currently made using a hand lay-up or spray-up process, or closed mold processes such as resin transfer molding (RTM) or vacuum infusion processes, such as vacuum assisted resin transfer molding (VARTM). Preform plus compression molding manufacturing is also used to manufacture these parts, and this type of manufacturing uses "molding compounds," which are highly filled thermoset formulations partly polymerized to obtain higher viscosity and a "putty-like" consistency. These molding compounds are pressed into their final shape via simple compression molding/tooling, and then heat is applied to finish polymerizing the material to its final rigid shape.

Currently in the European Union there is a harmonized fire standard which governs materials in mass transportation, and this standard is DIN EN 45545–1–7, "Fire protection on railway vehicles."[9] The flame spread test is based on the IMO radiant panel test ISO 5658–2. The heat release is measured using ISO 5660–2 cone calorimeter. Smoke opacity is measured by EN ISO 5659–2 and smoke toxicity is measured uses EN ISO 5659–2 with FT-IR used to measure toxic smoke components. These all go into giving the product a rating of HL1, HL2, or HL3. Each country can specify the rating level required for the applications in their country. The HL3 rating is the rating with the most stringent requirements which would probably require a phenolic resin or a highly filled modified acrylic resin to meet the requirements. The standard filled polyester resins in commercial use today likely can meet the HL1 and HL2 ratings. It will be difficult to meet even the HL1 rating with an unfilled halogenated polyester resin due to the smoke release from the halogenated FR.

25.1.3 Marine

Another form of mass transportation which uses polymer composites is marine applications such as ships that transport goods and people around the world. The size of the ship determines what level of fire protection is required for the polymer composite. For small ships, such as pleasure boat (small sailing ships, power boats, etc.), the resins used in hull manufacture do not require flame retardancy, with the exception of the fuel tanks. In the US, the fire code that governs these small ship fuel tanks is governed by the US Coast Guard under 33 CFR 183 Subpart J. In general, the fire requirements are not too difficult to pass with most polymer composite resins containing flame retardants, and instead the stricter part of the test is ensuring the polymer composite fuel tank can resist the fuel solvating/swelling and weakening the resin of the polymer composite. The introduction of ethanol into maritime fuels made this code harder to pass, and as maritime fuels continue to change (including desulfurization of diesel/maritime bunker fuel) fuel compatibility along with fire performance must be checked.

For larger vessels, the International Maritime Organization (IMO) has codes within the "Safety of Life at Sea" (SOLAS) codes that are used for any ocean going vessels or vessels traveling in international waters.[10,11] The requirement for wall ceiling and bulkhead linings in this code is very difficult to pass. This requires a resin with a very low flame spread, low smoke, and smoke toxicity. The applications are mostly for lavatories and passenger compartment walls rather than structural components. Next to the IMO SOLAS fire safety requirements, the US Navy has some very strict fire requirements for materials in general on ships, and especially so for FRP.[12,13] When structural properties need to be maintained in maritime applications, a high char yield material will normally do better in fire resistance tests like ASTM E119. The other way to meet the longer fire resistance times is by using a core material that can help insulate the back side of the composite. The core materials that have been used successfully are balsa wood and polyisocyanurate foam.[14–23] There are non-maritime commercial applications which also use this approach. One example is the Fire Safe House that is designed to protect miners in case of a fire or collapse inside of a mine. This incorporates a polymer composite made from a fire retardant vinyl ester resin with a core that allows this composite wall to obtain greater than a 1 hour fire rating on the ASTM E119 fire resistance test.

25.1.4 Offshore Platforms

Another example of a structural polymer composite maritime application is grating for offshore oil platforms. These composites are used as they can hold up to the corrosive seawater environment of offshore oil platforms, and greatly reduce maintenance cost (as well as ensure structural stability) for this grating used throughout the structure where water needs to drain as waves/storms go over the structure. The US Coast Guard and IMO have set up a test for qualifying composite grating for this application. The test requires the grating to be exposed to the hydrocarbon fire temperature profile in the ASTM E119 test apparatus. A weight is placed on the grating, and it cannot deflect more than the deflection of a steel grating under the same test conditions. The system currently used is a composite grating that is pultruded from a phenolic resin. Phenolic resins are lower heat release polymers which also form a great amount of char when they burn,[24–30] and this is the reason that phenolic resin based composites performs well in this application.

25.1.5 Building Materials

For building materials, one must look at both architectural and industrial applications. The use of polymer composites in architectural applications has been going for many years. They are used to produce exterior decorative items such as cornices, facades, sky lights, etc. and have also been used to make interior finish wall panels. An example of this is the composite façade used for the San Francisco Museum of Modern Art (Figure 25.1). The requirements for these applications are governed by the building codes, and the current building code is applied to polymer composites the way it is applied to any material. See Chapter 22 for more details on the International Building Code (IBC). The IBC has specific sections for polymer composites and describes how the codes should be applied to polymer composites parts in interior and exterior applications. For example, most materials over 40 ft in height would need to be rated "non-combustible" under tests in the IBC. However, for polymer composites, if the material is tested under ASTM E84 and has flame spread index of < 25 and it does not cover more than 20% of the exterior area of the building, then it can be used at any height.

The use of polymer composites is desired to replace older materials of construction such as concrete, stucco, etc. They have been used in many historical preservation projects. The reason they are desired is because of lightweight, durability ease of installation, and low maintenance (no dry rot, attack by insects, mold, etc.). These are typically nonstructural parts made by spray-up process or continuous panel process. The main resin used for these exterior architectural applications are halogenated polyester vinyl resins. Since there is not a smoke requirement, resins that meet

FIGURE 25.1 Composite façade at the San Francisco Museum of Modern Art Creative Commons License. 19 April 2017, Author: https://commons.wikimedia.org/wiki/User:Beyond_My_Ken.

the ASTM E84 flame spread requirement of < 25 can be used for this application. The other common material used for these applications is polyester resin or modified acrylic resins that are filled with aluminum trihydrate (ATH). The ATH filled resins can also be used in interior applications since they will typically also have low smoke values. Other resins can be used, but the low cost of vinyl ester resins has led to their extensive use in these applications.

The European Union has also implemented a new unified standard for building materials in EN 13501–2. This uses the single burning item (SBI) test EN 13823. [See Chapter 22 for more details on these fire codes.] The classifications under the EU are shown in Table 25.1. It should be noted that to date, there is no great correlation between materials that pass US building codes under ASTM E84, and those tested under the SBI.

Testing has been done that shows that a halogenated resin that can give a Class A classification by ASTM E84 gave a Euroclass D flame spread rating and a S3 smoke rating; rather poor performance in the Euroclass rating system but considered very good by ASTM E84. The level of halogen required to obtain a Euroclass B is much higher than what it takes to make a Class A flame spread in ASTM E84 test. It has also been found that the introduction of mineral fillers such as ATH will help give a better Euroclass rating. An all ATH filled modified acrylic resin system with 150 phr of ATH has been able to meet the Euroclass B,S1,d0 classification. Halogenated resin systems normally give an S3 smoke rating, even when ATH is included in the formulation. It is much easier to meet the S1 smoke rating with a halogen-free resin system.

The use of polymer composites in industrial applications has been around for over 70 years. Some typical applications where these are used include chemical storage tanks, scrubbers, carbon absorbers, stacks, stack liners, duct work, fan housings, fan blades, etc. They were first used in 1953 when a chlorinated polyester resin was used to make the cell covers and headers at a chlorine plant. The acceptance and use of these materials have only grown since then. The main reason that polymer composites are used in these applications is for their improved corrosion and chemical resistance in many environments. In some cases there is also a fire retardant requirement when FRP materials are used in chemical operations. The most common test for fire retardancy used in these applications is the ASTM E84 Steiner Tunnel test. Most of the time regulators are looking for a flame spread index of < 25 in these applications. The resin selection for these applications is first done on what environments it will be exposed to and at what temperatures. A brominated epoxy vinyl ester resin is typically the main resin used when both corrosion resistance and fire retardancy is required. These systems give very good chemical resistance to both acids and bases and can achieve the desired flame spread rating of < 25. The laminate will have a corrosion barrier that consists of a veil layer that is 90% resin and 10 fiber followed by chopped strand glass layer that is 70% resin and 30% glass fibers. The corrosion liner is not included in the structural wall calculations. The structural portion of the equipment is normally made by filament winding to obtain high strength and modulus in the structural wall. The normal resin content of these tanks is 30% with 70% glass. Fillers are not used in the corrosion liner because it can reduce the corrosion resistance of the resin. Fillers are also not generally used in the structural layer except for when antimony trioxide or antimony pentoxide is used to get improved flame spread results with

TABLE 25.1
European Classifications from EN 13501 for Construction Products

Euroclass	Typical Products Examples	Previous National Classifications		
		France	Germany	UK
A1	Stone, concrete	NC	A1	NC
A2	Gypsum boards	M0	A2	0/NC
B	Wood (treated)	M1		0
C	Wall covering on gypsum board	M2	B1	1
D	Wood (untreated)	M3		2
E	Low density wood fiber board	M4	B2	
E	Some plastics		B3	

the halogenated vinyl ester resin. The use of these resins will continue to grow as they replace metals in many applications because of longer life, lower maintenance and in many cases lower initial costs. Polymer composites give the engineer more design flexibility on the shape of the equipment and design the strength and stiffness where it is needed most. The use of finite element analysis has helped put more science into designing the strength and stiffness of the equipment instead of empirically experimenting and running tests to see what is obtained.

25.2 UNIQUE FIRE HAZARDS WITH COMPOSITES

Polymer composites present some additional fire hazards given their use in aerospace, transportation, and structural (building) applications, and the sometimes higher intensity fires present in those applications. Specifically, they can mechanically fail prior to their ignition, and they can present some challenges for firefighting. This is not to say that polymer composites are far more hazardous than other burning materials, but rather that due to their specific fire risk scenarios, there are some additional fire hazards that must be considered. Flame retardant use and fire protection barriers can lessen these hazards, but they will still be present given the inherent properties of polymer composites when they burn.

25.2.1 Mechanical Failure Prior to Ignition

Many polymeric materials have a glass transition temperature (T_g) which is the temperature at which the polymer goes from a rigid glassy state to a softer flexible state. For polymer composites, this is the upper use temperature for these materials when they are supporting a mechanical load. For most polymer composites, the T_g is below the decomposition temperature of the polymer, and this means that some polymer composites will deform and bend prior to igniting in a fire event. Usually 50% of the mechanical performance (flexural modulus) is lost as the temperature reaches T_g of the matrix.[31] If the material is a thermoplastic composite, the material can also melt during this pre-ignition event, leading to further mechanical failures as the resin matrix melts away. Depending upon the nature of the fire, this can result in catastrophic failures and fractures/breaks in the structure of the polymer composite early on in a fire event. If this happens, this will in turn lead to faster fire growth as fires penetrate through openings in the broken polymer composite structure, and there will also be higher surface area which can participate in the fire growth on the polymer composite surface. There have been numerous studies on mechanical properties of polymer composites under fire conditions, from low heat flux to high heat flux.[15–17,19–21,23,32–40] In general, if one wants to have a higher use temperature for a polymer composite, specialized crosslinking after the initial polymerization, or high cross-link density in the thermoset polymer, will be required. It should be noted that as crosslink density goes up, provided the crosslinks are sp^2 hybridized C-C bonds, heat release (flammability) of the polymer will tend to decrease. Therefore, pushing toward higher T_g with increased aromatic structure will also yield improved flame retardancy of the FRP.

25.2.2 Emissions and Firefighting Measures

Polymer composites burn differently when compared to cellulosic (wood) materials, and when carbon fibers are present, present some additional fire hazards. While some polymers rapidly melt and gasify (polyolefins, polystyrenes, selected polyurethanes), most polymers used in polymer composites are high char forming or are thermoset polymers, which don't go to a liquid form. Instead, they go from solid to gaseous state and, as mentioned above, can bend/deform under mechanical load which in turn changing the surface area of the sample when they burn. On top of the polymer contributing to the fire, the fiber needs to be considered. While fiberglass does not burn, natural fibers can burn and would contribute to the total fuel load available in a fire event. Even carbon fibers contribute to fire events even if they don't exactly "burn" with a flame but do suffer oxidative damage and consumption from oxidation. Carbon fibers conduct heat, and they can drive heat from the fire source deeper into the composite structure, causing the entire composite to heat up. Further, carbon fibers oxidize and release needle-like fragments in fires,[41–43] which would be a health concern for firefighters exposed to smoke from the fire and repair/salvage crews tending to the fire-damaged structure after the fire event. Self-contained breathing apparatus, such as that compliant with National Fire Protection Association 1971, is strongly recommended for firefighters involved in fighting polymer composite fires. Moreover, carbon fibers being electrically conductive, on exposure to the surrounding environment, can interfere with electrical systems.

Due to the abovementioned fire physics of polymer composites, firefighting measures when dealing with polymer composites are somewhat different than those used for putting out fires made from other materials. Typical firefighting extinguishments include water, foams, CO_2, halon, and powders for metal fires. Most aircraft fires with large jet fuel pool fires upon crashing utilize foam to smother the fuel pool fire that is formed post-crash so the fuel pool fire extinguishes. For metal-based aircraft, once the fuel pool fire is smothered with foam, the metal rapidly cools, and the entire fire is out. This is not the case for polymer composites. Once polymer composites ignite, the foam will simply put out the fuel pool fire, but the polymer composite will remain hot and will reignite (due to auto-ignition or entrained flames in structurally damaged composites), which may reignite the fuel pool fire. Although foam is needed for dealing with a fuel pool fire where the polymer composite may be exposed to flame, the foam has almost no effect on the polymer composite itself. Halon (now increasingly not in use due to Montreal Protocols on Ozone-depleting substances) also has little effect on polymer composites exposed to high heat sources. While the Halon will quickly extinguish the fire, the underlying polymer composite remains hot and will reignite once the Halon gas has dissipated. Effectively one has to use something to cool the polymer composite, which means either CO_2 or water. While water is quite common, a lot of it may need to be used, and when reconsidering the abovementioned fuel pool fire, the water will interfere with the foam and cause the foam to dissipate and the fuel pool fire to reignite. Therefore, polymer composites in aircraft crash fires present a particularly difficult fire to put out. Some research has been conducted on this by Boeing, and they have published specialized instructions for putting out composite fires when their aircraft (the Boeing 787) is involved in a large fire event.[44] For polymer composites used in new electrical vertical takeoff and landing (e-VTOL) aircraft, the fuel pool fire issue will not be present, but the large lithium ion battery pack in these aircraft will require large amounts of water to extinguish these fires. Therefore, for all aircraft fires involving polymer composites, water and cooling extinguishing methods will need to be present to put out these fires, rather than foam as is used today.

25.3 FLAME-RETARDANT DESIGN OF COMPOSITES

As has been discussed throughout this book, design for a material for flame retardancy and flame retardant selection are determined by the fire protection needed for the material. For polymer composites, this underlying rule is the same, the design and the flame retardant strategy chosen will be driven by the end use fire test. Once the test is known, then other requirements such as compatibility with polymer composite manufacture, matching flame retardant (FR) to polymer type, and balance of non-fire properties (mechanical, thermal, etc.) are considered. In general, for polymer composites, one can consider using traditional flame retardants or fire protection barriers, but one may select to use both of them at the same time, as will be discussed below.

25.3.1 Choice of Constituents and Geometry

The first step in designing any composite is to consider what it is to be made out of and what geometry it will have. The constituents of the composite (resin, fiber reinforcement, core material) make up the majority of the choices that must be considered in fire safe composite design.

25.3.1.1 Resins

For commercial applications there are four major classes of resins that are used for composites: phenolics, epoxies, unsaturated polyester, and vinyl ester resins, the choice of which though is application specific. The ranking of their major properties are summarized in Table 25.2,[45,46] which places epoxy as the material of choice for good mechanical properties, polyesters, and vinyl esters for the most cost-effective structural materials and phenolics for applications where products have to pass required fire regulatory tests.

Besides these, other resin systems such as cyanate esters, polyimides, furan, bismaleimides, and benzoxazines can be used. Their commercial use, however, is limited due to cost, high curing temperatures, or unfavorable mechanical properties, but they are all used commercially in selected applications. The inherent low flammability and high glass transition temperature of polyimides, bismaleimides, and benzoxazines lead to these materials often being used in aerospace applications. In order to compare different resins, a fire safety diagram (Figure 25.2(a)) based on cone calorimetric test results of different resins at 50 kW/m^2, performed at University of Bolton,[47–52] has been constructed. In Figure 25.2 total heat release (THR) has been plotted against the flashover propensity values calculated by dividing peak heat release rate (PHRR) by time-to-ignition (TTI).[53] The parameter, PHRR/TTI, is a measure of the severity of the fire, as a short TTI and a PHRR are considered to be necessary requirements for flashover to occur. The THR represents the total amount of heat that can be released by the fuel, that is, the larger the value of THR, the more heat the material releases as it burns. This ranking also places both formaldehyde based resins, that is, phenolics and melamine formaldehyde, in less hazardous category.

Although the mechanical performance of phenolics is inferior to epoxy, unsaturated polyester, and vinyl ester resins, their char-forming ability enables composites comprising them to retain mechanical strength for long times under fire conditions. In addition, such composites encapsulate

TABLE 25.2
Ranking of the Properties of Epoxy, Polyester, Vinyl Ester, and Phenolics

Property	Ranking
Mechanical Property	Epoxy > Polyester and Vinyl Ester > Phenolics
Processability	Polyester & Vinyl Ester > Epoxy > Phenolics
Cost-Effectiveness	Polyester & Vinyl Ester > Phenolics > Epoxy
Flammability	Phenolics > Epoxy > Polyester & Vinyl Ester

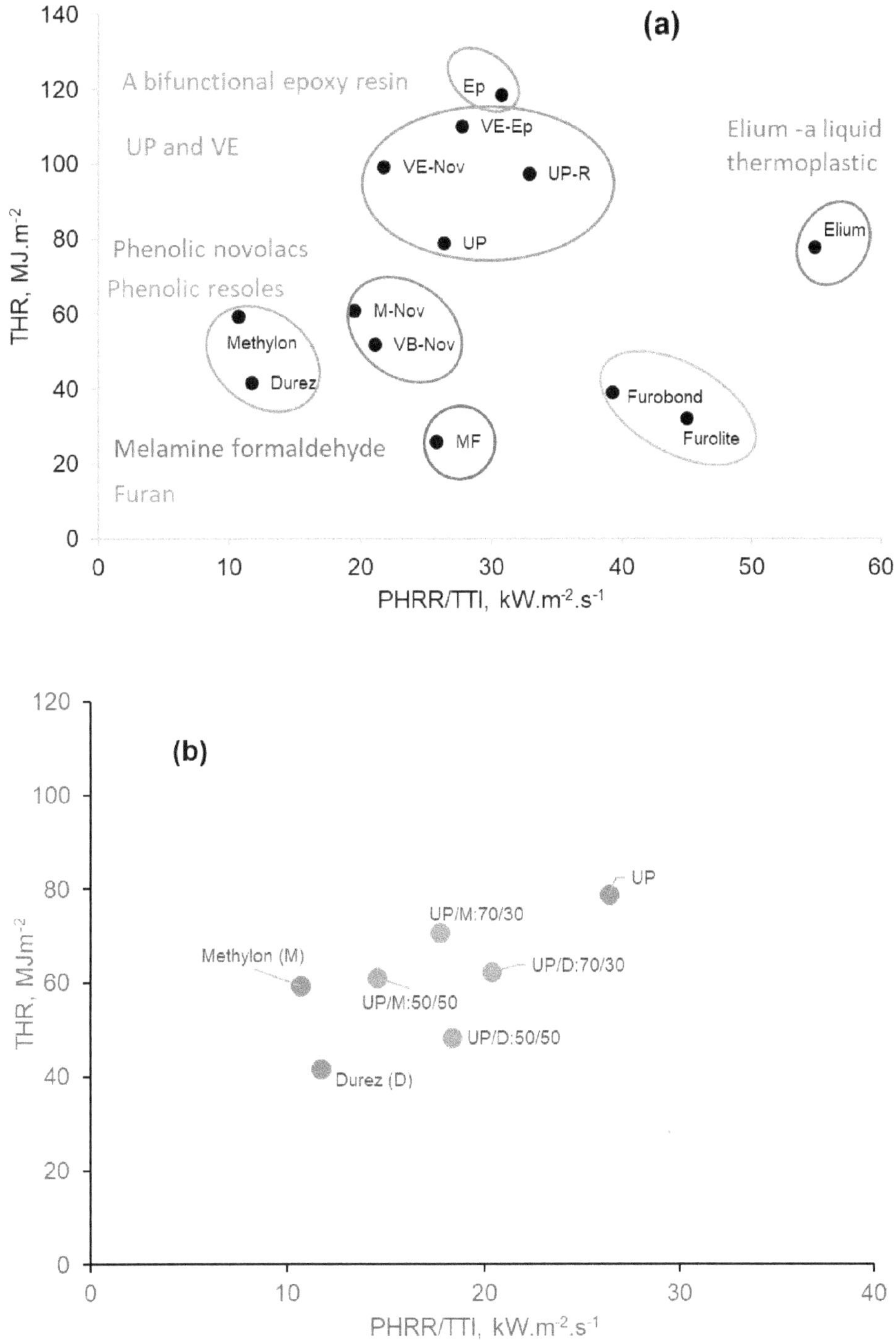

FIGURE 25.2 Fire safety diagram, that is, plot of total heat released (THR) vs. peak heat release rate (PHRR) divided by time to ignition (TTI), for (a) different resins and (b) UP, phenolic, and their 50/50 and 70/30 blended resins based on cone calorimetric data recorded with 50 kW/m^2 radiant heating.[47–52]

themselves in char, and they do not produce as much smoke when compared to epoxies and vinyl/polyester resins.

The thermal stability and flammability of an epoxy resin are determined by functionality of the aromatic resin (tetrafunctional epoxies are more thermally stable than tri- or bifunctional), and types of curing agent and hardener used.[54] Commonly used curing agents are amine, anhydride, and polymer- (polyamide, phenolic resin, *etc.*) based. Epoxy resins cured with amines tend to produce more char than anhydride-cured resins.[55] Curing agents containing phosphorus and nitrogen in their structure help in reducing flammability of the resin as they can have additional flame retardant effects when the polymer thermally decomposes.

Unsaturated polyesters display poor fire performances being composed of aromatics attached by relatively long aliphatic chains. The emission of styrene during the processing and the thermal combustion is a serious environmental issue due to the styrene toxicity. For fire retardant variants, the use of halogenated UP resin or replacement of curing agent from styrene to bromostyrene are common commercial solutions. Since the most combustible part of this formulation is styrene, it is desirable to replace it fully or partially with other reactive diluents such as maleic anhydride and dialkyl *p*-vinyl benzylphosphonates.[56]

Vinyl esters, as polyesters, display poor fire performance. As the amount of styrene in vinyl esters is generally higher than in polyesters, they are slightly more flammable and generate more smoke than unsaturated polyesters. However, brominated vinyl esters (which are commercially available materials through Ineos and other vendors) have much better fire performance but still tend to have high smoke release. Additional flame retardants can be added to make brominated vinyl esters have low flammability and flame spread, and with reductions in smoke release, and commercial versions of lower smoke/low smoke release vinyl esters are available.

25.3.1.2 Resin Blending

The blending of different polymers is a well-established and convenient method for preparing materials with improved properties, that is, the components are chosen so that the weaknesses of one polymer can, to a certain extent, be masked by the strengths of the other and vice versa. This is also true in case of resins as well. A flammable resin can be blended with another inherently flame-retardant and char-forming resin such as phenolic, polyimide, bismaleimide, melamine formaldehyde resin, etc. This approach is commonly used for chemically compatible resins with similar cure chemistry; some examples are unsaturated polyestesr (UP) blended with epoxy resins;[57] phenolics with epoxies.[58] Blending of incompatible resins such as UP with phenolics, however, can lead to phase separation. In such cases one of the resin needs modification or a compatibilizer to be used. In a recent research at Bolton, UP has been blended with resole phenolics, by using either a solvent with which both resins are compatible, or using a functional group in the phenolic resin so that it can either react with UP or cure in a similar manner as the UP.[47] With this approach, flammability of UP could be reduced to a significant extent (see Figure 25.1(b)). The curing temperatures of the blends, however, can be too high to be used for most commercial applications and processing equipment. This led to use of novolacs, which were either methacrylated[48] or vinylbenzylated[51] to be compatible with UP. UP-modified novolac blends can be cured at room temperature. The cured methacrylated, vinylbenzylated novolac, and co-cured blends of them with UP showed superior flame retardance to cured UP alone. Other resins that may be blended with unsaturated polyesters to improve their char-forming potential include urea-formaldehyde, melamine-formaldehyde,[50] and furan resins.[49]

25.3.1.3 Fibers

Similar to resins, the choice of fiber type is also application specific. Except for some particular fibers, such as ultra high molecular weight polyethylene (UHMWPE), most fibers, especially the commonly used ones (such as glass and carbon), add little to the fuel content of a composite. Fiberglass and other inorganic fibers (basalt, ceramic, etc.) are noncombustible. Carbon fiber, while it is not combustible and has very low heat release potential because of the thermally stable carbon forms present (often graphitized), oxidizes at temperatures > 500°C, decomposing and releasing needle-like particulates in fire, as discussed previously.[40–42]

The glass/carbon fiber-reinforced composites are less flammable than cast neat resins of similar thicknesses due to lower resin contents in the former and also the fibers present acting as fillers. Natural fibers (e.g., cellulosic based) or synthetic (polyolefins, polyamides, polyaramids) however are flammable and, hence, contribute to overall flammability of the composites, which though is still less than that of respective neat resins of similar thicknesses.[59] Specifically, fiber volume will affect flammability of the composite when it burns. Simply put, more nonflammable fiber (glass or inorganic fiber) or more low-flammability fiber (aramid, carbon, etc.) will replace flammable polymer in the composite, thus lowering aspects of its flammability.[60–67] Depending upon the chemistry of the polymer, higher fiber content may slow flame spread or lower total heat release. However, for some thermoplastics, fibers yield a fire behavior often referred to as "wicking," where the molten thermoplastic moves along the fibers via capillary action, similar to how molten wax moves along a wick in a candle. This ends up causing negative flame spread behaviors even if the fibers are preventing dripping of the thermoplastic.[68,69] Therefore, the material scientist must consider how the decomposing polymer behaves in the presence of fibers when considering how to balance both mechanical and fire properties of a polymer composite.

25.3.1.4 Core Materials

Core materials, usually in the form of honeycomb structures of a variety of base materials such as aluminum or aramid paper (Nomex®, DuPont) or balsa wood are used in composites to increase the flexural stiffness while keeping the overall weight low. The microstructure of wood is similar to the cellular hexagonal structure of synthetic honeycombs, thus included in the same category of honeycombs. The other category is of synthetic polymer foams of materials such as polystyrene (PS), polyvinyl chloride (PVC), polyurethane (PU), polyetherimide (PEI), etc. Usually thin laminate skins are glued on one or both sides of the core material. These core materials are fully encapsulated within the laminate, hence are not directly exposed to fire. However, when exposed to fire and/or heat, a foam will melt and burn, wood will burn and/or char depending upon the intensity of fire and aramid will char. Aluminum cores on the other hand will be initially unaffected in fire but will melt at very high temperatures and the mechanical properties of the sandwich structure will diminish as a function of time. In Figure 25.3(a) the burning behaviors of a thin glass/unsaturated polyester composite laminate (1 mm) thickness and composites laminate on one side of balsa wood core of two thicknesses (13 mm and 25 mm) are shown. As can be seen that TTI and PHRR of all samples are similar (within experimental error range) indicating no significant effect of balsa core, but due to continued slow burning/smoldering, heat transfer in sample with thinner core is higher than the one in thicker core, leading to sharper rise in back surface temperature in the former. In Figure 25.3(b) the effect of laminates on one or both sides is shown, where there is not much difference in HRR up to 300s, above that the sample with no laminate on the reverse surface shows more heat release and sharper temperature rise than the one with laminate on both sides. Cores which provide thermal insulation and have slow combustion will provide fire protection to the other side of the composite panel, and this may in turn delay the temperature rise to the T_g of the polymer and therefore slow structural failure of the composite in a fire event.

25.3.1.5 Geometry

Thickness of the composite, including due to the presence of the core, is an important factor in the overall flammability of a composite. Thin samples ignite early, show large peak heat release values, burn for shorter time, and produce less total heat release compared to thick samples, which burn slowly but for a longer time.[70,71] As an example, effect of thickness on selected cone calorimetric parameters of glass-fiber-reinforced polyester composite samples are shown in Figure 25.4. In these samples as the thickness of the composite increased, TTI and also duration of burning (flameout time) increased. When sample thickness became equivalent to being thermally thick (i.e., the heat wave penetration depth is less than the physical depth) stage, TTI was less affected, and burning slowed down. PHRR for thermally thin samples is also very high, which decreases until thermally thick behavior is achieved and then is less

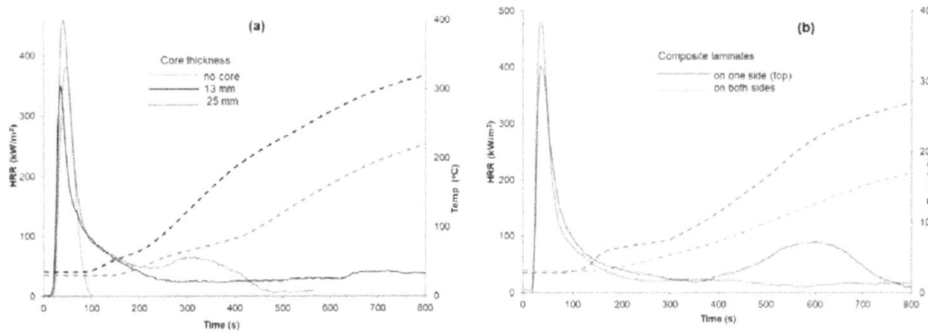

FIGURE 25.3 Heat release rate (solid lines) and temperature on the back surfaces of the sandwich structure (broken lines) versus time of (a) composite laminate of 1.1 mm thickness, sandwich composites with laminates on top surface only of balsa core of 13 mm and 25 mm thicknesses and (b) sandwich structures with balsa core of 25 mm thickness and laminates on one surface and both surfaces of the samples exposed to cone calorimeter at 50 kW/m^2 heat flux.[52]

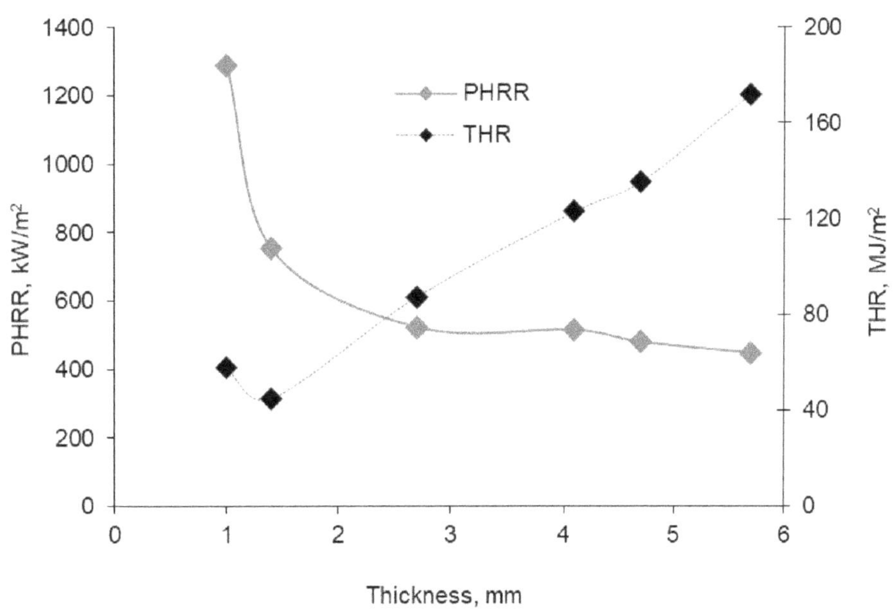

FIGURE 25.4 Effect of composite thickness on PHRR and THR at 50 kW/m^2 cone irradiance of glass/polyester composites.

Source: Reprinted from reference with permission from Elsevier[70]

affected. THR however, increases with thickness as there is more total polymer fuel available for burning. The exception to this would be cases where the composite forms enough char that the composite can self-extinguish and cease combustion. However, in cases of continuous heat flux exposure or very high heat fluxes, these protective chars can only slow combustion and cannot stop it completely. So for some fire safety engineering considerations, the total fuel load of the thick composite may be a detrimental feature, even though the PHRR may be lower. The effect of thickness due to the core in sandwich structures has already been discussed in the previous section, as well as its potential effects on heat release and thermal protection of the core on the rest of the composite structure

25.3.2 Use of Flame Retardants

Since this entire book discussed flame retardants of all types and chemistries, this part of the chapter will focus on some of the key selection criteria for a flame retardant to be used in a polymer composite. Firstly, there is the obvious selection choice of matching FR chemistry to compatibility with the polymer decomposition temperature and chemistry, which is covered in Chapter 2 of this book. But a highly important second criteria, which often can result in de-selection of flame retardants even if they are ideally suited for the application, is compatibility of the flame retardant with the polymer composite manufacturing process. This second criteria breaks down into two more sub-criteria: compatibility with the polymer chemistry and solubility/miscibility in the polymer matrix. These two sub-criteria seem similar, but they are quite different.

In regards to compatibility, one must consider if the flame retardant is reactive with the polymer matrix. Specifically, does it form covalent bonds with the polymer, either during polymerization with monomers (in the case of thermoset resins) or via grafting (in the case of thermoplastic resins). Ideally, reactive FRs are preferred since they cannot migrate out of the polymer over time, and therefore fire performance will be maintained over the lifetime of the polymer composite. Additionally, reactive FRs bring the benefit of benign bioaccumulation profiles from a regulatory perspective (see Chapter 23). However, reactive FRs can change polymer properties, including physical, mechanical, and thermal properties, and these changes may yield polymers that do not have the right properties for the end-use application or cause processing issues due to change in viscosity. In the case or thermoset resins, reactive FRs can change polymerization kinetics, which in turn can change manufacturing of the polymer composite. If the reactive FR speeds up polymerization, the resin may harden too quickly during polymer composite manufacture, and one may not fully "wet out" the fibers, or not fill the mold of the part made via RTM or VARTM processes. If the reactive FR slows polymerization, additional heating and molding/processing time may be required to get a good quality part, which can result in increased cost for the final polymer composite. The ability of the resin to fully wet out the fibers during polymer composite manufacture is critical to ensure good quality parts with thermoset resins. If polymerization kinetics are not well tuned to the manufacturing process, one can get voids, dry fiber areas, and delaminations in the final polymer composite that effectively turn it to scrap since thermoset resins cannot be "remelted" and processed again. Close attention to how reactive FRs work with the polymerization kinetics (and activation temperatures) is critical to ensure good quality parts. Details on reactive FRs with thermoset polymers are covered in detail in Chapters 4, 5, 8, 10, 11, and 13. Some examples of reactive flame retardants for epoxy resins include tetrabromobisphenol A (TBBPA) or its diglycidyl ether, 2,2-bis[3,5-dibromo-4-(2,3-epoxypropoxy) phenyl]propane and 9,10-dihydro-9-oxa-10-phosphaphenanthrene-10-oxide (DOPO). For unsaturated polyester and vinyl ester resins, use of bromostyrene has already mentioned in the previous section, and there is commercially available brominated acrylic monomers available as well for these systems.

Solubility/miscibility with the polymer matrix is another key sub-criteria to consider in polymer composite manufacture. Put simply, at the start of its manufacture, the polymer composite is nothing but a combination of fibers tightly woven together. If the polymer can infuse in between the fibers, but the flame retardant is not soluble/miscible in the molten polymer state, there is a possibility it will "filter out" as the resin infuses into the fibers. Typical methods of making polymer composite such as RTM, VARTM, and even compression molding/melt lamination of thermoplastic materials, all present scenarios where the FR can be filtered out by the fibers as the polymer composite is manufactured. Depending upon the specific fiber weave, the gaps between fibers may only be microns in width, and if the FR additive is an insoluble particle larger than that width, it will concentrate on the outer layers of the fiber. This will result in FR rich areas on the edges of the part, and areas with little to no FR deeper into the part. This often can result in resin not being able to make it past the areas of high FR concentration and one obtains a failed part. Or one ends up with a part that is highly filled on the outer layers with very little polymer present and therefore weakened mechanical properties. Good examples of insoluble/non-miscible FRs that are used in polymer composites are most additive type flame retardants such as mineral fillers and ammonium polyphosphate. If these FRs are not ground to very fine particle size, they will filter out when utilized. The exception to this would be chopped fiber molding compounds, where the fiber reinforcement is not continuous. In that case, the solubility/miscibility of the FR is not really important provided the polymer viscosity is sufficiently high enough to prevent the FRs from "settling out" over time. However, applications where hand lay-up technique is used, these additive type flame retardants are commonly used. Most commonly used flame retardant additives for unsaturated polyester, vinyl ester, and epoxy resins include aluminum trihydrate, magnesium hydroxide, expandable graphite, brominated compounds, ammonium polyphosphate, intumescent systems, organophosphates, phosphinates, etc.

25.3.3 Intumescent Fire Protection

When FR additives filtering out of the polymer composite cannot be resolved, and other FRs that are soluble/miscible/compatible with the polymer matrix are not suitable, the use of fire protection coatings such as intumescent fire protection is often employed.[72,73] Intumescent flame retardant chemistry and its fire protection mechanism is covered in Chapter 7 of this book, so it will not be repeated here. However, it is important to mention that intumescent fire protection only works if the intumescent "foam" is allowed to form. If the intumescent is incorporated into the composite in such a way that it cannot rise up/expand, the fire protection it would yield will be lessened or may be ineffective. Therefore, it is important to ensure that intumescent fire protection for composites is not restrained by the composite fiber structure, or other aspects of the composite construction.

Intumescent flame retardancy for composites is typically incorporated in one of two ways. The first way, which is simple to apply, is to use an intumescent paint/coating on the outer layer of the composite. Provided there is good adhesion between the paint/coating and the upper surface of the composite, this can be a cost-effective method of providing passive fire protection to a composite. However, if the coating should peel/flake off over time, or is scratched/damaged in such a way that the underlying composite is exposed, the fire protection will be greatly lessened, or may not be very effective at all in a fire event. For better durability, the binder used for the intumescent formulation should be compatible with the resin of the composite, as demonstrated in Figure 25.5 where the effectiveness of an epoxy based coating before and after impact and water soak teat has been compared with a water-based intumescent coating, the former maintaining fire barrier effectiveness after impact, water soak, impact plus water soak, whereas water-based coating, as expected, lost its effectiveness after water soak. Further, other sides of the composite that do not have the intumescent paint/coating will be fully vulnerable to fire damage. An intumescent fire barrier provides no protection if the fire starts on the other side of the composite where the protective coating is located. Therefore, understanding how the composite is to be used, and how the fire risk scenario will affect that composite, will guide the user on which side of the composite to apply the intumescent coating/paint. It must be borne in mind that traditionally intumescent coatings are used as "passive fire protection" systems for steel structures, which are nonflammable and the coatings are expected to maintain their structural integrity between 1 and 3 h when the temperature of the surroundings is in excess of 1100°C. For flammable substrates, such as polymer composites, these coatings have to be more efficient to further protect the resin from decomposition and/or ignition. Hence, to be effective, they have to be of particular thickness, usually of 1–5 mm, depending on the substrate geometry and level of fire protection required.[72,74,75] As an example the effect of thickness of an epoxy-based intumescent coating on a 3 mm thick glass epoxy composite, exposed to 50 kW/m^2 heat flux has been evaluated by measuring the surface and back surface temperatures of the composite by inserting thermocouples under the coating on top surface and under the back surface of the composite laminate.[75,76] In Figure 24.6 time to reach 250°C for control and coated with 1, 3, and 5 mm thick coatings have been plotted. As can be seen, for the coating to be effective for a considerable period of time, it has to be ≥ 3 mm thick, which for thin laminates (≤ 3 mm) is not practical. This approach is more commonly used in construction or trains/marine industries where thick composite laminates are used. Sorathia et al. however, have reported that these coatings are not sufficient to protect naval shipboard spacings during a fire and are not equivalent when used alone as direct replacement for blanket-type fibrous fire insulation installed aboard ships.[72] However, if combined with blanket-type, fibrous fire insulation, they can be effective in meeting required fire resistance criteria.

The other approach is to incorporate the intumescent in the upper layers of the composite so that it is fully integrated into the composite and cannot fall off/be removed over time.

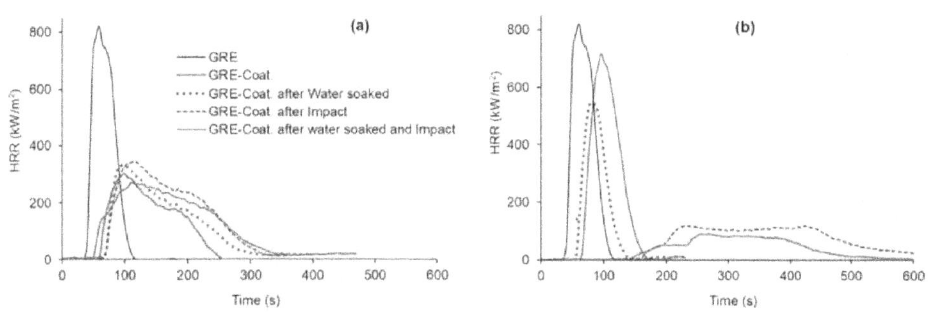

FIGURE 25.5 HRR *vs.* time curves of the glass fiber composite (GRE), intumescent-coated GRE samples before and after water-soak, impact, and water soak plus impact: (a) an epoxy-based and (b) water-based intumescent coating (1 mm thickness).[75]

Source: Taken from Ref.[75]

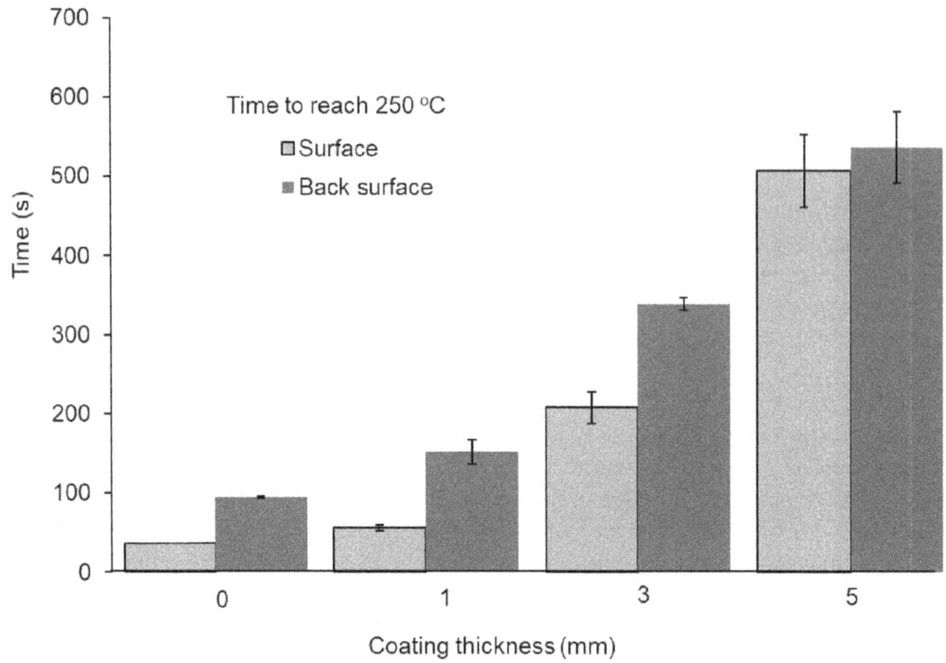

FIGURE 25.6 Thermal barrier effect (time taken to reach top and bottom surface temperatures 250°C) while exposed to a cone calorimetric test at 50 kW/m^2 heat flux.[76]

Source: Adapted from Ref.[76]

This can be accomplished with the use of intumescent mats/non-woven materials that are laid up into the upper layers of the composite during resin infusion/manufacture.[54,77–79] Quite often these mats contain a chemical, such as expandable graphite, that helps in generating intumescent char. As the thermoset resin polymerizes, the intumescent layer is covalently bonded into the overall composite structure, and therefore it is fully integral to the composite surface. As with the abovementioned intumescent coatings/paints, the intumescent barrier must be placed on the side of the composite where fire damage will be incurred. These mats though being very absorbent compared to unidirectional/woven reinforcements used for structural composites, surface becomes resin rich, leading to reduced time-to-ignition. Moreover, slow but prolonged burning, resulting in higher THR values for surface protected composites when compared to the control sample.[77] While intumescent char-forming mats do provide thermal insulation to the core composite, they however need to be heated first and fast enough in order to allow time for char expansion before the temperature is high enough to initiate depolymerization of the polymer resin in the underlying composite. However, if insulative mats purely as surface barriers (no infused resin) are used, this problem can be alleviated.[78]

Another approach to incorporate the intumescent FR system into the resin of the top layer of the composite, similar to the approach of using a resin-rich layer on the outside of a composite for aesthetic or environmental resistance reasons (a.k.a., a "Gel-Coat"). Again, the intumescent layer is covalently bonded to the outer layer of the composite, and it cannot break off/fall off during the lifetime of the composite.

As mentioned previously, if the intumescent layer is damaged, whether a paint or a covalently bonded layer, fire protection may be lost. Therefore, it is strongly encouraged that for severe fire threats/fire risk scenarios, that the user considers adding additional FR to the polymer underneath the intumescent layer. This "Defense-in-Depth" approach ensures that if the intumescent coating fails (or burns away) the inner FR system can further slow fire growth. While this approach will yield superior fire protection for a composite, it may not be usable if the flame retardants underneath the intumescent layer weaken the mechanical properties of the composite. Therefore, this approach is only viable if a balance of mechanical and fire properties is maintained.

25.3.4 Balance of Mechanical and Fire Properties

Polymer composites are typically designed for mechanical properties first, as that is the main reason why they get inserted into applications due to the benefits mentioned previously in this application. If the mechanical properties are not maintained, then very likely the composite will not be used. Therefore, maintaining a balance between mechanical and fire properties is essential. However, in many cases, adding flame retardants to the polymer matrix often lowers/interferes with the mechanical properties of the composite. If the flame retardant lowers T_g, then the use temperature of the composite is reduced. Flame retardants, particularly additive types, usually interfere with the cross-linking of the resin during curing stage, hence affecting the mechanical properties of the resin and composites. Moreover, flame retardant particles may interfere with the ability of the polymer matrix to stay fully adhered to fibers, or it may result in voids/resin-rich pockets that lead to composite delamination, cracks, or other mechanical flaw points. As an example, effect of flame retardant additives (ammonium polyphosphate (APP), alumina trihydrate (ATH), melamine phosphate (NH), and dipentaeryhtritol/melamine phosphate (NW)) at 20% loading on flammability and flexural mechanical performance of glass/unsaturated polyester composites is shown in Figure 25.7, where considerable decrease in flexural modulus can be seen.[80] Therefore, many composites are designed for mechanical properties first, and if fire performance is required, coatings/outer protection systems are used so that the mechanical properties are not compromised.

Another aspect is to maintain mechanical properties during or after exposure to extreme heat or fire. The loss of mechanical properties (tension, flexure, compression, and interlaminar shear properties) of polymeric composites occur mainly due to resin softening when heated, chemical degradation, thermal decomposition, combustion, and damage of the resin matrix that can form a char. When the resin part softens, up to 50% of the mechanical property of a composite laminate could be reduced at temperatures as low as 120°C (depending upon the glass transition temperature of the resin). On degradation/burning, the reduction could be up to 90% or completely lost depending on the resin type.[45,81–83] Thin laminates of glass/epoxy composites have been reported to have no post-fire strength.[74] The phenolic composites, despite having outstanding fire retardant properties, also have low retentions of mechanical properties,[83] which is because the char formed is usually very porous and brittle and provides minimal structural support to the fibers. In case of sandwich structure of glass/vinyl ester laminates as skin and balsa wood as core more reduction in impact properties (65–75% reduction) than the glass/vinyl ester laminate (20–30% reduction) has been reported by Ulven and Vaidya.[16] Greater reductions in post-fire properties in sandwich structures is because on exposure to heat/fire the phenolic skin and balsa wood char, the residual char being very porous and brittle, the structural integrity is also low. Similar behavior has been reported for glass/vinylester face sheets with poly (vinyl chloride) (PVC) and phenolic foam cores by Mouritz and Gardiner.[84]

Since flame-retardant additives are not effective at temperatures close to the glass transition temperature of the resin (usually < 200°C), they have no effect on mechanical properties of the composite, as shown in Figure 25.8(a), where the flame retardant used was an effective char former, but had no effect on mechanical performance of the composite on continued exposure to heat. However, if the flame retardant additives can help in forming oxidation resistant and coherent char, they can help in improving residual mechanical properties after heat/fire exposure of composites, as shown in Figure 25.8(b) where the flame retardant showed effectiveness for samples exposed to ≥ 350°C before being cooled to room temperature.

Thermal barrier surface coatings, on the other hand, help in maintaining the residual mechanical properties of the composites for a considerable period of time.[16,83,84] As an example the effect of different types of thermal barriers, that is, ceramic fabric, ceramic and intumescent coatings, hybrids of ceramic and intumescent coating, silicone foam, and phenolic skins on flammability of glass/vinyl ester composites, studied by Sorathia et al.,[85] is shown in Figure 25.9 where apart from providing fire protection to composites, the residual mechanical properties were much higher than that of the control sample without any surface protection.

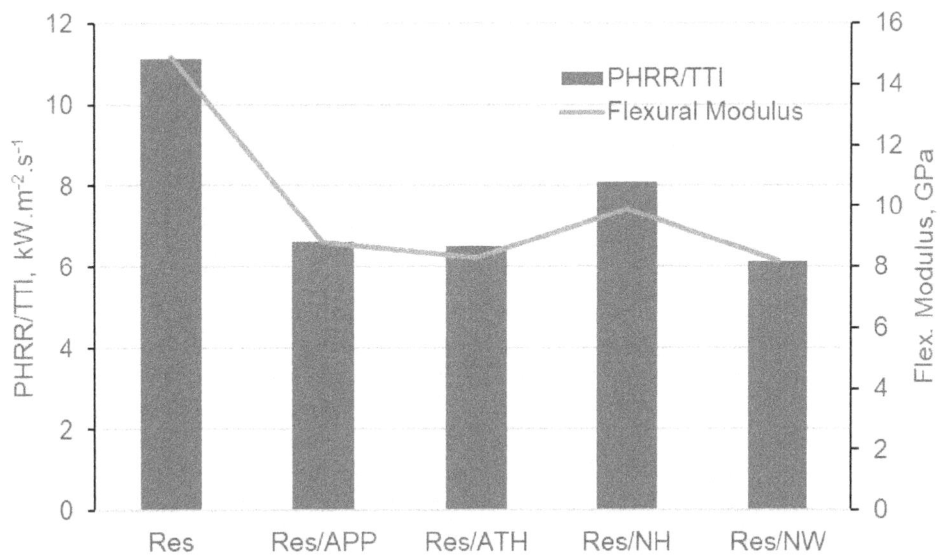

FIGURE 25.7 Effect of additives at 20% loading on flammability and flexural mechanical performance of glass/unsaturated polyester composites.[80]

Source: Adapted from Ref. 80

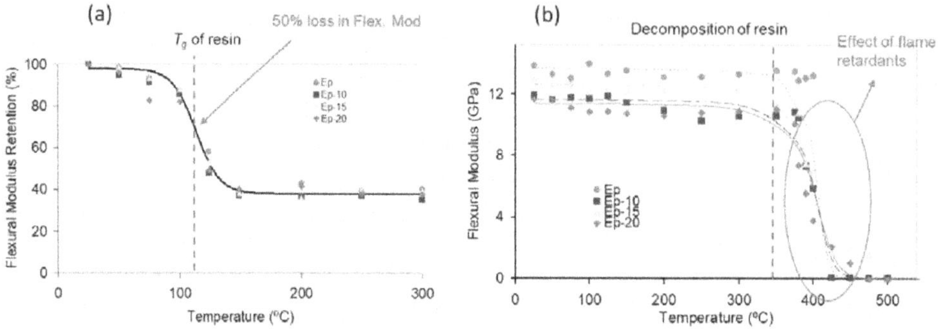

FIGURE 25.8 Residual flexural modulus retention of glass/epoxy composite without/with 10, 15, and 20 wt% of a char-forming flame retardant system (melamine phosphate plus pulverized Visil (a cellulosic fiber containing silica) fiber) as measured at (a) elevated temperatures in a fan-assisted furnace and (b) after exposure to elevated temperatures followed by rapid cooling to ambient temperature.

Source: Taken from ref.[31], reprinted with permission.[31]

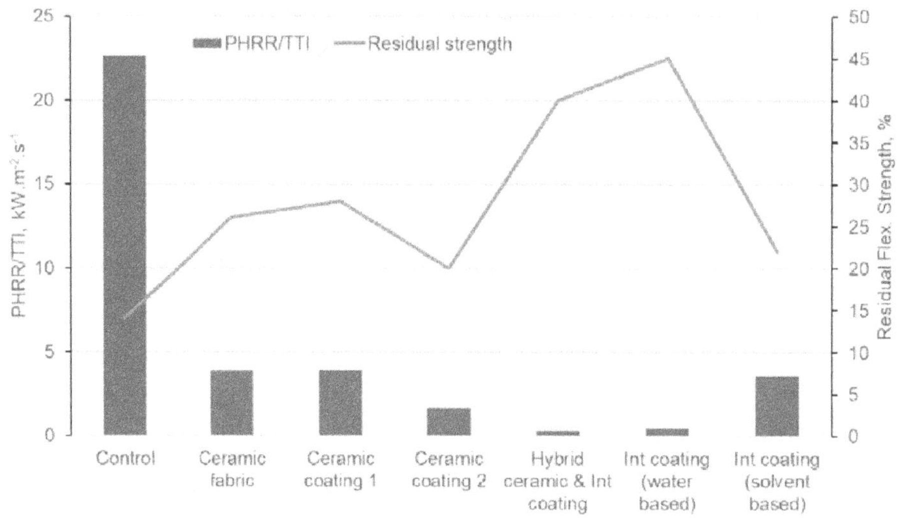

FIGURE 25.9 Effect of different fire barriers on flammability at 75 kW/m^2 heat flux in cone and residual flexural strength after exposure at 25 kW/m^2 for 20 min of glass/vinyl ester composites.[85]

Source: Adapted from Ref.[85]

25.4 THERMOPLASTIC COMPOSITES FOR STRUCTURAL APPLICATIONS

Thermoplastic composites for structural applications are reinforced with glass, carbon, aramid, or metal fibers. These composites can be divided into two classes: the engineering thermoplastics and high temperature thermoplastics. Engineering thermoplastic composites can use nearly any thermoplastic for the matrix, however, in practice choices have been limited to polyvinyl chloride, polypropylene, polyamide, polyesters, and polycarbonate, with polypropylene accounting for about 95% of commercial use. For advanced thermoplastic composites the matrix used could be polyphenylene sulfide (PPS), polyetherimide (PEI), and polyetheretherketone (PEEK).

The use of flame retardant additives is the most popular method of flame retarding thermoplastic composites containing short fibers, where FR chemicals can be added into the polymer and fiber prior to melt compounding. For long fiber composites, FR chemicals can be introduced into the polymer prior to forming sheets. The choice of flame retardant depends upon the type and processing temperature of the polymer used in the composite. The flame retardants used for different polymer types have been discussed in detail throughout this book, hence will not be repeated here. Use of inherently low flammability polymers (such as PPS, PEI, PEEK) in composite forms often removes the need for flame retardants, but additional fire protection systems may be needed to address the abovementioned mechanical failure of these composites when they go above their T_g. On final note, thermoplastic composites that do not char during a fire event can also drip and flow and so may present additional fire threats that need to be addressed depending upon the specific fire risk scenario they are exposed to. In those cases, flame retardants which inhibit dripping (provided they can be processed with the thermoplastic containing fibers) can address this issue in those fire risk scenarios.

25.5 NATURAL FIBER COMPOSITES: UNIQUE ISSUES

Natural fiber-reinforced composites are becoming more and more popular since the last two decades owing to recent environmental and regulatory pressures towards a circular economy, increased sustainability, and recyclability of the products. Amongst natural fibers, lignocellulosic fibers and in particular flax and jute are mostly used for reinforcement of polypropylene (thermoplastic) or epoxy (thermoset) resins for commercial applications.[59] However, to address the demand for fully bio-based composites, biodegradable thermoplastic matrices such as poly(lactic acid) (PLA), cellulose esters, starch plastic, poly(-caprolactone), and aliphatic polyester/co-polyester etc., and for thermosets, furan and bio-derived epoxy resins are potential candidates.

While in principle, the type of fiber-reinforcement (short fiber, nonwoven mat, woven structures etc.) and methods of composite preparation are similar to the respective synthetic fiber-reinforced composites, the choice of matrix material and processability parameters are limited by the thermal sensitivity of natural fibers at temperatures above 180–200°C.[59,86] Additionally, fabric preforms from natural-polymer fibers in co-woven,[59,86] woven from commingled fibers[87] or woven (natural fiber)-nonwoven (synthetic fibers)[88] forms can be prepared and melt pressed (compression molding) to make composites.

Natural fibers have poor compatibility with both synthetic and bio-polymers, leading to weak fiber-matrix interfacial adhesion, which though can be addressed by surface modification of fibers[89] or functionalization of the polymeric matrix.[90] Hydrophilicity of natural fibers leading to moisture absorption, environmental degradation, and their flammability are other major issues associated with their usage. The effect of ageing of jute/PP and jute/PLA composites in aqueous and chemical solutions for 21 days in a study conducted at University of Bolton[91] is shown in Figure 25.10 (a1). Chemical (salt, acidic, alkaline) solutions were selected to study the effect of different chemical environments as composites are commonly used in industries such as marine, oil and gas, chemical plants, offshore etc., where they are exposed to wet, aggressive environments. Both jute/PP and jute/PLA composites absorbed > 6 wt% water. The absorption behavior of each chemical solution was very similar to that for water for all composites. After drying in an oven at 50°C for 24 h, all the adsorbed water desorbed plus further small mass

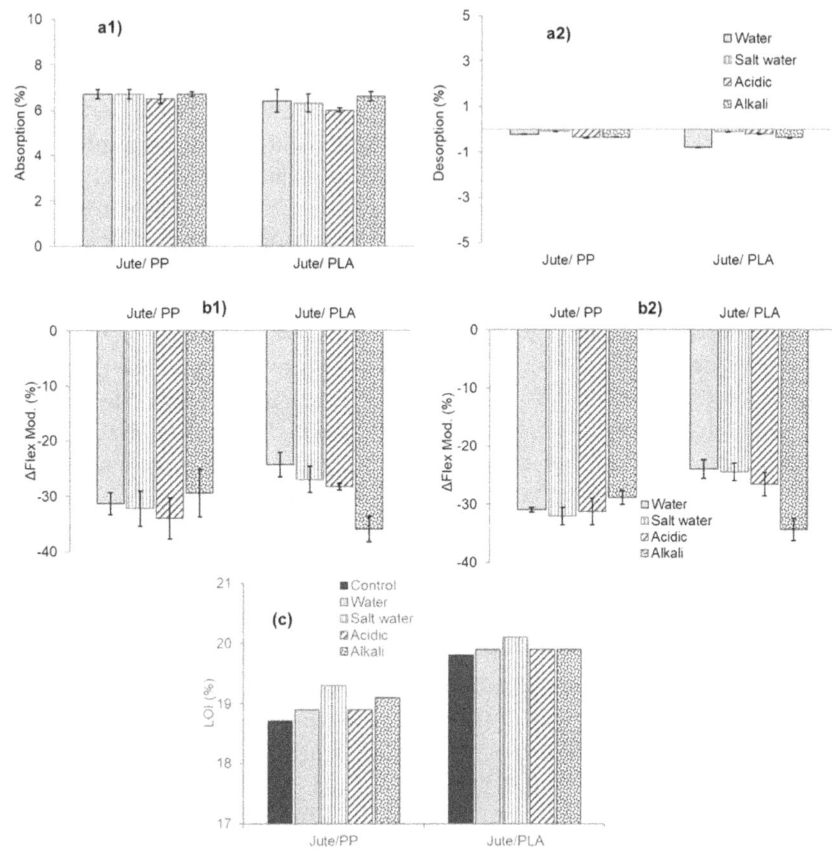

FIGURE 25.10 Effect of ageing Jute/PP and Jute/PLA solutions in water/chemical solutions on: (a1, a2) water/chemical solution absorption and desorption after drying at (a1) room temperature, (a2) at 50°C for 24 h; (b1, b2) change in flexural modulus between unaged and aged, ΔFlex: (b1) room temperature and (**b2**) at 50°C for 24 h; and (c) limiting oxygen index values after oven-drying at 50°C for 24 h.[91]

Source: Adapted from Ref.[91]

loss (< 1 wt%) for each exposed jute-reinforced composites was observed (Figure 25.10 (a2). The additional mass loss indicates loss of adsorbed water. Slightly more mass loss in acidic environment could be due to acid catalyzed degradation of cellulosic fiber and in alkaline environment due to dissolution of hemicellulose. Water absorption affects the mechanical properties of composites, as can be seen from reduced flexural moduli of all samples in Figure 25.10 (b1), which could not be restored even after desorption of water (Figure 25.10 (b2)). Water diffusion occurring through capillary action into fiber morphological voids promotes swelling and increases internal stresses, affecting the interfacial cohesion between the fiber surfaces and the matrix. Such swelling causes irreversible changes in fiber and fiber-matrix morphologies, thus, when the water is desorbed, the composite cannot simply return reversibly to its original physical form, hence explaining the non-restoration of flexural moduli.[91]

Water or aqueous chemical solutions' absorption on the other hand had no adverse effect on the flammability of the composites as seen from LOI values presented in Figure 25.10 (c). In all aged samples, on average LOI increased by 0.1 unit with respect to respective controls. This difference is, though very insignificant, significant in that it demonstrates an increase rather than a decrease. Salt solution exposure of jute/PP composite increased the LOI by about 0.4% units with respect to water-only exposed samples, which is possibly due to the residual Cl- ions present, which may react with the decomposing fiber and act in gas phase, reducing flammability. Acidic and alkaline environments had a little further effect than the water-only ageing.

For flame retardancy, all the strategies used for synthetic fiber-reinforced thermoplastic or thermoset composites discussed in above sections can also be used for natural fiber composites. In addition, fabrics from natural fibers or woven preforms from natural-polymer fibers can be treated with aqueous flame-retardant solution using the conventional textile processing technique of pad-dry and cure (if necessary). In the pad-dry method fabric is passed through the chemical formulation (mostly in aqueous form), passed through rollers to squeeze out the excess, and then dried in oven at 120°C. There are many well-established flame-retardant systems available, discussed in detail in Chapter 27. The FR treatment though weakens the fiber, affecting mechanical properties of the resultant composites, as shown in Figure 25.11, where flax/PP and flax/PLA woven fabrics from commingled fibers were treated with different commercial flame-retardant formulations prior to melt pressing in composites.[87] All flame retardants were very effective in reducing the flammability but had a detrimental effect on mechanical properties.

25.6 CONCLUSIONS AND GUIDANCE STATEMENTS

Polymer composites can be flame retarded like many polymers, but there are some notable restrictions which must be considered in their flame-retardant design:

- End use fire protection needs (including special firefighting measures)
- Compatibility of the flame retardant with the composite manufacturing process
- Mechanical property needs and whether flame retardants, protective coatings, or both should be used
- Effects of fiber types and interactions with the polymer during burning

This makes fire protection of composites more difficult than a thermoplastic molded part, but it can be achieved as there are many FR polymer composites in use today. Using lower heat release polymers when designing these polymer composites is an advisable starting point, but in the cases where this is not feasible from a cost or mechanical perspective, considering the above issues can lead to a successful FR polymer composite. There are additional requirements coming for polymer composites, including recycling needs (see Chapter 24 in this book), and some new fire threat scenarios, such as those seen in electric transportation (see Chapter 32 in this book), but these can be met as well with careful attention to chemistry and the above principles. Bio-based materials may also have a role in this space, as there are increasing amounts of natural fibers used to produce composites for transportation as well as building and construction needs. It's unlikely that bio-based materials will be used in electronic, maritime, or aerospace composite needs due to the performance requirements for those applications, but there may be some bio-based resin systems of low flammability and good performance that may yet find a place in these challenging applications.

The general approach to flame retarding a polymer composite should be as follows:

1) Determine if the polymer can be one with inherently low heat release. If not, due to mechanical or certification requirements, identify the polymer class and the flame retardants available for that polymer.
2) Consider the manufacturing process and if the flame retardants selected will be compatible with the manufacturing process. Start with reactive/polymeric flame retardants and if those cannot be used due to interference with manufacturing methods (or mechanical/certification requirements) then move to nonreactive additives. Nonreactive additives that are soluble in the polymer are preferred, and those which are insoluble should only be used if they will not filter out when infused into the fiber reinforcement, or will not result in manufacturing issues.
3) Consider the effects of fiber types and volume on the composite. This is often set by mechanical performance requirements, but where possible, determine if the fibers can play a role in the fire protection of the system, with higher fiber volumes of non- and low-flammability fibers typically making for a lower flammability composite.

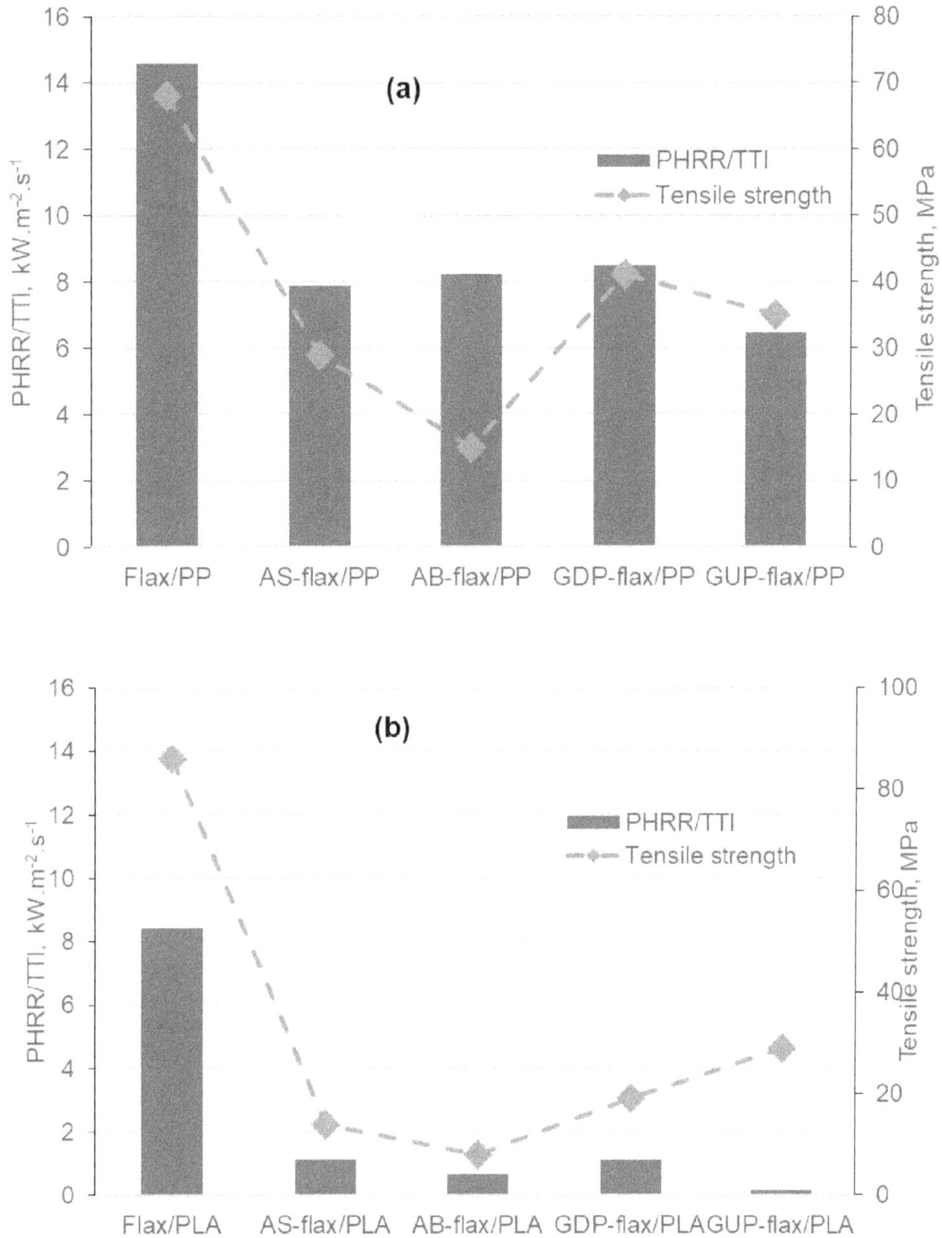

FIGURE 25.11 Effect of FR application on the fabrics from commingled fibers on flammability and flexural mechanical performance of (a) flax/PP and (b) flax/PLA composites. FRs are formulations with FR components as: AS = ammonium sulfamate, AB = ammonium bromide, GDP = guanidine dihydrogen phosphate, and GUP = guanylurea methylphosphonate.

Source: Adapted from the data taken from Ref.[87]

4) In the event flame-retardant additives cannot be put into the polymer matrix, investigate protective coatings. Protective systems which are fully integrated into the polymer composite surfaces (through co-lamination with the main composite) should be used first. If those cannot be used, then use paints/coatings that can be applied to the outer layer of the composite. This step should also be considered when the fire performance requirements are very strict and flame retardant additives were incorporated in step 2 above.

The above steps should yield a successful FR polymer composite, but it is expected that this may require iterative research to achieve the final desired balance of mechanical, thermal, and flammability properties.

REFERENCES

1. www.fire.tc.faa.gov/Handbook (accessed 01/11/23).
2. "Comparison of methods for thermal analysis: Application to PEEK and a composite PEEK+CF" Hache, F.; Delichatsios, M.; Fateh, T.; Zhang, J. *J. Fire Sci.* **2015**, *33*, 232–246.

3. "Investigation of the thermal decomposition and flammability of PEEK and its carbon and glass-fibre composites" Patel, P.; Hull, T. R.; Lyon, R. E.; Stoliarov, S. A.; Walters, R. N.; Crowley, S.; Safronava, N. *Polym. Degrad. Stab.* **2010**, *96*, 12–22.
4. "Heat release rate measurements of thin samples in the OSU apparatus and the cone calorimeter" Filipczak, R.; Crowley, S.; Lyon, R. E. *Fire Saf. J.* **2005**, *40*, 628–645.
5. www.ecfr.gov/current/title-49/subtitle-B/chapter-II/part-216 (accessed 01/11/23).
6. National Fire Prevention Association (NFPA) 130 "Standard for fixed guideway transit and passenger systems" www.nfpa.org/codes-and-standards/all-codes-and-standards/list-of-codes-and-standards/detail?code=130 (accessed 01/11/23).
7. "Standard test method for surface flammability of materials using a radiant heat source", ASTM E162, ASTM International, West Conshockent, PA.
8. "Standard test method for specific optical density of smoke generated by solid materials" ASTM E662, ASTM International, West Conshockent, PA.
9. www.en-standard.eu/set-din-en-45545-1-7-fire-protection-on-railway-vehicles/ (accessed 01/10/23).
10. https://en.wikipedia.org/wiki/SOLAS_Convention (accessed 01/11/23).
11. https://treaties.un.org/doc/Publication/UNTS/Volume%20 1184/volume-1184-I-18961-English.pdf (accessed 01/11/23).
12. "Fire safety of composites in the US Navy" Sorathia U, Ness J, Blum M. *Compos. Part A Appl. Sci. Manuf.* **1999**, *30*, 707–713.
13. "Screening tests for fire safety of composites for marine applications" Sorathia, U.; Long, G.; Gracik, T.; Blum, M.; Ness, J. *Fire Mater.* **2001**, *25*, 215–222.
14. "Cone calorimeter and room corner fire testing of balsa wood core/phenolic composite skin sandwich panels" Morgan, A. B.; Toubia, E. *J. Fire Sci.* **2014**, *32*, 328–345.
15. "Post-fire low velocity impact response of marine grade sandwich composites" Ulven, C. A.; Vaidya, U. K. *Compos. Part A Appl. Sci. Manuf.* **2006**, *37*, 997–1004.
16. "Impact response of fire damaged polymer-based composite materials" Ulven, C. A.; Vaidya, U. K. *Compos. Part B Eng.* **2008**, *39*, 92–107.
17. "Thermal response of composite materials to elevated temperatures" Lattimer, B. Y.; Ouelette, J.; Trelles, J. *Fire Technol.* **2011**, *47*, 823–850.
18. "Modelling decomposition and fire behaviour of small samples of a glass-fibre-reinforced polyester/balsa-cored sandwich material" Marquis, D. M.; Pavageau, M.; Guillaume, E.; Chivas-Joly, C. *Fire Mater.* **2013**, *37*, 413–439.
19. "Deterioration of the fire structural resistance of sandwich composite under tension due to water absorption" Anjang, A.; Chevali, V. S.; Feih, S.; Mouritz, A. P. *Compos. Part A Appl. Sci. Manuf.* **2016**, *87*, 263–270.
20. "Post-fire mechanical properties of sandwich composite structures" Anjang, A.; Chevali, V. S.; Lattimer, B. Y.; Case, S. W.; Feih, S.; Mouritz, A. P. *Compos. Struct.* **2015**, *132*, 1019–1028.
21. "Fire stability of glass-fibre sandwich panels: The influence of core materials and flame retardants" Horold, A.; Schartel, B.; Trappe, V.; Korzen, M.; Bunker, J. *Compos. Struct.* **2017**, *160*, 1310–1318.
22. "Fire resistance behaviour of GFRP-polyurethane composite sandwich panels for building floors" Proença, M.; Garrido, M.; Correia, J.R.; Gomes, M.G. *Compos. Part B Eng.* **2021**, *224*, https://doi.org/10.1016/j.compositesb.2021.109171
23. "Thermal response and edgewise compression failure of thermally degraded sandwich composite structures" Toubia, E. A.; Sihn, S.; Pitz, J.; Vernon, J. P.; Flores, M.; Parks, S. *Compos. Struct.* **2021**, *256*, 113070.
24. "Cardanol-based novolac resins as curing agents of epoxy resins" Campaner, P.; D'Amico, D.; Longo, L.; Stifani, C.; Tarzia, A. *J. Appl. Polym. Sci.* **2009**, *114*, 3585–3591.
25. "Blends of unsaturated polyester and phenolic resins for application as fire-resistant matrices in fibre-reinforced composites: Effects of added flame retardants" Kandola, B. K.; Krishnan, L.; Ebdon, J. R. *Polym. Degrad. Stab.* **2014**, *106*, 129–137.
26. "Fire performance of charring closed-cell polymeric insulation materials: Polyisocyanurate and phenolic foam" Hidalgo, J. P.; Torero, J. L.; Welch, S. *Fire Mater.* **2018**, *42*, 358–373.
27. "Impact of the formulation of biosourced phenolic foams on their fire properties" Delgado-Sanchez, C.; Sarazin, J.; Santiago-Medina, F. J.; Fierro, V.; Pizzi, A.; Bourbigot, S.; Celzard, A. *Polym. Degrad. Stab.* **2018**, *153*, 1–14.
28. "Cone calorimeter testing of S2 glass reinforced polymer composites" Morgan, A. B.; Gagliardi, N. A.; Price, W. A.; Galaska, M. L. *Fire Mater.* **2009**, *33*, 323–344.
29. "A study of the effect of thickness on the thermal degradation and flammability characteristics of some composite materials using a cone calorimeter" Fateh, T.; Kahanji, C.; Joseph, P.; Rogaume, T. *J. Fire Sci.* **2017**, *35*, 547–564.
30. "Phenolic foams: A review of mechanical properties, fire resistance and new trends in phenol substitution" Mougel, C.; Garnier, T.; Cassagnau, P.; Sintes-Zydowicz, N. *Polymer* **2019**, *164*, 86–117.
31. "Thermo-mechanical responses of fibre-reinforced epoxy composites exposed to high temperature environments. Part I: Experimental data acquisition" Kandare, E.; Kandola, B. K.; Myler, P.; Gerard E. *J. Compos. Mater.*, **2010**, *44* (26), 3093–3114. https://doi.org/10.1177/0021998310373511
32. "Mechanistic approach to structural fire modeling of composites" Boyd, S. E.; Bausano, J. V.; Case, S. W.; Lesko, J. J. *Fire Tech.* **2011**, *47*, 941–983.
33. "Compression creep rupture behavior of a glass/vinyl ester composite laminate subject to fire loading conditions" Boyd, S. E.; Lesko, J. J.; Case, S. W. *Compos. Sci. Technol.* **2007**, *67*, 3187–3195.
34. "Compressive response of composites under combined fire and compression loading" Liu, L.; Holmes, J. W.; Kardomateas, G. A.; Birman, V. *Fire Tech.* **2011**, *47*, 985–1016.
35. "Influence of fibre orientation on the tensile properties of sandwich composites in fire" Anjang, A.; Mouritz, A. P.; Feih, S. *Compos. Part A Appl. Sci. Manuf.* **2017**, *100*, 342–351.
36. "Influence of ply stacking pattern on the structural properties of quasi-isotropic carbon-epoxy laminates in fire" Grigoriou, K.; Mouritz, A. P. *Compos. Part A Appl. Sci. Manuf.* **2017**, *99*, 113–120.
37. "Evaluation of the residual strength of a polyester/E-glass composite tray after exposure to high temperatures due to a nearby fire event" Lopes, B. J.; da Silva, G. A. S.; d'Almeida, J. R. M. *Fire Technol.* **2018**, *54*, 853–865.
38. "Post-fire failure mechanisms of seawater-accelerated weathering composites for coastal and marine structures" Elmushyakhi, A.; Toubia, E. A.; Morgan, A. B. *Marine Struct.* **2018**, *63*, 304–317.
39. "Influence of water content on failure of phenolic composites in fire" Feih, S.; Mathys, Z.; Mathys, G.; Gibson, A. G.; Robinson, M.; Mouritz, A. P. *Polym. Degrad. Stab.* **2008**, *93*, 376–382.

40. "Flammability properties of aircraft carbon-fiber structural composite" Quintiere, J. G.; Walters, R. N.; Crowley, S. DOT/FAA/AR-07/57 Report, October 2007.
41. "Prevention of the formation of respirable fibers in carbon fiber reinforced epoxy resins during combustion by phosphorus or silicon containing flame retardants" Greiner, L.; Doring, M.; Eibly, S. *Polym. Degrad. Stab.* **2021**, *185*, 109497.
42. "Potential for the formation of respirable fibers in carbon fiber reinforced plastic materials after combustion" Eibl, S. *Fire Mater.* **2017**, *41*, 808–816.
43. "Smolder behavior and emissions byproducts of aircraft composite coupons" Hatch, J.; Wardall, A.; Jackson, J.; McNeilly, R.; Kirsh, J.; Parker, A.; Morgan, A.; Duran, C. *Fire Saf. J.* **2021**, *123*, 103366.
44. Boeing. "787 Aircraft Rescue & Firefighting Composite Structure April 2013." www.boeing.com/assets/pdf/commercial/airports/faqs/787_composite_arff_data.pdf, accessed on 21 January 2023.
45. "Composites having improved fire resistance" Kandola, B. K.; Kandare, E. *Advances in Fire Retardant Materials*, Horrocks, A. R.; Price, D., editors, Chapter 15. Cambridge: Woodhead Publishing Ltd., 2008, pp. 398–442.
46. "Flammability and fire resistance of composites", Horrocks A. R.; Kandola, B. K. *Design and Manufacture of Textile Composites*, Long, A. C., editor, Chapter 9. Cambridge: Woodhead Publishing Ltd., 2005.
47. "Blends of unsaturated polyester and phenolic resins for application as fire-resistant matrices in fibre—Reinforced composites. Part 2: Effects of resin structure, compatibility and composition of fire performance" Kandola, B. K.; Ebdon, J. R.; Krishnan, L.; Deli, D. *Polym, Deg Stab.* **2015,** *113*, 154–167.
48. "Fire and mechanical properties of a novel free-radically cured phenolic resin based on a methacrylate-functional novolac and of its blends with an unsaturated polyester resin" Kandola, B. K.; Krishnan, L.; Deli, D.; Luangtriratana, P.; Ebdon, J. R. *RSC Adv.* **2015,** *5*, 33772–33785.
49. "Flame retardance and physical properties of novel cured blends of unsaturated polyester and furan resins" Kandola, B. K.; Ebdon, J. R.; Chowdhury, K. P. *Polymers*, **2015,** *7*, 298–315. http://dx.doi.org/10.3390/polym7020298
50. "Thermal stability, flammability and mechanical performances of unsaturated polyester–melamine resin blends and of glass fibre-reinforced composites based on them" Krishnan, L.; Kandola, B. K.; Deli, D.; Ebdon, J. R. *Polymers* **2022**, 14, 4885.
51. "Novel flame retardant thermoset resin blends derived from a free-radically cured vinylbenzylated phenolic novolac and an unsaturated polyester for marine composites" Kandola, B. K.; Ebdon, J. R.; Luangtriratana, P.; Krishnan, L. *Polym. Degrdn. Stab.* **2016,** *127*, 56–64. http://dx.doi.org/10.1016/j.polymdegradstab.2016.01.005
52. "Fire resistant light weight sandwich composites for marine applications" Zhou, C. PhD Thesis, University of Bolton, 2020.
53. "The assessment of full-scale fire hazards from cone calorimeter data" Petrella, R. V. *J. Fire Sci.*, **1994**, *12*, 14–43.
54. "Flame retardants for epoxy resins: Application related challenges and solutions" Kandola, B. K.; Magnoni, F.; Ebdon, J. R. *J. Vinyl Addit. Techn.* **2022**, 28, 17–49.
55. "Combined pyrolysis and radiochemical gas chromatography for studying the thermal degradation of epoxy resins and polyimides, the degradation of epoxy resins in nitrogen between 400 and 700 °C" Bishop, D. P.; Smith, D. A. *J. Appl. Polym. Sci.*, **1970**, 14, 205.
56. "The effects of additional reactive diluents on flammability of unsaturated polyester/phenolic resin blends co-cured with styrene" Kandola, B. K.; Izunobi, J.; Krishnan, L.; Ebdon, J. R. Paper presented at the 14th European Meeting on Fire Retradant Polymers (FRPM13), Lille, June 2013.
57. "Epoxy-toughened, unsaturated polyester interpenetrating networks" Shaker, Z. G.; Browne, R. M.; Stretz, H. A.; Classidy, P. E.; Blanda, M. T. *J. Appl. Polym. Sci.* **2002**, *84*, 2283–2286.
58. "Structure–property relationships of void-free phenolic–epoxy matrix materials" Tyberg, C. S.; Bergeron, K.; Sankarapandian, M.; Shih, P.; Loos, A. C.; Dillard, D. A.; McGrath, J. E.; Riffle, J. S.; Sorathia, U. *Polymer* **2000**, *4*, 5053–5062.
59. "Flame retardant characteristics of natural fibre composites" Kandola, B. K. *Natural Polymers, Volume1: Composites.* John, M. J.; Thomas, S., editors. Chapter 5. Cambridge: RSC Publishing, 2012.
60. "Design and testing of innovative materials for passive fire protection" Landucci, G.; Rossi, F.; Nicolella, C.; Zanelli, S. *Fire Saf. J.* **2009**, *44*, 1103–1109.
61. "Carbon, glass and basalt fiber reinforced polybenzoxazine: The effects of fiber reinforcement on mechanical, fire, smoke and toxicity properties" Wolter, N.; Beber, V. C.; Sandine, A.; Blomqvist, P.; Goethals, F.; Van Hove, M.; Jubete, E.; Mayer, B.; Koscheck, K. *Polymers* **2020**, *12*, 2379.
62. "Thermal degradation of epoxy resin/carbon fiber composites: Influence of carbon fiber fraction on the fire reaction properties and on the gaseous species release" Dao, D. Q.; Rogaume, T.; Luche, J.; Richard, F.; Valencia, L. B.; Ruban, S. *Fire Mater.* **2015**, *40*, 27–47.
63. "Influence of unwoven roving and woven fabric carbon fiber reinforcements on reaction-to-fire properties of polymer matrix composites" Eibl, S. *Fire Mater.* **2020**, *44*, 557–572.
64. "Influence of out-of-plane fiber orientation on the reaction-to-fire properties of carbon fiber reinforced polymer matrix composites" Eibl, S.; Swanson, D. *Fire Mater* **2018**, *42*, 234–243.
65. "Behaviour of aeronotical polymer composite to flame: A comparative study of thermoset- and thermoplastic-based laminate" Schuhler, E.; Coppalle, A.; Vieille, B.; Yon, J.; Carpier, Y. *Polym. Degrad. Stab.* **2018**, *152*, 105–115.
66. "The effect of fibre content on the thermal and fire performance of polypropylene-glass composites" Kandola, B. K.; Toqueer-Ul-Haq, R. *Fire Mater.* **2012**, *36*, 603–613.
67. "Use of high-performance fibres and intumescents as char promoters in glass-reinforced polyester composites" Kandola, B. K.; Akonda, M. H.; Horrocks, A. R. *Polym. Degrad. Stab.* **2005**, *88*, 123–129.
68. "Recommendations on the scientific approach to polymer flame retardancy: Part 2—Concepts" Schartel, B.; Wilkie, C. A.; Camino, G. *J. Fire Sci.* **2017**, *35*, 3–20.
69. "Interfacial charring method to overcome the wicking action in glass fiber-reinforced polypropylene composite" Liu, L.; Liu, Y.; Han, Y.; Liu, Y.; Wang, Q. *Compos. Sci. Technol.* **2015**, *121*, 9–15.
70. "The effect of intumescents on the burning behaviour of polyester-resin-containing composite" Kandola, B. K.; Horrocks, A. R.; Myler, P.; Blair, D. *Compos. Part A Appl. Sci. Manuf.* **2002**, *33*, 805–817.
71. "Fire Properties of Polymer Composite Materials" Mouritz, A. P.; Gibson, A. G. Dordrecht: Springer, 2006.

72. "Evaluation of intumescent coatings for shipboard fire protection" Sorathia, U.; Gracik, T.; Ness, J.; Durkin, A.; Williams, F.; Hunstad, M.; Berry, F. *J. Fire Sci.* **2003**, *21*, 423–450.
73. "Navy R&D Programs for Improving the Fire Safety of Composite Materials" Sorathia, U.; Perez, I. *Fire and Polymers IV: Materials and Concepts for Hazard Prevention.* ACS Symposium Series #922, Wilkie, C. A.; Nelson, G. L., editor. Washington, DC: American Chemical Society, 2005, pp. 185–199.
74. "Polymers and composites—An examination of fire spread and generation of heat and fire products" Tewarson, A.; Macaione, D.P. *J. Fire Sci.* **1993,** *11*, 421–441.
75. "The effects of thermophysical properties and environmental conditions on fire performance of intumescent coatings on glass fibre-reinforced epoxy composites" Kandola, B. K.; Luangtriratana, P.; Duquesne, S.; Bourbigot, B. *Materials* **2015**, *8*, 5216–5237. https://doi.org/10.3390/ma8085216
76. "Quantification of thermal barrier efficiency of intumescent coatings on glass fibre-reinforced epoxy composites" Luangtriratana, P.; Kandola, B. K.; Duquesne, S.; Bourbigot, B. *Coatings* **2018**, *8*(10), 347. https://doi.org/10.3390/coatings8100347
77. "The use of fire-retardant intumescent mats for fire and heat protection of glass fibre-reinforced polyester composites: Thermal barrier properties" Kandare, E.; Chukwudole, C.; Kandola, B. K. *Fire Mater.* **2010**, *34*, 21–38.
78. "The effect of fire-retardant additives and a surface insulative fabric on fire performance and mechanical property retention of polyester composies" Kandare, E.; Chukwunonso, A. K.; Kandola, B. K. *Fire Mater.* **2011**, *35*, 143–155.
79. "New developments in flame retardancy of glass-reinforced epoxy composites" Kandola, B. K.; Horrocks, A. R.; Myler, P.; Blair, D. *J. Appl. Polym. Sci.* **2003**, *88*, 2511–2521.
80. "Flame-retardant unsaturated polyester resin incorporating functionalised nanoclays" Nazare, S.; Kandola, B. K.; Horrocks, A. R. *Polym. Adv Technol.* **2006**, *17*, 294–303.
81. "Post-fire mechanical properties of marine polymer composites" Mouritz, A. P.; Mathys, Z. *Compos. Struct.*, **1999**, *47*, 643–653.
82. "Post-fire mechanical properties of glass-reinforced polyester composites" Mouritz, A. P.; Mathys, Z. *Compos. Sci. Techno.*, **2001**, *61*, 475–490.
83. "Mechanical properties of fire-damaged glass-reinforced phenolic composites" Mouritz, A. P.; Mathys, Z. *Fire Mater.*, **2000**, *24*, 67–75.
84. "Compression properties of fire-damaged polymer sandwich composites" Mouritz, A. P.; Gardiner, C. P. *Compos. Part A Appl. Sci. Manuf.*, **2002**, **33**, 609–620.
85. "Improved fire safety of composites for naval applications" Sorathia, U.; Rollhauser, C. M.; Hughes, W. A. *Fire. Mater.*, **1992**, *16*, 119–125.
86. "Natural-Fiber Composites in the Automotive Sector" Huda, M. S.; Drzal, L. T.; Ray, D.; Mohanty, A. K.; Misra, M. *Properties and Performance of Natural-Fibre Composites*', K. L. Pickering, editor, Chapter 7. Cambridge: Woodhead Publication, 2008.
87. "Fire-resistant natural fibre-reinforced composites from flame retarded textiles", Pornwannachai, W.; Ebdon, J. R.; Kandola, B. K. *Polym. Degrdn. Stab.* **2018**, *154*, 115–123.
88. "Natural fibre-reinforced thermoplastic composites from woven-nonwoven textile preforms: mechanical and fire performance study" Kandola, B. K.; Mistik, S. I.; Pornwannachai, W.; Anand, S. C. *Compos. Part B. Eng.* **2018**, *153*, 456–464.
89. "Surface treatments of plant fibers and their effects on mechanical properties of fiber-reinforced composites: A review" Latif, R.; Wakeel, S.; Khan, N. Z.; Siddiquee, A. N.; Verma, S. L.; Khan, Z. A. *J. Reinf. Plast. Comp.* **2019**, *38*, 15–30.
90. "Interface modification on the properties of sisal fiber- reinforced polypropylene composites" Fung, K. L.; Li, R. K. Y.; Tjong, S. C. *J. Appl. Polym. Sci.* **2002**, *85*, 169–176. https://doi.org/10.1002/app.10584
91. "Effects of water and chemical solutions ageing on the physical, mechanical, thermal and flammability properties of natural fibre-reinforced thermoplastic composites', Kandola, B. K.; Mistik, S. I.; Pornwannachai, W.; Horrocks, A. R. *Molecules* **2021**, *26*, 4581. https://doi.org/10.3390/molecules26154581

26 Flame-Retardant Design and Protection for Wood and Wood Products

Laura E. Hasburgh, Grant T. Kirker, and Katie M. Ohno

26.1 INTRODUCTION

Two of the most important characteristics of wood, its combustibility and strength, have made it so that wood has served humankind both as fuel and as a building material for thousands of years. As a building material, wood is often used for structures, transportation networks, and infrastructure systems. Wood materials come in many forms, including dimensional lumber, heavy timber, and engineered wood products (e.g., plywood, orientated strand board, glulam, and cross-laminated timber). A more recent category of engineered wood products, known as mass timber, has increased the market and implementation of wood in both the building and infrastructure sectors. Mass timber products are thick, compressed layers of lumber that result in strong, structural load-bearing elements. However, as a building material, the combustibility of wood may increase fire risks, including fire development, fire propagation, and change in load bearing capacity (1). These risks are applicable to both interior building materials and materials exposed to the exterior fire hazards. Ultimately, to reduce these risks, fire protection of wood is often required by the model building codes.

To improve the overall fire performance of wood, it is commonly sheathed by a noncombustible product or treated or coated with fire retardants. Over the decades, fire retardants have been either additive or reactive and have traditionally relied on the combined effect between boron, phosphorus, and nitrogen to increase char formation (2, 3). With the increased use of wood-building products (1), increased wildfire exposure (4), and increased awareness of environmental issues and consumer safety for fire-retardant products (2, 5), these older technologies will need to be replaced or upgraded. Understanding wood chemistry and the combustion process is key to developing effective fire-retardant treatments since they must be compatible with the wood and they act by interfering with a particular stage of the wood combustion process (i.e., during heating, decomposition, ignition, or flame spread) (3). When developing new fire-retardant treatment (FRT) technologies, multiple factors must be assessed beyond the improved fire performance including health and environmental impacts, favorable application processes, limited negatives effects on strength and adhesion, increased durability, and end-of-life disposal.

26.1.1 Wood Structure and Properties

At the macroscale, wood is anisotropic, porous, and its complex material properties and chemical composition varies greatly between species and even within a single tree, ultimately affecting how it reacts to fire and its treatability. As a building product, wood is typically separated into two categories: hardwoods and softwoods. The wood obtained from the trunk of a tree is further divided into heartwood and sapwood. The longitudinal direction is parallel with the trunk, the radial direction is from the center of the trunk or branch to the outermost layers, and the tangential direction is perpendicular to both longitudinal and radial directions. Given these directions, the anisotropy of wood is usually referenced with respect to three different planes: the tangential-longitudinal, radial-longitudinal, and transverse (Figure 26.1a).

As a result of the orientation of the fibers within the wood, its properties, such as its strength, thermal conductivity, and ability to absorb treatments, are highly dependent on the direction that it is cut. To maximize the strength properties, wood-building products are typically cut parallel with the longitudinal direction. A thorough discussion detailing how wood strength is a function of structural scale (macro, micro, and polymeric) is provided in (8). When examining the transverse plane of a wood board, distinct growth rings are easily observed and consist of earlywood and latewood (Figure 26.1c). The latewood cells are denser, have thicker cell walls, and narrower lumens when compared to the earlywood cells. Because of this, the strength of the wood is largely derived from the latewood cells.

Both early and latewood cell walls are broken down into three main regions with varying amounts of the main polymeric constituents. Figure 26.1d illustrates the three main cell wall regions: the middle lamella, the primary wall, and the secondary wall. Each region contains varying amounts of the three main polymeric components: cellulose, hemicelluloses, and lignin. The dry weight of softwood species is made up of approximately 33–42% cellulose, 22–40% hemicelluloses, 27–32% lignin, and 2–3.5% extractives. The dry weight of hardwoods consists of approximately 38–51% cellulose, 17–38% hemicelluloses, 21–31% lignin, and 3% extractives (8–10). Since the chemical composition will affect both wood pyrolysis kinetics and its ability to be treated, a brief

DOI: 10.1201/9781003380689-26

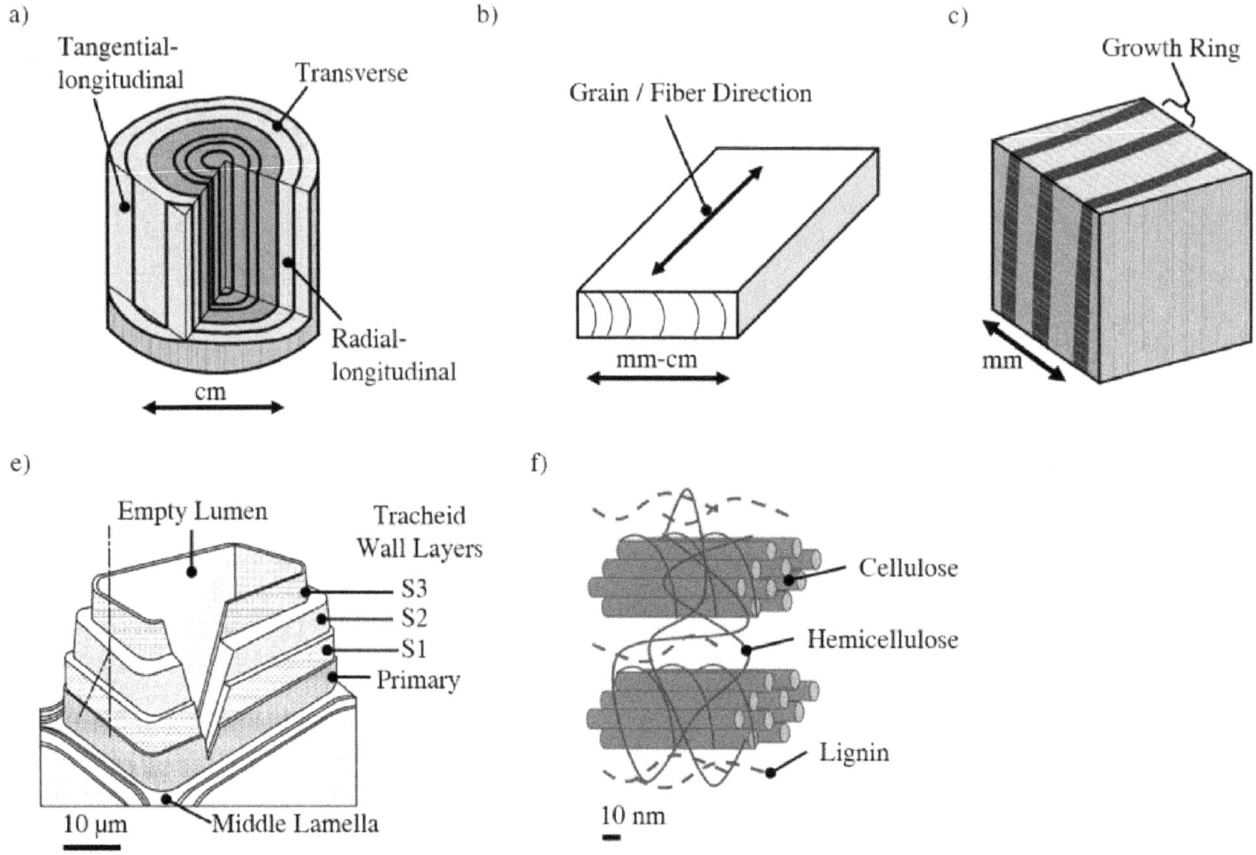

FIGURE 26.1 Graphical illustrations of the hierarchical structure of softwood from the (a) bulk to the (b) board of lumber to the (c) growth rings to the (d) cell wall to the (e) lignocellulose matrix.

Source: Adapted from (6, 7).

background on the chemistry of each wood polymer is discussed below with detailed reviews available elsewhere (8, 11).

Cellulose exists in wood in both crystalline (highly ordered) or amorphous (less ordered) phases (12). With a chemical formula of $(C_6H_{10}O_5)_n$, cellulose (Figure 26.2) is a linear polysaccharide whose primary unit is the cellobiose, which is comprised of two glucose units linked by a glycosidic bond (6, 8, 13). Additionally, each cellobiose unit has six hydroxyl (OH) groups that can build hydrogen bonds.

Hemicelluloses (Figure 26.3) are polysaccharides consisting of branched molecules that form hydrogen bonds with the cellulose (8). Several monomers are combined to make up hemicelluloses, such as mannose, arabinose, xylose, galactose, glucose, galacturonic acid, and glucuronic acid (8, 13). Common hemicellulose components found in the secondary cell wall are shown in Figure 26.3. Both hemicelluloses and cellulose are saturated with single carbon-oxygen bonds. They also contain a carbonyl bond, oxygen double bonded to carbon, which is not present in lignin. The presence of these types of bonds is important when evaluating the chemical changes caused by thermal decomposition.

Lignin (Figure 26.4) consists of cross-linked phenol polymers that bond cellulose and hemicelluloses within cell walls (14). The brittle matrix material of the middle lamella between the cells is also made of lignin. The structure of lignin is complex, consisting of a three-dimensional aromatic polymer network. Lignin is composed of phenylpropane units, including coniferyl alcohol, sinapyl alcohol, and p-coumaryl alcohol, that are linked by ether and carbon-carbon bonds (15).

Extractives account for a small amount (roughly 1 to 4%) of the dry mass of wood; they influence the appearance, odor, microbial characteristics, and reaction to fire (11). Extractives are a mixture of aromatic compounds, terpenes, and aliphatic acids and alcohols. The extractives can be beneficial but also can cause unsightly staining. For construction where the wood is visible, such as exterior siding, the wood is often treated (e.g., coated or heat-treated) to avoid bleeding of the extractives (17).

26.1.2 Wood Pyrolysis, Combustion, and Products

Traditionally, three elements have made up the fire triangle, which is the concept that oxygen, heat, and fuel are needed to sustain combustion. A fire cannot occur if there is not a material or fuel available to burn. Oxygen reacts with the

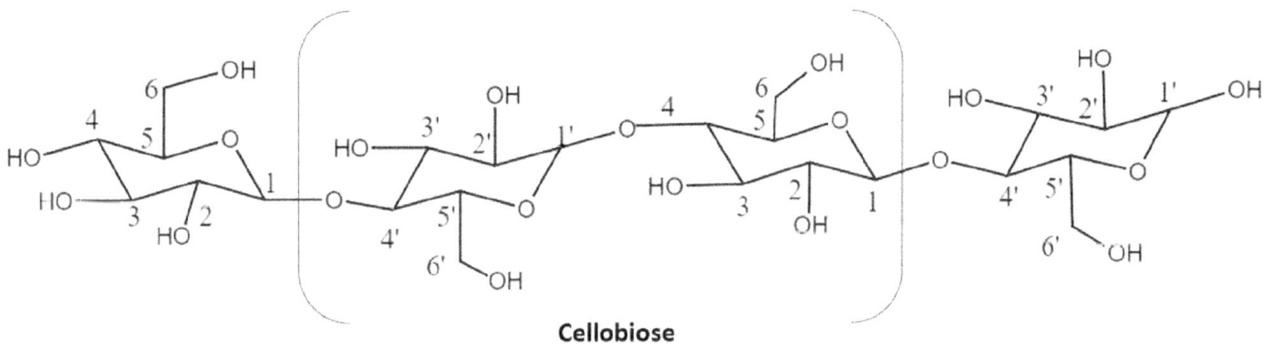

FIGURE 26.2 Molecular structure of cellulose showing the cellobiose unit, indicated by brackets.

Source: Adapted from (6, 8).

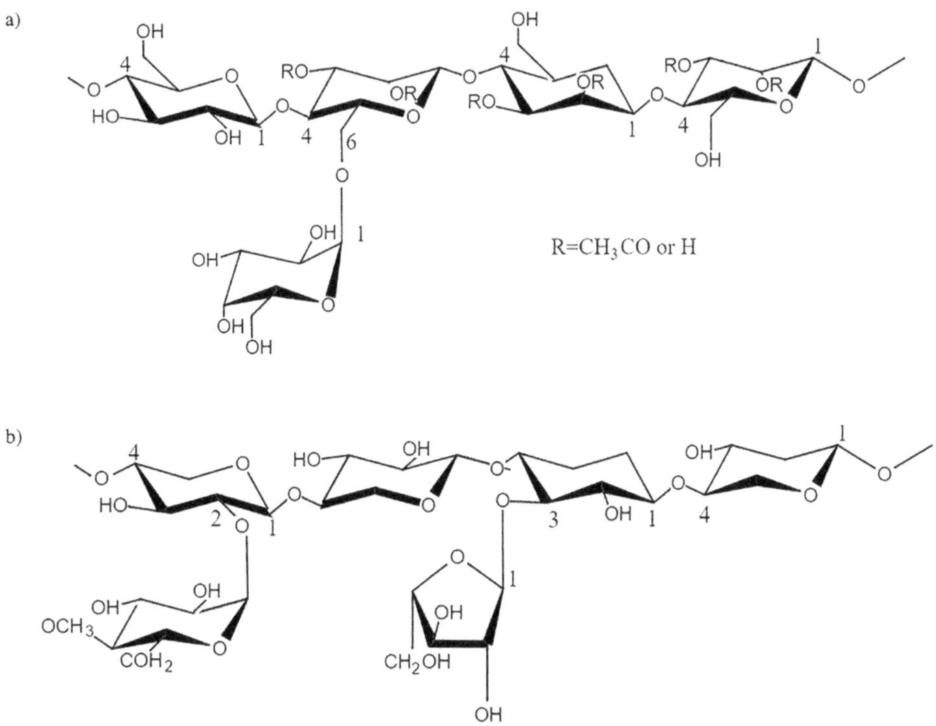

FIGURE 26.3 Structure of common hemicelluloses found in softwood: (a) glucomannan and (b) xylan.

Source: Redrawn from (8).

fuel to release carbon dioxide and heat. The heat is necessary to raise the temperature of the fuel to an ignition point and produce more reacting fuel via pyrolysis to sustain fire. Recently, "chemical chain reaction" was identified as the fourth necessary component of fire, and the concept has expanded from a triangle to a fire tetrahedron (18). If one or more of components are removed, then combustion will not begin, or an active fire will be extinguished.

As wood is exposed to high temperatures, a complex and interdependent series of chemical reactions and physical transformations occur. These begin with hydrolysis, then oxidation, dehydration, pyrolysis, and finally, combustion reactions are followed (19, 20). However, these reactions do not just start at high temperatures but are rather a continuum over time as the temperatures increase. Given enough time, the reactions can begin at temperatures lower than 100°C, as modeled in (21). Pyrolysis is the irreversible thermophysical decomposition of a material that results in changes in chemical composition and physical properties of the wood components. The thermal decomposition of wood has been extensively studied with more recent studies focused on modeling the kinetic reactions (22–26). Wood pyrolysis is often characterized by the thermal decomposition of the hemicelluloses, cellulose, and lignin (27–30). Meszaros et al. show that the extractives thermally decompose in the same temperature range as the three main wood

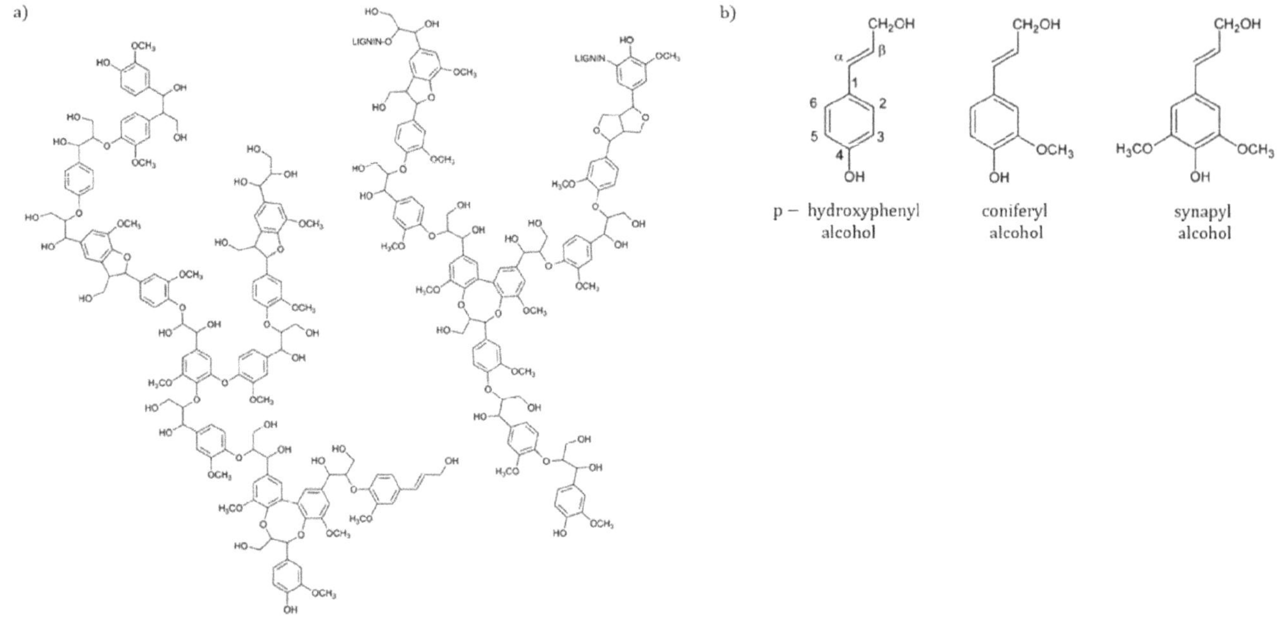

FIGURE 26.4 (a) Structure model of lignin in softwood and (b) monolignols, the primary precursors of lignin.

Source: Redrawn from (15, 16).

polymers (31). The overall effect of extractives on the thermal decomposition of wood depends on species, however, for all species studied to date, the removal of extractives resulted in improved thermal stability but decreased char yield (32–34). Above 200°C, the chemical structure of the polymeric components decomposes into light volatiles that are released from the surface. Hemicelluloses begin to break down between 200° and 260°C and contribute more to the production of volatiles and less to the formation of tar and char (29, 35). Cellulose will begin to decompose between 240 and 350°C with the formation of char, tar, and gaseous products particularly prevalent above 300°C (19). Lignin thermally decomposes between 280 and 500°C with the dominate result being the formation of char. At these higher temperatures, if the flammable volatiles that are produced by both the wood polymers and tar reach the lower flammability limit and react with oxygen, a diffusion flame is produced and known as flaming combustion (gas phase oxidation). Flaming combustion can be sustained by emanating sufficient heat (convective, conductive, and radiative) for pyrolyzing the virgin fuel to generate more flammable volatiles and, thus, a chain reaction. The reaction rates of flaming combustion must be sustained to overcome heat losses of a particular flow configuration, thus giving rise to fire retardant concepts for interfering with chain reactions within the combustion zone.

In addition to the gaseous products that result from wood pyrolysis, a carbon-rich residue, known as char, is produced. The structure of char varies widely and is dependent on heating rates and length of exposure to heat, wood species, material properties (such as density and porosity), and fire-retardant treatments (2). The char yield increases when wood is exposed to slower increases in heat flux and temperature rather than rapid increases. While several fire retardants will increase the char yield, they will not affect the char rate. The char rate is the rate at which a char front progresses into the depth of wood and has been shown to depend greatly on the external heat flux and oxygen concentration in addition to the density, moisture content, permeability of the wood (36–39). This char rate has been well characterized to be 0.6 mm min^{-1} for softwoods exposed to the standard time-temperature curve from the standard test from standards such as ASTM E119 (40) used to determine fire resistance for structural members. Once present, the char may undergo solid-phase oxidation, known as smoldering combustion, but will also act as an insulating material reducing the diffusion of heat into the wood, consequently decreasing the volatization rates that results in lowering the mass transport of volatiles released (41, 42). Compared to flaming combustion, wood smoldering combust is not as well understood, but a shift to focusing on understanding the smoldering combustion is occurring in conjunction with the increased use of mass timber structures. This is because these timber structures are susceptible to smoldering, associated with more amenable critical combustion conditions of oxygen concentrations, pyrolysis temperature, wood porosity, and moisture content that may be difficult for firefighters to observe, and may continue for hours or days after the end of flaming combustion (43, 44).

26.2 WOOD PROTECTION

Fire retardants are either chemical or physical treatments and the primary objectives of fire retardants for wood products are to hinder pyrolysis, reduce time to ignition, prevent flame spread, and suppress toxic smoke production (2, 45, 46). Flame spread is the movement of the flame front across the surface of a combustible material with the flame foot acting as the ignition source of the adjacent, preheated surface, and air/volatile mixture. Ignition refers to the visual confirmation of sustained combustion (flaming, smoldering, or glowing) on the surface of the material and can either be piloted or autoignition (47, 48). For wood, piloted ignition typically occurs around 300°C to 400°C while autoignition will occur at higher temperatures near 600°C. However, unpiloted ignition has been reported at temperatures as low as 200°C, and it is recommended that wood surfaces be maintained below 80°C to avoid incipient wood degradation that will result in lower ignition temperatures (47–49).

26.2.1 Wood Protection Mechanisms

Most fire-retardant treatments are effective because they combine two or more operational modes or mechanisms of action, which are categorized with examples in Table 26.1.

For wood, most commercial fire-retardant treatments include halogens, phosphorus, nitrogen, heavy metals, or boron and will result in systems that both increase the char production, reduce flammable volatiles, or form a protective layer.

TABLE 26.1
Mechanisms of Action for Wood Fire Retardant Treatments (From 2, 3, 45, 46, 50, 51)

Mechanism	Details
Chemical	Pyrolysis of wood directed towards the production of char and water rather than flammable volatiles. The increased char yield is due to an acid-catalyzed hydrolysis of the cellulose and hemicellulose, which dehydrates the wood and promotes the condensation and cross-linking of the carbon skeleton. *Examples: ammonium polyphosphates, guanidine phosphates*
Thermal	Altering the thermal properties to interfere with ignition or combustion can be accomplished by providing thermal insulation to the wood, absorbing heat via endothermal reactions, increasing the thermal conductivity of wood to dissipate the heat, or adding compounds that act as a chemical heat sink to lower the temperature in the wood near the flame. *Examples: acetate, magnesium hydroxide, cyanurates*
Gas	Production of inert or heavy gases that either dilute the flammable gases produced by wood to prevent the formation of a flammable mixture or inhibit chemical reactions in the flame so that combustion is incomplete, and less heat is available to sustain burning. *Examples: magnesium chloride, ammonium sulphamate, ammonium bicarbonate*
Coating/ Ceramification	Utilizing or producing glassy or foamy protective layers to prevent the escape of flammable vapors and access of oxygen or adding compounds that results in a wood-inorganic composite that is less flammable. *Examples: boric acid, silica, potassium silicate, aluminum oxide*

26.2.2 Wood Fire Retardant Treatment History

26.3.2.1 Pre-1950

Treating wood to increase fire retardancy goes back to the Romans using solutions of alum and vinegar to protect boats, but in 1820, Gay-Lussac introduced ammonium phosphate and borax as treatment options (51). The roots of Gay-Lussac's work are clearly visible in modern wood fire protection as variants of these compounds are still widely used today. The dawn of the 20th century saw new treatment formulations built on mineral constituents (silicates, sulfates, borates, phosphates, zinc, tin, or calcium), and by the World War I era, these were being used in combination with ammonia to produce ammonium chlorides, sulfates, and phosphates, which were found to be highly effective in wood (52). In the early 1930s the USDA-FS Forest Products Laboratory conducted extensive testing of one of the largest test groups of inorganic formulations to date (53). In that final report, over 130 formulations were tabulated and ranked based on their ability to reduce flame spread. Diammonium phosphate was found to be highly effective at reducing flame spread, and this marks the first mention of this compound as a highly effective fire retardant and would remain a popular choice up until today. This report also pointed out difficulties associated with several of the test compounds that were either too costly, highly corrosive to wood, imparting strength losses or other undesirable characteristics (54). Based on these findings, future endeavors would focus on in situ effects of fire retardants with the individual structural components of wood (55). Two of the earliest specifiers of fire-retardant treated wood were the US Navy, who were interested in fire proofing of deck and structural components of naval vessels (56) and the railroad, who frequently encounter increased fire hazard on trestle bridges via sparks generated by braking (57).

26.3.2.2 1950–2010

Towards the end of the 1960s, the focus on fire retardants shifted to engineered wood products such as plywood, particleboard, and oriented strand board due to their increasing use in both nonresidential (light frame) and residential construction. Analytical methods improved considerably during this time through the advent of thermogravimetric analysis (TGA) and differential thermal analysis (DTA) which complemented the other longstanding fire test methodologies (ASTM E84, ASTM E160) by providing information on heat release rate of fire-retardant treatments (29, 51, 57). Additionally, due to the leachable nature of most common fire retardants at this time, pressure treatments became a more standard method of application to ensure deep penetration of the retardant chemicals into the wood cellular matrix. In continuing the idea of *in situ* fire protection for wood and wood-based materials, there is proposed research involving the use of dual component or multi-component systems that instill both preservative benefits to the wood while also providing suitable fire protection (Figure 26.5). These combined treatment options were reviewed by White and Sweet (58), and a continuation of the discussion can also be found in Marney and Russel (59).

Bruce et al. (55) added zinc and phosphorus to chromated copper arsenate and found that it improved fire resistance considerably but was failed in market adoption due to concerns over the color and relative toxicity of the material. Jain and Cedercreutz (60) developed one of the first combined FRT systems, consisting of boric acid, zinc chloride, copper sulfate, and sodium bichromate. Although the material showed good fire resistance and promising activity against decay fungi, the material was prone to leach and therefore of limited use in this situation. Similarly, Purushtotam et al. (61) found similar outcomes with a similar mixture with the addition of monoammonium phosphate (not listed on timeline). Collister et al. (62) evaluated phosphorus as an additive to pentachlorophenol and creosote and this was abandoned due to concerns over penetration of triaryl phosphate. Dev and Kumar (63) took a similar approach to that of Purushtotam et al. (61), but instead used a gaseous carrier, but was lacking in fire performance. Goldstein and Drehrer's (64) initial work with amino resin formulations (not listed on timeline) found that aqueous solutions of dicyandiamide and phosphoric acid imparted fire resistance and the non-hygroscopic nature of this treatment reduced wood swelling by 40%, but leaching of the material was not addressed. Juneja (65) investigated the addition of urea, formaldehyde, phosphoric acid, and dicyanamide to overcome some of the issues with leaching, but the improvements were minimal. Follow-up work by Juneja and Calve (66) found that resin curing temperatures were insufficient to prevent leaching of phosphorus and nitrogen. Gyarmati et al. (67) evaluated the combination of ammonium sulfate, diammonium phosphate, borax, and sodium fluoride, but the combination again was subject to leaching. Ermus et al. (68) found that a mixture of cupric sulfate, borax, ammonium sulfate, and boric acid imparted both decay resistance and fire performance, presumably due to the formation of insoluble copper borates. Gardner et al. (69), evaluated a proprietary blend containing UDPF and other compounds and found phosphorus to be highly leachable. This period of time witnessed significant advancements in knowledge on leaching and adverse effects of fire retardants to wood, but limited progress had been made in advancing a marketable option. The mid to late 1980s also saw increased concern of fire-retardant treatment of trusses and plywood sheathing, particularly to first responders who rely on the structural integrity of these materials (Figure 26.6).

Alexiou et al. (70) published positive results using the same proprietary mixture used by Gardner et al. (69) but still did not address concerns over leaching of phosphorus, regardless of initial curing temps. Levan et al. (71) and Lilaro et al. (72) simultaneously developed a dual use FRT that consisted of resin impregnation, both of which were patented. The latter part of the 1980s saw a departure from the use of dual use treatments and more towards the use of insoluble organic compounds. Yasuda and Ota (73) used a multistep treatment of calcium chloride followed by sodium hydroxide, which resulted in the formation of insoluble calcium hydroxide. This was then impregnated with sodium aluminate and sodium sulfate to form ettringite, which is extensively reviewed in literature on the durability of concrete (74). Saka et al. (75–77) looked at the efficacy of sol gels on fire performance and biological durability and was found to be quite effective. Several other studies (Hirata et al. (78) = calcium carbonate (not listed on timeline), Ota et al. (79) = dibasic barium phosphate, and Ishikawa and Adachi (80) = diammonium phosphate + boric acid + adhesives) looked at various combinations of inorganics that formed insoluble end products in the wood and showed promising results.

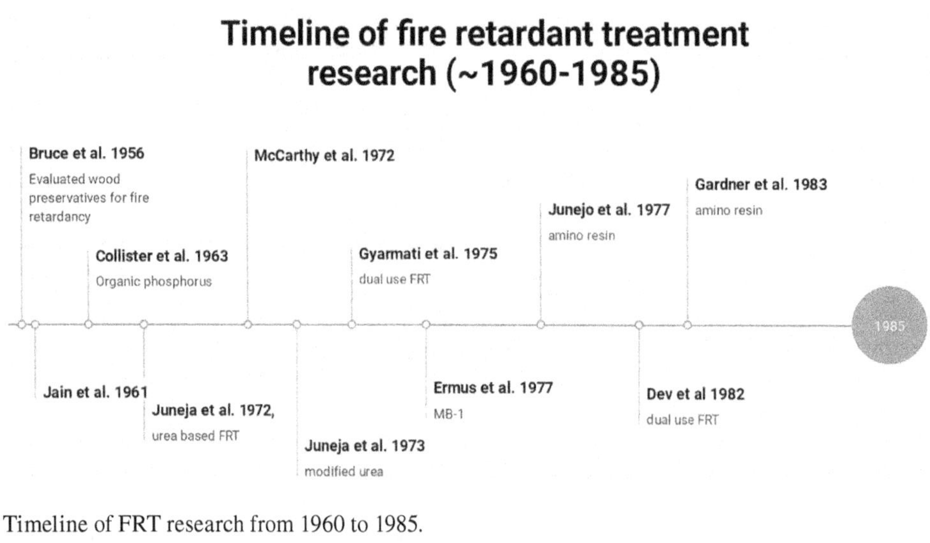

FIGURE 26.5 Timeline of FRT research from 1960 to 1985.

Source: Adapted from (58, 59).

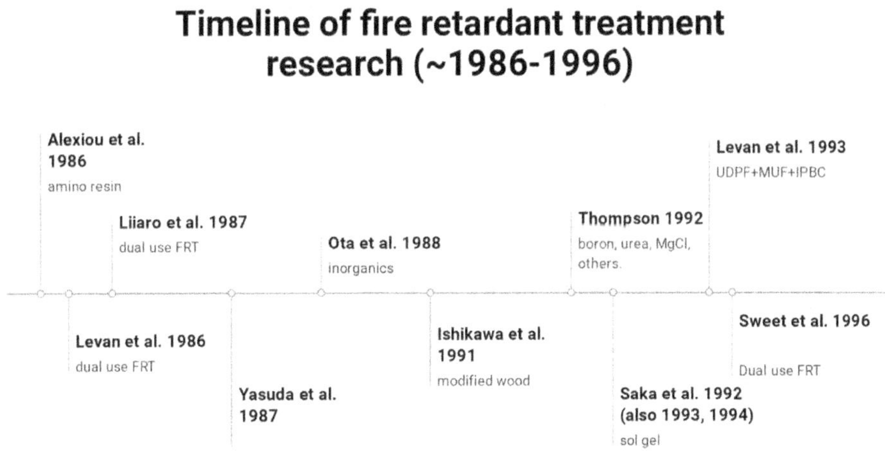

FIGURE 26.6 Timeline of fire-retardant treatment research from 1985 to 1996.

Source: Adapted from (58, 59).

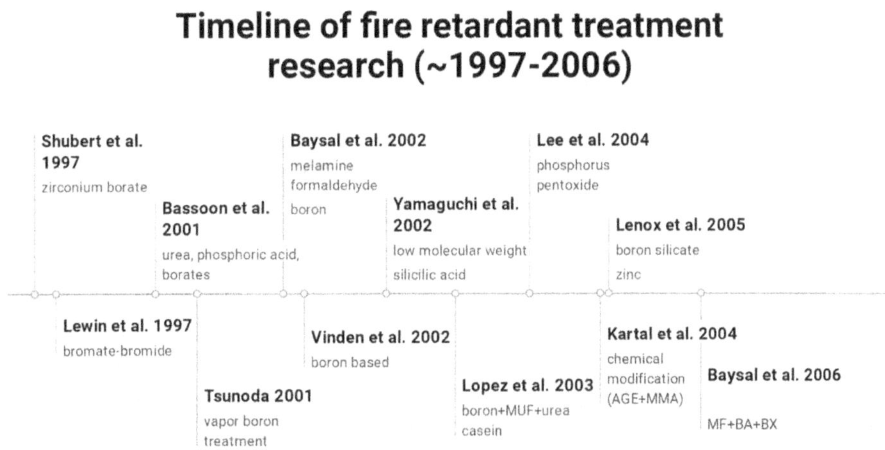

FIGURE 26.7 Timeline of fire-retardant treatment research from 1997 to 2006.

Source: Adapted from (58, 59).

The period from 1997 to 2006 saw renewed and increased usage of borates and chemically modified wood (Figure 26.7). Schubert et al. (81) evaluated zirconium borate as a fire retardant and found improved resistance to leaching, but little improvement in fire performance and biological durability. Lewin (82) evaluated the chemical modification of lignin with an acidified bromate bromine solution. The authors reported that at bromination levels less than 10%, a class 2 rating was achievable according to French standards and a class 1 rating according to the British standards applicable at that time. Basson et al. (83) combined common fire retardants (urea, phosphoric acid) with a waterborne solution of borates, but results were inconclusive. Tsunoda (84) evaluated vapor boron treatments and showed promising results based on laboratory tests, but concerns over leachability of these compounds prompted re-evaluation. Several other studies evaluated borates in combination with fire-retardant compounds, Baysal et al. (85), Vinden et al. (86), Lopez et al. (87), Lenox et al. (88), and Baysal et al. (89), and collectively these studies effectively demonstrate the lack of congruency between fire resistance and biological activity. The levels of borates typically needed to control termites and decay fungi are substantially lower than what is needed for adequate fire performance, which highlights the difficulties in the formulation of a borate-based dual component FRT treatment. Several studies were also conducted looking at chemical modification to existing structural components in the wood, Lee et al. (90) reacted phosphorous pentoxide loblolly pine and sweetgum, which improved decay resistance, but the fire performance was modeled using various methods and never scaled up to actual fire tests. Kartal et al. (91), evaluated chemical modification with allyl glycidyl ether in combination with methyl methacrylate to alleviate leaching. This approach showed promising biological data, but limited efficacy in relation to fire performance.

26.2.3 Present-Day Fire-Retardant Chemicals, Treatments, and Systems

Chemical fire-retardant systems may either be impregnated (pressure- or immersive-dip-treated) or additive in nature. Fire-retardant chemicals minimize ignition and flame spread and contain functional chemical additives like halogens, borates, phosphates, nitrogen, silicon, and nanomaterials. Physical fire-retardant treatments, also known as coatings, are applied to the surface of a material and can either be traditional or intumescent (45) [See Chapter 7 in this book for more information on intumescent materials]. These chemicals act to modify pyrolysis in untreated wood and can negatively affect initial and/or long-term strength of fire-retardant treated wood (45). Traditional coatings include paints, stains, oils, and varnishes that contain fire-retardant chemicals. These coatings delay ignition and slow the rate of burn because the fire-retardant chemicals are activated by ignition before the wood substrate is affected. Factors controlling flame-retardant surface coatings are ignition, flame spread, heat release, smoke, and gases; however, chemical makeup and pyrolysis behavior dictates modes of action in the gas or condensed phase. Intumescent coatings will form a layer of char by swelling to many times their original thickness when exposed to heat. This char acts as an insulation and slows the heating of the substrate material. Fire-retardant chemicals, treatments, and systems are discussed in further detail below.

26.2.3.1 Halogen-Based Fire-Retardant Chemicals

Chlorine and bromine are the most widely used halogen-based fire retardants (3); however, the use of these types of fire retardants are limited and even banned (92). The typical mode of action of halogen-based fire retardants are to create fire extinguisher properties in the gas phase to scavenge free radicals during combustion. Halogenated acid is typically liberated at the wood surface and helps to decrease temperature in the flame zone, which reduces via the formation of a halogen vapor blanket and interrupts reactive halogen-free radicals (3). The application of halogen-based flame retardants are restricted due to health impacts. Halogenated compounds release toxic and/or highly corrosive gases during combustion which are harmful to not only humans but the environment. Additionally, brominated fire retardants (i.e., poly brominated biphenyls/diphenyl esters, hexabromocyclododecanes, and tetrabromobisphenol) exposure in humans can be neurotoxic, affect thyroid function, and can lead to infertility all dependent upon the amount, type, and presence in the environment (92). See Chapter 4 of this book for more information on this class of flame retardants.

26.2.3.2 Inorganic Salts-Based Fire-Retardant Chemicals

Inorganic salt-based fire retardants consist of somewhat more environmentally friendly compounds (i.e., hydroxides, phosphates, carbonates, and sulfates) and are mostly used as co-additives. These types of fire retardants provide good thermal stability, release less smoke, and produce less toxic and/or corrosive gases. The mode of action of inorganic salt-based fire-retardant treatment systems are to cool the fuel source, increase char formation, lower surface thermal conductivity, reduce heat release rate, and delay ignition time (3). Additionally, these fire retardants undergo endothermic decomposition in which the metal hydrate acts as a chemical heat absorber which cools the wood by lowering flame temperature and increasing time to ignition (92). Aluminum and magnesium hydroxides are widely used in coating formulations (3). Other inorganic salts-based fire retardants include potassium carbonate, sodium tungstate, sodium tannate, sodium molybdate (92). Potassium carbonate has fungicidal properties and a high temperature of decomposition (800°C) and works as a dehydration catalyst during combustion which increases char formation, water vapor, and CO_2 gas on burning surfaces (92). Inorganic salts-based fire retardants are highly soluble in water and are subject to leaching due to salt migration with water movement in wood, thus are recommended for use in interior applications (3). See Chapters 6 and 8 of this book for more information on these types of flame retardants.

26.2.3.3 Boron-Based Fire-Retardant Chemicals

Boron-based fire retardants consist of boric acid, borax, and boron salts. These types of fire retardants are commonly used in wood preservation because they have high thermal resistance, biological resistance, nontoxic, cheap, easy to formulate, and can penetrate deep to give long lasting protection (3). Boron-based fire retardants have low combustion properties and low melting points and form glassy films during exposure to high temperatures (92, 93). These fire retardants act in the condensed phase during the decomposition of materials at high temperatures by forming carbon instead of CO/CO_2 gas. Additionally, boron-based fire retardants form a protective layer which prevents the access of oxygen by preventing carbon oxidation (92, 94). Boric acid promotes glassy micro surfaces and catalyzes dehydration reactions in cellulose which increases char formation, reduces volatile organic compound production, reduces smoldering and glowing combustion, but has little effect on flame spread (51). Borax reduces flame spread but promotes smoldering and glowing. For these reasons, boric acid and borax are typically used in combination with each other and/or other compounds (8, 95). The mode of action of boron compounds utilizes a combined effect of the formation of an intumescent, glassy, inorganic layer and an increase in char formation likely achieved though borate ester formation and the inhibition of volatile release (51). Additionally, borate compounds provide a heat sink by giving off water, fuel diluent, a propellant not in the flame zone, and a blowing agent (96, 97). Other common boron-based products include boric acid + trimethylolmelamine and dicyandiamide (resistant to decay after severe weathering), boron salts (catalyzes dehydration and esterification), boron salt + guanyl urea phosphate (improves thermal stability, increases char formation, lowers pyrolysis temperature), boron compounds + alumina trihydrate (forms porous and ceramic-like sintered layer at elevated temperatures), and boron salt + tannin-based formulations (3, 97–99). Boron-based fire retardants are subject to leaching and are not recommended for use in-ground contact (3). See Chapter 9 of this book for more information on boron-based flame retardants.

26.2.3.4 Phosphorus-Based Fire-Retardant Chemicals

Phosphorus-based fire retardants consist of phosphates, phosphines (and oxides), polyphosphonates, phosphonium compounds, and phosphonic acid salts (3, 18, 92). These types of fire retardants are used in wood preservation acting to reduce flammability and are generally produce less toxic gas and smoke compared to other fire-retardant treatments (3) and have replaced halogen-based fire retardants (18). Phosphorus-based fire retardants act in both the condensed phase and gas phase (3, 18). In the condensed phase, phosphoric acid, phosphonates, and phosphate promote char formation (18). In the gas phase, phosphorus-based radicals (PO, PO_2, HPO_2) inhibit chain reactions (18) and flame spread (92). These flame retardants are versatile and are customizable by adjusting the inorganic-organic content, phosphorus loading, and oxidation state of phosphorus derivatives (18). Phosphorus compounds increases the moisture content in the wood which can promote fungal growth, therefore, are suitable for indoor application (3). Additionally, some organophosphates can be hazardous for respiratory health in children (3). See Chapter 5 of this book for more information on boron-based flame retardants.

26.2.3.5 Nitrogen-Based Fire-Retardant Chemicals

Nitrogen-based fire retardants consist of melamine, melamine-formaldehyde resins, urea, and dicyandiamide (3). These fire retardants are environmentally friendly, non-toxic, recyclable, and have lower smoke emissions during combustion (92). Nitrogen-based fire retardants act in both the condensed phase and gas phase (3) to decrease oxygen saturation in the gas phase which suppresses polymer leakage (92). Nitrogen is typically combined with phosphorus in a 2:4 molar ratio (3) which promotes a highly stable crosslinked char layer that is effective in the condensed phase (92). Other common phosphorus-nitrogen-based fire retardants include mono/di ammonium phosphate, ammonium polyphosphate, polyol, amide, aziridinyl phosphazene, guanyl urea phosphate, phosphate melamine (3). See Chapter 10 of this book for more information on boron-based flame retardants.

26.2.3.6 Silicon Containing Fire-Retardant Chemicals

Silicon containing fire retardants consist of silicones, silicates, organosilanes, and silsesquioxanes (3). These fire retardants are environmentally friendly and pose no health threats to humans. The mode of action is to form silica layers within wood resulting in high thermal stability and promotes char formation (3). See Chapter 8 of this book for more information on boron-based flame retardants.

26.2.3.7 Nanocomposite Fire Retardants

Nanocomposite fire retardants are effective in low concentrations but must be used in combination with other fire-retardant materials. These typically consist of nanoclays (aluminosilicates), nano oxides (titanium and zinc), nano-silica sol/silicon compounds, and nanostructured (cellulose nanofiber) carbon materials (3). The mode of action for these fire retardants are to act as a barrier to protect oxidation of the char layer (3) until other fire-retardant chemistries (used in combination with nanocomposites) act to prevent flame spread (21). Additionally, silicates can act to reduce smoke generation (3). See Chapter 12 of this book for more information on boron-based flame retardants.

26.2.3.8 Fire-Retardant Treatments

26.2.3.8.1 Traditional Surface-Coating Systems

Flame-retardant surface coatings work by delaying time to ignition and slowing the burn rate (45). Traditional surface coatings are not used in place of fire-retardant coatings. Typically, long-term performance requires reapplication and durability in-service is often difficult to achieve (45). Surface coatings have shown longevity in small-scale laboratory assays but tend to underperform in large-scale testing (45). When applying surface coatings, wood surfaces must be free of paint, stain, and grease and must be applied properly (not layered on too thick), otherwise it can be subject to cracking.

26.2.3.8.2 Intumescent Fire-Retardant Coating Systems

Intumescent systems typically include a carbon source, an acid source, and a blowing agent. The carbon source tends to be in the form of polyhydric compounds (i.e., starch, pentaerythritol, phenol, formaldehyde, polyamines, polyurethanes, polycarbonates) (3). The acid source can be ammonium salts, phosphates, sulfates, amines/amides, and/or hydroxides and reacts with the carbon source during the condensed phase to form a char layer (18). Blowing agents can be urea, urea formaldehyde resins, melamine, and/or polyamides (3) and breaks down and dilutes combustible gases which helps to thicken/expand the char layer (2). The char layer that is formed by intumescent systems during exposure to high temperatures can swell and increase 50–200 times its original volume (3) which provides a protective barrier against heat and mass transfer between the condensed and gas phases (92). The success of an intumescent system relies on activation of components by the series of chemical reactions listed above (18). Coating failures can occur when coatings are improperly applied as well as when subject to certain environmental conditions, as outlined in Section 26.3.3.

In a study by Baena et al., nine formulations of water-borne intumescent coatings applied to radiata-pine timber were exposed to a heat flux equivalent to high-risk bush fire conditions using a cone calorimeter to provide insight into the level of protection provided (100). The formulations included varying combinations of expandable graphite, ammonium polyphosphate, and waterborne acrylic resin. Based on the results, most of the coated samples reduced the peak heat release rate and satisfied the requirements for bushfire-resistant timber.

26.2.3.8.3 Resin Types

The standard resin types used in coatings are summarized in Table 26.2. Lignin can also be used as a resin (92). Functional groups of lignin can be modified via phenolization, demethylation, and or/hydroxymethylation, which increase copolymerization with phenol and formaldehyde. Additionally, lignin contains phenolic groups which when depolymerized can be utilized in the preparation of phenolic resins. Advantages of using lignin-based phenolic resins include low cost, increased bonding strength, better gel time, faster curing speed, increased thermal and mechanical properties (101).

26.2.4 Modified Wood

Thermal modification typically involves incomplete combustion of wood where wood is heated to a range of temperatures between 170° and 220°C. Studies have shown that thermal modification does lead to increased flammability (111) and efforts to mitigate this involves the use of additives introduced posttreatment, these include: melamine impregnation (112), clay nanomaterials (113), silicon dioxide (113), and a host of others.

Several chemical wood modification technologies have been evaluated for improving fire resistance, all with varying degrees of success. The two most common examples of modified woods that have achieved commercialization are acetylated and furfurylation. Acetylated wood has been found to have diminished fire resistance compared to untreated wood. Furfurylated wood has been found to have similar performance to acetylated wood where flammability is elevated, but several additives have been evaluated for both acetylated and furfurylated wood that improve fire resistance. Similar to thermally modified wood, there are several options for improving fire resistance of chemically modified woods, as discussed in Spear et al. (114). There is also considerable overlap between the fire-retardant additives and wood modification.

26.2.5 Wood-Plastic Composites

Wood-plastic composites (WPC) are materials that are made of wood fiber and/or wood flour and thermoplastic(s). The thermoplastics used in WPCs are generally limited to those with melting temperatures below 200°C since wood thermally degrades at low temperatures and are typically either high-density polyethylene (HDPE), low-density polyethylene (LDPE), polypropylene (PP), or polyvinyl chloride (PVC) (115). A small amount (by weight) of additives may be included to improve properties, such as durability and fire performance. These additives are typically particles or powders that are added to the melted thermoplastic composite during processing.

Prior to 2010, a limited number of studies evaluated the fire performance of WPCs, and an even smaller number investigated the effect of fire-retardant additives. In 2010, Stark et al. compared an untreated HDPE WPC with treated HDPE WPC (116). They tested five different fire-retardant treatments compounded with HDPE including (1) decabromodiphenyl oxide and antimony trioxide, (2) magnesium hydroxide, (3) zinc borate, (4) melamine phosphate, and (5) ammonium polyphosphate. Using a cone calorimeter, the

TABLE 26.2
Resin Types, Properties, and Mode of Action Used in Coatings

Resin Types		Properties	Mechanisms	Reference
Solvent-based				
	Polyurethane	High energy absorption, high resistance, maximum elongation, thermal stability, chemical resistance, low cost	Char layer formation	92
	Epoxy	Compact, water repellant, high strength, strong adhesion	Char layer formation	92
	Alkyds	Synthetic	Increases hydrophobic layer	92
	Melamine-formaldehyde	Low cost, high performance, high reactivity, fills micro/macro wood pores, UV stable, water repellant	Enhances durability, dimensional and thermal stability	102–104
	Nitrocellulose	Easy application, rapid drying, polymer compatibility, high mechanical properties, high pigment dispersion rate, biodegradable	Lowers volatile emmisions	105–108
Water-based				
	Acrylic emulsions	Nontoxic, odorless, easy application	Char formation	109
	Polyurethane dispersions	Abrasion and chemical/solvent resistant, high hardness	Lowers volatile emmisions	92
	Hybrid emulsions	Gloss, hardness, weather, adhesion, and chemical resistant	Char formation, lowers volatile emmisions	92, 110

results showed that the magnesium hydroxide resulted in the greatest reductions in the peak heat release rate, while decabromodiphenyl oxide/antimony trioxide treatment resulted in the greatest reduction in average effective heat of combustion. Since this work, additional studies evaluating the flammability properties of various fire-retardant treatments for WPCs have been conducted. Bajwa et al. evaluated a green and safe fire-retardant system comprising of cellulose nanocrystals and zinc oxide nanoparticles added to HDPE and found the treatments decreased the average mass loss, peak heat release rate, and total smoke release compared to untreated HDPE (117). Others have also reported positive effects on the fire-retardant performance with the incorporation of ammonium polyphosphate (118–123), transition metal oxides (124), and borates (125).

26.2.6 Effects of Fire-Retardant Treatments on Strength

The effects of fire retardants on wood properties were a major focus of research beginning in the 1980s as reports of failing plywood and wood truss systems in North America increased (Figure 26.8). These failures triggered the collection of data to better understand and characterize the effects of fire retardants and/or elevated temperatures on the structural integrity of wood. This topic was extensively reviewed in 1990 (126) and is summarized as follows:

Levan and Winandy focused on five fundamental phenomena involved in the thermal degradation of wood: (1) chemistry of thermal degradation of wood, (2) mechanisms by which FRT alters the aforementioned chemistry, (3) effects of acids on wood strength, (4) effects of temperature on wood strength, and (5) the combined interaction of acid and elevated temperature on wood strength. The immediate and long-term effects of various FR treatment systems on the engineering properties of wood materials have been reviewed and modeled (127). Winandy et al., using actual roof temperature exposure data obtained in nearly identical structures in Mississippi and Wisconsin, derived comparable field-to-laboratory exposure factors for untreated and monoammonium phosphate-treated plywood and identified quantitative relationships between laboratory- and field-exposure conditions as a function of exposure temperature and duration on the rate of change in strength of plywood.

Fire retardant chemicals may cause wood to become acidic, resulting in strength reductions that can vary by species but are reported in the 2 to 10% range. Elevated temperatures, such as those found in roofs, have an additive effect on wood strength so that the compounded effects can be substantial. Acid hydrolysis via the combined action of acetic acid generating during the heating of wood and additional acidic groups when the fire retardants are heated affect glycosidic linkages in cellulose causing significant strength losses. For this reason, the National Design Standard specified strength reductions from 10–20% in their allowable design stresses (128). The effects of heat on wood constituents have been extensively reviewed and have been shown to impact individual structural components (129) that manifests into decreases in bending strength (126) and can also have detrimental impacts to internal bonding strength of engineered woods (130). The bending strengths were measured for untreated and treated southern pine exposed to three different temperatures for 160 days (Table 26.3) (131). The results showed that the modulus of rupture was more sensitive to fire-retardant treatments than the modulus of elasticity, and it was noted that there was a correlation between the treatment pH and the effect on mechanical properties such that lower pH treatments had the greatest effect on mechanical properties.

26.3 TESTING OF WOOD PROTECTION

Numerous test methods are available to evaluate the efficacy of wood fire retardant formulations and may be

FIGURE 26.8 Failed FRT plywood roof sheathing after removal.

Source: USDA Forest Products Laboratory.

TABLE 26.3
Reduction in Bending Properties of Treated Southern Pine over 160 Days (131)

Treatment and Exposure	Reduction in Property after 160 Days (Percent)			
	Modulus of Elasticity		Modulus of Rupture	
(°C)	Initial	Final	Initial	Final
Untreated control				
27	0	0	0	0
54	0	0	0	0
82	0	0	0	11
Phosphoric acid				
27	17	17	50	50
54	17	17	50	50
82	17	45	50	1
Monoammonium phosphate				
27	11	11	13	13
54	11	11	13	22
82	11	24	13	64
Borax-boric acid				
27	13	13	0	0
54	13	13	0	2
82	13	13	0	2
Guanylurea phosphate—boric acid				
27	13	13	8	8
54	13	13	8	8
82	13	13	8	44
Dicyandiamide phosphoric acid formaldehyde				
27	13	13	9	9
54	13	13	9	9
82	13	13	9	40
Diethyl-N, N-bis(2-hydroxyethyl) aminomethyl phosphate				
27	13	13	15	15
54	13	13	15	15
82	13	13	15	25

Notes: 1) Extreme degradation, specimens could not be tested. 2) Values increased by ~5%.

required by various regulating bodies or used for research and may be standard or not. This section highlights several tests methods that are applicable but is not all-encompassing. Additional test methods are discussed in 2, 20, 40.

26.3.1 Regulatory Fire Test Methods

The dominant purpose of regulating building materials and product location is to increase fire safety for first responders, building occupants, and adjacent structures (132). Specific requirements for the use and location of fire-retardant treated wood, including required standardized testing, can be found in the North American model building codes as well as in state and local modifications and vary based on interior or exterior use in a structure. Several of these codes and standards are discussed further in the chapter on "Regulations, Codes and Standards Relevant to Fire Issues in the United States." The required standardized tests are intended to confirm the fire-retardant capabilities and performance based on environmental conditions and typically result in a classification applicable to the end use. These tests must be conducted by qualified, certified laboratories, with periodic retesting and inspection by accredited quality control agencies. The main regulatory tests for fire-retardant treated wood are summarized below:

26.3.1.1 Flame Spread

Fire growth in a building or compartment is influenced by the rate at which flames will spread across the exposed surfaces. The main standardized test to evaluate flame spread is ASTM E84, *Standard Test Method for Surface Burning Characteristics of Building Materials* (133). ASTM E84 provides comparative measurements of surface flame spread and smoke density measurements to select grade red oak and fiber-cement board surfaces under the specified fire exposure conditions. This test utilizes a 7.6 m (25-foot) tunnel test chamber with two gas burners at one end used to expose materials to a 1.35 m (4.5 ft) long test flame with a heat release rate of approximately 88 kW. Measurements of surface flame propagation and optical smoke density during the test are used to calculate the flame spread index (FSI) and smoke developed index (SDI) as the area under the flame spread distance-time plot and the area under the smoke density-time plot, respectively. Three classes of interior finish are specified in the model building codes based on the FSI and SDI results from ASTM E84. These classes are A, B, and C with Class A having the lowest flame spread and C having the highest. The FSI for several species of solid wood and wood-based panel materials commonly used as interior finish are published by the American Wood Council (134). When tested to ASTM E84, lumber, plywood, and other wood-based materials typically achieve either a B or C classification. The inclusion of fire-retardant treatments for these materials improves the flame spread performance and wood-based products that achieve a Class A rating are commonly labelled with their classification.

An extended flame spread test, ASTM E2768, *Standard Test Method for Extended Duration Surface Burning Characteristics of Building Materials (30 min Tunnel Test)* (135), which is colloquially referred to as the 30-minute ASTM E84 test, is used to assess fire-retardant treated wood products. In this test, the same apparatus as ASTM E84 is used, but the burner is left on for 30 minutes. To pass ASTM E2768, the FSI determined for the initial ten-minute test period must be less than 25 and the flame front shall not progress more than 10.5 ft. (3.2 m) beyond the burners at any time during the 30-minute test period.

26.3.1.2 Fire Growth

As an alternative to ASTM E84, a full-size room burn test detailed in NFPA 286, *Standard Methods of Fire Tests for Evaluating Contribution of Wall and Ceiling Interior Finish to Room Fire Growth* (136) may be conducted. Similar full-scale room/corner test standards include ASTM 2257 (137) and ISO 9705 (138), with a comparison of the standard methods provided by White et al. (139). In NFPA 286, the material in question is applied to three walls and/or the ceiling, depending on the intended end use application, of a 2.4 m by 3.6 m by 2.4 m (8 ft by 12 ft by 8 ft) room with a single door opening for ventilation. A propane burner is placed in the back corner of the room and the burner protocol is 40 kW for five minutes followed by 160 kW for ten minutes. Per NFPA 286, flashover is considered to have occurred when two of the five parameters are attained. The five parameters are as follows: the heat release rate exceeds 1 MW, the heat flux at the floor exceeds 20 kW/m^2, the average upper layer temperature exceeds 600°C (1112°F), flames exit the doorway, and a paper target on the floor auto-ignites. During the 1990s, several series of these full-scale room/corner tests were conducted as part of an international effort to evaluate alternative reaction-to-fire tests and support modeling efforts (139–143). These test series evaluate two burner scenarios: (1) 40 kW for 300 seconds followed by a change to 160 kW for 300 seconds and (2) 100 kW for 600 seconds followed by an increase to 300 kW for 600 seconds. Some of the results from these test series are included in Table 26.4 and show that the fire-retardant treatments impact the fire growth in the room by either significantly delaying the time to flashover or flashover never attained.

26.3.2 Micro- and Bench-Scale Test Methods

Beyond the regulatory tests, developers and researchers employ a micro- and bench-scale tests to evaluate the effectiveness of a treatment. Since many fire-retardant formulations have encompassed one or more mechanism of action, a combination of these tests is typically required to fully characterize the reaction-to-fire performance. Summaries of the test methods commonly used to investigate the suitability of a wood fire-retardant treatment formulation for research and development purposes are provided in the next section (2, 20).

TABLE 26.4
Time to Flashover for Rooms with Walls Lined with Plywood

Material	Burner Scenario	Time to Flashover [s]
FRT Plywood N	40/160 kW	No Flashover
FRT Plywood N	100/300 kW	No Flashover
FRT Plywood M	40/160 kW	No Flashover
FRT Plywood M	100/300 kW	870
FRT Plywood F	40/160 kW	No Flashover
FRT Plywood F	100/300 kW	870
FRT Plywood R	40/160 kW	No Flashover
FRT Plywood R	100/300 kW	849
Douglas-fir plywood M	40/160 kW	378
Douglas-fir plywood M	100/300 kW	510
Douglas-fir plywood R	40/160 kW	391
Douglas-fir plywood R	100/300 kW	465
Southern pine plywood	40/160 kW	344
Southern pine plywood	100/300 kW	321

Source: Data from White et al. (139)

26.3.2.1 Thermogravimetric Analysis

Thermogravimetric analysis (TGA) is the most common method to quantitatively evaluate thermal decomposition, and the results are often used to identify the pyrolysis kinetics and residual mass (i.e., char formation). Samples are on the order of milligrams in size and are typically ground but may remain whole depending on the experimental design. The samples are heated in a chamber with a gas (nitrogen, air, or other gas). Different atmospheres will provide varying insight into the thermal decomposition and combustion of the material. As the sample is heated, the weight loss is recorded. The change in weight of the sample can be a function of time at a constant temperature (isothermal) or at different heating rates (non-isothermal). This method is frequently used to delineate between the mechanisms of action for wood fire retardants (20, 144).

26.3.2.2 Differential Scanning Calorimetry

Differential scanning calorimetry (DSC) is used to measure the specific heat capacity of a material and how it changes with temperature as well as identify when the material undergoes a phase transition. In DSC, both the sample (~milligrams) and a well-characterized reference are heated by separate heating elements. Endothermic reactions, such as a solid melting to a liquid, will require more heat to flow to the sample to increase its temperature. Crystallization is an exothermic process, and less heat will be required to raise the sample temperature. The amount of heat absorbed or released during a phase transition is quantified via the difference in heat flow between the sample and reference.

26.3.2.3 Cone Calorimeter

The cone calorimeter was developed in the 1980s and is used to measure material flammability based on heat release rate. The heat release rate is determined via the oxygen consumption method and other flammability parameters obtained include time to ignition, mass loss rate, smoke release rate, effective heat of combustion, and production rates of combustion gas (145). Samples are typically 100 mm by 100 mm and up to 50 mm thick and are exposed to a uniform radiant heat source and may be use a spark igniter for piloted ignition. For wood-based materials, the heat release rate curves as a function of time typically include two peaks, as shown in Figure 26.9. The first peak is attributed to the ignition of the volatiles produced. This peak is followed by a dip in the curve, which is when the insulating char layer is formed that restricts pyrolysis and production of flammable volatiles. The second peak is caused by sample burn through and the backside of the sample becoming involved in the combustion process (145–148). The results from the cone calorimeter have been used to develop a predictive model for flame spread (149) and have showed that the inclusion of fire-retardant treatments in wood will delay ignition, reduce peak heat release rates, and reduce the smoke production rates (150).

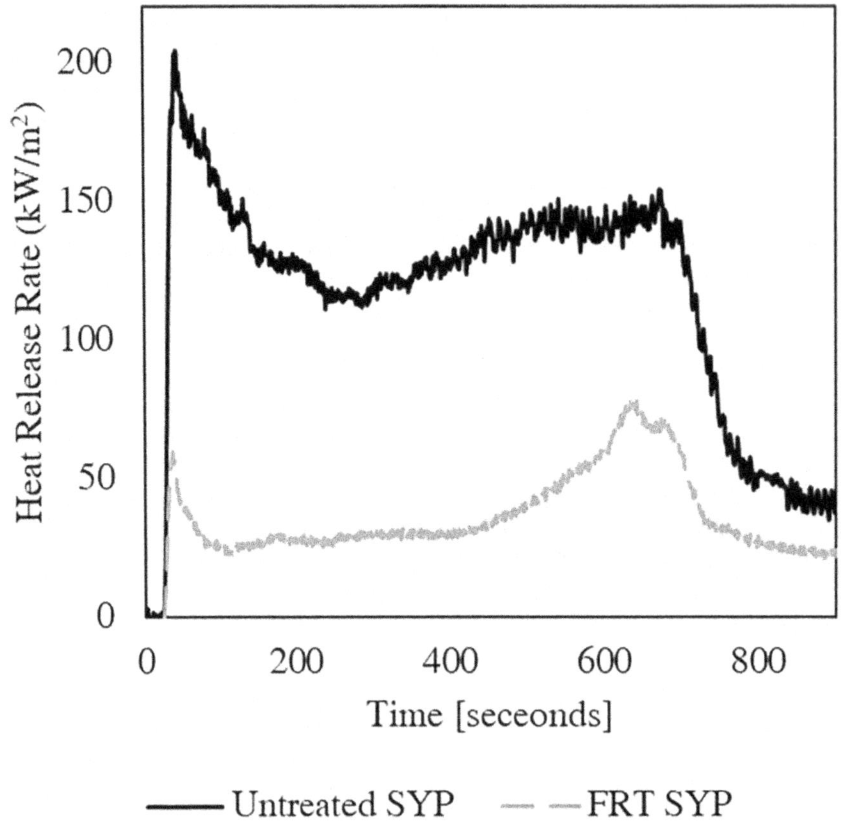

FIGURE 26.9 Heat release rate curves for untreated and treated southern yellow pine. Data plotted from Dietenberger, M. A.; White, R. H. 2001.

Source: Adapted from (150).

26.3.3 Durability/Impact to Wood Testing

Environmental conditions (i.e., moisture, UV, biological organisms) play a huge role in the degradation of wood products. Energy absorption from UV and visible light generates carbonyl and phenolic radicals that depolymerize wood components. Degradation due to UV-VIS can lead to discoloration, cracking, cupping, and splitting of wood products. Moisture fluctuations causes wood to swell, shrink, blister, erode, crack, and cup which can lead to biological degradation from mold, decay, and stain fungi. This is often more problematic for surface coating systems rather than more-deeply penetrated treatment systems. Figure 26.10 shows some common coating failures on wood products that were naturally and artificially weathered. To adequately predict service life, products are tested in laboratory and outdoor settings following ASTM standards. The typical durability standards are outlined below.

26.3.3.1 Weathering Tests

26.3.3.1.1 Laboratory Weathering

Laboratory weathering of fire-retardant treated wood follows ASTM D 2898, *Standard Practice for Accelerated Weathering of Fire-Retardant-Treated Wood for Fire Testing* (151). This method consists of exposure to accelerated cycles of wetting, drying, and UV prior to the evaluation of fire testing (ASTM E84). This standard consists of four evaluation methods (A–D). Method A has a total duration of 2016 h (84 days) with repeat cycles consisting of 96 h (4 days) of water exposure (0.30 L/min•m^2) and 72 h (3 days) of drying (57–60°C). Water recirculation is not permitted for this method, and specimens must change position in the weathering apparatus to ensure each specimen occupies all possible locations for an equal number of cycles. Method B has a total duration of 1,000 h (41 days, 16 h) with repeat cycles of 4 h wetting (12.2 ± 0.8 L/min•m^2), 4 h drying (66 ± 3°C), 4 h wetting (12.2 ± 0.8 L/min•m^2), 4 h drying (66 ± 3°C), 8 h dark. The last cycle is terminated prior to the 8 h dark period. Water can be recirculated after 72 h with a circulating volume of at least 19 L of fresh water through each spray nozzle for the 4 h wetting period. Specimens must change position to allow exposure to all locations in the apparatus. Method C has a total duration of 2016 h (84 days) with repeat cycles consisting of 4 h wetting (12.2 ± 0.8 L/min•m^2) and 4 h drying (63–68°C). Water recirculation is permitted after the first three wetting cycles, but before the last three wetting cycles. Specimens must change position. Method D has a total duration of 2,328 h (97 days) and consists of two repeat cycles of 168 h (7 days) of water exposure (0.30 L/min•m^2), 48 h (2 days) of rest, 120 h (5 days) of

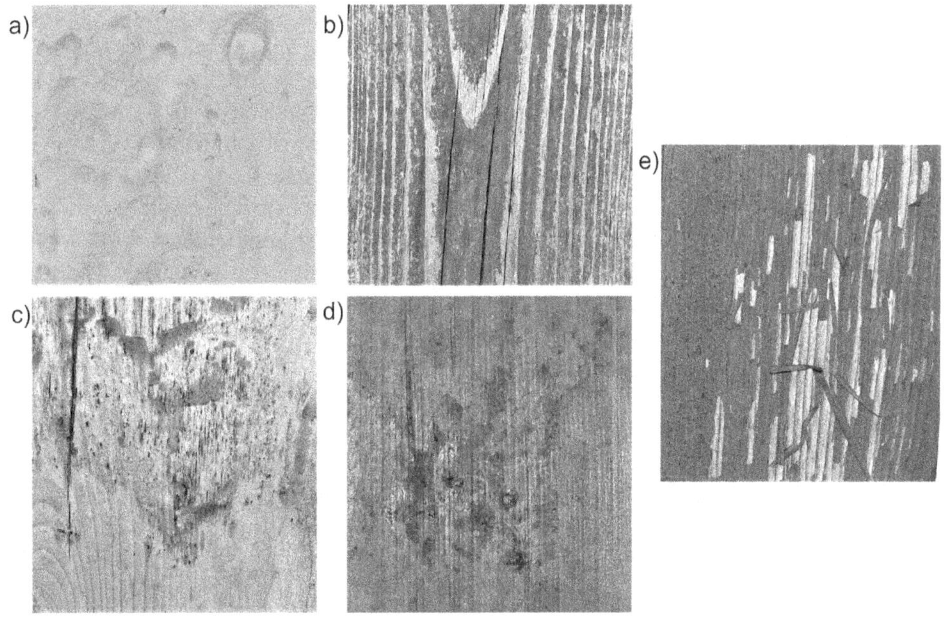

FIGURE 26.10 Common surface coating failures: (a) blistering, (b) erosion, (c) microcracking and flaking, (d) microbial growth, (e) flaking and microbial growth.

Source: Katie M. Ohno, USDA Forest Products Laboratory

drying (57–60°C). Water recirculation is not allowed, and specimens must change position. When using this standard, the reports must include the method used, description of the weathering apparatus, types of UV lamps, slope of the test specimen, water quality characteristics, amount of water recirculated when applicable, visual observations of the specimens pre and post testing, and any deviations from the specifications.

26.3.3.1.2 Outdoor Weathering

Outdoor weathering of fire-retardant treated wood follows ASTM G7/G7M, *Standard Practice for Natural Weathering of Materials* (152). This standard is used to assess the durability of materials in outdoor exposure, which can be variable due to differences in UV, relative humidity, time of wetness, temperature, wet-dry cycling, freeze-thaw cycling, pollutants, and other additional environmental factors. Typically, it is recommended to expose specimens in several locations representing different climates that represent a broad range of environmental conditions. Natural weathering tests are often conducted for at least two years in multiple exposure sites to accurately predict service life. Additionally, ASTM D3456, *Standard Practice for Determining by Exterior Exposure Tests the Susceptibility of Paint Films to Microbiological Attack* is also used to accurately predict service life of fire-retardant coatings (153).

26.3.3.1.3 Weathering Evaluations

There are several ASTM and ISO standards used to evaluate the service life of wood coatings. It is important to note that when performing these evaluations, the evaluator must visually rate the coating applied to the wood (not the wood itself). ASTM D3274, *Standard Test Method for Evaluating Degree of Surface Disfigurement of Paint Films by Fungal or Algal Growth, or Soil and Dirt Accumulation* is used to determine coating susceptibility to fungal growth when exposed outdoors (154). The rating scale is from 10 (no fungal growth) to 0 (complete coverage) and details the different microbial growth patterns. ASTM D4214, *Standard Test Method for Evaluating Degree of Chalking of Exterior Paint Films* is used to determine if there is formation of powder of a coating (155). The rating scale is from 10 (no chalking) to 0 (complete chalking). ASTM D660, *Standard Test Method for Evaluating Degree of Checking of Exterior Paints* is used to assess if there are slight breaks of the coating (that do not penetrate through to the surface of the wood) (156). The rating scale is from 10 (no checking) to 0 (complete checking) and details the types of checking. ASTM D661, *Standard Test Method for Evaluating Degree of Cracking of Exterior Paints* is used to report cracks in the coating (penetrate through to the wood surface) (157). The rating scale is from 10 (no cracking) to 0 (complete cracking) and details the types of cracks. ASTM D662, *Standard Test Method for Evaluating Degree of Erosion of Exterior Paints* is used to determine if the coatings are wearing away to expose the surface of the wood (158). The rating scale is from 10 (no erosion) to 0 (complete erosion). ASTM D714, *Standard Test Method for Evaluating Degree of Blistering of Paints* is used to evaluate if the coating forms blisters (159). The rating scale is from 10 (no blistering) to 0 (complete blistering) and details the density of the blistering. ASTM D772, *Standard Test Method for Evaluating Degree of Flaking (Scaling) of Exterior Paints* is used to assess if the coating has detached from the surface of the wood (160). The rating scale is from 10 (no flaking) to 0 (complete flaking). ASTM D523, *Standard Test Method for Specular Gloss* is used to measure gloss at 60°, 20°, and 85° (161). ISO 11664, *Colorimetry Part 4: CIE 1976 L*a*b* Colour Space* is used to evaluate color change, where L represents lightness from 0 (black) to 100 (white), a represents red (+) or green (-), and b represents yellow (+) or blue (-) (162). These evaluations should be obtained once a month for six months then yearly until the duration of the test is complete (2–5 y).

26.4 TECHNICAL RESOURCES

- International Organization for Standardization (ISO) is a non-profit organization that develops and publishes standards for an expansive range of materials and processes. ISO was founded in 1946 by an international delegation from 25 countries and now represents an international standardization body with membership with 168 national standards bodies.
- ASTM International (ASTM): ASTM, formerly known as American Society for Testing and Materials, is an international, consensus-based standards organization that develops and publishes technical standards for a wide range of materials, products, and services. Within ASTM, there are two committees that oversee the development of standards related to wood and fire: Committees D07 on Wood and E05 on Fire Standards. The scope for Committee D07 includes the development of standards and guides pertaining to forests, timber, wood, modified wood, veneer, wood-based structural panels, laminated wood, and other multi-material structural composites in which cellulosic materials are a significant component. D07 has the jurisdiction over the standards on moisture content determination, weathering, and strength properties. Committee E05 is responsible for developing and maintaining the fire standards intended for analysis and assessment of the fire performance of materials, products, and assemblies within their relevant environment. This includes standards on quantifying flame spread, heat release rate, and numerous flammability parameters.
- American Wood Protection Association (AWPA): AWPA is a leading organization that develops standards for wood preservation, including fire retardants. The association provides technical resources and publications related to pressure applied wood fire retardants which are accessible on awpa.com. To date, there are only two fire retardant formulations listed in the AWPA book of standards, while most of the remaining fire-retardant treatments are promulgated through ICC-ES (see below).
- Forest Products Laboratory (FPL): FPL is a research branch of the United States Forest Service. They conduct research on wood and wood products, including fire retardant treatments. FPL's website (https://www/fpl.fs.usda.gov) provides access to research papers, publications, and technical reports on wood fire retardants.
- International Code Council (ICC): ICC is an association responsible for developing and maintaining a set of building codes widely adopted in the United States and other countries. The ICC website (iccsafe.org) provides access to building codes, including provisions for fire-retardant-treated wood.
- National Fire Protection Association (NFPA): NFPA is a global nonprofit organization focused on fire safety. They develop codes and standards related to fire protection, including fire retardant treatments for wood. The NFPA website (nfpa.org) offers resources such as codes, standards, and research reports related to fire safety.
- Western Wood Preservers Institute (WWPI): WWPI is an industry trade group that deals primarily with Western wood species such as Douglas Fir. WWPI's website (fireresistantwood.org) contains a variety of technical resources on the proper use and specifications for fire retardant treated wood.
- U.S. Environmental Protection Agency (U.S. EPA): Flame retardants are subject to regulation by the U.S. EPA, multiple state agencies, and regulatory bodies in countries around the world. The U.S. EPA has authority to limit or prohibit a flame retardant's use if it presents an unreasonable risk of injury to human health or the environment. The Agency's authority applies to chemicals currently in commerce as well as new chemicals that manufacturers might wish to introduce into the marketplace. EPA regulates new and existing chemicals, including flame retardants, under the Toxic Substances Control Act (TSCA) and are subject to re-registration.
- Consumer Product Safety Commission (CPSC): CPSC has authority to regulate the safety of consumer products that may incorporate flame retardants to meet fire safety standards. As part of the Federal Hazardous Substances Act (FHSA), the CPSC can regulate chemicals in consumer products.
- North American Flame-Retardant Alliance (NAFRA): NAFRA members support a robust and transparent regulatory system that provides both strong fire protection and chemical safety. NAFRA was formed in 2011 and works collaboratively with several international trade organizations on the safe use of flame retardants. NAFRA also supports research on past, present, and future formulations of fire retardants.
- Southern Pine Inspection Bureau (SPIB), founded in 1915, SPIB maintains quality standards for the grading and manufacturing of softwood lumber products. The SPIB quality mark is commonly

used on end tags in accordance with the third-party quality assurance audit programs. Western Wood Services is a division of SPIB that deals exclusively with the western wood species (Douglas fir, cedars, etc.) and is based in British Columbia.

- Timber Products Inspection (TPI) is a third-party inspection and technical resource for the forest products industry. TPI offers contract testing and validation of lumber, auditing, engineered and technical services for clients in 16 countries covering over 20 different wood products.
- Underwriters Laboratory (UL) is a testing laboratory that provides vital testing for a wide range of appliances and materials. Founded in Chicago, UL has worked closely with the NFPA since 1914 to ensure all products adhere to acceptable performance standards, and this includes fire-retardant treated wood products.

26.5 DISPOSAL

As it does not contain active biocides unless part of a multi-component system, fire-retardant treated wood is not subject to Federal Insecticide, Fungicide, and Rodenticide Act (FIFRA) regulations and, therefore, there are fewer disposal stipulations than other chemically treated wood. Disposal routes for fire-retardant treated wood are similar to those for untreated wood. Fire-retardant treated wood can be incinerated using specialized boilers or incinerators. If not incinerated, fire-retardant treated wood is normally disposed in landfills, which can last for extended periods of time and exposure of breakdown products of fire retardants, particularly brominated materials, have been found to have significant ecological impacts (163).

REFERENCES

1. Sauerbier, P.; Mayer, A.K.; Emmerich, L.; Militz, H. Fire retardant treatment of wood–state of the art and future perspectives. In *Wood & Fire Safety: Proceedings of the 9th International Conference on Wood & Fire Safety*. Makovicka Osvaldova, L., Markert, F., Zelinka, S., Eds., **2020**, Vol. 9, pp. 97–102. DOI: 10.1007/978-3-030-41235-7_14
2. Lowden, L.A.; Hull, T.R. Flammability behaviour of wood and a review of the methods for its reduction. *Fire Sci Rev.* **2013**, *2*(4), 1–19. DOI: 10.1186/2193-0414-2-4
3. Popescu, C.; Pfriem, A. Treatments and modifications to improve the reaction to fire of wood and wood based products—an overview. *Fire Mater.* **2019**, 44, 100–111. DOI: 10.1002/fam.2779
4. Burke, M.; Driscoll, A.; Heft-Neal, S.; Xue, J.; Burney, J.; Wara, M. The changing risk and burden of wildfire in the United States. *Proc Natl Acad Sci.* **2021**, *118* (2). DOI: 10.1073/pnas.2011048118
5. Page, J.; Whaley, P.; Bellingham, M.; Birnbaum, L.S.; Cavoski, A.; Dilke, D.F.; Garside, R.; Harrad S.; Kelly, F.; Kortenkamp, A.; Martin, O.; Stec, A.; Woolley, T. A new consensus on reconciling fire safety with environmental & health impacts of chemical flame retardants. *Environ Int.* **2023**, *173*. DOI: 10.1016/j.envint.2023.107782
6. Moon, R.J.; Martini, A.; Nairn, J.; Simonsen, J.; Youngblood, J. Cellulose nanomaterials review: Structure, properties and nanocomposites. *Chem Soc Rev.* **2011**, *40*, 3941–3994. DOI: 10.1039/C0CS00108B
7. Jakes, J.E.; Hunt, C.G.; Zelinka, S.L.; Ciesielski, P.N.; Plaza, N.Z. Effects of moisture on diffusion in unmodified wood cell walls: A phenomenological polymer science approach. *Forests.* **2019**, *10*, 1084. DOI: 10.3390/f10121084
8. Rowell, R.M. Handbook of wood chemistry and wood composites. CRC Press, **2012**.
9. Sjostrom, E. Wood chemistry: Fundamentals and applications. 2nd Ed. Elsevier, **2013**.
10. Tarasov, D.; Leitch, M.; Fatehi, P. Lignin–carbohydrate complexes: Properties, applications, analyses, and methods of extraction: A review. *Biotechnol Biofuels.* **2018**, *11* (1), 1–28. DOI: 10.1186/s13068-018-1262-1
11. Fengel, D.; Wegener, G. Wood: Chemistry, ultrastructure, reactions. Walter de Gruyter, **1984**. DOI: 10.1002/pol.1985.130231112
12. Park, S.; Baker, J.O.; Himmel, M.E.; Parilla, P.A.; Johnson, D.K. Cellulose crystallinity index: Measurement techniques and their impact on interpreting cellulase performance. *Biotechnol Biofuels.* **2010**, (3), 10. DOI: 10.1186/1754-6834-3-10
13. Shafizadeh, F. The chemistry of pyrolysis and combustion. In The chemistry of solid wood. Rowell, R., Ed., Vol. 207; Advances in Chemistry, **1984**. DOI: 10.1021/ba-1984-0207.ch013
14. Chen, H. Biotechnology of lignocellulose: theory and practice. Springer, **2005**. DOI: 10.1007/978-94-007-6898-7
15. Duval, A.; Lawoko, M. A review on lignin-based polymeric, micro-and nano-structured materials. *React Funct Polym.* **2014**, *85*, 78–96. DOI: 10.1016/j.reactfunctpolym.2014.09.017
16. Crestini, C.; Crucianelli, M.; Orlandi, M.; Saladino, R. Oxidative strategies in lignin chemistry: a new environmentally friendly approach for the functionalisation of lignin and lignocellulosic fibers. *Catal Today.* **2010**, *156* (1–2), 8–22. DOI: 10.1016/j.cattod.2010.03.057
17. Hunt, C.G. Chapter 16: Finishing wood. In Wood handbook wood as an engineering material. General Technical Report FPL-GTR-282. U.S. Department of Agriculture, Forest Service, Forest Products Laboratory, **2021**, p. 36.
18. Lazar, S.T.; Kolibaba, T.J.; Grunlan, J.C. Flame-retardant surface treatments. *Nat Rev Mater.* **2020**, *5* (4), 259–275. DOI: 10.1038/s41578-019-0164-6.
19. Shafizadeh, F. Introduction to pyrolysis of biomass. *J Anal Appl Pyrolysis.* **1982**, *3* (4), 283–305. DOI: 10.1016/0165-2370(82)80017-X
20. Rowell, R.M.; Dietenberger, M.A. Thermal properties, combustion, and fire retardancy of wood. In Handbook of wood chemistry and wood composites, **2012**, pp. 121–151. DOI: 10.1201/b12487-11
21. Lebow, P.K.; Winandy, J.E. Verification of a kinetics-based model for long-term effects of fire retardants on bending strength at elevated temperatures. *Wood Fiber Sci.* **1999,** *31* (1), 49–61.
22. Park, W.C.; Atreya, A.; Baum, H.R. Experimental and theoretical investigation of heat and mass transfer processes during wood pyrolysis. *Combust Flame.* **2010**, *157* (3), 481–494. DOI: 10.1016/j.combustflame.2009.10.006
23. Grieco, E.; Baldi, G. Analysis and modelling of wood pyrolysis. *Chem Eng Sci.* **2011**, *66* (4), 650–660. DOI: 10.1016/j.ces.2010.11.018

24. Lautenberger, C.; Fernandez-Pello, C. A model for the oxidative pyrolysis of wood. *Combust Flame*. **2009**, *156* (8), 1503–1513. DOI: 10.1016/j.combustflame.2009.04.001
25. Kung, H.C. A mathematical model of wood pyrolysis. *Combust Flame*. **1972**, *18* (2), 185–195. DOI: 10.1016/S0010-2180(72)80134-2
26. Bryden, K.M.; Ragland, K.W.; Rutland, C.J. Modeling thermally thick pyrolysis of wood. *Biomass Bioenergy*. **2002**, *22* (1), 41–53. DOI: 10.1016/S0961-9534(01)00060-5
27. Parker, W.J. *Prediction of the heat release rate of wood*. PhD Dissertation, George Washington University, **1988**.
28. Parker, W.J.; LeVan, S.L. Kinetic properties of the components of douglas-fir and the heat of combustion of their volatile pyrolysis products. *Wood Fiber Sci*. **1989**, *21* (3), 289–305.
29. Tang, W.K. *Effect of inorganic salts on the pyrolysis, ignition, and combustion of wood, cellulose and lignin*. PhD Dissertation, University of Wisconsin, **1964**.
30. Dietenberger, M.A.; Hasburgh, L.E. Wood products: Thermal degradation and fire. In Reference module in materials science and materials engineering. Hashmi, S., Ed., **2016**, 1–8. DOI: 10.1016/B978-0-12-803581-8.03338-5
31. Mészáros, E.; Jakab, E.; Várhegyi, G. TG/MS, Py-GC/MS and THM-GC/MS study of the composition and thermal behavior of extractive components of Robinia pseudoacacia. *J Anal Appl Pyrolysis*. **2007**, *79* (1–2), 61–70. DOI: 10.1016/j.jaap.2006.12.007
32. Várhegyi, G.; Grønli, M.G.; Di Blasi, C. Effects of sample origin, extraction, and hot-water washing on the devolatilization kinetics of chestnut wood. *Ind Eng Chem Res*. **2004**, *43* (10), 2356–2367. DOI: 10.1021/ie034168f
33. Grønli, M.G.; Várhegyi, G.; Di Blasi, C. Thermogravimetric analysis and devolatilization kinetics of wood. *Ind Eng Chem Res*. **2002**, *41* (17), 4201–4208. DOI: 10.1021/ie0201157
34. Shebani, A.N.; Van Reenen, A.J.; Meincken, M. The effect of wood extractives on the thermal stability of different wood species. *Thermochim Acta*. **2008**, *471* (1–2), 43–50. DOI: 10.1016/j.tca.2008.02.020
35. Panshin, A.J.; De Zeeuw, C. Textbook of wood technology: Structure, identification, properties, and uses of the commercial woods of the United States and Canada. McGraw-Hill Book, **1980**.
36. Friquin, K.L. Material properties and external factors influencing the charring rate of solid wood and glue-laminated timber. *Fire Mater*. **2010**, *35* (5), 303–327. DOI: 10.1002/fam.1055
37. White, R.H.; Nordheim, E.V. Charring rate of wood for ASTM E 119 exposure. *Fire Technol*. **1992**, *28*, 5–30. DOI: 10.1007/BF01858049
38. Bartlett, A.I.; Hadden, R.M.; Bisby, L.A. A review of factors affecting the burning behaviour of wood for application to tall timber construction. *Fire Technol*. **2019**, *55*, 1–49. DOI: 10.1007/s10694-018-0787-y
39. White, R.H. Charring rate of composite timber products. In *Proceedings Wood and Fire Safety*, May 14–19, **2000**, Part I, pp. 353–363.
40. ASTM International. Standard Test Methods for Fire Tests of Building Construction and Materials; ASTM E 119, **2022**. DOI: 10.1520/E0119-22
41. White, R.H. *Charring rates of different wood species*. PhD Dissertation, University of Wisconsin–Madison, **1988**.
42. Dietenberger, M.A. Effect of backing board on the heat release rate of wood. In *Proceedings of the International Conference on Fire Safety*, Columbus, OH, **1999**.
43. McNamee, R.; Zehfuss, J.; Bartlett, A.I.; Heidari, M.; Robert, F.; Bisby, L.A. Enclosure fire dynamics with a cross-laminated timber ceiling. *Fire Mater*. **2021**, *45* (7), 847–857. DOI: 10.1002/fam.2904
44. Barber, D.; Rackauskaite, E.; Christensen, E.; Schulz, J. Exposed mass timber in high-rise structures: A practical discussion of a complex fire problem. *CTBUH J*. **2022**, *1*, 32–39.
45. Winandy, J.E.; Wiesner, F.; Hassan, B.; Morrell, J.J. Fire performance of timber: Review for use in wildland-urban interfaces. *Holzforschung*. **2022**, *76* (8), 679–698. DOI: 10.1515/hf-2022–0038
46. Green, J. Mechanisms for flame retardancy and smoke suppression-a review. *J Fire Sci*. **1996**, *14* (6), 426–442. DOI: 10.1177/073490419601400602
47. Babrauskas, V. Ignition of wood: A review of the state of the art. *J Fire Prot Eng*. **2002**, *12* (3), 163–189. DOI: 10.1177/10423910260620482
48. Dietenberger, M.A.; Hasburgh, L.E.; Yedinak, K.M. Fire safety of wood construction. In: Wood handbook wood as an engineering material. USDA Forest Service, Forest Products Laboratory, **2021**.
49. Hshieh, F.Y.; Richards, G.N. The effect of preheating of wood on ignition temperature of wood char. *Combust Flame*. **1990**, *80* (3–4), 395–398. DOI: 10.1016/0010-2180(90)90115-8
50. Browne, F.L. Theories of the combustion of wood and its control; Report No. 2136; USDA Forest Service, Forest Products Laboratory, **1958**.
51. LeVan, S.L. Chemistry of fire retardancy. In The chemistry of solid wood. Advances in Chemistry, **1984**, pp. 531–574. DOI: 10.1021/ba-1984-0207.ch014
52. Truax, T.R.; Harrison, C.A. Measuring fire resistance of wood. *Trans Am Soc Mech Eng*. **1930,** *52* (14), 33–40. DOI: 10.1115/1.4057987
53. Bescher, R.H.; Henry, W.T.; Dreher, W.A. A study of permanence of commercial fire retardants. In *Proceedings of American Wood-Preservers' Association Annual Meeting*, **1948**.
54. Van Kleeck, A. Corrosion studies with certain fire-retardant chemicals. In *Proceedings of American Wood-Preservers' Association Annual Meeting*, **1942.**
55. Bruce, H.D. Effects of some preservative treatments on the flammability of wood. In *Proceedings of American Wood-Preservers' Association Annual Meeting*, **1956**.
56. Ventriglio, D.R. Fire safe materials for navy ships. *Nav Eng J*. **1982**, *94* (5), 65–74. DOI: 10.1111/j.1559-3584.1982.tb02504.x
57. White, R.H. Ten decades of flame retardancy testing at the US Forest Products Laboratory. In *Proceedings of the 21st annual conference on recent advances on flame retardancy of polymeric materials*, Brooklyn, NY, May 24–26, **2010**, pp. 445–454.
58. White, R.H.; Sweet, M.S. Flame retardancy of wood: Present status, recent problems and future fields. Business Communications Company, **1992**.
59. Marney, D.; Russell, L. Combined fire retardant and wood preservative treatments for outdoor wood applications—a review of the literature. *Fire Technol*. **2008**, 44, 1–14. DOI: 10.1007/s10694-007-0016-6
60. Jain, J.C.; Cedercreutz, U. A study of the toxicity of some water soluble wood preservatives to wood destroying fungi. *Indian Forest*. **1961**, *87* (11), 678–691.
61. Purushotham, A.; Pande, J.N.; Sud, J.S. A note on fire resistive-cumantiseptic composition and fire resistive paint. *J. Timber Dryers Preserv Assoc*. **1963**, *9* (3), 27–28.

62. Collister, L.C. 1963. Fire-retardant treatments on the Santa Fe. *Proc Am Wood Preserv Assoc Ann Meet.* **1963**, *59*, 92–98.
63. Dev, I.; Kumar, S. Studies on the fire retardant cum preservative composition. *J Timber Develop Assoc.* **1982**, *28* (4), 30–33.
64. Goldstein, I.S.; Dreher, W.A. A non-hygroscopic fire-retardant treatment for wood. *For Prod J.* **1961**, *11* (5), 235–237.
65. Juneja, S.C. Urea based fire retardant formulations and products. Canadian Patent No. 917334, **1972**.
66. Juneja, S.C.; Calve L. Curing and fixation of amino-resin fire retardants. *J Fire Retard Chem.* **1977**, *4* (4), 235–240.
67. Gyarmati, B.; Kovacs, K.; Siklosi, M.; Varallyai, C.; Kovancsek, E. Combined wood-protecting compositions. Hungarian Patent No. 10350, **1975**.
68. Ermus, N.; Pastor, N.; Kalinis, A.; Luse, I.; Celmillere, M.; Kalachev, G.P.; Sukhanov V.A. Methods for the protective treatment of wood with the antiseptic MB-1. *Zashch Drev Biorazrusheniya.* **1977**, 49–59.
69. Gardner, W.D.; Alexiou, P.N.; Lind, P.; Butler, D. Evaluation of the fire retardant efficacy and leach resistance of an amino resin fire retardant. In *Proceedings of the International Research Group on Wood Preservation*, IRG/WP/3260, **1983**.
70. Alexiou, P.N.; Gardner, W.D.; Lind, P.; Butler, D. Efficacy of an amino resin fire retardant. *For Prod J.* **1986**, 36, 9–15.
71. LeVan, S.L.; Holmes, C.A. *Effectiveness of fire-retardant treatments for shingles after 10 years of outdoor weathering.* Research Paper FPL 474; USDA Forest Service, Forest Products Laboratory, Madison, WI, **1986**.
72. Lilaro, S.; Ota, Y.; Yasuda, A.; Nakai, T.; Seto, K. Fire retardants and preservatives for wood. German Patent no. 3630139, **1987**.
73. Yasuda, H.; Ota Y. Improved wood. Japanese Patent no. 62-116103, **1987**.
74. Diamond S. Delayed ettringite formation—processes and problems. *Cem Concr Compos.* **1996**, *18* (3), 205–215. DOI: 10.1016/0958-9465(96)00017-0
75. Saka, S.; Sasaki, M.; Tanahashi, M. Wood-inorganic composites prepared by solgel processing. I. Wood-inorganic composites with porous structure. *J Jpn Wood Res Soc.* **1992**, *38* (11), 1043–1049.
76. Saka, S.; Yakake, Y. Wood-inorganic composites prepared by Sol-gel process. III. Chemically-modified wood-inorganic composites. *J Jpn Wood Res Soc.* **1993**, *39* (3), 308–314.
77. Saka, S.; Tanno, F. Wood-inorganic composites prepared by Sol–gel process. VI. Effects of a property-enhancer on Fire-resistance in SiO_2–P_2O_5 and SiO_2–B_2O_3 wood inorganic composites. *J Jpn Wood Res Soc.* **1996,** *42* (1), 81–86.
78. Hirata, T.; Kawamoto, S. A conception for development of fire-retardant wood materials from research on thermal analysis of wood treated with water-insoluble retardants. In *Proceeding of XIX IUFRO World Congress* **1990,** 5–11.
79. Ota, Y.; Hirao, S.; Usui, H.; Nakai, T.; Ishikawa, H. Preservation of wood by impregnation and precipitation of inorganic compounds. Japanese Patent No. 63-132001, **1998**.
80. Ishikawa, H.; Adachi, A. Modified wood containing fireproofing agent for exterior wood. Japanese Patent No. 3-110103, **1991**.
81. Schubert, D.M.; Manning M.J. Compositions and methods for preserving wood products. U.S. Patent No. 5,612,094, **1997**. Washington, DC: U.S. Patent and Trademark Office.
82. Lewin, M. Flame retarding wood by chemical modification with bromate-bromide solutions. *J Fire Sci.* **1997**, *15*, 29–51. DOI: 10.1177/073490419701500103
83. Basson, G.R.; Conradie, W.E. Preservative and fire-retardant composition and combination and process. U. S. Patent No. 6,319,431, **2001**. Washington, DC: U.S. Patent and Trademark Office.
84. Tsunoda, K. Preservative properties of vapor-boron-treated wood and wood-based composites. *J Wood Sci.* **2001**, *47* (2), 149–153. DOI: 10.1007/bf00780565
85. Baysal, E. Determination of oxygen index levels and thermal analysis of scots pine (Pinus sylvestris L.) impregnated with melamine formaldehyde–boron combinations. *J Fire Sci.* **2002**, *20* (5), 373–389. DOI: 10.1177/0734904102020005485
86. Vinden, P.; Romero, F.J. Process for treating wood to protect from fungal decay, by applying boron based preservative and subjecting to moisture free environment to produce boron and alcohol by-product on reaction with moisture in wood. International Patent Application No. PCT/AU2002/00781, **2002**.
87. Lopez, R.A. Fire retardant compositions and methods for preserving wood products. U.S. Patent No. 6,620,349, **2003**. Washington, DC: U.S. Patent and Trademark Office.
88. Lenox, J.D.; Miller, N.T.; Rossi, M. Amino acid-solubilized borate silicate and zinc compositions, and methods for treating wood products. International Patent Application No. PCT/US05/010893, **2005**.
89. Baysal, E.; Sonmez, A.; Colak, M.; Toker, H. Amount of leachant and water absorption levels of wood treated with borates and water repellents. *Bioresour Technol.* **2006**, *97* (18), 2271–2279. DOI: 10.1016/j.biortech.2005.10.044
90. Lee, H.L.; Chen, G.C.; Rowell, R.M. Thermal properties of wood reacted with a phosphorus pentoxide-amine system. *J Appl Polym Sci.* **2004**, *91*, 2465–2481.
91. Kartal, S.N.; Yoshimura, T.; Imamura, Y. Decay and termite resistance of boron-treated and chemically modified wood by in situ co-polymerization of allyl glycidyl ether (AGE) with methyl methacrylate (MMA). *Int Biodeterior Biodegradation.* **2004**, *53* (2), 111–117. DOI: 10.1016/j.ibiod.2003.09.004
92. Mali, P.; Sonawane, N.S.; Patil, V.; Lokhande, G.; Mawale, R.; Pawar, N. Morphology of wood degradation and flame retardants wood coating technology: An overview. *Int Wood Prod J.* **2022**, *13* (1), 21–40. DOI: 10.1080/20426445.2021.2011552
93. Kartal, S.N. Boron-based wood preservatives and their use. In Handbook on borates: chemistry, production and applications. Chung, M.P., Ed., Nova Science, **2009**.
94. Wang, Q.; Li, J.; Winandy, J.E. Chemical mechanism of fire retardance of boric acid on wood. *Wood Sci Technol.* **2004**, *38*, 375–389. DOI: 10.1007/s00226-004-0246-4
95. Middleton, J.C.; Draganov, S.M.; Winters, F.T. Jr. Evaluation of borates and other inorganic salts as fire retardants for wood products. *For Prod J.* **1965**, *15* (12), 463–467.
96. Cullis, C.F.; Hirschler, M.M. The combustion of organic polymers. Clarendon Press, **1981,** 256–261.
97. Lewin, M.; Weil, E.D. Mechanisms and modes of action in flame retardancy of polymers. In Fire retardant materials. Horrocks, A.R.; Price, D., Eds. CRC Press, **2001**.
98. Shen K. Zinc borate as a flame retardant in halogen-free wire and cable systems. *Plast. Compd.* **1988,** *11* (7), 26–28.
99. Bourbigot, S.; LeBras, M.; Leeuwendal, R.; Schubert, D. In: Proceedings of 9th Annual Conference of Recent

Advances in Flame Retardancy of Polymeric Materials, Business Communications Co, Norwalk, **1998**.
100. Baena, J.C.; Wang, C.; Kabir, I.I.; Khalid, A.; Nazir, M.T.; Yuen, A.C.Y.; Ahmad, F.; Yeoh, G.H. Fire behaviour of waterborne intumescent coatings on timber substrate for bushfire exposure. *Fire Saf J.* **2023**, *140*. DOI: 10.1016/j.firesaf.2023.103836
101. Gao, Z.; Lang, X.; Chen, S.; Zhao, C. Mini-review on the synthesis of lignin-based phenolic resin. *Energy Fuels.* **2021**, *35* (22), 18385–18395. DOI: 10.1021/acs.energyfuels.1c03177
102. Pittman, C.U.; Kim, M.G.; Nicholas, D.D.; Wang, L.; Kabir, F.R.A.; Schultz, T.P.; Ingram, L.L. Wood enhancement treatments I. Impregnation of southern yellow pine with melamine-formaldehyde and melamine-ammeline-formaldehyde resins. *J Wood Chem Technol.* **1994**, *14* (4), 577–603. DOI: 10.1080/02773819408003114
103. Deka, M.; Saikia, C.B.K. Studies on thermal degradation and termite resistant properties of chemically modified wood. *Bioresour Technol.* **2002**, *84*, 151–157. DOI: 10.1016/s0960–8524(02)00016-0
104. Lin, X.; Zhang, S.; Qian, J. Synthesis and properties of a novel UV-curable waterborne hyperbranched polyurethane. *J Coat Technol Res.* **2014**, *11* (3), 319–328. DOI: 10.1007/s11998-013-9520-4
105. Winchester, C.M. Waterborne nitrocellulose wood lacquers with lower VOC. *J Coat Technol Res.* **1991**, *63* (803), 47–53.
106. Chen, C.; Webster, J.R.; Hong, R.I.; Tsai, Y.R.; Chen, C.C.; Tseng, K.Y.; Chang, P. U.S. Patent No. 7,235,307 B2, **2007**. Washington, DC: U.S. Patent and Trademark Office.
107. Javidnia, H.; Mohammadi, N.; Shooshtari, A.; Ghasemirad, S.; Farajpoor, T.; Ghanbari, M. The effect of butyl glycol acetate/ethyl acetate mixed solvents composition on nitrocellulose solution emulsification: The stability of resultant colloid and micro-filterability. *Iran Polym J.* **2010**, *19* (5), 323–332.
108. Ma, S.; Song, G.; Feng, N. Preparation and characterization of self-emulsified waterborne nitrocellulose. *Carbohydr Polym.* **2021**, *89* (1), 36–40. DOI: 10.1016/j.carbpol.2012.02.029
109. Weil, E. Fire-protective and flame-retardant coatings—a state-of-the-art review. *J Fire Sci.* **2011**, *29*, 259–296. DOI: 10.1177/0734904110395469
110. Zheng, J.; Fox, S.C.; Frazier, C. Rheological, wood penetration, and fracture performance studies of PF/PMDI hybrid resins. *For Prod J.* **2004**, *54* (10): 74–81.
111. Lahtela, V.; Kärki, T. The influence of melamine impregnation and heat treatment on the fire performance of scots pine (pinus sylvetris) wood. *Fire Mater.* **2015**, *40* (5), 731–737. DOI: 10.1002/fam.2338.
112. Kaya, A.I. Fore performance of thermally modified wood impregnated with clay nanomaterials. *Fresenius Environ Bull.* **2022**, *30* (5), 5292–5296.
113. Zhang, N.; Xu, M.; Cai, L. Improvement of mechanical, humidity resistance and thermal properties of heat-treated rubber wood by impregnation of SiO_2 precursor. *Sci Rep.* **2019**, *9* (1). DOI: 10.1038/s41598-018-37363-3
114. Spear, M.J.; Curling, S.F.; Dimitriou, A.; Ormondroyd, G.A. Review of functional treatments for modified wood. *Coatings* **2021**, *11* (3), 327. DOI: 10.3390/coatings11030327.
115. White, R.H.; Stark, N.M.; Ayrilmis, N. Recent activities in flame retardancy of wood–plastic composites at the forest products laboratory. In *Proceedings: 22nd BCC Annual Conference on Recent Advances in Flame Retardancy of Polymeric Materials*, Stamford, CT, **2011**, pp. 348–359.
116. Stark, N.M.; White, R.H.; Mueller, S.A.; Osswald, T.A. Evaluation of various fire retardants for use in wood flour–polyethylene composites. *Polym Degrad Stab.* **2010**, *95* (9), 1903–1910. DOI: 10.1016/j.polymdegradstab.2010.04.014
117. Bajwa, D.S.; Rehovsky, C.; Shojaeiarani, J.; Stark, N.; Bajwa, S.; Dietenberger, M.A. Functionalized cellulose nanocrystals: A potential fire retardant for polymer composites. *Polymers.* **2019**, *11* (8), 1361. DOI: 10.3390/polym11081361
118. Shen, K.K.; Olson, E. The use of borates and talc as fire retardants in wood plastic composite. In *Proceedings Seventeenth Annual BCC Conference on Flame Retardancy*, **2006**, pp. 289–296.
119. Abu Bakar, M.B.; Mohd, I.Z.A.; Taib, R.M.; Rozman, H.D.; Mohamad Jani, S. Mechanical and flame retardant properties of wood flour filled polypropylene composites. In *Proceedings 8th Pacific Rim Bio-Based Composites Symposium*, **2006**, pp. 349–356.
120. Schirp, A.; Hellmannn, A. Fire retardance improvement of high-density polyethylene composites based on thermo-mechanical pulp treated with ammonium polyphosphate. *Polym Comp* **2019**. DOI: 10:2410–2423.
121. Shao, B.; Zhang, Z.J.; Wang, Q.W.; Song, Y.M. 2008. Effects of APP on the fire retardant and mechanical properties of wood-flourHDPE composite. *Polym Mater Sci Eng.* **2008**, *24* (4), 93–96.
122. Dong, J.; Li, B. Study on the flame retardation and properties of PP-wood composite material by intumescent flame retardant. *Chem Adhesion.* **2007**, *29* (4), 269–273.
123. Bai, X.Y.; Wang, Q.W.; Fang, Y.Q.; Zhang, Z.J. Effects of transition metal oxides on the combustion properties of wood-flour/polyvinyl chloride composites. *Sci Silvae Sin.* **2008**, *44* (10), 103–108.
124. Kurt, R.; Mengeloglu, F. Utilization of boron compounds as synergists with ammonium polyphosphate for flame retardant wood-polymer composites. *Turk J Agric For.* **2001**, *35* (2), 155–163.
125. Ayrilmis, N.; Akbulut, T.; Dundar, T.; White, R.H.; Mengeloglu, F.; Buyuksari, U.; Candan, Z.; Avci, E. Effect of boron and phosphate compounds on physical, mechanical, and fire properties of wood–polypropylene composites. *Constr Build Mater.* **2012**, *33*, 63–69. DOI: 10.1016/j.conbuildmat.2012.01.013
126. LeVan, S.L.; Winandy, J.E. Effects of fire retardant treatments on wood strength: A review. *Wood Fiber Sci.* **1990**, *22* (1), 113–131.
127. Winandy, J.E.; Barnes, H.M.; Jones, P.D.; McIntyre, C.R. Laboratory and field exposures of fire retardant-treated plywood: Part 3—modeling exposure relationships. *Wood Fiber Sci.* **2014**, *46* (4), 563–572.
128. National Fire Protection Association. Standard on Respiratory Protection Equipment for Tactical and Technical Operations, NFPA 1986, **2023**.
129. Beall, F.C.; Eickner, H.W. Thermal degradation of wood components: A review of the literature. USDA Forest Service, Forest Products Laboratory, **1970**.
130. Ayrilmis N. Effect of fire retardants on internal bond strength and bond durability of structural fiberboard. *Build Environ.* **2007**, *42* (3), 1200–1206. DOI: 10.1016/j.buildenv.2005.11.017
131. LeVan, S.; Ross, R.J.; Winandy, J.E. Effects of fire retardant chemicals on the bending properties of wood at elevated temperatures. USDA Forest Service, Forest Products Laboratory, **1990**.
132. Hasburgh, L.E.; Sumathipala, K. Flammability tests for regulation of building and construction materials. In

Flammability testing of materials used in construction, transport and mining, 2nd ed., Apte, V., Ed.; Woodhead Publishing, **2022**, pp. 223–236. DOI: 10.1016/B978-0-08-102801-8.00004-1
133. ASTM International. Standard Test Method for Surface Burning Characteristics of Building Materials; ASTM E 84, **2021**. DOI: 10.1520/E0084-21A
134. AmericanWoodCouncil.Designforcodeacceptable1,Flame Spread Performance of Wood Products used for Interior Finish, **2019**. https://awc.org/wp-content/uploads/2021/12/AWC-DCA1-FlameSpreadPerformance-1906.pdf (accessed 2013-07-15).
135. ASTM International. Standard Test Method for Extended Duration Surface Burning Characteristics of Building Materials (30 min Tunnel Test); ASTM E 2768, **2018**. DOI: 10.1520/E2768-11R18
136. National Fire Protection Association. Standard Methods of Fire Tests for Evaluating Contribution of Wall and Ceiling Interior Finish to Room Fire Growth; NFPA 286, Quincy, MA, **2019**.
137. ASTM International. Standard Test Method for Room Fire Test of Wall and Ceiling Materials and Assemblies; ASTM E 2257, **2022**. DOI: 10.1520/E2257-22
138. International Organization for Standardization. Reaction to fire tests—room corner test for wall and ceiling lining products—Part 1: Test method for a small room configuration; ISO 9705-1:2016, **2016**.
139. White, R.H.; Dietenberger, M.A.; Tran, H.; Grexa, O.; Richardson, L.; Sumathipala, K.; Janssens, M. Comparison of test protocols for the standard room/corner test. *Fire Mater.* **1999**, *23* (3), 139–146. DOI: 10.1002/(sici)1099-1018(199905/06)23:3%3C139::aid-fam681%3E3.0.co;2-0
140. Grexa, O.; Dietenberger, M.A.; White, R.H. Reaction-to-fire of wood products and other building materials: Part 1, room/corner test performance. USDA Forest Service, Forest Products Laboratory, **2012**. DOI: 10.2737/fpl-rp-663
141. Dietenberger, M.A.; Grexa, O.; White, R.H. Reaction-to-fire of wood products and other building materials: Part II, cone calorimeter tests and fire growth models. USDA Forest Service, Forest Products Laboratory, **2012**.
142. Dietenberger, M.A.; Grexa, O.; Janssens, M.L.; Sweet, M.; White, R.H. Room/corner tests of wall linings with 100/300 kW burner. USDA Forest Service, Forest Products Laboratory, **1995**.
143. Tran, H.C.; Janssens, M.L. Wall and corner fire tests on selected wood products. *J Fire Sci.* **1991**, *9* (2), 106–124. DOI: 10.1177/073490419100900202
144. Hirata, T.; Kawamoto, S.; Nishimoto, T. Thermogravimetry of wood treated with water-insoluble retardants and a proposal for development of fire-retardant wood materials. *Fire Mater.* **1991**, *15* (1), 27–36. DOI: 10.1002/fam.810150106
145. Babrauskas, V. Development of the cone calorimeter—a bench-scale heat release rate apparatus based on oxygen consumption. *Fire Mater.* **1984**, *8* (2), 81–95. DOI: 10.1002/fam.810080206
146. Tran, H.C.; White, R.H. Burning rate of solid wood measured in a heat release rate calorimeter. *Fire Mater.* **1992**, *16* (4), 197–206. DOI: 10.1002/fam.810160406
147. Dietenberger, M. Effect of backing board on the heat release rate of wood. In *Proceedings of the International Conference on Fire Safety*, Columbus, OH, Product Safety Corporation, **1999**, Vol. 28, pp. 62–73.
148. Sanned, E.; Mensah, R. A.; Försth, M.; Das, O. The curious case of the second/end peak in the heat release rate of wood: A cone calorimeter investigation. *Fire Mater.* **2022**, *47* (4), 498–513. DOI: 10.1002/fam.3122.
149. White, R.H.; Dietenberger, M.A. Cone calorimeter evaluation of wood products. In *Fifteenth Annual BCC Conference on Flame Retardancy*, Jun 7–9, **2004**; Stamford, CT, pp. 331–342.
150. Dietenberger, M.A.; White, R.H. Reaction-to-fire testing and modeling for wood products. In *Proceedings of the 12th Annual BCC Conference on Flame Retardancy*, May 21–23, **2001**; Business Communications Co., Norwalk, CT, pp. 4–69.
151. ASTM International. Standard Practice for Accelerated Weathering of Fire-Retardant-Treated Wood for Fire Testing; ASTM D 2898, **2017**. DOI: 10.1520/D2898-10R17
152. ASTM International. Standard Practice for Natural Weathering of Materials; ASTM G7/G7M, **2021**. DOI: 10.1520/G0007_G0007M-21
153. ASTM International. Standard Practice for Determining by Exterior Exposure Tests the Susceptibility of Paint Films to Microbiological Attack; ASTM D3456, **2018**. DOI: 10.1520/D3456-18
154. ASTM International. Standard Test Method for Evaluating Degree of Surface Disfigurement of Paint Films by Fungal or Algal Growth, or Soil and Dirt Accumulation; ASTM D3274, **2021**. DOI: 10.1520/D3274-09R21
155. ASTM International. Standard Test Method for Evaluating Degree of Chalking of Exterior Paint Films; ASTM D4214, **2015**. DOI: 10.1520/D4214-07R15
156. ASTM International. Standard Test Method for Evaluating Degree of Checking of Exterior Paints; ASTM D660, **2019**. DOI: 10.1520/D0660-93R19
157. ASTM International. Standard Test Method for Evaluating Degree of Cracking of Exterior Paints; ASTM D661, **2005**. DOI: 10.1520/D0661-93R05
158. ASTM International. Standard Test Method for Evaluating Degree of Erosion of Exterior Paints; ASTM D662, **2019**. DOI: 10.1520/D0662-93R19
159. ASTM International. Standard Test Method for Evaluating Degree of Blistering of Paints; ASTM D714, **2017**. DOI: 10.1520/D0714-02R17
160. ASTM International. Standard Test Method for Evaluating Degree of Flaking (Scaling) of Exterior Paints; ASTM D772, **2018**. DOI: 10.1520/D0772-86R18
161. ASTM International. Standard Test Method for Specular Gloss; ASTM D523, **2018**. DOI: 10.1520/D0523-14R18
162. International Organization for Standardization. Colorimetry Part 4: CIE 1976 L*a*b* Colour Space; ISO/CIE 11664-4, **2019**.
163. Madyaratri, E.W.; Ridho, M.R.; Aristri, M.A.; Lubis, M.A.; Iswanto, A.H.; Nawawi, D.S.; Antov, P.; Kristak, L.; Majlingová, A.; Fatriasari, W. Recent advances in the development of fire-resistant biocomposites—a review. *Polymers.* **2022**, *14* (3), 362. DOI: 10.3390/polym14030362

27 Flame-Retardant Design and Protection for Textiles

Baljinder K. Kandola
Institute for Materials Research and Innovation, University of Bolton, Bolton, BL3 5AB

27.1 INTRODUCTION

Textile term more commonly used for fibers, yarns, and fabrics for apparel clothing and furnishing has expanded in meaning and applications since last century after the development of synthetic and high performance fiber-forming polymers. Until the 1940s, natural fibers like cotton, wool, silk have been used as apparel, curtains and furnishing fabrics; and flax, jute, sisal etc., as technical (e.g., ropes, tents) and geotextiles. The first commercially available synthetic fiber, viscose rayon, a regenerated cellulosic fiber was developed around 1910. In 1939 the first synthetic thermoplastic fiber forming polymer, nylon 6,6 (polyamide), was synthesized, followed by polyester in 1950s and polyolefins in 1960s. These developments revolutionized the textile world with new fibers offering several advantages over natural fibers and often at a competitive price. Their general popularity derives from improved durability, easy-care properties, and resistance to attack by biological agents such as insects and mildew. The application areas expanded from the traditional textiles areas like apparel, upholstery, furnishing, towels, curtains, carpets, etc., to technical (ropes, parachute fabrics), transport (reinforcement of tires, seat belts, brakes, interior furnishing fabrics), industrial products (packaging, filters in chemical plants), buildings (roofliners), geotextiles, and medical textiles. Thermoplasticity of these fibers, however, limits their applications in certain areas where high temperatures are achieved, such as protective clothing for firefighters and workers for oil and gas industry and molten metal workers. The real breakthrough came in the 1970s with the development of aramid fibers and other high-performance fibers in the 1980s, which could be used as protective and safety clothing. The development of inorganic fibers like glass and carbon expanded their use in fiber-reinforced composites for load-bearing structures in transportation and construction industry.

The hazard posed by burning textiles has also been realized for centuries, and salts like alum have been used since those times to reduce their flammability and so confer flame retardancy. With new fiber developments and application areas, fire hazards have also grown over the years. Synthetic fibers being thermoplastic will melt and drip upon heating and ignition, and tend to move away from the ignition source. When this happens, the synthetic fabrics do not burn further and, in some cases, pass certain flammability standards via this fire behavior. However, synthetic fibers/fabrics are not heat resistant and pose problems of melt dripping, as the molten drips are very hot and can cause severe burns to the wearer. These can also be burning drips, which, although take the flame away from the burning material, can fall on other combustible materials and, hence, become source of ignition. The development of inherently fire retardant aramid fibers has been a real breakthrough in the area of fire retardancy, establishing them as protective and safety clothing materials.

In different applications, while some materials must be fireproof, others can be fire resistant. The terms "resist," "retard," and "proof" have different meaning. While the words "resist" and "retard" imply a late ignition and/or low flame spread, "proof" is an absolute term meaning the fiber will not to be affected from the fire.[1] This chapter reviews the flammability of different fiber/fabric types used for different applications and fire-retardant strategies involved to make them fire retardant. This review is not exhaustive but is complimentary to other reviews written in the past.[2–10]

27.2 MEASUREMENT OF FABRIC FLAMMABILITY AND TESTING STANDARDS

Textile materials have high fiber surface to mass ratio and, hence, tend to ignite easily. As for most applications, the textiles are exposed to the environment and, hence, due to ease of accessibility of oxygen, burn faster than other bulk polymers. In general, flammable fabrics are those fabrics which ignite when subjected to a small flame for durations of up to 12 seconds and continue to burn after the source has been removed. Most of the work on flammable fabrics is therefore concerned and directed towards the observation and measurement of ease of ignition, the rate and extent of flame spread, the duration of flaming, measurement of heat release, and heat of combustion and quantitative description of burning debris, such as melt dripping. Rarely if ever does a single test method enable all these parameters to be measured. For self-extinguishing fabrics, such as flame retarded

DOI: 10.1201/9781003380689-27

fabrics, test methods include measurement of time of afterflame and afterglow and extent of fire damage in terms of char length, hole size, or weakened sample length.

No single laboratory test can determine the complete burning character of a particular textile.[11] There are two types of tests, scientific or research test methods, and the standard test methods. The research test methods help in understanding the burning behavior and are used to develop new products or fire-retardant finishes. The standard test methods are used to rate or regulate materials for sale but can also provide some understanding of burning behavior depending how the test measures flammability. For more insight on research vs. standard test methods, please see Chapter 17 of this book.

Amongst the research methods, most popular is the limiting oxygen index (LOI), also called oxygen index (OI) and defined as the minimum concentration of oxygen, expressed as volume percent, in a mixture of oxygen and nitrogen that will just support flaming combustion of a material.[12] Generally, textiles having LOI values of 21 vol% or less burn rapidly, those having values in the range 21 to 25 vol% burn slowly, and those with LOI ≥ 26 vol% exhibit some level of flame retardancy[11] and usually pass most small flame ignition tests in the horizontal and vertical orientations.[3] LOI tests are mainly used in determining the flammability of different fibers and effect of different flame retardant treatments and finishes. However, LOI value may be influenced by many fabric structural variables for the same fiber type,[11] hence is a relative rather than absolute value. Moreover, because the sample ignition occurs at the top and the sample burns vertically downward, this is not true representative of the burning of freely hanging material in the real world.

In the past different countries had their own textile standard testing methods.[11] Since 1990 within the EU, there has been an attempt to normalize these standards. In the UK for example, most new British Standards are prefixed by BS EN or BS EN ISO. Most of the test methods are based on principle that tests should be straightforward and easy to apply and are application specific. The standard test methods usually involve "pass-fail" and/or performance rating criteria[13] and are mostly product application specific. Some selective and more relevant test methods are discussed here in brief and for detailed information reader is referred to other reviews.[10,14]

27.2.1 Simple Ignition Tests

A simple ignition test includes a vertically oriented fabric subjected to a standard gas flame applied to the face or lower edge of the fabric specimen. Ignition is monitored by visual observations and the time taken to ignite the specimen is recorded. This test is used in many standards including BS 5438, EN ISO 6941, FAR (Federal Aviation Regulation) Part 25, etc. Many flammability standards like BS 5722 require simple vertical strip test, measured by BS5438 Test 2. This involves subjecting a vertically oriented fabric to a standard igniting flame source either at the edge or on the face of the fabric for specified time (10s). If after the removal of ignition source, the flame reaches either end of the fabric, it fails the test. If the flame is extinguished, the char length, size of hole if present, afterglow, and nature of any debris (molten drip, etc.) are noted.

27.2.2 Flame Spread

Rate of flame spread is usually calculated by measuring the distance and recording the time taken of the advancing flame front to reach defined distances. The upward fire spread is far more rapid than downward and horizontal flame spread and, hence, adopted as a better means of measuring the fire hazard of a fabric. Therefore, most standards, including BS 5438:1989,[15] standards for curtains and drapes, BS EN ISO 15025:2016, use this type of bench-scale test method for measuring vertical flame spread properties of fabrics in particular. In BS 5438 Test 3, for example, a sample size of 560 mm × 170 mm is used across which cotton trip threads connected to timers are placed, and the time taken to cut through each thread is used to measure the burning rate. For research purposes, sometimes samples of limited sizes are available, for which the test can be modified for indicative burning rate results. For example, in our laboratory we developed a number of synthetic fibers containing polymer layered silicate nanoclays and reactive flame retardants (discussed in detail in a later section). Due to limited sample sizes, sample holder, 190 × 70 mm, used in Test 1 of BS 5438 standard was used. Samples were marked at 60, 120, and 180 mm intervals. The first 10 mm of sample burning was not taken into account, and so times of burning were recorded once the flame had reached a line drawn 10 mm from the bottom edge against which the standard flame was applied for 10s as specified in the test. A video film was taken of the burning of each sample from which times to reach 60, 120, and 180 mm marks and/or to achieve flame-out were noted, and from this burning behavior of each sample was observed and rate of flame spread calculated.[16]

Measurement of flame spread under external heat flux is necessary where the thermal radiation is likely to impinge on the textile materials, for example the flooring material of the building or transport vehicles whose upper surfaces are heated by flames or hot gases, or both. The French test method, NF P 92–503 Brûleur Electrique, or "M" test, involves radiant panel for testing flame spread of flexible textile materials. This test method (flame spread under external heat flux) is the basis of that used by the FAA (Federal Aviation Administration) for assessing flammability of textile composites used in thermal/acoustic insulation materials (FAR 25.856 (a)) used in aircraft and has also been included by the EU for fire test approval of floorings such as EN ISO 9239.

For textile materials used as interior wall coverings in UK buildings including railway carriages, where the fabric could be in a vertical orientation attached to the wall panel, measurement of rate of flame spread under external heat flux is one of the requirements. For such applications, the test method (BS 476 Part 7) essentially requires a vertically oriented specimen exposed to gas fired radiant panel with incident heat flux of 32.5 kW/m^2 for 10 mins. In addition, a pilot flame is applied at the bottom corner of the specimen for 1 min 30 s, and rate of flame spread is measured. The same principle is used in the French test for carpets, NF P 92–506.

In the US, the National Fire Prevention Association (NFPA) 701 test is used for curtain materials to measure upward flame spread, and tests such as ASTM D6413 are used to measure vertical flame spread/burn length for textiles which are worn by people.

27.2.3 Heat Release Tests

When textiles comprise part of a building or transport structure, heat release rate measurement becomes an important criterion. One of the original, successful small-scale calorimeter was developed at Ohio State University (OSU), defined in the aviation standard FAR 25.853 Part IV Appendix F and ASTM E906–2009 for determining the heat release of internal structural materials in commercial aircraft originally in US and now worldwide. In this calorimeter a vertically oriented composite is exposed to a heat flux of 35 kW/m^2. A maximum peak heat release rate of ≤ 65 kW/m^2 and average heat release rate over the first two minutes of the test not exceeding 65 kW/m^2 are required for a textile composite to achieve a pass. In most cases the interiors are made of glass-fiber-reinforced phenolic composites, which are inherently flame retardant. The textile material (wool, silk, or wool/silk blend), if used, is usually glued on an aramid board. The textile materials used, although are inherently flame retardant, need flame-retardant back coating to pass the composite test discussed previously.[17] The greater availability and more scientific data output from the cone calorimeter has led a number of researchers including ourselves to compare heat release from cone (discussed below) and OSU calorimeters, so that cone calorimeter can be used to assess the flammability parameters of materials for aerospace. It was concluded that the cone and OSU calorimeters, used according to ISO 5660 and FAR 25.853 Part IV, respectively, do not give similar results at a particular heat flux. The main reasons for this are the different ignition sources and methods for measuring HRR in these two instruments. For samples that are difficult to ignite, it was observed that results at 35 kW/m^2 in the OSU correlate with those from cone calorimeter at 50 kW/m^2 for PHR values. However, the peak of heat release was delayed in the latter because of the spark ignition source used. Due to delayed ignition in the cone, THR for 2 min values were also different than in OSU calorimeter, however, all the values after 5 min exposure were similar.[17]

More sophisticated bench scale equipment, which measures heat release rate by oxygen consumption is the cone calorimeter. In this test, the fabric or composite specimen mounted over an insulating ceramic blanket is exposed to an external heat flux (0–75 kW/m^2). The volatiles released from the heated specimen are ignited using spark igniter and the time taken to ignite the gases is recorded as time-to-ignition of the specimen tested. Originally designed to study the fire characteristics of building materials which are physically and hence thermally thick, the cone calorimeter can now be used for thermally thin materials such as fabrics.[18] It has also been used for characterizing furnishing fires which incorporates the samples in a composite fabric/filling form (for example, an upholstery fabric on top of a polyurethane foam).[19] Work in our laboratory[13,18] has shown that when single layer of fabrics are tested with a cone calorimeter, fabrics curl, melt, and some char, thus changing the specimen configuration under the incident heat flux. Intumescing fiber containing samples such as silk and wool in particular formed a large dome-like char over the entire surface when heated prior to ignition. This dome structure sometimes reached a maximum height of 25 mm at the center, which resulted in a variation of distance between the sample surface and the cone heater. The specimen even sometimes touched the pilot ignition system, resulting in ignition by impingement, or fouling of the igniter. In case of acrylic and polyester-cotton, specimens shrank away rapidly from the radiant source, posing the problem of change in specimen configuration factor due to reduction in thickness and reduction in flux intensity imposed upon the surface. In order to stabilize fabrics and improve reproducibility, a cross-wire grid, as shown in Figure 27.1(a), was used across the surface of the fabric specimens. The design of grid has also an effect on time to ignition values and the PHRR values.[20] With the grid shown in Figure 27.1 (a), we tested lightweight cotton (87 g/m^2), heavyweight cotton (180 g/m^2), polyester-cotton (65:35,105 g/m^2), acrylic (118 g/m^2), lightweight silk (71 g/m^2), heavyweight silk (174 g/m^2), and wool (173 g/m^2) under 35 kW/m^2 external heat flux. In general and apart from silk and wool, single layer specimen results showed acceptable reproducibility in that CV ≤ 13%.[18] We also studied the effect of fabric layering (1 to 10 layers) on cone parameters. With increasing number of fabric layers, time-to-ignition (TTI) values increased as expected. The effect on peak heat release values plotted in Figure 27.1(b) shows that the increase in PHRR is significant when the number of layers increase from a single layer to 2–3 layers, stabilized or even decreases with a further increase in number of layers. The probable reason for this effect may be because the specimen becomes physically and thermally thick and that the diffusion of volatiles from beneath the surface is impeded by char formation in adjacent layers, which then reduces burning rates and respective PHRR values. This approach to thermally thick behavior is seen only for the char-forming fabrics such as cotton, silk, and wool (see Figure 27.1(b)). However, the total heat released (THR), which reflects the total specimen mass and fuel content, was higher for the multilayered specimen than the single layer specimens.

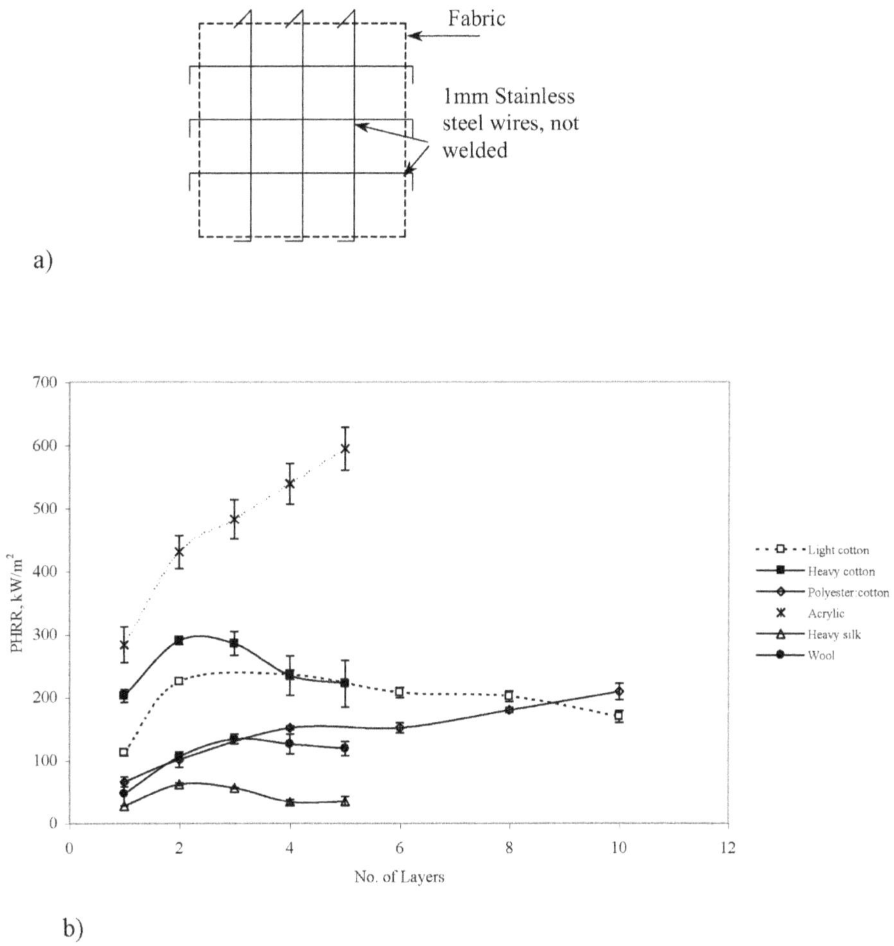

FIGURE 27.1 (a) Textile sample-retaining grid for the cone calorimeter and (b) effect of layers on PHRR values of all fabric samples.[18]
Source: From Nazaré S. et al, *Fire Materials*, 26, 191–199, 2002. With permission.

27.2.4 Mannequin Tests

The major means of determining the fabric burn hazard of textiles and more particularly, clothing assemblies, has been in the use of sensored mannequins. A typical mannequin is equipped with 100–122 individual heat-flux sensors distributed over the surface of the body. The test garment is placed on the mannequin at ambient temperature conditions and exposed to intense fire or flash fire simulation sources with controlled heat flux, duration, and flame distribution. The sensors measure incident heat flux upon the underlying mannequin surface during and after exposure. The changing temperature of the mannequin surface temperature simulates damage to human tissue at two skin thicknesses, one representing a second-degree burn injury point and the other a third-degree burn injury point is calculated. The computerized data acquisition system also calculates surface heat fluxes, skin temperature distribution histories, and predicts the skin burn damage for each sensor location. These tests offer a more analytical means of assessing apparel burning hazards and predict potential skin burn hazards of garments. They are useful in research but are very expensive and complex to be used in standard test procedures.[21] A major problem with this test is poor reproducibility because of fitting of the clothing assembly over the mannequin torso, even when garment sizes are identical. This is particularly the case in firefighter's clothing where the overlap between top coat and trousers may present a thermal weakness during flame exposure. Overall mannequins are considered to be important in simulating the burn injury severity in real-life accidents. More details about mannequin tests can be found in a review by Camenzind et al.[22] The American Society of Testing and Materials version of this full-scale manikin test is ASTM F1930, and it is often a required regulatory test for select military uniforms, firefighter turnout gear, and outer protective gear for workers in hot environments.

27.2.5 Full Product Tests

Products mostly comprise different materials. For example, upholstered furniture consists of textile material covering polyurethane foam on the supporting metal or wooden frame. Flammability tests for upholstered furniture using various ignition sources were first developed in 1979 as a British Standard BS 5852: Parts 1 and 2 in the United Kingdom. Source 0 is a smoldering cigarette and Source 1 a small butane flame which simulates a lighted match. Source 2 and 3 are more intense flame sources. The test specimen is a composite specimen consisting of fabric and filling material. A similar composite specimen assembly is used in the US standard Cal TB 116 for testing flame retardance of upholstered furniture. In addition to the smoldering cigarette ignition source, BS 5852:1979 and its subsequent variants (BS 5852:1990 (1998, 2006) replaces BS 5852–2) define use of a variety of pinewood cribs (Source 4–7) which match the calorific outputs of increasing numbers of full-size newspaper sheets. Development of BS 5852 as small-scale composite test was breakthrough in realistic model testing that cheaply and accurately indicated the ignition behavior of full-scale products of complex nature.[23] Good reproducibility, cost-effectiveness, and easy-to-use features of BS 5852 have led to the establishment of the concept which was further employed for flammability testing of bedding and mattresses.[3] For full product as a chair or mattress, furniture calorimeter is used. See Chapter 31 of this book for more information on fire testing of furniture items.

Other textile products that require composite flammability testing are protective clothing assemblies for firefighters' suits, military flight suits, etc. Flammability standards and test methods for textile components in protective clothing have previously been discussed by Bajaj[24] and Horrocks.[23,25] Their reviews suggest that standard bench scale tests for separate determination of flame resistance, thermal insulation/protection, and heat resistance may be undertaken on a single fabric or a composite form in a manner that reflects a real application or product requirement.

27.3 FLAMMABILITY OF DIFFERENT FIBER AND TEXTILE TYPES

The burning behavior of fibers depends upon their physical and chemical properties. All natural and synthetic fibers are organic polymers, and no organic material can withstand intense and prolonged heat without degradation, even in absence of oxygen. Given sufficient oxygen and energy input, these materials will burn. The burning of polymer/fiber is essentially a three-phase process—heating, thermo-pyrolytic or decomposition, and finally ignition. The behavior of a polymer/fiber during the initial or primary heating phase depends to a considerable extent upon the nature of its composition. Thermoplastic fibers, because of their linear molecular chains, will generally melt in the temperature range of 100–250°C and start to flow. Because of loss of rigidity, they move away from ignition source, hence further pyrolysis and ignition in some cases is prevented. Natural fibers like cellulose, wool, silk on the other hand remain unchanged during this stage. Their three-dimensional cross-linked molecular structure prevents softening or melting. Their decomposition or pyrolysis occurs between 250–500°C, depending upon the chemical composition of the fiber. Pyrolysis or decomposition involves unzipping of the polymer chains to yield flammable monomers or the random elimination of small chemical fragments. Both types of products can sustain gas phase flame reactions. The flammable gases formed by pyrolysis mix with atmospheric oxygen, reach the lower flash on limit, and are either ignited by external flame, or if temperature is sufficiently high, self-ignite. In some cases, however, recombination of some of these fragments also occur and leads to the formation of aromatic condensed ring system, called char, which is stable under the pyrolytic conditions.

The major difference between fibers and bulk polymers is the small thickness of individual fibers, typically being 15–30 μm in diameter, yielding yarns of 50–100 μm diameter and fabrics having thicknesses varying from as low as 100 μm to several mm. Due to very high fiber surface-to-mass ratios, the temperature gradient through the thickness of the fiber is very low, hence abovementioned reactions occur very fast, leading to easy ignition. Melting, pyrolysis, and combustion temperatures and flammability behavior of different fiber types taken from literature are given in Table 27.1.

TABLE 27.1
Thermal and Flammability Properties of Some Commonly Used Fibers[4,25]

Fiber	Melting Temp (°C)	Pyrolysis Temp (°C)	Ignition Temp (°C)	LOI (%)
Natural fibers				
Cotton	-	350	350	18.4
Wool	-	245	600	25
Silk	-	320	600	23
Synthetic—thermoplastic				
Nylon 6	215	431	450	20–21.5
Nylon 6,6	265	403	530	20–21
Polyester	255	420–477	480	20–21.5
Polypropylene	165	469	550	18/6
Poly vinyl chloride	> 180	> 180	450	37–39
Synthetic—char formers				
Viscose	-	350	420	18.9
Acrylic	> 320	290	> 250	18.2
Modacrylic	> 240	273	690	29–30
High-performance fibers				
Meta-aramid (e.g., Nomex)	375	310	500	28.5–30
Para-aramid (e.g., Kevlar)	560	590	> 550	29
Oxidized acrylic	-	> 640	-	55
Polybenzylimidazole (PBI)	-	> 500	> 500	40–42
Polytetrafluorethylene (PTFE)	> 327	400	560	95

The burning behavior of fabrics depends upon many factors, such as fiber type or fiber blend ratio, fabric area density, fabric structure (woven, knitted, or non-woven, open structure or closely woven etc.), finishes, and garment design, point of ignition and the intensity of ignition source, orientation of fabric (vertical or horizontal), etc. Due to their thermally thin character and the open structure of the textiles, it is easy for air to circulate between the burning fibers. In case of thermoplastics, some fabrics may cause shrinkage in one or more directions, curl (see Section 27.2.3), and hence change geometry. With thermoplastics or blends with higher ratio of thermoplastic fibers, melting occurs, giving rise to often flaming molten drips.

As discussed in the previous section, from burning hazard point of view the important parameters are time-to-ignition, rate of flame spread, and rate of heat release. In our laboratory we have studied these parameters for selective fabrics, most commonly used as apparels, and the results are given in Table 27.2.[13,18,26] Although the LOI values given in Table 27.1 are different for different fiber types, all fabrics ignited within 1–3 s, when studied by BS 5438 Test 1. Auto ignition temperatures for all fabrics were determined by placing the fabrics in a preheated tube furnace at different temperatures. Flame spread results under different orientations show that the fire hazard is maximum when the fabrics are free hanging in the vertical direction. Light cotton, as expected, has higher rate of flame spread than others. Acrylic has lower flame spread rate than other fabrics, whereas has highest peak and total heat release rate amongst all fabrics. These results show that the flame spread rate is inversely proportional to the fabric area density. Similar conclusion that "for a particular fiber type, the heavier the fabric the better it behaves in a fire" could also be drawn from a study by Hirschler et al.,[27] where they conducted a flammability study of 50 fabrics with different fiber types and different area densities using the NFPA 701 small-scale test.[28]

27.4 APPLICATION-BASED HAZARDS AND PERFORMANCE REQUIREMENTS

Designing flame-retardant textiles for required level of performance for specific end uses requires understanding of the end use conditions and specific flammability performance standards. Textiles in different applications are used either as a single component (e.g., apparel, curtains, bedsheets etc.) or a component of the composite structure (e.g., furniture). In the latter, components other than textile material and their layup will also influence the burning behavior of the whole structure. This section discusses different fiber and fabric types used for different applications, associated fire hazards, and most relevant flammability performance standards. In Section 27.2 testing standards were discussed, which are used for testing materials to comply with performance standards discussed briefly in this section and in detail elsewhere.[14]

27.4.1 Apparel Fabrics

Apparel fabrics for normal use generally include cotton, viscose, polyester, nylon, wool, silk fibers, and their blends. Adult apparel fabrics are generally not flame retarded because of lack of both consumer demand and stringent mandatory standards. The flammability behavior of some

TABLE 27.2
Ignition, Flame Spread, and Heat Release Properties of Some Commonly Used Fabrics as Apparel[13,18,26]

	Ignition Time Using BS 5438, Test 1(s)		Auto Ignition		Flame Spread Rate Using Modified BS 5438 Test 3 (m/s)			Cone Results at 35 kW/m² Heat Flux		
Fabric	Face Ign	Edge Ign	Ign. Temp (°C)	Ign time (s)	Vertical	45° angle	Horizontal	TTI (s)	PHRR (kW/m²)	THR (MJ/m²)
Lightweight cotton (87 g/m²)	2	1	480	16	57	37	6	9	94	1.0
Heavyweight cotton (180 g/m²)	4	1	480	35	27	18	3	14	128	3.2
Polyester-cotton (65:35, 105 g/m²)	2	1	574	20	37	24	9	10	154	1.9
Acrylic (118 g/m²)	*		-	-	23	13	6	17	292	4.5
Lightweight silk (71 g/m²)	2	2	909	3	***	***	***	***	***	***
Heavyweight silk (174 g/m²)	**	3	655	4	***	***	***	28	45	1.0
Wool (173 g/m²)	3	3	746	3	23	12	14	16	171	2.9

Notes: * Flames extinguished when the flame was moved away.

** Fabric melted away from the flame.

*** Test not performed.

fabrics commonly used for apparels has been discussed in previous section. In normal use apparel fabrics usually do not catch fire, but if they do, the risks of death/injury per incident are high (almost one in four accidents[29]), this is especially the case with loose clothing, for example, ladies nightwear[30] and saris worn by Indian women. The UK Nightwear (Safety) Regulations, 1985 requires the testing of all nightwear, including pajamas and dressing gowns, and demands that adult and children's nightwear carry a permanent label showing whether or not each item meets the requirements of BS 5722:1991[31] (which uses Test 3 of BS 5438: 1989, discussed in Section 24.2.2). This latter performance standard defines a maximum permissible burning rate of a vertically oriented fabric. European standard EN 1103 requires that all types of apparel exhibit a flame propagation rate (FPR) of at most 90 mm/s.[32] Both of these tests are good in predicting the potential hazard of fabrics made of natural fibers but do not take into account the thermoplasticity of synthetics, hence cannot relate the hazard to actual skin burns. There is no relationship between flame propagation rate and actual skin burns.[33] Fabrics made of blends with high percentages (> 50%) of synthetic fibers usually pass these tests but can present an increased burn risk due to the combined flame spread and melting effects of thermoplastics.[13]

27.4.2 Protective Clothing

Protective textiles often have to protect the wearer from more than one condition, hence need multifunctional properties. With regard to fire hazard, in addition to flame retardant properties, they should also be thermally resistant. Thermal protection relates to the ability of textile to resist conductive, convective, radiant thermal energy or a combination of two or more. Typical flame temperatures may range from 600 to 1,000°C, hence the textile structure should be able to withstand these temperatures without undergoing melting or degradation and thus loosing geometrical coherence. Synthetic thermoplastic fibers, hence, cannot be used for applications where very high performance is required. In most applications, multilayered clothing assemblies are used. The most important aspect of protective clothing testing is the evaluation of burn injury protection and thermal characteristics of clothing systems, which can be undertaken using an instrumented mannequin, as discussed in Section 27.2.4. Resistance to heat transfer by convective flame, radiant energy, or plasma energy sources is quantified in terms of thermal protective index (TPI) often related to the time taken for an underlying skin sample with or without an insulating air gap to achieve a minimum temperature or energy condition sufficient to generate a second-degree burn. This can be done by measuring the time for a thermocouple placed behind a fabric or assembly to reach a critical temperature equivalent to that causing a radiant energy source. Although this area is very vast and includes many applications, only selected examples are discussed here, and for detailed information refer to other reviews.[10,25]

27.4.2.1 Firefighters' Clothing

Common firefighters' protective clothings consist of a flame-resistant outer layer made from Nomex, Kevlar, or Zylon, and two inner layers.[34] The second layer is a moisture barrier and usually consists of aramid fiber. It increases firefighters' comfort and protection by preventing moisture, liquid, or vapor from reaching the skin when performing a task in a fire. Due to its insulation value and ability to block the passage of hot gas and steam, it can promote some burn protection. The inner layer is the thermal barrier that blocks the transfer of heat from the firefighting environment to the body of the wearer. The felt or batting is made from Nomex, Kevlar, a mixture of these two fibers, or polybenzimidazole (PBI).[34] Gloves are made from heavy-duty leather and Nomex liner. Relevant test methods for their performance testing include BS EN 469:2014, US NFPA 2112 standard for protection of personnel against flash fires.

27.4.2.2 Military Personnel Battlefield Clothing

Fireproof garments currently used in many armed forces are made of single layer aramid material (e.g., Nomex, DuPont), which is inherently fire retardant. However, due to the high cost of these and similar materials, they are used only for specialized applications such as coveralls for tank crewmen or to protect service personnel involved in military operations with potential for catastrophic fires. The low cost, lower performance flame-retardant treatments applied to more conventional fibers as finishes, flame retardant-modified synthetic (polyester and nylon) and regenerated fibers, use of fiber blends, and alteration of fabric construction and/or garment configuration are often preferred. For US Army use, outer garments are made from 50% nylon/50% cotton, whereas T-shirts are 100% cotton. Flammability standards for military textiles are defined in regards to performance criteria, but they do not necessarily dictate an approach and may not always be easy to find as this standard information can be considered to be controlled information. Different commercial suppliers to different armies across the globe use different standards and test methods to specify their products in meeting these standards, so readers are advised to take some time to research what is available before trying to "jump in" on research in this area. For detailed information, see other reviews written specifically on military clothing.[35,36]

27.4.2.3 Police Protective Garments or Systems

Only for first-response operations the armed police forces need protective clothing for protection against chemical, biological, radiological, and nuclear threats, termed as CBRN personal protective equipment. In a hostile situation the clothing may also be expected to protect against flame and the radiant heat caused by fire, ballistic impact, as well as resistance to shrapnel and slashing. The clothing is a multilayered assembly,[37] flame-retardant part of which is normally Proban-treated cotton, Kevlar® aramid (DuPont), or Twaron® (AKZO). There are no specific standard for their testing but use other methods for protective clothing, for example, BS EN 469:2014.

27.4.3 Curtains and Drapes

Curtains and drapes are vertically oriented fabrics used in interior furnishings and hence can ignite easily. The fibers involved are usually cotton, wool, silk, polyester, and nylon. The flammability testing of fabrics is often carried out by mounting the specimen vertically on the testing rig. The principal UK flammability test uses Test Method 3 described in BS 5438:1989 for curtains and drapes which measures the rate of flame spread of a vertically oriented fabric specimen. For curtains and drapes used for domestic usage, the flame application time is typically 10s whereas for contract furnishing, the test is more severe with longer flame application time of 15s. The test is even more stringent for the more hazardous applications such as hospitals, prison, etc. For these applications the fabric has to be tested with four flame application times of, 5, 15, 20, and 30 s. Other relevant standards are BS 5867: Part 2:1980, BS EN 13772, NFPA 701 (US), and NBC (National Building Code of Canada)—ULC-S 109.

27.4.4 Upholstery and Bedding Fabrics

In upholstered seating and mattresses and bedding, the textile material is component of a composite structure and is the exposed part. The statistics show that most domestic fires start by smoking in beds or upholstered furniture.[38,39] Some of the notorious fires caused by furniture in UK are Woolworths fire in 1979 (12 fatalities, 15 injuries), and Stardust in 1981 (10 fatalities, 53 injuries).[3] For upholstered furniture fabrics (usually polypropylene, cotton, wool, and blends) are used on a polyurethane foam, and the whole assembly is on a wooden or metallic frame. BS 5852 test method tests the whole piece of furniture and uses different ignition sources ranging from cigarette (Source 0) to a butane gas flame (Source 1) and to pinewood cribs of different sizes (Sources 3–7). The sources are selected based on application areas, for example, for domestic furniture, Source 1, and for military and certain sleeping wards in hospital, Source 7 is used. In California, the Cal TB 133 test method uses oxygen depletion calorimetry for testing flammability of seating used in public occupancies such as public auditoriums, hotels, hospitals, etc. This test method uses a 250 m^2 tube burner with heat generating capacity of 18 kW. The burner is placed on the seating area for 80 s. This full-scale furniture item is tested in a room, and the pass/fail criteria depends on the measurements used, that is, temperature increase at the ceiling thermocouple or peak heat release rate. Please see Chapter 31 of this book for more information on furniture flammability issues.

Bedding materials have traditionally been combustible materials. In the US the California Test Bulletin CAL TB 106 and Cal TB 603, both superseded by 16 CFR 1632 and 16 CFR 1633,[40] test methods are used for testing ignition resistance of mattresses uses a burning cigarette and large open flames from a gas burner, respectively. In the UK, the standard BS 6807:2006 is used to assess ignitability of mattresses and uses ignition sources specified in BS 5852.

27.4.5 Floor Coverings

Cotton, viscose, nylon, polypropylene, acrylic, polyester, or wool are generally used in carpets as the face or pile fabric over Neoprene latex. Carpets are usually flame retarded by back-coating with antimony-halogen system. The primary UK tests for carpet are BS 6307:1982[41] and BS 4790:1987,[42] which uses a methenamine tablet and hot metal nut, respectively as ignition sources placed in the center of the specimen. The damaged zone is measured, and the flaming and/or afterglow time is also recorded. In US ASTM E648[43] and NFPA 253[44] standards for floor covering textile materials are mainly applicable in the construction industry. These standards use the radiant panel test to evaluate the tendency of a flooring system to spread flame when exposed to the radiant heat flux. For transportation applications, carpets are tested according to the requirements of respective transportation departments. For floor coverings in motor vehicles the BS ISO 3795:1989 and FMVSS 302 (US Office of Vehicle Safety Compliance of the National Highway Traffic Safety Administration) are used, which measure horizontal rate of flame spread. For aerospace applications, the Federal Aviation Regulation (FAR) 14CFR25 (also known as FAR 25.853—Airworthiness Standards: Transport Category Airplanes) uses a radiant panel and a vertical flame test with carpet sample suspended vertically in a chamber. For naval applications the IMO (International Maritime Organization) Fire Test Procedures (FTP) for testing flammability of carpets (46CFR72.05–55) uses the radiant panel test method.[14]

27.4.6 Tents and Marquees

Tents, marquees, etc. use textile materials as single layer of fabrics (mostly cotton, cotton-synthetic blends, etc.) can be ignited easily as happened in Saudi Arabia where 10,000 tents caught fire during Haz in 2002, and 360 pilgrims died.[3] Hence, the fabrics used for such purposes should either be inherently flame retardant or have a durable flame-retardant treatment. Testing standards for tented structures involve The British standard BS 7837:1996 where fabric is tested in vertical orientation with bottom edge-ignition, BS 7157:1989 using the range of pinewood cribs of various dimensions as ignition sources and BS EN 14115:2002[45] with exposure of specimen to radiative heat. Other relevant tests are Swedish SIS 65 00 82, American NFPA-701,[46] ASTM E84[47] tunnel test, and Australian standard AS 1530.2.

27.4.7 Transportation

Textiles in transportation are used as seat coverings, carpets, wall coverings, battery separators, tires, belts, clutch lining, etc. Especially the seats, and sometimes other interior materials, foam materials are used beneath the covering fabric for more comfort. Although flame-retarded foams are used, fireblocker materials are used between the face fabric and the foam.[48] Fireblockers, made from inherently fire retardant fibers like oxidized acrylics and aramids, were first used for aircraft seats and now being increasingly used on trains, buses, and coaches. The performance requirements for textiles in transportation are more demanding compared to domestic applications as their use in the former is more rigorous, and they have to withstand daylight and UV radiations. From flammability point of view, the performance requirements are application specific, for example, for aerospace, requirements are more stringent than for motor vehicles. Fireblockers, also known as barrier fabrics, are discussed in more detail in Chapter 31 of this book for their use with upholstered furniture.

27.4.7.1 Motor Vehicles

Most of the passenger car interiors including car seats are made of polyester fiber (90% of the world market), and in some cases polypropylene fiber. The flammability testing of fabrics used in motor vehicles, in particular cars are not mandatory due to the fact that fire incidents in motor vehicles are rare, and moreover, fire spreads relatively slowly. Most manufacturers test seating covers and carpets conform to the US FMVSS (Federal Motor Vehicle Safety Standard) 302 test, which is a simple horizontal flame spread test. Other similar standards are German DIN 75 200, British, Australian BS AU 169, and Japanese JIS D 1201 automotive standards. The curtains and blinds are tested according to tests specific to them discussed earlier.

27.4.7.2 Rail Vehicles

Textile flammability standards are extremely severe because fires in railways can spread very quickly and can result in significant losses. Furnishing fabrics used in railway seatings, mainly 85% wool/15% nylon or polyester of FR grade (eg Trevira CS),[48] are tested in European countries as a complete assembly with UIC (Union Internationale des Chemins de. Fer.) 564.2. In the US, ASTM E1537–98, and in the UK, ceiling lining materials are tested according to BS 476.

27.4.7.3 Aircraft

Seats in the aeroplane need to conform to highest standards of fire safety, hence are usually made from wool (Zirpo treated), wool/nylon blends or leather. Flammability testing of all textile materials is regulated by the US Federal Aviation Administration (FAA) under Federal Aviation Regulations (FAR), and these latter extend to all commercial airliners operating across the world internationally. All textiles present in an aircraft have to pass the test requirements defined in FAR 25.253(b) in which a Bunsen burner flame impinges upon the bottom edge of a vertically oriented sample for 12 seconds. For seats, the complete seat assembly is tested. For textiles attached to internal wall and ceiling panels or partitions, they are tested as a composite for heat release using the OSU calorimeters, as discussed in Section 27.2.3.

27.4.7.4 Ships

Cruise ships use a large amount of furnishing fabrics, for example, bedding, curtains, carpets, etc., hence need to have durable high standards of flame retardancy. Standards required are DIN 4102 class B and BS476.[48] Flammability standards and test for furnishing fabrics, bedding, and draperies used in ships and submarine have been developed by the International Maritime Organization (IMO) and the National Fire Protection Association (NFPA), which include Safety of Life at Sea (SOLAS), High Speed Craft (HSC), and Fire Test Procedures (FTP) Codes. The Fire Test Procedures (FTP) code describes the flammability tests and performance criteria for combustible materials.[49] Bedding components including mattresses, pillows, blankets, quilts, and bedspreads used in ships are tested in accordance with IMO Resolution A.688 (17) according to SOLAS and HSC Code whereas NFPA 301 requires mattresses and mattress pads to comply with NFPA 267, ASTM E1590, and 16 CFR 1632 test methods. When tested according to the IMO Resolution A.688 (17), the bedding should not ignite readily or exhibit progressive smoldering when subjected to smoldering or flaming ignition.

27.4.8 Fiber-Reinforced Polymeric Composites

High-performance fibers like glass, carbon, and aramid are being increasingly used as reinforcing element in polymeric resins to form structural composites. Composites' application is increasing in all transport applications, in particular aerospace, trains, ships, and motor vehicles. The fiber-reinforced composites are generally less flammable than cast neat resins of similar thicknesses due to lower resin contents in the former, and also the fibers present act as fillers and thermal insulators.[50] Thick laminates with more fabric layers burn slowly, giving lower heat release than thin laminates with less fabric layers.[51] Amongst different composite types, there are different factors that can affect the flammability of the laminate and its residual mechanical property, namely: (i) fiber type (glass, carbon, aramid etc.), (ii) volume content of fiber, and (iii) reinforcement type (weave type, fiber/yarn arrangement, etc.). While glass fibers are noncombustible, exposure to high temperatures can result in softening and, finally, melting, which effectively lowers their mechanical properties. On the other hand, other commonly used fibers such as carbon and aramid are non-thermoplastic but, because they are organic in nature, will decompose and/or oxidize at elevated temperatures.

We have studied the effect of different fiber types on the burning behavior of epoxy composites.[52,53] Thermal decomposition behavior of glass, carbon, and aramid fibers in air is shown in Figure 27.2 (a) where glass shows inert behavior with no major mass loss up to 800°C, whereas, carbon fibers start losing mass at 280°C probably by physical desorption of volatiles and are fully oxidized at 800°C, leaving 5% charred residue. The aramid fiber starts losing mass earlier than carbon fiber. The curves for all fibers/resin combinations in 1:1 mass ratio shown in Figure 27.2 (b) indicate that

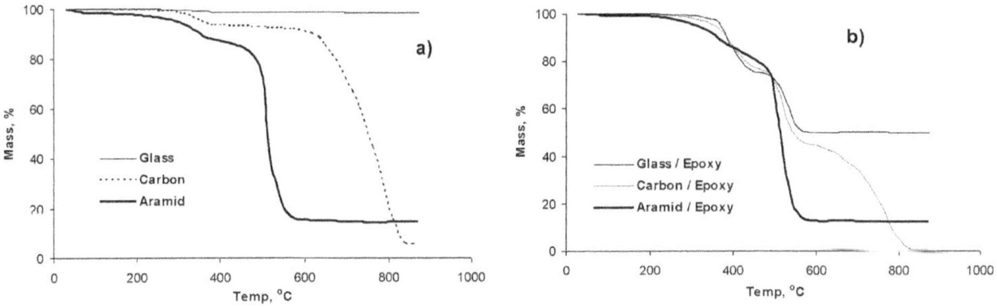

FIGURE 27.2 TGA curves in air of (a) glass, carbon, and aramid fibers and (b) fiber/epoxy mixtures in 1:1 mass ratio.[52]

Source: From Kandola, B.K. et al. 2006. In *Recent advances of flame retardancy of polymeric materials*, ed. Lewin, M., Proceedings of the 17th Conference, BCC, Stamford, Connecticut, USA. With permission.

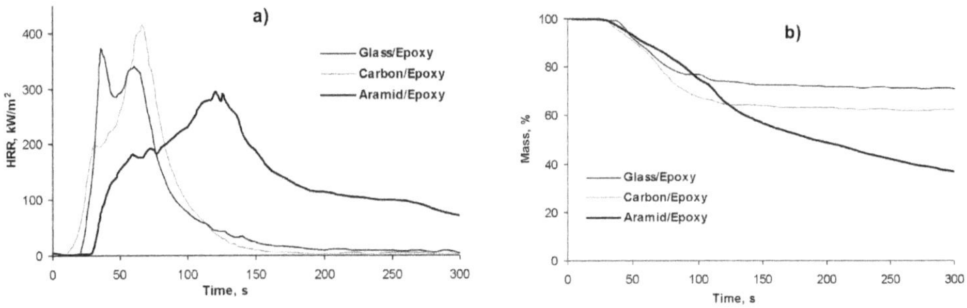

FIGURE 27.3 (a) Heat release rate and (b) mass loss versus time curves for different fiber/epoxy composites at 50 kW/m^2 heat flux.[53]

apart from glass fiber, both resin and fiber decompose or lose mass. This shows that carbon and aramid fibers may add to the fuel content, which in case of carbon is due to solid state oxidation. However, the TGA technique is not truly representative of actual fire conditions. The burning behavior of the laminates prepared from these fibers was studied by cone calorimetry at 50 kW/m^2 heat flux. Heat release curves in Figure 27.3 (a) show that the aramid-containing laminate has the lowest peak although the heat release curves are broader than other samples, and it burns for longer times. The aramid-containing composite had the highest total heat release (THR) value, which is a consequence of its decomposition and fuel-generating property as shown by the TGA results. However, contrary to the TGA results, the carbon fiber did not show any significant degradation (see Figure 27.3 (b)), which is not surprising since carbon does not decompose and the mass loss in TGA is due to solid state oxidation. For detailed information about this work, the reader is referred to our publications, where these results are also analyzed in terms of effect of resin content and fabric area density.[52,53] Flammability standards for composites are beyond the scope of this review, but are discussed elsewhere.[50]

27.5 FIRE-RETARDANT STRATEGIES (OR APPROACHES)

Similar to bulk polymers, textile fibers/fabrics also need certain elements such as phosphorus, nitrogen, halogen, sulfur, boron, metals, etc. to render them flame retardant. The flame-retardant elements containing chemicals can be either applied to existing fibers/fabrics on their surface as finishes, introduced as additives in polymer before fiber production (in synthetic fibers), or introduced into the polymer back-bone (inherently flame retardant fibers). Flame retardants functioning by physical means only, such as alumina trihydrate, magnesium hydroxide, etc., are not usually used for textiles, as for them to be effective as flame retardants large quantities (> 40 wt% of the polymer) are required, which is not possible to be introduced within the fiber or on the surface. The most effective flame retardants are those which function in condensed phase and promote char formation by converting the organic fiber structure to carbonaceous char and hence reduce volatile fuel formation.[10,54] Most often these flame retardants functioning mainly by chemical means require absorption of heat to operate and release nonflammable gases like CO_2, NH_3, and H_2O during chemical action, hence also function by physical means by cooling the substrate below the combustion temperature and diluting the flaming zone with nonflammable gases. These condensed phase active flame retardants are more effective in natural fibers such as cotton and wool with polymer backbone having reactive side groups, which on removal lead to unsaturated carbon bond formations and eventually a carbonaceous char following elimination of most of the non-carbon atoms present.[54] Synthetic fibers like polyester, polyamides, and polypropylene on the other hand tend to pyrolyze by chain scission or unzipping reactions and lack reactive groups and are not char-forming polymers and, hence, cannot be effectively flame retarded with these chemicals.[3,55] Vapor phase active halogenated flame retardants, however, will work with any fiber type, since they interfere with the flame chemistry by generating free radicals. The detailed mechanisms of action of these flame retardants on textiles are similar to the ones in bulk polymers and are beyond the scope of this review, for which readers are referred to other chapters in this book and other such reviews.[10,56] The main issues with flame-retarding textiles, however, are the method of application/introduction of these flame retardants on/in textiles, their wash durability and effect on other textile properties such as tensile and tear strength, handle, drape, appearance, etc. This section therefore reviews different fire-retardant strategies based on methods of application. Natural fibers/fabrics are generally treated with flame retardant finishes in the finished fabric form. In synthetics usually additives are added to the molten polymer/polymer dope before fiber extrusion or flame retardant copolymers/homopolymers are used. However, a coating with the help of resin binder can be applied to any fiber/fabric types. This section also reviews inherently-fire retardant fibers, blends, and some other novel approaches to flame-retard fibers.

27.5.1 Surface Treatments

Surface treatment is the oldest method of flame retarding the textiles, dating back to 1735 when borax, vitreol (a metal sulfate), and some other mineral substances were patented in England for flame-retarding canvas and linen.[57] There are two types of surface treatments, finishes, and coatings. Finish is applied by impregnating the fabrics in the aqueous solution of the chemical as a textile finishing process. Coating on the other hand is application of a continuous or discontinuous layer on the surface of the fabric generating heterogeneous fabric/polymer composite.[58] These are discussed in detail in this section. In textiles the main issue with any treatment is the durability to laundering. Some treatments are nondurable to laundering, others may be semi-durable and withstand a single water soak or dry cleaning process, and durable ones withstand 50 wash cycles. Durability requirements are usually determined by the application and specified by associated regulation.

27.5.1.1 Flame-Retardant Finishes

Most of the treatments and formulations were developed in 1950–1970 period[3,4,6] and in terms of new chemistry, nothing much has changed since then. Several interesting reviews have been published in this area.[3,4,6,21,29] Finishes are usually applied by pad-dry method where fabric is passed through the chemical formulation (mostly in aqueous form), passed through rollers to squeeze out the excess, and then dried in oven at 120°C. This gives a nondurable finish. To get a semi-durable or durable finish, fabric is passed through another oven set at higher temperature (usually 160°C) where a curing stage allows a degree of interaction between the finish and the fiber. In some cases curing stage might also involve chemical treatment. The best example is Proban process (discussed below) which requires ammonia gas curing process in order to polymerize the applied finish into the internal fiber voids. As discussed previously, finishes are applied to fabrics from natural fiber types, although there are some available commercially for synthetic fibers as well. Some generic types of finishes are discussed next; for more information and their commercial grades details, reader is referred to other reviews.[3,6,29,55]

27.5.1.1.1 Cotton

Ammonium phosphates are the most effective flame retardants for cotton as first identified by Gay-Lussac in 1821 and still widely used. All phosphates on heating release phosphoric acid, which catalyzes dehydration reactions of cellulose to yield char at the expense of volatiles formation reactions.[54] Ammonium phosphates like mono- or diammonium phosphates, however, are water soluble, hence applicable as nondurable treatments only. Ammonium bromide can be used in combination with ammonium phosphates to provide some vapor phase flame retardant action. Other examples include borax and boric acid, ammonium sulfamate, and sulfates. These nondurable finishes are useful for disposable fabrics, insulation, wall boards, theatrical scenery, packaging material, paper, etc. Ammonium polyphosphates are used in combination with urea to provide semi-durable finishes and by curing at 160°C, when some phosphorylation can occur. Semi-durable finishes are very useful for materials that may not need frequent washings, for example, mattresses, drapes, upholstery, carpets, etc. Some commercial examples of semi-durable finishes include Flammentin® FMB (Thor Specialities), Pyrovatim® PBS (Ciba, now marketed by Huntsman).

Amongst the commercially successful durable finishes, one type is based on N-methylol dialkyl phosphonopropionamides, from which well-known product Pyrovatex CP and CP-LF (Huntsman, formerly Ciba) is derived. Pyrovatex CP (N-methylol dimethyl phosphonopropionamide) is applied with a methylolated melamine resin in the presence of phosphoric acid, which catalyzes the formation of pyrovatex-resin-cellulose moieties via the C(6) OH group, represented by the reactions in Scheme 27.1. Another commercial product based on similar chemistry is Thor's Aflammit® KWB.

Other most successful durable treatment is based on tetrakis (hydroxymethyl) phosphonium derivatives. Very well-known brand marketed as Proban CC (Solvay, previously Rhodia and Albright & Wilson) involves padding of tetrakis (hydroxymethyl) phosphonium chloride (THPC) urea solution onto the cotton fabric, curing with ammonia in a specially designed reactor to generate a highly cross-linked three-dimensional polymer network. The fabric is then treated with hydrogen peroxide, which converts P^{3+} to the P^{5+} state. The reactions are shown in Scheme 27.2. Other similar commercial product is Thor's Aflammit® SAP. In literature many combinations of tetrakis (hydroxymethyl) phosphonium derivatives with other salts have

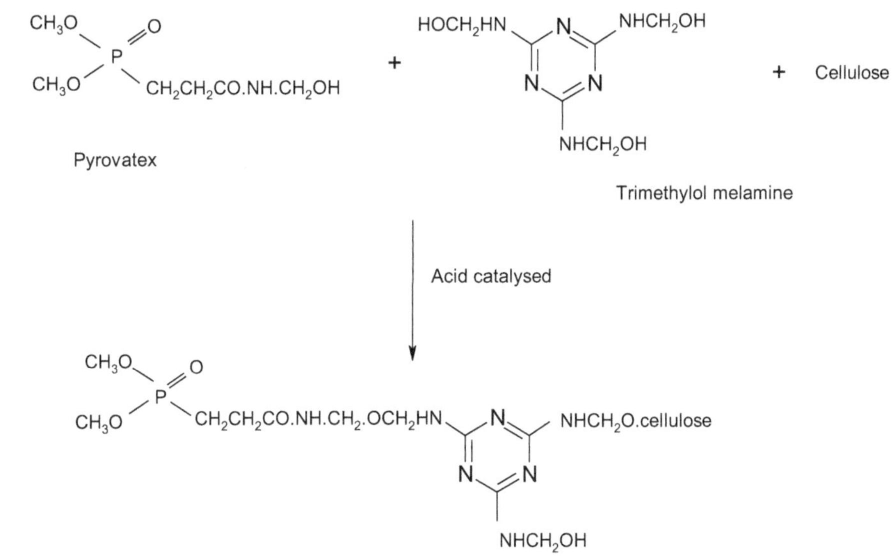

SCHEME 27.1 Chemistry of the Pyrovatex CP process.

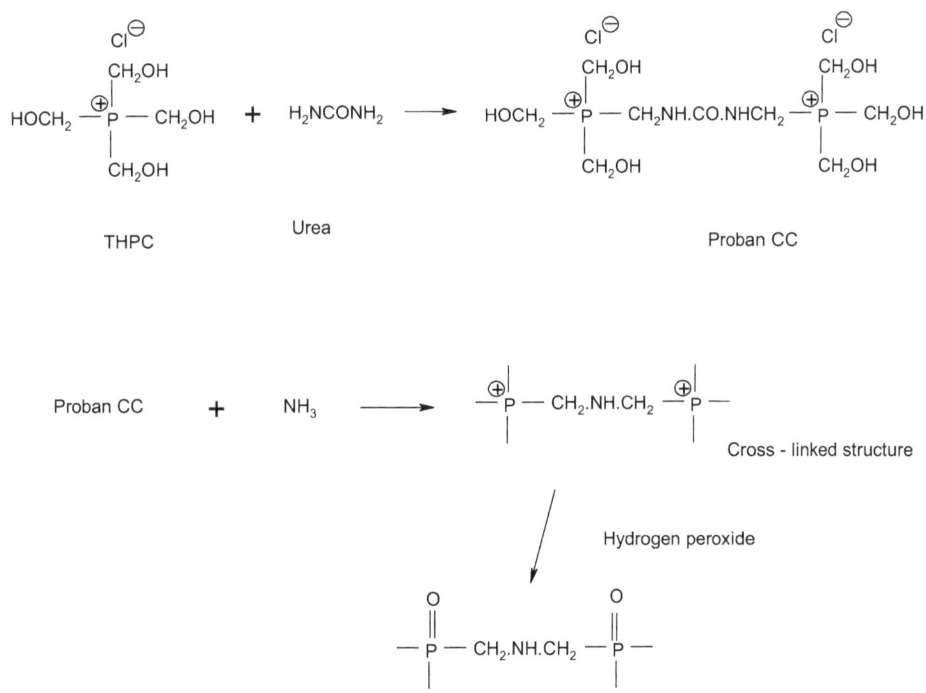

SCHEME 27.2 Chemistry of the Proban CC process.

been reported,[54] but the most successful so far has been the THPC-urea-NH_3 system discussed previously.

Pyrovatex CP type treatments have more dye affinity and, hence, are used for curtains, apparels (nightwear), etc., whereas Proban CC, which retains greater strength, is used for hospital bedsheets, military applications, hotels, nursing homes, etc.[4]

27.5.1.1.2 Viscose

In principle all the treatments effective on cotton should be effective on viscose as well. However, the delicate viscose fibers cannot withstand the harsh finishing processes and are rarely treated with a finish. Lyocell (developed by Coutaulds, now the term used as a genericized trademark to refer to Lyocell process), however can be treated with Pyrovatex CP (Huntsman, previously Ciba) to produce FR Lyocell.[59] The work at Bolton has shown that this fiber being more reactive than cotton needed only half the normal amount of this FR treatment applied to cotton cellulose to produces an equivalent degree of flame resistance.

27.5.1.1.3 Polyester

Tris(2,3-dibromopropyl) phosphate was the successful finish developed in 1970s for polyester fabrics but was withdrawn from the market after a short time due to being carcinogenic. Since then, the major product used in the thermosol dyeing treatment of polyester fabric has been Antiblaze® CU (formerly by Rhodia, now produced in China). Other similar commercial products include SMC Global's SMC 688, Thor's Afflamit® PE, and Schill and Seilacher's Flacavon® AZ.

27.5.1.1.4 Cotton-Polyester Blends

For blends sometimes finishes which are successful on one type of fiber may prove to have antagonistic effect on blends.[10] The general rule is to apply finish suitable to the majority fiber present or apply halogen-based coating (see section below). Most nondurable finishes for cotton will function on cotton-rich blends with polyester but not for the polyester-rich blend unless some bromine is present.[10] Durable treatments such as Proban and Pyrovatex type finishes can be applied on higher (> 55%) cotton content fabrics, for example, 70:20 cotton/polyester.

27.5.1.1.5 Wool

Wool, though not as flammable as cotton, still needs flame retardation for specific applications, for example, carpets, upholstered furniture in transport, etc. Ammonium phosphates and polyphosphate, boric acid-borax, ammonium bromide can be successfully used in nondurable flame-retardant finishes for wool. The most successful durable treatment for wool is Zirpro®, developed by Benisek, which involves exhaustion of negatively charged complexes of zirconium or titanium onto positively charged wool fibers under acidic conditions at 60°C. The treatment can be applied to wool at any processing stage from loose fiber to fabric using exhaustion techniques. Commercial products include Aflammit® ZR and ZAL (Thor), Cetaflam® PHFZ and ZAS (Avocet Chemicals).[10] For details of Zirpro and other finishes, reader is referred to review by Cardamone.[60]

27.5.1.1.6 Silk

Silk, although not very flammable (LOI 23%), needs flame-retardant treatment for certain applications, for example, nightwear garments, curtains, interior decoration of executive jet aeroplanes, etc. Nondurable finishes containing mixture of borax and boric acid,[61] inorganic salts, and quaternary ammonium salts, urea—phosphoric acid salt[62] have been reported to be effective on silk. Semi-durable finishes like organophosphorus flame retardant and trimethylolmelamine (TMM),[61] Pyrovatex CP (Huntsman, formerly Ciba)[63] are also reported to be effective on silk.

27.5.1.2 Coatings

As mentioned before flame retardant coatings can be applied on the surface of the fabric (including on the back) to confer flame retardancy to the overall fabric. Typical textile coating polymers include natural and synthetic rubbers (polyisobutylene, styrene butadiene, poly(butadiene-acrylonitrile), poly(chloroprene), etc.), poly(vinyl chloride) or PVC plastisols and emulsions, poly(vinyl alcohols), formaldehyde-based resins, acrylic copolymers, polyurethanes, silicones, and fluorocarbons, used for different applications such as water resistance, flexibility, moisture permeability, and flame retardancy. Some coatings have varying levels of inherent flame retardancy (e.g., PVC and other chlorine and fluorine-containing polymers), although the more commonly used polymers and copolymers are quite flammable and so presence of flame retardants additives is necessary to flame-retard both the coating matrix polymer and the underlying textile substrate. Flame retardants used for coatings include different phosphates and phosphonates, for example, triaryl phosphate, cresyl diphenyl phosphate, oligomeric phosphate-phosphonate, etc.

27.5.1.2.1 Back-Coating

Back-coating of textiles is a well-established application method where the flame retardant formulation containing brominated organic species and antimony trioxide is applied in a bonding resin to the reverse surface of the fabric. Application methods include doctor blade or the knife-coating methods, and the formulation is as a paste or foam. Most typical brominated derivatives are decabromodiphenyl oxide (DBDPO) and hexabromocyclododecane (HBCD), both are restricted now and largely replaced by either bis(pentabromophenyl) ethane or a brominated polymeric resin component such as pentabromobenzyl acrylate.[10] Such back-coatings are effective on a wide range of fabrics, including nylon, polypropylene, polyester-nylon blends, acrylics, and many blends. Important applications are in domestic, industrial, and automotive upholstered furniture, draperies for hotels and other public buildings. All halogen and more specifically, bromine-containing flame retardants, have come under scrutiny due to environmental issues associated with them, hence there is a pressure to replace bromine-containing flame retardants. Attempts to reduce the use of brominated flame retardants in textile back-coatings in the past have met with varying degrees of success.[64,65] In our laboratory phosphorus-based intumescent flame retardants have been tried in back-coating formulations on 100% cotton and 35% cotton–65% polyester blend fabrics. Out of many flame retardants, only ammonium polyphosphate-based formulations yielded simulated match passes but failed after the required 40°C water soak treatment.[66] To sensitize the decomposition or flame retarding efficiency of ammonium polyphosphate, metal ions as synergists for APP have also been used.[66] Metal ions promote thermal degradation of APP at lower temperatures than in their absence, and this enables flame retardant activity to commence at lower temperatures in the polymer matrix, thereby enhancing flame-retardant efficiency; however the problem of durability to water soaking still remains.

At Bolton, we also have attempted to introduce volatile and possible vapor phase-active, phosphorus-based flame-retardant components in back-coating formulations.[65,67]

The selected flame retardants included tributyl phosphate (TBP), a monomeric cyclic phosphate, and an oligomeric phosphate-phosphonate. When combined with an intumescent char-forming pentaerythritol derivative and applied as a back-coating on to cotton and polypropylene substrates, a significant improvement in overall flame retardancy was observed.

Attempts have also been made by Bourbigot et al.[68–70] for microencapsulation of ammonium phosphate with polyurethane and polyurea shells to make an intrinsic intumescent system compatible in normal polyurethane (PU) coating for textiles. Microencapsulation of ammonium phosphate helps in reducing the water solubility of the phosphate and, hence, increases the durability of the textile back-coating.

The inclusion of nano-particles in coating formulations has been investigated by Devaux et al.,[71] where nanoclay and polyhedral oligomeric silsesquioxanes (POSS) in polyurethane coatings applied to polyester and cotton fabrics were found to reduce peak heat release values of back-coated fabrics as determined by cone calorimetry. However, neither nano-species increased ignition times nor reduced extinction times. In fact, the reverse happened. Thus it was evident that the presence of nanoparticles alone could not impart a flame retarding effect. A similar study at Bolton by Horrocks et al.[65] has shown that the introduction of nanoparticulate clays has no beneficial effect to a back-coating polymeric film, and the introduction of fumed silica to a flame retarded back-coating formulation reduces its effectiveness.

27.5.1.2.2 Intumescent Coatings

Intumescent coatings have been around for nearly 70 years, normally used as surface treatments for structural materials and metals to protect them from heat and fire. Most intumescent formulations include ammonium phosphate and melamine, or melamine phosphate and a char former such as pentaerythritol or a polyol. On heating, they form a foamed char, which thermally insulates the underlying structure. In textiles although can be used as surface coating on the fabric,[72] but usually are applied in between different fabric layers, which are then needlebonded to consolidate the whole structure.[73,74] In carpets the intumescent layer can be applied between the tufting and the backing, or on the upper surface of the backing. The intumescing material also closes up the interstices of the fire-exposed fabric.

27.5.1.3 Sol-Gel Coatings

Sol-gel technique has been exploited in the last 15 years or so as an alternative method for providing flame retardant properties to textiles by synthesizing ceramic nanoparticles on the surfaces.[75–77] This method involves the hydrolysis and polycondensation of reactive (semi)metal oxide precursors, followed by gelation, ageing, and drying of the wet gel to form a fully inorganic or hybrid organic-inorganic 3D network (see Scheme 27.3).[78,79] Alongi et al. have been at the forefront of studying the flame retardancy of cotton fabric, using mainly silica precursors, namely, tetramethylorthosilicate (TMOS) and tetraethylorthosilicate (TEOS).[80,81] Using TEOS, the LOI values for cotton and cotton blends could be increased from 18 to 22 and 23 vol%, respectively and the amount of char increased due to the formation of silica film on the fabric surface. On the other hand, in the case of polyester fabric, the similarly developed silica films did not improve the thermal properties and LOI values remained similar to that of untreated polyester ~22 vol%.[82] As an alternative to silica oxide films, other oxide phases such as titania, zirconia, and alumina can also be synthesized by using tetraethylortho-titanate, zirconate, and aluminum isopropylate precursors,[76,77] the efficiency of these though are lower than that of silica.[78]

While with sol-gel treatment certain level of flame retardancy can be achieved, the inadequate thickness of the

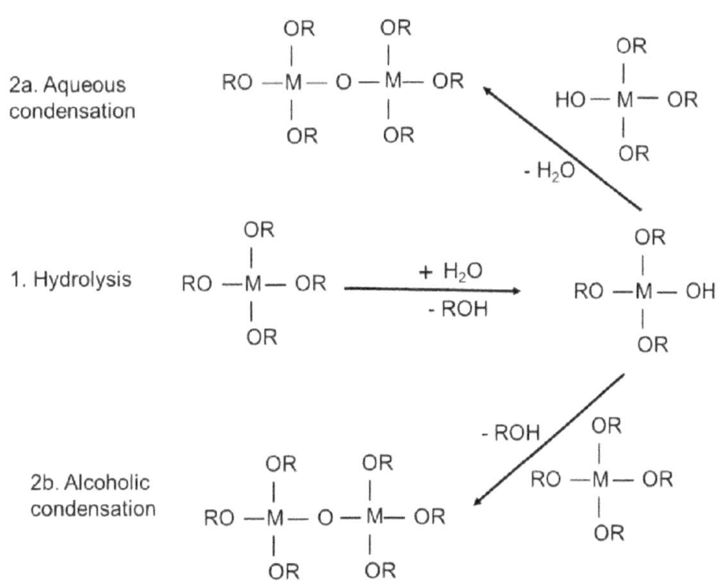

SCHEME 27.3 Schematic of the sol-gel process.

developed film may not be enough to provide the desired thermal barrier effect. Hence, quite often during sol-gel process, flame retardants are added, which become part of the 3-D structures and help in achieving required flame retardancy for the fabric.[76–80] Despite achieving certain levels of flame retardancy, this technique has certain limitations, particularly poor washing durability and damaging effects on mechanical properties of treated fabrics.[76] There have been a number of attempts to reduce wash durability of the coatings. Cireli et al. have reported that phosphorus-doped (as phosphoric acid) SiO_2 thin films on cotton could develop nonflammable property and some degree of durability up to ten wash cycles. However, the color of the P-doped- SiO_2 treated fabric changed from white to yellow after the treatment.[83] Przybylak et al.[84] developed a superhydrophobic flame retardant coating by adding ammonium dihydrogen phosphate and/or guanidine carbonate into an aminopropyltriethoxysilane based sol-gel coating for flame retardancy and using polysiloxane or a fluorofunctional silane for hydrophobicity. The latter also acted as a synergist for flame retardancy, raising the LOI of cotton from 18 to 71 vol%.

Lin et al. activated the cotton surface with oxygen plasma to increase the surface concentrations of carboxyl and hydroxyl groups.[85] Ammonium polyphosphate was then used as the flame-retardant with a mixture of TEOS and hydroxyl-terminated polydimethylsiloxane (HPDMS). The flame retarded fabric showed a degree of hydrophobicity and durability.[85]

In a more recent work at Bolton, we tested sol-gel treated cotton fabrics for their ability to act as a flame barrier over typical materials used as fillings in domestic furniture, such as polyurethane (PU) foam. This is an essential requirement of fabrics which have to fulfil the UK regulatory BS 5852 test standard requirements after the obligatory 30 min water-soak at 40°C. Sol-gel treatment involved combination of TEOS, and diammonium phosphate and urea (as a conventional phosphorus- and nitrogen-containing flame retardant system) on cotton fabric.[86] For testing, a simulated match test was developed, in which fabric pieces of 220 × 250 mm sizes were cut and sewn onto pieces of non-flame-retardant polyurethane foam (220 × 150 × 22 mm size; ~25 kg/m^3 density) to form a mock-up back of an upholstered chair (for details, see ref[86]). Using this test, the TEOS-diammonium phosphate-urea modified sol-gel treatment enabled fabric to pass the simulated match test before water-soaking, whereas, failed after water-soak test. However, by adding polydimethylsiloxane (20 wt% w.r.t. TEOS) to the formulation, the sample passed the SMT both before and after water-soaking.[86]

27.5.1.4 Layer-By-Layer Coatings

LbL is another method for the production of thin, multilayered films over textile surfaces through electrostatic attraction between cations and anions or nanoparticles, reviewed previously by many authors.[77,87,88] The process takes advantage of the weak electrical charge (usually negative) already present on the textile surfaces and involves deposition of several monomolecular layers alternatively, as alternating positively charged and negatively charged layers, by alternate immersion in or spraying on suitably charged polyelectrolyte solutions. The duration of each spraying or dipping step ranges from seconds to minutes, depending on the experimental settings, textile structure and its absorption capacity, ionic strength of polyelectrolytes, etc.[77] Quite often rinsing and drying operations are performed between each immersion step to remove excess of the deposited material and to prevent possible contamination that takes place between the oppositely charged solutions during each step. Other approaches utilizing attractive forces than electrostatic attraction, include hydrogen bonding,[89] donor/acceptor,[90] and covalent bond formation[91] to form LbL assembly. LbL assemblies could be fully inorganic (e.g., silica, polyhedral oligomeric silsesquioxane (POSS) based), or hybrid organic-inorganic (e.g., assemblies made of poly(acrylic acid) and amino-functionalized montmorillonite nanoplatelets).[92]

While LbL assembly was demonstrated for the first time in 1966,[93] its exploitation for flame retardancy was not conducted until ~2010. One of the first studies regarding the LbL deposition on cotton for flame reatrdancy was performed by Li et al., depositing 20 bilayers of cationic branched polyethyleneimine (BPEI) of different pH and anionic sodium montmorillonite (MMT) clay of different concentration on cotton fabrics, resulting in reduced afterglow time in vertical flame tests.[94] Li et al. in another research group conferred flame retardancy on cotton fabric by depositing bilayers of 3-aminopropyl triethoxysilane (APTS) hydrolysis product and sodium phytate through LbL assembly with drying at 100°C after each step. Fifteen bilayers could stop the fire after the removal of the ignition source.[95]

Pan et al. worked on cotton-polyester fabric (activated with NaOH), coating with positively charged polyethyleneimine (PEI) and negatively charged oxidized sodium alginate (OSA) followed by immersion in hypophosphorous acid to cross link PEI/OSA bilayers on to the fabric. The treated fabric was self-extinguishing in the horizontal flame test and could withstand 12 laundering cycles.[96] Carosio et. al. have used LbL assembly to flame retard polyester fabric by controlling the ionic strength of the repeated layers. The fabric surface was activated with poly(ethylene imine) then coated with the hybrid organic/inorganic flame-retardant coatings made of octa-ammonium POSS (polyhedral oligomeric silsesquioxane) and sodium montmorillonite clay. Washing the fabric with NaCl solution after the first BL coating increased the thickness and ionic strength of the deposited layers, which helped in stopping the melting dripping and reducing the rate of burning during the flammability testing.[97]

Biomolecules such as proteins,[98] chitosan,[99] starch,[100] hydrophobins,[101] phytic acid,[102] and DNA[103] have also been used for LbL: assemblies on cotton, some of which could impart self-extinguishing properties to the fabric. For detailed information see recent reviews in this area.[77,88] In a recent work, Vishwakarma et al. have used combination of egg white proteins and hypophosphorous acid for developing a flame-retardant coating for cotton fabrics by LbL technique.[104] Coated cotton fabric showed flame retardancy with 26% LOI value and self-extinguishing property with significant but fragile residual char structure, whereas uncoated fabric completely burnt with a little ash as residue. Wash durability of the coated fabrics, though showed a slight reduction in flame-retardant properties as a consequence of ~ 9.7 wt% coating loss during soaking, yet there was no after-glow.

An extensive discussion of LbL and other similar approaches for materials is covered extensively in Chapter 14 of this book.

27.5.1.5 Plasma Coating

Plasma technology can be used for surface modification (etching) of fibers and fabrics, functionalization of surface or surface coating by deposition of a thin layer on the surface followed by grafting. Cold (non-thermal) plasma technology can simultaneously graft and polymerize monomers onto the surface of the textile substrate. In the so-called plasma-induced-graft-polymerization process, the plasma is used to activate the surface and the plasma of an inert gas is used to initiate polymerization of the nonvolatile or solid monomer on the surface of the substrate. The most commonly studied cold plasma is of low-pressure (LP) type, obtained by ionizing gas at much lower pressure than the atmospheric pressure and requires both vacuum chamber and pump to create and contain the plasma. One of the pioneering works on flame retardant treatment of fabrics using LP cold plasma was reported in 1980s by Simonescu et al.,[105] where surface grafting of rayon fabrics was carried out with phosphorus containing polymers, and improvement in fire retardant properties was observed. Tsafack and Levalois-Grützmacher[106] achieved the argon plasma assisted grafting of phosphorus-containing acrylate monomers (diethyl(acryloyloxyethyl)phosphate (DEAEP), diethyl-2-(methacryloyloxyethyl)phosphate (DEMEP), diethyl(acryloyloxymethyl)phosphonate (DEAMP), and dimethyl(acryloyloxymethyl)phosphonate (DMAMP)) onto polyacrylonitrile fabric. Same authors also used low pressure argon plasma to graft DEAEP and perfluorinated acrylate monomer on cotton for flame retardant properties, followed by CF_4 plasma treatment for hydrophobicity.[107,108] In 2012 Paosawatyanyong et al.[109] used argon plasma to graft vinyl phosphonate in the presence of a cross-linking agent on cotton. Vannier et al.[110] grafted polyethylene methacrylate phosphate on cotton fabric. The LOI value rose from 21 vol% for the pure cotton to 32 vol% for the FR cotton and the treated fabric could achieve M2 rating, compared to M3 rating by pure cotton.

For polyester fabric, oxygen plasma has shown to be more effective, for example Omerogulları et al.[111] managed to reduce the amount of flame-retardant agent from 100 g/L to 50 g/L and enhance the wash durability of the flame retardant. In a similar work, Zhou et al.[112] used oxygen gas plasma on polyester fabric coated with maleic acid, the plasma activated fabric was then coated with pentaerythritol phosphate urea salt by pad-dry (at 80°C for 3 min)-cure (at 160°C for 3 min) method to achieve durable flame retardant treatment.

While above examples using LP cold plasma have shown some promising results, owing to the need of vacuum chamber for processing, the technology has limitations for industrial mass production, which is generally required for textiles. Since early 2000s atmospheric-pressure-glow or silent discharge technology involving continuous, open-width textile/polymer films processing has been developed, which has created a potential of commercializing some of the previously developed plasma assisted polymerization or grafting chemistries. Using atmospheric plasma, an earlier work at the University of Bolton has shown that deposition of silicon-based species on textile surfaces could significantly improve their flame retardancy defined in terms of improved flash fire resistance.[113] This improvement of flash fire resistance was observed on pure cotton, Proban®-treated cotton, and Nomex® aramid fabrics. More recent work by Tata et al.[114] showed that polyester fabrics could be etched initially by cold oxygen plasma and then finished with hydrotalcite, nanometric titania, and silica aqueous suspensions. This treatment gave improved fire performance levels, even after washing in demineralized water at 30°C for 30 min. Later research by Totolin et al.[115] reported grafting/cross-linking of sodium silicate layers onto viscose and cotton flannel substrates by using atmospheric pressure plasma, which increased fabric burning times during 45° testing. However, the presence of silicon-containing moieties alone on the cotton is insufficient to produce high levels of flame retardancy as observed from sol-gel and LbL treatments.

Lam et al.[116] coated He/O_2 plasma treated cotton fabric with Pyrovatex CP flame retardant and a melamine resin (cross-linking agent) with and without phosphoric acid and finally coated with zinc oxide. Vertical flammability test showed that plasma-treated samples reduced the burning rate and exhibited excellent flame-retardant performance even after three laundering cycles. Edwards et al.[117] applied a phosphoramidate onto cotton via an helium atmospheric plasma treatment. The results showed that while char levels were increased, flame self-extinguishability during vertical fabric testing was not achieved and correlated with the inability to apply the appropriate phosphorus levels into the grafted fabric.

The above examples demonstrate that atmospheric plasma, while having a capability of covalent grafting of precursor species via the formation of small functional groups on the polymer surfaces, is not very efficient due to (i) difficulty of achieving high enough add-ons sufficient to create self-extinguishability, (ii) weak physical and/or chemical bonding to the underlying fibers present which compromises wash durability, (iii) lack of adequate penetration below fiber surfaces of the flame retardant species or precursors applied which can allow increased add-ons to be achieved, and (iv) excitation states diminish over time. The solution to overcome these issues is to add a second high energy source to the plasma region. In that context MTIX Ltd. developed a novel, patented,[118] atmospheric plasma/UV laser facility for processing textiles. In this process, this latter is a 308 nm UV excimer laser able to break single covalent bonds (C-C, C-O, C-N, etc.) in both flame retardant precursor and fiber thereby increasing the chance of interaction. Using this system on cotton, polyester, and nylon fabrics, the recent work at Bolon has shown the generation of radical species and oxygen- (e.g., COOH) or nitrogen-containing functional groups (NH_2, only in nylon),[119,120] the concentrations of which depended on plasma power dosage, atmosphere composition, and the presence or absence of UV laser (308 nm XeCl) energy. The application of plasma/UV to cotton and polyester fabrics, pre-impregnated with a number of flame retardants comprising combinations of phosphorus-, nitrogen-, and silicon-containing species, increased the water-soak durability of otherwise water-soluble components.[121] In another work at Bolton, water soluble resorcinol bis(diphenyl phosphate), RDP, was applied to polyester and nylon6.6 by padding with 30 wt% solution in methanol. There was a 16% FR retention in polyester[122] and 3.4% in nylon 6.6 after the water/methanol soak test, for plasma pre-treated samples, in contrast to no FR retention in samples without plasma pretreatment (unpublished results). Ethyl diphenylphosphinite was also applied on both nylon and nylon-cotton blended fabrics. Plasma pretreatment with N_2/CO_2 gases helped in 6.7% retention of the FR after water-soak testing.

Based on the MLSE (MTIX) principle, a laboratory scale reel-to-reel dielectric barrier discharge (DBD) atmospheric pressure cold plasma was also constructed at Bolton and used to apply ethyl diphenylphosphinite on nylon and nyco fabrics through the plasma stream. This process could successfully deposit flame retardant on the fabric. The concentration of phosphorus on the fabric though was very low. To achieve required level of FR on the fabric surface, the optimization of equipment settings (fabric feed speed, type of plasma gas and power settings) and/or multiple treatments would be required. This work though has demonstrated proof of concept that volatile FRs can be durably applied on the textile surfaces using this technique.

27.5.2 Fire-Retardant Additives/Copolymers in Synthetic Fibers

As the title suggests, this approach is applicable to synthetic fibers only where either one of the monomer/homopolymer can be flame-retarded or the flame-retardant molecules can be attached to the polymer chain during polymerization, or flame-retardant additives can be in the polymer melt or in solution prior to extrusion. Synthetic fibers thus produced are also called inherently flame retardant. Some exemplar fibers are discussed here.

27.5.2.1 Viscose or Rayon

A well-known inherently flame retardant viscose fiber is Viscose FR, marketed by Lenzing. The fiber is produced by adding Viscofil® Exolit® 5060 (Clariant), which is bis(2-thio-5,5-dimethyl-1,3,2-dioxaphosphorinyl)oxide in the spinning dope before extrusion. As this additive is phosphorus based, it is similar to other phosphorus-based flame retardants in terms of mode of action (condensed phase). Other commercially available flame retardant viscose fibers include DFG (Daiwabo, Japan), White mountain FR (Jilin Chemical Group, China), Anti-fcel (Shangdong Helon Textile science &Technology Company, China), and Safewll (New Fibers Textile Corporation, Taiwan).

27.5.2.2 Polyester

One of the most successful commercially available flame retardant polyester fiber is Trevira CS, which is produced by incorporating a comonomeric phosphinic acid unit into the PET polymeric chain. Examples of flame retardant additives used in polyester are bisphenol-S-oligomer derivatives, for example, Heim/Toyobo GH (Toyobo, Japan), phosphorus-containing additive, e.g., Avora®Plus FR (KoSa), and phosphinate salts (Clariant). All these flame retardants do not promote char formation.[3]

27.5.2.3 Polypropylene

Although in principle, phosphorus-containing, halogen-containing, silicon-containing, metal hydrate, and oxide are effective in rendering PP flame retardant, invariably these flame retardants are required in high levels, typically > 20% w/w, to achieve specified level of flame retardancy. However, such high levels render the processing of compounded polymer into fibers difficult and the resulting fiber properties unsuitable for textile applications.[123] Apart from antimony-halogen, or in some cases, tin-halogen formulations, only one single flame retardant system, tris(tribomoneopentyl) phosphate (FR 370, ICL), is presently effective in polypropylene when required for fiber end uses. PP is very sensitive to UV radiations and normally requires UV-stabilizer for outdoor use. Flamestab® NOR 116 (BASF, formerly Ciba), a thermally stable hindered amine (an oligomer-linked N-alkoxy-2,2,6,6-tetramethyl-4-substituted-morpholine) used for UV stability, is also reported to have some flame retardant properties.

27.5.2.4 Polyamides

It is very difficult to incorporate additives in polyamides because of their melt reactivities. Phosphinate salts developed by Clariant, mainly aluminum diethyl phosphinates though have shown improvement in flame retardancy, achieving UL94 V0 ratings with < 20 wt% additive.[124] Lewin et al. showed that small weight percentage (< 2 wt%) of ammonium sulphamate and dipentaerythritol yield high levels flame retardancy of polyamide 6 with V-0 ratings and LOI values of 30–37 vol% being achieved even at such low concentrations.[125,126] These findings have also been supported by our work at Bolton.[127]

27.5.3 Nanocomposite-Based Flame Retardants in Textiles

While polymeric nanocomposites (nanoparticles or nanofillers dispersed in the polymer matrix) as flame retardant materials received extensive attention over last two decades, there has been a limited success in developing flame-retardant textile structures out of them. Amongst nanoparticles, most of the work has been on nanoclays. The major difference between fibers and bulk polymers is the small thickness of individual fibers, typically 15–30 μm in diameter, yielding yarns of 50–100 μm diameter and fabrics having thicknesses varying from as low as 100 μm to several mm. For fire-retardant properties, it is believed that the presence of clay in a polymer promotes carbonaceous-silicate buildup on the surface during burning, which insulates the underlying material. The accumulation of silicate layers on the surface is due to gradual degradation and gasification of the polymer. This mechanism works fine for thick polymer plaques. However, for physically and thermally thin samples like textiles, there is not enough time to make thermal barrier and hence the nanoclays in textiles are not as effective as seen in bulk polymers. The other important issue is that although nanoclays reduce peak heat release rate during cone calorimetric testing, they have no effect on time-to-ignition, which is important in textiles as most flammability tests require ignition resistance. Most often simple techniques such as limiting oxygen index show that the addition of nanoclays and other nanoparticles do not significantly increase LOI values unless their presence modifies the burning behavior by reducing melt dripping character of the polymer.[128] They, however, do promote char formation in otherwise non-charring polymers, which is an advantageous effect. While there are many academic research papers in this area where polymer nanocomposites as bulk materials have shown promising results, in most cases filaments could not be extruded from them. The work where filaments/fibers could be produced is summarized here.

Solarski et al. have developed polylactic acid (PLA)-clay nanocomposites fibers[129] by melt blending organomodified (OM)-MMT (1 to 4 wt%) with PLA and then melt spinning into multifilament yarns. The dispersion of the clay in the yarns was reported to be quite good. The multifilaments were knitted and the flammability studied using cone calorimetry at an external heat flux of 35 kW/m². Depending on the clay loading, the peak value of RHR decreased by up to 38% demonstrating improved fire performance of these PLA fibers.

In case of polypropylene, one of the difficulties of incorporating nano clays is the lack of polar groups in the polymer chain which makes direct intercalation or exfoliation almost impossible. Most often maleic anhydride-grafted polypropylene is used as a compatibilizer, which enhances the interaction between the clay and polymer with strong hydrogen bonding between -OH or -COOH and the oxygen groups of clay. Presence of such compatibilizers usually has adverse effects on fiber properties, however.[123] While this effect is concentration dependent, in order to keep the rheological properties necessary for extrusion into fiber, lower levels are preferred. Work in our laboratories at Bolton has shown that levels between 1–3% w/w of grafted PP along with sufficient compounding can improve the dispersion of the clay platelets in the PP matrix.[16,130] The clays used included a montmorillonite clay modified with dimethyl, dehydrogenated tallow quaternary ammonium ion, Bentone 107 (Elementis), Cloisite 20A and 30B (Southern Clay Products, USA), and montmorillonite modified with vinyltriphenyl phosphonium bromide. However, the clays were not exfoliated or intercalated, but dispersed at nanolevel as shown by TEM image of one sample[131] in Figure 27.4. From compounded nanocomposite pellets, filaments and tapes could be extruded. Filaments could also be knitted into fabric strips with a small, hand, circular knitting machine, gauge E7. The clays had minimal effect on LOI of the extruded tapes, which is not surprising as the clays reduce the ignition time. The clays also changed the burning behaviors of thermoplastics by holding them together rather than letting them melt away. The fabrics were tested for flame spread test on BS 5438 rig[15] used for flammability of textile materials, using sample holder used in Test 1 of this standard. The fabric containing highest level (5%) of compatibilizer and 3% clay self-extinguished, whereas the fabric with 1% compatibilser and 3% clay burned very slowly. It was seen that all clays reduced rate of burning.[131]

A subsequent study investigated the effect of introducing conventional flame retardants to PP-nanocomposites. The flame retardants were both phosphorus- and halogen-containing. Phosphorus-containing flame retardants included ammonium polyphosphate melamine phosphate and pentaerythritol phosphate, and the bromine-containing included tris (tribromopentyl) phosphate and tris(tribromophenyl)cyanurate species. Hindered amine stabilizer NOR 116 (BASF,

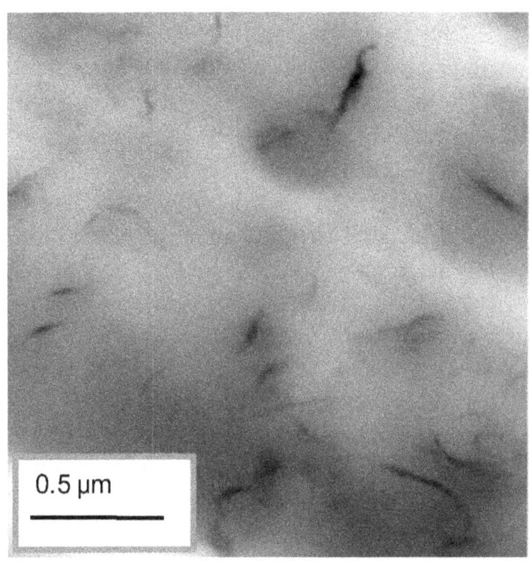

FIGURE 27.4 TEM of polypropylene containing 3% compatibilizer (Pb) and 2.5% Cloisite 20A clay.[131]

formerly Ciba) was also used as a UV stabilizer. The clay used for this set of samples was Elementis E. Extrusion into filaments proved to be challenging because of problems with optimizing clay and flame retardant dispersion, and this was especially the case when APP was present because of its very poor dispersion and relatively large particle size (25–30 μm). Hence, fabrics produced were of different area densities. In some cases fabrics could not be knitted. LOI values of strings were unaffected by either the presence of clays and/or flame retardant but then the low concentrations of flame retardants present (5 wt%) would not be expected to raise the LOI values significantly when present alone. Flame spread test with BS 5438 test rig however showed reduction in rate of flame spread.

Bourbigot et al.[132] have used poly(vinylsilsesquioxane) (POSS) in PP to melt spin filaments, which were then knitted into fabrics and tested using cone calorimetry. POSS presence had minimal effect on peak heat and total heat release values of PP fabric but delayed the time-to-ignition. This behavior of POSS is opposite to that of layered silicates, which have minimal effect on time-to-ignition but reduce peak heat release rate, indicating that POSS does not act as a flame retardant but only as a heat stabilizer via a decrease of the ignitability.

Lille group has also been successful in preparing yarns from polypropylene/multi-walled carbon nanotubes (MWNT) (1 and 2% by mass) nanocomposites.[6,133] Fabrics knitted from these yarns were tested by cone calorimetry. Peak heat rate reduced by 50% for a fraction of nanotubes of only 1 wt%, but the time to ignition of the nanocomposite was shorter.

Bourbigot et al. also reported the fire performance of nylon 6 or PA-6/clay hybrid fibers made by melt blending and by melt spinning.[134] PA-6 nanocomposite exhibited an exfoliated structure, and no degradation of the clay was observed after processing. However, the yarns were not very uniform due to reagglomeration of the MMT platelets, resulting in poor mechanic properties.[6] The fabrics prepared from these yarns were of 1020 g/m^2 area density and 2.5 mm thickness. These were tested by a cone calorimeter at 35 kW/m^2 heat flux, where ignition times of 70s and 20s and peak heat release rate (PHRR) values of 375 and 250 kW/m^2 respectively for the normal and nanocomposite polyamide 6 fabrics were recorded. While the latter represents a significant 33.3% reduction in PHRR, ignition resistance was significantly reduced, and total heat release was little affected. It must be noted that these fabrics have very high area density, hence this reduction in peak heat release might not be reflected by fabrics of area densities normally used as apparel fabrics, that is, 100–200 g/m^2.

At Bolton, we could also extrude filaments from nylon 6-nanocomposites.[135] Nylon 6 was compounded with Cloisite 15A and 25A nanoclays at different levels. Commercially available nylon 6 containing 2% nanoclay (UBE Industries Ltd) was also used. The dispersion of nanoclays was much easier for nylon 6 than in case of polypropylene, and filaments could be extruded without problem. Flame spread tests were done on strands of the compounded polymers by a modified UL-94 test to observe their burning behavior in both horizontal and vertical orientations. It was seen that presence of clays changes the burning behavior of nylon. It was observed that nylon melts and drips and hence does not burn for a longer time. With clay presence, the melting and dripping reduced, but as the clay was able to hold the polymer there, it burnt more than pure polymer. With very high clay levels, the char formation tendency increased, as can be seen from Figure 27.5, where a strand of nylon 6 is burning with flaming drips. In contrast, the sample containing clay burns steadily and after burning the char remains intact. In order to observe the effect of additional flame retardants, melamine phosphate, and tris (tribromopentyl) phosphate at 5 wt% level were also added. While tris (tribromopentyl) phosphate reduced the burning of nylon 6 nanocomposite, melamine phosphate showed antagonistic effect.[135] From these sample only nylon 6 nanocomposites could be extruded into filaments and knitted. The burning behavior was similar to that observed with strands, that the nanoclays reduce the thermoplasticity, hence change the burning behavior of nylon.[135]

In a subsequent work,[136] Cloisite 25A nanoclay was added along with two types of flame retardant formulations, (i) aluminum diethyl phosphinate (AlPi) at 10 wt%, known to work principally in the vapor phase and (ii) ammonium sulphamate (AS)/dipentaerythritol (DP) system present at 2.5 wt% and 1 wt% respectively, believed to be condensed phase active. None of the AlPi-containing formulations achieved self-extinguishability (see Figure 27.6), although the presence of nanoclay promoted lower burning and melt dripping rates. The AS/DP-containing formulations with total flame retardant levels of 5.5 wt% or less showed far superior properties and that with nanoclay showed fabric extinction times ≤ 39s and reduced melt dripping (Figure 27.7).

As a followup of this work, bicomponent filaments of nylon 6 were produced with core/sheath geometries containing aluminum diethyl phosphinate (AlPi), and/or nanoclay located individually or together in the sheath or core[137] (see Figure 27.8). Knitted fabrics were assessed for flammability by vertical strip testing, which demonstrated that the relative location of the clay and flame retardant had a significant effect on burning behavior. Samples containing flame retardant AlPi alone in the core or sheath appeared to burn more intensively than the clay-only-containing samples (see Figure 27.9), they showed significantly reduced times-to-burn and numbers of drops relative to these, although their burning length and mass loss values were similar. It was seen that when the AlPi was in the sheath, the total burning time was slightly less than when in the core, which might be explained in terms of the vapor phase activity of the AlPi sample being more effective when it is in the sheath. When the clay and flame retardant were present individually, in the sheath and core respectively and vice versa, their general burning properties were little changed from those of the AlPi-containing-only samples, although the number of

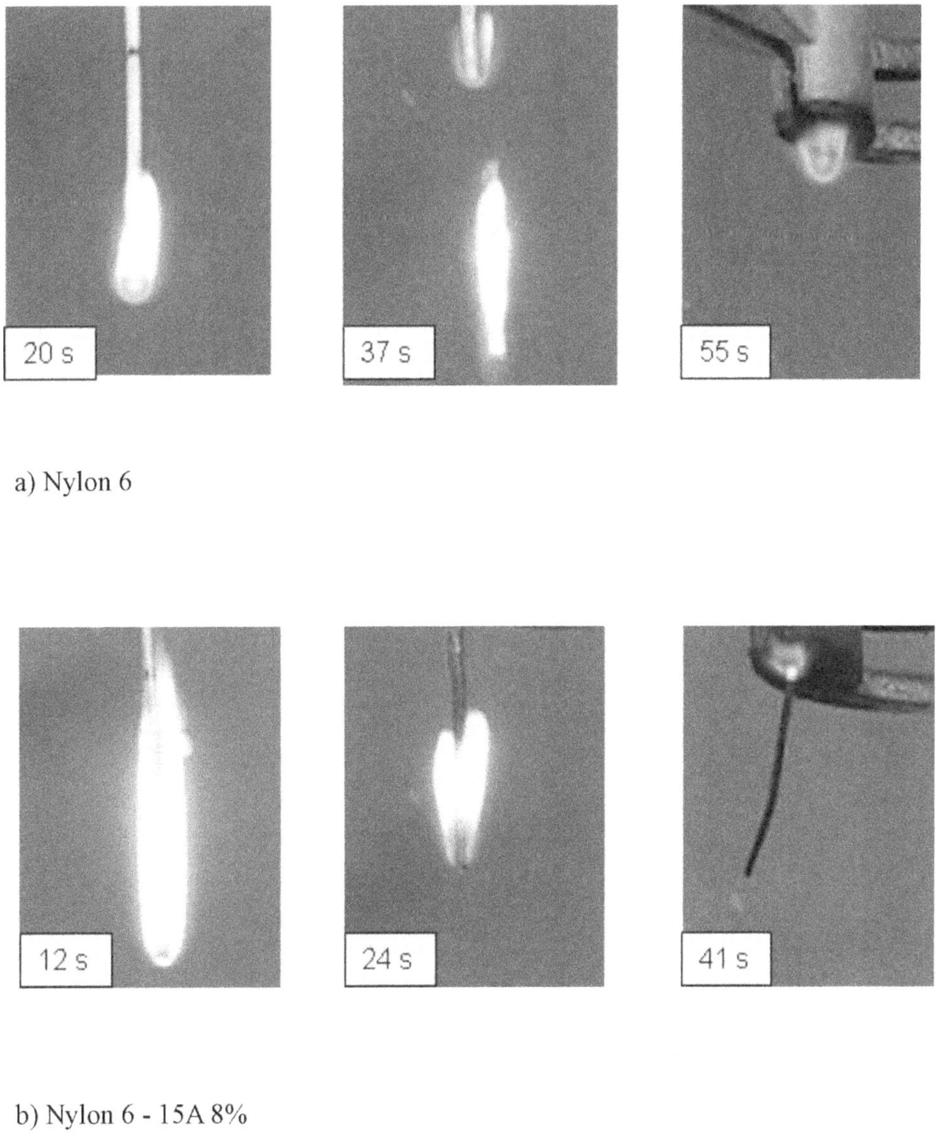

FIGURE 27.5 Vertical burning behavior of (a) nylon 6 and (b) nylon 6 strings containing 8% 15A nanoclay.[135]

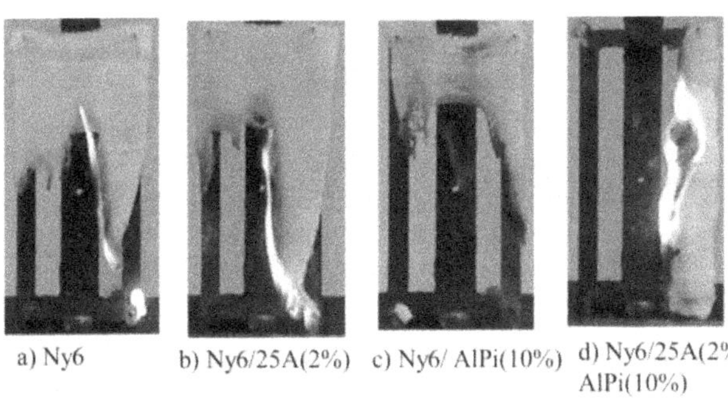

FIGURE 27.6 Selected images of burning Nylon 6 fabrics comprising AlPi and/or Cloisite 25A: 40 s after extinction of the igniting flame.[136]

drops formed had significantly reduced further by a factor of 2 or more. Finally, the presence of both nanoclay and AlPi together in the core had a negative effect in that dripping tendency was increased. The best results in terms of reduced burning time and reduced melt dripping were seen in the sample with both components present in the sheath, where the combined flame retarding effect appeared to protect the Nylon 6 core, thereby suggesting that the vapour phase active AlPi was at its most effective coupled with a possible tendency for the clay present to diffuse to the outer filament surface and so form a protective layer.

27.5.4 Heat Resistant and Inherently Fire-Retardant Fibers

Inherently flame and heat resistant fibers are either all-aromatic polymeric structures or inorganic and mineral based. The aromatic fibers are mostly used for apparel applications as protective clothing. Some commonly used fiber types are given in Table 24.1 and discussed later. These are non-thermoplastic, combustion-resistant with decomposition temperatures above 375°C[10] and with LOI values 30 vol% or

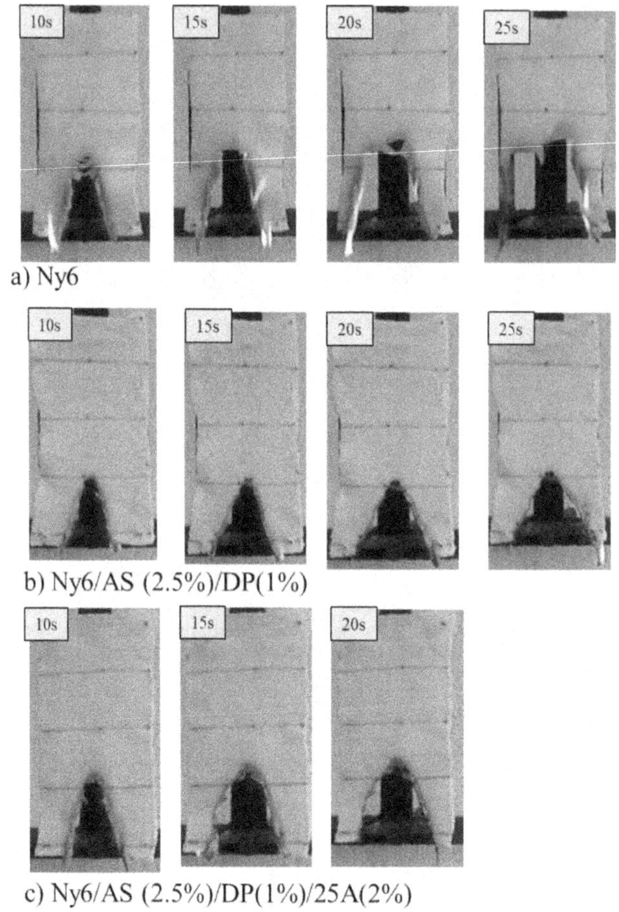

FIGURE 27.7 Burning and extinguishing behavior of ignited Nylon 6/DP/AS/clay-based fabric samples: 10, 15, 20, and 25s after extinction of the igniting flame.[136]

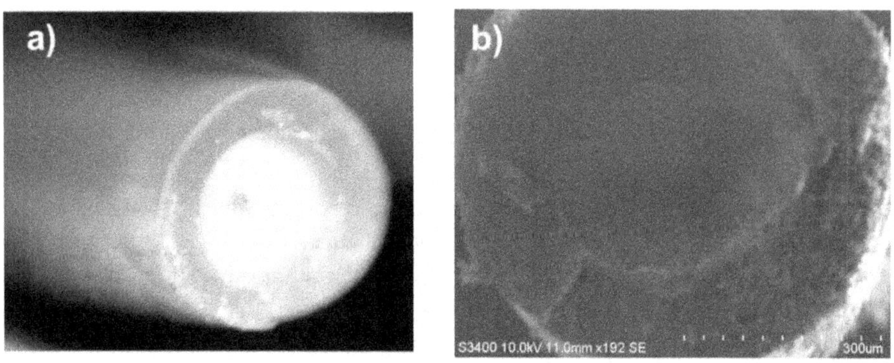

FIGURE 27.8 (a) Optical and (b) SEM images of Nylon 6 in sheath and Nylon 6/AlPi(10 wt%) core bicomponent fiber.[137]

more. Moreover, they have char-forming tendency. Detailed information about their chemical structures and mode of decomposition is provided elsewhere.[138]

First example of this type is aramids, which are most well-known and exploited for protective clothing. They have aromatic repeat units bonded together by strong amide—CO.NH-, imide—CO.N< or both in alternating manner. The most commonly used aramids are meta-aramids (Nomex, DuPont and Conex, Teijin), para-aramid (Kevlar, DuPont and Conex®, Teijin), poly-m-phenylenediamine isophthalamide (Twaron, Teijin) and poly(aramid-arimid) fiber (Kermel, Rhone-Poulenc). Nomex is used in firefighters and military protective apparel and accessories, as well as electrical insulation paper. Kevlar, the most fire-resistant of the aramids, has been much used in military protective garments and protective fabrics in vehicles. Aramid fibers however, have poor UV stability. The problem appears as discoloration or fading, cracking and sometimes, total product disintegration if cracking has proceeded far enough. This affects the fire performance of the product as well. Kevlar and Kermel have

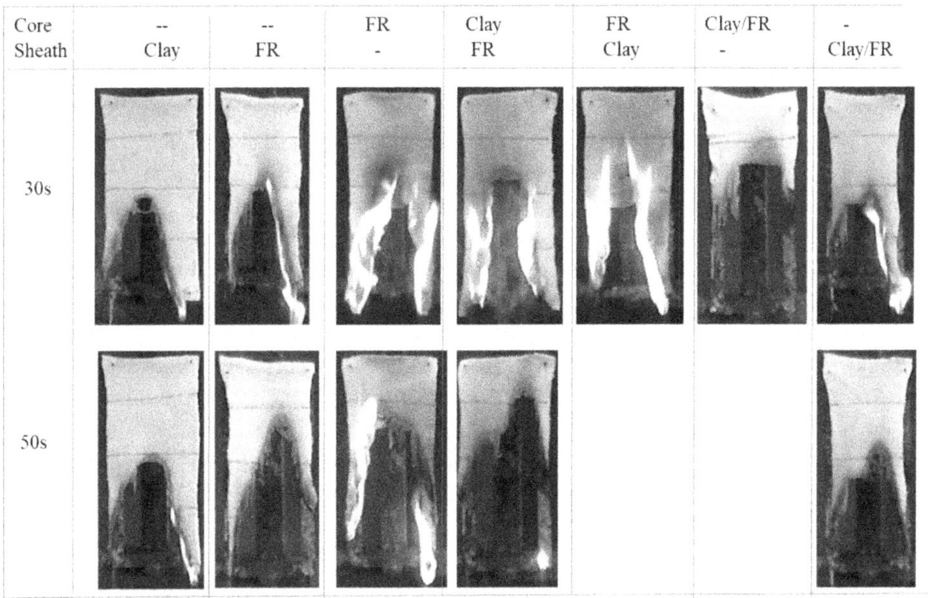

FIGURE 27.9 Images of burning Nylon 6 fabrics from bicomponent filaments following the extinction of the igniting flame. Bicomponent filaments with core/sheath geometries contain AlPi and Cloisite 25A nanoclay in different combinations.[137]

poor UV stability. Twaron ® HM (High modulus) has similar tensile strength as Kevlar 49 but better UV resistance. A fuller description of these fibers is given elsewhere.[139]

Polybenzazole group of fibers include polybenzimidazole (PBI, PBI Performance Products) and polybenzoxazole (Zylon, Toyobo). These are wholly aromatic structures. The applications for PBI include fire blocking fabrics for aircraft seats, firefighters suits, racing-car driver suit, etc. These fibers, however, have poor resistance to both UV and visible light, and so must be protected from intense radiation source.

Phenolic or novoloid fibers, produced by spinning and postcuring of phenol formaldehyde precondensate are noncombustible and char formers. However, due to poor fiber strength and abrasion properties, they are not suited for making apparel. Common example is Kynol, which can be blended with Nomex or FR viscose to produce protective clothing. Melamine fibers are other examples, which are produced by condensation reaction of formaldehyde with melamine or substituted melamines. Basofil (Basofil Fibers LLC, formerly BASF) is an example used in commercial aircraft seating, firefighter turnout gear, industrial protective clothing, friction parts, and automotive insulation.

Halogen-containing fibers, namely, modacrylics fibers are typically copolymers of vinyl chloride or vinylidene dichloride and acrylonitrile. Although these fibers do not melt-flow or drip, they shrink rapidly when exposed to the fire. Because of this, the best use of modacrylic fibers is in blends.

Polyphenylene sulfide fibers are other group of inherently flame retardant fibers. Example is Diofort® (Diolen Industrial Fibers, Netherlands).

Semicarbon or oxidized polyacrylonitrile fibers, produced by thermo-oxidative stabilization of either viscose or acrylic fibers, have excellent heat resistance, do not melt or burn, and have excellent resistance to molten metal splashes. Panox (SGL Group), Panotex (Universal Carbon Fibers) and Pyron® (Zoltek Corp) are some examples, produced from acrylic fibers.

27.5.4.1 Inorganic Fibers

This group of fibers includes glass, carbon and ceramic fibers. Glass and carbon fibers have already been mentioned in the "Composites" section (Section 27.4.8). Glass fiber is nonflammable but melts at temperatures above 600°C. Carbon fibers are extremely resistant to high temperatures: their melting temperature is 4,000°C.[6] They can also be considered as flame resistant since they will burn only at very high temperatures or in high oxygen-containing atmospheres. They hence are ideal choice for applications requiring extremely high temperatures, for example in the filtration of molten iron. A number of ceramic fibers, SiC, silicon or boron nitride, polycarbosilicones, alumina, etc., have been developed. Another example is Basalt, produced from relatively low-melting natural aluminosilicate rock. These ceramic fibers can withstand temperatures between 1,000–1,400°C. The fibers however, are very abrasive and may cause problems during processing. Their main application is in fire barriers and insulation materials.

27.5.5 Fiber Blending

Fiber blending is a very common method of reducing the flammability of flammable fibers. Polyester is usually blended with cotton and this poly-cotton, if has lower than 50% polyester content can pass the simple vertical strip flammability test. With higher polyester content, sometimes the blended fiber is more flammable than the individual components. This is called wicking effect, where the cotton acts like a wick, holding the polyester component together, which burns. Cotton-nylon blend are also quite commonly used to reduce flammability of cotton.

Cotton-wool blends are quite common as well. Aramids are blended with many fibers for different applications. Nomex can be blended with FR viscose and FR wool to produce fire-blocking fabric, for example, for aircraft seats.[4] Nomex blended with Kevlar shows better performance than 100% Nomex in firefighters' outer protective garments.[29] Various blends of glass fibers with aramids, melamine fibers, PVC fibers, and polyester have been reported for use in fire-protective non-woven veils for upholstery and mattresses.[140]

27.5.6 Composite Assemblies—Design Issues

In previous sections the flammability and fire resistance of individual fiber/fabric type are discussed. However, as also discussed before, the fire resistance of a fabric not only depends upon the nature of components and the flame retardant treatments applied, but also on fabric area density, construction, air permeability, and moisture content. Nonwovens for example will have superior properties to woven or knitted structure, even if all other variables are kept the same.[141] The air entrapped within the interstices of any fabric structure and between layers of fabrics within a garment assembly provides the real thermal insulation. For effective thermal and fire resistance in a fabric structure, these insulating air domains need to be maintained.[25] In general, for protective clothing and fire-block materials, for best performance multilayered fabric structures are employed. The assembly structures can be engineered to maximize their performance. It is beyond the scope of this review to go into details of these composite structures, hence the reader is referred to the literature on specified applications and products available.

27.6 FUTURE TRENDS

It is evident that the demand for fire and heat-resistant textiles at affordable price will sustain the research and development in this area. Development of high performance fibers with inherently high levels of fire and heat resistance is a major breakthrough in this area, but invariably these are expensive and used only when performance requirements justify cost. There is still need to flame retard conventional textiles such as cotton, wool, etc. For cotton there are well-established phosphorus- and nitrogen-containing commercial flame-retardant finishes, for example, Pyrovatex (Huntsman, previously Ciba) and Proban (Solvay, previously Rhodia, Albright & Wilson), which act in condensed phase and presently are not under threat from environmental point of view. However, for many applications using different fiber types, including cotton, antimony-halogen-based back-coatings are still being used, which are under pressure to be replaced by other environmentally friendly chemicals. The way forward is to introduce other volatile species, such as phosphorus-containing, in these coatings.

In recent years, there has been a big push towards using bio-derived flame retardants. While biomolecules such as proteins, chitosan, starch phytic acid, and DNA have shown promising results, their wash durability is not so good. New trends for surface modification of textiles include sol-gel, layer-by-layer, and plasma techniques. While with sol-gel and LbL treatments, certain level of flame retardancy can be achieved, the inadequate thickness of the developed film may not be enough to help textile material pass required fire standards, in particular vertical strip test. Moreover, their wash durability currently is also poor, but improvements are being made with the technology (see Chapter 14 of this book). Plasma technology involving modification of fiber and textile surface has a history spanning about 50 years, and although it has gained commercial significance within other sectors such as in improving paint/coating adhesion to plastics for automotive and other applications, its adoption by the textile industry has been slow. However, development of atmospheric plasma technique in last few years has enabled commercialization of this technique for surface treatments of textiles. With atmospheric plasma, functional groups can be created on the textile surfaces, and flame retardants can be covalently bonded with the developed functional groups, leading to better water soak durability. In addition, volatile flame retardants can be directly deposited on the surfaces of textiles using appropriate plasma gases.

REFERENCES

(1) "Introduction: An Overview", Lyons, J.W. In: *The Chemistry and Uses of Fire Retardancy*, Wiley Interscience, New York, 1970, p. 2.
(2) "Flame Retardant Finishes", Hauser, P.J.; Schindler, W.D. In: *Chemical Finishing of Textiles*, Schindler, W.D.; Hauser P.J., editors, Woodhead Publishing/CRC, Cambridge, 2004. pp. 98–116.
(3) "Textiles", Horrocks, A.R. In: *Fire Retardant Materials*, Horrocks, A.R.; Price, D., editors, Woodhead Publishing/CRC, Cambridge, 2001. pp. 128–181.
(4) "Heat and Flame Protection", Bajaj, P. In: *Handbook of Technical Textiles*, Horrocks, A.R.; Anand, S.C., editors, Woodhead Publishing, Cambridge, 2000. pp. 223–263.
(5) "Developments in Flame Retardant Textiles-A Review", Horrocks, A.R.; Kandola, B.K.; Davies, P.J.; Zhang, S.; Padbury S.A. *Polym. Degrad. Stab.* **2005**, *88*, 3–12.
(6) "Flame Retardancy of Textiles: New Approaches", Bourbigot, S. In: *Advances in Fire Retardant Materials*, Horrocks A.R.; Price, D., editors, Chapter 2. Woodhead Publishing, Cambridge, 2008. pp 9–40.
(7) "Handbook of Fire Resistant Textiles", Kilinc F., editor., Woodhead Publishing, Cambridge, 2013.
(8) "Flame Retardants for Textile Materials", Choudhury A.K.R., CRC Press, Routledge Taylor & Francis Group, Boca Raton, FL, 2021.
(9) "Update on Flame Retardant Textiles: State of the Art, Environmental Issues and Innovative Solutions", Alongi, J. Smithers Rapra Technology, London, 2013.
(10) "Textiles", Horrocks, A.R.; Kandola, B.K. In: *Plastics Flammability Handbook: Principles, Regulations, Testing,*

and Approval, Troitzsch, J.; Antonatus, E., editors, Chapter 4. Hanser Publishers, Munich, 2021. pp. 129–151.
(11) "The Burning Behaviour of Textiles and Its Assessment by Oxygen-Index Methods", Horrocks, A.R.; Tunc, M.; Price, D. *Tex. Prog.* **1989,** *18 (1/2/3)*, 1–186.
(12) "Standard Method for Measuring the Minimum Oxygen Concentration to Support Candle-Like Combustion of Plastics (Oxygen Index)", ASTM D2863-19.
(13) "Investigation and Prediction of Factors Influencing Flammability of Nightwear Fabrics", Gawande (Nazaré), PhD Thesis, University of Bolton, Bolton, 2002.
(14) "Flammability Testing of Fabrics", Nazare S.; Horrocks, A.R. In: *Fabric Testing*, Hu, J., editor, Chapter 12, Woodhead Publishing, Cambridge, 2008. pp 339–388.
(15) "Methods of Test for Flammability of Textile Fabrics When Subjected to a Small Igniting Flame Applied to the Face or Bottom Edge of Vertically Oriented Specimens", BS 5438:1989 British Standard.
(16) "Polypropylene Fibers Containing Dispersed Clays Having Improved Fire Performance Part II: Characterization of Fibers and Fabrics from PP—Nanoclay Blends", Smart, G.; Kandola, B.K.; Horrocks, A.R.; Marney, D. *Polym. Adv. Techn.* **2008,** *19*, 658–670.
(17) "Comparison of Cone and OSU Calorimetric Techniques to Assess the Flammability Behavior of Fabrics Used for Aircraft Interiors", Kandola, B.K.; Horrocks, A.R.; Padmore, K.; Dalton, J.; Owen, T., *Fire Mater.* **2006**, *30 (4)*, 241–255.
(18) "Use of Cone Calorimetry to Quantify the Burning Hazard of Apparel Fabrics", Nazaré, S.; Kandola, B.K.; Horrocks, A.R. *Fire Mater.* **2002**, *26*, 191–199.
(19) "Fire Safety of Upholstered Furniture—The Final Report on the CBUF Research Programme, EUR 16477EN", Sundstrom, B., editor, Interscience Communications, London, 1994.
(20) "Communication: Raised Grid for Ignitability and RHR Testing", Mikkola, E., *Fire Mater.* **1993**, *17 (1)*, 47–48.
(21) "Flame-Retardant Finishing of Textiles", Horrocks, A.R. *Rev. Prog. Coloration.* **1986**, *16*, 62–101.
(22) "Manikin Test for Flame Engulfment Evaluation of Protective Clothing: Historical Review and Development of a New ISO Standard", Camenzind, M.A.; Dale, D.J.; Rossi, R.M., *Fire Mater.* **2007**, *31 (5)*, 285–296.
(23) "Flammability Testing of Textiles", Horrocks, A.R.; Kandola, B.K. In: *Plastics Flammability Handbook*, Troitzsch J., editor, Chapter 6, Hanser Publication, Munich, 2004. pp. 173–188.
(24) "Finishing of Technical Textiles", Bajaj, P. In: *Handbook of Technical Textiles*, Horrocks, A.R.; Anand, S.C., editors, Woodhead Publishing, Cambridge, 2000. pp. 152–172.
(25) "Thermal (Heat and Fire) Protection", Horrocks, A.R. In: *Textiles for Protection*, Scott, R.A., editor, Chapter 15, Woodhead Publishing, Cambridge, 2005. pp. 398–440.
(26) "The Burning Hazard of Clothing—The Effect of Textile Structures and Burn Severity", Horrocks, A.R.; Gawande, S.; Kandola, B.K. In: *Recent Advances of Flame Retardancy of Polymeric Materials*, Volume XI, Lewin, M., editor. Proceedings of the 2000 conference, Business Communication Company, Stamford, CT, 2000.
(27) "Survey of Small—Scale Flame Spread Results of Modern Fabrics", Hirschler, M.M.; Piansay, T., *Fire Mater.* **2007**, *31 (6)*, 373–386.
(28) "Development of a Multiple Layer Test Procedure for Inclusion in NFPA 701: Initial Experiments", Davis, S.; Villa, K.M. 1989. NISTIR 89-4138. National Institute of Standards and Technology, Gaithersburg, MD, August 1989.
(29) "Flame Retardants in Commercial Use or Development for Textiles", Weil, E.D.; Levchik S. *J. Fire Sci.* **2008**. *26*, 243–281.
(30) "The Particular Flammability Hazards of Nightwear", Horrocks, A.R.; Nazare, S.; Kandola, B.K. *Fire Saf. J.* **2004**, *39 (4)*, 259–276.
(31) "Specification for Flammability Performance of Fabrics and Fabric Combinations Used in Nightwear Garments", BS 5722: 1991.
(32) "Textiles Burning Behaviour Fabrics for Apparel Detailed Procedure to Determine the Burning Behaviour of Fabrics for Apparel", EN 1103: 1995.
(33) "Comparison of Flame Spread of Textiles and Burn Injury Prediction with a Manikin", Rossi, R.M.; Bruggmann, G.; Stämpfli, R. *Fire Mater.* **2005**, *29 (6)*, 395–406.
(34) "Textiles for Survival", Holmes, D.A. In: *Handbook of Technical Textiles*, Horrocks, A.R.; Anand, S.C., editors, Woodhead Publishing, Cambridge, 2000. pp. 461–489.
(35) "Flight Suits for Military Aviators", Scott, R.A. In: *Textiles for Protection*, Scott, R.A., editor, Chapter 24, Woodhead Publishing, Cambridge, 2005. pp. 678–698.
(36) "Fire Protection in Military Textiles", Nazare, S. In: *Advances in Fire Retardant Materials*, Horrocks, A.R.; Price, D., editors, Chapter 19, Woodhead Publishing, Cambridge, 2008.
(37) "Chemical, Biological, Radiological and Nuclear Protection", Seed, M.; Anand, S.C.; Kandola, B.K.; Fulford, F. *Tech. Text. Int.* **2008**, 39–46.
(38) "Protective Clothing", Bajaj, P.; Sengupta, A.K. *Text. Progr.* **1992**, *22 (2–4)*, 1–110.
(39) "Flame Retardancy", Fung, W. In: *Coated and Laminated Textiles*, Woodhead Publishing, Cambridge, 2002. pp. 71–77.
(40) "Flame Retardancy Testing and Regulation of Soft Furnishings", Nazaré, S.; Davis, R.D. In: *Handbook of Fire Resistant Textiles (Woodhead Publishing Series in Textiles)*, Kilinc, F.S., editor, Chapter 17, Woodhead Publishing, Cambridge, 2013. pp. 456–498.
(41) "Method for Determination of the Effects of a Small Source of Ignition on Textile Floor Coverings (Methenamine Tablet Test)", BS 6307:1982, ISO 6925-1982.
(42) "Method for Determination of the Effects of a Small Source of Ignition on Textile Floor Coverings (Hot Metal Nut Method)", BS 470:1987.
(43) "Standard Test Method for Critical Radiant Flux of Floor-Covering Systems Using a Radiant Heat Energy Source", ASTM E648-99.
(44) "Standard Method of Test for Critical Radiant Flux for Floor Covering Systems Using a Radiant Heat Energy Source", NFPA 253.
(45) "British Standard for Burning Behavior of Materials for Marquees, Large Tents and Related Products", BS EN 14115:2002.
(46) "Standard Methods of Fire Tests for Flame Propagation of Textiles and Films", NFPA 701.
(47) "Standard Test Method for Surface Burning Characteristics of Building Materials", ASTM E84–03.
(48) "Textiles in Transportation", Fung W. In: *Handbook of Technical Textiles*, Horrocks, A.R.; Anand, S.C., editors, Chapter 18, Woodhead Publishing, Cambridge, 2000. pp. 490–528.
(49) "Flame Retardant Materials for Maritime and Naval Applications", Sorathia, U. In: *Advances in Flame Retardant*

Materials, Horrocks, A.R.; Price D., Chapter 19, Woodhead Publishing, Cambridge, 2008.
(50) "Composites Having Improved Fire Performance", Kandola, B.K.; Kandare, E. 2008. In: *Advances in Flame Retardant Materials*, Horrocks, A.R.; Price D., Chapter 16, Woodhead Publishing, Cambridge, 2008.
(51) "The Effect of Intumescents on the Burning Behavior of Polyester-Resin-Containing Composite", Kandola, B.K.; Horrocks, A.R.; Myler, P.; Blair, D. *Compos. Part A Appl. Sci. Manuf.* **2002,** *33*, 805–817.
(52) "Effect of Reinforcing Element on Burning Behavior of Fiber-Reinforced Epoxy Composites", Kandola, B.K.; Horrocks, A.R.; Rashid, M.R. In: *Recent Advances of Flame Retardancy of Polymeric Materials*, Lewin, M., editor, Proceedings of the 2006 Conference, Business Communication Company, Stamford, CT, 2006.
(53) "Effect of Fiber Type on Fire and Mechanical Behaviour of Hybrid Composite Laminates", Kandola, B.K.; Myler, P.; Horrocks, A.R.; Herbert, K.; Rashid, M.R. In: *Proceedings of the SAMPE Fall Technical Conference*, Dallas, TX, 2006.
(54) "Flame Retardant Treatments of Cellulose and Their Influence on the Mechanism of Cellulose Pyrolysis", Kandola, B.K.; Horrocks, A.R.; Price D.; Coleman, G.V. *J. Macromol. Sci.* **1996**, *C36 (4)*, 721–794.
(55) "Overview of Traditional Flame-Retardant Solutions", Horrocks A.R. In: *Update on Flame Retardant Textiles: State of the Art, Environmental Issues and Innovative Solutions*, Alongi J.; Horrocks A.R.; Carosio F.; Malucelli G., editors, Smithers Rapra, Shawbury, 2013. pp. 123–178.
(56) "Mechanism and Modes of Action in Flame Retardancy of Polymers", Lewin, M.; Weil, E.D. In: *Fire Retardant Materials*, Horrocks, A.R.; Price, D., editors, Woodhead Publishing/CRC, Cambridge, 2001. pp. 31–68.
(57) "Making or Preparing of Paper, Linen, Canvass and Such Like Substances, Which Will Neither Flame Nor Retain Fire, and Which Hath Also a Property in it of Resisting Moisture and Damps", Wyld, O. 1735. British Patent No. 551, London.
(58) "Flame Retardant Textile Coatings and Laminates", Horrocks, A.R. In: *Advances in Fire Retardant Materials*, editor, Horrocks, A.R.; Price, D., Chapter 7, Woodhead Publishing, Cambridge, 2008.
(59) "The Flammability of Lyocell", Hall, M.E.; Horrocks, A.R.; Seddon, H. *Polym. Degrad. Stab.* **1999**, *64*, 505–510.
(60) "Flame Resistant Wool and Wool Blends", Cardamone, J.M. In: *Handbook of Fire Resistant Textiles*, Kilinc, F., editor, Chapter 9, Woodhead Publishing, Cambridge, 2013. pp. 245–271.
(61) "Now and Future of Flame Retardancy of Fabrics", Zhou, H. *Guangxi Text. Sci. Technol.* **1995**, *24 (2)*, 35–39.
(62) "Flame Retardant Finishing of Cotton and Silk Fabrics", Achwal, W.B. *Colourage.* **1987,** June, 16–30.
(63) "Flame Retardancy Finish with an Organophosphorus Retardant on Silk Fabrics", Guan, J.P.; Chen, G.Q. *Fire Mater.* **2006**, *30 (6)*, 415–424.
(64) "The Sensitization of Thermal Decomposition of APP by Selected Metal Ions and Their Potential for Improved Cotton Fabric Flame Retardancy", Davies, P.J.; Horrocks, A.R.; Alderson, A. *Polym. Degrad. Stab.* **2005,** *88*, 114–122.
(65) "The Challenge of Replacing Halogen Flame Retardants in Textile Applications: Phosphorus Mobility in Back-Coating Formulations", Horrocks, A.R.; Davies, P.J.; Alderson, A.; Kandola, B.K. In: *Proceedings of 10th European Meeting of Fire Retardant Polymers, FRPM'05*, Berlin, 6–9 September 2005; Advances in the Flame Retardancy of Polymeric Materials: Current perspectives, Schartel, B., editor. Norderstedt, 2007, pp. 141–158.
(66) "Flame Retardant Textile Back-Coatings. Part 2. Effectiveness of Phosphorus-Containing Flame Retardants in Textile Back-Coating Formulations", Horrocks, A.R.; Wang, M.Y.; Hall, M.E.; Sunmonu, F.; Pearson, J.S. *Polym. Int.* **2000,** *49*, 1079–1091.
(67) "The Potential for Volatile Phosphorus-Containing Flame Retardants in Textile Back-Coatings", Horrocks, A.R.; Davies, P.; Alderson, A.; Kandola, B.K. *J Fire Sci.* **2007**, *25 (6)*, 523–540.
(68) "Flame Retarded Polyurea with Microencapsulated Ammonium Phosphate for Textile Coating", Giraud, S.; Bourbigot, S.; Rochery, M.; Vroman, I.; Tighzert, L.; Delobel, R.; Poutch, F. *Polym. Degrad. Stab.* **2005**, *88*, 106–113.
(69) "Microencapsulation of Ammonium Phosphate with a Polyurethane Shell Part I: Coacervation Technique", Saihi, D.; Vroman, I.; Giraud, S.; Bourbigot, S. *React. Funct. Polym.* **2005**, *64*, 127–138; **2006**, *66*, 1118–1125.
(70) "Recent Advances for Intumescent Polymers", Bourbigot, S.; Le Bras, M.; Duquesne, S.; Rochery, M. *Macromol. Mater. Eng.*, **2004**, *289 (6)*, 499–511.
(71) "Polyurethane/Clay and Polyurethane/POSS Nanocomposites as Flame Retarded Coating for Polyester and Cotton Fabrics". Devaux, E.; Rochery, M.; Bourbigot, S. *Fire Mater.* **2002**, *26 (4–5)*, 149–154.
(72) "Fire Barrier Fabrics", Tolbert, T.W.; Dugan, J.S.; Jaco, P.J.; Hendrix, J.E. Spring Industries Inc., US Patent 333174, 4 April 1989.
(73) Von Bonin, W.; Von Gizycki, U.; Bayer, A.G. Europe Patent 91121065.6, 9 December 1991.
(74) "Complex Char Formation in Flame-Retarded Fiber Intumescent Combinations—IV. Mass Loss Calorimetric and Thermal Barrier Properties", Kandola, B.K.; Horrocks, A.R. *Fire Mater.* **2000**, *24*, 265–275.
(75) "Functionalization of Textiles by Inorganic Sol-Gel Coatings", Mahltig, B.; Haufe, F.; Böttcher, H. *J. Mater. Chem.* **2005**, *15 (41)*, 4385–4389.
(76) "Muti-Component Flame Resistant Coating Techniques for Textiles", Alongi, J.; Frache, A.; Malucelli, G.; Camino, G. In: *Handbook of Fire Resistant Textiles*, Kilinc, F. editor, Chapter 4, Woodhead Publishing, Cambridge, 2013. pp. 68–91.
(77) "Sol-Gel and Layer-by-Layer Coatings for Flame-Retardant Cotton Fabrics: Recent Advances", Malucelli, G. *Coatings*. **2020**, *10*, 333. DOI:10.3390/coatings10040333
(78) "Thermal Stability, Flame Retardancy and Mechanical Properties of Cotton Fabrics Treated with Inorganic Coatings Synthesized Through Sol–Gel Processes", Alongi, J.; Ciobanu, M.; Malucelli, G., *Carbohydr. Polym.* **2007,** *87*, 2093–2099.
(79) "Novel Flame Retardant Finishing Systems for Cotton Fabrics Based on Phosphorus-Containing Compounds and Silica Derived from Sol-Gel Processes", Alongi, J.; Ciobanu, M.; Malucelli, G., *Carbohydr. Polym.* **2011**, *85 (3)*, 599–608.
(80) "Sol-Gel Derived Architectures for Enhancing Cotton Flame Retardancy: Effect of Pure and Phosphorus-Doped Silica Phases", Alongi, J.; Colleoni, C.; Rosace, G.; Malucelli, G., *Polym. Degrad. Stab.* **2014,** *99 (1)*, 92–98.

(81) "Sol–Gel Treatments on Cotton Fabrics for Improving Thermal and Flame Stability: Effect of the Structure of the Alkoxysilane Precursor", Alongi, J.; Ciobanu, M.; Malucelli, G., *Carbohydr. Polym.* **2012**, *87*, 627–635.
(82) "Thermal Stability and Flame Retardancy of Polyester, Cotton, and Relative Blend Textile Fabrics Subjected to Sol–Gel Treatments", Alongi, J.; Ciobanu, M.; Tata, J.; Carosio, F.; Malucelli, G., *J. Appl. Polym. Sci.* **2011**, *119*, 1961–1969.
(83) "Development of Flame Retardancy Properties of New Halogen-Free Phosphorous Doped SiO_2 Thin Films on Fabrics", Cireli, A.; Onar, N.; Ebeoglugil, M.F.; Kayatekin, I.; Kutlu, B.; Culha, O.; Celik, E., *J. Appl. Polym. Sci.*, **2007,** *105*, 3747–3756.
(84) "Multifunctional, Strongly Hydrophobic and Flame-Retarded Cotton Fabrics Modified with Flame Retardant Agents and Silicon Compounds", Przybylak, M.; Maciejewski, H.; Dutkiewicz, A.; Wesołek, D.; Władyka-Przybylak, M., *Polym. Degrad. Stab.* **2016**, *128*, 55–64.
(85) "One-Pot Fabrication of Superhydrophobic and Flame-Retardant Coatings on Cotton Fabrics via Sol-Gel Reaction", Lin, D.; Zeng, X.; Li, H.; Lai, X.; Wu, T.; *J. Colloid Interface Sci.* **2019,** *533*, 198–206.
(86) "Novel Environmentally Sustainable Surface Coatings for Textiles", Eivazi, PhD Thesis, University of Bolton, Bolton, 2018.
(87) "Flame Retardant Coatings Prepared Using Layer by Layer Assembly: A Review", Qiu, X.; Li, Z.; Li, X.; Zhang, Z., *Chem. Eng. J.* **2018**, *334*, 108–122. DOI:10.1016/j.cej.2017.09.194
(88) "Layer-by-Layer Deposition: A Promising Environmentally Benign Flame-Retardant Treatment for Cotton, Polyester, Polyamide and Blended Textiles", Magovac, E.; Vonˇcina, B.; Jordanov, I.; Grunlan, J.C.; Bischof, S., *Materials.* **2022**, *15*, 432. DOI:10.3390/ma150204
(89) "Polyvalent Hydrogen-Bonding Functionalization of Ultrathin Hyperbranched Films on Polyethylene and Gold", Bergbreiter, D.E.; Tao, G.; Franchina, J.G.; Sussman, L. *Macromolecules.* **2001,** *34*, 3018–3023. DOI:10.1021/ma000873d
(90) "Preparation of the Layer-by-Layer Deposited Ultrathin Film Based on the Charge-Transfer Interaction", Shimazaki, Y.; Mitsuishi, M.; Ito, S.; Yamamoto, M., *Langmuir.* **1997,** *13*, 1385–1387. DOI:10.1021/la9609579
(91) "Covalently Attached Multilayer Assemblies by Sequential Adsorption of Polycationic Diazo-Resins and Polyanionic Poly(Acrylic Acid)", Sun, J.; Wu, T.; Liu, F.; et al. *Langmuir.* **2000,** *16*, 4620–4624. DOI:10.1021/la991482z
(92) "Poly (Acrylic Acid)/Clay Thin Films Assembled by Layer-by-Layer Deposition for Improving the Flame Retardancy Properties of Cotton", Huang, G.; Liang, H.; Wang, X.; Gao, J. *Ind. Eng. Chem. Res.* **2012**, *51*, 12299–12309.
(93) "Multilayers of Colloidal Particles", Iler, R.K., *J. Colloid Interface Sci.* **1966**, *21*, 569–594.
(94) "Flame Retardant Behavior of Polyelectrolyte_Clay Thin Film Assemblies on Cotton Fabric", Li, Y.; Schulz, J.; Mannen, S.; Delhom, C.; Condon, B.; Chang, S.; Zammarano, M.; Grunlan, J.C., *Am. Chem. Soc.* **2010,** *4*, 3325–3337.
(95) "Fire Retardant and Thermal Degradation Properties of Cotton Fabrics Based on APTES and Sodium Phytate Through Layer-by-Layer Assembly", Li, Z.F.; Zhang, C.J.; Cui, L.; Zhu, P.; Yan, C.; Liu, Y., *J. Anal. Appl. Pyrolysis.* **2017**, *123*, 216–223.
(96) "Hypophosphorous Acid Cross-Linked Layer-by-Layer Assembly of Green Polyelectrolytes on Polyester-Cotton Blend Fabrics for Durable Flame-Retardant Treatment", Pan, Y.; Liu, L.; Wang, X.; Song, L.; Hu, Y., *Carbohydr. Polym.* **2018**, *201*, 1–8.
(97) "Controlling the Melt Dripping of Polyester Fabrics by Tuning the Ionic Strength of Polyhedral Oligomeric Silsesquioxane and Sodium Montmorillonite Coatings Assembled Through Layer by Layer", Carosio, F.; Di Pierro, A.; Alongi, J.; Fina, A.; Saracco, G., *J. Colloid Interface Sci.* **2018**, *510*, 142–151.
(98) "Flame Retardant Cellulosic Fabrics via Layer-by-Layer Self-Assembly Double Coating with Egg White Protein and Phytic Acid", Liu, X.; Zhang, Q.; Peng, B.; Ren, Y.; Cheng, B.; Su, X.; He, J.; Lin, S., *J. Clean. Prod.* **2020,** *243*, 118641. DOI:10.1016/j.jclepro.2019.118641
(99) "Intumescent Multilayer Nanocoating, Made with Renewable Polyelectrolytes, for Flame-Retardant Cotton", Laufer, G.; Kirkland, C.; Morgan, A.; Grunlan, J.C. *Biomacromolecules.* **2012**, *13*, 2843–2848.
(100) "Starch-Based Layer by Layer Assembly: Efficient and Sustainable Approach to Cotton Fire Protection", Carosio, F.; Fontaine, G.; Alongi, J.; Bourbigot, S., *ACS Appl. Mater. Interfaces.* **2015**, *7*, 12158–12167. https://doi.org/10.1021/acsami.5b02507
(101) "Green and Scalable Fabrication of Core-Shell Biobased Flame Retardants for Reducing Flammability of Polylactic Acid", Xiong, Z.; Zhang, Y.; Du, X.; Song, P.; Fang, Z., *ACS Sustain. Chem. Eng.* **2019**, *7*, 8954–8963. DOI:10.1021/acssuschemeng.9b01016
(102) "A Bio-Resourced Phytic Acid/Chitosan Polyelectrolyte Complex for the Flame Retardant Treatment of Wool Fabric", Cheng, X.-W.; Guan, J.-P.; Yang, X.-H.; Tang, R.-C.; Yao, F., *J. Clean. Prod.* **2019,** *223*, 342–349.
(103) "Green DNA-Based Flame Retardant Coatings Assembled Through Layer by Layer", Carosio, F.; Di Blasio, A.; Alongi, J.; Malucelli, G., *Polymer.* **2013,** *54*, 5148–5153. DOI:10.1016/j.polymer.2013.07.029
(104) Egg White Protein-Hypophosphorous Acid-Based Fire Retardant Single Bilayer Coating Assembly for Cotton Fabrics", Vishwakarma, A.; Reddy, V.J.; Kandola, B.K.; Kumar, V.; Dasari, A.; Chattopadhyay, S., *Cellulose.* **2021,** *28*, 10689–10705. DOI:10.1007/s10570-021-04208-8
(105) "Grafting of Rayon Fabrics with Phosphorus Containing Polymers in Cold Plasma in Order to Obtain Flame-Retardant Materials", Simionescu, C.I.; Dénes, F.; Macoveanu, M.M.; Cazacu, G.; Totolin, M.; Percec, S.; Balaur, D., *Cellul. Chem. Technol.* **1980,** *14*, 869–883.
(106) "Plasma-Induced Graft-Polymerisation of Flame Retardant Monomers onto PAN Fabrics", Tsafack, M.J.; Levalois-Grützmacher, J., *Surf. Coat. Technol.* **2006,** *200*, 3503–3510.
(107) "Towards Multifunctional Surfaces Using the Plasma-Induced Graft-Polymerization (PIGP) Process: Flame and Waterproof Cotton Textiles", Tsafack, M.; Levalois-Grützmacher, J., *Surf. Coat. Technol.* **2006,** *201 (6)*, 2599–2610.
(108) "Flame Retardancy of Cotton Textiles by Plasma-Induced Graft-Polymerization (PIGP)", Tsafack, M.; Levalois-Grützmacher, J., *Surf. Coat. Technol.* **2007**, *201 (12)*, 5789–5795.

(109) "Surface Nanomodification of Cotton Fiber for Flame Retardant Application", Paosawatyanyong, B.; Jermsutjarit, P.; Bhanthumnavin, W., *J. Nanosci. Nanotechnol.* **2012,** *12*, 748–775.
(110) "The Use of Plasma Induced Polymerization Technology to Develop Fire Retardant Textiles", Vannier, A.; Duquesne, S.; Bourbigot, S.; Delobel, R.; Magniez, C.; Vouters, M., In: *Proceeding of International Conference on Textile Coating & Laminating*, November 2006, pp. 8–29, Barcelona.
(111) "Application of Low-Frequency Oxygen Plasma Treatment to Polyester Fabric to Reduce the Amount of Flame Retardant Agent", Ömeroğulları, Z.; Kut, D., *Text. Res. J.* **2012,** *82*, 613–621.
(112) "Application of a Novel Plasma-Based Method of Introducing Maleic Acid and Pentaerythritol Phosphate Urea Salt Onto Polyester Fabrics for Durable Flame-Retardant Treatment", Zhou, T.; Wang, Y.; Xu, H.; Cai, L.; Liu, Y.; Zeng, C., *Text. Res. J.* **2019,** *89 (23–24)*, 4916–4928.
(113) "Surface Modification of Fabrics for Improved Flash-Fire Resistance Using Atmospheric Pressure Plasma", Horrocks, A.R.; Nazaré, S.; Masood, R.; Kandola, B.K.; Price, D., *Polym. Adv. Technol.* **2011**, *22*, 22–29.
(114) "Optimization of the Procedure to Burn Textile Fabrics by Cone Calorimeter: Part II. Results on Nanoparticle-Finished Polyester", Tata, J.; Alongi, J.; Frache, A., *Fire Mater.* **2012,** *36*, 527–537.
(115) "Environmentally Friendly Flame-Retardant Materials Produced by Atmospheric Pressure Plasma Modifications", Totolin, V.; Sarmadi, M.; Manolache, S.O.; Denes, F.S., *J. Appl. Polym. Sci.* **2012,** *124*, 116–122.
(116) "Effect of zinc oxide on flame retardant finishing of plasma pre-treated cotton fabric", Lam, Y.L.; Kan, C.W.; Yuen, C.W.M., *Cellulose.* **2011,** *18*, 151–165. DOI 10.1007/s10570–010–9466-y.
(117) "Towards Flame Retardant Cotton Fabrics by Atmospheric Pressure Plasma-Induced Graft Polymerization: Synthesis and Application of Novel Phosphoramidate Monomers", Edwards, B.; El-Shafei, A.; Hauser, P.; Malshe, P., *Surf. Coat. Technol.* **2012,** *209*, 73–79.
(118) "Method and Apparatus for Surface Treatment of Materials Utilizing Multiple Combined Energy Sources", Mistry, P.; Turchan, J., US Patent 9,309, 619 B2, 12 April 2016.
(119) "The Impact of Atmospheric Plasma/UV Laser Treatment on the Chemical and Physical Properties of Cotton and Polyester Fabrics", Ayesh, M.; Horrocks, A.R.; Kandola, B.K., *Fibers.* **2022,** *10*, 66. DOI:10.3390/fib10080066
(120) "Combined Atmospheric Pressure Plasma and UV Surface Functionalisation and Diagnostics of Nylon 6.6 Fabrics", Weclawski, B.T.; Horrocks, A.R.; Ebdon, J.R.; Mosurkal, R.; Kandola, B.K., *Appl. Surf. Sci.* **2021,** *562*, 150090. DOI:10.1016/j.apsusc.2021.150090
(121) "The Effect of Combined Atmospheric Plasma/UV Treatments on Improving the Durability of Flame Retardants Applied to Cotton", Ayesh, M.; Horrocks, A.R.; Kandola, B.K, *Molecules.* **2022,** 27, 8737. DOI:10.3390/molecules27248737
(122) "Novel Flame-Retardant Treatments for Cotton and Polyester Fabric Using Atmospheric Plasma Technology", Ayesh, M., PhD Thesis, University of Bolton, Bolton, 2020.
(123) "A Review of Flame Retardant Polypropylene Fibers", Zhang, S.; Horrocks, A.R., *Prog. Polym. Sci.* **2003,** *28*, 1517–1538.
(124) "Flame-Retarding Agents and Their Use in the Preparation of Fire-Proof Thermoplastic Polymers", Herwig, W.; Kleiner, H.J.; Sabel, H.D., European Patent 0006568, 1980.
(125) "The System Polyamide/Sulfamate/Dipentaerythritol: Flameretardancy and Chemical Reactions", Lewin, M.; Brozek, J.; Martens, M., *J. Polym. Adv. Technol.* **2003**, *13 (10–12)*, 1091–1102.
(126) "Flammability of Polyamide 6 Using the Sulfamate System and Organo-Layered Silicate", Lewin, M.; Zhang, J.; Pearce, E.; Gilman, J., *J. Polym. Adv. Technol.* **2007**, *18 (9)*, 737–745.
(127) "Effects of Nanoparticles on the Flame Retardancy of the Ammonium Sulphamate-Dipentaerythritol Flameretardant System in Polyamide 6", Dahiya, J.B.; Kandola, B.K.; Sitpalan, A.; Horrocks A.R., *Polym. Adv. Technol.* **2013**, *24 (4)*, 398–406.
(128) "Potential Applications of Nanocomposites for Flame Retardancy", Horrocks, A.R.; Kandola, B.K. In: *Flame Retardant Polymer Nanocomposites*, Morgan, A.B.; Wilkie, C.A., editors, Chapter 11, John Wiley & Sons, Inc., Hoboken, NJ, 2007.
(129) "(Plasticized) Polylactide/Clay Nanocomposite Textile: Thermal, Mechanical, Shrinkage and Fire Properties", Solarski, S.; Mahjoubi, F.; Ferreira, M.; Devaux, E.; Bachelet, P.; Bourbigot, S.; Delobel, R.; Coszach, P.; Murariu, M.; Da Silva Ferreira, A.; Alexandre, M.; Degée, P.; Dubois, P., *J. Mater. Sci.* **2007,** *42 (13)*, 5105–5117.
(130) "Effect of Different Compatibilizers on Nanoclay Dispersion, Thermal Stability and Burning Behavior of Polypropylene—Nanoclay Blends", Kandola, B.K.; Smart, G.; Horrocks, A.R.; Joseph, P.; Zhang, S.; Hull, T.R.; Ebdon, J.; Hunt, B.; Cook, A., *J. Appl. Polym. Sci.* **2008,** *108 (2)*, 816–824.
(131) "Polypropylene Fibers Containing Dispersed Clays Having Improved Fire Performance. I. Effect of Nanoclays on Processing Parameters and Fiber Properties", Horrocks, A.R.; Kandola, B.K.; Smart, G.; Zhang, S.; Hull, T.R., *J. Appl. Polym. Sci.* 2007, *106 (3)*, 1707–1717.
(132) "Polyhedral Oligomeric Silsesquioxanes: Application to Flame Retardant Textiles", Bourbigot, S.; Le Bras, M.; Flambard, X.; Rochery, M.; Devaux, E.; Lichtenhan, J. In: *Fire Retardancy of Polymers: New Applications of Mineral Fillers*, Le Bras, M.; Bourbigot, S.; Duquesne, S.; Jama, C.; Wilkie, C.A., editors, Royal Society of Chemistry (Pub), Cambridge, 2005. pp. 189–201.
(133) "Polymer/MWNTs Nanocomposite Yarns and Fabrics: Processing, Characterization and Flammability and Thermal Properties", Bellayer, S.; Bourbigot, S.; Flambard, X.; Rochery, M.; Gilman, J.W.; Devaux, E. In: *Proceedings of the 4th AUTEX Conference*, ENSAIT, Roubaix, 2004, O-3W1.
(134) "Flammability of Polyamide-6/Clay Hybrid Nanocomposite Textiles", Bourbigot, S.; Devaux, E.; Flambard, X., *Polym. Degrad. Stab.* **2001,** *75 (2)*, 397–402.
(135) "High Performance Fire Retardant Synthetic Fibers Incorporating Nanocomposites", Ratnayaka, A., MSc Thesis, University of Bolton, Bolton, May 2007.
(136) Flame Retardant Polyamide Fibers: The Challenge of Minimising Flame Retardant Additive Contents with Added Nanoclays", Horrocks, A.R.; Sitpalan, A.; Zhou, C.; Kandola, B.K., *Polymers.* **2016**, *8*, 288. DOI:10.3390/polym8080288
(137) "Design and Characterisation of Bicomponent Polyamide 6 Fibers with Specific Locations of Each Flame Retardant

Component for Enhanced Flame Retardancy", Horrocks, A.R.; Sitpalan, A.; Kandola, B.K., *Polym. Test.* **2019,** *79*, 106041. DOI:10.1016/j.polymertesting.2019.106041

(138) "Heat Resistance and Flammability of High Performance Fibers: A Review", Bourbigot, S.; Flambard, X., *Fire Mater.* **2002,** *26 (4–5)*, 155–168.

(139) "Technical Fibers for Heat and Flame Protection", Horrocks A.R. In: *Handbook of Technical Textiles* (2nd edition), Horrocks A.R.; Anand S.C., editors. Elsevier, Amsterdam, 2016. pp. 237–270.

(140) Weller, D.E. (to Owens Corning), European Patent Application 1,771,615, 2007.

(141) "Thermal Protective Performance of Heat-Resistance Fabrics in Various High Intensity Heat Exposures", Lee, Y.M.; Barker, R., *Text. Res. J.* **1987,** *57 (3)*, 123–132.

28 Flame-Retardant Design and Protection for Electrical and Electronic Systems

Sergei Levchik, Eyal Eden, and Lein Tange, ICL-IP

28.1 FIRE HAZARDS IN ELECTRICAL AND ELECTRONIC EQUIPMENT

Any electrical or electronic (E&E) device has circuit components in it. The electric power which can cause overheating, overloading, sparking, or arcing is the main ignition source in these devices. Under normal use conditions E&E devices are safe, however there is always a chance of manufacturing defects, loose connections, misuse, or abuse that can lead to a short circuit and ignition. Because thermoplastics and thermosets are excellent electric insulators, all domestic electronic devices, appliances, and electric tools as well as office equipment have printed circuits, connectors, cables, frame components, and housings made of polymeric materials. A fire can be caused internally within malfunctioning components, or it can be caused externally for example by a candle accidentally getting too close or directly contacting the equipment.[1] Both ignition sources are typically small and therefore the most cost-effective way to prevent accidental fires is by engineering design or chemical or physical ignition preventive measures. Use of flame-retardant polymers is a common passive fire protective practice[2] for E&E equipment. Significant use of polymeric materials has only been possible with simultaneous development and application of flame retardants.[3]

In a typical residential house or business office environment electronics are minor contributors to fuel load compared to furnishing and bedding. In general terms large E&E equipment and appliances should not contribute more to fire growth than other surrounding materials and structural components. Therefore, passive fire protection of E&E products is applied only to the early stages of fire, most commonly by preventing ignition or reducing flame propagation rate and rate of heat release if ignition occurs. On the other hand, E&E equipment is potentially the first item to ignite, and if not adequately protected, it can become a big enough ignition source to ignite other items in the room. For example, a recent computer modeling study[4] shows that a flat screen TV with a casing not protected by flame retardants is 4 to 15 times more likely to cause flashover in the living room or bedroom compared to a similar fire protected TV. Hong et al.[5] performed fire tests on eight halogen-free and one halogen-containing resins, some of which are used in computer monitors and others in TV housings that were rated V-0, V-1, V-2, and HB. They exposed 19 inch monitors and 25 inch TVs to the fire of methenamine tablet, candle, and paper ball. V-0 and V-1 rated items extinguished after short period of time whereas V-2 and HB rated items continue to burn and spread fire by dripping.

In 2000 a group of European researchers[6] expressed a concern over increasing number of fires caused by TV sets which they attributed to the relaxing requirements fire safety for TV enclosures due to switching from V-0 rated plastics to HB rated. The risk of use of flame retardants in consumer products like furniture and electronics due to the exposure to chemicals versus the benefits of a decreasing number of fire deaths, significant injuries, and monetary loses were reviewed by Stevens.[7,8] He concluded that many commercial flame retardants went through required human health and environmental testing before entering the market and therefore the balance of risk for death and injury favors the use of flame retardants for those compounds whose risks have been assessed. The monetary benefit of the use of flame retardants in TVs based on Life Cycle Analysis was presented by Simonson et al.[9]

28.2 FIRE TESTS

In this chapter the most common fire tests used in E&E devices are briefly reviewed. Although there are different levels of fire testing including final testing of the fully assembled devices, we will focus only on the fire tests which are important and widely used by the developers of polymeric materials for electronics. These tests intend to simulate as closely as possible real ignition risks. E&E equipment can operate only if it complies with the specific safety regulations and therefore polymeric material used as insulation or structural components should pass one or more fire tests. These are usually considered pre-selection reaction-to-flame tests which are performed on small scale standard size specimens. Although the tests are not performed on real polymeric parts, they are very useful because they allow the comparison of different materials and the performance of quality controls. Fire testing results along with mechanical and electrical properties allow engineers to design final devices.

As defined by Babrauskas and Simonson,[10] fire safety of E&E devices should meet three main objectives:

DOI: 10.1201/9781003380689-28

(1) Electrical components of the device should be designed to minimize the possibility of overheating or of starting a fire.
(2) Where there is no assurance that an electrical fault will not occur and lead to overheating or fire, effective barriers or enclosures must be provided so that fire will not propagate and ignite external objects.
(3) If the device has combustible external parts, small external ignition sources impinging on the appliance must not cause a large flaming fire.

Along with numerous national standardization bodies and organizations, there are two international organizations which are in charge of harmonization of national standards, thus helping to reduce trade barriers and promote safety on a global basis. The International Organization for Standardization (ISO) is responsible for all technical fields excluding electrical technology, on the other hand the International Electrotechnical Commission (IEC) is responsible for standards in the field of E&E engineering including fire safety. The key publication is IEC 60695 which outlines Combustion Characteristic Tests for Preselection of Materials in section 1–30. Please see Chapters 17 and 18 in this book for more information on fire test methods.

The most used ignition test is the 50 W vertical flame test IEC 60695–11–10 also known as the UL-94 vertical test developed in the late sixties by Underwriter's Laboratories Inc. in North America.[11] The flammability requirement classifications of UL-94 test were introduced in the 70s. UL fire safety standards have been developed with the participation of manufacturers, consumers, government officials, inspection authorities, and insuring companies. Because of this many UL standards have been adopted in the international standards and regulations. UL has testing laboratories and representative offices around the globe. The two most common ways UL certifies products and components are Listing and Recognition.[12] A UL listing means that UL has evaluated a representative sample of a product and found it to be in compliance with UL requirements. UL Component Recognition means that UL evaluated component parts, such as plastic materials, and certify that they can be used in a complete product or system. It should be noted that the Plastics Recognition Program has found worldwide acceptance.[13] Over 60,000 materials have been analyzed, not only for small open flame ignition resistance, but also for other electrical ignition scenarios.

The general layout of the UL-94 vertical test is shown in Figure 28.1. This test should be performed in an enclosure to minimize air drafts. It requires two sets of five specimens with a measurement of 125 × 13 mm. The thickness is defined by the thickness of the part in the final application. The most common thicknesses tested are 3.2, 1.6, 0.8, and 0.4 mm. Typically, the thinner the specimen, the more difficult it is to pass the test. One set of specimens is stored for at least 48 hours at 23°C and 50% relative humidity, whereas the second set is stored for 168 hours in an air-circulation oven at 70°C followed by four hours cooling to room temperature in a desiccator. Surgical cotton is placed 305 mm below the lower end of the vertically mounted specimen. A laboratory burner specified by UL uses propane gas to produce a 20 mm high blue flame which represents a 50 W ignition source. The flame is applied two times for a duration of 10 sec to each specimen. The second application occurs as soon as the specimen extinguishes after the first application. There are three classifications in the

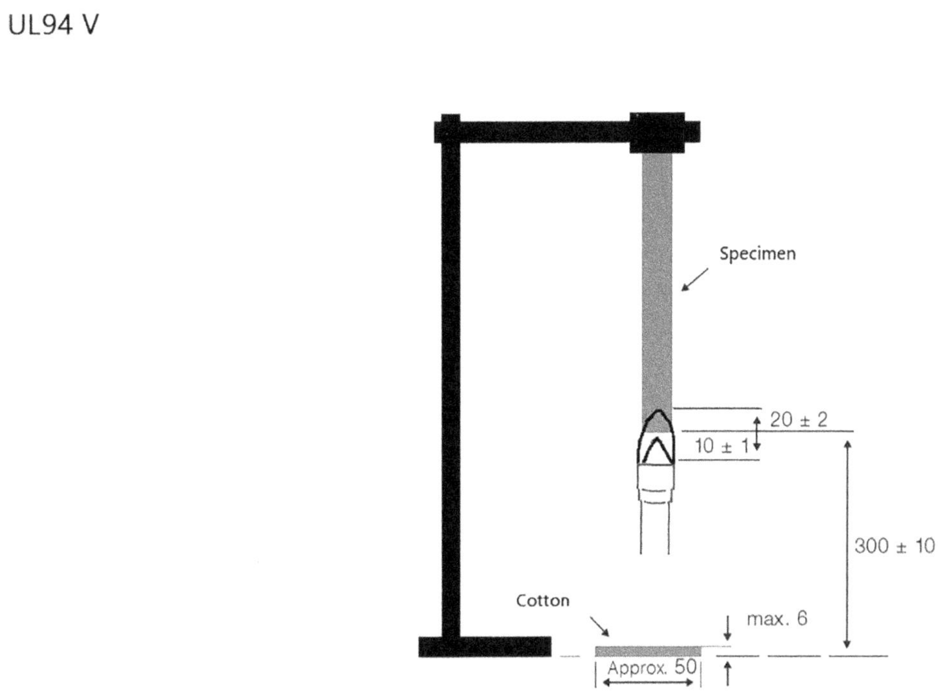

FIGURE 28.1 Layout of UL-94 vertical 50 W flame test.

UL-94 vertical test. The material is classified as V-0 if the single specimen afterflame time is ≤ 10 sec and the total set afterflame time ≤ 50 sec; the single specimen afterflame time and afterglow time is ≤ 30 sec; no specimens burn to the clamp, and there is no cotton ignition. The material is classified as V-1 if the single specimen afterflame time ≤ 30 sec and the total afterflame time ≤ 250 sec; the single specimen afterflame and afterglow time ≤ 50 sec; no specimens burn to the clamp, and there is no cotton ignition. The material is classified as V-2 if all of the criteria of the V-1 classification are met, with the exception of cotton ignition, which is permitted. In principle the UL-94 vertical test can be applied for the screening of solid plastic materials for any applications apart from building and construction and wires and cables.

It is important to keep in mind that UL-94 ratings were developed for a specific ignition scenario. For example, it was shown[14] that if ignition time is extended to 20 sec. some V-1 rated materials can perform better than some V-0 rated. Another point to make is that materials classified V-0, V-1, or V-2 should have a robust margin to achieve this rating. If afterflame or afterglow time and flammable dripping performance is borderline the rating becomes operator dependent. For example, shortening the flame application time by just 0.5 sec can significantly shorten afterflame time.[15] Faster than prescribed removal of the flame can cause immediate extinguishing in contrast with the development of a small continuous flame on the specimen.

Since the UL-94 vertical test takes into consideration dripping this phenomenon can be used as an advantage to pass the test. Wang et al.[16–18] studied in great detail the combustion of polymers in a UL-94 test configuration. By measuring the heat flux delivered by a Bunsen burner and taking into consideration the edge effect, they built a model to predict the ignition time, which correlated with other tests. They also modeled dripping as a competition between three forces, the gravity which pulls melted polymer to drip, and the melt viscosity and surface tension which prevent it from dripping. For small-size droplets, the surface melt flows downward due to gravity, and for large-size droplets, the bulk specimen bar is elongated by gravity. With some polymers, rapid decomposition can be promoted to quickly decrease the viscosity and to create very small drops which cool and extinguish before reaching the underlying cotton. A group of French researchers[19] created instrumentation which measures the weight of droplets, the weight loss of the specimen and the temperature gradients during the UL-94 test. Two flame retardant additives which promote dripping, and one intumescent flame retardant were tested in glass-reinforced and unreinforced polyamide 6. It was shown that unreinforced polyamide 6 with drip promoters loses its V-0 rating in reinforced polymer because the glass fibers delay dripping and serve as heat conductors to increase the temperature more quickly in the bulk of the polymer. Bertelli et al.[20] believe that detailed data on temperature distribution is helpful in eliminating the intrinsic ambiguity of the UL 94 classification in the case of flame retardant materials burning with dripping. Analysis of infrared camera profiles can help to gain insight in the flame retardant mechanism.[21] Every flame retarding mode of action has its own specific influence on the temperature profile.

Kempell et al.[22] identified four competing effects in UL-94 testing, for example, gasification, charring, flame inhibition, and melt flow/dripping. They used a particle finite element method model to predict the fire retardant performance of PC/ABS containing a phosphate ester flame retardant and an anti-dripping agent. The model predicted well the dripping and non-dripping phenomenon and the mass loss before extinguishment. To produce a more robust outcome from the UL-94 test apart from the pass/fail criteria Siripurapu and Agawal[23] added a statistical tool to the test, in particular such parameters as the probability of first-time pass and time-to-drip. This tool allowed them to optimize PC/ABS formulations for flammability and physical properties in a narrow interval of flame retardant loading. Interestingly, it was discovered[24] that UL-94 ratings also depended on the residual stress in the polycarbonate samples. In this study molding conditions, temperature, injection speed, and holding pressure and sample orientation (ignition from molding gate or no gate) affected UL-94 performance. More recently with the development of 3D printing techniques, it was found[25] that the dripping performance of polycarbonate depends very much on the direction of the strands laydown.

A common horizontal flammability test applied to E&E equipment is commonly known as UL-94 HB or IEC 60695–11–10. This test uses the same burner as the UL-94 V test with a 20 mm high flame corresponding to 50 W. The layout of the test is shown in Figure 28.2. The test is performed on three bars that are the same size as UL-94 V (125 × 13 mm) and the minimal thickness in the intended application. The bars have two reference marks at 25 and 100 mm. The specimens should be conditioned for 48 hours at 23°C and 50% relative humidity. The sample is mounted in a horizontal position inclined 45° in respect to the longitudinal axis. The burner is also tilted at 45°, and the burner is applied for 30 sec. If the 25 mm mark is achieved before 30 sec, the burner is removed earlier. The classification of HB 40 is achieved if one of three conditions are satisfied: (a) no visible flame after burner removal or (b) the specimen extinguishes before the 100 mm mark is achieved or (c) the burning rate between the reference marks is < 40 mm/min. The classification HB 75 is achieved if the burning rate between the reference marks is < 75 mm/min.

Testing for the classifications of 5V is the most stringent of all UL-94 tests because a more powerful 500 W burner is used. It is also known as the IEC 60695–11–20 test. A 5V rating is typically required for enclosures of large E&E weighing more than 18 kg.[26]The layout of the test is shown in Figure 28.3. The UL-94 5V test requires two sets of five bars of 125 × 13 mm size at the minimal thickness in the intended application. It also requires two sets of three plates of 150 × 150 mm size of minimal thickness. One set of specimens is conditioned for at least 48 hours at 23°C

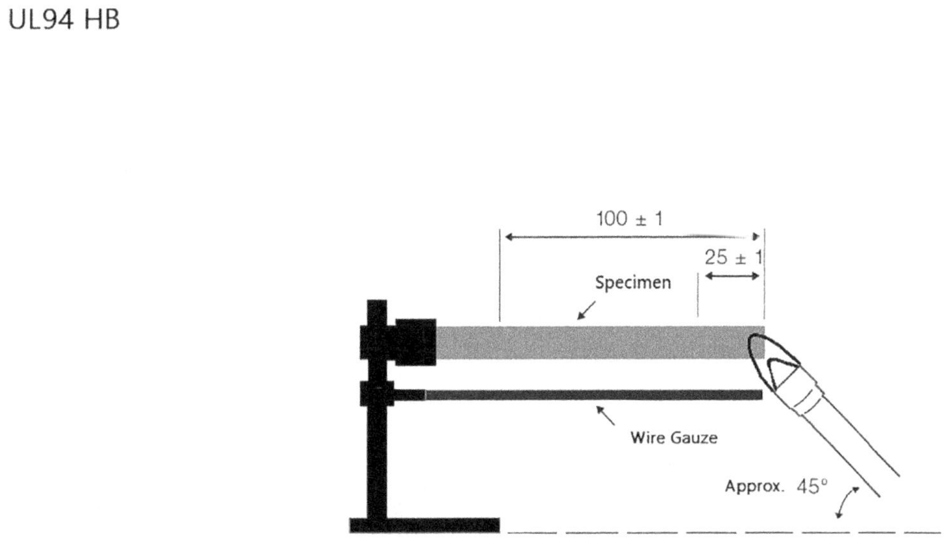

Figure 28.2 Layout of UL-94 horizontal 50 W flame test.

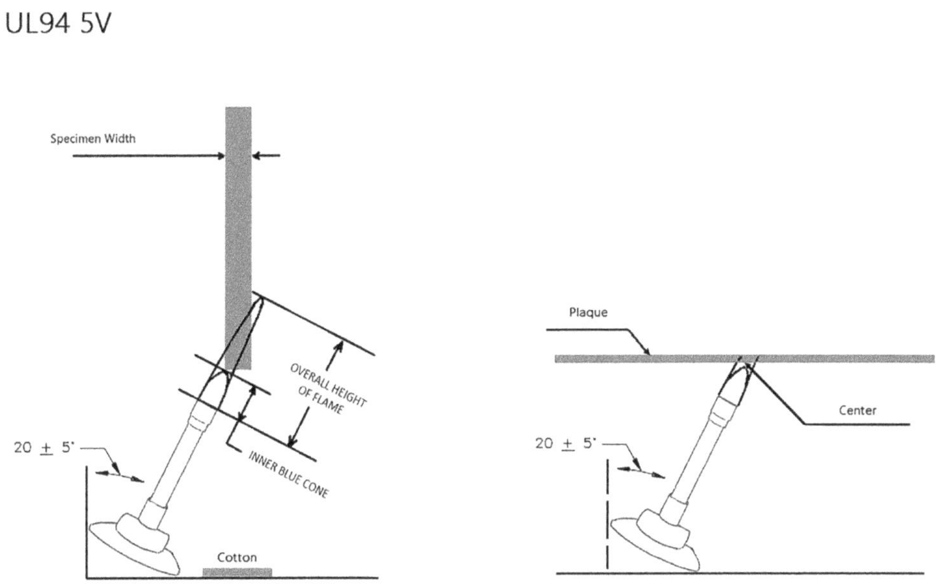

FIGURE 28.3 Layout of UL-94 5V vertical 500 W flame test.

and 50% relative humidity. The second set is stored in an air-circulating oven for 168 hours at 70°C followed by four hours cooling at room temperature. The burner is inclined at 20° to vertical with a 125 mm high and a 40 mm blue inner zone used. Surgical cotton is placed under the specimen, as shown in Figure 28.3. The flame is applied five times for five sec with five sec intervals. The material is classified as 5VA if the individual afterflame time is ≤ 60 sec after the fifth flame application on the vertical specimen, there is no ignition of the cotton, and no hole is developed in the plate specimens. The material is classified as 5VB if the individual afterflame time is ≤60 sec after the fifth flame application on the vertical specimen, there is no ignition of the cotton but a hole is developed in the plate specimens.

Another important fire test applied to E&E equipment is the Glow Wire Flammability Test Method for Materials (GWFI), IEC 60695–2–12. This test simulates the short-term effect of thermal stress which can be produced by glowing wires or overloaded resistors. The layout of the glow-wire apparatus is shown in Figure 28.4. The objective of this test is to ensure that the glow-wire does not ignite the plastic part, or if it ignites, the burn time should be short with no flame spread or falling burning or glowing debris. The glow-wire coil is 4 mm in diameter and made of a chromium-nickel alloy. The temperature of the wire is preset at 550 up to 900°C with 50°C intervals. The maximum temperature tested is 960° C. The size of the specimen is ≥ 60 mm length and ≥ 60 mm width (inside clamps).

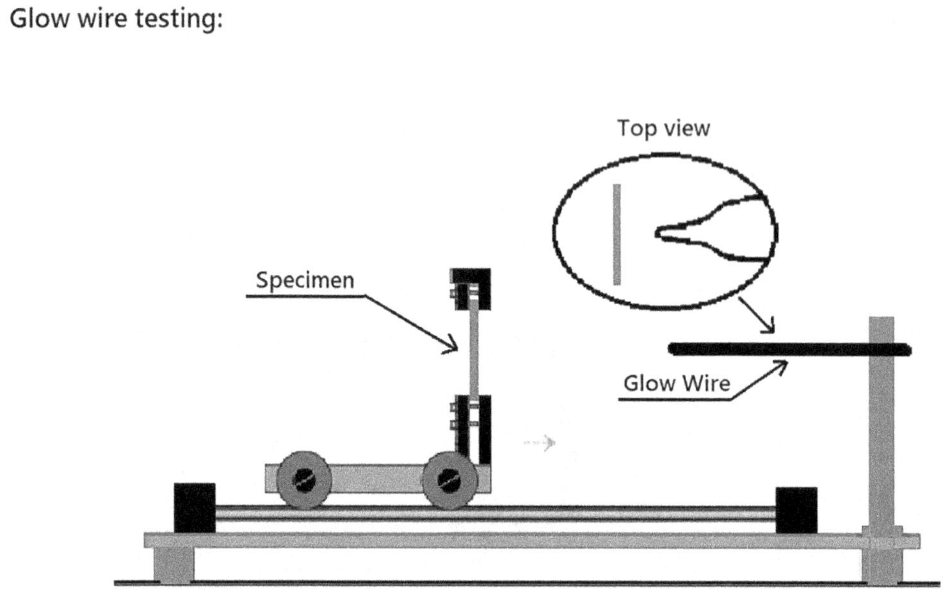

FIGURE 28.4 Layout of glow-wire test apparatus.

The thickness is not defined but preferred thicknesses are 0.1, 0.2, 0.4, 0.75, 1.5, and 3.0 mm. Typically, a set of 15 test specimens is used in vertical orientation. The glow wire is applied to the specimen at a force of 0.95 ± 0.10 N for 30 sec. The penetration of the wire into the specimen is limited to 7 mm. In order to evaluate the possibility of flame spread from falling debris, a tissue-covered wooden board is placed underneath the specimen. The test passes at a certain wire temperature if there is no ignition or (a) the flames or glowing of the specimen extinguishes < 30 sec after removal of the wire and no ignition of the wrapping tissue is observed. Then the temperature is increased by 50°C, and another specimen is tested. This is continued until the test fails or 960°C is reached.

As per UL 746 C there are three levels of fire safety of enclosures depending on the type of E&E devices.

(1) Portable attended appliances are required to have HB rated enclosure material. HB rated material will, once ignited, have a limited burning rate.
(2) Portable unattended appliances are required to have V rated enclosure material. V rated material will, once ignited, self-extinguish after a certain time.
(3) Other unattended appliances are required to have 5V-rated enclosure material. 5V material is also self-extinguishing and will additionally prevent a 500 W burner flame from penetrating the material.

As per UL 60335 some enclosures and internal parts of unattended household appliances need to pass the Glow Wire Ignition Test.

Every power cord used in E&E equipment needs to satisfy fire safety requirements. The most common test is the VW-1 (Vertical Wire) UL 1581 section 1080 test. This test is performed on 600 mm long single coated wire or small cable in a three walls sided enclosure. The cable or wire is fixed vertically. A 12.5 × 20 mm flag from Kraft paper is attached at the 425 mm mark from the lower end of the specimen. Surgical cotton is placed no further than 60 mm below the cable. A 9.5 mm in diameter Bunsen burner with additional air supply (Tirril gas burner) with a flame height of about 125 mm calibrated to a 500 W output is applied at a distance of 250 mm from the flag. The burner is inclined at 20° to the vertical axes and applied five times for 15 sec with 15 sec intervals. The wire or cable satisfies a VW-1 rating if less 25% of the flag is carbonized, the specimen burned < 60 sec after the flame application, and the cotton did not ignite.

28.3 MECHANICAL AND ELECTRICAL TESTS

Apart from flammability tests, plastic materials for E&E applications need to pass numerous mechanical and electrical tests. The addition of flame retardants to polymers can affect their mechanical and electrical properties. The changes in physical properties caused by flame retardants depend greatly on their characteristics. The most important ones are the following[27]:

(1) particle size and particle shape
(2) melting point
(3) compatibility with the polymer
(4) solubility in the polymer
(5) effective loading level

During the processing, the effects of flame retardants on physical properties can be related to whether they are: plasticizing, semiplasticizing, or nonplasticizing. Here we will briefly describe some of the most important properties which need to be maintained as close as possible to the original non-flame retardant materials after the required flammability properties have been achieved. Table 28.1 lists the most important mechanical property tests and references to the standards.

The most important mechanical properties for thermoplastics are ultimate tensile and flexural strengths, flexural modulus as a measure of stiffness, tensile elongation as a measure of flexibility, unnotched Izod impact strength (an indicator of material toughness), and notched Izod impact (a measure of notch sensitivity). In the design of the structural part, compromises between flammability resistance and mechanical properties may be necessary. In some cases, material weakness can be accommodated by changing the component thickness and geometry, but nowadays with the trend of miniaturization of electronics, this becomes difficult or impossible. Design engineers may prefer to just replace one plastic with another one with better mechanical properties even if it is more expensive. Short-term thermal stability is characterized by heat-deflection temperature (HDT). This test value is useful for selecting temperature limits on such post molding operations as annealing and soldering, but it does not necessarily relate to the long term thermal stability.

The comparative tracking index (CTI) test is used to estimate the sensitivity of the material with respect to the surface tracking. In essence it is an accelerated test to provide an indication of the performance of insulating materials under wet and contaminated conditions, where the voltage is between two electrodes on the surface of the plastic and a solution of ammonium chloride is dropped between electrodes. It is believed that there is instant conductivity between the electrodes. When the carbonaceous conductive tracks form, a continuous short conductivity heats up the polymer surface and causes charring up to the point where tracking breakdown occurs due to the carbonaceous conductive material.[28]

Low loss properties in printed wiring boards are assessed by the dielectric constant Dk (capacitance ratio) and the dissipation factor Df (tangent of the dielectric loss angle). The dielectric constant is the ratio of the capacitance with the material as the dielectric, to the capacitance of the same electrode in a vacuum as the dielectric. The dielectric constant is a measure of the ability of an insulating material to store electrical energy. A low dielectric constant value indicates that the material can store small amounts of energy. Since the loss of energy from a circuit into an insulator is generally undesirable, the best insulating materials are those having the lowest value of dielectric constant. Thermoset resins and the reinforcing material in printed wiring boards are high frequency sensitive and tend to resonate at specific frequencies. For this reason, electrical properties should always be tested over the frequency at which the material is expected to function.

The coefficient of thermal expansion (CTE) in Z-axes and the glass transition temperature (T_g) are two important parameters for the reliability of printed wiring boards (PWB). During use PWB heats and cools and as a result it expands and shrinks. Excessive expansion and shrinkage can cause the breakage of interconnections between different layers of the wiring board. So a lower CTE is beneficial for PWB reliability. The CTE increases significantly at temperatures above the T_g. Therefore, a high T_g is beneficial for PWB reliability. Another thermal stability parameter is T_d which is the temperature of 5% weight loss measured in a thermogravimetric analysis. This is the temperature at which the resin network undergoes irreversible decomposition. A high T_d is important for soldering and

TABLE 28.1
Mechanical and Electrical Tests Important for the Development of Flame Retardant Plastics and Printed Wiring Boards

Name	E&E Parts	Standard	Apparatus
Tensile strength	All solid plastic parts	ASTM D638	Tensile tester
Tensile module	All solid plastic parts	ASTM D638	Tensile tester
Elongation at break	All solid plastic parts	ASTM D638	Tensile tester
Izod impact strength	All solid plastic parts	ASTM D256	Pendulum type impact tester
Heat deflection temperature	All solid plastic parts	ASTM D648	Automatic deflection tester
Melt flow index	All solid plastic parts	ASTM D1238	Extrusion plastometer
Comparative tracing index	Plastics for connectors	ASTM D3638	Comparative tracking index tester
Dielectric constant and dissipation factor	Printed wiring boards	IPC-TM-650 2.5.5.2	Digital LCR meter
Glass transition temperature (T_g)	Printed wiring boards	IPC-TM-650 2.4.25	Differential scanning calorimeter (DSC)
Coefficient of thermal expansion (CTE)	Printed wiring boards	IPC-TM-650 2.4.24	Thermal mechanical analysis (TMA)
Decomposition temperature (T_d)	Printed wiring boards	IPC-TM-650 2.4.24.6	Thermogravimetric analyzer (TGA)
Pressure cooker test (PCT)	Printed wiring boards	IPC-TM-650 2.6.16	Autoclave and molten solder bath
Copper peel strength	Printed wiring boards	IPC-TM-650 2.4.8	Tensile tester

reworking operations where PWB can be locally exposed to high temperatures. A pressure cooker test is another test related to soldering. In this test a 10 × 10 cm fragment of PWB is exposed to saturated steam at 121°C in an autoclave for 30 min and longer and then immediately dipped into molten solder at 288°C. If the laminate absorbs moisture between plies it will pop up and blister. This test is critical for phosphorus-based flame retardants because they are more susceptible to moisture absorption compared to brominated flame retardants. Because some flame retardants especially the not reactive additive types can affect copper adhesion, copper peel strength is important for PWB reliability.

28.4 FLAME RETARDANTS IN DIFFERENT PARTS OF E&E EQUIPMENT

28.4.1 Enclosures

The main purpose of enclosures in E&E devices is the protection of users from electric power circuits inside the equipment. Because the enclosure is the object which the consumer sees and touches, it should have certain aesthetic features and good mechanical properties. The most common polymers used to manufacture the enclosures of audio, video and office machines are HIPS, ABS, PC/ABS and PPO/HIPS. Polypropylene can also be found in some home appliances.

HIPS and ABS are copolymers of styrene with other vinyl monomers such as acrylonitrile and butadiene and an impact modifier which is generally a grafted blend of a rubbery polymer. The polystyrene part of the copolymers has a tendency to depolymerize when exposed to fire temperatures, and the volatile products are materials of a high fuel value, namely, styrene monomer, styrene dimers, and related hydrocarbons such as benzene, lower-alkylbenzenes, and a few percent of oxygen-containing related aromatics.[29] HIPS and ABS unless blended with char-formers (PC or PPO), by themselves form little or no char.[30] Moreover, while depolymerization is taking place, melt flow and drip are also occurring. If flaming drips are allowed by the test standard, as in the UL-94 V-2 rating, then the dripping may be useful to meet flame retardancy requirements.

Approximately 10 wt% Br in combination with antimony trioxide is required to pass UL94 V0 requirements in HIPS and ABS. The exact amount will vary with the rubber content. Very high impact polymers have higher rubber content and may require more FR. Antimony trioxide is normally used as a synergist with bromine-based based flame retardants. The theoretical ratio would be three atoms of bromine to one atom of antimony, assuming that antimony tribromide is the desired active species but the optimum ratio is best determined experimentally.

For many years decabromodiphenyl ether (also known as decabromodiphenyl oxide) was the most widely used flame retardant for HIPS. It had great success because it was very rich in bromine (83%) and was very stable with a melting point of 305°C. However, it was banned in Europe in 2008 due to the potential risk of debromination and the formation of lower brominated diphenyl ethers which were found to be toxic. Later decabromodiphenyl ether was banned in other parts of the world. Since those times decabromodiphenylethane (Structure 28.1) replaced the ether. It has just marginally lower bromine content (82%), but a higher thermal stability with a melting point of 350°C and better UV stability.

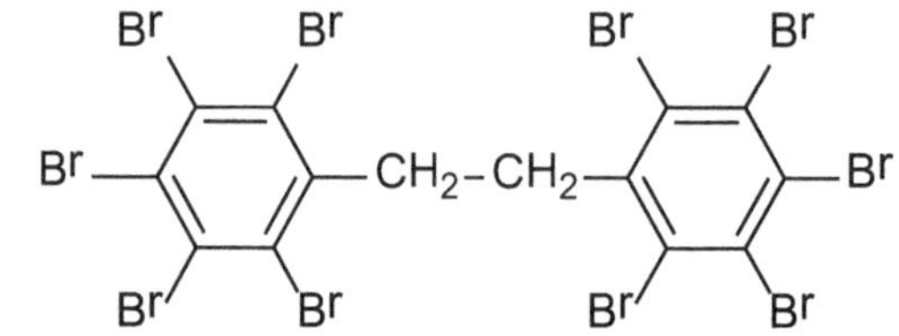

(28.1)

2,4,6-Tris(2,4,6-tribromophenoxy)-1,3,5-triazine (Structure 28.2) is another brominated flame retardant commonly used in HIPS and ABS. This compound is also high melting (230°C), but it is a melt-blendable solid with a pattern of utility rather similar to decabromodiphenylethane and with the advantages of better melt flow, impact strength and light stability.[31–33] The bromine content of tris(tribromophenoxy) triazine is 67%, which is lower than that of decabromodiphenylethane. An important feature of tris(tribromophenoxy) triazine is that it can be used to reach UL-94 V-0 in HIPS and ABS with little antimony trioxide, with the assistance of a very small amount of polytetrafluoroethylene (PTFE) to retard drips.[34,35] This becomes possible because tris(tribromophenoxy)triazine melts at the processing temperature and achieves better contact with antimony trioxide. A combination of this melt-blendable flame retardant (which aids melt flow) with a filler-like flame retardant such as decabromodiphenylethane (which can reduce melt flow) enables the molding of large dimension housings of electronic equipment made of HIPS, without the problems of melt flow and poor impact (Table 28.2). Tris(tribromophenoxy)triazine has a very favorable impact strength rating, over twice as high compared to decabromodiphenylethane by the falling ball test. Both good flow and high impact strength are probably due to its solubility in styrenics.

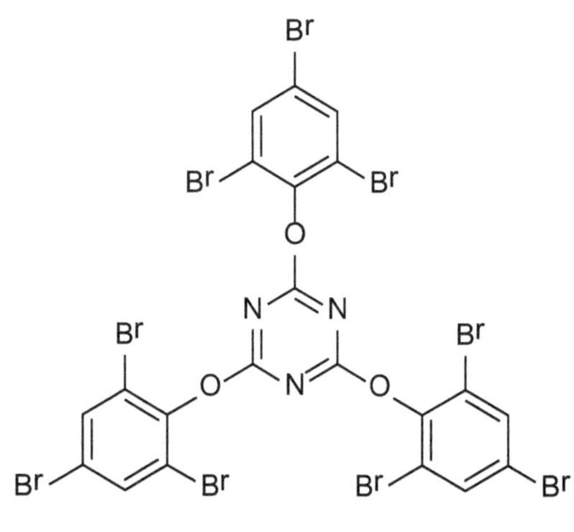

(28.2)

TABLE 28.2
Melt Flow and Impact Properties of 2,4,6-tris (2,4,6-tribromophenoxy)-1,3,5-triazine and Decabromodiphenyl Oxide Based HIPS

Flame Retardant	MFI at 200°C, 5 kg	Gardner Impact kg-cm
2,4,6-Tris(2,4,6-tribromophenoxy)-1,3,5-triazine	10	100
2,4,6-Tris(2,4,6-tribromophenoxy)-1,3,5-triazine and decabromodiphenylethane (1:1)	8	81
Decabromodiphenylethane	5	62

Tetrabromobisphenol A was another brominated flame retardant which was used in ABS because it improved resin flow. However, it was regulated recently in Europe in non-reactive applications and practically banned from use in ABS. Other geographical regions are considering a similar ban as well. However, tetrabromobisphenol A remains the predominant flame retardant in epoxy printed wiring boards (see next section).

Other brominated additives especially suitable for HIPS and ABS are brominated epoxy polymers which contain very little or no detectable levels of free tetrabromobisphenol A. There are two types of brominated epoxy polymers and oligomers: uncapped (Structure 28.4) or end-capped with 2,4,6-tribromophenol (Structure 28.5). These products are made from tetrabromobisphenol A diglycidyl ether and tetrabromobisphenol A via a chain extension process. They are made at different molecular weights from 700 up to 68,000 and depending on the molecular weight have different softening/melting temperatures and are suitable for different thermoplastics. Polymers with a molecular weight 1400–2000 and a softening temperature of about 100–120°C are especially suitable for HIPS and ABS. Uncapped polymers have 50–52% bromine content and end-capped have 56–60% bromine content. Uncapped low molecular weight polymers in addition to flame retardancy provide a heat stabilization effect due to acid scavenging by the epoxy groups. These polymers provide excellent flow properties, good thermal dimensional stability, and good light stability. Guidance for UL-94 V-0 formulations in HIPS including low antimony formulations are shown in Table 28.3 and for ABS in Table 28.4.

Where high light stability (color stability) is needed, ethylenebis(tetrabromophthalimide) (Structure 28.3) can be used. Besides good light stability, this additive resists bloom and plate-out, but it tends to reduce impact strength. Because of relatively higher cost, it is rarely used in HIPS and ABS but is used in more demanding applications (see next section).

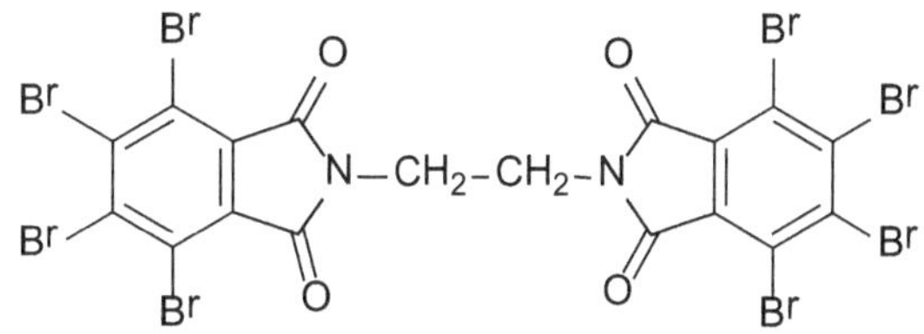

(28.3)

A decade ago, flame-retarded HIPS was mostly used for TV and monitor back panels and ABS seemed to be the choice for personal computers, printers, and scanners.[26] Keyboard housings/enclosures were being made from non-flame retarded HIPS, and the key buttons were made from non-flame retarded ABS. Flame retarded PC/ABS was mostly used for larger machines such as servers and main frames and some high-end laptops and display enclosures. Display housings/enclosures were produced from flame retarded ABS and PC/ABS. However, this started to change mostly because the cost gap between ABS and PC/ABS became much smaller. Another reason is because of eco-labeling standards, which have been adopted recently in Europe and are under discussion in some North American states. Although at a smaller scale compared to PC/ABS, PPO based alloys started to also increase their market share in E&E equipment.

Aromatic phosphates or aromatic phosphate oligomers (mostly diphosphates) are widely used in PC/ABS blends. Triphenyl phosphate (TPP) was the first phosphorus-based flame retardant used in PC/ABS. Although TPP is soluble in PC/ABS and doesn't bloom out at room temperature, it deposits on the mold surfaces during molding. Because of the low melting point of TPP (48°C), it leads to bridging at extrusion feeding ports. The next generation of aromatic phosphate FR in PC/ABS was isobutylated-phenyl phenyl phosphates (Structure 28.6). These are mixtures of triaryl phosphates of different degrees of butylation and also contain free triphenyl phosphate. These phosphates are liquid and somewhat less volatile than triphenyl phosphate.

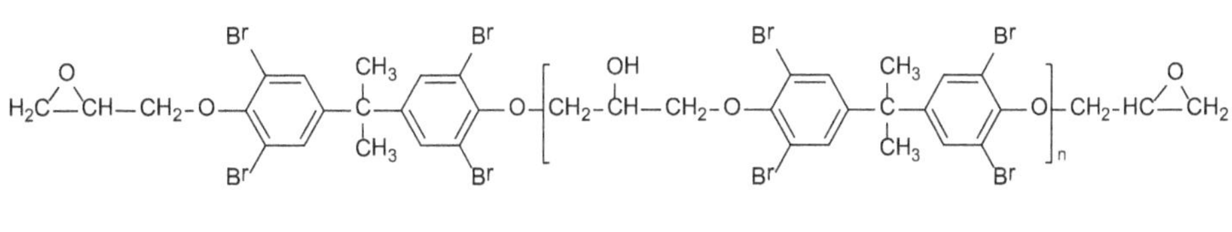

(28.4)

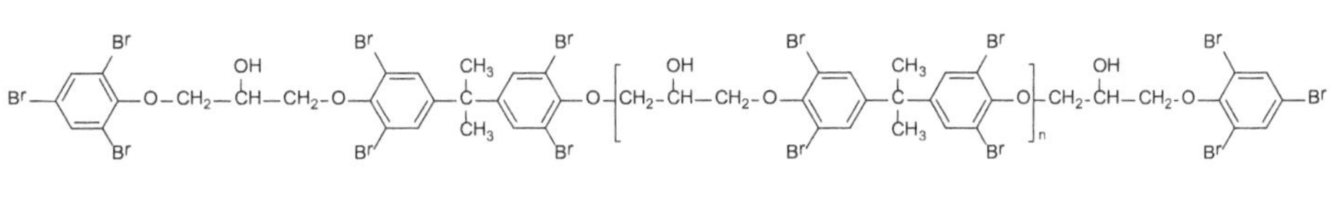

(28.5)

TABLE 28.3
Bromine-Based UL94 V-0 Formulations for HIPS

FR Type>	Units	Decabromodiphenyl-ethane	Tris(tribromophenoxy) triazine	End-Capped Brominated Epoxy Mw = 1400	End-Capped Brominated Epoxy Polymer Mw = 2000
HIPS	%	83.5	80.8	79.0	78.2
FR	%	12.2	14.9	16.7	17.5
Sb_2O_3	%	4	4	4	4
PTFE	%	0.1	0.1	0.1	0.1
Antioxidant	%	0.2	0.2	0.2	0.2
Sum	%	100.0	100.0	100.0	100.0
Br [calculated]	%	10	10	10	10
UL 94 1.6 mm rating		V-0	V-0	V-0	V-0
Izod notched impact	J/m	80–90	80–90	60–70	50–60
Tensile strength	MPa	24	27	26	26
Elongation at break	%	42	38	40	45
Tensile modulus	MPa	2,010	1,990	1,880	1,980
MFI at 200°C, 5 kg	g/10 min	6	10	14	17
HDT 1.8 MPa, 3.2 mm	°C	67	66	65	71
Initial color vs. QUV 500h	ΔE	26	21	29	17

TABLE 28.4
Bromine-Based UL94 V-0 Formulations for ABS

FR Type >	Units	Tetrabromobisphenol A	Tris(tribromophenoxy) triazine	End-Capped Brominated Epoxy Polymer Mw = 2000	Decabromodiphenyl-ethane
ABS	%	78.5	80.7	78.1	83.4
FR	%	17.1	14.9	17.5	12.2
Sb_2O_3	%	4	4	4	4
PTFE	%	0.2	0.2	0.2	0.2
Antioxidant	%	0.2	0.2	0.2	0.2
Sum	%	100.0	100.0	100.0	100.0
Br [calculated]	%	10	10	10	10
UL 94 1.6 mm rating		V-0	V-0	V-0	V-0
Izod notched impact	J/m	95	130	100	125
Tensile strength	MPa	1,910	1,960	2,000	2,050
Elongation at break	%	10	5	5	9
Tensile modulus	MPa	44.7	44.3	45.1	43.6
MFI at 220°C, 5 kg	g/10 min	8.3	4.2	6.8	2.4
HDT 1.8 MPa, 3.2 mm	°C	71	76	74	78
Gloss	60°	52	40	56	35
Initial color vs. QUV 168h	ΔE	37	33	33	30

Recently isobutylated-phenyl phenyl phosphates have been used mostly in PPE/HIPS and often as processing aids not as flame retardants.

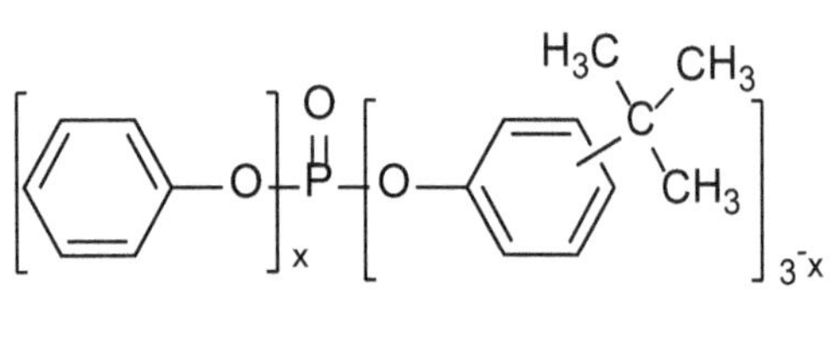

(28.6)

There is a large choice of oligomeric aromatic phenyl phosphates which are finding broader application than monophosphates because of better thermal stability and lower volatility.[36] The first product which became commercial was resorcinol bis(diphenyl phosphate) (RDP, Structure 28.7) which is a mixture of oligomers with two to five phosphorus atoms, but with the distribution heavily shifted towards the diphosphate.[37] In commercial PC/ABS blends where ABS content normally does not exceed 25%, RDP gives a V-0 rating at 8–12 wt% loading.[38] PTFE is usually added at < 0.5 wt% to retard dripping. RDP is somewhat less hydrolytically stable compared to other bisphosphates, which limits its application in humid environments and may cause a problem in recycling. Because of this shortcoming nowadays RDP is mostly used in PPE/HIPS which is not sensitive to acid formation.

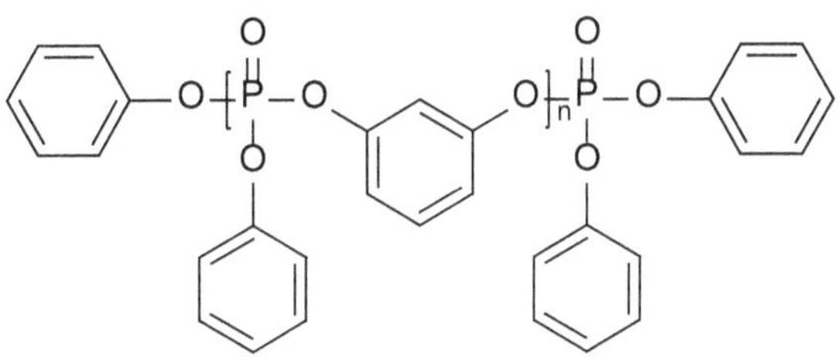

(28.7)

Bisphenol A bis(diphenyl phosphate) (BDP) (Structure 28.8) was introduced to the market as an alternative to RDP.[39] Since bisphenol A is less expensive that resorcinol, BDP is more cost efficient despite a lower phosphorus content (8.9% P for BDP vs. 10.7% P for RDP). On the negative side BDP is significantly more viscous than RDP (12500 cP for BDP vs. 600 cP for RDP at 25°C), and therefore it requires a heated storage tank and heated transfer lines. Because of the high viscosity, the oligomers content (n > 1, Structure 28.8) in the BDP mixture is usually limited to only 10–15% which creates a problem of potential crystallization in cold weather. Due to favorable cost-efficiency, BDP became the major product used in PC/ABS and the second largest phosphorus-based flame retardant produced. On the positive side BDP has better hydrolytic stability than RDP[40] and can be used in high humidity applications. PC/ABS with an ABS content less than 25 wt% usually needs more than 12 wt% BDP plus a small co-addition of PTFE to assure a V-0 rating.[41]

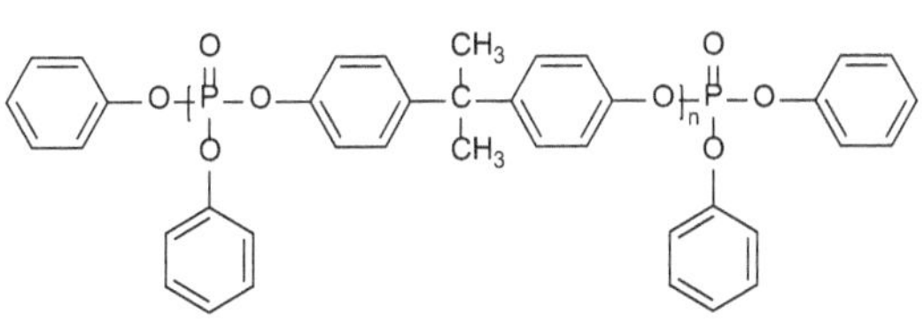

(28.8)

Although major compounders of PC-based blends are likely to be well equipped with liquid feeding systems, small- and medium-size compounders prefer to use solid bisphosphates even if they cost more than BDP. Very close to RDP, hydroquinone bis(diphenyl phosphate) (HDP, Structure 28.9) when made relatively pure with a low TPP content and a low oligomers content is a solid with a melting point of 105–108°C. It can be fed into the extruder without extensive cooling of the feeding zone, and therefore some large compounders also use this product where the flexibility of changing extrusion lines is desirable. HDP has a phosphorus content of 10.8%, is similar to BDP in hydrolytic stability, and requires 8–12 wt% loading in PC/ABS.[42]

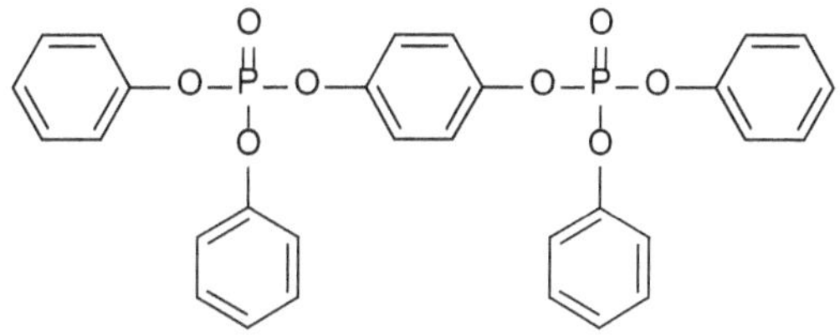

(28.9)

Resorcinol bis(di-2,6-xylyl phosphate) (RXP, Structure 28.10)[43] has been on the market for over 20 years, but mostly in Asia. Similar to HDP, RXP is mostly pure bisphosphate with very little oligomers present. Because of the specific chemical structure and high purity, RXP is a solid with a melting point of 95°C. The steric hindrance provided by the 2,6-xylyl groups makes this product more hydrolytically stable than BDP. RXP has a phosphorus content of 9.0%, and its fire-retardant efficiency is similar to that of BDP; it provides a V-0 rating in PC/ABS at 12–16 wt% loading.[44]

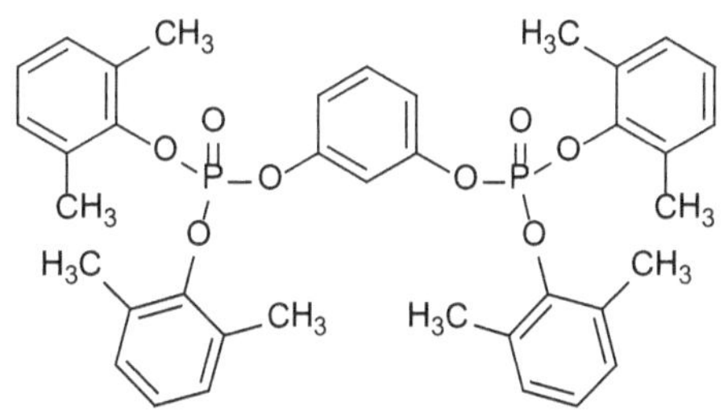

(28.10)

Generic formulations and physical properties for UL-94 V-0 PC/ABS are shown in Table 28.5.

Academic studies,[45,46] on the mechanism of the flame-retardant action of aromatic phosphates in a PC-based blend, revealed that BDP shows mostly condensed phase action, RDP shows a mixed condensed phase and gas phase, whereas TPP is mostly gas phase active. This was attributed to the temperature of decomposition of PC and phosphates, for example, TPP evaporates at a relatively low temperature and doesn't have a chance to react with PC, whereas bisphosphates react with PC,[47,48] RDP or BDP tend to cause PC to produce more char, decreasing the fuel supply to the flame and decreasing the flame temperature. TPP, which has gas phase activity, becomes more effective in the gas phase with a decrease in the flame temperature. HDP shows significant gas phase efficiency and when mixed with a mostly condensed phase active BDP exhibits a synergistic effect.[49] Another study showed[50] that the hindered structure of RXP slows down the reaction with PC, and therefore it shows less condensed phase action compared to RDP. Interestingly talc improves the flame retardant efficiency of bisphosphates because of the better protective properties of the char[51] and on the other hand glass fiber reinforcement deteriorates the flame retardant efficiency because it increases the combustion surface (candlewick effect).[52]

Another large application of aromatic bisphosphates and oligomers is in polyphenylene ether (PPE) based blends. Polyphenylene ether cannot be processed alone because of its very high melting temperature, but it is readily

TABLE 28.5
Phosphate Esters-Based UL94 V-0 Formulations for PC/ABS

FR Type >	Units	RDP	BDP	HDP	RXP
PC	%	73.4	70.9	72.6	69.9
ABS	%	18.3	17.7	18.1	17.5
FR	%	8	11.1	9	12.3
PTFE	%	0.3	0.3	0.3	0.3
Sum	%	100.0	100.0	100.0	100.0
P [calculated]	%	0.86	1	0.96	1.1
UL 94 1.6 [mm] rating		V-0	V-0	V-0	V-0
Izod notched impact	J/m	725	610	715	725
Flexural strength	MPa	83	83	78	80
Flexural modulus	MPa	1,520	1,590	1,520	1,520
Tensile strength	MPa	54	55	54	54
HDT 1.8 MPa, 3.2 mm	°C	84	85	88	88
Initial color vs. QUV 500h	ΔE	30	27	27	32

TABLE 28.6
Phosphate Esters-Based UL94 V-0 Formulations for PPE/HIPS

FR Type >	Units	RDP	BDP	HDP	RXP
PPO	%	53	51.7	52.9	51.8
HIPS	%	35.4	34.4	35.3	34.5
FR	%	11.2	13.5	11.4	13.3
PTFE	%	0.2	0.2	0.2	0.2
Antioxidant	%	0.2	0.2	0.2	0.2
Sum	%	100.0	100.0	100.0	100.0
P [calculated]	%	1.2	1.2	1.2	1.2
UL 94 1.6 [mm] rating		V-0	V-0	V-0	V-0
Izod notched impact	J/m	57	48	49	44
Tensile strength	MPa	53	56	53	55
Elongation at break	%	45	40	42	38
Tensile modulus	MPa	1,050	1,085	1,045	1,050
HDT 1.8 MPa, 3.2 mm	°C	97	96	95	98
MFI at 280°C, 5 kg	g/10 min	25	33	29	34
Initial color vs. QUV 500h	ΔE	9.2	7.5	8.6	9.1

compatible with many polymers and can be processed as a blend. Commercial PPE/HIPS blends, also known as modified PPE, contain from 35 to 65 wt% PPE. Typically, between 9 and 15 wt% of a phosphate ester is needed to achieve V-0 rating; the lower the PPE content in the blend, the higher the phosphate loading required. PTFE is required to prevent dripping. Because PPE is not sensitive to hydrolysis, any bisphosphate can be used in high humidity applications, as for example water pipes.[53] Generic formulations and physical properties of UL-94 V-0 PPE/HIPS are shown in Table 28.6.

Cyclic phenoxyphosphazenes are thermally and hydrolytically stable phosphorus-nitrogen products. Typically, these are mixtures of hexaphenoxytricyclophosphazene as a major component (Structure 28.11) with some octaphenoxytetraphosphazene and a smaller fraction of linear oligomers. Because cyclic phenoxyphosphazenes have low polarity and favorable electrical properties, they found use in high-end printed wiring boards.[54] Another commercial use of cyclic phenoxyphosphazene is in PC/ABS, where it is effective at 12–15 wt% loading.[55] Reportedly, it shows a higher heat distortion temperature compared to aromatic bisphosphates.[56] It also provides excellent hydrolytic stability and low temperature ductility. Despite numerous patents, real commercial use of phenoxyphosphazenes in E&E is unclear.

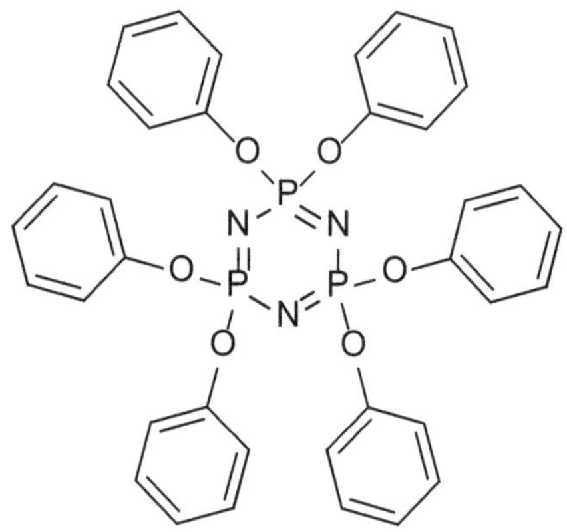

(28.11)

28.4.2 Printed Wiring Boards

Printed wiring board (PWB) is an essential component of any E&E device, and it always needs to be flame retarded to a minimum as V-1 (but practically to V-0) UL-94 rating. PWBs are typically produced by pressing copper clad laminates (CCL) and pre-pregs. The UL-94 test and other tests listed in Table 28.1 are performed on CCLs, not on the assembled PWBs. National Electrical Manufacturers Association (NEMA) assigns classifications to CCLs. In this chapter we will discuss only the CCL classes which are commonly used in consumer electronics and some industrial, medical, and automotive applications which require flame retardants. Ironically, most military and some avionic applications do not specify flame retardancy because there is a belief that tight standards of production control ensure high reliability and therefore the probability of failure is very low.

The FR-4 CCL grade is composed of epoxy resin reinforced with glass cloth and is a particularly important one because it is predominantly used in consumer electronics. Usually, FR-4 boards will be formulated to pass UL 94 V-0. Many consumer electric applications use CEM-1 which is epoxy resin, glass cloth shell, and paper core boards that are flame retardant, as required by UL. CEM-1 is easier to produce compared to FR-4 because trough holes in CEM-1 are punchable, whereas FR-4 trough holes are drilled which is the slowest operation in PWB production. FR-3 are epoxy resin paper board laminates which are punchable and used in lower end electronics.

For higher temperature performance above that of FR-4, other resins notably polyimides, cyanates, PTFE, other fluoropolymers, epoxy-PPE blends, and even ceramics are often used. Special high frequency requirements such as computer circuits operating in the Gigahertz range also may require particularly low dielectric constant materials, such as PTFE or primarily hydrocarbon type resins. Mineral fillers such as silica are often present, which is favorable to flame retardancy. Polyimides, cyanates, PTFE, and inorganic substrates are usually inherently flame retardant. Certain low dielectric constant crosslinked hydrocarbon-rich resins pose a more challenging flame-retardancy problem.

The phenolic paper FR-1 and FR-2 boards have long been flame retarded with triaryl phosphates (Structure 28.6) or oligomeric organic phosphates (Structures 28.7 to 28–10). In general, phenolic resins tend to help flame retardancy because of their char forming ability; on the other hand, the char may show glowing non-flame combustion (smoldering) and phosphorus flame retardants help deter that problem. Aryl phosphates and oligomeric phosphates are also good plasticizers and help to improve the flexibility of phenolic resin boards.

Encapsulated integrated circuits share many of the same considerations, but matching the thermal coefficient of expansion is particularly critical. Therefore, high levels of fillers, as high as 80% silica, are often used, as well as many of the same flame retardants used in PWB. A special test, UL 1694, suitable for small items is available using a needle flame. This is a test of the encapsulated package itself rather than just of the plastic material.[57]

At the present, tetrabromobisphenol A (TBBA, Structure 28.12) is the single largest flame retardant in terms of commercial value. It is used in the majority of all FR-4 epoxy printed wiring boards. There are three ways of incorporating TBBA in epoxy resin. One is called "chain extension" which is reacting TBBA with diglycidyl ether of bisphenol A in the presence of a catalyst. This process is performed in a reactor typically by epoxy resin manufacturers and the obtained brominated epoxy with 20–25% bromine content (Structure 28.13) is dissolved in acetone and delivered to CCL producers as an 80% solution (commonly known as A-80). The A-80 is added with dicyandiamide curing agent and methylimidazole catalyst, and this varnish is used to impregnate glass fiber mats, produce pre-pregs and press FR-4 laminates. Typical laminates made by this process do not exceed a glass transition temperature (T_g) of 130°C.

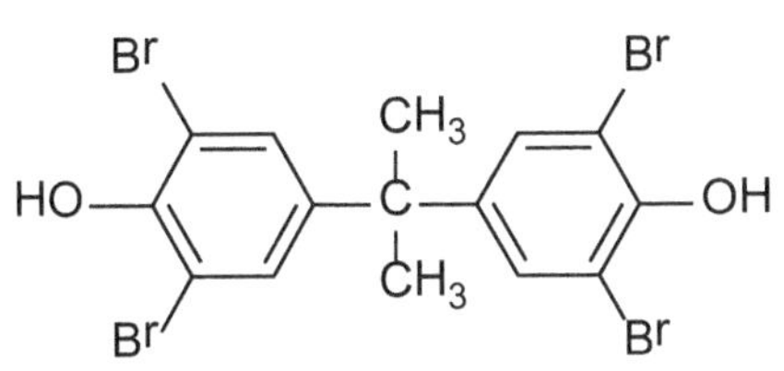

(28.12)

The chain extension process cannot be performed with multifunctional epoxies because the resin will gel in the reactor. As an alternative method of incorporating bromine into CCL, a preformed diglycidyl ether of tetrabromobisphenol A may be used (Structure 28.14), which is typically sold as a 60% solution in toluene, commonly known as T-60. It is suitable for use where a maximum level of bromine is needed to offset a large ratio of hardener or non-brominated resin. However, more often it is used in combination with multifunctional novolac epoxy resins which are cured with phenol novolacs. The laminates with such resins can achieve glass transition temperatures of 150°C or higher. Broad use of such FR-4 laminates began in early 2000 with the implementation of lead-free soldering which requires higher thermal stability.

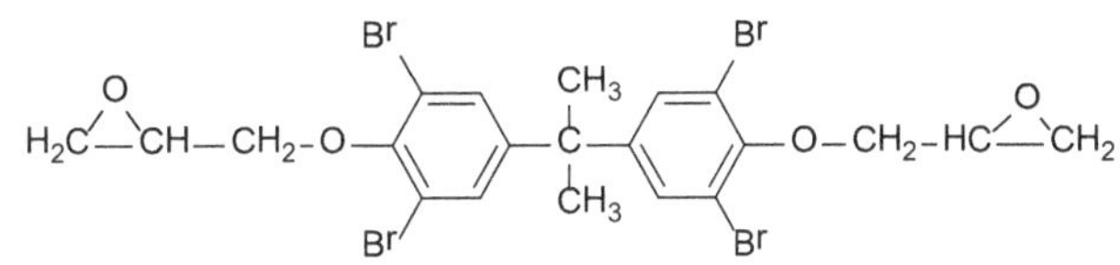

(28.14)

And finally, the third method of incorporating TBBA to epoxy is simply adding it to the varnish. In this case TBBA will react with epoxy during B-staging (prepreg production) laminating and pressing PWB. This method is typically used just to boost the bromine content in combination with the other two methods. Table 28.7 shows typical mechanical and electrical properties for TBBA based laminates of low, medium, and high T_g.

Recently unreacted TBBA uses were restricted in Europe and are under consideration in the USA and Asia. European law restricts levels of unreacted TBBA to < 1000 ppm. In a recent study seven TBBA based epoxy commercial laminates with a T_g range from 140 to 170°C, two high performance laminates with a T_g = 200°C with unknown FR and four halogen-free laminates as references were analyzed for free TBBA following EPA protocols for pressurized liquid extraction and HPLC analysis.[58] None of the laminates showed TBBA above the detection limit of 1 ppm.

Phosphorus-based FRs can be added to epoxy as an additive or can be incorporated in the epoxy network by phosphorylation of the epoxy resin or in the form of phosphorus-based cross-linking agents.[59] Reactive FRs are more preferred in epoxy because they show fewer negative effects on the physical properties, mostly glass-transition temperature and hydrolytic stability. The product of the reaction of o-phenyl phenol and phosphorus trichloride[60] followed by hydrolysis[61] resulted in a unique cyclic 9,10-dihydro-9-oxa-10-phosphaphenanthrene 10-oxide structure, also known as

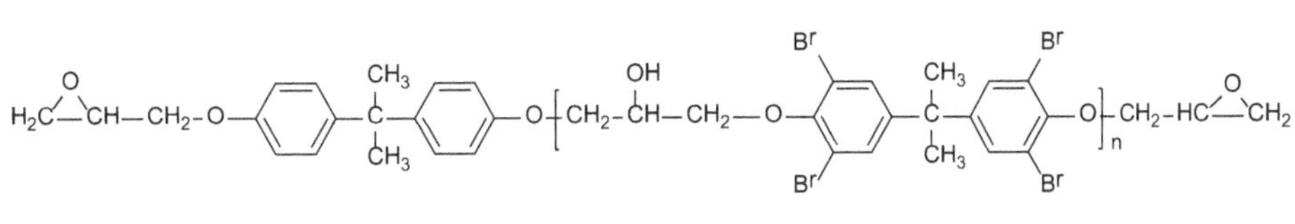

(28.13)

TABLE 28.7
Typical Properties of TBBA-Based Laminates

Property	Units	Low T_g	Medium T_g	High T_g
UL-94 rating		V-0	V-0	V-0
Glass transition temperature, T_g	°C	135	150	170
Decomposition temperature, T_d	°C	315	330	300
Coefficient of thermal expansion, CTE, pre T_g	ppm/°C	60–70	50–60	60–70
Coefficient of thermal expansion, CTE, post T_g	ppm/°C	240–250	240–250	240–250
Pressure cooker test, 30 min, 288°C		Pass	Pass	Pass
Dielectric constant, Dk, 1 GHz		4.20–4.35	3.70–3.80	3.80–3.95
Dissipation factor, Df, 1 GHz		0.020	0.025	0.015
Copper peel strength	MPa	1.20–1.35	1.10–1.20	1.20–1.30

DOPO (Structure 28.15). Its P-H bond is more reactive than a similar bond in many other phosphinates or phosphonates, and therefore DOPO can be reacted with epoxies. Although DOPO is monofunctional, it was adopted by the industry for use in PWB laminates[62] and now DOPO or its derivatives are the largest phosphorus FRs used in epoxy. One common practice developed early is reacting DOPO with a multifunctional novolac type epoxy[63] in order to achieve a phosphorus content of about 3 wt% (Structure 28.16). This still leaves on the average 2–4 epoxy functionalities unconsumed which allows further curing of the phosphorylated epoxy resin.

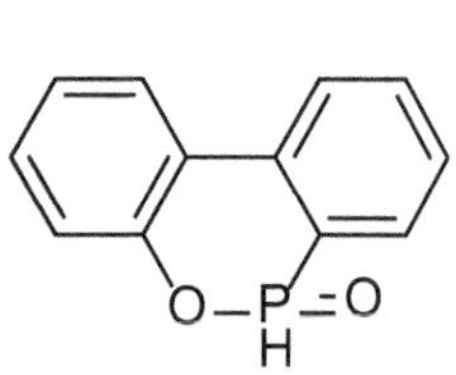

(28.15)

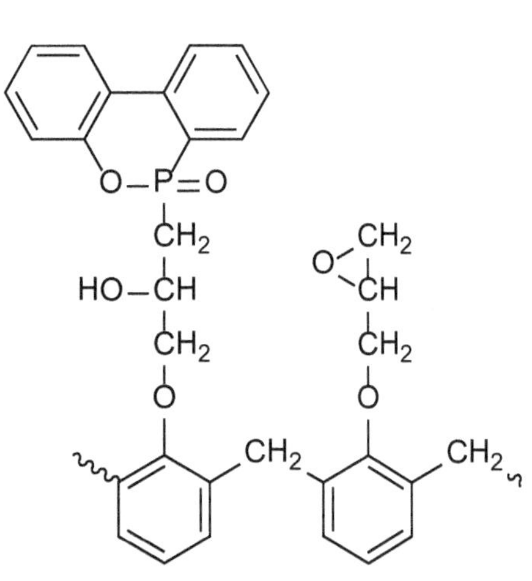

(28.16)

Because DOPO is monofunctional, it cannot be used with most common difunctional bisphenol A epoxy resins. In novolac-type epoxies DOPO provides V-0 at a relatively low phosphorus content of 2.0–2.5 wt%.[64] The high efficiency of DOPO compared to other phosphorus FRs is partially attributed to its gas phase action.[65] DOPO can be combined with ATH[66] which is normally not the case with many phosphorus FRs showing mostly condensed phase action. When combined with ATH or fine silica, DOPO-based laminates require only 1 wt% P or less to achieve a V-0 rating. The main disadvantage of DOPO modified epoxy is the challenge to achieve a high glass transition temperature T_g > 150°C even when combined with multifunctional epoxy.[67]

By the reaction of DOPO with butoxymethylated bisphenol A,[68] a mixture of phosphorylated bisphenols (DOPO-BPA) can be made with the major component presented in Structure 28.17. Because DOPO-BPA is a difunctional reactive FR and has a high phosphorus content of about 9% compared to phosphorylated epoxy of 3%, it allows the production of laminates with a T_g > 175°C and with good thermal stability as measured by a delamination test. It also has good solubility in solvents that are compatible with epoxy lamination processes. Another positive attribute of DOPO-BPA is good electrical properties in epoxy laminates.[69]

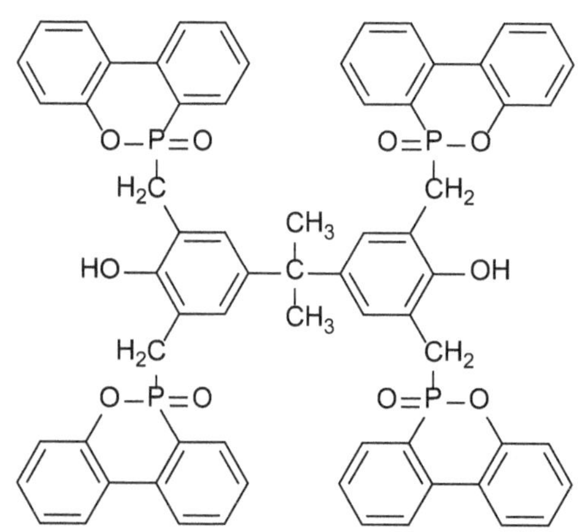

(28.17)

Typical properties of DOPO based epoxy laminates are listed in Table 28.8. Low T_g halogen-free laminates had poor performance and reliability compared to TBBA based counterparts and disappeared from the market. Therefore, we are listing the properties of only medium and high T_g laminates. Phosphorus-based laminates are more sensitive to moisture absorption and because of this the PCT test requirement is typically 60 min in an autoclave or longer.

With the development of personal computers and later cell phones emerged the need for high frequency low dielectric loss PWBs. Epoxy resins are too polar to satisfy some of these needs. The first move in such a direction was the partial replacement of epoxy resin with low molecular weight polyphenylene ether resin with a hydroxyl

TABLE 28.8
Typical Properties of DOPO-Based Laminates

Property	Units	Medium T_g	High T_g
UL-94 rating		V-0	V-0
Glass transition temperature, T_g	°C	150	170
Decomposition temperature, T_d	°C	350	360
Coefficient of thermal expansion, CTE, pre T_g	ppm/°C	30–50	30–50
Coefficient of thermal expansion, CTE, post T_g	ppm/°C	200–230	200–230
Pressure cooker test, 60 min, 288°C		Pass	Pass
Dielectric constant, Dk, 1 GHz		3.8–4.0	4.1–4.3
Dissipation factor, Df, 1 GHz		0.011–0.013	0.09–0.012
Copper peel strength	MPa	1.40–1.75	1.20–1.55

terminated chain end. This resin still reacted with epoxy containing TBBA. However, more demanding applications require complete replacement of the epoxy resin. The most common system for such laminates is polyphenylene resin with terminated acrylate or other unsaturated groups and butadiene rubber. Such resin is cured with a cross-linker and a free radical initiator. On the halogen side decabromodiphenylethane (Structure 28.1) or ethylen ebis(tetrabromophthalimide) (Structure 28.5) are used as nonreactive FRs.

On the phosphorus side hexaphenoxytricyclophosphazene (Structure 28.11) having low polarity and favorable electrical properties found use in epoxy-free PPE[70] based printed wiring boards. On the negative side the low melting point of cyclic phenoxyphosphazene (110°C) can cause excessive resin flow in the lamination process, and as a result its loading level in the formulation is limited. More recently p-xylenebis(diphenyl phosphine oxide) (Structure 28.18) was introduced to the market. Since phosphine oxide is not soluble in the common solvents used by PWB laminators, it is applied as a filler FR in high-frequency formulations.[71] Low dielectric loss laminates vary significantly between different manufacturers with a common narrative that the dissipation factor should be Df = 0.006–0.002.

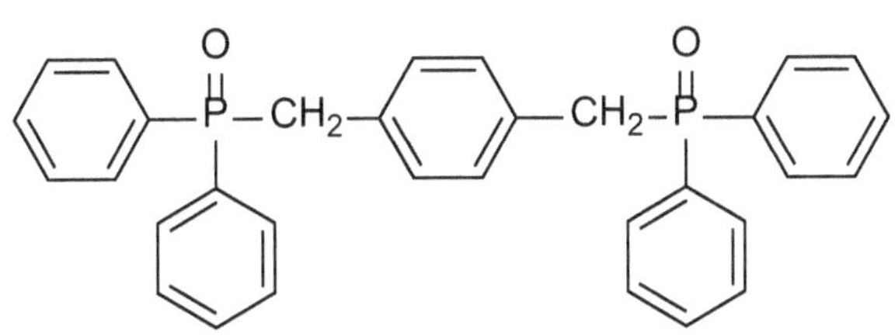

(28.18)

28.4.3 Connectors

All E&E connectors can be divided into internal and external. Their fire safety performance is defined by IEC 62368. All connectors are classified into three categories PS1, PS2, and PS3 depending on the power level. PS1 connectors are used for a power level of less than 15 Watt with no ignition source. They require only UL-94 HB rating. PS2 connectors with a power level between 15 and 100 Watt and PS3 with even higher power levels require as a minimum V-1 rating if the weight of the plastic is more than 4 grams. There are some specific designs where PS2 and PS3 connectors cannot be the ignition source. In this case a minimum V-2 rating is recommended if the connector is located at a distance of less than 50 mm from the potential ignition source. The cause of ignition can be arching between two pins or resistive ignition because of overloading due to the failure of other electronic components. Because of specific ignition scenarios the glow wire ignition test (GWIT) and the comparative tracking index (CTI) are important for connectors. Although tracking resistance is well-known for pure plastics, the addition of glass fibers and flame retardants can change the CTI significantly.[28,72] Therefore, it is imperative to measure the CTI for any specific formulation.

Polybutylene terephthalate (PBT), polyamide 6.6 (PA 6.6), and polyamide 6 (PA 6) are three predominant engineering thermoplastics used to produce connectors. PBT has an excellent processing profile, and its fast crystallization permits fast molding cycles.[73] The differences in the mold shrinkage of amorphous and crystalline regions are minimized by the presence of glass fibers and the high content of flame retardants.[74] Typical FR additives are only partially soluble or insoluble in the amorphous regions, and they act as fillers, reducing the mold shrinkage and increasing the modulus and specific gravity. The strength and stiffness of reinforced PBT allows its use in thinner wall sections than are possible with unreinforced materials.

Although some low molecular weight brominated FRs can be used in PBT and polyamides, polymeric FRs are more preferred for connectors because they provide better tensile strength and do not migrate to the surface. Terabromobisphenol A polycarbonate is one of the commonly used FRs in glass-filled PBT. TBBA polycarbonate can be end capped with phenol or tribromophenol (Structures 28–19 and 28–20). These polymers have a bromine content of 52 and 58% respectfully. They provide a good balance of flame retardancy with good physical properties, although with decreased melt flow. Another potential disadvantage of tetrabromobisphenol A polycarbonate is its tendency to transesterify with PBT, which is catalyzed by antimony trioxide. This is usually not a problem during processing and molding but can be seen during the heat aging test which is required for some applications. As patent literature shows,[75] a small amount such as 0.15% of calcium acid phosphate can inhibit the transesterification.

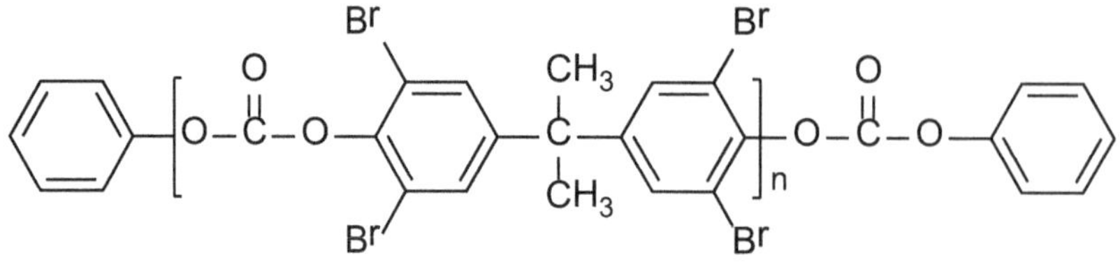

(28.19)

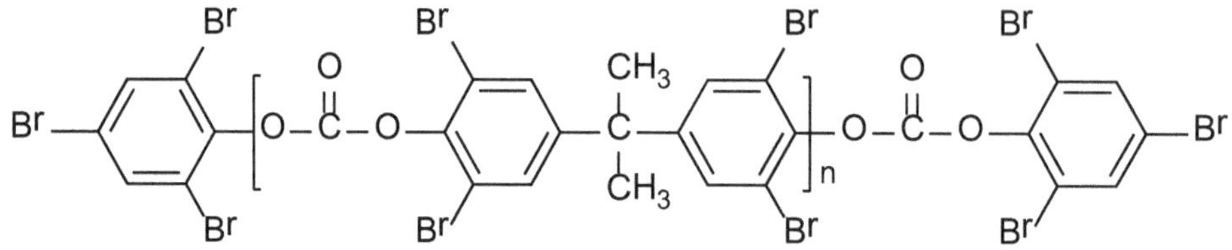

(28.20)

High molecular weight (Mw < 15000) TBBA epoxy polymers (Structures 28.4 and 28.5) are often used in PBT for connectors. The polymeric FRs of this series show good flame retardant performance, non-blooming behavior, and good UV stability in PBT. Thermal stability improves with increasing molecular weight, however for better resin flow the lower Mw version is preferable. The grade specifically recommended for glass-filled PBT is end capped epoxy with a Mw of about 15,000.[76] Corrosion of processing equipment is a disadvantage of brominated epoxy polymers, but this problem can be partially overcome with the use of acid scavengers and heat stabilizers.[77]

Another brominated polymer particularly suitable for PBT is poly(pentabromobenzy acrylate) (Structure 28.21). It is a very high molecular weight polymer (Mw ~ 600,000) with a bromine content of 71%. It shows advantages in impact strength, CTI and melt flow in glass-fiber-reinforced PBT.[78] Its stability in thermal aging under recycling conditions is quite good, and surprisingly the tensile strength is better retained than with non-flame retarded PBT.[79] It is believed that poly(pentabromobenzyl acrylate) serves as an adhesive promoter between glass-fibers and the polymer.[80]

High molecular weight (Mw ~ 600,000) brominated polystyrene (Structure 28.22) is mainly used in polyamides, but it also has some limited application in PBT where it affords both good CTI performance and good thermal aging properties.[81] It is produced by bromination of polystyrene, which puts two or three bromine atoms on each aromatic ring which results in a bromine content of about 66%. As a polymer with no vapor pressure, it is non-blooming. Because high molecular weight brominated polystyrene has relatively poor melt flow characteristics, there are also specialty versions with low molecular weight for the manufacture of thin wall parts.[82]

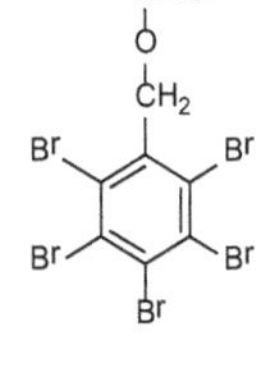

(28.21)

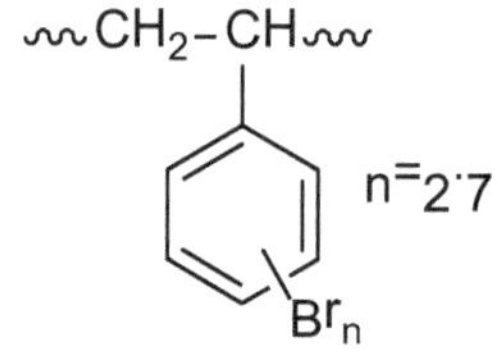

(28.22)

On the halogen-free side, stabilized and coated red phosphorus was used in glass-filled PBT and polyamides where it provided V-0 rating at only 6–12 wt% loading.[83] However, in contact with atmospheric moisture and air red phosphorus can disproportionate with slow long-term evolution of phosphine and the formation of electroconductive and corrosive phosphorous acids on the surface of the plastic. This caused problems in the use of red phosphorus in E&E applications; it is not used anymore in connectors.[84] Calcium and aluminum hypophosphites alone or in combination with melamine cyanurate are efficient in glass-filled PBT.[85–87] However, if unintentionally overheated, they can create a potential hazard during processing due to the evolution of phosphine[88,89] and can even catch fire. Because of this, actual commercial use of calcium and aluminum hypophosphites in PBT and polyamides for connectors is not clear.

In the late nineties aluminum diethyl phosphinate (DEPAL, Structure 28.23) was introduced to the market as a phosphorus-based FR for PBT and polyamides. In glass-filled PBT DEPAL can provide V-0 rating at about 18–22 wt% loading.[90,91] Although DEPAL is only moderately efficient in polyamides, it was found to be synergistic with melamine polyphosphate.[92] Further addition of a few percent of zinc borate as a stabilizer is required for high temperature processing of polyamide 6.6.[93] These products provide UL-94 V-0 ratings in glass-filled polyamides at 15–20 wt% loading. It seems there is no actual synergism between DEPAL and melamine salts in glass-filled PBT, but about 1/3 of DEPAL can be replaced with melamine cyanurate or melamine polyphosphate without the loss of the V-0 rating which is probably beneficial for cost saving. Compared to brominated flame retardants used in the same application, DEPAL allows a high Comparative Tracking Index (CTI) > 500 volt.[94,95] On the other hand, DEPAL and DEPAL-based synergistic combinations show significant wear (corrosion) of processing equipment,[96] which can be decreased by using acid scavengers.

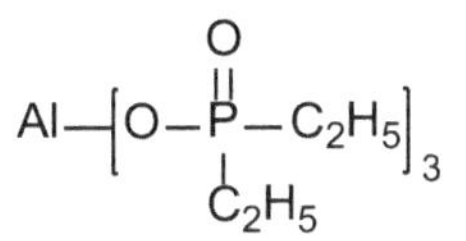

(28.23)

Based on academic studies, there is a strong indication that DEPAL mostly operates in the gas phase by a flame-inhibiting mechanism.[97,98] It was believed that DEPAL decomposes with the evolution of phosphinic acid, which evaporates to the flame, however there is other evidence showing that aluminum alkylphosphinates can volatilize without decomposition. Melamine polyphosphate provides a condensed phase action by increasing the charring of the polymer[99] and thus provides a synergistic action with DEPAL.[97] In contrast, melamine cyanurate mostly volatilizes and provides a cooling effect to the flame,[98] therefore its action is mostly adjunctive but not synergistic.

Interestingly melamine cyanurate alone can provide V-0 rating in unreinforced or mineral filled polyamides.[100] Unreinforced polyamides show better flexibility and elongation of break which is beneficial for very small connectors which can crack during installation. It appears that melamine cyanurate works by producing enhanced dripping (with non-burning drips) and also by dissociating endothermically to produce noncombustible vapor.[101–103] Ratings of UL 94 V-0 can be obtained at 5–10 wt% in unreinforced polyamide 66, at 8–15 wt% in unfilled polyamide 6, and at 13–16 wt% in some mineral-filled polyamide formulations. Polyamide 66 tends to pyrolyze to form cyclopentanone, which condenses with the melamine and/or cyanuric acid formed by the dissociation of melamine cyanurate causing cross-linking and residue formation. On the other hand, the pyrolysis of polyamide 6 yields mostly caprolactam, alkyl cyanides, and ammonia which do not interact much with the melamine cyanurate pyrolysis products.[104] This is the reason for melamine cyanurate to be somehow more effective in polyamide 6.6 compared to polyamide 6. In polyamide 6 melamine cyanurate decreases the yield of caprolactam, but increases the yield of oligomeric fragments, which are less flammable, but help to enhance melt-flow.[105]

Generic formulations and physical properties of UL-94 V-0 glass-filled PBT are shown in Table 28.9, polyamide 6.6 in Table 28.10 and polyamide 6 in Table 28.11. Due to the miniaturization of electronics, especially of laptops, tablets, and cell phones, some connectors used are small with very thin walls. Therefore a UL-94 V-0 rating is required for standard bars of 0.8 and even 0.4 mm thickness. The Glow Wire Ignition Test is often mandatory for thermoplastics used in the connectors. Halogen-based flame retardants typically show lower CTI compared to phosphorus-based FRs, however apparently CTI > 600 V is possible in halogen systems.[106]

Interestingly, a UL-94 V-0 rating can be achieved in glass-filled PBT without the use of antimony trioxide.[107,108] It requires an insignificant increase in the bromine content and one or two phosphorus synergists replacing the antimony trioxide. Table 28.12 shows some antimony trioxide-free formulations and their fire performance and physical properties. It should be noted that the removal of antimony from bromine-based formulations significantly increases CTI.

TABLE 28.9
Halogen and Halogen-Free UL94 V-0 Formulations for Glass-Filled PBT

FR Type >	Units	Tribromophenol End-Capped TBBA PolyCarbonate	Brominated Epoxy Polymer	Poly(penta-bromobenzyl acrylate)	Brominated PolyStyrene	DEPAL
PBT	%	50.5	48.9	53.0	52.2	49.3
Glass fiber	%	30	30.0	30.0	30.0	30.0
FR	%	13.8	15.4	11.3	12.1	20.0
Sb_2O_3	%	5.0	5.0	5.0	5.0	-
PTFE	%	0.5	0.5	0.5	0.5	0.5
Antioxidant	%	0.2	0.2	0.2	0.2	0.2
Sum	%	100.0	100.0	100.0	100.0	100.0
Br [calculated]	%	8	8	8	8	-
UL-94 @ 0.8 mm, rating		V-0	V-0	V-0	V-0	V-0
UL-94 @ 0.4 mm, rating		V-0	V-0	V-0	V-0	V-0
GWIT Rating @ 3.2 mm	°C	775	775	775	775	775
Izod notched impact	J/m	82	83	82	78	54
Tensile strength	MPa	135	129	115	114	88
Elongation at break	%	3.6	4.0	4.1	3.7	3.3
Tensile modulus	MPa	11.960	10,110	9,525	9,990	9,370
HDT 1.8 MPa, 3.2 mm	°C	203	205	208	206	206
MFI at 250°C, 2.16 kg	g/10 min	16	21	21	22	34
CTI	Volt	225	200	250	250	600

TABLE 28.10
Halogen and Halogen-Free UL94 V-0 Formulations for Glass-Filled Polyamide 6.6

FR Type >	Units	Brominated Polystyrene	Brominated Epoxy Polymer	DEPAL Melamine Polyphosphate
PA66	%	44.9	40.1	49.6
Glass fiber	%	30.0	30.0	30.0
FR	%	19.7	24.5	20.0
Sb_2O_3	%	5.0	5.0	-
Antioxidant	%	0.4	0.4	0.4
Sum	%	100.0	100.0	100.0
Br [calculated]	%	13	13	0
UL-94 @ 0.8 mm, rating		V-0	V-0	V-0
UL-94 @ 0.4 mm, rating		V-0	V-0	V-0
GWIT Rating @ 3.2 mm	°C	875	900	775
Izod notched impact	J/m	101	79	64
Tensile strength	MPa	150	168	121
Elongation at break	%	3.7	3.0	4.2
Tensile modulus	MPa	9,340	9,980	8,270
HDT 1.8 MPa, 3.2 mm	°C	243	242	240
MFI at 270°C, 1.2 kg	g/10 min	6.6	7.8	7.6
CTI	Volt	325	225	600

TABLE 28.11
Halogen and Halogen-Free UL94 V-0 Formulations for Glass-Filled Polyamide 6

FR Type >	Units	Brominated Polystyrene	Brominated Epoxy Polymer	Poly(pentabromo-benzyl acrylate)	DEPAL Melamine Polyphosphate
PA6	%	44.9	40.1	46.3	49.6
Glass fiber	%	30.0	30.0	30.0	30.0
FR	%	19.7	24.5	18.3	20.0
Sb_2O_3	%	5.0	5.0	5.0	0.0
Antioxidant	%	0.4	0.4	0.4	0.4
Sum	%	100.0	100.0	100.0	100.0
Br [calculated]	%	13	13	13	0
UL-94 @ 0.8 mm, rating		V-0	V-0	V-0	V-0
UL-94 @ 0.4 mm, rating		V-0	V-0	V-0	V-1
GWIT Rating @ 3.2 mm	°C	850	850	850	775
Izod notched impact	J/m	109	85	115	83
Tensile strength	MPa	139	160	148	114
Elongation at break	%	4.4	3.3	3.1	3.3
Tensile modulus	MPa	9,315	8,400	8.745	9,385
HDT 1.8 MPa, 3.2 mm	°C	201	199	204	203
MFI at 270°C, 1.2 kg	g/10 min	5.9	6.0	5.5	3.6
CTI	Volt	325	225	300	600

28.4.4 Wires

There are two types of wires in any E&E device, for example, external flexible cord and internal wires. According to UL (IEC) 60065 all external flexible cords connecting an AC outlet with an E&E device should be rated VW-1 according to UL 1581 section 1080. If the device has an external AC/DC convertor part of the cord connecting the convertor with the device should be rated VW-1 if peak DC voltage exceeds 35 V or carrying current exceeds 0.2 A. Internal wires should be rated VW-1 if the wire is located in a circuit that is considered a potential ignition source or is in contact with wiring located in a circuit that is a potential ignition source.

Despite the halogen-free policies of many E&E OEMs, flexible PVC is still the dominant polymer used for VW-1 power cords and wires. The fire properties of plasticized PVC are largely determined by the type and amount of

TABLE 28.12
Antimony Trioxide-Free UL94 V-0 Formulations for Glass-Filled PBT

Formulation #	Units	1	2	3
PBT	%	47.7	47.7	49.3
Glass fiber	%	30	30	30
Poly(pentabromobenzyl acrylate	%	14.1	14.1	10
HDP (Structure 28.9)	%	5		
Melamine polyphosphate	%	2.5		
Calcium hypophosphite			7.5	
DEPAL				10
PTFE	%	0.5	0.5	0.5
Antioxidant	%	0.2	0.2	0.2
Sum	%	100.0	100.0	100.0
Br	%	10	10	10
UL 94 0.8 [mm] rating		V-0	V-0	V-0
UL 94 0.4 [mm] rating		V-2	V-0	V-0
Izod notched impact	J/m	57	60	66
Tensile strength	MPa	107	115	114
Elongation at break	%	2	2	3
Tensile modulus	MPa	9,580	13,535	12,775
HDT	°C	191	199	198
MFI at 280°C, 5 kg	g/10 min	33	17	10
CTI	Volt	250	450	250

plasticizer. Typical plasticizers in the past were dialkyl phthalates, but their use is banned now in many geographical areas. They have been replaced with other esters; trimellitates, adipates, sebacates, citrates, and oligomeric or polymeric polyesters.[109] A traditional approach taken to reduce the flammability of plasticized PVC is the use of antimony trioxide.[110] It is typically used in the range of 3 to 7 phr (rarely as much as 12 phr). It is generally believed to act mainly by means of its reaction with HCl to form antimony trichloride, which is volatile at fire-exposure temperatures and which acts as an active flame inhibitor. A typical particle size range is 1.0–1.8 microns. The dust problem associated with antimony trioxide, occasionally causing respiratory or dermal irritation, has been solved by the addition of mineral oils or plasticizers to make dust-free grades or by compounding into pelletized antimony oxide concentrates.

Low smoke halogen-free (LSHF) cables and the materials used to produce them were known since the seventies. Typically, LSHF cables were installed in confined spaces where the toxicity and corrosivity of the smoke generated in a fire would be particularly problematic.[111] Basic polymers used in LSHF are polyethylene or polyethylene copolymers flame retarded with large loadings up to 70% of aluminum trihydrate or magnesium hydroxide. Similar formulations were tried in E&E power cords but were quickly abandoned due to the poor flexibility of the coated wires. Instead, other more flexible polymers or blends which can be flame retarded with significantly lower loadings of phosphorus or nitrogen flame retardants are used.

Most halogen-free formulations for power cords and wires in E&E devices are proprietary, but some generalities can be understood from the patent literature. For example, elastomeric polyurethane wire coatings have been flame retarded with melamine combined with resorcinol bis(diphenyl phosphate) (Structure 28.7). Melamine cyanurate may have some advantages over melamine[112] in combination with RDP. Many formulations based on RDP and ATH passing the VW-1 rating have been developed[113] and some were probably commercialized. Recently many patents appeared on formulations based on piperazine pyrophosphate and melamine pyrophosphates combined with bisphosphates.[114] About 30 wt% DEPAL (Structure 28.23) and some melamine salt allow the achievement of a VW-1 rating in TPU.[115] About 30 wt% DEPAL is required to achieve a VW-1 rating in thermoplastic elastomers (TPE) wire insulation,[116] and the same rating can be achieved if about 1/3 of the DEPAL is replaced with melamine polyphosphate[117] or melamine cyanurate.[118] Some of these technologies are probably commercial.[119]

Fire-retardant performance of styrene based thermoplastic elastomers, such as styrene-ethylene-butylene-styrene (SEBS) block copolymer can be enhanced by blending with polyphenylene resin (PPE). A copolymer of SEBS and maleic anhydride is used as a compatibilizer.[120] These blends are mostly used in electric wire coatings, and often polyolefins and HIPS are also included in the blends. As the patent literature indicates, aromatic bisphosphates RDP[121] and BDP[122] were originally used as flame retardants in PPE/SEBS blends. Bisphosphates are very compatible and soluble in PPE, but not in TPEs and polyolefins and therefore the total loading of bisphosphates is limited because of potential exudation. To overcome this problem in blends containing less than 50% PPE, solid flame retardants are added along with bisphosphates. The patent literature shows combinations of aromatic phosphates with magnesium hydroxide,[123] melamine phosphates,[124] ammonium polyphosphate,[125] or DEPAL.[126]

28.5 RECYCLING OF WEEE POLYMERS

28.5.1 Volumes of WEEE Plastics

Waste from Electrical and Electronic Equipment (WEEE) has been growing from 53.6 million tons in 2018, and it is forecasted to reach 74.7 million tons in 2030.[127] China is the biggest generator of WEEE waste followed by the USA and India.[128] WEEE is one of the challenges for plastics recyclers due to legacy additives in their input.[129] These additives include low molecular weight phthalates (such as diethylhehyl phthalate, benzy butyl phthalate, dibutyl phthalate and diisobutyl phthalate used as plasticizers), heavy metals (such as lead and cadmium compounds used as stabilizers) and some brominated flame retardants such as octaBDE and decaBDE. These legacy substances were added to plastics in the past but have been discontinued due to concerns to human health and environmental issues. If present above the Low Persistent Organic Pollutant (POP) Limit Concentration (LPLC) in the WEEE stream, such a stream is not allowed for direct mechanical recycling. Within WEEE plastics there are different categories with different amounts of legacy additives. The fractions of different plastic categories in European WEEE streams are shown in Table 28.13.

28.5.2 Collecting and Sorting

Within Europe a collecting system for WEEE is well established. Also, the sorting of the different plastics via density separation is a mature process. During the density sorting of the WEEE plastics, the legacy additives are concentrated in what is called the "heavy fraction" mainly above 1,150 kg/m^3 which contains flame-retarded ABS, HIPS, PC/ABS, and HIPS/PPO plus polymers like PVC and PMMA. In the sorted fraction between 1,050 and 1,150 kg/m^3 one can find polystyrene, HIPS, ABS, and PC/ABS without flame retardants and legacy substances. Figure 28.5 shows the plastics density overlap of the heavy and middle fractions.[131] Polyolefins are mostly found in the "light fraction" below 1000 kg/m^3. The middle and light fractions often undergo further tribo-electrostatic separation to concentrate specific plastics which can be mechanically recycled (remelting) into new products.

Today, the high-density fraction containing the flame retardants is not further sorted for mechanical recycling but rather sent to incineration, smelters, co-combustion, or landfill. There are a few reasons for this.

- The final volume of heavy fraction is smaller compared to the light and middle fractions, and it doesn't make economic sense to do further sorting. For comparison, the annual output of virgin ABS or HIPS in a full scale plant is about 100–300 Ktons, whereas the output of ABS or HIPS at a recycling plant is in the range of 5 KT up to 15 Kton.
- Another reason is the presence of black plastic which is common for E&E devices. The sorting of black plastics is challenging, and it requires significant investment. Although plastics containing brominated FRs can be separated using XRF the final quality of the further sorted out "heavy fraction" is not good enough for mechanical recycling and reuse in the same application.
- Full compliance with chemical regulations (e.g., REACH[132]) and RoHS Directive[133] is also challenging specifically for E&E plastics. Chemical analyses for the possible presence of legacy chemicals are prohibitive costly.

PC/ABS and PPO/HIPS are flame retarded with phosphorous flame retardants. These plastics also end up in the "heavy fraction" and, so far, are not separated due to difficulties in the process of sorting and detecting the low concentration of phosphorus (1% phosphorus compared

TABLE 28.13
European Amounts of WEEE Plastics from Four Categories in Tons per Year in 2020[130]

	WEEE Plastics	ABS	HIPS	PP	PC + PC/ABS	Others
Temperature exchange equipment	100,000	2,710	42,270	11,610	120	43,290
Large equipment	172,000	25,232	4,833	97,008	4,317	40,609
Screen and Monitors	80,000	17,512	32,496	9752	8,672	11,568
Small equipment/small IT and communication	379,000	135,000	60,299	55,145	44,191	84,365
Sum of four categories	731,000	180,454	139,898	173,515	57,300	179,832
Percentages	100%	25%	19%	24%	8%	25%

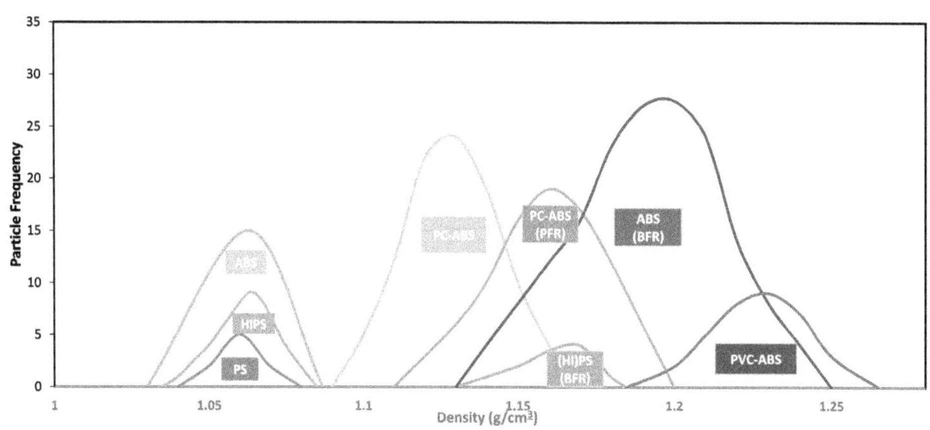

FIGURE 28.5 Overlap of plastic densities.

to 10–15% bromine). These plastics are not sorted out but are sent to incineration, co-combustion, or landfill. Other polymers containing FRs are present in WEEE plastics like polyesters and polyamides mostly used in connectors. Connectors are relatively small and contain metals, therefore they cannot be sorted technically and economically recycled. The fraction containing printed circuit boards and connectors goes to smelters as the main economical drive is the recovery of precious metals.

28.5.3 Recycling Processes

As Figure 28.6 shows recycling processes for thermoplastics which can be divided into two categories: (1) physical recycling and (2) chemical recovery.

- Physical Recycling (synonym: Material Recycling)—Separation of materials and substances. The polymers remain intact and can be reused.
- Chemical Recovery (synonym: Feedstock Recovery)—Decomposition of materials and substances. Polymer chains are always destroyed.

The physical recycling is the most attractive and economical method; however at this point of time it cannot be implemented on WEEE plastics because of incomplete sorting and the potential presence of POPs. Within the best available technology (BAT) solvent-based purification is suggested for the separation of legacy additives, which then must be followed by a thermal destruction step. POPs need to be treated according to the technical guidelines (TGs) developed under the Basel Convention.[134] The obligation is that the POPs are allowed to be separated but must be destroyed via high temperature treatment like incineration above 850°C. As pretreatment, the dissolution recycling purification process aimed at separating and concentrating (purifying) the POP substances is allowed whereafter they need to be destroyed via incineration or a high temperature hazardous waste incinerator.

Solvent-based purification (synonym: dissolution recycling) is based on a physical process also known as "selective extraction." This technology selectively dissolves the polymer and the brominated FRs. A filtration step is applied to remove the undissolved contaminants. The concentration of brominated FRs can be increased by using an antisolvent or ceramic filtration. Separation of the brominated FRs is done by distillation with solvent recovery or precipitation. If separated FRs contain POPs all sludge must be destroyed. The polymer is dried by removal of the solvent and granulated.

One of the known processes allowing the POPs to be separated from host thermoplastic is CreaSolv® Process[135] developed by Fraunhofer IVV and CreCycle GmbH shown in Figure 28.7. After completion of the European NONTOX[136] project on the separation of POPs from the polymers Fraunhofer-IVV has built a pilot plant at its premises in Freising Germany for further development of the CreaSolv® Process for recycling flame retardant ABS, HIPS, and PC/ABS. This unit operates at Technical Readiness Level (TRL)[137] 5–6, and it is ready for scale up into a demo plant. In the Netherlands a demo plant PSLoop B.V was built to handle BFR containing plastics including POPs.[138] So far, this plant has

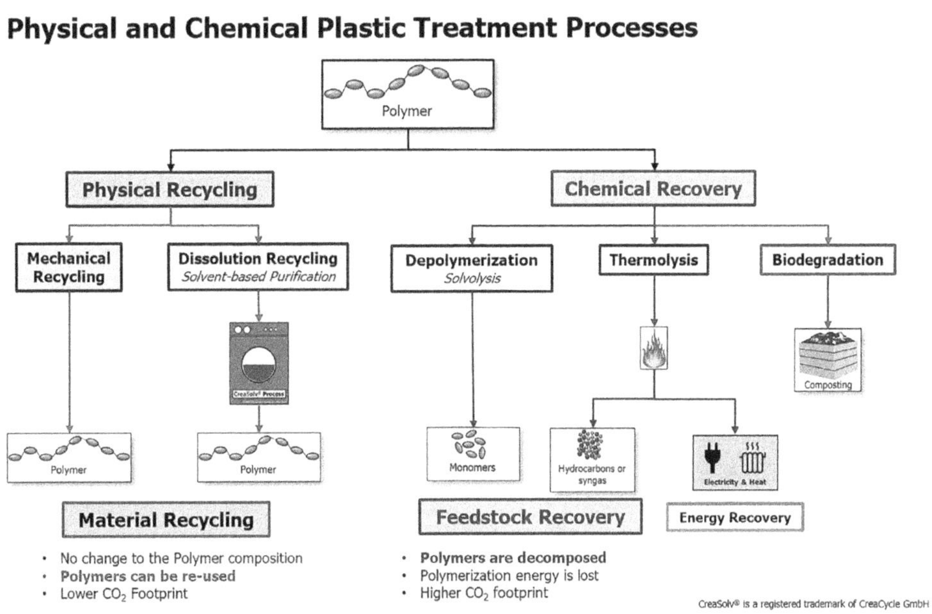

FIGURE 28.6 Recycling technologies for waste plastics.

Source: Reproduced with permission of CreaCycle GmbH www.creacycle.de

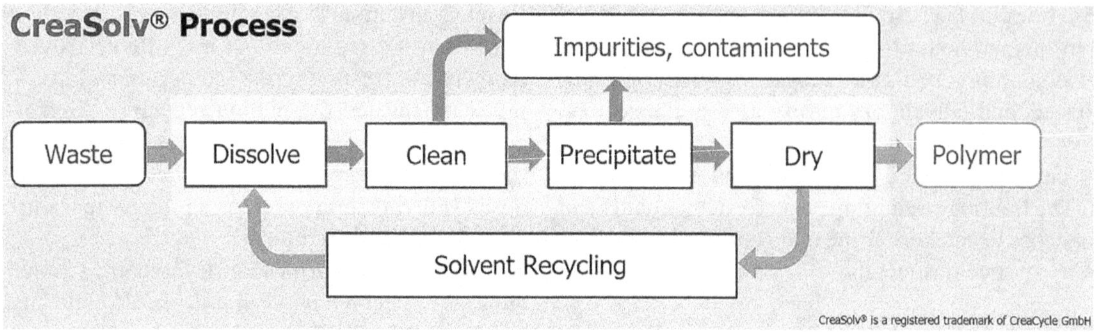

FIGURE 28.7 Plastics recycling using the solvent-based CreaSolv® Process.

Source: Reproduced with permission of Fraundhofer IVV www.ivv.fraunhofer.de/en/recycling-environment/recycling-plastics-creasolv.html

treated only polystyrene foam containing hexabromocyclododecane (POP, HBCD) but in principle it can treat flame retarded ABS and HIPS. The ABS or HIPS recyclate needs to have an unintentional trace contamination (UTC) limit in the recyclate for the POP PBDE's below 500 ppm, and for PS it is below 75 ppm for E&E applications. The separated BFR's sludge will be destroyed in the chemical treatment unit at 1,100° C followed by a bromine recovery step. Another pilot plant for WEEE HIPS and ABS is known to be operated by PolyStyvert[139] in Canada.

Currently, there are no demo plants treating the heavy fraction (FRs) from WEEE plastics in operation with a Technical Readiness level (TRL) above 6. The time needed for scaling up (5–7 years) from pilot up to the full scale plant is a major challenge. The main recycling technologies have been developed for packaging waste and are not directly applicable for plastics-waste-containing halogens including brominated flame retardants, which are mostly not used in packaging. The difficulties in the sorting technology of the heavy fraction and combining it with challenges in the scale of economy makes it very difficult to scale up recycling for most of the diverse WEEE plastics.

Chemical recycling and especially for plastics-containing flame retardants are under investigation at universities and European R&D projects. There are several claims that a certain conversion of the POP is possible but no scale up of a chemical process is in place which can handle flame retardant WEEE plastics containing POPs. There are several claims of projects being able to handle WEEE plastics by pyrolysis, but no commercial process exists today. The continuous change of legislation like the UTC for different POPs, makes it very difficult to comply with the ever-changing specifications. In addition, compliance with (ever stricter) chemicals regulations (e.g., REACH) makes it difficult to operate such plants in the EU. However, there are opportunities for the largest fractions like flame retarded ABS, HIPS, and PC/ABS to sort them out and to create an economical case for scaling up from a pilot plant via a demo plant into a full-scale plant.

28.6 CONCLUSIONS AND FUTURE TRENDS

Because electrical and electronic equipment always contains electrical circuits there will always be a need for some degree of fire safety. There are conflicting trends in the use of flame retardants in E&E devices. On the one hand, household video and audio equipment, computers and handhold devices now operate at lower voltage which means lower fire hazard and potential elimination of flame retardants in some parts. On the other hand, more and more electronics are now equipped with lithium-ion batteries which become a significant potential ignition source and the plastics surrounding the battery need more fire protection. Miniaturizations of E&E devices also bring the challenges of separating non-flame-retardant parts from potential ignition sources.

Although there is a general market trend to halogen-free flame retardants mostly dictated by original equipment manufacturers (OEMs) for their "green image," there are many areas where halogenated flame retardants cannot be replaced due to technical reasons. Phosphorus flame retardants have not escaped environmentalist concerns, and there are a growing number of academic publications studying the spread of organophosphorus flame retardants in the environment. However, it would be an error to say that phosphorus flame retardants are found more often because they replace halogens. Most of the relatively mobile organophosphorus flame retardants that were detected have been in use for half a century. Specific organophosphorus compounds raised sufficient concerns to warrant their discontinuance, but this was not generic to the entire class. Following the general trend of the flame-retardant industry towards polymeric and reactive products which show lower negative effects in the final products and minimal exposure to humans and the environment, both halogen and phosphorus flame retardants will follow this general strategy.

Plastics containing flame retardants from WEEE plastics typically end up in the heavy fraction during sorting combined with XRF separation of the BFRs. Until today, the only "approved" innovative pretreatment technology for POP containing WEEE plastics is a selective solvent-based purification process wherein the polymer is separated and reused while the extracted collected BFRs are destroyed in a high-temperature incinerator combined with a bromine recovery unit. Further incineration, co-combustion, and smelter processes are allowed. Existing processes are being developed until pilot scale and need more time to scale up to demo plant level and full scale plants.

REFERENCES

1. Sidiki, T.P., Hilty, R.D., and Shipe, J., Design for low-halogen green electronics, *IPC Apex Expo*, Las Vegas, NV, March 31—April 2, 2009, S32:01
2. Morgan, A.B., and Wilkie, C.A., An introduction to polymeric flame retardancy, its role in materials science, and the current state of the field, in *Fire Retardancy of Polymeric Materials*, Wilkie, C.A., and Morgan, A.B. (Eds), 2nd edition, CRC Press, Boca Raton, FL, 2009, pp. 1–14.
3. Sherman, L., The new breed of fire fighters, *Compounding World*, November 14–20, 2009.
4. Janssens, M.L., and Schluneker A.M., Evaluation of the likelihood of flashover in room fires involving a flat screen TV and the effect of flame retardant treatment of the casing, *Fire Mater.* 2022. https://doi.org/10.1002/fam.3090
5. Hong, S.; Yang, J.; Ahn, S.; Mun, Y.; Lee, G. Flame Retardancy Performance of Various UL94 Classified Materials Exposed to External Ignition Sources. *Fire Mater.* **2004**, *28*, 25–31. DOI:10.1002/fam.840
6. De Poortere, M.; Schonbach, C.; Simonson, M. The Fire Safety of TV Set Enclosure Materials, a Survey of European statistics. Fire Mater. 2000, 24, 53–60. https://doi.org/10.1002/(SICI)1099–1018(200001/02)24:13.0.CO;2-J
7. Stevenson, G.C., Countervailing risks and benefits in the use of flame retardants, in *Flame Retardants 2000*, Interscience Communications, London, 2000, pp. 131–145.
8. Emsley, A.M., and Stevens, G.C., The risks and benefits of flame retardants in consumer products, in *Advances in Fire Retardant Materials*, Horrocks, A.R. and Price, D., (Eds), Woodhead, Boca Raton, FL, 2008, pp. 363–397.
9. Simonson, M., Andersson, P., and Van den Berge, M., Cost benefit analysis model for fire safety and DecaBDE case study, in *Fire and Materials*, San Francisco, CA, January 29–31, 2007.
10. Babrauskas, V.; Simonson, M. Fire Behaviour of Plastic Parts in Electrical Appliances-Standards Versus Required Fire Safety Objectives. Fire Mater. 2007, 31, 83–96. https://doi.org/10.1002/fam.927
11. Fechtmann, G. J.; History of UL flammability testing of polymeric materials, *Annual BCC Conference on Recent Advances in Flame Retardancy of Polymeric Materials*, Stamford, CT, May 22–25, 2005.
12. Stimitz, J.S., Plastic-a UL perspective (UL 94, UL 746A, B, C, D), *Annual BCC Conference on Recent Advances in Flame Retardancy of Polymeric Materials*, Stamford, CT, May 21–23, 2001.
13. Chapin, J.T., and Backstrom, R., A primer of fire protection measures and the appropriate selection of flammability test methods for materials and products for specific end- use applications, *API 2004 Conference*, Las Vegas, 2004, pp. 144–150.
14. Schreiber, H. Vertical Burning Tests UL-94 and IEC 249–1 (4.3.4) Theoretical Considerations and Experimental Results. *Makromol. Chem. Macromol. Symp.* **1993**, *74*, 349–357.
15. Blaszkiewicz, M., Bowman, P., and Masciantonio, M., Understanding the repeatability and reproducibility of UL 94 testing, *Annual BCC Conference on Recent Advances in Flame Retardancy of Polymeric Materials*, Stamford, CT, 2007.
16. Wang, Y.; Zhang, F.; Chen, X.; Jin, Y.; Zhang, J. Burning and Dripping Behaviors of Polymers Under the UL94, Vertical Burning Test Conditions. Fire Mater. 2010, 34, 203–215. https://doi.org/10.1002/fam.1021
17. Wang, Y.; Zhang, F.; Jiao, C.; Jin, Y.; Zhang, J. Convective Heat Transfer of the Bunsen Flame in the UL94 Vertical Burning Test for Polymers. J. Fire Sci. 2010, 28, 337–356. https://doi.org/10.1177/0734904109351484
18. Wang, Y.; Kang, W.; Zhang, X.; Chen, C.; Sun, P.; Zhang, F.; Li, S.; Development of a Pendant Experiment using Melt Indexer for Correlation with the Large-Size Dripping in the UL-94 Test. Fire Mater. 2018, 42, 436–446. https://doi.org/10.1002/fam.2510
19. Dupretz, R.; Fontaine, G.; Duquesne, S.; Bourbigot, S. Instrumentation of UL-94 Test: Understanding of Mechanisms Involved in Fire Retardancy of Polymers. *Polym. Adv. Technol.* **2015**, *26*, 865–873. DOI:10.1002/pat.3507
20. Bertelli, G.; Benassi, R.; Marchini, R.; Camino, G.; Costa, L.; Luda, M. P. Evaluation of Polymer Combustion and Fire Retardance by using Thermography. *Fire Mater.* **1993,** *17*, 125–129.
21. Spiess, B.; Metzsch-Zilligen, E.; Pfaendner, R. Mechanistic Evaluation of Flame Retardants During UL94 Standard Testing via IR-Camera. Polym. Test. 2021, 103, 107320. https://doi.org/10.1016/j.polymertesting.2021.107320
22. Kempell, F., Schartel, B., Hofmann, A., Butler, K.M., Oriate, E., Idelsohn, S.R., Rossi, R., and Marti, J.M., Numerical simulation of polymeric materials in UL 94 test: Competition between gasification and melt flow/drip- ping, *Interflam 2010*, Interscience Communications, 2010, pp. 721–730.
23. Siripurapu, S., and Agarwal, N., Statistical tools for predicting flame resistance of PC/ABS blends, *ANTEC2006*, Charlotte, NC, 2006, Vol. 3, pp. 1018–1022.
24. Morizur, J.-F., and Asthana, H., Relationship between residual stress and flame resistance of polycarbonate, *ANTEC2012*, Orlando, FL, April 2–4, 2012.
25. Galucci, R.R., and Hocker, T., (Sabic) Additive manufactured items with flame resistance, process for making and process for testing their flame erformance, U.S. Patent, 2019: 10 434 705.
26. Wadehra, I. Designing Information Technology Products with Flame Retarded Plastics with Special Emphasis on current Flammability Safety and Environmental Concerns. Fire Mater. 2005, 29, 121–126. https://doi.org/10.1002/fam.863
27. Howarth, J. T.; Lindstrom, R. S.; Sheth, S. G.; Sidman, K. R. What Designers and Processors Should Know About Flame-Retardant Additives. *Plast World.* **1973**. 65–73.
28. Acquasanta, F.; Berti, C.; Colonna, M.; Fiorini, M.; Karanam, S. Study of Glow Wire Ignition Temperature (GWIT) and Comparative Tracking Index (CTI) Performances of Engineering Thermoplastics and Correlation with Material Properties. Polym. Degrad. Stab. 2011, 96, 566–573. https://doi.org/10.1016/j.polymdegradstab.2010.12.024
29. Levchik, S. V.; Weil, E. D. New Developments in Flame Retardancy of Styrene Thermoplastics and Foams, Polym. Int. 2008, 57, 431–448. https://doi.org/10.1002/pi.2282
30. Hirschler, M.M., Chemical aspects of thermal decomposition of polymeric materials, in *Fire Retardancy of Polymeric Materials*, Grand, A.F. and Wilkie, C.W. (Eds.), Marcel Dekker, New York, 2000, pp. 27–79.

31. Borms, R., Wilmer, R., Peled, M., Kornberg, N., Mazor, R., Bar Yaakov, Y., Scheinert, J., and Georlette, P., Improved and cost-efficient brominated fire retardant systems for plastics and textiles by reducing or eliminating antimony trioxide, in *Fire Retardancy of Polymers. New Applications of Mineral Fillers*, Le Bras, M., Wilkie, C.A., Bourbigot, S., Duquesne, S., and Jama, C. (Eds.), The Royal Society of Chemistry, Cambridge, 2005, pp. 401–411.
32. Squires, G., Thomas, R., Georlette, P., Finberg, I., and Reznick, G., Benefits of complex flame retardant systems on properties of plastic compositions, *Annual BCC Conference on Recent Advances in Flame Retardancy of Polymeric Materials*, Stamford, CT, May 12–23, 2007.
33. Bar-Yaakov, Y., and Finberg, I., (Bromine Compounds), Fire retardant formulations, U.S. Patent, 2010: 7 695 650.
34. Bar-Yaakov, Y., Finberg, I., Shteklcr, R. and Georlette, P., (Bromine Compounds) Flame-retardant compositions for styrene-containing polymers, U.S. Patent, 2013: 8 524 813.
35. Levchik, S., and Georlette, P., Low antimony trioxide and antimony-free flame retardant plastics for electrical and electronic applications, *Specialty Chem.*, September 31–33, 2013.
36. Aaronson, A.M., Oligomeric phosphate flame retar- dants, *Annual BCC Conference on Recent Advances in Flame Retardancy of Polymeric Materials*, Stamford, CT, May 20–22, 1996.
37. Bright, D. A.; Dashevsky, S.; Moy, P. Y.; Williams, B. Resorcinol Bis(Diphenyl Phosphate), a Non-Halogen Flame-Retardant Additive. *J. Vinyl Add. Technol.* **1997**, *3*, 170–174.
38. Gosens, J.C., Pratt, C.F., Savenije, H.B., and Claesen, C.A.A., (General Electric) Polymer mixture having aromatic polycarbonate, styrene-containing copolymer and/or graft polymer and a flame-retardant, articles formed therefrom, U.S. Patent, 1993: 5 204 394.
39. Kenichi, O., (Asahi chemicals) phosphoric ester flame retardant, Japanese Patent, 2000: 3 043 694.
40. Levchik, S. V.; Bright, D. A.; Moy, P.; Dashevsky, S. New Developments in Fire Retardant Non-Halogen Aromatic Phosphates. *J. Vinyl Additive Technol.* **2000**, *6*, 123–128.
41. Seidel, A., Eckel, T., Wittmann, D., and Kurzidim, D., (Bayer) Impact-modified polycarbonate blends, U.S. Patent, 2006: 7 067 567.
42. Moy, P.Y., Gregor, A.I., Pirelli, R.L., Bright, D.A., and Bright, L., (ICL-IP America) Oligomeric bis-phosphate flame retardants and compositions containing the same, European Patent, 2015: 2 089 402.
43. Matsumura, T., and Tanaka, Y., (Daihachi) Crystalline powders of aromatic diphosphates, their preparation and use, European Patent 1997: 0 509 506.
44. Katayama, M., Ito, M., and Otsuka, Y., (Daicel) Polycarbonate resin composition containing block copolymer, U.S. Patent, 2001: 6 316 579.
45. Pawlowski, K.H. and Schartel, B., Flame retardancy mechanisms of triphenyl phosphate, resorcinol bis(diphenyl phosphate) and bisphenol A bis(diphenyl phosphate) in polycarbonate/acrylonitrile-butadiene-styrene blends, *Polym. Int.*, 2007, 56, 1404–1414. DOI:10.1002/pi.2290
46. Schartel, B.; Pawlowski, K. H.; Perret, B. Flame Retardancy Mechanisms in Halogen-Free PC/ABS blends. *Polym. Mater, Sci. Eng.* **2008**, *98*, 245–246.
47. Murashko, E. A.; Levchik, G. F.; Levchik, S. V.; Bright, D. A.; Dashevsky, S. Fire Retardant Action of Resorcinol Bis(Diphenyl Phosphate) in PC-ABS Blend. II. Reactions in the Condensed Phase. *J. Appl. Polym. Sci.* **1999**, *71*, 1863–1872.
48. Perret, B.; Pawlowski, K. H.; Schartel, B. Fire Retardancy Mechanisms of Arylphosphates in Polycarbonate (PC) and PC/acrylonitrile-Butadiene-Styrene. The Key Role of Decomposition Temperature. J. Therm. Anal. Calorim. 2009, 97, 949–958. https://doi.org/10.1007/s10973-009-0379-7
49. Despinasse, M.-C.; Schartel, B. Aryl Phosphate-Aryl Phosphate Synergy in Flame-Retarded Bisphenol A Polycarbonate/acrylonitrile-butadiene-styrene. Thermochim Acta. 2013, 563, 51–61. https://doi.org/10.1016/j.tca.2013.04.006
50. Despinasse, M.-C.; Schartel, B. Influence of the Structure of Aryl Phosphates on the Flame Retardancy of Polycarbonate/acrylonitrile-butadiene-styrene. Polym Degrad. Stab. 2012, 97, 2571–2580. https://doi.org/10.1016/j.polymdegradstab.2012.07.005
51. Schartel, B.; Richter, K. H.; Boehning, M. Synergistic use of Talc in Halogen-Free Flame Retarded Polycarbonate/acrylonitrile-butadiene-styrene Blends, in *Fire and Polymers VI: New Advances in Flame Retardant Chemistry and Science*, ACS Symposium Series, Washington, DC, 2012, Vol. 1118, pp. 15–36.
52. Wang, W.; Zhao, G.; Wu, X.; Li, X.; Wang, C. Investigation on Phosphorus Halogen-Free Flame-Retardancy Systems in Short Glass Fiber-Reinforced PC/ABS Composites Under Rapid Thermal Cycle Molding Process Condition. Polym Compos. 2015, 36, 1653–1663. https://doi.org/10.1002/pc.23075
53. Chen, G., Na, Y., and Xu, Y., (Sabic), Water pipe for mining operations, European Patent, 2020: 3 362 517.
54. Taniguchi, M., Tada, Y., and Nishioka, Y., (Otsuka) Flame-retardant epoxy resin composition, molded article thereof, and electronic part, U.S. Patents, 2004: 6 797 750.
55. Maruyama, K., and Motoshige, R., (Mitsubishi), Flame retardant thermoplastic resin composition, European Patent, 2003: 0 728 811.
56. Yamamoto, Y., and Wursta, T., Phosphazene flame retardants as a green chemistry to replace halogen-type flame retardants, *ANTEC 2010*, Orlando, FL, May 16–20, 2010.
57. Moltzan, H., and Hall, L., UL1694 and plastic encapsulated integrated cirquits, *Annual BCC Conference on Recent Advances in Flame Retardancy of Polymeric Materials*, Stamford, CT, June 2–4, 1997.
58. Levchik, S.V., Analytical determination of unreacted TBBA flame retardant in printed wiring boards, *EIPC Winter Conference*, Rotterdam, February 13–14, 2020.
59. Levchik, S. V.; Weil, E. D. Thermal Decomposition, Combustion and Fire-Retardancy of Epoxy Resins—a Review of the Recent Literature. Polym. Int. 2004, 53, 1901–1929. https://doi.org/10.1002/pi.1473
60. Kleiner, H.J., (Hoechst), Process for preparing 6-chloro-(6H)-dibenzo[c,e][1,2]oxaphosphorin, U.S. Patent, 1995: 5481 017.
61. Kleiner, H.J., (Hoechst), Method of preparing 6H-dibenzo[c,e][1,2]oxaphosphorin-6-one, U.S. Patent, 1996: 5 391 798.
62. Ultz, R., and Sprenger, S., (Schill & Sellacher), Flame-retardant epoxy resins and flame retarders for epoxy resins, European Patent, 2001: 0 806 429.
63. Wang, C.-S., and Lin, C.H., (National Science Council of Taiwan), Epoxy resin rendered flame retardant by reaction with 9,10-dihydro-9-oxa-10-phosphaphenanthrene-10-oxide, U.S. Patent, 2001: 6 291 627.
64. Sprenger, S.; Utz, R. Reactive, Organophosphorus Flame Retardants for Epoxies. *J. Adv. Mater.* **2001**, *33*, 24–32.
65. Doering, M., Phosphorus containing flame retardants in epoxy resins, in *Flame Retardants 2008*, Interscience Communication, London, 2008, pp. 121–131.
66. Hirai, Y., Suzuki, H., Takeda, Y., Oohori, K., Kamoshida, S., Kakitani, M., and Abe, N., (Hitachi), Resin compo- sition,

and use and method for preparing the same, U.S. Patent, 2004: 6 720 077.

67. Rakotomalala, M.; Wagner, S.; Doering, M. Recent Developments in Halogen Free Flame Retardants for Epoxy Resins for Electrical and Electronic Applications. Materials. 2010, 3, 4300–4327. https://doi.org/10.3390/ma3084300
68. Gan, J., (Dow), Phosphorus-containing compounds useful for making halogen-free, ignition-resistant polymers, U.S. Patent, 2012: 8 143 357.
69. Wang, R.-T., Yu, L.-C., Lin, Y.-T., Chen, Y.-J., Tian, W., Ma, Z., and Lu, W., (Elite) Halogen-free resin composition, copper clad laminate using the same, and printed circuit board using the same, U.S. Patent, 2017: 9 650 512.
70. Sagara, T., Kakiuchi, H., Kashihara, K., Kitai, Y., Saito, H., Yokoyama, D., and Fujiwara, H., Resin composition, resin varnish, prepreg, metal-clad laminate, and printed wiring board, U.S. Patent, 2016: 9 528 026.
71. Gao, Y., Hsieh, C.-Y., Ma, Z., Wang, R., and Tian, W., Polyphenylene oxide resin, method of preparing polyphenylene oxide resin, polyphenylene oxide prepolymer and resin composition, U.S. Patent, 2016: 9 447 238.
72. Acquasanta, F.; Berti, C.; Colonna, M.; Fiorini, M.; Karanam, S. Glow Wire Ignition Temperature (GWIT) and Comparative Tracking Index (CTI) of Glass Fibre Filled Engineering Polymers, Blends and Flame Retarded Formulations. Polym. Degrad. Stab. 2011, 96, 2098–2103. https://doi.org/10.1016/j.polymdegradstab.2011.09.022
73. Gallucci, R.R., and Patel, B.R., Poly(butylene terephthalate), in *Modern Polyesters*, Scheirs, J., and Long, T., (Eds.), John Wiley, Chichester, 2003, pp. 293–321. DOI:10.1002/0470090685.ch8
74. Crosby, J. M.; Talley, K. L. Design in Reinforced Flame-Retardant Plastics. *Modern Plast.* **1982**, 66–70.
75. Gepraegs, M., (BASF) Flameproof thermoplastic molding material, e.g. for electrical plugs and lamp parts, contains polyester, halogenated fire retardant, antimony oxide, calcium phosphonate or aluminum phosphinate and anti-drip additive, German Patent Application, 2003: 10 132 058.
76. Reznick, G., Finberg, I., Staimetz, A., Bar Yaakov, Y., Georlette, P., and Geran, T., Tomorrow's fire retardants for technopolymers mean stricter fire safety and higher performance, *Annual BCC Conference on Recent Advances in Flame Retardancy of Polymeric Materials*, Stamford, CT, June 7–9, 2004.
77. Bar Yaakov, Y., Finberg, I., Peled, M., Georlette, P., Stainmetz, A., Plevinsky, D., and Geran, T., Development of brominated flame retardant systems with improved corrosion resistance for application in plastics, *FRCA Fall Conference*, Cleveland, OH, October 20–22, 2002, pp. 159–172.
78. Litzenburger, A. Criteria for, and Examples of Optimal Choice of Flame Retardants. *Polym. Polym. Compos.* **2000**, *8*, 581–592.
79. Borms, R., Reznick, G., Bar Yaakov, Y.Georlette, P., Polymeric flame retardant systems: A step forward to a safer environment, in *Flame Retardants 2006*, Interscience Communications, Greenwich, pp. 135–142.
80. Dvir, H.; Gottlieb, M. A Reactive Flame Retardant as A Poly(propylene)-Sized Glass Adhesion Promoter. Macromol. Mater. 2006, 291, 913–916. https://doi.org/10.1002/ mame.200600010
81. Reed, J.S., Landry, S.D., De Schryver, D.A., Dadgar, B.B., Balhoff, D.E., Kolich, C.H., Ao, M.-S., and Lin, H.C., (Albemarle) Polymer compositions containing brominated polystyrenic resins, U.S. Patent, 2001: 6 235 831.
82. De Schryver, D. Brominated Polystyrene for Engineering Thermoplastics. *Specialty Chem. Magazine.* **2006**, 32–33.
83. Gatti, N., New Red Phosphorus Masterbatches Find New Application Areas in Thermoplastics. *Plast. Add. Compound.* **2002**, 34–37.
84. Chen, Z.; Du, J.; Li, X.; Xie, Z.; Wang, Y.; Wang, H.; Zheng, J.; Yang, R. Failure Behavior of Nylon Products for Red Phosphorus Flame Retardant Electrical Connectors. RSC Adv. 2019, 9, 24935–24941. https://doi.org/10.1039/C9RA04027G
85. Engelmann, J., and Warting, D., (BASF), Halogen-free flameproof polyester, European Patent, 2005: 1 423 460.
86. Costanzi, S., and Leonardi, M., (Italmatch), Polyester compositions flame retarded with halogen-free additives, U.S. Patent, 2010: 7 700 680.
87. Roth, M., Uske, K., and Wolf, U., (BASF), Long-fiber- reinforced flame-retardant polyesters, U.S. Patent 2015: 8 962 717.
88. Yang, W.; Song, L.; Hu, Y.; Lu, H.; Yuen, R. K. K. Enhancement of Fire Retardancy Performance of Glass-Fibre Reinforced Poly(Ethylene Terephthalate) Composites with the Incorporation of Aluminum Hypophosphite and Melamine Cyanurate. Compos. B Eng. 2011, 42, 1057–1065. https://doi.org/10.1016/j.compositesb.2011.03.019
89. Tang, G.; Wang, X.; Xing, W.; Zhang, P.; Wang, B.; Hong, N.; Yang, W.; Hu, Y.; Song, L. Thermal Degradation and Flame Retardance of Biobased Polylactide Composites Based on Aluminum Hypophosphite. Ind. Eng. Chem. Res. 2012, 51, 12009–12016. https://doi.org/10.1021/ie3008133
90. Jenewein, E., Kleiner, H.-J., Wanzke, W., and Budinsky, W., (Clariant) Synergistic flame protection agent combination for thermoplastic polymers, U.S. Patent 2002: 6 365 071.
91. Kleiner, H.-J., and Budinsky, W., (Clariant) flameproof polymer moulding material, U.S. Patent, 2001: 6 270 560.
92. Schlosser, E., Nass, B., and Wanzke, W., (Clariant) flame- retardant combination, U.S. Patent, 2001: 6 255 371.
93. Dietz, M., Hoerold, S., Nass, B., Schacker, O., Schmitt, E., and Wanzke, W., New environmentally friendly phos- phorus based flame retardants for printed circuit boards as well as polyamides and polyesters in E&E applications, *Electronics Goes Green*, Berlin, September 6–8, 2004, 771–776.
94. Walz, R.; Baque, T. Metal Phosphinates in Industrial Applications, *Kunststoffe*. **2007**, December, 113–115.
95. Sullalti, S.; Colonna, M.; Berti C.; Fiorini M.; Karanam S. Effect of Phosphorus Based Flame Retardants on UL-94 and Comparative Tracking Index Properties of Poly(Butylene Terephthalate). Polym. Degrad. Stab., 2012, 97, 566–572. https://doi.org/10.1016/j.polymdegradstab.2012.01.015
96. Wanzke, W., Non halogen FR solutions for polyesters in E&E applications—the use of phosphinates, *AMI Conference Fire Retardants in Plastics*, Denver, CO, June 14–15, 2012.
97. Braun, U.; Schartel, B.; Fichera, M. A.; Jaeger, C. Flame Retardancy Mechanisms of Aluminium Phosphinate in Combination with Melamine Polyphosphate and Zinc Borate in Glass-Fibre Reinforced Polyamide 6,6. *Polym. Degrad. Stab.* **2007**, *92*, 1528–1545. DOI:10.1016/j.polymdegradstab.2007.05.007
98. Braun, U.; Schartel, B. Flame Retardancy Mechanisms of Aluminium Phosphinate in Combination with Melamine Cyanurate in Glass Fibre-Reinforced Poly(1,4-Butylene tere-Phthalate). Macromol. Mater. Eng. 2008, 293, 206–217. https://doi.org/10.1002/mame.200700330
99. Ding Y., Stoliarov S. I., Kraemer R. H. Pyrolysis Model Development for A Polymeric Material Containing Multiple Flame Retardants: Relationship between Heat Release Rate and Material Composition. Comb. Flame. 2019, 202, 43–57. https://doi.org/10.1016/j.combustflame.2019.01.003

100. Stern, G.; Horacek, H. Melamine Cyanurate. Halogen and Phosphorus Free Flame Retardant. *Int. J. Polym. Mater.* **1994**, *25*, 255–268.
101. Casu, A.; Camino, G.; De Giorgi, M.; Flath, D.; Morone, V.; Zenoni, R. Fire-Retardant Mechanistic Aspects of Cyanurate in Polyamide Copolymer. Polym. Degrad. Stab. 1997, 58, 297–302. https://doi.org/10.1016/S0141-3910(97)00061-X
102. Kaprinidis, N., and Zingg, J., Halogen free melamine based flame retardants for polyamides, *FRCA Spring Conference*, New Orleans, LA, March 9–12, 2003, pp. 167–175.
103. Nagasawa, Y.; Hotta, M.; Ozawa, K. Fast Thermoly-sis/FTIR Studies of Fire-Retardant Melamine-Cyanurate and Melamine-Cyanurate Containing Polymer. *J. Anal. Appl. Pyrolysis.* **1995**, *33*, 253–267.
104. Gijsman, P.; Steenbakker, R.; Furst, C.; Kersjes, J. Differences in the Flame Retardant Mechanism of Melamine Cyanurate in Polyamide 6 and Polyamide 66, Polym. Degrad. Stab. 2002, 78, 219–224. https://doi.org/10.1016/ S0141-3910(02)00136-2
105. Levchik, S. V.; Balabanovich, A. I.; Levchik, G. F.; Costa, L. Effect of Melamine and Its Salts on Combustion and Thermal Decomposition of Polyamide 6. *Fire Mater.* **1997**, *21*, 75–83.
106. Tjahjadi, M.; Gallucci, R. R. Tracking Resistance of Flame Retardant Glass Reinforced PBT. *J. Elastomers Plast.* **1999**, *31*, 206–212.
107. Hirchsohn, Y., Eden, E., and Epshtein Assor, Y., (ICL-IP America), flame retardant polyester composition, PCT Patent Application WO2018/026621: 2018.
108. Hirchsohn, Y., Eden, E., and Epshtein Assor, Y., (Bromine compounds), flame retardant polyester compositions, European Patent, 2022: 3 529 306.
109. Wypych, G., *PVC Formulary*, ChemTec Publishing, Toronto, 2009.
110. Weil, E. D.; Levchik, S. V.; Moy, P. Flame and Smoke Retardants in Vinyl Chloride Polymers—Commercial Usage and Current Developments. J. Fire Sci. 2006, 24, 211–236. https://doi.org/10.1177/0734904106057951
111. Bellasai, R., Finding clarity in LSHF cables, *Compounding Word*, 2016, December, 35–38.
112. Hackl, C., Lehrich, F., Bittner, G., (Elastogran), Self extinguishing thermoplastic polyurethanes and process for their preparation, European Patent 2004: 0 617 079.
113. Cogen, J.M., Chen, G., Tai X., and Brown, G.D., (Dow), Color-stable, halogen-free flame retardant thermoplastic polyurethane compositions, European Patent 2013: 2454316.
114. Makadia, C.M., (Lubrizol), Non halogen flame retardant thermoplastic polyurethane, U.S. Patent, 2015: 8 957 141.
115. Vanhalle, A., Verbeke, H., and Meynen, S.M.L., (Huntsman), Flame retardant thermoplastic composition, U.S. Patent, 2014: 8 872 034.
116. Karayianni, E., and Philippoz, J.-M., (DuPont), Flame retardant thermoplastic elastomer compositions, U.S. Patent, 2010: 7 790 790.
117. Fitie, C.F., (DSM), Flame retardant composition and insulated wires for use in electronic equipment, PCT Patent Application, 2019: WO 2019/063554.
118. Karayianni, E., Li, T., and Ni, Y., (DuPont), Fire-retardant copolyetherester composition and articles comprising the same, European Patent, 2017: 2 892 956.
119. Lee, B.-L., Worley, D., Scadding, P., Jones, B., Sakhalkar, S. and Giroux, W., Advances in TPE styrenic block copo- lymer compounding for UL flame retardant cable appli- cations, *57th International Wire and Cable Symposium*, Providence, RI, November 9–12, 2008, pp. 360–364.
120. Jiang, Z.; Liu, S.; Zhao, J.; Chen, X. Flame-Retarded Mechanism of SEBS/PPO Composites Modified with Mica and Resorcinol Bis(Diphenyl Phosphate). Polym. Degrad. Stab. 2013, 98, 2765–2773. https://doi.org/10.1016/j.polymdegradstab.2013.10.013
121. Kubo, H., and Sato, S., (GE), Resin composition for wire and cable covering material, European Patent, 2006: 1 563 008.
122. Kubo, H., Mhetar, V.R., Rajamani, V., Sato, S., Tai, X., and Yao, W., (GE) Flame retardant electrical wire, European Patent, 2008: 1 829 056.
123. Kosaka, K., Li, X., and Xiao, Y., (Sabic), Flame retardant thermoplastic composition and articles comprising the same, U.S. Patent, 2010: 7 799 854.
124. Klei, S.R., Pickett, J.E., and Xu, J.J., (Sabic), Poly(arylene ether) composition, method, and article, U.S. Patent, 2009: 7 576 150.
125. Harada, T., Rajamani, V., Sato, S., Tai, X., and Yao, W., (Sabic), Thermoplastic composition, coated conductor, and methods for making and testing the same, U.S. Patent, 2009: 7 504 585.
126. Gan, H., (Sabic), Poly(phenylene ether) composition and article, European Patent 2019: 3 317 350.
127. Forti, V., Balde, P.C., Kuehr, R., and Bel, G., The global E-waste monitor 2020. Quantities, flows, and the circular economy potential, United Nations University, Geneva, https://collections.unu.edu/view/UNU:7737
128. Ahirwar, R.; Tripathi, A. K. E-Waste Management: A Review of Recycling Process, Environmental and Occupational Health Hazards, and Potential Solutions. Environ. Nanotechnol. Monitor. Management. 2021, 15, 100409. https://doi.org/10.1016/j.enmm.2020.100409
129. Fjaeder, P., Turunen T., Rinne, P., Haekkinen, E, Kauppi, S., Sormunen, T., and Anderson, M., Harmful additives in WEEE plastics and the regulatory framework, Reports of the Finnish Environment Institute 33, Helsinki, 2022.
130. Cardamone, G. F.; Ardolino, F.; Arena, U.; About the Environmental Sustainability of the European Management of WEEE Plastics. Waste Manag. 2021, 126, 119–132. https://doi.org/10.1106/j.wasman.2021.02.040
131. Tange, L. A.; Van Houwelingen, J. A. B.; Peeters, J. R.; Vanegas, P. Recycling of Flame Retardant Plastics from WEEE, Technical and Environmental Challenges. *Adv. Prod. Eng. Manag.* **2013**, *8*(2), 67–77.
132. https://environment.ec.europa.eu/topics/chemicals/reach-regulation_en
133. www.rohsguide.com/
134. www.basel.int/Implementation/POPsWastes/TechnicalGuidelines/tabid/5052/Default.aspx
135. Schlummer, M., Fell, T., Maeurer, A., The role of chemistry in plastic recycling, *Kunststoffe Int.*, 2020, May, 34–37.
136. http://nontox-project.eu/
137. Debois, S., Hildebrandt, T.T., Marquard, M., and Slaats, T., Bridging the Valley of Death; A success story on Danish funding schemes paving a path from technology readiness level 1 to 9, *2015 IEEE/ACM 2nd International Workshop on Software Engineering Research and Industrial Practices*, Florence, May 15. DOI:10.1109/SERIP.2015.18
138. Demacsek, C.; Tange, L.; Reichenecker, A.; Altnau G. PolyStyreneLoop—the Circular Economy in Action. IOP Conf. Series Earth Environ. Sci. 2019, 323, 012149. https://doi.org/10.11088/1755–1315/323/1/012149
139. Tulo, A. H. Polystyvert to Build Recycling Plant. *Chem. Eng. News.* **2023**, *101*(20).

29 Flame-Retardant Design and Protection for Wire and Cable

Walid H. Awad

29.1 INTRODUCTION

One of the major flame-retardant (FR) market sectors is wire and cable. The unique thing about this market is the diversity of flame-retardant technologies employed, and the stringent requirements to meet the fire safety standards, on both the small- and large-scale fire testing. A broad spectrum of flame retardant materials are used in this market, ranging from the halogenated to the non-halogenated FR materials, and the combination of both types. The main theme in this industrial sector is to meet the fire safety criteria requirements needed by the electrical codes and standards. Meeting the fire safety standards should be done in parallel with keeping the other property space needed for wire and cable materials to function properly. In the following sections, we will explore the different aspects of this industry, starting from the polymers and flame retardants needed for building and designing the cables, the manufacturing process, and the fire testing to evaluate these materials.

29.2 PRODUCT DEVELOPMENT

To survive and stay competitive in the wire and cable business, developing new products should be an essential component in the business model. One of the main challenges facing the product design and development process for wire and cable is how to formulate and select the materials needed for the target application. Another challenge would be how to evaluate the progress made through this process and how to take advantage of the evaluation tools that are available at disposal for the scientist and engineer involved in the process, many publications have been dedicated to this important subject.[1,2]

Engaging in the product development process requires first knowledge about the market need and a plan to address this deficiency. Next step is designing and tailoring the new product that meets the target and, most of the time, the conflicting properties. Exploring the wide portfolio of polymer resins and FR technologies available would be an essential component in this cycle.

Developing a compound with robust performance should be a target in the design process. That means a compound that will process satisfactorily and yield acceptable properties even if processing or service conditions deviate from those anticipated or thought ideal. Therefore, the goal should be to develop the best compound that the application can afford.

29.3 POLYMERS USED IN WIRE AND CABLE

Cable technology is in principle a materials-based technology. The cable is basically composed of metallic conductor or optical fiber as the core element. Both copper and aluminum could be used as conductors, but copper is used predominantly for power/electrical transmission applications. Optical fiber is starting to be used more, but it is limited to communication systems. In order to have signal transmission across this metallic core, it should be isolated and surrounded with dielectric insulation.

A wide range of materials are used in the design of all cables categories, such as building, industrial power, and photovoltaic cables, (Figure 29.1). These materials in the context of fire performance can be classified as combustible and noncombustible materials. Polymers form the bulk of the combustible materials in these cables. Tables 29.1 and 29.2 show the chemical structure and the combustion properties for a number of polymeric materials.

Polymers offer a number of advantages in terms of cable manufacture and performance, generally exhibiting a very high level of electrical resistance (Tables 29.3) and ease of processing. On the other hand, these polymers, as organic materials, will eventually burn during the fire, but they need to be formulated in order to reduce their fire hazard. In the following sections, we will explore the different types of polymers used in wire and cable.[4–7] Understanding the polymers available helps one design flame retardancy for the polymer chosen.

29.3.1 Polyethylene

Polyethylene is a semicrystalline thermoplastic material, first developed in the early 1930s by British scientists at Imperial Chemicals Industries (ICI).[8] Since that time, polyethylene material has undergone a wide variety of changes, as scientists across the globe have been able to develop new polymeric structures by varying the type and amount of side-chain branching of the polymeric structure, as well

DOI: 10.1201/9781003380689-29

as the ability to control the polymer molecular weight and molecular weight distribution. This ability to tailor the molecular design of polyethylene was made possible using new generations of catalyst technologies, such as Chromium oxide, Ziegler-Natta, and Metallocene catalysts. These developments represented unprecedented success story and created a wide variety of commercial applications for polyethylene.[9–11]

The first application of polyethylene in wire and cable was for radar applications during World War II. Now, polyethylenes are used widely as insulating materials for power and communication cables due to their good electrical properties. The negligible polar character of the carbon hydrogen bond controls these electrical properties. Polyethylene exhibits low dielectric loss and great water resistance, which are important product attributes for wire and cable applications. Polyethylene also has high dielectric strength, low dielectric constant, low power factor, and low dissipation factor. All these electrical attributes secure a good position for polyethylene in the wire and cable market.

The three parameters that control the properties of polyethylene are density, melt index and molecular weight distribution. Polyethylene properties may be modified by incorporating additives and fillers, usually mixed with resins during the extrusion process. Antioxidants are important to prevent degradation of polymer properties during manufacture and in service. Degradation by ultraviolet light can be inhibited to a remarkable degree by adding carbon black or light stabilizers.

Based upon the density range and the comonomer content, polyethylene can be classified into high density (HDPE), medium density (MDPE), low density (LDPE), linear low density (LLDPE), and very low density (VLDPE). For wire and cable applications, ASTM D1248 classifies the polyethylene into different categories,[12] as shown in Table 29.4.

29.3.1.1 Cross-Linked Polyethylene

Among the family of polyethylene resins, cross-linked polyethylene (XLPE) has a unique position due to the enhanced thermo-mechanical deformation resistance, which enabled the material to be used at high temperature. Although additives can be added to polyethylene to improve the physical and mechanical properties, there is still limitation related to its inherent lower thermal stability, cross-linking is employed to overcome this limitation.

Because of its thermoset character, XLPE maintains its mechanical properties when exposed to temperatures that would cause linear polyethylene to melt, lose shape, and flow. For example, XLPE cables may be operated continuously at 90°C and intermittently at 130°C during fault conditions. Figure 29.2 shows typical construction of XLPE cable.

Cross-linking is a process where thermoplastic polyethylene converts into a three-dimensional network structure. This network structure renders high strength, toughness and crack resistance, high thermal stability, and better flame performance. These properties made it ideal for different applications, including wire and cable, especially as insulation for low voltage applications. The process of cross-linking can be classified into three different categories[13–15] (Figure 29.3):

- PEX-a is made using the Peroxide, or Engel method.
- PEX-b is made using Silane, or the moisture Cure method.
- PEX-c is made using the Electron beam method.

A brief description of each method will be covered later. These cross-linking processes must be considered in flame retardant selection, to ensure the flame retardants are compatible with the crosslinking process and will maintain their effectiveness after process exposure.

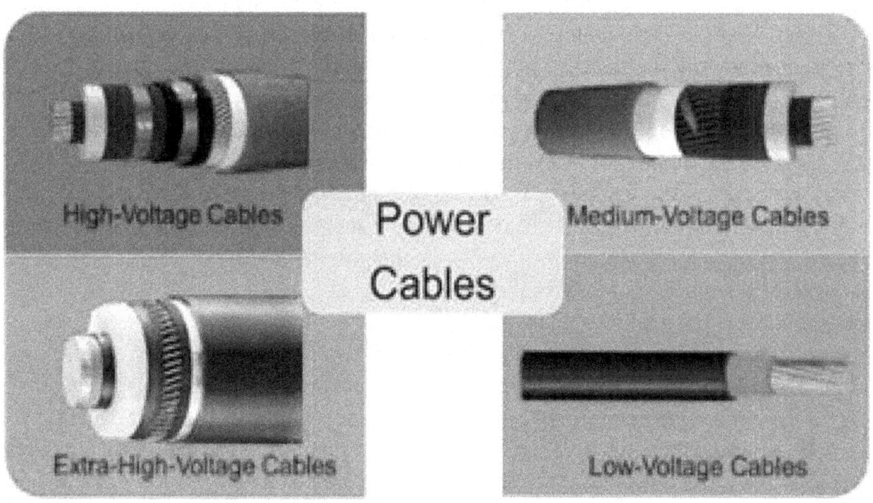

FIGURE 29.1 Different types of power cables.[3]

Source: Open access

29.3.1.1 Peroxide Cross-Linking (Continuous Vulcanization CV)

This is the most common method used for medium and high-voltage cable applications. A small amount of organic peroxide, the cross-linking agent, is added to the polymer—the most common one is dicumyl peroxide. During the extrusion process, which is the method used for cable manufacture, the peroxide remains stable due to its high decomposition temperature. After extrusion, the cable enters a long curing tube in the shape of a catenary, where the temperature is raised above the decomposition temperature of the peroxide. At this high temperature and pressure, the peroxide decomposes and triggers the cross-linking process.[16,17]

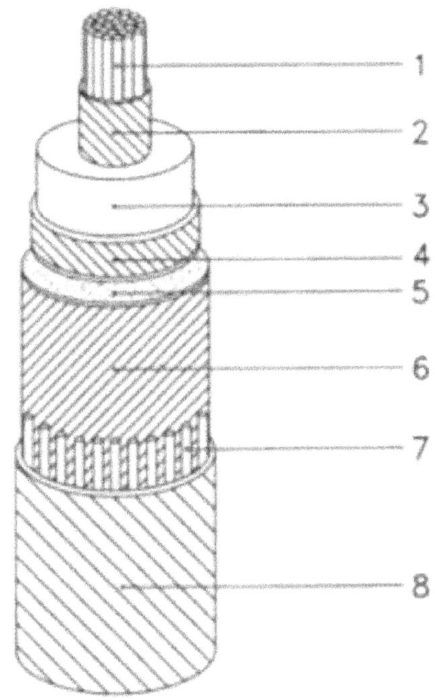

FIGURE 29.2 Typical construction of cross-linked XLPE-**insulated HV cable**: (1) stranded conductor, (2) semiconducting conductor shield, (3) XLPE insulation, (4) semiconducting insulation shield, (5) shielding copper tape, (6) inner sheath, (7) wire armoring, (8) outer sheath.

Source: From *Rubber Products Manufacturing Technology*, Bhowmick, A.K., Hall, M.M., Benarey, H.A., Eds., Marcel Dekker Inc, with permission.

Catenary CV lines are used in cable manufacture,[18] as shown in Figure 29.4. During the process, the core is prevented from touching the sides of the tube by accurately controlling the tension in the conductor. On completion of the crosslinking, the cable passes into the cooling section of the tube in which water or nitrogen gas is circulated. A key element in this process is to avoid the degradation of the peroxide during the extrusion process and the premature cross-linking. If this occurs, scorching (or gels of oxidized polymer) may form that can compromise the performance of the cable and leads to failure. Another challenge that needs to be addressed during manufacture is the deformation and sagging of the material due to the gravity effect. This can be properly controlled by proper control of the materials and processing conditions.

29.3.1.2 Silane Cross-Linking (Monosil and Sioplas)

Both peroxide and radiation cross-linking processes have some disadvantages such as high investment costs for the radiation equipment, the risk of scorching in peroxide cross-linking processes. Research efforts to improve cross-linking technology resulted in the development of silane cross-linking technique that was able to overcome these disadvantages.[19–21]

The silane-induced cross-linking, commonly known as "moisture curing," employs organic chemical agents called silanes that react with water and induce the cross-linking

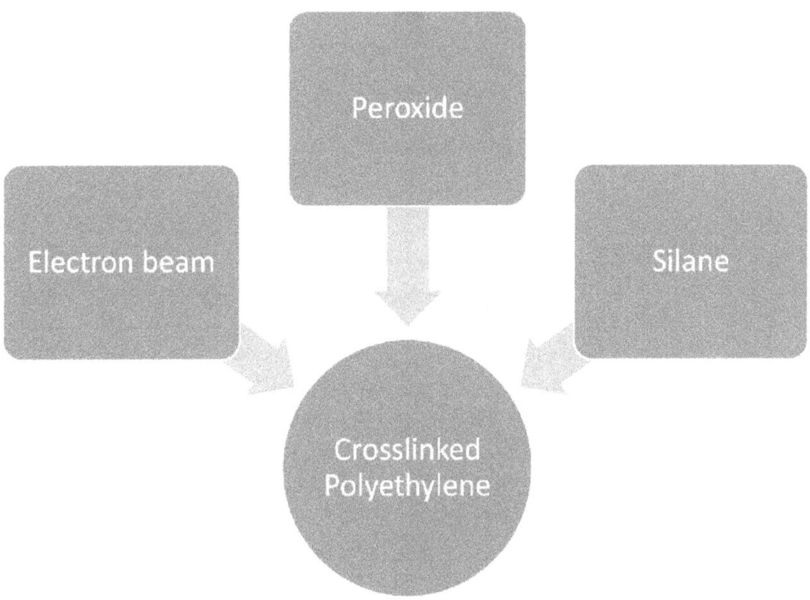

FIGURE 29.3 Different techniques for cross-linking polyethylene.

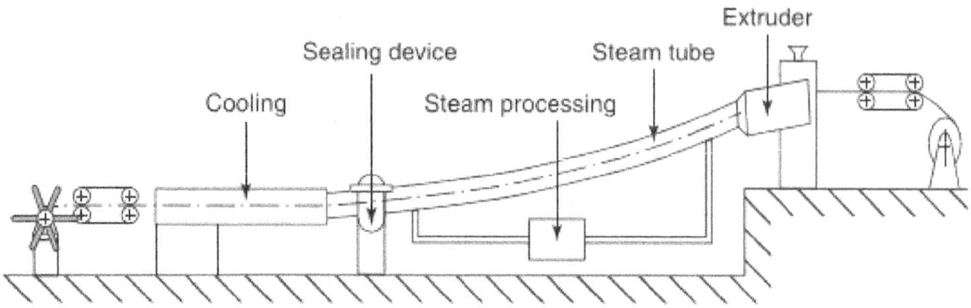

FIGURE 29.4 Catenary CV curing.

Source: From Herpich, B., in *The Global Cable Industry, Beyer,* G., Ed., Chapter 3, Wiley-VCH GmbH, 2021. With permission.

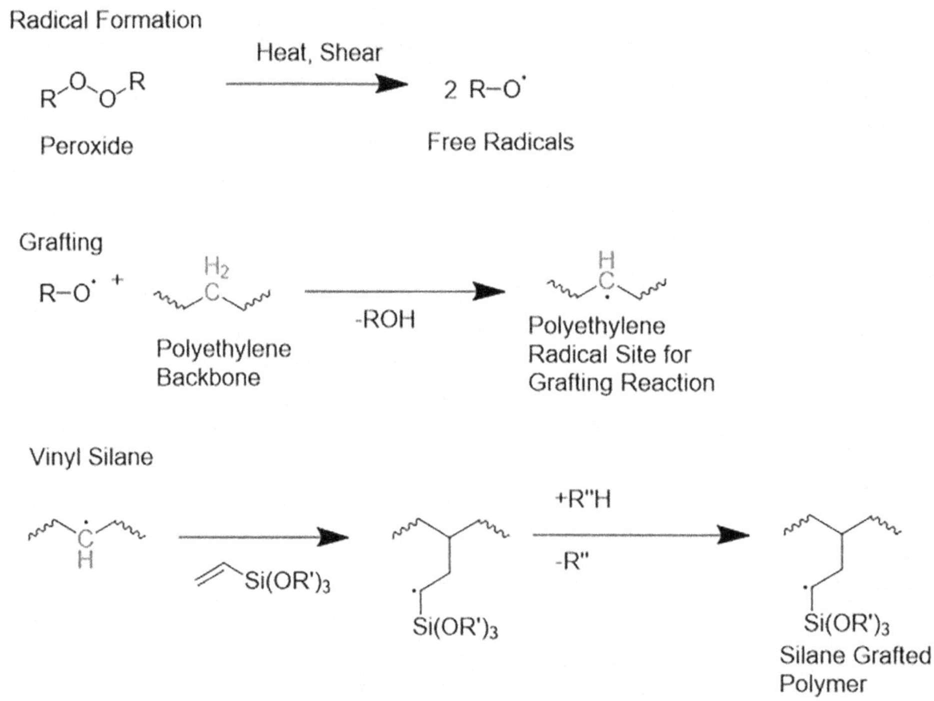

FIGURE 29.5 Formation of a silane grafted polymer.

process, as shown in Figures 29.5 and 29.6. In this category, the silane agent is included in the polyethylene backbone, as either grafting or copolymer. Two approaches can be employed for the silane cross-linking polyethylene known as Sioplas and Monosil.

Another way for the Sioplas process involves the use of a polyethylene-silane copolymer (rather than a grafted material), it allows the wire/cable producer to directly procure the silane-modified polyethylene; this simplifies the handling aspects.

The other approach called Monosil, developed by BICC-Maillefer, is a one-step process.[25,26] The silane grafting takes place during the crosshead extrusion on the conductor. It involves mixing all the ingredients, the polyethylene, the catalyst, and silane together during the extrusion process.

For both approaches, Sioplas and Monosil, the curing process involves subjecting the finished cable to water, the alkoxysilane groups hydrolyze to silanol groups, which further condense to form a three-dimensional network with siloxane cross-links between polyethylene chains.

Curing or cross-linking process takes place in the presence of water on the finished article, and there is no difference in mechanism between the graft or the copolymer. The curing proceeds at a rate dependent on (a) water diffusion and (b) the insulation wall thickness; the water must penetrate the wall for curing to occur. Raising the water

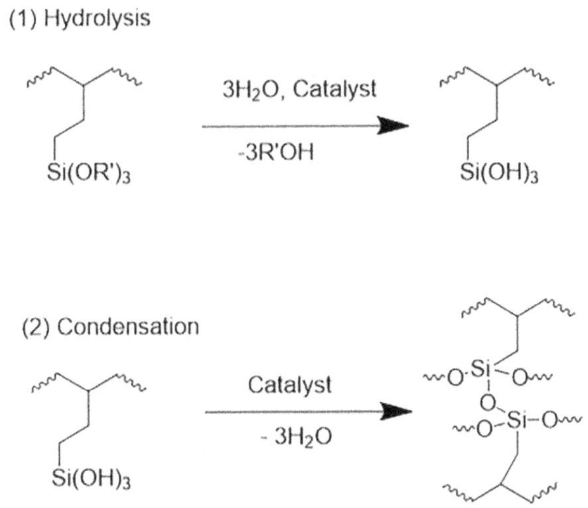

FIGURE 29.6 In Sioplas technology, first developed by Dow Corning,[22,23] the basic chemistry involves (a) grafting the alkoxy silane crosslinker onto the polymer chain via radical chemistry, (b) hydrolysis of the grafted alkoxy group, and (c) condensation of the silanol and the formation of three-dimensional network structure. In Sioplas production process, the silane grafted polyethylene combined with a catalyst masterbatch are extruded on the conductor to produce the cross-linked insulation. Many commercial products are available using this technology.[24]

temperature increases the water diffusion rate into the insulation and therefore the cross-linking (curing) rate; the thinner cable insulation walls will cure more rapidly.

29.3.1.3 Electron Beam Cross-Linking

Radiation processing of polymeric materials is the main step in certain modern technologies and has extensive applications. It involves mainly the use of either electron beams from electron accelerators or gamma radiation from ^{60}co-radiation facilities. Industrial applications of radiation processing of plastics include polymerization, grafting, and cross-linking. The major application of radiation processing for wire and cable is the cross-linking of polymeric insulation.[27,28]

In this process, a high-energy electron beam is used to trigger the cross-linking reaction, as shown in Figure 29.7. The electron beam creates radicals by abstracting hydrogen and the formation of carbon-carbon bond, hence inducing cross-linking, preferentially in the amorphous phase of the polymer structure. The radiation cross-linking process is performed at room temperature, which means, the polyethylene is in the solid state, which is different from other cross-linking techniques.

Because of the very short wavelength and high frequencies, electron beam has a strong penetrating power, and the depth of this penetration is very important for wire and cable. In general, the depth of penetration is directly proportional to the accelerator voltage. However, as the voltage increases, the potential for polymer degradation also increases.

The basic unit of radiation is called rad (equals to 100 ergs/gm); the commercial applications require higher doses and are termed grays (1 gray equivalent to 100 rads). The electron beam process needs no curing agent like sulfur or peroxides for vulcanization/cross-linking.

29.3.2 Polypropylene

Polypropylene is known to have good insulating properties such as high dielectric strength, low dielectric constant, low dissipation factor, and good moisture resistance, combined with high temperature resistance, but the conventional grade lacks the flexibility due to the crystal structure, however the recent catalyst technologies improved this deficiency.[29]

The Ziegler-Natta catalyst technology allowed manufacturing of commercially viable polypropylene. The resulting product was predominantly an isotactic polypropylene and could be manufactured at high molecular weight. Metallocene catalysts were the next generation catalysts and led to creating both isotactic and syndiotactic structure with highly controlled molecular weight distributions and tacticity.

As an advantage to thermoplastic insulation, polypropylene-based insulation can be used for conductor temperatures of 90°C, just like XLPE. In the cable industry, polypropylene is used in four applications: as insulation of twisted copper wires, in particular data cables, in buffer tubes in fiber optic cables, as an HFFR automotive cable, and as thermoplastic insulation of power cables.

29.3.3 Ethylene Vinyl Acetate

Ethylene vinyl acetate (EVA) is a type of copolymer that consists of ethylene and vinyl acetate units. It has many advantages for wire and cable applications, such as good flexibility, low flammability, high heat shrink ability (ability to shrink around and conform to underlying substrates), and excellent electrical properties.[30]

Ethylene vinyl acetate with vinyl acetate content in the range (18%–33%) is widely used for wire and cable, especially for jacket applications, the polarity of the resin makes

FIGURE 29.7 Electron beam processing of insulated wire and cable.

Source: From Goethals, R., in The Global Cable Industry, Beyer, G., Ed., Chapter 8, Wiley-VCH GmbH, 2021. With permission.

it difficult to use it as insulation. You can tailor the properties of EVA by varying the vinyl acetate content. EVA has better thermal stability and resistance to oxidation than polyethylene. The lower melting point gives it advantage in terms of processability. EVA is well-known for its oil resistance and synergistic flame retardant effect with hydrated mineral flame retardants, such as magnesium hydroxide and aluminum hydroxide (see Chapter 6 of this book).

The higher vinyl acetate content allows for high filler loadings, retention of properties and cold temperature flexibility. EVA is relatively amorphous and can tolerate the high filler loading typically required for high FR loading while providing reasonably good physical properties. EVA based wire and cable compounds can be cross-linked to improve their end use performance.

29.3.4 Ethylene Acrylate Copolymers (EAC)

These are copolymers of ethylene and acrylate esters such as methyl, ethyl, or butyl acrylates. They have higher softness, flexibility, and polarity than PE. They also have better oil resistance and flame retardancy than EVA. EAC can be used as a modifier for PE or as a base polymer for wire and cable compounds.[31]

29.3.5 Polyolefin Elastomers

These are copolymers of ethylene and alpha olefins such as butene, hexene, or octene. They exhibit extreme elastic elongation and flexibility when subjected to mechanical stress. They have low crystallinity, which allows for high filler loadings and their elastomeric characteristics allow them to better retain their properties, even with the high loading levels of fillers. They can be used as a modifier for PE or as a base polymer for wire and cable compounds.[32]

Both Ethylene Propylene Copolymer and Terpolymer rubber (EPR and EPDM) are widely used in wire and cable for high temperature applications. They have a combination of excellent electrical and mechanical properties. EPDM has unsaturation in the main chain, whereas EPR has a completely saturated structure. This difference gives EPR an advantage over EPDM in terms of heat aging, and so EPR is the preferred choice in high-voltage applications.

29.3.6 Chlorinated Polyethylene (CPE)

The manufacture of chlorinated polyethylene (CPE) is carried out by reacting the chlorine with high density polyethylene, and that creates a polar structure. The resulting polymer contains 25–45% chlorine by weight and has a higher density and crystallinity than polyethylene.[33] CPE can be cross-linked with peroxides to form an elastomer that has improved mechanical strength and elasticity. CPE can also be blended with other polymers, such as PVC, EVA, or EPDM, to modify its properties and reduce its cost.

CPE-based jacket materials have improved chemical and oil resistance in addition to improved flammability properties like PVC. A number of factors, such as the molecular weight and the molecular weight distribution, the chlorine content, and the chlorine distribution, influence the properties of CPE product.

The chlorination of polyethylene inhibits the crystallization process, and it determines how rubbery the modified polymer is. As you increase the chlorine content to more than 40% it becomes increasingly brittle. The most desirable CPE for wire and cable applications is obtained when the degree of chlorination is around 35%.

CPE is compatible with a wide range of additives, such as fillers, stabilizers, antioxidants, flame retardants, and colorants. These additives can enhance the properties of

CPE to meet specific requirements of different wire and cable applications. For example, CPE can be formulated with high chlorine content and flame retardants to achieve the desired rating for fire resistance. The main applications for CPE-based compounds are jackets for flexible power transmission, portable power, control, and mining cables.

29.3.7 Chlorosulfonated Polyethylene (CSM)

Chlorosulfonate polyethylene (CSM) is a synthetic rubber that has excellent resistance to heat, chemicals, ozone, UV, weathering, and abrasion. It is widely used as a jacketing and insulation material for wire and cable applications, especially in harsh environments where high temperature and oil resistance are required.[34] It has some inherent flame retardancy due to release of HCl and H_2SO_4 during combustion. However, these same gases released during burning are highly corrosive and can limit the use of this material to applications where high levels of flame retardancy are required, and there are no limitations on corrosive gas release.

CSM is produced by treating polyethylene with a mixture of chlorine and sulfur dioxide under UV-radiation. The resulting polymer contains 20–45% chlorine and a few percent of chlorosulfonyl groups that allow for vulcanization. Vulcanization is a process that creates cross-links between polymer chains, enhancing the strength and durability of the material. CSM can be vulcanized by various methods, such as peroxides, sulfur, or radiation.

CSM has several advantages over other types of polyethylene, such as low-density polyethylene (LDPE) or cross-linked polyethylene (XLPE). CSM does not melt when heated, unlike thermoplastic polyethylene. It can withstand temperatures up to 150°C and short-term peaks up to 250°C. CSM also has better oil resistance than LDPE and XLPE, making it suitable for applications where exposure to fluids and chemicals is expected. CSM does not require a paper separator between the conductor and the insulation, unlike some chemical cross-linked wires that use peroxides. CSM also has good electrical properties, such as low dielectric constant and high breakdown voltage.

CSM is compatible with a wide range of additives, such as fillers, stabilizers, antioxidants, flame retardants, and colorants. These additives can modify the properties of CSM to meet specific requirements of different wire and cable applications. For example, CSM can be blended with ethylene-propylene rubber (EPR) to improve flexibility and low-temperature performance. CSM can also be coated with other polymers, such as polyvinyl chloride (PVC) or polyurethane (PU), to enhance abrasion resistance or flame retardancy.

29.3.8 Polyvinyl Chloride

Polyvinyl chloride (PVC) is produced by polymerizing vinyl chloride monomer in a suspension or emulsion process. The resulting polymer has a linear structure with chlorine atoms attached to the carbon backbone. The degree of polymerization and the amount of chlorine determine the properties of PVC. PVC can be modified with various additives, such as plasticizers, stabilizers, fillers, pigments, and flame retardants, to achieve different performance and appearance.[35] The presence of chlorine in the polymer means the material is inherently flame retardant, due to the release of HCl during combustion of this polymer. This HCl release can be problematic for applications where corrosive gas release is prohibited.

A wide range of PVC resin properties may be obtained by controlling molecular weight, particle size, and particle configuration. High-molecular-weight resins are more difficult to process but exhibit superior strength and greater resistance to heat and chemicals. Particle configuration determines the degree of porosity and the ability to absorb plasticizers.

Plasticization of PVC, a process needed to improve low temperature flexibility, leads to a reduction in flame retardancy and increased smoke generation. To achieve the desired balance of properties, two or more plasticizers are usually required. Important considerations in the selection of a plasticizer are its compatibility, volatility, tendency to migrate to other materials, toxicity, odor, burning characteristics, dielectric properties, weatherability, and efficiency. Volatile plasticizers are less desirable as a result of their propensity for migration into the insulation layers when PVC is used as a jacketing material.

PVC is often the choice for low-voltage cable applications. It offers several unique properties that make it an attractive candidate for wire and cable applications. Along with excellent mechanical properties, PVC has good chemical and corrosion resistance. Furthermore, unplasticized PVC compounds are known for their inherent flame retardancy, requiring only antimony trioxide to provide exceptional flame resistance. These properties, coupled with low cost, make PVC materials useful for indoor applications.

The electrical properties of PVC insulation vary with formulation, but to match the capacitance of a polyethylene cable, a thicker layer of PVC insulation must be used. The dissipation factor for PVC limits its use as insulation in coaxial cable, but it may find use in jacketing compounds.

29.3.9 Fluorinated Polymers

Fluoropolymers are extensively used in cable applications, especially in North America. They demonstrate high permanent service temperatures, excellent dielectric performances, and chemical resistance; they are inherently flame retardant. These characteristics are attributable to the carbon-fluorine bond, among the strongest known in organic chemistry, and much stronger, for example, than the carbon-chlorine bond.[36]

The typical fluoropolymers used in cable applications are fluorinated ethylene propylene (FEP) polymer, perfluoroalkoxy (PFA), ethylene tetrafluoroethylene (ETFE) polymer, ethylene chlorotrifluoroethylene (ECTFE) polymer, and in some applications polytetrafluoroethylene (PTFE).

Fluoropolymers have been used as wire coatings in military and aerospace applications since their discovery. They have also found their way into wire and cable applications that required extreme performance based on thermal extremes (cryogenic and high temperature applications), chemical resistance, etc., but the major market expansion in fluoroplastic use in the wire and cable industry occurred in North America when the National Electric Code (NEC) allowed polymers with low smoke-generation and flame-spread to be installed in building plenums without metal conduits.

29.3.10 Silicone Polymers

Silicone rubber compounds are known for their wide operating temperature range.[37] They may be used for very low temperature applications, showing a brittle point at –118°C. Silicone elastomers may be used in applications up to 260°C, making them suitable for severe service applications. The cable industry utilizes silicone elastomers if high temperature resistance and high flexibility are needed. Silicones formulated with flame retardants show excellent char formation and low heat release values. By proper formulation of the silicones and the cable design, the char acts as a mechanically strong ceramic and can form an electrically insulating layer useful for circuit integrity applications.

The properties of the silicone rubber are determined by the length of the molecular chain and by the nature of the substituents of the silicon atoms. Silicone rubbers are composed of the dimethyl-siloxane units, $(CH_3)_2SiO$- as shown in Table 29.1. Silicone rubbers are prepared by the hydrolysis and subsequent condensation of lower molecular weight silicone fluids (also defined by the molecular structure).

For wire and cable insulation, some methyl groups are replaced with phenyl and/or vinyl groups to increase tensile strength, impart tear resistance, extend the useful temperature range, and improve resistance to gamma radiation.

Silicone elastomers are usually filled with silica to provide reinforcement and alumina trihydrate for arc resistance. Inexpensive fillers, such as diatomaceous earth or ground quartz may also be used to reduce the cost of the compound. Silica filled systems offer improved flame retardancy over the unfilled materials. This combination results in the formation of an insulating char layer and low smoke and toxic gas released.

29.3.11 Polyamides

Polyamide (PA), also known as nylon, is a synthetic polymer that is widely used for wire and cable applications.[38] Polyamide has many useful properties, such as high strength, toughness, elasticity, abrasion resistance, and chemical resistance. Polyamide is a thermoplastic material that can be melted and extruded or molded over the conductor or the cable core.

TABLE 29.1
Chemical Structure of Polymers Used in Wire and Cable

Material	Unit Chemical Structure
Polyethylene	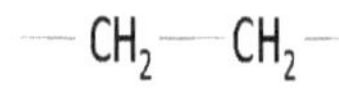
Polyethylene vinyl acetate	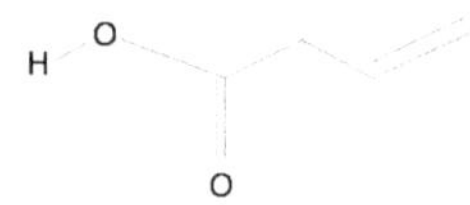
Polypropylene	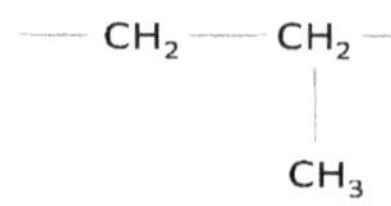
Chlorinated polyethylene	
Polyvinyl chloride	
Silicone rubber	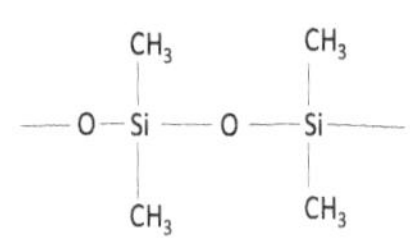
Nylon 6	
Chlorosulfonated polyethylene	

Polyamide is a thermoplastic with many properties that make it desirable for jacketing of wire and cable. Polyamides provide properties such as abrasion resistance, chemical resistance, high strength, dimensional stability, lightweight, and flexibility. Polyamides with long chains such as PA11 and PA12 have more advantage in terms of lower water absorption that make them good candidates for wire and cable. The most common use of polyamide in cable jacketing is the jacket materials in building wire, fiber optics for industrial application, transportation, and offshore applications.

Nylon can be customized with various additives, such as fillers, stabilizers, antioxidants, flame retardants, and colorants, to achieve different performance and appearance. Polyamide can also be blended with other polymers, such as PVC, EVA, or TPU, to modify its properties and reduce its cost. Polyamide is used for many types of wires and cables, such as power cables, control cables, communication cables, automotive cables, geophysical cables, industrial cables, and medical cables.

29.3.12 Thermoplastic Polyurethane (TPU)

TPU is produced by reacting diisocyanates with polyols and chain extenders. The diisocyanate and polyol form soft segments, while the diisocyanate and chain extender form hard segments. The hard segments are more polar and phase segregate from the soft segments, forming hard segment domains that act as physical cross-links. The soft segments provide elasticity and low-temperature performance, while the hard segments provide strength and rigidity.[39]

TPU can be used in wire and cable formulations and for those compounds that require excellent abrasion resistance, toughness, and weather resistance. Various grades of TPU, both polyether and polyester based, can be used to produce both soft and hard wire and cable formulations. Polyether-based TPUs are used when low temperature flexibility, good weather resistance, and resistance to wet environments is required. Polyester-based TPUs are used when applications require higher temperature resistance, chemical resistance, or higher mechanical properties.

TPU is used for many types of wires and cables, such as power cables, control cables, communication cables, automotive cables, geophysical cables, industrial cables, and medical cables.

29.4 FLAME RETARDANCY OF WIRE AND CABLES

Fire is a major concern to the wire and cable industry;[40] the burning cables will lead to substantial damage in different ways. This serious concern is reflected in the stringent fire safety requirements needed by the wire and cable codes and standards. For example, in the United States, the National Electrical Code (NEC) established a set of standards and requirements for flame spread and smoke generation of wires and cables installed in office building, residences, and factories.[41] These codes are prepared by the National Fire Protection Association (NFPA) and provide installation guidelines for flame-retardant cables.

There is a unique feature about fire in wire and cable, each cable will perform differently during a fire, depending upon its design, location, and its components. Therefore, to meet the target fire rating, the cable designer needs to consider all these factors. The component materials include the type of polymer resins and the associated flame retardants used for the insulation, bedding and the jacket, the metals used as electric conductors, or as layers to protect some cable components. The disparity in the heat conduction between the copper conductor and the polymeric materials is also a factor that has impact on the fire performance as it will affect the heat transfer process and the flame spread along the cables.

Given the serious fire risk posed by the wire and cable, there is a need to impart flame retardancy to these materials. The most common way to reduce the flammability of polymeric materials is by incorporating flame-retardant additives. Developing flame-retardant polymeric materials has gained a huge momentum over the past few decades, as evidenced by the number of publications dedicated to this subject,[42–47] and the commercialization of many flame-retarded systems.

In the following section, we will explore the principles of flame retardancy and the different types of flame retardants used in wire and cable.

29.4.1 Principles of Flame Retardancy

Polymer combustion, in principle, involves a multitude of steps, but essentially three elements are needed for a fire to happen: the combustible material (fuel), heat, and oxygen (Figure 29.8). These elements create what is known as the fire triangle; fire ceases to exist once one of these elements has been eliminated.

When a polymeric material is exposed to external heating source, and the heat energy exceeds the chemical bonding energy, the polymer starts to decompose, generating flammable gases. These gases, if mixed with air at stoichiometric ratio, will ignite if the temperature is high enough. If the heat evolved from this ignition is sufficient to make further degradation of the fuel, the combustion cycle will continue until all combustible materials are consumed. If the heat evolved from the ignition is below the amount needed to sustain the decomposition process of the fuel, self-extinguishment will occur.

The main function of flame retardant additive is to interrupt this fire cycle. The mechanism of this flame retardation is quite complex and has been reviewed extensively.[48,49] Generally speaking, there are a number of major mechanisms for flame retardant materials to interfere and interrupt this fire cycle, and they can be summarized as follows:

- *Gas phase mechanism*: This process involves inert gas dilution and chemical quenching of active radicals. The former (dilution effect) refers to the release of noncombustible vapors during combustion, diluting the oxygen supply to the flame or diluting the fuel concentration to below the flammability limit. Flame retardants or their

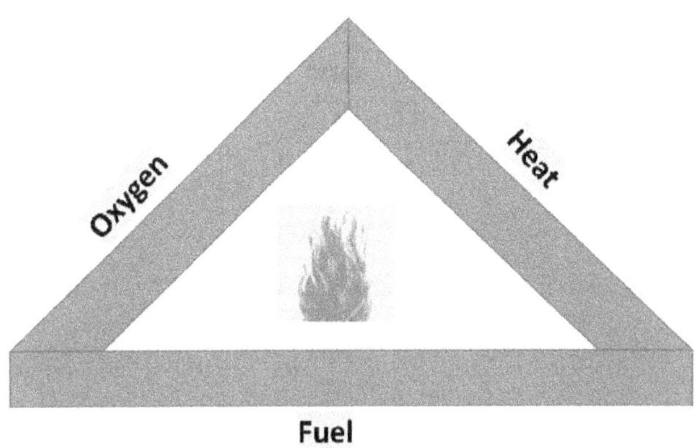

FIGURE 29.8 Fire triangle.

degradation products stop the radical mechanism of the combustion process that takes place in the gas phase, stopping the exothermic processes that occur in the flame and reducing the supply of flammable gases.

- *Condensed phase mechanism*: It is a process where flame retardants can act by forming a solid char (or a glassy layer), which interferes with the transfer of heat back from the gas phase to the condensed phase. This inhibits or prevents further thermal decomposition.
- *Removal of heat or cooling mechanism*: This represents a mineral-based flame retardant additives that have endothermic effect, such as aluminum trihydrate (ATH) and magnesium hydroxide (MDH). They release water and carbon dioxide, which dilute the fuel available for combustion. Endothermic decomposition cools down the condensation phase, slows degradation, and absorbs the fuel combustion heat.

Selecting the best flame retardants suitable for the wire and cable depends on several factors starting from the formulation design and ending with cable structure and environment. A number of requirements are needed for a good flame retardant, such as effectiveness at low loading, compatibility with the polymer resin, no detrimental effect on other physical, electrical, and mechanical properties, and last but not least, low cost. It is difficult to find flame retardant that meets all these demands, as some of these criteria are contradictory. The types of flame retardants used in wire and cable and their mechanism of action will be reviewed in the next section.

In the process of materials selection, it would be essential to have a good understanding of the mechanism of polymer thermal stability and degradation and the mechanism of how the selected flame retardant will operate in fire. This fundamental understanding will allow selecting the FR materials that have the optimum thermal and processing window with the polymer matrix. See Chapter 2 of this book for more details.

We should also keep in mind that it would be difficult to find a stand-alone flame retardant for a specific application. So taking advantage of the synergy between different FR mechanisms will be a must in order to find the sweet spot in the formulation that gives you adequate FR performance, in addition to the physical, rheological, electrical, and mechanical property space.

29.4.2 Halogenated Flame Retardants

Halogenated flame retardants are widely used in wire and cable, and this is attributed to the high effectiveness of these materials compared to other flame retardants. Early flame inhibition studies indicated that the effectiveness of the halogens decreases in the order: I > Br > Cl > F. The thermal stability of the iodine compound makes it difficult to survive the polymer processing conditions, whereas that of fluorine compound is highly stable. That makes both bromine and chlorine compounds the best choice for the halogen system.[50–52] Halogenated FRs are discussed more in Chapter 4 of this book.

Typically, the halogenated molecules (bromine or chlorine) are used in combination with antimony trioxide to deliver the synergy expected from their combination. Numerous studies indicated that the way halogenated flame retardants act during fire is by interfering with the combustion cycle in the gas phase. The very active combustion radicals (OH· and H·) that control the combustion process are removed by the decomposed halogenated species, thus effectively interfering in their oxidation as shown in equations 1 and 2. This disruption process leads to a significant reduction in the exothermicity of the reaction, thereby lowering the flame temperature.

$$H^{\bullet} + HX = H_2 + X^{\bullet} \quad (1)$$
$$OH^{\bullet} + HX = H_2O + X^{\bullet} \quad (2)$$

As mentioned above, one of the unique advantages of the halogen system is the high level of synergy exhibited between the halogen and the antimony. The antimony trioxide by itself has little or negligible flame retardant effect. However, by joining the halogenated molecules in the formulations, both act synergistically in the gas phase to extinguish the flame. One of the important factors that control this type of synergy is the molar ratio between the halogen and the antimony. Different ratios have been proposed in the literature, the most common one is 3:1 (halogen: antimony).

It should be noted that matching the thermal degradation of the halogenated molecules with that of the polymer resin is one aspect that needs to be considered in the selection of the best candidates. Halogen-containing species need to be selected to vaporize at the same temperature as the polymer fragments. Coexisting in the reactive area of the flame, the halogens are effective at decreasing the concentrations of the free radicals that propagate the flame, thus helping in making extinction to the flame. This type of assessment and preliminary information about the thermal degradation behavior of both the flame retarded molecule and the resin is important and could be obtained by thermal analysis techniques.

One of the factors for the dominant use of brominated flame retardants for wire and cable in the US, is the fact that with these types of flame retardants, it would be possible to pass the very stringent wet electrical test needed by UL standard. It would be difficult to pass this test with the non-halogenated flame retardants. This requirement is not needed by the European standards, and that enables the implementation of non-halogen flame retardants in Europe.

Even though the halogenated flame retardants, and specifically the brominated FR proved to be very effective in imparting flame retardancy to the wire and cable polymeric matrices, there is increasing pressure on these types of molecules due to environmental concern. This kind of pressure is pushing the wire and cable industrial sector to look for halogen-free FR as an alternative solution.

29.4.3 Non-Halogenated Flame Retardants

Hydrated mineral fillers represent the largest sector of this market,[53] where the term hydrated minerals include a wide range of materials, but the focus here would be more on magnesium hydroxide and aluminum trihydrate, because they play a major role in the wire and cable market.[54,55] See Chapter 6 of this book for more details on mineral filler flame retardants. Other classes of non-halogen flame retardants will be discussed as well in this section.

The use of these zero halogen FR materials has the potential to grow in the future due to the interest in developing "Low Smoke Zero Halogen" (LSZH) materials. The original concept of LSZH cables was identified through the requirements of underground railways in the 1970s to allow sufficient visibility for passengers' evacuation through a tunnel during a fire. Recently this concept is attracting more attention due to the need to use these types of cables in public buildings, transportation, and other spaces.

29.4.3.1 Aluminum Trihydrate

Aluminum trihydrate (ATH) is a white powder that is widely used in wire and cable for many reasons such as low cost, good flame retardant properties, and nontoxic. The typical particle size is about 1 micron or finer and must be uniformly dispersed throughout the resin matrix to ensure a good level of flame retardancy. Typical loading level in the formulation is about 60 wt% to achieve a decent flame retardancy. However, this high loading has negative effects in terms of degrading the mechanical and durability properties of the polymer composites. To minimize these adverse effects, ATH is often used in lower concentration in combination with another flame retardant.

ATH decomposes to form alumina and water through endothermic decomposition reaction as shown in Equation 3, where the main decomposition peak occurs at around 200°C. The removal of heat caused by the chemical reaction is about 1,300 joules per gram, and when combined with the production of water (about 34%) to dilute the combustible gases, this helps to cool down the system and prevent fire and smoke generation.

$$2Al(OH)_3 \rightarrow Al_2O_3 + 3H_2O \quad (3)$$

ATH operates in the condensed phase as a heat sink that extends the time it takes for the resin matrix to decompose. The nature of the residue, an inert oxide with a high heat capacity, is also a positive factor in slowing down the burning rate. The extent of its flammability performance depends on its particle size and the surface area. Another advantage of ATH that needs to be mentioned is its effectiveness in reducing the smoke evolution.

29.4.3.2 Magnesium Hydroxide

Magnesium hydroxide (MDH) is the second largest selling inorganic hydroxide flame retardant, next to ATH. It is a white powder that acts as a flame retardant in a similar manner to ATH, by undergoing through a highly endothermic decomposition reaction that slows the heating rate of the host material in a fire as shown in Equation 4.

$$Mg(OH)_2 \rightarrow MgO + H_2O \quad (4)$$

Magnesium hydroxide converts to magnesium oxide and water at about 340°C, which is a significantly higher temperature range compared with that of ATH. That means wide temperature processing window to the formulation. Besides, this is more endothermic reaction, with a higher enthalpy of decomposition (1450 joules per gram), hence, more efficient as a "heat sink." The hydroxyl groups bonded to the magnesium are converted into water vapor (31%), which dilutes the flammable species in the flame. The decomposition of magnesium compound yields magnesia that has good insulating properties. Also, magnesium hydroxide is generally more effective than ATH as smoke suppressant.

Surface treatment of MDH, ATH, and other mineral fillers is becoming a significant factor in making product differentiation. The basic value added from making surface treatment is improvement in the filler's dispersion quality, coupling with the matrix resin, and improvement in mechanical performance. Surface treatment doesn't make improvement in the flammability performance but is still needed in order to make balance in the property space given the fact that these fillers are used at high loading.

The chemistry used for mineral surface treatment is diverse, and there is significant competition in this market, but the dominant reactive modifiers are silanes and stearates-based coupling agents.

Dealing with hydrated minerals, mention should be made that, in addition to the precipitated grades of these materials, which will deliver better properties, especially the mechanical performance. There are also ground grades of these materials, such as brucite and bauxite materials. The obvious main driving force for these materials is the low cost, but that would be at the expense of the performance of the final product, as the ground grade will have a significant amount of impurities.

29.4.3.3 Boron-Based Flame Retardant

Boron-based flame retardants are used in wire and cable materials design, they can function as an effective charring promoter as well as a smoke suppressant.[56] Adding borates to hydrated minerals containing compounds can result in synergistic performance benefits, notably in the compound fire performance and rheological properties.[57] See Chapter 9 for more information on boron-based flame retardants.

29.4.3.4 Phosphorus Flame Retardants

Phosphorus as a unique flame retardant element can exist in different chemical forms, such as red phosphorus, phosphates, phosphate esters, phosphonates, and phosphinates. The phosphorus content in these compounds varies widely from 100% (red phosphorus) to 10% (phosphates). The main applications of the phosphorus flame retardants remain in oxygen-containing resin polymers. The literature indicates that the mode of action of the phosphorus compounds can be in the gas and condensed phase. In the gas phase, phosphorus mode of action is radical scavenger to inhibit the flame. In the condensed state, it is effective in forming char that promotes the flammability performance.[58] See Chapter 5 of this book for more information on phosphorus-based flame retardants.

29.4.3.5 Nitrogen-Based Flame Retardants

Nitrogen-based flame retardants are a class of various materials that can be used in different resins such as nylons, polyolefins, and polyurethane.[59] Some well-known examples of these materials include ammonium polyphosphate, melamine cyanurate, melamine borate, and melamine (poly) phosphate. Other nitrogen-based flame retardants include N-alkoxy hindered amines, azoalkanes, and phosphorus-nitrogen compounds such as phosphazenes and phospham. See Chapter 10 of this book for more information on nitrogen-based flame retardants.

29.4.3.6 Intumescent Flame Retardants

The intumescent system is a special case in flame retardancy and considered one of the most effective and environmentally friendly materials. They are typically comprised of three major components: an acid source (e.g., phosphoric acid), a carbonizing source (e.g., pentaerthyritol), and a foaming or blowing source (e.g., melamine), and their main mechanism is in the condensed phase, no gas phase involved here.[60,61]

So in a typical intumescent system, the first step is carbonization, which takes place around 300°C, followed by decomposition of the blowing source to release the gases that cause the char to swell and foam. This foam structure will eventually decompose at much higher temperature.

The advantages offered by using such a type of flame retardant system includes the lack of evolution of corrosive and toxic gases, the smoke, which is one of the main fire hazard, will also be reduced. However, the relatively high loading needed by this system in order to meet the safety requirements has deleterious effect on the physical and mechanical properties of the final system. Therefore, careful design of the final formulation is needed in order to take advantage of this system. See Chapter 7 of this book for more information on intumescent flame retardants.

29.4.3.7 Nanocomposites and Nanoparticle Flame Retardants

Polymer nanocomposite was developed as multiphase materials that incorporate particles on the nanoscale. These materials are considered a revolutionary type of flame retardancy, and their effectiveness is attributed to the ability to disperse the particulate additives in the resin matrix on the nanoscale. The initial phase in the development of these materials focused on the use of smectite nanoparticles such as montmorillonite materials. In this class, dispersion on the nano scale is an essential component in delivering the flame retardancy and was achieved using different processing techniques such as solution blending, in situ polymerization, and melt blending.[62]

Montmorillonite is a widely used class of nanoparticles, which are layered silicates possessing a crystalline structure made of two-dimensional sheets. These are formed by the sandwiching of an octahedral layer of aluminum hydroxide or magnesium hydroxide by two tetrahedral silica layers held by weak van der Waals or dipolar forces of attraction with a layer thickness of 1 nm. During processing, the polymer chains in the molten state enter into the interlayer space of clay platelets forming intercalate, and by applying more shear force during the melt blending, the montmorillonite will exfoliate into single layers.

Applying this technology to polyolefins was a challenge because montmorillonite is hydrophilic material, so in order to have a good dispersion in the hydrophobic polyolefins, surface coating of these materials with organic quaternary ammonium ions was essential. This type of surface treatment renders them hydrophobic, so they will be compatible with polyolefin resins. These types of clay materials became available commercially with different types of surface coatings, and the application of this new technology to wire and cable was cited in the literature.[63]

Later, other types of nanoparticles have been added to the list, such as carbon nanotubes, polyhedral oligomeric silsesquioxane (POSS), and inorganic nanoparticles. In all these systems, the main theme remains the same, the ability to disperse the particles on the nanoscale to take advantage of the high surface area contact with the resin matrix. See Chapter 12 of this book for more information on polymer nanocomposites used for flame-retardant applications.

29.4.3.8 Ceramifiable Materials

There has been a significant growth in demand for cables that are expected to continue to function for a period in a fire situation, enabling essential services to continue operation during the evacuation of buildings or during the initial firefighting stages.

That could be achieved by developing a robust char structure that protects the cable and enables it to do its function in delivering the electrical signals. Fire performance cables are required to continue to operate and provide circuit integrity when subjected to fire. To meet specific standards, cables must maintain circuit integrity when heated to a specified temperature (e.g., from 650°C to 1,000°C) for a specified time (from 15 minutes to 2 hours). According to these standards, the cables will be subjected to mechanical shocks and water spray.

The traditional way of meeting these requirements was to wrap the cables with tapes made of glass fibers and treated with mica. Recently, a new way to address these requirements was made possible by developing ceramifibale materials.

Polymer-based ceramifiable composites generally consist of polymer matrix with refractory silicate minerals that form the ceramic framework in combination with a flux system. A key characteristic of ceramifying polymers is their ability to form a self-supporting structure throughout the temperature range from ambient service temperature to over 1,000°C.

The ceramification in these materials is not simply the bonding or fusing of the silicate particles by a viscous liquid phase, such as with relatively high levels of low melting point glasses. Such materials tend to collapse at high temperatures and are not self-supporting. Ceramification involves reaction sintering assisted by the controlled level of liquid phase.[64–66]

29.4.3.9 Smoke Suppressants

One of the major fire risks for wire and cable, especially inside buildings, is the smoke production. The smoke density produced will be dependent upon the formulation components, especially the resin matrix and the type of flame retardants used. For the resins, PVC is well-known for its high smoke production. Halogenated flame retardants release high smoke volume during fire. Non-halogenated flame retardants such as ATH and MDH are much better in terms of smoke production, but typically high loading will be needed. So it depends on the formulation design in order to balance all the needed properties. A number of materials technologies are available that can be employed in order to mitigate the smoke production during fire, such as molybdenum and boron-based compound.[67]

29.5 CABLE MANUFACTURING: COMPOUNDING AND COATING PROCESS

Cable manufacturing is a complex and well-established industry. It covers a broad range of cable designs, and it has been discussed in numerous publications.[68,69] We will limit ourselves here to discuss the two basic elements of this industry, the compounding and the coating process. As stated previously regarding cross-linking, the flame retardants to be used must be compatible with the compounding and coating processes selected. Therefore, thermal stability of the flame retardant and its ability to remain in the polymer during coating onto the wire (and if the flame retardant will affect adhesion to the wire) must be considered.

29.5.1 Extrusion Compounding

Wire and cable are used to transmit electrical signals or power from one point to another. They require polymers that can provide insulation, protection, flexibility, durability, and flame retardancy. Polymer extrusion compounding is the processing technique used to make the insulations materials. It involves mixing, melting, and shaping polymers with various additives to achieve the desired properties for wire and cable applications.

Two principal mechanisms are used to describe the type of mixing configuration during the compounding: the distributive mixing, which involves the even distribution of the components throughout the volume, and the dispersive mixing, which involves reducing the size of the particles.

Polymer extrusion compounding for wire and cable is a pelletizing process. It is the process of producing uniform pellets or granules of polymer compounds that can be stored, transported, and fed into the wire coating extruder. The compound could be thermoplastic or cross-linkable according to the application and the specification.

A wide range of additives and fillers will be incorporated into the compounding process, including the flame retardant additives, in order to regulate the behavior of polymers when in use. Antioxidants and UV stabilizers will be needed in order to stabilize the compound against thermal and UV aging. The resultant compound must meet a blend of characteristics and specifications, the most important of which are FR properties, mechanical and electrical properties. These properties tend to be competitive rather than complementary and therefore present the formulator with a significant challenge.

The resultant compound can be difficult to manufacture and produce consistently. So it requires more than just formulating skills in order to successfully and reliably balance these competing parameters, and between formulation and conversion onto cable lies the extremely important step of mixing and compounding, where the most common compounding technique employed is the co-rotating twin screw extrusion.[70]

29.5.2 Reactive Extrusion Compounding (Cross-linked HFFR)

As mentioned before, due to the environmental concern on the halogenated materials, halogen-free flame retardant (HFFR) compounds started to get traction in the cable industry. Thermoplastic jacket application is the target for most of these HFFR compounds. However, in terms of standards, there is increasing demand for high performance in both mechanical and thermal properties. These kinds of stringent requirements can be addressed by employing cross-linking technology to the polyolefin compounds, which seems to be an attractive approach. In principle, there are three ways of cross-linking polyethylene, as mentioned previously:

- (PEX-a) or Peroxide cross-linking
- (PEX-b) or Silane cross-linking
- (PEX-c) or Electron-beam cross-linking

These HFFR formulations are known to be shear sensitive, so the use of BUSS Kneader seems to be appropriate choice. This mixing technology employs low shear force where the reciprocating screw simultaneously executes an axial oscillating motion with every rotation. Such reactive extrusion technique provides efficiency in both the distributive and dispersive mixing caused by the combination of the radial and longitudinal actions at moderate shear rates.[71,72]

29.5.3 Wire and Cable Coating

Wire and cable coating is the process of applying a thin layer of polymer compound over a metal conductor or a core wire by passing it through a heated die. Coating is carried out by employing an extrusion process, where the polymer in its melt state is extruded continuously over the axial copper conductor, or the conductor is pulled through the extruded polymer melt. A typical wire coating line consists of a pay-off device, a wire preheater, an extruder equipped with a crosshead die, a cooling trough, and a take-off device. Various control and measuring instruments are utilized in the commercial line. The two basic wire-coating dies are pressure-type dies and tubing-type dies. Most wires and cables used in electrical and telecommunication applications and the electronics industry are coated this way.[73,74]

It should be pointed that one of the factors that affect the performance and safety of the cables is the size of the conductor AWG (American Wire Gauge) used in the coating process. AWG is a standardized system that specifies the diameter and cross-sectional area of round, solid, or stranded wires. The AWG size is inversely proportional to the diameter and directly proportional to the cross-sectional area. That means a smaller AWG number indicates a larger diameter and a larger cross-sectional area, and vice versa.

29.6 FLAMMABILITY TESTING FOR WIRE AND CABLE

Due to the dangers associated with the cables, many national and international testing methods have been proposed to evaluate the fire safety of these materials.[75] Historically, assessment of the fire hazard of electrical cables was based on the traditional approach where the flame spread over the cable surface, and the smoke density was used as criteria for evaluation. Many of these testing methods were developed in the 1970s, meant to meet the regulatory requirements of pass and fail criteria, but they are not updated to match the huge development that was made in fire science over the last decades.

More recently, a new approach was established that is based on an assessment of the reaction to fire class. This testing enables the measurements of heat release rate by oxygen consumption technique. Parameters such as flame spread (FS), peak heat release rate (PHRR), total heat released (THR), and fire growth rate (FIGRA) can be measured. Based upon these measurements, the cables can be classified into different classes.

Some of the fire testing used for evaluating the wire and cable will be presented in the following section. It includes both test methods used by R&D for product development along with other tests needed by different standards and regulations. Table 29.5 gives a brief comparison of the different types of testing, and the intensity hierarchy of these tests are shown in Figure 29.9.

TABLE 29.2
Basic Flammability Properties for a Number of Polymers[76,77]

Material	Limiting Oxygen Index	Ignition Temperature (°C)	Average Effective Heat of Combustion ΔH_{ch} (MJ/Kg)
Polyethylene	17–18	349	43.4
Cross-linked polyethylene (XLPE)	17–18	-	23.8
Polyethylene vinyl acetate	18–19	-	-
Polypropylene	17–18	350	44
Chlorinated polyethylene	21	-	-
Polyvinyl chloride (plasticized)	40–45	454	7.7
Silicone rubber	23	550	25
Nylon 6	23–26	-	28.8
Chlorosulfonated polyethylene	25	-	-
Polyurethane	16.5	416	28
Polytetrafluoroethylene (PTFE)	95	580	Endothermic

TABLE 29.3
Electrical Properties of Selected Polymers

Material	Dielectric Constant	Dielectric Strength (KV mm)
Polyethylene (LDPE)	2.3 (50–60 Hz)	200
Cross-linked polyethylene (XLPE)	2.3 (1 MHz)	220
Polypropylene (PP)	2.27 (50–60 Hz)	200
Ethylene-propylene diene rubber (EPDM)	2.5–3.5 (50–60 Hz)	20
Silicone rubber	2.5–3.2 (50–60 Hz)	20–30
Polyvinyl chloride (PVC)	2.9–3.1	140–200

TABLE 29.4
Classification of Polyethylene Plastics Extrusion Materials for Wire and Cable According to Type[9]

Polyethylene Type	Density Range (g/cm^3)	Type (ASTM D1248)
Elastomer	0.855–0.885	0
Plastomer	0.885–0.915	0
VLDPE or ULDPE	0.890–0.915	0 or I
LDPE	0.910–0.925	I
LLDPE	0.915–0.925	I
MDPE	0.926–0.940	II
HDPE	0.941–0.966	III and IV

TABLE 29.5
Cable Fire Testing

Fire Test	Test Chamber	Sample Orientation	Gas Burner and Flame Size	Duration	Pass/Fail Requirements
FT-2 UL44 & UL2556	Small scale (4 m^3 chamber)	Horizontal	Methane 500W	30 seconds	The total length of char shall not exceed 100 mm (4 in). No ignition for the cotton.
VW-1 UL44 & UL2556	Small scale (4 m^3 chamber)	Vertical	Methane 500W	15 seconds × 5 applications	Specimen does not continue to flame longer than 60 seconds. The specimen does not show more than 25% of the indicator flag burned away
FT4 (IEEE1202) Tray Cable	Large scale	Vertical	Propane 20.6 KW (70,000 BTU/hr.)	20 minutes	The cable char height shall be less than 4 ft 11 in. Total smoke released shall not exceed 150 m2. Peak smoke release rate shall not exceed 0.4 m2/s.
UL1685 Tray Cable	Large scale	Vertical	Propane 20.6 KW (70,000 BTU/hr.)	20 minutes	The cable char height shall be less than 4 ft 11 in. Total smoke released shall not exceed 150 m2. Peak smoke release rate shall not exceed 0.4 m2/s.
UL1666 Riser Test	Large scale	Vertical	Propane 154.5 KW (527,000 BTU/ hr.)	30 minutes	Flame propagation height must not equal or exceed 12 ft. Maximum temperature must not exceed 850°F.
UL910 & NFPA262 Plenum Test	Large scale	Horizontal	Methane 88 KW (300,000 BTU/ hr.)	20 minutes	Maximum Flame Propagation ≤ 5 (ft). Optical density: O Peak Optical Density ≤ 0.5 O Average Optical Density ≤ 0.15.
EN50399	Large scale	Vertical	20/30 KW	40 minutes	

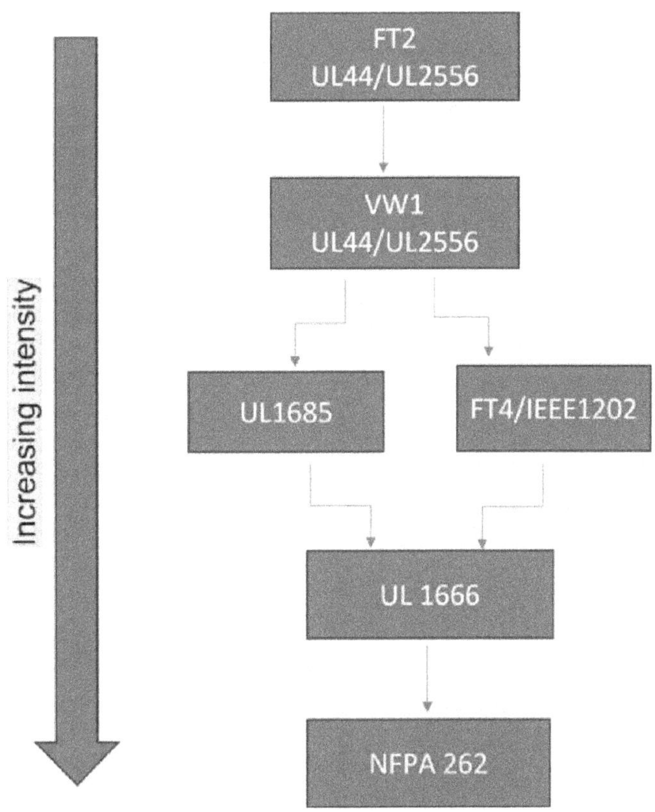

FIGURE 29.9 Hierarchy of cable fire testing.

29.6.1 Small-Scale Fire Test

29.6.1.1 Limiting Oxygen Index

The limiting oxygen index (LOI) is a reliable method for measuring the minimum oxygen concentration needed to sustain ignition.[78] Even though this method is used frequently in industry as a quality control, it is difficult to make prediction for the fire behavior based upon this test. The method is described in various United States and European test standards, including ASTM D2863, ISO 4589–2, and NES 714. The LOI apparatus consists of a vertical chimney standing 450 mm high with an internal diameter of 75 mm. Oxygen and nitrogen are pumped in at the base of the chimney where they pass through a layer of glass beads that ensures even mixing before entering the main test chamber. A needle valve or paramagnetic oxygen cell inside the chimney is used to measure the oxygen concentration. The gas mixture flows upwards to a vertical test specimen, a small gas flame is used to ignite the upper end of the specimen, and the subsequent burning behavior is monitored.

29.6.1.2 Cone Calorimeter

The most widely known instrument for measuring the fire reaction properties of combustible materials is the cone calorimeter.[79] It is based upon the oxygen consumption principle, and it was developed at the National Bureau of Standards (now known as the National Institute of Standards and Technology). The popularity of the cone calorimeter comes from its capability to measure many reactions to fire properties by burning a small sample (Figure 29.10).

Materials can be fire tested at incident heat fluxes ranging from 0 to 100 kW/m^2, thereby simulating pre-flashover fires from low intensity up to high temperature fuel fires.

FIGURE 29.10 Cone calorimeter.

Source: From Saco AEI Polymers lab.

The properties measured by the cone calorimeter includes peak and average heat release rate, time-to ignition, time of sustained flaming, effective heat of combustion, smoke density, soot yield, mass loss rate, and yields of CO, CO_2, and other combustion gases.

29.6.1.3 FT-2 Horizontal Flame Test for Cable (Single Wire Testing)

This test is required for thermoset insulated wire and cable based upon UL44 and UL2556 standards.[80,81] The burner is placed on the floor of the chamber. A 10-inch-long specimen of the wire or cable is secure in a horizontal position by a lab stand or other apparatus. The burner is positioned such that the tip of the flame will just touch the center of the specimen at a right angle to the specimen. The cotton is placed under the specimen and on the top of the burner base. The test procedure is to tilt the burner forward to apply the test flame for 30 seconds and then to quickly tilt the flame away. To meet the requirement of this test, the specimen must not have been burned more than four inches, and the cotton must not have been ignited.

29.6.1.4 VW-1 Vertical Flame Test for Cable (Single Wire Testing)

This test is also required for thermoset insulated wire and cable based upon UL44 and UL2556 standards.[80,81] To prepare for this test the burner is placed on the floor of the chamber. An 18 inch long specimen of the wire or cable is secured in a vertical position by a lab stand or other apparatus. The burner is positioned such that the tip of the flame will just touch the specimen. The cotton is placed under the specimen and on the top of the burner base. A strip of Kraft paper is wrapped around the cable to create a flag 10 inches above the flame impingement point.

The test procedure is to tilt the burner forward to apply the test flame for 15 seconds and then to quickly tilt the flame away for 15 seconds or until the flaming of the specimen ceases, whichever happens last. This procedure is repeated four times for a total of five flame applications.

To meet the requirements of this test, the paper flag must not have been burned more than 25%, the cotton must not have been ignited, and the flame must not have continued more than 60 seconds after any application of flame.

29.6.1.5 FV-1/Vertical Flame (Single Wire testing)

This test is for photovoltaic wire based upon UL4703 standard.[82] It is identical to VW-1 test mentioned above except that during the test procedure, the flame shall be reapplied after 15 seconds, regardless of whether flaming of the specimen persists longer than 15 seconds. The time for flaming of the specimen to self-extinguish shall be recorded after removal of the burner flame following the fifth application.

To meet the requirements of this test, the paper flag must not have been burned more than 25%, the cotton must not have been ignited, and the flame must not have continued more than 60 seconds after any application of flame.

29.6.2 Large-Scale Fire Test

29.6.2.1 Tray Cable

This is a large-scale vertical fire test referenced in IEEE 1202 and UL 1685 standards.[83] A set of defined cables is installed in a steel ladder type cable tray that is 12" wide, 3" deep and 8 feet long, half a diameter apart, and filling 75% of the tray width. The cables are submitted to a 70,000 BTU/hour (20 kW) flame for 20 minutes. The 10" wide flame is applied at an angle of 20°, placed in front of the cables and 1 foot from the bottom of the cables. The rating requires the set of cables to self-extinguish before reaching the top of the tray (Figure 29.11).

29.6.2.2 Riser Test

A different type of vertical flame spread test for cables is the riser test based upon UL1666 standard.[84] Riser cables, as their name implies, rise between floors. The test simulates a cable installation in a vertical building riser shaft which penetrates more than one floor. The cable samples are 17.5" long and are mounted in a single layer 12" wide, vertically passing through a slot in a floor at 12" high and in a test chamber 19" high. They are exposed to a large flame source of 527,500 BTU/hour (154.5 kW) from a 12" square burner that impinges a flame for 30 min onto the cables on the tray. The test requires the cables to pass a flame spread requirement with the burned length of less than 12" and a requirement on the temperature of any of the thermocouples in the tray with values not to exceed 850°F (Figure 29.12).

FIGURE 29.11 UL1685 vertical tray test.

FIGURE 29.12 UL1666 riser test.

29.6.2.3 Plenum Test

The plenum cable test is a version of the Steiner Tunnel, but with some special requirements and criteria for passing. According to NFPA262/UL 910 standard,[85] horizontally installed bunched cables are tested in a preheated tunnel, more than 70 m in length, in which 7.3 m long test samples are installed on a horizontal ladder. The cables are ignited for 20 min by an 88 KW (300,000 BTU/hr.) gas burner, and the tunnel is flowed through by an air stream. At the end of the tunnel the smoke density is measured. During the 20-minute test, the flame spread is observed through small windows spaced one foot apart. Smoke is measured by a photocell installed in the exhaust duct. To qualify, cables must have a flame spread of fewer than 5 feet beyond the end of the 4–1/2 foot ignition flame, a peak optical density of 0.5 maximum (33% light transmission), and a maximum average optical density of 0.15 (70% light transmission) (Figure 29.13).

29.6.2.4 Fire Resistance Test

The circuit integrity of the cable under fire can be tested according to IEC 60331–11 standard,[86] which is a test method for fire with shock at a temperature of at least 830°C for cables of rated voltage up to and including 0.6/1.0 kV and with an overall diameter exceeding 20 mm. The test apparatus used for testing cables required to maintain circuit integrity when subject to fire alone where the test condition is based upon a flame with a controlled heat output corresponding to a temperature of at least 750°C.

29.6.2.5 CPR Testing

The Construction Products Regulations (CPR) of the European Union lays down the essential requirements for building products. CPR has been set up to facilitate the free circulation of construction products in the EU, removing trade barriers eventually created by technical specification.

FIGURE 29.13 Plenum test.

Manufacturers must test the cables based upon EN 50399 standard.[87] The following are the measured quantities:

- Vertical flame spread (m)
- Total heat released during test (MJ)
- Peak heat release rate (kW)
- Smoke production rate (m^2/s)
- Total smoke production (m^2)
- Fire growth rate index (FIGRA) (Ws^{-1})

The performance rating for the different cables can be rated into different classes (Figure 29.14 and Table 29.6): Class Aca, Class B1ca, Class B2ca, Class Cca, Class Dca, and Class Eca.

FIGURE 29.14 External view of EN 50399 standard test chamber.[88]

Source: Open access

29.6.2.6 EN 60332-1-2

The vertical flame propagation of a single wire is evaluated based upon EN 60332–1–2 standard.[89] The cable sample (length: 600 mm) will be burned by applying 1 KW flame for a time that depends upon the cable diameter, and then remove the flame. The cable will pass the test if the distance between the lower edge of the top support and the char onset is higher than 50 mm.

29.6.2.7 Smoke Measurement

Test methods have been developed specifically to measure smoke production during the combustion process; best known example is the NBS smoke chamber. This method is described in ASTM E662, Standard Test Method for Specific Optical Density of Smoke Generated by Solid Materials.[90] The test is conducted at heat flux of 25 Kw/m^2 and can be conducted with and without pilot flames and measures the reduction in the visibility within the chamber by light beam. However, the test has been subjected to criticism because the smoke generated by the specimen accumulates inside the chamber and eventually affects combustion.

29.6.3 Correlation between Small- and Large-Scale Fire Testing

One of the major challenges dealing with product development for flame-retardant materials is to establish a correlation between the small-scale and large-scale testing. Typically, during the product development process, it would be more feasible and practical to evaluate the progress made using small-scale testing. However, translating the data collected to reliable scenario for the large-scale fire testing proved to be a daunting challenge.

Early attempt was made by Weil et al. using the limiting oxygen index to predict other fire testing.[91] A statement was

TABLE 29.6
European CPR Classification

Class	Test Method(s)	Classification Criteria	Additional Requirements
A_{ca}	EN ISO 1716	PCS ≤ 2.0 MJ/kg PCS ≤ 2.0 MJ/kg	
$B1_{ca}$	EN 50399 (30 kW flame source)	FS ≤ 1.75 m THR1200s ≤ 10 MJ Peak HRR ≤ 20 kW FIGRA ≤ 120 W/s	Smoke production and acidity
$B2_{ca}$	EN 60332–1–2 EN 50399 (20.5 kW flame source)	H ≤ 425 mm FS ≤ 1.5 m THR_{1200s} ≤ 15 MJ Peak HRR ≤ 30 kW FIGRA ≤ 150W/s	Smoke production and acidity
C_{ca}	EN 60332–1–2 EN 50399 (20.5 kW flame source)	H ≤ 425 mm FS ≤ 2.0 m THR_{1200s} ≤ 30 MJ Peak HRR ≤ 60 kW FIGRA ≤ 300W/s	Smoke production and acidity
D_{ca}	EN 60332–1–2 EN 50399 (20.5 kW flame source)	THR_{1200s} ≤ 70 MJ Peak HRR ≤ 400 kW FIGRA ≤ 1300W/s	Smoke production and acidity
E_{ca}	EN 60332–1–2	H ≤ 425 mm	
F_{ca}	EN 60332–1–2	H > 425 mm	

Source: PCS: gross heat of combustion
H: vertical flame spread
Fs: vertical flame spread
PHRR: peak heat release rate
THR: total heat release
FIGRA: Fire Growth Rate

made that LOI value of 30 was needed to pass IEEE 383 and IEC 332 fire testing for sheathing materials. However, it was concluded that it is difficult to make such prediction given the fact that the fire behavior of the cable depends not only on the sheathing material, but other factors need to be considered such as the cable construction and the heat sink effect of the copper conductor.

Given the fact that the cone calorimeter is one of the most significant tool for investigating fire behavior, the heat release rate was cited as the most important indicator on the fire intensity.[92,93] A comparison was made between small- and large-scale testing for cable materials using the cone calorimeter. In this study the construction design for all cables was constant, but the chemical composition of the materials was different. The large-scale test used was the tray cable CSA FT-4, and it was found that the peak of heat release can be used as indictive for the fire performance on large scale.[94] However, it was indicated that it is difficult to use simple correlation and more detailed modeling will be needed.

An extensive research program to study the power cables used in nuclear power plants was made under the name CHRISTIFIRE program.[95,96] The main goal of this investigation was to study and quantify the burning characteristics of the electrical cables. Testing was done on bench-scale using the cone calorimeter and large-scale tray cable in horizontal and vertical orientation. Based upon this investigation, FLASH-CAT model was developed to predict the fire scenario of the cables.

Recently, an attempt was made to predict the European CPR large-scale testing that is based upon EN50399 standard.[97] In this study a bench scale testing using a practical module of the cone calorimeter was used to simulate the large-scale testing. The setup of the cone typically used for collecting the heat release data was maintained, the cable samples tested in the vertical orientation, but the length was reduced to a fifth of the original length used in the CPR. The module was able to link the key properties such as the peak heat release, the fire growth rate, and the flame spread to that in CPR by a factor of around 5.

Under the same trend of predicting the flammability properties of materials, an investigation was made using the small-scale Pyrolysis-combustion flow calorimetry to calculate the main flammability properties of the materials. The method is simply based upon molar group contribution, and it gives a quantitative way to rate each group from the flammability standpoint.[98,99]

29.7 FUTURE TRENDS IN WIRE AND CABLE

The global market for wire and cable and the associated flame retardants additives that will be needed are expected to grow in the future.[100,101] This growth is driven by the increasing demand for sophisticated wiring and cabling solutions across various industries, such as automotive, telecommunications, construction, and renewable energy. The rising need for efficient and reliable data transmission, better connectivity, and improved electrical performance are some of the factors that fuel the demand for high-performance wire and cable compounds.

Some emerging trends and technologies will shape the future of wire and cable:

5G Technology: 5G technology is expected to revolutionize the telecommunication industry as it provides faster data speed, higher bandwidth, and a more reliable network. This technology will rely on wire and cable materials in its expansion. Polymers that can provide excellent dielectric properties, low dissipation factor, high heat resistance, and low smoke emission are in high demand for 5G applications. Fluoropolymers, such as polytetrafluoroethylene (PTFE), perfluoroalkoxy (PFA), and fluorinated ethylene propylene (FEP), are some of the preferred polymers for 5G wire and cable.

Electric Vehicles (EV): The market for electric vehicles is growing rapidly. EVs require more complex wiring and cabling systems to handle higher voltages, currents, temperatures, and mechanical stresses. Polymers that can offer unique properties such as high electrical insulation, thermal stability, flame retardancy, abrasion resistance, and chemical resistance are essential for EV wire and cable. Some of the polymers that are suitable for EV applications are cross-linked polyethylene (XLPE), silicone rubber, thermoplastic elastomers, and polyphenylene sulfide.

Renewable Energy: Governments and organizations are investing in clean and sustainable energy sources such as solar, wind, hydro, and biomass. Reliable and durable wire and cable solutions that can withstand harsh environments will be needed. Polymers that can provide high weatherability, UV resistance, moisture resistance, corrosion resistance, and flame retardancy are ideal for renewable energy wire and cable. Some of the polymers that are widely used in renewable energy applications are ethylene propylene diene monomer (EPDM), polyolefin elastomers, polyvinylidene fluoride, and ethylene tetrafluoroethylene.

29.8 CONCLUSIONS

This chapter provided a brief introduction to the diverse fields of wire and cable industry, where flame retardancy plays a major role. Electrical cables are indispensable components in almost everything we encounter in our daily life, such as buildings, vehicles, electrical equipment, etc. So it is no wonder that fire safety is the number one priority in this industry.

The major components in this industry have been covered in this chapter, starting from the materials used to build and manufacture the cables, how to impart flame retardancy to these materials, and how to test and evaluate the final products.

Cross-linked polyethylene plays a major role in wire and cable due to the unique properties such as dielectric constant, dielectric strength, volume resistivity, mechanical properties in addition to excellent processability.

PVC continues to play a major role in this market due to its inherent flame retardancy, excellent electrical and mechanical properties.

Other polymers and rubber materials have also been covered as they still hold a significant market share due to their outstanding properties.

Halogenated flame retardants are still a major FR additive, especially in North America. The profound effectiveness of these materials makes it difficult to find non halogenated alternatives with the same effectiveness.

The environmental concern is pushing the FR market to the non-halogenated flame retardant, and materials such as magnesium hydroxide and aluminum trihydrate are the most important candidates of this market.

Manufacturing and getting the cables ready to the market includes a wide range of processes, starting from compounding the pellets with all functional additives, and ending with coating these compounds to make the final cable article.

Testing and meeting the standard requirements and regulations is a major concern, and the major testing for this industry has been covered. Efforts are still underway to develop a reliable methodology to predict the fire behavior of the cables from small-scale testing.

The global wire and cable market will continue to grow, and some emerging trends and technologies will have their impact on this market in terms of materials selection, cable design and fire testing, such as the 5G, renewable energy, and the electric vehicles industries.

REFERENCES

1. Kahn, K.B., Castellion, G., and Griffin, A., Eds., *The PDMA Handbook of New Product Development*, John Wiley & Sons, 2005.
2. Gessinger, G.H., Ed., *Materials and Innovative Product Development*, Butterworth-Heinemann, Elsevier, 2009.
3. Plesa, I.; Notingher, P. V.; Stancu, C.; Wiesbrock, F.; Schlogl, S. Polyethylene Nanocomposites for Power Cable Insulations. *Polymers*. **2019**, *11*(24).
4. Cousins, K., *Polymers for Wire and Cable*, RAPRA, 2000.
5. Sen, A.K., Cable Technology, in *Rubber Products Manufacturing Technology*, Bhowmick, A.K., Hall, M.M., and Benarey, H.A., Eds., Marcel Dekker, Chapter 19, 1994.
6. Bernstein, B.S., Fundamentals of Electrical Insulation Materials, in *Electrical Power Cable Engineering*, Thue, W., Ed., CRC Press, Chapter 5, 2012.
7. Geussens, T., Thermoplastics for Cables, in *The Global Cable Industry*, Beyer, G., Ed., Wiley-VCH GmbH, Chapter 2, 2021.

8. Fawcett, E.W., Gibson, R.O., and Perrin, M.W., Polymerizing Ethylene, US Patent Appl. 2153553, assigned to Imperial Chemical Industries, 1939.
9. Kissin, Y.V., Catalysts for the Manufacture of Polyethylene, in *Handbook of Industrial Polyethylene and Technology*, Spalding, M.A., and Chatterjee, A.M., Eds., John Wiley & Sons, Chapter 2, 2018.
10. Nowlin, T.E., *Business and Technology of the Global Polyethylene Industry*, Scrivener Publishing, 2014.
11. Hoff, R., and Mathers, R.T., Eds., *Handbook of Transition Metal Polymerization Catalysts*, John Wiley & Sons, 2009.
12. ASTM D1248-16, Standard Specification for Polyethylene Plastics Extrusion Materials for Wire and Cable.
13. Sen, A., In Cable Technology, in *Polymer Grafting and Crosslinking*, Bhattcharya, A., Rawlins, J.W., and Ray, P., Eds., John Wiley & Sons, Chapter 9, 2009.
14. Wasserman, S.H., Chaudhary, B.I., Cogen, J.M., Esseghir, M., and Person, T.J., Wire and Cable Applications of Polyethylene, in *Handbook of Industrial Polyethylene and Technology*, Spalding, M.A., and Chatterjee, A.M., Eds., John Wiley & Sons, Chapter 43, 2018.
15. Thomas, J., Thomas, M.E., and Thomas, S., Crosslinked Polyethylene: State-of-the-Art and New Challenges, in *Crosslinkable Polyethylene*, Thomas, J., Thomas, S., and Ahmad, Z., Eds., Springer, Chapter 1, 2021.
16. Sultan, B.-A., Crosslinking of Polyolefins, in *Polymeric Materials Encyclopedia*, Vol. 2C, Salamone, J.C., Ed., CRC Press, p. 1552, 1996.
17. Ahmad, H.; Rodrigue, D. Crosslinked Polyethylene: A Review on the Crosslinking Techniques, Manufacturing Methods, Applications, and Recycling. *Polym. Sci. Eng.* **2022**, *62*(8), 2376–2401.
18. Herpich, B., Elastomers for Cables, in *The Global Cable Industry*, Beyer, G., Ed., Wiley-VCH GmbH, Chapter 3, 2021.
19. Munteanu, D., Moisture-Crosslinkable Silane Grafted Polyolefins, in *Metal-Containing Polymeric Systems*, Sheats, J.E. Carraher, C.E. Jr., and Pittman, C.U. Jr., Eds., Plenum Press, p. 479, 1985.
20. Munteanu, D., Moisture-Crosslinkable Silane-Modified Polyolefins, in *Reactive Modifiers for Polymers*, Al-Malika, S., Ed., Blackie Academic & Professional, Chapter 5, 1997.
21. Morshedian, J.; Hoseinpour, P. M. Polyethylene Cross-Linking by Two-Step Silane Method: A Review. *Iran. Polym. J.* **2009**, *18*(2), 2009, 103.
22. Scott, H.G. (to Midland Silicones), Crosslinking of a Polyolefin with a Silane, U.S. Patent 3,646,155, 1968.
23. Scott, H.G. (to Dow Corning Ltd.), British Patent 1,286,460, 1972.
24. Pexidan, Moisture Cure Technology, www.sacoaei.com/products.
25. Swarbrick, P., Green, W.J., and Maillefer, C. (to BICC Ltd. and Maillefer, S.A.), Extruding Articles from Crosslinked Polymers—Mixing Polymer and Additives in the Extruder Cylinder, U.S. Patent 4117195, 1974.
26. Swarbrick, P., Green, W.J., and Maillefer, C. (to BICC Ltd. and Maillefer, S.A.), British Patent 1,526,398, 1978.
27. Dole, M. The History of the Crosslinking of Polyethylene. *J. Macromol. Sci.* **1981**, *A15*, 1403.
28. Makuuchi, K., and Cheng, S., *Radiation Processing of Polymer Materials and Its Industrial Applications*, John Wiley & Sons, Inc., Chapter 4, 2012.
29. Albizzati, E., Giannini, U., Collina, G., Noristi, L., and Resconi, L., *Catalysts and Polymerizations, Polypropylene Handbook*, Carl Hanser Verlag, 1996.
30. Meisenheimer, H., and Zens, A., Ethylene Vinyl Acetate Elastomers, in *Handbook of Specialty Elastomers*, Klingender, R.C., Ed., Taylor & Francis Group, Chapter 10, 2008.
31. Muschiatti, L.C., Wu, Y.T., McBride, E., and Kammerer, K., Ethylene/Acrylic Elastomer Formulation Design, in *Handbook of Specialty Elastomers*, Klingender, R.C., Ed., Taylor & Francis Group, Chapter 6, 2008.
32. Walton, K.L., Clayfield, T., Hemphill, J., and Madenjian, L., Applications of Polyethylene Elastomers and Plastomers, in *Handbook of Industrial Polyethylene and Technology*, Spalding, M.A., and Chatterjee, A.M., Eds., John Wiley & Sons, Chapter 47, 2018.
33. Laakso, R., Compounding with Chlorinated Polyethylene, in *Handbook of Specialty Elastomers*, Klingender, R.C., Ed., Taylor & Francis Group, Chapter 8, 2008.
34. Klingender, R.C., Chlorosulfonated Polyethylene and Alkylated Chlorosulfonated Polyethylene, in *Handbook of Specialty Elastomers*, Klingender, R.C., Ed., Taylor & Francis Group, Chapter 9, 2008.
35. Titow, W.V., *PVC Technology*, Elsevier Applied Science Publisher, 1984.
36. Ferrandez, P., Fluoroelastomers, FKM, and FEPM, in *Handbook of Specialty Elastomers*, Klingender, R.C., Ed., Taylor & Francis Group, Chapter 4, 2008.
37. Noll, W., *Chemistry and Technology of Silicones*, Academic Press, 1978.
38. Melvin, K., *Nylon Plastics Handbook*, Carl Hanser Verlag, 1995.
39. Grigoryeva, O., Fainleib, A., and Sergeeva, L., Polyurethane-Based Thermoplastic Elastomers, in *Handbook of Condensation Thermoplastic Elastomer*, Fakirov, S., Ed., Wiley-VCH Verlag, Chapter 12, 2005.
40. Babrauskas, V., Peacock, R., Braun, E., Bukowski, R.W., and Jones, W., Fire Performance of Wire and Cable: Reaction-to-Fire Tests, NIST Technical Note 1291, 1991.
41. Earley, M.W., Coache, C.D., Cloutier, M., Moniz, G., and Vigstol, D., Eds., *National Electrical Code Handbook*, National Fire Protection Association, 2016.
42. Lyons, J.W., *The Chemistry and Uses of Fire-Retardants*, 16. Wiley Interscience, 1970.
43. Lewin, M., Atlas, S.M., and Pearce, E.M., Eds., *Flame Retardant Polymeric Materials*, Plenum Press, Volume 3, 1982.
44. Horrock, A.R., and Price, D., Eds., *Fire Retardant Materials*, CRC Press, 2001.
45. Wilkie, C.A., and Morgan, A.B., Eds., *Fire Retardancy of Polymeric Materials*, CRC Press, 2010.
46. Papaspyrides, C.D., and Kiliaris, P., Eds., *Polymer Green Flame Retardants*, Elsevier, 2014.
47. Morgan, A.B., Ed., *Non-Halogenated Flame Retardant Handbook*, Scrivener Publishing, Wiley, 2022.
48. Lewin, M., and Weil, E.D., Mechanisms and Modes of Action in Flame Retardancy of Polymers, in *Fire Retardant Materials*, Horrock, A.R., and Price, D., Eds., CRC Press, Chapter 2, 2001.
49. Wypych, G., *Handbook of Flame Retardants*, ChemTec Publishing, Chapter 5, 2021.
50. Larsen, E.R., Fire Retardants (Halogenated), in *Kirk-Othmer Encyclopedia of Chemical Technology*, 3rd edn, Wiley-Interscience, Vol. 10, 373–95, 1980.
51. Georlette, P., Applications of Halogen Flame Retardants, in *Fire Retardant Materials*, Horrock, A.R., and Price, D., Eds., CRC Press, Chapter 8, 2001.
52. Bocchini, B., and Camino, G., Halogen-Containing Flame Retardants, in *Fire Retardancy of Polymeric Materials*,

Wilkie, C.A., and Morgan, A.B., Eds., CRC Press, Chapter 4, 2010.

53. Sauerwein, R., Mineral Filler Flame Retardants, in *Non-Halogenated Flame Retardant Handbook*, Alexander, B.M., Ed., Scrivener Publishing, Wiley, Chapter 3, 2022.
54. Sauerwein, R., Application of Submicron Metal Hydrate Fillers in Flame Retardant Cables, in *Proceedings of the 55th IWCS/Focus International Wire & Cable Symposium*, p. 399, 2006.
55. Chen, T., Isarov, A., New Magnesium Hydroxides Enabling Low-smoke Cable Compounds, in *Proceedings of the 56th IWCS/Focus International Wire & Cable Symposium*, p. 190, 2007.
56. Shen, K.K., Boron-Based Flame Retardants in Non-Halogen Based Polymers, in *Non-Halogenated Flame Retardant Handbook*, Morgan, A.B., Ed., Scrivener Publishing, Wiley, Chapter 7, 2022.
57. Chen, T., and Shen, K., Synergistic Benefits of Metal Hydroxides and Zinc Borate in Flame Retardant Wire and Cable Performance Compounds, in *Proceedings of the 54th IWCS/Focus International Wire & Cable Symposium*, p. 217, 2005.
58. Levchik, S., Phosphorus-Based Flame Retardants, in *Non-Halogenated Flame Retardant Handbook*, Morgan, A.B., Ed., Scrivener Publishing, Wiley, Chapter 2, 2022.
59. Morgan, A.B., and Klatt, M., Nitrogen-Based Flame Retardants, in *Non-Halogenated Flame Retardant Handbook*, Morgan, A.B., Ed., Scrivener Publishing, Wiley, Chapter 5, 2022.
60. Le Bras, M., Camino, G., Bourbigot, S., and Delobel, R., Eds., *Fire Retardancy of Polymers: The Use of Intumescence*, The Royal Society of Chemistry, 1998.
61. Bourbigot, S., Intumescence-Based Flame Retardant, in *Non-Halogenated Flame Retardant Handbook*, Morgan, A.B., Ed., Scrivener Publishing, Wiley, Chapter 4, 2022.
62. Morgan, A.B., and Wilkie, C.A., Eds., *Flame Retardant Polymer Nanocomposites*, John Wiley & Sons, 2007.
63. Beyer, G. Flame Retardancy of Nanocomposites: From Research to Technical Products. *J. Fire Sci.* **2005**, *23*, 75–87.
64. Hanu, L. G.; Simon, G. P.; Mansouri, J.; Burford, R. P.; Cheng, Y. B. Development of Polymer-Ceramic Composites for Improved fire Resistance. *J. Mater. Process. Technol.* **2004**, *153–154*(1), 401–407.
65. Mansouri, J.; Burford, R. P.; Cheng, Y.B.; Hanu, L. Formation of Strong Ceramified Ash from Silicone-Based Compositions. *J. Mater. Sci.* **2005**, *40*, 5741.
66. Alexander, G., Cheng, Y.B., Burford, R.P., Shanks, R., Mansouri, J., Barber, K.W., Rodrigo, D.P.D., and Preston, C., Ceramifying Composition For Fire Protection, US Patent 8,409,479 B2, 2013.
67. Levchik, S., Hirschler M., and Weil, E., *Practical Guide to Smoke and Combustion Products from Burning Polymers – Generation, Assessment and Control*, iSmithers, 2011.
68. Moore, G.F., Ed., *Electric Cables Handbook*, BICC Cables, Blackwell Publishing Company, 1997.
69. Thue, W., Ed., *Electrical Power Cable Engineering*, CRC Press, 2012.
70. Giles, H.F., and Wagner, J.R., *Extrusion: The Definitive Processing Guide and Handbook*, William Andrew, 2005.
71. Beyer, G., and Hopmann, C., Eds., *Reactive Extrusion Principles and Applications*, Wiley-VCH Verlag GmbH & Co., 2018.
72. Holmes, M., Wired for Performance, in *AMI Compounding World Magazine*, p. 43, May 2019. ISSN 2053-7174.
73. Moore, G.F., Ed., *Electric Cables Handbook/BICC Cables*, Blackwell Publishing Company, Chapter 14, 1997.
74. Thue, W., Ed., *Electrical Power Cable Engineering*, CRC Press, Chapter 11, 2012.
75. Muller, B., Electrical Engineering, in *Plastics Flammability Handbook*, Troitzsch, J., Ed., Hanser Publishing, Chapter 12, 2004.
76. Tewarson, R., Flammability, in *Physical Properties of Polymers Handbook*, 2nd ed., Mark, J.E., Ed., Springer Science Business Media, 2007.
77. Tewarson, R., Flammability of Polymers, in *Plastics and the Environment*, John Wiley & Sons, Chapter 11, 2003.
78. ASTM D2863, Standard Test Method for Measuring the Minimum Oxygen Concentration to Support Candle-Like Combustion of Plastics (Oxygen Index).
79. ASTM E1354, Standard Test Method for Heat and Visible Smoke Release Rates for Materials and Products Using an Oxygen Consumption Calorimeter.
80. UL44, Thermoset-Insulated Wires and Cables, 2018.
81. UL2556, Wire and Cable Test Methods, 2015.
82. UL4703, Photovoltaic Wire, 2014.
83. UL1685, Vertical-Tray Fire-Propagation and Smoke-Release Test for Electrical and Optical-Fiber Cables, 2007.
84. UL1666, Test for Flame Propagation Height of Electrical and Optical-Fiber Cables Installed Vertically in Shafts, 2012.
85. NFPA262, Standard Method of Test for Flame Travel and Smoke of Wires and Cables for Use in Air-Handling Spaces, 2023.
86. IEC 60331-11, Tests for Electric Cables Under Fire Conditions—Circuit Integrity, International Electro- technical Commission, 1999.
87. EN 50399, Test for Flame Spread on Vertically Mounted Bunched Wires or Cables.
88. Kaczorek-Chrobak, K., Fangrat, J., and Papis, B.K., Calorimetric Behavior of Electric Cables, *Energies* 14, 2021, 1007.
89. EN 60332-1-2, Tests on Electric and Optical Fiber Cables Under Fire Conditions-Part 1-2.
90. ASTM E662, Standard Test Method for Specific Optical Density of Smoke Generated by Solid Materials.
91. Weil, E. D.; Patel, N. G.; Said, M.M.; Hirschler, M. M.; Shakir, S. Oxygen Index: Correlations to Other Fire Tests. *Fire Mater.* **1992**, *16*, 159–167.
92. Babrauskas, V., Effective Measurement Techniques for Heat, Smoke, and Toxic Fire Gases, in *Int. Conf. FIRE: Control the Heat-Reduce the Hazard, Fire Research Station*, paper 4, London, 24–25 October 1988.
93. Babrauskas, V., and Peacock, R.D., Heat Release Rate: The Single Most Important Variable in Fire Hazard, in *Fire Safety Developments and Testing: Toxicity, Heat Release, Product Development, Combustion Corrosivity*, pp. 67–80, Fire Retardant Chemicals Association, Fall 1990 Meeting, FRCA, Lancaster, PA, 1990.
94. Hirschler, M. M. Comparison of Large-and Small-Scale Heat Release Tests with Electrical Cables. *Fire Mater.* **1994**, *18*, 61–76.
95. McGrattan, K., Lock, A., Marsh, N., Nyden, M., Bareham, S., Price, M., Morgan, A., Galaska, M Schenck, K., and Stroup, D., Cable Heat Release, Ignition, and Spread in Tray Installations During Fire (CHRISTIFIRE). Phase 1: Horizontal Trays, NUREG/CR-7010, U.S. NRC, 2012.
96. McGrattan, K., and Bareham, S., Cable Heat Release, Ignition, and Spread in Tray Installations During Fire

(CHRISTIFIRE). Phase 2: Vertical Shafts and Corridors, NUREG/CR-7010, U.S. NRC, 2013.

97. Gallo, E.; Stöcklein, W.; Klack, P.; Schartel, B. Assessing the Reaction to Fire of Cables by a New Bench-Scale Method. *Fire Mater.* **2017**, *41*, 768.
98. Walters, R.; Lyon, R. Molar Group Contributions to Polymer Flammability. *J. Appl. Polym. Sci.* **2003**, *87,* 548–563.
99. Lyon, R.; Takemori, M.; Safronava, N.; Stoliarov, S.; Walters, R.; A Molecular Basis for Polymer Flammability. Polymer. 2009, 50, 2608–2617.
100. Wire and Cable Polymers Market Reports, Forecast, Polaris Market Research, New York City, NY, 2022–2030.
101. Aupetit, A., Overview of the Global Cable Industry—Markets and Materials, in *The Global Cable Industry*, Beyer, G., Ed., Wiley-VCH GmbH, Chapter 1, 2021.

30 Flame-Retardant Design and Protection for Building Materials

Alexander B. Morgan[1], *Alessandra Lorenzetti*[2]
[1] University of Dayton Center for Flame Retardant Materials Science, Dayton, OH 45469
[2] University of Padova, Department of Industrial Engineering, Padova (Italy),

30.1 INTRODUCTION

Fire protection of materials used in structures is one of the oldest fields of fire protection engineering, as protecting homes and buildings against fire is as old as recorded history. The earliest applications of flame retardant chemistry were for natural polymeric materials (wood products) where vinegar (acetic acid) and alum (hydrates of aluminum sulfate) were used. Plaster of Paris and Gypsum were also used as early as the 17th century because of their ability to contain fire within the room of origin and prevent its spread to other rooms. The use of flammable synthetic polymeric materials in building materials is a recent phenomena of the past 60–70 years. Houses built in the past 30 years have large volumes of synthetic materials such that they burn very differently than houses built in earlier times.[1] Due to the benefits that polymeric materials can bring to modern houses, including good durability against the elements, improved energy efficiency, resistance to fungal rot/insect attack, and speed of assembly, they will continue to be used in increasing amounts despite their increased fire hazard. As alluded to throughout the book, this means that there is an increased need for fire protection of these polymeric materials, and building materials definitely require constant attention to ensure the polymers used meet the fire protection targets that have been set by various codes and standards. To help understand where flame retardants can be used in building materials, and some of the way in which they are successfully applied, it is important to begin by studying a typical building structure and the polymeric materials used in its construction.

30.1.1 Building Components That Require Fire Protection

Shown in Figure 30.1 is a schematic of a modern residential building and some of the typical polymeric materials used that require fire protection. It's important to note that this figure does not show some of the other notable polymeric components in the home, namely, interior furnishings and wire and cable. Wire and cable materials that run through a building do require fire protection and are covered in Chapter 29 of this book. Interior furnishings typically do not have fire protection requirements, even though they can be a notable contribution to fire risk scenarios.[2–7] The largest interior fuel load in a building, namely, flexible polyurethane foam, is covered in Chapter 31 of this book. Wood materials used in building structures are covered in more detail in Chapter 26, but fire protection of wood + polymer composites is covered in this chapter, along with polymeric insulation, roofing materials, and siding. As will be discussed later in this chapter, the materials used in residential buildings/homes vs. mass occupancy buildings or industrial buildings can vary as the fire risk scenarios are different for each of those buildings. However, mass occupancy buildings and industrial buildings also have polymeric materials present, but they are inserted/protected in different ways as per building codes, as will be discussed next.

Before giving more details on how these polymeric materials are flame retarded, it is important to note that polymeric building materials are used as parts of overall assemblies. Specifically, they work with other building materials to yield the final structure/building sub-assembly that serves a specific function. For example, polymeric foam insulation is not used by itself as it is placed between interior walls and outer siding/roofing materials. Polymeric roofing membranes are used with outer shingles and underlying structural roofing members. Polymeric window frames are combined with glass and framing structural members. As such, the flammability of the polymeric material may be studied both by itself, as well as in combination with other materials. Very often, the polymeric components are flame retarded factoring in the overall assembly in the final structure. This results in fire risk scenarios that occur during construction of a building that do not occur after the building is constructed as the fire performance of the final assembly in use is typically regulated, not the assembly as it is being put together. This will be elaborated in more detail below, but realizing that polymeric materials are used in combination with other materials which are often non-polymeric helps understand why specific flame retardant approaches are used and why the building codes and their fire tests factor this "assembly approach" into their application.

30.1.2 Fire Risk during and after Construction

As mentioned in the previous section, the fire risk scenario for a building in construction is different than that

DOI: 10.1201/9781003380689-30

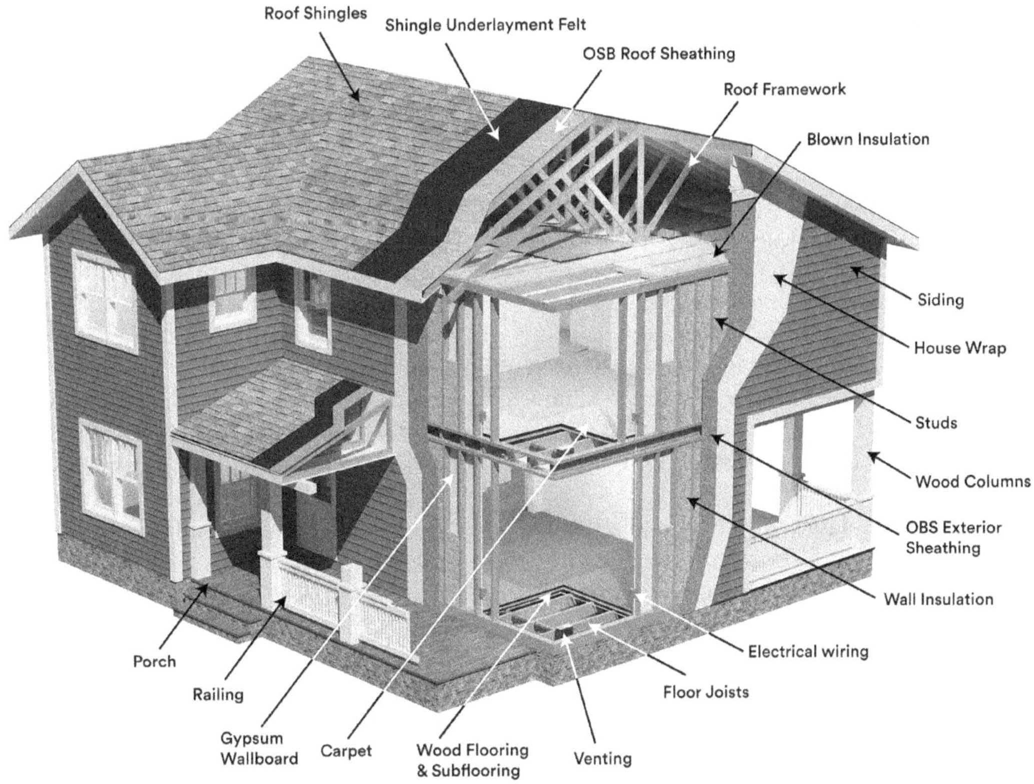

FIGURE 30.1 A modern residential structure and its polymeric components.

of a fully constructed building. The overall fire protection of a building must be assembled like the rest of the building, and at various points in the construction, the fire protection system(s) are not yet ready to protect the structure. Therefore, during construction of a building is when fires can occur and can cause significant damage because the fire protection is not present. For example, if a fire protection engineer has decided that an industrial manufacturing building requires a sprinkler system, up to the point where the sprinkler system is installed, the fire protection of the building is quite limited, especially if flammable polymeric materials, such as insulation, flooring, or other components are installed prior to the sprinkler system being put in place. Sparks from welding or hot metal from sealing pipes during welding, as well as open flame sources, can cause fires that quickly get out of control when a building is still being constructed and its full fire protection system is not yet in place. There are many reported examples in the press/media of large buildings, including apartment complexes, burning up completely due to exposed polymeric components catching fire since it was not fully fire protected by its final assembly.[8–10] One can argue that this may mean that building materials require fire protection not just in their final use but also during construction, but the codes and standards consider the final building, not the manner or timeline for its construction. Fire protection of buildings as they are assembled is best accomplished through diligence of keeping hot/spark-generating operations away from flammable components and considering how the building is to be constructed to minimize exposure of flammable materials. The international building code (IBC) talks about fire protection safeguards during construction, and Chapter 33 of the IBC only requires fire extinguishers be nearby and the ability to access water for firefighting activities. Other fire protection measures are not required. Still, depending upon the value of the building and insurance needs, there may be future requirements for polymeric materials to have fire protection both during construction and after construction is complete.

30.1.3 Building Codes—Local vs. International

In this book, Chapter 22 focuses on building codes and standards, but some discussion of codes and standards are needed in this chapter as well when considering flame-retardant materials for building and construction. It is important to realize that building codes and standards are different within each country around the world, and even within a country there can be local differences. Some areas have harmonized building codes (example, European Union), but even in those areas, there can be specific local rules and requirements that may mandate additional fire protection or prevent the use of specific flame retardants due to environmental regulations. It is important for the fire safety scientist designing a flame retardant material to be aware of what the codes and standards are when designing their material, as these codes and standards will set the performance targets that have to be met to sell the product into that market. While this book cannot be exhaustive in looking at all known exceptions and specific local codes and standards, there are some examples worth discussing here relevant to flame-retardant material design.

Firstly, there is the International Building Code (IBC).[11] These codes tell how buildings should be built to maximize safety and durability and have entire sections that describe what fire performance requirements exist for building materials as a function of building type. While these codes set the minimum requirements for building materials and assemblies, they must be adopted by a particular country/jurisdiction to be utilized. Very often, the IBC serves as a foundation for fire performance of materials used in building and constructions, and then other countries will adopt aspects of these codes into their own localized fire performance requirements.

One country which builds upon IBC requirements and then requires even higher levels of fire safety performance is Japan. Japan is an earthquake-prone zone due to its location along various tectonic plates. Further, Japan has regions of high population density and was the location of the world's largest earthquake + fire event, the Great Kanto Earthquake.[12] The earthquake led to major fire losses, with estimates of 105,000–142,000 dead from earthquake and fire. As Japan rebuilt from WWII, it implemented strict building codes that ensured buildings could withstand earthquakes and the secondary fire events that often happen from earthquakes.[13] Specifically, in an earthquake, gas and electrical lines can break and lead to ignition events, and therefore building materials must be fire hardened so that if they survive the earthquake, they will also resist further ignition and fire damage. Japan's fire codes remain some of the strongest in the world given their location and desire to prevent fire losses in earthquake events, and as such, this country maintains a high level of fire safety in its buildings, even when polymeric materials are used.

In the European Union (EU), the IBC is used, but commonly one finds the Eurocodes, which are harmonized building codes to be used across the EU to facilitate commerce and ease of construction.[14] Prior to this harmonization, each country of the EU had its own unique fire test requirements for building materials, with a wide range of performance existing between countries. This made it difficult for building materials in one country to be used in the adjacent country, but with the Eurocodes, this has mostly been addressed. However, some countries still require their own tests on top of the Eurocodes to be performed to ensure fire safety of the products in that country, and so some local autonomy still complicates the fire standards in the EU. Indeed, when considering Eurocodes, IBC, and country requirements, it can require some careful attention to the codes prior to designing a fire-safe polymeric material for building use. Despite this potential for multiple requirements, the Euroclass rating system for materials has proven effective,[15] and it uses the following test methods to set the rating for a building material.

- The Single Burning Item (SBI) test.[16] This heat release based test is used for all materials except flooring. Materials are put into a room corner configuration and ignited with an open flame source, simulating a "single burning item" radiating heat and impinging flame onto the materials which make up the room corner. Fire growth and smoke growth rates are measured through oxygen consumption calorimetry.
- Non-combustibility test (EN ISO 1182). This test inserts a material into a 750°C tube furnace, and temperature rise is measured by thermocouples. The test is similar to the US/IBC Non-combustibility test (ASTM E136), but there are some differences in the test, which is why ASTM E2652 exists as well. Effectively the test measures the potential heat output from a thermally decomposing material that could ignite other objects. The test has been in use for a long time, and if used correctly, ensures materials are used which cannot contribute to fire growth. The test is difficult for polymeric materials to pass, but not impossible for these materials to pass.
- Total heat release via oxygen bomb calorimetry (EN ISO 1716). This measures the maximum heat release of a material when combusted completely and can be envisioned as the total heat release/fuel load of a material in a fire event.
- Radiant Panel Test (EN ISO 9239–1). This test is similar to the US radiant panel test (ASTM E648 and E162) which studies flame spread and critical heat fluxes for ignition. Smoke measurements can also be measured in this test.
- Ignitability (EN ISO 11925–2). This test exposes a vertical oriented material to a small 20 mm gas flame and is a simple vertical flame spread and ignition resistance test. One can view this test as a way of verifying the material can resist the smallest minimum flames that may be encountered in a building fire event.

It should be noted that passing results in any of these tests does not confirm that a pass will occur in the other tests. As this book has explained many times, each polymeric material must be flame retarded/fire protected for each test. There may be cases where something that is flame retarded to a very strict fire test can pass a weaker fire test, but this is not always ensured, and testing verification must be accomplished. Of final note, the EU also has harmonized Euroclasses for wire and cable materials used in buildings. This is not discussed in this chapter and is instead covered in Chapter 29 of this book.

One organization which has a heavy role in codes and fire safety requirements for building materials in the United States, which in turn generates specialized requirements for the US in concert with the IBC, is the US National Fire Prevention Association (NFPA). NFPA codes and standard tests are covered in detail in Chapter 22 of this book but are worth mentioning again here so the reader is aware that NFPA standards are used in building materials. In some cases, NFPA standards are used verbatim in the IBC and have equivalents in other standard organizations (ASTM, ISO), but not always. Sometimes the NFPA codes have their own fire requirements which are strictly voluntary to implement or are used by select organizations for select building applications. Still, most NFPA codes look at specific building materials in specific applications (roofing, walls, electrical systems, etc.) and prescribe specific fire test methods for the safe installation and use of polymeric materials.

Finally, there is the International Wildland-Urban Interface Code (IWUIC) which provides details on how to fire harden buildings against wildfires that release burning embers/ashes.[17] As with the IBC, this standard must be adopted by various countries, but it is increasingly being used, and it provides guidance on test methods that building materials must pass to be fire resistant in these areas. Resistance to burning embers is a bit of a unique fire risk scenario as the single embers may not ignite something, but a large pile of collected embers could serve as a significant ignition source. Likewise, if embers ignite something else near the house (a dry shrub/bush or dry mulch by the walls of the house), then these secondary fire sources can grow into large primary fire sources which can be difficult to protect against.

Ultimately, the material scientist seeking to develop and implement flame-retarded polymeric materials for building and construction applications should pay attention to the codes and standards to ensure their product can be sold. These codes and standards describe the tests and performance to design to. It's important to realize that these tests describe the minimum fire performance required in these applications, and it is up to the fire safety engineer to decide if higher performance above and beyond that described in the codes is required. Therefore, material scientists designing fire-safe materials for building and construction should not only pay attention to the tests but also be aware of how fire safety engineers will prescribe fire protection in these applications. It is strongly encouraged for those producing fire-safe materials for building and construction to be aware of codes and standards so they can ensure their polymeric materials maintain performance over time as the codes and standards change. Codes and standards are not static and do change as new fire risk scenarios are discovered, or as the new non-fire requirements (energy efficiency, environmental requirements) are implemented.

30.1.4 Fire Protection of Structures

One key aspect of fire protection of buildings is the fire protection of the building support structure. This is done to prevent collapse of the building in a fire event, ensure sufficient time to escape, and for the fire service to arrive and extinguish the fire. Building structural fire protection is often discussed in regards to fire exposure times, and discussions of fire protection ratings for half hours and hours is common. Further, the fire protection times also consider mechanical load on the structure to ensure that the building material holds up the entire structure while heat soaks into the structure. Some materials can maintain their structure just fine in a fire if not under mechanical load, while others will quickly buckle and collapse when load is applied. It's important to understand this aspect of load and fire exposure time when designing materials for building and construction.

One common example of fire protections of building structure is the use of intumescent coatings for concrete and steel. These intumescent coatings are covered in more detail in Chapter 7 of this book, but they basically provide thermal protection to prevent softening/melting of steel, and structural failure of the concrete in fire events. Intumescent coatings can be used for many other building materials where appropriate, including wood framing systems, insulation, and gaps around utility pipes/conduits that penetrate fire rated walls and substructures. Notionally, intumescent coatings could be used for exterior siding and roofing, but currently are not used for those applications since they tend not to be that durable against continual exposure to UV (sunlight) and water (rain, snow, humidity). Another way to protect structural assemblies is with the use of plaster/gypsum wallboard to compartmentalize fire and prevent it from spreading to the structure, but they can only be used inside the structure and not exposed to the elements. Sprinkler systems can also be used to again attack the fire and keep heat low during a fire event so the structure remains intact.

Structural fire protection tends to rely more on polymeric materials to enhance protection of the structure, as the polymeric materials are rarely ever used for structure. Many polymeric materials mechanically fail well before they ignite due to their glass transition temperatures (T_g) occurring well below their thermal decomposition temperatures. Therefore, polymeric materials tend to not be tested in structural fire tests, but they can be part of the overall protective system if designed and installed correctly.

30.2 FIRE PROTECTION OF ROOFING AND SIDING

As discussed previously, when considering fire protection of roofing and siding of buildings, the fire risk scenario must be considered first. Notionally, if the roofing or siding of the building is made from nonflammable materials (example—metal or shale roofing, metal or brick walls), then fire protection may be easily obtained. If the roofing or siding is made out of polymeric materials, then one must consider the test and the flame retardant approaches for it. Typical polymers used for roofing and siding including tar-paper singles, elastomeric (rubber) roofing membranes, polyolefin water/weather seals, polyvinylchloride siding and wood siding. Polymeric-wood composites may also be part of the exterior systems depending upon what the building codes allow and how the composite is installed.

Testing of building materials is covered in some detail in Chapter 18 of this book, but it's worthwhile to talk about these tests here as well, especially those used for roofing and siding. Many materials used in building and construction, including those for siding, use some sort of flame spread test. In the EU, the SBI test is most often used, but in the US, the "Steiner Tunnel" test (ASTM E84) is most commonly used. Other tests that may be required include the abovementioned non-combustibility test (EN ISO 1182/ASTM E136), and radiant panel flame spread tests (EN ISO 9239–1, ASTM E162). Roofing materials often need to pass tests such as UL 790 (ASTM E108), which places a large burning wood source on the roof, meant to reflect a burning branch landing on a roof, and measures flame spread and burn through from that heat source. Roofing shingles in the US can have three different ratings—Class A (least flammable), Class B, and Class C (minimum level of fire protection) depending upon what is allowed in the building codes. More severe tests that look at true fire resistance of roofing materials would require testing such as UL 263 (ASTM E119), which is a large furnace tests that assigns fire resistance times.

Typical roofing shingles are composite materials, containing not just flammable materials (such as tars or polymers), but also fillers, grit for traction on the roof, and fiberglass reinforcements/mats. The tar/pitch used to make roofing shingles is best flame retarded with mineral fillers (aluminum hydroxide, calcium carbonate), but other inorganic fillers can be used as well (silicas, wollastonite, etc.) that will lower the overall flammability of the shingle and help it meet flame spread requirements. See Chapters 6 and 8 for more details on the types of mineral fillers that can be used. Halogenated flame retardants can also be used, but some UV stabilizers would likely be needed to maintain performance over time in the elements. Notionally, intumescent flame retardants could be used as well, but they would need to be water/weather durable, and this is not always easy to achieve with the typical chemistry used in intumescent flame-retardant additives/packages. High levels of flame retardant may be required to get satisfactory fire performance in tests like UL790.

Polymeric siding is more complex to flame retard because of the complex geometry of the siding and the myriad components that may be adjacent/in contact with the siding. Specifically, the polymeric siding may be directly in contact with building insulation or other components that may affect the flame spread rating of the siding. ASTM E84 is most commonly used to measure flame spread of polymeric siding, and depending upon how the material sags and melts during the test, specialized testing methods/annexes within ASTM E84 may be required to be used to ensure valid results. The most common polymer used for siding is vinyl, or poly(vinyl chloride) (PVC) siding, which has some inherent flame retardancy to it due to the presence of chlorine in the polymer backbone. However, high use of plasticizer to get good mechanical properties can cause the PVC siding to have higher flammability. Therefore, typical flame retardants like mineral fillers (calcium carbonates, huntite, aluminum/magnesium hydroxide) are used (see Chapter 6), with halogenated flame retardants also used in some cases.[18,19] Silicas and other inorganic flame retardants (Chapter 8) can also be used to address some of the melting action seen in PVC siding. Of final note, wood siding is still used in some places and can be flame retarded via various pressure treatments that penetrate the flame retardant into the wood so that it can meet the flame spread tests (see Chapter 26). Wood does not have the melting/dripping problem that polymeric siding has, although wood does have issues with smolder since it tends to char as it burns.[20–28]

Wood-polymer composites used in decking are not used for roofing or siding, but as exterior components that come into contact with siding, they merit some discussion on fire protection here. Most commonly, these wood-polymer composites are extruded from polyolefin + wood dust (also known as wood flour)/wood chip/wood fiber mixtures, but versions with other polymers and fiber reinforcement exist as well. Tests for these composites as exterior decking materials include both flame spread and tests that simulate "burning brands" from WUI type fires. In the US, deck assemblies are built, and a fire put underneath the deck to see how firebrands/fire sources starting underneath the deck may result in ignition and flame spread of the decking material.[29] Additionally, there are some jurisdictions in the US which use burning brand tests on top of the deck, such as ASTM E108[30] or ASTM E2726,[31] the latter standard which is only used in California. Polymeric materials do not perform as well as wood decking in flame spread and burn through tests unless flame retardants are involved, as the commonly used polyolefins are far more flammable than wood, and they melt and flow during fire/heat exposure.[32–36] There are a wide range of flame retardants that can work for polyolefin decking materials.[37–46] Mineral fillers, intumescent additives, and halogens can be used, although UV stabilizers will be needed for halogenated FR additives to prevent discoloration. Additionally, high mineral filler loading levels will begin to color the decking white, and pigments may be needed to restore a more "wood-like" appearance. As higher flame retardant loading levels are used, the balance of mechanical properties will be an issue.

30.3 FIRE PROTECTION OF INSULATION SYSTEMS

Insulation systems can be built by using foamed materials which can be inorganic (like rockwool) or organic like polymeric foams (e.g., polyurethane or polystyrene foams). Rockwool, because of its inorganic nature, is incombustible and reaches A1 Euroclass. Polymeric foams are a two-phase system built up by many cells, comprising solid walls and struts, filled with a gaseous phase (mainly blowing agent or air) which may be flammable (e.g., pentane). This particular structure is responsible of the low thermal conductivity, vital for the insulating properties. Moreover, polymeric foams are organic materials, and thus they are combustible in nature, but their burning behavior differs from that of the bulk polymer because of their particular structure. In general, for common thermally thick combustible materials (greater than a few millimeters), the time to ignition is proportional to the thermal inertia that can be expressed as $\lambda*\rho*c$ (where λ is the thermal conductivity, ρ is the density, and c is the heat capacity). Thermal inertia characterizes the rate of surface temperature rise of the material when exposed to heat. Low values of thermal inertia lead to a rapid temperature rise for a given applied heat flux and, hence, to a rapid ignition.[47] Polymeric foams have much lower thermal inertia than the corresponding solid materials, because of their low thermal conductivity and density,[48] and thus the surface temperature of the first heats up more rapidly than that of the latter. Foam surface may reach the ignition temperature ten times faster than the solid polymer, and this, obviously, contributes significantly to pyrolysis.[49] Thus, fire involving polymeric cellular materials develops in extremely rapid way when compared with solid polymer, giving rise to high temperatures and producing large amount of smoke in a short period of time. Moreover, during pyrolysis of closed cell foams, a huge amount of a flammable compound may be released due to the potential presence of flammable blowing agent inside the foam cells.

In order to increase the understanding of the fire phenomena of rigid polyurethane foams (RPUFs) used in thermal insulation, which is the foundation for developing flame-retardant approaches,[50] Günther et al. studied[51] a set of polymeric foams with systematically varied properties. In particular, the effect of density (which has been varied from 30 to 100 kg/m^3) and blowing agent (which were water or pentane) has been analyzed. The variation of density has been obtained by varying the amount of blowing agent used, so without affecting the chemical composition of the material. Besides RPUF, pentane blown rigid polyisocyanurate (PIR) foams were examined, since these foams are known for their higher stability and improved fire behavior from a thermodynamic point of view.[52] Preliminarily, it has been verified that foams have comparable cell morphology at similar densities, rendering them comparable when referring to their burning behavior. It has been verified that the density of the foams has a significant effect on both the limiting oxygen index (LOI) as well as under forced flaming conditions. The LOI increased with increasing density, although the effect is small. Indeed, the influence of density is limited to 1.2 up to 1.6 vol% and was related to the increased thermal inertia and the better char yield of higher-density foams.[53] Water-blown foams showed slightly higher LOI with respect to pentane blown foams with the same density, probably because of the nonflammable blowing agent.

During cone calorimeter testing, all the samples underwent distinct charring, and none exhibited structural collapse nor formed a pool fire of liquid pyrolysis products, as instead is known for flexible PUF.[54] They also showed very short time of ignition (t_{ig}), that is 4 s or lower. An increase in t_{ig} was measured for higher densities, although it was not substantial; this is because the thermal inertia ($\lambda\rho c$) of foams rises with density. Moreover, the ignition was followed immediately by the peak of heat release rate (PHRR) and the formation of a carbonaceous protective layer. Indeed, due to the low heat conduction into the bulk of foams, cellular polymers generally respond very quickly to an imposed heat flux. Therefore, the surface of all tested samples heated up promptly once the specimen was exposed to the external heat flux of the cone calorimeter. Then, in a second stage, plateau-like behavior of the HRR occurred, with less intense burning. PHRR and average HRR are considerably lower for PIR compared to all tested RPUF samples, but PIR exhibited a prolonged time to flameout, so the fire load was very similar for the same density. The length of the second stage plateau depended on the density of the materials for both RPUF and PIR, since this affects the amount of combustible material, being the chemical composition of the foam constant. The same behavior has been observed for solid polymers with increasing thickness and therefore amount of combustible material. The phenomenon of increased steady burning for foams is based on the same effect, but instead of increasing the sample thickness for a higher amount of combustible material, the density is the parameter varied. While for different sample thicknesses the pyrolysis front consuming the sample is constant but the burning time increases because the front must travel longer distance, for samples with varying densities, the pyrolysis front must travel the same distance, but its velocity decreases with increasing density.[51] A different peak of heat release rate (PHRR) trend has been shown for RPUF and PIR. A higher PHRR has been revealed for RPUF with increasing density samples because the higher the amount of pyrolysis product in the top layer, the more heat can be released, thus leading to an increased PHRR for higher-density materials. On the contrary, the PHRR of PIR samples dropped slightly with increasing density, since for these materials the effect of more distinct charring prevailed over the effect of the higher amount of combustible material on the surface for higher-density materials, as instead observed for RPUF samples. In general, the residue formed by the PIR foam was more stable, denser, and thicker compared to PUR foams. Difference in charring behavior can also be confirmed considering the residues obtained by quenching with liquid nitrogen at 50 wt% of total mass loss (TML) (Figure 30.2). It is evident that RPUFs developed a residue in which holes were visible, because of the bubbling during combustion, while the residues of PIR had a closed and compact surface.

A further confirmation of the better charring behavior of PIR with respect to RPUF has been obtained by monitoring the temperature development inside burning specimens, during the cone calorimeter test, at depths of 10, 20, and 30 mm (Figure 30.3). The $T_{95\%}$ obtained from TGA (i.e., the temperature of 5 wt % mass loss) was consulted

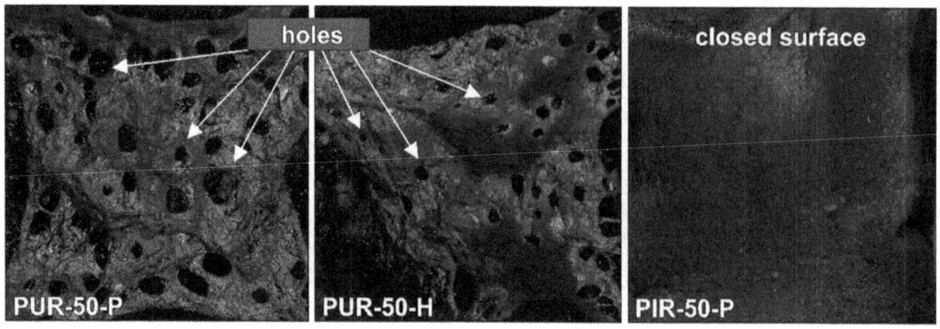

FIGURE 30.2 Macro-photograph of residue surfaces of RPUF blown with pentane (PUR-50-P), with water (PUR-50-H) and PIR (PIR-50-P), with 50kg/m^3 density.

Source: Taken from ref.[51]

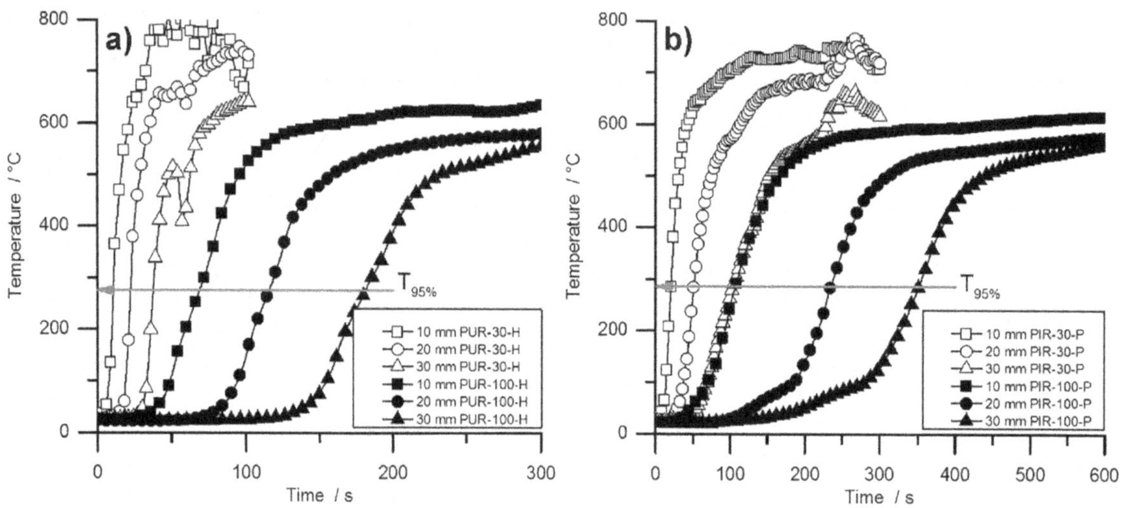

FIGURE 30.3 Temperature development inside (a) RPUF and (b) PIR foams at densities of 30 and 100 kg/m^3

Source: Taken from ref.[51]

to define the time at which the pyrolysis front reached the depth of temperature measurement. The graphs confirmed the good insulating properties of the foams: even though the PHRR of the burning top layer was already reached before 20s for all the materials, the temperature measured at a depth of 10 mm remained low. By the time the $T_{95\%}$ was reached at a depth of 10 mm, the protective layer on the surface had already formed, and HRR began to drop to the steady burning HRR. In any case, it can be clearly seen that the times to reach $T_{95\%}$ at each depth of measurement increased significantly for PIR with respect to RPUF, while the slope of temperature rise (i.e., the rate) is lower for PIR than for RPUF, as a consequence of the better fire performance of PIR foams under forced flaming conditions. Moreover, it has also been noted that the slope of the temperature curves is steeper with lower density, for both PIR and RPUF, and the onset of temperature rise is shorter for lower density, meaning that the increasing density of the foam leads to an earlier formation of char. In other words, for low-density foams the temperature signal exhibited a steep increase as soon as the pyrolysis front approached the thermocouple. In contrast, for high-density foams a more moderate rise was observed, including a minor rise in the temperature inside the specimen prior to pyrolysis, caused by thermal conduction through the bulk of the foam. This implies a shift of the burning behavior toward the behavior of a solid, non-cellular material with increasing density.

No significant effect has been revealed in the cone test related to the different blowing agent used: even though the content of the closed cells is either highly flammable (iso and cyclo-pentane for pentane-blown foams) or inert (CO_2 for water-blown foams), the peak of heat release rate (PHRR), average HRR, and fire load differs from each other only slightly. This is because the pentane adds no more than 3–5 wt% to the overall weight and thus barely any fuel or increased effective heat of combustion (EHC), the fire behaviour of a RPUF is determined mainly by the solid polymer and not by the gas phase.[55] The only difference that has been revealed is the slightly higher maximum heating rate (MHR, i.e., the maximum slope of the plot of the temperature inside the specimen versus time) for pentane blown RPUF, which is probably an effect of the highly combustible blowing agent, although the overall fire performance determined with the cone calorimeter was not affected compared to water-blown RPUF.

Therefore, we can state that PIR foams have advantageous fire behavior with respect RPUF, whatever the density considered. However, to meet strict fire regulation requirements (e.g., B s1 d0 according to the Euroclasses), flame retardants must be used for both PIR and RPUF.

Several reviews on flame retardants for polyurethane foams have been recently reported (e.g., by Yadav et al.,[56] Zhu et al.,[57] Muhammed Raji et al.,[58] Levchik and Weil,[59] and by Wang[60]). Flame retardancy of PU foam is commonly achieved through use of reactive or additive flame-retardants: the first are chemically linked to the polymer backbone, while the latter are added to foam formulations as separate compounds.

Reactive flame retardants with multi-hydroxyl, multi-amino, or multi-epoxy groups can serve as polyols to participate in the curing process of RPUF. Their flame-retardant function is mainly derived from the phosphorus, nitrogen, silicon, or sulfur elements in the structure.[57] For example, 9,10-Dihydro-9-oxa-10-phosphaphenanthrene-10-oxide (DOPO) is commonly used precursor to prepare P-containing reactive flame retardants. While it is nowadays declining, large use of brominated polyols have also been reported in the past,[58] and they are still commercially available.[61] In general, reactive flame retardants exhibit good interfacial compatibility with the matrix because of their chemical bonding interactions, and thus negligible effects on mechanical properties of RPUF. Meanwhile, the chemical bond makes the reactive flame retardants unable to migrate from the matrix, bringing about great durability in industrial applications.[57] Unfortunately, the reactive flame retardants usually require a high loading level to enhance the flame retardancy of RPUF. Moreover, for most of the reactive flame retardants, they cannot guarantee high level of flame retardancy: for example the foams cannot be rated UL-94 V-0 when reactive FRs are used alone or low LOI values (< 31.0%) are obtained.[57] It should be noted that since UL-94 V-0 ratings are not used for building applications, some care should be used in connecting the UL-94 V test to building material requirements, and other testing should be done to confirm if a "fail" in the UL-94 V test also equals a fail in a building material test.

The additive FRs can be physically blended into the PU at the mixing phase, which usually accounts for a relatively less expensive and more effective way to introduce a wider number of FRs. However, in some cases (e.g., inorganic hydroxides), the need for large amounts of FRs to achieve satisfactory fire performance leads to mechanical properties deterioration due to incompatibility with PU matrix, contributed by the microphase separation.[56] Additive FRs are generally based on halogen, phosphorus, nitrogen, inorganic hydroxides, and carbon; compounds based on halogen and phosphorus (e.g., chlorinated organophosphates like TCPP, tris(1-chloro-2-propyl)phosphate) or nitrogen and phosphorus are very often used because of their synergy.

Despite the effectiveness, the negative persistence, bioaccumulation, and toxicity of select halogenated FRs limit their wide usage and efforts were dedicated to develop non-halogenated FRs. For example, TCPP, which is very effective in both gas and condensed phase, has been assessed very recently by the European Chemicals Agency (ECHA) for regulatory needs due to its toxicity[62] and is being phased out of use for insulation materials in the EU and US.

Phosphorus-based FRs used in RPUF are based on phosphates (e.g., triethyl phosphate, TEP), phosphonates (e.g., diethyl ethylphosphonate, DEEP), or phosphinates (both organic or inorganic ones, like aluminum phosphinate). Phosphorus FRs are very effective since they can act in both solid and gas phase, although it has been shown that their main mechanism of action depends on the phosphorus valency.[63] Recently, 9,10-dihydro-9-oxa-10-phosphaphen-anth-rene-10-oxide (DOPO) and its derivatives have been introduced as phosphorus FRs for PU foams.[64,65] Nitrogen FRs (e.g., melamine) are seldomly used alone because of their ineffectiveness at low nitrogen content, poor compatibility, and deterioration in mechanical properties of PU foams.[56] It must also be underlined that in Europe a decrease in the use of melamine can be expected since melamine has been recently classified as a suspected human reproductive toxicant while melamine cyanurate is already classified H373 (specific organ toxicity repeated exposure STOT re.2 kidney).[66] Indeed, although classification of melamine is not relevant for melamine as used as a flame retardant, that is, after inclusion in a material, it is relevant to compounders and processers handling melamine and the new classification is expected to affect foam producers choice. Most commonly, nitrogen FRs are used in combination with phosphorus FRs to promote synergy: the well-known and highly effective ammonium polyphosphate (APP) is an example (see Chapter 10 of this book for more detail). However, when high APP content (> 15%) is incorporated as a standalone additive, possible breakdown in mechanical properties and FR efficiency, mostly during extended use, high temperature and humid environments, can be experienced. Nowadays, microencapsulation technique is used as popular surface modification approach of APP to fulfil the need for flame retardancy of polymers by using diverse surface material and overcome the drawbacks listed before.[58] Very commonly, the phosphorus and nitrogen-based FRs are intumescent systems and they are primarily the mixture of three components: inorganic acid sources (e.g., APP), char former (e.g., pentaerythritol or the polymer itself), and blowing agent (e.g., melamine). An intumescent system can be obtained also by using expandable graphite (EG) which use in RPUF has been reported since 2002[67]: EG leads to the formation of an intumescent and compact char layer characterized by the presence of "worms," deriving from its expansion. The expansion of EG is due to a redox process between H_2SO_4, intercalated between graphite layers, and the graphite itself that originates the blowing gases. By comparing several EGs with the same mean particle size, with different expansion volumes and intercalants (sulfuric or phosphoric acid),[68] it has been shown that EGs intercalated with sulfur showed better flame retardancy than the one with phosphorus in both LOI and cone calorimeter. The expansion volume seemed to play a limited role in small-scale tests to assess flammability, like the LOI, whereas no effect of this parameter was revealed under forced flaming conditions in the cone test. Above a critical level, increasing EG loadings deteriorated the mechanical properties and thermal conductivity of foams. Furthermore, PU foams modified using large particle size expandable graphite demonstrated better flame retardancy than those prepared with small particle size.[56]

More recently, besides reactive and additive FR technologies, some coating technologies have also been proposed[57] although to a greatly limited extent when compared to the layer-by-layer (LBL) technique for flexible PUF. The flame-retardant coatings on the RPUF surface can efficiently protect the underlying foam during combustion.[69] Compared to additive flame retardants, surface-coating treatments exhibit, in general, overwhelming advantages because they do not damage bulk properties of RPUF. However, there still remains some challenges for their practical application, and the major one is the unsatisfactory interface adhesion between RPUF and the coating, resulting in the decreased flame-retardant efficiency and durability. In addition, they also suffer from the relatively low flame-retardant efficiency and negative impact on the thermal insulation of RPUF.[57] Due to the closed-cell structure of RPUF, the coatings need to have a higher flame-retardant efficiency or greater thickness to achieve a significant flame-retardant level, which is the main problem of the current surface flame-retardant RPUF (SFR/RPUF) system.[70] Flame-retardant coatings (FRC) are divided into non-intumescent (NIFRC) and intumescent (IFRC) ones, depending on whether the flame-retardant coating expands when heated or not. NIFRCs for RPUF mainly include hydrogel/sol, aerogel, and ceramic materials.[69] The intumescent flame-retardant coating can be prepared by adding conventional intumescent FRs (such as EG) to the coating substrate or by reacting the flame retardant with the coating substrate (all-in-one).[69] In general, the interfacial adhesion performance and hydrophobicity need to be further addressed.

Along with the above mentioned polyurethane insulation materials, mineral wool insulation is used as well. This includes insulation such as fiberglass and mineral wool fibers spun from various molten minerals and ceramic slags. These inorganic fiber insulation materials are often installed as batting with various wrappers to enable ease of installation. While the fibers themselves are nonflammable, the overall insulation batting assembly may have some flammable components, including adhesives, binders, fiber sizings, and the abovementioned wrappers. These would be flame retarded as needed as the overall assembly is often tested for its flame-retardant performance, rather than testing of the individual components that make up the fiber batting assemblies.

Finally, there should be some discussion on the other major polymeric insulation used in building materials, namely, polystyrene-based foams. This is often sold and installed in two forms: expanded polystyrene bead foam (EPS) and extruded polystyrene foam (XPS). EPS many people would recognize as polystyrene foam beads that have been pressed into board shapes that can be cut and shaped to fit into various building cavities. XPS is extruded as a continuous foam board which is then cut and shaped to fit into various building cavities. Both EPS and XPS often need to meet fire tests, but since polystyrene is a high heat release material, EPS and XPS use an unusual method of passing the fire test. Unlike the abovementioned PU foams which can be heavily flame retarded and can char in place, the loading levels required for polystyrene foams to char and resist flame result in a product which is highly dense and cannot be foamed. Therefore, EPS and XPS drip away from the fire source by using flame retardants which induce depolymerization of the polystyrene backbone. Essentially, the melt viscosity of the EPS and XPS decreases rapidly upon exposure to heat, and the material drips away from the flame and creates an "air gap" between the heat source and the remaining EPS/XPS.[71,72] This melt-back behavior does result in an increase in heat release for EPS and XPS,[73] and so this limits applications for EPS and XPS in some fire-risk scenarios.

30.4 FIRE PROTECTION OF STRUCTURAL SYSTEMS

As discussed previously in this chapter, structural frames of buildings require fire protection. Steel needs fire protection so that it does not exceed its softening/melting temperatures that will cause it to buckle/deform under mechanical load when holding up a structure. Concrete also requires fire protection as it will explode during a fire event. Specifically, the water in concrete becomes superheated in a fire event and boils out with explosive force, causing some of the concrete to crack and "pop off" the concrete surface, which is described as "spalling." With more heat, more concrete cracks and spalls off, and eventually parts of the concrete structure start to give way. Additionally, steel reinforcement (rebar) in the concrete structure will now be exposed, and it also will in turn start to heat up and structurally fail in a fire. So while concrete itself isn't flammable, something else of high heat release will damage the concrete and cause it to structurally fail. Most often, steel and concrete are protected with intumescent coatings, examples of which can be seen in Figure 30.4 below. These coatings come in paint and spray-on forms depending upon how the coating will need to interact with the environment. These coatings must often remain in place over the lifetime of the building and so should have good adhesive properties to the steel/concrete they are applied to. If the coating should fall off, then the underlying material will have lost its fire

FIGURE 30.4 Example of spray-on intumescent coatings for applied to I-beams.[76]

protection and will structurally fail prematurely in a fire event. Periodic inspections of coatings are recommended, especially in places of high humidity where corrosion of the underlying metal may occur. The chemistries of these intumescent coatings are highly varied depending upon the end use requirements.[74,75] For more details on the range of approaches and chemistries available, please see Chapter 7 of this book.

Another structural member which can use intumescent coatings as well, but can also benefit from other chemical approaches, is wood. Wood is being used increasingly as a structural member in multi-level structure, sometimes referred to as "Mass Timber" buildings. Because wood will char, this char protects the underlying unburned wood, thus slowing the fire burning through the structure. Mass Timber does require some different fire protection systems and specific fasteners vs. steel and wood, but wood can be protected with flame retardants (see Chapter 26 of this book) as well as traditional fire protection methods (sprinklers, gypsum wallboard) and the abovementioned intumescent coatings. Depending upon how the mass timber structure is engineered, there may be potential for flame retardants to be added into adhesive/polymeric components as well, assuming the addition of said flame retardants does not weaken the mechanical properties of the structural wood member.

30.5 CONCLUSIONS AND OUTLOOK

Fire protection of buildings is as old as recorded history and will continue to be required for as long as there are fire risk scenarios in buildings. Since sprinklers cannot be inserted into every structure, the fire protection must be tailored to the specific building type (industrial, residential, medical) as per the building codes, and wherever there are polymeric building materials, flame retardants may be a necessary part of the building material.

As discussed in other parts of the book (Chapters 22 and 23), changing fire requirements and flame-retardant regulations can be in conflict with one another, with entire classes of flame retardants under sufficient regulatory scrutiny such that they cannot be used for fire protection. This means that the material scientist must pay attention to not only the needed fire protection for their material, but also the regulatory issues about which flame retardant chemistry may be allowable in their product. Changing fire-risk scenarios (energy storage systems, WUI) further mean that new fire tests will be developed and polymeric materials may not always be able to withstand severe fire hazards, even with the use of flame retardants. However, increased demand for energy efficiency will require more insulation in buildings, and polymeric insulation can be flame retarded to high levels of fire safety for this need.

To conclude, flame retardants remain a vital part of building materials with the increased use of polymeric materials in buildings. This trend of increased use of polymeric materials is unlikely to change, as polymeric building materials tend to have lower cost than stone, brick, and steel, and wood is not always an appropriate material in all building applications.

REFERENCES

1. www.nfpa.org/News-and-Research/Publications-and-media/NFPA-Journal/2021/Fall-2021/Features/1700 (accessed 11/22/22).
2. "Revisiting flexible polyurethane foam flammability in furniture and bedding in the United States." Morgan, A. B. *Fire Mater.* **2021**, *45*, 68–80.
3. "Combustion characteristics of flat panel televisions with and without flame retardants in the casing." Blais, M.; Carpenter, K. *Fire Technol.* **2015**, *51*, 19–40.
4. "Comparative room burn study of furnished rooms from the United Kingdom, France and the United States." Blais, M. S.; Carpenter, K.; Fernandez, K. *Fire Technol.* **2020**, *56*, 489–514.
5. "Fire-LCA model: TV case study." Simonson, M.; Blomqvist, P.; Boldizar, A.; Möller, K.; Rosell, L.; Tullin, C.; Stripple, H.; Sundqvist, J. O. *SP Report 2000:13*. Printed in 2000.
6. "Fire performance of selected IT-equipment." Simonson, M.; Andersson, P.; Bliss, D. *Fire Tech.* **2004**, *40*, 27–37.
7. "Fire behaviour of plastic parts in electrical appliances—standards versus required fire safety objectives." Babrauskas, V.; Simonson, M. *Fire Mater.* **2007**, *31*, 83–96.
8. www.latimes.com/local/california/la-me-downtown-fire-20141209-story.html (accessed 12/19/22).
9. www.enr.com/articles/53600-blaze-leaves-nearly-finished-65m-apartment-project-a-total-loss (accessed 12/19/22).
10. www.nfpa.org/News-and-Research/Publications-and-media/NFPA-Journal/2019/November-December-2019/Features/Large-Loss/Large-Loss-incidents (accessed 12/19/22).
11. www.iccsafe.org/products-and-services/i-codes/2018-i-codes/ibc/ (accessed 12/11/22).
12. https://en.wikipedia.org/wiki/1923_Great_Kant%C5%8D_earthquake (accessed 12/11/22).
13. www.bcj.or.jp/upload/international/baseline/BSL Introduction201307_e.pdf (accessed 12/11/22).
14. www.en-standard.eu/eurocodes/ (accessed 12/11/22).
15. http://virtual.vtt.fi/virtual/innofirewood/stateoftheart/database/euroclass/euroclass.html (accessed 12/11/22).

16. www.en-standard.eu/csn-en-13823-reaction-to-fire-tests-for-building-products-building-products-excluding-floorings-exposed-to-the-thermal-attack-by-a-single-burning-item/ (accessed 12/11/22).
17. https://codes.iccsafe.org/content/IWUIC2021P1 (accessed 12/11/22).
18. "Flame and smoke retardants in vinyl chloride polymers—commercial usage and current developments." Weil, E. D.; Levchik, S.; Moy, P. *J. Fire Sci.* **2006**, *24*, 211–236.
19. "A targeted review of bio-derived plasticizers with flame retardant functionality used in PVC." Morgan, A. B.; Mukhopadhyay, P. *J. Mater. Sci.* **2022**, *57*, 7155–7172.
20. "Review of large-scale fire tests on cross-laminated timber." Ronquillo, G.; Hopkin, D.; Spearpoint, M. *J. Fire Sci.* **2021**, *39*, 327–369.
21. "Thermal analysis and cone calorimeter study of engineered wood with an emphasis on fire modelling." Ira, J.; Salek, V.; Jahoda, M.; Vystrcil, V.; Hasalova, L. *Fire Technol.* **2020**, *56*, 1099–1132.
22. "Experimental research on temperature distribution and charring rate of typical components of wood structure building." Wei, S.; Yang, H.; Gao, B.; Cheng, H.; et al. *J. Fire Sci.* **2022**, *40*, 134–152.
23. "Flammability properties for charring materials." Delichatsios, M.; Paroz, B.; Bhargava, A. *Fire Saf. J.* **2003**, *38*, 219–228.
24. "Fire performance properties of commonly used South African hardwood." Maake, T.; Asante, J.; Mwakikunga, B. *J. Fire Sci.* **2020**, *38*, 415–432.
25. "Performance of type X gypsum board on timber to non-standard fire exposure." Chorlton, B.; Forrest, B.; Gales, J.; Weckman, B. *Fire Mater.* **2021**, *45*, 885–900.
26. "Tensile strength of wood in high temperatures before charring." Kuronen, H.; Mikkola, E.; Hostikka, S. *Fire Mater.* **2021**, *45*, 858–865.
27. "Experimental study of spruce wood reaction to fire in single burning item test." Terrei, L.; Zeinali, D.; Acem, Z.; Marchetti, V.; Lardet, P.; Boulet, P.; Parent, G. *J. Fire Sci.* **2022**, *40*, 293–310.
28. "A review of factors affecting the burning behaviour of wood for application to timber construction." Bartlett, A. I.; Hadden, R. M.; Bisby, L. A. *Fire Technol.* **2019**, *55*, 1–49.
29. ASTM E2632/E2632M-20. "Standard Test Method for Evaluating the Under-Deck Fire Test Response of Deck Materials."
30. ASTM E108-20a. "Standard Test Methods for Fire Tests of Roof Coverings."
31. ASTM E2726/E2726M-12a. "Standard Test Method for Evaluating the Fire-Test-Response of Deck Structures to Burning Brands."
32. "Reaction-to-fire properties of wood–polypropylene composites containing different fire retardants." Nikolaeva, M.; Kärki, T. *Fire Technol.* **2015**, *51*, 53–65.
33. "Flame-retardant wood polymer composites (WPCs) as potential fire safe bio-based materials for building products: Preparation, flammability and mechanical properties." Kalali, E. N.; Zhang, L.; Shabestari, M. E.; Croyal, J.; Wang, D.-Y. *Fire Saf. J.* **2019**, *107*, 210–216.
34. "Flammability of wood–polypropylene composites." Borysiak, S.; Paukszta, D.; Helwig, M. *Polym. Degrad. Stab.* **2006**, *91*, 3339–3343.
35. "Flammability and thermal stability of thermoplastic-based composites filled with natural carbon." Al-Majali, Y. T.; Trembly, J. P. *J. Fire Sci.* **2022**, *40*, 175–193.
36. "Fire performance properties of solid wood and lignocellulose-plastic composite deck boards." Fabian, T. Z. *Fire Technol.* **2014**, *50*, 125–141.
37. "A strategy to improve the compatibility of carboxyl methyl cellulose with silica fume-based geopolymer inorganic siliceous coatings for flame-retarding plywood." Wang, Y.; Kou, X.; Zhao, J.; Deng, J. *J. Appl. Polym. Sci.* **2022**, *139*, e53090.
38. "Investigation of mechanical property, flame retardancy and thermal degradation of LLDPE-wood-fibre composites." Li, B.; He, J. M. *Polym. Degrad. Stab.* **2004**, *83*, 241–246.
39. "Evaluating the use of calcium hydrogen phosphate dihydrate as a mineral-based fire retardant for application in melamine-urea-formaldehyde (MUF)-bonded wood-based composite materials." Ozyhar, T.; Tschannen, C.; Thoemen, H.; Zoppe, J. O. *Fire Mater.* **2022**, *46*, 595–604.
40. "A new flame retardant for wood materials tested in wood-plastic composites." Seefeldt, H.; Braun, U. *Macromol. Mater. Eng.* **2012**, *297*, 814–820.
41. "Flame retarded polyethylene/wood flour composites with high performances: Satisfying both sides with intumescent flame retardants and synergistic compatibilizers, respectively." Hong, H.; Liu, H.; Zhang, H.; He, H.; et al. *Polym. Compos.* **2018**, *39*, 569–579.
42. "Flammability and flame-retardant mechanism of high density polyethylene/wood fiber/modified ammonium polyphosphate composite." Liu, W.; Wang, D.; Zhang, Y.; Bai, T.; et al. *Polym. Compos.* **2018**, *39*, 1192–1199.
43. "Routes to halogen-free flame-retardant polypropylene wood plastic composites." Yin, H.; Sypaseuth, F. D.; Schubert, M.; Schoch, R.; Bastian, M.; Schartel, B. *Polym. Adv. Technol.* **2019**, *30*, 187–202.
44. "Intumescence in fire retardancy of lignocellulosic panels." Grexa, O.; Poutch, F.; Mainkova, D.; Martvonova, H.; Bartekova, A. *Polym. Degrad. Stab.* **2003**, *82*, 373–377.
45. "Influence of mineral fillers on the fire retardant properties of wood-polypropylene composites." Nikolaeva, M.; Kärki, T. *Fire Mater.* **2013**, *37*, 612–620.
46. "Evaluation of various fire retardants for use in wood flour-polyethylene composites." Stark, N. M.; White, R. H.; Mueller, S. A.; Osswald, T. A. *Polym. Degrad. Stab.* **2010**, *95*, 1903–1910.
47. "Flammability hazard of materials." Madrzykowski, D.; Stroup, D. W. In *Fire Protection Handbook*, 20th edn, Cote, A. E., Grant, C. C., Hall, C. C. et al., Eds; National Fire Protection Association: Quincy, MA, **2008**; vol. 1, 31–48.
48. "How do foams perform under fire conditions?" Troitzsch, J. In *Fire and Cellular Polymers*, Buist, J. M., Grayson, S. J., Woolley, W. D., Eds.; Elsevier Applied Science: London, **1986**, 77–91.
49. "Fundamentals of the fire behaviour of cellular polymers." Drysdale, D. D. In *Fire and Cellular Polymers*, Buist, J. M., Grayson, S. J., Woolley, W. D., Eds.; Elsevier Applied Science: London, **1986**, 61–75.
50. "Polymer combustion and flammability—Role of the condensed phase." Kashiwagi, T. *Symp. Int. Combust.* **1994**, *25*, 1423–1437.
51. "Fire phenomena of rigid polyurethane foams." Günther, M.; Lorenzetti, A.; Schartel, B. *Polymers.* **2018**, *10*, 1166.

52. "Improvement on fire behaviour of water blown PIR–PUR foams: Use of an halogen-free flame retardant." Modesti, M.; Lorenzetti, A. *Eur. Polym. J.* **2003**, *39*, 263–268.
53. "From cells to residues: Flame-retarded rigid polyurethane foams." Günther, M.; Lorenzetti, A.; Schartel, B. *Combust. Sci. Technol.* **2020**, *192*(12), 2209–2237.
54. "Heat release and structural collapse of flexible polyurethane foam." Krämer, R. H.; Zammarano, M.; Linteris, G. T.; Gedde, U. W.; Gilman, J. W. *Polym. Degrad. Stab.* **2010**, *95*(6), 1115–1122.
55. "Flammability characteristics of cellular plastics." Hilado, C. J. *Ind. Eng. Chem. Prod. Res. Dev.* **1967**, *6*, 154–166.
56. "Recent advancements in flame-retardant polyurethane foams: A review." Yadav, A.; De Souza, F. M.; Dawsey, T.; Gupta, R. K. *Ind. Eng. Chem. Res.* **2022**, *61*(41), 15046–15065.
57. "Recent advances in fire-retardant rigid polyurethane foam." Zhu, M.; Ma, Z.; Liu, L.; Zhang, J.; Huo, S.; Song, P. *J. Mater. Sci. Technol.* **2022**, *112*, 315–328.
58. "Emerging trends in flame retardancy of rigid polyurethane foam and its composites: A review." Muhammed Raji, A.; Hambali, H. U.; Khan, Z. I.; Binti Mohamad, Z.; Azman, H.; Ogabi, R. *J. Cell. Plast.* **2023**, *59*(1), 65–122.
59. "Commercial flame retardancy of polyurethanes." Weil, E. D.; Levchik, S. V. *J. Fire Sci.* **2004**, *22*(3), 183–210.
60. "A brief review on fire retardants for polymeric foams." Wang, J. Q.; Chow, W. K. *J. Appl. Polym. Sci.* **2005**, *97*(1), 366–376.
61. www.solvay.com/en/product/polyol-ixol-b-251 (accessed on 31st July 2023).
62. https://echa.europa.eu/assessment-regulatory-needs?p_p_id=disslists_WAR_disslistsportlet&p_p_lifecycle=1&p_p_state=normal&p_p_mode=view&_disslists_WAR_disslistsportlet_javax.portlet.action=searchDissLists (accessed on 31st July 2023).
63. "Influence of phosphorus valency on thermal behaviour of flame retarded polyurethane foams." Lorenzetti, A.; Modesti, M.; Besco, S.; Hrelja, D.; Donadi, S. *Polym. Degrad. Stab.* **2011**, *96*(8), 1455–1461.
64. "The preparation and properties of flame-retardant polyisocyanurate-polyurethane foams based on two DOPO derivatives." Liu, Y.; He, J.; Yang, R. *J. Fire Sci.* **2016**, *34*(5), 431–444.
65. "Synthesis of a DOPO-triazine additive and its flame-retardant effect in rigid polyurethane foam." Liu, L.; Lv, R. *e-Polymers.* **2019**, *19*(1), 235–243.
66. www.pinfa.eu/mediaroom/information-on-melamine-reach-dossier/#gsc.tab=0 (accessed on 31st July 2023).
67. "Flame retardancy of polyisocyanurate-polyurethane foams: Use of different charring agents." Modesti, M.; Lorenzetti, A. *Polym. Degrad. Stab.* **2002**, *78*(2), 341–347.
68. "Expandable graphite in polyurethane foams: The effect of expansion volume and intercalants on flame retardancy." Lorenzetti, A.; Dittrich, B.; Schartel, B.; Roso, M.; Modesti, M. *J. Appl. Polym. Sci.* **2017**, *134*(31), 1–8.
69. "A green highly-effective surface flame-retardant strategy for rigid polyurethane foam: Transforming UV-cured coating into intumescent self-extinguishing layer." Huang, Y.; Jiang, S.; Liang, R.; Liao, Z.; You, G. *Compos. Part A Appl. Sci. Manuf.* **2019**, *125*(July), 105534.
70. "Surface flame-retardant systems of rigid polyurethane foams: An overview." Jiang, Y.; Yang, H.; Lin, X.; Xiang, S.; Feng, X.; Wan, C. *Materials (Basel).* **2023**, *16*(7).
71. "Fire retardant mechanism of aliphatic bromine compounds in polystyrene and polypropylene." Kaspersma, J.; Doumen, C.; Munro, S.; Prins, A.-M. *Polym. Degrad. Stab.* **2002**, *77*, 325–331.
72. "Flame retardants for polystyrenes in commercial use or development." Weil, E. D.; Levchik, S. V. *J. Fire Sci.* **2007**, *25*, 241–264.
73. "Characterisation of smoke in expanded polystyrene combustion." Rossi, M.; Camino, G.; Luda, M.-P. *Polym. Degrad. Stab.* **2001**, *74*, 507–512.
74. "Fire-protective flame-retardant coatings—a state-of-the-art review." Weil, E. D. *J. Fire Sci.* **2011**, *29*, 259–296.
75. "Recent developments of intumescent fire protection coatings for structural steel: A review." Mariappan, T. *J. Fire Sci.* **2016**, *34*, 120–163.
76. https://commons.wikimedia.org/wiki/User:Quadell Creative Commons License; http://creativecommons.org/licenses/by-sa/3.0/

31 Flame-Retardant Design and Protection for Furniture and Furnishings

Mauro Zammarano,[1] *Alessandra Lorenzetti,*[2]
Michele Modesti,[2] *Alexander B. Morgan*[3]
[1] Fire Research Division, Engineering Laboratory, National Institute of Standards and Technology, Gaithersburg, Maryland, USA.
[2] University of Padova, Department of Industrial Engineering, Padova (Italy)
[3] University of Dayton Center for Flame Retardant Materials Science, Dayton, OH 45469 Alexander B. Morgan

31.1 INTRODUCTION TO THE FIRE/FURNITURE PROBLEM

There are many parts of a building that require fire protection as per building codes and standards (see Chapter 22 which covers building codes and Chapter 30 which covers building materials, both within this book), but it is also important to consider the materials inside the final structure given how they can and will affect fire growth. The severity of a fire strongly depends on a wide range of factors including ignition source, fuel load, fuel type, room of origin, layout, and fire protection used. Fire codes and regulations for interior components are primarily focused on electrical systems and components, appliances, heating ventilation, and air conditioning (HVAC), and soft interior components, such as carpeting, drapes/window coverings, and items containing flexible polyurethane foam (FPUF). According to the National Fire Protection Agency (NFPA)—FPUF persists as the leading item first ignited causing 28% of residential fire deaths.[1–3] This chapter will focus on the fire threat associated with FPUF and the products that contain FPUF (soft furnishing—mattress and upholstered furniture) since these products present the greatest fire threat. Namely, mattresses and upholstered furniture present a major fire hazard if ignited due to fast fire growth.[2–4]

This chapter includes detailed discussions on the soft furnishing fire problem, with an emphasis on the role of FPUF. This will include descriptions of why FPUF is a notable fire hazard as well as discussion of its fire physics and thermal decomposition chemistry. Aspects of the fire risk scenario, and what is or isn't covered under existing regulatory methods will also be discussed in detail in this chapter, including the important distinction between pass/fail smolder-based testing, and open flame/heat release-based testing. Regulatory and experimental test methods used to assess the flammability of soft furnishings in use today will also be discussed. Most of the chapter will focus on flame retardant chemistry that can be applied to FPUF, with detailed discussions on how FPUF decomposes, and how its chemical structure interacts with commercial and non-commercial flame-retardant systems. Finally, some discussion of the expected future of fire safety for furniture will be discussed, as the regulatory changes in furniture also have broad implications for material fire safety and flame-retardant chemical use. In particular, fire barrier fabrics are an emerging technology with the potential to act as an alternative to or in combination with flame retardants. The impact of fire barrier fabrics on fire safety, their mechanisms of action, failure mechanisms, and test methodology will be discussed.

31.2 FIRE CODES AND REGULATIONS FOR FURNITURE AND FURNISHINGS

Before discussing the current fire codes and regulations for furniture and furnishings, it is important to discuss the history of the known fire hazards of these items, since flame retardancy and fire protection are reactive to known/studied hazards. The fire hazards of upholstered furniture and mattresses containing FPUF have been known since the 1970s in the US.[5–7] Since this hazard has been known for some time, regulations have been enacted regarding the fire safety of mattresses and furniture, with standards initially focused on cigarette ignition or very small flame tests. Until 2013, California Technical Bulletin #117 (TB117–2000) required compliance with a cigarette ignition test based on the Upholstered Furniture Action Council (UFAC) test and a small open-flame scale test. Outside the US, in 1988, the United Kingdom (UK) implemented the British Standard (BS) #5852, which utilized large-intensity flaming ignition sources to test furniture mockups (simple combinations of fabrics/interlayers and padding).[8,9] Research continued even after the implementation of these methods, and detailed studies on furniture flammability have continued, with major studies occurring in the US[10] and EU[11] in the mid-1980s and mid-1990s, respectively.

DOI: 10.1201/9781003380689-31

Research into mattress/bedding fires began more recently, starting in the 1990s, with a subsequent more rapid progression from acknowledgement of the hazard to promulgation of regulatory tests than occurred for FPUF. Research conducted by the US National Institute of Standards and Technology (NIST) studied not just mattresses but also bedding (blankets, sheets, comforters) used on top of the mattress. This research concluded that mattresses present a genuine fire hazard, with the FPUF component being the largest fuel source for mattress fires.[12] Cigarette-based fire testing for mattresses had been in place for some time via 16 CFR 1632,[13] but open flame tests were only recently established. The abovementioned NIST studies, along with further research carried out by the US Consumer Product Safety Commission (CPSC), yielded a new regulation for open ignition of mattresses—16 CFR 1633.[14] What was novel about the 16 CFR 1633 mattress standard was that it focused on heat release rate and used a large ignition source including a side burner of about 9 kW applied for 50 s and a top burner of about 18 kW applied for 70 s. The test required a heat release rate below 200 kW for the first 30 min and a total heat release of less than 15 MJ for the first 10 min. No matter what the actual mattress manufacture/assembly might be, by testing the entire mattress against a large open flame source, and measuring heat release, the test ensures that the mattress will prevent a rapid fire growth, allow for prolonged time to escape and for the fire services to arrive, reduce risk of flashover and reduce emission rate of toxic gases.[15–17] 16 CFR 1633 was, and is, a revolutionary fire safety standard with its emphasis on heat release rate, and it has ensured significant fire safety for mattresses in the US. A 2020 NIST study by Gilbert et al. suggested that—for bed fires started by a flaming ignition source—the standard successfully reduced the number of deaths from bed fires by approximately 70% with only a 50% to 80% replacement of older mattresses.[18] These data unequivocally demonstrate the effectiveness of a well-designed fire standard. Mattresses meet the requirements of 16 CFR 1633 by using specialty fabrics that are designed to slow down the burning rate of the padding material in soft furnishings and are known as fire barrier fabrics. Noticeably, when fire barrier fabrics are used, compliance with 16 CFR 1633 can be met without the use of flame retardants (FRs).

Returning to the present day, the only open flame test requirement in US was removed with the adoption of the 2013 edition of California Technical Bulletin # 117 (TB117–2013), which is a cigarette-based fire test method.[19] The move to smolder-based testing may have been driven by concerns about the use of FRs and their potential environmental/health hazards with arguable improvements in fire safety (see Chapter 22 of this book). In 2022, TB117–2013 was adopted in 16 CFR 1640 federal standard and became a federal regulation applied to the entire US. The fire risk scenario addressed by the regulation is the smoldering ignition of upholstered furniture by a lit cigarette. Smoldering constitutes a potential threat to human life and property safety in two ways. First, it produces potentially deadly amounts of carbon monoxide at a higher yield (even though at a slower rate) as compared to flaming combustion and can lead to incapacitation or death by asphyxiation.[20] Second, smoldering can induce flaming ignition via smoldering-to-flaming transition with heat sources otherwise too weak to directly cause flaming ignition.[21]

Smoldering ignition of upholstered furniture due to cigarettes or other hot/glowing objects has been reported as the leading ignition source in deadly fires where upholstered furniture is the first item ignited, and these fires account for 55% of home fire deaths.[22,23] For this reason, regulators have been paying particular attention to smoldering ignition of upholstered furniture. However, most fire deaths (about 95%) occur when fire spreads beyond the upholstery furniture that was the first item to ignite, and about two-thirds of fire deaths occur when fire spreads beyond the room of origin, presumably after flashover.[24] These scenarios imply the occurrence of flaming combustion occurs in furniture either by direct open flame ignition of the furniture or transition from smoldering to flaming of the FPUF in the furniture.[25] Hence, independent of the type of ignition source (i.e., smoldering vs. flaming), it is likely that most fire deaths occur due to flaming rather than smoldering.

TB117–2013 is a smoldering ignition test that does not provide any insight into the flammability of upholstered furniture. Further, it favors high heat release synthetic materials that drip/melt away from a heat source and disfavors natural fibers with lower heat release which tend to char. Put more plainly, high heat release synthetic materials are far more likely to pass TB117–2013 than lower heat release natural fibers.

TB117–2013 was based on a voluntary industry test for cigarette ignition developed by UFAC in the 1970s. According to CPSC staff, conformance with the UFAC cigarette ignition test was already over 90% in 2008, thus, the adoption of TB117–2013 in 2021 as a federal regulation was not expected to result in a significant reduction in fire losses.[26] CPSC staff also concluded that the materials prescribed by TB117–2013 for testing, and in particular the FPUF called for in this test method, provided inconsistent smoldering performance. Therefore, there was a lack of confidence in the TB117–2013 method as the unpredictable smoldering behavior of FPUF may be influencing the conclusions on fabrics' smoldering performance. NIST showed that this inconsistency was due to the inability during mass production foaming to tightly control the morphological parameters of the foam that affects smoldering (i.e., pore size, air-flow rate, and foam mass density).[27,28]

Ultimately, CPSC staff concluded that the bench-scale test in TB117–2013 did not adequately predict the fire performance of full-scale/finished production furniture.[25] Additionally, an NIST study found that TB117–2013 systematically underestimated smoldering in upholstered furniture. The study concluded that this may be due to the TB117–2013 test configuration limiting oxygen to the smoldering front, thus decreasing the severity of smoldering and a small modification to the TB117–2013 test configuration would improve oxygen supply and correlation with the prediction of finished product performance.[29] For all the aforementioned reasons, no significant reductions in fire losses are likely to be realized by the introduction of 16 CFR 1640 and furniture fire safety remains a major issue.[30]

Going back to the changes in fire safety standards for FPUF in furniture, some background about the regulation of FRs in FPUF is needed. In the US, regulations that impose severe limitations on the use of FRs in upholstered furniture have been introduced or are under consideration by legislators in multiple states. For example, in 2017 the state of Maine banned the use of any chemical compound for which "fire retardant" appears on the material safety data sheet for use in furniture.[31] In 2018, California enacted Assembly Bill 2998 (AB2998), which banned the use in upholstered furniture, mattresses, and juvenile products of the following: (1) FRs based on halogenated, organo-phosphorous, or organo-nitrogen chemicals; (2) nanofillers; (3) FRsor synergists to FRs listed as chemicals of high concern (see Section 173–334–130 of Title 173 of the Washington Administrative Code); and (4) "designated chemicals" (as defined in Section 105440 of the Health and Safety Code).[32] Effectively, most FRs or FRs systems have been banned in California, while only a few inorganic FRs, such as ammonium polyphosphate and aluminum hydroxide, are still allowed. Due to the size of the California market and the rising number of states banning the use of FRs, furniture manufacturers have a strong incentive to avoid the use of FRsin the entire US. At the federal level, CPSC recommended refraining from intentionally adding hazardous additives such as nonpolymeric organo-halogen FRs and has an ongoing project to evaluate additive organo-halogenated FRs.[33,34]

At the time of writing this chapter, United Kingdom (UK) BS5852 test is the only open-flame ignition fire safety test for residential upholstered furniture, and even that test method is under scrutiny in the UK for the same reasons that drove the US switch to smolder-based testing for furniture—namely, the use of FRs in BS5852 compliant furniture.[35]

There is a perception by some that technologies promoting fire safety and health/environment safety are mutually exclusive. This is leading to a gradual withdrawal of preexisting open flaming testing and is halting the development of new standards and regulations. As demonstrated by 16 CFR 1633, not only can properly designed open flame testing have a major impact on fire safety, but products can meet stringent open flame resistance requirements with fire commercially available barrier fabrics that do not pose a health or environmental hazards.

31.3 FIRE BARRIER FABRICS

Fire barrier fabrics (referred to simply as barriers in the remainder of this chapter) are specialty fabrics used in soft furnishings to prevent ignition of the padding material or delay its contribution to the overall heat release rate once ignition has occurred. Barriers can act by limiting heat transfer (e.g., from a flame or an external heat source) to a flammable substrate, thus limiting the rate at which combustible pyrolizates are generated, and by controlling the rate and location at which these pyrolizates are released and able to burn. These two mechanisms of action are physically based (i.e., not requiring chemical FRs) and can be referred to as heat and mass transfer effects, respectively. Additionally, barriers can also utilize chemical mechanisms of action by including FRs. Based on their mechanisms of action, barriers can be classified as passive barriers (i.e., using physical mechanisms only, with no FRs), which rely on inherently fire-retardant materials (e.g., glass, carbon, and silicic-acid/rayon fibers), and active barriers (i.e., using a combination of chemical mechanisms, such as FRs based on free-radical trapping or promoting charring, and physical mechanisms).[36,37]

Noticeably, California's AB2998 applies to all soft furnishing components except barriers used in mattresses. This is a temporary exemption intended to provide sufficient time for furniture manufacturers to transition from active barriers to passive barriers while complying with 16 CFR 1633.

31.4 TEST METHODS

FPUF is subject to both flaming and smoldering combustion (discussed in section 31.5). Hence, the soft furnishing fire hazard is a dual problem, and both forms of combustions need to be properly screened and tested. Therefore, some discussion of tests methods is needed. As mentioned previously, TB117–2013 is the only test used to regulate residential upholstered furniture in US. TB117–2013 measures the smoldering ignition propensity of cover fabrics and padding materials in a mock-up assembly within a plywood supporting frame. Pass/fail criteria are based on qualitative metrics—the occurrence of persistent smoldering and transition to flaming and a quantitate metric—the char length measured on the cover fabric. If either the cover or the padding material fails the test, a smoldering barrier (i.e., a thermoplastic polyester batting that stops the smoldering wave by melting away from the heat source) must be placed between the cover and the padding to comply with TB117–2013. National Fire Prevention Association (NFPA) 260 is another cigarette ignition propensity test which uses the same test setup as TB-117–2013, except the foam thickness is two inches instead of three inches and concealed smoldering (i.e., presence of sustained smoldering in the crevice that is detected by separating the panels of the mock-up at the end of the test) is included as an additional failure criterion. Both differences address or mitigate concerns on oxygen depletion and concealed smoldering that were demonstrated by NIST studies.[28] Mattress fire performance in terms of cigarette ignition resistance is regulated in US by federal laws 16 CFR 1632.

In Europe, the British cigarette ignition standard for seats BS 5852–1 has been adopted as EN1021–1 and ISO 8191–1. Similar to TB117–2013, the test is based on a mock-up of upholstery materials, but the dimensions are significantly larger (approaching actual couch cushion size), and the frame is built with a metal wire mesh that does not affect oxygen supply like the plywood sample holder of TB117–2013. For bedding, cigarette ignition resistance in Europe is carried out according to EN ISO 12952–1.

For flaming combustion, there was no US open flaming requirement for upholstered furniture since 2013 when TB117–2000 was superseded by TB117–2013. The only flammability standard test method for upholstered furniture still in use in US is ASTM E1537, which is based on the now withdrawn California technical bulletin #133.[38] Mattress flammability is regulated by federal regulation—16 CFR 1633, which had a major impact on fire safety.[7] ASTM E1474 and ASTM E3367 are two test methods that use the cone calorimeter to assess the performance of upholstered furniture and mattress components or composites with regard to open flame resistance. ASTM E1474 is based on the European Combustion Behavior of Upholstered Furniture (CBUF) research program, and it will referred to as CBUF test in the reminder of the chapter.[10] ASTM E3367 is a recent standard published on July 2023 that is based on NIST research.[39] This standard is sometimes referred to as the "Cube test" because it uses a special cube-shaped specimen holder. The CBUF test and Cube test will be further discussed in the following paragraphs.

In Europe, outside of UK and Ireland where the open flame performance is regulated by BS 5852–2, there is no requirement for flaming performance of residential upholstered furniture and mattresses.

31.5 FLEXIBLE POLYURETHANE FOAM FIRE PHYSICS AND FIRE GROWTH[1]

The thermal degradation of FPUF in air is the result of two competitive decomposition pathways: (1) an oxidative decomposition pathway leading to char, which is promoted by an increase in oxygen availability and surface area; and (2) a non-oxidative decomposition pathway leading to a liquid residue, which is promoted by an increase in heating rate.[40]

Typically, the first mechanism (charring) dominates for heating rates in the order of 10°C/min or less.[41] This is why the combustion of FPUF is strongly affected by the intensity of the ignition source. Weak ignition sources like cigarettes will likely lead to char formation and smoldering combustion, while stronger ignition sources will likely "liquify" and collapse the foam at a temperature below 300°C and lead to the release of liquid and gaseous pyrolyzates that are highly flammable and ignitable.[42]

During foam collapse, the gaseous products are mainly thermal decomposition byproducts of isocyanates, which can sustain flaming, while the liquid products are usually referred to as "regenerated polyols" because their composition is similar to the one of the starting polyols used in the foaming process.[43] Regenerated polyols partially accumulate on the surface of the collapsing foam and, as shown by micro combustion calorimetry data, start contributing to the combustion process at temperatures above 300°C when the foam had already fully collapsed.[40]

FPUFs pose a very high fire hazard because they are easy to ignite (i.e., via both flaming and smoldering sources) and lead to rapid-fire growth. The relatively low density, thermal conductivity, and thermal stability of FPUFs contribute to their high ignitability. The rapid-fire growth seen in PFUFs is attributed to the formation of regenerated polyols, which account for about 70% of the total heat release of a FPUF.[40] The accumulation of regenerated polyols can result in a pool fire that facilitates an increase in burning surface area, flame spread to adjacent items, and a flame feedback loop between the pool fire and the initial burning item.[44,45] Several studies have also shown that the heat release rate from combusting FPUF increases extremely rapidly after the release of regenerated polyols underneath the burning item. A comprehensive study which included an overview of upholstered furniture fire testing and the role of pool fire has been published recently.[46]

Based on this observation, NIST developed the Cube test which has been recently standardized as ASTM E3367.[38] This new standard uses a cone calorimeter to evaluate the tendency of multilayered products to release flammable liquid products and assess the ability of barriers to inhibit pool formation.[38]

The ability of the Cube test and the CBUF test (both bench-scale) to assess the performance of barriers in soft furnishing has been discussed in a NIST report.[47] In the CBUF test the gaseous pyrolyzates can escape from the edge of the barrier in the test specimen and cause a drastic increase in heat release rate. This is clearly an artefact that does not that affect the actual performance of a barrier in a finished product. This issue has been addressed in the Cube test by compressing and sealing the barrier to the sample holder (see ASTM E3367 for details).[48] The specimen in the Cube test is intended to represent a cross section of the upholstered furniture cushion or mattress.

Figure 31.1 shows a snapshot from a typical Cube test at the time when a pool fire is generated for an upholstery material assembly including a fire barrier at the top and bottom of the specimen. The sides of the specimen are thermally insulated and sealed with a metal liner to generate—as much as possible—a one-dimensional heat/mass transfer. It should be noted that Figure 31.1 shows a prototype sample holder which is slightly different from the one described in ASTM E3367.

Figure 31.1 includes a wide-view image of the entire setup on the left, a detail of the top surface on the top right, a detail of the bottom surface on the bottom right and the heat release curve as an overlay. The wide-view image shows the cone heater on the top of the specimen and an aluminum catch pan underneath the specimen. Note that the specimen holder is open on the bottom (94 mm × 94 mm opening, identical to the top opening) so regenerated polyols can drip and accumulate into the catch pan. The images of

FIGURE 31.1 Composite image from a Cube test including an image of the entire setup (left), a detail of the top surface (top right), a detail of the bottom surface (bottom right), and the heat release curve as an overlay.

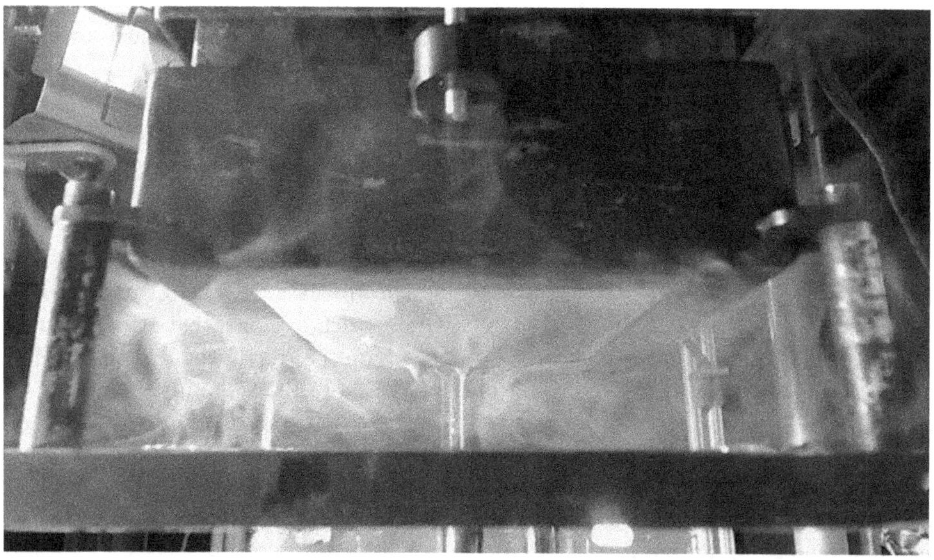

FIGURE 31.2 Image of the bottom surface of the upholstery specimen captured by an air-cooled camera during a Cube test. The image shows regenerated polyols percolating through the barrier and then dripping into the catch pan.

Figure 31.1 are captured by three dedicated air-cooled cameras in a custom-made aluminum enclosure. These cameras provide information for analyzing the response to fire of the specimen.

For example, Figure 31.2 shows very clearly the release of regenerated polyols from the bottom of the specimen. The time at which regenerated polyols are first observed on the bottom surface of the specimen is defined as "time to wetting," t_w.

This parameter has been proved to be extremely useful in understanding the full-scale performance of soft furnishing. From a physical point of view, t_w is providing (1) the time required for the FPUF to liquify, plus (2) the time required for the regenerated polyols to percolate through the barrier, if present, at the bottom of the specimen.

The Cube test has been designed based on the fundamental hypothesis that fire growth in soft furnishings is largely controlled by FPUF pyrolysis and liquid products percolation through the barrier. This hypothesis was validated by running Cube tests and open-flame full-scale tests on chair mock-ups as described in two NIST publications.[45,49] In the first publication, a combination of one padding material, one cover fabric and 7 passive barriers (B1 to B6) plus a control specimen without barrier (C0) were run in triplicates in the Cube test and in the chair mock-up test. An external heat flux of 75 kW/m^2 was required to

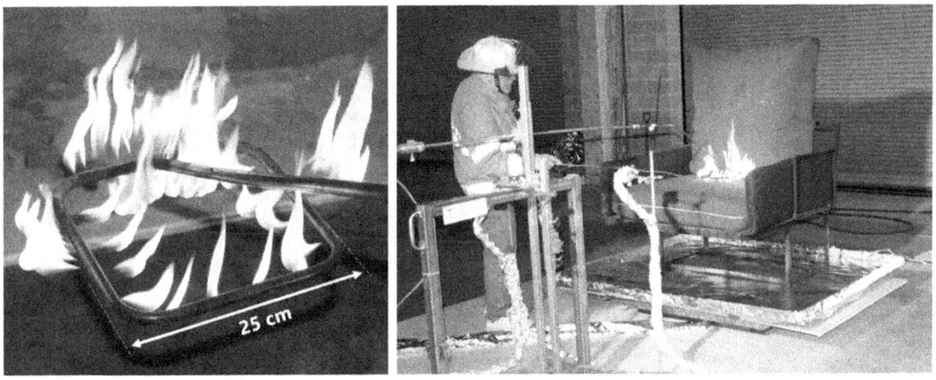

FIGURE 31.3 Pictures showing the ignition source (on the left) and the setup used in the full-scale chair mock-ups (on the right).

"liquify" the FPUF in the Cube test without significant char formation, thus allowing for measurement of the time to wetting. Videos and data for the Cube tests are available here.[50] In the second publication, a burner (as described in ASTM E1537) with a heat release rate of about 18 kW was applied for 80 s over the top of the seat cushion and about one inch from the back cushion for 80 s. Figure 31.3 shows the setup used for the chair mock-up tests. Videos and data for the chair mock-up tests are available from NIST's Fire Calorimetry Database.[51]

The videos for chair mock-up tests without a barrier show that flame jets from the burner simultaneously ignited the top surface of the seat cushion and the front bottom of the back cushion within a few seconds. Flames started spreading rapidly upwards on the vertical back cushion and, at a lower rate, over the horizontal seat cushion from the center toward the sides and front of the cushion. By the time the ignition source was extinguished at t = 80 s, the padding material was heavily involved in the burning process and shortly after the removal of the burner, the back cushion completely collapsed. Intense flaming over the seat cushion eventually generated a hole in the center of the cushion, then regenerated polyols poured through the hole onto the catch pan and quickly formed a pool fire. Heat flux feedback from the pool fire to the seat cushion increased the generation and release rate of flaming liquid, which in turn produced a larger pool fire and stronger heat flux feedback over the seat cushion and pool fire. This was a self-accelerating process that led to an abrupt increase in heat release rate and continued until the FPUF and regenerated polyol in the seat cushion and the liquid pool were depleted. The peak heat release was reached approximately one minute from the initial pool fire formation.

The videos for chair mock-up tests with a barrier show that the burner ignited the cover fabric over the seat cushion and along the lower front bottom of the back cushion within a few seconds, similar to the chair mock-up without a barrier, but the rate of flame spread over the back cushion and the seat cushion was slower and varied depending on the type of barrier.[50] The polypropylene cover fabric quickly melted/pyrolyzed along the flame front exposing the barrier, which remained intact and protected the padding material by decreasing the heat transfer from the flames. While the barrier was still intact, the back cushion collapsed and tipped forward. A quasi-steady burning phase ("plateau") was characterized by flaming over the entire top surface of the seat cushion and insides of the armrests was observed after cushion collapse. During this plateau, flame spread propagated very slowly from the inside to the outside of the mock-up arms and then to the bottom of the mock-up. Meanwhile, regenerated polyols started percolating through the barrier at the bottom of the seat cushion. Similarly, to what observed in the Cube test, the regenerated polyols became visible due to wetting of the barrier after the cover fabrics burned and pulled away. The plateau phase continued until the flames on the cover fabric reached the bottom of the chair and ignited the regenerated polyols. This phenomenon will be referred to as "bottom ignition" in the remainder of the chapter. Bottom ignition was revealed by prolonged and localized flaming on the bottom surface of the chair mock-up; before bottom ignition, flaming was characterized by a weak moving flaming front accompanied by melting and dripping of the cover fabric.

Figure 31.4 shows pictures of the bottom of a chair mock-up at the time of "wetting" and the time of bottom ignition.[48] Bottom ignition always quickly followed wetting when polyols become available for ignition on the bottom surface of the seat cushion. For example, in the test shown in Figure 31.4 bottom ignition occurred about 23 min after ignition and about 30 s after wetting. Shortly after bottom ignition, flaming drips accumulated under the chair, generating a pool fire. The flame size increased until the peak of heat release rate was reached about two to three minutes after bottom ignition.

Figure 31.5 shows the heat release curve measured in the same chair mock-up test of Figure 31.4. The heat release rate increased up to a plateau value after about 3 min and then stabilized until the time to bottom ignition (t_{BI}). Figure 31.5 also shows average heat release rate measured between test start and bottom ignition ($AHRR_{BI}$). This is an important

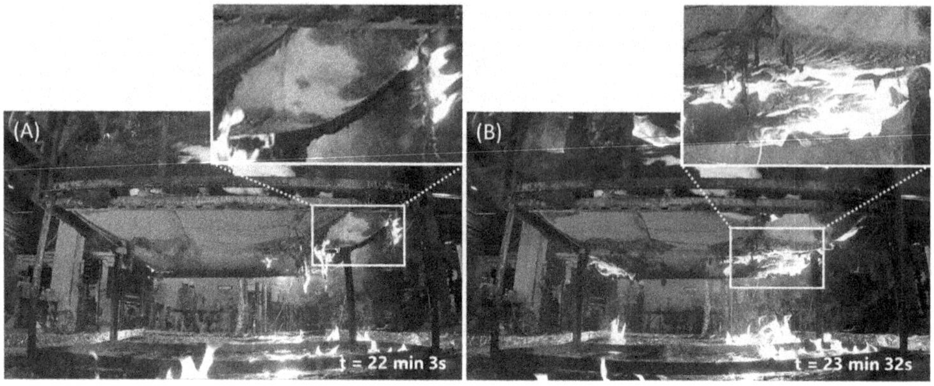

FIGURE 31.4 Video frame grabs for a representative chair mock-up with barrier B6 illustrating: (a) "wetting" of the barrier due to the percolation of regenerated polyols through the barrier (see inset); (b) "bottom ignition" due to the ignition of liquid products percolating through the barrier (see inset).

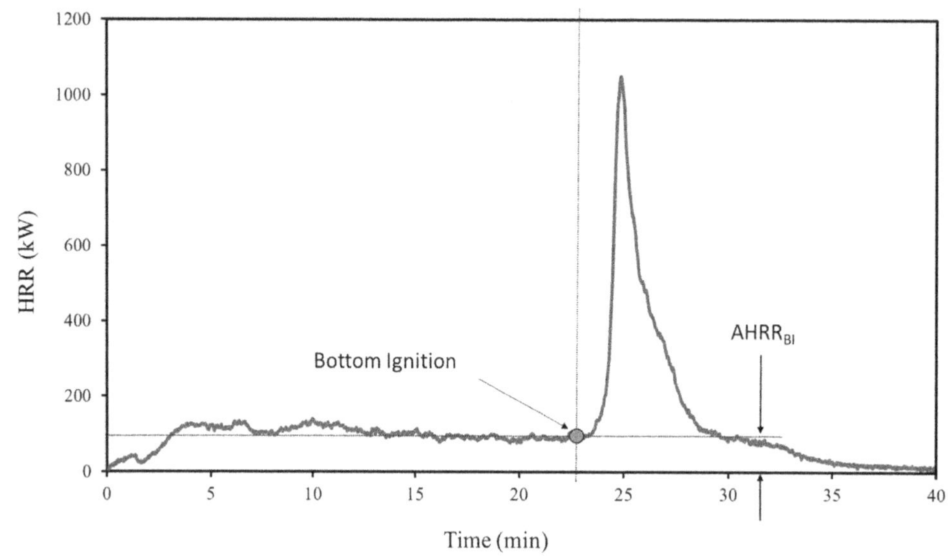

FIGURE 31.5 Representative heat release curve measured in a chair mock-up test with barrier B6 (same test of Figure 31.4). $AHRR_{BI}$ is the average heat release rate measured between test start and bottom ignition.

parameter because it defines the performance of a barrier before bottom ignition, which in turn affects tenability and likelihood to observe second item ignition.

After bottom ignition, the heat release rate abruptly increased up to the peak of heat release rate (see Figure 31.5). A similar behavior was observed in all chair mock-up tests except—to a certain extent—chairs B3 due to tearing of the barrier during chair construction.[48] Because the peak of heat release rate always closely followed bottom ignition, a strong correlation between these two parameters was observed (see Figure 31.6).[46,48] Data point B3 in Figure 31.6 was considered an outlier. Linear fit and R^2 were calculated using the method described by York and assuming no correlation between x and y error.[52] Uncertainties for all data were reported as standard deviation of the mean ($\sigma/\sqrt{3}$), where σ is the standard deviation calculated over three independent observations.

Based on these data, the "failure" mechanism of the barrier, which led to a sharp increase in heat release rate at the end of the plateau in all chair mock-ups, was attributed to wetting and consequent bottom ignition rather than a mechanical failure associated with bursting of the barrier.

A strong correlation was also identified between time to wetting in the Cube test and bottom ignition in the chair mock-up tests (see Figure 31.7). This correlation is extremely important because it allows to estimate the time to bottom ignition (i.e., time to failure of a barrier) using a bench-scale test. This strong correlation supports the fundamental hypothesis at the basis of the Cube test design, that is, fire growth in the chair mock-up is largely controlled by FPUF pyrolysis and liquid products percolation through the barrier.

The time to wetting in the Cube test and chair mock-up tests are related to the same phenomena (namely,

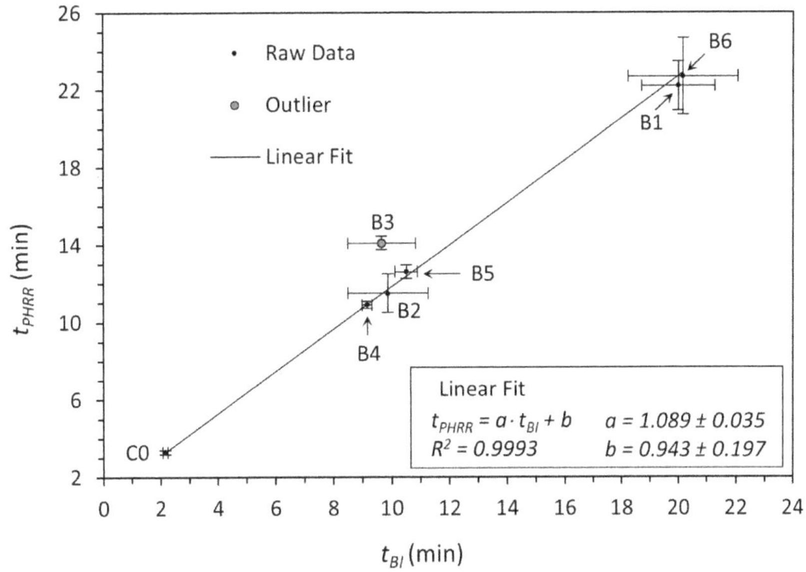

FIGURE 31.6 Scatterplot of time to peak of heat release rate (t_{PHRR}) vs. time to bottom ignition (t_{BI}).

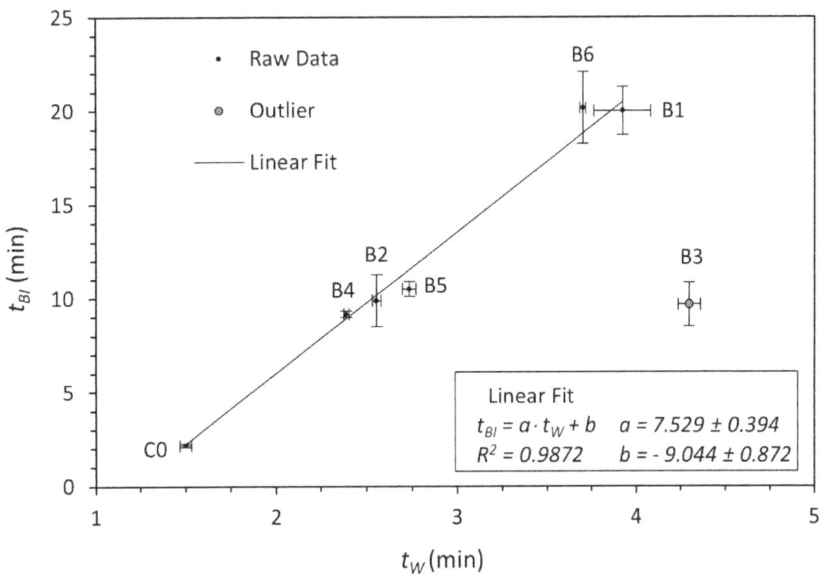

FIGURE 31.7 Scatterplot of time to bottom ignition (t_{BI}) in the chair mock-up tests vs. time to wetting in the Cube test (t_W). Linear fit, R^2, and uncertainties are calculated with the same method used for Figure 31.6.

liquefaction and percolation of regenerated polyols through the barrier) so a correlation between them is expected. In this case, we also observed a correlation between time to wetting in the Cube test and time to bottom ignition in the chair mock-up tests because bottom ignition readily followed wetting due to flaming of the cover fabric. In general, time to wetting in the Cube test provides a conservative minimum value for bottom ignition because it is assumed that bottom ignition and wetting occur simultaneously.

Remarkably, t_w shows a strong correlation also with other performance parameters of the chair mock-up tests, namely, the peak of heat release rate (PHRR)[53] and the average heat release rate measured between test start and bottom ignition ($AHRR_{BI}$) (see Figure 31.8). This finding is somewhat surprising, in fact, full-scale fire growth is always affected by flame spread, which the Cube test is not designed to capture. These results seem to indicate that fire growth is predominantly affected by burn through and release of regenerated polyols and that flame spread plays a second-order effect. This hypothesis is supported by the fact that, when a barrier is adopted, flame spread over the cover fabric can occur without significant involvement of the padding material, in which case the effect on heat release rate might be marginal.

The aforementioned data indicated that a simple bench-scale test like the Cube test, designed to mimic FPUF

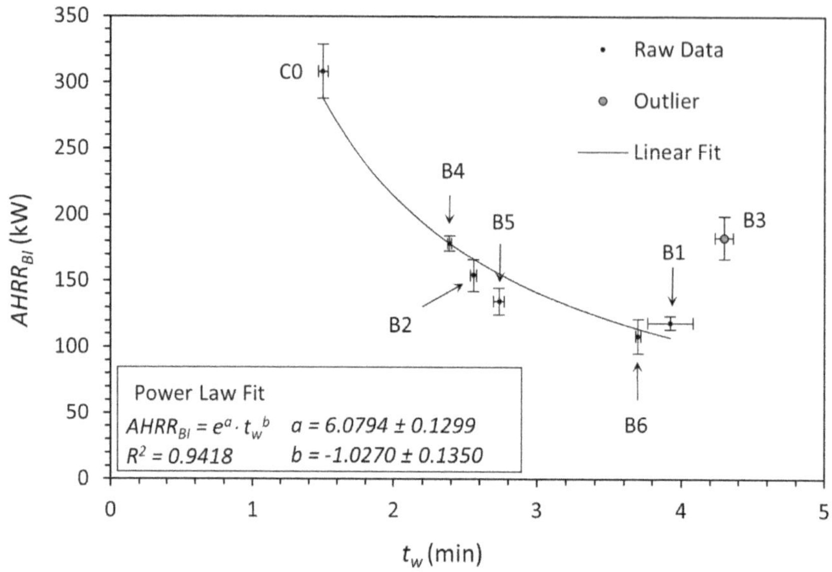

FIGURE 31.8 Scatterplot of $AHRR_{BI}$ in the chair mock-up tests vs. time to wetting in the Cube test (t_W). Linear fit, R^2, and uncertainties are calculated with the same method used for Figure 31.6.

pyrolysis and liquid products percolation in the chair seat cushion without accounting for flame spread might prove to be useful as a predictive tool for soft furnishing fire performance incorporating barriers. Further testing with different soft furnishing items is required to validate these findings.

The next important question to answer is: how effective are passive barriers in a realistic fire scenario? For mattresses, compliance with 16 CFR 1633 basically allows to fully prevent fire deaths due to flaming ignition.[17] Compliance with 16 CFR 1633 can be achieved with passive barriers, however, the use of active barriers is still allowed.[31]

For upholstered furniture, a study from Underwriter Laboratories (UL) showed that with a 35 mm flame ignition source (match equivalent) a chair including a glass barrier quickly extinguished after ignition, whereas the chairs with FPUFs containing 5.2% by weight of tris (1,3-dichloro isopropyl) phosphate (TDCPP) or 2.9% by weight of a reactive organo-phosphate blend ((4-tert-butylphenyl) diphenyl phosphate, (2,4-di-tert-butylphenyl) diphenylphosphate (B4tBPPP), and tris (4-tert-butylphenyl)phosphate) did not show a significant reduction in fire growth compared to the control chairs with no barrier or FRs.

With a larger ignition source (i.e., 18 kW square burner), NIST showed that FPUF ignition could not be prevented in upholstered furniture in the presence of passive barriers but the PHRR could be delayed by about 19 min.[46] More recent unpublished data showed that by selecting higher performance passive barriers it is possible to delay the PHRR by over 30 min.[54] The impact of barriers on residential upholstered furniture fires was also assessed by utilizing the CFAST Fire Data Generator (CData) for a Monte Carlo analysis.[55] The results showed that for most passive barriers the likelihood of observing flashover or second item ignition is negligible and that the use of passive barriers in residential upholstered furniture would result in a reduction of over 150 fire deaths per year in US.[55]

NIST also investigated the effect of passive barriers in fully furnished living rooms containing a sofa and found that the addition of a barrier was able to delay flashover by over 15 min from about 7 min to 22 min. Such a delay can lead to a substantial saving of human lives escaping a burning structure. This can be contextualized by analyzing the objective (target) fire service response times in NFPA 1720.[56] According to NFPA 1720, first responders are expected to intervene within 12 min in urban areas, 13 min in suburban areas, and 17 min in rural areas. Considering target fire service response times, a 15 min delay (in these specific room tests leading to an increase in time to flashover from about 7 min to 22 min) is important because it likely allows first responders to intervene before flashover and potentially prevent flashover occurrence. This is important because about two-thirds of fire deaths presumably occur after flash-over.[57]

31.6 FLEXIBLE POLYURETHANE FOAM FORMULATIONS AND CHEMISTRY

While materials used in furniture over time have changed, this section of the chapter will focus on FPUFs only. Interactions between materials in furniture affect fire growth and risk, but since the FPUF is typically the largest fuel load in furniture, it makes sense to focus on that product here. Readers should be aware that other materials in older furniture (e.g., horsehair, straw, wood, metal springs) have lower heat release than materials used in modern furniture (e.g., polyester, polyurethane, polyolefins), and so the complexities of FPUF burning with the other materials have changed over time.[1]

As told before in this chapter, the "soft furnishings," namely, mattresses and upholstered furniture (sofas, chairs, and ottomans) containing FPUF, present a major fire hazard if they are allowed to ignite and lead to notable fire growth. Flexible polyurethane foam, in contrast to rigid types, are open-cell materials that allow free movement of air throughout the materials when flexed. Even if it is not easy to give a universal definition of flexible foam, it may be preferable to describe these foams in terms of stress-strain relationships, considering that a flexible foam shows relatively low load-bearing properties with high recovery properties, similar to a coiled metal spring,[58] in contrast with both semi-flexible (a.k.a. semi-rigid) foams that are used for automobile dashboard panels or some under-the-hood applications, and rigid foams used for thermal insulation (see Chapter 30 of this book). FPUF can be further classified as conventional, viscoelastic (VE), and high resilience (HR). VE foams, a.k.a. memory foams, are characterized by slow recovery after compression. They were developed by NASA in the early 1960s, and they are now extensively employed for mattresses and pillows. HR foams are so called because they have, indeed, a good resilience, meaning they quickly regain their original shape after compression, ensuring prolonged comfort and durability. HR foams have improved hysteresis curves and higher resiliency than conventional flexible foams: for example, the sag factor of HR foams ranges from 2.4 to 3.0 and the percent ball rebound is about 70%. These properties make HR foams the choice for many seating applications.[21]

31.6.1 Isocyanate and Polyol Chemistry Effects

The main raw materials for FPUF, whether conventional, VE, HR, or polyurethane foams in general, are isocyanate and polyols.[59] The isocyanate is used as the source of the functional groups (NCO) that react with the hydroxyl groups from the polyol and water in the formulation. For flexible foams, the most commonly used isocyanate is toluene di-isocyanate (TDI). In the United States, the 80/20 isomer ratio of 2,4-/2,6-TDI is by far the most commonly used. In contrast, the 65/35 TDI is used for some specific applications and is the more widely used in Europe. 65/35 TDI will give foams with higher load-bearing characteristics, but significant changes are normally required in catalysis, since the reactivity of this isomer ratio is less than that of the 80/20 TDI. Diphenylmethane diisocyanate (MDI), in various forms, has found increasing use with time, because of the lower volatility with respect to TDI, making its use safer than TDI. Since pure MDI is solid at room temperature, it is used either as a crude product (also referred to as polymeric MDI) or as modified products that are liquid and offer processing and/or property enhancements.

In general, FPUF are produced from polyols of moderately high molecular weight and low degree of branching. The mostly used polyols are polyethers; although polyoxypropylene diols have some application, the polyethers primarily used are triols or polyols or higher functionality.[21] The use of polyester polyols in FPUF is quite limited (they are most extensively used in rigid polyurethane foams). Polyethers give softer, more resilient foams with better hydrolysis resistance than polyesters, while the polyester-based foams may be expected to have greater tensile strength and better resistance to oils, solvents, and oxidation. With either type, an increase in the degree of branching results in increased load-bearing capacity, lower elongation, and a higher glass transition temperature.

VE foam is prepared using a polyol blend of high- and low-molecular weight polyols; the isocyanate can be TDI or polymeric MDI.[60] TDI-based VE foams are soft to the touch and have very low resilience and slow recovery. MDI-based VE foams are easier to process and can be produced in a wider range of isocyanate indexes. However, physical strengths of MDI-based foams are lower, resilience is higher, and compression set is better than TDI-based foams.[22]

The formulation principle of HR foams is to reduce intermolecular interactions such as hydrogen bonding of urethane linkages and irregular structures of polymeric molecule. Examples of polyether polyol used for HR foams include high-molecular weight polyether polyols while isocyanates include blends of TDI-8020/polymeric MDI in an 80/20 weight ratio and the single use of polymeric MDI.[22]

Mattresses can also be manufactured by partly using polymeric polyols, to enhance the load-bearing capability. Polymeric polyols are defined as very fine and stable dispersions of solid polymers (vinylic polymers, like SAN, and copolymers or polyurea) in liquid polyethers[21]; they contain up to 40 wt% of solid particles.

The flammability of FPUF strongly depends on the physical properties of the foams. These foams burn when exposed to an ignition source and may burn quite rapidly, because of the combination of high surface area (from its cellular structure), ready access of oxygen (due to its open cellular structure) and low heat capacity (thermal inertia).[61] Thus, FPUF density and openness of the cells (air flow) are the most important physical properties in determining flammability.[62–65] For example, the density of the FPUF strongly influences the first HRR peak in cone calorimeter test: the lower the density, the faster the step of liquefaction, the higher flame propagation rate,[25] the higher the intensity of the first peak of heat release rate (PHRR), and the production of smoke during the first stage of combustion.[26] Also, the open cell content influences flammability since a higher air permeability leads to a better access of oxygen from air to the flame.[27] A FPUF with a lower open cells content (i.e., with more closed cells) is less permeable and heat conductive, and this reduction in the permeability slows the flow of combustible products of degradation towards the flame and so the time to ignition in the cone calorimeter increases.[26] The physical properties of an FPUF are dictated by its final application.[66] It is also important to note that since physical properties are determined by the foam's formulation (i.e., polyol, isocyanate, blowing agent, catalysts, etc.), the effect of raw materials on flammability of FPUF is worth mentioning. Indeed, it is known that variations in the composition of polymeric materials can alter the combustion characteristics of the polymer, leading to enhanced char and altered decomposition kinetics.[67] It has been shown that the fire behavior of an FPUF is weakly related to the isocyanate index (which is in any case low, less than 115, for an FPUF), as well as the surfactant and the catalyst contents.[26] The blowing agent level (and so the density) is the major factor modifying the combustion of a FPUF. Besides the aforementioned physical effects, foam density might also have a chemical effect on flammability.[26,29] Indeed, FPUFs contain a complex mixture of polyol rich soft domain with polyurea hard domains. The polyurea content is controlled by the amount of water in the formulation and, thus, increasing the water content (used as a blowing agent to reduce density) leads to an increase in hard polyurea domains and a reduction in crosslinking. Hence, additional isocyanate is required when increasing the water content to keep the required cross-linking. These effects on foam chemistry are likely to affect the combustion of the foam. Cone calorimetry data showed that when FPUF density was decreased by increasing water and TDI content, the first peak of HRR and smoke production increased but the second peak of HRR decreased.[26] This second peak is related to the quantity of polyol used for the synthesis of the FPUF: an increase in the water amount used necessarily requires the use of higher quantities of isocyanate and consequently a reduction in the percentage by weight of polyol in FPUF.[26] Schrock et al.[29] decoupled foam density from polyurea content by using not only water but also a physical blowing agent. They showed that an increase in polyurea content and cross-link density in high resiliency foams, apparently are not causing these foams to perform differently in tests like BS 5852 and the OSU calorimeter. This was quite surprising because the polyurea in the foam samples changed from a low of 2 meq/g polymer to a high of 3 meq/gram polymer. The fraction of aromatic polyurea changing over this magnitude would be expected to have dramatic effects on burn properties owing to the low flammability of these aromatic structures. Moreover, higher polyurea content would correlate with lower cross-link density which would affect foam collapse and liquefaction from thermally induced decomposition.

Höhne et al.[68] analysed the effect of isocyanate (TDI or MDI) and polyol (polyether or polyester) on fire behavior of FPUF by comparing two virgin water-blown polyether-polyol-based TDI- or MDI and one polyether-polyester-polyol-based MDI FPUFs. It has been shown that MDI-based FPUFs perform better than TDI-based foam in fire test like limiting oxygen index, cone calorimeter, FMVSS 302, and FAR 25.853. Besides the type of isocyanate, the specific polyols also have a significant influence on the flammability of FPUF. For example, a polyether-polyester foam performs better than a polyether one. However, polyester polyols make the FPUF prone to scorching (see later in this chapter) and therefore are often mixed with polyether polyols. Indeed, polyester polyols tend to degrade under humid conditions and to degrade quickly when thermal decomposition or hydrolysis products of FRs are present.[21] In another work, the superior fire performance in the cone calorimeter of MDI foams compared to TDI foams was confirmed. The same work also reported a better fire behavior of foams based on SAN-graft polyol as compared to foam based on polyether polyol.[69] Graft-polyurea polyols (PHD) have even better fire performance than SAN polyols (e.g., in TB117–2000) as PHD based foam can pass this test without the need for FRs.[70]

Sometimes, the partial substitution of polyols with halogens- and/or phosphorus-based hydroxyl terminated compounds (the so-called reactive approach in flame retardancy) is proposed to increase the flame retardancy of FPUF, although it is not so common as in rigid polyurethane foams (see Chapter 30 in this book). This is because of several drawbacks on mechanical properties from including these compounds in foams; moreover, these reactive materials are generally short-chain products, and they cause reactivity problems that are not easily overcome.[21] It has been reported that the use of such compounds deteriorates the cell microstructure of foams: in particular, the number of open cells and the strut joints decreased and the diameters of the cells increased.[71,72] Since it is well-known that the cell structure of a foam can greatly affect its mechanical properties and that the tensile strength and elongation at break of FPUF decreased as the cell diameters increased, it is not surprising that worse mechanical properties are reported when using these reactive FRs, especially for high content (> 20 php, parts per hundred polyol).[28] Moreover, very often these compounds have higher hydroxyl number than the polyol they are substituting, so they consumed more isocyanate. Therefore, more hard segments were formed and, as a result, the ratio of soft hard segments in FPUF decreased, causing the foam to have a high compression set at 50%.[33]

For the above reasons, only low amount of such reactive FRs can be used in FPUF; though even at these low concentrations they may provide flame retardancy since they are generally more effective than additive P based FRs. There are several commercial reactive polyol-type FRs, like Exolit OP550 (ethyl phosphate oligomer diol)[73] and OP560 (oligomeric phosphonate polyol[74]) (Clariant) and VeriQuel-F (ICL-IP) which are phosphorus-based or FR-513 (ICL-IP) which is based on tribromoneopentyl alcohol. The use of such reactive FRs is very useful when fogging and emission are relevant to the application (e.g., in the automotive sector), such as in children's car seats.[21,75] Content of less than 10 php are suggested to meet automotive standard FMVSS 302 (SE), thus showing an up to three times higher efficiency than Tris(2-chloropropyl)phosphate (TCPP).[35]

31.6.2 Relevant Flame Retardant Chemistry

One of the primary approaches to add flame retardancy to FPUF has been through the use of liquid FRs containing combinations of phosphorus, chlorine, and/or bromine. Halogen-phosphate compounds are becoming less commonly used as some have been banned in US and Europe or, like TCPP, are under scrutiny by the European Chemicals Agency (ECHA) due to their toxicity.[76] Therefore, more attention will be devoted in this paragraph to phosphorus-based compounds, which are under increasing scrutiny, but are not (as of yet) being as widely banned or restricted beyond some selective bans on specific compounds and states (see Chapter 23 for information on flame-retardant regulations).

Liquid phosphates are largely used as FRs. However, some of these compounds have been recognized as responsible for scorching, a phenomenon that occurs when foam overheats and degrade during foaming. Scorching lead to discoloration of the interior of the foam slab and loss of mechanical properties. Therefore, scorching should be avoided during foaming.[35] Scorching is attributed to thermal degradation and oxidation and is particularly visible in some flame retarded foams. However, not all fire-retardant foams undergo scorching.[77] It has been shown that scorching is related to the presence of aromatic amines inside the foams and to their reactions with some phosphates.[35,39,78] Indeed, there are some free isocyanate groups in the foam because of their entrapment in the cross-linked network where they cannot come into proximity with reactive groups due to steric hindrance. These isolated isocyanate groups can generate free amino groups. Once amines are formed by hydrolysis, they can react with free isocyanate forming ureas, or with oxygen in a reaction that is faster than urea formation for secondary amines.[39] The reaction between amines and oxygen forms oxidized compounds (e.g., quinoneimine structures) that are extremely chromophoric with very strong visible light absorption bands. The presence of some chloroalkyl (e.g., TCPP) or alkyl phosphates aggravate scorching because these groups can alkylate an aminoaryl group, forming alkylaminophenyl groups, which are more readily oxidized than the unalkylated aminophenyl groups.[35] Moreover, the alkylation reaction also adds to the exotherm during foaming. Conversely, phosphorus compounds that cannot alkylate amino groups are those which do not aggravate scorch; examples are aryl phosphates, such as isopropyl phenyl diphenyl phosphate.

Liquid phosphate FRs can cause not only scorching but also fogging problems because they can migrate out of the foam. Although fogging is more relevant for automotive standards, which consider accelerated aging in some material specifications, migration of the FRs may also be relevant for furnishings.[21] Some of the liquid phosphates commonly used (e.g., tricresyl phosphate) are volatile enough to be released into the atmosphere and stable enough to remain in the environment for years without any significant degradation (see Chapter 23 of this book). Therefore, they have been detected in indoor atmospheres and/or dust. Several studies observe correlations between the concentrations of certain FRs in dust from bedroom floors and concentrations in mattresses used in these rooms.[79]

Migration and volatilization can be hindered by using reactive FRs or by using oligomeric phosphates (See Chapters 5 and 23 of this book). Many examples of oligomeric phosphates are currently offered as commercial products. For example, Emerald® NH-1 (Lanxess) is a low viscosity liquid, probably a member of the aryl phosphate family, designed to meet, in combination with melamine, BS 5852 for upholstered furniture in the UK.[80] Lanxess's Disflamol DPK is also a member of the triaryl phosphate family and is recommended for both furniture and automotive low fogging applications.[42] Fyrol PNX (by ICL-IP), an oligomeric ethyl ethylene glycol phosphate, outperforms traditional products in TB117–2000. PNX is 50% to 60% more effective than traditional chlorinated alkyl phosphates.[35] Fyrol HF-5 (ICL-IP) is an oligomeric highly efficient phosphate ester flame retardant for upholstered furniture specially designed to satisfy older IKEA requirements related to very low emissions and absence of triphenyl phosphate and chloroalkyl phosphates; HF-9 (ICL-IP) is a highly efficient low scorch FR for broad upholstered furniture market.[42]

To substitute halogen-phosphate FRs, several authors studied the use of DOPO (9,10-dihydro-9-oxa-phosphaphenthrene-10-oxide) derivates as FRs for FPUF, although DOPO derivates are currently not in use on industrial scale for FPUF.[27,81] For example, it has been shown that methyl-DOPO is quite effective because it is released in the gas phase in the same temperature range as the decomposition products of FPUF, thus assuring an optimal interaction in the gas phase. DOPO mainly acts by diluting the flammable gaseous pyrolyzates and by a radical scavenger effect.[27]

Phosphate FRs were generally sufficient to meet the withdrawn flaming requirements of TB117–2000, while in the UK, the more stringent fire safety standard (BS 5852) must still be met by upholstered furniture. In UK, melamine is generally used.[82] Melamine has been extensively used in FPUF, although until 2009 its main application as FR in FPUF was limited to higher density foams (> 50kg/m^3).[83] König et al.[84] were the first to apply melamine in lower density foams (i.e., ≈ 25 kg/m^3), and they attributed its flame retardant efficiency to dilution of the flame by the sublimed melamine or its decomposition products rather than to the heat sink effect. It was also shown that 20 php melamine acted as a very efficient smoke suppressant. Especially in Europe, melamine has been often used in combination with a haloalkyl phosphate, such as tris(1,3-dichloro-2-propyl) phosphate. A synergistic interaction noted with melamine and TDCPP (but not with the more volatile TCPP in combustion-modified high resilience foam has been recently explained by Lefebvre et al.[47] on the basis of a chemical interaction producing char or a hard to ignite low-melting semi-solid. The more volatile TCPP escapes before this interaction can take place.[35,85] Therefore, if melamine-free foams are required, it is better to choose a liquid flame retardant which tends to decompose earlier (like TCPP). However, if melamine containing foams are to be produced, it is advisable to choose a liquid flame retardant which enables melamine to undergo char formation.[47] Denecker et al.[86] reported that the interaction between melamine and TCPP takes place because a cross-linking reaction between melamine, the matrix, and phosphoric acid.

A combination of melamine and chloroalkyl phosphate FRs is generally sufficient to meet TB117–2000 flaming tests, as well as BS5852 with Crib 5. In some foam formulations, aluminum trihydrate (ATH) was also added, but it should be noted that the addition of a solid such as ATH increases the viscosity of regenerated polyols and, hence, hinders the ability of FPUF to flow away from the heat source.[44] A similar effect has also been reported for phosphoramidates in melamine+chloroalkyl phosphate FPUF.[45]

Although melamine is quite extensively used in FPUF, a decrease in its use in Europe can be expected since it has been recently classified as a suspected human reproductive toxicant.[87] Although this classification is not relevant for melamine when used as an FR (i.e., once embedded in a polymer matrix), it is relevant to compounders handling melamine.

ATH is not generally used in FPUF for furniture and mattresses as the only FR. In fact, due to its low effectiveness, high loadings would increase the viscosity and affect foaming. To overcome this problem, ATH can be used as a surface treatment,[45] but this is mostly limited to fabric back-coatings that protect foam from fire damage, as high loads of ATH, even in surface treated foam, inhibit FPUF manufacture.

Other FRs for FPUF, like expandable graphite (EG), are used in aviation seating foams to meet strict oil burner tests but are not used in consumer furniture and mattress applications due to appearance, cost, and performance issues.[1] Another FR used in minor amounts in FPUF is ammonium polyphosphate (APP). In polyester-polyol-based FPUF, which have a greater charring propensity than polyether-polyol-based FPUFs, insoluble high molecular weight APP, crystal type II, is especially useful as a low smoke non-migrating flame retardant additive.[44]

Beside reactive and additive flame retardant approaches, another approach for FPUF is the use of nanoparticles. When embedded in foam, nanoparticles cannot be released unless the foam is ignited; however the foaming process is affected by the inclusion of nanoparticles which requires a complete reformulation of the foaming process.[88,89] Carbon nanofibers at a concentration of 4% by mass and organo-modified montmorillonite at a concentration of 5% by mass generated a percolated network of nanoparticles, however, only carbon nanofibers created an entangled fiber network that was thermally stable through the combustion. A carbon nanofiber entangled network prevented the collapse of the foam and led to the formation of a low density near net shape residue that shielded the underlying material from the incident heat flux and adsorbed the regenerated polyols preventing dripping. The net result was a reduction of about 35% in PHRR compared to the control foam containing 4% by mass of talc, as measured in a modified cone calorimeter test configuration accounting for the effects of pool formation.

Nanoparticles can be also used in FPUF to generate conformal coatings, sometimes called polyelectrolyte coatings or layer-by-layer (LbL) coatings, that are applied to the final foam, rather than being incorporated into the foam itself. Because this is a post-foaming process, there is no effect of the coating on the foaming process. These thin films were commonly fabricated through alternating deposition of a positively charged layer and negatively charged layer (called a bilayer, BL). This is noteworthy because it applies the flame retardant protection on every wall, cell, and strut of the foam, rather than attempting to incorporate itself into the polyurethane foam structure.[1] Moreover, the flame retardant multilayered films are constructed between a substrate and the exterior which can directly interfere with the combustion process and eliminate the challenges associated with processing foam.[90] In addition, the LbL assembly technique relies on mild processing conditions such as room temperature, atmospheric pressure and a low concentration of to-be-assembled materials (below 1 wt%), which often means it is an easy route for fabricating coatings. These types of coatings have been studied extensively for fire protection of polyurethane foam.[50,91–93] Open cell FPUFs exhibit a high surface to bulk ratio; thus, the FR properties can be enhanced by the LbL deposition that is able to penetrate inside the foam and homogeneously covers the foam surface.[53] Nanomaterials such as polymer carbon nanotubes and montmorillonite can all be assembled on the surface of FPUF.[52] The first attempt to improve fire behavior of FPUF by LbL were made by Grunlan and coworkers[94] who assembled polyethyleneimine stabilized carbon nano-fibres (CNF; cationic layer) with poly(acrylic acid) (anionic layer). They showed that even though the CNF distribution is not completely uniform and the CNFs are not completely embedded in the polymer coating, the coating significantly reduces the flammability of foam. In fact, the LbL coating significantly reduces the heat release rate, total heat released, and total burn time of the FPUF with just four bilayers (e.g., about 40% reduction in the peak of heat release rate, PHRR), at significantly lower additive concentration than traditional FRs (e.g., 1.6 wt% CNF coating on FPUF yields a 23% lower PHRR than a 20 wt% brominated FR in FPUF). Later, Grunlan et al.[95] developed an LbL assembly for enhancing PU flame retardancy without using nanoparticles by using positively charged chitosan and anionic polyvinyl sulfonic acid sodium salt. This coating system completely stops foam melt dripping upon exposure to the direct flame from a butane torch. Furthermore, a 30 nm thick coating adds only 5.5% to the foam's weight and completely stop flame propagating on the foam. Cone calorimetry reveals that this coated foam has a 52% reduction in PHRR relative to an uncoated control. Also intumescent LbL coatings have been proposed: phosphorus-containing polymer electrolytes are widely used as an acid source while nitrogen-containing charring polymer electrolytes serve as both the gas source and the charring agent.[51] For example, Carosio et al.[96] developed adaptive hybrid organic-inorganic nanoarchitectures capable of withstanding the exposure to extreme heat, by exploiting a chemical reaction, which leads to the formation of thermally-stable carbon-based structures (char). The chosen LbL architecture consists of a four-layer repetitive unit made of poly(acrylic acid) (-), chitosan (+), poly(phosphoric acid) (-), and chitosan (+). In the coating composition, chitosan and poly(acrylic acid) exhibit natural char-forming characteristics that can be further enhanced by the dehydration abilities of poly(phosphoric acid). In another work,[97] the same authors used a similar system to improve the traditional LbL assembly method to make it faster (three to 4 orders of magnitude faster than conventional LbL treatments) in view of a potential industrial application. This approach is based on the standard LbL adsorption process, significantly accelerating the deposition times from some minutes to less than a second. It consists in imparting strong deformation to the substrate in order to remove the excess of polyelectrolyte/water before moving to the next deposition step. In this method, a highly efficient LbL formulation, capable of reaching high performances at low deposition steps (2 bilayers) while withstanding the deformation imparted to the sample during the process, is needed. To meet these requirements Carosio et al. used chitosan and poly(phosphoric acid) and found that the structure of FPUF, protected by the thermally stable carbonaceous structures produced by the coating, slowly shrank as it is slowly consumed by the fire without collapsing and resulted in a 33% reduction in PHRR.

Despite many advantages, the durability of FR coatings remains a major issue to be addressed. The coatings are usually deposited on substrates that often undergo significant wear-and-tear in service, and thus the coatings should possess excellent durability to survive these external damaging forces and retain high fire retardant efficiency.[51]

31.7 CONCLUSIONS AND FUTURE DIRECTIONS

The flammability of furniture and furnishings is a well-studied phenomenon, and FPUF present in these items remains the main fire hazard that must be protected from transitioning to flaming combustion.

In the past, once a basic understanding of polyurethane foam smoldering was acquired and strategies to mitigate the smoldering hazard were developed, the upholstered furniture manufacturer voluntarily adopted the UFAC cigarette ignition test that eventually evolved as a federal standard for smoldering ignition resistance. Hopefully, a similar path will be followed by upholstery manufacturers in terms of flaming performance now that, as summarized in this chapter, we might have all the necessary knowledge and technology to do so. New regulatory requirements that strongly discourage flame retardant chemical use in upholstered furniture in the US lead to consideration of new fire protection systems, namely, fire barriers. In fact, (1) fire barriers that can drastically increase fire safety while complying with the regulatory restrictions on FRsare now commercially available, (2) the mechanisms of action/failure are well understood, and (3) a bench-scale tool to assess barrier fabric performance exists in the form of the abovementioned Cube test (ASTM E3367). However, the current regulatory fire tests in the US (namely 16 CFR 1640) can be met without the use of barriers, so it remains to be seen if barriers will be used by furniture manufacturers. The implementation of barriers would have a significant cost in terms of materials (commercial barriers typically cost between $1 and $10 per square yard) and labor (by laminating the barrier to a polyester batting layer which is already commonly used in upholstered furniture would largely decrease this cost) and might prevent barriers from being used. In absence of a voluntary or regulatory standard, the only way to drive demand towards fire-safe products is via education and awareness of the consumer. Barriers are commonly used in mattresses to provide fire protection for the FPUF under US regulatory tests like 16 CFR 1633, but their use might remain limited to mattresses even though barriers are expected to be equally effective in upholstered furniture.

If flame retardant chemicals are so desired to be used in FPUF, there is a wealth of known chemistry to choose from as was outlined in this chapter. However, due to the concerns on flame retardant chemicals migrating out of foam over time (which led to the changes in regulatory fire tests in the US), any future flame retardant chemical put into FPUF *must be* a reactive-based FR that covalently bonds with the polyurethane during foam manufacture. By using a reactive FR, the FR cannot get into the environment (see Chapters 23 and 24 in this book), and further, it ensures fire protection of the product over its lifetime. This requirement to use reactive-based FRs applies to any FRs put into fire barrier materials as well.

Future directions to address the flammability of furniture and furnishings are not at all clear at the moment. While the fire hazard remains in place as FPUF did not become less flammable with time, the regulatory change in the US is unlikely to change anytime soon, despite the known fire hazards. Therefore, in the US the future direction to address the flammability of these items may be around hazard awareness education for the public. More specifically, to focus on educational campaigns on keeping ignition sources away from furniture.

The situation regarding the fire safety of soft furnishing in other parts of the world seems to be equally unclear. The EU is not moving to change its fire safety requirements for furniture, and the UK is currently (as of the writing of this chapter) considering legislation to withdraw the open flame testing of its furniture. Hopefully fire losses will not have to go up again before fire tests are implemented with meaningful fire safety, as it is possible and probable to have fire protection without negative environmental and public health effects. Flame retardant chemicals don't always have to be used to fire protect furniture, and one can use reactive FRs as well. For now, one should assume that new furniture items sold in the US will have no FRs and will not be able to withstand open flame ignition, while mattresses in the US will continue to have strong fire safety performance. For furniture items in the rest of the world, it is likely that the regulations will remain unchanged as well. Therefore, future flame-retardant chemistry research into FPUFs seems to be limited to academic studies for the near future.

NOTE

1. Part of this work was carried out by the National Institute of Standards and Technology (NIST), an agency of the U.S. government, and by statute is not subject to copyright in the United States. Certain commercial equipment, instruments, materials, services, or companies are identified in this chapter in order to specify adequately the experimental procedure. This in no way implies endorsement or recommendation by NIST. The policy of NIST is to use metric units of measurements in all its publications, and to provide statements of uncertainty for all original measurements. In this document, however, data from organizations outside NIST are shown, which may include measurements in nonmetric units or measurements without uncertainty statements. wt% is used throughout this manuscript and is identical to mass fraction %.

REFERENCES

1. Home Structure Fires, Hall, Shelby, National Fire Protection Agency, April 2023. www.nfpa.org/-/media/files/news-and-research/fire-statistics-and-reports/building-and-life-safety/oshomes.pdf (accessed on 09/22/2023).
2. "Revisiting flexible polyurethane foam flammability in furniture and bedding in the United States." Morgan, A. B. *Fire Mater.* **2021**, *45*, 68–80.
3. "Comparative room burn study of furnished rooms from the United Kingdom, France and the United States." Blais, M. S.; Carpenter, K.; Fernandez, K. *Fire Technol.* **2020**, *56*, 489–514.
4. "Bedding ignition, vertical flame spread and subsequent thermal impact on underlying mattress." Gratkowski, M. T. *Fire Mater.* **2021**, *45*, 28–36.
5. "A bibliography of published information on furniture flammability, part I." Damant, G. H. *J. Consum. Prod. Flammab.* **1978**, *5*, 245–248.
6. "A bibliography of published information on furniture flammability, part II." Damant, G. H. *J. Consum. Prod. Flammab.* **1980**, *7*, 127–129.
7. "A bibliography of published information on furniture flammability, part III." Damant, G. H.; Wortman, P. S. *J. Consum. Prod. Flammab.* **1981**, *8*, 235–238.
8. "The Furniture and Furnishings (Fire) (Safety) Regulations 1988." www.legislation.gov.uk/uksi/1988/1324/contents/made (accessed 04/14/20).
9. "An overview and experimental analysis of furniture fire safety regulations in Europe." Guillaume, E.; de Feijter, R.; van Gelderen, L. *Fire Mater.* **2020**, *44*, 624–639.
10. "Fire Behavior of Upholstered Furniture." Babrauskas, V.; Krasny, J. F. *NBS Monograph 173*, US National Bureau of Standards, **1985**, Gaithersburg, MD. www.nist.gov/publications/fire-behavior-upholstered-furniture-nbs-monograph-173 (accessed 04/14/20).
11. "Fire Safety of Upholstered Furniture—The Full Report of the European Commission Research Programme CBUF." Sundström, B. Editor, Interscience Communications Ltd, **1995**, London.
12. "Flammability Assessment Methodology for Mattresses." Ohlemiller, T. J.; Shields, J. R.; McLane, R. A.; Gann, R. G. *NISTIR 6497*, US National Institute of Standards and Technology, **2000**, Gaithersburg, MD. https://doi.org/10.6028/NIST.IR.6497
13. www.cpsc.gov/s3fs-public/pdfs/blk_media_testmatt.pdf (accessed 04/14/20).
14. www.cpsc.gov/s3fs-public/pdfs/blk_media_labmanual.pdf (accessed 04/14/20).
15. "Heat release rate: The single most important variable in fire hazard." Babrauskas, V.; Peacock, R. D. *Fire Saf. J.* **1992**, *18*, 255–272.
16. "Combustion toxicity regulations for combustion products." Babrauskas, V. *J. Fire Sci.* **2020**. https://doi.org/10.1177/0734904119880472
17. "Fire conditions for smoke toxicity measurement." Gann, R. G.; Babrauskas, V.; Peacock, R. D.; Hall, J. R. *Fire Mater.* **1994**, *18*, 193–199.
18. "Estimating the impact of the mattress fire safety Standard 16 CFR Part 1633 on bed fire outcomes." Gilbert, S. W.; Butry, D. T.; Davis, R. D.; Gann, R. G. *Fire Mater.* **2021,** 45, 17–27.
19. "Technical Bulletin 117-2013, Requirements, Test Procedure and Apparatus for Testing the Smolder Resistance of Materials Used in Upholstered Furniture." State of California, Department of Consumer Affairs, January **2019**. https://bhgs.dca.ca.gov/laws/tb117_2013.pdf (accessed 08/30/23).
20. "Toxic product yields and hazard assessment for fully enclosed design fires." Purser, D. A. *Polym Int.* **2000**, 49(10), 1232–1255.
21. "Smoldering Combustion." Ohlemiller, T. J. *NBSIR 85-3294*, **1986**, Gaithersburg MD. https://nvlpubs.nist.gov/nistpubs/Legacy/IR/nbsir85-3294.pdf (accessed on 08/31/2023).
22. "Home Fires That Began with Upholstered Furniture." Ahrens, M. National Fire Protection Association Research, **2017**. www.nfpa.org//-/media/Files/News-and-Research/Fire-statistics-and-reports/US-Fire-Problem/Fire-causes/osupholstered.pdf (accessed 08/30/23).
23. "Estimating fires when a product is the primary fuel but not the first fuel, with an application to upholstered furniture." Hall, J. R.; *Fire Technol.* **2015**; *51*(2), 381–391.
24. "The Challenge of Realizing Low Flammability Home Furnishings." Gann, R. G., Interflam, **2013**, London.
25. "Summary and Conclusions of a Workshop on Quantifying the Contribution of Flaming Residential Upholstered Furniture to Fire Losses in the United States." Pitts, W.M. *NIST Technical Note 1757*, National Institute of Standards and Technology, **2012**, Gaithersburg, MD. https://nvlpubs.nist.gov/nistpubs/TechnicalNotes/NIST.TN.1757.pdf (accessed on 08/30/23).
26. "Upholstered Furniture Flammability; Staff Activities and Recommendation." U.S. Consumer Product Safety Commission, **2019**. https://cpsc.gov/s3fs-public/Upholstered%20Furniture%20Update%202019.pdf (accessed 09/14/23).
27. "Factors Influencing the Smoldering Performance of Polyurethane Foam." Zammarano, M.; Krämer, R. H.; Matko, S.; Mehta, S.; Gilman, J. W.; Davis, R. D., NIST TN 1747, **2012**, Gaithersburg, MD. www.nist.gov/customcf/get_pdf.cfm?pub_id=911753 (accessed on 08/31/23).
28. "Smoldering in Flexible Polyurethane Foams: The Effect of Foam Morphology." Zammarano, M.; Matko, S.; Krämer, R. H.; Davis, R. D. *ACS Symposium Series 1118. Fire and Polymers VI: New Advances in Flame Retardant Chemistry.* ACS, **2013**, pp. 459–479. DOI: 10.1021/bk-2012-1118.ch029
29. "Smoldering propensity in upholstered furniture: Effects of mock-up configuration and foam thickness." Yang, J.; Rein, G.; Chen, H.; Zammarano, M. *Appl. Therm. Eng.* **2020**, 181.
30. "Soft furnishing fires: They're still a problem." Ahrens, M. *Fire Mater.***2021**, *45*, 8–16.

31. "An Act to Protect Firefighters by Establishing a Prohibition on the Sale and Distribution of New Upholstered Furniture Containing Certain Flame-Retardant Chemicals". 128th Maine Legislature House of Representatives 182; 2017. www.mainelegislature.org/legis/bills/bills_128th/chapters/PUBLIC311.asp (accessed on 09/01/23).
32. "Consumer Products: Flame Retardant Materials." State of California Office of Legislative Counsel Bloom 2998; 2018. https://leginfo.legislature.ca.gov/faces/billTextClient.xhtml?bill_id=201720180AB2998 (accessed on 09/01/23).
33. "Project Plan: Organohalogen Flame Retardant Chemicals Assessment." Babich, M., Bevington, C., Chen, X., et al. Consumer Product Safety Commission, **2020**. www.cpsc.gov/s3fs-public/OFR-plan-report-package-final.pdf?TDF6XKQhE3NgKpZaUZuTqVXXUW32E.kh.
34. "Guidance Document on Hazardous Additive, Non-Polymeric Organohalogen Flame Retardants in Certain Consumer Products." CPSC-2015-0022 CDN. Commission CPS, Editor. Vol 82: Federal Register, **2017**. www.federalregister.gov/documents/2017/09/28/2017-20733/guidance-document-on-hazardous-additive-non-polymeric-organohalogen-flame-retardants-in-certain.
35. "Open Letter on FR Chemicals in Furniture." 2 April **2022**. https://openletter.earth/open-letter-on-fr-chemicals-in-furniture-b4a072ae (accessed on 09/05/2023).
36. "Flame retardants in commercial use or development for textiles." Weil, E.; Levchik, S. *J. Fire Sci.* **2008**, *26*, 243.
37. "A review of fire blocking technologies for soft furnishings." Nazare, S.; Davis, R. *Fire Sci. Rev.* **2012**, *1*(1), 1.
38. California Bureau of Household Goods and Services (BHGS), State of California, Department of Consumer Affairs, North Highlands, CA. https://bhgs.dca.ca.gov/industry/tb129.pdf and https://bhgs.dca.ca.gov/industry/tb133.pdf
39. "Reduced-scale test to assess the effect of fire barriers on the flaming combustion of cored composites: An upholstery-material case study." Zammarano, M.; Shields, J. R.; Leventon, I.; Kim, I.; Nazare, S.; Thompson, A.; Davis, R. R., Chernovsky, A.; Bundy, M. *Fire Mater.* **2020**, *45*(1), 114–126.
40. "Smolder characteristics of flexible polyurethane foams." Rogers, F.; Ohlemiller, T. *J. Fire Flammab.* **1980**, *11*(1), 32–44.
41. "NIST TN 1747 – Factors influencing the smoldering performance of polyurethane foam." Zammarano, M.; Krämer, R. H.; Matko, S.; Mehta, S.; Gilman, J. W.; Davis, R. D. **2012**.
42. "Heat release and structural collapse of flexible polyurethane foam." Kramer, R. H.; Zammarano, M.; Linteris, G. T.; Gedde, U. W.; Gilman, J. W. *Polym. Degrad. Stab.* **2010**, *95*, 1115–1122.
43. "Flexible polyurethane foam. I. Thermal decomposition of a polyether-based, water-blown commercial type of flexible polyurethane foam." Ravey, M.; Pearce, E. M. *J. Appl. Polym. Sci.* **1997**, *63*(1), 47–74. DOI: 10.1002/(sici)1097-4628(19970103)63:1<47::Aid-app7>3.0.Co;2-s.
44. "Fundamentals of the Fire Behaviour of Cellular Polymers." Drysdale, D. D. *Fire and Cellular Polymers*. Buist, J. M., Grayson, S. J., Woolley, W. D. Editors, Springer, Dordrecht. **1986**, pp. 61–75. https://doi.org/10.1007/978-94-009-3443-6_4
45. "NIST Technical Note 1493 – Aspects of the fire behavior of thermoplastic materials." Ohlemiller, T. J.; Shields, J. R. *NIST Tech. Note* 1493, National Institute of Standards and Technology, 2008, Gaithersburg, MD. https://doi.org/10.6028/NIST.tn.1493.
46. "Assessing the predictive capability for real-scale residential upholstered furniture mock-up fires using cone calorimeter measurements. Part 1: Real-scale experiments." Pitts, W.; Werrel, M.; Fernandez, M.; Long, A. M.; Eisenberg, E. A.; Runyon, C. D., **2020**. DOI: 10.6028/NIST.SP.1246.
47. "Reduced-Scale Test to Assess the Effect of Fire Barriers on the Flaming Combustion of Cored Composites." National Institute of Standards and Technology, **2020**. www.nist.gov/el/fire-research-division-73300/flammability-reduction-73304/low-heat-release-upholstered-0.
48. "Standard Test Method for Determining the Combustion Behavior of Layered Assemblies using a Cone Calorimeter." ASTM E3367-23, ASTM International, **2023**, West Conshohocken, PA.
49. "Performance and failure mechanism of fire barriers in full-scale chair mock-ups." Thompson, A.; Kim, I.; Hamins, A.; Mundy, M.; Zammerano, M. *Fire Mater.* **2022**, *46*(1), 329–346. DOI: 10.1002/fam.3007.
50. www.nist.gov/el/fire-research-division-73300/flammability-reduction-73304/low-heat-release-upholstered-0 (accessed on 09/03/23).
51. www.nist.gov/el/fcd/fire-barriers-full-scale-chair-mock-ups (accessed on 09/03/23).
52. "Least-squares fitting of a straight line." York, D. *Can. J. Phys.* **1966**, *44*(5), 1079–1086.
53. US National Institute of Standards and Technology unpublished data.
54. "High Performance Fire Barrier Fabrics in Chair Mock-ups." Zammarano, M.; Kim, I.; Bundy, M. NIST Technote, National Institute of Standards and Technology, **2024** (in preparation).
55. "Impact of Barrier Layers on Hazardous Conditions in Residential Upholstered Furniture Fires." Reneke, P. A.; Gilbert, S. W.; Zammarano, M; Peacock, R. D.; Cleary, T. G. NIST Technical Note 2264, National Institute of Standards and Technology, 2024, Gaithersburg, MD.
56. NFPA. *NFPA 1720, Standard for the Organization and Deployment of Fire Suppression Operations, Emergency Medical Operations, and Special Operations to the Public by Volunteer Fire Departments*, **2010**. www.nfpa.org/codes-and-standards/1/7/2/nfpa-1720 (accessed on 01/27/2024).
57. "The Challenge of Realizing Low Flammability Home Furnishings." Gann, R. G., Interflam, **2013**, London.
58. "Flexible and Semi-Flexible Foams." Szycher, M. *Szycher's Handbook of Polyurethanes*. Szycher, M., Editor, CRC Press, **2012**, Boca Raton, FL, pp. 181–255.
59. "Flexible Polyurethane Foams: A Practical Guide." Defonseka, C. Walter de Gruyter, Boston, USA. DOI: https://doi.org/10.1515/9783110643183, **2019**.
60. "Polyurethane Foams." Kaneyoshi A. *Polyurethane and Related Foams Chemistry and Technology*, CRC Press, **2007**, Boca Raton, FL, pp. 65–99.
61. "The effect of foam density on combustion characteristics of flexible polyurethane foam." Stone, H.; Pcolinsky, M.; Parrish, D. B.; Beal, G. E. *J. Cell. Plast.* **1991**, *27*(1), 78–79.
62. "Recent progress in flame retardancy of polyurethane foam." Weil, E. D., Ravey, M.; Gertner, D. In *Proceedings of the BCC Conf. Recent Advances in Flame Retardancy of Polymeric Materials*, Stamford, CT, **1997**, *7*, 191–200.
63. "Flame spread of flexible polyurethane foam: Comprehensive study." Lefebvre, J.; Bastin, B.; Le Bras, M.; Duquesne, S.; Ritter, C.; Paleja, R.; Poutch, F. *Polym. Test.* **2004**, *23*(3), 281–290.
64. "Thermal stability and fire properties of conventional flexible polyurethane foam formulations." Lefebvre, J.; Bastin, B.; Le Bras, M.; Duquesne, S.; Paleja, R.; Delobel, R. *Polym. Degrad. Stab.* **2005**, *88*(1), 28–34.

65. "Methyl-DOPO-a new flame retardant for flexible polyurethane foam." König, A.; Kroke, E. *Polym. Adv. Technol.* **2011**, *22*(1), 5–13.
66. "Foamed polymers. Cellular structure and properties." Shutov, F. A. *Ind. Dev.* **2005**, 155–218.
67. "The influence of polymer morphology on the combustion of melamine filled flexible foams." Schrock, A. K.; Solis, R.; Beal, G. E.; Skorpenske, R. G.; Parrish, D. P. *J. Fire Sci.* **1990**, *8*(3), 174–193.
68. "Intrinsic flame resistance of polyurethane flexible foams: Unexpectedly low flammability without any flame retardant." Höhne, C. C.; Hanich, R.; Kroke, E. *Fire Mater.* **2018**, *42*(4), 394–402.
69. "Combustion behaviour of polyurethane flexible foams under cone calorimetry test conditions." Vanspeybroeck, R.; Van Hees, P.; Vandevelde, P. *Fire Mater.* **1993**, *17*(4), 155–166.
70. "Performance of HS-HR flexible foams in small open-flame combustibility tests." Hager, S.; Skorpenske, R.; Blaszkiewicz, M.; Masciantonio, M. *J. Cell. Plast.* **2004**, *40*(4), 315–340.
71. "Inherently flame-retardant flexible polyurethane foam with low content of phosphorus-containing cross-linking agent." Chen, M. J.; Chen, C. R.; Tan, Y.; Huang, J. Q.; Wang, X. L.; Chen, L.; Wang, Y. Z. *Ind. Eng. Chem. Res.* **2014**, *53*(3), 1160–1171.
72. "Persistently flame-retardant flexible polyurethane foams by a novel phosphorus-containing polyol." Rao, W. H.; Xu, H. X.; Xu, Y. J.; Qi, M.; Liao, W.; Xu, S.; Wang, Y. Z. *Chem. Eng. J.* **2018**, *343*, 198–206.
73. "Recent Progress in Flame Retardancy of Polyurethane and Polyisocyanurate Foams." Levchik, S. V.; Weil, E. D. *Fire and Polymers IV, Materials and Concepts for Hazard Prevention.* Wilkie, C. A.; Nelson, G. L., Editors, American Chemical Society, Washington, D.C., USA. **2005**, 922, pp. 280–290.
74. www.clariant.com/de/Corporate/News/2014/10/Clariant-s-Exolit--OP-560-confirmed-as-safer-flame-retardant-for-PU-foam (accessed on 31st July 2023).
75. www.clariant.com/en/Solutions/Products/2014/03/18/16/31/Exolit-OP-560 (accessed on 3rd August 2023).
76. https://echa.europa.eu/assessment-regulatory-needs?p_p_id=disslists_WAR_disslistsportlet&p_p_lifecycle=1&p_p_state=normal&p_p_mode=view&_disslists_WAR_disslistsportlet_javax.portlet.action=searchDissLists (accessed on 31st July 2023).
77. "Discolouration in fire retardant flexible polyurethane foams. Part I. Characterisation." Luda, M. P.; Bracco, P.; Costa, L.; Levchik, S. V. *Polym. Degrad. Stab.* **2004**, *83*(2), 215–220.
78. "Relevant factors in scorch generation in fire retarded flexible polyurethane foams II reactivity of isocyanate, urea and urethane groups." Luda, M. P.; Costa, L.; Bracco, P.; Levchik, S. V. *Polym. Degrad. Stab.* **2004**, *86*(1), 43–50.
79. "Fireproofing of domestic upholstered furniture: Migration of flame retardants and potential risks." Lounis, M.; Leconte, S.; Rousselle, C.; Belzunces, L. P.; Desauziers, V.; Lopez-Cuesta, J. M.; Julien, J. M.; Guenot, D.; Bourgeois, D. *J. Hazard. Mater.* **2019**, *366*, 556–562.
80. "Phosphorus-Based FRs." Levchik, S. *Non-Halogenated Flame Retardant Handbook.* Morgan, A. B.; Wilkie, C. A., Editors, John Wiley & Sons, **2014**, Hoboken, NJ, pp. 17–74.
81. "Flame retardant flexible polyurethane foams from novel DOPO-phosphonamidate additives." Gaan, S.; Liang, S.; Mispreuve, H.; Perler, H.; Naescher, R.; Neisius, M. *Polym. Degrad. Stab.* **2015**, *113*, 180–188.
82. "Commercial flame retardancy of polyurethanes." Weil, E. D.; Levchik, S. V. *J. Fire Sci.* **2004**, *22*(3), 183–210.
83. "Recent Developments in Flame Retardancy of Flexible Polyurethane Foams." Neisius, N. M.; Liang, S.; Mispreuve, H.; Gaan, S. *Fire and Polymers VI: New Advances in Flame Retardant Chemistry and Science.* Morgan, B. A.; Wilkie, C. A.; Nelson, G. L., Editors, Vol. 1118, American Chemical Society, Washington D.C. **2012**, pp. 251–270.
84. "Thermal decomposition behavior of the flame retardant melamine in slabstock flexible polyurethane foams." König, A.; Fehrenbacher, U.; Kroke, E.; Hirth, T. *J. Fire Sci.* **2009**, *27*(3), 187–211.
85. "Flexible polyurethane foams: Flammability." Lefebvre, J.; Le Bras, M.; Bastin, B.; Paleja, R.; Delobel, R. *J. Fire Sci.* **2003**, *21*(5), 343–367.
86. "Relationship between the thermal degradation chemistry and flammability of commercial flexible polyurethane foams." Denecker, C.; Liggat, J. J.; Snape, C. E. *J. Appl. Polym. Sci.* **2006**, *100*(4), 3024–3033.
87. www.pinfa.eu/mediaroom/information-on-melamine-reach-dossier/#gsc.tab=0 (accessed on 31st July 2023).
88. "Flammability reduction of flexible polyurethane foams via carbon nanofiber network formation." Zammarano, M.; Krämer, R. H.; Harris Jr., R.; Ohlemiller, T. J.; Shields, J. R.; Rahatekar, S. S.; Lacerda, S.; Gilman, J. W. *Polym. Adv. Technol.* **2008**, *19*(6), 588–595. DOI: 10.1002/pat.1111.
89. "Characterizing particle emissions from burning polymer nanocomposites." Nyden, M. R.; Harris, R. H.; Kim, Y. S.; Davis, R. D.; Marsh, N. D.; Zammarano, M. In *Proceedings of the 21th BCC Conference on Flame Retardation*, Vol. 1, National Institute of Standards and Technology-Nanotech, **2010**, Gaithersburg, MD, pp. 717–719.
90. "Flame retardant coatings prepared using layer by layer assembly: A review." Qiu, X.; Li, Z.; Li, X.; Zhang, Z. *Chem. Eng. J.* **2018**, *334*, 108–122.
91. "Surface-coating engineering for flame retardant flexible polyurethane foams: A critical review." Yang, H.; Yu, B.; Song, P.; Maluk, C.; Wang, H. *Compos. Part B Eng.* **2019**, *176*, 107185.
92. "Review of layer-by-layer self-assembly technology for fire protection of flexible polyurethane foam." Liu, Q.; Gao, S.; Zhao, Y.; Tao, W.; Yu, X.; Zhi, M. *J. Mater. Sci.* **2021**, *56*(16), 9605–9643.
93. "Layer-by-Layer Nanostructured Coatings for Flame Retardancy." Carosio, F.; Alongi, J.; Malucelli, G.; Applicata, S.; Torino, P. *Comprehensive Guide for Nanocoatings Technology.* Aliofkhazraei, M., Editor, Vol. 4, Nova Science, **2015**, New York, pp. 235–258.
94. "Development of layer-by-layer assembled carbon nanofiber-filled coatings to reduce polyurethane foam flammability." Kim, Y. S.; Davis, R.; Cain, A. A.; Grunlan, J. C. *Polymer.* **2011**, *52*(13), 2847–2855.
95. "Exceptionally flame retardant sulfur-based multilayer nanocoating for polyurethane prepared from aqueous polyelectrolyte solutions." Laufer, G.; Kirkland, C.; Morgan, A. B.; Grunlan, J. C. *ACS Macro Lett.* **2013**, *2*(5), 361–365.
96. "Self-assembled hybrid nanoarchitectures deposited on poly(urethane) foams capable of chemically adapting to extreme heat." Carosio, F.; Di Blasio, A.; Cuttica, F.; Alongi, J.; Malucelli, G. *RSC Adv.* **2014**, *4*(32), 16674–16680.
97. "Ultra-fast layer-by-layer approach for depositing flame retardant coatings on flexible PU foams within seconds." Carosio, F.; Alongi, J. *ACS Appl. Mater. Interfaces* **2016**, *8*(10), 6315–6319.

32 The Future of Material Fire Protection—Unmet Needs, New Fire Risk Scenarios, and New Flame Retardant Approaches

Alexander B. Morgan
University of Dayton Center for Flame Retardant Materials Science, Dayton, OH 45469

32.1 INTRODUCTION AND BRIEF HISTORY OF REACTION-TO-FIRE EVENTS DRIVING FIRE PROTECTION NEEDS

As indicated throughout this book, fire protection has been predominantly a reactive field of science and engineering. Specifically, it means that fire protection is studied and put in place (via codes and standards) *after* a major fire event occurs. Effectively, the learnings from that fire event put in place new fire protection to prevent the fire event that occurred from happening again. Because one cannot predict the future, fire protection is rarely proactive, although one can start to see fire threats that may occur in the near future by applying the fundamentals of fire protection engineering, and material flammability, to envision new fire risk scenarios. In general, as new technologies and approaches are inserted into society, very often fire protection is playing "catch up" to these trends and working to provide protection after technologies are inserted. One would think that with all the emphasis on total life cycle analysis design that these things would start to be considered in new design work, and sometimes they are considered, as will be discussed later. However, the complicating factor is very often fire protection engineers cannot envision all the ways that new technologies will be inserted into use, and therefore, it becomes very difficult to be fully proactive in fire protection, especially when commonly used flammable polymeric materials are part of the technological insertion. A good example of this is changes in firefighting measures for modern vs. older homes. Due to the large increase in polymeric materials in a modern home, the modern home requires different firefighting tactics because the room fires become oxygen starved very quickly due to faster heat release/fire growth rates, vs. older homes which have less polymeric materials and slower heat release/fire growth rates.[1] While individual components in these newer houses all meet building code, the large increase in polymeric materials for furniture and the contents of the room/house have yielded increased fire hazard faster than the codes and standards can keep up. This has happened before in human history as new materials are inserted into use and will likely continue to happen. While this historical observation should drive further design work to consider fire safety in new building materials, one cannot tell (or enforce) the owner of the home how to do things. If a home owner decides to make a sound-proofing room on their own with flexible polyurethane foam (ideal for this purpose), and then inserts a lot of electrical equipment into the room, and something short circuits and catches the foam on fire, this is going to result in a loss of the room and likely the home due to the high heat release of polyurethane foam. Fire marshals and fire protection engineers cannot predict this type of owner behavior, nor can they really prevent it from happening either. With this in mind, what one can do to be proactive on fire safety is look at current and likely trends of technology insertion, and use fire safety principles and known material flammability behavior to predict the fire protection for the future. This then guides the fire safety engineer and the fire safety science on where to focus their efforts.

This chapter considers the future of fire protection needs but also considers the current unmet needs in polymeric material fire protection, especially in regards to specific chemical approaches and design considerations about when flame retardants should be used, and how they should be incorporated. Also, this chapter discusses and where there are needs for new methods and methodology for fire safe material development. Finally, discussion around current growing fire risks/fire hazards is included, along with some projection on how these needs will be met.

32.2 CURRENT UNMET NEEDS FOR POLYMERIC MATERIAL FIRE PROTECTION

32.2.1 Reactive and Polymeric Additives

In regards to unmet needs, the first topic is one that could be argued is already being met. Specifically, the need for

DOI: 10.1201/9781003380689-32

flame retardant additives that cannot get into the environment. Most of the new flame retardants released commercially from the main flame retardant manufacturers (such as ICL, Lanxess, Clariant, Albemarle) are usually polymeric in structure, or they have the potential to react (covalently bond) with the polymer they are added to. The reasons for this are covered in detail in Chapters 23 and 24 (as well as a few other places throughout the book), but they are worth repeating here. Polymeric and reactive flame retardants have lower chances of migrating out of a polymeric product over its lifetime, and therefore the flame retardant stays in place where it is needed, and environmental impact of the additives is lessened. This further results in low potential for bioaccumulation of these additives over their lifetime, and so gives these polymeric and reactive flame retardants more positive "persistence, bioaccumulation, toxicity" (PBT) profiles that should further enable their use. However, despite the increase in available polymeric and reactive flame retardants, what is available is rather limited in utility in polymers when you look at the chemical structures available. When studying the phosphorus-based reactive flame retardants available (Chapters 5, 24, 28, 31), it's quite clear that what is available is limited to thermoset polymers. There is a limited number of reactive halogenated flame retardants (Chapter 4) available to be copolymerized with thermoset polymers, and again, most of the reactive halogenated structures are more suited for thermoset polymers.

There is a definite need for more reactive flame retardants for thermoplastic polymers, such as condensation polymers (polyesters), but it is admittedly more difficult to generate reactive flame retardants for some thermoplastics, such as polyolefins and nylons. Polyolefins tend to use catalysts that interact poorly with monomers containing heteroatoms, and polyamides have particular polymerization processes that do not always lend themselves to other monomer chemistries. Potentially, free-radical initiated thermoplastics would work well with a wide range of alkene functionalized flame retardant monomers, but this isn't always the case. Indeed, reactive flame retardants are not easy to implement because of the reactivity of the flame retardant to polymerization conditions and potential differences in reactivity between the monomer of the polymer to be synthesized and the reactive flame retardant.

The chemical structure of a monomer will determine its reactivity to polymerization reactions, whether they be condensation or free radical polymerizations. The reactivity can result in very different kinetics of polymerization or require more activation energy to get polymerization to occur. To further complicate matters, the flame-retardant chemical structure must be compatible with the chemical environment it is polymerized into. For example, in an epoxy polymerization, a phosphate ester functionalized monomer could effectively react with alcohol groups from an epoxide polymerization, or could react with amine curing agents instead, thus changing the flame retardant functional group and the final polymer structure. Effectively, using a flame-retardant monomer which is copolymerized with other monomers yields a brand-new polymer that has never before been made, and one must work to achieve balance of thermal, mechanical, and fire properties for that new polymer. Many of the polymers in common use today took years of work to successfully scale up and commercialize, and so it does require notable research, development, testing, and evaluation (RDT&E) to enable a new reactive flame retardant to enter the market. Part of the attractiveness of "nonreactive" additives is that all you have to do (in theory) is just add them in and you don't have to worry about changing the polymerization kinetics and properties of the final polymer, as the properties of the base polymer are known at the start of the process. There is still a need for balance of properties even with this approach, but it is easier if you already know the base properties of the starting polymer. However, given the environmental issues associated with these nonreactive additives, it makes more sense to start investing the RDT&E into reactive flame retardants for more than just thermoset polymers, and stop investing in nonreactive/non-polymeric flame retardants.

Polymeric flame retardant additives in theory are more user friendly to the materials scientist trying to make a new flame retardant materials with a good balance of properties. A polymeric additive blended into another polymer is a polymer/polymer blend, which can be a lot easier to make than starting with a reactive approach. Interestingly, polymeric flame retardant additives lend themselves more for thermoplastic systems than thermoset, because of the polymer/polymer blend aspect to the project. One can incorporate a polymeric additive into a thermoset monomer, but it requires a lot of attention to miscibility of the additive in the monomer, and ensuring the polymeric additive does not phase separate as the thermoset monomers polymerize. So while polymeric additives are easier to incorporate into thermoplastic materials, they still require RDT&E to ensure that the polymers don't phase separate, and they maintain a good balance of properties. This is particularly true for thermoplastic polymers with a high degree of crystallinity (such as polyolefins and polyamides) and also for polymers with acid/base sensitivity that could cross-link/copolymerize with one another when mixed (such as polycarbonates). Attention to chemical structure interactions is required prior to mixing two polymers together, and indeed for thermoplastic polymers, quality of both dispersive and distributive mixing is important to get a good flame retardant polymer blend at the end of the mixing process.

32.2.2 Recyclable/End-of-Life Compatible Flame-Retardant Chemistries

Chapter 24 of this book covers this concept in detail, but some discussion again on the importance of this for the future needs to be repeated here. In general, most current polymer materials are not designed for end-of-life, nor are they designed for traditional recycling methods (e.g., regrind/melt/reuse for thermoplastics). While thermal recycling methods of converting used polymers into energy/raw materials exist (i.e., combustion vs. pyrolysis), even then, if one considers how flame retardants often inhibit heat release or promote char formation, thermal recycling may be problematic for flame retardant materials of the future. Flame retardant effects/capabilities of a future polymeric material may be in direct conflict with recycling/energy capture requirements for future materials at end-of-life for the product. It's unclear how to provide further guidance on this future requirement other than if there are easy ways to depolymerize a polymer at end-of-life and return to monomers via a chemical switch (rather than a thermal one), these should be considered for future material RDT&E.

Another option to consider here would instead be to focus on bio-based/bio-derived flame retardant materials. Chapter 13 covers this topic, but if the final chemical structure is bio-derived, then it may be possible for the flame retardant product to be bio-degradable at its end-of-life. However, then the issues of favorable PBT requirements for the final polymeric material come back into play as the flame retardant structure may not be something that can be allowed to release into the environment. If one is lucky and all aspects of the future flame retardant material have favorable PBT profiles and is all bio-derived/bio-compatible, then the problem would be solved, but since polymeric material often have to resist the elements/biological exposure, this seems unlikely to be achieved across a wide-range of applications. It's one thing to have a short-lifetime bio-compostable consumable flame retardant electronic device, but it's another to assume the same could be achieved for an electrically powered car, or a power system inside a house that is regularly getting exposed to heat, humidity, and biological organisms and is meant to last for a decade or more. Designing for life cycle is something that is very challenging, but it must be considered for all flame retardant materials of the future. The design targets will likely have to be compromises, but should inform future RDT&E targets for new materials to meet this need.

32.2.3 Test Methods

The importance of test methods to assessing material flammability and flame retardant performance is covered in detail in Chapters 17 and 18 of this book and has been published on extensively.[2–4] Further, new test methods are being developed for emerging fire risk scenarios, although this is not a fast process (see Chapter 22 for more details). What continues to be an unmet need in regards to test methods is the ability of test methods to measure flammability at the small scale and have those results correlate to larger scale flammability testing. The reasons for needing smaller scale test methods that correlate to larger scale are time and money. Specifically, scaling up enough flame retardant material for full-scale testing requires notable amounts of resources (money) and time to accomplish, and this slows down the development of new fire safe materials.

There are two small scale methods that show much promise to correlation to larger tests, namely, the microscale combustion calorimeter (MCC)[5,6] and the microscale flaming combustion calorimeter (MFCC),[7,8] but even these two techniques have their limitations.[9,10] The MCC is the more advanced of the two techniques in its potential correlations to other methods and has shown some good ability to correlate results from milligrams of material to other bench-scale tests. Specifically, it has shown some relationship to cone calorimeter results and for probabilistic prediction of UL94 V performance for pure polymers (no additives present).[11–15] Further, it is being considered by the US Federal Aviation Administration as a tool for material requalification, which suggests that it can provide meaningful assessments of material fire safety.[16] However, MCC has yet to show the ability to predict to larger scale fire tests that involve specific configurations of flame spread and sample orientation, such as ASTM E84 or large scale vertical burn tests. This is where the real need for new test methods is located: being able to take bench-scale results and use them to predict (with medium to high probability) full-scale test results. There is notable difficulty in achieving this when going from milligrams/grams of materials to large assemblies of materials in specific fire risk configurations. As discussed in Chapters 3, 17, 18, 20, and 21 of this book, how materials interact with one another during burning is highly complex, and combining multiple materials together in a large assembly with gas fed burners or other heat/flame sources results in a highly complex burning scenario that is tough for any one single test to correlate to. There is some good reason to believe that cone calorimeter testing could be one of the techniques to help provide this correlation,[17–21] but in some cases, notable modifications to the cone calorimeter and sample holders need to be made to get good correlation.[22–27]

While the MCC, MFCC, and cone calorimeter show the most promise of correlation to larger scale tests, there is more work that needs to be done, and other test methods may be needed. MCC, MFCC, and cone calorimeter have limited ability to capture vertical flame spread and flame spread caused by dripping of burning polymers, and this is important for a variety of fire risk scenarios. While cone calorimeter does have a vertical orientation mode, this testing setup is not easy to use and does have scale limitations. Therefore, MCC, MFCC, and cone calorimeter may need to be combined with small-scale vertical tests that capture these effects. It is currently unclear what test methods should be combined with MCC, MFCC, and cone calorimeter to properly assess aspects of vertical flame spread, although it seems logical that measuring rheology up to the temperatures of thermal decomposition may give some insight into heat release. There also been some work taking simple bench-scale tests like UL94 V testing and adding further instrumentation to either measure mass loss rate or heat release,[28–31] but the low intensity of the UL94 V test likely limits the ability of this test to be predictive for larger scale tests. Further work by fire safety scientists, engineers, and material scientists is needed to tackle this problem.

Of final note in regards to new test methods, there is a need for modernized test methods that more accurately capture modern fire risk scenarios. As discussed in Chapter 22, regulations/codes determine what test method gets used for fire testing of materials, and changing these tests can be very difficult for a variety of political, economic, and conservative fire protection reasons. Regardless of the reasons, there is benefit to fire safety scientists and engineers reconsidering existing regulatory test methods and looking to see if there is a better way of running the test that results in improved reproducibility, or if it yields more realistic results vs. the fire risk scenario of interest. For example, lots of building and construction materials in the US use ASTM E84 as a default test method for verification of flame spread. While this test has served fire protection needs very well, the test has lots of issues with it, including how materials are mounted, smoke measurement consistency, and how one assesses flame spread visually during testing when the material releases lots of smoke. Further, one can argue that flame spread of a panel put on the ceiling of a tunnel may not capture all aspects of flame spread/fire performance correctly for vertically oriented materials, or complex assemblies containing new materials, electrical sub-systems (powered during a fire event), or materials with air-gaps/direct contact with other objects not rated to E84 testing. More funding for fire test method development is needed, and hopefully there will be continued work in this area in the future.

32.2.4 Cost-Benefit Analysis—When Should We Use FR Additives?

While flame retardants do work and provide meaningful fire safety benefits,[32] we should consider when we need to use them, and confirm when the cost associated with their use is justified. While there has been some life cycle analysis and studies on the broader benefits vs. drawbacks of flame retardant use,[33–39] there is more to be done. To be more specific, one should step back and look at the fire risk scenario and quantify if flame retardants are truly the best solution when considering the actual fire hazard, fire probability, and the holistic life cycle assessment of the flame retardant and the product it is put into. Going back to the discussions of regulatory issues (Chapters 22 and 23), the need for such studies is made moot if the laws require fire protection even if the life cycle analysis should show more drawbacks than benefits to fire protection of a particular product. More plainly said, it really doesn't matter if the cost-benefit analysis suggests fire protection isn't needed if the codes and standards say fire protection is needed—one will have to fire protect the product because it is mandated. The life cycle analysis may provide impetus to change the codes and standards, but until the codes are changed, they must be followed. Likewise, if the life cycle analysis indicates that use of flame retardants far outweigh their potential drawbacks, if the laws ban their use, then one cannot use the flame retardants. Still, with good data, it becomes possible to make informed decisions and propose changes to codes and standards.

The other benefit of these analyses lies in identifying places where we should be focusing our efforts on next generation flame retardants. If the cost-benefit analysis finds that the flame retardant is providing meaningful benefit but has some significant drawbacks (environmental, cost, etc.), then that becomes a good scientific target for focused research. As stated earlier, fire safety science is reactionary in that it reacts to fire threats as they are found. Combining this reactionary approach with life cycle assessments allows us to better focus our efforts to not just put out the immediate fires, but also ensure the reduction of fire losses in the future is done in lasting and resilient way. Ulimately, considering cost-benefit analysis in fire protection will require fire safety engineers, flame-retardant chemists, and economists to work together to sort this all out, and come up with useful models for future flame-retardant research.

32.3 NEW FIRE RISK SCENARIOS

While it has been stated several times that fire protection is reactionary in nature, we can still use fire protection principles, known material flammability, and our insight into technology trends and human behavior to predict near term fire risk scenarios of concern that are not yet being met, and need notable amounts of fire protection, and perhaps new flame retardant polymeric materials/chemicals. Listed below are five near-term fire risk scenario areas to be aware of. While there is some similarity in regards to some of them (namely transportation focused and electrical issues), they are still organized separately to allow for detailed discussion and explanation of the fire risk scenario.

32.3.1 Electric Transportation

Fire risk and hazards have been in place for transportation for a long time and are often revisited when a new form of fuel or propulsion is put into use. The emerging fire risk scenario in more recent times is that of electric propulsion systems, whether they be for automotive or aviation. Electric ground transportation is not new, but what is new is the energy storage systems for these vehicles, namely, lithium ion batteries (LIB). This same technology (LIB) is being used to spur a potential revolution in aviation, namely, to power unmanned aerial systems (UAS) and electric vertical takeoff and landing (e-VTOL) aircraft. While a ground vehicle with a LIB doesn't look that different than an internal combustion engine ground vehicle, UAS and e-VTOL look very different than traditional fixed wing combustion engine aircraft (Figure 32–1), and as such, bring different fire risk scenarios. For our discussion of changing fire hazards, this particular section will focus on electric ground transportation using LIBs first, and e-VTOL aircraft will be covered in the next section. While both use LIBs, the fire risk scenarios are different enough that they merit separate sections of discussion.

Electric cars with LIBs present two notable fire hazards: increased total fuel load and the fire hazard of the LIB battery itself. In regards to total fuel load, due to a desire to increase the range of the electric car between charging of the LIB, the cars are increasingly replacing metal with lighter weight composites and plastics. This trend has been happening with gasoline/diesel powered vehicles as well. The net effect of this replacement of metal with polymer composites and thermoplastics is that the overall fuel load/flammability of these parts and components, and of the car itself, is greatly increased.[40–42] Since the flame retardant standard for polymers use in ground vehicles, Federal Motor Vehicle Safety Standard 302 (FMVSS30) has no heat release testing requirements associated with it and is a simplistic horizontal flame spread test, this means that the plastics used in ground vehicles are quite high in heat release and are easily ignited. The net effect is that these cars give off a lot of heat when involved in a fire. The intensity of the car fire itself is enough to cause notable damage to concrete parking garages,[43,44] and inside a house, is very likely to ignite the garage and the rest of the house as well. For traditional fueled cars, the plastic will ignite when exposed to an external flame source, or in a post-crash scenario, where fuel is spilled and an ignition occurs. For LIB powered cars, the ignition scenario is more intense, as the LIB presents its own fire hazard which can occur when the vehicle is not in motion.

LIB fires and their intensity have been well documented.[45–58] LIBs give off high levels of energy in very short time when the battery pack is mechanically damaged, thermally overloaded, or electrically overloaded. The fast discharge of electrical power and heat, combined with reactions between battery components, results in what is often called "thermal runaway" as all that energy is released which in turn easily ignites other objects. The heat levels from the failing LIB easily ignite plastics and are difficult to protect against. What makes LIBs further different from fuel fires is that they can occur not just when the car is involved in a crash, but they can occur when the LIB is charging or discharging. While most LIBs have thermal and battery management systems to prevent overheating and overcharging that lead to failures, these systems can fail, and the LIB charging in a garage (or outside) can go into thermal runaway and result in total loss of the vehicle. The high energy output of the LIB, combined with the high heat release of automotive plastics makes for a very difficult fire to put out. Indeed, LIBs require specialized extinguishing systems to put out, as foam used for fuel pool fires does nothing to cool a LIB, and instead large amounts of water must be used. Further, LIBs can continue to react/burn as the thermal runaway propagates from one battery cell to another, resulting in fires that can keep igniting long after the initial fire has been put out.

The underlying fire hazards of automotive plastics and LIBs present some challenging fire risk scenarios to design protection for. For parking garages, mandatory use of sprinklers seems to be a reasonable way forward, but lowering the heat release of automotive plastics would also go a long way toward improving the fire hazard that burning cars present to buildings.[59] This can be achieved by selecting lower heat release polymers, or considering the addition of relevant flame retardants. For the LIBs, however, a combination of engineering measures and very low-level heat release polymers may be required. So far, it's proven highly challenging to develop a LIB with built-in fire protection *and* good energy density. Attempts have been made at putting flame retardants directly into the electrolyte, with mixed success.[60–64] Fire protection of the battery case, along with mechanical reinforcement should also be considered. In the future, there is discussion of next generation solid-state LIBs that use ceramic/inorganic electrolytes and separators,[65,66] and these batteries should have inherently low heat release in a thermal runaway event, but more work is needed on these materials to confirm this potential.

FIGURE 32.1 Electric car example (left—By Vauxford—Own work, CC BY-SA 4.0, https://commons.wikimedia.org/w/index.php?curid=70077203) and e-VTOL example (right—By Matti Blume—Own work, CC BY-SA 4.0, https://commons.wikimedia.org/w/index.php?curid=62464463).

32.3.2 Aviation

Aviation has had a lot of change in recent years, especially with concerns over high fuel prices and the above mentioned e-VTOL/UAS revolutions in aviation. The Federal Aviation Administration (FAA) continues to study changes in aircraft design and shape and continues to put forth research to ensure the safety of these vehicles. The FAA sets regulations for materials used in aircraft, and what level of fire safety they need to meet, but it is reactive, not proactive in its approach for all the same reasons mentioned earlier in this chapter. Currently, the FAA is developing fire safety requirements and other regulations for electric aircraft, but as of the writing of this chapter, nothing has been definitively published or put forth for public comment other than to apply existing fire requirements for traditional fueled aircraft to electric aircraft carrying people. Fire requirements for UAS has not been decided upon. While putting low heat release materials currently used in combustible fueled aircraft will lower fire hazard for e-VTOLs, some discussion of the difference in fire hazard for e-VTOLs vs. traditional aircraft is needed.

When a traditional Jet A fueled aircraft crashes, there will be a "jet fuel pool fire," which is of great intensity and will burn its way into the fuselage. Most aircraft out there are aluminum skinned, but an increasing number of carbon-fiber + polymer composite skinned aircraft are being used to increase aircraft range by making the aircraft lighter. Carbon fiber + polymer composites burn more intensely than aluminum metal (aluminum melts, but can oxidize exothermically in a fire event), and these polymer composites can also release carbon fibers when they burn, presenting additional hazards in a post-crash fire event.[67–69] Despite this challenging fire risk scenario, existing fire safety regulations for aircraft under FAR 25.853 have been successful in giving about five minutes of time to escape the burning aircraft.

When an electric-powered aircraft crashes that contains a LIB, as mentioned above in regards to ground vehicles, the high intensity of the LIB fire is very hard to protect against. There will not be a "fuel pool fire" in an e-VTOL crash, but rather a LIB fire instead. The high intensity heat release from a LIB may overwhelm existing FAR 25.853 rated materials, but this needs to be verified. Where the LIB is located in the aircraft also complicates fire protection, as LIBs needed for electric aviation can be quite large, and if located inside the cabin to keep them at optimal temperatures, this could result in rapid heat injury to passengers. Further, LIBs may ignite in flight if they fail due to thermal or charging issues when in use during flight. LIBs are temperature sensitive and lose the ability to charge and discharge readily when they get cold in the upper atmosphere and so will require some insulation. Likewise, if the e-VTOL has a sudden surge in need for power while in flight, this rapid draw against the batteries may increase the chances of overheating the battery and causing a thermal runaway event while in flight.[70] Finally, in cases where a potential crash could occur, the pilots of a traditional aircraft can dump fuel prior to impact, but currently there is no way to "dump charge" off the battery prior to an impact event. Therefore, in a crash today, the LIB will go into thermal runaway unless it has a highly hardened outer shell to prevent damage to LIB on impact. Needless to say, LIB fires for aircraft are very different than they are for jet-fueled aircraft, and they require notable protection systems that have yet to be invented.

The other potential fire hazard for aircraft returns to combustion engines, and that is the fire hazard associated with different types of fuel. A new fuel being considered for aviation is liquefied natural gas (LNG) because of its energy density, lower cost, and potential to help serve as a cooling fluid in modern aircraft engines. LNG requires low temperature and pressure to remain liquid but can combust cleanly in a traditional turbine engine. However, the hazard of LNG in a post-crash scenario is very different than that of fuel pool fire. Specifically, when a potential future aircraft fueled by LNG crashes, all that pressurized fuel will rapidly convert to gas, which may result in explosion events rather than post-crash fuel pool fires. Currently, there have been no definitive studies on how LNG will affect aviation fire safety, but LNG is used safely in turbine engines for power plants, so it stands to reason there can be ways of considering LNG as a fuel for aircraft with the correct engineering safety measures. Depending upon the rate of release of LNG and its heat output intensity (namely, slow release vs. fast release), existing fire safe materials used in aircraft should perform well against small fire events, but it is unlikely they will survive fast release explosion events.

32.3.3 Wildland Urban Interface (WUI)

Forest fires and fires in the natural environment that spread to the built environment where humans live is not a new fire risk scenario. However, the increased population density moving out of classic "bricks and mortar" cities into sprawling suburbs and houses intertwined with forest/wildland areas in recent years is making WUI more of a fire risk scenario to pay attention to.[71–74] What makes WUI fires somewhat unique is the nature of the burning vegetation and how it impacts/interacts with flammable materials used in housing today. As discussed in Chapter 30, the amount of polymeric materials being used in building and construction is greatly increasing, which is changing how these materials burn, and it changes how buildings respond to WUI fires. An older house of brick/metal siding can still get involved in a fire in a WUI setting, but a more modern house with polymeric siding, energy-efficient insulation, and other polymeric components making up more of the structure will respond very differently and likely will burn up faster in a WUI fire. There are attempts to improve cities and towns in the WUI to make them more fire hardened/fire ready, as can be seen in a public domain image generated by the US National Fire Prevention Association (Figure 32.2) below.

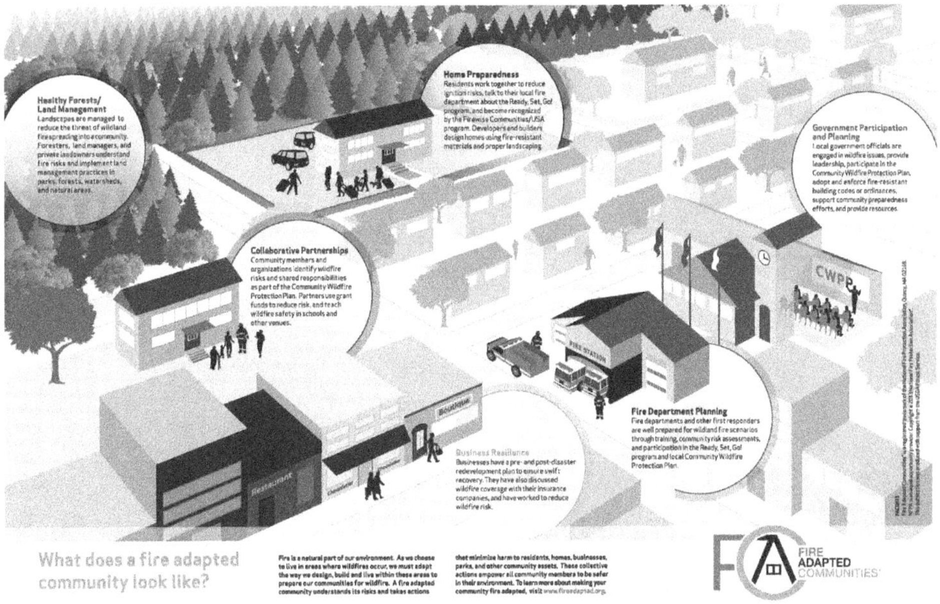

FIGURE 32.2 Fire adapted communities image showing WUI and changes to community to help it resist fire better.

Source: National Fire Protection Association—National Fire Protection Association, Public Domain, https://commons.wikimedia.org/w/index.php?curid=42616523

What makes WUI fires more of a novel fire risk scenario is that WUI fires are a combination of burning brands/embers and radiant heat ignition sources. To put it more plainly, building materials in WUI situations must not only resist the heat coming off a burning tree/bush near the house which is radiating heat onto the structure, but also resist the buildup of embers/small burning plant matter that blow onto the house, impact it, and lead to spot ignition sources depending upon how the wind is blowing. Effectively a house can get hit with lots of spot ignition sources and radiant heat at the same time depending upon where it is located in a WUI setting and what the vegetation near the house is. For the fire protection engineer and flame retardant chemist, this means new tests that have to be met to show sufficient fire protection. As outlined in Chapters 18 and 30, tests that meet radiant heat sources that affect walls or roofs are well-known. Tests that simulate "burning embers" are still in development[75–77] and should be finalized in the codes and standards soon to help determine what materials are appropriate for use in WUI settings. The author suspects that many materials which meet radiant heat fire tests by not dripping away from the heat source will perform well in these wind-blown burning ember tests, but this needs to be verified, and so this is hopefully something that will be researched in the near future by regulatory agencies.

32.3.4 Additive Manufacturing

Additive manufacturing has enabled all sorts of new applications and is capable of producing one-off replacement parts as well as novel shapes and forms that could not be achieved with traditional manufacturing methods. In regards to polymeric materials, some truly novel shapes and forms can be made depending upon the additive manufacturing method used (Figure 32.3).

Stereolithography (SLA) or fused deposition modeling (FDM) are two commonly used techniques. SLA uses light based focused polymerization (lasers typically) to control the polymerization and build the part one layer at a time. FDM also builds the part one layer at a time but lays down a strand of thermoplastic material which fuses together with the previous layer. The relevance of additive manufacturing to fire threats is that it is very easy to build a part out of a non-flame retardant polymer and insert it into an application that requires a flame retardant polymer. For example, a plastic case on a power supply cracks when it is dropped, and a consumer prints themselves a replacement case. If the power supply should short out or overheat in the future, the plastic used for the additive manufactured case is highly likely to not be able to resist this ignition threat. While there are some FDM plastics of inherently low heat release/low flammability (such as thermoplastic polyetherimide and polyphenylsulfone) many of the common plastics used for FDM are made from high heat release styrenic polymers such as high impact polystyrene and poly(acrylonitrile-butadiene-styrene), or semi-crystalline flammable polymers like poly(lactic acid) and polyamides. SLA polymers are difficult to flame retard as one must incorporate flame retardants that do not interfere with the light-based polymerization process. Effectively, any flame retardant for SLA processes must be a "Reactive" flame retardant.

So one must ask what the fire risk scenario is here, and it's often one of unintended flammable material insertion. This particular fire risk scenario may actually be best

FIGURE 32.3 Examples of parts made via Stereolithography (SLA).

Source: Creative Commons Attribution-Share Alike 4.0 International License, https://3dprintingcolorado.com/expanded-sla-3d-printing-services/

served through education, rather than more use of flame retardants. Specifically, education to users of 3D printers that the parts they make, unless made out of the right type of plastic, are inherently flammable and should not be used with electrical/electronic components unless rated for that application. Otherwise, cost-effective flame retardants for FDM and SLA will be needed, as there is continued discussion of using additive manufacturing for actual part production, not just part replacement, and this would put additively manufactured parts directly into applications with fire hazard. Therefore, flame retardants for additively manufactured parts will need to be not only compatible with the polymer they are put into, but they must be compatible with the additively manufactured process and allow for the parts to built layer-by-layer, without part deformation during printing. For thermoplastics, this is not that different than optimizing a flame retardant formulation for extrusion or injection molding, but for polymers used in SLA, understanding polymerization kinetics will be critical to flame retardant reactive monomers used for SLA processes.

32.3.5 Changing Power Systems

With an increased emphasis on renewable power to address climate change, there has been notable change in power systems. Rather than a reliance on alternating current (AC), there has been increased use of direct current (DC) which is often generated by renewable power systems like wind and solar. While fire protection against electrical systems is pretty well understood, the changes of more DC power being used, and uses of "smart grid" technology to moderate/control the follow of this power, does change the fire risk scenario a bit. The main changes to the fire risk scenario move from just a short circuit/electrical arc igniting something, to large power storage units failing (LIB systems for home power storage) or DC power from a solar panel/wind system continuing to produce power during the fire, thus creating additional hazard in the fire event. The first change to the fire risk scenario is easy to understand when one considers LIB batteries and how they fail. However, LIB batteries for home use are quite large, and present some unique fire hazards that could burn down an entire structure if something that large should go into thermal runaway. Currently, there is ongoing work by Underwriter's Laboratories (UL) in the US to implement UL 9540 to develop safe standards for these large batteries, but there is still more work to be done in the codes and standards.[78] While this standard will call out where to place the battery and how to address some aspects of thermal runaway, the fire potential remains, and fire protection for the building materials near the battery may require fire performance above and beyond what is used for most building materials today. If the codes and standards move these types of batteries far from the structure so that it cannot involve the house in the fire, then there may not be a need for new fire safety standards for materials placed near/adjacent to the large format battery pack.

Regarding the fact that renewable power is constantly "on," in a fire event, if the sun is still shining/wind blowing in a fire, all the power must go somewhere, and potential this could result in additional overheating, ignitions, or electrical shock hazards to those escaping from the fire. More study is needed on how renewable power systems change the fire risk scenario, and if additional fire hazards are created by the power still being generated during a fire event. Additional fire hardening may be needed for wires/roofing/building materials adjacent to these power generation units, but it stands to reason that existing materials currently meeting existing fire protection tests should be able to inserted into this application. This could mean that the codes and standards insist upon higher rated flame-retardant/fire-resistant materials be used when near these systems, or customized tests are to be developed and followed. In any case, close attention to this new fire risk scenario is needed, and research is needed to determine what the real hazards are and what the probability is of them occurring.

32.4 CONCLUSIONS

Fire protection of materials is a very old field of science, and the need to fire protect these materials will continue for as long as there are fire risk scenarios and polymers involved in those scenarios. So to some extent, this field of research is eternal, and short of very novel fire extinguishing systems or fire extinguishing robots always being available in our lives, flame retardancy and fire protection will need to be studied, applied, and optimized for future use. This chapter has focused on why fire protection is reactive, rather than proactive, and what the new needs for flame retardants are. Specifically, new flame retardants must have good environmental profiles, and flame retardants with polymeric and reactive structures are the way to proceed here. The days of using small molecule flame retardants which can migrate out of polymers are done, and continued research in this area, other than for basic research and flame retardant discovery, should be discouraged.

There are some notable unmet needs which need to be focused on, including figuring out how to make flame retardants of the future more recyclable and compatible with end-of-life business models for those polymeric products, whether it be physical/chemical recycling, or waste-to-energy processes. Test methods to improve prediction between bench-scale and full-scale testing is needed, especially for complex fire risk scenarios that involve flame spread and unique configurations only seen at full scale and not easily scaled down. It's not practical to burn down an entire house just to prove new flame-retardant materials can meet WUI-type fires or resist a LIB-powered car going into thermal runaway, and so new test methods are sorely needed. Most importantly, scientific analysis combined with cost-benefit analysis to determine when flame retardants should be used in a fire risk scenario is greatly needed. These types of analyses can help us focus our limited resources on targeted research that the cost-benefit analysis shows the most impact per currency spent, or where there is the greatest unmet need that is leading to major fire losses that needs to be addressed now.

New fire risk scenarios including changes in transportation, building designs, additive manufacturing, and the entire changing landscape of power generation give us new applications to provide fire protection for polymeric materials. The intense fire threats of LIB power systems require close attention, and fire hardening of polymeric materials against LIB fires will require notable resources for fire protection to be successfully obtained. It is highly likely that yet more fire risk scenarios will be discovered as further developments in building materials and power generation systems come online, and as the codes and standards attempt to keep up with all this change. While regulations do tell the fire protection engineer and fire safety scientist what minimum level of fire safety is needed, a newly emerged fire risk scenario that overwhelms the fire safety of our products/materials does not give us an excuse to not be vigilant against these new fire risk scenarios. Fire safety engineers and scientists should pay close attention to technological and socio-economic trends to find the fire risk scenarios of the future, and plan accordingly to ensure these future scenarios do not cause major amounts of fire damage before they are stopped.

To conclude, there is still much that needs to be done in flame retardancy of polymeric materials, and this chapter gives some insight into future needs, while the rest of the book provides practical information on how to protect polymeric materials against fire. Reading is just part of the solution—experimentation and application of the information in this book is required, and new research (both in funding and discovery) is needed if we are to stay ahead of fire losses and keep polymers from burning.

REFERENCES

1. www.nfpa.org/News-and-Research/Publications-and-media/NFPA-Journal/2021/Fall-2021/Features/1700 (accessed 11/22/22).
2. "Recommendations on the scientific approach to polymer flame retardancy: Part 1—Scientific terms and methods." Schartel, B.; Wilkie, C. A.; Camino, G. *J. Fire Sci.* **2016**, *34*, 447–467.
3. "Recommendations on the scientific approach to polymer flame retardancy: Part 2—Concepts." Schartel, B.; Wilkie, C. A.; Camino, G. *J. Fire Sci.* **2017**, *35*, 3–20.
4. "Development of fire-retarded materials—Interpretation of cone calorimeter data." Schartel, B.; Hull, T. R. *Fire Mater.* **2007**, *31*, 327–354.
5. "Pyrolysis combustion flow calorimetry." Lyon, R. E.; Walters, R. N. *J. Anal. Appl. Pyrolysis* **2004**, *71*, 27–46.
6. "Standard test method for determining flammability characteristics of plastics and other solid materials using

microscale combustion calorimetry." ASTM D7309-21b. www.astm.org/d7309-21b.html
7. "A milligram-scale flame calorimeter pyrolyzer system used to emulate burning of nonthermally thin solid samples." De Beer, J. A.; Raffan-Montoya, F.; Stoliarov, S. I. *Fire Mater.* **2022**, *46*, 302–312.
8. "Screening flame retardants using milligram-scale flame calorimetry." Raffan-Montoya, F.; Stoliarov, S. I.; Levchik, S.; Eden, E. *Polym. Degrad. Stab.* **2018**, *151*, 12–24.
9. "Pyrolysis combustion flow calorimeter: A tool to assess flame retarded PC/ABS materials?" Schartel, B.; Pawlowski, K. H.; Lyon, R. E. *Thermochim. Acta* **2007**, *462*, 1–14.
10. "Microcombustion calorimetry as a tool for screening flame retardancy in epoxy." Morgan, A. B.; Galaska, M. *Polym. Adv. Technol.* **2008**, *19*, 530–546.
11. "Molar group contributions to polymer flammability." Walters, R. N.; Lyon, R. E. *J. App. Polym. Sci.* **2002**, *87*, 548–563.
12. "Prediction of the burning rates of charring polymers." Stoliarov, S. I.; Crowley, S.; Walters, R. N.; Lyon, R. E. *Combust. Flame* **2010**, *157*, 2024–2034.
13. "A molecular basis for polymer flammability." Lyon, R. E.; Takemori, M. T.; Safronava, N.; Stoliarov, S. I.; Walters, R. N. *Polymer* **2009**, *50*, 2608–2617.
14. "Material properties and fire test results." Lyon, R. E.; Safronava, N.; Quintiere, J. G.; Stoliarov, S. I.; Walters, R. N.; Crowley, S. *Fire Mater.* **2014**, *38*, 264–278.
15. "A molecular-level fire growth parameter." Lyon, R. E.; Safronava, N.; Crowley, S.; Walters, R. N. *Polym. Degrad. Stab.* **2021**, 1009478.
16. "Microscale flammability criterion for constituents of aircraft cabin materials." Lyon, R. E.; Safronava, N. US Department of Transportation, Federal Aviation Administration. Report # DOT/FAA/TC-22/22, August 2022. www.fire.tc.faa.gov/pdf/TC22-22.pdf (accessed 11/22/22).
17. "Mechanistic study of the combustion behavior of polymeric materials in Bench-scale tests. I. Comparison between cone calorimeter and traditional tests." Costa, L.; Camino, G.; Bertelli, G.; Borsini, G. *Fire Mater.* **1995**, *19*, 133–142.
18. "Cone calorimeter analysis of UL-94 V-rated plastics." Morgan, A. B.; Bundy, M. *Fire Mater.* **2007**, *31*, 257–283.
19. "Heat release rate measurements of thin samples in the OSU apparatus and the cone calorimeter." Filipczak, R.; Crowley, S.; Lyon, R. E. *Fire Saf. J.* **2005**, *40*, 628–645.
20. "Flammability testing of wool/cellulosic and wool/synthetic fiber blends: Vertical flame spread and heat release results." Morgan, A. B.; Galaska, M. L. *J. Fire Sci.* **2020**, *38*, 522–551.
21. "Cone calorimeter and room corner fire testing of balsa wood core/phenolic composite skin sandwich panels." Morgan, A. B.; Toubia, E. *J. Fire Sci.* **2014**, *32*, 328–345.
22. "Reduced-scale test to assess the effect of fire barriers on the flaming combustion of cored composites: An upholster-material case study." Zammarano, M.; Shields, J. R.; Leventon, I.; Kim, I.; Nazare, S.; Thompson, A.; Davis, R. D.; Chernovsky, A.; Bundy, M. *Fire Mater.* **2021**, *45*, 114–126.
23. "Optimization of the procedure to burn textile fabrics by cone calorimeter: Part I. Combustion behavior of polyester." Tata, J.; Alongi, J.; Carosio, F.; Frache, A. *Fire Mater.* **2011**, *35*, 397–409.
24. "Apparatus for the vertical orientation cone calorimeter testing of flexible polyurethane foams" Galaska, M. L.; Morgan, A. B.; Schenck, K. A.; Stalter, J. G. *Fire Mater.* **2015**, *40*, 158–176.
25. "Assessing the reaction to fire of cables by a new bench-scale method." Gallo, E.; Stocklein, W.; Klack, P.; Schartel, B. *Fire Mater.* **2017**, *41*, 768–778.
26. "Considerations on reaction to fire tests of polyethylene foam with a cone calorimeter apparatus." Apostolopoulou, N.; Romeos, A.; Hinopoulos, G.; Perrakis, K.; Panidis, T. *J. Fire Sci.* **2018**, *36*, 240–255.
27. "Testing of liquids with the cone calorimeter." DiDomizio, M. J.; Ibrahimli, V.; Weckmen, E. J. *Fire Saf. J.* **2021**, *126*, 103449.
28. "Burning and dripping behaviors of polymers under the UL94 vertical burning test conditions." Wang, Y.; Zhang, F.; Chen, X.; Jin, Y.; Zhang, J. *Fire Mater.* **2010**, *34*, 203–215.
29. "Instrumentation of UL-94 test: Understanding of mechanisms involved in fire retardancy of polymers." Dupretz, R.; Fontaine, G.; Duquesne, S.; Bourbigot, S. *Polym. Adv. Technol.* **2015**, *26*, 865–873.
30. "In-situ investigation of temperature evolution of drippings via an optimized UL-94 instrumentation: Application to flame retarded polybutylene succinate." Hu, C.; Fontaine, G.; Tranchard, P.; Delaunay, T.; Collinet, M.; Marcille, S.; Bourbigot, S. *Polym. Degrad. Stab.* **2018**, *155*, 145–152.
31. "Quantification of polymer degradation during melt dripping of thermoplastic polymers." Kandola, B. K.; Ndiaye, M.; Price, D. *Polym. Degrad. Stab.* **2014**, *106*, 16–25.
32. "The well-meaning but misguided rollback of fire safety in the United States." Morgan, A. B. *J. Fire Sci.* **2022**, *40*, 249–253.
33. "Fire-LCA model: TV case study." Simonson, M.; Blomqvist, P.; Boldizar, A.; Möller, K.; Rosell, L.; Tullin, C.; Stripple, H.; Sundqvist, J. O. SP Report 2000:13, 2000.
34. "Fire performance of selected IT-equipment." Simonson, M.; Andersson, P.; Bliss, D. *Fire Tech.* **2004**, *40*, 27–37.
35. "Emissions from fires Part I: Fire retarded and non-fire retarded TV-sets." Blomqvist, P.; Rosell, L.; Simonson, M. *Fire Tech.* **2004**, *40*, 39–58.
36. "Emissions from fires Part II: Fire retarded and non-fire retarded TV-sets." Blomqvist, P.; Rosell, L.; Simonson, M. *Fire Tech.* **2004**, *40*, 59–73.
37. "Fire behaviour of plastic parts in electrical appliances—standards versus required fire safety objectives." Babrauskas, V.; Simonson, M. *Fire Mater.* **2007**, *31*, 83–96.
38. "Fire emissions of organics into the atmosphere." Blomqvist, P.; Persson, B.; Simonson, M. *Fire Technol.* **2007**, *43*, 213–231.
39. "Application of a cost-benefit analysis model to the use of flame retardants." Simonson-McNamee, M.; Andersson, P. *Fire Technol.* **2015**, *51*, 67–83.
40. "Human survivability in motor vehicle fires." Digges, K. H.; Gann, R. G.; Grayson, S. J.; Hirschler, M. M.; Lyon, R. E.; Purser, D. A.; Quintiere, J. G.; Stephenson, R. R.; Tewarson, A. *Fire Mater.* **2008**, *32*, 249–258.
41. "Investigation of smoke gases and temperatures during car fire—large-scale and small-scale tests and numerical investigations." Krüger, S.; Hofmann, A.; Berger, A.; Gude, N. *Fire Mater.* **2016**, *40*, 785–799.
42. "Post-collision fires in road vehicles between 2002 and 2015." Otxoterena, P.; Bjornstig, U.; Lindkvist, M. *Fire Mater.* **2020**, *44*, 767–775.
43. "Fire hazard analysis of modern vehicles in parking facilities." Boehmer, H. R.; Klassen, M. S.; Olenick, S. M. *Fire Technol.* **2021**, *57*, 2097–2127.
44. "Fire safety assessment of sprinkler systems for car parks using the J-value methodology." Alimzhanova, M.; Spearpoint, M.; Jomaas, G. *Fire Saf. J.* **2022**, *129*, 103565.

45. "Experimental study on the combustion characteristics of primary lithium batteries fire." Chen, M.; DeZhou, C.; Wang, J.; He, Y.; Chen, M.; Yuen, R. *Fire Technol.* **2016**, *52*, 365–385.
46. "The impact of thermal runaway on sprinkler protection recommendations for warehouse storage of cartoned lithium-ion batteries." Ditch, B. *Fire Technol.* **2018**, *54*, 359–377.
47. "On a method to mitigate thermal runaway and propagation in packages of lithium ion batteries." Quintiere, J. G. *Fire Saf. J.* **2022**, *130*, 103573.
48. "Heat release during thermally-induced failure of a lithium ion battery: Impact of cathode composition." Liu, X.; Wu, Z.; Stoliarov, S. I.; Denlinger, M.; Masias, A.; Snyder, K. *Fire Saf. J.* **2016**, *85*, 10–22.
49. "Evaluating the thermal failure risk of large-format lithium-ion batteries using a cone calorimeter." Wang, Z.; Ning, X.; Zhu, K.; Hu, J.; Yang, H.; Wang, J. *J. Fire Sci.* **2019**, *37*, 81–95.
50. "Standard Guide for shipboard use of lithium-ion (Li-Ion) batteries." ASTM F3353-19.
51. "Handling lithium-ion batteries in electric vehicles: Preventing and recovering from hazardous events." Bisschop, R.; Willstrand, O.; Rosengren, M. *Fire Technol.* **2020**, *56*, 2671–2694.
52. "Analysis of effectiveness of suppression of lithium ion battery fires with a clean agent." Said, A. O.; Stoliarov, S. I. *Fire Saf. J.* **2021**, *121*, 103296.
53. "The implications of post-fire physical features of cylindrical 18650 lithium-ion battery cells." Nagourney, T.; Jordan J.; Marsh, L.; Scardino, D.; May, B. M. *Fire Technol.* **2021**, *57*, 1707–1722.
54. "More on methods to measure the energetics of lithium ion batteries in thermal runaway." Quintiere, J. G. *Fire Saf. J.* **2021**, *124*, 103382.
55. "Experimental comparison of Oxygen Consumption Calorimetry and Sensible Enthalpy Rise Approach for determining the heat release rate of large-scale lithium-ion battery fires." Voight, S.; Straubig, F.; Palis, S.; Kwade, A.; Knaust, C. *Fire Saf. J.* **2021**, *126*, 103447.
56. "Experimental investigation of suppression of 18650 lithium ion cell array fires with water mist." Said, A. O.; Garber, A.; Peng, Y.; Stoliarov, S. I. *Fire Technol.* **2022**, *58*, 523–551.
57. "Full-scale experimental study on suppressing lithium-ion battery pack fires from electric vehicles." Cui, Y.; Liu, J.; Han, X.; Sun, S.; Cong, B. *Fire Saf. J.* **2022**, *129*, 103562.
58. "Review-meta-review of fire safety of lithium-ion batteries: Industry challenges and research contributions." Diaz, L. B.; He, X.; Hu, Z.; Restuccia, F.; Marinescu, M.; Barreras, J. V.; Patel, Y.; Offer, G.; Rein, G. *J. Electrochem. Soc.* **2020**, *167*, 090559.
59. Guide on methods for evaluating fire hazard to occupants of passenger road vehicles, national fire prevention association standard # 556. www.nfpa.org/codes-and-standards/all-codes-and-standards/list-of-codes-and-standards/detail?code=556 (accessed 11/20/2022).
60. "An attempt to formulate nonflammable lithium ion electrolytes with alkyl phosphates and phosphazenes." Xu, K.; Ding, M. S.; Zhang, S.; Allen, J. L.; Jow, T. R. *J. Electrochem. Soc.* **2002**, *149*, A622–A626.
61. "Experimental study on flame stability limits of lithium ion battery electrolyte solvents with organophosphorus compounds addition using a candle-like wick combustion system." Guo, F.; Ozaki, Y.; Nishimura, K.; Hashimoto, N.; Fujita, O. *Combust. Flame* **2019**, *207*, 63–70.
62. "Thermal effect and mechanism analysis of flame-retardant modified polymer electrolyte for lithium-ion battery." Wu, Z.-H.; Huang, A. C.; Tang, Y.; Yang, Y.-P.; Liu, Y.-C.; Li, Z.-P.; Zhou, H.-L.; Huang, C.-F.; Xing, Z.-X.; Shu, C.-M.; Jiang, J.-C. *Polymers* **2021**, *13*, 1675.
63. "Recent advances in non-flammable electrolytes for safer lithium-ion batteries." Chawla, N.; Bharti, N.; Singh, S. *Batteries* **2019**, *5*, 19. DOI: 10.3390/batteries5010019.
64. "Phosphoryl-rich flame retardant ions (FRIONs): Toward safer lithium-ion batteries." Rectenwald, M. F.; Gaffen, J. R.; Rheingold, A. L.; Morgan, A. B.; Protasiewicz, J. D. *Angew. Chem.* **2014**, *53*, 4173–4176.
65. "A solid-state, rechargeable, long cycle life lithium-air battery." Kumar, B.; Kumar, J.; Leese, R.; Fellner, J. P.; Rodrigues, S. J.; Abraham, K. M. *J. Electrochem. Soc.* **2010**, *157*, A50–A54.
66. "Kinetics of all-solid-state sulfur cathodes." Wang, Y.; Liu, T.; Estevez, L.; Kumar, J. *Energy Storage Mater.* **2020**, *27*, 232–243.
67. "Prevention of the formation of respirable fibers in carbon fiber reinforced epoxy resins during combustion by phosphorus or silicon containing flame retardants." Greiner, L.; Doring, M.; Eibl, S. *Polym. Degrad. Stab.* **2021**, *185*, 109497.
68. "Potential for the formation of respirable fibers in carbon fiber reinforced plastic materials after combustion." Eibl, S. *Fire Mater.* **2017**, *41*, 808–816.
69. "Smolder behavior and emissions byproducts of aircraft composite coupons." Hatch, J.; Wardall, A.; Jackson, J.; McNeilly, R.; Kirsh, J.; Parker, A.; Morgan, A.; Duran, C. *Fire Saf. J.* **2021**, *123*, 103366.
70. www.ainonline.com/aviation-news/business-aviation/2020-01-24/eviations-electric-alice-aircraft-catches-fire-during-ground-tests (accessed 11/22/22).
71. "Trends in fire danger and population exposure along the wildland-urban interface." Peterson, G. C. L.; Prince, S. E.; Rappold, A. G. *Environ. Sci. Technol.* **2021**, *55*, 16257–16265.
72. "Wildland–urban interface fires in Spain: Summary of the policy framework and recommendations for improvement." Pastor, E.; Muñoz, J. A.; Caballero, D.; Àgueda, A.; Delmau, F.; Planas, E. *Fire Technol.* **2020**, *56*, 1831–1851.
73. "Regulatory controls for buildings in wildfire-prone areas of Australia." Baker, G.; Webb, A.; Whiting, P. *Fire Technol.* **2020**, *56*, 1903–1935.
74. "Review of the transition from smouldering to flaming combustion in wildfires." Santoso, M. A.; Christensen, E. G.; Yang, J.; Rein, G. *Front. Mech. Eng.* **2019**. DOI: 10.3389/fmech.2019.00049.
75. https://tsapps.nist.gov/publication/get_pdf.cfm?pub_id=900080 (accessed 11/22/22).
76. www.nist.gov/publications/investigating-ignitions-fuel-beds-using-reduced-scale-firebrand-generator (accessed 11/22/22).
77. "Firebrand generator." ISO/DIS 6021(en). www.iso.org/obp/ui/#iso:std:81912:en (accessed 11/22/22).
78. "Energy storage systems and equipment." Underwriter's Laboratory Standard # UL9540. www.shopulstandards.com/ProductDetail.aspx?UniqueKey=36788 (accessed 11/20/2022).

Index

Note: Page numbers in *italics* indicate figures, and page numbers in **bold** indicate tables in the text.

G

H

I

T

www.ingramcontent.com/pod-product-compliance
Lightning Source LLC
LaVergne TN
LVHW080105011026
849160LV00013B/1031
* 9 7 8 1 0 3 2 4 6 2 3 3 2 *